The Illustrated Moss Flora of Antarctica

Mosses are a major component of the vegetation in ice-free coastal regions of Antarctica. They play an important role in the colonisation of ice-free terrain, accumulation of organic matter, release of organic exudates, and also provide a food and habitat resource for invertebrates. They serve as model organisms for physiological experiments designed to elucidate problems of plant cold tolerance and survival mechanisms and for monitoring biological responses to climate change.

This Flora provides the first comprehensive description, with keys, of all known species and varieties of moss in the Antarctic biome. It has involved microscopic examination of around 10,000 specimens from Antarctica and, for comparison, from other continents. All species are illustrated by detailed line drawings, alongside information about their reproductive status, ecology, and distribution. This is an invaluable resource for bryologists worldwide, as well as to Antarctic botanists and other terrestrial biologists.

*Detailed keys to all species allow the user to make an initial assessment of taxa

*Provides a comprehensive taxonomic account of every family, genus and species of Antarctic moss, making this an invaluable identification guide

*Extensive bibliography provides a comprehensive list of publications on Antarctic mosses

RYSZARD OCHYRA is Head of the Laboratory of Bryology at the Institute of Botany, Polish Academy of Sciences in Kraków. He has undertaken fieldwork in Tanzania, South Africa, Antarctica, Tierra del Fuego, the Falkland Islands, Prince Edward Islands, Îles Crozet and Îles Kerguelen. He has received numerous awards including the prize of the Prime Minister of the Polish Government in 2000 for outstanding scientific achievement.

RONALD I. LEWIS SMITH joined the British Antarctic Survey in 1964 as a cryptogamic ecologist, and in 1975 he was appointed as head of the BAS Plant Biology Section. He worked extensively throughout South Georgia, South Orkney and South Shetland Islands, and the Antarctic Peninsula and continental Antarctica, before retiring from BAS in 2002. He was awarded the British Polar Medal in 1977 and a clasp in 2002 in recognition of his contribution to Antarctic science.

HALINA BEDNAREK-OCHYRA is based at the Institute of Botany, Polish Academy of Sciences in Kraków, where she became Curator of the Bryophyte Herbarium in 1993. She is a skilled botanical illustrator, and in 1996 and 2004 was awarded a prize by the Second Division of the Polish Academy of Sciences for outstanding scientific achievements.

Schistidium urnulaceum

The Illustrated Moss Flora of Antarctica

Ryszard OCHYRA
Institute of Botany
Polish Academy of Sciences, Kraków

Ronald I. LEWIS SMITH
Centre for Antarctic Plant Ecology and Diversity, Moffat, Dumfriesshire, Scotland

Halina BEDNAREK-OCHYRA
Institute of Botany
Polish Academy of Sciences, Kraków

CAMBRIDGE
UNIVERSITY PRESS

CAMBRIDGE UNIVERSITY PRESS
Cambridge, New York, Melbourne, Madrid, Cape Town, Singapore, São Paulo, Delhi

Cambridge University Press
The Edinburgh Building, Cambridge CB2 8RU, UK

Published in the United States of America by Cambridge University Press, New York

www.cambridge.org
Information on this title: www.cambridge.org/9780521814027

First published 2008

Printed in the United Kingdom at the University Press, Cambridge

A catalogue record for this publication is available from the British Library

ISBN 978-0-521-81402-7 hardback

This book is dedicated to those early explorers of Antarctica in appreciation of their courageous voyages and who, despite often enduring appalling hardships throughout their search for new lands, succeeded in making valuable botanical collections and observations which have served as the basis for future generations of botanists to venture south of research the plant life of the 'great white continent'.

CONTENTS

Taxonomic and Nomenclatural Novelties

NEW STATUSES AND COMBINATIONS

Cratoneuropsis relaxa (Hook.f. & Wilson) Broth. subsp. *minor* (Wilson & Hook.f.) Ochyra – p. 517
Ditrichum ditrichoideum (Cardot) Ochyra – p. 142
Pohlia wilsonii (Mitt.) Ochyra – p. 433
Splachnidae Ochyra – p. 415

NEW SYNONYMS

Amblystegium chilense Lorentz = *Cratoneuropsis relaxa* subsp. *minor* – p. 517
A. excurrens Cardot & Broth. = *Cratoneuropsis relaxa* subsp. *minor* – p. 519
A. kerguelense Mitt. = *Cratoneuropsis relaxa* subsp. *minor* – p. 517
Amblystegium philonotis Paris = *Cratoneuropsis relaxa* subsp. *minor* – p. 518
Amblystegium tenellum Cardot & Broth. = *Cratoneuropsis relaxa* subsp. *minor* – p. 519
Argyro-bryum subrotundum Hampe = *Bryum argenteum* var. *muticum* – p. 458
Austrophilibertiella Ochyra = *Ditrichum* – p. 138
Bartramia acicularis Müll.Hal. = *Philonotis polymorpha* – p. 504
Blindia austrocrispula Müll.Hal. = *Hymenoloma crispula* – p. 210
B. brevipes Müll.Hal. = *Hymenoloma antarcticum* – p. 214
B. pallidifolia Müll.Hal. = *Hymenoloma crispula* – p. 210
Brachythecium majusculum M.E.Newton = *Sciuro-hypnum fuegianum* – p. 563
Bryoxiphiales H.A.Crum & L.E.Anderson = Dicranales M.Fleisch. – p. 129
Bryum amblyolepis Cardot = *Bryum argenteum* var. *muticum* – p. 460
B. anomobryoides Cardot & Broth. = *B. dichotomum* – p. 463
B. argenteum Hedw. var. *obtusifolium* Broth. = *B. argenteum* var. *muticum* – p. 458
B. argenteum var. *subamblyolepis* Cardot & Broth. = *B. argenteum* – p. 454
B. cephalozioides Cardot = *Bryum argenteum* var. *muticum* – p. 458
B. confusum Lewis Smith = *B. pseudotriquetrum* – p. 474
B. crateris Dixon = *B. archangelicum* – p. 482
B. ellipsifolium Müll.Hal. = *Bryum argenteum* var. *muticum* – p. 458
B. filicaule Broth. = *B. pseudotriquetrum* – p. 473
B. imperfectum Cardot = *B. archangelicum* – p. 482
B. inclinatum (Brid.) Turton var. *magellanicum* Cardot = *B. archangelicum* – p. 482
B. inconnexum Cardot = *B. pallescens* – p. 486
B. inconnexum Cardot var. *tomentosum* Cardot = *B. pallescens* – p. 486
B. laetevirens Cardot & Broth. = *B. dichotomum* – p. 463
B. maceratum Müll.Hal. = *Bryum argenteum* var. *muticum* – p. 458
B. myurella Dusén = *Bryum argenteum* var. *muticum* – p. 458
Brachythecium turquetii Cardot = *Warnstorfia fontinaliopsis* – p. 546
Calliergonaceae Vanderpoorten, Hedenäs, C.J.Cox & A.J.Shaw = Amblystegiaceae – p. 514
Cochlearidiaceae Ochyra & Mahu = Andreaeaceae, p. 74
Conostomum antarcticum F.Weber & D.Mohr = *C. pentastichum* – p. 497
Dicranella fuegiana Cardot & Broth. = *Ditrichum heteromallum* – p. 139
D. recurvata Ochyra, Arts & Lewis Smith = *D. varia* – p. 165
Dicranoweisia breviseta Cardot = *Hymenoloma antarcticum* – p. 214
D. grimmiacea (Müll.Hal.) Broth. fo. *robusta* Cardot = *Hymenoloma grimmiaceum* – p. 206
Dicranum Hedw. subgen. *Trichodontium* Dixon = *Kiaeria* I.Hagen – p. 177
Ditrichum gemmiferum Ochyra & Lewis Smith = *Ditrichum ditrichoideum* – p. 142
D. lewis-smithii Ochyra = *Ditrichum hyalinocuspidatum* – p. 145
Drepanocladus austrostramineus (Müll.Hal.) Paris. var. *tenuissimus* Cardot & Broth. = *Warnstorfia fontinaliopsis* – p. 546
D. uncinatus (Hedw.) Warnst. fo. *gracillimus* Cardot = *Sanionia uncinata* – p. 527
Gemmabryum J.R.Spence & H.P.Ramsay = *Bryum* – p. 450
Grimmia abscondita Cardot = *Schistidium praemorsum* – p. 238
G. antarctici Cardot var. *pilifera* E.B.Bartram = *Schistidium urnulaceum* – p. 247

G. chubutensis Cardot & Broth. = *Schistidium praemorsum* – p. 238
G. defoliata Müll.Hal. = *Bucklandiella didyma* – p. 306
G. donatii Herzog & Thér. = *Schistdium amblyophyllum* – p. 263
G. donniana Sm. fo. *antarctica* Kuc = *Orthogrimmia sessitana* – p. 279
G. fallax Dusén = *Schistidium andinum* – p. 241
G. hyalinocuspidata Müll.Hal. var. *mutica* Cardot & Broth. = *Schistdium amblyophyllum* – p. 263
G. occulta Müll.Hal. = *Schistidium andinum* – p. 241
G. suborthotrichacea Müll.Hal. var. *robustissima* Müll.Hal. = *Bucklandiella pachydictyon* – p. 298
G. syntrichiacea Müll.Hal. = *Schistidium andinum* – p. 241
Guembelia lorentziana Müll.Hal. = *Schistidium andinum* – p. 241
Hennedia intermedia R.Br.bis = *Hennediella antarctica* – p. 365
H. microphylla R.Br.bis = *Hennediella antarctica* – p. 365
Hygroamblystegium fuegianum (Besch.) Reimers var. *compactum* E.B.Bartram = *Cratoneuropsis relaxa* subsp. *minor* – p. 519
H. fuegianum (Besch.) Reimers var. *secundum* E.B.Bartram = *Cratoneuropsis relaxa* subsp. *minor* – p. 519
Hymenoloma nordenskjoeldii Dusén = *H. antarcticum* – p. 214
Hyophilales M.Fleisch. = Phascales – p. 313
Hypnum Hedw. sect. *Orthotheciella* Müll.Hal. = *Cratoneuropsis* – p. 516
H. austroaduncum Müll.Hal. = *Sanionia uncinata* – p. 527
H. austrocatenulatum Müll.Hal. = *Cratoneuropsis relaxa* subsp. *minor* – p. 518
H. austrofluviatile Müll.Hal. = *Cratoneuropsis relaxa* subsp. *minor* – p. 518
H. austropulchellum Müll.Hal. = *Isopterygiopsis pulchella* – p. 587
H. austrostramineum Müll.Hal. var. *gracillimum* Müll.Hal. = *Warnstorfia fontinaliopsis* – p. 544
H. austrostramineum Müll.Hal. var. *subfluitans* Müll.Hal. = *Warnstorfia fontinaliopsis* – p. 544
H. chalarocladum Müll.Hal. = *Cratoneuropsis relaxa* subsp. *minor* – p. 518
H. desmiocladum Müll.Hal. = *Cratoneuropsis relaxa* subsp. *minor* – p. 518
H. filum Müll.Hal. = *Cratoneuropsis relaxa* subsp. *minor* – p. 518
H. georgicoantarcticum Müll.Hal. = *Plagiothecium orthocarpum* – p. 574
H. uncinatum Hedw. var. *polare* Renauld = *Sanionia georgicouncinata* – p. 531
Imbribryum N.Pedersen = *Bryum* – p. 450
Lescuraea chilensis Herzog = *Cratoneuropsis relaxa* subsp. *minor* – p. 519
Leskea fuegiana Besch. = *Cratoneuropsis relaxa* subsp. *minor* – p. 518
Meesia austrogeorgica Müll.Hal. = *Philonotis polymorpha* – p. 504
Mielichhoferia austrogeorgica Müll.Hal. = *Schizymenium pusillum* – p. 424
Mitteniales A.J.Shaw = Dicranales M.Fleisch. – p. 129
Ochiobryum J.R.Spence & H.P.Ramsay = *Bryum* – p. 450
Orthodontiaceae Goffinet = Bryaceae – p. 416
Philibertiella Cardot = *Ditrichum* – p. 138
Philonotis gourdonii Cardot = *Ph. polymorpha* – p. 504
Ph. luteola E.B.Bartram = *Ph. polymorpha* – p. 504
Ph. varians Cardot = *Ph. polymorpha* – p. 504
Ph. varians Cardot fo. *gracilis* Cardot = *Ph. polymorpha* – p. 504
Ph. varians Cardot fo. *robusta* Cardot = *Ph. polymorpha* – p. 504
Plagiothecium antarcticum Mitt. = *Isopterygiopsis pulchella* – p. 587
P. conostegium Herzog = *P. orthocarpum* – p. 574
P. simonovii L.I.Savicz & Smirnova = *P. orthocarpum* – p. 574
Pottia acutidentata Cardot & Thér. = *Hennediella antarctica* – p. 365
P. austrogeorgica Cardot var. *microphylla* Cardot & Broth. = *Hennediella antarctica* – p. 365
P. fuscomucronata Müll.Hal. = *Hennediella antarctica* – p. 365
P. grata R.Br.bis = *Hennediella antarctica* – p. 365
P. naumanii Müll.Hal. = *Hennediella heimii* – p. 359
Pottia oedipodioides Müll.Hal. = *Hennediella heimii* – p. 359
Pottiales M.Fleisch. = Phascales – p. 313
Pseudoleskea antarctica Cardot = *Cratoneuropsis relaxa* subsp. *minor* – p. 518
P. calochroa Cardot = *Cratoneuropsis relaxa* subsp. *minor* – p. 518
P. fuegiana (Besch.) Cardot var. *gracilis* Cardot & Broth. = *Cratoneuropsis relaxa* subsp. *minor* – p. 518
P. fuegiana (Besch.) Cardot var. *skottsbergii* Cardot = *Cratoneuropsis relaxa* subsp. *minor* – p. 518
P. lurida Cardot = *Cratoneuropsis relaxa* subsp. *minor* – p. 518
P. platyphylla Cardot = *Cratoneuropsis relaxa* subsp. *minor* – p. 518
P. sordidoviridis Cardot & Broth. = *Cratoneuropsis relaxa* subsp. *minor* – p. 519
P. strictula Cardot = *Cratoneuropsis relaxa* subsp. *minor* – p. 518
Racomitrium convolutum Mont. = *Bucklandiella didyma* – p. 306
R. flavescens Cardot = *Bucklandiella didyma* – p. 306
R. subrupestre Roiv. = *Bucklandiella didyma* – p. 306
Schistidium subpraemorsum Broth. = *S. amblyophyllum* – p. 263
Schistostegales M.Fleisch. = Dicranales M.Fleisch. – p. 129
Sciaromium gracile Dusén = *Cratoneuropsis relaxa* subsp. *minor* – p. 518
Scorpidiaceae Ignatov & Ignatova = Amblystegiaceae – p. 514
Tortula conferta E.B.Bartram = *Syntrichia magellanica* – p. 398
T. grossiretis Cardot = *Syntrichia magellanica* – p. 396
T. grossiretis Cardot var. *atrata* Cardot = *Syntrichia magellanica* – p. 398
T. heteroneura Cardot = *Syntrichia magellanica* – p. 398
Trichostomales M.Fleisch. = Phascales – p. 313
Weissia crispula Hedw. var. *ambigua* Wilson & Hook.f. = *Hymenoloma antarcticum* – p. 214
Willia senilis Müll.Hal. = *W. austroleucophaea* – p. 371
Wollnya Herzog = *Pohlia* – p. 427

NEW EPITYPES

Leskea pulchella Hedw. – p. 587
Pohlia inclinata Brid. – p. 482
Polytrichum juniperinum Hedw. – p. 105
P. longisetum (Sw.) *ex* Brid. – p. 98
P. piliferum Hedw. – p. 101
P. strictum Brid. – p. 108
Webera nutans Hedw. – p. 444
Weissia heteromalla Hedw. – p. 139

NEW LECTOTYPES

Gymnostomum heimii Hedw. – p. 359
G. ovatum Hedw. – p. 345
Hennedia intermedia R.Br.bis – p. 365
Hennedia microphylla R.Br.bis – p. 365
Hygroamblystegium fuegianum (Besch.) Reimers var. *compactum* E.B.Bartram – p. 519
H. fuegianum (Besch.) Reimers var. *secundum* E.B.Bartram – p. 519
Hymenoloma nordenskjoeldii Dusén = *H. antarcticum* – p. 214
Hypnum austroaduncum Müll.Hal. – p. 527
H. austrocatenulatum Müll.Hal. – p. 518
H. austrofluviatile Müll.Hal. – p. 518
H. austroglareosum Müll.Hal. – p. 551
H. austropulchellum (Müll.Hal.) Kindb. – p. 587
H. austrosalebrosum (Müll.Hal.) Kindb. – p. 555
H. austrostramineum Müll.Hal. var. *gracillimum* Müll.Hal. – p. 544
H. austrostramineum Müll.Hal. var. *subfluitans* Müll.Hal. – p. 544
H. austrouncinatum Müll.Hal. – p. 527
H. chalarocladum Müll.Hal. – p. 518
H. desmiocladum Müll.Hal. – p. 518
H. euryodictyon Müll.Hal. – p. 555
H. filum Müll.Hal. – p. 518
H. fontinaliopsis Müll.Hal. – p. 544
H. georgicoglareosum Müll.Hal. – p. 555
H. subpilosum Hook.f. & Wilson – p. 559
H. uncinatum Hedw. *var. polare* Renauld – p. 531
H. wahlenbergii F.Weber & D.Mohr – p. 430
Isopterygium fuegianum Besch. – p. 587
Kiaeria I.Hagen – p. 177
Leptostomum pusillum Hook. & Wilson – p. 424
Leptotrichum glaciale Müll.Hal. – p. 377
Lescuraea chilensis Herzog – p. 519
Leskea fuegiana Besch. – p. 518
L. pulchella Hedw. – p. 587
Meesia austrogeorgica Müll.Hal. – p. 504
M. uliginosa Hedw. – p. 492
Mielichhoferia austrogeorgica Müll.Hal. – p. 424
M. microstoma Hampe – p. 424
Mnium crudum Hedw. – p. 440
Orthotrichum antarcticum Cardot – p. 411
Pachyneurum J.J.Amann – p. 353
Philibertiella ditrichoidea Cardot – p. 142
Philonotis gourdonii Cardot – p. 504
Ph. varians Cardot – p. 504
Plagiotheciumm antarcticum Mitt. – p. 587
P. leptoplumulosum Dusén – p. 587
P. orthocarpum Mitt. – p. 574
Pogonatum alpinum (Hedw.) Röhl. fo. *elatum* Cardot – p. 94
Pohlia inclinata Brid. – p. 482
Polytrichum alpinum Hedw. – p. 94
P. alpestre Hoppe – p. 108
P. antarcticum Cardot – p. 101
P. brevifolium R.Br. – p. 94
P. juniperinum Hedw. – p. 105
P. longisetum (Sw.) *ex* Brid. – p. 98
P. piliferum Hedw. – p. 101
P. strictum Brid. – p. 108
P. trichodon Hook. & Wilson – p. 90, 92
Pottia acutidentata Cardot & Thér. – p. 365
P. austrogeorgica Cardot – p. 365
P. austrogeorgica Cardot var. *microphylla* Cardot & Broth. – p. 366
P. charcottii Cardot – p. 359
P. fuscomucronata Müll.Hal. – p. 365
P. naumanii Müll.Hal. – p. 359
Pottia oedipodioides Müll.Hal. – p. 359
Pseudodistichium austrogeorgicum Cardot – p. 151
P. calochroa Cardot – p. 518
Pseudoleskea fuegiana (Besch.) Cardot var. *gracilis* Cardot & Broth. – p. 518
P. fuegiana (Besch.) Cardot var. *skottsbergii* Cardot – p. 518
P. lurida Cardot – p. 518
P. platyphylla Cardot – p. 518
P. sordidoviridis Cardot & Broth. – p. 519
P. strictula Cardot – p. 518
Racomitrium Brid. [unranked] *Curviseta* Huebener – p. 283
R. flavescens Cardot – p. 306
R. nigritum A.Jaeger – p. 306
R. symphyodontum (Müll.Hal.) Paris var. *dusenii* Cardot – p. 306
R. symphyodontum (Müll.Hal.) Paris var. *muticum* Cardot – p. 298
Sarconeurum antarcticum Bryhn – p. 377
Schistidium subpraemorsum Broth. – p. 263
Stereodon revolutus Mitt. – p. 579
Tortula conferta E.B.Bartram – p. 398
T. didymodontoides Broth. – p. 383
T. fuscoviridis Cardot – p. 383
T. grossiretis Cardot var. *atrata* Cardot – p. 398
T. magellanica Mont. – p. 396
T. princeps De Not. – p. 399
T. ruralis (Hedw.) P.Gaertn., B.Mey. & Scherb. var. *gypsophila* G.Roth – p. 394
Webera gerlachei Cardot – p. 473
W. nutans Hedw. – p. 444
W. pyriformis Hedw. – p. 419
W. racovitzae Cardot – p. 444
W. racovitzae Cardot var. *laxiretis* Cardot – p. 473
Weissia crispula Hedw. – p. 210
W. crispula var. *ambigua* Wilson & Hook.f. – p. 214
W. heteromalla Hedw. – p. 139
W. latifolia Schwägr. – p. 340
W. recurvirostra Hedw. – p. 329
Willia grimmioides Müll.Hal. – p. 371
Wollnya stellata Herzog – p. 433

NEW NEOTYPES

Bryum inflexum Müll.Hal. – p. 436
Pottia hemii (Hedw.) Hampe var. *brevinervis* L.I.Savicz & Smirnova – p. 359

Foreword

By Sir Martin Holdgate CB

former Chief Biologist of the British Antarctic Survey
and past Director General of IUCN – The World Conservation Union

Fifty years ago, the late Dr Stanley Greene visited South Georgia and made a comprehensive collection of both vascular plants and bryophytes. His visit was supported by the Falklands Islands Dependencies Survey (FIDS), whose geological unit was then based at Birmingham University. Greene and his wife and professional assistant, Dorothy, also worked at Birmingham, in the Department of Botany, and it was there that they embarked on the sorting and taxonomic description of their collections. In 1964 Stanley Greene published an account of the vascular flora of South Georgia and in 1970 the first of a planned series on the Antarctic Moss Flora appeared.

In 1961 I became involved with FIDS (re-named British Antarctic Survey, BAS, in 1962), and between 1962 and 1966 I directed its biological research programme. We decided to focus our studies on Signy Island in the South Orkney Islands (60°43′ S, 45°35′ W). Looking back, it is striking to recall how little was known at that time about the biota of this or any other part of the maritime Antarctic region. Our plan was to examine the principal components of the terrestrial ecosystem alongside those of the freshwater lakes and inshore seas, and to define the nutrient flows and other interactions between them. It was at once apparent that the characterisation of the soils, microclimates and cryptogamic vegetation would be an essential prerequisite for the terrestrial studies. Signy Island is diverse topographically and geologically, and relatively low-lying, rising to no more than 300 m, and about half of its surface is free of ice and snow in summer. Many low-lying areas are heavily influenced by penguins, other seabirds and seals. Radiative warming is extremely important, raising the temperature within moss carpets to over 30° C on still, sunny summer days and hence slope aspect is a critical factor. The result is a mosaic of habitats, reflected in a complex pattern of plant communities and their associated soil fauna.

Preliminary soil and vegetation surveys were carried out in the summers of 1961–62 and 1963–64, and these were extended by more thorough research between 1964 and 1967. In this period other locations along the Antarctic Peninsula and in the nearby South Shetland, South Orkney and South Sandwich island groups were also examined. Material collected in the field was transferred to Birmingham for taxonomic investigation by Dr Greene and his colleagues, who provided initial identifications validating field names. The resulting classification of bryophyte and lichen communities in the Maritime Antarctic region, in which the co-author of the present volume, Dr Ronald Lewis Smith, played a central part, has stood the test of time.

As the authors of this book indicate in their Preface, Stanley Greene planned a definitive taxonomic account of the Antarctic moss flora as a whole. The work was supported, initially, through BAS as a component of the Natural Environment Research Council and then through the NERC Institute of Terrestrial Ecology (ITE) to which Dr Greene's Bryophyte Project Group was transferred in 1975. Unfortunately, as the years went by taxonomy was seen as of lower priority than more 'cutting edge' aspects of ecology and experimental biology, and support for it and for long-term monitoring waned. I will not linger on the arguments for and against this change in priorities except to comment that it is surely obvious that a sound taxonomic foundation is essential for reliable ecological field work, and that in a world where so much store is set on the conservation of biodiversity we cannot afford to ignore the discipline essential for the characterisation of that diversity.

Dr Greene served for some years as Head of a Plant Biology subdivision in ITE, but he left the Institute, retired and died before his planned work on Antarctic mosses was anywhere near completion. Plant ecology and physiology, however, continued to thrive in the BAS Plant Biology and Environment Section, led by Dr Lewis Smith and in the past two decades he has been the driving force behind the reinstatement of bryophyte and lichen systematics as a component of research on Antarctic biodiversity. This has in turn led to the completion of volumes on Antarctic lichens and hepatics, the former through a partnership with Professor Øvstedal and the latter with Professor Ochyra and Dr Bednarek-Ochyra, co-authors of the present volume, and Professor Váňa. The series now reaches completion in the present monograph, which is the first complete treatment of the moss flora of the Antarctic. It is timely, for ecological change is gaining momentum, especially in the Antarctic Peninsula where most of the species described in these pages are concentrated, and monitoring of their shifting patterns of distribution – and of the invasion of further species from lands to the north

– should be an essential element in research on climate change and its consequences.

It gives me the greatest pleasure to commend this important book. It is the culmination of fifty years of endeavour, against all manner of odds. The Antarctic – indeed world – science community is fortunate that it has been led by Dr Lewis Smith, who has been involved in the research from almost its beginning and has pursued it with single-minded purpose. He has been fortunate to collaborate with Professor Ochyra, a renowned moss taxonomist, and his wife Dr Bednarek-Ochyra, an equally renowned illustrator and bryophyte taxonomist. I congratulate the authors on their achievement, and am sure that this volume will not only be an invaluable aid to future generations of polar botanists, but a landmark in the history of Antarctic plant science.

December 2007

Martin Holdgate

Preface

This is the first comprehensive systematic treatment of the moss flora of Antarctica. Mosses are an important component of the Antarctic flora and vegetation, and the terrestrial ecosystem in general, especially in the more northerly coastal regions; here, they often predominate over lichens, which constitute the main life form in drier and inland habitats. Although Antarctica is the last continent to receive a full descriptive moss Flora, it is at the same time the first one to get such modern, comprehensive and complete treatment. To achieve this for an entire and vast biome (albeit with $<1\%$ of its area available for colonisation) has been possible because it comprises a discrete and isolated geographical unit with simple ecosystems and relatively low biodiversity. The task of elaborating this Flora might be construed as relatively easy because of the low species richness. However, the lack of any modern regional moss Flora in the cool-temperate and cold zones of the Southern Hemisphere, which could have greatly benefitted the development of the taxonomic concepts of the Antarctic taxa, was a major constraint in the progress of this work. This is particularly true for southern South America with which West Antarctica, the bryologically richest region in the Antarctic biome, exhibits the greatest floristic affinity.

While Antarctic terrestrial biological research has progressed and diversified rapidly since the early 1960s, scientists in the field have often been constrained by ever-changing and frequently inaccurate taxonomic identity, particularly with regard to the bryoflora. This has inevitably led to nomenclatural confusion in the literature and frequent misdeterminations of specimens. Not until now has an exhaustive revision of the biome's moss flora been undertaken. The only reasonably detailed account of Antarctic mosses has long been *La flore bryologique des terres magellaniques, de la Géorgie du Sud et de l'Antarctide* by Jules Cardot, published in 1908. However, it was not a classical Flora but an annotated catalogue of all mosses discovered in Antarctica prior to 1905.

The great majority of specimens on which the present Flora is based are deposited in the herbarium of the British Antarctic Survey, Cambridge (AAS), but we have obtained and studied specimens from most of the world's herbaria with collections of mosses made in Antarctica and conterminous regions of the Subantarctic and southern cool-temperate zone, to allow critical comparisons of taxa from within and between localities and to establish the accurate identity and status of the species concerned. Furthermore, type specimens of all species and infraspecific taxa described here have been examined, together with those of many related taxa, and all taxon names have been carefully typified. Consequently, it is hoped that this Flora will serve as a useful guide to help researchers in their understanding of ecological and physiological processes in the Antarctic environment and to provide useful phytosociological descriptions of plant communities in which mosses play an important role. Nevertheless, it is appreciated that many of the species described here are difficult to determine reliably in the field. The systematic accounts presented here should also provide a major contribution to global bryophyte taxonomy, nomenclature and biogeography.

Recent taxonomic treatments of Antarctic liverworts (Bednarek-Ochyra *et al.* 2000) and lichens (Øvstedal & Lewis Smith 2001) are testimony to the relatively comprehensive widespread bryological and lichenological survey of this remote part of the world, although many areas have still to be visited by professional botanists. However, unlike the far greater diversity and morphological complexity of lichens throughout the biome, we believe that collections made in new areas are unlikely to yield many taxa as yet unrecorded for the biome or new to science. Certainly, additions to distribution patterns may be expected.

As emphasised by Øvstedal & Lewis Smith (2001) regarding Antarctic lichens, we also appreciate that bryophyte taxonomy, like that of other phyla, is entering a new era in which DNA sequencing and other molecular techniques may revolutionise biological systematics in the near future (Bopp & Capesius 1998). Indeed, publication of this volume coincides with a major scientific initiative, the International Polar Year (IPY) Science Programme, 2007–2009. Although the present work was not formally one of the many research themes of IPY, it will complement the *Evolutionary Biology in Antarctica* programme. This proposes to explore the evolutionary history of selected modern biota, examine how biological diversity in the Antarctic influences the way present-day ecosystems function, and thereby predict how the biota may respond to future environmental change. While taxonomy is specified as a component of this programme, this will be largely at the molecular level and concentrate on marine organisms. It is hoped that the present Flora, based on traditional taxonomic concepts, will serve a useful practical purpose in today's more technological approach to the identification of and relationships between the moss components of the Antarctic terrestrial ecosystem.

It was the International Geophysical Year (IGY) of 1957–1958 which initiated a new era in the biological exploration of Antarctica. Unlike the early days of Antarctic exploration, when collections of mosses were rarely made by professional botanists and bryologists, in the post-IGY period collections have often been made by specialists. This resulted in the rapid expansion of information on Antarctic mosses and yielded many new records for this biome. Although many bryophyte collections have been made throughout Antarctica during the past half century, taxonomic progress has been slow. As a result, ecological research has been hindered by often unreliable and uncritical taxonomy. Meaningful phytosociological and ecosystem studies rely on accurate identification of species, as do comparative physiological studies of individual species. Many national research programmes in Antarctica have provided valuable data on the role of mosses in the composition of plant communities, the structure and functioning of terrestrial ecosystems and the physiological behaviour of species, but few have been based on the same taxonomic concepts. A major aim of the present Flora is to consolidate these concepts to provide a source of textual and illustrative information which will serve all bryologists, ecologists and conservationists working in the Antarctic biome.

It was the initiative of the late Stanley W. Greene, formerly of the British Antarctic Survey, that inspired the present work. He, and his co-workers, began the revision of the mosses of subantarctic South Georgia and the Antarctic in the early 1960s. However, due to the rapid decline in support for traditional taxonomic research by funding agencies, this endevour was terminated in the mid-1980s.

The challenge of producing a complete flora of the mosses of Antarctica was taken up by the present authors, and work commenced in 1995 when *The Illustrated Moss Flora of Antarctica* was conceived. The objective of this work is to provide a comprehensive up-to-date account of all taxa of mosses recorded in Antarctica by examining specimens from the majority of the world's herbaria holding Antarctic collections. We have taken a monographic approach. The type of every taxon described from the study area has been examined, and we also tried to investigate all historical collections and, for the most part, we have succeeded. Much of the work has been based on the detailed examination of the large collection of specimens in AAS, as well as those in several other major herbaria willing to loan material for the purpose. International support to achieve this work has been provided by the Scientific Committee on Antarctic Research (SCAR) Working Group on Biology, particularly through the former BIOTAS (*Biological Investigations of Terrestrial Antarctic Systems*) Programme which had expressed the urgent need for such Floras. We hope that this volume will prove informative and useful to all researchers with an interest in Antarctic mosses.

In the introductory part of this Flora we have deliberately replicated some of the text used in the lichen volume because it is equally relevant to the bryoflora, and including appropriate or modified sections here will save the reader consulting the complementary volume. We consider it important to familiarise potential users of the Flora with this vast and, for many, largely unknown region because there appear to be many interpretations of the 'Antarctic' biogeographical element in the bryological literature, and this biome is often neglected in global treatments of bryogeography.

ACKNOWLEDGEMENTS

During many years of taxonomic research on austral mosses and work of this Flora we received generous help from many individuals and institutions, without whose cooperation completion of this volume would not have been possible. In particular, we wish to express our sincere gratitude to the British Antarctic Survey, Cambridge (RILS) and Institute of Botany of the Polish Academy of Sciences, Kraków (RO and HBO) which supported and encouraged research throughout its prolonged gestation. The project has gained financial support from the British Antarctic Survey (for the colour plates), Polish State Committee for Scientific Research and the Polish Ministry of Science and High Education through grants No. 6P204 02006 and 2 P04G 043 29 for RO. In retirement RILS was awarded a Leverhulme Emeritus Fellowship grant to allow completion of his work. Christine Phillips, BAS librarian, willingly offered assistance with difficult Antarctic literature and her cooperation is gratefully acknowledged. Special thanks go to Helen Peat, Curator of the BAS herbarium (AAS), for her frequent and prompt despatch of specimens to RO in Kraków and provision of information from the AAS Antarctic Plant Database.

RILS is also indebted to the captains and officers of the BAS ships RRS *Shackleton*, *John Biscoe*, *Bransfield*, *James Clark Ross* and *Ernest Shackleton*, and pilots of the BAS Twin Otter aircraft for support in the field, throughout the maritime Antarctic and on South Georgia; grateful thanks are also due to the Directors of the Argentine, Australian and Italian Antarctic Programmes for their invitations to undertake research at their stations, and to the helicopter pilots serving the Australian and Italian Programmes for field support. RILS also expresses his deepest gratitude to his wife, Elinor, for her unceasing support and patience during his many and prolonged absences in the Antarctic and for providing a steady supply of coffee during his periods of writing and editing this volume in his retirement.

RO is indebted to the late Celina M. Matteri, Buenos Aires, Valdon R. Smith, Stellenbosch, and Marc Lebouvier and Yves Frenot, Paimpont, for arranging his field work in Tierra del Fuego, Prince Edward Is. and Îles Crozet and Îles Kerguelen, respectively, which markedly contributed to increasing the distributional data of many species and a better understanding of the biology and taxonomy of many austral moss taxa.

RO and HBO are very grateful to Len T. Ellis, Alan Harrington and Brian O'Shea (BM), Denis Lamy, Hélène Bischler and Catherine Rausch (PC), Olga M. Afonina (LE), Lars Hedenäs and Thor-Björn Engelmark (S), Timo Koponen, Sinikka Piippo and Johannes Enroth (H), and Herman Stieperaere (BR) for assistance and hospitality during their personal visits to these herbaria which possess exceptional and invaluable collections of Antarctic mosses, especially from the early expeditions. Their gratitude is extended to the Directors and Curators of the herbaria at ADT, ALTA, B, BG, CHR, COLO, E, FH, G, GRO, H, HBG, HO, HIRO, JE, LD, MANCH, MEL, MO, NY, NIPR, O, PC, PRE, U, US, UPS and W, who arranged loans of thousands of moss specimens from their herbaria, as well as to Manuel Mahu, Santiago, the late Celina M. Matteri, Buenos Aires, and the late Vera Komárková, Leysin, for sending numerous Antarctic and South American specimens. Bill Buck, Len Ellis, Denis Lamy, Brian O'Shea, Herman Stieperaere, and the late Stanley W. Greene, have often willingly provided copies of many taxonomic papers not readily accessible. Rod Seppelt, Kingston, Niek Gremmen, Diever, Valdon R. Smith, Stellenbosch and Tamás Pócs, Eger, provided critical comment on some parts of the manuscript; Misha Ignatov, Moscow, confirmed the distribution of many species in Russia; and María T. Gallego, Murcia, confirmed the determination of *Syntrichia caninervis* var. *gypsophila*; Werner Greuter, Berlin, Gea Zijlstra, Utrecht, John McNeill, Edinburgh, and the late Riclef Grolle, Jena, have always offered their valuable advice and comments on nomenclatural issues. We are very thankful for their kind help and collaboration. We are also much indebted to the Publishers and Editors of the journals indicated in the text for kind permission to reproduce some illustrations and distribution maps, to Denis Lamy for providing the portraits of Jules Cardot and Joseph D. Hooker and to Len Ellis and Herman Stieperaere for the images of the title pages of some books reproduced in the present Flora.

Very special thanks are due to Robin Stevenson, King's Lynn, and Howard Matcham, Chichester, for valuable comment on and very careful proof reading of Chapter 6 – *Systematic Accounts of the Taxa*. We are also pleased to acknowledge the meticulous assistance of Katarzyna Biłyk for generating the distribution maps and incorporating corrections in the formatted text. The line drawings are by HBO and the colour photographs by RILS, unless otherwise stated.

Finally, we pay tribute to Marian Wysocki for his invaluable cooperation (with RO) in elaborating the layout of the text, scanning, inserting and arranging the Figures and photographic Plates, as well as formatting the book.

Kraków – Moffat
22 February 2008

Ryszard Ochyra
Ronald I. Lewis Smith
Halina Bednarek-Ochyra

1. *Introduction*

The Antarctic is unique amongst the biomes of the world in having a terrestrial biota comprising almost exclusively lower organisms (microorganisms, lichenised fungi, algae, bryophytes, invertebrates). There are only two native phanerogams, namely Antarctic hairgrass (*Deschampsia antarctica* Desv.) and Antarctic pearlwort (*Colobanthus quitensis* (Kunth) Bartl.), and these occur only on the islands of the Scotia Ridge and along the western coast of the Antarctic Peninsula. While the greatest diversity of bryophytes and lichens also occurs in this region, lichens are the dominant life form throughout much of it and in virtually all of continental Antarctica. However, mosses predominate wherever there is a regular supply of water. This restricts the great majority of species and development of distinct moss-dominated communities to lowland coastal habitats, especially in the relatively milder and wetter maritime Antarctic (see Table 1). Species diversity, and extent and complexity of moss communities decrease wherever conditions become colder and drier, for instance in the coastal continental Antarctic, while only a very few species can survive the extreme conditions of inland sites.

Increasing scientific interest in Antarctica since about 1980, much of it focusing on the climate change scenario, has changed perceptions of the continent. From being seen as remote and disconnected from most general fields of science it is now regarded as commanding a key position in many research disciplines. In addition, growing tourism, interest in global biodiversity and a need for more active conservation have led to the realisation that modern identification manuals are essential for Antarctica.

This is the first comprehensive and critical taxonomic account of the mosses of the Antarctic biome. It will contribute to a better understanding of the flora, provide more accurate distributional data and permit more reliable assessments of biogeographical affinities, dispersal routes and evolutionary trends. Although bryophytes play an insignificant role in the Antarctic terrestrial trophic web (there is no published evidence of them being grazed by invertebrates), they are the predominant components of many communities. In experimental bryology there is increasing interest in their physiological and biochemical responses to climate change, increasing potentially damaging UV-B levels and other stress factors, and to their synthesis of both cryo- and photoprotectant organic compounds.

We have deliberately provided an extended account of biogeographic zones and bryological history, and of the environment and moss-dominated communities of the Antarctic, because such information is often difficult to obtain for this biome which is still the least researched in the world. Because the flora of Antarctica has low species diversity it is possible to provide probably the most comprehensive assessment of the major terrestrial components (mosses, liverworts, lichens, invertebrate fauna) of any biome or continent on the planet. Collections of moss have been made at most localities where vegetation is well developed and usually supports a higher species diversity.

The objective of this work is to provide a comprehensive account of the mosses of Antarctica based on critical assessment of specimens in most of the world's major herbaria with Antarctic holdings, and on ecological data for habitats and communities. Individual taxa have been compared with type specimens wherever possible, and with relevant material from other parts of the world. We have sought the expertise of many bryologists with specialist knowledge in particular genera.

BIOGEOGRAPHIC ZONES

The Antarctic (Fig. 1) is defined geopolitically, for the purposes of the internationally adopted Antarctic Treaty (Triggs 1987; Heap 1994), as all land and adjoining ice shelves south of latitude 60° S (Article VI). In accordance with the Antarctic Treaty, there are no national territorial claims recognised within this region, and all pre-Antarctic Treaty claims, often still indicated on maps, have been waived (Anonymous 1961). Although this landmass covers about 14 million sq. km (about 1.5 times the area of Europe), only *ca* 0.3% is ice-free in summer (Fox & Cooper 1994), and much of that is inaccessible precipitous inland rock. South Georgia and the South Sandwich Is. (both lying north of 60° S and therefore excluded from the Antarctic Treaty Area) are a dependent territory of the United Kingdom

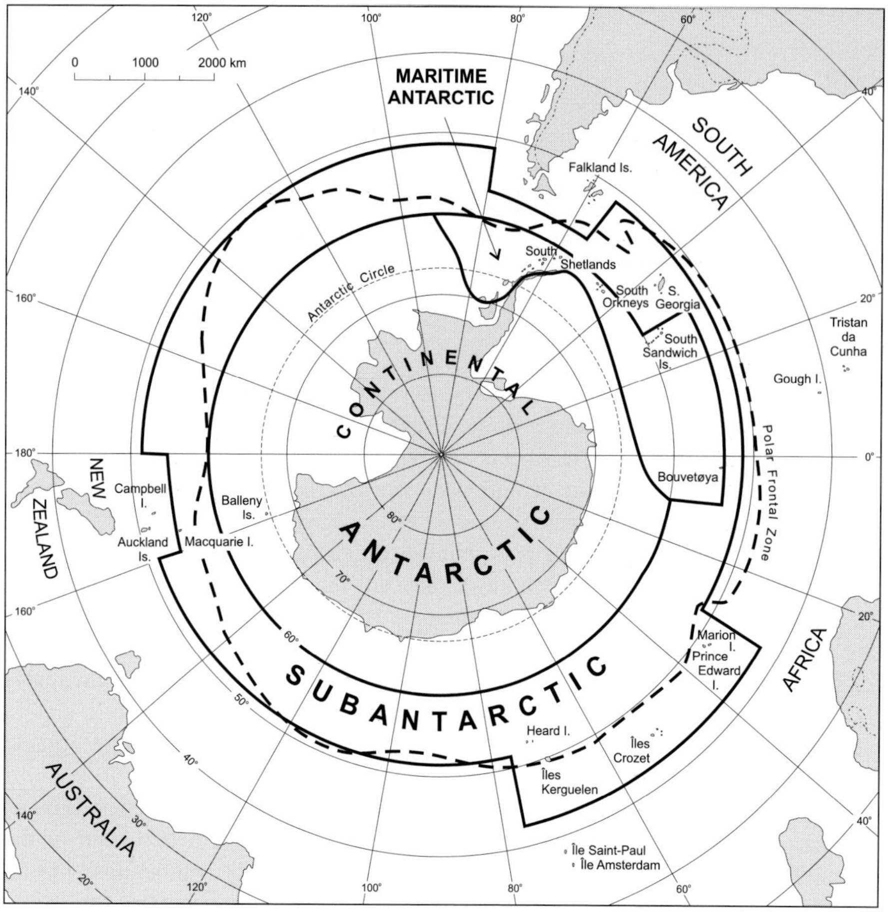

Figure 1. Antarctica, the Southern Ocean and subantarctic islands and conterminous land masses, showing the Polar Frontal Zone (broken line, also known as the Antarctic Convergence in older literature) and subdivision of the Antarctic into biogeographic zones (after Ochyra 1998*a*, modified).

situated on the Scotia Ridge, a submarine mountain range extending eastwards from the Andes Mountains in southern South America to Trinity Peninsula in the northern Antarctic Peninsula. Also on this arc rise the archipelagoes of the South Orkney Is. and South Shetland Is. Bouvetøya, a Norwegian territory, lies in the mid-South Atlantic Ocean and is also outwith the Antarctic Treaty Area.

Biogeographically, the Antarctic biome may be divided into zones which correspond to distinct climatic regions. These broadly latitudinal zones have been variously defined in several phytogeographic accounts (e.g. Skottsberg 1912*a*, 1960; Godley 1960; Wace 1960, 1965, 1969; Greene 1964*a*; Holdgate 1977; Lewis Smith 1984*a*; Pickard & Seppelt 1984*a*; Longton 1988). Greene (1968*a*) divided the 'Antarctic botanical zone' (i.e. all land south of 60° S, together with the South Sandwich Is. and Bouvetøya) into six longitudinal sectors, named after predominant geographical features. These are: Maud (from 30° W to 30° E), Enderby (30–90° E), Wilkes (90–150° E), Ross (150° E–150° W, Byrd (150–90° W), Scotia (90–30° W). The system adopted here (Table 1; Fig. 2) is taken from Lewis Smith (1984*a*). The South Sandwich Is. and Bouvetøya, although lying outside the Antarctic Treaty Area, are included in this biome because of their climatic and biological affinity with the northern maritime Antarctic botanical zone (see below).

An alternative view of the biogeographic subdivision of the Antarctic biome was presented by Bliss (1979) who developed a general classification system which embraces comparable Northern and Southern Hemisphere vegetation. He combined alpine, arctic, and antarctic vegetation into one classification system and defined the Antarctic as including closed vegetation of grass-forb, herbfields, dwarf shrub, mires and open cushion fellfields found at higher elevations. According to this system Antarctica comprises southern islands including Macquarie I., Heard I., Îles Kerguelen, Îles Crozet, Prince Edward Is. (Marion I. and Prince Edward I.), South Georgia, South Orkney Is. and South Shetland Is. Thus, in this system Antarctica and the Subantarctic are merged into a single unit. This system does not seem to be satisfactory because the impoverished southern flora has a different balance of life forms and has not gained acceptance (Lewis Smith 1984*a*).

Continental Antarctic zone

Continental Antarctica (Fig. 2) includes all of the continent and its offshore islands (including central and southern Alexander I.), but excluding the western side of the Antarctic Peninsula north of 70° S. The climate in this zone is severe with the average monthly temperature never exceeding freezing point and mean annual precipitation is *ca* 10–15 cm (water equivalent).

Because the climate of the east coast of the Antarctic Peninsula is much colder and drier than the west side, the

Table 1. Geobotanical zones of the Antarctic (modified from Lewis Smith 1984*a*).

Localities	Province	Region	Climatic features	Biotic features
Bouvetøya, South Sandwich, South Orkney and South Shetland Islands, west coast of Antarctic Peninsula to *ca* 66° S.	Maritime Antarctic	Northern	Cold moist maritime climate, mean monthly temperatures >0° C for 3–4 months in summer but rarely < –10 to –12° C in winter; precipitation 350–500 mm per annum with much as rain in summer. Snow-free period at sea level (SFP) *ca* 15–20 weeks.	Fellfield dominated by cryptogams but including small closed stands of *Deschampsia antarctica* and associated *Colobanthus quitensis*; locally diverse flora near coast; mosses form closed stands in wetter habitats, locally accumulating peat; liverworts frequent; lichens predominate in exposed drier situations and inland; snow algae and macrofungi frequent in summer. Abundant marine bird and mammal fauna; substantial invertebrate fauna dominated by mites and springtails and including two native higher insects (Diptera).
West coast of Antarctic Peninsula from *ca* 66 to 70° S, and offshore islands; also north-east coast to *ca* 64° S.	Maritime Antarctic	Southern	Cold dry maritime climate, mean monthly temperatures > 0° C for 1–3 months in summer but rarely < –15° C in winter; precipitation < 350 mm water equivalent with occasional rain. SFP *ca* 12–18 weeks.	As for Northern Region but cryptogamic diversity less and closed stands infrequent; two phanerogams relatively common and one Diptera occasional at west coast sites only; moss peat accumulation minimal.
Coastal fringe of East Antarctica and West Antarctica south of 70° S, and east coast of Antarctic Peninsula south of 64° S; includes Peter I Øy, Scott and Balleny Is.	Continental Antarctic	Coastal	Cold arid climate, mean monthly temperatures >0° C for 0–2 months in summer; winter monthly means from –5 to –25° C but some maritime influence narrowing temperature range; precipitation *c.* 100–150 mm water equivalent; rain very rare. SFP *ca* 10–15 weeks.	Semi-desert with moss and alga vegetation on moist ahumic soil; bryophyte diversity very low; lichens relatively diverse, locally forming extensive stands; snow algae occasional in some localities. Invertebrates locally diverse and abundant. Seabird colonies locally frequent and large; marine mammals abundant offshore.
Mountain and glacier zone inland from coast encircling the central ice plateaux; includes isolated nunataks in ice plateau and ice-free 'dry valleys'.	Continental Antarctica	Slope	Extremely cold and arid climate, mean monthly temperatures < –5° C in summer and < –30° C in winter; precipitation < 100 mm water equivalent; no rain. 1–3 months continual daylight in summer.	Desert with very occasional discontinuous lichen stands; mosses very rare; most vegetation associated with snow petrel colonies and microclimatically favourable oases. Some mites and nematodes; springtails very rare. Endolithic microorganisms in suitable sandstones.
Continental interior, generally >2000 m altitude.	Continental Antarctica	Ice Plateau	Extreme continental climate, all mean monthly temperatures < –15° C, < –50° C in winter; slight precipitation. Daylight as Slope Province.	Macroscopic life restricted to very occasional lichens, algae and cyanobacteria; some microorganisms. Vagrant birds very rare.

former region is considered to be transitional between maritime and continental Antarctic. Within this region are isolated small areas with a relatively diverse terrestrial biota e.g. on Joinville I., Dundee I., Vega I., James Ross I., and at Seal Nunataks, Cabinet Inlet, parts of eastern and north-western Alexander I. (attached to south-western Antarctic Peninsula by the George VI Sound Ice Shelf) and Charcot I. Parts of the coastal region of Eights Coast (notably Thurston I.) have probably the most diverse biota in continental Antarctica, and represent an important transition between those of the tectonically distinct plates of East and West Antarctica (Gilbert *et al.* 1969). There are several small areas of coastal Victoria Land which have a relatively diverse moss flora (e.g. Crater Cirque, Kay I., Harrow Peaks). Continental Antarctica may be subdivided into coastal, slope and inland provinces, although the latter is almost devoid of vegetation. The climatically more favourable coastal province has the greatest biotic diversity, while the slope and inland provinces, much of them at high altitude and with a very dry cold climate, have a macrobiota restricted almost exclusively to a few lichen taxa. Within the inland province, mosses occur only very rarely on nunataks (e.g. Siple 1938; Cameron *et al.* 1971; Ryan *et al.* 1989).

Maritime Antarctic zone

The maritime Antarctic includes the South Sandwich Is. (Fig. 3), South Orkney Is. (Fig. 4) and South Shetland Is. (Fig. 5), and the west coast of the Antarctic Peninsula and its offshore islands, to about 70° S in southern Marguerite Bay (Fig. 6). Floristically, this biological zone possesses the two vascular plants (to about 69°20′ S) (P. Convey, personal communication) and has a relatively diverse bryophyte and lichen biota (Gimingham & Lewis Smith 1970; Lewis Smith 1984*a*, 1996). It includes the climatically similar but extremely isolated small glacierised island Bouvetøya (southern Atlantic Ocean) which has an entirely cryptogamic flora. The region may be divided into northern and southern sub-zones or provinces.

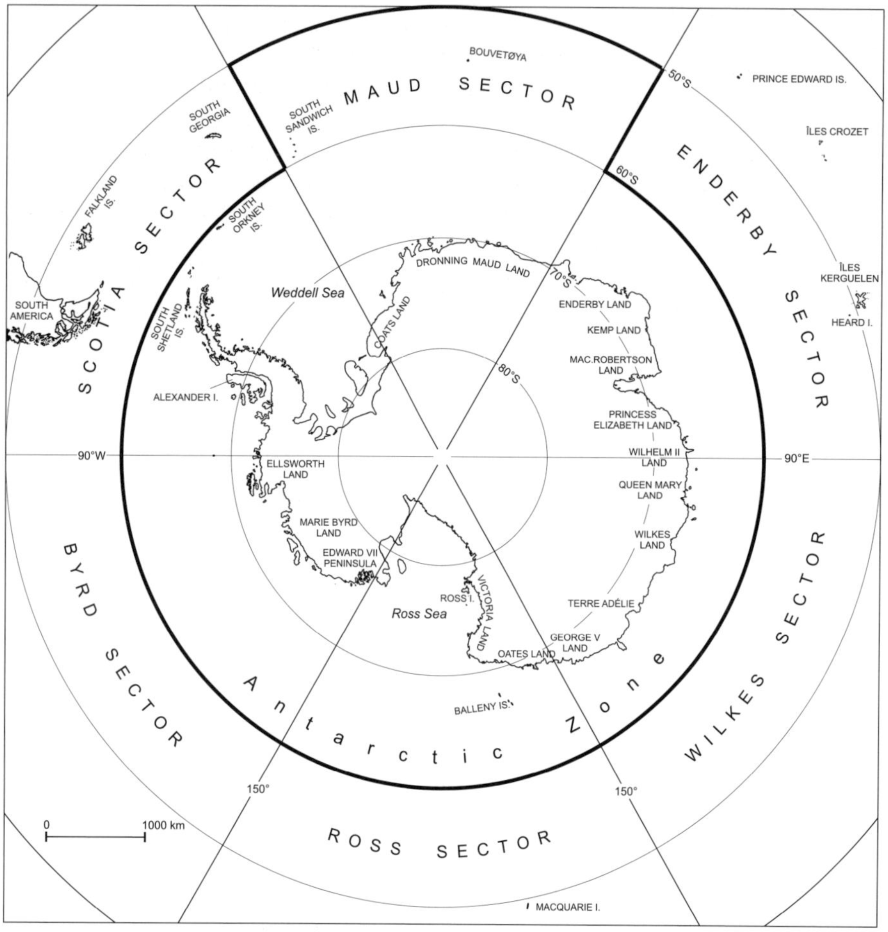

Figure 2. Map of Antarctica, showing the sectors and major regions cited in the text.

CLIMATE

Unlike the Arctic biome, the Antarctic has very cold summers which is a major factor determining the survival of terrestrial biota in the south polar regions. Like lichens, mosses are poikilohydric/poikilothermic organisms with a remarkable capacity to tolerate extreme conditions, notably low and rapidly fluctuating temperatures and tissue water contents (see Longton & Holdgate 1967; Matsuda 1968; Bayfield 1973; Proctor 1979, 1980, 1981, 1982; Longton 1980; Noakes & Longton 1988, 1989; Kappen & Valladares 1999; Schlensog & Schroeter 2000), and are consequently one of the dominant terrestrial life forms. There is a very wide latitudinal range of climatic conditions, especially of temperature, precipitation and day length, throughout the Antarctic (see Table 2). However, it should be stressed that for organisms such as mosses, the climate of the more northerly and coastal parts of the Antarctic biome, as in high Arctic and alpine regions, is not particularly stressful as they are well adapted to contend with the prevailing conditions.

Within the Antarctic, air temperatures at lowland coastal sites often exceed 0° C during the day in the short summer (generally considered as December to February), but typically drop below freezing at night, thereby frequently subjecting the biota to a diurnal freeze–thaw cycle during its most physiologically active period. Winter temperatures rarely exceed 0° C and are generally far below freezing point but, since terrestrial organisms are usually covered by an insulating blanket of snow, they remain physiologically dormant and unaffected. However, some remain physiologically active and capable of growth beneath winter snow (Collins 1973, 1977; J. Fowbert & Lewis Smith, unpublished). Some Antarctic mosses have been shown to continue photosynthesis *in situ* to –8° C and dark respiration to –8° to –10° C, beyond which both processes cease (e.g. Davis & Harrisson 1982, Schroeter *et al.* 1997). Preliminary analyses on a number of Antarctic mosses (Lewis Smith, unpublished), showed that in hydrated moss shoots within-tissue ice formation occurs between about –5 and –8° C, thereby preventing any further gas exchange. Furthermore, Kennedy (1993) showed for *Polytrichum strictum* that plants with a low tissue water content can withstand freeze–thaw cycles to –20° C better than plants with a high tissue water content, suggesting that desiccation may improve survival at low temperatures. This may also imply that exposure to sub-zero temperatures and freeze–thaw cycles may act as limiting factors on

Table 2. Approximate mean summer (December–February) and mean annual air temperatures in major bryophyte localities in Antarctica (data taken from Lewis Smith 1993*c*,*d*; King & Turner 1997; J. King, personal communication).

Region	Mean temperature (° C)	
	Summer	Year
Maritime Antarctic		
Northern		
Bouvetøya	1	−2
South Sandwich Is.	1	−2
South Orkney Is.	1	−3 to −4
South Shetland Is.	1 to 2	−2 to −3
North-western Antarctic Peninsula	0 to 1	−4 to −5
Mid-western Antarctic Peninsula	0 to 1	−3 to −4
Southern		
South-western Antarctic Peninsula	0	−5 to −7
North-eastern Antarctic Peninsula	−2	−9
Mid-eastern Antarctic Peninsula	−3	−10 to −12
South-eastern Antarctic Peninsula	−5	−16 to −20
Alexander I.	−2	−15 to −20
Continental Antarctic		
Coastal (near sea level)		
Edward VII Peninsula	−4 to −9	−13 to −20
Northern Victoria Land	−2 to −5	−12 to −15
Southern Victoria Land	−5 to −7	−17 to −20
Terre Adélie	−2	−10
Wilkes Land	−1	−8 to −9
Queen Mary Land	−3	−11
Princess Elizabeth Land	0 to −1	−10
Mac.Robertson Land	−2	−11 to −12
Enderby Land	−2	−10 to −11
Dronning Maud Land (Schirmacher Oasis)	−1 to −2	−10
Inland		
Southern Antarctic Peninsula	−10	−25
Ellsworth Mountains	−12 to −15	−25 to −30
McMurdo Dry Valleys	−5 to −10	−20 to −30
Southern Transantarctic Mountains	−10 to −20	−35 to −50

species' distribution and viability, particularly when the insulating effect of snow cover is small.

In summer, during periods of sunshine, moss cushions and turves become much warmer than the ambient air temperature and, for short spells at least, may exceed 35–40° C (Lewis Smith 1988*a*, 1999). With night temperatures often dropping to −10° C or less, such plants may be exposed to diurnal amplitudes in excess of 50° C, and annual amplitudes of perhaps 60° C to 70° C, which must

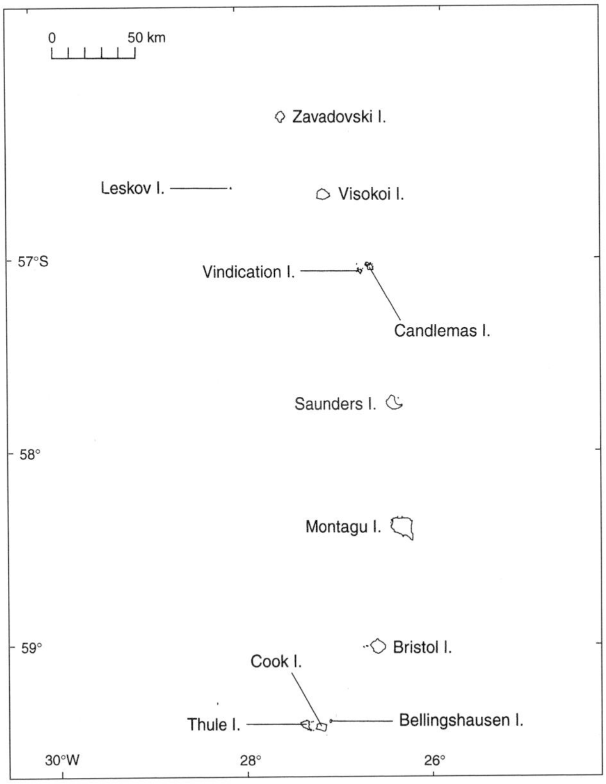

Figure 3. South Sandwich Islands.

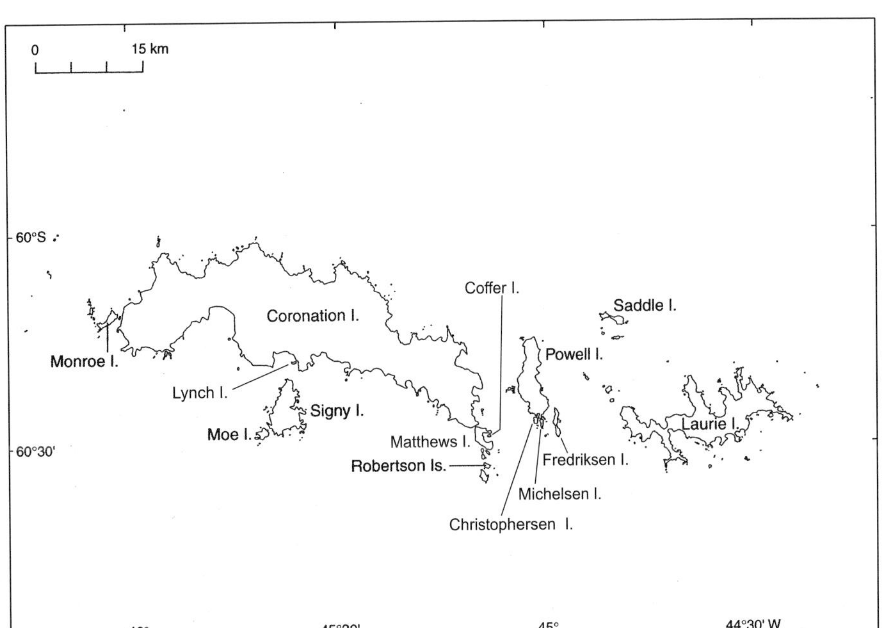

Figure 4. South Orkney Islands.

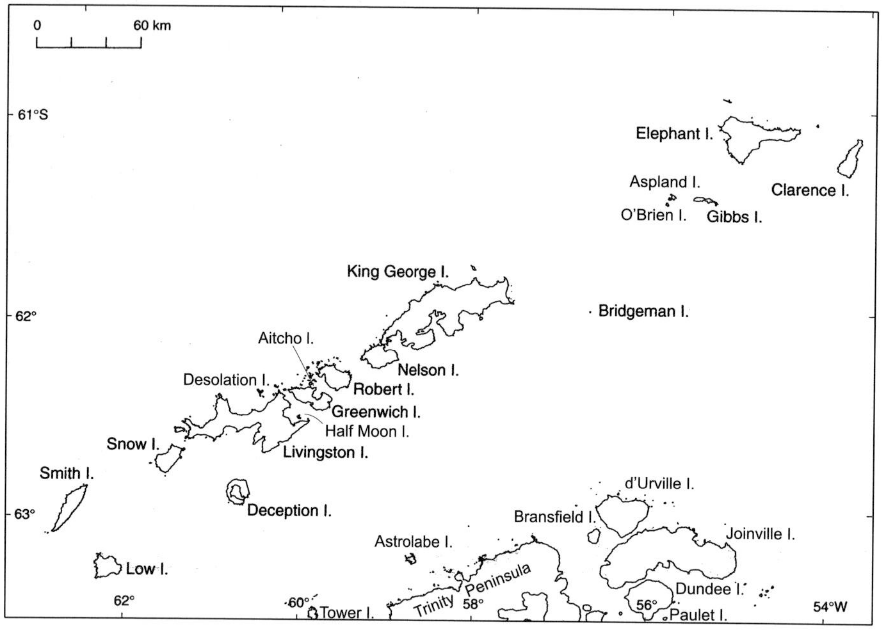

Figure 5. South Shetland Islands.

create considerable physiological stress within their tissues. Extreme air temperatures during summer may very briefly (a few hours at the most and usually during föhn wind events) reach about 12° C at coastal sites, even as far south as central Victoria Land. The approximate mean summer and annual air temperatures for most regions of the Antarctic where mosses predominate are given in Table 2. At most coastal stations with long-term meteorological records there has been a steady increase in air temperature of between 0.02° and 0.05° C per year since the 1950s (King & Turner 1997). This trend is most pronounced in the maritime Antarctic.

Precipitation in the maritime Antarctic is frequently as rain during summer, but snow can fall at any time of the year; there is deep snow cover for up to eight months, even at sea level. Farther south, rain becomes an increasingly rare phenomenon, and winter snow cover can last for eight to ten months of the year. Rainfall equivalent ranges from about 400–500 mm in the northern maritime Antarctic to 50–100 mm in the southern Antarctic Peninsula and much of coastal continental Antarctica. With increasing latitude precipitation declines, and in the *ca* 2500 sq. km McMurdo Dry Valleys region there is no precipitation for long periods, and here mosses occur only very rarely, usually around the edge of boulders where moisture is retained. Here, too, the very low temperatures and low precipitation, strong solar irradiance, and strong winds subject these plants to extremely desiccating and physiologically stressful conditions. In such environments several mosses often exist in a subpsammic habitat buried by up to a centimetre or more of wind-blown fine mineral particles.

GEOLOGY

Antarctic mosses, much more than lichens, are quite strongly influenced by geochemistry, with the exception of a rather exclusive group of species occurring on highly calcareous rock (e.g. on Signy I., South Orkney Is.). The following broad account is taken from Tingey (1991), Lewis Smith (1993*c*,*d*) and Øvstedal & Lewis Smith (2001).

Antarctica consists of two geological provinces: the ancient continental shield of East or Greater Antarctica, and the mainly magmatic arc of West or Lesser Antarctica. These are geographically separated by the 3500 km long Transantarctic Mountains. East Antarctica is composed largely of crystalline basement rocks which are intruded by a variety of igneous rocks. Almost all are Precambrian in age. By contrast, the Transantarctic Mountains contain a variety of sedimentary rocks of mainly Cambrian to Triassic age. The older Cambrian to Ordovician rocks are strongly deformed, whereas the younger Devonian to Triassic sediments of the Beacon Supergroup (Transantarctic Mountains and Prince Charles Mountains) are usually flat-lying. With the exception of limestones in the Cambrian rocks, the whole sedimentary sequence is dominated by sandstones and conglomerates, with less frequent occurrences of shales. In the coastal areas of Victoria Land and Ross I. there are exposures of late Cenozoic basaltic volcanic rocks; these include the active volcanoes of Mt. Erebus, Mt. Melbourne and Mt. Rittmann. Similar rocks are present on the coast at Gaussberg (90° E). The spectacular mountain ranges in Dronning Maud Land are mainly of granitic gneisses, migmatite, charnockite and diorites.

West Antarctica is much younger and more complex than East Antarctica, and geologically related to southern

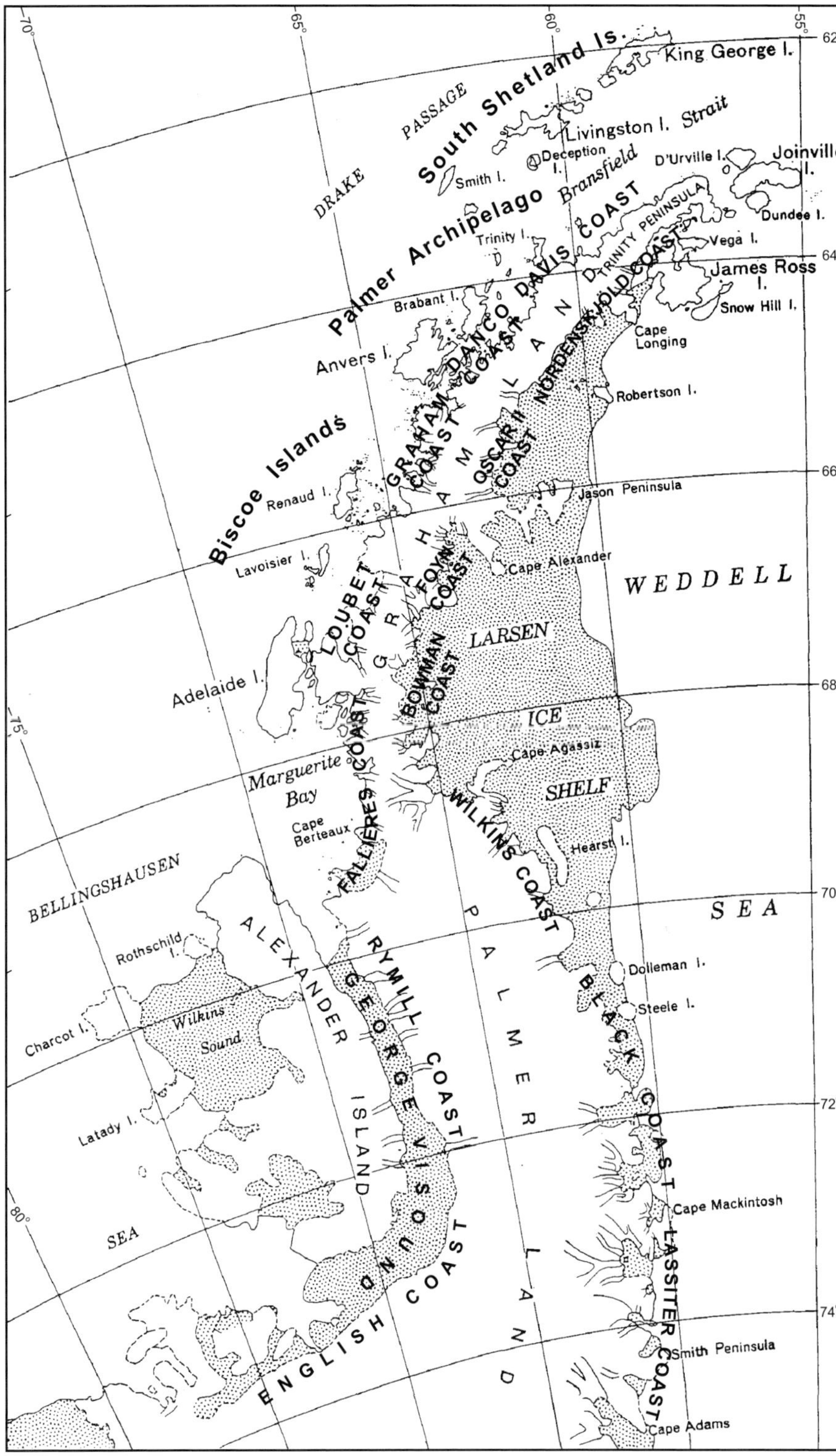

Figure 6. Antarctic Peninsula, showing the major areas cited in the text.

South America and New Zealand. It comprises five crustal blocks or provinces, of which the Antarctic Peninsula is the largest. Most of this province consists of the remnants of a long phase of magmatic arc activity. Metamorphic rocks (schists and gneisses) occur at many places. Most of the rocks forming the Antarctic Peninsula and farther south (e.g. Thurston I. crustal block) consist either of the volcanic and plutonic products of the magmatic activity or of the sedimentary rocks which accumulated in the seas around its margins. From at least Jurassic until early Cenozoic times the volcanism was dominated by andesitic rocks, with lesser amounts of rhyolites and basalts; plutonic rocks include granite, granodiorite, diorite and gabbro. Sedimentary rocks include Late Palaeozoic-Triassic shales and greywackes and Jurassic-Cretaceous fossiliferous marine shales and sandstones. Outcrops of marble (limestone) are frequent on Signy I., but very rare elsewhere in the Antarctic Peninsula region, e.g. isolated patches on the east coast of Graham Land and south-west coast of Palmer Land; calcareous nodules and concretions occur in some Antarctic Peninsula localities (e.g. James Ross I.). The Ellsworth-Whitmore Mountains crustal block is mainly sedimentary and includes limestones, sandstones and shales of Cambrian to Permian age, which had originally been deposited by rivers or in shallow seas. Late Cenozoic basaltic and other volcanic rocks occur in the James Ross I. area, the South Shetland Is., the western seaboard of the northern Antarctic Peninsula, Peter I Øy and Marie Byrd Land. The volcano of Deception I. is still active, and there is fumarolic activity here and on some of the South Sandwich Is. and Bouvetøya; all are composed of basalts and lavas.

2. History of Muscological Investigations in Antarctica

Antarctica was the final continent to be discovered and explored and, for a long time it, together with other southern polar regions, had attracted little attention from professional systematic botanists working *in situ* because of its relative inaccessibility and inhospitable environment. The responsibility for making and recording natural history observations was usually undertaken by the expedition surgeon or medical officer, some of whom made collections of and illustrated the flora and fauna during the voyages. However, botanical collections and observations were often minimal because of the nature of the landing sites and the paucity of vegetation. This also dissuaded future expeditions from sending professional botanists to make first-hand investigations.

Several phases of exploration, discovery and exploitation of the southern polar regions have been proposed (Headland 1989; Senchina 2005). Bryological exploration of Antarctica comprises four distinct phases related to the discovery and subsequent expansion of human activity in this continent.

EARLY EXPLORATION (1820–1896)

Exploration of the southern polar lands commenced around 1820 when sealing expeditions from the United States of America and Great Britain made landings in the then newly discovered South Shetland Is. off the Antarctic Peninsula. A few plants and lichens were collected and casual observations made on the nature of the vegetation and documented in the diaries of some of the ships' personnel. There were several rather vague references to the existence of moss at sites in the South Shetland Is. reported in the ships' logs and observers' accounts of the earliest sealing expeditions to that region. Unfortunately, no specimens from these landings are known to exist.

The first observation of a moss from the Antarctic was made during the third voyage of William Smith on the merchant ship *Williams* from England. On 16 October 1819 he landed on King George I., South Shetland Is., and took possession of it for King George III. Adam Young, a surgeon aboard this ship, made a short note on the island's vegetation:

> ... we met with no sort of vegetation except here and there small patches of stunted grass growing upon the surface of the thick coat of dung which the sea-fowls left in the crevices of the rocks, and a species of a moss, which occasionally we met with adhering to the rocks themselves (Young 1821).

Also in that summer another report of moss in the South Shetland Is. was made by Thomas Maine Bone, midshipman on Edward Bransfield's Antarctic voyage, 1819–1820 (Anonymous 1821, 1946). On 23 January 1820 on Penguin I., in King George Bay, King George I., it was reported that '... some mosses were found and taken on board'. Again, there is no record of these specimens ever reaching a herbarium or being identified. A few days later in this general area, Bone also observed:

> The swampy land, the lowest of all, was covered with a sort of grass and moss, nourished by the dung of several oceanic birds; this moss and grass abounds in great quantities, and is all that deserves to be called vegetation (Anonymous 1821).

There is also reference to brown skuas building a 'nest of moss'.

During the sealing voyage of the *Lady Trowbridge* the vessel was wrecked off Cape Melville in the north-east of King George I. on 25 December 1820 and the captain, Richard Sherratt, made some observations on the island's vegetation:

> I wish I could say something in favour of its vegetable productions; but, alas! little or nothing of that can be said. There is not a tree, not a bush, not a shrub, not a flower, in all the islands. There is a little coarse moss here and there, and in Potter's Cove there is a small plot of land with a little grass of a small kind, and very short (Sherratt 1821; Anonymous 1952).

In January 1821 the Russian Antarctic Expedition of 1819–1821 with the ships *Vostok* and *Mirnyy*, led by Thaddeus Th. Bellingshausen, made a landing at North Foreland, in the north-east of King George I. Bellingshausen stated in the narrative of the expedition, published

Figure 7. *Sanionia uncinata*, one of the two oldest extant moss specimens collected in the South Shetland Is. in the Antarctic by J. Eights and deposited at NY. The other, from the same location, is *Polytrichastrum alpinum*.

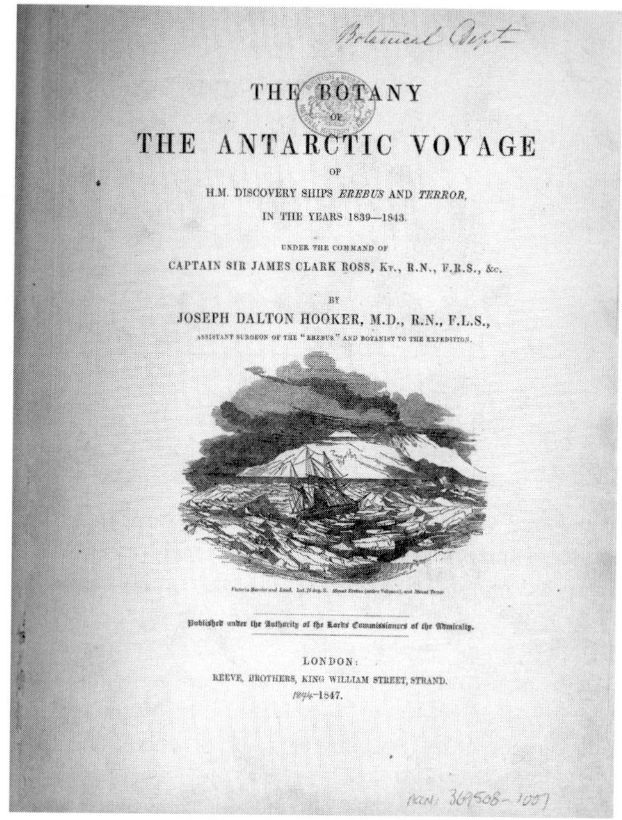
THE BOTANY
OF
THE ANTARCTIC VOYAGE
OF
H.M. DISCOVERY SHIPS *EREBUS* AND *TERROR*,
IN THE YEARS 1839—1843.
UNDER THE COMMAND OF
CAPTAIN SIR JAMES CLARK ROSS, Kt., R.N., F.R.S., &c.
BY
JOSEPH DALTON HOOKER, M.D., R.N., F.L.S.,
ASSISTANT SURGEON OF THE "EREBUS" AND BOTANIST TO THE EXPEDITION.

Published under the Authority of the Lords Commissioners of the Admiralty.

LONDON:
REEVE, BROTHERS, KING WILLIAM STREET, STRAND.
1844–1847.

Figure 9. Title page of Hooker's (1847) *The botany. The Antarctic voyage of the H.M. discovery ships* Erebus *and* Terror *in the years 1839–1843. Part I. Flora Antarctica.*

Figure 8. Joseph Dalton Hooker (1817–1911), the first professional botanist who collected mosses in the Antarctic. Courtesy of Gestion Fonds Photos (Cryptogamie), Département Systématique et Evolution, Muséum National d'Histoire Naturelle, Paris.

in 1831, that he had sent ashore three shipmates who returned with moss specimens, noting that:

> The shore consisted of rocks covered with crumbling soil and moss; they observed no flora except the moss (Bellingshausen 1831; see Debenham 1945).

Some years later, an unnamed moss was observed on Deception I. by W. H. B. Webster during the British Naval Expedition of 1828–1831 on the *Chanticleer* captained by Henry Foster. He described the vegetation of this island as including:

> ... only the growth of diminutive moss, and a striped coralloid lichen [= *Usnea* sp.], identically the same as that which is found on the lofty hills of Cape Horn and Tierra del Fuego (Webster 1834).

Despite its generality, this is the first important phytogeographical observation recognising an affinity of the flora and vegetation of the maritime Antarctic with those of the southernmost tip of South America.

The earliest account which refers to a moss by name was by James Eights (1798–1882), naturalist to the U.S. Exploring Expedition of 1829–1831 (Hedgpeth 1971):

> A species of Polytrichum resembling the alpinum of Lin. one or two lichens and a fucus found in the sea, along the shores – when you add to these an occasional plant of a small species of avena, you complete the botanical catalogue of the islands (Eights 1833).

On King George I. Eights collected *Polytrichastrum alpinum* and this specimen is now housed in the National

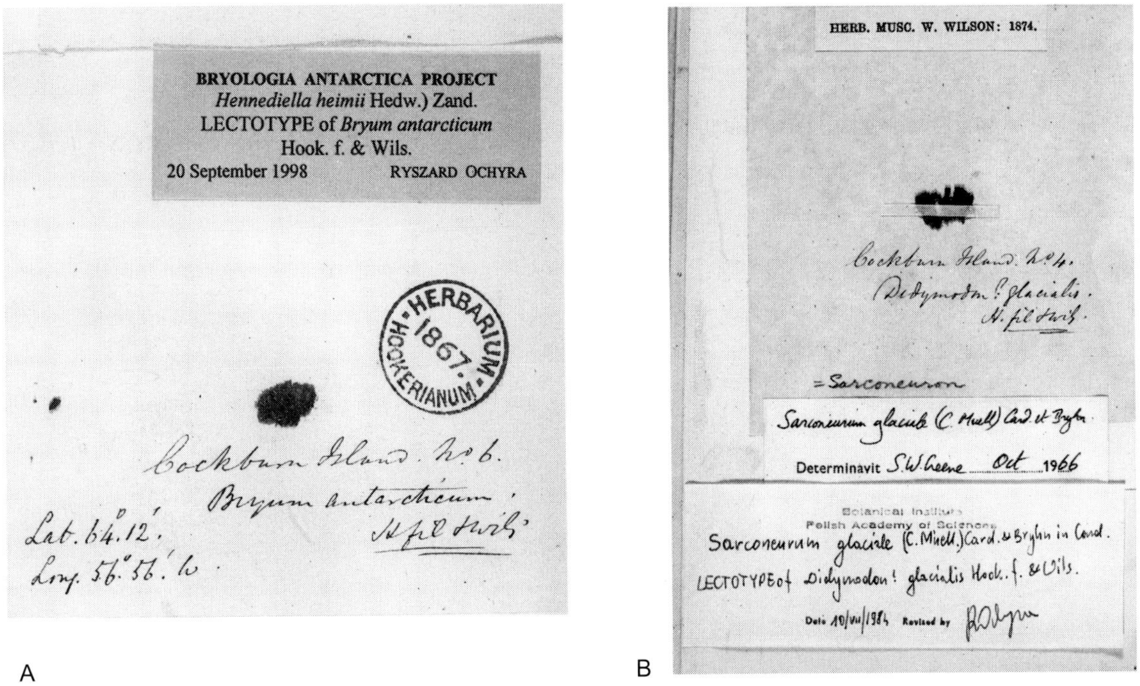

Figure 10. The type specimens of *Bryum antarcticum* (A) and *Didymodon glacialis* (B) collected by J. D. Hooker on Cockburn I. in January 1843 during the voyage of the Antarctic Expedition 1839–1843 under the command of James Clark Ross (see Plate 6).

Museum of Natural History in Washington (US) and a duplicate is in the New York Botanical Garden Herbarium (NY). Additionally, he collected a specimen of *Sanionia uncinata* (Fig. 7) which is also preserved in the latter herbarium.

The first professional botanist to set foot on land south of lat. 60° S was Joseph Dalton Hooker (1817–1911) (Fig. 8), surgeon-botanist to James Clark Ross' British expedition on the *Terror* and *Erebus* in 1839–1843. Hooker made the first comprehensive collection of mosses and lichens from an Antarctic locality. His collection included five mosses gathered on Cockburn I. (lat. 64°12′S, long. 56°51′W) off north-east Antarctic Peninsula (Plate 6), and these were reported in *Flora Antarctica* (Wilson & Hooker 1847) (Fig. 9). Specimens included (Table 4) *Hennediella heimii* (Hedw.) R.H.Zander (as *Bryum antarcticum* Hook.f. & Wilson), *Syntrichia sarconeurum* Ochyra & R.H.Zander (as *Didymodon ? glacialis* Hook.f. & Wilson), *S. magellanica* (Mont.) R.H.Zander (as *Tortula laevipila* (Brid.) Schwägr. var. 3 and 4), *Didymodon brachyphyllus* (Sull.) R.H.Zander (as *Tortula gracilis* Hook. & Grev.) and *Bryum argenteum* Hedw. var. *argenteum*. Of these *Bryum antarcticum* (Fig. 10A) and *Didymodon glacialis* (Fig. 10B) were described as new to science.

The very small number of mosses discovered during these early expeditions was a reflection of the emphasis on geographical exploration and discovery, and exploitation of the marine resources, particularly seals, rather than a biological discovery. Also, at this time (early-mid 1800s) weather and ice conditions at sea were much more severe than today as these southern regions were still emerging from the 'Little Ice Age'. This had a major effect on where expeditions in sail ships could reach and, even when landings were possible, any vegetation may have been covered by snow as a result of severe summer weather.

HEROIC AGE (1897–1918)

For over half a century after Hooker's visit to the Antarctic there were no scientific expeditions, and no major bryophyte discoveries were made during this time. However, in 1895 the International Geographical Congress, London, announced that the Antarctic regions remained the greatest challenge to geographical exploration, and adopted a resolution recommending that scientific societies throughout the world should urge their governments in whatever way would be most effective to undertake such exploration, for the benefit of all branches of science, before the close of the nineteenth century. This resulted in a number of national expeditions to the Antarctic, most of which made small, but significant, contributions to bryological knowledge of the biome. This became known as the 'Heroic Age' of Antarctic exploration and accounts of these were published by several of the leading bryophyte taxonomists of the time (see Tables 3–4), of which the most significant were those by the French bryologist Jules Cardot (1860–1934) (Fig. 11).

The first large collection of mosses in Antarctica was

Table 3. Synopsis of published moss collections made in the Antarctic between 1829 and 1941.

No.	Expedition and years	Ship and Captain or Leader	Total number of species	Total number of infraspecific taxa	Number of species new to science	Number of infraspecific taxa new to science	Number of species new to Antarctica	Publication(s)
1.	U.S. sealing voyage 1829–1831	B. Pendleton (*Seraph*) & N. B. Palmer (*Annawan*)	2	–	–	–	1	Eights (1833)
2.	British naval expedition 1839–1843	J. C. Ross (*Terror* & *Erebus*)	5	–	2	–	5	Wilson & Hooker (1847)
3.	Belgian Antarctic Expedition 1897–1899	A. V. de Gerlache de Gomery (*Belgica*)	27	6	16	3	21 (+ 1 variety)	Cardot (1900, 1901)
4.	British Antarctic Expedition 1898–1900	C. E. Borchgrevink & B. Jensen (*Southern Cross*)	3	–	1	–	–	Gepp (1902); Bryhn (1902)
5.	Swedish South Polar Expedition 1901–1904	N. O. G. Nordenskjöld (*Antarctic*)	23	8	5	8	6	Skottberg (1905); Cardot (1906*a*, 1908*a*)
6.	German South Polar Expedition 1901–1903	E. D. von Drygalski (*Gauss*)	1	–	1	–	–	Brotherus (1906*a*)
7.	British National Antarctic Expedition 1901–1904	R. F. Scott (*Discovery*)	7	–	2	–	–	Cardot (1907*b*)
8.	Scottish National Antarctic Expedition 1902–1904	W. S. Bruce (*Scotia*)	13	–	–	–	1	Wright (1905*b*); Cardot (1911*c*)
9.	French Antarctic Expedition 1903–1905	J.-B. E. A. Charcot (*Français*)	18	7	1	2	2	Cardot (1906*b*, 1907*a*)
10.	British Antarctic Expedition 1907–1909	E. H. Shackleton (*Nimrod*)	4	–	–	–	1	Cardot (1910*b*)
11.	French Antarctic Expedition 1908–1910	J.-B. E. A. Charcot (*Pourquoi Pas?*)	34	2	7	1	4	Cardot (1911*a*,*b*, 1913)
12.	Australasian Antarctic Expedition 1911–1914	D. Mawson (*Aurora*)	5	–	–	–	–	Dixon & Watts (1918)
13.	Imperial Trans-Antarctic Expedition 1914–1916	E. H. Shackleton (*Endurance*)	2	–	–	–	–	Darbishire (1923)
14.	Naval work for the Falkland Islands Dependencies 1917–1919	J. C. Robins	8	–	1	–	–	Dixon (1920)
15.	Discovery Expedition 1933–1935	N. A. Mackintosh (*Discovery II*)	10	–	–	–	1	Dixon (1935)
16.	U.S. Antarctic Expedition 1933–1935	R. E. Byrd (*Bear of Oakland*)	5	2	2	2	1	Bartram (1938)
17.	U.S. Antarctic Service Expedition (1939–1941)	R. E. Byrd (*North Star* & *Bear*)	24	4	1	1	2	Bartram (1957)

Figure 11. Jules Cardot (1860–1934), the eminent French bryologist who named many early collections of mosses from Antarctica. Courtesy of Gestion Fonds Photos (Cryptogamie), Département Systématique et Evolution, Muséum National d'Histoire Naturelle, Paris.

made in 1898 in the Gerlache Strait region on the western Antarctic Peninsula by Émile G. Racovitza (1868–1947), naturalist to the Belgian (*Belgica*) Expedition, 1897–1899. It was studied by Cardot (1900, 1901) who named 27 species and six varieties, including 16 species and three varieties new to science (Fig. 12). These were described in detail and fully illustrated in coloured plates (Fig. 13). According to modern taxonomic concepts, this work contributed in total 20 species and one variety of the muscoflora of Antarctica, but of the newly described species only one, *Andreaea depressinervis*, is currently accepted as a distinct species.

The British (*Southern Cross*) Expedition, 1898–1900, explored the bryologically depauperate region of northern Victoria Land on the Antarctic continent. Only three species were gathered by Carstens Borchgrevink (1864–1934), leader of the expedition, namely *Syntrichia magellanica*, *S. sarconeurum* and *Bryum argenteum* (Gepp 1902), all having been formerly recorded from the Antarctic Peninsula region by different names. However, on the basis of this small collection Bryhn (1902) described the monotypic genus *Sarconeurum*, with a single species *S. antarcticum*, which has long been considered endemic to the Antarctic biome.

In the period before and during World War I eleven national expeditions worked in the Antarctic and made moss collections (Table 3). Of these, the most remarkable results were achieved by the Swedish South Polar Expedition, 1901–1903, during which mosses were collected by Carl Skottsberg (1880–1963) in the South Shetland Is.

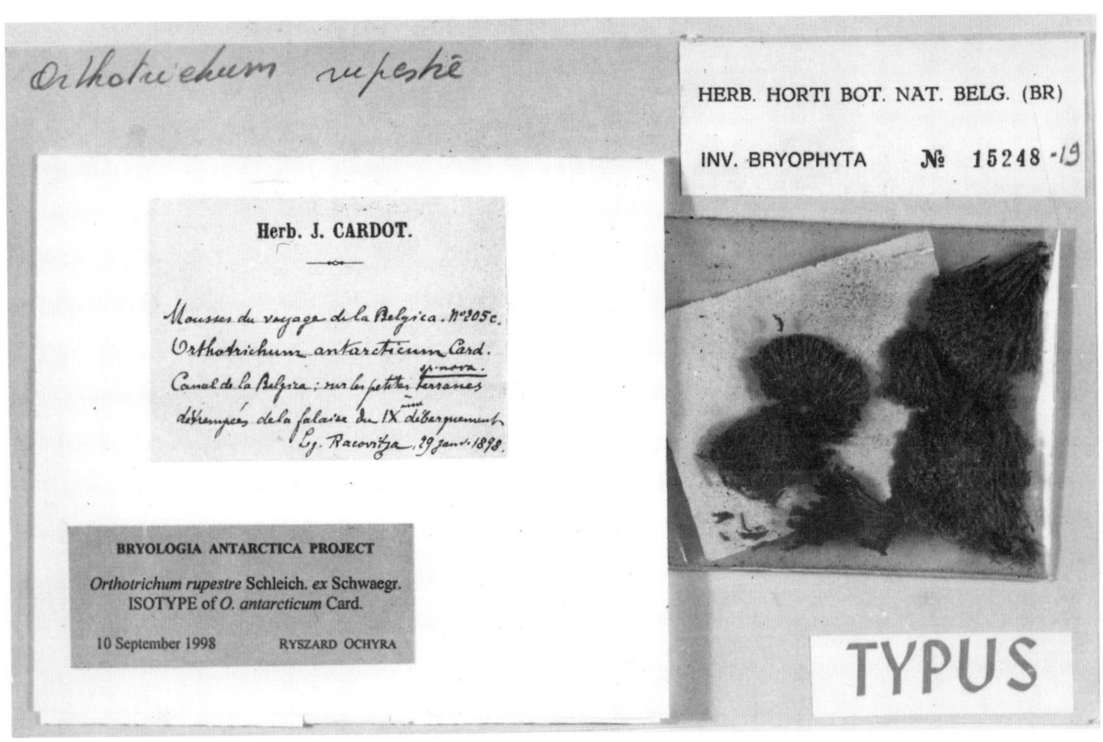

Figure 12. The isotype specimen of *Orthotrichum antarcticum* collected by É. Racovitza in January 1898 on Cape Anna on the Antarctic Peninsula during the voyage of the Belgian Antarctic Expedition 1897–1899.

Table 4. Chronological synopsis of the moss taxa published from the Antarctic between 1833 and 2007. Names in capital letters refer to currently accepted taxa; names in boldface refer to taxa described from the Antarctic material; names in capital letters and boldface refer to taxa described from the Antarctic and accepted in modern taxonomy; underlined names refer to currently accepted species and indicate their first Antarctic report under this name, though sometimes in a different genus.

No.	Recorded species	Publication	Current status	First synonymisation or determination
		EARLY EXPLORATION (1829–1896)		
1.	*Polytrichum* cf. *alpinum* Hedw.	Eights (1833: p. 65)	<u>*POLYTRICHASTRUM ALPINUM*</u> (Hedw.) G.L.Sm.	———
2.	***Didymodon glacialis*** Hook.f. & Wilson, *hom. illeg.* [*non* (Brid.) Wallr. 1827]	Wilson & Hooker (1847: p. 408)	*SYNTRICHIA SARCONEURUM* Ochyra & R.H.Zander	———
3.	*Tortula laevipila* (Brid.) Schwägr. var. 3 & 4	Wilson & Hooker (1847: p. 408)	*SYNTRICHIA MAGELLANICA* (Mont.) R.H.Zander	Present work
4.	*Tortula gracilis* of Wilson & Hook.f. [*non* Hook. & Grev. 1824]	Wilson & Hooker (1847: p. 408)	*DIDYMODON BRACHYPHYLLUS* (Sull.) R.H.Zander	Ochyra & Zander (2002: p. 38)
5.	<u>*BRYUM ARGENTEUM*</u> Hedw.	Wilson & Hooker (1847: p. 413)	———	———
6.	***Bryum antarcticum*** Hook.f. & Wilson	Wilson & Hooker (1847: p. 414)	*HENNEDIELLA HEIMII* (Hedw.) R.H.Zander	Kanda (1981*b*: p. 102)
		HEROIC AGE (1897–1918)		
7.	***Andreaea pycnotyla*** Cardot	Cardot (1900: p. 42)	*ANDREAEA REGULARIS* Müll.Hal.	Greene (1968*b*: p. 143)
8.	***Andreaea pygmaea*** Cardot	Cardot (1900: p. 43)	*Andreaea regularis* Müll.Hal.	Greene (1968*b*: p. 143)
9.	***ANDREAEA DEPRESSINERVIS*** Cardot	Cardot (1900: p. 43)	———	———
10.	<u>*CERATODON PURPUREUS*</u> (Hedw.) Brid.	Cardot (1900: p. 43)	———	———
11.	***Ceratodon antarcticus*** Cardot	Cardot (1900: p. 43)	*Ceratodon purpureus* (Hedw.) Brid.	Ochyra (1998*a*: p. 103)
12.	*Distichium capillaceum* (Hedw.) Bruch & Schimp. var. *brevifolium* Bruch & Schimp.	Cardot (1900: p. 43)	*DISTICHIUM CAPILLACEUM* (Hedw.) Bruch & Schimp.	Arnott (1825: p. 36)
13.	***Orthotrichum antarcticum*** Cardot	Cardot (1900: p. 43)	*ORTHOTRICHUM RUPESTRE* Schwägr.	Ochyra (1986: p. 144)
14.	*Orthotrichum rupicola* Müll.Hal.	Cardot (1900: p. 43)	*Orthotrichum rupestre* Schwägr.	Lewinsky (1978: p. 63)
15.	***Webera cruda*** (Hedw.) Schwägr. var. ***imbricata*** Cardot	Cardot (1900: p. 43)	*POHLIA CRUDA* (Hedw.) Lindb.	Ochyra (1998*a*: p. 198)
16.	***Webera racovitzae*** Cardot	Cardot (1900: p. 44)	*POHLIA NUTANS* (Hedw.) Lindb.	Greene *et al.* (1970: p. 19)
17.	***Webera gerlachei*** Cardot	Cardot (1900: p. 44)	*BRYUM PSEUDOTRIQUETRUM* (Hedw.) P.Gaertn., B.Mey. & Scherb.	Ochi (1970: p. 44)
18.	***Bryum imperfectum*** Cardot	Cardot (1900: p. 44)	*BRYUM ARCHANGELICUM* Bruch & Schimp.	Present work
19.	***Bryum inconnexum*** Cardot	Cardot (1900: p. 44)	*BRYUM PALLESCENS* Schwägr.	Present work
20.	***Bryum austropolare*** Cardot	Cardot (1900: p. 45)	*Bryum pseudotriquetrum* (Hedw.) P.Gaertn., B.Mey. & Scherb.	Ochi (1970: p. 44)
21.	***Bryum amblyolepis*** Cardot	Cardot (1900: p. 45)	*BRYUM ARGENTEUM* Hedw. var. *MUTICUM* Brid.	Present work
22.	*Polytrichum alpinum* var. *brevifolium*	Cardot (1900: p. 45)	*Polytrichastrum alpinum* (Hedw.) G.L.Sm.	Greene *et al.* (1970: p. 26, 50)
23.	***Polytrichum subpiliferum*** Cardot	Cardot (1900: p. 45)	*POLYTRICHUM JUNIPERINUM* Hedw.	Greene *et al.* (1970: p. 26, 50)
24.	***Polytrichum antarcticum*** Cardot	Cardot (1900: p. 45)	*POLYTRICHUM PILIFERUM* Hedw.	Greene *et al.* (1970: p. 29, 52)
25.	<u>*POLYTRICHUM STRICTUM*</u> Brid.	Cardot (1900: p. 45)	———	———
26.	***Pseudoleskea antarctica*** Cardot	Cardot (1900: p. 45)	*CRATONEUROPSIS RELAXA* (Hook.f. & Wilson) Broth. subsp. *MINOR* (Wilson & Hook.f.) Ochyra	Present work
27.	*Brachythecium georgicoglareosum* Müll.Hal.	Cardot (1900: p. 46)	Erroneous report	Cardot (1901: p. 42)
28.	***Brachythecium antarcticum*** Cardot	Cardot (1900: p. 46)	*BRACHYTHECIUM AUSTROSALEBROSUM* (Müll.Hal.) Kindb.	Newton (1979*a*: p. 138)
29.	***Brachythecium antarcticum*** var. ***cavifolium*** Cardot	Cardot (1900: p. 46)	*Brachythecium austrosalebrosum* (Müll.Hal.) Kindb.	Newton (1979*a*: p. 138)
30.	***Amblystegium densissimum*** Cardot	Cardot (1900: p. 46)	*PLATYDICTYA JUNGERMANNIOIDES* (Brid.) Berk.	Ochyra (1999*k*: p. 562)
31.	*Hypnum uncinatum* Hedw.	Cardot (1900: p. 46)	<u>*SANIONIA UNCINATA*</u> (Hedw.) Loeske	———
32.	*Hypnum austrostramineum* Müll.Hal.	Cardot (1900: p. 46)	*WARNSTORFIA FONTINALIOPSIS* (Müll.Hal.) Ochyra	Ochyra & Matteri (2001: p. 72)

33.	*HYPNUM REVOLUTUM* (Mitt.) Lindb.	Cardot (1900: p. 46)	———	———
34.	***Andreaea depressinervis*** var. ***compacta*** Cardot	Cardot (1901: p. 23)	*Andreaea depressinervis* Cardot	Greene (1968*b*: p. 110)
35.	*Grimmia donniana* of Cardot [*non* Sm. 1804]	Cardot (1901: p. 30)	*ORTHOGRIMMIA SESSITANA* (De Not.) Ochyra & Żarnowiec	Present work
36.	*Webera nutans* Hedw.	Cardot (1901: p. 34)	*Pohlia nutans* (Hedw.) Lindb.	———
37.	*Hypnum uncinatum* Hedw. fo. *gracile* Renauld, *nom. inval.*	Cardot (1901: p. 44)	*Sanionia uncinata* (Hedw.) Loeske	Present work
38.	*Hypnum austrostramineum* Müll.Hal. var. *gracillimum* Müll.Hal.	Cardot (1901: p. 44)	*Warnstorfia fontinaliopsis* (Müll.Hal.) Ochyra	Ochyra & Matteri (2001: p. 72)
39.	*Barbula fuegiana* (Mitt.) A.Jaeger	Gepp (1902: p. 319)	*Syntrichia magellanica* (Mont.) R.H.Zander	Lightowlers (1986*b*: p. 61)
40.	***Sarconeurum antarcticum*** Bryhn	Bryhn (1902: p. 205)	*Syntrichia sarconeurum* Ochyra & R.H.Zander	Cardot (1907*b*: p. 3)
41.	*Campylopus introflexus* of Wright [*non* Hedw.) Brid.]	Wright (1905*b*: 104)	*CHORISODONTIUM ACIPHYLLUM* (Hook.f. & Wilson) Broth.	Cardot (1908*a*: p. 265)
42.	*Campylopus vesticaulis* of Wright [*non* Mitt.]	Wright (1905*b*: 104)	*Chorisodontium aciphyllum* (Hook.f. & Wilson) Broth.	Cardot (1908*a*: p. 266)
43.	*Grimmia amblyophylla* of Wright [*non* Müll.Hal.]	Wright (1905*b*: 104)	*Schistidium antarctici* (Cardot) L.I.Savicz & Smirnova	Cardot (1908*a*: p. 271)
44.	*Grimmia* cf. *apocarpa* of Wright [*non* Hedw.]	Wright (1905*b*: 104)	*Hymenoloma grimmiaceum* (Müll.Hal.) Ochyra	Present work
45.	***Brachythecium antarcticum*** Cardot fo. ***robustum*** Skottsb.	Skottsberg (1905: p. 413)	*Brachythecium austrosalebrosum* (Müll.Hal.) Kindb.	Present work
46.	*Andreaea regularis* Müll.Hal.	Cardot (1906*a*: p. 28)	———	———
47.	*Dicranoweisia grimmiacea* (Müll.Hal.) Broth.	Cardot (1906*a*: p. 29)	*HYMENOLOMA GRIMMIACEUM* (Müll.Hal.) Ochyra	———
48.	***Dicranoweisia grimmiacea*** (Müll.Hal.) Broth. fo. ***brevifolia*** Cardot	Cardot (1906*a*: p. 29)	*Hymenoloma grimmiaceum* (Müll.Hal.) Ochyra	Ochyra (1998*a*: p. 128)
49.	***Dicranoweisia grimmiacea*** (Müll.Hal.) Broth. fo. ***latifolia*** Cardot	Cardot (1906*a*: p. 29)	*Hymenoloma grimmiaceum* (Müll.Hal.) Ochyra	Ochyra (1998*a*: p. 128)
50.	***Dicranum nordenskioldii*** Cardot	Cardot (1906*a*: p. 29)	*Chorisodontium aciphyllum* (Hook.f. & Wilson) Broth.	Bell (1973*a*: p. 34)
51.	***Dicranum nordenskioldii*** fo. ***minus*** Cardot	Cardot (1906*a*: p. 29)	*Chorisodontium aciphyllum* (Hook.f. & Wilson) Broth.	Ochyra (1999*f*: p. 512)
52.	***Blindia skottsbergii*** Cardot fo. ***robusta*** Cardot	Cardot (1906*a*: p. 29)	*Chorisodontium aciphyllum* (Hook.f. & Wilson) Broth.	Ochyra (1993*a*: p. 507)
53.	***Ceratodon grossiretis*** Cardot	Cardot (1906*a*: p. 29)	*Ceratodon purpureus* (Hedw.) Brid.	Ochyra (1998*a*: p. 103)
54.	***Ceratodon grossiretis*** var. ***validus*** Cardot		*Ceratodon purpureus* (Hedw.) Brid	Ochyra (1998*a*: p. 103)
55.	***Tortula excelsa*** Cardot	Cardot (1906*a*: p. 30)	*SYNTRICHIA FILARIS* (Müll.Hal.) R.H.Zander	Lightowlers (1985*b*: p. 49, 53)
56.	*Tortula monoica* Cardot	Cardot (1906*a*: p. 30)	*Syntrichia magellanica* (Mont.) R.H.Zander	Lightowlers (1986*b*: p. 61)
57.	***Grimmia antarctici*** Cardot	Cardot (1906*a*: p. 30)	*SCHISTIDIUM ANTARCTICI* (Cardot) L.I.Savicz & Smirnova	———
58.	*Bartramia pycnocolea* Müll.Hal.	Cardot (1906*a*: p. 31)	*BARTRAMIA PATENS* Brid.	Robinson (1972*a*: p. 174)
59.	***Bryum cephalozioides*** Cardot	Cardot (1906*a*: p. 31)	*Bryum argenteum* var. *muticum* Brid.	Present work
60.	***Pogonatum alpinum*** (Hedw.) Röhl. fo. ***elatum*** Cardot	Cardot (1906*a*: p. 31)	*Polytrichastrum alpinum* (Hedw.) G.L.Sm.	Greene *et al.* (1970: p. 26, 50)
61.	*BRACHYTHECIUM SUBPILOSUM* (Hook.f. & Wilson) A.Jaeger	Cardot (1906*a*: p. 32)	———	———
62.	*Hypnum sarmentosum* Wahlenb.	Cardot (1906*a*: p. 32)	*WARNSTORFIA SARMENTOSA* (Wahlenb.) Hedenäs	———
63.	***Webera racovitzae*** Cardot var. ***laxiretis*** Cardot	Cardot (1906*a*: p. 34)	*Bryum pseudotriquetrum* (Hedw.) P.Gaertn., B.Mey. & Scherb.	Ochyra (1998*a*: p. 194)
64.	***Bryum inconnexum*** Cardot var. ***tomentosum*** Cardot	Cardot (1906*b*: p. 34)	*Bryum pallescens* Schwägr.	Present work
65.	*BRACHYTHECIUM AUSTROGLAREOSUM* (Müll.Hal.) Kindb.	Cardot (1906*b*: p. 34)	———	———
66.	***Brachythecium turquetii*** Cardot	Cardot (1906*b*: p. 34)	*Warnstorfia fontinaliopsis* (Müll.Hal.) Ochyra	Present work
67.	*Hypnum uncinatum* Hedw. fo. *polare* (Renauld) Cardot	Cardot (1906*b*: p. 35)	*SANIONIA GEORGICOUNCINATA* (Müll.Hal.) Ochyra & Hedenäs	Present work
68.	***Bryum filicaule*** Broth.	Brotherus (1906: p. 91)	*Bryum pseudotriquetrum* (Hedw.) P.Gaertn., B.Mey. & Scherb.	Present work
69.	*Bartramia diminutiva* Müll.Hal.	Cardot (1907*a*: p. 3)	*Bartramia patens* Brid.	Robinson (1972*a*: p. 174)
70.	*Brachythecium georgicoglareosum* (Müll.Hal.) Kindb.	Cardot (1907*a*: p. 7)	*Brachythecium austrosalebrosum* (Müll.Hal.) Kindb.	Newton (1979*a*: p. 138)
71.	***Didymodon gelidus*** Cardot	Cardot (1907*b*: p. 4)	*Didymodon brachyphyllus* (Sull.) R.H.Zander	Zander & Ochyra (2001: p. 374)

(cont.)

Table 4. Continued.

No.	Recorded species	Publication	Current status	First synonymisation or determination
72.	***Bryum algens*** Cardot	Cardot (1907*b*: p. 5)	*Bryum pseudotriquetrum* (Hedw.) P.Gaertn., B.Mey. & Scherb.	Ochyra (1998*a*: p. 194)
73.	***Andreaea depressinervis*** Cardot var. ***robusta*** Cardot	Cardot (1908*a*: p. 263)	*Andreaea depressinervis* Cardot	Greene (1968*b*: p. 140)
74.	*Blindia skottsbergii* Cardot	Cardot (1908*a*: p. 266)	*Chorisodontium aciphyllum* (Hook.f. & Wilson) Broth.	Ochyra (1993*a*: p. 507)
74.	*Grimmia apocarpa* of Cardot [*non* Hedw.]	Cardot (1908*a*: p. 270)	*Schistidium antarctici* (Cardot) L.I.Savicz & Smirnova	Present work
76.	*Polytrichum strictum* Brid. var. *alpestre* (Hoppe) Rabenh.	Cardot (1908*a*: p. 278)	*Polytrichum strictum* Brid.	Host (1831: p. 718)
77.	*DICRANELLA HOOKERI* (Müll.Hal.) Cardot	Cardot (1910*b*: p. 77)	———	———
78.	***Philonotis gourdonii*** Cardot	Cardot (1911*a*: p. 600)	*PHILONOTIS POLYMORPHA* (Müll.Hal.) Broth.	Present work
79.	***Tortula heteroneura*** Cardot	Cardot (1911*a*: p. 601)	*Syntrichia magellanica* (Mont.) R.H.Zander	Present work
80.	***Bryum perangustidens*** Cardot	Cardot (1911*a*: p. 601)	*Bryum pseudotriquetrum* (Hedw.) P.Gaertn., B.Mey. & Scherb.	Ochi (1979: p. 77)
81.	***ANDREAEA GAINII*** Cardot	Cardot (1911*a*: p. 601)	———	———
82.	***Racomitrium substenocladum*** Cardot	Cardot (1911*a*: p. 601)	*BUCKLANDIELLA SUDETICA* (Funck) Bednarek-Ochyra & Ochyra	Frisvoll (1986: p. 344)
83.	***Pottia charcotii*** Cardot	Cardot (1911*a*: p. 601)	*Hennediella heimii* (Hedw.) R.H.Zander	Matteri (1977*a*: p. 141)
84.	*Pseudoleskea calochroa* Cardot	Cardot (1911*a*: p. 601)	*Cratoneuropsis relaxa* (Hook.f. & Wilson) Broth. subsp. *minor* (Wilson & Hook.f.) Ochyra	Present work
85.	*Bartramia patens* Brid. fo. *austrogeorgica* Paris	Cardot (1911*a*: p. 601)	*Bartramia patens* Brid.	Cardot (1906*a*: p. 23)
86.	*Dicranoweisia subinclinata* (Müll.Hal.) Broth.	Cardot (1911*a*: p. 601)	*HYMENOLOMA CRISPULUM* (Hedw.) Ochyra	Ochyra (1998*a*: p. 125)
87.	***Ceratodon minutifolius*** Cardot	Cardot (1911*a*: p. 602)	*Ceratodon purpureus* (Hedw.) Brid.	Ochyra (1998*a*: p. 103)
88.	*Bartramia oreadella* Müll.Hal.	Cardot (1911*a*: p. 602)	*Bartramia patens* Brid.	Robinson (1972*a*: p. 174)
89.	*Bartramia oreadella* Müll.Hal. var. *microphylla* Cardot	Cardot (1911*a*: p. 602)	*Bartramia patens* Brid.	Newton (1973: p. 13)
90.	***Brachythecium austroglareosum*** (Müll.Hal.) Kindb. var. ***diffusum*** Cardot	Cardot (1911*a*: p. 602)	*Brachythecium austroglareosum* (Müll.Hal.) Kindb.	Newton (1979*a*: p. 137)
91.	*Dicranum aciphyllum* Hook.f. & Wilson	Cardot (1911*c*: p. 68)	*Chorisodontium aciphyllum* (Hook.f. & Wilson) Broth.	———
92.	*Drepanocladus uncinatus* (Hedw.) Warnst. fo. *gracillimus* Cardot, *nom. nud.*	Cardot (1913: p. 28)	*Sanionia uncinata* (Hedw.) Loeske	Present work
93.	*Grimmia fastigiata* of Dixon & Watts [*non* Cardot]	Dixon & Watts (1918: p. 6)	*Orthogrimmia sessitana* (De Not.) Ochyra & Żarnowiec	Present work
94.	*Grimmia stolonifera* of Dixon & Watts [*non* Cardot]	Dixon & Watts (1918: p. 6)	*Schistidium antarctici* (Cardot) L.I.Savicz & Smirnova	Present work
95.	*Tortula grossiretis* Cardot	Dixon (1920: p. 66)	*Syntrichia magellanica* (Mont.) R.H.Zander	Present work
96.	***Bryum crateris*** Dixon	Dixon (1920: p. 67)	*Bryum archangelicum* Bruch & Schimp.	Present work
97.	*Brachythecium skottsbergii* Cardot	Dixon (1920: p. 67)	*Brachythecium austrosalebrosum* (Müll.Hal.) Kindb.	Newton (1979*a*: p. 138)
98.	*Amblystegium subvarium* of Darb. [*non* Broth.]	Darbishire (1923: p. 107)	*Cratoneuropsis relaxa* (Hook.f. & Wilson) Broth. subsp. *minor* (Wilson & Hook.f.) Ochyra	Present work
	PRE-ANTARCTIC TREATY PERIOD (1919–1960)			
99.	*Racomitrium lanuginosum* (Hedw.) Brid.	Dixon (1935: p. 381)	Excluded	Bell (1973*c*: p. 93)
100.	*Drepanocladus brachiatus* of Dixon [*non* (Mitt.) Paris]	Dixon (1935: p. 382)	*Sanionia georgicouncinata* (Müll.Hal.) Ochyra & Hedenäs	Present work
101.	***Barbula byrdii*** E.B.Bartram	Bartram (1938: p. 720)	*Didymodon brachyphyllus* (Sull.) R.H.Zander	Zander & Ochyra (2001: p. 375)
102.	***Grimmia antarctici*** Cardot var. ***pilifera*** E.B.Bartram	Bartram (1938: p. 722)	*SCHISTIDIUM URNULACEUM* (Müll.Hal.) B.G.Bell	Present work
103.	***Grimmia antarctici*** Cardot var. ***percompacta*** E.B.Bartram	Bartram (1938: p. 723)	*Schistidium antarctici* (Cardot) L.I.Savicz & Smirnova	Ochyra (1998*a*: p. 165)
104.	***Bryum siplei*** E.B.Bartram	Bartram (1938: p. 723)	*Bryum argenteum* Hedw.	Greene (1967*b*: p. 12)
105.	*Andreaea rupestris* of E.B.Bartram [*non* Hedw.]	Bartram (1957: p. 139)	*Andreaea regularis* Müll.Hal.	Greene *et al.* (1970: p. 69)

106.	***Tortula conferta*** E.B.Bartram	Bartram (1957: p. 140)	*Syntrichia magellanica* (Mont.) R.H.Zander	Present work
107.	*GRIMMIA PLAGIOPODIA* Hedw.	Bartram (1957: p. 141)	———	———
108.	***Grimmia plagiopodia*** Hedw. var. ***antarctica*** E.B.Bartram	Bartram (1957: p. 141)	*Grimmia plagiopodia* Hedw.	Ochyra (1993*c*: p. 22)
109.	*Grimmia alpicola* Hedw. var. *rivularis* (Brid.) Wahlenb.	Bartram (1957: p. 141)	*SCHISTIDIUM RIVULARE* (Brid.) Podp.	———
110.	*Pohlia cruda* (Hedw.) Lindb.	Bartram (1957: p. 141)	———	———
111.	*Bryum inclinatum* (Brid.) Turton var. *magellanicum* Cardot	Bartram (1957: p. 142)	*Bryum archangelicum* Bruch & Schimp.	Present work
112.	*Bartramia patens* Brid.	Bartram (1957: p. 142)	———	———
113.	***Bryum korotkevicziae*** L.I.Savicz & Smirnova	Savicz-Lyubitskaya & Smirnova (1959: p. 34)	*Bryum pseudotriquetrum* (Hedw.) P.Gaertn., B.Mey. & Scherb.	Ochyra (1998*a*: p. 194)
114.	***Bryum korotkevicziae*** L.I.Savicz & Smirnova var. ***hollerbachii*** L.I.Savicz & Smirnova	Savicz-Lyubitskaya & Smirnova (1960: p. 25)	*Bryum pseudotriquetrum* (Hedw.) P.Gaertn., B.Mey. & Scherb.	Ochyra (1998*a*: p. 194)
	POST-ANTARCTIC TREATY PERIOD (1961 – PRESENT)			
115.	***Bryum ongulense*** Horik. & Ando	Horikawa & Ando (1961: p. 166)	*Bryum pseudotriquetrum* (Hedw.) P.Gaertn., B.Mey. & Scherb.	Ochyra (1998*a*: p. 194)
116.	***Bryum inconnexum*** Cardot var. ***fragile*** Horik. & Ando	Horikawa & Ando (1961: p. 166)	*Bryum pseudotriquetrum* (Hedw.) P.Gaertn., B.Mey. & Scherb.	Ochyra (1998*a*: p. 196)
117.	***Bryoerythrophyllum recurvirostrum*** (Hedw.) P.C.Chen var. ***antarcticum*** L.I.Savicz & Smirnova	Savicz-Lyubitskaya & Smirnova (1963*a*: p. 356)	*BRYOERYTHROPHYLLUM RECURVIROSTRUM* (Hedw.) P.C.Chen	Seppelt (1986: p. 230)
118.	***Pottia heimii*** (Hedw.) Hampe var. ***brevinervis*** L.I.Savicz & Smirnova	Savicz-Lyubitskaya & Smirnova (1963*b*: p. 193)	*Hennediella heimii* (Hedw.) R.H.Zander	Kanda (1981*b*: p. 102)
119.	***Plagiothecium simonovii*** L.I.Savicz & Smirnova	Savicz-Lyubitskaya & Smirnova (1964*c*: p. 34)	*PLAGIOTHECIUM ORTHOCARPUM* Mitt.	Present work
120.	***Grimmia lawiana*** J.H.Willis	Filson (1966: p. 148)	*COSCINODON LAWIANUS* (J.H.Willis) Ochyra	———
121.	*Psilopilum antarcticum* (Müll.Hal.) Paris	Longton (1966: p. 91)	*NOTOLIGOTRICHUM TRICHODON* (Hook. & Wilson) G.L.Sm.	Smith (1969: p. 67)
122.	*Tortula fuscoviridis* Cardot	Gimingham (1967: p. 252)	*SYNTRICHIA SAXICOLA* (Cardot) R.H.Zander	Lightowlers (1985*b*: p. 72, 75)
123.	*Brachythecium subplicatum* of Gimingham [*non* (Hampe) A.Jaeger]	Gimingham (1967: p. 252)	*Brachythecium austrosalebrosum* (Müll.Hal.) Kindb.	Present work
124.	*CAMPYLOPUS INTROFLEXUS* (Hedw.) Brid.	Longton & Holdgate (1967: p. 243) & Longton (1967: p. 229)	———	———
125.	***Grimmia donniana*** Sm. fo. ***antarctica*** Kuc	Kuc (1969: p. 659)	*Orthogrimmia sessitana* (De Not.) Ochyra & Żarnowiec	Muñoz (1998*c*: p. 395)
126.	*Tortula densifolia* Hook.f. & Wilson	Wijk *et al.* (1969: p. 80)	Erroneous report	Present work
127.	*Andreaea gainii* Cardot var. *parallela* (Müll.Hal.) S.W.Greene	Greene *et al.* (1970: p. 11)	*Andreaea gainii* Cardot	Ochyra (1996*b*: p. 479)
128.	***Sarconeurum tortelloides*** S.W.Greene	Greene *et al.* (1970: p. 54)	*TORTELLA ALPICOLA* Dixon	Zander (1993: p. 322)
129.	*Calliergon* cf. *stramineum* of Gimingham & Lewis Smith [*non* (Brid.) Kindb.]	Gimingham & Lewis Smith (1971: p. 4)	*Warnstorfia sarmentosa* (Wahlenb.) Hedenäs	Present work
130.	*Tortula* cf. *rubra* of Zanten [*non* Mitt.]	Zanten (1971: p. 193)	*Syntrichia filaris* (Müll.Hal.) R.H.Zander	Present work
131.	*Racomitrium crispulum* of Zanten [*non* (Hook.f. & Wilson) Hook.f. & Wilson]	Zanten (1971: p. 198)	*Schistidium antarctici* (Cardot) L.I.Savicz & Smirnova	Present work
132.	*Brachythecium austrosalebrosum* (Müll.Hal.) Kindb.	Zanten (1971: p. 214)	———	———
133.	*Pseudodistichium fuegianum* of H.Rob. [*non* Roiv.]	Robinson (1972*a*: p. 168)	*DITRICHUM HYALINUM* (Mitt.) Kuntze	Ochyra & Lewis Smith (1998: p. 51)
134.	*Pottia heimii* (Hedw.) Hampe	Robinson (1972*a*: p. 171)	*Hennediella heimii* (Hedw.) R.H.Zander	———
135.	*CONOSTOMUM PENTASTICHUM* (Brid.) Lindb.	Lewis Smith (1972: p. 18)	———	———
136.	*Conostomum perangulatum* Cardot	Lewis Smith (1972: p. 18, 52)	*Conostomum magellanicum* Sull.	Bell (1973*a*: p. 42)
137.	*Dicranum oleodictyon* Dixon	Lewis Smith (1972: p. 18)	*KIAERIA PUMILA* (Mitt.) Ochyra	Ochyra (1998*a*: p. 121)
138.	*Distichium* cf. *capillaceum* (Hedw.) Bruch & Schimp.	Lewis Smith (1972: p. 18, 27, 30, 59, 62)	*Distichium capillaceum* (Hedw.) Bruch & Schimp.	———

(cont.)

Table 4. Continued.

No.	Recorded species	Publication	Current status	First synonymisation or determination
139.	*Pottia austrogeorgica* Cardot	Lewis Smith (1972: p. 27, 30, 33, 37, 42, 54, 59, 62)	*HENNEDIELLA ANTARCTICA* (Ångstr.) Ochyra & Matteri	Ochyra (1998*a*: p. 139)
140.	*Drepanocladus* cf. *plicatus* Dixon	Lewis Smith (1972: p. 30, 33, 54, 62)	*Sanionia uncinata* (Hedw.) Loeske	Ochyra & Matteri (2001: p. 46)
141.	*STEGONIA LATIFOLIA* (Schwägr.) Broth.	Lewis Smith (1972: p. 30, 33, 37, 59)	———	———
142.	*Orthotrichum crassifolium* Hook.f. & Wilson	Lewis Smith (1972: p. 42)	*MUELLERIELLA CRASSIFOLIA* (Hook.f. & Wilson) Dusén	———
143.	*CONOSTOMUM MAGELLANICUM* Sull.	Bell (1973*b*: p. 131)	———	———
144.	*Mielichhoferia austrogeorgica* Müll.Hal.	Lewis Smith & Corner (1973: table 1)	*SCHIZYMENIUM PUSILLUM* (Hook. & Wilson) A.J.Shaw	Present work
145.	*Pohlia inflexa* (Müll.Hal.) Wijk & Margad.	Allison & Lewis Smith (1973: p. 187)	*POHLIA DRUMMONDII* (Müll.Hal.) A.L.Andrews	Ochyra (1998*a*: p. 201)
146.	*Racomitrium austrogeorgicum* Paris	Bell (1973*c*: p. 91)	*Bucklandiella sudetica* (Funck) Bednarek-Ochyra & Ochyra	Frisvoll (1986: p. 344)
147.	*Dicranoweisia* cf. *immersa* of R.Webb [*non* Broth.]	Webb (1973: table 1)	*DITRICHUM HYALINOCUSPIDATUM* Cardot	Present work
148.	*Dicranoweisia mackayi* (Broth. & Dixon) Broth.	Webb (1973: table 1)	*Hymenoloma antarcticum* (Müll.Hal.) Ochyra	Present work
149.	*Encalypta* cf. *patagonica* Broth.	Webb (1973: table 1)	*ENCALYPTA RHAPTOCARPA* Schwägr.	Ochyra (1998*a*: p. 134)
150.	*Encalypta* cf. *peruviana* of R.Webb [*non* Broth.]	Webb (1973: table 1)	*Encalypta rhaptocarpa* Schwägr.	Present work
151.	*Grimmia* cf. *alpicola* Hedw.	Webb (1973: table 1)	*Schistidium rivulare* (Brid.) Podp.	Present work
152.	*Pohlia wahlenbergii* (F.Weber & D.Mohr) A.L.Andrews var. *glacialis* (Brid.) E.F.Warb.	Webb (1973: table 2)	*POHLIA WAHLENBERGII* (F.Weber & D.Mohr) A.L.Andrews	Shaw (1982*b*: p. 272)
153.	*Racomitrium austrogeorgicum* Paris	Bell (1973*c*: p. 91)	*Bucklandiella sudetica* (Funck) Bednarek-Ochyra & Ochyra	Frisvoll (1986: p. 344)
154.	*Drepanocladus* cf. *aduncus* of J.J.Light & R.B.Heywood [non (Hedw.) Warnst.]	Light & Heywood (1973: p. 535)	*DREPANOCLADUS LONGIFOLIUS* (Mitt.) Paris	Present work
155.	*Hypnum revolutum* (Mitt.) Lindb. var. *dolomiticum* (Milde) Mönk.	Ando (1973: p. 206)	*Hypnum revolutum* (Mitt.) Lindb.	Present work
156.	*Hypnum revolutum* (Mitt.) Lindb. fo. *pumilum* (Husn.) Ando	Ando (1973: p. 201–202)	*Hypnum revolutum* (Mitt.) Lindb.	Present work
157.	*ENCALYPTA PROCERA* Bruch	Newton (1974*a*: p. 3)	———	———
158.	*CAMPYLIUM POLYGAMUM* (Schimp.) Lange & C.E.O.Jensen	Light & Heywood (1975: p. 200)	———	———
159.	*Dicranoweisia antarctica* (Müll.Hal.) Kindb.	Bell (1976: p. 97)	*HYMENOLOMA ANTARCTICUM* (Müll.Hal.) Ochyra	———
160.	*Tortula grossiretis* Cardot var. *atrata* Cardot	Pizarro & Sáiz (1977: p. 86)	*Syntrichia magellanica* (Mont.) R.H.Zander	Present work
161.	*Tortula tenella* Broth.	Pizarro & Sáiz (1977: p. 86)	Excluded	Present work
162.	*Bryum argenteum* Hedw. var. *obtusifolium* Broth.	Savicz-Lyubitskaya & Smirnova (1978: p. 277)	*Bryum argenteum* var. *muticum* Brid.	Present work
163.	*Bryum pseudotriquetrum* (Hedw.) P.Gaertn., B.Mey. & Scherb.	Ochi (1979: p. 77)	———	———
164.	*Campylopus canescens* (Müll.Hal.) A.Jaeger	Longton & Holdgate (1979: p. 9)	*Campylopus introflexus* (Hedw.) Brid.	Frahm (1978: p. 847)
165.	*Campylopus flavoviridis* Dusén	Longton & Holdgate (1979: p. 9)	Excluded	Present work
166.	*Campylopus incrassatus* Müll.Hal.	Longton & Holdgate (1979: p. 9)	Excluded	Present work
167.	*CAMPYLOPUS SPIRALIS* Dusén	Longton & Holdgate (1979: p. 9)	———	———
168.	*Bryum billardierei* Schwägr.	Haji Mohamed (1991: p. 411)	Excluded	Present work
169.	*Grimmia grisea* Cardot	Gray *et al.* (1982: p. 197)	Excluded	Present work

170.	*Psilopilum trichodon* (Hook. & Wilson) Mitt.	Lewis Smith (1984*b*: 29, 32; 1984*c*: 56)	*Notoligotrichum trichodon* (Hook. & Wilson) G.L.Sm.	—
171.	*Funaria* cf. *hygrometrica* Hedw.	Lewis Smith (1984*b*: 29, 42; 1984*c*: 56)	*FUNARIA HYGROMETRICA* Hedw.	—
172.	*Leptobryum* cf. *pyriforme* (Hedw.) Wilson	Lewis Smith (1984*b*: 29, 44; 1984*c*: 56, 59)	*LEPTOBRYUM PYRIFORME* (Hedw.) Wilson	—
173.	*Grimmia nordenskioldii* Cardot	Seppelt & Selkirk (1984: p. 278)	Excluded	Present work
174.	*Grimmia trichophylla* Grev.	Seppelt & Selkirk (1984: p. 278)	Excluded	Present work
175.	*Encalypta rhaptocarpa* Schwägr.	Ochyra (1984: No. 30)	—	—
176.	*Ditrichum austrogeorgicum* (Cardot) Seppelt	Ochyra (1984: No. 105, 181)	*Ditrichum hyalinum* (Mitt.) Kuntze	Ochyra (1999*a*: p. 250)
177.	*Schistidium* cf. *hyalinocuspidatum* (Müll.Hal.) B.G.Bell	Ochyra (1984: No. 164)	***SCHISTIDIUM HALINAE*** Ochyra	Ochyra (1998*c*: p. 268)
178.	*Bryum dichotomum* Hedw.	Ochyra (1984: No. 116, 165)	Erroneous record	Ochyra (1998*a*: p. 208)
179.	*Pohlia wahlenbergii* of Ochyra [*non* (F.Weber & D.Mohr) A.L.Andrews]	Ochyra (1984: No. 117)	*Pohlia cruda* (Hedw.) Lindb.	Present work
180.	*Bryum urbanskyi* Broth.	Ochyra (1984: No. 193)	Erroneous record	Ochyra (1998*a*: p. 208)
181.	*SCHISTIDIUM FALCATUM* (Hook.f. & Wilson) B.Bremer	Ochyra & Bell (1984: p. 77)	—	—
182.	*Pterygoneurum* cf. *ovatum* (Hedw.) Dixon	Lewis Smith (1985: p. 512)	*PTERYGONEURUM OVATUM* (Hedw.) Dixon	
183.	*Tortula saxicola* Cardot	Lewis Smith (1985: p. 512)	*Syntrichia saxicola* (Cardot) R.H.Zander	—
184.	*Tortula princeps* of Lewis Smith [*non* De Not.]	Lewis Smith (1985: p. 512)	*Syntrichia magellanica* (Mont.) R.H.Zander	Present work
185.	***Ceratodon kinggeorgicus*** Kanda	Kanda (1986*a*: p. 324)	*MEESIA ULIGINOSA* Hedw.	Ochyra & Lewis Smith (1998: p. 6)
186.	*Bryoerythrophyllum recurvirostrum* (Hedw.) P.C.Chen	Seppelt (1986: p. 229)	—	—
187.	*Tortula filaris* (Müll.Hal.) Broth.	Lightowlers (1986*b*: p. 71)	*Syntrichia filaris* (Müll.Hal.) R.H.Zander	—
188.	*Schistidium* cf. *celatum* of Bell & Blom [*non* (Cardot) B.G.Bell]	Bell & Blom (1986: p. 20)	*Schistidium antarctici* (Cardot) L.I.Savicz & Smirnova	Present work
189.	*Schistidium syntrichiaceum* (Müll.Hal.) B.G.Bell	Bell & Blom (1986: p. 20)	*SCHISTIDIUM ANDINUM* Mitt.) Herzog	Present work
190.	*Racomitrium sudeticum* (Funck) Bruch & Schimp.	Frisvoll (1986: p. 345)	*Bucklandiella sudetica* (Funck) Bednarek-Ochyra & Ochyra	
191.	*Dicranella cardotii* (R.Br.bis) Dixon	Ochyra & Newton (1985: p. 94)	*DICRANELLA CAMPYLOPHYLLA* (Taylor) A.Jaeger	Robinson (1986: p. 10)
192.	*Schistidium angustifolium* (Mitt.) Herzog	Ochyra & Afonina (1986: p. 324)	*SCHISTIDIUM AMBLYOPHYLLUM* (Müll.Hal.) Ochyra & Hertel	Ochyra (1998*a*: p. 163)
193.	*Schistidium chrysoneurum* of Ochyra *et al.* [*non* (Müll.Hal.) Ochyra]	Ochyra *et al.* (1986: p. 56, 59)	*Schistidium antarctici* (Cardot) L.I.Savicz & Smirnova	Present work
194.	*Bryum orbiculatifolium* of Ochyra & Ochi [*non* Cardot & Broth.]	Ochyra & Ochi (1986: p. 212)	*Bryum pseudotriquetrum* (Hedw.) P.Gaertn., B.Mey. & Scherb.	Present work
195.	*Bryum amblyodon* of Ochyra & Ochi [*non* Müll.Hal.]	Ochyra & Ochi (1986: p. 214)	*Bryum archangelicum* Bruch & Schimp.	Present work
196.	*CAMPYLOPUS PYRIFORMIS* (Schultz) Brid.	Broady *et al.* (1987: p. 103)	—	—
197.	***SCHISTIDIUM STEEREI*** Ochyra	Ochyra (1987*a*: p. 609–611)	—	—
198.	*Brachythecium glaciale* Schimp.	Lightowlers (1987: p. 91)	*SCIURO-HYPNUM GLACIALE* (Schimp.) Ignatov & Huttunen	—
199.	*Brachythecium majusculum* M.E.Newton	Lightowlers (1987: p. 92)	*SCIURO-HYPNUM FUEGIANUM* (Broth.) Ochyra & Żarnowiec	Present work
200.	*Orthotrichum rupestre* Schwägr.	Lightowlers (1987: p. 92)	—	—
201.	*Blindia dicranellacea* Müll.Hal.	Kanda (1987*b*: p. 76)	Excluded	Present work
202.	*Philonotis acicularis* (Müll.Hal.) Broth.	Lewis Smith (1988*d*: p. 132)	*Philonotis polymorpha* (Müll.Hal.) Broth.	Present work
203.	*RACOMITRIUM LANUGINOSUM* (Hedw.) Brid.	Lewis Smith (1988*d*: p. 132)	—	—
204.	*Racomitrium heterostichum* (Hedw.) Brid.	Lewis Smith (1988*d*: p. 132)	Excluded	Present work
205.	*Brachythecium populeum* (Hedw.) Schimp.	Zhou *et al.* (1989: p. 351)	Excluded	Present work

(cont.)

Table 4. Continued.

No.	Recorded species	Publication	Current status	First synonymisation or determination
206.	*Ceratodon purpureus* (Hedw.) Brid. subsp. *convolutus* (Reichard) Burley & N.M.Pritsch.	Burley & Pritchard (1990: p. 58)	Excluded	Present work
207.	*Andreaea obovata* of Wu & Hu [*non* Thed.]	Wu & Hu (1990: p. 478)	*Andreaea gainii* Cardot	Ochyra (1998*a*: p. 87)
208	*Barbula unguiculata* of Wu & Hu [*non* Hedw.]	Wu & Hu (1990: p. 479)	*Syntrichia saxicola* (Cardot) R.H.Zander	Ochyra (1998*a*: p. 155)
209.	*Tortula ruralis* of Wu & Hu [*non* Hedw.]	Wu & Hu (1990: p. 479)	*Syntrichia magellanica* (Mont.) R.H.Zander	Present work
210.	*Schistidium urnulaceum* (Müll.Hal.) B.G.Bell	Ochyra (1990*a*: p. 141–142)	—	—
211.	*Andreaea acutifolia* Hook.f. & Wilson	Ochyra (1990*a*: p. 141)	Excluded	Present work
212.	*Bryum pallescens* Schwägr.	Ochi & Kanda (1991: p. 123)	—	—
213.	*Bryum* cf. *gemmiferum* R.Wilczek & Demaret	Lewis Smith (1991: p. 322)	Excluded	Present work
214.	*Racomitrium heterostichoides* Cardot	Lewis Smith (1991: p. 322)	Excluded	Present work
215.	*Hygrohypnum palustre* Loeske var. *polare* (Lindb.) Mönk.	Linskens *et al.* (1991: p. 236)	Excluded	Present work
216.	*Cratoneuropsis relaxa* (Hook.f. & Wilson) Broth. subsp. *minor* (Wilson & Hook.f.) Myrcha, Ochyra & Tatur, *comb. inval.*	Myrcha *et al.* (1991: p. 163)	—	—
217.	*HOLODONTIUM STRICTUM* (Hook.f. & Wilson) Ochyra	Ochyra (1993*b*: p. 93)	—	—
218.	*Racomitrium striatipilum* Cardot	Chen *et al.* (1993: p. 49)	Excluded	Ochyra (1998*a*: p. 183)
219.	*Bryum muehlenbecki* Bruch & Schimp.	Chen *et al.* (1993: p. 49)	Excluded	Present work
220.	*Drepanocladus exannulatus* (Schimp.) Warnst.	Chen *et al.* (1993: p. 50)	Excluded	Ochyra (1998*a*: p. 241)
221.	*Hypnum recurvatum* (Lindb. & Arnell) Kindb.	Chen *et al.* (1993: p. 50)	Excluded	Ochyra (1998*a*: p. 252)
222.	*Aloina* cf. *brevirostris* (Hook. & Grev.) Kindb.	Convey & Lewis Smith (1993: p. 299)	*ALOINA BREVIROSTRIS* (Hook. & Grev.) Kindb.	—
223.	*DISTICHIUM INCLINATUM* (Hedw.) Schimp.	Convey & Lewis Smith (1993: p. 299)	—	—
224.	*Grimmia anodon* Bruch & Schimp.	Convey & Lewis Smith (1993: p. 299)	Excluded	Present work
225.	*Polytrichum longisetum* Brid.	Lewis Smith (1993*a*: p. 72)	*POLYTRICHASTRUM LONGISETUM* (Brid.) G.L.Sm.	—
226.	*Hypnum recurvatum* (Lindb. & Arnell) Kindb. fo. *strictum* (Meylan) Ando *ex* F.D.Chen, X.D.Li & J.X. Liu	Chen *et al.* (1993: p. 122)	Excluded	Present work
227.	*Tortula geheebiaeopsis* (Müll.Hal.) Broth.	Chen *et al.* (1995: p. 53)	Excluded	Present work
228.	*Hygroamblystegium austrofluviatile* (Müll.Hal.) Broth.	Lewis Smith (1996: p. 32)	*Cratoneuropsis relaxa* (Hook.f. & Wilson) Broth. subsp. *minor* (Wilson & Hook.f.) Ochyra	Present work
229.	*BRYUM DICHOTOMUM* Hedw.	Lewis Smith (1996: p. 32)	—	—
230.	*Ditrichum brotherusii* (R.Br.bis) Seppelt	Ochyra (1996*d*: p. 1019)	Excluded	Present work
231.	***Ditrichum lewis-smithii*** Ochyra	Ochyra (1996*e*: p. 304)	*Ditrichum hyalinocuspidatum* Cardot	Present work
232.	*Dicranoweisia dryptodontoides* (Müll.Hal.) Broth.	Marshall & Convey (1997: p. 380)	Excluded	Present work
233.	*Dicranella recurvata* Ochyra, Arts & Lewis Smith, *nom. nud.*	Ochyra *et al.* (1998: p. 166)	*DICRANELLA VARIA* (Hedw.) Schimp.	Present work
234.	*DITRICHUM HETEROMALLUM* (Hedw.) E.B.Britton	Ochyra & Lewis Smith (1998: p. 38)	—	—
235.	*DITRICHUM CONICUM* (Mont.) Mitt.	Ochyra & Lewis Smith (1998: p. 42)	—	—
236.	***Ditrichum gemmiferum*** Ochyra & Lewis Smith	Ochyra & Lewis Smith (1998: p. 48–50)	*DITRICHUM DITRICHOIDEUM* (Cardot) Ochyra	Present work

237.	*Ditrichum hyalinum* (Mitt.) Kuntze	Ochyra (1998*a*: 109)	——	——
238.	*Kiaeria pumila* (Mitt.) Ochyra	Ochyra (1998*a*: 119)	——	——
239.	*Dicranoweisia brevipes* (Müll.Hal.) Cardot	Ochyra (1998*a*: 123)	*Hymenoloma antarcticum* (Müll.Hal.) Ochyra	Present work
240.	*Dicranoweisia crispula* (Hedw.) Milde	Ochyra (1998*a*: 125)	*HYMENOLOMA CRISPULUM* (Hedw.) Ochyra	
241.	*Hennediella antarctica* (Ångstr.) Ochyra & Matteri	Ochyra (1998*a*: 139)	——	——
242.	*Grimmia reflexidens* of Ochyra [*non* Müll.Hal.]	Ochyra (1998*a*: 157)	*Orthogrimmia sessitana* (De Not.) Ochyra & Żarnowiec	Present work
243.	*Schistidium amblyophyllum* (Müll.Hal.) Ochyra & Hertel	Ochyra (1998*a*: 163)	——	——
244.	*SCHISTIDIUM CUPULARE* (Müll.Hal.) Ochyra	Ochyra (1998*a*: 169)	——	——
245.	*Schistidium occultum* (Müll.Hal.) Ochyra & Matteri	Ochyra (1998*a*: 174)	*Schistidium andinum* (Mitt.) Herzog	——
246.	***Schistidium halinae*** **Ochyra**	Ochyra (1998*a*: 174; 1998*c*: p. 268)	——	——
247.	*Pohlia drummondii* (Müll.Hal.) A.L.Andrews	Ochyra (1998*a*: 201)	——	——
248.	*Pohlia wahlenbergii* (F.Weber & D.Mohr) A.L.Andrews	Ochyra (1998*a*: 205)	——	——
249.	*Meesia uliginosa* Hedw.	Ochyra (1998*a*: 209)	——	——
250.	*Orthotheciella varia* of Ochyra [*non* (Hedw.) Ochyra]	Ochyra (1998*a*: 224)	*Cratoneuropsis relaxa* (Hook.f. & Wilson) Broth. subsp. *minor* (Wilson & Hook.f.) Ochyra	Present work
251.	*Sanionia georgicouncinata* (Müll.Hal.) Ochyra & Hedenäs	Ochyra (1998*a*: 227)	——	——
252.	*Warnstorfia laculosa* (Müll.Hal.) Ochyra & Matteri	Ochyra (1998*a*: 234)	*Warnstorfia fontinaliopsis* (Müll.Hal.) Ochyra	Ochyra & Matteri (2001: p. 72)
253.	*Platydictya jungermannioides* (Brid.) H.A.Crum	Ochyra (1998*a*: 250)	——	——
254.	*WILLIA AUSTROLEUCOPHAEA* (Besch.) Broth.	Ochyra *et al.* (1998: p. 167)	——	——
255.	*Tortella alpicola* Dixon	Ochyra *et al.* (1998: p. 167)	——	——
256.	*TORTELLA FRAGILIS* (Drumm.) Limpr.	Ochyra *et al.* (1998: p. 167)	——	——
257.	*Racomitrium orthotrichaceum* (Müll.Hal.) Paris	Ochyra *et al.* (1998: p. 167)	Excluded	Present work
258.	*Grimmia fuscolutea* Hook.	Ochyra *et al.* (1998: p. 167)	*DRYPTODON FUSCOLUTEUS* (Hook.) Ochyra & Żarnowiec	——
259.	*Schistidium chubutense* Cardot & Broth.	Ochyra *et al.* (1998: p. 167)	*SCHISTIDIUM PRAEMORSUM* (Müll.Hal.) Herzog	Present work
260.	*Leptobryum wilsonii* (Mitt.) Broth.	Ochyra *et al.* (1998: p. 167)	*POHLIA WILSONII* (Mitt.) Ochyra	——
261.	*Schizymenium pusillum* (Hook. & Wilson) A.J.Shaw	Ochyra *et al.* (1998: p. 167)	——	——
262.	*Bryum subrotundifolium* A.Jaeger.	Ochyra *et al.* (1998: p. 167)	*Bryum argenteum* var. *muticum* Brid.	Present work
263.	*Philonotis polymorpha* (Müll.Hal.) Broth.	Ochyra *et al.* (1998: p. 167)	——	——
264.	*Drepanocladus longifolius* (Mitt.) Paris	Ochyra *et al.* (1998: p. 167)	——	——
265.	*Brachythecium fuegianum* Broth.	Ochyra *et al.* (1998: p. 168)	*Sciuro-hypnum fuegianum* (Broth.) Ochyra & Żarnowiec	——
266.	*PLAGIOTHECIUM OVALIFOLIUM* Cardot	Ochyra *et al.* (1998: p. 168)	——	——
267.	*Plagiothecium georgicoantarcticum* (Müll.Hal.) Kindb.	Ochyra *et al.* (1998: p. 168)	*PLAGIOTHECIUM ORTHOCARPUM* Mitt.	Present work
268.	*ISOPTERYGIOPSIS PULCHELLA* (Hedw.) Z.Iwats.	Ochyra *et al.* (1998: p. 168)	——	——
269.	*Bryum* cf. *atropurpureum* Bruch & Schimp.	Singh & Semwal (2000: p. 176)	*Bryum dichotomum* Hedw.	Lindberg (1879: p. 16)
270.	*Bryum* cf. *bornholmense* Wink. & R.Ruthe	Singh & Semwal (2000: p. 176)	Excluded	Present work
271.	*Bryum* cf. *mildeanum* Jur.	Singh & Semwal (2000: p. 176)	Excluded	Present work
272.	*Grimmia kidderi* James	Muñoz & Pando (2000: p. 39)	Excluded	Present work
273.	*Dicranella hilariana* of Ochyra *et al.* [*non* (Mont.) Mitt.]	Ochyra *et al.* (2000: p. 155)	*Dicranella varia* (Hedw.) Schimp.	Present work
274.	*BRYUM ORBICULATIFOLIUM* Cardot & Broth.	Convey *et al.* (2001: p. 1283)	——	——
275.	*Bryum pseudomicron* Müll.Hal.	Convey *et al.* (2001: p. 1283)	Excluded	Present work
276.	*Schizymenium austrogeorgicum* (Müll.Hal.) A.J.Shaw	Convey *et al.* (2001: p. 1283)	*Schizymenium pusillum* (Hook. & Wilson) A.J.Shaw	Present work
277.	*Andreaea acuminata* Mitt.	Putzke & Pereira (2001: p. 56)	Excluded	Present work

(cont.)

Table 4. Continued.

No.	Recorded species	Publication	Current status	First synonymisation or determination
278.	*BUCKLANDIELLA PACHYDICTYON* (Cardot) Bednarek-Ochyra & Ochyra	Putzke & Pereira (2001: p. 56)	———	———
279.	*Polytrichum commune* Hedw.	Kappen & Schroeter (2002: p. 337)	Excluded	Present work
280.	*Didymodon brachyphyllus* (Sull.) R.H.Zander	Ochyra & Zander (2002: p. 35)	———	———
281.	***SCHISTIDIUM LEWIS-SMITHII*** Ochyra	Ochyra (2003*b*: p. 366)	———	———
282.	***SCHISTIDIUM DECEPTIONENSE*** Ochyra, Bednarek-Ochyra & Ochyra	Ochyra *et al*. 2003: p. 570)	———	———
283.	***SCHISTIDIUM LEPTONEURUM*** Ochyra	Ochyra (2004*b*: p. 126, 128)	———	———
284.	*TORTULA HOPPEANA* (Schultz) Ochyra	Ochyra (2004*d*: p. 500)	———	———
285.	*BUCKLANDIELLA DIDYMA* (Mont.) Bednarek-Ochyra & Ochyra	Lewis Smith (2005*a*: p. 238)	———	———
286.	*BUCKLANDIELLA SUBSECUNDA* (Harv.) Bednarek-Ochyra & Ochyra	Lewis Smith (2005*a*: p. 238)	———	———
287.	*Ditrichum ditrichoideum* (Cardot) Ochyra *ex* Lewis Smith, *comb. inval.*	Lewis Smith (2005*a*: p. 238)	———	———
288.	*Ditrichum hyalinocuspidatum* Cardot	Lewis Smith (2005*a*: p. 238)	———	———
289.	*Schistidium andinum* (Mitt.) Herzog	Lewis Smith (2005*a*: p. 239)	———	———
290.	*Schistidium praemorsum* (Müll.Hal.) Herzog	Lewis Smith (2005*a*: p. 239)	———	———
291.	*Syntrichia andersonii* (Ångstr.) R.H.Zander	Lewis Smith (2005*c*: p. 199)	Excluded	Present work
292.	*Syntrichia magellanica* (Mont.) R.H.Zander	Lewis Smith (2005*c*: p. 199)	———	———
293.	*Bryum confusum* Ochyra *ex* Lewis Smith, *nom. nud.*	Lewis Smith (2005*c*: p. 200)	*Bryum pseudotriquetrum* (Hedw.) P.Gaertn., B.Mey. & Scherb.	Present work
294.	*Warnstorfia fontinaliopsis* (Müll.Hal.) Ochyra	Lewis Smith (2005*c*: p. 200)	———	———
295.	*Dicranella varia* (Hedw.) Schimp.	Convey & Lewis Smith (2006: p. 531)	———	———
296.	*Ditrichum ditrichoideum* (Cardot) Ochyra	Present work	———	———
297.	*Dicranella campylophylla* (Taylor) A.Jaeger	Present work	———	———
298.	*AMPHIDIUM LAPPONICUM* (Hedw.) Bruch & Schimp.	Present work	———	———
299.	*CAMPYLOPUS VESTICAULIS* Mitt.	Present work	———	———
300.	*GUEMBELIA LONGIROSTRIS* (Hook.) Ochyra & Żarnowiec	Present work	———	———
301.	*Orthogrimmia sessitana* (De Not.) Ochyra & Żarnowiec	Present work	———	———
302.	*SYNTRICHIA CANINERVIS* Mitt.	Present work	———	———
303.	*SYNTRICHIA CANINERVIS* Mitt. var. *GYPSOPHILA* (G.Roth) Ochyra	Present work	———	———
304.	*Bryum argenteum* var. *muticum* Brid.	Present work	———	———
305.	*BRYUM NIVALE* Müll.Hal.	Present work	———	———
306.	*Bryum archangelicum* Bruch & Schimp.	Present work	———	———
307.	*Cratoneuropsis relaxa* (Hook.f. & Wilson) Broth. subsp. *minor* (Wilson & Hook.f.) Ochyra	Present work	———	———
308.	*Plagiothecium orthocarpum* Mitt.	Present work	———	———

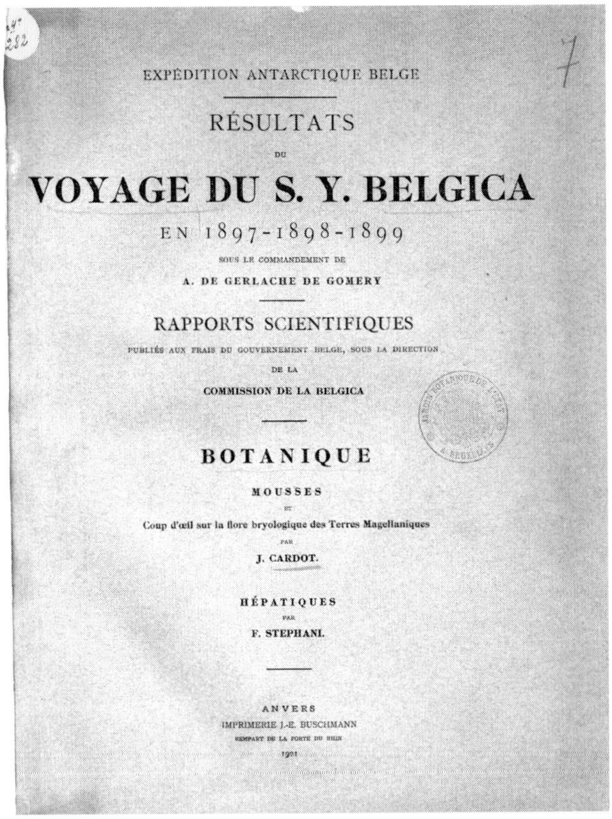
EXPÉDITION ANTARCTIQUE BELGE
RÉSULTATS
DU
VOYAGE DU S. Y. BELGICA
EN 1897-1898-1899
SOUS LE COMMANDEMENT DE
A. DE GERLACHE DE GOMERY
RAPPORTS SCIENTIFIQUES
PUBLIÉS AUX FRAIS DU GOUVERNEMENT BELGE, SOUS LA DIRECTION
DE LA
COMMISSION DE LA BELGICA
BOTANIQUE
MOUSSES
ET
Coup d'œil sur la flore bryologique des Terres Magellaniques
PAR
J. CARDOT.
HÉPATIQUES
PAR
F. STEPHANI.
ANVERS
IMPRIMERIE J.-E. BUSCHMANN
REMPART DE LA PORTE DU RHIN
1901

Figure 13. Title page of Cardot's (1901) *Mousses et coup d'oeil sur la flore bryologique des Terres Magellaniques* based upon collections made during the voyage of the Belgian Antarctic Expedition 1897–1899 under the command of Adrién de Gerlache de Gomery.

and on the north-western Antarctic Peninsula. Altogether 23 species were discovered, five of which were described as new (Cardot 1906*a*). Of these, only one species, *Schistidium antarctici*, is still accepted as a species in its own right. Of these, *Schistidium antarctici* is still accepted as a species in its own right, endemic to the Antarctic.

Two French expeditions, of 1903–1905 and 1908–1910, led by Jean-Baptiste A. E. Charcot (1867–1935), also proved to be bryologically important. They explored areas on the western coast of the Antarctic Peninsula, from the South Shetland Is. to Marguerite Bay, and yielded important moss collections. Although both expeditions contributed only six species new to the Antarctic, including *Andreaea gainii* which is currently accepted as a valid species, they provided a lot of valuable distributional data for a number of species (Cardot 1906*b*,*c*, 1907*a*, 1908*b*, 1911*a*,*b*, 1913).

Cardot (1908*a*) collated and evaluated the early moss collections made by various expeditions to the Antarctic, subantarctic South Georgia as well as Tierra del Fuego and the Falkland Is. in his well known opus *La flore bryologique des terres magellaniques, de la Géorgie du Sud et de l'Antarctide* (Fig. 14). It included the results of the Swedish South Polar Expedition of 1901–1903 as well as all earlier moss collections from these regions up to 1905 most of which Cardot had personally examined. Until the 1980s Cardot's annotated catalogue was the basic source of information on mosses of Antarctica and conterminous regions of the Subantarctic and South America. Of the later expeditions of the Heroic Age of exploration, only the Australasian Antarctic Expedition of 1911–1914, under the leadership of Douglas Mawson (1882–1958) made significant moss collections in Queen Mary Land (Dixon & Watts 1918).

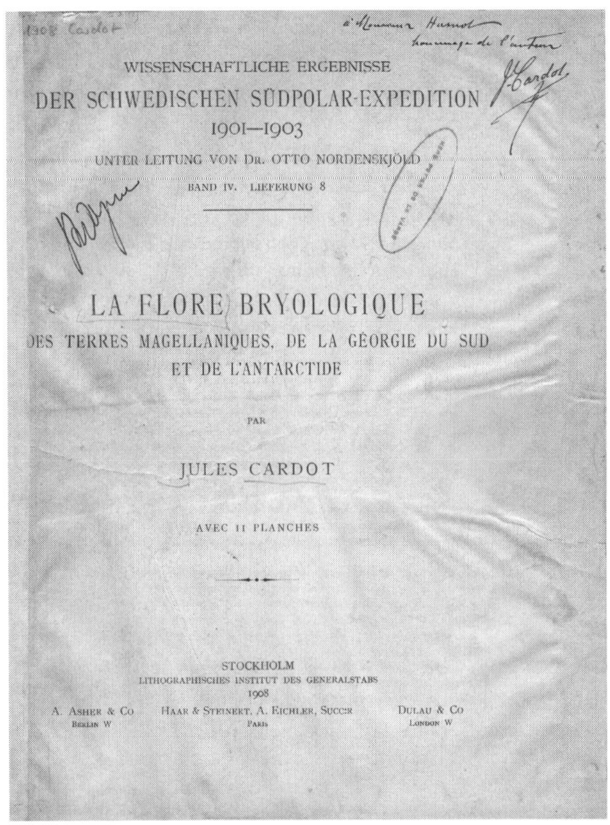
WISSENSCHAFTLICHE ERGEBNISSE
DER SCHWEDISCHEN SÜDPOLAR-EXPEDITION
1901—1903
UNTER LEITUNG VON DR. OTTO NORDENSKJÖLD
BAND IV. LIEFERUNG 8
LA FLORE BRYOLOGIQUE
DES TERRES MAGELLANIQUES, DE LA GÉORGIE DU SUD
ET DE L'ANTARCTIDE
PAR
JULES CARDOT
AVEC 11 PLANCHES
STOCKHOLM
LITHOGRAPHISCHES INSTITUT DES GENERALSTABS
1908
A. ASHER & CO — BERLIN W
HAAR & STEINERT, A. EICHLER, SUCC'R — PARIS
DULAU & CO — LONDON W

Figure 14. Title page of Cardot's (1908) *La flore bryologique des Terres Magellaniques, de la Géorgie du Sud et de l'Antarctide.*

PRE-ANTARCTIC TREATY PERIOD (1919–1960)

After the World War I the nature of Antarctic exploration changed. Expeditions tended to have greater scientific aims, with particular emphasis on earth sciences and zoology, while botany was largely neglected. Only three expeditions provided significant plant collections between the World Wars. Although primarily concerned with the biology of whales, whale stock management and oceanography, the British *Discovery Investigations*, 1918–1951, made a minor contribution to bryology during the cruise of the *Discovery II* (1933–1935). A small collection of mosses was made in the South Orkney Is. and on Wiencke I. in the Palmer Archipelago (Dixon 1935). No new regional discoveries were published, except for *Drepanocladus brachiatus* which, however, proved to be a misdetermination.

Bryologically, the most prolific were two American Expeditions led by Richard E. Byrd (1888–1957). The second Byrd Antarctic Expedition, 1933–1935, to Marie Byrd Land yielded five species of moss (Bartram 1938). One newly described variety, *Grimmia antarctici* var. *pilifera*, represents, in fact, *Schistidium urnulaceum*. In addition, Siple (1938), who collected the moss specimens, gave an excellent first-hand account of their habitats and environmental conditions under which they exist. The third Byrd Antarctic Expedition (1940–1941) provided important moss collections from both the maritime and continental Antarctic, including two additions to the moss flora, *Grimmia plagiopodia* and *Schistidium rivulare* (Bartram 1957).

Concurrent with the second Byrd Expedition, the British Graham Land (*Penola*) Expedition of 1934–1937 conducted scientific exploration along the western coast of the Antarctic Peninsula, from the Palmer Archipelago to southern Alexander I. The main achievements concerned mapping and aerial photography of the region, but some preliminary scientific findings were published (Fleming *et al.* 1938), including a general account of the bryophyte flora of the Argentine Is. and adjacent areas of the Graham Coast. G. C. L. Bertram (in Fleming *et al.* 1938) described Berthelot I. as 'botanically the richest' in Antarctica. He noted that 'moss colonies on this island were an acre in size, sometimes forming three-foot-deep bogs which were perennially frozen save for a couple of inches on the top which thaw during the summer'. These were formed by *Chorisodontium aciphyllum* and *Polytrichum strictum* (RILS, personal observation) and this is the first report of such moss peat-banks which are a characteristic feature of the landscape in the northern maritime Antarctic (Fenton & Lewis Smith 1982; Lewis Smith & Corner 1973). No bryological account of mosses collected by this expedition was published, but these collections are deposited in the Natural History Museum in London (BM) and have been examined for the present Flora.

During the World War II (1943–1945) the British government established several small bases in the South Shetland Is. and western Antarctic Peninsula under the code name 'Operation Tabarin'. From 1944 to 1946 I. Mackenzie Lamb (1911–1990) made the most intensive botanical survey ever undertaken up to then, at Deception I. in the South Shetland Is., and at Hope Bay and Wiencke I. on the Antarctic Peninsula. His huge collection included many mosses. In 1946 Operation Tabarin became the Falkland Islands Dependencies Survey (FIDS), still coordinated by the British government, and several new bases were established throughout the maritime Antarctic for the specific purpose of mapping the region and conducting geological surveys. During the 1950s biology became increasingly important. Numerous small collections of plants were made in the vicinity of these bases, and by teams making long dog-sledge journeys along the Antarctic Peninsula. These collections are also housed at BM, and have been examined in the present work.

Between 1956 and 1958 the International Geophysical Year (IGY) Programme promoted a wide range of scientific studies by 12 nations operating throughout Antarctica, and the establishment of many more bases. However, biology was not one of the major themes and few plant collections and observations of merit were made. Remarkably, only two bryological papers resulted from the IGY, namely Savicz-Lyubitskaya & Smirnova's (1959, 1960) descriptions of a new species, *Bryum korotkevicziae*, and its variety var. *hollerbachii*, from a continental Antarctic lake.

A valuable catalogue of the mosses recorded in the first three periods of bryological investigations in Antarctica was presented by Steere (1961), and subsequently supplemented by Greene (1968*a*). Altogether, between 1829 and 1960, 112 taxa (83 species, 18 varieties and eight forms) were recorded in the literature from the Antarctic. Of these, 58 taxa (40 species, 12 varieties and six forms) were described as new to science. A critical taxonomic assessment of the type specimens and voucher collections revealed that a considerable number of these taxa proved to be either identical with other species or misdeterminations. Accordingly, the real number of moss taxa discovered in the first three phases of the Antarctic exploration has been reduced in the present work to 45 species and one variety. This group includes almost all of the most common and widespread taxa known throughout the Antarctic biome.

POST-ANTARCTIC TREATY PERIOD (1961 – PRESENT)

As a result of the scientific and political success of the IGY, participating governments formulated the Antarctic Treaty and laid the foundations for the future use and management of the continent (indeed, of all land south of 60° S, referred to as the Antarctic Treaty Area) for the purpose of peaceful scientific endeavour. It came into force in 1961 (Triggs 1987; Ensminger *et al.* 1999) and has subsequently been the major influence on Antarctic affairs. Since then, many more nations have established research stations throughout the Antarctic and undertaken ever-expanding science programmes, many involving significant botanical research. Greene (1964*b*), at the first Scientific Committee for Antarctic Research (SCAR) Symposium devoted to Antarctic biology, outlined problems and progress towards the study of Antarctic bryophytes. This initiated major advances in Antarctic bryophyte taxonomy by specialists working with collections of specimens made by ecologists and other scientists (especially geologists visiting remote areas). Greene (1967*a*) discussed developments in bryological research in the decade after the IGY, while Ando (1979) and Seppelt *et al.* (1998) summarised achievements in Antarctic bryology, respectively, in the first two and four decades of the Antarctic Treaty period.

With the adoption of the Antarctic Treaty in 1961 the FIDS was renamed British Antarctic Survey (BAS). Most of the Antarctic botanical research undertaken by BAS has been concentrated at Signy station (Signy I., South Orkney Is.) and, since 1994, at Rothera station (Adelaide I.). However, botanical investigations have been carried out by BAS personnel throughout the maritime Antarctic region, from the South Sandwich Is. to Alexander I., and in international collaborative projects at numerous locations in continental Antarctica. These have resulted in the largest and most diverse collections of Antarctic cryptogams; these are housed in the herbarium of this institute (AAS) and details of all specimens may be accessed on the Antarctic Plant Database (APD) (Greene 1972; Greene & Greene 1975; Lightowlers 1988; Peat 1998). All data concerning material preserved in AAS are available on the BAS Internet web site (http://www.antarctica.ac.uk/bas_research/data/access/plantdatabase/index.php).

Records of mosses were published in many general and ecological accounts on Antarctic vegetation (e.g. Greene 1964*b*; Longton 1966, 1967; Longton & Holdgate 1967, 1979; Collins 1969; Gimingham & Lewis Smith 1970; Lewis Smith 1972, 1993*c*, 1996; Allison & Lewis Smith 1973; Lewis Smith & Corner 1973), but in many cases definitive determinations of species were lacking. This was because of the general state of the taxonomy of austral mosses, which was then in its infancy, and no keys to determine species were available. For nearly 180 years of botanical exploration of Antarctica much confusion has existed regarding the precise status of many species which were uncritically reported in the literature. Steere (1965) commented on this situation: 'The real problem ... is not to find in the published literature what names of what bryophytes have been reported, but to find out what these names mean'.

During the 1960s detailed ecological and phytosociological studies were being undertaken for the first time in the Antarctic biome, thus increasing the number of moss collections being made. These stimulated the need for improved methods of identification, and the first detailed taxonomic studies on Antarctic mosses commenced. Greene (1968*b*) presented a taxonomic assessment of the difficult genus *Andreaea*, three species of which play an important role in maritime Antarctic vegetation. Furthermore, at his suggestion, an ambitious project of the revision of Antarctic mosses was then initiated. However, this resulted in only the first of a proposed series entitled *Antarctic Moss Flora* (Greene *et al.* 1970) (Fig. 15). It comprised detailed treatments of five genera and 11 species, including distribution maps and meticulous listing of all known specimens from Antarctica. Unfortunately, this work was terminated owing to the decline of financial support for taxonomic investigations. In the mid-1960s another project was initiated to examine the geographical distribution of various elements of the indigenous flora and fauna of Antarctica, including six species of moss; these were the first distribution maps for bryophytes in this biome (Greene 1967*b*).

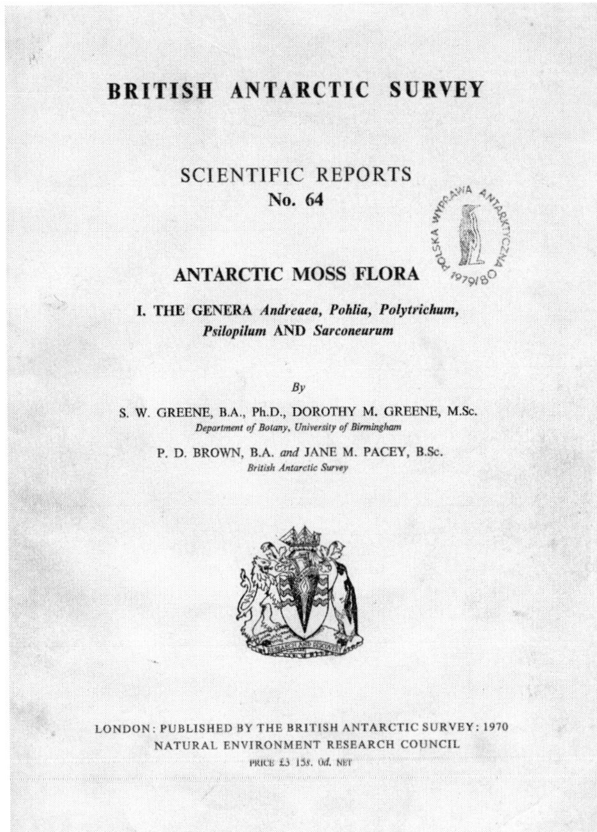

BRITISH ANTARCTIC SURVEY

SCIENTIFIC REPORTS
No. 64

ANTARCTIC MOSS FLORA

I. THE GENERA *Andreaea*, *Pohlia*, *Polytrichum*, *Psilopilum* AND *Sarconeurum*

By

S. W. GREENE, B.A., Ph.D., DOROTHY M. GREENE, M.Sc.
Department of Botany, University of Birmingham

P. D. BROWN, B.A. *and* JANE M. PACEY, B.Sc.
British Antarctic Survey

LONDON: PUBLISHED BY THE BRITISH ANTARCTIC SURVEY: 1970
NATURAL ENVIRONMENT RESEARCH COUNCIL
PRICE £3 15*s*. 0*d*. NET

Figure 15. Title page of Greene *et al.*'s (1970) *Antarctic moss flora. Part I.*

Since the early 1960s, intensive botanical investigations have also been made by scientists at, or working from, several national stations. They resulted in new approaches to the study of Antarctic mosses and contributed much to knowledge of the moss flora and distribution of taxa. Notable collections and published accounts have been made by bryologists from Australia, Brazil, Chile, China, India, Japan, Poland, Russia, USA, and UK and the specimens of Antarctic mosses and liverworts have become widely available in major world herbaria through *Bryophyta Antarctica Exsiccata*, a series consisting of 200 numbers (Ochyra 1984; Ochyra *et al.* 1986).

A number of ecological and taxonomic accounts and regional listings of species have been published in that time. For the maritime Antarctic, these include Bouvetøya (Bell & Blom 1986; Engelskjøn & Jørgensen 1986), the South Sandwich Is. (Longton & Holdgate 1979; Convey *et al.* 2001), Signy I. (Lewis Smith 1972, 2005*c*), Elephant I. (Allison & Lewis Smith 1973), King George I. (Putzke & Pereira 1990; Chen *et al.* 1995; Ochyra 1998*a*), Deception I. (Lewis Smith 1984*b*,*c*, 2005*a*,*b*), Argentine Is. (Lewis Smith & Corner 1973), and Charcot I. (Convey *et al.* 2000). For continental Antarctica, important publications include: Schirmacher Oasis (Richter 1990, 1995; Singh & Semwal 2000; Lal 2004; Ochyra & Singh 2008), Ongul Is. (Horikawa & Ando 1961, 1967*a*; Kanda 1981*a*, 1986*b*, 1987*c*), Kemp Land (Seppelt & Ashton 1978), Mac.Robertson Land (Filson 1978), Larsemann Hills

Figure 16. Title page of Ochyra's (1998) *The moss flora of King George Island, Antarctica.*

(Seppelt 1984, 1986), Bunger Oasis (Kuc 1969), Wilkes Land (Lewis Smith 1988*b*) and southern Victoria Land (Seppelt & Green 1998). In addition to these local treatments covering relatively small areas, there have been some attempts to summarise the moss flora of continental Antarctica (Savicz-Lyubitskaya & Smirnova 1978; Kanda 1987*a*) and the maritime Antarctic (Robinson 1972*a*). A checklist of mosses of Antarctica produced by Ochyra *et al.* (1998) served as the origin of the present Flora.

There have been several taxonomic studies which attempted to clarify the status of some of the Antarctic genera, for example *Ceratodon* (Horikawa & Ando 1963; Kanda 1986*a*), *Andreaea*, *Polytrichum* and *Pohlia* (Greene *et al.* 1970), *Encalypta* (Newton 1974*a*), *Bryum* (Ochi 1979; Seppelt & Kanda 1986), *Pottia* (Kanda 1981*b*; Matteri 1977*a*), *Orthotrichum* (Ochyra 1986), *Tortula* (Lightowlers 1986*b*), *Ditrichum* (Ochyra & Lewis Smith 1998), *Meesia* (Ochyra & Lewis Smith 1999). Special attention was paid to *Schistidium* which proved to be the largest genus in the Antarctic, and includes six endemic or sub-endemic species, five of which were described as new to science (Ochyra & Bell 1984; Ochyra 1987*a*, 1990*a*, 1998*c*, 2003*b*, 2004*b*; Ochyra *et al.* 2003). The most comprehensive moss flora prior to the present volume was *The moss flora of King George Island, Antarctica* (Ochyra 1998*a*) (Fig. 16). This is the largest island in the South Shetland Is. archipelago and has the most diverse moss flora of any site of similar size in the Antarctic. However, its 63 species compares with 59 recorded in the much smaller area of Signy I. in the South Orkney Is., with most of these species occurring within *ca* 1 sq. km (Lewis Smith, personal observation).

During the 1970s there was rapid development of experimental research into plant ecological and physiological processes by several nations operating in Antarctica. Unfortunately, this coincided with a global recession in cryptogamic taxonomy. It was not until the Convention on Biological Diversity in Rio de Janeiro (UNEP 1992) that biologists began to recognise the importance of mosses as a major terrestrial life-form in Antarctica. This led to a rapid interest in assessing the plant and lichen diversity of the biome and their role in terrestrial ecosystems. However, there remain relatively few specialists working on the taxonomy and biology of Antarctic mosses. A developing line of research is addressing problems concerning the provenance of the moss flora and its relationship to that of other biomes and continental landmasses, evolutionary traits of some genera or species within the continent and their potential to have survived in glacial refugia since before the Pleistocene.

Physiologists, biochemists and molecular biologists are investigating the response of some taxa to increasing UV-B irradiance resulting from the annual depletion of stratospheric ozone, or to increasing summer temperatures as a result of the current trend of global warming (Robinson *et al.* 2003). Such research has prompted a fundamental requirement for accurate identification of the taxa being studied, and this has provided a stimulus to bryologists to re-examine collections and produce taxonomic revisions of well documented collections from throughout the Antarctic biome. In conjunction with this it is crucial to collate and standardise nomenclature from the many published accounts so as to avoid confusion wherever possible. This is a primary objective of the present Flora.

During nearly 50 years of the intensive exploration and taxonomic investigations, 196 moss taxa have been recorded from the Antarctic during the 'Post-Antarctic Treaty' period, including *nomina nuda* and invalidly published names. Of these, 16 taxa were described as new to science, including 12 species, three varieties and one form, of which only six species are accepted in the Flora. Very many of these records have proved to be erroneous and have been excluded after a careful assessment of the voucher collections. Nonetheless, the list of accepted taxa increased nearly two and a half-fold in this period, from 46 in 1957 to 113 in 2008.

Altogether, during all periods of the exploration of the moss flora 74 taxa were described as new from the Antarctic, including 52 species, 16 varieties and six forms. Of these, only ten species are accepted in modern moss taxonomy, namely *Syntrichia sarconeurum*, *Andreaea depressinervis*, *A. gainii*, *Schistidium antarctici*, *Coscinodon lawianus*, *S. steerei*, *S. halinae*, *S. lewis-smithii*, *S. deceptionense* and *S. leptoneurum*.

3. Terrestrial Environment and Moss Ecology of Antarctica

The vegetation of Antarctica is strongly affected by the geographical isolation of the continent as well as its climatic and environmental conditions. The Antarctic terrestrial ecosystems have only limited extent and the vegetation is usually a poorly developed tundra, which is dominated by cryptogams, with lichens and mosses being the principal constituents of the plant communities (Plates 1–11). The flowering plant flora is very scanty and it is represented by only two native species, namely the Antarctic hair grass *Deschampsia antarctica* Desv. (Poaceae) and the Antarctic pearlwort *Colobanthus quitensis* (Kunth) Bartl. (Caryophyllaceae). They occur only in the maritime Antarctic (Greene & Greene 1963; Greene 1970; Greene & Holtom 1971; Lewis Smith 1982; Komárková 1984; Lewis Smith & Poncet 1985, 1987; Komárková *et al.* 1985, 1990), reaching their southern limit on north-west Alexander I. at lat. 69°20′ S (P. Convey, personal communication, March 2008).

In the past half century a great body of information has accumulated, relating to the description of plant communities and to the physical and environmental factors which are responsible for their development. Comprehensive summaries have been presented in a number of synthetic treatments (e.g. Mackenzie Lamb 1970; Gimingham & Lewis Smith 1970; Lewis Smith 1984*a*, 1993*c*,*d*, 1996, 2001; Longton 1988; Kanda & Komárková 1997; Beyer & Bölter 2002), and detailed descriptions of the vegetation are available for most major regions of both maritime and continental Antarctica. This brief account is not intended to be a comprehensive assessment of the ecology and phytosociology of moss-dominated communities as these may be found in many publications elsewhere, but it provides a background for the better understanding of distribution patterns, ecology and role of mosses in the Antarctic terrestrial environment.

GENERAL ECOLOGICAL FEATURES OF THE MOSS-DOMINATED COMMUNITIES

The first attempt to describe the moss-dominated communities was by Skottsberg (1912) who provided annotated lists of species for a range of habitats in the South Shetland Is. and some locations along the north-western Antarctic Peninsula. He recognised several assemblages of moss and lichen tundra vegetation, including *Polytrichum*-tundra and *Andreaea*-tundra on Moss I., Danco Coast, and moss mats formed by *Brachythecium antarcticum* (= *B. austrosalebrosum*) at Beatrice Point (= Cape Andreas) in the same area. However, it was not until the early 1960s that a more formalised classification was proposed, albeit based on Skottsberg's (1912) floristic units. Holdgate (1964) recognised three vegetation 'formations' on Signy I., South Orkney Is., each with several component communities:

1. The *Andreaea–Usnea* formation of rocky knolls and exposed ridges.
2. The *Polytrichum–Dicranum* (= *Chorisodontium aciphyllum*) formation of well drained favourable situations.
3. The *Drepanocladus* (= *Sanionia*)*–Acrocladium* (= *Warnstorfia*)*–Brachythecium* formation of wet habitats, runnels and snow-patches.

This simplistic system was further developed by Gimingham (1967) and Longton (1967), culminating in a subjective classification of discrete communities of the maritime Antarctic based on growth-form (subformations), dominance (associations) and constancy (sociations) by Gimingham & Lewis Smith (1970) and Lewis Smith (1972). The scheme was adopted and augmented by many other workers for various regions throughout the biome (e.g. Lindsay 1971; Allison & Lewis Smith 1973; Lewis Smith & Corner 1973; Longton 1973, 1979; Nakanishi 1977; Seppelt & Ashton 1978; Ochyra & Furmańczyk 1982; Engelskjøn 1986, 1987; Hu 1998; Hu & Seppelt 1989).

Relatively few classification and ordination schemes have been derived from multivariate analyses (e.g. Lewis Smith & Gimingham 1976; Lewis Smith 1988*b*). Major works dealing with Antarctic bryophyte ecology and

Table 5. Major publications on bryophyte ecology and bryophyte-dominated communities in the Antarctic (excluding taxonomic, biogeographic, physiological and species list accounts).

Sector	Localities	References
Maud	Bouvetøya	Engelskjøn (1981, 1987)
	South Sandwich Is.	Longton & Holdgate (1979), Convey *et al.* (2001); Convey & Lewis Smith (2006)
	Dronning Maud Land	Engelskjøn (1986), Kanda *et al.* (2002)
Enderby	Sôya Coast	Nakanishi (1977), Kanda (1981*a*, 1986*b*, 1987*c*)
	Mawson Rock	Seppelt & Ashton (1978)
	Vestfold Hills	Pickard (1986)
Wilkes	Windmill Is.	Lewis Smith (1988*a*, 1990*b*), Hu & Seppelt (1989), Melick *et al.* (1994), Melick & Seppelt (1997), Seppelt (2002)
Ross	Victoria Land	Longton (1973), Kappen (1985), Lewis Smith (1999), Schwarz *et al.* (1992)
Byrd	Marie Byrd Land, Edward VII Land	Siple (1938)
Scotia	South Orkney Is.	Lewis Smith (1972, 1985, 2005*c*)
	South Shetland Is.	Collins (1969), Lindsay (1971), Allison & Lewis Smith (1973), Furmańczyk & Ochyra (1982), Lewis Smith (1984*b*,*c*, 1988*d*, 2005*a*,*b*), Myrcha *et al.* (1991), Pereira & Putzke (1994), Putzke *et al.* (1998), Hu (1998), Olech (2002), Kim *et al.* (2007)
	Antarctic Peninsula	Lewis Smith & Corner (1973), Lewis Smith (1988*c*, 1993*b*, 1996), Leeuw *et al.* (1998)

communities are given in Table 5 and the simplified classification of principal Antarctic moss-dominated communities is shown in Table 6.

Throughout the Antarctic biome mosses dominate, or are major components of the terrestrial biota, in moist, low altitude, sheltered coastal habitats with a northerly aspect. However, their occurrence and the development of communities are strongly influenced by local environmental conditions, notably instability of the substratum (e.g. cryoturbation of soil and stones, exfoliation of rock surfaces), rock type and surface texture, rock and soil chemistry, permanent shading from direct solar radiation, excessive exposure to strong wind, the abrasive action of wind-blown mineral particles and ice crystals and, in the close proximity of bird and seal colonies, the impact of disturbance and toxic levels of certain chemical elements. Some inland sites with breeding colonies of snow petrels (*Pagodroma nivea*) and Antarctic petrels (*Thalassoica antarctica*), often several hundred kilometres from the sea (notably in Dronning Maud Land), support sparse growth of mosses, often encrusted by several lichen species and black cyanobacteria (Siple 1938; Bowra *et al.* 1966; Engelskjøn 1985, 1986; Ryan *et al.* 1989; Richter 1995; Kanda *et al.* 2002).

The most extensive stands of moss, sometimes covering up to several hectares, occur in the maritime Antarctic region (Plates 1–7). Here, depending on habitat features, communities dominated by one or more species predominate. Habitats with relatively stable conditions tend to have low species diversity, but increasing habitat complexity, such as typify fellfields and sites of variable small-scale geology, lead to increased species diversity both in bryophytes and lichens and create the most heterogeneous communities of the biome. Several moss species have wide ecological amplitudes and, in the absence of serious competition from other bryophytes and vascular plants, attain virtually monospecific status and develop almost pure closed stands of several hundred square metres; elsewhere, two or more species may co-dominate. It is a feature of the Antarctic moss flora that few species frequently attain widespread dominance over more than about 25 sq. m. However, many species may do so on a small scale by forming almost pure patches of a few square metres within more complex communities.

Wet habitats with impeded drainage in the maritime Antarctic are typically dominated by four carpet-forming pleurocarpous species (*Sanionia georgicouncinata*, *S. uncinata*, *Warnstorfia fontinaliopsis*, *W. sarmentosa*) (Plates 16–17, 22), while mildly calcareous wet soils support *Campylium polygamum*. Habitats subjected to moving water (margins of melt runnels, wet rocks, etc.) support large loose hummocks of *Brachythecium austrosalebrosum* and *Bryum pseudotriquetrum* (Plates 15, 18, 25) and, occasionally in the more northerly region, *Syntrichia filaris*. Seepage areas in coastal continental Antarctica sometimes support large stands of *Bryum argenteum* var. *muticum* (Plates 19–21).

A unique community type of moist, well drained rocky hillsides in the northern maritime Antarctic is formed by the tall turf-forming mosses *Chorisodontium aciphyllum* and *Polytrichum strictum* (Plates 13–14). These species have developed deep accumulations of peat rising 1–2.5 m above the surface of the terrain, and which possess permafrost 20–30 cm below the living moss surface and active layer. The integrity of these moss banks is maintained by the dense rhizoid systems of the two mosses, but particularly by the wick-like tomentum of the *Polytrichum*. These moss banks often stand as isolated features with eroding vertical edges 1–2 m high. Samples of peat from the base of banks on Signy I. and Elephant I. have provided radiocarbon ages of about 5500 years (Fenton & Lewis Smith 1982; Björck *et al.* 1991) and 4090 (Fabiszewski & Wojtuń 1997) to *ca* 5000 years (Birkenmajer *et al.* 1985) from King George I.

The most diverse and disparate moss-dominated communities are those of fellfields where environmental conditions fluctuate greatly both diurnally and seasonally. Fellfield or feldmark is the most typical ecosystem in

Table 6. Simplified subjective classification of the principal Antarctic moss-dominated communities of the Antarctic Non-vascular Cryptogam Tundra Formation (modified from Lewis Smith 1972, 1996; Longton 1979, 1988). Growth-form terminology is taken largely from Gimingham (1967) and Longton (1988, p. 35), and the phytosociological terminology from Gimingham & Lewis Smith (1970). Where only the generic name is given two or more species are involved. Each association may contain several sociations in which one or more species (including lichens or liverworts) attain dominance; these may include species otherwise occurring as subordinate associates.

Short moss turf and cushion subformation

Assemblages of mosses with predominantly short cushion and turf growth-form which, in terricolous communities, may coalesce to form dense or closed stands with a smooth to highly convoluted surface, often encrusted with lichens and cyanobacteria but with fruticose lichens often associated; typical of fellfields on drier stony substrata; lithophyte communities comprise scattered moss cushions and many lichens.

Andreaea association
Bryum association
Bryum–Ceratodon–Schistidium association
Ceratodon purpureus–Pohlia nutans association
Coscinodon lawianus–Syntrichia sarconeurum association
Hennediella association
Syntrichia sarconeurum association
Schistidium–Syntrichia association

Tall moss turf subformation

Assemblages of mosses with tall turf growth-form, some communities of which accumulate peat to >1.5 m depth and have permafrost about 20–30 cm below the surface; usually colonised by various crustose and fruticose lichens. On well drained soils and slopes.

Campylopus association (geothermal soils)
Dicranella hookeri–Philonotis polymorpha association (geothermal soil)
Polytrichum strictum–Chorisodontium aciphyllum association
Polytrichastrum alpinum–Pohlia nutans association

Bryophyte carpet and mat subformation

Assemblages of predominantly pleurocarpous mosses of shallow turf-like carpet and prostrate mat growth-forms, and similar forms of foliose and thallose liverworts on moist to wet soils; lichens sparse or absent.

Campylium polygamum association
Sanionia–Warnstorfia association
[Several liverwort associations]

Tall moss cushion (hummock) subformation

Assemblages of predominantly tall moss cushions and occasionally deep undulating carpet growth-forms on wet soil and rock and along melt-stream courses; lichens sparse or absent.

Brachythecium austrosalebrosum association
Bryum pseudotriquetrum–Sanionia association

the Antarctic biome, comprising windswept dry to moist glacial debris and screes or talus covering most ice-free terrain and frequently rendered unstable by cryoturbic activity. The entire Antarctic terrestrial biome could be considered as fellfield, with relatively small areas affording stability, a regular source of moisture, nutrient input, shelter and a northerly aspect supporting the other principal community types (Plates 1–3, 12, 23–28). The predominant species are mainly those of short cushion and turf growth-forms, with occasional small mats of liverworts in moister niches.

There is considerable variation in the fellfield communities of the maritime Antarctic, usually dominated by species of *Andreaea*, but with numerous associated short acrocarpous species, including *Polytrichastrum alpinum* (Plate 39), *Bartramia patens* (Plate 37), *Bryum* spp., *Bucklandiella sudetica*, *Orthogrimmia sessitana* (Plate 36), *Encalypta rhaptocarpa* (Plate 41), *Ceratodon purpureus*, *Hymenoloma* spp., *Kiaeria pumila*, *Pohlia nutans*, several polytrichoid species and others (Plates 3, 23–24). On calcareous, sandstone and volcanic fellfield soils and gravels, physiognomically similar but floristically different communties occur. These are usually dominated by species of *Schistidium* and *Syntrichia*, with *Bryum* spp., *Ceratodon purpureus*, *Distichium capillaceum* (Plate 38), *Ditrichum* spp., *Hennediella heimii*, *H. antarctica* (Plate 42), *Pohlia cruda*, etc. as subordinate associates. Fellfield communities on continental Antarctica are basically similar, but with restricted species composition, for instance *Bryoerythrophyllum recurvirostrum*, *Bryum* spp., *Ceratodon purpureus*, *Coscinodon lawianus*, *Didymodon brachyphyllus*, *Hennediella heimii*, *Schistidium antarctici*, *Syntrichia sarconeurum*.

Throughout the northern maritime Antarctic the current trend of regional climate warming has caused substantial recession of ice field and glacier margins and thickness since the mid-20th century. This has revealed many sub-fossil moss turf banks and patches of other moribund mosses radiocarbon dated at several thousand years old (Ingólfsson *et al.* 1992; Lewis Smith 1996); these have retained their original structure and the component species are still readily identifiable. However, there is no evidence that any species existed earlier in the Holocene which do not occur at present. Lower Cretaceous (100–120 m. y. ago) fossil mosses are known from the South Shetland Is., and at least three species of Pleistocene (2–3 m. y. ago) unidentified mosses have been excavated from sub-fossil sediment at Oliver Bluffs, Transantarctic Mountains (D. Harwood & R. Hill, personal communication, and examination by RO and RILS).

MOSSES IN EXTREME HABITATS

Most habitats in interior continental Antarctica can be regarded as extreme, in terms of low temperature, prolonged drought, rapidly fluctuating thermal and hydric regimes, high and continuous summer solar irradiance, high UV-A and B radiation levels, and prolonged winter darkness. The first reports of mosses in extreme habitats were that of Siple (1938) and Bartram (1938) in western Marie Byrd Land and Edward VII Peninsula. Details of those species known from high altitudes and high latitudes are given in Table 7 (see also Lewis Smith 2000).

Epilithic mosses are absent from continental Antarctica, except in the Antarctic Peninsula region. At high elevations and at far southern latitudes mosses occur only as chasmoliths or on soil adjacent to rock or stones where there is slight moisture retention. In such habitats, they are exposed for long periods to temperatures as low as −40 to −50° C in winter (no sunlight to warm the substratum, and often only a thin insulating cover of snow) and, more critically, a summer diurnal range of from 15–20° C (colony and substratum microclimate) to −30° C. As a response to extreme conditions, especially high UV-A and B levels, most of the few moss species growing in open exposed habitats have dark photoprotective pigments (sun form) compared to those in shaded habitats which are largely unpigmented (shade form) (e.g. Newsham 2003; Green *et al.* 2005).

High latitude habitats

Only five bryophyte species have been reported beyond 80° S, all in the Transantarctic Mountains, southern Victoria Land (Table 7). Wise & Gressitt (1965) noted 'moss' on a ridge east of Massam and Barrett Glaciers (84°35′ S), but also referred to moss collected at Cape Smith (84°42′ S). Seppelt & Green (1998) reported *Bryum pseudotriquetrum* from Black Ridge, Commonwealth Range (84°15′ S), *Schistidium antarctici* and *Grimmia plagiopodia* from Mt. Harcourt (83°49′ S), and *Ceratodon purpureus* and *Syntrichia sarconeurum* from Mt. Kyffin (83°48′ S), all in the southern Transantarctic Mountains. *Bryum argenteum* and *Schistidium antarctici* were found by Siple (1938 as *Bryum siplei* and *Grimmia antarctici* var. *percompacta*, respectively) on Mt. Washington, Edward VII Peninsula (78°05′ S). Some other records for these mosses at around 78°S are listed by Seppelt & Green (1998).

High altitude habitats

Five mosses have been collected above 1500 m and, of these, two occur in the unique habitat provided by geothermal activity near the summits (>2500 m) of the three active volcanoes in Victoria Land (Mt. Erebus, Mt. Melbourne and Mt. Rittmann, see below). The highest recorded occurrences of mosses in unheated habitats are in several of the mountain ranges in Dronning Maud Land. Engelskjøn (1986) found *Coscinodon lawianus* at 1800 m, and this species, with *Syntrichia sarconeurum*, formed a distinctive community at 1500 m in the Gjelsvikfjella-Mühlig-Hofmannfjella area. However, G. Thor (unpublished and personal communication) has recorded these two species at *ca* 2220 m at a site in Wrighthamaren, Heimefrontfjella. Further collections of *Coscinodon lawianus* have been made in the Tottanfjella at *ca* 2015 m (Bowra *et al.* 1966) and in the Filchnerfjella at *ca* 1650 and 1700–1800 m (Table 7). *Syntrichia sarconeurum* has also been recorded in the Tottanfjella at 1550 m (Bowra *et al.* 1966), and this species, and *Bryum argenteum*, were collected at 1540 m on Mt. McGee in Victoria Land (Lewis Smith, personal observation). There appear to be few, if any, other published records of mosses occurring above 1000 m, although several species have been found above this altitudinal limit (Table 7).

Geothermal habitats

Active volcanic areas in Antarctica are few, but all possess a unique habitat for bryophyte colonisation. The porous ashy soil and scoria, in places directly influenced by fumaroles, are kept permanently warm and moist in an otherwise inhospitable environment, thereby permitting some immigrant plant propagules to become established which are unable to do so if they are deposited on unheated substrata (see Lewis Smith 1993*a,d*). Soil temperatures beneath geothermal bryophyte communities are frequently within the range of 25–50° C, and some show a diurnal thermal cycle. Three areas in the northern maritime Antarctic possess volcanically heated substrata. These occur between sea level and *ca* 500 m altitude on Bouvetøya (54°25′ S), several of the South Sandwich Is. archipelago (56–60° S), and Deception I. in the South Shetland Is. (62°57′ S) (Plates 29–30).

The only vegetated geothermal site reported on Bouvetøya is at Nyrøysa (*ca* 35 m alt.), in the north-west of the island, where *Brachythecium austrosalebrosum* is the principal bryophyte (Engelskjøn 1987). Associated species included *Bryum* sp., *Ceratodon purpureus*, *Dicranella hookeri* and *Syntrichia filaris*. Although soil temperatures were not high, there was little diurnal fluctuation, remaining between *ca* 10° C and 18° C when air temperatures were around 0° C, day and night (Engelskjøn 1981).

In the South Sandwich Is. geothermal habitats occur up to *ca* 350 m altitude on Bellingshausen (Plates 31–32), Candlemas, Leskov and Visokoi Is. Around many of the fumarole vents, bryophytes have developed distinct zones, each usually dominated by one to several mosses and liverworts with numerous associated species. Species composition of these zones appears to be dependent on the temperature of the substratum, thereby creating a series of communities along thermal gradients (Longton & Holdgate 1979; Convey *et al.* 2001; Convey & Lewis Smith 2006). When ambient air temperatures have been *ca* 0–2° C, temperatures of 50° C within turves

Table 7. Altitudinal and latitudinal limits of mosses in Antarctica. All moss records above 1000 m are included.

Species	Altitude (m) and latitude (°S)	Location	Reference
Campylopus pyriformis (protonemata)	*ca* 3500 m 77°32′S	Mt. Erebus, Ross I., Victoria Land	Broady (1984), Skotnicki *et al.* (2000)
Campylopus pyriformis	*ca* 2700 m 74°21′S	Mt. Melbourne, Victoria Land	Broady *et al.* (1987), Skotnicki *et al.* (2000)
Pohlia nutans	*ca* 2600 m 73°28′S	Mt. Rittmann, Victoria Land	Bargagli *et al.* (1996), Lewis Smith (personal observation)
Coscinodon lawianus	*ca* 2200 m	Wrighthamaren, Heimefrontfjella, Dronning Maud Land	G. Thor (personal communication)
Coscinodon lawianus	*ca* 2015 m	Tottanfjella, Dronning Maud Land	Bowra et al. (1966 as *Grimmia lawiana*)
Orthogrimmia sessitana	1800 m	Drygalskifjella, Prinsesse Astrid Kyst	Present work
Coscinodon lawianus	1800 m	Gjelsvikfjella-Mühlig- Hofmannfjella, Dronning Maud Land	Engelskjøn (1986 as *Grimmia lawiana*)
Coscinodon lawianus	1700–1800 m	Filschnerfjella, Dronning Maud Land	Present work
Syntrichia sarconeurum	1550 m	Tottanfjella, Dronning Maud Land	Bowra *et al.* (1966 as *Sarconeurum glaciale*)
Syntrichia sarconeurum *Bryum argenteum*	1540 m	Mt. McGee, Victoria Land	Present work (Lewis Smith, personal observation)
Syntrichia sarconeurum	1500 m	Gjelsvikfjella-Mühlig- Hofmannfjella, Dronning Maud Land	Engelskjøn (1986 as *Sarconeurum glaciale*)
Andreaea regularis	1360 m	Engel Peak, east Palmer Land, Antarctic Peninsula	Present work
Bucklandiella sudetica	1350 m	Engel Peak, east Palmer Land, Antarctic Peninsula	Present work
Schistidium antarctici	1379 m	Sør Rondane, Prinsesse Ragnhild Kyst	Present work
Tortella fragilis	1150 m	Engel Peak, east Palmer Land, Antarctic Peninsula	Present work
Ceratodon puerpureus	1347 m	Mt. Wakefield, Wilkins Coast	Present work
Schistidium urnulaceum	*ca* 1100 m	Ford Range, Marie Byrd Land	Present work
Syntrichia magellanica	1100 m	Mt. McGee, Victoria Land	Present work
Bryum argenteum var. *muticum*	1050 m	Mt. England, Granite Harbour, Victoria Land	Present work
Pohlia nutans	1015 m	Engel Peak, east Palmer Land, Antarctic Peninsula	Present work
'Moss'	84°42′S	Cape Smith, Transantarctic Mountains	Wise & Gressitt (1965)
'Moss'	84°35′S	Ridge east of Massam and Barrett Glaciers, Transantarctic Mountains	Wise & Gressitt (1965)
Bryum pseudotriquetrum	84°15′S	Black Ridge, Commonwealth Range, Transantarctic Mountains	Seppelt & Green (1998)
Schistidium antartcici *Grimmia plagiopodia*	83°49′S	Mt. Harcourt, Transantarctic Mountains	Seppelt & Green (1998)
Ceratodon purpureus *Syntrichia sarconeurum*	83°48′S	Mt. Kyffin, Transantarctic Mountains	Seppelt & Green (1998)
Bryum argenteum *Schistidium antarctici*	78°05′S	Mt. Washington, Edward VII Land	Siple (1938 as *Bryum siplei* and *Grimmia antarctici* var. *percompacta*)

of *Campylopus spiralis* and *Sanionia uncinata* have been recorded, while at the moss–soil interface (*ca* 2.5 cm beneath the moss surface), they reached 85° C, rising to over 90° C at 5 cm depth. Those bryophytes most commonly or typically associated with heated ground include the mosses *Bryum dichotomum*, *Campylopus introflexus* (Plate 33), *C. pyriformis*, *C. spiralis*, *Dicranella hookeri*, *Ditrichum hyalinocuspidatum*, *D. heteromallum*, *D. ditrichoideum*, *Notoligitrichum trichodon* and *Pohlia nutans*, and several liverworts.

Geothermal areas on Deception I. also possess distinct assemblages of mosses, but generally liverworts are rare (Lewis Smith 1984*b*,*c*, 1988*d*, 2005*a*,*b*). Following the 1967 eruption on the island, Collins (1969) reported an abundance of fruiting *Funaria hygrometrica* on steaming ashy soil. This was its first known occurrence in the Antarctic, although it disappeared after a few years when the ground cooled. Zones of mosses occur around series of fumarole vents at *ca* 450–500 m alt. on a ridge north of Mt. Pond and, more generally, scattered amongst the warm moist scoria at a number of other geothermal sites. Although several mosses form thick turves on heated ground centred on several vents on Mt. Pond, the dominant species are *Dicranella hookeri* and *Philonotis polymorpha*

(Plates 29–30). The latter species is unknown elsewhere in the Antarctic. Temperatures of 85–90° C were recorded at a fumarole vent, while 10 cm from the vent the surface of a 5 cm thick turf of these mosses reached 25–40° C, and 35–50° C 3 cm beneath the base of the moss. At 40 cm from the vent the soil temperature below the moss was *ca* 30–40° C. The air temperature during the recording period ranged from –5° to 5° C (Lewis Smith 2005*a*,*b*).

A unique habitat occurs at high altitude on three volcanoes on continental Antarctica. These are the geothermal soils near the summits of Mt. Erebus, Ross I., Mt. Melbourne and Mt. Rittmann, all in Victoria Land. Here, at altitudes of over 2500 m, ambient conditions are amongst the most severe for life on the planet, yet volcanic activity provides a microenvironment favourable for the development of a very few bryophytes and numerous microalgae. The substratum is maintained continually at temperatures well above freezing, and usually >25° C. With air temperatures permanently below about –20° C, the temperature differential at the soil–air interface creates high humidity, often in the form of steam, which condenses to provide a source of warm liquid water; any snowfall immediately melts to provide liquid water. Steam, often sulphurous or with high levels of CO_2, also emanates from fumarole vents. Although these sites are in total darkness for many months in winter, they either remain snow- and ice-free or are covered by hollow ice-towers up to 3 m high, formed by rime frost crystals, and serving as heated 'greenhouses' (see cover illustration).

At *ca* 3500 m near the summit of Mt. Erebus (77°32′ S), a protonemal moss was reported by Broady (1984), but various attempts at culturing this failed to stimulate gametophyte development, and hence the possibility of identification. However, Skotnicki *et al.* (2001) have identified it, by genetic fingerprinting, as *Campylopus pyriformis*, a species (with others of this genus) typical of geothermally heated soil elsewhere in the world. On Cryptogam Ridge, Mt. Melbourne (74°21′ S), at *ca* 2700 m, *C. pyriformis* and *Cephaloziella varians* are frequent on heated soil and, in places, have accumulated up to 10 cm of organic matter (Broady *et al.* 1987). Within-vegetation temperatures ranged from 14° C to 31° C; at 15 cm depth they increased to 40°–60° C. These are the only sites where *C. pyriformis* has been recorded in continental Antarctica. The most recent discovery of a thermophilic moss occurring at high altitude is at *ca* 2600 m below the crater rim of Mt. Rittmann (73°28′ S) (Bargagli *et al.* 1996). *Pohlia nutans* forms several discrete populations on heated compact clay. The growth-form is unusual, each plant comprising a rosette of very short yellow-green shoots. The temperature within the vegetation ranged from 25° C to 45° C (Bargagli *et al.* 1996; Lewis Smith, unpublished).

Subpsammic habitats

In many continental Antarctic areas several mosses exist in a highly unstable edaphic environment. They are capable of colonising loose, porous sandy soils during early summer when melt water moistens the substratum sufficiently to afford short-term stability and establishment. The successful species of such subpsammic communities (including *Bryoerythrophyllum recurvirostrum*, *Bryum argenteum*, *B. dichotomum*, *B. pseudotriquetrum*, *Ceratodon purpureus*, *Hennediella heimii*) are often buried by water-borne and wind-blown silt and sand to a depth of several centimetres for much of each year (Lewis Smith 1999). Etiolated growth and photosynthesis continue during summer beneath this protective and insulating mineral layer, as very low levels of light can penetrate the loose crystalline structure of the sand.

Submerged habitats

Mosses have long been known to inhabit the stable environment of freshwater lakes in Antarctica (see summaries by Light & Heywood 1973; Kaspar *et al.* 1982; Seppelt 1983; Kanda & Mochida 1992; Wagner & Seppelt 2006) and, indeed, Light & Heywood (1975) posed the question 'is the vegetation of continental Antarctica predominantly aquatic?', an hypothesis that has never been challenged, but is clearly spurious. Several species of moss are now known to be exclusively aquatic in Antarctica, although they exist as terrestrial plants elsewhere in the world. These include *Plagiothecium orthocarpum* (= *P. simonovii*), which has its only known continental Antarctic station in several lakes in the Schirmacher Oasis, Dronning Maud Land, where it grows to depths of 32.5 m (Savicz-Lyubitskaya & Smirnova 1964*c*, 1965*c*; Richter 1995; Bednarek-Ochyra *et al.* 1999). *Pohlia wilsonii* is another exclusively benthic moss reported from lakes near the Japanese Sôya station, Enderby Land, where it forms unique 'towers' up to 60 cm high and 40 cm wide at depths of 3–5 m (Kanda & Mochida 1992 as *Leptobryum pyriforme*; Imura *et al.* 1999). What is probably the same taxon has been reported from lakes in Schirmacher Oasis by Tewari & Pant (1996). Both of these mosses occur as terrestrial species in South America.

Bryum pseudotriquetrum (as *B. algens* and *B. korotkeviczaie*, see Kaspar *et al.* 1982; Seppelt 1983; Pickard & Seppelt 1984*b*) is known from many lakes around coastal continental and maritime Antarctica to considerable depths (e.g. Savicz-Lyubitskaya & Smirnova 1959, 1960; Kanda & Iwatsuki 1989; Kanda & Mochida 1992). It has also been recorded from the permanently ice-covered Lake Vanda (77°32′ S) in the Dry Valleys of southern Victoria Land from 9–31 m depth (Kaspar *et al.* 1982), and from the exceptional depth of 81 m in Radok Lake in Amery Oasis, Mac.Robertson Land (Wagner & Seppelt 2006). A recent survey using a remotely operated vehicle in a lake in Bunger Oasis sampled a protonemal moss growing in large masses at depths of 70–90 (possibly to 100) m (D. Anderson, personal communication). Attempts by J. Duckett and RILS to culture samples failed to produce gametophytes, but it is suspected that the moss is likely to be *B. pseudotriquetrum*, which grows in some of the neighbouring lakes. This is a very common terrestrial moss throughout Antarctica and also the most frequently encountered aquatic moss. At Signy I., South Orkney Is., the typically hydric moss *Warnstorfia sarmentosa* and aquatic *Drepanocladus longifolius* are frequent in many lakes, notably Moss Lake, to 9 m depth (Light & Heywood 1973; Priddle 1979 as *Calliergon sarmentosum* and *Drepanocladus* sp. or *Drepanocladus* cf. *aduncus*). Here, and in many lakes in the South Shetland Is. and a few on James Ross I. these have been recorded in shallow lakes to about 9 m depth. *Campylium polygamum* occurs in several lakes on Alexander I. (Light & Heywood 1975; Lewis Smith, personal observation).

CONSERVATION OF BRYOPHYTES

On the basis of current knowledge, no moss (or liverwort) taxon anywhere in Antarctica is endangered, and therefore none is deemed to require special protection. Several are known from single collections or apparently restricted to single sites, but human impact is on such a small and limited scale that none is considered to be threatened. However, to a degree, all bryophytes, indeed all plants and lichens, within the Antarctic Treaty Area are protected by the *Protocol on Environmental Protection to the Antarctic Treaty* (Anonymous 1993), Annex II, Conservation of Antarctic Fauna and Flora (Article 3, Protection of Native Fauna and Flora). This states that no native plants (including fungi and lichens), or their propagules, may be removed or damaged in 'such quantities that their local distribution or abundance would be significantly affected'. Harmful interference is not permitted, i.e. by 'significantly damaging concentrations of native terrestrial plants by landing aircraft, driving vehicles, or walking on them, or by other means' and 'any activity that results in the significant adverse modification of habitats of any species or population of native … plant …'. To achieve anything more than very small-scale collecting of samples or specimens requires the issue of a permit by the national agency to whom the collector is responsible.

Table 8. Antarctic Specially Protected Areas (ASPA) for which mosses are specified as a major criterion for designation (from Antarctic Protected Areas Information Archive 2006).

ASPA No.	Location	Reference to moss vegetation as a reason for site designation
105	Beaufort I., Ross Sea	'… the most extensive, continuous area of mosses yet known for the McMurdo Sound region … the moss community [*ca* 2.5 ha] is dominated by a single species, *Bryum argenteum* [= *B. argenteum* var. *muticum*]'
106	Cape Hallett, northern Victoria Land	'… contains a variety of habitats with communities that are considered important as the most extensive, representative, examples known at the northern end of … Victoria Land … vegetation surveys have recorded five species of moss, dominated by *Bryum subrotundifolium* [= *B. argenteum* var. *muticum*]'
108	Green I., Berthelot Is., Antarctic Peninsula	'… well developed continuous banks of moss turf formed by *Chorisodontium aciphyllum* and *Polytrichum strictum* … overlie peat of more than one metre in depth … considered to be the most important examples of their vegetation feature in the west Antarctic Peninsula region, occupying an area of over 0.5 ha.'
109	Moe I., South Orkney Is.	'… contains the greatest continuous expanses of *Chorisodontium-Polytrichum* moss turf found in the Antarctic.'
110	Lynch I., South Orkney Is.	'… several species of moss found on the island (*Polytrichastrum alpinum* and *Muelleriella crassifolia*) are unusually fertile for their southerly location … The moss *Plagiothecium ovalifolium* occurs in moist shaded rock crevices near the shore …'
112	Coppermine Cove, Robert I., South Shetland Is.	'… outstanding feature of the vegetation is a 1.5 ha closed carpet of the mosses *Calliergidium austro-stramineum* [= *Warnstorfia fontinaliopsis*], *Calliergon sarmentosum* [= *Warnstorfia sarmentosa*] and *Drepanocladus uncinatus* [= *Sanionia uncinata* and *S. georgicouncinata*] representing one of the largest continuous moss stands in the Antarctic.'
113	Litchfield I., Arthur Harbour, Anvers I., Palmer Archipelago, Antarctic Peninsula	'… supported one of the most extensive moss carpets in the Antarctic Peninsula region … several banks of *Chorisodontium aciphyllum* and *Polytrichum strictum* of up to 1.2 m in depth … Some areas previously richly carpeted by mosses have been completely destroyed [by fur seals], while others have suffered moderate to severe damage.'
115	Lagotellerie I., Marguerite Bay, Antarctic Peninsula	'… numerous mosses and lichens also form well developed communities … a few of the mosses are fertile, a rare phenomenon in most Antarctic localities.'
116	New College Valley, Caughley Beach, Cape Bird, Ross I.	'… contains some of the richest stands of mosses … in the Ross Sea region …'
117	Avian I., off Adelaide I., Marguerite Bay, Antarctic Peninsula	'… the moss *Warnstorfia laculosa* [=*W. fontinaliopsis*] … is at the southern limit of its known range.'

(cont.)

Table 8. Continued.

ASPA No.	Location	Reference to moss vegetation as a reason for site designation
118	Cryptogam Ridge, Mt. Melbourne, Victoria Land	'The warmest areas of ground created by fumaroles support patches of moss, liverwort and algae … Mt. Melbourne has the only known leafy examples of the moss *Campylopus pyriformis* on the Antarctic continent.'
126	Byers Peninsula, Livingston I., South Shetland Is.	'… inland lakes … with a particularly important biota, including aquatic mosses …'
130	Tramway Ridge, Mt. Erebus., Ross I.	'The single … moss species … is unusual in that it persists in the protonematal stage.'
131	Canada Glacier, Lake Fryxell, Taylor Valley, Victoria Land	'… some of the richest plant growth (bryophytes and algae) in the southern Victoria Land Dry Valleys.'
135	North-east Bailey Peninsula, Budd Coast, Wilkes Land	'Moss and lichen communities are used to monitor the environmental impacts of Casey Station … contrasting moss fields that are the subject of taxonomic, ecological and physiological studies … monitoring annual growth increments in mosses.'
136	Clark Peninsula, Budd Coast, Wilkes Land	'Moss and lichen communities … are used as control plots to monitor the environmental impacts of Casey Station … rich associations of macrolichens and bryophytes that occupy very specific ecological niches.'
139	Biscoe Pt., Anvers I., Palmer Archipelago, Antarctic Peninsula	'One stand of mosses … extends almost continuously for 150 m along the valley floor, covering an area of approximately 6500 m^2.'
140	Parts of Deception I., South Shetland Is.	'11 sites of unique botanical interest … The island has the greatest number of rare and extremely rare plant species of any site in the Antarctic. 28 of the 54 mosses recorded on the island … are considered to be rare or extremely rare [these numbers have since been reduced following the present authors' revision] … The more stable geothermal areas … have developed bryophyte communities … each with a distinct and unique flora.'
141	Yukidori Valley, Langhovde, Lützow Bay, Enderby Land	'… a typical continental fellfield ecosystem … Permanent quadrats for monitoring lichen and moss vegetation have been established …'
147	Ablation Pt. – Ganymede Heights, Alexander I.	'This Area has the greatest bryophyte diversity of any site at this latitude in Antarctica (at least 21 species) … Many of the bryophytes and lichens are at the southern limit of their known distribution … Several mosses occur in lakes and ponds to depths of 9 m … These are the best examples of aquatic vegetation in the Antarctic Peninsula region. Several bryophyte species … are fertile … and some of these are not known or are very rare in this condition elsewhere in the Antarctic.'
154	Botany Bay, Cape Geology, Victoria Land	'… extremely rich botanical refuge for such a high latitude location, with a lichen and moss species diversity and abundance that is unique for southern Victoria Land …'
163	Dakshin Gangotri Glacier, Dronning Maud Land	'Terrestrial mosses are quite widespread in the Schirmacher Oasis colonising a range of habitats.'
165	Edmonson Pt., Wood Bay, Victoria Land	'… moss dominated communities that favour continually moist and wet habitats …'

The Antarctic Protected Areas System, including Sites of Special Scientific Interest (SSSI) and Specially Protected Areas (SPA) (Bonner & Lewis Smith 1985; Clark & Dingwall 1985; Anonymous 1997) has recently been revised. Both categories of protected area are now classed as Antarctic Specially Protected Areas (ASPA) and, for many, the importance of bryophytes or vegetation in general (including lichens) is stated as a primary criterion for their protected status (Table 8). Each has its own management plan (currently being revised) which strictly controls activities so as to afford maximum protection to the values for which the sites have been designated.

A unique situation exists at several localities in the northern maritime Antarctic where populations of Antarctic fur seals (*Arctocephalus gazella*) have been increasing rapidly since the mid-1970s. At Signy I. the annual summer influx of non-breeding male animals has increased from a few dozen in the early 1970s to over 20,000 from about 1990. Their impact on the lowland terrestrial ecosystem has been devastating and large expanses (*ca* 80 ha) of bryophyte and lichen communities have been destroyed (Lewis Smith 1988*e*, 1990*b*, 1997*a*). Of particular concern has been their destruction of localised calcicolous communities, unknown elsewhere in the Antarctic. Fortunately, all bryophyte species lost in these habitats are known on marble outcrops at higher altitudes which remain inaccessible to the seals.

A single place-name recognises the prominence of a moss. In 1990 The British Antarctic Place-names Committee officially named 'Andreaea Plateau' on Signy I. because 'the feature is notable for the largest known stand in the Antarctic of the black-brown moss *Andreaea* spp.'. This plateau, at *ca* 180 m altitude, is covered by a stand of about 1 ha of almost continuous mixed *Andreaea regularis* and *A. gainii*. It is situated above an extensively moss-dominated hillside named 'Moss Braes' (Plate 24).

4. *Diversity and Phytogeography of the Moss Flora*

Mosses were first collected in the far southern latitudes in the last decades of the eighteenth century, during various voyages to this remote part of the globe, and were subsequently described in the classical works of Hedwig (1801) and Bridel (1826–1827). However, it was the British Antarctic Expedition of James C. Ross in 1839–1843 which was crucial to the study of antipodal moss floras. The young Joseph D. Hooker made substantial bryophyte collections in widely separated areas of the Southern Hemisphere. These were the basis of the first comprehensive treatments of the mosses of southern South America (including Tierra del Fuego and the Falkland Is.) and Îles Kerguelen (Wilson & Hooker 1847), the Auckland Is. and Campbell I., off New Zealand, in the southern-cool-temperate zone (Wilson & Hooker 1845), as well as New Zealand (Wilson 1854) and Tasmania (Wilson 1859).

The moss floras of New Zealand and southern South America were subsequently revisited by Mitten (1867, 1869). In addition, numerous accounts have been published in which many new species were described from the insular and continental regions of the Antipodes. This period ended with significant taxonomic contributions to the moss flora of southern South America by Dusén (1903*a,b*, 1905*a,b,c*, 1906, 1907), Australia (Watts & Whiteledge 1902, 1906), Tasmania (Rodway 1914) and New Zealand (Dixon 1913, 1914, 1923*a*, 1926, 1927, 1929). A characteristic feature of these early works was an exclusive concentration on local and regional taxa, with only rudimentary (or no) references to, or comparative studies on, taxa in the widely separated land masses in the Southern Hemisphere. However, much later, floristic relationships and affinities alluded to by Hooker were interpreted and documented by Turrill (1953) and Affolter (1980).

During the next half-century, roughly until the 1970s, the Musci of the cool and cold austral regions were almost completely neglected, with the notable exception of a catalogue of the Argentine mosses (Kühnemann 1938) and a modern moss flora of New Zealand (Sainsbury 1955). Since then, however, significant progress has been made in the knowledge of these plants in various parts of this vast region. A number of catalogues, checklists, floras and floristic accounts of the mosses have been published for various areas in the cool-temperate and cold zones of the Southern Hemisphere (Table 9).

In contrast to the fairly extensive floristic literature, the number of taxonomic treatments of the mosses of the cool Antipodes is disproportionately low, only a few moss genera having been surveyed globally or regionally in the cool-temperate part of the Southern Hemisphere (Table 10). Many large and critical genera which play an important role in the vegetation, for example *Andreaea*, *Bryum*, *Bucklandiella*, *Dicranoloma*, *Ditrichum*, *Hymenoloma*, and *Schistidium*, still remain underworked, despite the publication of many miscellaneous accounts on the taxonomy and nomenclature of austral mosses. (Amongst these may be cited: Ochyra 1988*c*, 1989*c*, 1990*a,b,c,d*, 1991*a,b*, 1992*a,b*, 1993*a,b,c,e,f*, 1994*a,c*, 1996*a,b,c,d,e,f,g,h,i*, 1997*a,c,d,e*, 1999*a,b,c,d,e,g,k,l,m,n*, 2003*c*; Ochyra & Bednarek-Ochyra 1997, 1998*d,e,f*, 1999; Bednarek-Ochyra & Ochyra 1992, 1994, 1996, 1998; Ochyra & Buck 2003; Ochyra & Lewis Smith 1996, 1999; Ochyra & Mahu 1990; Ochyra & Matteri 1996*a,b*, 1997; Ochyra *et al.* 1996).

The imperfect and defective taxonomy of very many genera and families of mosses in the austral regions obscures salient and important phytogeographical facts, and the lack of taxonomic 'cleansing', especially of large and critical genera, is a major factor inhabiting a meaningful phytogeographical analysis of the flora and evaluation of the diversity of the area concerned.

Sound biogeographical interpretation of antipodal moss floras is also bedevilled by the inadequate exploration of many areas. This is especially true for some subantarctic islands, including Îles Crozet and Îles Kerguelen, and even smaller remote islands in the cool-temperate zone, namely Gough I. and Tristan da Cunha (in the South Atlantic), as well as Île Amsterdam and Île St. Paul (in the South Indian Ocean). Exploration of these small specks of land should yield interesting discoveries which change, or allow more accurate interpretation of, the distribution patterns of many species. This supposition is based upon recent significant discoveries of a number of species on subantarctic Marion I. in the Prince Edward Is. group, including *Dicranella gremmenii* (Ochyra 1999*d*) and *Bucklandiella valdon-smithii* (Ochyra & Bednarek-Ochyra

Table 9. Selected modern taxonomic and floristic treatments of the mosses in various areas of the cool-temperate and cold regions in the Southern Hemisphere.

Area	Authors	Category
Antarctica	Greene *et al.* (1970) Ochyra *et al.* (1998)	Part one of a Flora Checklist
Argentina	Matteri (2003)	Checklist
Auckland Is.	Vitt (1979*a*)	Detailed taxonomic and floristic account
Australia	Streimann & Curnow (1989), Streimann & Klanzenga 2002) McCarthy (2006)	Catalogue First volume of a complete moss Flora
Aysén Region (Chile)	Seki (1974)	Floristic account
Bío-Bío Region (Chile)	Ireland *et al.* (2006)	Floristic account
Campbell I.	Vitt (1974*a*)	Detailed taxonomic and floristic account
Chile	He (1998)	Catalogue
Falkland Is.	Matteri (1986),	Catalogue
Heard I.	Bergstrom & Selkirk (1997)	Checklist
Juan Fernandez	Robinson (1975)	Detailed taxonomic and floristic account
King George I.	Ochyra (1998*a*)	Complete moss Flora
Los Glaciares National Park (Argentina)	Matteri & Schiavone (2004)	Floristic account
Macquarie I.	Seppelt (2004)	Complete moss Flora
Nahuel Huapi National Park (Argentina)	Calabrese & Matteri (1999)	Floristic account
New Zealand	Fife (1995)	Checklist
Prince Edward Is.	Zanten (1971) Ochyra (2008)	A detailed taxonomic and floristic account Checklist
South Georgia	Greene (1973), Bell (1973*a*, 1974, 1984), Clarke (1973*a*), Newton (1974*b*, 1977*a*, 1979*a*, 1983*a*), Bell & Greene (1975), Matteri (1977*b*), Lightowlers (1985*b*), Ochyra & Lightowlers (1988)	Incomplete synoptic Flora covering many moss genera (see Ochyra *et al.* 2002 for detailed listing)
Southern Patagonia	Hässel de Menendez *et al.* (1984), Matteri (1985*a*)	Detailed floristic accounts
Southern South America, South Georgia and Antarctica	Greene (1986)	Catalogue
Tasmania	Dalton *et al.* (1991)	Checklist
Tierra del Fuego	Matteri (1975, 1985*b*), Schiavone (1993), Ochyra & Matteri (2001)	Four fascicles of a cryptogamic Flora
Tierra del Fuego and the Falkland Is.	Matteri & Schiavone (2002)	Checklist

1999), which are new to science, as well as *Entosthodon productus* (Ochyra & Smith 2003) and *Thuidium delicatulum* (Ochyra *et al.* 2003) which are new to the Subantarctic. A recent bryological survey of the moss flora of Îles Kerguelen yielded no less than 30 species new to the archipelago and at least two species new to science (Ochyra, unpublished).

These data clearly demonstrate that an accurate evaluation of the biodiversity of mosses (and other groups of plants) in the cold Antipodes requires, on the one hand, critical taxonomic assessment of all historical collections and, on the other hand, intensive and extensive exploration, especially of under-investigated regions. In general, as elsewhere, the majority of taxonomic work carried out on the mosses of the cool Antipodes is ancillary to phytogeographic and biodiversity studies, as well as to ecological surveys. Interestingly, a comparison of taxonomic estimates of diversity patterns, molecular estimates based on standing nucleotide diversity, and phylogenetic diversity, the latter taking into account the historical information contained in a molecular phylogenetic tree for the mosses, suggests that moss diversity is highest in the Southern Hemisphere and lowest in the Northern Hemisphere, with the tropics having an intermediate level (Shaw *et al.* 2005). However, these are only preliminary results and statistical analysis of phylogenetic diversity should be

Table 10. Selected taxonomic monographs, revisions and surveys of the moss genera and families in the cool-temperate and cold parts of the Southern Hemisphere. NB: some regional treatments in the Floras are included in Table 9.

Genus	Reference	Area covered
Bartramia	Matteri (1984) Fransén (2004*a*, 2005)	Southern South America, Subantarctica, Antarctica Global
Bartramidula	Matteri (1984)	Southern South America
Blindia	Bartlett & Vitt (1986)	Global
Breutelia	Matteri (1973*a*)	Southern South America
Bryum	Ochi (1970) Ochi (1972) Ochi (1982)	Australasia Subantarctica (Kerguelen Province) Southern South America
Camptochaete	Tangney (1997)	Global
Campylopodium	Giese & Frahm (1985)	Global
Campylopus	Frahm (1975, 1976, 2005) Frahm (1987*b*) Frahm (1985)	Southern South America Australia Subantarctica (Kerguelen Province)
Ceratodon	Burley & Pritchard (1990)	Global
Chorisodontium	Frahm (1981), Hyvönen (1991*b*)	Southern South America
Conostomum	Matteri (1984) Fife (1998)	Southern South America, Subantarctica, Antarctica New Zealand
Crosbya	Vitt (1977)	Global
Dicranoloma	Klanzenga (2003)	Australasia
Ditrichum	Seppelt (1982) Ochyra & Lewis Smith (1998)	Australasia Antarctica
Fissidens	Beever *et al.* (2002) Müller & Pursell (2003)	New Zealand Southern South America (Chile)
Funariaceae	Fife & Seppelt (2001)	Australia
Grimmia	Greven (1998*a*) Greven (2000*b*) Muñoz (1999) Greven (2003)	New Zealand Australia Southern South America Global
Holodontium	Ochyra (1993*b*)	Global
Hookeriaceae	Streimann (1997, 1999, 2000, 2001)	Australia
Hypopterygiaceae	Matteri (1973*c*) Kruijer (2002)	Southern South America Global
Leptostomum	Matteri (1991)	Southern South America
Leptotheca	Churchill & Buck 1982	Global
Lepyrodon	Allen (1999)	Global
Macromitrium	Vitt & Ramsay (1985*a*,*b*, 1986)	Australasia
Muelleriella	Vitt (1976*b*)	Global
Orthotrichum	Lewinsky (1984) Lewinsky (1985, 1987)	Australasia South America
Philonotis	Matteri (1968)	Southern South America
Pleurophascum	Fife & Dalton (2005)	Global
Pohlia	Shaw (2006)	Australia
Ptychomniaceae	Kühnemann & Gonçalves-Carralves (1975)	Southern South America
Rhacocarpus	Frahm (1996)	Global
Rigodium	Zomlefer (1993)	Global
Sphagnum	Fife (1996)	New Zealand
Valdonia	Ochyra (2003*c*)	Global
Vittia	Ochyra (1987*c*)	Global
Zygodon	Calabrese (2006)	Southern South America

interpreted with caution because the Southern Hemisphere estimate did not incorporate taxon subsampling. Also, the degree of molecular differentiation among regions is very small, only about 1% of the total diversity among mosses at the two loci sampled in the study.

In recent decades only liverworts have received the considerable taxonomic attention which now permits a meaningful phytogeographical analysis. They have been treated in detail, with numerous supporting distribution maps by Schuster (e.g. 1963, 1969, 1972, 1976, 1979, 1981, 1982, 1983*a*, 2000, 2002), Grolle (e.g. 1963, 1965, 1966, 1969, 2002), Engel (e.g. 1978, 1980, 1990), Bednarek-Ochyra *et al.* (2000), and Engel & Smith Merrill (2004). Nonetheless, exploration of subantarctic islands still yields remarkable collections which contain species new to science, for example *Pachyglossa grolleana* Váňa (Váňa & Gremmen 2005) from Heard I. and Îles Kerguelen (Váňa & Gremmen 2006) and species of *Riccardia* from the Prince Edward Is. (Schuster 1989), or species of great phytogeographical interest, for instance *Prasanthus suecicus* (Gottsche) Lindb. on Prince Edward I. (Váňa *et al.*, unpublished), a High Arctic species which has so far never been recorded beyond the Holarctic.

As shown above, the information for mosses is more scattered and less comprehensive because generally this group has been less studied taxonomically than liverworts in the austral cool-temperate regions. However, with the completion of the present work, Antarctica will undoubtedly have the best known muscoflora in the Southern Hemisphere. In addition, good collecting coverage of the biome and distributional data from almost all ice-free areas (Peat *et al.* 2006) make possible a critical assessment of the diversity and biogeography of the Antarctic moss flora.

DIVERSITY AND SPECIES RICHNESS

The flora of the far southern regions of the globe is renowned for its relatively low species diversity in all groups of plants and fungi, in comparison to their northern counterparts (Bliss 1979). For example, the Subantarctic and Antarctic flora of vascular plants consists of about 70 species of native pteridophytes and spermatophytes (Greene & Greene 1963; Greene & Walton 1975, although there have been a few more recent additions), and an additional 108 species are regarded as naturalised or transient aliens (Frenot *et al.* 2005). The Falkland Is. possess about 170 species (Moore 1968) and the Auckland Is. 187 species (Johnson & Campbell 1975). By way of contrast, the Arctic vascular plant flora comprises approximately 900 native species (Polunin 1959). While no vascular plants exist beyond 70° S, the High Arctic boasts very many between 68° and 80–83° N, for example 173 species in Spitsbergen (Elven & Elvenbakk 1996), more than 340 in the Canadian Arctic Archipelago (Porsild 1964) and 435 in northern Alaska (Wigins & Thomas 1962).

A remarkable disparity in the number of species also exists amongst lichens, which are considered to be the organisms best adapted to living in polar deserts. The lichen flora of Antarctica comprises *ca* 386 species (471, if South Georgia is included) (Øvstedal & Lewis Smith 2001; Kappen 2004), whereas that of Spitsbergen in the High Arctic consists of 597 species (Elvebakk & Hertel 1996).

As noted above, an evaluation of the richness of austral moss floras is hampered by taxonomic uncertainties and inadequate exploration. Nevertheless, the differences in species diversity between northern and southern moss floras are also great (Table 11). By austral standards of diversity the moss flora of Antarctica appears to be fairly rich. It consists of 111 species and two varieties, which belong within 55 genera and 17 families and is much richer than the moss floras of small subantarctic islands such as the Prince Edward Is., Îles Crozet, Heard I. and Macquarie I., despite their position in a zone where the climate is more favourable.

Although the intra-Antarctic distributions given here provide an indication of broad patterns, for many species these will almost certainly change as more collections are made from un- or under-explored areas. The observed taxonomic richness of a region depends very much on the sampling intensity, especially by experts in a given group of organisms. To place the present work in the context of current taxonomic and floristic research the numbers of moss taxa recorded for each of the major geographical regions considered in this Flora are given in Table 12.

As would be expected, the species richness of the Antarctic moss flora declines significantly with latitude. This is very obvious in comparisons of species richness between the maritime and continental Antarctic. The former is confined to the Antarctic Peninsula region, especially to its western coast and conterminous archipelagoes, and has the greatest diversity of all the major regions, but it does extend over a much greater latitudinal and climatic gradient than the other regions. In total, 109 of the total 113 taxa of moss, including 107 species and two varieties, have been recorded in this region.

Here, the South Shetland Is. exhibit the greatest diversity and have the richest muscoflora in terms of extent of moss-dominated communities, consisting of 87 species and one variety, of which 12 species have not been collected elsewhere in Antarctica. The archipelago has a relatively large area, covering 3824 sq. km including 11% ice-free land (Peat *et al.* 2006). It has been extensively and intensively studied by both professional bryologists and well trained botanists and ecologists. However, the exceptional richness of the moss taxa should primarily be attributed to the wide range of habitats, including sites with heated ground associated with volcanic activity on Deception I. where a number

Table 11. Some numerical comparisons between the moss floras of selected northern polar areas and southern cool-temperate and polar areas. NB: The areas listed vary considerably in area; all northern areas are situated above the Arctic Circle (lat. 67° N).

North Polar regions			South Polar regions			
Location	Number of species	Reference	Location	Latitude	Number of species	Reference
Chukotka	467	Afonina (2004)	Magallanes Region	*ca* 48° – 56° S	450	He (1998)
Arctic Alaska	415	Steere (1978)	Tierra del Fuego	52°30′–56° S	*ca* 340	Matteri & Schiavone (2002)
Beringian Chukotka	378	Afonina & Czernyadjeva (1995)	Aisén Region	*ca* 47° S	310	Seki (1974), He (1998)
Canadian Arctic Archipelago	*ca* 350	Ireland *et al.* (1987)	Southern Patagonia	51°–52° S	206	Hässel de Menendez *et al.* (1984), Matteri (1985*a*)
Polar Ural	329	Afonina & Czernyadjeva (1995)	Auckland Is.	50°30′–51° S	145	Vitt (1979*a*)
Canadian Eastern Arctic	304	Steere (1946)	Falkland Is.	51°–52°30′ S	141	Matteri (1986)
Yamal-Gydan Region	247	Afonina & Czernyadjeva (1995)	Îles Kerguelen	48°30′–50° S	*ca* 140	Ochyra (unpublished data)
Svalbard	288	Frisvoll & Elvebakk (1996)	Tristan da Cunha	37°02′–37°24′ S	130	Wace & Dickson (1965)
Taimyr Peninsula	292	Afonina & Czernyadjeva (1995)	South Georgia	53°30′–55° S	*ca* 125	Ochyra *et al.* (2002, 2003*b*), Ochyra (2005), Ochyra & Bednarek-Ochyra (2007*a*), and unpublished data
Wrangel I.	238	Afonina & Czernyadjeva (1995)	Campbell I.	52°33′ S	119	Vitt (1974*a*)
Novaya Zemlya	190	Afonina & Czernyadjeva (1995)	Antarctica	South of 60° S	111	Present work
Novosibirskiye Is.	163	Afonina & Czernyadjeva (1995)	Prince Edward Is.	46°30′–47° S	94	Ochyra (2008)
Ellesmere I. (northern)	151	Brassard (1971)	Macquarie I.	54°30′ S	91 (incl. four unnamed species)	Seppelt 2004; Bednarek-Ochyra & Ochyra 2007; Ochyra & Seppelt 2008*a*,*b*; Ochyra *et al.* 2008
Axel Heiberg I.	131	Kuc (1973*a*)	Gough I.	40° S	66	Wace & Dickson (1965)
Franz Josef Land	103	Afonina & Czernyadjeva (1995)	Îles Crozet	46°–46°30′ S	60	Ochyra (unpublished data)
Severnaya Zemlya	94	Afonina & Czernyadjeva (1995)	Heard I.	53° S	49	Bergstrom & Seppelt (1988), Bergstrom & Selkirk (1997)

Table 12. Occurrence, frequency, fertility, altitudinal limits and geographical elements of mosses in peri-Antarctic islands and archipelagoes and sectors of the continental and the maritime Antarctic.

Species	Region														Altitudinal range [m]	Fre-quency	Fertility	Geographical element
	BOU	SSI	SOI	SHI	NWP	CWP	SWP	NEP	SEP	MAU	END	WIL	ROS	BYR				
ANDREAEACEAE																		
Andreaea depressinervis			•	•	•	•		•							0–450	C	ST	A–SG
A. gainii	•	•	•	•	•	•	•	•	•					•	0–550	C	C	T–AP
A. regularis	•	•	•	•	•	•	•	•	•					•	0–500	C	C	S–AA
POLYTRICHACEAE																		
Notoligotrichum trichodon		•	•	•											5–200	O	ST	S–AA
Polytrichastrum alpinum	•	•	•	•	•	•	•	•							0–800	C	O	BI–I
P. longisetum			•	•											0–50	VR	ST	BI–I
Polytrichum piliferum		•	•	•	•	•	•	•	•						0–600	C	VR	BI–I
P. juniperinum		•	•	•	•	•		•							0–200	C	ST	BI–I
P. strictum		•	•	•	•	•		•							0–250	C	VR	BI–S
FUNARIACEAE																		
Funaria hygrometrica				•								•			5	VR	C	COS
ENCALYPTACEAE																		
Encalypta procera			•	•		•	•	•	•						0–600	F	VR	BI–S
E. rhaptocarpa			•	•			•	•	•						0–650	C	C	BI–I
DITRICHACEAE																		
Distichium capillaceum			•	•	•	•	•	•	•						0–375	C	O	BI–I
D. inclinatum								•							50	VR	VR	BI–S
Ditrichum heteromallum		•		•											35–500	O	VR	BI–I
D. ditrichoideum		•		•											2–200	R	VR	T–AA
D. hyalinocuspidatum		•	•	•		•		•							0–600	F	R	S–SA
D. conicum				•											2–350	R	VR	T–AA
D. hyalinum			•	•		•									0–300	F	VR	T–AP
Ceratodon purpureus	•	•	•	•	•	•	•	•	•	•	•	•	•		0–1347	C	VR	COS
DICRANACEAE																		
Dicranella varia		•													170	VR	ST	BI–I
D. hookeri	•			•									•?		0–480	R	VR	T–PA
D. campylophylla				•			•								5–70	R	ST	T–PA
Amphidium lapponicum						•									100	VR	ST	BI–I
Kiaeria pumila			•	•	•	•		•							0–238	F	VR	T–PA
Chorisodontium aciphyllum			•	•	•	•									0–400	C	ST	T–PA
Campylopus introflexus		•													0–225	R	VR	BI–S
C. pyriformis													•		2700 (3500)	VR	ST	BI–I
C. vesticaulis		•													50–200	VR	ST	T–PA
C. spiralis		•													50–200	VR	ST	T–SA
SELIGERIACEAE																		
Holodontium strictum				•											*ca* 15	VR	F	S–AA
Hymenoloma grimmiaceum	•	•	•	•	•	•									0–350	C	C	A–SG
H. crispulum				•	•	•	•	•	•						25–480	F	C	BI–I
H. antarcticum	•	•	•	•	•	•		•	•						0–460	C	C	T–PA
GRIMMIACEAE																		
Coscinodon lawianus										•	•				20–2200	F	ST	A–CO
Schistidium falcatum				•											35–220	VR	ST	S–AA
S. lewis-smithii				•											35–60	VR	F	A–MA
S. rivulare		•	•	•	•	•									0–120	F	C	BI–I
S. praemorsum				•	•	•									150–525	R	C	T–SA
S. andinum	•			•		•	•	•							0–500	F	C	T–SA
S. steerei				•											30–110	VR	C	A–MA
S. urnulaceum				•	•								•	•	130–1100	O	C	A–SG

Table 12. Continued.

Species	Region														Altitudinal range [m]	Fre-quency	Fertility	Geographical element
	BOU	SSI	SOI	SHI	NWP	CWP	SWP	NEP	SEP	MAU	END	WIL	ROS	BYR				
S. halinae				•				•							20–150	O	C	A–MA
S. antarctici	•	•	•	•	•	•	•	•	•	•	•	•	•	•	0–1379	C	C	A–SG
S. leptoneurum				•											5–220	R	C	A–MA
S. deceptionense				•											100–200	VR	C	A–MA
S. amblyophyllum		•		•			•	•							40–500	O	C	T–AA
S. cupulare				•											35–60	VR	C	S–AA
Grimmia plagiopodia				•		•	•	•		•	•	•	•	•	0–900	C	VR	BI–S
Guembelia longirostris									•						200	VR	ST	COS
Orthogrimmia sessitana				•	•	•	•	•	•	•	•	•	•		0–1800	C	F	BI–I
Dryptodon fuscoluteus								•				•			250	VR	F	BI–I
Racomitrium lanuginosum				•											100–130	VR	ST	BI–I
Bucklandiella subsecunda				•											125	VR	ST	PT–O
B. pachydictyon				•											100–350	VR	ST	T–AA
B. sudetica		•	•	•	•	•			•						0–1350	F	ST	BI-S
B. didyma		•		•											75–130	VR	ST	T-PA
POTTIACEAE																		
Tortella alpicola						•	•								165–717	O	ST	BI–I
T. fragilis					•	•	•	•	•						12–1150	F	ST	BI–I
Bryoerythrophyllum recurvirostrum			•	•			•	•	•		•	•			0–250	F	VR	BI–I
Didymodon brachyphyllus			•	•		•	•	•					•	•	0–540	C	ST	BI–I
Stegonia latifolia			•	•		•									15–125	R	R	BI–S
Pterygoneurum ovatum			•					•							8–150	VR	F	BI–I
Aloina brevirostris							•	•							0–500	R	F	BI–I
Tortula hoppeana					•										5–270	VR	C	BI–S
Hennediella heimii		•	•	•	•	•	•	•		•	•	•	•	•	0–540	C	C	BI–S
H. antarctica		•	•	•	•						•				0–265	C	C	BI–S
Willia austroleucophaea						•									5–300	VR	ST	S–SA
Syntrichia sarconeurum			•	•	•	•	•	•		•	•	•	•	•	0–1550	C	ST	A–PA
S. saxicola			•	•				•							0–450	C	VR	S–AA
S. filaris	•		•	•				•							0–340	C	ST	S–AA
S. caninervis var. *caninervis*								•							10–65	VR	ST	BI–S
S. caninervis var. *gypsophila*								•							830	VR	ST	BI–S
S. magellanica	•	•	•	•	•	•	•	•				•	•	•	0–1100	C	F	S–AA
ORTHOTRICHACEAE																		
Muelleriella crassifiolia			•	•	•	•									0–100	F	VR	S–PA
Orthotrichum rupestre					•	•									0–100	O	VR	BI–I
BRYACEAE																		
Leptobryum pyriforme				•		•									16–50	VR	C	COS
Schizymenium pusillum				•	•	•									2–25	R	C	S–SA
Pohlia wahlenbergii				•											0–500	R	ST	BI–I
P. wilsonii										•?	•				70	VR	ST	T–AA
P. drummondii		•	•	•	•	•									0–238	F	ST	BI–S
P. cruda		•	•	•	•	•	•	•							0–830	C	VR	BI–I
P. nutans	•	•	•	•	•	•	•	•	•			•	•	•	0–2800	C	F	BI–I
Bryum argenteum var. *argenteum*		•	•	•	•	•	•	•	•				•	•	0–750	C	ST	COS
B. argenteum var. *muticum*		•		•	•	•	•	•	•	•	•		•		0–1050	C	ST	COS
B. dichotomum	•	•	•	•	•		•								0–600	F	ST	BI–I
B. nivale							•								??	VR	ST	S–AA
B. orbiculatifolium		•		•						•					0–500	R	ST	T–AA
B. pseudotriquetrum	•	•	•	•	•	•	•	•	•	•	•	•	•		0–2600	C	F	BI–I
B. archangelicum			•	•	•	•	•								0–220	C	F	BI–I
B. pallescems		•	•	•	•	•	•								0–300	F	VR	BI–I

(cont.)

Table 12. Continued.

Species	Region														Altitudinal range [m]	Frequency	Fertility	Geographical element
	BOU	SSI	SOI	SHI	NWP	CWP	SWP	NEP	SEP	MAU	END	WIL	ROS	BYR				
								MEESIACEAE										
Meesia uliginosa			•	•				•							0–200	O	ST	BI–S
								BARTRAMIACEAE										
Conostomum pentastichum			•												20–200	R	ST	T–PA
C. magellanicum			•	•											0–275	R	ST	S–SA
Philonotis polymorpha				•											20–525	VR	ST	S–PA
Bartramia patens		•	•	•	•	•	•	•							0–700	C	C	T–AA
								AMBLYSTEGIACEAE										
Cratoneuropsis relaxa subsp. *minor*			•	•	•	•									0–500	C	ST	T–PA
Campylium polygamum			•	•			•								15–420	F	ST	BI-I
Sanionia uncinata		•	•	•	•	•									0–400	C	VR	BI–I
Sanionia georgicouncinata		•	•	•	•	•									0–340	C	VR	BI–S
Drepanocladus longifolius			•	•				•							0–50	R	ST	T–PA
Warnstorfia sarmentosa			•	•	•			•							0–275	C	ST	BI–I
W. fontinaliopsis			•	•	•	•									0–100	C	ST	S–PA
								BRACHYTHECIACEAE										
Brachythecium austroglareosum			•	•	•	•			•						0–330	F	ST	T–PA
B. austrosalebrosum	•	•	•	•	•	•	•	•							0–600	C	ST	S–PA
B. subpilosum					•										?	VR	ST	T–AA
Sciuro-hypnum fuegianum		•		•											75–350	R	ST	T–SA
S. glaciale			•	•	•										0–80	O	ST	BI–S
								PLAGIOTHECIACEAE										
Plagiothecium ovalifolium			•			•									0–100	O	ST	T–AA
P. orthocarpum										•					?	VR	ST	T–AA
								HYPNACEAE										
Hypnum revolutum			•	•	•	•	•	•							0–515	C	ST	BI–S
Platydictya jungermannioides			•	•	•	•	•	•	•						0–600	F	ST	BI–S
Isopterygiopsis pulchella			•	•		•									0–85	O	R	BI–I
TOTAL NUMBER OF TAXA (Species and varieties)	15	39	59	87 +1	49 +1	54 +1	36 +1	45 +2	20 +1	10 (11) +1	11 +1	11 +1	13 (14) +1	11				
	Antarctic Peninsula region 109 (107 species + 2 varieties)									Continental Antarctic region 24 (23 species + 1 variety)								

EXPLANATION:

REGION: **BOU** – Bouvetøya; **SSI** – South Sandwich Is.; **SOI** – South Orkney Is.; **SHI** – South Shetland Is.; **NWP** – north-western Antarctic Peninsula (Davis + Danco Coasts, western Trinity Peninsula – south to lat. *ca* 65° S); **CWP** – central Antarctic Peninsula (Graham and Loubet Coasts – lat. 65°–70° S); **SWP** – south-western Antarctic Peninsula (Fallières, Rymill and English Coasts, George VI Sound, Alexander I., Charcot I. – lat. 70°–75° S); **NEP** – north-eastern Antarctic Peninsula (eastern Trinity Peninsula, Joinville Island and James Ross Island groups – south to lat. 64° S); **SEP** – eastern and south-eastern Antarctic Peninsula (Oscar II, Foyn, Bowman, Wilkins, Black, Lassiter and Orville Coasts – lat. 64°–77° S); **MAU** – Dronning Maud sector (long. 30° W – 30° E); **END** – Enderby sector (long. 30°–90° E); **WIL** – Wilkes sector (long. 90°–150° E); **ROS** – Ross sector (long. 150° E – 150° W); **BYR** – Marie Byrd sector (long. 150° W – 90° W).

FREQUENCY: **VR** – very rare (1–2 records); **R** – rare (3–5 records); **O** – occasional (6–10 records); **F** – frequent (11–20 records); **C** – common (21 or more records).

FERTILITY: **ST** – Sterile; **VR** – very rare (1–2 records); **R** – rare (3–5 records); **O** – occasional (6–10 records); **F** – frequent (11–25 records); **C** – common (26 or more records).

GEOGRAPHICAL ELEMENT:

Antarctic Endemic: **A–PA** – pan-Antarctic endemic; **A–CO**: continental Antarctic endemic; **A–MA**: maritime Antarctic endemic; **A–SA** – Antarctic–South American subantarctic (Antarctic–South Georgian);
Subantarctic: **S–SA** – South American; **S–AA** – amphiatlantic; **S–AP** – amphipacific; **S–PA** – circumsubantarctic;
Temperate: **T–SA** – South American; **T–AA** – amphiatlantic; **T–AP** – amphipacific; **T–PA** – pan-temperate;
Bipolar: BI–S – without intermediate station; **BI–I** – with intermediate station in the tropics;
Cosmopolitan: **COS**;
Tropical: **PT–O** – pantropical oreophyte.

of species unknown elsewhere in Antarctica have been recorded (e.g. *Ditrichum conicum*, *Racomitrium lanuginosum*, *Bucklandiella pachydictyon*, *B. subsecunda*, *Schistidium deceptionense* and *Philonotis polymorpha*). Besides these, an additional six species have been found exclusively in the South Shetland Is., namely *Holodontium strictum*, *Schistidium cupulare*, *S. lewis-smithii*, *S. falcatum*, *S. steerei* and *S. leptoneurum*, the last three being Antarctic endemics.

The much smaller South Orkney Is. archipelago, occupying 626 sq. km with a 15.8% ice-free surface, has the second richest moss flora consisting of 59 species. Only one species, *Conostomum pentastichum*, is exclusively restricted in its Antarctic distribution to this area. All of the mosses in the South Orkney Is. are known from the small Signy I. (*ca* 25 sq. km) which has a complex geology, being especially rich in basic rocks, thus resulting in a broader range of habitats for mosses. However, the South Orkney Is. are colder and slightly drier than the South Shetland Is., despite being farther north, due to the Weddell Sea circulation effect. Uniquely, about 85% of these species, i.e. about 50% of the entire Antarctic moss flora, have been recorded in an area of about 1 sq. km at the northern end of the island (Lewis Smith 2005*c*).

The existence of latitudinal and, implicitly at least, environmental gradients in species richness is well demonstrated along the Antarctic Peninsula. This is a very large area which covers a wide range of climatic environments, and this has a marked effect on its biota. The western coast of the Antarctic Peninsula may be divided into three distinct zones:

(*a*) northern (lat. 62–65° S) which covers *ca* 26,400 sq. km of which 6.5% is ice-free;

(*b*) central (lat. 65–70° S) which occupies *ca* 89,700 sq. km, with 2.3% of ice-free area;

(*c*) southern (lat. 70–*ca* 77° S) which covers *ca* 1,052,400 sq. km of which only 0.65% is ice-free.

In contrast, the eastern coast of the Antarctic Peninsula may be divided into northern (lat. 62–64° S) and southern zones (lat. 64–77° S).

In total, 67 species and two varieties of moss have been discovered in the Antarctic Peninsula region and nine of these occur exclusively in this region. Of these, only *Tortella fragilis* is present in all the zones on the western and eastern coasts and three species are known from two zones, namely *Orthotrichum rupestre* in the northern and central zones of the western Antarctic Peninsula, *Tortella alpicola* in its central and southern zones and *Aloina brevirostris* in the southern zone of the western and in the northern zone of the eastern coast of the Antarctic Peninsula. The remaining five species are known to occur in single zones of the western Antarctic Peninsula, namely *Tortula hoppeana* and *Brachythecium subpilosum* (northern), *Amphidium lapponicum* and *Willia austroleucophaea* (central) and *Bryum nivale* (southern). All but one of the Peninsula species and a single variety are known to occur on the western coast, i.e. 66 species and one variety, including 49, 54 and 36 species in its northern, central and southern parts, respectively, and with *Bryum argenteum* var. *muticum* occurring in every part.

The moss flora of the eastern coast comprises rather less, 47 species and two varieties, but the extent of moss-dominated communities is very much smaller. Of these, only two taxa, *Syntrichia caninervis* var. *caninervis* and var. *gypsophila* has its only Antarctic occurrence in this region and, intriguingly, its only known occurrence in the Southern Hemisphere. The northern part of this coast has a fairly rich biota, with 45 species and two varieties of moss, whereas the southern part has a significantly reduced moss flora; only 20 species and one variety have been recorded south of lat. 64° S, on scattered nunataks.

The very remote, small and volcanic South Sandwich Is. and Bouvetøya are generally included as part of the maritime Antarctic, on account of their climate and terrestrial biotas, although they are situated a little to the north of lat. 60° S. They have restricted moss floras which, however, are not distinctly different from those elsewhere on the Antarctic Peninsula. The former group consists of 11 islands, most with recent volcanic activity, which together have 38 species and one variety. Four of these are associated with fumaroles or other geothermal ground and are unique to the Antarctic, notably *Dicranella varia*, *Campylopus introflexus*, *C. spiralis* and *C. vesticaulis*, while another three (*Ditrichum heteromallum*, *D. ditrichoideum* and *Sciuro-hypnum fuegianum*) are known additionally only from similar habitats on Deception I., a small island in the South Shetland Is. group, also with recent volcanic activity. Bouvetøya, on the other hand, has a very reduced biota and so far only 15 species of moss have been recorded on this extremely isolated volcanic island in the South Atlantic Ocean. These are mostly ubiquitous species widespread throughout the maritime and continental Antarctic, with the exception of *Dicranella hookeri* which is a very rare species in this biome, being otherwise only associated with geothermal sites on Deception I.

The moss flora of the continental Antarctic contrasts markedly in its paucity with that of the maritime Antarctic, despite the huge disparity in size between them. The Antarctic continent, except for the Scotia Sector of the Antarctic Peninsula south of lat. 70° S, covers *ca* 10,797,500 sq. km, but the percentage of ice-free areas is exceptionally low. The Wilkes Sector is the largest on the continent (*ca* 3,422,400 sq. km) but only 0.02% of its area is ice-free, whereas in the Enderby Sector (*ca* 3,001,800 sq. km), which is the second largest, 0.24% is occupied by ice-free ground. The Maud and Byrd Sectors cover *ca* 2,027,000 and 1,468,900 sq. km, with 0.18% and 0.15% of ice-free ground, respectively. The Ross Sector (*ca* 877,200 sq. km) is the smallest but has the largest expanses of ice-free land (2.4%). Most of the ice-free areas available for the colonisation by mosses are situated between lat. 65° and 70° S and only in the Ross Sector do they extend well beyond lat. 80° S, although here moss diversity and cover are exceptionally low.

Altogether 24 species and one variety of moss have been discovered in continental Antarctica. Of these, the record of *Dicranella hookeri* from southern Victoria Land (Cardot 1910*b*) has never been confirmed, nor has the voucher collection ever been located. *Funaria hygrometrica* was collected once, from soil in a horticultural greenhouse in Casey Station in Wilkes Land, and is thus a typical adventive element in the moss flora of the continental Antarctic. Realistically, therefore only 22 species may be considered as native to continental Antarctica (as defined here).

Only four species of moss are known to occur exclusively on the Antarctic continent: *Campylopus pyriformis*, *Coscinodon lawianus*, *Plagiothecium orthocarpum* and *Pohlia wilsonii*. Except for *C. lawianus*, they have been collected from extreme habitats, namely deep lakes (*Plagiothecium orthocarpum* and *Pohlia wilsonii*) and geothermal ground (*Campylopus pyriformis*), and are known from very few sites.

The species diversity of the six continental Antarctic sectors is very similar, and varies from ten species in the Maud Sector to 13 species in the Ross Sector (Table 12). Only four species (*Grimmia plagiopodia*, *Hennediella heimii*, *Schistidium antarctici* and *Syntrichia sarconeurum*) have been discovered in all sectors and an additional three species (*Bryum pseudotriquetrum*, *Ceratodon purpureus* and *Orthogrimmia sessitana*) are known in all but the Byrd Sector. Eight species are known from single sectors, namely *Bryum orbiculatifolium* and *Plagiothecium orthocarpum* (Maud Sector), *Hennediella antarctica* and *Pohlia wilsonii* (Enderby Sector), *Dryptodon fuscoluteus* (Wilkes Sector), *Campylopus pyriformis* (Ross Sector) and *Andreaea gainii* and *A. regularis* (Byrd Sector). The other seven species have been recorded from two adjoining sectors, namely *Coscinodon lawianus* (Maud and Enderby Sectors), *Bryoerythrophyllum recurvirostrum* (Enderby and Wilkes Sectors), *Bryum argenteum* var. *argenteum*, *Didymodon brachyphyllus* and *Schistidium urnulaceum* (Ross and Byrd Sectors), whilst *Pohlia nutans* and *Syntrichia magellanica* are known from three adjacent sectors (Wilkes, Ross and Byrd Sectors). The only taxon which shows a disjunct distribution pattern on the continent is *Bryum argenteum* var. *muticum*, which is frequent in the Ross and Maud Sectors, very rare in the Enderby Sector, and so far has not been recorded from the Wilkes and Byrd Sectors.

The fact that ice-free areas are restricted very largely to a narrow latitudinal span of coastline between lat. *ca* 65° and 70° S in most parts of the continental Antarctic markedly restricts any evidence of a latitudinal gradient in the distribution of species. The only area where there is a slight trend in declining species southwards is along the Victoria Land coast and the Transantarctic Mountains, where isolated ice-free areas occur from *ca* 72° to 86° S. This area is relatively well collected bryologically, and has revealed a cline from 13 species in northern Victoria Land to eight in southern Victoria Land, beyond lat. 76° S. Beyond 80° S there have been scattered collections of only five species (*Syntrichia sarconeurum*, *Bryum pseudotriquetrum*, *Schistidium antarctici*, *Grimmia plagiopodia* and *Ceratodon purpureus*) (Table 7).

Evaluation of the species richness and diversity of individual areas in the Antarctic is rather difficult, mainly because of the unequal stage of their bryological exploration. The present state of knowledge indicates that the richest moss flora, comprising 63 species, occurs on King George I. in the maritime Antarctic. This is the largest island in the South Shetland Is., covering over 1,300 sq. km with nearly 10% ice-free ground and a relatively wide range of habitats. In addition, it has been intensively studied since the early days of Antarctic exploration (Ochyra 1998*a*). Only four species fewer have been recorded on the tiny Signy I. (*ca* 25 sq. km) in the South Orkney Is. In any ranking of the bryologically richest sites in Antarctica the next two positions are occupied by two islands of the South Shetland Is., namely the large Livingston I. (56 species) and the small volcanic Deception I. (54 species and one variety), both intensively explored bryologically in the Post-Antarctic Treaty period. The true diversity of other sites may be underestimated as some are certainly under-collected. For example, on Coronation I. in the South Orkney I. 38 species have been detected, James Ross I. – 33 species and one variety, Anvers I. – 33, the Argentine Is. – 32, Adelaide I. – 31, Cape Tuxen – 30, Cuverville I. – 28, Alexander I. – 27, Joinville I. – 25, Petermann I. – 25 and Elephant I. – 24.

Frequency analysis of the Antarctic moss flora shows that the bulk of it (66 species, or 59.5%) falls into two clearly demarcated groups. Of these, the largest is a group of 'common' species (36) which have been recorded at 21 or more sites. These are usually ubiquitous mosses with a wide ecological amplitude growing, often in great abundance, in a wide range of habitats throughout the whole of the Antarctic, for example *Bryum pseudotriquetrum*, *Ceratodon purpureus*, *Schistidium antarctici*, *Bryum argenteum*, *Hennediella heimii*, *Syntrichia magellanica* and *Pohlia nutans*, or in most regions of the maritime Antarctic only, for instance *Andreaea gainii*, *A. regularis*, *Polytrichastrum alpinum*, *Hymenoloma antarcticum*, *Pohlia nutans*, *Bartramia patens*, *Sanionia georgicouncinata* and *S. uncinata*.

The other group consists of very rare species which have been collected at only one or two sites. It comprises 30 species which are distributed primarily in the maritime Antarctic, whereas on the continent only *Campylopus pyriformis*, *Plagiothecium orthocarpum* and *Pohlia wilsonii* are known from single sites. The Grimmiaceae, with 24 Antarctic species, contains most of the very rare moss species. These include three endemics (*Schistidium deceptionense*, *S. lewis-smithii* and *S. steerei*). The Dicranaceae, with 10 species in the genus, and the Pottiaceae, with 17, contribute five very rare species each. Some, including *Dicranella varia*, *Distichium inclinatum*, *Syntrichia caninervis* and *Tortula hoppeana* are bipolar taxa, having their only known occurrences in the Southern Hemisphere in Antarctica.

The aforementioned frequency groups are supplemented by the next two categories of 'rare' and 'frequent' species which have been collected at 3–5 and 11–20 localities, respectively. In the 'rare' category are 16 species that are scattered or localised, or restricted to one or a few areas in the maritime Antarctic, for instance *Campylopus introflexus*, *Ditrichum ditrichoideum* and *D. conicum*, found on islands with recent volcanic activity in the South Sandwich Is. and Deception I. The 'frequent' category consists of 18 species which are confined in distribution to the Antarctic Peninsula region, for example *Bucklandiella sudetica*, *Ditrichum hyalinocuspidatum*, *Encalypta procera*, *Kiaeria pumila* and *Schistidium andinum*, whereas in the continental Antarctic only *Bryoerythrophyllum recurvirostrum* and *Coscinodon lawianus* may be included within this category.

Sporophytes are known in more than 50% of Antarctic mosses (61 species), but relatively few are commonly fertile. For 21 species sporophytes have been observed in only one or two populations. In 13 of 51 barren species sometimes male, female or both gametangia have been occasionally observed, but in four species of Antarctic mosses sporophytes have never been recorded although very occasionally gametangia have been detected, namely in *Andreaea depressinervis*, *Coscinodon lawianus*, *Syntrichia sarconeurum* (all Antarctic endemics) and *Tortella alpicola*. In some species sporophytes may be produced in profusion, but they often remain immature and underdeveloped or produce aborted spores which certainly do not serve for effective reproduction. Examples include *Polytrichum piliferum*, *P. strictum*, *Polytrichastrum alpinum*, *Pohlia nutans* and *Bryum pseudotriquetrum*, although the latter often produces mature sporophytes too. The extreme southern location reported for moss sporophytes is for *Hennediella heimii* which was found with immature capsules from the Scott Coast in southern Victoria Land at lat. 77°55′S (Seppelt *et al.* 1992).

Only 22 species commonly and abundantly produce sporophytes, although some of these are relatively rare. This group includes, *inter alia*, all species of *Hymenoloma* and *Schistidium* (except for *S. falcatum*), *Andreaea gainii*, *A. regularis*, *Bartramia patens*, *Encalypta rhaptocarpa*, *Hennediella antarctica*, *H. heimii*. Of the few pleurocarpous mosses, most are sterile and only some populations of *Isopterygiopsis pulchella* have been found in fine fruiting condition with abundant and fully mature sporophytes. Mature sporophytes have been found in *Sanionia uncinata*, although only rarely, whereas in *S. georgicouncinata* immature sporophytes have been observed only once.

A number of species produce gemmae which serve as an effective mode of asexual reproduction. These are usually either rhizoidal tubers, which are sometimes produced in abundance (e.g. *Coscinodon lawianus*, *Ditrichum ditrichoideum*, *Dicranella* spp., *Leptobryum pyriforme*, *Pohlia wilsonii*), or axillary brood-bodies (*Didymodon brachyphyllus*, *Encalypta procera* and *Platydictya jungermannioides*). Some species reproduce by caducous leaf apices, for instance *Syntrichia sarconeurum* and *Tortella* spp.

PHYTOGEOGRAPHICAL ELEMENTS

A precise definition of phytogeographical elements is crucial for a proper understanding of the regional affinities of the flora of a given area. Only if they are well defined can they be meaningfully compared with the floras of different geographical regions, and thus enable evaluation of their affinities and provenances. The main difficulty in assessing the phytogeographical relationships of Antarctic mosses is defining the limits of south-temperate and subantarctic distribution patterns. These elements merge, sometimes imperceptibly, and become confused. Therefore, some authors have ignored them and have united them into a single category, alternatively termed 'antipodal' (Schuster 1963, 1969), 'subantarctic' (Grolle 1969) or simply 'antarctic' (Fulford 1963, 1966).

A detailed analysis of distribution patterns of liverwort species occurring on the Brunswick Peninsula, in southern South America, and the Falkland Is. led Engel (1978, 1990) to propose the first comprehensive system of phytogeographical units for the austral regions. This recognised, and justified, the existence of distinct austral temperate and subantarctic units, some of which are quite small. This system was successfully applied to the hepatic flora of Antarctica (Ochyra & Váňa 1989; Bednarek-Ochyra *et al.* 2000) and, with some minor modifications, also to the mosses of the continent (Ochyra 1998*a*; Ochyra *et al.* 1998). It has also been adopted in the present work, with some changes which reflect significant progress in the increasing knowledge of the distributions and taxonomy of austral mosses, especially in hitherto under-explored areas such as the Prince Edward Is., Îles Kerguelen, Heard I. and Macquarie I. in the Subantarctic. It is to be expected that our understanding of the distribution patterns of the species concerned will undergo further changes as new data become available, especially from southern South America and Tristan da Cunha, as well as Île Amsterdam and Île St. Paul in the southern Indian Ocean.

The moss flora of Antarctica is divided into six major phytogeographical elements, further subdivided into 16 sub-elements (Table 13). They define the various distribution patterns of species within the regions concerned and reflect the current knowledge of their geographical ranges (which are indicated on the distribution maps in the systematic part of the Flora). Distribution ranges were obtained primarily from specimens which we personally examined. However, data from taxonomic monographs, revisions and surveys as well as floristic treatments and contributions, were, after critical evaluation, also incorporated.

A striking feature of the Antarctic moss flora is the high proportion of bipolar species (50 species: 45.1%) which, together with cosmopolitan species (5 species: 4.5%),

Table 13. Conspectus of geographical elements of Antarctic mosses.

Category	Number of species	Percentage of total flora
Antarctic Endemic	**11**	**9.9**
Pan-Antarctic	1	0.9
Continental Antarctic	1	0.9
Maritime Antarctic	5	4.5
Antarctic-South Georgian	4	3.6
Subantarctic	**18**	**16.2**
Circumsubantarctic	4	3.6
Amphipacific	1	0.9
Amphiatlantic	9	8.1
South American	4	3.6
South-temperate	**26**	**23.4**
Pan-temperate	11	9.9
Amphipacific	1	0.9
Amphiatlantic	10	9.0
South American	4	3.6
Bipolar	**50**	**45.1**
Strict	18	16.2
Intermediate	32	28.9
Cosmopolitan	**5**	**4.5**
Tropical	**1**	**0.9**
Pan-tropical oreophyte	1	0.9

constitute almost half of the total number of species. The second largest group is composed of southern temperate species (26 species: 23.4%) and the third consists of subantarctic species (18: 16.2%). These two groups together contribute 44 species (39.6%) and comprise a fairly large number which have their ranges centred about the cool, moist southern temperate and subantarctic regions. Some of them may be 'relict' taxa, but others may be relatively new and recently evolving. However, this is in contrast to the liverworts which have a high proportion of 'relict' groups in the antipodal regions (Schuster 1969; Engel 1978, 1990).

The smallest group is composed of 11 (9.9%) strictly antarctic species. The total number of strictly austral species, i.e. those with south-temperate, subantarctic and antarctic distributions, is balanced by bipolar and cosmopolitan species, each grouping consisting of 55 species. The tropical element of the flora is represented by a single species, *Bucklandiella subsecunda*.

A. **Antarctic Endemic element**

Antarctica is a biome with a low level of moss endemism. The continent, together with its associated archipelagoes, was almost completely covered by continental ice during the Pleistocene, so that putative endemics must either have immigrated from farther north or survived on glacial refugia (nunataks) where, in addition to moss species, some mites and lichens are believed to have survived. Some may also have evolved in this biome.

In the course of bryological exploration 52 species and 22 infraspecific taxa have been described as new to science from the Antarctic. The majority of these 'endemics' proved after critical examination to be identical to taxa known from more northerly areas, and only ten species are currently accepted as Antarctic endemics. Seven of these are strict Antarctic endemics whilst an additional three penetrate weakly into subantarctic South Georgia, but having their maximum occurrence in the Antarctic. Two of these Antarctic–South Georgian endemics, *Hymenoloma grimmiaceum* and *Schistidium urnulaceum*, were originally described from South Georgia, but they now appear to have their main centre of distribution in Antarctica. One species originally described from Antarctica, *Andreaea gainii*, has long been considered an Antarctic endemic. However, it has since proved to have a wider range in the Holantarctic and is now considered not to have this phytogeographical status.

At present 11 species (9.9% of the flora) are considered to have an Antarctic distribution. They show four distinct distribution patterns within the Antarctic; these are here considered as geographical subelements. The low level of endemism of the moss flora contrasts significantly with that of the lichens, which are the other major component of the terrestrial biota of Antarctica. According to Øvstedal & Lewis Smith (2001 and unpublished data) the lichen flora consists of 101 endemics in the maritime Antarctic and 50 in the continental Antarctic. Conversely, the hepatic flora of this biome is very poor and contains no endemic taxa (Bednarek-Ochyra *et al.* 2000).

(*1*) *Pan-Antarctic subelement* — This category comprises taxa which are widespread in both West and East Antarctica. Only one species, *Syntrichia sarconeurum*, belongs to this subelement. It is widely distributed throughout the continental and maritime Antarctic (Fig. 171–172). It is an good example of a moss for which its phytogeographical status has been interpreted differently due to taxonomic confusion. Its range was at one time expanded to include southern South America (Greene 1975; Matteri 1982, 1985*a*; Lightowlers 1985*c*) but Ochyra & Zander (2007) provided evidence that these southern South American plants in fact belong to a different, though closely related species, *S. lithophila*.

Another three species, *Didymodon gelidus*, *Andreaea gainii* and *Schistidium antarctici*, were formerly considered as pan-Antarctic endemics (Ochyra *et al.* 1998) but have changed their phytogeographical status, though they are widespread in both West and East Antarctica. *Didymodon gelidus* proved to be conspecific with *D. brachyphyllus* and is now established as a bipolar species, whereas the geographical ranges of the other two have been extended to subantarctic South Georgia and other regions of the southern cool-temperate zone.

(*2*) *Continental Antarctic subelement* — One species, *Coscinodon lawianus*, is restricted to the continent. It is known from many localities on the periphery of the Maud and Enderby Sectors (Fig. 95). It is similar to some South American species of *Coscinodon* but, until fruiting material is found, any discussion of its affinities is inappropriate at present.

Plate 1–2. Limestone outcrops provide a habitat for unique moss communities dominated by species of *Bryum, Schistidium* and *Syntrichia.* Between the small hills the rock is acidic quartz-mica-schist and the vegetation dominated by *Andreaea* spp. and lichens, especially *Usnea* spp.
MARBLE KNOLLS, SIGNY I., SOUTH ORKNEY IS.

Plate 3. Stable, moist hillsides with a northerly and north-westerly aspect in the northern maritime Antarctic often support extensive moss-dominated vegetation. In the foreground *Andreaea*-dominated fellfield covers the drier ground; in the middle ground a wet seepage area is dominated by a continuous carpet of *Sanionia georgicouncinata, Warnstorfia fontinaliopsis* and *W. sarmentosa,* and in the background *Chorisodontium aciphyllum* and *Polytrichum strictum* have developed thick banks of moss peat.
SKUA TERRACE, NORTH-WEST SIGNY I., SOUTH ORKNEY IS.

Plate 4. Coastal hillsides in the northern maritime Antarctic often support moss communities, especially down-wind of penguin and other seabird colonies. These produce nitrogenous compounds which are deposited over the nearby landscape as a nutrient-rich aerosol.
AITCHO I., SOUTH SHETLAND IS.

Plate 5. Deeply-eroded ice-free hills of unstable dry sedimentary deposits, largely devoid of vegetation, but the only known habitat of *Syntrichia caninervis* var. *caninervis.* [*Syntrichia caninervis* var. *gypsophila* is known from a single locality at nearby Larsen Inlet on the Antarctic Peninsula mainland.]
NEAR CAPE WIMAN, SEYMOUR I., NORTH-EAST ANTARCTIC PENINSULA.

Plate 6. The light-coloured area is the Adelie penguin and blue-eyed cormorant colony where J. D. Hooker landed from HMS *Erebus* in 1844 and made the first collection of mosses in Antarctica by a professional botanist.
NORTH-EASTERN COCKBURN I., JAMES ROSS ISLAND GROUP, NORTH-EAST ANTARCTIC PENINSULA.

Plate 7. Coastal hillsides in the northern maritime Antarctic often support locally extensive moss communities, especially down-wind of penguin and other seabird colonies. These produce nitrogenous compounds which are deposited over the nearby landscape as a nutrient-rich aerosol.
BAILEY HEAD, DECEPTION I., SOUTH SHETLAND IS.

Plate 8. Dissected ice-free plateau of unstable dry sedimentary deposits, devoid of vegetation.
SEYMOUR I., NORTH-EAST ANTARCTIC PENINSULA.

Plate 9. Undulating ice-free topography with freshwater streams and lakes, but with very sparse moss vegetation.
VESTFOLD HILLS, PRINCESS ELIZABETH LAND.

Plate 10. One of many extensive inland ice-free areas in the McMurdo Dry Valleys region of southern Victoria Land. Mosses are restricted mainly to seepage areas below glaciers.
LOWER WRIGHT VALLEY, SOUTHERN VICTORIA LAND [Photo D. D. Wynn-Williams].

Plate 11. One of several high diversity botanical sites in continental Antarctica. Most moss species known in East Antarctica have been recorded here.
CRATER CIRQUE, NORTHERN VICTORIA LAND.

Plate 12. In some ice-free areas seals have penetrated many kilometres inland and died. Their mummified carcasses and skeletons occasionally serve as a nutrient-rich oasis for mosses and lichens in an otherwise infertile and barren landscape. This crabeater seal skeleton is probably several centuries old and the soil around it is colonised by *Bryum argenteum, B. pseudotriquetrum, Ceratodon purpureus, Hennediella antarctica, Schistidium antarctici* and *Syntrichia magellanica.*
SAN JOSÉ VALLEY, JAMES ROSS I., NORTH-EAST ANTARCTIC PENINSULA.

Plate 13. The most extensive almost-continuous community of *Chorisodontium aciphyllum* and *Polytrichum strictum* in Antarctica. The black areas are where the living moss surface has been killed and eroded by a snow field that developed and later disappeared during a period when the climate became colder for a period of decades or centuries, possibly during the "Little Ice Age".
MOE I., SOUTH ORKNEY IS.

Plate 14. A 2-metre deep moss turf bank formed exclusively by *Chorisodontium aciphyllum* and *Polytrichum strictum*. Beneath the living surface the mosses have accumulated a largely undecomposed peat reaching 2 m depth (3 m in one exceptional bank on Elephant I., South Shetland Is.) which has a permafrost below about 30–35 cm. Such communities are frequent in the South Orkney Is. And some sites on the western coast of the northern Antarctic Peninsula, but rare in the South Shetland Is. The moss peat at the base of these ancient features has been radiocarbon dated at *ca* 5500 years old.
NORTH-WEST COAST, SIGNY I., SOUTH ORKNEY IS.

Plate 15. Seepage area dominated by *Brachythecium austrosalebrosum* (fore- and middle-ground) with *Sanionia georgicouncinata* and *Warnstorfia fontinaliopsis* in background. A few years after this photograph was taken the entire area was destroyed by fur seals.
MIROUNGA FLATS, SIGNY I., SOUTH ORKNEY IS.

Plate 16. Melt water flushes on a limestone slope dominated by *Bryum pseudotriquetrum* and *Sanionia georgicouncinata.*
THULLA PT., SIGNY I., SOUTH ORKNEY IS.

Plate 17. Probably the most extensive (*ca* 4 ha) continuous stand of moss in Antarctica, dominated by *Sanionia georgicouncinata.* The small island benefits from a relatively high summer rainfall and nutrient input from large penguin colonies nearby.
AITCHO I., SOUTH SHETLAND IS.

Plate 18. Seepage area below a receding icefield, dominated by *Bryum pseudotriquetrum*.
TERRAPIN HILL, JAMES ROSS I., NORTH-EAST ANTARCTIC PENINSULA.

Plate 19. Melt runnel bordered by a thick (to *ca* 15 cm) carpet of *Bryum argenteum* var. *muticum*.
SOUTHERN END OF EDMONSON POINT, VICTORIA LAND.

Plate 20. Melt stream and seepage area dominated by *Bryum argenteum* var. *muticum*.
EDMONSON PT., VICTORIA LAND WITH MT. MELBOURNE IN BACKGROUND.

Plate 21. Boundary between a thick stand of *Bryum argenteum* var. *muticum* (left) at the moist margin of a melt stream and *Ceratodon purpureus* (right) on dry porous ash.
EDMONSON PT., VICTORIA LAND.

Plate 22. Carpet of *Warnstorfia fontinaliopsis* in a wet hollow with rounded turves of *Polytrichum strictum* at the drier margin. The light brown circles and spots are formed by *Thyronectria hyperantarctica,* an ascomycete fungus which commonly infects mosses of wet habitats in the maritime and, occasionally, continental Antarctic.

Plate 23. *Andreaea* spp. and *Bucklandiella sudetica* in fellfield on a moist hillside with dense moss cover.
MORAINE VALLEY, SIGNY I., SOUTH ORKNEY IS.

Plate 24. The largest known closed stand of *Andreaea* in Antarctica covering *ca* 1 ha, although broken by a few small soil boils and occasional turves of *Chorisodontium aciphyllum.* The community comprises mainly *A. gainii* and *A. regularis,* but *A. depressinervis* is also present.
ANDREAEA PLATEAU, SIGNY I., SOUTH ORKNEY IS.

Plate 25. *Brachythecium austrosalebrosum, Bryum pseudotriquetrum, Schistidium antarctici, Syntrichia saxicola, S. magellanica* in a community on moist calcareous soil (pH 6.5–8.5). Such calcicolous communities on Signy I. can have up to 15 moss species within 1 sq. m.
MARBLE KNOLLS, SIGNY I., SOUTH ORKNEY IS.

Plate 26. Cushions-shaped turves of *Ceratodon purpureus* and *Syntrichia magellanica* emerging from a melting snow drift. The mosses are growing on coarse sand which often buries the cushions for much of the growing season. – Crater Cirque, northern Victoria Land.

Plate 27. In continental Antarctica the convoluted turf of *Ceratodon purpureus* becomes extremely dry and cracked during periods of drought, and may remain in this condition throughout most of the summer.
CLARK PENINSULA, WILKES LAND.

Plate 28. Convoluted turf formed by *Schistidium antarctici* (dark green) with *Bryum pseudotriquetrum* (yellow-green) colonising the moist troughs between the ridges.
BAILEY PENINSULA, WILKES LAND.

Plate 29. Thick turf of *Dicranella hookeri* and *Philonotis polymorpha* on soil warmed by geothermal activity. The steam issuing from the fumarole vents often exceeds 90°C and condenses as a film of water droplets on the moss.
MT. POND AT *CA* 500 M ALTITUDE ON DECEPTION I., SOUTH SHETLAND IS.

Plate 30. Fumaroles emitting steam in a geothermal area. Such areas are colonised by several moss species, most of which are unique to this habitat, notably *Dicranella hookeri, Ditrichum* spp. and *Philonotis polymorpha.* The soil surface below the moss turf reaches 25–40°C, and locally exceeds 70°C at 5 cm below the surface.
MT. POND AT *CA* 500 M ALTITUDE ON DECEPTION I., SOUTH SHETLAND IS.

Plate 31. Unique bryophyte community centred on a steam-emitting fumarole. The principal species are *Ditrichum* spp. and *Dicranella hookeri* (dark green, left), *Campylopus* spp. (green, upper), *Notoligotrichum trichodon* (green scattered shoots, lower right), and the liverwort *Cryptochila grandiflora* (brown, lower left). BELLINGSHAUSEN I., SOUTH SANDWICH IS. [Photo P. Convey]

Plate 32. Line of fumaroles with almost concentric zones of bryophytes radiating from the vents. The inner zone is dominated by *Campylopus introflexus* and *C. spiralis,* the next zone by *Marchantia berteroana,* and the outer zone by *Bucklandiella didyma, Notoligotrichum trichodon, Pohlia nutans* and *Sanionia uncinata.* Within the *Campylopus* zone the temperature of the soil surface beneath the moss ranged from 15 to 25°C, at 2.5 cm depth from 20 to 35°C, and at 5 cm depth from 25 to >50°C.
BELLINGSHAUSEN I., SOUTH SANDWICH IS. [Photo P. Convey]

Plate 33. *Campylopus introflexus* on geothermal ground.
CANDLEMAS I., SOUTH SANDWICH IS.

Plate 34. *Racomitrium lanuginosum* on coarse scoria. This species is known from only one small site where it appears to be a recent colonist.
PENDULUM COVE, DECEPTION I., SOUTH SHETLAND IS.

Plate 35. *Orthogrimmia sessitana* on metamorphic rock.
ROTHERA PT., ADELAIDE I., MARGUERITE BAY, SOUTHERN ANTARCTIC PENINSULA.

Plate 36. *Encalypta procera* and *Bryum argenteum* var. *argenteum* on moist silt.
MARS OASIS, ALEXANDER I.

Plate 37. *Bartramia patens* in diorite rock crevice.
ROTHERA PT., ADELAIDE I., MARGUERITE BAY, SOUTHERN ANTARCTIC PENINSULA.

Plate 38. *Distichium capillaceum* with exceptionally abundant sporophytes.
LÉONIE I., MARGUERITE BAY, SOUTHERN ANTARCTIC PENINSULA.

Plate 39. The only known population of *Polytrichastrum alpinum* producing abundant sporophytes; throughout its Antarctic range it is very rarely fertile. The site is very sheltered and has an unusually favourable microclimate.
CRATER LAKE, DECEPTION I., SOUTH SHETLAND IS.

Plate 40. *Pterygoneurum ovatum* colonising moist silt on a moraine.
ORWELL GLACIER, MORAINE VALLEY, SIGNY I., SOUTH ORKNEY IS.

Plate 41. *Encalypta rhaptocarpa* on a pocket of fine soil amongst scree.
LÉONIE I., MARGUERITE BAY, SOUTHERN ANTARCTIC PENINSULA.

Plate 42. *Hennediella antarctica* colonising an area of ground, *ca* 15 sq. m, that had been severely disturbed by human activity close to a research station. A few years later the population was destroyed by fur seals.
FACTORY COVE, SIGNY I., SOUTH ORKNEY IS.

(*3*) *Maritime Antarctic subelement* — This group consists of five species confined to the maritime Antarctic, and it is remarkable that they are all in the genus *Schistidium*. They occur exclusively in the South Shetland Is. archipelago, except for *S. halinae* which has also been recorded from nearby James Ross I. off the north-eastern tip of the Antarctic Peninsula. Until recently *Schistidium* has been one of the most taxonomically neglected of all moss genera and many distinct species have been merged into *S. apocarpum*, making it both an artificial and unrealistic 'species'. This makes any assessment of the phytogeographical status of the maritime Antarctic *Schistidium* endemics difficult. On the one hand, some are very distinct and taxonomically isolated species, especially *S. lewis-smithii*, *S. halinae* and *S. steerei*. These have very distinctive morphologies which it is hard to visualise having evolved rapidly in the recently deglaciated areas on King George I. and Livingston I. in the South Shetland Is. archipelago and on James Ross I. On the other hand, it is possible that they could be recent immigrants from more northerly areas where they are as yet unknown. This hypothesis could only be verified by a critical monographic evaluation of *Schistidium* in the cool-temperate and cold regions of the Southern Hemisphere. This would need to be preceded by intensive field work since, until recently, the genus seems to have been undercollected, especially in southern South America. What is remarkable is that the genus is well represented on Îles Kerguelen, both in terms of frequency and species diversity, but no species with morphologies similar to those in Antarctic taxa have been recorded (Ochyra, personal observations), nor have recent collections from the Fuegian region yielded similar plants.

Schistidium deceptionense
Schistidium halinae
Schistidium leptoneurum
Schistidium lewis-smithii
Schistidium steerei

(*4*) *Antarctic–South Georgian subelement* — This category contains four species which are restricted in occurrence to the Antarctic and subantarctic South Georgia. Despite their occurrence in two phytogeographical regions, they are here considered to belong to the Antarctic endemic group, rather than the subantarctic, because they appear to have their main centre of occurrence in the Antarctic biome, being either restricted to the Antarctic Peninsula region (*Andreaea depressinervis* and *Hymenoloma grimmiaceum*) or having pan-continental Antarctic distribution (*Schistidium antarctici* and *S. urnulaceum*). They occur only rarely on some of the subantarctic islands.

It is possible that *Andreaea drepressinervis* represents a palaeoantarctic element which survived the Pleistocene glaciation *in situ* because it is a barren plant with very limited modes of dispersal. *Schistidium antarctici* may also possibly be an Antarctic palaeoendemic since it is very scarce on South Georgia and Bouvetøya, to which it might only be only a recent immigrant, reaching these isolated islands via long distance dispersal of the spores which are produced by most populations.

In contrast to the two previous species, *Hymenoloma grimmiaceum* and *Schistidium urnulaceum* were originally described from South Georgia but only recently discovered in the Antarctic. The former is much more widespread and abundant in Antarctica than on South Georgia and also produces sporophytes in great profusion, so it is possible that it survived the Pleistocene climatic catastrophe in the Antarctic and subsequently spread to South Georgia, though it may be a survivor on both areas. *Schistidium urnulaceum* is more frequent on South Georgia, but in the Antarctic occurs in the Scotia, Ross and Byrd Sectors (Fig. 109) which may indicate its survival *in situ*.

Andreaea depressinervis
Hymenoloma grimmiaceum
Schistidium antarctici
Schistidium urnulaceum

B. **Subantarctic element**

Separation of the subantarctic and south-temperate elements presents a major problem in the evaluation of the phytogeographical status of antipodal moss species. This is because they overlap and may be easily confused, and the status of the species concerned depends heavily on the available distributional data from under-worked areas. For example, *Andreaea regularis* was previously considered as a temperate species (Ochyra 1998*a*; Ochyra *et al.* 1998) but detailed analysis of its current distribution pattern, with consideration of many recent records, has resulted in the reinterpretation of this species as subantarctic, since it is abundant in Antarctica and the subantarctic islands, whilst its distribution in southern South America is limited to alpine elevations. Conversely, *Plagiothecium ovalifolium* is now considered as a south-temperate species, having been recorded at many low altitude stations in southern South America. The subantarctic element is here interpreted as comprising species having an optimum occurrence and abundance on subantarctic islands though occasionally penetrating into more northerly areas (mostly at high elevations), as well as into the Antarctic.

Eighteen species (16.2%) have the major part of their range on subantarctic islands as defined by Greene (1964*a*) and Lewis Smith (1984*a*). Four distinct subelements can be recognised.

(*1*) *Circumsubantarctic subelement* — The ranges of species demonstrating this distribution pattern cover all, or most, subantarctic islands and include the southernmost extremities of South America and Australasia. Only four species (3.6%) occurring in the Antarctic belong within this subelement.

Brachythecium austrosalebrosum
Muelleriella crassifolia
Philonotis polymorpha
Warnstorfia fontinaliopsis

(*2*) *Amphipacific Subantarctic subelement* — This is a very rare distribution pattern and consists of species occurring on the islands in the Australasian and South American sectors of the Subantarctic, i.e. on Macquarie I. and South Georgia, respectively. At present it is represented by a sole species, *Andreaea gainii*, which until recently was classified as having a West Antarctic (Ochyra 1998*a*) or pan-Antarctic (Ochyra *et al.* 1998) element. However, this status was changed as a result of its recent discovery on Macquarie I. and Tasmania (Fig. 21). It is likely that further taxonomic and floristic studies on *Andreaea* will contribute to a refinement of its geographical range.

(*3*) *Amphiatlantic Subantarctic subelement* — This category comprises species occurring within the South American and Atlantic sectors of the Subantarctic and extending to the Kerguelen Province in the South Indian Ocean. It is represented by nine species (8.1%):

Andreaea regularis
Bryum nivale
Holodontium strictum
Notoligotrichum trichodon
Schistidium cupulare
Schistidium falcatum
Syntrichia filaris
Syntrichia magellanica
Syntrichia saxicola

(*4*) *South American Subantarctic subelement* — This category consists of species which are confined to the South American subantarctic island of South Georgia and which also occasionally extend northwards in the South American Cordillera, occurring at scattered alpine sites from Tierra del Fuego to Colombia, very rarely in the mountains in south-eastern Brazil, and southwards to the northern maritime Antarctic.

Conostomum magellanicum
Ditrichum hyalinocuspidatum
Schizymenium pusillum
Willia austroleucophaea

C. **South-temperate element**

This element comprises species which are restricted in distribution to the southern temperate zone. These are generally found in New Zealand and Australia, disjunct in southern South America and/or South Africa, and occasionally extending to more northerly latitudes in tropical mountains. This is the largest group amongst antipodal mosses in the Antarctic, consisting of 26 species (23.4%). Within this broad phytogeographical category, four subelements can be recognised.

(*1*) *Pan-Temperate subelement* — This unit comprises species which are generally widespread and locally frequent and abundant throughout all or most temperate regions of the Southern Hemisphere. They occur in southern South America, New Zealand, south-eastern Australia, Tasmania, southern Africa, as well as on remote islands in the South Atlantic Ocean (Tristan da Cunha and Gough I.) and South Indian Ocean (Île Amsterdam and Île Saint-Paul). In addition, they often penetrate into the tropical zone where they occur at high elevations in the mountains. Eleven species (9.9%) belong within this category.

Brachythecium austroglareosum
Bucklandiella didyma
Campylopus vesticaulis
Chorisodontium aciphyllum
Conostomum pentastichum
Cratoneuropsis relaxa
Dicranella hookeri
Dicranella campylophylla
Drepanocladus longifolius
Hymenoloma antarcticum
Kiaeria pumila

(*2*) *Amphipacific Temperate subelement* — Only one species (0.9%), *Ditrichum hyalinum*, is distributed in the temperate regions of southern South America, and disjunct on Campbell I. in the Australasian sector (Fig. 60).

(*3*) *Amphiatlantic Temperate subelement* — This subelement consists of ten species (9.0%) which are distributed mainly in the temperate regions of southern South America and South Africa, including Tristan da Cunha and Gough I., and which extend to the subantarctic islands of the Kerguelen Province of the South Indian Ocean. They are not considered as subantarctic species because their northward extensions from the Subantarctic are not restricted to high elevations.

Bartramia patens
Brachythecium subpilosum
Bryum orbiculatifolium
Bucklandiella pachydictyon
Ditrichum conicum
Ditrichum ditrichoideum
Plagiothecium orthocarpum
Plagiothecium ovalifolium
Pohlia wilsonii
Schistidium amblyophyllum

(*4*) *South American Temperate subelement* — This subelement consists of species having their main centre of occurrence in southern South America, including the Falkland Is. and South Georgia. They also penetrate northward to the Valdivian region and, occasionally, to the northern Andes where they occur mainly in the alpine zone. There are four species (3.6%) in this subelement.

Campylopus spiralis
Schistidium andinum
Schistidium praemorsum
Sciuro-hypnum fuegianum

D. **Bipolar element**

This element includes the most widespread and common species of Antarctic mosses, which are also widely distributed in the Northern Hemisphere and disjunct in Australasia and/or in southern South America or Antarctica. These taxa are not particularly primitive and it is likely that they reached their southern ranges by long-distance dispersal along mountain chains, rather than as a result of continental break-up. In recent decades the problem of bipolar taxa in bryophytes has received much attention (e.g. Schuster 1969, 1983*a*; Schofield & Crum 1972; Schofield 1974; Ochyra 1992*b*; Ochyra & Buck 2003). These authors accepted the classic and rather broad definition of bipolar plants formulated by Du Rietz (1940), which considered as bipolar those taxa which occur in the polar and cool-temperate regions of both hemispheres, but which are absent from the tropical lowlands. They may, or may not, have intermediate stations in mountains in the tropical regions.

Because the majority of bipolar bryophytes are clearly of Holarctic origin, it is generally postulated that they reached their present austral locations by long-distance dispersal from the Northern to the Southern Hemisphere along one or more of the three main trans-tropical routes, namely (*1*) the American pathway across the Neotropics to Patagonia and the maritime Antarctic; (*2*) the African pathway along the East and South African mountains to Subantarctic islands in the Kerguelen Province; (*3*) the Indomalayan-Melanesian pathway, across the chain of Malesian islands with high mountains to southeastern Australia and New Zealand and their associated nearby islands.

Bipolar species are the largest group in the moss flora of Antarctica, consisting of 50 species (45.1%). They are markedly diverse, and they are here subdivided into two distinct subelements.

(*1*) *Strict Bipolar subelement* — This contains species which occur in the cold or temperate regions of both hemispheres, without any transitional localities at higher elevations in tropical and equatorial areas. The number of bipolar moss taxa (in the strict sense) has changed in recent decades through increased exploration of the tropics and austral polar and cool-temperate regions, as well as through progress in the taxonomic study of many groups. Accordingly, the early lists of bipolar species have been substantially modified. For example, Schofield (1974) listed 24 trans-American bipolar moss species without intermediate occurrences in the Neotropics. However, as the result of a careful examination of the global ranges of the species concerned, Ochyra (1992*b*) excluded 17 species from this list, but at the same time added five species to it, so the final list consisted of only 12 species. After another decade of field work the list was re-examined and four further species were deleted, but were replaced by another 12 species. Thus, the current list of trans-American bipolar moss disjuncts in the strict sense consists of 20 species. However, the complete list of strict bipolar moss species is longer, because some species also occur at bipolar stations in Australasia and on some subantarctic islands.

In the Antarctic this subelement is represented by 18 species (16.2%). Most of these are rare or occasional, sometimes known only from solitary stations (e.g. *Distichium inclinatum*, *Syntrichia caninervis* and *Tortula hoppeana*), and only *Sanionia georgicouncinata* appears to be a common species of this category in the maritime Antarctic.

Bucklandiella sudetica
Campylopus introflexus
Distichium inclinatum
Encalypta procera
Grimmia plagiopodia
Hennediella antarctica
Hennediella heimii
Hypnum revolutum
Meesia uliginosa
Platydictya jungermannioides
Pohlia drummondii
Polytrichum strictum
Pterygoneurum ovatum
Sanionia georgicouncinata
Sciuro-hypnum glaciale
Stegonia latifolia
Syntrichia caninervis
Tortula hoppeana

(*2*) *Transitional Bipolar subelement* — Species showing this distribution pattern usually have some transitional localities at higher elevations in tropical and equatorial areas. This is the largest subelement in the Antarctic muscoflora, consisting of 32 species (28.9%). Some bipolar species with transitional occurrences in the tropics are common and widespread in the Antarctic, for example *Bryum pseudotriquetrum*, *Pohlia cruda*, *P. nutans*, *Polytrichastrum alpinum* and *Polytrichum piliferum*, but some are very rare and known from single records, for instance *Amphidium lapponicum*, *Campylopus pyriformis*, *Dicranella varia*, *Dryptodon fuscoluteus* and *Racomitrium lanuginosum*.

Aloina brevirostris
Amphidium lapponicum
Bryoerythrophyllum recurvirostrum
Bryum archangelicum
Bryum dichotomum
Bryum pseudotriquetrum
Bryum pallescens
Campylium polygamum
Campylopus pyriformis
Dicranella varia
Didymodon brachyphyllus
Distichium capillaceum
Ditrichum heteromallum
Dryptodon fuscoluteus
Encalypta rhaptocarpa
Hymenoloma crispulum
Isopterygiopsis pulchella
Orthogrimmia sessitana
Orthotrichum rupestre
Pohlia cruda

Pohlia nutans
Pohlia wahlenbergii
Polytrichastrum alpinum
Polytrichastrum longisetum
Polytrichum piliferum
Polytrichum juniperinum
Racomitrium lanuginosum
Sanionia uncinata
Schistidium rivulare
Tortella alpicola
Tortella fragilis
Warnstorfia sarmentosa

E. **Cosmopolitan element**

This element consists of very widespread and 'weedy' species, often growing on disturbed and open sites, which are not restricted to higher elevations in the tropics. Five species (4.5%) of the Antarctic moss flora have a wide pan-continental distribution. Sometimes species such as *Polytrichum juniperinum* and *P. piliferum* are considered to be cosmopolitan, but a careful analysis of their ranges shows that they are rather infrequent in the tropics and occur only at high elevations in the mountains. Therefore they are interpreted here as bipolar species with intermediate stations in the tropical regions.

Bryum argenteum
Ceratodon purpureus
Funaria hygrometrica
Guembelia longirostris
Leptobryum pyriforme

F. **Tropical element**

Species with their main centre of occurrence in tropical regions are generally absent from the polar and cool-temperate regions of the Southern Hemisphere. The only exception in the Antarctic moss flora is *Bucklandiella subsecunda*, which has been recorded once, on heated ground on the volcanic Deception I. in the South Shetland Is. This is an oreophyte occurring in tropical regions of all three sectors, i.e. America, Africa and Asia-Australasia, and therefore it is included in the pan-tropical subelement. The record of the neotropical *Dicranella hilariana* from the volcanic Bellingshausen I. in the South Sandwich Is. archipelago (Ochyra *et al.* 2000) has proved to be a misdetermination and the voucher material actually belongs to *D. varia*.

SOME REMARKS ON THE ORIGIN OF THE MOSS FLORA OF ANTARCTICA

The fossil records of bryophytes are poor compared to those of vascular plants, and the Antarctic record is particularly sparse. Bryophytes are rarely fossilised because they lack the resistant structures typically found in vascular plants, for example thick cuticles and lignified vascular tissues. Moreover, fossil bryophytes may also have been easily overlooked because of their small size, or they may have been destroyed by many of the commonly used methods of extracting vascular plant fossils from sediments. Fossil bryophytes are much less common in the Southern Hemisphere than in the Northern Hemisphere where the oldest fossil moss, *Muscites plumatus*, is known from the Lower Carboniferous (Thomas 1972).

Very little is known about the pre-Quaternary moss flora of Antarctica. The oldest reported Antarctic moss is the permineralised *Merceria augustica* from the Transantarctic Mountains (Smoot & Taylor 1986); it is also the oldest Southern Hemisphere moss and the oldest bryophyte with anatomical preservation of the gametophyte. This is an exceptional record from this continent and only a few unpublished records from the Lower Cretaceous (100–120 m. y. ago) and the Pleistocene (2–3 m. y. ago) are known from the Antarctic (see p. 29). In contrast, fossil liverworts are comparatively more frequent in the Antarctic and a number of species of the Lower Cretaceous period are known from Alexander I. (Cantrill 1997) and the South Shetland Is. (Cantrill 2000), as well as from the Early Jurassic from Trinity Peninsula (Halle 1913; Gee 1989; Rees & Cleal 2004; Ociepa 2007). These few data strongly support the hypothesis promoted by Schuster (1972, 1976, 1979, 1981, 1982, 1983*a*) that a specialised flora of *Glossopteris* and associated plants which had occurred in the Southern Hemisphere in the later Palaeozoic, and subsequent araucarian floras of the Mesozoic, were rich in liverworts and mosses. Descendants of these ancient bryophytes persist in the present subantarctic and cool-temperate flora of various groups of liverworts and, very probably, mosses. Unfortunately, palaeobotanical data do not throw much light on the problem of the origin and composition of the present-day moss flora of Antarctica. Therefore, in any discussion of this issue bryogeographers must speculate on possible phytogeographical affinities of the species, taking into consideration their biology and ecology as well as modes and effectiveness of dispersal.

A striking feature of the moss flora of Antarctica is the equally-matched proportion of broadly conceived Holantarctic and Holarctic taxa which contain 55 species each. They constitute 99.1% of the total flora. The first of these groups contains south-temperate, subantarctic and Antarctic endemic species, while the second is composed of bipolar and cosmopolitan species which reach their maximum occurrence in the Northern Hemisphere. This phytogeographical composition of the moss flora of Antarctica is in sharp contrast with the liverwort flora of this biome in which the antipodal element comprises about 75% of the total number of species. However, despite the lower percentage, bipolar species of hepatics are dominant in the Antarctic, whereas most of the antipodal taxa appear to be rare and sparse (Bednarek-Ochyra *et al.* 2000). A similar situation also exists with mosses, the bipolar taxa predominating in the vegetation of Antarctica, both in terms of frequency and cover.

The cosmopolitan nature of the Antarctic moss flora is best evidenced at the generic level. Of 55 moss genera currently recognised in this biome, the overwhelming majority have worldwide or predominantly Holarctic distribution. Only seven genera can be considered as Holantarctic or Southern Hemisphere taxa. These are *Cratoneuropsis* and *Muelleriella*, which have a strictly pan-south-temperate distribution, and *Chorisodontium* and *Notoligotrichum*, which additionally penetrate the Neotropics, occurring at high elevations in the mountains. Two additional genera, *Holodontium* and *Willia*, have amphiatlantic subantarctic ranges, and *Schizymenium* is a widely distributed pan-tropical oreophyte. All these genera are represented by single species in Antarctica which, except for *Chorisodontium aciphyllum*, are very rare or occasional and play an inconspicuous role in the vegetation cover.

In general, most of the Antarctic moss species have a strong affinity with their nearest continental land masses, notably southern South America and New Zealand. The similarities of the floras of southern South America, south-eastern Australia, Tasmania and New Zealand result from the land connection between these lands via Antarctica about 65 m. y. ago. At that time there was a temperate climate throughout the region and there is palaeobotanical evidence that, during the Oligocene (*ca* 60 m. y. ago), forest existed in Antarctica. Indeed, probably until the Miocene (23–12 m. y. ago), relatively complex tundra communities must have persisted around the moister coastal fringes of the continent. Cold-adapted taxa might have survived well into the Pliocene in the coastal plains (Raven & Axelrod 1974). An exchange of biota would therefore have been possible without the necessity for long-range aerial dispersal. South America became isolated from Antarctica *ca* 60 m. y. ago and from Australia *ca* 45 m. y. ago. During the initial phase of these separations an exchange of biota was probably possible via short-range aerial transport. However, with increasing distance between the land masses such a chance migration became less feasible and, with progressing glaciation, the phanerogamic flora of Antarctica became totally obliterated, although a few cryptomagic taxa probably persisted on nunataks. Schuster (1969, 1972, 1976, 1979, 1981, 1982, 1983*a*) discussed the causes of disjunct liverwort ranges and concluded that long-range dispersal was mainly operative in phylogenetically young and ruderal species, whereas in phylogenetically old species, continental drift may provide the satisfactory explanation for disjunct ranges.

Various hypotheses have been proposed to elucidate the cause of distributional patterns and the origin of Antarctic mosses. Most favour the concept of Holocene immigration by long-distance dispersal of diaspores. The nearest likely source of immigrant species that recolonised the Antarctic continent is South America, some 700–1000 km distant. It could not have been South Georgia because this island was covered by its own ice sheet extending 60 km offshore (Sudgen & Clapperton 1977). The chance of bird dispersal is rather slight because, except for the sheathbill, there are no true land birds in the Antarctic and sea birds come ashore only to breed. However, skuas, gulls, terns and sheathbills migrate to the Fuegian area and Patagonia but rarely visit inland sites. They probably do occasionally settle on bog or marsh habitats near the coast from where vegetative propagules or spores of mosses could become attached to their feet or feathers.

A more probable hypothesis is that wind is the most likely dispersal agent (Winkworth *et al.* 2002; Muñoz *et al.* 2004). Although the effectiveness of aerial long-range dispersal is limited by many factors, it is, at least in mosses, probably more important than is assumed because of the very long period of time which has been available for this mechanism to operate. According to experiments carried out by Zanten (1976, 1978, 1983, 1984) and Zanten & Gradstein (1988), the success of long-range dispersal is higher in the temperate than in the tropical regions. Zanten & Pócs (1981) and Longton & Schuster (1983) discussed at length the various factors determining the dispersability of bryophytes and the conditions for effective long-distance dispersal and emphasised its importance in the establishment of present day species distributions.

There is ample evidence of trans-oceanic dispersal by wind by the entrapment of exotic diaspores and pollen at sites in the Antarctic (Kappen & Straka 1988; Lewis Smith 1991; Linskens *et al.* 1993; Marshall 1996), as well as local dissemination of moss propagules (Rudolph 1970; Marshall & Convey 1997). Several studies of the composition and dynamics of soil propagule banks on Signy I. and James Ross I. in the maritime Antarctic (Lewis Smith 1987, 1991, 1993*a*, 1997*b*; Lewis Smith & Coupar 1987) and in Victoria Land in the continental Antarctic (Lewis Smith & Ochyra 2006) have revealed a wide range of bryophyte species which developed from buried propagules when the soils were incubated under a range of temperature and nutrient conditions. Virtually all of the species were typical of the local floras and presumed to have originated from these (see also Ayukawa *et al.* 2001). This was verified by spore trapping experiments (Chalmers *et al.* 1996; Marshall & Convey 1997). With the current trend in regional warming in the maritime Antarctic (Convey 2003; King *et al.* 2003), there is an increasing probability that propagules in the soil, previously unable to develop into plants because of adverse climatic conditions, will be able to become established (Lewis Smith 1990*b*, 1993*a*). During the course of a long-term study of soil propagule banks at several Antarctic sites by RILS, soil samples from Signy I. cultured under a range of temperatures have yielded two mosses, *Campylopus introflexus* and *Leptobryum pyriforme*, a basidiomycete fungus

(*Coprinus* sp.) and a fern prothallus not known in the Antarctic (Lewis Smith 1993*a*, unpublished). Likewise, cultured soil samples from Coulman I. in northern Victoria Land have produced sterile shoots of a funarialean moss which was identified, with some reservation, as *Entosthodon subnudus* (Taylor) Fife, which is unknown in the Antarctic (Lewis Smith & Ochyra 2006).

Evidence for persistent survival of mosses in the Antarctic through the Pleistocene is weak but some believe that pre-glacial survival on ice-free nunataks and oases may have played an important role as dispersal centres. Endemic species are strong candidates for pre-Pleistocene survivors *in situ*. This may be especially true for *Andreaea depressinervis* which is a taxonomically isolated species that propagates only asexually and its limited mode of dispersal is probably responsible for its restricted distribution in the northern maritime Antarctic, with only very occasional occurrences on South Georgia. The same is probably true for *Syntrichia sarconeurum* which is widespread in the maritime and continental Antarctic, probably because of its very effective means of asexual reproduction by caducous leaf apices. Nevertheless, it has never crossed the oceanic barrier separating Antarctica from southern South America where its sibling species, *S. lithophila*, occurs rarely in Patagonia. Actually, all Antarctic and Antarctic-South Georgian endemics are very distinct taxa and, until they are discovered in more northerly regions, they should be interpreted as pre-Pleistocene survivors.

5. Background to the Flora

AIMS AND OBJECTIVES

Knowledge of the composition of the moss flora of the Antarctic biome has remained inadequate until recently, and the taxonomic status of many taxa and collections was very imprecise and controversial. As a result, most published ecological, biogeographic or comparative studies contain misleading or inaccurate information. This is a major problem for ecological studies because they rely on accurate identification of the floristic components of communities and ecosystems. Likewise, the identities of some of the commonest species of moss were suspect, while for others it was difficult or virtually impossible to distinguish one taxon from another in the field, for example species of *Bryum*, *Schistidium*, *Hymenoloma*, *Warnstorfia*, *Sanionia* and *Brachythecium*. Therefore, with increasing research on terrestrial and aquatic ecosystems and their biota in Antarctica, especially with regard to physiological tolerances and responses to changing climate, there is an urgent need for the moss flora to be revised.

The need for a thoroughly revised monograph of Antarctic mosses is particularly justified because, after lichens, mosses are the second largest group of terrestrial macro-organisms. They play an important and crucial role in the impoverished Antarctic tundra vegetation and dominate many plant communities, especially those associated with moist and wet habitats. Antarctic lichens, together with liverworts, another small but biologically important group of terrestrial plants, have recently been taxonomically revised in modern descriptive Floras (Øvstedal & Lewis Smith 2001; Bednarek-Ochyra *et al.* 2000). They cover the entire Antarctic continent and offshore archipelagoes, and the former deals additionally with lichens of South Georgia.

Hitherto, no modern moss flora covering the entire Antarctic biome has been published. An attempt to produce such a treatment was undertaken in the late 1960s by British botanists under the leadership of the late Stanley W. Greene but, regrettably, this project was never completed. By the late 1970s funding for traditional taxonomic studies declined significantly and, in some countries, notably Great Britain, research institutes and universities terminated virtually all such research. Only the first of several planned volumes of the Antarctic Moss Flora was published, comprising treatments of 12 species and five genera (Greene *et al.* 1970). Additionally, S.W. Greene and his collaborators also published a dozen parts of a Synoptic Flora of South Georgian mosses (see Ochyra *et al.* 2002 for a detailed list), but this incomplete project ceased in the mid-1980s.

In general, there have been only a few descriptive, ecological and distributional accounts published, including illustrative material, of mosses of some regions in Antarctica. Kanda (1987*a*) provided an overview of 11 widespread species of moss and one liverwort in East Antarctica. This important treatment is written in Japanese, and is thus mostly unavailable to most bryologists, although line drawings, colour photographs and distribution maps of species are certainly valuable and useful. Detailed taxonomic accounts of the mosses, with excellent illustrations, have been completed for Vestfold Hills in Princess Elizabeth Land (Seppelt 1984, 1986) and southern Victoria Land (Seppelt & Green 1998), but these areas have exceedingly impoverished moss floras consisting of only six and seven species, respectively.

Taxonomic treatments of mosses in the maritime Antarctic are similarly scarce and all refer to the South Shetland Is., specifically King George I., the largest island in this archipelago. The first survey of the island's mosses was published by Putzke & Pereira (1990) and it contained keys to determinations, short descriptions and illustrations of 43 species encountered by these authors. It was followed by a similar treatment of 48 moss species occurring on Fildes Peninsula in the south-western part of the island published, in Chinese, by Chen *et al.* (1995). A comprehensive revised descriptive moss flora of King George I. was published by Ochyra (1998*a*) and it comprised detailed taxonomic and phytogeographic treatment of 61 species which were collected by this author in this area. Finally, Putzke & Pereira (2001) published an account of Antarctic mosses with particular reference to the South Shetland Is. These authors adopted taxonomic concepts largely taken from Ochyra's (1998*a*) moss flora of King George I., including textual use of descriptions of many species, although the latter source received no mention in this work (Ochyra 2002*c*).

The concept of preparing a comprehensive moss Flora for a distinct region of Antarctica was conceived during a visit of the first author to the South Shetland Is. in 1979–1980. Then, it was obvious that the preparation of such a treatment would require prolonged and laborious taxonomic studies in the broadest geographical scope in order to resolve the problems of the identities and the status of moss taxa described or reported from Antarctica. Such revisionary studies were conducted, with various intensity, throughout the next 18 years, culminating in the publication of the moss Flora of King George I. (Ochyra 1998*a*). During the same period, RILS had been seeking support to resurrect S.W. Greene's vision of an Antarctic Moss Flora. In 1995 the project of producing such a Flora was formally taken up by the present authors, with the objective of preparing a modern monographic treatment of all taxa of mosses recorded in Antarctica. In this phase of the research a total of 104 species of moss had been reported from within the Antarctic biome (Ochyra *et al.* 1998). Concurrently, taxonomic studies continued on critical genera, for instance *Ditrichum* (Ochyra & Lewis Smith 1998; Ochyra 1999*a*), *Platydictya* (Ochyra 1999*k*), *Meesia* (Ochyra & Lewis Smith 1999; Ochyra 2001*a*), *Bryoerythrophyllum* and *Didymodon* (Ochyra & Zander 2002), *Grimmia* (Ochyra 2004*a*) and *Schistidium* (Ochyra 2003*b*, 2004*c*, Ochyra *et al.* 2003*a*), while studies of some obscure collections revealed interesting taxa which have not hitherto been reported from Antarctica (e.g. *Bryum nivale*, *Bucklandiella subsecunda*, *Syntrichia caninervis*) or even previously undescribed species (e.g. in *Schistidium*).

As a result of these studies, the number of taxa for Antarctica currently stands at 111 species and two varieties. The primary aim of this treatment is to provide detailed descriptions and illustrations of all taxa of mosses now known in the Antarctic (as defined here), as well as keys for their identification. Furthermore, details of habitat and reproduction of each species are provided and their regional distribution is mapped. In addition, to illustrate the phytogeographical affinity of the muscoflora of Antarctica, the global distribution of each species has been carefully reviewed and, for almost all, these are also mapped.

SOURCE OF SPECIMENS

During the preparation of this flora most of the historical collections made by early Antarctic expeditions, and deposited in several major world herbaria, were located and examined (Table 14). Additionally, in very many cases Subantarctic, South American and Australasian taxa have been studied because some taxonomic problems could be satisfactorily resolved only after examination of the taxa in question in the broadest geographical terms. Following normal practice in this type of taxonomic work, as far as possible the relevant type collections of taxon names have been located and investigated. This was of particular importance since it allowed the true identities of the taxa concerned to be determined.

Apart from historical collections a large number of recent collections have been examined. The vast majority of them are deposited in the herbarium of the British Antarctic Survey in Cambridge, UK (AAS) and they are the basis of the present treatment. Although the AAS herbarium is currently secure within BAS, it has been threatened with closure and disposal on several occasions. For this reason, duplicates of all specimens examined in the present study have been donated to KRAM. They were made predominantly by RILS and other scientific staff of this institution, mainly since the early 1960s, in a wide range of localities and habitats throughout the maritime Antarctic, from Bouvetøya and the South Sandwich Is. to Alexander I. and some remote inland mountain ranges, and also from continental Antarctica, especially from Victoria Land, Princess Elizabeth Land and Wilkes Land (Table 15). These collections are of special importance because they yielded many distributional data and contributed many new regional moss records, as well as providing valuable ecological data.

Another valuable collection of Antarctic mosses was made by RO on King George I. and it was the basis of the Flora of this island (Ochyra 1998*a*). Some 180 specimens of moss from this island and Deception I. collected by RO have been distributed in the *Bryophyta Antarctica Exsiccata* (Ochyra 1984; Ochyra *et al.* 1986) and these are the first Antarctic specimens of mosses which are widely available in numerous world herbaria. In general, but with some exceptions, most voucher collections of mosses cited in the Antarctic botanical literature have been examined during the course of the present work (Table 14).

In total, approximately 10,000 specimens of moss collected in the Antarctic botanical zone have been investigated and used for completing the distribution maps of the species.

ARRANGEMENT OF TEXT

The Illustrated Moss Flora of Antarctica provides descriptions and illustrations, as well as ecological and phytogeographical information for all taxa recorded in the Antarctic botanical zone. Detailed descriptions are provided of the taxa of all ranks, from the division Bryophyta to species and varieties, if recognised. The systematic treatment of the taxa is preceded by a dichotomous generic key which is a guide to the allocation of specimens to genera. It should be emphasised that the keys are primarily constructed for the identification of the Antarctic genera and

Table 14. Principal bryological herbaria housing the most important moss collections studied during the work on this flora. Acronyms of herbaria adopted from *Index Herbariorum* (Holmgren *et al.* 1990) with subsequent modifications available on the NYBG internet web site http://sciweb.nybg.org/science2/IndexHerbariorum.asp.

Acronym	Herbarium name	Voucher or unpublished collections
AAS	British Antarctic Survey, Cambridge, UK	Allison & Lewis Smith (1973), Bell (1973*a,b,c*, 1974, 1976, 1984), Bell & Greene (1975), Clarke (1973*a*), Clarke & Lightowlers (1983), Convey *et al.* (2000, 2001), Greene (1973), Greene *et al.* (1970), Lewis Smith (1972, 1984*b,c*, 1985, 1988*d*, 1990*a*, 1993*b*, 2005*a,b,c*), Lewis Smith & Corner (1973), Light & Heywood (1973), Lightowlers (1984*b*, 1985*b*, 1986*a,b*, 1987), Lindsay (1971), Longton (1966), Longton & Holdgate (1979), Matteri (1977*a,b*, 1985*a*), Newton (1972, 1973, 1974*a,b,c*, 1977*a,b*, 1979*a,b*, 1980, 1983*a,b*), Webb (1973); also many collectors listed in Table 15
ADT	Antarctic Division, Kingston, Australia	Seppelt (1984, 1986, 2004), Seppelt & Green (1998)
ALTA	University of Alberta, Edmonton, Canada	Vitt (1974*a*, 1979*a*)
BG	Museum of Natural History, University of Bergen, Norway	Bell & Blom (1986)
BM	Natural History Museum (formerly British Museum, Natural History), London, UK	Darbishire (1923), Dixon (1912, 1913, 1914, 1920, 1923*a*, 1926, 1927, 1929, 1930, 1034*b,c*, 1935, 1937, 1939, 1960), Dixon & Watts (1918), Hooker & Wilson (1844), Mitten (1851, 1859, 1867, 1869, 1876*a,b*, 1879), Wilson (1854, 1859), Wilson & Hooker (1845, 1847); also unpublished collections made by the British Graham Land Expedition (1934–1937) and Operation Tabarin (1944–1946)
BR	National Botanic Garden, Meise, Belgium	Cardot (1900, 1901),
FH	Farlow Herbarium, Harvard University, Cambridge, Massachusetts, USA	Bartram (1938, 1957), Roivainen & Bartram (1937)
GRO	State University of Groningen, Holland	Zanten (1971)
H	Botanical Museum, University of Helsinki, Finland	Brotherus (1906*a*), Cardot & Brotherus (1923), Müller (1883, 1885, 1889, 1890*a*), Roivainen & Bartram (1937)
HBG	Herbarium Hamburgense, Biozentrum Klein-Flottbek, Hamburg, Germany	Müller (1883, 1889, 1890*a*)
HIRO	Hiroshima University, Japan	Horikawa & Ando (1961, 1963, 1967*a*)
KRAM	Institute of Botany, Polish Academy of Sciences, Poland	Kuc (1969), Ochyra (1998*a*), Przywara *et al.* (1983, 1984), V. Komárková (present work); unpublished collections of R. Ochyra from the Prince Edward Is., Îles Crozet and Îles Kerguelen and N. Gremmen from Heard I., Gough I. and Tristan da Cunha
LD	Botanical Museum, Lund, Sweden	Cardot (1905*a*, 1906*a*, 1908*a*), Cardot & Brotherus (1923)
LE	Komarov Botanical Institute, Russian Academy of Sciences, Sankt Petersburg, Russia	Savicz-Lyubitskaya & Smirnova (1959, 1960, 1961, 1962, 1963*a,b*, 1964*a,b,c,d*, 1965*a,b,c*, 1966, 1970*a,b*, 1971, 1972, 1973, 1978)
M	Botanische Staatssammlung, Munich, Germany	Müller (1890*a*)
MEL	National Herbarium of Victoria, Royal Botanic Gardens, South Yarra (Melbourne), Australia	Dixon & Watts (1918), Filson (1966), Filson & Willis (1975)
NIPR	National Institute of Polar Research, Tokyo, Japan	Kanda (1986*a*, 1987*a,b*), Kanda & Iwatsuki (1989)
NY	New York Botanical Garden, USA	Bartram (1938, 1957), Eights (1833), Mitten (1851, 1859, 1867, 1869, 1876*a,b*, 1879); unpublished collection of W. C. Steere from Victoria Land
PC	Muséum National d'Histoire Naturelle, Paris, France	Cardot (1900, 1901, 1905*a*, 1906*a,b*, 1907*a,b*, 1908*a*, 1910*b*, 1911*a,b,c*, 1913), Cardot & Brotherus (1923), Müller (1883, 1885, 1889, 1890*a*)
S	Swedish Museum of Natural History, Stockholm, Sweden	Cardot (1905*a*, 1906*a*, 1908*a*), Cardot & Brotherus (1923), Dusén (1903*a,b*, 1905*a,b,c*, 1906, 1907)
UPS	University of Uppsala, Sweden	Cardot (1905*a*, 1906*a*, 1908*a*), Cardot & Brotherus (1923)
US	National Herbarium, Smithsonian Institution, Washington, DC, USA	Bartram (1938, 1957), Eights (1833), Robinson (1972*a*)

Table 15. Principal collectors of Antarctic mosses in AAS and other herbaria used in the text under 'Specimen examined'.

Abbreviation	Name of collector	Herbarium	Region
AC	A. Cameron	AAS	Western Antarctic Peninsula
ADK	A. D. Kennedy	AAS	Princess Elizabeth Land
AM	A. Moffat	AAS	Brabant I.
AS	A. Scott	AAS	Argentine Is.
BASM	B.A.S. Miscellaneous Collections	AAS	Antarctic Peninsula region and continental Antarctica
BGLE	British Graham Land Expedition (1934–1937)	BM	Argentine Is.
BJ	B. Jabłoński	KRAM	King George I.
BJT	B. J. Taylor	AAS	King George I., Alexander I.
CGB	C. G. Brading	AAS	Trinity Peninsula
CS	C. J. F. Skottsberg	PC, S	Antarctic Peninsula, South Georgia, Tierra del Fuego
DCL	D. C. Lindsay	AAS	South Shetland Is., Signy I.
DM	D. Mason	AAS	Deception I.
DWB	D. W. Brown	AAS	Marguerite Bay
DDWW	D. D. Wynn-Williams	AAS	Marguerite Bay, Alexander I.
DI	Discovery Investigation (1935)	BM	South Orkney Is., Palmer Archipelago
DWHW	D. W. H. Walton	AAS	Victoria Land
ER	É. Racovitza	BR, PC	Danco Coast
FIDS	Falkland Islands Dependencies Survey (1947–1961)	BM	Antarctic Peninsula
GAL	G. A. Llano	NY	Victoria Land, Palmer Archipelago
GCSC	G. C. S. Clarke	AAS	South Georgia
IML	I. Mackenzie Lamb	BM	Northern and mid-western Antarctic Peninsula
JF	J. Fenton	AAS	Western Antarctic Peninsula, Adelaide I.
JJL	J. J. Light	AAS	Signy I., Alexander I.
JK	J. Killingbeck	AAS	Adelaide I.
JMG	J. M. Gnau	US	Palmer Archipelago
JPB	J. P. Baylis	AAS	Elephant I. group
JRR	J. R. Rastorfer	US	Palmer Archipelago
JS	J. Smith	AAS	Western Antarctic Peninsula
JSA	J. S. Allison	AAS	Elephant I. group
KA	K. Archibald	AAS	Western Antarctic Penisula
KR	K. Richard	AAS	Livingston I.
MJC	M. J. Cousins	AAS	Eastern Antarctic Peninsula
MMM	M. McManmon	AAS	South Orkney Is.
MWH	M. W. Holdgate	AAS	South Sandwich Is., Signy I., Bouvetøya
NJC	N. J. Collins	AAS	Deception I.
OB	O. Benitez	AAS	Danco Coast
OG	Operation Gooseflesh (1964–1965)	BM	Palmer Archipelago
OT	Operation Tabarin (1944–1946)	BM	Northern and mid-western Antarctic Peninsula
PAS	P. A. Siple	FH, NY, US	Marie Byrd Land, Palmer Archipelago
PC	P. Convey	AAS	South Sandwich Is., Charcot I., Ellsworth Land
PK	P. Kennett	AAS	Eastern Antarctic Peninsula
RBH	R. B. Heywood	AAS	Signy I., Alexander I.
RDS	R. D. Seppelt	ADT	Victoria Land, Princess Elizabeth Land, Macquarie I.
REL	R. E. Longton	AAS	South Georgia, South Sandwich Is., Signy I., Deception I., Argentine Is., Victoria Land
RF	R. Filson	MEL	Mac.Robertson Land
RILS	R. I. Lewis Smith	AAS	South Georgia, South Orkney Is., South Shetland Is., James Ross I., western Antarctic Peninsula, Alexander I., Wilkes Land, Princess Elizabeth Land, Victoria Land
RO	R. Ochyra	KRAM	King George I., Deception I.
RMS	R. M. Schuster	US	Livingston I., Anvers I., Danco Coast, Argentine Is., Hook I.
RT	R. Tindall	AAS	Eastern Antarctic Peninsula
RW	R. Webb	AAS	South Orkney Is.
RWMC	R. W. M. Corner	AAS	Argentine Is.
SWG	S. W. Greene	AAS	South Georgia, Victoria Land
TE	T. Engelskjøn	BG	Bouvetøya
VK	V. Komárková	KRAM	Kintg George I., Anvers I., western Antarctic Peninsula
WCS	W. C. Steere	NY	Victoria Land
WH	W. Hankinson	AAS	Brabant I.

may not always be applicable elsewhere. Those having knowledge of the familial placement of specimens may commence the determination of a taxon from the key to the genera within a given family. In addition, keys to the classes of the Bryophyta, the subclasses of the Bryopsida, and the orders of the Dicranidae are provided.

This is a specimen-based Flora and, whenever possible, type specimens have been studied. Type information is given for all taxon names and, according to normal practice, the type is cited from the published protologue, followed by data taken from the herbarium labels. For each genus and species name all homotypic synonyms are given and, additionally, some heterotypic synonyms are provided if they are used for taxa described or reported for material collected from Antarctica. In many cases taxa described from the conterminous territories of the Subantarctic, especially South Georgia and Îles Kerguelen, southern South America or, occasionally, other parts of the austral region, are discussed in detail.

Moreover, all taxon names are carefully typified, following the procedures suggested by the current Code of Botanical Nomenclature (Vienna Code) (McNeill *et al.* 2006). Because a remarkable proportion of the Antarctic moss flora comprises bipolar species, which were largely described in the early classical works of Hedwig (1801) and Bridel (1826–1827) from material collected in Europe and other parts of the Northern Hemisphere, names of these taxa are also lectotypified in this Flora. Examination of type material is indicated within the text by the citation of an acronym of a herbarium (after Holmgren *et al.* 1990), accompanied by an exclamation mark (!).

For each name full bibliographic data are cited, following abbreviations of publications suggested by Crosby (2003). The name citations of authors of taxonomic names strictly follow the protocol of Brummitt & Powell (1992). If some author's names are missing from this work, abbreviations or citations of them are proposed here following the general spirit of their formation in this compendium.

In order to avoid repetition of the formal taxonomic descriptions of families and suprafamilial taxa, which are readily available in many modern Floras, they are omitted from the present account. Also, the generic and familial composition of many families and orders has undergone significant modification as a result of recent molecular studies, and these new taxonomic concepts still require confirmation before they are likely to be universally accepted. Instead, these taxa are characterised in a descriptive form, with special emphasis on essential and taxonomically important traits. The nomenclature of suprageneric names is carefully revised and the rule of priority has been generally accepted following Recommendation 16.B of the Vienna Code. Accordingly, the well known and widely accepted order Pottiales has been changed to the largely neglected Phascales. Likewise, authors stated in parenthesis are not consistently cited for suprageneric names after Article 49.2 of the Code but, for these names, epithet-bringing synonyms are quoted, if necessary (Ochyra *et al.* 2006).

Descriptions of genera are detailed and consider a full spectrum of character states, thus making them universal. On the other hand, descriptions of species, as well as both recognised varieties, are also exhaustive but are based only on Antarctic material. For species which are known only from their gametophytic state or the material is otherwise incomplete in Antarctica, characteristics that occur only in plants from outside the biome in the conterminous regions of the Subantarctic and/or southern cool-temperate zones are also included, but these parts of the descriptions are given within square brackets. In monotypic genera, the generic description is given as usual, but the species description is condensed and usually restricted to the measurements of the plants, and morphological and anatomical characters.

All the species and varieties are illustrated comprehensively with full-page plates of line drawings, allowing those which are more easily recognised to be more readily compared with morphologically similar species that are less easily identifiable. The figures depict habit sketches, leaves, stem and leaf sections, leaf areolation, gemmae if present, and perichaetial leaves, as well as sporophyte details, including exothecial cells, peristomes and spores. As a rule, a number of leaves are illustrated to show their variability, and leaf sections present a full spectrum of variation in the leaf anatomy from the base to the apex. Characteristics that are unknown in Antarctic plants, for example sporophytes and peristomes, are also often illustrated, usually from specimens from the Southern Hemisphere. Detailed legends below the plates are given, with indication of the specimens used for the illustrations. Unless otherwise stated, specimens and microscopic slides used for illustrations are preserved in KRAM.

In addition, for each species and variety, the following data are given:

(*1*) *Discussion* — This includes information on diagnostic characters, peculiarities of morphology, taxonomic problems, sources of confusion during identification, comparison with closely related or otherwise similar species, information on the first discovery in Antarctica, and interesting biological phenomena, if available. Furthermore, a taxonomic history of species if discussed and nomenclatural problems relating to accepted and synonymous names are clarified, including typification of these names.

(*2*) *Reproduction* — Detailed information on sexual or asexual reproduction is given for each species, together with notes on frequency and abundance of sporophytes, if produced.

(*3*) *Habitat* — This section provides detailed ecological notes on each species based on our field experience or taken from labels of the herbarium specimens. Apart from habitat details, information on the occurrence in the principal plant communities and associated species is usually given.

(*4*) *Antarctic distribution* — This is presented as dot maps. In the case of pan-Antarctic species separate maps have been compiled for the Antarctic Peninsula region and continental Antarctica, whereas for some very rare or occasional antipodal species their Antarctic distribution is presented on more generalised maps which include South Georgia and/or the Fuegian region. The distribution of all species and varieties is based on the examination of specimens. However, in a few cases some literature reports are also considered, if they constitute phytogeographically important records, but they are marked by different symbols on their respective maps. For each species available altitudinal limits are provided, with particular consideration of the highest known records, and the southernmost occurrences are also indicated.

(*5*) *Global distribution* — This part of the species presentation deviates markedly from the generally accepted standards in most Floras and taxonomic treatments in which only generalised information is usually provided, and sometimes repeats the same errors and inconsistencies. Here, global ranges of species are carefully revised and distributions of almost all species are presented on separate maps, usually as dots and/or delineated areas. They are based on an examination of specimens, especially in the Southern Hemisphere, or extracted from reliable literature reports, usually monographic or revisionary studies. In many cases local distribution, especially in Eurasia, was determined in consultation with residentiary bryologists.

(*6*) *Specimens examined* — For very rare, rare and occasional species usually all available specimens are quoted, whereas for frequent, common and very common species only selected specimens are listed which, however, represent all major geographical areas within Antarctica from which they are known. They are always arranged according the same scheme and the first major division refers to West and East Antarctica. Then, for the former region specimens are first listed for the peri-Antarctic islands and archipelagoes (Bouvetøya, South Sandwich Is., South Orkney Is. and South Shetland Is.) and then for the archipelagoes, islands and coasts of the western and eastern Antarctic Peninsula in a north–south direction. For the continental Antarctic, specimens are listed according to lands and coasts starting with the Maud Sector and progressing clockwise to the Byrd Sector.

All species reported from Antarctica on the basis of misdeterminations or otherwise erroneous reports are placed in a 'Species excluded' section at the end of each family. Technical terms in the keys and descriptions that may be unfamiliar to the user are explained in the glossary that follows the systematic chapter.

6. *Systematic Accounts of the Taxa*

CONSPECTUS OF CLASSIFICATION

The systematic arrangement used in the present Flora reflects our current state of understanding of moss diversity and is primarily based on morphology. It is derived from the latest systems of Buck & Goffinet (2000) and Goffinet & Buck (2004), with some modifications and changes resulting from our own ideas on evolution and phylogeny. Following current tendencies, mosses or Musci are classified into the separate phylum Bryophyta which is equiponderant to liverworts or Hepaticae (Marchantiophyta) and hornworts or Anthocerotae (Anthocerotophyta). The classes and subclasses are designated sequentially by capital letters and a combination of letters and Arabic numerals, respectively, whereas orders and families are numbered consecutively by Roman numerals. The genera are numbered sequentially by Arabic numerals and only species and infraspecific taxa are numbered consecutively by Arabic letters and a combination of Arabic numerals and capital letters, respectively, within a genus.

Division **BRYOPHYTA** Schimp.
Class A. **Andreaeopsida** J.H.Schaffn.
Subclass A1. **Andreaeidae** Engl.
Order I. **Andreaeales** Limpr.
Family I. Andreaeaceae Dumort.
1. *Andreaea* Hedw.
1. *A. depressinervis* Cardot
2. *A. gainii* Cardot
3. *A. regularis* Müll.Hal.

Class B. **Polytrichopsida** Doweld
Subclass B1. **Polytrichidae** Ochyra
Order II. **Polytrichales** M.Fleisch.
Family II. Polytrichaceae Schwägr.
2. *Polytrichastrum* G.L.Sm.
1. *P. alpinum* (Hedw.) G.L.Sm.
2. *P. longisetum* (Brid.) G.L.Sm.
3. *Polytrichum* Hedw.
1. *P. piliferum* Hedw.
2. *P. juniperinum* Hedw.
3. *P. strictum* Brid.
4. *Notoligotrichum* G.L.Sm.
1. *N. trichodon* (Hook. & Wilson) G.L.Sm.

Class C. **Bryopsida** Pax
Subclass C1. **Funariidae** Ochyra
Order III. **Funariales** M.Fleisch.
Family III. Funariaceae Schwägr.
5. *Funaria* Hedw.
1. *F. hygrometrica* Hedw.

Subclass C2. **Encalyptidae** Ochyra, Żarnowiec & Bednarek-Ochyra
Order IV. **Encalyptales** Dixon
Family IV. Encalyptaceae Schimp.

6. *Encalypta* Hedw.
    1. *E. procera* Bruch
    2. *E. rhaptocarpa* Schwägr.

Subclass C3. **Dicranidae** Doweld

Order V. **Dicranales** M.Fleisch.

Family V. Ditrichaceae Limpr.

7. *Distichium* Bruch & Schimp.
    1. *D. capillaceum* (Hedw.) Bruch & Schimp.
    2. *D. inclinatum* (Hedw.) Bruch & Schimp.
8. *Ditrichum* Hampe
    1. *D. heteromallum* (Hedw.) E.Britton
    2. *D. ditrichoideum* (Cardot) Ochyra
    3. *D. hyalinocuspidatum* Cardot
    4. *D. conicum* (Mont.) Mitt.
    5. *D. hyalinum* (Mitt.) Kuntze
9. *Ceratodon* Brid.
    1. *C. purpureus* (Hedw.) Brid.

Family VI. Dicranaceae Schimp.

10. *Dicranella* (Müll.Hal.) Schimp.
    1. *D. varia* (Hedw.) Schimp.
    2. *D. hookeri* (Müll.Hal.) Cardot
    3. *D. campylophylla* (Taylor) A.Jaeger
11. *Amphidium* Schimp.
    1. *A. lapponicum* (Hedw.) Bruch & Schimp.
12. *Kiaeria* I.Hagen
    1. *K. pumila* (Mitt.) Ochyra
13. *Chorisodontium* (Mitt.) Broth.
    1. *Ch. aciphyllum* (Hook.f. & Wilson) Broth.
14. *Campylopus* Brid.
    1. *C. introflexus* (Hedw.) Brid.
    2. *C. pyriformis* (Schultz) Brid.
    3. *C. vesticaulis* Mitt.
    4. *C. spiralis* Dusén

Order VI. **Grimmiales** M.Fleisch.

Family VII. Seligeriaceae Schimp.

15. *Holodontium* (Mitt.) Broth.
    1. *H. strictum* (Hook.f. & Wilson) Ochyra
16. *Hymenoloma* Dusén
    1. *H. grimmiaceum* (Müll.Hal.) Ochyra
    2. *H. crispulum* (Hedw.) Ochyra
    3. *H. antarcticum* (Müll.Hal.) Ochyra

Family VIII. Grimmiaceae Arn.

17. *Coscinodon* Spreng.
    1. *C. lawianus* (J.H.Willis) Ochyra
18. *Schistidium* Bruch & Schimp.
    1. *S. falcatum* (Hook.f. & Wilson) B.Bremer
    2. *S. lewis-smithii* Ochyra
    3. *S. rivulare* (Brid.) Podp.
    4. *S. praemorsum* (Müll.Hal.) Herzog
    5. *S. andinum* (Mitt.) Herzog
    6. *S. steerei* Ochyra
    7. *S. urnulaceum* (Müll.Hal.) B.G.Bell
    8. *S. halinae* Ochyra
    9. *S. antarctici* (Cardot) L.I.Savicz & Smirnova
    10. *S. leptoneurum* Ochyra
    11. *S. deceptionense* Ochyra, Bednarek-Ochyra & Lewis Smith

12. *S. amblyophyllum* (Müll.Hal.) Ochyra & Hertel
13. *S. cupulare* (Müll.Hal.) Ochyra

19. *Grimmia* Hedw.
1. *G. plagiopodia* Hedw.

20. *Guembelia* Hampe
1. *G. longirostris* (Hook.) Ochyra & Żarnowiec

21. *Orthogrimmia* (Schimp.) Ochyra & Żarnowiec
1. *O. sessitana* (De Not.) Ochyra & Żarnowiec

22. *Dryptodon* Brid.
1. *D. fuscoluteus* (Hook.) Ochyra & Żarnowiec

23. *Racomitrium* Brid.
1. *R. lanuginosum* (Hedw.) Brid.

24. *Bucklandiella* Roiv.
1. *B. subsecunda* (Harv.) Bednarek-Ochyra & Ochyra
2. *B. pachydictyon* (Cardot) Bednarek-Ochyra & Ochyra
3. *B. sudetica* (Funck) Bednarek-Ochyra & Ochyra
4. *B. didyma* (Mont.) Bednarek-Ochyra & Ochyra

Order VII. **Phascales** Limpr.

Family IX. Pottiaceae Schimp.

25. *Tortella* (Müll.Hal.) Limpr.
1. *T. alpicola* Dixon
2. *T. fragilis* (Drumm.) Limpr.

26. *Bryoerythrophyllum* P.C.Chen
1. *B. recurvirostrum* (Hedw.) P.C.Chen

27. *Didymodon* Hedw.
1. *D. brachyphyllus* (Sull.) R.H.Zander

28. *Stegonia* Venturi
1. *S. latifolia* (Schwägr.) Broth.

29. *Pterygoneurum* Jur.
1. *P. ovatum* (Hedw.) Dixon

30. *Aloina* Kindb.
1. *A. brevirostris* (Hook. & Grev.) Kindb.

31. *Tortula* Hedw.
1. *T. hoppeana* (Schultz) Ochyra

32. *Hennediella* Paris
1. *H. heimii* (Hedw.) R.H.Zander
2. *H. antarctica* (Ångstr.) Ochyra & Matteri

33. *Willia* Müll.Hal.
1. *W. austroleucophaea* (Besch.) Broth.

34. *Syntrichia* Brid.
1. *S. sarconeurum* Ochyra & R.H.Zander
2. *S. saxicola* (Cardot) R.H.Zander
3. *S. filaris* (Müll.Hal.) R.H.Zander
4. *S. caninervis* Mitt.
   4A. var. *caninervis*
   4B. var. *gypsophila* (G.Roth) Ochyra
5. *S. magellanica* (Mont.) R.H.Zander

Subclass C4. **Orthotrichidae** Ochyra

Order VIII. **Orthotrichales** Dixon

Family X. Orthotrichaceae Arn.

35. *Muelleriella* Dusén
1. *M. crassifolia* (Hook.f. & Wilson) Dusén

36. *Orthotrichum* Hedw.
1. *O. rupestre* Schwägr.

Subclass C5. **Bryidae** Engl.

Order IX. **Bryales** Limpr.

Family XI. BRYACEAE Schwägr.
37. *Leptobryum* (Schimp.) Wilson
1. *L. pyriforme* (Hedw.) Wilson
38. *Schizymenium* Harv.
1. *S. pusillum* (Hook. & Wilson) A.J.Shaw
39. *Pohlia* Hedw.
1. *P. wahlenbergii* (F.Weber & D.Mohr) A.L.Andrews
2. *P. wilsonii* (Mitt.) Ochyra
3. *P. drummondii* (Müll.Hal.) A.L.Andrews
4. *P. cruda* (Hedw.) Lindb.
5. *P. nutans* (Hedw.) Lindb.
40. *Bryum* Hedw.
1. *B. argenteum* Hedw.
1A. var. *argenteum*
1B. var. *muticum* Brid.
2. *B. dichotomum* Hedw.
3. *B. nivale* Müll.Hal.
4. *B. orbiculatifolium* Cardot & Broth.
5. *B. pseudotriquetrum* (Hedw.) P.Gaertn., B.Mey. & Scherb.
6. *B. archangelicum* Bruch & Schimp.
7. *B. pallescens* Schwägr.
Family XII. MEESIACEAE Schimp.
41. *Meesia* Hedw.
1. *M. uliginosa* Hedw.
Family XIII. BARTRAMIACEAE Schwägr.
42. *Conostomum* F.Weber & D.Mohr
1. *C. pentastichum* (Brid.) Lindb.
2. *C. magellanicum* Sull.
43. *Philonotis* Brid.
1. *Ph. polymorpha* (Müll.Hal.) Broth.
44. *Bartramia* Hedw.
1. *B. patens* Brid.

Subclass C6. **HYPNIDAE** W.R.Buck, Goffinet & A.J.Shaw

Order X. **HYPNALES** W.R.Buck & Vitt

Family XIV. AMBLYSTEGIACEAE Kindb.
45. *Cratoneuropsis* (Broth.) M.Fleisch.
1. *C. relaxa* (Hook.f. & Wilson) Broth. subsp. *minor* (Wilson & Hook.f.) Ochyra
46. *Campylium* (Sull.) Mitt.
1. *C. polygamum* (Schimp.) Lange & C.E.O.Jensen
47. *Sanionia* Loeske
1. *S. uncinata* (Hedw.) Loeske
2. *S. georgicouncinata* (Müll.Hal.) Ochyra & Hedenäs
48. *Drepanocladus* (Müll.Hal.) G.Roth
1. *D. longifolius* (Mitt.) Paris
49. *Warnstorfia* Loeske
1. *W. sarmentosa* (Wahlenb.) Hedenäs
2. *W. fontinaliopsis* (Müll.Hal.) Ochyra
Family XV. BRACHYTHECIACEAE Schimp.
50. *Brachythecium* Schimp.
1. *B. austroglareosum* (Müll.Hal.) Kindb.
2. *B. austrosalebrosum* (Müll.Hal.) Kindb.
3. *B subpilosum* (Hook.f. & Wilson) Kindb.
51. *Sciuro-hypnum* Hampe
1. *S. fuegianum* (Broth.) Ochyra & Żarnowiec
2. *S. glaciale* Schimp.
Family XVI. PLAGIOTHECIACEAE M.Fleisch.
52. *Plagiothecium* Schimp.

1. *P. ovalifolium* Cardot
2. *P. orthocarpum* Mitt.

Family XVII. HYPNACEAE Schimp.

53. *Hypnum* Hedw.
    1. *H. revolutum* (Mitt.) Lindb.
54. *Platydictya* Berk.
    1. *P. jungermannioides* (Brid.) H.A.Crum
55. *Isopterygiopsis* Z.Iwats.
    1. *I. pulchella* (Hedw.) Z.Iwats.

KEY TO THE GENERA OF ANTARCTIC MOSSES

Antarctic species of moss are classified into 55 genera and 17 families. Of these, only *Chorisodontium*, *Cratoneuropsis*, *Holodontium*, *Muelleriella*, *Notoligotrichum* and *Willia* may be considered as antipodal, i.e. restricted in their distribution to the broadly understood Gondwanaland, primarily the cool and cold regions of the Southern Hemisphere and occasionally penetrating into the tropics along high mountain chains. The majority of the Antarctic moss genera are either cosmopolitan or Holarctic (Laurasian) taxa which have a bipolar distribution owing to the presence of some representatives at sites in the Antarctic, Subantarctic or in southern cool-temperate regions, and usually without intermediate sites in the tropical regions. The latter group, i.e. without tropical occurrences, is quite small and encompasses only three strictly bipolar genera, namely *Kiaeria*, *Platydictya* and *Stegonia*, while three other genera, *Amphidium*, *Coscinodon* and *Isopterygiopsis*, have some species occurring also in the tropics. All of these genera are largely distinct and precisely defined taxa and should not pose serious problems for their identification.

In most cases recognition of the genera of mosses is based upon characters of both gametophyte and sporophyte generations. Unfortunately, mosses occurring in the severe climatic conditions of the Antarctic biome are often sterile and/or only seldom produce fully mature sporophytes. Therefore, the key to the genera is mostly artificial and refers rather to local species without consideration of the whole spectrum of morphological variability of the genera concerned, and it is constructed primarily with special emphasis on gametophyte features which are much easier to observe and examine. Sporophyte characters are also used, especially if the genera concerned profusely and frequently produce mature capsules but, as a rule, in order to avoid miscomprehension and confusion, such genera are keyed several times to allow users a choice if sterile and/or fertile material is available. Where species of a genus may exhibit both alternatives of a dichotomy, or interpretation of some characters may be difficult owing to their variability, for example the papillosity and length of the laminal cells, excurrence and width of the costa and leaf shape, allowance has been made as far as possible in the key. Nevertheless, in doubtful cases it may be necessary to try both alternatives of the dichotomy because allowing for every eventuality would markedly extend the key, making it unintelligible and impractical. If the natural sequence of couplets in the key is interrupted by new general dichotomies, the numbers of these couplets are printed in bold-face and groups of dichotomies referring to the same general character are separated by a broader space for the sake of clarity.

We declined from constructing a key to families because these are large taxa which are often diagnosed by sporophyte characters and show great morphological diversity that can often not be observed in the Antarctic moss flora.

As is the case with liverworts (Bednarek-Ochyra *et al.* 2000), it is very likely that, with progress in exploration of hitherto not or under-investigated areas, further species and genera of mosses may be discovered in the Antarctic biome, particularly in the maritime Antarctic. They may be recruited from taxa occurring in southern South America, especially in the *Nothofagus* zone at the western fringes of the continent and Tierra del Fuego, as well as from the Northern Hemisphere, and the discovery of species such as *Campylopus spiralis*, *C. pyriformis*, *Ditrichum ditrichoideum*, *D. conicum*, *Dicranella varia*, *D. hookeri* and *Racomitrium lanuginosum* in volcanically active areas following recent eruptions indicates that these are perhaps the most promising sites for new species becoming established. Likewise, Antarctic soil seems to hold a remarkable reservoir of diaspores which can germinate in favourable ecological conditions, as has been shown in response to regional climate change (Lewis Smith 1987, 2001; Lewis Smith & Ochyra 2006). Finally, discovery of species new to science is both possible and likely in Antarctica, especially in some under-investigated genera, such as *Schistidium* which includes five endemics in this biome.

1. Leaves distichous, i.e. arranged in two rows on the stem .... ***Distichium*** (p. 130)
1. Leaves arranged all around the stem (foliate stems occasionally flattened) 2
    2. Leaves with green longitudinal photosynthetic lamellae or filaments on the ventral (upper) surface of the costa **3**

2. Leaves without green photosynthetic lamellae or filaments on the ventral (upper) surface of the costa (but propagules sometimes present) **7**

3(2). Leaves with filaments on the ventral surface of the costa .... ***Aloina*** (p. 348)
3. Leaves with longitudinally oriented lamellae on the ventral surface of the costa 4

4. Ventral lamellae 2–4; plants, less than 0.5 cm; peristome absent .... ***Pterygoneurum*** (p. 343)
4. Ventral lamellae 10–40; plants mostly larger; peristome present 5

5. Leaves not or only moderately differentiated into sheath and limb; costa narrow, flanked by a broad lamina; ventral lamellae segmented, undulate; calyptra naked or sparsely hairy at the apex .... ***Notoligotrichum*** (p. 90)
5. Leaves with a distinct membraneous and elamellose sheathing base and a firm, broadly lamellate, divergent limb; costa broad, occupying most of the width of limb, flanked by a narrow lamina, 2–12 cells wide; ventral lamellae, straight, not segmented; calyptra densely pilose 6

6. Leaf margin dentate or serrate, plane or incurved but never involute; capsule terete or rounded to 4–6-angled, without a distinct hypophysis; exothecial cells smooth, not pitted; peristome teeth not keeled; spores large, 13–26 μm wide, warty papillose .... ***Polytrichastrum*** (p. 93)
6. Leaf margin entire, broadly involute and enclosing the upper surface of the limb; capsules sharply 4(–5)-angled, with a distinct hypophysis; exothecial cells mammillose, pitted; peristome teeth keeled at back; spores small, 6–17 μm wide, finely papillose .... ***Polytrichum*** (p. 100)

7(2). Leaves ecostate or costa short and double, double, or single with 2–3 lateral spurs or branches **8**
7. Leaves with a single costa extending to at least mid-leaf... **13**

8. Plants lithophilic, deep red-brown, brown to blackish-brown, forming short tufts or mats; laminal cells with strongly incrassate and often porose walls; capsules, if present, terminal on the tips of the stem or branches, splitting by longitudinal slits into 4(–10) valves fused above and below .... ***Andreaea*** (p. 74)
8. Plants of various habitats, forming tufts, mats or wefts, not deep red-brown, brown or blackish-brown; laminal cells thin- to thick-walled, not or slightly pitted; capsules, when present, lateral on the stems or branches, dehiscent by means of a lid-like operculum 9

9. Stems and branches complanate-foliate; leaves distinctly decurrent, usually asymmetric .... ***Plagiothecium*** (p. 571)
9. Stems and branches not flattened; leaves non-decurrent, symmetric 10

10. Plants large; leaf acumen channelled; costa very variable, often single; alar cells large, inflated, thin-walled and hyaline, becoming thicker-walled and brownish with age, forming large, transversely triangular auricles often reaching the costa .... ***Campylium*** (p. 523)
10. Plants small; leaf acumen plane; costa lacking or short and double; alar cells undifferentiated or small, thick-walled and subquadrate, forming a small group in the basal angles 11

11. Leaves falcate-secund; alar cells well differentiated, small and subquadrate; rhizoids arising below the leaf insertion, smooth; capsules inclined to pendulous, asymmetric .... ***Hypnum*** (p. 578)
11. Leaves straight, occasionally somewhat secund; alar cells not or scarcely differentiated; rhizoids in the leaf axils, papillose; capsules ± erect, symmetric................................ 12

12. Plants very small and slender, thread-like; leaves minute, 0.15–0.3 mm long; pseudoparaphyllia present, filiform; laminal cells oblong-rhomboidal to rhombic; perichaetial leaves spinosely dentate at margins .... ***Platydictya*** (p. 582)
12. Plants larger, not filamentous; leaves more than 0.5 mm long; pseudoparaphyllia always lacking; laminal cells linear-flexuose to linear-fusiform; margins of the perichaetial leaves entire .... ***Isopterygiopsis*** (p. 586)

13(7). Stems procumbent or prostrate, usually much branched; plants forming mats or wefts; sporophytes, if present, arising laterally on shoots **14**

13. Stems erect, not or sparingly branched; plants forming tufts or extensive patches; sporophytes, when present, arising terminally on main shoot or branches **21**

14. Costa extending to the leaf apex to long excurrent, very seldom subpercurrent.. .... **15**

14. Costa always vanishing well below the leaf apex **17**

15(14). Laminal cells with distinctly sinuose lateral walls .... ***Bucklandiella*** (p. 293)

15. Laminal cells with straight, smooth or somewhat pitted but never corrugately thickened lateral walls.. 16

16. Leaves ovate-lanceolate, shortly acuminate; alar cells small, ± isodiametric, thick-walled, forming not or scarcely differentiated auricles; paraphyllia sometimes present on the stems .... ***Cratoneuropsis*** (p. 516)

16. Leaves narrowly oblong-lanceolate, gradually tapering to a long-acuminate, often filiform acumen; alar cells large, short-rectangular, forming small, inflated auricles; paraphyllia always wanting .... ***Drepanocladus*** (p. 535)

17(14). Leaves falcate-secund, rarely only weakly curved to straight; stem with a distinct unistratose hyalodermis in cross-section.. ***Sanionia*** (p. 526)

17. Leaves straight, occasionally slightly curved; stem without a hyalodermis in transverse section 18

18. Leaf acumen distinctly canaliculate; costa variable, often branched .... ***Campylium*** (p. 523)

18. Leaf acumen plane; costa unbranched or only occasionally spurred 19

19. Leaves with a group of hyaline, thin-walled rhizoid-initial cells at the apex .... ***Warnstorfia*** (p. 539)

19. Leaves lacking a group of thin-walled, rhizoid-initial cells at the apex 20

20. Leaves distinctly vertically plicate, not or scarcely decurrent; stem leaves ovate-lanceolate, broadly ovate to elongate-triangular, gradually or, seldom, more abruptly acuminate .... ***Brachythecium*** (p. 550)

20. Leaves smooth or hardly plicate, longly or shortly decurrent with tapering auricles; stem leaves, cordate-triangular to broadly ovate-cordate abruptly narrowed to a filiform acumen .... ***Sciuro-hypnum*** (p. 562)

21(13). Leaves with a broad, sheathing base, abruptly contracting to a long subula **22**

21. Leaves lacking a sheathing base, subulate or not **25**

22. Leaves 2-ranked on the opposite sides of the stem .... ***Distichium*** (p. 130)

22. Leaves multi-ranked, spirally arranged on the stem 23

23. Leaf base with a distinct border of narrow and elongate, thin-walled, hyaline cells .... ***Ditrichum*** (p. 137)

23. Basal marginal leaf cells undifferentiated or eventually somewhat narrower............. 24

24. Basal cells hyaline throughout, thin-walled, elongate-linear; leaf margins serrulate to the shoulders; leaf subula 3 or more times as long as the base; stem with a unistratose hyalodermis, with collapsed outer walls .... ***Bartramia*** (p. 507)

24. Basal cells chlorophyllose throughout, thin- to thick-walled, rectangular; leaf margin entire throughout or only roughened in the subula; leaf subula less than 3 times as long as the base; stem lacking a hyalodermis ..... ***Dicranella*** (p. 164)

25(21). Costa very broad, occupying ⅓–½ the leaf base **26**

25. Costa narrower, occupying ¼ or less the leaf base **32**

26. Costa not clearly demarcated and gradually merging with the laminal cells..... **27**

26. Costa clearly separated from the laminal cells, at least in the proximal part..... **29**

27. Costa bistratose throughout, composed of undifferentiated cells .... ***Andreaea*** (p. 74)

27. Costa multistratose, with a distinct dorsal stereid band and a row of guide cells 28

28. Leaves ovate to oblong, broadly rounded at the apex, closely imbricate, 5-ranked .... ***Conostomum*** (p. 496)

28. Leaves lanceolate, tapering to a long subula, unranked .... ***Ditrichum*** (p. 137)

29(26). Laminal cells appearing papillose in cross-section owing to the longitudinal cuticular ridges covering the lumina; leaves falcate-secund to curved .... ***Holodontium*** (p. 201)

29. Laminal cells smooth; leaves straight or with a flexuose acumen 30

30. Laminal cells linear to linear-rhomboidal throughout, undifferentiated in the basal angles .... ***Leptobryum*** (p. 418)

30. Laminal cells isodiametric to short-rectangular at the shoulders, short and hyaline at the base and usually forming distinct auricles 31

31. Costa in cross-section with a few rows of ventral stereids ..... ***Chorisodontium*** (p. 181)

31. Costa in cross-section with a ventral layer of hyalocysts .... ***Campylopus*** (p. 186)

32(25). Upper laminal cells smooth **33**

32. Upper laminal cells papillose, pseudopapillose, mammillose or prorate **72**

33. Costa bistratose **34**

33. Costa 3- to multistratose **35**

34. Costa broad, occupying ⅓–¼ the leaf base, diffuse .... ***Andreaea*** (p. 74)

34. Costa narrow, occupying ⅕ or less the leaf base, clearly demarcated .... ***Schistidium*** (p. 226)

35(33). Leaves broadly ovate to orbicular, rounded-obtuse to apiculate at the apex **36**

35. Leaves lanceolate to ovate lanceolate, acute, acuminate or subulate **40**

36. Plants silvery, silver-green to whitish-green **37**

36. Plants not silvery or whitish **38**

37. Plants small, gemmiform, with the stem less than 3 mm and closely imbricate leaves; laminal cells with dorsal superficial cell walls flat and thickened and ventral superficial cell walls bulging and thin .... ***Stegonia*** (p. 339)

37. Plants larger, with the stem 4–25 mm long and upper leaves forming a distinct capitulum and lower leaves laxly erect-imbricate; laminal cells with evenly thickened dorsal and ventral superficial cell walls .... ***Bryum*** (p. 449)

38(36). Upper cells thin-walled, rhombic, rhomboidal to shortly oblong-hexagonal; leaf base often red or brown pigmented .... ***Bryum*** (p. 449)

38. Upper laminal cells quadrate, rounded-quadrate or rhomboidal, firm- to thick-walled; leaf base not pigmented 39

39. Basal leaf cells thin-walled, lax, hyaline; leaf margins plane to incurved often roughly serrate or toothed near the apex by projecting cells .... ***Hennediella*** (p. 358)

39. Basal leaf cells firm-walled, sturdy, concolorous with the upper cells; leaf margins plane to recurved, entire.......... ***Ceratodon*** (p. 154)

40(35). Leaves subulate **41**

40. Leaves acute to acuminate, not subulate.......... **45**

41. Alar cells differentiated, inflated or flat, usually coloured **42**

41. Alar cells not differentiated **43**

42. Leaves curved to straight on drying; laminal cells thin-walled, entirely smooth .... ***Kiaeria*** (p. 177)

42. Leaves crisped when dry; laminal cells thick-walled, pseudopapillose owing to the lengthwise cuticular lamellae.......... ***Hymenoloma*** (p. 203)

43. Laminal cells isodiametric to short rectangular at the shoulders ...... ***Dicranella*** (p. 164)

43. Laminal cells linear to linear-rhomboidal throughout 44

44. Leaf margin serrulate in the subula .... ***Leptobryum*** (p. 418)
44. Leaf margin entire throughout .... ***Ditrichum*** (p. 137)

45(40). Laminal cells with thin transverse walls and thick and sinuose to sinuose-nodulose longitudinal walls **46**
45. Laminal cells with straight, evenly thickened walls or with transverse walls thicker than vertical walls **50**

46. Basal cells short- or long-rectangular, mostly with straight and uniformly thickened walls or with transverse walls thicker than vertical walls, always without spiral thickenings, concolorous with the upper cells; costa semi-terete or rarely reniform, not situated in a furrow, with 2 cells at the ventral side in cross-section, usually not markedly larger than central and dorsal cells; plants acrocarpous; stems erect, often with forked branches; central strand present or absent **47**
46. Basal cells long-rectangular to linear, with strongly nodose longitudinal walls and often with spiral thickenings, usually forming a coloured strip in 1–3 rows along the insertion; costa reniform, elliptical to semi-terete, often lying at the bottom of a furrow, in cross section in the proximal part with 2–5 large ventral cells markedly different from central and dorsal cells; plants cladocarpous; stems often prostrate, usually with abbreviated branches; central strand always absent **49**

47. Capsules immersed, symmetric, systylious; seta very short, straight; calyptra minute, covering only the beak of the operculum; costa in transverse section scarcely differentiated, 2–3(–5)-stratose, without a central stereid band .... ***Schistidium*** (p. 226)
47. Capsules emergent to exserted, with deciduous operculum; seta longer, arcuate; columella remaining in the urn at dehiscence; calyptra larger, covering the capsule mouth; costa in transverse section usually with a distinct central band of small hydroid or stereid cells 48

48. Leaves canaliculate and narrowly U-shaped distally to broadly U-shaped throughout; costa semi-elliptic to reniform and curved in transverse section, not clearly delimited from the laminal cells and scarcely prominent on the dorsal surface..... .... ***Guembelia*** (p. 275)
48. Leaves sharply keeled and V-shaped distally; costa semi-terete in transverse section, clearly delimited from the laminal cells and prominent on the dorsal surface .... ***Dryptodon*** (p. 283)

49(46). Laminal cells with large flat cuticular thickenings with secondary papillulae over the longitudinal walls and almost the whole lumina, except for a narrow slit in the middle; hyaline hair-points long, strongly papillose and erose-dentate, long decurrent down the leaf margins .... ***Racomitrium*** (p. 288)
49. Laminal cells smooth or pseudopapillose owing to longitudinal cuticular ridges; hyaline hair-point, if present, smooth or denticulate, never papillose, not or short-decurrent .... ***Bucklandiella*** (p. 293)

50(45). At least the uppermost leaves with a hyaline hair-point **51**
50. Leaves lacking a hyaline hair-point **56**

51. Leaves distinctly plicate on both sides of the costa in the distal part; plications often tristratose; plants .... ***Coscinodon*** (p. 223)
51. Leaves not plicate; plants hoary or not 52

52. Costa in transverse section homogeneous, 2–3(–5)-stratose, without a central stereid band; capsules immersed, symmetric, systylious; seta very short, straight; calyptra minute, covering only the beak of the operculum .... ***Schistidium*** (p. 226)
52. Costa in transverse section usually with a distinct central band of small hydroid or stereid cells; capsules emergent to long exserted or if immersed then asymmetric, ventricose at the base; seta short, cygneous or long; columella remaining in the urn at dehiscence; calyptra larger, covering the capsule mouth 53

53. Leaves deeply concave, unistratose throughout or occasionally with bistratose marginal spots near the apex; capsules immersed, asymmetric and ventricose at the base; setae arcuate, eccentrically attached, shorter than the urn .... ***Grimmia*** (p. 270)

53. Leaves canaliculate to sharply keeled in the distal half; lamina unistratose with bistratose patches; margins bistratose for one to several rows of cells; capsules emergent to long exserted, symmetric, not ventricose at the base; setae straight or curved, centrally attached, longer than the urn 54

54. Leaves canaliculate and narrowly U-shaped distally to broadly U-shaped throughout; costa semi-elliptic to reniform and curved in transverse section, not clearly delimited from the laminal cells and scarcely prominent on the dorsal surface .... ***Guembelia*** (p. 275)

54. Leaves sharply keeled and V-shaped in the distall part; costa semi-terete in cross-section, clearly delimited from the laminal cells and prominent on the dorsal surface 55

55. Leaf margins plane throughout or incurved above, occasionally narrowly recurved proximally on one side to the mid-leaf; basal marginal cells of the leaves with transverse walls much thicker than the longitudinal walls; seta straight; capsule smooth .... ***Orthogrimmia*** (p. 279)

55. Leaf margins recurved on one or both sides from the base to ⅔ way up the leaf; basal marginal cells of the leaves with all walls equally thin and straight; seta arcuate or flexuose; capsule striate-ribbed to rugose-plicate .... ***Dryptodon*** (p. 283)

56(50). Upper laminal cells short, with a 2 : 1 or less length/breadth ratio, quadrate, rounded, oblate, isodiametric or rounded-hexagonal **57**

56. Upper laminal cells long, with a (2–)3 : 1 or greater length/breadth, rhombic- or oblong-hexagonal, elongate, rectangular or linear **65**

57. Costa undifferentiated in transverse section, comprising cells of more or less similar size and shape **58**

57. Costa differentiated in transverse section, consisting of 1 or 2 stereid bands, a row of guide cells and an epidermis, at least on the ventral surface **59**

58. Leaves bistratose distally, bordered with a broad, strong, (2–)3–4-layered swollen limbidium; costa ending below the apex; basal cells quadrate to short-rectangular, about 1 : 2; capsule exserted; spores multicellular; blackish-green, exclusively a near-shore plant .... ***Muelleriella*** (p. 406)

58. Leaves 1–2-stratose distally, with a narrow 1–2-stratose marginal border; costa percurrent to long excurrent; basal cells rectangular, 1 : 3 or more, if shorter the leaf margin bluntly denticulate at the apex; capsule immersed; spores unicellular; coastal or inland plant .... ***Schistidium*** (p. 226)

59(57). Leaves closely appressed, arranged in five ranks, giving the shoot a ridged appearance .... ***Conostomum*** (p. 496)

59. Leaves not ranked; shoots terete 60

60. Leaf margin recurved .... ***Ceratodon*** (p. 154)

60. Leaf margin plane 61

61. Costa with a row of large guide cells **62**

61. Costa lacking guide cells **63**

62. Basal leaf cells thin-walled, lax, hyaline; costa with a dorsal stereid band lacking epidermis; leaf margins roughly serrate or toothed near the apex by projecting cells .... ***Hennediella*** (p. 358)

62. Basal leaf cells firm-walled, sturdy, concolorous with the upper cells; costa mostly with dorsal and ventral stereid bands and with a distinct epidermis on the dorsal surface; leaf margins entire .... ***Ceratodon*** (p. 154)

63(61). Leaves oblong-lanceolate to lingulate; costa occupying about ¼ or more the leaf base .... ***Meesia*** (p. 491)

90. Leaves lanceolate, oblong-lanceolate, oblong- to ovate-lingulate, not or occasionally slightly constricted at or below the middle, not or hair-pointed, with recurved margins; dorsal stereid band crescent-shaped; capsules long-exserted; peristome present; calyptra cucullate; inner perichaetial leaves not or only slightly differentiated 91

91. Basal leaf cells evenly thin-walled, lax; calyptra cucullate, covering only the operculum and the capsule apex .... ***Syntrichia*** (p. 375)

91. Basal leaf cells with markedly thickened transverse walls; calyptra campanulate-mitrate, covering the entire capsule .... ***Encalypta*** (p. 120)

Division BRYOPHYTA Schimp.

Bryophyta Schimp. *in* Zitt., Handb. Palaeont. Abt. 2 (Palaeophyt.): 1. 1879 ['Stamm Bryophyta'].

The division Bryophyta has traditionally been considered to comprise three major classes of land plants called bryophytes (Braun 1864; Schimper 1879), namely Musci or mosses, Hepaticae or liverworts and Anthocerotae or hornworts. The basic differences between the mosses and liverworts were defined at the end of the 18th century by Hedwig (1784), but the hornworts were separated from the Hepaticae a century later (Howe 1899; Campbell 1905). Schuster (1977) argued that hornworts should be placed in their own division and a monophyletic concept of bryophytes has been formally rejected by Mishler & Churchill (1984) who consider, on the basis of morphological, physiological and biochemical characters, that the Bryophyta are composed of a paraphyletic group comprising three major lineages which are remotely related to each other. This is supported by molecular analyses (Mishler *et al.* 1994; Lewis *et al.* 1997; Bopp & Capesius 1996, 1998; Hedderson *et al.* 1996, 1998). Accordingly, the former classes of Bryophyta have been raised to the rank of separate divisions or phyla (e.g. Renzaglia & Vaughn 2000; Crandall-Stotler & Stotler 2000; Buck & Goffinet 2000; Crum 2001). Consequently, the term 'bryophyte' has lost its original meaning, as did 'thallophyte' and 'pteridophyte'. Its current use is justified only because mosses, liverworts and hornworts show similarities in structure, reproduction, development, and ecology.

Mosses, liverworts and hornworts are small, relatively simple plants but may form a major component of the vegetation in cool temperate and humid tropical regions. They are essentially terrestrial plants that achieve their greatest diversity as the gametophyte stage. They share similar life cycles in which the gamete-bearing, free-living and photosynthetic gametophytes are dominant and the spore-bearing sporophyte remains attached to the gametophyte and is photosynthetic for only a short time. Additionally, the gametophytes develop from single cell initials, the sex organs are similar, sperm cells are biflagellate and conducting tissue is virtually absent. Other than these important and significant similarities mosses, liverworts and hornworts have little in common. They exhibit important morphological dissimilarities in their gametophyte and sporophyte generations which outweigh the similarities. The differences are coupled with major disparities in morphogenesis, including spore germination and ontogeny of both generations as well as gametangial development and structure, cellular organisation and biochemistry. The similarities and differences between the main lineages of 'bryophytes' were summarised by Chopra (1967) and Crandall-Stotler (1984). They proposed that mosses, liverworts and hornworts are distantly related but independently established groups of land plants. Following these modern concepts, the division Bryophyta in the present treatment is restricted only to mosses, or Musci.

Mosses differ from liverworts and hornworts in structure and ontogeny. The gametophyte is sharply divided into protonema and gametophore. The protonema arises from the germinating spore and is chlorophyllose, either flat and thallose or (more often) extensively branched and filiform. The protonema gives rise to the gametophore which is usually a perennial leafy shoot. The leaves are undivided, mostly spirally arranged (occasionally in two or three ranks) and have single or double, multistratose costa, although this is sometimes absent. The leaves do not have stomata. The rhizoids are most often filamentous, much-branched, linearly multicellular with oblique cross-walls. Occasionally they are biseriate and plate-like or cylindrical. They resemble the protonema but lack chlorophyll and usually have brown walls. The sex organs are intermixed with sterile, hair-like paraphyses and the antheridia have a cap of differentiated cells that serve in dehiscence. The sporophyte is long-lived, leafless and unbranched and it is intimately associated with the gametophyte, being attached by a pointed basal part (the foot). It bears the sporangium (capsule) that is either elevated on a seta which elongates before the capsule differentiates or is extended on a pseudopodium (a leafless extension of the stem). The capsule walls are always multistratose, photosynthetic and usually have stomata. The capsules usually dehisce by an operculum or lid, but some genera exhibit irregular dehiscence (e.g. in *Acaulon*, *Pseudephemerum* and *Pleuridium*) or open by longitudinal slits (in *Andreaea* and *Acroschisma*). Within the capsule there is a spore sac with the archesporium which is endothecial in origin and gives rise to spores and a central column of sterile tissue (the columella) running most of the length through the capsule. Rarely the archesporium is absent (in *Archidium*). The developing capsule is protected by a cap (the calyptra) which is formed from the expanded upper portion of the archegonium when the growing embryo breaks through its neck. Comprehensive accounts of the structure, life-cycles and phylogeny of the Musci can be found in Parihar (1965), Watson (1967), Schuster (1983–1984), Puri (1985), Schofield (1985), Chopra & Kumra (1988), Shaw & Goffinet (2000) and Crum (2001). Important earlier works are those of Haberlandt (1886), Müller (1894–1900), Cavers (1911), Campbell (1895, 1918), Ruhland (1924*a*,*b*,*c*), Goebel (1930), Lorch (1931), Verdoorn (1932), Bower (1935) and Smith (1955).

The Bryophyta are the second largest division of land plants, after angiosperms. In the two centuries since publication of the fundamental opus *Species muscorum frondosorum* by Hedwig (1801), which is generally accepted as the starting point of moss nomenclature (except for *Sphagnum*), some 29,100 species of moss have been described. However, a recent world checklist of the mosses (Crosby *et al.* 2000) recognises only 12,760 species belonging to 901 genera. The remaining species have been considered as superfluously described and their names have been reduced to synonymy of accepted species names. A realistic estimation of the number of mosses is difficult because of the slow progress in taxonomic overviews of some large genera and extra-Holarctic moss floras. After all necessary reductions, the division Bryophyta is likely to consist of no more than 8500–9000 species worldwide. They are currently classified in 864 genera which are assigned to 116 families (Buck & Goffinet 2000; Goffinet & Buck 2004). Only 111 species and two varieties, in 55 genera and 17 families, have been recorded from Antarctica.

The classification of mosses has undergone major modifications over the past two centuries since Hedwig (1782, 1784) and Bridel (1826–1827) laid down the foundations of modern muscology (Vitt 2000) and certainly it will continue to be revised, especially with the increasing use of molecular techniques in moss phylogenetic and taxonomic studies (e.g. Capesius & Stech 1997; Cox & Hedderson 1999; Newton 2000; Buck *et al.* 2000*a*,*b*). At present, the general tendency is to give higher ranks to different lineages within the traditionally defined Musci. Until recently only the single class Bryopsida was recognised, subdivided into seven subclasses, Sphagnidae, Andreaeidae and Bryidae (Walther 1983; Vitt 1984), Tetraphididae, Polytrichidae and Buxbaumiidae (Crosby 1980) and Archidiidae (Schofield 1985).

By the end of the 20th century two 'new generation' classifications of the mosses were proposed on the basis of peristomial characters. Both are supported by molecular data which, though far from complete, provide sound evidence of separate origins and lines of evolution of the main groups of 'bryophytes'. Vitt *et al.* (1998) divided the Bryophyta into four classes, namely Sphagnopsida, Andreaeopsida, Polytrichopsida and Bryopsida, while Buck & Goffinet (2000) and Goffinet & Buck (2004) developed this classification further and recognised two additional monotypic classes, Takakiopsida and Andreaeobryopsida, for the highly isolated and bizarre moss genera, *Takakia* and *Andreaeobryum*, respectively. Most recently, Crum (2001) segregated Takakiopsida and Sphagnopsida into separate divisions, Takakiophyta and Sphagnophyta, parallel to Bryophyta, Hepatophyta and Anthocerotophyta, leaving the Bryophyta composed of two classes, Bryopsida or true mosses with six subclasses (i.e. Bryidae, Buxbaumiidae, Tetraphididae, Polytrichidae, Dawsoniidae and Archidiidae), and Andreaeopsida or lantern mosses with two orders (i.e. Andreaeales and Andreaeobryales).

The great majority (102) of species occurring in Antarctica belong to the true mosses (Bryopsida), three to the lantern mosses (Andreaeopsida) and six to the hair-cap mosses (Polytrichopsida). The Bryopsida are subsequently divided into six subclasses and eight orders.

KEY TO THE CLASSES OF THE ANTARCTIC BRYOPHYTA

1. Protonema thallose, ribbon-like; rhizoids filamentous or plate-like, uni- or biseriate with straight cross-walls; capsules elevated shortly beyond perichaetial leaves on a pseudopodium or a short and stout seta, dehiscing by means of 4(–6–10), lengthwise valves, free or attached at apex; dark brown, reddish or blackish plants mostly on acidic rocks .... **Andreaeopsida** (p. 73)
1. Protonema filamentous, extensively branched; rhizoids hair-like, filiform, mostly branched with oblique cross-walls; pseudopodium absent; capsules elevated on setae of variable length, dehiscing by means of an operculum or indehiscent; plants of variable size and colour growing on different substrata 2
    2. Peristome nematodontous; leaves typically with longitudinal ventral and occasionally dorsal chlorophyllose lamellae; apex of the columella forming an expanded epiphragm; conducting tissues composed of hydroids and leptoids present in stem .... **Polytrichopsida** (p. 87)
    2. Peristome arthrodontous; leaves mostly lacking lamellae; epiphragm absent; hydroids and leptoids lacking in stem .... **Bryopsida** (p. 112)

Class A. **ANDREAEOPSIDA** J.H.Schaffn.

Andreaeopsida J.H.Schaffn., Ohio Nat. 9: 451. 1909 ['Andreaeae'].

The class Andreaeopsida is a small and highly isolated group of mosses which retains many primitive features indicating its ancient origin. Members of this group are predominantly epilithic mosses growing mostly on acidic rocks and for that reason this group is sometimes called 'granite mosses'. However, this is a misleading term because the monotypic genus *Andreaeobryum* is sometimes included in this class, but it has an ecological preference for limestone and, furthermore, many acidic rocks are not granite. Hence, a more appropriate name is 'lantern mosses' referring to the resemblance of the sporangium to a Chinese lantern.

The gametophyte of the lantern mosses resembles that of the true mosses of the class Bryopsida, but the sporophyte is quite different. The lantern mosses are easily recognised by the capsules with solid walls, lacking stomata, dehiscing along longitudinal valves and having epidermal cells with nodose, secondary thickenings on radial walls. The protonema is generally persistent, branched, thallose, ribbon-like or cylindrically thickened and functions as rhizoids. It often begins development inside the germinating exospore. The sporogenous tissue and columella are derived from the endothecium, and a dome-shaped spore-sac overarches the persistent, thick, rod-like columella and it is not separated from the capsule wall by air spaces as in the Bryopsida.

The name Andreaeopsida was first proposed by Schaffner (1909), although it is often ascribed to Rothmaler (1951) who elevated the order Andreaeales, described by Limpricht (1876, 1885 as Ordnung Andreaeaceae), to the rank of class. The taxonomic and phylogenetic problems in this class have been discussed in detail by Murray (1982, 1988*a*). The Andreaeopsida are considered to be the closest relatives of the even more primitive Takakiopsida and, additionally, some links with liverworts are also possible (through *Haplomitrium*), but not with other mosses. The class consists of four genera referred to two separate orders, Andreaeales (*Andreaea*, *Acroschisma*, *Bicosta*) and Andreaeobryales (*Andreaeobryum*). Buck & Goffinet (2000) placed these orders in two separate classes, Andreaeopsida and Andreaeobryopsida, but it is doubtful whether the differences in the ecological preferences, sexuality, capsule shape and the presence or absence of the seta are sufficient to maintain Andreaeobryopsida as a class separate from Andreaeopsida. It is more reasonable to retain *Andreaeobryum* in a relationship with *Andreaea* and therefore both genera are placed in two subclasses within the Andreaeopsida, namely Andreaeidae and Andreaeobryoidae (Ochyra *et al.* 2003). The latter subclass is a monotypic taxon consisting of a single species, *A. macrosporum*, belonging to a monotypic family Andreaeobryaceae, widely distributed but sporadic in the mountains of north-west North America from Alaska to northern British Columbia (Steere 1978; Murray 1987*a*, 1988*a*).

Subclass A1. **ANDREAEIDAE** Engl.

Andreaeidae Engl., Syllabus Grosse Ausg.: 47. 1892 ['Unterklasse Andreaeales'].

The subclass Andreaeidae is diagnosed primarily by a set of sporophyte characters. The perichaetial leaves are concave or convolute and develop partly prior to fertilisation enclosing the aggregate archegonia. The seta is not developed and the capsule is elevated after maturity on the pseudopodium. The capsules are symmetric and usually dehisce regularly along the sutures of thin-walled cells into 4(–6–10) valves which are usually attached at the apex. The epidermal cells of the valves are long-rectangular, thick-walled and oval in transverse section. There are no stomata. The calyptra is small, mitrate and apical, usually bistratose in cross-section. The mucilage papillae are present only in the leaf axils. They are straight, cylindric or clavate and lack an apical beak, and the mucilage is secreted through the lateral walls. The plants are commonly cladautoicous, more rarely dioicous or gonioautoicous.

A subclass with one order, the Andreaeales.

Order I. **ANDREAEALES** Limpr.

Andreaeales Limpr. *in* Cohn, Krypt.-Fl. Schlesien 1(1): 217. 1876 ['Ordnung Andreaeceae'].

The Andreaeales were first recognised in the present sense by Campbell (1895). Murray (1988*a*) ascribed the first recognition of this order to Milde (1869), but the 'ordo' of this author should rather be interpreted as a family, not as an order in the present sense according to Art. 18.2 of the Code (McNeill *et al.* 2006). It is characterised by the features of the only family Andreaeaceae which is well represented in the maritime Antarctic.

Family I. ANDREAEACEAE DUMORT.

Andreaeaceae Dumort., Anal. Fam. Pl.: 68. 1829. — Type: *Andreaea* Hedw.

Cochlearidiaceae Ochyra & Mahu, J. Bryol. 16: 221. 1990, ***syn. nov.*** — Type: *Cochlearidium* Ochyra & Mahu.

The Andreaeaceae are one of the most distinctive moss families, which is well characterised by the general features of the subclass Andreaeidae. It comprises short, acrocarpous mosses growing in cushions or short turves on non-calcareous rocks, especially in mountainous areas. The plants are usually dark brown, reddish or blackish and mostly rigid and brittle when dry. The stems are simple or branched, lack a central strand and are attached to the substratum by rhizoids. These are confined to the basal part of the stem only and are of four types (Redhead 1973), uni- or biseriate, plate-like or cylindrical, depending on whether the moss is growing in crevices or on the surface of rocks. The leaves are spirally arranged and very variable in shape, from lanceolate and subulate to broadly ovate and shortly apiculate. They are ecostate or have a costa which is single or branched, distinctly or indistinctly delineated, bi- to multistratose, broad or narrow and short to excurrent. The laminal cells are short throughout or only in the upper part, becoming rectangular below, and they are smooth or papillose at back. The cell walls are usually strongly incrassate and the longitudinal walls are often variously porose or sinuose. The leaf lamina is generally unistratose, except for the costal area, or bi- to multistratose in small patches. Very occasionally (in *Bicosta fuegiana*), two submarginal multistratose, costa-like thickenings are present. The perichaetial leaves do not markedly differ from the vegetative leaves or are enlarged, convolute and sheathe the pseudopodium. The capsules are spindle-shaped or ellipsoid and dehisce by 4 or sometimes 6–10 longitudinal valves which are connate at the apex and strongly bulging when dry because of the contraction of the capsule in length, or sometimes (in *Acroschisma wilsonii*) the capsule is cylindrical with only 4–8, free apical valves. The spores are variously papillose to smooth, small to large, often with a large percentage aborted and shrivelled, and their size ranges from 13 to 60 μm, although in extreme cases they may be as small as 10 μm or as large as 110 μm.

The family has a worldwide distribution, but is particularly abundant in polar and cold temperate regions, whereas in the tropics its members are confined to altimontane areas. Generally, the Andreaeaceae are considered to be a monotypic family consisting of a single genus, *Andreaea*, but two other genera, *Acroschisma* and *Bicosta*, have been established to accommodate two rare, antipodal, cool-adapted species. Ochyra & Mahu (1990) described the monotypic genus *Cochlearidium* from northern Chile, but its only species, *C. aymaranum*, is conspecific with *Andreaea nitida* Hook.f. & Wilson and, accordingly, this generic name is a synonym of *Andreaea*, an amphipacific south-temperate species (Murray 2006). Although the Andreaeaceae have not been globally monographed, recognition of these genera is based on a suite of distinct characters. It is noteworthy that many accepted genera of the true mosses (Bryopsida) are distinguished on the basis of single characters. The family is well represented in the cool temperate zone in the Southern Hemisphere and in the Subantarctic, and the genus *Andreaea* is also prominent in the maritime Antarctic, occasionally extending to the Antarctic continent.

1. ANDREAEA HEDW.

Andreaea Hedw., Spec. Musc. Frond.: 47. 1801. — Lectotype (*vide* Schultze-Motel 1970: p. 27): *Andreaea rupestris* Hedw.

Cochlearidium Ochyra & Mahu, J. Bryol. 16: 215. 1990. — Type: *Cochlearidium aymaranum* Ochyra & Mahu. First synonymised by Churchill *et al.* (1995: p. 102).

Plants small to large and robust, perennial, growing in brittle short turfs or cushions, dark brown to blackish, often tinged with red, purple or dark green. *Stems* erect, variously branched, densely leafy, in cross-section composed of ± uniform, thick-walled cells or outer cortical cells smaller and thicker-walled than medullary cells, central strand absent; *axillary hairs* with 1–4 short, brown basal cells and an elongate, cylindrical or clavate terminal cell. *Leaves* crowded, erect and appressed or falcato-secund when dry, spreading to squarrose when wet, symmetric or asymmetric, plane or concave to cochleariform, narrowed at base or sometimes auriculate, lanceolate, linear-lanceolate or panduriform to ovate or obovate, rarely ovate-elliptical, broadly acute to slenderly acuminate, sometimes subulate or obtuse and minutely apiculate, mostly clasping or sheathing at base; *margins* plane to incurved, sometimes reflexed, entire or crenulate to denticulate; *costa* absent, single or sometimes branched, distinct or indistinctly defined, narrow to very broad and stout, short to long-excurrent; *laminal cells* unistratose, sometimes bistratose, dorsally smooth or papillose, very variable, ± short throughout or short, quadrate, rounded-quadrate or irregular to short-rectangular, with strongly incrassate walls above, becoming rectangular, with conspicuously thickened, often pitted and sinuose longitudinal walls below. *Cladautoicous*, rarely *dioicous*, *gonioautoicous* or *synoicous*. *Perigonia* lateral or terminal, small, bulbiform, usually comprising 4–10 elliptic antheridia on biseriate stalks, with few or lacking paraphyses. *Perichaetia* terminal; *perichaetial leaves* undifferentiated from the vegetative leaves or enlarged, mostly 4–6, convolute and sheathing the pseudopodium. *Capsules* shortly exserted, spindle-shaped to ellipsoid, splitting by longitudinal slits into 4, sometimes 6–10 valves remaining fused above and below. *Stomata* absent. *Spores* frequently aborted, small to very large, (10–)13–60(–110) µm in diameter, papillose to reticulate-papillose, sometimes almost smooth. *Calyptra* very small, apical, campanulate-mitrate, fugacious or not, smooth, laciniate at the base.

A relatively large genus comprising, realistically, 50–70 species. However, about 175 species have been validly described (Wijk *et al*. 1959, 1969; Crosby *et al*. 1992; Crosby 2000), of which more than 90 species are still recognised (Crosby *et al*. 2000). The actual number of the species cannot be precisely established because the original survey of the genus by Roth (1911, 1913, 1916) is now out of date and, recently, *Andreaea* has been revised taxonomically only in part, mostly in the Northern Hemisphere, including the Nearctic (Murray 1987*b*), Italy (Raffaelli 1973), the British Isles (Murray 1988*b*), Taiwan (Chiang 1998) and China (Cao & Gao 1995, 1999), and in sub-Saharan Africa (De Sloover 1977; Magill 1981), while Schultze-Motel (1970) presented a world revision of the costate species.

In the Southern Hemisphere the genus exhibits the greatest diversity but it has not been studied thoroughly here, except for Australia (Murray 2006) and subantarctic Macquarie I. (Seppelt 2004). The early treatment of the Patagonian species by Dusén (1903*a*) provides fine and still useful illustrations of some species, but the species concept needs a modern revision. Antarctic species have been studied by Greene *et al*. (1970), Ochyra (1998*a*) and Liu & Li (1998*a*); in addition, Greene (1968*b*) surveyed some South Georgian and southern South American species, Seppelt (2004) studied the genus on Macquarie I., Vitt (1980) revised the *A. acutifolia* complex, and some miscellaneous studies on *Andreaea* have been published by Ochyra (1996*a*,*b*) and Matteri & Farias (1999).

Britton & Emerson (1913) proposed *Andreaea petrophila* (Ehrh.) *ex* Fürnr. as the lectotype of the generic name *Andreaea*, but it cannot be accepted. However, this species is missing from the description of the genus (Hedwig 1801) which originally consisted only of *A. rupestris* and *A. alpina*. Schultze-Motel (1970) selected the former species as lectotype and this lectotypification meets the requirements of the current Code (McNeill *et al*. 2006).

In the Antarctic *Andreaea* is represented by three species which are important constituents of several communities, both in terms of frequency and cover, and are pioneers of exposed rocks and fellfields. The genus may be confused with *Schistidium* or *Grimmia* but it is most readily recognised by the 4 or more long vertical splits along the sides of the capsules. In the absence of capsules, which are a constant feature of *A. depressinervis*, its distinctive reddish-brown to dark purplish-green coloration, ecostate or having broad and indistinct costa leaves and moderately to strongly thick-walled, rounded cells, usually papillose on the back, aid the identification of *Andreaea*.

The infrageneric classifications of *Andreaea* proposed by Brotherus (1901*a*, 1924*a*) are in need of revision because the subgeneric and sectional boundaries seem to be in flux (Murray 1986, 1988*b*). Two Antarctic species, *A. gainii* and *A. regularis*, clearly belong within the type subgenus and type section encompassing ecostate and denticulate and non-denticulate species with the enlarged and convolute perichaetial bracts. The subgeneric placement of *A. depressinervis* is still debatable since the species is invariably sterile. It was initially placed by Brotherus (1901*a*) in the type subgenus of *Andreaea* but later transferred to subgen. *Chasmocalyx* Lindb. which consists of species with large but undifferentiated and non-tubular perichaetial bracts. Greene *et al*. (1970) found and described one perichaetium in this species consisting of differentiated and convolute bracts which may suggest its position in the type subgenus. Nonetheless, because *A. depressinervis* occupies an isolated position in the genus its placement in the separate sect. *Depressinerves* W.Schultze-Motel is fully justified.

1. Leaves with a central bistratose strip of elongate, mostly porose cells, forming a broad, indistinctly delineated costa extending into an acumen .... 1. ***A. depressinervis***
1. Leaves ecostate, unistratose throughout or with occasional bistratose patches above 2
    2. Most leaves panduriform, broadest at or above constriction, abruptly narrowed to triangular apices; leaf margins crenate to dentate below; upper laminal cells smooth or inconspicuously papillose on the dorsal side .... 2. ***A. gainii***
    2. Leaves lanceolate or ovate-lanceolate, broadest near base, gradually tapering into a narrow long acumen; leaf margins entire throughout; upper laminal cells distinctly papillose on the dorsal side .... 3. ***A. regularis***

1. **Andreaea depressinervis** Cardot Fig. 17

Andreaea depressinervis Cardot, Rev. Bryol. 27: 43. 1900. — Type citation: ... détroit de Gerlache, XIIme débarquement; later Cardot (1901) wrote: Détroit de Gerlache: XIIème débarquement, île de Cavelier de Cuverville, sur les corniches de la falaise. (Nos. 239 et 240c; 2 février 1898). [Lectotype (*vide* Schultze-Motel 1970: p. 95): 'Herb. J. Cardot. Mousses du voyage de la Belgica. N° 239. Andreaea depressinervis Card. sp. nova. Canal de la Belgica: XIIème débarquement, sur les corniches de la falaise. Leg. Racovitza, 2 févr. 1898' – BR!; isolectotype: FH, NY!, PC!, S!; syntype: same label data, only the collecting number is 'Racovitza 240^{c}' – PC!; isosyntype: BR!].

Andreaea depressinervis var. *compacta* Cardot, Voy. Belgica Mousses: 23. 1901. — Type citation: Détroit de Gerlache: XXème débarquement, Terre de Danco, à l'entrée du chenal de Lemaire, sur des roches isolées au milieu d'un glacier, à 50 métres au-dessus du niveau de la mer. (N° 477, 12 février 1898). [Lectotype (*selected here*) {or perhaps holotype}: 'Herb. J. Cardot. Mousses du voyage de la Belgica. N° 477. Andreaea depressinervis Card.var. compacta Card. Detroit de Gerlache: XXème débarquement, rochers alt. 50 m Leg. Racovitza, 12 févr. 1898' – PC!]. First synonimised by Greene (1968*b*: p. 140).

Andreaea depressinervis fo. *robusta* Cardot, Wiss. Erg. Schwed. Südpolar-Exp. 4(8): 263. 1908 [fo. *robusta* Cardot, Bull. Herb.

Boissier Sér. 2, 6: 28. 1906, *nom. nud.*]. — TYPE CITATION: Région de Graham: terre Louis-Philippe: pointe Béatrice (Skottsberg, nº 430; île des Mousses (Skottsberg nº 431). [Lectotype (*selected here*): 'Herb. J. Cardot. Svenska Sydpolarexped. 1901–03. nº 430. Andreaea depressinervis Card. f. robusta Card. Terre Louis Philippe: pointe Béatrice. Leg. C. Skottsberg, 29/11 1902'– PC!; isolectotype: S!; syntype: 'Svenska Sydpolarexped. 1901–03. N:r 431. Andreaea depressinervis Card. forma robusta Card. Terre Louis-Philippe: île des Mousses. 1/12 1902 C. Skottsberg. Det. J. Cardot.' – PC!; isosyntypes: LD!, S!]. First synonymised by Greene (1968*b*: p. 140).

Plants slender, medium-sized to large, not or somewhat lustrous, in loose or compact, hummocky cushions or low tufts, usually loosely appressed to the substratum, sometimes scattered among other mosses, light to dark brown, olive-green, olive-brown, blackish-green to reddish- or blackish-brown. *Stems* rather stiff, 1–4 or occasionally 5–8 cm long, erect to ascending, usually profusely irregularly branched, sometimes simple or forked, in cross-section round, consisting of a 2(–3)-layered cortex of small cells with strongly thickened walls surrounding a medulla of large, firm-walled cells, central strand absent; *axillary hairs* with 1–2 small, quadrate, brown basal cells and a single, elongate apical cell. *Leaves* densely set throughout, non-decurrent, straight or sometimes somewhat secund towards the stem or branch tips, erect and imbricate to loosely appressed when dry, erecto-patent from appressed bases when moist, 1.0–1.4(–1.6) mm long, 0.3–0.7 mm wide, slightly to strongly concave, ovate, ovate-lanceolate to elliptical, gradually tapering into a short or long, broad or slender, acute or obtuse acumen; *margin* plane, entire; *costa* single, bistratose, usually poorly delineated, especially in fresh leaves, occupying ⅓–¼ the leaf base and extending up to the acumen and imperceptibly merging into the laminal cells, in the lower part composed of elongate cells with incrassate, pitted longitudinal walls, 25–50 µm long, 6–10 µm wide, becoming short rectangular in mid-leaf to elliptical in the apex; *laminal cells* unistratose; *upper cells* 8–25(–30) µm long, 5–15 µm wide, mostly isodiametric, rounded or elliptic, with rather thick and sometimes porose walls, usually with distinct collenchymatous, corner thickenings, smooth or moderately to strongly papillose on the dorsal side, smooth or faintly papillose ventrally, becoming short rectangular, 20–30 µm long, 8–10 µm wide in the juxtacostal regions below; *basal marginal cells* rounded, elliptical to oblate-subquadrate in 4–7 rows, 6–8 µm wide, with strong corner thickenings. *Perichaetia* [according to Greene *et al.* 1970] terminal, little longer than the vegetative leaves; *perichaetial bracts* broadly ovate to elliptical, rounded at the apex, deeply concave to convolute, with an indistinct costa or nerveless. *Sporophyte* unknown.

Chromosome number: $n = 10$ – King George I. (Kuta *et al.* 1982).

Discussion. *Andreaea depressinervis* differs most obviously, and is immediately recognised from other species of the genus in the Antarctic, by the single costa that forms a wide, bistratose and not sharply delimited but otherwise distinct strip in the centre of the leaf lamina. It extends to the apex where it imperceptibly merges with the laminal cells and fills it as a bistratose area. The costa is present in all but the youngest apical leaves and the shape of the costa is unique amongst Antarctic mosses. The laminal cells are smooth or weakly to strongly papillose distally on the dorsal surface and smooth or weakly papillose on the ventral surface and this is also a unique feature of this species, all others having leaf cells that are papillose only dorsally. *Andreaea depressinervis* is the largest and greenest of all *Andreaea* species in the Antarctic and therefore it often bears a superficial similarity to species of *Schistidium*. However, the characteristic leaf areolation of rounded cells with strong collenchymatous thickenings and elongate and porose supracostal cells in the basal part distinguish this species from all other mosses known in this biome. *Andreaea depressinervis* exhibits some variation in the stature of the plants and leaf shape, and this prompted Cardot (1901, 1906*a*, 1908*a*) to recognise two infraspecific taxa, viz. *A. depressinervis* var. *compacta* and *A. depressinervis* fo. *robusta*. Phenotypic variation in this species is certainly influenced by habitat, but microscopic details of the leaves remain stable and constant. Accordingly, these infraspecific taxa do not merit any taxonomic distinction and therefore their names have been reduced to synonymy (Greene 1968*b*).

Roth (1911) mentioned the presence of 'rounded or elliptical gemmae'. If true, this would be another unique feature of this species because no propagulae have ever been found in *Andreaea*. However, this observation cannot be confirmed and we agree with Schultze-Motel (1970) that young and underdeveloped leaves could have been misinterpreted by Roth (1911) as gemmae.

Andreaea depressinervis was originally described by Cardot (1900, 1901) from material collected by the *Belgica* expedition on Cuverville I. off the Danco Coast on the west coast of the Antarctic Peninsula, and is one of very few species of moss described from Antarctica which is accepted in modern moss taxonomy.

Reproduction in Antarctica. It is always sterile and neither sporophytes nor sexual organs have been detected. Greene *et al.* (1970) reported a single perichaetium but this material could not be traced. Liu & Li (1998*a*) provided a description and illustration of the sporophyte in *Andreaea depressinervis* from Fildes Peninsula, King George I., South Shetland Is. However, because these authors consider sporophytes in this species to be of frequent occurrence, we believe that they probably belong to *A. regularis* which also has a strongly papillose dorsal leaf surface.

Habitat. It occurs most commonly in dry to damp, exposed to sheltered conditions on acidic rock outcrops, boulders, ledges,

⟶

Figure 17. *Andreaea depressinervis.* — **1–2**. Habit. **3**. Portion of branch, dry. **4**. Portion of branch, wet. **5**. Cross-section of stem. **6–9**. Axillary hairs. **10–19**. Leaves. **20**. Leaf apex. **21**. Upper supracostal cells. **22**. Mid-leaf cells at margin. **23**. Angular leaf cells. **24**. Basal leaf cells. **25**. Lower supracostal cells. **26**. Basal juxtacostal cells. **27–33**. Cross-section of leaves. — [1, 6–10, 13 from *CS 430* (lectotype of *A. depressinervis* fo. *robusta*), PC; 2, 18 from *JPB 48A*; 3–5, 14, 20–21, 25–27, 29–33 from *RO 5149/79*; 11, 28 from *RO 5074/79*; 12, 23 from *BJT 446*; 15–16, 24 from *ER 477* (lectotype of *A. depressinervis* fo. *compacta*), PC; 17, 19 from *ER 239* (lectotype of *A. depressinervis*), PC; 22 from *RWMC 731*]. — Scale bars: a – 50 µm (6–9, 20–21, 23, 25–26); b – 100 µm (5, 22, 24, 27–33); c – 1 mm (3–4); d – 0.5 cm (1–2); e – 1 mm (10–19).

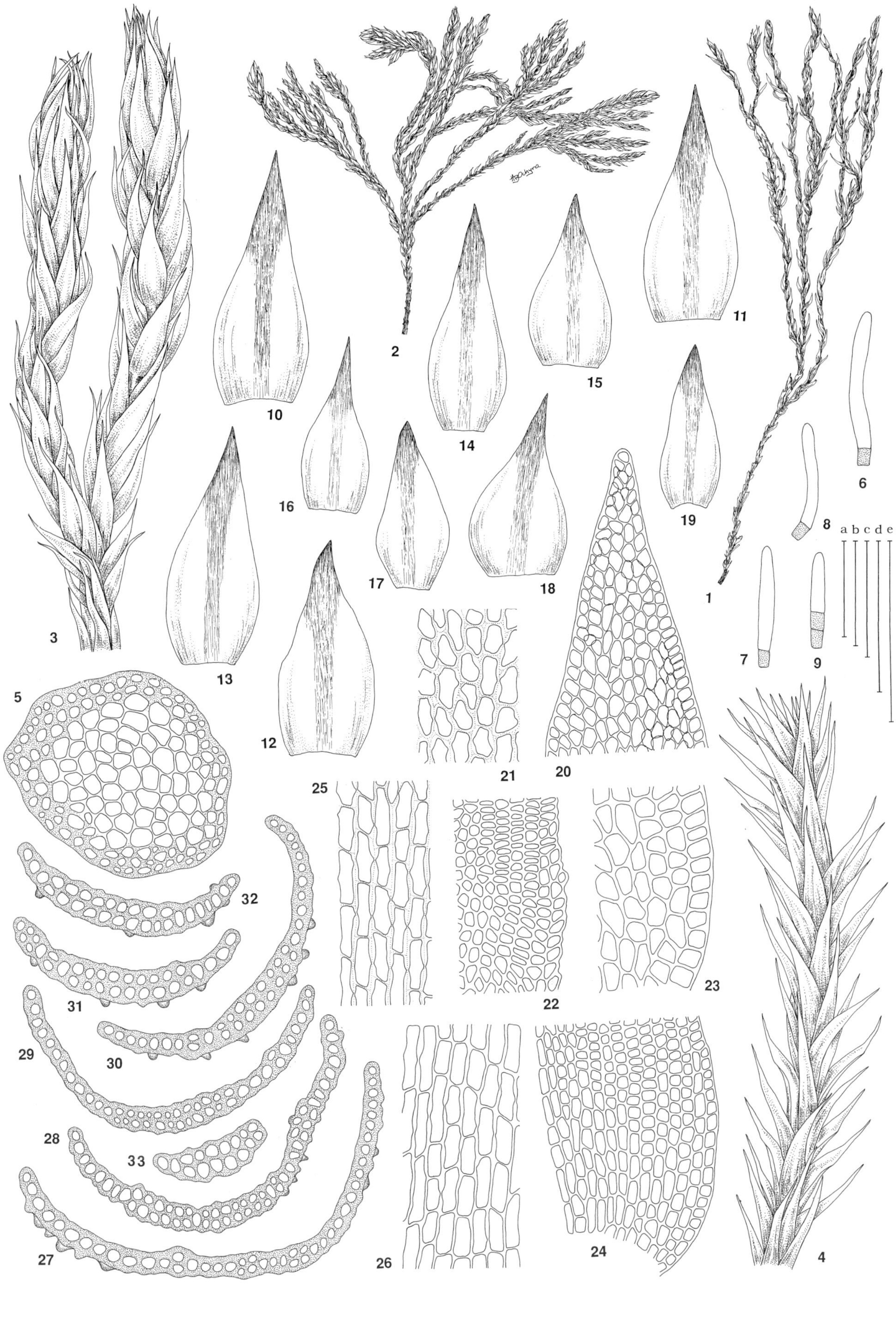
1
2
3
4
5
6
7
8
9
10
11
12
13
14
15
16
17
18
19
20
21
22
23
24
25
26
27
28
29
30
31
32
33
a b c d e

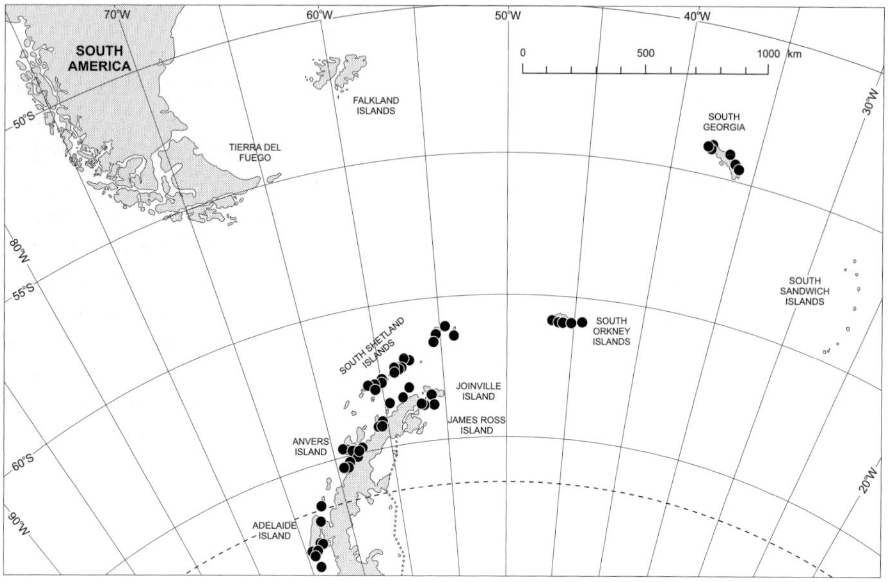

Figure 18. Global distribution of *Andreaea depressinervis*.

cliffs, scree, stony and gravelly ground, on pockets of soil or humus and, rarely, on eroding moss peat banks. The species is a common constituent of the fellfield vegetation dominated by various communities in the *Andreaea-Usnea* association of the fruticose lichen and moss cushion subformation (Gimingham & Lewis Smith 1970; Lewis Smith 1972).

World range. An Antarctic species, having the main centre of its occurrence in the maritime Antarctic and extending northward to South Georgia where it appears to be rare and localised (Fig. 18).

Distribution in Antarctica. Widely distributed and locally common and abundant in the South Orkney and South Shetland Is. and the west coast of the Antarctic Peninsula, south to the Faure Is., northern Marguerite Bay, at lat. *ca* 68°06′ S. Also sparsely distributed in the northernmost part of the east coast of the Antarctic Peninsula in the Hope Bay area and on Joinville I. and Paulet I. (Fig. 19). It usually occurs at elevations from sea level to about 250 m, decreasing in abundance at higher altitudes, occasionally above 300 m and exceptionally at 450 m on Powell I., South Orkney Is.

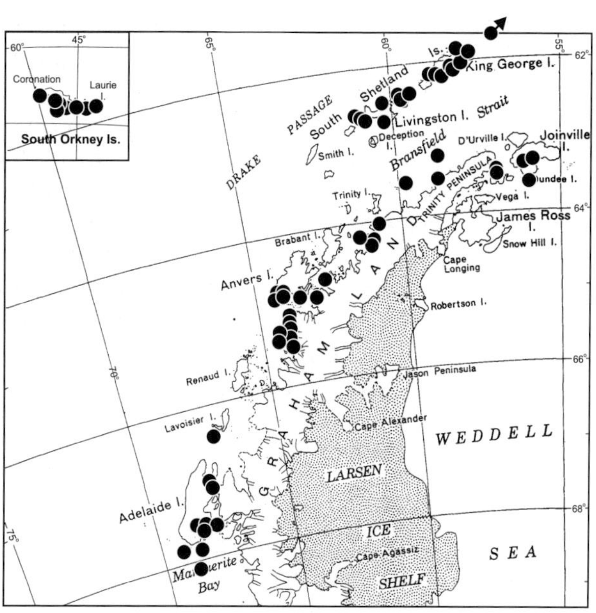

Figure 19. Distribution of *Andreaea depressinervis* in the Antarctic. The localities in the northern group of the South Shetland Is. (Elephant I., O'Brien I. and Clarence I.) are indicated by an arrow.

Selected specimens examined. WEST ANTARCTICA. SOUTH ORKNEY IS. **Coronation I.** Mansfield Pt., 10–20 m, *MMM 3 & 17* (AAS, KRAM); Cape Hansen, 5–60 m, *RILS 3222B* (AAS, KRAM); Saunders Pt., *RILS 22* (AAS, KRAM). **Lynch I.** *Ca* 15 m, *DCL 944* (AAS, KRAM). **Matthews I.** Near The Divide, 100 m, *RILS 160* (AAS, KRAM); NE peninsula, 10 m, *MMM 91* (AAS, KRAM). **Coffer I.** Alt. 12–15 m, *AC & PK 9A* (AAS, KRAM). **Powell I.** Michelsen I., *ca* 15 m, *RILS 41A & 42* (AAS, KRAM); Falkland Harbour, *ca* 15 m, *RILS 50* (AAS, KRAM); John Peaks, 450 m, *RILS 271B* (AAS, KRAM). **Christoffersen I.** S end, 50 m, *KR 242A & 243A* (AAS, KRAM). **Laurie I.** Pt. Rae, 15–65 m, *MMM 109* (AAS, KRAM); Cape Dundas, 65 m, *MMM 176* (AAS, KRAM); Cabo Moneta, 150 m, *Bellisio 11764* (AAS, KRAM, NY); Cape Geddes, 15–60 m, *MMM 152* (AAS, KRAM). **Signy I.** Jebsen Pt., *ca* 25 m, *MWH 775A* (AAS, KRAM); Robin Peak, 190 m, *RILS 7010B* (AAS, KRAM); Factory Cove, *ca* 25 m, *RILS 686B* (AAS, KRAM) and 80 m, *RILS 1A* (AAS, KRAM); Berntsen Pt., *ca* 30 m, *BJT 348* (AAS, KRAM). **Moe I.** 10–200 m, *RILS 5260* (AAS, KRAM).

SOUTH SHETLAND IS. **Elephant I.** SW of Endurance Glacier, 244 m, *JSA 63 & 178* (AAS, KRAM); Walker Pt., 138 m, *JSA 202A* (AAS, KRAM). **O'Brien I.** North-east coast, 150–180 m, *JPB 48A* (AAS, KRAM). **Clarence I.** Southern end, 152 m, *JSA 1G* (AAS, KRAM). **King George I.** ***Fildes Peninsula***: Ardley I., 25 m, *RO 2490/80* (KRAM). ***Admiralty Bay***: Sphinx Hill, 70 m, *RO 4834/79* (KRAM); Vauréal Peak, 40 m, *RO 5260/79* (AAS, KRAM). **Robert I.** Edwards Pt., 50 m, *DCL 606* (AAS, KRAM). **Aitcho Is.** *RMS 69-924* (US) and 50–80 m, *RILS 11455* (AAS, KRAM). **Heywood I.** Penguin rookery, 10 m, *DCL 646* (AAS, KRAM). **Greenwich I.** Discovery Bay, Spark Pt., 20 m, *DCL 628* (AAS KRAM); Mt. Plymouth, *RMS 69-967 & 68-1802* (US). **Desolation I.** *DI 1487* (S). **Livingston I.** ***Byers Peninsula***: Barnard Pt., 25 m, *DCL 28* (AAS, KRAM); South Beaches, Vietor Rock, 10 m, *DCL 470* (AAS, KRAM); between Smellie Pt. and Devils Pt., 5 m, *KR 28B* (AAS, KRAM); Lair Pt., 5 m, *KR 59A* (AAS, KRAM).

WEST ANTARCTIC PENINSULA. **Trinity Peninsula.** Cape Roquemaurel, 10–50 m, *RILS 3926* (AAS, KRAM). **Davis Coast.** Astrolabe I., 10–30 m, *RILS 4054 & 4055* (AAS, KRAM); Tower I., *RILS 702* (AAS, KRAM); Cape Andreas, *CS 429 & 430* (PC, S – type of *A. depressinervis* fo. *robusta*). **Danco Coast.** Cierva Pt., *OB 21* (AAS, KRAM); Moss I., *CS 431* (PC, S – type of *A. depressinervis* fo. *robusta*); Spring Pt., *VK 912, 1212 & 1263* (KRAM); Cuverville I., *ER 239* (BR, FH, NY, PC, S – lectotype of *A. depressinervis*) and 10–20 m, *RILS 11943 & 11954B* (AAS, KRAM); Coughtrey Peninsula, *RMS 69-352 & 69-387* (US). **Palmer Archipelago.** ***Anvers I.***: Joubin Is., *JR & JMG 13, 14 & 15* (US); Cape

Monaco, *VK 796* (KRAM); Dream I., *VK 922, 931 & 1008* (KRAM); Litchfield I., *VK 1083* (KRAM) and 30 m, *RWMC 386* (AAS, KRAM); Cormorant I., 20 m, *AS 17* (AAS, KRAM) and *RILS 4182* (AAS, KRAM); Laggard I., *ca* 15 m, *REL 1301* (AAS, KRAM). **GRAHAM COAST.** Rasmussen I., *AS 77A* (AAS, KRAM) and *RILS 1922* (AAS, KRAM); Cape Tuxen, 15 m, *RILS 10667* (AAS, KRAM); Mt. Demaria, 150 m, *RWMC 682A* (AAS, KRAM); Cape Pérez, 10 m, *RWMC 634* (AAS, KRAM). **WILHELM ARCHIPELAGO.** ***Lemaire Channel***: entrance to the channel, *ER 477* (BR, PC – type of *A. depressinervis* var. *compacta*); Booth I., *Turquet 288b* (PC) and *Gain 130a, 151a, 152 & 264* (PC). ***Penola Strait***: Petermann I., *ca* 20 m, *RWMC 731* (AAS, KRAM, PC), 50 m, *AS 110* (AAS, KRAM), 55 m, *AS 121* (AAS, KRAM), 70 m, *AS 116* (AAS, KRAM) and 89 m, *AS 133* (AAS, KRAM). ***Argentine Is.***: Uruguay I., *RWMC 818A* (AAS, KRAM) and *RILS 4273* (AAS, KRAM); Corner Is., *JR & JMG 56 & 118* (US); Galindez I., *RWMC 485* (AAS, KRAM); The Barchans, 5 m, *RWMC 752* (AAS, KRAM). ***Collins Bay***: Berthelot Is., largest island, 100 m, *RWMC 623* (AAS, KRAM); Green I., 20–60 m, *RILS 3375* (AAS, KRAM); Darboux I., 10–20 m, *RILS 3366A* (AAS, KRAM). ***Grandidier Channel***: largest island off Takaki Pt., 10–35 m, *RILS 4347* (AAS, KRAM). **BISCOE IS.** Between Watkins I. and Belding I., *RILS 5320A* (AAS, KRAM). **LOUBET COAST.** ***Bennett Is.***: Liard I., Tutton Pt., 230 m, *JK 156A* (AAS, KRAM); Pfaff I., *JK 175* (AAS, KRAM); Megaw I., *JK 167* (AAS, KRAM, PC). ***Laubeuf Fjord***: Webb I., *JK 253* (AAS, KRAM); Piñero I., 10–25 m, *RILS 4830* (AAS, KRAM). **FALLIÈRES COAST.** ***Adelaide I.***: Rothera Pt., 30 m, 29 Dec 1984, *Birkenmajer s.n.* (KRAM) and *JF 41A* (AAS, KRAM). ***Léonie Is.***: Lagoon I., *VK 268* (KRAM); Anchorage I., 25–35 m, *RILS 9205* (AAS, KRAM); Léonie I., 5 m, *RILS 8932* (AAS, KRAM). ***Marguerite Bay***: Avian I., *ca* 15 m, *REL 937* (KRAM, LE) and *JR & JMG 131, 132 & 133* (US); Guébriant Is., *ca* 85 m, *BJT 446* (AAS, KRAM, LE); Faure Is., *MJC 72* (AAS, KRAM, PC).

EAST ANTARCTIC PENINSULA. **TRINITY PENINSULA.** Hope Bay, *CS 427* (BM, PC, S); Nobby Nunatak, 295 m, *CGB 4* (AAS, KRAM); Mt. Flora, 25–50 m, *RILS 3848* (AAS, KRAM). **JOINVILLE ISLAND GROUP.** ***Joinville I.***: Mt. Alexander, 250 m, *RILS 3673* (AAS, KRAM); Active Sound, 200 m, *RILS 3729A* (AAS, KRAM). ***Paulet I.***: *CS 428* (H, PC, S).

2. **Andreaea gainii** Cardot — Fig. 20

Andreaea gainii Cardot, Compt. Rend. Hebd. Séances Acad. Sci. 153: 602. 1911. ≡ *A. parallela* Müll.Hal. var. *gainii* (Cardot) Ochyra, Fragm. Florist. Geobot. 41: 479. 1996. — TYPE CITATION: [Antarctica] Terre de Graham: Cap Tuxen. [Lectotype (*selected here*): 'Herb. Mus. Paris. 209. Andreaea Gainii sp. nova. Terre du Graham: cap Tuxen, rochers, 2ème expéd. Charcot. Leg. Gain, 8/1 1909' – PC!; isotypes: BM!, NY!].

Plants medium-sized to large, stiff and brittle, with a rather coarse appearance, dark, reddish-brown to purplish-black, sometimes brownish- or olive-green above, not or slightly glistening, in small to large, loosely adhering cushions or short turfs. *Stems* up to 5 cm long, repeatedly branched, erect to ascending, in cross-section consisting of a 2–3-stratose cortex of thick-walled cells with small lumina, surrounding a relatively small medulla of large, thin-walled cells, central strand absent; *axillary hairs* with 1–2 small, quadrate, brown basal cells and elongate, club-shaped apical cell. *Leaves* crowded, often eroded below, stiff, erect and imbricate or sometimes loosely appressed, straight or with inflexed apices when dry, erect-spreading to widely spreading from appressed bases, often with apices becoming recurved when moist, panduriform to obovate-spathulate or oblong, with a strongly contracted sinus, broadest at or above middle, (0.6–)0.8–1.2(1.3) mm long, 0.2–0.6 mm wide, symmetric, moderately concave to weakly or strongly ventricose below, gradually or abruptly narrowed to a short or long, broad or narrow, triangular apex, 0.2–0.4 the length of leaves; *margins* plane, entire distally and at the extreme base, variably denticulate to crenate from above a short distance from the base to mid-leaf constriction due to prominently projecting cell ends, appearing bordered by several rows of oblique, thinner-walled cells, entire in the most proximal 4–16 cells; *costa* absent; *laminal cells* unistratose with occasional bistratose patches apically, short, rounded to elliptical, in ± radiating rows, about 5–7 μm wide, 6–15 μm long, thick-walled, mostly with large collenchymatous corner thickenings, smooth or weakly papillose dorsally in the upper half, ± abruptly passing into basal cells, rectangular to long-rectangular, 15–35 μm long, 6–10 μm wide, with longitudinal walls heavily incrassate and usually pitted or nodose in almost the entire width of the base, becoming similar to upper cells or oblate in a few marginal rows; *basal marginal cells* short-rectangular to quadrate or rhombic. *Cladautoicous*. *Perichaetial bracts* differentiated, convolute and sheathing the pseudopodium, much larger than the vegetative leaves, 2.0–2.5 mm long, obovate, acute, reaching the base of the capsule or nearly so. *Perigonia* terminal or lateral, bulbiform, comprising 4–10 elliptic antheridia on biseriate stalks. *Capsules* exserted 0.5–1.0 mm above the bracts, spindle-shaped, 0.5–1.3 mm long, dark brown, 4-valved. *Spores* frequently aborted, shrivelled and brown, 17–30(–50) μm in diameter, and turgid and green, 28–50(–60) μm in diameter.

Chromosome number: $n = 10$ – King George I. (Kuta *et al.* 1982).

Discussion. *Andreaea gainii* is a very distinctive species which can be distinguished readily from all other species in the Antarctic by its rather coarse appearance, usually dark, reddish-brown to purplish-black coloration and mostly pandurate leaves with dentate to crenate margins in the lower half, from the shoulders to a short distance from the leaf base consisting of 4–16 cells. The marginal dentation is variable and on some leaves it may be obscure, but otherwise distinct as blunt or wavy protrusions. The leaves are broadest at or above the middle and rapidly or gradually taper to triangular, broad and short or narrower and longer apices, rarely occupying more than 20 per cent of the length of the leaf. Greene (1968*b*) and Greene *et al.* (1970) recognised such phenotypes as separate varieties, respectively, var. *gainii* and var. *parallela* (Müll.Hal.) S.W.Greene, but the taxonomic value of these taxa is doubtful as there is often a range of shapes on the same plant.

The recognition of these varieties has resulted in some nomenclatural problems. *Andreaea gainii* var. *parallela* was invalidly published because no basionym was cited. Also, this name is illegitimate because the resulting infraspecific name has been transferred from an older one, *A. parallela* Müll.Hal., introduced by Müller (1883, 1889) for a new species described from Îles Kerguelen, to the more recently named *A. gainii*, which had been used for a new species described by Cardot (1911*a*,*b*, 1913) from the Antarctic. Ochyra (1996*b*) introduced nomenclatural changes in order to preserve the usage proposed by Greene (1968*b*) and Greene *et al.* (1970), but here *A. gainii* is considered as a species distinct from *A. parallela*. The latter is clearly conspecific with the bipolar *A. alpina* Hedw. and *A. gainii*, though closely related to the latter, differs from it in one fundamental character.

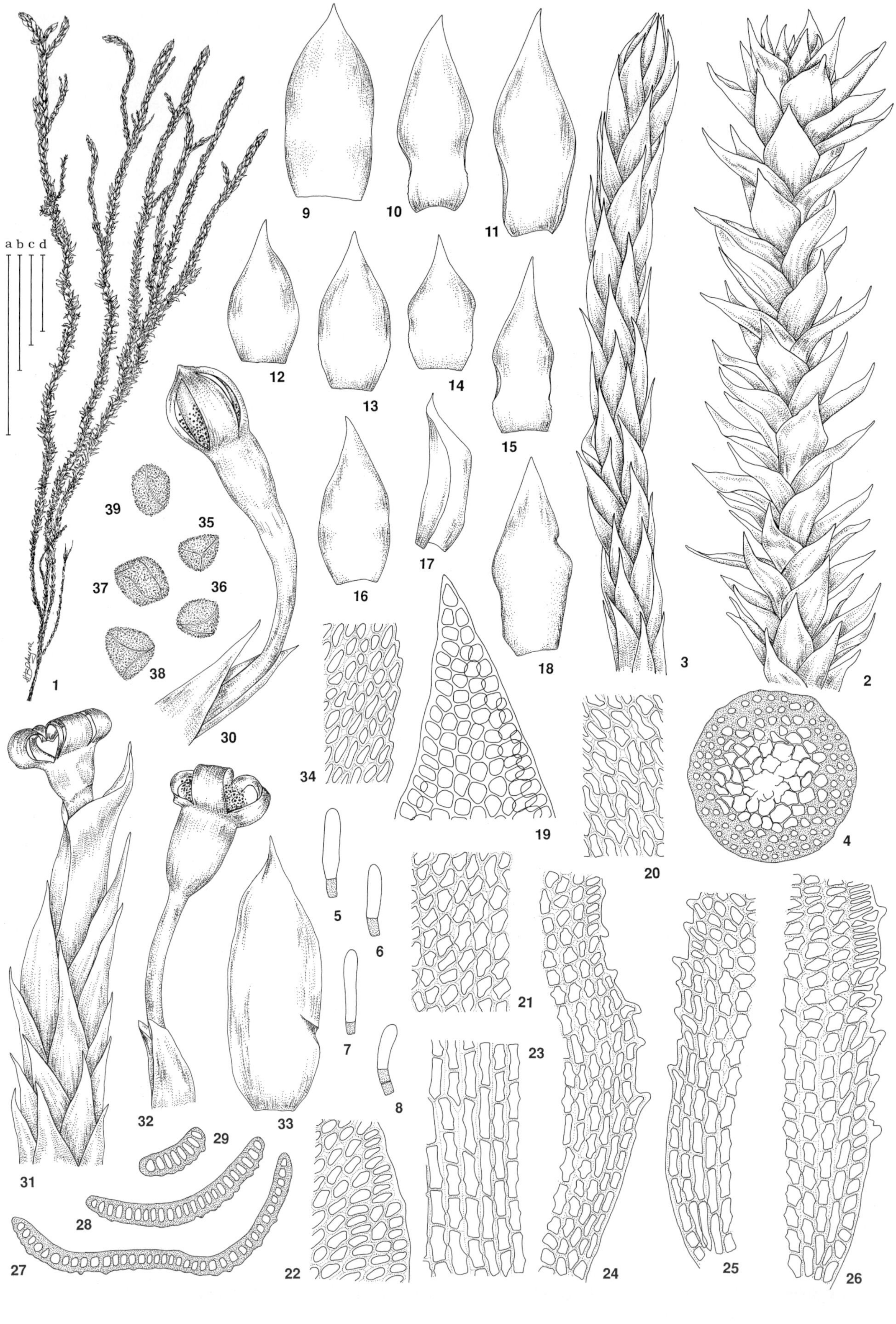
a b c d
1
2
3
4
5
6
7
8
9
10
11
12
13
14
15
16
17
18
19
20
21
22
23
24
25
26
27
28
29
30
31
32
33
34
35
36
37
38
39

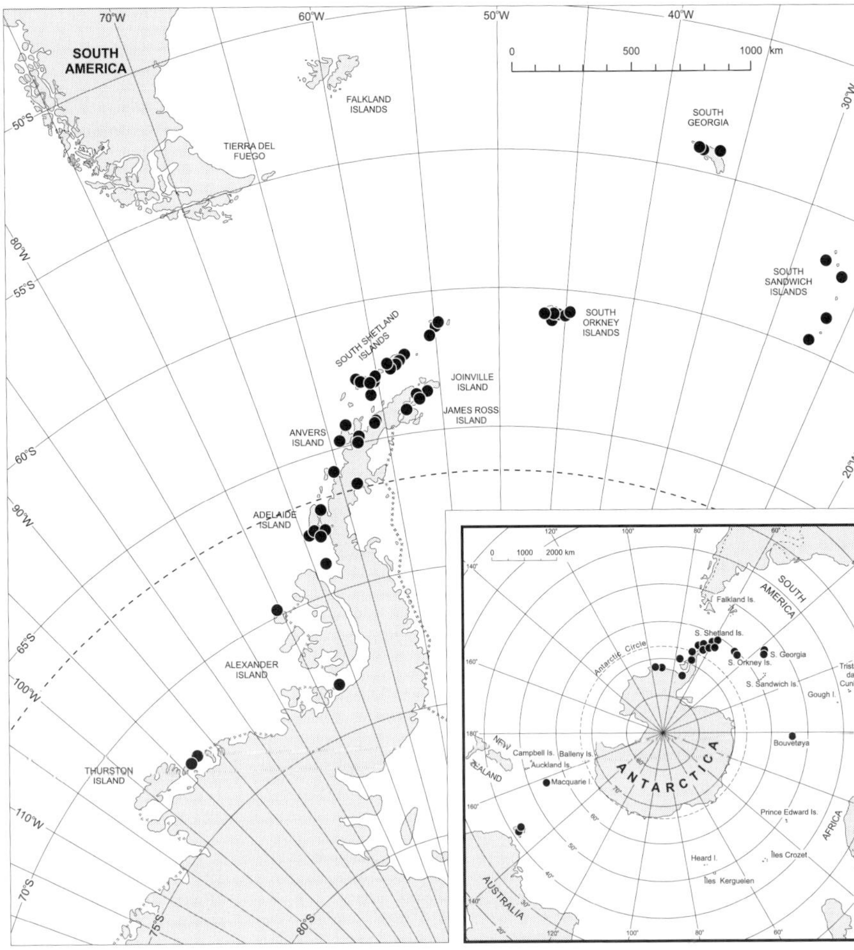

Figure 21. Distribution of *Andreaea gainii* in South Georgia, the South Sandwich Is. and Antarctic Peninsula region. Inset: global distribution; the unconfirmed literature records from southern South America and Bolivia are not indicated.

In *A. alpina* the leaf margins are denticulate or crenate to the extreme base, whereas in *A. gainii* the dentation starts at some distance from the base, usually between 4–16 cells up.

In some Antarctic specimens, even from the type locality at Cape Tuxen on the Graham Coast (e.g. *RWMC 845*, AAS, KRAM, LE, PC, S) some leaves are somewhat sinuate near the leaf base and B. M. Murray annotated the specimen in S as possibly belonging within *A. alpina* but still requiring further study. Such plants are interpreted here as *A. gainii* and, so far, typical material of *A. alpina* has not been discovered in the Antarctic.

When describing *Andreaea gainii* Cardot (1911*a*,*b*, 1913) compared it to *A. verruculosa* Cardot from Tierra del Fuego from which it differed in its coarser stature and the leaves being twice as long and broad, and less acuminate. *Andreaea verruculosa* is conspecific with *A. regularis* (Greene 1968*b*) and has nothing in common with *A. gainii*. Both species belong within the type subgenus and section of *Andreaea* characterised by the ecostate leaves, but *A. gainii* represents a group of dentate-margined species, whereas *A. regularis* clearly belongs within an entire-margined group of species. *Andreaea gainii* is very closely related to *A. pseudomutabilis*, a species described from western Patagonia (Dusén 1903*a*), and to the bipolar *A. alpina*. These species constitute a critical complex which needs a careful global assessment. Until such study is completed *A. gainii* is best treated as a distinct species.

To the group of dentate-margined species also belong two closely related species, *Andreaea marginata* Hook.f. & Wilson from southern South America (Greene 1986) and the pan-south-temperate *A. acuminata* Mitt. (Vitt 1980). These species share with *A. gainii* crenate, but not denticulate, leaf margins in the basal part but differ in having slenderly long-acuminate leaves, and elongate and strongly thick-walled basal marginal cells.

Reproduction in Antarctica. Gametangia and sporophytes common throughout its range.

Habitat. As with other species of the genus, *Andreaea gainii* is a pioneer of exposed or sheltered, dry and wet rock faces, boulders, cliffs, rock ledges, rocky knolls, scree and overhanging rocks. Also, it grows on stony or gravelly ground, on pockets of mineral soil between rocks and boulders, in rock crevices, sometimes on volcanic scoria, on humus over rock surfaces, and on rocks and gravel moistened by melt water. On volcanic islands of the South Sandwich Is. it occurs on rocks near fumaroles moistened by steam but not warmed. The species is one of the commonest components of the moss cushion and fruticose lichen

←

Figure 20. *Andreaea gainii*. — **1**. Habit. **2**. Portion of branch, wet. **3**. Portion of branch, dry. **4**. Cross-section of stem. **5–8**. Axillary hairs. **9–18**. Leaves. **19**. Leaf apex. **20**. Upper cells of perichaetial leaf. **21**. Mid-leaf cells. **22**. Marginal cells at the widest part of leaf. **23**. Cells in the middle of leaf base. **24–26**. Basal cells at margin; **27–29**. Cross-sections of leaves. **30**. Young sporophyte. **31–32**. Mature sporophytes. **33**. Perichaetial leaf. **34**. Upper marginal cells of perichaetial leaf. **35–39**. Spores. — [1–7, 10–11, 18–19, 21–23, 26–29 from *RO 30/80*; 8–9, 13, 17 from *Gain 209* (type), PC; 12, 14, 25, 31 from *RO 1964/80*; 15 from *MWH 437*; 16, 24 from *Schoeman 40*; 20, 30, 32–39 from *RO 596/80*]. — Scale bars: a – 1 mm (9–18, 30–32); b – 1 mm (2–3, 33); c – 100 µm (4, 27–29); d – 0.5 cm (1) and 50 µm (5–8, 19–26, 34–39).

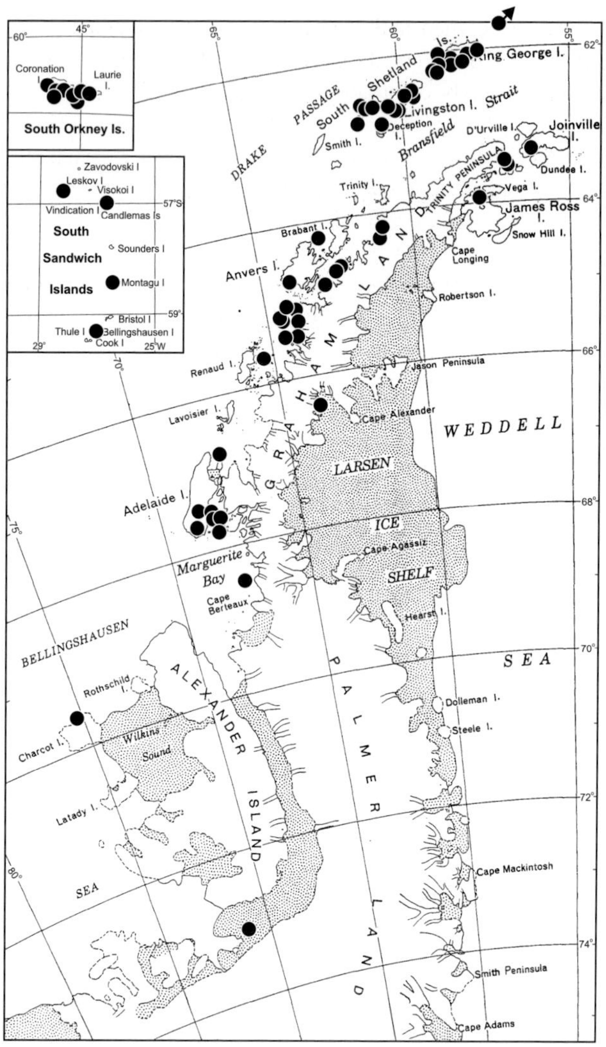

Figure 22. Distribution of *Andreaea gainii* in the Antarctic Peninsula area. The localities in the northern group of the South Shetland Is. (Elephant I. and Gibbs I.) are indicated by an arrow.

subformation (Lewis Smith 1972) and in suitable habitats may be dominant over large areas, for example on Andreaea Plateau on Signy I., South Orkney Is., where it forms, with *A. regularis*, an almost close stand of more than one hectare.

World range. *Andreaea gainii* was long considered to be an Antarctic endemic species (Ochyra 1998*a*), but it has recently been confirmed from South Georgia (Ochyra *et al.* 2002), Macquarie I. (Seppelt 2004) and Tasmania and southern South America (Murray 2006) (Fig. 21, inset), although no specimens have been seen from the two latter regions. The species was reported from subantarctic Prince Edward Is. (Zanten 1971) but the voucher collections are *A. alpina*. Also, it has been reported from Bolivia (Churchill *et al.* 2000) but the relevant voucher collections have not been available for examination. Until the *A. alpina* complex is taxonomically assessed in the Southern Hemisphere, it is best to consider *A. gainii* as an amphipacific Subantarctic–Antarctic species attaining maximum abundance in the maritime Antarctic.

Distribution in Antarctica. A widespread and locally common and abundant species, known from Bouvetøya, the South Sandwich, South Orkney and South Shetland Is. and occurring along the west coast of the Antarctic Peninsula from Joinville I. off Trinity Peninsula to Mushroom I. at lat. 68°54′ S in southern Marguerite Bay, and scattered on the east coast on James Ross I. and extending to lat. 66°22′ S on the Foyn Coast. The southernmost records are from very isolated sites, namely Charcot I., off northwest Alexander I. (lat. 69°45′ S), Eklund I., English Coast (lat. 73°14′ S) and Dustin I. and McNamara I., Eights Coast (72°34′ S and 72°36′ S, respectively, the latter having abundant sporophytes) (Figs 21–22). Typically, *Andreaea gainii* occurs from sea level to about 300 m and occasionally on higher summits, for instance at 550 m on a ridge on Cape Casey, Foyn Coast, east coast of the Antarctic Peninsula (*RT 5*, AAS, KRAM).

Selected specimens examined. WEST ANTARCTICA. BOUVETØYA. Top of Rustadkollen, *Schoeman 40* (KRAM, LE, PC) and 300–450 m, *TE 222* (BG); Moseryggen, 260–287 m, *TE 268* (BG).

SOUTH SANDWICH IS. **Leskov I.** Below the summit ridge, 70 m, *PC 255C* (AAS, KRAM). **Vindication I.** NE corner, *ca* 180 m, *MWH 437* (AAS, KRAM, LE); E side, *ca* 135 m, *REL 745 & 747* (AAS, KRAM). **Montagu I.** Allen Pt., 70 m, *PC 93 & 100C* (AAS, KRAM) and 150 m, *MWH 804A* (AAS, KRAM). **Bellingshausen I.** Eastern crater, 150 m, *PC 45A* (AAS, KRAM) and *ca* 200 m, *MWH 825A* (AAS, KRAM).

SOUTH ORKNEY IS. **Monroe I.** W coast, 35–100 m, *MMM 78* (AAS, KRAM). **Coronation I.** Meier Pt., *ca* 35 m, *RILS 578* (AAS, KRAM, PC); Wave Peak, *ca* 135 m, *FIDS H 609/1* (AAS, KRAM); Shingle Cove, *ca* 135 m, *RILS 445* (AAS, KRAM, PC). **Matthews I.** Near The Divide, *ca* 35 m, *RILS 159* (AAS, KRAM, LE). **Powell I.** John Peaks, 455 m, *RILS 271A* (AAS, KRAM); W coast, *RILS 239a* (AAS, KRAM). **Laurie I.** Cape Geddes, *ca* 330 m, *MMM 201* (AAS, KRAM); Cape Dundas, *ca* 65 m, *MMM 192* (KRAM). **Signy I.** Foca Pt., 30 m, *RW 45* (AAS, KRAM); Jane Peak, *ca* 170 m, *MWH 764a* (AAS, KRAM, PC); Factory Cove, 50 m, *REL 850* (AAS, KRAM, PC); Gneiss Hills, 225 m, *RILS 8106* (AAS, KRAM).

SOUTH SHETLAND IS. **Elephant I.** SW of Endurance Glacier, 167 m, *JSA 21B* (AAS, KRAM); Emma Cove, 138 m, *JSA 126* (AAS, KRAM). **Gibbs I.** The Spit, 30–120 m, *JPB 217B* (AAS, KRAM). **King George I.** ***Fildes Peninsula***: Suffield Pt., 30 m, *RO 2448/80* (KRAM). ***Admiralty Bay***: Agat Pt., 25 m, *RO 352/80* (KRAM); ***Bransfield Strait***: Red Hill, 100 m, *RO 1107/80* (KRAM). **Nelson I.** Rip Pt., 50 m, *DCL 793A* (AAS, KRAM); Strachan Hill, 50 m, *DCL 853* (AAS, KRAM). **Greenwich I.** ***Discovery Bay***: Spark Pt., 20 m, *DCL 633* (AAS, KRAM); Mt. Plymouth, *RMS 69-987* (US). **Half Moon I.** S of huts, 50 m, *DCL 579A* (AAS, KRAM). **Meade Is.** Cave I., 6 m, *DCL 672* (AAS, KRAM). **Livingston I.** ***Byers Peninsula***: Clark Nunatak, 180 m, *DCL 209* (AAS, KRAM, LE, PC); Charity Glacier, 70 m, *DCL 60* (AAS, KRAM, PC). ***Hurd Peninsula***: Punta Hesperides, *Pandurski 92* (KRAM); Miers Bluff, *RMS 69-814* (US); False Bay, *RMS 69-814* (US); Johnsons Dock, 20 m, *Redon 110B* (AAS, KRAM). **Deception I.** NW of Kroner Lake, *ca* 10 m, *RILS 61* (AAS, KRAM). **Snow I.** Plateau President Head, *KR 2C & 3B* (AAS, KRAM).

WEST ANTARCTIC PENINSULA. **Danco Coast.** Cierva Pt., *VK 520* (KRAM); Spring Pt., *VK 1201* (KRAM); Cuverville I., 20 m, *de Leeuw 28A* (AAS, KRAM); Birdsend Bluff, 5–35 m, *RILS 4235* (AAS, KRAM). **Palmer Archipelago.** ***Melchior Is.***: Lystad I., *PAS 376.3* (FH). ***Anvers I.***: Norsel Pt., *RMS 69-846, 69-855 & 69-858* (US). **Graham Coast.** Edge Hill, 30 m, *RWMC 662A* (AAS, KRAM); Cape Tuxen, *Gain 209* (BM, FH, NY, PC – type of *A. gainii*) and 370 m, *RWMC 845* (AAS, KRAM, LE, PC, S); Mt. Demaria, *ca* 160 m, *RWMC 682B* (AAS, KRAM); Cape Pérez, *ca* 12 m, *RWMC 641B* (AAS, KRAM). **Wilhelm Archipelago.** ***Penola Strait***: Petermann I., 50 m, *AS 129* (AAS, KRAM). ***Argentine Is.***: Irizar I., 8 m, *RWMC 606* (AAS, KRAM); Uruguay I., 8 m, *RWMC 597B* (AAS, KRAM); Skua I., *RWMC 505A* (AAS, KRAM). ***Collins Bay***: Berthelot Is., *ca* 65 m, *RWMC 622A* (AAS, KRAM). ***Grandidier Channel***: Hook I., *RMS 69-513, 69-710, 69-718, 69-884, 69-895 & 69-902* (US). **Loubet Coast.** ***Laubeuf Fjord***: Piñero I., 10–25 m, *RILS 4832* (AAS, KRAM). ***Bigourdan Fjord***: Blaiklock I., *ca* 15 m, *RILS 836* (AAS, KRAM). **Fallières Coast.** ***Adelaide I.***: Stork Ridge, *ca* 420 m, *PC 264, 265C & 266C* (AAS, KRAM); Reptile Ridge, 300 m, *RILS 3421B* (AAS, KRAM). ***Léonie Is.***: Léonie I., *ca* 135 m, *JK 212B* (AAS, KRAM). ***Marguerite Bay***: unnamed island near Pourquoi Pas I., *DDWW 166A* (AAS, KRAM); Mushroom I., *Poncet 1A* (KRAM). **George VI Sound.** ***Charcot I.***: Rils Nunatak, 125 m, *RILS 10537, 10472 & 10710* (AAS, KRAM). **English Coast.** Eklund Is., 240 m, *RILS 5531A* (AAS, KRAM).

EAST ANTARCTIC PENINSULA. **Trinity Peninsula.** Hope Bay, Scar Hills, *ca* 110 m, *CGB 19* (AAS, KRAM); Nobby Nunatak, *ca* 295 m, *CGB 31* (AAS, KRAM, PC); Mt. Flora. 25–50 m, *RILS 3846* (AAS, KRAM) and 115 m, *REL 37A* (AAS, KRAM). **Joinville Island group.** ***Joinville I.***: Active Sound, 200 m, *RILS 3729B* (AAS, KRAM). **James Ross Island group.** ***James Ross I.***: Martha Cove, *ca* 150 m, *RILS 7426, 7533 & 7547* (AAS, KRAM). **Foyn Coast.** Cabinet Inlet, Cape Casey, 550 m, *RT 5* (AAS, KRAM).

ELLSWORTH LAND. **Eights Coast.** Dustin I., 100 m, 16 Feb 1960, *Leech s.n.* (AAS, KRAM); McNamara I., Peeler Bluff, *ca* 265 m, *RILS 5422* (AAS, KRAM).

3. **Andreaea regularis** Müll.Hal. Fig. 23

Andreaea regularis Müll.Hal. *in* Neumayer, Int. Polarforsch. Deutsch. Exped. 2: 286. 1890. — TYPE CITATION: Austro-Georgia, Ostseite des Vexirberges, 17. Februar 1883. [Lectotype (*selected here*): 'Andreaea regularis C. Muell. nsp. Austro-Georgia, Ostseite des Vexirberges, 17.II.83. Südpolarexpedition' – HBG!; isotypes: M! (2 specimens), PC!].

Andreaea pycnotyla Cardot, Rev. Bryol. 27: 42. 1900. ≡ *A. regularis* var. *pycnotyla* (Cardot) Cardot, Wiss. Erg. Schwed. Südpolar-Exp. 4(8): 262. 1908. — TYPE CITATION: [...] détroit de Gerlache, XXème débarquement; later Cardot (1901: p. 22) wrote: Détroit de Gerlache: XXème débarquement, Terre de Danco, à l'entrée du chenal de Lemaire, sur des roches isolées au milieu d'un glacier, à 50 métres au-dessus du niveau de la mer. (Nos. 267a et 270; 12 février 1898). [Lectotype (*selected here*): 'Herb. J. Cardot. Mousses du voyage de la Belgica. n° 270. Andreaea pycnotyla Card. sp. nova. Canal de la Belgica: XXème débarquement, sur des roches isolées au milieu d'un glacier, à 50 m d'alt. Leg. Racovitza, 12 févr. 1898' – PC!; isolectotype: BR!; syntype: same label data only the collecting number 'Racovitza 267^a^' – PC!; isosyntype: BR!]. First synonymised by Greene (1968*b*: p. 144).

Andreaea pygmaea Cardot, Rev. Bryol. 27: 43. 1900. — TYPE CITATION: [...] détroit de Gerlache, Xème débarquement; later Cardot (1901: p. 22) wrote: Détroit de Gerlache: Xème débarquement, île Brabant, sur des roches complétement entourées de glaces, à 350 métres au-dessus du niveau de la mer. (N° 252d; 30 janvier 1898). [Lectotype (*selected here*): 'Herb. J. Cardot. Mousses du voyage de la Belgica. n° 252^d^. Andreaea pygmaea Card. sp. nova. Canal de la Belgica: Xème débarquement, sur des roches completément entourées de glaces, alt. 350 m. Leg. Racovitza, 30 janv. 1898' – PC!; isotype: BR!]. First synonymised by Greene (1968*b*: p. 143).

Plants small to medium-sized, slender, stiff and brittle, dull or slightly lustrous, brown, reddish- to blackish-brown or ferruginous, in low, small to large, compact cushions or short turfs. *Stems* 0.5–2.5, occasionally up to 5.0 cm tall, erect to ascending, fastigiately to repeatedly branched, with scattered rhizoids at base, in cross-section round, consisting of 2–3 layers of lacunose cortical cells with small lumina and brown and very incrassate walls surrounding a medulla of larger cells with strongly thickened walls; *axillary hairs* relatively short, composed of 1–4 brown, short, basal cells and long, claviform apical cell. *Leaves* crowded, tightly or sometimes loosely imbricate, erect, straight or with apices somewhat spreading when dry, erect-spreading to spreading from appressed bases, sometimes with recurved apices when wet, symmetric, ovate to ovate-lanceolate, gradually or sometimes abruptly tapering to a long or shorter, narrow or broader acumen, rounded at base, non-decurrent, 0.5–1.1 mm long, 0.2–0.5 mm wide, concave, often distinctly channelled in acumen and ventricose at base, *margins* plane throughout or inflexed above, entire throughout or occasionally weakly crenulate or sinuolate at shoulders; *costa* lacking; *laminal cells* unistratose throughout, usually strongly papillose or mammillose dorsally in the upper half; *upper cells* rounded-quadrate, oval, short-rectangular to hexagonal above, 5–14 μm wide, 6–25 μm long, with moderately to strongly thickened and porose walls, often with large corner thickenings; *lower cells* short- to long-rectangular towards the base, 20–40 μm long, 5–10 μm wide, with strongly incrassate and mostly porose longitudinal walls, sometimes becoming similar to upper cells or transversely rectangular towards margins; *basal marginal cells* generally rectangular and similar to mid-base cells, sometimes mixed with a few rounded-quadrate to rhombic cells. *Cladautoecious*. *Perigonia* small, with few paraphyses. *Perichaetial leaves* differentiated, sheathing and convolute, 2–4 times as long as the vegetative leaves, obovate or oblong-obovate, abruptly shortly and broadly acuminate, nearly reaching the base of the capsule. *Capsules* medium-sized, about 1 mm long, dehisced by 4 or rarely 6 valves, elliptic and with closed valves when wet, contracted in length and with bulging valves when dry. *Spores* frequently aborted, shrivelled and brown, 15–22 μm in diameter, and turgid and green, 26–34 μm in diameter.

Discussion. *Andreaea regularis* is readily identified. The ecostate leaves distinguish it from *A. depressinervis*, and the ovate to ovate-lanceolate leaves which are broadest near the base, separate it from the pandurate leaves of *A. gainii* which are broadest at or above the middle. Additionally, the leaf margins in the latter are distinctly dentate and/or crenate in the lower half to near the base, whereas in *A. regularis* they are entire throughout or only occasionally wavy or indistinctly crenulate near the insertion. *Andreaea regularis* is the smallest and most slender of all species of the genus in Antarctica but the stature of the plants and leaf size are variable. The leaf shape also varies from ovate to ovate-lanceolate and in the degree of the acumination but, in general, the leaves are short acuminate. This character distinguishes it from the closely related *A. acutifolia* Hook.f. & Wilson and *A. flabellata* Müll.Hal., pan-south-temperate species which, however, have longly and narrowly acuminate leaves.

Andreaea regularis was originally described from South Georgia (Müller 1890*a*). It was reported for the first time from the Antarctic by Cardot (1906*a*,*b*, 1907*a*, 1908*a*) from a few localities in the South Shetland Is. and on the west coast of the Antarctic Peninsula. However, this species was collected several years earlier in this region by É. Racovitza during the Belgian 1897–1899 expedition and described as *A. pygmaea* and *A. pycnotyla* by Cardot (1900, 1901). The latter species was subsequently reduced to a variety of *A. regularis* (Cardot 1908*a*), but neither taxon merits taxonomic recognition, as discussed by Greene (1968*b*), who considered both species to be conspecific with *A. regularis*. We confirm this, following examination of the relevant type collection.

Reproduction in Antarctica. Gametangia and sporophytes frequent throughout its range.

Habitat. In the Antarctic, *Andreaea regularis* is ecologically very similar to other species of this genus, notably *A. gainii*, with which it usually grows as a component of communities of the fruticose lichen and moss cushion subformation which most

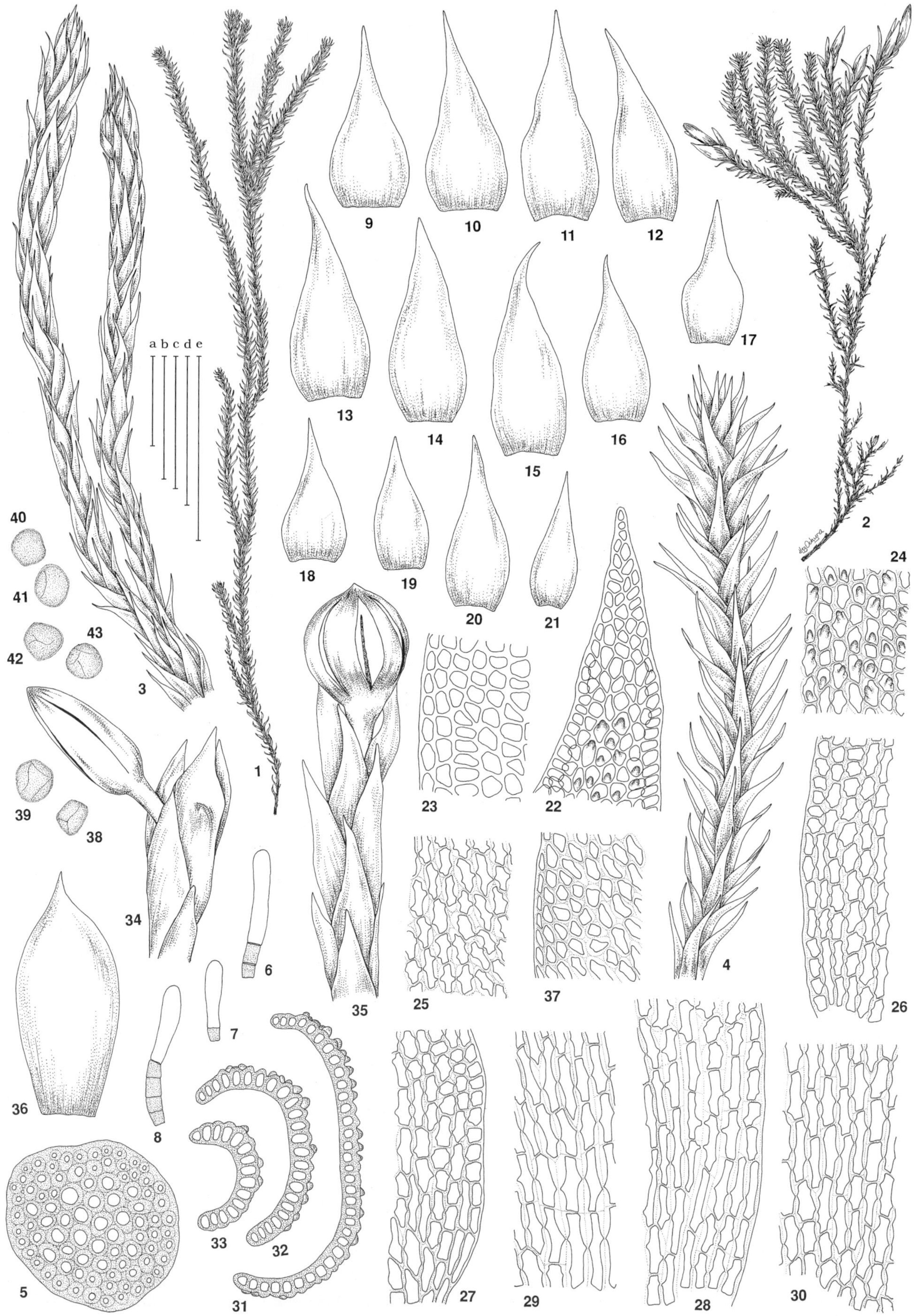
a b c d e
1
2
3
4
5
6
7
8
9
10
11
12
13
14
15
16
17
18
19
20
21
22
23
24
25
26
27
28
29
30
31
32
33
34
35
36
37
38
39
40
41
42
43

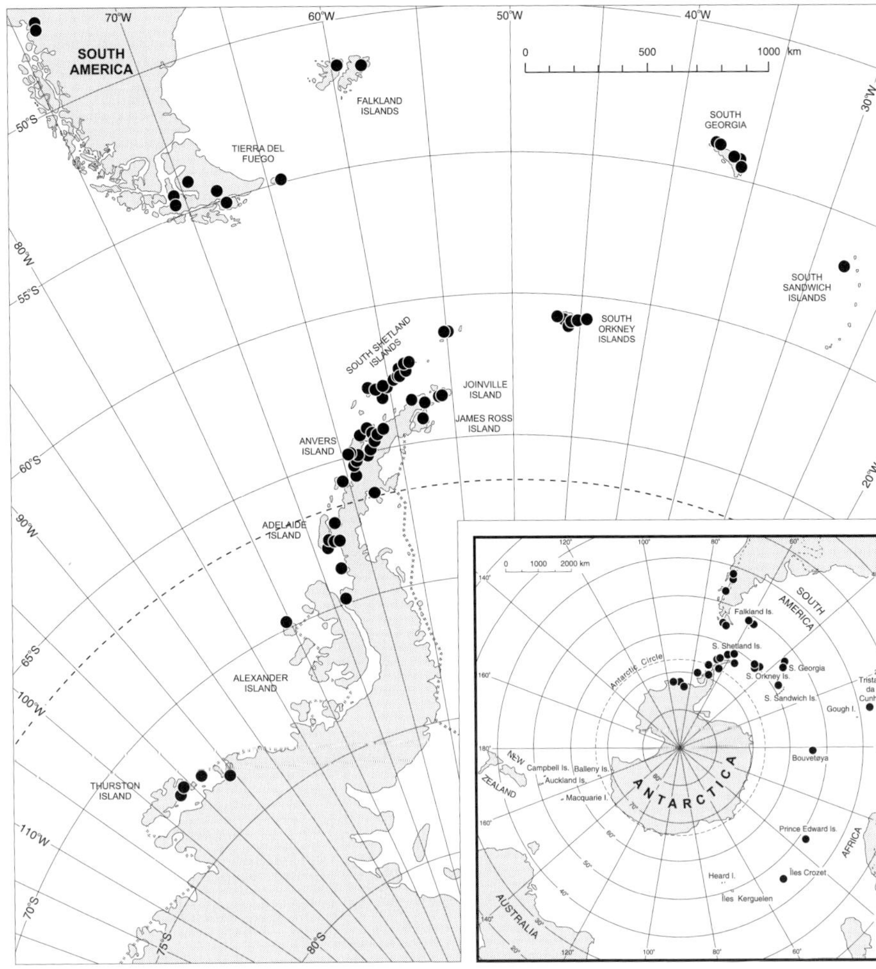

Figure 24. Distribution of *Andreaea regularis* in the Fuegian region and Antarctica. Inset: global distribution; the literature record from Bolivia is not indicated.

commonly occupies fellfields. It grows as cushions or continuous turves on dry or seasonally wet and exposed or sheltered, acidic boulders, stones, cliffs, scree, rock ledges of basalt, mica-schist, granite, etc. On volcanic islands it also occurs on coarse ash and scoria. Additionally, it grows on pockets of mineral soil between boulders and pebbles, on temporarily wet glacial soil and on dry sandy glacial debris, and occasionally also on eroding moss peat. Like other species of the genus it has a broad tolerance of various ecological conditions including aspect, elevation, substrate and moisture, and it occasionally occupies intermittently wet seepage areas.

World range. *Andreaea regularis* is here considered to be an amphiatlantic subantarctic species, rather than an amphiatlantic south-temperate one as suggested by Ochyra (1998*a*), because its northern stations in the cool temperate zone are few and restricted to high elevations (Fig. 24, inset). The species is most common and abundant in the maritime Antarctic and on South Georgia, and it is less frequent on other subantarctic islands including Prince Edward Is. (Zanten 1971; Ochyra, personal observations) and Îles Crozet (*Bell 1534*, AAS, KRAM). Additionally, it was recorded at 1900 m from Tristan da Cunha (Dixon 1960), at 1270 m from Tierra del Fuego (as *A. verruculosa* Cardot – Cardot 1905*a*, 1908*a*; Greene 1968*b*), as well as from the Falkland Is. and Isla de los Estados (Greene 1968*b*), the Fuegian region (Bartram 1952*a*), western Patagonia as far north as lat. *ca* 40°S in the Valdivian Province (Seki 1974; Greene 1986; Deguchi 1991*a*; He 1998), and from an isolated station in Bolivia (Churchill *et al.* 2000).

Distribution in Antarctica. A widespread and locally common and abundant species, especially in the South Orkney and South Shetland Is. as well as in the central part of the west coast of the Antarctic Peninsula (Danco Coast, Anvers I., northern part of the Graham Coast), declining in abundance southwards and extending to the entrance of George VI Sound at 69°31′S and Charcot I. at 69°45′S. It is rare and localised elsewhere including Bouvetøya, South Sandwich Is., northern and eastern Antarctic Peninsula, and is also known from several stations in continental Antarctica on the Eights Coast where it extends to

◄——

Figure 23. *Andreaea regularis*. — **1–2**. Habit. **3**. Portion of branch, dry. **4**. Portion of branch, wet. **5**. Cross-section of stem. **6–8**. Axillary hairs. **9–21**. Leaves. **22**. Leaf apex. **23**. Mid-leaf cells at margin. **24–25**. Mid-leaf cells. **26–28**. Basal cells at margin. **29–30**. Leaf cells in the middle of base. **31–33**. Cross-sections of leaves. **34**. Young sporophyte. **35**. Mature sporophyte. **36**. Perichaetial leaf. **37**. Upper cells of perichaetial leaf at margin. **38–43**. Spores. — [1 from *JS 219*; 2, 11–12, 24, 26, 30 from *Will s.n.*, 17.ii.1883 (lectotype of *A. regularis*), HBG; 3–4 from *RO 537/80*; 5, 36–43 from *RO 418/80*; 13, 29 from *RO 1266/80*; 18 from *RO 1926/80*; 19 from *RO 464/80*; 21 from *RO 1386/80*; 34–35 from *622/80*; 6–8, 14–15, 22–23, 31–33 from *ER 267A* (syntype of *A. pycnotyla*), BR; 9–10, 27 from *ER 270A* (lectotype of *A. pycnotyla*), BR; 16–17, 25, 28 from *ER 252A* (lectotype of *A. pygmaea*), BR; 20 from *MMM 199*]. — Scale bars: a – 50 µm (6–8, 22–30, 37–43); b – 0.5 cm (1–4); c – 100 µm (5, 31–33); d – 1 mm (36); e – 1 mm (9–21, 34–35).

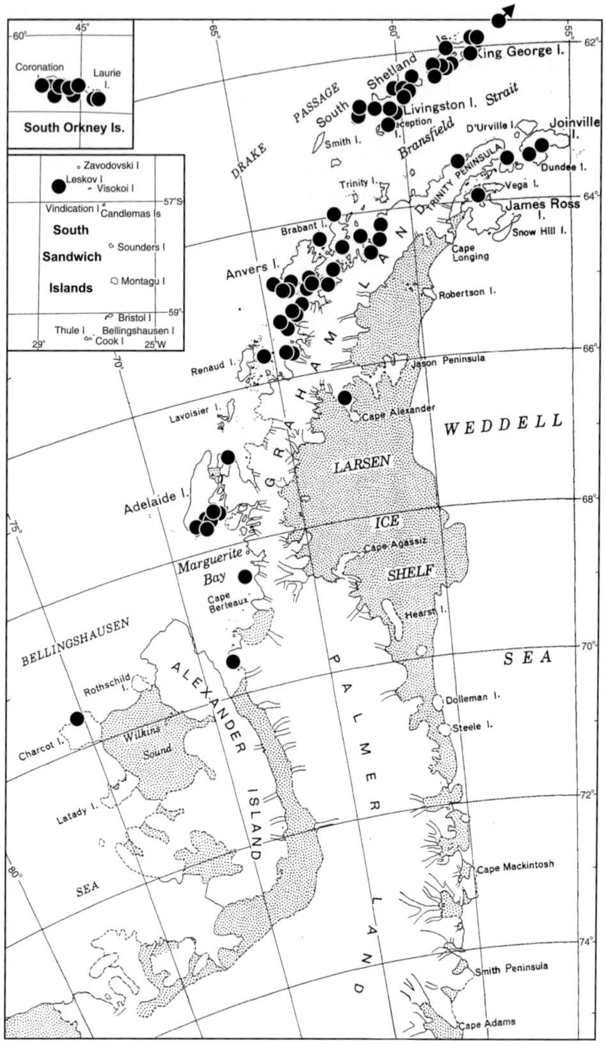

Figure 25. Distribution of *Andreaea regularis* in the Antarctic Peninsula area. The localities in the northern group of the South Shetland Is. (Elephant I.) are indicated by an arrow.

73°37′S, this being its southernmost known occurrence (Figs 24–25). The species occurs from sea level to 300 m, occasionally ascending to 500 m on Trinity Peninsula, Adelaide I. and on the Oscar II Coast.

Selected specimens examined. WEST ANTARCTICA. BOUVETØYA. Moserygen, 270–287 m, *TE 278, 279 & 316* (BG).

SOUTH SANDWICH IS. **Leskov I.** Summit ridge, 100 m, *PC 254D* (AAS, KRAM) and 200 m, *REL 530* (AAS, KRAM, PC).

SOUTH ORKNEY IS. **Monroe I.** Plateau, *RILS 8006* (AAS, KRAM). **Coronation I.** Mansfield Pt., 15–35 m, *RILS 297* (AAS, KRAM, LE); Laws Glacier, *ca* 15 m, *RW 134B* (AAS, KRAM); Olivine Pt., 230 m, *RILS 118* (AAS, KRAM, PC). **Powell I.** Eastern coast, *ca* 135 m, *RILS 219* (BM, KRAM, PC) *& 197C* (AAS, KRAM); western side, *ca* 300 m, *RILS 230* (AAS, KRAM). **Laurie I.** Scotia Bay, Pt. Davis, 7–35 m, *MMM 122* (KRAM); Cape Dundas, *ca* 335 m, *MMM 199* (KRAM). **Signy I.** Andreaea Plateau, 185 m, *RILS 8029* (AAS, KRAM); Moraine Valley, 35 m, *RILS 369* (AAS, KRAM); Factory Cove, 10 m, *RW 1B* (AAS, KRAM); Observation Bluff, ca 85 m, *MWH 136A* (AAS, KRAM); Paal Harbour, *ca* 35 m, *MWH 752B* (AAS, KRAM); Rusty Bluff, 220 m, *MWH 738A & 741A* (AAS, KRAM, PC); between Rethval Pt. and Lenton Pt., *ca* 50 m, *REL 1158* (AAS, KRAM); Berntsen Pt., *BJT 394* (AAS, KRAM); between Rethval Pt., and Lenton Pt., 50 m, *REL 1158* (AAS, KRAM). **Moe I.** 5–210 m, *RILS 5265* (AAS, KRAM).

SOUTH SHETLAND IS. **Elephant I.** Endurance Glacier, 259 m, *JSA 59A* (AAS, KRAM); Walker Pt., 158 m, *JSA 203* (AAS, KRAM); Stinker Pt., 107 m, *JSA 84* (AAS, KRAM); Muckle Bluff, 122 m, *JSA 197A* (AAS, KRAM). **King George I.** ***Fildes Peninsula***: Ardley I., 5 m, *RO 2481/80* (AAS, KRAM); Bellinghausen Station, 10 m, *RO 2426/80* (KRAM); Lake Kitezh, 25 m, *RO 2467/80 & 2468/80* (KRAM). ***Drake Passage***: Davey Pt., *BJ 45* (KRAM). ***Destruction Bay***: Faraway Nunataks, *BJ 172* (KRAM); North Foreland, *BJ 102* (KRAM); Pyrites I., *BJ 36* (KRAM). ***King George Bay***: Lions Rump, *BJ 95* (KRAM); Turret Pt., *BJ 139* (KRAM). ***Bransfield Strait***. Red Hill, 100 m, *RO 477/80 & 1094/80* (KRAM); The Tower, 340 m, *RO 811/98* (KRAM); Demay Pt., 55 m, *RO 1386/80* (AAS, KRAM); Creeping Slopes, 170 m, *RO 1296/80* (KRAM); Zamek, 335 m, *RO 21/80* (KRAM); Penguin Ridge, 85 m, *RO 4963/79* (KRAM); Hala, 30 m, *RO 5208/79* (KRAM); Skua Cliff, 45 m, *RO 622/80* (KRAM); Ubocz, 110 m, *RO 4941/79 & 4961/79* (KRAM); Jersak Hills, 200 m, *RO 5093/79* (AAS, KRAM). ***Ezcurra Inlet***: Jardine Peak, 275 m, *RO 5168/79* (KRAM); Italia Valley, 70 m, *RO 42/80* (KRAM); Urbanek Crag, 130 m, *RO 2336/80* (KRAM). ***MacKellar Inlet***: Klekowski Crag, 230 m, *RO 2264/80* (KRAM); Misty Nunatak, 280 m, *RO 2032/80* (KRAM); Admiralen Peak, 305 m, *RO 2023/80* (KRAM); Komandor Peak, 270 m, *RO 929/80* (KRAM); Kapitan Peak, 200 m, *RO 1926/80* (KRAM); Wegger Peak, 320 m, *RO 1985/80* (KRAM); Garnuszewski Peak, 300 m, *RO 2018/80* (KRAM). ***Keller Peninsula***: Ore Pt., 4 m, *RO 503/80 & 505/80* (KRAM); Speil Pt., 15 m, *RO 537/80* (AAS, KRAM); Round Hill, 50 m, *RO 546/80* (KRAM); British Pt., 70 m, *558/80* (AAS, KRAM); Moraine Pt., 60 m, *RO 475/80* (KRAM); Mt. Flagstaff, 230 m, *RO 418/80* (KRAM); Mt. Birkenmajer, 300 m, *RO 440/80* (KRAM). ***Martel Inlet***: Ullman Spur, 70 m, *RO 579/80* (KRAM); Ternyck Needle, 433 m, *RO 1834/80* (KRAM); Point Hennequin, 15 m, *RO 2238/80* (KRAM); Basalt Pt., 15 m, *2227/80* (KRAM); Mt. Wawel, 90 m, *RO 2179/80* (KRAM); Bell Zygmunt, 300 m, *RO 2676/80* (KRAM). ***Marian Cove***: North Spit, *BJ 147* (KRAM) and *DCL 782* (AAS, KRAM). ***Barton Peninsula***. Noel Hill, *BJ 204* (AAS); Winship Pt., *BJ 192* (KRAM). **Nelson I.** Harmony Cove, *CS 425* (PC, S); Moss Bay, 26.2.1989, *Komárek s.n.* (KRAM). **Robert I.** Coppermine Cove, 15–35 m, *RILS 908* (KRAM, PC). **Greenwich I.** Mt. Plymouth, *RMS 69-992* (US); Duff Pt., 15 m, *DCL 558* (AAS, KRAM). **Half Moon I.** S of huts, 50 m, *DCL 585* (AAS, KRAM). **Livingston I.** ***Byers Peninsula***: Lair Pt., 10 m, *KR 58C* (AAS, KRAM); Robbery Beaches, 10 m, *KR 54C* (AAS, KRAM); Negro Hill, 100 m, *KR 14C* (AAS, KRAM); between Chester Cone and Sealers Hill, 25–30 m, *RILS 3819* (AAS, KRAM); South Beaches, 6 m, *DCL 85* (AAS, KRAM). ***Hurd Peninsula***: near Spanish base, *Acuña 7002* (KRAM); Johnsons Dock, 80 m, *Redon 114* (AAS, KRAM). ***False Bay***: Barnard Pt., 10 m, *DCL 19* (AAS, KRAM). **Deception I.** Between Kroner Lake and Ronald Hill, *ca* 15 m, *REL 903* (AAS, KRAM); Stonethrow Ridge, 60 m, *RILS 5806C* (AAS, KRAM). **Snow I.** President Head, *KR 5F* (AAS, KRAM).

WEST ANTARCTIC PENINSULA. **Trinity Peninsula.** Marescot Ridge, 450 m, *CGB 34* (AAS, KRAM). **Danco Coast.** Cierva Pt., *OB 22* (AAS, KRAM); Moss I., *CS 432* (LD, PC, S); Spring Pt., *VK 1244* (KRAM); Challenger I., *CS 426* (PC, S); Charlotte Bay, Recess Cove, 2–10 m, *RILS 3991* (AAS, KRAM); Cuverville I., *ca* 25 m, *de Leeuw 25* (AAS, KRAM); Coughtrey Peninsula, *RMS 69-355* (US). **Palmer Archipelago.** ***Brabant I.***: Cape Roux, 10 m, *AM 11* (AAS, KRAM); NE of Metchnikoff Pt., 80 m, *AM 71B* (AAS, KRAM); Hill Bay, 5 m, *AM 55A & 56A* (AAS, KRAM). ***Melchior Is.***: Lysted I., 7–10 m, *PAS 371.6 & 376* (FH); Lambda I., *PAS 345.3 & 347.7* (FH); Omega I., 2–10 m, *RILS 4123* (AAS, KRAM); Kappa I., 12 m, *Lamb & Waterhouse 7932* (NY); Gamma I., 5–10 m, *RILS 4147 & 4148* (AAS, KRAM). ***Anvers I.***: Cape Monaco, *VK 746 & 780* (KRAM); Norsel Pt., *JR & JMG 68* (US); Litchfield I., *JR & JMG 37 & 38* (US); Cormorant I., 8 m, *AS 57* (AAS, KRAM); Jubin Is., 5–10 m, *RILS 1993* (AAS, KRAM); Laggard I., 15 m, *REL 1302* (AAS, KRAM); Doumer I., *ca* 100 m, *FIDS Misc. 2* (AAS, KRAM); Wiencke I., *ca* 7 m, *FIDS Misc. 11* (AAS, KRAM). **Graham Coast.** Rasmussen I., 20 m, *AS 85* (AAS, KRAM) and *RILS 1923* (AAS, KRAM); Cape Tuxen, 15 m, *RILS 10671* (AAS, KRAM). **Wilhelm Archipelago.** ***Lemaire Channel***: Wandel I., *Turquet 237B & 288C* (PC); Booth I., *Gain 153* (PC). ***Penola Strait***: Petermann I., *Gain 167* (PC) and 70 m, *AS 112A* (AAS, KRAM). ***Argentine Is.***: Corner Is., *JR & JMG 51, 115 & 117* (US); Galindez I., Stella Creek. *ca* 15 m, *RWMC 447* (AAS, KRAM); Black I., 2–10 m, *RILS 4296* (AAS, KRAM). ***Grandidier Channel***: Lahille I., 2–5 m, *RILS 4332* (AAS, KRAM); Takaki Pt., 10–35 m, *RILS 4365C* (AAS, KRAM); Hook I., 15–40 m, *RMS 69-711* (US). **Loubet Coast.** ***Matha Strait***: nunatak on mainland facing Hanusse Bay, 233 m, *Lamb & Waterhouse 8069* (AAS, NY). ***Laubeuf Fjord***: Piñero I., 10–25 m, *RILS 4831* (AAS, KRAM); Webb I., *DWB 3A* (AAS, KRAM). **Fallières Coast.** ***Adelaide I.***: Reptile Ridge above Rothera Station, 140 m, *RILS*

10579 (AAS, KRAM), 205 m, 30.12.1984, *Birkenmajer s.n.* (KRAM); NW of Cape Alexandra, *JBK 200* (AAS, KRAM); Cape Alexandra, 220–250 m, *RILS 3593* (AAS, KRAM). ***Léonie Is.***: Léonie I., 30 m, *RILS 8931* (AAS, KRAM); Lagoon I., 15–35 m, *RILS 9175* (AAS, KRAM). ***Marguerite Bay***: Mushroon I., 60 m, *Poncet 1A* (AAS, KRAM). **GEORGE VI SOUND.** Föhn Bastion, *DDWW 205B* (AAS, KRAM). ***Charcot I.***: Rils Nunatak, 125 m, *RILS 10471 & 10713B* (AAS, KRAM).

EAST ANTARCTIC PENINSULA. **TRINITY PENINSULA.** Hope Bay, Summit Pass, 400–430 m, *OT 2571* (AAS, KRAM). **JOINVILLE ISLAND GROUP.** ***Joinville I.***: Mt. Alexander, 250 m, *RILS 3674 & 3676* (AAS, KRAM); Active Sound, 200 m, *RILS 3729A* (AAS, KRAM). **JAMES ROSS ISLAND GROUP.** ***James Ross I.***: Martha Cove, *ca* 150 m, *RILS 7427* (AAS, KRAM). **OSCAR II COAST.** Churchill Peninsula, 330–500 m, *RT 20* (AAS, KRAM). **WILKINSON COAST.** Engel Peaks, 1280 m, *Lovegrove 192* (AAS, KRAM).

ELLSWORTH LAND. **EIGHTS COAST.** Lepley Nunatak, 200 m, *RILS 5419* (AAS, KRAM); McNamara I., Peeler Bluff, *ca* 265 m, *RILS 5422A* (AAS, KRAM). ***Thurston I.***: unnamed locality, 72°18.7′ S, 96°16.2′ W, 17.2.1960, *Wagoner s.n.* (AAS, KRAM, NY); S of Bel Knap Nunatak, *RILS 5430A* (AAS, KRAM).

EXCLUDED SPECIES OF THE ANDREAEACEAE

Andreaea acuminata Mitt. — This species was reported from Nelson I., South Shetland Is., by Putzke & Pereira (2001) but no voucher collection was available for examination. It is a widely distributed south-cool-temperate species occurring in New Zealand, southern South America and some subantarctic islands, but so far no specimens of *Andreaea acuminata* from the Antarctic botanical zone have been seen.

A. gainii Cardot var. ***parallela*** (Müll.Hal.) S.W.Greene — All specimens reported under this name from Antarctica (Greene *et al.* 1970) are *Andreaea gainii* or *A. regularis*. See also the note under *A. gainii*.

A. obovata Thed. — This species was reported for King George I., South Shetland Is., by Wu & Hu (1990) but the voucher specimen is *Andreaea gainii* (Ochyra 1998*a*). So far, *A. obovata* has only once been recorded in the Southern Hemisphere on subantarctic South Georgia (Ochyra *et al.* 2002) and, additionally, this species occurs at high elevations in tropical East Africa (De Sloover 1977).

A. rupestris Hedw. — This Northern Hemisphere species was reported from several stations on Lysted I. and Lambda I. in the Melchior Is. archipelago off the western coast of the Antarctic Peninsula (Bartram 1957) but all specimens are *A. regularis* (Greene 1968*b*).

Class B. **POLYTRICHOPSIDA** Doweld

Polytrichopsida Doweld, Prosyll. Tracheophyt.: i. 2001.

Polytrichopsida Vitt, Goffinet & Hedd. *in* Bates, Ashton & Duckett, Bryol. Twenty-First Cent.: 119. 1998, *nom. nud.*

Polytrichopsida W.R.Buck & Goffinet *in* A.J.Shaw & Goffinet, Bryoph. Biol.: 96. 2000, *nom. inval. descr. angl.*

The class Polytrichopsida forms a relatively small and isolated evolutionary line. It shares many general features with other mosses, but a suite of characters, especially in the structure of the sporophytes, is essentially restricted to this group. In the older classification schemes of mosses it was usually considered as a subclass within the broadly understood class Musci or Bryopsida, and Mitten (1859) was the first to recognise it as a separate group, Nematodonti, to accommodate two families, Polytrichaceae and Buxbaumiaceae. According to recent phylogenetic studies and molecular data (Mishler & Churchill 1984; Vitt *et al.* 1998; Hyvönen *et al.* 1998, 2004; Newton *et al.* 2000) all mosses with a single nematodontous peristome, as well as the buxbaumiaceous mosses with a complex double peristome, are closely related and constitute a separate lineage of mosses which deserves to be recognised as a separate class. This concept has subsequently been accepted by Buck & Goffinet (2000) and Goffinet & Buck (2004) in their latest classification systems of mosses, and the Polytrichopsida was formally validated as a new class by Doweld (2001) which antedates validation of this class name made by Ochyra *et al.* (2003) (see Ochyra *et al.* 2006).

The Polytrichopsida are very diversified gametophytically and comprise species with the smallest and most reduced (in *Buxbaumia*) and the largest and most robust (in *Dawsonia*) gametophores known in the bryophytes. The class is essentially founded on structure of the sporophyte, especially the sporangium and the peristome. All taxa placed in it are united by the nematodontous peristome which is composed of whole, elongate cells with walls thickened all around. The peristome is mostly single but occasionally it is lacking (in *Alophozia*, *Bartramiopsis*, *Lyellia* and *Oedipodium*) or double (in *Buxbaumia*). In the latter case it is formed by 3–6, occasionally as many as 10, concentric cell layers and consists of an inner, pleated, truncate cone and an outer row or rows of rudimentary teeth. The peristome teeth are solid or articulate and their number varies from 4 to 32 or 64.

Mosses belonging within the Polytrichopsida show great structural diversity, both in gametophytes and sporophytes, which is distinctly reflected in three major evolutionary lines.

Consequently, three subclasses are recognised to accommodate members of these lineages, namely Buxbaumiidae, Tetraphidiidae and Polytrichidae. It is possible that the monotypic, highly isolated, gymnostomous genus *Oedipodium* also represents a separate lineage which merits recognition into a separate subclass. These subclasses have already been distinguished as separate taxa within the class Musci or Bryopsida in earlier classifications proposed, for example, by Crosby (1980), Walther (1983) and Schofield (1985). Although the names of these subclasses have been widely used in bryological literature, they have only recently been validly published (Doweld 2001; Ochyra *et al.* 2003).

The Polytrichopsida comprise about 220 species which grow mostly on mineral soil or humus, less often on decaying wood or rocks. They have a worldwide distribution from the High Arctic to the Antarctic and from sea level to high alpine elevations. In the Antarctic biome the class is represented only by members of the subclass Polytrichidae.

Subclass B1. **POLYTRICHIDAE** OCHYRA

Polytrichidae Ochyra *in* Ochyra, Żarnowiec & Bednarek-Ochyra, Cens. Cat. Polish Mosses: 155. 2003. Validly published by reference to Bryidae Engl. superorder Polytrichanae W.Frey *in* W.Frey, Hurka & Oberw., Beitr. Biol. Nied. Pflanz.: 132. 1977.

Musci Endl. subclass Polytrichidae Crosby *in* R.J.Taylor & A.E.Leviton, Moss. N. Am.: 120. 1980, *nom. inval. descr. angl.*

Tribus Archidontei M.Fleisch., Musci Buitenzorg 1: xxv. 1904, *nom. illeg.*

The subclass Polytrichidae or hair-cap mosses encompasses mostly large and robust, rhizomatous mosses with perennial gametophores which reach the greatest structural complexity among mosses. The stems and the leaves show considerable tissue differentiation. They have a well developed water-conducting system unknown in any other mosses and composed of hydroids and leptoids which are analogous, respectively, to the tracheid and sieve elements of vascular plants (Kawai 1969; Hébant 1974, 1977). Another peculiar feature of the gametophores in this subclass is the common occurrence of parallel lamellae along the ventral surface of the costa; these are densely filled with chlorophyll and markedly increase the photosynthetic area of the leaf (Proctor 2005). They are reduced only in a few members of this subclass (*Alophozia*, *Atrichopsis*, a few *Pogonatum* species, one *Notoligotrichum* species), while analogous structures occur only very rarely in the Bryopsida, for instance in the Pottiaceae (*Aloina*, *Pterygoneurum*) and Grimmiaceae (*Aligrimmia*).

In addition to the gametophyte characters, the subclass is characterised by its peculiar peristome which is composed of a single row of 32 or 64, rarely 16 or none, solid, lingulate teeth which are typically attached to an epiphragm that covers the mouth of the capsule and represents the expanded apex of the columella (Smith 1974). This 'polytrichoid' peristome differs fundamentally from that of arthrodontous mosses in being composed of fascicles of crescent or horseshoe-shaped cells. These attain their final shape by a process of symplastic growth and elongation similar to that of libriform fibres (Wijk 1929; Smith 1971).

Because of their widespread and often abundant occurrence, relatively large size and high visibility, hair-cap mosses attracted more attention from muscologists than any other moss group, and *Polytrichum commune* is often used in botanical textbooks as the representative moss. Since the classic monograph of this group by Lorch (1908) there have been numerous treatments dealing with various aspects of these mosses. Smith (1971) described in considerable detail the structure, diversity, generic classification and interrelationships of the Polytrichidae, and recently Hyvönen *et al.* (1998, 2004) discussed thoroughly their phylogeny.

Fleischer (1908) recognised the hair-cap mosses as a separate taxon intermediate between the Ordnung and the Reihe called 'Unterordnung Polytrichineae', and later Fleischer (1920, 1923) named it 'Reihengruppe Polytrichinales', whereas Frey (1977) considered them as a superorder Polytrichanae. These taxa are all equivalents of the subclass Polytrichidae in the present sense and Ochyra *et al.* (2003) validly published the subclass Polytrichidae by direct reference to the name of the latter superorder. Following the original concept of Fleischer (1920, 1923) this subclass is subdivided into two orders, Dawsoniales and Polytrichales, the latter being represented in the Antarctic.

Order II. **POLYTRICHALES** M.FLEISCH.

Polytrichales M.Fleisch., Hedwigia 61: 400. 1920.

The Polytrichales are defined by their peristome which, if present, comprises 32 or 64, rarely 16, solid, lingulate teeth. They are derived from a restricted number (usually 4) of concentric layers of cells originating through periclinal divisions of cells in two layers of the inner amphithecium and becoming U-shaped owing to growth pressures. These peristome teeth are markedly different from those in the Dawsoniales which are long and filiform and initially spiralled but becoming spread out and brush-like with age. Additionally, they are formed by asymmetric cell divisions in about eight concentric layers of cells in radial rows and many tiers of the inner amphithecium, each cell cutting off 1–2 small cells at its corner. The calyptra in the Polytrichales is cucullate and it is often densely covered with branched hairs consisting of uniseriate cells.

The order consists of a single family, Polytrichaceae.

Family II. **POLYTRICHACEAE** Schwägr.

Polytrichaceae Schwägr. *in* Willd., Spec. Pl. Ed. 4, 5(2): 1. 1830 ['Familia Polytrichii']. — Type: *Polytrichum* Hedw.

The Polytrichaceae comprise very distinctive acrocarpous mosses which are readily recognised by their characteristic habit. They are terrestricolous plants growing as a short or tall turf or sometimes scattered on a persistent protonema, and which are very variable in size but mostly coarse and robust. The stem is simple or branched and arises from a horizontal, subterranean rhizome densely coated by rhizoids. It has a complex anatomical structure and consists of a broad central cylinder, composed of hydrome and leptome with traces extending to the leaves, which forms a water-conducting system. The leaves are usually differentiated into a unistratose, membranous, elamellose and mostly sheathing base and a divergent, firm-textured limb or, less often (in *Atrichum*, *Psilopilum*, *Notoligotrichum*), the leaf lamina is not or scarcely differentiated. The costa is single and well developed, greatly broadened in the limb and usually occupying most of its width. It consists of a single row of large guide cells surrounded by ventral and dorsal stereid bands as seen in transverse section. The ventral and occasionally also the dorsal surface of the costa and the adjacent lamina are covered with few to many longitudinal lamellae and only occasionally they are absent. The marginal cells of the lamellae are, in profile, entire, crenulate or serrate and, in section, undifferentiated or sharply distinct in size and shape from those beneath. The laminal cells are smooth and mostly isodiametric and the margins of the lamina are entire, dentate or serrate, sometimes bordered by linear and uni- to polystratose cells. The cells in the sheath are elongate-rectangular to linear and the transition in areolation between the sheath and the limb is gradual or, most often, abrupt through a group of incrassate, transversely elongate cells forming a hinge-tissue. Asexual reproduction occurs exceedingly rarely (only in *Alophozia*) by stalked, thickened-discoid brood bodies.

The polytrichalean plants are mostly dioicous and infrequently monoicous. The perigonia are discoid or cupulate and they are indeterminate and give rise to innovations which continue the growth of the stem. The perichaetia are terminal and the perichaetial bracts are strongly or not differentiated. The setae are elongate, solitary or sometimes several in a single perichaetium. The capsules are erect to horizontal, terete or angled, ovoid, obloid, cylindrical, campanulate, radial or bilateral. The neck is short and scarcely differentiated or distinct, hemispheric and abruptly constricted from the urn as a hypophysis. The exothecial cells are smooth, mammillose or papillose, sometimes pitted. Stomata are absent (in *Atrichum*, *Itatiella* and *Pogonatum*) or present and their number per capsule ranges from several hundred in some taxa to fewer than a dozen in others. They are superficial, with the guard cells often not completely separated but consisting at maturity of a single torus-like cell and with the central pore normally thickened or, rarely, with one or two additional cell divisions, resulting in an abnormal stoma with three of four guard cells or a 3–4-armed pore. The annulus is lacking or uniseriate and the calyptra is cucullate, densely pilose or rarely naked.

The Polytrichaceae are a medium-sized family consisting of 18 extant genera and approximately 160 species (Hyvönen *et al.* 1998). Of these, seven genera are monotypes, three comprise 2–4 species, five of about 9–10, one of about 15 and only *Polytrichum* and *Pogonatum* have greater diversity with about 30 and 50 species, respectively. The family has a cosmopolitan distribution and some species are common and important constituents of the vegetation in temperate, cool and polar regions of both hemispheres. It is well represented in the cool temperate zone in the Southern Hemisphere and in the Subantarctic, and Smith (1972) postulated that it had a Gondwanan origin.

The Polytrichaceae have been monographed for Tierra del Fuego (Schiavone 1993), South Georgia (Greene 1973) and Australia (Hyvönen 2006). Modern treatments of the Polytrichaceae with excellent illustrations have been presented for Arctic North America and Greenland (Long 1985), Central Siberia (Ignatov & Smith Merrill 1995), Japan (Osada 1965, 1966), Papua New Guinea (Hyvönen 1986) and sub-Saharan Africa (De Sloover 1986). In the Antarctic the family has been surveyed taxonomically by Greene *et al.* (1970). It is represented here by three genera and six species.

KEY TO THE ANTARCTIC GENERA OF THE POLYTRICHACEAE

1. Leaves not or only moderately differentiated into sheath and limb; costa narrow flanked by a broad lamina; ventral lamellae segmented, undulate; calyptra naked or sparsely hairy at the apex .... 2. ***Notoligotrichum***

1. Leaves with a distinct sheathing base and a firm, broadly lamellate, divergent limb; costa broad, occupying most of the width of limb, flanked by a narrow lamina, 2–12 cells wide; calyptra densely pilose 2
   2. Leaf margin dentate or serrate, plane or incurved but never involute; capsule terete or rounded to 4–6-angled, without a distinct hypophysis; exothecial cells smooth, not pitted; peristome teeth not keeled; spores large, 13–26 μm wide, warty papillose .... 3. ***Polytrichastrum***
   2. Leaf margin entire, broadly involute and enclosing the upper surface of the limb; capsules sharply 4(–5)-angled, with a distinct hypophysis; exothecial cells mammillose, pitted; peristome teeth keeled at back; spores small, 6–17 μm wide, finely papillose .... 4. ***Polytrichum***

2. NOTOLIGOTRICHUM G.L.SM.

Notoligotrichum G.L.Sm., Mem. New York Bot. Gard. 21(3): 50. 1971. — Type: *Notoligotrichum australe* (Hook.f. & Wilson) G.L Sm. (*Polytrichum australe* Hook.f. & Wilson).

Plants small and slender to large and robust, gregarious or loosely caespitose to densely tufted, dull green, yellowish- to brownish-green or brown. *Stems* erect, simple or rarely branched, densely leafy. *Leaves* rigid to rather soft-textured, closely imbricate or curled, crisped to falcato-secund and spreading when dry, erecto-patent to patent when wet, ovate to oblong-elliptic, ovate- or oblong-lanceolate from a scarcely or distinctly expanded, not or slightly sheathing base, concave, obtuse and apiculate to acute or acuminate, cucullate or flat at apex; *margin* entire, crenate-dentate to coarsely serrate, erect, flat to broadly inrolled above; *costa* narrow to broad, well defined throughout, ending below the apex to shortly excurrent, dorsally smooth, toothed or low lamellose near the apex, in transverse section with numerous dorsal stereids in a broad band and few ventral stereids forming a small and greatly reduced ventral band; *ventral lamellae* few to many, mostly tightly set on both costa and lamina or only on costa, straight or undulate, entire or laciniate-dissected into short segments, 1–10 cells high, in lateral view entire to retuse, shallowly lobed or dentate, with smooth or papillose but otherwise not differentiated end-cells; *lamina* usually broad, involute or rarely flat, unistratose; *upper laminal cells* subquadrate to hexagonal, smooth to striate-papillose on the lower surface, moderately thick-walled, with marginal cells like the others or somewhat elongate, narrower and thicker-walled forming an indistinct border; *lower cells* short rectangular, wider, thinner-walled, somewhat enlarged, thin-walled and inflated at basal angles. *Dioicous*. *Perigonia* discoid; *perigonial bracts* short, broad, spreading. *Perichaetial leaves* not differentiated. *Setae* elongate, erect, straight or somewhat flexuose. *Capsules* erect or inclined to horizontal, symmetric to gibbous, terete to laterally compressed, cylindrical but broadest near bases, obliquely ovoid to obloid, narrowed at oblique mouth, without apophyses; *exothecial cells* polygonal, thin- to firm-walled; *stomata* conspicuous, cryptoporous, restricted to capsule base; *peristome teeth* 16 or 32, tapering, sharp-pointed to obtuse, unequal or rarely absent. *Operculum* short to long rostrate or subulate, with a slanted beak. *Calyptra* glabrous or with few scattered hairs at the apex.

A small antipodal genus consisting of nine or fewer species widely distributed in southern temperate areas and in the Subantarctic, and extending north to Mexico along the Andean cordillera (Smith Merrill 1994) and to New Guinea (Hyvönen 1986). It is the austral counterpart of the Arctic genus *Psilopilum* with which it had been formerly merged. However, Smith (1971) segregated it in a genus of its own on account of its peristome which is composed of triangular, acute and distant teeth, whereas *Psilopilum* has linear, blunt and crowded teeth. The genus has not yet been revised taxonomically and only southern South American species have been studied by Schiavone (1993). A single species, *Notoligotrichum trichodon*, is known from the Antarctic.

1. **Notoligotrichum trichodon** (Hook. & Wilson) G.L.Sm. Fig. 26

Notoligotricum trichodon (Hook. & Wilson) G.L.Sm., Mem. New York Bot. Gard. 21(3): 51. 1971. ≡ *Polytrichum trichodon* Hook. & Wilson, London J. Bot. 6: 289, *t. 10, f. 1–10.* 1847. ≡ *Psilopilum trichodon* (Hook. & Wilson) Mitt., J. Linn. Soc. Bot. 12: 607. 1869. — TYPE CITATION: [Colombia] Paramo of Ruiz, on volcanic ashes near the line of perpetual snow. July, 1846. (*Purdie*). [Lectotype (*selected*

⟶

Figure 26. *Notoligotrichum trichodon*. — **1**. Habit, dry. **2**. Portion of shoot, dry. **3**. Plant with sporophyte, dry. **4**. Portion of shoot, wet. **5**. Portion of cross-section of stem. **6–12**. Leaves. **13–14**. Lower leaves. **15**. Upper laminal cells at margin. **16**. Mid-leaf cells at margin. **17**. Mid-leaf cells. **18**. Lower laminal cells at margin. **19**. Lower laminal cells. **20**. Basal leaf cells. **21**. Lateral view of lamella. **22**. Cross-section of leaf lamina about 5/6 up the leaf. **23**. Cross-section of costa in the lower part of leaf. **24**. Plant with perigonium, wet. **25–26**. Outer perigonial bracts. **27**. Inner perigonial bract. — [1–2, 10–21, 24–27 from *MWH 828A*; 3 from *RO 487/95*; 4, 6–9 from *RILS 458*; 5, 22–23 from *MWH 405A*]. — Scale bars: a – 2 mm (3); b – 1 mm (6–14, 25–27); c – 1 mm (24) and 2 mm (1); d – 2 mm (2, 4); e – 100 μm (5, 15–20, 22–23); f – 100 μm (21).

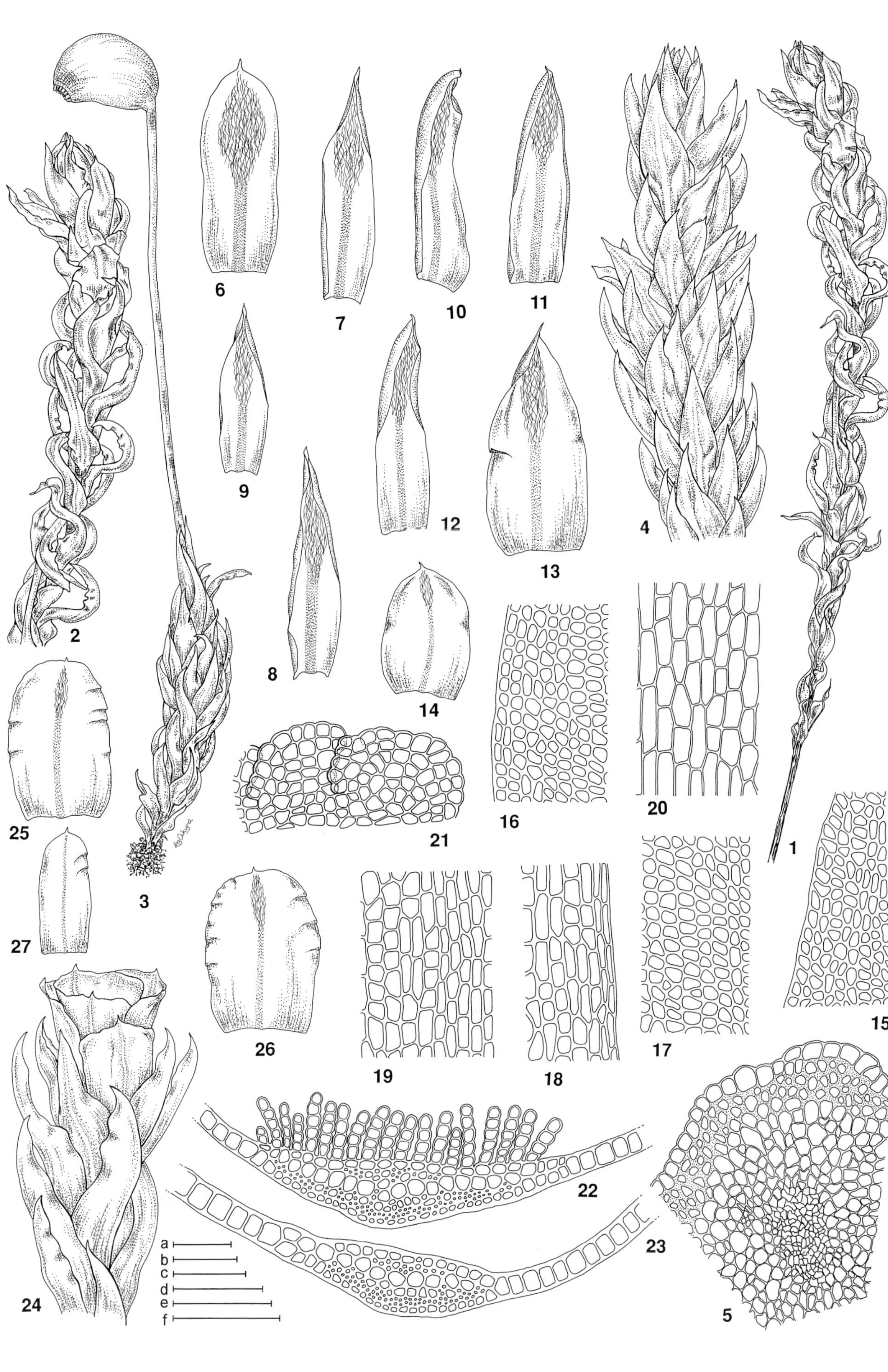
1
2
3
4
5
6
7
8
9
10
11
12
13
14
15
16
17
18
19
20
21
22
23
24
25
26
27
a
b
c
d
e
f

here): 'Paramo of Ruiz, N. Granada N[r] perpetual snow on volcanic ashes, rare July 1846 W. Purdie Polytrichum trichodon HfWils L. J. Bot 6 t. 10 Wm Purdie' – BM-Hooker!; isotypes: BM-Hooker!, NY-Mitten!].

Catharinea antarctica Müll.Hal., Bot. Jahrb. Syst. 5: 77. 1883. ≡ *Psilopilum antarcticum* (Müll.Hal.) Paris, Ind. Bryol.: 1039. 1898. ≡ *P. trichodon* (Hook. & Wilson) Mitt. var. *antarcticum* (Müll.Hal.) Mitt. *ex* Paris, Ind. Bryol.: 1039. 1898, *nom. inval. err. pro P. antarcticum*. — TYPE CITATION: all new species described by Müller (1883) are included in the section 'I. Bryologia Kerguelensis'; later Müller (1889: p. 11) wrote: 'Ins. Kerguelen, Betsy Cove (12.74)'. [Lectotype (*selected here*): 'Psilopilum antarcticum mihi n. sp. Kerguelen Insula Decbr 1874 Hb. C. Müll. Gazelle-Expedition' – BM!; isotype: U (not seen)]. First synonymised by Smith (1969: p. 67).

Plants small to moderately large and robust, rigid, forming short turves, densely to loosely caespitose or gregarious and intermixed amongst other bryophytes, dull to slightly glistening, greenish- or yellow-brown to brown, sometimes rusty- to blackish-brown or sometimes dull green. *Stems* erect, mostly simple, whitish-radiculose at base, (0.5–)1.0–2.0(–4.5) cm tall, usually with a comal appearance due to the dense foliage in the upper part or leafy for most of its length, in cross-section with a typical polytrichoid central strand. *Leaves* scattered along the stem or, more often, crowded and somewhat larger towards the stem apex, erecto-patent to loosely imbricate, curled and incurved, usually with tubular limbs on drying, erecto-patent to spreading on wetting, rather stiff and rigid, oval-oblong or ovate- to oblong-lanceolate from a broad, clasping base, gradually tapering from about mid-leaf to a cucullate, broadly acute or mucronate apex, concave, (1.8–)2.2–4.0(–5.2) mm long, 0.8–1.6(–1.9) mm wide; *margins* entire, plain, unistratose, broadly involute in the upper half; *costa* relatively narrow below, usually ⅕ or less of the width of lamina, 80–110 µm wide, broadening above, distinct but not sharply delimited from the laminal cells, ending in the apex or excurrent as a very short mucro, dorsally smooth or faintly roughened near the apex, in transverse section composed of large dorsal and small ventral stereid bands separated by a single row of large guide cells; *ventral lamellae* 18–30 borne on broadened costa and adjacent lamina in the upper half, wavy, with a continuous basal portion 2–3 cells high and upper segmented portion 4–6 cells high, in side-view entire but crenulate with bulging, smooth and undifferentiated end-cells. *Lamina* 25–35 cells broad on each side in the lamellose part of limb, unistratose; *upper laminal cells* irregular, mostly rounded-quadrate, elliptical to transversely rectangular, (11–)14–20(–25) µm wide, thicker-walled with distinct corner thickenings, becoming smaller, thinner-walled and obliquely rhombic at margins but not forming a distinct border or sometimes 2–3 rows of

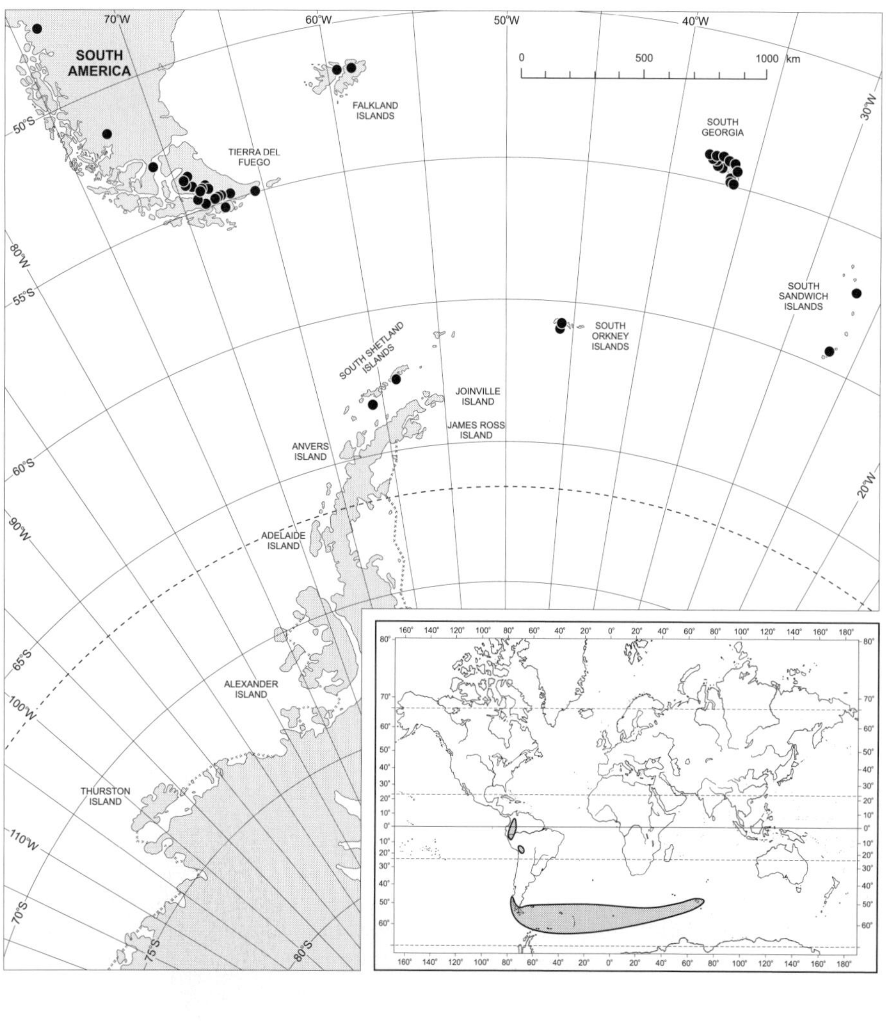

Figure 27. Distribution of *Notoligotrichum trichodon* in the Fuegian and Antarctic Peninsula region. Inset: global distribution.

marginal cells hyaline; *lower cells* subquadrate to rectangular, thinner-walled, 20–30 µm wide, 25–70(–100) µm long, becoming narrower towards the margins and thin-walled, lax and somewhat bulging at base. *Dioicous*. *Perigonia* cupuliform; *perigonial bracts* oblong, concave, external broad, truncate and shortly apiculate, entire-margined. [*Setae* elongate, erect, straight or flexuose, 1.3–3.5 cm long. *Capsules* inclined to horizontal, laterally compressed, obloid, gibbous, light to dark brown, 3–5 mm long, 2–3 mm wide; *exothecial cells* thin-walled; *peristome teeth* unequal, pale, 180–200 µm long. *Operculum* subulate with a slanted rostrum. *Spores* reniform, 20 × 25 µm].

Chromosome numbers: $n = 7$ – South Georgia (Newton 1980).

Discussion. With its polytrichoid habit the species is readily distinguished from all other regional moss species and from other members of the family by its curled and characteristically incurved, often glossy, leaves lacking a distinct sheathing base. Microscopically, it has distinct undulate and segmented ventral lamellae which occupy a relatively small area flanked on both sides by a broad lamina. The plants are variable in general stature and in the acuteness of the leaf apex, but they fall well within the broad range of variation of these features observed throughout its wide geographical range.

Greene *et al.* (1970) reported this species from the Antarctic as *Psilopilum antarcticum*. This species was described from Îles Kerguelen (Müller 1883, 1889) but Smith (1969) showed its conspecificity with *P. trichodon*, which was originally described from the Andean paramos in Colombia (Hooker & Wilson 1847). This conclusion is here confirmed by examination of original collections and, additionally, both names are typified. However, the authority of *Polytrichum trichodon* was erroneously ascribed to J. D. Hooker & Wilson (i.e. 'Hook.f. & Wilson') but Ochyra (2003*e*) proved that in fact this species was correctly described by W. J. Hooker & Wilson (i.e. 'Hook. & Wilson').

Reproduction in Antarctica. Sterile, but perigonia occasionally observed.

Habitat. On volcanic islands the species grows predominantly in moist and heated places near fumaroles, on ash, scoria and cinder as well as on stony flats and knolls, often near vents emitting steam. Elsewhere, it grows mostly on acidic soil in moist situations and on rocky slopes. In the South Orkney Is. its typical habitat is late snow beds, associating with *Andreaea* spp., *Bucklandiella sudetica*, *Polytrichastrum alpinum* and several liverworts.

World range. An amphiatlantic subantarctic species (Fig. 27, inset), most abundantly and widely distributed on South Georgia. Additionally, it is known from Îles Kerguelen, northern maritime Antarctic, Falkland Is. (Cardot & Brotherus 1923; Matteri 1986), Tierra del Fuego (Schiavone & Hyvönen 1993), Brunswick Peninsula (Deguchi 1991*b*), western Patagonia (Hässel de Menendez *et al.* 1984; Matteri 1985*a*) and disjunct and isolated stations at altimontane elevations of 3600–4600 m in the Andes of Bolivia, Ecuador and Colombia (Churchill *et al.* 2000). Information on its occurrence on the Juan Fernandez Is. (He 1998) is incorrect (see Robinson 1975).

Distribution in Antarctica. Relatively frequent only on Bellingshausen I. in the South Sandwich Is., occasional in the South Orkney Is. and rare in the South Shetland Is., extending southwards to Deception I. at lat. 62°58′S (Fig. 27). On the latter island the site was buried by a lahar from the 1970 eruption. The species was not seen again until 2002 when a small clump of colonising shoots was found at the same site (Lewis Smith 2005*a*,*b*). Its altitudinal range is from (5–)50 to 200 m.

Selected specimens examined. WEST ANTARCTICA. SOUTH SANDWICH IS. **Candlemas I.** N of eastern lagoon, 50 m, *REL 782* (AAS, KRAM); Lucifer Hill, above Clinker Gulch, *PC 136A* (AAS, KRAM). **Bellingshausen I.** S flanks of crater rim, *ca* 200 m, *MWH 814A* (AAS, BM, KRAM) & *824A* (AAS, KRAM); S slope of the cone, *ca* 75 m and *ca* 85 m, *MWH 418* (AAS, KRAM); inner southern slopes of crater, 120 m, *PC 66E* (AAS, KRAM), 130 m, *PC 63D*, *64A* & *65A* (AAS, KRAM), 150 m, *PC 41C*, *42B*, *44A*, *50B*, *51C*, *56A* & *74C* (AAS, KRAM) and 170 m, *PC 32C* (AAS, KRAM); western side of the main crater, *ca* 135 m, *MWH 429B* & *480* (AAS, BM, KRAM); SE rim of crater, *ca* 65 m, *MWH 824A* (AAS, KRAM).

SOUTH ORKNEY IS. **Coronation I.** Shingle Cove, 200 m, *RILS 4968* (AAS, KRAM); Cape Hansen, *RILS 8684* (AAS, KRAM). **Signy I.** Rusty Bluff, *ca* 135 m, *RILS 458* (AAS, BM, KRAM); Moraine Valley, *ca* 35 m, *REL 1088* (AAS, KRAM, PC), *ca* 60 m, *RILS 305* & *372* (AAS, KRAM) and *ca* 65 m, *RW 121* (AAS, KRAM).

SOUTH SHETLAND IS. **King George I.** ***Potter Peninsula***: Three Brothers Hill, 90 m, 22.12.1993, *Schulz s.n.* (KRAM). **Deception I.** SSW of Ronald Hill, 8–17 m, *REL 33* (AAS, KRAM) and *RILS 69B* (AAS, KRAM); between Kroner Lake and Ronald Hill, 2–4 m, *RILS 11246* (AAS, KRAM).

3. POLYTRICHASTRUM G.L.Sm.

Polytrichastrum G.L.Sm., Mem. New York Bot. Gard. 21(3): 50. 1971. — Type: *Polytrichastrum alpinum* (Hedw.) G.L.Sm. (*Polytrichum alpinum* Hedw.).

Plants medium to large and coarse, loosely tufted or gregarious. *Stems* short or tall, simple or branched, erect. *Leaves* crowded, with a well-defined, thin, broad, oblong to obovate, sheathing base abruptly contracted into lanceolate to linear-lanceolate limbs, erect to somewhat spreading, straight or weakly, widely patent when wet, opaque, acute or blunt, sometimes cucullate; *margins* erect, plain, serrate to sharply dentate, sometimes entire, not bordered; *costa* narrow and sharply delimited in sheathing base, becoming much widened and poorly defined in limbs, short- to long-excurrent, often scabrous to denticulate dorsally near apex, in transverse section with well-developed dorsal and ventral stereid bands; *ventral lamellae* numerous, closely spaced, straight, covering costa and most of lamina, in side-view entire to crenulate, with end-cells undifferentiated to enlarged, thin- to thick-walled, smooth or finely papillose; *basal cells* unistratose, hyaline, long-rectangular to linear; *limb* unistratose forming narrow wings on both sides of costa, with rounded-quadrate to transversely rectangular, thick-walled cells, mostly with collenchymatous corner thickenings; *marginal teeth* single, 1–4-celled. *Dioicous*. *Setae* erect, elongate, solitary in perichaetium. *Capsules* inclined to horizontal, terete or obtusely 4–6-angled, symmetric or nearly so, with not or weakly differentiated hypophyses; *exothecial cells* smooth, not pitted, polygonal; *stomata* numerous in the neck region, consisting of a single slit in a single guard cell; *peristome teeth* 64 or fewer by fusion, simple, leiodontous, not keeled at back; *epiphragm* readily detached, thick and entire or with erect, tooth-like processes opposite the peristome teeth and attached to their inner surface. *Operculum* long-rostrate. *Spores* warty papillose, 13–26 µm wide. *Calyptra* cucullate, densely long-pilose.

A small genus segregated from the traditionally understood *Polytrichum* (Smith 1971) and characterised by the terete or bluntly 4–6-angled capsule without a differentiated hypophysis, the smooth and unpitted exothecial cells, the leiodontous peristome teeth, the epiphragm thick and entire or with marginal tooth-like processes, and relatively large, warty papillose spores. It comprises about 13 species widely distributed in the Northern and Southern Hemispheres. Two species occur in the Antarctic.

1. Marginal cells of ventral lamellae strongly differentiated, narrowly ovate to pyriform in cross-section, with upper walls strongly incrassate, densely and finely papillose; lamina narrow, 2–6 cells wide .... 1. ***P. alpinum***
1. Marginal cells of ventral lamellae ± undifferentiated, rounded or elliptical in cross-section, with smooth, thin and even walls; lamina broader, 5–15 or more cells wide .... 2. ***P. longisetum***

1. **Polytrichastrum alpinum** (Hedw.) G.L.Sm.

Fig. 28, pl. 39

Polytrichastrum alpinum (Hedw.) G.L.Sm., Mem. New York Bot. Gard. 21(3): 37. 1971. ≡ *Polytrichum alpinum* Hedw., Spec. Musc. Frond.: 92. 1801. ≡ *Pogonatum alpinum* (Hedw.) Röhl., Ann. Wetterauischen Ges. Gesammte Naturk. 3(2): 226. 1814. — TYPE CITATION: Habitat in alpibus Scoticis, Scandinavicis, Tyrolensibus, Helveticis, in Sudetibus, Bructero Hercyniae etc.; also references to Dillenius (1741: t. 55, f. 4), Linnaeus (1763: p. 1112), Oeder (1766: *pl. 296*), Murray (1784: p. 947), Gmelin (1791: p. 1327) and Bridel (1798: p. 99) are provided. [Lectotype (*selected here*): The central specimen in the upper row with four mature capsules on the sheet labelled 'Polytrichum alpinum e Bruct' – G-Hedwig/Schwägrichen!].

Polytrichum brevifolium R.Br., Suppl. App. Capt. Parry's Voy.: 294. 1824. ≡ *Pogonatum alpinum* var. *brevifolium* (R.Br.) Brid., Bryol. Univ. 2: 131. 1827. ≡ *Polytrichum alpinum* var. *brevifolium* (R.Br.) Müll.Hal., Syn. Musc. Frond. 1: 211. 1848. ≡ *Polytrichastrum alpinum* var. *brevifolium* (R.Br.) Brassard, Bryologist **75**: 531. 1972. — TYPE CITATION: Melville Island, leg. Ross. [Lectotype (*selected here*) {or perhaps holotype}: '70. Polytrichum brevifolium Mr James Ross' – BM-Brown!]. First synonymised with *Polytrichum alpinum* by Greene *et al.* (1970: p. 26, 50).

Pogonatum alpinum fo. *elatum* Cardot, Bull. Herb. Boissier Sér. 2, 6: 31. 1906. — TYPE CITATION: Terre-Louis-Philippe: île Challenger. [Lectotype (*selected here*): 'Svenska Sydpolarexpeditionen 1901–03. N:r 463. Pogonatum alpinum Röhl. var. brevifolium Brid. f. elata Card. Terre Louis-Philippe, île Challenger, 2/12 1902, Carl Skottsberg, det. J. Cardot' – PC!; isotype: S!]. First synonymised with *Polytrichum alpinum* by Greene *et al.* (1970: p. 26, 50).

Plants medium to fairly tall, coarse, loosely to densely caespitose or gregarious, forming loose short to tall turves, dull to slightly lustrous, brown, yellow- to reddish-brown, sometimes blackish or glaucous-green. *Stems* erect, often from a decumbent base, simple or often forked, varying in size from very short in exposed habitats, 0.5 cm tall, to 4–6 or occasionally up to 13 cm, often leafless, slender and threadlike below, not or weakly radiculose only at base with hyaline rhizoids, in transverse section obtusely pentagonal in leafy part to triangular in lower leafless part, with a typical polytrichoid central strand. *Leaves* loosely or densely set, erect and appressed to erect-spreading, not or little twisted when dry, erect- to wide-spreading or recurved when wet; *sheath* elliptic, ovate to obovate, 1.5–3.0 mm long, 1.3–1.9 mm wide, with tapering shoulders, hyaline, ± lustrous, entire-margined, closely clasping the stem; *limb* linear-lanceolate, long acuminate to narrowly acute, (2.5–)3.5–9.5 mm long, 0.5–07(–0.8) mm wide, concave and often subtubulose when dry, but not cucullate; *margins* erect and inflexed, conspicuously and coarsely serrate or serrulate to mid-limb or nearly to the shoulders to almost entire; *costa* narrow in sheath, becoming strongly widened above and occupying almost the whole limb, excurrent as a brown, serrate to entire point, smooth or toothed at back near the apex; *ventral lamellae* (11–)20–25(–40) straight, 5–9 cells high, entire in lateral view, with marginal cells in cross-section ovate to narrowly ovate with strongly thickened, often yellowish to dark brown, densely and finely papillose outer wall; *basal cells* narrowly rectangular to linear, thin-walled, 6–10 µm wide, 35–65(–80) µm long, becoming short and oblate, to

Figure 28. *Polytrichastrum alpinum*. — **1**. Habit, wet. **2**. Portion of shoot, dry. **3**. Portion of shoot, wet. **4**. Portion of cross-section of stem. **5–11**. Leaves. **12–14**. Cells about two-thirds up leaf margin. **15**. Laminal cells at the leaf shoulder. **16**. Upper cells of sheath below shoulders; **17**. Laminal cells in mid-sheath. **18–21**. Cross-sections of leaf, a sequence from apex of limb to mid-sheath; **22**. Side-view of lamella. **23**. Antheridial head. **24–25**. Perigonial bracts. **26–27**. Perichaetial leaves. **28**. Deoperculate capsule. **29**. Operculate capsule. **30**. Calyptra. **31**. Portion of peristome and spores. **32**. Mid-exothecial cells. **33**. Stomata. — [1–4, 18–21 *RO 2375/80*; 5–6, 12–13 from *BJT 39*; 7–8, 14–17 & 22 from *RO 1381/80*; 9 from *CS 469* (type of *Pogonatum alpinum* fo. *elatum*), S; 10 from *ER 270B*, BR; 11, 23–27 from *RO 516/80*; 28, 31–33 from *RILS 3215*; 29 from *RO 1397/80*; 30 from *BJ s.n.*, 23.12.1979]. — Scale bars: a – 1 mm (2); b – 1 mm (28–30; c – 1 mm (3, 5–11, 23–27); d – 100 mm (4, 12–22, 31–33); e – 0.5 cm (1).

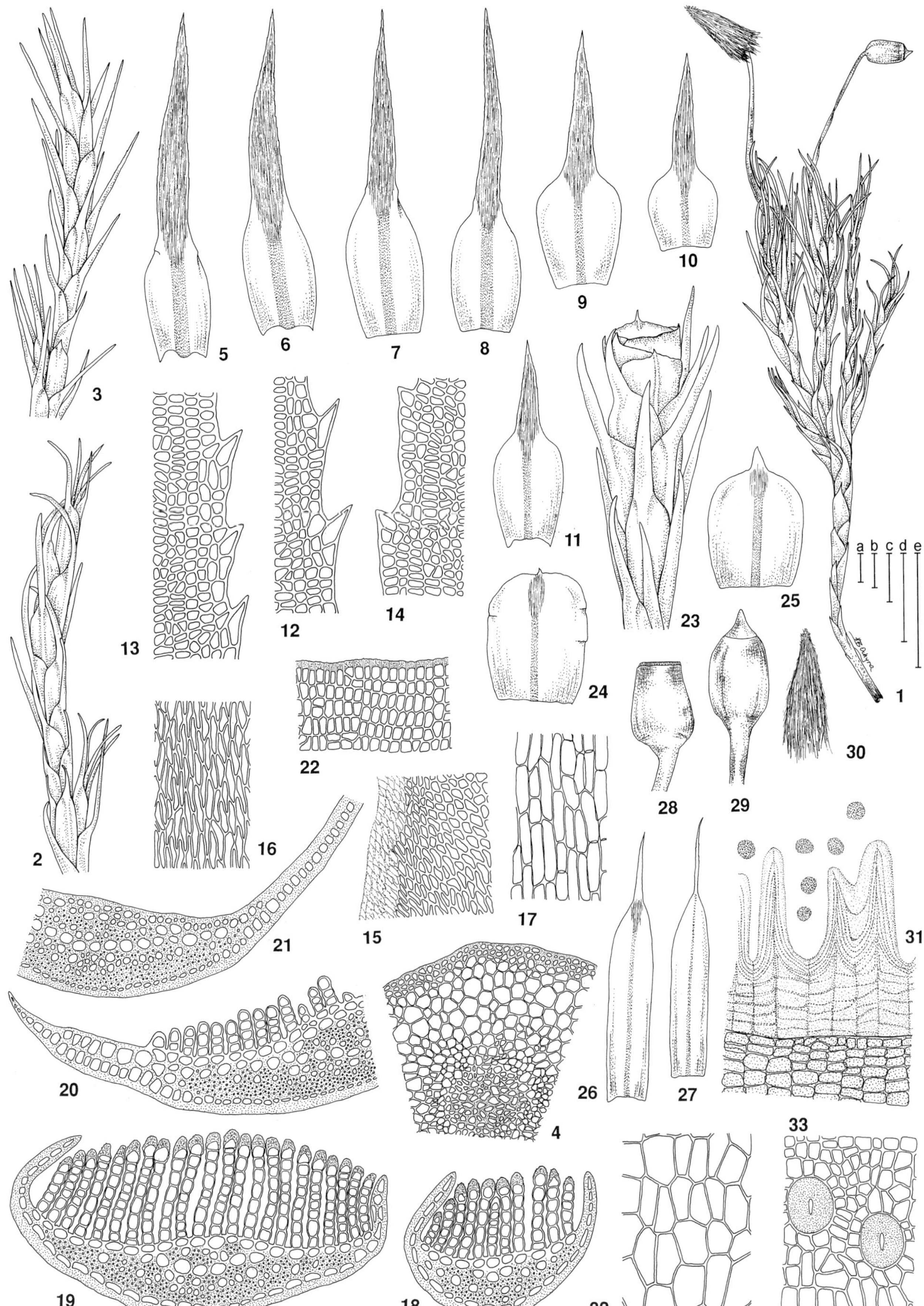
1
2
3
4
5
6
7
8
9
10
11
12
13
14
15
16
17
18
19
20
21
22
23
24
25
26
27
28
29
30
31
32
33
a b c d e

12 µm wide, at the shoulders and forming a pale border at margins; *upper cells* quadrate to transversely rectangular, 9–16(–25) µm wide, evenly thick-walled, forming narrow, 2–6 rows of cells broad, unistratose wings on both sides of costa; *marginal teeth*, if present, variable, directed upwards, 1–3 celled, with larger, acute to blunt, apical cell. *Dioicous*. *Setae* erect, flexuose, brown, 1.2–2.5 cm long. *Capsules* suberect to inclined, terete, slightly asymmetric, subglobose, ovoid to short-cylindric, 2–4 mm long, yellow-green to light brown, becoming dark brown with age, with indistinct hypophysis; *exothecial cells* quadrate to rectangular or irregular, 20–40 µm wide in mid-urn; *stomata* bicellular or with a central slit in a circular cell; *peristome teeth* 35–50, irregular, narrowly lanceolate, 140–230 µm long arising from a low basal membrane, pale to somewhat coloured in the median part. *Operculum* rostrate from a convex base. *Spores* with warty papillar projections, 16–20 µm wide. *Calyptra* hairy, covering the upper portion of the capsule.

Chromosome numbers: $n = 7$ – King George I. (Kuta *et al.* 1982; Zhou *et al.* 1995) and South Georgia (Newton 1980); the same number was recorded 13 times for the Holarctic (Fritsch 1991).

Discussion. *Polytrichastrum alpinum* was the first moss species reported from the Antarctic (Eights 1833). The species is easily distinguished in the field by the typically tall, branched plants forming wide and extensive tufts, the denticulate leaf margins and, when fertile, by the terete capsules. The papillose and strongly thickened outer walls of the uppermost border cells of the ventral lamellae are clearly visible in microscopic transverse section. These are unknown in any other regional species of the Polytrichaceae.

Polytrichastrum alpinum is a very variable species, especially in size and compactness of the turf, the extent of branching, and the size and dentation of the leaves. The leaf blades are often short, stiff and rigid, erect and appressed, with reduced dentation of the margins which may be sometimes almost entire. In contrast, large and robust plants have long, erect-spreading leaves with sharply dentate margins. However, all possible intergradations occur between these extremes and, in fact, this phenotypic variation is continuous as evidenced by worldwide biometric studies of this species (Yli-Rekola 1980). The former phenotypes have been recognised taxonomically in the Arctic as var. *brevifolium*, and some of the earliest Antarctic collections of this species were also so-named (Cardot 1901, 1906*a*, 1908*a*). In northern polar regions many environmentally induced phenotypes have been recognised taxonomically (Long 1985). In contrast, in the Antarctic only one such modification was described by Cardot (1906*a*, 1908*a*), namely *Pogonatum alpinum* fo. *elatum* which was a typical short-leaved expression of the species with relatively long stems, up to 8 cm. However, Müller (1890*a*) described five new species of *Polytrichum* from South Georgia, all of which proved to be conspecific with *P. alpinum* (Greene 1973).

The generic position of this species has long been debatable and it was usually placed either in *Pogonatum* on the basis of its terete capsules or *Polytrichum* on account of the presence of stomata and smooth exothecial cells. This confusing problem was satisfactorily resolved when Smith (1971) established the genus *Polytrichastrum* for which *P. alpinum* serves as the generitype.

Reproduction in Antarctica. Mostly sterile but capsules are produced occasionally. They are mostly immature and misshaped or malformed with aborted spores and normally dehisced capsules are exceedingly rare. Although usually of very restricted size, sexually reproducing populations appear to produce sporophytes (often only 3–5) each year, e.g. over 35 years at Signy I. and Lynch I., South Orkney Is. (Lewis Smith, personal observation). However, a single population within an area of several hundred square metres at Crater Lake, Deception I., South Shetland Is., has an exceptional abundance of sporophytes (Lewis Smith 2005*a*,*b*).

Habitat. A weedy species growing in a wide range of habitats, from dry to moist and exposed to sheltered. It grows most typically on mineral, clay and gravelly soil, on rocky or stony ground, scree, well drained slopes and moraines, in rock crevices and on ledges, on sandy glacial debris, at margins of melt-water runnels, and sometimes on humus or peat. In volcanically active areas it often grows at geothermal sites near fumaroles as well as on lava rocks, ash, lapilli and scoria. *Polytrichastrum alpinum* occurs as an associate in many different communities but it has its optimum of frequency and abundance in the moss turf subformation within which it forms a separate association occurring commonly at low altitudes in coastal areas. The species commonly forms circular colonies, formed by the radiating rhizome system. These often coalesce to form almost pure stands covering several square metres, especially on moist organic soil close to elephant seal wallows and breeding colonies of giant petrels, and around the periphery of penguin colonies. The principal bryophyte associates are *Sanionia georgicouncinata*, *Cephaloziella varians*, *Pohlia nutans*, *Ceratodon purpureus* and sometimes *Andreaea depressinevis* and, occasionally near the shore, also the vascular plants *Deschampsia antarctica* and *Colobanthus quitensis*. The species is also a frequent but seldom abundant associate in fellfield communities of the fruticose lichen and moss cushion subformation, and drier margins of the moss carpet subformation.

World range. A bipolar species (Fig. 29). In the Northern Hemisphere it has a continuous arctic-boreal-alpine range, reaching the highest possible latitudes in the Canadian Arctic Archipelago and on Greenland (Long 1985), and on Svalbard (Kuc 1973*b*), becoming rarer southwards and extending to central California (Koch 1950) and North Carolina (Ireland 1960) in North America, to the Iberian Peninsula, Sicily, northern Greece, Turkey and the Caucasus in Europe, and to the Himalayas,

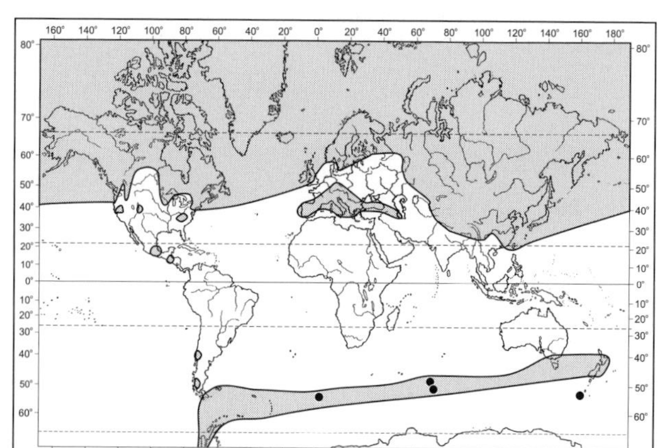

Figure 29. Global distribution of *Polytrichastrum alpinum*. The dots on the Southern Ocean indicate localities on Bouvetøya, Îles Kerguelen, Heard I. and Macquarie I.

China and Taiwan in Asia. In the Southern Hemisphere it is common on South Georgia (Greene 1973) and in the maritime Antarctic, scarcer elsewhere including the Falkland Is. (Ochyra & Broughton 2004), Tierra del Fuego (Schiavone & Hyvönen 1993), western Patagonia (Matteri 1985*a*) extending as far north as the Valdivian region in Chile (Thériot 1935; He 1998). In addition, it is known from the subantarctic islands of Kerguelen, Heard and Macquarie as well as from southwest Australia and Tasmania (Streimann & Curnow 1989; Hyvönen 2006) and both islands of New Zealand (Sainsbury 1955). It is known from intermediate stations in the neotropics in Mexico (Merrill 1994) and Guatemala (Delgadillo *et al.* 1995).

Distribution in Antarctica. One of the commonest and most widely distributed species throughout the maritime Antarctic from Bouvetøya and the South Sandwich Is. to Mica Is. at lat. 69°20′S in southern Marguerite Bay on the west coast of the Antarctic Peninsula; exceedingly rare on the east coast, extending south to James Ross I. at lat. 63°54′S (Fig. 30). It extends from sea level to about 300 m; the highest recorded collection is at 800 m on Mt. Liotard in south-eastern Adelaide I., near its southern limit.

Selected specimens examined. WEST ANTARCTICA. BOUVETØYA. Rustadkollen, 300–340 m, *TE 226, 244 & 309* (BG); Moseryggen, 260–287 m, *TE 270, 306.1–4 & 405* (BG).

SOUTH SANDWICH IS. **Leskov I.** S end of summit ridge, *ca* 65 m, *MWH 835A* (AAS, KRAM); N end of summit ridge, 100 m, *PC 254A* (AAS, KRAM); E slope of summit ridge, 100 m, *PC 274B* (AAS, KRAM). **Visokoi I.** Wordie Pt., *PC 104* (AAS, KRAM). **Vindication I.** NE coast, *ca* 100 m, *MWH 851A* (AAS, KRAM, PC). **Candlemas I.** W shore, *ca* 35 m, *REL 590* (KRAM, PC); N end of ice-cap, 50 m, *REL 683* (KRAM, PC); Clinker Gulch, *PC 158 & 182C* (AAS, KRAM); Lucifer Hill, *PC 133A* (AAS, KRAM); Breakbones Plateau, *PC 120B* (AAS, KRAM). **Montagu I.** Allen Pt., 70 m, *PC 100B* (AAS, KRAM). **Bellingshausen I.** S slopes of main cone, 100 m, *MWH 404* (KRAM, PC) and *PC 36A* (AAS, KRAM).

SOUTH ORKNEY IS. **Monroe I.** SW part, 35 m, *MMM 68A* (AAS, KRAM). **Coronation I.** Meier Pt., 5–20 m, *MMM 27* (AAS, KRAM); Mansfield Pt., *ca* 35 m, *RILS 285* (AAS, KRAM, PC); Cape Hansen, 5 m, *RILS 3215* (AAS, KRAM); Saunders Pt., *ca* 25 m, *RILS 20* (KRAM, PC). **Lynch I.** N facing slope, 10 m, *RILS 514* (AAS, KRAM). **Matthews I.** The Divide, 65–135 m, *RILS 177* (AAS, KRAM). **Coffer I.** 10–15 m, *AC & PK 9B* (AAS, KRAM). **Powell I.** Michelsen I., *ca* 23 m, *RILS 46* (AAS, KRAM). **Fredriksen I.** W face, 50 m, *DI 1090/1* (AAS, KRAM). **Saddle I.** E part, 50–150 m, *MMM 165* (AAS, KRAM). **Laurie I.** Pt. Davis, 5–35 m, *MMM 119* (KRAM); Cape Dundas, 65 m, *MMM 183* (AAS, KRAM). **Signy I.** Factory Cove, 8–15 m, *REL 826* (AAS, KRAM, PC); Berntsen Pt., *BJT 355* (KRAM, PC).

SOUTH SHETLAND IS. **Elephant I.** Endurance Glacier, 122 m, *JSA 54* (AAS, KRAM). **O'Brien I.** NE coast, 60–75 m, *JPB 29* (AAS, KRAM). **King George I.** ***Fildes Peninsula***: Ardley I., 10 m, *RO 2483/80* (KRAM). ***King George Bay***: Penguin I., *BJ 62 & 67* (KRAM); Turret Pt., *BJ127* (KRAM). ***Admiralty Bay***: Rescuers Hills, 35 m, *RO 4883/79* (KRAM); Vauréal Peak, 35 m, *RO 5237/79 & 5245/79* (KRAM). **Nelson I.** Harmony Cove, *CS 461 & 462* (PC, S); Moss Bay, 262.1989, *Komárek s.n.* (KRAM); Strachan Hill, 40 m, *DCL 855A* (AAS, KRAM). **Robert I.** Coppermine Cove, 10–30 m, *RILS 921* (KRAM, PC). **Greenwich I.** Discovery Bay, Mt. Plymouth, *RMS 69-998* (US); Yankee Harbour, 150 m, *Bellisio 14070A* (AAS, KRAM). **Aitcho Is.** *RMS 69-929 & 935* (US). **Half Moon I.** Hill S of huts, 50 m, *DCL 588A & 592* (AAS, KRAM). **Desolation I.** Near gull rookery, 20 m, *DI 1487/4* (KRAM, PC). **Livingston I.** ***Byers Peninsula***: Ocoa Pt., 20–75 m, *KR 44C* (AAS, KRAM); Sealer Hill, 10–15 m, *RILS 3812* (AAS, KRAM); South Beaches, 8 m, *DCL 436* (AAS, KRAM). ***Hurd Peninsula***: False Bay, *RMS 69-580* (US); 'Marine Lion Bay', *Pandurski 118* (KRAM). ***False Bay***: Barnard Pt., 30 m, *DCL 6* (KRAM, PC). **Deception I.** Neptunes Bellows, *ca* 15 m, *AC & PK 36* (AAS, KRAM, PC); Mt. Pond, 5 m, *BJT 14* (AAS, KRAM, PC); Collins Pt., *BJT 39* (AAS, KRAM, PC) and 20–30 m, *RILS 3620A* (AAS, KRAM); Vapour Col, 130 m, *RILS 5633A* (AAS, KRAM); Stonethrow Ridge towards Wensleydale Beacon, 75–105 m, *RILS 5798D* (AAS, KRAM);

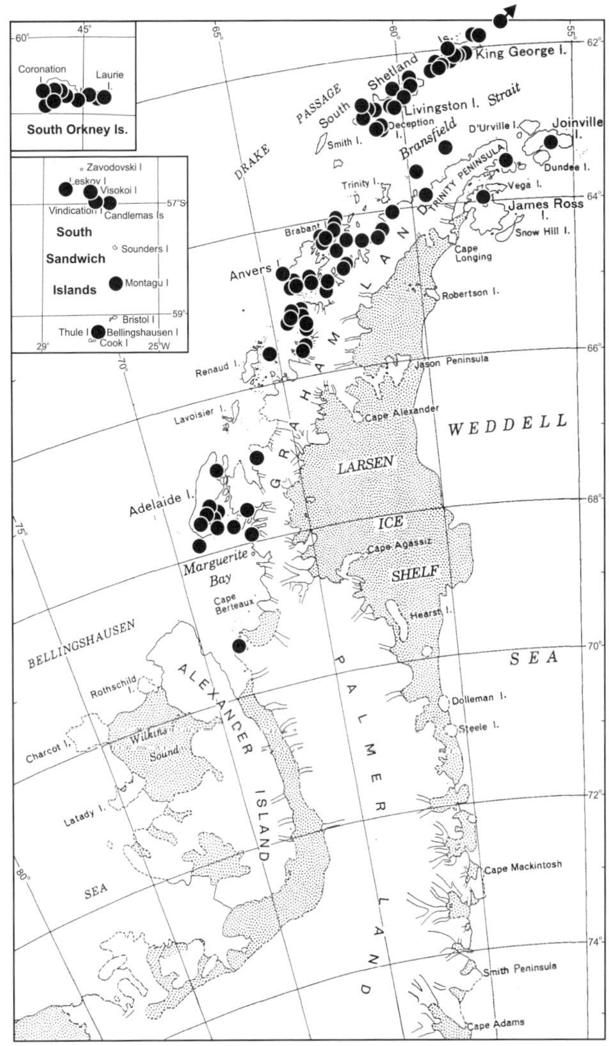

Figure 30. Distribution of *Polytrichastrum alpinum* in the Antarctic Peninsula region. The occurrence in the northern group of the South Shetland Is. (Elephant I.) is indicated by an arrow.

Crater Lake, 15 m, *RILS 5651A, 5659B & 11413* (AAS, KRAM). **Snow I.** Plateau President Head, *KR 2D* (AAS, KRAM).

WEST ANTARCTIC PENINSULA. **Davis Coast.** Astrolabe I., 10–30 m, *RILS 4092* (AAS, KRAM); Tower I., *RILS 701A* (AAS, KRAM); Almond Pt., 10–20 m, *AC & PK 77A* (AAS, KRAM); Cape Andreas, *CS 467 & 468* (PC, S). **Danco Coast.** Cierva Pt., *VK 618, 656 & 931* (KRAM); Spring Pt., *VK 913, 968, 1191, 1210, 1221, 1232 & 1258* (KRAM); Challenger I., *CS 463* (PC, S – type of fo. *elata*); Recess Cove, 2–10 m, *RILS 3983* (AAS, KRAM); Cuverville I., *ER 241* (BR, PC) and *RILS 4210* (AAS, KRAM); Beneden Head, *ER 235* (BR, PC) and 15–30 m, *AC & PK 53* (KRAM, PC); Coughtrey Peninsula, Paradise Harbour, 14.1.1985, *Birkenmajer s.n.* (KRAM). **Palmer Archipelago.** ***Melchior Is.***: Omega I., 2–10 m, *RILS 4124* (AAS, KRAM); Gamma I., 5–10 m, *RILS 4149* (AAS, KRAM). ***Brabant I.***: Metchnikoff Pt., 100 m, *WH 18A* (AAS, KRAM); Astrolabe Needle, *VK 889* (KRAM); Driencourt Pt., *AM 13A* (AAS, KRAM); Pinel Pt., 350 m, *AM 48B* (AAS, KRAM); Koch Glacier, 400 m, *AM 116* (AAS, KRAM). ***Anvers I.***: Cape Monaco, *VK 822* (KRAM); Norsel Pt., *RMS 69-848* (US); Litchfield I., *JR & JMG 25 & 28* (US); Cormorant I., 15 m, *AS 50* (AAS, KRAM) and 20 m, *AS 18* (AAS, KRAM); Wiencke I., *Turquet 339b* (PC); Goudier I., Port Lockroy, *Gain 107* (PC). **Graham Coast.** Cape Tuxen, *Gain 197A* (PC) and 300 m, *RILS 10650* (AAS, KRAM); Cape Pérez, *ca* 10 m, *RWMC 641A* (AAS, KRAM). **Wilhelm Archipelago.** ***Lemaire Channel***: entrance to the channel, *ER 270B* (BR, PC); Booth I., *Gain 126, 129, 132, 135 & 138* (PC) and *Turquet 275* (PC); Hovgaard I., 22.2.1905, *Turquet s.n.* (PC). ***Penola Strait***: Petermann I., 70 m, *AS 120B* (KRAM); ***Argentine Is.***: Galindez I., above Stella Creek, 18 m, *AS 64*

(KRAM); Corner Is., *JR & JMG 111* (US). ***Grandidier Channel***: Lahille I., 2–5 m, *RILS 4333* (AAS, KRAM). **BISCOE IS.** Hook I., 15–40 m, *RMS 69-522, 530 & 706A* (US). **LOUBET COAST.** Orford Cliff, *RILS 5283C* (AAS, KRAM). ***Bennett Is.***: Pfaff I., *JK 181* (AAS, KRAM). **FALLIÈRES COAST.** Camp Pt., 10–15 m, *RILS 4621* (AAS, KRAM). ***Adelaide I.***: Reptile Ridge, 140 m, *RILS 10576* (AAS, KRAM) and 175 m, *RILS 10624* (AAS, KRAM); Stork Ridge, *PC 269 & 271* (AAS, KRAM); Mt. Gaudry, *ca* 330 m, *JK 142A* (AAS, KRAM); Mt. Mangin, *ca* 260 m, *JK 147* (AAS, KRAM); Mt. Liotard, *ca* 830 m, *JK 189A* (AAS, KRAM). ***Léonie Is.***: Lagoon I., *VK 269* (KRAM) and 35 m, *RILS 19597* (AAS, KRAM). ***Marguerite Bay***: Blaiklock I., *Poncet 6C* (AAS, KRAM); Pourquoi Pas I., Dalgliesh Bay, *BAS Misc. 43* (AAS, KRAM); Emperor I., Dion Islets, 30 m, *DWB 17B* (AAS, KRAM); Mica Is., *DDWW 144A* (AAS, KRAM).

EAST ANTARCTIC PENINSULA. **TRINITY PENINSULA.** Hope Bay, Base Hut, *ca* 25 m, *RWMC 870B* (AAS, KRAM). **JOINVILLE ISLAND GROUP.** ***Joinville I.***: Mt. Alexander, 250 m, *RILS 3679* (AAS, KRAM); to NE of Mt. Alexander, 15–50 m, *RILS 3699* (AAS, KRAM). **JAMES ROSS ISLAND GROUP.** ***James Ross I.***: Andreassen Pt., 10 m, *RILS 7609 & 7611D* (AAS, KRAM).

2. **Polytrichastrum longisetum** (Brid.) G.L.Sm.

Fig. 31

Polytrichastrum longisetum (Brid.) G.L.Sm., Mem. New York Bot. Gard. 21(3): 35. 1971. ≡ *Polytrichum longisetum* (Sw.) *ex* Brid., J. Bot. (Schrader) 1800(1–2): 286. 1801. — TYPE CITATION: In alpibus Sueciae et Scotiae. [Lectotype (*selected here*): [icon in] Swartz (1799): f. 16a–d on pl. 8; epitype (*selected here*): '1356 Polytrichum (gracile var.) longisetum Mus. pal. bot. Stockholm Herb. bryol. Swartzii' – S-Swartz! (No. B 97703)].

Plants coarse, medium-sized to fairly large, dull to light green or brownish-green. *Stems* simple, erect, 1–6 cm high, not or moderately radiculose at or near the base with long, whitish-grey rhizoids, in transverse section rounded-pentagonal, with large, typically polytrichoid central duct. *Leaves* crowded towards the stem apex, erect and appressed, usually somewhat flexuose or curled when dry, erect-spreading to spreading or recurved when wet, from an oblong or oblong-ovate sheathing base, 1.1–1.4 mm wide, 2.0–2.5 mm long gradually or abruptly constricted into a linear-lanceolate, long-acuminate limb, 2.5–4.5 mm long, 0.5–0.7 mm wide, canaliculate when dry, flat when wet; *margins* incurved when dry, strongly serrate throughout by unicellular, ascending, spinose teeth; *costa* narrow, almost concolorous to dark brownish-yellow and often poorly delimited in sheath, widening above and occupying a major part of the leaf blade, shortly excurrent as a serrate arista, with a few scattered teeth on the back, in transverse section with two large stereid bands separated by 1–2 rows of enlarged guide cells; *ventral lamellae* variable in number from 20 to 40, straight, 3–6 cells high, in lateral view flat-topped, with end-cells undifferentiated, rounded to elliptical, smooth, thin-walled; *mid-sheath cells* rectangular, thin-walled, (30–)50–80(–90) µm long, 10–20 µm wide, becoming narrower towards the margins and short and oblate at the shoulders, to 20 µm wide; *upper cells* unistratose forming wings on both sides of costa, 5–15 cells wide, rounded-quadrate to irregular, 12–20 µm wide, with moderately incrassate walls. [*Dioicous*. *Setae* slightly flexuose, yellow to reddish, (2.0–)4.5–6.5 mm long. *Capsules* long or short, obloid-cylindrical, obtusely 5–6-angled when dry, almost terete when wet, 3–4 mm long, with distinct hypophyses; *exothecial cells* smooth, isodiametric to short-rectangular, 20–40 µm wide; *stomata* bi- or unicellular; *peristome teeth* 64 or fewer through confluence, irregularly developed, 190–230 µm long arising from a low basal membrane, pale yellowish to hyaline. *Operculum* slanted-subulate from a convex base. *Spores*, 18–21 µm wide, finely warty papillose. *Calyptra* cucullate, long-pilose, yellowish to golden-brown or brown.]

Chromosome numbers: unknown in Antarctic and subantarctic plants; $n = 7$ and $n = 14$ reported three and nine times, respectively, from Europe (Fritsch 1991).

Discussion. This species closely resembles a long-leaved form of *Polytrichastrum alpinum* in the field but, microscopically, *P. longisetum* is recognisable by the strongly serrate laminal wings which are 5–15 or more cells wide, and the entirely smooth and thin-walled, undifferentiated marginal cells of the ventral lamellae. Sporophyte-bearing plants are common in Tierra del Fuego, and these have distinctive rounded-angulate capsules which are almost terete when wet and have no constriction between the urn and hypophysis.

The species was originally described as *Polytrichum longisetum* by Swartz (1799) from Sweden and this pre-Hedwigian name was subsequently validated by Bridel (1801). Because the Bridel herbarium at B holds no material eligible for typification of this species name and there is no evidence that this author could have seen any original material in the Swartz herbarium, the illustration of this species in Swartz's (1799) work is here chosen as a lectotype and the corresponding specimen from the Swartz herbarium at S is selected as an epitype.

Reproduction in Antarctica. Sterile.

Habitat. On recently deglaciated, moist, gravelly base-rich soil at the margin of boulders, associated with *Ceratodon purpureus*, *Bryum pseudotriquetrum*, *Sanionia georgicouncinata* and *Polytrichum juniperinum*.

World range. A bipolar species (Fig. 32, inset), widespread throughout much of the Holarctic, especially frequent in the boreal zone and extending northwards to the Low Arctic, with an isolated station on Spitsbergen in the High Arctic (Kuc 1963; Bednarek-Ochyra 1993*a*), becoming rarer southwards, extending at high altitudes to California, Colorado and Arizona (Lawton 1971), the Pyrenees (Casas Sicart 1986), northern Italy (Cortini Pedrotti 2001*a*), the Caucasus and Central Siberia (Abramova *et al.* 1954), northern Mongolia (Tsegmed 2001*a,b*), north-east China, Yunnan and Sichuan (Wu & Wang 2005), Korea and Japan (Osada 1966), with some isolated stations in Sicily (Dia *et al.* 1987), the Lebanon (Arzeni 1965), China and Taiwan (Redfearn *et al.* 1996). In the Southern Hemisphere

→

Figure 31. *Polytrichastrum longisetum*. — **1**. Portion of shoot, wet. **2**. Habit, dry. **3**. Portion of cross-section of stem. **4–9**. Leaves. **10**. Marginal cells about two-thirds up leaf. **11–12**. Marginal cells in the lower part of limb. **13**. Inner laminal cells at shoulder. **14**. Laminal cells in mid-sheath. **15**. Side-view of lamella. **16**. Cross-section of lamina about two-third up the leaf. — [All from *RILS 8697*]. — Scale bars: a – 1 mm (3–9) and 50 µm (10–16); b – 2 mm (1); c – 2 mm (2).

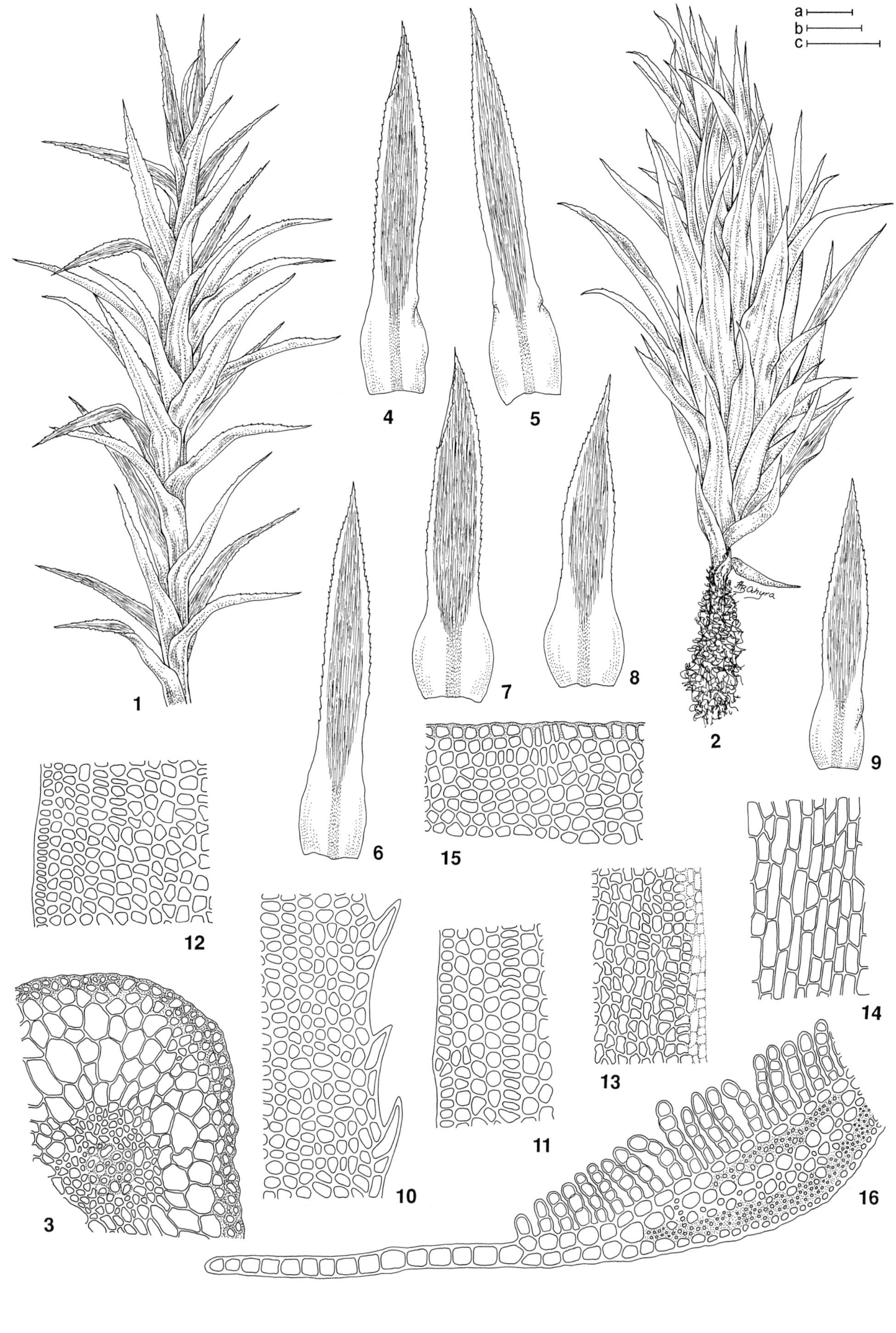
a
b
c
1
2
3
4
5
6
7
8
9
10
11
12
13
14
15
16

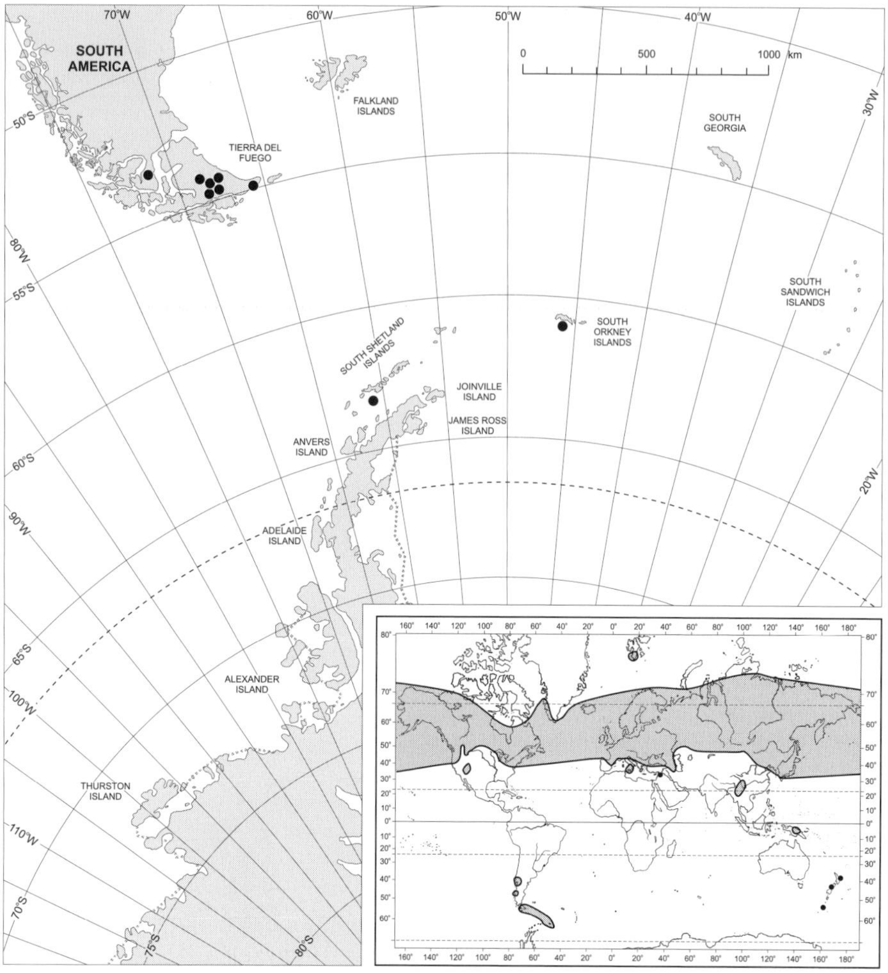

Figure 32. Distribution of *Polytrichastrum longisetum* in the Fuegian and Antarctic Peninsula region. Inset: global distribution.

Polytrichastrum longisetum is bicentric in distribution. It is extremely rare on the North (Sainsbury 1955) and South Islands (Bartlett 1985) of New Zealand and on the adjacent Auckland Is. (Sainsbury & Allison 1962; Vitt 1979*a*) in Australasia, whereas in the Western Hemisphere it is quite common and locally abundant on Tierra del Fuego (Schiavone & Hyvönen 1991) and rare and localised on the Brunswick Peninsula (Deguchi 1991*b*) and in western Patagonia extending northwards to *ca* 40° S in the Valdivian region (Herzog 1939; Deguchi 1991*b*). In the tropics the species is exceedingly rare at high elevations of New Guinea (Hyvönen 1986).

Distribution in Antarctica. Known from only two sites, and probably a recent immigrant at both (Fig. 32). On Signy I., South Orkney Is., several small colonies appeared on gravelly soil when it became exposed by a receding ice field in the 1980s above Changing Lake. In 2002 a small population was discovered on fine ashy gravel on Ronald Hill on Deception I., South Shetland Is., deposited by a lahar during an eruption in 1970 (Lewis Smith 1993*a*, 2005*a,b,c*). Because of its morphological similarity to large forms of *Polytrichastrum alpinum*, this species may be overlooked in the field.

Specimens examined. WEST ANTARCTICA. SOUTH ORKNEY IS. **SIGNY I.** Between Changing Lake and Sprindrift Col, 50 m, *RILS 8695, 8697 & 10770A* (AAS, KRAM).

SOUTH SHETLAND IS. **DECEPTION I.** Between Kroner Lake and Ronald Hill, 2–4 m, *RILS 11247* (AAS, KRAM).

4. **POLYTRICHUM** HEDW.

Polytrichum Hedw., Spec. Musc. Frond.: 88. 1801. — Lectotype (*vide* Abramov & Savicz-Lyubitskaya 1963: p. 382): *Polytrichum commune* Hedw.

Plants perennial, rigid, coarse and often robust, medium-sized to large, loosely to densely tufted or gregarious, dull, dark green to yellowish-, brownish- or bluish-green. *Stems* simple or sometimes branched, erect, radiculose only at base or sometimes densely tomentose to near the tip with whitish to whitish-grey rhizoids, in transverse section with typical polytrichoid structure. *Leaves* distant and scale-like below, crowded above; *upper leaves* from a broad, enlarged, erect, ± membranous and closely sheathing base, suddenly or gradually constricted into a limb, with a well developed hinge tissue at the junction to the limb; *limb* linear- to subulate-lanceolate, gradually long-acuminate, usually awned, erect and appressed to erect-spreading, straight or flexuose, flat or canaliculate when dry, erect-spreading, widely spreading-recurved when wet, opaque, almost totally covered on the ventral surface from the shoulders to the apex by densely packed lamellae; *margins* erect or strongly infolded and overlapping the ventral face of the limb, unbordered, entire or coarsely toothed; *costa* narrow, flat and well defined in sheath, broadened and dorsally convex in limb, usually denticulate at back toward the apex, short excurrent as a denticulate, reddish arista or spinulose, hyaline hair-point; *ventral lamellae* numerous, closely spaced, green, variously crenulate to crenate in lateral view, with end-cells sharply differentiated from lower lamellar cells, rounded, flat-topped or grooved, thin- or thick-walled, smooth; *lower cells* thin-walled, rectangular to linear, unistratose, becoming shortly oblate-oblong to irregular, strongly thick-walled at the shoulders; *upper cells* bistratose except the 2–15 cells wide margin, isodiametric to short-rectangular, smooth, thick-walled, usually oblate to transversely rectangular in infolded wings. *Dioicous* or very rarely *autoicous*. *Perigonia* conspicuous, terminal, rosulate, innovating by central proliferation. *Perichaetial leaves* differentiated, with a longer sheath. *Setae* solitary, elongate, erect to flexuose. *Capsules* suberect to inclined or horizontal, cubic or prismatic, sharply 4–6-angular, symmetric, with sharply differentiated, discoid hypophyses, separated by a deep constriction from the urn; *exothecial cells* bulging-mammillose, centrally pitted on the outer walls; *stomata* somewhat numerous at hypophysis, immersed, large, usually unicellular and consisting of a central slit; *annulus* none or uniseriate; *peristome teeth* 64, simple, vertically keeled at back, with spur-like projections; *epiphragm* remaining attached to the peristome, thin and delicate except the margin dissected into pendent sac-like lobes alternating with the peristome teeth. *Operculum* short-beaked from a flat or umbonate base. *Spores* spherical, punctate but appearing smooth, small, usually less than 10 μm wide. *Calyptra* cucullate, enclosing the whole capsule, densely pilose by a matted felt of rusty brown hairs.

A cosmopolitan genus of about 30 species, closely related to *Polytrichastrum* but differing in its sharply 4(–6)-angled capsules, deeply constricted between the urn and the hypophysis, the pitted exothecial cells, the peristome teeth keeled at back and the thin epiphragm with sac-like ventral marginal lobes. Three species occur in the maritime Antarctic, all belonging to sect. *Juniperina* I.Hagen.

1. Limb suddenly constricted at the apex into a long, flexuose hair-point, hyaline throughout or sometimes brownish at base, usually more than 0.8 mm long; sheathing base ovate........ 1. ***P. piliferum***
1. Limb gradually narrowed at the apex into a short arista, usually less then 0.8 mm long, brown throughout or sometimes hyaline at the apex; sheathing base rectangular 2
    2. Stem tomentose only in lower part; leafy shoots mostly more than 8 mm thick when moist; leaves often slightly curved, erect to erecto-patent, 4–7 mm long ... . 2. ***P. juniperinum***
    2. Stem densely matted with a tomentum of rhizoids nearly to the tip; leafy shoots usually less than 7 mm thick when moist; leaves straight and appressed, 2.5–4.5 mm long .... 3. ***P. strictum***

1. **Polytrichum piliferum** Hedw. Fig. 33

Polytrichum piliferum Hedw., Spec. Musc. Frond.: 90. 1801. — TYPE CITATION: [Europe] Locis editioribus siccioribus, in ericetis, muris argillaceis. Lipsiae ad Bieniz, Sct. Theclae, in muris pagorum, Schoenefeld. etc. also 'prope Sattelpass in Coburgo – Saxonia observavi S[chwägrichen]; in addition, references to Vaillant (1727: p. 131, *pl. 23*, *f. 7*), Dillenius (1741: p. 426, *pl. 54*, *f. 3*), Linnaeus (1763: p. 1573), Schreber (1771: p. 74) and Bridel (1798: p. 85). No material suitable for typification of this name is present in G-Hedwig/Schwägrichen. [Lectotype (*selected here*): [icon in] Dillenius (1741): fig. 3 on pl. 54; epitype (*selected here*): 'E. Bauer: Musci Europaei Exsiccati No. 1199: Polytrichum piliferum Schreb., fr. Frankreich: Le Havre, Wald von Montgeon, April 1905 legit I. Theriot' – KRAM!].

Polytrichum antarcticum Cardot, Rev. Bryol. 27: 45. 1900. — TYPE CITATION: Détroit de Gerlache: XIème débarquement; later Cardot (1901: p. 40) wrote: Détroit de Gerlache: XIème débarquement, Terre de Danco, cap Van Beneden, sur les corniches de la falaise. (No 236; 1 février 1898. [Lectotype (*selected here*): 'Herb. J. Cardot. Mousses du voyage de la Belgica. no 236. Polytrichum antarcticum Card. sp. nova Canal de la Belgica: XIème débarquement, sur les corniches de la falaise. Leg. Racovitza, 1 févr. 1898' – PC!; isotypes: BR!, PC! (2 specimens), S!]. First synonymised by Greene *et al.* (1970: p. 31, 51).

Plants loosely caespitose or gregarious, relatively short, rigid, dull, brown to bluish-green. *Stem* erect, unbranched, prominently reddish, 1–5 or sometimes more cm high, usually densely leafy upwards and forming a comose shoot, seldom regularly leafy throughout, not or moderately radiculose only at base, in cross-section obtusely 5-angled, with a large central strand and 1–2-stratose cortex of small cells with reddish and strongly incrassate walls. *Leaves* straight to slightly incurved, erect and tightly appressed when dry, erect-spreading when wet, (2.5–)3.0–4.5 mm long (without hair-points), from a broader, ovate, hyaline, closely sheathing base, 0.8–1.2 mm wide, 1.5–2.2 mm long, quite suddenly contracted into a lanceolate to oblong- or linear-lanceolate limb, 0.5–0.7 mm wide, rounded at apices and abruptly long-aristate; *margins* flat, forming broadly inflexed wings, 5–12 cells wide, enclosing totally or nearly so the costa covered with ventral lamellae, minutely crenulate to entire; *costa* broad above and occupying the entire width of limb, smooth dorsally, excurrent in a long, 0.6–1.3 mm, sharply spinulose awn, hyaline throughout or reddish-brown at the extreme base; *ventral lamellae* 25–32 covering the whole width of costa, straight, regularly crenate to irregularly crenulate-dentate in profile, 6–9 cells high, with marginal cells conic or pyriform in the median lamellae, mammillate in section and with lumina smaller than those of adjacent lower cells because of the thicker outer cell walls, forming knob-like apical thickenings, in lateral lamellae end-cells often undifferentiated, elliptical and with thin or slightly thickened walls; *sheath cells* thin-walled, quadrate or short- to elongate-rectangular, 10–15 μm wide, (15–)25–60(–80) μm long, becoming subquadrate, elliptical to transversely rectangular, thick-walled at the shoulders and hyaline at margins forming a distinct border; *limb cells* transversely rectangular with incrassate upper and lower walls resulting in narrow, oblate lumina as evident in cross-section, 10–20 μm wide, 25–35 μm long. *Dioicous*. *Setae* yellow-brown to brown, 1.5–3.0 cm long, erect, straight. *Capsules* erect and glaucous when young, becoming inclined to horizontal, reddish-brown when mature, cubic to short-prismatic, 1.5–3.5 mm long, 1.5–2.0 mm wide, sharply 4(–5)-angled, strongly constricted above a short, discoid hypophysis; *exothecial cells* short-rectangular to hexagonal, mammillose with a large, rounded or elliptical central pit; *stomata* with 2 guard cells situated in the constriction between urn and hypophysis; *annulus* present, uniseriate; *peristome teeth* 60–64, oblong, obtuse, equal, whitish, 120–150 μm long. *Operculum* stoutly beaked from a conical base. *Spores* 9–12 μm wide, echinulate but appearing smooth under the light microscope. *Calyptra* sordid-white to light-brown, enclosing the capsule.

Chromosome numbers: $n = 7$ – King George I. (Kuta *et al.* 1982) and South Georgia (Newton 1972); the same number was recorded 13 times for the Holarctic (Fritsch 1991).

Discussion. *Polytrichum piliferum* is quite variable with regard to the stature and size of the plants and leaves (Walther 1940) but its diagnostic characters remain very stable. Its short rosulate shoots with long, conspicuous, hyaline and strongly spinulose awns produced from the obtuse leaf apices allow it to be readily identified in the field. Short hair-pointed forms may be confused with *P. juniperinum* and *P. strictum* which have similar leaf characters, notably the strongly infolded margins entirely obscuring their ventral surface. Also, in some populations, for example in the type material of *P. antarcticum*, the leaf bases may be markedly reddish-brown, not hyaline as with the typical plants of the species. However, in these species the limb is acute at the apex and ends with a relatively short, stout and entirely brown arista.

Cardot (1900, 1901) described the plants from the Danco Coast as a separate species, *Polytrichum antarcticum*, on the basis of the quadrate basal sheath cells and weakly differentiated end-cells of the ventral lamellae. Later, having examined more Antarctic material, Cardot (1906*a,b*, 1907*a*, 1913) suggested it may be only a polar variety of *P. piliferum*. Greene *et al.* (1970) considered them to be conspecific, and here this taxonomic concept is confirmed since the characters shown to be diagnostic for *P. antarcticum* fall well within the range of variation of *P. piliferum*. The undifferentiated, thin- to slightly thicker-walled end cells on the lateral ventral lamellae and the brownish bases of the hair-points resemble those in the High Arctic *P. hyperboreum* R. Br. However, the latter species usually has branched stems (like those in *Polytrichastrum alpinum*), acute and channelled leaf apices, sparsely papillose costa at the back near the apex and rectangular sheathing bases with narrower and longer cells.

Polytrichum piliferum is a moss species known from the end of the 17th century when it was designated by the phrase name *Muscus trichodes minor, foliis oblongis angustis obscure viridibus in longum et praeteneum mucronem desinentibus* (Ray 1696) and *Muscus capillaceus stellatus prolifer* (Tournefort 1700). The species was given two binomial names at the same time, namely *Polytrichum piliferum* (Schreber 1771) and *P. pilosum* (Necker 1771), and the first of these was validated by Hedwig (1801). In the validating description this author gave references to some pre-1801 publications, including Vaillant (1727), Dillenius (1741), Linnaeus (1763), Schreber (1771) and Bridel (1798) and cited several specimens from Germany. Because no specimens eligible for typification of *P. piliferum* are available in the Hedwig herbarium at G (Price 2005), the illustration of this species in *Historia muscorum* of Dillenius (1741) which is cited by Hedwig (1801) in the validating diagnosis and certainly studied by this author is here selected as a lectotype of this species name. In order to dispel any doubts regarding the interpretation of this species, the specimen from France which represents a typical phenotype of *P. piliferum* is selected as an epitype.

Reproduction in Antarctica. Mostly sterile, although male plants are occasionally found, but female plants are exceedingly rare. Plants with immature sporophytes have been found only in populations from Cierva Pt. on the Danco Coast (*OB 8B & 15*, AAS, KRAM).

Habitat. Often a pioneer species of exposed, dry and damp habitats growing on stony and gravelly ground, especially on raised beaches, hillsides and moraines, sometimes on stony humus and on rock ledges, predominantly on acidic substrata, but occasionally (on the only known station in the South Orkney Is.) also on base-rich gravelly soil. At the latter site, which became deglaciated around 1980, *Polytrichum piliferum* has established several small colonies which are spreading quite rapidly (Lewis Smith 2005*c*). In general, it is mostly associated with various communities of the moss turf and fruticose lichen and moss cushion subformations, but sometimes it grows in association with *Deschampsia antarctica* in the herb tundra formation.

→

Figure 33. *Polytrichum piliferum*. — **1**. Habit. **2**. Portion of shoot, wet. **3**. Portion of cross-section of stem. **4–8**. Leaves. **9**. Stipe leaf. **10**. Base of hyaline hair-point. **11**. Apex of hyaline hair-point. **12**. Marginal cells of the infolded part of leaf lamina. **13**. Marginal cells at the leaf shoulder. **14**. Lamina cells in the inner part of the leaf shoulder. **15**. Lamina cells in the middle of sheath. **16**. Lower cells of leaf sheath at margin. **17**. An outline of cross-section of lamina about two-thirds up the leaf. **18**. Cross-section of the portion of the leaf at the junction between limb and sheath. **19–20**. Cross-sections of the portions of limb in the middle part. **21**. Side-view of lamella. **22**. Perigonium. **23–25**. Perigonial bracts. **26**. Plants with immature capsule. **27–28**. Perichaetial leaves. **29**. Calyptra. **30**. Capsule. — [1–2, 5–6, 10–16, 18, 21 from *RO 1466/80*; 3, 7, 9, 17, 19–20 from *RO 543/80*; 4 from *ER 236* (type of *P. antarcticum*), PC; 8 from *RO 5206/79*; 22–25 from *OB 20*; 26–28 from *OB 15*; 29–30 from *Roivainen 680*]. — Scale bars: a – 0.5 cm (23–25); b – 0.5 cm (30) and 1 mm (29); c – 100 μm (3), 0.5 cm (26), 2 mm (27–28) and 0.25 cm (22); d – 100 μm (10–16, 18–21) and 200 μm (17); e – 1 mm (2, 4–9) and 2 mm (1).

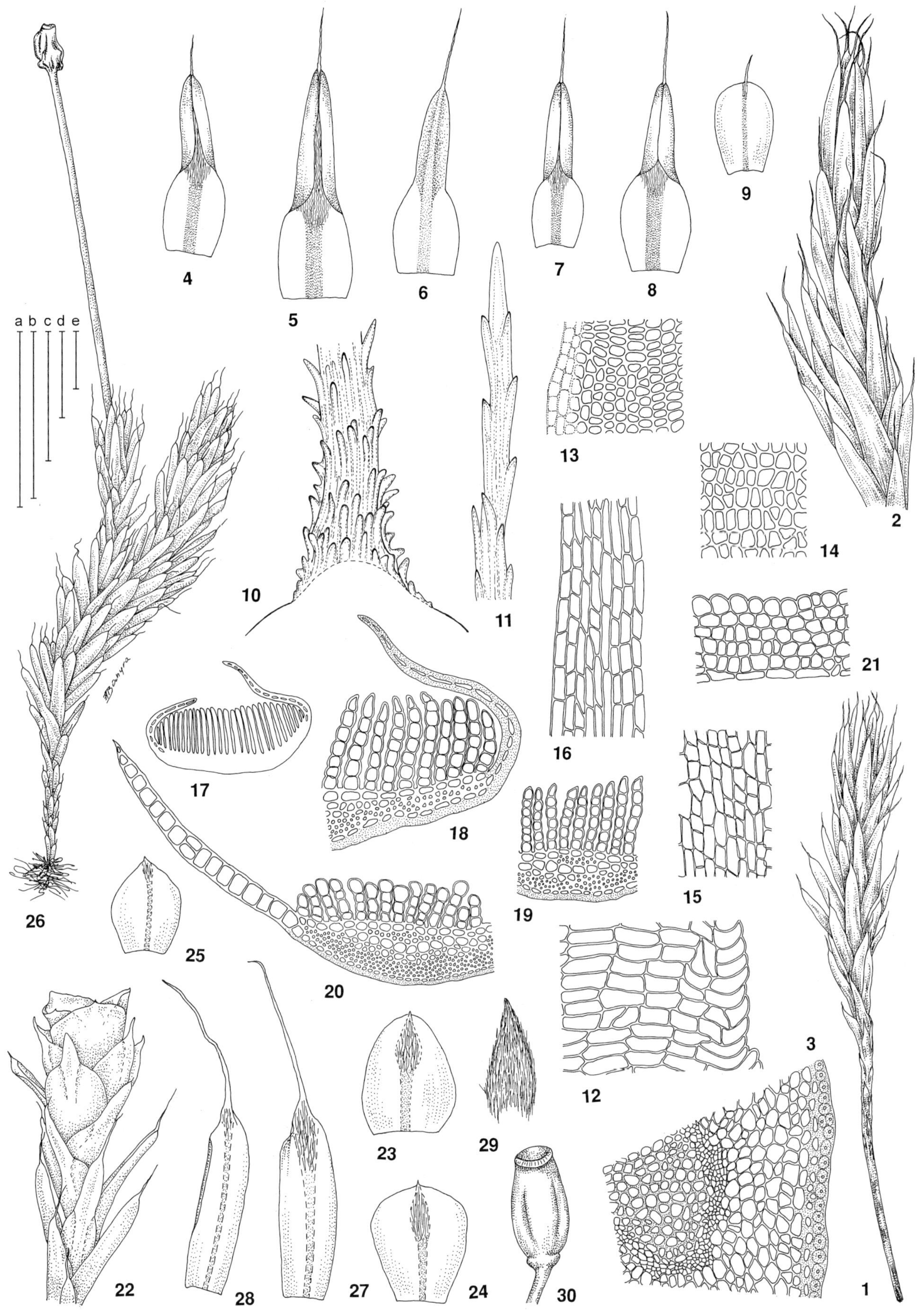
a b c d e
1
2
3
4
5
6
7
8
9
10
11
12
13
14
15
16
17
18
19
20
21
22
23
24
25
26
27
28
29
30

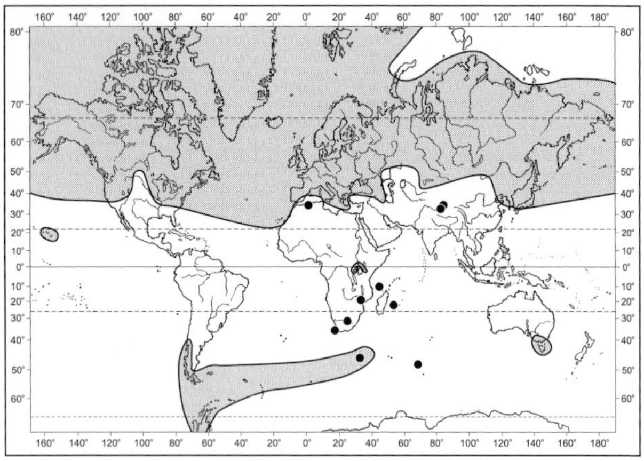

Figure 34. Global distribution of *Polytrichum piliferum*. The dots on the Southern Ocean indicate localities on Marion I. and Îles Kerguelen.

World range. Although many authors consider it to be a cosmopolitan (e.g. Dierßen 2001) or subcosmopolitan (Long 1985) species, *Polytrichum piliferum* is a strictly bipolar species with restricted occurrence in the tropics and the Southern Hemisphere (Fig. 34). It is very widespread and locally common throughout much of the Holarctic, reaching high latitudes in the Arctic and extending south to California and North Carolina in North America (with a gap in central states of the United States), the Azores and the Canary Is., the Mediterranean, northern Turkey, the Caucasus, northern Mongolia (Tsegmed 2001*a*,*b*), northwest and northeast China (Wu & Wang 2005) and central Japan (Osada 1966) with some isolated stations in Algeria in North Africa (Jelenc 1955) and Xizang (Li 1985). In the Southern Hemisphere it is frequent and has a continuous range only in southern South America along the western fringes of the continent from Bio-Bio Province to Tierra del Fuego (Schiavone & Hyvönen 1991; Schiavone 1993; He 1998), on South Georgia (Greene 1973) and in the maritime Antarctic; also on subantarctic Marion I. (Zanten 1971) and in southeast Australia and Tasmania (Streimann & Curnow 1989; Hyvönen 2006). In the tropics it is scattered only in South and East Africa from the Cape to the Ruwenzori, and in the Mascarenes (De Sloover 1986), on Réunion I. (Müller & Schäfer-Verwimp 1999). It is absent from tropical Asia and Central and South America, but has been recorded from the Hawaiian Is. (Hoe 1974).

Distribution in Antarctica. A widely distributed but scattered species in the maritime Antarctic, becoming increasingly rare towards the south and reaching its southernmost locality on the Rymill Coast at 71°40′S (Fig. 35). It is exceedingly rare on the South Sandwich and South Orkney Is. but relatively frequent on some islands in the South Shetlands archipelago and on the Danco and northern Graham Coasts; it is rare in the northern Marguerite Bay area, yet locally common at a site on Charcot I. (Convey *et al.* 2000). Moreover, it is widely scattered on the east coast of the Antarctic Peninsula, extending as far south as Francis I. off the Bowman Coast at 67°37′S. Generally, the species is most frequent and abundant at lower elevations, from sea level to about 200 m, but there are several records from 350–400 m on Brabant I. and its highest recorded locality is at about 600 m on Cape Casey on the Foyn Coast.

Selected specimens examined. WEST ANTARCTICA. SOUTH SANDWICH IS. **CANDLEMAS I.** Breakbones Plateau, *PC 110* (AAS, KRAM); between Chaimaera Flats and Breakbones Plateau, *PC 193A* (AAS, KRAM); N of western lagoon, *REL 701* (AAS, KRAM).

SOUTH ORKNEY IS. **SIGNY I.** Between Spindrift Col and Changing Lake, 80 m, *RILS 8140* (AAS, KRAM).

SOUTH SHETLAND IS. **KING GEORGE I.** ***Fildes Peninsula***: Suffield Pt., 10 m, *RO 2438A/96* (KRAM). ***King George Bay***: Turret Pt., *BJ 128* (KRAM). ***Admiralty Bay***: Llano Pt., 10 m, *RO 4864/79* (KRAM); Penguin Ridge, 30 m, *RO 1436/80* (KRAM); Stenhouse Bluff, 30 m, *RO 2581/80* (KRAM). **ROBERT I.** Coppermine Cove, 15–30 m, *RILS 922* (BM, KRAM, S). **GREENWICH I.** Discovery Bay, Spark Pt., *DCL 645A* (AAS, KRAM). **AITCHO IS.** 25–80 m, *RILS 11454* (AAS, KRAM). **LIVINGSTON I.** ***Byers Peninsula***: South Beaches, 10 m, *DCL 183* (AAS, KRAM). ***Hurd Peninsula***: Johnsons Dock, 2 m, *Redon 121A* (AAS, KRAM). **DECEPTION I.** Ronald Hill, *ca* 10 m, *RILS 67* (AAS, KRAM). **SNOW I.** President Head, *KR 5A* (AAS, KRAM).

WEST ANTARCTIC PENINSULA. **DAVIS COAST.** Almond Pt., *AC & PK 77C* (AAS, KRAM). **DANCO COAST.** Cierva Pt., *OB 8B, 15 & 20* (AAS, KRAM); Moss I., *CS 466* (PC, S, UPS); Spring Pt., *VK 1168, 1229 & 1287* (KRAM); Cuverville I., 20 m, *de Leeuw 10A* (AAS, KRAM); Beneden Head, *ER 236* (BR, PC – type of *P. antarcticum*). **PALMER ARCHIPELAGO.** ***Brabant I.***: Hunt I., *AM 51A* (AAS, KRAM); Pinel Pt., 350 m, *AM 46A & 49* (AAS, KRAM); Koch Glacier, 400 m, *AM 115* (AAS, KRAM); Mt. Bulcke, 300 m, *AM 101* (AAS, KRAM). ***Melchior Is.***: Pi Is., 15–25 m, *OG 7972A* (AAS, KRAM, NY). ***Anvers I.***: Norsel Pt., *Lippert 12* (AAS, NY); Bonaparte Pt., *VK 125* (KRAM); Litchfield I., *VK 70* (KRAM). **GRAHAM COAST.** Edge Hill, *ca* 65 m, *RWMC 648A* (AAS, KRAM); Rasmussen I., 20 m, *AS 83B* (AAS, KRAM); Cape Tuxen, *Gain 198* (PC); Mt. Demaria, 30 m, *RILS 10654* (AAS, KRAM); Cape Pérez, *ca* 25 m, *RWMC 633A* (AAS, KRAM). **WILHELM ARCHIPELAGO.** ***Collins Bay***: largest of the Berthelot Is., *ca* 130 m, *RWMC 612A* (AAS, KRAM); Darboux I., 8–150 m, *RILS*

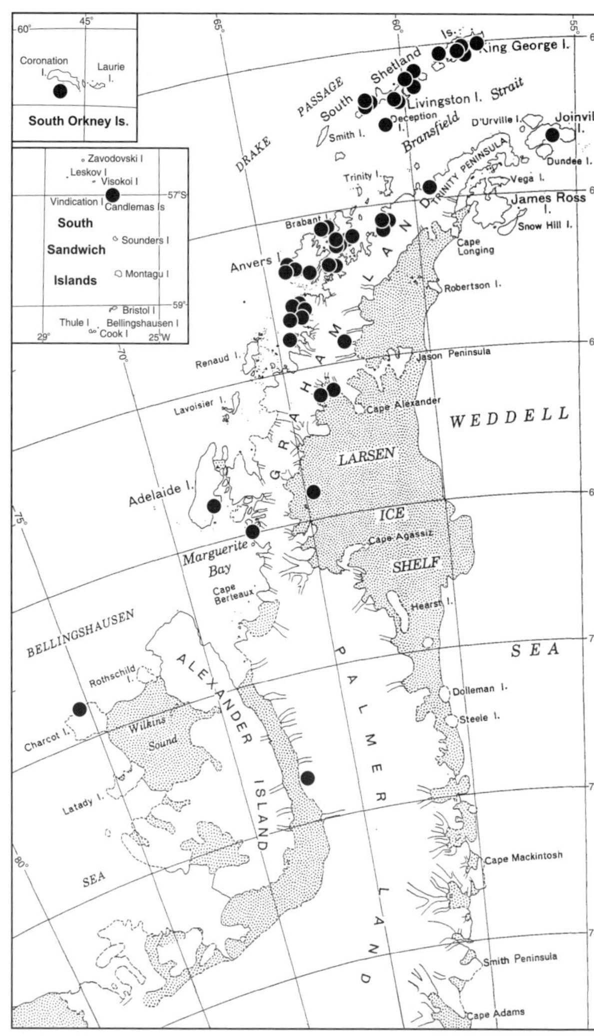

Figure 35. Distribution of *Polytrichum piliferum* in the Antarctic.

3332 (AAS, KRAM). ***Grandidier Channel***: Lahille I., 2–5 m, *RILS 4326* (AAS, KRAM). **FALLIÈRES COAST.** Camp Pt., 10–15 m, *RILS 4622A* (AAS, KRAM). ***Adelaide I.***: Stork Ridge, *PC 261A* (AAS, KRAM). **GEORGE VI SOUND.** ***Charcot I.***: Rils Nunatak, 125 m, *RILS 10473, 10712 & 10715A* (AAS, KRAM). **RYMILL COAST.** Steeple Ridge, *ca* 400 m, *BAS Misc. 220* (AAS, KRAM).

EAST ANTARCTIC PENINSULA. **JOINVILLE ISLAND GROUP.** ***Joinville I.***: NE of Mt. Alexander, 15–50 m, *RILS 3700* (AAS, KRAM). **OSCAR II COAST.** Bildad Peak, *RILS 2768A* (AAS, KRAM). **FOYN COAST.** Cabinet Inlet, Cape Casey, 330–650 m, *RT 1 & 3C* (AAS). **BOWMAN COAST.** Francis I., 325 m, *RILS 2771* (AAS, KRAM).

2. **Polytrichum juniperinum** Hedw. Fig. 36

Polytrichum juniperinum Hedw., Spec. Musc. Frond.: 89. 1801. — TYPE CITATION: European [Locus montosus, subalpestris, delineata exempla ex Helvetia; also cited Vaillant (1727: p. 130, *pl. 23, f. 6*), Buxbaum (1728: p. 42, *pl. 62, f. 1–4*), Dillenius (1741: p. 424, *pl. 54, f. 2*), Linnaeus (1763: p. 1573) and Bridel (1798: p. 83)]. No material suitable for typification of this name is present in G-Hedwig/ Schwägrichen). [Lectotype (*selected here*): [icon in] Hedwig (1801): figs 6–10 on pl. 18; epitype (*selected here*): 'E. Bauer: Musci Europaei Exsiccati No. 1549b: Polytrichum juniperinum Willd., c. fr. Et masc. Voralberg: Bregenz, linker oberer Rand des Schleifertobers auf trockenen Waldboden, 550 m s.m., 6. Mai 1916 legit Joseph Blumrich' – KRAM!].

Polytrichum subpiliferum Card., Rev. Bryol. 27: 42. 1900. — TYPE CITATION: Détroit de Magellan: environs de Punta Arenas and [...] dètroit de Gerlache: XIIème débarquement; later Cardot (1901) wrote: 'Détroit de Magellan: environs de Punta Arenas, en mélange avec l'espèce précédente; fertile. (N° 60^{a}; 9 décembre 1897). Détroit de Gerlache: XII^{ème}débarquement, Terre de Danco, à l'entrée du chenal de Lemaire, sur des roches isolées au milieu d'un glacier, à 50 mètres au-dessus de niveau de la mer; stérile. (N° 268^{c}; 12 février 1898)'. [Lectotype (*vide* Ochyra 1997*a*: p. 176): 'Herb. J. Cardot. Mousses du voyage de la Belgica n° 60a Polytrichum subpiliferum Card. sp. nova Détroit de Magellan: environs de Punta Arenas. Leg. Racovitza, 9 Déc^{bre} 1897' – PC-Cardot!; isolectotype: BR!], *pro parte* excluding syntype (*Racovitza 268C* – BR!, PC-Card!) which is *Polytrichum strictum* (*vide* Ochyra 1997*a*: p. 176). First synonymised by Greene *et al.* (1970: p. 29, 52).

Plants coarse, small or medium-sized to fairly large and robust, gregarious or in loose tufts or patches, not or somewhat lustrous, red-brown or brown throughout or only below and green to glaucous- or greyish-green above. *Stems* erect, rigid, stiff, simple or rarely forked, 1–6 or rarely up to 10 cm high, densely and evenly leafy above, with leafy shoots 8–11 mm wide when wet, 2.2–4.0 mm when dry, bracteate to naked towards the base, rhizoidous only at base or rarely with a tomentum of pale brownish rhizoids extending into the lower part of the leafy region, in cross-section obscurely pentagonal above, with typical polytrichoid central strand and 2–3-stratose cortex of small cells with brown, strongly incrassate walls and small rounded lumina. *Leaves* crowded in upper ¾ of stem, 4–6 mm long, mostly erect and appressed, straight or slightly curved, sometimes widely spreading to recurved when dry, erecto-patent to patent when wet, consisting of a well defined, erect, oblong, membranous, hyaline, yellowish to reddish-brown and hyaline-margined sheath, 1.8–2.5 mm long, 1.5–1.8 mm wide, closely clasping stem and rather abruptly narrowed into a stout, thick, narrowly lanceolate limb, 2.5–4.0 mm long, 0.5–0.8 mm wide, becoming canaliculate near the subulate apex and gradually tapering into a stout, short, toothed arista; *margins* broadly inflexed from the shoulders to the base of the awn, finely serrulate to crenulate, 6–9 cells wide, membranous, transparent, overlapping or nearly so and almost entirely covering the ventral surface of the limb; *costa* sharply denticulate dorsally towards the apex, excurrent as a short, stout hair-point, 0.4–0.9 mm long, reddish-brown throughout or pale to hyaline at the extreme apex, denticulate to nearly smooth; *ventral lamellae* 25–40, arising only from the broadened costa, straight, notched to crenulate in profile, 6–8(–10) equal-sized cells high, terminating with smooth, ovoid-conic to pyriform cells as seen in transverse section and ending as a distinct, incrassate knob in central lamellae and ovoid and thinner-walled in the lateral lamellae; *lower cells* in mid-sheath elongate-rectangular, thin-walled, 7–12 μm wide, 50–100 μm long, becoming isodiametric to transversely rectangular at the shoulder and hyaline at the margins, forming a distinct border of 4–8 rows of cells; *upper cells* unistratose forming infolded wings, 5–8 cells wide, oblate to irregular at margins, obliquely oriented, 10–15 μm wide, 30–45 μm long, with strongly incrassate walls. *Dioicous. Perigonia* terminal; *perigonial bracts* with broad sheaths and very shortened limbs. [*Setae* stout, red to orange-brown, smooth, erect, 2.3–5.5 cm long. *Capsules* suberect to horizontal, 3–5 mm long, 1.5–2 times longer than broad, prismatic, sharply 4-angled, reddish-brown when mature with a glaucous hue when fresh, deeply constricted between the urn and the hypophysis; *exothecial cells* not or somewhat mammillose, with a rounded or elliptical central spot; *annulus* none; *peristome teeth* 64, oblong, obtuse, 210–230 μm long, divided to 0.8 and arising from a low basal membrane, keeled at back. *Operculum* stoutly rostrate to subulate from a plano-convex base. *Spores* 8–12 μm in diameter, smooth. *Calyptra* white to light brown, pilose.]

Chromosome numbers: $n = 7$ – King George I. (Kuta *et al.* 1982); the same number was recorded 16 times for the Holarctic (Fritsch 1991) and, additionally, $n = 14$ was once reported from Japan and three old counts $n = 6$ are known from Japan and Europe, respectively (Fritsch 1991).

Discussion. *Polytrichum juniperinum* is very variable throughout its wide geographical range and this prompted Papp (1933) and Walther (1934) to distinguish several infraspecific taxa. They are no longer accepted in modern moss taxonomy since they refer to such characters as the habit and colour of the plants, the dentation of the hair-point, the leaf shape and arrangement and the shape of the capsule, all of which are subject to variation induced by environmental factors. The broadly inflexed leaf margins covering the whole ventral surface of the limbs and the stout, short, entirely reddish-brown and toothed hair-points are immediately diagnostic for *Polytrichum juniperinum*.

The only regional species for which *Polytrichum juniperinum* is likely to be mistaken is *P. strictum*. Typical plants of *P. juniperinum* differ in the broader, shorter and loosely aggregated shoots which are only brownish radiculose at the bases and lack a wooly tomentum of greyish-white rhizoids in the leafy parts. Additionally, the ventral lamellae are more numerous,

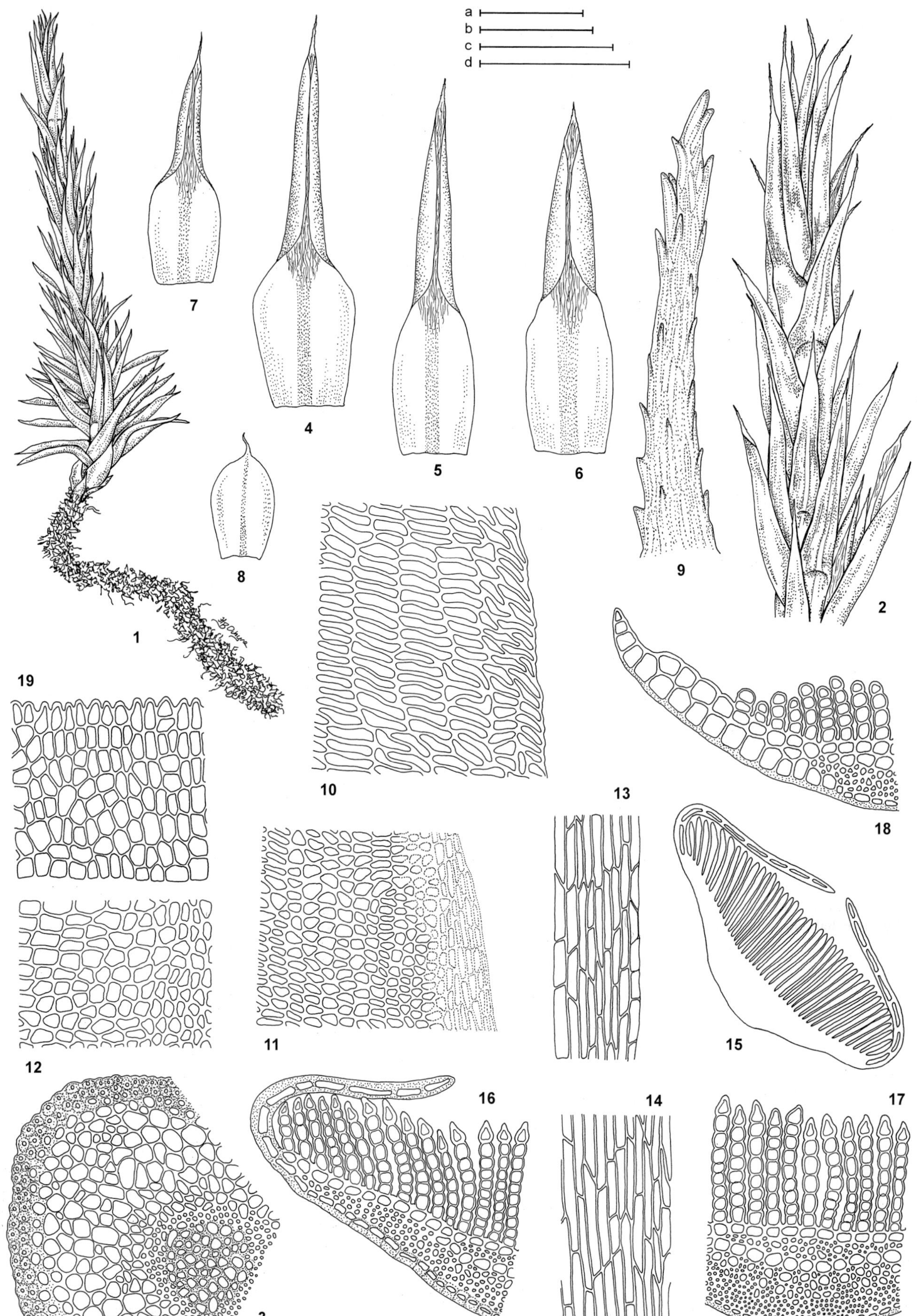
a
b
c
d
1
2
3
4
5
6
7
8
9
10
11
12
13
14
15
16
17
18
19

the leaves are longer and occasionally curved, the capsules are longer and the spores smaller, but the sporophyte characters cannot be tested in the Antarctic due to the sterile condition of the plants.

Although the extreme phenotypes of both species do not pose serious problems in identification, the determination of intermediate plants may sometimes be difficult. The is exampliﬁed by the type material of *Polytrichum subpiliferum*. Cardot (1900, 1901) described this species on the basis of the specimens from Tierra del Fuego and the Antarctic and compared them with *P. piliferum* with which it shared the hyaline uppermost part of the leaf arista. The fertile Fuegian plants represent *P. juniperinum*, whereas the Antarctic plants clearly belong within *P. strictum*, as correctly observed by Greene *et al.* (1970) and, in order to preserve the synonymy proposed by these authors, Ochyra (1997*a*) selected the Fuegian plants as lectotype. The presence of the hyaline uppermost parts in the hair-points is sometimes observed in other Antarctic mosses, for example in *Ceratodon purpureus*, *Hymenoloma antarcticum* and *Bryum pseudotriquetrum*. This tendency to 'hyalinisation' of leaf apices may be interpreted as a response to severe climatic conditions.

Polytrichum juniperinum has been known since the earliest days of modern botany. It was first distinguished by Thal (1588) as *Adiantum aphyllon medium* and later by Bauhin (1623) as *Polytrichum aureum medium* and a full list of early synonyms of this species name was presented by Dillenius (1741). The species was given the present binomial name by Willdenow (1787) and it was subsequently validated by Hedwig (1801). In the Hedwig herbarium at G there is no material of this species which could be seen by this author. The only existing specimen is labeled by F. Schwägrichen (Price 2005) and it is not eligible for typification of *Polytrichum juniperinum*. Therefore, the illustration of this species in *Species Muscorum Frondosorum* of Hedwig (1801) is here chosen as a lectotype and a specimen from Voralberg in Austria is here selected as an epitype. It originates from the conterminous territory of Switzerland from where originated the material which was used by Hedwig (1801) to illustrate his *P. juniperinum*.

Reproduction in Antarctica. Only perigonia occasionally observed; perichaetia and sporophytes unknown.

Habitat. A pioneer species growing on gravelly soil and humus, glacial debris, volcanic ash and cinders in dry or moist and generally open habitats, sometimes at margins of melt water rills and in seepage areas. On volcanic islands it grows most often in geothermal areas around fumaroles and steaming vents. It is an occasional associate of *Deschampsia antarctica* swards and is a colonizer of fellfield communities in the fruticose lichen and moss cushion subformation.

World range. As with *Polytrichum piliferum*, *P. juniperinum* is generally considered to be a cosmopolitan species, but it is actually a bipolar species, common in polar to temperate areas of both hemispheres with some intermediate montane stations in the tropics (Fig. 37). It has a continuous range throughout the

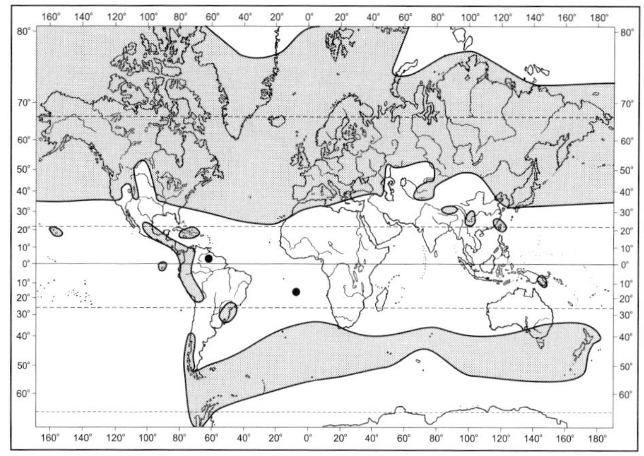

Figure 37. Global distribution of *Polytrichum juniperinum*.

Holarctic extending as far south as central California, Arizona and Georgia in North America (Crum & Anderson 1981), to Macaronesia (Eggers 1982), North Africa (Ros *et al.* 1999), northern Turkey (Henderson & Prentice 1969), the Caucasus and mountainous areas of Central Asia and Central Siberia (Abramova *et al.* 1954), northern Mongolia (Tsegmed 2001*a*,*b*), Jilin in northeast China (Redfearn *et al.* 1996; Wu & Wang 2005) to Japan (Osada 1966), with some isolated stations in the Himalayas (Gangulee 1969), Xizang (Li 1985), Yunnan (Wang *et al.* 2000), as well as Sichuan and Xinjiang (Wu & Wang 2005). In the Southern Hemisphere it is widespread in the temperate areas of southern South America from central Chile to Tierra del Fuego (He 1998; Schiavone 1993), in southwest Australia, Tasmania and Lord Howe I. (Streimann & Curnow 1989; Hyvönen 2006) and throughout New Zealand and adjacent islands (Sainsbury 1955), on some islands of the Southern Ocean including South Georgia (Greene 1973), Tristan da Cunha (Dixon 1960), Gough I. (Wace & Dickson 1965), Marion I. (Zanten (1971), Îles Crozet (Hébrard 1970*a*,*b*), Île Amsterdam (Tixier 1980), Campbell I. (Vitt 1974*a*), Auckland Is. (Vitt 1979*a*) and Macquarie I. (Seppelt 2004) and extending to the maritime Antarctic. In the tropical zone *P. juniperinum* occurs only in the mountains and it is only widespread in the Neotropics along the Andean Cordillera from central Mexico to Bolivia and on some islands in the West Indies including Cuba, Jamaica and Hispaniola (Delgadillo *et al.* 1995) as well as in southeast Brazil (Yano 1981), whereas in Africa it is rare in Cape Province (Sim 1926; De Sloover 1986) and in Asia it occurs only in New Guinea (Hyvönen 1986). Additionally, it is known from the Hawaiian Is. (Bartram 1933*a*; Hoe 1974) and the Galapagos Is. (Gradstein & Weber 1982) in the Pacific Ocean, and also from St. Helena in the Atlantic Ocean (Geheeb 1910).

Distribution in Antarctica. This is the least common of all species of the genus, being largely confined to the northern maritime Antarctic (Fig. 38). It is locally frequent on Candlemas I. and Bellingshausen I. in the South Sandwich Is. and on King George I. in the South Shetland Is., but scarce elsewhere in that archipelago and in the South Orkney Is. (mainly Signy I.).

◄——

Figure 36. *Polytrichum juniperinum.* — **1**. Habit. **2**. Portion of stem, wet. **3**. Portion of cross-section of stem. **4–7**. Leaves. **8**. Scale-like leaf. **9**. Apex of leaf hair-point. **10**. Marginal cells of infolded part of leaf lamina. **11**. Marginal leaf cells at shoulder. **12**. Laminal cells in inner part of leaf shoulder. **13**. Lower cells of leaf sheath at margin. **14**. Laminal cells in mid-sheath. **15**. An outline of cross-section of lamina about ⅔ up the leaf. **16–17**. Portions of cross-sections of mid-leaf. **18**. Cross-section of portion of leaf at junction between limb and sheath. **19**. Side-view of lamella. — [1–2 from *RO 1521/80*, 3, 7–9, 15–19 from *RO 4995/79*, 4 from *RO 5220/79* and 5–6, 10–14 from *RO 1826/80*]. — Scale bars: a – 100 μm (9); b – 100 μm (10–14, 16–18) and 200 μm (15); c – 0.5 cm (1), 2 mm (2, 4–8) and 50 μm (19); d – 100 μm (3).

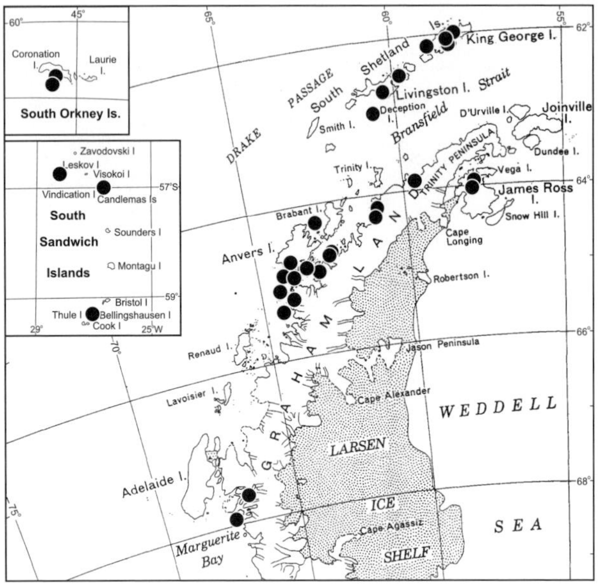

Figure 38. Distribution of *Polytrichum juniperinum* in the Antarctic.

Additionally, it occurs at scattered sites on the north-west coast of the Antarctic Peninsula and offshore islands from Charcot to Beascochea Bays, with two isolated stations on the Fallières Coast. It extends south to 67°58′ S at Camp Pt. and, on the east coast of the Peninsula, it occurs on James Ross I. It is a species of low elevations, from sea level to about 50 m, occasionally ascending to about 200 m, for example at Signy I. and at Cape Tuxen on the Graham Coast.

Selected specimens examined. WEST ANTARCTICA. SOUTH SANDWICH IS. **Leskov I.** Summit ridge, 100 m, *PC 2451 & 249B* (AAS, KRAM). **Candlemas I.** N part, *ca* 35 m, *MWH 443* (AAS, BM, KRAM, S); Lucifer Hill, *PC 127C* (AAS, KRAM); Breakbones Plateau, *PC 194, 196B, 200C & 209A* (AAS, KRAM); Clinker Gulch, *PC 162B* (AAS, KRAM). **Bellingshausen I.** Active fumaroles, 150 m, *PC 41B, 53C & 73A* (AAS, KRAM).

SOUTH ORKNEY IS. **Coronation I.** Cape Hansen, 20–200 m, *RILS 3234* (AAS, KRAM). **Signy I.** Between Spindrift Col and Changing Lake, 80 m, *RILS 8139* (AAS, KRAM); Thulla Pt., 65 m, *MWH 154* (AAS, KRAM).

SOUTH SHETLAND IS. **King George I. *Fildes Peninsula*:** Ardley I., 5 m, *RO 2482B/80* (KRAM). ***Admiralty Bay*:** Penguin Ridge, 40 m, *RO 1647/80* (KRAM); Crépin Pt., 25 m, *RO 2039/80* (KRAM); Keller Peninsula, Speil Pt., 2 m, *RO 536/80* (KRAM). **Half Moon I.** 12 m, *RILS 11936* (AAS, KRAM). **Livingston I. *Hurd Peninsula*:** South Bay, *Schulz 182* (KRAM). **Deception I.** Ronald Hill, 10 m, *RILS 3639* (AAS, KRAM); abandoned British Station, 4 m, *RO 2764/80* (KRAM); Crater Lake, 30 m, *RILS 11414* (AAS, KRAM).

WEST ANTARCTIC PENINSULA. **Davis Coast.** Almond Pt., 5–20 m, *AC & PK 77B* (AAS, KRAM). **Danco Coast.** Spring Pt., *VK 1276* (KRAM); Cuverville I., *ca* 25 m, *de Leeuw 13B* (AAS, GRO, KRAM); Birdsend Bluff, 5–35 m, *RILS 4217B & 4231* (AAS, KRAM); Coughtrey Peninsula, Paradise Harbour, *VK 344, 1013 & 1024* (KRAM). **Palmer Archipelago. *Melchior Is.*:** Omega I., 2–10 m, *RILS 4125* (AAS, KRAM). ***Anvers I.*:** Norsel Pt., *VK 954, 986 & 1071* (KRAM) and 20 m, *RILS 1938* (KRAM); Shortcut I., *VK 61, 1044 & 1047* (KRAM); Litchfield I., *VK 1158* (KRAM); Wiencke I., Damoy Pt., 10 m, 1985, *Birkenmajer s.n.* (KRAM). **Graham Coast.** Cape Tuxen, *ca* 200 m, *RILS 4897* (AAS, KRAM). **Wilhelm Archipelago. *Argentine Is.*:** Uruguay I., 10–25 m, *RILS 4269* (AAS, KRAM) and *REL 1370* (AAS, KRAM, LE, NY, PC). ***Collins Bay*:** Darboux I., 8–150 m, *RILS 3331* (AAS, KRAM). **Fallières Coast.** Camp Pt., 10–15 m, *RILS 4640* (AAS, KRAM).

EAST ANTARCTIC PENINSULA. **James Ross Island group. *James Ross I.*:** S side of St. Martha Cove, 1–2 m, *RILS 7475* (AAS, KRAM); valley to N of St. Martha Cove, 50 m, *RILS 7510A* (AAS, KRAM), 60 m, *RILS 7570* (AAS, KRAM) and 80 m, *RILS 7645* (AAS, KRAM).

3. **Polytrichum strictum** Brid. Fig. 39

Polytrichum strictum (Menzies) *ex* Brid., J. Bot. (Schrader) 1800(1–2): 286. 1801 (IV). ≡ *P. juniperinum* Hedw. var. *strictum* (Brid.) Röhl., Deutschl. Fl. (Ed. 2) Krypt. Gew. 3: 58. 1813. ≡ *P. juniperinum* subsp. *strictum* (Brid.) Nyl. & Saelán, Herb. Mus. Fenn. (Ed. 2) Musci: 65. 1859. — Type citation: In America septentrionali et Scotia. No material suitable for typification of this name is present in the Bridel herbarium at B. [Lectotype (*selected here*): [icon in] Menzies (1798): fig. 1 on pl. 7; epitype (*selected here*): 'Newfoundland: Logy Bay, Avalon Peninsula, just north of St. John's, lat. 47°38′ N, long. 52°40′ W, shrub tundra with fen seeps and acidic rock outcrops, 3 July 1984, leg. D. H. Vitt No 31505' – KRAM!].

Polytrichum alpestre Hoppe, Bot. Taschenb. 1801: 198. 1801 (IV). ≡ *P. strictum* Brid. var. *alpestre* (Hoppe) Rabenh., Deutschl. Krypt. Fl.: 238. 1848. — Type citation: An felsigen Orten auf dem Untersberge bei Salzburg. [Lectotype (*selected here*): 'Polytrichum alpestre Hopp. in sumo jugo montis Untersberg' – M!]. First synonymised by Host (1831: p. 718).

Plants rather slender but tall and rigid, in deep, dense and usually extensive turves, brownish to yellowish-green. *Stems* erect, simple or sparingly branched, 2–20 or more cm high, densely matted with a thick, woolly, whitish to brownish tomentum of rhizoids extending well up among the leaves, sometimes to within about 1 cm of the stem tip, leafy shoots 1–3 mm wide when dry, 2–7 mm when wet, with a dense foliage for most of their length, 0.3–04 mm thick, in cross-section rounded, with a large, polytrichoid central strand and a poorly defined cortex composed of relatively large, thin-walled cells. *Leaves* densely set throughout, straight, stiffly erect to closely imbricate, rarely erect-spreading when dry, erecto-patent when wet, (2.0–)2.5–4.5(–6.0) mm long, from an oblong or ovate-oblong, erect, brownish, hyaline-margined base closely sheathing the stem, 1.1–2.2 mm long, 0.7–1.2 mm wide, ± abruptly constricted into a narrowly lanceolate limb, 0.2–0.5 mm wide, flat below and becoming channelled towards the apex, gradually tapering into a short, subulate arista, with broadly infolded, membranous laminal wings, overlapping or nearly so and obscuring the ventral lamellae; *margins* minutely crenulate to entire;

→

Figure 39. *Polytrichum strictum*. — **1**. Habit. **2**. Portion of stem. **3**. Cross-section of stem. **4–10**. Leaves. **11**. Apex of leaf hairpoint. **12**. Marginal cells of infolded part of leaf lamina. **13**. Marginal leaf cells at shoulder. **14**. Laminal cells in mid-sheath. **15**. Cross-section of portion of leaf at junction between limb and sheath. **16**. Cross-section of mid-leaf. **17**. Side-view of lamella. **18–19**. Perichaetial leaves. **20**. Capsule. **21**. Exothecial cells in mid-urn. **22**. Peristome teeth, spores and cells at orifice. — [1–2 from *Sladen 1091/19*; 3–6, 11–13, 15, 17 from *BJT 130*; 7, 16 from *ER 238C* (syntype of *P. subpiliferum*), BR; 8 from *ER 234A*, BR; 9 from *RO 2494/80*; 10, 14 from *VK 1078*; 18–22 from *CS 144*, PC]. — Scale bars: a – 1 mm (2); b – 1 mm (20); c – 1 mm (4–10); d – 100 μm (3, 11–17, 21–22); e – 0.5 cm (1) and 2 mm (18–19).

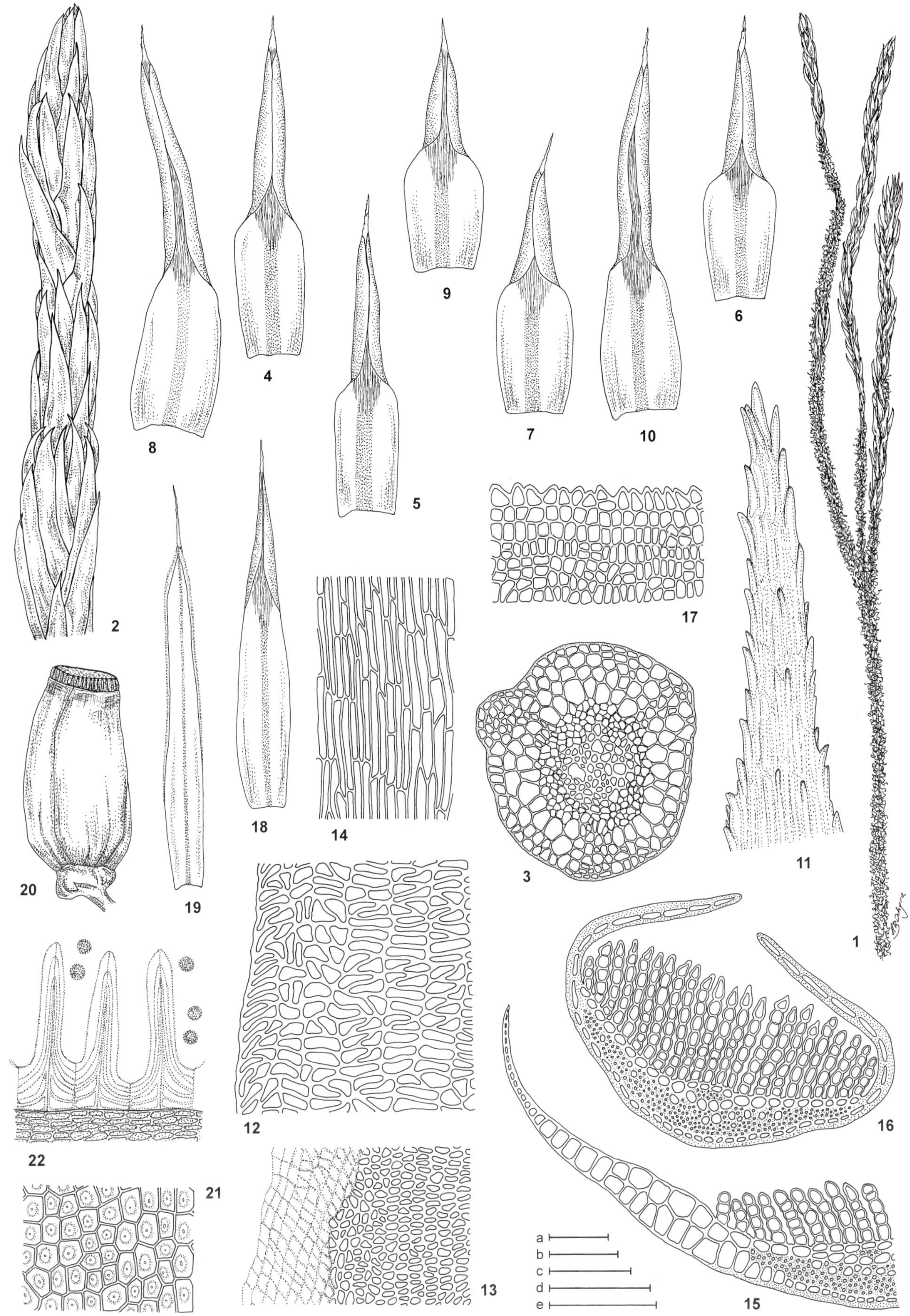
1
2
3
4
5
6
7
8
9
10
11
12
13
14
15
16
17
18
19
20
21
22
a
b
c
d
e

costa denticulate dorsally in the upper half or only near the apex, excurrent as a short, stout hair-point, 0.3–0.6 or rarely more mm long, entirely reddish-brown or sometimes pale or hyaline at the extreme apex, with short teeth throughout to almost smooth; *ventral lamellae* 21–35, borne only over the broadened costa, 6–8 cells high, bluntly and irregularly crenate in side-view, with marginal cells of median lamellae pyriform in cross-section and ending with a distinct, knob-like apical papilla, and ovate to elliptical, thin-walled in lateral lamellae; *mid-sheath cells* rectangular to linear-rectangular, 40–85 µm long, 7–12 µm wide, becoming short, isodiametric, thick-walled at the shoulders and hyaline towards the margins, forming 5–7-seriate borders; *limb cells* unistratose, forming membranous, infolded wings, 6–8 cells wide, irregular to oblate, 8–12 µm wide, 30–45 µm long. *Dioicous*. *Perigonial bracts* with broad sheaths and short limbs, forming conspicuous rosettes. *Perichaetial leaves* elongate with long, narrow sheaths passing gradually into short, aristate limbs. *Setae* stout, erect, straight, yellowish-brown, 0.5–2.2 cm long. *Capsules* inclined to horizontal, shortly obloid to almost cubic, 2–3 mm long, 1.2–1.5 mm wide, reddish-brown, sharply 4-angled, with distinct hypophyses separated by a deep constriction from the urns; *exothecial cells* isodiametric, firm-walled, subquadrate to hexagonal, mammillose, with a central, rounded or elliptical pore; *peristome teeth* 64, narrow, oblong, obtuse, equal, whitish, 110–150 µm long, arising from a low basal membrane. *Operculum* stoutly beaked from a convex base. *Spores* smooth, (8–)10–12 µm in diameter. *Calyptra* pilose, dirty white to light brown, enclosing the capsule.

Chromosome numbers: unknown in the Antarctic and Subantarctic; $n = 7$ and $n = 14$ have been found three times in Europe and Japan and once in Europe, respectively (Fritsch 1991).

Discussion. When typically developed, *Polytrichum strictum* does not pose any problems with identification by its characteristic growth form. The plants form tall and compact turves in which the stems are tightly bound together by the thick felt of rhizoids covering the greater part of their length. However, in some depauperate or juvenile plants the tomentum may be underdeveloped or reduced, and such plants may be confused with *P. juniperinum*. In those, the leaves have shorter limbs and broader sheaths and the imbricate foliage makes the leafy shoots generally more slender and less than 7 mm wide on wetting. The sporophytes, if present, have relatively stable and constant characters, with small and almost cubic capsules. Unfortunately, this character cannot often be verified in Antarctic material as it is very rarely fertile and, even then the capsules are usually under-developed.

Polytrichum strictum is a sibling species of *P. juniperinum* and often it is considered to be merely a well marked ecological variant of the latter. The problem of the status of this species needs further investigation, including genetic and molecular techniques because traditional methods do not seem to be satisfactory. For example, on the basis of multiannual observations on permanent plots Abolin' (1985) concluded that *P. strictum* is merely a phenotype of *P. juniperinum* conditioned by changes in habitat moisture. The problem seems to be more complex and sometimes plants growing in drier conditions, including in the Antarctic, retain their characteristic features. Therefore, *P. strictum* is accepted here as a species in its own right since it is well separated both morphologically and ecologically from *P. juniperinum*.

Besides the taxonomic problems, there also exists a nomenclatural problem associated with *Polytrichum strictum*, for which another frequently used name is *P. alpestre*. By an unusual coincidence both names were published simultaneously on 19 April 1801 (Sayre 1959) by Bridel (1801) and Hoppe (1801), respectively. Host (1831) was the first to choose *P. strictum* as the correct name for this species and placed *P. alpestre* in this taxon. Hence, according to Article 11.5 of the current Code (McNeill *et al.* 2006) *P. strictum* has priority.

Polytrichum strictum was described by Menzies (1798) on the basis of specimens from England, Scotland and Nova Scotia in Canada and the latter was illustrated in his paper. In fact, the species was first collected by J. Banks in Newfoundland, and he introduced this epithet in his unpublished manuscript (Menzies 1798). This name was subsequently validated by Bridel (1801) who apparently accepted this species on the basis of its description and illustration since no pre-1801 material is present in his personal herbarium at B. Therefore, the illustration of this species presented in the work of Menzies (1798) is here selected as a lectotype and a modern specimen from Newfoundland is here chosen as an epitype.

Reproduction in Antarctica. Gametangia occur occasionally but often remain in the immature stage. Sporophytes are known from a single population at Norsel Pt., Anvers I., where it was noted with aborted sporophytes in 1963 by Longton (1966), and in 1967 and again in 1981 by Lewis Smith (1996). Detailed observations of reproduction of *Polytrichum strictum* and other species of this family in Antarctica and on South Georgia have been presented by Longton (1972). In the latter area sporophytes are produced in profusion.

Habitat. *Polytrichum strictum* generally grows on acidic mineral soils. However, it also colonises thick turves of *Chorisodontium aciphyllum* on well drained slopes or gravelly porous ground. As the cushion-shaped turves coalesce, they form extensive thick turf banks of semi-ombrogeneous peat which accumulate organic matter often up to 1–2 m in depth (Lewis Smith 1972, 1979, 1996; Collins 1976*a*,*b*; Fenton 1980, 1982; Fenton & Lewis Smith 1982). In moister situations it forms turf mounds up to 15 cm high which usually coalesce to form small undulating stands. In the South Sandwich Is. it is a rare component of vegetation around fumaroles.

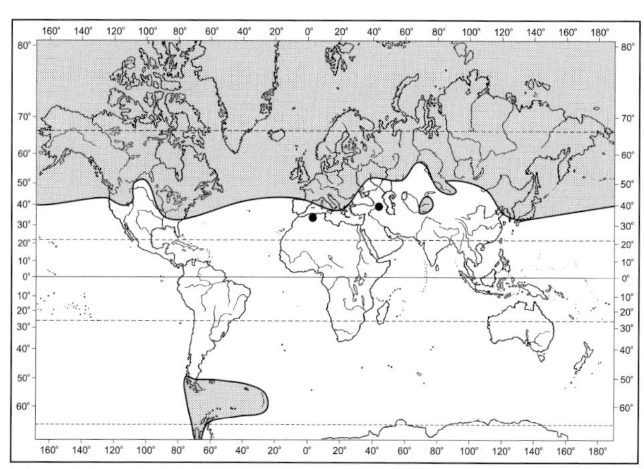

Figure 40. Global distribution of *Polytrichum strictum*.

The thick moss turf banks formed by *Polytrichum strictum* and *Chorisodontium aciphyllum* are a unique feature of the northern Maritime Antarctic vegetation. The largely undecomposed moss accumulates to depths of 1–2 m, exceptionally to almost 3 m on Elephant I., South Shetland Is. (Lewis Smith 1972; Fenton & Lewis Smith 1982). These banks stand above the level of the terrain, rather than infilling depressions, and possess a permafrost below the upper active layer of *ca* 25 cm depth. These are ancient features, and peat from the base of 1.5 deep banks on Signy I. and Elephant I. has been radiocarbon dated at *ca* 5000 and 5500 years old, respectively (Fenton & Lewis Smith 1982; Björck *et al.* 1991).

World range. A bipolar species, widespread throughout much of the Holarctic, especially in the arctic and boreal zones, becoming rarer southwards and roughly extending as far as the boundary of *Sphagnum*-dominated peat bogs with which it is mainly associated (Fig. 40). In North America its southern boundary extends to Washington, Colorado, Ohio, Illinois and Georgia (Smith Merrill 2007) and in Eurasia to the Pyrenees (Casas Sicart 1986), Sicily (Dia *et al.* 1987), Macedonia (Pavletić 1955), Bulgaria (Petrov 1975) and northern Greece (Duell 1995), the steppe zone of the Ukraine (Bachurina & Mel'nichuk 1987) and Russia (Abramova *et al.* 1954), northern Mongolia (Tsegmed 2001*a*,*b*), northwest and northeast China (Redfearn *et al.* 1996; Wu & Wang 2005) to Japan (Osada 1966), with some isolated stations in Algeria in North Africa (Ros *et al.* 1999), the Caucasus (Ignatov & Afonina 1992) and Central Asia (Mamatkulov 1990). In the Southern Hemisphere *Polytrichum strictum* occurs in the southernmost tip of South America including western Patagonia (Cardot & Brotherus 1923; Matteri 1985*a*), Tierra del Fuego (Schiavone 1993), Falkland Is. (Cardot & Brotherus 1923), South Georgia (Greene 1973) and Antarctica.

Distribution in Antarctica. A widely distributed though scattered species of acidic soils, ranging from the South Sandwich Is. to northern Marguerite Bay (Fig. 41). It is common and abundant in the South Orkney Is., in the southern part of the Danco Coast (Arctowski Peninsula area, Paradise Harbour, southern Anvers I. and some offshore islands) and on the northern Graham Coast and offshore islands (Petermann I., Argentine Is., Berthelot Is., Cape Tuxen). It is rare in the South Sandwich and South Shetland Is. and in the northernmost Antarctic Peninsula. It reaches its southernmost station on Jenny I. in northern Marguerite Bay at lat. 67°44′ S [but discovered in February 2008 at lat. *ca* 69°20′ S on NW Alexander I. (P. Convey, personal communication)]. *Polytrichum strictum* occurs mainly at low elevations, from sea level to about 200–250 m, but it reaches 400 m on Mt. Demaria above Cape Tuxen.

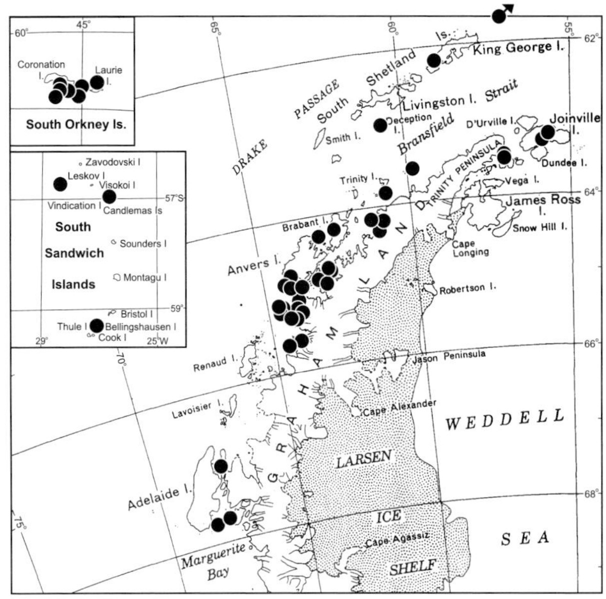

Figure 41. Distribution of *Polytrichum strictum* in the Antarctic. The occurrence in the northern group of the South Shetland Is. (Elephant and Clarence Is.) is indicated by an arrow.

Selected specimens examined. WEST ANTARCTICA. SOUTH SANDWICH IS. **Leskov I.** Summit ridge, *REL 512* (AAS, KRAM). **Candlemas I.** Clinker Gulch, *PC 163A & 166A* (AAS, KRAM); Lucifer Hill, *PC 126A, 129A & 228B* (AAS, KRAM). **Bellingshausen I.** S slopes of main cone, *ca* 95 m, *MWH 412A* (AAS) and 150 m, *PC 51A* (AAS, KRAM).

SOUTH ORKNEY IS. **Coronation I.** Mansfield Pt., *ca* 35 m, *RILS 286* (AAS, KRAM); Laws Glacier, 100 m, *RW 132 & 133B* (AAS, KRAM); Shingle Cove, *ca* 15 m, *BASM 7* (AAS, KRAM); Olivine Pt., 15–35 m, *RILS 134A & 136* (AAS, KRAM); Saunders Pt., *ca* 25 m, *RILS 21* (AAS, KRAM). **Lynch I.** *Ca* 15 m, *DCL 942A* (AAS, KRAM). **Robertson Is.** Steepholm I., 75 m, *RILS 3245* (AAS, KRAM). **Powell I.** SW end, 35–65 m, *RILS 242* (AAS, KRAM). **Laurie I.** Cape Geddes, 15–70 m, *MMM 151* (AAS, KRAM). **Signy I.** Factory Cove, *ca* 30 m, *MWH 3* (AAS, KRAM); Berntsen Pt., *ca* 20 m, *FIDS* H621/1 (KRAM, PC). **Moe I.** 10–200 m, *RILS 5258* (AAS, KRAM).

SOUTH SHETLAND IS. **Elephant I.** Walker Pt., 150 m, *JPB 240* (KRAM); SW of Endurance Glacier, 152 m, *JSA 32A* (AAS, KRAM) and 305 m, *JSA 241A* (AAS, KRAM); 3 km SW of Walker Pt., *ca* 50 m, *JSA 184A* (AAS, KRAM); Muckle Bluff, *ca* 50 m, *JSA 200A* (AAS, KRAM). **Clarence I.** South end, *ca.* 51 m, *JSA 1J* (AAS, KRAM). **King George I.** Ardley I., 15 m, *RO 2494/80* (AAS, KRAM). **Deception I.** Pendulum Cove, 100 m, *RILS 5750A* (AAS, KRAM).

WEST ANTARCTIC PENINSULA. **Davis Coast.** Tower I., *DCL 1690* (AAS, KRAM). **Danco Coast.** Cierva Pt., *VK 612 & 655* (KRAM); Moss I., *CS 465* (S, PC); Spring Pt., 10–30 m, *RILS 3882* (AAS, KRAM); Cuverville I., *ER 243A* (BR, PC) and 10–20 m, *RILS 11937* (AAS, KRAM); Birdsend Bluff, *ca* 100 m, *RILS 4229* (AAS, KRAM); Beneden Head, *ER 234A* (BR, PC); Coughtrey Peninsula, Paradise Harbour, *RMS 69-353* (US) and 65 m, 7 Jan 1984, *Birkenmajer s.n.* (KRAM); Bryde I., *ca* 15 m, *JS 179A* (AAS, KRAM). **Palmer Archipelago.** ***Brabant I.***: Minot Pt., 100 m, *AM 27 & 28A* (AAS, KRAM). ***Anvers I.***: Norsel Pt., *VK 1063& 1094* (KRAM), *RMS 69-847 & 856* (US) and 10–20 m, *RILS 1958* (AAS, KRAM); Bonaparte Pt, *VK 67* (KRAM); Shortcut I., *VK 1120* (KRAM); Litchfield I., *VK 1078, 1097, 1105, 1134 & 1146* (KRAM) and *JRR & JMG 26, 27 & 29* (US); Laggard I., 5–15 m, *REL 825* (AAS, KRAM). **Graham Coast.** Loubat Pt., 330 m, *PK 53* (AAS, KRAM); Edge Hill, *ca* 65 m, *RWMC 648B* (AAS, KRAM); Rasmussen I., 15 m, *AS 89* (KRAM) and 19 m, *AS 76B* (KRAM) and *Gain 283A* (PC); Cape Tuxen, *Gain 196 & 204* (PC); Mt. Demaria, 400 m, *RILS 10657* (AAS, KRAM). **Wilhelm Archipelago.** ***Lemaire Channel***: entrance to the channel, *ER 268C* (BR, PC – syntype of *P. subpiliferum*). ***Penola Strait***: Petermann I., 71 m, *AS 114A* (KRAM) and *Gain 147, 155, 162, 166A & 270A* (PC). ***Argentine Is.***: Corner Is., *JRR 54, 59 & 79* (US); Galindez I.: Woozle Hill, *BJT 50 & 51* (KRAM, PC), Stella Creek, 18 m, *AS 60* (AAS, KRAM); Winter I., 3 m, *BJT 124* (AAS, KRAM, PC) and 5 m, *BJT 130* (AAS, KRAM, PC); The Barchans, *ca* 5 m, *RWMC 749A* (AAS, KRAM). ***Collins Bay***: Berthelot Is., *Gourdon 180A & 181* (PC); Darboux I., 8–150 m, *RILS 3329* (AAS, KRAM). ***Grandidier Channel***: largest island off Takaki Pt., 10–35 m, *RILS 4341* (AAS, KRAM). **Loubet Coast.** Bennett Is., Pfaff I., *JK 180* (AAS, KRAM). **Fallières Coast.** Pourqui Pas I., Lainez Pt., *S. & J. Poncet 11 & 13* (KRAM); Jenny I., *Gain 237A, 238A, 239, 240, 241A & 242A* (PC), *JBK 226 & 229A* (AAS, KRAM, PC) and *BJT 427* (AAS, KRAM, PC).

EAST ANTARCTIC PENINSULA. **Trinity Peninsula.** Hope Bay, *CS 464* (PC, S) and *ca* 225 m, *BJT 232* (KRAM, PC); Lake Boeckella, *ca* 50 m, *CGB 29* (AAS, KRAM); Mt. Flora, 35–65 m, *FIDS D1091/19* (KRAM, PC), *ca* 85 m, *CGB 18* (KRAM, PC) and 250 m, *BJT 326* (AAS, KRAM, PC). **Joinville Island group.** ***Joinville I.***: NE of Mt. Alexander, 15–50 m, *RILS 3698* (AAS, KRAM); Mt. Alexander, 250 m, *RILS 3678* (AAS, KRAM).

EXCLUDED SPECIES OF THE POLYTRICHACEAE

Polytrichum commune Hedw. — This species is mentioned as an error by Kappen & Schroeter (2002). They ascribed this record to Lewis Smith (1972) who was to show it as an example of a polytrichoid moss forming shoots from a single subterranean system of radiating stems. Indeed, the latter author presented a diagram of a circular moss colony, but of *Polytrichastrum alpinum*, not *Polytrichum commune*. This last species does not occur in the austral cool-temperate and polar zones, although it was once reported from Gough I. (Wright 1905a) but this record needs verification. However, in the moss flora of Macquarie I. Seppelt (2004) presented an SEM micrograph of a transverse section of the leaf of the moss designated as *Polytrichastrum alpinum*. The apical cells of the lamellae are distinctly and deeply grooved in a manner typical of *Polytrichum commune*. This would suggest that this species does occur in the Australasian sector of the Subantarctic. Unfortunately, no details of the photographed specimen are available to confirm its origin.

Class C. **BRYOPSIDA** PAX

Bryopsida Pax, Prantl's Lehrb. Bot. Ed. 11: 220. 1900 ['Bryophyta'].

The class Bryopsida includes the great majority (*ca* 97–98%) of all known moss species, and are commonly called the true mosses. They show the greatest morphological diversity and their gametophores reach a high degree of specialisation. These are mostly perennial plants which represent two basic growth-forms: (*a*) acrocarpous with stems erect, sparsely branched from subfloral innovations and usually crowded together to form a turf or cushion with terminal perichaetia and sporophytes which limit each year's growth; (*b*) pleurocarpous with stems mainly prostrate, freely branched and interwoven in mats, bearing perichaetia and sporophytes laterally and consequently growing indefinitely from an apical cell. However, the growth-form is not necessarily correlated with systematic position (Mägdefrau 1982; La Farge-England 1996). The branches are exogenous and the leaves are undivided and most often spirally inserted due to the activity of the tetrahedral apical cell. Only occasionally the apical cell is bifacial, resulting in a two-ranked leaf arrangement. The gametophores never have leptoids and also often lack hydroids. Mosses of this class are typically loosely attached to the substrate by rhizoids, which are much-branched filaments of uniseriately arranged cells with diagonal cross-walls, although many pleurocarpous mosses do not have rhizoids.

The sporophyte is long-lived and never elevated at maturity on a pseudopodium, but the capsule is usually raised by a gradually elongated seta. The endothecium usually forms the columella and the archesporium. All cells of the sporogeneous tissue undergo meiosis to form spores. The columella penetrates the cylindrical spore-bearing layer and continues up to the apex of the capsule and only very rarely is it absent (in *Archidium*). The spore sac is usually separated from the wall of the capsule by an intercellular space.

The fundamental diagnostic character of the Bryopsida is the structure and development of the peristome. About 90% of the genera in this class are ochyrostomous (having a peristome), whereas the remaining *ca* 10% are either gymnostomous (lacking a peristome) or cleistocarpous (lacking an operculum), but they are usually clearly related to peristomate taxa by other gametophyte and/or sporophyte characters. The peristome is a single or double ring of articulate teeth around the orifice of a dehiscent capsule.

The peristome teeth develop from the two or three innermost concentric layers of the amphithecium which are called the outer (OPL), primary (PPL) and inner (IPL) peristomial layers. Each of these contributes different numbers of cells to the peristome and is expressed in the peristomial formula proposed by Edwards (1977, 1979, 1984). To ease comparison, the number of cells in ⅛ of the capsule circumference is used in the peristomal formula. For example, a 4:2:8 formula means that in a whole capsule there are 32, 16 and 64 cells in the OPL, PPL and IPL, respectively. Depending on the number of cell layers that contribute to the formation of the peristome teeth, two basic types of the arthrodontous peristome are recognised.

The diplolepidous peristome is typically composed of two rings of teeth, an outer exostome and an inner endostome, derived from three concentric layers of the amphithecial cells. The exostome consists typically of 16 teeth, which may sometimes be reduced to being entirely absent or fused into eight pairs with distinct median lines on the outer surface which mark the junction of two rows of cells. They are usually trabeculate on the inner surface by projecting remnants of cell walls. The endostome also has 16 segments arising from a high or low basal membrane and most often alternating in position with the exostome teeth or sometimes lying opposite them. The segments are more delicate than the teeth, are often keeled and are usually separated by groups of cilia which originate as the result of additional divisions in the IPL.

By contrast, the haplolepidous peristome has a single ring of teeth derived from the abutting tangential walls of two concentric cell layers of the amphithecium. It is homologous to the endostome of the diplolepidous peristome and the outer surface of the teeth lacks a median line because they are elaborated from a single column of cells. This peristome also consists of 16 teeth, but sometimes they are deeply split nearly to the base so that there appear to be 32 present, but sometimes the teeth may be fused into an extensive tube or may be vestigial or entirely absent.

The teeth are usually highly hygroscopic at maturity and assist in gradual dissemination of the spores. The hygroscopic nature of the peristome is variable and its behaviour, when it is moistened, is correlated with the degree of thickening on the outer or inner surface of the tooth.

The peristomial features have long been the most important character used to construct classifications of moss taxa which could reflect natural relationships and evolutionary processes in this group. Mitten (1859) was the first to challenge the traditional views on the moss classifications based on the position of sex organs, and used the structure of the peristome to distinguish two basic groups of mosses, Arthrodonti and Nematodonti, which correspond to the modern classes Bryopsida and Polytrichopsida. This concept was greatly developed by H. Philibert who, between 1884–1902, published ten pioneer articles on the structure of the peristome in various groups of mosses (Taylor 1962). These investigations were of fundamental importance for understanding relationships of mosses, and heavily influenced Fleischer (1904–1923) who constructed the first phylogenetic system of classification of mosses based on the sporophyte characters. It was popularised by Brotherus (1924*b*, 1925) in the second edition of the treatment of the mosses in Engler's *Die Natürlichen Pflanzenfamilien*, and since then it has been widely known as the Fleischer-Brotherus system (Miller 1979). With some minor modifications it was accepted in the subsequent systems of classification of mosses in the 20th century (e.g. Dixon 1932*a*; Reimers 1954; Robinson 1971; Crosby 1980; Walther 1983).

The peristome offers a wide array of characters, some of which are of paramount importance for modern classification systems of higher taxa of mosses. This refers, in particular, to the pattern of cell division and peristome architecture. Since the early 1970s many studies on peristome architecture and development have been carried out (e.g. Robinson 1972*b*; Vitt & Hamilton 1974; Vitt 1981, 1984; Shaw & Crum 1984; Shaw & Fife 1984; Shaw & Robinson 1984; Shaw & Rohrer 1984; Shaw 1984*a*, 1985*a,b,c*, 1986; Shaw & Allen 1985; Magill 1987*a*; Shaw *et al.* 1987, 1989*a,b*; Shaw & Anderson 1986, 1988; Lewinsky 1989; Schwartz 1994). These have helped to reveal the close relationships between various groups of mosses with similar peristomes, regardless of the morphology of their gametophores.

These studies resulted in recognition of four basic peristome types: (*a*) *Funaria*-type – unfused, opposite peristome with symmetric divisions in the IPL and lacking cilia, defined by the formula 4 : 2 : 4; (*b*) *Orthotrichum*-type – mostly alternate peristome with the OPL thickened or fused and with symmetric division in the IPL, also defined by the formula 4 : 2 : 4; (*c*) *Bryum*-type – ciliate alternate peristome with asymmetric late stage division in the IPL, defined by the formula 4 : 2 : 8; (*d*) *Dicranum*-type – opposite peristome with no OPL/PPL development and with asymmetric division in the IPL, defined by the formula [4] : 2 : 3.

In early classifications groups with haplolepidous peristomes preceded those with the diplolepidous peristomes. However, Crosby (1980) and Vitt (1981) suggested that the haplolepidous peristome may be derived from the diplolepidous type because the main trend of evolution in the sporophyte has been reduction and simplification. Acceptance of this concept implies that diplolepidous groups should come first in the classification scheme, and such a sequence of higher taxa is visible in the systems proposed by Vitt (1984) and Vitt *et al.* (1998).

In the latest classification schemes of the Bryopsida Buck & Goffinet (2000) and Goffinet & Buck (2004) largely followed Crosby's (1980) suggestion, but they placed the diplolepidous subclass Bryidae at the end of the sequence, thus retaining some concepts from the Fleischer-Brotherus system. They subdivided the Bryopsida into four subclasses, Diphysciidae, Funariidae, Dicranidae and Bryidae, and this sequence of higher taxa is also adopted in the present treatment, although no representatives of the first subclass are known in Antarctica. On the other hand, considering the essential peculiarities in the peristome structure in the genera *Encalypta* and *Orthotrichum* and their allies, two subclasses, Encalyptidae and Orthotrichidae, are recognised here following the suggestion of Vitt *et al.* (1998).

KEY TO THE SUBCLASSES OF THE ANTARCTIC BRYOPSIDA

1. Peristome haplolepidous, derived from two layers of the amphithecium, when well developed containing a single row of 16 teeth composed of fused cell plates from two cell layers, with 16 cell plates in the outermost layer and 24 plates in the innermost layer, thus showing the peristomial formula [4] : 2 : 3; teeth lacking a median line; acrocarpous mosses .... C3. **Dicranidae**

1. Peristome diplolepidous, derived from three layers of the amphithecium, when well developed containing two concentric rows of teeth, with the outer row (exostome) consisting of 16 teeth composed of fused cell plates from two cell layers, with 32 cell plates

in the outermost layer and 16 plates in the innermost layer; the peristomial formula ranging from 4 : 2 : 4 to 4 : 2 : 14; exostome teeth with a distinct median line; acrocarpous or pleurocarpous mosses 2

2. Exostome teeth positioned opposite the endostome segments 3
2. Exostome teeth positioned alternate to the endostome segments 4

3. Endostome, when present, not fused with the exostome; leaf cells smooth throughout, those at base similar or somewhat larger than upper cells .... C1. **Funariidae**
3. Endostome, when present, partly or wholly adherent to the exostome; basal leaf cells hyaline, thin-walled and long-rectangular, sharply differentiated from small, bulging, densely papillose or rarely smooth upper cells .... C2. **Encalyptidae**

4. Inner layer of the exostome teeth much thinner than outer layer, not or weakly trabeculate; preperistome often present .... C4. **Orthotrichidae**
4. Inner layer of the exostome teeth markedly thicker than outer layer, usually strongly trabeculate; preperistome absent 5

5. Plants erect, mostly unbranched, acrocarpous, very seldom pleurocarpous ... C5. **Bryidae**
5. Plants with creeping stems, typically freely branching, pleurocarpous ... C6. **Hypnidae**

Subclass C1. **FUNARIIDAE** OCHYRA

Funariidae Ochyra *in* Ochyra, Żarnowiec & Bednarek-Ochyra, Cens. Cat. Polish Mosses: 19, 100. 2003. Validly published by reference to Bryidae Limpr. superorder Funarianae W.Frey *in* W.Frey, Hurka & Oberw., Beitr. Biol. Nied. Pflanz.: 134. 1977.

Funariidae Vitt, Goffinet & Hedd. *in* Bates, Ashton & Duckett, Bryol. Twenty-First Cent.: 119. 1998 ['Funaridae'], *nom. nud.*

Funariidae W.R.Buck & Goffinet *in* A.J.Shaw & Goffinet, Bryoph. Biol.: 97. 2000, *nom. inval. descr. angl.*

The subclass Funariidae is primarily characterised by its peristomial characters. Although the peristome is very variable in this group, sometimes being absent, single and rudimentary, or some taxa being cleistocarpous, when well developed it is typically diplolepidous and double and represents a *Funaria*-type. It clearly follows a 4 : 2 : 4 pattern with symmetric divisions in the IPL. It consists of 16 exostome teeth which are papillose on the outer surface in the upper and median parts and sometimes cohering in pairs or tetrads. They are often united at the tips by a small perforated epiphragm and are often spirally twisted in an anticlockwise direction. The teeth lie opposite the endostome segments, but are not fused with each other. The endostome lacks a basal membrane and cilia. On the basis of these features, the funariaceous peristome has often been interpreted as the most primitive among extant mosses (Vitt 1981, 1982*a*). However, the reticulate pattern of the exostomial ornamentation at the extreme base of the teeth in some species, for example *Funaria hygrometrica*, is commonly encountered in many taxa with the diplolepidous bryaceous peristome (Shaw 1985*a*,c), a feature which is considered to be evolutionarily advanced.

The peristomial features are coupled with a suite of other sporophyte and gametophyte characters, making the Funariidae a distinctive taxon. It comprises exclusively small acrocarpous or cladocarpous mosses with short erect stems or sometimes with creeping, fleshy, rhizomatous, subterranean stems and erect short branches. The setae are terminal, mostly elongate and bear exserted, erect and symmetric capsules which, less commonly, are inclined and asymmetric. The calyptra is either small and mitriform or larger, cucullate and, occasionally, inflated at the base, while the operculum is conic or convex.

The subclass also exhibits a remarkable gametophyte uniformity. The leaves are inserted in many rows around the stem, but often a few larger, mostly broad and obovate, leaves are crowded at the stem tips. They have a single costa which, however, is sometimes indistinct or wanting. The laminal cells are large, thin-walled, lax, hexagonal to oblong-hexagonal and entirely smooth. It is also remarkable that, in general, the representatives of the Funariidae are characterised by an ephemeral life cycle. They are usually annual or biennial plants, growing gregariously or forming loose turves, mostly on soil in disturbed habitats which are at least periodically available for colonisation.

Buck & Goffinet (2000) and Goffinet & Buck (2004) placed three orders in the Funariidae, namely Encalyptales, Timmiales and Funariales, but here this subclass is interpreted in a strict sense to encompass only the single order Funariales. Following the suggestion of Vitt *et al.* (1998), Ochyra *et al.* (2003) segregated the Funariales into a subclass of its own by reference to superorder Funarianae of Frey (1977). Also, the placement of the unusual and phylogenetically isolated genus *Timmia*, which does not occur in the Antarctic, in this subclass is questionable as evidenced by many peculiarities of its peristome as well as gameophyte (Allen 1998) and, accordingly, Ochyra *et al.* (2003) established the separate subclass Timmiidae to accommodate this genus.

Order III. **FUNARIALES** M.Fleisch.

Funariales M.Fleisch., Musci Buitenzorg 2: xii, 381. 1904 ['Reihe Funaroideae'].

Epicranoideae M.Fleisch., Musci Buitenzorg 2: 466. 1904, *nom. illeg.*

The distinctive features of the order are those of the subclass, as given above. The Funariales consist of three families, Gigaspermaceae, Disceliaceae and Funariaceae, although originally Fleischer (1904) also placed here the Splachnaceae (as a separate suborder) and the Ephemeraceae. Recent molecular data clearly indicate that the former family is sister to the orthotrichalean clade, not to the funarialean clade and, accordingly, it is segregated into a separate subclass (Vitt *et al.* 1998) or a separate order within the Bryidae (Buck & Goffinet 2000; Goffinet & Buck 2004). Also, the cleistocarpous family Ephemeraceae, comprising minute mosses developing from persistent protonemata, appears to be closer to the Pottiaceae in its gametophyte characters, including leaf areolation with occasionally papillose leaf cells, and consequently it was transferred to the order Phascales (Buck & Goffinet 2000; Ochyra *et al.* 2003). The relationship of this family with the phascalean mosses is also corroborated by molecular data (Werner *et al.* 2007).

In the Antarctic the Funariales are represented by a single species of the Funariaceae which is not a native plant on the continent.

Family III. **FUNARIACEAE** Schwägr.

Funariaceae Schwägr. *in* Willd., Spec. Pl. Ed. 4, 5(2): 43. 1830 ['Familia Funariae']. — Type: *Funaria* Hedw.

The Funariaceae are one of very few families comprising mosses which exhibit a remarkable uniformity in gametophyte morphology. The majority of species of this family are small to medium-sized, pale green, epigean, gregarious or loosely tufted, annual or ephemeral plants. The stem is usually sympodially branched by subperigonial innovation, less often unbranched or forked and its internal structure is strikingly uniform. It consists of a central strand surrounded by a medulla of large thin-walled cells and a cortex of 1–4 layers of smaller, firm-walled, reddish-brown cells, with the outermost row of large, thin-walled cells forming a hyalodermis. The leaves are soft, erect to spreading, sometimes twisted, larger and more crowded above, smaller and distant below, mostly oblong-obovate, less often broadly elliptic, ovate to lanceolate, concave, entire or serrate by projecting cell ends, sometimes ciliate, usually acute to acuminate, rarely obtuse. The costa is single and ceases in mid-leaf or is percurrent to excurrent and in cross-section it consists of a small central stereid band surrounded by one dorsal and 1–2 ventral layers of larger, thin-walled cells. The laminal cells are large, pale, lax, glabrous, thin-walled, oblong-hexagonal above, rectangular below, sometimes becoming linear at the margin. The majority of species are autoicous, less often paroicous or synoicous, whereas dioicy is unknown in the family. The antheridia are intermixed with paraphyses with a terminal pyriform to globose cell; this type of paraphysis is unique to the Funariaceae.

In contrast to the gametophyte uniformity, the Funariaceae exhibit a high degree of variation in sporophyte morphology. Loeske (1914–1929) stated that, on the basis of gametophyte characters, all species of the family may be placed in a single genus, but that the family includes what he called 'Sporophyten-Genera'. Apart from the funariaceous type of the peristome, the other peculiar sporophyte feature of the Funariaceae are the 'doughnut' stomata. These are restricted to the neck and consist of an elongated slit enclosed by a single, superficial or immersed guard cell.

The sporophyte in this family exhibits a strong tendency towards progressive simplification. The genera present the full range of capsule morphology, from stegocarpous to cleistocarpous and from gymnostomous to ochyrostomous, the latter also showing a complete gradation in peristome development, from rudimentary to well developed and from single to double. Additionally, the distinct evolutionary tendencies include the enlargement of the spores from very small (*ca* 10 μm diameter) to very large (*ca* 90 μm) and the development of thin-walled and isodiametric exothecial cells. The radial walls of the exothecial cells are often cuneate in transverse section (Fife 1985). The capsules are immersed to long-exserted, erect or curved to strongly asymmetric, operculate to inoperculate and usually they have distinct necks. The calyptra is persistent or deciduous, mitrate or cucullate, usually long-rostrate and often inflated at the base.

In addition to their ephemeral nature, the funarialean mosses show a remarkable ability to hybridise and develop polyploidic races. Because they are also easily cultured, they have received much intensive physiological, cytological and genetic attention (Wettstein 1932) but, taxonomically, the family has attracted less attention. Fife (1985) presented a generic revision of the Funariaceae, and Holarctic taxa have often been studied and described in many local and regional Floras. However, far fewer tropical and austral taxa have been studied (e.g. Fife 1982, 1986, 1987; Magill 1987*b*; Enroth 1991; Eddy 1996; Fife & Seppelt 2001).

The Funariaceae comprise 16 genera (Crosby *et al.* 2000) and approximately 250 species, although their final number may be lower when the largest genera of the family, including *Funaria*, *Entosthodon* and *Physcomitrium*, are critically revised on a global scale. These large genera are surrounded by a number of small 'satellite' genera, 11 of which are monotypic, one bitypic (*Physcomitrella*) and one (*Goniomitrium*) consists of five species. Species of the family are widely distributed in temperate and tropical regions and exhibit less diversity in polar regions of both hemispheres. In the southern cool temperate and polar regions the Funariaceae are poorly represented and only five or six species of *Entosthodon*, two of *Funaria* and three of *Physcomitrium* have been recorded and, in general, they are not abundant. Of these, a single species of *Funaria* has been recorded from the Antarctic. In addition, *Entosthodon* cf. *subnudus* (Taylor) Fife germinated from spores deposited in soil from Coulman I., Ross Sea (Lewis Smith & Ochyra 2006).

5. FUNARIA HEDW.

Funaria Hedw., Spec. Musc. Frond.: 172. 1801. — Type: *Funaria hygrometrica* Hedw.

Plants soft, gregarious or caespitose, small to medium-sized, bright green to yellowish. *Stems* short, mainly to 1 cm tall, erect, simple or with a short, basal antheridial branch, with smooth, reddish-brown rhizoids at base, in cross-section with a distinct central strand and a unistratose, thin-walled hyalodermis. *Leaves* increasingly larger acropetally and rosulate at stem apices, slightly contorted on drying, patent to wide-spreading when wet, mostly concave, elliptical, spathulate or oblong-ovate to broadly ovate, acute to acuminate, sometimes rounded, apiculate, mucronate to piliferous or, rarely, obtuse at the apex; *margins* plane or sometimes broadly incurved, entire, sinuate, serrate to crenulate-dentate, infrequently weakly bordered; *costa* single, ending well below the apex to percurrent, occasionally long-excurrent; *laminal cells* smooth, large, lax and thin-walled, rectangular below, oblong-hexagonal to rhomboidal above, sometimes becoming narrower towards the margins and forming an indistinct border. *Autoicous. Perigonia* borne terminally on a slender male branch arising at the base of the stem; *paraphyses* multicellular, with enlarged, globose or pyriform apical cell, apparently none in the perichaetia. *Perichaetial leaves* not differentiated. *Setae* elongate, slender, erect and flexuose to strongly and irregularly curved and twisted. *Capsules* exserted, erect, inclined to pendent, elongated-pyriform, strongly asymmetric and usually curved, small-mouthed, often deeply sulcate, with a distinct neck, often shrivelled when dry; *annulus* large, compound and revoluble or none; *operculum* convex or conic with cells in obliquely radiating rows; *exothecial cells* rectangular to oblong-hexagonal, with incrassate walls, often stronger toward the inner tangential walls than toward the outer surface, resulting in a cuneate shape in cross-section; *stomata* present on the neck; *peristome* double; *exostome teeth* 16, lanceolate-subulate, filiform and often appendiculate at the tips and united into a latticed disc, often sigmoid, obliquely arranged, trabeculate at back, papillose-striate on the outer surface; *endostome segments* ½ to ⅔ or sometimes about as long as the teeth, occasionally more reduced, papillose to rarely papillose-striate. *Spores* spherical to angled, finely papillose to verrucate, sometimes appearing smooth owing to small and uniform verrucae. *Calyptra* cucullate and inflated, long-beaked.

Funaria has often been considered to be congeneric with *Entosthodon* and, indeed, the separation of these genera is not always clear, especially because an apparent gradation in peristome reduction is evident in both genera. However, species of *Funaria* are united by their capsules which are always deeply furrowed when dry, due to vertical bands of strongly cuneate exothecial cells alternating with non-cuneate but otherwise firm-walled cells. Moreover, the exostome teeth are obliquely directed and strongly sigmoid and they are fused at the tips to form a perforated epiphragm, and the annulus, if present, is always compound and revoluble. Additionally, Fife (1985) found that the verrucate and verrucate-lyrate ornamentation of the pores occurs throughout *Funaria* and this was interpreted as another distinguishing feature of the genus.

The number of species of *Funaria* cannot be evaluated because the genus needs a modern taxonomic overview. Crosby *et al.* (2000) listed 86 species in *Funaria* but, of these, 68 have never been studied since their inception. A single, ubiquitous species, *F. hygrometrica*, is known from the Antarctic biome.

1. **Funaria hygrometrica** Hedw. Fig. 42

Funaria hygrometrica Hedw., Spec. Musc. Frond.: 172. 1801. ≡ *Mnium hygrometricum* (Hedw.) With., Syst. Arr. Brit. Pl. Ed. 4, 3: 787. 1801. ≡ *Strephedium hygrometricum* (Hedw.) P.Beauv. *in* Brid., Bryol. Univ. 2: 51. 1827, *nom. inval. in synon.* — TYPE CITATION: Locis subhumidis umbrosis aut apertis, in rupibus terra tenuis tectis per omnem Germaniam; also cited Vaillant (1727), Dillenius (1741), Haller (1742, 1768), Linnaeus (1763), Hedwig (1782) and Murray (1784). [Lectotype (*selected here*): the second specimen from the left in the upper row forming a tuft with many mature capsules on the sheet labeled 'Bryum bulbiforme aureum, calyptra quadrangulari,

→

Figure 42. *Funaria hygrometrica*. — **1**. Sterile plant, wet. **2**. Fertile plant, dry. **3**. Fertile plant, wet. **4**. Cross-section of stem. **5–7**. Axillary hairs. **8–12**. Leaves. **13**. Leaf apex. **14**. Upper cells at margin. **15**. Mid-leaf cells. **16**. Basal cells. **17–21**. Cross-sections of leaves, a sequence from apex to base. **22–23**. Operculate capsules, dry. **24**. Operculate capsule, wet. **25**. Exothecial cells in middle of urn. **26**. Lower exothecial cells and stomata. **27**. Peristome teeth, spores, annulus and rim cells of capsule. — [1, 3–4, 8–18 from *RWMC 44*; 2, 5–7, 19–27 from *SWG 3171A*]. — Scale bars: a – 100 µm (5–7); b – 1 mm (8–12) and 200 µm (28); c – 100 µm (4, 13–21, 25–27); d – 1 mm (1–3, 22–24).

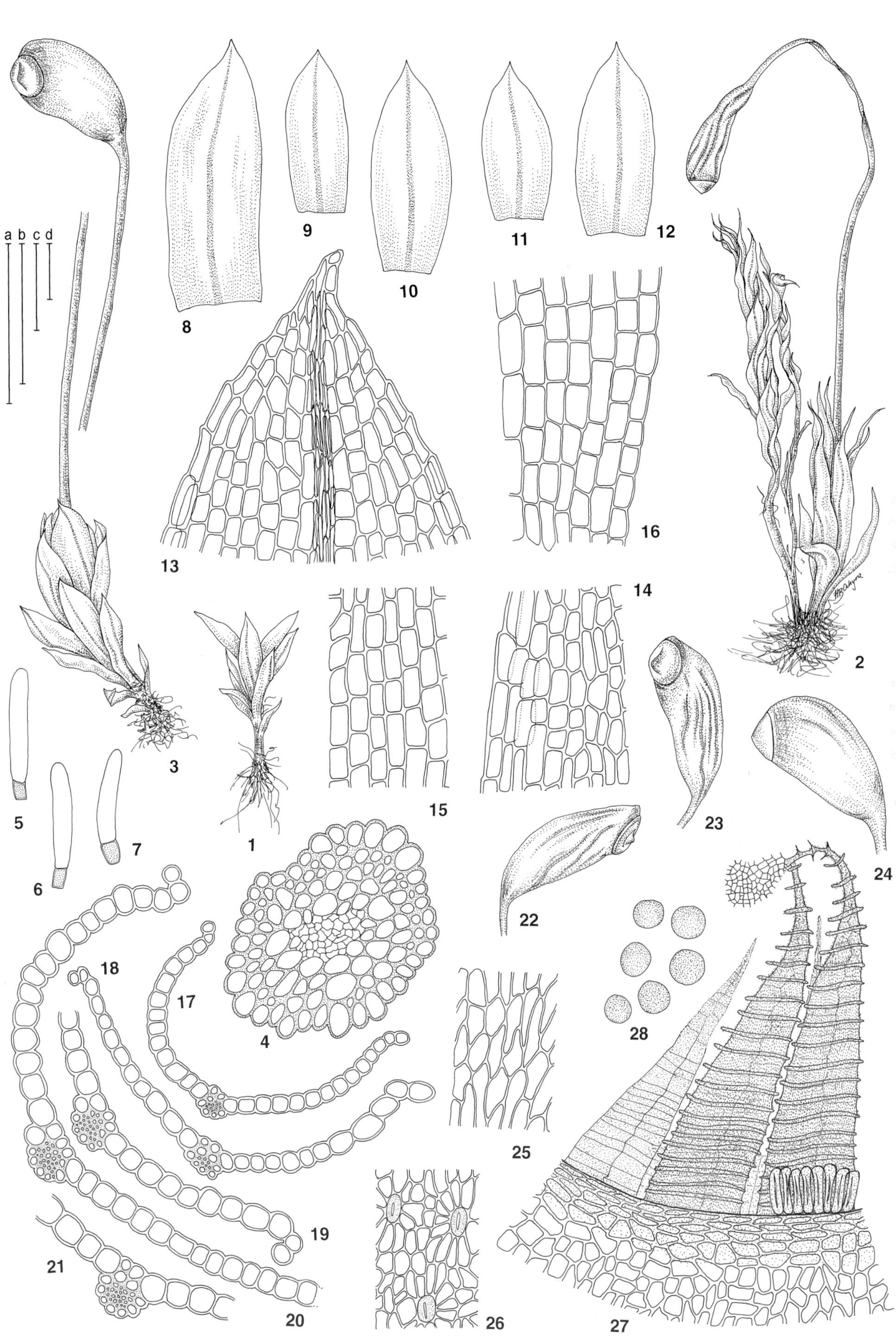
a b c d
1
2
3
4
5
6
7
8
9
10
11
12
13
14
15
16
17
18
19
20
21
22
23
24
25
26
27
28

capsules pyriformibus nutantinus. Dill. H. Musc. 407. t. 52. f. 75. Mnium hygrometricum acaule anthera nutante calyptra reflexa tetragona Linn. Spec. pl. 2. p. 1575. n. 5.' – G-Hedwig/Schwägrichen!].

Plants small, forming brownish- or yellowish-green patches. *Stems* to 0.5 cm tall, erect, simple or with male basal branches, usually densely matted with smooth brown rhizoids at base, nearly naked below and leafy and bulb-like above or laxly foliate throughout; *axillary hairs* club-like, bicellular, consisting of a short, brown basal cell and elongate club-shaped upper cell. *Leaves* smaller below and increasing in size acropetally, erect and contorted when dry, patent when wet, concave, broadly elliptical to oblong-ovate, 1.2–2.3 mm long, 0.5–0.8 mm wide, acute to short acuminate; *margins* plane, entire or slightly sinuate-serrulate above; *costa* single, ending in or just below the leaf tip, in cross-section round, consisting of a small central stereid band surrounded by one layer of larger, thin-walled cells; *laminal cells* lax, thin-walled, smooth, rectangular and bulging at base, short-rectangular to hexagonal above, 30–40 µm wide, 40–100 µm long in mid-leaf, often becoming narrower in 1–2 marginal rows. *Autoicous. Perigonia* and *perichaetia* on separate branches; *antheridia* in large, yellowish discoid heads, becoming brown with age. *Setae* 1.2–1.5 cm long, yellowish, straight for most of its length but arched above. *Capsules* horizontal to pendent, long-pyriform, gibbous, strongly asymmetric, obliquely mouthed, to 2.5 mm long, reddish-brown, strongly sulcate on drying, with distinct neck, furrowed when dry; *annulus* large, compound and revoluble; *operculum* slightly convex; *exothecial cells* irregularly rectangular, firm-walled to thicker-walled on ribs, becoming transversely rectangular in 4–6 rows at mouth and smaller, subquadrate to short-rectangular on neck; *stomata* numerous on neck; *peristome* perfectly developed, double; *exostome teeth* sigmoid, lanceolate-subulate, yellowish-brown, papillose and strongly appendiculate above, papillose-striate below on the outer surface, strongly transversely lamellate on the inner surface, joined at the apex to a fragile, ephemeral meshwork; *exostome segments* lanceolate, slenderly long-acuminate, about ⅔ or more the length of the teeth, pale yellow, finely papillose-striate. *Spores* spherical, yellowish, 12–15 µm in diameter, finely verrucate. *Calyptra* cucullate, strongly inflated at base, long-rostrate.

Chromosome numbers: not studied in the Antarctic; $n = 14$, $n = 28$ and $n = 56$ were recorded at least 45 times from various parts of the world (Fritsch 1991).

Discussion. *Funaria hygrometrica* is one of the best known of all mosses and has gained more investigative attention than any other single species of moss. Despite remarkable phenotypic variation (Weitz & Heyn 1981), it is unlikely to be confused with any other moss, especially because it commonly produces very characteristic and unique sporophytes which readily enable its placement. The strongly asymmetric, horizontal to pendulous, deeply furrowed capsules are very distinctive. They are exserted on long, cygneous setae which twist and curl with changes in humidity, and the specific epithet refers to this feature. The peculiar shape of the capsule is coupled with its oblique mouth, well developed and revoluble annulus and double, sigmoid peristome. The large capsule markedly contrasts with a short-stemmed gametophyte having the leaves crowded in a bulb-like cluster. Its ecological preference for disturbed soil, often at the site of a former fire, is also quite unusual amongst mosses.

Reproduction in Antarctica. The naturally occurring population from Deception I. produced abundant sporophytes, some of which were mature, whereas the plants found in the Casey Station, Wilkes Land, 'greenhouse' were sterile.

Habitat. A weed, growing on soil in man-made or otherwise disturbed, ephemeral habitats. Although no ecological data appear on the label of the specimen from Deception I., Collins (1969) and Lewis Smith (1984*b*, 1991) considered *Funaria hygrometrica* as a moss 'associated only with fumaroles'. This site was created by an eruption in 1967 and colonised by *F. hygrometrica* within one year. Collins (1969) described in detail the habitat and stated that the moss colonised extensive areas where geothermal activity occurred. It grew on ash and lapilli, the surface of which had been consolidated by fumarolic sublimates from the continuous emission of steam from the vents and on a more level area where the bombs were being broken down to a fine grit. Temperatures recorded 0.5 cm below the surface of the moss at 3–15 cm from the fumarole vents on an overcast windy day were 27–36°C. In 2006, fruiting specimens were found at Rothera Station on Adelaide I., having been imported by attachement to vehicles brought from the Falkland Is. The continental plants grew in a typical anthropogenic habitat for this species on hydroponic trays of vermiculite and horticultural compost and gravel used to grow lettuce in a greenhouse.

World range. A truly cosmopolitan weedy species, widespread in northern and temperate latitudes throughout the Holarctic and widely scattered but generally infrequent in the tropics and in the temperate zone in the Southern Hemisphere. It is rare and scattered in southern South America (Greene 1986). It appears to be absent from the Subantarctic, although it has been discovered on South Georgia (Ochyra *et al.* 2002) and Macquarie I. (Seppelt 2004) but these populations represent recent and temporary introductions, as is the case with those in the Antarctic.

Distribution in Antarctica. The three records of *Funaria hygrometrica* represent recent, ephemeral introductions and the species has never become firmly established in this region. On the volcanic Deception I. it was only once observed in a geothermal area north of Telefon Bay created by an eruption in 1967 (Collins 1969; Young & Kläy 1971; Lewis Smith 1984*b*,*c*). This site was destroyed by the 1970 eruption. Later, Lewis Smith (1988*d*) reported that in 1981 moist heated ground at a nearby site at Pendulum Cove on this island was sparsely colonised by several species of moss including a *Funaria*-like plant, but its occurrence was not recorded in 2002 (Lewis Smith 2005*a*,*b*,*c*). The moss may have been a natural immigrant, but it may have been introduced accidentally by humans from spores or shoot fragments adhering to clothing or footwear. This is certainly the case with the population on Adelaide I. where the species was evidently introduced with vehicles imported from the Falkland Is. in 2006. In contrast to the *in situ* occurrences in the maritime Antarctic, the continental record of *F. hygrometrica* came from imported soil in a horticultural greenhouse at a research station.

Specimens examined. WEST ANTARCTICA. SOUTH SHETLAND IS. **DECEPTION I.** Valley to the N of Telefon Bay, *NJC 44* (AAS, KRAM).

WEST ANTARCTIC PENINSULA. **FALLIÈRES COAST.** ***Adelaide I.***: Rothera Pt., 5 m, *Hughes 1 & 3B* (AAS, KRAM).

EAST ANTARCTICA. WILKES LAND. **BUDD COAST.** Bailey Peninsula, Casey Station, *RILS 6372* (AAS).

Subclass C2. ENCALYPTIDAE Ochyra, Żarnowiec & Bednarek-Ochyra

Encalyptidae Vitt, Goffinet & Hedd. *ex* Ochyra, Żarnowiec & Bednarek-Ochyra, Cens. Cat. Polish Mosses: 19, 103. 2003.

Encalyptidae Vitt, Goffinet & Hedd. *in* Bates, Ashton & Duckett, Bryol. Twenty-First Cent.: 119. 1998, *nom. nud.*

Arthrodontei Mitt. [unranked group] Heterolepideae M. Fleisch., Musci Buitenzorg 1: xxiv. 1904, *nom. illeg.*

The Encalyptidae have formally been recognised as a separate subclass of the Bryopsida by Ochyra *et al.* (2003) following the concept suggested by Vitt *et al.* (1998). Although it is a small group, it has long received special interest in phylogenetic speculations on mosses because the peristome of the eponymous genus, *Encalypta*, shows greater variation than in any other genus of moss. Apart from some gymnostomous species, there are others with single and double peristomes and, additionally, in some species that generally have gymnostomous capsules, there are rare populations with a vestigial peristome.

The isolated phylogenetic position of the 'extinguisher' mosses (named after their large campanulate-cylindrical calyptra, resembling a candle-snuffer) is primarily associated with an array of peculiarities in the structure of the peristome. According to Philibert (1889) *Encalypta* contains nematodontous, haplolepidous and diplolepidous peristomes. Therefore, many former authors considered this genus to be a living remnant of an ancient group ancestral to both arthrodontous and nematodontous mosses, and Fleischer (1904) created a separate group, Heterolepideae, equivalent to Haplolepideae and Diplolepideae, to accommodate it. However, recent detailed studies on the peristome in *Encalypta* (Horton 1982*a*; Edwards 1984; Vitt 1984) clearly showed that all species in this genus are arthrodontous and at least some of them have a complex diplolepidous, opposite and often flanged peristome.

When the peristome is double it is composed of 16 exostome teeth and 16 endostome segments which are partly or completely fused and consequently such a peristome appears to be single. The exostome teeth are long and linear with a vertical median line on the outer surface, which is missing when the peristome is truly single. The endostome segments are shorter than the teeth and arise from a distinct, pleated basal membrane. The outer surface of this membrane contains two fused, revolute flanges attached to the joint between adjoining segments which alternate with the exostome teeth. According to Edwards (1984) they are derived from the same layers as the exostome.

The peristome in *Encalypta* is highly modified and it can be interpreted as an extant connecting link along the evolutionary lineage from the diplolepidous to haplolepidous peristome. In many details it is very similar to the peristomes in *Buxbaumia* and *Diphyscium*, although these taxa are sufficiently distant to place them in different subclasses.

The Encalyptidae consist of a single order, Encalyptales.

Order IV. ENCALYPTALES Dixon

Encalyptales Dixon *in* Verd., Man. Bryol.: 409. 1932. Validly published by reference to Pottiales M.Fleisch. suborder Encalyptineae M.Fleisch., Hedwigia 61: 393. 1920.

Bryidae order Heterolepideae M.Fleisch., Musci Buitenzorg 3: xiii. 1908, *nom. illeg.*

The order consists of the single family Encalyptaceae and is diagnosed by its characters. All species of this family look very alike in habit, with a cylindrical erect capsule, large mitriform and rostrate calyptra, and elongate leaves with more or less the same width in the upper and lower parts. All grow in tufts or mats, sometimes as solitary plants with other mosses on basic or siliceous soil or soil-covered rocks.

Family IV. ENCALYPTACEAE Schimp.

Encalyptaceae Schimp., Coroll. Bryol. Eur.: 38. 1856. — Type: *Encalypta* Hedw.

Pottiaceae Schimp. subfam. Encalyptoideae Broth. *in* Engl. & Prantl, Nat. Pflanzenfam. 1(3): 381. 1902. — Type: *Encalypta* Hedw.

Apart from the peculiar peristomial features, which are characteristic of the subclass Encalyptidae and the order Encalyptales, the most prominent character of the Encalyptaceae providing a high degree of recognition is their large, long-mitrate calyptra. It is campanulate-cylindrical, abruptly or gradually contracted to a long, straight or slightly curved rostrum, usually extending well below the capsule, erose, lacerate or fringed at the base, non-plicate, smooth to papillose but never hairy, brown to golden and lustrous or dull. In addition to its characteristic calyptra, the family has distinctive spores which are subject to considerable interspecific variation in size, structure and ornamentation, unknown in any other family of moss (Vitt & Hamilton 1974; Járai-Komlódi & Orbán 1975; Horton 1982*a*). The capsules are cylindrical, smooth or longitudinally or spirally ribbed.

The exceptional variability in sporophyte characters in the Encalyptaceae contrasts with a high uniformity in the structure of the gametophytes. Vegetative plants show the greatest similarity to some genera of the large family Pottiaceae, in particular to *Syntrichia* and *Pottia*. The elliptic to oblong-obovate or lingulate leaves are muticous or hyaline hair-pointed at the apex and their margins are plane to recurved and crenulate-papillose to bluntly serrulate. The upper laminal cells are chlorophyllose, firm-walled, subquadrate to short-oblong and they are prominently mammillose and smooth (in *Bryobrittonia*) or mammillose-papillose on both the abaxial and adaxial surfaces (in all species of *Encalypta*). The papillae are compound, forked, rounded-quadrate or irregularly lunate in surface view and, because they densely cover the lumina, the laminae are distally opaque. The basal cells are large, oblong, smooth, lax and hyaline (only exceptionally chlorophyllose in *Bryobrittonia*) and they have thin and hyaline vertical walls and strongly thickened, reddish-brown to dark yellow transverse walls. Additionally, they are often perforated by a single large pore.

Species of the Encalyptaceae are gonioautoicous, except for *Bryobrittonia longipes* and *Encalypta streptocarpa* which are dioicous. Also, only two species (*E. streptocarpa* and *E. procera*) produce abundantly filiform brood bodies on the stems, whereas all other species lack any specialised means of asexual reproduction.

The Encalyptaceae are a small family consisting of two genera, *Bryobrittonia* and *Encalypta*. The former is a monotype, consisting of a single species, *B. longipes*, which has circumpolar, arctic-montane distribution in the Holarctic, while *Encalypta* has a worldwide distribution, including the Antarctic. Horton (1982*a*, 1983) presented a detailed taxonomic account of the family, with particular reference to North American taxa, but southern South American taxa still need a taxonomic overview.

6. **ENCALYPTA** HEDW.

Encalypta Hedw., Spec. Musc. Frond.: 60. 1801. — Lectotype (*vide* Horton 1983: p. 374): *Encalypta ciliata* Hedw.

Plants small to rather large, forming dense, dull, green, sordid- to brownish-green or glaucous tufts. *Stems* erect, forked, densely radiculose below with two types of rhizoids, large and thick, brown, shiny and much-branched giving rise to minor paler ones, in cross-section without or, rarely, with central strand, occasionally bearing clusters of brown filiform brood bodies in the leaf axils; *axillary hairs* filiform, multicellular, hyaline or brownish. *Leaves* little altered and variously twisted at the tips when dry, erect-spreading when moist, concave or somewhat keeled, elliptic, oblong, lingulate to spathulate, sometimes ovate-lanceolate, obtuse, acute to short-acuminate, occasionally cucullate, muticous, mucronate, apiculate or shortly hyaline-awned; *margins* plane throughout or recurved below, crenulate-papillose above, entire below; *costa* single, strong, prominently convex dorsally, subpercurrent to short-excurrent, in cross-section with a large dorsal stereid band and smaller ventral band of 2–3 rows of large cells including a chlorophyllose and papillose epidermis; *laminal cells* quadrate to hexagonal or irregular, thick-walled, bulging and strongly papillose with C-shaped papillae above, much enlarged, oblong, smooth, hyaline, lax, with cross walls yellow, orange to red-brown, noticeably thicker than vertical walls at the base. *Gonioautoicous*, very rarely *dioicous*. *Setae* terminal on stem or branches, erect to flexuose, dextrorse above, sinistrorse below. *Capsules* exserted, erect and symmetric, long-cylindrical, smooth or furrowed, with an inconspicuous neck; *annulus* present, broad and ± deciduous or none; *operculum* straight, long-rostrate; *exothecial cells* oblong to long-rectangular, thin- to thick-walled, in ± regular, longitudinal or spiral rows, becoming quadrate to short-rectangular in several rows at the orifice; *stomata* present at the base of the urn, superficial, consisting of 2 guard cells with a long pore; *peristome* single, double or lacking, with lanceolate to filiform, entire or variously split, papillose exostome teeth and pale, delicate, papillose segments usually adhering wholly or partly to the teeth, arising from a low basal membrane. *Spores* ± smooth to papillose, generally large. *Calyptra* campanulate-cylindrical, long-rostrate, smooth to scindulose, erose, fringed or lobed at the base.

Encalypta is well characterised by the features of the subclass and the family. The genus consists of about 25 species divided into five sections (Horton 1983) and has a worldwide distribution. It has greatest diversity in the Northern Hemisphere, where most species are restricted to montane or arctic-montane habitats, and it is scattered in tropical mountains and in the Southern Hemisphere, where only northern species occur. In the southern cool-temperate and polar regions only five or six species are known to occur (Greene 1986), but the genus is in need of taxonomic revision. Only one species, *E. procera*, is recorded from South Georgia (Ochyra *et al.* 2002) and two species are known from the Antarctic and they have been surveyed taxonomically by Newton (1974*a*). They represent two different sections of the genus, namely sect. *Streptotheca* (Kindb.) Broth. (*E. procera*) and sect. *Rhabdotheca* Müll.Hal. (*E. rhaptocarpa*).

1. Plants relatively large and robust, mostly sterile, commonly with prominent clusters of brown, filiform propagules in the leaf axils; capsules spirally ribbed; base of the calyptra lacerate; peristome double, 0.8–0.9 mm long; spores small, isopolar, 14–24 µm wide ... . 1. ***E. procera***
1. Plants smaller, usually fruiting, without propagules; capsules longitudinally ribbed; base of the calyptra entire to erose; peristome single, rudimentary or none, about 0.25 mm long; spores large, heteropolar, 30–50 µm wide .... 2. ***E. rhaptocarpa***

1. **Encalypta procera** Bruch Fig. 43, pl. 36

Encalypta procera Bruch, Abh. Math.-Phys. Cl. Königl. Bayer. Akad. Wiss. 1: 283, *pl. 11, f. 1–18.* 1832. — Type citation: An Felsen auf Alpen in Norwegen in dichten Rasen. (Kurr). [Lectotype (*vide* Horton 1983: p. 386): 'Encalypta procera aus Norwegen Hubner Kurr 1828' – BM-Schimper!].

Plants medium-sized to relatively large and robust, densely or occasionally loosely tufted, usually interwoven by copious, reddish-brown, smooth rhizoids below, dull, green, olive- or sordid-green, often with a glaucous tinge above, brown below. *Stems* erect, dichotomously branched, 1.5–4.5 or, occasionally, to 7 cm long, in cross-section obscurely triangular, consisting of a distinct central duct surrounded by a 3–4-stratose medulla of large, thin-walled, hyaline cells and an outer cortex of 1–2(–3) layers of smaller cells with thicker, brown walls; *axillary hairs* sparse, filiform, composed of 5–7 elongate, hyaline upper cells and 1–2 shorter, brownish basal cells; *brood bodies* copious, in dense clusters in leaf axils or attached to the base of the costa, to 3 mm long, filamentous, profusely branched in the upper part, brown, finely papillose. *Leaves* incurved and almost unaltered to twisted and ± folded when dry, spreading when wet, oblong, oblong-elliptic to obovate-oblong, 2.5–4.5(–5.5) mm long, 0.7–1.2 mm wide, obtuse, rounded-obtuse to broadly acute, ± cucullate, muticous or shortly hyaline-awned in perichaetial and uppermost leaves; *margins* papillose-crenulate above, entire below, unistratose, plane or sometimes weakly recurved below; *costa* single, subpercurrent to percurrent, prominently convex dorsally, usually dark red to golden-brown, densely papillose from the apex to the transitional cells on ventral side, smooth distally or with low, rounded scindulae on the dorsal surface and ± densely papillose from near the base to just above the transitional cells with O-shaped papillae, in cross-section with a 3–5-stratose dorsal stereid band, 2–3 rows of large, ventral cells and without guide cells; *laminal cells* quadrate, hexagonal to irregularly angular above, 10–16 µm wide, thick-walled, bulging on both surfaces, densely papillose with 2–5 per cell, compound papillae, irregularly C-shaped in surface view, quite abruptly passing into oblong, hyaline, usually irregularly perforated basal cells, 30–80 µm long, 10–25 µm wide, with cross walls orange, prominently thickened at corners. *Gonioautoicous. Setae* terminal, reddish-brown, straight to flexuose, 10–15 mm long, dextrorse above. *Capsules* cylindrical, 2–3 mm long, pale brown, dextrorsely ribbed; *annulus* conspicuous, deciduous in fragments, composed of 2–3 rows of enlarged, vesiculose cells; *operculum* rostrate, 1.2–1.5 mm long; *exothecial cells* elongate-rectangular, thick-walled, in regular spiral to vertical rows, becoming scarcely differentiated in 1–2 rows at the mouth; *stomata* present at the base of the urn; *peristome* double, inserted slightly below the mouth, the exostome teeth and the endostome segments fused in lower ½–⅔; *exostome teeth* filiform, 0.8–0.9 mm long, recurved when dry, dark orange below, paler above, densely finely papillose, ± distinctly trabeculate on the inner surface, with a distinct median line on the outer surface, joined at the base by a low, fenestrate membrane; *endostome segments* linear, about ⅓ shorter than the teeth, yellow to orange-brown, densely finely papillose on the outer surface, arising from a low, yellow, papillose basal membrane. *Spores* spherical to elliptical, isopolar, 15–25 µm wide, finely papillose. *Calyptra* 6–8 mm long, shiny, golden-brown, finely papillose above, lacerate at the base.

Chromosome number: unknown in the Antarctic and Subantarctic; $n = 26 + m$ reported from Alaska (Steere 1954).

Discussion. The most conspicuous vegetative character of *Encalypta procera* is the numerous clusters of brood bodies which occur in the axils of leaves. They arise from a stalk consisting of unbranched or forked filaments of oblong cells with oblique cross walls which are joined laterally to form a uniseriate mat. The principal propagules are a richly branched mass of filaments which develop from the upper part of the stalk and are composed of shortly oblong or quadrate cells with perpendicular cross walls and with papillose superficial walls. Such propagules are unknown in any other moss species in the Antarctic, making *E. procera* easily identifiable. Most leaves in this species are muticous, narrowly obtuse to short-apiculate, and ± cucullate at the apex, but the perichaetial and uppermost vegetative leaves are shortly hair-pointed and this character may lead to confusion with *E. rhaptocarpa* which, however, lacks brood bodies. Most descriptions and illustrations of *E. procera* emphasize that the margins at the leaf base are narrowly recurved but in most collections from the Antarctic they are proximally mostly plane or narrowly deflexed to occasionally recurved.

Newton (1974*a*) indicated that *Encalypta procera* differs from *E. rhaptocarpa* in having a well defined basal marginal border of thick-walled and conspicuously narrowed cells, serrulate to serrate proximal leaf margins and papillose upper cells versus a poorly defined basal marginal border of thin-walled and slightly narrowed cells, ± entire proximal leaf margins and mammillose upper cells. Examination of several dozens of specimens from the Antarctic revealed that these characters have no potential taxonomic significance because there is considerable overlap, and certainly the upper laminal cells are distinctly mammillose-papillose in both species.

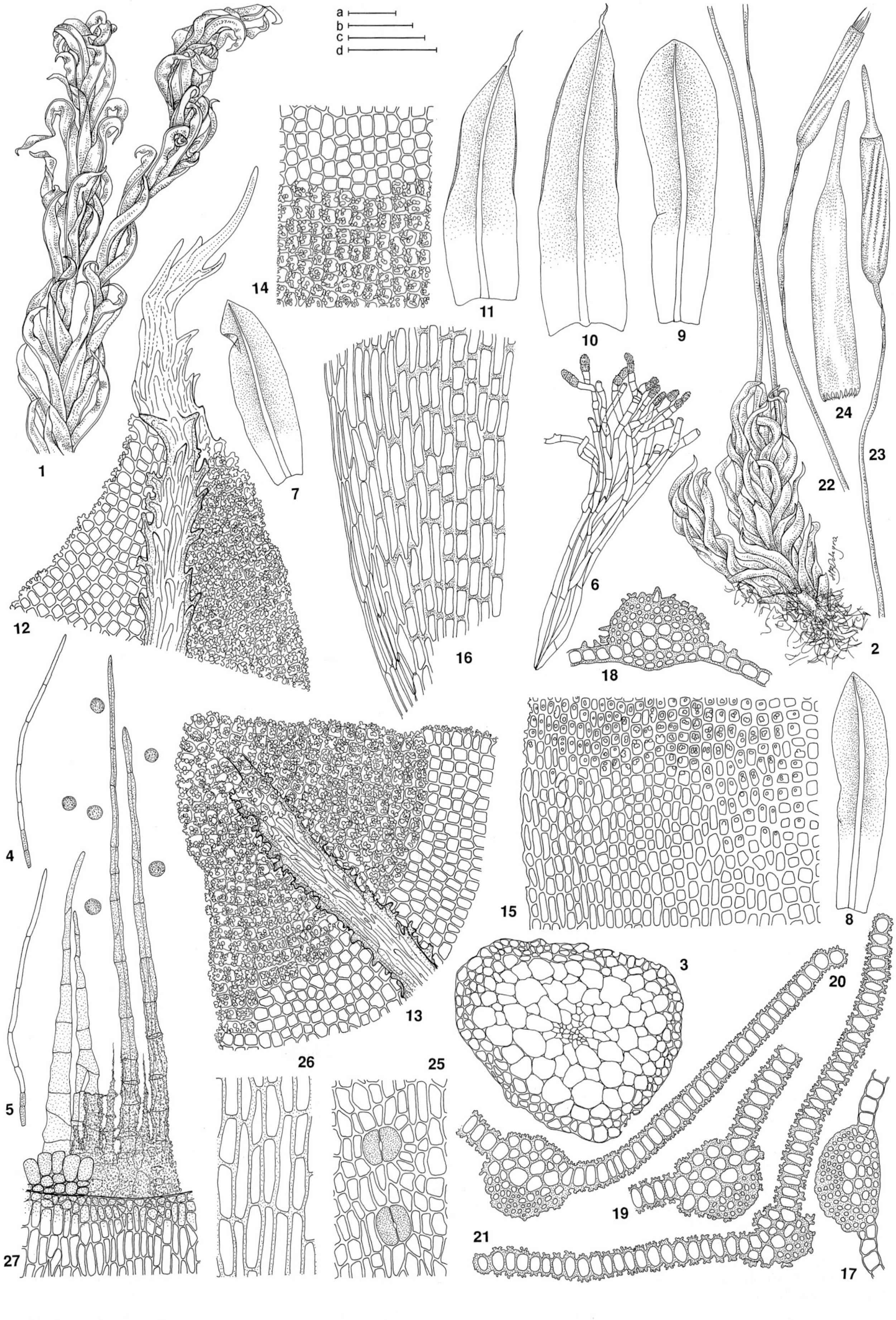
a
b
c
d
1
2
3
4
5
6
7
8
9
10
11
12
13
14
15
16
17
18
19
20
21
22
23
24
25
26
27

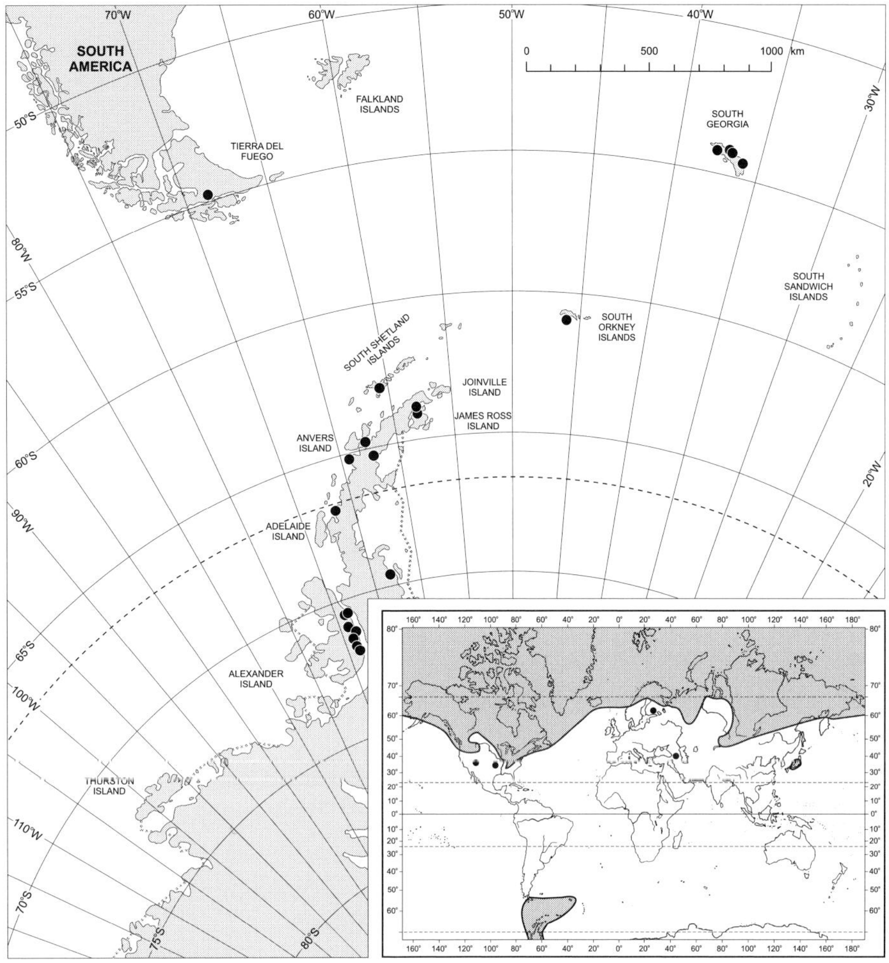

Figure 44. Distribution of *Encalypta procera* in the Fuegian region and Antarctica. Inset: global distribution.

Reproduction in Antarctica. All regional populations reproduce asexually by brood bodies which are abundantly produced in leaf axils. Sporophytes are known only in a single collection from Alexander I. (*RBH & JJL 35*, AAS, KRAM) but the peristomes are malformed and the spores are aborted. However, other occurrences have been observed on Alexander I. by RILS, these representing the southern limit of the species (Lewis Smith & Convey 2002).

Habitat. A pioneer moss of exposed or sheltered, dry or moist habitats, colonising gravelly or sandy soil, crevices in rock outcrops, rock ledges, overhangs, cliffs and scree. It is often associated with soil along melt-water runnels and at the margins of solifluction lobes subjected to early season flooding and consequently temporary burial by fine soil. The species is primarily associated with base-rich habitats, but it can also tolerate neutral to slightly acidic conditions.

World range. A bipolar species, having a continuous arctic-boreal range in the Northern Hemisphere (Horton 1983) and extending to mountains in the southern part of the Holarctic in North America (Horton 1983), the Caucasus (Ignatov & Afonina 1992), Central Asia (Bardunov 1974) and Japan (Horton 1983) (Fig. 44, inset). In the Southern Hemisphere it has been recorded, apart from in the Antarctic, in South Georgia (Newton 1974*a*; Ochyra *et al.* 2002) and in Tierra del Fuego (Ochyra & Matteri 2007).

Distribution in Antarctica. Relatively frequent and locally abundant only in central and south-eastern coastal areas on Alexander I., where it reaches its southernmost locality at lat. 72°05′ S. Widely scattered and sparse elsewhere, including the South Orkney Is. and South Shetland Is. and the Antarctic Peninsula region, notably James Ross I. (Fig. 44). It grows from sea level to about 200 m, reaching its highest locality at 600 m on Natal Ridge on Alexander I.

Selected specimens examined. WEST ANTARCTICA. SOUTH ORKNEY IS. **SIGNY I.** Jebsen Pt., *ca* 200 m, *RILS 594A* (AAS, KRAM, NY); above Paal Harbour, 80 m, *RILS 678* (AAS, KRAM); Everson Ridge, 200 m, *RILS 8166 & 8169* (AAS, KRAM).

SOUTH SHETLAND IS. **LIVINGSTON I. *Hurd Peninsula***: Johnsons Dock, 25 m, *Schulz 25* (KRAM); near Spanish Station, 20 m, *Schulz 180* (KRAM).

◄———

Figure 43. *Encalypta procera*. — **1**. Sterile plant, dry. **2**. Plant with sporophytes, dry. **3**. Cross-section of stem. **4–5**. Axillary hairs. **6**. Cluster of brood-bodies. **7–11**. Leaves. **12–13**. Leaf apices (papillae partly not indicated). **14**. Mid-leaf cells (papillae partly not indicated). **15**. Laminal cells in transition zone to basal part. **16**. Basal leaf cells. **17–19**. Cross-sections costa in lower part. **20–21**. Cross-section of leaves. **22**. Deoperculate capsule. **23**. Operculate capsule. **24**. Calyptra. **25**. Basal exothecial cells and stomata. **26**. Exothecial cells in middle of urn. **27**. Exothecial cells at mouth, portion of peristome and spores. — [1, 3–8, 17–21 from *RILS 3314*; 2, 22–26 from *Nilsson-Ehle s.n.*, 19.7.1898; 9, 13–16 from *Schulz 25*; 10–12 from *BJT 522*; 27 from *Czernadjeva 84*]. — Scale bars: a – 1 mm (1–2, 22–24); b – 100 µm (25–27); c – 100 µm (3, 12–21) and 200 µm (4–6); d – 1 mm (7–11).

WEST ANTARCTIC PENINSULA. **Danco Coast.** Cuverville I., 10–20 m, *RILS 11951* (AAS, KRAM). **Graham Coast.** ***Collins Bay***: Darboux I., 10–15 m, *RILS 3314* (AAS, KRAM). **Loubet Coast.** Orford Cliff, *RILS 5270A* (AAS, KRAM). **George VI Sound.** ***Alexander I.***: Ablation Pt., 10 m, *RBH & JJL 35* (AAS, KRAM); Moutonnée Valley, 6 m, *RBH & JJL 64* (AAS, KRAM); Callisto Cliffs, *BJT 203A* (AAS, KRAM); Succession Cliffs, *ca* 20 m, *BJT 531* (AAS, KRAM); Eros Glacier, *ca* 65 m, *BJT 525* (AAS, KRAM); Ablation Valley, *RBH & JJL 26, 33A & 56* (AAS, KRAM); Scarab Ridge Plateau, 400 m, *RILS 9357 & 9358* (AAS, KRAM); Fossil Bluff, summit of Khufu, 400–500 m, *RILS 9332* (AAS, KRAM); Uranus Glacier, Waitabit Cliffs, *ca* 20 m, *BJT 522* (AAS, BM, H, KRAM) and *ca* 170 m, *BJT 513* (AAS, KRAM); Natal Ridge, 600 m, *RILS 10418* (AAS, KRAM); Ares Oasis, 50 m, *RILS 9457C* (AAS, KRAM); Mars Oasis, 30 m, *RILS 10373* (AAS, KRAM), 150 m, *RILS 10375* (AAS, KRAM) and 175 m, *RILS 9424A* (AAS, KRAM); Ganymede Heights, 2–10 m, *RILS 11110* (AAS, KRAM); upper Flatiron Valley, 500 m, *RILS 11089B* (AAS, KRAM); Two Step Moraine, 30 m, *RILS 9492* (AAS, KRAM); Coal Nunatak, 100 m, *RILS 10443* (AAS, KRAM).

EAST ANTARCTIC PENINSULA. **James Ross Island group.** ***James Ross I.***: Lachman Crags, 35 m, *RILS 7381B & 7382B* (AAS, KRAM); St. Martha Cove, 150 m, *RILS 7537* (AAS, KRAM); Terrapin Hill, 85 m, *RILS 7742* (AAS, KRAM); Hidden Lake, 180 m, *RILS 5941A* (AAS, KRAM). **Oscar II Coast.** Starbuck Glacier, *ca* 400 m, *PK 39* (AAS, KRAM). **Wilkins Coast.** Finlay Heights, *ca* 230 m, *MJC 2 & 7* (AAS, KRAM).

2. **Encalypta rhaptocarpa** Schwägr. Fig. 45

Encalypta rhaptocarpa Schwägr., Spec. Musc. Suppl. 1(1): 56, *pl. 16*. 1811. ≡ *E. vulgaris* Hedw. var. *rhaptocarpa* (Schwägr.) E.Lawton, Moss Fl. Pacific Northwest: 118. 1971. — TYPE CITATION: ... ad nives alpis Carinthiae Heiligenbluter Tauern cum Hypno iulaceo, aliis locis similibus Gebhard invenit, augusto fructiferum. [Lectotype (*vide* Horton 1983: p. 413): G-Hedwig/Schwägrichen (*non vidi*)].

Encalypta patagonica Broth. *in* Dusén, Bot. Not. 1905: 301. 1905. — TYPE CITATION: Fretum magellanicum, Punta Arenas in terra. [Lectotype (*vide* Ochyra 1998*a*: p. 134): 'Patagonia australis ad Punta Arenas emporium in terra, Dusén 31' – S!; isotypes: BM!, H-Brotherus!]. First synonymised by Ochyra (1998*a*: p. 134).

Plants medium-sized, densely or loosely caespitose, dull, dark or bright green to glaucous-green or sometimes yellowish to brownish above, brown or orange-brown below. *Stems* 0.5–3.0 cm tall, erect, forked, radiculose below or sometimes near to the tips, in transverse section rounded, with a distinct central strand surrounded by a medulla of large, thin-walled, hyaline cells and a cortex of 1–2 layers of smaller cells with thicker, brown walls; *axillary hairs* hyaline, filiform, 10–15-celled. *Leaves* closely appressed, irregularly spirally twisted and incurved, somewhat keeled when dry, erect-spreading when moist, oblong to oblong-elliptic, 1.5–3.5 mm long, 0.5–1.0 mm wide, broadly short-acuminate, acute to obtuse and muticous to long- piliferous; *hair-point* hyaline, flexuose, denticulate; *margins* unistratose, plane throughout or erect near the apex, entire to bluntly serrulate below to crenulate-papillose above; *costa* ceasing below the apex to percurrent, often disappearing in the awn, golden or golden-brown, smooth or distally papillose at back; *laminal cells* rounded-hexagonal to subquadrate, 12–16 μm wide, thick-walled, opaque, chlorophyllose, bulging and densely papillose with lunate papillae; *basal cells* hyaline, rectangular, with thick, pale yellow cross walls, often becoming narrower and elongate towards the margins forming a well defined hyaline or yellowish border of 3–4 rows of cells. *Autoicous. Setae* erect, orange-brown to dark red-brown, 5–10 mm long, smooth, dextrorse above. *Capsules* erect, cylindrical, 2–3 mm long, about 1 mm wide, abruptly narrowed at the base, with strong or weaker, dark reddish-brown, perpendicular ribs, wrinkled at the short neck when dry; *annulus* none or usually persistent, consisting of 2 rows of small, orange-brown, quadrate cells; *operculum* 0.8–1.3 mm long, consisting of a long, straight rostrum arising from a plano-convex base with red edge; *exothecial cells* oblong-hexagonal to rectangular, thin-walled in grooves between ridges and thicker-walled on ribs, scarcely differentiated at the orifice; *stomata* superficial, bicellular, long-pored, scattered in the lower half of the urn; *peristome* absent or single, inserted somewhat below the mouth, consisting of 16, short, to 250 μm long, lanceolate, truncate, dark yellow to orange, rather coarsely papillose teeth. *Spores* 30–50 μm wide, spherical or subreniform, warty-papillose. *Calyptra* 3.5–4.5 mm long, shiny, consisting of a golden cylinder, entire or slightly erose at the basal margin, gradually or abruptly narrowed rostrum, 1.1–1.5 mm long, smooth to slightly scabrous, fuscous at the tip.

Chromosome numbers: $n = 13$ – King George I. (Kuta *et al.* 1982); the same number was recorded from Poland (Przywara *et al.* 1984) and $n = 26$ from Alaska (Steere (1954).

Discussion. Fertile plants of *Encalypta rhaptocarpa* with normally-developed capsules do not pose any problems with identification and the most distinctive features of the species are prominently straight-ribbed and fluted capsules, the peristome of which is often lacking in Antarctic populations, and large, warty-papillose spores. The latter are distinctly heteropolar with sculpture of proximal and distal surfaces markedly differentiated. The proximal surface is narrowly, radially plicate with some small central warts, whereas the distal surface is ornamented by many large verrucae. Sterile plants of *E. rhaptocarpa*, especially those lacking long piliferous leaves and having obtuse and muticous leaf apices, may be confused with *E. procera*. However, the absence of axillary propagules in *E. rhaptocarpa* enables distinction between the species.

Encalypta rhaptocarpa was reported for the first time from the Antarctic by Newton (1974*a*) as *E. patagonica*. However, both Longton (1967) and Lewis Smith (1972) reported it, as *Encalypta* sp., and Webb (1973) as *E.* cf. *peruviana*, on Signy I. This species

→

Figure 45. *Encalypta rhaptocarpa*. — **1**. Plant with sporophytes, dry. **2**. Plant with sporophyte, wet. **3**. Cross-section of stem. **4–5**. Axillary hairs. **6–12**. Leaves. **13–14**. Leaf apices (papillae partly not indicated). **15**. Mid-leaf cells (papillae partly not indicated). **16**. Lamina cells in transition zone to basal part. **17**. Basal leaf cells. **18–20**. Cross-sections of leaves. **21**. Cross-section of costa in lower part. **22**. Basal exothecial cells. **23**. Exothecial cells in middle of urn. **24**. Exothecial cells at mouth, portion of peristome and spores. — [1–2, 11–12, 14–15 from *RO 2286/80*; 3–5 from *RO 1161/80*; 6–7, 9–10, 16–19 from *RW 127*; 8, 13, 20–21 from *REL 1178*; 22–24 from *REL 1098*]. — Scale bars: a – 100 μm (4–5); b – 100 μm (17, 22–24); c – 1 mm (1–2) and 100 μm (3, 13–16, 18–21); d – 1 mm (6–12).

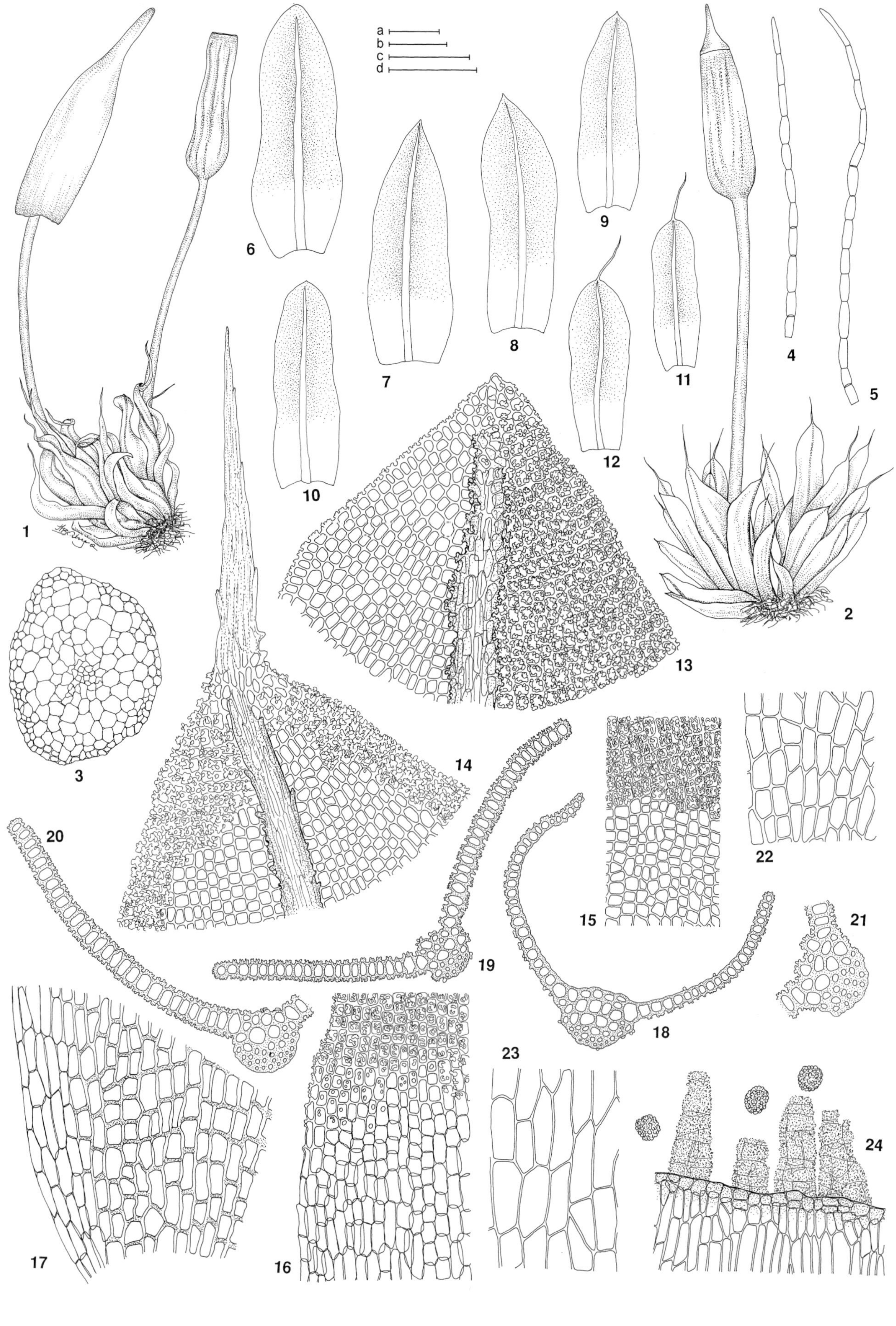
a
b
c
d
1
2
3
4
5
6
7
8
9
10
11
12
13
14
15
16
17
18
19
20
21
22
23
24

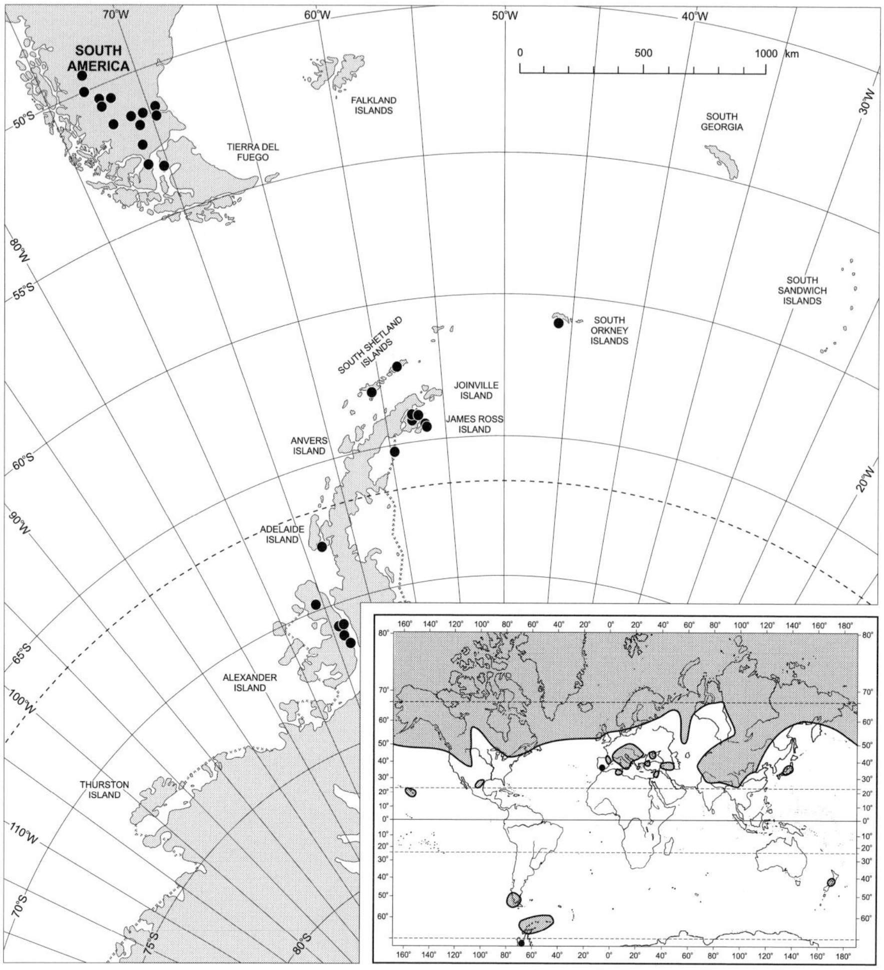

Figure 46. Distribution of *Encalypta rhaptocarpa* in the Fuegian region and Antarctica. Inset: global distribution.

was described twice by Dusén (1905*c*, 1906) from southern Patagonia, but Ochyra (1998*a*) reduced this name to synonymy of *E. rhaptocarpa*. In general, this is a highly protean species subject to considerable geographical variation. Horton (1982*b*, 1983) discussed variability within this species in detail and stated that there exist numerous populations showing features intermediate between *E. rhaptocarpa* and *E. vulgaris*, including the loss of ribs on the capsules and various reductions of the peristome to its total absence in the utmost stage. Gymnostomous capsules are frequently observed in Antarctic populations, but the grooves of the capsule are mostly well developed. Although populations with such intermediate features were often recognised taxonomically as distinct species or infraspecific taxa, until a thorough taxonomic study of this complex is carried out the broad concept of *E. rhaptocarpa* is accepted in the present treatment.

Reproduction in Antarctica. Commonly with sporophytes. In exposed sites the capsules are sometimes misshaped or malformed with aborted spores, but normally dehisced capsules with well developed spores are very frequent.

Habitat. Populations of *Encalypta rhaptocarpa* occur in a wide variety of base-rich or neutral, sheltered or windswept habitats including ledges and crevices in rock outcrops, soil amongst stones (Pl. 41) as well as silty, gravelly or sandy soil and morainic detritus in fellfields. It can tolerate xeric and more mesic sites, and colonies of *E. rhaptocarpa* are often found in temporary seepage areas, at the margins of melt-water runnels, drainage channels and solufluction stripes, colonising soil disturbed by frost action and forming clay puddles and frost boils. On Signy I. it is a frequent component of the community dominated by *Syntrichia saxicola*, *S. magellanica* and *Schistidium antarctici*, typically developed on soils derived from marble and amphibolite outcrops (Lewis Smith 1972, 2005*c*; Lewis Smith & Gimingham 1976).

World range. A bipolar species (Fig. 46, inset), having a continuous range in the Northern Hemisphere across North America and Eurasia in boreal and arctic regions, becoming increasingly rare southwards and extending to Colorado in North America (Horton 1983), the Mediterranean including North Africa (Ros *et al.* 1999, 2000), the Caucasus and the Near East (Frey & Kürschner 1991), Central Asia (Mamatkulov 1990), Central Siberia (Bardunov 1974), Mongolia (Tsegmed 2001*a*,*b*), China (Cao *et al.* 1992) and Japan (Noguchi 1988). Outside the Holarctic it is known only from north-east Mexico (Horton 1994) and the Hawaiian Is. (Bartram 1952*b*) and from the Southern Hemisphere, including New Zealand (Bartlett 1984), southern South America (Matteri 1984 as *E. patagonica*; Deguchi 1991*a*) and in the maritime Antarctic.

Distribution in Antarctica. A widely scattered, occasional to locally frequent species in the maritime Antarctic (Fig. 46). It is infrequent and localised on the South Orkney Is. (with the exception of Signy I. where it is quite common on calcareous soils) and South Shetland Is. and rare on the west coast of the Antarctic Peninsula. However, it is locally frequent in the James Ross I. group off the north-east coast of the Antarctic Peninsula and on Alexander I. where it extends as far south as lat. 71°20′ S. The species grows most frequently from sea level to about 200 m and attains its highest known altitude at 650 m on Alexander I.

Selected specimens examined. WEST ANTARCTICA. SOUTH ORKNEY IS. **Signy I.** Between Thulla Pt. and Jebsen Pt., *ca* 50 m, *MWH 163* (AAS, KRAM); ridge above cove south of Foca Pt., *ca* 15 m, *REL 1098* (AAS, KRAM) and *ca* 30 m, *RW 127* (AAS, KRAM); Foca Pt., *ca* 30 m, *RILS 334 & 335* (AAS, KRAM); North Pt., *ca* 165 m, *REL 1201 & 1202* (AAS, KRAM, NY); Three Lakes Valley, W of Knob Lake, *ca* 30 m, *RW 88A* (AAS, KRAM); Three Lakes Valley, between Pumphouse Lake and Waterpipe Beach, 15 m, *MWH 144A* (AAS, KRAM); Elephant Flats, 20 m, *RILS 377* (AAS, KRAM) and *ca* 140 m, *RILS 597* (AAS, KRAM); Knife Pt., *RW 15* (AAS, KRAM); Factory Cove, 1–39 m, *REL 839* (BM, KRAM, PC); McLeod Glacier, 104 m, *RILS 7049* (AAS, KRAM). **Moe I.** Landing Cove, 50–100 m, *RILS 5238A* (AAS, KRAM).

SOUTH SHETLAND IS. **King George I.** ***Bransfield Strait***: Blue Dyke, 125 m, *RO 1161/80* (KRAM). ***Admiralty Bay***: Krzesanica, 80 m, *RO 4970/79* (KRAM); Klekowski Crag, 110 m, *RO 2286/80* (KRAM). **Livingston I.** ***Hurd Peninsula***: South Bay, *Schulz 156* (KRAM).

WEST ANTARCTIC PENINSULA. **Fallières Coast.** ***Léonie Is.***: Léonie I., 20 m, *RILS 10644* (AAS, KRAM) and 30 m, *RILS 8927A* (AAS, KRAM). **George VI Sound.** ***Alexander I.***: ridge running W from Douglas Range to Hampton Glacier, *ca* 650 m, *RILS 6027* (AAS, KRAM); Ablation Pt., *RBH & JJL 30* (AAS, KRAM); Ablation Valley, 2–3 m, *RILS 9273A* (AAS, KRAM) and 5 m, *RILS 2715* (AAS, KRAM); Ablation Lake, 2–3 m, *RILS 9272B, 9277E & 9317A* (AAS, KRAM); Moutonnée Valley, 5 m, *RBH & JJL 29* (AAS, KRAM); Fossil Bluff, Scarab Ridge Plateau, 400 m, *RILS 9355B & 9359B* (AAS, KRAM).

EAST ANTARCTIC PENINSULA. **James Ross Island group.** ***Vega I.***: False Island Pt., 25–30 m, *RILS 7830* (AAS, KRAM). ***James Ross I.***: Lachman Crags, 3–50 m, *RILS 7340* (AAS, KRAM); Brandy Bay, 5–35 m, *RILS 7327C* (AAS, KRAM); Fortress Hill, 25 m, *RILS 7810* (AAS, KRAM) and 150 m, *RILS 7752* (AAS, KRAM); San José Pass, 165 m, *RILS 7529* (AAS, KRAM); St. Martha Cove, 15 m, *RILS 7479* (AAS, KRAM), 50 m, *RILS 7464* (AAS, KRAM) and 75 m, *RILS 7438A* (AAS, KRAM); Terrapin Hill, 85 m, *RILS 7741* (AAS, KRAM), 200 m, *RILS 7733* (AAS, KRAM), 250 m, *RILS 7727* (AAS, KRAM) and 540 m, *RILS 7713* (AAS, KRAM); Tumbledown Cliffs, 100 m, *RILS 5491B* (AAS, KRAM); Blancmange Hill, 12–15 m, *RILS 7687* (AAS, KRAM); Dreadnought Pt., 3 m, *RILS 7493* (AAS, KRAM). ***Cockburn I.***: plateau, *ca* 400 m, *RILS 7970A* (AAS, KRAM). ***Seymour I.***: Cape Wiman, 20 m, *RILS 7876* (AAS, KRAM). **Nordenskjöld Coast.** Robertson I., Seal Nunataks, Oceana Nunataks, 50 m, *RILS 6416D* (AAS, KRAM) and 100 m, *RILS 6430A* (AAS, KRAM).

Subclass C3. **DICRANIDAE** Doweld

Dicranidae Doweld, Prosyll. Tracheophyt.: i. 2001.

Dicranidae Vitt, Goffinet & Hedd. *in* Bates, Ashton & Duckett, Bryol. Twenty-First Cent.: 119. 1998, *nom. nud.*

Dicranidae W.R.Buck & Goffinet *in* A.J.Shaw & Goffinet, Bryoph. Biol.: 98. 2000, *nom. inval. descr. angl.*

Bryidae Engl. order Haplolepideae M.Fleisch., Musci Buitenzorg 3: xii. 1908 ['Reihe Haplolepideae'], *nom. illeg.*

The subclass Dicranidae corresponds to the group of mosses which was first designated by Philibert (1884) as 'Aplolépidées'. This term was subsequently latinised by Fleischer (1904, 1908) as the Haplolepideae. He considered it as one of three major groups of the tribus Arthrodontei, parallel to Heterolepideae and Diplolepideae which were initially considered as 'Subtribus' (Fleischer 1904) and later as 'Reihes' i.e. orders (Fleischer 1908). Dixon (1932*a*) designated it as a sub-clan within the clan Arthrodonteae which is equivalent to the Bryopsida as defined in the present treatment.

The subclass Dicranidae is a monophyletic group of mosses which is primarily characterised by a single haplolepidous peristome. The exostome and endostome components in this type of the peristome are opposite. The exostome is typically absent, but sometimes it is present as a very reduced, vestigial or rudimentary structure called a preperistome or prostome which is situated outside, and generally adhering to, the base of the main peristome teeth. The haplolepidous peristome teeth develop from periclinal wall pairs of the primary (PPL) and inner (IPL) peristomial layers. They are homologous to the endostome of the diplolepidous peristome and more precisely to the cilia of the *Bryum*-type endostomes (Shaw *et al.* 1989*b*). They are not vertically keeled as is usually the case with the *Bryum*-type because the PPL cells do not usually become enlarged or convex.

The haplolepidous peristome consists of a ring of 16 teeth, each tooth having horizontal lines across the outer face, but without vertical lines on the outer surface. In typical situations this peristome exhibits a sesquilepidous pattern – 16 PPL cells opposite 24 IPL cells (2 : 3 ratio). Therefore, for each pair of PPL cell columns there are three columns of IPL cells. Thus, the inner surface of each tooth is made up of columns of one and a half plates in a staggered right–left arrangement. The internal plates of each tooth are very unequal due to the asymmetric first secondary divisions in the IPL. Only every third anticlinal wall in the IPL is aligned with an anticlinal wall in the PPL, whereas the intervening IPL walls are not aligned with PPL or OPL walls. Therefore the teeth are asymmetric and form pairs. This is a characteristic feature of the haplolepidous peristome as is the total absence of cilia.

The majority of haplolepidous taxa complete development with a [4] : 2 : 3 peristomial formula, but some taxa deviate from this pattern. Sometimes the IPL does not undergo the final asymmetric divisions and consequently it consists of only 16 or 8 cells resulting in the [4] : 2 : 2 or [4] : 2 : 1 peristomial formulae, respectively. Examples include *Lorentziella* (Rushing & Snider 1980), *Bruchia* (Shaw *et al.* 2000) and the Rhachitheciaceae (Goffinet 1997). Sometimes, deviant peristomial patterns have been observed, for example 2 : 5 in some species of *Dicranum* and *Ditrichum pallidum* (Shaw *et al.* 1989*b*) and 2 : 6 at the base of the teeth in *Eucamptodon* (Allen 1987*a*).

The classic haplolepidous peristome has many varied expressions which are generally characteristic for different orders, families and/or genera of this subclass. The teeth remain free or are exserted on a low or high basal membrane. They are either undivided and entire or perforated, or are variously forked at the tips or deeply split near to the base into filiform segments. Both faces of the teeth may be smooth or ornamented and variously thickened. Consequently, they may be not or distinctly trabeculate on the outer or inner face. In some genera various peculiarities may occur in the peristome structure, but nonetheless these can be interpreted as deviations from the basic pattern which is called the *Dicranum*-type peristome (Vitt *et al.* 1998; Vitt 2000). Edwards (1979) described the separate *Seligeria*-type peristome and designated it as diplolepidous, but it is actually a haplolepidous peristome often with a well developed prostome (Crum 1994*b*; Allen 2001).

The Dicranidae comprise exclusively acrocarpous or cladocarpous mosses which exhibit considerable variation in gametophytes but consistently lack paraphyllia and pseudoparaphyllia. This is a very large and markedly diverse group consisting of nearly 4000 species, i.e. about one third of all known moss species, which are classified into 25 families and nearly 220 genera. Fleischer (1920, 1923) and Brotherus (1924*b*) divided the haplolepidous mosses into four orders, namely Fissidentales, Dicranales, Pottiales and Grimmiales. On the basis of his studies on the peristomes, Edwards (1979, 1984) recognised two additional orders, Syrrhopodontales and Seligeriales, of which only the latter has been accepted in the latest classifications proposed by Vitt *et al.* (1998) and Buck & Goffinet (2000). On the other hand, these authors merged the Fissidentales and the Dicranales on account of the identical features of the peristomes. Additionally, Buck & Goffinet (2000) placed the Archidiales in the Dicranidae, but this monotypic taxon, consisting of the single genus *Archidium*, represents a distinctive evolutionary line among the true mosses characterised by the absence of a columella, unusually large spores with a unique mode of germination which, additionally, are very few per capsule, as well as the enormous foot of the sporophyte (Snider 1975*a*,*b*). These characteristics certainly warrant the status of the separate subclass Archidiidae for the Archidiales, and it has been recognised in the recent moss classifications by Crum & Anderson (1981), Schofield (1985), Snider (1994*a*), Gradstein *et al.* (2001) and Crum (2001).

Acceptance of the *Seligeria*-type peristome as haplolepidous implies that recognition of the separate order Seligeriales is unwarranted and the Seligeriaceae fall well within the concept of the Grimmiales. Such a relationship to this family has already been suggested by Fleischer (1904) and Walther (1983). Thus, the subclass Dicranidae is here accepted to comprise only three large orders, Dicranales, Grimmiales and Phascales, each of which is represented in the Antarctic and may be differentiated in the following key.

KEY TO THE ORDERS OF THE DICRANIDAE

1. Inner layer of the peristome teeth thicker than outer, with well developed transverse walls forming distinct trabeculae or ridges; peristome teeth typically forked into 2 prongs or undivided, lanceolate in shape, often vertically striate on the outer surface; basal membrane absent or low, very rarely high; alar cells often differentiated .... V. **Dicranales**

1. Inner layer of the peristome teeth thinner or equal to outer, the latter often distinctly trabeculate or ridged; peristome teeth undivided or deeply split into 2 filamentous segments, less often divided into 2 prongs, entire or papillose, very occasionally striate; basal membrane absent, low or very high; alar cells mostly undifferentiated 2

   2. Peristome teeth straight, undivided, entire or perforated, less often forked or split into 2 filiform segments, smooth or papillose on one or both surfaces; basal membrane absent or low, very rarely high; leaves narrow, usually ovate-lanceolate, sometimes subulate, very rarely broadly lingulate; upper leaf cells short and ± isodiametric to elongate, smooth or papillose; alar cells sometimes differentiated; lower leaf cells not markedly different from upper leaf cells ..... VI. **Grimmiales**

   2. Peristome teeth straight or often spirally twisted, frequently split into 2 filiform segments to a low or high basal membrane, papillose on one or both surfaces, rarely smooth or spirally striate; leaves often broad, spathulate, ovate to lanceolate, never subulate; upper leaf cells mostly isodiametric, densely pluripapillose, rarely smooth; alar cells always absent; lower leaf cells hyaline, somewhat enlarged and sharply set off from upper cells or more or less intergrade in size and degree of papillosity with upper cells .... VII. **Phascales**

Order V. DICRANALES H.Philib. *ex* M.Fleisch.

Dicranales H.Philib. *ex* M.Fleisch., Musci Buitenzorg 1: xxvi, 11–12. 1904 ['Reihe Dicranoideae'].

Fissidentales M.Fleisch., Hedwigia 61: 391. 1920. First synonymised by Vitt *et al.* (1998: p. 119).

Schistostegales M.Fleisch., Hedwigia 61: 394. 1920, ***syn. nov.***

Bryoxiphiales Á.Löve & D.Löve *ex* H.A.Crum & L.E.Anderson, Moss. East. N. Am. 1: 113. 1981, ***syn. nov.***

Mitteniales A.J.Shaw, Syst. Bot. 10: 228. 1985, ***syn. nov.***

This order is primarily characterised by the classic haplolepidous peristome which is sometimes referred to as dicranoid because the teeth are often forked to about halfway into two branches or sometimes split nearly to the base into two filaments. When typically developed it has the ventral plates of the teeth much thicker than the dorsal plates and they are usually distinctly trabeculate, whereas the latter are most often vertically striate. In addition, the basal membrane is lacking or low and the preperistome is only occasionally visible. In some genera the peristomes are variously reduced or irregular, for instance with the ventral and dorsal plates of the teeth equally thickened, with the reduced trabeculation and/or weak striation, but they remain essentially similar in the general pattern to the normal dicranoid peristome. However, in many genera of the Dicranales the capsules are cleistocarpous or gymnostomous.

The Dicranales exhibit considerable gametophyte diversity and all major features of the leaves and leaf cells are present in this order in various combinations, although differentiated alar cells, often forming large and inflated auricles, occur in this group more frequently than in other orders of the acrocarps. Conversely, papillose laminal cells and hyaline awns are rare. The plants vary in size from minute to fairly large and robust and the leaves are usually narrow, elongate, narrowly-lanceolate and often subulate. The foliage is sometimes distichous and complanate.

The Dicranales are interpreted here in the broader sense to encompass the large genus *Fissidens* which, apart from the unusual leaves differentiated into two sheathing laminae clasping the stem, share with this order the perfect dicranoid peristome. Also, three monotypic orders, Bryoxiphiales, Mitteniales and Schistostegales, are merged with the Dicranales. Such interpretation makes the Dicranales the largest order of the haplolepidous mosses consisting of about 2000 species classified into 13 families and over 100 genera. In the Antarctic the order is represented by two families, Ditrichaceae and Dicranaceae, comprising 18 species belonging to eight genera.

Family V. DITRICHACEAE Limpr.

Ditrichaceae Limpr., Laubm. Deutschl. 1: 482. 1887, *nom. cons.* — Type: *Ditrichum* Hampe (*nom. cons.*).

Distichiaceae Schimp., Syn. Musc. Eur.: cxxx, 135. 1860 ['Distichieae], *nom. rej.* — Type: *Distichium* Bruch & Schimp.

Ceratodontaceae Schimp., Syn. Musc. Eur.: cxxx, 138. 1860 ['Ceratodonteae'], *nom. rej.* — Type: *Ceratodon* Brid.

Dicranaceae Schimp. subfam. Ditrichoideae Lindb., Utkast Eur. Bladmoss.: 34. 1878 ['Ditricheae']. — Type: *Ditrichum* Hampe (*nom. cons.*).

The modern circumscription of the Ditrichaceae was presented by Limpricht (1887) and his concept was subsequently adopted in the treatment of the Musci in the second edition of *Die Natürlichen Pflanzenfamilien* by Brotherus (1924*b*). Since then the Ditrichaceae have gained almost universal acceptance among bryologists and their recognition as a family distinct from the Dicranaceae has had only a few opponents (e.g. Smith & Newton 1968; Newton 1977*a*; Eddy 1988). They view the latter family very broadly, following the initial concept of Brotherus (1901*b*) presented in the treatment of the Musci in the first edition of *Die Natürlichen Pflanzenfamilien*. Gametophytically it is nearly impossible to distinguish the members of the Ditrichaceae and the Dicranaceae, but many sporophyte characters, especially the architecture of the peristome teeth, separate them.

In the Ditrichaceae the teeth, when present, are linear and deeply divided into two terete, distinctly barred prongs which are smooth or variously papillose but never vertically pit-striated. Conversely, within the Dicranaceae they are mostly flat and lanceolate, never barred, most often split to about half-way into two divisions and have a characteristic vertical pit-striated ornamentation. A characteristic feature of the Ditrichaceae is a strong tendency to cleistocarpy, which is practically unknown in the Dicranaceae, and all genera are, without exception, stegocarpous. However, the monotypic genus *Pseudephemerum*, traditionally placed in this family, has an inoperculate capsule. It has recently been transferred to the Ditrichaceae by Smith (1978, 2004) and Matsui & Iwatsuki (1990) who pointed out its affinity to *Eccremidium* subgen. *Pseudopleuridium*. In the Ditrichaceae six of 24 genera are cleistocarpous (Goffinet 1998). Additionally, there is a distinct trend in this family to shortening the seta and at least 11 genera exclusively have the capsules immersed in the leaves, including all cleistocarpous genera. However, in some typically stegocarpous genera, for example *Ditrichum*, there are also species with immersed capsules. On the other hand, in the Dicranaceae the capsules are mostly exserted on long setae and only occasionally are they immersed to short-exserted.

Gametophytically the Ditrichaceae are nondescript, making this family difficult to characterise. Its members completely lack the differentiated alar cells which are so diagnostic of many of the dicranaceous genera, but this distinction is not absolute. The family comprises mainly small to minute mosses, usually growing in tufts or gregarious on bare, often disturbed soil. The leaves are arranged in many rows but sometimes they are distichous or trifarious. They are lanceolate, acuminate or subulate, rarely ovate or obovate, not or sheathing at the base. The costa is single, broad or narrow, subpercurrent to long-excurrent, very rarely absent. The laminal cells are short to linear, thin- to thick-walled, uni- to bistratose and smooth or sometimes densely mammillose.

Cytological evidence suggested by Bryan (1956) in support of the separation of the Ditrichaceae and Dicranaceae seems to be insufficient, mostly due to the scant cytological data available for other members of these families and, additionally, the ditrichaceous and dicranaceous mosses exhibit the great cytological heterogeneity (Smith & Newton 1968). Because of some sporophyte differences, it seems reasonable to separate the Ditrichaceae from the Dicranaceae. This separation, though based on minor differences, is less for convenience, as suggested by Crum & Anderson (1981), than for the structural dissimilarities in sporophytes.

The Ditrichaceae are very diverse both gametophytically and sporophytically and Limpricht (1887), when establishing the present concept of the family, subdivided it into three 'Gruppen' which were later given the rank of subfamilies (Brotherus 1924*b*), namely Ditrichoideae, Distichioideae and Ceratodontoideae. In fact, the latter two subfamilies were recognised by Schimper (1860) as families in their own rights, Distichiaceae and Ceratodontaceae and, because these names antedate the better known and widely used name Ditrichaceae, Magill (1977) proposed conserving the latter name and since the Sydney Code (Voss *et al.* 1983) it is on the list of *Nomina familiarum conservanda.*

The Ditrichaceae are the third largest family of the Dicranales, consisting of about 170 species and 24 genera. It has a worldwide distribution, with most taxa occurring in temperate and subtropical regions. There are five genera known in the south polar and cool temperate regions, namely *Ceratodon*, *Distichium*, *Ditrichum*, *Chrysoblastella* and *Skottsbergia*, the latter being a monotype endemic to South Georgia and Tierra del Fuego (Ochyra *et al.* 2002). The first three genera are well represented in the Antarctic, with the number of species ranging from one to five.

KEY TO THE GENERA OF THE ANTARCTIC DITRICHACEAE

1. Leaves distichous; plants strongly flattened .... 7. ***Distichium***
1. Leaves in many rows; plants not flattened 2
   2. Leaves from a lanceolate, oblong- or ovate-lanceolate and often sheathing base abruptly narrowed to a long subula; margins plane .... 8. ***Ditrichum***
   2. Leaves ovate to lanceolate, acute to acuminate, not sheathing at the base and never setaceous at the apex; margins mostly strongly recurved ...... 9. ***Ceratodon***

7. **DISTICHIUM** BRUCH & SCHIMP., *nom. cons.*

Distichium Bruch & Schimp. *in* Bruch, Schimp. & W.Gümbel, Bryol. Eur. 2: 153. 1846 [Fasc. 29–30 Mon.: 1], *nom. cons.* ≡ *Leptotrichum* Hampe *ex* Müll.Hal. sect. *Distichium* (Bruch & Schimp.) Mitt., J. Linn. Soc. Bot. 7: 148. 1863. — TYPE: *Distichium capillaceum* (Hedw.) Bruch & Schimp. (*Cynontodium capillaceum* Hedw.) (*typ. cons.*).

Cynontodium Hedw., Spec. Musc. Frond.: 57. 1801, *nom. rej.* — TYPE: *non designatus.*

Plants small to large but slender, strongly flattened, bright to dark green or yellowish-green, in dense, glossy or dull tufts. *Stems* erect, often elongated, simple or dichotomously branched, densely tomentose below, with a large central strand; *axillary hairs* filiform, composed of 8–10 elongate cells, hyaline throughout or with 2–3 brownish basal cells. *Leaves* distichous, inserted at right angles to the long axis of the stem, neither crispate nor plicate, non-decurrent, from a concave, conduplicate, erect, tightly clasping, oblong or oblong-ovate, sheathing base abruptly narrowed to a short or long, erect-spreading to spreading, subulate or setaceous, rough acumen; *margins* plane, somewhat convolute below, entire to weakly crenulate at the shoulders; *costa* single, strong, filling most of the subula, excurrent, rough, in cross-section with a central row of large guide cells separating two stereid bands, sometimes with the ventral band lacking or poorly developed; *laminal cells* smooth, small, subquadrate to irregularly hexagonal, thick-walled, obscure at the shoulders, becoming long-rectangular to linear, thin-walled and pellucid below; *alar cells* not differentiated. *Autoicous*, *paroicous* or *synoicous*. *Perichaetial leaves* not differentiated. *Setae* terminal, straight, erect, elongate. *Capsules* long-exserted, erect or inclined, cylindrical to ovoid-cylindrical, symmetric to inclined and asymmetric, not sulcate or strumose; *annulus* large, compound, deciduous to ± persistent; *operculum* low-conical; *exothecial cells* not in distinct rows, irregularly rectangular to hexagonal, thin- to firm-walled; *stomata* present at the base of the urn, phaneroporous, bicellular; *peristome* single, consisting of 16 shortly lanceolate teeth, free or sometimes paired, variously divided or perforated nearly to the base, papillose-roughened or smooth with oblique striations, without or with a low basal membrane. *Spores* spherical, densely papillose. *Calyptra* cucullate, smooth.

The two-ranked leaves consisting of a strongly clasping base and a slender, distinctly mammillose subula give the stems a very marked, flabellate, flattened, distichous appearance which is unique among Antarctic mosses and makes *Distichium* a distinct and unmistakable genus. However, taxonomically *Distichium* is a difficult genus because of the lack of reliable gametophyte characters which could be useful for separation of species, so it is based mostly upon sporophyte characters. This is especially the case in Antarctica where the two species of *Distichium* very rarely produce sporophytes.

Distichium is a small bipolar genus consisting probably of no more than five species widely distributed throughout the Holarctic, especially in northern and montane regions, with some isolated occurrences in tropical mountains. In the Southern Hemisphere it occurs mainly in cool and cold temperate and polar regions where it is represented by two species, of which *D. inclinatum* is here reported for the first time from the Southern Hemisphere. *Distichium austrogeorgicum* Müll.Hal. from South Georgia is conspecific with *Chorisodontium aciphyllum* (Ochyra 1996*c*). Useful treatments of *Distichium* are available for Australasia (Sainsbury 1955; Seppelt 1982), South Georgia (Newton 1977*a*) and Antarctica (Newton 1977*b*).

1. Capsules erect, cylindrical and symmetric or almost so; spores 15–25 µm wide; plants usually tall in bright to dark green, glossy tufts .... 1. ***D. capillaceum***
1. Capsules inclined to horizontal, ovoid or oblong-ovoid, asymmetric; spores 30–45 µm wide; plants shorter, dull green, less shining .... 2. ***D. inclinatum***

1. **Distichium capillaceum** (Hedw.) Bruch & Schimp. Fig. 47, pl. 38

Distichium capillaceum (Hedw.) Bruch & Schimp. *in* Bruch, Schimp. & W.Gümbel, Bryol. Eur. 2: 156, *tab. 193*. 1846 [Fasc. 29–30 Mon.: 4, *tab. I*]. ≡ *Cynontodium capillaceum* Hedw., Spec. Musc. Frond.: 57. 1801. ≡ *Swartzia capillacea* (Hedw.) Brid., J. f. Bot. 1800(2): 289. 1801. ≡ *Bryum capillaceum* (Hedw.) Dicks., Fasc. Pl. Crypt. Brit. 4: index [i]. 1801. ≡ *Didymodon capillaceus* (Hedw.) F.Weber & D.Mohr, Ind. Mus. Pl. Crypt.: 2. 1803. ≡ *Trichostomum capillaceum* (Hedw.) Sm., Fl. Brit. 3: 1236. 1804. ≡ *Cynodontium capillaceum* (Hedw.) Brid., Spec. Musc. 1: 156. 1806. ≡ *Leptotrichum capillaceum* (Hedw.) Mitt., J. Proc. Linn. Soc. Suppl. Bot. 1: 10. 1859. — Type citation: Saxonia inferior, Niedersachswerfen et Neuhof, in montibus gypsaceis prope Steigerthal, ad ripas fluvii Bode, (in calcareis Franconiae, alestribus Austriae,) Suecia, Islandia in fissuris rupium; also cited Hedwig (1789: p. 76, *pl. 28*) and Bridel (1798: p. 117). [Lectotype (*selected here*): The right specimen in the lower row, on the sheet labelled 'Swartzia capillacea Hedw. St. crypt. Vol. II. p. 72. t. 26' – G-Hedwig/Schwägrichen!].

Didymodon capillaceus var. *compactus* Huebener, Musc. Germ.: 282. 1833. ≡ *Swartzia montana* Lindb. var. *compacta* (Huebener) Braithw., Brit. Moss Fl. 1: 103. 1881. ≡ *Distichium capillaceum* var. *compactum* (Huebener) Dalla Torre & Sarnth., Fl. Tirol. 5: 195. 1904. ≡ *Didymodon subulatus* Schkuhr, Deutschl. Krypt. Gew. 2(2): 65, *t. 28*. 1811. ≡ *Didymodon distichus* Brid., Muscol. Recent. Suppl. 4: 101. 1819. ≡ *D. capillaceus* var. *distichus* (Brid.) Hampe, Flora 20: 282. 1837. — Type citation: No type material is provided by Huebener (1833) in the protologue but the types of two species names are included, *Didymodon subulatus* (*1*) and *D. distichus* (*2*), and they are eligible for typification of *D. capillaceus* var. *compactus*: (*1*) Habitat in Carinth.; (*2*) in alpibus Sanensibus anno 1796 Augusto legi et pro *Didymodonte capillaceo* habui. *Schleicher* et *Thomas* serius eodem nomine instructum miserunt. [Lectotype (*selected here*): 'Didymodon distichus Cynodontium distichum Bryol. Univ In Alpibus Sanensibus 1806. Aestate Thomas etiam serius misit' – B-Bridel!]. First synonymised with *Didymodon capillaceus* by Arnott (1825: p. 36).

Distichium capillaceum var. *brevifolium* Bruch & Schimp. *in* Bruch, Schimp. & W.Gümbel, Bryol. Eur. 2: 156, *pl. 193*. 1846 [Fasc. 29–30 Mon.: 4, *pl. I*]. ≡ *Swartzia montana* var. *brevifolia* C.E.O.Jensen, Medd. Grønland 15: 418. 1898, *nom. illeg. incl. var. prior.* [Didymodon capillaceus var. compactus Huebener, 1833].

Plants slender, small to medium-sized, in compact, rarely loose tufts, lustrous, often silky, bright or dark green to yellowish-green, becoming brownish in older parts. *Stems* simple or forked, erect, to 6.5 cm tall, densely intertwined below with the tomentum of red-brown, smooth and extensively branched rhizoids, in cross-section oval, consisting of a 1–3-stratose cortex of small cells, with brown, incrassate walls surrounding a medulla of 3–4 layers of large, thin-walled, hyaline cells and a large central strand. *Leaves* closely imbricate and markedly distichous, 1.5–3.5(–5.2) mm long, consisting of a conduplicate, oblong-ovate to oblong, hyaline, membraneous, sheathing base, tightly appressed to the stem, rather abruptly constricted into a setaceous subula of about ½–¾ or, occasionally, only ¼ of the total length of leaf, erect to horizontally spreading and flexuose when dry, erecto-patent to sharply squarrose when wet, coarsely mammillose throughout, acute and denticulate at the apex, composed mainly of the costa, lunate to terete in cross-section; *margins* plane, erect, hyaline-bordered at base, entire below to lightly crenulate at the shoulders and low denticulate at the acumen; *costa* broad, occupying to ¼ of the width of base and rather diffuse, well defined in the subula, percurrent to shortly excurrent in an awn, in cross-section at middle of leaf base about 2–3 times as wide as deep, strongly convex dorsally, with a strong dorsal stereid band and a thinner or occasionally reduced ventral stereid band separated by a single row of large guide cells; *basal cells* rectangular to linear, thin-walled, not pitted, unistratose, pellucid, 5–10(–14) µm wide, (17–)20–45(–75) µm long, becoming narrowed and thinner-walled towards the margins and usually forming a hyaline border and shorter and more variable in shape in the mid-sheath; *shoulder cells* variable, square, short-rectangular to irregular in outline, obscure, moderately thick-walled, 6–9 µm wide, 6–15 µm long; *limb cells* similar in size and shape to the shoulder cells, bistratose and appearing papillose due to the prominent, protruding end of cross-walls separated by a collapsed, thin surface wall. *Paroicous* or *gonioautoicous*. *Antheridia* naked in the axils of the uppermost leaves or protected by the perigonial bracts and forming the small, axillary, gemmiform inflorescences. *Setae* 0.8–1.5(–2.0) cm long, yellow-brown to reddish, smooth, straight, flexuose and sinistrorse above when dry. *Capsules* erect to slightly inclined, cylindrical to ovoid-cylindrical, symmetric or nearly so, 1.0–1.6(–2.0) mm long, 0.4–0.7 mm wide, light to darker brown, lustrous, with a dark orange-red mouth; *annulus* well differentiated, large, biseriate, not revoluble; *operculum* conic, blunt, to ¼–⅕ of the urn length; *exothecial cells* rectangular to irregularly hexagonal, rather thin-walled, 15–25(–30) µm wide, 40–70 µm long in the mid-urn, becoming shorter, subquadrate towards the neck and smaller, rounded-quadrate in 1–3 rows forming an orange-red strip at the orifice; *stomata* few, superficial, restricted to the neck; *peristome* of 16 lanceolate teeth inserted somewhat below the mouth, to 220 µm long, pale orange- or yellowish-brown, bifid or irregularly perforated nearly to the base into 2 nearly equal branches, finely papillose-roughened or very faintly obliquely striate. *Spores* spherical, 15–20(–25) µm wide, yellowish-brown, densely papillose. *Calyptra* cucullate, smooth.

Chromosome numbers: unknown in the Antarctic and Subantarctic; $n = 7$ (1 record), $n = 14$ (9 records) and $n = 28$ (1 record) from the Holarctic (Fritsch 1991).

Discussion. When fertile, *Distichium capillaceum* is readily distinguished by its distichous habit together with the erect, cylindrical and symmetric capsules. Additionally, the spores are relatively small and the typically long-subulate plants are strongly shining and silky and have the leaf subulae widely spreading to reflexed. However, the leaf subulae show a considerable range of variation, their length being nearly twice that of the leaf base or accounting for only one-quarter of the total leaf length. Depauperate plants with short-subulate, stiff, erect leaves predominate in polar and altimontane regions, and most of the Antarctic populations represent such phenotypes. They have been taxonomically distinguished as var. *brevifolium* or var. *compactum* but they do not merit any taxonomic recognition because the length of the subulae shows continuous variation and many intermediates occur within the same colony. Therefore, the names of these varieties are here formally reduced to synonymy.

Newton (1977*b*) assessed variation in the size of the cells at the sheath base and found that in some populations they are markedly shorter (to 17 µm) and wider (to 13 µm). Such forms with aberrant areolation occasionally occur in Fuegian plants and were described by Roivainen & Bartram (1937) as fo. *subquadratum* which otherwise has erect and symmetric capsules, 1.3–1.5 mm long. These cells are usually confined to the juxtacostal area, the outer cells being more rectangular. It is possible that such forms are only modifications caused by severe climatic conditions and, because the short cells gradually merge with more rectangular cells, this taxon is not recognised here. The consistently short, subquadrate basal sheath cells, to 11 µm long and to 8 µm wide, are known only in the Chinese endemic *Distichium bryoxiphioidium* (Cao & He 1999).

Owing to the shortage of reliable gametophyte characters, numerous Antarctic populations lacking fruiting plants are named as *Distichium capillaceum* by convention on the basis of their general similarity to the remaining fruiting austral plants as did Newton (1977*b*). However, the discovery of fruiting *D. inclinatum* on James Ross I. indicate that this species may have wider distribution in the Antarctic. However, sterile plants of both species are almost indistinguishable from each other, especially if the plants have short subulate leaves.

The roughened, mammillose leaf subula of *Distichium capillaceum* is very similar to *Ditrichum ditrichoideum*, and Newton (1977*b*) so named the plants from the South Sandwich Is. However, *D. ditrichoideum* has distinctive leaves which are arranged in many rows and also produces copious gemmae.

Reproduction in Antarctica. Mostly sterile, but fully mature sporophytes have been found, occasionally in great abundance, in about ten populations. *Distichium capillaceum* does not produce gemmae and information on the presence in this species of 4-celled clavate gemmae borne in clusters below leaf insertions refers to *Ditrichum ditrichoideum*.

Habitat. *Distichium capillaceum* is a characteristic species of damp or dry, shaded, sheltered cliffs, hillsides, scree and moraines growing in crevices, beneath overhangs and on ledges with other chasmophytes such as *Didymodon brachyphyllus*, *Encalypta rhaptocarpa*, *Bartramia patens*, *Syntrichia magellanica* and *Pohlia cruda*. In addition, it also grows on soil, amongst gravel and rock debris as well as in seepage areas, and at the margins of melt streams. It is typically associated with base-rich and, less frequently, siliceous and other mildly acidic rocks. It is a prominent species of mildly calcareous soils derived from sedimentary rocks in the James Ross Island group and on Alexander I.

→

Figure 47. *Distichium capillaceum.* — **1**. Plant with sporophyte. **2–3**. Portions of shoots. **4–5**. Axillary hairs. **6**. Cross-section of stem. **7–17**. Vegetative leaves. **18**. Cells at leaf shoulder. **19**. Mid-leaf cells. **20**. Mid-leaf cells at margin. **21**. Basal leaf cells. **22–28**. Cross-sections of leaves, a sequence from base to apex. **29–30**. Perichaetial leaves. **31**. Operculate capsule. **32**. Exothecial cells at base of urn and stoma. **33**. Exothecial cells in middle of urn. **34**. Exothecial cells at orifice, annulus, portion of peristome and spores. — [1–2, 14, 18–19, 21, 23, 28–34 from *RO 1775/80*; 3, 17 from *VK 1240*; 4–5, 9, 20 from *RO 730/80*; 6, 8, 22, 24–27 from *RO 810/80*; 7 from *VK 1243*; 10–11 from *RO 1789/80*; 12–13 from *RO 1152/80*; 15 from *Lind s.n.*, ii.1907, s; 16 from *BJT 439*]. — Scale bars: a – 1 mm (1); b – 50 µm (18–28); c – 100 µm (4–5); d – 1 mm (2–3, 7–17, 29–31); e – 100 µm (6, 32–34).

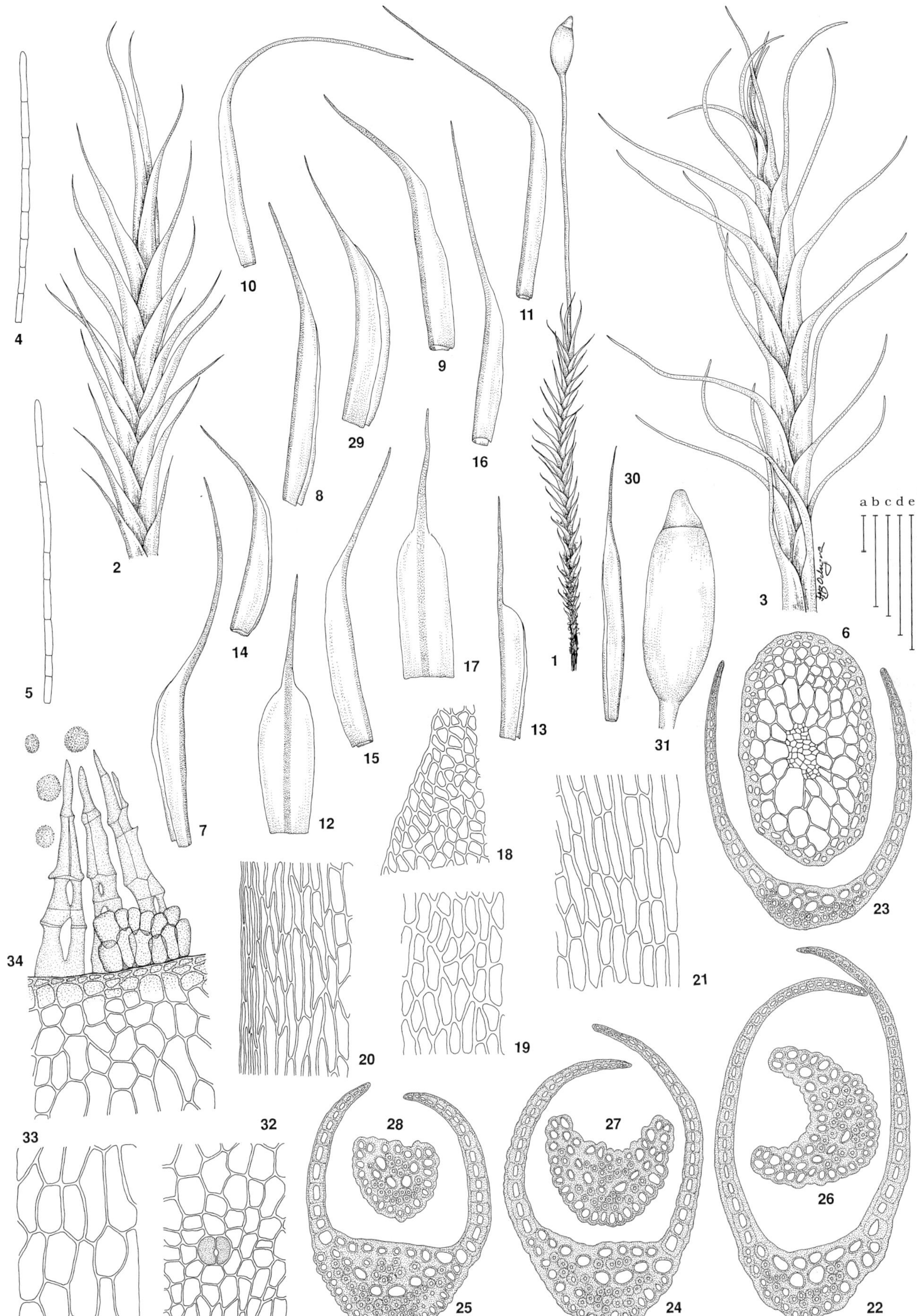
1
2
3
4
5
6
7
8
9
10
11
12
13
14
15
16
17
18
19
20
21
22
23
24
25
26
27
28
29
30
31
32
33
34
a b c d e

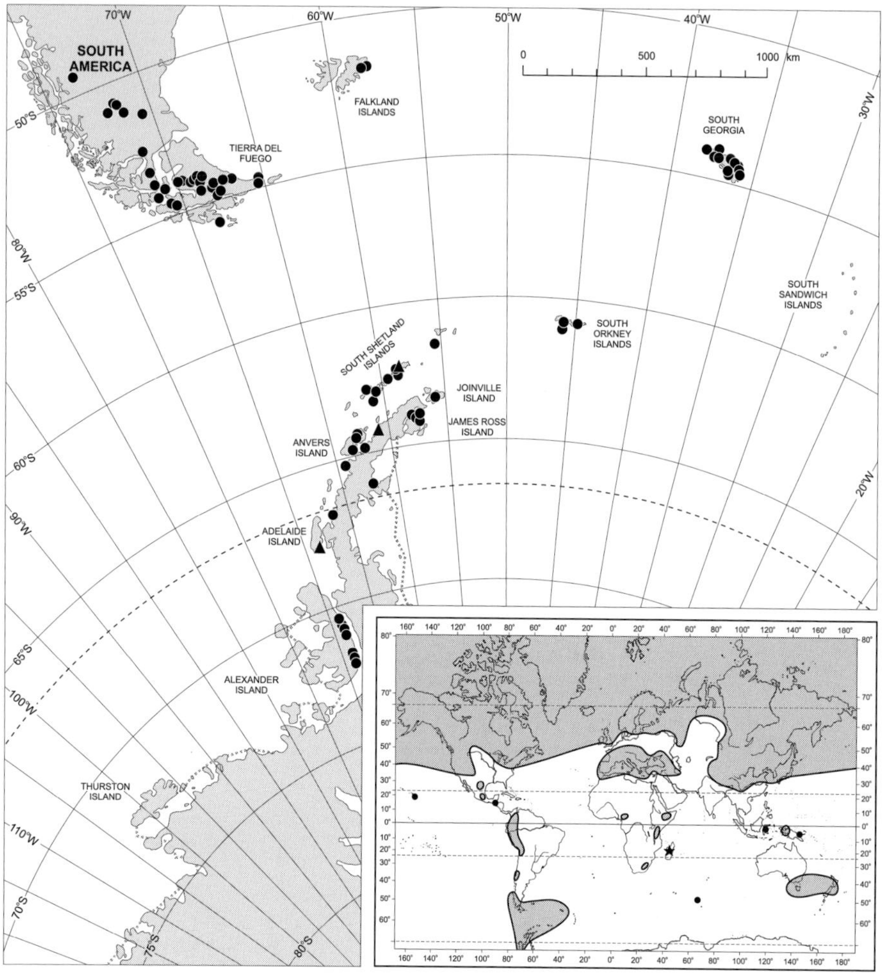

Figure 48. Distribution of *Distichium capillaceum* in the Fuegian region and Antarctica. Fruiting populations are marked with a triangle. Inset: global distribution. Inexact locality indicated by the star.

World range. A bipolar species with frequent intermediate stations at high elevations in the tropics (Fig. 48, inset). It has a wide pan-Holarctic distribution in northern latitudes and occurs mostly in highlands and mountains towards the south, extending to California, Arizona, New Mexico and New York in North America (Seppelt 2007), Morocco and Algeria in North Africa (Ros *et al.* 1999), Sicily (Dia *et al.* 1987), Crete (Duell 1995) and central Turkey (Henderson & Prentice 1969) in the Mediterranean, northern Iraq, Iran and Afghanistan (Frey & Kürchner 1991), the Himalayas (Gangulee 1969), Central China (Cao & He 1999) and Japan (Matsui & Iwatsuki 1990) in Asia. In the tropics it is widely distributed but generally scattered and confined to high mountain habitats in the Neotropics, ranging from Mexico (Snider 1994*b*) through the Central American isthmus (Allen 1994*a*) to Bolivia (Churchill *et al.* 2000), in Cameroon and East Africa (O'Shea 1999) and Madagascar (Dixon 1942), Sulawesi (Eddy 1988) and New Guinea (Norris & Koponen 1988) in Malesia and in the Hawaiian Is. (Bartram 1952*b*). Until recently, *Distichium capillaceum* appeared to be amphipacific in distribution, since the reports from Marion I. and Îles Crozet refer to *Ditrichum ditrichoideum*, but in 2007 it was discovered on Îles Kerguelen (Ochyra, unpublished) and, accordingly, the species should be considered as circumsubantarctic in the austral region. In Australasia the species is scattered in New Zealand, SE Australia and on Tasmania (Seppelt 1982), and in southern South America it is scattered in the Valdivian region (He 1998), western Patagonia and Tierra del Fuego (Greene 1986), on the Falkland Is. (Matteri 1986) and on subantarctic South Georgia (Newton 1977*a*).

Distribution in Antarctica. It is widely distributed throughout the maritime Antarctic, but seldom very frequent, reaching its southernmost limit on Alexander I. at lat. 72°05′S (Fig. 48). The species occurs from sea level to 375 m. Newton (1977*b*) reported *Distichium capillaceum* from the South Sandwich Is. but these plants were of *Ditrichum ditrichoideum* (Ochyra & Lewis Smith 1998). In addition, she mentioned this species from Victoria Land, Hope Bay on Trinity Peninsula, and Moss I. on the Danco Coast, attributing these reports to Cardot (1907*b*, 1908*a*). However, *D. capillaceum* was not reported from these areas by that author and these reports are certainly erroneous.

Selected specimens examined. WEST ANTARCTICA. SOUTH ORKNEY IS. **Coronation I.** Island opposite Saunders Pt., 5 m, *RILS 34* (AAS, KRAM). **Laurie I.** Paso de la Bellavista, 100 m, *Bellisio 11791* (NY). **Signy I.** Foca Pt., *ca* 30 m, *RILS 342* (BM, KRAM, LE, NY, S); Stygian Cove, *RILS 489* (AAS, KRAM, NY, S); Elephant Flats, *ca* 15 m, *RILS 390* (KRAM, NY, PC); Paal Harbour, 75 m, *RILS 10869* (AAS, KRAM).

SOUTH SHETLAND IS. **Gibbs I.** W of the Spit, 120 m, *JPB 190* (AAS, KRAM). **King George I.** ***Fildes Peninsula***: Gemmel Peaks, 145 m, *RO 2475B/80* (KRAM). ***Bransfield Strait***: Blue Dyke, 120 m, *RO 1152/80* (KRAM) and 135 m, *RO 1147/80* (KRAM); The Tower, 350 m, *RO 812/80* (KRAM); Bastion, 240 m, *RO 983/80* (KRAM); Creeping Slopes, 100 m, *RO 1297/80* (KRAM). ***Ezcurra Inlet***: Belweder, 145 m, *RO 1594/80* (KRAM); Dufayel I., Gdynia Pt, 80 m, *RO 1760/80* (KRAM) and 85 m, *RO 1772/80* (KRAM). ***Mackellar Inlet***: Klekowski Crag, 120 m, *RO 2288/80* (KRAM) and 250 m, *RO 2259/80* (KRAM). ***Keller Peninsula***: Ore Pt., 30 m, *RO 528/80* (KRAM), 40 m, *RO 519/80* (KRAM) and 50 m, *RO 517/80* (KRAM); Mt. Flagstaff, 275 m, *RO 414/80* (KRAM); Moraine Pt, 30 m, *BJT 272B* (KRAM, PC). **Robert I.** Coppermine Cove, 25–30 m, *RILS 926* (AAS, KRAM). **Livingston I.** ***Byers Peninsula***: Pt. Smellie,

10 m, *KR 35A* (AAS, KRAM); Devil's Pt., 20 m, *KR 24B* (AAS, KRAM). ***Hurd Peninsula***: bay off New Plymouth, *DCL 379* (BM, LE). **DECEPTION I.** Cliffs above Fumarole Bay, 50 m, *RILS 11352* (AAS, KRAM) and 75 m, *RILS 11360 & 11362* (AAS, KRAM).

WEST ANTARCTIC PENINSULA. **DANCO COAST.** Spring Pt., *VK 901, 1240, 1242 & 1243* (KRAM); Cape Anna, *ER 202A* (BR, PC). **PALMER ARCHIPELAGO. *Brabant I.***: Claude Pt., 300 m, *WH 87* (AAS, KRAM). ***Melchior Is.***: Kappa I., *ML 7904* (NY). **GRAHAM COAST.** Cape Tuxen, *ca* 15 m, *RWMC 668* (AAS, KRAM). **LOUBET COAST.** Orford Cliff, *RILS 5269A, 5275B & 5288B* (AAS, KRAM). **FALLIÈRES COAST.** Jenny I., *ca* 15 m, *BJT 439* (KRAM, PC) and *ca* 155 m, *BJT 426* (KRAM, NY, PC). ***Adelaide I.***: Reptile Ridge, 200 m, *RILS 3455* (AAS, KRAM); Stork Ridge, *PC 263* (AAS, KRAM). ***Léonie Is.***: Léonie I., 30 m, *RILS 8927H & 8928*, 50 m, *RILS 8947* (AAS, KRAM) and 150–175 m, *RILS 9130 & 9131* (AAS, KRAM). **GEORGE VI SOUND. *Alexander I.***: Ablation Pt., *RBH & JJL 2, 28, 85* (AAS, KRAM); Ablation Valley, 2–3 m, *RILS 9270A, 9273B, 9277A & 9315A* (AAS, KRAM); Moutonnée Lake, *RILS 11122* (AAS, KRAM); Succession Cliffs, 20 m, *BJT 532* (AAS, KRAM); Eros Glacier, 65 m, *BJT 526* (AAS, KRAM, PC); Uranus Glacier, *BJT 519B* (AAS, S); Waitabit Cliff, *BJT 514* (BM, LE); Viking Valley, 300 m, *RILS 9467 & 9470* (AAS, KRAM); Mars Oasis, 30 m, *RILS 10420* (AAS, KRAM), 32 m, *RILS 10439* (AAS, KRAM), 35 m, *RILS 10422* (AAS, KRAM), 125 m, *RILS 9415* (AAS, KRAM) and 150 m, *RILS 10374* (AAS, KRAM); Two-Step Cliffs, *RILS 9507* (AAS, KRAM); Coal Nunatak, *RILS 10317B* (AAS, KRAM).

EAST ANTARCTIC PENINSULA. **JOINVILLE ISLAND GROUP. *Joinville I.***: Mt. Alexander, 60–75 m, *RILS 3667B* (AAS, KRAM). **JAMES ROSS ISLAND GROUP. *Vega I.***: False Island Pt., 25–40 m, *RILS 7828B & 7839A* (AAS, KRAM). ***James Ross I.***: Lachman Crags, 25 m, *RILS 7333A* (AAS, KRAM) and 35 m, *RILS 7324 & 7381A* (AAS, KRAM); Lagrelius Pt., 10 m, *RILS 5962A* (AAS, KRAM); Fortress Hill, 25 m, *RILS 7813* (AAS, KRAM) and 150 m, *RILS 7773A* (AAS, KRAM); St. Martha Cove, 75 m, *RILS 7434B* (AAS, KRAM), 150 m, *RILS 7534* (AAS, KRAM) and 200 m, *RILS 7476B & 7477* (AAS, KRAM); Terrapin Hill, 85 m, *RILS 7743* (AAS, KRAM) and 200 m, *RILS 7734A & 7735B* (AAS, KRAM); Blancmange Hill, 12–15 m, *RILS 7674A* (AAS, KRAM); Dreadnought Pt., 3 m, *RILS 7493 & 7496B* (AAS, KRAM). **OSCAR II COAST.** Jason Peninsula, 375 m, *RILS 2760B* (AAS, KRAM).

2. **Distichium inclinatum** (Hedw.) Bruch & Schimp. Fig. 49

Distichium inclinatum (Hedw.) Bruch & Schimp. *in* Bruch, Schimp. & W.Gümbel, Bryol. Eur. 2: 157, *tab. 194.* 1846 [Fasc. 29–30 Mon.: 5, *tab. II*]. ≡ *Cynontodium inclinatum* Hedw., Spec. Musc. Frond.: 58. 1801. ≡ *Bryum inclinatum* (Hedw.) Dicks., Fasc. Pl. Crypt. Brit. 4: index [i]. 1801. ≡ *Didymodon inclinatus* (Hedw.) F.Weber & D.Mohr, Ind. Mus. Pl. Crypt.: 2. 1803. ≡ *Grimmia inclinata* (Hedw.) Sm., Fl. Brit. 3: 1193. 1804. ≡ *Swartzia inclinata* (Hedw.) P.Beauv., Prodr.: 90. 1805. ≡ *Cynodontium inclinatum* (Hedw.) Brid., Spec. Musc. 1: 155. 1806. ≡ *Cynodon inclinatus* (Hedw.) Brid., Mant. Musc.: 99. 1819. ≡ *Ceratodon inclinatus* (Hedw.) Huebener, Musc. Germ.: 273. 1833. ≡ *Leptotrichum inclinatum* (Hedw.) Mitt., J. Proc. Linn. Soc. Suppl. Bot. 1: 10. 1859. — TYPE CITATION: Upsaliae legit Cl. Ehrhart (Hedwig 1789: 75) and [....] in paludibus turfosis Scotiae, humidis Austriae et Alpibus Sanensibus Helvetiae habitat (Bridel 1798: 119). [Lectotype (*selected here*): The middle specimen with the distinctly inclined capsule, on the sheet labelled 'Swartzia inclinata Hedw. St. crypt. Vol. II. p. 74. t. 27' – G-Hedwig/Schwägrichen!].

Plants slender, small, densely caespitose, somewhat lustrous, light yellowish- or brownish-green above, brown below. *Stems* simple or branched, erect, 1.0–1.5 cm tall, densely tomentose below with brown, smooth and strongly branched rhizoids, in cross-section oval, with a large central strand surrounded by large, hyaline, thin-walled medullary cells in 2–3 layers and a cortex of 1–2 layers of small cells with brown and thick walls. *Leaves* crowded, imbricate, 2-ranked, 1.1–1.9 mm long, from a conduplicate, convolute, oblong- to oblong-ovate, hyaline, sheathing base, ± rapidly narrowed to an erect to erect-spreading, rigid, mammillose subula, as long as the base or only about ⅓–¼ of its length, acute and bluntly denticulate at the apex, mostly filled by the costa; *margins* plane, erect, entire throughout; *costa* relatively weak and narrow in the sheath, usually occupying ⅕ of its width, in cross-section not prominent dorsally, with a relatively weak dorsal stereid band and none or very weak and reduced to a few cells ventral stereid band, more distinct and broader in the subula, percurrent to shortly excurrent, and consisting of two stereid bands separated by a row of large guide cells; *basal cells* shortly to long-rectangular, occasionally short and subquadrate, thin- to moderately firm-walled, not pitted, unistratose, pellucid, 6–10(–13) µm wide, (15–)20–40(–60) µm long, narrowed and thinner-walled towards the margins, becoming shorter and more variable in shape in the mid-sheath; *shoulder* and *limb cells* variable, quadrate to short-rectangular, rather thick-walled, 5–9 µm wide, 5–20 µm long, obscure, bistratose and appearing mammillose in the limbs. *Gonioautoicous. Antheridia* in the axillary gemmiform inflorescences with 1–3 short, subulate bracts below the female inflorescences.

Setae 0.8–1.0 cm long, reddish-brown, smooth, straight, flexuose and sinistrorse above when dry. *Capsules* inclined, ovoid, asymmetric, about 1 mm long, 0.6 mm wide, light brown; *operculum* conic, blunt, about ⅕ of the urn length; *annulus* well differentiated, large, 2–3-seriate, not rolling away; *exothecial cells* rectangular to oblong-hexagonal, thin-walled, 20–30 µm wide, 30–85 µm long in the mid-urn, becoming shorter and narrower towards the neck and orifice; *stomata* not many in the neck; [*peristome* of 16 relatively broad, lanceolate teeth inserted somewhat below the mouth, to 360 µm long, yellowish-brown to brownish-red, irregularly cleft by perforations nearly to the base into 2–3 prongs, obliquely striolate. *Spores* spherical, 30–45 µm wide, yellowish-brown, densely papillose. *Calyptra* cucullate, smooth.]

Chromosome numbers: unknown in the Antarctic and Subantarctic; $n = 13$ (1 record) and $n = 14$ (2 records) from the Holarctic (Fritsch 1991).

Discussion. Fertile and typically developed plants of *Distichium inclinatum* can be distinguished by their short, ovoid, asymmetric and inclined to horizontal capsules as well as the large spores which are 30–45 µm in diameter. Unfortunately, the critical sporophyte characters are usually unavailable and cannot be verified with certainty in most Antarctic material. Its determination relies on the three plants with better developed capsules in which at least the shape characteristic for *D. inclinatum* is preserved and in a single capsule a few spores have been found which are 30–35 µm in diameter. Also, a few underdeveloped and partly broken peristome teeth seem to be equidistant; this is important because another species, *D. hagenii*, may also be involved here. This is a predominantly High Arctic species which, in many respects, is similar to *D. inclinatum* but differs by its peristome teeth which are united into eight blocks and by its obloid, inclined and asymmetric capsules. Until now this is

a
b
c
1
2
3
4
5
6
7
8
9
10
11
12
13
14
15
16
17
18
19
20
21
22
23
24
25
26

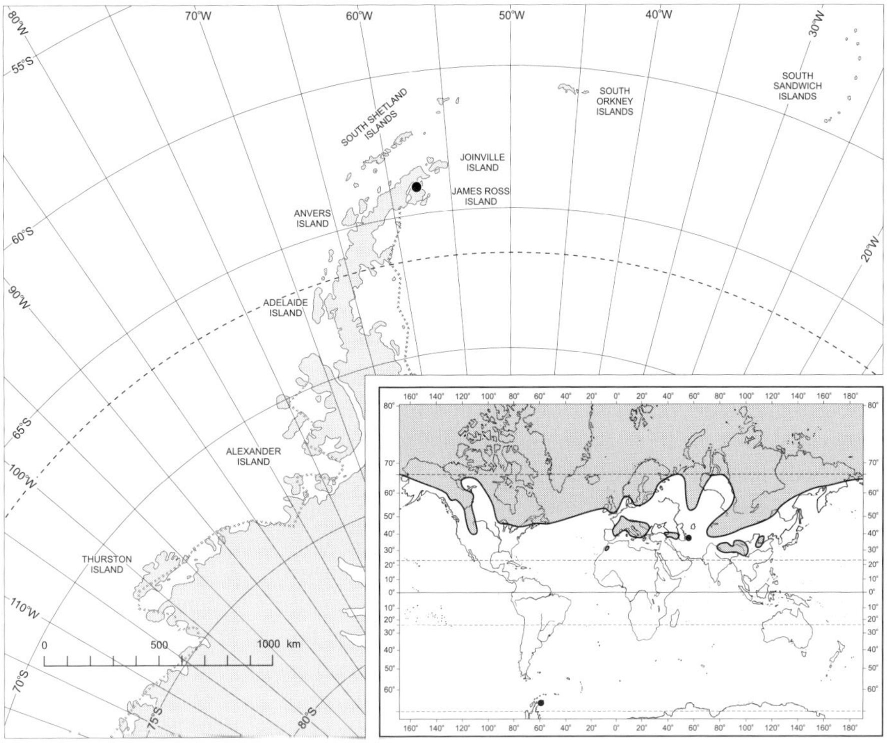

Figure 50. Distribution of *Distichium inclinatum* in Antarctica. Inset: global distribution.

the only record of *D. inclinatum* from the Southern Hemisphere but the genus needs revision in the austral regions. Sainsbury (1955) and Seppelt (1982) reported some New Zealand plants with suberect capsules and large spores, to 36 μm in diameter, so it is possible these may represent *D. inclinatum*.

Reproduction in Antarctica. The sole Antarctic collection bears several immature sporophytes and only three capsules which appear to be mature. Of these, only one capsule had a few spores, and several underdeveloped, hyaline but otherwise distinct peristome teeth could be detected.

Habitat. The only known Antarctic population was on the moist, stony margin of a melt stream. The plants were almost totally buried in a fine gravel and grew in association with *Cephaloziella varians*, *Bryum pseudotriquetrum* and cyanobacteria.

World range. A bipolar species, having a continuous range throughout the arctic and boreal zones in the Northern Hemisphere, common in the High Arctic and becoming more scattered and restricted mostly to the mountains in the south of the Holarctic (Fig. 50, inset). In North America *Distichium inclinatum* extends along the Rocky Mountains to Colorado and California in the west (Vitt *et al.* 1988) and to the Great Lakes (Ireland & Ley 1992) and Newfoundland (Brassard 1983*a*) in the east. In Eurasia it is scattered in the British Isles (Pitkin 1992), Fennoscandia (Söderström 1996) and across northern Russia to the Chukchi Peninsula (Ignatov *et al.* 2006). In southern Eurasia it is widespread in all mountain ranges in Europe, from the Pyrenees to the Caucasus, and scattered in Middle Asia (Mamatkulov 1990), Central Siberia (Bardunov 1974), the Himalayas (Gangulee 1969) and China (Cao & He 1999), with isolated stations in the High Atlas of Morocco (Ros *et al.* 1999, 2000).

Distribution in Antarctica. Known only from James Ross I. off the NE coast of the Antarctic Peninsula (Fig. 50).

Specimen examined. **WEST ANTARCTICA.** EAST ANTARCTIC PENINSULA. **James Ross Island group.** ***James Ross I.***: Solorina Valley to N of St. Martha Cove, 50 m, *RILS 7513* (AAS, KRAM).

8. DITRICHUM Hampe, *nom. cons.*

Ditrichum Hampe, Flora 50: 181. 1867. ≡ *Leptotrichum* Hampe subgen. *Ditrichum* (Hampe) Besch., Mem. Soc. Sc. Nat. Cherbourg 16: 174. 1872. ≡ *Leptotrichum* Hampe sect. *Ditrichum* (Hampe) Müll.Hal., Gen. Musc. Frond.: 312. 1900. — Type: *Ditrichum homomallum* (Hedw.) Hampe (*Didymodon homomallus* Hedw.) (*typ. cons.*) [= *Ditrichum heteromallum* (Hedw.) E. Britton (*Weissia heteromalla* Hedw.)].

Aschistodon Mont., Ann. Sci. Nat. Bot. Sér. 3, 4: 109. 1845, *nom. rej.* ≡ *Leptotrichum* Hampe sect. *Aschistodon* (Hampe) Müll.Hal., Linnaea 38: 627. 1874. — Type: *Aschistodon conicus* Mont. [≡ *Ditrichum conicum* (Mont.) Mitt.)].

◄———

Figure 49. *Distichium inclinatum*. — **1**. Plant with sporophyte. **2**. Portion of shoot. **3**. Cross-section of stem. **4–5**. Axillary hairs. **6–10**. Vegetative leaves. **11**. Cells at leaf shoulder. **12**. Mid-leaf cells. **13**. Basal leaf cells. **14–19**. Cross-sections of leaves, a sequence from base to apex. **20**. Deperculate capsule with destroyed peristome. **21–23**. Operculate capsules. **24**. Exothecial cells in middle of urn. **25**. Exothecial cells at base of urn and stomata. **26**. Exothecial cells at orifice, annulus, portion of peristome and spores. — [1–21 from *RILS 7513*; 22–26 from *Bauer 735*]. — Scale bars: a – 1 mm (6–10) and 2 mm (1–2, 20–23); b – 100 μm (3–5, 11–19); c – 100 μm (24–26).

Diaphanophyllum Lindb., Öfvers. Förh. Kongl. Svenska Vetensk.-Akad. 19: 605. 1863, *nom. rej.*

Lophiodon Hook.f. & Wilson, London J. Bot. 3: 543. 1844, *nom. rej.* — Type: *Lophiodon strictus* Hook.f. & Wilson [*Ditrichum strictum* (Hook.f. & Wilson) Hampe (*Lophiodon strictus* Hook.f. & Wilson)].

Pseudodistichium Cardot, Rev. Bryol. 32: 45. 1905. — Type: *Pseudodistichium austrogeorgicum* Cardot [=*Ditrichum hyalinum* (Mitt.) Kuntze (*Leptotrichum hyalinum* Mitt.]. First synonymised by Seppelt (1980: p. 128).

Philibertiella Cardot, Rev. Bryol., 41: 37. 1914, *hom. illeg.* [*non* Vail, Bull. Torrey Bot. Cl. 26: 305. 1897]. ≡ *Austrophilibertiella* Ochyra, Fragm. Flor. Geobot. 41: 478. 1996, ***syn. nov.*** — Type: *Austrophilibertiella ditrichoidea* (Cardot) Ochyra (*Philibertiella ditrichoidea* Cardot) [*Ditrichum ditrichoideum* (Cardot) Ochyra].

Plants small to medium-sized, less often large but slender, gregarious or loosely tufted to densely caespitose, variously coloured, green, yellowish-, brownish- to blackish-green or yellowish- to reddish-brown, dull or lustrous. *Stems* erect or ascending, simple or sparingly branched, sparsely or densely radiculose with branched, brown, reddish or yellow rhizoids, in cross-section round, with a distinct central strand; *axillary hairs* filiform, 4–10-celled, hyaline throughout or with 1–2 short, brown basal cells. *Leaves* densely set in many rows, erect, erect-spreading to spreading, sometimes homomallous or falcate, non-decurrent, not altered or somewhat contorted on drying, from a lanceolate, ovate, oblong-ovate or elliptic, concave or convolute, often clasping base, gradually acuminate or suddenly narrowed to a slender or stout, short or long, setaceous, sometimes twisted, stiff and rigid or flexuose, smooth or strongly roughened, channelled acumen, acute or obtuse, entire or denticulate at the apex; *margins* plane, inflexed to tubulose or recurved, entire or denticulate at the apex of the subula; *costa* single, strong, mostly diffuse in the sheath, distinct above and usually occupying all or major part of the subula, percurrent to excurrent, not prominent and lunate or strongly convex dorsally, in cross-section with stereid bands differentiated on both sides of guide cells or only on the dorsal side; *laminal cells* smooth, thin-walled and lacking pits or thick-walled with pits, unistratose throughout or sometimes 2–3-stratose at the shoulders and in the subulae, rectangular throughout or rectangular to linear below, subquadrate, oval to elongate above; *alar cells* not differentiated. *Asexual reproduction* lacking or by ovoid, 3–5-cellular rhizoidal tubers. *Monoicous* or *dioicous*. *Perichaetial leaves* not or strongly differentiated, enlarged and clasping at the base. *Setae* elongate or occasionally very short, erect and flexuose, straight or sometimes bent. *Capsules* erect or inclined, cylindrical, obloid to ellipsoidal, often curved and asymmetric, not or distinctly narrowed at the mouth, smooth or rarely furrowed or striate when dry and empty; *annulus* 2–3-seriate of large cells, compound, deciduous or ± persistent; *operculum* conic or conic-rostrate with a straight or slanted beak; *exothecial cells* rectangular to oblong-hexagonal, rather thin-walled; *stomata* present, bicellular; *peristome* present, very rarely absent, single, inserted below the orifice, the teeth 16, short or long, straight to slightly twisted, split to the base into 2 threadlike, terete, papillose or spirally striate divisions or sometimes undivided. *Spores* spherical, smooth or papillose, uni- or sometimes multicellular. *Calyptra* cucullate, naked.

Differing facies associated with leaf shape make *Ditrichum* quite difficult to define, yet it is characterised well enough by peristome teeth architecture which are deeply divided into two filiform segments, and the name of the genus refers to this feature. In some species, for instance *D. strictum*, *D. punctulatum* and *D. buchananii*, the teeth are undivided but are otherwise filiform and terete. Gametophytically, it is difficult to satisfactorily delimit some species of *Ditrichum* from *Dicranella* with which they share the absence of differentiated alar cells, often sheathing leaves and similar costae. Despite these external similarities, *Ditrichum* is unrelated to the dicranaceous genera but it shows rather close genetic affinity to the cleistocarpous genus *Pleuridium* which is attested by the existence of frequent intergeneric hybrids, including the hybrid genus ×*Pleuriditrichum* (Andrews & Hermann 1959).

Due to the diversity in both gametophyte and sporophyte details, some austral species which are currently placed in *Ditrichum* had initially been placed in separate genera, including *Aschistodon*, *Lophiodon* and *Philibertiella*. The first two antedate the name *Ditrichum* but they are now considered as rejected names and *Ditrichum* is conserved over them (Magill 1993). In fact, species currently placed in *Ditrichum* were at first placed by Hampe (1847) in the genus *Leptotrichum* which, however, proved to be an illegitimate homonym of a fungal generic name. Lindberg (1863*a*) and Hampe (1867) introduced the names *Diaphanophyllum* and *Ditrichum*, respectively, as replacements of the nomenclaturally illegitimate name *Leptotrichum* and the latter was retained as a conserved name on the suggestion of Cardot (1910*a*).

Ditrichum is a cosmopolitan genus consisting of approximately 50–55 species. It is represented by at least 15 species in the cool and cold regions in the Southern Hemisphere, some of which play a significant role in the tundra vegetation. The genus has been revised for Australia and New Zealand (Seppelt 1982) but it still needs a monographic study in southern South America and in the Subantarctic, although many taxonomic and phytogeographical papers have been published dealing with particular species of this genus (e.g. Ochyra 1994*a*, 1996*d,e*, 1999*a,b,c*; Matteri 1996; Seppelt 1996; Ochyra & Deguchi 1999). Five species of *Ditrichum* are known to occur in the Antarctic and their detailed overview was presented by Ochyra & Lewis Smith (1998).

1. Cells elongate throughout the lamina; leaf base non-sheathing, without hyaline margin; costa with or without a weak ventral stereid band .... 1. ***D. heteromallum***
1. Cells of the upper leaf base subquadrate to shortly rectangular; leaf base vaginate with distinct hyaline margin; costa with two strong stereid bands 2

2. Leaf subula strongly mammillose; rhizoidal tubers, 3–5-celled, common .... 2. ***D. ditrichoideum***
2. Leaf subula smooth; rhizoidal tubers absent 3
3. Costa narrow, 60–70 µm wide at the base; leaf subula flexuose; capsule immersed in the perichaetial leaves; peristome absent .... 3. ***D. hyalinocuspidatum***
3. Costa broad, 90–160 µm wide at the base; leaf subula mostly straight and rigid or somewhat crisped; capsule exserted on a long seta; peristome present 4
4. Cells irregular in outline, isodiametric to shortly elongate throughout the leaf sheath (except for its base), 3–5 : 1; costa 130–180 µm wide at base; capsule symmetric, straight; peristome teeth strongly papillose, without striolations; spores 10–14 µm wide; dioicous .... 4. ***D. conicum***
4. Cells subquadrate at the shoulders becoming elongate rectangular in the mid-sheath and below, 6–10 : 1; costa 100–120 µm wide at base; capsule asymmetric, slightly curved; peristome teeth papillose with oblique striolations; spores 22–34 µm wide; monoicous... 5. ***D. hyalinum***

1. **Ditrichum heteromallum** (Hedw.) E.Britton Fig. 51

Ditrichum heteromallum (Hedw.) E.Britton, N. Am. Fl. 15: 64. 1913. ≡ *Weissia heteromalla* Hedw., Spec. Musc. Frond.: 71. 1801. ≡ *Grimmia heteromalla* (Hedw.) F.Weber & D.Mohr, Ind. Mus. Pl. Crypt.: 2. 1803. ≡ *Didymodon heteromallus* (Hedw.) Hook. & Taylor, Musc. Brit.: 68, *t. 20.* 1818. ≡ *Trichostomum heteromallum* (Hedw.) Lindb. *ex* Austin, Bull. Torrey Bot. Cl. 6: 74. 1876. — Type citation: In montibus Saxoniae silvis, ad vias cavas, udas et earum orbitas pfofundiores [sic!]. (In Hercynia, Sudetibus, Austria.); also a reference to Hedwig (1787: p. 22, *pl. 8*) is provided who subsequently provided the specimen from 'Chemnitii in sylva Zeissigwald' (no material suitable for typification of this name is present in G-Hedwig/Schwägrichen). [Lectotype (*selected here*): Tab. VIII in Hedwig (1787); epitype (*selected here*): 'A. Kopsch: Bryotheca Saxonica. No. 13: Ditrichum homomallum (Hedw.) Hpe. Erzgebirge: an Waldrändern bei Oberwiesenthal. 2.II.1919. leg. Alfr. Kopsch' – KRAM!].

Dicranella fuegiana Cardot & Broth., Kongl. Svenska Vetensk. Acad. Handl. 63(10): 12. 1923, ***syn. nov***. — Type citation: Fuegia: Sierra Valdivieso, in trajectu Paso de las Lagunas, alt. c. 700 m. s. m. (747). [Isotypes: 'Herb. J. Cardot Dicranella fuegiana Card. sp. nova Terre de Feu: Sierra Valdivieso, paso de las Lagunas reg. Alpina Leg. C. Skottsberg, 1908, nº 747' – H-Brotherus!, BM-Dixon!, S-Roth!].

Plants small, in loose tufts or seldom gregarious, lustrous, somewhat silky, green or yellowish-green. *Stems* usually branched or less often simple, with straight innovations, 5–15 mm high, erect, sparsely radiculose below with long, reddish-brown, smooth, branched rhizoids, in cross-section rounded or elliptic to bluntly triangular, consisting of a uni- to bistratose in places cortex composed of small cells with moderately thickened walls, brown to yellow-brown cells, a 3–5-stratose medulla of large, hyaline, thick-walled cells and a large and distinct central strand; *axillary hairs* 4–7-cellular, composed of 2–3 small, brown basal cells and 2–4 elongate, hyaline upper cells. *Leaves* erect to erect-spreading, straight to secund, mostly rigid, (1.3–)1.8–2.5(–3.0) mm long, from a non-sheathing, concave, short, ovate-lanceolate to lanceolate base, 0.25–0.5 mm wide, gradually narrowed into a rigid, straight or falcate, channelled subula, entire or indistinctly sinuate and bluntly rounded or acute at the apex, 2–3 times exceeding the leaf base length; *margins* entire throughout, plane; *costa* single, yellow to pale yellow-brown or sometimes concolorous with the laminal cells, rather diffuse because of several rows of bistratose juxtacostal cells, 110–140 µm wide at the base, excurrent, filling most of the width of the subula, smooth on both surfaces, in cross-section very flattened, with large guide cells in a single row, a strong dorsal stereid band, 2–4-stratose and a very weak ventral stereid band, reduced to a few cells above, completely missing in the base; *laminal cells* smooth, transparent, often yellowish at the base, with thin to moderately thick, not pitted walls, narrowly rectangular to linear throughout, (4–)6–10 µm wide, (25–)40–70(–80) µm long, unistratose and bistratose in several rows near the costa, becoming linear and thinner-walled towards the margins but not forming a distinct band; *alar cells* undifferentiated. *Dioicous*. *Perichaetia* terminal, with single sporophytes; *outer perichaetial leaves* similar to the vegetative leaves; *inner perichaetial leaves* up to 3.3 mm long from an oblong-ovate, base abruptly long subulate. *Setae* erect, straight, 1.0–1.5 cm long, somewhat flexuose and dextrorse above when dry, yellowish-brown to reddish-brown with age. *Capsules* long exserted, erect, symmetric to somewhat gibbous or curved, abruptly narrowed into the seta and narrowed at the mouth, smooth, reddish-brown to blackish-brown with age, mostly with a shiny lustre, narrowly obloid-ovoid or ellipsoid, 1.0–1.5 mm long, 0.5–0.6 mm wide; *operculum* short, *ca* ¼ of the capsule length, conical, with a slanted beak; *annulus* large, consisting of 2 rows of cells, revoluble; *exothecial cells* rectangular to oblong-hexagonal, thin-walled in the mid-urn, 20–30 µm wide, 40–90 µm long, becoming regularly rectangular and thicker-walled at the base of the urn; *cells at the mouth* rounded to hexagonal, rather firm-walled and concolorous with the other exothecial cells; *stomata* 6–8 per capsule at the extreme base of the urn, superficial, rounded, *ca* 30 µm in diameter, bicellular, variously oriented, with rounded pori; *peristome teeth* 16, yellow-brown, 240–280 µm long, finely papillose, with a very low basal membrane, usually divided to the base into two filiform, mostly closely joined prongs or with elongate gaps between them along the midline. *Spores* globose,

a
b
c
d
1
2
3
4
5
6
7
8
9
10
11
12
13
14
15
16
17
18
19
20
21
22
23
24
25
26
27
28
29
30
31
32
33
34
35
36
37

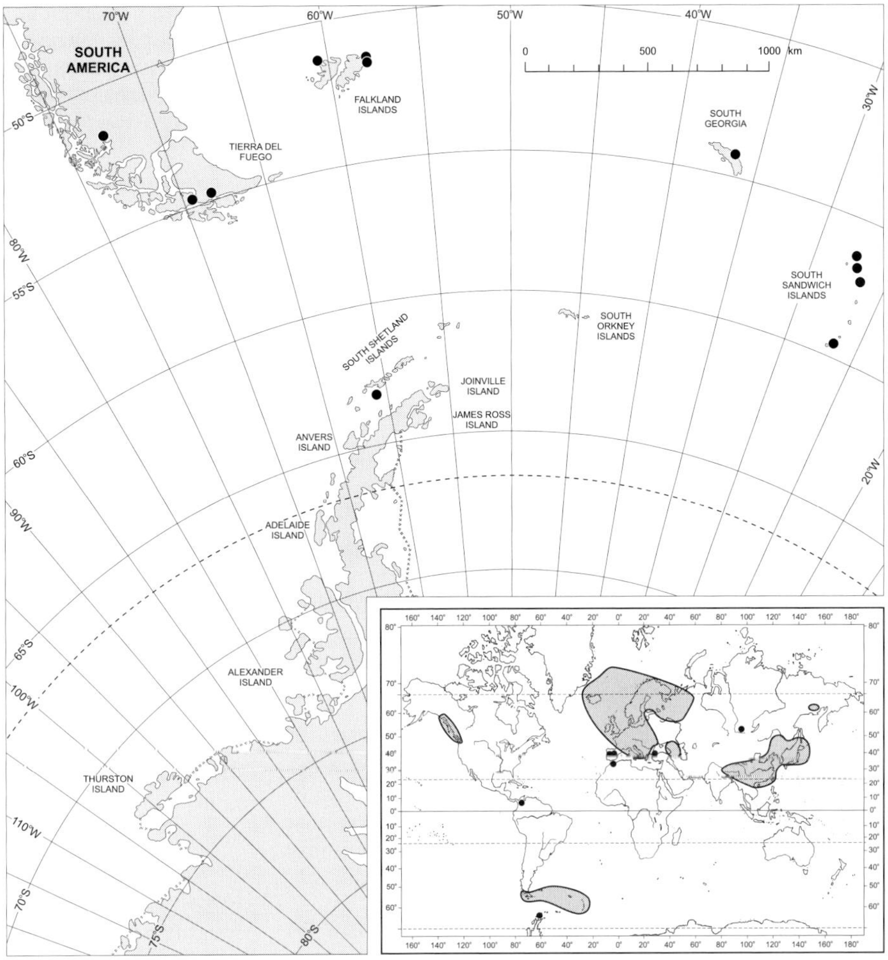

Figure 52. Distribution of *Ditrichum heteromallum* in the Fuegian region and Antarctica. Inset: global distribution. (Reproduced from *Annales Botanici Fennici*, 35: p. 37, 1998, expanded, with permission.)

pale, 11–14 μm in diameter, smooth. *Calyptra* cucullate, smooth, covering only the upper half of the urn.

Chromosome numbers: unknown in the Antarctic and Subantarctic; $n = 13$ (5 records) and $n = 13 + \mathrm{m}$ (1 record) from the Holarctic (Fritsch 1991).

Discussion. *Ditrichum heteromallum* is a distinct species which is distinguished from all other local species by its leaves lacking the sheathing base, its leaf areolation of long-rectangular to linear cells throughout the lamina but not forming a hyaline marginal border at the leaf base and, anatomically, by the absence or poor development of the ventral stereid band in the costa. The often abundantly produced sporophytes make this species readily identifiable from all other Antarctic acrocarps by its reddish-brown, at least below, setae and more or less symmetric, erect capsules with short, conical lids with slanted beaks. Sterile plants might be mistaken for *Kiaeria pumila*, but the narrow costa and the presence of the well developed alar cells should aid in the identification of the material.

In its habit, leaf shape and especially anatomical structure of the costa *Ditrichum heteromallum* is also similar to *Campylopus pyriformis*. However, this species is recognised by its slightly flexuose leaves which are serrate at the apex, as well as its leaf areolation of quadrate, rhomboidal or short rectangular upper leaf cells and short rectangular, widenened and relatively lax basal cells. In its broad costa and long subulate leaves it may be confused with *Leptobryum pyriforme*, which has been recorded in the Antarctic from heated ground on the volcanic Deception I. and may be expected to occur in similar habitats on the South Sandwich Is. However, the leaves of *L. pyriforme* are flexuose when dry and have long setaceous subulae, whereas in *D. heteromallum* these are stiff and straight or secund.

Reproduction in Antarctica. Sporophytes with fully mature capsules produced in great profusion have been seen in one or two fertile populations from the South Sandwich Is., whereas all plants from Deception I. are sterile. In addition, some populations from the South Sandwich Is. consisted of abundant male plants with numerous perigonia.

Habitat. A pioneer species colonising bare heated ground associated with fumaroles. It grows on cemented soil, crusty ash slopes and on moist fine ash, sometimes around violently active steam

◄———

Figure 51. *Ditrichum heteromallum*. — **1**. Plant with mature sporophytes, dry. **2**. Male plant with perigonium, dry. **3**. Cross-section of stem. **4–6**. Axillary hairs. **7–13**. Vegetative leaves. **14–15**. Leaf apices. **16**. Cells at leaf shoulder. **17–18**. Mid-leaf cells. **19**. Basal leaf cells. **20–28**. Cross-sections of leaves, a sequence from base to apex. **29–30**. Perichaetial leaves. **31**. Cells at shoulder of perichaetial leaf. **32**. Capsule with calyptra. **33–34**. Operculate capsules, dry. **35**. Exothecial cells at base of urn and stoma. **36**. Exothecial cells in middle of urn. **37**. Exothecial cells at orifice, annulus, portion of peristome and spores. — [1, 3–37 from *PC 180*; 2 from *PC 213A*]. — Scale bars: a – 1 mm (7–13, 29–30, 32–34); b – 2 mm (1) and 100 μm (3, 20–28); c – 1 mm (2) and 100 μm (4–6, 35–37); d – 100 μm (14–19, 31). (Reproduced from *Annales Botanici Fennici*, 35: p. 36, 1998, with permission.)

and sulphur vents with the temperature of the ground reaching 90–95°C at 10 cm beneath its surface. It forms usually monospecific loose patches or dense tufts, and occasional associates are *Bryum orbiculatifolium*, *Pohlia wahlenbergii*, *Campylopus spiralis*, *Cryptochila grandiflora* and *Lophozia excisa*.

World range. *Ditrichum heteromallum* is a bipolar species with a wide but strongly dissected, boreal-temperate range in the Northern Hemisphere (Fig. 52, inset). It is widespread throughout much of Europe where its seems to have its main centre of occurrence, ranging from Iceland and northern Fennoscandia to the Mediterranean. In Asia it is rare in Central Siberia (Bardunov 1974), in the Russian Far East (Bardunov & Cherdantseva 1982; Blagodatskikh 1984), the Himalayas and northen India (Gangulee 1969), China (Cao & He 1999), Taiwan (Kuo & Chiang 1987) and Japan (Matsui & Iwatsuki 1990). In North America it occurs only on the Pacific coast extending from Alaska to Oregon (Lawton 1971). In the Southern Hemisphere the species is scattered in Tierra del Fuego (Cardot & Brotherus 1923; Ochyra 1999*c*), western Patagonia (Matteri 1985*a*), on the Falkland Is. (Matteri 1986) and South Georgia (Ochyra *et al.* 2002) as well as in the northern maritime Antarctic. Additionally, *D. heteromallum* occurs at high elevations in the northern Andes of Colombia (Florschütz-de Waard & Florschütz 1979; Churchill *et al.* 1995, 2000).

Distribution in Antarctica. The species is known exclusively from geothermal habitats on islands with recent volcanic activity (Fig. 52). It occurs on four islands in the South Sandwich Is. archipelago and on Deception I. It was not observed on the latter island prior to 1967–1970 eruptions (Collins 1969), so it may represent a recent introduction via long distance dispersal.

Specimens examined. WEST ANTARCTICA. SOUTH SANDWICH IS. **ZAVODOVSKI I.** Central ash plain, *PC 259A & 260A, C–D* (AAS, KRAM). **VISOKOI I.** Finger Pt., *MWH 861* (AAS, KRAM). **CANDLEMAS I.** W shore, *ca* 35 m, *REL 632* (AAS, KRAM); S-facing slope of main cone, 200 m, *REL 621* (AAS, KRAM); Lucifer Hill, *PC 180 & 213A* (AAS, KRAM). **BELLINGSHAUSEN I.** Floor of crater, 50 m, *PC 67, 68, 70B, 71A & 76C* (AAS, KRAM); S slopes, *ca* 75 m, *MWH 813* (AAS, KRAM) and 100 m, *MWH 411B* (AAS, KRAM).

SOUTH SHETLAND IS. **DECEPTION I.** Mt. Pond, 480 m, *RILS 5819A, 5822B & 5825C* (AAS, KRAM), 485–500 M, *RILS 11165* (AAS, KRAM) and 500 m, *RILS 11148 & 11149* (AAS, KRAM); 1.5 km NNW of Mt. Pond, 410–525 m, *RILS 6897* (AAS, KRAM); SW Fumarole Bay, Lake Irizar, 100–125 m, *RILS 6908, 6911 & 6918* (AAS, KRAM); south ened of Fumarole Bay, *ca.* 125 m, *RILS 11272C, 11273B, 11279A, 11280, 11282, 11283B & 11284B* (AAS, KRAM); ridge above Fumarole Bay, 100 m, *RILS 5563 & 5564* (AAS, KRAM).

2. **Ditrichum ditrichoideum** (Cardot) Ochyra Fig. 53

Ditrichum ditrichoideum (Cardot) Ochyra, *comb. nov.* ≡ *Philibertiella ditrichoidea* Cardot, Rev. Bryol., 41: 37. 1914. ≡ *Austrophilibertiella ditrichoidea* (Cardot) Ochyra, Fragm. Flor. Geobot. 41: 478. 1996. — TYPE CITATION: Chili: île Chiloé, rochers maritimes: Ancud (C. Skottsberg, 1908), Quellon (Halle et Skottsberg, 1909). – Environs de Valparaiso (Porter; herb. Thériot). [Lectotype (*selected here*): 'Chili leg Porter Philibertiella ditrichoidea Card. In sched. Rev. bryol. 1914, nº 3' – PC-Thériot!; isolectotype: H-Brotherus!].

Ditrichum gemmiferum Ochyra & Lewis Smith, Ann. Bot. Fennici 35: 46, *f. 7.* 1998, ***syn. nov.*** — TYPE: Antarctica, South Shetland Islands, Deception Island: Port Foster, geothermal area on slopes to E. and S.E. of Pendulum Cove, lat. 62°57′S, long. 60°38′W, plants on rocks with no soil, 50–100 m; 1987–1988, R. I. Lewis-Smith 6921. [Holotype: 'Herbarium of the British Antarctic Survey (AAS) Flora of South Shetland Islands Ditrichum austro-shetlandicum Ochyra & Lewis-Smith Plants on rocks. No soil. Alt. 50–100 m. Geothermal area on slopes to E. and S.E. of Pendulum Cove. Deception Island. lat = 6257 S long = 6038 W Leg. Greenfield, L. 1987–88 Det. R. Ochyra, 22 May 1997 No. Smith, R. I. L. 06921' – AAS!; isotype: KRAM!].

Plants small to medium-sized, loosely or densely caespitose, dull or weakly glossy, olive-green to light olivaceous or dark green above and dark brown to blackish-green below, sometimes light olive-brown throughout. *Stems* erect, usually sparingly branched, 0.5–1.5 tall, in cross-section rounded, consisting of 2(–3) rows of smaller cortical cells with moderately thickened, yellowish-brown walls surrounding 3(–4) layers of large, hyaline and thin-walled medullary cells, central strand relatively large and conspicuous; *rhizoids* common in the lower part of the stem, long and sparsely branched, light brown, not forming a tomentum; *rhizoidal tubers* common, relatively small, pale yellowish to brownish, consisting of (2–)3–4 large, bulging, roughly spherical cells, from 25–50 μm in diameter, arranged in a short row and attached to the rhizoid by a much smaller cell; *axillary hairs* numerous in the apical part of the stem and branches, hyaline throughout, composed of 7–10 elongate cells. *Leaves* (1.5–)1.8–2.5(–3.0) mm long, (0.4–)0.6–0.8 mm wide, from an ovate to oblong-ovate, concave, vaginate base abruptly constricted into the subula, erect and more or less contorted in the upper part when dry, erect-spreading when wet; *subula* rigid, straight to curved when dry, about as long as the sheathing base, channelled, coarsely papillose throughout, usually sharply denticulate or sometimes entire and rounded at the apex; *margins* plane, entire throughout or slightly crenulate at the shoulders; *costa* single, yellowish-brown, often concolorous with the lamina cells, diffuse, subpercurrent to percurrent, strongly flattened, 100–160 μm wide at the base, occupying most of the width of the subula and invisibly merging into the multistratose lamina cells, usually ridged on the abaxial surface, in transverse section consisting of

→

Figure 53. *Ditrichum ditrichoideum*. — **1**. Plant with mature sporophytes, dry. **2**. Sterile plant, dry. **3**. Male plant with perigonium. **4**. Cross-section of stem. **5–6**. Axillary hairs. **7–11**. Rhizoidal tubers. **12–18**. Vegetative leaves. **19–20**. Leaf apices. **21**. Papillae of subula in profile. **22**. Cells at leaf shoulder. **23**. Lamina cells above mid-sheath at margin. **24–25**. Lamina cells just below mid-sheath at margin. **26–27**. Basal leaf cells. **28–38**. Cross-sections of leaves, a sequence from base to apex. **39–40**. Outer perichaetial leaves. **41**. Inner perichaetial leaf. **42**. Young operculate capsule. **43–44**. Deoperculate capsules. **45**. Exothecial cells at base of urn and stoma. **46**. Exothecial cells in middle of urn. **47**. Exothecial cells at orifice, peristome and spores. — [1, 5–6, 12–13, 20–21, 23–24, 28, 36, 39, 41–43, 45–47 from *RILS 6921* (isotype of *Ditrichum gemmiferum*); 2 from *RILS 6925* (paratype of *D. gemmiferum*); 3, 19, 25–26, 31–32, 38 from *RILS 6926* (paratype of *D. gemmiferum*); 4, 7–11, 14–16, 22, 27, 29, 33–35, 37 from *RO 2792/80* (paratype *D. gemmiferum*); 18, 40, 44 from *Mahu 22834* (paratype of *D. gemmiferum*); 17, 30 from *REL 572* (paratype of *D. gemmiferum*)]. — Scale bars: a – 2 mm (1); b – 2 mm (3), 50 μm (21) and 100 μm (5–6); c – 1 mm (12–18, 39–41); d – 100 μm (4, 19–20, 22–38); e – 1 mm (42–44), 4 mm (2) and 100 μm (7–11, 45–47). (Reproduced from *Annales Botanici Fennici*, 35: p. 47, 1998, with permission.)

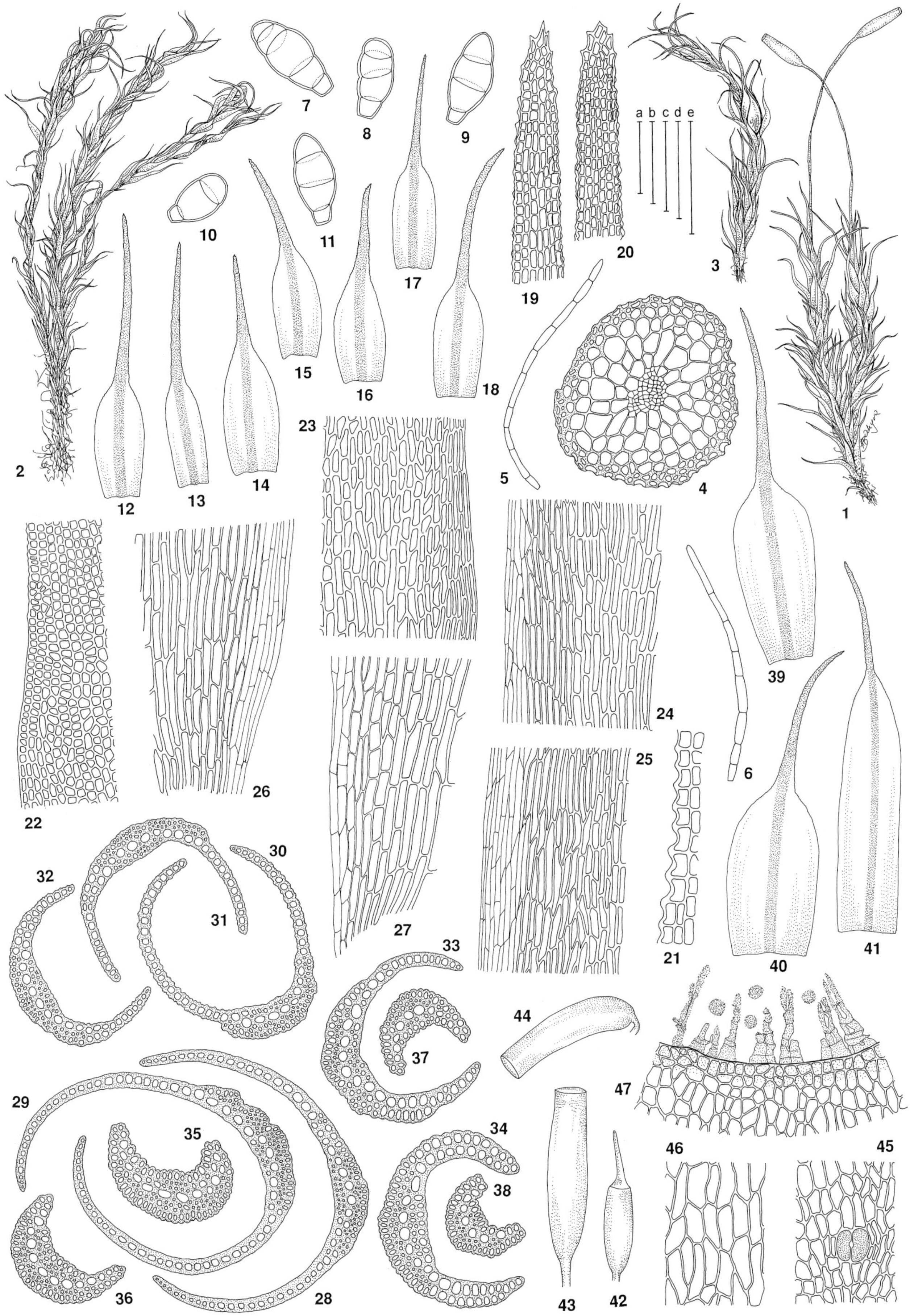
a b c d e
1
2
3
4
5
6
7
8
9
10
11
12
13
14
15
16
17
18
19
20
21
22
23
24
25
26
27
28
29
30
31
32
33
34
35
36
37
38
39
40
41
42
43
44
45
46
47

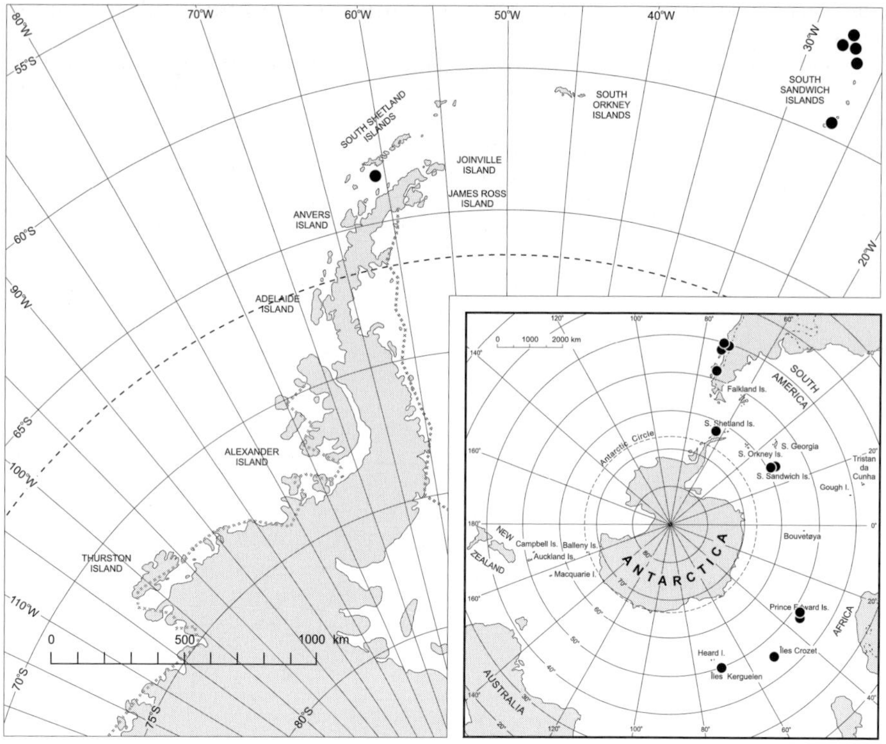

Figure 54. Distribution of *Ditrichum ditrichoideum* in Antarctica. Inset: global distribution.

a median row of large guide cells separating strong dorsal stereid band and a usually weaker ventral stereid band; *laminal cells* thick-walled and not porose throughout, short to linear-rectangular in the mid-sheath to the base, transparent, (25–)40–60(–80) µm long, (4–)6–8(–10) µm wide, unistratose and distinctly thinner-walled towards the margins, usually forming a distinct, often hyaline border extending from the mid-sheath to the base; *angular cells* not differentiated; *shoulder cells* mostly regular in outline, varying from quadrate or rounded to elliptic, 8–18 µm long, 6–8(–10) µm wide, unistratose to variously bistratose, translucent; *subula cells* strongly papillose due to the collapsed thin outer cell walls and protuberant thickened ends of cell cross-walls at the point of junction, giving the subula a papillose appearance, short rectangular over the costa, isodiametric in the lamina, 10–20 µm long, 8–10 µm wide. *Dioicous. Perichaetial leaves* similar in shape to the vegetative leaves only larger, 3.5–4.0 mm long, 1.0–1.2 mm wide. *Setae* erect or somewhat bent, pale yellow-brown, not twisted when dry, 6–7 mm long. *Capsules* exserted, erect to somewhat inclined, symmetric or slightly gibbous, smooth, brown, usually with a varnish lustre, cylindrical, 1.0–1.5 mm long, *ca* 0.5 mm wide, not or slightly narrowed at the mouth; *operculum* conical, long rostrate, with a straight or slightly slanted beak, *ca* ½ length of the urn; *annulus* present, persistent; *exothecial cells* thin-walled, oblong-hexagonal to hexagonal throughout the urn, except its base, 40–60 µm long, 15–25 µm wide, decreasing in size towards the mouth, cells at the rim somewhat smaller, forming a narrow deeper brown strip composed of 2–3 cell rows, cells at the urn base hexagonal, 20–30 µm long, 10–15 µm wide; *stomata* few at the base of the urn, bicellular, superficial, variously oriented with rounded pori; *peristome* single, composed of 16, pale yellowish teeth, to 100 µm long, irregularly split nearly to the base into 2 filiform, terete, papillose divisions. *Spores* 14–18 µm wide, brownish, finely papillose-verrucose. *Calyptra* cucullate, smooth, brown.

Chromosome number: unknown.

Discussion. *Ditrichum ditrichoideum* is recognised from all other local congeners by its strongly and coarsely mammillose leaf subulae and the presence of the copious rhizoidal gemmae. Finely mammillose leaf subulae occur in some species of *Ditrichum*, for example in *D. punctulatum* and *D. brachycarpum* from Australasia (Seppelt 1982) and in Asian *D. divaricatum* (Matsui & Iwatsuki 1990), but in each species the mammillosity is not so strong and coarse as in *D. ditrichoideum* and, additionally, these species differ in other structural characters. In the shape of the leaf subulae *D. ditrichoideum* closely resembles *Distichium capillaceum*, and Newton (1977*b*), relying on this character, so named all specimens from the South Sandwich Is. However, *D. capillaceum* is distinct in its strongly distichous leaf arrangement and the leaf bases in this species are sharply conduplicate and almost impossible to spread out, whereas the vaginate leaf bases in *D. ditrichoideum* are concave and easy to unfold.

The mammillose leaf subulae are also known in two Antarctic species of *Dicranella* which, additionally, have clasping leaf bases but in this genus these are short and almost square, not oblong, and widest at the shoulders. Furthermore, the laminal cells in the sheath are much wider in *Dicranella*.

Another characteristic feature of *Ditrichum ditrichoideum* is the common occurrence of copious rhizoidal tubers. These asexual propagules have been described for most European species of *Ditrichum* (Risse 1987; Arts 1994), but among the austral species of this genus these are known only in the present species, although originally the material was erroneously ascribed to *D. brotherusii* (Ochyra 1996*d*). They occur rarely in

D. heteromallum, but represent a quite different type of tuber, consisting of a spirally twisted uniseriate row of 3–9 cells (Risse 1985; Arts 1994) and are unlike those of *D. ditrichoideum* which consist of a short row of 3–5 cells.

Ditrichum ditrichoideum was initially described from the Antarctic as the new species *D. gemmiferum* (Ochyra & Lewis Smith 1998) which proved to be inseparable from *Austrophilibertiella ditrichoidea*. This taxon was originally described as *Philibertiella ditrichoidea* (Cardot 1914; Ochyra 1996*f*) and has long remained obscure and poorly known, mostly because of the difficulty in locating the original material.

Reproduction in Antarctica. Fruiting specimens with mature capsules have been found on only two specimens from Deception I., but in all populations copious rhizoidal tubers have been observed (Ochyra 1996*d*).

Habitat. In the Antarctic *Ditrichum ditrichoideum* grows mainly on heated ground, most often on dry scoria amongst fumarole vegetation, and its most frequent associates are the mosses *Pohlia nutans*, *Ceratodon purpureus*, *Dicranella hookeri*, *Campylopus introflexus*, *C. vesticaulis*, *Polytrichum juniperinum* and the liverworts *Cryptochila grandiflora*, *Cephaloziella varians* and *Lophozia excisa*.

World range. An amphiatlantic south-temperate species (Fig. 54, inset), scattered in western Patagonia (Ochyra & Deguchi 1999) and extending to Kerguelen Province in the Subantarctic, where it is quite common on Prince Edward I. and less frequent on Marion I. in the Prince Edward Is. archipelago; some records are also known from Îles Crozet and Îles Kerguelen (Ochyra & Poulsen 2003; Ochyra, personal observations), as well as from the maritime Antarctic.

Distribution in Antarctica. Restricted in its occurrence to areas with recent volcanic activity in the South Sandwich Is. (Zavodowski I., Leskov I., Candlemas I., Bellingshausen I.) and South Shetland Is. (Deception I.) (Fig. 54). Until recently, it seemed that its main centre of occurrence lay in the South Sandwich Is., at least judging from the number of the herbarium specimens, but it is now known to be very common and abundant on Prince Edward I. and less frequent on Marion I. in the Prince Edward Is. archipelago (Ochyra, personal observations).

Selected specimens examined. WEST ANTARCTICA. SOUTH SANDWICH IS. **Zavodovski I.** Ash plain above NW coast, *ca* 20 m, *PC 258* (AAS, KRAM). **Leskov I.** Summit ridge, 100 m, *PC 245A, 246D & 249I* (AAS, KRAM) and *ca* 200 m, *REL 509b* (AAS, KRAM). **Visokoi I.** Finger Pt., *ca* 35 m, *REL 758 & 759* (AAS, KRAM). **Candlemas I.** Between lagoons and most southerly exposed lava ridge, 65–100 m, *REL 572* (AAS, KRAM); Lucifer Hill, top of SE slope, *PC 117A & 215A* (AAS, KRAM); same locality, 20 m below crater rim, *PC 214D* (AAS, KRAM); same locality, northern crater, *PC 119B* (AAS, KRAM); same locality, E rim of SW crater, *PC 107A & 185A* (AAS, KRAM); same locality, NW ridge, *PC 241D* (AAS, KRAM); Breakbones Plateau, *PC 197B, 199, 200B, 201A, 208B & 209D* (AAS, KRAM); Clinker Gulch, *PC 151B & 155C* (AAS, KRAM); southern edge of shallow gully above Clinker Gulch, *ca* 50 m below summit, *PC 124A, 127A, 128, 130A, 133C & 136C* (AAS, KRAM). **Bellingshausen I.** Near fumarole at S lip of crater, 150 m, *PC 21 & 46C* (AAS, KRAM).

SOUTH SHETLAND IS. **Deception I.** South East Pt., 150 m, *NJC 48* (AAS, KRAM); Whalers Bay, 'Perchuč Cone', 240 m, *RILS 11238* (AAS, KRAM); Pendulum Cove, near destroyed Chilean Station, 2–5 m, *RO 2792/80* (AAS, KRAM) and *RILS 6933, 6936 & 11311A1* (AAS, KRAM); slopes to east and south-east of Pendulum Cove, 50–100 m, *RILS 6919, 6921, 6922, 6923, 6925 & 6926* (AAS, KRAM); southern part of Pendulum Cove, 100 m, *RILS 5751A* (AAS, KRAM); east of Crimson Hill, 130 m, *RILS 11208A* (AAS, KRAM); Telefon Bay, *Mahu 22834* (AAS, KRAM); southern end of Fumarole Bay, *ca* 125 m, *RILS11278A* (AAS, KRAM).

3. **Ditrichum hyalinocuspidatum** Cardot Fig. 55

Ditrichum hyalinocuspidatum Cardot, Bull. Herb. Boissier Sér. 2, 6: 20. 1906. — Type citation: [South Georgia] Cumberland bay: Pot harbour, mt. Duse. [Lectotype (*selected here*): 'Svenska Sydpolarexpeditionen 1901–03. N:r 286 in parte. Ditrichum hyalinocuspidatum Card. sp. nova. South Georgia: Cumberland Bay. Pot Harbour, ad saxa 15/5 1902 Carl Skottsberg. Det. J. Cardot' – PC!; isolectotype: S!, S-Möller!; syntype: 'Svenska Sydpolarexpeditionen 1901–03. N:r 312 in parte. Ditrichum hyalinocuspidatum Card. sp. nova South Georgia: Cumberland Bay. Mt. Duse 2/5 1902 Carl Skottsberg. Det. J. Cardot' – PC!; isosyntype: S!].

Ditrichum lewis-smithii Ochyra, Ann. Bot. Fennici 33: 304, *f. 1.* 1996, ***syn. nov.*** — Type citation: Antarctica, South Shetland Islands, Admiralty Bay, King George Island: Admiralty Bay, Ezcurra Inlet, Pond Hill, lat. 62°10′25″S, long. 58°26′W, elev. 160 m; in moist situation on bare ground on NW-facing slope; 11.II.1980, Ochyra 1829/80. [Holotype: 'Herbarium Instituti Botanici Academiae Scientiarum Polonae – Cracoviae Bryophyta Ditrichum lewis-smithii Ochyra, sp. nov. Antarctica. South Shetland Islands – King George Island: Admiralty Bay, Ezcurra Inlet, Pond Hill, 62°10′25″S, 58°26′W, elev. 160 m; in moist situation on bare ground on NW-facing slope. 11.II.1980, Coll. R Ochyra 1829/80 Det. R. Ochyra, 1996' – KRAM!; isotypes: AAS!, BM!, H!].

Plants small to medium-sized, forming low or, occasionally, tall turves, usually strongly shining, yellowish-brown to olive-green throughout or blackish-brown above and brown below. *Stems* erect or ascending, often repeatedly branched, 1.0–1.5(–4.0) cm tall, sparsely radiculose at base to densely matted with light brown, smooth and branched rhizoids, in cross-section oval or rounded, consisting of 2(–3) rows of smaller cortical cells with moderately thickened, brown walls surrounding 3–4 layers of large, hyaline and thin-walled medullary cells with distinct corner thickenings and a large and distinct central strand; *axillary hairs* frequent at the apex of the stem and branches, filiform, hyaline throughout, 5–7-celled. *Leaves* (1.5–)2.0–2.5(–3.2) mm long, 0.4–0.7 mm wide at the base, from a concave, oblong-lanceolate, non-decurrent and sheathing base tapering quickly into the subula, erect-spreading when dry, sharply spreading when wet, 1.5 to 2 times as long as the sheath, narrow, strongly flexuose when dry, denticulate near the apex, subtubulose, smooth at the back; *margins* entire, erect to inflexed; *costa* single, yellow-brown, subpercurrent to percurrent, 60–70 µm wide at the base, occupying most of the width of the subula and invisibly merging into the multistratose lamina cells, smooth dorsally, in cross-section consisting of a median row of large guide cells between the ventral and dorsal stereid bands; *laminal cells* smooth, thick-walled and non-porose, unistratose throughout or variously 2–3-stratose at the shoulders and in the subula; *sheath cells* rectangular to linear-rectangular, transparent, up to 90 µm long, 6–8(–10) µm wide, becoming distinctly narrower and thinner-walled towards the margins, forming a distinct hyaline border extending from the mid-leaf to the base; *angular cells* not differentiated; *shoulder cells* mostly irregular, subquadrate, rounded, rhombic or short rectangular, 4–20 µm long, 4–8(–12) µm wide, obscure; *subula cells* mostly isodiametric, rounded or elliptic,

6–16 μm long, 4–6 μm wide. *Autoicous*. *Perichaetia* terminal with solitary sporophytes; *outer perichaetial leaves* similar to the vegetative leaves; *innermost perichaetial leaves* strongly modified, enlarged and clasping at the base, 6.0–7.5 mm long, 1.0–1.2 mm wide, from a broad, convolute, oblong-lanceolate base gradually narrowed into a short subula, much shorter than the lamina. *Setae* erect, yellow-brown, not twisted when dry, about 2 mm long. *Capsules* immersed, erect to somewhat inclined, symmetric or slightly gibbous, smooth to slightly wrinkled with age, brown, lustrous, short cylindrical, 1.3–1.5 mm long, slightly narrowed at the mouth and abruptly constricted into the seta; *annulus* present, biseriate, compound, revoluble; *operculum* conic-rostrate, with a straight or somewhat slanted beak, about half of the urn length; *exothecial cells* thin-walled, oblong-hexagonal to hexagonal, 40–60 μm long, 15–25 μm wide, becoming smaller at the urn rim and forming a distinct, orange-brown strip composed of 4–6 rows of cells; *stomata* few at the base of the urn, bicellular, superficial, with rounded pori; *peristome* none. *Spores* spherical, 20–26 μm wide, brownish, finely papillose-verrucose. *Calyptra* cucullate, smooth.

Chromosome number: unknown.

Discussion. *Ditrichum hyalinocuspidatum* has been a long-forgotten and poorly known species, mainly because it was described from sterile material (Cardot 1906*a*, 1908*a*). Consequently, fertile specimens of this species found in the Antarctic were described as a new species, *D. lewis-smithii*, because they showed some differences in gametophyte characters, including the lack of distinctly hyaline leaf apices (Ochyra 1996*e*). However, when more material of this species became available, including fertile specimens from South Georgia, it was clear that hyalinised leaf apices are not a structural character but merely a response of the plants to harsh climatic conditions, a phenomenon frequently observed in Antarctic and Subantarctic mosses. Because other gametophyte and sporophyte characters of the type and non-type Antarctic and South Georgian material of *D. lewis-smithii* were identical, the species names have to be considered synonymous.

Ditrichum hyalinocuspidatum is otherwise a peculiar and distinct species which occupies an isolated position in the genus *Ditrichum*. It is characterised by the absence of the peristome, a unique feature in this genus, with all other congeners normally having ochyrostomous capsules. Additionally, the capsules are deeply immersed in the perichaetial leaves on very short setae, about 2 mm long. The only other member of this genus with similar capsules is *D. subaustrale* Broth., which is better known as *D. immersum* Zanten, a species known from the subantarctic islands of the Kerguelenian Province including the Prince Edward Is., Îles Crozet, Îles Kerguelen and Heard I. (Zanten 1971; Seppelt 1996; Ochyra personal observations). However, this species, except for having a normal peristome, is a large and robust plant, with rigid, erect leaves 4–7 mm long. In contrast, the plants of *D. hyalinocuspidatum* are smaller and the leaves are 2.0–2.5 mm long on the average, with flexuose, erecto-patent to widespread subulae.

Sterile plants of *Ditrichum hyalinocuspidatum* are likely to be confused with other local species of the genus, especially *D. hyalinum*, but they have generally much broader costae, usually well over 100 μm, whereas in *D. hyalinocuspidatum* the costa is narrow, mostly 60–70 μm wide at the base. In addition, the plants of *D. hyalinocuspidatum* are strongly lustrous and mostly yellowish-brown, whereas those of *D. hyalinum* are generally dull and most often dark yellowish- or olive-green.

There may be some confusion between *Ditrichum hyalinocuspidatum* and some species of *Hymenoloma*, and some South Georgian and Antarctic plants have been misnamed as *H. grimmiaceum*. However, the distinct alar cells, mostly crisped leaves and pseudopapillose laminal cells should preclude misdetermination. In fact, this species was initially reported from the Antarctic from Signy I. (*RILS 689*, AAS, KRAM), with some reservation, as *Hymenoloma immersum* (Bell 1976 as *Dicranoweisia immersa*).

Reproduction in Antarctica. Mostly sterile, but some specimens have abundant sporophytes.

Habitat. A chasmophytic and epigean moss growing on soil covering rock ledges, on humus in rock fissures and crevices as well as on sandy or peaty soil, on stony ground, fine gravel and lava flows. It forms almost monospecific cushions or short turves, sometimes intermingled with liverworts such as *Cephaloziella varians*, *Herzogobryum teres* and *Pachyglossa dissitifolia*. Occasionally, it grows in moss carpets associated with *Polytrichastrum alpinum*, *Andreaea gainii*, *Sanionia georgicouncinata* and *Hymenoloma grimmiaceum*. Generally, the species occupies dry habitats, although some specimens have been collected in moist sites and on heated ground on volcanic islands.

World range. An American subantarctic species, having its main centre of distribution on South Georgia (Ochyra *et al.* 2002) and in the northern maritime Antarctic, and extending also to the Falkland Is., Tierra del Fuego and western Patagonia (Cardot & Brotherus 1923), but all known South American localities are restricted to high alpine elevations (Fig. 56). *Ditrichum hyalinocuspidatum* appears to produce fully mature sporophytes only in the Antarctic and on South Georgia, and all other populations are sterile.

Distribution in Antarctica. An infrequent species restricted to some islands in the northern maritime Antarctic including the South Sandwich, South Orkney and South Shetland Is. as well as Joinville I. with an isolated station on Adelaide I. in northern Marguerite Bay at lat. 67°33′ S (Fig. 56). So far, it has been recorded only once on the Antarctic Peninsula mainland. It usually forms rather small, inconspicuous populations from sea level to 600 m on the Graham Coast, but is most frequent at elevations below 150 m.

→

Figure 55. *Ditrichum hyalinocuspidatum*. — **1**. Habit, dry. **2**. Habit, wet. **3**. Cross-section of stem. **4–5**. Axillary hairs. **6–11**. Vegetative leaves. **12**. Leaf apex. **13**. Cells at leaf shoulder. **14**. Mid-leaf cells. **15**. Mid-leaf cells at margin. **16**. Basal leaf cells at margin. **17**. Basal leaf cells near costa. **18–30**. Cross-sections of leaves, a sequence from base to apex. **31**. Capsule immersed in perichaetial leaves. **32–33**. Perichaetial leaves. **34**. Cells at shoulder of perichaetial leaf. **35**. Capsule with seta. **36**. Exothecial cells in middle of urn. **37**. Exothecial cells at base of urn and stoma. **38**. Exothecial cells at orifice, annulus and spores. — [1, 3–5, 8–9, 12–19, 25–26, 31, 34–38 from *RO 1829/80* (holotype of *Ditrichum lewis-smithii*); 6–7, 24, 27, 32–33 from *RO 1994/80* (paratype of *Ditrichum lewis-smithii*); 10–11, 20–23, 28–30 from *RILS 689* (paratype of *Ditrichum lewis-smithii*); 2 from *RO 2009/80* (paratype of *Ditrichum lewis-smithii*)]. — Scale bars: a – 1 mm (32–33, 35); b – 1 mm (31); c – 1 mm (6–11) and 2 mm (1–2); d – 100 μm (3–5, 12–30, 34, 36–38). (Reproduced from *Annales Botanici Fennici*, 33: p. 305, 1996, with permission.)

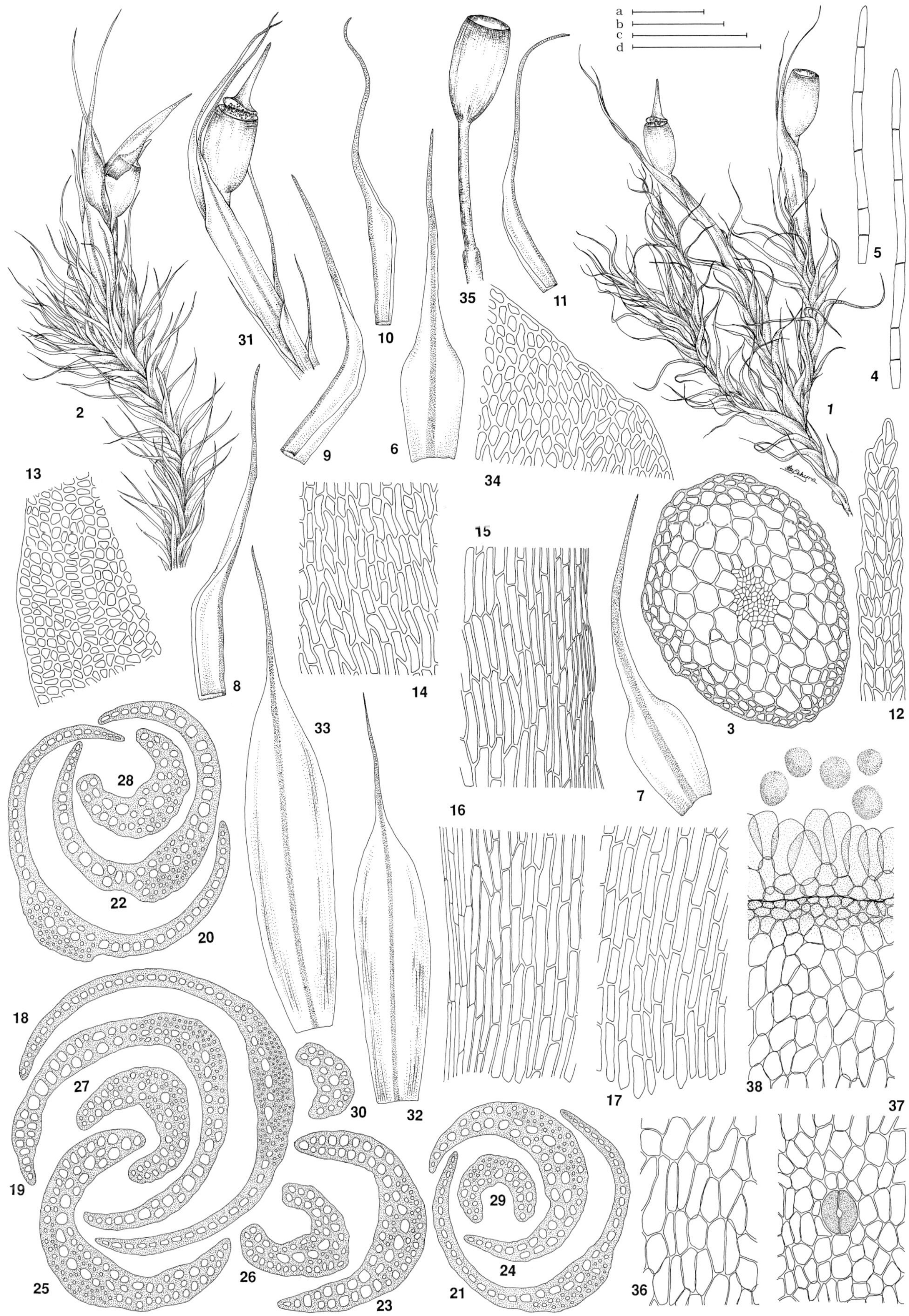
a
b
c
d
1
2
3
4
5
6
7
8
9
10
11
12
13
14
15
16
17
18
19
20
21
22
23
24
25
26
27
28
29
30
31
32
33
34
35
36
37
38

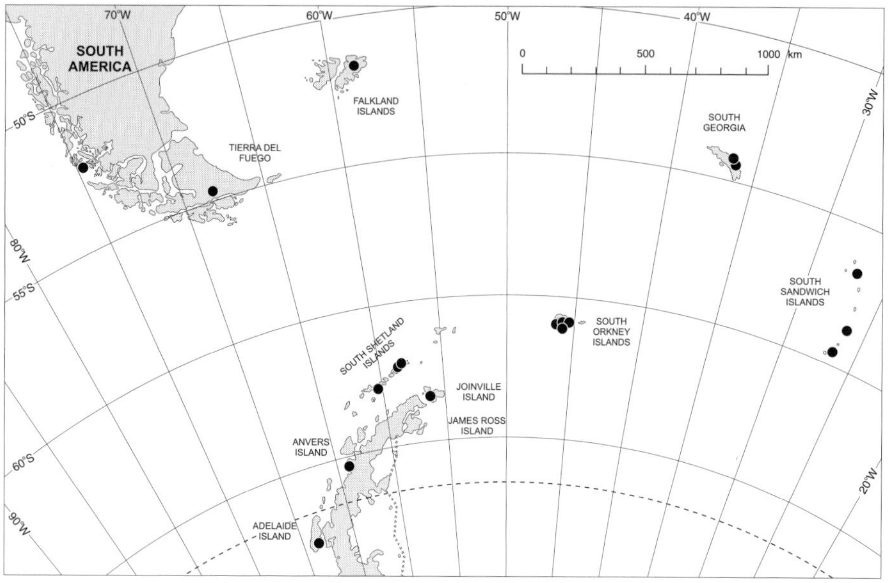

Figure 56. Global distribution of *Ditrichum hyalinocuspidatum*.

Specimens examined. WEST ANTARCTICA. SOUTH SANDWICH IS. **Visokoi I.** Finger Pt., *ca* 6 m, *KA 1* (AAS, KRAM); NE coast, 20–35 m, *REL 773* (AAS, KRAM); Wordie Pt., *PC 102A* (AAS, KRAM). **Montagu I.** Allen Pt., 70 m, *PC 100A* (AAS, KRAM). **Bellingshausen I.** E end of crater rim, *ca* 200 m, *MWH 825F* (AAS, KRAM).

SOUTH ORKNEY IS. **Coronation I.** E side of Laws Glacier, 100 m, *RW 130* (AAS, KRAM); Cape Hansen, 60 m, *RILS 5186* (AAS, KRAM). **Powell I.** SE coast, *ca* 135 m, *RILS 202B* (AAS, KRAM). **Signy I.** Waterpipe Beach, *ca* 15 m, *RILS 505A, C* (AAS, KRAM); between Spindrift Col and Changing Lake, 80 m, *RILS 8141* (AAS, KRAM); Factory Cove. 35 m, *REL* 1062 (AAS, KRAM); Factory Bluffs, *ca* 65 m, *RILS 675A & 869* (AAS, KRAM); Observation Bluff, *ca* 65 m, *MWH 114B* (AAS, KRAM); Moraine Valley, 100 m, *RILS 10844B* (AAS, KRAM); Rusty Bluff, 65 m, *RILS 414* (AAS, KRAM), 70 m, *RILS 3125* (AAS, KRAM), 75 m, *RILS 3130 & 3131* (AAS, KRAM), *ca* 115 m, *RILS 674* (AAS, KRAM), 200 m, *RILS 8628* (AAS, KRAM) and 220 m, *MWH 737B* (AAS, KRAM).

SOUTH SHETLAND IS. **King George I.** ***Barton Peninsula***: N side, 10 m, *DCL 764* (AAS, KRAM). ***Admiralty Bay***: Pond Hill, 160 m, *RO 1829/80* (AAS, BM, H, KRAM); Wegger Peak, 310 m, *RO 1994/80* (KRAM) and 320 m, *RO 2006/80 & 2009/80* (KRAM); Crépin Pt., 10 m, *RO 2042/80* (KRAM). **Livingston I.** ***Hurd Peninsula***: East of Juan Carlos I Station, 50 m, *Schulz 133* (KRAM). **Deception I.** Pendulum Cove, 2 m, *RILS 11325B* (AAS, KRAM); E of Crimson Hill, 130 m, *RILS 11209* (AAS, KRAM); Vapour Col, 150 m, *DM 14* (AAS, KRAM).

WEST ANTARCTIC PENINSULA. **Graham Coast.** Mt. Demaria, 600 m, *RILS 10653* (AAS, KRAM). **Fallières Coast.** ***Adelaide I.***: Reptile Ridge, 250 m, *RILS 1885* (AAS, KRAM).

EAST ANTARCTIC PENINSULA. **Joinville Island group.** ***Joinville I.***: Mt. Alexander, 60–75 m, *RILS 3668A–B & 3672* (AAS, KRAM); to NE of Mt. Alexander, 15–50 m, *RILS 3706* (AAS, KRAM).

4. **Ditrichum conicum** (Mont.) Mitt. Fig. 57

Ditrichum conicum (Mont.) Mitt., Phil. Trans. R. Soc. London 168: 25. 1879. ≡ *Aschistodon conicus* Mont., Ann. Sci. Nat. Bot. Sér. 3, 4: 110. 1845. ≡ *Cynodontium conicum* (Mont.) Mitt., J. Linn. Soc. Bot. 15: 528. 1877. — TYPE CITATION: In provinciis meridionalibus Chiles à cl. C. Gay lectus. [Lectotype (*selected here*): 'Herb. Mus. Paris Herbier du Chili austral envoyé par M. Gay (3e envoi) Aschistodon nov. gen. Mont. A. conicum Montag.' – PC-Montagne!; isotypes: BM-Hooker!, BM-Schimper!, BM-Dixon!, PC-Montagne! (2 specimens)].

Plants medium-sized, rather delicate, dull to lustrous, tall, rather densely tufted, olivaceous, yellow or light green above, brown to blackish-brown below. *Stems* simple or often repeatedly forked, 2.0–3.5 cm tall, erect or sometimes ascending, radiculose below with scattered, pale brownish or yellowish, smooth, branched rhizoids, in cross-section rounded, consisting of (1–)2(–3) rows of smaller, moderately thick-walled, brownish cells surrounding 3–4 layers of large, hyaline and thin-walled medullary cells and a large, conspicuous central strand; *axillary hairs* hyaline throughout, composed of 4–6, elongate cells. *Leaves* (2.0–)2.5–3.5(–4.0) mm long, (0.4–)0.6–0.7(–0.9) mm wide in the middle of the base, erect, appressed, straight or ± contorted in the upper part when dry, erect-spreading when moist, from an oblong-ovate to oblong-lanceolate, concave, non-decurrent and sheathing base gradually narrowed into a channelled, smooth or weakly mammillose subula; *subula* rigid, strict or curled on the upper leaves, erect, usually as long as or up to 1.5 times longer than the leaf base, narrow, acute, entire or more or less denticulate at the apex; *margins* entire, plane in the sheath, inflexed in the subula; *costa* single, pale yellowish-brown to brown, usually not sharply separated from the lamina cells, percurrent to excurrent, 130–180 μm wide at the base, flattened, filling most of the width of the subula, in cross-section consisting of

⟶

Figure 57. *Ditrichum conicum.* — **1**. Plant with mature sporophytes, dry. **2**. Sterile plant, dry. **3**. Cross-section of stem. **4–5**. Axillary hairs. **6–12**. Vegetative leaves. **13–16**. Leaf apices. **17**. Cells at leaf shoulder. **18**. Mid-leaf cells. **19**. Basal leaf cells. **20–29**. Cross-sections of leaves, a sequence from base to apex. **30**. Perigonium. **31–32**. Perichaetial leaves. **33–34**. Operculate capsules, wet. **35**. Exothecial cells at base of urn and stoma. **36**. Exothecial cells in middle of urn. **37**. Exothecial cells at orifice, annulus, portion of peristome and spores. — [1, 11–13, 31–34 from *Gay s.n.* (isotype of *Aschistodon conicus*), BM; 2–10, 14–30 from *RILS 3642*; 35–37 from *Lobb 23*, BM]. — Scale bars: a – 1 mm (1–2); b – 1 mm (6–12, 31–32) and 50 μm (13–16); c – 1 mm (30, 33–34); d – 100 μm (17–19); e – 100 μm (3, 20–29); f – 100 μm (4–5, 35–37). (Reproduced from *Annales Botanici Fennici*, 35: p. 39, 1998, with permission.)

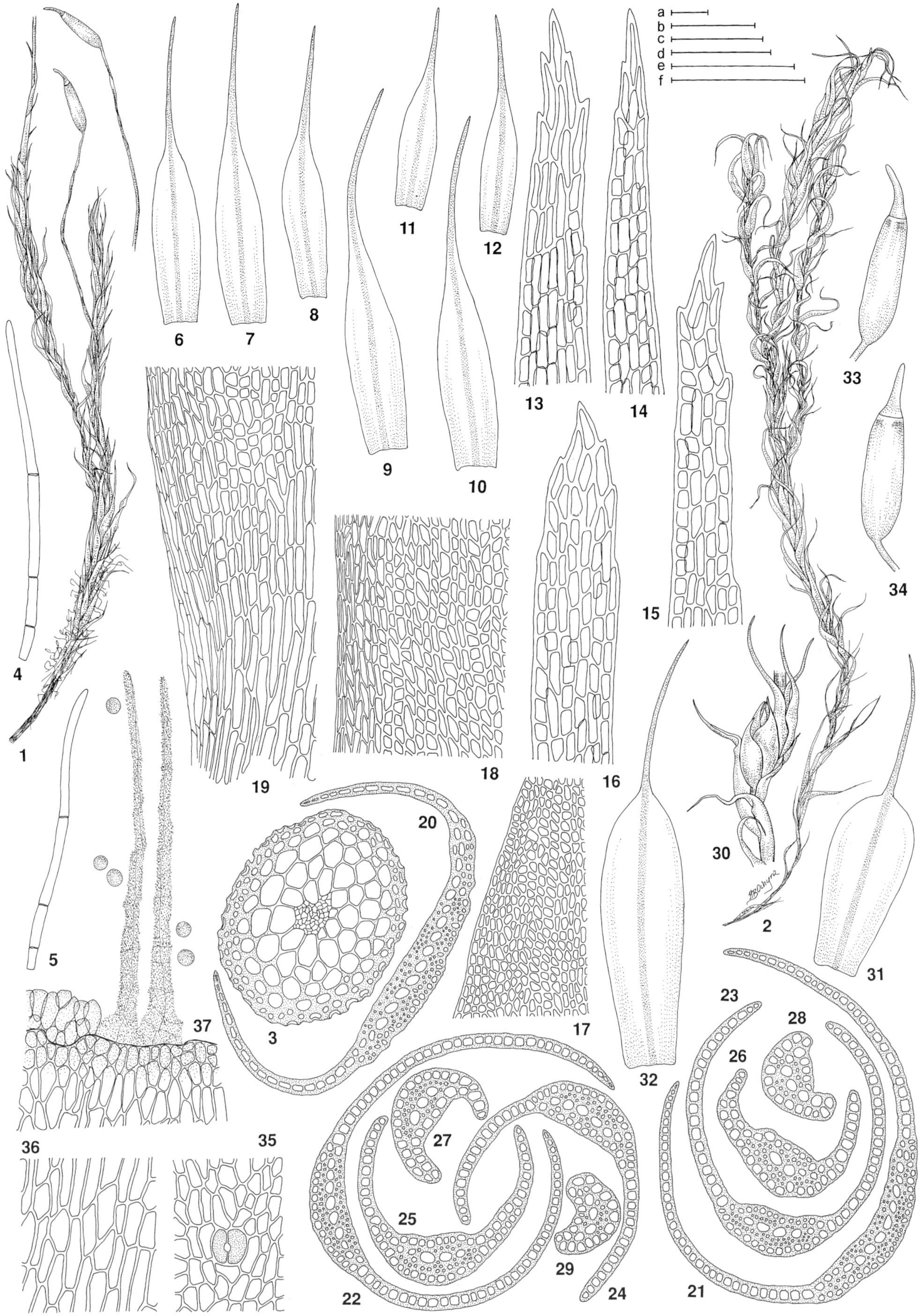
a
b
c
d
e
f
1
2
3
4
5
6
7
8
9
10
11
12
13
14
15
16
17
18
19
20
21
22
23
24
25
26
27
28
29
30
31
32
33
34
35
36
37

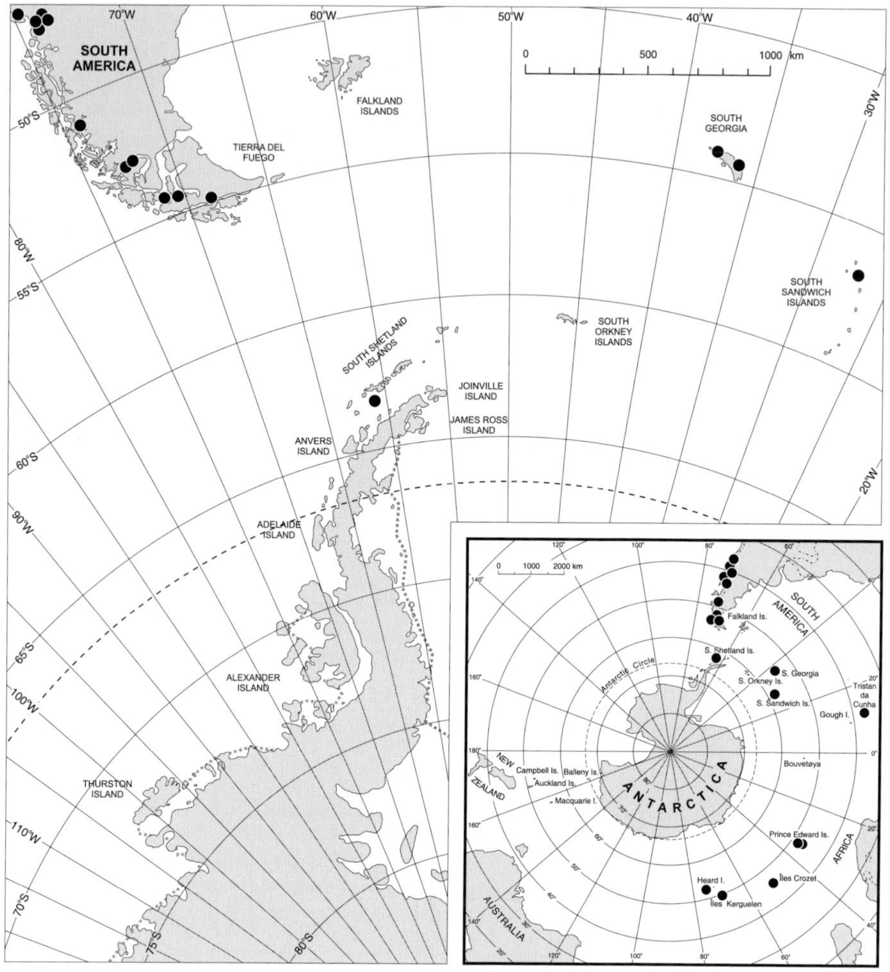

Figure 58. Distribution of *Ditrichum conicum* in the Fuegian region and Antarctica. Inset: global distribution.

a median row of large guide cells and 2 stereid bands; *laminal cells* smooth, non-porose, transparent, thin- to moderately thick-walled, randomly arranged, unistratose throughout and distinctly thinner towards the margins, *sheath cells* variable in shape below, irregular, elliptical, obliquely rhomboidal to short rectangular, 7–10 μm wide, 10–35 μm long, becoming thin-walled and narrower towards the margin, oblong-rhomboidal to linear-rectangular, 4–6 μm wide, up to 50 μm long, usually forming a distinct basal hyaline border, often extending as high as the mid-sheath; *shoulder cells* rounded, subquadrate to transversely elliptical, 6–10 μm wide, 6–12 μm long; *subula cells* isodiametric to short rectangular, rhombic, rounded or elliptic, 5–15 μm long, 4–8 μm wide, unistratose; *alar cells* not differentiated. *Dioicous. Perichaetia* terminal with single sporophytes; *outer perichaetial leaves* 3.0–4.0 mm long, from an ovate or oblong-ovate sheathing base immediately narrowed into the subula, with elliptic to elongate-elliptic, thick-walled cells at the shoulders and rectangular to linear-rectangular, regularly disposed cells below; *innermost perichaetial leaves* strongly modified, enlarged, up to 6.0 mm long, 1.1–1.3 mm wide, from a broad, convolute, oblong-lanceolate base gradually tapering into a short subula, much shorter than the lamina and with areolation of linear-rectangular cells throughout becoming short rectangular to elliptic at the shoulder. *Setae* erect, yellow or pale brownish, not twisted when dry, 0.7–1.5 cm long. [*Capsules* long exserted, erect, symmetric, smooth, light brown to dark brown with age, more or less glossy, short cylindrical, 1.3–1.5 mm long, 0.6–0.7 mm wide, slightly narrowed at the orifice, abruptly constricted into the seta; *annulus* present, persistent; *operculum* conical, with a straight beak; *exothecial cells* thin-walled, oblong-hexagonal to hexagonal throughout the urn, except its base, 40–60 μm long, 10–20 μm wide, decreasing in size towards the rim and forming a distinct, orange-brown strip of 3–4 rows of isodiametric cells; *stomata* few at the base of the urn, bicellular; *peristome* of 16 yellow or pale, papillose teeth, split to the base into 2 filiform branches, 240–300 μm long. *Spores* unicellular, 10–14 μm wide, brownish, finely papillose. *Calyptra* cucullate, smooth.]

Chromosome number: unknown.

Discussion. *Ditrichum conicum* is distinct from all other species in the austral region by its leaf areolation. The laminal cells are irregular in outline and short throughout the leaf, usually only three times as long as wide, occasionally slightly longer, but the length/width ratio never exceeds five and, additionally, they are usually arranged irregularly on the vegetative leaves. However, the cells of the perichaetial leaves are rectangular to linear-rectangular and arranged in quite regular lengthwise rows. Such leaf areolation is not unique in the genus *Ditrichum* and some northern species, for example *D. zonatum* and *D. sekii*, have short laminal cells but are obviously different in certain other characters.

In the leaf areolation *Ditrichum conicum* approaches *Kiaeria pumila* and *Hymenoloma grimmiaceum* but the lack of hyaline basal marginal border, which is replaced by more or less developed alar cells, as well as a much narrower costa and non-vaginate leaves, separate these species. In its vaginate leaves *D. conicum* is likely to be confused in Antarctica with *Dicranella* species, namely *D. campylophylla* and *D. hookeri*. However, they can be distinguished with a hand-lens by their erect-spreading to squarrose limbs and short, more or less square leaf bases. Microscopically, they are distinguished by their bistratose and/or highly mammillose cells of the leaf subula.

Reproduction in Antarctica. No mature sporophytes have been seen on any Antarctic plants, but they possess sexual organs in abundance and some specimens possess young setae without developed capsules.

Habitat. On heated ground, especially on gravelly ash, near fumaroles.

World range. An amphiatlantic temperate species widespread in the *Nothofagus* zone along the western fringe of southern South America from the Valdivian region at *ca* 40°S to Tierra del Fuego, and known from most Subantarctic islands including South Georgia (Ochyra *et al.* 2002), Tristan da Cunha (Dixon 1960), Prince Edward Is. (Zanten 1971; Ochyra, personal observations), Heard I. (Bergstrom & Seppelt 1988; Bergstrom & Selkirk 1997) and Îles Kerguelen (Hébrard 1970*a*; Ochyra 1992*a*; Ochyra, personal observations) (Fig. 58, inset).

Distribution in Antarctica. The species is known only from a single collection from Visokoi I. in the South Sandwich Is. and from a few localities on the volcanic Deception I. in the South Shetland Is. (Fig. 58). On the latter it was not observed prior to the 1967–1970 eruptions of the volcano, but the island was rather undercollected bryologically before then.

Specimens examined. WEST ANTARCTICA. WEST ANTARCTICA. SOUTH SANDWICH IS. **Visokoi I.** Finger Pt., *ca* 35 m, *REL 760* (AAS, KRAM).

SOUTH SHETLAND IS. **Deception I.** Pendulum Cove, behind the ruined Chilean station, 50 m, *RILS 3642* (AAS, KRAM); at back of beach about 100 m from shore, 2 m, *RILS 11322, 11323, 11324, 11325A & 11326* (AAS, KRAM); Vapour Col, 400 m, *DM 31A, 33B & 36B* (AAS, KRAM); 2 km N of Vapour Col, *ca* 175 m, *RILS 11255, 11256A, 11257A, 11258A, 11261D, 11266B, 11268A & 11270* (AAS, KRAM); Mt. Kirkwood, 350 m, *DM 119D* (AAS, KRAM).

5. **Ditrichum hyalinum** (Mitt.) Kuntze Fig. 59

Ditrichum hyalinum (Mitt.) Kuntze, Rev. Gen. Pl. 2: 835. 1891. ≡ *Leptotrichum hyalinum* Mitt., J. Linn. Soc. Bot. 4: 67. 1859. ≡ *Cynontodium hyalinum* (Mitt.) Mitt., J. Linn. Soc. Bot. 12: 43. 1869. — Type citation: On the ground on the hills, Hermite Island, Cape Horn, and in the Falkland Islands, Dr. J. D. Hooker. [Lectotype (*vide* Ochyra 1999*a*: p. 250): 'No. 204. Didymodon longifolius Brid.? var. 2. Falkland Islands 50. Antarct. Exp. 1839–1843. J.D.H.' – BM-Hooker!; isolectotypes: BM-Wilson!, NY-Mitten!; syntype: 'No. 133b Didymodon on ground on the hills Hermite Island, Cape Horn Antarct. Exp. 1839–1843. J.D.H. – NY-Mitten!; isosyntypes: PC!, S-Roth/Möller].

Pseudodistichium austrogeorgicum Cardot, Rev. Bryol. 32: 45. 1905 ['austro-georgicum']. ≡ *Ditrichum austrogeorgicum* (Cardot) Seppelt, J. Hattori Bot. Lab. 51: 142. 1982. – Type citation: Géorgie du Sud. [Lectotype (*selected here*): 'Svenska Sydpolarexpeditionen 1901–03. Nr. 287. Pseudodistichium austrogeorgicum Card. gen. et spec. nov. + Philonotis scabrifolia (HfW) Broth.) South Georgia, Cumberland Bay Bore valley 4/5 1902 Carl Skottsberg Det. J. Cardot' – PC!; isotype: S!]. First synonymised by Ochyra (1999*a*: p. 250).

Plants small to medium-sized, densely caespitose or occasionally loosely gregarious, dull or somewhat glossy, dark olive-green or yellowish- to brownish-green above, dark brown to blackish-brown below. *Stems* 0.5–1.5(–4.5) cm high, erect or ascending, sparsely branched, moderately radiculose below with light brown, smooth rhizoids, in transverse section rounded, consisting of a uni- or incompletely bistratose cortex composed of smaller cells with moderately thickened, brown walls surrounding 3–4-layered medulla of large, hyaline and thin-walled cells and a large and very distinct central strand; *axillary hairs* frequent, hyaline throughout, composed of 5–7 elongate cells. *Leaves* (1.2–)2.5–3.5(–5.0) mm long, (0.3–)0.4–0.5(–0.7) mm wide, usually tightly appressed and not altered when dry, from an oblong or oblong-ovate, concave, non-decurrent and sheathing base rather abruptly tapering to a channelled, smooth or slightly roughened subula, erect when dry, erect-spreading when wet; *subula* strict or sometimes weakly flexuose, erect, usually as long as or to 1.5 times the leaf base length, narrow, acute and more or less denticulate at the apex; *margins* entire, plane in the vaginate leaf base, inflexed in the subula; *costa* single, pale yellowish-brown, usually not sharply set off from the laminal cells, percurrent to short excurrent, flattened dorsally, (90–)100–120 μm wide at the base, occupying most of the width of the subula, smooth on the abaxial surface, in transverse section consisting of a median row of large guide cells between 2 stereid bands; *laminal cells* smooth, non-porose, transparent, unistratose throughout and distinctly thinner towards the margins; *sheath cells* moderately thick- to thin-walled, elongate from the mid-sheath to the base, (20–)30–60 μm long, 8–14 μm wide, with oblique or truncate ends, arranged in more or less regular longitudinal rows, becoming thin-walled and narrower towards the margin, usually forming a distinct hyaline border extending from the mid-sheath to the base; *shoulders* and *subula cells* mostly irregular in outline, varying from subquadrate, rounded or elliptic to short rectangular, (4–)8–15(–20) μm long, (4–)8–10(–12) μm wide, thick-walled; *angular cells* not differentiated. *Autoicous. Perichaetia* terminal with solitary sporophytes; *perichaetial leaves* similar to the vegetative leaves, only somewhat larger and more abruptly narrowed into the subula. *Setae* erect, yellow or pale reddish-brown to brown with age, not twisted when dry, about 1 cm long. *Capsules* exserted, erect to somewhat inclined, asymmetric, smooth to slightly wrinkled with age, brown, lustrous, short cylindrical, 1.3–1.5(–2.0) mm long, 0.5–0.8 mm wide, slightly narrowed at the mouth and abruptly constricted into the seta; *annulus* present, compound, revoluble, composed of a single outer row of large, thin-walled cells and 1 row of smaller basal cells; *operculum* low-conical, with a straight or slanted beak; *exothecial cells* thin-walled, oblong hexagonal to

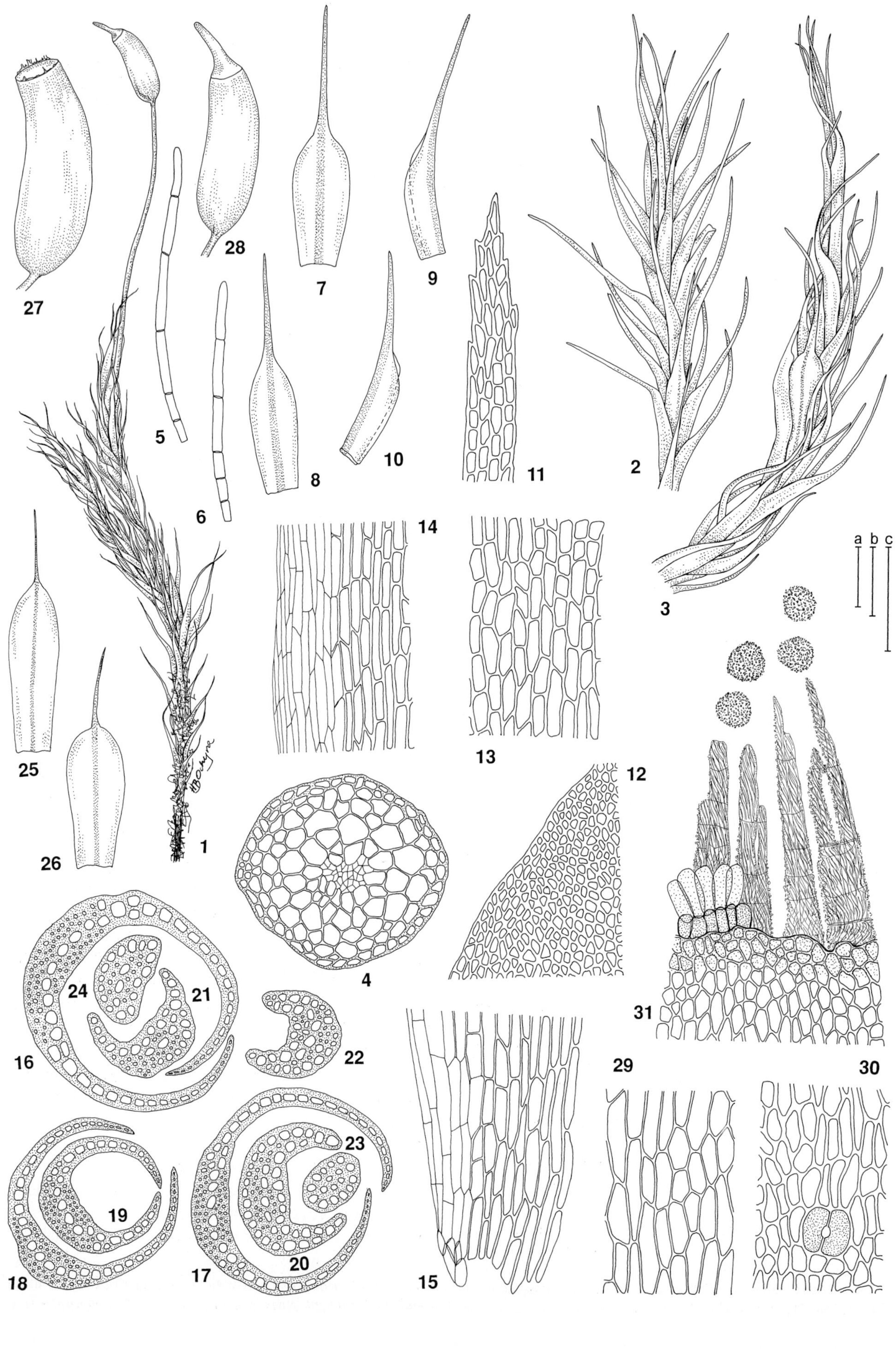
27
28
7
9
5
6
8
10
11
2
3
14
13
12
a
b
c
25
26
1
4
31
16
24
21
22
29
30
18
19
17
20
23
15

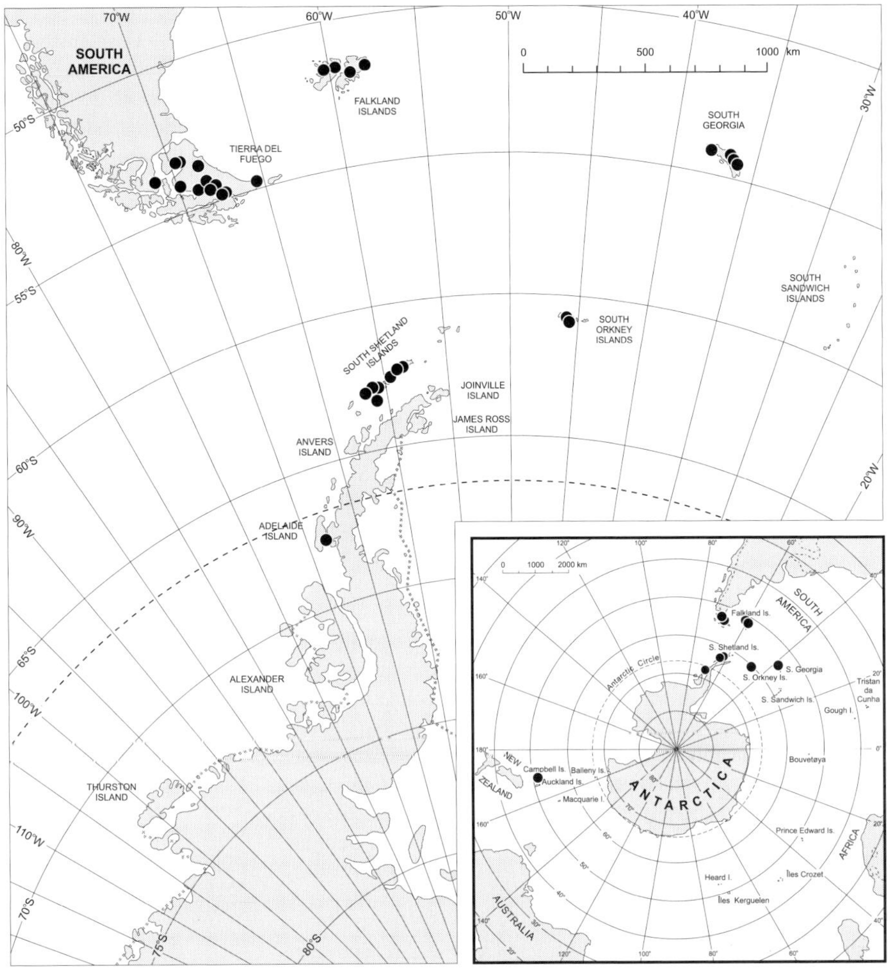

Figure 60. Distribution of *Ditrichum hyalinum* in the Fuegian region and Antarctica. Inset: global distribution.

hexagonal throughout the urn, except its base, 40–90 µm long, 10–20 µm wide, becoming smaller towards the rim, forming a distinct, orange-brown strip composed of 4–5 rows of isodiametric cells; *stomata* few at the base of the urn, bicellular, superficial, with rounded pori; *peristome* of 16 reddish or yellowish teeth, unequally split longitudinally, papillose, obliquely striolate. *Spores* spherical, (22–)25–30(–34) µm wide, brownish, finely papillose. *Calyptra* cucullate, smooth.

Chromosome number: $n = 14$ – South Georgia (Newton 1972 as *Distichium* cf. *austrogeorgicum*).

Discussion. *Ditrichum hyalinum* is better known in regional treatments as *D. austrogeorgicum* but they are clearly conspecific (Ochyra 1999*a*). When describing *D. austrogeorgicum*, Cardot (1905*b*) established the separate genus *Pseudodistichium* to accommodate it, but Seppelt (1980) correctly pointed out its congenericity with *Ditrichum*. *Ditrichum hyalinum* is identified by its low dense tufts and erect and straight leaf subulae which are almost unaltered in the dry state and microscopically by its rectangular cells from mid-sheath to the base. The length of the subulae varies considerably in this species. In the South Georgian type material the subulae slightly exceed the sheath length. However, Cardot (1908*a*) described phenotypes, also from South Georgia, with shorter subulae (fo. *brevifolium*) and later Cardot & Brotherus (1923) recognised, from the Falkland Is., var. *longifolium* with extremely long subulate leaves. These taxa are doubtless habitat responses caused by local climatic or edaphic conditions and do not deserve any taxonomic recognition (Ochyra & Lewis Smith 1998; Ochyra 1999*a*). The majority of Antarctic populations have short subulate leaves, probably a response to the severe climatic conditions. This view is supported by the occurrence of phenotypes matching the South Georgian type in the length of the leaf subulae on Deception I., where geothermal areas create favourable conditions for the development of vegetation (Lewis Smith 1988*d*, 2005*a*,*b*).

Ditrichum hyalinum is likely to be mistaken in Antarctica for *Kiaeria pumila* with which it shares a similar leaf shape. However, *K. pumila* is distinguished by the presence of more or less distinct angular cells and a costa which is composed of almost homogeneous cells in transverse section. Some large plants of

←

Figure 59. *Ditrichum hyalinum*. — **1**. Habit. **2**. Portion of shoot, wet. **3**. Portion of shoot, dry. **4**. Cross-section of stem. **5–6**. Axillary hairs. **7–10**. Vegetative leaves. **11**. Leaf apex. **12**. Cells at leaf shoulder. **13**. Mid-leaf cells. **14**. Mid-leaf cells at margin. **15**. Basal leaf cells at margin. **16–24**. Cross-sections of leaves, a sequence from base to apex. **25–26**. Perichaetial leaves. **27**. Deoperculate capsule, wet. **28**. Operculate capsule, wet. **29**. Exothecial cells in middle of urn. **30**. Exothecial cells at base of urn and stoma. **31**. Exothecial cells at orifice, annulus, portion of peristome and spores. — [1–26, 28 from *RO 590/80*; 27, 29–31 from *Roivainen s.n.*, 10.iii.1929, H]. — Scale bars: a – 1 mm (1) and 0.5 mm (27–28); b – 1 mm (25–26) and 50 µm (4–6, 11–24, 29–31); c – 1 mm (2–3, 7–10). (Reproduced from *Annales Botanici Fennici*, 35: p. 43, 1998, with permission.)

D. hyalinum may be confused with depauperate phenotypes of *Chorisodontium aciphyllum* which were described in the Antarctic as *Blindia skottsbergii* Card. (Ochyra 1993*a*). The two have similarly subulate leaves and broad costae with identical internal structure. However, *Ch. aciphyllum* is distinguished by its lax hyaline alar cells.

Reproduction in Antarctica. Mostly sterile, but plants with fully mature sporophytes have occasionally been collected.

Habitat. In the Antarctic *Ditrichum austrogeorgicum* is an epigean moss usually growing on dry mineral soil amongst stones and boulders, on soil and humus over rock outcrops and in rock crevices, as well as on volcanic ash. Occasionally it occurs on moist clayey soil. The species is most frequently associated with various communities of the fruticose lichen and moss cushion subformation typical of sparsely vegetated fellfield dominated by moss species of the genera *Andreaea*, *Hymenoloma*, *Schistidium* and *Hennediella*, as well as the lichens *Usnea* spp. and many crustose taxa (Lewis Smith 1972). Less frequently *D. hyalinum* is found in the association of *Polytrichastrum alpinum* and *Sanionia georgicouncinata*.

World range. *Ditrichum hyalinum* is an amphipacific south-temperate species (Fig. 60, inset). It is very rare in the eastern austral region where it is known from a single sterile collection from Campbell I. (Seppelt 1982), but frequent and locally abundant in the Fuegian–Antarctic region in the western sector of the austral region. It occurs on the Brunswick Peninsula, on Isla Grande de Tierra del Fuego where it seems to have its main centre of occurrence, on the Falkland Is. (Ochyra 1999*a*) and South Georgia (Cardot 1908*a*; Cardot & Brotherus 1923; Ochyra *et al.* 2002), as well as in the maritime Antarctic.

Distribution in Antarctica. The species is restricted in its occurrence to the South Orkney and South Shetland Is. archipelagoes in the northern maritime Antarctic, with a single isolated station on southern Adelaide I. at lat. 67°31′S; hitherto, it has not been recorded from the Antarctic Peninsula mainland (Fig. 60). It is frequent on King George I. (Ochyra 1998*a*), and occasional on other islands. The reports of *Ditrichum hyalinum* from the South Sandwich Is. (Convey *et al.* 2001) are based upon misdetermination, and these specimens should refer to *D. hyalinocuspidatum* (Ochyra *et al.* 2002). The altitudinal range of *D. hyalinum* in the Antarctic extends from sea level to almost 300 m, although it appears to occur most frequently at elevations below 150 m.

Specimens examined. WEST ANTARCTICA. SOUTH ORKNEY IS. **CORONATION I.** Cape Hansen, 20–200 m, *RILS 3237 & 3244* (AAS, KRAM). **LYNCH I.** Summit plateau, 50 m, *RILS 5183 & 7029* (AAS, KRAM). **POWELL I.** NW coast, 300 m, *RILS 240B* (AAS, KRAM). **SIGNY I.** Spindrift Col, 90 m, *RILS 3124* (AAS, KRAM); western coast opposite Spindrift Rocks, 65 m, *RILS 1144* (AAS, KRAM); Khyber Pass, 85 m, *RILS 10892* (AAS, KRAM); Jane Peak, 150 m, *RILS 3147* (AAS, KRAM); McLeod Glacier, 104 m, *RILS 7050* (AAS, KRAM); Factory Bluffs, 80 m, *RILS 3133 & 8617* (AAS, KRAM), 110 m, *RILS 3126, 10890A & 10891A* (AAS, KRAM), 115 m, *RILS 472* (AAS, KRAM) and 120 m, *RILS 8699* (AAS, KRAM); Polynesia Pt., 50 m, *RILS 8616* (AAS, KRAM); Moraine Valley, 50 m, *RILS 5077 & 7103* (AAS, KRAM), 65 m, *RILS 612* (AAS, KRAM) and 100 m, *RILS 10847* (AAS, KRAM); Gneiss Hills, 200 m, *RILS 8110* (AAS, KRAM).

SOUTH SHETLAND IS. **KING GEORGE I.** ***Fildes Peninsula***: Ardley I., 25 m, *RO 2492/80* (KRAM). ***Barton Peninsula***: Noel Hill, *BJ 202* (KRAM). ***Potter Cove***: Three Brothers Hill, *BJ 228* (KRAM). ***Bransfield Strait***: Red Hill, 100 m, *RO 1086/80, 1097/80 & 1099/80* (KRAM); Demay Pt., 40 m, *RO 1369/80* (KRAM). ***Admiralty Bay***: Sphinx Hill, 100 m, *RO 227/80 & 241/80* (KRAM); Jersak Hills, 100 m, *RO 5068/79, 5070/79, 5072/79 & 5073/79* (KRAM); Jardine Peak, 285 m, *RO 5190/79* (KRAM); Italia Valley, 100 m, *RO 106/80 & 107/80* (KRAM); Breccia Crag, 140 m, *RO 932/80* (KRAM); Kapitan Peak, 200 m, *RO 1924/80* (KRAM); Wegger Peak, 295 m, *RO 2000/80* (KRAM); Ullman Spur, 40 m, *RO 590/80* (KRAM). ***King George Bay***: Turret Pt., *BJ 129* (KRAM). **ROBERT I.** Coppermine Cove, 25–30 m, *RILS 926* (AAS, KRAM). **LIVINGSTON I.** ***Byers Peninsula***: Sealers Hill, 10–15 m, *RILS 3814* (AAS, KRAM); Lair Pt., 10 m, *KR 58D* (AAS, KRAM); Clark Nunatak, 180 m, *DCL 219 & 238* (AAS, KRAM); South Beaches, 20 m, *DCL 504* (AAS, KRAM); Robbery Beaches, 10 m, *KR 52E* (AAS, KRAM). ***Hurd Peninsula***: vicinity of the Spanish Station, *Acuña 7006* (KRAM); Miers Bluff, *RMS 804* (US). **SNOW I.** President Head, *KR 1B & 5D* (AAS, KRAM). **DECEPTION I.** Ronald Hill, 10–15 m, *REL 32A & 34* (AAS, KRAM); Kroner Lake, 10 m, *RWMC 872* (AAS, KRAM); east of Crimson Hill, 130 m, *RILS 11208B* (AAS, KRAM).

WEST ANTARCTIC PENINSULA. **FALLIÈRES COAST.** ***Adelaide I.***: Stork Ridge, 420 m, *PC 268* (AAS, KRAM).

3. **CERATODON** BRID.

Ceratodon Brid, Bryol. Univ. 1: 480. 1826. — Lectotype (*vide* Lindberg 1878: p. 34): *Ceratodon purpureus* (Hedw.) Brid. (*Dicranum purpureum* Hedw.).

Plants small to medium, in short or tall, loose to dense tufts, dark green to yellowish green, often with a reddish or purplish tinge. *Stems* erect, simple or sparingly branched, radiculose at the base, with a distinct central strand. *Leaves* appressed to imbricate, straight or somewhat contorted when dry, erect-spreading to spreading when moist, lanceolate, ovate-lanceolate to broadly ovate, keeled, acute to gradually acuminate, sometimes short awned, to obtuse and rounded at the apex, non-decurrent; *margins* recurved to revolute, occasionally plane, notched or irregularly serrate near the apex to entire throughout; *costa* percurrent to excurrent or ending below the apex, prominent at back, in cross-section with 2 stereid bands separated by a row of guide cells; *laminal cells* smooth, unistratose, thick- to thin-walled, subquadrate, rounded to short-rectangular above, short rectangular and rather thin-walled below; *alar cells* absent. *Dioicous. Perichaetial leaves* terminal, differentiated, consisting of an oblong sheathing, convolute base abruptly shortly pointed or rounded and crenulate at the apex. *Setae* erect, elongate, straight, dark red to orange or yellowish when old. *Capsules* ellipsoid to cylindrical, sometimes ovoid, sulcate when dry, erect or suberect, symmetric or nearly so and non-strumose or inclined to horizontal, curved, distinctly asymmetric and strumose at the base; *annulus* large, revoluble, 2–3-seriate, composed of large cells in the outer row; *operculum* high-conic, straight; *exothecial cells* oblong-hexagonal to long-rectangular, thin- to moderately thick-walled; *stomata* present, few at the neck, superficial, bicellular; *peristome* single, consisting of 16, dark or pale, yellow, orange-reddish to red teeth, arising from a low basal membrane, linear to linear-lanceolate, distinctly articulated, papillose above, cross-striolate below, without or with distinct, pale border, bifid nearly to the base into 2 filiform, terete prongs, united at the nodes. *Spores* yellow to yellowish-green, globose, smooth or minutely roughened. *Calyptra* cucullate, naked, yellowish with a reddish tip, fugacious.

Ceratodon is a natural and well delimited genus which is primarily characterised by sporophyte features. The most distinctive are the peristome teeth which are deeply split to near the base into two terete, filamentous segments. The tips of these turn inward towards the centre of the capsule and resemble the horns of a goat, and the generic name alludes to the shape of the peristome teeth. The teeth have a very distinct articulate appearance due to many trabeculae and the prongs of teeth are distinctly united at their nodes. In addition, the capsule in *Ceratodon* is strongly furrowed and is usually strumose at the neck, a unique feature in the Ditrichaceae, and the innermost perichaetial leaves are convolute and short-pointed to obtuse. Gametophytically, *Ceratodon* is less distinct and may be confused with genera of distantly related families. Typically, the leaves are clearly recurved to revolute at the margins, the upper and median laminal cells are mostly isodiametric, subquadrate, elliptical to short-rectangular, and the costa in transverse section has two stereid bands separated by a median row of large guide cells. However, this typical expression is often masked and altered by considerable phenotypic plasticity of the plants.

Taxonomically, *Ceratodon* is one of the most difficult of all moss genera to determine, especially when sterile. During two centuries of the history of modern bryology approximately 70 species and nearly 100 infraspecific taxa have been recognised under this generic name (Wijk *et al.* 1959, 1969), but currently *Ceratodon* is accepted as consisting of only two or three species (Crosby *et al.* 2000). However, the most radical option treats *Ceratodon* as a monotype and this is sometimes accepted when *C. conicus* is considered as a subspecies or a variety of *C. purpureus* (e.g. Husnot 1884–1890; Dixon & Jameson 1896; Mönkemeyer 1927; Smith 1978). *Ceratodon* has recently been revised on a worldwide scale by Burley & Pritchard (1990) who recognised four species in this genus but this concept has not gained wide acceptance.

Ceratodon is one of the commonest and most widespread taxa in the impoverished moss flora in the Antarctic. Due to the considerable phenotypic plasticity of the plants, apart from the cosmopolitan *C. purpureus*, five species in this genus were described from local material, including *C. antarcticus* (Cardot 1900, 1901), *C. grossiretis* (Cardot 1906*a*, 1908*a*), *C. minutifolius* (Cardot 1911*a,b*, 1913), *C. validus* (Cardot 1906*a*; Horikawa & Ando 1963) and *C. kinggeorgicus* (Kanda 1986*a*). Of these, the first four species truly belong within *Ceratodon*, whereas the latter proved to be *Meesia uliginosa* (Ochyra 1998*a*; Ochyra & Lewis Smith 1999). They were mostly accepted in the former taxonomic treatments of *Ceratodon* in the Antarctic biome (Horikawa & Ando 1963; Robinson 1972*a*; Kanda 1986*a*), but here *Ceratodon* is considered to be a monotypic genus consisting only of the extremely protean species *C. purpureus*.

1. **Ceratodon purpureus** (Hedw.) Brid. Figs 61–62

Ceratodon purpureus (Hedw.) Brid., Bryol. Univ. 1: 480. 1826. ≡ *Dicranum purpureum* Hedw., Spec. Musc. Frond.: 136, *f. 36.* 1801. ≡ *Didymodon purpureus* (Hedw.) Hook. & Taylor, Musc. Brit.: 65, *f. 20.* 1818. ≡ *Trichostomum purpureum* (Hedw.) De Not., Syll.: 189. 1838. ≡ *Desmatodon purpureus* Hassk., Fl. Regist. 1–25: 428. 1852 *err. pro* Didymodon purpureus. — Type citation: [...] per omnem Europam vulgatissima species, inque America septentionali. [Lectotype (*selected here*): The left specimen on the sheet labelled 'Bryum tenue stellatum setis purpureis Dill. Musc. 386. t. 49. f. 51. Mnium purpureum caule dichotomo, axillis pedunculiferis, anthera erecta, foliis carinatis Linn. Spec. ed. II 1575.6' – G-Hedwig/Schwägrichen!].

Ceratodon antarcticus Cardot, Rev. Bryol. 27: 43. 1900. — Type citation: [...] détroit de Gerlache: XIIème débarquement; later Cardot (1901) wrote: Détroit de Gerlache: XIIème débarquement, île de Cavalier de Cuverville, sur les corniches de la falaise. (N° 240b; 2 février 1898. [Lectotype (*selected here*): 'Herb. J. Cardot. Mousses du voyage de la Belgica. N° 240b. Ceratodon antarcticus Card. sp. nova. Canal de la Belgica: XIIème débarquement, sur les corniches de la falaise. Leg. Racovitza, 2 févr. 1898' – PC!; isotype: BR!]. First synonymised by Ochyra (1998*a*: p. 103).

Ceratodon grossiretis Cardot, Bull. Herb. Boiss. Sér. 2: 6: 29. 1906. — Type citation: Terre Louis-Philippe: débarquement *c*; later Cardot (1908*a*) identified this locality as Moss Island. [Lectotype (*selected here*): 'Herb. J. Cardot. Svenska Sydpolarexped. 1901–03. n° 445. Ceratodon grossiretis Card. sp. nova.Terre Louis Philippe: débarquement c. 1/12 1902 Leg. C. Skottsberg Det. J. Cardot' – PC!; isotypes: LD!, S!, UPS!]. First synonymised by Ochyra (1998*a*: p. 103).

Ceratodon grossiretis var. *validus* Cardot, Bull. Herb. Boiss. Sér. 2: 6: 29. 1906. ≡ *C. validus* (Cardot) Horik. & Ando, Hikobia 3: 278. 1963. — Type citation: Terre Louis-Philippe: débarquement *c*; later Cardot (1908*a*) identified this locality as Moss Island. [Lectotype (*selected here*): 'Herb. J. Cardot. Svenska Sydpolarexped. 1901–03. n° 446. Ceratodon grossiretis Card. sp. nova. var. validus Card. var. nova Terre Louis-Philippe: débarquement c. 1/12 1902 Leg. C. Skottsberg Det. J. Cardot' – PC!; isotype: S!, S-Roth!]. First synonymised by Ochyra (1998*a*: p. 103).

Ceratodon minutifolius Cardot, Compt. Rend. Hebd. Séances Acad. Sci. 153: 602. 1911. — Type citation: Baie Marguerite: île Jenny. [Lectotype (*selected here*): 'Herb. Mus. Paris. 258a. Ceratodon minutifolius Card. sp. nova. Baie Marguerite: île Jenny, le long de petites cascades produites par la fonte des neiges, 70–80 m. 2ème expéd. Charcot Leg. Gain, 30/1 1909' – PC! isotypes: H-Brotherus!, PC-Thériot!]. First synonymised by Ochyra (1998*a*: p. 103).

Plants small, medium-sized to relatively large but slender, mostly stiff, sometimes fragile, densely or, less often, loosely caespitose, sometimes forming extensive patches or short turves by coalescing cushions or robust turves with convoluted sinuous ridges, to 10–12 cm thick, separated by troughs up to 5 cm deep, occasionally growing in small, isolated hemispherical cushions, partially buried in soil, dull, very variable in colour, dark green, sordid- or brownish-green, yellowish- to blackish-green, reddish to purplish-brown above, brown below or brown throughout. *Stems* (0.5–)0.8–2.0 or sometimes to 4.5 cm tall, simple or sparsely branched, sometimes julaceous, dark to light brown, erect, sometimes bearing slender, flagelliform innovations with much reduced leaves, variously tomentose below with clusters of brown, smooth, branched rhizoids, in cross-section round, composed of 1–2 layers of small, thicker- and brown-walled cortical cells surrounding 2–4 layers of large, thin-walled, hyaline medullary cells and a distinct, large or small central strand; *axillary hairs* filiform, 4–10-celled, usually with 2–4 short, brown basal cells and rectangular to cylindrical upper cells. *Leaves* crowded or sometimes distant, loosely imbricate to appressed, straight or slightly twisted and flexuose when dry, erecto-patent to patent when wet, ovate, ovate-lanceolate to lanceolate, sometimes triangular to elliptical, gradually short-acuminate to acute, sometimes obtuse, rounded to subacute or cuspidate at the apex, (0.5–)0.9–1.8(–2.4) mm long, 0.4–0.8(–1.0) mm wide, carinate to concave, straight or somewhat falcate at the tips, occasionally distant, concave and curved when dry giving the shoots a catenulate appearance; *margins* recurved to revolute nearly to the apex, sometimes plane to weakly deflexed, unistratose, entire or slightly denticulate, crenulate or irregularly notched near the apex; *costa* single, stout, (40–)60–120(–150) µm wide at the base, reddish, reddish-brown, brown or yellow-brown, straight or sometimes flexuose, occasionally spurred, percurrent to short-excurrent, occasionally ending shortly below the apex or long-excurrent as a stout cuspidate point, prominently convex and smooth dorsally, in cross-section typically composed of a strong dorsal stereid band and weaker, sometimes reduced to a few cells or lacking ventral stereid group and a row of large guide cells; *laminal cells* unistratose, pellucid; *upper laminal cells* quadrate, round, irregularly polygonal to very short-rectangular, (7–)9–19(–22) µm wide, thick- to thin-walled, often with distinct corner thickenings; *basal cells* short-rectangular, 20–40(–50) µm long, 12–18 µm wide, moderately thick- to thin-walled, sometimes sublax, not differentiated in the basal angles. *Asexual reproduction* occasional by filamentous gemmae, 100–200 µm long, 25–40 µm wide, composed of 1–2 rows of 3–20 smooth, somewhat bulging, pale brown, slightly thick-walled cells, borne from axillary rhizoids. *Dioicous. Perichaetial leaves* elongate, 1.5–2.0 mm long, the outer leaves abruptly narrowed to a short point from a tall sheathing base, the inner leaves convolute, truncate, rounded or crenulate at the apex or shortly and broadly pointed. *Setae* erect, 10–25 mm long, straight, dark red to orange or yellowish. *Capsules* ellipsoid to cylindrical, 1–2 mm long, 0.5–0.7 mm wide, deeply sulcate when dry, inclined to horizontal, curved, distinctly asymmetric and strumose at the base; *annulus* large, revoluble, 2–3-seriate; *operculum* conical, straight, to 0.5 mm long; *exothecial cells* oblong-hexagonal to long-rectangular, thin- to moderately thick-walled, 60–110 µm long, 8–10 µm wide, becoming rounded in 1–2 tiers at the mouth; *stomata* 8–10 in one row at the extreme base of the urn, superficial, bicellular; *peristome* single, consisting of 16 teeth, pale yellow above, orange-brown below, linear-lanceolate from a low basal membrane, (250–)350–450(–600) µm long, distinctly articulated, finely papillose throughout, bifid nearly to the base into 2 filiform, terete prongs, united at the nodes. *Spores* yellowish-green, 10–14 µm in diameter, smooth or nearly so. *Calyptra* cucullate, naked, yellowish with a reddish tip.

Chromosome number: unknown in the Antarctic and Subantarctic; *n* = 13 – King George I. (Przywara *et al.* 1984) and Ongul I., Enderby Land (Tatuno 1963; Inoue 1976); the same chromosome count was detected in over 30 studies from throughout the world (Fritsch 1991).

Discussion. Fruiting plants of *Ceratodon purpureus* are readily identified by their dark red, asymmetric, subinclined and strumose capsules that are markedly furrowed when dry. However, the sporophyte features cannot be verified in Antarctic material owing to its sterility and therefore the moss can often be difficult to determine. Gametophytically it may be easily confused with some pottialean mosses because of the mostly isodiametric upper laminal cells which are suggestive of *Didymodon* or *Barbula*. Further confusion may arise from the unusual plasticity and polymorphism of this species resulting in the creation of many phenotypic expressions which, in extreme situations, differ considerably and may imitate entirely different species. The typical expressions of *C. purpureus* are readily recognised by the strongly recurved to revolute leaf margins, often irregularly notched or serrulate near the apex and the firm- or thick-walled, entirely non-papillose, subquadrate to rounded upper laminal cells. In addition, plants, especially those growing in exposed, dry situations are reddish-brown, being strongly pigmented with anthocyanins and carotenoids (Post 1990). In contrast, plants growing in shaded sites are often bright green.

Antarctic plants of *Ceratodon purpureus* exhibit considerable diversity in size, leaf shape and areolation, as well as recurvature of leaf margins, and this may warrant recognition of some distinct phenotypes as species in their own rights. However, it is very easy to demonstrate a morphological continuity between these taxa, and even in a single collection there can exist the full spectrum of leaves from ovate-lanceolate and acuminate with a long excurrent costa and recurved margins, through ovate and acute leaves with a percurrent costa and weakly recurved margins to broadly ovate, concave and rounded at the apex with a costa ceasing slightly below the apex and plane leaf margins, as clearly illustrated by Burley & Pritchard (1990: f. 7). Consequently, these authors included all Antarctic plants in the single species *C. antarcticus* and considered all regional species to be conspecific with it. Also, they considered *C. antarcticus* to be a bipolar species because *C. purpureus* subsp. *arcticus* from Spitsbergen was placed in its synonymy. According to these authors *C. antarcticus* differs from *C. purpureus* on the basis of differences in cell size. In many Antarctic populations the mid-leaf cells are (12–)14–18(–22) µm wide on average and are larger than those in the plants from temperate latitudes in which the mid-leaf cells tend to be generally smaller, (7–)9–12(–14) µm on average. However, no clear difference in cell size is visible as the dimensions intergrade, and consequently many populations could be randomly classified into either species. Accordingly, *C. antarcticus* is here treated as conspecific with *C. purpureus*.

→

Figure 61. *Ceratodon purpureus*. — **1**. Habit. **2–4**. Portions of shoots. **5**. Cross-section of the stem. **6–8**. Axillary hairs. **9–28**. Leaves. **29–31**. Leaf apices. **32**. Mid-leaf cells. **33**. Basal leaf cells. **34–43**. Cross-sections of the leaves, a sequence from the base to the apex. — [1 from *RO 452/80*; 2 from *RO 966/80*; 5–6, 11–12, 32–33, 36, 41 from *RO 774/80*; 7–8, 20–21, 29, 37, 43 from *RO 2728/80*; 40, 42 from *RO 1424/80*; 10 from *RO 841/80*; 3, 31 from *VK 1010*; 4, 22–23 from *VK 1297*; 9, 30, 35, 38 from *TE 97*, BG; 13, 18–19, 39 from *CS 445* (isotype of *C. grossiretis*), S; 14 from *CS 446* (isotype of *C. grossiretis* var. *validus*), S; 15–17 from *Gain 258a* (lectotype of *C. minutifolius*), PC; 24–27, 34 from *ER 240B* (isotype of *C. antarcticus*), BR; 28 from *Nakano A.B. 2*, HIRO]. — Scale bars: a – 1 mm (3); b – 100 µm (5–8, 29–43); c – 1 mm (2, 4, 24–27); d – 1 mm (9–23, 28); e – 0.5 cm (1).

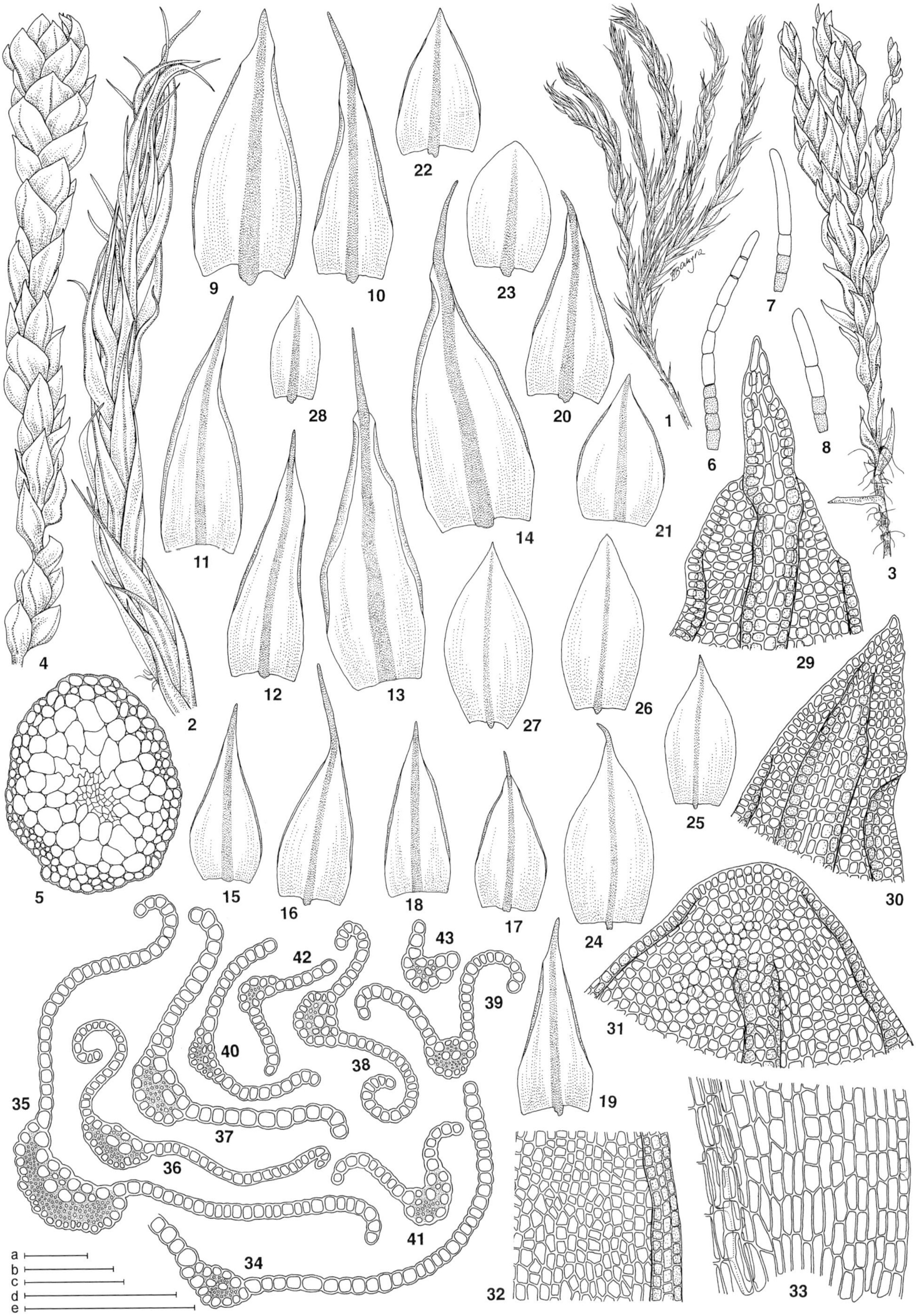
1
2
3
4
5
6
7
8
9
10
11
12
13
14
15
16
17
18
19
20
21
22
23
24
25
26
27
28
29
30
31
32
33
34
35
36
37
38
39
40
41
42
43
a
b
c
d
e

It seems that the enlargement of the leaf cells is a characteristic feature of the polar morphs of *Ceratodon purpureus*, and a similar pattern of leaf areolation is also known in *C. heterophyllus* from the Arctic. This species is dubiously distinct from *C. purpureus* and is gametophytically inseparable from many Antarctic populations with rounded, concave and flat-margined leaves and having costae ending at some distance from the apex as emphasised by Cardot (1907*b*). Burley & Pritchard (1990) showed some differences in the sporophyte features between *C. heterophyllus* and *C. purpureus* in the number of the articulations on the peristome teeth and spore size but due to the sterility of the Antarctic plants testing them is impossible. Preliminary DNA studies (Skotnicki *et al.* 1998; Shaw & Beer 1999) showed that *C. purpureus* exhibits considerable infra- and interpopulational genetic variation, and some experiments (Jules & Shaw 1994) showed that it can undergo relatively rapid evolution in response to strong selective pressures. This may possibly explain some of its protean nature, especially in polar regions. Until more biosystematic studies, including molecular analyses, are performed it is practical to refer all Antarctic material to the single, highly polymorphous species *C. purpureus*.

Some Antarctic phenotypes, especially those described as *Ceratodon antarcticus*, are likely to be mistaken for some smooth-celled morphs of *Hennediella heimii* which were recognised in the Antarctic as *Pottia charcotii*. However, the latter have much longer, thin-walled and lax cells in the lower half of the leaves and at least on some juvenile leaves the weak papillosity of the cells can be traced. In addition, the leaf margins are always distinctly inflexed and the costa is clearly biconvex in section. In the field *C. purpureus* may be confused with *Didymodon brachyphyllus* and *Bryoerythrophyllum recurvirostrum*, but they have opaque and papillose upper leaf cells, as well as with *Pohlia nutans* which has elongate-rhomboidal laminal cells and plane leaf margins, usually serrate distally.

Morphologically, *Ceratodon purpureus* may also be confused with some small plants of *Bryum pseudotriquetrum* and *Schistidium antarctici* but they can be distinguished by the leaf areolation and the costal anatomy which lacks the median row of large guide cells. In the former species the basal leaf cells are always reddish-tinged and the upper cells are thin-walled and oblong-hexagonal becoming narrower and longer at the margins and forming ± distinct border, whereas in the latter the laminal cells are small, usually 7–9 µm wide, often have sinuose walls and are bistratose at the margins.

Reproduction in Antarctica. Only a single population of *Ceratodon purpureus* from Deception I. has been found with fully mature sporophytes (*RILS 3644B*, AAS, KRAM) after the 1970 eruption, and possibly this population represented a recent introduction to the geothermal site.

However, gametangia are occasionally observed. The species readily regerates from detached shoot apices, and soil propagule bank culture studies at Signy I., South Orkney Is. almost always yielded shoots of this species, even though the moss is not always present in the immediate sampling area (Lewis Smith 1987, 1993*a*, 1997*a*,*b*; Lewis Smith & Coupar 1987). These propagules are particularly abundant in soil adjacent to receding icefields. Antarctic plants of *C. purpureus* cultivated under controlled conditions show stronger potential for regeneration in stem cells than plants from temperate regions (Kanda 1979) although, according to Seppelt & Hancock (1991), such shoots do not appear to be especially effective in establishing new populations. In addition, the species frequently produces rhizoidal, filamentous gemmae (Imura & Kanda 1986) which are probably the principal mode of dissemination of this moss in this biome. Vegetative reproduction is very rarely found in this species elsewhere, and similar gemmae were only rarely reported for European (Loeske 1931) and North American populations (Zander & Ireland 1979).

Habitat. The species has a very wide ecological amplitude, growing in most habitats, both open and sheltered, often on disturbed soil and stony ground, on scree, moraines, on rock ledges, and in crevices in cliffs. Although it often grows in dry and exposed sites, it tends to occupy the moist or intermediate zone in hydrological gradients, in particular on well drained slopes as well as below late snow patches, and melt stream channels which quickly become dry after inundation. In areas with recent volcanic activity it is a frequent coloniser of ash, scoria and lapilli but is seldom a component of fumarole vegetation. In the maritime Antarctic *Ceratodon purpureus* is a frequent component of various communities, including those dominated by *Deschampsia antarctica* and *Colobanthus quitensis* which develop on nutrient-rich soil near penguin rookeries and in bird nesting areas. It is commonly associated with *Polytrichastrum alpinum*, *Bryum pseudotriquetrum*, *Sanionia georgicouncinata* and *Syntrichia filaris* in communities of the moss hummock subformation. Although it is scarcer above *ca* 250 m altitude, it is frequently found in lower altitude fellfield communities dominated by fruticose lichens and short cushion mosses. In coastal continental Antarctic sites it is often associated with *Bryum pseudotriquetrum* and *Schistidium antarctici*.

World range. *Ceratodon purpureus* is usually considered as a cosmopolitan weedy moss. It is very common in the temperate latitudes of both hemispheres, but in the tropics it is largely confined to montane regions, mostly above 2000 m, and it is absent from most lowland regions, for example Amazonia (Churchill 1998) in the Neotropics and from the African lowlands (Kis 1985). In addition, the tropical plants are often distinguished as subsp. *stenocarpus* (Burley & Pritchard 1990; Allen 1994*a*) which is sometimes given the status of a separate species (e.g. Norris & Koponen 1988; Churchill *et al.* 1995, 2000). In the austral cool temperate and polar regions, *C. purpureus* is widespread and common, for example in the maritime Antarctic, although it is absent or sparse on some Subantarctic islands, for instance on the Prince Edward Islands (Ochyra, personal observations).

Distribution in Antarctica. Together with *Bryum pseudotriquetrum*, *Ceratodon purpureus* is the most widespread pan-Antarctic moss species (Figs 63–64). In the maritime Antarctic it is common in all offshore archipelagoes and widespread throughout the west coast of the Antarctic Peninsula, from Joinville I. to Alexander I. where it extends to lat. 72°05′S. It decreases in frequency southwards and on Alexander I. it is infrequent and localised. It is rare on the east coast of the Antarctic Peninsula, but is quite frequent on James Ross I. in the northern part of the region and very rare elsewhere, extending to lat. 70°30′S on the Wilkins Coast. On the continent *C. purpureus* is widely distributed but scattered in coastal areas. It is the most frequent

→

Figure 62. *Ceratodon purpureus*. — **1**. Plant with mature sporophyte. **2–3**. Operculate capsules, wet. **4–5**. Deoperculate capsules, dry. **6–7**. Outer perichaetial leaves. **8–11**. Inner perichaetial leaves. **12**. Upper cells of inner perichaetial leaf. **13**. Exothecial cells in middle of urn. **14**. Exothecial cells at base of urn and stomata. **15**. Exothecial cells at orifice, annulus, portion of peristome and spores. — [All from *RILS 3644B*]. — Scale bars: a – 1 mm (6–11); b – 100 µm (12); c – 100 µm (13–15); d – 1 mm (2–5); e – 1 mm (1).

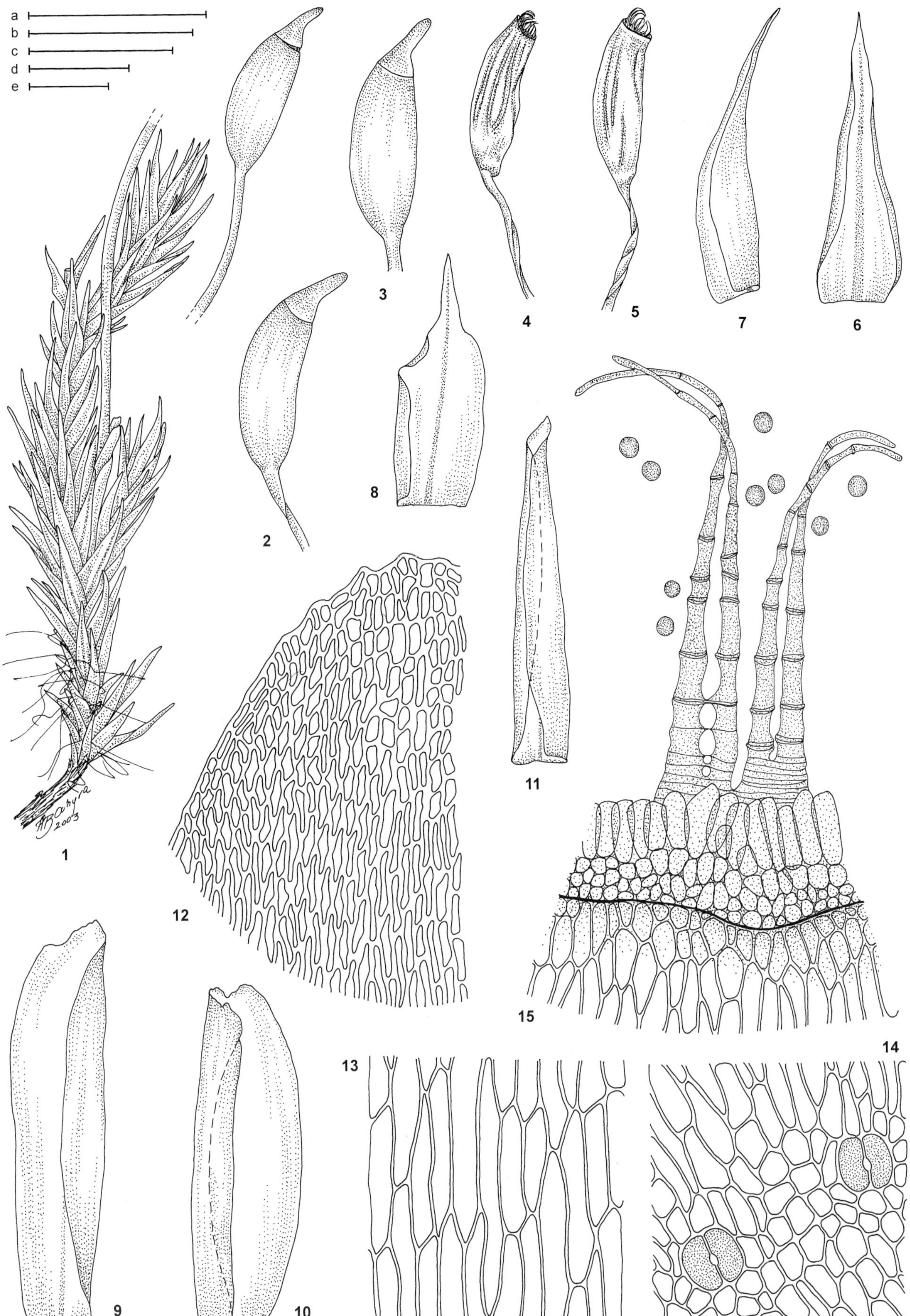
a
b
c
d
e
1
2
3
4
5
6
7
8
9
10
11
12
13
14
15

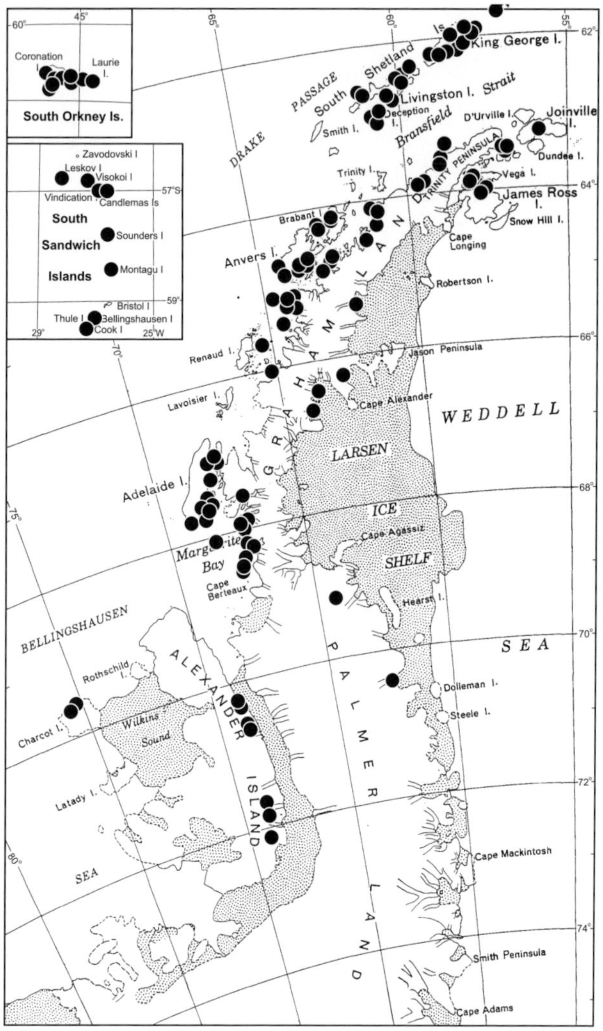

Figure 63. Distribution of *Ceratodon purpureus* in the Antarctic Peninsula region.

in Victoria Land, where it reaches its southernmost occurrence on Mt. Kyffin, Dufek Coast, at lat. 83°48′ S (Seppelt & Green 1998). In the Northern Hemisphere the species reaches its northernmost locality at lat. 83°09′ N in Peary Land in North Greenland (Hesselbo 1923; Holmen 1960). Moreover, the species is known elsewhere from Queen Maud Land (Ryan *et al.* 1989; Richter 1995; Singh & Semwal 2000), Enderby Land (Horikawa & Ando 1961, 1963, 1967*a,b*; Matsuda 1964, 1968; Kanda 1979, 1981*a*, 1986*a*, 1987*b*), Princess Elizabeth Land (Seppelt 1984, 1986), Queen Mary Land (Dixon & Watts 1918; Kuc 1969) and Wilkes Land (Lewis Smith 1989, 1990*a*; Melick *et al.* 1994). *C. purpureus* is a moss of lower elevations, growing mainly from sea level to *ca* 250 m, but it has been recorded at about 600 m on the Foyn Coast, and 1347 m in the Wakefield Highlands in south Central Graham Land.

Selected specimens examined. WEST ANTARCTICA. BOUVETØYA. Kapp Valdivia, 60–90 m, *TE 281, 282, 283.1* (BG); Nyrøysa, *TE 97* (BG).

SOUTH SANDWICH IS. **LESKOV I.** Summit ridge, 100 m, *PC 245D & 249H* (AAS, KRAM); N end, 135 m, *REL 522B* (AAS, KRAM). **VISOKOI I.** Finger Pt., 6 m, *KA 2 & 36* (AAS, KRAM) and 35 m, *REL 752, 755 & 766* (AAS, KRAM); Wordie Pt., *PC 102B* (AAS, KRAM); Irving Pt., *PC 242 & 244* (AAS, KRAM). **CANDLEMAS I.** Lucifer Hill, *PC 214E & 215B* (AAS, KRAM); between Chimera Flats and Breakbones Plateau, *PC 191B* (AAS, KRAM); SW region, 35–100 m, *MWH 448A* (AAS, KRAM); N end of ice cap, 135 m, *REL 715* (AAS, KRAM). **VINDICATION I.** S slopes, *ca* 200 m, *MWH 856A* (AAS, KRAM); NE corner, *ca* 180 m, *MWH 438* (AAS, KRAM). **SAUNDERS I.** Harper Pt., *ca* 20 m, *MWH 846A* (AAS, KRAM). **MONTAGU I.** Allen Pt., 50 m, *PC 96* (AAS, KRAM). **BELLINGSHAUSEN I.** Crater rim, 170 m, *PC 32A & 35B* (AAS, KRAM); W side of crater, *ca* 130 m, *MWH 430D* (AAS, KRAM). **COOK I.** Reef Pt., 30 m, *PC 14B* (AAS, KRAM).

SOUTH ORKNEY IS. **MONROE I.** SW part, 35 m, *MMM 71* (AAS, KRAM). **CORONATION I.** Meier Pt., 10 m, *RILS 8026B* (AAS, KRAM); Mansfield Pt., 10–20 m, *MMM 14* (AAS, KRAM); Cape Vik, 35–100 m, *RILS 571* (AAS, KRAM); Saunders Pt., *ca* 6 m, *RILS 30* (AAS, KRAM). **GOSLING IS.** Centre, 35–50 m, *MMM 37* (AAS, KRAM). **LYNCH I.** N-facing bank, *ca* 13 m, *DCL 949* (AAS, KRAM). **MATTHEWS I.** NW coast, *ca* 135 m, *RILS 157D* (AAS, KRAM). **POWELL I.** Falkland Harbour, *ca* 15 m, *RILS 53* (AAS, KRAM); S tip, 35 m, *MMM 107* (AAS, KRAM); Ellefsen Harbour, 10 m, *RILS 49* (AAS, KRAM). **MICHELSEN I.** E side, *ca* 15 m, *RILS 44* (AAS, KRAM). **FREDRIKSEN I.** NE side, 100 m, *NJC 14* (AAS, KRAM). **SADDLE I.** E part, 50–150 m, *MMM 168* (AAS, KRAM). **LAURIE I.** Scotia Bay, Pt. Davis, 5–35 m, *MMM 120 & 124* (AAS, KRAM); Cape Dundas, *MMM 185* (AAS, KRAM). **SIGNY I.** North Pt., 50 m, *RILS 8056A* (AAS, KRAM); Jane Col, 160 m, *RILS 8070* (AAS, KRAM); Drying Pt., *ca* 3 m, *RILS 691G* (AAS, KRAM); Factory Cove, *ca* 15 m, *RILS 8* (AAS, KRAM) and 35 m, *REL 7 & 1064* (AAS, KRAM); Factory Bluffs, 110 m, *RILS 10890D* (AAS, KRAM); Moraine Valley, 50 m, *RILS 413* (AAS, KRAM). **MOE I.** Landing Cove, 50 m, *RILS 5237* (AAS, KRAM).

SOUTH SHETLAND IS. **ELEPHANT I.** Stinker Pt., 77 m, *JSA 82D* (AAS, KRAM). **GIBBS I.** The Spit, 165 m, *JPB 135* (AAS, KRAM). **O'BRIEN I.** NE coast, 90 m, *JPB 22A & 50A* (AAS, KRAM). **KING GEORGE I.** ***Fildes Peninsula***: Bellinghausen Station, 15 m, *RO 2404/80 & 2422/80* (KRAM); Green Pt., *BJ 20 & BJ 41* (KRAM). ***Drake Passage***: False Round, *BJ76* (KRAM); Ridley I., *BJ 27* (KRAM); Davey Pt., *BJ 46* (KRAM). ***Destruction Bay***: Cape Melville, *BJ 78* (KRAM); Wrona Buttress, *BJ 176* (KRAM); Jenny Buttress, 270 m, *BJ 169* (KRAM); North Foreland, *BJ 104* (KRAM). ***Sherrat Bay***: Three Sisters Pt., *BJ 119 & 120* (KRAM). ***King George Bay***: Turret Pt., *BJ 134 & 137* (KRAM); Penguin I., *BJ 60* (KRAM). ***Legru Bay***: Low Head, *BJ 90* (KRAM). ***Admiralty Bay***: Uchatka Pt., 5 m, *RO 1048/80* (KRAM), 25 m, *RO 1042/80* (KRAM); Demay Pt., 50 m, *RO 1349/80* (KRAM); Creeping Slopes, 120 m, *RO 1288/80* (KRAM); Bastion, 250 m, *1010/80* (KRAM); Sphinx Hill, 6 m, *RO 412A/80* (KRAM) and 95 m, *RO 238/80* (KRAM); Czajkowski Needle, 230 m, *RO 684/80* (KRAM); Penguin Ridge, 45 m, *RO 1424/80* (KRAM); Jasnorzewski Gardens, 4 m, *RO 1455/80, 1465/80 & 1489/80* (KRAM). ***Ezcurra Inlet***: Dutkiewicz Cliff, 10 m, *RO 5083/79* (KRAM) and 15 m, *RO 5081/79* (KRAM); Krzymiński

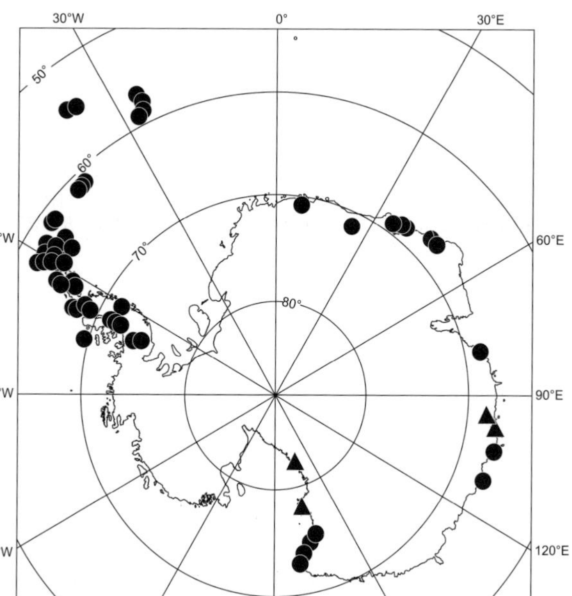

Figure 64. Distribution of *Ceratodon purpureus* in Antarctica. Literature records marked with the triangle.

Pt., 30 m, *RO 973/80* (KRAM). ***MacKellar Inlet***: Klekowski Crag, 220 m, *RO 2302/80* (KRAM); Komandor Peak, 180 m, *RO 1940/80* (KRAM); Crépin Pt., 5 m, *RO 2040/80* (KRAM). ***Keller Peninsula***: Ore Pt., 30 m, *RO 452/80* (KRAM); Speil Pt., 3 m, *RO 530/80* (KRAM); Round Hill, 60 m, *RO 542/80* (KRAM); Moraine Pt., 40 m, *RO 481/80* (KRAM). ***Martel Inlet***: Stenhouse Bluff, 4 m, *RO 2616/80 & 2617A/80* (KRAM); Ullman Spur, 2 m, *RO 583/80* (KRAM); Warkocz, 300 m, *RO 2551/80* (KRAM). ***Bransfield Strait***: Red Hill, 90 m, *RO 1088/80* (KRAM); Blue Dyke, 4 m, *RO 1205/80* (KRAM). ***Marian Cove***: North Spit, *BJ 154* (KRAM). ***Barton Peninsula***: Winship Pt., *BJ 185* (KRAM). ***Potter Peninsula***: Three Brothers Hill, *BJ 163* (KRAM); Stranger Pt., *BJ 108 &110* (KRAM). **NELSON I.** Moss Bay, 26.2.1989, *Komárek s.n.* (KRAM). **ROBERT I.** Coppermine Cove, 15–35 m, *RILS 930* (AAS, KRAM). **GREENWICH I.** Duff Pt., 15 m, *DCL 563* (AAS, KRAM). **HALF MOON I.** S of huts, 50 m, *DCL 580* (AAS, KRAM). **AITCHO IS.** *RMS 69-937* (US). **LIVINGSTON I.** ***Byers Peninsula***: South Beaches, 6 m, *DCL 188* (AAS, KRAM). ***Hurd Peninsula***: False Bay, *RMS 69-412, 570 & 698* (US). **DECEPTION I.** Baily Head, 35 m, *DM 1* (AAS, KRAM); Ronald Hill, 60 m, *RO 2782/80* (KRAM); Lake Irizar, 100–125 m, *RILS 6909A* (AAS, KRAM); Vapour Col, 30 m, *DM 24A* (AAS, KRAM) and 100 m, *RILS 5609* (AAS, KRAM); Kendall Terrace, 65–85 m, *DN 100B* (AAS, KRAM).

WEST ANTARCTIC PENINSULA. **DAVIS COAST.** Astrolabe I., *RILS 6974A, 6976A & 6981* (AAS, KRAM); Cape Roquemaurel, 10–50 m, *RILS 3930* (AAS, KRAM); Bone Bay, Young Pt., 5–15 m, *RILS 3904* (AAS, KRAM); Almond Pt., 5–20 m, *AC & PK 73* (AAS, KRAM); Auster Pt., 15–20 m, *AC & PK 81A* (AAS, KRAM). **DANCO COAST.** Cierva Pt., *VK 186, 294, 513, 611A & 666B* (KRAM); Moss I., *CS 444 & 455* (PC, S), *CS 445* (LD, PC, S, UPS – type of *C. grossiretis*) and *CS 446* (PC, S – type of *C. grossiretis* var. *validus*); Spring Pt., *VK 1231, 1288, 1289 & 1297* (KRAM); Andrée I., 25 m, *RILS 4016 & 4039* (AAS, KRAM); Recess Cove, 2–10 m, *RILS 3988* (AAS, KRAM); Cuverville I., *ER 240B* (BR, PC – type of *C. antarcticus*), 2–10 m, *RILS 4173A* (AAS, KRAM) and 10–20 m, *RILS 11956 & 11960* (AAS, KRAM); Birdsend Bluff, 5–35 m, *RILS 4219A & 4226* (AAS, KRAM); Beneden Head, 15–35 m, *AC & PK 56* (AAS, KRAM); Coughtrey Peninsula, Paradise Harbour, *RMS 69-390* (US). **PALMER ARCHIPELAGO.** ***Brabant I.***: Astrolabe Needle, *VK 365 & 888* (KRAM); Driencourt Pt., 10 m, *AM 14A* (AAS, KRAM). ***Melchior Is.***: Omega I., 2–10 m, *RILS 4120* (AAS, KRAM); Gamma I., 5–10 m, *RILS 4145 & 4157B* (AAS, KRAM). ***Anvers I.***: Cape Monaco, *VK 537* (KRAM); Norsel Pt., *VK 136, 150, 155, 212, 291, 298, 372, 904 & 999* (KRAM); Bonaparte Pt., 10 m, *RILS 1932* (AAS, KRAM); Marr Ice Piedmont, 10 m, *RILS 1930* (AAS, KRAM); Copper Peak, 2–3 m, *RILS 1909* (AAS, KRAM); Joubin Is., 5–10 m, *RILS 2008* (AAS, KRAM); Dream I., *VK 907, 963 & 1010* (KRAM); Litchfield I., *JRR & JMG PAL-42* (US); Cormorant I., 22 m, *AS 36* (AAS, KRAM); Biscoe Bay, *VK 958* (KRAM). ***Wiencke I.***: Goudier I. in Port Lockroy, *VK 99, 108 & 115* (KRAM); Damoy Pt., 5–15 m, *RILS 1912* (AAS, KRAM). **GRAHAM COAST.** Edge Hill, 70 m, *RWMC 658* (AAS, KRAM); Rasmussen I., 20 m, *AS 81* (AAS, KRAM); Cape Tuxen, 25 m, *RILS 10665* (AAS, KRAM) and 100 m, *RWMC 851D* (AAS, KRAM); Cape Pérez, *ca* 25 m, *RWMC 633B* (AAS, KRAM). **WILHELM ARCHIPELAGO.** ***Lemaire Channel***: without details, *ER 268B* (BR, PC). ***Penola Strait***: Petermann I., 10 m, *RWMC 784A & 786* (AAS, KRAM) and 73 m, *AS 115A* (AAS, KRAM). ***Argentine Is.***: Uruguay I., 35 m, *RWMC 820A* (AAS, KRAM); Corner Is., 35 m, *REL 1339 & 1341* (AAS, KRAM); Galindez I., Stella Creek, 15 m, *REL 1323* (AAS, KRAM); Skua I., 5 m, *RWMC 759* (AAS, KRAM). ***Collins Bay***: Berthelot Is., *Gourdon 182A* (H, PC) and *RWMC 621* (AAS, KRAM); Green I., 15–20 m, *RILS 3411* (AAS, KRAM); Sommerville I., *ca* 25 m, *BJT 122C* (AAS, KRAM). ***Grandidier Channel***: Lahille I., 2–5 m, *RILS 4324* (AAS, KRAM); Hook I., *RMS 69-514, 69-523, 69-527, 69-712, 69-825 & 69-890* (US). ***Holtedahl Bay***: Prospect Pt., 2 m, *RILS 1904 & 1906* (AAS, KRAM). **LOUBET COAST.** ***Bennett Is.***: Liard I., Tutton Pt., *ca* 250 m, *JK 156B* (AAS, KRAM); Pfaff I., *JK 156B* (AAS, KRAM); Megaw I., 15 m, *JK 156B* (AAS, KRAM). ***Laubeuf Fjord***: Webb I., *JK 249B* (AAS, KRAM). **FALLIÈRES COAST.** ***Adelaide I.***: E of Mt. Bouvier, 15 m, *JK 236B* (AAS, KRAM); Rothera Pt., 3 m, *JF 12* (AAS, KRAM) and 15 m, *REL 910* (AAS, KRAM); Reptile Ridge, 140 m, *RILS 10588 & 10599* (AAS, KRAM); Jenny I., *Gain 238B, 245B* (PC) *& 258A* (H, PC – type of *C. minutifolius*) and 15–50 m, *RILS 4715* (AAS, KRAM); Avian I., *JRR & JMG PAL-126, 134 & 135* (US) and 15 m, *REL 938A* (AAS, KRAM). ***Léonie Is.***: Lagoon I., *RILS 8828E* (AAS, KRAM); Léonie I., 20 m, *RILS 844* (AAS, KRAM); Anchorage I., *RILS 9212 & 9213* (AAS, KRAM). ***Marguerite Bay***: Piñero I., 10–25 m, *RILS 10665* (AAS, KRAM); Blaiklock I., 8 m, *RILS 844* (AAS, KRAM); Horseshoe I., Lystad Bay, 10–15 m, *RILS 1898 & 1901* (AAS, KRAM); Lagotellerie I., 50 m, *RILS 4842* (AAS, KRAM); Line Is., *ca* 5 m, *RILS 4604* AAS, KRAM); Camp Pt., 10–15 m, *RILS 4631* (AAS, KRAM); Faure Is., Dismal I., 10–25 m, *RILS 3476 & 3545* (AAS, KRAM); Debenham Is., Barbara I., 10 m, *Collinge & Jennings 42* (AAS, KRAM); Neny I., 5 m, *Collinge & Jennings 38* (AAS, KRAM); Roman Four Promontory, 10–30 m, *RILS 4566* (AAS, KRAM); Refuge Is., 10–25 m, *RILS 4522* (AAS, KRAM); Terra Firma Is., Barn Rock, *Poncet 6* (COLO, KRAM); same group, Alamode I., *Poncet 3J* (KRAM); Mushroom I., *Poncet 1C* (KRAM). **GEORGE VI SOUND.** ***Charcot I.***: central N-facing slope, 85 m, *RILS 10495, 10496B, 10507B, 10528 & 10529B* (AAS, KRAM); Rils Nunatak, *ca* 100 m, *PC 368* and 125 m, *RILS 10476* (AAS, KRAM). ***Alexander I.***: Ablation Valley, 1–3 m, *RILS 9235 & 9238* (AAS, KRAM) and 10 m, *RILS 9308* (AAS, KRAM); Moutonnée Valley, 77 m, *RILS 11123A* (AAS, KRAM); Upper Flatiron Valley, 515 m, *RILS 11094* (AAS, KRAM); Horrocks Block, *RILS 8853C* (AAS, KRAM); Ares Oasis, 30 m, *RILS 10440C* (AAS, KRAM) and 50 m, *RILS 9462* (AAS, KRAM).; Coal Nunatak *PC 393* (AAS, KRAM).

EAST ANTARCTIC PENINSULA. **TRINITY PENINSULA.** Hope Bay, *CS 443* (PC, S) and *RWMC 870C* (AAS, KRAM); Scar Hills, *ca* 110 m, *CGB 35* (AAS, KRAM); Mt. Flora. 25–50 m, *RILS 3847* (AAS, KRAM); Lake Boeckella, 45 m, *CGB 9* (AAS, KRAM). **JOINVILLE ISLAND GROUP.** ***Joinville I.***: to NE of Mt. Alexander, 15–50 m, *RILS 3708* (AAS, KRAM). **JAMES ROSS ISLAND GROUP.** ***James Ross I.***: Abernethy Flats, 100 m, *RILS 7625 & 7632B* (AAS, KRAM); Lachman Crags, 25 m, *RILS 7333B* (AAS, KRAM) and 30 m, *RILS 7374A* (AAS, KRAM); St. Martha Cove, 15 m, *RILS 7481A & 7482A* (AAS, KRAM), 50 m, *RILS 7514* (AAS, KRAM) and *ca* 150 m, *RILS 7430B* (AAS, KRAM); Terrapin Hill, 250 m, *RILS 7722A* (AAS, KRAM); Blancmange Hill, 12–15 m, *RILS 7676B & 7677A* (AAS, KRAM); Dreadnought Pt., 3 m, *RILS 7494A* (AAS, KRAM). **OSCAR II COAST.** Starbuck Glacier, 40 m, *PK 41B* (AAS, KRAM). **FOYN COAST.** Jason Peninsula, *ca* 400 m, *RILS 2752 & 2758* (AAS, KRAM); Cabinet Inlet, Cape Casey, 330 m, *RT 2* (AAS, KRAM) and *ca* 600 m, *RT 57C* (AAS, KRAM); Mt. Denucé, *ca* 130 m, *RILS 2769B* (AAS, KRAM). **WILKINS COAST.** Mt. Wakefield, 1347 m, *PC 657* (AAS, KRAM); near Eielson Peninsula, *Bryant 59* (FH).

ELLSWORTH LAND. Behrendt Mountains, 801 m, *PC 565A* (AAS, KRAM).

EAST ANTARCTICA. DRONNING MAUD LAND. **PRINSESSE ASTRID KYST.** Schirmacher Oasis, 92 m, *Singh 1* (KRAM), 124 m, *Singh 12 & 13* (KRAM) and 133 m, *Singh 21* (KRAM). **PRINCE HARALD COAST.** Robertskollen, 'Petrel's Rest', 326 m, *Watkins 25A* (AAS, KRAM); same locality, 'Ice Axe Peak', 480 m, *Watkins 23A* (AAS, KRAM).

ENDERBY LAND. **SÔYA COAST.** Olav, *Nakano A.B. 1* (HIRO, KRAM, PC); Laye Vole, *Nakano A.B.23* (HIRO, KRAM); Utre, *Nakano A.B. 3* (HIRO); East Ongul I., *Matsuda A.B. 28* (HIRO) and *Fukushima A.B. 5 & 6* (HIRO); West Ongul I., *Matsuda 103* (HIRO) and *Fukushima 39* (HIRO, KRAM, PC) *& 40* (AAS, HIRO, KRAM, LE). **PRINCE OLAV COAST.** ***Molodezhnaya Station***: Ozernaya Hill, 60 m, 6.3.1967, *Alexandrov s.n.* (KRAM, LE); Polcanova Oasis, *Alexandrov 28* (KRAM, LE); Mt. Vechernaya, 25 m, *MacNamara 3* (US); Alasheeva Cove, *Korotkevich 12 & 21* (KRAM, LE).

PRINCESS ELIZABETH LAND. **INGRID CHRISTENSEN COAST.** ***Vestfold Hills***: Taleg Gorge, 30 m, *RILS 6393* (AAS, KRAM).

QUEEN MARY LAND. **KNOX COAST.** Delta Bluff, *Harrison 2* (NSW); Watson Bluff, *Harrison 8A & 9A* (NSW). ***Bunger Oasis***: near the Polish Base, *Różycki 4* (KRAM); 0.5 km N of the Base, *Hollerbach 12/397* (KRAM, LE).

WILKES LAND. **BUDD COAST.** ***Windmill Is.***: Clark Peninsula, 20 m, *RILS 6209A & 6213B* (AAS, KRAM); Browning Peninsula, 60 m, *RILS 6348 & 6368* (AAS, KRAM); Bailey Peninsula, *ca* 30 m, *RILS 6261* (AAS, KRAM).

VICTORIA LAND. **BORCHGREVINK COAST.** Hallett Base, Crater Hill, *ca* 530 m, 11–12.1.1958, *Croll et al.* (Hb. K. W. Allison 6) (AAS); Crater Cirque, 100 m, *RILS 9957 & 10172* (AAS, KRAM); Daniell Peninsula, Cape Philipps, 250 m, *RILS 10167* (AAS, KRAM); Apostrophe I., 25 m, *RILS 9790 & 9791* (AAS, KRAM); Cape King, 100–125 m, *RILS 10258* (AAS, KRAM); Tinker Glacier, *RILS 9656* (AAS, KRAM); W side of Whitehall Glacier, 14.11.1964, *Wise s.n.* (US); Harrow Peaks, 75 m, *RILS 9644A, 9875 & 9876B* (AAS, KRAM); Kay I., 130 m, *RILS 10218 & 10222* (AAS, KRAM); North Valley, Edmondson Pt., 8–10 m, *RILS 9530* (AAS, KRAM), 20 m, *RILS 9775* (AAS, KRAM) and 50 m, *RILS 9628* (AAS, KRAM); Terra Nova Bay Station, *DWHW 200B* (AAS, KRAM). **SCOTT COAST.** ***Granite Harbour***: Cuff Cape, *ca* 25 m, *WCS & SWG* 65/50 (AAS, KRAM); The Flatiron, *WCS & SWG* 65/21 (AAS, KRAM); Bay of Sails, *ca* 100 m, *WCSS & SWG* 65/68 (AAS, KRAM).

EXCLUDED TAXA OF THE DITRICHACEAE

Ceratodon purpureus subsp. ***convolutus*** (Reichardt) Burley & N.M.Pritch. — This taxon was reported by Burley & Pritchard (1990) from an obscure single collection (*A. Knudsen*, OXF) from the South Sandwich Is. This is one of three subspecies recognised by these authors within the exceedingly polymorphous *Ceratodon purpureus* and its differentiating characters refer exclusively to the sporophyte. It has an erect to suberect capsule lacking a struma with strongly bordered peristome teeth with 5–10 trabeculae. According to Burley & Pritchard (1990) this is a pan-south-temperate taxon, distributed in New Zealand, Australia, Tasmania, southern South America, including the Falkland Is., Madagascar, Réunion, as well as on the Southern Ocean islands, including Tristan da Cunha, Îles Kerguelen and Heard I. The concept of this taxon is very strange and obscure because the oldest synonym of its name is *Trichostomum chilense* Mont. which actually refers to a very sound and distinct moss *Chrysoblastella chilensis* (Mont.) Reimers. Also, the placement of *Ceratodon planomarginatus* Dixon from Tristan da Cunha in synonymy with *C. purpureus* subsp. *convolutus* is strange since this is a pottialean moss having nothing in common with *Ceratodon*. Furthermore, numerous collections of *Ceratodon* from the islands of the Southern Ocean are sterile, for example on Îles Kerguelen no fertile specimens of this genus are known at all (Ochyra, personal observations), whereas the sporophytes are profusely produced on Île de la Possession, Îles Crozet, and they do not deviate from typical *C. purpureus* (Ochyra, personal observation). All specimens examined of *Ceratodon* from the South Sandwich Is. are sterile and they do not differ materially from those collected in other parts of Antarctica which are here considered as *C. purpureus*. Recognition of *C. purpureus* subsp. *convolutus* on practically unverifiable characters would be very impractical and the reliability of this taxon should be confirmed by biosystematic studies including molecular methods.

Ditrichum brotherusii (R.Br.bis) Broth. — This species was reported from several collections from Livingston I. by Robinson (1972*a*) as *Pseudodistichium fuegianum*, which is conspecific with *Ditrichum brotherusii* (Seppelt 1982) and from a single collection from Deception I. by Ochyra (1996*d*). Examination of the voucher specimens revealed that all Livingston I. specimens are *D. hyalinum*, whereas the Deception I. specimen is *D. ditrichoideum*. Thus, *D. brotherusii* must be withdrawn from the Antarctic moss flora and must still be considered as an amphipacific temperate species.

Ditrichum strictum (Hook.f. & Wilson) Hampe — The occurrence of this species in the maritime Antarctic was mentioned by Kappen & Schroeter (2002: p. 339). However, this is evidently an error as these authors were referring to the Subantarctic Prince Edward Is. when citing the paper of Huntley (1971). *Ditrichum strictum* is a temperate/subantarctic species of the Eastern Hemisphere, common on Macquarie I. and on islands in the Kerguelen Province, as well as on offshore islands of New Zealand (Auckland Is., Campbell I.) and extending to Tasmania and South Africa (Seppelt 1982), but never correctly recorded in the Western Hemisphere.

Family VI. **DICRANACEAE** SCHIMP.

Dicranaceae Schimp., Coroll. Bryol. Eur.: 11. 1856. — Type: *Dicranum* Hedw.

The circumscription and delimitation of the Dicranaceae is one of the most difficult among the acrocarpous mosses due to their remarkable internal morphological diversity and transitions to neighbouring families of the Dicranales. The family is best characterised by the combination of the haplolepidous dicranoid peristome and the falcate-secund leaves giving the plants a characteristic 'dicranaceous' aspect. The peristome is single and consists of 16 lanceolate, flat, often reddish teeth which are divided about one third to two third their length into two or, rarely, three non-filiform segments or, less often, remain undivided, solid to variously fenestrate. They are usually distinctly transversely barred on the inner surface and papillose above and vertically pitted-striolate below on the outer face, although sometimes ventral trabeculae are reduced or absent. In some species or genera the peristome is lacking or variously reduced, with both dorsal and ventral plates equally thin and/or the dicranoid shape and the 2 : 3 pattern irregular or almost lost, but its structure remains essentially similar to that of the normal dicranoid peristome.

The leaves are frequently falcato-secund, sometimes variously twisted or straight, and mostly lanceolate to subulate. They are mostly unicostate, only rarely ecostate (*Braunfelsia*, *Parisia*), and very often have differentiated, large, usually inflated, hyaline or coloured, uni- or multistratose alar cells. However, undifferentiated angular cells are also encountered and in some genera both character states are found. The costa is strong, narrow to very broad, ending near the apex or is excurrent, sometimes as a hyaline awn and is smooth or lamellose dorsally. It is internally undifferentiated or consists of two stereid bands separated by a median row of large guide cells. In some genera (*Campylopodiella*, *Brothera*, *Paraleucobryum*, some species of *Campylopus*) guide cells and stereids are replaced by chlorocysts and hyalocysts. Other gametophyte characters of the Dicranaceae show considerable variation. The leaf margins are entire to variously toothed, uni- to bistratose and erect to inrolled or sometimes recurved. The laminal cells vary from isodiametric to elongate above and rectangular to linear below. They are smooth or rarely papillose or mammillose, uni- to bistratose and the cell walls are often incrassate and porose.

The Dicranaceae comprise small and tiny, epigean and ephemeral plants to large and robust plants growing in extensive mats, tuft or turves on rock, humus, soil or as epiphytes. The stems are erect or very rarely creeping (*Mesotus*), not much branched and have a well developed central strand. The sporophytes are terminal but may become laterally displaced by branch innovations. The perichaetial leaves are similar to the vegetative leaves or markedly differentiated, often being strongly sheathing and convolute. In dioicous species the male plants are often dwarfed and grow on the tomentum of the female plants. The capsules are mostly long-exserted on elongate, straight or cygneous setae and only very rarely they are immersed on a short seta (*Mesotus*). They are cylindrical, obloid to ovoid, erect to inclined, smooth to furrowed, symmetric to asymmetric and strumose. The operculum is usually obliquely long rostrate from a conic base or high conic. The calyptra is cucullate, smooth, naked, sometimes fringed at the base or, exceptionally, mitrate and deeply lobed (in *Mesotus*).

The gametophyte and sporophyte features found in the Dicranaceae, taken together, make the family distinct, but considered individually all can be found in other families of the Dicranales, including Leucobryaceae, Dicnemonaceae and Ditrichaceae. The family still awaits a thorough evaluation and possibly some discordant elements should be segregated into families in their own right. Buck (1979) resurrected the Bruchiaceae for *Bruchia* and *Trematodon* and this concept has gained acceptance (Allen 1994*a*); also Stech (1999*a*) proposed reinstating the Aongstroemiaceae for *Aongstroemia*, *Polymerodon* and *Aongstroemiopsis*. In addition, the Rhabdoweisiaceae are often recognised for *Rhabdoweisia* and *Amphidium*, but recently Buck & Goffinet (2000) expanded the concept of this family and transferred to it several genera including those which were originally positioned in it by Limpricht (1886), namely *Dichodontium*, *Cynodontium*, *Oreas* and *Oreoweisia*, but other genera including *Dicranoweisia* and *Holodontium* have different affinities. This family seems poorly founded and cannot be explicitly diagnosed, even by a single character, and molecular data preclude a close relationship between *Rhabdoweisia* and *Amphidium* (Stech 1999*b*).

The Dicranaceae are the largest and most diverse family within the Dicranales comprising 54 genera and around 1000 species. It has worldwide distribution, with most genera occurring in the temperate regions of both hemispheres, although a few genera are found mostly in the tropics. The present infrafamilial classification of the Dicranaceae was outlined by Hagen (1915) who subdivided the family into five subfamilies. It was subsequently accepted and slightly modified by Brotherus (1924*b*) who accommodated within it all known world genera. He distinguished six subfamilies, of which the Trematodontoideae are now treated as the separate family Bruchiaceae (Buck 1979; Allen 1994*a*), and the Paraleucobryoideae are merged with the Campylopodioideae (Allen 1994*a*). This conclusion was partly supported by molecular studies (Stech 1999*a*). Instead, the separate monotypic subfamily Mesotoideae was established for the isolated Australasian genus *Mesotus* (Allen 1987*b*), and the isolated genus *Amphidium* was placed in the separate subfamily Amphidioideae (Ochyra *et al.* 2003). Thus, six subfamilies are currently recognised within the Dicranaceae and the nomenclatural issues associated with these subfamilies have been clarified by Ochyra (2002*a*). They are quite distinct taxa with regard to the core genera but grade into one another via peripheral genera. The subfamilial classification of the Dicranaceae still needs revision and recent molecular data (Stech 1999*a*) have yielded promising results.

In the Antarctic, the Dicranaceae are well represented, but most species are not prominent components of the tundra vegetation and some of them are restricted to areas with current volcanic activity. Ten species have been discovered in this biome; they belong within five genera which can be classified into currently recognised subfamilies as follows: Dicranelloideae (*Dicranella*), Campylopoideae (*Campylopus*), Dicranoideae (*Kiaeria*, *Chorisodontium*) and Amphidioideae (*Amphidium*).

KEY TO THE GENERA OF THE ANTARCTIC DICRANACEAE

1. Alar cells not differentiated 2
1. Alar cells clearly differentiated 3
    2. Laminal cells smooth and transparent or mammillose in the limb; leaves ovate-lanceolate, erect to falcato-secund or differentiated into sheathing base and subulate, squarrose limb .... 10. ***Dicranella***
    2. Upper laminal cells densely warty pluripapillose, opaque; leaves oblong- to linear-lanceolate, curled-contorted .... 11. ***Amphidium***
3. Costa narrow, less than ⅓ the leaf base width, in cross-section rounded, with stereids and median guide cells not or poorly differentiated .... 12. ***Kiaeria***

3. Costa broad, ⅓ or more the leaf base width, in cross-section strongly flattened, with well developed stereids, guide cells and hyalocysts 5
    4. Costa in cross-section with several layers of ventral stereids ... 13. ***Chorisodontium***
    4. Costa in cross-section with a single ventral layer of hyalocysts .... 14. ***Campylopus***

10. **DICRANELLA** (MÜLL.HAL.) SCHIMP., *nom. cons.*

Dicranella (Müll.Hal.) Schimp., Coroll. Bryol. Eur.: 13. 1856, *nom. cons.* ≡ *Aongstroemia* Bruch & Schimp. sect. *Dicranella* Müll.Hal., Syn. Musc. Frond. 1: 430. 1848. — Type: *Dicranella heteromalla* (Hedw.) Schimp. (*Dicranum heteromallum* Hedw.) (*typ. cons.*).

Anisothecium Mitt., J. Linn. Soc. Bot. 12: 39. 1869, *nom. illeg. incl. lectotyp. gen.*

Plants small to medium, occasionally large but slender, in lax to dense tufts, turves and patches or scattered on soil, somewhat lustrous, dark or bright green to yellowish-green, often brown below, sometimes reddish-tinged. *Stems* erect, simple or branched, sparsely radiculose at base to tomentose below with brown, branched, smooth rhizoids, in cross-section rounded to obtusely triangular, with a distinct, small to large central strand and 1–2-stratose cortex of small, thick-walled, brown cells or larger, thinner-walled cells only slightly different from medullary cells; *rhizoidal tubers* frequent, relatively small, 1–3-celled to multicellular, spherical or irregularly shaped; *axillary hairs* filiform, 3–7-celled, hyaline throughout or with 1–3 short, brown basal cells and elongate upper cells. *Leaves* non-decurrent, lanceolate, lanceolate-triangular to linear-lanceolate, erect to flexuose-erect, often secund, from an ovate-lanceolate, non-vaginate base gradually acuminate to subulate or from an ovate, square or oblong sheathing base clasping the stem abruptly narrowed to a channelled, subulate to linear-subulate, erect-spreading to squarrose-flexuose subula; *margins* plane or recurved, uni- to bistratose, entire below, serrulate towards the apex; *costa* single, stout, sometimes not well defined, usually occupying ⅓–⅕ the leaf base width, often filling virtually all of the limb, percurrent to excurrent, in cross-section with ventral and dorsal stereid bands and a median row of guide cells or sometimes ventral band highly reduced to lacking; *laminal cells* smooth, unistratose, rectangular to linear-rectangular, becoming shorter and irregular at the shoulders and short-rectangular to quadrate, bistratose and mammillose in the limb, not pitted, thin-, firm- to moderately thick-walled; *alar cells* not differentiated or occasionally somewhat shorter and widened. *Dioicous* or *autoicous*. *Perichaetial leaves* not differentiated. *Setae* erect, elongate, straight to weakly flexuose, deep red to purplish-brown. *Capsules* globose, ovoid, erect to inclined, symmetric to asymmetric and gibbous, not or strumose, smooth or faintly striate; *annulus* none or present, complex and revoluble or persistent with 1 to several rows of small, thick-walled cells; *operculum* conical, with a short or long, slanted beak; *exothecial cells* short-rectangular, thick-walled; *stomata* absent or present at the urn base, superficial, bicellular, with rounded pori; *peristome* single, consisting of 16 lanceolate, dark red teeth arising from a low basal membrane or inserted at the mouth, forked to ⅓ or more than halfway to the base, densely papillose to papillose-striate. *Spores* globose, yellow to greenish-brown, smooth to papillose. *Calyptra* cucullate, naked, fugacious, covering about half of the capsule.

Dicranella is here interpreted in the broader sense to include *Anisothecium* which is sometimes recognised as a separate genus, although nomenclaturally this name is not available because it is illegitimate under Art. 6.4 of the current Code (McNeill *et al.* 2006). These genera, together with the mostly tropical genus *Leptotrichella*, which is better known as *Microdus* (Ochyra 1997*b*), constitute a difficult complex diagnosed exclusively by sporophyte characters. Allen (1994*a*) stated that *Anisothecium* is characterised by the combination of three features, namely the presence of stomata, the exannulate capsules and the presence of the low basal membrane of the peristome. Of these, only the first character is consistently lacking in *Dicranella s. str.* but its value seems to be overemphasised and in other well defined genera, for example in *Seligeria*, *Blindia*, *Hymenoloma* or *Dicranoweisia*, both character states are frequent. Because the other two character states are also present in various species of *Dicranella s. str.*, in practice *Anisothecium* could be based upon a single sporophyte character of doubtful value. Hence, the separation of *Anisothecium* seems to be unnatural and, additionally, it could pose some nomenclatural problems because the lectotype species of *Dicranella* (*D. varia*) is a member of *Anisothecium*. However, this has been successfully resolved by conserving the generic name *Dicranella* with the conserved type *D. heteromalla* (Margadant & Geissler 1995).

As presently understood *Dicranella* is a large worldwide genus comprising approximately 70 well-known species, but about 135 additional obscure species (Crosby *et al.* 2000) still need careful taxonomic assessment. It is certainly a polyphyletic and unnatural group which can be characterised only in general terms by undifferentiated alar cells, narrow costae occupying less than ⅕ of the leaf base (although it is much broader in the type species itself), straight to flexuose setae and erect to erect-spreading, flexuose or falcato-secund leaves. Some species of *Dicranella* resemble diminutive species of *Dicranum* in general aspect, for example *D. heteromalla* or *D. cerviculata*. However, there are many species in this genus in which the vaginate leaves tightly clasp the stems, and this character state is unknown in *Dicranum*; these species closely resemble species of *Ditrichum*.

The greatest diversity of *Dicranella* species is in warm and temperate regions. In polar regions the genus is poorly represented and its species are not prominent components of the bryoflora. In the austral polar regions only six or seven species are known, but they await taxonomic revision; a useful treatment of the genus is available only for two South Georgian species (Newton 1977*a*). From other regions mostly individual species have been studied taxonomically (e.g. Matteri & Calabrese 1999; Ochyra 1999*d*). Three species, all belonging within sect. *Anisothecium* (Kindb.) Ochyra, have been found rarely in the Antarctic (Cardot 1910*b*; Newton 1977*b*; Ochyra & Newton 1985; Lewis Smith 1988*d*; Ochyra *et al.* 2000).

1. Leaves erect, erect-flexuose to subsecund, lacking an enlarged sheathing and clasping base, ± gradually tapering from the insertion; leaf margin narrowly recurved throughout .... 1. ***D. varia***
1. Leaves from an enlarged, sheathing and usually clasping base abruptly or gradually narrowed into an erect, erecto-patent or squarrose limb; leaf margin plane throughout 2
   2. Laminal cells smooth throughout .... 2. ***D. hookeri***
   2. Laminal cells in the limb highly mammillose on the dorsal surface .... 3. ***D. campylophylla***

1. **Dicranella varia** (Hedw.) Schimp. Fig. 65

Dicranella varia (Hedw.) Schimp., Coroll. Bryol. Eur.: 13. 1855 [1856]. ≡ *Dicranum varium* Hedw., Spec. Musc. Frond.: 133. 1801. ≡ *Aongstroemia varia* (Hedw.) Müll. Hal., Syn. Musc. Frond. 1: 435. 1849. ≡ *Anisothecium varium* (Hedw.) Mitt., J. Linn. Soc. Bot. 12: 40. 1869. — Type citation: In Anglia, Dania, Gallia, Pedemontia; also cited Dillenius (1741), Linnaeus (1763), Hudson (1762), Murray (1784), Hedwig (1787), Dickson (1793) and Bridel (1798). [Lectotype (*vide* Geissler *in* Margadant & Geissler 1995: p. 616): 'Dicranum varium Hedw. St. Crypt. II. p. 93. t. 34. Bryum simpl. S. V. p. 949 Lips. in solo graminoso lectum' – G-Hedwig/Schwägrichen!].

Dicranella recurvata Ochyra, Arts & Lewis Smith *in* Głowacki & Bednarek, Polish Polar Stud. 25 Int. Polar Symp.: 166. 1998, *nom. nud.*, ***syn. nov.*** — Chirotype: Antarctica, South Sandwich Islands, Bellingshausen Island: small circular heated area, knoll on crater rim W of access point, alt. 170 m., lat. 59°26′ S, long. 27°05′ W, 21 Jan 1997, leg. P. Convey 31 – AAS!, KRAM!

Plants small, light to yellowish-green, dull, in cushions 5–13 mm high. *Stems* simple or occasionally forked, brown, in cross-section rounded or oval, with a large and distinct central duct and consisting of an indistinct cortex of relatively large cells with slightly thickened, brown or yellowish-brown walls and 3–4 layers of large, thin-walled medullary cells; *rhizoids* scattered at stem base, pale to dark brown, smooth, branched; *rhizoidal tubers* infrequent, pale brown to reddish-brown, consisting of a straight or bent uniseriate row of 1–5 large subsphaerical swollen cells, 50–90 µm long, usually with a small lenticular initiating cell at the distal end of large cells, 25–35 µm in diameter; *axillary hairs* filiform, 3–4-celled, with 2–3 basal cells brown and elongate hyaline apical cell. *Leaves* narrowly lanceolate, erect to erect-flexuose or subsecund, 1–2 mm long, 0.3–0.5 mm wide at base, non-decurrent, gradually tapered to a narrow, blunt, subentire or bluntly serrulate apex; *margins* recurved on both sides from near the base to the apex, sometimes strongly so, bistratose almost throughout in one row of cells; *costa* strong, 40–60 µm wide at the insertion, dark yellow to yellow-brownish, subpercurrent, in cross-section strongly convex dorsally, consisting of many substereid cells surrounded by larger epidermal cells; *laminal cells* variable in shape, short-rectangular to linear, straight with truncate apices, non-porose, pale, 5–8(–10) µm wide, (15–)20–80(–150) µm long, becoming shorter towards the apex, thin- to thick-walled, unistratose except for the margins, not differentiated at the basal angles. [*Dioicous. Perichaetial leaves* not different from the upper vegetative leaves. *Setae* red, erect, 5–8(–15) mm long, straight. *Capsules* ovoid, inclined to horizontal, asymmetric, gibbous, 0.7–1.0 mm long, smooth, not strumose, usually contracted at the mouth when dry and empty; *annulus* absent; *operculum* obliquely long-rostrate, nearly as long as the urn; *exothecial cells* short-rectangular to subquadrate, thick-walled; *stomata* at the neck, bicellular, superficial; *peristome* bifid in the upper ⅓, 350–400 µm long, papillose above, red and pitted-striolate below. *Spores* globose, light brown, 15–18 µm wide, finely papillose. *Calyptra* cucullate, naked.]

Chromosome numbers: unknown in the Antarctic and Subantarctic; $n = 7$ (1 record), $n = 14$ (9 records) and $n = 15$ (2 records) from the Holarctic (Fritsch 1991).

Discussion. The determination of sterile specimens of *Dicranella* is often difficult because the diagnostic characters usually refer to mature sporophytes. Initially, the present material from the South Sandwich Is. was suggested as a new species, *D. recurvata* (Ochyra *et al.* 1998) but subsequently it was re-determined as *D. hilariana* (Ochyra *et al.* 2000; Convey *et al.* 2001). A critical re-examination of the voucher collection led to the conclusion that it is *D. varia*, a species hitherto unknown in the austral region. Although both species are very similar gametophytically, they are placed in two different subgenera of *Dicranella* on account of their sporophyte features. *Dicranella hilariana* lacks stomata and has a large, revoluble annulus, whereas *D. varia* has well developed stomata but the annulus is missing. Therefore, the former is positioned in the type subgenus of *Dicranella* and the latter is a representative of subgen. *Anisothecium*. However, the gametophytes of both species share variously recurved leaf margins, a rare feature in this genus. Additionally, their leaf areolation is similar, although in *D. varia* the upper median laminal cells are oblong-linear to linear, usually >40 µm long, whereas in *D. hilariana* they are rectangular to short-rectangular, usually <40 µm long. The only character which enables accurate distinction of sterile plants is the bistratose leaf margin in *D. varia* which, in contrast, is unistratose in *D. hilariana*.

The Antarctic material of *Dicranella varia* fits quite well the type of var. *obtusifolia* from Spitsbergen (Berggren 1875) in its lanceolate leaves gradually tapering to a short, blunt point with broadly recurved margins on both sides from the base to the apex, but this taxon seems to have a weak taxonomic value.

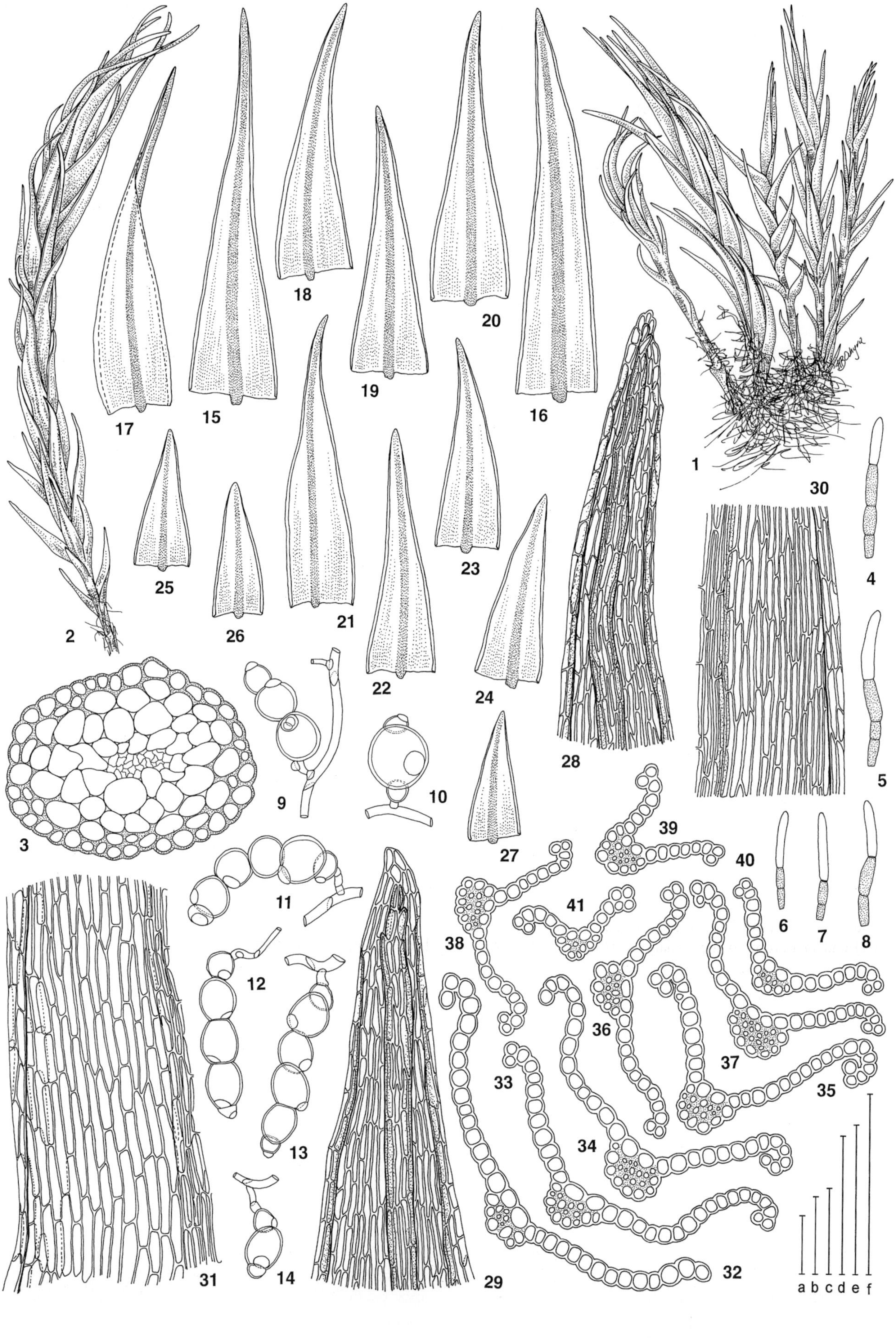
1
2
3
4
5
6
7
8
9
10
11
12
13
14
15
16
17
18
19
20
21
22
23
24
25
26
27
28
29
30
31
32
33
34
35
36
37
38
39
40
41
a b c d e f

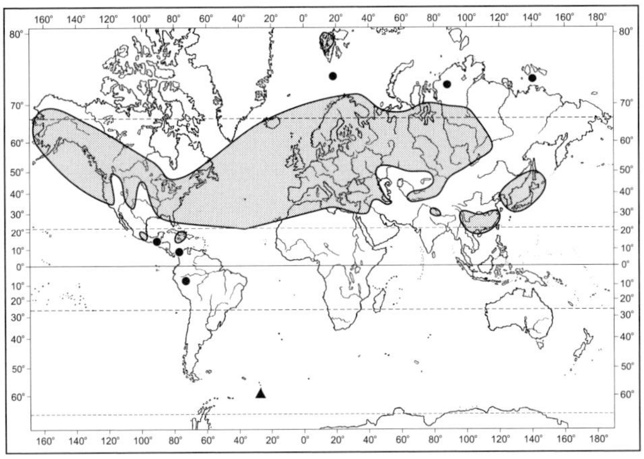

Figure 66. Global distribution of *Dicranella varia*. Antarctic locality marked by the triangle.

Dicranella varia may be confused with juvenile plants of *D. hookeri*, having an oblong base gradually tapering to the limb and not tightly clasping the stem. Nonetheless, the leaf margins in the latter are always flat, the limb is mostly bistratose and the cells in the limb are subquadrate to short-rectangular.

Reproduction in Antarctica. Invariably sterile but infrequently producing rhizoidal tubers. These are not unusual in *Dicranella* and have been described and illustrated in most European species and often they have diagnostic value (Whitehouse 1966, 1969; Risse 1986; Ochyra 1988*a*).

Habitat. In the Antarctic the single record of *Dicranella varia* was in a small volcanically heated area where it grew on clayey soil in association with *Ditrichum ditrichoideum*, *Ceratodon purpureus*, *Cryptochila grandiflora* and *Cephaloziella varians*.

World range. *Dicranella varia* is a bipolar species which is here reported for the first time from the polar regions of the Southern Hemisphere (Fig. 66). It has an almost continuous range in boreal and temperate regions in the Northern Hemisphere, only rarely extending into the Arctic in Alaska (Steere 1978), Svalbard (Frisvoll & Elvebakk 1996) and Siberia (Afonina & Czernyadjeva 1995). Its southern limit may be obscure in some areas because the species has often been confused with the closely related but otherwise distinct *D. howei* (Crunwell & Nyholm 1977), so some literature data certainly need verification. In North America *D. varia* extends to southern California, Arizona, eastern Texas and Florida (Reese 1984; Crum 2007) and in Europe to the Canary Is. (Eggers 1982), North Africa (Jelenc 1955; Ros *et al.* 1999) and the Caucasus (Manakyan 1989). In Asia it is relatively frequent in the Near East (Frey & Kürschner 1991), Middle Asia including Tadzhikistan (Mamatkulov 1990), Pakistan and Afghanistan (Crundwell & Nyholm 1977), and very rare in Siberia and the Russian Far East (Bardunov 1974; Bardunov & Cherdantseva 1982; Ignatov *et al.* 2006) as well as scattered in Japan (Noguchi 1987), south-eastern China (Gao *et al.* 1999) and in northern India (Crunwell & Nyholm 1977). Additionally, *D. varia* is scattered in the Neotropics at high elevations in Mexico (Crum 1994*a*), Guatemala (Bartram 1949; Allen 1994*a*), southern Cuba (Duarte Bello 1997), Haiti (Crum & Steere 1958) and Jamaica (Bartram 1928; Crum & Bartram 1958) in the West Indies, as well as in the Andes of Colombia (Churchill *et al.* 2000) and Peru (Robinson 1967).

Distribution in Antarctica. *Dicranella varia* is known from a single collection from the volcanic Bellingshausen I. in the South Sandwich Is. archipelago at an elevation of 170 m (Fig. 66).

Specimen examined. WEST ANTARCTICA. SOUTH SANDWICH IS. **BELLINGSHAUSEN I.** Knoll on heated crater rim, 170 m, *PC 31* (AAS, KRAM).

2. **Dicranella hookeri** (Müll.Hal.) Cardot Fig. 67

Dicranella hookeri (Müll.Hal.) Cardot, Bull. Herb. Boissier Sér. 2, 6: 19. 1906. ≡ *Aongstroemia hookeri* Müll.Hal., Syn. Musc. Frond. 2: 607. 1851. ≡ *Anisothecium hookeri* (Müll.Hal.) Broth. *in* Engl., Nat. Pflanzenfam. Ed. 2, 10: 178. 1924. — TYPE CITATION: Insula Eremitae ad Cap. Horn: J. D. Hooker. [Lectotype (*selected here*): 'N° 129 Dicranum vaginatum Hook. Hermite Island, Cape Horn Antarct. Exped. 1839–1843 J. D. H.' – BM-Wilson!; isotype: FH-Taylor!].

Plants small to medium, slender to moderately robust, in low to deeper, dense tufts or extensive patches, sometimes gregarious, sordid- or yellowish- to bright-green above, brownish below, dull or slightly glistening. *Stems* erect, to 10 cm tall, simple or sparingly branched by subterminal innovations, usually light- to reddish-brown and lustrous, rather sparsely radiculose below with light brown, lustrous, smooth, branched rhizoids, in cross-section rounded with a large central strand, 3–4-stratose medulla of large, yellowish-hyaline cells with thin, brownish walls and a unistratose or incompletely bistratose cortex of somewhat smaller cells with thicker, brown to reddish-brown walls; *axillary hairs* filiform, consisting of 5–7 elongate cells, hyaline except for 2–3 shorter, brownish basal cells; *rhizoidal gemmae* occasional, spherical, multicellular, reddish-brown, 90–140 µm in diameter. *Leaves* smaller below, becoming progressively larger upwards, (0.9–)1.3–2.5(–3.1) mm long, (0.3–)0.5–0.7(–0.9) mm wide, widely spaced, erect and appressed or erecto-patent to patent, altered or sometimes straight when dry, from a broad, ovate or quadrate, semi-amplexicaulous base, becoming often non-sheathing on the lower leaves or on those on innovations, abruptly or rather gradually tapered to a channelled, erect to erect-spreading limb, straight or usually altered on drying, entire, obtuse to subacute at the apex, usually as long as or to twice exceeding the leaf base; *margins* plane, entire throughout or somewhat crenulate at the shoulders; *costa* single, subpercurrent, 60–120 µm wide at base, smooth at back, sharply delimited to somewhat diffuse in the sheath, gradually merging into the bistratose laminal cells at the shoulders and in the limb and often seemingly filling up the whole of the subula, in cross-section variable, flattened below, more convex

◄——

Figure 65. *Dicranella varia*. — **1**. Habit. **2**. Sterile shoot. **3**. Cross-section of stem. **4–8**. Axillary hairs. **9–14**. Tubers. **15–27**. Leaves, a sequence from base to apex of shoots. **28–29**. Leaf apices. **30**. Mid-leaf cells. **31**. Basal cells. **32–41**. Cross-sections of leaves, a sequence from base to apex. — [All from *PC 31*]. — Scale bars: a – 100 µm (9–14); b – 1 mm (15–26); c – 100 µm (3, 32–41); d – 1 mm (2); e – 1 mm (1) and 100 µm (4–8); f – 100 µm (28–31).

dorsally in the subula, typically consisting of larger dorsal and smaller ventral bands of stereids, the latter being often reduced or lacking, separated by a row of large guide cells; *laminal cells* smooth throughout; *basal cells* thin- to firm-walled, short- to long-rectangular, (20–)25–60(–85) µm long, (5–)8–14(–18) µm wide, unistratose, translucent, not differentiated at the basal angles; *shoulder cells* short-rectangular to subquadrate, unistratose or variously bistratose towards the costa and then relatively obscure, (10–)15–40(–45) µm long, 8–13(–15) µm wide; *limb cells* short-rectangular to subquadrate, (8–)15–30(–40) µm long, 8–15 µm wide, uni- to bistratose except for 1–3 marginal rows. *Dioicous*. *Perigonia* terminal, gemmiform; *perigonial bracts* from a strongly concave, ovate base abruptly subulate. *Setae* erect, somewhat flexuose, *ca* 1 cm long, brown. [*Capsules* erect to inclined, obloid, 0.7–1.1 mm long, slightly asymmetrical, urceolate when dry and empty, dark brown, lustrous; *annulus* absent; *operculum* stoutly rostrate with a slanted beak, equal to the urn in length; *exothecial cells* short-rectangular to subquadrate, 20–30 µm long, 7–10 µm wide, with markedly incrassate longitudinal walls and much thinner transverse walls, becoming isodiametric to oblate in 2–3 rows at the orifice; *stomata* few at the base of the urn, bicellular; *peristome* teeth 16, arising from a low basal membrane consisting of 2–5 rows of cells, 350–390 µm long, broadly triangular, divided nearly halfway to the base, occasionally perforated, vertically pitted-striolate on the outer surface. *Spores* globose, 24–32 µm in diameter, pale brownish, minutely papillose. *Calyptra* cucullate.]

Chromosome number: $n = 26$ – South Georgia (Newton 1972).

Discussion. *Dicranella hookeri* is a distinct species recognised by its long, widely spaced and erecto-patent leaves composed of a more or less quadrate sheathing base which is abruptly to gradually narrowed to a channelled subula terminating in an acute or obtuse apex. The laminal cells are smooth throughout, unistratose throughout to partly or entirely bistratose at the shoulders and in the subula. In the Antarctic it may be confused with *D. campylophylla* which it closely resembles, but the latter is easily separated by the laminal cells in the limb which are highly mammillose, often with double mammillae. *Dicranella hookeri* is very variable with regard to the size of the plants and the leaf length (Ochyra 1999*e*). Kaalaas (1912) described *D. hookeri* var. *elongata* from specimens collected on Île de la Possession in the Îles Crozet group. A characteristic feature of this taxon is its exceedingly elongate shoots, up to 20 cm in length, and on Marion I. this taxon forms extensive carpets with stems reaching 25 cm in length in habitats permanently inundated by water from a nearby waterfall (Ochyra, personal observations). Some populations from the South Sandwich Is. growing on heated ground near fumaroles produce a profusion of subterminal innovations with the leaves having weakly developed sheathing bases, erect and appressed to the axis, but the leaves on shoots buried in ash are of typical shape. There may also be confusion between *D. hookeri* and some species of *Ditrichum*, notably *D. conicum* which, however, is distinguished by its oblong leaf base with a broad hyaline border and markedly shorter cells in the sheath.

Dicranella hookeri is closely related to the neotropical *D. vaginata*, and its type collection from Hermite I. near Cape Horn, was initially referred to this species (Wilson & Hooker 1847), but Müller (1851) described it as a separate species. Mitten (1869) introduced taxonomic and nomenclatural confusion by considering it to be conspecific with *D. jamesonii*, a species described from Ecuador by Taylor (1847) as *Dicranum jamesonii*. However, this name was an illegitimate homonym, and the epithet *hookeri* had priority. Furthermore, *Dicranum jamesonii* is conspecific with *Dicranella vaginata* as initially stated by Mitten (1851). Although the differences between *D. hookeri* and *D. jamesonii* were later confirmed by Cardot (1908*a*), the epithet *jamesonii* has remained in use, especially in Australian and New Zealand literature (e.g. Dixon 1914; Sainsbury 1955; Streimann & Curnow 1989; Beever *et al.* 1992; Fife 1995).

Reproduction in Antarctica. Plants are almost always sterile, but a single specimen with old setae has been collected on a plant from Deception I. (*RILS 5829B*, AAS, KRAM). Gametangia are frequently formed, sometimes in great profusion (e.g. *REL 632*, AAS, KRAM), and fertilised archegonia have occasionally been observed. In addition, rhizoidal gemmae are occasionally produced in some populations.

Habitat. The species is exclusively associated with geothermal areas, growing on moist to wet, ashy soil, on fine ash and scoria around fumarole vents, sometimes violently active, among boulders with steam rising between them, occasionally under them in strongly shaded niches. It grows as a short turf associated with species of *Campylopus*, *Bryum pseudotriquetrum*, *Brachythecium austrosalebrosum*, *Notoligotrichum trichodon*, *Marchantia berteroana*, *Cephaloziella varians* and *Clasmatocolea rigens*; in sheltered sites shoots attain 10 cm in length.

World range. A pan-south-temperate species (Fig. 68, inset) widely distributed in southern South America, including Tierra del Fuego, southern and western Patagonia (Greene 1986) and the Falkland Is. (Matteri 1986), and extending along the Andean chain from the Valdivian region in Chile (He 1998), through Bolivia and Ecuador to Colombia and Venezuela where it occurs at high elevations of 3800–4100 m (Churchill *et al.* 2000). It also occurs on Tristan da Cunha (Dixon 1960), Gough I. (Bartram 1959), in South Africa (Ochyra, unpublished), south-eastern Australia (Streimann & Curnow 1989), Tasmania (Dalton *et al.* 1991) and New Zealand (Dixon 1914; Sainsbury 1955) from where it is reported as *D. jamesonii*. Additionally, *Dicranella hookeri* penetrates into the Subantarctic and West Antarctic. It has been recorded recorded from the Prince Edward Is. (Zanten 1971; Ochyra 1999*e*) and Îles Crozet and Kerguelen (Brotherus 1906*a*), and from some islands in the northern maritime Antarctic.

→

Figure 67. *Dicranella hookeri.* — **1.** Plant with sporophyte. **2–4.** Habit, dry (2 & 4) and wet (3). **5.** Cross-section of stem. **6–8.** Axillary hairs. **9–17.** Leaves. **18.** Leaf apex. **19–20.** Cells at leaf shoulder. **21.** Basal cells. **22–33.** Cross-sections of leaves, a sequence from base to apex. **34.** Outer perigonial bract. **35.** Inner perigonial bract with antheridia. **36.** Deoperculate capsule. **37.** Calyptra. **38.** Operculate capsule. **39.** Exothecial cells in middle of urn. **40.** Exothecial cells at base of urn and stoma. **41.** Exothecial cells at orifice, peristome and spores. — [1–4, 6–8, 13–15, 20, 22, 24, 28, 34–36, 39–41 from *RILS 1165*; 5, 9–12, 18–19, 21, 23, 25–27, 29–33 from *RILS 5821B*; 16–17 *RILS 5667A*; 37–38 from *Hässel de Menendez s.n.*, 5.ii.1957]. — Scale bars: a – 1 mm (1); b – 1 mm (3), 50 µm (5–8, 18–33, 39–41) and 0.5 mm (37–38); c – 1 mm (2, 4, 34–35) and 0.5 mm (36); d – 1 mm (9–17).

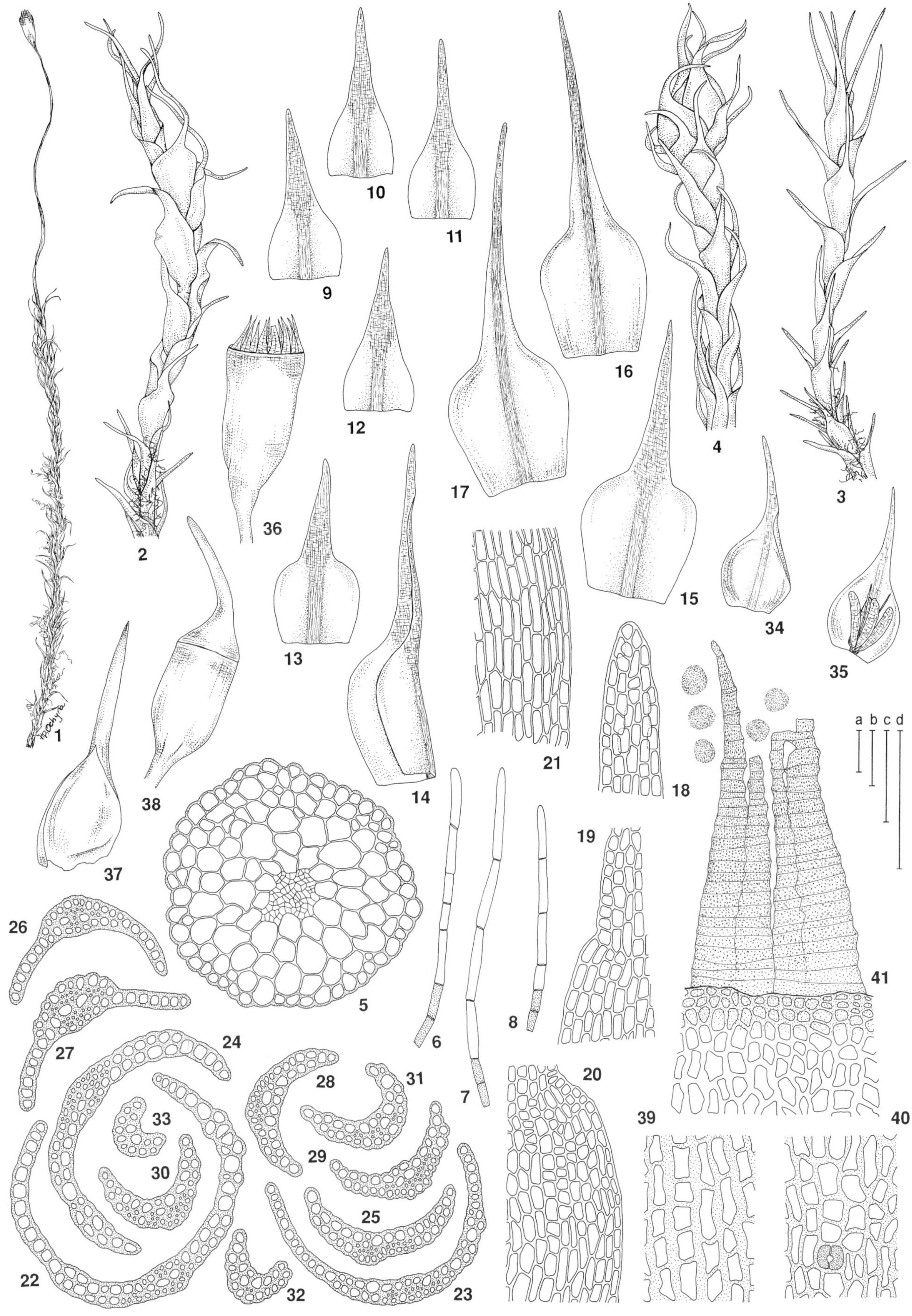
1
2
3
4
5
6
7
8
9
10
11
12
13
14
15
16
17
18
19
20
21
22
23
24
25
26
27
28
29
30
31
32
33
34
35
36
37
38
39
40
41
a
b
c
d

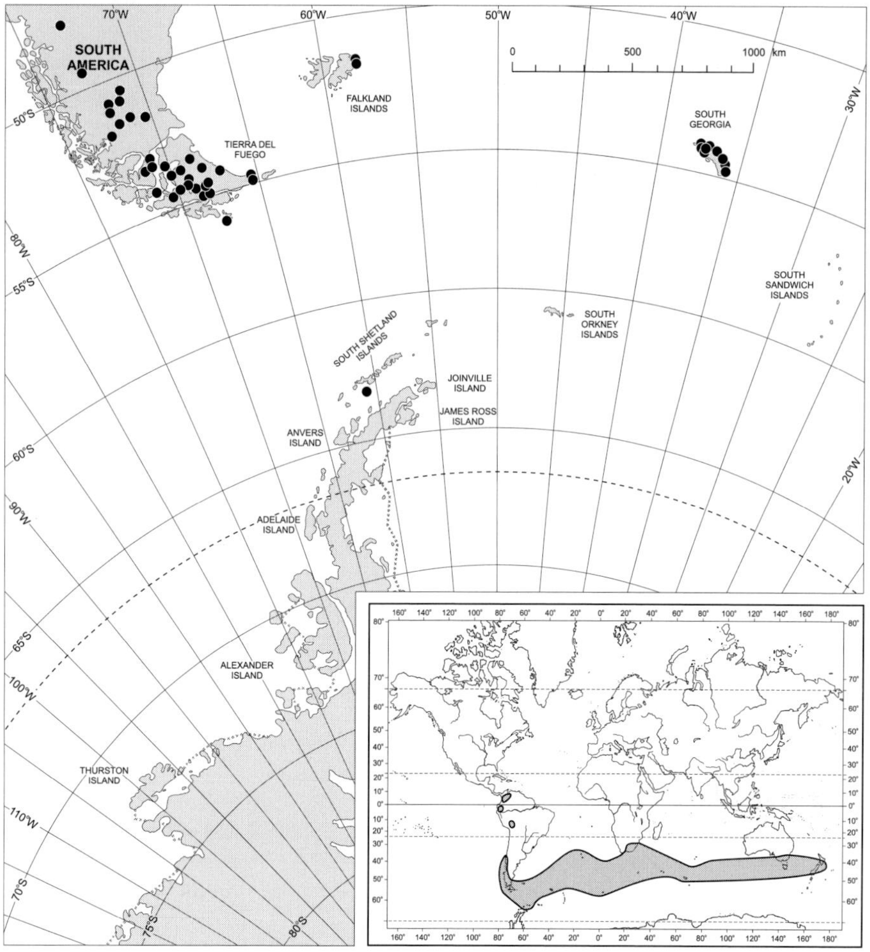

Figure 68. Distribution of *Dicranella hookeri* in the Fuegian region and Antarctica. Inset: global distribution.

Distribution in Antarctica. The species is known only from two islands in the West Antarctic with recent volcanic activity, namely the isolated island of Bouvetøya and Deception I. in the South Shetland Is. (Fig. 68), occurring from near sea level to 480 m. Cardot (1910*b*) reported *Dicranella hookeri* from Victoria Land but the voucher collection has never been located (Newton 1977*b*; Ochyra, personal observations), nor has the species been rediscovered by subsequent collectors (Seppelt & Green 1998; Lewis Smith, personal observations). The records from the South Sandwich Is. (Convey *et al.* 2001) are based on misdeterminations.

Selected specimens examined. WEST ANTARCTICA. BOUVETØYA. Nyrøysa, N border, *TE 209* (BG) and western part, 15 m, *TE 321 & 322* (BG).

SOUTH SHETLAND IS. **DECEPTION I.** ENE of Ronald Hill, 80–100 m, *RILS 6873B, 6874 & 6876* (AAS, KRAM); ridge to N of Mt. Pond, 380 m, *RILS 5812A* (AAS, KRAM), 430 m, *RILS 5816A* (AAS, KRAM), 480 m, *RILS 5821B, 5822A, 5823A-B, 5824A, 5826B, 5827C, 5829B, 5830B, 5831B, 5833BB & 5834A* (AAS, KRAM), 480–500 m, *RILS 11157A, 11162 & 11163A* (AAS, KRAM) and 500 m, *RILS 11146, 11147, 11151B, 11153 & 11154A* (AAS, KRAM); Fumarole Bay, 100 m, *RILS 5667A-C* (AAS, KRAM); same locality, N of lagoon, *RILS 11272A, 11274B, 11278B &11284A* (AAS, KRAM); Lake Irizar, *RILS 6912* (AAS, KRAM); 2 km N of Vapour Col, *RILS 11266A* (AAS, KRAM).

3. **Dicranella campylophylla** (Taylor) A.Jaeger Fig. 69

Dicranella campylophylla (Taylor) A.Jaeger, Ber. Thätigk. S. Gallischen Naturwiss. Ges. 1870–71: 382. 1872. ≡ *Dicranum campylophyllum* Taylor, London J. Bot. 7: 281. 1848. ≡ *Aongstroemia campylophylla* (Taylor) Müll.Hal., Syn. Musc. Frond. 2: 608. 1851. ≡ *Anisothecium campylophyllum* (Taylor) Mitt., J. Linn. Soc. Bot. 12: 40. 1869. — TYPE CITATION: On Pichincha. Prof. W. Jameson. 8th Aug., 1847. [Lectotype (*selected here*): 'Dicranum campylophyllum Tayl. Mss On Pichincha Prof. W. Jameson 8 August 1847' – FH-Taylor!; isotype: BM-Wilson!].

Dicranum cardotii R.Br.bis, Trans. & Proc. New Zealand Inst. 35: 329, *pl. 36*, *f. 9.* 1903. ≡ *Dicranella cardotii* (R.Br.bis) Dixon, New Zealand Inst. Bull. 3(3): 77. 1923. ≡ *Anisothecium cardotii* (R.Br.bis) Ochyra, Moss Fl. King George Island Antarctica: 114. 1998. — TYPE CITATION: [New Zealand] On damp banks, tributary of the River Hapuka, near Koikoura. Collected by Robert Brown. [Holotype (*vide* Ochyra & Newton 1985: p. 97): BM-Dixon (*non vidi*)]. First synonymised by Robinson (1986: p. 10).

Plants small to medium-sized, slender, densely caespitose or, less often, in scattered mats, occasionally forming loose submerged mats, bright- to yellowish- or

⟶

Figure 69. *Dicranella campylophylla.* — **1–3**. Habit, wet (1 & 2) and dry (3). **4**. Portion of cross-section of stem. **5–6**. Axillary hairs. **7–14**. Leaves. **15**. Leaf apex. **16–17**. Cells at leaf shoulder. **18**. Mid-leaf cells. **19**. Mid-leaf cells at margin. **20**. Basal cells. **21–30**. Cross-sections of leaves, a sequence from base to apex. **31**. Tuber on rhizoids. — [1, 4–6, 11–15, 19–20, 26–30 from *RO 889/80*; 2–3, 7–10, 16–18, 22, 24–25 from *Bell 1388*; 21, 23, 31 from *Greene 2763*]. — Scale bars: a – 1 mm (2) and 50 µm (4–6, 15–30); b – 100 µm (31); c – 1 mm (1, 3); d – 1 mm (7–14).

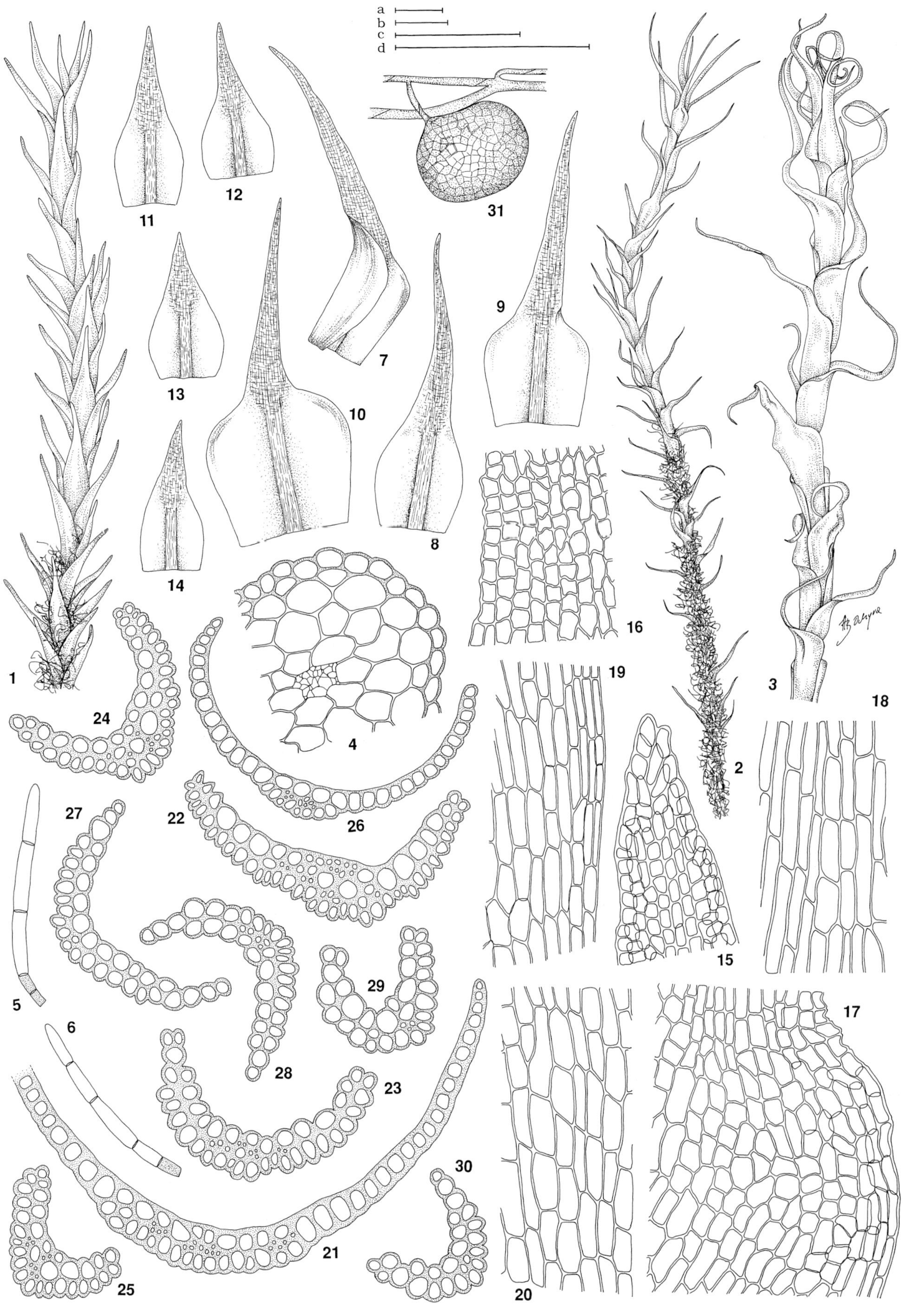
a
b
c
d
1
2
3
4
5
6
7
8
9
10
11
12
13
14
15
16
17
18
19
20
21
22
23
24
25
26
27
28
29
30
31

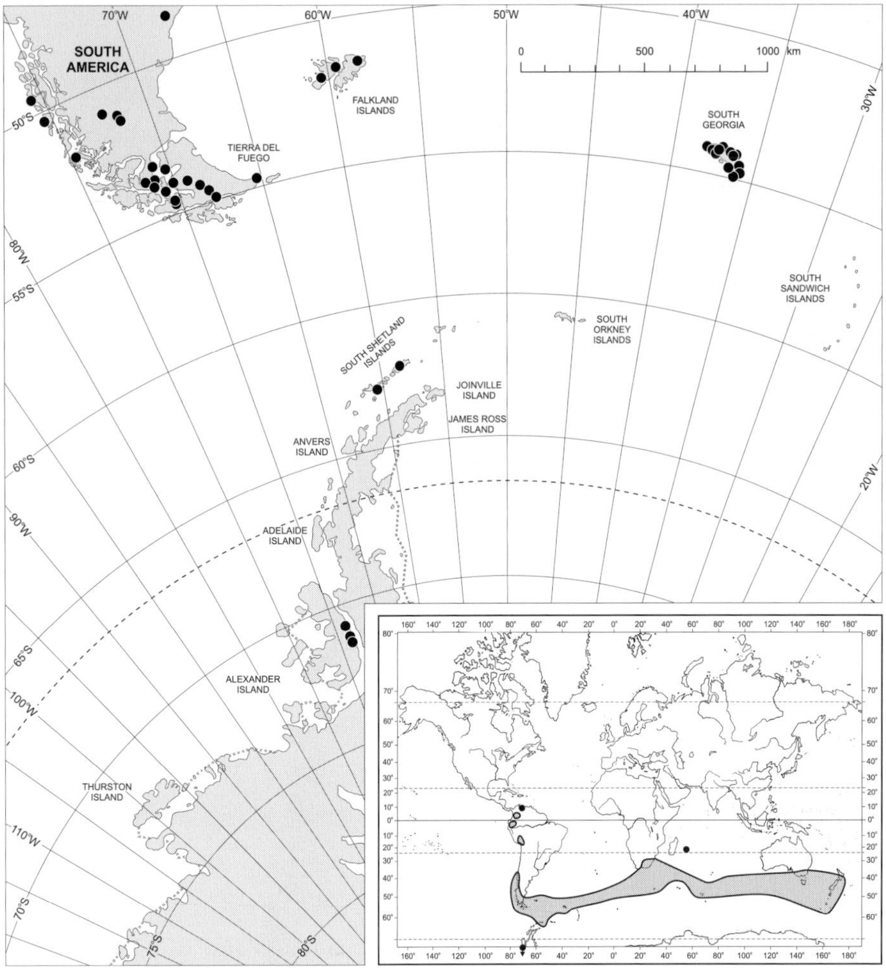

Figure 70. Distribution of *Dicranella campylophylla* in the Fuegian region and Antarctica. Inset: global distribution. The locality on Alexander I. is indicated by an arrow.

sometimes sordid-green, dull to weakly lustrous. *Stems* erect, mostly simple or with ascending lateral branches, 1–2(–4) cm high, usually richly radiculose below with long, brownish, smooth, glistening, not much branched rhizoids, in cross-section round, consisting of a distinct central strand surrounded by a 3–4-stratose medulla of large, thin-walled, hyaline cells and a cortex of 1 or incompletely 2 somewhat smaller cells with brownish, thinner walls; *axillary hairs* filiform, 5–6-celled, consisting of 1–2 short, brownish basal cells and 3–5 elongate, hyaline upper cells; *rhizoidal gemmae* copious, globose, multicellular with slightly protuberant cells, 150–190 µm in diameter. *Leaves* distantly set, (0.6–)0.8–1.6 mm long, (0.3–)0.4–0.6 mm wide, from an ovate, oblong-ovate to square vaginate base, loosely clasping the stem, gradually narrowed to erecto-patent, flexuose-spreading to recurved, channelled, distinctly roughened limb, subacute at the apex, usually as long as or up to twice exceeding the leaf base; *margins* plane, entire or obscurely denticulate at the apex; *costa* relatively narrow, 60–70 µm wide at the base, subpercurrent, usually well demarcated in the vaginant base but gradually merging into the laminal cells in the limb, in cross-section flattened below and more convex dorsally above, often only bistratose in the leaf base and in the lower part of the subula, with poorly developed or absent stereid bands; *laminal cells* of the sheath unistratose, smooth, thin- to firm-walled, rectangular, 25–60 µm long, 6–14 µm wide, becoming short-rectangular to subquadrate at the shoulders, 10–20 µm long, 6–10 µm wide, unistratose; *limb cells* bistratose, quadrate to shortly rectangular and distinctly mammillose on the dorsal surface, 10–20 µm long, 6–10 µm wide and rectangular, smooth over the ventral surface of the costa, 20–50(–70) µm long, 8–14 µm wide. [*Dioicous*. *Setae* erect, brown, 1–2 cm long, flexuose when dry. *Capsules* erect to inclined, obloid, turbinate when dry and empty, to 1 mm long, brown, symmetric to slightly asymmetric and gibbous, non-strumose, exannulate; *operculum* rostrate, with an oblique beak as long as the urn; *exothecial cells* rectangular to quadrate, 40–60 µm long, 10–20 µm wide, with strongly thickened longitudinal walls and much thinner transverse walls, becoming short, quadrate, rounded or transversely rectangular, thick-walled in 2–3 rows at the mouth; *stomata* at the base of the urn, superficial, bicellular; *peristome* teeth 16, reddish-orange below, yellow-orange above, arising from a low basal membrane consisting of 2–5 rows of cells, 350–410 µm long, broadly triangular, divided nearly halfway to the base, sometimes perforated, vertically pitted-striolate on the outer surface. *Spores* globose, pale brown, 20–30 µm wide, papillose. *Calyptra* cucullate, smooth.]

Chromosome number: unknown.

Discussion. Typically developed *Dicranella campylophylla* is readily identified, the leaves with vaginate base, tightly clasping the stem and rapidly tapered to a long, squarrose and strongly papillose subula due to the mammillose laminal cells of the dorsal surface being immediately diagnostic. In Antarctic species similarly papillose leaf subulae are known in *Distichium* species and *Ditrichum ditrichoideum*. However, the former are distinguished by their perfectly distichous leaves, whereas the latter has an oblong leaf base, not widened at the shoulders, which is distinctly bordered below by elongate, hyaline cells and has generally narrower cells. The Antarctic plants of *D. campylophylla* growing in severe climatic conditions or submerged in water differ markedly from those from lower latitudes in their smaller stature and leaf bases gradually tapered into the limbs, and consequently they may be mistaken for *D. hookeri*. However, the distinctly mammillose cells of the dorsal surface of the subula afford a clear means of discrimination. Also, due to the small size and/or submersion of the Antarctic plants, they have a markedly less differentiated costal anatomy with highly reduced or lacking stereid bands, whereas typically in this species dorsal and ventral stereid bands separated by a row of large guide cells are developed.

Dicranella campylophylla is a variable species, mostly with regard to the shape of the leaf subula, and accordingly various phenotypes were given taxonomic recognition in the past. The species has long been known as *D. cardotii*, until Robinson (1986) suggested it should be replaced by the older and neglected name *D. campylophylla*, which was used for a species described by Taylor (1848) from Ecuador as *Dicranum campylophyllum*. This taxonomic conclusion has been confirmed after examination of the relevant type specimens (Ochyra, unpublished observation).

Reproduction in Antarctica. Fertile plants unknown, but tubers have frequently been encountered in some populations.

Habitat. Plants from the South Shetland Is. are epigean and grow on wet soil or in crevices on steep slopes or cliffs, these also being typical habitats for *Dicranella campylophylla* in more northerly latitudes (Dusén 1905*a*; Newton 1977*a*). It is associated with *Bryum pseudotriquetrum*, *Ceratodon purpureus* and *Polytrichum juniperinum*. By contrast, plants from Alexander I. were collected in a freshwater pond at a depth of 2–3 m together with *Campylium polygamum*, *Encalypta procera*, *Hypnum revolutum*, *Distichium capillaceum* and *Bryum pseudotriquetrum* (Light & Heywood 1975) and on fine gravelly soil in an alluvial outwash fan.

World range. A pan-south-temperate species (Fig. 70, inset) widely distributed but generally scattered in the *Nothofagus* zone along the western fringes of South America from central Chile to Tierra del Fuego, and infrequent on the Falkland Is.; in South Africa, south-eastern Australia, on Tasmania and in New Zealand and its offshore islands, then extending southwards to the Subantarctic, including South Georgia, Marion I., Îles Kerguelen and Macquarie I., as well as to the maritime Antarctic, and penetrating into the tropics along the high elevations of the Andes north to Venezuela, reaching its highest elevation of 4500 m in Colombia (Ochyra & Newton 1985; Churchill *et al.* 2000).

Distribution in Antarctica. A very rare and localised species, known only from two sites in the South Shetland Is. (Ochyra & Newton 1985; Sancho *et al.* 1999) and from several sites on Alexander I. at lat. 70°53′S (Fig. 70). It was found only at low elevations, to 70 m.

Specimens examined. WEST ANTARCTICA. SOUTH SHETLAND IS. **King George I.** ***Admiralty Bay***: Ezcurra Inlet, Pond Hill, 70 m, *RO 889/80* (KRAM). **Livingston I.** ***Hurd Peninsula***: South Bay, *Schulz 117* (HBG, KRAM).

WEST ANTARCTIC PENINSULA.**George VI Sound.** ***Alexander I.***: Ablation Pt., 10 m, *JJL & RBH 7, 8 & 11* (AAS, KRAM); Ganymede Heights, 5 m, *RILS 11075B* (AAS, KRAM); Striation Valley, *RILS 11079, 11081B & 11119* (AAS, KRAM).

11. AMPHIDIUM Schimp., *nom. cons.*

Amphidium Schimp., Coroll. Bryol. Eur.: 39. 1856, *nom. cons.* — Type: *Amphidium lapponicum* (Hedw.) Schimp. (*Anictangium lapponicum* Hedw.).

Amphoridium Schimp., Syn. Musc. Eur.: 247. 1860, *hom. illeg.* — Type: *Amphoridium lapponicum* (Hedw.) Schimp. (*Anictangium lapponicum* Hedw.).

Cylindrocarpus Lindb., Öfvers. Förh. Kongl. Svenska. Vetensk.-Akad. 19: 605. 1863, *nom. illeg. incl. typ. gen.*

Plants small to medium-sized, erect, rather soft, densely caespitose, dull to slightly lustrous, yellowish-green above, brownish below. *Stems* erect, repeatedly branched, variously radiculose below with clusters of smooth, light brown, branched rhizoids, in cross-section triangular, lacking central strand, with large, thin-walled medullary cells and a 1–2-stratose cortex of smaller cells, with thickened, brown walls, densely and minutely papillose on the outer surface. *Leaves* crowded, evenly spaced, tristichous, appressed below, flexuose, curled or contorted distally when dry, erect- to wide-spreading when wet, narrowly linear- to oblong-lanceolate from a scarcely differentiated, sheathing base, not or slightly decurrent, narrowly acute and apiculate; *margins* plane above, narrowly recurved on 1 or both sides below, entire or sinuouse to irregularly serrate in the upper half; *costa* single, subpercurrent, convex, smooth or papillose dorsally, in cross-section with median guide cells and well developed dorsal and weaker ventral stereid or substereid bands; *laminal cells* unistratose, rounded, rounded-quadrate to rounded-hexagonal, thick-walled, pluripapillose on both sides and opaque above, becoming rectangular, thin-to thick-walled, faintly and minutely papillose-striate in the lower half; *alar cells* not differentiated. *Autoicous* or *dioicous*. *Perichaetial leaves* strongly or weakly differentiated, broad, short-acuminate and entirely sheathing or elongate, long-acuminate from ± sheathing base, with areolation of elongate, smooth or somewhat papillose cells. *Setae* short, erect to slightly curved. *Capsules* erect and symmetric, immersed to short-exserted, pyriform from a broad neck gradually tapered to the seta, distinctly 8-ribbed and often urceolate and constricted below the mouth when dry and empty, light brown and dark brown on the ribs, lustrous, exannulate; *operculum* conic-apiculate to conic-rostellate, with a slanted or rarely straight beak; *exothecial cells* irregularly rectangular, ± thick-walled, usually differentiated along the ribs, becoming smaller in several rows at the orifice; *stomata* superficial, bicellular in the neck; *peristome* absent. *Calyptra* cucullate, naked or slightly roughened at the apex, fugacious. *Spores* subglobose to angular, brownish, papillose.

Amphidium is recognised by a combination of both gametophyte and sporophyte characters including the linear-lanceolate leaves which are tristichous and curled-contorted on drying; isodiametric, densely pluripapillose distal laminal cells and rectangular and distinctly papillose-striolate lower leaf cells; the minutely but otherwise distinctly papillose stem which is triangular in transverse section and lacks a central strand; the strongly differentiated perichaetial leaves; and the gymnostomous and exannulate, strongly 8-ribbed capsules which are immersed to slightly exserted. The relationships of this genus have been a subject of continual controversy, aggravated by the lack of a peristome. Some authors (e.g. Schimper 1856; Dixon & Jameson 1924; Crum & Anderson 1981; Magill 1998) have incorporated *Amphidium* into the Orthotrichaceae, especially near *Zygodon*, on account of the similarity of the leaves, calyptra, antheridial buds and, particularly, in the shape and the development of the young capsules (Lewinsky 1976). On the other hand, the costal anatomy, leaf areolation and general habit of the plants indicate the alliance of *Amphidium* with the dicranaceous mosses, especially with *Rhabdoweisia* and, accordingly, many authors placed it in the Dicranaceae (e.g. Hagen 1915; Brotherus 1924*b*; Malta 1926).

The placement of *Amphidium* in the Dicranaceae has been strongly supported by molecular data (Stech 1999*b*), although they indicate that this genus is unrelated to *Rhabdoweisia*, as often suggested (e.g. Anderson & Crum 1959). Because *Amphidium* seems to be a discordant element in either subfamily of the Dicranaceae, the most practical solution seems to be to establish the separate subfamily Amphidioideae to accommodate this genus (Ochyra *et al.* 2003).

Amphidium is a pancontinental oreophyte which has hitherto been known from all continents except Antarctica. The genus has recently been revised by Frahm *et al.* (2000) who concluded that it consists of only three species, one of which, *A. lapponicum*, has been recorded at a single station on the Antarctic Peninsula.

1. **Amphidium lapponicum** (Hedw.) Schimp.

Fig. 71

Amphidium lapponicum (Hedw.) Schimp., Coroll. Bryol. Eur.: 40. 1855 [1856]. ≡ *Anictangium lapponicum* Hedw., Spec. Musc. Frond.: 40. 1801. ≡ *Bryum lapponicum* (Hedw.) Dicks., Fasc. Pl. Crypt. Brit. 4: 10. 1801. ≡ *Gymnostomum lapponicum* (Hedw.) F.Weber & D.Mohr, Ind. Mus. Pl. Crypt.: 2. 1803. ≡ *Hymenostylium lapponicum* (Hedw.) Fürnr *ex* Hampe, Flora 20: 280. 1837. ≡ *Zygodon lapponicus* (Hedw.) Bruch & Schimp. *in* Bruch, Schimp. & W.Gümbel, Bryol. Eur. 3: 38, *tab. 206*. 1838 [Fasc. 4 Mon.: 6, *tab. 1*]. ≡ *Amphoridium lapponicum* (Hedw.) Schimp., Syn. Musc. Eur.: 247. 1860. ≡ *Cylindrocarpus lapponicus* (Hedw.) Lindb., Öfvers. Förh. Kongl. Svenska. Vetensk.-Akad. 19: 605. 1863. ≡ *Didymodon lapponicus* (Hedw.) Mitt., J. Linn. Soc. Bot. 8: 18. 1864. — TYPE CITATION: In alpibus Lapponiae. (In Hercynia Ehrh. Schrad.). [Lectotype (*selected here*): The first specimen from the right, on the sheet labelled 'Hedwigia lapponica Hedw. St. crypt. v. 3. p. t. nomine Gymnost. Exempl. a Cl. Swartz' – G-Hedwig/Schwägrichen!].

Plants medium-sized, in erect, compact tufts, greenish- or yellowish-brown above, brown below, slightly lustrous. *Stems* to 3 cm high, branched, profusely radiculose at the base and among the leaves, in section triangular, minutely but distinctly papillose, without central strand, consisting of a 1(–2)-stratose cortex of small, brown, thick-walled cells and a medulla of large, thin-walled cells. *Leaves* densely set, ± distinctly tristichous, erect and appressed below, erect and moderately crisped above on drying, erect-spreading on wetting, oblong- to linear-lanceolate, 1–2 mm long, 0.25–0.35 mm wide, acute, keeled above, non-decurrent; *margins* entire, unistratose, plane distally, recurved below; *costa* single, sharply set off from the laminal cells, ceasing somewhat below the apex, 40–60 µm wide, minutely papillose and strongly convex on the dorsal side, with linear, minutely roughened cells on the ventral side, in cross-section subround or crescent-shaped dorsally, with a row of median guide cells, a single row of ventral substereids or stereids and 1–2 rows of dorsal stereids or substereids; *laminal cells* unistratose throughout, opaque; *upper cells* irregularly rounded, rounded-quadrate to rounded-hexagonal, 8–10(–12) µm wide, with incrassate walls, densely pluripapillose with rounded, ± flat papillae, more than 8 per cell; *lower cells* short- to long-rectangular, 20–45 µm long, (4–)6–8(–10) µm wide, thin- to moderately thick-walled, translucent, ± hyaline, finely papillose-striate; *alar cells* undifferentiated. [*Autoicous*. *Perchaetial leaves* strongly sheathing, elliptical to oblong, 1.2–1.8 mm long, short-pointed, enclosing the seta along its whole length, with smooth laminal cells. *Setae* short, erect, straight, to 1.5 mm long, yellow-green, broader and sinistrorse above. *Capsules* erect, emergent to short-exserted, brownish, 0.8–1.3 mm long, pyriform, becoming urceolate and widened at the mouth when dry and empty, strongly 8-furrowed; *annulus* absent; *operculum* conic-rostellate, obliquely beaked; *exothecial cells*, irregularly rectangulat to hexagonal, somewhat broader, more thicker-walled, reddish-brown on the ribs, becoming smaller and flattened in 2–5 rows at the orifice; *stomata* bicellular, superficial in the capsule neck; *peristome*

→

Figure 71. *Amphidium lapponicum*. — **1**. Habit. **2**. Portion of shoot, dry. **3**. Portion of shoot, wet. **4**. Cross-section of stem. **5–7**. Axillary hairs. **8–15**. Leaves. **16**. Leaf apex (papillae partly removed to show areolation). **17**. Upper cells (papillae partly removed to show areolation). **18**. Midleaf cells(papillae partly removed to show areolation). **19**. Cells in transition zone between median and basal parts of leaf. **20**. Basal cells. **21–26**. Cross-sections of leaves. — [All from *RWMC 851C*]. — Scale bars: a – 0.5 cm (1); b – 1mm (8–15), 2 mm (2–3) and 100 µm (5–7); c – 100 µm (4, 21–26); d – 100 µm (16–20).

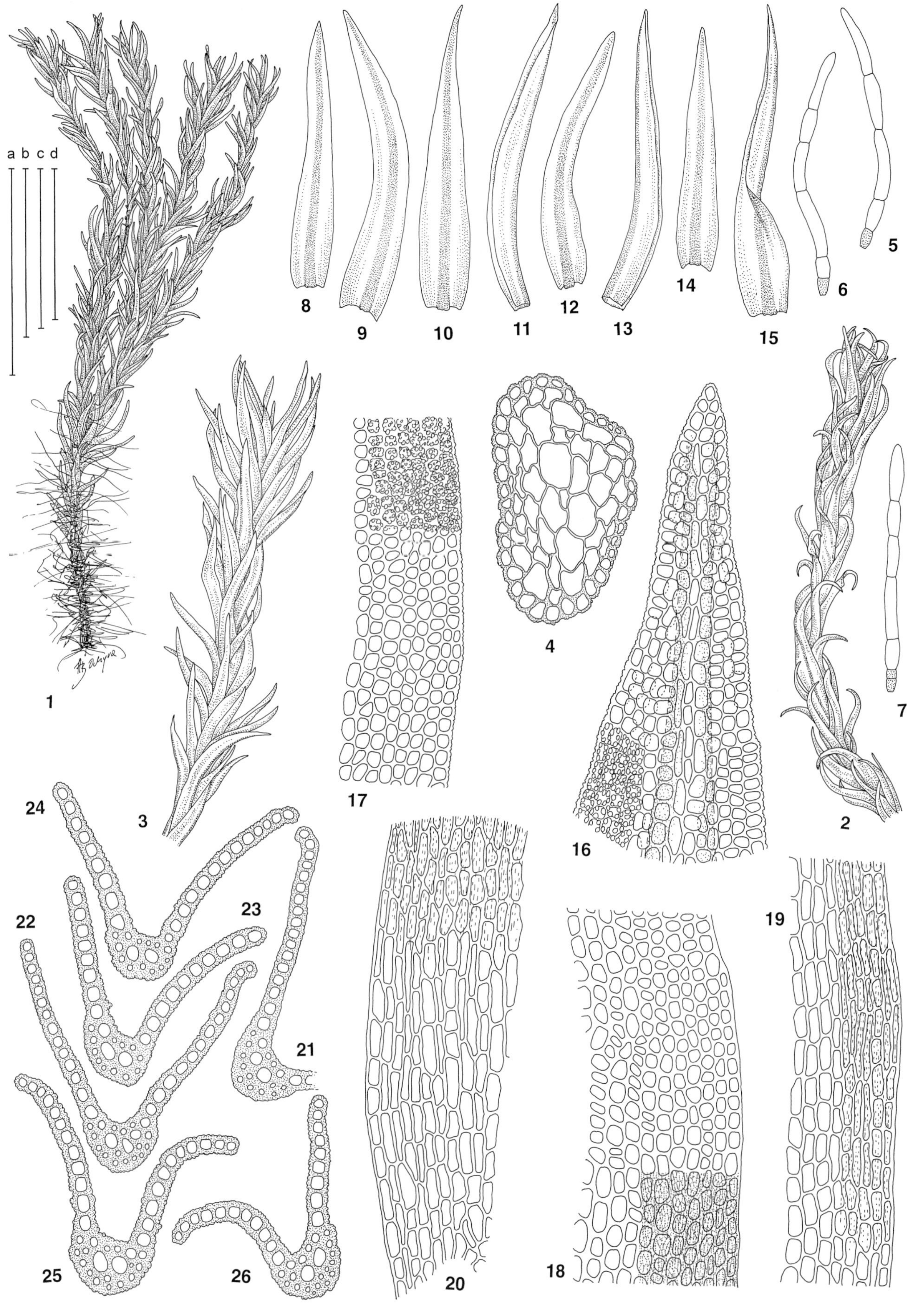
a b c d
1
2
3
4
5
6
7
8
9
10
11
12
13
14
15
16
17
18
19
20
21
22
23
24
25
26

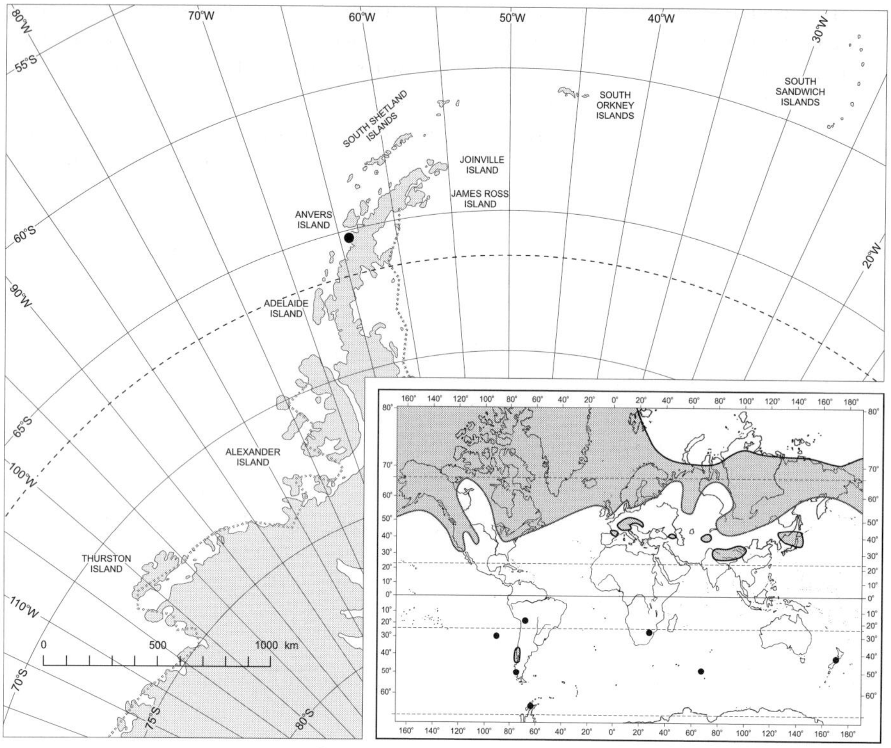

Figure 72. Distribution of *Amphidium lapponicum* in Antarctica. Inset: global distribution.

absent. *Spores* subglobose, 10–12 μm in diameter, light reddish-brown, finely roughened. *Calyptra* cucullate, naked, yellowish with a reddish, papillose tip, covering half the capsule.]

Chromosome numbers: unknown in the Antarctic and Subantarctic; *n* = 13 (2 records), *n* = 13 + 3m (1 record) and *n* = 16 (3 records) from the Holarctic (Fritsch 1991).

Discussion. *Amphidium lapponicum*, as an autoicous species, often produces sporophytes in lower latitudes allowing the genus and species to be easily recognised by its emergent to shortly exserted, gymnostomous capsules which are distinctly urceolate and strongly ribbed when dry and dehiscent. In addition, the plants grow in dense tufts and the leaves are strongly curled and contorted when dry. However, no sporophyte characters are available in Antarctic plants due to their sterility and the upper leaves are usually eroded and less contorted. The strongly papillose and opaque upper laminal cells are very deceptive and may resemble a pottiaceous taxon. The Antarctic material was originally misnamed as *Bryoerythrophyllum recurvirostrum*. However, *A. lapponicum* is immediately distinct from all pottiaceous mosses in its finely but otherwise distinctly papillose-striate basiscopic laminal cells and minutely papillose stem surface. Moreover, the more or less distinctly 3-ranked, oblong- to linear-lanceolate leaves, which are somewhat contorted when dry, and the triangular stem lacking a central strand are unknown in any local species of the Pottiaceae and should aid in unmistakable identification.

The specimens on the sheet in the Hedwig/Schwägrichen herbarium were selected by Rooy (1992) and Magill (1998) as holotype since they argued that Hedwig (1801) cited only one locality 'in alpibus Lapponiae' in the protologue. However, he actually also quoted, in parantheses, another locality 'in Hercynia' where the species was collected by Ehrhart and Schrader. Thus, correctly lectotypification of *Amphidium lapponicum* is necessary. There are four specimens affixed in one row on the sheet in the Hedwig/Schwägrichen herbarium (Price 2005). As indicated on the label they originated from Lappland (hence the species name) and were donated by Olof Swartz. On the label there is additionally a reference to Hedwig's (1792) earlier work where the species is illustrated, but none of the specimens corresponds to the illustration. The first specimen from the right, consisting of four capsules and three perichaetia, is here selected as lectotype.

Reproduction in Antarctica. Sterile.

Habitat. The single Antarctic specimen is a saxicole, collected in 'rocky areas'.

World range. A bipolar species, having a continuous, arctic-alpine range in the Northern Hemisphere (Ochyra & Szmajda 1983) and strongly dissected in the Southern Hemisphere with some intermediate stations in tropical mountains (Fig. 72, inset). In the Holarctic it is widespread throughout much of the Arctic and boreal zone, although its absence in some northern regions of Arctic Siberia and offshore archipelagoes (Afonina & Czernyadjeva 1995) may be due to undercollecting. It reaches its highest latitudes in Greenland (Lewinsky & Mogensen 1977) and on Spitsbergen (Frisvoll & Elvebakk 1996). It becomes rarer southwards and is restricted mostly to alpine elevations in California, Arizona and Colorado in western North America (Vitt *et al.* 1988) and extends to Minnesota, Wisconsin, Michigan and New Jersey in the east of North America (Crum & Anderson 1981). In Europe it is widespread in Iceland (Jóhannsson 1990) and Fennoscandia (Söderström 1996), rarer in Britain (Blockeel 1994) and reappearing in the Pyrenees, Alps, Sudetes, Carpathians and Caucasus with some isolated stations in the Harz and in Bulgaria (Ochyra & Szmajda 1983). In Asia it is scattered in Tadzhikistan in Middle Asia (Mamatkulov 1990), Central Siberia including the Western Sayan (Bardunov 1965, 1969, 1974) and mountains of northern Lake Baikal (Bardunov 1961), northern Mongolia (Tsegmed 2001*a*,*b*), the Russian Far East (Bardunov & Cherdantseva 1982), Japan (Noguchi 1989), extending southwards to Xizang (Gao *et al.* 1999) and Kashmir (Frahm *et al.* 2000). Reports from North Africa and the Hawaiian Is. are doubtful (Ros *et al.* 1999; Frahm *et al.* 2000). In the Southern Hemisphere it has hitherto been recorded in South Africa (Rooy 1992), New Zealand, Bolivia, Juan Fernandez Is. (Frahm *et al.* 2000), western Patagonia (Müller 2000*a*, 2002), Îles Kerguelen (Ochyra & Poulsen 2003) and from the Antarctic.

Distribution in Antarctica. *Amphidium lapponicum* is known from a single collection made at an elevation of about 100 m on Cape Tuxen on the Graham Coast on the Antarctic Peninsula at lat. 65°16′ S (Fig. 72).

Specimen examined. WEST ANTARCTICA. WEST ANTARCTIC PENINSULA. GRAHAM COAST. Cape Tuxen, *ca* 100 m, *RWMC 851C* (AAS, KRAM).

12. KIAERIA I.HAGEN

Kiaeria I.Hagen, Kongel. Norske Vidensk. Selsk. Skr. (Trondheim) 1914(1): 109. 1915. — Lectotype (*selected here*): *Kiaeria starkei* (F.Weber & D.Mohr) I.Hagen (*Dicranum starkei* F.Weber & D.Mohr).

Trichodontium (Dixon) Fife, Bryologist 98: 315. 1995. ≡ *Dicranum* Hedw. subgen. *Trichodontium* Dixon, Bull. New Zealand Inst. 3(3): 81. 1923, ***syn. nov.*** — Type: *Dicranum aucklandicum* Dixon [= *Trichodontium falcatum* (R.Br.bis) Fife].

Plants medium-sized and slender to rather large and robust, in erect, dull to somewhat lustrous, loose to fairly dense tufts or extensive patches, green, yellowish-, brownish- or occasionally blackish-green. *Stems* erect to ascending, simple or sparingly branched, thin, sparsely radiculose with brown, smooth, branched rhizoids scattered among the leaves, in cross-section rounded to obscurely rounded 3–5-angled, without or with a distinct central strand, consisting of a hyaline medulla of large, thin-walled cells and a weak, unistratose cortex of somewhat smaller and rather thin-walled cells; *axillary hairs* filiform, hyaline, 6–9-celled, composed of mostly elongate cells. *Leaves* erect-spreading to falcato-secund, flexuose to somewhat crisped when dry, straight to falcate, from a narrowly ovate or lanceolate, sometimes subsheathing base, gradually tapered to a concave or subtubulose, short or long subula, acute to narrowly obtuse at the apex, non-decurrent; *margins* plane, entire throughout or finely serrulate towards the apex, unistratose or occasionally bistratose above; *costa* single, narrow, percurrent to long-excurrent, in cross-section (2–)3–5-stratose, nearly homogeneous, without stereids or occasionally with a few dorsal substereids, smooth or papillose to serrulate distally on the dorsal surface; *laminal cells* unistratose, transparent, with thin to moderately thick, non-pitted to porose walls, smooth or low mammillose on the dorsal surface above, rectangular to long-rectangular throughout or becoming short-rectangular to subquadrate distally; *alar cells* weakly to well differentiated, forming a flat to ventricose group, uni- to partially bistratose. *Autoicous* or sometimes *polygamous*. *Perigonia* gemmiform below the perichaetium or apical in the stem or branches. *Perichaetia* terminal; *perichaetial leaves* differentiated, larger than the vegetative leaves, from a convolute, sheathing base abruptly short subulate. *Setae* straight to flexuose, elongate, smooth, yellow, brown to reddish-brown, variously twisted. *Capsules* erect to inclined, brown to light brown, obloid to short-cylindrical, symmetric or asymmetric, straight to somewhat curved, not or strumose, smooth or ribbed and often contracted below the mouth when dry; *annulus* compound, of 2–3 rows of large cells, revoluble or simple and persistent; *operculum* from a conical base obliquely long-rostrate, smooth or roughened near the tip; *exothecial cells* thin-walled, subquadrate to rectangular or irregularly hexagonal; *stomata* phaneroporous, bicellular, in 1 row at the base of the capsule; *peristome* single, consisting of 16, red, purple to yellow-red teeth arising from a low basal membrane, divided about halfway down into 2 prongs, vertically striolate or pitted-striolate below, papillose above or papillose-striate throughout on the outer surface. *Spores* spherical to ellipsoidal, yellow to greenish-yellow, finely papillose. *Calyptra* cucullate, fugacious, covering about half of the capsule, naked, yellowish or with a dark, reddish-brown tip.

Kiaeria is a small pancontinental genus of five species which was segregated by Hagen (1915) from the large and all-encompassing genus *Dicranum* on account of its nearly homogeneous costal anatomy and autoicous sex condition. Three species were originally placed in the genus but none of them was indicated as type and therefore *K. starkei* is here selected as lectotype of *Kiaeria*. It is principally a Northern Hemisphere genus where four species having an arctic-alpine distribution occur. The only Southern Hemisphere species *K. pumila*, which occurs also in the Antarctic, has been variously classified. Most often it was placed in the genus *Holodontium* (Ochyra 1993*b*) but Dixon (1923*a*) retained it in *Dicranum* (as *D. aucklandicum*) in the separate subgenus *Trichodontium*. This subgenus has subsequently been elevated to genus by Fife (1995) who placed in it a single species, *T. falcatum*, which is an older name for *D. aucklandicum*. *Trichodontium falcatum* is conspecific with *Kiaeria pumila* which perfectly fits the concept of the genus *Kiaeria* in the structure of the costa and monoicy. Also, shorter and papillose-striate peristome teeth are not exceptional in *Kiaeria* and are known to occur in *K. falcata*. This species also lacks a deciduous, compound annulus which is present in all remaining species of the genus including *K. pumila*. Accordingly, *Trichodontium* has to be considered a taxonomic synonym of *Kiaeria*.

1. **Kiaeria pumila** (Mitt.) Ochyra   Fig. 73

Kiaeria pumila (Mitt.) Ochyra, Fragm. Florist. Geobot. 38: 79. 1993. ≡ *Symblepharis pumila* Mitt. *in* Hook.f., Handb. New Zealand Fl. 2: 405. 1867. ≡ *Holomitrium pumilum* (Mitt.) A.Jaeger, Ber. Thätigk. St. Gallischen Naturwiss. Ges. 1870–1871: 449. 1872. — TYPE CITATION: [New Zealand] Middle Island: dry ground, Otago, Hector and Buchanan. [Lectotype (*vide* Ochyra 1993*b*: p. 79): 'N° 10.

D. pumilum m. dry ground Otago, 1863, Hector' – NY-Mitten!; isolectotypes: BM-Dixon!, BM-Hooker!, H-Brotherus!; syntype: 'Weissia microcarpa Hf & W. N Zeal Buch[n] 55' – BM-Wilson!].

Dicranum oleodictyon Dixon, Kongel. Norske Vidensk. Selsk. Forh. 7: 1. 1934. — TYPE CITATION: [South Georgia] 'Albatros', Grytviken, 300 m alt., 13 March 1933 (125, 136). [Lectotype (*vide* Ochyra 1998*a*: p. 121): 'Herb. H. N. Dixon. Ref. No. 125 Dicranum oleodictyon Dix. sp. nov. K. Norske Vidensk. Selsk Forh. vii, Nr. 1 (1934) 'Albatros', Grytviken, S. Georgia 300 m alt. 13 Mar 1933 T. Tröim Herb. Mus. Nidoros. Type' – BM-Dixon!; syntype: same label data but the collecting number 'Tröim 136' – BM-Dixon!]. First synonymised by Ochyra (1998*a*: p. 121).

Plants medium-sized, slender to fairly robust, densely caespitose, dull to somewhat glistening, yellow- to brownish-green or sometimes dark to blackish-green above, brown to light brown below. *Stems* short to moderately tall, 1–3(–4) cm high, erect, sparingly branched or less often simple, variously radiculose with brown, lustrous, smooth rhizoids scattered among the leaves, in cross-section with a small but distinct central duct surrounded by a 3–4-stratose medulla of large, thin-walled, hyaline cells and a uni- or partially bistratose cortex of smaller cells with weakly thickened, brown or pale brownish walls; *axillary hairs* 6–7-celled, hyaline throughout. *Leaves* straight to falcate, sometimes subcircinate at the stem tips, non-decurrent, erect, erect-spreading to falcato-secund, often flexuose and somewhat crispate when dry, (1.5–)1.8–3.5(–4.1) mm long, (0.25–)0.3–0.5(–0.7) mm wide, lanceolate to ovate-lanceolate or from a ± oblong, convolute and loosely clasping base tapering to a long, narrow, channelled, smooth or finely mammillose subula, acute to subacute at the apex; *margins* plane, entire or bluntly serrulate at the apex, unistratose or with bistratose spots at the extreme apex; *costa* narrow, 40–60 μm wide at the base, yellow but often nearly concolorous with the laminal cells and weakly demarcated, percurrent, in cross-section 2–3-stratose, convex dorsally, composed of ± undifferentiated cells, smooth or scabrous at back near the apex; *laminal cells* thin- to somewhat thick-walled, non-porose, uni- to partially bistratose at the apex of the subula; *upper cells* subquadrate to short-rectangular, (6–)8–11(–13) μm wide, (10–)20–35(–50) μm long, smooth or weakly mammillose on the dorsal surface; *lower cells* short- to long-rectangular, 7–16 μm wide, (25–)30–75(–80) μm long; *alar cells* quadrate to short-rectangular, usually 15–25 μm wide, 15–30 μm long, thin- to firm-walled, partly bistratose, forming mostly a sharply delimited, conspicuous, flat or convex, hyaline-brownish to yellow- or golden-brown group occupying to about half of the leaf base. *Autoicous*. *Perigonia* sessile, close to the perichaetia, bud-like. *Perichaetia* terminal; *perichaetial leaves* somewhat larger, 2.2–3.2 mm long, from a convolute, sheathing base abruptly constricted into the erect to erect-spreading, flexuose subula, as long as or only to ½–⅓ exceeding the leaf base length. *Setae* straight, erect, 7–12 mm long, yellowish-brown to brown, smooth, lustrous, sinistrorse above. *Capsules* ovoid to obloid, (0.8–)1.0–1.3 mm long, *ca* 0.5 mm wide, erect to slightly inclined, straight, symmetric to weakly gibbous, non-strumose, brown, lustrous, somewhat furrowed when dry; *annulus* compound, biseriate, revoluble; *operculum* conic-rostrate with a slanted rostrum nearly as long as the urn; *exothecial cells* rectangular to oblong-hexagonal, 40–70 μm long, 15–25 μm wide, thin- to firm-walled, becoming short, rounded, thick-walled and orange-brown in 1–2 rows at the mouth; *stomata* few in a single row at the base of the urn, superficial, bicellular, round-pored; *peristome* single, consisting of 16 narrowly lanceolate teeth arising from a low basal membrane, irregularly split to about ⅔ way down into 2, mostly unequal prongs, 250–290 μm long, intensely yellow to orange-brown below, pale to hyaline above, variously ornamented on the outer surface, mostly with dense, oblique papillose-striation below intermixed with tall conical papillae, occasionally regularly pitted-striolate below, densely papillose above. *Spores* spherical, pale brownish, 12–14 μm in diameter, papillose. *Calyptra* cucullate, fugacious, naked, yellow with a dark tip, covering half of the urn.

Chromosome number: unknown.

Discussion. *Kiaeria pumila* is a very distinct species and unlikely to be confused with any other local species, although the habit of the plants may be quite variable. Quite often, the plants have a typical 'dicranoid' aspect with the regularly falcato-secund leaves, which are occasionally almost circinate at the stem apices, and they are easily identifiable even in the field. In some populations the falcation of the leaves is less evident or absent, but microscopically the leaf areolation and especially the prominent, often ventricose and coloured alar cells are diagnostic. In these characters *K. pumila* is similar to *Hymenoloma grimmiaceum* which, however, has strongly crispate and contorted leaves on drying and the laminal cells are densely covered by lengthwise cuticular ridges giving the cells a papillose appearance which is clearly visible in leaf cross-sections. Additionally, the costal anatomy in *H. grimmiaceum* is remarkably different because of the presence of the dorsal and ventral stereid bands separated by a row of large guide cells. In the Antarctic, *K. pumila* may also be mistaken for short subulate plants of *Chorisodontium aciphyllum* which were described as *Blindia skottsbergii* (Ochyra 1993*a*) with which it shares a similar leaf

Figure 73. *Kiaeria pumila*. — **1**. Plant with sporophyte. **2**. Portion of shoot. **3**. Cross-section of stem. **4–5**. Axillary hairs. **6–13**. Leaves. **14–15**. Leaf apices. **16–17**. Upper lamina cells at leaf shoulder. **18**. Upper mid-leaf cells. **19**. Lower mid-leaf cells. **20–21**. Alar cells. **22**. Cross-section of alar region. **23–30**. Cross-sections of leaves, a sequence from base to apex. **31**.Inner perichaetial leaf. **32**. Outer perichaetial leaf. **33**. Operculate capsule. **34**. Exothecial cells in middle of urn. **35**. Exothecial cells at base of urn and stoma. **36**. Exothecial cells at orifice, annulus, peristome teeth and spores. — [1, 33 from *Greene 1827*; 2, 12, 14, 27 from *REL 1055*; 3, 6–7, 16, 18, 23–26, 28, 30 from *RO 533/80*; 4–5, 8 from *Clarke & Greene 316*; 9, 15, 19, 21–22, 29 from *JPB 253*; 10, 17, 20 from *Tröim 125* (lectotype of *Dicranum oleodictyon*), BM; 11 from *Hamilton 532* (holotype of *Verrucidens intermedius*), BM; 13 from *Will s.n.*, 23.i.1883 (lectotype of *Blindia dicranellacea*), HBG; 31–32, 34–36 from *Matteri 2239*]. — Scale bars: a – 1 mm (1); b – 1 mm (2); c – 1 mm (6–13, 31–33); d – 100 μm (3–5, 14–30, 34–36).

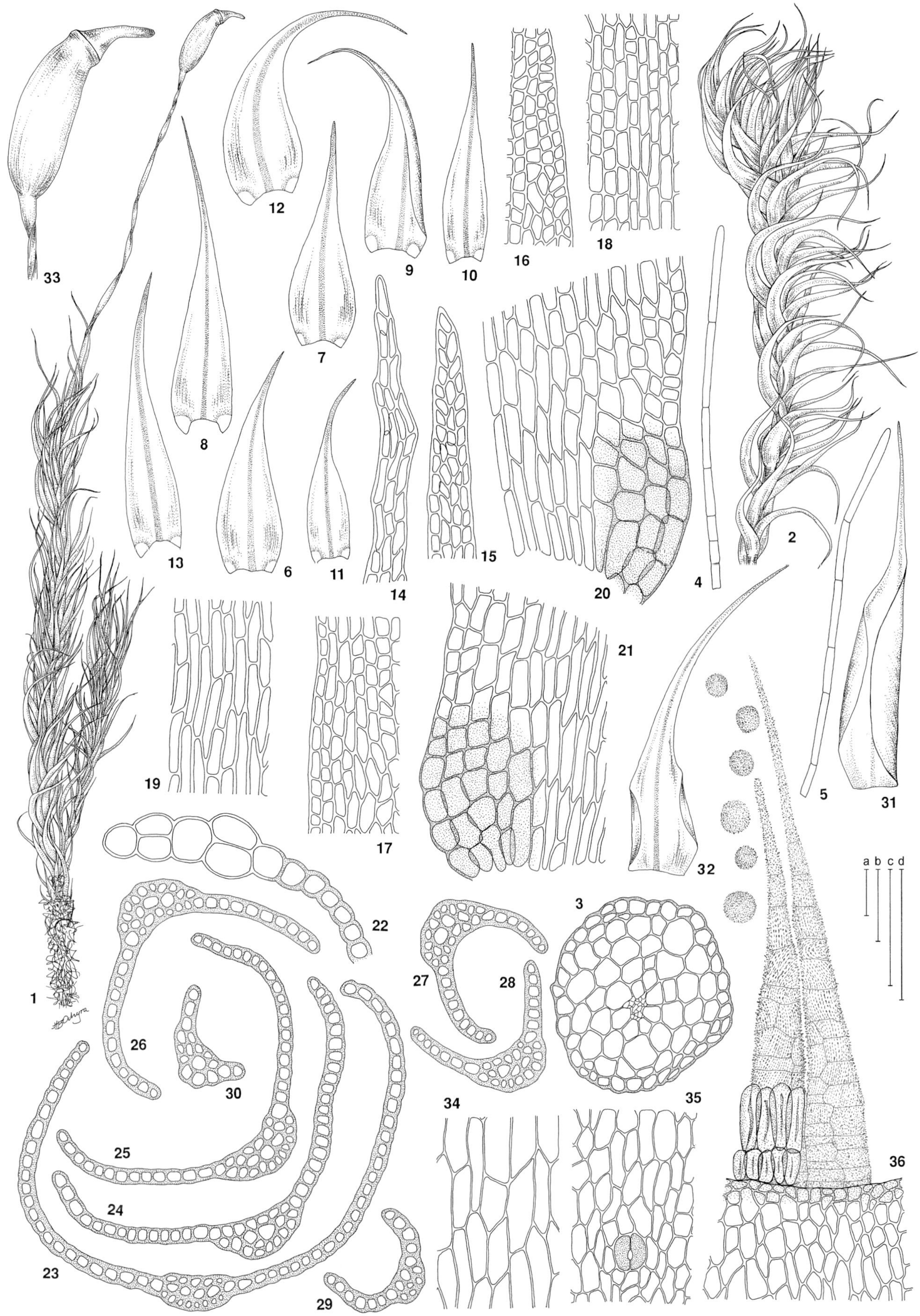
33
12
9
10
16
18
7
8
13
6
11
14
15
20
4
2
19
17
21
32
5
31
22
3
a b c d
27
28
1
26
30
35
34
25
24
23
29
36

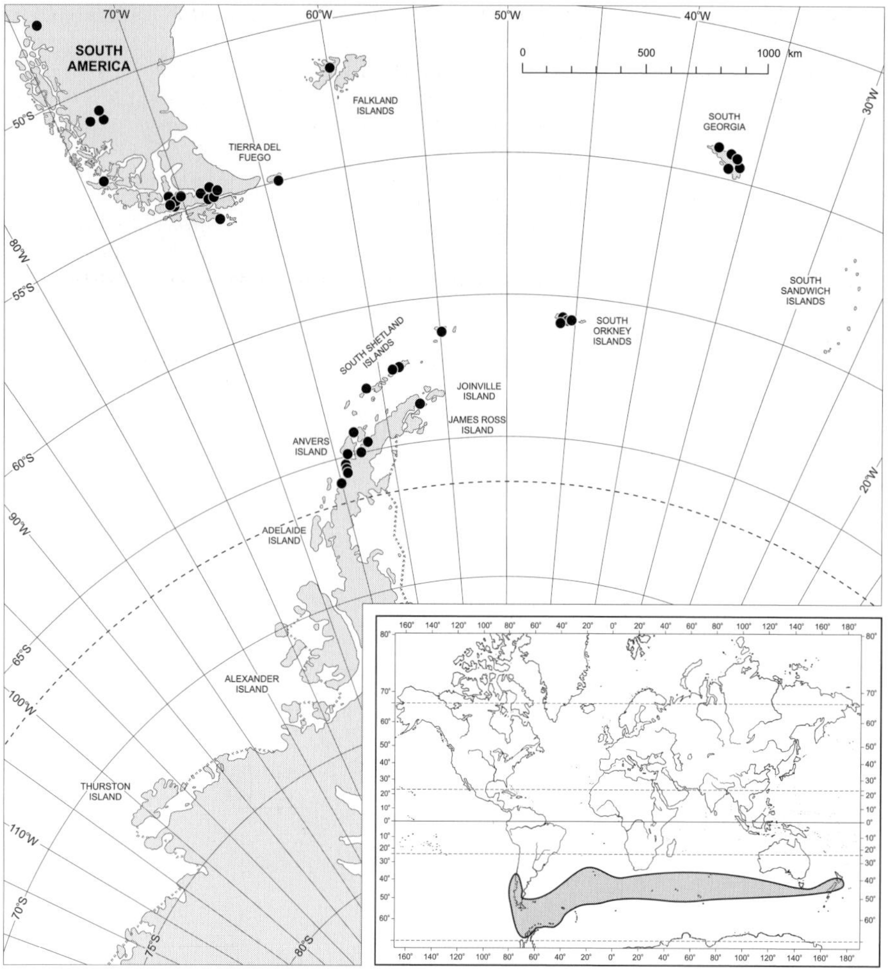

Figure 74. Distribution of *Kiaeria pumila* in the Fuegian region and Antarctica. Inset: global distribution.

areolation, the presence of the alar cells and subtubulose subula. However, *Ch. aciphyllum* is distinct in its broad costa which occupies about one third of the leaf base and shows internal differentiation into stereid groups and guide cells.

Kiaeria pumila is the only member of this genus in the Southern Hemisphere and it is closely related to the northern *K. starkei*, and this species was first recognised as an unnamed variety of *Dicranum starkei* (≡ *K. starkei*), material of which was subsequently described by Mitten (1869) as two separate species, *Dicranum pumilum* and *D. ramulosum*. By coincidence, the epithet *pumilum* was used two years earlier by Mitten (1867) for another new species, *Symblepharis pumila*, from New Zealand and both species are conspecific, the latter name having priority (Ochyra 1993*b*). *Kiaeria pumila* is otherwise markedly distinct from its northern counterpart *K. starkei* in its symmetric or nearly symmetric capsules which lack the struma, ornamentation of the outer surface of the peristome teeth and smaller spores.

Reproduction in Antarctica. Only two populations, from Rasmussen I. off the Graham Coast (*RILS 1916B & 1919*, AAS, KRAM), have been found with fully mature sporophytes, all other collections being sterile, although in some gametangia have been abundantly produced.

Habitat. *Kiaeria pumila* is associated primarily with various communities belonging to the fruticose lichen and moss cushion subformation. It grows mostly on non-calcareous substrata, in dry to moist and open to sheltered habitats, rock ledges and outcrops, scree, in crevices and rock depressions, often on gravelly or stony soil in late snow patches and at the edge of melt streams and in temporarily flushed areas. Occasionally, it grows on peaty soil colonising re-exposed *Chorisodontium aciphyllum* banks. *Kiaeria pumila* is most often associated with species of *Andreaea*, *Bartramia patens*, *Ceratodon purpureus*, *Hymenoloma grimmiaceum*, *Pohlia cruda* and *Sanionia uncinata*, and its short turf colonies are often interwoven with various species of liverworts including *Barbilophozia hatcheri*, *Cephaloziella varians*, *Herzogobryum teres* and *Lophozia excisa*.

World range. An infrequent pan-south-temperate species (Fig. 74, inset) widely scattered in the *Nothofagus* zone in southern South America, ranging from the Valdivian region at lat. *ca* 40°S to Cape Horn (Ochyra 1983*b* – type of *Dicranum pumilum*), and rare on Isla de Los Estados (Müller 1885 as *Blindia humilis*) and the Falkland Is. (Cardot & Brotherus 1923) in the western hemisphere. It is widespread on both islands of New Zealand in the eastern hemisphere (Dixon 1923*a*), scattered on some islands in the Southern Ocean including Île Amsterdam (Tixier 1980) and Tristan da Cunha (Dixon 1960, as *Dicranoweisia falcifolia*). It extends into the Subantarctic on the Prince Edward Is. (Zanten 1971, as *Holodontium pumilum*) and on South Georgia (Bell 1973*a*, 1983, as *Dicranum oleodictyon*), and into the northern maritime Antarctic.

Distribution in Antarctica. Generally, *Kiaeria pumila* is an infrequent species, although locally it may occur in profusion. It occurs most frequently on the South Orkney Is. where it was recorded for the first time by Bell (1973*b*) and is very rare and localised on the South Shetland Is. and scattered along the western

Antarctic Peninsula from Trinity Peninsula to the Graham Coast, extending southwards to lat. 65°33′S (Fig. 74). It is a species of low and medium elevations, occurring from near sea level to 150 m, and exceptionally to 238 m on Elephant I. A report of the species from the South Sandwich Is. (Convey *et al.* 2001) is based upon misidentification.

Selected specimens examined. WEST ANTARCTICA. SOUTH ORKNEY IS. **Coronation I.** Cape Hansen, 25 m, *RILS 3235, 3236 & 8689* (AAS, KRAM); Sunshine Glacier, 175 m, *RW 150A* (AAS, KRAM); Olivine Pt., 15–30 m, *RILS 122* (AAS, KRAM). **Powell I.** Middle of E coast, 135 m, *RILS 223B & 227* (AAS, KRAM). **Signy I.** Coast opposite Spindrift Rocks, *ca* 65 m, *REL 1143* (AAS, KRAM); ridge E of Spindrift Rocks, *ca* 80 m, *MWH 765H* (AAS, KRAM); Stygian Cove, *RILS 1808* (AAS, KRAM); between the Wallows and Stygian Cove, 25 m, *RILS 5156 & 5157* (AAS, KRAM); Knob Lake, 30–50 m, *RILS 619 & 628* (AAS, KRAM); Knife Pt., 30 m, *RILS 10772* (AAS, KRAM); Factory Cove, 6 m, *RILS 8694* (AAS, KRAM); Factory Bluffs, 65 m, *REL 1055* (AAS, KRAM) and 80 m, *RILS 10775* (AAS, KRAM); Rusty Bluff, 135 m, *RILS 604 & 1795* (AAS, KRAM); Green Gable, 180 m, *RILS 8627* (AAS, KRAM); Khyber Pass, 85 m, *RILS 8678* (AAS, KRAM). **Moe I.** Alt. 35–100 m, *RILS 5261B* (AAS, KRAM).

SOUTH SHETLAND IS. **Elephant I.** Walker Pt., *ca* 120 m, *JPB 253* (AAS, KRAM); Endurance Glacier, 238 m, *JSA 171AA* (AAS, KRAM). **King George I.** ***Fildes Peninsula***: Ardley I., *Li 890117* (AAS, KRAM). ***Admiralty Bay***: Speil Pt., 4 m, *RO 533/80* (KRAM). **Livingston I.** ***Byers Peninsula***: between Chester Cone and Sealers Hill, 25–30 m, *RILS 3820A* (AAS, KRAM); south beach below Negro Hill, 5 m, *KR 18A* (AAS, KRAM).

WEST ANTARCTIC PENINSULA. **Danco Coast.** Charlotte Bay, Andrée I., 25 m, *RILS 4040A* (AAS, KRAM); Coughtrey Peninsula, Almirante Brown, *RMS 69-359* (US). **Palmer Archipelago.** ***Melchior Is.***: Gamma I., 5–10 m, *RILS 4151 & 4152* (AAS, KRAM). ***Anvers I.***: Litchfield I., *ca* 15 m, *RWMC 393B* (AAS, KRAM). **Graham Coast.** Rasmussen I., *RILS 1916B & 1919* (AAS, KRAM); Cape Pérez, *ca* 12 m, *RWMC 638* (AAS, KRAM). **Wilhelm Archipelago.** ***Penola Strait***: Petermann I., 20 m, *RWMC 792C* (AAS, KRAM). ***Collins Bay***: largest of Berthelot Is., *RWMC 625* (AAS, KRAM). ***Grandidier Channel***: largest island off Takaki Pt., 10–35 m, *RILS 4342B, 4343, 4348B & 4365A* (AAS, KRAM).

EAST ANTARCTIC PENINSULA. **Trinity Peninsula.** Mt. Flora. *ca* 115 m, *REL 39* (AAS, KRAM).

13. **CHORISODONTIUM** (Mitt.) Broth.

Chorisodontium (Mitt.) Broth. *in* Engl., Nat. Pflanzenfam. Ed. 2, 10: 204. 1924. ≡ *Dicranum* Hedw. sect. *Chorisodontium* Mitt., J. Linn. Soc. Bot. 12: 62. 1869. ≡ *Dicranum* Hedw. subgen. *Chorisodontium* (Mitt.) Broth. *in* Engl. & Prantl, Nat. Pflanzenfam. 1(3): 328. 1901. — Lectotype (*selected here*): *Chorisodontium aciphyllum* (Hook.f. & Wilson) Broth. (*Dicranum aciphyllum* Hook.f. & Wilson).

Plants moderate-sized to large, occasionally small, slender to robust, loosely or densely caespitose, glossy, dark-, olive- to yellowish-green or yellowish- to golden-brown above, light to dark brown below. *Stems* erect, sparingly branched, weakly radiculose below to densely tomentose near to the apex with tawny to rusty-brown, occasionally whitish, branched, smooth rhizoids, in cross-section round, with a small but otherwise distinct central strand surrounded by a multilayered medulla of large, thin-walled, hyaline cells and a 1–4-stratose cortex of small, thick-walled, brown cells; *axillary hairs* hyaline throughout, composed of 8–15, mostly elongate cells. *Leaves* evenly spaced, crowded to fairly distant, straight to falcate, erect and appressed to erect-spreading when dry, erect-spreading when wet, sometimes secund to curved- or falcate-secund, somewhat decurrent, from an ovate, oblong-ovate or lanceolate, broadly concave, often sheathing and clasping base, gradually to quite abruptly tapering into a long setaceous, tubulose subula, straight or flexuose when dry, acute at the apex, as long as or exceeding up to 5 times the length of the base; *margin* plane, entire throughout or bluntly denticulate to sharply, remotely serrate near the apex; *costa* single, narrow to very broad, occupying ⅕–⅓ the width of the leaf base, long-excurrent, clearly delimited or diffuse and concolorous with the laminal cells below, flaring outwards above and gradually merging with the bistratose laminal cells at the shoulders, filling most of the subula and imperceptibly fusing with the bistratose lamina in the subula, in cross-section strongly flattened below, lunate in the subula, smooth or scabrous dorsally, with a distinct median row of enlarged guide cells and conspicuous stereid bands on the ventral and dorsal sides; *laminal cells* smooth throughout, unistratose below, becoming partly bistratose at the shoulders and in the subula; *basal cells* rectangular, oblong-hexagonal to linear, with thin or moderately to strongly incrassate, distinctly porose to unpitted walls, often becoming narrower towards the margins; *shoulder* and *subula cells* shorter, irregularly rhomboidal, subquadrate, oval to short-rectangular, not or variously pitted; *alar cells* quadrate to short-rectangular, thin- to thick-walled, forming flat to swollen, hyaline, hyaline-brownish to orange-brown group, often easily detached, extending to the costa or occupying much of the leaf base. *Dioicous*. *Perichaetia* terminal; *outer perichaetial leaves* consisting of a short, quadrate base abruptly narrowed into a much longer awn-like subula; *inner percihaetial leaves* comprising a long, vaginate, convolute and strongly clasping base rapidly constricted into a short or long setaceous subula. *Setae* elongate, erect, straight, rather stout, yellowish to light brown, dextrorse above when dry. *Capsules* erect, cylindrical, symmetric to slightly asymmetric, dark brown, somewhat glossy, smooth; *annulus* compound, 2–3-seriate, deciduous or simple and persistent; *operculum* long-rostrate, with a rostrum to as long as the urn; *exothecial cells* short- to long-rectangular, thick-walled, becoming isodiametric, orange-brown in 2–3 tiers at the mouth; *stomata*, few in a single row in the neck, bicellular, superficial; *peristome teeth* 16, lanceolate, split in the upper half into 2, often unequal segments, orange- or yellow-brown below, hyaline above, vertically striolate throughout or papillose at the pale tips. *Spores* globose, brownish, unicellular or ellipsoidal, green and 2–4-cellular. *Calyptra* cucullate, smooth, naked, light brown, dark brown at the tip.

The only reliable generic diagnostic character of *Chorisodontium* is its bistratose laminal cells at the leaf shoulders and in the leaf subula which imperceptibly merge with the costa. Accordingly, it gives the impression that the costa, which is often diffuse near the leaf base, is markedly flaring outwards at the junction of the leaf base and the subula. In general aspect of the often large and robust plants and the structure of the costa which is long-excurrent, strongly flattened below and broad to very broad, at least in the cool-adapted austral species, *Chorisodontium* may resemble some species of *Campylopus*, with which it additionally shares the leaf areolation of the cells with the incrassate and strongly porose walls and prominent angular cells. However, the genus *Campylopus* entirely lacks stomata which are always present in *Chorisodontium*, and therefore the genus is placed in the subfamily Dicranoideae rather than the Campylopodioideae. Moreover, in all Antarctic species of *Campylopus* the costa has ventral hyalocysts in transverse section, whereas in *Chorisodontium* it is internally differentiated into a median row of guide cells separating two stereid bands on either side, and this should prevent any confusion.

Chorisodontium was originally established by Mitten (1869) as a section within the then all-encompassing genus *Dicranum*, which was subsequently raised to genus by Brotherus (1924*b*). Five species were originally placed in section *Chorisodontium* but none of them was indicated as type and therefore lectotypification of this generic name is needed. This is particularly important because the genus, as currently understood, comprises two distinct morphological groups which have different geographical ranges and, to some extent, ecological preferences. The austral group comprises six species, widespread in the south cool-temperate zone in southern South America and, less often, in New Zealand, Subantarctica, Antarctica, occasionally extending northwards to Bolivia and Peru. These are characterised by a very broad costa, shorter leaves with entire or nearly so leaf margins, less distinct alar cells, a deciduous annulus and small unicellular spores, 18–30 µm wide and they are often peat-forming plants. The neotropical group consists of three species, occurring on the Central American Cordillera and in the northern Andes, extending to northern Peru and occasionally to Bolivia. They differ from the austral species in their larger leaves, distinctly serrate at the apex, prominent, often thick-walled and coloured alar cells, simple, persistent annulus and large, often multicellular spores, 20–80 µm in diameter, germinating precociously in the capsules. *Dicranum aciphyllum* (≡ *Chorisodontium aciphyllum*) is here selected as lectotype of *Chorisodontium* since it represents the larger group of species within the genus and fits well the the original diagnosis of this taxon.

The austral cool-adapted *Chorisodontium* species were reviewed by Roivainen & Bartram (1937) and Hyvönen (1991*a*), and the neotropical species were overviewed by Frahm (1989). Additionally, Frahm (1981) provided illustrations of the type specimens and a key to the species of *Chorisodontium*. At present, the genus is considered to consist of nine species. One of these, *Ch. aciphyllum*, is widespread in the northern maritime Antarctic.

1. **Chorisodontium aciphyllum** (Hook.f. & Wilson) Broth. Fig. 75

Chorisodontium aciphyllum (Hook.f. & Wilson) Broth. *in* Engl., Nat. Pflanzenfam. Ed. 2, 10: 204. 1924. ≡ *Dicranum aciphyllum* Hook.f. & Wilson, London J. Bot. 3: 541. 1844. — TYPE CITATION: Hermite Island, Cape Horn, and Falkland Islands. [Lectotype (*vide* Bell 1973*a*: p. 35): 'N°. 123. Dicranum aciphyllum H.fil. & Wils. Hermite Island, Cape Horn Antarct. Exp. 1839–1843 J.D.H.' – BM-Wilson!; isolectotypes: BM-Wilson!, E!, MANCH!; syntype: '52 Falkl. Islands W. 200 Dicranum aciphyllum Wils.' – BM-Wilson!, isosyntypes: E!, MANCH!].

Dicranum nordenskioldii Cardot, Bull. Herb. Boissier Sér. 2, 6: 29. 1906. ≡ *Chorisodontium nordenskioldii* (Cardot) Roiv. *in* Roiv. & E.B.Bartram, Ann. Bot. Soc. Zool. Bot. Fenn. Vanamo 9(2): 42. 1937. — TYPE CITATION: Terre Louis-Philippe: débarquement c. [Lectotype (*vide* Bell 1973*a*: p. 35): 'Svenska Sydpolarexpeditionen 1901–03. N:r 440. Dicranum nordenskjöldii Card. sp. nova Terre Louis-Philippe: debarquement c, 1/12 1902 Carl Skottsberg det. J. Cardot' – UPS!; isotypes: PC!, S!]. First synonymised by Bell (1973: p. 34).

Dicranum nordenskioldii fo. *minus* Cardot, Bull. Herb. Boissier Sér. 2, 6: 29. 1906. — TYPE CITATION: Terre Louis-Philippe: débarquement c. [Lectotype (*selected here*): 'Svenska Sydpolarexpeditionen 1901–03. N:r 441. Dicranum nordenskjöldii Card. fo. minus Card. Terre Louis-Philippe: île des Mousses 1/12 1902 Carl Skottsberg det. J. Cardot' – PC!; isotype: S!]. First synonymised by Ochyra (1999*f*: p. 512).

Blindia skottsbergii Cardot fo. *robusta* Cardot, Bull. Herb. Boissier Sér. 2, 6: 29. 1906. — TYPE CITATION: Terre Louis-Philippe: débarquement c. [Lectotype (*selected here*): 'Svenska Sydpolar-expeditionen 1901–03. N:r 442. Blindia Skottsbergii Card. forma robusta Card. Antarctide: Terre Louis-Philippe: île des Mousses 1/12 1902 Carl Skottsberg. Det. J. Cardot' – PC!; isotypes: H-Brotherus!, S!]. First synonymised by Ochyra (1993*a*: p. 507).

Plants medium to large, generally slender, growing in loose to compact turves, lustrous, yellowish- to golden-brown or yellowish-green to dark green, conspicuously interwoven by numerous, reddish-brown to brown rhizoids. *Stems* reddish, erect, sparsely branched, in cross-section consisting of a unistratose cortex of small, thick-walled, reddish-brown cells, a thick, 5–7-stratose medulla of large, hyaline, thin-walled cells and a distinct, small central strand; *axillary hairs* filiform, hyaline, 9–12 celled. *Leaves* crowded, closely appressed to erect-spreading, straight to flexuose when dry, erect-spreading when moist, 5–10(–12) mm long, 0.5–1.0(–1.1) mm wide, from a lanceolate to oblong-lanceolate, concave, sheathing and

→

Figure 75. *Chorisodontium aciphyllum*. — **1**. Plant with sporophytes. **2**. Habit, dry. **3**. Portion of shoot, dry. **4**. Portion of cross-section of stem. **5–6**. Axillary hairs. **7–15**. Leaves. **16**. Leaf apex. **17–18**. Cells at leaf shoulder. **19**. Mid-leaf cells at margin. **20–22**. Mid-leaf cells. **23**. Basal cells. **24–29**. Cross-sections of leaves, a sequence from base to apex. **30**. Outer perichaetial leaf. **31**. Inner perichaetial leaf. **32**. Upper cells of inner perichaetial leaf. **33**. Deoperculate capsule. **34**. Operculate capsule. **35**. Calyptra. **36**. Exothecial cells in middle of urn. **37**. Exothecial cells at base of urn and stoma. **38**. Exothecial cells at orifice, annulus, peristome and spores. — [1, 30–32, 36–38 from *Schäfer-Verwimp & Verwimp 10808*; 2, 7, 24, 29 from *RO 2163/80*; 3, 9, 22 from *RO 1247/80*; 4 from *Will s.n.*, 14.vii.1883 (lectotype of *Distichium austrogeorgicum*), H; 5–6, 10, 23, 26, 28 from *RO 594/80*; 8, 16–19, 21 from *CS 440* (isotype of *Dicranum nordenskioldii*), S; 11, 27 from *CS 441* (isotype of *Dicranum nordenskioldii* fo. *minus*), S; 12 from *RO 2401/80*; 13, 25 from *CS 442* (isotype of *Blindia skottsbergii* fo. *robusta*), S; 14–15, 20 from *CS 283* (isotype of *Blindia skottsbergii*), S; 33–35 from *Poelt s.n.*, 8.i.1989]. — Scale bars: a – 1 mm (3, 7–15); b – 100 µm (5–6) and 50 µm (4, 24–29); c – 1 mm (30–31, 33–35); d – 0.5 cm (1–2) and 50 µm (16–22, 32); e – 100 µm (23, 36–38).

a
b
c
d
e
33
34
35
12
20
14
15
13
11
17
32
1
31
30
18
19
8
9
10
2
3
7
6
5
21
16
23
22
38
27
29
26
25
24
28
4
36
37

loosely clasping base, gradually narrowing to a long, awn-like, channelled, straight to flexuose subula, acute, as long as or somewhat longer than the base; *margin* plane, entire to bluntly, sparingly denticulate towards the apex; *costa* very broad, 200–320 μm wide in mid-leaf, usually occupying more than one-third of the leaf base, excurrent, dorsally smooth or papillose, mostly diffuse and concolorous, gradually merging into the lamina cells at the shoulders and in the subula, in cross-section with a median row of enlarged guide cells and conspicuous ventral and dorsal stereid bands; *laminal cells* smooth throughout; *lower cells* unistratose, short- to long-rectangular, thin- to thick-walled, with markedly porose to unpitted walls, 8–16 μm wide, 30–100 μm long, becoming narrower, 4–6 μm wide, oblong-hexagonal, rhomboidal to linear towards the margin and forming an ill-defined border; *upper cells* variable, short-rectangular, rhomboidal, rhombic, rounded-quadrate to elliptical, thick-walled, 7–20 μm long, 5–8 μm wide, partly bistratose at the shoulders and in the subula and imperceptibly grading into the costa; *alar cells* large, brownish, numerous, thin-walled, forming distinct, rather plane auricles extending to the costa and often easily detached. [*Dioicous*. *Setae* erect, straight, 1.5–3.0 cm long, yellow to light brown, twisted to the right above. *Capsules* erect, cylindrical, 1.5–3. mm long, 0.5–0.7 mm wide, dark brown, weakly lustrous, smooth; *annulus* compound, 2–3-seriate, revoluble; *operculum* with a long, slanted rostrum, as long as the urn; *exothecial cells* short- to long-rectangular, 30–80 μm long, 10–20 μm wide, moderately to strongly thick-walled, becoming rounded-quadrate in 2–3 rows at the mouth and forming an orange-brown strip; *stomata* superficial, bicellular in the neck; *peristome teeth*, lanceolate, 400–550 μm long, divided to the middle or below into 2, usually unequal prongs, yellow-brown below, hyaline above, vertically striolate below, papillose above. *Spores* spherical, 23–30 μm wide, papillose. *Calyptra* cucullate, naked, light brown becoming dark at the tip.]

Chromosome numbers: $n = 10$ (Zhou *et al.* 1995) and $n = 12$ (Kuta *et al.* 1982) from King George I.

Discussion. Amongst the Antarctic mosses, *Chorisodontium aciphyllum* is usually an unmistakable and easily identified species in the field by its peculiar growth form. It typically grows in deep, compact and usually extensive turves in which the shoots are tightly bound to one another by the dense tomentum of reddish-brown to brown rhizoids covering much of the stems. Additionally, the erect and long-subulate leaves are very distinctive. Microscopically, the species is distinguished by its very broad costa which, though usually diffuse, occupy about one third of the leaf base. In the Antarctic moss flora such exceptionally broad costae are known only in species of *Campylopus* and *Holodontium strictum* and, moreover, all these taxa share the prominent alar cells. *Chorisodontium aciphyllum* differs from all Antarctic species of *Campylopus* by the lack of hyalocysts in the costal anatomy, whereas *H. strictum* has the laminal cells covered with longitudinal cuticular ridges giving them a papillose appearance in transverse section.

Chorisodontium aciphyllum is a very variable species, especially with respect to the size of the plants, the length of the leaves, the thickness and porosity of the laminal cells as well as the papillosity of the dorsal surface of the costa. This prompted different authors to recognise its various phenotypes as separate taxa. All these characters are extremely variable and they are conditioned by different environmental factors and, as such, cannot be accepted for delimiting separate taxa. Therefore, numerous taxa described within this complex have been considered conspecific with the polymorphous *Ch. aciphyllum*. Hyvönen (1991*a*) recognised eight synonymous names of *Ch. aciphyllum*, but other studies by Bell (1973*a*, 1977) and Ochyra (1993*a*, 1994*a*, 1996*c*) have increased this to 13 synonyms at the species level and an additional two synonyms referred to the names of the infraspecific taxa (Ochyra 1993*a*, 1999*f*).

The plants from Moss I. were originally described by Cardot (1906*a*, 1908*a*) as *Dicranum nordenskioldii* on the basis of a narrower and better delimited costa and thinner and less porose laminal cells, characters which are particularly liable to considerable variability in *Chorisodontium aciphyllum* and therefore Bell (1973*a*) rightly considered them to be conspecific. Additionally, Cardot (1906*a*, 1908*a*) recognised another specimen from Moss I. as *Blindia skottsbergii* fo. *robusta* on account of its weakly developed alar cells and poorly developed ventral stereid band in the costa, but this taxon also represented *Ch. aciphyllum* (Ochyra 1993*a*).

Reproduction in Antarctica. Sterile. However, on Subantarctic South Georgia sporophytes are locally produced in abundance.

Habitat. Together with *Polytrichum strictum*, *Chorisodontium aciphyllum* is the principal component of the community within the moss turf subformation (Gimingham & Lewis Smith 1970; Lewis Smith 1972; Lewis Smith & Gimingham 1976) which forms extensive banks of semi-ombrogenous peat on gentle, well drained slopes or gravelly porous ground. These reach 1–2 (exceptionally to 3) m in depth and are a unique feature of the vegetation in some regions of the northern maritime Antarctic (Lewis Smith 1972, 1979, 1981, 1996; Collins 1976*a*,*b*; Fenton 1980, 1982; Fenton & Lewis Smith 1982; Lewis Smith & Corner 1973). The formation of the peat banks is possible, in part because of the high water content (Gimingham & Lewis Smith 1971) and the high net annual production of *Ch. aciphyllum* and low rate of decomposition of peat (Baker 1972; Fenton 1980, 1982). Additionally, the species occasionally grows on ledges, rock faces and slabs and outcrops, as well as on mineral soil between rocks and stones, on humus and stony soil, generally in relatively dry to moist and sheltered situations. It has also been collected on timber from derelict huts; at lower latitudes in southern South America *Ch. aciphyllum* often grows on the lower part of tree trunks.

World range. *Chorisodontium aciphyllum* is a pan-south-temperate species with its most frequent occurrence in the American sector (Fig. 76, inset). It is common and locally abundant on the Juan Fernandez Is. (Robinson 1975) and in the *Nothofagus* zone in Patagonia and Tierra del Fuego, the Valdivian region in Chile at lat. *ca* 38° S to Cape Horn, and on the Falkland Is. (Hyvönen 1991*a*); it is common on the Subantarctic island of South Georgia (Bell 1973*a*) and in the northen maritime Antarctic. Elsewhere, it is rare and occasional on Tristan da Cunha (Dixon 1960), Île Amsterdam (Hyvönen 1991*a*), Îles Kerguelen (Cardot 1916; Ochyra, personal observations) and on the South Island of New Zealand (Bartlett & Frahm 1983).

Distribution in Antarctica. A frequent species, locally growing in great profusion in the northern maritime Antarctic, south-

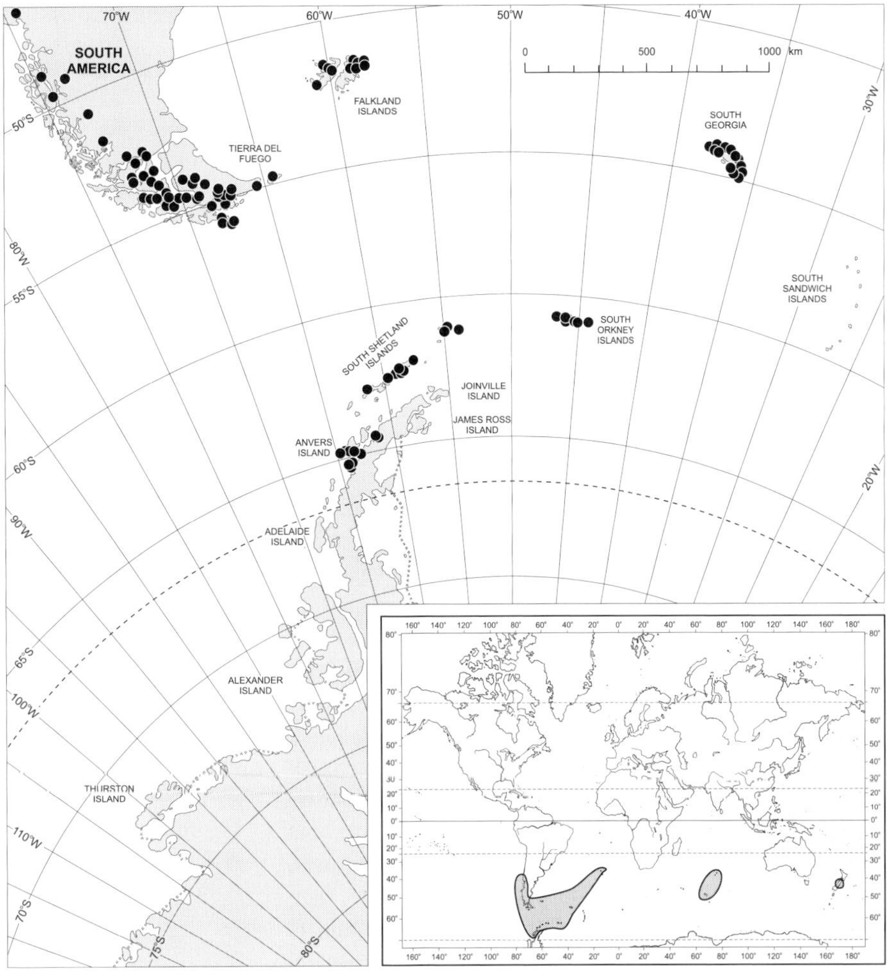

Figure 76. Distribution of *Chorisodontium aciphyllum* in the Fuegian region and Antarctica. Inset: global distribution.

wards to lat. 65°19′ S on Green I. in the Berthelot Is. archipelago off the Graham Coast (Fig. 76). It is most common in the South Orkney Is. and the mid-west Antarctic Peninsula including its offshore islands (Danco and Graham Coasts), and rare and localised in the South Shetland Is. It is a species of lower elevations, occurring most often from sea level to about 250 m and is much rarer at higher elevations, ascending to 400 m on Mt. Demaria above Cape Tuxen on the Graham Coast.

Specimens examined. WEST ANTARCTICA. SOUTH ORKNEY IS. **Monroe I.** W coast, 35–100 m, *MMM 79* (AAS, KRAM); plateau, 80 m, *RILS 8007* (AAS, KRAM). **Coronation I.** Cape Hansen, 20–200 m, *RILS 3217* (AAS, KRAM); Saunders Pt., *ca* 30 m, *RILS 18B* (AAS, KRAM). **Lynch I.** N-facing slope, 10–30 m, *RILS 521* (AAS, KRAM). **Matthews I.** NE peninsula, 5–20 m, *MMM 103* (AAS, KRAM). **Robertson Is.** Steepholm I., 75 m, *RILS 3247* (AAS, KRAM). **Powell I.** W of John Peaks, 15–65 m, *RILS 248A* (AAS, KRAM); SE coast, 135 m, *RILS 206A* (AAS, KRAM). **Christoffersen I.** Falkland Harbour, *ca* 15 m, *RILS 54A* (AAS, KRAM) and 60 m, *KR 221A* (AAS, KRAM); S end, 50 m, *KR 172A & 197A* (AAS, KRAM). **Laurie I.** Cape Valavielle, *MMM 134* (AAS, KRAM); Cape Dundas, 330 m, *MMM 198* (AAS, KRAM); Cerro Moneta, 70 m, *Bellisio 11744* (AAS, KRAM); Scotia Bay, x.1903, *Rudmose Brown s.n.* (BM, E, PC). **Signy I.** Port Jebsen, *ca* 15 m, *MWH 771B* (AAS, KRAM); Polynesia Pt., 50 m, *RILS 10750* (AAS, KRAM); Clowes Bay, *ca* 10 m, *MWH 680, 693B & 695* (AAS, KRAM). **Moe I.** 10–200 m, *RILS 5251* (AAS, KRAM).

SOUTH SHETLAND IS. **Elephant I.** Walker Pt., 120 m, *JPB 241* (AAS, KRAM); Endurance Glacier, 167 m, *JSA 22A* (AAS, KRAM). **Clarence I.** S end, 152 m, *JSA 1B* (AAS, KRAM). **King George I.** ***Fildes Peninsula***: Ardley I., 15 m, *RO 2493/80* (KRAM); Bellinghausen Station, 15 m, *RO 2401/80* (KRAM). ***Marian Cove***: North Spit, *BJ 141* (KRAM). ***Barton Peninsula***: Narębski Pt., *BJ 179* (KRAM); Noel Hill, *BJ 195* (KRAM); Winship Pt., *BJ 180* (KRAM). ***Potter Cove***: Stranger Pt., *BJ 107B* (KRAM). ***Admiralty Bay***: Creeping Slopes, 90 m, *RO 1247/80* (KRAM); Jersak Hills, 170 m, *RO 5064/79* (KRAM); Ore Pt., 3 m, *RO 521/80* (KRAM); Speil Pt., 50 m, *RO 544/80* (KRAM); Ullman Spur, 60 m, *RO 594/80* (KRAM); Mt. Wawel, 40 m, *RO 2163/80* (KRAM). ***Destruction Bay***: Faraway Nunataks, *BJ 17* (KRAM). **Robert I.** Coppermine Cove, 15 m, *RILS 935* (AAS, KRAM). **Livingston I.** South Beaches, Clark Nunatak, 30 m, *DCL 254* (AAS, KRAM); between Smellie Pt. and Devil's Pt., 5 m, *KR 29A* (AAS, KRAM). **Snow I.** President Head, *KR 7D* (AAS, KRAM).

WEST ANTARCTIC PENINSULA. **Danco Coast.** Cierva Pt., *VK 620* (KRAM) and *OB 10* (AAS, KRAM); Moss I., *CS 440* (PC, S, UPS – type of *Dicranum nordenskioldii*), *441* (LD, PC, S – type of *D. nordenskioldii* fo. *minus*) *& 442* (H, PC, S – type of *Blindia skottsbergii* fo. *robusta*); Cuverville I., 10–20 m, *RILS 11938A* (AAS, KRAM) and 100 m, *RILS 8801* (AAS, KRAM). **Palmer Archipelago.** ***Anvers I.***: Cape Monaco, *VK 620* (KRAM); Norsel Pt., *VK 66, 1065 & 1095* (KRAM), *RILS 1937* (AAS, KRAM) and *REL 1230* (AAS, KRAM); Bonaparte Pt., *VK 45, 47, 1038 & 1165* (KRAM) and *Llano & Flint 3024* (NY); Joubin Is., *JRS & JMG PAL-18 & 24* (US); Dream I., *VK 577, 923, 935 & 991* (KRAM); Humbe I., 6 m, *RWMC 860* (AAS, KRAM); Litchfield I., *VK 73, 85, 500, 1076, 1096, 1108, 1123, 1135, 1147 & 1159* (KRAM) and *JRS & JMG PAL-30, 32, 33, 34 & 154* (US); Shortcut I., *VK 59, 1040, 1046, 1057 & 1109* (KRAM); Hermit I., *RILS 891A* (AAS, KRAM); Cormorant I., 10 m, *AS 59* (AAS, KRAM); Laggard I., 15 m, *REL 1298* (AAS, KRAM). **Graham Coast.** Edge Hill, 35 M, *RWMC 656A* (AAS, KRAM); Rasmussen I., 10 m, *AS 73* (AAS, KRAM); Cape Tuxen, 100 m, *Gain 194, 195A & 202a* (PC); Mt. Demaria, 400 m, *RILS 10661* (AAS, KRAM). **Wilhelm Archipelago.** ***Penola Strait***: Petermann I., 60 m, *AS 125A* (AAS, KRAM) and 70 m, *AS 117* (AAS, KRAM). ***Argentine Is.***: Uruguay I., *RWMC 816A* (AAS, KRAM); Corner Is., *JRS & JMG PAL-55, 94, 101 & 112* (US); Galindez I., *DWB 22& 23* (AAS, KRAM) and 15 m, *REL 1248* (AAS, KRAM). ***Collins Bay***: Green I., *BGLE 1074C* (BM) and 20–60 m, *RILS 3368B* (AAS, KRAM).

14. CAMPYLOPUS BRID.

Campylopus Brid., Musc. Recent. Suppl. 4: 71. 1819. — Lectotype (*vide* Pfeiffer 1873: p. 572): *Campylopus flexuosus* (Hedw.) Brid. (*Dicranum flexuosum* Hedw.).

Plants varying in size from very small and slender or filiform to large and robust, loosely to densely caespitose, soft or coarse and rigid, erect, glossy, green, olive-, yellowish-, whitish- to golden-green, brown, reddish- to blackish-brown or blackish. *Stems* erect, simple or variously forked or branched, sometimes by innovations, usually loosely or densely, evenly foliate, sometimes interruptedly comose or verticillate foliate or with a terminal comal tuft, usually reddish-, reddish-brown or sometimes whitish radiculose at the base or almost throughout, with a tomentum of smooth or occasionally papillose rhizoids, frequently arising from cells on the lower dorsal surface of the costa, variable in abundance, usually dense but sometimes scanty or absent, in cross-section with a 2–4-stratose cortex of small cells with variously thickened, brown cells surrounding a medulla of large, thin-to firm-walled cells and a small or large, distinct central strand, occasionally with a unistratose hyalodermis of large, thin-walled cells; *axillary hairs* filiform, hyaline, 5–10-celled. *Leaves* tightly appressed, erect, erect- to flexuose-spreading, straight to slightly curved, not or somewhat altered when dry, erect to erect-spreading when wet, broadly or narrowly lanceolate or ovate-lanceolate, usually gradually tapering to short or long, lance-acuminate to setaceous, channelled to subtubulose subula, usually denticulate towards the apex and with concolorous or hyaline tips, occasionally ovate, broadly acute and cucullate at the apex; *margin* more or less incurved throughout, entire below, variable serrate towards the apex, nearly always denticulate at the extreme tip; *costa* strong, broad, occupying one third or more of the leaf base width, occasionally narrower, entire or rarely spurred, percurrent to long-excurrent, sometimes as a long, hyaline, straight or recurved hair-point, filling most of the subula, smooth, denticulate to furrowed at the back distally, sometimes with high, multicellular lamellae, in cross-section varying in structure, either with dorsal and ventral stereid bands separated by a row of large guide cells, or with a row of large ventral hyalocysts and dorsal stereid band, occasionally interrupted by large cells, or sometimes with a dorsal group of substereids weakly differentiated from the ventral hyalocysts; *laminal cells* unistratose throughout, smooth, pellucid; *basal cells* short- to long-rectangular, thin-walled, lax and hyaline or thick-walled, porose or unpitted, often wider beside the costa than at the margin and forming a distinct border, gradually or abruptly passing into the upper cells, often higher up along the margins and making a distinct V-shaped group; *distal cells* variously shaped and sized, oval, quadrate, rhomboidal, elongate-rhomboidal, short- to long-rectangular, sometimes vermicular, mostly with incrassate, unpitted walls; *alar cells* mostly differentiated, enlarged, hyaline, forming distinct, inflated, hyaline or rusty-red auricles extending to the costa. *Asexual reproduction* occasional by anisophyllous brood leaves or small, microphyllous branches in the upper leaf axils or by rhizoidal gemmae. *Dioicous*. *Perichaetia* terminal, bud-like; *perichaetial leaves* widened at the base, suddenly constricted to a long subula. *Setae* elongate, cygneous, sinuose or rarely erect-flexuose when moist, smooth, yellow to brownish, dextrorse above or untwisted when dry. *Capsules* exserted, erect, inclined to horizontal or pendulous, ovoid to ellipsoid, straight and symmetric to curved and asymmetric with a slightly oblique mouth, brown, smooth or scabrous at the base, usually furrowed when dry, not or strumose; *annulus* compound and revoluble or simple and persistent; *operculum* conic-rostrate; *exothecial cells* rectangular, thick-walled, uniform or differentiated in corresponding longitudinal bands of alternating ribs and furrows; *stomata* absent; *peristome teeth* 16, erect, curved inwards when dry, lanceolate, split to the middle or a little over halfway, orange-brown and vertically striolate below, hyaline and papillose above. *Spores* spherical, yellow or brown, papillose. *Calyptra* cucullate, usually fringed at the base but with the fringes often caducous.

Campylopus is one of several genera including *Dicranodontium*, *Pilopogon*, *Bryohumbertia*, *Metzleria* and *Microcampylopus* belonging to the large complex characterised by 'perfect dicranoid' peristomes and very broad and excurrent costae which constitutes the separate subfamily Campylopoideae within the Dicranaceae (Frahm 1991*a*). Its generic limits are not always clearly and precisely defined and *Campylopus* intergrades with the aforementioned genera by peripheral groups of species which exhibit various combinations of characters. As presently understood, the characters common to all species of *Campylopus* are dioicy of the plants, cygneous or sinuous setae on wetting, broad and excurrent costae and the lack of stomata. However, these character states are rather generalised and are of questionable value in defining this genus. This is especially true for the sinuous setae which have evolved several times in the Campylopoideae (Frahm 1983, 1986).

Despite many intergradations with other genera, a major group of species of *Campylopus* is readily assigned to this genus by a combination of both gametophyte and sporophyte features. This group comprises moderate to large, dioicous mosses with rigid, erect to spreading linear-lanceolate leaves. They are mainly composed of a broad costa which occupies approximately ½–⅔ the leaf base width and exhibits extreme diversity of anatomical differentiation, unknown in any other moss genus. Apart from the typical dicranaceous costal anatomy with two outer stereid bands and a median band of guide cells, in many species the costa consists of a ventral layer of enlarged, hyaline cells (hyalocysts) and a well developed dorsal stereid band. The major diagnostic character of *Campylopus* is the seta which is sigmoidally curved or cygneous in the middle and twisted in the upper and lower parts and due to the special twist mechanism (Frey & Frahm 1987), the seta remains uncoiled when moistened.

Nonetheless, a similar state of the seta is known in all adjacent genera of the Campylopoideae.

The infrageneric classification of *Campylopus* is still poorly developed. Traditionally, the genus was subdivided into three subgenera, subgen. *Palinocraspis*, subgen. *Pseudocampylopus* and subgen. *Campylopus*, by Limpricht (1886) who used it for European taxa. It was further developed by Brotherus (1901*b*, 1924*b*), who expanded it for all species, although subsequent authors recognised additional subgenera, namely subgen. *Leucocampylopus* (Herzog 1916) and subgen. *Lucidi* (Bizot & Kilbertus 1979). This classification system has utilised the anatomical structure of the costa and it proved to be impractical and artificial because of the considerable structural plasticity of the costa, even within a single leaf and/or plant (Florschütz & Florschütz-de Waard 1974; Frahm 1982). Therefore, the system based on the sporophyte characters (Frahm 1983, 1991*b*; Vital 1984) in which *Campylopus* is subdivided into three subgenera, subgen. *Thysanomitrion*, subgen. *Campylopidulum* and subgen. *Campylopus* (with two sections, sect. *Campylopus* and sect. *Homalocarpus*), appears to be natural.

Campylopus is one of the largest genera of mosses and over 1000 species have been described in or transferred to this genus (Frahm 1987*a*, 1999). This number proved to be greatly exaggerated and, after many revisionary studies of taxa from all parts of the world, Frahm (1999 and cited references) markedly reduced it to about 150 species. The genus is abundantly represented in the subtropics and tropics where over 90% of all known species occur from lowland to alpine regions. In the austral cool-temperate and polar regions only 14 species have been detected (Frahm 1988). Yet, Frahm (1988, 1990*a*) postulates that *Campylopus* could have originated in the relatively moist and cool climatic conditions in the southern peripheries of the ancient Gondwanaland. From here, the ancestors of the tropical and the Northern Hemisphere species could have migrated northwards and spread throughout the tropical mountain systems, especially in the Neotropics where 65 species are known (Frahm 1991*b*) and Africa where 50 species have been recorded (Frahm 1985). Four species of *Campylopus* occur sparsely in areas with recent volcanic activity in the Antarctic and all belong to the type subspecies and type section of the genus.

1. Leaves ending in a spinose, hyaline hair-point; basal cells extending along the margins, forming a clearly demarcated, V-shaped group .... 1. ***Campylopus introflexus***
1. Leaves without hyaline hair-point; basal cells gradually passing into the upper cells, not extending along the margins 2
    2. Upper lamina cells rectangular; costa with a relatively weak dorsal substereid band; alar cells not or slightly differentiated; leaves usually to 2.5 mm long .... 2. ***Campylopus pyriformis***
    2. Upper laminal cells oval, rhombic, quadrate to short-rectangular; costa with a strong dorsal stereid band; alar cells distinct; leaves generally more than 3 mm long 3
3. Upper laminal cells oval; costa deeply furrowed distally....... 3. ***Campylopus vesticaulis***
3. Upper laminal cells rhombic, quadrate to short-rectangular; costa irregularly shallowly ribbed distally .... 4. ***Campylopus spiralis***

1. **Campylopus introflexus** (Hedw.) Brid.

Fig. 77, pl. 33

Campylopus introflexus (Hedw.) Brid., Mant. Musc.: 72. 1818 [1819]. ≡ *Dicranum introflexum* Hedw., Spec. Musc. Frond.: 147. 1801. — Type citation: Insularum meridionalium incola. [Lectotype (*selected here*): the second specimen from the left specimen on the sheet labelled 'Dicranum introflexum Spec. Musc. p. 147 t. 29. f. 1–7 Nova Holl. (ver. Hedwig)' – G-Hedwig/Schwägrichen!].

Campylopus canescens (Müll.Hal.) Schimp. *ex* A.Jaeger, Ber. Thätigk. S. Gallischen Naturwiss. Ges. 1870–71: 435. 1872. ≡ *Dicranum canescens* Müll.Hal., Bot. Zeit. (Berlin) 17: 220. 1859. — Type citation: Insulae Maclovianae, in ins. orient. monte Mt. William, 800 ped. alt. inter Cyperaceas, Sept. 1850: W. Lechler. [Lectotype (*selected here*): 'W. Lechler pl. ins. Maclovian. Ed. M. F. Hohenacker 90. Campylopus canescens W. P. Schimper ad ins. Maclov. orient. sinum Port William Stanley Sept. m.' – BM!; isotypes: JE!, LE!]. First synonymised by Frahm (1978: p. 847).

Plants moderate to occasionally large, slender to relatively robust, in wide, rigid, sometimes deep turves, weakly lustrous, green, yellow-, olive- to brownish-green above, brown to blackish-brown below. *Stems* simple or branched, 1–5, occasionally to 11 cm tall, usually appearing layered with the swollen gametangial nodes and slender internodes because of the repeated annual growth by subfloral innovations stretching in the same direction as the main axis, erect, evenly foliate, with comose terminal tufts of perichaetial leaves on fertile shoots, rather sparsely reddish to reddish-brown tomentose with long, branched, smooth, glossy rhizoids, in transverse section round, with a large central strand surrounded by 3–5 layers of firm-walled, brownish medullary cells and a 1–3-layered cortex of smaller, substereidal, red-brown cells. *Leaves* tightly appressed, not or slightly altered when dry, erecto-patent when wet, straight to slightly curved, 2.5–5.5 mm long, 0.7–1.2 mm wide, elliptical, oblong-lanceolate or lanceolate, gradually or rather abruptly long acuminate, concave below, subtubulose above; *margin* plane, incurved, entire; *costa* broad, ⅓–⅗ of the leaf base width, usually long excurrent as a hyaline, to 2 mm long, dentate, frequently reflexed hair-point, usually at right angles, or less often percurrent, apex with a short hyaline tip or without it, on the dorsal, distal surface with lamellae 1(–2) cells high, in proximal section with 25–30 guide cells, large, weakly thickened ventral cells in a single layer and a strong dorsal band of stereid cells or stereid groups of 3–4 cells separated by larger substereid cells, in distal section ventral and dorsal cells similar, generally substereids throughout; *laminal cells* unistratose throughout; *basal cells* rectangular to linear rectangular, becoming oblong-hexagonal above, 40–120 µm long, 8–16 µm wide, becoming narrower towards the margin, lax, thin-walled, hyaline, rarely slightly chlorophyllose, extending higher up along the margins and forming a V-shaped, mostly clearly demarcated group; *upper cells* oval, rhombic to rhomboidal, 6–8 µm wide, 6–16 µm long, thick-walled; *alar cells* thin-walled, hyaline to dark red-brown, not or scarcely differentiated, forming convex auricles, varying in size and extending to the costa. *Dioicous. Setae* 7–11 mm long, yellow-brown, erect, cygneous to sigmoid, smooth, lustrous. *Capsules* erect, horizontal to cernuous, ellipsoid to cylindrical-ellipsoid, 1.0–1.3 mm long, slightly asymmetric, brown-blackish, with a varnish lustre, smooth above, somewhat scabrous in the neck, strongly ribbed, estomatose, not strumose; *annulus* biseriate, compound, deciduous; *operculum* conic-rostrate, oblique; *exothecial cells* rectangular, 40–60 µm long, 12–16 µm wide, with thin and porose walls along the furrows and with unpitted and strongly incrassate longitudinal walls over the ribs, becoming short-rectangular to isodiametric, thick-walled in 1–2 rows at the mouth; *peristome teeth* 16, lanceolate, 480–500 µm long, 50–60 µm wide at base, arising from a low basal membrane, divided to about the middle into 2, often unequal prongs, orange-brown and regularly vertically striolate below, hyaline and spiculose above. *Spores* spherical, pale brownish, 12–13 µm wide, minutely roughened. *Calyptra* cucullate, smooth, naked, brown below, blackish at the tip, fringed at the base.

Chromosome number: unknown in the Antarctic and Subantarctic; $n = 12$ from Poland (Ochyra & Kuta 1990) and Australia (Ramsay 1974).

Discussion. *Campylopus introflexus* is a distinct species which is unlikely to be confused with any other species in the Antarctic. When typically developed, it is recognised in the field by its long, hyaline hair-points which are usually strongly recurved when dry, often at right angle or more. However, the species is very variable and quite often the leaves are either muticous or have a very short hyaline tip. Such plants are identified microscopically by the combination of oval upper laminal cells and distinctly lamellose costa on the dorsal surface, down to the leaf shoulders. The upper and basal laminal cells are usually sharply demarcated along the V-shaped transition because of the extension of the basal hyaline cells upwards along the margins. Some specimens from the South Sandwich Is. were determined as *Campylopus canescens* (Longton & Holdgate 1979) which species is considered to be conspecific with *C. introflexus* (Frahm 1978).

In the leaf areolation *Campylopus introflexus* resembles some populations of *C. incrassatus*, especially the Chilean type of this species (Müller 1845; Frahm 1975). Some specimens from Candlemas I. (*REL 622*, *706*, *712B*, AAS, KRAM) and Leskov I. (*REL 520*, AAS, KRAM) have been determined as *C. flavoviridis* (Longton & Holdgate 1979; Longton 1979), which is conspecific with *C. incrassatus* (Frahm 1985, 1987*a*). Some of these specimens (e.g. *REL 622*, AAS, KRAM) are in fine fruiting condition and their asymmetric capsules unquestionably indicate *C. introflexus* along with other gametophyte characters, whereas *C. incrassatus* has symmetric capsules which clearly place it in sect. *Holamocarpus*. The latter has narrowly lanceolate and long-setaceous leaves, ending with a concolorous or short, hyaline hair-point, which is always straight, never recurved and, additionally, smooth to weakly ridged costa distally, and subquadrate, rhombic to short-rectangular laminal cells at the shoulders (Frahm 1976, 1987*b*, 1994*a*; Bartlett & Frahm 1983; Catcheside & Frahm 1985). One Antarctic population (*REL 520*, AAS, KRAM) comprises very tall, luxuriant plants which resemble *C. incrassatus* in habit and, moreover, some leaves have weakly lamellose to shallowly ridged costa on the dorsal surface, but this feature is subject to considerable variation in this species (Giacomini 1955; Frahm 1974). Nonetheless, the very long, recurved to recurved-squarrose hair-points as well as the oblong-lanceolate leaves with the clearly oval laminal cells at the shoulders are all typical of *C. introflexus*.

Campylopus introflexus was originally described by Hedwig (1801) from '*insulae meridionales*'. Although the label on the type collection indicates 'Nova Hollandia' [Australia] as the *locus classicus* of this species, the original material, as pointed out by Touw (1971), most probably was collected in New Zealand by J. Banks during Cook's first voyage round the world of 1768–1771. The original material of this species in the Hedwig herbarium consists of seven specimens permanently attached to the sheet in a single row. Although the label indicates that the specimens come from Australia, the identity of particular shoots clearly indicate that the material could not originate from this region because four shoots (No. 1, 3, 5 & 7) represent *Campylopus aureonitens* (Müll.Hal.) A.Jaeger, and one shoot (No. 4) is *C. nivalis* (Brid.) Brid. These species do not occur in Australasia but are known to occur in East and South Africa and Indian Ocean islands off East Africa. If it is accepted that these shoots were collected in one place, the only 'austral island' where they grow together is Réunion and, in all likelihood, the original material was collected on this island. The shoot of *C. introflexus* exhibits all typical features of the species.

Reproduction in Antarctica. Mostly sterile but fully mature sporophytes are occasionally produced.

Habitat. Antarctic populations of *Campylopus introflexus* are mostly associated with fumarole vegetation where they grow on moist, warmed ground, ash and lapilli among lava blocks around active vents emitting steam and sulphur. The dense, luxuriant turves of this species often form an extensive zone around these vents and completely cover large areas, except on the steep sides

Figure 77. *Campylopus introflexus*. — **1–2**. Plants with sporophytes. **3**. Sterile plant, dry. **4**. Portion of shoot, dry. **5**. Portion of cross-section of stem. **6–12**. Leaves. **13–14**. Leaf apices. **15–16**. Upper lamina cells at margin. **17**. Mid-leaf cells. **18**. Marginal cells in middle of leaf. **19**. Lamina cells in transitional zone between basal and upper parts of lamina. **20**. Alar cells. **21–26**. Cross-sections of leaves, a sequence from base to apex. **27**. Plant with perigonium. **28**. Deoperculate capsule, dry. **29**. Exothecial cells in middle of urn. **30**. Peristome teeth, spores and exothecial cells at mouth. — [1–2, 5, 13, 15–17, 21, 23–24, 28–30 from *REL 706*; 3, 11–12, 14, 18, 22, 25–26 from *REL 626*; 4, 6–8, 20 from *REL 513*; 9–10 from *MWH 822A*; 19 from *REL 622*; 27 from *REL 712B*]. — Scale bars: a – 1 mm (2–4); b – 1 mm (27); c – 0.5 cm (1); d – 1 mm (28) and 2 mm (6–12); e – 100 µm (5, 15–26) and 200 µm (13–14, 29–30).

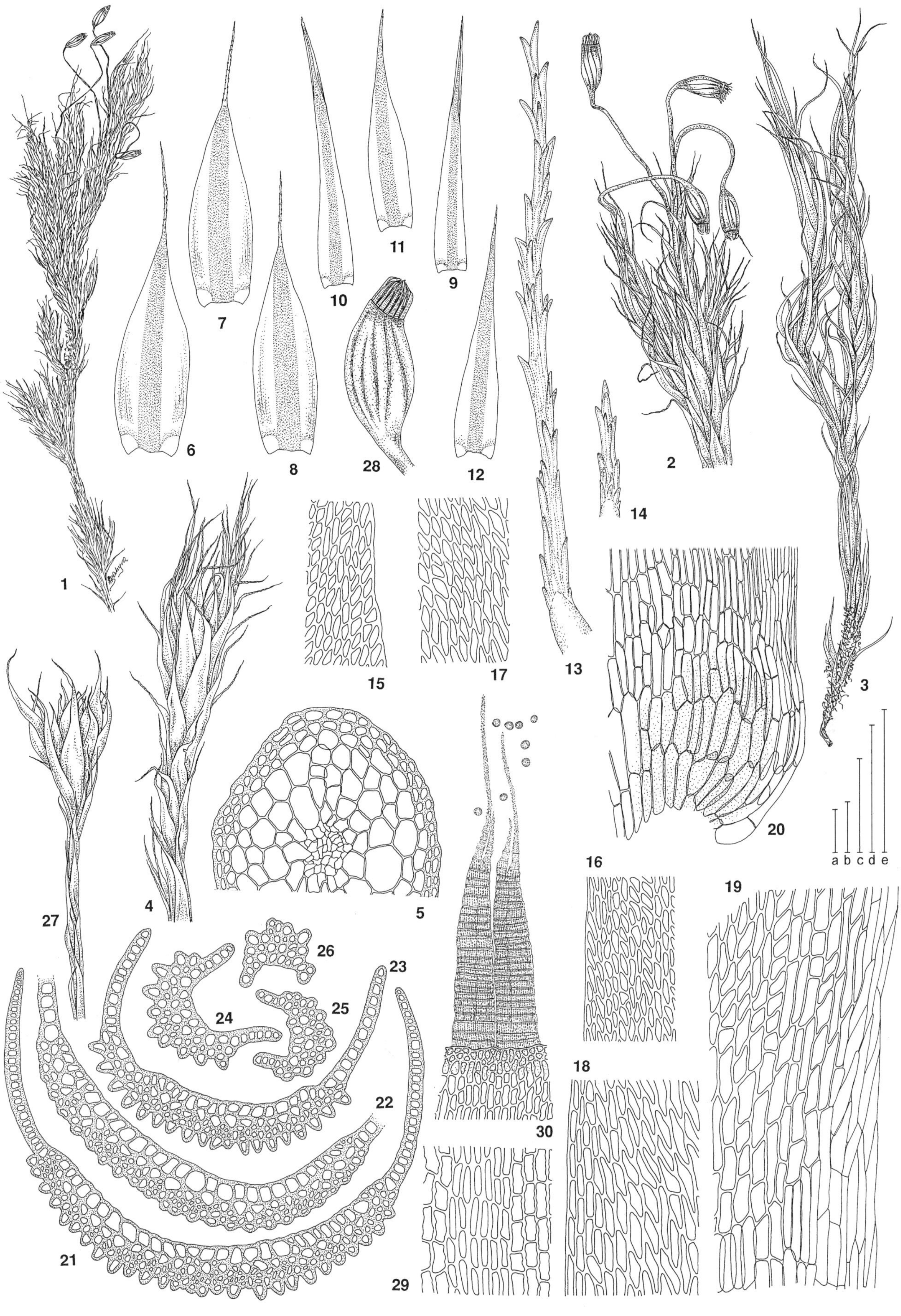
1
2
3
4
5
6
7
8
9
10
11
12
13
14
15
16
17
18
19
20
21
22
23
24
25
26
27
28
29
30
a b c d e

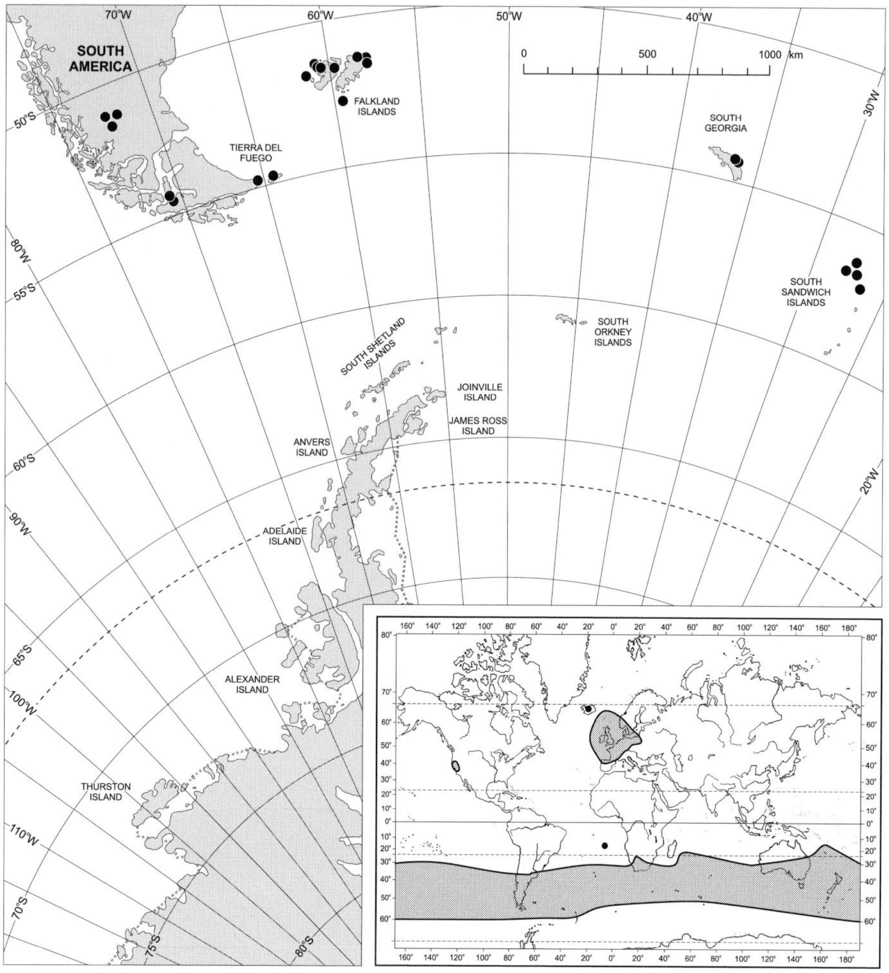

Figure 78. Distribution of *Campylopus introflexus* in the Fuegian region and Antarctica. Inset: global distribution.

of projecting lava boulders (Longton & Holdgate 1979). Its most frequent associates are *Cephalozia badia*, *Cephaloziella varians*, *Ceratodon purpureus*, *Cryptochila grandiflora*, *Ditrichum ditrichoideum*, *Marchantia berteroana*, *Pohlia nutans* and *Polytrichastrum alpinum*.

World range. A bipolar species indigenous to the Southern Hemisphere where it has a pan-temperate range (Fig. 78, inset). It covers New Zealand and southern Australia with Tasmania, southern Africa and southern South America (Gradstein & Sipman 1978; Ochyra 1983; Frahm 1988), extending north to New Caledonia (Gradstein & Sipman 1978; Frahm 1990*b*), Réunion and St. Helena (Frahm 1985). It has been recorded from almost all islands in the Southern Ocean including Campbell I. (Vitt 1974*a*), Auckland Is. (Vitt 1979*a*), Île Amsterdam (Tixier 1980), Îles Kerguelen (Frahm 1985), Îles Crozet (Hebrard 1970*a*), Prince Edward Is. (Zanten 1971), Tristan da Cunha (Dixon 1960), Gough I. (Wace & Dickson 1965), South Georgia (Ochyra *et al.* 2002) and the South Sandwich Is. (Longton & Holdgate 1979; Convey & Lewis Smith 2006). In 1941 it was discovered in England and in 1942 in Ireland, and since then it has spread throughout the British Isles (Richards 1963; Richards & Smith 1975; Corley 1992). In the early 1950s it invaded mainland Europe and it has subsequently spread eastwards (Frahm 1972; Ochyra 1983). At present its easternmost localities are in the eastern Czech Republic (Soldan 1996; Mikulášková 2006), western Poland (Fudali 1992), the Polish Carpathians (Stebel 1995) and Kotlina Sandomierska in eastern Poland (Fojcik & Gumieniak 1999) as well as in Kaliningrad Province in Russia (Razgulyaeva *et al.* 2001). Moreover, it is known from southern Norway and Sweden (Söderström 1996) and northern Iceland (Jóhannsson 1991) as well as from California in western North America (Frahm 1980; Showers 1982; Shevock & Toren 2001; Norris & Shevock 2004).

Distribution in Antarctica. The species is known only from the northern island group of the South Sandwich Is. archipelago (Fig. 78). It occurs sparsely on Zavodovski, Leskov and Visokoi Is. (known collectively as Traversay Is.), and is relatively frequent and abundant on Candlemas I. (forming with Vindication I. the group of Candlemas Is.) where it reaches its southernmost locality in the West Antarctic at lat. 57°03′S. It occurs from near sea level to the summits of the highest volcanic cones at an altitude of about 225 m. The reports of *Campylopus introflexus* from Deception I. in the South Shetland Is. (Lewis Smith 1984*b*,*c*, 1988*d*) are based upon misidentification.

Selected specimens examined. WEST ANTARCTICA. SOUTH SANDWICH IS. **ZAVODOVSKI I.** Central N ash plain, *PC 259B & 260B* (AAS, KRAM). **LESKOV I.** N end, *ca* 130 m, *REL 518B & 520* (AAS, KRAM); summit ridge, 100 m, *PC 246F* (AAS, KRAM), *ca* 200 m, *REL 508B, 513 & 514* (AAS, KRAM) and *ca* 225 m, *MWH 833B* (AAS, KRAM). **VISOKOI I.** Finger Pt., *ca* 30 m, *REL 757* (AAS, KRAM). **CANDLEMAS I.** N of Sea Serpent Cove, *KA 4* (AAS, KRAM); eastern lagoon, 50 m, *REL 779* (AAS, KRAM); western lagoon, 50 m, *REL 706 & 712B* (AAS, KRAM); between lagoons, 65–100 m, *REL 568* (AAS, KRAM); W shore, *ca* 30 m, *REL 622, 626 & 631* (AAS, KRAM); Lucifer Hill, *PC 129C, 140, 142B, 145A, 227A, 230D, 231A, 234A* (AAS, KRAM); Clinker Gulch, *PC 152, 156B & 238B* (AAS, KRAM).

2. **Campylopus pyriformis** (Schultz) Brid.

Fig. 79

Campylopus pyriformis (Schultz) Brid., Bryol. Univ. 1: 469. 1826. ≡ *Dicranum pyriforme* Schultz, Prodr. Fl. Starg. Suppl. 73. 1819. ≡ *Campylopus flexuosus* (Hedw.) Brid. subsp. *pyriformis* (Schultz) Dixon *in* Dixon & Jameson, Stud. Handb. Brit. Moss.: 95. 1896. ≡ *C. fragilis* (Brid.) Bruch & Schimp. var. *pyriformis* (Schultz) Agst., Ned. Kruidk. Arch. 57: 332. 1950. — TYPE: Ad pagum Ballin in dem Holz hinter dem Hofgarten. Aug. [Lectotype (*selected here*): 'Dicranum pyriforme S Fl. Starg. Schultz 1819' – B-Bridel!].

Plants small, to 5 mm tall, in dense, weakly glossy, yellowish-green tufts. *Stems* erect, to 3 mm long, simple or branched by innovation, not comose foliate, densely tomented at the base with reddish to colourless, branched, glossy rhizoids, in cross-section rounded, with a very weak central strand, a 3–4-stratose medulla of large, hyaline to yellowish-brown cells and 1–2-layered cortex of relatively large, thicker- and brown-walled cells; *axillary hairs* filiform, hyaline throughout, 6–7-celled. *Leaves* erecto-patent when moist, becoming more appressed on drying, often homomallous and flexuose at the stem tips, straight to somewhat falcate, (1.5–)1.8–2.5(–2.7) mm long, 0.25–0.45 mm wide, from an ovate, elliptic to ovate-lanceolate, concave base, widest at or very near the insertion, gradually or rather abruptly contracted to a long, slender, channelled subula, 3 or more times as long as the base; *margin* plane, entire below, weakly serrate towards the apex; *costa* broad, occupying ⅓–⅗ of the leaf base width and a major part of the subula, percurrent to short excurrent, never hyaline-tipped, grooved or scabrid dorsally, in cross-section with a row of ventral hyalocysts as large as or larger than and about equal in number to median guide cells and a dorsal band of substereid and stereid cells, occupying ⅓–½ of the costa thickness; *laminal cells* rectangular, sometimes intemixed with short, subquadrate to irregular cells, moderately thick-walled above, (10–)16–40(–45) μm long, 3–8 μm wide, becoming rectangular to oblong-hexagonal, (20–)30–60 μm long, 10–15 μm wide, usually narrower towards the margins, thin-walled, hyaline and usually lax; *angular cells* variable, usually undifferentiated, hyaline or pale brown, occasionally forming more distinct auricles reaching the costa. *Asexual reproduction* by ovoid, cylindrical to vermiform, brownish, multicellular tubers, 100–600 μm long, 100–150 μm wide, occurring at the end of large main rhizoid branches. *Dioicous*. *Male plants* with bud-like perigonia, consisting of strongly concave, shortly pointed innermost bracts. [*Setae* erect, sigmoid to cygneous, yellow to yellow-brownish, 6–9 mm long; *Capsules* ellipsoid, nearly symmetric, erect to slightly inclined, olive-green to light brown, glossy, distinctly ribbed, estomatose; *annulus* compound, deciduous; *operculum* conic-rostrate, brown, darker than the urn, with a slanted, long rostrum; *exothecial cells* rectangular, in alternating thin- and thick-walled bands in furrows and over ribs, respectively; *peristome teeth* 16, lanceolate, 240–350 μm long, divided somewhat more halfway to the base, reddish-brown and vertically striolate below, yellowish-hyaline and papillose above. *Spores* globose, olive, finely papillose, 8–16 μm wide. *Calyptra* cucullate, brownish, fringed or not at the base.]

Chromosome number: unknown in the Antarctic and Subantarctic; $n = 11$ (2 records) from India and the British Isles (Fritsch 1991).

Discussion. With the exception of *Dicranella*, *Campylopus pyriformis* is unlikely to be confused with any other species of the genus in Antarctica. The leaves are very narrow, long-subulate and distinctly canaliculate at the apex and indeed this species may resemble species of *Dicranella*, especially some phenotypes of *D. hookeri* with a somewhat broadened and diffuse costa, from the South Sandwich Is. However, the very broad costa with characteristic large ventral hyalocysts and dorsal stereids and/or substereids immediately excludes any alliance with this genus. Also, the leaf areolation of the rectangular cells at the shoulders is unique amongst the Antarctic representatives of *Campylopus*. A valuable diagnostic feature of *C. pyriformis* is the rhizoidal gemmae. They are large, pale brownish, multicellular, ovoid, long-cylindrical to vermiform tubers arising on the main rhizoid branches (Arts 1986, 1989, 1993; Risse 1988).

Although *Campylopus pyriformis* is a distinct and sound species, it may pose problems with identification due to the high degree of variability, mostly depending on habitat, shown by many of its populations from various parts of its wide geographical range (Corley 1976). The species varies primarily in the size of the plants which may attain only a few millimetres and are nearly stemless, consisting of only a comal tuft, as is the case with the Antarctic plants, or are up to 3 cm high, for example on peaty soil on the Falkland Is. (Ochyra 2003*a*). As a result, numerous phenotypes of this species were given taxonomic recognition as species in their own right and *C. pyriformis* is accompanied by nearly 50 heterotypic synonyms (Corley & Frahm 1982; Frahm 1987*a*, 1994*b*).

Reproduction in Antarctica. Sterile, but male plants with perigonia have been recorded. Antarctic plants produce copious multicellular rhizoidal gemmae (Skotnicki *et al.* 2001) which represent the type 1 as described by Arts (1986).

Habitat. Growing exclusively on geothermally heated fine ash soil, or beneath snow hummocks formed by rime ice over fumaroles.

World range. *Campylopus pyriformis* is a bipolar temperate species (Fig. 80). For a long time it was considered to be a European endemic species (Frahm 1984), but the conspecificity of this species with many austral species (Corley & Frahm 1982; Frahm 1987*a*) extended its range to the austral region. In the Southern Hemisphere, *C. pyriformis* has a pan-temperate range, including South America where it extends to the Juan Fernandez Is. and central Chile (He 1998) and Rio de Janeiro in south-east Brazil (Frahm 1991*b*), the Falkland Is. (Ochyra 2003*a*), Tristan da Cunha, Gough I., southern Africa, Île Amsterdam, Île Saint-Paul (Frahm 1985) and Réunion (Risse 1988) in the Southern Indian Ocean, as well as Tasmania, southern and eastern Australia north to Queensland (Frahm 1994*a*), New Caledonia (Frahm 1990*b*) and New Zealand (Bartlett & Frahm 1983), south to Campbell I. (Vitt 1974*a*) and Auckland Is. (Vitt 1979*a*) in the south cool-temperate zone, Marion I. in the Subantarctic (Frahm 1985; Ochyra, personal observations) and Continental Antarctica from where it was reported by Broady *et al.* (1987), Meurk (1989), Given *et al.* (1991) and Linskens *et al.* (1991). Additionally, it is scattered in the East African mountains to Malawi (Frahm 1994*b*). In the

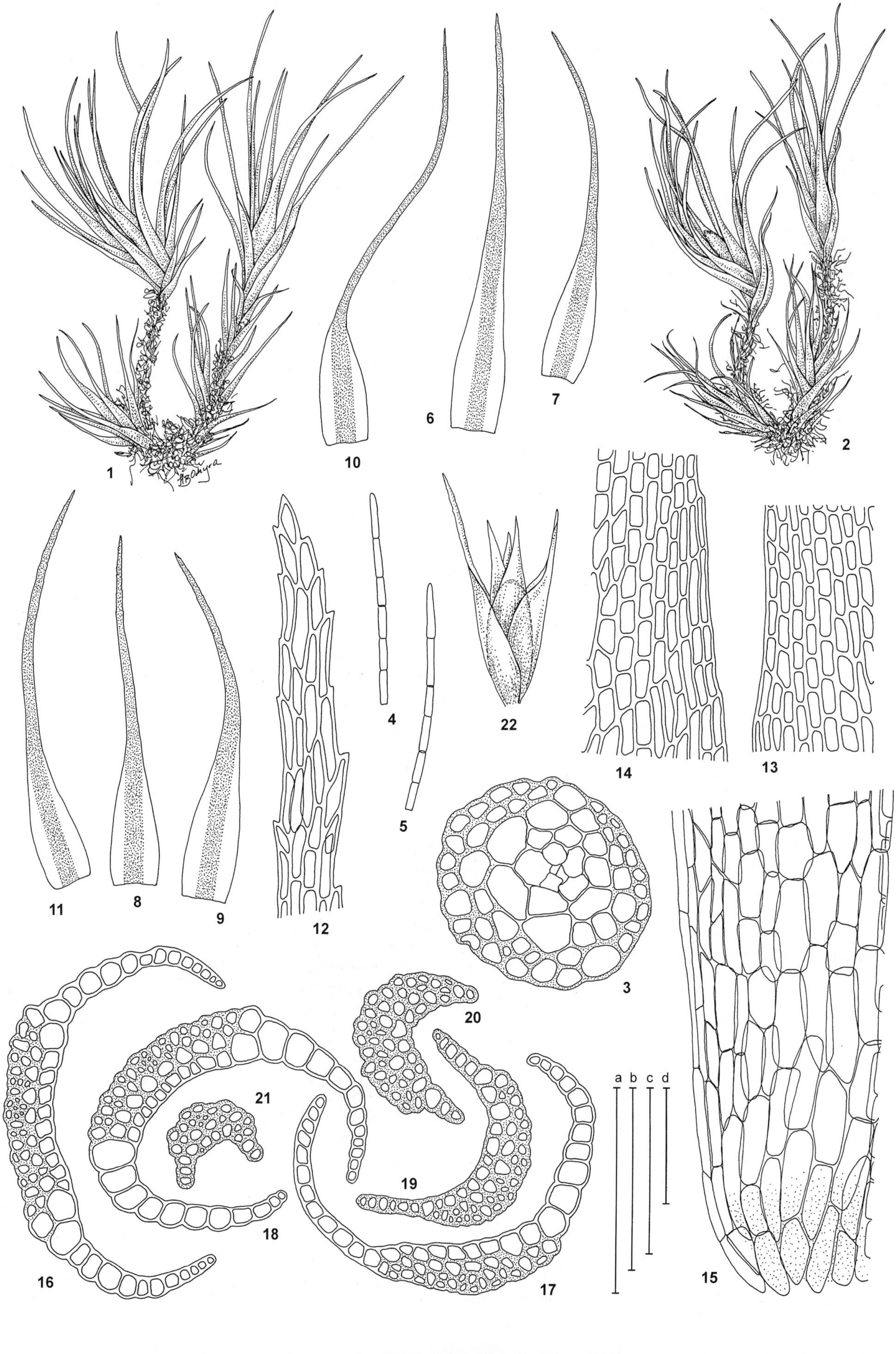
1
2
3
4
5
6
7
8
9
10
11
12
13
14
15
16
17
18
19
20
21
22
a
b
c
d

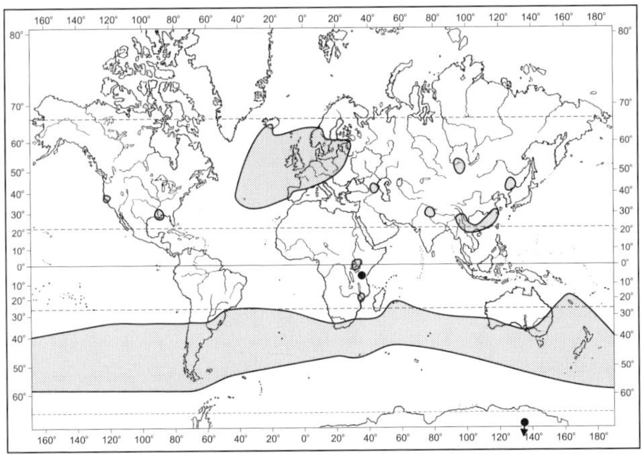

Figure 80. Global distribution of *Campylopus pyriformis*. Antarctic occurrence in Victoria Land is indicated by the arrow.

Northern Hemisphere it was long known only from Western and Central Europe (Frahm 1984), extending to Iceland (Jóhannsson 1991) and southern Scandinavia (Söderström 1996) in the north, to Estonia (Ingerpuu *et al.* 1994), north-eastern Poland (Szafran 1957) and Romania (Mohan 1998) in the east and to northern Italy (Cortini Pedrotti 2001*a*) in the south. Recently, it has been recorded from the Caucasus, Central Siberia and Russian Far East (Ignatov *et al.* 2006), Mongolia (Tsegmed 2001*a,b*), northern India (Gangulee 1971) and China (Frahm 1992; Gao *et al.* 1999) in Asia and also from North America including Louisiana, Mississippi and Florida (Arts & Frahm 1990) and California (Shevock & Toren 2001). The distribution pattern of *C. pyriformis* shows many analogies with that of *C. introflexus* and possibly it is also an adventive species which could have been introduced to Europe in the early years of the colonial expansion in the 16th or 17th centuries. This is especially likely because tuber-bearing mosses may be readily distributed by human activity (Crundwell & Whitehouse 1978).

Distribution in Antarctica. Known only from a single locality at an elevation of about 2700 m on the summit of Mt. Melbourne in Victoria Land at lat. 74°21′S. In addition, an initially unidentified moss protonema reported from Mt. Erebus on Ross I. (Lyon & Giggenbach 1974) was determined as belonging to this species by the use of the RAPD technique (Skotnicki *et al.* 2001) (Fig. 80).

Specimens examined. EAST ANTARCTICA. VICTORIA LAND. BORCHGREVINK COAST. Mt. Melbourne, Cryptogam Ridge immediately S of the main crater, 2700 m, *Given 13747* (AAS, CHR, KRAM) and *DWHW 301* (AAS, KRAM).

3. **Campylopus vesticaulis** Mitt. Fig. 81

Campylopus vesticaulis Mitt. *in* Melliss, St. Helena: 359. 1875. — TYPE CITATION: Tristan d'Acunha, Milne and M'Gillivray. [Lectotype (*selected here*): 'Tristan d'Acunha Milne Denham's Voyage' – NY-Mitten!; isolectotype: BM-Hooker!, NY-Mitten! (2 specimens); syntype: 'Tristan d'Acunha Mc Gillivray' – NY-Mitt!; isosyntype: BM-Hooker!].

Plants medium-sized, rather robust, occasionally slender, densely tufted, often buried in soil, weakly lustrous, light- or yellowish-green above, brownish to blackish-brown below. *Stems* erect, mostly branched, 0.8–2.5 cm tall, densely, evenly foliate, sparsely radiculose with reddish-brown to colourless, branched rhizoids, in cross-section round, with a medium and distinct central strand surrounded by a medulla of large, thick-walled hyaline-yellowish cells in 5–6 layers and an outer 1–2-stratose cortex of smaller cells with brown, thick walls; *axillary hairs* filiform, 5–6-celled with an intensively brown basal cell and hyaline upper cells. *Leaves* erect-spreading when wet, more appressed and not or slightly altered when dry, rigid, straight or somewhat curved, 4–5 mm long, 0.6–0.8 mm wide, narrowly lanceolate, gradually tapering to a relatively short, canaliculate subula, concolorous and distinctly toothed at the extreme apex; *margin* plane, incurved, entire except for the tip of the subula; *costa* proportionately broad, more than half of the leaf base width, 350–420 µm wide, filling a major part of the subula, percurrent to excurrent as a short, concolorous, toothed awn, in cross-section with a ventral layer of large, thin-to moderately thick-walled hyalocysts, overlying the central row of large guide cells, and a dorsal band of small stereid cells alternating with larger substereid cells protruding distally as low ridges; *laminal cells* oval, rhombic, oblong-rhomboidal, trapezoid to fusiform above, 14–35 µm long, 5–10 µm wide, thick-walled, transparent, becoming rectangular, 30–90 µm long, (6–)9–14(–18) µm wide, thin- to firm-walled below, distinctly narrowed at the margins; *alar cells* enlarged, thin-walled, forming hyaline to hyaline-brownish dilated, somewhat convex auricles. [*Dioicous*. *Setae* erect, sigmoid to cygneous, yellow to blackish-brown with age, smooth, dextrorse above. *Capsules* oblong-ellipsoid to cylindrical, 1.2–1.5 mm long, gradually narrowed towards the neck, constricted above and strongly furrowed when dry, glossy, light brown; *exothecial cells* rectangular in regular vertical rows, 35–50 µm long, 14–20 µm wide, thin-walled and unpitted over grooves to thick-walled and porose over ribs, markedly thick-walled near the orifice; *peristome teeth* lanceolate, split halfway down the base, 400–500 µm long, dark orange-brown and densely vertically striolate below, yellowish-hyaline and papillose above. *Spores* globose, yellow-brownish, 15–16 µm wide, papillose. *Operculum* and *calyptra* unknown.]

Chromosome number: unknown.

Discussion. *Campylopus vesticaulis* is a distinct species with its thin-walled basal laminal cells and oval to fusiform, thick-walled upper leaf cells coupled with a conspicuous dorsal stereid band and lack of a hyaline awn. It may be confused with *C. spiralis* with which grows around fumaroles, but this species

←

Figure 79. *Campylopus pyriformis*. — **1**. Habit, wet. **2**. Habit, dry. **3**. Cross-section of stem. **4–5**. Axillary hairs. **6–11**. Leaves. **12**. Leaf apex. **13**. Upper laminal cells at margin. **14**. Marginal cells in middle of leaf at margin. **15**. Alar cells. **16–21**. Cross-sections of leaves, a sequence from base to apex. **22**. Perigonium. — [1–6, 8–10, 12–22 from *DWHW 3001*; 7, 11 from *Given 13747*]. — Scale bars: a – 100 µm (3–5, 12–21); b – 1 mm (22); c – 1 mm (6–11); d – 1 mm (1–2).

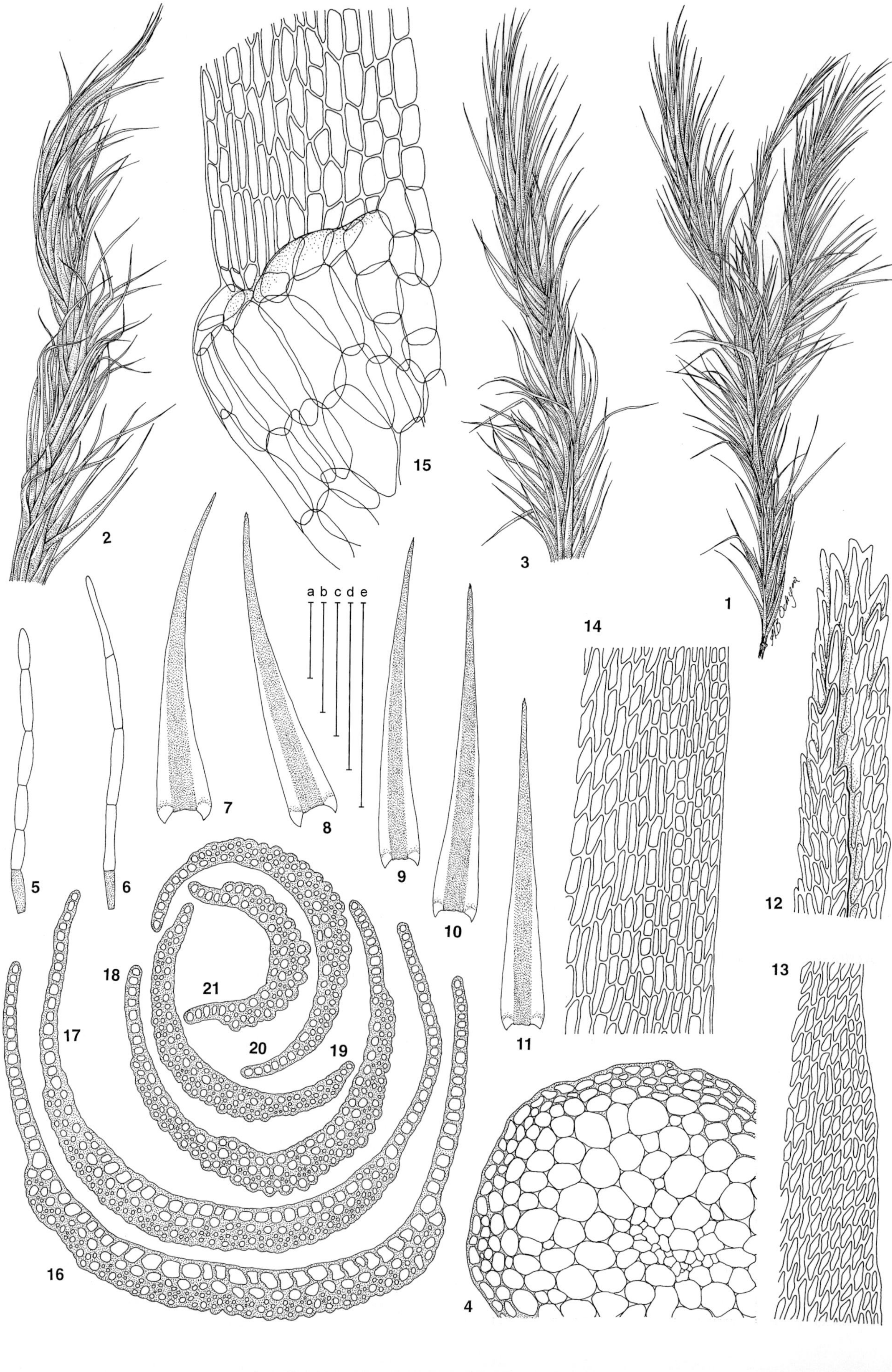
1
2
3
4
5
6
7
8
9
10
11
12
13
14
15
16
17
18
19
20
21
a b c d e

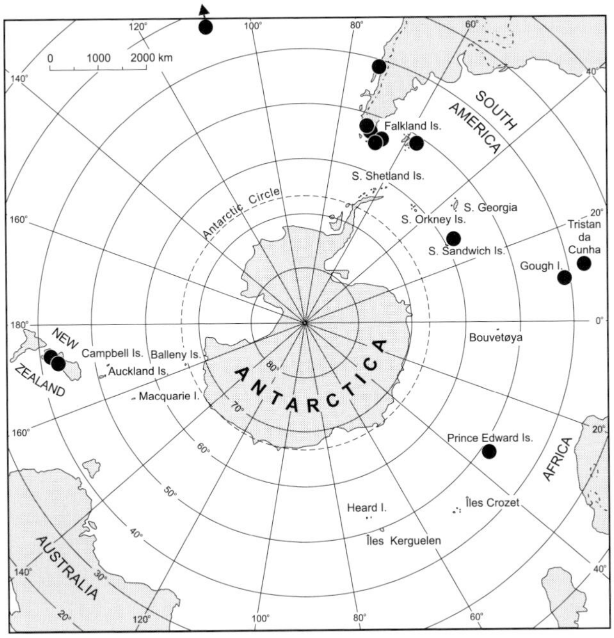

Figure 82. Global distribution of *Campylopus vesticaulis*. Occurrence on Easter I. is indicated by the arrow.

has distinctly short-rectangular to subquadrate, never oval upper laminal cells. The Antarctic plants are remarkably uniform morphologically and have well developed alar cells which are, however, subject to considerable variation in this species. For example, all collections from Easter I., which were described as four separate species (Brotherus 1924*c*), have prominent, swollen auricles, whereas the type material from Tristan da Cunha has weak alar cells and in the type material of Patagonian *C. guaitecae* (Dusén 1905*b*) this character varies from auriculate to exauriculate states.

When describing *Campylopus vesticaulis*, Mitten (1875) cited in the protologue two specimens from Tristan da Cunha collected by Milne and McGillivray, but neither of these was indicated as the holotype. Frahm (1999) indicated the specimen collected by Milne and preserved in the Mitten herbarium at NY as holotype, but this typification was apparently incorrect since lectotypification is here necessary. In the Mitten herbarium are preserved three duplicates of Milne's collection (Slavick 1992) and one of these, being in fine fruiting condition, is here chosen as lectotype.

Reproduction in Antarctica. Sterile.

Habitat. On heated ground and damp rock surfaces near fumaroles, usually around active steam vents and vigorous sulphur-emitting fumaroles, growing in pure tufts or sometimes mixed with other mosses, of which the most frequent associates are *Ceratodon purpureus*, *Cryptochila grandiflora*, *Dicranella hookeri*, *Pohlia nutans* and *Sanionia georgicouncinata*.

World range. A rare and widely scattered pan-south-temperate species (Fig. 82), known from western Patagonia, Tierra del Fuego and the Falkland Is. (Greene 1986; Frahm 1976, 1988), then recurring in the Tristan da Cunha group (Frahm 1985) from where it was originally described (Mitten 1875), Subantarctic Marion I. (Ochyra & Hertel 1990) and South Island of New Zealand, with an isolated occurrence on Easter I. in the southern Pacific Ocean, and extending southwards to the South Sandwich Is. in the maritime Antarctic.

Distribution in Antarctica. Known from a single collection from Candlemas I. but frequently found on the small volcanic Bellingshausen I. in the Southern Thule group of the South Sandwich Is. at lat. 56°26′S (Fig. 82), occurring from about 50 m to the summit of the volcanic cone at *ca* 200 m. *Campylopus vesticaulis* seems to have increased over the island in the last four decades because during the former survey of Bellingshausen I. in the early 1960s (Longton & Holdgate 1979) only two specimens were collected, whereas in the 1997 survey (Convey *et al.* 2001) the species was collected as a frequent component of the fumarole vegetation.

Specimens examined. WEST ANTARCTICA. SOUTH SANDWICH IS. **CANDLEMAS I.** Clinker Gulch, *PC 165B* (AAS, KRAM). **BELLINGSHAUSEN I.** S slopes of cone, *ca* 75 m, *MWH 830D* (AAS, KRAM), 130 m, *PC 63B & 65C* (AAS, KRAM); S crater lip access point, 150 m, *PC 46E, 47, 52B, 54B & 55* (AAS, KRAM); W of S access to crater lip, 150 m, *PC 39E, 41D, 56C, 57D & 58A* (AAS, KRAM); SE edge of main crater, 150 m, *PC 22* (AAS, KRAM); S wall of crater, *ca* 200 m, *MWH 810A* (AAS, KRAM); at floor of crater, 50 m, *PC 69 & 70A* (AAS, KRAM).

4. **Campylopus spiralis** Dusén Fig. 83

Campylopus spiralis Dusén, Ark. Bot. 4(13): 4, *pl. 3, f. 7–16 & pl. 3, f. 1–2.* 1905. — TYPE CITATION: Patagonia australis ad oppidum Punta Arenas in fruticetis, ubi speciem fructibus parum evolutis praeditis mense Majo legi; Patagonia occidentalis in insulis Guiatecas in terra aperta, subgraminosa, uni species et fructibus parum evolutis et fructibus nonnulis fere maturis ornata mense Majo inventa fuit. [Lectotype (*selected here*): '617. Campylopus spiralis Dus. Patagonia occ. in insulis Guaiatecas in terra Majo 5 a. 1897 P. Dusén' – BM!; isolectotype: NY! (2 specimens); syntype: '12. Campylopus spiralis Dus. Fretum Mag, Punta Arenas in fruticetis M. Majo a. 1896. P. Dusén' – NY!].

Plants moderate, rather slender, densely caespitose, somewhat glossy, yellow-green above, brown to yellowish-brown below. *Stems* erect or subascending, simply or dichotomously branched, to 6 cm high, densely and evenly foliate, densely coated from shortly below the apex to the base with a tomentum of reddish-brown, branched, glossy rhizoids, in cross-section round, consisting of a distinct central strand surrounded by a 3–5-stratose medulla of large, hyaline, thin-walled cells and a 2–3-stratose cortex of small, thick-walled cells; *axillary hairs* filiform, 5–6-cellular, with a shorter, brown basal cell and elongate upper cells. *Leaves* erect, appressed, not or slightly altered when dry, erect-spreading when moist, 2.5–4.2 mm long, 0.6–0.8 mm wide, straight to somewhat falcate at the stem apex, rigid, from a narrowly lanceolate or ovate-lanceolate, concave base gradually canaliculate-subulate, lacking a hyaline hair-point; *margin* plane, incurved and

Figure 81. *Campylopus vesticaulis*. — **1**. Habit. **2**. Portion of shoot, dry. **3**. Portion of shoot, wet. **4**. Portion of cross-section of stem. **5–6**. Axillary hairs. **7–11**. Leaves. **12**. Leaf apex. **13**. Upper laminal cells at margin. **14**. Mid-leaf cells at margin. **15**. Alar cells. **16–21**. Cross-sections of leaves, a sequence from base to apex. — [1–3 from *PC 47*; 4–8, 12–18, 21 from *PC 69*; 9–11, 19–20 from *PC 70A*]. — Scale bars: a – 1 mm (7–11); b – 0.5 cm (1); c – 100 µm (4, 12–21) and 0.5 cm (3); d – 100 µm (5–6); e – 0.5 cm (2).

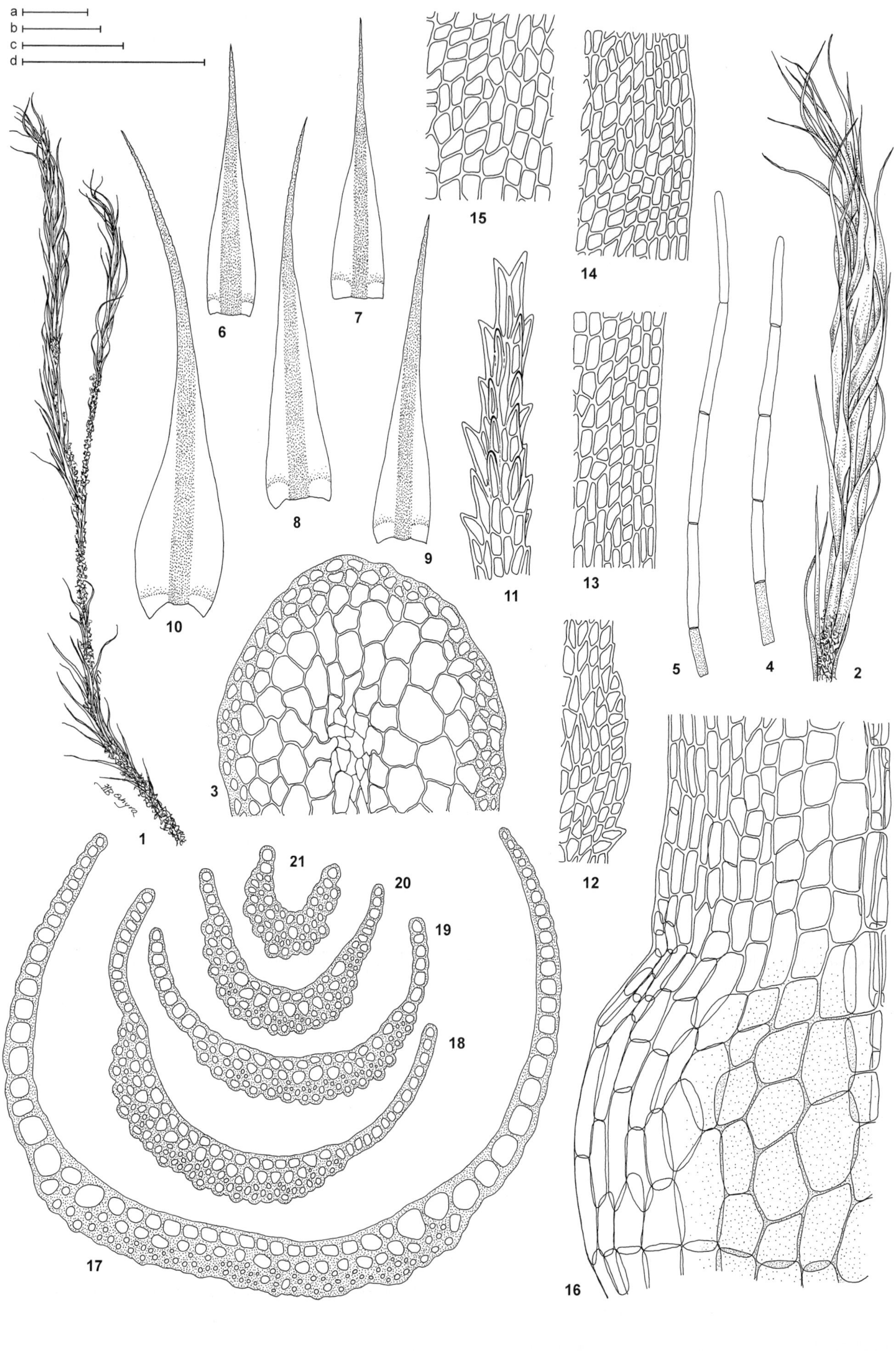
a
b
c
d
1
2
3
4
5
6
7
8
9
10
11
12
13
14
15
16
17
18
19
20
21

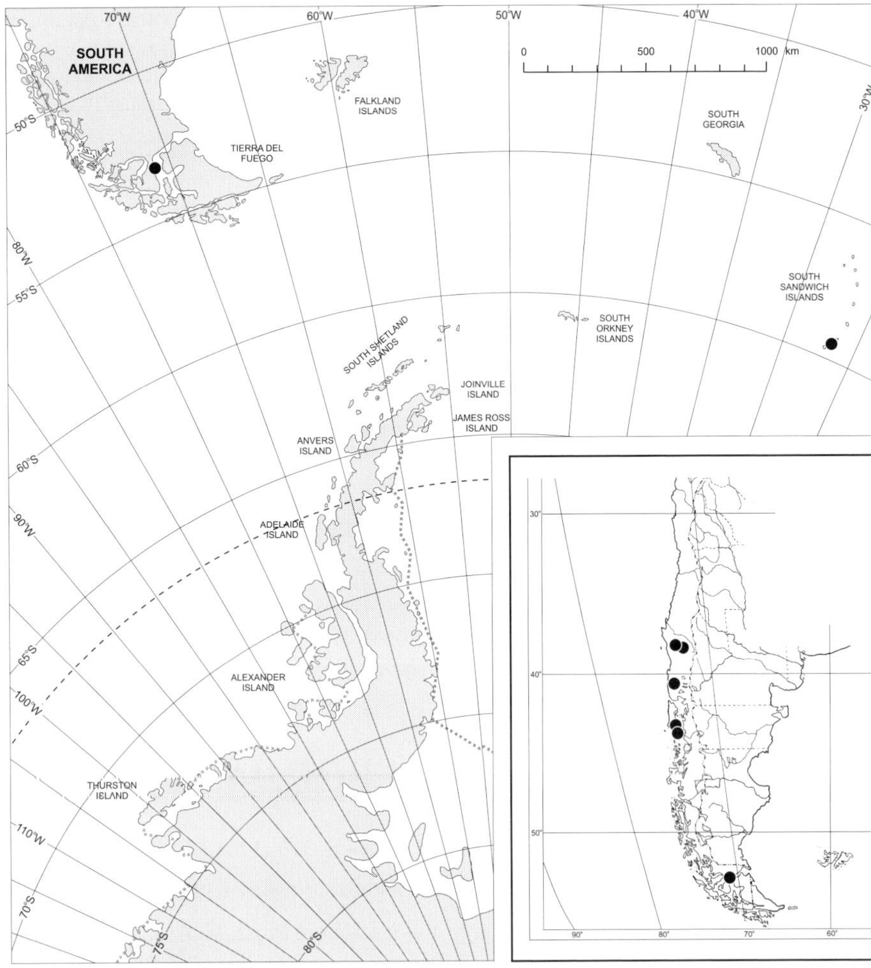

Figure 84. Distribution of *Campylopus spiralis* in the Fuegian region and Antarctica. Inset: South American distribution.

entire to the mid-subula or sometimes to the shoulders only, sharply, rather remotely serrulate to serrate above; *costa* relatively narrow, occupying about ⅓ of the leaf base width, 170–190 µm wide, percurrent to short excurrent as a concolorous, sharply toothed awn, winged by a relatively broad lamella in the subula, in cross-section with a ventral layer of large hyalocysts, a median row of guide cells and dorsal stereid band, distinctly grooved or ridged dorsally in the subula; *laminal cells* pellucid, short-rectangular, subquadrate to trapezoid, rhombic or rhomboidal at the shoulders and in the subula and extending down to the mid-leaf along the margins, 10–20(–25) µm long, 5–8 µm wide, with moderately thickened but not incrassate walls, becoming rectangular to quadrate and markedly widened, 12–16 µm wide in the mid-leaf near the costa and larger, short- to long-rectangular, 20–50 µm long, 14–20 µm wide near the basal; *alar cells* large, thin-walled, quadrate to short-rectangular, forming distinct, flat or swollen, hyaline to yellow-brownish auricles. [*Dioicous*. *Setae* erect, sigmoid and distinctly dextrorse when dry, cygneous when wet, *ca* 10 mm long, yellow to brownish with age, smooth. *Capsules* obloid, erect, slightly asymmetric, 8-ribbed when dry, lustrous, light brown, *ca* 1.2 mm long; *exothecial cells* rectangular, thin-walled in the furrows and thicker-walled over the ribs, 30–45 µm long, 15–20 µm wide; *annulus* compound, biseriate, deciduous; *opeculum* convex-conic, with a short, slanted beak; *peristome teeth* 16, lanceolate, forked halfway to the base, *ca* 420 µm long, deeply red and vertically striolate below, hyaline and papillose above. *Spores* globose, 13–16 µm wide, yellow-brownish, minutely papillose. *Calyptra* cucullate, fringed at the base.]

Chromosome number: unknown.

Discussion. Although Frahm (1988) considered *Campylopus spiralis* to be a doubtful species, a number of specimens from Bellingshausen I. match perfectly the South American type collections and, consequently, it is recognised here as a good, though still poorly and inadequately known species. In some details of the costal anatomy it resembles *C. vesticaulis*, but the costa is much narrower and occupies only about one third of the leaf base

←

Figure 83. *Campylopus spiralis*. — **1**. Habit. **2**. Portion of shoot, dry. **3**. Portion of cross-section of stem. **4–5**. Axillary hairs. **6–10**. Leaves. **11**. Leaf apex. **12**. Lamina cells in upper part of subula. **13**. Upper lamina cells at margin. **14**. Marginal cells in middle of leaf. **15**. Mid-leaf cells. **16**. Alar cells. **17–21**. Cross-sections of leaves, a sequence from base to apex. — [1–2 from *MWH 841b*; 3–5, 7, 11–12, 16–18, 21 from *MWH 827C*; 6, 19–20 from *MWH 822*; 8–9, 14–15 from *MWH 409B*; 10, 13 from *Dusén 617* (isotype of *C. spiralis*), BM]. — Scale bars: a – 1 mm (2); b – 3 mm (1); c – 1 mm (6–10); d – 100 µm (3–5, 11–21).

width and also does not fill the entire subula, leaving a relatively broad lamina nearly to the apex. Additionally, the leaf subula is distinctly serrate to serrulate from the apex down to the middle or sometimes to the leaf shoulders and the leaf areolation of rectangular to subquadrate, rather thin-walled upper laminal cells is unlike that of *C. vesticaulis*. The Antarctic plants are clearly auriculate, often with coloured and swollen alar cells. However, it differs in this character from the type material in which mostly hyaline alar cells are present, forming flat auricles; Dusén (1905*b*) described the leaves in this species as exauriculate.

Reproduction in Antarctica. Sterile.

Habitat. On warm, wet ground in the inner zone of fumaroles, usually near vents vigorously emitting steam and sulphur, sparsely associated with *Cryptochila grandiflora*, *Marchantia berteroana*, *Dicranella hookeri* and *Ceratodon purpureus*.

World range. An imperfectly known, American south-temperate species, widely scattered in western Patagonia of Chile from the Valdivian region to Brunswick Peninsula (Dusén 1905*b*; Thériot 1934, 1935; Herzog 1939), and extending to the maritime Antarctic (Fig. 84, inset).

Distribution in Antarctica. Known only from the volcanic Bellingshausen I. in the South Sandwich Is. archipelago (Fig. 84) where it occurs together with *Campylopus vesticaulis* at moderate altitudes (50–200 m).

Specimens examined. WEST ANTARCTICA. SOUTH SANDWICH IS. **BELLINGSHAUSEN I.** S slopes of cone, *ca* 75 m, *MWH 821B*, *827C & 841B* (AAS, KRAM), *ca* 100 m, *MWH 409B*, *412B & 415* (AAS, KRAM) and 130 m, *PC 61*, *61A & 62C* (AAS, KRAM); W of access point, 150 m, *PC 40* (AAS, KRAM); SE rim of crater, *ca* 200 m, *MWH 822* (AAS, KRAM); at floor of crater, 50 m, *PC 71B* (AAS, KRAM).

EXCLUDED SPECIES OF THE DICRANACEAE

Campylopus incrassatus Müll.Hal. — This species was reported from the South Sandwich Is. by Longton & Holdgate (1979) and Longton (1979) as *Campylopus flavoviridis* Dusén but examination of all voucher collections revealed that they represent *C. introflexus* (see discussion under this species).

Order VI. GRIMMIALES M.FLEISCH.

Grimmiales M.Fleisch., Musci Buitenzorg 1: xxviii, 370. 1904 ['Reihe Grimmioideae']. — Type: *Grimmia* Hedw.

Platycranoideae M.Fleisch., Musci Buitenzorg 1: xxviii. 1904 ['Reihe Platycranoideae'], *nom. illeg.*

Seligeriales Crosby *in* R.J.Taylor & Leviton, Moss. N. Am.: 125. 1980, *nom. inval. desc. angl.*

The Grimmiales are recognised as a separate order by the combination of both sporophyte and gametophyte characters. They present a broad range of variation in particular families and genera and to some extent make this group intermediate between the Dicranales and Phascales, combining many characters of both orders. Sporophytically, the Grimmiales are well diagnosed by their haplolepidous peristome which is usually present and only in some species is it absent, variously reduced or vestigial. When present, the peristome consists of 16 teeth arising from a low or very occasionally high basal membrane which, however, is often lacking. The teeth are entire to variously perforated and/or forked above into two flat prongs, but in some genera they are divided nearly to the base into two filamentous, terete segments which are somewhat paired and give the impression that the peristome is composed of 32 teeth. A fundamental difference of the peristome type in the Grimmiales in comparison to that of the dicranoid type lies in the outer face of the teeth which is more strongly developed and trabeculate than the inner face or, occasionally, both faces may be equally thickened. Additionally, the teeth may be smooth or variously ornamented, papillose or coarsely granulose on both faces, and the prostome is often present.

Different patterns of the relative thickness of the cell wall remnants in a peristome appears to have adaptive significance and is probably correlated with hygroscopic motion of the teeth which, in turn, affects the timing of spore dispersal (Shaw & Robinson 1984). In both the Grimmiales and Phascales the stronger development of the outer face of the peristome teeth may be associated with the xeric nature of most representatives of these orders in contrast to the Dicranales which are mostly mosses of mesic habitats. Edwards (1979) recognised the seligerioid peristome type and interpreted it as diplolepidous, and this prompted Crosby (1980) and Vitt (1984) to distinguish a separate order and suborder, respectively, to accommodate *Seligeria* and its allies. The seligerioid peristome represents a reversal of the dicranoid type with the distinctly trabeculate dorsal, not ventral side, although sometimes the dorsal trabeculation is indistinct or missing. Therefore, taxa belonging to the Seligeriaceae are included in the Grimmiales following the suggestions of Fleischer (1904) and Walther (1983).

The perichaetia and sporophytes in the Grimmiales are terminal or lateral. The setae are straight or curved, long or very short and, accordingly, the capsules are either distinctly exserted over the perichaetial leaves or deeply immersed within them. A remarkable feature of this order is the constant presence of stegocarpous capsules, and cleistocarpy has never been observed in any taxon within this group, in contrast to the Dicranales or Phascales where cleistocarpous genera and species are quite frequent.

Gametophytically, mosses belonging within the Grimmiales present considerable variation in all features of leaf and stem morphology and anatomy. The size of plants ranges from very small to large and robust. The leaves are very rarely subulate, but usually they are lanceolate and, less often, ovate-lanceolate, ovate to lingulate. The foliage is never distichous and complanate, although occasionally the leaves are arranged in distinct spiral rows. In many species the leaves are pointed with a hyaline awn which may be an adaptation to dry habitats. The alar cells are occasionally developed and the laminal cells are mostly smooth, although in some taxa they are variously papillose or pseudopapillose, and the cell walls are usually incrassate.

The Grimmiales were initially recognised by Fleischer (1904) as the order Platycranoideae, which is a nomenclaturally illegitimate name, and comprised two families, Grimmiaceae and Seligeriaceae, but the former was broadly interpreted to include Ptychomitriaceae and Scouleriaceae in the current comprehension. Hedderson *et al.* (1999) indicated the close affinity of the monotypic South African genus *Wardia*, representing the monotypic family Wardiaceae (Welch 1943), to Seligeriaceae, and subsequently Buck & Goffinet (2000) transferred this family to the Seligeriales. These authors also transferred the small monogeneric family Drummondiaceae to the Grimmiales (Buck & Goffinet 2000).

The Grimmiales are the smallest order of the Dicranidae, consisting of about 400 species belonging to six families and about 30 genera. Members of two families, Seligeriaceae and Grimmiaceae, are present in the Antarctic and comprise 27 species classified in ten genera.

Family VII. **SELIGERIACEAE** SCHIMP.

Seligeriaceae Schimp., Coroll. Bryol. Eur.: 22. 1856. — Type: *Seligeria* Hedw.

Blindiaceae Schimp, Syn. Musc. Eur.: 117. 1860 ['Blindieae']. — Type: *Blindia* Bruch & Schimp.

Brachyodontaceae Schimp., Syn. Musc. Eur.: 116. 1860 ['Brachyodonteae'], *nom. illeg.* — Type: *Brachyodus* Fürnr. *ex* Nees & Hornsch. (*nom. illeg.*).

The Seligeriaceae are generally understood to be minute to moderately sized, acrocarpous, gregarious or tufted, rupestral mosses widely distributed at high latitudes and in the mountains in both hemispheres, but which also occur rarely at high elevations in the tropics. They belong to several genera of which the most prominent are *Seligeria*, with a predominantly Holarctic distribution, and the bipolar *Blindia* with a marked species diversity in the Southern Hemisphere. Taxonomically, the family was usually classified within the order Dicranales as a close neighbour of the Dicranaceae (Fleischer 1920, 1923; Brotherus 1924*b*) or included in this family as a separate subfamily (Brotherus 1901*b*; Reimers 1936).

The connection of the Seligeriaceae with the Dicranaceae was mostly owing to the typically dicranaceous habit of the plants. The stems are usually erect or occasionally descending, simple or few-branched, not tomentose and mostly radiculose at the base. The leaves are erect-appressed, erect-spreading to flexuose, sometimes falcato-secund or crispate when dry, arranged in many rows but occasionally they are distinctly trifarious. They are lanceolate, linear to long-subulate from a concave base. The margins are plane or sometimes recurved and they are most often entire throughout and only denticulate at the extreme apex, but very seldom they are serrulate below. The costa is single, often filling most of the subula, terete, percurrent to excurrent, sometimes as a long subula. It may lack internal differentiation or have two stereid bands separated by a single row of large guide cells. The laminal cells are variable, short and isodiametric or rectangular to linear, with thin to moderately thick, non-porose walls. They often become narrower and elongated near the margins but they do not form a distinct border. The leaf cells are unistratose or sometimes bistratose at the margins, smooth and pellucid or sometimes they are densely covered with lengthwise cuticular ridges. These give them a papillose appearance when viewed in transverse sections and in surface view, with careful focusing and adjustment of the light, these structures appear as faint longitudinal striations. The alar cells are undifferentiated or well developed, often forming coloured, enlarged, convex and decurrent auricles which extend nearly to the costa and are sharply separated from the adjacent laminal cells.

The Seligeriaceae also share many of the sporophyte features with the Dicranaceae, except for the peristome type. The perichaetial leaves are either similar to the vegetative leaves or markedly differentiated and sometimes they are convolute-sheathing. The setae are very short to moderately long, often stout and thick, not or variously twisted when dry, straight to cygneous. The capsules are globose, obovoid to obloid, often becoming turbinate to pyriform with age and sometimes they are strongly widened at the mouth when dry and empty, a feature which has not been observed in the dicranaceous taxa. They are either deeply immersed or exserted, erect, inclined to horizontal or pendulous in taxa with strongly curved setae. The capsule walls are smooth or ribbed and pachydermous or leptodermous. The exothecial cells are thin- to more often thick-walled, isodiametric to rectangular and usually become oblate at the rim of the capsule. The stomata are usually few at the base of the urn, bicellular, superficial or sometimes they are lacking. The capsules are sometimes systylious when the columella lengthens markedly after dehiscence and the operculum remains attached to its tip. This feature is also not found in dicranaceous mosses, but is frequent in some grimmiaceous mosses. The operculum is rostrate, straight or slanted and the calyptra is cucullate, smooth and naked. The spores are spherical, small to large, smooth or papillose. The annulus is absent, simple and adherent or complex and deciduous.

The fundamental difference between the Seligeriaceae and Dicranaceae lies in the distinctive peristome type of the Seligeriaceae. In this family the peristome consists of 16 teeth which are relatively broad at the base, triangular, undivided and deeply inserted below the mouth. They are often truncate above and thus markedly reduced in length or sometimes reduced to 1–3 hyaline cells, and in a few cases they are entirely absent (*Seligeria donniana*, *Blindia caespiticia*). The teeth are smooth or coarsely granulose to papillose and are considerably more thickened and trabeculate on the dorsal than on the ventral side, in contrast to that in the dicranoid peristome type. Edwards (1979) described the seligerioid peristome as having a very reduced 2:3 haplolepidous peristomial pattern, a thin membrane that adheres to the margins of the teeth and the prostome at the base of the teeth. He interpreted both the thin membrane and the prostome as an exostome and consequently considered the seligerioid peristome as diplolepidous.

In many species of *Seligeria* and *Blindia* the peristomes are so considerably reduced that they are not particularly useful in any accurate interpretation of the seligerioid peristome because they have only an aberrant 2:2 peristomial pattern and lack either a marginal membrane or a basal prostome. However, in species with well developed peristomes, for example in *Seligeria acutifolia*, *S. polaris* or *Blindia acuta*, the teeth have very prominent dorsal trabeculae and a distinct marginal membrane which is thin, hyaline, and appears to represent a primary cell wall. Most of the membrane consists of PPL wall remnants, although in some places OPL wall remnants are also present (Allen 2001 and personal observations). The unusual features of the seligerioid peristome type were overestimated by Edwards (1979) and its interpretation as diplolepidous is probably unjustified. The reduced size of the seligerioid peristome and its reversed type of thickening may be associated with a pattern of spore dispersal that has adaptive significance.

The Seligeriaceae have traditionally been interpreted by Brotherus (1924*b*) to include six genera, namely *Brachydontium*, *Seligeria*, *Blindia*, *Trochobryum*, *Stylostegium* and *Hymenoloma*. The first five genera constitute the core of the family, although *Trochobryum* and *Stylostegium* are usually merged with, respectively, *Seligeria* (Nyholm 1954, 1987) and *Blindia* (Bartlett & Vitt 1986). On the other hand, Brotherus (1924*b*) broadly interpreted the genus *Hymenoloma* to include *Verrucidens*. Ochyra (1993*b*) expanded the original concept of the Seligeriaceae and transferred to it *Dicranoweisia* and *Holodontium*, genera traditionally associated with the Dicranaceae because they also have seligerioid peristomes. The distant relationship of *Dicranoweisia* to the type subfamily of the Dicranaceae was confirmed by molecular data (Stech 1999*a*). However, the placement of this genus in the Rhabdoweisioideae (Buck & Goffinet 2000; Goffinet & Buck 2004) does not seem to be a well founded solution because it makes this subfamily a very heterogeneous taxon, comprising genera with two fundamentally different peristome types, dicranoid and seligerioid.

Allen (1994*b*, 2000, 2001, 2005) argued for retaining *Dicranoweisia* in the Dicranaceae because of the internal costal differentiation. However, some dicranaceous genera such as *Arctoa* and *Kiaeria* also have an undifferentiated costal anatomy. Therefore, the structure of the peristome should be of critical importance in the classification of genera with otherwise dicranaceous aspect to the gametophytes.

Considering the differences in the costal anatomy as well as in the ornamentation of the peristome teeth Ochyra (1993*b*, 1994*b*) subdivided the Seligeriaceae into two subfamilies. The Seligerioideae comprise genera with a homogeneous costal anatomy and smooth peristome teeth on the outer surface, namely *Seligeria* (incl. *Trochobryum*), *Blindia* and *Brachydontium*. The second subfamily, Dicranoweisioideae, consists of genera with heterogeneous costal elements, coarsely granulose to strongly papillose teeth on the outer surface of the peristome and often with longitudinal ridges on the laminal cells, a peculiar feature of some species of *Dicranoweisia*, in the broad sense, first observed by Flowers (1956).

Two genera, *Dicranoweisia* and *Holodontium*, have initially been placed in this subfamily but several others have been added. However, *Dicranoweisia* itself is a heterogeneous taxon and, for a group of species centred around *D. crispula*, it was necessary to reinstate the genus *Hymenoloma* (Ochyra *et al.* 2003). This subfamily is particularly well represented in the Southern Hemisphere and recently a new genus, *Valdonia*, was recognised for a taxonomically isolated species from the Kerguelen Province of the Subantarctic (Ochyra 2003*c*). A monotypic South American genus *Camptodontium* (Dusén 1905*c*, 1907; Cardot 1908*a*), which is closely related to *Dicranoweisia* but lacks internal costal differentiation and has entirely smooth peristome teeth and laminal cells, also belongs in this family. In general, the division of the Seligeriaceae into subfamilies is only provisional and the family urgently needs a generic revision.

Particular genera of the family also require taxonomic revision. So far, only the genus *Blindia* has been revised worldwide (Bartlett & Vitt 1986) and another large genus, *Seligeria*, has been revised for North America (Vitt 1976*a*) and New Zealand (Vitt & Bartlett 1983).

The Seligeriaceae are a rather small family comprising about 13–14 genera, eight or nine of which are monotypic, and approximately 60 species. Six monotypic and taxonomically isolated genera are endemic to cool temperate and polar regions in the Southern Hemisphere, and here *Blindia* and *Hymenoloma* have much greater species diversity than in the Northern Hemisphere. Two genera, *Holodontium* and *Hymenoloma*, with one and three species, respectively, are known in the maritime Antarctic, and some species of the latter genus are prominent constituents of fellfield vegetation.

KEY TO THE GENERA OF THE ANTARCTIC SELIGERIACEAE

1. Leaves not or only little altered when dry, rigid, straight to falcato-secund, long-subulate; costa strong, long excurrent, broad, occupying about ⅓ or more the width of the leaf base and filling the entire subula .... 15. ***Holodontium***
1. Leaves flexuose, strongly crisped when dry, lanceolate, ovate-lanceolate to short-subulate; costa narrow, subpercurrent, occupying ¼ or less the width of the leaf base; leaf subula with distinct lamina on both sides of the costa .... 16. ***Hymenoloma***

15. **HOLODONTIUM** (Mitt.) Broth.

Holodontium Broth. *in* Engl., Nat. Pflanzenfam. Ed. 2, 10: 198. 1924. — Lectotype (*vide* Ochyra 1993*b*: p. 82): *Holodontium inerme* (Mitt.) Broth. (*Dicranum inerme* Mitt.) [= *Holodontium strictum* (Hook.f. & Wilson) Ochyra].

Plants medium-sized to large and robust, dull, loosely or densely tufted, olive-green or less often golden- to brownish-green, sometimes blackish. *Stems* stiff and rigid, erect, simple or irregularly and sparsely branched, radiculose at the base with smooth, branched, brown rhizoids or rhizoids scattered in the lower part of the stem, densely covered with many longitudinal cuticular ridges numerous on the surface and appearing papillose in transverse section, rounded or oval, consisting of a medulla composed of large, thin-walled cells with clear corner thickenings sharply grading to a bi- or occasionally tristratose cortex of small, thick-walled cells, central strand distinct but small or lacking; *axillary hairs* infrequent, long, hyaline. *Leaves* rigid and stiff, crowded, appressed, erect-flexuose to erect-spreading or falcato-secund, not much altered when wet, often almost circinate and forming an erect, penicillate tuft at stem or branch tips, abruptly or gradually long-subulate from a broader, concave, ovate-lanceolate to oblong-lanceolate base, subulae semi-lunar to canaliculate in transverse section, up to twice the lamina length, narrowly obtuse at the apex; *margins* plane, entire; *costa* single, strong and distinct, smooth, occupying about ⅓ or more the width of the leaf base, filling the subula, in transverse section, with 2 stereid bands and a row of large guide cells and continuous layers of large dorsal and ventral epidermal cells densely covered with cuticular thickenings; *laminal cells* unistratose throughout, transparrent, smooth but appearing papillose on both sides in transverse section owing to numerous lengthwise cuticular ridges, non-porose, thin- to moderately thick-walled, usually in more or less regular longitudinal rows, quadrate to short rectangular above, becoming long rectangular below, often much narrowed towards the margins and forming an indistinct border; *alar cells* sharply differentiated, large, varying from thin-walled and hyaline to thick-walled and reddish-brown, forming distinct, usually inflated, decurrent auricles extending to the costa, occasionally poorly differentiated. *Autoicous*. *Perichaetia* terminal; *perichaetial leaves* similar in shape to vegetative leaves, but somewhat larger and sheathing to convolute at the base with areolation of large and thin-walled, hyaline to reddish cells in the lower part. *Perigonia* sessile, bud-like, lateral just below the perichaetia or sometimes terminal on separate branches, with numerous antheridia intermixed with many filiform, yellow paraphyses, somewhat exceeding the antheridia. *Setae* single per perichaetium, erect, elongate, stout, twisted to the right above, yellowish-brown to brown. *Capsules* exserted, erect and symmetric, smooth, globose or obloid to short cylindrical, without neck, usually narrowed at the mouth, light to dark brown, dull or occasionally with a varnish lustre; *annulus* absent, rudimentary and persistent or occasionally revoluble and deciduous; *operculum* conical, obliquely long-rostrate; *exothecial cells* isodiametric, mostly hexagonal to rounded, moderately thick-walled, becoming oblate to quadrate, with brown, incrassate walls, in 5–8 rows at the orifice; *stomata* not numerous at the extreme base of the urn, bicellular, phaneroporous, with rounded pori; *peristome* consisting of 16 undivided, lanceolate teeth, yellow-brown to pale yellowish at the tips, coarsely papillose on both dorsal and ventral surfaces, strongly trabeculate on the dorsal face; *preperistome* distinct, hyaline, smooth, extending ⅓ the length of the teeth. *Spores* globose, papillose. *Calyptra* naked, smooth, cucullate, entire or sometimes split at the base.

Holodontium is an amphiatlantic subantarctic genus extending northward to Ecuador along the Andean chain but only at high elevations. It was originally established as a section of *Dicranum* (Mitten 1869) which was subsequently elevated to genus (Brotherus 1924*b*). A taxonomic revision of *Holodontium* (Ochyra 1993*b*) revealed that the genus consisted of an unnatural assemblage of five species, four of which were conspecific with *Kaeria pumila*. Exclusion of these species left *Holodontium* a monotypic genus, comprising only *H. strictum*. In its overall appearance *Holodontium* closely resembles dicranaceous genera, and its very broad costa, occupying one third or more the leaf lamina, suggests its close affinity to *Campylopus* and *Chorisodontium*. The seligerioid peristome precludes any alliance of this genus with the Dicranaceae and *Holodontium* is distinguished by its rigid, long-subulate leaves and numerous longitudinal cuticular ridges which densely cover both surfaces of the leaf lamina

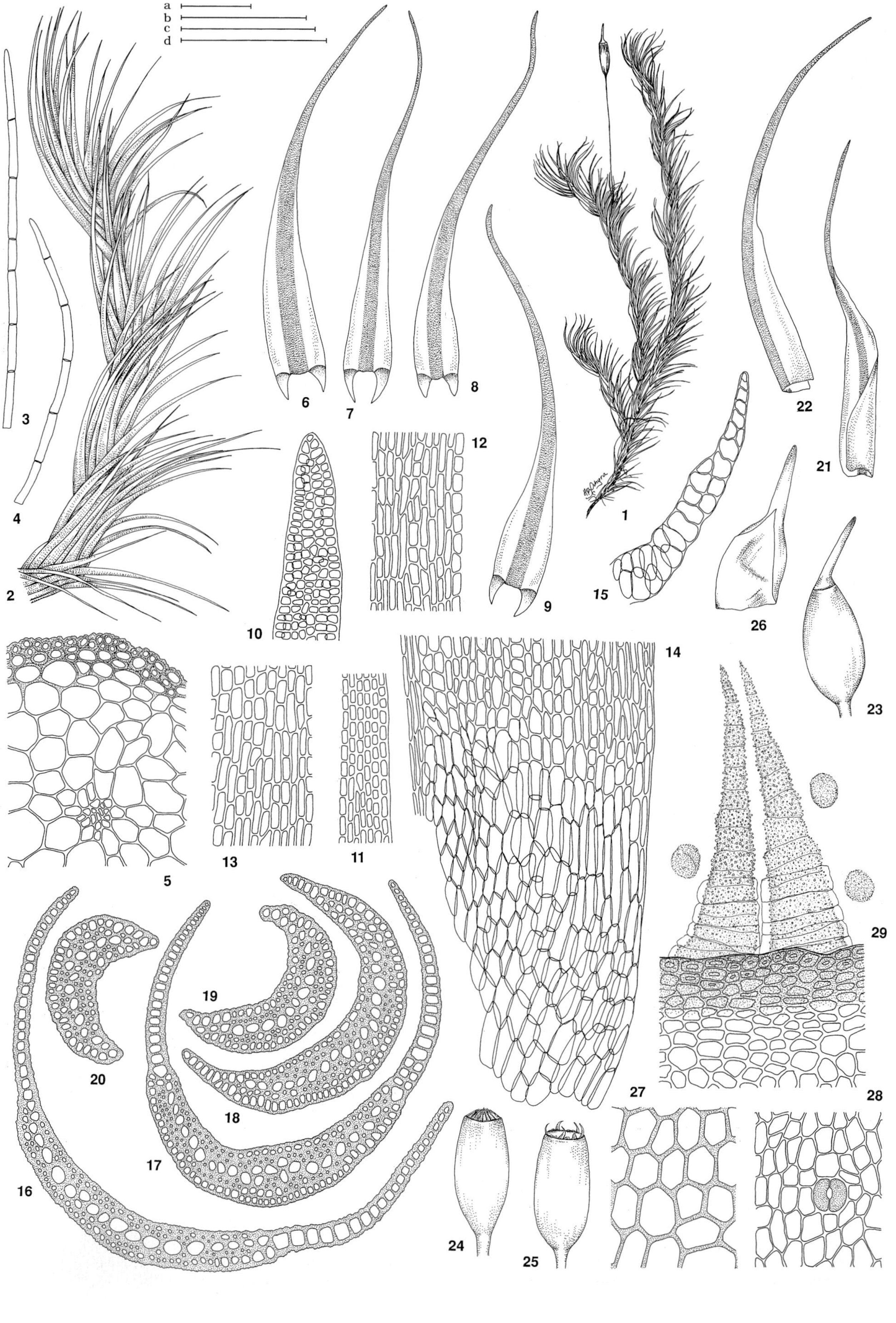
a
b
c
d
1
2
3
4
5
6
7
8
9
10
11
12
13
14
15
16
17
18
19
20
21
22
23
24
25
26
27
28
29

and the stems and branches and give them a papillose appearance in cross-section. Similar cuticular ridges are characteristic only for species of *Dicranoweisia* which, however, are recognised by the strongly crispate leaves and much narrower and weaker costae, never excurrent as a stout, long subula. *Holodontium* is closely related to the other monotypic antipodal genus *Valdonia* from the Kerguelen Province (Ochyra 2003*c*).

1. **Holodontium strictum** (Hook.f. & Wilson) Ochyra Fig. 85

Holodontium strictum (Hook.f. & Wilson) Ochyra, Fragm. Flor. Geobot. 38: 84, *f. 1–3*. 1993. ≡ *Weissia stricta* Hook.f. & Wilson, London J. Bot. 3: 540. 1844. ≡ *Blindia stricta* (Hook.f. & Wilson) Müll.Hal., Syn. Musc. Frond. 1: 343. 1848. ≡ *Leptotrichum strictum* (Hook.f. & Wilson) Mitt. *ex* Paris, Ind. Bryol.: 119. 1894, *nom. inval. in synon.* — Type citation: Kerguelen's Island. [Lectotype (*vide* Ochyra 1993*b*: p. 84): 'N° 243. Weissia stricta H.fil. & Wils. Kerguelen's Land. Antarct. Exped. 1839–1843. J.D.H.' – BM-Hooker!; isotypes: BM-Dixon! (2 specimens), BM-Hooker! (2 specimens), BM-Hampe! (2 specimens), BM-Lyon!, BM-Wilson! (9 specimens), BR!, E!].

Plants medium-sized to relatively large, 2–7 cm tall. *Leaves* (3.5–)4.0–5.0(–5.5) mm long, 0.6–0.9 mm wide; *costa* 200–300 μm wide at the base; *upper laminal cells* 7–18 μm long, 7–8 μm wide, becoming more elongate below, up to 50–60 μm, and much narrower and linear in 3–8 rows at the margins, 5–6 μm wide, 80–120 μm long; *alar cells* 50–75 μm long, 20–30 μm wide, varying from thin-walled and hyaline to thick-walled and reddish-brown and forming a distinct, usually inflated, decurrent group extending to the costa. *Perichaetial leaves* 5.5–7.0 mm long, 1.1–1.3 mm wide. *Setae* 1–2 mm long. *Capsules* 1.2–1.4 mm long; *exothecial cells* 25–45 μm wide at the mid-urn; *operculum* with a slanted beak 0.8–1.0 mm long; *peristome teeth* 230–280 μm long. *Spores* spherical, light brown, papillose, 20–25 μm in diameter.

Discussion. *Holodontium strictum* is a prominent species in the Antarctic moss flora, and its relatively large plants with long-subulate, falcato-secund leaves, often forming penicillate tufts at the tips of shoots are very distinctive. The leaves are prominently auriculate and have a very broad costa which occupies about one-third or more of the leaf base and is excurrent as a stout, long subula; similarly broad costae are present only in species of *Campylopus* and *Chorisodontium*. However, in these genera the laminal cells are smooth, whereas those in *H. strictum* are prominently papillose due to numerous cuticular thickenings covering the leaf lamina on both surfaces which, in transverse section, look like prominent papillae.

As with all mosses associated with wet habitats, *Holodontium strictum* exhibits considerable variation of some gametophyte characters, especially alar cells and papillosity of the laminal cells (Ochyra 1993*b*). However, Antarctic plants represent a typical expression of the species and quite well match the type material from Îles Kerguelen originally described as *Weissia stricta* (Hooker & Wilson 1844), but also the Ecuadorean type of *Dicranum inerme* (Mitten 1869) under which name this species was better known until recently.

Reproduction in Antarctica. In both Antarctic populations capsules have been occasionally observed, those from Bukhta Korablev being fully mature.

Habitat. A hydrophilous species forming large, rather loose almost pure colonies or tufts on damp or wet gravelly soil by streams and lakes. A small admixture in one specimen includes *Sciuro-hypnum glaciale*.

World range. An amphiatlantic subantarctic species, widespread on South Georgia, where it seems to have its main centre of occurrence, occasional on Îles Crozet and Îles Kerguelen and penetrating the maritime Antarctic. Additionally, it is infrequently found in Tierra del Fuego and in the Valdivian region in western Patagonia at elevations above 500 m, and was once recorded at a remote and highly disjunct locality in the Northern Andes of Ecuador at an elevation of 4200 m (Ochyra, 1993*b*) (Fig. 86, inset).

Distribution in Antarctica. Exceedingly rare, known only from two collections from Fildes Peninsula on King George I., South Shetland Is., its southernmost locality at lat. 62°12′S (Fig. 86).

Specimens examined. WEST ANTARCTICA. SOUTH SHETLAND IS. **King George I.** ***Fildes Peninsula***: Bukhta Korablev between Suffield Pt. and Vent Pt. on Maxwell Bay, 28 Mar 1976, *Petrov s.n.* (AAS, F, H, KRAM, LE); Lake Kitezh, *Mahu 21925* (KRAM).

16. HYMENOLOMA Dusén

Hymenoloma Dusén, Ark. Bot. 4(1): 19. 1905. — Type: *H. nordenskjoeldii* Dusén [= *H. antarcticum* (Müll.Hal.) Ochyra (*Blindia antarctica* Müll.Hal.)].

Verrucidens Cardot, Wiss. Erg. Schwed. Südpolar-Exp. 4(8): 85. 1908. — Type: *V. turpis* (Cardot) Cardot (*Blindia turpis* Cardot) [≡ *Hymenoloma turpe* (Cardot) Cardot]. First synonymised by Cardot & Brotherus (1923: p. 10).

Plants small to medium-sized, sometimes moderately large, in short or tall, dense tufts or cushions, occasionally forming extensive patches, dull to lustrous, bright yellow, yellowish- to dark or blackish-green above, brown or

◄———

Figure 85. *Holodontium strictum*. — **1**. Habit. **2**. Portion of shoot, wet. **3–4**. Axillary hairs. **5**. Portion of cross-section of stem. **6–9**. Vegetative leaves. **10**. Leaf apex. **11**. Cells at leaf shoulder. **12–13**. Mid-leaf cells. **14**. Basal leaf cells. **15**. Cross-section of alar region. **16–20**. Cross-sections of leaves, a sequence from base to apex. **21–22**. Perichaetial leaves. **23**. Operculate capsule. **24–25**. Deoperculate capsules. **26**. Calyptra. **27**. Exothecial cells in middle of urn. **28**. Exothecial cells at base of urn and stoma. **29**. Exothecial cells at orifice, annulus, peristome and spores. — [1–9, 12–13, 15–16, 19–20, 23 from *Petrov s.n.*, 28.iii.1976; 10, 17–18 from *Hébrard s.n.*, 10.ii.1969 (holotype of *Campylopus quezelii*), hb. Hébrard; 11 from *Jameson 129* (holotype of *Dicranum inerme*), NY; 14, 21–22, 24–25, 27–29 from *Greene 2523*; 26 from *Hooker 243* (lectotype of *Weissia stricta*), BM]. — Scale bars: a – 1 mm (6–9, 21–26); b – 100 μm (3–5, 10–20, 27–29); c – 1 cm (1); d – 0.5 cm (2).

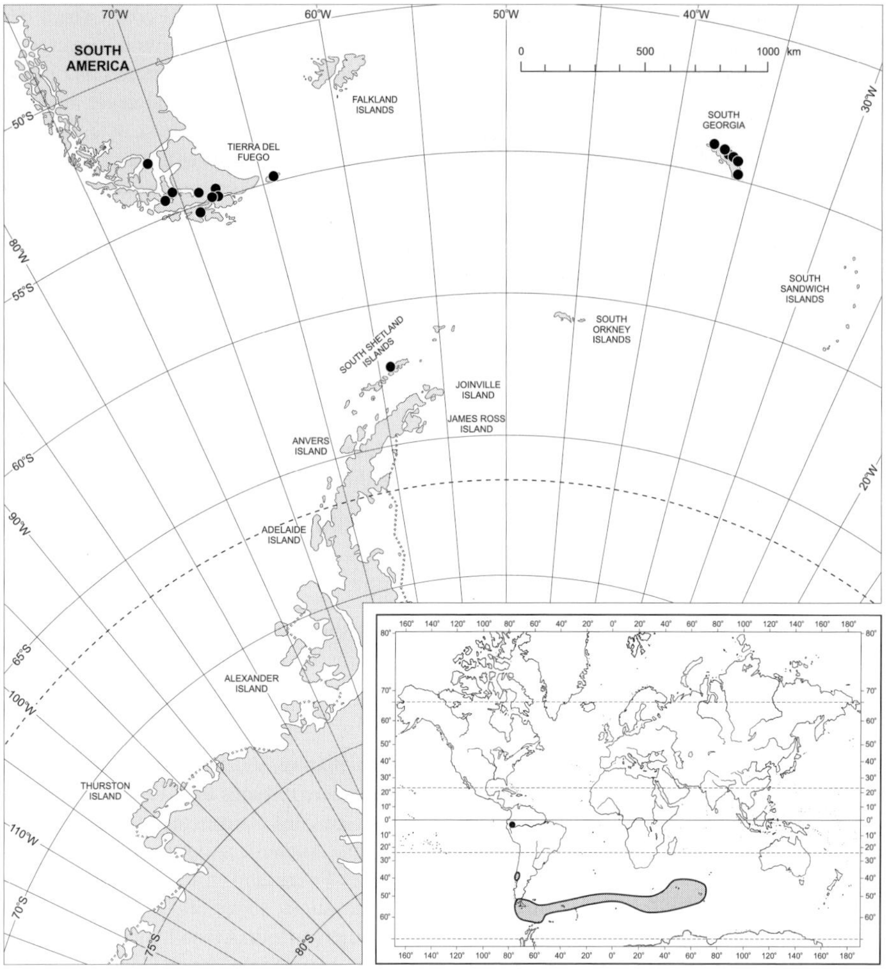

Figure 86. Distribution of *Holodontium strictum* in the Fuegian region and Antarctica. Inset: global distribution.

blackish-brown below. *Stems* erect or ascending, sparingly dichotomously branched, usually with short, fastigiate branches, sometimes simple, in cross-section rounded, to oval, often appearing papillose due to many longitudinal cuticular thickenings, consisting of a distinct, small or large central strand, 1–3-stratose cortex of small, thick-walled, brown cells and 3–6-stratose medulla of large, thin-walled, hyaline to yellowish-hyaline cells; *rhizoids* brown to reddish-brown, lustrous, smooth, either small, slender and branched, clustered at the stem base, or relatively thick, long and mostly unbranched, scattered in the leaf axils in the lower part of the stem; *axillary hairs* multicellular, filiform, often flexuose, hyaline throughout or with 2–3 short, brown basal cells and elongate, hyaline upper cells. *Leaves* densely or quite distantly set, flexuose to somewhat arcuate, strongly crisped, curled to tortuose when dry above, scarcely altered below, erect to erecto-patent dry, erect-spreading, erect-flexuose to spreading when wet, from a concave, ovate, lanceolate or oblong, non-decurrent base gradually tapering to a short or elongate, lance-acuminate or subulate, channelled to subtubulose point, bluntly acute to obtuse at the apex; *margin* entire to weakly crenulate by projecting cell ends, plane, distally erect to incurved, smooth or appearing papillose owing to cuticular ridges over cells, unistratose; *costa* single, subpercurrent to percurrent, narrow, occupying about ⅕ width of the leaf base, usually flat and not prominent or more convex dorsally, especially above, with short, quadrate or rectangular to linear epidermal cells on the ventral surface, in cross-section appearing papillose on both surfaces owing to cuticular ridges covering epidermal cells, consisting of a single row of large guide cells separating usually weaker ventral and stronger dorsal stereid bands or sometimes a ventral band poorly developed to lacking, often with somewhat enlarged ventral and dorsal epidermal cells; *laminal cells* usually in distinct longitudinal rows, unistratose, with incrassate to moderately thick walls, densely covered with lengthwise cuticular ridges on both ventral and dorsal surfaces, giving them a papillose appearance in transverse section, quadrate, rounded-quadrate, irregular to short rectangular at the shoulders, becoming subquadrate to short-rectangular or long-rectangular to linear below, with strongly thickened walls and truncate to slanted ends, often markedly narrowed towards the margins and forming an ill-defined border; *alar cells* often enlarged, thin- to moderately thick-walled, brown, reddish-brown to yellowish-hyaline forming prominent, bulging or flat auricles occupying about half of the leaf base, or sometimes weakly or not differentiated. *Autoicous*. *Perigonia* gemmiform, just below the perichaetia. *Perichaetia* terminal or appearing lateral due to subfloral innovations; *perichaetial leaves* convolute-sheathing, oblong-ovate, shortly, bluntly acuminate or sometimes truncate or abruptly short- or long-subulate, with areolation similar to that of the vegetative leaves but lacking cuticular ridges. *Setae* straight, short to elongate, slender to stout, smooth, yellow to reddish-brown, not twisted or sinistrorse above. *Capsules* long exserted, emergent to immersed, erect to subinclinate, symmetric, globose, ovoid, obloid to cylindrical, lacking neck, smooth or occasionally wrinkled with age, lepto- or pachydermous, yellowish-brown, light brown to castaneous, often with a varnish lustre, sometimes systylous; *annulus* absent or indistinct, persistent; *operculum* conic with a long, slanted rostrum, often attached after dehiscence to the elongate columella; *exothecial cells* isodiametric to short-rectangular, rounded-quadrate, irregular to hexagonal, thin- to moderately thicker-walled, becoming transversely rectangular, thicker-walled in 2–4 tiers at the mouth and forming a coloured strip; *stomata* at the neck of the urn, bicellular, superficial, sometimes absent; *peristome teeth* 16, deeply inserted below the mouth, lanceolate, triangular or dagger-shaped, acute, obtuse or often truncate, entire or slightly forked at the apex, with basal 3–4 lamellae smooth, hyaline, fragile and reddish-orange above or hyaline and reddish- or yellowish-orange throughout, densely trabeculate and variously low papillose, papillose-striate or with scattered coarse papillae on the dorsal surface, ventral trabeculae weakly thickened. *Spores* spherical, yellowish to brownish, smooth or faintly papillose to verrucose. *Calyptra* cucullate, smooth, naked, entire at the base.

Hymenoloma is a genus recognised by the combination of the seligerioid peristome, strongly crispate, curled to tortuous leaves on drying, entirely unistratose and seemingly papillose laminal cells when viewed in transverse section due to numerous cuticular lamellae densely covering both surfaces of the leaf and making them rugged, plane to distally incurved leaf margins and internal differentiation of the costa. This set of characters is unknown in any other local moss genus and makes *Hymenoloma* an unmistakable genus in the Antarctic.

Species of *Hymenoloma* have traditionally been placed in *Dicranoweisia*, but this proved to be a heterogeneous taxon comprising at least two distinct groups of species deserving separate generic status (Ochyra *et al.* 2003). The Holarctic *Dicranoweisia cirrata* was a discordant element in the genus due, among other characters, to its entirely smooth laminal cells, lack of longitudinal cuticular ridges, strongly and widely recurved bistratose leaf margins, undifferentiated perichaetial leaves, as well as the presence of the compound, deciduous annulus and multicellular brood-bodies in the axils and on the costae of the upper leaves. On the other hand, several species belonging to the *D. crispula* complex constitute a separate genus. Unfortunately, due to strictly nomenclatural reasons the generic name *Dicranoweisia* has to be retained for the genus comprising only *D. cirrata* and *D. africana*. When establishing this genus, Milde (1869) placed both *D. cirrata* and *D. crispula* in it but neither was indicated as the type. For a long time *D. crispula* was considered to be a lectotype of *Dicranoweisia* (Wijk *et al.* 1962; Farr *et al.* 1979) and this lectotypification was ascribed to Kindberg (1882). However, this choice had to be superseded because *D. crispula* was a typical ‘residue lectotype’ and such lectotypification was in conflict with Art. 7.11 of the current Code (Greuter *et al.* 2000). Therefore, the lectotypification of *Dicranoweisia* with *D. cirrata* proposed by Williams (1913*a*) is now accepted (Greuter *et al.* 1993).

The oldest available generic name for species of the *Dicranowieisa crispula* complex is *Hymenoloma* and it was used for a monotypic genus described by Dusén (1905*a*) from western Patagonia. This genus was considered to be closely related to *Dicranoweisia* and the only distinctive difference was the presence of a prominent prostome. Reimers (1936) discussed the status of *Hymenoloma* in detail and merged it with *Dicranoweisia*, transferring its only species, *H. nordenskjoeldii*, to this genus. The aptness of this conclusion is confirmed here and the generitype of *Hymenoloma* is actually conspecific with *H. antarcticum*.

Cardot (1908*a*) established the monotypic austral genus *Verrucidens*, typified by the Fuegian *V. turpis*, and later consisting of six species distributed in southern South America and on sub-antarctic islands (Reimers 1936; Ochyra & Matteri 1996*a*). *Verrucidens* was merged with *Hymenoloma* by Cardot & Brotherus (1923) who transferred its generitype to this genus and Ochyra (1999*g*) found that *Verrucidens* shared all diagnostic features of *Dicranoweisia*, and therefore considered these genera to be congeneric.

Hymenoloma is a small genus of about eight species but the final species composition may be established after a thorough taxonomic assessment of all species which still remain in *Dicranoweisia*. Sometimes the peristomial features are disregarded and various species with strongly curled leaves have been placed in *Dicranoweisia*, for example *D. papuana* from Papua New Guinea (Norris & Koponen 1990) which correctly belongs to *Holomitrium* (Eddy 1988).

Six species of *Hymenoloma* are known in austral polar region, five of which are endemics and one is a bipolar species. Four of these are widely distributed throughout the maritime Antarctic and some of them play a prominent role in various plant communities, especially in fellfield vegetation. The genus has not been investigated taxonomically in the austral polar region and only a brief, preliminary account of the species on Signy I., South Orkney Is., is available (Bell 1976). Although it is easily distinguished, taxonomically *Hymenoloma* is a very difficult genus, mostly because in some species important diagnostic characters refer to the sporophytes, whereas the gametophyte features, especially the leaf areolation, are subject to considerable variability, even within the same shoot. Fortunately, in harsh polar conditions, most populations of *Hymenoloma* species produce sporophytes in profusion.

1. Basal cells mostly uniform, 8–14 μm wide, irregular, quadrate, short-rectangular, oblong to hexagonal, usually less than 5 times as long as wide, moderately thick-walled, forming lax areolation; epidermal cells on the ventral surface of the costa in and just below the subula, short, rounded-quadrate to short-rectangular .... 1. ***H. grimmiaceum***

1. Basal juxtacostal cells 3–6(–8) μm wide, showing a strong tendency to elongation, long-rectangular to linear, well spaced due to strongly thickened longitudinal walls, usually 6–20 times as long as wide; epidermal cells on the ventral surface of the costa in and just below the subula, short, rectangular to linear 2

2. Inner perichaetial leaves broadly obtuse to acute, convolute; spores 14–20 µm wide, nearly smooth to finely papillose; setae mostly slender, sinistrorse above, yellow-brownish, 5–12 mm long; capsules long-exserted; peristome teeth densely irregularly striate-papillose; capsules obloid to short-cylindrical, leptodermous, light brown, wrinkled with age; plants often bright yellow-green; leaves crisped .... 2. ***H. crispulum***

2. Inner perichaetial leaves subulate, often with the subula as long as the base, scarcely convolute but ± sheathing; spores 20–35 µm wide; setae stout, dark yellow- to reddish-brown, 3–5 mm long; capsules short-exserted, globose to ovoid, rather pachydermous, castaneous, smooth; peristome teeth often coarsely papillose; plants usually dark to sordid-green; leaves often tortuous .... 3. ***H antarcticum***

1. **Hymenoloma grimmiaceum** (Müll.Hal.) Ochyra Fig. 87

Hymenoloma grimmiaceum (Müll.Hal.) Ochyra *in* Ochyra, Żarnowiec & Bednarek-Ochyra, Cens. Cat. Polish Mosses: 115. 2003. ≡ *Blindia grimmiacea* Müll.Hal. *in* Neumayer, Int. Polarforsch. Deutsch. Exp. 2: 299. 1890. ≡ *Dicranoweisia grimmiacea* (Müll.Hal.) Broth. *in* Engl. & Prantl, Nat. Pflanzenfam. 1(3): 318. 1901. — TYPE CITATION: Austro-Georgia, ad rupes 'am Ausgange des Brocken-Thales' in pulvinulis vigens, 23. Januario 1883. [Lectotype (*selected here*): 'Blindia grimmiacea C. Müll. nsp. Am Ausgange des Brockenthales in kleinen Polstern 23.I.83 Sub. No. 7. Südpolarexpedition' – HBG!; isotype: H-Brotherus!].

Dicranoweisia grimmiacea fo. *brevifolia* Cardot, Bull. Herb. Boissier Sér. 2, 6(1): 29. 1906. — TYPE CITATION: Terre Louis-Philippe: île Challenger. [Lectotype (*selected here*): 'Svenska Sydpolarexpeditionen 1901–03. N:r 438. Dicranoweisia grimmiacea (C. M.) Broth. forma brevifolia Card. Terre Louis Philippe: île Challenger 2/12 1902 Carl Skottsberg Det. J. Cardot' – PC-Cardot!; isotype: S!]. First synonymised by Ochyra (1998*a*: p. 128).

Dicranoweisia grimmiacea fo. *latifolia* Cardot, Bull. Herb. Boiss. Sér. 2, 6(1): 29. 1906. — TYPE CITATION: Terre Louis-Philippe: île Challenger. [Lectotype (*selected here*): 'Svenska Sydpolarexpeditionen 1901–03. N:r 439. Dicranoweisia grimmiacea (C. M.) Broth. forma latifolia Card. Terre Louis Philippe: ïle Challenger 2/12 1902 Carl Skottsberg Det. J. Cardot' – PC-Cardot!; isotype: S!]. First synonymised by Ochyra (1998*a*: p. 128).

Dicranoweisia grimmiacea fo. *robusta* Cardot, Wiss. Erg. Schwed. Südpolar-Exp. 4(8): 202. 1908, ***syn. nov.*** — TYPE CITATION: Baie Cumberland: Jason-Harbour (Skottsberg, n° 282). [Lectotype (*selected here*): 'Svenska Sydpolarexpeditionen 1901–03. N:r 282. Dicranoweisia grimmiacea (C. M.) Broth. forma robusta South Georgia, Cumberland Bay Jason Harbour 23/4 1902 Carl Skottsberg Det. J. Cardot' – PC-Cardot!; isotype: S! (2 specimens)].

Plants small to medium-sized, densely caespitose or in compact, hemispheric cushions, sometimes in patches, dull to somewhat lustrous, yellow-, olive-, sordid-, or brownish-green or dark green to blackish-green above, brown to blackish-brown below, sometimes blackish throughout. *Stems* usually not tightly coherent in tufts or cushions, rather slender, (0.7–)1.0–2.0(–3.0) cm tall, sparingly dichotomously branched, radiculose at base with short, branched, brown rhizoids, often with large, thick, sparingly branched, castaneous rhizoids scattered in the leaf axils in the lower part, in cross-section round, consisting of a small, rather weak central strand, 2–4-stratose medulla of large, thin-walled, hyaline or yellowish-brownish cells and a 1–2-stratose cortex of relatively large, substereid cells with moderately thickened, brown walls; *axillary hairs* 8–12-celled, filiform, with 2–3 basal short, brown cells and 6–10 elongate, hyaline upper cells. *Leaves* moderately crisped at the shoot apex, sometimes only arcuate and flexuose, mostly straight below, rather laxly set, erect and appressed on drying, erect-spreading to spreading on wetting, (1.2–)1.5–2.2(–2.5) mm long, (0.45–)0.5–0.7(–0.75) mm wide, from an ovate- or oblong-lanceolate, concave, non-decurrent base gradually narrowed to a short or long, straight or somewhat curved, subtubulose subula, mostly bluntly acute at the tip; *margin* plane, incurved distally, entire; *costa* single, well defined, 60–120 µm wide at base, yellow-brown, subpercurrent, not filling the entire subula, on the ventral surface in and somewhat below the subula covered with rounded-quadrate to short-rectangular cells, becoming long-rectangular to linear in the basal part, rather flat and not prominent dorsally below, more convex at the back distally, in cross-section comprising a median row of large guide cells separating rather weakly developed, ventral and dorsal stereid or substereid bands, often with epidermal cells larger on both surfaces and distinctly papillose by cuticular ridges; *laminal cells* unistratose throughout, very occasionally with bistratose patches in the subula, translucent, arranged in regular longitudinal rows, moderately thick-walled to thin-walled below, almost uniform in shape throughout, quadrate, rounded-quadrate, short

→

Figure 87. *Hymenoloma grimmiaceum*. — **1**. Habit, dry. **2**. Portion of shoot, wet. **3**. Cross-section of stem. **4–5**. Axillary hairs. **6–13**. Leaves. **14**. Cells on ventral surface of costa and laminal cells in leaf subula. **15–16**. Mid-leaf cells at margin. **17**. Cells at leaf shoulder. **18**. Mid-leaf cells. **19**. Alar cells. **20–24**. Cross-sections of leaves, a sequence from base to apex. **25–26**. Perichaetial leaves. **27**. Cells at shoulder of perichaetial leaf. **28**. Operculate capsule. **29–30**. Deoperculate capsules. **31**. Calyptra. **32**. Exothecial cells in middle of urn. **33**. Exothecial cells at base of urn. **34**. Exothecial cells at orifice, peristome teeth and spores. — [1, 17 from *RW 102*; 2, 4–5, 25–31, 33 from *PC 148A*; 32, 34 from *PC 192*; 3, 6, 18, 20, 22–23 from *RO 826/80*; 7 from *BJ 69*; 8–9, 14 from *BJT 360*; 10, 19 from *RW 4*; 11, 21, 24 from *Will 7* (lectotype of *Blindia grimmiacea*), HBG; 12, 16 from *CS 439* (isotype of *Dicranowesia grimmiacea* fo. *latifolia*), S; 13, 15 from *CS 438* (isotype of *D. grimmiacea* fo. *brevifolia*), S]. — Scale bars: a – 1 mm (28–31); b – 2 mm (1–2) & 100 µm (14–19, 27, 32–34); c – 1 mm (6–13, 25–26) and 100 µm (3–5, 20–24).

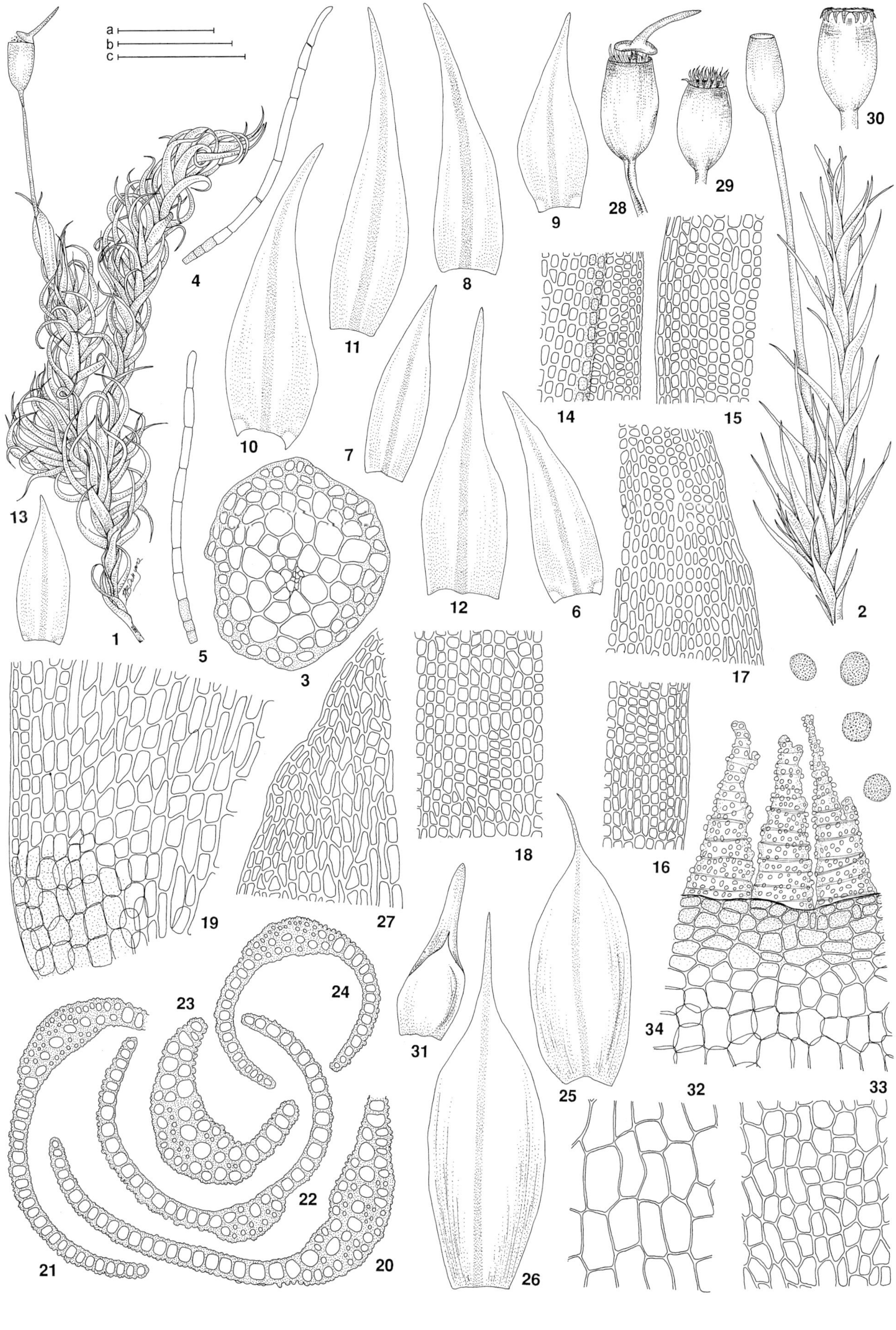

a
b
c
1
2
3
4
5
6
7
8
9
10
11
12
13
14
15
16
17
18
19
20
21
22
23
24
25
26
27
28
29
30
31
32
33
34

rectangular to transversely rectangular in the subula and at the shoulders, 7–10 μm wide, (5–)8–15(–20) μm long, somewhat wider and thinner-walled near the base, 8–13 μm wide, quadrate, short rectangular, oblong to hexagonal, with the basal juxtacostal cells not showing a tendency for elongation, usually less than 5 times as long as wide, becoming somewhat narrower in a few rows at the margins but not forming a distinct border and gradually passing into wider median cells; *alar cells* not differentiated or larger, thin-walled and more lax than the adjacent basal cells, forming more or less distinct auricles, flat or convex, hyaline, yellowish-hyaline to brownish auricles, extending about halfway to the costa. *Autoicous*. *Perichaetial leaves* 1.8–3.1 mm long, from an ovate to elliptical, strongly concave, vaginate, base abruptly constricted into a short, straight or curved acumen, with an areolation of irregular cells with heavily incrassate walls above and rectangular, thin-walled and lax cells in the lower half. *Setae* short, 4–5 mm long, yellow to brown with age, stout or occasionally once twisted to the left, not twisted, straight or somewhat curved. *Capsules* emergent to short-exserted, erect to inclined, subglobose to ovoid, 0.9–1.2 mm long, 0.8–0.9 mm wide, leptodermous, stramineous to light brown, smooth or wrinkled with age, somewhat lustrous, often systylious; *annulus* persistent or lacking; *operculum* conic, obliquely long-rostrate, with a rostrum, about as long as the urn; *exothecial cells* thin-walled, rounded to oblong-hexagonal, 30–60 μm long, 20–40 μm wide, becoming somewhat smaller, thicker-walled and often oblate near the orifice and forming a distinct, yellow to yellow-brownish strip of 3–5 tiers; *stomata* absent; *peristome* consisting of 16 teeth, deeply inserted below the mouth, erect to recurved when dry, delicate and fragile and mostly disintegrated in old capsules, hyaline, yellow-hyaline to brownish, triangular, entire or variously erose above and often truncated at the apex, with dense or scattered, coarse verrucae on both surfaces. *Spores* spherical, brown, densely papillose, 20–32 μm in diameter. *Calyptra* cucullate, smooth, naked, entire at the base.

Discussion. *Hymenoloma grimmiaceum* is easy to identify, the crisped leaves and a very characteristic leaf areolation of relatively broad and short cells arranged in regular longitudinal rows being distinctive. The leaf areolation is somewhat similar to that in *Kiaeria pumila* and, additionally, both species share similar alar cells, but the crispate leaves and pseudopapillose laminal cells should prevent confusion. The leaf areolation and relatively less crispate leaves differ *H. grimmiaceum* from all other congeners which, in addition, have elongate, long-rectangular to linear epidermal cells over the ventral surface of the costa in the leaf subula. A remarkable feature of *H. grimmiaceum* is the uniform leaf areolation near the leaf base, whereas in many species of the genus the basal juxtacostal cells show a tendency to be elongate. Bell (1976) stated that *H. grimmiaceum* differs from *H. antarcticum*, among others, by its smooth peristome teeth, but actually they are distinctly papillose to verrucose in this species. This author also misspelled the epithet of this species as *grimmiaceae* which spelling was later sometimes used in the Antarctic botanical literature (e.g. Longton & Holdgate 1979; Lewis Smith 1984*b*–*c*).

The species exhibits relatively little variation, but there is some in the crispness and the size of the leaves. Cardot (1906*a*, 1908*a*) recognised two forms in the material collected from Challenger I. off the western Antarctic Peninsula, fo. *brevifolia* and fo. *latifolia*. They evidently represent ecological variants, conditioned by different environmental conditions, which do not merit taxonomic recognition because anatomical details of the leaves are exceptionally stable. Generally, in contrast to other species of the genus, the leaf areolation in *H. grimmiaceum* is not very variable and only the angular cells are subject to greater variation, either forming prominent, coloured and convex auricles or being not much different from the adjacent laminal cells. Cardot (1908*a*) suggested that the length of the seta varies markedly in this species, from 1.5 to 5 mm. He studied several specimens from South Georgia with many immature sporophytes and such underdeveloped setae are indeed very short, but in mature sporophytes their length is quite constant and varies between 4 and 5 mm.

Reproduction in Antarctica. Fully mature sporophytes are profusely produced in most populations.

Habitat. A predominantly saxicolous moss growing in a wide range of habitats, mostly occupied by various associations of the fruticose lichen and moss cushion subformation. It is also locally abundant on dry gravelly soil, forming relatively large patches due to the coalescence of many small cushions and tufts of the moss, but frequently it also thrives on sheltered rock faces supplied with trickling melt water. It frequently grows on scree, rock ledges and outcrops, in sheltered and shaded rock crevices and clefts, and under overhangs. It usually prefers dry and sheltered sites, but it also grows in moist habitats including margins of runnels and moist cliffs. In areas with volcanic activity it occasionally grows near fumarole vents, on lava blocks moistened by steam, but usually in the outer zone of the fumarole vegetation. Occasionally, it grows on peaty soil and rock detritus and in drier moss carpets dominated by *Sanionia georgicouncinata*. Its most frequent associates include species of *Andreaea*, *Kiaeria pumila*, *Polytrichastrum alpinum*, *Schistidium antarctici*, *Pohlia nutans*, *Bartramia patens*, *Ceratodon purpureus* and the liverworts *Pachyglossa dissitifolia*, *Barbilophozia hatcheri* and *Cephaloziella varians*.

World range. An Antarctic–South Georgian species, known only from South Georgia, from whence it was originally described (Müller 1890*a*), and from the maritime Antarctic (Fig. 88). Brotherus (1906*a*) reported it from Heard I. but the material represents *Hymenoloma antarcticum*. Cardot (1908*a*) reported it from Îles Kerguelen but this record has not been confirmed by any voucher specimens.

Distribution in Antarctica. A widely distributed species which is locally common and abundant in the archipelagoes of the South Sandwich and South Orkney Is., becoming progressively less frequent southwards in the South Shetland Is. and on the west coast of the Antarctic Peninsula, where it ranges from Astrolabe I. to Hook I. at lat. 65°38′ S near the Graham Coast (Fig. 88). It is a species of lower elevations, occurring most frequently from sea level to about 100 m, and only occasionally extends above 300 m (e.g. at 340 m on Bouvetøya and 350 m on King George I.).

Hymenoloma grimmiaceum was first reported from the Antarctic by Cardot (1906*a*) from Challenger I., Danco Coast, on the basis of material collected by C. Skottsberg during the Swedish South Polar Expedition of 1902–1903. However, in the Brotherus herbarium in Helsinki there is an earlier specimen

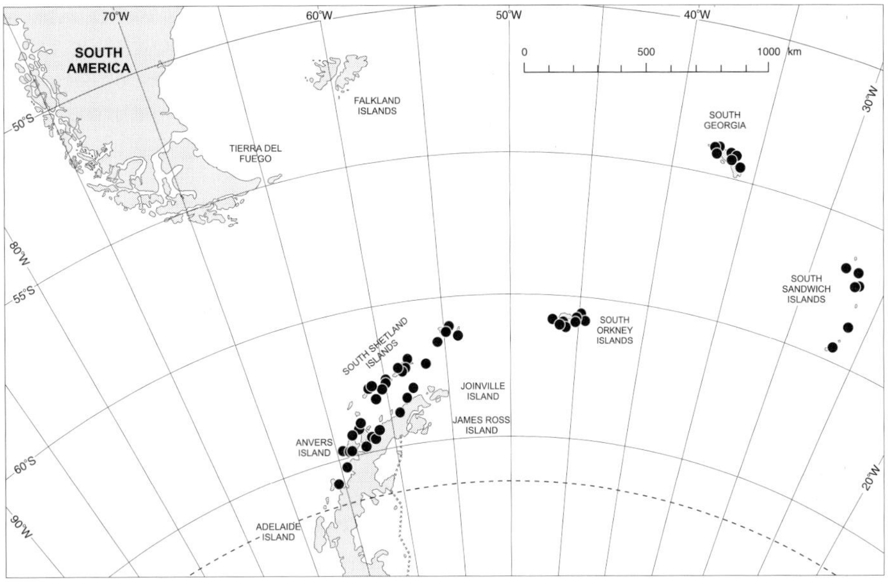

Figure 88. Global distribution of *Hymenoloma grimmiaceum*. The locality on Bouvetøya not indicated.

collected in the Antarctic in 1899 by Dr Cook, a member of the Belgian Antarctic Expedition of 1897–1899 from an unspecified locality; it was recognised by V. F. Brotherus and N. Bryhn as a new species of *Dicranoweisia* which, however, remained undescribed.

Specimens examined. WEST ANTARCTICA. BOUVETØYA. Moseryggen, 260–287 m, *TE 265, 311, 391.2 & 395.1* (BG); Rustadkollen, 300–340 m, *TE 243, 258, 375, 400 & 408.1–4* (BG); nunatak ridge E of Rustadkollen, *ca* 290 m, *TE* 200 (BG).

SOUTH SANDWICH IS. **Leskov I.** N end, *ca* 130 m, *REL 521, 522A & 523* (AAS, KRAM); N-facing slopes, *ca* 100 m, *MWH 836B & 837C* (AAS, KRAM); summit ridge, 100 m, *PC 245J, 248A, 252B, 254E & 257* (AAS, KRAM) and *ca* 200 m, *REL 483, 528 & 534* (AAS, KRAM); W slopes, 70 m, *PC 255A* (AAS, KRAM); near centre of W side, *ca* 135 m, *REL 526* (AAS, KRAM) and *REL 523* (AAS, KRAM). **Visokoi I.** Finger Pt., *ca* 35 m, *REL 761* (AAS, KRAM). **Candlemas I.** Clinker Gulch, *PC 148A* (AAS, KRAM); between Chimaera Flats and Breakbones Plateau, *PC 192* (AAS, KRAM); E lagoon, *ca* 35 m, *REL 603* (AAS, KRAM). **Vindication I.** N end, 100 m, *REL 719* (AAS, KRAM). **Montagu I.** Allen Pt., 70 m, *PC 100F* (AAS, KRAM) and 150 m, *MWH 804B* (AAS, KRAM); Scarlett Pt., 50 m, *PC 98B & 99A* (AAS, KRAM). **Bellingshausen I.** E end of crater rim, *ca* 200 m, *MWH 825D* (AAS, KRAM); W side of crater rim, *ca* 130 m, *MWH 430B* (AAS, KRAM) and *PC 28A & 29A* (AAS, KRAM).

SOUTH ORKNEY IS. **Monroe I.** W coast, 35–100 m, *MMM 75 & 86* (AAS, KRAM); SE coast, 35–65 m, *MMM 63* (AAS, KRAM); S coast, 80 m, *RILS 8008* (AAS, KRAM). **Coronation I.** Mansfield Pt., 10–20 m, *MMM 13* (AAS, KRAM) and 35 m, *RILS 289* (AAS, KRAM); Cape Vik, 100 m, *RILS 562* (AAS, KRAM); Cape Hansen, 20–200 m, *RILS 3243* (AAS, KRAM); Olivine Pt., 15-60 m, *RILS 144* (AAS, KRAM). **Lynch I.** N-facing slope, 3 m, *RW 109* (AAS, KRAM). **Matthews I.** Promontory N of Coffer I., 35 m, *RILS 187* (AAS, KRAM). **Robertson Is.** Steepholm I., 75 m, *RILS 3251* (AAS, KRAM). **Powell I.** W of John Peaks, 50 m, *RILS 258* (AAS, KRAM); Falkland Harbour, *ca* 15 m, *RILS 52* (AAS, KRAM); Michelsen I., *ca* 15 m, *RILS 41B* (AAS, KRAM). **Saddle I.** E part, 50–150 m, *MMM 171* (AAS, KRAM). **Laurie I.** Scotia Bay, x.1903, *Rudmose Brown s.n.* (E, PC); Cape Geddes, 15–65 m, *MMM 7 & 155 m* (AAS, KRAM). **Signy I.** Foca Pt., 50 m, *REL 1118* (AAS, KRAM); coast opposite Spindrift Rocks, *ca* 65 m, *REL 1142* (AAS, KRAM); Heywood Lake, 15 m, *RILS 536* (AAS, KRAM); between Starfish Cove and The Wallows, 35 m, *RILS 474* (AAS, KRAM); Factory Cove, 65 m, *RW 35* (AAS, KRAM); Berntsen Pt., 50 m, *BJT 360* (AAS, KRAM); Observation Bluff, 60 m, *RILS 8663* (AAS, KRAM); Polynesia Pt., 50 m, *RILS 10749* (AAS, KRAM). **Moe I.** 25–200 m, *RILS 5259* (AAS, KRAM).

SOUTH SHETLAND IS. **Elephant I.** Stinker Pt., 107 m, *JSA 87* (AAS, KRAM); Endurance Glacier, 238 m, *JSA 1712A* (AAS, KRAM). **Clarence I.** S end, 135 m, *JSA 226A* (AAS, KRAM). **Eadie I.** *JPB 61* (AAS, KRAM). **Bridgeman I.** 135 m, *JSA 240* (AAS, KRAM). **King George I.** ***Bransfield Strait***: Red Hill, 100 m, *RO 1111/80* (KRAM). ***Admiralty Bay***: Creeping Slopes, 110 m, *RO 1270/80* (KRAM); The Tower, 350 m, *RO 1031/80* (KRAM); Emerald Pt., 30 m, *RO 826/80* (KRAM); Urbanek Crag, 100 m, *RO 2318/80* (KRAM). ***Penguin I.***: crater rim, *BJ 66 & 69* (KRAM). ***Destruction Bay***: North Foreland, 20 m, *DCL 658* (AAS, KRAM). **Robert I.** Coppermine Cove, 15 m, *RILS 925* (AAS, KRAM). **Greenwich I.** Duff Pt., 15 m, *DCL 555* (AAS, KRAM); Spark Pt., 8 m, *DCL 645B* (AAS, KRAM) and 20 m, *DCL 630* (AAS, KRAM); NE of Mt. Plymouth, *RMS 69-979, 989, 999 & 1801* (US). **Half Moon I.** S of Huts, 50 m, *DCL 579B & 594* (AAS, KRAM). **Livingston I.** ***Byers Peninsula***: Vietor Rock, 8 m, *DCL 464* (AAS, KRAM); Devils Pt., 100 m, *KR 21A & 22C* (AAS, KRAM); E end of South Beaches, 6 m, *DCL 99* (AAS, KRAM) and 60 m, *DCL 161A & 164* (AAS, KRAM); Clark Nunatak, 30 m, *DCL 126 & 256* (AAS, KRAM); Lair Pt., 2 m, *KR 61A* (AAS, KRAM) and 17 m, *DCL 525* (AAS, KRAM); Robbery Beaches, 10 m, *KR 54A & 56D* (AAS, KRAM); Laager Pt., 20 m, *KR 40C* (AAS, KRAM). ***Hurd Peninsula***: Miers Bluff, *RMS 69-810, 818 & 819* (US). **Deception I.** Headland of entrance, *Dixon 1358* (BM). Stonethrow Ridge, *RILS 5798C* (AAS, KRAM).

WEST ANTARCTIC PENINSULA. **Davis Coast.** Astrolabe I., 5–25 m, *RILS 4075* (AAS, KRAM) and 10–30 m, *RILS 4058* (AAS, KRAM); Cape Roquemaurel, 10–50 m, *RILS 3922, 3923, 3924, 3927, 3928 & 3933* (AAS, KRAM); Cape Kjellman, *CS 433* (PC, S). **Danco Coast.** Cierva Pt., *VK 526 & 603* (KRAM) and *OB 14* (AAS, KRAM); Valdivia Pt., *CS 435* (PC, S); Challenger I., *CS 436, 348 & 439* (PC, S); Cuverville I., 15 m, *de Leeuw 17B* (AAS, KRAM) and 10–20 m, *RILS 11942* (AAS, KRAM). **Palmer Archipelago.** ***Brabant I.***: Cape Roux, 10 m, *AM 10B* (AAS, KRAM); Metchnikoff Pt., 30 m, *AM 60C* (AAS, KRAM); Claude Pt., 25 m, *WH 117* (AAS, KRAM). ***Melchior Is.***: Gamma I., 5–10 m, *RILS 4150* (AAS, KRAM); Lysted I., *PAS 344* (FH). ***Anvers I.***: Joubin Is., *RILS 1988* (AAS, KRAM); Dream I., *VK 934* (KRAM); Litchfield I., *VK 86* (KRAM), *JRS & JMG PAL-43 & 44* (US), 10–25 m, *RILS 1963 & 1964* (AAS, KRAM) and *RWMC 369* (AAS, KRAM). **Wilhelm Archipelago.** ***Argentine Is.***: Irizar I., *RWMC 608B* (AAS, KRAM). ***Grandidier Channel***: Hook I., *RMS 69-505 & 69-710* (US).

2. **Hymenoloma crispulum** (Hedw.) Ochyra

Fig. 89

Hymenoloma crispulum (Hedw.) Ochyra *in* Ochyra, Żarnowiec & Bednarek-Ochyra, Cens. Cat. Polish Mosses: 114. 2003. ≡ *Weissia crispula* Hedw., Spec. Musc. Frond.: 68, *t. 12, f. 1–6*. 1801. ≡ *Grimmia crispula* (Hedw.) Sm., Fl. Brit. 3: 1192. 1804. ≡ *Weissia cirrata* Hedw. var. *crispula* (Hedw.) Wahlenb., Fl. Lapp.: 323. 1812. ≡ *Blindia crispula* (Hedw.) Müll. Hal., Syn. Musc. Frond. 2: 584. 1851. ≡ *Holomitrium crispulum* (Hedw.) Mitt., J. Linn. Soc. Bot.

Suppl. 1: 25. 1859, *hom. illeg.* ≡ *Dicranoweisia crispula* (Hedw.) Milde, Milde, Bryol. Siles.: 49. 1869. — TYPE CITATION: Fagimonti (Buchberg) in Bohemia legit Ludwig. [Lectotype (*selected here*): The second specimen from the left on a sheet labelled 'Weissia crispula sp. Musc. p. 68 t. 12. f. 1.2–6 Fagimonti (Buchberg) in Bohemia Ludw.' – G-Hedwig/Schwägrichen!].

Blindia austrocrispula Müll.Hal., Flora 68: 404. 1885 ['austro-crispula']. ≡ *Dicranoweisia austrocrispula* (Müll.Hal.) Paris, Ind. Bryol.: 339. 1895, ***syn. nov.*** — TYPE CITATION: Fuegia, Staten Island, Port Vancouver, ad rupes montanas, Martio 1882: Spegazzini. [Lectotype (*selected here*): 'Ex Museo botanico Berolinensi. Blindia austro-crispula C. Müll. Original! Fuegia, Staten Island Port Vancouver leg. Spegazzini' – PC-Cardot!].

Blindia subinclinata Müll.Hal. *in* Neumayer, Int. Polarforsch. Deutsch. Exp. 2: 301. 1890. ≡ *Dicranoweisia subinclinata* (Müll.Hal.) Broth. *in* Engl. & Prantl, Nat. Pflanzenfam. 1(3): 318. 1901. — TYPE CITATION: Austro-Georgia, ad rupes der Ostseite des Vexirberges, 17. Januario 1883, c. fr. maturis; am Ausgange des Brockenthales 23. Januario 1883 c. fr. mat. [Lectotype (*vide* Ochyra 1998*a*: 125): 'Nº 7. Blindia subinclinata C Müll. nsp. Fundort: Felsen am Ausgang des Brockenthales in kleinen Polstern. Süd-Georgia, 23/I 83 Will' – HBG!; syntype: 'Ex Museo botanico Berolinensi. Blindia subinclinata C. Müll. Dicranoweisia Austro-Georgia Vexirberg 23.I.1883' – PC-Cardot!; isosyntypes: M!]. First synonymised by Ochyra (1998*a*: p. 125).

Blindia pallidifolia Müll.Hal. *in* Neumayer, Int. Polarforsch. Deutsch. Exp. 2: 301. 1890. ≡ *Dicranoweisia subinclinata* (Müll.Hal.) Paris, Ind. Bryol.: 342. 1895, ***syn. nov.*** — TYPE CITATION: Austro-Georgia, Felsblöcke des südlichen Ufers der Landzunge, 13. Octobri 1882 c. fr. juvenilibus. [Lectotype (*selected here*): 'Nº 7. Blindia pallidifolia C Müll. nsp. Fundort: Felsblöcke des südlichen Ufers der Landzunge. Kleine (3–4 cm im Durchmessen) Rasen bildens. Süd-Georgien. 13. October 1882. Will Südpolar-expedition' – HBG!; isotypes: M! (2 specimens), PC!].

Plants rather small and slender, in dense tufts and cushions, mostly glistening, bright to pale yellow-green above, pale brown to blackish-brown below. *Stems* erect, thin, 1–3 cm tall, dichotomously branched, in cross-section rounded, with a cortex of 1–2(–3) layers of cells with thick, brown walls, a 3–4-stratose medulla of large, thin-walled cells and a small but distinct central duct; *rhizoids* branched, smooth, slender, brown in clusters at the stem base and thick, dark brown, lustrous, sparingly branched scattered in the leaf axils along the stem; *axillary hairs* filiform, hyaline, 8–10-celled, with short basal and elongate upper cells. *Leaves* strongly crisped, curled and incurved when dry, straight to somewhat curved, erect-spreading when moist, (1.7–)2.0–3.0(–3.3) mm long, (0.25–)0.3–0.5(–0.6) mm wide, from a concave, erect, ovate- to oblong-lanceolate, subclasping base gradually or sometimes more rapidly tapering to a long, slender, linear, straight or flexuose, subtubulose subula, acute to obtusely acute and entire at the apex; *margins* entire to crenulate above by projecting cuticular thickenings, plane to distinctly incurved distally; *costa* single, percurrent, relatively slender, 40–60 μm wide, yellow to yellow-brownish, sometimes nearly concolorous and with the laminal cells and appearing diffuse, filling about 1/3 of the subula, on the ventral surface covered with long-rectangular to linear epidermal cells, in cross-section rather flat and not prominent at the back, flat on the ventral side, consisting of a distinct median row of large guide cells and better developed dorsal and weaker ventral stereid bands, the latter often becoming reduced or lacking above, sometimes with rather larger ventral and dorsal epidermal cells; *laminal cells* unistratose throughout, translucent, appearing papillose in transverse section because of the long, cuticular ridges, subquadrate, rounded-quadrate to short-rectangular, sometimes oblate to irregular, arranged in regular, well spaced rows owing to the thicker lateral walls, 6–20 μm long, 4–8 μm wide, becoming very variable in size and shape in the lower half of the base, mostly with strongly incrassate longitudinal walls and truncate at the apices, long-rectangular to linear near the costa or throughout the base, 30–100 μm long, 4–6 μm wide, or becoming shorter in several rows towards the margin; *alar cells* sometimes enlarged, quadrate to hexagonal, thin-walled forming small, yellow-brownish, somewhat bulging auricles or most often thicker-walled, short-rectangular, sometimes laxer, forming a flat, oval, yellowish group not markedly differentiated from the adjacent laminal cells. *Autoicous. Perichaetia* terminal; *outer perichaetial leaves* similar to the vegetative leaves, often with a recurved subula; *inner perichaetial leaves* oblong to oblong-ovate, convolute, concave, truncate, obtuse or broadly acute at the apex, sometimes abruptly contracted into a short, blunt point. *Setae* slender, erect, straight or somewhat flexuose, 7–10 mm long, sinistrorse above, yellow becoming yellow-brown with age. *Capsules* long-exserted, erect to somewhat inclined, ovoid, obloid to short-cylindrical, symmetric, 1.5–2.0 mm long, 0.6–0.7 mm wide, leptodermous, light brown, mostly wrinkled when dry and empty; *annulus* persistent; *operculum* with a long, slanted beak; *exothecial cells* quadrate to hexagonal, thin-walled, 20–40 μm long, 15–25 μm wide, becoming somewhat smaller in 1–2 rows at the mouth; *stomata* few at the base of the urn, bicellular, superficial; *peristome teeth* deeply inserted below the mouth, entire, 160–180 μm long, with 3–5 basal, hyaline, smooth lamellae, orange to

→

Figure 89. *Hymenoloma crispulum.* — **1**. Plant with capsule, dry **2**. Portion of cross-section of stem. **3–5**. Axillary hairs. **6–11**. Leaves. **12**. Leaf apex. **13**. Cells on ventral surface of costa and laminal cells in leaf subula. **14**. Cells at leaf shoulder. **15–16**. Mid-leaf cells. **17–18**. Basal cells. **19–25**. Cross-sections of leaves, a sequence from base to apex. **26**. Perichaetium and mature capsule, wet. **27–28**. Perichaetial leaves. **29**. Cells at shoulder of perichaetial leaf. **30**. Deoperculate capsule with whole seta and vaginula, wet. **31**. Operculate capsule, wet. **32**. Deoperculate capsule, dry. **33**. Operculate capsule with calyptra. **34**. Exothecial cells at base of urn and stoma. **35**. Exothecial cells in middle of urn. **36**. Exothecial cells at orifice, preperistome, peristome teeth and spores. — [1, 11, 18, 30 from *Acuña 7044*; 2–6, 14, 16, 20, 23–24 from *RO 501/80*; 7, 12, 15, 19, 22, 25, 33 from *RO 2123/80*; 8 from *RO 479/80*; 9, 13, 21, 26–29, 34–35 from *RO 2022/80*; 10, 17 from *RO 599/80*; 32 from *RO 578/80*; 31 from *RMS 69–599a*, US; 36 from *Roivainen 2580*]. — Scale bars: a – 1 mm (30); b – 1 mm (6–11, 26–28, 31–33) and 2 mm (1); c – 100 μm (2–5, 12–25, 29, 34–36).

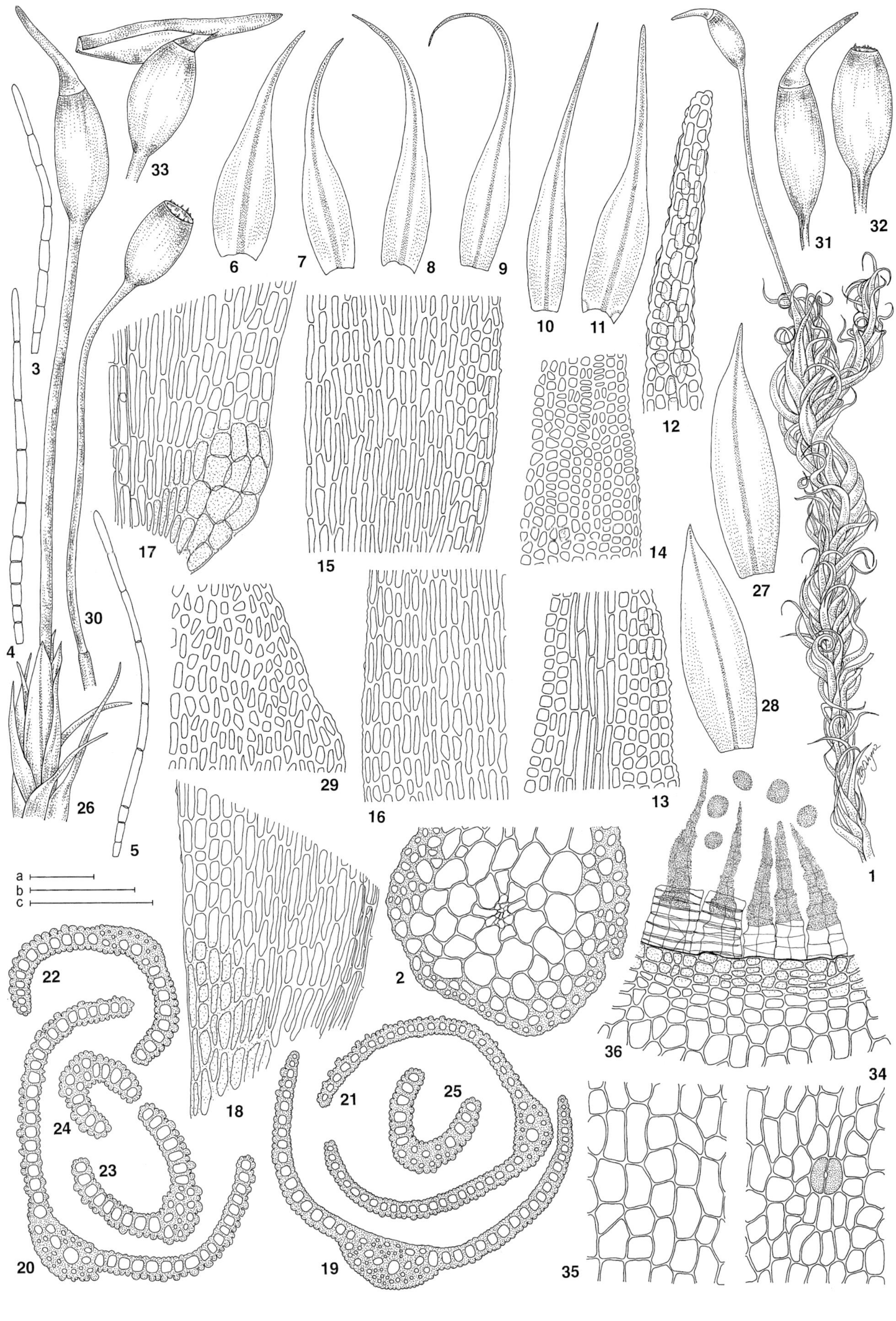

33
6
7
8
9
10
11
12
31
32
3
4
30
17
15
14
27
29
16
13
28
26
5
1
a
b
c
18
2
36
34
22
20
24
23
21
25
19
35

orange-brownish, papillose to obscurely papillose-striate above on both outer and inner surfaces; *prostome* sometimes present, hyaline. *Spores* spherical, finely papillose to nearly smooth, 14–20 μm wide. *Calyptra* cucullate, smooth.

Discussion. *Hymenoloma crispulum* is primarily defined by sporophyte characters and therefore the safest determination is warranted if fertile plants are available. The species is diagnosed by the convolute innermost perichaetial leaves which are truncate, broadly rounded to obtusely acute at the apex or, occasionally, terminated with a short, blunt point. Additionally, the spores are smaller, 14–20 μm in diameter, usually finely papillose, sometimes nearly smooth. In practice, these are the only characters which separate *H. crispulum* from its bipolar counterpart, *H. antarcticum*, in which the innermost perichaetial leaves are subulate, sometimes with a flexuose or tortuous subula as long as the sheathing base, and the spores are generally larger, 20–35 μm in diameter, and finely granulate to verrucose. *Hymenoloma crispulum* has rather slender, yellow setae, 7–10 mm long, and admittedly this length overlaps that of the setae in *H. antarcticum* which ranges from 3–18 mm. However, Antarctic populations of the latter species have very short setae, 3–5 mm long, so the seta length is a useful character for discrimination of the local populations. Moreover, *H. antarcticum* lacks stomata.

The gametophytes of both species are quite similar and the population from Campbell I., which was later described as *Blindia antarctica* (Müller 1848), had initially been recognised by Wilson & Hooker (1847) as a variety, *Weissia crispula* var. *ambigua*. These authors commented: 'After a very careful comparison of the Campbell's Island plant with specimens of *W. crispula* from Snowdon, we are at length satisfied of the specific identity of the two'.

Gametophytically, *Hymenoloma crispulum* is very variable, especially with regard to the robustness of the plants and leaf areolation. Some Arctic populations consist of large and rather robust plants with the leaves showing somewhat peculiar torsion of the subula when dry, with spirally curled tips as well as having prominent, bulging auricles, resembling in this respect some populations of *H. antarcticum*. However, the austral populations of *H. crispulum* have remarkably uniform gametophytes and the plants are generally rather small and slender, glossy, bright yellow to pale yellow-green and the leaves have a narrowly oblong-lanceolate base gradually tapering to a long, flexuose subula. In addition, the alar cells are scarcely differentiated and only very seldom form somewhat bulging brownish auricles. This set of gametophyte characters may be locally useful for determination of sterile plants of *H. crispulum*, and some populations from the Antarctic were so-named on account of the perfect correspondence of their gametophyte features with those found in fertile plants.

Hymenoloma crispulum was reported from the Southern Hemisphere only recently (Ochyra 1998*a*), but it had been described earlier from Isla de los Estados as *Blindia austrocrispula* (Müller 1885) and South Georgia as *B. pallidifolia* and *B. subinclinata* (Müller 1890*a*) but the corresponding type specimens of these species do not differ materially from *H. crispulum* as defined by the European type and accordingly all these taxa are considered to be conspecific with it.

Reproduction in Antarctica. Of the 58 specimens available for examination, 27 were in a fertile state. However, only very few fully mature capsules have been detected in which peristomial features could be observed. Most fertile plants possess either immature sporophytes or only old capsules with destroyed peristomes which, in general, are very fragile and are soon lost.

Habitat. A species associated with a wide range of rock habitats occupied by various sociations of the fruticose lichen and moss cushion subformation. It grows both in dry, damp or wet and exposed or sheltered situations on mostly acidic substrata, including rock crevices, rock faces and ledges, outcrops, scree, as well as on gravelly, rocky or sandy soil between stones, boulders and pebbles, occasionally also on ash soil and coarse scoria. Sometimes it occurs on rock surfaces moistened by trickling water. It is associated with a wide array of lichen, moss and hepatic species, of which the most important are species of *Usnea* and numerous crustose lichens, *Andreaea*, *Schistidium*, *Sanionia*, *Bartramia patens*, *Barbilophozia hatcheri* and *Cephaloziella varians*.

World range. A bipolar species having a pan-Holarctic, arctic-boreal-montane distribution in the Northern Hemisphere, and scattered in southern South America and Antarctica, with some intermediate altimontane stations in the northern Andean countries of Colombia (Churchill *et al.* 1995) and Peru (Menzel 1986) (Fig. 90, inset). Until the genus *Hymenoloma* is revised taxonomically, the geographical distribution of *H. crispulum* cannot be precisely evaluated. For example, a careful taxonomic assessment is needed to determine the status of the northern populations with distinctly pointed inner perichaetial leaves which have been described as *Dicranoweisia contermina*, and which are common in some regions of North America, for example in California (Lawton 1971). Likewise, western North American populations with distinctly bistratose laminae (Flowers 1956; Schofield 2007) do not belong to *H. crispulum* but represent *H. mulahacenii*, the taxonomic status of which also requires a detailed evaluation. The typical phenotypes of *H. crispulum* are common throughout the Arctic, reaching the highest latitudes, as well as in the boreal zone, and decrease markedly southwards where they are restricted mostly to higher elevations in the mountains. In North America it is reported to extend to Oregon, Michigan, Maine and New Hampshire, with some isolated occurrences in Arizona and New Mexico (Crum & Anderson 1981; Vitt *et al.* 1988). In Europe it extends to the Mediterranean including Sicily (Cortini Pedrotti 2001*a*,*b*), Crete (Duell 1995) and Cyprus (Koppe 1976) as well as Morocco and Algeria in North Africa (Ros *et al.* 1999). In Asia, *H. crispulum* extends to the Caucasus (Manakyan 1989), Central Asia (Mamatkulov 1990), Central Siberia (Bardunov 1974) and northern Mongolia (Tsegmed 2001*a*,*b*) and the Russian Far East (Ignatov & Afonina 1992), with some disjunct stations in the Western Himalayas and Kashmir (Chopra 1975), China (Gao *et al.* 1999) and Japan (Noguchi 1987). In southern South America the species is scattered in western Patagonia and Tierra del Fuego, and most collections reported as *Dicranoweisia austrocrispula*, *D. perpulvinata* and *D. subinclinata* (Greene 1986) refer to *Hymenoloma crispulum*.

Distribution in Antarctica. An infrequent and occasional species, widely scattered from the South Shetland Is. to Charcot I. where it reaches its southernmost limit at lat. 69°45′ S and the plants are still fertile (Fig. 90). It is most frequent in the South Shetland Is. where it seems to have the main centre of its regional occurrence and where it quite frequently produces sporophytes; it is rare and localised along the western coast of the Antarctic Peninsula, but with a particular concentration of localities on the Graham Coast, although all collections from this region are sterile. Sterile but otherwise typical plants have occasionally been recorded on Joinville I. and Francis I. on the east coast of the Antarctic Peninsula. The species has not yet been recorded from the South Orkney Is. *Hymenoloma crispulum* is a moss of higher elevations, quite often collected above 250 m, although some specimens originate from

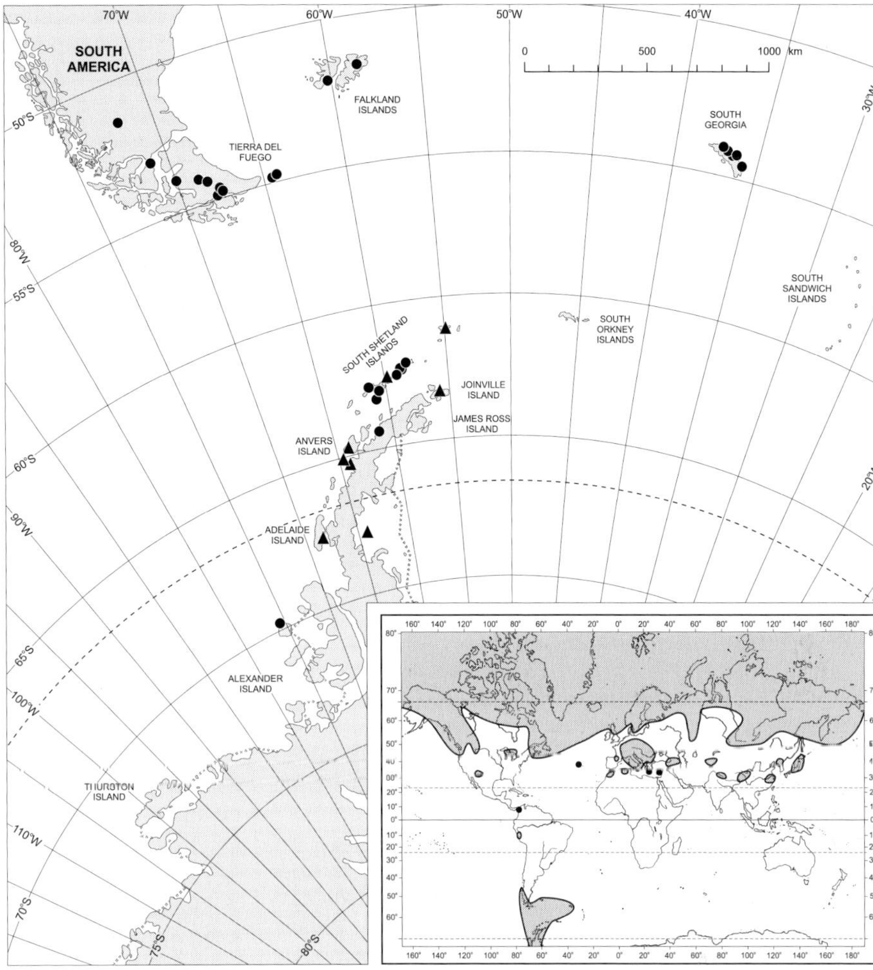

Figure 90. Distribution of *Hymenoloma crispulum* in the Fuegian region and Antarctica. Inset: global distribution.

near sea level. Cardot (1913) reported this species for the first time from the Antarctic as *Dicranoweisia subinclinata* from an elevation of 480 m on 'Mont de Tranchant' on the Graham Coast. This information is somewhat confusing because this mountain, now known as Edge Hill (Hattersley-Smith 1991), rises to only 285 m. It is possible that the moss was collected on the nearby Cape Tuxen which reaches to 735 m and where Charcot's expedition did land.

Specimens examined. WEST ANTARCTICA. SOUTH SHETLAND IS. **Elephant I.** 2 km SW of Endurance Glacier, 167 m, *JSA 21A* (AAS, KRAM) (ster!); 0.5 km SW of Endurance Glacier, 183 m, *JSA 61 & 147* (AAS, KRAM) (both ster!) and 244 m, *JSA 176 & 179A* (AAS, KRAM) (both ster!). **King George I.** ***Fildes Peninsula***: Ardley I., *Kühnemann 145* (AAS, KRAM) (ster!). ***Barton Peninsula***: N side of the peninsula, 10 m, *DCL 765* (AAS, KRAM); Noel Hill, *BJ 200* (KRAM). ***Admiralty Bay***: Point Thomas, Skua Cliff, 60 m, *RO 635/80* (KRAM); Misty Nunatak, 260 m, *RO 2031/80* (KRAM) (ster!); Admiralen Peak, 305 m, *RO 2022/80 & 2029/80* (ster!) (KRAM); Kapitan Peak, 195 m, *RO 1917/80* (KRAM) (ster!); Wegger Peak, 310 m, *RO 2003/80* (KRAM); Garnuszewski Peak, 300 m, *RO 2016/80* (KRAM); Ore Point, 10 m, *RO 501/80* (KRAM); near old British base, 10–15 m, *RILS 715* (AAS, KRAM); Mt. Flagstaff, 100 m, *RO 419/80* (KRAM); Moraine Point, 25 m, *RO 483/80* (KRAM) and 40 m, *RO 479/80* (KRAM); Keller Peninsula, without specific locality, *RMS 69-946 pp.* (US); Martel Inlet, Ullman Spur, 60 m, *RO 578/80* (KRAM) and 80 m, *RO 599/80* (KRAM); Mt. Wawel, 60 m, *RO 2123/80* (KRAM). **Half Moon I.** S of Huts, 50 m, *DCL 593* (AAS, KRAM) (ster!). **Livingston I.** ***Byers Peninsula***: West end opposite Astor I., 21 m, *DCL 361* (AAS, KRAM) (ster!); valley between Chester Cone and Sealer Hill, 25–30 m, *RILS 3821* (AAS, KRAM) (ster!). ***Hurd Peninsula***: Johnsons Dock, 35 m, *DCL 70 & 73* (AAS, KRAM) (both ster!) and 80 m, *Guzmán 111* (AAS, KRAM); eastern side of False Bay, *RMS 69-533* ster!), *69-560 & 69-599A* (US); near Spanish base, *Acuña 7044* (KRAM). **Deception I.** Pendulum Cove, E of Crimson Hill, 130 m, *RILS 11203* (AAS, KRAM); between Kroner Lake and Ronald Hill, *ca* 15 m, *REL 901* (AAS, KRAM); Crater Lake, 70–200 m, *RILS 6849 & 6854* (AAS, KRAM); Mt. Kirkwood, 250 m, *RILS 5715C* (AAS, KRAM).

WEST ANTARCTIC PENINSULA. **Danco Coast.** Cierva Pt., 5 m, *OB 26* (AAS, KRAM). **Graham Coast.** Edge Hill, *Gain 275* (PC) (ster!); Cape Tuxen, 15 m, *RWMC 671* (AAS, KRAM) and 100 m, *RWMC 851B* (AAS, KRAM) (both ster!); Mt. Demaria, *ca* 270 m, *RWMC 691* (AAS, KRAM) (ster!). **Wilhelm Archipelago.** ***Penola Strait***: Petermann I., *ca* 25 m, *RWMC 793* (AAS, KRAM) (ster!). ***Argentine Is.***: Galindez I., Stella Creek, 15 m, *REL 1326* (AAS, KRAM) (ster!) and 20 m, *RWMC 415* (AAS, KRAM) (ster!); Skua I., Skua Creek, 15 m, *RWMC 461C* (AAS, KRAM) (ster!). ***Collins Bay***: Berthelot Is., *ca* 65 m, *RWMC 622C* (AAS, KRAM) (ster!). **Fallières Coast.** ***Adelaide I.***: E side of of Mt. Mangin, 270 m, *JK 149B* (AAS, KRAM) (ster!). **George VI Sound.** ***Charcot I.***: Rils Nunatak, 85 m, *RILS 10491* (AAS, KRAM) and 125 m, *RILS 10475 & 10538* (ster!) (AAS, KRAM).

EAST ANTARCTIC PENINSULA. **Joinville Island group.** ***Joinville I.***: Mt. Alexander, 60–75 m, *RILS 3696* (AAS, KRAM) (ster!). **Bowman Coast.** NW corner of Francis I., *ca* 300 m, *RILS 2770B & 2773AA* (AAS, KRAM) (ster!).

3. **Hymenoloma antarcticum** (Müll.Hal.) Ochyra

Figs 91–92

Hymenoloma antarcticum (Müll. Hal.) Ochyra *in* Ochyra, Żarnowiec & Bednarek-Ochyra, Cens. Cat. Polish Mosses: 114. 2003. ≡ *Blindia antarctica* Müll.Hal., Syn. Musc. Frond. 1: 344. 1848. ≡ *Dicranum antarcticum* (Müll.Hal.) Mitt., J. Linn. Soc. Bot. 12: 63. 1869. ≡ *Dicranoweisia antarctica* (Müll.Hal.) Kindb., Enum. Bryin. Exot.: 54. 1888. — Type citation: Insula Campbelli et Eremitae ad Cap. Horn: J. Hooker. [Lectotype (*selected here*): 'W. 25 Weissia crispula var. β. ambigua C. Island – Campbells Island. Antarct. Exp. 1839–1843. J.D.H.' – BM!; isolectotypes: BM!, BM-Wilson!, BM-Hampe! (2 specimens), BM-Shuttleworth!; syntype: 'N°. 99. W. 126. Blindia antarctica C. Müller l.344 Hermite Island C. Horn' – BM-Wilson!; isosyntypes: BM!, BM-Hampe!].

Weissia crispula Hedw. var. *ambigua* Wilson & Hook.f. *in* Hook. f., Fl. Antarct. 1: 127, *pl. 58, f. 2.* 1847, ***syn. nov.*** — TYPE CITATION: Campbell's Island; on exposed rocks from 600–1200 feet above the level of the sea, abundant. [Lectotype (*selected here*): 'W. 25 Weissia crispula var. β. ambigua C. Island – Campbells Island. Antarct. Exp. 1839–1843. J.D.H.' – BM!; isolectotypes: BM!, BM-Wilson!, BM-Hampe! (2 specimens), BM-Shuttleworth!].

Blindia brevipes Müll.Hal. *in* Neumayer, Int. Polarforsch. Deutsch. Exp. 2: 300. 1890. ≡ *Dicranoweisia brevipes* (Müll.Hal.) Cardot, Bull. Herb. Boiss. Sér. 2, 5: 998. 1905, ***syn. nov.*** — TYPE CITATION: Austro-Georgia, ad rupes des Köppenberges, 19. Majo 1883. [Lectotype (*selected here*): 'Südpolarexpedition Blindia brevipes C.Müll. nsp. Austro-Georgia, ad rupes des Köppenberges, 19.V.1883 No. 7' – HBG!; isotype: PC-Cardot!].

Hymenoloma nordenskjoeldii Dusén, Ark. Bot. 4(1): 20, *pl. 5, f. 16–20* & *pl. 6, f. 1–13.* 1905. ['nordenskjöldii'] ≡ *Dicranoweisia nordenskjoeldii* (Dusén) Reimers, Notizbl. Bot. Mus. Berlin-Dahlem 13: 44. 1936, ***syn. nov.*** — TYPE CITATION: Patagonia occidentalis in valle fluminis Rio Aysen, ubi speciem in truncis arborum putrescentibus rare crescentem et fructibus operculatis praeditam mense Januario legi. [Lectotype (*selected here*): '506. Hymenoloma Nordenskjöldii Dus. Patagonia occ. in valle fluminis Aysén in truncis putridis. 30.I.1897. P. Dusén' – S-Dusén!; isotypes: PC-Cardot!, S!].

Dicranoweisia breviseta Cardot, Bull. Herb. Boissier Sér. 2, 5: 998. 1905, ***syn. nov.*** — TYPE CITATION: Terre-de-Feu: Ushuaia. [Lectotype (*selected here*): 'Svenska Sydpolarexpeditionen 1901–03. N:r 15. Dicranoweisia breviseta Card. sp. nova Tierra del Fuego, Monte Martial supra Ushuaia, 1135 m s.m. 22/3 1902 Carl Skottsberg. Det. J. Cardot' – PC-Cardot!; isotype: S!].

Dicranum mackayi Broth. & Dixon, J. Linn. Soc. Bot. 40: 437, *pl. 20, f. 6–11.* 1912. ≡ *Dicranoweisia mackayi* (Broth. & Dixon) Broth. *in* Engl., Nat. Pflanzenfam. ed. 2, 10: 198. 1924. ≡ *Holodontium mackayi* (Broth. & Dixon) Broth. *in* Engl., Nat. Pflanzenfam. ed. 2, 10: 199. 1924. — TYPE CITATION: [New Zealand] Nun's Veil, Mt. Cook district, alt. 6000 ft.; leg. Dr Mackay (No. 130). [Lectotype (*selected here*): 'Herb. H. N. Dixon Ref. No. 130 Dicranum Mackayi Broth. & Dixon sp. nov. {Journ. Linn. Soc., XL, 437 (1912)} Nun's Veil, Mt. Cook district, South I., New Zealand, alt. 6000 ft. Coll. Dr. Mackay Dec. 1907 Comm. Brit. Antarct. Exped. Type' – BM-Dixon!; isotype: H-Broth!]. First synonymised by Sainsbury (1948*a*: p. 80).

Plants small to moderately sized, forming dense but rather loosely coherent tufts or cushions, occasionally in extensive patches, dull to lustrous, mostly dark to sordid-green, or yellow-, olive- to brownish-green above, brown to blackish-brown below. *Stems* erect, 0.5–2.5, sometimes up to 4.5 cm tall, dichotomously branched, slender, variously radiculose at the base with brown, smooth, branched rhizoids, often with large, thick, sparingly branched, dark brown rhizoids scattered in the leaf axils, in cross-section with a 1–2(–3)-stratose cortex of small, stereid cells with brown walls, 3–5-stratose medulla of large, thin-walled, hyaline cells and small to medium-sized central strand; *axillary hairs* filiform, 5–10-celled, hyaline throughout or with 2–3 short brownish basal cells. *Leaves* crisped to tortuous at the stem apex, less altered, erect to erect-spreading below, often falcato-secund to hamate on drying, erecto-patent to spreading, flexuose to subhomomallous when moist, 2.0–4.5 mm long, 0.3–0.8 mm wide, from a concave, ovate, ovate- or oblong-lanceolate base, gradually or sometimes more abruptly narrowed to a long, straight or flexuose, subtubulose subula, obtusely acute at the apex; *margins* entire to crenulate, plane below, incurved above; *costa* single, percurrent, 30–100 μm wide, intensively yellow to brownish-yellow, filling about ⅓ of the subula, distinctly set off from the laminal cells or sometimes diffuse and subconcolorous in the base, on the ventral surface covered with long-rectangular to linear epidermal cells, in cross-section flat dorsally below, more prominent and convex at the back in the subula, with a stronger dorsal stereid band and weaker or sometimes absent ventral stereid or substereid band separated by a median row of large guide cells; *laminal cells* unistratose, transparent, appearing papillose in cross-section owing to the cuticular thickenings covering many cells in length and only one cell in width over the lumen; *upper cells* quadrate, rounded-quadrate to oblong, thick-walled, in regular, longitudinal, well spaced rows, 4–8 μm wide, 5–20 μm long; *basal cells* very variable in size and shape, short- or long-rectangular to linear, often mixed with rounded-quadrate to oblong, 3–8 μm wide, 10–105 μm long, mostly in regular, well spaced, longitudinal rows, often becoming shorter and narrower towards the margins but not forming distinct borders; *alar cells* large, quadrate to hexagonal, thin- to thicker-walled, brownish to yellowish-brown, forming prominent, rounded or oval, bulging auricles or not or weakly differentiated, forming flat, often yellowish group imperceptibly merging into the basal laminal cells. *Autoicous. Inner perichaetial leaves* 2.0–2.5 mm long, from a concave, oblong-ovate, sheathing base, rapidly constricted into a short or long, straight or flexuose subula, sometimes as long as or longer than the sheathing base. *Setae* straight, stout, dark yellow- to reddish-brown, 3–5 mm long, not twisted. *Capsules* short-exserted, erect to inclined, subglobose to ovoid, 0.9–1.2 mm long, 0.7–0.9 mm wide, rather pachydermous, usually lustrous, castaneous, smooth; *annulus* persistent; *operculum* conical, obliquely long-rostrate, with a beak almost as long as the urn; *exothecial cells* rounded

⟶

Figure 91. *Hymenoloma antarcticum* (typical phenotype). — **1**. Portion of shoot with capsule, dry **2**. Sterile plant, dry. **3**. Cross-section of stem. **4–5**. Axillary hairs. **6–10**. Leaves. **11**. Leaf apex. **12**. Cells on ventral surface of costa and laminal cells in leaf subula. **13**. Cells at leaf shoulder. **14**. Mid-leaf cells at margin. **15**. Mid-leaf cells. **16**. Alar cells. **17–23**. Cross-sections of leaves, a sequence from base to apex. **24–25**. Perichaetial leaves. **26**. Cells at shoulder of perichaetial leaf. **27**. Operculate capsule. **28–29**. Deoperculate capsules. **30**. Exothecial cells at base of urn. **31**. Exothecial cells in middle of urn. **32**. Exothecial cells at orifice, peristome teeth and spores. — [1, 25 from *Hooker W.25* (lectotype of *Blindia antarctica*), BM; 2–23 from *RT 56*; 24, 26–32 from *Seppelt 14435*]. — Scale bars: a – 2 mm (2), 100 μm (32) and 200 μm (30–31); b – 100 μm (3, 11–23, 26) & 200 μm (4–5); c – 1 mm (6–10, 24–25, 27–29); d – 1 mm (1).

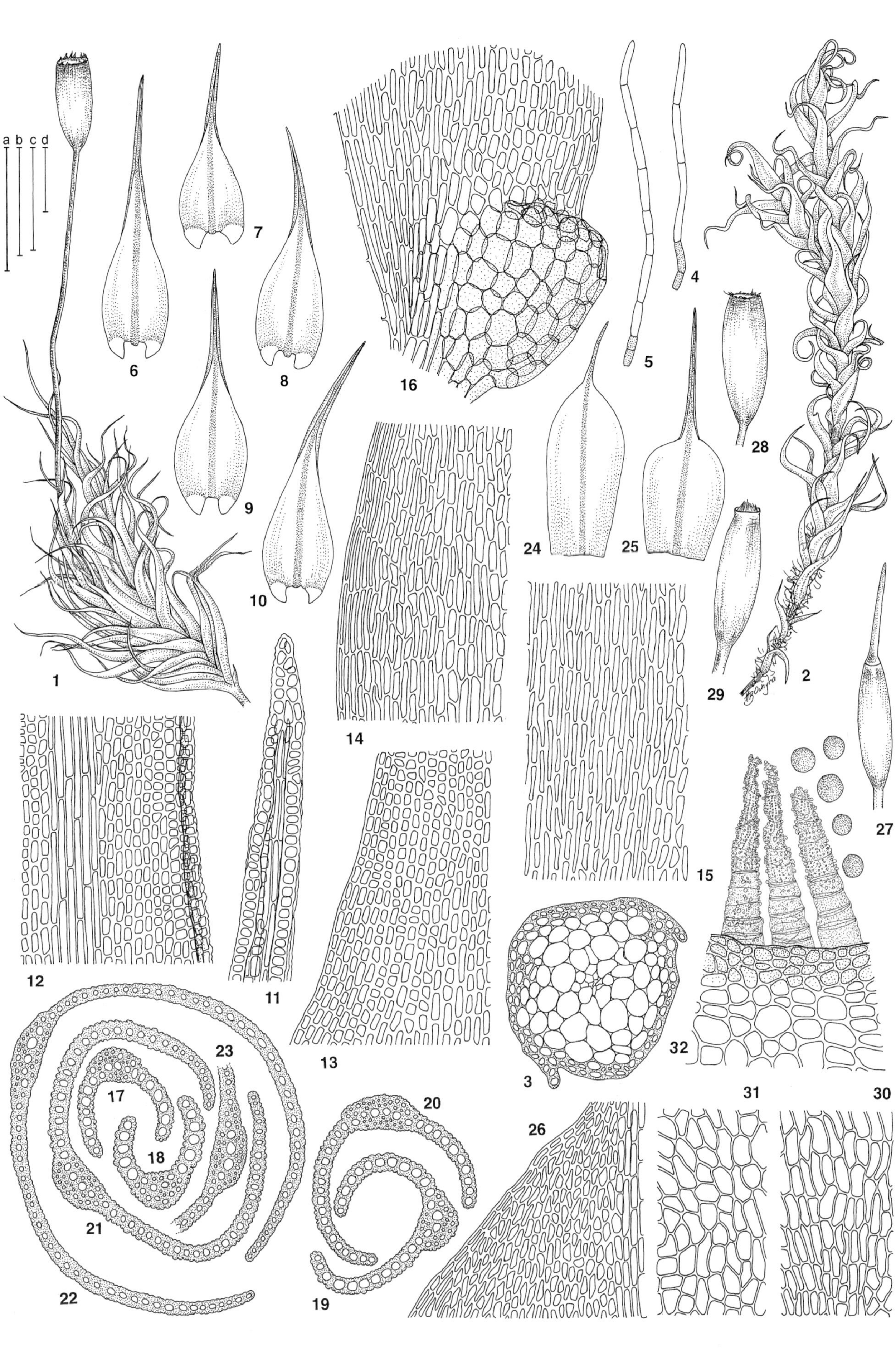
a b c d
1
2
3
4
5
6
7
8
9
10
11
12
13
14
15
16
17
18
19
20
21
22
23
24
25
26
27
28
29
30
31
32

to hexagonal, thin-walled, 25–40 µm wide, 40–70 µm long, becoming smaller, elliptical, thicker-walled at the mouth, forming an orange-brown strip of 3–5 rows of cells; *stomata* absent; *peristome teeth* triangular, entire or variously eroded at the apex, 180–240 µm long, erect or sometimes recurved, orange-brown, reddish-orange to hyaline-yellow, often soon becoming hyaline and fragile, densely papillose throughout or coarsely verrucose distally and finely papillose at the base. *Spores* globose, brownish, densely papillose to granulose, (18–)20–35 µm in diameter. *Calyptra* cucullate, smooth, naked.

Discussion. *Hymenoloma antarcticum* is closely related to *H. crispulum*, and the differences between them are discussed under the latter species. Throughout its wide geographical range *H. antarcticum* is very variable, especially in the length of the seta. In this character the Antarctic populations are much closer to the lectotype material from Campbell I. in which the seta is about 6 mm long (Fig. 123), whereas in the Antarctic plants it varies from 3 to 5 mm.

Gametophytically, the Antarctic populations clearly represent two phenotypes. Plants from the South Shetland Is., especially King George I. and Livingston I., as well as from the western coast of the Antarctic Peninsula fit perfectly the type of the South Georgian *Blindia brevipes* (Müller 1890*a*) and the almost identical Fuegian *Dicranoweisia breviseta* (Cardot 1905*a*, 1908*a*). Plants in these populations are slender and have relatively short basal laminal cells and lack convex auricles. Conversely, many populations from the South Orkney Is., especially Signy I., as well as from the volcanic Deception I. in the South Shetland Is., comprise larger plants with distinctly tortuous upper leaves, broader and distinctly auriculate leaves and very long, often linear, basal laminal cells. They approach the type material of *Blindia antarctica* from Campbell I. and also Hermit I. in the Fuegian region. However, all intermediate characters occur between these extremes and therefore recognition of either phenotype as a separate taxon would lead to serious confusion, and some populations could be named only randomly. There is considerable variation of some structural characters even within a single shoot or leaf. For example, on a single leaf an auricle in one leaf angle may be prominent but with completely undifferentiated cells in the opposite one. The same is also true for the leaf areolation. Therefore, *Dicranoweisia brevipes* and *D. breviseta* are here reduced to synonymy with *Hymenoloma antarcticum*. The conspecificity of *Dicranoweisia mackayi* from New Zealand (Dixon 1912) with *H. antarcticum*, first suggested by Sainsbury (1948*a*), is also confirmed and the generitype of *Hymenoloma*, *H. nordenskjoeldii* from western Patagonia (Dusén 1905*a*), also represents this species, and therefore both names are reduced to synonymy.

A feature of the austral species of the *Hymenoloma crispulum–H. antarcticum* complex is a progressive reduction of the length of the seta. In *H. dryptodontoides* from Îles Kerguelen the seta is only 1–2 mm long and the perichaetial leaves reach the capsules which are barely emergent. The reduction of the seta length is most advanced in *H. insulare* from the Kerguelen Province in which the seta is less than 1 mm long and consequently the capsules are concealed within the perichaetial leaves and the plants resemble superficially species of *Schistidium*.

Reproduction in Antarctica. About one third of the 162 specimens examined are sterile, but in most fertile populations the sporophytes are produced in great abundance. This species is one of the most frequent producers of sporophytes amongst the Antarctic mosses, together with species of *Schistidium*, *Andreaea gainii*, *A. regularis*, *Bartramia patens*, and *Hennediella antarctica*.

Habitat. The ecological requirements of *Hymenoloma antarcticum* are identical to those of its closely related *H. crispulum*. It grows most often on rocks and cliffs, both dry and with a plentiful water supply, in rock fissures and crevices, and on scree, gravelly and stony ground, on shingle, pebbles, rocky slopes as well as on scoria and volcanic ash. Occasionally, it colonises re-exposed peat, and on islands with recent volcanic activity it is a component of the fumarole vegetation, but usually at the unheated outer edge. In general, it is associated with fellfield vegetation belonging to the fruticose lichen and moss cushion subformation.

World range. A pan-south-temperate species (Fig. 93, inset), scattered throughout both islands of New Zealand (Dixon 1914; Sainsbury 1955), in south-eastern Australia and on Tasmania (Streimann & Curnow 1989) and along the western coast of southern South America from the Valdivian region to Cape Horn (Greene 1986) as well as occurring on all Subantarctic islands, from South Georgia to Macquarie I. and extending to the maritime Antarctic. As with *Hymenoloma crispulum*, the global range of *H. antarcticum* cannot be precisely designated until a global taxonomic revision of this critical complex has been undertaken. For example, *H. antarcticum* is known from the northern Andes of Ecuador, where it is represented by the original material of *Dicranoweisia fastigiata*.

Distribution in Antarctica. *Hymenoloma antarcticum* is a widespread and locally common and abundant species, especially on Signy I., King George I. and Deception I. Unfortunately, its local range cannot be strictly determined because many populations are sterile and their identity is uncertain, especially on Bouvetøya, the South Sandwich Is. and on the Antarctic Peninsula (Fig. 93), although some gametophyte features, as well as the presence in critical areas of fertile plants, seems to indicate that the local populations represent *H. antarcticum* rather than *H. crispulum*. The three southernmost populations of the species are sterile. The furthest south occurrence is at lat. 69°14′ S on the Wilkins Coast on the eastern coast of the Antarctic Peninsula. Its elevational range extends from sea level to 460 m, although it seems to prefer the higher elevations.

Selected specimens examined. WEST ANTARCTICA. BOUVETØYA. Rustadkollen, 290 m, 1 Jan 1979, *TE s.n.* (BG).

→

Figure 92. *Hymenoloma antarcticum* (*brevipes* phenotype). — **1**. Plant with capsule, dry **2**. Cross-section of stem. **3–6**. Axillary hairs. **7–13**. Leaves. **14**. Leaf apex. **15**. Cells on ventral surface of costa and laminal cells in leaf subula. **16**. Cells at leaf shoulder. **17**. Mid-leaf cells. **18–19**. Basal cells. **20–28**. Cross-sections of leaves, a sequence from base to apex. **29**. Perichaetium and mature capsule, dry. **30–32**. Perichaetial leaves. **33**. Cells at shoulder of perichaetial leaf. **34**. Deoperculate capsule with whole seta and vaginula, wet. **35–36**. Operculate systylious capsules, wet. **37**. Exothecial cells at base of urn. **38**. Exothecial cells in middle of urn. **39**. Exothecial cells at orifice, peristome teeth and spores. — [1, 7, 20 from *RO 2177/80*; 2, 32 from *RO 61/80*; 3–6, 12–13 from *RO 1819/80*; 9–10, 14–17, 19, 23, 26–28, 34 from *RO 689/80*; 22, 24–25, 39 from *RO 105/80*; 29 from *RO 2651/80*; 30–31, 33, 36–38 from *RO 4797/79*; 35 from *RO 5126/79*; 8 from *BJ 165*; 11, 18, 21 from *RW 93*]. — Scale bars: a – 1 mm (7–13); b – 2 mm (1), 100 µm (2–6, 20–28) & 200 mm (37–38); c – 1 mm (29–32, 34–36) & 100 µm (14–19, 33, 39).

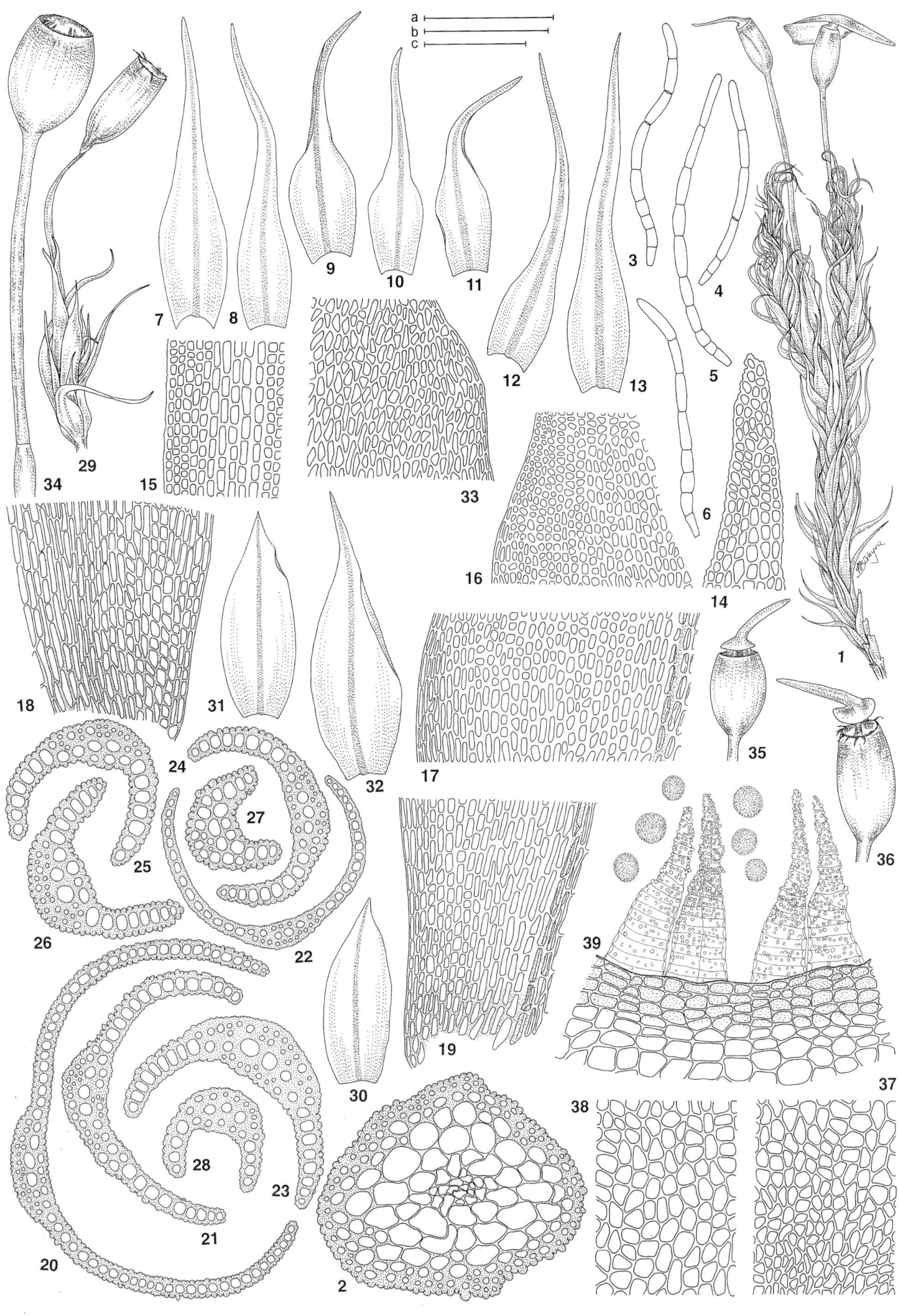
a
b
c
1
2
3
4
5
6
7
8
9
10
11
12
13
14
15
16
17
18
19
20
21
22
23
24
25
26
27
28
29
30
31
32
33
34
35
36
37
38
39

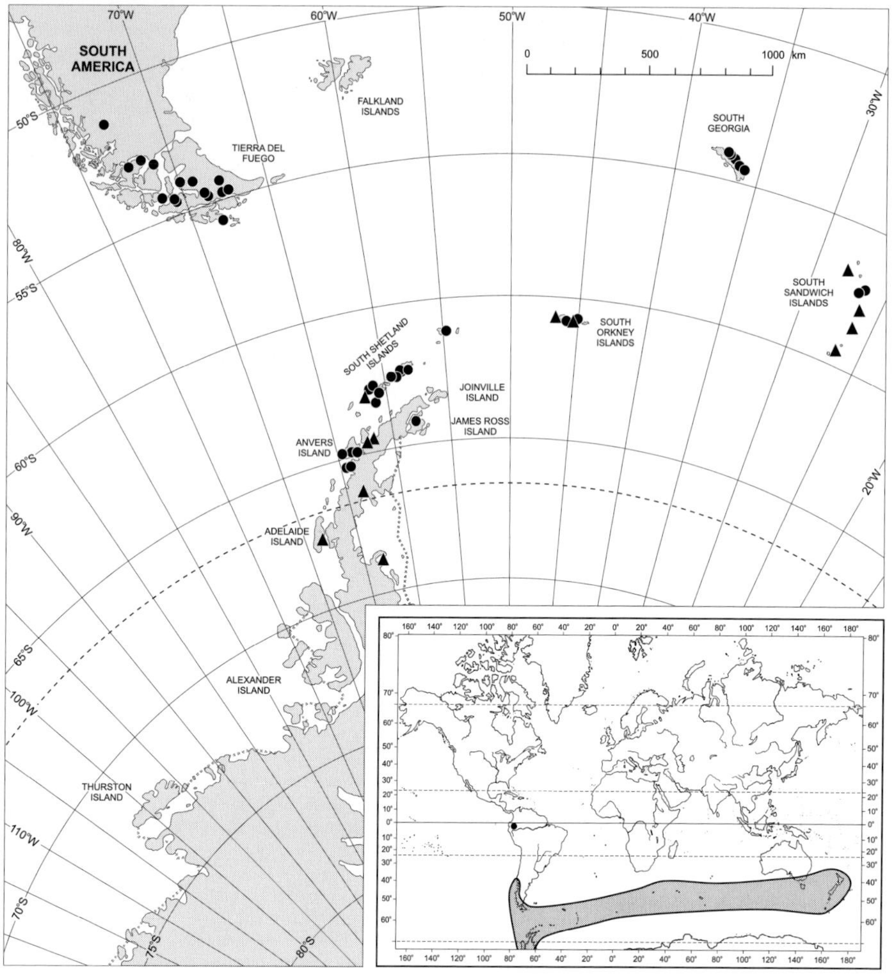

Figure 93. Distribution of *Hymenoloma antarcticum* in the Fuegian region and Antarctica. Inset: global distribution.

SOUTH SANDWICH IS. **Leskov I.** Summit ridge, 200 m, *REL 510* (AAS, KRAM) (ster!). **Candlemas I.** Sea Serpent Cove, *KA 11B* (AAS, KRAM). **Vindication I.** NE corner of the island, 180 m, *MWH 439* (AAS, KRAM). **Saunders I.** Harper Pt., *ca* 20 m, *MWH 846AA* (AAS, KRAM) (ster!). **Montagu I.** Scarlett Pt., 50 m, *PC 97A* (AAS, KRAM) (ster!). **Bellingshausen I.** W side of main crater, 135 m, *MWH 478* (AAS, KRAM); S slopes of main cone, 100 m, *MWH 406* (AAS, KRAM), 120 m, *PC 66A* (AAS, KRAM) and 150 m, *PC 45C* (AAS, KRAM) (all ster!).

SOUTH ORKNEY IS. **Monroe I.** S part, 80 m, *RILS 8005* (AAS, KRAM) (ster!). **Coronation I.** Mansfield Pt., 20 m, *MMM 2* (AAS, KRAM) (ster!); Olivine Pt., 1 m, *RILS 8680* (AAS, KRAM); Saunders Pt., 35 m, *RILS 32* (AAS, KRAM). **Lynch I.** 15 m, *DCL 966 & 967B* (AAS, KRAM). **Signy I.** Knob Lake, *ca* 65 m, *RW 11, 18, 21, 26, 100, 113A & 122* (AAS, KRAM); between Knob Lake and Heywood Lake, *ca* 65 m, *RW 2 & 82* (AAS, KRAM); Changing Lake, 40 m, *RILS 10792 & 10794B* (AAS, KRAM); between Moss Lake and Jane Peak, *ca* 65 m, *RILS 599* (AAS, KRAM); Factory Cove, *ca* 35 m, *RW 93* (AAS, KRAM); Observation Bluff, 60 m, *RILS 8662* (AAS, KRAM); Rusty Bluff, 80 m, *RILS 5153* (AAS, KRAM); above Paal Harbour, *ca* 80 m, *RILS 667* (AAS, KRAM); Gneiss Hills, 200 m, *RILS 8098* (AAS, KRAM). **Moe I.** 35–100 m, *RILS 5261A* (AAS, KRAM) (ster!). **Matthews I.** The Divide, *ca* 65 m, *RILS 173* (AAS, KRAM) (ster!). **Robertson Is.** Steepholm I., 75 m, *RILS 3252* (AAS, KRAM) (ster!). **Powell I.** NW coast, 300 m, *RILS 233* (AAS, KRAM); John Peaks, 460 m, *RILS 272* (AAS, KRAM).

SOUTH SHETLAND IS. **Elephant I.** Stinker Pt., 107 m, *JSA 86* (AAS, KRAM). **King George I.** ***Fildes Pneinsula***: Gemmel Peaks, *ca* 100 m, *RO 2476/80* (KRAM) and 120 m, *RO 2477/80* (KRAM). ***Barton Peninsula***: Noel Hill, *BJ 201* (KRAM). ***Marian Cove***: North Spit, *BJ 156* (KRAM). ***Bransfield Strait***: Red Hill, 100 m, *RO 1101/80* (KRAM); Blue Dyke, 120 m, *RO 1183/80* (KRAM); The Tower, 340 m, *RO 814/80* (KRAM). ***Admiralty Bay***: Sphinx Hill, 100 m, *RO 193/80* (KRAM); Czajkowski Needle, 230 m, *RO 689/80* (KRAM); Skua Cliff, 105 m, *RO 623/80* (KRAM); Jersak Hills, 140 m, *RO 5059/79* (KRAM); Jardine Peak, 280 m, *RO 5176/79* (KRAM); Emerald Pt., 45 m, *RO 821/80* (KRAM); Keller Peninsula, Ore Point, 90 m, *RO 450/80* (KRAM); Shark Fin, 150 m, *RO 2644/80 & 2651/80* (KRAM); Precious Peaks, 100 m, *RO 2662/80* (KRAM); Mt. Wawel, 40 m, *RO 2141/80* (KRAM); Harnasie Hill, 140 m, *RO 1845/80* (KRAM). **Nelson I.** Moss Bay, 26.ii.1989, *Komárek s.n.* (KRAM) (ster!); Rip Pt., 80 m, *DCL 793B* (AAS, KRAM) (ster!). **Greenwich I.** Spark Pt., 20 m, *DCL 639* (AAS, KRAM). **Livingston I.** ***Byers Peninsula***: Clark Nunataks, 180 m, *DCL 235* (AAS, KRAM); between Smellie Pt. and Chester Cone, 35–50 m, *RILS 3802* (AAS, KRAM) (ster!); Negro Hill, 100 m, *KR 14D* (AAS, KRAM) (ster!). ***Hurd Peninsula***: False Bay, *RMS 69-545* (US); Barnard Pt., 80 m, *DCL 43* (AAS, KRAM) (ster!). **Deception I.** South-East Pt., 250 m, *REL* (AAS, KRAM); Mt. Pond, *ca* 250 m, *RILS 74A* (AAS, KRAM); between Kroner Lake and Whalers Bay, 3–5 m, *CGB 40* (AAS, KRAM) and *RILS 11242, 11243 & 11244* (AAS, KRAM); Crimson Hill, 115 m, *RILS 11214A* (AAS, KRAM) and 130 m, *RILS 11200* (AAS, KRAM); Telefon Ridge, 30–60 m, *RILS 5761A & 5757* (AAS, KRAM); Cross Hill, 60 m, *RILS 5795A* (AAS, KRAM); Stonethrow Ridge, 60 m, *RILS 5804A* (AAS, KRAM) and 104–240 m, *RILS 5802A* (AAS, KRAM); Vapour Col, *ca* 175 m, *RILS 11267 & 11271* (AAS, KRAM); Wensleydale Gap, 300 m, *NJC 28* (AAS, KRAM); Wensleydale Beacon, 30–100 m, *RILS 6667B, 6669 & 6674A* (AAS, KRAM); between Telefon Ridge and Stonethrow Ridge, 150 m, *RILS 5733B, 5734B, 5735B & 5736B* (AAS, KRAM); Fumarole Bay, 10 m, *RILS 10308 & 10309* (AAS, KRAM), 50 m, *RILS 11349, 11536 & 11537* (AAS, KRAM) and 100 m, *RILS 11434, 11507 & 11511* (AAS, KRAM); Crater Lake, 30 m, *RILS 11416* (AAS, KRAM); Mt. Kirkwood, 250 m, *RILS 5715A & 5721C* (AAS, KRAM). **Snow I.** President Head, *KR 13A* (AAS, KRAM) (ster!).

WEST ANTARCTIC PENINSULA. **Danco Coast.** Birdsend Bluff, 5–35 m, *RILS 4224* (AAS, KRAM) (ster!); Coughtrey Peninsula, Paradise Harbour, *RMS 69-356, 69-367 & 69-370* (US) (ster!). **Palmer Archipelago.** ***Brabant I.***: Metchnikoff Pt., 50 m, *AM 68C* (AAS, KRAM); island S of Metchnikoff Pt., 50 m, *WH 45* (AAS, KRAM). ***Melchior Is.***:

Omega I., 2–10 m, *RILS 4129* (AAS, KRAM) (ster!). ***Anvers I.***: Norsel Pt., 5–10 m, *RILS 1948* (AAS, KRAM); Jubin Is., middle island, 5–10 m, *RILS 1994* (AAS, KRAM) (ster!); Jubin Is., largest island, 20–50 m, *RILS 1978* (AAS, KRAM); Hermit I., *ca* 15 m, *RILS 874 & 875* (AAS, KRAM) (ster!); Doumer I., *ca* 65 m, *FIDS Misc. 4* (AAS, KRAM). **GRAHAM COAST.** Cape Tuxen, 200 m, *RILS 4905 & 4906* (AAS, KRAM). **WILHELM ARCHIPELAGO.** ***Argentine Is.***: largest island between Irizar I. and Uruguay I., *ca* 15 m, *RWMC 435A* (AAS, KRAM). **FALLIÈRES COAST.** ***Adelaide I.***: Stork Ridge, *PC 266A* (AAS, KRAM) (ster!).

EAST ANTARCTIC PENINSULA. **JAMES ROSS ISLAND GROUP.** ***James Ross I.***: St. Martha Cove, 75 m, *RILS 7432* (AAS, KRAM) and 150 m, *RILS 7534* (AAS, KRAM). **FOYN COAST.** Cabinet Inlet, Cape Casey, *ca* 600 m, *RT 56* (AAS, KRAM) (ster!). **WILKINS COAST.** Finley Heights, *ca* 235 m, *MJC 5A* (AAS, KRAM) (ster!).

EXCLUDED SPECIES OF THE SELIGERIACEAE

Blindia dicranellacea Müll.Hal. — Kanda (1987*b*) reported this species from King George I., South Shetland Is. However, the true identity of this material remains unknown because the relevant herbarium collection has not been available for study. The type material of *Blindia dicranellacea* from South Georgia actually belongs within *Kiaeria pumila* (Ochyra 1998*a*).

Dicranoweisia dryptodontoides (Müll.Hal.) Broth. — This endemic species of the Kerguelen Province, correctly belonging to *Hymenoloma* as *H. dryptodontoides* (Müll.Hal.) Ochyra, is closely related to *H. antarcticum* but otherwise is a distinct species, so the suggestion of Marshall & Convey (1997) that some Antarctic populations named *Dicranoweisia antarctica* may belong to this species should be rejected.

D. immersa Broth. — This endemic species of the Kerguelen Province, correctly belonging within *Hymenoloma* as *H. immersum* (Broth.) Ochyra, was reported, with some reservation, from a single collection (*RILS 689*, AAS, KRAM) from Signy I., South Orkney Is., by Bell (1976) but the voucher collection represents *Ditrichum hyalinocuspidatum* (see p. 146).

Family VIII. GRIMMIACEAE ARN.

Grimmiaceae Arn., Disp. Méth. Mousses: 19. 1825 [‘Ordre Grimmoideae’]. — TYPE: *Grimmia* Hedw.

The Grimmiaceae are one of the most distinctive moss families. The great majority of its members exhibit marked adaptation to xerophytic habitats and they show a superficial resemblance to the diplolepidous family Orthotrichaceae. They are primary colonizers which are mostly found on bare, usually dry and exposed rocks and stones, forming predominantly dark green to blackish cushions or tufts, although many species also occur on wet or damp rocks and boulders along the streams and rivers. However, *Schistidium maritimum* is one of very few moss halophytes restricted to the seashore where it grows within the salt-water spray zone. Some species of the Grimmiaceae are highly specialised and confined to a specific rock type, including sedimentary, igneous or metamorphic rocks, while others are generalists that grow on a variety of rock types. Epigean habitats are not frequently occupied grimmialean mosses and very few are epiphytes.

Although members of the Grimmiaceae are rather easily discernible from other acrocarps, the family is difficult to characterize in simple and unequivocal terms, mostly because the essential diagnostic characters are variously expressed throughout it. This refers especially to the hyaline hair-points of the leaves and sinuose laminal cells. Hyaline awns are present in most species of this family, but their shape and length vary markedly, even on the same plant, from a very short, translucent tip to an arista exceeding the length of the lamina, and on some lower leaves they may be absent. They are flat or terete, smooth, denticulate, spinose or papillose, and not or long decurrent.

The leaves are typically lanceolate to ovate-lanceolate, less often ovate or linear. They are highly hygroscopic, being tightly appressed or crisped about the stem on drying and erect-spreading to patent on wetting. The laminal cells are short to elongate, smooth, pseudopapillose or papillose and uni- to polystratose. The cell walls are generally incrassate, often variously sinuose to strongly nodose, and this expression of the leaf areolation has been considered an adaptation to xeric conditions (Lorch 1931; Tallis 1959). The leaves always have a single costa which is usually strongly convex dorsally, semi-terete, semi-elliptic or reniform in cross-section, but sometimes it is scarcely prominent on the abaxial surface. Its internal anatomy is differentiated into three cell layers, including ventral (adaxial) and dorsal (abaxial) epidermis and a central stereid or substereid band. On the basis of the morphology and differentiation of these layers Kawai (1965, 1968) distinguished five types of costal architecture, of which the type C, characterised by a more or less clear differentiation of the three layers is typical and most common in this family; however, all of Kawai's (1968) costal types can be found throughout Grimmiaceae. For instance, in *Schistidium maritimum* the costal type E occurs in which the central stereid band is divided by a median row of large guide cells into two groups, dorsal and ventral. Costal anatomy is of considerable taxonomic importance in most genera of the family. The leaf margins are plane, inflexed, recurved to revolute, uni- to variously polystratose. The alar cells are not developed or sometimes they form distinct hyaline or coloured auricles.

The Grimmiaceae have a haplolepidous seligerioid peristome but its structure is extremely variable (Deguchi 1979; Maier 1999). In general, two extreme peristome types can be distinguished which are linked by a continuum of intermediate forms. At one extreme the peristome teeth are lanceolate, entire or perforate to variously split into two or three unequal prongs and smooth or variously ornamented, lack a basal membrane and have a distinctly thicker and trabeculate dorsal side and thin ventral side (*Schistidium* type). At the other extreme the teeth are linear, divided nearly to the base into two filiform somewhat paired segments, have a distinct, usually low, basal membrane and often prostome and are equally thickened and less prominently trabeculate on both dorsal and ventral sides (*Racomitrium* type). A few species in this family have gymnostomous capsules but the great majority of taxa possess ochyrostomous capsules, although in some species the peristome is vestigial and/or variously reduced.

Other sporophyte characters are also subject to considerable variation in the Grimmiaceae. The perichaetial leaves are either similar to the vegetative leaves or much enlarged and strongly modified. The capsules are ovoid, obloid, cyathiform or cylindrical, symmetrical or strongly ventricose at the base and gibbous, smooth or sulcate. They are long exserted to emergent on the straight or curved setae or deeply immersed and concealed within the perichaetial leaves on the short, straight or curved setae. The annulus is absent or well developed and deciduous. The stomata are superficial and bicellular and confined to the base of the urn. The calyptra is mitrate or cucullate, naked or scabrous, 2–4-stratose in transverse section, smooth or plicate, entire, erose or lobate at the base and completely covers the urn or barely the operculum. The operculum is umbonate to long rostrate and sometimes attached to the columella after dehiscence. The spores are unicellular, globose, nearly smooth or variously papillose and the pattern of ornamentation is essentially similar to that observed in the Pottiaceae, Dicranaceae and Encalyptaceae (Hirohama 1978).

The Grimmiaceae comprise small to relatively large and robust plants which grow in small cushions, tufts or large mats or patches. The stems are sympodially or monopodially branched, erect or ascending to prostrate, with acrocarpous or cladocarpous gametangia. In transverse section they consist of a distinct cortex surrounding a central medulla of large cells, whereas the central strand is present or absent. Vegetative reproduction in the Grimmiaceae is by means of gemmae, but is restricted to a few species only.

Despite considerable variation of gametophyte and sporophyte morphologies, the Grimmiaceae as a whole are a very distinct moss family and some recent molecular investigations have provided evidence that it is monophyletic (Tsubota *et al.* 2003). It was the first family of the Bryophyta to be described (Arnott 1825). However, Crosby & Magill (1981) stated that the family name Splachnaceae is older by one year, but, in fact, the part of the paper by Greville & Arnott (1822–1826) describing this family was published in 1826, not 1824 as indicated by these authors.

The preference for arid conditions has been the dominant evolutionary leitmotiv within the Grimmiaceae and consequently various morphological structures in many of these mosses are considered as xeromorphic adaptations. Many of the gametophyte peculiarities that so readily distinguish members of this family are convergent and are also found in other genera and species from distantly related moss families, for example hyaline hair-points, multistratose leaves and strongly incrassate cell walls. Interestingly, some peripheral genera of the Grimmiaceae, namely *Aligrimmia*, *Coscinodontella* and *Indusiella*, share some convergent xeromorphological adaptations such as chlorophyllose costal lamellae, specialised laminal chlorophyllose tissue and bistratose semi-succulent leaves with some genera of the Pottiaceae, for instance *Pseudocrossidium*, *Aloina*, *Crossidium*, *Pterygoneurum* and *Tortula* (Murray 1984).

The Grimmiaceae seem to occupy an intermediate position between the Seligeriaceae and the Pottiaceae. Some authors (e.g. Jones 1933; Churchill 1981) have considered this family to be most closely related to the Dicranaceae because they regarded the *Schistidium* peristome type to be primitive in this family. However, this suggestion is not strictly valid because the peristome type in *Schistidium*, *Grimmia* and related genera is typically seligerioid, not dicranoid, and this may be strong evidence of derivation of the Grimmiaceae and the Seligeriaceae from common ancestral stock. Relationships to the Pottiaceae appear to be real but rather distant. Similarities of the peristome in *Racomitrium* and some species of *Syntrichia*, *Pleurochaete* and *Dolotortula* are possibly in part due to parallelism in development. On the other hand, taxa having the *Racomitrium* peristome type show a marked similarity to the Ptychomitriaceae, which are sometimes treated as a subfamily of the Grimmiaceae (Brotherus 1902; Jones 1933; Churchill 1981, 1987).

The Grimmiaceae are usually regarded as a notoriously difficult family, especially its largest genera, including *Grimmia* and *Racomitrium* in the traditional sense. Many species of these genera display considerable morphological variability and as a result numerous phenotypes and habitat modifications of the same species were given taxonomic recognition in various parts of their broad geographical ranges. This led to description of an excessive number of poorly defined species and infraspecific taxa, often known only from the type collections. The most instructive examples here are *Guembelia longirostris* which name has no less than 55 validly published heterotypic synonymous names and ten *nomina nuda* (Muñoz 1998*a*), and *Bucklandiella sudetica* with 32 heterotypic synonyms including one *nomen nudum* (Frisvoll 1988). However, the converse is true for a number of good and well defined species which, in the last half-century, have been lumped into a single species. Accordingly, the number of species decreased markedly in some genera, but some unnatural and artificial species have been created, for instance *Bucklandiella crispula* (Clifford 1955) and *Schistidium apocarpum* (Bremer 1980, 1981, 1982), which are difficult to define and defend.

The subfamilial and generic classification of the Grimmiaceae has long been unresolved and has been in a considerable flux. In the first edition of the Musci in *Die Natürlichen Pflanzenfamilien* Brotherus (1902) divided the family into three subfamilies, Scoulerioideae, Grimmioideae and Ptychomitrioideae, but in the second edition of the Musci in this series the latter subfamily was excluded from the Grimmiaceae and elevated to family rank and placed in the remote order Isobryales (Brotherus 1925). The initial concept of Brotherus (1902) was resurrected by Churchill (1981), but he considered the Scoulerioideae as a family in its own right. In addition, he recognised the new subfamily Coscinodointoideae for several small genera, namely *Coscinodon*, *Jaffueliobryum*, *Coscinodontella* and *Aligrimmia*, and transferred *Racomitrium* to the Ptychomitrioideae.

Finally, Ochyra *et al.* (2003) divided the Grimmiacae into two subfamilies, Grimmioideae and Racomitrioideae, which correspond well to the two extreme peristome types present in this family and, additionally, they placed *Coscinodon* in the former subfamily. The only diagnostic character of the subfamily Coscinodontoideae is the plicate calyptra, whereas all gametophyte characters considered to be diagnostic of this subfamily *Coscinodon* and its allies share with *Grimmia* and its segregates, so this subfamily seems to be weakly founded. Though closely related, it seems that the Ptychomitriaceae deserve to be kept apart from the Grimmiaceae, among others, on account of the lack of the distinct dorsal trabeculae on the teeth, the presence of a characteristic air-gap at the base of the teeth, the internal differentiation of the costa (which is of the E type, Kawai 1968), the strongly plicate calyptra, the presence of the small male branch within the perichaetium (Deguchi 1977), and on the different pattern of spore ornamentation (Hirohama 1978). The last two character states are known also in some genera of the Grimmiaceae, at least in part, namely *Coscinodon*, *Jaffueliobryum*, *Coscinodontella* and *Aligrimmia* (Murray 1984; Churchill 1987), and this clearly confirms the close relationships of both families.

As currently understood, the Grimmiaceae consist of 17 genera disposed in two subfamilies. The Grimmioideae are primarily characterised by having the *Schistidium* peristome type and comprise 13 genera, namely *Leucoperichaetium*, *Coscinodontella*, *Indusiella*, *Aligrimmia*, *Hydrogrimmia* and *Streptocolea* (all monotypic), *Jaffueliobryum* (with four species), *Coscinodon* (with eight species) as well as *Grimmia*, *Schistidium*, *Orthogrimmia*, *Guembelia* and *Dryptodon*. The last five genera along with *Hydrogrimmia* and *Streptocolea* have traditionally been united into a very large, single genus *Grimmia*. However, it proved to be a very heterogeneous and unnatural taxon and Ochyra *et al.* (2003) split it into smaller and better defined genera which appear to be natural taxa.

The second subfamily, Racomitrioideae, is diagnosed by the *Racomitrium* peristome type, the thick and sinuose walls of the epidermal cells of the vaginula, the lack of a central strand and the frequent presence of a preperistome. It consists of four genera which, until recently, have been included into the large and very heterogeneous *Racomitrium* which was split into four smaller but natural and well defined genera (Ochyra *et al.* 2003), namely *Racomitrium* (with three species), *Niphotrichum* (with eight species), *Codriophorus* (with 15 species) and *Bucklandiella* (with about 45 species).

In total, the Grimmiaceae comprise about 220 species and are cosmopolitan in distribution. The definite number of species cannot be established until its largest and taxonomically most neglected genera, including *Schistidium*, *Bucklandiella* and *Codriophorus*, are globally revised. Some segregates of the large genus *Grimmia*, namely *Guembelia* and *Dryptodon*, still need critical revisions, although most of the type collections of its species and infraspecific taxon names have been assessed (Geissler & Maier 1995; Muñoz & Pando 2000) and a world synopsis of species is available (Greven 2003), so at least the approximate number of species in this genus can be given. Nonetheless, new species continue to be discovered (e.g. Greven 2002*a*, 2003; Muñoz 2002; Muñoz *et al.* 2002) or resurrected (e.g. Greven 2000*a*,*b*,*c*; Delgadillo 2002; Delgadillo & Villaseñor 2002; Hastings & Greven 2007*a*). Of other genera, only the small genera *Jaffueliobryum* (Churchill 1987) and *Coscinodon* (Hastings 1996, 1999) have been revised, and the three monotypic genera, *Coscinodontella*, *Indusiella* and *Aligrimmia*, have been assessed (Murray 1984). There are many regional treatments of the Grimmiaceae which are mostly out of date, although they provide substantial information about the classification and taxonomy of genera and species, for example for Norway (Hagen 1909), Europe (Loeske 1913, 1930), the former Czechoslovakia (Vilhelm 1924), North America (Jones 1933), Japan (Deguchi 1979; Noguchi 1974) and the Altai Mountains (Ignatov & Cao 1994).

In the austral cool temperate and polar regions the Grimmiaceae are well represented, both in terms of the number of species and their frequency and abundance. In the Antarctic they are the richest moss family, consisting of 23 species classified into eight genera. However, the Grimmiaceae, and especially the genera *Schistidium* and *Bucklandiella*, are taxonomically underworked in the Southern Hemisphere. The major problem in identification of the genera in this family is that some of them are best or exclusively defined by sporophyte characters, especially *Schistidium*, *Grimmia*, *Dryptodon*, *Guembelia*, *Orthogrimmia* and *Coscinodon*. Although most species of the two first genera produce sporophytes in profusion and frequently in the Antarctic, recognition of the genera may pose problems. Therefore, the key to the genera is highly artificial and considers mostly characters found in local species, with special emphasis on the gametophyte traits because quite often only sterile material is available for examination.

KEY TO THE GENERA OF THE ANTARCTIC GRIMMIACEAE

1. Leaf cell walls straight or sinuose; basal cells short- or long-rectangular, mostly with straight and uniformly thickened walls or with transverse walls thicker than vertical walls, always without spiral thickenings, concolorous with the upper cells; costa semi-terete, occasionally elliptical to reniform, not situated in a furrow, 2 cells wide at the ventral side in cross-section, usually not markedly larger than central and dorsal cells; plants acrocarpous; stems erect, often with forked branches; central strand present or absent 2
1. Leaf cell walls strongly sinuose to sinuose-nodulose throughout; basal cells long-rectangular to linear, with strongly nodose longitudinal walls and often with spiral thickenings, usually forming a coloured strip in 1–3 rows along the insertion; costa reniform, elliptical to semi-terete, often lying at the bottom of a furrow, in cross section in the basal part with 2–5 large ventral cells markedly different from central and dorsal cells; plants cladocarpous; stems often prostrate, usually with abbreviated branches; central strand always absent 7
   2. Leaves distinctly plicate on both sides of the costa in the distal part; plications often tristratose; plants hoary owing to the presence of the long, flat hyaline hair-points; calyptrae large, campanulate, longitudinally plicate, covering most of the urn .... 17. ***Coscinodon***
   2. Leaves not plicate; plants hoary or not; calyptrae small, mitriform or cucullate, hardly reaching the urn or slightly covering it, smooth 3
3. Capsules immersed, symmetric, systylious; seta very short, straight; calyptrae minute, covering only the beak of the operculum; costa in transverse section scarcely differentiated, 2–3(–5)-stratose, without a central stereid band .... 18. ***Schistidium***
3. Capsules emergent to long exserted or if immersed then asymmetric, ventricose at the base; seta short, cygneous; columella remaining in the urn at dehiscence; calyptrae larger, covering the capsule mouth; costa in transverse section usually with a distinct central band of small hydroid or stereid cells 4
   4. Leaves deeply concave, unistratose throughout or occasionally with bistratose marginal spots near the apex; capsules immersed, asymmetric and ventricose at the base; setae arcuate, eccentrically attached, shorter than the urn .... 19. ***Grimmia***
   4. Leaves canaliculate to sharply keeled in the distal half; lamina unistratose with bistratose patches; margins bistratose for one to several rows of cells; capsules emergent to long exserted, symmetric, not ventricose at the base; setae straight or curved, centrally attached, longer than the urn 5
5. Leaves canaliculate and narrowly U-shaped distally to broadly U-shaped throughout; costa semi-elliptic to reniform and curved in transverse section, not clearly delimited from the laminal cells and scarcely prominent on the dorsal surface................... 20. ***Guembelia***
5. Leaves sharply keeled and V-shaped distally; costa semi-terete in transverse section, clearly delimited from the laminal cells and prominent on the dorsal surface 6
   6. Setae straight; capsules smooth; margins plane throughout or incurved above, occasionally narrowly recurved proximally on one side to the mid-leaf; basal marginal cells of the leaves with transverse walls much thicker than the longitudinal walls .... 21. ***Orthogrimmia***
   6. Setae arcuate or flexuose; capsules striate-ribbed to rugose-plicate; margins recurved on one or both sides from the base up to ⅔ way up the leaf; basal marginal cells of the leaves with all walls equally thin and straight .... 22. ***Dryptodon***
7. Laminal cells with large flat papillae with small secondary papillulae over the longitudinal walls and almost the whole lumina, except for a narrow slit in the middle; hyaline hair-points long, strongly papillose and erose-dentate, long decurrent down the leaf margins; setae sinistrorse when dry, strongly papillose; capsule slightly ventricose at the base; peristome teeth long-linear, divided to the base into 2 filiform regular branches .... 23. ***Racomitrium***
7. Laminal cells smooth or pseudopapillose owing to longitudinal cuticular ridges; hyaline hair-point, if present, smooth or denticulate, never papillose, not or short-decurrent; setae dextrorse when dry, smooth; capsule not bulging at the base; peristome teeth short, ovate-triangular to lanceolate, irregularly divided in the upper half into 2–4 prongs....... .... 24. ***Bucklandiella***

17. COSCINODON Spreng.

Coscinodon Spreng., Anleit. Kenntn. Gew. 3: 281. 1804. — Type: *Coscinodon pulvinatus* Spreng. [=*Coscinodon cribrosus* (Hedw.) Spruce].

Apocarpum P.Beauv., Mém. Soc. Linn. Paris 1 (Fasc. planch.): *t. 4, f. 2.* 1822. — Type: *Apocarpum commune* P.Beauv. [= *Coscinodon cribrosus* (Hedw.) Spruce]. First synonymised by Ochyra & Lamy (2004: p. 187).

Plants small to medium-sized, occasionally larger, in dense or loose cushions, sometimes coalescent and forming wide mats or patches, yellow-, brown- to blackish-green above, olive-green to dark brown below, often hoary due to long hair-points of leaves. *Stems* erect, stiff, few to moderately short-branched, sometimes fastigiate, sparsely radiculose at base with brownish, smooth, simple or sparsely dichotomously branched rhizoids, in cross-section without or with a small and indistinct or larger and distinct central strand, composed of nearly uniform, large, hyaline, thin-walled cells becoming only somewhat smaller and thicker-walled towards the epidermis and forming an indistinct unistratose cortex, or with 1–2 rows of smaller, thick-walled and brownish cortical cells; *axillary hairs* filiform, composed of 3–6 uniseriate, hyaline cells. *Leaves* crowded, imbricate, erect or rarely erecto-patent when dry, erect-spreading to spreading when wet, lanceolate to ovate-lanceolate, narrowly to broadly keeled above, concave below, biplicate throughout or only distally, gradually tapering to acute, rounded-acuminate or obtuse, cucullate or flat, awned apex; *hair-point* short or long, often longer than the lamina, flat, hyaline, smooth to sparsely denticulate, not or slightly decurrent, straight or somewhat flexuose; *margins* plane and unistratose below, incurved or very seldom recurved or plane, bi- or rarely 3–4-stratose distally, entire or appearing crenulate by bulging cells; *costa* single, percurrent to shortly excurrent, semi-terete to reniform, clearly delimited, prominent on the dorsal surface, consisting of 2 ventral epidermal cells in transverse section; *laminal cells* fully to partially 2–3(–4)-stratose distally, rounded, irregularly rounded-quadrate, oblate to short-rectangular, not much differentiated at plications, thick-walled, with straight or sinuose walls; *basal cells* uni- or bistratose with unistratose patches, rectangular to long-rectangular near the costa and forming a distinct translucent area, becoming quadrate to rectangular or transversely rectangular with thin longitudinal and incrassate transverse walls at the margin, forming a 3–5-seriate border. *Dioicous*, exceptionally *autoicous*. *Perichaetia* terminal; *perichaetial leaves* markedly or only slightly larger than the vegetative leaves, oblong- or ovate-lanceolate to oblong-ovate, ending in a long awn equalling the lamina, pleated throughout or only distally, with areolation similar to the vegetative leaves or cells longer in all parts. *Setae* straight, short, 0.3–2.4 mm long, centrally attached to the capsule. *Capsules* immersed to emergent or exserted, cylindrical, ovoid or sometimes cupulate, 0.5–1.7 mm long, symmetric, straight, smooth, often constricted below the mouth; *annulus* persistent, composed of 2–3(–4) rows of thin-walled, red-brown to pale orange cells; *operculum* funnel-like with a short or long, straight beak; *exothecial cells* thick-walled, oblong-hexagonal, becoming quadrate to oblate in 2–3 rows at the rim; *stomata* superficial, bicellular, few at the base of the urn; *peristome* rarely absent or single, consisting of 16 triangular, orange-brown teeth, entire or with a few splittings at the tips, erect, hygrocastique or xerocastique, densely low or coarsely papillose on both surfaces, solid or cribrose. *Spores* spherical, densely papillose, pale yellow-orange. *Calyptra* campanulate, with the base erose and expanded at maturity, straw-gold, covering ⅓ to entire mature capsule, strongly plicate.

Coscinodon is a small genus of seven species which is best diagnosed and immediately separated from all other genera of the Grimmiaceae by its strongly plicate calyptra which usually covers the whole capsule or only to the middle. Other character states which are considered to be diagnostic to this genus are also widely present in *Grimmia*, *Orthogrimmia*, *Dryptodon*, *Guembelia* and *Schistidium*; for example the strongly biplicate leaves are typical of *Orthogrimmia caespiticia*; the immersed capsules in enlarged perichaetial leaves occur in all species of *Grimmia*, in *Guembelia involucrata* and *G. kidderi* as well as in *Dryptodon pseudanodon*; and the cribrose peristome teeth are found in some species of *Schistidium*, for example in *S. cribrodontium* from East Africa. Therefore recognition of a separate subfamily for this genus (Churchill 1981), which is based on a single character state, seems to be doubtful and *Coscinodon* certainly show much closer alliance to *Grimmia* than to *Racomitrium* and its allies which constitute the separate subfamily Racomitrioideae.

Coscinodon is a predominantly Northern Hemisphere genus where five species and one subspecies are known (Hastings & Deguchi 1997; Hastings 1999), of which only *C. cribrosus* and *C. calyptratus* seem to have a widely disjunct, pan-Holarctic distribution, and the other taxa are narrow endemics. Two species occur at high elevations in the Andes of Bolivia (Hastings 1996), and *C. calyptratus* has been recorded in New Zealand (Muñoz 1998*b*) but this record is based upon misdeterminations of the specimens which correctly belong within *Guembelia longirostris* (Hastings 1999). Sometimes a third Andean species, *C. trinervis*, from Bolivia and Peru is placed in *Coscinodon* (Hastings 1996), but the small mitrate and non-plicate calyptra covering only the operculum excludes this species from this genus. On account of its sigmoid to coiled setae, eccentrically attached to the ventricose capsule, it is best placed in the type subgenus of *Grimmia* (Muñoz 1999). However, exclusion of this species is balanced by *C. reflexidens* (Müll.Hal.) Ochyra from southern Chile. Maier (2002) provided convincing evidence that the type material of *Grimmia reflexidens* is in fact a *Coscinodon*. Although the type material lacks the calyptra, the strongly plicate leaves are very suggestive of *Coscinodon* and, accordingly, Ochyra (2004*a*) formally transferred *G. reflexidens* to this genus as *Coscinodon reflexidens*. This species seems to be closely related to the High Arctic *C. hartzii* C.E.O.Jensen in which the leaf margins are plane throughout and the upper lamina usually has tristratose patches. In contrast, in *C. reflexidens* the proximal leaf margins are narrowly recurved on one side and its upper lamina is mostly bistratose, with only occasional tristratose streaks.

Until recently, *Coscinodon* has not been recorded from the austral polar region but, after a careful taxonomic assessment, Ochyra (2004*a*) considered that the Antarctic endemic *Grimmia lawiana* should be best placed in that genus.

1. **Coscinodon lawianus** (J.H.Willis) Ochyra

Fig. 94

Coscinodon lawianus (J.H.Willis) Ochyra, Polish Pol. Res. 25: 112. 2004. ≡ *Grimmia lawiana* J.H.Willis *in* Filson, ANARE Sci. Rep. Ser. B (II) Bot. 82: 148, *pl. 41, f. a–f.* 1966. — TYPE CITATION: Ring Rocks, Mac.Robertson Land, Antarctica, leg. R. Filson No. 4225, 16 September 1962. [Holotype: 'National Herbarium of Victoria (MEL), Australia MEL 1048478 Grimmia lawiana Loc.: Antarctica – Ring Rocks, MacRobertson Land. Coll. Rex Filson, no. 4225 Date 16th September 1962 Notes: Abundant in cracks of boulders on the larger rock areas' – MEL!; isotypes: BM!, H!, S!].

Plants small to medium-sized, in dense, fine-textured cushions or forming small mats owing to coalescing cushions, not or hoary, light-, dark- to pale olive-green above, blackish-brown to dark brown below. *Stems* slender, erect, 0.5–1.3 or, occasionally, to 3.5 cm tall, sparingly branched, radiculose at the base with smooth, light brown, sparingly branched rhizoids, in cross-section rounded, with a small but distinct central strand surrounded by 3–4-stratose medulla of large, hyaline, thin-walled cells and 1–2-layered cortex of somewhat smaller, brownish, thicker-walled cells; *axillary hairs* sparse, filiform, hyaline, 3–4-celled. *Leaves* densely set, imbricate, erect-appressed when dry, erecto-patent when moist, ovate to broadly ovate-lanceolate, 0.7–0.9(–1.1) mm long (without the hair-point), 0.25–0.6 mm wide, acute to rounded-acute, biplicate and strongly keeled distally, smooth and deeply and broadly concave below, epilose or mostly terminated with a stout, flat, not or slightly decurrent, entire to obscurely denticulate, hyaline awn; *margins* plane and unistratose in the proximal half, erect to manifestly incurved in the distal half, 40–50 μm wide at the base, prominently convex dorsally, semi-terete, internally differentiated in cross-section, with 2 ventral epidermal cells, distinct central stereid band and a dorsal epidermis; *laminal cells* bistratose throughout the distal part or occasionally tristratose at plications, unistratose in the middle and below with frequent bistratose streaks; *upper cells* irregularly quadrate, rounded-quadrate to short-rectangular, becoming often oblate along the margins, 3–10 μm wide, 5–13 μm long, thick-walled, not differentiated over plications; *basal juxtacostal cells* short- to long-rectangular, with evenly thickened walls, 8–12 μm wide, 15–35 μm long, becoming quadrate to rectangular, 8–12 μm wide, 15–40 μm long at the margins, with evenly thickened or with transverse walls thicker than the longitudinal walls, sometimes thin-walled and forming 2–3-seriate marginal border. *Asexual reproduction* occasional by filamentous protonemal gemmae, composed of uniseriate chains of chlorophyllose cells 15–40 μm long, 15–25 μm wide, with somewhat thickened, hyaline walls. Fertile plants unknown.

Chromosome number: unknown.

Discussion. *Coscinodon lawianus* is relatively easy to identify, the small imbricate leaves which are distinctly biplicate and carinate in the distal part being immediately diagnostic. The leaf plicae are restricted to the upper half of the leaves and are clearly visible in transverse section as bi- or occasionally tristratose spots. The leaves are strongly pilose with long hyaline hair-points which give the plants a hoary appearance. In addition, they are ovate to broadly ovate-lanceolate and very small, rarely longer than 1 mm (without the hair-point), this making them one of the smallest leaves in Antarctic moss species.

In the Antarctic *Coscinodon lawianus* is likely to be mistaken only for *Orthogrimmia sessitana*, especially on the continent where only sterile populations of this species are known. However, the latter species has strongly keeled but otherwise plane leaves and the lamina is invariably bistratose with frequent unistratose streaks distally. The hoary phenotypes of *Grimmia plagiopodia* are also superficially similar to *C. lawianus* but it has broadly concave, non-plicate leaves with unistratose lamina, and fertile plants are distinct at a glance by their asymmetric, ventricose capsules.

Because all of the Antarctic populations of *Coscinodon lawianus* are sterile, establishing an accurate generic placement for this species has been one of the most difficult decisions made during the present taxonomic work on the Antarctic mosses. The structure of the gametophytes may be highly deceptive and misleading as diagnostic features at the generic level owing to convergent evolution of species living in similar environmental conditions. Leaf plication has developed independently in several lineages of the Grimmiaceae including, besides *Coscinodon*, also *Orthogrimmia* and *Grimmia*, so it must be considered with caution as a diagnostic generic trait. This is especially so in two species of the genus *Coscinodon* itself, *C. yukonensis* and *C. calyptratus*, in which the leaves are not plicate. Therefore, the only reliable character supporting the recognition of *Coscinodon* as a genus in its own right is the shape of the calyptra and, until fertile plants of *C. lawianus* are found, the generic placement of this species is uncertain. The present decision is based mainly on the overall similarity of the plants and leaves of the Antarctic species with some Andean species of *Coscinodon*, namely *C. pseudocribrosus* and *C. bolivianus* (Hastings 1996). The latter species shares with *C. lawianus* the size of the plants, the size and shape of the leaves and the restriction of the plication to the upper part of the leaves. However, it is distinct in having frequent 3–4-stratose strands beyond the plication. The present placement of *C. lawianus* was also suggested by Muñoz & Pando (2000) but Greven (2003) retained it in *Grimmia*. However, the latter author studied very few specimens of this species and his drawings, especially of transverse sections

Figure 94. *Coscinodon lawianus*. — **1**. Habit. **2**. Portion of branch. **3**. Cross-section of stem. **4–6**. Axillary hairs. **7–13**. Leaves. **14**. Leaf apex. **15**. Upper mid-leaf cells at margin. **16**. Upper mid-leaf cells. **17**. Lower mid-leaf cells. **18–19**. Basal cells. **20–26**. Cross-sections of leaves, a sequence from base to apex. — [1, 10–12, 17, 21, 25 from *RF 4225* (type of *Grimmia lawiana*); 2, 7 from *RF 4171*; 3–4, 6, 20, 23–24, 26 from *RF s.n.*, 1.11.1973; 5, 15, 18, 22 from *RF 4182*; 9, 14, 19 from *RF 4503*; 8, 13, 16 from *RF 14930*; all in MEL]. — Scale bars: a – 1mm (7–13); b – 1 mm (2); c – 100 μm (3–6, 14–26); d – 1 mm (1). (Reproduced from *Polish Polar Research*, 25: p. 113, 2004, with permission.)

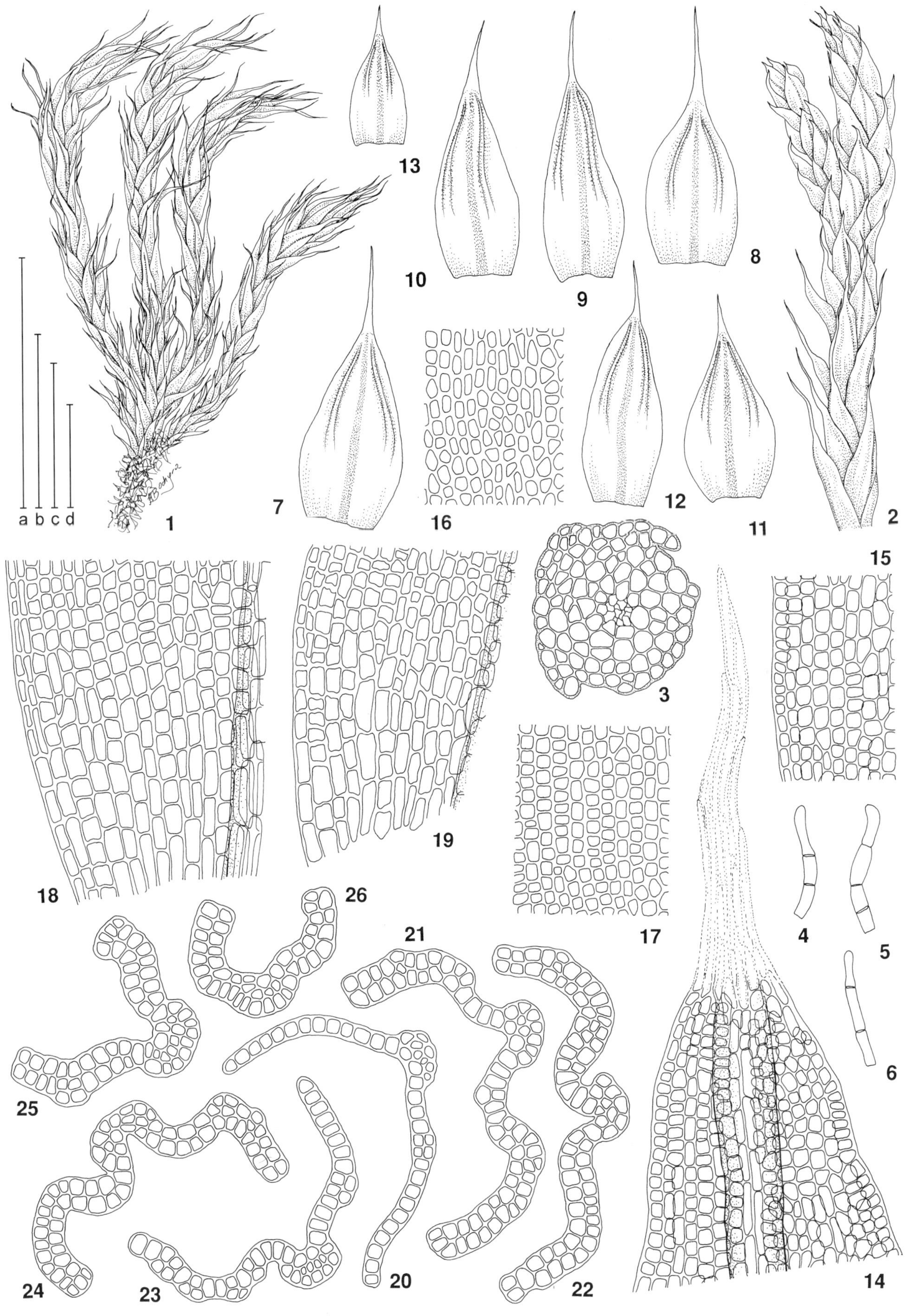
a
b
c
d
1
2
3
4
5
6
7
8
9
10
11
12
13
14
15
16
17
18
19
20
21
22
23
24
25
26

of the leaf, are very incorrect and he makes no mention of the leaf plication.

Coscinodon lawianus was originally described as *Grimmia lawiana* from several specimens from Mac.Robertson Land (Filson 1966). The strongly biplicate leaves make it a very distinct species which does not fit any known species of this genus in the world (Muñoz & Pando 2000), especially in southern South America (Muñoz 1998*a*), or the Subantarctic and Australasia (Greven 2003). Thus, it is the only endemic species of moss which is restricted to continental Antarctica. As is the case with *Syntrichia sarconeurum*, it may be considered as a relict of the ancient Antarctic flora existing on this continent in the early Tertiary which could have survived the climatic catastrophe *in situ* on nunataks and other ice-free refuges. Having limited modes of vegetative reproduction, *C. lawianus* is restricted in its present distribution to a few regions on the continent, in contrast to *S. sarconeurum* which produces vegetative propagulae in masses and these have facilitated its widespread distribution around the continent. *Coscinodon hartzii*, the northern counterpart of *C. lawianus*, is also very rare and widely scattered in the American High Arctic and Greenland (Jensen 1898; Murray 1992; Hastings 1999).

Reproduction in Antarctica. Always sterile, but female plants occasionally observed. However, some colonies of *Coscinodon lawianus*, especially if colonised by algae and/or lichens, produce copious protonemal gemmae which are easily broken into 1–2-cellular fragments (Imura & Kanda 1986).

Habitat. *Coscinodon lawianus* is a saxicole associated with a variety of acidic rocks, growing in crevices, cracks and fissures, on talus slopes, in hollows amongst larger boulders in moraines and patterned ground. It grows predominantly in dry habitats, but has also been found on wet rocks partially covered with fine mineral soil, beside swift flowing streams running from glaciers (Kanda 1986*b*, 1987*d*), on sand and gravel subject to flooding near small melt pools, as well as in seepage areas (Seppelt & Ashton 1978). It avoids areas affected by sea birds or marine salts and appears to prefer nutrient-deficient soils at higher elevations lacking marine and biotic influence. It forms either poor monospecific sociations or is a main component of simple plant communities, with *Ceratodon purpureus* (Kanda 1981*a*, 1986*b*, 1987*b,d*) or *Bryum pseudotriquetrum* (Seppelt & Ashton 1978), belonging to the short moss turf and cushion subformation which are often invaded by imperfect lichens and/or algae.

World range. An Antarctic endemic, so far known only from East Antarctica (Fig. 95).

Distribution in Antarctica. *Coscinodon lawianus* is restricted to the continental Antarctic where it is widespread but never abundant. It is frequent in Dronning Maud Land where, in addition to the specimens cited below, it is known from Schirmacher Oasis (Richter 1990, 1995) and nunataks of Gjelsvikfjella and Mühlig-Hofmannfjella (Engelskjøn 1985), as well as from the Sôya Coast in Enderby Land (Nakanishi 1977; Kanda 1981*a*, 1986*b*, 1987*a,b,d*). All known localities of this species are situated south

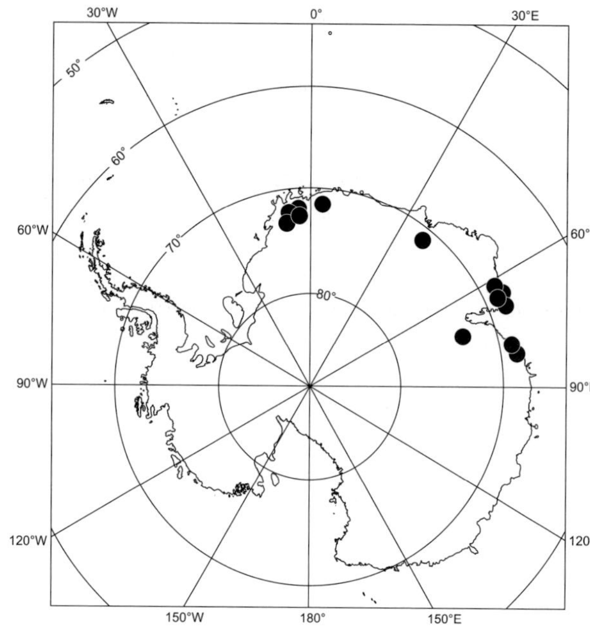

Figure 95. Global distribution of *Coscinodon lawianus*.

of the Antarctic Circle, and the southernmost are in Tottanfjella in Dronning Maud Land at lat. *ca* 75° S (Bowra *et al.* 1966; Thor 1995). In the coastal region of Enderby Land and Mac.Robertson Land it occurs near sea level at 20–30 m, but on the inland nunataks in Tottanfjella it occurs at elevations of 1650–2160 m.

Specimens examined. EAST ANTARCTICA. DRONNING MAUD LAND. **KRONPRINSESSE MÄRTHA KYST.** ***Heimerfrontfjella***: Tottanfjella, Johnsonhogna, *ca* 1650 m, *GTB 20A & 28A* (AAS, KRAM); unnamed nunatak NW of Cottontoppen, 1667 m, *GTB 7* (AAS, KRAM); Blades Ridge, *ca* 2000 m, *GTB 2 & 3* (AAS, KRAM); Sivorgfjella, Mathisenskaget, *ca* 2160 m, *GTB 4* (AAS, KRAM). **PRINSESSE ASTRID KYST.** Henriksenskjera, 1200 m, *RILS 10068* (AAS, KRAM), 1296 m, *RILS 10108* (AAS, KRAM), *ca* 1300 m, 'Wind-Scoop Peak', *RILS 10077A, 10078 & 10080* (AAS, KRAM) and 1363 m, *RILS 11138* (AAS, KRAM); Filchnerfjella, Rakenkniven, 1700 m, *RILS 10061 & 10062A* (AAS, KRAM) and 1800 m, *RILS 10065B* (AAS, KRAM).

ENDERBY LAND. **SÔYA COAST.** Langhovde, 20 m, *Kashiwandani 4006* (KRAM, LE).

MAC.ROBERTSON LAND. **MAWSON COAST.** Chapman Ridge, *Seppelt 1506* (KRAM); Hayes Peak, *RF 4378* (MEL); Cape Bruce, *RF 4335* (MEL); Ufs Is., Melt-Lake Valley, *RF 14930* (MEL); Ring Rock, *RF 4378* (MEL); *RF 4325* (MEL – type of *Grimmia lawiana*); Mawson Rock, *RF 4171* (MEL); West Bay, *RF 4069* (MEL); North Masson Range, *RF 4092 & 4121* (MEL); Telurometer Peak, *RF 4417* (MEL); South Masson Range, Peak 1070 S of Trost Peak, 1030 m, *RF 4503* (MEL); Departure Rocks Group, *RF 4157* (MEL); Goldsworth Range, Mt. Henderson, *RF 4185* (MEL); Fisher Nunatak, *RF 4182* (MEL). **PRINCE CHARLES MOUNTAINS.** Mawson Escarpment, north-west 'J', 1 Nov 1973, *RF s.n.* (MEL 1012186 & 1012187); Radok Lake, 150 m, *Adamson 19050* (KRAM).

PRINCESS ELIZABETH LAND. **INGRID CHRISTENSEN COAST.** Larsemann Hills, *RILS 10330 & 10339* (AAS, KRAM) and *Riddle 3* (KRAM); Rauer Is., Stonewall Peninsula, *Cracknell R195* (KRAM).

18. SCHISTIDIUM BRUCH & SCHIMP., *nom. cons.*

Schistidium Bruch & Schimp. *in* Bruch, Schimp. & W.Gümbel, Bryol. Eur. 3: 93. 1845 [Fasc. 25–28, Mon.: 1]. — Type: *Schistidium maritimum* (Turner) Bruch & Schimp. (*Grimmia maritima* Turner) (*typ. cons.*).

Plants small, medium-sized to large, slender to robust, stiff, forming loose or compact, low or deep cushions, tufts or extensive mats, dull, glistening or with a greasy lustre, variegated above, with all possible tones and combinations of colours including green, brown, orange-red, orange-brown, ferrugineous, reddish, ochraceous, yellowish, olivaceous, vinaceous, greyish-green, brownish- or purplish-black to jet-black, brown, blackish-brown to black below, not or hoary. *Stems* erect or creeping to decumbent, sparingly to profusely, irregularly, dichotomously or fastigiately branched, rarely simple, not or julaceous, sparsely radiculose at the base with reddish or brownish, smooth, branched rhizoids, in cross-section round, consisting of 1–4(–5) layers of small cortical cells with brown, strongly thickened walls and narrow lumina, 3–6 layers of large, hyaline, thin- to moderately thick-walled medullary cells clearly delimited from the cortex, central strand absent or present, narrow to broad; *axillary hairs* filiform, hyaline throughout, 4–10-celled. *Leaves* crowded or more distant, straight, erect, appressed and ± closely imbricate or from a subsheathing, appressed base recurved, falcate to falcate-secund, not or sometimes altered, contorted to curved when dry, erect-spreading when moist, sometimes arranged in distinct spiral rows, not or distinctly short-decurrent, ovate, ovate-lanceolate, ovate-triangular, lanceolate, elliptical or oblong-ligulate, mostly gradually acuminate, acute, rounded to broadly obtuse, mostly hyaline awned, sometimes epilose or very shortly hyaline tipped, concave, not plicate, mostly keeled and V-shaped in cross-section, sometimes narrowly to broadly canaliculate and U-shaped in transverse section in the distal portion; *hair-point* hyaline, yellowish-hyaline to yellowish, freely variable in length, flat, canaliculate or terete, smooth to coarsely denticulate, dentate or spinulose, straight or irregularly bent to flexuose; *margins* mostly asymmetrically recurved, sometimes ± throughout at least on one side, most often only in the broadest part of the leaf, occasionally revolute or plane, entire or repand-denticulate near the apex, uni- to bistratose in 1 to several rows of cells, sometimes 3–5-stratose; *costa* subpercurrent to long excurrent, clearly differentiated, broadest in the central and upper parts, strongly convex dorsally, smooth or papillose on the dorsal surface, in cross-section semi-terete, elliptical, trapezoid or subrectangular, sometimes flattened dorsally, bi- to multistratose, nearly homogeneous or with ventral and dorsal epidermal cells larger than the central stereids or substereids, very rarely consisting of a central row of large guide cells separating dorsal and ventral stereid bands; *laminal cells* unistratose or partly to completely bistratose in the distal part, smooth or papillose, usually more densely on the dorsal than on the ventral side and in the upper and central than in the lower part of the leaf, with thickened, straight or sinuose walls, square, rounded, oval, ovate, oblate or subtriangular above becoming somewhat elongate in the central and lower parts of the leaf; *basal juxtacostal cells* short- to long-rectangular, thick- to thin-walled, hyaline or yellowish, mostly with straight and evenly thickened walls, rarely with nodulose longitudinal walls; *basal marginal cells* short, quadrate, oval to oblate, esinuose, with cross-walls thicker than the longitudinal walls, forming a hyaline border or otherwise distinct group or elongate, sinuose, chlorophyllose with evenly thickened walls. *Autoicous* or very rarely *dioicous*. *Perigonia* bud-like, axillary below the perichaetia. *Perichaetia* terminal but appearing lateral due to subfloral innovation stretching to nearly the same direction as that of the preceding stem; *perichaetial leaves* markedly larger and different in shape from the vegetative leaves, with the innermost pair usually ligulate and hyaline and the remaining ones elliptical to lingulate, chrorophyllose, concave, smooth or occasionally plicate, short- or long-acuminate, straight and symmetric or sometimes curved and asymmetric. *Setae* straight, shorter or as long as the urn, yellowish, orange to reddish- or orange-brown; *vaginula* very short, with thin and smooth epidermal cells. *Capsules* deeply immersed to emergent, erect, smooth or faintly striate, ovoid, cupulate, globose, cyathiform to obloid or cylindrical, often wide-mouthed, stramineous, red, brown, orange or golden-brown, systylious; *annulus* absent; *operculum* conic with a slanted or erect beak, occasionally mammillate; *columella* attached to the operculum after dehiscence; *exothecial cells* oblong, rectangular, isodiametric to transversely rectangular, gradually becoming oblate towards the orifice, with thin to incrassate walls; *stomata* present at the base of the urn, unicellular, superficial, with rounded pori or absent; *peristome teeth* narrowly triangular to linear-lanceolate, sometimes truncate, vestigial or absent, undivided, entire or variously perforate to cribrose, erect to squarrose-recurved, red, orange-red to reddish-brown, finely to coarsely papillose on both surfaces. *Spores* spherical, small to large, smooth to verrucose. *Calyptra* small, fugacious, not extending to the capsule mouth, cucullate to mitriform.

Schistidium is closely related to *Grimmia* and gametophytically both genera are difficult to separate, but the scarcely differentiated costa, lacking a central band of small stereids or hydroids, as clearly seen in transverse section, may be considered as a reliable character separating *Schistidium* from *Grimmia* and its segregates. Nonetheless, these genera basically differ in two sporophyte characters. Although the reference to sporophytes may appear to be impractical, *Schistidium* is generally 'over-fertile' and the majority of its species and populations produce capsules frequently and in profusion. The genus has consistently systylious capsules, i.e. with the operculum remaining permanently attached to the tip of columella after dehiscence (the only exception is *S. trichodon*), and small, mitrate to cucullate calyptrae which do not reach the capsule mouth. These characters coupled with the short, straight and centrally attached setae, resulting in the immersion of the symmetric capsules in the enlarged perichaetial leaves, make *Schistidium* a distinct and natural genus. The capsules usually appear to be lateral because of the subfloral innovations which continue their growth nearly in the same direction as the preceding stems. The characteristic appearance of the sporophytes makes *Schistidium* readily recognisable from other grimmiaceous mosses.

Systylious capsules have evolved several times in the mosses and they are found in various, distantly related moss families, for example in *Tayloria* and *Splachnum* in the Splachnaceae, *Hennediella*, *Hymenostylium* and *Gymnostomum* in the Pottiaceae, as well as in *Seligeria*, *Blindia*, *Valdonia* and *Hymenoloma* in the Seligeriaceae but, except for *Hymenostylium*, they are known only in selected species within genera concerned and have no diagnostic value. In the Grimmiaceae the exserted columella is known only in the monotypic genus *Streptocolea*. The function of systylious capsules is still unknown but most probably the attachment of the columella to the operculum may regulate spore dispersal by releasing them in times of more optimal conditions for their germination.

Schistidium is a globally distributed genus which is widespread in cold, cool and temperate regions of both hemispheres and at high elevations on tropical mountains. It has had a chequered nomenclatural and taxonomic history. Although it was established relatively early (Bridel 1819) it has not gained wide acceptance and has usually been merged with the then all-encompassing *Grimmia*. In addition, its generitype, *S. pulvinatum*, proved to be conspecific with *Grimmia anodon* (Ochyra 1989*a*) and accordingly the generic name *Schistidium* had to be conserved (Ochyra & Isoviita 1989) with the conserved type *S. maritimum*. About 130 species have been described in or transferred to this genus (Wijk *et al.* 1967; Crosby *et al.* 1992; Blom 1996; Ochyra 1998*b*; Ochyra & Muñoz 2000). Unfortunately, the real number of species in *Schistidium* cannot be established at present because the genus needs a modern global taxonomic revision, although it is likely that the genus consists of about 100 species. Bremer (1980, 1981, 1982) presented a worldwide revision of *Schistidium* in which she recognised 12 species, but she accepted a very broad and unreliable concept of *S. apocarpum* with which she had lumped a great number of distinct species from throughout the world.

In contrast, Blom (1996, 1998) proposed a totally different concept of species in the *Schistidium apocarpum* complex and in his revision of this group in Scandinavia he distinguished no less than 31 species and several infraspecific taxa, many of which were described as new or reinstated as good species. These are mostly distinct and clearly delimited taxa and many of them have wide European and/or pan-Holarctic distribution. The same situation also appears to be true in other parts of the world and, for instance, many austral species considered by Bremer (1981) to be conspecific with *S. apocarpum* proved to be well defined species (Bell 1984; Ochyra 1998*a*).

Owing to the imperfect state of the taxonomy, the infrageneric classification of *Schistidium* is practically non-existent. Kindberg (1898) considered this genus as a rankless subdivision within *Grimmia* and subdivided it into two unranked groups, *Platyphylloideae* and *Apocarpiformes*, and Vilhelm (1922) divided the genus into two subgenera, *Apocarpa* and *Conferta*. On the other hand, Bremer (1980, 1981, 1982) in her global revision of *Schistidium* did not propose any internal classification of the genus. Blom (1996) recognised five groups within the *S. apocarpum* complex, *Robusta*, *Conferta*, *Tenera*, *Atrofusca* and *Apocarpa*, the latter with three subgroups *Stricta*, *Umbrosa* and *Apocarpa*. Finally, Ochyra *et al.* (2003) presented the first infrageneric classification of *Schistidium* and recognised three subgenera and six sections within this genus but certainly this classification will require refinement by increased taxonomic study.

In many informal classifications and keys to the determination species of *Schistidium* are divided into two groups on the basis of spore size (Blom 1998), namely *Rivulare* and *Apocarpum* groups, having the spores 14–25(–36) µm and 8–15 µm wide in diameter, respectively. This criterion seems to have limited and doubtful value at the subgenus level since some species of the *Apocarpum* group have consistently larger spores (e.g. *S. grandirete* 15–21 µm) and some morphologically closely related species differ markedly in spore size (e.g. *S. cinclidodonteum*, *S. occidentale* and *S. falcatum* having spores 9–10, 9–11 and 20–28 µm wide, respectively). Thus, until the genus is revised taxonomically and the relationships between the species carefully assessed any classification of the genus is tentative.

In the Southern Hemisphere, species of *Schistidium* are very widespread and common in temperate and polar regions and, in some areas, play an important role in plant communities. However, the taxonomy of this genus has not been thoroughly studied in the austral region and the only available treatments are for South Georgia (Bell 1984), King George Island in the maritime Antarctic (Ochyra 1998*a*) and New Zealand (Fife 2000). Nonetheless, many species described from Îles Kerguelen, South Africa and New Zealand still await a careful taxonomic assessment, but at present one can state that species diversity of *Schistidium* in the Southern Hemisphere is quite similar to that in the Holarctic, at least judging from preliminary and as yet unpublished taxonomic studies, especially in southern South America, as well as from discovery of some new species in the Antarctic (e.g. Ochyra 1987*a*, 1998*c*, 2003*b*).

In the Antarctic *Schistidium* consists of 13 species and is the largest moss genus. Some of its species are very common and abundant components of the tundra vegetation and belong to the most frequent mosses in this biome.

1. Leaf margin plane or inflexed; leaves concave to canaliculate throughout, U-shaped in cross-section; costa often diffuse, elliptical to semi-elliptical in cross-section, scarcely prominent on the dorsal surface in the distal part (subgen. *Canalicularia*) 2

1. Leaf margin recurved to revolute for varying lengths; leaves keeled, V-shaped in cross-section, at least in the distal portion; costa semi-terete to subrectangular in cross-section, mostly strongly convex and prominent on the dorsal surface (subgen. *Apocarpa*) 3

2. Leaves lanceolate, mostly falcato-secund; costa excurrent as a cuspidate point; peristome well developed; spores 15–28 µm wide .... 1. ***S. falcatum***

2. Leaves ovate, ovate-lanceolate to lanceolate, straight or recurved, never secund; costa percurrent; peristome very rudimentary or absent; spores 9–11 µm .... 2. ***S. lewis-smithii***

3. Spores large, 15–25 µm wide; leaf margins sometimes distantly denticulate to serrulate at the apex; plants of very moist to wet habitats; leaves broadly ovate to broadly ovate-lanceolate, subobtuse, always lacking hair-point; costa semi-terete, 4–5-stratose in cross-section (sect. *Rivularia*) .... 3. ***S. rivulare***

3. Spores small, 7–13 µm wide; leaf margins always entire; plants of dry habitats; leaves ovate-lanceolate to lanceolate, often piliferous; costa semi-terete to subrectangular, 2–3-stratose in cross-section 4

4. Lamina irregularly bistratose in the distal part of the leaf; leaf margin recurved on one or both sides for 3/4–4/5 of the leaf length; exothecial cells predominantly oblong to rectangular in central and lower parts of the urn (sect. *Atrofusca*) .... 4. ***S. praemorsum***

4. Lamina unistratose throughout, with occasional bistratose strands distally; leaf margin recurved to revolute to the apex; exothecial cells predominantly transversely elongate and isodiametrical in central and lower parts of the urn 5

5. Basal marginal cells chlorophyllose, small, isodiametrical, not forming a distinct group or border; plants red to red-brown, sometimes with blackish, greyish or green-yellowish tones; reaction of the upper leaves with KOH red or orange; peristome teeth curved, entire (sect. *Apocarpiformia*) .... 5. ***S. andinum***

5. Basal marginal cells hyaline, isodiametrical to elongate with thickened transverse walls, forming a rectangular group or a leaf border; plants olivaceous, often with yellowish tones; reaction of the upper leaves with KOH yellow; peristome teeth straight, often irregular and strongly perforated, sometimes reduced to rudimentary (sect. *Conferta*) ... 6

6. Leaf margins revolute throughout; peristome teeth vestigial or absent .... 6. ***S. steerei***

6. Leaf margins narrowly recurved for varying lengths; peristome teeth well developed to variously reduced 7

7. Plants hoary; leaves with spinulose-denticulate hair-points, longer than 0.4 mm 8

7. Plants not hoary; leaves epilose or with smooth or denticulate hair-points, shorter than 0.4 mm 9

8. Exothecial cells with strongly incrassate walls, 4–8 µm thick; upper laminal cells rectangular, with sinuose, strongly thickened walls; central strand absent .... 7. ***S. urnulaceum***

8. Exothecial cells thin-walled, usually less than 1 µm thick; upper laminal cells isodiametric to oblong, with straight, moderately evenly thickened walls; central strand present.. 8. ***S. halinae***

9. Peristome teeth reduced, mostly truncate in the middle; leaves usually arranged in distinct spiral rows .... 9. ***S. antarctici***

9. Peristome teeth well developed, occasionally eroded at the apex; leaves not spirally ranked 10

10. Costa weak, to 40 µm wide, scarcely prominent on the dorsal surface, often nearly of the same thickness as the lamina .... 10. ***S. leptoneurum***

10. Costa strong, 50 or more µm wide, prominently convex on the dorsal surface, much thicker than the laminal cells 11

11. Capsules cupulate, 0.5–0.6 mm long; exothecial cells with incrassate walls; costa semi-terete to elliptic, with undifferentiated ventral cells; leaves sharply keeled throughout, narrowly V-shaped, with blades of the lamina in transverse section forming a $<30°$ angle, usually parallel at the junction with the costa; furrow very narrow .... 11. ***S. deceptionense***

11. Capsules ovoid to obloid, 0.7–1.1 mm long; exothecial cells thin-walled; costa subrectangular to semi-terete, usually with 2 ventral cells somewhat larger than central and dorsal cells; leaves concave below, keeled in the distal portion, broadly V- or U-shaped, with blades of the lamina in transverse section forming a $>45°$ angle, divergent at base; furrow more open 12

12. Upper laminal cells long-rectangular, with strongly sinuose-nodulose walls; basal cells never lax, merging into long-rectangular sinuose-walled cells extending along the margin nearly to the base; central strand absent .... 12. ***S. amblyophyllum***

12. Upper laminal cells quadrate to shortly rectangular, with straight or weakly sinuate walls; basal cells often lax, straight-walled, merging into rectangular cells with sinuose walls not extending to the base along the margins; central strand present .... 13. ***S. cupulare***

1. **Schistidium falcatum** (Hook.f. & Wilson) B.Bremer Fig. 96

Schistidium falcatum (Hook.f. & Wilson) B.Bremer, Lindbergia 6: 110. 1980. ≡ *Grimmia falcata* Hook.f. & Wilson, London J. Bot. 3: 539. 1844. — TYPE CITATION: Kerguelen's Land. [Lectotype (*vide* Bell 1984: p. 89): 'Grimmia Scouleria falcata H. fil. & Wils. in the waterfall on the S-side of Christmas Harbour, Kerguelen's Land. May 1840. Antart. [sic!] Exp. 1839–1843. J. D. Hoker 245' – BM-Hooker!; isotypes: BM!, BM-Hooker!].

Plants medium-sized to large, rigid, in loose or dense ascending tufts, forming extensive coalescing mats, dull, dark or red-green to rufous above, dark or blackish-brown below. *Stems* prostrate to ascending or suberect, 1.5–4.5 cm tall, stiff, copiously fastigiately branched, appearing bristly below due to erosion of the leaf laminae except for the stout costae, occasionally radiculose at the base with brown, lustrous, smooth, sparsely branched rhizoids, in cross-section round, lacking a central strand and with a distinct, 1–3-stratose cortex of small, substereid cells with strongly incrassate, dark to reddish-brown walls, sharply delimited from large, hyaline medullary cells with firm, yellowish walls; *axillary hairs* hyaline, filiform, 4–5-celled. *Leaves* rather distant below, becoming densely set toward the apex, imbricate, erect to erect-spreading, recurved to falcato-secund toward the apex, less often straight when dry, erecto-patent when moist, lanceolate to oblong-lanceolate, gradually tapering to a narrow, obtuse or subacute apex, epilose, 2–3 mm long, (0.4–)0.5–0.8 mm wide, concave, broadly canaliculate, distinctly decurrent; *margins* entire, erect to incurved, unistratose or with bistratose spots distally; *costa* strong, usually ill-defined and diffuse, 85–120 μm wide, percurrent to shortly excurrent, at least in the upper leaves, in transverse section semi-elliptical, plano-convex, 3–5-stratose, with somewhat larger dorsal and ventral epidermal cells and smaller median cells; *laminal cells* unistratose throughout or partly bistratose above, smooth, not bulging, arranged in regular longitudinal rows, quadrate, rounded-quadrate, oval to shortly rectangular above, 6–16 μm long, 6–12 μm wide, with straight, moderately thickened walls, becoming isodiametric to shortly rectangular below, 10–25 μm long, 8–12 μm wide; *basal juxtacostal cells* short-rectangular, 14–28 μm long, 8–12 μm wide, thick- and straight-walled; *basal marginal cells* similar, mostly shorter, often with slightly sinuose walls, sometimes oblate at the margins. [*Monoicous*. *Perichaetial leaves* oblong-lanceolate, concave, 2.8–3.4 mm long, 0.9–1.0 mm wide. *Setae* short, 0.3–0.5 mm long. *Capsules* immersed, subhemispheric to obloid, 0.8–1.2 mm long, 0.7–1.0 mm wide, light to dark brown, usually with varnish luster, pachydermous; *operculum* conic, with a short, straight or oblique rostrum; *exothecial cells* irregular, collenchymatous, with moderately thickened walls, rounded, elliptical to short-rectangular, 10–20 μm long, 8–20 μm wide, becoming oblate, thick-walled in 5–7 tiers at the mouth, forming dark reddish-brown strip; *stomata* absent; *peristome teeth* lanceolate, 200–250 μm long, 60–80 μm wide at the base, orange-reddish to dark reddish-brown, entire or perforate above, papillose-striate. *Spores* globose, pale brownish, minutely roughened, (15–)20–28 μm wide. *Calyptra* mitrate.]

Chromosome number: unknown.

Discussion. *Schistidium falcatum* is a distinct and impressive species which can be readily distinguished from all other congeners by its lanceolate to oblong-lanceolate, strongly falcato-secund, decurrent leaves with erect to incurved margins and stout, mostly excurrent costa as well as a characteristic areolation of rounded, thick-walled cells, nearly homogeneous throughout the lamina. As with most aquatic mosses, the species is very variable, and in most populations from outside the Antarctic the costa is very strong, 100–125 μm wide, often diffuse and usually long excurrent as a stout cuspidate point.

In the Antarctic the species may only be confused with *Schistidium lewis-smithii*, and the material from King George I. which represents this species was originally misdetermined as *S. falcatum* (Ochyra 1998*a*). Both species share similar leaf areolation and structure of the costa, but the fundamental difference is in the capsules which are perfectly gymnostomous in *S. lewis-smithii*. Additionally, this species has ovate to ovate-lanceolate and never falcato-secund leaves and the leaf shape should enable safe discrimination of the species because *S. falcatum* is not known to fruit in the Antarctic.

Schistidium falcatum together with *S. lewis-smithii* from the Antarctic, *S. agassizii*, *S. cinclidodonteum*, *S. occidentale* and *S. pacificum* from the Northern Hemisphere, as well as the Australasian *S. flexifolium*, constitute a distinct group within the genus which is best characterised by erect to incurved leaf margins and broadly canaliculate leaves. This group was distinguished as a separate subgenus, subgen. *Canalicularia*, which differs from all other species of the genus which have variously recurved leaf margins and mostly distinctly keeled leaves (Ochyra 2003*b*).

Reproduction in Antarctica. Neither gametangia nor sporophytes have been observed, although outside the Antarctic the species is commonly found in fruiting condition.

Habitat. Outside the Antarctic *Schistidium falcatum* is a typical hydrophyte growing, often submerged, on rocks, boulders and stones in swiftly flowing streams and in waterfalls. The Antarctic plants were found on moist, bare gravelly or ashy soil among boulders on scree. The only associates of the species are *Leptogium puberulum* and *Sanionia georgicouncinata*.

World range. An amphiatlantic subantarctic species (Fig. 97, inset) widespread on South Georgia (Bell 1984), the Prince Edward Is. (Zanten 1971; Ochyra, personal observations), Îles Crozet (Ochyra & Bell 1984) and Îles Kerguelen from where it was originally described (Hooker & Wilson 1844; Wilson & Hooker 1847). In addition, it occurs at high elevations in

→

Figure 96. *Schistidium falcatum*. — **1**. Habit, dry. **2**. Portion of branch and perichaetium, wet. **3**. Cross-section of stem. **4–8**. Leaves. **9**. Upper laminal cells at margin. **10**. Mid-leaf cells at margin. **11**. Basal juxtacostal cells. **12**. Angular cells. **13–18**. Cross-sections of leaves, a sequence from base to apex. **19–21**. Perichaetial leaves. **22**. Operculate capsule, wet. **23**. Operculum with attached columella. **24**. Calyptra. **25**. Exothecial cells in middle of urn. **26**. Lower exothecial cells. **27**. Peristome teeth, spores and rim cells of capsule. — [1, 3–4, 7–18 from *RO 437/80*; 2, 5–6, 19–27 from *Bell 143A*]. — Scale bars: a – 1 mm (2); b – 1 mm (1, 4–8, 19–23); c – 100 μm (3, 9–18) and 200 μm (25–27); d – 1 mm (24).

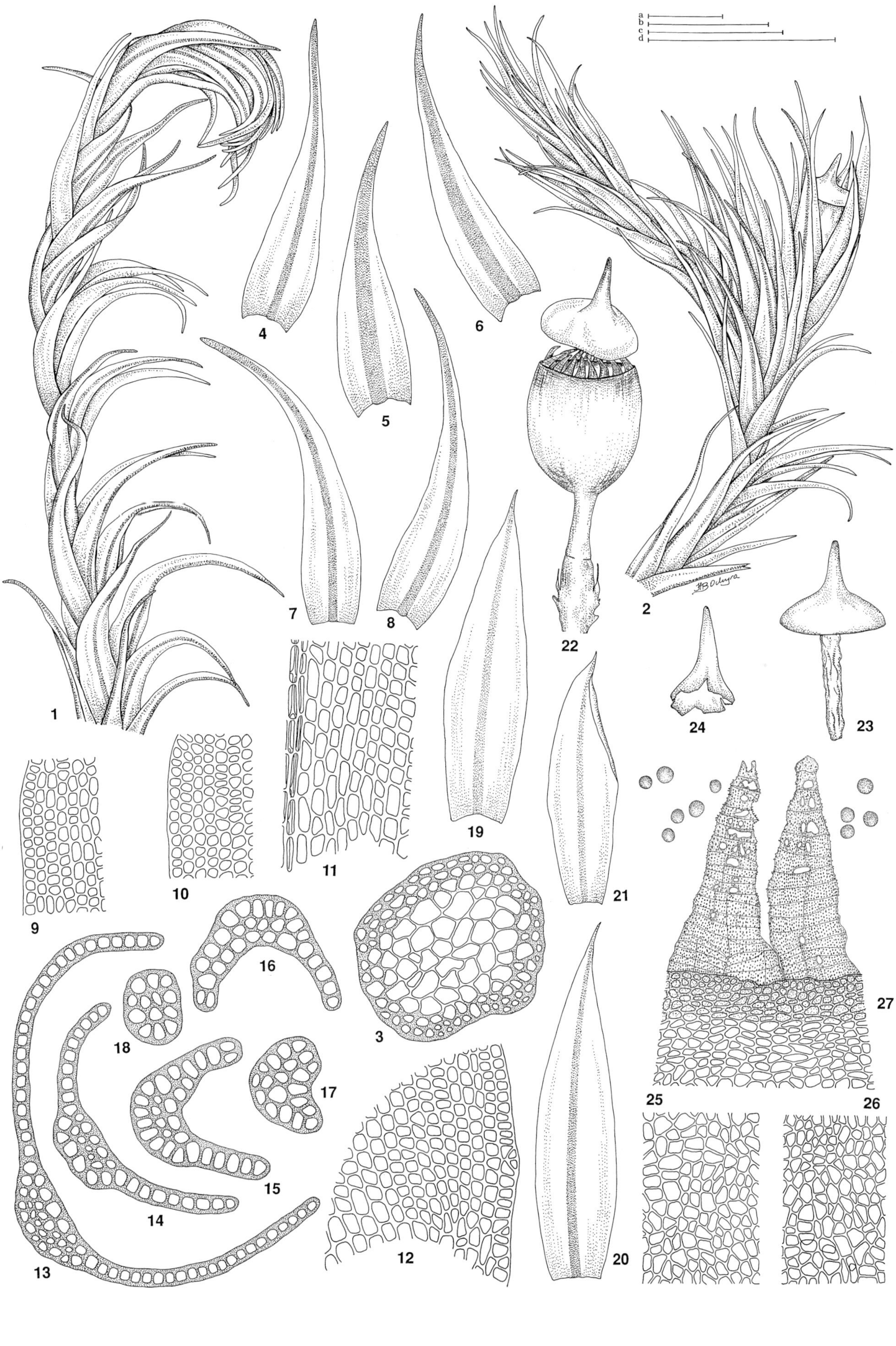
a
b
c
d
1
2
3
4
5
6
7
8
9
10
11
12
13
14
15
16
17
18
19
20
21
22
23
24
25
26
27

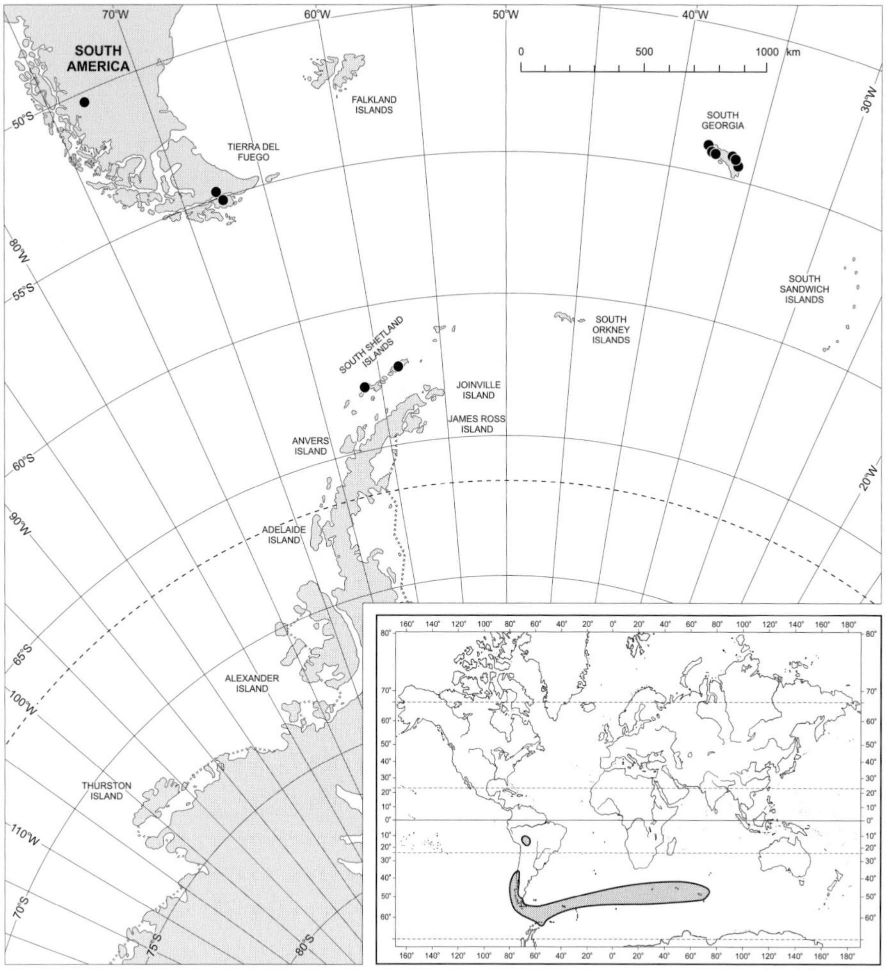

Figure 97. Distribution of *Schistidium falcatum* in the Fuegian region and Antarctica. Inset: global distribution.

Tierra del Fuego and in western Patagonia, extending to central Chile, with some isolated stations in the Andes of Bolivia at elevations above 4300 m.

Distribution in Antarctica. Known only from two localities, one on King George I. (Ochyra & Bell 1984) and the other on Livingston I., both in the South Shetland Is., reaching its southernmost occurrence at the latter site (lat. 62°32′S) (Fig. 97). It occurs at elevations of 35–220 m. A second specimen reported from King George I. (Ochyra 1998*a*) proved to be a misdetermination and represents a new species, *S. lewis-smithii* (Ochyra 2003*b*).

Specimens examined. WEST ANTARCTICA. SOUTH SHETLAND IS. **KING GEORGE I. *Admiralty Bay***: Keller Peninsula, Tyrell Ridge, 220 m, *RO 437/80* (AAS, B, BA, BM, C, CANM, CHR, F, FH, G, GL, H, KRAM, NY, PC, S, U). **LIVINGSTON I. *Byers Peninsula***: between Smellie Pt. and Chester Cone, 35–50 m, *RILS 3790B* (AAS, KRAM).

2. **Schistidium lewis-smithii** Ochyra Fig. 98

Schistidium lewis-smithii Ochyra, Nova Hedwigia 77: 364, *f. 1.* 2003. — TYPE CITATION: Antarctica, South Shetland Islands, King George Island: Admiralty Bay, Creeping Slopes above Paradise Cove, 62°12′30″S, 58°26′30″W, alt. 60 m, on bare earth among boulders on scree slope with easterly aspect, 2 Feb 1980, Ochyra 1380/80. [Holotype: 'Herbarium Instituti Botanici Academiae Scientiarum Polonae – Cracoviae Bryophyta. ~~Schistidium antarctici (Card.) Sav.-Lub. et Smirn.~~ = Schistidium lewis-smithii Ochyra, sp. nov. Antarctica, South Shetland Islands – King George Island: Admiralty Bay, Creeping Slopes above Paradise Cove, 62°12′30″S, 58°26′30″W, elev. 60 m, on bare earth among boulders on scree slope with easterly aspect, 2/II/1980 Coll. R. Ochyra No. 1380/80 Det. R. Ochyra' – KRAM!; isotype: NY!].

Plants medium-sized, stiff and rigid, densely tufted, dull to weakly glistening, reddish-brown to dark olive-green above, brown below. *Stems* 0.8–2.5 cm tall, erect, much irregularly branched, rather thick, brown, in transverse section rounded, consisting of 1–2-layered cortex of small to medium cells with brown, thickened walls sharply delimited from a medulla of 3–6 layers of large, hyaline, thin- to moderately thick-walled cells, central strand small

→

Figure 98. *Schistidium lewis-smithii.* — **1**. Habit, dry. **2**. Habit, wet. **3**. Portion of branch and perichaetium, wet. **4**. Cross-section of stem. **5–6**. Axillary hairs. **7–11**. Leaves. **12**. Leaf apex. **13**. Mid-leaf cells at margin. **14**. Basal juxtacostal and angular cells. **15–23**. Cross-sections of leaves, a sequence from base to apex. **24–26**. Perichaetial leaves. **27–28**. Operculate capsules, wet. **29**. Operculum with attached columella. **30**. Lower exothecial cells and stoma. **31**. Exothecial cells in middle of urn. **32**. Exothecial cells at mouth, showing a vestigial peristome tooth and spores. **33**. Two vestigial peristome teeth. — [All from *RO 1369/80* (holotype)]. — Scale bars: a – 1 mm (3); b – 1 mm (27–29); c – 100 μm (4, 30–33); d – 1 mm (24–26); e – 2 mm (1–2) and 100 μm (15–23); f – 1 mm (7–14); g – 100 μm (5–6). (Reproduced from *Nova Hedwigia*, 77: p. 365, 2003, with permission.)

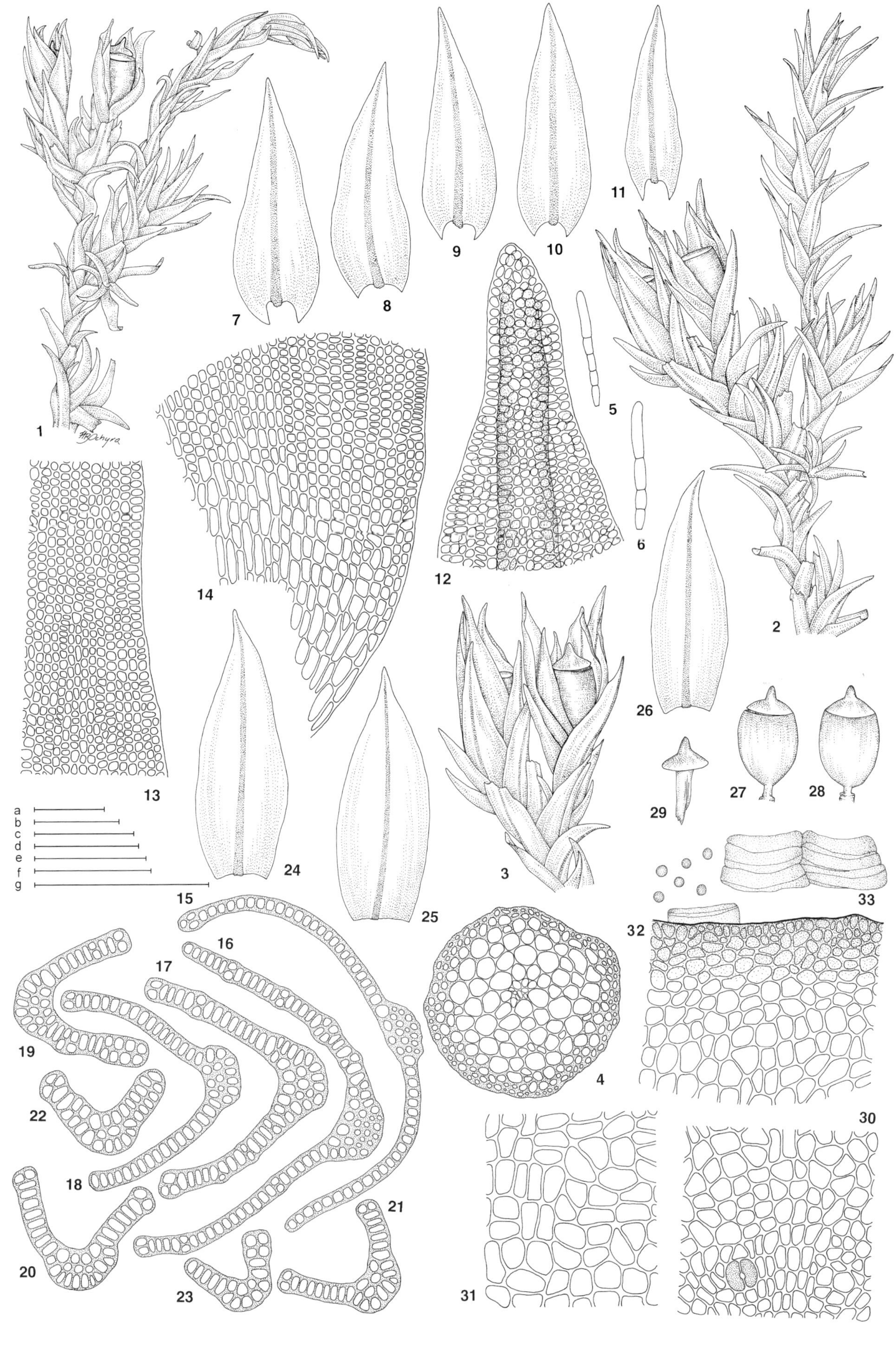
1
2
3
4
5
6
7
8
9
10
11
12
13
14
15
16
17
18
19
20
21
22
23
24
25
26
27
28
29
30
31
32
33
a
b
c
d
e
f
g

but distinct; *rhizoids* few, scattered at the base of the stem, brown, lustrous, smooth, sparsely branched; *axillary hairs* filiform, hyaline, 50–60 µm long, composed of 3–4 elongate cells. *Leaves* crowded, erect and appressed at the stem tips, becoming erecto-patent, often with recurved apices below, not or weakly altered on drying, erect-spreading on wetting, ovate to ovate-lanceolate or lanceolate, straight or slightly curved, gradually narrowed to an acute or narrowly obtuse apex, always lacking a hair-point, (1.5–)1.8–2.1 mm long, 0.6–0.9 mm wide, concave, distinctly decurrent; *margins* erect to incurved, entire, unistratose with frequent uniseriate bistratose patches near the apex; *costa* percurrent, well defined to somewhat diffuse near the base, 50–80 µm wide, orange-yellow, plano-convex, not prominent dorsally, in cross-section semi-elliptical to semi-terete, 3–4-stratose, with somewhat larger ventral and dorsal epidermal cells and smaller central substereid cells; *laminal cells* unistratose throughout, with frequent bistratose streaks or occasionally almost entirely bistratose in the distal part, smooth, not bulging, nearly homogeneous throughout the lamina, rounded, rounded-quadrate, oblate to shortly rectangular, 4–15 µm long, 4–10 µm wide, thick- and straight-walled, becoming short-rectangular in a small area near the base of the costa, 25–35 µm long, 8–10 µm wide. *Monoicous. Perigonia* on separate branches below the perichaetia; *perigonia bracts* ovate with percurrent to subpercurrent costa, with many club-shaped antheridia, paraphyses wanting. *Perichaetial leaves* variable in size, the outer ones oblong-lanceolate, 1.9–2.1 mm long, 0.7–0.8 mm wide, the inner larger, 2.1–2.3 mm long, about 1 mm wide, oblong-ovate, gradually or abruptly shortly acuminate, narrowly obtuse, epilose; *costa* percurrent or ending below the apex, weak, sometimes of only 2 layers of cells; *laminal cells* similar to those in the vegetative leaves above, becoming rectangular, hyaline, not lax, with thin to moderately thickened walls in the proximal part. *Setae* short, 0.2–0.3 mm long. *Capsules* deeply immersed, subglobose, 0.8–0.9 mm long, 0.7–0.8 mm wide, pachydermous, yellow-brown, smooth; *operculum* conic, shortly straight-beaked; *exothecial cells* variable, mostly rounded to shortly rectangular, 20–45 µm long, 15–30 µm wide, thick-walled, collenchymatous, becoming somewhat smaller, rounded to oblate, with distinctly incrassate walls at the orifice, forming 5–7-seriate, intensively orange-brown strip; *stomata* absent; *peristome* absent or very rudimentary, to 15 µm long, consisting of short basal segments, not or very slightly exceeding beyond the rim of the urn, pale yellow, nearly smooth. *Spores* spherical, pale, 8–11 µm in diameter, smooth to minutely roughened. *Calyptra* not seen.

Chromosome number: unknown.

Discussion. *Schistidium lewis-smithii* is clearly related to *S. falcatum* but differs in its gymnostomous capsules and much smaller, more or less smooth spores. The essential differences in leaf shape between these species produce various habits which are discussed under *S. falcatum*. However, both species share broadly canaliculate leaves, erect to incurved leaf margins, leaf areolation, as well as elliptical to semi-elliptical costa well visible in transverse section. These characters serve to link a small group of species which, apart from the two austral species, comprises *S. pacificum*, *S. occidentale*, *S. cinclidodonteum* and *S. agassizii* (it has the leaf margins narrowly recurved in the proximal part for a short distance from the base or, exceptionally, narrowly deflexed higher up) from the Northern Hemisphere, into a separate subgenus, subgen. *Canalicularia* (Ochyra *et al.* 2003; Ochyra 2003*b*). Species of this section differ from other species of *Schistidium* which are characterised by carinate leaves and variously recurved leaf margins.

Apart from *Schistidium steerei*, *S. lewis-smithii* is the only other gymnostomous species of the genus in the Antarctic. This state of the capsules is generally rare in *Schistidium*, with the majority of species having ochyrostomous capsules, the only three exceptions being the Holarctic *S. flaccidum*, *S. cryptocarpum* and *S. atrofuscum*. In *S. antarctici* and in some populations of *S. praemorsum* the peristome teeth are markedly reduced and mostly truncate in the middle, although it is worth noting that very rarely the gymnostomous *S. flaccidum* forms very fragile peristome teeth, as reported in a single population from the Alps (Blom 1996).

Reproduction in Antarctica. Fertile, rather frequently producing fully mature sporophytes.

Habitat. The only known specimens of the species were collected from bare soil between boulders on a scree slope which was temporarily wet judging from the dense coating of the lower parts of the plants with silt, and on moist ashy soil dominated by *Leptogium puberulum*.

World range. Antarctic endemic, so far recorded only in the South Shetland Is. (Fig. 99).

Distribution in Antarctica. Known only from two collections from King George I. and Livingston I., South Shetland Is., extending to lat. 62°32′ S.

Specimens examined. WEST ANTARCTICA. SOUTH SHETLAND IS. KING GEORGE I. ***Admiralty Bay***: Creeping Slopes, 60 m, *RO 1380/80* (KRAM – type). LIVINGSTON I. ***Byers Peninsula***: between Smellie Pt. and Chester Cone, 35–50 m, *RILS 3787* (AAS, KRAM).

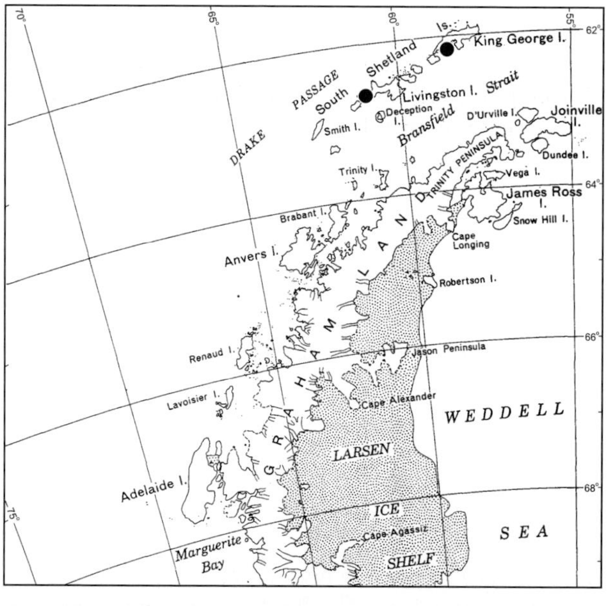

Figure 99. Global distribution of *Schistidium lewis-smithii*. (Reproduced from *Nova Hedwigia*, 77: p. 367, 2003, with permission.)

3. **Schistidium rivulare** (Brid.) Podp. Fig. 100

Schistidium rivulare (Brid.) Podp., Beih. Bot. Centralbl. 28: 207. 1911. ≡ *Grimmia rivularis* Brid., J. Bot. (Schrader) 1800(1): 276. 1801. ≡ *G. alpicola* Hedw. var. *rivularis* (Brid.) Wahlenb., Fl. Lapp.: 32. 1812. ≡ *G. apocarpa* Hedw. var. *rivularis* (Brid.) Nees & Hornsch., Bryol. Germ. 2(1): 101. 1827. ≡ *Schistidium apocarpum* (Hedw.) Bruch & Schimp. var. *rivulare* (Brid.) Bruch & Schimp., Bryol. Eur. 3: 100, *pl. 234γ*. 1845. ≡ *S. alpicola* var. *rivulare* (Brid.) Limpr., Laubm. Deutschl. 1: 708. 1889. — Type citation: [...] ad Thuringici saltus rivulos frigidos, ex gr. prope Klein Schmalkalden. [Lectotype (*selected here*) {or perhaps holotype}: 'Grimmia apocarpa v. aquatica. Grimmia rivularis N. Schmalkalden .... (illegible) 1799' – B-Bridel!].

Plants medium-sized to large, rather robust, in loose or dense, rigid tufts or forming extensive straggling tufts, somewhat lustrous to dull, dark green, yellow-brown to olive-green or blackish-brown above, brown to brown-black below. *Stems* rather stiff, erect to ascending, 1–5 cm tall, usually strongly and irregularly branched above, sparsely so below, occasionally almost unbranched, sparsely radiculose below with long, brown, smooth, weakly branched rhizoids, in cross-section consisting of 1–3-stratose cortex of small, brown cells with strongly incrassate walls surrounding 4–5 rows of large, thin-walled, hyaline medullary cells, central strand small and indistinct to lacking; *axillary hairs* filiform, hyaline throughout, 6–8-celled. *Leaves* crowded, loosely or more closely imbricate, erect or occasionally somewhat secund, not much altered on drying, erecto-patent on wetting, broadly ovate-triangular to ovate-lanceolate, sometimes slightly asymmetric, 1.8–2.4(–3.2) mm long, 0.7–1.3 mm wide, subobtuse to bluntly acute at the apex, always without a hair-point, sharply carinate distally, broadly concave in the proximal part, somewhat decurrent; *margins* narrowly recurved on both sides for varying distances, usually up ⅔–¾ the leaf length or sometimes only in the broadest part of the leaf, entire or distantly repand-denticulate to serrulate at the apex, bistratose in 2–4 rows of cells in the distal part; *costa* strongly projecting at the dorsal side, subpercurrent to percurrent, well defined, very strong, reddish-brown, 60–100 μm wide, in transverse section plano-convex, semi-terete to rectangular, 4–5-stratose, composed of 2–4 larger ventral and many dorsal epidermal cells and somewhat smaller, central cells; *laminal cells* smooth, not bulging, unistratose or with bistratose strands distally, with thin to moderately evenly thickend, esinuose or slightly sinuose walls, nearly homogeneous except at the base, rounded, rounded-quadrate to oval, 6–12 μm long, 5–12 μm wide; *basal cells* shortly rectangular, 20–45 μm long, 6–15 μm wide; *basal marginal cells* isodiametric, chlorophyllose, esinuose, not differentiated from the upper laminal cells. *Autoicous. Perichaetial leaves* yellowish- or olive-green to brown, much larger than the vegetative leaves, broadly ovate, oblong-lanceolate to elliptical, 2.9–3.5 mm long, 0.9–1.5(–1.8) mm wide, epilose, deeply concave, bluntly acute to broadly obtuse, with the areolation similar to that in the vegetative leaves in the distal part and composed of rectangular, thin-walled cells in the lower half, costa subpercurrent to ceasing at some distance from the apex. *Setae* short, 0.2–0.5 mm long, straight, centrally attached. *Capsules* deeply immersed in perichaetial leaves, brown to dark reddish-brown with age, subglobose to obovoid, 0.8–1.1 mm long, 0.7–1.1 mm wide, wide-mouthed; *operculum* rostrate to rostellate, with a straight or oblique beak; *exothecial cells* irregular, isodiametric, rounded-quadrate to irregularly angular, or shortly oblong, 20–40 μm long, 15–35 μm wide, thick-walled with collenchymatous corner thickenings, becoming oblate in 4–6 rows at the orifice and forming a distinct, reddish strip; *stomata* few at the extreme base of the urn, superficial, bicellular, round-pored; *peristome teeth* lanceolate, revolute, 400–520 μm long, 110–160 μm wide at the base, gradually long-acuminate or broadly short-pointed, sometimes irregularly truncate, entire or cribrose above, often with a few narrow slits in the upper part, dark purple-red to orange- or reddish-brown, densely, coarsely or finely papillose on both surfaces. *Spores* spherical, brownish, papillose, (15–)19–24 μm in diamter. *Calyptra* conical, smooth, mitriform, lobed at base, not extending to the urn.

Chromosome number: unknown in the Antarctic and Subantarctic; n = 13 (3 records) from the Holarctic (Fritsch 1991).

Discussion. In the Antarctic, *Schistidium rivulare* is an almost unmistakable species which is readily distinguished by a set of both sporophyte and gametophyte characters. The most important diagnostic features of the species relate to the capsules which are subglobose and hemispherical when empty. The exothecial cells are thick-walled and have distinct collenchymatous thickenings, and the spores are papillose and large, mostly 19–23 μm in diameter. Sterile plants are best recognised by their broadly ovate-triangular to ovate-lanceolate, sometimes slightly asymmetric leaves which are subobtuse to bluntly acute at the apex, always lack a hair-point and are sharply carinate in the upper half and broadly concave below. The leaf margins are bluntly repand-denticulate to distantly serrulate at the apex and this feature is best observed on young upper leaves, whereas older, lower leaves usually have entire margins. The costa is very strong, usually subpercurrent, 4–5-stratose and strongly projected on the dorsal side. In contrast to the Holarctic populations, the laminal cells are unistratose distally in the Antarctic plants or exceptionally they are arranged in bistratose strands. Also, the habitat of *S. rivulare* is unusual since very few other acrocarpous mosses grow submerged on rocks in streams in the Antarctic.

The only species in the Antarctic with which *Schistidium rivulare* is likely to be confused is *S. praemorsum*. Both species share a similar shape, size and colour of the leaves, areolation and the structure of the costa but in the latter the uppermost leaves are usually hyaline tipped and the upper laminal cells are irregularly, sometimes almost entirely, bistratose. However, the most important difference is in the spores which are much smaller (10–13 μm wide) and nearly smooth. Moreover, its peristome teeth are orange and somewhat eroded and reduced at the apex, at least in the Antarctic plants.

Together with two Holarctic species, *Schistidium platyphyllum* and *S. sordidum*, *S. rivulare* belongs to the small sect. *Rivularia*. They are all associated with aquatic habitats, have epilose to shortly piliferous leaves, a strong, dorsally prominent costa and large and granulose spores.

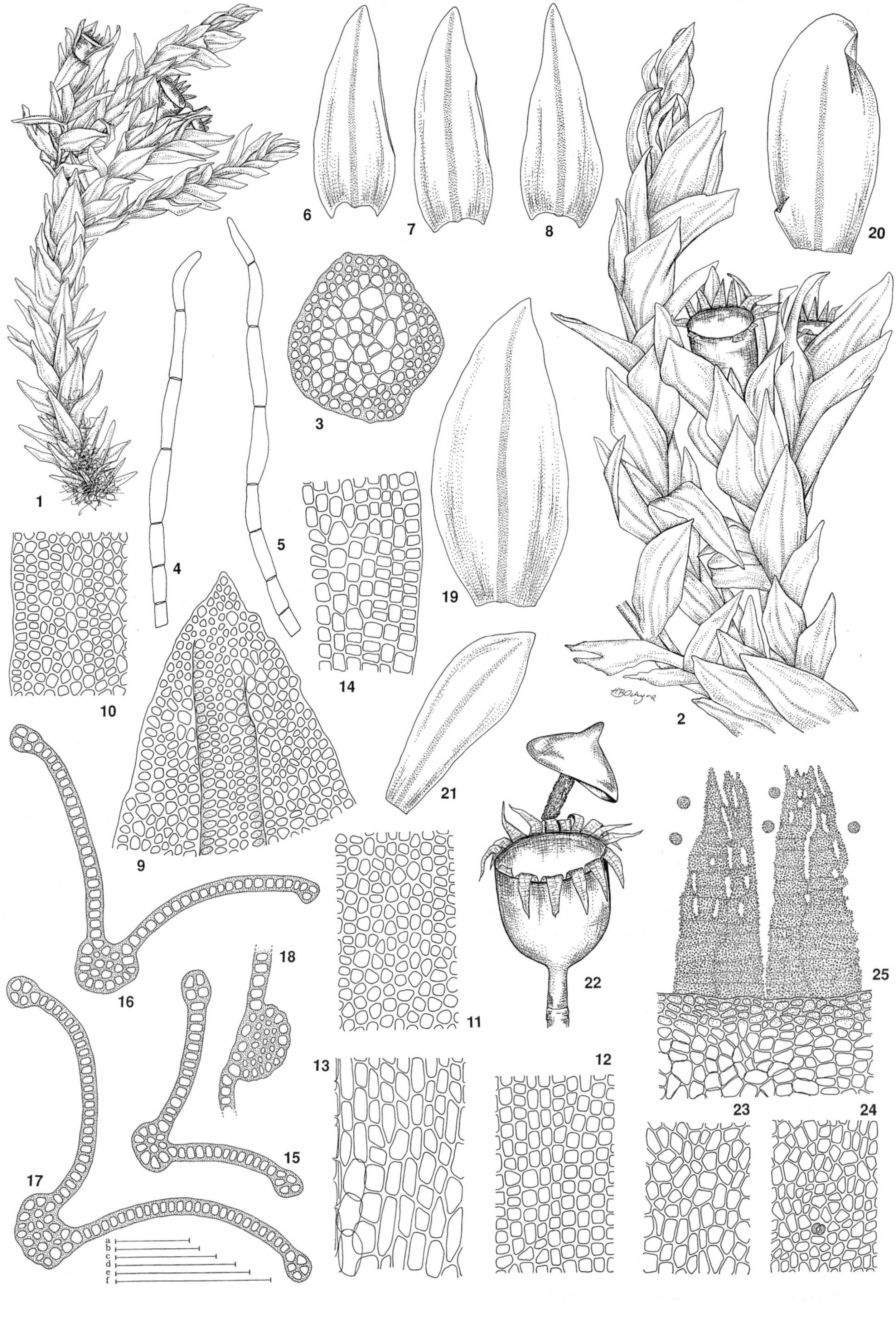
1
2
3
4
5
6
7
8
9
10
11
12
13
14
15
16
17
18
19
20
21
22
23
24
25
a
b
c
d
e
f

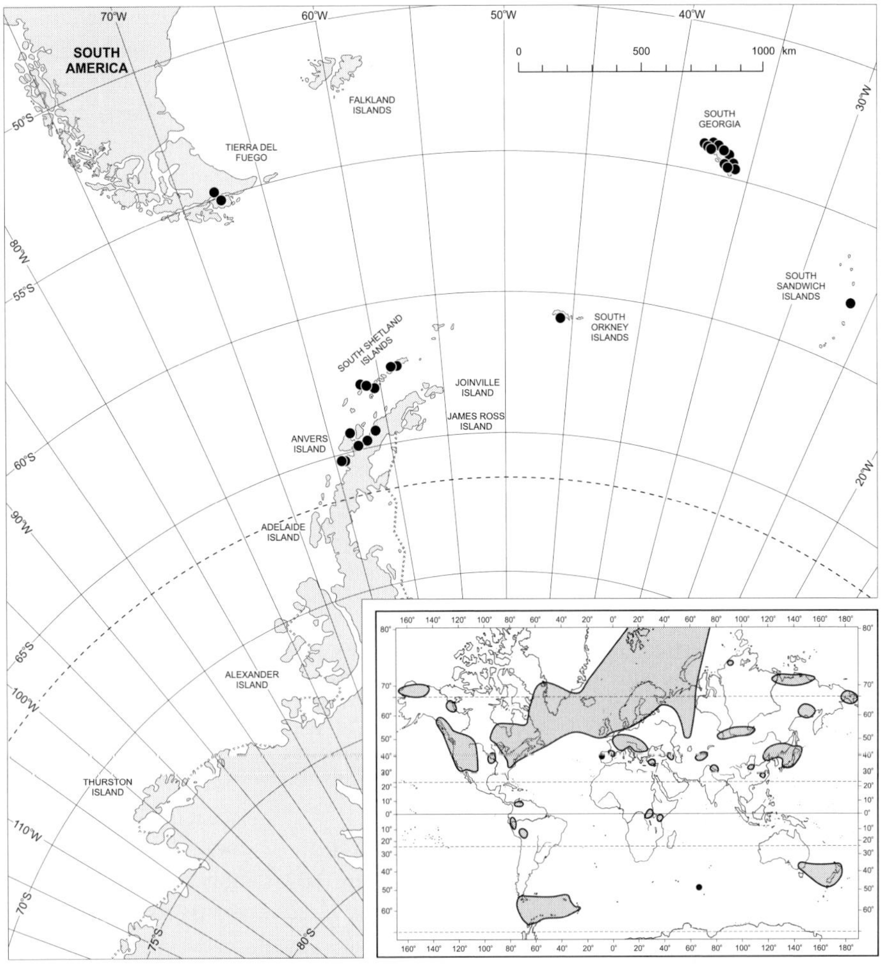

Figure 101. Distribution of *Schistidium rivulare* in the Fuegian region and Antarctica. Inset: global distribution.

Reproduction in Antarctica. Most populations are found in fruiting condition and sporophytes are usually produced in abundance.

Habitat. A hydrophytic moss, usually forming extensive dark green to brown patches or isolated turves on wet, calcareous and non-calcareous rocks and boulders in and along streams, by waterfalls, in flood zone of lakes, on moist to wet soil and in wet flush areas, sometimes totally submerged in swiftly flowing water. It usually grows in pure stands and is occasionally associated with *Sanionia georgicouncinata*, *Bryum pseudotriquetrum*, *Brachythecium austrosalebrosum*, *Ceratodon purpureus* and *Pohlia cruda*.

World range. A bipolar species with some transitional stations in tropical mountains (Fig. 101, inset). Its global range is still inadequately known due to unresolved taxonomic problems and unclear relationships between various aquatic and large-spored taxa of *Schistidium*. In the Northern Hemisphere it has a boreal montane range with a limited penetration into the high Arctic in Alaska (Steere 1978), Svalbard (Kuc 1973*b*) and Russian Arctic (Afonina & Czernyadjeva 1995) and extending as far south as California, Oklahoma, Tennessee and North Carolina in North America (McIntosh 2007) and southern Turkey, Central Asia (Mamatkulov 1990), Central Siberia (Ignatov & Cao 1994), the Himalayas (Chopra 1975) and Central China (Cao & Vitt 1986; Cao *et al.* 2003). Outside the Holarctic, typical populations of *S. rivulare* appear to occur only in the Fuegian region, on South Georgia, in the maritime Antarctic and on Îles Kerguelen (Ochyra, personal observations). The tropical American and African plants possibly belong to different species, at least judging from some specimens examined from East Africa (Ochyra 1993*d*) and the Central Andes (Deguchi 1997) which have much smaller, smooth or only minutely roughened spores (10–13 μm wide). The same may also be true for the Australian and New Zealand populations (Bremer 1980; Fife 2000) because the specimens described as *Grimmia flexifolia*, *G. subflexifolia* or *G. aquatica* and considered to be identical to *S. rivulare* certainly represent a distinct species.

Distribution in Antarctica. A rare but widely scattered and locally common species, very occasional in the South Sandwich Is. (Saunders I.) and infrequent in the South Orkney Is. (Signy I.) and South Shetland Is. (King George I., Livingston I.)

←

Figure 100. *Schistidium rivulare*. — **1**. Habit. **2**. Portion of the stem with perichaetium, wet. **3**. Cross-section of stem. **4–5**. Axillary hairs. **6–8**. Leaves. **9**. Leaf apex. **10**. Upper lamina cells at margin. **11**. Mid-leaf cells. **12**. Lower lamina cells. **13**. Basal juxtacostal cells. **14**. Angular cells. **15–17**. Cross-sections of leaves, a sequence from apex to base. **18**. Cross-section of costa in lower part. **19–21**. Perichaetial leaves. **22**. Dehisced capsule and operculum with attached columella, wet. **23**. Exothecial cells in middle of urn. **24**. Lower exothecial cells and stoma. **25**. Peristome teeth, spores and rim cells of capsule. — [All from *RO 637/80*]. — Scale bars: a – 1 mm (2); b – 1 mm (6–8, 19–21); c – 100 μm (3, 15–18); d – 1 mm (22); e – 100 μm (4–5, 9–14) and 50 μm (23–25); f – 0.5 cm (1).

and on the Danco and Graham Coasts on the western shore of the Antarctic Peninsula, extending south to Cape Tuxen at lat. 65°16′ S (Fig. 101). It occurs mostly at low elevations from near sea level to about 50 m, and only rarely has it been observed at higher elevations of 100–120 m.

Specimens examined. WEST ANTARCTICA. SOUTH SANDWICH IS. SAUNDERS I. Northern end, *ca* 15 m, *MWH 434B* (AAS, KRAM).

SOUTH ORKNEY IS. SIGNY I. Between Jane Col and Express Cove, 100 m, *RILS 10906* (AAS, KRAM); Tranquil Lake, 25 m, *RILS 8187* (AAS, KRAM); between Knob and Heywood Lakes, 20 m, *RILS 534, 1447, 1813 & 1814* (AAS, KRAM) and *REL 1172* (AAS, KRAM); Heywood Lake, 20 m, *RILS 535* (AAS, KRAM); Changing Lake, 40 m, *RILS 10796* (AAS, KRAM); Andreaea Plateau, 100–120 m, *RILS 8630* (AAS, KRAM).

SOUTH SHETLAND IS. KING GEORGE I. ***Fildes Peninsula***: 1.5 km W of Half Three Pt., 10 m, *Schulz 1* (KRAM); banks of a little river, 35 m, *Schulz 20* (KRAM). ***Admiralty Bay***: Petrified Forest Creek, 25 m, *RO 160/80* (KRAM) and 30 m, *RO 637/80* (AAS, KRAM). LIVINGSTON I. ***Byers Peninsula***: between Smellie Pt. and Chester Cone, 35–50 m, *RILS 3776 & 3801* (AAS, KRAM); Negro Hill, 100 m, *KR 16A* (AAS, KRAM); wide valley between Chester Cone and Sealers Hill, 25–30 m, *RILS 3817* (AAS, KRAM); Smellie Pt., *KR 38C* (AAS, KRAM). ***Hurd Peninsula***: in the vicinity of the Spanish Base, ii.1992, *Schulz s.n.* (KRAM).

WEST ANTARCTIC PENINSULA. DANCO COAST. Spring Pt., *VK 472* (KRAM); Charlotte Bay, Andrée I., 20–50 m, *RILS 4001* (AAS, KRAM); Birdsend Bluff, 5–35 m, *RILS 4232 & 4233* (AAS, KRAM). **PALMER ARCHIPELAGO.** ***Melchior Is.***: Lysted I., Water Landing, *PAS 360.3* (FH). **GRAHAM COAST.** Cape Tuxen, 50–100 m, *RILS 4903* (AAS, KRAM). **WILHELM ARCHIPELAGO.** ***Penola Strait***: Petermann I., 20 m, *RWMC 723 & 789* (AAS, KRAM) and *REL 1357* (AAS, KRAM).

4. **Schistidium praemorsum** (Müll.Hal.) Herzog Fig. 102

Schistidium praemorsum (Müll.Hal.) Herzog, Biblioth. Bot. 87: 53. 1916. ≡ *Guembelia praemorsa* Müll.Hal., Linnaea 43: 452. 1882. ≡ *Grimmia praemorsa* (Müll.Hal.) Paris, Ind. Bryol.: 534. 1896. — TYPE CITATION: Argentinia Catamarcana, supra Yakutula in convallis altae Granadillas, in Vayas altas alpina, ineunte Februarii 1872, c. fr. vetustis et junioribus. [Lectotype (*selected here*): 'Herbarium Haussknecht , Jena Gümbelia praemorsa CM Argentinia, in alpibus prope Catamarca, Yakutula, in Vayas altas, convallis altae Grandillas Febr. 1872 leg P. G. Lorentz com. Schliephacke' – JE-Schliephacke/Winter!; isotype: S!].

Grimmia abscondita Cardot, Wiss. Erg. Schwed. Südpolar-Exp. 4(8): 107, *f. 19*. 1908. ≡ *Schistidium absconditum* (Cardot) Ochyra, Fragm. Florist. Geobot. 43: 104. 1923, ***syn. nov.*** — TYPE CITATION: Patagonie australe; lac San Martin, Rio Fosiles, vers 1000 m. (Dusén, nº 5944). [Lectotype (*selected here*): '5944. G. abscondita Card. sp. nova Grimmia celata Card. pseudo-celata Card. Patagonia australis Lago San Martin Rio Fosiles in saxis c. 1000 m s.m. Apr. 1 a. 1905 P. Dusén' – PC-Cardot!; isotype: S!].

Grimmia chubutensis Cardot & Broth., K. Svensk. Vet. Akad. Handl 63(10): 26, *t. 2, f. 6a–e*. 1923. ≡ *Schistidium chubutense* (Cardot & Broth.) Ochyra, Fragm. Florist. Geobot. 43: 105. 1998, ***syn. nov.*** — TYPE CITATION: Patagonia andina: Territ. Chubut, Valle 16. de Octubre, ad saxa: Futulaefú, Casa Rees. [Lectotype (*selected here*): 'Herb. J. Cardot. Grimmia chubutensis Card. sp. nova. Patagonie: Territoire Chubut, vallée 16 de Octubre, pierres Leg. C. Skottsberg, 1908 nº 432' – PC-Cardot!; isotype: H-Brotherus!].

Plants medium-sized, forming compact but rather easily crumbling, slightly glistening, dark green, brownish- or blackish-green to reddish-brown above, brown below. *Stems* 1.5–2.5 cm tall, rigid, rather robust, erect, much branched and fertile stems usually dichotomously branched in the upper part, densely radiculose at the base with brown, lustrous, smooth, branched rhizoids, in transverse section rounded to elliptical, consisting of a 2–3-stratose cortex of small cells with orange-brown, incrassate walls and small lumina surrounding 3–4-stratose medulla of large, hyaline to yellowish-hyaline, moderately thick-walled cells and a distinct central strand. *Leaves* crowded, erect and closely imbricate when dry, erecto-patent to recurved when moist, sharply carinate above, broadly concave below, straight to slightly curved, ovate, broadly ovate-lanceolate to lanceolate, 1.3–1.8 mm long, 0.5–0.9 mm wide, gradually shortly acuminate, bluntly acute to narrowly obtuse, muticous or some uppermost leaves ending with a coarse, denticulate hyaline hair-point, mostly terete in its upper part, to 0.4 mm long; *margins* entire, narrowly revolute or recurved on both sides to ¾–⅘ way up the leaf, plane or sometimes slightly deflexed close to the apex, bistratose almost throughout, except for the extreme base, below usually bistratose in one row of bulging cells; *costa* percurrent or ceasing somewhat below the apex and merging with the bistratose laminal cells, strong above, 60–70 µm wide, becoming weaker towards the base, strongly convex dorsally, smooth, in transverse section semi-terete to subrectangular, tristratose in the upper and median parts, 4-stratose towards the base, bistratose at the extreme apex; *laminal cells* regularly bistratose distally, often with unistratose patches, unistratose in the median part but with bistratose streaks, entirely unistratose in the proximal part; *upper cells* rounded to transversely elliptical, 4–7(–9) µm long, 6–8(–10) µm wide, thick-walled; *median cells* rounded-quadrate to short-rectangular, with evenly thickened, straight or often slightly sinuose walls, 10–16 µm long, 7–8 µm wide; *basal juxtacostal cells* rectangular, with evenly moderately thickened, straight walls forming a small area extending to ¼–⅓ way up the costa, becoming shorter and similar to the median cells towards the margins; *basal marginal cells* quadrate to short-rectangular, pellucid, with straight and thickened walls. *Autoicous*. *Perichaetial leaves* elliptical to oblong-lanceolate, 2.5–2.9 mm long, 0.9–1.2 mm wide, shortly acuminate, acute to narrowly obtuse, epilose or with a short, flat, serrulate hyaline awn, to 0.2 mm long; *margins* recurved in the distal half, plane below. *Setae* straight, short, 3.5–4.0 mm long, thick, centrally attached. *Capsules* shallowly immersed, spherical, 1.0–1.1 mm long, *ca* 1 mm wide, wide-mouthed, light

→

Figure 102. *Schistidium praemorsum*. — **1**. Habit. **2**. Portion of branch, dry. **3**. Cross-section of stem. **4–6**. Axillary hairs. **7–16**. Leaves. **17–18**. Leaf apices. **19**. Upper laminal cells. **20**. Mid-leaf cells. **21**. Basal juxtacostal cells. **22**. Angular cells. **23–30**. Cross-sections of leaves, a sequence from base to apex. **31**. Perichaetium with systylious capsule. **32**. Perichaetium with operculate capsule and calyptra. **33–34**. Perichaetial leaves. **35**. Operculate capsule, wet. **36**. Exothecial cells at urn base and stoma. **37**. Exothecial cells in middle of urn. **38**. Peristome teeth, spores and rim cells of capsule. — [All from *VK 596*]. — Scale bars: a – 1 mm (7–16, 33–34) and 100 µm (36–38); b – 1 mm (31–32, 35); c – 100 µm (3, 23–30); d – 2 mm (1) and 100 µm (4–6, 17–22); e – 1 mm (2).

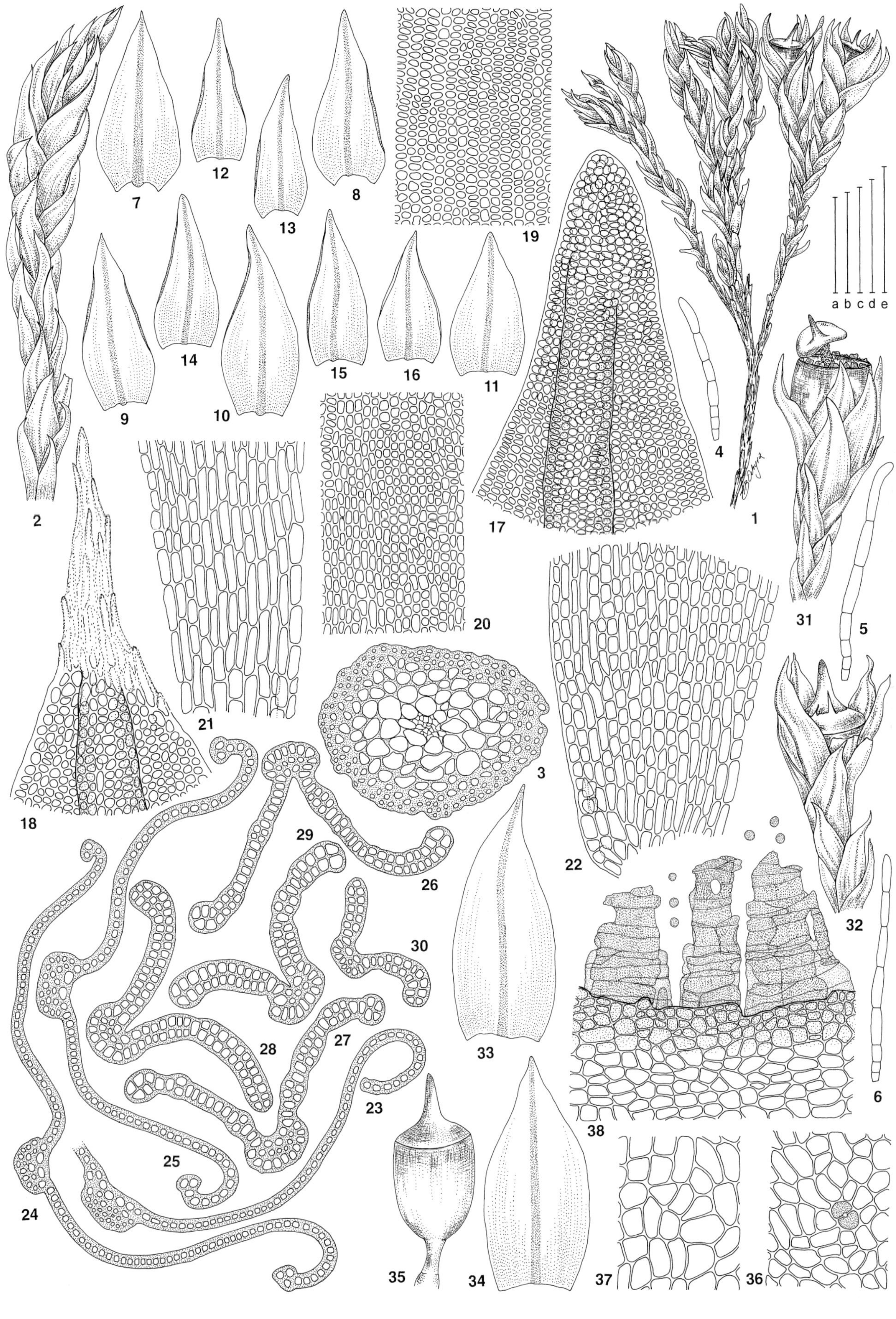

2 7 12 13 8 19 17 1 4 9 14 10 15 16 11 5 31 18 21 20 3 22 32 29 26 30 33 28 27 23 38 6 24 25 35 34 37 36

brown to yellowish-brown, leptodermous; *operculum* flat, short- and straight-beaked; *exothecial cells* predominantly oblong, mixed with quadrate, rounded-quadrate or irregular cells, 25–50 µm long, 15–30 µm wide, forming an irregular pattern, with slightly and evenly thickened walls, becoming oblate to rounded at the mouth in 6–8 tiers and forming a bright orange-brown strip; *stomata* few at the base of the urn, superficial, bicellular, variously oriented, with round pori; *peristome teeth* erect to patent, usually broken with age, lanceolate but often with truncate apices to short and rudimentary, irregular in outline with very uneven margins, 120–260 µm long, 65–80 µm wide at the base, truncate to variously eroded at the apex, entire to somewhat perforate, orange to reddish-brown, finely striate-papillose on both ventral and dorsal surfaces. *Spores* globose, pale, smooth, 10–12 µm in diameter. *Calyptra* mitrate, lacerate at the base, smooth, naked, brownish at the tip, stramineous at the base.

Chromosome number: unknown.

Discussion. Like most southern South American species of this genus, *Schistidium praemorsum* is a very poorly known and obscure species. It was originally described by Müller (1882) as *Guembelia praemorsa* from the Andes of Prov. Catamarca in NE Argentina. Later it was reported only once from Bolivia by Herzog (1916), who transferred it to *Schistidium*. Bremer (1981) considered this species, without any comment, to be conspecific with *S. apocarpum*.

Schistidium praemorsum is a distinct species recognised by its entirely or nearly so bistratose laminal cells in the distal part, the presence of the coarse, stiff, denticulate and terete hair-point which, however, is often missing or very short on some leaves. In addition, the leaf margins are plane or occasionally only somewhat deflexed close to the apex, the exothecial cells are oblong to rectangular and the peristome teeth show a strong tendency to reduction. They are finely papillose-striolate, usually irregularly eroded or truncate at the apex, have very uneven and variously eroded margins and sometimes they are markedly reduced and rudimentary, for example in plants from the Danco Coast (*VK 596*, KRAM). The plane leaf margins in the upper ¼–⅕ of the leaf, the oblong to rectangular exothecial cells in the middle ofthe urn and the bistratose lamina in the distal portion clearly indicate the placement of *S. praemorsum* in sect. *Atrofusca*. This species is superficially similar to *S. andinum* with which it often shares reddish-brown coloration of the plants. However, the latter is distinct in having only bistratose leaf margins in 1–2 rows and unistratose lamina with occasional small, irregular bistratose streaks in the distal portion, coarsely papillose peristome teeth and quadrate to transversely rectangular exothecial cells in the middle of the urn.

In practice, the bistratose laminal cells in the distal portion distinguish *Schistidium praemorsum* from all local congeners. Amongst the southern South American species a similar leaf condition is exhibited by only two poorly known species, *S. absconditum* and *S. chubutense*. The former has so far been known only from the type collection from the vicinity of Lago San Matin in southern Patagonia in the Argentine province of Santa Cruz. This material was initially reported by Dusén (1907) as *Grimmia occulta* but Cardot (1908*a*) showed its distinctness in leaf shape, areolation and anatomy from both *G. occulta* and *G. celata* and described it as a new species, *Grimmia abscondita*, which was transferred to *Schistidium* by Ochyra (1998*b*). Bremer (1981) reduced this name, without any comment, to synonymy with *S. apocarpum* and this taxonomic conclusion was subsequently approved by Bell (1984). Cardot (1908*a*) described the spores in this species as very small, 6–8 µm in diameter, but, in fact they are larger, 9–12 µm wide. All other character states of *S. absconditum* agree perfectly with those of *S. praemorsum* and the species are considered to be conspecific. The second species, *S. chubutense* was described from western Patagonia as *Grimmia chubutensis* (Cardot & Brotherus 1923). Though sterile, the type material of this species perfectly matches the type collection of *S. praemorsum*, and the names are here considered synonymous.

The partially bistratose uppermost laminal cells are also known only in *Schistidium falcatum* and *S. lewis-smithii* which, however, have the leaves broadly canaliculate with plane to inflexed but never recurved margins. A similar state of the distal laminal cells is also known in many populations of *S. rivulare*, but all known Antarctic populations have consistently unistratose upper cells. Nonetheless, the large, papillose spores and the broadly ovate-triangular to ovate-lanceolate leaves, always lacking hair-points and often denticulate at the apex, immediately preclude confusion.

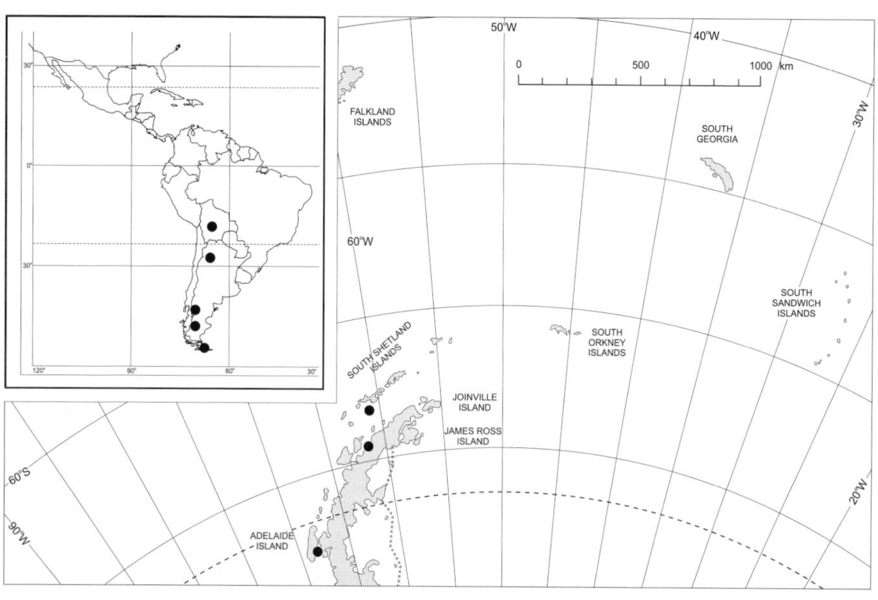

Figure 103. Distribution of *Schistidium praemorsum* in Antarctica and South America (inset).

Amongst the Antarctic species some populations of *Schistidium praemorsum* with reduced peristome teeth are likely to be mistaken for *S. antarctici*. However, the broadly ovate-lanceolate, unranked leaves with bistratose laminal cells above are the safest distinguishing characters.

Reproduction in Antarctica. Fully mature sporophytes produced in profusion.

Habitat. On Deception I. the species grows in a geothermal area but on unheated yellowish tephra rock outcrops immediately adjacent to fumaroles in association with *Schistidium amblyophyllum*. The specimen from the Danco Coast lacks closer ecological data but it was collected during a survey of the two native Antarctic vascular plants and the collector described this locality as 'on rock ledges' (Komárková *et al.* 1985). This type of habitat is confirmed by the associated mosses in the sample, namely *Pohlia cruda*, *Brachythecium austrosalebrosum* and *Sanionia georgicouncinata*.

World range. A poorly known American south-temperate species which has hitherto been known only from a few collections from the Central Andes of Bolivia and NW Argentina, western and southern Patagonia (the type collections of *Grimmia chubutensis* and *G. abscondita*) and recently discovered on Isla Navarino in the Fuegian region (*Buck 40759*, KRAM, NY) (Fig. 103, inset).

Distribution in Antarctica. Known only from three localities, at Mt. Pond on Deception I. in the South Shetland Is. where it grows at an elevation of about 500 m, at Cierva Pt. on the Danco Coast and extending to Léonie I. on the Fallières Coast on the Antarctic Peninsula at lat. 67°36′S (Fig. 103). The Cierva Pt. site is remarkable as it also has the only Antarctic record for *Tortula hoppeana* and from where the only Antarctic fruiting plants of *Polytrichum piliferum* are known.

Specimens examined. WEST ANTARCTICA. SOUTH SHETLAND IS. **DECEPTION I.** Ridge of hill 1.5 km NNW of Mt. Pond, 410–525 m, *RILS 6920B* (AAS, KRAM); fumarole area N of Mt. Pond, 500 m, *RILS 11172B* (AAS, KRAM).

WEST ANTARCTIC PENINSULA. **DANCO COAST.** Cierva Pt., *VK 596* (KRAM). **FALLIÈRES COAST.** ***Léonie Is.***: Léonie I., 150–175 m, *RILS 9124B* (AAS, KRAM).

5. **Schistidium andinum** (Mitt.) Herzog Fig. 104

Schistidium andinum (Mitt.) Herzog, Biblioth. Bot. 87: 53. 1916. ≡ *Grimmia andina* Mitt., J. Linn. Soc. Bot. 12: 97. 1869. — TYPE CITATION: Andes Quitenses, in monte Carguairazo in rivulis, Spruce, n. 132c. [Lectotype (*vide* Bremer 1981: p. 93): '[Grimmia andina] Carguairazo Spruce' – NY!; isotypes: 'Musci Amazonici et Andini No. 132c Schistidium andinum Mitt. Legit Ric. Spruce And. Quit. Determinavit W. Mitten' – BM-Hooker!, H-Lindberg!, NY-Mitten!, S!].

Guembelia lorentziana Müll.Hal., Linnaea 43: 450. 1882. ≡ *Grimmia lorentziana* (Müll.Hal.) Paris, Ind. Bryol.: 529. 1896. ≡ *Schistidium lorentzianum* (Müll.Hal.) Ochyra & Matteri, Fragm. Florist. Feobot. 41: 1008. 1996, ***syn. nov.*** — TYPE CITATION: Argentinia subtropica Tucumanensis, in Sierra de Aconquija prope Tafi et in via ab Tafi per Quebrada de Monteros, ineunte Aprilis 1872. [Lectotype (*vide* Ochyra & Matteri 1996*a*: p. 1008): 'Herbarium Haussknecht, Jena Gümbelia lorentziana ssp. Tafi Argentinia leg. Lorentz' – JE-Schliephacke/Winter!].

Grimmia occulta Müll.Hal. *in* Neumayer, Int. Polarforsch. Deutsch. Exp. 2: 314. 1890. ≡ *Schistidium occultum* (Müll.Hal.) Ochyra & Matteri, Fragm. Florist. Feobot. 41: 1008. 1996, ***syn. nov.*** — TYPE CITATION: Austro-Georgia, inter muscos, 6. Febr. 1883. [Lectotype (*vide* Ochyra & Matteri 1996*a*: p. 1008): 'Grimmia (Platystoma) occulta C. Müll. nsp. Austro-Georgia, Drehkuppel, zwischen Moos 6/II 83 leg. Will Südpolarexpedition' – HBG!; isotype: HBG!].

Grimmia syntrichiacea Müll.Hal. *in* Neumayer, Int. Polarforsch. Deutsch. Exp. 2: 315. 1890. ≡ *Schistidium syntrichiaceum* (Müll.Hal.) B.G.Bell, Br. Antarct. Surv. Bull. 63: 97. 1984, ***syn. nov.*** — TYPE CITATION: Austro-Georgia, Felsblöcke des südlichen Ufers des Landzunge, inter Blindiam pallidifoliam n. sp. specimina perpauca, 13. Oct. 1882. [Lectotype (*selected here*): 'Grimmia syntrichiacea C. Müll. n.sp. Inter Blindiam pallidifoliam von Süd-Georgien auf Felsblöcken des südl. Landzunge Südpolarexpedition' – HBG!; isotype: S!].

Grimmia fallax Dusén *in* Scott, Rep. Princeton Univ. Exp. Patagonia Botany 8: 77, *f. 7 & pl. 8, f. 5–6*. 1903. ≡ *Schistidium fallax* (Dusén) Ochyra & Matteri, Fragm. Florist. Feobot. 41: 1008. 1996, ***syn. nov.*** — TYPE CITATION: Patagonia australis, ubi verisimilariter in territorio fontinali fluminis Rio Chico in saxis lecta est. [Lectotype (*selected here*) {or perhaps holotype}: 'Princeton Scientific Expedition to Patagonia 1896–1897. P. 141. Grimmia fallax Dusén. Patag. austr. [Verisimilariter] In territorio fontinali fluminis Rio Chico in saxis lecta est J. B. Hatcher' – S!].

Plants medium-sized, rarely small or large, rather coarse and rigid, forming loose to moderately compact tufts or patches, dull or weakly glistening, red-brown to brown, sometimes with yellowish-green, blackish or greyish tones above, brown to black-brownish below, occasionally brown throughout. *Stems* rigid, erect, 1–6 cm long, much irregularly branched with short to elongate branches, less often dichotomously or fastigiately branched, usually sparsely radiculose with long, branched, brown, smooth rhizoids, confined to the base or scattered in the lower part, in transverse section round, consisting of a small but distinct or sometimes absent central strand and 2–3-stratose cortex of strongly thick-walled, brown cells surrounding 3–4-stratose medulla of large cells with moderately thickened walls; *axillary hairs* filiform, hyaline, 4–6-celled, mostly with elongate distal cells. *Leaves* straight to slightly curved, erect, closely imbricate when dry, erect-spreading when moist, ovate-lanceolate, 1.0–1.8(–2.1) mm long, (0.4–)0.5–0.8(–0.9) mm wide, carinate, gradually acuminate, acute to narrowly obtuse, epilose to piliferous; *hair-point* absent to 0.5 mm long, hyaline to yellowish-hyaline, rather coarse and stiff, straight to flexuose, usually with clearly visible remnants of cell walls, somewhat decurrent, not flattened, usually sharply to bluntly denticulate, sometimes nearly smooth; *margins* entire, narrowly to broadly recurved throughout, bistratose in the upper and central parts for 1–2 rows, forming somewhat bulging thickenings, unistratose below with frequent bistratose patches; *costa* smooth dorsally or rarely with a few low papillae in the distal portion, percurrent, clearly delimited from the laminal cells, 40–55 µm wide below, intensively yellow- or orange-brown to brown, in cross-section strongly convex dorsally, terete to subrectangular, 2–3-stratose in the distal half, 3(–4)-stratose in the proximal part; *laminal cells* unistratose throughout with occasional bistratose patches in the distal part, irregular in size and shape, sometimes guttulate, with thick, straight or slightly to distinctly sinuose walls, rounded-quadrate, oval to short-rectangular in the upper and central parts, (3–)5–13(–15) µm long, (3–)4–8(–9) µm wide, becoming short-rectangular in the lower part, (8–)10–20(–23) µm long, (3–)5–9(–11) µm wide; *basal juxtacostal cells* short- to long-rectangular, 15–30 µm long, 4–6 µm wide, thick- and straight-walled, forming a narrow chlorophyllose area, gradually passing into shorter, sinuose-walled cells; *basal marginal cells* chlorophyllose, quadrate to shortly oblong, mostly with markedly thickened transverse walls. *Autoicous. Perichaetial leaves* much larger than the vegetative leaves, ovate- to oblong-lanceolate, 2.5–3.4 mm long (with a hair-point), 0.8–1.1 mm wide, concave, gradually long-acuminate, acute and usually long-piliferous, rarely epilose, with margins strongly recurved in the distal part, plane below. *Setae* short, 0.35–0.5 mm long, yellow, thick. *Capsules* deeply immersed, obloid, 0.8–1.2 mm long, 0.5–0.7 mm wide, about 1.2–1.5 times as long as wide, stramineous to light brown, smooth, leptodermous; *operculum* flat with a short and straight beak; *exothecial cells* in patches of quadrate to irregularly isodiametric and transversely rectangular cells, occasionally mixed with short-oblong cells, 15–30 µm long, 20–40 µm wide, thin-walled, becoming oval to oblate in 4–6 tiers and forming bright orange-brown strip at the orifice; *stomata* at the extreme base of the urn, bicellular, round-pored, 5–8 per urn; *peristome teeth* lanceolate, usually gradually tapering to a fine point, 320–400 µm long, 100–120 µm wide at the base, squarrose-recurved, curved, reddish- or orange-brown, entire or irregularly perforate in the upper part, symmetric or with uneven margins, densely coarsely papillose thoughout with short, narrow papillae. *Spores* globose, pale brownish, smooth to finely roughened, 10–13 µm in diameter. *Calyptra* mitriform, wide, mostly split into 3–4 lobes at the base extending nearly to the margin of the urn.

Chromosome number: unknown.

Discussion. In the Antarctric *Schistidium andinum* is rather easy to distinguish in the field by the coarse reddish-brown plants and the well developed, never reduced, reddish- to orange-brown and coarsely papillose peristome teeth which are easily observable with a hand-lens. Microscopically, the irregular, unistratose laminal cells with markedly thickened and variously sinuose walls, bistratose leaf margins in the distal portion, undifferentiated, chlorophyllose basal marginal cells and short-rectangular, straight- and thick-walled, chlorophyllose basal juxtacostal cells and the transversely rectangular mixed with isodiametric exothecial cells are characteristic of this species. The species shows rather narrow interpopulational variation and it refers mostly to the size and coloration of the plants and the presence and the length of the hair-point but all microscopic characters are rather invariable. In the Antractic *S. andinum* may possibly be mistaken for *S. antarctici* but the distinctly 5-ranked leaves, markedly reduced peristome teeth and the hyaline basal marginal border preclude confusion. The entire, lanceolate peristome teeth are also typical of *S. amblyophyllum* but the elongate and strongly sinuose-walled laminal cells and long-rectangular basal juxtacostal cells forming characteristic well defined groups at the leaf base are immediately diagnostic. In habit *S. andinum* is quite similar to *S. praemorsum* but the bistratose distal laminal cells and more or less regular leaf areolation enable safe distinction between the species.

Schistidium andinum was described as *Grimmia andina* by Mitten (1869) from the Andes of Ecuador, but the species has remained poorly known and misunderstood since its description. It was reported only once from Bolivia by Herzog (1916), who correctly transferred it to *Schistidium*. The main reason for neglecting this and many other species belonging within the *S. apocarpum* complex has been the poor state of taxonomy of this group which appeared to be one of the most difficult and misinterpreted amongst mosses. Until recently, it has been the common practice to consider most species of *Schistidium* to be conspecific with *S. apocarpum* and, accordingly, this species has become a highly artificial and broadly interpreted taxon. Bremer (1981) lectotypified *Grimmia apocarpa* with the specimen from the Hedwig herbarium but, at the same time, she has lumped the majority of Holarctic and extra-Holarctic species, including *S. andinum*, with *S. apocarpum*, making it an incomprehensible and obscure species which could hardly be defined and defended. Finally, Blom (1996) precisely defined and described *S. apocarpum* and his work represents a very important advance towards understanding this difficult complex.

Having defined the central species of this complex it is much easier to describe other morphologically distinct entities in this group in other geographical regions. *Schistidium andinum* certainly belongs to subgen. *Apocarpa* and sect. *Apocarpiformia* and a characteristic feature of this group is the orange-red or red reaction of the upper leaves with KOH. The species is closely related to *S. apocarpum*, but the latter is easily distinguished by its denticulate upper leaf margins and low papillose costa on the dorsal surface in the distal part. Nonetheless, *S. apocarpum* is a readily modified and variable species, and occasionally its distal margins and costa on the dorsal surface are smooth. Much taxonomic work and revision of more South American specimens is needed before *S. andinum* is fully understood with regard to its variation and relationships. As with other extra-Holarctic moss species, *S. andinum* has been described several times under different names from widely separated geographical areas. Some of these species have been examined during the course of the present study and their names have been reduced to synonymy with *S. andinum*, including *Grimmia occulta* and *G. syntrichiacea* from South Georgia (Müller 1890*a*), *G. fallax* from southern Patagonia (Dusén 1903*b*) and *Guembelia lorentziana* from the Andes of north-western Argentina (Müller 1882).

Reproduction in Antarctica. Commonly fruiting.

Habitat. Usually growing in dry to moist situations on rocks, rock faces, cliff ledges and in rock clefts and crevices, also on

→

Figure 104. *Schistidium andinum*. — **1**. Habit. **2**. Portion of branch, dry. **3**. Cross-section of stem. **4–6**. Axillary hairs. **7–15**. Leaves. **16**. Leaf apex. **17**. Upper laminal cells at margin. **18**. Mid-leaf cells. **19**. Lower mid-leaf cells. **20**. Basal juxtacostal cells. **21**. Angular cells. **22**. Basal cells. **23–31**. Cross-sections of leaves, a sequence from base to apex. **32**. Perichaetium with systylious capsule. **33–34**. Perichaetial leaves. **35**. Operculate capsule. **36**. Deoperculate capsule, wet. **37**. Operculum with attached columella. **38**. Exothecial cells at base of urn and stomata. **39**. Exothecial cells in middle of urn. **40**. Peristome teeth, spores and rim cells of capsule. — [1–2, 8–10, 17–21, 32, 35–37 *RO 189/80*; 3–6, 11–15, 22–31, 33–34, 38–40 from *RILS 7469*; 7, 16 from *Will s.n.*, 6.ii.1883, lectotype of *Grimmia occulta*, HBG]. — Scale bars: a – 1 mm (7–15), 100 µm (4–6, 16–22) and 200 µm (38–40); b – 0.5 cm (1); c – 100 µm (3, 23–31); d – 1 mm (2, 32, 35–37); e – 1 mm (33–34).

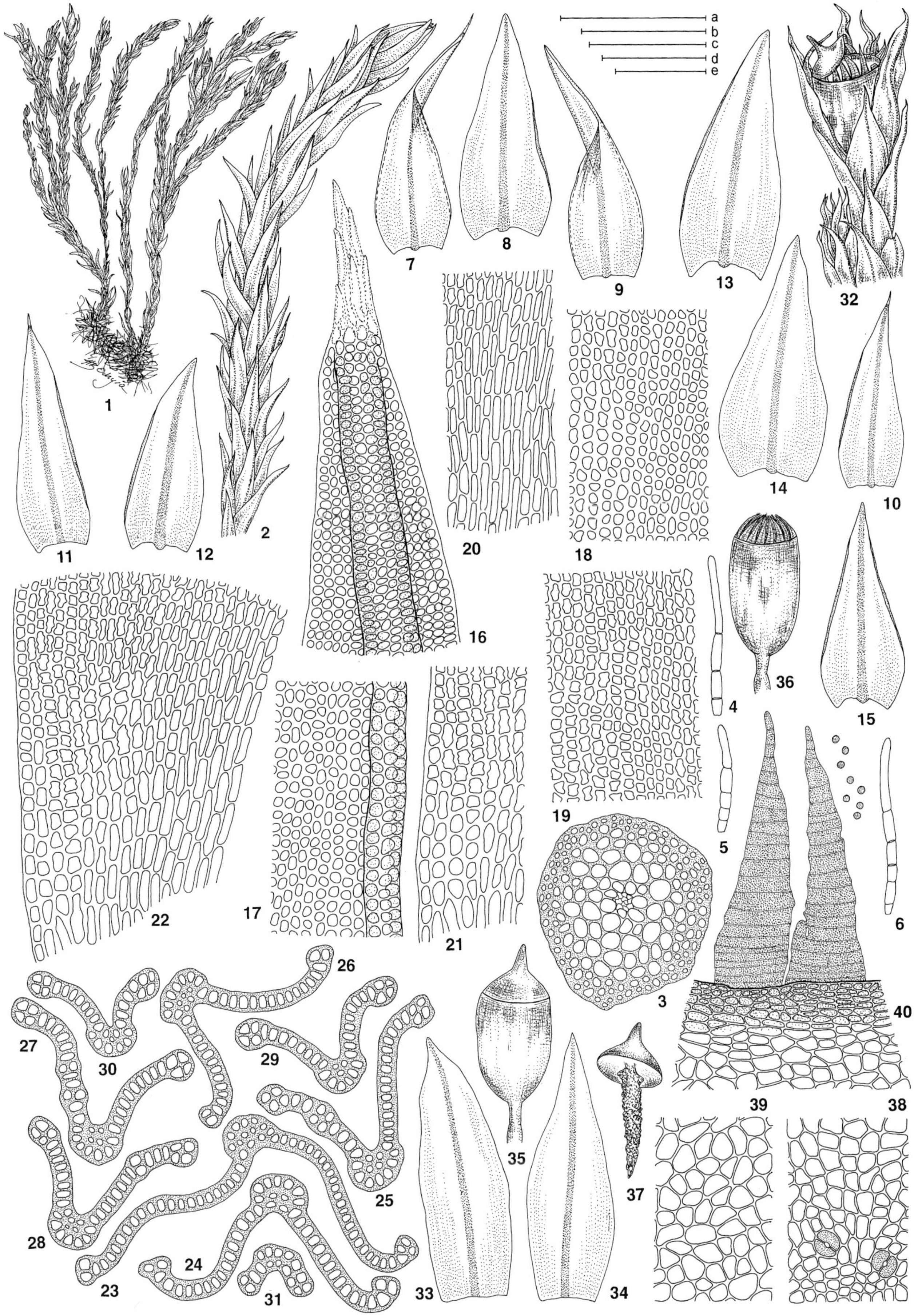
a
b
c
d
e
1
2
3
4
5
6
7
8
9
10
11
12
13
14
15
16
17
18
19
20
21
22
23
24
25
26
27
28
29
30
31
32
33
34
35
36
37
38
39
40

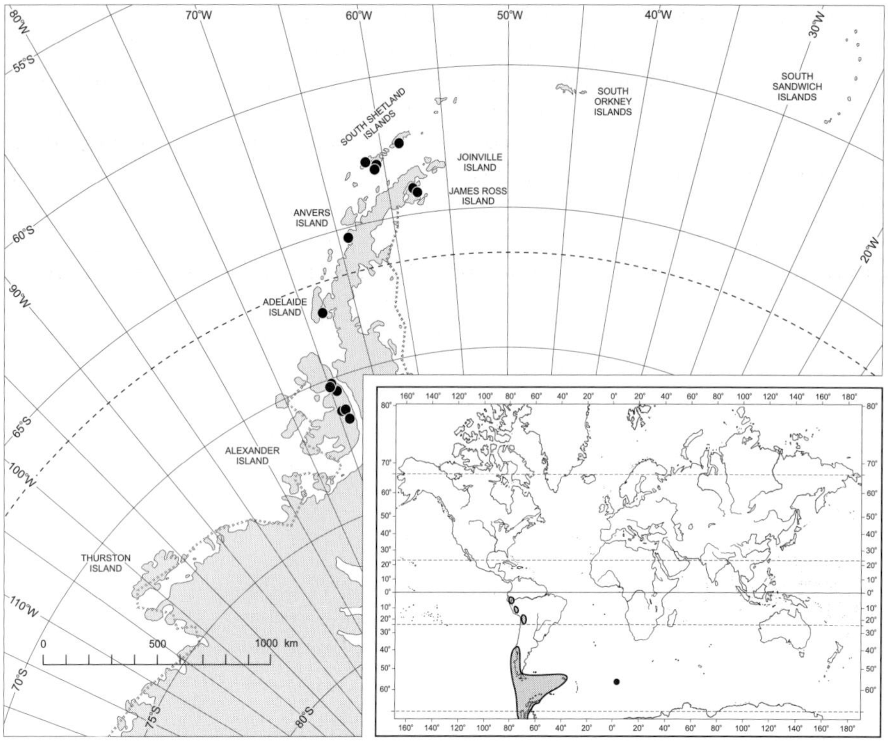

Figure 105. Distribution of *Schistidium andinum* in Antarctica. Inset: global distribution.

gravelly or sandy soil, on soil over ledges of rock outcrops, in dry stable channels of stones at the margin of solifluction lobes and stripes, occasionally also in moist seepage areas below snow patches, in damp watercourses, on scree slopes and in moist habitats beside streams. The data on collection labels of *Schistidium andinum* are meagre with regard to information on accompanying species and plant communities or formations in which it has been collected. The species usually forms monospecific cushions or patches, with an occasional admixture of other species, for example *Hypnum revolutum*, *Bryoerythrophyllum recurvirostrum* and *Tortella alpicola* on Alexander I. and *Hypnum revolutum*, *Bryum pseudotriquetrum*, *Distichium capillaceum*, *Encalypta procera* and *Leptogium puberulum* on James Ross I., whereas on King George I. the species was found in a community dominated by *Polytrichastrum alpinum*.

World range. Apparently an American south-temperate species but its true phytogeographical status cannot be established until the genus *Schistidium* has been thoroughly revised in the tropics and austral region. The species seems to be scattered at alpine elevations along the Andean chain from Ecuador to western Patagonia, then it appears to be quite frequent in the Tierra del Fuego archipelago and common on subantarctic South Georgia, becoming widely distributed but occasional in the maritime Antarctic (Fig. 105, inset).

Distribution in Antarctica. An infrequent species, rare and occasional on the South Shetland Is., on the Graham and Fallières Coasts on the western Antarctic Peninsula, and on Alexander I., extending southwards to Mars Glacier at lat. 71°50′S. In addition, it is occasional on James Ross I. on the north-eastern coast of the Antarctic Peninsula and on a highly isolation station on Bouvetøya (Fig. 105). Its altitudinal range extends from near sea level to 500 m.

Specimens examined. WEST ANTARCTICA. BOUVETØYA. Nyrøysa, *ca* 25 m, *TE 293* (BG).

SOUTH SHETLAND IS. **KING GEORGE I.** ***Admiralty Bay***: Rescuers Hills, 45 m, *RO 189/80* (KRAM). **LIVINGSTON I.** ***Byers Peninsula***: between Chester Cone and Sealers Hill, 25–30 m, *RILS 3818 & 3827* (AAS, KRAM). ***Hurd Peninsula***: South Bay, Johnsons Dock, *ca* 45 m, *Schulz 115* (KRAM). **DECEPTION I.** N summit of Mt. Pond, 480 m, *RILS 5828A* (AAS, KRAM); fumarole area N of Mt. Pond, 500 m, *RILS 11168, 11169, 11170, 11172B, 11173, 11175, 11176 & 11182B* (AAS, KRAM); Wensleydale Gap, *ca* 300 m, *NJC 30* (AAS, KRAM); N end of cliffs above Fumarole Bay, 50 m, *RILS 11297B, 11298C &11534* (AAS, KRAM); same locality, W end of cliffs, 75 m, *RILS 11361* (AAS, KRAM).

WEST ANTARCTIC PENINSULA. **WILHELM ARCHIPELAGO.** ***Collins Bay***: Darboux I., *RILS 3317* (AAS, KRAM). **FALLIÈRES COAST.** ***Léonie Is.***: Léonie I., 30 m, *RILS 8970B* (AAS, KRAM) and 150–175 m, *RILS 9140B* (AAS, KRAM). **GEORGE VI SOUND.** ***Alexander I.***: Middle Flatiron Valley, 315 m, *RILS 11103A* (AAS, KRAM), 317 m, *RILS 11102* (AAS, KRAM) and 420 m, *RILS 11096A & 11097A* (AAS, KRAM); Ablation Valley, near W shore of Ablation Lake, 2–3 m, *RILS 9220A, 9221A, 9222A, 9257A, 9267A & 9268A* (AAS, KRAM) and 5 m, *RILS 2707 & 2708B* (AAS, KRAM); Fossil Bluff, Khufu, 300–400 m, *RILS 9325A, 9347A & 9348* (AAS, KRAM); Mars Glacier, Ares Ridge, *RILS 9501* (AAS, KRAM).

EAST ANTARCTIC PENINSULA. **JAMES ROSS ISLAND GROUP.** ***James Ross I.***: Cape Lachman, 15–35 m, *RILS 7323 & 7337A* (AAS, KRAM); St. Martha Cove, 50 m, *RILS 7468A & 7469* (AAS, KRAM) and 80 m, *RILS 7446* (AAS, KRAM); Terrapin Hill, 250 m, *RILS 7728* (AAS, KRAM); Tumbledown Cliffs, 30 m, *RILS 5480* (AAS, KRAM).

6. **Schistidium steerei** Ochyra Fig. 106

Schistidium steerei Ochyra, Mem. New York Bot. Gard. 45: 609, *f. 1–21.* 1987. — TYPE CITATION: Antarctica, South Shetland Islands: King George Island, Admiralty Bay, Ezcurra Inlet, Golden Cove, Cytadela Cliff, 62°11′S, 58°35′W, elev. 30 m; on banks of a small stream on bare, gravelly soil on slope with northerly aspect, 20 Jan 1980, Ochyra 946/80. [Holotype: 'Herbarium Instituti Botanici Academiae Scientiarum Polonae – Cracoviae Bryophyta. Schistidium steerei Ochyra, sp. nova Antarctica, South Shetland Islands: King George Island, Admiralty Bay, Ezcurra Inlet, Golden Cove, Cytadela Cliff, 62°11′S, 58°35′W, elev. 30 m; on banks of a small stream on bare, earth on slope with northerly aspect, Jan. 20, 1980, Coll. R. Ochyra No 946/80 Det. R. Ochyra' – KRAM!; isotypes: AAS!, ALTA!, BM!, CANM!, F!, H!, NY!].

Plants medium-sized in relatively large, compact cushions, usually half-buried in soil, rather dull, yellowish- to brownish-green above, brown below. *Stems* 1.5–2.5 long, erect, freely branched, with scattered, brown, smooth, branched rhizoids in the lower part, in cross-section with 1–2 layers of smaller, brown, moderately thick-walled cortical cells surrounding 4–6 rows of large, thin-walled medullary cells, central strand lacking or very inconspicuous. *Leaves* densely set, straight or sometimes subsecund at the tips, closely imbricate, erect when dry, erect-spreading when moist, 1.3–2.0 mm long, 0.4–0.8 mm wide at the widest part, lanceolate to ovate-lanceolate, gradually acuminate, obtuse to bluntly acute, muticous or sometimes the uppermost leaves ending in a very short, to 0.4 mm long, hyaline, denticulate awn, sharply keeled in the upper half, somewhat concave below; *margins* entire, strongly revolute on both sides from about ⅕ the leaf length above the base to the apex, unistratose below, in upper part with frequent bistratose spots; *costa* percurrent, channelled above, strongly convex at the dorsal side, 50–60 μm wide above, becoming narrower towards the base, smooth throughout on the dorsal surface, in cross-section bistratose in the upper part and composed of uniform, large cells, tristratose below including one row of smaller inner cells and ventral and dorsal rows of larger epidermal cells; *laminal cells* unistratose throughout excluding bistratose margins in the upper part, totally smooth, in regular, longitudinal rows, upper and median cells irregularly rounded-quadrate to transversely short rectangular, (5.5–)7.5–11.0(–13.0) μm long, 7.5–11.0 μm wide, with straight or slightly sinuose, moderately incrassate walls; *basal marginal cells* quadrate to short-rectangular with transverse walls much thicker than longitudinal walls, hyaline to yellowish-hyaline, forming a distinct multiseriate border; *basal juxtacostal cells* long-rectangular, 20–40 μm long, 5–11 μm wide, rather thin-walled, sometimes somewhat lax. *Monoicous*. *Perichaetial leaves* yellowish- to olive-green, almost twice as long as sporophytes, 3.0–3.5 mm long, 1.5–1.8 mm wide, broadly elliptical, suddenly contracted into a stout, cuspidate point, muticous or ending with very short, hyaline, denticulate hair-point, deeply concave, strongly flaring when dry, erect and domed because of the incurved tips, loosely wrapped around the sporophyte; *margins* strongly revolute from the apex to well below mid-leaf; *upper laminal cells* like those in vegetative leaves, but in less regular rows, becoming long rectangular in mid-leaf and passing into lax-rectangular, thin-walled, hyaline cells in the basal part. *Setae* short, 0.3–0.6 mm long, straight, centrally attached. *Capsules* deeply immersed in perichaetial leaves, yellowish-brown to brown with age, ovoid or obloid, wide-mouthed, thin-walled, 1.0–1.2 mm long, 0.8–1.1 mm wide at mouth; operculum conical, rostellate, orange-brown, the beak short, straight; *exothecial cells* mostly rounded-quadrate, occasionally shortly oblong or transversely rectangular, (15–)20–40(–60) μm long, (10–)15–20(–25) μm wide, thin-walled with distinct collenchymatous thickenings in the middle of the urn, becoming transversely rectangular below the mouth, those at mouth smaller, 8–13 μm wide, orange-brownish, in 1–3 transparent, thin-walled tiers, directly juxtaposed with the marginal cells of the operculum; *stomata* 5–8, bicellular, at the extreme base of the urn; *peristome teeth* lacking in places or vestigially developed, papillose on both surfaces, yellowish-orange, composed of 1–3 eroded segments slightly extending beyond the capsule rim. *Spores* spherical, smooth to very finely papillose, 7–9 μm wide. *Calyptra* conical, smooth, mitriform, lobed at base, about 0.7 mm long.

Chromosome number: unknown.

Discussion. *Schistidium steerei* is one of the most distinct and sharply defined species in the genus *Schistidium*, characterised chiefly by its strongly revolute leaf margins from near the base to the extreme apex, gymnostomous capsules, nearly always muticous leaves, excepting only the innermost perichaetial leaves that sometimes have a very short, less than 0.4 mm long, hyaline, denticulate hair-points and bud-like perichaetia not closely wrapped around the capsule, composed of leaves with typically incurved tips forming a dome over the deeply immersed capsules. The revolute leaf margins in *S. steerei* are a unique feature in the genus, the majority of species having only variously recurved margins, sometimes broadly so, for example *S. poeltii*, *S. umbrosum* and *S. pulchrum* from the Northern Hemisphere, but they are never revolute from near the base to the extreme apex. In the rare *S. crenatum* which was only recently described from Fennoscandia, Siberia and northern North America the margins are more or less broadly recurved to revolute towards the leaf apex, but is is at once distinct from *S. steerei* in having coarsely papillose-denticulate margins in the distal part of the leaf and short, thin, flattened and coarsely and irregularly spinulose hair-points.

Schistidium steerei clearly belongs within sect. *Conferta* which is characterised by the olivaceous coloration of the plants, smooth leaves which show a yellow reaction with KOH and more or less hyaline basal marginal cells that have strongly incrassate transverse walls and form a distinct rectangular alar group or a marginal border (Blom 1996, 1998; Ochyra *et al.* 2003). This is the richest section of the genus in Antarctica, consisting of eight of 13 species now known from this biome. The gymnostomous state of the capsules in this section is known in *S. flaccidum* and *S. cryptocarpum* which, however, are immediately distinct in having narrowly recurved leaf margins that are almost plane close to the apex, piliferous to pilose leaves and hemispherical to widely cyathiform capsules. Except for *S. lewis-smithii*, *S. steerei* is the only regional species with a strongly vestigial or almost absent peristome. However, these species are unlikely to be mistaken because they belong within different subgenera of *Schistidium* and *S. lewis-smithii* is at a glance distinct in its plane leaf margins. As is the case with the latter species, *S. steerei* has been discovered and described only recently (Ochyra 1987*a*).

Reproduction in Antarctica. Fertile, producing an abundance of mature sporophytes.

Habitat. A species of acidic substrata, growing on gravelly soil or basalt rubble, but apparently never truly saxicolous; usually in moist to dry situations. It often forms extensive, compact cushions, usually half-buried in soil.

World range. An Antarctic endemic, so far known only from the South Shetland Is. (Fig. 107).

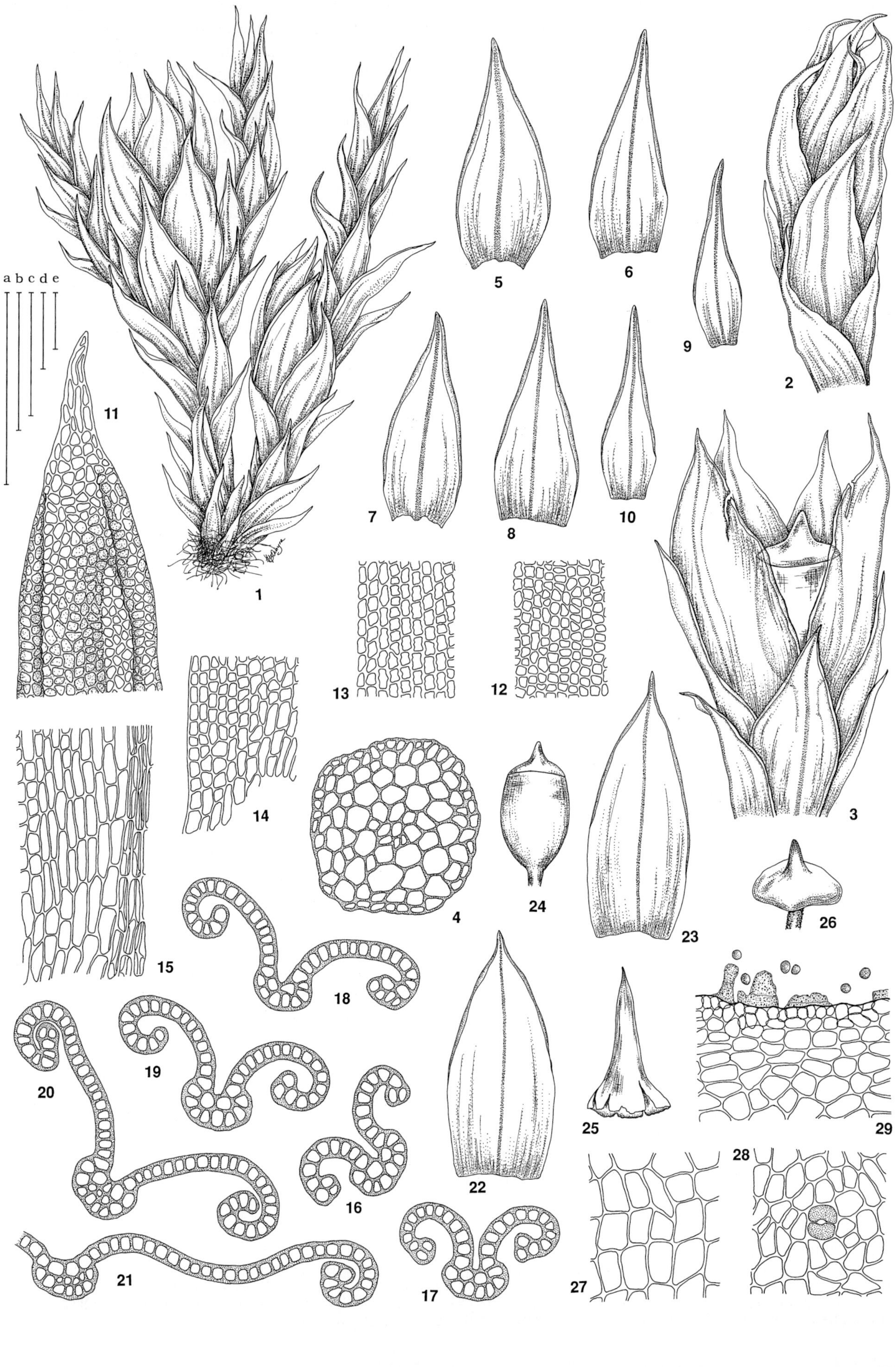
a b c d e
1
2
3
4
5
6
7
8
9
10
11
12
13
14
15
16
17
18
19
20
21
22
23
24
25
26
27
28
29

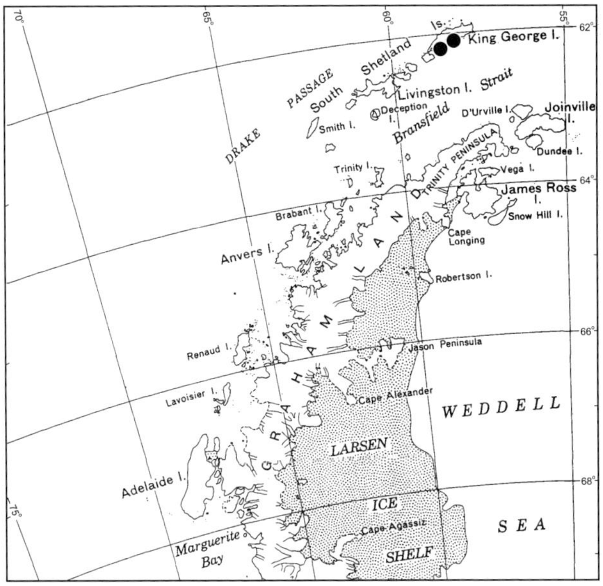

Figure 107. Global distribution of *Schistidium steerei.*

Distribution in Antarctica. Known only from two localities in the Admiralty Bay area on King George I., at low elevation between 30 and 110 m.

Specimens examined. WEST ANTARCTICA. SOUTH SHETLAND IS. **King George I.** ***Admiralty Bay***: Ezcurra Inlet, Cytadela, 30 m, *RO 946/80* (AAS, ALTA, BM, CANM, F, H, KRAM, NY); Martel Inlet, Mt. Wawel, 110 m, *RO 2176/80* (KRAM, NY).

7. **Schistidium urnulaceum** (Müll.Hal.) B.G.Bell Fig. 108

Schistidium urnulaceum (Müll.Hal.) B.G.Bell, Br. Antarct. Surv. Bull. 63: 99. 1984. ≡ *Grimmia urnulacea* Müll.Hal. *in* Neumayer, Int. Polarforsch. Deutsch. Exp. 2: 313. 1890. — Type citation: Austro-Georgia, ad rupes am Ausgange des Brockenthales, cum *Gümbelia immerso-leucophaea* consociata, 23. Januario 1883. [Lectotype (*vide* Ochyra 1990*a*: p. 135): 'Grimmia urnulacea C. Müll. n. sp. Austro-Georgia, ad rupes am Ausgange des Brockenthales. Südpolar-expeditionen' – HBG!; isotype: S!].

Grimmia antarctici Cardot var. *pilifera* E.B.Bartram, Ann. Missouri Bot. Gard. 25: 722. 1938, ***syn. nov.*** — Type: Marie Byrd Land, Edsel Ford Range, Mt. Rea-Cooper, P. Siple, F. A. Wade, S. Corey & O. D. Standcliff 27. [Lectotype (*selected here*) {or perhaps holotype}: 'Byrd Antarctic Expedition II. Mosses of Marie Byrd Land, Antarctica. No. 27 Grimmia Antarctici Card. var. pilifera Bartr. Locality: Mt. Rea Collected by Paul A. Siple on granite sand' – FH-Bartram!].

Plants small, in dense, yellow-or olive-green, somewhat lustrous cushions, weakly hoary above, brown to blackish-brown below. *Stems* 0.5–2.0 cm long, erect, regulary repeatedly branched, with scattered clusters of smooth, brown rhizoids, in cross-section without a central strand and composed of nearly uniform cells, or with the outer row consisting of slightly smaller, weakly thickened and brownish cells surrounding 4–5 rows of large, thin-walled medullary cells; *axillary hairs* filiform, hyaline throughout, 5–7-celled. *Leaves* densely set, straight to somewhat curved, keeled, closely imbricate, erect to erecto-patent when dry, erect-spreading when moist, (1.0–)1.2–1.6(–2.0) mm long, 0.2–0.4 mm wide, narrow and lanceolate, gradually acuminate, acute, ending in a rather short, about 0.5 mm long, hyaline, denticulate to nearly smooth hair-point; *margins* entire, plane or recurved on one or both sides for varying distances, unistratose below, in upper part with infrequent bistratose spots; *costa* weakly convex at the dorsal side, smooth and flattened at back, 35–45 µm wide above, becoming thinner towards the base, ending near but usually not in the hyaline point, in cross-section bistratose throughout; *laminal cells* unistratose throughout, quadrate to short rectangular in upper and median parts, 8–20 µm long, (3–)5–9(–11) µm wide with sinuose and strongly incrassate walls; *basal juxtacostal cells* short- to long-rectangular, not lax, straight-walled, 20–60 µm long, 5–12 µm wide; *basal marginal cells* pellucid, hyaline to yellowish-hyaline, quadrate to rectangular, forming a distinct multiseriate rectangular group. *Autoicous. Perichaetial leaves* yellowish-green, somewhat larger than the vegetative leaves but otherwise similar in shape, lanceolate or ovate- to oblong-lanceolate, 1.9–2.1 mm long (without hairpoint), 0.55–0.65 mm wide, terminated with a hyaline, 0.3–0.5 mm long, denticulate to almost smooth awn, weakly concave, straight, erect; *margins* plane below, recurved in the upper half; *upper laminal cells* similar to those of the vegetative leaves, subquadrate to short rectangular, sinuose and thick-walled, becoming in mid-leaf short to long rectangular, not lax, straight-walled; *costa* 35–45 µm wide above, channelled, becoming thinner, 22–25 µm wide, at the base. *Setae* short, 0.2–0.3 mm long straight, centrally attached. *Capsules* deeply immersed in the perichaetial leaves, light brown or yellowish-brown, 0.6–0.8 mm long, globose or cyathiform to obloid, wide-mouthed when dry and empty; *operculum* conic, rostellate, with attached columella after dehiscence; *exothecial cells* irregular, isodiametric to short-rectangular, 10–30 µm long, 8–15 µm wide, with strongly incrassate, 4–8 µm thick, straight walls, becoming somewhat smaller, isodiametric to transversely rectangular, orange-brown, in 2–4 tiers at the capsule rim; *stomata* few at the extreme base of the urn, bicellular, superficial; *peristome teeth* lanceolate, 250–300 µm long, 50–60 µm wide at the base, widely patent or squarrose, yellow- to reddish-brown, irregularly perforate to cracked above, roughly papillose on both surfaces throughout. *Spores* spherical, smooth,

←

Figure 106. *Schistidium steerei.* — **1**. Habit, dry. **2**. Perichaetium, dry. **3**. Perichaetium, wet. **4**. Cross-section of stem. **5–10**. Leaves. **11**. Leaf apex. **12**. Upper lamina cells. **13**. Mid-leaf cells. **14**. Angular cells. **15**. Basal juxtacostal cells. **16–21**. Cross-sections of leaves, a sequence from apex to base. **22–23**. Perichaetial leaves. **24**. Operculate capsule, wet. **25**. Calyptra. **26**. Operculum with attached columella. **27**. Exothecial cells in middle of urn. **28**. Lower exothecial cells and stoma. **29**. Vestigial peristome teeth, spores and rim cells of capsule. — [All from *RO 2176/80* (holotype)]. — Scale bars: a – 1 mm (25); b – 100 µm (4, 11–21, 27–29); c – 1 mm (2–3, 5–10, 26); d – 1 mm (22–24); e – 1 mm (1).

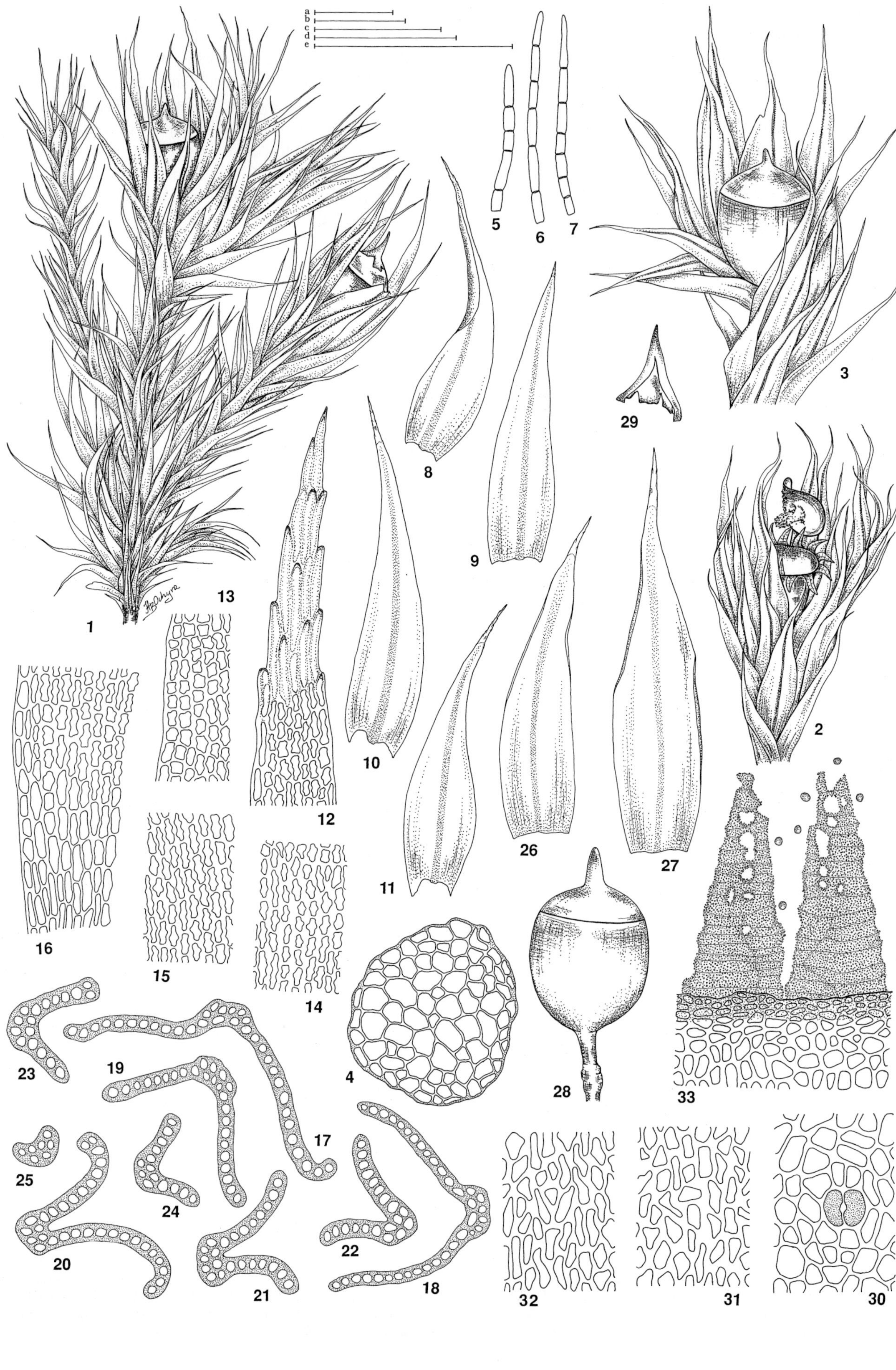
a
b
c
d
e
1
2
3
4
5
6
7
8
9
10
11
12
13
14
15
16
17
18
19
20
21
22
23
24
25
26
27
28
29
30
31
32
33

7–9 µm wide. *Calyptra* conical, smooth, mitriform, lobed at base, not extending the urn.

Chromosome number: unknown.

Discussion. *Schistidium urnulaceum* is a distinct, stenotypic species which is remarkably uniform in character states of the capsules, peristome, spores and leaves. The somewhat hoary appearance enables its separation in the field from all other congeners, except for *S. halinae* from which, however, it differs in a wide array of microscopic characters. There is some variation in form of the hyaline hair-point in *S. urnulaceum* whose length may vary from 0.1 to 0.5 mm both within and between populations. Additionally, lower leaves are muticous in most populations and hair-pointed in others. These differences represent a spectrum of variability inherent in populations of most species of the Grimmiaceae and are influenced by such ecological factors as the humidity and insolation of the substratum and exposure to sunlight.

Schistidium urnulaceum is most easily identified by its oddly shaped exothecial cells which are isodiametric to shortly oblong and have strongly incrassate walls, 4–8 µm thick. This shape is exceptional in *Schistidium* and is known also only in *S. deceptionense* which, however, differs in a number of characters that are discussed under this species. Exothecial cells with strongly thickened walls are exceptional in the Grimmiaceae generally, being known only in *Orthogrimmia asperitricha*, *Dryptodon tortuosus* and *Grimmia incrassicapsulis* from the austral region. The subglobose or cyathiform capsules which are wide-mouthed when dry and empty, unistratose leaf margins with infrequent bistratose spots in the distal part and the narrow and weak costa which is bistratose throughout in transverse section and scarcely convex on the dorsal surface are other important sporophyte features of this species.

Leaf areolation of cells with strongly incrassate and sinuose walls is another unusual feature of *Schistidium urnulaceum*. This leaf cell pattern is known in a number of *Schistidium* species, but in the Antarctic it is most evident in *S. amblyophyllum*. However, this species differs from *S. urnulaceum* by its epilose or very shortly hyaline-tipped leaves. *Schistidium cupulare* also has sinuose-walled laminal cells but they are short-rectangular, with less incrassate walls and are confined to the transitional zone between the base and mid-leaf. In addition, this species differs immediately from *S. urnulaceum* in its markedly thickened, 2–3-stratose leaf margins and a strong costa which is semi-terete in shape, prominent on the dorsal surface and tristratose below.

Grimmia antarctici var. *pilifera* was described from sterile material from Marie Byrd Land (Bartram 1938, 1957). This taxon was initially interpreted as a modification of *Schistidium antarctici* (Ochyra 1998*a*) but a reconsideration of the type and non-type material revealed that it fits much better into the concept of *S. urnulaceum*, mostly on account of the identical leaf areolation and hoary plants.

Schistidium urnulaceum was described from South Georgia as *Grimmia urnulacea* by Müller (1890*a*) and Bell (1984) correctly transferred it to the proper genus. For a long time this species was considered to be endemic for South Georgia until it was recorded from King George I. in the Antarctic by Ochyra (1990*a*). Additional reports of *Schistidium urnulaceum* from this island (Chen *et al.* 1995; Liu & Li 1998*b*) were based upon misidentifications of *S. antarctici*.

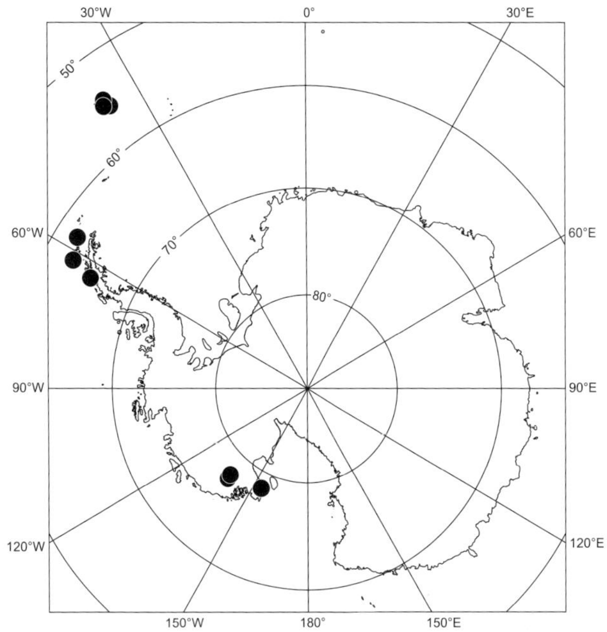

Figure 109. Global distribution of *Schistidium urnulaceum*.

Reproduction in Antarctica. Commonly in fine fruiting condition.

Habitat. A saxicolous species growing on acidic rocks, on weathered andesite and basalt rocks, on scree as well as in the crevices of rock outcrops. It is usually associated with exposed rock ledges covered with thin soil and scarcely vegetated, growing with *Schistidium antarctici*, *Andreaea regularis*, *A. gainii*, *Syntrichia saxicola* and *Hymenoloma antarcticum*.

World range. An American Subantarctic species, widely distributed on South Georgia and occasionally extending southwards to the northern maritime Antarctic, and also recorded from Marie Byrd Land on the continent (Fig. 109).

Distribution in Antarctica. A rare, though locally frequent species. In the maritime Antarctic, known only from the South Shetland Is. and Anvers I. off the Danco Coast (Fig. 109). Additionally, it has been recorded from Marie Byrd Land where it reaches its southernmost locality at lat. *ca* 78°S. The maritime Antarctic sites are near sea level to 130 m, but the continental stations are at altitudes above 1100 m.

Specimens examined. WEST ANTARCTICA. SOUTH SHETLAND IS. **King George I.** ***Admiralty Bay***: Ezcurra Inlet, Scalpel Pt., 160 m, *RO 1816/80* (H, KRAM, NY); Martel Inlet, Mt. Wawel, 80 m, *RO 2127/80* (KRAM), 120 m, *RO 2183/80* (H, KRAM) and 130 m, *RO 2180/80* (KRAM). **Livingston I.** ***Hurd Peninsula***: False Bay, *RMS 69-542, 550, 557 & 561* (US).

WEST ANTARCTIC PENINSULA. **Palmer Archipelago.** ***Anvers I.***: Norsel Pt., 5–10 m, *RILS 1946* (AAS, KRAM).

EAST ANTARCTICA. MARIE BYRD LAND. **Ruppert Coast.**

←

Figure 108. *Schistidium urnulaceum*. — **1**. Habit, dry. **2**. Perichaetium, dry. **3**. Perichaetium, wet. **4**. Cross-section of stem. **5–7**. Axillary hairs. **8–11**. Leaves. **12**. Leaf apex. **13**. Upper lamina cells at margin. **14–15**. Mid-leaf cells. **16**. Angular cells. **17–25**. Cross-sections of leaves, a sequence from apex to base. **26–27**. Perichaetial leaves. **28**. Operculate capsule, wet. **29**. Calyptra. **30**. Lower exothecial cells and stoma. **31–32**. Exothecial cells in middle of urn. **33**. Peristome teeth, spores and rim cells of capsule. — [All from *RO 2180/80*]. — Scale bars: a – 1 mm (1); b – 100 µm (33); c – 1 mm (2–3, 8–11, 26–28); d – 100 µm (4–7, 12–25, 30–32); e – 1 mm (29).

Ford Range: Mt Rea, *PAS 27* (FH – type of *Grimmia antarctici* var. *pilifera*) & *98a* (FH); Mt. Donald Woodward, *PAS 97* (FH). **SHIRASE COAST.** Rockefeller Mountains: Mt. Tenant, *Frazier & Wade 304* (FH).

8. **Schistidium halinae** Ochyra Fig. 110

Schistidium halinae Ochyra, Ann. Bot. Fenn. **35**: 268, *f. 1.* 1998. — TYPE CITATION: Antarctica, South Shetland Islands, King George Island: Admiralty Bay, Puchalski Peak, lat. 62°10′S, long. 58°17′30″W, elev. 140 m; on dry, bare andesite rocks on a west-facing slope; 16.III.1980, Ochyra 2711/80. [Holotype: 'Herbarium Instituti Botanici Academiae Scientiarum Polonae – Cracoviae Bryophyta. Schistidium ~~cf. hyalino-cuspidatum (C. M.) Bell~~ halinae Ochyra, sp. nov. Antarctica, South Shetland Islands – King George Island: Admiralty Bay, Puchalski Peak between Nature Conservation Glacier and Rybak Glacier, 62°10′S, long. 58°17′30″W, elev. 140 m; on dry bare andesite rocks on western slope 16/III/1980 Coll. R. Ochyra No. 2711/80 Det. R. Ochyra' – KRAM!; isotypes: AAS, ALA, ALTA, B, BA, BM, BP, BR, C, CANM, CHR, COLO, DUIS, DUKE, F, FH, FLAS, G, GJO, GL, H, HIRO, HO, HSC, IFP, JE, KRAM, L, M, MEXU, MICH, MIN, MO, NAM, NFLD, NIPR, NICH, NSW, NY, PC, PE, PRE, S, TENN, TNS, TRH, U, UBC, US – all isotypes were distributed in *Bryophyta Antarctica Exsiccata* No. 164 as *Schistidium* cf. *hyalinocuspidatum*).

Plants medium-sized, rarely smaller, in relatively large, compact cushions, dull, olivaceous to brown or yellow-brown above, brown below, rarely brown or blackish throughout. *Stems* (1.0–)1.5–2.5(–3.0) cm long, erect, much branched, in cross-section with 1–2 layers of smaller, brown to orange-brown, moderately thick-walled, cortical cells surrounding 3–4 layers of large, hyaline, thicker-walled medullary cells with distinct collenchymatous thickenings; *central strand* large and conspicuous; *rhizoids* scattered in the lower part of the stem, brown and smooth, weakly branched; *axillary hairs* hyaline throughout, 5–7-celled, composed of short rectangular cells or with elongated upper cells. *Leaves* crowded, arranged in 5 distinct, spiral rows giving the stem a ridged appearance, straight, closely imbricate, erect when dry, erect-spreading when moist, (1.3–)1.5–1.8(–2.0) mm long, 0.4–0.7 mm wide at the widest part, lanceolate to ovate-lanceolate, gradually acuminate, variable on the same shoot as regards the presence and length of the hyaline hair-points, the lower ones acute to obtuse and muticous, the uppermost acute and pilose, sharply keeled in the upper half, obtusely keeled to canaliculate below; *hair-points* hyaline, 0.2–1.0 mm long, becoming progressively longer up the stem, rather stiff, flattened, membraneous, narrow above and becoming widened towards the insertion and embracing a small part of the upper lamina, not or shortly decurrent down the margins, finely and irregularly spinulose-denticulate with short narrow spicules; *margins* entire, plane in lower half or more of the leaf length on one side and in lower ¼–⅓ on the other side, narrowly recurved above almost to the apex, at least on one side, irregularly bistratose in upper and central part for 1–2 rows of cells, in the lower half unistratose; *costa* long-excurrent, channelled above, strongly convex dorsally, 40–65 μm wide above, becoming narrower towards the base, smooth throughout on the dorsal surface, in cross-section hemispherical to subrectangular in outline, bistratose throughout, occasionally with tristratose spots below, composed of uniform, large cells; *laminal cells* unistratose throughout with occcasional bistratose spots in the upper half, entirely smooth, thick-walled; *upper* and *median cells* rounded, rounded-quadrate to oblong, sometimes transversely short rectangular, (4.5–)6.0–10.0 μm long, 6–8 μm wide, straight-sided to slightly sinuose, becoming short rectangular in the lower mid-leaf, 10–14 μm long, 7–8 μm wide; *basal cells* chlorophyllose to subhyaline, forming a large juxtacostal band of short- to long-rectangular cells with straight and strongly incrassate longitudinal walls, 20–50 μm long, 2.5–4.5(–5.0) μm wide; *basal marginal cells* thick-walled with transverse walls thicker than longitudinal walls, 7–8 μm wide, 6–20 μm long, forming a group of 4–6 rows of cells wide, imperceptibly merging into the juxtacostal group. *Autoicous. Perichaetial leaves* yellowish- to olive-green, about twice as long as sporophytes, 2.0–2.5(–3.0) mm long, 1.1–1.3 mm wide, concave, sometimes faintly plicate, ovate, gradually acuminate, ending with a very long, 0.6–1.4 mm, hyaline, flat, spinulose-denticulate hair-point, non-decurrent down the margins but down the dorsal side of the costa resulting in a U- or V-shaped hyaline portion of the upper part of the perichaetial leaves. *Setae* short, 0.5–0.6 mm long, straight, centrally attached; *vaginula* very short, with thin and smooth epidermal cells. *Capsules* deeply immersed in the perichaetial leaves, pale brown, shortly oblong-cylindrical, thin-walled, 0.9–1.0 mm long, 0.7–0.8 mm wide at mouth; *operculum* conic, with a straight beak, orange-brown; *exothecial cells* mostly isodiametric, rounded-quadrate to short-oblong, 20–35 μm long, 20–30 μm wide, thin-walled, with clear corner thickenings, becoming transversely rectangular below the mouth, those at the urn rim markedly smaller, 8–15 μm wide, thick-walled, orange-brown, in 2–3 tiers; *stomata* 7–10 at the extreme base of the urn, bicellular, superficial, round-pored, variously oriented; *peristome teeth* 280–350 μm long, orange, erect, straight, lanceolate or irregular in outline, with uneven margins, acute to obtuse, irregularly perforated, often with irregular cracks, usually eroded at

⟶

Figure 110. *Schistidium halinae.* — **1**. Habit. **2**. Portion of branch, dry. **3**. Perichaetium with deoperculate capsule. **4**. Cross-section of stem. **5–7**. Axillary hairs. **8–12**. Leaves. **13**. Leaf apex. **14**. Upper laminal cells at margin. **15**. Mid-leaf cells. **16**. Lower mid-leaf cells. **17**. Basal cells. **18**. Angular cells. **19–26**. Cross-sections of leaves, a sequence from base to apex. **27–28**. Perichaetial leaves. **29**. Deoperculate capsule, dry. **30**. Deoperculate capsule, wet. **31**. Operculate capsule, wet. **32**. Operculum with attached columella, dry. **33**. Calyptra. **34**. Exothecial cells at urn base and stomata. **35**. Exothecial cells in middle of urn. **36**. Peristome teeth, spores and rim cells of capsule. — [1, 4–10, 13–14, 16–17, 19–36 from *RO 2711/80*, holotype; 2–3, 11–12, 15, 18 from *937/80*]. — Scale bars: a – 1 mm (8–12) and 100 μm (4–7); b – 100 μm (34–35); c – 1 mm (2) and 3 mm (1); d – 1 mm (29–33) and 100 μm (13–18); e – 1 mm (3, 27–28) and 100 μm (19–26); f – 100 μm (36). (Reproduced from *Annales Botanici Fennici*, 35: p. 269, 1998, with permission.)

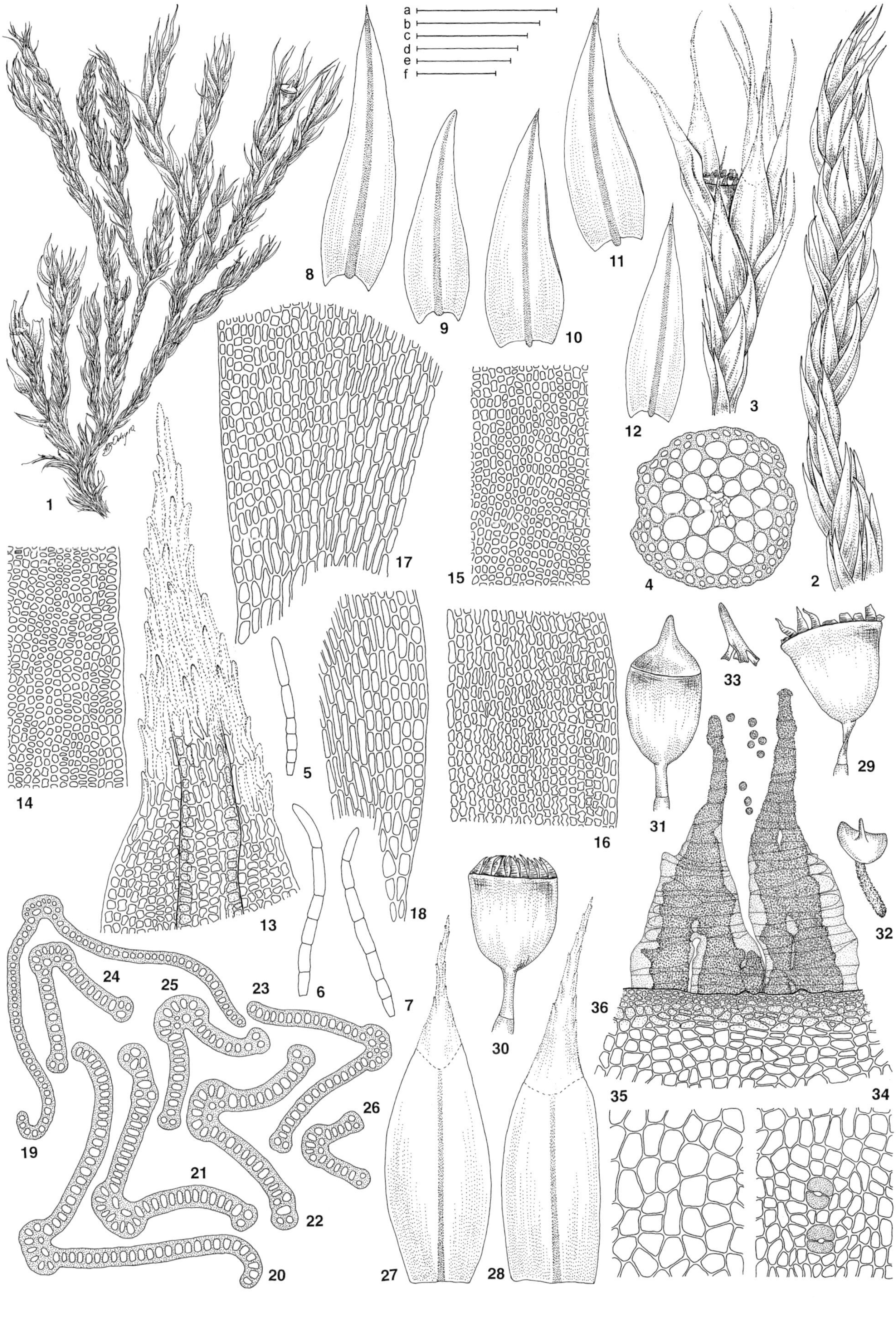
a
b
c
d
e
f
1
2
3
4
5
6
7
8
9
10
11
12
13
14
15
16
17
18
19
20
21
22
23
24
25
26
27
28
29
30
31
32
33
34
35
36

margins, densely papillose. *Spores* spherical, pale brown, smooth to very finely papillose, 10–12 μm wide. *Calyptra* conical, smooth, mitriform, lobed at base, not extending to the urn.

Chromosome number: unknown.

Discussion. *Schistidium halinae* can be recognised *in situ* by its hoary appearance because of the long, hyaline hair-points on the uppermost vegetative and perichaetial leaves, which is the key character of the species. They are broad, flattened and membraneous and tend to be very long and pronounced in the upper vegetative leaves, reaching to 1.4 mm on the perichaetial leaves and it are unusual in that the cell wall remnants are clearly visible in them. Such hair-points have evolved in several species growing in dry, often sunny and exposed habitats and they occur in distantly related species such as *S. flaccidum*, *S. frisvollianum*, *S. bryhnii*, *S. venetum* and *S. scandicum*. However, the hair-point in *S. halinae* is much stiffer because of the strong and long-excurrent costa and does not tend to be as flexuose as that in the aforementioned species. This is reminiscent of the hyaline awn in some species of *Grimmia*, for example *G. anodon*.

The only other Antarctic species with long-pilose leaves is *Schistidium urnulaceum* which can be recognised microscopically by its rectangular and strongly sinuose-walled laminal cells, the central strand, the very short setae, 0.2–0.3 mm long, the small subglobose or cyathiform capsules, and the isodiametric shortly oblong exothecial cells with strongly incrassate walls, 4–8 μm thick.

Additionally, the species is recognised by a combination of gametophyte and sporophyte characters including the rounded to rounded-quadrate upper and median laminal cells becoming short-rectangular and sinuose in the lower mid-leaf, the long-rectangular and straight-walled basal juxtacostal cells and quadrate to short-rectangular basal marginal cells forming a broad band, the presence of the central strand, the shortly oblong-cylindrical or cyathiform capsules and the thin-walled, isodiametric to short-oblong exothecial cells.

Schistidium halinae shows the closest relationship to *S. frigidum* and *S. cryptocarpum* (and possibly *S. venetum*) from the Northern Hemisphere with which it belongs in sect. *Conferta* (Ochyra 1998*c*). The northern species may be similar to *S. halinae* in habit, but the hair-points, if not encrusted by calcareous mineral matter, are much softer. *S. halinae* also bears some similarity to the Scandinavian *S. scandicum* which is distinct in its shallowly immersed capsules, red peristome teeth, narrowly recurved margins of the perichaetial leaves in the upper half, larger laminal cells, and granulose to verrucose, somewhat larger spores.

Reproduction in Antarctica. Frequently producing fully mature sporophytes.

Habitat. In exposed places on stones and gravelly ground, mostly in dry situations, occasionally in moist sites below late snow patches.

World range. An Antarctic endemic, so far known only in the northern maritime Antarctic region (Fig. 111).

Distribution in Antarctica. *Schistidium halinae* is so far known from two widely separated centres, namely the South Shetland Is. and especially the James Ross Island group, where it is locally abundant, in the northern part of the East Antarctic Peninsula, both in the northern maritime Antarctic (Fig. 111). In the South Shetland Is. the species is known only from King George I. and Livingston I. Its southernmost station is at lat. 63°56′S and the altitudinal range extends from 20 to 150 m.

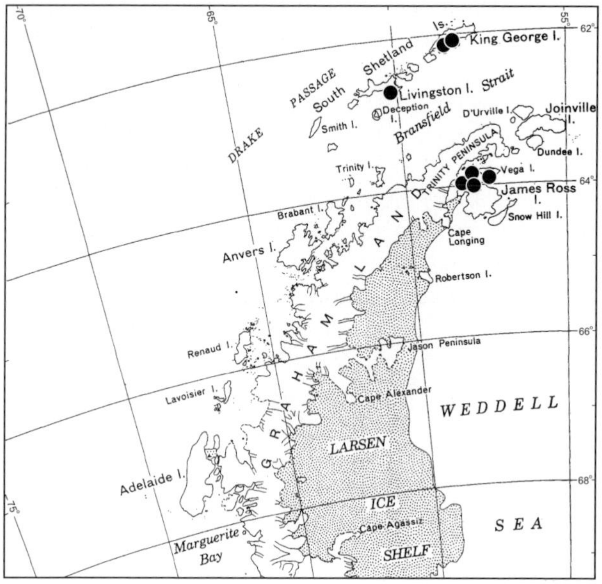

Figure 111. Global distribution of *Schistidium halinae*.

Specimens examined. WEST ANTARCTICA. SOUTH SHETLAND IS. **KING GEORGE I.** ***Admiralty Bay***: Ezurra Inlet, Cytadela between Monsimet Cove and Zalewski Glacier, 20 m, *RO 937/80* (KRAM) and 80 m, *RO 947/80* (KRAM). **LIVINGSTON I.** ***Hurd Peninsula***: near a melt stream near Juan Carlos I Station, *Schulz 97* (KRAM).

EAST ANTARCTIC PENINSULA. **JAMES ROSS ISLAND GROUP.** ***Vega I.***: Pastorizo Bay, 50 m, *RILS 7855* (AAS, KRAM, TRH). ***James Ross I.***: Abernethy Flats, near Monolith Lake, 100 m, *RILS 7635* (AAS, KRAM, TRH); Lachman Crags, 25–50 m, *RILS 7332* (AAS, KRAM); Crame Col, 100–120 m, *RILS 7408* (AAS, KRAM); Fortress Hill, 150 m, *RILS 7777 & 7781C* (AAS, KRAM); St. Martha Cove, 35 m, *RILS 7455A* (AAS, KRAM, TRH), 50 m, *RILS 7469* (AAS, KRAM), 80 m, *RILS 7648* (AAS, KRAM) and 150 m, *RILS 7541A* (AAS, KRAM).

9. **Schistidium antarctici** (Cardot) L.I.Savicz & Smirnova — Fig. 112

Schistidium antarctici (Cardot) L.I.Savicz & Smirnova, Novosti Syst. Nizš. Rast. 1965: 252. 1965. ≡ *Grimmia antarctici* Cardot, Bull. Herb. Boissier Sér. 2, 6: 30. 1906. — TYPE: Shetland du Sud: d'île Nelson, Harmony Cove. Terre Louis-Philippe: débarquement *a*. Archipel de Graham: d'île Paulet. [Lectotype (*vide* Ochyra *et al.* 1986: p. 59): 'Herb. J. Cardot. Svenska Sydpolarexped. 1901–03. nº 425 in parte. Grimmia antarctici Card. sp. nova Shetland méridionales: d'île Nelson, Harmony Cove. Leg. C. Skottsberg 11/1 1902' – PC-Cardot!; isolectotype: S!; syntypes: (1) 'Herb. J. Cardot. Svenska Sydpolarexped. 1901–03. nº 448. Grimmia antarctici Card. sp. nova Terre Louis-Philippe: cap Kjellman. Leg. C. Skottsberg 28/11 1902' – PC-Cardot!; isosyntype: S! (2) Svenska Sydpolarexpeditionen 1901–03. N:r 449. Grimmia antarctici Card. sp. nova. Graham Archipelago, Paulet I. /10 1903 C. Skottsberg 28/11 1902 Det. J. Cardot' – PC-Cardot!; isosyntypes: FH-Bartram!, LE!, S!].

Grimmia antarctici var. *percompacta* E.B.Bartram, Ann. Missouri Bot. Gard. 25: 723. 1938. — TYPE: King Edward VII Land: Rockefeller Mts., Mt. Helen Washington, P. Siple, F. A. Wade, S. Corey & O. D. Standcliff 68a. [Lectotype (*selected here*) {or perhaps holotype}: 'Byrd Antarctic Expedition II. Mosses of King Edward VII Land, Antarctica. No. 68a Grimmia Antarctici Card. var. percompacta Bartr. Locality: Mt. Helen Washington Collected by Paul A. Siple' – FH-Bartram!]. First synonymised by Ochyra (1998*a*: p. 165).

Plants small to medium-sized, in small to large, dense or looser cushions, often coalescing and forming large patches or mats, weakly lustrous, brown, reddish-brown, olivaceous, dark green, olive-, yellowish-, blackish- or brownish-green to blackish-brown above, brown to blackish-brown below. *Stems* 0.3–5.5 cm tall, erect, straight, slender, much fastigiately branched, in cross-section rounded to obtusely rounded-angulate, consisting of 1–2(–3) layers of small, brown to orange-brown, thick-walled cortical cells sharply delimited from 4–5-layered medulla of large, hyaline or yellowish-hyaline, thin-walled cells, often with distinct collenchymatous thickenings; *central strand* small and often indistinct, occasionally absent; *rhizoids* light to dark brown, smooth, branched, scattered in the lower part of the stem or nearly absent, sometimes repeatedly branched and chlorophyllose at the apices and becoming protonemata forming multicellular protonemal gemmae; *axillary hairs* filiform, hyaline throughout, 4–5-celled. *Leaves* crowded, usually disposed in 3–4 distinct spiral rows, erect and closely imbricate, rarely erect-spreading when dry, erecto-patent when moist, straight to slightly secund, ovate, ovate-triangular, ovate- or oblong-lanceolate to lanceolate, 0.8–2.9(–2.5) mm long, 0.25–0.6 mm wide, keeled, gradually acuminate, acute to narrowly obtuse; *hair-point* absent to 0.5 mm long, hyaline to yellowish, flattened, non-decurrent but embracing parts of the upper lamina, smooth or nearly so; *margins* entire, narrowly to broadly recurved on one or both sides throughout or for varying length, unistratose to variously bistratose and bulging distally in 1 row; *costa* percurrent or ending just below the apex, 40–60 μm wide, smooth, in cross-section somewhat flattened, rectangular to elliptical above or semi-terete throughout, bistratose above, 3(–4)-stratose in the middle and below; *laminal cells* unistratose throughout, smooth, not bulging, with thick, straight or sinuose walls, variable in shape, mostly quadrate to short-rectangular above, often oblate at the margins, 8–16(–20) μm long, (5–)7–10(–12) μm wide, usually becoming elongate below the middle; *basal juxtacostal cells* short-rectangular, 20–40 μm long, 10–13 μm wide, with thick, straight or sinuose walls, forming a small group, usually becoming shorter and narrower towards the margins; *basal marginal cells* quadrate to short-rectangular, hyaline, pellucid, with markedly thickened cross-walls, sometimes forming a distinct border of 3–5 short rows. *Autoicous*. *Perichaetial leaves* broadly ovate, ovate-triangular to oblong- or ovate-lanceolate, 1.7–2.2 mm long, 0.9–1.2 mm wide, acute to broadly acute, muticous to shortly hyaline tipped, narrowly recurved on both sides in the upper half, plane below. *Setae* short, 3.5–4.5 mm long, straight. *Capsules* shallowly or deeply immersed in the perichaetial leaves, subglobose, ovoid to obloid, 0.9–1.1 mm long, *ca* 0.8 mm wide, wide-mouthed, smooth, rather leptodermous, yellow-brownish to pale brown; *operculum* conical, with a straight beak, 4–6 mm long; *exothecial cells* variable, predominantly oblong to short-rectangular but usually mixed with rounded, rounded-square or less often transversely rectangular cells, 20–45 μm long, 15–30 μm wide, with moderately, evenly thickened walls, clearly collenchymatous, becoming smaller, rounded to oblate at the mouth and forming a flaring, orange-brown 5–7-seriate strip at the mouth; *stomata* few at the extreme base of the urn, bicellular, superficial, with rounded pori; *peristome teeth* short or rudimentary, 25–220 μm long, patent to squarrose or revolute with age, straight, lanceolate to irregular in outline, truncate, usually cleft at the apex, variously eroded resulting in very uneven margins, entire or perforated in central and upper parts, orange to reddish-orange, densely, finely papillose throughout. *Spores* spherical, pale, smooth, 8–10 μm in diameter. *Calyptra* mitrate, small, smooth, lacerate at the base.

Chromosome number: $n = 13$ – King George I. (Przywara *et al*. 1984).

Discussion. *Schistidium antarctici* is one of the most distinctive species of the genus, being primarily characterised by its rudimentary, highly reduced peristome teeth. This feature is readily observed in plants from the maritime Antarctic where they usually produce sporophytes in profusion. The peculiar shape of the peristome is shared only with some Antarctic populations of *S. praemorsum*, to which it is remotely related. *Schistidium antarctici* is clearly a member of sect. *Conferta* on account of, among others, the presence of the distinct basal marginal border of hyaline cells, whereas *S. praemorsum* belongs to sect. *Atrofusca* and has chlorophyllose basal leaf cells. The latter is also distinct by its almost entirely bistratose leaf lamina in the distal half as well as by its stronger costa which is semi-terete and tristratose distally.

Sterile plants of *Schistidium antarctici*, which are common on the continent, are also easy to distinguish by their characteristic arrangement of closely imbricate leaves in 3–4 distinct spiral rows giving the shoots a ridged appearance. When sterile, the species may be also confused with *S. andinum* in which, however, the leaves are not closely imbricate and spirally arranged, the costa is stronger, semi-terete and tristratose above, the leaf margin is often bistratose above in 2–3 rows of cells and the leaves lack the hyaline basal marginal border.

Gametophytically, *Schistidium antarctici* is quite variable, especially with regard to the size of the plants, growth form and the presence of hyaline hair-points, but various phenotypes showing different combinations of these characters are clearly habitat modifications. For example, the stunted habit of the plants described from the continent as var. *percompacta* (Bartram 1938) is doubtless a result of living in extremely severe climatic conditions and similar modifications are exhibited in practice by most, if not all, other species. Also, piliferous and epilose leaves may be often observed in the same population, usually depending on the position of plants in the cushion. This character is usually subject to considerable variation in all grimmiaceous mosses.

Schistidium antarctici was originally described by Cardot (1906*a*) as *Grimmia antarctici* from three specimens from the South Shetland Is. and the Antarctic Peninsula region, and Savicz-Lyubitskaya & Smirnova (1965*a*) transferred it to *Schistidium* on account of its systylious capsules. Ochyra *et al*. (1986) considered this species to be identical to *S. chrysoneurum*, a poorly known species from Îles Kerguelen. Closer examination of the scanty type material of the latter revealed that, despite the similar leaf areolation, *S. chrysoneurum* differs markedly in the anatomical structure of the leaves which are additionally papillose in the upper part and, accordingly, *S. antarctici* has been reinstated as a good species (Ochyra 1998*a*).

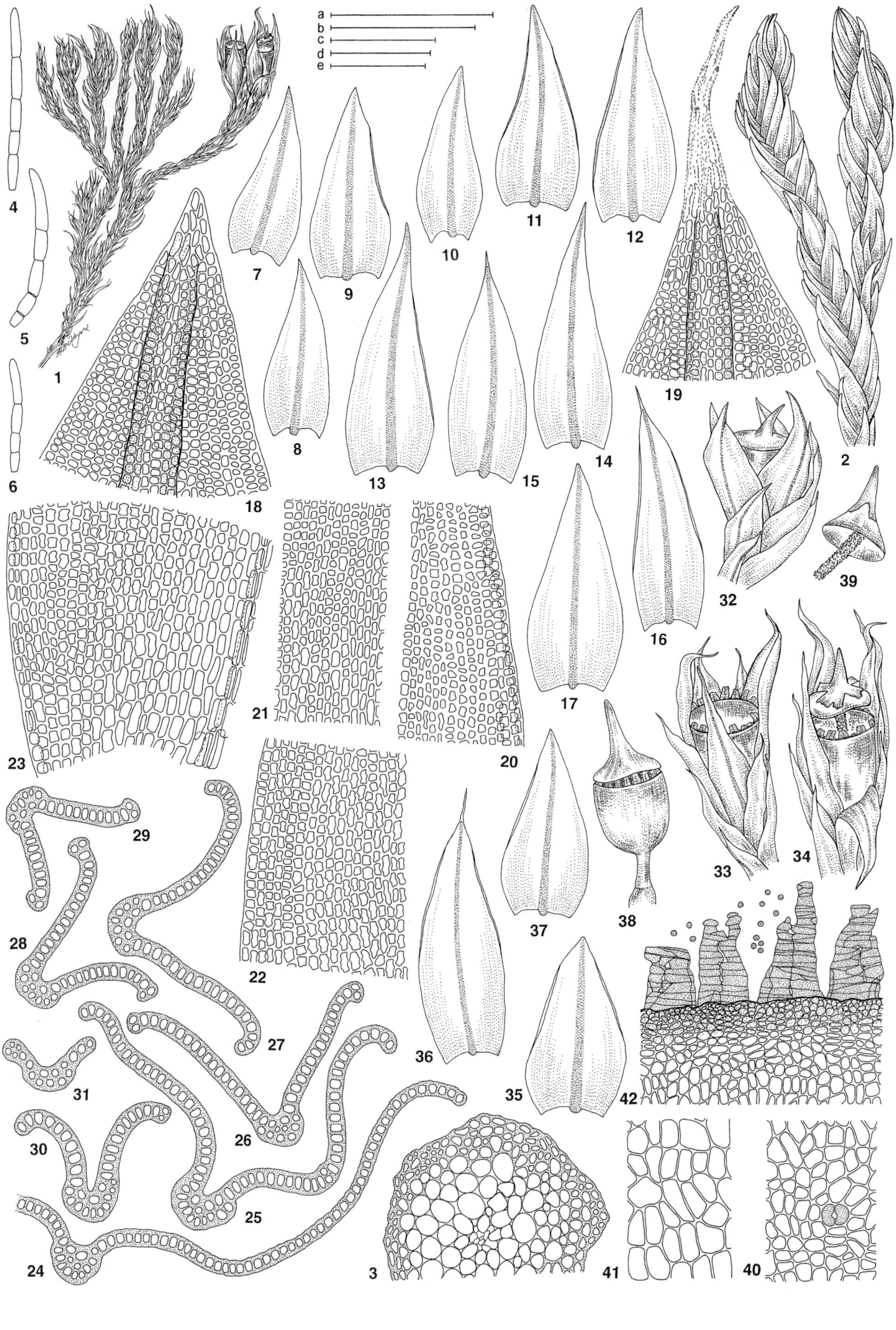
a
b
c
d
e
1
2
3
4
5
6
7
8
9
10
11
12
13
14
15
16
17
18
19
20
21
22
23
24
25
26
27
28
29
30
31
32
33
34
35
36
37
38
39
40
41
42

Reproduction in Antarctica. Commonly fruiting in the maritime Antarctic, although it is seldom fertile when growing on soil. On the continent sporophytes have been observed only occasionally in specimens from Marie Byrd and Ellsworth Lands. Sterile plants, especially from the continent, often reproduce vegetatively by means of protonemal gemmae which develop on the chlorophyllous apices of rhizoids (Savicz-Lyubitskaya & Smirnova 1965*a*).

Habitat. A rupicolous and epigean species growing in a wide range of habitats, both dry and wet as well as exposed and sheltered, forming small to large cushions or tufts which may coalesce to form large turves or patches which are frequently broken by frost action, leaving unstable clay puddles and stone polygons. It grows on rock faces, ledges, erratics, outcrops and knolls, in rock cracks and fissures, on cliffs and outcrops as well as on scree, sandy or gravelly ground, on coarse scoria and volcanic ash and on pockets of mineral soil between blocks and boulders, on thin soil covering stones, on weathered debris and morainic detritus. *Schistidium antarctici* in the maritime Antarctic is a constant component of various communities of the fruticose lichen and moss cushion subformation which cover large expanses on fellfields and moraines. It has its ecological optimum on base-rich substrata and forms a separate community in which it co-dominates with *Syntrichia saxicola*, for instance on Signy I. (Lewis Smith 1972), but it appears to grow equally extensively on acidic substrata, for example on King George I. (Ochyra 1998*a*). In seepage areas at continental sites it frequently forms an undulating turf, admixed with *Bryum pseudotriquetrum* and *Ceratodon purpureus* (Kappen *et al.* 1989; Lewis Smith 1990*a*). Having a wide ecological amplitude it grows in association with many species of moss, although some short cushion- and turf-forming species occur only sparsely with *S. antarctici*, e.g. *Bartramia patens*, *Bryum pseudotriquetrum* and *Ceratodon purpureus*. Also, some species of lichens are frequent associates of *S. antarctici*, for example *Ochrolechia frigida*, *Psoroma hypnorum* and *Leptogium puberulum*.

World range. An Antarctic endemic, penetrating into the Subantarctic on South Georgia where it is very rare and not abundant (Ochyra *et al.* 2002) (Fig. 113).

Distribution in Antarctica. Together with *Ceratodon purpureus* and *Bryum pseudotriquetrum*, *Schistidium antarctici* is one of the most widespread and frequently collected moss species in the Antarctic (Figs 113–114). In the maritime Antarctic it is particularly common in the South Orkney and South Shetland Is., for example on Signy I., King George I. and Deception I., but becomes less abundant along the west coast of the Antarctic Peninsula, extending southwards to central Alexander I. on its east coast where it reaches its southernmost latitude at 71°31′S. On the eastern coast of the Antarctic Peninsula it is frequent only on James Ross I. but extends to lat. 65°12′S on the Oscar II Coast. On the continent *S. antarctici* is widely scattered in all sectors, and locally abundant, with particular concentrations of localities in Queen Mary Land, Wilkes Land (Savicz-Lyubitskaya & Smirnova 1965*a*), although no material has been seen from Enderby, Kemp and Mac.Robertson Land. However, Seppelt & Green (1998) reported this species from Enderby and Mac.Robertson Lands ascribing the record from the latter region to Filson (1966) but in his Flora this species is not mentioned. The southernmost locality of *S. antarctici* was reported from lat. 83°49′S in southern Victoria Land (Seppelt & Green 1998), but this material was not available for study. The altitudinal range extends from sea level to about 900 m on James Ross I. in West Antarctica and 1379 m in East Antarctica, although the majority of collections come from elevations below 200 m.

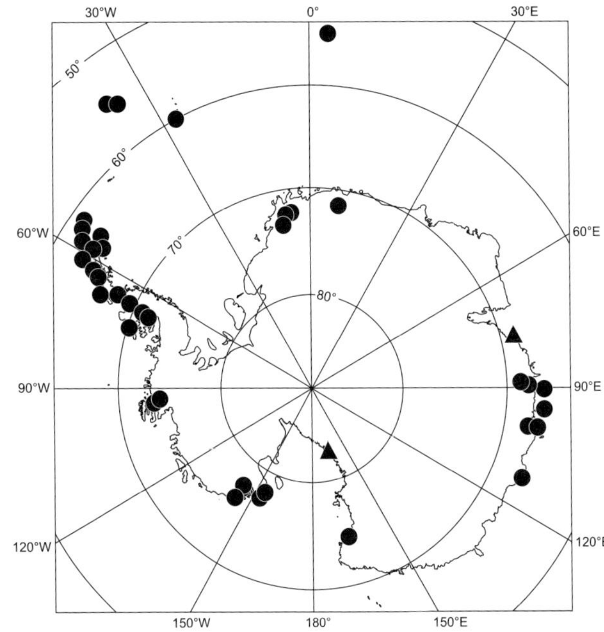

Figure 113. Global distribution of *Schistidium antarctici*. Literature records marked with the triangle.

Selected specimens examined. WEST ANTARCTICA. BOUVETØYA. Kapp Circoncision, *ca* 15 m, *MWH 868B* (AAS, KRAM) and 40 m, *MWH 864B* (AAS, KRAM); Nyrøysa, 20 m, *TE 111* (BG), 25 m, *TE 297* (BG) and 30 m, *TE 363* (BG); Rustadkollen, *ca* 300 m, *TE 231* (BG), 320 m, *TE 31.2* (BG), 323 m, *TE 41.2 & 114* (BG) and 320–340 m, *TE 401.3* (BG).

SOUTH SANDWICH IS. **Bellingshausen I.** W side of main crater, *ca* 130 m, *MWH 431C* (AAS, KRAM).

SOUTH ORKNEY IS. **Coronation I.** Opposite Gosling Is., 10–15 m, *MMM 52* (AAS, KRAM); Meier Pt., 10–20 m, *MMM 26 & 53* (AAS, KRAM) & *RILS 579 & 8026A* (AAS, KRAM); Cape Vik, 100 m, *RILS 558* (AAS, KRAM); Shingle Cove, 20 m, *RILS 450* (AAS, KRAM); Windscoop Buttress east of Sunshine Glacier, *ca* 175 m, *RILS 48, 63A, 64A & 171* (AAS, KRAM); Olivine Pt., 5–15 m, *RILS 114* (AAS, KRAM) and 30 m, *RILS 125* (AAS, KRAM); opposite Saunders Pt., *ca* 8 m, *RILS 36* (AAS, KRAM). **Lynch I.** South side, 35 m, *RILS 7027* (AAS, KRAM). **Powell I.** Promontory west of John Peak, *ca* 8 m, *RILS 254 & 256* (AAS, KRAM); promontory west of Falkland Harbour, *ca* 15m, *RILS 273* (AAS, KRAM); promontory

◄

Figure 112. *Schistidium antarctici*. — **1**. Habit, dry. **2**. Portion of branch, dry. **3**. Portion of cross-section of stem. **4–6**. Axillary hairs. **7–17**. Leaves. **18–19**. Leaf apices. **20**. Upper lamina cells at margin. **21**. Mid-leaf cells. **22**. Lower mid-leaf cells. **23**. Basal cells. **24–31**. Cross-sections of leaves, a sequence from base to apex. **32**. Perichaetium with operculate capsule. **33**. Perichaetium with deoperculate capsule. **34**. Perichaetium with systylious capsule and calyptra. **35–37**. Perichaetial leaves. **38**. Operculate capsule, wet. **39**. Calyptra and operculum with attached columella. **40**. Exothecial cells at base of urn and stoma. **41**. Exothecial cells in middle of urn. **42**. Peristome teeth, spores and rim cells of capsule. — [1, 3–6, 13–14, 28–29, 31, 33–34, 36, 40–42 from *RO 952/80*; 2, 11–12 from *RO 441A/80*; 9–10, 18, 23, 25, 27, 30 from *RO 2452/80*; 15–17, 19–20, 22, 24, 35, 38–39 from *RO 5055/79*; 26 from *RO 4982/79*; 32 from *RO 2789/80*; 7–8, 21, 37 from *CS 425* (lectotype of *Grimmia antarctici*), PC]. — Scale bars: a – 0.5 cm (1) and 100 µm (4–6); b – 1 mm (7–17); c – 100 µm (3, 18–23, 40–41) and 200 µm (42); d – 1 mm (2, 32–34, 38–39); e – 1 mm (35–37) and 100 µm (24–31).

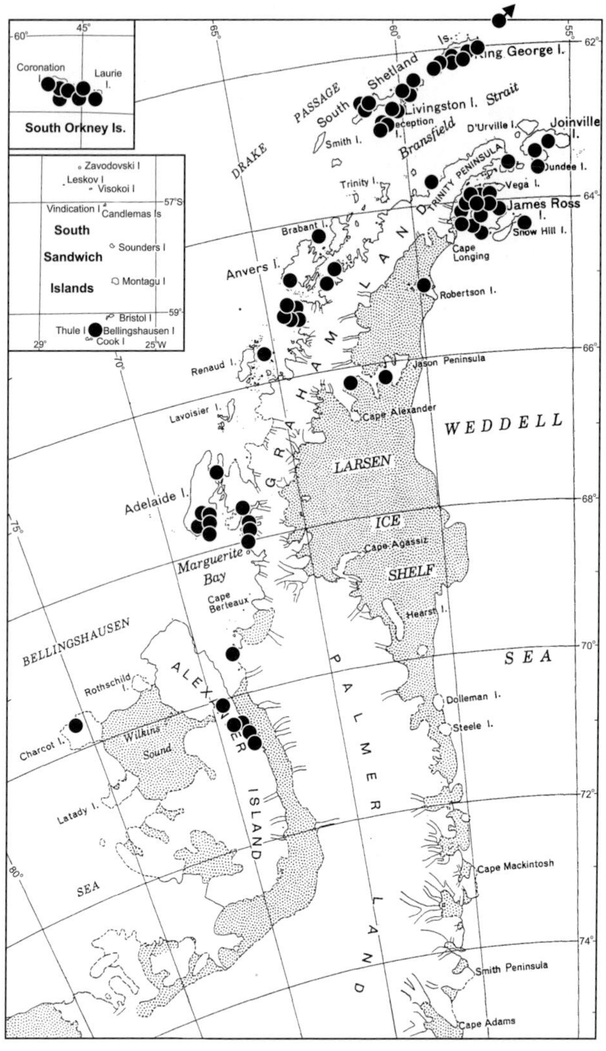

Figure 114. Distribution of *Schistidium antarctici* in the Antarctic Peninsula area. Localities in the northern group of the South Shetland Is. (Elephant I. and Gibbs I.) are indicated by an arrow.

in south-east coast of the island, 130 m, *RILS 207* (AAS, KRAM); north-west coast, 300 m, *RILS 236* (AAS, KRAM). **MATTHEWS I.** The Divide, *ca* 35 m, *RILS 171* (AAS, KRAM). **ROBERTSON IS.** Steepholm I., 10–80 m, *RILS 3248* (AAS, KRAM). **LAURIE I.** Mackenzie Peninsula, 150 m, *Bellisio 11698* (AAS, KRAM); Scotia Bay, 8 Dec 1903, *Rudmose Brown s.n.* (BM, E, PC). **SIGNY I.** Cummings Cove, 150 m, *RILS 8125* (AAS, KRAM); Thulla Pt., *ca* 30 m, *REL 1130* (AAS, KRAM); Foca Pt., 30 m, *RW 44* (AAS, KRAM); ridge opposite Spindrift Rocks, 3–6 m, *RILS 8060* (AAS, KRAM) and 65 m, *REL 1134B* (AAS, KRAM); Stygian Cove, 65 m, *RILS 662* (AAS, KRAM); S end of Three Lakes Valley, *ca* 15 m, *REL 1190* (AAS, KRAM); Jane Peak, 220 m, *RW 31* (AAS, KRAM); Mooring Pt., 15 m, *RW 43* (AAS, KRAM); Knife Pt., 25 m, *RW 116* (AAS, KRAM), 30 m, *RW 47 & 123* (AAS, KRAM) and 130 m, *RW 16, 22A & 30* (AAS, KRAM); Factory Cove, 10 m, *RW 17 & 34* (AAS, KRAM), 50 m, *MWH 152A* (AAS, KRAM) and 65 m, *RW 38* (AAS, KRAM); Paal Harbour, 80 m, *RILS 672* (AAS, KRAM); Cemetery Bay, 50 m, *RILS 5555* (AAS, KRAM); Orwell Glacier, 30 m, *RILS 10823* (AAS, KRAM); Factory Cove, 50 m, *REL 1162* (AAS, KRAM); Observation Bluff, 65 m, *MWH 113* (AAS, KRAM). **MOE I.** Landing Cove, 30–50 m, *RILS 5236* (AAS, KRAM).

SOUTH SHETLAND IS. **ELEPHANT I.** Emma Cove, 138 m, *JSA 141* (AAS, KRAM); 3 km SW of Walker Pt., 92 m, *JSA 223A* (AAS, KRAM); 2 km SW of Endurance Glacier, 152 m, *JSA 25, 31 & 44A* (AAS, KRAM). **CLARENCE I.** 480 m, *Chuter 7 & 8* (AAS, KRAM). **KING GEORGE I.** ***Fildes Peninsula***: Bellinghausen Station, 10 m, *RO 2415/80* and 15 m, *RO 2407/80* (KRAM); Suffield Pt., 50 m, *RO 2444/80* (KRAM). ***Marian Cove.*** North Spit, *BJ 151* (KRAM). ***Barton Peninsula.*** Winship Pt., *BJ 183* (KRAM). ***Potter Peninsula.*** Three Brothers Hill, *BJ 158* (KRAM); Florence Nunatak, 340 m, *RO 1314/80* (KRAM). ***Admiralty Bay***: Red Hill, 100 m, *RO 1124/80* (KRAM); Blue Dyke, 4 m, *RO 1209/80 & RO 1211/80* (KRAM); The Tower, 310 m, *RO 1041/80* (KRAM) and 340 m, *RO 813/80* (KRAM); Creeping Slopes, 60 m, *RO 1380/80* (KRAM); Sphinx Hill, 120 m, *RO 648/80* (KRAM); Rescurs Hills, 90 m, *RO 169/80* (KRAM); Llano Pt., 16 m, *RO 4891/79* (KRAM); Ubocz, 90 m, *RO 5014/79* (KRAM); Italia Valley, 60 m, *RO 54/80* (KRAM); Emerald Pt., 50 m, *RO 820/80* (KRAM); Keller Peninsula, Ore Point, 70 m, *RO 449/80* (KRAM); Stenhouse Bluff, 30 m, *RO 2593/80* (KRAM); Tern Nunatak, 255 m, *RO 2514/80* (KRAM); Mt. Wawel, 80 m, *RO 2172/80* (KRAM); Vauréal Peak, 40 m, *RO 5236/79* (KRAM). ***King George Bay.*** Low Head, *BJ 89* (KRAM); Penguin I., *BJ 72* (KRAM). ***Destruction Bay***: Cape Melville, *BJ 80* (KRAM). **NELSON I.** Rip Pt., 80 m, *DCL 797* (AAS, KRAM); Harmony Cove, *CS 225* (PC, S – lectotype of *Grimmia antarctici*). **ROBERT I.** Coppermine Cove, 15–30 m, *RILS 918* (AAS, KRAM). **GREENWICH I.** Ash Pt., 10 m, *DCL 706* (AAS, KRAM); Spit Pt., 40 m, *DCL 547* (AAS, KRAM). **LIVINGSTON I.** ***Byers Peninsula***: Sayer Nunatak, 35 m, *RILS 5867, 5876A & 5879* (AAS, KRAM); Devils Pt., 100 m, *KR 22A* (AAS, KRAM); Pt. Smellie, 10 m, *KR 32B, 33A & 35D* (AAS, KRAM); Barnard Pt., 100 m, *DCL 46B* (AAS, KRAM); South Beaches, 6 m, *DCL 90B* (AAS, KRAM). ***Hurd Peninsula***: Johnsons Dock, 20 m, *Redon 100A* (AAS, KRAM); SE end of Johnsons Dock, 35 m, *DCL 75* (AAS, KRAM); Miers Bluff, *RMS 69-800* (US). **DECEPTION I.** Baily Head, 75 m, *DM 70C* (AAS, KRAM); Macaroni Pt., 70–100 m, *DM 128* (AAS, KRAM); Cathedral Crags, 80 m, *RILS 76* (AAS, KRAM); Mt. Pond, 500 m, *RILS 11178* (AAS, KRAM); Ronald Hill, 65 m, *BJT 250* (AAS, KRAM); Mt.Pond, 410–525 m, *RILS 6895* (AAS, KRAM); Crimson Hill, 130 m, *RILS 11201 & 11204* (AAS, KRAM); Telefon Ridge, 30–60 m, *RILS 5758B* (AAS, KRAM); Kendall Terrace, *JBK 74A* (AAS, KRAM); Wensleydale Beacon, 9 m, *RILS 5598B* (AAS, KRAM); Stonethrow Ridge, 75–105 m, *RILS 5800A* (AAS, KRAM); Vapour Col, 340 m, *DM 116* (AAS, KRAM); Fumarole Bay, 10 m, *RILS 11310* (AAS, KRAM); Crater Lake, 110 m, *RILS 5728D* (AAS, KRAM); Mt. Kirkwood. 250 m, *RILS 5717A* (AAS, KRAM); South Pt., 155 m, *DM 128* (AAS, KRAM); Collins Pt., 50–70 m, *RILS 6754* (AAS, KRAM).

WEST ANTARCTIC PENINSULA. **DAVIS COAST.** Cape Kjelmann, *CS 448* (PC, S – syntype of *Grimmia antactici*). **DANCO COAST.** Cuverville I., 2–20 M, *RILS 4178 & 4179* (AAS, KRAM). **PALMER ARCHIPELAGO.** ***Melchior Is.***: Omega I., *RILS 4121* (AAS, KRAM). ***Anvers I.***: Norsel Pt., 5–10 m, *RILS 1939* (AAS, KRAM); LitchfieldI., 10–25m, *RILS 1962* (AAS, KRAM); Hermit I., *RILS 885* (AAS, KRAM). **GRAHAM COAST.** Edge Hill, 20 m, *RWMC 654B* (AAS, KRAM) and 35 m, *RWMC 649* (AAS, KRAM); Cape Tuxen, 15 m, *RWMC 677B* (AAS, KRAM) and 20 m, *RILS 10663* (AAS, KRAM) and 30–300 m, *RILS 4911* (AAS, KRAM); Cape Pérez, *Gain 272a* (PC) and 6 m, *RWMC 635C* (AAS, KRAM). **WILHELM ARCHIPELAGO.** ***Lemaire Channel***: Booth I., *Expéd. Charcot 164A & 234C* (PC). ***Penola Strait***: Petermann I., 15 m, *REL 1345A & 1355A* (AAS, KRAM). ***Argentine Is.***: largest island between Irizar I. and Uruguay I., 15 m, *RWMC 433* (AAS, KRAM); Corner Is., 10 m, *RWMC 582* (AAS, KRAM); Galindez I., Stella Creek, 10 m, *RWMC 407A, 464, 487A & 590* (AAS, KRAM) and 35 m, *REL 1320* (AAS, KRAM); Black I., 2–10 m, *RILS 4293* (AAS, KRAM) and 10 m, *RWMC 583* (AAS, KRAM). ***Collins Bay***: Berthelot Is., 35 m, *RWMC 628A* (AAS, KRAM); Darboux I., *RILS 3316, 3318 & 3366B* (AAS, KRAM). ***Grandidier Channel***: Hook I., *RMS 69-510, 512 & 888* (US). **LOUBET COAST.** ***Bennett Is.***: Pfaff I., *JK 173B & 178A* (AAS, KRAM). **FALLIÈRES COAST.** ***Adelaide I.***: Mt. Mangin, 600 m, *JK 150B* (AAS, KRAM); Mt. Gaudry, 330 m, *JK 142C* (AAS, KRAM) and 465 m, *JK 145* (AAS, KRAM); Rothera Pt., 20–30 m, *RILS 1868, 1873& 8878* (AAS, KRAM); Reptile Ridge, 350–500 m, *RILS 1880, 1881, 1884B, 1887, 3422, 3436 & 3459* (AAS, KRAM); Jenny I., *Gain 252B & 257* (PC), 15–50 m, *RILS 4716 & 4754* (AAS, KRAM) and 150 m, *BJT 424A* (AAS, KRAM). ***Léonie Is.***: Lagoon I., 25 m, *RILS 9176* (AAS, KRAM); Léonie I., 15 m, *RILS 8986* (AAS, KRAM), 25 m, *RILS 8935* (AAS, KRAM) and 150–175 m, *RILS 9123, 9124C & 9125* (AAS, KRAM). ***Marguerite Bay***: Blaiklock I., 15–20 m, *RILS 847* (AAS, KRAM); Pourquoi Pas I., 5–20 m, *RILS 4772* (AAS, KRAM); Horseshoe I., *REL 1268 & 1269* (AAS, KRAM) and *RILS 1752* (AAS, KRAM); Camp Pt., 5–15 m, *RILS 4673* (AAS, KRAM); Neny I., *ca* 65 m, *RILS 798* (AAS, KRAM). **RYMILL COAST.** Zonda Towers, *DDWW 204* (AAS, KRAM). **GEORGE VI SOUND.** ***Charcot I.***: central N-facing slope, 85 m, *RILS 10493* (AAS, KRAM) and 150 m, *PC 325* (AAS, KRAM). ***Alexander I.***: Zebra Ridge, 600 m, *RILS 10025 & 10027* (AAS, KRAM); Ablation Pt., *RBH & JJL 48 & 84* (AAS, KRAM); Ablation Valley, 1–3 m, *RILS 9247, 9255A, 9256, 9265 & 9278* (AAS, KRAM) and 75 m, *RILS 9288 & 9289* (AAS, KRAM); Moutonée

Valley, *RBH & JJL 65 & 66* (AAS, KRAM) and 183 m, *RILS 11082B* (AAS, KRAM); Fossil Bluff, 100 m, *BJT 508* (AAS, KRAM), 300–400 m, *RILS 9318B, 9322, 9323B, 9326A, 9346 & 9347B* (AAS, KRAM) and 500–600 m, *RILS 9337B* (AAS, KRAM); Waitabit Cliffs, 235 m, *BJT 516* (AAS, KRAM); Eros Glacier, *BJT 5* (AAS, KRAM).

EAST ANTARCTIC PENINSULA. **Trinity Peninsula.** Hope Bay, between Scar Hills and Mt. Flora, 85 m, *CGB 36* (AAS, KRAM). **Joinville Island group.** ***Joinville I.***: Mt. Alexander, 60–75 m, *RILS 3669B* (AAS, KRAM); Active Sound, 200 m, *RILS 3730* (AAS, KRAM); Paulet I., *CS 449* (FH, LE, PC, S – syntype of *Grimmia antarctici*). **James Ross Island group.** ***Vega I.***: Pastorizo Bay, 50 m, *RILS 7854* (AAS, KRAM); False Island Pt., 25–40 m, *RILS 7841* (AAS, KRAM). ***James Ross I.***: Abernethy Flats, 100 m, *RILS 7624* (AAS, KRAM); Cape Lachman/Bibby Pt., *RILS 5311A* (AAS, KRAM); Brandy Bay, 250 m, *RILS 5385D* (AAS, KRAM); Rink Pt., 500 m, *RILS 5968* (AAS, KRAM); Kotick Pt., 100 m, *RILS 5472A* (AAS, KRAM); St. Martha Cove, 15 m, *RILS 11136B* (AAS, KRAM), 50 m, *RILS 7467* (AAS, KRAM), 75 m, *RILS 7436* (AAS, KRAM), 100 m, *RILS 5377* (AAS, KRAM) and 150 m, *RILS 7532 & 7549* (AAS, KRAM); Terrapin Hill, 85 m, *RILS 7746* (AAS, KRAM) and 175 m, *RILS 7738 & 7739A* (AAS, KRAM); Tumbledown Cliffs, 20 m, *RILS 5487A* (AAS, KRAM), 30 m, *RILS 5483* (AAS, KRAM) and 100 m, *RILS 5491A* (AAS, KRAM); Nygren Pt., 30 m, *RILS 5478A* (AAS, KRAM); Hamilton Pt., 500 m, *RILS 5381A* (AAS, KRAM); James Ross Ice Cap, 905 m, *RILS 5981A* (AAS, KRAM). ***Cockburn I.***: *ca* 400 m, *RILS 7964 & 7965* (AAS, KRAM). **Oscar II Coast.** Seal Nunataks, Castor Nunatak, 150 m, *RILS 6435* (AAS, KRAM); Jason Peninsula, Stratton Inlet, 280 m, *RILS 2737C* (AAS, KRAM); Gulliver Nunatak, 265 m, *PK 12* (AAS, KRAM).

ELLSWORTH LAND. **Eights Coast.** ***Thurston I.***: Mt. Dowling, 500 m, *RILS 5415 & 5417* (AAS, KRAM); Morgan Inlet, 100 m, *RILS 5408* (AAS, KRAM). ***Jones Mountains***: north of Pillsbury Tower, *ca* 200 m, *RILS 5400A & 5401* (AAS, KRAM).

EAST ANTARCTICA. DRONNING MAUD LAND. **Prince Harald Coast.** ***Vestfjella***: lokalitet PL, 1968–1969, *Winsnes s.n.* (KRAM, O); same area, lokalitet 9, 1968–1969, *Winsnes s.n.* (KRAM, O); same area, lokalitet 15, 1968–1969, *Hjelle s.n.* (KRAM, O). **Prinsesse Astrid Kyst.** Novolazarevskaya Station, *Meyer 3* (KRAM, NY). **Prinsesse Ragnhild Kyst.** Sør Rondane, Utsteinen, 1379 m, *Ertz 10220* (KRAM).

QUEEN MARY LAND. **Knox Coast.** Watson Bluff, *Harrison 5 & 7* (BM). ***Davis Sea***: Haswell I., 19 Jan 1956, *Markov s.n.* (KRAM, LE); Mirny Station, *Markov 1* (KRAM, LE) and *Meyer 14* (KRAM, NY). ***Shackleton Ice Shelf***: Chugonov I., 1961, *Konovalov s.n.* (KRAM, LE). ***Bunger Oasis***: Lake Figurnoye, *Lishkin & Zhivago 3* (KRAM, LE).

WILKES LAND. **Budd Coast.** ***Windmill Is.***: Clark Peninsula, Whitney Pt., 25 m, *RILS 6198A & 6203* (AAS, KRAM); NE of Stevenson Cove, 25 m, *RILS 6218, 6219 & 6225* (AAS, KRAM); Shirley I., *RILS 6302* (AAS, KRAM); Browning Peninsula, *GAL 273 & 2773A* (KRAM, NY) and 50 m, *RILS 6347 & 6359* (AAS, KRAM); Bailey Peninsula, *ca* 30 m, *RILS 6241, 6255B & 6257* (AAS, KRAM) and 40 m, *RILS 6157* (AAS, KRAM); Haupt Nunataks, 90–100 m, *RILS 6325* (AAS, KRAM).

VICTORIA LAND. **Borchgrevink Coast.** Harrow Peaks, 75 m, *RILS 9640 & 9642* (AAS, KRAM) and 85 m, *RILS 9876A* (AAS, KRAM).

MARIE BYRD LAND. **Edward VII Peninsula.** ***Rockefeller Mountains***: northern group, *Frazier & Wade 323* (FH); Mt. Franklin, *Fitzimmons 291* (NY); Mt. Helen, *PAS 68 & 100A* (NY), *68A* (FH – type of *Grimmia antarctici* var. *percompacta*) *& 76* (FH, NY) and *Fitzsimmons 48* (NY). **Ruppert Coast.** ***Edsel Ford Ranges***: Mt. Rea, *PAS 2 &98A* (NY); Skua Gull Peak, *PAS 1, 49, 75 & 99* (NY) *& 7* (FH, NY); Marujupu, *Perkins 136B* (NY); Mt. Corey, *PAS 43* (FH, NY); Mt. Donald Woodward, *PAS 97* (NY); Chester Mountains, *PAS 40 & 44* (FH, NY).

10. **Schistidium leptoneurum** Ochyra Fig. 115

Schistidium leptoneurum Ochyra, Cryptogamie, Bryol. 25: 126, *f. 1.* 2004. — Type citation: Antarctica, South Shetland Islands, King George Island: Admiralty Bay, Keller Peninsula, Tyrell Ridge between Mt. Flagstaff and Piasecki Pass, lat. 62°04′S, long. 58°24′30″W, alt. 220 m; on bare earth in crevices of dry andesite rocks, 9 Jan 1980, Ochyra 429/80. [Holotype: 'Herbarium Instituti Botanici Academiae Scientiarum Polonae – Cracoviae Bryophyta. ~~Schistidium antarctici (Card.) Sav.-Lub. et Smirn.~~ Antarctica, South Shetland Islands – King George Island: Admiralty Bay, Keller Peninsula, Tyrell Ridge between Mt Flagstaff and Piasecki Pass, alt. 220 m; in crevices of dry andesite rocks, on bare earth = Schistidium lewis-smithii Ochyra, sp. nov. Jan. 9, 1980 Coll. R. Ochyra No 429/80 Det. R. Ochyra' – KRAM!].

Plants small to very small, forming compact cushions or growing in loose tufts, mostly with a waxy lustre to dull, green, olive-green to yellowish-brown in the upper part, brown to blackish-brown in the lower portion. *Stems* erect, 0.4–1.3 mm tall, slender, irregularly branched, sparsely radiculose at the base or in the lower part buried in soil with brown, smooth, branched rhizoids, in transverse section round, consisting of a large and distinct central strand, a unistratose cortex of relatively large cells with moderately thickened, brown walls and large lumina, rather sharply set off from 3–4 layers of large, hyaline medullary cells with thin to moderately thickened, yellow to yellowish-brownish walls; *axillary hairs* filiform, hyaline throughout, 5–7-celled, composed of elongate cells in the distal part and shorter basal cells or all cells elongate. *Leaves* small, erect and loosely imbricate when dry, erect-spreading when moist, straight to slightly curved, ovate to shortly ovate-lanceolate, 0.8–1.3(–1.6) mm long, 0.25–0.4 mm wide, gradually acuminate, acute, epilose, somewhat decurrent, obtusely keeled to narrowly canaliculate above, broadly concave below; *margins* entire, narrowly recurved on one side up to ¾–⅘ of the leaf length, occasionally higher, plane or narrowly recurved in the mid-leaf portion on the other side, plain or deflexed on one side near the apex, 1–2-stratose in 1 row of cells in the distal portion, unistratose in the proximal part; *costa* percurrent, very narrow, (32–)35–40(–43) µm wide in the upper and median parts, becoming weaker downwards, yellowish-brownish, poorly delimited and nearly concolorous with the laminal cells, in transverse section flattened, smooth but appearing uneven owing to bulging dorsal epidermal cells, subrectangular, scarcely prominent on the dorsal surface to nearly of the same thickness as the lamina, especially below, bistratose throughout, occasionally with additional 1–2 central cells in the upper part; *laminal cells* smooth, unistratose throughout, slightly bulging below, thick-walled, with sinuose walls throughout the lamina or mostly straight-walled above and sinuose-walled in the central and lower parts, occasionally nearly straight-walled throughout, very variable in size and shape, isodiametric to shortly oblong, rounded, rounded-quadrate to elliptical, sometimes transversely oblong, (4–)6–16(–20) µm long, 6–8 µm wide in the upper, central and suprabasal parts; *basal juxtacostal cells* short rectangular, chlorophyllose, thick- and straight-walled, (15–)20–35(–40) µm long, 4–8(–10) µm wide, yellowish, yellowish-hyaline to subhyaline; *basal marginal cells* quadrate to oblong, 10–22 µm long, 5–10 µm wide, with straight, thin to moderately thickened longitudinal walls and markedly incrassate transverse walls, forming a rectangular, subhyaline, 1–3-seriate border composed of up to 12 cells. *Autoicous. Perichaetial leaves* ovate, broadly ovate to oblong-ovate, 1.9–2.5 mm long, 0.8–1.2 mm wide,

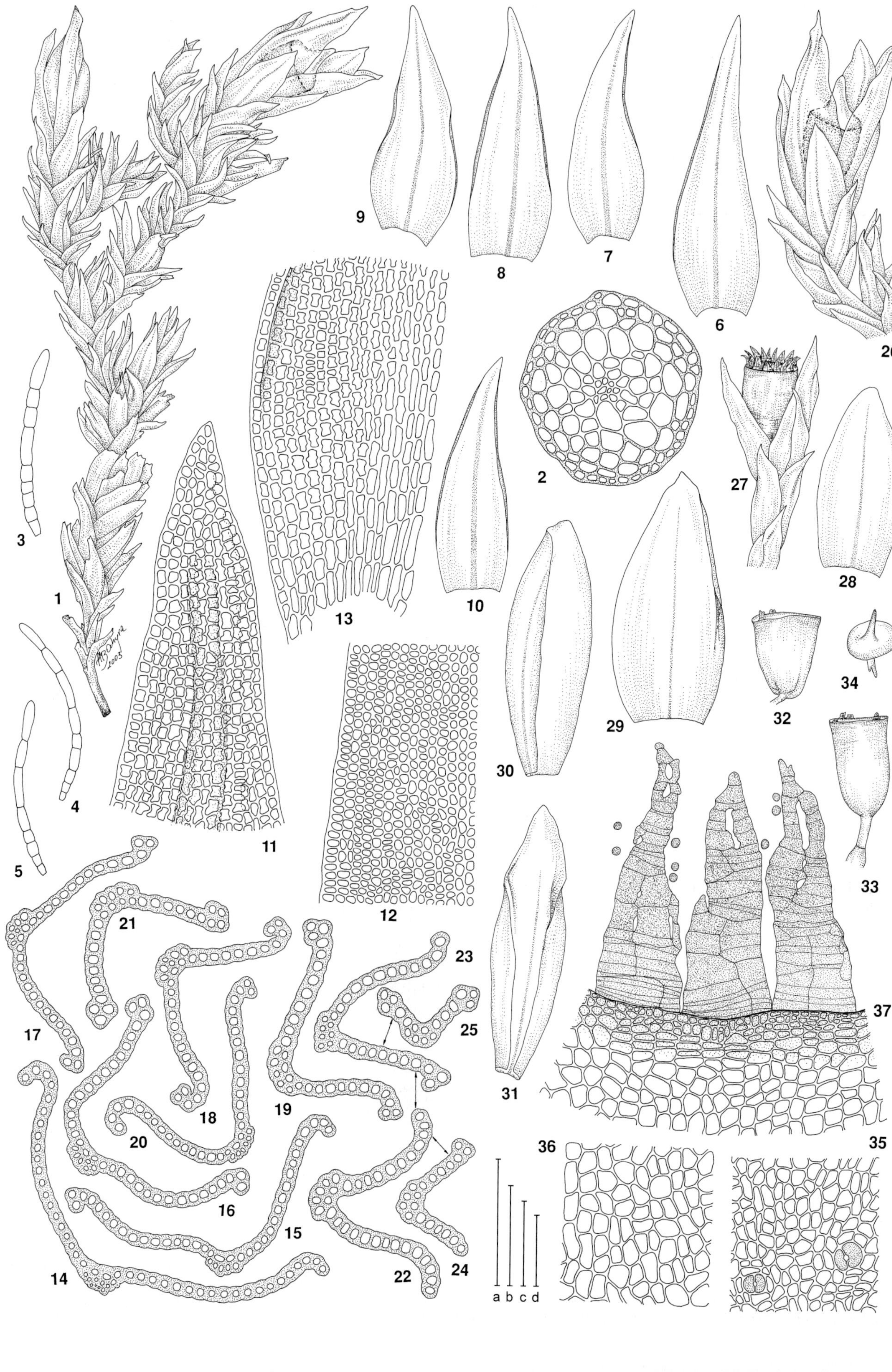
1
2
3
4
5
6
7
8
9
10
11
12
13
14
15
16
17
18
19
20
21
22
23
24
25
26
27
28
29
30
31
32
33
34
35
36
37
a b c d

broadly obtuse to shortly apiculate, epilose or terminated with a short, to 0.25 mm long, flat, decurrent, denticulate hyaline awn when young, straight, concave, with a very narrow costa, nearly concolorous with the laminal cells, ceasing well below the apex and margins plane below and narrowly recurved above. *Setae* stout, 0.35–0.4 mm long, yellowish-brown to pale brown. *Capsules* shallowly immersed, ovoid to obloid, often somewhat campanulate when old, 0.9–1.1 mm long, 0.7–0.8 mm wide, light brown to yellowish-brown, somewhat lustrous, leptodermous; *operculum* distinctly convex, conical, with a straight, acute rostrum, to 0.3 mm long; *exothecial cells* predominantly isodiametric mixed with scattered shortly oblong and transversely oblong cells, (15–)20–40(–55) µm long, 16–38(–45) µm wide, thin-walled but with clear collenchymatous thickenings, becoming oblate to rounded at the orifice in 3–5, orange-brown tiers; *stomata* at the base of the urn, 5–8 per urn, superficial, bicellular, round-pored; *peristome teeth* lanceolate, usually gradually narrowed to a broad and obtuse or narrow and acute apex, orange to reddish-orange, 250–320 µm long, 80–115 µm wide at the base, straight, erect to erecto-patent, becoming recurved with age, variously perforate in the upper part, entire or with somewhat uneven margins, densely striate-papillose throughout. *Spores* globose, pale brownish, smooth to minutely roughened, 10–13 µm in diameter. *Calyptra* very small and narrow, mitriform, split into 4 lobes at the base, extending nearly to the base of the urn rim.

Chromosome number: unknown.

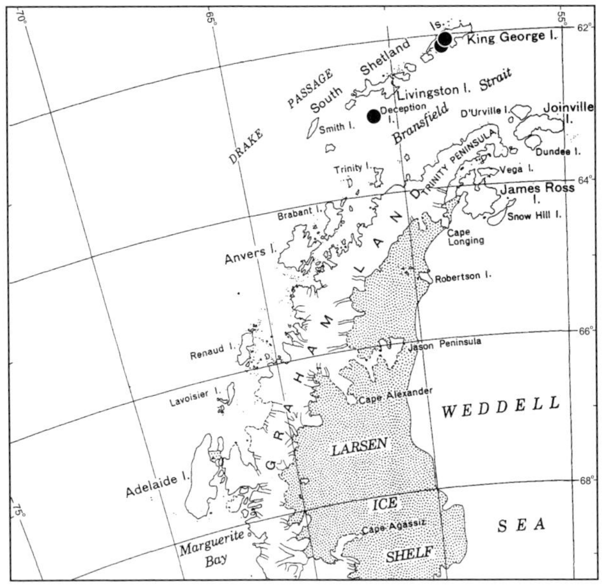

Figure 116. Global distribution of *Schistidium leptoneurum*.

Discussion. Superficially, *Schistidium leptoneurum* is a very inconspicuous species which may be easily overlooked *in situ* and confused with other species, and therefore it is possibly undercollected in the Antarctic. It was only recently described as a new species (Ochyra 2004*b*) and formerly the plants were confused with both *S. antarctici* and *S. amblyophyllum*. It is easily distinguished from the former by having normally developed peristome teeth, and from the latter in its very weak costa which is scarcely prominent on the dorsal surface.

The species is primarily characterised by its peculiar costa which is the weakest one not only amongst regional species but in the entire genus *Schistidium*. On average, the costa is only 35–40 µm wide in the upper part, and only occasionally is it narrower or broader, to 32 or 43 µm, respectively. The costa is entirely bistratose and only sometimes in the proximal part can one or two central cells be observed. It is scarcely prominent dorsally and in the lower part is nearly as thick as the lamina. A similarly narrow costa, 35–45 µm wide, is known only in *S. urnulaceum* which species is, however, immediately distinct in its hoary appearance, whereas *S. leptoneurum* entirely lacks hyaline hair-points or they rarely occur only on the juvenile perichaetial leaves. In addition, the strongly incrassate walls of the exothecial cells, the elongate laminal cells and unistratose leaf margins preclude any confusion of the species. In the perichaetial leaves the costa ends far below the apex and this is also a unique feature in this genus.

Schistidium leptoneurum is the smallest species of all Antarctic members of this genus and its leaves do not exceed 1.3 mm. They are entirely epilose and only some juvenile perichaetial leaves bear short, inconspicuous hyaline hair-points which, however, disappear from the older leaves. In some plants (*RO 2323/80*, KRAM) the upper leaves are whitish, with seemingly hyaline leaf apices but in the Antarctic such 'hyalinisation' of upper leaves is a relatively frequent effect of living in severe climatic conditions and it is observed in species which consistently lack hyaline hair-points, for example in *Hymenoloma antarcticum*.

The species shows some variation in leaf areolation, especially with regard to the sinuosity of the cell walls which is especially distinct on plants from Deception I. On the same shoot there are leaves with markedly sinuose cell walls and with nearly straight walls. Laminal cells with variously sinuose, 'racomitrioid' walls are present in several Antarctic species of *Schistidium*, including *S. urnulaceum*, *S. deceptionense*, *S. cupulare*, *S. halinae* and *S. amblyophyllum*, but they are distinct in having much broader costae which are prominently convex on the dorsal side. The racomitrioid laminal cells occur often in some populations of *S. antarctici* but this species is readily recognised by its 5-ranked leaves and variously reduced peristome teeth which are mostly truncate in the middle. In contrast to the upper and mid-leaf cells, the areolation in the basal part in *S. leptoneurum* is very constant. The short juxtacostal cells and the subhyaline marginal cells with distinctly thickened transverse walls indicate the affinity of the species to *S. antarctici* from which it is distinct in its well developed peristome and unranked leaves.

◄——

Figure 115. *Schistidium leptoneurum*. — **1**. Habit. **2**. Cross-section of stem. **3–5**. Axillary hairs. **6–10**. Leaves. **11**. Leaf apex. **12**. Mid-leaf cells at margin. **13**. Basal cells. **14–25**. Cross-sections of leaves, a sequence from base to apex. **26**. Perichaetium with operculate capsule, dry. **27**. Perichaetium with deoperculate and well developed peristome teeth, wet. **28–31**. Perichaetial leaves. **32**. Old operculate capsule with missing peristome teeth, dry. **33**. Same, wet. **34**. Operculum. **35**. Exothecial cells at base of urn and stomata. **36**. Exothecial cells in middle of urn. **37**. Peristome teeth, spores and rim cells of capsule. — [1–5, 7–11, 13, 23–35 from *RO 429/80* (holotype); 6, 12, 14–22, 36–37 from *RILS 11296A* (paratype)]. — Scale bars: a – 1 µm (2–5, 11–25); b – 1 mm (6–10, 28–31); c – 1 mm (26–27, 32–34) and 100 µm (35–37); d – 1 mm (1). (Reproduced from *Cryptogamie, Bryologie*, 25: p. 127, 2004, with permission.)

Nonetheless, the presence of the basal marginal border clearly indicates the affiliation of *S. leptoneurum* to sect. *Conferta*.

Reproduction in Antarctica. In all known populations of the species fully mature sporophytes are commonly produced in great abundance.

Habitat. A rupicolous species growing on weathered volcanic rocks including a yellowish tephra, as well as on soil in crevices of dry andesite rocks.

World range. Antarctic endemic.

Distribution in Antarctica. A rare and poorly known species, possibly undercollected, known only from a few localities on King George I. and Deception I. in the South Shetland Is. (Fig. 116), occurring at elevations of 50–220 m.

Specimens examined. WEST ANTARCTICA. SOUTH SHETLAND IS. **KING GEORGE I.** ***Admiralty Bay***: Urbanek Crag, 100 m, *RO 2323/100* (KRAM); Keller Peninsula, Tyrell Ridge, 220 m, *RO 429/80* (KRAM – type); without closer locality data, *ca* 210 m, *Bennett 1404 & 1405* (KRAM, LD). **DECEPTION I.** Ronald Hill, 60 m, *RO 2773/80* (KRAM); Telefon Ridge, 210 m, *RILS 5744E* (AAS, KRAM); Fumarole Bay, 50 m, *RILS 11296A & 11298B* (AAS, KRAM).

11. **Schistidium deceptionense** Ochyra, Bednarek-Ochyra & Lewis Smith Fig. 117

Schistidium deceptionense Ochyra, Bednarek-Ochyra & Lewis Smith, The Bryologist 106: 570, *f. 1–35*. 2003. — TYPE CITATION: Antarctica, South Shetland Islands, Deception Island: Fumarole Bay, north side of lagoon, lat. 62°58′ S, long. 60°42′ W, alt. 100–200 m; on coarse yellow tephra cliffs and outcrops, 5 Feb 2002, leg. Lewis Smith 11435. [Holotype: 'British Antarctic Survey Collection R. I. L. Smith No. 11435 Flora of South Shetland Islands Deception Island, Fumarole Bay Schistidium deceptionense Ochyra, Bednarek-Ochyra & Lewis Smith North side of lagoon on coarse yellow tephra rock of cliffs and outcrops. Altitude 100–200 m Date 5/2/2002' – AAS!; isotype: KRAM!].

Plants small, coarse, forming compact cushions, dull to weakly lustrous, olive-green to dark green above, blackish-brown to dark brown below. *Stems* erect, 0.5–1.2 mm long, much irregularly branched, not or sparsely radiculose at the base with brown, somewhat branched, smooth rhizoids, in transverse section round, lacking a central strand and consisting of a relatively thin, 1–2-stratose cortex of small, substereid cells with small lumina and strongly thickened, brown walls surrounding 3–4 layers of large, hyaline medullary cells with moderately thickened walls; *axillary hairs* filiform, 5–6-celled, hyaline throughout, consisting of 1–2 short basal cells and elongate distal cells. *Leaves* loosely imbricate, erect when dry, erecto-patent when moist, not ranked, straight to slightly curved, narrowly lanceolate, gradually tapering to an acute or subacute apex, epilose to piliferous, 1.0–1.7 mm long, 0.3–0.5 mm wide, sharply carinate; *hair-point* absent on the lower leaves, mostly present on the uppermost and perichaetial leaves, to 0.4 mm long, hyaline, pellucid, with well preserved areolation of elongate cells, straight and stiff, occasionally somewhat flexuose, flat below, terete above, distinctly decurrent, distantly bluntly or sharply denticulate throughout to nearly smooth; *margins* entire, mostly plane in the basal part of the leaf, narrowly recurved in the lower ¾–⅘ of the leaf length for varying distances on one or both sides, mostly flat to weakly deflexed near the apex, 2–3-stratose for 1–2 rows of cells in the upper and central part pf the leaf, mostly unistratose with bistratose patches in the proximal part; *costa* percurrent to subpercurrent, smooth, 40–60 µm wide above, becoming weaker towards the base, mostly dark brown and clearly delimited from the laminal cells, in transverse section strongly dorsally convex, terete to elliptical, 2–3-stratose above, mostly bistratose below; *laminal cells* unistratose throughout, only with occasional bistratose patches in the distal part, smooth, with thick and variously sinuose walls, very variable in size and shape, isodiametric to shortly oblong, rounded to angular, often transversely oblong, (3–)5–12 µm long, 5–8(–10) µm wide from the upper to suprabasal part; *basal juxtacostal cells* short rectangular, 20–40 µm long, 6–8 µm wide, mostly yellowish, thick- and straight-walled, chlorophyllose; *basal marginal cells* mostly oblong, 15–25 µm long, 6–8 µm wide, with straight, moderately thickened longitudinal walls and markedly incrassate cross-walls, forming a rectangular, subhyaline, 2–6-seriate alar group. *Autoicous. Perichaetial leaves* lanceolate, ovate-lanceolate to broadly ovate, 1.7–2.0 mm long (without hair-point), 0.6–0.7 mm wide, gradually acuminate, mostly piliferous, erect, usually with incurved apices when dry forming a bud-like, dome-shaped perichaetium, with plane margins in the lower half and variously recurved in the upper half. *Setae* relatively stout, yellow-green becoming brownish with age, 2–3 mm long. *Capsules* deeply immersed, cupulate, 0.5–0.6 mm, as long as wide, not widened at the mouth, stramineous to light yellowish-brown, brown when old, moderately pachydermous; *operculum* slightly convex, short- and straight-beaked; *exothecial cells* oblong in mid-urn, mixed with some isodiametric and irregular cells, (12–)20–35(–40) µm long, (5–)8–15(–18) µm wide, with intensively yellow, evenly, strongly incrassate walls, 3.5–5.5 µm thick, becoming smaller and rounded towards the mouth and forming a bright orange-brown strip of 3–4 tiers of rounded, thick-walled cells the the capsule rim; *stomata* at the extreme base of the urn, 2–4 per urn, superficial, bicellular, round-pored; *peristome teeth* lanceolate, usually gradually tapering to a broad point, 240–270 µm long, 100–110 µm wide at the base, bright orange, straight, erect

⟶

Figure 117. *Schistidium deceptionense.* — **1**. Habit. **2**. Portion of branch, dry. **3**. Cross-section of stem. **4**. Axillary hair. **5–11**. Leaves. **12**. Leaf apex. **13**. Upper laminal cells. **14**. Mid-leaf cells at margin. **15–17**. Basal cells. **18–25**. Cross-sections of leaves, a sequence from base to apex. **26–28**. Perichaetial leaves. **29**. Perichaetium with deoperculate capsule, wet. **30**. Operculate capsule, wet. **31**. Deoperculate capsule, wet. **32**. Calyptra. **33**. Exothecial cells in middle of urn. **34**. Exothecial cells at base of urn and stomata. **35**. Peristome teeth, spores and rim cells of capsule. — [All from *RILS 11435* (isotype)]. — Scale bars: a – 1 mm (1); b – 1 mm (2); c – 1 mm (29–32) and 100 µm (33–35); d – 100 µm (3, 12–25); e – 1 mm (5–11, 26–28); f – 100 µm (4). (Reproduced from *The Bryologist*, 106: p. 571, 2003, with permission.)

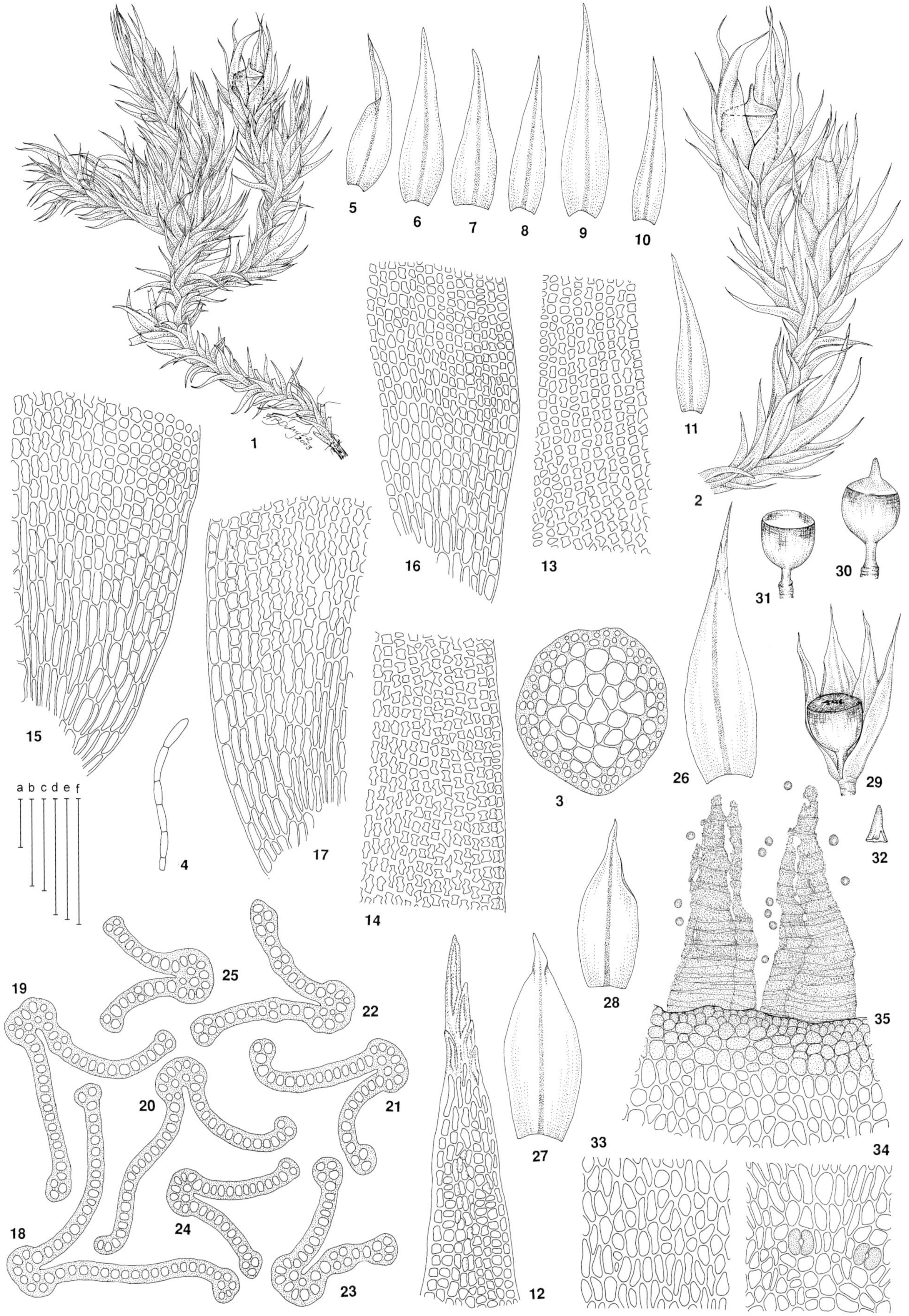
1
2
3
4
5
6
7
8
9
10
11
12
13
14
15
16
17
18
19
20
21
22
23
24
25
26
27
28
29
30
31
32
33
34
35
a b c d e f

to erecto-patent, becoming squarrose with age, strongly perforate with more or less wide cracks along the median line, sometimes with prongs split off at margins or rarely almost entire, mostly asymmetric with uneven margins, densely striate-papillose below, coarsely papillose in the upper part with short, rather distant pebble-like papillae. *Spores* globose, pale brownish, smooth to minutely roughened, 8–11 μm in diameter. *Calyptra* very small and narrow, mitriform, mostly split into 3–4 lobes at the base, extending mostly to the base of the rostrum.

Chromosome number: unknown.

Discussion. Although known only from the single type collection (Ochyra *et al.* 2003*a*), *Schistidium deceptionense* is a distinct species which is distinguished by its odd sporophyte characters. It has the smallest capsules amongst the Antarctic species of the genus, which are deeply immersed in the perichaetial leaves. They are perfectly cupulate and are 0.5–0.6 mm in length and width. The exothecial cells of the urn have strongly incrassate walls, usually 0.35–0.55 μm thick, and this is rather unique in *Schistidium*.

The similarly greatly incrassate walls of the exothecial cells, mostly 4–8 μm thick, are known only in *Schistidium urnulaceum*, another endemic of the maritime Antarctic and Subantarctic South Georgia, and *S. crassithecium* from eastern North America (Allen 2005). *Schistidium deceptionense* and *S. urnulaceum* appear to be closely related and share a number of characters including the lack of a central strand, the thick- and sinuose-walled laminal cells, the presence of a basal marginal border of pellucid cells with markedly thickened transverse walls, the small pachydermous capsules, the small and almost smooth spores and the small calyptrae not extending over the urn. However, *S. deceptionense* differs in having strongly carinate leaves that are narrowly V-shaped, with blades of the lamina in transverse section forming a < 30° angle, usually parallel at the junction with the costa, a stout costa which is semi-terete in transverse section and strongly convex dorsally, bi- or occasionally tristratose leaf margins forming distinct, 1–2-seriate, bulging thickenings in the distal part of the leaf as well as shorter upper and median laminal cells which are generally 5–12 μm long (vs. 8–20 μm long in *S. urnulaceum*). In contrast, *S. urnulaceum* has the leaves concave below and broadly keeled, V- or U-shaped in the distal portion, with the blades of the lamina in transverse section forming a > 45° angle, divergent at base, much weaker costa, 35–45 μm wide above, which is subrectangular in transverse section, flattened and weakly convex on the dorsal side, unistratose leaf margins with only occasional bistratose patches in the upper half and elongate upper and median laminal cells, 8–20 μm long. In addition, *S. urnulaceum* is a hoary plant due to the constant presence of the hyaline hair-points, to 0.5 mm long, on the uppermost vegetative and perichaetial leaves as well as on many lower leaves, whereas in *S. deceptionense* the lower leaves are generally epilose and only some uppermost vegetative leaves and perichaetial leaves are piliferous, with shorter hair-points, to 0.4 mm long.

Schistidium deceptionense and *S. urnulaceum* constitute an isolated group within the genus on account of the odd shape of their exothecial cells. Nonetheless they show the closest affinity to sect. *Conferta*, as defined by Blom (1996) and Ochyra *et al.* (2003*a*), with which they share the pellucid alar region of the hyaline cells with cross-walls much thicker than longitudinal walls, the narrow upper laminal cells, the irregular and mostly strongly perforated peristome teeth; and the yellow reaction of the upper leaves and leaf cells with KOH. It is worth noting that *S. confertum*, the type species of the section, often has thickened walls to the exothecial cells, although not so much as those in *S. deceptionense* and *S. urnulaceum*.

Reproduction in Antarctica. Sporophytes are produced in great profusion.

Habitat. A saxicolous moss growing in a dry situation on coarse yellow tephra rock of cliffs and rock outcrops.

World range. Endemic to Antarctica.

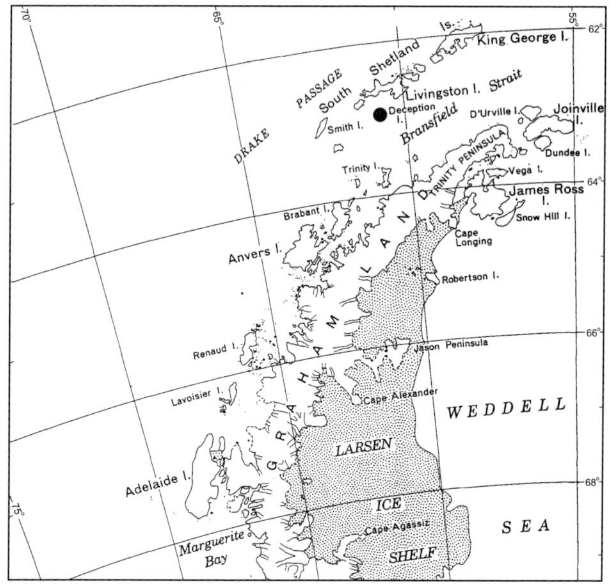

Figure 118. Global distribution of *Schistidium deceptionense*.

Distribution in Antarctica. Known only from a single collection on the volcanic Deception I. in the South Shetland Is. at lat. 62°58′S (Fig. 118).

Specimen examined. WEST ANTARCTICA. SOUTH SHETLAND IS. **DECEPTION I.** Fumarole Bay, 100–200 m, *RILS 11435* (AAS, KRAM).

12. **Schistidium amblyophyllum** (Müll.Hal.) Ochyra & Hertel Fig. 119

Schistidium amblyophyllum (Müll.Hal.) Ochyra & Hertel, Polish Bot. Stud. 1: 26. 1990. ≡ *Grimmia amblyophylla* Müll.Hal., Syn. Musc. Frond. 1: 779. 1849. — TYPE CITATION: Insula Eremitae et Kerguelen ad Cap Horn [sic!]: J. D. Hooker. [Lectotype (*vide* Ochyra & Hertel 1991: p. 26): 'No. 122. Grimmia apocarpa Linn. var. 3. Hermite Island, Cape Horn. Antarct. Exp. 1839–1843. J. D. H.' – BM-Hooker!; isotype: E!, NY-Mitten!; syntype: 'No. 246. Grimmia apocarpa var. Kerguelen's Land 730. Antarct. Exp. 1839–1843. J.D.H.' – BM-Hooker!; isosyntypes: E!, NY-Mitten!].

Grimmia angustifolia Mitt., J. Linn. Soc. Bot. 12: 97. 1869. ≡ *Schistidium angustifolium* (Mitt.) Herzog, Biblioth. Bot. 87: 52. 1916. — TYPE CITATION: Andes Quitenses, in summo monte Pichincha, Jameson. [Lectotype (*selected here*) {or perhaps holotype}: 'Grimmia angustifolia m. rocks on the summit of Pichincha Jameson 92' – NY-Mitten!]. First synonymised by Ochyra (1998*a*: p. 163).

Grimmia stylostegia Müll.Hal., Bot. Jahrb. Syst. 5: 80. 1883. ≡ *Schistidium stylostegium* (Müll.Hal.) Zanten, Marion and Prince Edward Islands: 193. 1971, *non rite publ. basion. non citat.* — TYPE CITATION: all species described by Müller (1883) are included in the section 'I. Bryologia Kerguelensis'; later Müller (1889) wrote: 'Ins. Kerguelen, Foundery branch, in rupibus siccis (11.1874)'. [Lectotype (*selected here*): 'Grimmia stylostegia C. Müller n. sp. Kerguelen, Foundery branch Nov 1874, C. Naumann ex Herb. Prahl' – HBG!; isotypes: BM-Bescherelle!, H-Brotherus!, PC!]. First synonymised by Ochyra and Hertel (1990: p. 26).

Grimmia hyalinocuspidata Müll.Hal. *in* Neumayer, Int. Polarforsch. Deutsch. Exp. 2: 315. 1890. ≡ *Schistidium hyalinocuspidatum* (Müll.Hal.) B.G.Bell, Br. Antarct. Surv. Bull. 63: 90. 1984. — TYPE CITATION: Austro-Georgia, ad rupes des Köppenberges 19. Majo 1883; Südwest-Gletscherthal, 7. Majo 1883. [Lectotype (*selected here*): 'Dryptodon hyalino-cuspidatus C. Müll. n. sp. Austro-Georgien, Köppenberg an Felsen, 9.vi.83 leg. Will Südpolarexpedition' – HBG!; syntype: 'Dryptodon hyalino-cuspidatus C. Müll. Austro-Georgien, SW Gletscherthal, 9/v.83 leg. Will Südpolarexpedition' – HBG!]. First synonymised by Ochyra (1998*a*: p. 163).

Schistidium subpraemorsum Broth. *in* Herzog, Biblioth. Bot. 87: 53. 1916, ***syn. nov.*** — TYPE CITATION: Oberes Llavetal gene Tunarisee, alt. 4300 m, No. 4867. [Lectotype (*selected here*): 'Plantae in itinero secundo per Bolivian lectae. No. 4867. Grimmia subpraemorsa Broth. n. sp. (Schistidium) An Felsen in einer Bachrinne: oberes Llavetal ca. 4300 m, leg. Th. Herzog Mai 1911' – H-Brotherus!; isotype: JE-Herzog!].

Grimmia hyalinocuspidata var. *mutica* Cardot & Broth., Kongl. Svenska Vet. Ak. Handl. 63(10): 26. 1923, ***syn. nov.*** — TYPE CITATION: Patagonia, Punta Arenas, Rio de las Minas, ad saxa (368). [Lectotype (*selected here*): 'Expeditio suecica 1907–1909. 368. Grimmia hyalinocuspidata C.M. forma Patag. austr.: Punta Arenas, Rio de las Minas, pierres. 16.2.1908 T. Halle et C. Skottsberg' – PC-Cardot!; isotypes: BM!, S!].

Grimmia donatii Herzog & Thér. *in* Herzog, Rep. Spec. Nov. Regni Veg. 41: 290, *f. 236e–n*. 1937 ['donati']. ≡ *Schistidium donatii* (Herzog & Thér.) Ochyra & Matteri, Fragm. Florist. Geobot. 41: 1008. 1996, ***syn. nov.*** — TYPE CITATION: Territoria Aisen. Regio Lago San Martin. Ventisquero, L. 3, ca. 1500 m, leg. A. Donat, 13.V.33 (no. 48). [Lectotype (*selected here*): 'Dr. A. Donat Plantae Patagonicae. No.: 48 Grimmia (Schistidium) Donati Herz. et Thér. n. sp. Terr. Aisen. Reg.: L. S. M. Ventisqueros, L. 3 Alt. ca 1500 m 13.II.33 det.Thériot, X.1936' – JE-Herzog!; isotype: PC-Thériot!].

Plants small, medium-sized to relatively large, dull to weakly glistening, densely caespitose, olivaceous, yellow-, brownish or sordid-green in the upper part, brown to blackish-brown in the lower part, sometimes blackish throughout. *Stems* erect, 0.5–2.0 cm tall, much branched, in cross-section rounded, consisting of a distinct, (1–)2–3-stratose cortex of small, brown, stereid or substereid cells with strongly incrassate walls and small lumina, rather sharply set off from 3–4 rows of large, hyaline medullary cells with moderately thickened walls, central strand absent; *rhizoids* few to many in the lower part of the stem, short or long, brown, branched, smooth; *axillary hairs* filiform, 5–9-celled, hyaline throughout, comprising 2–4 short basal and elongate distal cells. *Leaves* crowded, erect and imbricate to erecto-patent when dry, erect-spreading when wet, not ranked, straight to slightly curved, narrowly lanceolate or oblong- to ovate-lanceolate, (1.1–)1.6–2.1(–2.5) mm long, 0.3–0.7 mm wide, not or short-decurrent, carinate above, narrowly canaliculate below, gradually long-acuminate, acute, muticous or the uppermost leaves piliferous; *hair-point* absent or short, to 0.4 mm, exceptionally to 0.6 mm long, stiff, straight, flattened, bluntly denticulate, yellowish, yellowish-brownish or hyaline to yellowish-hyaline, decurrent and occasionally making up the greater part of the lamina, usually with well preserved areolation of elongate cells; *margins* entire, recurved on both sides in the lower ⅚ of the leaf length, plane somewhat below the apex, in the basal part unistratose with occasional bistratose spots, usually recurved on one side and plane on the other side, in the upper and central parts mostly bistratose for 1–2 rows with unistratose spots or sometimes entirely unistratose throughout; *costa* smooth, narrow and relatively weak, 40–60 μm wide above becoming thinner towards the base, ceasing in the hair-point, distinctly convex dorsally, except in the basal part, subrectangular to semi-terete in transverse section, 2–3-stratose above, tristratose or occasionally bistratose in the lower part, with 2 somewhat larger ventral epidermal cells; *laminal cells* unistratose throughout, smooth, with strongly thickened, sinuose to strongly sinuose-nodulose, yellowish to yellow-whitish walls, shortly oblong to rectangular in the distal and median parts, 10–25 μm long, 5–10 μm wide, becoming long-rectangular below, 15–30(–40) μm long, 5–6 μm wide; *basal juxtacostal cells* long-rectangular to linear, 35–90 μm long, 4–7 μm wide, with strongly thickened, straight walls, yellowish or yellowish-hyaline, forming a large, prominent, sharply delimited group, separated from the margin by a group of shorter cells with strongly sinuose walls, often extending to the leaf insertion; *basal marginal cells* hyaline, subquadrate to oblong, translucent, with straight walls, with thickened cross-walls, forming ± distinct, 1(2–3)-seriate marginal border composed of 10–25 cells. *Autoicous*. *Perichaetial leaves* larger than the vegetative leaves, erect to erecto-patent, straight and rather stiff, concave, oblong- to broadly ovate-lanceolate, 3.0–3.5 mm long, 0.8–1.2 mm wide, shortly acuminate, acute to rarely broadly obtuse, recurved on both sides in the upper half, epilose or terminated with a hyaline, not or weakly decurrent, slightly denticulate hair-point, to 0.4 mm long, with an areolation of strongly sinuose cells in the distal part and long-rectangular, thin-walled in the proximal part. *Setae* short, 0.2–0.4 mm long, yellow to light brown, straight. *Capsules* deeply immersed in the perichaetial leaves, stramineous to light brownish, 0.7–0.9(–1.1) mm long, 0.5–0.7 mm wide, smooth, ovoid to obloid, usually cyathiform or campanulate when dry and empty; *operculum* conic, with a straight, erect to slanted beak; *exothecial cells* in mixed, irregular patches of quadrate, rounded-quadrate to oblong, up to 20–40 μm long, 15–25 μm wide, thin-walled but usually with small collenchymatous corner thickenings, becoming oblate to quadrate, orange-brownish pigmented in 5–7 tiers at the orifice; *stomata* 5–8 per urn at its extreme base, bicellular, round-pored; *peristome teeth* straight and erect, 240–360 μm long, 100–120 μm wide at the base, orange-brown to orange-red, erect-spreading to recurved with age, lanceolate, tapering to a rather broad obtuse apex, sometimes with uneven margins, entire or with narrow submarginal perforations, sometimes cleft at the apex, finely warty papillose. *Spores* globose, pale brownish, smooth to minutely roughened, 9–11 µm in diameter. *Calyptra* small, subcucullate, irregularly split at the base.

Chromosome number: unknown.

Discussion. *Schistidium amblyophyllum* is a distinct and easily defined species of sect. *Conferta* which is recognised amongst Antarctic species of this genus by the following combination of characters: (*1*) a very characteristic leaf areolation of elongate, racomitrioid cells with strongly incrassate and sinuose to sinuose-nodulose walls; (*2*) a large group of yellow to yellowish-hyaline, rectangular to linear basal cells forming prominent juxtacostal 'windows' separated from the margins by distinctly sinuose-walled cells; and (*3*) a well developed basal marginal border composed of cells with markedly thickened cross-walls; (*4*) well developed peristome teeth. The presence of strongly sinuose-walled cells in this species may cause confusion with species of *Bucklandiella* but the short compact growth form and normally present immersed capsules contrast with the more robust growth forms and exserted capsules of taxa in that genus.

Schistidium amblyophyllum is closely related to *S. cupulare* and the distinguishing characters between these species have been thoroughly discussed by Ochyra (2004*b*). The species are most safely distinguished by their leaf areolation. The leaf cells of *S. amblyophyllum* are elongate throughout the lamina, whereas in *S. cupulare* mid-leaf and distal cells are short, mostly isodiametric, irregularly quadrate to, rarely, short-rectangular and sometimes oblate near the apex, with straight or weakly sinuate, but not distinctly sinuose or nodulose walls. The basal cells are long rectangular throughout the base, becoming shorter towards the margins, with straight, thick or sometimes thinner walls which make the basal areolation lax. They are usually yellowish-hyaline and sharply set off from the supra-basal cells which are intensely yellow, long-rectangular and weakly to strongly sinuose-walled. In addition, the costa in *S. amblyophyllum* is weaker, 40–60 µm wide in upper and median parts, subrectangular to semi-terete in transverse section, 2–3-stratose above, tristratose or occasionally bistratose in the lower part, with 2 somewhat larger ventral epidermal cells. Conversely, in *S. cupulare* the costa is stronger, 50–75 µm wide in the distal and median part, semi-terete to subrectangular in transverse section, (2–)3-stratose above, occasionally also 4-stratose below, strongly convex on the dorsal surface. The leaf margins are unistratose throughout or variously bistratose in one row of cells in *S. amblyophyllum*, whereas those in *S. cupulare* are usually regularly bi- or occasionally tristratose in (1–)2(–3) rows of cells forming fleshy, bulging limbidia. Moreover, *S. amblyophyllum* lacks the central duct and its peristome teeth are finely warty papillose, whereas *S. cupulare* has a distinct and usually large central strand and smooth or only finely roughened peristome teeth.

Schistidium amblyophyllum is a variable species. Many Antarctic populations lack hair-points to the leaves and in piliferous plants they are short, not exceeding 0.4 mm, and are confined to the uppermost, mostly perichaetial leaves. The hyaline hair-point is occasionally to 0.6 mm long, for example in the plants from the Eklund Is. near the English Coast in George VI Sound (*RILS 5531*, AAS, KRAM), and it makes up the greater part of the distal portion of the lamina. The local populations of the species are also quite variable with regard to the structure of the costa and the leaf margin. Typically, the costa is 2–3-stratose in the central and lower parts but sometimes it is entirely bistratose throughout. Likewise, in some plants the leaf margins are unistratose throughout, but normally they are bistratose in the distal half with occasional unistratose patches.

The species forms various modifications which are related to habitat. Plants growing in dry and exposed conditions are shorter, have smaller vegetative leaves and ovate-lanceolate, acuminate perichaetial leaves with hyaline hair-point shorter than the hair-point in less demanding niches, whereas plants growing in moist and sheltered conditions are more robust, with larger vegetative leaves and large, mostly ovate, broadly obtuse and epilose perichaetial leaves. On the other hand, there is little variation in sporophyte characters in *S. amblyophyllum*, especially in the size of the capsule, peristome teeth, exothecial cells and spore size. Nonetheless, the broad range of morphological variation exhibited throughout its relatively large geographical area of occurrence resulted in the recognition of several phenotypes as species in their own rights. Some of them, including *Grimmia angustifolia* from Ecuador (Mitten 1869), *G. stylostegia* from Îles Kerguelen (Müller 1883, 1889), *G. hyalinocuspidata* from South Georgia (Müller 1890*a*) and its variety from southern Patagonia var. *mutica* (Cardot & Brotherus 1923), *G. donatii* from western Patagonia (Herzog 1937) and *Schistidium subpraemorsum* from Bolivia (Herzog 1916) are here considered to be identical to *S. amblyophyllum*. In addition, *Grimmia perichaetialis* from the Ruwenzori Mountains and Mt. Kenya in East Africa (Potier de la Varde 1955) also belong to this species, considerably extending its geographical range (Ochyra, unpublished).

Laminal cells with variously sinuose walls are present in several Antarctic species of *Schistidium*. *Schistidium urnulaceum* and *S. deceptionense* have a more or less similar leaf areolation to that in *S. amblyophyllum* but they are distinct in having exothecial cells with strongly incrassate walls. In addition, the former has a hair-point on almost all leaves and is usually a hoary plant, whereas in the latter the supra-basal cells are shorter, irregularly quadrate to shortly rectangular. Of the other two species with sinuose laminal cell walls, *S. halinae* is a strongly hoary plant and in *S. leptoneurum* the costa is exceptionally weak and narrow and these characters immediately separate them from *S. amblyophyllum*. Racomitrioid laminal cells often occur in some populations of *S. antarctici* but this species is recognised by its 5-ranked leaves and very small basal juxtacostal groups of short oblong cells.

Schistidium amblyophyllum was originally described by Müller (1849) as *Grimmia amblyophylla* from material collected by J. D. Hooker from Hermite I. in the Tierra del Fuego archipelago and Îles Kerguelen during the British Antarctic voyage of 1839–1843 and reported as an unnamed variety of *Grimmia apocarpa* (Wilson & Hooker 1847). The species was subsequently given a name under *Schistidium* by Ochyra & Hertel (1990). It was first reported from the Antarctic from King George Island,

→

Figure 119. *Schistidium amblyophyllum*. — **1**. Habit. **2**. Perichaetium with operculate capsule. **3**. Cross-section of stem. **4–6**. Axillary hairs. **7–12**. Leaves. **13**. Leaf apex. **14**. Upper laminal cells at margin. **15**. Mid-leaf cells. **16**. Alar cells. **17**. Basal cells. **18–25**. Cross-sections of leaves, a sequence from base to apex. **26–27**. Perichaetial leaves. **28–30**. Deoperculate capsules. **31**. Operculum with attached columella. **32**. Exothecial cells at base of urn and stoma. **33**. Exothecial cells in middle of urn. **34**. Peristome teeth, spores and rim cells of capsule. — [1–2 from *RILS 5597A*; 9–10, 15 from *RILS 5674*; 28–30 from *RILS 6917*; 34 from *RILS 5669*; 3–6, 11–14, 16–27, 31–33 from *RO 913/80*; 7–8 from *DM 42* (7–8)]. — Scale bars: a – 100 µm (4–6); b – 1 mm (26–27); c – 1 mm (2, 28–30) and 100 µm (13–17, 32–33); d – 100 µm (18–25); e – 100 µm (3) and 200 µm (34); f – 1 mm (31) and 2 mm (1); g – 1 mm (7–12).

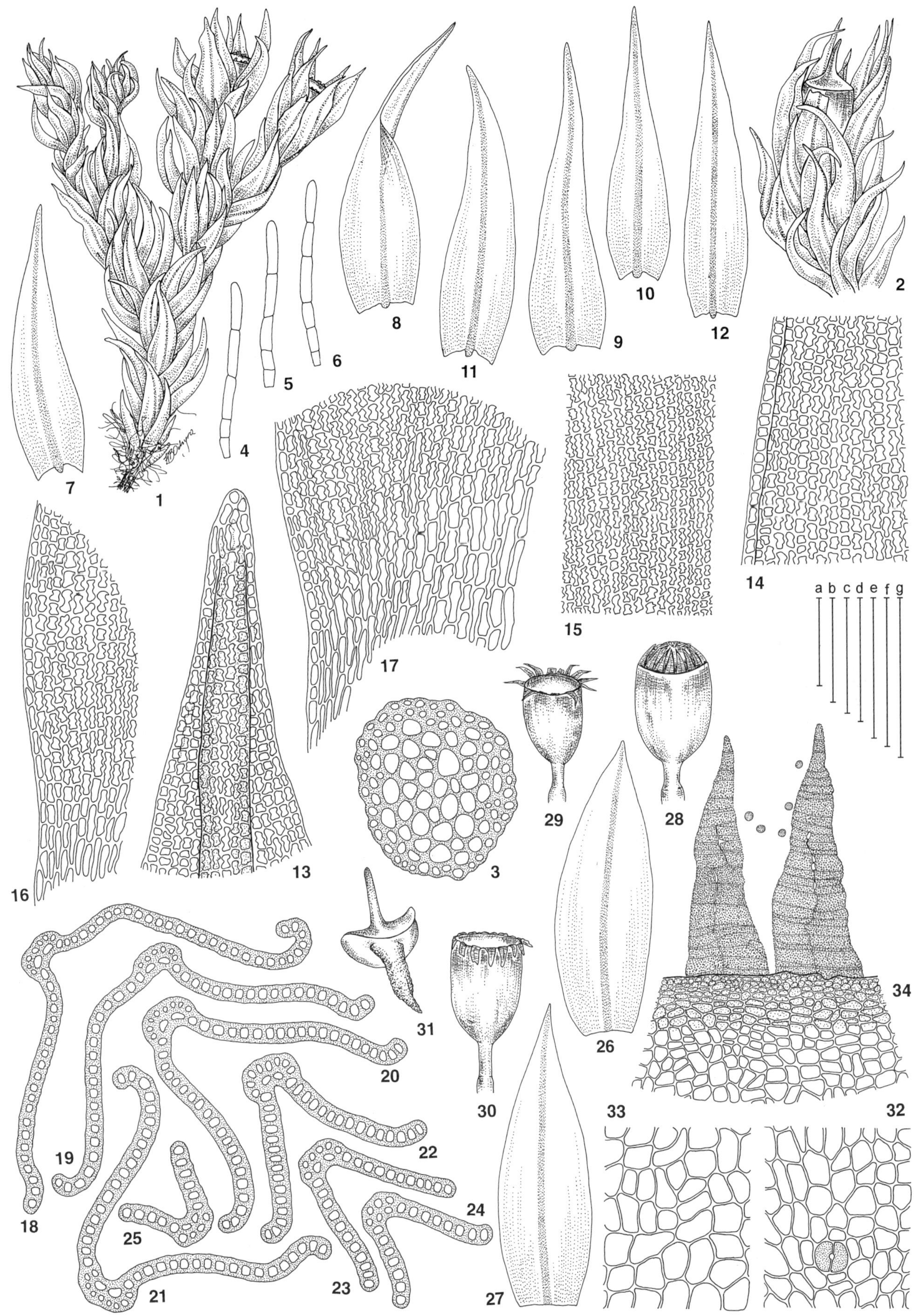
1
2
3
4
5
6
7
8
9
10
11
12
13
14
15
16
17
18
19
20
21
22
23
24
25
26
27
28
29
30
31
32
33
34
a b c d e f g

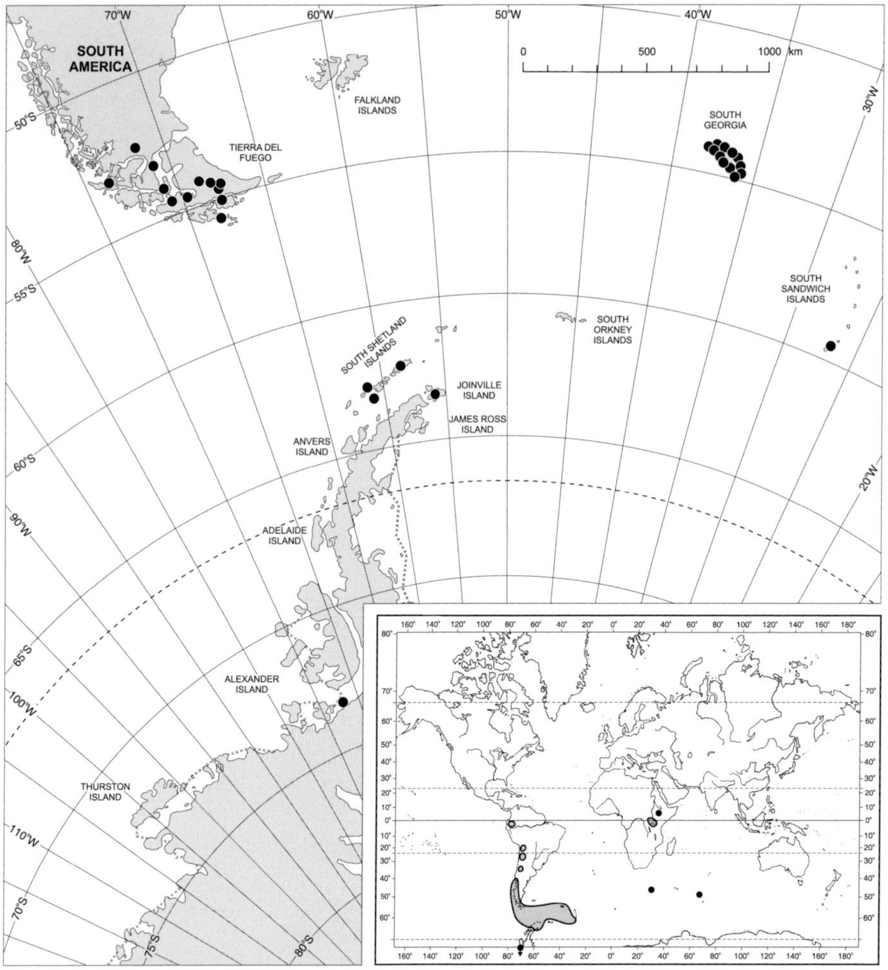

Figure 120. Distribution of *Schistidium amblyophyllum* in the Fuegian region and Antarctica. Inset: global distribution.

South Shetland Is., as *S. angustifolium* (Ochyra & Afonina 1986) with which it is here considered to be conspecific.

Schistidium amblyophyllum is a bipolar counterpart of *S. frigidum*, a widespread arctic-alpine species in the Northern Hemisphere which was only recently described as a separate species (Blom 1996). Both species share a number of gametophyte and sporophyte characters and their mutual relationships need careful assessment, especially because *S. frigidum* is a very variable species. It has distinctly imbricate leaves arranged in spiral rows which are mostly longipilose, especially in the Arctic, whereas in epilose or brevipilose ecads the laminal cells are more thin-walled and less sinuose. In addition, the peristome teeth are smooth or finely papillose in the basal part and the urn is obloid to obloid-cylindrical with the length/width ratio (1.1–)1.3–1.5(–1.7). However, in var. *haavaasii*, rarely found in Norway and Britain, the hair-point is absent or short, to 0.4 mm, the laminal cells have strongly incrassate and sinuose walls, the perichaetial leaves are widely ovate-lanceolate, shortly pointed and deeply concave and the capsule is widely cyathiform, obovoid to shortly obloid and nearly as long as wide. In these characters this variety is very close to *S. amblyophyllum*.

Reproduction in Antarctica. Commonly fruiting.

Habitat. A saxicolous moss growing in dry and exposed situations on andesite and basalt, and soil on rock ledges, in rock crevices of sedimentary rocks, on volcanic ash and scoria, sometimes close to fumaroles but on unheated ground and occasionally, on Bellingshausen I., on moist volcanic rocks emitting steam from fissures. Generally, it is a component of various associations in the fruticose lichen and moss cushion subformation dominated by *Usnea antarctica*, *Bartramia patens*, *Andreaea* spp., *Ditrichum hyalinum*, *Schistidium antarctici* and *Hymenoloma antarcticum*.

World range. An amphiatlantic south-temperate species, widespread in southern South America and extending to the subantarctic islands of South Georgia, Prince Edward Is. and Îles Kerguelen and southwards to the maritime Antarctic and northwards to the Valdivian region of Chile, then recurring at high elevations in the Andes of northern Argentina, Bolivia and Ecuador and appearing at high elevations in the East African mountains including the Ruwenzori and Kilimanjaro Mountains and Mt. Kenya (Fig. 120, inset).

Distribution in Antarctica. A very rare and occasional, though locally quite abundant species, known from the South Shetland Is., including King George I., Livingston I. and Deception I. and from a highly isolated and disjunct station on the Eklund Is. at Ronne Entrance to George VI Sound at lat. 73°14′ S where it reaches its southernmost occurrence. Additionally, it has been found on Joinville I. at the northernmost tip of the Antarctic Peninsula (Fig. 120) and on Bellingshausen I. in the South Sandwich Is. It grows at various elevations, from 40 to 500 m.

Selected specimens examined. WEST ANTARCTICA. SOUTH SANDWICH IS. **BELLINGSHAUSEN I.** Main crater, *ca* 135 m, *MWH 430C* (AAS, KRAM).

SOUTH SHETLAND IS. **KING GEORGE I.** ***Admiralty Bay***: The Tower, 360 m, *RO 1026/80* (KRAM); Italia Valley, 80 m, *RO 87/80* (KRAM);

Breccia Crag, 100 m, *RO 913/80* (KRAM); Shark Fin, 210 m, *RO 2634/80* (KRAM); Precious Peaks, 120 m, *RO 2660/80* (KRAM). **LIVINGSTON I.** ***Byers Peninsula***: between Smellie Pt. and Chester Cove, 35–50 m, *RILS 3782* (AAS, KRAM). **DECEPTION I.** Fumarole area N of Mt. Pond, 500 m, *DM 42 & RILS 11172A & 11180* (AAS, KRAM); hill 1.5 km NNW of Mt. Pond, 410–525 m, *RILS 6899 & 6902B* (AAS, KRAM); Mt. Pond, 500 m, *RILS 11174 & 11177* (AAS, KRAM); Pendulum Cove, east of Crimson Hill, *RILS 11205B* (AAS, KRAM); southern end of Fumarole Bay, *ca* 125 m, *RILS 11279C & 11288* (AAS, KRAM); ridge above Fumarole Bay, opposite Argentine Base, 100 m, *RILS 5669 & 5674* (AAS, KRAM); Lake Irizar at SW Fumarole Bay, 100–125 m, *RILS 6910, 6913 & 6917* (AAS, KRAM).

WEST ANTARCTIC PENINSULA. **ENGLISH COAST.** Eklund Is., 240 m, *RILS 5531* (AAS, KRAM).

EAST ANTARCTIC PENINSULA. **JOINVILLE ISLAND GROUP.** ***Joinville I.***: Mt. Alexander, 60–75 m, *RILS 3669A* (AAS, KRAM).

13. **Schistidium cupulare** (Müll.Hal.) Ochyra

Fig. 121

Schistidium cupulare (Müll.Hal.) Ochyra, Fragm. Florist. Geobot. 43: 105. 1998. ≡ *Grimmia cupularis* Müll.Hal., Bot. Jahrb. Syst. 5: 80. 1883. — TYPE CITATION: all species described as new by Müller (1883) are included in the section 'I. Bryologia Kerguelensis'; later Müller (1889) wrote: 'Ins. Kerguelen, Foundery branch, in rupibus siccis (11.1874) cum Andreaeis'. [Lectotype (*vide* Ochyra 1998*a*: 169): 'Ex Museo botanico Berolinensi. Grimmia cupularis C. Müll. n. sp. Kerguelenland. Foundery branch in rupib. siccis Nov. 1874 Dr. Naumann' – S-Dusén!; isotypes: H-Brotherus!, PC-Cardot!, PC-Thériot!].

Grimmia celata Cardot, Bull. Herb. Boissier Sér. 2, 6: 22. 1906. ≡ *Schistidium celatum* (Cardot) B.G.Bell, Brit. Antarct. Surv. Bull. 63: 86. 1984. — TYPE CITATION: [South Georgia] Cumberland bay: connecting valley between S. and W. fiords. [Lectotype (*selected here*): 'Svenska Sydpolarexpeditionen 1901–03. Ser. N:r 318 in parte. Grimmia celata Card. sp. nova Géorgie du Sud: baie Cumberland, vallée entre les fjords S. et O. D. 21./5 1902 Carl Skottsberg. Det. J. Cardot' – PC-Cardot!; isotypes: S!, S-Roth/Möller!]. First synonymised by Ochyra (2004*c*: p. 125).

Plants small to medium-sized, forming short, rigid, compact cushions, not or somewhat lustrous, yellow- or olive-green above, brown to blackish-brown below, sometimes blackish throughout. *Stems* erect, 0.5–2.0 cm tall, copiously dichotomously branched, in transverse section rounded, consisting of 2–3 layers of small, brown, stereid or substereid cortical cells with strongly thickened walls and small lumina, rather sharply demarcated from 3–4 rows of large, hyaline to yellowish-hyaline medullary cells with moderately thickened walls, central strand present, usually large; *rhizoids* usually copious at the base of the stem, long, brown, branched, smooth; *axillary hairs* filiform, composed of 6–9 cells, hyaline throughout, composed of 2–4 short basal and elongate distal cells. *Leaves* densely set, closely imbricate and erect on drying, erecto-patent on wetting, not ranked, straight to somewhat curved, narrowly ovate-lanceolate to more widely ovate- to oblong-lanceolate, (0.9–)1.3–2.1(–2.4) mm long, 0.3–0.6 mm wide, non-decurrent, sharply keeled in the distal part, narrowly canaliculate in the proximal portion, gradually tapering to an acute or sometimes rounded and obtuse apex, epilose, or the uppermost leaves shortly piliferous, with a short awn, to 0.4 mm long; *hair-point* short, to 0.4 mm, straight, flattened, bluntly denticulate, yellowish-brownish to yellowish-hyaline, mostly non-decurrent; *margins* entire, recurved on both sides for varying distances, usually to ⅘–⅚ way up the leaf, plane or somewhat deflexed near the apex, 2(–3)-stratose in the distal portion forming 1–2(–3)-seriate fleshy, bulging limbidia, becoming unistratose with frequent bistratose patches in the proximal portion; *costa* ending in the apex or somewhat below, smooth, strong, 50–75 µm wide above, becoming thinner towards the base, distinctly convex on the dorsal surface, in transverse section semi-terete to subrectangular 2–3-stratose above, 3(–4)-stratose in the proximal portion; *laminal cells* unistratose throughout or with occasional bistratose streaks, patches or short striae below the apex, smooth, mostly isodiametrical, irregularly quadrate to shortly oblong in the upper and median parts, sometimes oblate near the apex, 4–9(–12) µm long, 3–8(–12) µm wide, with straight or weakly sinuate, but not distinctly sinuose or nodulose walls, becoming long-rectangular, 20–40 µm long, 5–10(–13) µm wide, with weakly to strongly sinuose walls in the supra-basal area; *basal juxtacostal cells* long-rectangular to linear, 35–90 µm long, 5–10 µm wide, becoming shorter towards the margins, hyaline or yellowish-hyaline and usually sharply demarcated from intensely yellow supra-basal cells, with thick or often thin walls, making the basal areolation lax, and consistently straight walls throughout the base; *basal marginal cells* hyaline, subquadrate to oblong, translucent, straight-walled, with markedly thickened transverse walls, forming a more or less distinct, 1(2–3)-seriate marginal border composed of 10–25 cells. *Autoicous*. *Perichaetial leaves* very large, up to twice the size of vegetative leaves, erect, straight and rather stiff, concave, widely oblong- or ovate-lanceolate to ovate, 2.0–3.2 mm long, 0.8–1.2 mm wide, gradually acuminate to a short, acute or less often broadly obtuse cuspidate point, recurved on both sides in the upper half, epilose or ending with a short, hyaline or yellowish-hyaline, smooth hair-point, to 0.3 mm long, with an areolation of irregular cells with strongly incrassate cells in the upper half, becoming rectangular and sinuose-walled in the median and lower parts and thinner- and straight-walled in the proximal part. *Setae* short, 0.25–0.4 mm long, pale yellowish-brownish, straight, thick. *Capsules* deeply immersed in the perichaetial leaves, ovoid to obloid, sometimes suburceolate, wide-mouthed, 0.7–1.0 mm long, 0.7–0.9 mm wide, pale yellowish to light brownish, smooth; *operculum* convex, rostellate, with a slanted or straight rostrum; *exothecial cells* irregular, mostly rounded or rounded-quadrate mixed with oblong, 18–40 µm long, 10–20(–25) µm wide, moderately thick-walled with distinct corner thickenings, becoming transversely rectangular, oblate to quadrate, orange-brownish in 5–7 layers at the mouth; *stomata* few, 3–5 per urn at the extreme base, bicellular, round-pored; *peristome teeth* straight and erect, 280–320 µm long, 110–130 µm wide at the base, yellow-orange to orange-brown, erect-spreading to recurved with age, lanceolate, narrowed to a rather broad obtuse apex, sometimes with uneven margins, entire or perforated, sometimes cleft

a b c d
1
2
3
4
5
6
7
8
9
10
11
12
13
14
15
16
17
18
19
20
21
22
23
24
25
26
27
28
29
30
31
32
33
34
35

at the apex, finely roughened to almost smooth. *Spores* globose, pale brownish, smooth to minutely papillose, 9–12 µm wide. *Calyptra* small, subcucullate, irregularly split at the base, covering only the rostrum.

Chromosome number: unknown.

Discussion. *Schistidium cupulare* has long remained a poorly known and obscure species. It was originally described as *Grimmia cupularis* by Müller (1883, 1889) from a single collection from Îles Kerguelen. Later, it was reported only once from this archipelago (Thériot 1924), although it is a very common and locally very abundant species (Ochyra, personal observations). The species was transferred to *Schistidium* by Ochyra (1998*b*) and, subsequently, it was recorded from King George I. in the maritime Antarctic (Ochyra 1998*a*). Detailed taxonomic studies of austral species of *Schistidium* showed its conspecificity with the South Georgian *Schistidium celatum* (Müll.Hal.) B.G.Bell (Ochyra 2004*c*).

Schistidium cupulare is primarily distinguished by its leaf areolation. The basal cells are elongate, moderately thick- to thin- and consistently straight-walled, from the costa to the leaf margin. The basal marginal cells are hyaline, subquadrate to oblong, transparent and have markedly thickened transverse walls. They form a more or less distinct, 1(–3)-seriate marginal border composed of 10–25 cells which warrant the placement of the species in sect. *Conferta*. The basal cells grade into short-rectangular supra-basal cells with distinctly sinuose walls and the cells in the median and upper parts of the leaf are mostly isodiametrical, irregularly quadrate to shortly oblong, sometimes oblate near the apex and have straight or weakly sinuate, but not distinctly sinuose or nodulose walls. Additional diagnostic traits of *S. cupulare* are (*1*) the stronger costa which is semi-terete to subrectangular and strongly convex on the dorsal surface in cross-section, (2–)3-stratose distally and sometimes 4-stratose in the proximal part; (*2*) the regularly 2(–3)-stratose leaf margins in 1–3 rows of cells forming a bulging border; (*3*) the presence of a distinct central strand in the stem; and (*4*) the smooth to finely roughened peristome teeth.

The short dense cushions or tufts of *Schistidium cupulare* are easily confused with forms of several other species, especially with *S. urnulaceum* and *S. amblyophyllum*. The former species is readily distinguishable from *S. cupulare* by its regularly hair-pointed narrower leaves, much weaker and narrower costae, small, globose or cyathiform to obloid capsules, exothecial cells with strongly incrassate walls and densely papillose peristome teeth. The separation from *S. amblyophyllum* is often difficult and depends on close examination of all leaves on individual stems and discriminatory notes are given under that species.

In leaf areolation *Schistidium cupulare* is somewhat similar to *S. halinae* but this species is distinguished at a glance by its pilose leaves terminated with long hyaline hair-points which give the plants a remarkable hoary appearance, whereas the leaves in *S. cupulare* are mostly epilose or have a short to 0.4 mm long, yellowish-brown to yellowish-hyaline point and consequently the plants are never hoary. *Schistidium deceptionense* also has a similar leaf areolation but it is distinct in its

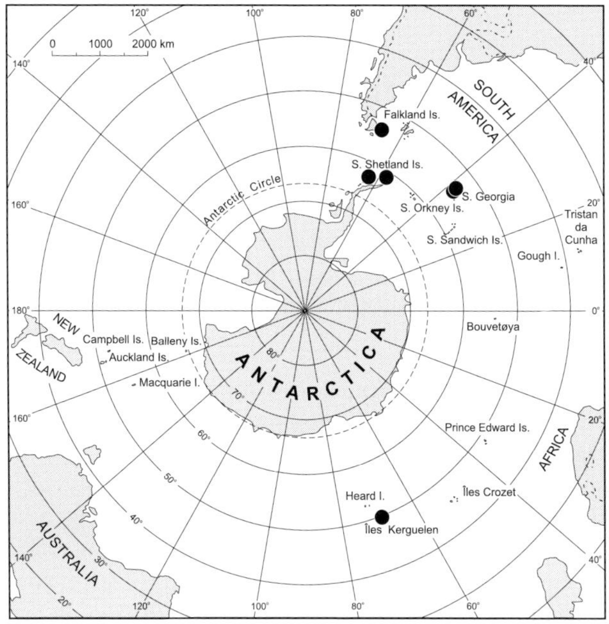

Figure 122. Global distribution of *Schistidium cupulare*.

small capsules with strongly thick-walled exothecial cells, as well as strongly carinate leaves with costae which are mostly bistratose in transverse sections. Finally, *S. leptoneurum* is easy to separate from *S. cupulare* by its very weak costae which are bistratose throughout in cross-section and scarcely prominent on the dorsal surface.

Reproduction in Antarctica. All known populations produce fully mature sporophytes.

Habitat. A saxicole showing a relatively broad ecological amplitude, growing on dry and exposed andesite on a rock outcrop, on vertical rock faces, as well as on moist ashy soil. It is associated with communities of the fruticose lichen and moss cushion subformation and often grows together with *Usnea antarctica*, *Bartramia patens*, *Hymenoloma antarcticum*, *Schistidium antarctici* and *Leptogium puberulum*.

World range. An amphiatlantic subantarctic species (Fig. 122), occurring on Îles Kerguelen where it is locally very common (Ochyra, personal observations) and South Georgia (Ochyra 2004*c*), extending southwards to the northern maritime Antarctic and, additionally, known from a single station on Isla Grande de Tierra del Fuego where it was collected at a high elevation in Rio Olivia near Ushuaia (Cardot & Brotherus 1923).

Distribution in Antarctica. A very rare species, possibly under-collected, known only from King George I. and Livingston I. in the South Shetland Is., reaching its southernmost locality at lat. 62°41′S. It grows at low elevations between 35 and 60 m a.s.l.

Specimens examined. WEST ANTARCTICA. SOUTH SHETLAND IS. **King George I.** ***Admiralty Bay***: Upłaz, 40 m, *RO 1536/80* (KRAM). **Livingston I.** ***Byers Peninsula***: between Smellie Pt. and Chester Cone, 35–50 m, *RILS 3788* (AAS, KRAM).

←

Figure 121. *Schistidium cupulare*. — **1**. Habit. **2**. Cross-section of stem. **3–4**. Axillary hairs. **5–10**. Leaves. **11–12**. Leaf apices. **13**. Upper laminal cells. **14**. Mid-leaf cells. **15**. Lower mid-leaf cells. **16–17**. Basal cells. **18–26**. Cross-sections of leaves, a sequence from base to apex. **27**. Perichaetium with deoperculate capsule. **28–30**. Perichaetial leaves. **31**. Deoperculate capsule, wet. **32**. Calyptra and operculum with attached columella, dry. **33**. Exothecial cells at urn base and stomata. **34**. Exothecial cells in middle of urn. **35**. Peristome teeth, spores and rim cells of capsule. — [All from *RO 1536/80*]. — Scale bars: a – 1 mm (1); b – 1 mm (5–10, 27–32); c – 100 µm (2, 11–26); d – 100 µm (3–4) and 200 µm (33–35).

19. GRIMMIA HEDW.

Grimmia Hedw., Spec. Musc. Frond.: 75. 1801. — Lectotype (*vide* Mårtensson 1956: p. 106): *Grimmia plagiopodia* Hedw.

Grimmia subgen. *Gasterogrimmia* Schimp., Coroll. Bryol. Eur.: 46. 1856. ≡ *Gasterogrimmia* (Schimp.) Buyss., Feuille Jeunes Naturalistes 13: 63. 1883. ≡ *Schistidium* Bruch & Schimp. subgen. *Gasterogrimmia* (Schimp.) C.E.O.Jensen ['Gastro-Grimmia'], *nom. illeg. incl. lectotyp. gen.*

Grimmia subgen. *Gastrogrimmia* Schimp. *ex* I.Hagen, Kongel. Norske Vidensk. Selsk. Skr. (Trondheim) 1909(5): 68. 1909, *orthogr. pro* Grimmia *subgen.* Gasterogrimmia.

Usmania Laz., Dopov. Akad. Nauk Ukrajinsk. RSR 11: 1039. 1970. — Type: *Usmania campylopoda* Laz. [= *Grimmia pitardii* Corb.]. First synonymised by Saito (1973*a*: p. 163).

Plants small, rather slender, forming short, compact cushions, grey, dark green to brown or olive-, reddish- to blackish-brown, occasionally glaucous or rufous, dull or somewhat lustrous, mostly strongly hoary. *Stems* erect, sparingly branched, rather sparsely radiculose at the base with smooth, brown, sparingly branched rhizoids, in cross-section with a large central strand, consisting of large, thin-walled, hyaline to yellowish-hyaline medullary cells and a uni- to bistratose cortex of smaller cells with moderately thickened, brown walls; *axillary hairs* 3–8-celled, filiform, hyaline throughout. *Leaves* crowded, tightly appressed to imbricate, erect when dry, erecto-patent when moist, ovate, oblong-ovate, oblong-lanceolate, muticous below, becoming somewhat enlarged and pilose upwards, broadly concave to narrowly canaliculate, occasionally carinate in the distal half, gradually to abruptly tapering to an obtuse or acute, epilose to piliferous, plane or sometimes cucullate apex, plane or sometimes deeply plicate; *hair-point* hyaline to yellowish-hyaline, very variable in length, often as long as or longer than the lamina on the uppermost leaves, flat or terete, often decurrent down the margins, smooth or denticulate to spinose, straight; *margins* entire, uni- to bistratose in one to several rows of cells, plane below, somewhat deflexed or inflexed in the distal portion; *costa* single, percurrent to subpercurrent, well differentiated from the laminal cells, weaker in the proximal part, becoming stronger in the apical part, in cross-section semi-terete, distinctly convex on the dorsal side, consisting of 2–4 ventral epidermal cells and a distinct group of small median stereid cells; *laminal cells* unistratose throughout to entirely bistratose or, most often, with alternating uni- and bistratose patches in the distal part, smooth and not bulging throughout, mostly isodiametric to short-rectangular or oblate, with thick, straight or faintly sinuose walls in the upper part, becoming elongate-rectangular in the basal part, usually with transverse walls thicker than longitudinal walls at the basal margins and straight to faintly nodulose walls near the costa, occasionally thin-walled and lax throughout the base. *Autoicous* or *dioicous*. *Perichaetial leaves* wider and longer than the vegetative leaves, concave and sheathing in the lower part, sometimes strongly modified and scarcely discernible, hyaline and filmy except for a small patch of chlorophyllous cells at the apex. *Setae* curved, arcuate to coiled, of capsule length or shorter than the urn, eccentrically attached. *Capsules* immersed or emergent, asymmetric and distinctly gibbose at the base, globose, ovoid or obloid, erect to slightly inclined or horizontal, smooth or striate when dry, wide-mouthed or constricted at the orifice; *annulus* compound, 1–4-seriate, often tardily deciduous; *operculum* plano-convex, umbonate with a low central wart or conic and long-rostellate; *exothecial cells* variable in size and shape, irregular, isodiametric to short-rectangular, thin-walled, occasionally with strongly incrassate walls and very narrow lumina; *stomata* in the neck of the urn, bicellular, superficial, round-pored; *peristome* absent or present, single, consisting of 16 teeth without basal membrane, triangular or lanceolate, orange to reddish, entire or irregularly split at the apex into 2–3 prongs, variously perforate to cribrose, finely to densely papillose; *columella* persistent. *Spores* spherical, smooth or minutely papillose. *Calyptra* smooth, cucullate or mitrate, deeply lobed, covering the entire operculum at least to the capsule mouth.

For a long time *Grimmia* was considered to be the largest genus within the Grimmiaceae which realistically consists of about 70–90 species distributed worldwide, mostly in montane areas. It is a genus of saxicolous mosses growing primarily on acidic rocks, less often on basic substrata, and only occasionally are some species epigean or aquatic plants. They are usually pioneers and primary colonisers on bare rock, forming small, compact, rigid, dark brown or green to blackish cushions or short turves, or sometimes growing in larger mats of coalesced colonies. The genus has always had a reputation as one of the most difficult of all moss genera, mostly because its species are often highly polymorphous and protean in nature. This resulted in the recognition of an excessive number of species and infraspecific taxa which usually proved to be merely habitat modifications and/or geographical races of more widely distributed species. Apart from some unrelated moss species which were given names under *Grimmia*, over 700 species and infraspecific taxa definitely belonging to this genus or closely related genera have been described in this genus. An excellent synopsis of these is presented by Muñoz & Pando (2000) who reduced the great majority of their names to synonymy and/or transferred many to other genera of the Grimmiaceae, leaving only 71 species as still residual in *Grimmia s. lato*.

In the traditional circumscription of Hagen (1909) and Brotherus (1924*b*) *Grimmia* was a typical polythetic taxon that could not be defined on the basis of any single character, but it was given cohesion by a specific combination of characters. Therefore, many bryologists strongly argued that *Grimmia* was an unnatural and polyphyletic genus consisting of several distinct monophyletic assemblages, well defined by sets of sporophyte and gametophyte characters (Buysson 1883; Churchill

1981; Cao & Vitt 1986; Muñoz 1998*c*). This is also confirmed by results of molecular investigations (Streiff 2006). They were mostly considered as subgenera (Schimper 1856; Limpricht 1889; Brotherus 1902, 1924*b*; Hagen 1909; Loeske 1913; Jones 1933; Hastings & Greven 2007*b*), sections (Nyholm 1956, 1998) or unranked groups (Loeske 1930), but recently Ochyra *et al.* (2003) split the traditional genus *Grimmia* into seven segregates. Five of these had already been recognised as separate genera and have been in use during various historical periods, so they have been resurrected.

According to the currently accepted nomenclatural rules, these genera are named and diagnosed as follows: (*a*) *Grimmia s. str.* (seta cygneous and asymmetrically attached, capsule asymmetric, ventricose at the base); (*b*) *Guembelia* Hampe (seta straight, capsule smooth and symmetric, leaves concave with plane margins and costa semi-elliptic and indistinct, weakly convex on the dorsal side); (*c*) *Orthogrimmia* (Schimp.) Ochyra & Żarnowiec (seta straight, capsule smooth and symmetric, leaves canaliculate to strongly keeled with plane margins and costa semi-terete and clearly delimited, strongly convex on the dorsal side); (*d*) *Dryptodon* Brid. (seta variously curved, capsule symmetric and mostly ribbed, leaves canaliculate to strongly keeled with recurved margins and costa semi-terete to reniform and distinctly differentiated); (*e*) *Schistidium* Bruch & Schimp. (capsule systylious, calyptra very small, covering in practice only the beak of the operculum); (*f*) *Hydrogrimmia* (I.Hagen) Loeske (leaves soft with plane margins, deeply concave and broadly rounded at the apex, costa narrow, subpercurrent, semi-rounded and weakly convex on the dorsal side); (*g*) *Streptocolea* (I.Hagen) Ochyra & Żarnowiec (alar cells differentiated, hyaline and convex, vaginula spirally twisted, capsule systylious).

Some authors (e.g. Brotherus 1902, 1924*b*; Jones 1933) merged *Schistidium* with *Grimmia* and considered it as a separate subgenus, but recently *Schistidium* has gained wide acceptance as a genus in its own right. Moreover, Crum & Anderson (1981) also 'lumped' into *Grimmia* species which are normally placed in *Coscinodon* and *Jaffueliobryum*, a usage reminiscent of the early days of moss taxonomy, when *Grimmia* was used as a catch-all genus.

The greatest concentration of species of *Grimmia s. lato* is in the Northern Hemisphere, with 40 species in Europe, 37 in North America and 28 in temperate and polar Asia (Muñoz & Pando 2000). The number of species decreases markedly southwards and in the Subantarctic only six species are known, while in the Antarctic there are only four, each representing a separate genus of the recently recognised segregates of *Grimmia s. lato*. In both biomes they are not significantly important constituents of the vegetation and are restricted to rocks and gravelly soil in various fellfield communities. *Grimmia s. lato* has received much attention from bryologists and there are many taxonomic treatments of this genus, both regional and worldwide. The most important are global revisions of *Grimmia s. lato* (Greven 2004), with 93 accepted species, and of *Orthogrimmia* (Muñoz 1998*c* as subgen. *Orthogrimmia*), as well as a world synopsis of *Grimmia s. lato* (Muñoz & Pando 2000). Additionally, in the last three decades there have been important treatments of this genus published for Japan (Deguchi 1979), South Georgia (Bell 1984), China (Cao & Vitt 1986), Central Siberia (Ignatov & Cao 1994), Europe (Geissler & Maier 1995; Maier & Geissler 1995; Greven 1995, 1999), New Zealand (Greven 1998*a*), Central and South America (Deguchi 1984, 1987; Muñoz 1999), Australia (Greven 2000*b*) and North America (Hastings & Greven 2007).

The generic name *Grimmia* is lectotypified by *G. plagiopodia* Hedw. (Mårtensson 1956). Consequently, as a result of splitting into the aforementioned segregates, *Grimmia* has become a small, oligotypic genus characterised by cygneous and asymmetrically attached setae and asymmetric capsules due to the ventricose base on one side. It corresponds to subgen. *Gasterogrimmia* of Bruch *et al.* (1845*a*), Schimper (1856), Hagen (1909) and Brotherus (1924*b*) and genus *Gasterogrimmia* of Buysson (1883), but both names are illegitimate because they include the lectotype-species of the generic name *Grimmia*.

As currently understood, *Grimmia s. str.* consists of only eight species, namely *G. plagiopodia*, *G. anodon* Bruch & Schimp., *G. crinita* Hedw., *G. americana* E.B.Bartram, *G. poecilostoma* Cardot & Sebille, *G. incrassicapsulis* B.G.Bell, *G. pitardii* Corb. and *G. trinervis* R.S.Williams, although the taxonomic position of the last two species is debatable and some authors place them, respectively, in the genera *Campylostelium* (Maier 1998) and *Coscinodon* (Hastings 1996). Only *G. plagiopodia*, occurs in the Antarctic.

1. **Grimmia plagiopodia** Hedw. Fig. 123

Grimmia plagiopodia Hedw., Spec. Musc. Frond.: 78, *pl. 15, f. 6–13.* 1801. ≡ *Weissia plagiopodia* (Hedw.) Poir., Enc. Meth. Bot. 8: 795. 1808. ≡ *Gasterogrimmia plagiopodia* (Hedw.) Buyss., Feuille Jeunes Naturalistes 13: 64. 1883, *comb. inval.* ≡ *Schistidium plagiopodium* (Hedw.) Loeske, Laubm. Eur. 1: 22, 52, *f. 13.* 1913. — TYPE CITATION: In Saxoniae rupibus detexit Cl. Flügge. [Lectotype (*selected here*; *vide* also Geissler & Maier 1995: p. 502): The right specimen in the upper row on the sheet labelled 'Grimmia plagiopus. a Hg s.n. b. a Flörkio' – G-Hedwig/Schwägrichen!].

Grimmia plagiopodia var. *antarctica* E.B.Bartram, Bryologist 60: 141. 1957. — TYPE CITATION: [Antarctica] Mt. Marujupu: in crevices on granite near rookery, B37.1. [Lectotype (*selected here*): 'Plants of Antarctica. United States of America Service Expedition of 1940–41. No. B37.1. Grimmia plagiopodia Hedw. var. antarctica Bartr. var. nov. Det. E. B. Bartram Mt. Marujupu in crevices on granite near rookery. J. E. Perkins, Collector. Nov. 27, 1940' – FH-Bartram!; isotype: US!]. First synonymised by Ochyra (1993*c*: p. 22).

Plants small, 3–10 mm high, in dense, compact mats or short turves, olive-brown to rufous or yellowish-green to dark green, usually hoary owing to the long hyaline awns on the uppermost leaves. *Stems* erect, 2–6 mm high, simple or sparsely branched, radiculose at the base with light brown, smooth rhizoids, in cross-section with a distinct or indistinct central strand, 2–4-layered medulla of large hyaline cells and 1–2-layered cortex of small, thick-walled, brown cells; *axillary hairs* filiform, hyaline throughout, consisting of 6–9 short cells. *Leaves* crowded, distinctly larger towards the shoot apex or ± uniform in size throughout, straight, closely imbricate when dry, erect to patent or homomallous when wet, concave or canaliculate-concave, (0.5–)1.0–1.5(–1.8) mm long (without the hair-point), 0.25–0.4(–0.6) mm wide, oval, widely ovate- or oblong-lanceolate to cymbiform, obtuse or bluntly acute, the lower leaves epilose, the upper leaves usually ending with a short or long, flat, straight or flexuose, sharply denticulate awn, often decurrent along the lamina, rarely with a short hyaline tip or muticous; *margins* entire, erect or occasionally weakly recurved distally, unistratose throughout or sometimes with occasional bistratose spots in the distal two thirds; *costa* single, ceasing below the apex, flat or somewhat canaliculate,

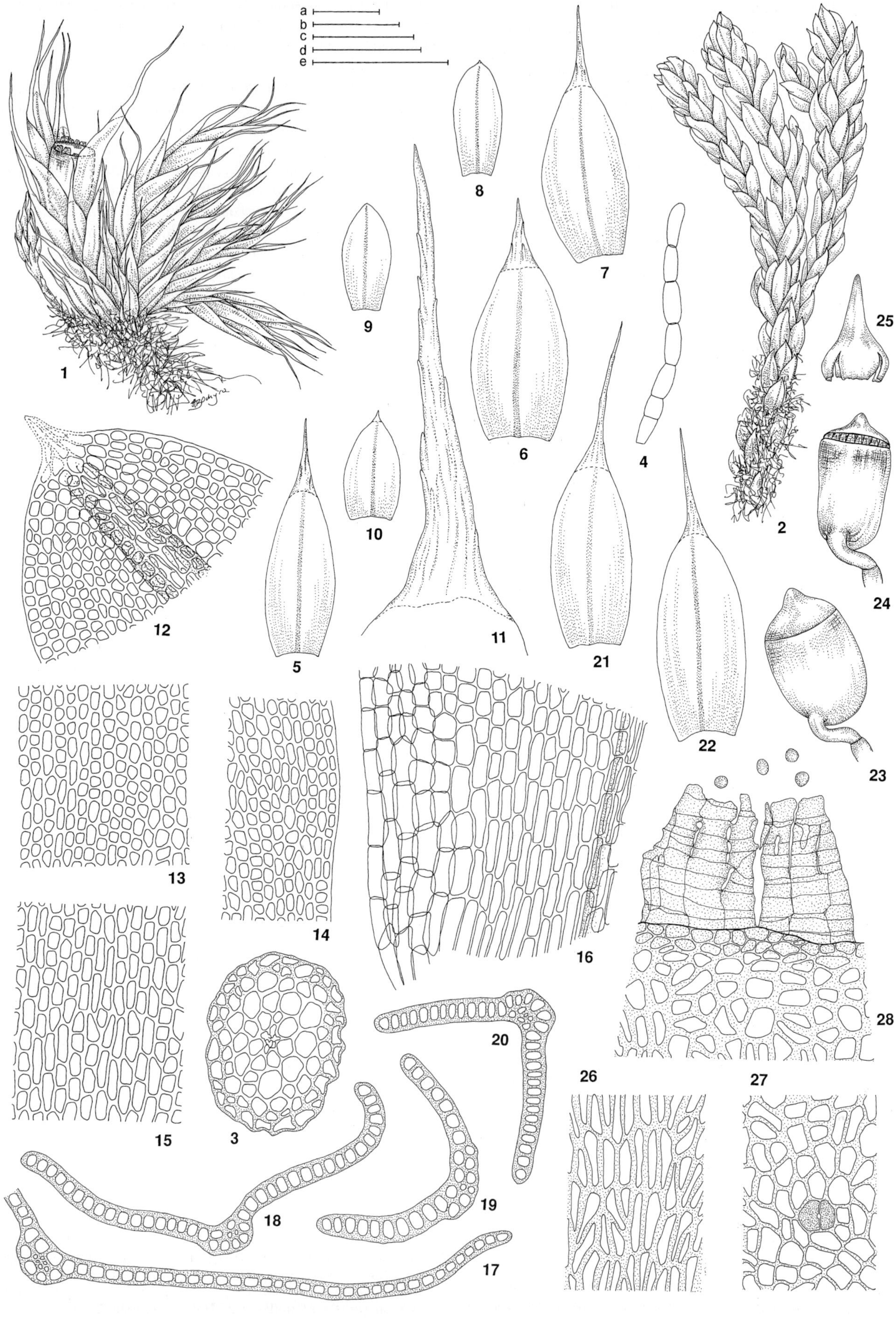
a
b
c
d
e
1
2
3
4
5
6
7
8
9
10
11
12
13
14
15
16
17
18
19
20
21
22
23
24
25
26
27
28

becoming wider towards the apex, in cross-section semi-terete, differentiated, tristratose, with 2 large ventral and one row of large dorsal epidermal cells surrounding a median stereid band of a few very small and incrassate cells or occasionally only bistratose without the central stereid band; *laminal cells* unistratose throughout, except for occasional bistratose spots at the distal margins, not bulging, rounded-quadrate to short-rectangular, with incrassate, straight to somewhat sinuose walls, 7–20 µm long, in the upper and median part, becoming isodiametric to short-rectangular, 12–40 µm long, 12–15 µm wide, with straight, thin- to moderately thick walls in the basal juxtacostal area and isodiametric to short-rectangular, 20–27 µm long, 10–15 µm wide, with transverse walls thicker than the longitudinal walls in the basal marginal area. *Autoicous*. *Perigonium* bud-like, terminal or placed below the perichaetium, without paraphyses and with obtuse perigonial bracts. *Perichaetial leaves* similar to the uppermost vegetative leaves but somewhat larger, convolute, 1.3–1.8 mm long, 0.7–0.8 mm wide. *Setae* curved and attached off-centre to the capsule, 0.6–1.2 mm long. *Capsules* immersed in the perichaetial leaves, ovoid to subglobose, 1.0–1.2 mm long, asymmetric, swollen on one side at the base, wide-mouthed, smooth, stramineous; *annulus* simple and persistent; *operculum* plano-convex with a stout, blunt, erect apiculus; *exothecial cells* irregular, elongate, 30–50 µm long, 12–25 µm wide, thick-walled, becoming transversely rectangular at the mouth in 3–4 tiers; *stomata* large, few, at the extreme base of the urn; *peristome teeth* 16, erect, orange-brown to light brown, lanceolate to truncate, 250–300 µm long, perforate or irregularly cleft at the apex, papillose throughout on the ventral surface and smooth proximally and papillose distally on the dorsal surface. *Spores* spherical, 11–13 µm in diameter, smooth. *Calyptra* mitrate, deeply lobed, slightly longer than the operculum.

Chromosome number: unknown in the Antarctic and Subantarctic; $n = 26$ (2 records) from Central Asia (Fritsch 1991).

Discussion. *Grimmia plagiopodia* is a distinct and characteristic species which is unlikely to be mistaken for any local moss species. It has immersed, ovoid to subglobose capsules which are distinctly ventricose at the base and are eccentrically attached to the sigmoid seta that is shorter than the urn. These features are all diagnostic for the type subgenus. The capsules concealed within the perichaetial leaves are also typical for *Schistidium* in which, however, they are always systylious, whereas in *G. plagiopodia* the columella is never attached to the operculum after dehiscence. Sterile plants are readily recognised by oval to oblong-ovate, concave and obtuse leaves which most often, especially in the distal part of shoots, are terminated by long, flat, denticulate and somewhat decurrent hair-points, giving the plants a hoary appearance. In transverse section the leaves are broadly concave and U-shaped, never keeled, and this is the most reliable character distinguishing *G. plagiopodia* from the closely related *G. anodon*.

Hair-point presence and length are subject to considerable variation in *Grimmia plagiopodia* (Ochyra & Bednarek-Ochyra 1994). In some continental plants the leaves have very short hyaline tips and in extreme cases they are totally epilose. Such plants were recognised as var. *antarctica* (Bartram 1957; Savicz-Lyubitskaya & Smirnova 1970*a*) and they have a characteristic appearance due to their cymbiform epilose leaves with cucullate apices. Epilose ecads are frequently observed in other species of *Grimmia*, for example in *G. anodon*, but they do not differ materially in other structural characters from normally piliferous plants of these species. Except for the lack of the hyaline hair-point *G. plagiopodia* var. *antarctica* shares all other characters with typical expressions of *G. plagiopodia*, and Ochyra (1993*c*) found it does not merit taxonomic recognition since all possible integradation to normally pilose plants can be observed.

The leaf lamina in *Grimmia plagiopodia* is entirely unistratose, and sometimes this feature is emphasised as a diagnostic character for this species. However, sometimes the leaf margins have scattered bistratose patches in the distal part, as clearly visible on the type material of *G. plagiopodia* (Geissler & Maier 1995). Seppelt & Green (1998) illustrated plant(s) from Southern Victoria Land (but without indication of the specimen(s) from which the drawings were made) and showed variously bistratose patches in the upper part of the lamina, occasionally tristratose leaf margins and, additionally, keeled leaves. These characters are strange for *G. plagiopodia* and suggest a rather different identity. We studied two specimens from this region, and they have the typical anatomical structure of *G. plagiopodia*, with U-shaped, unistratose leaf lamina, although one specimen (*Seppelt 19775*, AAD, KRAM) has quite regularly bistratose leaf margins in the distal one third. The leaves of this plant are almost epilose but they are distinctly cucullate at the apex, unlike *Schistidium antarctici*.

Reproduction in Antarctica. Only two of the continental populations examined were fertile, i.e. those from Dustin I. on the Eights Coast (*RILS 5407*, AAS, KRAM) and northern Victoria Land (*RILS 9654*, AAS, KRAM), whereas in the plants from the maritime Antarctic sporophytes are encountered more frequently, though they are not abundant.

Habitat. A saxicole associated with various types of acidic rocks including granite, basalt, diorite, arkosic sandstone, rarely occurring also on volcanic tufa and breccia. It grows mostly in dry habitats on rocks, boulders, ledges, in rock crevices, and occasionally on gravel and soil; less often it occurs on moist to wet habitats on rocks, at the edge of late snow beds and in niches fed by melt water. Occasionally, it grows on rocks used as bird perches which are heavily affected by nitrogen, and consequently the moss is sometimes overgrown by epiphytic lichens, e.g. *Physcia caesia* and *Xanthoria candelaria*.

←

Figure 123. *Grimmia plagiopodia*. — **1**. Habit, dry, piliferous plant. **2**. Habit, dry, epilose plant. **3**. Cross-section of the stem. **4**. Axillary hair. **5–10**. Leaves. **11**. Hyaline hair-point. **12**. Leaf apex. **13**. Upper mid-leaf cells. **14**. Mid-leaf cells at the margin. **15**. Lower mid-leaf cells. **16**. Basal cells. **17–20**. Cross-sections of the leaves, a sequence from the base to the apex. **21–22**. Perichaetial leaves. **23–24**. Operculate capsules. **25**. Calyptra. **26**. Exothecial cells in the middle of the urn. **27**. Lower exothecial cells and stoma. **28**. Portion of the peristome, exothecial cells at the orifice and spores. — [1, 17–18, 24–25 from *Ruiz Leal-Carretero 13230*, FH; 2, 8–10, 12 from *Perkins B37* (holotype of *G. plagiopodia* var. *antarctica*), FH; 4, 6, 19–20 from *Perkins B37*.1, FH; 3, 5, 7, 11, 13–16 from *RILS 9335A*; 21–23, 26–28 from *RILS 3653*]. — Scale bars: a – 1 mm (1); b – 1 mm (21–22); c – 100 µm (17–20); d – 100 µm (4, 12–16, 26–28), 200 µm (11) and 1 mm (2); e – 1 mm (5–10, 23–25) and 100 µm (3).

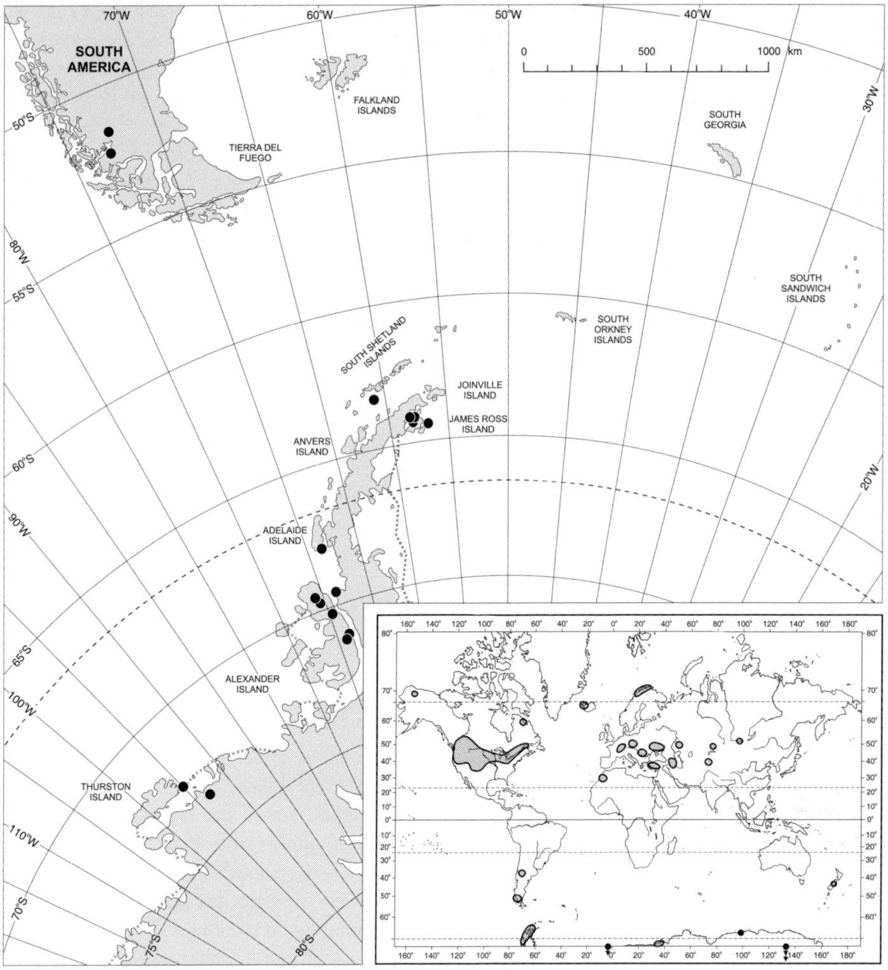

Figure 124. Distribution of *Grimmia plagiopodia* in the Antarctic Peninsula and Fuegian regions. Inset: global distribution (localities in Dronning Maud Land and Coats Land are indicated by an arrow and in Victoria Land by double arrow).

World range. A bipolar species (Fig. 124, inset) which is highly disjunct in temperate regions of the Northern Hemisphere, only occasionally penetrating the Nearctic in Alaska (Steere 1978) and the Eastern Canadian Archipelago (Steere 1946). In North America *Grimmia plagiopodia* extends throughout the continent from British Columbia, Oregon and California in the west to Ontario, Ohio and Maryland in the east and Colorado and New Mexico in the south (Stewart 1988). In Europe it ranges from France across Germany (Duell 1994), Hungary (Boros 1968), Slovakia (Greven 1995), Romania (Mohan 1975), the southern Ukraine (Bachurina & Mel'nichuk 1988), southern Russia (Mel'nichuk 1970; Ignatov & Ignatova 2003) to the Southern Ural Mountains (Ignatov & Afonina 1992), Caucasus Mountains (Manakyan 1989) and Western Anatolia in Turkey (Walther 1967), and extending in the north to Iceland and northernmost Scandinavia (Söderström 1996). In Asia the species is rare in the Altai and NW Mongolia (Ignatov & Cao 1994), Central Asia (Mamatkulov 1990; Ignatov & Afonina 1992, Ignatov *et al.* 2006) and Afghanistan (Kilbertus & Bizot 1971), and an isolated station is known from the Anti-Atlas of Morocco (Jelenc 1955). In the Southern Hemisphere it is rare in southern South America, South Island of New Zealand (Greven 1998*a*) and in the Antarctic (Ochyra 1993*c*).

Distribution in Antarctica. *Grimmia plagiopodia* is widely distributed but scattered and localised throughout the Antarctic. In the maritime Antarctic it is rare on Deception I. in the South Shetland Is. and on Léonie I. in northern Marguerite Bay, but more frequent in the northern part of Alexander I. where it extends to lat. 71°21′S (Fig. 124). On the continent it is present in all sectors but, in general, it is very rare everywhere. The southernmost locality from which a specimen has been exam-

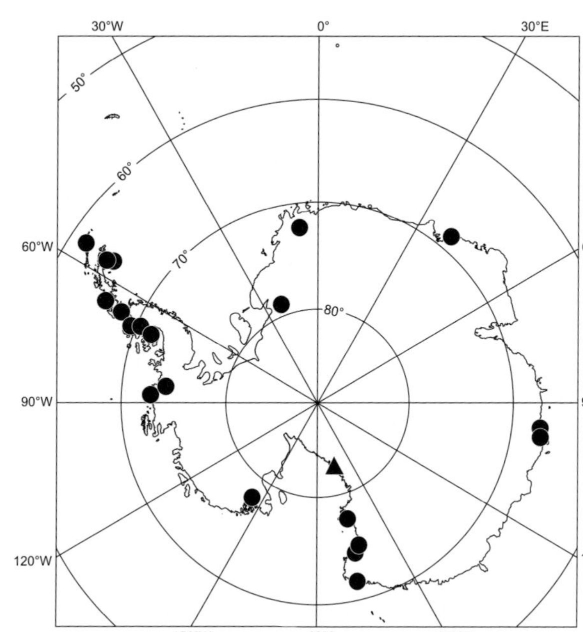

Figure 125. Distribution of *Grimmia plagiopodia* in Antarctica. Literature record marked by the triangle.

ined in this study is in the Theron Mountains in Coats Land at lat. 79°04′ S, but Seppelt & Green (1998) reported it from lat. 83°49′ S in the Transantarctic Mountains (Fig. 125). The altitudinal range of the species extends from near sea level to 900 m in the Theron Mountains and Marguerite Bay area.

Specimens examined. WEST ANTARCTICA. SOUTH SHETLAND IS. DECEPTION I. Pendulum Cove, Crimson Hill, 10 m, *RILS 11524* (AAS, KRAM); Fumarole Bay, 8 m, *RILS 3653* (AAS, KRAM) and 50 m, *RILS 11296B, 11299B, 11300 & 11535* (AAS, KRAM); Wensleydale Beacon, 30–100 m, *RILS 6661A & 6678A* (AAS, KRAM); without exact locality, *Ruiz Leal-Carretero 13230* (FH).

WEST ANTARCTIC PENINSULA. **FALLIÈRES COAST.** ***Léonie Is.***: Léonie I., 150–175 m, *RILS 9124A & 9126* (AAS, KRAM). ***Marguerite Bay***: Föhn Bastion, *ca* 915 m, *DDWW 205A* (AAS, KRAM). **GEORGE VI SOUND.** ***Alexander I.***: Lassus Mountains, 750 m, *BASM 208A* (AAS, KRAM); S of Bartok Glacier, *ca* 500 m, *BASM 205* (AAS, KRAM); Zebra Ridge at S side of Tumble Glacier, 600 m, *RILS 10026* (AAS, KRAM); Fossil Bluff, Khufu, 400–500 m, *RILS 9319 & 9335A* (AAS, KRAM); Mt. Ariel, *ca* 135 m, *BJT 505* (AAS, KRAM).

EAST ANTARCTIC PENINSULA. **JAMES ROSS ISLAND GROUP.** ***James Ross I.***: Fortress Hill, 25 m, *RILS 7815* (AAS, KRAM) and 150 m, *RILS 7798* (AAS, KRAM); Terrapin Hill, 250 m, *RILS 7717 & 7739B* (AAS, KRAM); Dreadnought Pt., 5 m, *RILS 7503* (AAS, KRAM). ***Seymour I.***: Cape Wiman, 10 m, *RILS 7889B* (AAS, KRAM).

ELLSWORTH LAND. Mt. Murphy Range, near Turtle Peak, *ca* 700 m, *RILS 9991A* (AAS, KRAM). **EIGHTS COAST.** Dustin I., Ehlers Knob, 100 m, *RILS 5407* (AAS, KRAM); Jones Mountains, W of Pillsbury Tower, *ca* 235 m, *RILS 5403A & 5405A* (AAS, KRAM).

EAST ANTARCTICA. COATS LAND. **THERON MOUNTAINS.** Marø Cliffs, *ca* 900 m, *Brook 22* (AAS, KRAM).

DRONNING MAUD LAND. **KRONPRINSESSE MÄRTHA KYST.** 'Grunehogna – 1285', *ca* 800 m, *Watkins 26A* (AAS, KRAM). **SÔYA COAST.** Skarvsnes, Oyako Lake, 130 m, *Nakanishi C72* (NIPR).

QUEEN MARY LAND. **KNOX COAST.** Bunger Oasis, Mt. Vyalov, *Korotkevich 5* (KRAM, LE) and *Hollerbach 12/381* (KRAM, LE); Obruchev Oasis, *Hollerbach 15/442* (KRAM, LE).

VICTORIA LAND. **PENNELL COAST.** Morozumi Range, *Champness MCM-53* (US). **BORCHGREVINK COAST.** Cape King, 100–125 m, *RILS 10256B & 102262A* (AAS, KRAM); moraine of Tinker Glacier-Burns Glacier, *RILS 9654* (AAS, KRAM). **SCOTT COAST.** Granite Harbour, Kar Plateau, *RDS 19775* (ADT, KRAM).

MARIE BYRD LAND. **SAUNDERS COAST.** ***Edsel Ford Range***: Mt. Majurupu, *Perkins B37.1* (FH, US – type of *G. plagiopodia* var. *antarctica*) & *B37* (FH).

20. GUEMBELIA HAMPE

Guembelia Hampe *in* Müll.Hal., Bot. Zeit. (Berlin) 4: 124. 1846. ≡ *Grimmia* subgen. *Guembelia* (Müll.Hal.) Schimp., Coroll. Bryol. Eur.: 48. 1856. ≡ *Grimmia* sect. *Guembelia* (Müll.Hal.) Mitt., J. Linn. Soc. Bot. 12: 96, 101. 1869. — Lectotype (*vide* Pfeiffer 1874: p. 1511): *Guembelia elliptica* (Funck) Hampe (*Grimmia elliptica* Funck) [= *Guembelia ovalis* (Hedw.) Müll.Hal.].

Grimmia Hedw. subgen. *Litoneurum* I.Hagen, Kongel. Norske Vidensk. Selsk. Skr. (Trondheim) 1909(5): 6, 10. 1909. ≡ *Grimmia* subsect. *Litoneurum* (I.Hagen) G.N.Jones *in* Grout, Moss Fl. N. Amer. 2: 5, 24. 1933. ≡ *Grimmia* sect. *Litoneurum* (I.Hagen) Nyholm, Ill. Fl. Nord. Mosses 4: 335. 1998. — LECTOTYPE (*vide* Ochyra *et al.* 2003: 127): *Grimmia commutata* Huebener, *nom. illeg.* [= *Grimmia ovalis* (Hedw.) Lindb.]. First synonymised by Ochyra *et al.* (2003: p. 127).

Plants small to medium-sized, sometimes quite large, slender to robust, rigid, densely caespitose, sometimes in compact but readily disintegrating turves or flat patches, dull to shiny, often hoary, dark green, yellowish-, olive-, grayish- or blackish-green, golden, rusty, red-brown, to black above, brownish, blackish-brown to black in the internal part. *Stems* erect to ascending, simple or sparsely branched, occasionally repeatedly forked by subfloral innovations, slightly radiculose at the base with brown, lustrous, smooth rhizoids, in transverse section with or without a central strand, a medulla of large, thin-walled cells and a uni- or bistratose cortex of small, substereid cells with brown, thickened walls; *axillary hairs* filiform, hyaline throughout, 5–10-celled. *Leaves* small, scale-like and tightly appressed to the stem below and usually progressively larger towards the stem tips, crowded, erect to appressed or imbricate when dry, erect-spreading to patent when wet, oblong-ovate, oblong-lanceolate, triangular, ovate to elliptic, gradually or abruptly narrowed to an acute, obtuse or rounded-obtuse apex, epilose and cucullate or usually terminated with a hyaline hair-point, sometimes homomallous, falcato-secund from a clasping base tapering to a long, subulate, uncinate point, concave to canaliculate, not plicate; *hair-point* absent or hyaline, brittle, straight to moderately flexuose, variable in length, flattened or terete, smooth, serrate, strongly denticulate to spinulose, often decurrent down the margins; *margins* entire, plane throughout or occasionally recurved proximally, erect to incurved distally, unistratose or multistratose in the upper two thirds, occasionally forming bulging, fleshy, marginal or inframarginal limbidia; *costa* single, indistinct, undifferentiated to weakly differentiated, mostly flattened, semi-elliptic to reniform in transverse section, scarcely prominent dorsally; *laminal cells* usually 1–2-stratose, sometimes with 3–4-layered strands in the distal part, smooth or sometimes strongly mammillose, mostly isodiametric to short-rectangular, rounded-quadrate, oval or oblate, obscure, with thick, straight or sinuose walls in the upper part, becoming shortly oblong to long-rectangular near the costa in the basal part, with thin or thick and nodulose walls; *basal marginal cells* short, transversely rectangular, short-rectangular to subquadrate, obscure to hyaline, straight-walled, with the transverse walls much thicker than the longitudinal walls or rectangular with slightly thickened walls. *Dioicous* or less often *gonioautoicous*. *Perichaetial leaves* larger than vegetative leaves, slightly sheathing to convolute, sometimes hyaline and filmy, scarcely discernible. *Setae* erect and straight, elongate, sometimes stout, sometimes flexuose-curved when dry. *Capsules* emergent to exserted, rarely immersed, symmetrical, erect, ovoid, obloid to cylindrical, narrowed at the mouth, stramineous to castaneous, dull or shiny, smooth or weakly striate at the base; *annulus* compound and revoluble, of 1–4 rows of cells, detaching singly or as spirals; *operculum* conical, with a straight or slanted, short or long, blunt or rostellate beak; *exothecial cells* variable in shape, mostly oblong-hexagonal, thin- to thick-walled; *stomata* present at the base of the urn, bicellular, superficial, with rounded pori; *peristome teeth* lanceolate, entire to variously split and cribrose above, reddish-brown, variously papillose on both surfaces. *Spores* small, brownish, globose, smooth to minutely granulose. *Calyptra* mitrate, often with a single slit, usually covering the upper third of the capsule.

Guembelia is a distinct genus which is diagnosed primarily by its poorly differentiated costae which, in transverse section, are semi-elliptic or reniform and scarcely prominent on the dorsal surface, very often being nearly of the same thickness as the lamina. As a result, the costa is indistinct and usually poorly defined and imperceptibly merges with the laminal cells. Additionally, the leaves are deeply concave, with plane, erect to incurved margins, and the setae are short. This combination of characters, especially the costal structure and the broadly concave leaves, separates *Guembelia* from *Orthogrimmia* and other segregates of *Grimmia s. lato*.

Guembelia was recognised as a separate genus by E. Hampe (Müller 1846) who placed in it eight species which, in terms of modern taxonomy, were rather remotely related and actually belong to other subgenera of *Grimmia sensu* Hagen (1909). Of these, only one species was given a name under *Guembelia*, namely *G. elliptica* which was selected as lectotype of this generic name (Pfeiffer 1874). The genus was adopted by Müller (1849) who placed in it ten species, and subsequently this author expanded the concept of *Guembelia* and added to it further species, amongst them species of the genus *Cinclidotus* (Müller 1851), thus making it a very unnatural assemblage of species. Schimper (1856) reduced *Guembelia* to a subgenus of *Grimmia* and in this form the name has remained, especially after Hagen (1909) and Brotherus (1924*b*) approved it. However, it should be stressed that the two latter authors used this name incorrectly because, for strictly nomenclatural reasons, the name should be applied to subgen. *Litoneurum*, whereas subgen. *Guembelia* in the sense of these authors should be called subgen. *Orthogrimmia*. *Guembelia* is a cosmopolitan genus, consisting of about 20 species but the placement of some of these in this genus still needs careful evaluation. One of these, *G. longirostris*, is known from a single collection in the Antarctic.

1. **Guembelia longirostris** (Hook.) Ochyra & Żarnowiec Fig. 126

Guembelia longirostris (Hook.) Ochyra & Żarnowiec *in* Ochyra, Żarnowiec & Bednarek-Ochyra, Cens. Cat. Polish Mosses: 127. 2003. ≡ *Grimmia longirostris* Hook., Musci Exot. 1: *tab. 62*. 1818. — TYPE CITATION: Locis scopulosis, apricis, in declivitate montis Chimborazo, altit. 2380 hexapod. super porphyrite (Regnum Quitense). Humboldt et Bonpland. [Lectotype (*vide* Deguchi 1984: p. 21): 'N° 76. M. de Humboldt' – BM-Wilson!; isotypes: BM-Hampe! (2 specimens), BM-Hooker!; BM-Wilson!, PC!].

Plants small, densely caespitose, somewhat glistening, not hoary, olive-brown above, brown to blackish-brown below. *Stems* erect, sparingly branched, to 1 cm high, sparsely radiculose below with smooth, branched, brownish rhizoids, in cross-section round, with a small but distinct central strand, 1(–2)-stratose cortex of small, thick-walled, brownish cells and 3–5-layered medulla composed of large, thin-walled, hyaline cells; *axillary hairs* filiform, 5–6-celled, hyaline throughout. *Leaves* erect and appressed, not or only slightly altered on drying, erect to erect-spreading, rigid on wetting, ovate-lanceolate to lanceolate, 1.5–2.1 mm long (excluding hair-point), 0.3–0.5 mm wide, larger and hyaline-awned towards the stem tips, acute, not plicate, canaliculate, not or weakly keeled distally, mostly epilose below; *hair-point* hyaline, terete, erect and rigid, smooth to slightly denticulate; *margins* entire, recurved on one or both sides to ½–⅔ up the leaf length, plane above, bistratose; *costa* percurrent, 50–60 µm wide at the base, weakly to clearly delimited from the laminal cells, scarcely prominent on the dorsal surface in the distal part, in transverse section reniform, U-shaped on the ventral surface, bistratose above, tristratose below, with 2–4 large ventral epidermal cells, a median layer of substereid cells and a row of larger dorsal epidermal cells; *laminal cells* regularly bistratose in the distal half, often with unistratose strands toward the costa, unistratose in the proximal half; *upper cells* isodiametric, short-rectangular to oblate, 4–10 µm long, not bulging, with thickened, often sinuose walls; *basal juxtacostal cells* rectangular, 25–80 µm long, 7–12 µm wide, with incrassate, nodulose vertical walls; *basal marginal cells* short-rectangular to subquadrate, 10–20 µm long, 7–10 µm wide, hyaline, straight-walled, with the transverse walls markedly thicker than the longitudinal walls. [*Autoicous*. *Perigonia* bud-like on short shoots arising from the base of shoots producing perichaetia. *Perichaetial leaves* much larger than vegetative leaves, convolute, with a long hyaline hair-point. *Setae* erect, straight, sinistrorse, 1–4 mm long. *Capsules* emergent to long-exserted, ovoid to cylindrical, 1.0–2.5 mm long, 0.5–0.7 mm wide, smooth, stramineous, becoming brown when empty; *annulus* compound, 2–4-seriate, revoluble; *operculum* conic, short-beaked to long- and straight-rostrate; *exothecial cells* irregular, isodiametric to rectangular, 20–50 µm long, 15–20 µm wide, thin-walled, becoming short in 2–3 tiers at the orifice; *stomata* at the base of the urn, few, superficial, bicellular, round-pored; *peristome teeth* lanceolate, entire to irregularly cleft above, 220–300 µm long, 50–80 µm wide at the base, orange- to reddish-brown, densely papillose on both surfaces or smooth in the lower ⅓ on the

→

Figure 126. *Guembelia longirostris*. — **1**. Habit, dry. **2**. Portion of branch, wet. **3**. Cross-section of stem. **4–7**. Axillary hairs. **8–14**. Leaves. **15**. Leaf apex with hyaline hair-point. **16**. Mid-leaf cells at margin. **17**. Basal cells. **18–27**. Cross-sections of leaves, a sequence from base to apex. **28–29**. Perichaetial leaves. **30**. Exothecial cells at base of urn and stomata. **31**. Mid-urn exothecial cells. **32**. Portion of peristome, annulus, exothecial cells at orifice and spores. — [1–7, 15–27 from *CGB 15*; 8–14, 28–32 from *Churchill, Franco & Parra 18812*]. — Scale bars: a – 1 mm (1, 28–29); b – 1 mm (2); c – 100 µm (18–27, 30–32); d – 100 µm (3–7, 15–17); e – 1 mm (8–14).

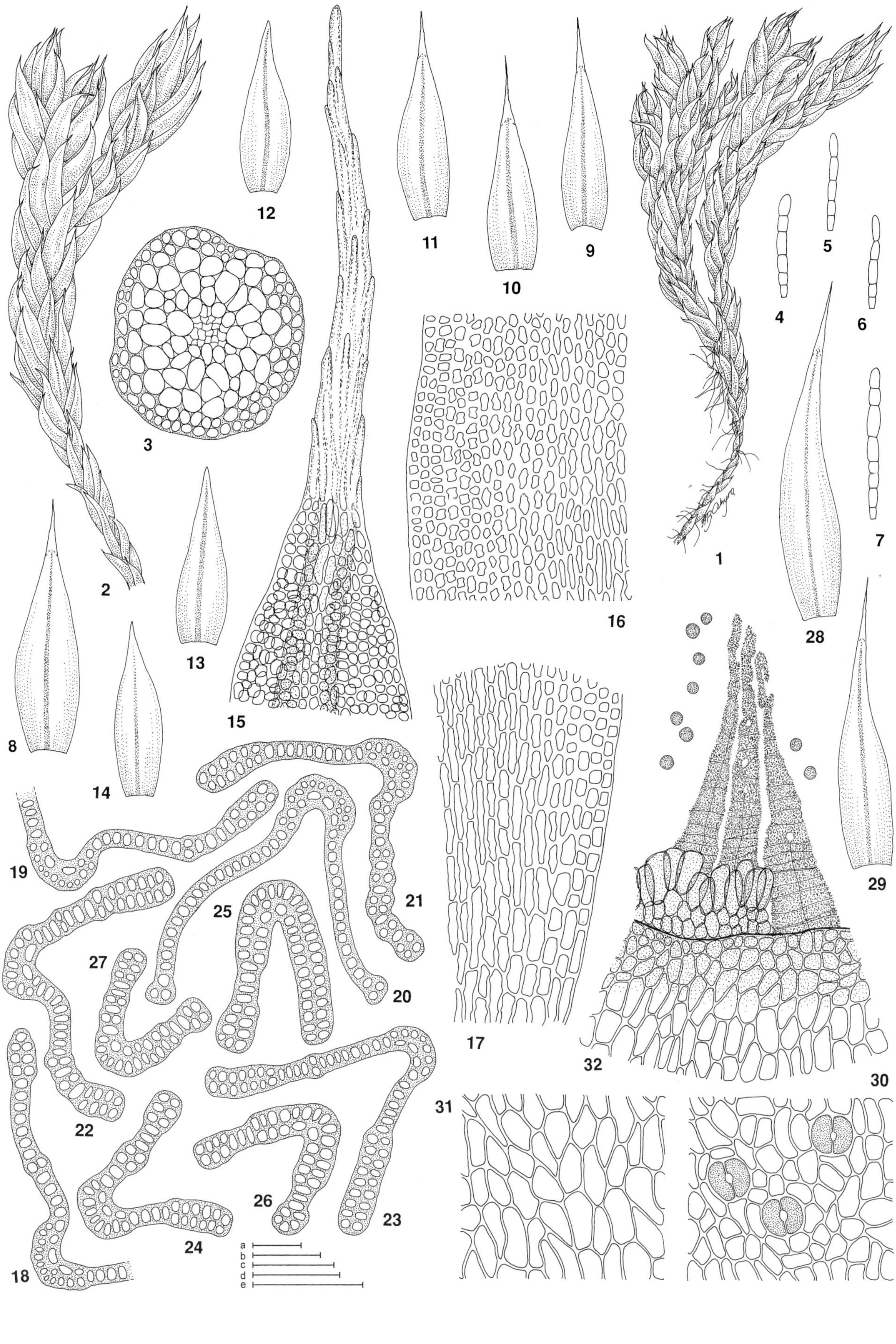
1
2
3
4
5
6
7
8
9
10
11
12
13
14
15
16
17
18
19
20
21
22
23
24
25
26
27
28
29
30
31
32
a
b
c
d
e

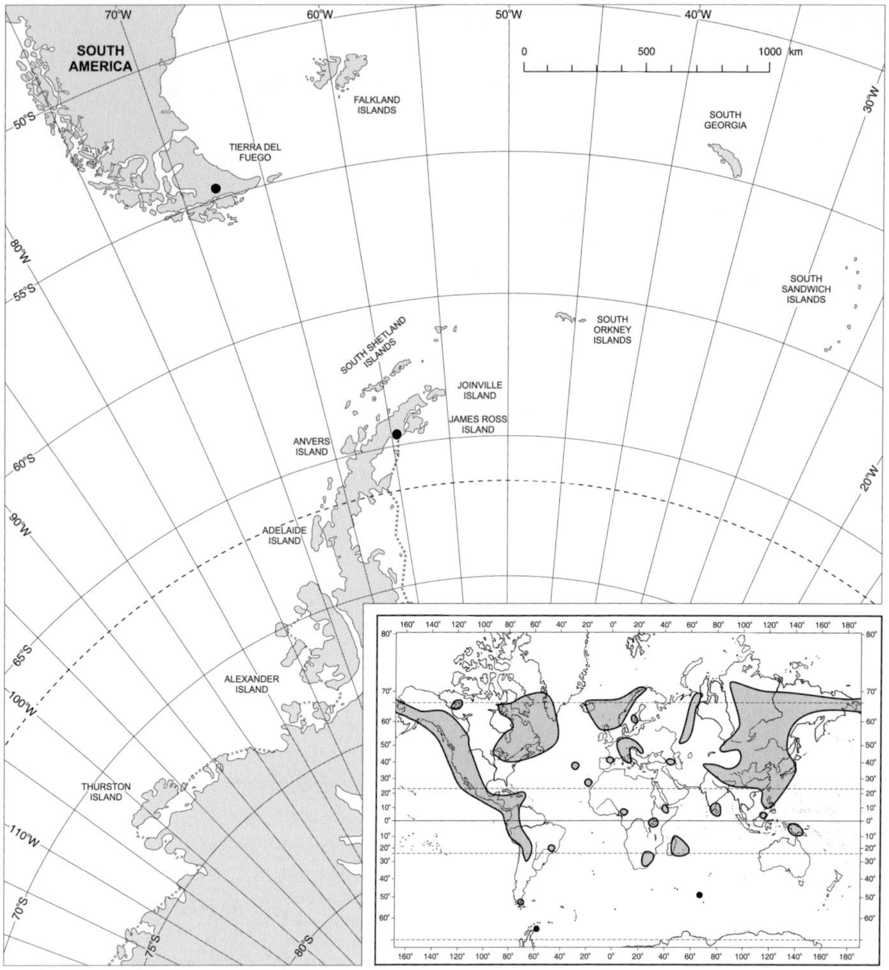

Figure 127. Distribution of *Guembelia longirostris* in the Fuegian region and Antarctica. Inset: global distribution.

outer surface. *Spores* spherical, brownish, 8–12 μm wide. *Calyptra* mitrate or rarely cucullate, naked, covering only the operculum.]

Chromosome numbers: unknown in the Antarctic and Subantarctic; n = 13 (3 records) and n = 14 (1 record) from the Holarctic (Fritsch 1991 as *Grimmia affinis*).

Discussion. *Guembelia longirostris* is distinguished by its canaliculate, U-shaped leaves in the distal part and semi-elliptic to reniform and curved costa, both characters clearly visible in transverse leaf sections. The costa has 2–4 ventral epidermal cells which are markedly larger than the dorsal ones. Additional characters that support accurate identification of *G. longirostris* are the recurved leaf margins, at least on one side in the proximal half, the elongate basal juxtacostal cells with strongly thickened and nodulose longitudinal walls and the shorter, hyaline basal marginal cells with strongly incrassate transverse walls. This combination of characters separates *G. longirostris* from all other species of *Guembelia* and confirms its isolated taxonomic position in the genus.

Guembelia longirostris is a very variable species but most of its variation involves quantitative characters such as the length of the seta, capsule, operculum and hair-point. This prompted many workers to recognise various phenotypes of this species as separate taxa. A detailed assessment of the taxonomy and variability of *G. longirostris* has been presented by Muñoz (1998*a*), who recognised 55 synonyms of this name from all parts of its global range.

Reproduction in Antarctica. Sterile.

Habitat. In crevices associated with grit and soil on a small cliff.

World range. A nearly cosmopolitan oreophyte known from all continents except Australia (Fig. 127, inset). In the Northern Hemisphere it is widespread in most mountains in the temperate zone, but only rarely penetrating into the Arctic in West Greenland and the Chukchi Peninsula. In the Americas it ranges from central Alaska to northern Argentina, and in tropical Africa it is frequent in the East African mountains from Ethiopia and Lesotho, in Madagascar and on Réunion. In tropical Asia it is known from the southernmost tip of India and Sri Lanka, the Philippines, Borneo and New Guinea. In the cool temperate region in the Southern Hemisphere it is exceedingly rare in Tierra del Fuego (Muñoz 1998*a*), while in the Subantarctic region it is known only from Îles Kerguelen where it was described by Müller (1883, 1889) as a separate species, *Grimmia rufa* (Muñoz 1999). It is known from a single location in the Antarctic (Ochyra 2004*a*).

Distribution in Antarctica. Known only from a single collection on the east coast of the Antarctic Peninsula at an elevation of about 200 m. The locality is situated at lat. 64°26′ S which is the southernmost occurrence of *Guembelia longirostris* (Fig. 127).

Specimen examined. WEST ANTARCTICA. EAST ANTARCTIC PENINSULA. **NORDENSKJÖLD COAST.** SW corner of Larsen Inlet, *ca* 200 m, *CGB 15* (AAS, KRAM).

21. **ORTHOGRIMMIA** (SCHIMP.) OCHYRA & ŻARNOWIEC

Orthogrimmia (Schimp.) Ochyra & Żarnowiec *in* Ochyra, Żarnowiec & Bednarek-Ochyra, Cens. Cat. Polish Mosses: 118. 2003. ≡ *Grimmia* Hedw. subgen. *Orthogrimmia* Schimp., Coroll. Bryol. Eur.: 48. 1856. — Lectotype (*vide* Muñoz 1998*c*: p. 375): *Orthogrimmia donniana* (Sm.) Ochyra & Żarnowiec (*Grimmia donniana* Sm.).

Plants small and rather delicate, rigid, forming short, dense, compact or less cohesive, flat or hemispheric cushions or tufts, mostly hoary, dull or shiny, grey, dark green, yellow- to olive-green, glaucous or blackish at the tips, dark green, brownish-green, brown or blackish-brown to blackish below. *Stems* erect, sparsely to strongly branched, adhering to the substrate by clusters of smooth, brown, branched rhizoids, occasionally with rhizoids scattered throughout the stem length, in transverse section with a weakly or well developed central strand and a poor differentiation between cortical and medullary cells; *axillary cells* filiform uniseriate, composed of 3–8 cells, with 1–2 shorter basal cells. *Leaves* densely set, in the lower part of the stem smaller, muticous or shortly hair-pointed, becoming gradually larger towards the stem tips, straight, erect and appressed, with incurved or flexuous, occasionally homomallous tips when dry, erect-spreading to spreading, rigid or flaccid when moist, ovate, ovate-lanceolate to lanceolate, acute to acuminate, keeled, plane or plicate, typically ending in a fairly long, slender, flat or terete, not or weakly decurrent, straight or flexuous, hyaline or yellowish, smooth or denticulate to spinulose hair-point, but the awn sometimes very short or even lacking; *margins* entire, bi- to polystratose in one to several rows of cells, plane throughout or incurved in the distal half and sometimes cucullate at the apex, rarely recurved on one or both sides in the proximal part; *costa* single, percurrent, well differentiated from the laminal cells, mostly strongly projecting on the dorsal side, in transverse section semi-terete, consisting of large dorsal and ventral epidermal cells and a distinct internal band of stereids or substereids; *laminal cells* variously 2–3(–4)-stratose, smooth or pseudopapillose, plane or bulging, isodiametric to short-rectangular or oblate, with thick, straight or sinuose walls in the distal portion; *basal cells* uniform throughout, hyaline, narrowly rectangular, with even, very thin, straight, scarcely discernible walls, making the basal leaf areolation lax or proximal juxtacostal cells isodiametric to rectangular, with straight or nodulose walls, uniformly thickened or with cross walls thicker than longitudinal walls and basal marginal cells quadrate to rectangular, with transverse walls thicker than longitudinal walls. *Autoicous* or *dioicous*. *Perichaetial leaves* similar to vegetative leaves, only slightly larger, or concave and convolute in the lower part and 2–3 times as large as vegetative leaves. *Setae* straight and erect, occasionally curved, twisted anticlockwise when dry. *Capsules* emergent to exserted, occasionally immersed, ovoid, ellipsoid or obloid, not or somewhat narrowed at the mouth, abruptly or gradually connected with the seta, symmetric, erect or sometimes inclined to horizontal, stramineous to castaneous, smooth; *annulus* simple and persistent or deciduous and compound, composed of 2–3 rows of isodiametric, inflated and hyaline cells; *operculum* conical, with a short, blunt beak or slanted-rostrate; *exothecial cells* thin- to thick-walled, isodiametric rectangular to transversely rectangular, becoming small and oblate, usually reddish- or brownish-pigmented in several rows at the mouth; *stomata* absent or present in the neck of the urn, bicellular, superficial, with rounded pori; *peristome* single, consisting of 16, lanceolate, ± entire, irregularly divided or cribrose teeth, densely papillose on the inner surface, smooth or slightly papillose proximally and distantly papillose distally on the outer surface. *Spores* spherical, minutely granulose. *Calyptra* mitrate or cucullate, smooth, covering the entire operculum.

The genus *Orthogrimmia* corresponds to *Grimmia* subgen. *Guembelia sensu* Hagen (1909) and Brotherus (1924*b*). It was originally recognised as a subgenus of *Grimmia* by Schimper (1856) and was only recently raised to genus by Ochyra & Żarnowiec (in Ochyra *et al.* 2003). The genus is characterised by straight or only occasionally curved setae, symmetric and smooth capsules, canaliculate to strongly carinate leaves with plane, erect to incurved margins (only very seldom narrowly recurved in the proximal half) and sharply delimited costae which are semi-terete in transverse section and project prominently on the dorsal surface. The structure of the costa and the carinate leaves readily separate *Orthogrimmia* from the closely related *Guembelia*.

A worldwide revision of *Orthogrimmia* (as a subgenus of *Grimmia*) was presented by Muñoz (1998*c*) who recognised eight species in the genus, and Ochyra *et al.* (2003) increased the number of species to 14. Ten of these are distributed mostly in the Northern Hemisphere, occasionally penetrating into the Neotropics in Central America, one in Africa, one in New Zealand and two species are bipolar, including *O. sessitana* which has a wide range in the Antarctic. *Orthogrimmia* is subdivided into two sections, the type section which comprises species with a simple and persistent peristome and sect. *Montanae* containing species with a compound and revoluble peristome.

1. **Orthogrimmia sessitana** (De Not.) Ochyra & Żarnowiec Fig. 128, pl. 35

Orthogrimmia sessitana (De Not.) Ochyra & Żarnowiec *in* Ochyra, Żarnowiec & Bednarek-Ochyra, Cens. Cat. Polish Mosses: 118. 2003. ≡ *Grimmia sessitana* De Not., Atti Reale Univ. Genova 1: 704. 1869. ≡ *G. alpestris* (Schleich. *ex* F.Weber & D.Mohr) Schleich. var. *sessitana* (De Not.) I.Hagen, K. Norsk. Vid. Selsk. Skrift. 1905(5): 22. 1909. ≡ *G. alpestris* fo. *sessitana* (De Not.) Loeske, Laubm. Eur. 1: 104. 1913. — TYPE CITATION: In praeruptis infra Hospitium della Valdobbia, in valle Sessitana, legit Carestia. [Lectotype (*vide* Cao & Vitt 1986: p. 164): RO (*non vidi*); isotypes: 'Grimmia sessitana DNtrs ad Hospitium della Valdobbia legit Carestia. DNtrs' – BM-Davies!, BM!, WRSL! – both Erbario Crittogamico Italiano – Ser. II, No. 55)].

Grimmia donniana Sm. fo. *antarctica* Kuc, Rev. Bryol. Lichénol. 36: 659, *f. 1–2, 3d–e, 4–5.* 1969, ***syn. nov.*** — TYPE CITATION: Antarctic, «the Bunger Oasis», ad stationem nomine Dobrowolski, leg.

S. Rózycki, 25.VI.1959. [Holotype: 'Grimmia Doniana Sm. f. antarctica Kuc okazy karłowate Antarktyda Oaza Bungera. Koło Polskiej Stacji Na gołoborzu, między kamieniami. Na piasku 25.VI.1959. Leg. S. Różycki Det. M. Kuc' – KRAM!]. First synonymised with *Grimmia reflexidens* by Muñoz (1998*c*: p. 395).

Plants small to medium-sized, slender, rigid, in compact, usually very hoary tufts or cushions, slightly lustrous, olive-, brownish- or sordid-green above, brown to blackish-green below. *Stems* erect, sparsely branched, 0.4–1.5 cm tall, radiculose at the base with light brown, smooth, branched rhizoids or sometimes rhizoids scattered throughout most of the stem, in cross-section rounded with central strand surrounded by 4–5 layers of large, thin- to thicker-walled, hyaline medullary cells and 1(–2)-layered cortex of smaller, thick-walled cells with brown walls; *axillary hairs* short, filiform, 3–4-celled, hyaline. *Leaves* rigid, erect and imbricate and not altered when dry, erect-spreading to patent, rigid to somewhat flaccid when moist, 0.8–1.8 mm long (without hair-point), 0.25–0.5 mm wide, ovate, ovate- to oblong-lanceolate or lanceolate, acute, sharply keeled, plane, not or very weakly plicate; *margins* plane or incurved, occasionally recurved on one side below, entire, 2–3(–4)-stratose above, unistratose proximally; *costa* subpercurrent, semiterete, prominently convex on the dorsal surface, sharply differentiated from the laminal cells, in cross-section with 2 ventral epidermal cells, a distinct central stereid band and a row of larger dorsal epidermal cells; *laminal cells* entirely or partly bistratose distally, unistratose with bistratose streaks below, plane or somewhat bulging, smooth or slightly pseudopapillose, isodiametric, rounded-quadrate to short rectangular, thick-walled, 8–12 μm wide in the distal half, becoming elongate downwards, up to 6 times as long as wide near the costa, 20–55 μm long, 8–13 μm wide, with straight, uniformly thickened walls in the juxtacostal area and with transverse walls thicker than the longitudinal walls near the margins. *Autoicous*. *Perigonia* terminal or axillary. *Perichaetial leaves* larger than the vegetative leaves, 1.6–2.7 mm long, 0.5–0.8 mm wide, broadly oblong-ovate or oblong-lanceolate, concave, long-piliferous. *Setae* erect, straight, 1.1–2.5 mm long, sinistrorse when dry. *Capsules* emergent to exserted, ovoid to ellipsoid, 1.1–1.8 mm long, 0.6–0.8 mm wide, erect to slightly inclined, smooth, yellowish-brown, leptodermous; *annulus* simple, persistent; *operculum* conic, with a straight, blunt or acute beak; *exothecial cells* rectangular to oblong-hexagonal, thin-walled, 30–70 μm long, 10–30 μm wide; *stomata* few at the base of the urn, superficial, bicellular, with rounded pori; *peristome teeth* lanceolate, to 180 μm long, 50–70 μm wide at the mouth, entire or irregularly cleft at the apex, often variously perforate, orange- to yellowish-brown below, somewhat paler above, papillose throughout on both ventral and dorsal surfaces. *Spores* globose, papillose, 9–14 μm wide. *Calyptra* cucullate, naked, lobed at the base.

Chromosome number: unknown.

Discussion. *Orthogrimmia sessitana* is a distinct species which is unlikely to be mistaken in Antarctica for any other, especially when fertile. The hoary plants with exserted capsules and plane or incurved leaf margins distinguish it from *Schistidium* species, and the sharply keeled leaves in the distal half, with the mostly bistratose laminal cells, separate it from *Grimmia plagiopodia*. In its keeled leaves and leaf areolation it resembles *Coscinodon lawianus*, but the latter has distinctly plicate leaf lamina with occasional tristratose cells at the plicae, whereas the leaves in the Antarctic plants of *O. sessitana* are not plicate, and beyond the Antarctic they are only occasionally very weakly plicate.

The basal marginal cells are always distinct and never hyaline, lax and thin-walled. They are subquadrate to short-rectangular, up to six times as long as wide, and have the transverse walls thicker than the longitudinal walls. The shape of the basal cells distinguishes this species from *O. donniana* in which the proximal marginal cells are more or less hyaline and inflated, lax-walled, elongate, and extend upwards along the margin in a V-shaped pattern farther than the juxtacostal cells. The emphasis on this character is of special importance because all early collections of *O. sessitana* in the Antarctic were misnamed as *O. donniana* (Cardot 1901, 1908*a*, 1911*a*,*b*, 1913; Bartram 1957; Kuc 1969; Savicz-Lyubitskaya & Smirnova 1970*b*, 1978; Robinson 1972*a*; Lewis Smith & Corner 1973); recently Ochyra (1998*a*) and Muñoz (1998*c*) used the name *Grimmia reflexidens* Müll.Hal. for these plants. However, Maier (2002) and Ochyra (2004*a*) showed that the type material of the latter species correctly belongs within *Coscinodon*. The problem of the correct name for the plants named by Muñoz (1998*c*) as *G. reflexidens* was thoroughly discussed by Maier (2002) who convincingly showed that the oldest available name for them is *G. sessitana*, whereas another potential candidate, *G. lamellosa* Müll.Hal. was identical to *G. alpestris* F.Weber & D.Mohr [= *Orthogrimmia alpestris* (F.Weber & D.Mohr) Ochyra & Żarnowiec].

Orthogrimmia sessitana is gametophytically very variable throughout its wide geographical range, but Antarctic plants growing in more or less similar habitat and climatic conditions are not markedly differentiated morphologically, except only for their size. Some specimens collected in extremely harsh environments on the continent were described as *Grimmia donniana* fo. *antarctica* (Kuc 1969) which, however, except for being dwarf, do not differ materially in their structural characters from more luxuriant plants.

Reproduction in Antarctica. All continental populations are sterile, but in most populations from the maritime Antarctic fully mature sporophytes are produced in profusion. The southernmost known population with perfectly developed sporophytes is on Charcot I. at lat. 69°45′ S.

→

Figure 128. *Orthogrimmia sessitana.* — **1**. Habit. **2**. Portion of shoot with sporophyte. **3**. Cross-section of stem. **4–6**. Axillary hairs. **7–14**. Leaves. **15**. Upper lamina cells. **16**. Mid-leaf cells at margin. **17**. Mid-leaf cells. **18–20**. Basal cells. **21–28**. Cross-sections of leaves, a sequence from base to apex. **29–30**. Perichaetial leaves. **31–32**. Operculate capsules. **33**. Deoperculate capsule. **34**. Calyptra. **35**. Lower exothecial cells and stoma. **36**. Exothecial cells in middle of urn. **37**. Portion of peristome, exothecial cells at orifice and spores. — [1, 34 from *Fukuchi 81*, NIPR; 2–6, 8, 16, 21–28, 30 from *Fukuchi 34*, NIPR; 7, 15, 19, 31–32 from *RO 627/80*; 9, 29, 33 from *Schulz 112*; 10 from *ER 475*, PC; 11 from *Różycki 6*; 12, 17 from *Gain 231C*, PC; 13, 20 from *Różycki 8*, holotype of *Grimmia donniana* fo. *antarctica*; 14, 18 from *Ledingham s.n.*, ii.1986; 35–37 from *RILS 8987A*]. — Scale bars: a – 1 mm (31–34) and 100 μm (15–28, 35–37); b – 1 mm (7–14, 29–30), 2 mm (1–2) and 50 μm (4–6); c – 100 μm (3).

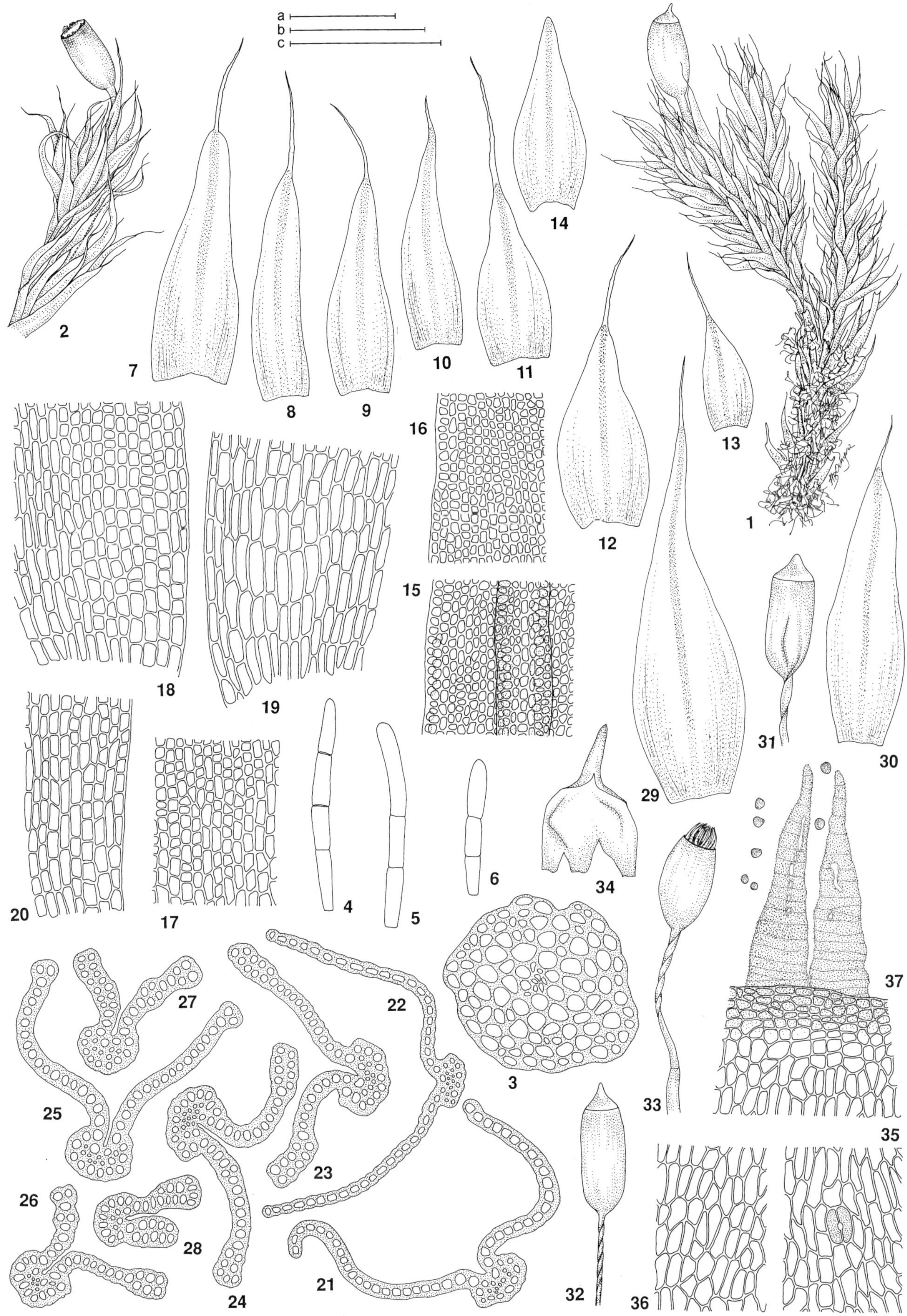
a
b
c
1
2
3
4
5
6
7
8
9
10
11
12
13
14
15
16
17
18
19
20
21
22
23
24
25
26
27
28
29
30
31
32
33
34
35
36
37

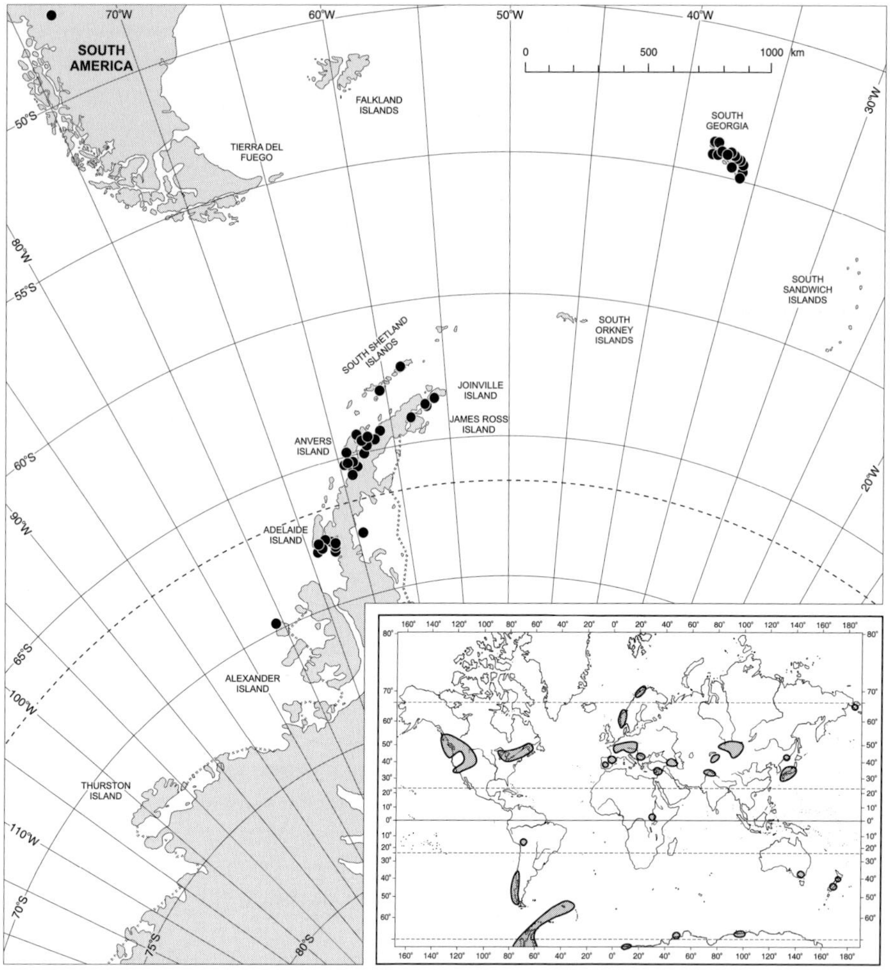

Figure 129. Distribution of *Orthogrimmia sessitana* in the Antarctic Peninsula and Fuegian regions. Inset: global distribution (localities in Victoria Land not indicated).

Habitat. An epilithic moss predominantly of dry habitats and only occasionally found in moist or wet situations. It grows on rock faces, ledges, in cracks and fissures, on boulders in scree, stones on raised beaches, beneath overhangs, and rarely also on stony or gravelly soil (e.g. on Lagoon I. in Marguerite Bay). It is associated with acid rocks including andesite, basalt, granite and greywacke, and some populations have been found on volcanic tuff. It forms pure cushions or tufts without any admixture of other moss species, and its most frequent associates are species of other saxicolous mosses, especially *Andreaea* and *Schistidium*.

World range. A bipolar species with some isolated stations at altimontane elevations in the Andes of Bolivia and the Ruwenzori Mountains of Uganda in the tropics (Fig. 129, inset). In the Northern Hemisphere it is a pan-Holarctic, disjunct oreophyte known from all major mountain ranges in the temperate region and very rarely extending beyond the Arctic Circle in Scandinavia and on the Chukchi Peninsula, growing at high altitudes from 850 to 3500 m (Muñoz 1998*c*). In the Southern Hemisphere it is scattered in the southern Andes of Chile and Argentina in the *Nothofagus* zone, SE Australia and New Zealand, and is quite widespread in the Antarctic.

Distribution in Antarctica. A pan-Antarctic species, widely distributed but scattered and localised and never in great abundance. In the maritime Antarctic it is scattered mainly on the western coast of the Antarctic Peninsula, from Cierva Pt. on the Danco Coast to Charcot I. at lat. 69°45′ S, very rare in the South Shetland Is. but absent from the South Orkney Is., and occasional on the east coast of the Antarctic Peninsula (Fig. 129). On the continent it is known from all sectors except Marie Byrd sector, with the greatest concentration of localities in Dronning Maud Land and Queen Mary Land. It reaches its southernmost known locality in

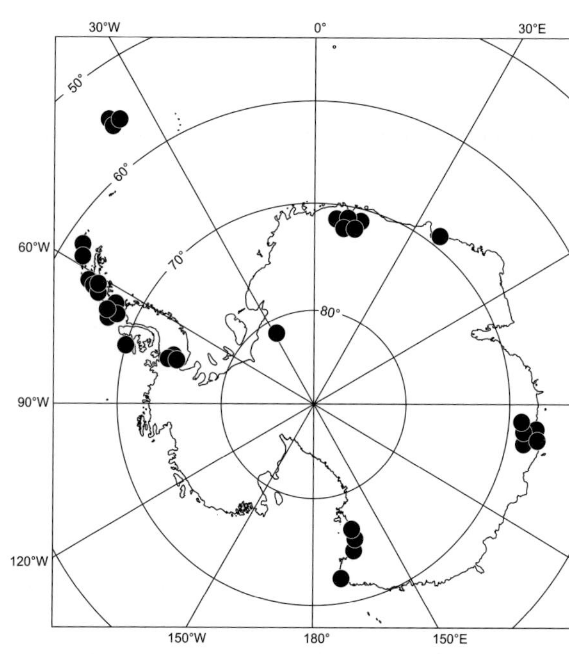

Figure 130. Distribution of *Orthogrimmia sessitana* in Antarctica.

the Shackleton Range at lat. 80°27′ S (Fig. 130). It grows from sea level to 1800 m in the Drygalski Mountains.

Specimens examined. WEST ANTARCTICA. SOUTH SHETLAND IS. King George I. ***Admiralty Bay***: Skua Cliff, 45 m, *RO 627/80* (KRAM). Livingston I. ***Hurd Peninsula***: South Bay, 115 m, *Schulz 112* (KRAM); False Bay, *RMS 69-537 & 69-575* (US).

WEST ANTARCTIC PENINSULA. Danco Coast. Cierva Pt., 10 m, *OB 16* (AAS, KRAM) and 200 m, *OB 24* (AAS, KRAM); Charlotte Bay, Andrée I., 25 m, *RILS 4018* (AAS, KRAM); Cuverville I., 100 m, *RILS 8816* (AAS, KRAM); Birdsend Bluff, 5–35 m, *RILS 4234* (AAS, KRAM); Coughtrey Peninsula, Almirante Brown, *RMS 69-358, 386, 391 & 392* (US) and *Fukuchi 14, 34, 77 & 81* (KRAM, NIPR). Palmer Archipelago. ***Brabant I.***: Mt. Bulcke, 350 m, *AM 102* (AAS, KRAM); eastern coast, 300 m, *ER 475* (BR, PC). ***Melchior Is.*** Gamma I., 5–10 m, *RILS 4158* (AAS, KRAM); Omega I., *RILS 4122* (AAS, KRAM); Lystad I., *PAS 376.2* (FH). ***Anvers I.***: Norsel Pt., 5–10 m, *RILS 1941* (AAS, KRAM). Graham Coast. Rasmussen I., 20 m, *AS 83A* (AAS, KRAM) and *RILS 1920* (AAS, KRAM); Edge Hill, *ca* 35 m, *RWMC 661* (AAS, KRAM); Cape Tuxen, 100–200 m, *RILS 4904* (AAS, KRAM) and *ca* 350 m, *RWMC 685* (AAS, KRAM); Mt. Demaria, 65 m, *RWMC 693* (AAS, KRAM) and *ca* 500 m, *RILS 10651* (AAS, KRAM). Wilhelm Archipelago. ***Argentine Is.***: Uruguay I., *ca* 15 m, *RILS 954 & 955* (AAS, KRAM); Galindez I., Stella Creek, 10 m, *RWMC 405* (AAS, KRAM), 15 m, *RWMC 445* (AAS, KRAM) and 35 m, *REL 1322* (AAS, KRAM); N side of Galindez I., 3–5 m, *RILS 3281* (AAS, KRAM); Woozle Hill, 25 m, *RILS 3282* (AAS, KRAM); Skua I., *RWMC 460* (AAS, KRAM). ***Collins Bay***: largest of Berthelot Is., 100 m, *RWMC 630* (AAS, KRAM). ***Grandidier Channel***: Lahille I., 2–5 m, *RILS 4327* (AAS, KRAM). Loubet Coast. ***Laubeuf Fjord***: Webb I., 100–135 m, *DWB 3A* (AAS, KRAM). Fallières Coast. ***Adelaide I.***: Rothera Pt., 25 m, *RILS 3421* (AAS, KRAM); Reptile Ridge, 175 m, *RILS 10628* (AAS, KRAM) and 300 m, *RILS 3421* (AAS, KRAM); Stork Ridge, 350–500 m, *RILS 1882 & 1891* (AAS, KRAM); Mt. Ditte, *JK 201* (AAS, KRAM). ***Léonie Is.***: Léonie I., 5–50 m, *RILS 8987A* (AAS, KRAM); Lagoon I., 5 m, *RILS 9165* (AAS, KRAM). ***Marguerite Bay***: Jenny I., *Gain 231C & 252A* (PC) and 15–50 m, *RILS 4717* (AAS, KRAM); Blaiklock I., *ca* 12 m, *RILS 848* (AAS, KRAM); Horsehoe I., Lystad Bay, 5–8 m, *RILS 1751* (AAS, KRAM); Camp Pt., 10–15 m, *RILS 4624 & 4671* (AAS, KRAM); Stonington I., *ca* 50 m above glacial ice, *Healy 65* (FH). George VI Sound. ***Charcot I.***: Rils Nunatak, 125 m, *RILS 10474, 10540 & 10708* (AAS, KRAM) and *PC 316 & 326A* (AAS, KRAM).

EAST ANTARCTIC PENINSULA. Trinity Peninsula. Scar Hills, *ca* 85 m, *CGB 213* (AAS, KRAM); Mt. Flora. 115 m, *REL 37B & 40B* (AAS, KRAM). ***Prince Gustav Channel***: Alectoria I., *ca* 50 m, *CGB 13* (AAS, KRAM). Joinville Island group. ***Joinville I.***: N end of Active Sound, 200 m, *RILS 3731* (AAS, KRAM). Bowman Coast. Francis I., 325 m, *RILS 2773A–B* (AAS, KRAM).

ELSWORTH LAND. Behrendt Mountains, 801 m, *PC 564A, 566B, 567A, 568B, 569B, 575B & 578B* (AAS, KRAM), 849 m, *PC 9B & 116* (AAS, KRAM) and 949 m, *PC 591A* (AAS, KRAM); Quilty Nunataks, 856 m, *PC 500C, 501C, 502, 503A, 506B, 508A & 509A* (AAS, KRAM), 860 m, *PC 494 & 498C* (AAS, KRAM) and 870 m, *PC 524, 525, 527 & 529B 533* (AAS, KRAM).

EAST ANTARCTICA. COATS LAND. Shackleton Range. Flat Top, *ca* 1000 m, *REL 1387* (AAS, KRAM).

DRONNING MAUD LAND. Kronprinsesse Märtha Kyst. Veten Nunatak, 1966–1967, *Poole s.n.* (US). Prinsesse Astrid Kyst. ***Schirmacher Oasis***: Posadka I, Feb 1959, *Voronov s.n.* (KRAM, LE). ***Drygalskifjella***: Nunatak 'Yunye', 1400 m, 12 Jan 1960, *Konovalov s.n.* (KRAM, LE) and 1800 m, 1960, *Konovalov s.n.* (KRAM, LE); nunatak 'Malenkiy', 12 Jan 1960, *Konovalov s.n.* (KRAM, LE).

ENDERBY LAND. Kronprins Olav Kyst. Molodezhnaya Station, Feb 1965, *Simonov s.n.* (KRAM, LE).

QUEEN MARY LAND. Knox Coast. Watson's Nunatak, *Harrison 4, 4A, 6, 8 & 9* (BM). ***Bunger Oasis***: Cape Moyes, *Hollerbach 20/612* (AAS, KRAM); Taylor I., *RDS 17324* (KRAM); Transkriptskiy Inlet, 10 m, Feb 1906, *Ledingham s.n.* (KRAM); 'Kholmy Ozernye', *Hollerbach 15/439* (KRAM, LE); Polish Base, 25 June 1959, *Różycki s.n.* (KRAM – type of *G. donniana* fo. *antarctica*); Chugunov I., 1961, *Konovalov s.n.* (KRAM); Hippo I. off the coast of David I., *RDS 17232* (KRAM).

WILKES LAND. Budd Coast. Windmill Is.: Mitchell Peninsula, *RDS 19844* (KRAM).

VICTORIA LAND. Pennell Coast. Cape Adare, *Terra Nova Expedition s.n.* (BM). Borchgrevink Coast. Harrow Peaks, 175 m, *RILS 9554 & 9641* (AAS, KRAM); 'Gondwana' area, 150 m, *RILS 9733, 9739 & 9741* (AAS, KRAM); Cape Sastrugi, *DWHW 225 & 227* (AAS, KRAM) and 100 m, *RILS 10128* (AAS, KRAM); Vegetation I., 125 m, *DWHW 207* (AAS, KRAM) and *RILS 9702* (AAS, KRAM); Star Nunatak, *DWHW 205* (AAS, KRAM); Snowy Pt., *Bargagli 32* (AAS, KRAM).

22. **DRYPTODON** Brid.

Dryptodon Brid., Bryol. Univ. 1: 191. 1826. ≡ *Grimmia* Hedw. subgen. *Dryptodon* (Brid.) Bruch & Schimp. *in* Bruch, Schimp. & W.Gümbel, Bryol. Eur. 3: 139. 1845 [Fasc. 25–28 Mon.: 5]. ≡ *Grimmia* sect. *Dryptodon* (Brid.) Sull *in* Gray, Manual Ed. 2: 638. 1856 (IX). — Lectotype: *Dryptodon patens* (Dicks. *ex* Hedw.) Brid. (*Bryum patens* Dicks. *ex* Hedw.) (*vide* Ochyra *et al.* 1988: p. 264).

Racomitrium Brid. [unranked] *Curviseta* Huebener, Muscol. Germ.: 138. 1833 ['Curviseti']. — Lectotype (*selected here*): *Racomitrium patens* (Dicks. *ex* Hedw.) Huebener (*Bryum patens* Dicks. *ex* Hedw.).

Grimmia Hedw. subgen. *Rhabdogrimmia* Limpr., Laubm. Deutschl. 1: 759. 1889. — Lectotype (*vide* Ochyra *et al.* 2003: p. 121): *Grimmia pulvinata* (Hedw.) Sm. (*Fissidens pulvinatus* Hedw.). First synonymised by Ochyra *et al.* (2003: p. 121).

Trigonodictyon Dixon & P.de la Varde, Ann. Cryptog. Exot. 1: 40. 1928. — Type: *Trigonodictyon indicum* Dixon & P.de la Varde [≡ *Grimmia indica* (Dixon & P.de la Varde) Goffinet & Greven]. First synonymised by Ochyra *et al.* (2003: p. 121).

Plants small and slender, moderate to large and fairly robust, growing in low, dense rounded cushions or forming low or deep, large, loose to dense turves or mats, soft or coarse and rigid, dull or lustrous, often hoary, generally dark, olive- to yellowish-green or yellowish-, reddish- to blackish-brown above, brown to blackish-brown below. *Stems* erect to ascending, occasionally procumbent, rather rigid, repeatedly forked to fasciculately branched, rather sparsely radiculose at the base with smooth, brown, branched rhizoids, less often with rhizoids scattered in clusters throughout the stem length, in transverse section with or without a central strand, consisting of large, thin-walled, hyaline to yellowish-hyaline medullary cells and a uni- to few-layered cortex of smaller cells with incrassate, brown walls, or the cortex poorly differentiated from the medullary cells; *axillary hairs* filiform, uniseriate, hyaline, composed of 3–9 cells. *Leaves* crowded or occasionally loosely disposed on the stem, tightly or loosely appressed, sometimes imbricate, erect, straight to twisted, contorted, curled or curved, sometimes with secund or flexuose apices when dry, erecto-patent to spreading, occasionally recurved to squarrose on wetting, lanceolate, oblong, ovate- or oblong-lanceolate, ligulate to linear, gradually or rapidly tapering to an obtuse or acute, plane or cucullate apex, usually sharply keeled or, occasionally, canaliculate to nearly flat in the distal half, broadly carinate to concave in the proximal part, epilose or piliferous; *hair-point* hyaline to yellowish-hyaline, very variable in length, often as long as or much longer than the lamina on the uppermost leaves, flat or terete, usually brittle and rigid, straight or flexuose, often decurrent along the margin, smooth to sharply denticulate or spinulose; *margins* entire, recurved on one or both sides for varying distances below, plane, deflexed or recurved above, uni- to few-stratose in one to many rows of cells, sometimes forming fleshy limbidia in the upper part; *costa* single, percurrent or ending somewhat below the apex, well differentiated and sharply delimited from the laminal cells, sometimes situated on the bottom of a broad- or narrow-angled fold, in transverse section semi-terete to reniform, prominently convex on the dorsal side, consisting of 2–6 ventral epidermal cells, enlarged dorsal epidermal cells and a prominent central stereid band, occasionally distinctly winged dorsally; *laminal cells* unistratose throughout to variously bi- to polystratose beyond the margins in the distal half, plane or bulging, smooth or sometimes pseudopapillose, transparent to opaque, mostly isodiametric to short-rectangular or oblate, with thick, straight or sinuose walls in the upper half, becoming rectangular to linear, with uniformly thickened walls or with transverse walls thicker than the longitudinal walls, the latter being straight or distinctly sinuose to nodulose, undifferentiated throughout the base or with the basal juxtacostal cells elongate and markedly differentiated from the marginal cells, the latter being quadrate to short-rectangular, with thicker cross walls or sometimes elongate, thin-walled and lax. *Asexual reproduction* occasional by multicellular gemmae produced in conspicuous clusters at the tips of the leaves or on the dorsal side at the base of the costa or the upper lamina. *Gonioautoicous* or *dioicous*. *Perichaetial leaves* similar to but slightly larger than the vegetative leaves, usually convolute in the basal part and prolonged in the narrow acuminate distal part. *Seta* longer than the urn, flexuose and sinistrorse when dry, distinctly arcuate when wet or sometimes straight and much shorter than the urn; *capsule* emergent to exserted, occasionally immersed, symmetrical or sometimes slightly asymmetrical, horizontal to pendent, sometimes erect, ovoid, obloid to ellipsoid, occasionally cyathiform, abruptly narrowed to the seta, very seldom with a long, distinct apophysis, narrowed at the orifice, stramineous, dull or shiny, smooth or distinctly ribbed; *annulus* compound and revoluble, consisting of 3–4 rows of cells, detaching spirally or as groups, sometimes tardily deciduous; *operculum* conical, rostrate, with a straight or slanted, short or long, mostly blunt beak; *exothecial cells* isodiametric to rectangular, thin-walled, sometimes with collenchymatous thickening, occasionally irregular with strongly incrassate walls; *stomata* present at the base of the urn, bicellular, superficial, with rounded pori; *peristome teeth* lanceolate, entire or irregularly cleft at the apex, sometimes cribrose, orange, with the external surface nearly smooth proximally and papillose distally, rarely papillose throughout, and the internal surface papillose throughout, sometimes peristome absent. *Spores* small, brownish, globose, minutely or seldom coarsely granulose. *Calyptra* mitrate or rarely cucullate, covering the operculum.

Dryptodon is the largest segregate of the genus *Grimmia* and is primarily characterised by variously curved setae; symmetric, mostly ribbed and exserted capsules; canaliculate to carinate leaves with margins recurved to varying degrees; and semi-terete to reniform and distinctly differentiated costae that are strongly projected on the dorsal side. However, this combination of characters does not occur in all species of the genus which also contains some with straight setae and smooth capsules, for example *D. elongatus* (Kaulf.) Hartm. and *D. lesherae* (Greven) Ochyra & Żarnowiec, placed here on account of their recurved leaf margins. There are also some species with very short straight setae resulting in immersed capsules, for instance *D. molestus* (J.Muñoz) Ochyra & Żarnowiec and *D. pseudoanodon* (Deguchi) Ochyra & Żarnowiec which, however, have recurved and plane leaf margins, respectively, but otherwise appear to be closely related taxa. These species are in need of further study to elaborate their relationships.

Dryptodon was established as a separate genus by Bridel (1826–1827) who initially placed in it 17 species, but in the supplement to Vol. 1 of *Bryologia universa* (which is bound with Vol. 1 but published in 1827) he refined the concept of this genus when removing from it *D. fontinaloides* (Hook.) Brid., which was transferred to the newly described genus *Hydropogon*. At the same time, he added a further eight species to the genus. Bridel (1826–1827) subsequently divided *Dryptodon* into two groups, *Curviseti* and *Rectiseti*, which were not formally described, although it is assumed that they refer to groups of species with curved and straight setae. The majority of species originally placed in *Dryptodon* belonged to *Grimmia* subgen. *Rhabdogrimmia sensu* Hagen (1909), and only relatively few species were representatives of modern genera such as *Bucklandiella*, *Coscinodon*, *Guembelia*, *Campylostelium* and *Cynodontium*.

The generic name *Dryptodon* has been variously interpreted by bryologists, mostly because for a long time it was not lectotypified. Quite often *Grimmia atrata* Hornsch. was included in this genus (e.g. Limpricht 1889; Roth 1904; Loeske 1930; Jensen 1939; Podpěra 1954; Savicz-Ljubitskaya & Smirnova 1970*c*), although nomenclaturally this was inadmissible as Bridel (1826–1827) did not place this species in the original concept of *Dryptodon* and retained it in *Grimmia*. This species is now placed in the separate monotypic genus *Streptocolea* (I.Hagen) Ochyra & Żarnowiec (Ochyra *et al.* 2003; Ochyra & Bednarek-Ochyra 2004). The other frequent option has been the consideration of *Dryptodon* as a monotypic genus for *D. patens* (e.g. Ochyra & Szmajda 1978; Corley *et al.* 1981; Duell 1984; Ignatov & Afonina 1992), and this interpretation was corroborated by formal lectotypification of the generic name with this species (Ochyra *et al.* 1998). Because there has been a considerable divergence of opinion about the relationship of *D. patens* (for a review see Deguchi 1979), *Dryptodon* as thus understood was considered to be an intermediate genus between *Grimmia* and *Racomitrium* (Crundwell 1971). However, it appears that the *Racomitrium*-like characters of *D. patens* including the presence of a basal membrane on the peristome, deeply divided peristome teeth and general appearance of the plants can also be easily found in *Grimmia s. lato*, so their diagnostic value is doubtful. Deguchi (1979) stated that *D. patens* has nodulose walls of the epidermal cells of the vaginula and this would be a strong indicator of an affinity of the species to *Racomitrium*. However, the epidermal cells of the vaginula are thick but straight-walled and never sinuose-walled like those in species of *Racomitrium s. lato*, so this character also does not provide a strong argument for the unique position of this species. One of the most peculiar features of *D. patens* which is often stressed is the presence of wings on the dorsal surface of the costa. However, weakly winged costae can also be observed in some phenotypes of *Grimmia hartmanii* (Loeske 1930; Bednarek-Ochyra 1996) with which *D. patens* seems to be most closely related. They also share ribbed capsules, curved setae and divided peristome teeth. Therefore, the arguments for recognising *Dryptodon* as a monotypic genus are weak since its generitype shares all diagnostic characters with other species which are currently placed in it (Ochyra *et al.* 2003).

As presently understood *Dryptodon* is a large genus of about 45–50 species distributed worldwide, but in the austral polar region it is very poorly represented. So far only one species, *D. fuscoluteus*, has been discovered at a single station in the Antarctic, and another species, *D. orbicularis* (Wilson) Ochyra & Żarnowiec, is known on Îles Kerguelen whence it was described as *Grimmia stolonifera* Müll.Hal. (Muñoz & Pando 2000).

1. **Dryptodon fuscoluteus** (Hook.) Ochyra & Żarnowiec Fig. 131

Dryptodon fuscoluteus (Hook.) Ochyra & Żarnowiec *in* Ochyra, Żarnowiec & Bednarek-Ochyra, Cens. Cat. Polish Mosses: 122. 2003. ≡ *Grimmia fuscolutea* Hook., Musci Exot. 1: *tab. 63*. 1818 ['fusco-lutea']. — Type citation: In frigidis Tobeccae et Islahuacae, altitudine 1330 hexapod. (Regnum Mexicanum). Humboldt et Bonpland. [Lectotype (*vide* Muñoz & Pando 2000: p. 31): 'N° 2673. S. Am. Humboldt 50' – BM-Hooker!].

Grimmia fastigiata Cardot, Bull. Herb. Boissier Sér. 2, 5: 1003. 1906. — Type citation: Terre-de-Feu: Ushuaia, mont Martial. [Lectotype (*selected here*): 'Svenska Sydpolarexpeditionen 1901–03. Nr 70. Grimmia fastigiata Card. sp. nova Terre-de-Feu: Ushuaia, mt. Martial, alt. 1275 m, 11/3 1902 Carl Skottsberg Det. J. Cardot' – PC-Cardot!; isotype: S!]. First synonymised by Muñoz (1999: p. 138).

Plants small, rather slender but rigid, densely caespitose, somewhat lustrous, yellowish- to brown-green above, blackish-brown to dark brown below, mostly hoary because of long hyaline awns. *Stems* erect, about 1 cm tall, fastigiately branched, sparsely radiculose below with brownish, smooth, branched rhizoids, in cross-section with a 1(–2)-stratose cortex of small cells with thick, brown walls, 3–5 stratose medulla of large, thin- to firm-walled, hyaline cells and small but distinct central strand; *axillary hairs* filiform, hyaline throughout, 4–6-celled. *Leaves* erect, appressed and slightly altered when dry, erect when moist, 1.3–2.3 mm long, 0.3–0.5 mm wide, lanceolate, gradually long acuminate, not plicate, keeled above, ending in a terete, flexuose, denticulate hair-point, to 2 mm long; *margins* entire, recurved on one or both sides up to ⅔ the leaf length, mostly bistratose distally in 1–2 rows of cells; *costa* percurrent, 40–50 μm wide at the base, sharply differentiated, in cross-section semi-terete, prominent on the dorsal surface, consisting of 2 ventral epidermal cells, a distinct median stereid band in the lower ⅔ and a row of larger dorsal epidermal cells; *laminal cells* unistratose throughout, except for bistratose margins, isodiametric, subquadrate, rounded-quadrate or oblate to short rectangular above, 10–30 μm long, 6–10 μm wide, with thick and sinuose walls; *basal juxtacostal cells* rectangular, 40–100 μm long, 7–12 μm wide, with moderately thickened, nodulose walls, becoming rectangular, 30–80 μm long, 7–12 μm wide, thin- and straight-walled towards the margins. *Autoicous*. *Perigonium* bud-like, positioned below the perichaetium. *Perichaetial leaves* larger than the vegetative leaves, 2.5–3.5 mm long, 0.6–0.7 mm wide, long-piliferous, convolute. *Setae* arcuate, 1.3–2.5 mm long, yellow-brownish, sinistrorse. *Capsules* emergent or exserted, ovoid to obloid, 0.7–1.2 mm long, 0.5–0.6 mm wide, symmetric, stramineous, smooth or faintly striate; *annulus* compound, biseriate, deciduous; *operculum* conic, mammillate or obtusely beaked; *exothecial cells* irregular, oblong-hexagonal to rectangular, 25–60 μm long, 10–20 μm wide, thin-walled, becoming oblate in 3–4 tiers at the mouth and forming a golden-orange strip; *stomata* few at the base of the urn, bicellular, superficial, rounded-pored; *peristome teeth* 16, lanceolate, 170–200 μm long, 50–60 μm wide at the base, orange, moderately perforated, densely papillose on both ventral and dorsal surfaces. *Spores* globose, brownish, 9–12 μm wide, minutely papillose. *Calyptra* cucullate, naked.

Chromosome number: unknown.

Discussion. *Dryptodon fuscoluteus* can be recognised by the combination of the recurved leaf margins, emergent to exserted capsules attached to the distinctly arcuate setae and uniformly thin-walled and elongate basal marginal cells. Muñoz (1999) stressed the yellowish shine of the leaf bases to be diagnostic for this species, but this feature is less obvious in Antarctic material although correspondence in other taxonomically essential characters is perfect. In the Antarctic the species may be mistaken only for *Guembelia longirostris* due to its recurved leaf margins and nodulose basal juxtacostal cells, but in the latter the proximal marginal cells are short and have much more thickened transverse than longitudinal walls.

Dryptodon fuscoluteus is a poorly known species which has long been known in the Northern Hemisphere under the name *Grimmia apiculata*. As emphasised by several authors (Mårtensson 1956; Deguchi 1979; Muñoz 1998*a*) this name has been the source of permanent nomenclatural confusion which was introduced by Hornschuch (1819*a–b*). However, this problem was resolved by Crum (1994*c*) who pointed out that the oldest name for the plants commonly named *G. apiculata* is *G. fuscolutea*, which name was used by Hooker (1818) for a Mexican species and recently was transferred to *Dryptodon* (Ochyra *et al.* 2003).

Because *Dryptodon fuscoluteus* varies markedly in both gametophyte and sporophyte characters, its name has about a dozen heterotypic synonyms used for species described from various parts of its wide geographical range (Muñoz & Pando 2000). Material from Tierra del Fuego was described as *Grimmia fastigiata* (Cardot 1905*a*), and Dixon & Watts (1918) subsequently reported this species from Queen Mary Land on the Antarctic continent. However, this material actually belongs to *Orthogrimmia sessitana*.

The relationships of *Dryptodon fuscoluteus* have been subject to some controversy. Hagen (1909) and Brotherus (1924*b*) placed this species in subgen. *Guembelia* which is characterised by having an undifferentiated costa, semi-elliptic in transverse

a
b
c
1
2
3
4
5
6
7
8
9
10
11
12
13
14
15
16
17
18
19
20
21
22
23
24
25
26
27
28
29
30
31
32
33
34
35
36

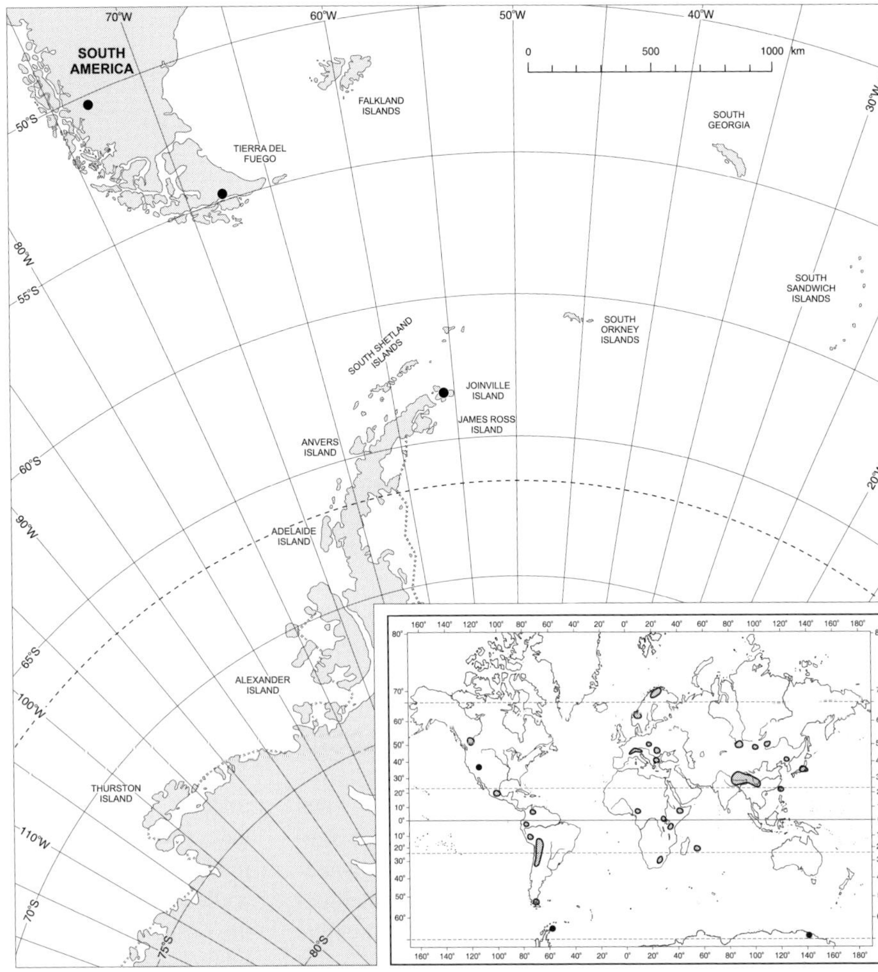

Figure 132. Distribution of *Dryptodon fuscoluteus* in the Fuegian region and Antarctica. Inset: global distribution.

section, plane to incurved leaf margins, concave leaves and straight seta. None of these characters is present in this species, which fits the concept of *Dryptodon* on account of its distinct costa that is semi-terete in cross-section, recurved leaf margins, keeled leaves and curved seta. The same affinity for this species was also suggested by Schimper (1856) and Buysson (1883).

Reproduction in Antarctica. Both Antarctic populations are in fine fruiting condition and have a number of well developed and fully mature sporophytes.

Habitat. On dry, exposed rock face at an elevation of 250 m, with a small admixture of *Andreaea regularis*. The cushions are also colonised by some crustose lichens.

World range. A bipolar species with intermediate occurrences in the tropical regions (Fig. 132, inset). In the Northern Hemisphere *Dryptodon fuscoluteus* is a highly disjunct oreophyte with the main centre of its occurrence in Europe where it is relatively frequent in the Alps and occasional in the Carpathians, Bulgaria and Scandinavia (Mohan 1975; Duell 1984; Bachurina & Mel'nichuk 1988; Greven 1995; Söderström 1996). In Asia it is scattered in mountains of the temperate and subtropical zone including the Altai (Muñoz & Pando 2000), northen Mongolia (Tsegmed 2000*a*,*b*), Honshu in Japan (Deguchi 1979), the Himalayas, Xizang and Sichuan in China (Cao & Vitt 1986; Cao *et al.* 2003) and Taiwan (Muñoz & Pando 2000). The species is relatively frequent in the Neotropics, especially in the Andes, ranging from Venezuela to Bolivia (Muñoz 1999), with an isolated occurrence in Mexico (Crum 1994*c*), and in sub-Saharan Africa where it extends from Ethiopia to Lesotho in East Africa, with some isolated localities in Cameroon in West Africa and on Réunion in the South Indian Ocean (Muñoz & Pando 2000). In the cool temperate and polar regions of the Southern Hemisphere the species is exceedingly rare and so far it has been recorded only from Tierra del Fuego and Antarctica.

Distribution in Antarctica. *Dryptodon fuscoluteus* is known from single collections from Joinville I. at the Antarctic Peninsula and Terre Adélie on the continent (Fig. 132).

Specimen examined. WEST ANTARCTICA. EAST ANTARCTIC PENINSULA. **Joinville Island group.** ***Joinville I.***: Mt. Alexander, 250 m, *RILS 3677* (AAS, KRAM).

EAST ANTARCTICA. Terre Adélie. No locality and collector, Aug 1912 (HO).

◄——

Figure 131. *Dryptodon fuscoluteus.* — **1**. Habit. **2**. Portion of branch. **3**. Cross-section of stem. **4–7**. Axillary hairs. **8–13**. Leaves. **14**. Leaf apex. **15**. Upper cells at margin. **16**. Mid-leaf cells at margin. **17**. Lower mid-leaf cells. **18–19**. Basal cells. **20–27**. Cross-sections of leaves, a sequence from base to apex. **28–30**. Perichaetial leaves. **31**. Deoperculate capsule, seta and vaginula, wet. **32**. Operculate capsule, dry. **33**. Deoperculate capsule, dry. **34**. Lower exothecial cells. **35**. Exothecial cells in the middle of the urn. **36**. Portion of peristome, exothecial cells at orifice and spores. — [All from *RILS 3677*]. — Scale bars: a – 1mm (2, 8–13, 28–33); b – 100 µm (3–7, 14–27, 34–36); c – 1 mm (1).

23. RACOMITRIUM Brid.

Racomitrium Brid., Muscol. Recent. Suppl. 4: 78. 1819. ≡ *Trichostomum* Hedw. sect. *Racomitrium* (Brid.) Duby, Bot. Gall. 2: 573. 1830. ≡ *Grimmia* Hedw. sect. *Racomitrium* (Brid.) Müll.Hal., Syn. Musc. Frond. 1: 806. 1849 ['Rhacomitrium']. — Lectotype (*vide* Schimper 1860: p. 730): *Racomitrium lanuginosum* (Hedw.) Brid. (*Trichostomum lanuginosum* Hedw.).

Rhacomitrium Brid. *ex* Lorentz, Bryol. Notizb.: 94. 1864, *orthogr. pro* Racomitrium Brid.

Racomitrium Brid. 1 [unranked] *Lanuginosa* Kindb., Eur. N. Am. Bryin. 2: 235. 1898. ≡ *Racomitrium* Brid. sect. *Lanuginosa* (Kindb.) Nog., J. Hattori Bot. Lab. 38: 346. 1974. — Type: *Racomitrium hypnoides* Lindb. [= *R. lanuginosum* (Hedw.) Brid.].

Plants medium-sized and relatively slender to large and robust, coarse, rigid, somewhat lustrous, mostly hoary, in dense, deep, often extensive patches or mats, greyish-, brownish- or yellowish-green, yellow or yellow-, grey- to blackish-brown. *Stems* decumbent to erect, elongate, stiff, tumid and brittle, variously but mostly pinnately branched, with many short, lateral, horizontal, tuft-like branchlets, lacking rhizoids or somewhat radiculose at the base with light brown, smooth rhizoids, in transverse section without a central strand and with a thick, multilayered cortex of small, stereid cells with brown to reddish-brown, incrassate walls, surrounding a central medulla of large, hyaline, thick-walled cells; *axillary hairs* filiform, hyaline throughout, composed of 8–12, relatively short cells. *Leaves* crowded, loosely imbricate, erect to slightly secund, with flexuose or, rarely, crisped points when dry, loosely erect to erect-spreading or widespreading-recurved when moist, narrowly long-lanceolate to ovate-lanceolate, concave below, broadly carinate to canaliculate above, gradually tapering to a long, slender, hyaline acumen; *hair-point* hyaline, densely papillose, eroso-dentate, long-decurrent, sometimes to ⅔ down the leaf margins, with the decurrencies flat or ruffled to wavy, with the teeth spreading at 20°–90°, distinctly papillose throughout to nearly smooth in the distal half; *margins* unistratose throughout, variably recurved to revolute, entire below, coarsely dentate along the hyaline border formed by decurrencies from the hyaline hair-point; *costa* single, percurrent, yellow-brownish, sharply demarcated from the laminal cells, in transverse section strongly convex on the dorsal surface, semi-terete to short-rectangular, bistratose, with 2 large ventral epidermal cells in the upper half, becoming reniform to long-rectangular and markedly flattened, tristratose in the proximal portion, with 3–4 large ventral cells and small stereid median and dorsal cells; *laminal cells* unistratose throughout, dull and opaque, with strongly incrassate and sinuose longitudinal walls, distinctly papillose with large, flat papillae covering the longitudinal walls and almost the whole of the lumina except for a narrow central slit, with small, rounded secondary papillulae over both ventral and dorsal surfaces, short-rectangular in the distal part, becoming long-rectangular in the median and lower parts; *basal cells* long-rectangular to linear with incrassate and nodulose longitudinal walls, forming a broad orange strip along the insertion; *alar cells* undifferentiated; *basal marginal cells* yellowish-hyaline, rectangular, translucent, with straight or occasionally weakly sinuose, thin to moderately thickened walls, forming 1(–2)-seriate band, consisting of to 30 cells. *Dioicous*. *Inner perichaetial leaves* oblong, oblong-lanceolate to elliptical, abruptly constricted into an awn terminated with a short, hyaline, weakly serrulate to fairly smooth, papillose hair-point, convolute and lax-areolated in the proximal half. *Setae* singly or often 2–3 per perichaetium, sinistrorse and mostly flexuose on drying, blackish-red with age, strongly papillose. *Capsules* exserted, erect, straight, symmetric, ovoid to ovoid-cylindrical, distinctly narrowed at the orifice, somewhat ventricose in the base, lacking a neck and rapidly constricted into the seta, smooth, brown, lustrous; *annulus* compound, revoluble, composed of 3–4 rows of thick-walled, transparent cells, progressively larger from the lower to the upper row; *operculum* straight, erect, conic-rostrate; *exothecial cells* variable, isodiametric to elongate, irregular to oblong or oblong-hexagonal, with thin to moderately thickened walls, becoming quadrate, thick-walled and forming an orange-brown, 5–6-seriate band at the orifice; *stomata* numerous at the base of the urn, bicellular, superficial, with round pori; *peristome* single, consisting of 16, long, reddish-brown teeth arising from a low basal membrane, split nearly to the base into 2 filiform, terete, strongly papillose branches, with a distinct prostome. *Spores* spherical, pale yellowish, finely roughened to nearly smooth. *Calyptra* cucullate, smooth to slightly scabrous above, split at the base into a few lobes.

Until recently, apart from *Grimmia s. lato*, *Racomitrium* was the largest genus of the Grimmiaceae. Wijk *et al.* (1967, 1969) listed under this name over 200 species and nearly the same number of various infraspecific taxa, but in effect they accepted only 81 species. However, this number did not reflect the real number of species, and subsequent taxonomic studies of some complexes in this genus (e.g. Frisvoll 1983*a*, 1988; Bednarek-Ochyra 2006) yielded some new species or reinstated species which had previously been 'sunk'. The best example of the latter was provided by Clifford (1955) who lumped many austral species with *Racomitrium crispulum* (Hook.f. & Wilson) Hook.f. & Wilson. These represented distinct species, as subsequent investigations convincingly showed (e.g. Roivainen 1955*a*; Bell 1974; Deguchi 1984, 1987). In her world synopsis of *Racomitrium*, Bednarek-Ochyra (1995*a*) accepted 61 species in this genus but since then several new species have been described (e.g. Buck 1997; Ochyra & Bednarek-Ochyra 1999; Bednarek-Ochyra 1999, 2000, 2004*a*; Bednarek-Ochyra & Ochyra 1996, 2000, 2003). Thus, it may be assumed that the real number of taxa in *Racomitrium s. lato* is approximately 80 species. *Racomitrium* is a genus of fairly large and robust mosses growing in a wide range of lithic habitats, but primarily associated with various types of acidic, non-calcareous rocks.

Moreover, some species are pioneers growing on sandy or gravelly soil or on soil among rocks, and a few species are typical aquatic and rheophytic mosses thriving in swiftly flowing water in streams and waterfalls.

In the traditional circumscription (e.g. Bridel 1819; Bruch *et al.* 1845*b*; Schimper 1860, 1876; Limpricht 1889; Brotherus 1902, 1924*b*; Loeske 1913, 1930), *Racomitrium* was considered as a distinct genus which was easy to characterise and to distinguish on the basis of its leaf areolation of thick and strongly sinuose to nodulose longitudinal cell walls, coupled with the special peristome type which, in its typical form, has linear teeth arising from a low or high basal membrane and divided nearly to the base into two filiform, somewhat paired segments that are equally thickened and less prominently trabeculate on both dorsal and ventral sides. In addition, the genus was characterised by the consistent lack of central strand, the usual presence of a prostome, the sinuose-walled epidermal cells of the vaginula, and a cladocarpous sex condition. This combination of characters made *Racomitrium* a readily recognisable genus and any new species were accurately placed in it. Hence, this generic name almost entirely lacks any heterotypic synonyms, the only exception being *Bucklandiella* Roiv. which was originally used for a peculiar rheophytic species from the Fuegian region (Roivainen 1955*b*, 1972).

The infrageneric classification of *Racomitrium* was quite poorly developed, and Kindberg (1898) was the first to divide the genus into four unranked groups, namely *Lanuginosa*, *Papillosa*, *Canescentia* and *Laevifolia*, which were subsequently validated as sections by Noguchi (1974). Bednarek-Ochyra (1995*a*) presented a detailed infrageneric classification of *Racomitrium* which was largely based on the peristome structure, the shape of the perichaetial leaves and the papillosity of the laminal cells. The genus was divided into four subgenera, namely subgen. *Racomitrium*, subgen. *Cataractarum* Vilh., subgen. *Niphotrichum* Bednarek-Ochyra and subgen. *Ellipticodryptodon* (Vilh.) Bednarek-Ochyra & Ochyra, which corresponded well to the four aforementioned groups recognised within this genus by Kindberg (1898).

A subsequent revisionary study of *Racomitrium* showed that the genus in its traditional circumscription was an artificial heterogeneous taxon and, as a result, it was split into four genera, *Racomitrium s. str.*, *Codriophorus* P.Beauv., *Niphotrichum* (Bednarek-Ochyra) Bednarek-Ochyra & Ochyra and *Bucklandiella* Roiv. (Ochyra *et al.* 2003). They are considered to be distinctive taxa through various combinations of unique or otherwise presumably advanced characteristics for the Grimmiaceae, such as presence of pellucid alar cells, variously papillose laminal cells, papillosity of seta and calyptra, torsion of seta, shape of peristome teeth, anatomy of costa, and shape of hair-point.

The segregates of the former large and artificial genus *Racomitrium* are natural and well established taxa which are easily distinguished from one another. Nonetheless, these segregates still constitute a distinctive and isolated group of genera, well diagnosed by a set of characters of the former aggregate genus *Racomitrium* that warrant recognition as a separate subfamily Racomitrioideae in the Grimmiaceae. *Racomitrium s. lato* has sometimes been associated with *Ptychomitrium* and *Campylostelium* and placed in the subfamily Ptychomitrioideae in the Grimmiaceae (Brotherus 1902; Churchill 1981), and usually treated as a separate family (Brotherus 1925). Despite their overall morphological similarity these taxa seem to be only remotely related and we prefer to keep *Racomitrium* and its allied genera apart. The Racomitrioideae are well characterised by their consistently sinuose-nodulose walls of the leaf cells and epidermal cells of the vaginula and non-plicate calyptrae, and the plants are cladocarpous. By contrast, in *Ptychomitrium* the leaf cells are straight or only weakly sinuose, the calyptrae deeply plicate and the plants acrocarpous. In addition, a preperistome is lacking, the peristome teeth lack trabeculae and have a characteristic air gap at the base of the teeth (Edwards 1979). Moreover, the plants are cryptoicous because the male branches are small and arise from the base of vaginula inside the perichaetial leaf circle (Deguchi 1977).

The nomenclature of the infrageneric taxa and segregates of the former *Racomitrium* was strongly influenced by typification of this name. Until the mid-1990s *R. canescens* was commonly accepted as lectotype and this lectotypification was made by Pfeiffer (1874). However, a much earlier and long-overlooked lectotypification of *Racomitrium* was made by Schimper (1860) who designated *R. lanuginosum* as lectotype. This lectotypification was unfortunate because *Racomitrium s. str.* is the smallest segregate of the broadly conceived genus *Racomitrium* but it is was in accordance with the rules of the Code (McNeill *et al.* 2006).

Racomitrium s. str. is a distinct genus which is characterised by the following combination of characteristics: strongly papillose setae twisted to the left; long hyaline, strongly papillose hair-points which are long decurrent along the leaf margins and eroso-dentate; large flat papillae with small secondary papillulae densely covering the longitudinal cell walls and almost the whole lumina, except for a narrow slit in the middle; very long peristome teeth divided to the base into 2(–3) filiform branches and slightly ventricose capsules at the base. The genus consists of only three species, namely *R. lanuginosum* (Hedw.) Brid. with two subspecies, subsp. *lanuginosum* and subsp. *geronticum* (Müll.Hal.) Vitt & Marsh, *R. pruinosum* (Wilson) Müll.Hal. and *R. patagonicum* Bednarek-Ochyra & Ochyra, and a key for their determination has been presented by Bednarek-Ochyra & Ochyra (2003). Of these, only the type subspecies of *R. lanuginosum* is distributed worldwide, whereas the remaining taxa are strictly austral in distribution, widespread and locally common, and very abundant in the southern cool-temperate region and in Subantarctica. In the Antarctic, the genus is represented only by *R. lanuginosum* subsp. *lanuginosum*, which is exceedingly rare and known only on the volcanic Deception I. in the South Shetland Is. archipelago.

1. **Racomitrium lanuginosum** (Hedw.) Brid.

Fig. 133, pl. 34

Racomitrium lanuginosum (Hedw.) Brid., Muscol. Recent. Suppl. 4: 79. 1819. ≡ *Trichostomum lanuginosum* Hedw., Sp. Musc. Frond.: 109. 1801. — Type citation: Montana, arida, arenosa, lapidosa loca; (maxime in frustis granitis montium subalpinorum, Bructeri, Sudetum.). [Lectotype (*vide* Frisvoll 1987: p. 83): the left specimen on the sheet labelled 'Trichostomum lanuginosum Hedw. St. Crypt. Vol. III. p. 3. t. 2' corresponding to fig. 2 on table 2 in Hedwig (1792) – G-Hedwig/Schwägrichen!].

Grimmia glacialis Müll.Hal. *in* Neumayer, Int. Polarforsch. Deutsch. Exped. 2: 317. 1890, *hom. illeg.* ≡ *Racomitrium glaciale* Kindb., Enum. Bryin. Exot.: 108. 1891. — Type citation: Austro-Georgia, Brockenthal ad rupes cespites magnos sistens; Whalerthal und am grossen Gletscher, 10. Febr. 1883. [Lectotype (*vide* Bednarek-Ochyra & Ochyra 2003: p. 187): '№ 45 Rhacomitrium glaciale CMüll. nsp. Fundort: Brockenthal. An Felasen in grossen Polstern. Findet sich ausserdem im unt. Whalerthal u. am grossen Gletscher Süd-Georgien 10/II 83 Will' – HBG!; isotype: M!]. First synonymised by Cardot (1908*a*: p. 219).

Plants small to medium-sized, fairly slender, coarse and rigid, growing singly or in small, loose, hoary patches, olive- to yellowish-green at the tips, black to brownish-black below. *Stems* erect, 0.5–1.5 cm long, stiff, variously branched, usually with elongate lateral branches or with short tuft-like horizontal branchlets, exceptionally small, underdeveloped plants unbranched, radiculose at the base with pale brown, smooth, poorly branched rhizoids, in transverse section without a central strand and with a 2(–3)-stratose cortex of small, brown stereids and a central medulla of large, hyaline, thick-walled cells; *axillary hairs* filiform, hyaline throughout, consisting of 9–10 relatively short, barrel-shaped cells. *Leaves* crowded, erect to slightly falcato-secund and appressed when dry, erecto-patent when moist, 2.5–3.5 mm long, 0.6–0.9 mm wide, narrowly lanceolate to oblong-lanceolate, gradually narrowed to a long slender hair-point, non-decurrent, canaliculate distally, concave to broadly carinate in the proximal part; *hair-point* hyaline, flat to canaliculate, densely papillose, broadly evenly decurrent to ¼–⅓ along the margins, eroso-dentate to nearly smooth, with the teeth short, to 30 μm long, sharp or blunt, spreading at the angle 30°–90°, moderately papillose throughout; *margins* unistratose all around, recurved on one side to the middle, plane to erect elsewhere, entire below, toothed to sinuate in echlorophyllose part formed by decurrencies from the hyaline hair-point; *costa* single, percurrent, yellow-brownish, distinct, 60–80 μm wide at the base, in transverse section rectangular to reniform, strongly convex on the dorsal surface, bistratose with two large ventral epidermal cells in the distal half, becoming partially tristratose, with 2–3 large ventral epidermal cells towards the base; *laminal cells* unistratose throughout, dull, with thickened, sinuose longitudinal walls, distinctly papillose with large flat papillae over these walls, covering almost the whole of the lumina except for a narrow central slit, with rounded secondary papillae over both ventral and dorsal surfaces, short-rectangular, (10–)15–40 μm long, 7–8 μm wide in the distal part, becoming long-rectangular, 50–60 μm long, 6–8 μm wide in the median and lower parts; *basal cells* long-rectangular, to 80 μm long, 7–8 μm wide, strongly nodulose, forming a broad orange strip along the insertion; *alar cells* not differentiated; *basal marginal cells* rectangular, 20–40 μm long, 7–9 μm wide, pellucid, with straight or weakly nodulose walls, forming a uniseriate translucent band, consisting of up to 20 cells. [*Dioicous*. *Inner perichaetial leaves* oblong-lanceolate to elliptical, 1.5–3.5 mm long, 0.4–0.7 mm wide, concave and sheathing at the base, abruptly contracted into a short, hyaline-tipped, smooth or nearly so acumen. *Setae* 1–3 per perichaetium, erect, flexuose, brown to reddish-brown, papillose, 3–7 mm long, twisted to the left. *Capsules* exserted, erect, straight, symmetric, ovoid to obloid, 1.2–1.5 mm long, 0.6–0.7 mm wide, smooth, brown, glistening; *annulus* compound, revoluble, 3–4-seriate; *operculum* straight, erect, conic-rostrate; *exothecial cells* isodiametric to rectangular, 20–50 μm long, 15–20(–25) μm wide, thin- to moderately thick-walled, becoming quadrate, thick-walled at the orifice, forming an orange-brown, 5–6-seriate band; *stomata* many at the base of the urn, bicellular, superficial, with round pori; *peristome* single, consisting of 16 teeth, 500–700 μm long, arising from a low basal membrane, divided almost to the base into 2 filiform, terete, strongly papillose branches. *Spores* spherical, 9–12 μm in diameter, pale yellowish, finely roughened to nearly smooth. *Calyptra* cucullate, smooth to slightly scabrous above.]

Chromosome numbers: unknown in the Antarctic and Subantarctic; n = 12 (1 record), n = 13 (2 records) and n = 14 (2 records) from the Holarctic (Fritsch 1991).

Discussion. *Racomitrium lanuginosum* is one of the most distinctive and easily known of all mosses, its hyaline, eroso-dentate, strongly papillose hair-points, long decurrent as a white border along the upper portion of the leaf, being the most characteristic feature of this species, easy to observe in the field with a hand-lens. Other characteristic traits of *R. lanuginosum* are large, flat papillae densely covering the longitudinal walls of the leaf cells with additional secondary small papillulae over both the dorsal and ventral surfaces; the lack of alar cells; the presence of the long basal marginal border of yellowish-hyaline, translucent, straight-walled cells; papillose setae; barrel-shaped capsules distinctly ventricose at the base; and long peristome teeth, divided to the base into two filiform segments. The last three characters cannot be observed in Antarctic material since the plants are always sterile.

Racomitrium lanuginosum exhibits remarkable variability throughout its global range especially with regard to the length, papillosity and angle of the awn teeth, and two subspecies have been distinguished on this basis (Vitt & Marsh 1988). The type subspecies has shorter teeth, less than 30 μm, that spread at 55°–90°, mostly more than 70°, whereas, in contrast, subsp. *geronticum* (Müll.Hal.) Vitt & Marsh has longer (40–65 μm) distal awn teeth, which spread at 20°–50°, mostly less than 45°. The latter is a South American cool-temperate species, occurring in the *Nothofagus* zone from the Valdivian region to Tierra del Fuego and the Falkland Is. and extending eastwards to subantarctic South Georgia (Vitt & Marsh 1988; Ochyra *et al.* 2002), while the type subspecies is worldwide in distribution. The Antarctic plants have mostly poorly developed dentation of the hyaline awn and some hair-points appear to be nearly smooth. The teeth are quite short, mostly blunt and they often appear as low triangular protrusions spreading at right angles or nearly so. The two other species of the genus, *R. pruinosum* and *R. patagonicum*, are also austral in distribution but are unlikely to be mistaken for *R. lanuginosum*. The former species is known by its awn teeth that are weakly papillose to nearly smooth in the distal portion and spread at 45°–60°. On the other hand, the latter has the decurrencies of the hair-points distinctly ruffled and long, extending to ½–⅔ along the leaf margins (Bednarek-Ochyra & Ochyra 2003).

Figure 133. *Racomitrium lanuginosum*. — **1**. Habit. **2**. Portion of branch, wet. **3**. Cross-section of stem. **4–5**. Axillary hairs. **6–9**. Leaves. **10–11**. Margins of hyaline hair-points. **12**. Mid-leaf cells. **13**. Alar and supra-alar cells. **14**. Basal leaf cells. **15–19**. Cross-sections of leaves, a sequence from base to apex. — [All from *RILS 5749*]. — Scale bars: a – 1 mm (1–2, 6–9); b – 100 μm (3–5, 10–19).

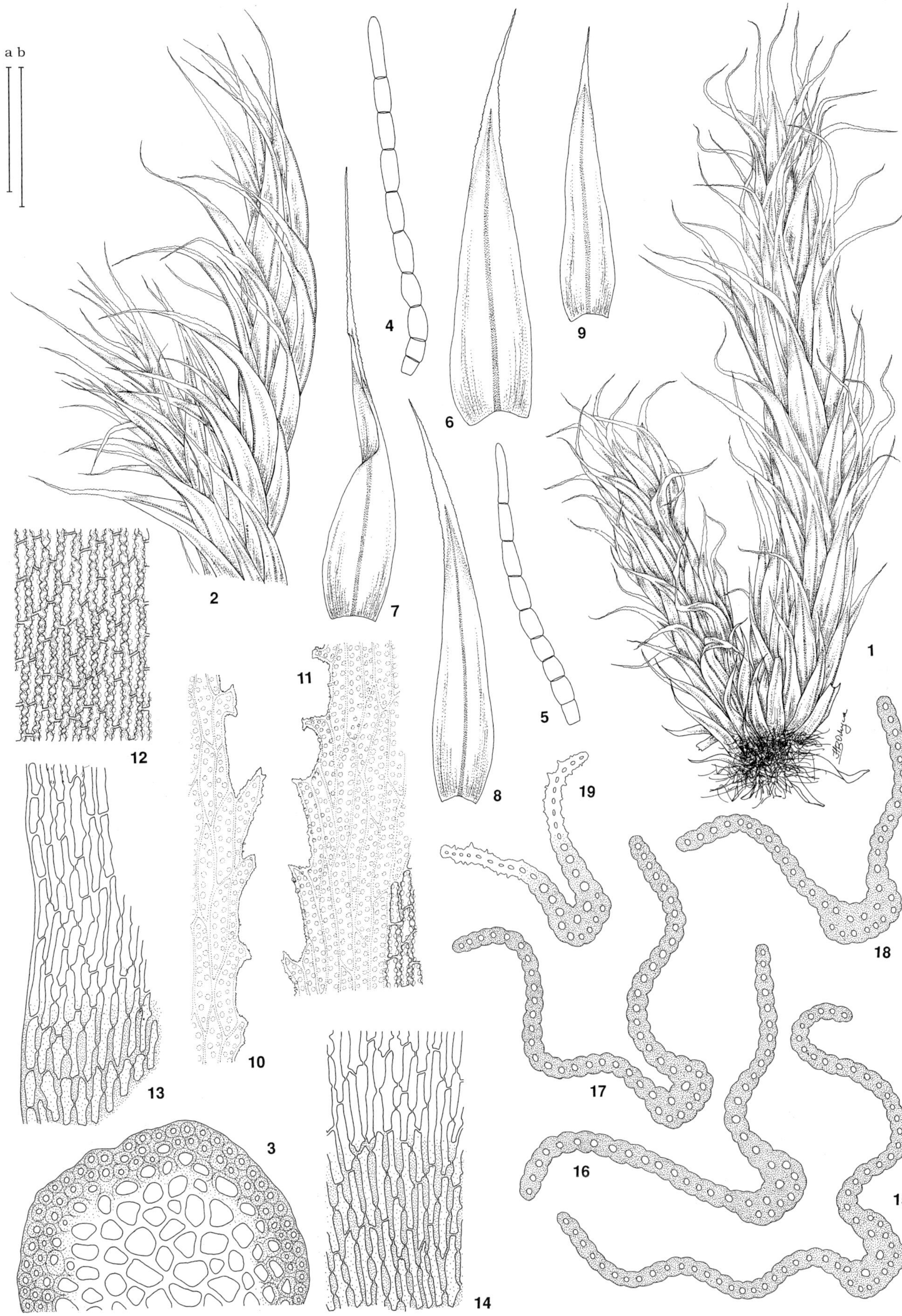
a b
1
2
3
4
5
6
7
8
9
10
11
12
13
14
15
16
17
18
19

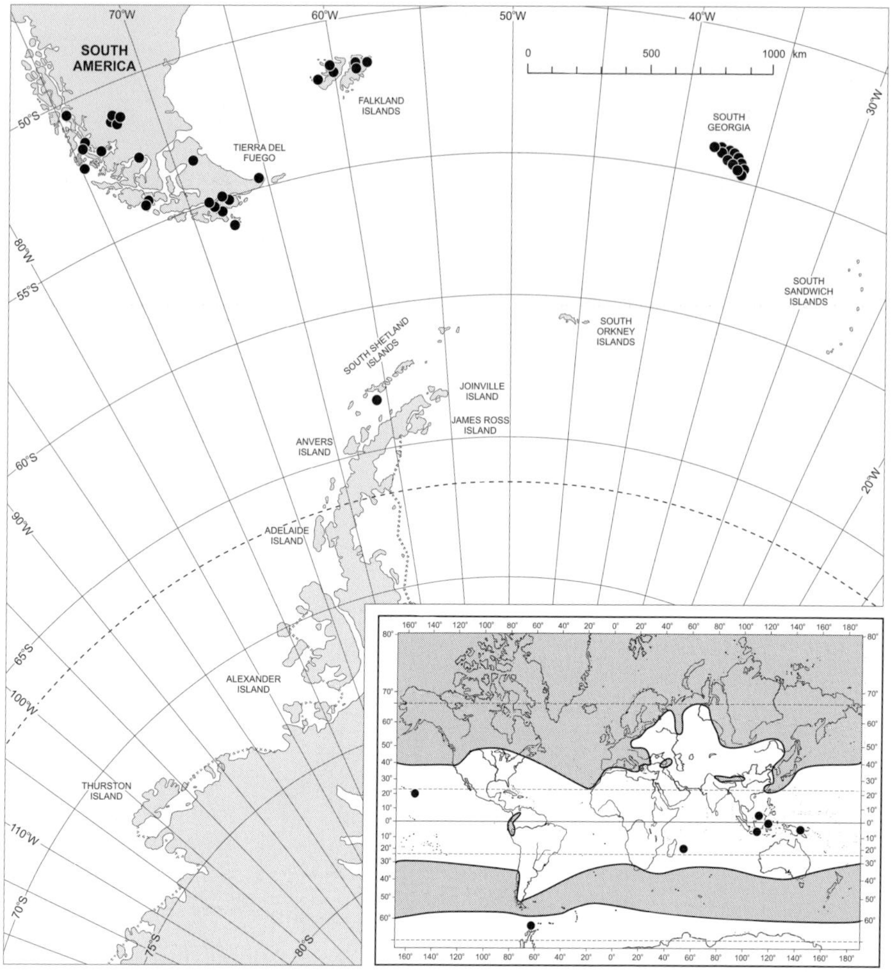

Figure 134. Distribution of *Racomitrium lanuginosum* in the Fuegian region and Antarctica. Inset: global distribution.

Reproduction in Antarctica. Consistently sterile.

Habitat. On dry very coarse scoriaceous boulders in a former geothermal area.

World range. *Racomitrium lanuginosum* is a bipolar species with several intermediate stations at altimontane elevations in the tropics (Ochyra *et al.* 1990; Bednarek-Ochyra 1995*a*) (Fig. 134, inset). In the Northern Hemisphere it is a circumpolar arctic-boreal-montane species that is widespread throughout much of the Holarctic, reaching its farthest north in Spitsbergen (Frisvoll 1982) and Greenland (Mogensen & Lewinsky 1987). The southern boundary of its continuous range in North America is in California and Wyoming, through the southern fringes of Hudson Bay to Québec and Maine in the eastern part of the continent. In Europe, it is widespread from Iceland and the British Isles (Tallis 1958) to the Ural Mountains (Jalas 1955; Ignatov & Ignatova 2003), as well as in the Alps, where it reaches an elevation of 3500 m (Loeske 1930) and in the Carpathians. In the south of the continent it is scattered in the mountains from the Iberian Peninsula to the Balkans, extending as far south as Sicily, from where it was described as *R. marginatum* (Dia & Raimondo 1984; Dia *et al.* 1987), and Macaronesia, and to Asia Minor and the Caucasus in the east. In Asia, *R. lanuginosum* is widespread in Arctic Siberia, extending southwards to northern Mongolia and Japan and occurring disjunctively at high elevations in the Himalayas and on Taiwan (Cao *et al.* 2003). In the Southern Hemisphere, the species has a circumpolar range that is seemingly disjunct because of the discontinuity of land masses in this part of the world. It is common and abundant in suitable sites in New Zealand, south-eastern Australia, Tasmania, southern Africa and in the cool-temperate zone in southern South America, as well as on all Subantarctic islands from South Georgia to Macquarie I. In the tropical zone, there are several stations of *R. lanuginosum* at high elevations in the Andes of Colombia, Ecuador and Peru (Churchill *et al.* 2000), on the Hawaiian Is., as well as in Java, Sumatra, Borneo and New Guinea (Vitt & Marsh 1988).

Distribution in Antarctica. *Racomitrium lanuginosum* is known from a very small area on the volcanic Deception I. in the South Shetland Is. (lat. 62°56′ S), at an elevation of 100–130 m where a few small plants occur. This is the southernmost locality of this moss (Lewis Smith 1988*d*, 2005*a*,*b*,*c*) (Fig. 134). The alleged record of the species from Michelsen I. in the South Orkney Is. (Dixon 1935; Greene 1968*b*) is incorrect since the plant morphology and associated species with the specimen (including *Syntrichia robusta* and *Acaena magellanica*, neither of which is known from the Antarctic) suggest a South Georgian origin for the voucher collection (Lewis Smith 1972; Bell 1973*c*). The record of *Racomitrium lanuginosum* from Astrolabe I. off the Davis Coast (Greenfield 2000) cannot be confirmed because no voucher specimen has been available for examination. [It is believed to have originated from Deception I. where Greenfield also visited (pers. comm.)].

Specimens examined. WEST ANTARCTICA. SOUTH SHETLAND IS. **DECEPTION I.** East and south-east of Pendulum Cove, *ca* 100 m, *RILS 6027, 6029 & 6930* (AAS, KRAM); southern Pendulum Cove, 100 m, *RILS 5749 & 5754D* (AAS, KRAM); Pendulum Cove, east of Crimson Hill, 130 m, *RILS 11196* (AAS, KRAM).

24. BUCKLANDIELLA Roiv.

Bucklandiella Roiv., Ann. Bot. Fenn. 9: 116. 1972. — Type: *Bucklandiella bartramii* (Roiv.) Roiv. (*Bucklandia bartramii* Roiv.).

Bucklandia Roiv., Arch. Soc. Zool. Bot. Fenn. 'Vanamo' 9: 98. 1955, *hom. illeg.* [non *Bucklandia* Sternb. 1825; nec Brongn. 1828; nec R.Br. *ex* Griff. 1836]. — Type: *Bucklandia bartramii* Roiv. [=*Bucklandiella bartramii* (Roiv.) Roiv.].

Dryptodon Brid. subgen. *Ellipticodryptodon* Vilh., Acta Bot. Bohem. 2: 53. 1923. ≡ *Racomitrium* Brid. sect. *Ellipticodryptodon* (Vilh.) Ochyra, Sérgio & Schumacker, Bull. Jard. Bot. Nat. Belg. 58: 246. 1988. ≡ *Racomitrium* subgen. *Ellipticodryptodon* (Vilh.) Bednarek-Ochyra & Ochyra, Fragm. Florist. Geobot. 39: 362. 1994. — Type: *Dryptodon ellipticus* Turner [*Bucklandiella elliptica* (Turner) Bednarek-Ochyra & Ochyra]. First synonymised by Ochyra *et al.* (2003: p. 142).

Racomitrium Brid. subgen. *Microcarpa* Vilh., Věstn. Král. Ceské Společn. Nauk Tř. Mat.-Přír. 1925(5): 22. 1926 ['Microcarpae']. — Type: *Racomitrium microcarpon* (Hedw.) Brid. [*Bucklandiella microcarpa* (Hedw.) Bednarek-Ochyra & Ochyra]. First synonymised by Ochyra *et al.* (2003: p. 142).

Plants small, medium-sized or large, gracile or robust, stiff, often coarse and rigid, in loose or compact, often deep small or large tufts or often forming extensive paches or mats, dull or somewhat lustrous, not or hoary and then appearing greyish, green, brown, yellowish, dark or light olive-green, less often rusty, blackish-brown or occasionally jet-black. *Stems* erect, decumbent or creeping, mostly elongate, much dichotomously to irregularly, occasionally subpinnately to pinnately branched, rarely almost simple and unbranched, sparsely radiculose at the base or lacking rhizoids when erect, or with abundant bundles of brownish-red, smooth, branched rhizoids scattered along creeping stems, in transverse section round, consisting of an outer cortex of 2–3 or more layers of small, stereid cells with brown to reddish-brown, incrassate walls and very small lumina, sharply or indistinctly grading into a medulla of large, hyaline or yellowish-hyaline, thin- or thick-walled cells, central strand lacking; *axillary hairs* usually numerous, filiform, hyaline throughout, composed of 6–15, relatively short cells or with some cylindrical cells in the distal part. *Leaves* crowded or occasionally distant, loosely or closely appressed to imbricate, erect, secund to falcato-secund, rarely contorted when dry, erecto-patent and often recurved at the tips when wet, narrowly to broadly lanceolate, long-, oblong or ovate-lanceolate, rarely ovate-triangular, gradually long-acuminate, acute or occasionally narrowly obtuse at the apex, long-piliferous to short-hyaline-tipped or sometimes epilose, smooth or plicate, concave below, canaliculate to carinate; *hair-point*, if present, hyaline to yellowish-hyaline, erect to squarrose, straight to strongly flexuose, flat to canaliculate, thin and delicate to terete and stout, never papillose, not or short-decurrent down the leaf margins, sharply or bluntly denticulate to strongly spinulose or entire; *margins* recurved to revolute on one or both sides for varying distances, entire or sometimes with some small teeth or crenulations made up by protruding cell ends at the apex, unistratose throughout or entirely to variously bistratose in 1 to several rows of cells, sometimes polystratose and forming multiseriate, strong, fleshy limbidia confluent at the apex and extending to the base; *costa* single, never branched, sharply delimited from the laminal cells, percurrent, yellow-brownish to brown, relatively slender and narrow to strong and broad, in transverse section strongly convex on the dorsal surface, semi-terete, short-rectangular to elliptical in the upper and median parts, reniform to strongly dorsally flattened and often situated at the bottom of a shallow or deep, wide-angled groove in the proximal portion, entire or occasionally with wings and furrows on the dorsal surface, 2- to multistratose, with 2 to many enlarged, ventral epidermal cells and small, substereid or stereid central and dorsal cells; *laminal cells* unistratose throughout or variously bi- to polystratose in the distal half, dull and opaque to transparent, with moderately to strongly incrassate, sinuose to nodulose lateral walls, smooth or often pseudopapillose due to many lengthwise cuticular ridges covering both ventral and dorsal surfaces and giving the cells a papillose appearance in transverse section, quadrate, rounded-quadrate, oblate to rectangular in the upper part, becoming elongate, long-rectangular to linear towards the base, sometimes less sinuose-walled in the proximal portion with the lateral walls strongly porose and nodulose, usually somewhat shorter and wider in 1–3 rows at the insertion, forming a yellow or orange-yellow strip; *alar cells* undifferentiated or isodiametric, thick-walled, forming decurrent, flat or convex, brown to orange- or yellow-brown, decurrent auricles; *basal marginal cells* not differentiated from the laminal cells or hyaline to yellowish-hyaline, quadrate to rectangular, straight- or weakly sinuose-walled, with thin to moderately thickened walls, forming translucent 1(–2)-seriate border, consisting of up to 30(–40) cells extending along the leaf margins. *Asexual reproduction* very rare by spherical or elliptical gemmae produced on the leaves. *Dioicous*. *Outer perichaetial leaves* similar to vegetative leaves, only somewhat larger, with erect to recurved apex on wetting; *inner perichaetial leaves* sheathing at the base but otherwise similar to vegetative leaves, or strongly modified, ovate, ovate-lanceolate to lingulate, epilose to short-piliferous, entirely hyaline to yellowish-hyaline, with a weak, short costa ceasing far below the apex and a lax areolation of very thin-walled cells throughout or cells in the distal part with strongly incrassate, porose or sinuose walls. *Setae* 1–3 per perichaetium, erect, straight, slender or stout, smooth, yellow or orange-yellow, becoming brown with age, twisted clockwise above. *Capsules* exserted, erect, straight, symmetric, ovoid, ellipsoid, obloid to narrowly cylindrical, narrowed at the mouth, without a neck and abruptly narrowed into the seta or with an indistinct apophysis, smooth, pachydermous, dull or glistening, sometimes with a varnish lustre, light or dark brown, castaneous or blackish-brown; *annulus* compound, abruptly or tardily deciduous, 2–3-seriate, consisting of large, rectangular, vesiculose, thick-walled, transparent cells in the upper row; *operculum* conic-rostrate or long-rostellate from a conic base, with a straight or oblique, slender or stout rostrum; *exothecial cells* variable in size and shape, subquadrate to oblong, with thin to thick, straight or flexuose walls, becoming quadrate or oblate, thick-walled at the mouth in 1–7 rows, forming a reddish-brown band; *stomata* many in 1–2 rows at the extreme base of the urn, bicellular, superficial, with round pori; *peristome* single, composed of 16, short to relatively long, lanceolate to triangular, reddish-brown or brown teeth, with many articulations, arising from a low or sometimes high basal membrane, seldom basal membrane lacking, cleft into 2–3 filiform, terete prongs in the distal half, occasionally bifid nearly to the base, often irregularly divided or connate at the joints to variously perforated, papillose to densely spiculose, with a distinct prostome. *Spores* spherical, pale yellowish to brown, finely granular to densely coarsely papillose. *Calyptra* conic-mitrate, 4–5-lobed at the base, smooth, naked, light brown below, becoming dark brown at the tip.

Bucklandiella is the largest and taxonomically most difficult segregate of the former genus *Racomitrium*. The genus is a clearly defined and easily recognised taxon. In contrast to the other three segregates of *Racomitrium*, it comprises taxa with smooth laminal cells. They may eventually be pseudopapillose due to numerous longitudinal cuticular ridges which make the leaf surface rugged and appearing papillose in leaf cross-section. The smooth laminal cells are coupled with relatively short, lanceolate or triangular peristome teeth which are irregularly divided in the upper half into 2–3 or occasionally 4 branches, or sometimes they are only irregularly perforate along the median line or, in some cases, the teeth remain undivided. Additionally, the following combination of characters is characteristic of this genus: narrowly lanceolate to ovate-triangular leaves; unbranched, percurrent costa, hair-point, if present, smooth, denticulate or spinulose but never papillose; leaf margins 1–2(–4)-stratose in one to several rows; seta always smooth, twisted to the right when dry and smooth calyptra.

Bucklandiella is the only available name for this genus which at the infrageneric level in *Racomitrium* was called subgen. *Ellipticodryptodon* (Vilh.) Bednarek-Ochyra & Ochyra or sect. *Laevifolia* (Kindb.) Nog. (Bednarek-Ochyra 1995*a*). The name was originally used for a monotypic genus from the Fuegian region, *Bucklandia bartramii* Roiv. (Roivainen 1955*b*), which proved to be a later homonym of at least three identical names and accordingly Roivainen (1972) renamed it *Bucklandiella*. Robinson (1974) showed that this species belongs to the group of *Racomitrium* species with a broad costa, and subsequent studies (Deguchi 1987; Ochyra *et al.* 1988) confirmed this to be correct. *Bucklandiella bartramii* along with two other species, *B. lamprocarpa* and *B. elliptica*, and possibly the Brazilian *B. visnadiae* which is known only in the sterile condition (Buck 1997; Bednarek-Ochyra *et al.* 1999), constitute an isolated group within the genus characterised by large, coarsely papillose spores, 18–32 µm in diameter, but otherwise exhibits all the typical features of the remaining species.

Bucklandiella comprises about 50 species distributed worldwide, but only some of the Holarctic species belonging to sect. *Laevifoliae* have been revised taxonomically (Frisvoll 1988). The genus shows a greater diversity in both sporophyte and gametophyte characters, and Bednarek-Ochyra (1995*a*) provisionally recognised eight sections to accommodate species formerly placed in *Racomitrium* subgen. *Ellipticodryptodon* which is the taxonomic equivalent of *Bucklandiella*. This classification is primarily based upon the shape and areolation of the perichaetial leaves, as well as on some peculiarities in the structure of sporophytes and gametophytes, and it was accommodated in *Bucklandiella* by Ochyra *et al.* (2003). Further revisions of *Bucklandiella* will refine the taxonomy of this genus. In addition, a taxonomic revision of the genus *Codriophorus* has shown that some species formerly placed in *Racomitrium* subgen. *Cataractarum*, which is the taxonomic equivalent of *Codriophorus*, correctly belong within *Bucklandiella*, including two Andean species, *B. cucullatifolia* (Hampe) Bednarek-Ochyra & Ochyra and *B. grimmioides* (Herzog) Bednarek-Ochyra & Ochyra, and *B. lepervanchei* (Besch.) Bednarek-Ochyra & Ochyra from Réunion (Bednarek-Ochyra 2006).

Bucklandiella is well represented in the austral polar region and some species are abundant constitutents of various communities in the fellfield vegetation (Gremmen 1982). A large group of austral species centred around *B. crispula* needs critical taxonomic assessment, so the real number of species cannot be definitely established at present. It was unfortunate that most of these species had once been lumped with *B. crispula* (Clifford 1955), making it an unnatural and artificial taxon. Although many species from this region uncritically described by Müller (1849, 1883, 1885, 1889, 1890*a*) and other workers (e.g. Cardot 1900, 1901, 1905*a*, 1908*a*, 1911*a*,*b*; Dusén 1907; Cardot & Brotherus 1923; Bartram 1946; Roivainen 1955*a*) need reduction, subsequent taxonomic studies (e.g. Roivainen 1955*a*; Lawton 1972, 1973; Bell 1974; Deguchi 1984, 1987; Ochyra & Bednarek-Ochyra 1999) confirmed the distinctness of several species. Moreover, Frisvoll (1984*a*) lectotypified and precisely defined *B. crispula* itself.

Following careful evaluation of a large number of records, at least 12 species of the genus occur throughout the Subantarctic, and four penetrate into the Antarctic biome. One of these, *B. subsecunda*, which is unknown in the Subantarctic, is known from a single location in Antarctica. Apart from *B. sudetica* they have a very narrow distribution here and are restricted to areas with recent volcanic activity. According to the classification scheme proposed by Bednarek-Ochyra (1995*a*) they represent three different sections, namely sect. *Subsecundae* (Bednarek-Ochyra) Bednarek-Ochyra & Ochyra (*B. subsecunda*), sect. *Sudeticae* (Bednarek-Ochyra) Bednarek-Ochyra & Ochyra (*B. sudetica*) and sect. *Marginatae* (Bednarek-Ochyra) Bednarek-Ochyra & Ochyra (*B. pachydictyon* and *B. didyma*).

1. Leaf margins unistratose, only occasionally with bistratose patches above 2
1. Leaf margins at least partly bistratose for 1–9 cell rows in the distal half 3
    2. Costa strong, 60–85 µm wide and distinctly flattened below, consisting of 4–6 enlarged ventral cells, symmetric throughout; basal marginal border always absent .... 1. ***B. subsecunda***

2. Costa weak, 35–60 µm wide at the base, semi-terete and usually asymmetric in outline, consisting of 2–3 enlarged ventral cells in cross-section in the proximal part; basal marginal border usually present, consisting of 10–15 pellucid, esinuose cells .... 2. ***B. pachydictyon***

3. Basal marginal border absent or consisting of 2–5 short, hyaline or yellowish, often sinuose-walled cells; costa narrower, 60–80 µm wide near the base, strongly convex and semi-terete on the dorsal surface, consisting of 3(–4) enlarged ventral cells in cross-section in the proximal part; alar cells undifferentiated .... 3. ***B. sudetica***

3. Basal marginal border distinct, often biseriate, consisting of 17–25 or more hyaline, pellucid, esinuose cells; costa broad, 90–105 µm wide near the base, reniform to distinctly flattened dorsally, consisting of 4 enlarged ventral cells in cross-section in the proximal part; alar cells differentiated .... 4. ***B. didyma***

1. **Bucklandiella subsecunda** (Harv.) Bednarek-Ochyra & Ochyra Fig. 135

Bucklandiella subsecunda (Harv.) Bednarek-Ochyra & Ochyra *in* Ochyra, Żarnowiec & Bednarek-Ochyra, Cens. Cat. Polish Mosses: 147. 2003. ≡ *Trichostomum subsecundum* Hook. & Grev. *ex* Harv. *in* Hook., Icon. Pl. 1: *pl. 17, f. 5.* 1836. ≡ *Racomitrium subsecundum* (Harv.) Wilson *in* Mitt. & Wilson, Hooker's J. Bot. Kew Gard. Misc. 9: 324. 1857. ≡ *Grimmia subsecunda* (Harv.) Mitt., J. Proc. Linn. Soc. Bot. Suppl. 1: 45. 1859. — Type citation: All new species illustrated by Harvey (1836) and accompanied by analytical captions were included in the section entitled 'Musci indici'; later Harvey & Hooker (1840) showed that the original material was collected in Nepal. [Lectotype (*vide* Frisvoll 1988: p. 207, 257–258 and Long 1995: p. 24–25): BM-Hooker!; isolectotypes: BM-Hooker! (2 specimens), BM-Wilson! (3 specimens), NY-Mitten!].

Plants small, in medium-sized, dense tufts, largely buried in soil, dull, olive-green at the tips, brown below. *Stems* ascending to erect, to 1.5 cm long, sparingly irregularly branched or almost unbranched, rather slender but stiff, dark brown, densely foliate throughout, with the lower leaves usually eroded; in cross-section rounded, composed of a 1–2-stratose cortex of relatively large substereid cells with strongly thickened, dark orange-brown walls and fairly large lumina surrounding a 4–5-layered medulla of large, hyaline, thin-walled cells with intensely yellow walls, central strand absent; *rhizoids* frequent, scattered in clusters along the stem and not forming a tomentum, long and sparsely branched, dark reddish-brown, smooth; *axillary hairs* numerous and frequent in the upper part of the stem, filiform, hyaline throughout, 8–10-celled, composed of short, barrel-shaped lower cells and elongate upper cells. *Leaves* crowded, erect, straight to falcato-secund, loosely appressed to the stem and mostly with more or less reflexed apices when dry, erecto-patent when moist, not plicate, concave, narrowly to broadly canaliculate throughout, lanceolate to ovate-lanceolate, 2.1–2.8 mm long, 0.4–0.6 mm wide, with distinct, shortly decurrent auricles, gradually tapering to a long, canaliculate acumen, with a blunt and narrowly obtuse or acute apex lacking a hair-point; *margins* entire, unistratose throughout, variously recurved on one or both sides in the proximal part to about the mid-leaf, plane or nearly so above; *costa* single, concolorous with the laminal cells but otherwise clearly separated from them, light yellow to yellow-brown, 60–80 µm wide at base, gradually tapering upwards, ceasing well below the apex and imperceptibly merging with apical cells, prominent and strongly flattened dorsally and usually situated at the bottom of a shallow to fairly deep groove below, semi-terete to elliptical and less convex on the dorsal surface above, ventrally canaliculate throughout, in transverse section bistratose above with 2 ventral and 3–4 dorsal cells, bistratose with occasional tristratose patches and 2–3 ventral and 7–12 dorsal cells in the median part, bi- to incompletely tristratose below, with 4–6 large ventral cells and 12–15 small dorsal cells below; *laminal cells* unistratose, smooth, with very incrassate and nodulose lateral walls and thinner, oblique or truncate transverse walls, short- to long-rectangular throughout the lamina, (15–)20–40(–50) long, 6–8(–10) µm wide, occasionally subquadrate at margins in the apical part; *basal cells* rectangular, with strongly thickened and porose walls, 30–70 µm long, 10–12 µm wide, dark yellow, forming a narrow, 1–2-seriate strip along the insertion; *alar cells* larger, subquadrate to oblong, with thick, porose or straight walls, yellow, pale yellowish or orange-yellow, forming plane or somewhat inflated, decurrent auricles or occasionally undifferentiated and similar to the adjacent basal cells; *supra-alar marginal cells* undifferentiated from the laminal cells or 2–3(–5) cells slightly differentiated, pellucid, yellowish-hyaline to yellowish, with moderately thickened, sinuose or esinuose walls. [*Dioicous. Outer perichaetial leaves* ovate- to oblong-lanceolate, gradually or abruptly narrowed to a short, obtuse, epilose acumen, recurved on wetting, 3.0–3.5 mm long, 0.7–0.9 mm wide, with a relatively slender, concolorous, percurrent costa and a strongly modified areolation, composed of rectangular to linear cells, with strongly incrassate and porose cells; *inner perichaetial leaves* convolute and strongly sheathing, oblong-ovate, 2.5–3.0 mm long, broadly obtuse, with an areolation in the distal half similar to that in the outer bracts, and lax, yellowish-hyaline to hyaline at margins in the proximal part. *Setae* singly in perichaetium, erect, straight to flexuose, dextrorse when dry, brown to reddish-brown, smooth, 5–15 mm long. *Capsules* exserted, erect, straight, symmetric, ovoid, obloid to narrowly obloid-cylindrical, 1.7–3.5 mm long, 0.6–0.8 mm wide, smooth, pachydermous, brown to castaneous, lustrous; *annulus* compound, revoluble, 2–3-seriate, with the upper row of large, thick-walled vesiculose cells; *operculum* conic-rostrate, with a long, straight rostrum; *exothecial cells* varying from isodiametric to oblong or rectangular, 15–60 µm long, 5–10(–20) µm wide, thick-walled, becoming rounded to elliptical, strongly thick-walled in 5–7 rows at the orifice, forming a reddish-brown strip; *stomata* numerous, up to 15 in a single row at the extreme base of the urn, bicellular, superficial, with round pori; *peristome* single, consisting of 16 lanceolate, yellow-brown teeth, 350–600 µm long, lacking a basal membrane, split nearly to the base into 2 filiform, terete, prongs, strongly warty papillose on both ventral and dorsal surfaces. *Spores* spherical, yellowish-green, 12–15 µm in diameter, finely papillose. *Calyptra* mitrate, smooth.]

Chromosome number: unknown.

Discussion. Because of the sterile condition of the material from Deception I., there was an initial problem with its correct assigation to species. The combination of some gametophyte characters, including the lack of basal marginal border, elongate laminal cells and very broad and flattened costa, seemed to be unique and unknown in any austral species, but subsequent comparison of the only Antarctic collection with extra-antipodal species revealed that it falls quite well within the range of variation of *Bucklandiella subsecunda*. This pantropical species exhibits considerable variability throughout its range, in the robustness of plants and the leaf size, as well as in the presence of the hair-point, leaf margins and internal structure of the costa. It is best characterised by its squarrose, not hyaline perichaetial leaves with strongly modified areolation; usually secund vegetative leaves; the strong costa which is quite flat dorsally in the lower part and consists of 5–9 large ventral cells; entirely unistratose laminal cells; absence of a basal marginal border; and large and inflated auricles composed of cells with strongly incrassate walls.

Typical phenotypes of *Bucklaneidlla subsecunda* which are common in the Himalayan–Yunnan region in Asia whence the species was originally described (Harvey 1836; Harvey & Hooker 1840; Long 1995), are characterised by having a broad costa, 80–110 µm wide, which is flat below and consists of 6–9 large ventral cells; entirely unistratose leaves; broadly recurved leaf margin on one side; and enlarged, auriculate alar cells with strongly incrassate walls giving them a stellate appearance. The Antarctic plants represent a smaller and epilose phenotype which is characterised by having the leaves with almost flat margins in the distal half, and the costa with 4–6 enlarged ventral cells in the basal part, as well as flat to weakly inflated auricles. In the number of ventral cells in the costa they approach south-east Asian plants (once recognised as *Racomitrium javanicum* Doz. & Molk.) which are especially frequent in altimontane sites in New Guinea (Frisvoll 1988). Similar populations are also quite widespread in tropical Africa where they were described as *Racomitrium alare* (Broth.) Paris (Ochyra 1993*d*) and in Brazil (Bednarek-Ochyra *et al.* 1999).

In the austral region *Bucklandiella subsecunda* may be confused with *B. striatipila* (Cardot) Bednarek-Ochyra & Ochyra in its strongly flattened costa in the lower part, elongate laminal cells and unistratose leaf margins, but the long basal marginal border of hyaline and pellucid cells of the latter species should prevent confusion. In its muticous leaves the Antarctic plants of *B. subsecunda* may be mistaken with some epilose ecads of *B. didyma* (Mont.) Bednarek-Ochyra & Ochyra and *B. crispula*, austral temperate species which are most probably conspecific taxa, but the wide bistratose margins in those species should prevent misidentification. The mostly muticous leaves and unistratose laminal cells resemble *B. rupestris* (Hook.f. & Wilson) Bednarek-Ochyra & Ochyra, an antipodal pan-temperate species, but it can be recognised with a hand-lens by its catenulate leaves, and microscopically by its short upper laminal cells and tristratose costa in the upper part and narrow and semi-terete costa in the proximal portion. *Bucklandiella subsecunda*, especially its epilose phenotypes, is likely to be mistaken for some phenotypes of *B. crispipila* (Taylor) Bednarek-Ochyra & Ochyra with entirely muticous leaves which had previously been given the status of separate taxa. This is especially true for Bolivian plants described as *Racomitrium dimorphum* (Müll.Hal. *ex* E.Britton) Paris which is considered to be conspecific with *Bucklandiella crispipila* (Deguchi 1987). However, these phenotypes can be differentiated by the presence of a long basal marginal border consisting of 10–25(–30) hyaline, transparent cells and an entirely bistratose costa in the median and lower parts.

Along with seven other Asiatic species that are restricted in their distribution to the Himalayan-Yunnan region and East Asia, *Bucklandiella subsecunda* belongs to sect. *Subsecundae* as defined by Bednarek-Ochyra (1995*a*). It is a distinct group that is well diagnosed by having an unusual form of the perichaetial bracts that are consistently epilose. The outer ones are squarrose on wetting and the 2–3 innermost ones are strongly modified, convolute, hyaline-yellowish to hyaline below and chlorophyllose above, with an areolation of cells with strongly incrassate and porose walls. Additionally, the costa is 2–3-stratose and the distal laminal cells tend to be elongate.

Reproduction in Antarctica. Sterile.

Habitat. As is the case with other species of the genus, *Bucklandiella subsecunda* grows on dry or wet rocks and gravel but in alpine regions it also thrives on soil. The only known Antarctic population of this species was found on soil close to fumaroles. The plants were almost totally buried in soil and grew in association with *Schistidium amblyophyllum*.

World range. *Bucklandiella subsecunda* is a pantropical montane species (Fig. 136) occurring mainly at high elevations, between 1100 and 4500 m, with maximum occurrence in the Himalayan–Yunnan region, but in Asia it is also widely scattered in the mountains of southern India, Sri Lanka, Taiwan, and Malesia including Sumatra, Java, Borneo, the Philippines and New Guinea (Frisvoll 1988). It is widespread and locally common on most volcanic mountains at elevations of (2200) 3000–4500 m in Kenya, Rwanda, Uganda, Democratic Republic of Congo and Tanzania, with an isolated station in Cameroon in West Africa, whence it was described as *Grimmia dura* Müll.Hal. *ex* Broth., and on Madagascar and Réunion in the Mascarenes in the southern Indian Ocean (De Sloover 1977; Ochyra 1993*d*). In the Neotropics, the species is widely scattered in the Central

→

Figure 135. *Bucklandiella subsecunda*. — **1**. Habit. **2**. Cross-section of stem. **3–4**. Axillary hairs. **5–8**. Leaves. **9–10**. Leaf apices. **11**. Upper laminal cells. **12**. Mid-leaf cells at margin. **13–14**. Alar and basal cells. **15–22**. Cross-sections of leaves and costae, a sequence from base to apex. — [All from *RILS 11287*]. — Scale bars: a – 0.5 cm (1); b – 1 mm (5–8) and 100 µm (2–4, 9–22).

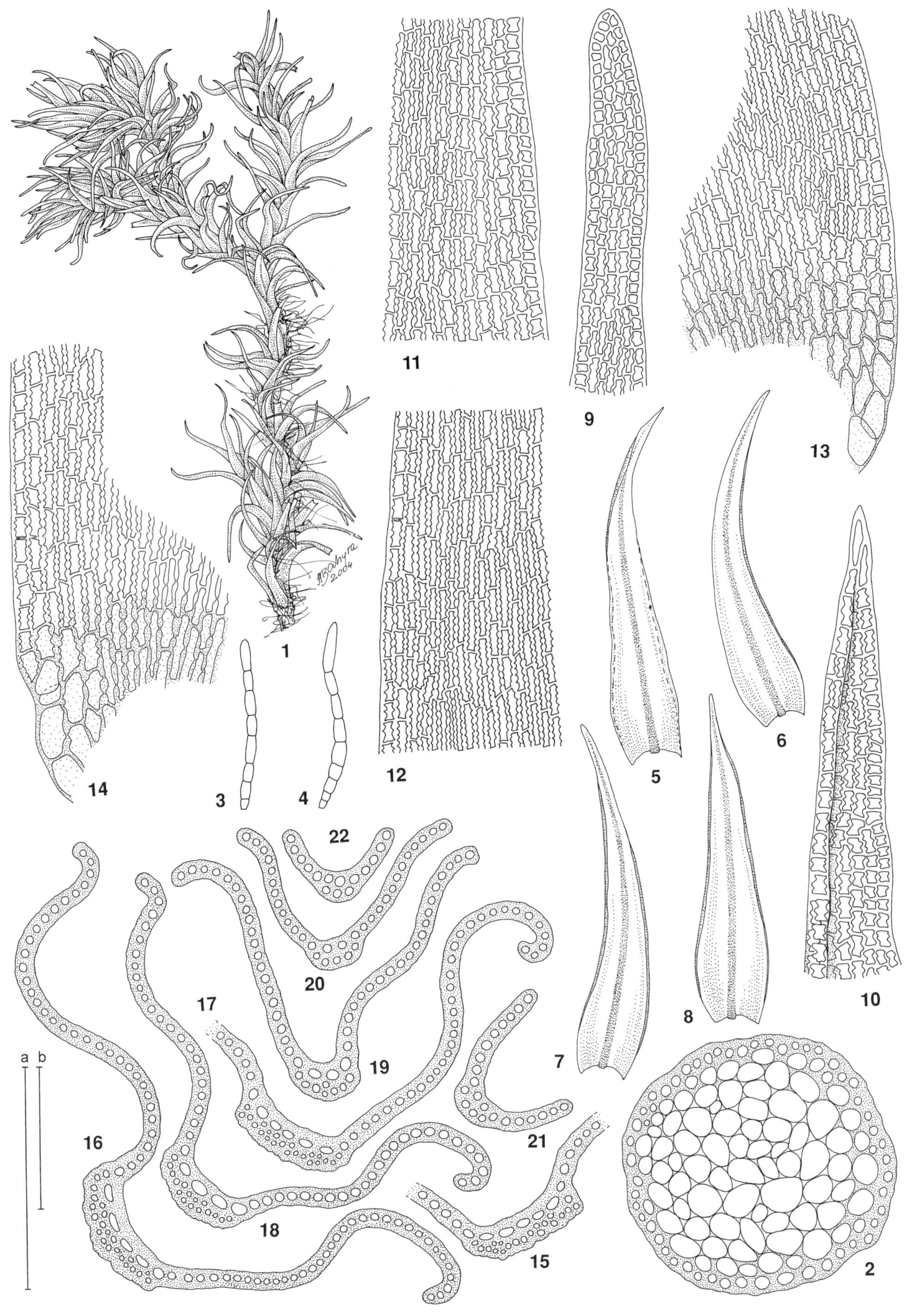
1
2
3
4
5
6
7
8
9
10
11
12
13
14
15
16
17
18
19
20
21
22
a
b

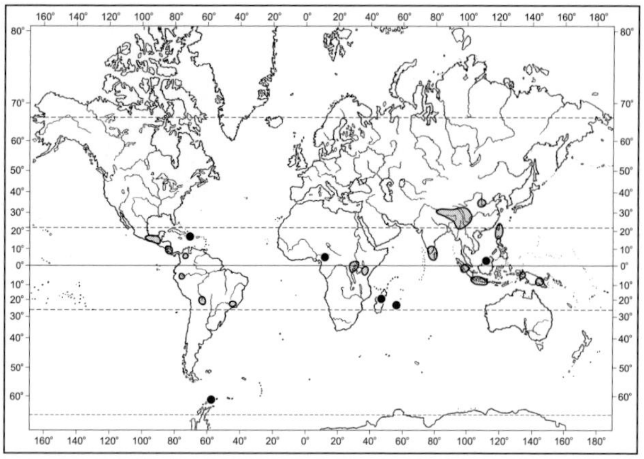

Figure 136. Global distribution of *Bucklandiella subsecunda*.

American isthmus from southern Mexico to Panama, with an isolated station in the Dominican Republic in the West Indies (Frisvoll 1988; Allen 2002), then recurring in the Andes of Colombia, Ecuador and Bolivia and in the states of Espírito Santo and Minas Gerais in south-eastern Brazil (Bednarek-Ochyra *et al.* 1999; Bednarek-Ochyra 2001), and at a highly disjunct and isolated station in the Antarctic. The latter occurrence of *B. subsecunda* is unusual and it was probably established only recently as the result of long distance dispersal, possibly from the Brazilian populations since the Antarctic and Brazilian plants share many morphological and anatomical features. It also represents the lowest altitude record for this species.

Distribution in Antarctica. Known from a single collection on the volcanic Deception I. in the South Shetland Is. at lat. 62°58′S (Fig. 136).

Specimens examined. WEST ANTARCTICA. SOUTH SHETLAND IS. **DECEPTION I.** Active fumarole at S end of Fumarole Bay, north of lagoon. *ca* 125 m, *RILS 11287* (AAS, KRAM).

2. Bucklandiella pachydictyon (Cardot) Bednarek-Ochyra & Ochyra Fig. 137

Bucklandiella pachydictyon (Cardot) Bednarek-Ochyra & Ochyra *in* Ochyra, Żarnowiec & Bednarek-Ochyra, Cens. Cat. Polish Mosses: 147. 2003. ≡ *Racomitrium pachydictyon* Cardot, Wiss. Erg. Schwed. Südpolar-Exp. 4(8): 113, *f.* 22. 1908. ≡ *Racomitrium symphyodontum* (Müll.Hal.) Paris var. *muticum* Cardot, Bull. Herb. Boissier Sér. 2, 5: 1004. 1905. — TYPE CITATION: Patagonie occidentale: Puerto Bueno (Dusén). Terre-de-Feu: Rio Azopardo (Dusén). Ile des Etats: Port-Cook (Skottsberg, n°s 78, 80). Géorgie du Sud. [Lectotype (*vide* Bell 1974: p. 87): 'Svenska Sydpolarexpeditionen 1901–03. N:r 78. Rhacomitrium pachydictyon Card. sp.nova. Ile des Etats: Port Cook D. 18/11 1903 Carl Skottsberg. Det. J. Cardot' – PC-Cardot!; isotypes: H-Brotherus!, PC!, S! Syntypes: (1) 'Herb. J. Cardot. Svenska Sydpol. Exped. 1901–03. n° 80. Rhacomitrium pachydictyon Card. sp.nova. Ile des Etats: Port Cook Leg. C. Skottsberg., 18/11 1903' – PC!, S!; (2) '266. Rhacomitrium pachydictyon Card. Fuegia austr., Rio Azopardo, ad marginem paludosis 29.II.1896 P. Dusén' – H-Brotherus!, PC!, S!; (3) 'Herb. J. Cardot. Svenska Sydpol. Exped. 1901–03. n° 312 in parte. Rhacomitrium pachydictyon Card. sp.nova. Géorgie du Sud: Baie Cumberland, mt. Duse. Leg. C. Skottsberg., 2/5 1902' – PC!, S!). Note: The fourth syntype cited in the protologue from Puerto Bueno (*Dusén 34*, PC, S) was recognised by Cardot (1908*a*: p. 285) as a separate taxon, *Racomitrium symphyodontum* (Müll.Hal.) Paris var. *dusenii* Cardot; it is considered identical to *Racomitrium didymum* (Mont.) Lorentz by Deguchi (1984: p. 30).

Grimmia suborthotrichacea Müll.Hal. var. *robustissima* Müll. Hal., Forschungsr. Gazelle Bot. 4: 30. 1889, ***syn. nov.*** — TYPE CITATION: Kerguelen (J. D. Hooker, Dr. Naumann). [Lectotype (*selected here*): 'No. 732 Kerguelen Land W. 247. Trichostomum Racomitr. cataractarum Brid. var. 3' – BM-Hooker!; isolectotype: BM-Hampe!].

Racomitrium symphyodontum (Müll.Hal.) Paris var. *muticum* Cardot, Bull. Herb. Boissier Sér. 2, 5: 1004. 1905. — TYPE CITATION: Ile des Etats: Port-Cook. [Lectotype (*selected here*): 'Svenska Sydpolarexpeditionen 1901–03. Nr 78. Rhacomitrium pachydictyon Card. sp.nova. Ile des Etats: Port Cook D. 18/11 1903 Carl Skottsberg. Det. J. Cardot' – PC-Cardot!; isotypes: H-Brotherus!, PC!, S!]. Homotypic with *Racomitrium pachydictyon* (*vide* Cardot 1908*a*: p. 113).

Plants occasionally small, mostly medium-sized to large, slender to fairly robust, densely caespitose or growing as individual shoots in hollows of the porous and rough surface of scoria, stiff and rigid, somewhat glossy, yellow-green to olive-brown or brown above, blackish-brown below. *Stems* short to elongate, (0.5–)1.0–8.5 cm tall, erect, sparingly fastigiately branched, usually bearing a few lateral branchlets above, densely radiculose at the base on juvenile, less on adult plants with brown, glossy, smooth, sparingly branched rhizoids, in transverse section rounded, consisting of 2–3-layered cortex of small, brown, stereid cells with small lumina, abruptly passing into a 3–5-stratose medulla of large, hyaline cells with thin to moderately thickened walls, central strand wanting; *axillary hairs* filiform, hyaline throughout, composed of 6–7 mostly elongate cells. *Leaves* crowded, stiff, erect and closely appressed when dry, erecto-patent when wet, straight, shortly decurrent, narrowly lanceolate to oblong-lanceolate, (1.5–)1.9–2.5 mm long, 0.3–0.5(–0.6) mm wide, gradually long-acuminate, epilose, obtuse to subacute at the apex, narrowly canaliculate, smooth on the left side of the costa and with a prominent fold on the right side, stretching from the base up to the mid-leaf or higher, partially overlapping the costa and resulting in its asymmetric outline as seen in transverse section; *margins*

→

Figure 137. *Bucklandiella pachydictyon*. — **1**. Habit, wet. **2**. Habit, dry. **3**. Portion of shoot, dry. **4**. Portion of shoot, wet. **5**. Cross-section of stem. **6–8**. Axillary hairs. **9–15**. Leaves. **16**. Leaf apex. **17–18**. Upper leaf cells at margin. **19**. Mid-leaf cells. **20**. Supra-basal leaf cells. **21–22**. Alar and supra-alar cells. **23–33**. Cross-sections of leaves, a sequence from base to apex. **34–37**. Perichaetial leaves, from outermost to innermost. **38**. Apex of innermost perichaetial leaf. **39–40**. Old deoperculate capsules with destroyed peristome teeth, dry. **41**. Operculate capsule, wet. **42–43**. Old deoperculate capsules with peristomes destroyed, wet. **44**. Calyptra. **45**. Exothecial cells in middle of urn. **46**. Exothecial cells at base of urn and stomata. **47**. Peristome teeth, spores and rim cells of capsule. — [1–4, 9, 22, 34–37 from *SWG 2016*; 5, 25–28 from *RILS 11205A*; 6–8 from *SWG 3943*; 10 from *CS 80* (syntype of *Racomitrium pachydictyon*), PC; 11, 18–19, 23–24 from *CS 78* (lectotype of *R. pachydictyon*), PC; 12, 44–47 from *RILS 1303*; 13–15, 29–33 from *RILS 5749C*; 16–17, 20–21 from *RILS 5749C*; 38–43 from *Halle & CS 403*, PC]. — Scale bars: a – 1 mm (3–4, 39–44); b – 100 µm (45–47); c – 1 cm (1), 1 mm (9–15, 34–37) and 100 µm (5–8); d – 100 µm (16–33, 38); e – 0.5 cm (2).

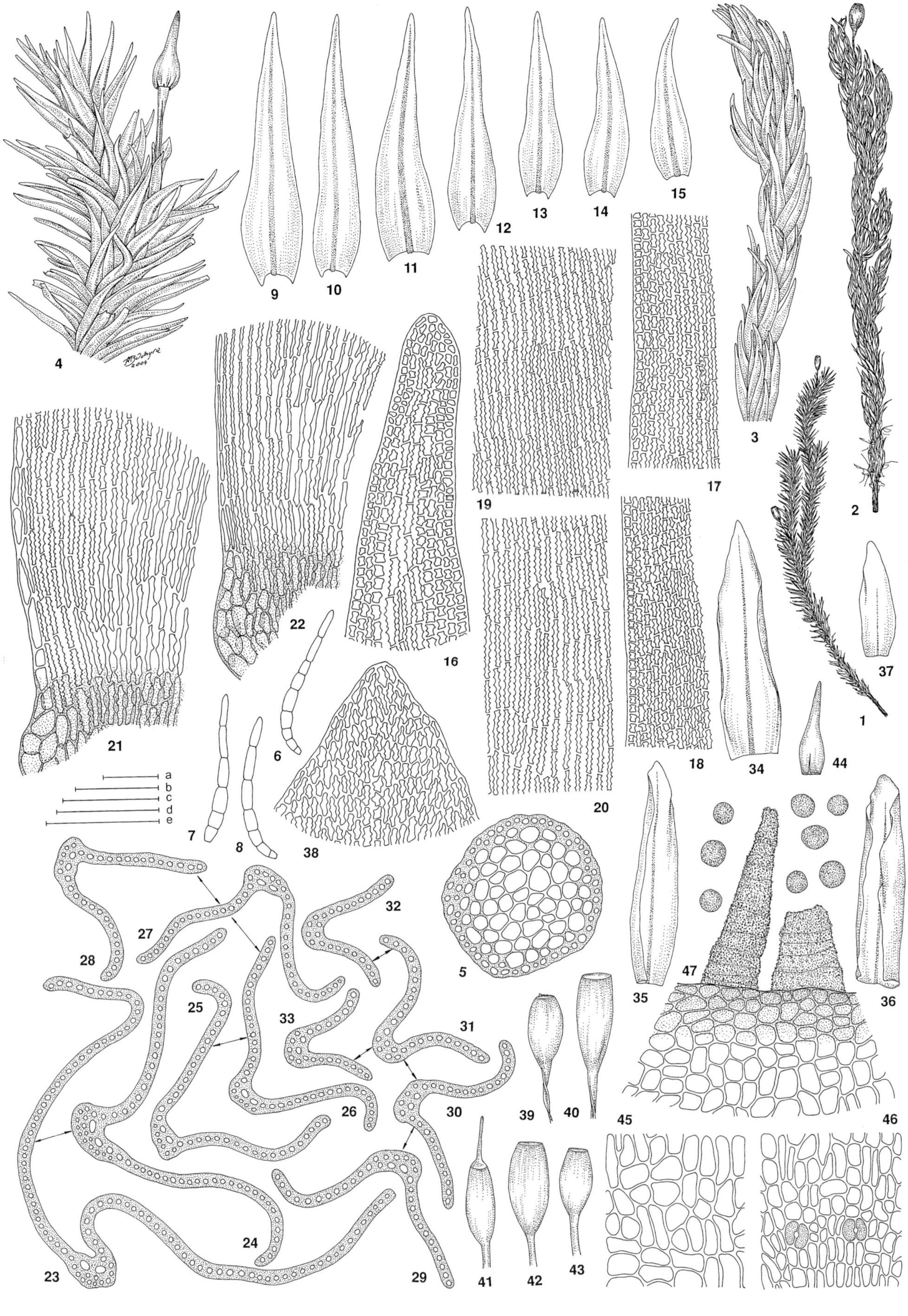
1
2
3
4
5
6
7
8
9
10
11
12
13
14
15
16
17
18
19
20
21
22
23
24
25
26
27
28
29
30
31
32
33
34
35
36
37
38
39
40
41
42
43
44
45
46
47
a
b
c
d
e

entire, unistratose throughout, plane or nearly so on the right side, recurved for varying length on the other; *costa* single, light yellow to yellow-brown, usually concolorous, not distinctly separated from the laminal cells being partially covered by the longitudinal fold of the lamina, narrow, 35–60 μm wide at the base, gradually tapering upwards, percurrent or usually ending well below the apex, narrowly canaliculate throughout, distinctly convex dorsally, in transverse section semi-terete with 2 ventral epidermal cells and bi- or very seldom tristratose in the upper and median parts, becoming distinctly asymmetric and irregular in outline, with 2 or 3 enlarged ventral cells in the proximal part; *laminal cells* unistratose throughout, smooth, thick-walled with strongly incrassate and sinuose lateral walls, elongate throughout, long-rectangular, 14–30(–40) μm long, 4–8 μm wide, in the upper part, becoming long-rectangular to linear in the lower half, 40–70 μm long, 4–5 μm wide, with strongly incrassate, 6–8 μm wide, sinuose-nodulose longitudinal walls; *basal cells* rectangular, shorter and wider than supra-basal cells, 20–40 μm long, 6–8 μm wide, strongly porose, forming a yellow-brown, 1–3-seriate strip along the insertion; *alar cells* quadrate, rounded to shortly oblong, often with strongly incrassate, dark brown or orange-brown walls, forming flat or swollen, decurrent auricles; *basal marginal cells* short- to long-rectangular, pellucid, hyaline, with straight and thin to moderately thickened walls, forming a distinct border consisting of (5–)10–15 cells extending along the leaf margin, occasionally on some leaves the marginal cells poorly differentiated or undifferentiated. [*Dioicous. Perichaetial leaves* differentiated, markedly different from vegetative leaves, epilose; *outer perichaetial leaves* ovate-lanceolate, 2.0–2.3 mm long, secund, with areolation similar to that of vegetative leaves; *inner perichaetial leaves* oblong-lanceolate to oblong-ovate, to 2 mm long, 0.7–0.8 mm wide, sheathing, abruptly cuspidate, chlorophyllose, with a remarkably modified areolation, upper cells short-rectangular, strongly porose, becoming long-rectangular to linear with nodulose and porose walls, and several rows of hyaline cells at the margins in the proximal portion, 1–2 innermost bracts much smaller than others to 1.5 mm long, chlorophyllose only at the apex, consisting of cells with strongly thickened walls, hyaline and lax below. *Setae* single in perichaetium, erect, 1.5–2.5 mm long, straight, dextrorse when dry, yellow when young, becoming brown with age. *Capsules* exserted, straight, symmetric, ovoid to short-obloid, occasionally subspherical, 1.1–1.5 mm long, 0.7–0.8 μm wide, smooth, yellow-brown becoming brown with age, glossy; *annulus* compound, revoluble, 2–3-seriate, consisting of a row of upper, enlarged, vesiculose cells with strongly incrassate walls; *operculum* long-rostrate, with a straight beak a little longer than the urn; *exothecial cells* of varying size and shape, mostly isodiametric to oblong, 10–50 μm long, 10–20 μm wide, thick-walled, becoming rounded to oblate, with incrassate walls in 3–4 tiers at the mouth and forming a yellow-brown strip; *stomata* few at the base of the urn, bicellular, superficial, with rounded pori; *peristome* single, consisting of 16 entire, red teeth, to 300 μm long, papillose throughout on both surfaces. *Spores* spherical, brownish, 22–30 μm in diameter, roughly papillose. *Calyptra* conic-mitrate, smooth, naked, 4–5-lobed at the base.]

Chromosome number: unknown.

Discussion. The above description of *Bucklandiella pachydictyon*, especially that concerning the habit of the plants and sporophyte details, is based upon specimens from outside the Antarctic because the plants collected on Deception I. are very small and stunted as the result of growing in an extreme environment. Consequently, they do not possess a characteristic habit which allows them to be easily recognisable in the field. Yet, even highly depauperate plants of *B. pachydictyon* retain the typical features of the leaf structure which make it a distinct species. It is diagnosed by the following set of traits: (*1*) leaves narrowly lanceolate to oblong-lanceolate, gradually long acuminate; (*2*) leaf apex obtuse to subacute, always lacking a hair-point; (*3*) leaf lamina with a prominent fold on the right side of the costa, extending from near the leaf base to at least mid-leaf; (*4*) laminal cells entirely unistratose; (*5*) costa narrow, often ending well below the apex, bistratose throughout or rarely tristratose below, with two or three larger ventral cells in the proximal portion, in transverse section distinctly asymmetric and irregular in outline in the lower half; (*6*) elongate upper laminal cells, becoming linear with very incrassate lateral walls in the basal part; (*7*) basal marginal border mostly present, composed of (5–)10–15 pellucid, hyaline cells; (*8*) alar cells usually differentiated, often forming swollen, decurrent auricles.

The most characteristic, distinctive and constant feature of *Bucklandiella pachydictyon* is a markedly asymmetric costa, as clearly visible in its transverse section. This shape results from a distinct plica on the right side of the costa which stretches from near the base upwards to the middle of the leaf or somewhat higher. This plication of the leaf lamina on one side of the costa is found in some other species of *Bucklandiella*, for instance *B. emersa* (Müll.Hal.) Bednarek-Ochyra & Ochyra and *B. membranacea* (Mitt.) Bednarek-Ochyra & Ochyra, but in neither of these is the shape of the costa markedly affected and it is symmetric in transverse section. In addition, the costa usually ceases well below the leaf apex, is narrowly canaliculate and narrow, and in transverse section it is semi-terete and distinctly convex on the dorsal side, mostly bistratose throughout and has 2 or 3 enlarged ventral cells in the proximal portion. The structure of the costa permits separation of *B. pachydictyon* from other congeners. In *B. subsecunda* and *B. didyma* the costa is very broad and always symmetric, markedly flattened and consisting of at least 4–6 enlarged ventral cells in the basal part in transverse section.

Another distinctive feature of *Bucklandiella pachydictyon* is its leaf areolation. The laminal cells are elongate throughout and in the proximal part they are linear, with strongly incrassate longitudinal walls, much thicker than the lumina, and from this peculiar feature the specific name was coined. The basal marginal border is quite variable in this species; sometimes it is absent from some leaves and present on others on the same shoot. Nevertheless, it is most often present and very prominent, composed of up to 15 pellucid cells which vary from subquadrate to linear and very narrow. The swollen, brown and decurrent auricles are a distinct feature of many South Georgian and South American plants, whereas in the Antarctic plants the angular cells are few but otherwise distinctly larger and form flat, shortly decurrent auricles.

Originally, the specimens considered in the present treatment as *Bucklandiella pachydictyon* were named *B. heterostichoides* (Cardot) Bednarek-Ochyra & Ochyra. This is an American Subantarctic species having its maximum occurrence on Subantarctic South Georgia (Bell 1974) and occurring rarely at alpine elevations in the Fuegian region (Roivainen 1955*a*) and on Tristan da Cunha (Dixon 1960). Both *B. pachydictyon* and *B. heterostichoides* share the presence of the hyaline basal marginal border, narrow costa, elongate laminal cells and the form of the inner perichaetial leaves, but the latter is immediately recognised by its distinctive branching pattern, the regularly ascending branches from the main prostrate stem with many short lateral branchlets being very characteristic and giving the species a peculiar ctenoid habit. In addition, the leaves are consistently piliferous with a short, smooth hyaline hair-point on most upper and lower leaves, and the leaves are not auriculate.

Bucklandiella pachydictyon clearly belongs within sect. *Marginatae* of *Bucklandiella* which is best characterised by the peculiar areolation of the innermost perichaetial leaves. The cholorophyllose upper leaf cells of these have strongly incrassate walls and are unlike those in the vegetative leaves. In general, *B. pachydictyon* seldom produces sporophytes, but they exhibit several distinct features, including thick-walled exothecial cells, short and entire peristome teeth and large, papillose spores, 22–30 µm in diameter.

Bucklandiella pachydictyon was distinguished for the first time as a separate taxon, *Grimmia suborthotrichacea* var. *robustissima*, from material collected from Îles Kerguelen (Müller (1889) and reported by Wilson & Hooker (1847) as an unnamed variety of *Racomitrium protensum* var. 3. Likewise, Cardot (1905*a*) initially recognised this species as a variety, *Racomitrium symphyodontum* var. *muticum*, from material collected by C. Skottsberg from Isla de los Estados, but it was subsequently raised to *Racomitrium pachydictyon* (Cardot 1908*a*). He based the concept of this new species not only on the type of var. *muticum* but additional specimens from Isla de Los Estados, Tierra del Fuego, South Georgia and western Patagonia were also used for its description. Of these, the material from Puerto Bueno in western Patagonia was erroneously placed in this species, and Cardot (1908*a*) himself corrected this when recognising this specimen as a separate variety, *Racomitrium symphyodontum* var. *dusenii*. It was later considered identical to *Racomitrium didymum* (Deguchi 1984) which is here treated as conspecific with *Bucklandiella crispula*.

Reproduction in Antarctica. Sterile.

Habitat. Small populations usually consisting of a few short and scattered shoots were found on very coarse scoria in geothermal areas, close to permanent ice.

World range. An amphiatlantic south-temperate species (Fig. 138). It is widely distributed but scattered in the Fuegian region, western Patagonia (Cardot & Brotherus 1923; Matteri 1985*a*) and extending northwards to lat. *ca* 39° S in Neuquén Province of Argentina (*Moreau 4719*, H, KRAM) and Cautin Province of Chile (Herzog 1938) in the Valdivian region in the *Nothofagus* zone in southern South America. It occurs on the Subantarctic islands of South Georgia (Bell 1974), Marion I. in the Prince Edward Is. and Île de la Possession in the Îles Crozet archipelago (Bednarek-Ochyra & Ochyra 2008*a*) and Îles Kerguelen (Cardot 1916; Ochyra, personal observations), and penetrates southwards into the maritime Antarctic. An isolated station of *Bucklandiella pachydictyon* is reported from the Central Andes of Bolivia from an elevation of 3600 m (Churchill *et al.* 2000).

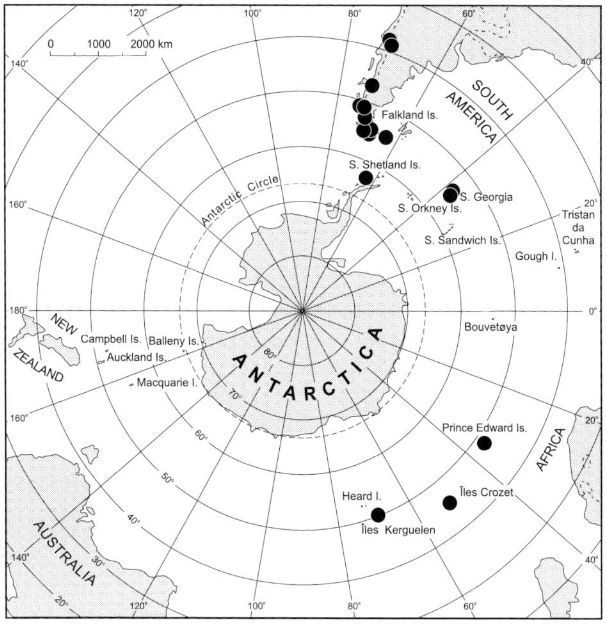

Figure 138. Global distribution of *Bucklandiella pachydictyon*. Literature record from Bolivia is not indicated.

Distribution in Antarctica. A few scanty collections of *Bucklandiella pachydictyon* are known from the volcanic Deception I. in the South Shetland Is. at lat. 62°58′ S (Fig. 170), where the species grows at elevations from *ca* 100 to 130 m, and, at one site, at 350 m (Lewis Smith, 2005*a*,*b*). The species was also reported from Nelson I. (Putzke & Pereira 2001) but no voucher specimens have been available for examination.

Specimens examined. WEST ANTARCTICA. SOUTH SHETLAND IS. **Deception I.** E of Crimson Hill, 130 m, *RILS 11205A* (AAS, KRAM); southern Pendulum Cove, 100 m, *RILS 5749C & 5754D* (AAS, KRAM); slopes to E and SE of Pendulum Cove, *ca* 100 m, *RILS 6924* (AAS, KRAM); summit of peak 2.2 km WNW of Mt. Kirkwood, 350 m, *DM 125* (AAS, KRAM).

3. **Bucklandiella sudetica** (Funck) Bednarek-Ochyra & Ochyra Fig. 139

Bucklandiella sudetica (Funck) Bednarek-Ochyra & Ochyra *in* Ochyra, Żarnowiec & Bednarek-Ochyra, Cens. Cat. Polish Mosses: 147. 2003. ≡ *Trichostomum sudeticum* Funck, Deutschl. Moose: 26, *t. 18*. 1820. ≡ *Racomitrium sudeticum* (Funck) Bruch & Schimp. *in* Bruch, Schimp. & W.Gümbel, Bryol. Eur. 3: 141, *f. 264*. 1845 [Fasc. 25–28 Mon.: 7, *f. 1*]. ≡ *Dryptodon sudeticus* (Funck) Brid., Bryol. Univ. 1: 195. 1826. ≡ *Campylopus sudeticus* (Funck) Fürnr., Flora 12: 595. 1829. ≡ *Racomitrium microcarpon* (Hedw.) Brid. var. [β] *sudeticum* (Funck) Huebener, Muscol. Germ.: 202. 1833. ≡ *R. heterostichum* (Hedw.) Brid. subsp. *sudeticum* (Funck) Dixon *in* Dixon & H.Jameson, Stud. Handb. Brit. Mosses: 154. 1896. ≡ *R. heterostichum* (Hedw.) Brid. var. *sudeticum* (Funck) Dixon *ex* E.Bauer, Musci Eur. & Amer. Exs. 41: no. 2023. 1929. — Type citation: [Czech Republic] Ad saxa in Sudetis (am Weisswasser im Teufelgrund) Jun. [Lectotype (*vide* Frisvoll 1984*b*: p. 314): '[Heinrich Christian Funck Deutschlands Moose] ? 15 [Trichostomum] sudeticum F., caule procumbente subramoso: ramis simplicibus, foliis lanceolatis erecto-patulis, piliferis, setis lateralibus rectis, thecis ovatis, operculo rostrato. Ad saxa in Sudetis (an Weisswasser im Teufelgrund.) Jun.' – M!; isotype: H! (2 specimens)].

Grimmia austropatens Müll.Hal. *in* Neumayer, Int. Polarforsch. Deutsch. Exped. 2: 316. 1890 ['austro-patens']. ≡ *Racomitrium austrogeorgicum* Paris, Index Bryol.: 519. 1895 ['austro-georgicum'], *nom. illeg.* — TYPE CITATION: Austro-Georgia, Bach-Grund oberhalb des Pinguin-Thales, 26 Januario 1883. [Lectotype (*vide* Frisvoll 1986: p. 340): '№ 53 Dryptodon austro-patens CMüll nsp. Fundort: Brachgrund oberhalb des Pinguinbay Süd-Georgian, 26/I 83 Will Südpolarexpeditionen' – HBG!; isotype: PC-Cardot!]. First synonymised with *Racomitrium sudeticum* by Frisvoll (1986: p. 340, 344).

Racomitrium substenocladum Cardot, Compt. Rend. Hebd. Séances Acad. Sci. 153: 603. 1911. — TYPE CITATION: [Antarctica] Terre de Graham: cap Tuxen [p. 601] (Je possède également cette espèce de la Terre de Feu) [p. 603]. [Lectotype (*vide* Frisvoll 1986: p. 342): 'No. 200 Racomitrium substenocladum Card. sp.nova Terre de Graham: cap Tuxen, lieux humides, alt. env. 100 m 2éme expéd. Charcot leg. Gain 8.01.1909' – PC!; isolectotype: H-Brotherus!; syntype: 'Expeditio suecica 1907–1909. 404. Rhacomitrium substenocladum nsp. Fuegia, supra ostium fluminis Azopardo, ad saxa 600 m s.m. 3/3 08 Det. J. Cardot. Leg. T. Halle et C. Skottsberg.' – H-Brotherus!, PC!, S!, UPS!]. First synonymised with *Racomitrium austrogeorgicum* by Roivainen (1955*a*: p. 86) and subsequently with *Racomitrium sudeticum* by Frisvoll (1986: p. 344).

Plants small and very short to medium-sized, rather slender but stiff and rigid, forming loosely coherent or dense cushions, tufts or mats, dull, very variable in colour, sordid-, olive-, yellowish- or drab-green, yellowish- or orange-brown to olivaceous above, brown to blackish-brown below, occasionally yellowish to jet-black throughout. *Stems* erect to ascending, occasionally prostrate, (0.5–)1.5–4.5(–7.5) cm tall, usually sparingly dichotomously branched to almost unbranched, or with more lateral branchlets, mostly sparsely radiculose below with brown, smooth, branched rhizoids, in transverse section round, lacking a central strand and consisting of a 2–3-layered cortex of small, stereid cells with strongly incrassate, brown walls and small lumina, rather gradually merging into a 5–6-stratose medulla of large, yellowish-hyaline, thick-walled cells; *axillary hairs* filiform, hyaline throughout, composed of 6–9 cells, short below, elongate in the distal part. *Leaves* crowded, appressed to imbricate, erect and often slightly curved-secund when dry, erecto-patent to patent and often recurved at the tips when moist, lanceolate to narrowly ovate-lanceolate, (1.2–)1.5–2.3(–2.6) mm long, 0.4–0.7 mm wide, gradually acuminate, usually ending in a hyaline hair-point, sometimes epilose and narrowly obtuse at the apex, broadly canaliculate to carinate, non-decurrent; *hair-point* hyaline, short, usually to 0.15 mm but occasionally to 0.35 mm, stout, non-decurrent along the margins of the lamina, denticulate to crenate, erect to erect-recurved or recurved, not or slightly flexuose when dry; *margins* entire, broadly recurved to ½–¾ the leaf length on one side, more narrowly and shortly recurved or sometimes plane on the other side, bistratose for 1 or, in patches, for 2 rows of cells and somewhat bulging to unistratose or unistratose with bistratose patches in the distal half, unistratose with bistratose spots in the proximal portion; *costa* single, subpercurrent to percurrent, intensely yellow to yellow-brown, sharply set off from the laminal cells, unbranched, gradually tapering upwards, (55–)60–80(–90) μm wide at the base, strongly convex on the dorsal surface, in transverse section semi-terete, 2(–3)-stratose above with 2–3 ventral cells, 3(–4)-stratose in the middle and at the base, with 3(–4) enlarged ventral cells; *laminal cells* unistratose with occasional bistratose strands above, smooth to pseudopapillose due to the lengthwise cuticular ridges appearing papillose in cross-section, with thick and sinuose walls, quadrate, rounded-quadrate to short-rectangular, mixed with transversely rectangular above, 5–15(–25) μm long, 5–8 μm wide, becoming gradually elongate towards the base; *basal cells* rectangular, (15–)25–50(–60) μm long, 8–10 μm wide, with thickened, sinuose-nodulose to nodulose lateral walls, intensely yellow in 1–2 rows at the insertion; *alar cells* not or only slightly differentiated, usually yellowish; *basal marginal cells* weakly differentiated, quadrate to short-rectangular, pellucid, with moderately thickened and mostly sinuose walls, forming an indistinct border of 2–5(–10) cells. *Dioicous*. [*Perichaetial leaves* similar to vegetative leaves, only somewhat larger but with a similar areolation, usually piliferous, the innermost 1–3 leaves ovate-lanceolate with a subhyaline base. *Setae* singly per perichaetium, 2.5–3.5 mm long, straight or sometimes arcuate, yellow to brown with age, smooth, dextrorse when dry. *Capsules* exserted, erect or occasionally inclined, straight, symmetric, ovoid to obloid, occasionally subspherical to obloid-cylindrical, (0.7–)1.1–1.5 mm long, 0.4–0.65 mm wide, pachydermous, brown, dull to glistening, smooth; *annulus* 2–3-seriate, compound, deciduous, consisting of an upper row of large, vesiculose cells with strongly incrassate walls; *operculum* conic convex, with a straight or somewhat slanted beak, reaching to ½ the urn length; *exothecial cells* variable in shape, mostly isodiametric, rounded-quadrate to quadrate or short-rectangular to hexagonal, thin-walled; *stomata* numerous in 1–2 rows at the base of the urn, superficial, bicellular, with rounded pori; *peristome* single, consisting of 16, orange-yellow to reddish-brown, papillose teeth, 280–410 μm long, arising from a low basal membrane, 35–50 μm high, usually variously cleft into 2–3 prongs in the upper half, sometimes nearly to the base, sometimes only variably perforated or undivided. *Spores* spherical, yellowish, almost smooth,

→

Figure 139. *Bucklandiella sudetica*. — **1–2**. Plants with sporophytes. **3**. Sterile plant. **4**. Portion of branch, wet. **5**. Portion of cross-section of stem. **6–8**. Axillary hairs. **9–14**. Leaves. **15**. Leaf apex. **16**. Upper lamina cells at margin. **17–19**. Mid-leaf cells. **20**. Basal leaf cells. **21–22**. Alar and supra-alar cells. **23–30**. Cross-sections of leaves, a sequence from base to apex. **31**. Operculate capsule, wet. **32**. Deoperculate capsule, dry. **33**. Exothecial cells in middle of urn. **34**. Lower exothecial cells and stomata. **35**. Peristome teeth, spores and exothecial cells at orifice. — [1–2, 31–35 from *Bell 169*; 3, 17, 21, 24, 27, 29 from *RO 535/80*; 4–5, 9, 18, 23 from *Gain 200* (lectotype of *R. substenocladum*), PC; 6–8, 12 from *RO 507/80*; 10, 15, 19, 30 from *RO 506/80*; 11, 26, 28 from *RO 599/80*; 13 from *RWMC 465*; 14, 16, 20, 22, 25 from *Will 53* (lectotype of *Grimmia austropatens*), HBG]. — Scale bars: a – 100 μm (16–22); b – 1 mm (2); c – 1 mm (31–32) and 2 mm (4); d – 1 mm (9–14) and 100 μm (5–8, 15, 23–30, 33–35); e – 0.5 cm (1, 3).

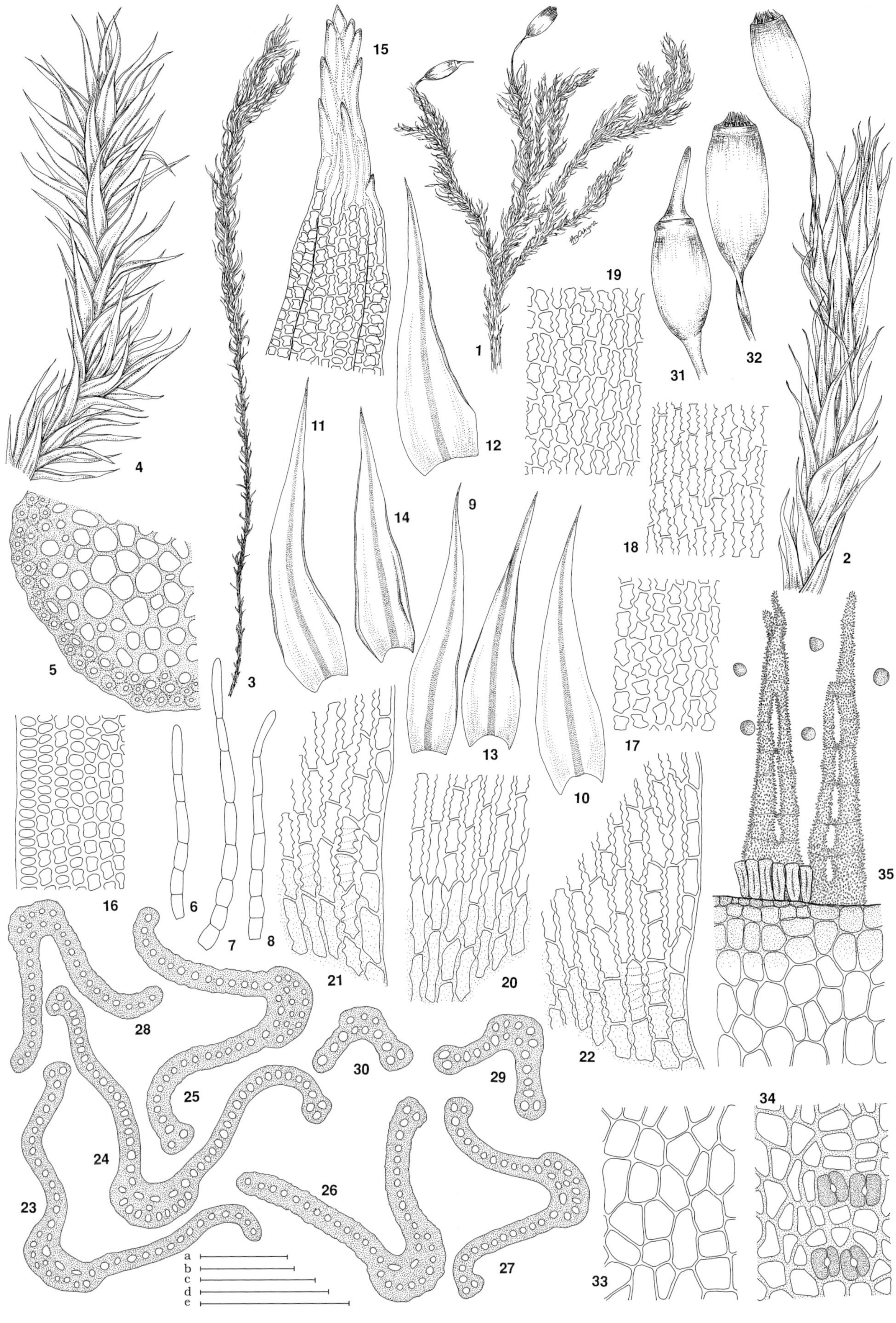
1
2
3
4
5
6
7
8
9
10
11
12
13
14
15
16
17
18
19
20
21
22
23
24
25
26
27
28
29
30
31
32
33
34
35
a
b
c
d
e

(10–)12–16(–18) µm in diameter. *Calyptra* conic-mitrate, smooth, naked, 4–5-lobed at the base.]

Chromosome number: unknown in the Antarctic and Subantarctic; $n = 14$ – a single record from the Ukraine (Vysotska 1975).

Discussion. Throughout its wide geographical range *Bucklandiella sudetica* is a very variable taxon, forming a number of different morphotypes which, in the past, were given taxonomic status. Some of them are possibly conditioned genetically but there is considerable variation, with many combinations of morphological characteristics. However, some morphological and anatomical structures exhibit constancy, making *B. sudetica* a fairly well defined species and difficult to confuse with other species of the genus. It is best characterised by the following set of traits: (*1*) plants usually growing in compact tufts; (*2*) stems slender, sparingly branched; (*3*) leaves short and narrow; (*4*) hyaline hair-point short, to 0.4 mm, or absent; (*5*) leaf margins recurved, bistratose in the distal portion with frequent unistratose patches; (*6*) costa relatively narrow, distinctly convex on the dorsal surface and semi-terete in cross-section, 2–3-stratose above, 3–4-layered in the proximal part, with 3–4 enlarged ventral cells; (*7*) upper laminal cells tending to be short or very short, quadrate to short-rectangular; (*8*) basal cells mostly sinuose-walled to the insertion; (*9*) basal marginal border indistinct, composed of 2–5 pellucid cells, with moderately thickened and partly sinuose walls; (*10*) all perichaetial leaves undifferentiated, similar to vegetative leaves in shape and areolation; (*11*) setae and capsules short; (*12*) basal membrane of the peristome low, 35–50 µm.

Bucklandiella sudetica was first described in the Southern Hemisphere as *Grimmia austropatens* from South Georgia (Müller 1890*a*). Paris (1895, 1898) transferred this species to *Racomitrium* and, by mischance, changed the specific epithet *austropatens* to *austrogeorgicum*. It was because he assumed that *Grimmia austropatens* of Müller (1890*a*) was a later homonym of the South African *G. austropatens* Müll.Hal. *ex* Geh. (Geheeb 1878). However, the latter name was then a *nomen nudum*, so the change of the name *G. austropatens* to *Racomitrium austrogeorgicum* was unjustified, and therefore it has to be considered illegitimate. Nevertheless, the name *Racomitrium austrogeorgicum* has been widely used (e.g. Cardot 1906*a*, 1908*a*; Cardot & Brotherus 1923; Roivainen 1955*a*; Bell 1973*c*, 1974; Greene 1986; Ochyra *et al.* 1986), until it was reduced to synonymy with *Racomitrium sudeticum* (Frisvoll 1986). Clifford (1955) had suggested the conspecificity of *R. austrogeorgicum* with *R. crispulum*, but his imperfect taxonomic concept of the latter species has not gained wide approval, although it is still accepted in *Index muscorum* (Wijk *et al.* 1967).

Antarctic plants of *Bucklandiella sudetica* were originally recognised by Cardot (1911*a,b*) as a separate species, *Racomitrium substenocladum*, that was also partly based on Fuegian material, and later Cardot & Brotherus (1923) described another new species *R. skottsbergii* on the basis of material from South Georgia and Tierra del Fuego. Roivainen (1955*a*) convincingly showed that both species differed only slightly from *R. austrogeorgicum*, reducing their names to synonymy, and Frisvoll (1986) confirmed the conspecificity of these species with *R. sudeticum*. In fact *R. substenocladum* should be the correct name for *R. austrogeorgicum*, if considered as a distinct species.

The morphological variation of the austral plants of *Bucklandiella sudetica* falls entirely in the range of variation of the species in the Northern Hemisphere, and the lectotype material of *Racomitrium austropatens* does not differ much from the lectotype of *Bucklandiella sudetica* itself which originated from the Czech sector of the Sudetes (Funck 1820). The leaves of the former are only slightly shorter, the costa is somewhat broader and the leaf margin is less recurved on one side, but these features are subject to considerable variation in this species. On the other hand, the costa in *Racomitrium substenocladum* is identical in its size and anatomy to that of the lectotype of *Bucklandiella sudetica*.

In contrast to the Northern Hemisphere where *Bucklandiella sudetica* has often been mistaken for various species of the genus, in the Southern Hemisphere the possibility of its confusion with other Antarctic and Subantarctic species is minimal. It can be distinguished from *B. subsecunda* by its semi-terete costa in the proximal portion and short upper laminal cells, and from *B. didyma* and *B. pachydictyon* by the absence of the basal marginal border which eventually may be indistinct and reduced to 2–5 partly sinuose cells, and by the sinuosity of the laminal cell walls to the extreme leaf base. The absence of the basal marginal border and the semi-terete costa is shared by *B. sudetica* and *B. rupestris*, but the entirely unistratose leaf margins, epilose leaves and rope-like appearance of the latter species should prevent confusion.

Reproduction in Antarctica. Always sterile, although both male and female inflorescences have been observed on specimens from various sites, for example on Signy I. in the South Orkney Is. and King George I. in the South Shetland Is.

Habitat. A rupicolous moss growing primarily in dry, sheltered fellfield habitats, less often in moist places, on rock surfaces, ledges and fissures, on stony or gravelly ground among boulders and pebbles on raised beaches, scree, well drained stony slopes and hillsides, sometimes on pockets of soil and humus. It is mostly associated with communities of the fruticose lichen and moss cushion subformation, especially within the *Andreaea–Usnea* association, occasionally forming small, almost pure stands of coalesced cushions. It is usually associated with *Andreaea* spp., *Sanionia uncinata*, *Hymenoloma antarcticum*, *H. grimmiaceum*, *Polytrichastrum alpinum*, *Barbilophozia hatcheri*, as well as fruticose lichens such as *Himantormia lugubris*, *Cladonia* spp., *Stereocaulon* spp. and *Usnea* spp.

World range. *Bucklandiella sudetica* is a bipolar species (Fig. 140, inset). In the Holarctic it is generally a boreal-montane moss, occasionally penetrating into the Arctic, which has a markedly discontinuous range with the main centres of its occurrence in Europe and North America (Frisvoll 1988; Bednarek-Ochyra *et al.* 1990; Bednarek-Ochyra 1995*a*). In northern Europe it occurs frequently and abundantly in western Fennoscandia (extending eastwards to Lake Ladoga), in Britain (Scotland, Wales to central England) and rarely in Ireland, on the Faröes and Iceland and rarely penetrating into the Arctic on Jan Mayen (Watson 1964; Frisvoll 1983*b*), Bjørnøya (Frisvoll 1988) and Spitsbergen (Wegener 1993). On the continent, *B. sudetica* occurs commonly in all mountain ranges from the Pyrenees and the West Caucasus, extending in the south to Corsica and the Rila massif in Bulgaria, with some relictual stations in the Central European lowlands in Germany (Koperski 1989) and Poland (Bednarek-Ochyra *et al.* 1990; Bednarek-Ochyra 1995*a*). In North America, the species is bicentric in distribution (Ochyra & Bednarek-Ochyra 2007*b*). In the eastern part of the continent it is scattered throughout much of Labrador, Newfoundland south to New York, with some isolated and relictual stations in northern Michigan in the Great Lakes Basin. In western North America it is frequent in the coastal region from central Alaska

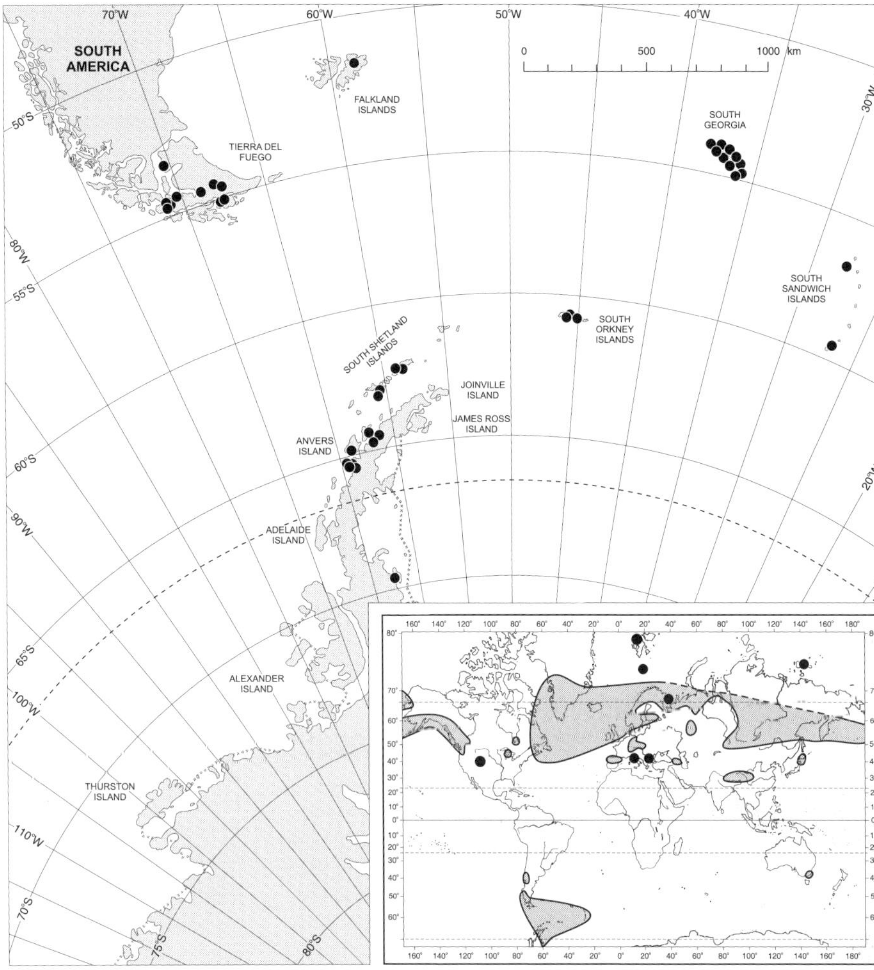

Figure 140. Distribution of *Bucklandiella sudetica* in the Fuegian region and Antarctica. Inset: global distribution.

to northern California, with some isolated stations in the Rocky Mountains of Montana and Colorado and, additionally, it is scattered throughout the Aleutian Is. In Greenland, the species is known only from the southern part, extending as far north as lat. 70° N. In Asia, *B. sudetica* is rare and scattered, but possibly this is due to undercollecting in that region. It is scattered in Central Siberia, the Russian Far East, Chukotka and Kamchatka (Ignatov & Cao 1994; Bednarek-Ochyra 2004*b*), Japan (Noguchi 1974) and the Himalayan region and China (Cao *et al.* 2003), with isolated stations on the New Siberian Is. in the Arctic (Afonina & Czernadjeva 1995). In the Southern Hemisphere, *B. sudetica* is uncommon but widely scattered in the Valdivian region of Chile, southern Patagonia and Tierra del Fuego (Roivainen 1955*a*; Frisvoll 1986; Bednarek-Ochyra 1993*b*), and on the Falkland Is. (Matteri 1986). Additionally, it is relatively common and abundant on Subantarctic South Georgia (Bell 1974), and penetrates into the northern maritime Antarctic from where it was first recorded by Cardot (1911*a*,*b*) as *Racomitrium substenocladum* and subsequently reported by Bell (1973*c*) as *R. austrogeorgicum*. Moreover, the species is very rare in the Australian Alps from where it was originally described as *Grimmia amoena* Broth. (Frisvoll 1986).

Distribution in Antarctica. *Bucklandiella sudetica* is widely distributed but generally an infrequent and localised species (Fig. 140), although is some places it is fairly abundant, for example on Signy I. and the Keller Peninsula on King George I. It is very rare on the South Sandwich Is. where is known only from two collections on Leskov and Bellinghausen Is., infrequent on the South Orkney Is. (except Signy I.) and on the South Shetland Is. where it is most frequent on King George I. and very rare on Livingston I. and Deception I. Additionally, the species occurs at a few sites on the western coast of the Antarctic Peninsula, including the Danco Coast and adjacent islands (Brabant I. and Anvers I.) and the Graham Coast and its offshore islands, reaching lat. 65°33′ S. On the eastern coast of the Antarctic Peninsula *B. sudetica* is exceedingly rare and is known from a single locality on the Wilkins Coast at lat. 69°32′ S, its southernmost occurrence. The majority of Antarctic stations of *B. sudetica* are at lower elevations, usually below 100 m, and only occasionally does it grow at higher altitudes; its highest known record at 1000–1350 m is from Engel Peaks on the Wilkins Coast.

Specimens examined. WEST ANTARCTICA. SOUTH SANDWICH IS. **Leskov I.** North end, *ca* 135 m, *REL 524* (AAS, KRAM). **Bellingshausen I.** Fumarole, *PC 29B* (AAS, KRAM).

SOUTH ORKNEY IS. **Coronation I.** Cape Hansen, 25 m, *RILS 8686* (AAS, KRAM) and 20–200 m, *RILS 3221*(AAS, KRAM); moraine below Windscoop Buttress on E side of Sunshine Glacier, *RW 149* (AAS, KRAM); Olivine Pt., *ca* 30 m, *RILS 126* (AAS, KRAM, LE, NY, PC). **Powell I.** SW coast, W of John Peak, 15–80 m, *RILS 244* (AAS, KRAM). **Signy I.** Gneiss Hills towards Porteous Pt., 150 m, *RILS 8124* (AAS, KRAM); ridge E of Spindrift Rocks, *ca* 80 m, *MWH 765J* (AAS, KRAM); col between Changing Lake and Knob Lake, *ca* 65 m, *RW 89* (AAS, KRAM); Changing Lake, *ca* 65 m, *RW 113B & 144* (AAS, KRAM); Factory Cove, slope behind the station, *ca* 35 m, *MWH 48* (AAS, KRAM) and 80 m, *RILS 12* (AAS, KRAM); top of hill behind the station, *ca* 100 m, *REL 1052* (AAS, KRAM, PC); Observation Bluff, *ca* 85 m, *MWH 136B* (AAS, KRAM, NY).

SOUTH SHETLAND IS. KING GEORGE I. ***Admiralty Bay***: Skua Cliff, 105 m, *RO 621/80* (KRAM); Pond Hill, 30 m, *RO 862/80* (KRAM) and 40 m, *RO 871/80* (KRAM); Ore Pt, 4 m, *RO 504/80 & 506/80* (KRAM); Speil Pt, 2.5 m, *RO 535/80* (KRAM) and 3 m, *RO 507/80* (KRAM); Round Hill, 50 m, *RO 541/80* (KRAM); Ullman Spur, 80 m, *RO 599/80* (KRAM). ***Fildes Peninsula***: Bukhta Gidrografov, *Li GWS28* (AAS, KRAM). **LIVINGSTON I.** E side of False Bay, *RMS 69-535, 69-574, 69-576 & 69-587* (US). **DECEPTION I.** Whalers Bay, near the Hanger, 10 m, *RWMC 873* (AAS, KRAM).

WEST ANTARCTIC PENINSULA. **DANCO COAST.** Spring Pt., *VK 1211* (KRAM); Recess Cove, Chartlotte Bay, 2–10 m, *RILS 3984 & 3989* (AAS, KRAM); Charlotte Bay, Andrée I., 25 m, *RILS 4040B* (AAS, KRAM). **PALMER ARCHIPELAGO.** ***Brabant I.***: point to SE of Hill Bay, 5 m, *AM 55C & 57* (AAS, KRAM). ***Anvers I.***: Cape Monaco, *VK 820* (AAS, KRAM). **GRAHAM COAST.** Cape Tuxen, *ca* 100 m, *Gain 200* (H, PC – lectotype of *Racomitrium substenocladum*) and 200–300 m, *RILS 4910* (AAS, KRAM). **WILHELM ARCHIPELAGO.** ***Argentine Is.***: Galindez I., overlooking Stella Creek, 5 m, *RWMC 465* (AAS, KRAM); same island, overlooking Meek Channel, *ca* 8 m, *RWMC 472 & 448* (AAS, KRAM); same island, Stella Creek, *ca* 15 m, *REL 1327* (AAS, KRAM). ***Collins Bay***: Berthelot Is., *ca* 35 m, *RWMC 619F* (AAS, KRAM). ***Grandidier Channel***: Lahille I., 2–5 m, *RILS 4325A* (AAS, KRAM); Takaki Pt., 5–20 m, *RILS 4388 & 4390* (AAS, KRAM); largest island off Takaki Pt., 10–35 m, *RILS 4348* (AAS, KRAM).

EAST ANTARCTIC PENINSULA. **WILKINS COAST.** Engel Peaks, 1000–1350 m, *MJC 59* (AAS, KRAM).

4. **Bucklandiella didyma** (Mont.) Bednarek-Ochyra & Ochyra   Fig. 141

Bucklandiella didyma (Mont.) Bednarek-Ochyra & Ochyra *in* Ochyra, Żarnowiec & Bednarek-Ochyra, Cens. Cat. Polish Mosses: 145. 2003. ≡ *Grimmia didyma* Mont., Ann. Sci. Nat. Bot. Sér. 3, 4: 122. 1845. ≡ *Racomitrium didymum* (Mont.) Lorentz, Bot. Zeit. (Berlin) 24: 187. 1866. — TYPE CITATION: Hab. in Chile australiori legit cl. C. Gay. Herb. Mus. Par. [Lectotype (*selected here*): 'Herb. Mus. Paris. Herbier du Chili austral envoyé par M. Gay. (3me. envoi) Grimmia didyma Montag. n. sp.' – PC-Montagne!; isotypes: BM-Bescherelle!, BM-Hampe! (2 specimens), PC-Montagne!].

Racomitrium convolutum Mont., Ann. Sci. Nat. Bot. Sér. 3, 4: 122. 1845. ≡ *Grimmia convoluta* (Mont.) Müll.Hal., Syn. Musc. Frond. 1: 802. 1849, ***syn. nov.*** — TYPE CITATION: Hab. in truncis arborum reipublicae chilensis a cl. C. Gray [sic!] lecta. Herb. Mus. Par. [Lectotype (*vide* Bednarek-Ochyra *et al.* 1999: p. 99): 'Herb. Mus. Paris. Herbier du Chili austral envoyé par M. Gay. (3me. envoi) Racomitrium convolutum Montag. n. sp. absq. fr. ad truncos Mali Los Angeles Concepcion Chili' – PC-Montagne!; isotypes: BM-Bescherelle!, BM-Hooker!, PC-Montagne!].

Grimmia nigrita Müll.Hal., Syn. Musc. Frond. 1: 801. 1849, *hom. illeg.* ≡ *Racomitrium nigritum* A.Jaeger, Ber. Thätigk. St. Gallischen Naturwiss. Ges. 1872–1873: 90. 1874. — TYPE CITATION: Insula Eremitae ad Cap. Horn: J. D. Hooker. [Lectotype (*selected here*): the left specimens in the bottom row on the sheet labelled as follows: 'N°. 135. Racomitrium protensum A. Braun R. cataractarum (Brid.) var. 1. Hermite Island, Cape Horn Antarct. Exped. 1839–1843. J. D. H.' – BM-Hooker!; isotypes: BM-Hooker!, BM-Wilson! (3 specimens), BR!, MICH!, NY!, S!]. First synonymised with *Racomitrium didymum* by Deguchi (1984: p. 30).

Grimmia symphyodonta Müll.Hal., Syn. Musc. Frond. 1: 809. 1849. ≡ *Racomitrium symphyodontum* (Müll.Hal.) Wilson *in* Hook.f., Fl. Tasman. 2: 181. 1859 ['symphyodon']. — TYPE CITATION: Insula Eremitae ad Cap. Horn: J. D. Hooker. [Lectotype (*selected here*): 'N°. 138. Racomitrium fasciculare Dill. var. 3 tenue Hermite Island, Cape Horn Antarct. Exped. 1839–1843. J. D. H.' – BM-Hooker!; isotypes: BM-Hooker!, MICH!; syntype: N°. 138.b. Racomitrium fasciculare (Dill.) var. 2 Hermite Island, Cape Horn Antarct. Exped. 1839–1843. J. D. H.' – BM-Hooker!; isosyntype: BM-Hooker!, BR!, F!, MICH!, NY-Mitten!].]. First synonymised by Deguchi (1984: p. 30).

Grimmia defoliata Müll.Hal., Bot. Jahrb. Syst. 5: 81. 1883. ≡ *Racomitrium kerguelense* Ochyra & Matteri, Fragm. Florist. Geobot. 41: 1009. 1996 (non *Racomitrium defoliatum* Dixon, Smithsonian Misc. Collect. 69(2): 16, *f. 4.* 1918). ≡ *Bucklandiella defoliata* (Müll.Hal.) Bednarek-Ochyra & Ochyra *in* Ochyra, Żarnowiec & Bednarek-Ochyra, Cens. Cat. Polish Mosses: 144. 2003, ***syn. nov.*** — TYPE CITATION: all new species described by Müller (1883) are included in the section 'I. Bryologia Kerguelensis'; later Müller (1889) provided details of the type collection as follows: 'Ins. Kerguelen, Betsy Cove 12.I.1875'. [Lectotype (*vide* Ochyra & Matteri (1996*a*: p. 1009): 'Ex Museo botanico Berolinensi. Grimmia defoliata C. Müll. n. sp. Kerguelensland Betsy Cove. 74. Dr. Naumann' – S-Dusén!; isotype: PC-Cardot!].

Racomitrium flavescens Cardot, Rev. Bryol. 27: 41. 1900. ≡ *R. symphyodontum* (Müll.Hal.) Wilson fo. *flavescens* (Cardot) Cardot *in* Cardot & Broth., Kongl. Svenska Vetensk. Akad. Handl. 63(10): 29. 1923, ***syn. nov.*** — TYPE CITATION: Détroit de Magellan: île Londonderry, Whale Boat Sund, baie des Astéries. [Lectotype (*selected here*): 'Herb. J. Cardot. Mousses du voyage de la Belgica. n° 177 Rhacomitrium flavescens Card. sp. nova détroit de Magellan: île Londonderry, Whale Boat Sound, baie des Astéries Leg. Racovitza 17 déc^{bre} 1897' – PC-Cardot!; isotype: BR!].

Racomitrium symphyodontum (Müll.Hal.) Wilson var. *dusenii* Cardot, Wiss. Erg. Schwed. Südpolar-Exp. 4(8): 285. 1908. — TYPE CITATION: Patagonie occidentale: Puerto Bueno (Dusén). [Lectotype (*selected here*): 'Herb. J. Cardot. Rhacomitrium symphyodontum var. Dusenii Card. (var. muticum Dus. non Card.) Patag. occid.: Puerto Bueno. in saxis rupturque littoreis. Leg. P. Dusén, 1896. n° 34' – PC!; isotype: S!]. First synonymised with *Racomitrium didymum* by Deguchi (1984: p. 30).

Racomitrium subrupestre Roiv., Arch. Soc. Zool. Bot. Fenn. 'Vanamo' 9: 94, *fig. 1, 4.* 1955, ***syn. nov.*** — TYPE CITATION: Type no. 2214: Fuegia occ.: Fjordo de Agostini, Bahia Groth-Hansen, on micaceous schist blocks not far from the shore, with young and old capsules. [Holotype: 'Bryophyta Fuegiana 2214. Rhacomitrium subrupestre Roiv. n. sp. Fuegia occidentalis, Fjordo de Agostini, Bahia Groth Hansen, micaceous schist blocks not far from the shore. 22.2.1929 Leg. H. Roivainen' – H!].

Plants medium-sized and slender to large and fairly robust, in loose to moderately dense tufts or extensive patches or mats, dull or weakly glossy, sordid-, olive- or yellowish-green, sometimes with the blackish tinge to light brown or light olivaceous above, brown, dark brown

→

Figure 141. *Bucklandiella didyma.* — **1**. Habit, dry. **2**. Portion of shoot, dry. **3**. Portion of cross-section of stem. **4–6**. Axillary hairs. **7–11**. Leaves. **12**. Leaf apex. **13**. Upper leaf cells. **14**. Mid-leaf cells. **15**. Basal leaf cells. **16**. Alar and supra-alar cells. **17–26**. Cross-sections of leaves, a sequence from base to apex. **27–30**. Perichaetial leaves, from outermost to innermost. **31**. Apex of innermost perichaetial leaf. **32**. Basal cells of innermost perichaetial leaf. **33**. Apex of median perichaetial leaf. **34**. Lower cells of median perichaetial leaf. **35**. Calyptra. **36**. Capsule with calyptra, wet. **37**. Capsule with operculum, wet. **38**. Deoperculate capsules with peristome. **39–40**. Deoperculate capsules with peristome teeth destroyed, dry. **41**. Exothecial cells in middle of urn. **42**. Exothecial cells at base of urn and stoma. **43**. Peristome teeth, spores and rim cells of capsule. — [1, 3, 25–32 from *Bell 490*; 2, 12, 15–16, 20–21 from *MWH 829B*; 4–9, 17–19, 22, 24 from *PC 38*; 10, 14 from *MWH 410B*; 11, 13, 23 from *MWH 824E*]. — Scale bars: a – 1 mm (35–40) and 2 mm (1); b – 1 mm (2, 7–11); c – 100 µm (41–43); d – 1 mm (27–30); e – 100 µm (3–6, 16); f – 100 µm (17–26, 31–34); g – 100 µm (12–15).

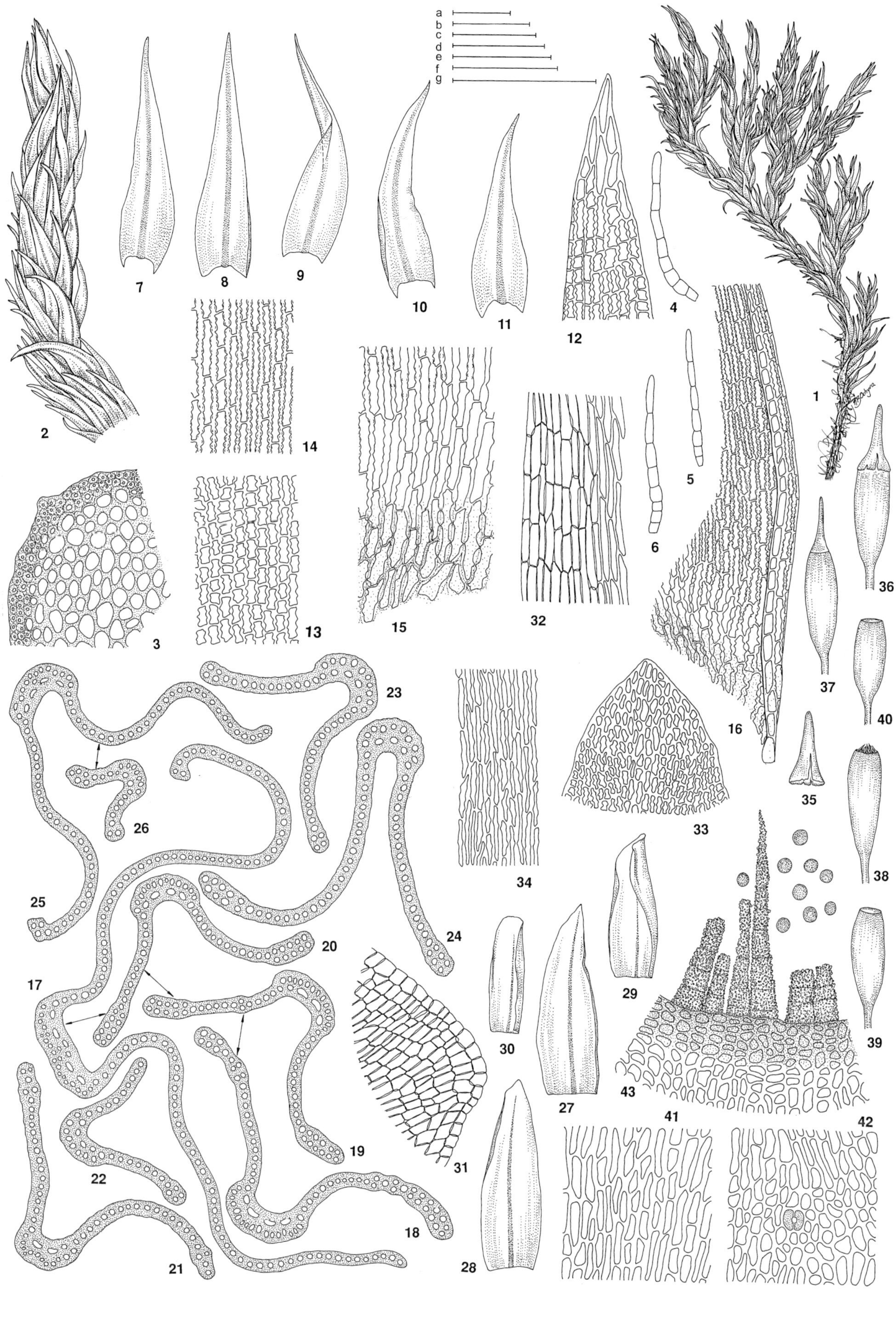
a
b
c
d
e
f
g
1
2
3
4
5
6
7
8
9
10
11
12
13
14
15
16
17
18
19
20
21
22
23
24
25
26
27
28
29
30
31
32
33
34
35
36
37
38
39
40
41
42
43

to blackish-brown below, occasionally olive-brown to light brown throughout. *Stems* erect to ascending, 1.5–6.5 cm long, sparsely to abundantly fastigiately or irregularly branched, occasionally almost unbranched to dichotomously branched, rather thin but stiff and coarse, dark brown, densely foliate, with the lower leaves often strongly eroded, sparingly radiculose at the base or with reddish-brown, smooth, glossy and sparsely branched rhizoids or rhizoids scattered in the lower part of the stem buried in soil and not forming a tomentum, in transverse section rounded, composed of 2–3 layers of small, stereid cortical cells with strongly incrassate, dark brown cells surrounding 4–6 rows of large, thick-walled medullary cells with intensely yellow walls and distinct corner thickenings, central strand absent; *axillary hairs* numerous and frequent in the upper part of the stem, filiform, hyaline throughout, 7–9-celled, composed of short, barrel-shaped lower cells and elongate upper cells. *Leaves* crowded, loosely or closely imbricate, erect when dry, erect-spreading when moist, straight to falcate, concave, canaliculate throughout, not or weakly plicate below, ovate-lanceolate to lanceolate, 2.1–3.1 mm long, (0.5–)0.6–0.8(–1.0) mm wide, distinctly short- to long-decurrent, gradually tapering to an obtuse or acute apex; *hair-point* absent or short, to 150 μm, erect, mucronate, edenticulate to faintly denticulate, yellow-brownish throughout or hyalinised at the extreme tip, indistinctly grading into the similarly coloured lamina, with a distinct and clearly visible areolation of thick- and straight-walled, elongate cells; *margins* entire throughout, broadly recurved on one side to about ¾ the leaf length or higher, and on the other side plane to very narrowly recurved to about ⅓–½ the leaf length, in the distal half bistratose in (1–)2–9 cell rows or occasionally unistratose on one or both sides, in the proximal half unistratose or bistratose on one or few cell rows; *costa* single, percurrent to short-excurrent, brown or yellow brown and sharply set off from the laminal cells, 90–105 μm wide at the base, gradually tapering upwards and about 60 μm wide below the apex, strongly flattened to reniform dorsally and situated at the base of a deep groove in the lower half, semi-terete to reniform and distinctly convex on the dorsal side in the upper half, canaliculate throughout on the ventral surface, in transverse section 2(–3)-stratose above with 3–4 ventral and 8–14 dorsal cells, becoming (2–)3(–4)-stratose in the median and lower part with 4 enlarged ventral cells and 15–20 dorsal cells; *laminal cells* unistratose with occasional bistratose strands in the upper part, smooth or sometimes weakly pseudopapillose due to longitudinal cuticular thickenings, with strongly incrassate and sinuose walls, isodiametric, rounded-quadrate or irregular, especially in the bistratosely thickened portions, to rectangular, 6–25(–30) μm long in the upper and median portions, becoming long-rectangular to linear in the lower part, (25–)30–55(–70) μm long, 4–7 μm wide, with sinuose-nodulose walls, *basal cells* intensively yellow- or orange-brown, rectangular but shorter and wider than the adjacent cells above, 20–50 μm long, 9–14 μm wide, with strongly thickened and porose walls, forming a coloured, 1–2(–3)-seriate strip along the insertion; *alar cells* differentiated, isodiametric to short rectangular, with thick and mostly strongly porose walls, yellow to orange- or yellow-brown, forming flat or somewhat swollen, short- or long-decurrent auricles; *basal marginal cells* differentiated, quadrate to long-rectangular, yellowish-hyaline, esinuose, moderately thick-walled, forming translucent, 1–2-seriate border, of 17–25(–35) cells in the marginal row and to 10 cells in the second row. *Dioicous. Outer perichaetial leaves* oblong, shortly broadly acute, obtuse at the apex or mucronate, 1.7–2.0 mm long, 0.7–0.8 mm wide, with broad, concolorous, subpercurrent costa and areolation of elongate, strongly thick-walled cells in the upper half, becoming long-rectangular below with strongly porose lateral walls and subhyaline in several cell rows towards the margins; *inner perichaetial leaves* oblong, 1.7–1.9 mm long, 0.7–0.9 mm wide, convolute, hyaline throughout, with a weak, indistinct costa extending to ½–¾ way up the leaf, with thin-walled, lax areolation of hexagonal cells in the distal portion, becoming long-rectangular, not lax and thin-walled in the proximal portion. *Setae* singly or geminate per perichaetium, erect, straight, 2–6 mm long, dextrorse when dry, brown. [*Capsules* erect, exserted, straight, symmetric, ovoid, obloid to obloid-cylindrical, (1.5–)1.6–2.1 mm long, 0.6–0.8 mm wide, smooth, brown, lustrous, pachydermous, gradually tapering into the seta; *annulus* compound, revoluble, composed of 2–3 cell rows, with the upper row of large, thick-walled, transparent, vesiculose cells; *operculum* conic-rostrate, with a straight beak, as long as half or more of the urn length; *exothecial cells* varying from rounded-quadrate to oblong, 18–60 μm long, 6–18 μm wide, with strongly incrassate walls, becoming smaller, rounded to oblate in 4–5 tiers at the urn mouth and forming a bright yellow-brown strip of 2–3 rows of cells at the orifice; *stomata* not many in a single row at the extreme base of the urn, superficial, bicellular; *peristome* single, consisting of 16 yellow-brown teeth, to 250 μm long, deeply cleft into two terete, irregular prongs to well below the middle, spiculosely papillose throughout on both dorsal and ventral surfaces. *Spores* spherical, 14–20 μm in diameter, pale brownish, finely papillose. *Calyptra* mitrate, naked, smooth, 4–5-lobed at the base.

Chromosome number: unknown

Discussion. Specimens determined and reported as *Racomitrium crispulum* (Hook.f. & Wilson) Hook.f. & Wilson [≡ *Bucklandiella crispula* (Hook.f. & Wilson) Bednarek-Ochyra & Ochyra] (Bell 1973*c*; Longton & Holdgate 1979) and *R. orthotrichaceum* (Müll.Hal.) Paris [=*Bucklandiella orthotrichacea* (Müll.Hal.) Bednarek-Ochyra & Ochyra] (Ochyra *et al.* 1998; Convey *et al.* 2001) are here interpreted as *Bucklandiella didyma*. These three species are closely related and their relationships have been the subject of considerable confusion. They belong within sect. *Emersa sensu* Bednarek-Ochyra (1995*a*) and, together with several other austral species that are still poorly known taxonomically, are very common members of the moss flora in the temperate and polar zones in the Southern Hemisphere, both in terms of the number of species and their abundance in the vegetation cover. To some extent, the problem with this group was similar to that in the *Racomitrium heterostichum* complex in the Northern Hemisphere, until Frisvoll (1984*b*, 1988) clearly defined and circumscribed its species and proved they were distinct taxa on account of several gametophyte and sporophyte characters. Interestingly, some authors (e.g. Clifford 1955; Lawton 1972, 1973) suggested a close relationship between the principal species of both northern and southern complexes which, however, are only remotely related.

Although there have been many species described within the *Bucklandiella crispula* complex in various parts of the austral region, they long remained understudied and were usually considered as independent species in South America (Mitten 1869; Cardot 1908*a*; Cardot & Brotherus 1923), New Zealand (Wilson 1854; Mitten 1867), Tasmania (Wilson 1859) and Australia (Mitten 1883). The situation changed radically when Dixon (1926) found that the status of the austral plants of the *B. crispula* complex, which he considered as a part of the genus *Racomitrium*, is very difficult to unravel and stated that numerous species from the Southern Hemisphere were 'too closely intergrading to be regarded as specific'. As a result he reduced several species names from Australia, New Zealand and South America to synonymy with *Racomitrium crispulum*, eventually recognising one variety within it, var. *rupestre* (Hook.f. & Wilson) Dixon. This concept was developed by Clifford (1955) who listed 18 species names supposedly synonymised with *R. crispulum* by former authors, and also suggested 20 taxon names as putative new synonyms of the latter. Accordingly, the majority of *Racomitrium* taxa formerly described from throughout the Southern Hemisphere have been lumped with *R. crispulum*. The broad and unrealistic concept of this species, additionally supported by the taxonomic studies of Lawton (1972, 1973), has been widely adopted in all major Floras and catalogues of mosses in the Australasian region (e.g. Sainsbury 1955; Scott & Stone 1976; Streimann & Curnow 1989; Beever *et al.* 1992; Streimann & Klanzenga 2002; Seppelt 2004) and, more importantly, also in *Index muscorum* (Wijk *et al.* 1967, 1969).

In the western part of the Southern Hemisphere Roivainen (1955*a*) accepted most species described in the *Bucklandiella crispula* complex by Cardot (1908*a*) and added several for the flora of Tierra del Fuego. Similarly, Bell (1974), in his treatment of the genus *Racomitrium* for a synoptic moss Flora of South Georgia, adopted the name *R. crispulum* for some populations and recognised many species in this complex. Deguchi (1984) was the first to make a careful comparison of the type collections of the two oldest species names in the *B. crispula* complex, i.e. *Dryptodon crispulus* from Campbell I. (Hooker & Wilson 1844) and *Grimmia didyma* from southern Chile (Montagne 1845*a*) and decided to keep them as distinct, through closely related species. This conclusion agrees with the description and illustrations of the lectotype of *Dryptodon crispulus* by Frisvoll (1984*a*) and it was subsequently approved by Bednarek-Ochyra *et al.* (1999) in the treatment of the genus *Racomitrium* in Brazil.

Because the shape of the capsule is often subject to variation in different species of *Bucklandiella*, we studied the larger material of this species from the *locus classicus* on Campbell I. in order to evaluate the reliability of the capsule size and shape as a taxonomic character in this species. Campbell I. is a small isolated island in the South Pacific Ocean lying about 600 km south of the South Island, New Zealand. Although it has had a long history of exploration, having been visited by numerous expeditions in the 19th and 20th centuries, only a few brought back collections of cryptogams. Vitt (1974*a*) provided a comprehensive treatment of the island's moss flora. Alas, none of the nine specimens collected by this author, and reported as *Racomitrium crispulum*, fits the type of this species, and all belong within other species of the *Bucklandiella crispula* complex. He also carried out field work on the adjacent Auckland Is., which is phytogeographically very close to Campbell I. (Vitt 1979*a*). Of 45 specimens of *B. crispula*, four perfectly fit the type of this species name, and a thorough examination of them confirmed that it is a distinct species well characterised by both gametophyte and sporophyte traits. All specimens examined of this species, including the type material (Frisvoll 1984*a*), have exceedingly small capsules, 0.8–1.2 mm long, 0.6–0.7 mm wide, which are subglobose to shortly ovoid and abruptly narrowed into the seta. Moreover, the leaves are terminated by a yellow-brownish hair-point, 0.3–0.4 mm long, and the alar cells are not differentiated. In contrast, the type and non-type specimens of *B. didyma* have obloid-cylindrical capsules, (1.5–)1.7–2.0(–2.2) mm long, 0.6–0.8 mm wide, gradually tapering into the seta. In addition, the yellow-brownish hair-point is very short, to 0.15 mm, and the alar cells are usually differentiated, forming swollen auricles. Bell (1974) gave the measurements 1.2–1.5 mm for the capsules of *B. crispula* from South Georgia, but these specimens refer to *B. orthotrichacea*, a closely related species from this complex characterised by large spores, 20–25 μm in diameter, and strongly pseudopapillose laminal cells.

Our preliminary studies of the specimens of the *Bucklandiella crispula* complex in the Australasian region have not revealed the occurrence of this species in New Zealand, Tasmania and south-eastern Australia, and therefore it is tentatively considered a narrow endemic of the Auckland Is. and Campbell I. These tiny islands are situated on the Campbell Plateau and are the eroded remnants of basaltic volcanoes which underwent a major eruptive phase in the Oligocene/Miocene. These ancient islands possess several moss endemics, including *Crosbya nervosa* (Hook.f. & Wilson) Vitt (Vitt 1977), *Muelleriella angustifolia* (Hook.f. & Wilson) Dusén, *M. aucklandica* Vitt (Vitt 1976*b*), *Pohlia ochii* Vitt, *Trichostomum imshaugii* (Vitt) R.H.Zander (Vitt 1971*a*), *Blindia contecta* (Hook.f. & Wilson) Müll.Hal. (Ochyra 2003*c*), and *Ditrichum maxwellii* (Vitt) Ochyra & Lewis Smith (Ochyra & Lewis Smith 1998).

During the course of the present study we located and examined all relevant type collections of the *Bucklandiella crispula* complex from southern South America and Subantarctica. As a result we confirmed three former synonymisations of taxon names proposed by Deguchi (1984), and four new synonyms of *B. didyma* are here added, including *Racomitrium convolutum* Mont. from southern Chile (Montagne 1845*a*), *R. flavescens* Cardot (Cardot 1900, 1901) and *R. subrupestre* Roiv. (Roivainen 1955*a*) from the Fuegian region, and *Grimmia defoliata* Müll.Hal. from Îles Kerguelen (Müller 1883, 1889).

In the Antarctic *Bucklandiella didyma* is easily distinguished from other congeners by its multiseriate leaf margins in the distal part and a broad, flattened costa which is distinctly grooved in the proximal part and consists of 3–4 enlarged ventral cells in transverse section of leaves. Although all Antarctic specimens are consistently sterile, the species is distinguished from the closely related *B. orthotrichacea* by its completely smooth or only somewhat roughened laminal cells and more flattened and thinner costa which is mostly bistratose in the distal part. Conversely, in *B. orthotrichacea* the laminal cells are densely covered on both the dorsal and ventral surfaces by lengthwise cuticular thickenings which give them a papillose appearance in cross-section, and the costa is mostly tristratose in the upper part.

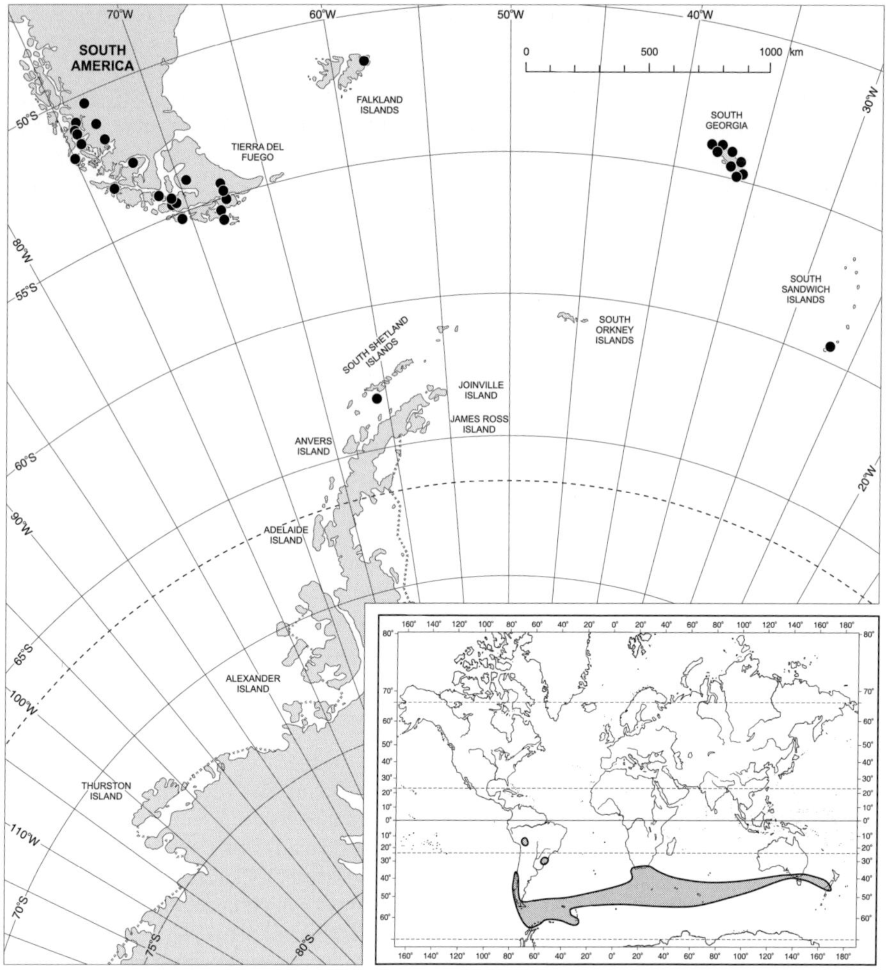

Figure 142. Distribution of *Bucklandiella didyma* in the Fuegian region and Antarctica. Inset: global distribution.

Reproduction in Antarctica. Consistently sterile, although one specimen from Bellingshausen I. (*MWH 811E*) possessed both male and female inflorescences and a solitary seta.

Habitat. On Bellingshausen I. *Bucklandiella didyma* is strictly associated with the fumarole vegetation, growing in mixed mats on moist heated ground within 45–570 cm of fumarole vents (Longton & Holdgate 1979). It thrives in a community dominated by *Cryptochila grandiflora*, and is interspersed with *Sanionia uncinata*, *Polytrichastrum alpinum*, *Notoligotrichum trichodon*, *Pohlia nutans*, *Campylopus spiralis*, *Ceratodon purpureus* and *Usnea antarctica*. Occasionally, it grows on cooler ground and rocks above fumaroles. On Deception I. the species was also found in a former geothermal area on very coarse scoria.

World range. Following the current taxonomic concept *Bucklandiella didyma* is a pan-south-temperate species (Fig. 142, inset). It is widely distributed in the *Nothofagus* zone at the western fringes of southern South America from the Valdivian region to Tierra del Fuego, on the Falkland Is. and South Georgia, and extending to the southern fringes of South Africa, as well as Prince Edward Is., Îles Crozet, Îles Kerguelen and Heard I. in the Subantarctic, south-eastern Australia and the South Island of New Zealand (Bednarek-Ochyra & Ochyra 2008*b*) and the northern maritime Antarctic, with some isolated stations in Bolivia at an elevation of 4000 m (*Lewis 84479*, KRAM) and in Santa Catarina Province in south-eastern Brazil (Bednarek-Ochyra *et al.* 1999). The species is likely to occur also on Tristan da Cunha but many specimens recorded from this archipelago as *Racomitrium crispulum* (Dixon 1939, 1960; Wace & Holdgate 1958; Wace & Dickson 1965) and *R. nigritum* (Dixon 1960) represent other species of *Bucklandiella*. Also, the specimens reported from Gough I. as *Racomitrium flavescens* (Wright 1905*a*) and *R. symphyodontum* (Cardot 1911*c*) are actually *Bucklandiella lamprocarpa* (Müll.Hal.) Bednarek-Ochyra & Ochyra.

Distribution in Antarctica. *Bucklandiella didyma* is an exceedingly rare species in the Antarctic which is restricted to two volcanic islands, namely Bellingshausen I. in the South Sandwich Is. archipelago and Deception I. in the South Shetland Is. (Fig. 142). It appears to be relatively frequent on the former island where it was collected for the first time in the early 1960s (Bell 1973*c*; Longton & Holdgate 1979, both as *Racomitrium crispulum*) and subsequently rediscovered at the same sites after a quarter of a century (Convey *et al.* 2001, as *Racomitrium orthotrichaceum*). In contrast, the species is very rare on the second island where only a few shoots have been found. In both areas, *B. didyma* occurs at moderate elevations from 75 to 130 m.

Specimens examined. WEST ANTARCTICA. SOUTH SANDWICH IS. **BELLINGSHAUSEN I.** South slopes, around fumarole, *ca* 75 m, *MWH 811E & 829B* (AAS, KRAM), 100 m, *MWH 410B* (AAS, KRAM), 120 m, *PC 66C* (AAS, KRAM), 130 m, *PC 65B* (AAS, KRAM) and 150 m, *PC 38, 39B, 42E, 45B, 46G & 53B* (AAS, KRAM); east end of the crater rim, *ca* 200 m, *MWH 811E & 824E* (AAS, KRAM) *& 825E* (KRAM, NY, PC).

SOUTH SHETLAND IS. **DECEPTION I.** Pendulum Cove, east of Crimson Hill, 130 m, *RILS 11197* (AAS, KRAM).

Excluded species of the Grimmiaceae

Bucklandiella crispula (Hook.f. & Wilson) Bednarek-Ochyra & Ochyra — This species has been recorded several times as *Racomitrium crispulum* from Bouvetøya (Zanten 1971), the South Sandwich Is. (Bell 1973*b*; Longton & Holdgate 1979), Signy I. in the South Orkney Is. (Gimingham & Lewis Smith 1971; Lewis Smith 1972), and Livingston I. in the South Shetland Is. (Robinson 1972*a*). The specimens from the former archipelago are actually *Bucklandiella didyma* and they are discussed under this species. Specimens recorded from Signy I. as *Racomitrium* cf. *crispulum* were subsequently redetermined as *R. austrogeorgicum* which is currently considered conspecific with *Bucklandiella sudetica*. Finally, all specimens of *Racomitrium crispulum* collected by R. M. Schuster on Livingston I. are also *Bucklandiella sudetica*.

B. heterosticha (Hedw.) Bednarek-Ochyra & Ochyra — This species was erroneously reported as *Racomitrium heterostichum* (Hedw.) Brid. by Lewis Smith (1988*d*) from Deception I. in the South Shetland Is., but this author was, in fact, referring to *R. heterostichoides* (see Lewis Smith 2005*a*,*b*). *Bucklandiella heterosticha* is a widespread Euro-American oceanic species (Frisvoll 1988; Bednarek-Ochyra 1995*a*) and its reports from southern South America are doubtless erroneous. The specimen collected on the Falkland Is. by J. D. Hooker during the Antarctic voyage 1839–1843 (Wilson & Hooker 1847; see also Dusén 1903*b* and Cardot 1908*a*) is *B. striatipila* (Cardot) Bednarek-Ochyra & Ochyra (*Hooker 206*, BM), and this is the first known record of this Patagonian-Fuegian species. The only other regional records of *B. heterosticha* are those from the Valdivian region in western Patagonia by Herzog (1938), but they need verification.

B. heterostichoides (Cardot) Bednarek-Ochyra & Ochyra — It was reported from Deception I. in the South Shetlands (Lewis Smith 1991; Ochyra *et al.* 1998) but the specimens proved to be misnamed, and in the present Flora they are referred to as *Bucklandiella pachydictyon*.

B. orthotrichacea (Müll.Hal.) Bednarek-Ochyra & Ochyra — The specimens reported as *Racomitrium orthotrichaceum* (Müll.Hal.) Paris from Bellingshausen I. in the South Sandwich Is. (Ochyra *et al.* 1998; Convey *et al.* 2001) correctly belong within *Bucklandiella didyma*.

B. striatipila (Cardot) Bednarek-Ochyra & Ochyra — This species was reported as *Racomitrium striatipilum* Cardot from a single locality on Fildes Peninsula (Chen *et al.* 1993, 1995; Li *et al.* 1998) but the voucher specimen is *Bucklandiella sudetica*. *Bucklandiella striatipila* is a Patagonian-Fuegian-South Georgian species (Cardot 1905*a*, 1908*a*) which is considered by Deguchi (1984) to be identical to the Afro-American *R. crispipilum* (Taylor) A.Jaeger (cf. also discussion in Bednarek-Ochyra 1995*b*). Notwithstanding the definite status of *B. striatipila* this species is quite different from *R. sudeticum* in the internal structure of the costa, as well as the presence of a very long hyaline hair-point and a basal marginal border of hyaline pellucid cells.

Dryptodon trichophyllus (Grev.) Brid. — Seppelt & Selkirk (1984) stated that '... other species such as *G.*[*rimmia*] *donniana* Sm., *G. nordenskioldii* Card., *G. grisea* Card., and *G. trichophylla* Grev., all of which have been reported from continental Antarctica'. Unfortunately, we have been unable to trace any published record of *Grimmia trichophylla* or *Dryptodon trichophyllus*, and it is likely that this information is erroneous. This is inferred by the erroneous placement of *Grimmia nordenskioldii* and *G. grisea* in this list of species (see discussion under the respective species). *Dryptodon trichophyllus* is otherwise a bipolar species, widespread in the Fuegian region (Muñoz 1999), Australia (Greven 2002*b*) and New Zealand (Greven 1998*a*).

Grimmia anodon Bruch & Schimp. — Convey & Lewis Smith (1993) reported *Grimmia anodon* from the maritime Antarctic without specifying localities, and Lewis Smith (1996) cited it from Alexander I. This name was initially used for the specimens which in the present treatment are referred to as *G. plagiopodia*. *Grimmia anodon* is a northern species, widely scattered throughout the Holarctic, extending southwards to North Africa and Mexico, with highly isolated and disjunct stations in Bolivia (Muñoz 1999; Muñoz & Pando 2000), New Zealand (Greven 1998*a*) and Australia (Greven 2002*b*).

G. fastigiata Cardot — This species was reported by Dixon & Watts (1918) from Queen Mary Land on the Antarctic continent, but the voucher specimens represent *Orthogrimmia sessitana*. The type material of *Grimmia fastigiata* itself, originating from Tierra del Fuego (Cardot 1905*a*), is identical to *Dryptodon fuscoluteus* (Muñoz 1999).

G. nordenskioldii Cardot — This species was mentioned by Seppelt & Selkirk (1984) as a single record from continental Antarctica. This appears to be an error since we could not find any pre-1984 record of *Grimmia nordenskioldii* from this region. However, this species was referred to by Kuc (1969) but, in the context of a discussion of the taxonomic status of *G. donniana* fo. *antarctica*, it may be misconstrued that the species occurred in the Antarctic. *Grimmia nordenskioldii* is currently conspecific with *Guembelia kidderi* (James) Ochyra & Żarnowiec (Ochyra & Hertel 1990), an amphiatlantic subantarctic species, occurring on South Georgia and all islands in the Kerguelen Province, and extending to Tristan da Cunha and a few sites in western Patagonia (see also a note under *Guembelia kidderi*).

G. stolonifera Müll.Hal. — Two specimens of this species were reported from Queen Mary Land by Dixon & Watts (1918) but they are *Schistidium antarctici*. The type material of *Grimmia stolonifera* from Îles Kerguelen (Müller 1883, 1889) belongs within *Dryptodon orbicularis* (Wilson) Ochyra & Żarnowiec (Muñoz & Pando 2000), although Greven (2003) maintained it as a species in its own right.

Guembelia kidderi (James) Ochyra & Żarnowiec — This species was reported by Muñoz & Pando (2000) from the Antarctic, although without closer locality data, and it is probable that the information on its Antarctic occurrence in Greven's (2003) handbook of *Grimmia* is taken from this source. These authors provided only the collector name (J. Healy) and collecting number (No. 65) and showed that the voucher specimen was housed at FH. From these data we were able to identify this specimen as collected during the United States Antarctic Service Expedition of 1940–41 and reported by Bartram (1957) as *Grimmia donniana*. The specimen bears the following label data: 'No 65. Grimmia donniana Sm. East Base: top of nunatak, 150 ft above glacial ice, 68°00′ S, 67°47′ W, growing deep in crack. 9 Febr 1941, J. Healy'. The locality 'East Base' is situated on Stonington I. (lat. 68°11′ S, long. 67°00′ W), off the south side of the terminus of Northeast Glacier, Neny Bay, Marguerite Bay, Fallières Coast, and was the site of the U.S. Antarctic Service Expedition (Hattersley-Smith 1991). Although the material is quite scanty, the plants are otherwise well developed. The older stem leaves have a semi-terete and sharply demarcated costa which is strongly convex on the dorsal side, are sharply keeled in the distal part and the margin is narrowly recurved on one side in the proximal part. These features are typical of all Antarctic populations of *Orthogrimmia sessitana* and exclude any close relationship with *Guembelia kidderi*. The latter species has an elliptical and indistinct costa which is only slightly projecting on the dorsal side, the leaves are deeply concave and the leaf margins are plane throughout.

Orthogrimmia donniana (Sm.) Ochyra & Żarnowiec — This species was reported as *Grimmia donniana* Sm. many times from both the continental (Kuc 1969; Savicz-Lyubitskaya & Smirnova 1970*b*, 1978) and maritime Antarctic, including the South Shetland Is. (Robinson 1972*a*; Myrcha *et al.* 1991) and the Antarctic Peninsula region (Cardot 1901, 1908*a*, 1911*a,b*, 1913; Bartram 1957; Lewis Smith & Corner 1973). All the voucher collections proved to be *Orthogrimmia sessitana*. *O. donniana* is a Northern Hemisphere species, occasionally extending to Mexico (Muñoz 1998*c*).

O. grisea (Cardot) Ochyra & Żarnowiec — This species was reported as *Grimmia grisea* Cardot from Galindez I. in the Argentine Is. archipelago off the Graham Coast on the west Antarctic Peninsula (Gray *et al.* 1982; Lewis Smith 1996). The voucher material is *Orthogrimmia sessitana*. Seppelt & Selkirk (1984) stated that this species was recorded in the literature from continental Antarctica but we have been unable to find any report of this species from there. Kuc (1969) did discuss and illustrate *Orthogrimmia grisea* (as *Grimmia grisea*) in his treatment of the mosses of Bunger Hills, but perusal of this paper clearly indicates that it was only a part of his taxonomic discussion of the status of the local populations which were recognised and described by this author as *Grimmia donniana* fo. *antarctica*. Although Muñoz (1998*c*) considered *O. grisea* to be conspecific with *O. sessitana*, it appears to be a distinct species which was originally described from South Georgia as *Grimmia grisea* (Cardot 1906*a*) and subsequently recorded from New Zealand (Greven 1998*a*) and from stations in Iceland (Greven 1998*b*) and Newfoundland (Greven 2003). Recently, Hastings & Greven (2007*b*) reduced *G. grisea* to synonymy with *G. reflexidens* which is also considered as *Coscinodon reflexidens* (Müll.Hal.) Ochyra (Ochyra 2004*a*).

Schistidium agassizii Sull. & Lesq. — This species was reported, with some reservation, as *Grimmia* cf. *alpicola* Hedw. from Signy I., South Orkney Is. (Webb 1973) and the voucher material correctly represents *Schistidium rivulare*. *Schistidium agassizii* is a Northern Hemisphere species and all records from outside the Holarctic are apparently based on misdeterminations.

S. apocarpum (Hedw.) Bruch & Schimp. — This species has been recorded from several sites in the maritime Antarctic, also as *Grimmia apocarpa* Hedw. The problems associated with its correct taxonomic interpretation are discussed in detail in relevant entries under *Schistidium* and *S. andinum*. A re-examination of most voucher collections from Antarctica showed that most of them correctly belonged to *S. antarctici*, for instance those from Bouvetøya (Bell & Blom 1986; Engelskjøn & Jørgensen 1986), the South Orkney Is. (Wright 1905*b*; Cardot 1911*c*), and the South Shetland Is. and the Antarctic Peninsula area (Robinson 1972*a*). Some others were *S. andinum*, for example from Livingston I. (Sancho *et al.* 1999) and Deception I. (Lewis Smith 1984*b,c*) in the South Shetland Is.

S. chrysoneurum (Müll.Hal.) Ochyra — The record of this species from within the Antarctic was based upon an erroneous taxonomic conclusion on the conspecificity of this species with *Grimmia antarctici* (Ochyra *et al.* 1986). This species was described by Müller (1883, 1889) as *Grimmia chrysoneura* from material collected from Îles Kerguelen by F. C. Naumann and remained a poorly known species, but it was rediscovered by RO at many locations during a bryophyte survey in 2006–2007. Closer examination of the type material of *G. chrysoneura* (for lectotypification of this species name see Ochyra 1998*a*) revealed that this species is distinct in its variously bistratose laminal cells in the upper half, strongly thickened, 3–4-stratose in several rows of cells, bulging margins in the upper part and sinuose laminal cells nearly to the base.

Order VII. **PHASCALES** Limpr.

Phascales Limpr. *in* Cohn, Krypt-Fl. Schlesien 1: 52, 210. 1876 ['Ordnung Phascaceae']. — Type: *Phascum* Hedw.

Trichostomales M.Fleisch., Musci Buitenzorg 1: 11. 1904 ['Reihe Trichostomoideae'], ***syn. nov.*** — Type: *Trichostomum* Hedw.

Hyophilales M.Fleisch., Musci Buitenzorg 1: 11. 1904 ['Reihe Hyophiloideae'], ***syn. nov.*** — Type: *Hyophila* Brid.

Pottiales M.Fleisch., Hedwigia 61: 392. 1920, ***syn. nov.*** — Type: *Pottia* Fürnr.

Monocranoideae M.Fleisch., Musci Buitenzorg 1: xxvi. 1904 ['Reihe Monocranoideae'], *nom. illeg.*

Ditrichocranoideae M.Fleisch., Musci Buitenzorg 1: xxvii. 1904 ['Reihe Ditrichocranoideae'], *nom. illeg.*

Ditrichostomoideae M.Fleisch., Musci Buitenzorg 1: 278. 1904 ['Reihe Ditrichostomoideae'], *nom. illeg.*

Syrrhopodontales S.R.Edwards *in* G.C.S.Clarke & Duckett, Syst. Assoc. Spec. Vol. 14: 328. 1979, *nom. inval. descr. angl.*

The modern concept of the Phascales was outlined by Fleischer (1920), who used the name Pottiales for this order, and subsequently it has gained wide acceptance when popularised by Brotherus (1924*b*) in the second edition of the treatment of the Musci in *Die Natürlichen Pflanzenfamilien*. The order contains mosses that have a haplolepidous peristome with a distinct basal membrane. This may be short and hidden within the capsule mouth or high and forming a distinct tube giving rise to the segments of the peristome. When present, the peristome consists basically of 16 teeth which are straight or spirally twisted. They are frequently forked or divided to the base into two filamentous branches arising more or less equally from the basal membrane and giving the impression that the peristome is composed of 32 teeth. Only occasionally the teeth are undivided and/or paired. The segments of the teeth may be thickened and variously ornamented on both faces but usually the PPL pattern on the outer face may be particularly prominent in the form of ridges and trabecules. The **2**:3 pattern appears as tesselations in those species which have a tall tubular basal membrane. The sporophytes in the Phascales are acrocarpous and only seldom cladocarpous. The capsules are mostly upright and smooth and frequently cleistocarpous and the calyptra may be deciduous or persistent and cucullate or mitriform. Vegetative propagation by various types of gemmae and tubers is also frequent.

Gametophytically, taxa of the Phascales exhibit considerable diversity. The plants vary in size from very small to large and from slender to robust, although occasionally they are minute and delicate ephemerals growing from sparse or abundant protonemata. The foliage is always multifarious, but no taxa are known with the distichous leaf arrangement in this group. Because the majority of taxa of the Phascales are associated with dry habitats, various adaptative xeromorphic characters often exist, including crisped and contorted leaves with incurved to involute margins; short and isodiametric, usually thick-walled and densely pluripapillose laminal cells; strong costae with two stereid bands separated by a median layer of enlarged guide cells; and the presence of hyaline awns. Alar cells are entirely lacking in the Phascales. The enlarged, banana-shaped 'dicranaceous' angular cells are present only in the monotypic genus *Weissiodicranum insularum* Reese from the Galapagos, Jamaica and Puerto Rico, but its taxonomic placement is still uncertain owing to the lack of sporophytes (Reese & Buck 1991; Zander 1993).

When establishing the order Pottiales (now Phascales), Fleischer (1920) divided it into three suborders with four families, namely Syrrhopodontineae (Syrrhopodontaceae, Calymperaceae), Encalyptineae (Encalyptaceae) and Pottiineae (Pottiaceae). Of these, the Encalyptineae are currently recognised as a separate order that is placed in the separate subclass Encalyptidae. Edwards (1979) suggested recognition of the separate order Syrrhopodontales for the Calymperaceae, mostly on the basis of examination of the peristome in the South African genus *Hypodontium* Müll.Hal. In this genus the peristome is composed of a prostome of 32 segments at the base of the teeth and 16 simple teeth with a poorly developed **2**:3 pattern. Besides the presence of the prostome, the structure of the peristome in *Hypodontium* does not differ essentially from that in other taxa of the Pottiaceae and Calymperaceae (Salazar Allen 1993), so recognition of the separate order for the Calymperaceae does not seem to be justified. Additionally, because *Hypodontium* shares many gametophyte and sporophyte characters with the Pottiaceae, Zander (1993) transferred that genus to this family. In the present circumscription the Phascales comprise, apart from the Pottiaceae and Calymperaceae, three small families, namely Cinclidotaceae which are an often neglected segregate of the Pottiaceae (Schimper 1860), Bryobartramiaceae (Sainsbury 1948*b*), and Serpotortellaceae (Reese & Zander 1987). Also, the cleistocarpous family Ephemeraceae is here placed on account of molecular data (Goffinet *et al.* 2001; Werner *et al.* 2007). However, on the basis of molecular studies LaFarge *et al.* (2000) showed that the Calymperaceae (and its segregate Octoblepharaceae) are cladistically distant from the Phascales and should be transferred to the Dicranales. These workers also suggested that the Rhabdoweisiaceae and Ditrichaceae, which have previously been considered 'dicranealean' families, are more closely related to the Phascales than the Dicranales. On the other hand, Robinson (1971) and Ignatov & Ignatova (2003) placed the genera of the Phascales in the much larger Dicranales, making it a very large and quite impractical order. Therefore, retaining the Phascales in the traditional circumscription seems to be the practical solution.

Apart from the uncertainties of classification, the Phascales also offer some nomenclatural problems because this oldest available name for this order has long been forgotten. It was used for an order described by Limpricht (1876) as 'Ordnung Phascaceae' to accommodate two cleistocarpous moss families, Archidiaceae and Phascaceae. The latter was a heterogeneous taxon according to modern moss taxonomy which also comprised, apart from typical pottiaceous genera (*Acaulon*, *Microbryum* and *Phascum*), *Pleuridium*, *Cleistocarpidium* and *Pseudephemerum* of the Ditrichaceae, *Physcomitrella* of the Funariaceae and *Ephemerum* of the Ephemeraceae. Yet, because this ordinal name is typified by a name of a pottiaceous genus, *Phascum* (= *Tortula*), it is the oldest available name for an order encompassing the Pottiaceae, and it should be used following Recommendation 16B of the current Code (McNeill *et al.* 2006). Nevertheless, according to Art. 11.10 of the same Code, the principle of priority is not mandatory for names of taxa above the rank of family. However, the neglected name Phascales is here accepted in favour of the Pottiales, although the latter is a widely used and well established name in bryological literature. This is because the name Pottiales is also antedated by two other names introduced by Fleischer (1904), the Trichostomales and Hyophilales, which are typified by *Trichostomum* and *Hyophila*, two well-known names of pottialean genera. Fleischer (1904) himself rejected the Trichostomales in favour of the illegitimate descriptive names Ditrichostomoideae and Ditrichocranoideae, and he replaced the Hyophilales with the illegitimate Monocranoideae. Interestingly, in the latter order, this author placed three predominantly tropical families, Calymperaceae, Leucophanaceae and Syrrhopodontaceae, which constitute the suborder Syrrhopodontineae, whereas the type genus *Hyophila* was positioned in the Trichostomales.

After the Dicranales, the Phascales are the second largest order of the Dicranidae, consisting of about 1550 species. Of these, the overwhelming majority of species (*ca* 1300) belong within the single family Pottiaceae, whereas four smaller satellite families contribute only about 250. In the Antarctic the Phascales are well represented, exclusively by the members of the Pottiaceae which often play an important role in the vegetation cover.

Family IX. **POTTIACEAE** SCHIMP.

Pottiaceae Schimp., Coroll. Bryol. Eur.: 24. 1856, *nom. fam. cons.* — Type: *Pottia* Fürnr.

Hyophilaceae Hampe, Bot. Zeit. (Berlin) 11: 303. 1953, *nom. fam. rej.* — Type: *Hyophila* Brid.

Phascaceae Schimp., Coroll. Bryol. Eur.: 4. 1856. — Type: *Phascum* Hedw.

Astomaceae Schimp., Coroll. Bryol. Eur.: 7. 1856. — Type: *Astomum* Hampe.

Weissiaceae Schimp., Coroll. Bryol. Eur.: 7. 1856. — Type: *Weissia* Hedw.

Anoectangiaceae Schimp., Coroll. Bryol. Eur.: 11. 1856. — Type: *Anoectangium* Schwägr.

Trichostomaceae Schimp., Syn. Musc. Eur.: 141. 1860 ['Trichostomeae']. — Type: *Trichostomum* Bruch (*nom. cons.*).

Systegiaceae De Not., Atti Reale Univ. Genova 1: 33. 1869 ['Systegiacei'], *nom. illeg.* — Type: *Systegium* Schimp. (*nom. illeg.*).

The Pottiaceae are one of the most taxonomically difficult families in the Bryopsida. The delimitation and circumsciption of this taxon have been a source of controversy among bryologists and a long list of heterotypic synonyms of this family name provides evidence of discrepancies in the disposition of genera. The main reason for these uncertainties is the considerable variability of taxonomically important characters, and their convergence and morphological reductions, as well as the reticulating nature of character variation. The morphological characters of the gametophores in the Pottiaceae are mostly not correlated with each other and vary independently. The close association of the gametophyte characters is found only in the curvature of the leaf margin, the shape of the hyaline area in the leaf base, and the occurrence of gemmae. However, the peristome variations are usually correlated with the size and shape of the capsule, annular cells, lids and setae. Consequently, depending upon a subjective weighting of the various features, the same genus may be classified in different ways. For example, the genera *Stegonia* and *Pterygoneurum* are usually aligned with *Tortula* (Brotherus 1924*b*; Savicz-Lyubitskaya & Smirnova 1970*c*; Nyholm 1990) when the features of the leaf shape and large upper laminal cells are emphasised. However, Zander (1993) placed these genera in a group centred around *Hyophila* by stressing the presence of upper laminal cells that are bulging ventrally and weakly convex dorsally. Likewise, Saito (1973*b*, 1975) transferred the genus *Scopelophila* to the Pottieae, close to *Pottia* and *Tortula* in the subfamily Pottioideae, on account of the absence of the ventral stereid band in the costa, whereas Zander (1993) positioned this genus in a group centred around *Gymnostomum* in the tribe Barbuleae of the subfamily Merceyoideae by emphasising the nearly homogeneous anatomical structure of the stem which is composed of a parenchymatous tissue lacking a distinct cortex, medulla and central strand. In general terms, species of the Pottiaceae appear to exhibit every possible variation and combination of variations in both gametophyte and sporophyte characters, and this precludes more detailed definition of the family.

Members of the Pottiaceae are small to large plants which grow in dense tufts or short turves, but sometimes are loosely caespitose or gregarious. The stem is very short to elongate, mostly rounded-pentagonal in cross-section, with a cortex that is well or poorly developed, or even lacking, and with or without a central strand. The axillary hairs are filiform, uniseriate, 2–15-celled, hyaline throughout or with 1–2 yellowish or brownish basal cells. When dry, the leaves are crisped, contorted, curled, involute, but also more or less straight and appressed or erecto-patent, whereas they are recurved, reflexed, squarrose or secund when wet. They vary markedly in shape from ovate, circular, elliptical to lanceolate, linear, oblong, lingulate or spathulate, and sometimes they are differentiated into a limb and a vaginate base. The apex is commonly narrowly acute, but rounded-obtuse to broadly rounded apices are sometimes present and it is usually flat or sometimes strongly cucullate. The leaf margins are plane, recurved, revolute or incurved to involute, entire, serrulate to dentate, uni- to polystratose in one to several rows. The costa is always present and single, narrow or wide, percurrent or subpercurrent to short-excurrent or often long excurrent as a hyaline or brownish hair-point, sometimes it ceases well below the apex. Anatomically, the costa is usually differentiated and consists of one or two stereid bands which are separated by one or, occasionally, two rows of large guide cells and often also an epidermis on ventral or both ventral and dorsal sides. Occasionally, the costa is covered with ventral outgrowths which are differentiated as filaments, lamellae or bulging pads of cells. The leaf lamina is unistratose, except for the margins, or sometimes bistratose throughout. The upper laminal cells are mostly isodiametric, less often hexagonal to short-rectangular, usually evenly thick-walled, superficially flat to bulging on both dorsal and ventral surfaces or bulging only ventrally. They are usually densely pluripapillose and obscured by papillae which may be solid or pitted, simple or bifid and flattened or compound. In contrast, the basal leaf cells are usually differentiated, rectangular, often lax and hyaline, transparent, smooth or weakly papillose. Asexual reproduction is relatively frequent in the Pottiaceae, usually by multicellular gemmae borne in the leaf axils or on leaves, less often by rhizoidal tubers. The plants are auto-, di- or, rarely, rhizoautoicous. The perichaetial leaves are larger than the cauline leaves, long-ovate to long-lanceolate and often convolute or not much differentiated. The sporophytes are usually long-exserted, but immersed or emergent capsules are also present. The capsules are ovoid to cylindrical and then stegocarpous or sometimes spherical and then cleistocarpous and rhexolytic. The annulus is usually present, 1–4-seriate, deciduous or persistent, or rarely absent. The operculum is short-beaked and the calyptra is smooth, mostly cucullate and rarely mitrate. The peristome is sometimes lacking or rudimentary, but mostly present and well developed. There are 16 teeth, arising from a low or high basal membrane, which is sometimes lacking, usually deeply cleft into two filamentous branches, erect or twisted clockwise, yellow, orange or red, spiculose, striolate or papillose. The spores are globose, mostly small, 8–15 µm in diameter, occasionally large, 20–40 µm in diameter.

The Pottiaceae were first described by Schimper (1856), although this name itself appeared as a *nomen nudum* much earlier in the treatments of *Pottia* and *Anacalypta* in *Bryologia europaea* (Bruch *et al.* 1843) and in the work of Hampe (1853). This name was proposed for conservation (Magill 1977) against the Hyophilaceae and, since the Berlin Code (Greuter *et al.* 1988), the Pottiaceae are on a list of conserved familial names. Zander (1993) claimed that the conservation of the Pottiaceae was redundant because the name Hyophilaceae was invalidly published. Indeed, the proposal for conservation of the Pottiaceae was rather carelessly prepared. The authorship of the Hyophilaceae was incorrectly ascribed to Karl Müller of Halle, whereas the true author of this name was Hampe (1847). Also, more importantly, in Hampe's (1847) publication this name was a *nomen nudum*, as pointed out by Zander (1993), so the conservation of the Hyophilaceae would have been unnecessary. However, a year earlier, Hampe (1846) recognised the family 'Hyophilaceen' and provided its description ['Die Hyophilaceen haben flache Blätter, niemals einen zurückgeschlagenen Rand, welcher bei den Pottiaceen häufig ist, der runde Nerv (costa) ist auf beiden Seiten der Blattflächengleich vorstehend (bei den Pottiaceen gewöhnlich nach aussen stärker, daher mehr oder weniger die folia carinata häufig vorkommen). Die Costa is hornartig durchscheinend, ...']. In this article the name Hyophilaceae is published with a non-Latin ending and is thus invalid according to Art. 18.4 of the current Code (McNeill *et al.* 2006). However, in his subsequent publication Hampe (1853) used the family name Hyophilaceae with a proper Latin termination and gave a direct reference to the diagnosis published in the 1846 work, thus clearly validating this familial name. Because the Hyophilaceae of Hampe (1853) antedate the Pottiaceae of Schimper (1856), the conservation of the latter name is justified.

Species of the Pottiaceae are often designated as mosses of variable or harsh habitats since they are frequently conspicuous constituents of the vegetation of arid, disturbed, ruderal, alpine and polar areas. They grow mostly on soil, humus and rocks, especially on calcareous substrates, especially on disturbed ground, sometimes epiphytically on tree trunks and branches, mostly in dry situations. A few species are associated with moist and wet habitats, growing on wet or damp soil and rocks beside streams or waterfalls or in flushes, bogs and wet hollows in arctic and subantarctic tundra. The predominantly Mediterranean *Dialytrichia* (Schimp.) Limpr. and the New Zealand *Tetracoscinodon* R.Br.bis seem to be the only genera of the family adapted to wet habitats – *D. mucronata* (Brid.) Broth., sometimes grows submerged in streams and brooks and *T. irroratus* (Mitt.) R.H.Zander is commonly found on dripping limestone rocks.

The Pottiaceae are the largest family in the Bryophyta in number of species (about 1300) and genera (79), and consist of an aggregation of many small, mono- or oligotypic, fairly well defined genera and a few large, poorly defined genera which are in need of taxonomic revision. The family is cosmopolitan in distribution, but it is much more abundant in temperate and cold than tropical regions.

Various classifications of the Pottiaceae have been published which divided this large family into a number of subfamilies and tribes of varying degrees of homogeneity (Brotherus 1924*b*; Hilpert 1933; Chen 1941; Saito 1975). Zander (1993) published a monumental generic monograph of the Pottiaceae which is a fundamental work for all attempting to understand this complex family. He recognised seven subfamilies, four of which (Timmielloideae, Gertrudielloideae, Erythrophyllopsideae and Chionolomoideae) are small, mono- or bigeneric taxa, while the remaining three subfamilies (Trichostomoideae, Merceyoideae and Pottioideae) comprise the bulk of genera and species and are the most complex taxonomically. Zander (1993) accepted 76 genera in the Pottiaceae but since then five genera have been added, namely *Hyophiladelphus* (Müll.Hal.) R.H.Zander (Zander 1995), *Microcrossidium* J.Guerra & Cano (Cano *et al.* 1993), *Neophoenix* R.H.Zander & During (Zander & During 1999), *Pottiopsis* Blockeel & A.J.E.Sm. (Blockeel & Smith 1998) and *Protobryum* J.Guerra & Cano (Guerra & Cano 2000). However, *Chionoloma* Dixon is considered congeneric with *Pseudosymblepharis* Broth. (Sollman 2000) and *Sarconeurum* Bryhn, long considered to be the only endemic Antarctic moss genus, is now united with *Syntrichia* Brid. (Ochyra & Zander 2007). Thus, at present the family Pottiaceae comprises 79 genera. The results of recent molecular studies suggest (Werner *et al.* 2004, 2005) that further modifications of the present classification of the family are inevitable, and Zander (2006) introduced some changes in his former system. For example, he suggested the transfer of *Sarconeurum* to the tribe Bryoerythrophylleae.

The subfamily Trichostomoideae is comparatively small and consists of only eight genera focused around the large genus *Trichostomum*. Compared with the Merceyoideae and Pottioideae, this subfamily appears to be more abundant in tropical and warm temperate latitudes. Recurvature of the leaf margin is almost unknown in the Trichostomoideae, the stem cortex is poorly differentiated from the medulla and the hyalodermis is commonly present. Other features of importance are a yellow colour of the upper lamina cell walls when treated with KOH, strongly crowded papillae on the upper laminal cells and the lack of a differentiated costal epidermis on the dorsal side. Perhaps the most immediately obvious character of this family is the 'tortelloid' leaf base – elongate, hyaline basal cells extend along the margins farther than along the costa to form a differentiated V-shaped area.

The subfamily Merceyoideae has a well developed stem cortex which is usually sharply differentiated from the medulla. In addition, recurvature of the leaf is present, upper laminal cells are relatively small, the costa usually has two stereid bands and is often grooved along the adaxial surface, the basal portion of the leaf is commonly differentiated, and the axillary rhizoids are frequent and they usually produce propagula. Zander (1993) recognised four tribes in the Merceyoideae, two of which are represented in the Antarctic by a single genus. The tribe Bryoerythrophylleae consists mostly of taxa in which the upper laminal cell walls are red in reaction with KOH, the dorsal stereid band of the costa is usually reniform in section, on the dorsal side the costa is flattened and the leaf margins are recurved or revolute and often denticulate or dentate. This tribe comprises five genera centred around the largest genus *Bryoerythrophyllum* which is present in the Antarctic. The second tribe Barbuleae can, in part, be recognised by having the dorsal stereid band of the costa round or semi-terete in cross-section and the KOH colour reaction of the upper laminal cell walls are yellow or yellowish-orange, very seldom red. This tribe is also worldwide in distribution and comprises nine genera aggregated around the largest genus *Barbula*. In the Antarctic, it is represented by a single species of the genus *Didymodon*.

The Pottioideae is the most diverse and abundant subfamily of the Pottiaceae and comprises 13 species distributed in seven genera. It is especially defined by its lack of a ventral stereid band in the costa and lack or poor differentiation of the stem cortex which gradually merges into enlarged medullary cells. Additionally, the leaves are often broadly lingulate to spathulate and the leaf base is usually little differentiated in shape. Zander (1993) divided this subfamily into two tribes which are both represented in the Antarctic. The tribe Hyophileae is primarily discernible by its upper laminal cells that are usually bulging on the ventral side and are weakly convex on the dorsal side. In the Antarctic this tribe consists of three genera, each with a single species, namely *Stegonia*, *Pterygoneurum* and *Aloina*, all of which are very rare and local taxa. Conversely, the tribe Pottieae is prominent in the Antarctic and is comprised of four genera with ten species, namely *Hennediella*, *Syntrichia*, *Tortula* and *Willia*. This tribe differs from the Hyophileae in its upper laminal cells which are equally bulging on both adaxial and abaxial surfaces.

In the Antarctic, the Pottiaceae are the second largest moss family, after the Grimmiaceae, consisting of 17 species belonging within 10 genera and three subfamilies. Yet, despite considerable taxonomic diversity, most species of the Pottiaceae are rare and inconspicuous constituents of the Antarctic fellfield vegetation and have restricted and localised ranges, although locally some may be quite abundant within a few square metres, notably *Stegonia latifolia* and *Pterygoneurum ovatum*. There are only a few species which have a wide distribution throughout the West and East Antarctic, for example *Syntrichia magellanica*, *S. sarconeurum*, *Didymodon brachyphyllus* and *Hennediella heimii*. In general, *Syntrichia* is one of the most diverse and widespread of moss genera in the Antarctic; this is also true for all Subantarctic islands (Lightowlers 1986*a*).

KEY TO THE GENERA OF THE ANTARCTIC POTTIACEAE

1. Leaves narrow, lanceolate, ovate-lanceolate to ovate; costa with two stereid bands in transverse section 2
1. Leaves broad, broadly ovate to ovate-lanceolate, lingulate, spathulate to oblong-ovate; costa with one stereid band in transverse section 4
 2. Hyaline basal leaf cells abruptly differentiated from the chlorophyllose upper cells, ascending along the margins and forming a V-shaped area; central strand absent; leaf margins plane or incurved .... 25. ***Tortella***
 2. Hyaline basal leaf cells gradually merging into the chlorophyllose upper cells, not ascending along the margins and not forming a V-shaped pattern; central strand distinct; leaf margin recurved in the distal part 3
3. Plants reddish-brown to rusty; smooth pellucid apical leaf cells present, at least on some young leaves; axillary hairs hyaline throughout; gemmae absent .... 26. ***Bryoerythrophyllum***
3. Plants green to brownish-green; smooth pellucid apical leaf cells always absent; axillary hairs with brown basal cell; axillary gemmae present .... 27. ***Didymodon***
 4. Leaves broadly ovate to circular 5
 4. Leaves ovate-lanceolate, lingulate or spathulate 6
5. Costa without lamellae on the dorsal surface .... 28. ***Stegonia***
5. Costa with 2–4 chlorophyllose lamellae on the ventral surface .... 29. ***Pterygoneurum***
 6. Leaves thick, cucullate at the apex; costa very broad with many chlorophyllose filaments on the ventral surface .... 30. ***Aloina***
 6. Leaves thin, flat at the apex; costa narrow lacking filaments on the ventral surface ... 7
7. Junction between the upper chlorophyllose cells and the basal hyaline cells gradual, the former not descending along the margins; basal cells without a distinct border and not forming a fenestrate pattern 8
7. Junction between the upper chlorophyllose cells and basal hyaline cells abrupt, the former distinctly descending along the margins and passing into a clear multiseriate basal marginal border of narrow cells, resulting in a fenestrate pattern of the basal hyaline cells, if not or weakly so, then the leaf lamina constricted below the apex and ending in a caducous, cylindrical, solid propagulum 9
 8. Leaves broadly acute to rounded at the apex, ending in a smooth yellowish-brown to brownish awn; margins recurved; capsule not systylious, ochyrostomous .... 31. ***Tortula***
 8. Leaves shortly acuminate, without a hyaline awn; margins plane; capsule systylious, gymnostomous .... 32. ***Hennediella***
9. Leaves distinctly panduriform, with a long hyaline hair-point and plane margins; dorsal stereid band semicircular in shape; capsules immersed to emergent; peristome absent; calyptra mitriform; inner perichaetial leaves markedly differentiated from the cauline leaves .... 33. ***Willia***
9. Leaves lanceolate, oblong-lanceolate, oblong- to ovate-lingulate, not or occasionally slightly constricted at or below the middle, not or hair-pointed, with recurved margins; dorsal stereid band crescent-shaped; capsules long-exserted; peristome present; calyptra cucullate; inner perichaetial leaves not or only slightly differentiated .... 34. ***Syntrichia***

25. TORTELLA (MÜLL.HAL.) LIMPR., *nom. cons.*

Tortella (Müll.Hal.) Limpr., Laubm. Deutschl. 1: 599. 1888, *nom. cons.* ≡ *Barbula* Hedw. sect. *Tortella* Müll.Hal., Syn. Musc. Frond. 1: 599. 1849. ≡ *Mollia* Schrank *ex* Lindb. [unranked] *Tortella* (Müll.Hal.) Lindb., Musci Scand.: 21. 1879. ≡ *Mollia* sect. *Tortella* (Müll.Hal.) Braithw., Brit. Moss Fl. 1: 230. 1885. ≡ *Barbula* [unranked] *Tortella* (Müll.Hal.) Kindb., Eur. N. Amer. Bryin. 2: 245. 1898. ≡ *Barbula* subgen. *Tortella* (Müll.Hal.) Kindb., Rev. Bryol. 36: 117. 1909. — Type: *Tortella caespitosa* (Schwägr.) Limpr. (*Barbula caespitosa* Schwägr.).

Plants small to medium-sized or, occasionally, fairly large, slender to rather robust, loosely to densely caespitose or forming expanded turves, sometimes cohering below by dense, felt-like rhizoids, generally stiff and rigid, dull but appearing lustrous owing to the basal hyaline cells and prominent shining costae, green, yellow-green, yellowish or brown above, brown, tan or blackish below. *Stems* erect, simple or branched, densely or sparsely radiculose at the base with long, smooth, brown, lustrous rhizoids, occasionally tomentose, in transverse section rounded-pentagonal, consisting of a poorly defined cortex of small to relatively large, substereid cells with moderately thickened walls, not sharply demarcated from large, hyaline medullary cells, often with thickened walls, central strand absent, hyalodermis present, only seldom with collapsed outer cell walls; *axillary hairs* long, 10–20-celled, hyaline throughout. *Leaves* densely set, erect, cirrose, crispate or contorted, occasionally only erect-incurved and somewhat twisted around the stem on drying, straight to flexuose, erect-spreading, patent to recurved on wetting, lanceolate to linear- or oblong-lanceolate, sometimes with a somewhat differentiated, sheathing base, gradually tapering to a long acuminate, acute, flat or sometimes cucullate apex or abruptly narrowed to an obtuse, mucronate point, occasionally forming easily detached apical propagula; *margins* plane to erect, often incurved in the distal part, entire or papillose-crenulate, sometimes irregularly notched, flat or sometimes strongly undulate, not or sometimes distinctly bordered by elongate, pellucid, smooth cells above the mid-leaf; *costa* strong, subpercurrent, percurrent to excurrent, smooth at the back, shining and papillose at both surfaces, in transverse section semi-terete to elliptical in outline, with a median row of enlarged guide cells and two stereid bands, with ventral epidermis present and dorsal present, weak or absent; *upper laminal cells* chlorophyllose, unistratose or occasionally bistratose, rounded-quadrate to rounded-hexagonal, evenly thick-walled, usually obscure owing to the dense papillosity with spiculose, bifid or occasionally simple conical papillae, 3–6 per lumen, occasionally obscurely unipapillose and pellucid; *basal leaf cells* enlarged, laxly long-rectangular, smooth, thin-walled, hyaline or occasionally brown and thick-walled, usually sharply demarcated from the chlorophyllose distal cells, extending above the shoulders along the margins as a short or fairly elongate border and forming a more or less distinct V-shaped region, occasionally the transition to the chlorophyllose cells gradual; *laminal KOH colour reaction* usually yellow, occasionally yellowish-orange to reddish-brown. *Asexual reproduction* occasional by caducous or fragile propaguloid leaf apices or fragments of the lamina. *Dioicous* or seldom *monoicous*. *Perichaetia* terminal; *perichaetial leaves* undifferentiated or the innermost leaves long-lanceolate, sheathing the seta. *Setae* single in perichaetium, long, erect, yellowish, reddish or light brown, smooth, dextrorse below when dry. *Capsules* exserted, erect and symmetric, ocasionally somewhat curved, cylindrical to ellipsoid, smooth or wrinkled when dry, yellowish to reddish-brown, stegocarpous or very seldom cleistocarpous; *annulus* absent or present, persistent, 1–4-seriate, composed of weakly to distinctly vesiculose cells; *operculum* conic to long-rostrate, with a straight or slanted rostrum, half as long as the urn or longer; *exothecial cells* rectangular to oblong-hexagonal, thin-walled, becoming short at the mouth; *stomata* few in one row at the urn base, bicellular, phaneroporous; *peristome* mostly present, rarely rudimentary or absent, single, consisting of 16 teeth deeply bifid to a low basal membrane into long, filiform, orange-red, densely spiculose-papillose segments, wound together into 1–3 counter-clockwise segments, sometimes only obliquely inclined. *Spores* spherical, yellowish-brown, finely to moderately coarsely papillose or nearly smooth. *Calyptra* cucullate, smooth, naked.

The most distinctive feature of the genus *Tortella* is the abrupt differentiation of the distal and proximal laminal cells along a V-shaped line because elongated, hyaline basal cells ascend along the leaf margins above the shoulders and make the basal cells patterned as a 'V'. Occasionally, however, the junction between the chlorophyllose upper cells and the hyaline basal cells is gradual or the basal cells do not rise very high along the margins and form a rather low 'V'. These species resemble very much the closely related genus *Trichostomum* and the limits between these genera need a thorough re-assessment. Apart from the features of the laminal cells, *Tortella* is also distinguished by its distinct stem hyalodermis composed of rectangular cells, usually with a few collapsed outer walls, incurved-contorted leaves when dry, and plane to erect leaf margins. The costa is mostly short-excurrent as a smooth mucro and it has a distinctive anatomy with two large stereid groups separated by a row of enlarged guide cells. In addition, the back of the costa has a shining appearance and sharply contrasts with the greenness of the lamina which is dulled by the dense papillosity of the cells. The peristome in *Tortella* exhibits remarkable variation and, although 32 twisted filamentous segments are predominant, there are also some gymnostomous and cleistocarpous species in this genus.

Although *Tortella* is a sound and undisputably accepted genus it has had a rather chequered nomenclatural history. This name was introduced by Müller (1849) as a section of *Barbula* but now it is generally considered to be illegitimate because it includes the type of an earlier sectional name, *Barbula* sect. *Tortuosae* Bruch & Schimp. (Wijk *et al.* 1959). This is not acceptable because Bruch *et al.* (1842) divided the genus *Barbula* into two sections for which no names were proposed. The first section comprised all but one species of the genus and was subdivided into several 'Untergruppen' which were named after the type species ['Bei den Untergruppen haben wir immer gut bekannte Art als Grundform angenommen...']. One such group was 'Untergruppe *Tortuosae*' which was based on *Barbula tortuosa* but it should not be designated a section. Because 'Untergruppe *Tortuosae*' does not affect sect. *Tortella* of Müller (1849), the latter is a legitimate name, and Lindberg (1879) transferred it to the genus *Mollia* as an unranked taxon. Limpricht (1888) raised sect. *Tortella* to genus as *Tortella* (Müll.Hal.) Limpr. Unfortunately, when establishishing the genus, this author also included *Tortella squarrosa* (Brid.) Limpr. in it which was recognised much earlier by Lindberg (1864) as a separate genus, *Pleurochaete* Lindb. The name *Tortella* was saved by Cardot (1910*a*) who proposed it to conserve against *Pleurochaete* and this was approved and effected in 1932 (Dixon 1932*b*; Magill 1993). Nonetheless, both *Tortella* and *Pleurochaete* are very similar morphologically and their close relationship is also confirmed by molecular studies (Spagnuolo *et al.* 1999).

Tortella is a medium-sized genus comprising 53 species which are distributed worldwide (Zander 1993). It has not been taxonomically overviewed and, apart from its treatments in various regional Floras, only a few species complexes of *Tortella* were carefully studied taxonomically in Europe and North America (Crundwell & Nyholm 1962, 1963; Eckel 1998) and Macaronesia (May 1986). In the austral polar region, including the Antarctic, the genus is represented by only two species, *T. fragilis* and *T. alpicola*, which have undistorted and fragile leaves; the others have leaves so crispate that they become entangled. Additionally, Jenkin (1972) reported *T. calycina* (Schwägr.) Broth. from Macquarie I. but Seppelt (2004) excluded this species from the island's moss flora, and Eckel (1997*a*) predicted the occurrence of *T. knightii* (Mitt.) Broth., a high-latitude Australasian species recently discovered in Tierra del Fuego, in frigid South America, and in Antarctica but no material matching this species has hitherto been found in this biome, including South Georgia. The genus *Tortella* is unmistakable in Antarctica and, apart from the distinct structural characters discussed above, it is readily known by its regularly broken leaf apices which form distinctive propagulae. In this character both species are similar to *Syntrichia sarconeurum* in which, however, the chlorophyllose upper cells are not sharply set off from the basal cells and are separated by an indistinct straight line running across the leaf base.

1. Plants small, fairly slender; stems 0.5–1.5(–2.0), very occasionally to 5 cm tall, scarcely or not tomentose; distal leaves 1.5–2.0 mm long; leaf apices, when undetached, incurved-circinate, weakly contorted when dry, forming apical propagula, 0.2–0.7 mm long, articulated by a series of barrel-shaped constrictions fracturing in several pieces; marginal cells of leaf tips undifferentiated; laminal cells unistratose throughout; leaf bases strictly hyaline, without a yellowish hue; leafy stems with rhizoids not formed into a dense tomentum .... 1. ***T. alpicola***

1. Plants medium-sized to fairly large, coarse and rigid; stems (0.5–)1.0–5.5 cm tall, densely tomentose along the entire stem length; distal leaves to 5.5 mm long; leaf apices, when undetached, usually twisted around the stem when dry, not forming articulate apical propagula, 0.5–1.0 mm long, fracturing as a single rigid unit; marginal cells of leaf tips with patches of elongate, translucent, smooth cells; distal laminal cells often bistratose; leaf bases usually with a yellowish hue; leafy stems often densely compacted into axillary tomentum .... 2. ***T. fragilis***

1. **Tortella alpicola** Dixon Fig. 143

Tortella alpicola Dixon, Ann. Bryol. 3: 54. 1930. — TYPE CITATION: [India] Bhutna Valley, Ladakh, alt. inter 9000 et 17,000 ped., 1928, coll. Kerr & Lillie (2209*b*). [Holotype: 'Herb. H. N. Dixon Ref. No. 2209b Tortella alpicola Dix. sp. nov. {Ann. Bryologici III, 54 (1930)} Bhutna Valley, Ladakh, India, alt. 9000–17000 ft. Coll. Kerr & Lillie, 1928 Comm. D. Lillie Type' – BM-Dixon!].

Sarconeurum tortelloides S.W.Greene *in* S.W.Greene, D.M. Greene, P.D.Br. & Pacey, Brit. Antarct. Surv. Sci. Rep. 64: 38, 54, *f. 28.* 1970. ≡ *Tortella tortelloides* (S.W.Greene) H.Rob., Antarct. Res. Ser. Amer. Geophys. Union 20: 170. 1972. ≡ *Tortella fragilis* (Drumm.) Limpr. var. *tortelloides* (S.W.Greene) R.H.Zander & Hoe, Bryologist 82: 84. 1979. — TYPE CITATION: [Antarctica] NNE facing rock niches Alt. 200 ft above glacier. Spur at head of Eros Glacier, Alexander Island. Leg. B. J. Taylor 2.12.1962. No. Taylor 529. [Holotype: 'British Antarctic Survey Flora of Antarctic Peninsula Sarconeurum tortelloides S. W. Greene N.N.E. facing rock niches Alt. 200 ft above glacier. Spur at head of Eros Glacier, Alexander Island. Leg. B. J. Taylor 2.12.1962. Det. S. W. Greene March 1967 No. Taylor 529' – AAS!; isotypes: BM!, KRAM!, LE!]. First synonymised by Zander (1993: p. 322).

Plants small to medium, fairly slender, in soft, low or very seldom deeper, compact tufts, dull but with shining white leaf bases, light to dark green above, pale buff-brown below. *Stems* erect, simple or sparingly branched, 0.5–1.5 or, occasionally, to 5.5 cm tall, usually densely radiculose at the base with light brown to rufous, smooth, branched rhizoids but often rhizoids extending high among the leaves, in transverse section rounded or obtusely pentagonal, with a distinct, complete or incomplete hyalodermis, 1–2-stratose cortex of larger cells with thicker, brownish walls not sharply set off from large, hyaline, thin-walled medullary cells and with a distinct central strand or central strand indstinct to absent; *axillary hairs* composed of 10–15 hyaline, elongate cells. *Leaves* crowded, rather rigid but appearing soft, erect, loosely to tightly appressed, incurved-circinate and weakly contorted when dry, erect-spreading to, occasionally, widely patent when moist, uniform in size, 1.5–2.0 mm long (without a caducous apex), 0.5–0.7 mm wide, from an elliptical, oblong to ovate-oblong base, long-lanceolate, gradually or abruptly tapered to a narrow, channelled, straight or most often recurved subula, very fragile and mostly incomplete except in the comal group; *apex* readily detached, narrowly acuminate, sometimes sharply contracted into a subula, fleshy, solid, polystratose, tubulose to nearly cylindrical, 0.2–0.5 mm long, consisting of a series of regular, barrel-shaped constrictions disarticulating in several pieces, densely papillose, except for an acute apiculus of 1 or 2, elongate, sharply pointed, pellucid cells on young leaves; *margins* entire, erect to incurved, plane to weakly undulate, lobed or with scalloped indentations in the distal portion; *costa* single, rather strong, pale yellowish and well demarcated from the laminal cells in the base, (25–)30–50(–60) μm wide near the base, gradually tapering upwards, vanishing in the apex, smooth on the dorsal side throughout the leaf length, densely papillose on the ventral surface in the upper subulate region, covered distally by an epidermis of quadrate cells on the ventral side, in transverse section weakly differentiated, with a ventral epidermis always present, rather thin-walled ventral and dorsal substereid bands present, separated by one row of large, often chlorophyllose guide cells, the ventral stereid cells usually disappearing in the distal portion of the leaf; *laminal cells* abruptly differentiated between the proximal and distal parts; *basal cells* hyaline, lax, thin-walled, rectangular, 60–90 µm long, 15–20 µm wide, ascending the margins and forming a V-shaped, pellucid area; *upper cells* chlorophyllose, unistratose, only in the solid region of the apex 2–3-layered, mostly isodiametric, quadrate to rounded-quadrate, (8–)10–14 µm wide, with not or only little thickened walls, heavily papillose and obscured due to an abundance of knob-shaped papillae throughout the limb, undifferentiated at the margins. *Asexual reproduction* common by propaguloid apices of the youngest leaves articulated into several barrel-shaped segments. [Apparently *dioicous*. *Perichaetia* terminal on the successive perichaetial innovations; *outer perichaetial leaves* differentiated, longer than the stem leaves, elongate- to linear-lanceolate, fragile, terminated with a long, rigid, subulate awn, rising above the contorted stem leaves on drying, with a broad basal region concolorous with the lamina, the latter being unbordered, not propaguloid and quickly disappearing into the costa; *inner perichaetial leaves* long-triangular. *Sporophytes* unknown.]

Chromosome number: unknown.

Discussion. *Tortella alpicola* has long remained an obscure and poorly known species which only recently has been reconfirmed taxonomically as a species in its own right. It was first described from the north-western Himalayas (Dixon 1930) but since then it was mentioned only occasionally by Indian bryologists (Gangulee 1972; Chopra 1975). The species was resurrected by Zander (1993) who considered it to be conspecific with *T. tortelloides* from the Antarctic and Hawaiian Is. The latter species was described from the Antarctic by S. W. Greene (in Greene *et al.* 1970) as *Sarconeurum tortelloides*. Although no more detailed taxonomic discussion of the species was provided, its placement in *Sarconeurum*, then considered an Antarctic endemic genus, was probably on the basis of the form of its caducous leaf apices. These were emphasised by Greene *et al.* (1970) as the best diagnostic character for *Sarconeurum* rather than the costal anatomy which has less differentiation throughout and no differentiation in the cells adaxially. Robinson (1972*a*) transferred *S. tortelloides* to *Tortella* on account of its 'tortelloid' basal leaf areolation forming a V-shaped area. The discovery of material, perfectly matching the Antarctic plants, in the Hawaiian Is. prompted Zander & Hoe (1979) to make a detailed study of *Tortella tortelloides*. They emphasised the close relationship of this species with *T. fragilis* and reduced it to its variety, considering it to be an ecomorph of the type variety, adapted to the stress of low temperatures which are typical of the Antarctic region and high elevations in tropical mountains. These authors indicated a number of structural characters differentiating var. *fragilis* and var. *tortelloides*.

Eckel (1991) reinstated *Tortella tortelloides* as a distinct species and reported it from the High Nearctic and the Rocky Mountains of Wyoming, and this status has subsequently been accepted by Zander (1993) who stated its conspecificity with the older Himalayan *T. alpicola*. The taxonomy of *T. alpicola* and its affinities with *T. fragilis* have been detailed by Eckel (1998) who noted that the chief morphological distinction between these species is the nature of the propaguloid, fragile and deciduous leaf apices. In *T. alpicola* they are articulated by regular periodic constrictions, and consequently they break into several barrel-shaped pieces. In contrast, in *T. fragilis* the apical propagule is not articulated and breaks as a single unit. Additionally, the leaves in *T. alpicola* are generally much smaller, 1.5–2.0 mm long without deciduous apices which may be up to 0.7 mm and their bases are strictly hyaline, whereas in *T. fragilis* the leaves are much longer in the Antarctic, to 6 mm, with leaf apices to 1 mm long and leaf bases usually with a yellowish hue. When describing *Sarconeurum tortelloides*, Greene *et al.* (1970) gave the leaf length in this species as 2.4–5.2 mm but it is because the description was based upon heterogeneous material. Examination of all paratypes revealed that two of them from the east coast of the Antarctic Peninsula (*MJC 3*, AAS, KRAM; *FIDS D746/3*, BM) are actually *T. fragilis*. According to Eckel (1998) *T. alpicola* consistently has a central strand, which is said to be entirely lacking in *T. fragilis*. However, in the Antarctic populations the central strand is observed only in the type material of *Sarconeurum tortelloides* whereas paratypes and all examined non-type collections lack a central strand. An additional character which is helpful in discriminating *T. alpicola* is the cell size. They are 10–14 µm wide and are undifferentiated at the margin of the limb, whereas in *T. fragilis* the distal marginal cells of young leaves in the comal tufts are elongate and transparent and form a distinct border. Moreover, the upper laminal cells are always entirely unistratose and the basal cells are hyaline in *T. alpicola*, whereas in *T. fragilis* the upper cells are often bistratose in the distal lamina, although in many Antarctic populations of this species it is consistently unistratose, and therefore the use of this character for separation of these species is limited; also, the basal leaf cells usually have a yellowish tinge in *T. fragilis*.

Tortella alpicola shows little variation in the Antarctic with regard to its structural characters, but there are some differences in the size of the plants. Generally, it is a small species, with the stem length to 1.5 cm, but occasionally the plants may reach to 3.5 cm (*RILS 9349*, AAS, KRAM) or even to 5 cm (*RILS 11102B*, AAS, KRAM). The latter plant was collected on a wet stream bank and the plants are densely encrusted with silt in the lower part. Nevertheless, they retain their slenderness and appear soft, although they are actually rather rigid.

Reproduction in Antarctica. The species is known only in the barren state and, so far, no plants with sporophytes have been

→

Figure 143. *Tortella alpicola*. — **1**. Habit, wet. **2**. Portion of branch, wet. **3**. Cross-section of stem. **4–5**. Axillary hairs. **6–12**. Leaves. **13–14**. Propaguloid leaf apices (papillae partly removed). **15**. Upper leaf cells (papillae partly removed). **16**. Cells at the junction of basal and upper parts of leaf. **17–24**. Cross-sections of leaves, a sequence from mid-leaf to apex. — [1–2 from *RILS 9349*; 3, 17–18 from *BJT 529*, isotype of *Sarconeurum tortelloides*; 4–9 from *RILS 9355A*; 10–16, 19–24 from *RILS 11102B*]. — Scale bars: a – 1 mm (1); b – 1 mm (2); c – 100 µm (3–5, 17–24); d – 1 mm (6–12) and 100 µm (13–16).

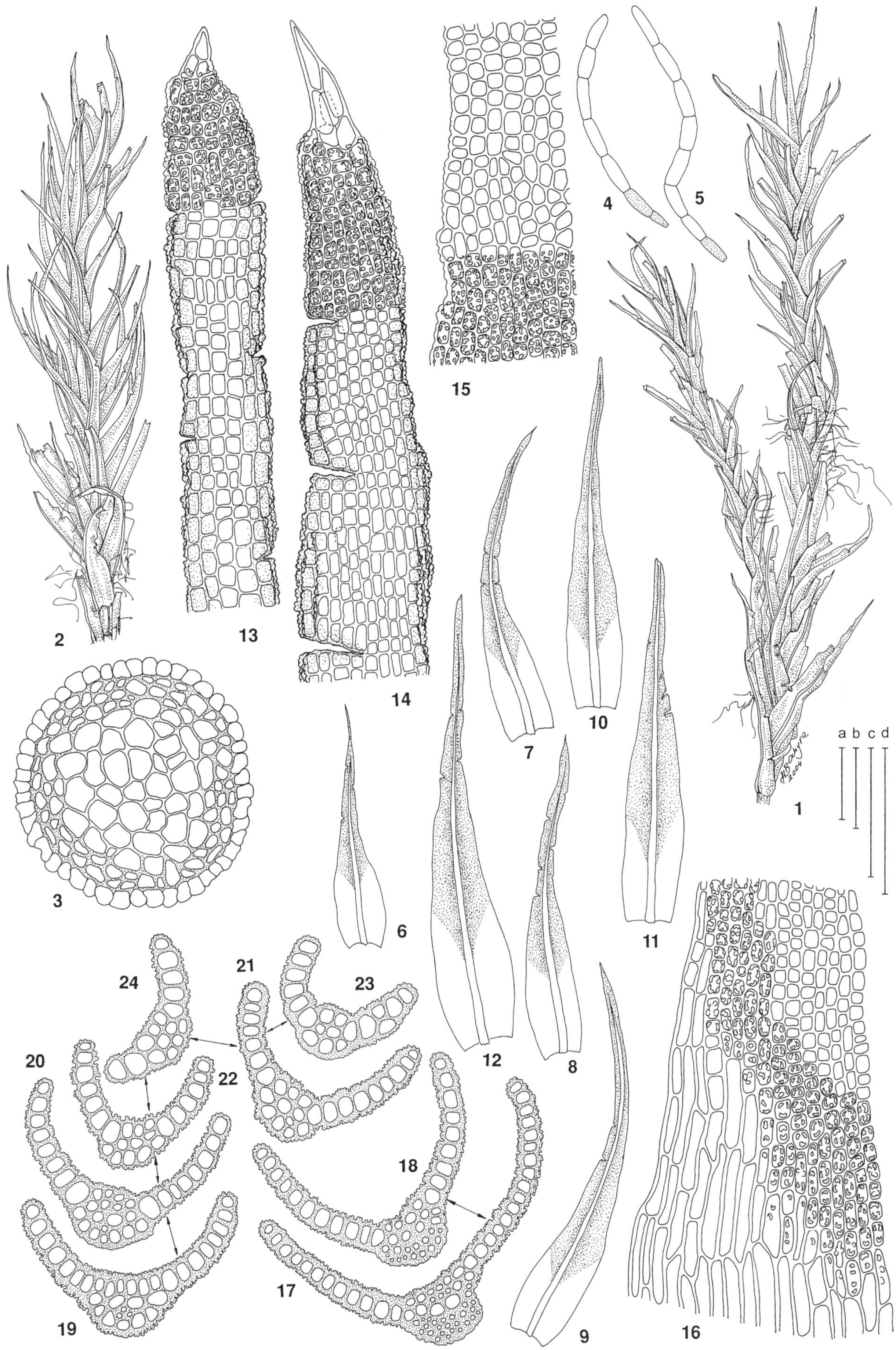
1
2
3
4
5
6
7
8
9
10
11
12
13
14
15
16
17
18
19
20
21
22
23
24
a b c d

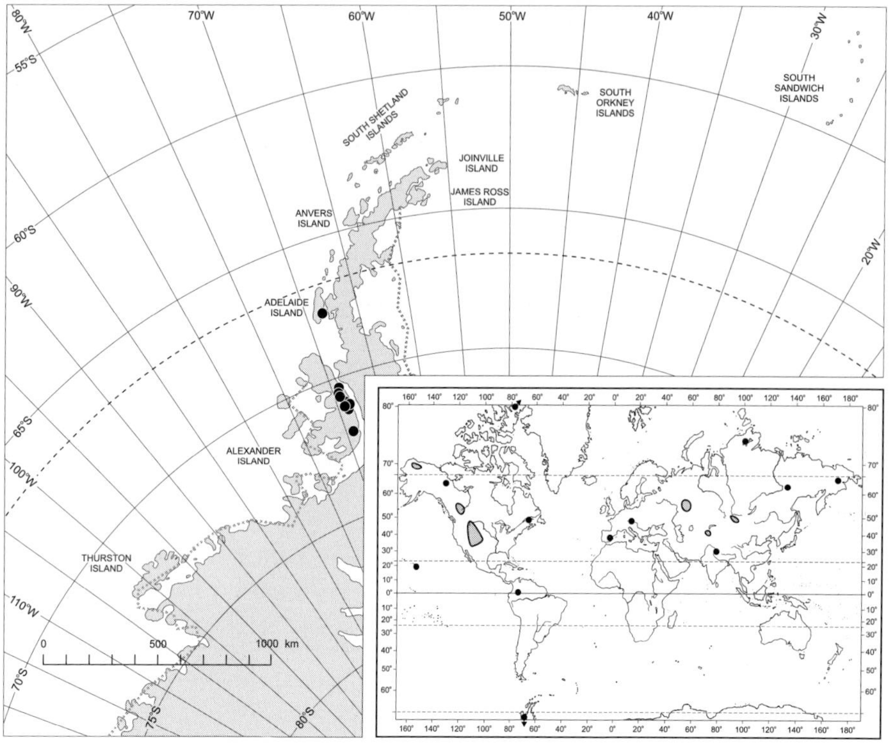

Figure 144. Distribution of *Tortella alpicola* in Antarctica. Inset: global distribution. The localities on Alexander I. are indicated by an arrow.

detected in any part of its wide geographical range. It commonly reproduces asexually by numerous propaguloid leaf apices which readily detach when the young leaves are formed.

Habitat. *Tortella alpicola* is associated with sheltered or exposed and wet or dry habitats, growing in rock crevices, on soil on rock ledges and among boulders, and on moist gravelly soil in depressions and on moist stream banks. It occurs mainly in base-rich habitats but it also tolerates neutral or slightly acidic conditions on sandstone and siltstone. The species most frequently forms monospecific aggregations and only sporadically *Bryum pseudotriquetrum*, *Distichium capillaceum*, *Encalypta procera*, *Hypnum revolutum* and *Pohlia nutans* are found as occasional associates; turves growing in exposed and well illuminated habitats are sometimes encrusted with lichens.

World range. A bipolar species, with some intermediate stations in the tropics in the Hawaiian Is. (Zander & Hoe 1979) and in the northern Andes of Colombia (Churchill *et al.* 1995) (Fig. 144, inset). In the Northern Hemisphere it is arctic-alpine in distribution but its geographical range is still poorly and imperfectly known owing to its obscure taxonomic status. It shows a heavily disjunct distribution pattern in the north, including North America, Asia and the easternmost part of Europe. In North America it is bicentric in distribution, being decidedly western cordilleran with an outlier in the Gaspé Peninsula, Quebec (Eckel 1998). In the Nearctic, *Tortella alpicola* is very rare and widely scattered in the Yukon, Alaska and northern Ellesmere I. (Eckel 1991, 1997*b*), whereas in the south it occurs throughout the eastern Rocky Mountains, from Alberta to Colorado and Arizona, with some extensions to the Great Plains of Nebraska (Eckel 1998) and South Dakota (Eckel 1999) as well as to California (Norris & Shevock 2004). In Asia, *T. alpicola* is so far known only from the Western Himalayas of India from where it was originally described by Dixon (1930), but recently a number of localities have been discovered in Siberia, Mongolia and Middle Asia (Uzbekistan and Kyrgyzstan) as well as in the Ural Mountains in eastern Europe (Otnyukova *et al.* 2004). In Europe, the species was additionally detected in the Sierra Nevada in southern Spain (Rams *et al.* 2006), and several unpublished records are known from the Eastern Alps (H. Köckinger, personal communication). In the Southern Hemisphere, *T. alpicola* is known only from the southern part of the maritime Antarctic from where it was originally described as *Sarconeurum tortelloides* (Greene *et al.* 1970).

Distribution in Antarctica. *Tortella alpicola* is restricted in its distribution to the far south-western Antarctic Peninsula. On eastern Alexander I. it ranges from Ablation Pt. to Coal Nunatak at lat. 72°05′S. Outside this area it is known from a single collection from Adelaide I. on Reptile Ridge, Wright Peninsula (Fig. 144). Moreover, there is a single specimen without locality data collected by the United States Antarctic Service Expedition of 1940–41 (*Bryant 84*, FH, NY) which was misdetermined by Bartram (1957) as *Sarconeurum glaciale*. It is likely that it originates from the south-western coast of the Antarctic Peninsula region where this expedition operated, having its main base at Stonington I. in Marguerite Bay. Greene *et al.* (1970) reported this species, as *Sarconeurum tortelloides*, from two localities on the east coast of the Antarctic Peninsula but the voucher collections are actually *T. fragilis*. *Tortella alpicola* is a species of moderate elevations, occurring from 165 to 717 m, and particularly between 300–500 m.

Specimens examined. WEST ANTARCTICA. WEST ANTARCTIC PENINSULA. **FALLIÈRES COAST.** ***Adelaide I.***: Reptile Ridge, 350–500 m, *RILS 1884A* (AAS, KRAM). **GEORGE VI SOUND.** ***Alexander I.***: 'Ablation Peak' south-west of Ablation Valley, 717 m, *RILS 11114B* (AAS, KRAM); Ablation Pt., *ca* 530 m, *JJL & RBH 52* (AAS, KRAM); middle Flatiron Valley, 317 m, *RILS 11102B* (AAS, KRAM); upper Flatiron Valley, 500 m, *RILS 11089C* (AAS, KRAM); Succession Cliffs, *BASM 209* (AAS, KRAM); Scarab Ridge Plateau, 400 m, *RILS 9355A, 9359A& 9361* (AAS, KRAM); Fossil Bluff, *ca* 360 m, *BJT 509B* (AAS, KRAM); Fossil Bluff, cliffs below summit of Khufu Peak, 300–400 m, *RILS 9324, 9327 & 9349* (AAS, KRAM); Eros Glacier, *BJT 529* (AAS, KRAM); Mt. Ariel, *ca* 135 m, *BJT 506* (AAS, KRAM); Coal Nunatak, *PC 380A* (AAS, KRAM).

2. **Tortella fragilis** (Drumm.) Limpr.

Figs 145–146

Tortella fragilis (Drumm.) Limpr., Laubm. Deutschl. 1: 606. 1888. ≡ *Didymodon fragilis* Drumm., Musci Amer. Drummond: no. 127. 1828 ['fragile']. ≡ *Trichostomum fragile* (Drumm.) Müll.Hal., Syn. Musc. Frond. 1: 586. 1849. ≡ *Barbula fragilis* (Drumm.) Schimp., Bryol. Eur. 6: 159, *pl. 639*. 1855 [Fasc. 62–64 Mon.: Suppl. 4: 1, *pl. 1*], *hom. illeg.* ≡ *Tortula fragilis* (Drumm.) C.Hartm., Handb. Skand. Fl. Ed. 7: 377. 1858, *hom. illeg.* ≡ *Tortula drummondii* Mitt., J. Linn. Soc. Bot. Suppl. 1: 27. 1859. ≡ *Barbula drummondii* (Mitt.) Milde, Bryol. Siles.: 124. 1869. ≡ *Mollia fragilis* (Drumm.) Lindb., Musci Scand.: 21. 1879. ≡ *Tortella tortuosa* (Hedw.) Limpr. subsp. *fragilis* (Drumm.) J.J.Amann, Fl. Mouss. Suisse 2: 100. 1918. — TYPE CITATION: [North America, Rocky Mountains] Margins of alpine lakes; fruit very rare. [Lectotype (*selected here*): '127 Drummond. Musci Am. Is certainly a Tortula 'Didymodon fragile' Hook. Tortula fragilis Wils. Mss. R. Mounts. Alpine Lakes – fruits very rare' – BM-Wilson!; isotypes: BM-Hooker!, BM-Lyon!, BM-Schimp!, BM-Shuttleworth!, BM-Wilson!].

Plants medium-sized to fairly large, occasionally small, coarse, in rigid, deep, dense tufts, basically lustreless but appearing brilliantly shining owing to whitish leaf bases and smooth costae prominently jutted dorsally, yellow-green to light or dark green above, pale brown to blackish-brown below. *Stems* erect, simple or sparsely branched, (0.5–)1.0–5.5 cm high, usually densely tomentose throughout most of its length or, occasionally, less radiculose when small with rigid, brown to rusty, smooth, branched rhizoids, in cross-section rounded or indistinctly obtusely angled, comprising a complete or incomplete hyalodermis, 1–2 layers of relatively large cortical cells with thicker, brownish walls, not sharply separated from a medulla composed of large, hyaline, thin-walled cells and lacking a central strand; *axillary hairs* long, consisting of 10–20 cells, shorter below and elongate distally, hyaline throughout. *Leaves* densely set, firm, coarse, rigid and brittle, uniform in size and shape throughout the stem, not appreciably enlarged at the stem apex, typically fragile only at base of the deciduous apical subula, stiffly erect-appressed, rarely contorted-circinate, especially at the tips of innovations when dry, erecto-patent and mostly straight when moist, (2.8–)3.5–5.5 mm long, 0.5–0.7 mm wide, narrowly lanceolate to linear, gradually tapering above the broadly oblong base until gradually or quickly contracting into a narrower propaguloid region up to about half of the distal leaf length, narrowly to broadly concave, carinate toward the apex in non-propaguloid leaves; *apex* easily detached, narrowly acute, not cucullate, long-subulate, fragile, with a subula obtusely thickened, multistratose, solid and nearly cylindrical in the distal portion, 0.5–1.0 mm long; *margins* entire, erect to inflexed, plane or occasionally somewhat undulate; *costa* single, strong, yellowish-brown to brown, sharply differentiated from the laminal cells in the proximal portion, 80–120 μm wide near the base, gradually narrowing upwards, short excurrent as a smooth mucro of several rhomboidal cells, smooth on the dorsal side throughout the length of the leaf, densely papillose on the ventral surface in the upper subulate region, covered distally on the ventral side by an epidermis of quadrate cells except for a narrow median groove of smooth, elongate cells, in transverse section strongly convex out the dorsal surface, crescent-shaped, with two prominent stereid bands separated by a row of large guide cells, becoming undifferentiated in the subula; *laminal cells* sharply differentiated between the basal and subulate parts, usually without a distinct transitional zone; *basal cells* hyaline, lax, thin-walled, rectangular, 50–100 μm long, 10–20 μm wide, extending a short way up the lamina or to the mid-leaf, forming a V-shaped pellucid area, usually with a yellowish hue; *upper cells* chlorophyllose, frequently bistratose on one or both sides beside the costa and often equally extending toward the margins in various degrees, making the lamina completely 2-layered, rounded-quadrate to hexagonal, (9–)10–12(–13) μm wide, strongly papillose with knob-shaped papillae, thin- to slightly thicker-walled, forming a more or less distinct areolation; *marginal cells* undifferentiated in the median leaf region, papillose-crenulate, becoming differentiated as 1 or 2 rows of smooth, thick-walled, short-rectangular, epapillose, yellowish cells along the distal median margin and in the subula, extending to the apex. *Asexual reproduction* common by caducous, subulate, smooth-sided leaf apices, breaking as a single, rigid unit. [*Dioicous*. *Perichaetia* terminal; *perichaetial leaves* fragile, setaceous, with propaguloid apices as in vegetative leaves. *Setae* singly per perichaetium, occasionally geminate, erect, straight, 1.5–3.0 cm long, smooth, red-orange and dextrorse below, yellow and sinistrorse above. *Capsules* long exserted, erect, symmetric, cylindrical, 1.8–3.0 mm long, straight or slightly curved, leptodermous, pale yellow-brown, reddish at the mouth; *annulus* none or 1–2-seriate, composed of weakly vesiculose cells; *operculum* reddish, high-conical, *ca* 1.5 mm long, with a blunt, somewhat oblique beak; *exothecial cells* thin-walled, rectangular, becoming small, rounded and thick-walled in 4–6 rows at the orifice; *stomata* few in a single row at the base of the urn, phaneroporous, pale, with an elliptical perforation; *peristome* single, consisting of 16 red, long teeth divided into 32 filamentous, densely spiculose-papillose segments, spirally wound together, twisted anticlockwise 1–3 times, arising from a low basal membrane, to 40 μm tall. *Spores* globose, 8–10 μm in diameter, yellowish-green, almost smooth. *Calyptra* cucullate, naked, nearly smooth.]

Chromosome number: unknown.

Discussion. The long, subulate, coarse and rigid, as well as usually fragile leaves with caducous propaguloid apices that are not or only slightly contorted on drying are distinctive for *Tortella fragilis*. In this latter character the species is similar to *T. alpicola* and the traits distinguishing both species are detailed in the discussion under this species. In ambiguous cases, the presence of a marginal border of the smooth, pellucid, elongate and thicker-walled cells in the distal leaf regions and in the propagules is a characteristic feature in *T. fragilis* and, so far, no specimens of *T. alpicola* have been found with at least some leaves displaying differentiated distal marginal cells. Eckel (1998) speculates that the presence of a strong marginal border may contribute to the maintenance of the structural integrity of the propagules when they detach.

In the Antarctic *Tortella fragilis* exhibits a range of variation similar to that portrayed by Eckel (1998) for northern populations. In most populations the leaves are stiff and merely arcuate, but there are a few populations (e.g. *RILS 9393*, *10429*, AAS, KRAM) in which the upper leaves are strongly contorted-circinate, but not crisped, with some undulations of the lamina. These occur mostly in innovating branches which are found in the plants buried in fine soil. Such populations usually show a distinctive dimorphism in the leaves, and in some plants from the central part of a large compact tuft, there occur typically subulate-propaguloid leaves, whereas the peripheral plants have more strongly twisted to contorted-circinate leaves. Caducous leaf apices are present in the majority of Antarctic plants but occasionally they are rare or almost lacking (e.g. *RILS 10417*, *11072*, AAS, KRAM). Identification of such plants may be problematic since the leaf shape, usually coupled with the unistratose or only partially bistratose upper lamina, may suggest another species of *Tortella*, but the overall rigidity of the leaves and the elongate marginal cells in the distal lamina are distinctive. Because the Antarctic plants of *T. fragilis*, as well as those from other sites in the Southern Hemisphere perfectly match those from the Northern Hemisphere, the suggestion of Hyvönen (1991*b*) of the provisional assignment of the austral plants to this species cannot be accepted.

Tortella fragilis was originally described as *Didymodon fragilis* from the material collected in the Rocky Mountains in North America and distributed by Drummond (1828) in his *Musci americani* as No. 127. Although this species name is designated only as 'nov. sp.' and is not accompanied by any author's name, the authorship of *D. fragilis* is often erroneously ascribed to W. J. Hooker & Wilson ('Hook. & Wilson') (e.g. Haring 1938; Nyholm 1956, 1990; Wijk *et al.* 1962; Magill 1981; Frahm 1995; Smith 2004), W. J. Hooker alone ('Hook.') (Müller 1849) or even to J. D. Hooker & Wilson ('Hook.f. & Wilson') (Smith 1978). The erroneous ascription of this species name to W. J. Hooker & Wilson should apparently be credited to Schimper (1855, 1860, 1876) who used it in his major works on European mosses. However, a careful analysis of the literature concerning the first series of Drummond's exsiccata of North American mosses (e.g. Hooker 1834, 1835, 1836; Wilson 1841; Bruch & Schimper 1843) shows that he was himself responsible for the naming of mosses distributed in these exsiccata, so the authorship of new taxa appearing in this series, unless specifically credited to W. J. Hooker, should be accredited to Drummond, as correctly suggested by Andrews (1922) and used in most recent publications (e.g. Crum & Anderson 1981; Corley *et al.* 1981; Zander 1993; Eckel 1998; Cortini Pedrotti 2001*b*; Ochyra *et al.* 2003). A number of sets of *Musci americani* were distributed to major world herbaria (Sayre 1971) and here the specimen from the personal Wilson herbarium at BM, which is accompanied by analytical drawings on the sheet, is selected as a lectotype of *Didymodon fragilis*. William Wilson was the first author to critically study Drummond's exsiccata (Wilson 1841) and he was the author of the second, posthumously published, edition of *Musci americani* (Wilson & Hooker 1841) which comprised the specimens collected by Drummond in 1830–1835 during his second voyage to North America under the patronage of W. J. Hooker.

Reproduction in Antarctica. As elsewhere in the Southern Hemisphere, the species is sterile in the Antarctic, but it effectively reproduces by asexual propagulae which are formed in great profusion by deciduous leaf apices.

Habitat. *Tortella fragilis* grows on various substrates, predominantly on calcareous rocks, but it does not avoid acidic rocks, both in open and sheltered and dry to wet sites, on soil, often on rock ledges and cliffs or in rock crevices, at margins of melt runnels and solifluction lobes, and on pockets of soil amongst boulders. It usually forms pure compact tufts, but is sometimes interwoven with *Cephaloziella varians*; in looser tufts species such as *Distichium capillaceum*, *Encalypta procera*, *E. rhaptocarpa*, *Hypnum revolutum*, *Schistidium andinum* and *Syntrichia magellanica* may be closely associated.

World range. *Tortella fragilis* is a bipolar species in the strict sense, i.e. having no intermediate stations in the tropics (Fig. 147). In the Northern Hemisphere it has a continuous pan-Holarctic range throughout the Arctic and boreal zone, becoming rarer towards the south where it is mostly restricted to montane regions. In North America, it extends to central California (Norris & Shevock 2004), Nevada, Colorado, South Dakota, Iowa, the Great Lakes Basin, New York with some extensions to Tennessee and North Carolina along the Appalachians (Eckel 1998) reaching eastwards to the maritime provinces of Canada (Ireland 1982), Newfoundland (Brassard 1983*b*) and Labrador (Favreau & Brassard 1988), as well as Greenland where it is widespread (Holmen 1960). In northern Europe the species is common in Iceland (Jóhannsson 1992), Fennoscandia (Söderström 1998) and northern Russia south to Bashkiria (Ignatov & Ignatova 2003) and rare in Scotland in the British Isles (Blockeel 1992*a*). In mainland Europe it is widespread in the Alps (Amann 1918; Braunmiller *et al.* 1971; Meinunger & Schröder 2007; Grims 1999) and rare in the Pyrenees (Casas *et al.* 2001), the Carpathians of Poland (Chałubiński 1886; Šmarda 1948) and Romania (Mohan 1998) and in the Caucasus (Manakyan 1989), reaching the southernmost localities in Tuscany in Italy (Cortini Pedrotti 2001*a*,*b*), Montenegro (Pavletić 1955) and in the Pirin in Bulgaria (Petrov 1975). An isolated station of *T. fragilis* is known from the Middle Atlas of Morocco (Jelenc 1955). In Asia, outside the Arctic the species is known from Central Siberia (Bardunov 1974), Central Asia (Lazarenko 1938; Mamatkulov 1990), north-western Mongolia (Tsegmed 2001*a*,*b*), China (Li *et al.* 2001), Taiwan (*Si He 36107*, *36207* & *36508*, MO), Japan (Saito 1975), Kashmir and the Himalayas (Gangulee 1972) and Thailand (*Larsen et al. 746*, *2173*, *2704* & *3159*, MO). In the Southern Hemisphere, *T. fragilis* is very rare but widely scattered, especially in the Antarctic. Outside this biome, it is known from Tierra del Fuego (Hyvönen 1991*b*), Lesotho in South Africa (Magill 1981), the South Island of New Zealand (Fife 1984) and Campbell I. (Vitt 1974*a*), as well as in South Georgia (Ochyra *et al.* 2002) and Îles Kerguelen (Ochyra & Poulsen 2003) in the Subantarctic.

Distribution in Antarctica. *Tortella fragilis* is widely distributed but generally rare and scattered along both the west and east coasts of the Antarctic Peninsula (Fig. 147). On the

⟶

Figure 145. *Tortella fragilis*. — **1**. Habit, dry. **2**. Portion of branch. **3**. Cross-section of the stem. **4–5**. Axillary hairs. **6–8**. Leaves with caducous apices. **9**. Leaf with propaguloid apex. **10**. Leaf apex. **11**. Cells below the leaf apex; papillae removed on one side to show better the areolation. **12**. Mid-leaf cells (papillae partly not indicated). **13**. Mid-leaf cells at the margin (papillae partly removed). **14**. Lamina cells in the transition zone to the hyaline base. **15–17**. Cross-sections of the leaf, a sequence from the base to the apex. — [All from *VK 1241*]. — Scale bars: a – 100 μm (10–14); b – 100 μm (3, 15–17); c – 100 μm (4–5) & 2 mm (2); d – 1 mm (6–9); e – 1 mm (1).

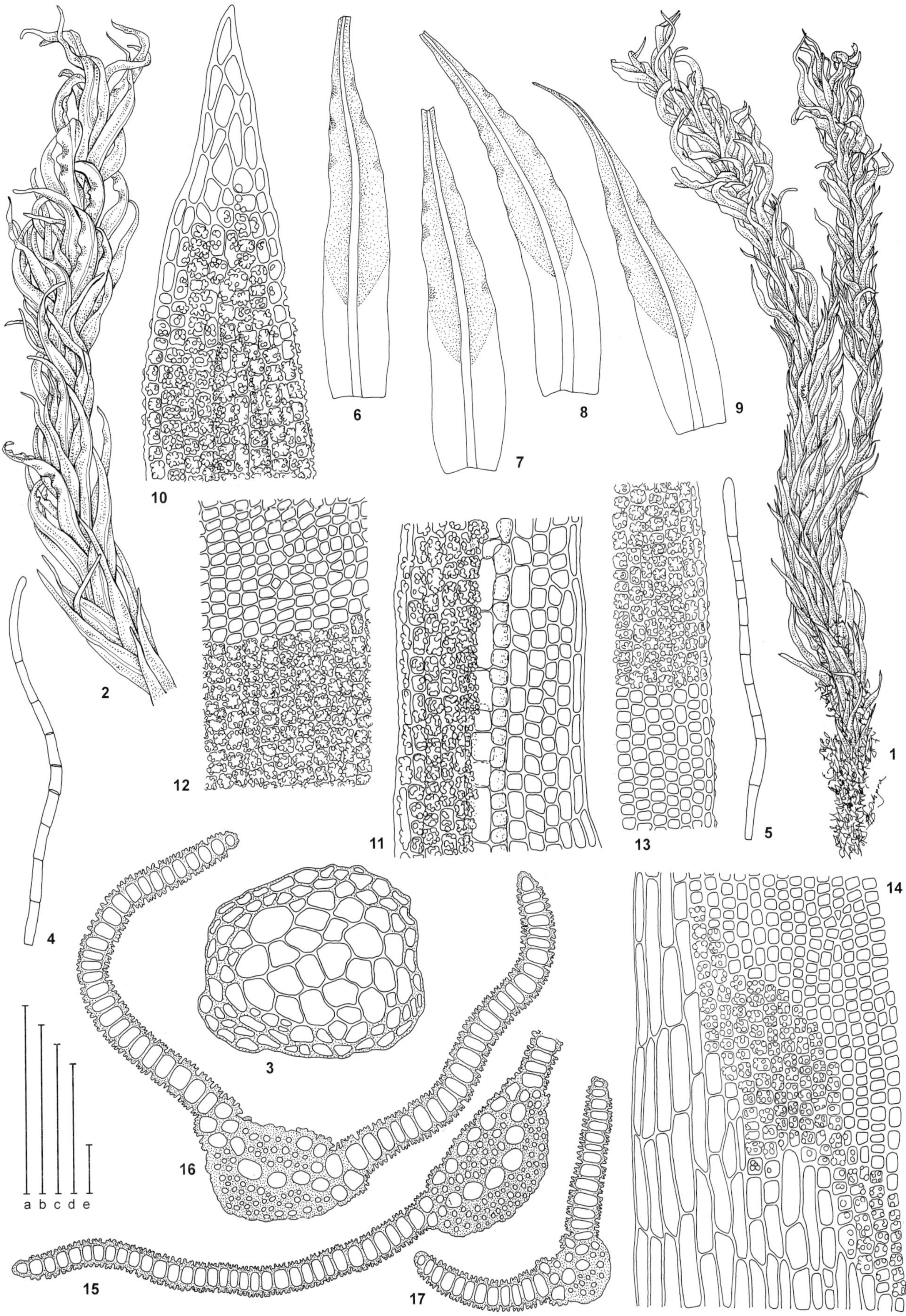
1
2
3
4
5
6
7
8
9
10
11
12
13
14
15
16
17
a b c d e

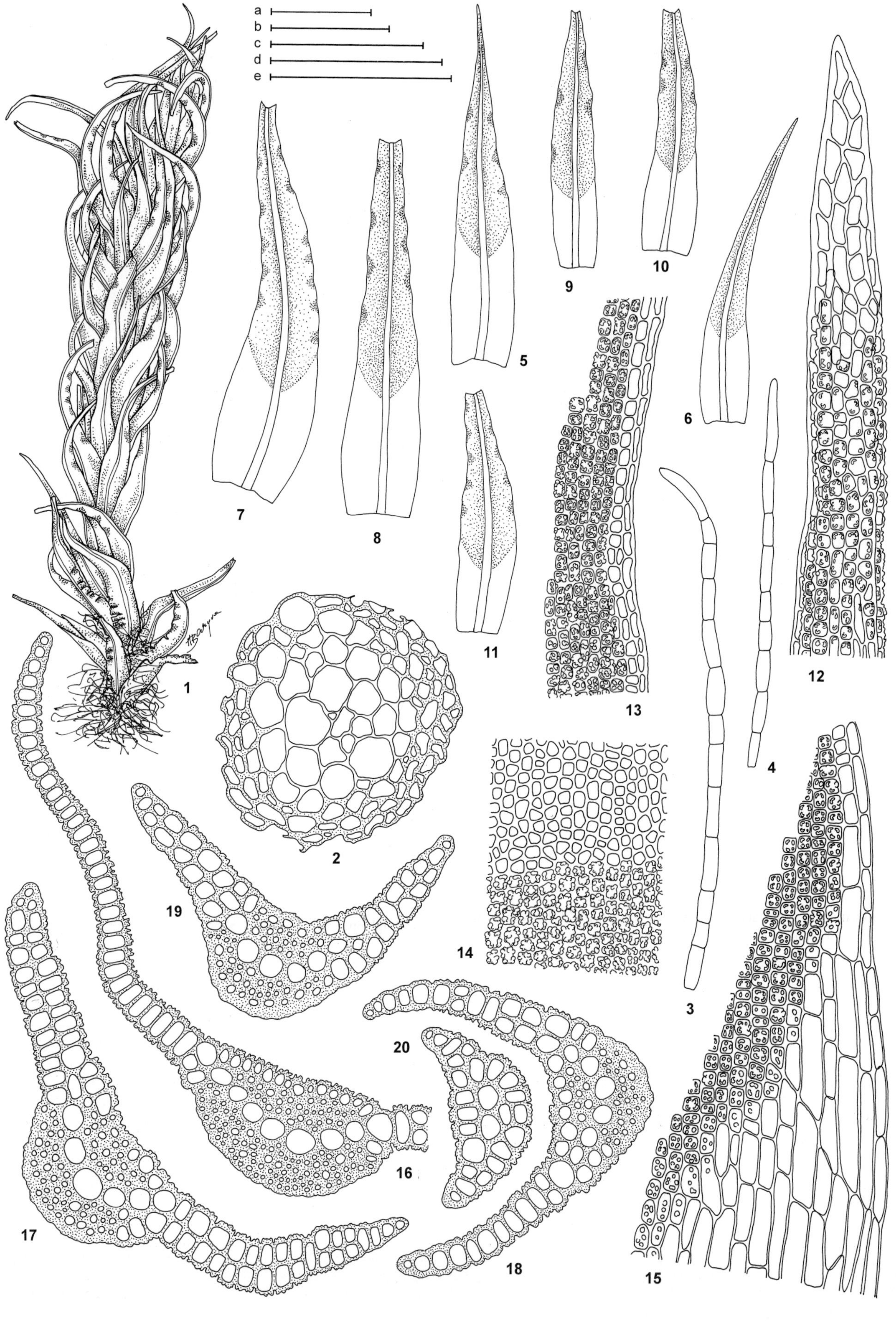
a
b
c
d
e
1
2
3
4
5
6
7
8
9
10
11
12
13
14
15
16
17
18
19
20

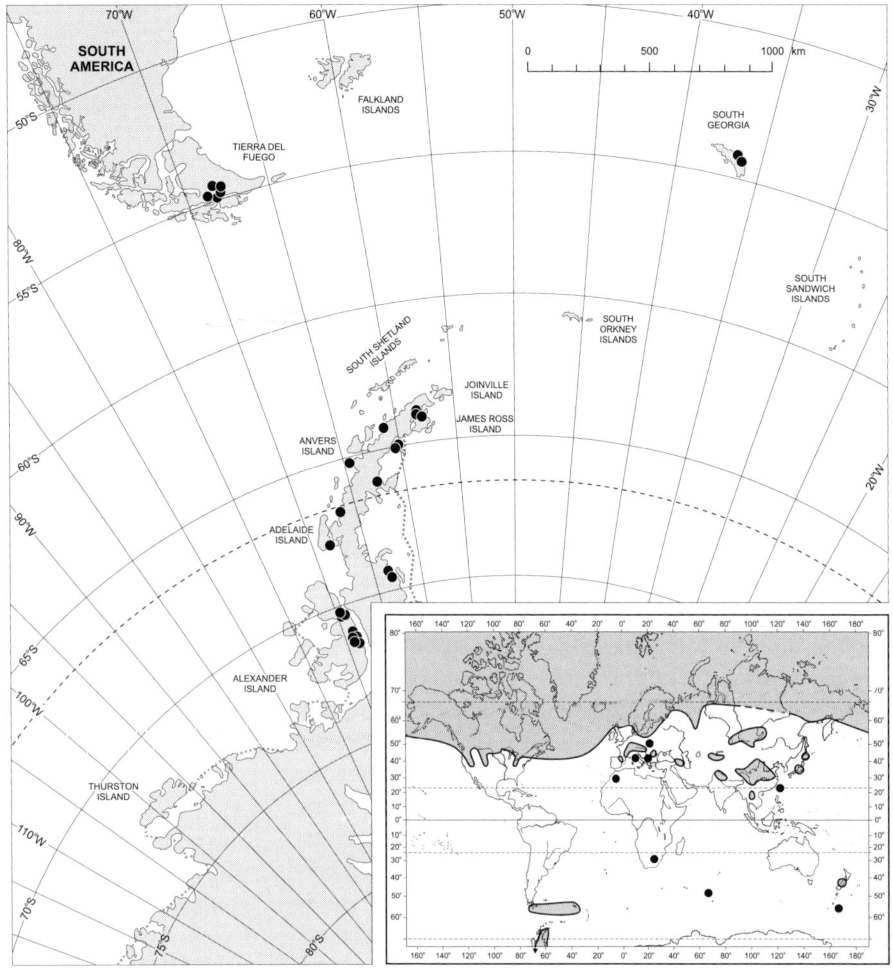

Figure 147. Distribution of *Tortella fragilis* in the Fuegian region and Antarctica. Inset: global distribution.

west Antarctic Peninsula it tends to occur at high latitudes and it is most common and locally abundant on eastern Alexander I. where it reaches its southernmost latitude of 71°52′ S, whereas in most northerly parts of the Antarctic Peninsula it is very rare and occasional at isolated stations on the Danco, Graham and Loubet Coasts. Likewise, the species occurs on the east Antarctic Peninsula, where it ranges from James Ross I. to Engel Peaks on the Wilkins Coast at lat. 69°32′ S, but appears to be relatively abundant only on James Ross I. *Tortella fragilis* occurs mainly from sea level to 610 m, but has been collected at 1150 m at its southernmost locality on the east Antarctic Peninsula.

Specimens examined. WEST ANTARCTICA. WEST ANTARCTIC PENINSULA. **Danco Coast.** Spring Pt., *VK 1241* (KRAM). **Wilhelm Archipelago.** ***Penola Strait***: Peterman I., *ca* 25 m, *RWMC 730A* (AAS, KRAM). **Loubet Coast.** Orford Cliff, *RILS 5275A & 5288A* (AAS, KRAM). **Fallières Coast.** ***Léonie Is.***: Léonie I., 150–175 m, *RILS 9139* (AAS, KRAM). **George VI Sound.** ***Alexander I.***: Ablation Valley, Ablation Lake, 2–3 m, *RILS 9224D, 9251C, 9259, 9267D, 9270B, 9271A, 9273C, 9274B & 9277D* (AAS, KRAM) and *ca* 5 m, *RILS 2711 & 2714* (AAS, KRAM); Ablation Pt., *JJL & RBH 42 & 86* (AAS, KRAM); Flatiron Valley, 510 m, *RILS 11072* (AAS, KRAM); Moutonée Valley, 6 m, *JJL & RBH 68, 69 & 73* (AAS, KRAM); Fossil Bluff, *ca* 350 m, *BJT 509C* (AAS, KRAM) and 600 m, *RILS 9392 & 9393* (AAS, KRAM); Natal Ridge, 600 m, *RILS 10417* (AAS, KRAM); Mars Glacier, Ares Ridge, *RILS 9392 9498* (AAS, KRAM); Mars Oasis, 70–75 m, *RILS 9403 & 10429* (AAS, KRAM), 125 m, *RILS 9420* (AAS, KRAM) and 175 m, *RILS 9441 & 9445* (AAS, KRAM); Two Step Moraine, *RILS 9505* (AAS, KRAM).

EAST ANTARCTIC PENINSULA. **James Ross Island group.** ***James Ross I.***: Cape Lachman, 20–35 m, *RILS 7328D & 7335* (AAS, KRAM); hill on south side of St. Martha Cove, 75 m, *RILS 7431 & 7438B* (AAS, KRAM) and 150 m, *RILS 7428 & 7542A, D* (AAS, KRAM); valley on north side of St. Martha Cove, 80 m, *RILS 7651* (AAS, KRAM); Blancmange Hill, 12–15 m, *RILS 7674B* (AAS, KRAM); hill north-east of Blancmange Hill, 300 m, *RILS 7823* (AAS, KRAM). **Nordenskjöld Coast.** ***Seal Nunataks***: Evensen Nunatak, *ca* 135 m, *FIDS 17C* (BM – paratype of *Sarconeurum tortelloides*); Bruce Nunatak, 50 m, *RILS 6416C* (AAS, KRAM). **Oscar II Coast.** Unnamed nunatak on Bruce Plateau, lat. 66°15′ S, long. 62°55′ W, *RILS 2760A* (AAS, KRAM). **Wilkins Coast.** Finley Heights, *ca* 230 m, *MJC 3* (AAS, KRAM); Engel Peaks, 1000–1150 m, *MJC 60* (AAS, KRAM).

◄

Figure 146. *Tortella fragilis*. — **1**. Habit, dry. **2**. Cross-section of the stem. **3–4**. Axillary hairs. **5–6**. Leaves with unbroken apices. **7–11**. Caducous leaves. **12**. Leaf apex. **13**. Cells below the leaf apex. **14**. Mid-leaf cells (papillae partly not indicated). **15**. Lamina cells in the transition zone to the hyaline base. **16–20**. Cross-sections of the leaf, a sequence from the base to the apex. — [All from *RILS 9259*]. — Scale bars: a – 1 mm (1); b – 1 mm (5–11); c – 100 µm (12–15); d – 100 µm (16–20); e – 100 µm (2–4).

26. BRYOERYTHROPHYLLUM P.C.CHEN

Bryoerythrophyllum P.C.Chen, Hedwigia 80: 4. 1941. ≡ *Barbula* Hedw. sect. *Erythrophyllum* Lindb. *ex* Braithw., Brit. Moss Fl. 1: 260. 1887 [*Trichostomum* Bruch e. [unranked] *Erythrophyllum* Lindb., Öfvers. Förh. Kongl. Svenska Vetensk.-Akad. 21: 213. 1864, *nom. nud.*; *Barbula* A. [unranked] *Erythrophyllum* Lindb., Musci Scand.: 22. 1879, *nom. nud.*]. ≡ *Didymodon* Hedw. [unranked] *Erythrophyllum* (Lindb. *ex* Braithw.) Limpr., Laubm. Deutsch. 1: 544. 1888. ≡ *Erythrophyllum* (Lindb. *ex* Braithw.) Loeske, Hedwigia 47: 175. 1908, *hom. illeg.* [*non Erythrophyllum* J.Agardh 1872 (Algae)]. ≡ *Barbula* subgen. *Erythrophyllum* (Lindb. *ex* Braithw.) C.E.O.Jensen, Skand. Bladmossfl.: 242. 1939. — Type: *Barbula rubella* Mitt. *in* Lindb., *nom. illeg. incl. spec. prior.* (*Anacalypta rubella* Huebener, *nom. illeg. incl. spec. prior.*) [= *Bryoerythrophyllum recurvirostrum* (Hedw.) P.C.Chen (*Weissia recurvirostra* Hedw.)].

Erythrobarbula Steere, Bryologist 54: 191. 1951, *nom. illeg. incl. gen. prior.*

Plants small to medium-sized, rather rigid and stiff, forming dense tufts or loosely caespitose, green, reddish-green or reddish-yellow above, reddish-brown, rusty to brick-red below. *Stems* erect, simple or irregularly branched, in transverse section rounded-pentagonal, mostly without a hyalodermis, with a distinct cortex of small, stereid or substereid cells with brown, thickened walls, grading into a medulla of large, hyaline, thin-walled cells and usually with a distinct central strand; *axillary hairs* filiform, consisting of up to 10 cells, hyaline or occasionally with 1–3 brown basal cells; *rhizoids* rather sparse, brown to red-brown, branched, lustrous, smooth. *Leaves* erect usually sheathing at base, appressed to appressed-incurved, incurved-tubulose or twisted when dry, erecto-patent to widespreading when moist, ovate, ovate-triangular, ovate-lanceolate, oblong-lanceolate, lanceolate, elliptical or oblong-ligulate, ventrally broadly concave or channelled to carinate, rounded-obtuse to acute and apiculate at the apex, shortly to non-decurrent; *margins* recurved throughout or recurved only in the lower ½–⅔, plane above, unistratose, entire or usually dentate above with irregular, pellucid teeth, unbordered or with a 3–4-seriate border of thick-walled cells; *costa* single, strong, ending a few cells below the apex, percurrent to short-excurrent as a sharp, smooth apiculus, with ventral superficial cells quadrate to short-rectangular and papillose and dorsal cells quadrate and papillose to elongate and smooth, differentiated as an epidermis ventrally or on both sides, in cross-section with strong, reniform in outline dorsal and usually weak ventral stereid bands, and 2–5 large guide cells in 1 or occasionally 2 layers; *laminal cells* differentiated, showing red to orange-red colour reaction with 2% KOH, unistratose or very seldom bistratose in patches or sometimes nearly continuously bistratose near the apex; *upper cells* irregularly quadrate, rounded-hexagonal, oblate to short-rectangular, evenly thick-walled, obscure and densely pluripapillose with solid or occasionally hollow, bifid or flattened papillae, 4–6 per lumen, almost flat or sometimes bulging in transverse section, gradual in transition to the proximal cells; *basal cells* conspicuously enlarged, oblong to rectangular, rarely quadrate, thin-walled and lax, hyaline or red-brown, smooth to weakly papillose, transparent; *alar cells* not differentiated. *Asexual reproduction* occasional by multicellular axillary gemmae or multicellular spherical or obovoid tubers borne on rhizoids. *Dioicous* or occasionally *monoicous* (*synoicous* or *paroicous*). *Perichaetia* terminal; *perichaetial leaves* larger than the cauline leaves, usually sheathing in the lower ⅓–¾, long-oval to long-lanceolate, with a prosenchymatous areolation in the proximal portion. *Setae* elongate, erect, red, mainly singly per perichaetium, dextrorse below and usually sinistrorse above. *Capsules* long-exserted, ellipsoid to cylindrical, straight to somewhat arcuate, smooth, brown to reddish-brown, leptodermous; *annulus* present, 1–2-seriate, composed of usually vesiculous cells, often revoluble or deciduous in pieces; *operculum* conical to conical-rostrate, with a straight or slanted, stout beak; *exothecial cells* thin-walled, rectangular, becoming oblate to quadrate, firm-walled in a few coloured rows at the mouth; *stomata* in 1 row at the base of the urn, superficial, bicellular, with elliptical pori; *peristome teeth* variable, none, rudimentary or well developed, consisting of 16 teeth arising from a low basal membrane, perforated along the median line or deeply bifid, red, finely papillose to spiculose, erect to twisted to 2–4 times counterclockwise. *Spores* globose, light brown, finely papillose. *Calyptra* cucullate, naked, smooth.

The generic name *Bryoerythrophyllum*, meaning red-leaved, refers to the red coloration of the plants, which is a distinctive character of the genus. It is especially strong brick-red in its most widespread species, *B. recurvirostrum*, whereas in others the leaves are usually reddish-brown to reddish-green but the red colour is much enhanced by placing a small drop of 10% KOH solution on the leafy plant. The genus is similar and closely related to *Barbula*, *Didymodon* and *Pseudocrossidium* and they form a rather nondescript assemblage of pottiaceous genera. In some recent treatments (e.g. Smith 1978; Nyholm 1990) *Bryoerythrophyllum* is merged with a broadly conceived *Barbula*. In a series of papers Zander (1978*a*, 1980, 1981*a*, 1993) refined the concept of *Bryoerythrophyllum* and distinguished it from the genera *Barbula* and *Didymodon* by the combination of the reddish coloration of the plants, massive papillae on the upper leaf cells, well differentiated leaf bases, a costa that has two well developed stereid bands, the dorsal one being reniform in outline, as well as enlarged ventral epidermal cells as clearly seen in leaf cross-section. Zander (1993) positioned *Bryoerythrophyllum*, along with *Barbula* and *Didymodon*, in the subfamily Merceyoideae but placed them in two different tribes: Bryoerythrophylleae and Barbuleae. The former tribe consists of five genera (*Bryoerythrophyllum*, *Dialytrichia*, *Mironia*, *Pseudocrossidium*, *Rhexophyllum*) that have leaf laminae red in KOH, commonly denticulate upper leaf margins, usually absent dorsal epidermis and a reniform dorsal stereid band in the costa. In contrast, the tribe Barbuleae lacks this combination of characters and its most obvious feature is the shape of the dorsal stereid band that is round or semicircular in outline in transverse section.

Bryoerythrophyllum is most closely related to two neotropical genera, *Mironia* and *Rhexophyllum*, which differ from it in having bistratose leaf margins (*Mironia*) or irregularly bistratose upper leaf lamina (*Rhexophyllum*). The distinctions between these three genera are obscured by the presence of the consistently bistratose upper cells in the Mexican *Bryoerythrophyllum sharpii* (Zander 1986, 1994*a*) and occasional bistratose patches in *B. ferruginascens*, although in all other species of *Bryoerythrophyllum* the laminal cells, including leaf margins, are unistratose. In the Antarctic, *Bryoerythrophyllum* is most likely to be confused with *Didymodon* but each has the radically different leaf papillosity. The former genus has hollow papillae, whereas the latter has solid papillae. This character can best be seen in a leafy stem whole-mount manipulated so that a leaf surface is seen in profile. The cell lumina of *Bryoerythrophyllum* can then be seen to enter the bases of the papillae, while the papillae of *Didymodon* can be seen to be formed entirely of wall material. In addition, the papillae in *Didymodon* are fairly low, making the leaves rather pellucid, whereas in *Bryoerythrophyllum* the upper leaf laminae are much obscured and opaque due to the dense concentration of papillae.

Bryoerythrophyllum has had a complex nomenclatural history. This generic name itself was proposed by Chen (1941) as a replacement of *Erythrophyllum*, a name introduced by Loeske (1908), because it proved to be a later illegitimate homonym of the algal genus *Erythrophyllum* J.Agardh of 1872. Loeske (1908) transferred three species which had usually been placed in *Didymodon* to his new genus, namely *E. recurvirostrum*, *E. alpigenum* and *E. rubrum*. When establishing this genus, Loeske (1908) gave a direct reference to the rankless taxon *Erythrophyllum* of S. O. Lindberg. This name was introduced by Lindberg (1864) for an unranked taxon within *Trichostomum* to accommodate *T. rubellum*. Because no description of this taxon was provided by Lindberg, this name was a typical *nomen nudum*. Later, Lindberg (1879) again used this name as a rankless taxon within *Barbula* for the same species but again no description was available. It was Braithwaite (1887) who validated Lindberg's unranked taxon as section *Erythrophyllum* within *Barbula* and, a year later, Limpricht (1888) transferred this section to *Didymodon* as an unranked taxon. Since the reference in Loeske's (1908) work is not restricted to a definite Lindberg publication, it can be considered as an indirect reference to Braithwaite's (1887) work in which sect. *Erythrophyllum* was validated and this name itself was ascribed to Lindberg. Steere (1951) overlooked the name *Bryoerythrophyllum* proposed ten years earlier by Chen (1941) and introduced the name *Erythrobarbula* to replace the illegitimate *Erythrophyllum* of Loeske (1908), but this name is also illegitimate.

Typification of the generic name *Bryoerythrophyllum* also needs clarification. Wijk *et al.* (1959) indicated *Bryoerythrophyllum recurvirostrum* as the lectotype of this generic name but lectotypification is not necessary in this case. When validating *Barbula* sect. *Erythrophyllum*, Braithwaite (1887) placed in it only a single species, *Barbula rubella* Mitt., which has to be considered as a holotype of the generic name *Erythrophyllum* Loeske and its substitute *Bryoerythrophyllum*. However, this name is illegitimate since it was nomenclaturally superfluous when published because *Barbula rubella* Mitt., based on *Anacalypta rubella* Huebener, included the type of *Weissia recurvirostra* which ought to have been adopted under the rules. Hence, *Bryoerythrophyllum recurvirostrum* is the only resulting type species of the generic name *Bryoerythrophyllum*.

Bryoerythrophyllum is a medium-sized genus which consists of 27 species. Except for a few species having wide pan-Holarctic or pantropical distributions, the genus has two distinct centres of diversity in the Neotropics and Asia. In the austral cool-temperate and polar regions, including Antarctica, only the bipolar *B. recurvirostrum* is widely distributed but scattered. Zander (1993) placed the Antarctic *Barbula byrdii* in *Bryoerythrophyllum* but actually this species is conspecific with *Didymodon brachyphyllus* (Zander & Ochyra 2001; Ochyra & Zander 2002).

1. **Bryoerythrophyllum recurvirostrum** (Hedw.) P.C.Chen Fig. 148

Bryoerythrophyllum recurvirostrum (Hedw.) P.C.Chen, Hedwigia 80: 255, *f. 52.1–2.* 1941. ≡ *Weissia recurvirostra* Hedw., Sp. Musc. Frond.: 71. 1801. ≡ *Bryum recurvirostrum* (Hedw.) With., Syst. Arr. Brit. Pl. Ed. 4, 3: 812. 1801. ≡ *Grimmia recurvirostra* (Hedw.) Sm., Fl. Brit. 3: 1190. 1804. ≡ *Anacalypta recurvirostra* (Hedw.) Fürnr., Flora 12(2) Erg.: 25. 1829. ≡ *Trichostomum recurvirostrum* (Hedw.) Lindb., Öfvers. Förh. Kongl. Svenska Vetensk.-Akad. 20: 416. 1863 ['recurvirostre']. ≡ *Erythrophyllum recurvirostrum* (Hedw.) Loeske, Hedwigia 47: 175. 1908. ≡ *Didymodon recurvirostrus* (Hedw.) Jenn., Man. Mosses W. Pennsylvania: 97. 1913 ['recurvirostre']. ≡ *Erythrobarbula recurvirostra* (Hedw.) Steere, Bryologist 54: 191. 1951. ≡ *Barbula recurvirostra* (Hedw.) Nyholm, Ill. Moss Fl. Fennoscandia. II. Musci 2: 119. 1956, *comb. inval. basion. non cit.* — TYPE CITATION: Locis lapidosis sabulosis; in addition, Hedwig (1801) gave a direct reference to his own earlier work (Hedwig 1785–1787: p. 19, tab. 7) in which the following localities are cited: 'Copiose eundem inveni olim ad radices faxorum vicinorum furnis calcinatoriis Rottlusiensium prope Chemnitium Saxoniae, nec non hic Lipsiae ad interstitia baseos Columnae, quae proxime ad pagum Rückmarsdorf militaria Martisburgum versus indicat'. [Lectotype (*selected here*): due to the lack of any type material in the Hedwig herbarium at G, the plate 7 in Hedwig (1785–1787) is selected as lectotype of *Weissia recurvirostra*; epitype (*selected here*): '[Deutschland] Kgr. Sachsen: Am Elbufer bei Königstein, etwa 120 m s.m., 26. Sept. 1905 legit W. Krieger' (E. Bauer, *Musci europaei exsiccati* No. 273) – KRAM!; isoepitypes: BM, FH, G, NY, PC; for details see Sayre (1971)].

Bryoerythrophyllum recurvirostrum var. *antarcticum* L.I.Savicz & Smirnova, Bot. Žurn. (Moscow & Leningrad) 48: 356, *f. 1–2.* 1963. — TYPE CITATION: Antarctida Orientalis, Bunger Hills, pars centralis, ad W ab statione 'Oasis', fundus lapidosus vallis cum rivulo, **o**, 24 I 1956, legit E. S. Korotkevicz. [Holotype: 'Sectio Cryptogamica Instituti Botanici Academiae Sc. URSS Bryoerythrophyllum recurvirostrum (Hedw.) Chen var. antarcticum L. Savicz & Z. Smirn. Antarctida Orientalis, Bunger Hills, pars centralis, ad W ab statione 'Oasis', fundus lapidosus vallis cum rivulo, 1956. I. 24 Leg. E. S. Korotkievicz 2/3 1960 Det. L. L. Savicz et Z. Smirnova' – LE!; isotype: KRAM!]. First synonymised by Seppelt (1986: p. 230).

Plants small to medium-sized, in dense tufts, rarely loosely caespitose, green, yellow-green, brown or yellow- to reddish-brown above, reddish-brown, brick-red to pale rusty-red below. *Stems* erect, irregularly branched, rarely simple, (0.5–)1.0–2.5(–4.0) cm tall, with clusters of red to red-brownish, smooth, branched rhizoids scattered throughout the stem length or only at the base, in transverse section rounded or rounded-pentagonal, consisting of a large central strand, 3–4 layers of large, hyaline, thin-walled medullary cells, and a 1–3-layered cortex of relatively large cells with brown and moderately thickened walls and lacking a hyalodermis; *axillary hairs* filiform, hyaline throughout or with 1–2 basal cells slightly brownish, 7–8-celled, with short basal cells becoming progressively elongate upwards. *Leaves* ovate- or oblong-lanceolate, lanceolate to linear-lanceolate from an oblong, erect sheathing base, erect-incurved to somewhat spreading and usually twisted to tightly appressed when dry, erect-spreading to spreading at 45° when wet, (0.4–)1.3–2.2 mm long, (0.2–)0.4–0.5 mm wide, acute or bluntly and broadly acute to apiculate, usually ending in a smooth, pellucid conical apical cell, at least on young leaves, plane or broadly concave; *margins* recurved to the apex or in the lower ¾, unistratose, not bordered, entire or with 1–3 irregular, sharp to obsolete teeth at the extreme apex; *costa* subpercurrent to percurrent, 60–80 μm wide, gradually tapering upwards, covered ventrally with quadrate papillose cells, in transverse section strongly convex dorsally and flat to weakly convex ventrally, consisting of a median row of 2–4 large guide cells, large, reniform in outline dorsal and weakly developed or often reduced ventral stereid bands and a well differentiated epidermis of large cells on ventral side; *upper laminal cells* subquadrate, quadrate to irregular, 9–12 μm in diameter, superficially flat, with evenly moderately thickened walls, opaque, densely multipapillose with hollow, bifid papillae, mostly 4–6 per lumen; *basal cells* enlarged, narrowly rectangular, hyaline, smooth, thin-walled, bulging and lax throughout the sheathing base. *Synoicous* or *paroicous*. *Perichaetia* terminal; *perichaetial leaves* elongate, sheathing, somewhat larger than the vegetative leaves but otherwise similar in shape. *Setae* erect, straight, 7–10 mm long, smooth, reddish-brown, twisted to the right when dry. *Capsules* exserted, erect and symmetric to slightly asymmetric, straight, cylindrical, 1.2–1.8 mm long, smooth, leptodermous, reddish-brown; *annulus* 1–3-seriate, revoluble or deciduous, composed of thick-walled, vesiculose cells; *operculum* stout, conical, 0.5–0.7 mm long, straight or obliquely short-rostrate; *exothecial cells* irregular, rectangular to isodiametric, 15–30 μm wide, 30–70 μm long, thin-walled, becoming smaller but not otherwise different at the capsule mouth, forming a differentiated, 1–2-seriate, brownish suboral ring; *stomata* at the extreme base of the urn, superficial, bicellular, with elliptical pori; *peristome teeth* 16, arising from a low basal membrane or basal membrane absent, linear-lanceolate to filamentous, 120–150 μm long, nodose-articulate, pale brown to yellowish, entire or irregularly cleft, densely finely papillose. *Spores* globose, brownish, 14–16 μm in diameter, lightly papillose. *Calyptra* cucullate, naked, smooth.

Chromosome numbers: unknown in the Antarctic and Subantarctic; $n = 12 + m$ (1 record), $n = 13$ (14 records), $n = 13 + m$ (3 records) and $n = 28$ (1 record) from the Holarctic (Fritsch 1991).

Discussion. *Bryoerythrophyllum recurvirostrum* is easy to recognize, even when sterile, by the rusty- or brick-red tinge of the older parts of the plants which may be markedly amplified when they are treated with a 10% solution of KOH. This characteristic colour, together with the lanceolate to linear-lanceolate leaves, readily aid positive determination. Microscopically, the leaf apex is unmistakable because of the several well developed to nearly obsolete teeth and the minute, smooth and pellucid apiculus which presents a sharp contrast to the densely papillose and almost opaque upper laminal cells. Additionally, the leaf margins are strongly recurved and the basal leaf cells are rectangular, hyaline and lax. The species is exceedingly variable in Antarctica. Plants from the maritime Antarctic represent typical phenotypes of the species which do not pose any problems in identification. In contrast, plants from the east coast of the Antarctic Peninsula and the continent are very different in habit. These grow in low, very dense tufts and have shorter, ovate-lanceolate leaves, usually with obtuse and entire apices. These continental forms are likely to be confused with *Didymodon brachyphyllus*, especially because in some plants described as *Barbula byrdii* the basal cells are more elongate and lax and the upper cells are highly papillose. However, the pellucid apiculus is seen at least on some young leaves and the axillary hairs are hyaline throughout, whereas in *D. brachyphyllus* the basal cell is brown. Additionally, plants of *D. brachyphyllus* are green to brown-green, the leaf apex is often cucullate, the pellucid, smooth apiculus is always absent and axillary gemmae are often present. The continental populations were recognised taxonomically by Savicz-Lyubitskaya & Smirnova (1963*a*) as *B. recurvirostrum* var. *antarcticum*, which appears to be merely a habitat modification of the otherwise typical plants induced by the severe environmental conditions. Therefore, *B. recurvirostrum* var. *antarcticum* was reduced to synonymy with var. *recurvirostrum* (Seppelt 1986; Ochyra & Zander 2002).

The overall appearance and colour of Antarctic plants is reminiscent of *Ceratodon purpureus*, one of the commonest species of moss in this region, but which is readily distinguished by its smooth laminal cells.

Bryoerythrophyllum recurvirostrum was described by Hedwig (1785–1787) as *Weissia recurvirostra* and this pre-starting point name was subsequently validated in his *Species muscorum frondosorum* (Hedwig 1801). Nevertheless, for a long time this specific epithet was neglected in favour of the epithets *curvirostrum* and *rubellum* which were introduced by Dickson (1790 as *Bryum curvirostrum*) and Hoffmann (1796 as *Bryum rubellum*). These names were subsequently validated, respectively,

→

Figure 148. *Bryoerythrophyllum recurvirostrum.* — **1**. Plant with sporophytes, wet. **2**. Sterile plant, dry. **3**. Portion of branch, dry. **4–11**. Leaves. **12**. Leaf apex of the uppermost young leaf. **13–14**. Leaf apices (papillae partly removed to show areolation). **15**. Mid-leaf cells at margin (papillae partly not indicated). **16**. Mid-leaf cells (papillae partly removed). **17**. Lamina cells in transition zone to basal part. **18**. Basal leaf cells. **19**. Cross-section of stem. **20–24**. Cross-sections of leaves, a sequence from base to apex. **25–27**. Axillary hairs. **28–29**. Operculate capsules, wet. **30–32**. Deoperculate capsules with distroyed peristomes, wet. **33**. Exothecial cells in middle of urn. **34**. Exothecial cells at base of urn and stomata. **35**. Exothecial cells at orifice, annulus, peristome and spores. — [1, 4–6, 14, 28–35 from *BJT 527*; 2–3, 7, 16, 23 from *Korotkevich L6* (paratype of *B. recurvirostrum* var. *antarcticum*); 8, 12, 18 from *Korotkevich 2/3* (isotype of *B. recurvirostrum* var. *antarcticum*); 9–11, 13, 15, 17, 19–21, 24–27 from *RILS 7714*; 22 from *Kanda 969*]. — Scale bars: a – 0.5 cm (1); b – 1 mm (4–11); c – 100 μm (12, 20–24); d – 1 mm (28–32); e – 1 mm (3), 2 mm (2) and 100 μm (13–19, 25–27, 33–35).

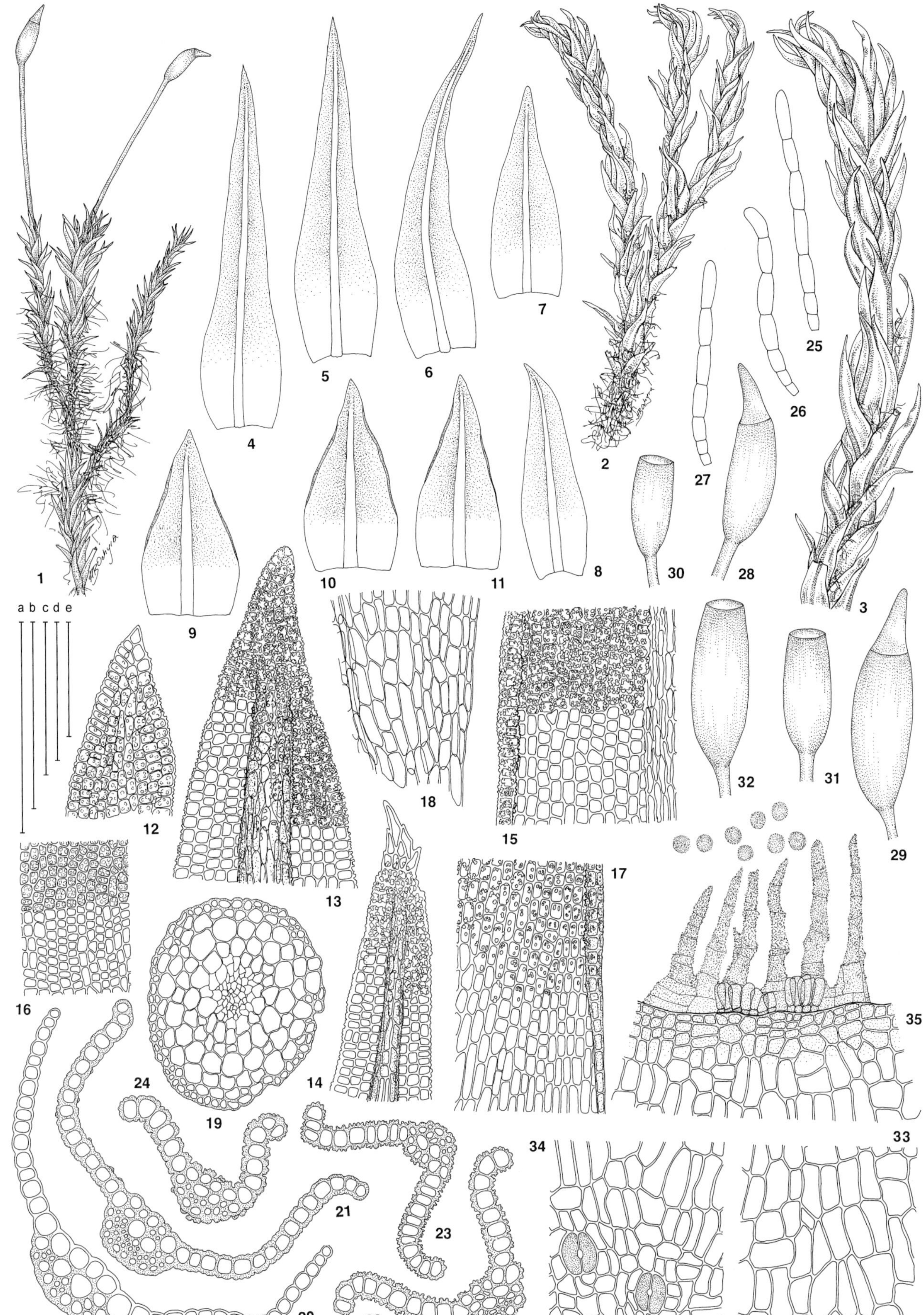
1
2
3
4
5
6
7
8
9
10
11
12
13
14
15
16
17
18
19
20
21
22
23
24
25
26
27
28
29
30
31
32
33
34
35
a b c d e

by Dickson (1801 as *Bryum curvirostrum*) and Hübener (1833 as *Anacalypta rubella*). The latter epithet was given by Bruch & Schimper (1846) as *Didymodon rubellus* in *Bryologia europaea*, and since then it was firmly rooted in bryology. Although these authors ignored the principle of priority, their opus was the oracle as far as the taxonomic and nomenclatural questions in European mosses were concerned. The Hedwigian epithet *recurvirostrum* was resurrected by Loeske (1908) who used it for this species under *Erythrophyllum*. Subsequently, Jennings (1913) and Steere (1938) adopted it for this species under *Didymodon* and, finally, Chen (1941) transferred *Weissia recurvirostra* to *Bryoerythrophyllum* and this name has gained wide acceptance.

Another problem is typification of *Weissia recurvirostra*. Steere (1938) discussed this issue in detail and showed some discrepancies between the description and illustrations of this species in Hedwig's (1785–1787, 1801) works and the true characters of the species in its present circumscription, and adopted the Hedwigian name for his *Didymodon recurvirostrus* (≡ *Bryoerythrophyllum recurvirostrum*). However, he did not study the original material in the Hedwig/Schwägrichen herbarium at G, and only Zander (1978*a*) indicated the specimen from this herbarium as holotype. Unfortunately, this specimen cannot be accepted as the original collection since it originates from 'Gottingae' (= Göttingen) and this locality is not mentioned in Hedwig's (1785–1787) work, nor are date or collector given. Moreover, the label is handwritten by Ch. Schwägrichen, not J. Hedwig. In the absence of suitable original material in the Hedwig/Schwägrichen herbarium, the plate with an illustration of *Weissia recurvirostra* in Hedwig's (1785–1787) *Descriptio et adumbratio microscopico-analytica muscorum frondosorum* is here selected as iconotype, and the specimen distributed by Bauer (1908) in his *Musci europaei exsiccati* as No. 273 is chosen as epitype. It was collected from Sachsen in Germany, a region from which Hedwig (1785–1787) collected, described and illustrated this species. The epitype material is in perfect fruiting condition and matches exactly the current concept of *Bryoerythrophyllum recurvirostrum*.

The spelling of the specific epithet of *Bryoerythrophyllum recurvirostrum* is confusing because there exist two adjectival versions of it. It is a compound word composed of two Latin elements: *recurvus* (= recurved) and *rostrum* (= beak). The latter component may have different forms: *-rostris*, *-rostre* and *-rostrus* (or *-roster*), *-rostra* and *-rostrum*, and both *-rostris* and *-rostrus* alternatives are admissible (Crundwell 1970). Some authors, including the compilers of *Index muscorum* (Wijk *et al.* 1959), use the former. However, because Hedwig (1801) used the latter form as *Weissia recurvirostra*, it should consequently be retained in nomenclatural combinations, as it resulted from transfers of the species to various genera. Thus, depending on gender of the generic name, it should be spelled as *Didymodon recurvirostrus*, *Barbula recurvirostra* and *Bryoerythrophyllum recurvirostrum*.

Reproduction in Antarctica. The majority of Antarctic populations are sterile, although gametangia are occasionally present. Only a few populations from Alexander I. (e.g. *BJT 527 & 528* and *RILS 9234*, AAS, KRAM) were found with sporophytes, with

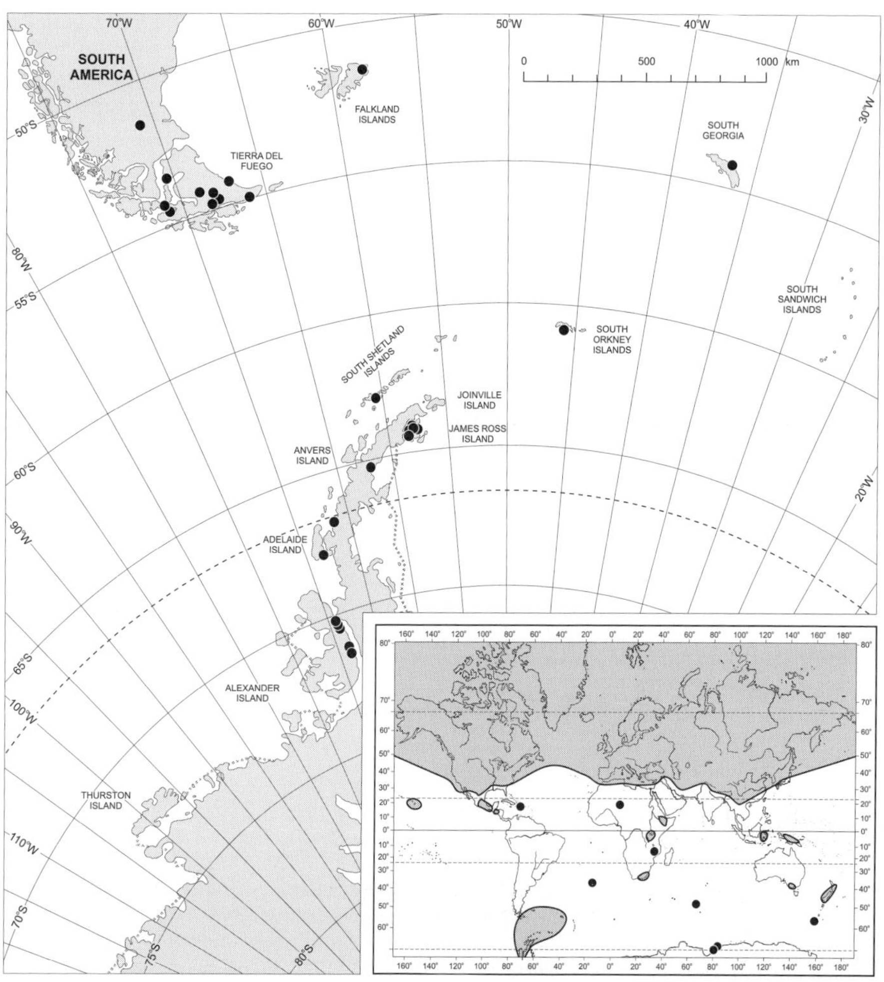

Figure 149. Distribution of *Bryoerythrophyllum recurvirostrum* in the Antarctic Peninsula and Fuegian regions. Inset: global distribution.

few fully mature capsules, but many immature and aborted sporophytes were also present (Lewis Smith & Convey 2002). Savicz-Lyubitskaya & Smirnova (1963*a*) described peculiar clavate hairs among young sexual organs and considered them to be propagula. Indeed, in the axils of the upper leaves there are frequently observed hyaline or hyaline-brownish filaments that appear to be merely aborted antheridia or archegonia, but their interpretation as propagula is doubtful.

Habitat. In Antarctica, *Bryoerythrophyllum recurvirostrum* occurs in a wide range of habitats, both in dry and moist situations. It frequently grows on soil, silt and gravel in wet places, in seepage areas, on margins of melt streams below late snow patches and occasionally in habitats subjected to early season flooding. In such habitats it is frequently associated with *Bryum pseudotriquetrum*, *Encalypta procera* and *Cephaloziella varians*. Additionally, this species occurs frequently in dry or moist rock fissures, on soil on rock ledges, in cliff faces and amongst boulders. It is quite frequent in such habitats on Signy I. where it grows on limestone and marble rocks together with such chasmophytes as *Distichium capillaceum*, *Platydictya jungermannioides*, *Encalypta rhaptocarpa*, *Brachythecium austrosalebrosum*, *Hypnum revolutum*, and *Tortella alpicola*.

World range. *Bryoerythrophyllum recurvirostrum* is considered by bryogeographers as a cosmopolitan species but this status requires verification. A detailed analysis of its global range (Fig. 149, inset) indicates that it is a typical bipolar species having scattered intermediate stations at high elevations in the tropics. In the Northern Hemisphere, it is a weedy species widespread throughout much of the arctic, boreal and temperate zones, extending southwards to Baja California, northern Mexico, Arizona, Colorado, Nebraska, Michigan and New York in North America (Zander 1994*a*, 2007), to the Mediterranean region in Europe and North Africa (Ros *et al.* 1999; Jiménez *et al.* 2002), Iran (Størmer 1963), Afghanistan (Froehlich 1964), the Himalayas (Gangulee 1972), southern China (Li *et al.* 2001) and central Honshu of Japan (Saito 1975). In the Southern Hemisphere, *B. recurvirostrum* is most frequent in the West Antarctic, but also occurs in East Antarctica, then recurring on some subantarctic islands including South Georgia (Ochyra *et al.* 2002), Îles Kerguelen (Savicz-Lyubitskaya & Smirnova 1963*a*; Ochyra, personal observation) and Macquarie I. (Seppelt 2004), in Tierra del Fuego (Cardot 1908*a*; Matteri & Schiavone 1991), Tristan da Cunha (Dixon 1960 as *Didymodon austroalpigenus*; Savicz-Lyubitskaya & Smirnova 1963*a*), Lesotho and KwaZulu-Natal in South Africa (Magill 1981), Victoria in south-eastern Australia (Streimann & Klanzenga 2002) and New Zealand (Sainsbury 1955). In the tropics the species is generally rare and not abundant. In the Neotropics it is known only from Mexico (Zander 1994*a*) and Guatemala (Allen 2002) and the Dominican Republic in the West Indies (Buck & Steere 1983), whereas records from the Andes of South America (Delgadillo *et al.* 1995; Churchill *et al.* 2000) refer to *B. aeneum* (Müll.Hal.) B.H.Allen, a species closely related to *B. recurvirostrum* and sometimes considered as its variety (Zander 1978*a*). Likewise, the species is rare and occasional in Malesia including Sulawesi and New Guinea (Norris & Koponen 1989; Eddy 1990) and the Hawaiian Is. (Bartram 1933*a*). It appears to be relatively frequent in sub-Saharan Africa including Chad (Ros *et al.* 1999), East African mountains from Ethiopia to Tanzania and the Democratic Republic of Congo (O'Shea 2003) and Mozambique (Ochyra & Pócs 1987).

Distribution in Antarctica. *Bryoerythrophyllum recurvirostrum* is a widely distributed but scattered pan-Antarctic species

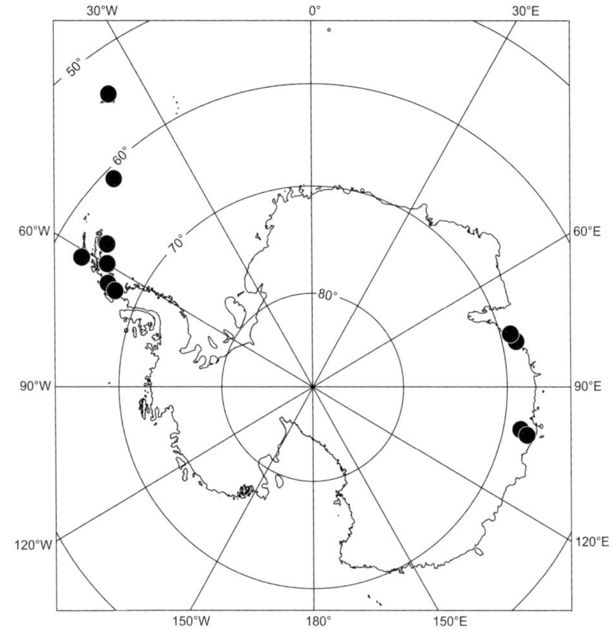

Figure 150. Distribution of *Bryoerythrophyllum recurvirostrum* in Antarctica.

(Figs 149–150). It occurs in the maritime Antarctic from the South Orkney Is. to Alexander I. where it reaches lat. 71°50′S, this being its southernmost station. On the eastern coast of the Antarctic Peninsula the species is relatively frequent on James Ross I. and has been recorded from the Oscar II Coast at lat. 65°05′S. On the continent it has been recorded only at two oases, namely Vestfold and Bunger Hills. Its altitudinal range extends in Antarctica from sea level to 250 m on James Ross I.

Selected specimens examined. WEST ANTARCTICA. SOUTH ORKNEY IS. **Signy I.** Cove S of Foca Pt., 15–35 m, *REL 1101 & 1110* (AAS, KRAM); Foca Cove, *RILS 1798* (AAS, KRAM); cliffs above Stygian Cove on Robin Peak, 25 m, *RILS 8148* (AAS, KRAM); Moraine Valley, *RILS 1799* (AAS, KRAM); Paal Harbour, *ca* 85 m, *RILS 667 & 668* (AAS, KRAM).

SOUTH SHETLAND IS. **Livingston I.** Hurd Peninsula, Juan Carlos I Station region, Feb 1992, *Schulz s.n.* (KRAM).

WEST ANTARCTIC PENINSULA. **Loubet Coast.** Orford Cliff, *RILS 5287B* (AAS, KRAM). ***Léonie Is.***: Léonie I., 20 m, *RILS 10642* (AAS, KRAM) and 150–175 m, *RILS 9136* (AAS, KRAM). **George VI Sound.** ***Alexander I.***: Ablation Pt., 10 m, *RBH & JJL 36 & 37* (AAS, KRAM); Ablation Valley, Ablation Lake, *RBH & JJL 41* (AAS, KRAM) and *RILS 8834E, 8842G, 9234, 9274A, 9277B & 9280* (AAS, KRAM); Moutonnée Valley, *RBH & JJL 75* (AAS, KRAM); Eros Glacier, *BJT 524A, 527 & 528* (AAS, KRAM); Waitabit Cliffs, *ca* 25 m, *BJT 521* (AAS, KRAM); Mars Oasis, 70–175 m, *RILS 9404, 9407 & 9430B* (AAS, KRAM).

EAST ANTARCTIC PENINSULA. **James Ross Island group.** ***Vega I.***: N end of False Island Pt., 25–40 m, *RILS 7833 & 7834* (AAS, KRAM). ***James Ross I.***: between Cape Lachman and Bibby Pt., *RILS 5308C* (AAS, KRAM); Lagrelius Pt., 10 m, *RILS 5962B* (AAS, KRAM); Fortress Hill, 150 m, *RILS 7776* (AAS, KRAM); S side of Martha Cove, 50 m, *RILS 7671B* (AAS, KRAM); valley S of Martha Cove, 15 m, *RILS 7481B* (AAS, KRAM); Terrapin Hill, 85 m, *RILS 7744 & 7745B* (AAS, KRAM) and 250 m, *RILS 7721* (AAS, KRAM); plateau S of Terrapin Hill, 250 m, *RILS 7715* (AAS, KRAM); N side of Blancmange Hill, 12–15 m, *RILS 7675, 7676A & 7686* (AAS, KRAM); valley below the glacier NW of Dreadnought Pt., 3–10 m, *RILS 7495 & 7500* (AAS, KRAM). **Oscar II Coast.** Seal Nunataks, Bruce Nunatak, 50 m, *RILS 6415B* (AAS, KRAM).

EAST ANTARCTICA. PRINCESS ELIZABETH LAND. **Ingrid Christensen Coast.** Vestfold Hills, Learner's Way, *Kanda 949, 951,*

952, 953, 967, 968, 969 (NIPR) & *969* (KRAM, NIPR); Davis Station, 1974, *Seppelt s.n.* (NY); Mossel Hut area, 10–30 m, *RILS 6374 & 6376* (AAS, KRAM); Langneset Peninsula, 15 Dec 1956, *leg. Korotkevich s.n.* (KRAM, LE).

QUEEN MARY LAND. **KNOX COAST.** Bunger Oasis, west of the Station, *Korotkevich 2/3* (KRAM, LE – type of *B. recurvirostrum* var. *antarcticum*) & *L6* (KRAM, LE); same locality, northernmost part towards the Highjump Achipelago, *Hollerbach 8/256* (KRAM, LE); same locality, at foot of hill, 36 m, *Vinogradov 12/640* (KRAM, LE); same locality, without further data, *Vjalov 4A* (AAS, KRAM, LE).

27. DIDYMODON HEDW.

Didymodon Hedw., Sp. Musc. Frond.: 104. 1801. ≡ *Barbula* Hedw. sect. *Didymodon* (Hedw.) Giacom., Atti Ist. Bot. Univ. Pavia Ser. 5, 4: 208. 1947. — Lectotype (*vide* Steere 1938: p. 186): *Didymodon rigidulus* Hedw.

Geheebia Schimp., Syn. Musc. Eur. Ed. 2, 2: 233. 1876. ≡ *Barbula* Hedw. subgen. *Geheebia* (Schimp.) Szafran, Fl. Polsk. Mchy 1: 213, 225. 1958, *comb. inval. basion. non citat.* — Type: *Geheebia cataractarum* Schimp. [= *Didymodon giganteus* (Funck) Jur.]. First synonymised by Juratzka (1882: p. 102).

Husnotiella Cardot, Rev. Bryol. 36: 71. 1909. — Type: *Husnotiella revoluta* Cardot [= *Didymodon revolutus* (Cardot) R.S.Williams]. First synonymised by Williams (1913*b*: p. 25).

Dactylhymenium Cardot, Rev. Bryol. 36: 72. 1909. — Type: *Dactylhymenium pringlei* Cardot [= *Didymodon tophaceus* (Brid.) Lisa]. First synonymised by Zander (1993: p. 157).

Trichostomopsis Cardot, Rev. Bryol. 36: 73. 1909. — Type: *Trichostomopsis crispifolia* Cardot [= *Didymodon umbrosus* (Müll.Hal.) R.H.Zander]. First synonymised by Zander (1981*b*: p. 384).

Limneria Stirt., Trans. Bot. Soc. Edinburgh 26: 428. 1915. — Type: *Limneria viridula* Stirt. [= *Didymodon spadiceus* (Mitt.) Limpr.]. First synonymised by Dixon (1923*b*: p. 51).

Astericium (Müll.Hal.) Hilp., Beih. Bot. Centralbl. 50(2): 618. 1933, *hom. illeg.* ≡ *Barbula* Hedw. [unranked] *Astericium* Müll.Hal., Linnaea 42: 342. 1879. ≡ *Didymodon* Hedw. sect. *Astericium* (Müll.Hal.) R.H.Zander, Cryptogamie Bryol. Lichénol. 2: 383. 1982. — Lectotype (*vide* Robinson 1970: p. 184): *Barbula umbrosa* Müll.Hal. [= *Didymodon umbrosus* (Müll.Hal.) R.H.Zander]. First synonymised by Zander (1981*b*: p. 383).

Prionidium Hilp., Beih. Bot. Centralbl. 50(2): 640. 1933. — Type: *Prionidium setschwanicum* (Broth.) Hilp. [= *Didymodon erosodenticulatus* (Müll.Hal.) K.Saito]. First synonymised by Saito (1975: p. 500].

Plants small, medium-sized to large, slender to fairly robust, rigid, in dull, loose or compact tufts, occasionally gregarious, green, olive-green, brown to red-brown above, brown, reddish-brown or tan below. *Stems* erect, simple or sparsely, irregularly branched, radiculose at base or with scarce clusters of reddish-brown, smooth rhizoids scattered along the stem, in cross-section obscurely rounded-pentagonal or occasionally rounded-triangular, usually with a distinct central strand and a few-layered cortex of small, brown-walled sclerenchymatous cells, mostly lacking a hyalodermis; *axillary hairs* filiform, short, 5–6 celled, consisting of 1–2 brown subquadrate stalk cells and hyaline elongate upper cells. *Leaves* crowded, occasionally somewhat distant, erect-appressed, incurved or twisted to contorted, sometimes soiled round the stem on drying, erect-spreading to squarrose-recurved when moist, ovate, ovate-lanceolate, lanceolate to linear-lanceolate, usually subulate from a weakly differentiated to broader, oblong and semi-sheathing, not or decurrent base, rounded, obtusely or narrowly acute to acuminate, plane or cucullate at the apex, sometimes with apices abruptly differentiated from the lamina, very fragile and caducous as swollen, thickened and fleshy propagula, broadly concave or sometimes channelled to carinate; *margins* entire or occasionally slightly denticulate to crenulate by projecting cells, recurved or rarely revolute or plane, unistratose or bistratose in patches or entirely so; *costa* single, stout, vanishing several cells below the apex, percurrent or often nearly filling the subula and excurrent as a fleshy, blunt cusp, with quadrate or elongate superficial cells on the ventral side and usually with short-rectangular dorsal cells in the distal portion, smooth or sometimes papillose ventrally at the mid-leaf, in transverse section terete or reniform, plane or convex ventrally, strongly convex dorsally, with a weak or occasionally no ventral stereid band and a strong, semicircular dorsal stereid band, separated by a row of 2–6(–8) large guide cells in 1 or occasionally 2 layers, with or without a ventral epidermis and with a usually weak dorsal epidermis; *laminal cells* unistratose or occasionally bistratose in patches or entirely, showing yellow or red KOH reaction; *upper cells* isodiametric, subquadrate to rounded-angular, sometimes rhombic or short-rectangular, thin- to evenly thick-walled, usually bulging on both surfaces, mostly bluntly uni- to pluripapillose, with low, simple or bifid, solid papillae, not strongly obscuring the lumina or smooth and translucent; *basal cells* short and firm or rectangular, smooth or weakly papillose. *Asexual reproduction* occasionally present, by green, spherical or ellipsoid, uni- or multicellular gemmae, borne in the leaf axils on the ventral side of the costa or on rhizoids, sometimes by fragile, caducous leaf apices. *Dioicous*. *Perichaetia* terminal; *inner perichaetial leaves* ovate to elongate-lanceolate, not or occasionally convolute-sheathing, often enlarged and commonly loosely sheathing in the proximal portion, with a rhomboidal areolation in the lower half. *Setae* elongate, smooth, yellowish to reddish-brown, dextrorse below, occasionally sinistrorse above. *Capsules* exserted, erect, straight, symmetrical, ellipsoid to oblong-cylindrical, smooth, leptodermous, yellowish- to reddish-brown; *annulus* none or present, 1–3-seriate, composed of hexagonal, often vesiculose cells, revoluble or deciduous in pieces; *operculum* short- to long-conic or conic-rostrate, with a straight or somewhat slanted beak; *exothecial cells* rectangular, thin-walled, becoming shorter, quadrate to transversely rectangular in a few rows at the mouth but not forming a colourful strip; *stomata* at the base of the urn, superficial, bicellular; *peristome* mostly present, sometimes rudimentary or absent, consisting of 16 teeth, free or joined in a low basal membrane, oblong, linear to elongate-triangular, entire or variously perforate to sometimes split nearly to the base into 2 filiform branches, straight or oblique and erect or long and weakly spiralled anticlockwise, hyaline, yellow, yellow-brown, reddish-yellow to red-orange, densely papillose to spiculose, occasionally spiral-striate. *Spores* globose, light brown, smooth or roughened. *Calyptra* cucullate, smooth, naked.

Although *Didymodon* is a Hedwigian genus, it has not been properly understood and consequently it has often been confused with related genera of pottialean mosses, especially with *Barbula*, with which it has usually been merged. The genus is difficult to characterise owing to the variability of all features. According to Saito (1975) the only distinct character separating *Didymodon* and *Barbula* is the form of axillary hairs: in *Barbula* they are entirely hyaline, whereas in *Didymodon* one or two short stalk cells are brown. Additionally, *Didymodon* is distinguished by quadrate to shortly oblong cells on the dorsal surface of the costa, although *D. japonicus* (Broth.) K.Saito with its linear abaxial cells is an exception. However, this feature is sometimes difficult to observe in many species because of the channelling of the apex of the leaves. These two characters are quite effective in distinguishing *Didymodon* from *Barbula* and they correlate well with leaf areolation. In general terms *Didymodon* has very low papillose cells with rounded and clearly outlined lumina, making the leaf lamina rather transparent, whereas *Barbula* has cells so densely papillose that the walls are obscured and the leaf lamina is markedly opaque; the leaves are mostly lanceolate to elongate-lanceolate. Moreover, *Didymodon* has mostly lanceolate to elongate-lanceolate leaves versus triangular to lingulate in *Barbula*; relative proximal and distal leaf cell differentiation vs. strong diversification; quadrate cells on the ventral side of the costa in the distal portion vs. elongate; and short and erect to weakly twisted peristome teeth vs. long and strongly twisted (Zander 1978*b*, 1981*b*).

Didymodon is extremely diverse in morphology and this prompted Steere (1947) to designate it as 'a synthetic genus composed of discordant elements originating in several genera of the Pottiaceae, agreeing among themselves only in their (supposedly!) untwisted peristome teeth'. This statement remains valid, and Zander (1993) subdivided *Didymodon* into four sections that represent four evolutionary lines in the genus and suggested that possibly they deserved recognition as independent genera. However, in a more recent evolutionary analysis of *Didymodon* Zander (1998*a*) recognised only two sections in the genus, sect. *Didymodon* and sect. *Fallaces* Steere. The heterogeneity of *Didymodon* is supported by its heterogeneous colour reaction in KOH. Among the unusual features that are encountered in some species, for instance in *D. vinealis* (Brid.) R.H.Zander, when immersed in 10% KOH solution is the bright to dark red coloration of the laminal cells, similar to that in species of *Bryoerythrophyllum*, whereas many other species exhibit a typically yellow reaction. The characters separating *Didymodon* from *Bryoerythrophyllum* are discussed under the latter genus. Together with other eight genera *Didymodon* belongs to the tribe Barbuleae which is very close to the tribe Bryoerythrophylleae and, in practice, they differ from each other in the shape of the dorsal stereid band of the costa, reniform in the Bryoerythrophylleae versus round to semicircular in the Barbuleae.

As presently circumscribed *Didymodon* is a large genus that consists of over 120 species distributed worldwide (Crosby *et al.* 2000). It exhibits greatest diversity in the warm temperate and subtropical regions in the Northern Hemisphere, but some species occur in cool-temperate and polar regions, although a few species are predominantly distributed in the Arctic. In the Southern Hemisphere the genus is not well represented and is poorly known taxonomically. In Australasia only six species are known (Fife 1995; Streimann & Klanzenga 2002), whereas in southern South America over 20 species have been recorded (Greene 1986) but most of them have never been critically examined; consequently the status of many species remains dubious. In the Subantarctic, the genus is practically absent and only recently *D. brachyphyllus* was encountered on South Georgia (Ochyra 2005) and Îles Kerguelen (Ochyra, unpublished). This species is also widely distributed throughout the Antarctic and is the only species of the genus in this biome (Ochyra & Zander 2002).

1. **Didymodon brachyphyllus** (Sull.) R.H.Zander Fig. 151

Didymodon brachyphyllus (Sull.) R.H.Zander, Phytologia 41: 24. 1978. ≡ *Barbula brachyphylla* Sull. *in* Sull. & Lesq., Musci Bor.-Amer.: No. 100^c. 1857 (II) and Sull. *in* Whipple & Ives, Expl. Railroad Mississippi Pacific Descr. Moss. Liverw. 4(2): 186. 1857 (VIII). ≡ *Didymodon vinealis* (Brid.) Zander var. *brachyphyllus* (Sull.) R.H.Zander, Cryptogamie Bryol. Lichénol. 2: 411. 1981. — Type citation: In California (legit Bigelow). [Lectotype (*vide* Zander & Ochyra 2001: p. 374): 'Barbula brachyphylla California. Bigelow. 1854' – FH-Sullivant!, isotypes: CU!, FH-Bartram!, and Musci Boreali-Americani ed. 1, No. 100^c – FH! and Musci Boreali-Americani Ed. 2: No. 135 – FH!).

Didymodon gelidus Cardot, Nat. Antarct. Exped. 1901–04, Nat. Hist. 4: 4, *pl. 1, fig. 1–11*. 1907. — Type citation: [Antarctica, southern Victoria Land] Granite Harbour, Jan. 20, 1902. [Lectotype (*selected here*): 'Herb. J. Cardot. Didymodon (?) gelidus Card. sp. nova. Terre Victoria: Granite Harbour, L. S. 77°. Expéd. de la Discovery, 20 janvier 1902' – PC!: isotypes: BM!, S!]. First synonymised by Zander & Ochyra (2001: p. 374).

Barbula byrdii E.B.Bartram, Ann. Missouri Bot. Gard. 25: 720. 1938. ≡ *Bryoerythrophyllum byrdii* (E.B.Bartram) R.H.Zander, Bull. Buffalo Soc. Nat. Sci. 32: 115. 1993. — Type citation: [Antarctica] Marie Byrd Land, Edsel Ford Range, Lichen Peak [76°55′S, 145°20′W], P. Siple & S. Corey G. [Lectotype (*selected here*): 'Byrd Antarctic Expedition II Mosses of Marie Byrd Land, Antarctica No. G Type Barbula Byrdii Bartr. sp. nov. Locality: Mt. 73 Collected by: Paul A. Siple Determined by: Edwin B. Bartram' – FH-Bartram!; isotypes: AAS!, BM!). First synonymised by Zander & Ochyra (2001: p. 375).

Plants small, slender, rather rigid, in compact tufts, dull, light to dark green above, brownish-green to brown below, usually with a reddish cast. *Stems* erect, to 2 cm tall, simple or sparingly branched, matted together below with light brown, lustrous, smooth rhizoids, in transverse section rounded, consisting of a large central strand, 3–4 layers of large, thin-walled and hyaline medullary cells and a (1–)2(–3)-stratose cortex of small cells with thick and reddish-brown walls; *axillary hairs* filiform, 4–6-celled with a short, brown basal cell and elongate hyaline upper cells. *Leaves* erect, appressed or closely imbricate to somewhat patent, concave and bent when dry giving the shoots a catenulate appearance, stiffly erecto-patent when moist, ovate to ovate-lanceolate, (0.5–)0.7–1.0(–1.2) mm long, (0.35–)0.4–0.5(–0.55) mm wide, widest below the middle, abruptly short-acuminate, broadly acute or blunt and often cucullate near the apex, without a smooth, pellucid apical cell, non-decurrent; *margins* unistratose, undifferentiated, entire to papillose-crenulate, plane below, weakly recurved above; *costa* yellowish to yellow-brownish, strong, 50–60 µm wide, slightly tapering upwards, subpercurrent and usually ending a few cells before the apex, with papillose, quadrate to short-rectangular adaxial costal cells in 4(–5) rows, in transverse section biconvex, with a median layer of 4–5 large guide cells, 2–3-stratose dorsal band of small substereid cells and 2 layers of 4–6 large ventral cells at mid-leaf; *laminal cells* unistratose throughout, undifferentiated at the margin, yellow to red-brown with KOH; *upper cells* irregularly rounded-quadrate to quadrate, 7–10 µm in diameter, with walls moderately thickened, weakly bulging on both sides, mostly obscured by simple or bifid papillae, 2–4 per lumen, sometimes nearly smooth; *basal cells* short-rectangular to quadrate, smooth, thin-walled, not differentiated at margins. *Asexual propagation* by multicellular gemmae borne in clusters arising from a short basal appendage in the leaf axils, consisting of a straight or weakly curving uniseriate row of 2–6 cells. [*Dioicous*. *Perichaetia* terminal; *perichaetial leaves* somewhat longer and wider than the cauline leaves, sheathing at the base. *Setae* 10–15 mm, straight, smooth, reddish-brown, twisted clockwise. *Capsules* exserted, erect, symmetric, cylindrical, 1.5–2.5 mm long, brown, smooth, leptodermous; *annulus* 1–2-seriate, revoluble or deciduous in pieces, composed of vesiculose cells; *operculum* long-rostrate, reaching half the length of the urn, obliquely beaked; *exothecial cells* rectangular, thin- to firm-walled; *peristome teeth* 16 arising from a very low basal membrane, red becoming pale with age, papillose, 200–250 µm long, erect, deeply bifid into linear prongs, twisted half-way around or scarcely so much. *Spores* spherical, 10–13 µm in diameter, pale brown, smooth. *Calyptra* cucullate, naked, smooth.]

Chromosome number: unknown.

Discussion. *Didymodon brachyphyllus* is a distinct species showing little variation. It may be confused in Antarctica with *Bryoerythrophyllum recurvirostrum* and the characters differentiating these species are discussed in detail under the latter. A problem with their misdetermination may occur in some areas in West Antarctica where both species occur together, namely Signy I., Livingston I., James Ross I., Léonie I. and Alexander I., although the plants of *B. recurvirostrum* are typically developed in this region and a characteristic pellucid and smooth mucro on the cauline leaves should warrant safe determination. In addition, the plants of *D. brachyphyllus* often produce multicellular gemmae which are absent in *B. recurvirostrum*. In habit, *D. brachyphyllus* may be similar to some species of *Schistidium* and *Ceratodon purpureus*, but its papillose laminal cells enable its recognition.

Didymodon brachyphyllus was described from California as *Barbula brachyphylla* and only recently it was transferred to *Didymodon* (Zander 1978*b*). Its diagnosis was published twice almost simultaneously by Sullivant (1856*a*,*b*), and the problem of priority of the publication of this species names was discussed in detail by Zander & Ochyra (2001). The species was long neglected and poorly known until Zander (1981*b*) considered it as a variety of *Didymodon vinealis*. However, it has subsequently been reinstated as a separate species (Zander 1994*b*; Zander & Ochyra 2001).

Interestingly, the first collection of *Didymodon brachyphyllus* was made in the Antarctic 11 years earlier than the type collection of the species was gathered in California. On 6 January 1843 J. D. Hooker collected this species on the small Cockburn I. off the north-east coast of the Antarctic Peninsula during his Antarctic voyage of 1839–1843 under the command of Captain James Ross. The material was determined (and illustrated) as *Tortula gracilis* Hook. & Grev. (Wilson & Hooker 1847). After 146 years, *D. brachyphyllus* was rediscovered on Cockburn I. in 1989 by Lewis Smith (1993*b*). This species was also discovered in the Antarctic in southern Victoria Land in 1902 by the *Discovery* Expedition, and Seppelt & Green (1998) pointed out that the plants may have been collected near Discovery Bluff and/or Botany Bay in Granite Harbour. This material was described by Cardot (1907*b*) as a distinct species, *Didymodon gelidus*, which was subsequently recorded from Deception I. in the maritime Antarctic (Dixon 1920). Later, several specimens of this species were discovered by the Second Byrd Antarctic Expedition on Edward VII Land, and Bartram (1938) described them as *Barbula byrdii*. However, Zander (1993) transferred this species to *Bryoerythrophyllum*, but Zander & Ochyra (2001) and Ochyra & Zander (2002) reduced both *Didymodon gelidus* and *Barbula byrdii* to synonymy with *Didymodon brachyphyllus* and, accordingly, this species was established as a new bipolar moss taxon.

Reproduction in Antarctica. Sterile, but occasionally female plants have been observed. The species produces multicellular axillary gemmae in great profusion.

Habitat. This largely xeric species thrives in Antarctica in a wide range of habitats, both in dry and damp situations. Most frequently it inhabits rock crevices and ledges, sheltered niches under overhangs, and volcanic ash, but also grows on pockets of soil amidst boulders, on silty soil, thriving on both acidic

→

Figure 151. *Didymodon brachyphyllus*. — **1**. Habit, wet. **2**. Habit, dry. **3**. Cross-section of stem. **4–6**. Axillary hairs. **7**. Propagules. **8–13**. Leaves. **14–15**. Leaf apices (papillae partly removed). **16**. Upper lamina cells at margin (papillae partly removed). **17**. Mid-leaf cells over ventral side of costa. **18**. Lamina cells in lower part at margin. **19–20**. Basal leaf cells. **21–24**. Cross-sections of leaves, a sequence from base to apex. — [1–6, 8–11, 14, 16–19, 21–24 from *RO 2286a/80*; 7, 12–13, 15, 20 from *Discovery Expedition s.n.*, 20.1.1902, (holotype of *Didymodon gelidus*), PC]. — Scale bars: a – 1 mm (1); b – 1 mm (2) & 100 µm (3); c – 100 µm (4–6, 14, 16–19); d – 100 µm (15, 20–24) and 200 µm (7); e – 1 mm (8–13). (Reproduced from *Journal of Bryology*, 24: p. 36, 2002, with permission.)

8
9
11
12
13
10
17
4
5
6
7
14
15
2
1
18
3
16
19
24
22
23
21
20
a
b
c
d
e

and base-rich rocks. *Didymodon brachyphyllus* is a component of various moss assemblages and frequently occurs with *Distichium capillaceum*, *Bartramia patens*, *Syntrichia saxicola*, *Pohlia cruda*, *Platydictya jungermannioides*, *Encalypta rhaptocarpa* and *Brachythecium austrosalebrosum*.

World range. *Didymodon brachyphyllus* is a bipolar species (Fig. 152, inset). In the Northern Hemisphere it is arctic-alpine in distribution (Zander & Ochyra 2001) with its main centre of occurrence in western North America, from Oregon to Baja California Norte in Mexico east to Colorado and New Mexico, with some isolated stations in British Columbia and in south-west Alaska on the Alaska Peninsula. Additionally, it is known from West Greenland (Mogensen & Zander 1999), Iceland (Jóhannsson 2003) and central Mexico (Zander 1994*b*). In the Southern Hemisphere the species is known only from the Antarctic and from a single station on South Georgia (Ochyra 2005) and Îles Kerguelen (Ochyra, unpublished).

Distribution in Antarctica. *Didymodon brachyphyllus* is a widely distributed but scattered pan-Antarctic species (Figs 152–153). In the West Antarctic it occurs on the South Orkney and South Shetland Is., along the west coast of the Antarctic Peninsula from the Graham Coast through Marguerite Bay to Alexander I. reaching its southernmost station at lat. 71°31′S. On the east coast of the Antarctic Peninsula the species is relatively frequent on James Ross, Vega and Cockburn Islands, and on the continent it occurs on Edward VII Land and southern Victoria Land where it reaches its southernmost occurrence at lat. *ca* 77°02′S at Cape Roberts. Its atitudinal range extends from sea level to 540 m on James Ross I., although most of its stations are at lower elevations, below 150 m.

Specimens examined. WEST ANTARCTICA. SOUTH ORKNEY IS. **MATTHEWS I.** The Divide, *ca* 35 m, *RILS 166A–B* (AAS, KRAM). **LAURIE I.** Mackenzie Peninsula, *Bellisio 11699* (KRAM, NY). **SIGNY I.** Jebsen Pt., *ca* 65 m, *RILS 426* (AAS, KRAM); Foca Pt., *ca* 65 m, *RILS 501* (AAS, KRAM); north of Spindrift Rocks, *ca* 700 m, *RILS 8062B* (AAS, KRAM); between North Pt. and coast opposite Spindrift Rocks, *ca* 65 m, *REL 1135 & 1136* (AAS, KRAM); North Pt., 50 m, *RILS 655* (AAS, KRAM); bluffs to S of North Pt., 50 m, *RILS 8056B* (AAS, KRAM); between North Pt. and Robin Peak, 100 m, *RILS 3169* (AAS, KRAM); Robin Peak, 25 m, *RILS 1800 & 8147* (AAS, KRAM); west side of Stygian Cove, 35–65 m, *RILS 491, 492 & 661* (AAS, KRAM); north side of Observation Bluff, *ca* 15 m, *RILS 17B* (AAS, KRAM); Limestone Valley, Snow Hills, 35 m, *RILS 8038* (AAS, KRAM).

SOUTH SHETLAND IS. **KING GEORGE I.** ***Admiralty Bay***: Klekowski Crag, 110 m, *RO 2286A/80* (KRAM). ***Fildes Peninsula***: Two Summit I., *Li S16* (AAS); Great Wall Station, 25 m, *Schulz 3* (KRAM). **LIVINGSTON I.** ***Hurd Peninsula***: Juan Carlos I Station region, Feb 1992, *Schulz s.n.* (KRAM). **DECEPTION I.** Whalers Bay, *Robins 447* (BM), *Bennett 1874b* (BM) and *RILS 75B* (AAS, KRAM); Cathedral Crags, *ca* 50 m, *Leech 165A* (KRAM, NY), 65 m, *RILS 70B* (AAS, KRAM) and 85 m, *RILS 79* (AAS, KRAM); Neptunes Windows, *ca* 65 m, *REL & RILS 868* (AAS, KRAM); Collins Pt., 21 Jan 1961, *R. E. & T. S. Leech s.n.* (FH) and *RILS 6766* (AAS, KRAM); Wensleydale Beacon, 2–15 m, *RILS 5599, 5600C, 5601B & 6619* (AAS, KRAM); Whalers Bay, E of whaling station, 10–25 m, *RILS 6553 & 6555* (AAS, KRAM); Whalers Bay, Ronald Hill, 25 m, *RILS 5778B & 5781B* (AAS, KRAM), 35 m, *BJT 252B* (AAS, KRAM) and 40 m, *RILS 3772* (AAS, KRAM) and *RO 2785/80* (KRAM); Pendulum Cove, 2–4 m, *RILS 3279A & 5746A* (AAS, KRAM); Mooring Post, 10 m, *RO 2790/80* (KRAM);

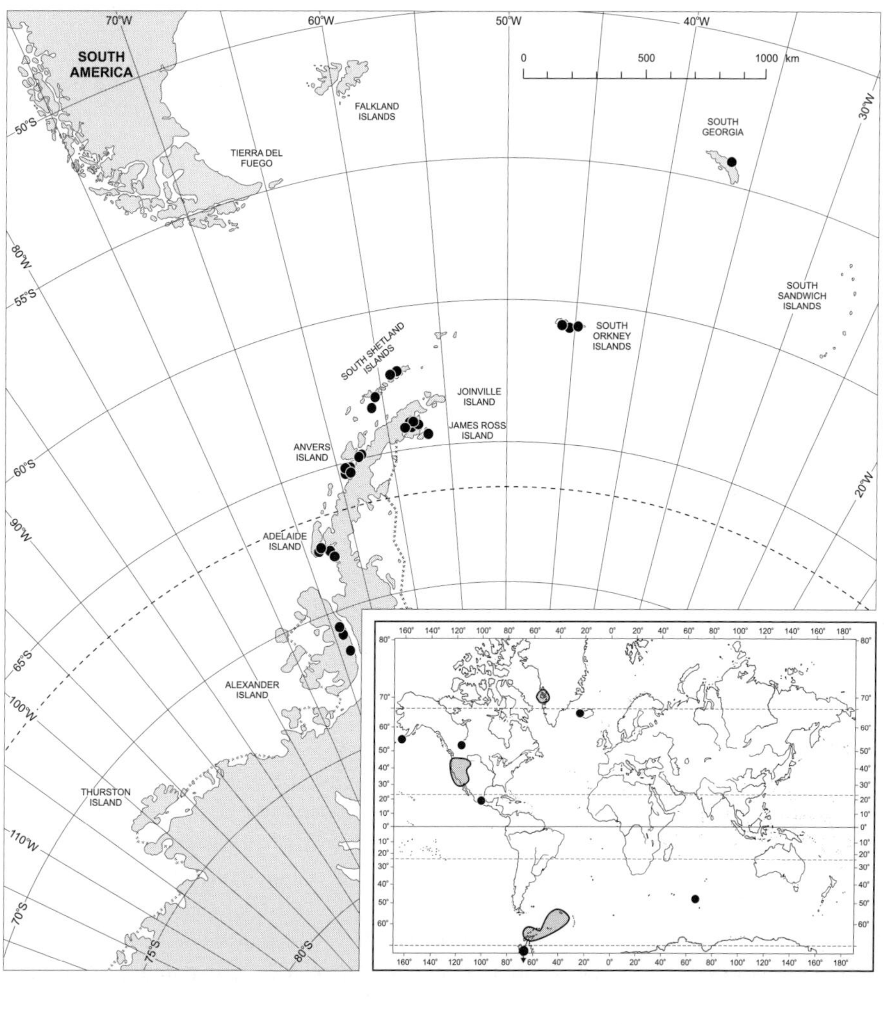

Figure 152. Distribution of *Didymodon brachyphyllus* in the Antarctic Peninsula and Fuegian regions. Inset: global distribution (localities in Victoria Land and Marie Byrd Land are not indicated).

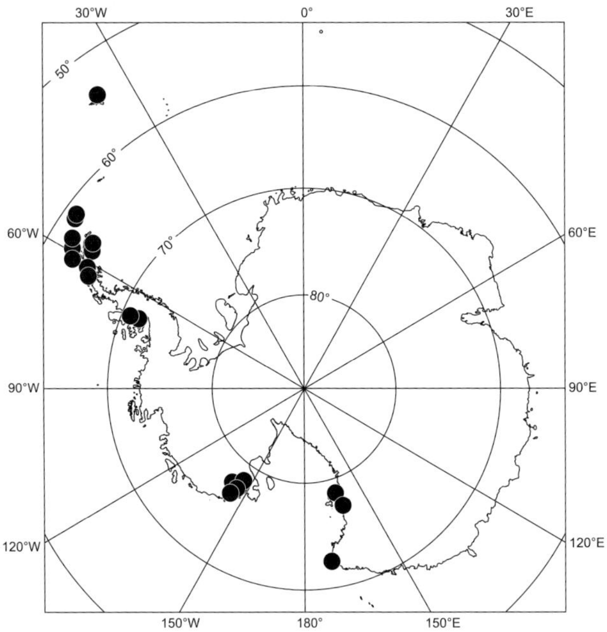

Figure 153. Distribution of *Didymodon brachyphyllus* in the Antarctic and Subantarctic.

1.5 km N of summit of Mt. Pond, 500 m, *DM 41* (AAS, KRAM); E face of N summit of Mt. Pond, 480 m, *RILS 5820B & 5827D* (AAS, KRAM); ridge 1.5 km NNW of Mt. Pond, 410–525 m, *RILS 6880, 6881, 6883, 6889, 6890, 6898, 6902A & 6905* (AAS, KRAM); Telefon Ridge, 60–150 m, *RILS 5796E & 5811B* (AAS, KRAM); between Stonethrow Ridge and Telefon Ridge, 45–230 m, *RILS 5734E & 5807G* (AAS, KRAM).

WEST ANTARCTIC PENINSULA. **Danco Coast.** Cuverville I., north side, 10–20 m, *RILS 11946B & 11952* (AAS, KRAM); Coughtrey Peninsula, Paradise Harbour, 100 m, *RILS 11622A* (AAS, KRAM). **Graham Coast.** Edge Hill, *ca* 15 m, *RWMC 650* (AAS, KRAM); Cape Tuxen, 10–15 m, *RWMC 676 & 846* (AAS, KRAM). **Wilhelm Archipelago.** ***Penola Strait***: Petermann I., *ca* 15 m, *RWMC 725 & 795* (AAS, KRAM) and *REL 1349* (AAS, KRAM). ***Argentine Is.***: Corner Is., 10 m. *RWMC 570* (AAS, KRAM); Galindez I., overlooking Stella Creek, 5 m, *RWMC 466B* (AAS, KRAM); same island, Meek Channel, *ca* 15 m, *RWMC 462* (AAS, KRAM); Skua I., north coast, *ca* 8 m, *REL 1335* (AAS, KRAM); Skua Creek, 10 m, *RWMC 455* (AAS, KRAM). ***Collins Bay***: Darboux I., 10–15 m, *RILS 3313* (AAS, KRAM). **Loubet Coast.** ***Adelaide I.***: Reptile Ridge, 350–500 m, *RILS 1889* (AAS, KRAM). ***Léonie Is.***: Léonie I., 25–30 m, *RILS 8927B & 8940* (AAS, KRAM). **Fallières Coast.** ***Marguerite Bay***: Camp Pt., 10–15 m, *RILS 4630 & 4675* (AAS, KRAM); Millerand I., 5–15 m, *RILS 3762B* (AAS, KRAM). **George VI Sound.** ***Alexander I.***: Succession Clifffs, *ca* 300 m, *BJT 534* (AAS, KRAM); Waitabit Cliffs, *ca* 165 m, *BJT 512* (AAS, KRAM).

EAST ANTARCTIC PENINSULA. **James Ross Island group.** ***Vega I.***: N end of False Island Pt., 25–40 m, *RILS 7842* (AAS, KRAM). ***James Ross I.***: Brandy Bay, *RILS 5303A* (AAS, KRAM); E side of Lachman Crags, 3–50 m, *RILS 7328b, 7329, 7330, 7380B, 7381C, 7384 & 7385B* (AAS, KRAM); N of Andreassen Pt., 10 m, *RILS 7607* (AAS, KRAM); Fortress Hill, 150 m, *RILS 7751* (AAS, KRAM); valley SW of Fortress Hill, 25 m, *RILS 7811* (AAS, KRAM); valley S of Martha Cove, 50 m, *RILS 7463* (AAS, KRAM); hill S of Martha Cove, 75 m, *RILS 7435 & 7437* (AAS, KRAM) and 150 m, *RILS 7542B–C* (AAS, KRAM); Terrapin Hill, 85 m, *RILS 7744 & 7745B* (AAS, KRAM) and 540 m, *RILS 7714* (AAS, KRAM); Humps I., 2–50 m, *RILS 7399 & 7401* (AAS, KRAM). ***Cockburn I.***: without closer locality data, *Hooker 3* (BM); between Adélie colony and foot of cliff, 120 m, *RILS 7914* (AAS, KRAM).

VICTORIA LAND. **Borchgrevink Coast.** Hallett Base, Crater Hill, *New Zealand Geol. Surv. Party Antarctc. Exped. 1957–58* (*Allison 7119 & 7121*) (AAS, KRAM). **Scott Coast.** Granite Harbour, 20 Jan 1902, *Discovery Expedition s.n.* (BM, PC, S – type of *Didymodon gelidus*); Cape Roberts, *Steere & Greene 65020* (AAS, KRAM).

MARIE BYRD LAND. **Saunders Coast.** ***Edsel Ford Ranges***: Lichen Peak, *Siple & Corey G* (FH – type of *Barbula byrdii*); Skua Gull Peak, *Siple & Corey 1A, 4, 8 & 99A* (FH, NY); Mt. Donald Woodward, *Siple, Wade, Corey & Standcliff 97* (FH, NY); Mt. Marujupu, *Perkins 129A* (FH, NY).

28. STEGONIA Venturi

Stegonia Venturi, Rev. Bryol. 10: 96. 1883. ≡ *Didymodon* Hedw. [unranked] *Stegonia* (Venturi) Kindb., Eur. N. Amer. Bryin. 2: 272. 1898. ≡ *Pottia* Fürnr. sect. *Stegonia* (Venturi) Müll.Hal., Gen. Musc. Frond.: 386. 1900. ≡ *Didymodon* subgen. *Stegonia* (Venturi) Kindb., Rev. Bryol. 36: 117. 1909. — Type: *Stegonia latifolia* (Schwägr.) Venturi *ex* Broth. (*Weissia latifolia* Schwägr.).

Hyalophyllum Lindb. *ex* Warnst., Hedwigia 53: 284. 1913, *nom. illeg. incl. gen. prior.*

Plants small, bulbiform, forming loose or dense patches or clusters, sometimes gregarious, pale green, silvery-green, white to whitish-green above, brownish below. *Stems* short, unbranched, scarcely radiculose at the base with brownish, smooth, sparingly branched rhizoids, in transverse section lacking a cortex and hyalodermis, with a central strand and more or less uniform tissue of large, thin-walled, hyaline cells; *axillary hairs* filiform, hyaline throughout, consisting of 4–7 cells, with 2 short stalk cells and elongate distal cells. *Leaves* imbricate and appressed when dry, little altered when moist, loosely involute above and weakly spreading below, nearly circular to obovate or broadly obovate-spathulate, sometimes broadly elliptical, non-decurrent, not differentiated at the base, deeply concave, usually upper ⅓ to ¼ of the leaf hyaline, rounded to broadly acute at the apex, mostly with a short, acute or obtuse apiculus or terminated with a long, flexuose, flat, faintly denticulate to subentire hyaline hair-point; *margins* plane, unistratose, usually broadly involute, faintly serrulate or notched above to nearly entire; *costa* single, thin but well defined, usually ending some distance below the apex or percurrent to long excurrent as a hyaline hair-point, with short-rectangular, smooth and bulging superficial cells on the ventral surface in 3–4 rows across the costa, rectangular and smooth dorsally, in transverse section round, with a distinct dorsal stereid band, lacking dorsally an epidermis in the distal portion and with an epidermis in the proximal part, with 2 large guide cells in 1 row and a distinct ventral epidermis forming a bulging ridge; *laminal cells* unistratose throughout, smooth, showing yellow colour in KOH; *upper cells* rhombic, rhomboidal or hexagonal, often hyaline in upper ⅓ to ¼ of the lamina, thin-walled medially and becoming thicker-walled towards the apex, superficially thick-walled on the dorsal surface and thin-walled and bulging on the ventral side; *basal cells* differentiated, rectangular, occasionally bulging, thin-walled. *Monoicous* (*paroicous* or *autoicous*). *Perichaetia* terminal; *perichaetial leaves* little differentiated. *Setae* very short to elongate, singly per perichaetium, rarely geminate, straight, light yellow-brown to reddish-brown, smooth, twisted to the right below and occasionally to the left above. *Capsules* immersed and cleistocarpous, ovoid and shortly apiculate or exserted and stegocarpous, ellipsoid to cylindrical, straight or somewhat bent, brown, smooth, leptodermous; *annulus* persistent, biseriate, composed of thick-walled, vesiculose cells; *operculum* conical, stout, obliquely rostrate; *exothecial cells* oblong-hexagonal, thin-walled, becoming smaller, isodiametric in 2–5 rows at the mouth and forming a dark brown band; *stomata* superficial in the base of the urn, large, bicellular, with elliptical pori; *peristome* well developed, rudimentary to absent in stegocarpous capsules, consisting of 16 teeth, erect and straight, lingulate to lanceolate, entire and perforate to irregularly cleft into 2 or 3 prongs, with a few articulations, pale red, yellowish or orange, densely spiculose. *Spores* spherical, light to dark brown, papillose. *Calyptra* cucullate, smooth, naked.

Stegonia is a small oligotypic genus which is distinct in its bulbiform habit; nearly circular to obovate or broadly obovate-spathulate, cochleariform leaves lacking differentiated bases; incurved leaf margins; smooth laminal cells which are rhomboidal to hexagonal, hyaline and thick-walled in the upper third; very thick-walled dorsal and thin-walled and bulging ventral laminal cells as clearly seen in a leaf cross-section; and anatomical structure of the costa which consists of a dorsal stereid band, two enlarged guide cells and a strongly bulging ventral epidermis. These characters, especially the incurved leaf margins and the laminal cells ventrally bulging and dorsally flat walls, confirm the placement of *Stegonia* in subfam. Pottioideae and tribe Hyophileae.

Stegonia was established as a separate genus by Venturi (1883) for a species that had long been known as *Pottia latifolia* (Schwägr.) Müll.Hal., but formally this species was given a name under *Stegonia* 40 years later (Brotherus 1923) and since then it has gained wide acceptance. Warnstorf (1913) established the genus *Hyalophyllum* to accommodate *P. latifolia*, but this name is superfluous and illegitimate because it comprises the type of an earlier generic name. *Stegonia* is a small genus of three species. Two of these, *S. hyalinotricha* (Cardot & Thér.) R.H.Zander and *S. mouretii* (Corb.) Broth., are species endemic to small areas of, respectively, California (Norris & Shevock 2004) and Mexico (Zander 1994*c* as *Phascum hyalinotrichum*), and Morocco (Corbière 1913 as *Pottia mouretii*; Jelenc 1955, 1967; Ros *et al.* 1999). The third species, *S. latifolia*, is a pan-Holarctic species, but also discovered recently from the Antarctic (Ochyra & Lewis Smith 1996), thus establishing *Stegonia* as a bipolar moss genus. A fourth species listed in the genus in *Index muscorum* (Wijk *et al.* 1967), *S. lorentzii* (Müll.Hal.) I.Hagen from the Andes of Argentina, is a member of the monotypic genus *Saitobryum* R.H.Zander (Ochyra 1999*h*).

1. **Stegonia latifolia** (Schwägr.) Venturi *ex* Broth. Fig. 154

Stegonia latifolia (Schwägr.) Venturi *ex* Broth., Fl. Fennica 1: 145. 1923. ≡ *Weissia latifolia* Schwägr. *in* Schult., Reise Glockner 2: 362. 1804. ≡ *Grimmia latifolia* (Schwägr.) F.Weber & D.Mohr, Bot. Taschenb.: 147. 1807. ≡ *Anacalypta latifolia* (Schwägr.) Fürnr., Flora 12(2) Erg.: 25. 1829. ≡ *Dermatodon latifolius* (Schwägr.) Huebener, Musc. Germ.: 112. 1833. ≡ *Pottia latifolia* (Schwägr.) Müll.Hal., Syn. Musc. Frond. 1: 549. 1849. — TYPE CITATION: [Austria] Heiligenbluter Tauern. [Lectotype (*selected here*) {or perhaps holotype}: 'Herb. Duby junct. Herb. Boissier 1886 Weissia latifolia Hohes Thor Heilblutes 1800 (Hohes Thor, Heiligenbluts 1800. Tauern) Scripsit: Schwaegrichen. teste. G. Colomb-Duplan Pottia latifolia (Schwgr.) C. M. Paris Index bryologicus, 1, p. 1025' – G-Duby/Boissier!].

Plants very small, simple, bud-like in loose or dense, pale green to silvery tufts. *Stems* simple, unbranched, to 2.5 mm long, with clusters of light brown, smooth and branched rhizoids at the base, in transverse section rounded, composed of 5–6 layers of relatively large, thin-walled hyaline cells, lacking sclerodermis or cortical cells with somewhat brownish walls but otherwise similar to internal cells, and with a distinct, medium-sized central strand; *axillary hairs* numerous, hyaline throughout, composed of 4–7 cells. *Leaves* closely imbricate when dry, unchanged to weakly spreading when moist, very concave, obovate, broadly ovate to nearly circular, 1.5–2.0 mm long, rounded to broadly acute at the apex, without a hyaline hair-point; *margins* plane, incurved, unistratose, entire or faintly to distinctly serrulate above due to protruding cell ends; *laminal cells* smooth, unistratose throughout, frequently hyaline in upper ⅓–¼ of the leaf and giving the plants a silvery appearance; *upper cells* rhomboidal to narrowly rectangular, usually smaller, rounded-quadrate and thicker-walled towards the margin and apex, 8–22 µm long, 8–10 µm wide, in transverse section with dorsal superficial cell walls thick and flat throughout and ventral superficial cell walls thin and bulging; *basal cells* oblong-hexagonal to rectangular, occasionally swollen, thin-walled, 60–80(–100) µm long, (10–)15–20(–25) µm wide, becoming shorter toward the mid-leaf; *costa* weak, well demarcated from the laminal cells, pale yellowish-brown, ending below the apex, to 40 µm wide at the base, in transverse section biconvex and round in outline, with a dorsal stereid band, lacking an epidermis in the distal portion and with an epidermis in the proximal part, 2 large guide cells and a distinct ventral epidermis composed of 3–4 rows of large,

⟶

Figure 154. *Stegonia latifolia.* — **1**. Immature plant with very young sporophyte. **2**. Mature plant. **3**. Cross-section of stem. **4–5**. Axillary hairs. **6–11**. Leaves. **12**. Cells at leaf apex. **13**. Upper lamina cells at margin. **14**. Mid-leaf cells at margin. **15**. Mid-leaf cells. **16**. Basal leaf cells. **17–18**. Cross-sections of leaf. **19**. Calyptra. **20**. Exothecial cells in middle of urn. **21**. Stoma and basal exothecial cells. **22**. Peristome, spores and exothecial cells at mouth. — [1, 3–18 from *RO 1161/80*; 2, 19–22 *Lisowski 1806*]. — Scale bars: a – 1 mm (1–2, 19) and 0.5 mm (6–11); b – 100 µm (3–5, 12–18, 20–22). (Reproduced from *Fragmenta Floristica et Geobotanica*, 41: p. 988, 1996, with permission.)

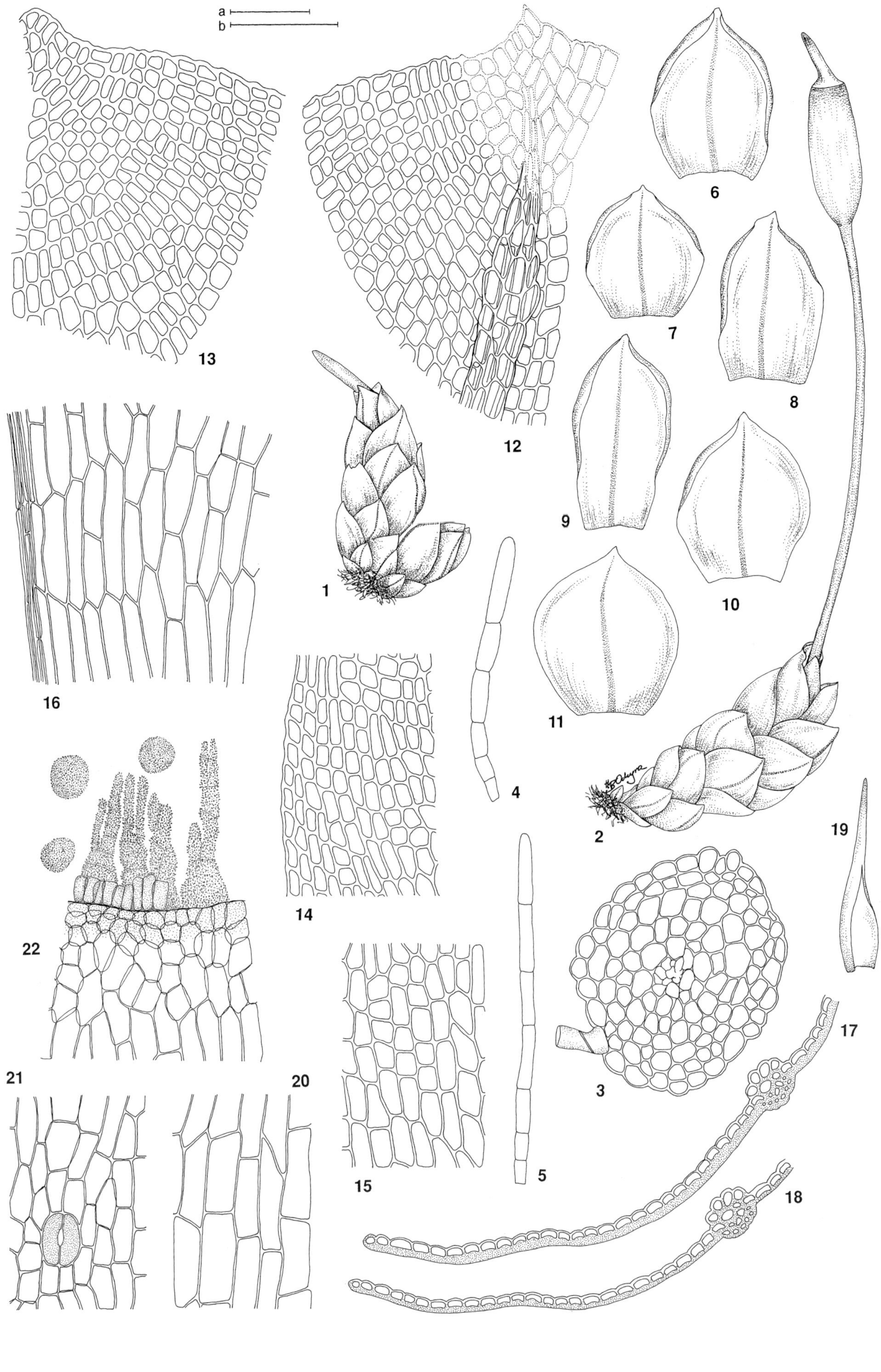
a
b
1
2
3
4
5
6
7
8
9
10
11
12
13
14
15
16
17
18
19
20
21
22

short-rectangular, smooth, bulging cells forming a swollen ridge over the costa. *Monoicous*. *Perichaetia* terminal; *perichaetial leaves* not much different from the cauline leaves. *Setae* singly per perichaetium, 6–10 mm long, straight, fairly stout, smooth, dextrorse below, light yellow-brown, becoming reddish-brown with age. *Capsules* exserted, ellipsoid to short-cylindrical, 1.2–1.5 mm long, 0.5–0.6 mm wide, straight or somewhat curved, smooth, castaneous, somewhat lustrous, leptodermous; *annulus* persistent, biseriate, composed of thick-walled, vesiculose cells; *operculum* conical, stout, obliquely rostrate; *exothecial cells* rectangular to oblong-hexagonal, thin-walled, 70–125 µm long, 20–40 µm wide, in regular longitudinal rows, becoming smaller, isodiametric in 2–5 rows at the mouth and forming a darker brownish strip; *stomata* superficial in the base of the urn, large, bicellular, with elliptical pori; *peristome* absent, rudimentary or consisting of 16 teeth, erect, straight, lanceolate, 120–160 µm long, irregularly divided into 2–3 branches, with a few articulations, pale yellowish-brown to brownish, densely papillose. *Spores* spherical, light to dark brown, 30–40 µm in diameter, densely papillose. *Calyptra* cucullate, smooth, naked.

Chromosome number: unknown in the Antarctic and Subantarctic; $n = 26$ – a single record from Canada (Anderson & Crum 1959).

Discussion. *Stegonia latifolia* is readily recognised by its gemmiform habit, silvery to pale green colour of the plants and nearly circular and cochleariform leaves. Microscopically, the species has distinct leaf areolation of entirely smooth, thick-walled, rhombic, rhomboidal to oblong-hexagonal, and usually hyaline upper laminal cells which are distinctly bulging ventrally and very thick-walled on the dorsal surface as evidenced by leaf cross-sections. Additionally, the anatomical structure of the costa is distinct by having a protuberant ventral epidermis, two large guide cells and dorsal stereid band lacking an epidermis in the distal portion.

Stegonia latifolia is a variable species with regard to the excurrence of the costa and the development of the peristome. In the Northern Hemisphere, the costa is often long excurrent as a hyaline awn and such populations are often recognised as a separate taxon, var. *pilifera* (Brid.) Broth., which occasionally is given the status of a species in its own right (Crum & Anderson 1989; Anderson *et al.* 1990), although the value of this character is disputable. All Antarctic populations are consistently epilose and the costa always vanishes several cells below the apiculus. Likewise, the species exhibits a broad range of variability of the peristome, and apart from normal ochyrostomous capsules with fully developed or variously reduced peristomes, gymnostomous capsules rarely occur which entirely lack a peristome. The Antarctic specimens from Signy I. and Camp Pt. produce mature sporophytes with gymnostomous capsules, but this character state cannot be verified in the plants from King George I. since they have only immature sporophytes. The gymnostomous capsules are a diagnostic trait of *S. mouretii* but this poorly known species is distinct in its subsessile capsules on a short seta, to about 1.5 mm, and small spores, 22–25 µm in diameter. Besides the lack of the peristome, the Antarctic plants perfectly match the northern plants of *S. latifolia* including the type from the Alps of Austria (Schwägrichen 1804).

Stegonia latifolia was first reported from the Antarctic by Lewis Smith (1972) from Signy I. but Lightowlers (1987) redetermined the material as *Pterygoneurum ovatum*. These species may be easily confused, especially in the Antarctic, where *S. latifolia* produces gymnostomous capsules and this state is typical of *P. ovatum*. Additionally, both species share the leaf shape and areolation as well as the anatomical structure of the costa and leaf lamina. *P. ovatum* may be distinguished by the presence of longitudinally elongated lamellae covering a major part of the ventral side of the costa, whereas *S. latifolia* lacks any ventral costal outgrowths. Additionally, the leaves of *P. ovatum* are long piliferous and the spores are much smaller, only 22–23 µm in diameter.

In the Antarctic, *Stegonia latifolia* can also be mistaken for some phenotypes of *Hennediella heimii* having the ovate and imbricate leaves. However, the leaves of this species are distinctly short- to long-acuminate, much smaller (usually about 1 mm long) and often keeled. The distal laminal cells in *H. heimii* are mostly papillose, although in some populations they are nearly smooth, but in this species 2–4 marginal rows of cells are somewhat elongate and form a more or less distinct, yellowish border. The safest character separating both species is the cross-section of the leaf cells: in *H. heimii* the outer cell walls are thin and uniform in the thickness, whereas in *S. latifolia* the dorsal superficial cell walls are much thicker than the ventral ones.

The only other moss species in the Antarctic with leaves of comparable size and shape to those in *Stegonia latifolia* are *Bryum orbiculatifolium* and *B. argenteum* var. *muticum*, the latter additionally being a silvery to whitish plant. However, both species are relatively large plants with elongate stems, usually more than 1 cm long. In doubtful cases, the cross-section of the leaves should separate these genera since in *Bryum* both ventral and dorsal cell walls are equally thin and never thickened. Moreover, the axillary hairs are very short, usually 3-celled with a pigmented basal cell.

Reproduction in Antarctica. Most populations are found in fine fruiting condition and sporophytes are usually produced in abundance.

Habitat. A pioneer colonist of bare silt and fine, fairly calcareous soil amongst stones, and on soil-covered ledges, growing in dry or moist sites which are often influenced by Antarctic terns and Dominican gulls which nest nearby. On Signy I. *Stegonia latifolia* is a constituent of a characteristic ephemeral community of short acrocarpous mosses, such as *Encalypta rhaptocarpa*, *Hennediella antarctica*, *Pterygoneurum ovatum* and *Bryum argenteum* var. *muticum*. These appear in early summer, with all but *Bryum* usually producing sporophytes in profusion, and then apparently disappearing during the winter. Their minute shoots often become buried by fine clay during the frequent freezing and thawing of the soil surface at the onset of winter and during the spring melt (Lewis Smith 1985). On King George I. *Stegonia latifolia* forms small, monospecific patches which grow in association, but not intermingled, with *Encalypta rhaptocarpa*, *Syntrichia magellanica*, *Schistidium antarctici*, *Hennediella antarctica*, *Bartramia patens*, *Polytrichastrum alpinum*, *Andreaea gainii*, *A. regularis* and *Hymenoloma grimmiaceum*.

World range. *Stegonia latifolia* is a rare bipolar species having a strongly discontinuous, arctic-alpine range in the Northern Hemisphere (Fig. 155, inset). In Europe, it has a typically bicentric distribution, being widely distributed in the north throughout the Scandes (Mårtensson 1956; Nyholm 1990), scattered in Scotland (Blockeel 1992*b*), Iceland (Jóhannsson 1992) and Spitsbergen (Kuc 1973*b*) as well as in the Russian Arctic including Novaya Zemlya, Franz Joseph Land, Vaygach I. and the North Ural Mountains (Abramova *et al.* 1961; Afonina

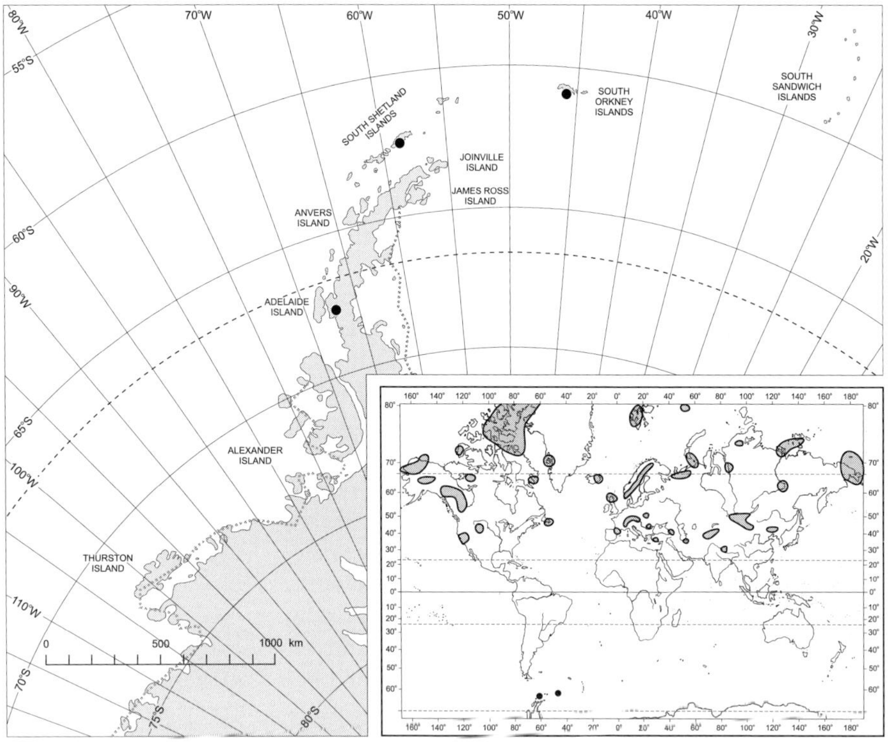

Figure 155. Distribution of *Stegonia latifolia* in Antarctica. Inset: global distribution.

& Czernyadjeva 1995) and then recurring in mountains in southern and central parts of the continent including the Pyrenees (Casas Sicart 1986), the Alps (Amann 1918; Duell & Meininger 1989; Grimms 1999), the Western (Chałubiński 1886; Šmarda 1948) and Southern Carpathians (Mohan 1998), Turkey (*Erdag 847*, KRAM) and the Caucasus (Manakyan 1989). In Asia *S. latifolia* is widely scattered throughout the Russian Arctic, from the Taymyr Peninsula to Chukotka (Abramova *et al.* 1961; Stepanova 1986; Afonina & Czernyadjeva 1995; Afonina 2004) and, disjunctly, in the Elburz Mountains of North Iran (Schiffner 1908), the Hindukush Mountains of NW Pakistan (Størmer 1954), Middle Asia (Pamir, Tien-Shan, Pamiro-Alai) (Mamatkulov 1990), the Altai Mountains in Central Siberia (Bardunov 1974; Ignatov 1994) and the Khangai and Mongolian Altai Mountains in Mongolia (Abramova & Abramov 1983; Tsegmed 2001*a*,*b*), reaching its highest elevation of *ca* 4300 m in Pamir (Brotherus 1906*a*). Additionally, the species is known from various provinces of China (Li *et al.* 2001). In North America, *S. latifolia* is a rare and bicentric species which is widely scattered throughout the Canadian Arctic Archipelago and in West Greenland from Baffin I. (Steere 1946; Brassard *et al.* 1979) to northern Ellesmere I. (Brassard 1971). On mainland North America it is common in Arctic Alaska (Steere 1978), scattered in southern and Central Alaska (Persson 1963), the Yukon Territory (Vitt 1974*b*) and Northwest Territories (Steere 1977), extending through the Canadian Rocky Mountains (Schofield 1976) to Wyoming and Colorado (Lawton 1971) and southern California (Norris & Shevock 2004). In eastern North America *S. latifolia* is known only from Newfoundland (Tuomikoski *et al.* 1973). Outside the Holarctic, the species is known only from the maritime Antarctic (Ochyra & Lewis Smith 1996).

Distribution in Antarctica. An apparently very rare and localised species, forming small populations in three widely separated centres in the maritime Antarctic, namely on Signy I. in the South Orkney Is., King George I. in the South Shetland Is. (Myrcha *et al.* 1991; Ochyra 1998*a*) and on the Fallières Coast on the western coast of the Antarctic Peninsula in Marguerite Bay at lat. 67°58′ S (Fig. 155). The species grows at low elevations from 15 to 125 m. Ochyra & Lewis Smith (1996) stated that the site of *Stegonia latifolia* described by Lewis Smith (1985) has been destroyed by the increasing population of fur seals which have affected much of lowland Signy I., including the lower part of the Orwell Glacier moraine. However, the species was rediscovered in 2000 at the north end of this moraine, at a slightly higher elevation of 30 m, and this finding supports the assumption that the species is firmly established in the Antarctic and effectively reproduces and spreads on Signy I.

Specimens examined. WEST ANTARCTICA. SOUTH ORKNEY IS. **SIGNY I.** Moraine Valley, below Orwell Glacier, 15 m, *RILS 546* (AAS, BM, KRAM) & *3155A* (AAS, KRAM); same locality, north end of Orwell Glacier moraine, 30 m, *RILS 10825A* (AAS, KRAM).

SOUTH SHETLAND IS. **KING GEORGE I.** ***Bransfield Strait***: Blue Dyke, 125 m, *RO 1161/80* (KRAM).

WEST ANTARCTIC PENINSULA. **FALLIÈRES COAST.** ***Marguerite Bay***: Camp Pt., 10–20 m, *RILS 4684* (AAS, KRAM).

29. PTERYGONEURUM JUR., *nom. et orth. cons.*

Pterygoneurum Jur., Laubm-Fl. Oesterr.-Ung: 95. 1882 ['Pterigoneurum']. ≡ *Pottia* Fürnr. subgen. *Pterygoneurum* (Jur.) Boulay, Musc. France 1: 469. 1884. ≡ *Tortula* Hedw. [unranked] *Pterygoneurum* (Jur.) Braithw., Brit. Moss Fl. 1: 206. 1885. ≡ *Tortula* sect. *Pterygoneurum* (Jur.) Dixon, Stud. Handb. Brit. Mosses: 176. 1896. — Type: *Pterygoneurum cavifolium* Jur., *nom. illeg.* (*Pottia cavifolia* Fürnr., *nom. illeg.*) [= *Pterygoneurum ovatum* (Hedw.) Jur. (*Gymnostomum ovatum* Hedw.)].

Fiedleria Rabenh., Flora 31: 252. 1848, *hom. illeg.* [*non* Rchb. 1844]. — Type: *Fiedleria subsessilis* (Brid.) Rabenh. (*Gymnostomum subsessile* Brid.).

Pharomitrium Schimp., Syn. Musc. Eur.: 120. 1860, *nom. rej.* — Type: *Pharomitrium subsessile* (Brid.) Schimp. (*Gymnostomum subsessile* Brid.).

Plants small, often bud-like, loosely to densely caespitose or gregarious, greyish-green to green above, pale brown below. *Stems* short to very short, simple or occasionally sparingly branched, often buried in soil and densely coated with pale brown, smooth, branched rhizoids, in transverse section obtusely angled, rounded-pentagonal, comprising a central strand and a more or less homogeneous tissue of large, hyaline to yellowish-hyaline, thin-walled cells, without differentiated cortex and hyalodermis; *axillary hairs* filiform, short, 4–7-celled, hyaline throughout or with 1–2 brown stalk cells. *Leaves* imbricate, erect to erect-incurved when dry, little altered when moist, ovate, oblong-ovate, elliptic to ligulate, obtuse to acute and often weakly cucullate at the apex, terminated with a short or long hyaline awn, broadly concave, not differentiated at the base, non-decurrent; *margins* weakly to broadly incurved distally, plane below, sometimes recurved, entire to serrulate or crenulate above, occasionally dentate or erose near the costa, unistratose throughout; *costa* single, well demarcated from the laminal cells, short- to long-excurrent as a smooth or denticulate, flexuose, hyaline hair-point, sometimes as long as or longer than the lamina or occasionally absent in lower cauline leaves, in transverse section circular, with a relatively weak, circular dorsal stereid band, with or lacking a dorsal epidermis, with 2 guide cells and 2–4 lamellae longitudinally inserted on the ventral surface, extending from the apex to the mid-leaf or reaching the base, widest near the apex and usually gradually tapered to 1 cell high at their base, occasionally wider at their bases; *lamellae* chlorophyllose, to 12 cells in height, with entire or crenulate margins, thin-walled, smooth or papillose, bulging superficially, sometimes with short lateral, simple or branched filaments, usually gradually increasing in degree of branching towards the leaf apex and forming dense clusters; *laminal cells* unistratose throughout, showing yellow coloration in KOH or occasionally with red or orange patches medially near the apex or the leaf base; *upper cells* quadrate, rounded-quadrate, rhomboidal to short-rectangular, pellucid, thin- to moderately thick-walled, sometimes with weak collenchymatous thickenings, superficially flat or somewhat bulging on both sides or only ventrally, with superficial walls thin and even on both sides or occasionally with dorsal superficial walls exceedingly crassate, smooth or papillose on the dorsal surface with small, simple, solid or hollow papillae, 2–3 per cell; *basal cells* differentiated, subquadrate, oblong to rectangular, weakly inflated, smooth, thin-walled, little wider than the upper cells, occasionally rising higher medially. *Autoicous* (*paroicous* or *cladautoicous*). *Perichaetia* terminal; *perichaetial leaves* little differentiated. *Setae* singly per perichaetium, short to long, erect, straight, rigid, smooth, red-brown to yellow-brown, twisted to the left. *Capsules* stegocarpous or occasionally cleistocarpous, immersed to exserted, erect, symmetric, ovoid, subhemispherical to obloid-cylindrical, leptodermous, brown, irregularly wrinkled, ridged or wrinkled-plicate when dry, macrostomous; *annulus* absent or present, persistent, consisting of 1–3 somewhat vesiculose cells; *operculum* obliquely long-rostrate from a convex conic base; *exothecial cells* rectangular, oblong-hexagonal to subquadrate, thin-walled, in fairly regular longitudinal rows; *stomata* in a single row at the base of the urn, superficial, with 2 yellow guard cells and narrowly elliptical perforation; *peristome* absent or rudimentary, fragile and usually adherent to the operculum as anastomosing elliptical fragments forming a lattice-work borne on a conical hyaline membrane, yellowish, finely papillose, with several articulations, slightly twisted clockwise. *Spores* large, globose or slightly ellipsoid, brown or yellow-brown, verrucose. *Calyptra* cucullate or mitrate and lobed at the base, smooth, naked.

Pterygoneurum is a distinct and easily identifiable genus by its deeply concave, ovate, elliptical to ligulate leaves with 2–4 photosynthetic lamellae on the ventral surface of the costa, and the name alludes to this distinctive character. In its short, often bulbiform habit, it is similar to *Stegonia* but that genus entirely lacks any supracostal outgrowths. In addition, the leaves in *Pterygoneurum* are always piliferous, giving the plants a greyish and hoary appearance, whereas the leaves in *Stegonia latifolia*, the only representative of the genus in the Antarctic, are only shortly mucronate, never ending with a hyaline awn. *Pterygoneurum* is apparently most closely related to the genera *Crossidium* Jur. and *Microcrossidium* J.Guerra & Cano in which the ventral costal outgrowths are differentiated as separate filaments, not lamellae, the spores are small, 8–15 μm in diameter, the capsules are without exception ochyrostomous and the annulus is well differentiated, composed of vesiculose cells (Delgadillo 1973*a*,*b*, 1975; Cano *et al.* 1993). Amongst the Antarctic genera, the green costal outgrowths are known only in *Aloina* as well as in all genera of the Polytrichopsida, but they are unlikely to be confused with *Pterygoneurum*. In *Aloina* the leaves are fleshy, incurved and tubulose, with strongly infolded margins, and the costal outgrowths are always formed as filaments, whereas in the Polytrichaceae the leaves are distinctly differentiated into a photosynthetic limb, with a broad costa densely covered with numerous longitudinal chlorophyllose lamellae and a vaginant, echlorophyllose base.

Pterygoneurum is a small genus that consists of nine to twelve species depending on different taxonomic interpretations (Wijk *et al.* 1967; Crosby *et al.* 2000), but this number may be reduced after critical studies of some obscure and poorly known species, such as the South African *P. macleanum* and Australian *P. kemsleyi* (Guerra *et al.* 1995*a*). However, these potential reductions may be compensated by the discovery of new species, for example, four have recently been described, namely *P. crossidioides* from the Dead Sea area in the Near East (Frey *et al.* 1990), *P. compactum* from Spain (Cano *et al.* 1994), *P. squamosum* from Spain (Segarra *et al.* 1998) and *P. papillosum* from

Germany (Oesau 2003). So far, *Pterygoneurum* has not been revised taxonomically on a global scale but an interesting account of the spore morphology in all species of this genus was published by Carrión *et al.* (1995). Additionally, a useful survey of *Pterygoneurum* is available for the Iberian Peninsula, where it shows remarkable diversity, being represented by six species (Guerra *et al.* 1995*b*), and a key to the determination of nine European species of the genus is presented by Oesau (2003).

The genus *Pterygoneurum* is predominantly distributed in the arid and semiarid regions in the Northern Hemisphere and only one species, *P. macleanum*, is known from semiarid shrubland in Cape Province in South Africa and from southern Australia (Guerra *et al.* 1995*a*), exhibiting an ancient Circum-Tethyan range related to the Permo-Triassic configuration of the Pangaean continents (Frey & Kürschner 1988). In addition, *P. ovatum* is a bipolar species, recorded also from Australasia and Antarctica.

1. **Pterygoneurum ovatum** (Hedw.) Dixon Fig. 156, pl. 40

Pterygoneurum ovatum (Hedw.) Dixon, Rev. Bryol. Lichénol. 6: 96. 1934. ≡ *Gymnostomum ovatum* Hedw., Sp. Musc. Frond.: 31, 2 *f. 1–2*. 1801. ≡ *Bryum ovatum* (Hedw.) Dicks., Fasc. Pl. Crypt. Brit. 4: Index 1. 1801. ≡ *Pottia ovata* (Hedw.) Fürnr. *ex* Lindb., Öfvers. Förh. Kongl. Svenska Vetensk.-Akad. 21: 218. 1864, *nom. inval. in synon.* ≡ *Pterygoneurum cavifolium* Jur., Laubm-Fl. Oesterr.-Ung: 95. 1882, *nom. illeg. incl. spec. prior.* [Gymnostomum ovatum Hedw., 1801]. — Type citation: In muris argillaceis circa Lipsiam aliisque regionibus frequens; additionally, references to Morison (1699), Hedwig (1785–1787), Bridel (1798) are given. [Lectotype (*selected here*): The specimen with five mature sporophytes on the sheet labelled 'Gymnostomum ovatum Hedw. Crypt. 1, 15, 6' – G-Hedwig/Schwägrichen!].

Pottia cavifolia (Ehrh.) *ex* Fürnr., Flora 12 (Erg. 2): 13. 1829 [*P. cavifolia* Ehrh., Beitr. Naturk. 2: 187. 1788, *nom. inval.*], *nom. illeg. incl. spec. prior.* [Gymnostomum ovatum Hedw., 1801].

Pottia pusilla Lindb., Öfvers. Förh. Kongl. Svenska Vetensk.-Acad. 20: 410. 1864 [*Bryum pusillum* Hedw., Fund. Hist. Nat. Musc. Frond. 2: 32, 87. 1782, *nom. inval.*]. ≡ *Pterygoneurum pusillum* (Lindb.) C.E.O.Jensen, Danmarks Mosser 2: 345, *f. 17*. 1923, *nom. illeg. incl. spec. prior.* [Gymnostomum ovatum Hedw., 1801].

Tortula pusilla Mitt., J. Linn. Soc. Bot. 12: 165. 1869, *nom. illeg. incl. spec. prior.* [Pottia cavifolia (Ehrh.) *ex* Fürnr. = Gymnostomum ovatum Hedw., 1801].

Plants small, mostly gemmiform, in dense, broad, low hoary tufts, sometimes gregarious, light green to yellow-brown or brownish. *Stems* short, erect, (1–)3–5 mm long, simply or sparingly branched, usually buried in soil and with a very strong and extensive rhizoidal system, often much longer than the whole plant including the sporophyte, composed of light brown, lustrous, smooth, sparsely branched rhizoids, in cross-section rounded-pentagonal, with a fairly large central and a hyaline to yellowish-hyaline tissue of large, thin-walled cells, lacking a differentiated cortex and hyalodermis; *axillary hairs* filiform, hyaline throughout, composed of 4–5 cells, short at the base and elongate in the distal part. *Leaves* imbricate, erect and incurved on drying, somewhat erect-spreading on wetting, deeply concave, broadly ovate, 0.7–1.0 mm long, 0.7–0.8 mm wide, obtuse to broadly acute at the apex, long piliferous with a hyaline or pale yellowish, smooth or faintly serrulate, flexuose hair-point, usually as long as or shorter than the lamina; *margins* entire to serrulate or crenulate above, plane, incurved, unistratose, with undifferentiated cells; *costa* single, well defined, yellow-brown, 35–50 µm wide at the base, reaching the apex and long excurrent as a hyaline awn, bearing on the upper surface green lamellae with entire margins, composed of 2–7 rows of thin-walled, somewhat bulging cells, in transverse section not prominent dorsally, with only a single, dorsal stereid band and with 2 or no guide cells; *laminal cells* unistratose, smooth, transparent, often hyaline in the upper third; *upper cells* rounded-quadrate, quadrate to rhombic, 15–20(–25) µm long, 10–15(–18) µm wide, moderately thick-walled, usually with distinct corner thickenings, in cross-section usually with much more thickened and flat dorsal superficial cell walls and thin and bulging ventral superficial cell walls; *basal cells* subquadrate, short-rectangular to oblong, (20–)30–50(–65) µm long, 20–25 µm wide, thin-walled, lax and somewhat bulging to firm-walled. *Autoicous*. *Perichaetia* terminal, with the interior perichaetial bracts having shorter hyaline awn than the cauline leaves. *Setae* erect, straight, fairly stout, 2.0–3.5 mm long, smooth, yellow-brown to reddish-brown, twisted anticlockwise when dry. *Capsules* shortly exserted, ovoid to obloid, 1.0–1.5 mm long, 0.5–0.7 mm wide, straight, erect, symmetric to slightly asymmetric with a somewhat oblique mouth, brown to castaneous, lustrous, wrinkled-plicate when dry, leptodermous; *annulus* none or persistent, uniseriate, composed of thick-walled, somewhat vesiculose cells; *operculum* long-rostrate, with a stout, slanted rostrum; *exothecial cells* variable, mostly oblong to rectangular in fairly regular longitudinal rows, usually mixed with irregular, subquadrate to hexagonal, (20–)40–90(–110) µm long, 20–35 µm wide, thin-walled, becoming rounded to oblate, of the same colour as lower cells but thicker-walled in 2–4 rows at the orifice; *stomata* 7–8 in 1 row at the base of the urn, phaneropore, elliptical, bicellular relatively large, 55–65 µm long, 35–40 µm wide, with a narrow, elongate slit; *peristome* none. *Spores* globose, brown, 22–35(–40) µm in diameter, papillose. *Calyptra* cucullate, naked, smooth.

Chromosome number: unknown in the Antarctic and Subantarctic; $n = 26$ – nine records from the Holarctic (Fritsch 1991).

Discussion. The lamellae on the upper surface of the costa coupled with piliferous leaves are sufficient basis for separating *Pterygoneurum ovatum* from any other moss species in the Antarctic. In addition, the plants are generally small to minute and usually produce an abundance of fully mature gymnostomous capsules. Members of the Polytrichopsida also possess supracostal chlorophyllose lamellae but these are generally large and coarse plants with a unique habit resulting from a differentiation of the leaves into a limb and clasping base.

The Antarctic plants of *Pterygoneurum ovatum* exhibit some minor differences in comparison with the northern plants, including a more lax foliage, enlarged upper laminal cells averaging 15–25 µm in width, thickened dorsal superficial cell walls in the distal part, and somewhat curved and asymmetric capsules. Also, the leaf lamellae are usually rather small and often reduced to only one to three rows of cells. Accordingly, some workers suggested initially they could be a basis for their recognition as a separate variety or subspecies (Lewis Smith 1985). Enlarged cells and more thickened cell walls are known in other Antarctic taxa, for example *Ceratodon purpureus* and *Syntrichia magellanica*, and these characters may be interpreted as a response to severe climatic conditions in the polar regions. However, because the Antarctic plants of *P. ovatum* match perfectly the northern ones in all other characters, their recognition as a separate taxon would be unjustified.

Pterygoneurum ovatum belongs to a small group of species within the genus which is characterised by having smooth lamellae which lack branched filaments on their sides. This group also comprises *P. macleanum* from South Africa and southern Australia (Magill 1981; Guerra *et al.* 1995*a*) but this species is distinct from *P. ovatum* by the lack of the hyaline hair-point (instead the costa is short-excurrent and the leaves are mucronate) and undulate lamellae.

Pterygoneurum ovatum has a long and quite complex nomenclatural history. It was recognised, but not described, for the first time by Hedwig (1782) as *Bryum pusillum* and placed in the genus *Gymnostomum*. Later, Hedwig (1785–1787) rejected the epithet *pusillum* and described it as *Gymnostomum ovatum*, and this name was validated in his *Species muscorum frondosorum* (Hedwig 1801). In the pre-starting point period for moss nomenclature, this species was also recognised by Ehrhart (1787) as *Pottia cavifolia*. Both *Bryum pusillum* and *Pottia cavifolia* were validated by Lindberg (1863*b*) and Fürnrohr (1829*a*) and were widely used by bryologists, especially when Juratzka (1882) accepted *Pterygoneurum cavifolium* when establishing the genus *Pterygoneurum*. However, when in 1910 at the Third International Botanical Congress it was agreed to arbitrarily change the stating point for the nomenclature of mosses (*Sphagnum* excepted) from Linnaeus' (1753) *Species plantarum* to Hedwig's (1801) *Species muscorum frondosurum*, *Gymnostomum ovatum* was resurrected as the oldest available name for this species, and it was given a name under *Pterygoneurum* by Dixon (1934*a*).

Reproduction in Antarctica. In most populations fully mature capsules have been detected and in some samples (e.g. *RILS 10826*, AAS, KRAM) the sporophytes are produced in great profusion, but sometimes the capsules are underdeveloped with aborted spores.

Habitat. A pioneer moss growing on fine base-rich soil, glacial till, silty clay in frost-heaved soil circles, amongst stones and marble outcrops. On Signy I. *Pterygoneurum ovatum* along with some short acrocarps such as *Stegonia latifolia*, *Hennediella antarctica*, *Encalypta rhaptocarpa* and *Bryum argenteum* var. *muticum*, forms a unique pioneer community colonising young lateral moraines (Lewis Smith 1985). On James Ross I. the species grows on pockets of soil among boulders and on soil on patterned ground in a fellfield community dominated by *Usnea sphacelata*.

World range. *Pterygoneurum ovatum* is a bipolar species (Fig. 157, inset). In the Northern Hemisphere it has a strongly discontinuous range with its greatest occurrence in Europe. Here, it is widely scattered but locally common and abundant throughout the temperate part of the continent, ranging from southern Scandinavia (Söderström 1998) and Britain (Blockeel 1992*c*) to the Iberian Peninsula and Madeira (Casas *et al.* 1989), and Morocco, Algeria and Tunisia in North Africa (Jelenc 1955; Ros *et al.* 1999; Cano *et al.* 2002), eastwards to the Ural Mountains (Ignatov & Ignatova 2003), the Caucasus (Manakyan 1989) and Turkey (Henderson & Prentice 1969). In Asia, the species is rare and widely scattered, mostly in the Near East including Syria, Israel and Jordan (Frey & Kürschner 1991), Iraq (Agnew & Vondraček 1975), Iran (Schiffner 1908) and Central Asia (Mamatkulov 1990), with some isolated stations in Mongolia (Tsegmed 2001*a*,*b*) and in China (Xinjiang and Inner Mongolia) (Li *et al.* 2001). A highly disjunct station of *P. ovatum* is known from Wrangel I. and it is the only known record of the species in the Arctic (Afonina 2004). In North America, it is predominantly distributed in the western part of the continent, extending from southern British Columbia (Schofield 1976) to southern California (Koch 1955; Norris & Shevock 2004) and Arizona (Williams 1959), in the Rocky Mountains south to Colorado and in the Great Plains of North Dakota, South Dakota and Kansas with isolated stations in the prairies of Saskatchewan or Alberta (Lawton 1971; Churchill 1985). In eastern North America, *P. ovatum* is exceedingly rare in West Virginia (Williams 1959), Ontario (Kucyniak1958; Ireland & Ley 1992) and Quebec (Favreau & Brassard 1988). Vitt (1974*b*) reported *P. ovatum* from the Yukon Territory but it is evidently due to misinterpretation of Persson's (1952) record of *Bryum ovatum* Jur., a species usually considered to be conspecific with *B. neodamense* Itzigs. An isolated, and the southernmost, locality of *P. ovatum* in the Northern Hemisphere is in central Mexico (Delgadillo & Cárdenas 1979; Crum 1994*d*). In the Southern Hemisphere, *P. ovatum* is known from the South Island, New Zealand, where it represents a recent introduction (Allison 1971). It also occurs in southern Australia (Scott & Stone 1976), but its status needs clarification because Scott & Stone (1976) and Catcheside (1980) stated that it shows considerable variation and possibly more than one species may be included. In the western part of the Southern Hemisphere, *P. ovatum* is currently known only from Antarctica and southern Patagonia (Matcham & Duckett 2008*a*). Greene (1986) also reported this species from southern South America and credited this record to Dusén (1906) and Warnstorf (1916). In fact, the former author did not mention this species, but did report *Tortula pusilla* Ångstr., a species which has little in common with *P. ovatum* but belongs within the *Syntrichia antarctica* complex (Kramer 1988). This confusion is probably due to the existence of *Tortula pusilla* Mitt. which is an illegitimate name synonymous with *P. ovatum*. This name was used by Mitten (1869) for a collection made by Gillies in Ecuador and, if its identity is confirmed, it could be the only known record of *P. ovatum* from the Northern Andes. However, the compilers of various checklists of Andean mosses (Steere 1948; Delgadillo *et al.* 1995; Churchill *et al.* 2000) have not commented on this record, although Warnstorf (1916) referred to it when mentioning the occurrence of *P. ovatum* in South America.

Distribution in Antarctica. An exceedingly rare species, known only from Signy I. in the South Orkney Is. and from

→

Figure 156. *Pterygoneurum ovatum*. — **1**. Habit. **2**. Cross-section of stem. **3–6**. Axillary hairs. **7–10**. Leaves. **11**. Upper lamina cells and hyaline hair-point. **12**. Basal leaf cells. **13–15**. Cross-sections of leaves. **16**. Cross-section of costa. **17**. Capsule, wet. **18**. Calyptra. **19**. Stoma and basal exothecial cells. **20**. Exothecial cells in middle of urn. **21**. Exothecial cells at mouth and spores. — [All from *RILS 3164*]. — Scale bars: a – 100 μm (11); b – 100 μm (2–10, 12–16, 19–21); c – 1 mm (1, 17–18).

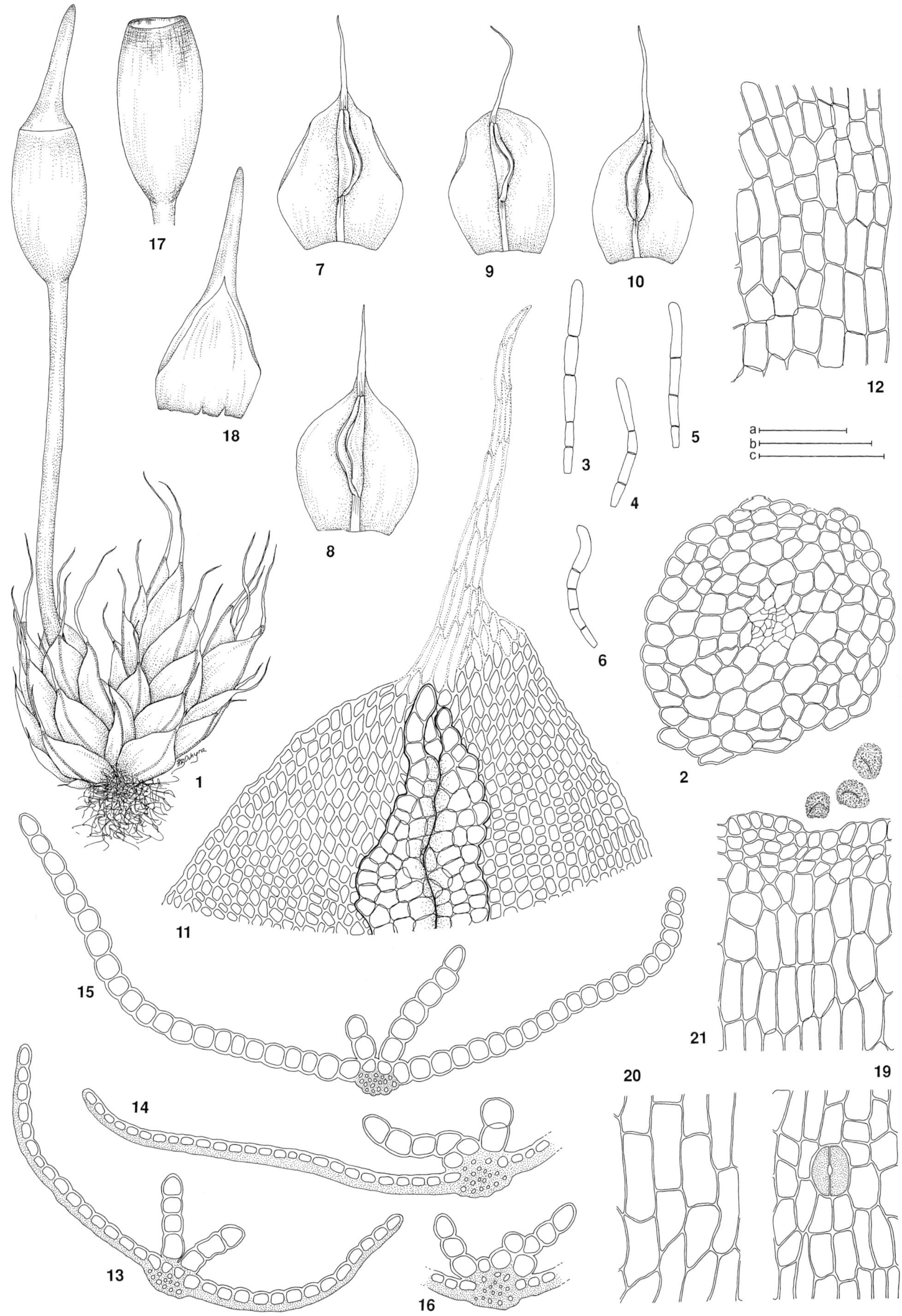
17
7
9
10
12
18
3
4
5
a
b
c
8
6
2
1
11
15
21
20
19
14
13
16

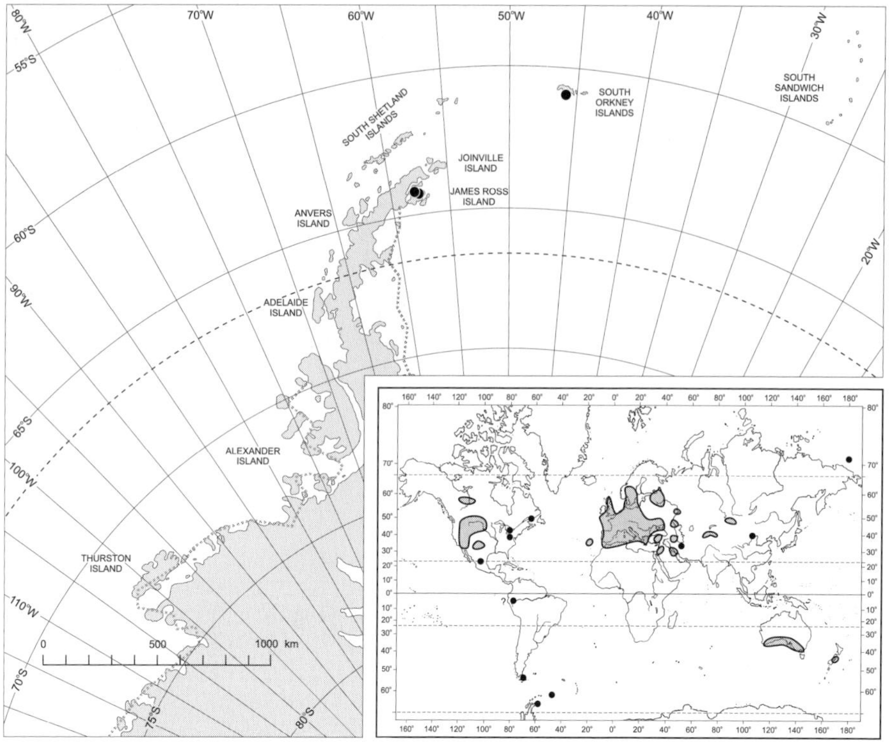

Figure 157. Distribution of *Pterygoneurum ovatum* in Antarctica. Inset: global distribution.

James Ross I. off the east coast of Trinity Peninsula (Fig. 189), although it is usually adundant on a very local scale. On the former island it occurs at an altitude from 8 to 35 m and on the latter at 50–150 m.

Specimens examined. WEST ANTARCTICA. SOUTH ORKNEY IS. **SIGNY I.** Between Marble Knolls and Knob Lake, 35 m, *RILS 3164* (AAS, KRAM); north of Elephant Flats, *ca* 30 m, *RILS 673* (AAS, KRAM); Moraine Valley, below Orwell Glacier, *ca* 8 m, *RILS 5013* (AAS, BM, KRAM), 10 m, *RILS 8096* (AAS, KRAM), 15 m, *RILS 3155* (AAS, BM, KRAM); same locality, north end of Orwell Glacier moraine, 30 m, *RILS 10826* (AAS, KRAM).

EAST ANTARCTIC PENINSULA. **JAMES ROSS ISLAND GROUP.** ***James Ross I.***: hill top on south side of St. Martha Cove, 150 m, *RILS 7540* (AAS, KRAM); in valley south of St. Martha Cove, 50 m, *RILS 7462 & 7462A* (AAS, KRAM).

30. ALOINA KINDB., *nom. cons.*

Aloina Kindb., Bih. Kongl. Svenska Vetensk.-Akad. Handl. 6(19): 22. 1882, *nom. cons.* ≡ *Barbula* Hedw. sect. *Aloina* Müll.Hal., Syn. Musc. Frond. 1: 592. 1849, *nom. illeg. incl. sect. prior*. [Tortula sect. Aloideae De Not. 1869]. ≡ *Tortula* Hedw. sect. *Aloina* Mitt., J. Linn. Soc. Bot. 12: 143, 157. 1869, *nom. illeg. incl. sect. prior.* — Lectotype (*vide* Craig 1939: p. 212): *Aloina aloides* (W.D.J.Koch *ex* Schultz) Kindb. (*Trichostomum aloides* W.D.J.Koch *ex* Schultz).

Aloidella (De Not.) Venturi, Comment. Fauna Veneto Trentino 1: 124. 1868, *nom.rej.* ≡ *Tortula* Hedw. [unranked] *Aloidella* De Not., Musci Ital.: 3, 14. 1862. ≡ *Barbula* Hedw. subgen. *Aloidella* (De Not.) Schimp., Syn. Musc. Eur. 2: 188. 1876. ≡ *Barbula* Hedw. sect. *Aloidella* (De Not.) Lesq. & James, Man. Mosses N. America: 115. 1884, *nom. illeg. incl. sect. prior.* — Type: *not designated.*

Plants small, bud-like, few-leaved, rigid and stiff, growing in short tufts or gregarious, dark green, brown, reddish- or greenish-brown above, brown below. *Stems* short, simple or seldom branched, usually buried in soil and densely covered with long, pale, smooth, branched rhizoids, in transverse section rounded-pentagonal, without or with a small central strand, composed of a uniform tissue of large, thin-walled cells, not differentiated into a cortex or hyalodermis; *axillary hairs* filiform, hyaline throughout, consisting of 6–7 elongate cells. *Leaves* stiff and rigid, thick, alutaceous, erect-appressed to erecto-patent, incurved and often tubulose when dry, erect-spreading to spreading when moist, usually with a well differentiated, broadly rectangular and auriculate, pale, sheathing base, often with distinct shoulders, broadly elliptical to subcircular, lingulate to lingulate-spathulate or lingulate-lanceolate, very concave above the base, cucullate and blunt, obtuse and rounded to mucronate or less often abruptly awned, apiculate to acute or acuminate at the apex; *margins* strongly infolded and covering a major part or almost the whole upper surface of the leaf blade, entire to weakly sinuose or denticulate above, unistratose throughout; *costa* distinctly developed, occasionally scarcely differentiated, flat, broad, vanishing below the apex, percurrent to long excurrent as a smooth or denticulate, terete hyaline hair-point, smooth or coarsely papillose on the dorsal surface near the apex, densely covered ventrally in the distal portion with chlorophyllose, uniseriate filaments, consisting of 2–6, barrel-shaped, thin-walled cells, with terminal cell somewhat enlarged and conical, thin or thick-walled and occasionally papillose, also extending beyond the stereid group, in transverse section broadly reniform, usually with a dorsal, crescent-shaped stereid band, with or without a weakly differentiated dorsal epidermis of elongate, rarely short-rectangular to quadrate cells and 6–12 guide cells in 1(–2) layers; *laminal cells* smooth throughout, showing red reaction with KOH; *upper cells* oblate-rectangular to elliptical, with incrassate walls, medially bistratose and unistratose towards the margins; *basal cells* at margins oblong, hyaline or yellow to yellow-brown, thin-walled, forming a broad border extending to the shoulders, those at or near the shoulders quadrate and thick-walled. *Dioicous* or *monoicous*. *Perichaetia* terminal; *perichaetial bracts* not or only little differentiated from the cauline leaves, occasionally with plane and not infolded margins. *Setae* elongate, straight, smooth, brown to reddish-brown, dextrorse below, sinistrorse above. *Capsules* exserted, erect or inclined, ellipsoid to cylindrical, straight or slightly curved, brown to yellow-brown, lustrous, smooth, leptodermous; *annulus* persistent or deciduous, 1–2-seriate, composed of thick-walled, vesiculose cells; *operculum* conic to rostrate, with a straight or slanted beak; *exothecial cells* rectangular to irregular, thin-walled, becoming short, isodiametric, rounded-quadrate, angular to oblate at the orifice in 2–5 rows, sometimes only deeper brown; *stomata* in a single row at the base of the urn, superficial, bicellular, pale, with rounded holes; *peristome* teeth 16, linear, usually deeply split near the basal membrane into 2 filamentous branches, spiculose, with several articulations, twisted to the left. *Spores* small, spherical, light brown, granulate. *Calyptra* cucullate, smooth, naked, large, covering about ⅔ the length of the urn.

Aloina is one of the most distinctive and characteristic of all moss genera, readily known by its fleshy leaves giving the plants the succulent aspect of an *Aloë* from which its names derives. It can be distinguished by its leaves that are differentiated into an auriculate, vaginant base and a deeply concave blade with strongly inflexed margins that cover about one third of its surface on both sides or sometimes meet over its middle. The upper laminal cells are thick-walled and mostly oblate and bistratose in the portion adjacent to the costa, and thin-walled, rectangular and hyaline at the margins in the basal part in some species, often forming a distinct, membraneous border. Moreover, the costa is very broad, only very rarely is it almost absent or poorly differentiated from the laminal cells. In the limb it is densely covered with filiform, unbranched filaments that also extend beyond the costa and cover the bistratose parts of the limb. The genus is closely related to *Aloinella* Cardot and *Crossidium* Jur., with which it shares the presence of a pad of the photosynthetic filaments, but in these genera it is restricted only to the supracostal region and does not extend beyond it. Additionally, in these genera the leaf margins are not infolded and the leaf lamina is entirely unistratose.

The group of species which constitute the genus *Aloina* was recognised relatively early by Fürnrohr (1829*b*) as the subdivision Aloideae within *Barbula*. Wijk *et al.* (1959) interpreted it as a section but it cannot be accepted because this author variously designated it as 'Reihe', 'Gruppe' and 'Rotte', but not as a section. However, it was given the status of section by De Notaris (1836) as *Tortula* sect. *Aloideae*. This group of species was subsequently recognised as section *Aloina* within *Barbula* by Müller (1849), but this name is illegitimate because it comprises the type of sect. *Aloideae*. Kindberg (1882) used this sectional name *Aloina* for his newly recognised genus and this generic name has to be ascribed only to him, not to '(Müll.Hal.) Kindb.'. In fact, such authority for this generic name was cited by Kindberg (1883) in his subsequent work and more recently it has been incorrectly used by, among others, Brotherus (1924*b*), Delgadillo (1973*a*, 1975), Gallego *et al.* (1999) and Gallego & Cano (2000). Following Cardot's (1910*a*) proposal, which was approved and effected in 1932 (Dixon 1932*b*; Magill 1993), *Aloina* is considered as a conserved name against *Aloidella* (De Not.) Venturi. This conservation is redundant since the generic name *Aloidella*, introduced by Venturi (1868), is illegitimate because it is based upon sect. *Aloidella*. When introducing this name, De Notaris (1862) placed in its synonymy sect. *Aloina* of Müller (1849) but, as a name including the type of an earlier sectional name, it is illegitimate.

Aloina is a small genus consisting of from nine to twelve species depending on different taxonomic interpretations. Delgadillo (1975), in his worldwide revision of the genus, listed nine species and three varieties, two of which are often considered as distinct species (Gallego *et al.* 1999). Recently, a further distinct species was discovered on the Canary Is. (Gallego *et al.* 1998), while the status of two species from the Andes of Argentina still needs elucidation (Delgadillo 1975; Crosby *et al.* 2000). All species of the genus are adapted to xerophytic conditions (Watson 1914) and are distributed in regions with arid and semiarid climates on all continents, although most species are known to occur in the Northern Hemisphere, especially in the Mediterranean region (Gallego *et al.* 1999). In the Southern Hemisphere the genus is poorly represented, both in terms of frequency and cover, although the total number of species is relatively high. Two species, *A. catillum* (Müll.Hal.) Broth. and *A. roseae* (R.S.Williams) Delgad., are narrow Andean endemics (Delgadillo & Schiavone 2004) and four species are bipolar in distibution, including *A. bifrons* (De Not.) Delgad. in Australia and South Africa, *A. ambigua* (Bruch & Schimp.) Limpr. in Australia, *A. rigida* (Hedw.) Limpr. in the Andes of Ecuador, Peru and Bolivia, and *A. brevirostris* in the Antarctic.

1. **Aloina brevirostris** (Hook. & Grev.) Kindb. Fig. 158

Aloina brevirostris (Hook. & Grev.) Kindb., Bih. Kongl. Svenska Vetensk.-Akad. Handl. 7(9): 137. 1883. ≡ *Tortula brevirostris* Hook. & Grev., Edinburgh J. Sci. 1: 289, *pl. 12*. 1824. ≡ *Barbula rigida* Hedw. var. *brevirostris* (Hook. & Grev.) Brid., Bryol. Univ. 1: 824. 1827. ≡ *B. brevirostris* (Hook. & Grev.) Bruch *in* F.A.Müll., Flora 12: 406. 1829. — TYPE CITATION: Walls and clay banks, Sweden; whence it has been received by Mr. Turner, Sir J. E. Smith, and ourselves. Germany, Funck. Le Vallais, Hooker. On an old wall, near Edinburgh, D. Stewart, Esq. [Lectotype (*vide* Gallego *et al.* 1999: p. 184): 'Sweden, same as H-1674. Herb. Hort. Kew., Swartz, 11-d' – BM (not seen); isolectotype: 'H. 1674. Sweden T. rigida of Swartz from Dr. Swartz different from our T. rigida (see fig. in Fl. Brit.)' – BM!].

Plants small to very small, 1–2 mm tall, rigid, gemmiform when dry and moist, in loose clusters, scattered thin tufts or gregarious, dark lurid, greenish-brown or brown to reddish-brown with age. *Stems* very short, 0.5–1.0 mm, simple, mostly buried in soil and densely radiculose at the base with clusters of long, pale brown to whitish, smooth, sparingly branched rhizoids, in transverse section rounded-pentagonal, with a small but distinct central strand and an undifferentiated parenchymatous tissue of large, thin-walled, hyaline cells, in places sometimes becoming smaller in 1 or 2 layers at the outer surface and imitating a cortex; *axillary hairs* thread-like, hyaline throughout, composed of 6–7 elongate cells. *Leaves* incurved or erect-appressed, slightly tubulose when dry, erect to spreading when moist, (0.5–)1.0–2.0 mm long, 0.3–0.7 mm wide, smaller at the stem base and becoming appreciably larger towards the stem apex, from a broadly oblong, clasping base grading to a broadly lingulate, suborbicular or ellipsoid, very concave blade, as long as or shorter than the base, cucullate, rounded or obtuse at the apex; *margins* entire throughout, broadly involute in the blade, becoming erect to incurved, membranous in the base owing to several rows of hyaline marginal cells; *costa* single, ending at the apex, very broad and flat below, becoming narrowed upwards, poorly delimited, 70–250 µm wide, smooth on the dorsal surface, in transverse section strongly flattened, consisting of 1–3 layers of dorsal stereids forming a flat or crescent-shaped band, lacking a dorsal epidermis, 9–20 large guide cells and ventral epidermis differentiated as a series of uniseriate photosynthetic filaments, inserted also on the adjacent portions of the lamina, composed of 3–5 short, thin-walled, barrel-shaped cells and terminated by an ovoid-conical to subspherical cell; *laminal cells* smooth, bistratose in the juxtacostal region in the blade; *upper cells* variable in size and shape, mostly transversely elliptical to oblate-oblong, 10–18 µm in longer and 2–6 µm in shorter diameters, mixed with rounded-quadrate cells, 6–10 µm wide, thick-walled and with markedly thickened dorsal superficial cell walls, becoming quadrate medially at the shoulders, 12–16 µm wide; *inner basal cells* laxly rectangular to oblong-hexagonal, 20–60 µm long, 8–15 µm wide, moderately thick- to thin-walled, yellowish- or brownish-hyaline, becoming thinner-walled towards the margins and forming a membranous, hyaline border of 3–7 rows of cells, extending up to the shoulders. *Synoicous*. *Perichaetia* terminal; *inner perichaetial bracts* little differentiated from the cauline leaves, with a reduced blade and costa. *Setae* erect, straight, elongate, smooth, 3–7 mm long, reddish-brown, sinistrorse below, dextrorse above. *Capsules* exserted, erect, oblong-ovoid to short-cylindrical, symmetric, 1.4–1.6 mm long, 0.5–0.6 mm wide, straight or somewhat arched, gradually or abruptly tapering at the neck, not much narrowed at the mouth, brown, pachydermous, lustrous, smooth; *annulus* biseriate, composed of large, vesiculose cells, fugacious; *operculum* stoutly short-rostrate from a conic base, about ¼–⅓ the length of the urn, straight or oblique, often bluntly apiculate; *exothecial cells* mostly rectangular to subquadrate, 20–55 µm long, 10–15 µm wide, becoming isodiametric in 2–4 tiers at the orifice; *stomata* in a single row at the base of the urn, phaneropore, bicellular, pale, with round perforation; *peristome teeth* 16 arising from a short basal membrane projecting above the capsule mouth, split nearly to the base into two filamentous branches, 400–600 µm long, yellow-orange to orange, papillose, rather indistinctly articulated, strongly twisted in one spiral anticlockwise. *Spores* spherical, yellowish-green to pale brownish, 16–23 µm in diameter, finely papillose. *Calyptra* cucullate, smooth, naked.

Chromosome numbers: unknown in the Antarctic and Subantarctic; $n = 24$ (1 record) and $n = 28$ (1 record) from Alaska (Steere 1954; Inoue 1979).

Discussion. Because *Aloina brevirostris* is the only representative of the genus in the Antarctic, it is unlikely to be mistaken for any other moss species in this biome. The characteristic shape of the cauline leaves which are differentiated into more or less hyaline, sheathing bases and broadly lingulate to elliptical, very concave blades with margins strongly inflexed over it, as well as obtuse to rounded and cucullate apices are unknown in any other local moss species, except for three species of *Polytrichum*. However, these are coarse, large and robust mosses with lanceolate to linear-lanceolate limbs ending with long, brown or hyaline aristae. In the general gemmiform aspect of the plants, *A. brevirostris* may be confused with *Pterygoneurum ovatum* and *Stegonia latifolia* which occupy similar habitats, but they lack inflexed leaf margins. In addition, the former has two or three lengthwise lamellae on the adaxial surface of the costa, whereas the latter lacks any supracostal outgrowths.

Aloina brevirostris shows considerable variation of most structural characters of the gametophyte and sporophyte, including shape and size of the leaves, capsules, lids, as well as spore size and seta length. Yet, the Antarctic plants exhibit little interpopulational variability and this may suggest that they could be established as the result of a single introduction. Pettersson (1940) showed experimentally that this species is capable of long distance dispersal, and although Persson (1944) and Bergeron (1944) questioned the provenance of the spores which had reached southern Finland, their transportation by wind is not in doubt (Koponen & Oittinen 1967). *A. brevirostris* is closely related to *A. rigida* (Hedw.) Limpr. and both species were not differentiated from each other until Hooker & Greville (1824) drew attention to the short and blunt operculum in some plants from Sweden, Britain and continental Europe, and recognised them as separate species. It differs from *A. rigida* in its synoicous inflorescences and mostly bistratose stereid band in the costa, although the latter may vary from one to three layers. In Antarctic plants the stereids are mostly arranged in

⟶

Figure 158. *Aloina brevirostris.* — **1**. Habit. **2**. Portion of cross-section of stem. **3–4**. Axillary hairs. **5–9**. Leaves. **10**. Cells at leaf apex. **11**. Lamina cells at leaf shoulder. **12**. Basal cells at leaf margin. **13**. Basal leaf cells over costa. **14**. Cross-section of leaf in the upper part. **15**. Cross-section of leaf at junction of limb and sheath. **16**. Perichaetial leaf. **17**. Operculate capsule. **18**. Exothecial cells in middle of urn. **19**. Exothecial cells at base of urn and stoma. **20**. Exothecial cells at mouth, annulus, peristome and spores. — [All from *RILS 935*]. — Scale bars: a – 100 µm (18–20; b – 100 µm (2–4, 10–15); c – 1 mm (1, 17); d – 1 mm (5–9, 16).

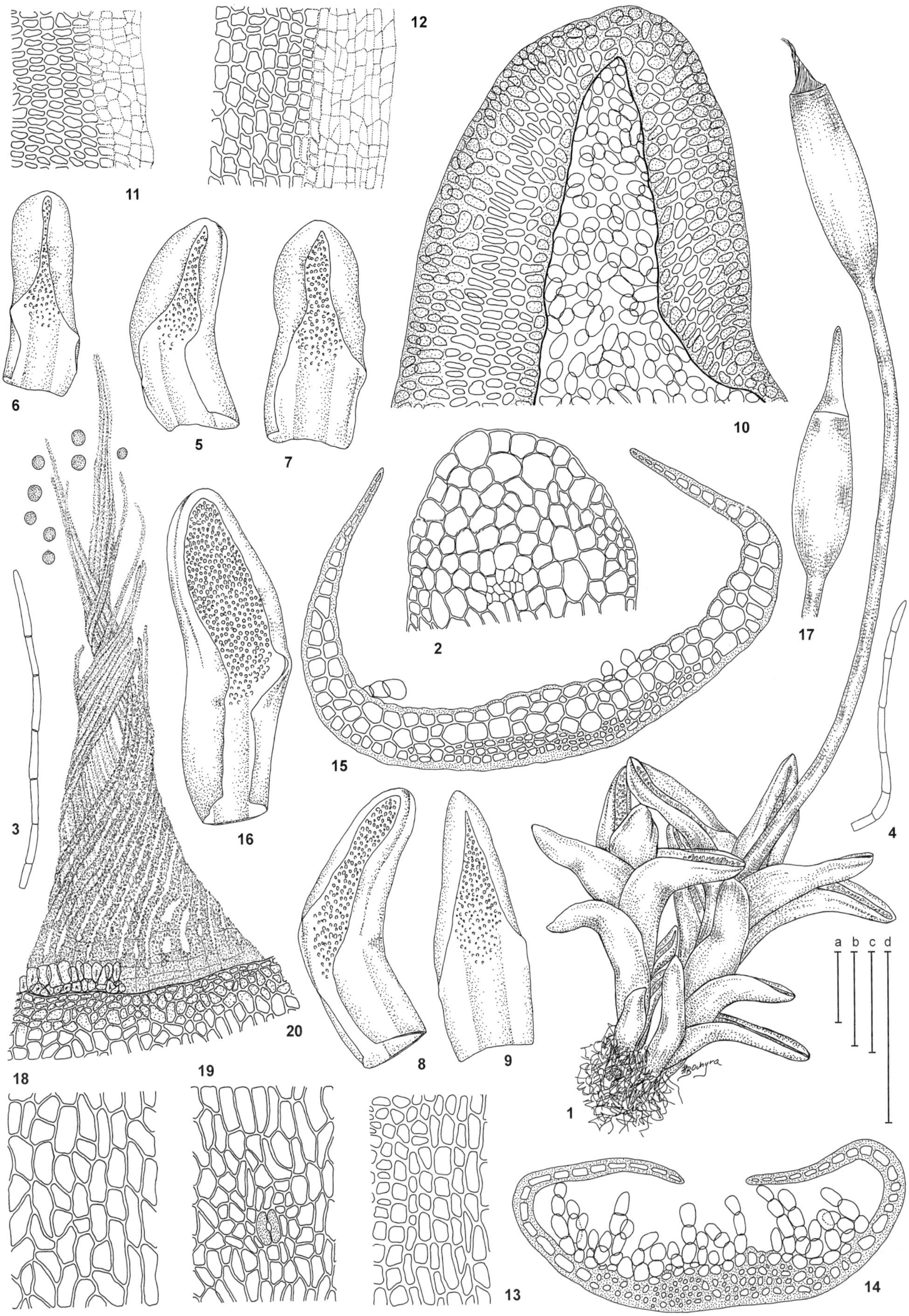
1
2
3
4
5
6
7
8
9
10
11
12
13
14
15
16
17
18
19
20
a
b
c
d
Bachyra

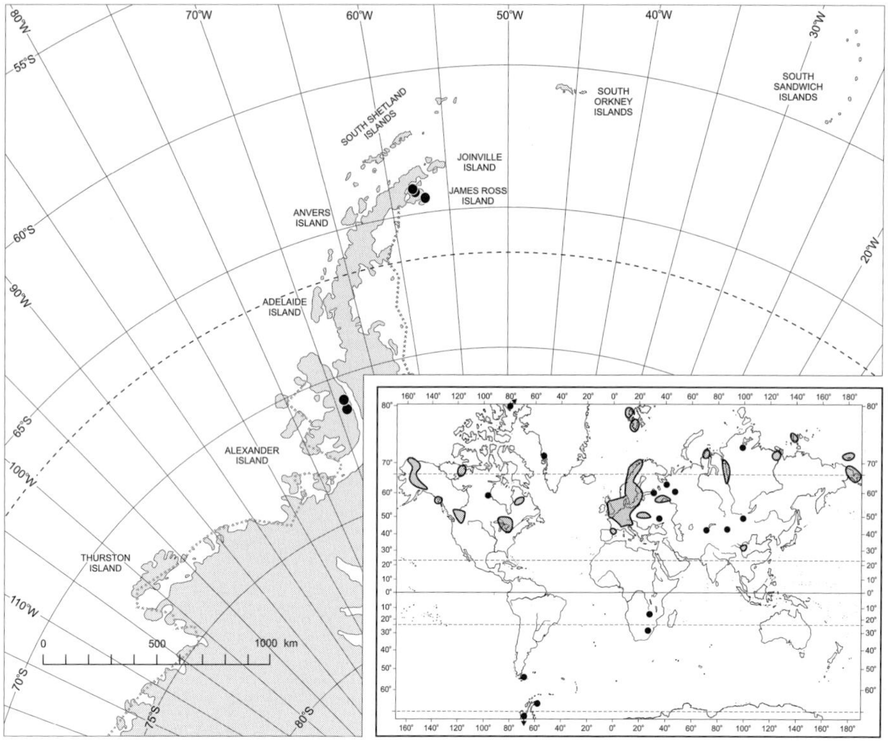

Figure 159. Distribution of *Aloina brevirostris* in Antarctica. Inset: global distribution. The localities on Alexander I. are indicated by an arrow.

two layers, and only occasionally on some leaves on the plants from Alexander I. (*RILS 9351*, AAS, KRAM) is the stereid band composed of three and, irregularly, in places four layers. Smith (1978, 2004) used the spore size for discriminating *A. rigida* from *A. brevirostris*, recording 14–16 μm in diameter for the former and 18–22 μm for the latter. However, spore size in both species overlaps markedly, although *A. rigida* shows a distinct tendency for having smaller spores, 11–22 μm in diameter, and *A. brevirostris* for having larger spores, 15.4–33.0 μm (Delgadillo 1975). In Antarctic plants the spores are often aborted, but those which are well developed are rather large, 20–23 μm in diameter, and this confirms their identity as *A. brevirostris*.

Reproduction in Antarctica. Almost all populations have been found with mature sporophytes, but the spores are often aborted.

Habitat. The species grows in dry or temporarily moist habitats, on fine silty soil amongst stones and boulders or gravelly silt on stream banks, as well as on soil on ledges and in rock crevices. It is occasionally associated with *Bryum dichotomum* and *Bryoerythrophyllum recurvirostrum*, but its colonies are often encrusted with lichens.

World range. A bipolar species, with its main centre of occurrence in the Northern Hemisphere, and known only from the Antarctic Peninsula region and southern Africa in the Southern Hemisphere (Fig. 159, inset). In the Holarctic it has a wide but strongly discontinuous range. In North America it is scattered in Alaska, the Yukon and Northwest Territories, reaching its highest known latitude at *ca* 83° N at northern Ellesmere I. in the Canadian Arctic Archipelago (Brassard 1971), Ontario (Ireland & Ley 1992) and Labrador (Favreau & Brassard 1988), southward to Montana and New York and with a single record in West Greenland (Delgadillo 1975). *Aloina brevirostris* is most widespread in Europe where it is scattered in Spitsbergen in the Arctic (Kuc 1973*b*), Fennoscandia (Söderström 1996) and Britain (Blockeel 1992*d*) in the north, and on the continent its range extends from the Netherlands (Bijlsma *et al.* 1985; Touw & Rubers 1989), across Germany (Duell & Meinunger 1989) and Poland (Kuc 1964) to the Western Ukraine (Bachurina & Mel'nichuk 1988) and middle Russia (Ignatov & Ignatova 2003) south to northern Italy (Cortini Pedrotti 2001*a*,*b*) and Spain (Gallego & Cano 1998, 2000). In Asia, *Aloina brevirostris* is rare and scattered in the Russian Arctic and conterminous territories, including the Yamal Peninsula (Czernyadjeva 2001), West Siberia (Lindberg & Arnell 1890), the Taymyr Peninsula (Kannukene & Matveyeva 1996), Yakutia (Stepanova 1986; Ignatov *et al.* 2001) and Chukotka and Wrangel I. (Afonina 2004), and at highly disjunct and isolated stations in Tadzhikistan in Central Asia (Mamatkulov 1990), Khangai Mountains in Mongolia (Tsegmed 2001*a*,*b*) and in China (Li *et al.* 2001). Outside the Holarctic *A. brevirostris* is known only from Zimbabwe (Best 1990; Manyanga & Perold 2004) and Lesotho (Hodgetts *et al.* 1999) in southern Africa and is now recorded from southern Patagonia (Matcham & Duckett 2008*b*) and Antarctica.

Distribution in Antarctica. *Aloina brevirostris* is bicentric in its known distribution in the Antarctic where it is restricted to arid areas on James Ross I. and Seymour I. on the east coast of the Antarctic Peninsula, and on Alexander I. where it reaches its southernmost locality at lat. 71°24′ S (Fig. 159). It also occurs at a considerable altitudinal range, growing from near sea level to 25 m in the James Ross Island group to 300–500 m on Alexander I.

Specimens examined. WEST ANTARCTICA. WEST ANTARCTIC PENINSULA. **GEORGE VI SOUND.** ***Alexander I.***: Fossil Bluff, below Khufu, 300–400 m, *RILS 9318A & 9351* (AAS, KRAM) and 400–500 m, *RILS 9334* (AAS, KRAM); edge of Uranus Glacier, northern part of Waitabit Cliff, *ca* 15 m above glacier level, *BJT 519A* (AAS, KRAM).

EAST ANTARCTIC PENINSULA. **JAMES ROSS ISLAND GROUP.** ***James Ross I.***: to south of St. Martha Cove, 25 m, *RILS 7615* (AAS, KRAM); Dreadnought Pt., 3 m, *RILS 7496A* (AAS, KRAM). ***Seymour I.***: Cape Wiman, 20 m, *RILS 7877B* (AAS, KRAM).

31. **TORTULA** Hedw., *nom. cons.*

Tortula Hedw., Sp. Musc. Frond.: 122. 1801, *nom. cons.* ≡ *Barbula* Hedw. [unranked] *Tortula* (Hedw.) Rebent., Prodr. Fl. Neomarch.: 257. 1804 ['Tortulae']. — Type: *Tortula subulata* Hedw. (*typ. cons.*).

Phascum Hedw., Sp. Musc. Frond.: 19. 1801. — Lectotype (*vide* Grout 1939: p. 195): *Phascum cuspidatum* Hedw. [= *Tortula acaulon* (With.) R.H.Zander].

Anacalypta Röhl. *ex* Léman, Dict. Sci. Nat. Ed. 2, 2 Suppl.: 38. 1816, *nom. rej.* ≡ *Pottia* Fürnr. sect. *Anacalypta* (Röhl. *ex* Léman) Müll.Hal., Syn. Musc. Frond. 1: 547. 1849, *nom. illeg. prior. ut genus.* ≡ *Pottia* subgen. *Anacalypta* (Röhl. *ex* Léman) Boulay, Fl. Crypt. Est Muscin.: 509. 1872, *nom. illeg. prior. ut genus.* ≡ *Trichostomum* Hedw. [unranked] *Anacalypta* (Röhl. *ex* Léman) Müll.Hal., Linnaea 42: 316. 1879, *nom. illeg. prior. ut genus.* — Type: *Anacalypta lanceolata* (Hedw.) Nees & Hornsch. (*Encalypta lanceolata* Hedw.). [= *Tortula lanceola* R.H.Zander].

Desmatodon Brid., Muscol. Recent. Suppl. 4: 86. 1819. ≡ *Trichostomum* Hedw. sect. *Desmatodon* (Brid.) Müll.Hal., Syn. Musc. Frond. 1: 588. 1849. ≡ *Tortula* Hedw. sect. *Desmatodon* (Hedw.) Mitt., J. Linn. Soc. Bot. 12: 145, 164. 1869. ≡ *Tortula* [unranked] *Desmatodon* (Brid.) Lindb., Musci Scand.: 20. 1879. ≡ *Tortula* subsect. *Desmatodon* (Brid.) Braithw., Brit. Moss Fl. 3: 212. 1885. ≡ *Didymodon* Hedw. [unranked] *Desmatodon* (Brid.) Kindb., Eur. N. Amer. Bryin. 2: 273. 1898. ≡ *Tortula* subgen. *Desmatodon* (Brid.) Kindb., Rev. Bryol. 36: 117. 1909. — Lectotype (*vide* Venturi 1868: p. 123): *Desmatodon latifolius* (Hedw.) Brid. (*Dicranum latifolium* Hedw.) [= *Tortula hoppeana* (Schultz) Ochyra].

Physedium Brid., Bryol. Univ. 1: 51. 1826. — Type: *Physedium splachnoides* (Hornsch.) Brid. (*Phascum splachnoides* Hornsch.) [= *Tortula splachnoides* (Hornsch.) R.H.Zander].

Zygotrichia Brid., Bryol. Univ. 1: 520. 1826. — Type: *Zygotrichia leucostoma* (R.Br.) Brid. (*Barbula leucostoma* R.Br.) [= *Tortula leucostoma* (R.Br.) Hook. & Grev.].

Pottia (Ehrh) *ex* Fürnr., Flora 12(2) Ergänzungsbl.: 10. 1829, *nom. cons.* (*Gymnostomum* Nees & Hornsch. subgen. *Pottia* Ehrh. *ex* Rchb., Consp. Regn. Veg. 1: 33. 1828, *nom. nud.*). ≡ *Tortula* Hedw. [unranked] *Pottia* (Fürnr.) Kindb., Bih. Kongl. Svenska Vetensk.-Akad. Handl. 7(9): 131. 1882. — Lectotype (*vide* Wareham 1939: p. 197): *Pottia truncata* (Hedw.) Bruch & Schimp. (*Gymnostomum truncatum* Hedw.) [= *Tortula truncata* (Hedw.) Mitt.].

Beccaria Müll.Hal. *in* Venturi, Nuovo Giorn. Bot. Ital. 4: 11. 1872. — *Pottia* Fürnr. sect. *Beccaria* (Müll.Hal.) Müll.Hal., Gen. Musc. Frond.: 389. 1900. — Lectotype (*selected here*): *Beccaria minor* Müll.Hal. [= *Tortula minor* (Müll.Hal.) R.H.Zander].

Dendia R.Br.ter., Trans. New Zealand Inst. 30: 411. 1898. — Type: *Dendia maritima* R.Br.bis [= *Tortula maritima* (R.Br.bis) R.H.Zander].

Hyophilopsis Cardot & Dixon, J. Bot. 49: 137. 1911. — Type: *Hyophilopsis entosthodontacea* Cardot & Dixon [= *Tortula entosthodontacea* (Cardot & Dixon) R.H.Zander].

Pachyneurum J.J.Amann, Fl. Mouss. Suisse 2: 112. 1918. ≡ *Tortula* subgen. *Pachyneurum* (J.J.Amann) C.E.O.Jensen, Skand. Bladmossfl.: 200. 1939. — Lectotype (*selected here*): *Pachyneurum atrovirens* (Sm.) J.J.Amann (*Grimmia atrovirens* Sm.) [= *Tortula atrovirens* (Sm.) Lindb.].

Plants very small, small to medium-sized, fairly coarse and rigid, scattered, gregarious or loosely caespitose or forming dense, low tufts or patches, not or sometimes hoary, yellow, brownish, green, yellowish- or blackish-green, sometimes reddish-brown, bluish-green to grayish above, yellow-brown to dark brown below. *Stems* erect, simple or occasionally branched, very short, to about 2 cm in length, usually densely radiculose to the base with brown, smooth, branched rhizoids, in transverse section rounded-pentagonal, with or very seldom without a central strand, hyalodermis and sclerodermis, consisting of nearly homogeneous, parenchymatous cells, large and thin-walled around the central strand and becoming somewhat smaller and brownish in 1–3 outer layers but not forming a distinct cortex; *axillary hairs* filiform, composed of 5–9 hyaline cells, thin-walled throughout or with 1–3 basal cells somewhat thicker-walled. *Leaves* few to many, crowded, equidistant and uniform or sometimes larger and more crowded at the stem tips, erect, appressed-incurved to lax, often flexuose or contorted, sometimes spirally twisted around the stem when dry, erect-spreading to widely patent when moist, nearly flat, broadly channelled to concave or concave-keeled distally, occasionally grooved along the costa, usually spathulate to obovate, sometimes oblong- to ovate-lanceolate, ligulate, ovate to elliptical, non-decurrent, with the base not or scarcely differentiated in shape to elliptical, seldom slightly auricled, acuminate, acute to broadly acute or rounder and obtuse at the apex, usually ending with a short, stout apiculus or hyaline, flexuose, smooth or serrulate hair-point; *margins* entire or occasionally slightly serrulate to crenulate near the apex, recurved to revolute below or nearly to the apex, rarely almost plane throughout, uni- or bistratose, often with 1–4 rows of marginal cells differentiated from the medial cells, less papillose and smaller, thicker-walled or elongate; *costa* single, usually short- to long-excurrent, less often percurrent to subpercurrent, with ventral superficial cells quadrate or occasionally short-rectangular, papillose or smooth and dorsal epidermal cells short-rectangular to elongate and papillose or smooth, rarely with an ellipsoid pad of cells bulging from the ventral surface of the costa, in transverse section circular or semicircular, strongly to weakly convex dorsally, consisting of a round, elliptical or semicircular dorsal stereid band, 2(–3) guide cells in 1(–2) layers, ventral stereid band lacking or occasionally small and represented by a few cells, epidermis present on both ventral and dorsal sides or occasionally absent or incomplete on the dorsal and well developed on the ventral surfaces; *laminal cells* unistratose throughout, except for occasional bistratose margins, usually showing yellow colour reaction with KOH, only seldom red or reddish-orange; *upper cells* rounded-quadrate to hexagonal or seldom rhomboidal, with thin or evenly thickened walls, convex superficially, mostly papillose with simple or bifid, usually hollow and C-shaped, occasionally conical papillae, 4–6 per lumen, occasionally smooth; *basal cells* differentiated, often lax, rectangular, thin-walled, hyaline, seldom only little differentiated. *Asexual reproduction* absent. *Dioicous* or *monoicous* (*autoicous* or *paroicous*). *Perichaetia* terminal, with inner bracts little differentiated or somewhat larger than the vegetative leaves. *Setae* very short to elongate, erect, straight, singly per perichaetium, very rarely geminate, yellowish-brown to brown, dextrorse, sinistrorse or not twisted. *Capsules* cleistocarpous or stegocarpous, sessile and immersed to long-exserted, globose, ovoid, ellipsoid, obovoid to cylindrical, erect or occasionally inclined to horizontal, symmetric or seldom asymmetric, curved or gibbous, usually macrostomous, leptodermous, pale brownish, brown to dark red-brown, smooth or finely wrinkled- or shriveled-striate; *annulus* 1–2-seriate, composed of thick-walled, vesiculose cells, persistent or revoluble, occasionally missing; *operculum* long-conic, short- or long-rostrate, usually obliquely so, with the cells arranged in oblique, anticlockwise rows; *exothecial cells* rectangular, thin- to evenly thick-walled, becoming isodiametric in 3–6 rows at the mouth, concolorous with mid-urn cells; *stomata* phaneroporous in a single row at the base of the urn, bicellular with elliptical openings; *peristome* present, rudimentary or absent, inserted somewhat below the mouth, consisting of 16 teeth, entire, short and triangular or deeply bifid near the base into long filaments, spiculose, straight or twisted anticlockwise, separate or arising from a low or high, tessellated and spiculose basal membrane. *Spores* globose to ellipsoid, light brown, papillose to densely spiculose. *Calyptra* cucullate, smooth, naked.

Tortula is a Hedwigian genus which has gained universal acceptance since its inception, although it has been interpreted in several ways during two centuries of its history. Of three species originally placed in *Tortula* by Hedwig (1801), only two (*T. subulata* and *T. muralis*) are still residual in the genus, whereas the third (*T. tortuosa*) is a now treated as a member of the genus *Tortella*. The basic diagnostic character which cemented the genus was the long peristome with the teeth twisted together in a screw-like fashion, and the generic name *Tortula* alludes to this. Consequently, for a long time *Tortula* served as a catch-all genus in which were placed numerous species of the Pottiaceae with twisted peristomes (e.g. De Notaris 1836, 1862; Lindberg 1864, 1879). The long list of heterotypic synonyms of *Tortula* at the genus level clearly confirms the divergent interpretation of this genus. The concept of *Tortula* adopted in the present Flora follows Zander (1989, 1993) who segregated several peripheral species in genera of their own (*Chenia*, *Dolotortula*, *Hilpertia*, *Sagenotortula*, *Stonea*), and defined precisely some closely related genera (*Syntrichia*, *Hennediella*, *Willia*), making *Tortula* a fairly homogeneous taxon which is remarkably cohesive in its gametophytes and exhibits a characteristic tendency to reduction of the sporophytes.

The major diagnostic features of the genus *Tortula* include the absence of sclerodermis and hyalodermis and the presence of the central strand in the stem anatomy; obovate to spathulate leaves with entire and proximally narrowly recurved margins; the presence of the semicircular to round dorsal stereid band and epidermis and the hydroid strand in the costal anatomy; large and relatively clear upper laminal cells with the walls relatively unobscured with papillae; the absence of propagula; and the yellow KOH reaction of the distal leaf lamina. In the present circumscription *Tortula* includes species traditionally arranged in the separate and widely accepted genera *Desmatodon*, *Phascum* and *Tortula*. Zander (1993) placed the last two genera in the separate section *Pottia*, although nomenclaturally this name cannot be accepted and, strictly, the section consisting of the genera *Phascum* and *Pottia* should be named sect. *Cuneifoliae* (Schimp.) Ochyra (Ochyra *et al.* 2003). However, *Desmatodon* was placed by Zander (1993) in the type section of *Tortula*. *Desmatodon* was usually distinguished as a genus in its own right on the basis of some slight differences in the peristome structure (Chen 1941). In contrast to the typical '*Tortula*' peristome which has 32 subterete, equally spaced identical branchlets wound in a spiral, the '*Desmatodon*' peristome has 16 teeth approximate in pairs which are split to near the base or a basal membrane into 2–3 flattened branchlets, erect or only slightly wound together. Zander (1993) interpreted this difference as a result of reduction of the peristome having paired teeth.

Tortula is closely related to *Syntrichia*, *Hennediella* and *Willia* and in the Antarctic these genera may be easily confused, especially in the sterile state. *Syntrichia* is readily distinguishable by its crescent-shaped dorsal costal stereid band lacking epidermis and the red KOH reaction of the distal laminal cells. *Willia* shares gross characteristics with *Syntrichia* but has highly differentiated perichaetial leaves, a strong tendency towards reduced sporophytes, as well as commonly panduriform cauline leaves. Finally, *Hennediella* is distinct from *Tortula* in its red KOH reaction of the upper part of the leaves and commonly dentate or serrate and plane upper leaf margins, as well as superficially flattened distal laminal cells.

In the austral polar region and cool-temperate zone *Tortula* is poorly represented and all species described previously in this area under *Tortula*, *Pottia* or *Desmatodon* are now placed in *Syntrichia*, *Hennediella* or *Willia*, although some species have not yet been taxonomically assessed, especially in southern South America (Greene 1986). In the Antarctic biome, the genus is represented by a single species, *Tortula hoppeana*, which is known only from a solitary record from the maritime Antarctic (Ochyra 2004*d*).

1. **Tortula hoppeana** (Schultz) Ochyra Fig. 160

Tortula hoppeana (Schultz) Ochyra, Bryologist 107: 499. 2004. ≡ *Trichostomum hoppeanum* Schultz *in* Hornsch., Syll. Pl. Nov. 2: 140. 1828. — Type citation: Habitat in rupibus editissimis Carinthiae (*Hopp.* et *Hornsch.*) et in alpibus Salisburg. (*Funck*). [Lectotype (*vide* Ochyra 2004*d*: p. 499): 'Desmatodon latifolius Brid. Brid. Bryol. 1. 524. Trichostomum latifolium Hedw. musc. frond. tab. xxxiii Auf der Pasterze bei Glockn.[er]', GZU-Hoppe].

For a full list of synonyms see Ochyra (2004*d*: 499).

Plants medium-sized, densely caespitose, dull, yellowish-green above, brownish-green below. *Stems* simple or forked, to 4 cm tall, erect, moderately stiff, in transverse section rounded to elliptical, consisting of a distinct and relatively large central strand, 5–6 layers of large, thin-walled, hyaline medullary cells and a poorly differentiated cortex of smaller, thicker-walled and brownish cells; *rhizoids* numerous scattered in clusters along the stem, long, sparingly branched, pale brownish to reddish-brown, smooth, lustrous; *axillary hairs* filiform, hyaline throughout, composed of 8–10 cells, elongate above and shorter in the basal part. *Leaves* erect-appressed to erecto-patent below, becoming larger and somewhat rosulate at the stem tip, contorted and erect when dry, erect to erect-spreading when moist, oblong to oblong-lanceolate or elliptical to oblong-elliptical, (1.5–)1.7–1.9(–2.1) mm long, 0.8–0.9 mm wide, concave, broadly short-acuminate to acute, ending in a short, brownish or yellowish-brown awn, smooth or somewhat papillose near the base; *margins* entire, finely crenulate above with projecting papillae, unistratose throughout, broadly recurved to revolute to near the leaf apex; *costa* short- to long-excurrent as a hair-point, rarely percurrent in the lower leaves, 60–75 µm wide near the base, becoming progressively narrowed upwards, yellowish-brown, sharply demarcated from the laminal cells, inserted laterally or to 35°, with quadrate and papillose dorsal and ventral epidermal cells in the distal half, 3–4 cells across the costa ventrally, in transverse section terete, prominent dorsally, consisting of a semicircular to elliptical dorsal stereid band, 2–3 large guide cells and lacking ventral stereids; *laminal cells* unistratose throughout, not bulging in cross-section; *upper cells* rounded-quadrate, hexagonal, rhomboidal to somewhat elongate, 13–20 µm long, 10–18 µm wide, thin- to moderately thick-walled, densely covered but not obscured with hollow, knob- or crescent-shaped, simple or bifid papillae, 4–6 per cell; *basal cells* rectangular to oblong-hexagonal, 35–85 µm long, 15–22 µm wide, thin-walled, rather lax, hyaline, gradually passing into chlorophyllose, papillose upper cells along an ill-defined line at ± the same level throughout or a little lower at the leaf margin than at the costa, not differentiated across the leaf base or somewhat narrower at the margins but otherwise similar to inner basal cells. *Autoicous*. *Perigonia* bud-like situated near the perichaetium. *Perichaetia* terminal; *perichaetial leaves* undifferentiated. *Setae* elongate, straight, erect, 0.9–1.2 mm long, yellow, becoming reddish-brown with age, lustrous, twisted to the right below, sinistrorse above. *Capsules* exserted, erect, symmetric or nearly so, straight or somewhat curved, obloid-cylindrical, 1.5–1.7 mm long, 0.6–0.7 mm wide, rather abruptly narrowed to the seta, not narrowed at the mouth, rarely systylious, rather leptodermous, castaneous, lustrous, smooth to slightly striate when dry and empty; *annulus* persistent, consisting of 2(–3) rows of small cells; *operculum* obliquely rostrate from a short conical base, with a stout beak about ⅓ the length of the urn, mostly deciduous but rarely attached to the extending columella; *exothecial cells* thin-walled, mostly elongate, rectangular to oblong-hexagonal, sometimes mixed with shorter and irregular in outline, 30–110 µm long, 15–25(–30) µm wide, becoming isodiametric, quadrate to hexagonal in 2–3 rows at the orifice; *stomata* in one row at the base of the urn, relatively large, 40–50 µm long, 35–40 µm wide, pale yellowish to subhyaline, superficial, bicellular, with an elongate slit; *peristome* single, inserted below the orifice, consisting of 16 teeth, erect, curved inwards and forming a dome above the urn mouth when dry, to 400 µm long, pale yellow to yellow-brownish, irregularly divided into 2–3 flat prongs or connate at the joints, sometimes variously perforate, densely spiculosely papillose, arising from a low basal membrane, to 50 µm high. *Spores* globose, 18–24 µm in diameter or ellipsoid, 25–28 µm long, 18–22 µm wide, olive-brown, densely papillose. *Calyptra* cucullate, smooth, naked, reaching the base of the urn.

Chromosome numbers: unknown in the Antarctic and Subantarctic; $n = 26$ (2 records) and $n = 28$ (1 record) from the Holarctic (Fritsch 1991 as *Desmatodon latifolius*).

Discussion. *Tortula hoppeana* has traditionally been placed in the genus *Desmatodon* and under this generic name the oldest available name for it is *D. latifolius* (Hedw.) Brid. This is a long-established name which was originally introduced by Hedwig (1785–1787, 1801) as *Dicranum latifolium*. However, this epithet is not available for this species in *Tortula* if *Desmatodon* is considered to be congeneric with it (Lindberg 1879; Zander 1993). Ochyra (2004*d*) found that the correct name for this species in *Tortula* is *T. hoppeana*. It is based on *Trichostomum hoppeanum* Schultz, a name introduced by Schultz (1828) for a species described from the Alps of Austria which antedates recently used names such as *T. euryphylla* R.H.Zander (Zander 1993) and *T. eucalyptrata* Lindb. (Cano & Gallego 2003).

Tortula hoppeana is a very variable species as regards the size of the plants, shape and papillosity of the leaves, length of the costa and excurrent hair-point, length and shape of the capsule, as well as length of the columella which may be exserted and sometimes the operculum remains temporarily attached to it, making the capsules systylious. This prompted many authors to recognize various phenotypes of this protean species as separate species and varieties. There are many heterotypic synonyms of *T. hoppeana* and a full list of them is presented by Podpěra (1954) and Ochyra (2004*d*). Generally, two varieties are now distinguished within it which, under *Desmatodon latifolius*, are designated as var. *pilifer* (Dicks.) Rabenh. and var. *muticus* (Brid.) Brid. The former comprises the smaller and densely tufted plants which have broadly spathulate to oblong-lanceolate, acuminate leaves with the costa long excurrent as a piliferous hyaline or yellowish hair-point. In fact, this variety represents the typical expression of the species, as represented by the types of *Dicranum latifolium* and *Trichostomum hoppeanum* and consequently var. *pilifer*, if recognised at all, should be designated as the type variety. In contrast, var. *muticus* consists of taller plants growing in looser mats which are characterised by oblong leaves with an obtuse or short-apiculate apex and the costa ceasing below the apex. Both varieties are recognised upon characters which vary continuously and they do not deserve taxonomic recognition. Rather, they may be considered habitat modifications of plants growing in drier (var. *hoppeana*) and moister (var. *mutica*) habitats. The only known Antarctic population of *T. hoppeana* represents the typical piliferous modification.

Tortula hoppeana is characterised by the oblong-lanceolate to spathulate, unbordered and awned leaves with relatively large, papillose cells, recurved leaf margins to the apex, ill-defined transition zone between thin-walled, lax and hyaline basal cells and chlorophyllose, papillose upper cells. In addition, the stem

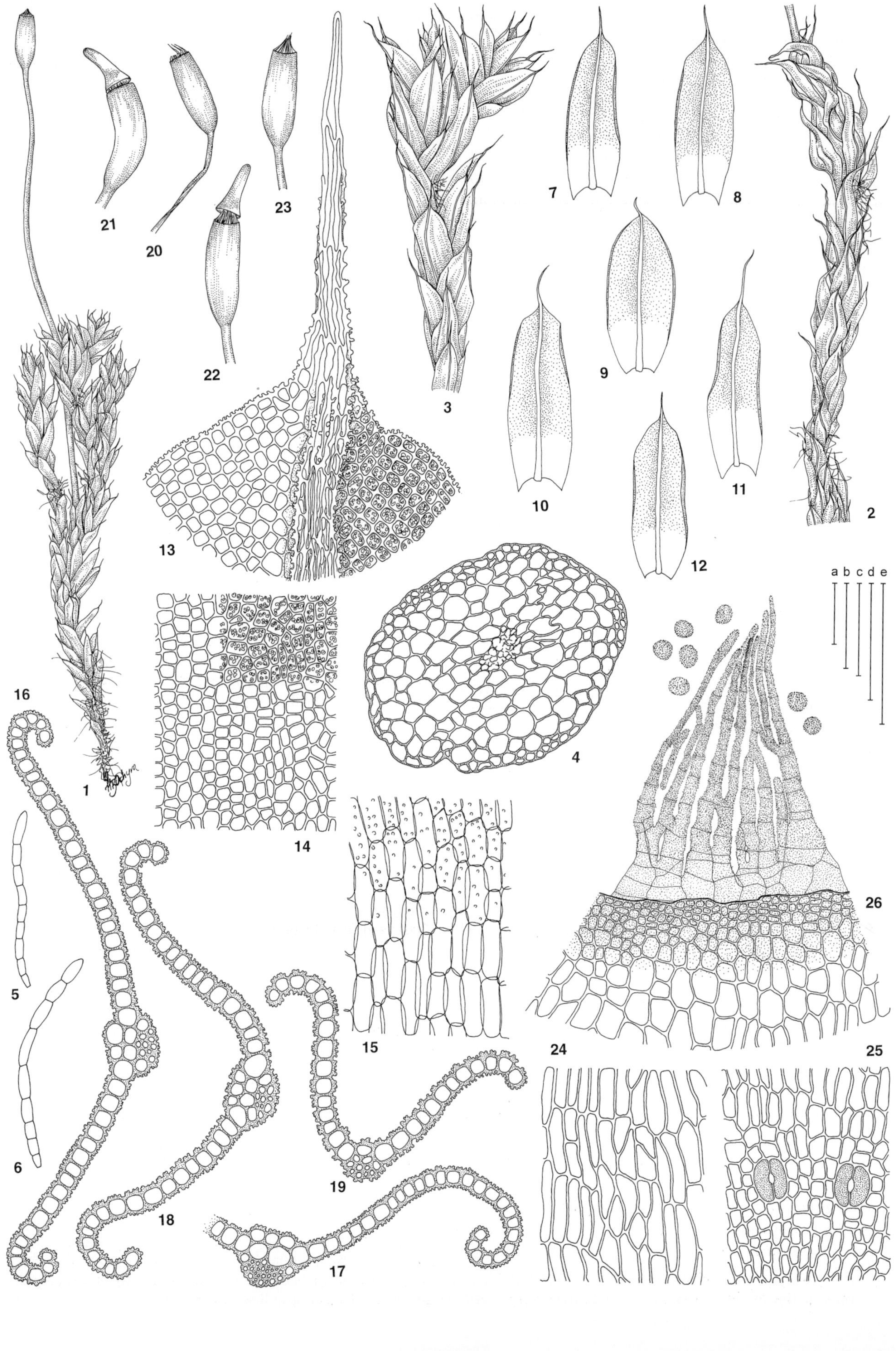
1
2
3
4
5
6
7
8
9
10
11
12
13
14
15
16
17
18
19
20
21
22
23
24
25
26
a b c d e

has a distinct central strand, the peristome teeth are irregularly divided into 2–3 flat branches arising from a low basal membrane. In the Antarctic the species may be confused only with *Syntrichia magellanica* and *Willia austroleucophaea* with which it shares the leaf shape with a similar hair-point. In contrast to the yellowish and smooth leaf awn in *T. hoppeana*, the hair-point in these species is hyaline and denticulate. Moreover, in *S. magellanica* the leaf margins are recurved at mid-leaf for up to two-thirds of the leaf length, the stem central strand is absent, the chlorophyllose and papillose upper laminal cells descend along the leaf margins and abruptly set off from the lax, hyaline basal cells which form a regular fenestrate pattern, and the peristome teeth are split into 32 terete branchlets arising from a high basal tube and twisted into a dextrorse helix. Sterile plants of *Willia austroleucophaea* are readily distinguishable by having pandurate leaves with plane margins, as well as a similar pattern of the leaf areolation. Fertile plants of this species, which have not hitherto been found in the Antarctic, have immersed to emergent capsules, mitriform calyptra, markedly differentiated inner perichaetial leaves from the cauline leaves and lack the peristome.

Reproduction in Antarctica. The only Antarctic population of *Tortula hoppeana* produces fertile plants with mature sporophytes in great abundance.

Habitat. *Tortula hoppeana* grows in the Antarctic in damp rock crevices, which is a typical habitat for this species. It forms small monospecific colonies, although some scattered shoots of *Cephaloziella varians* are concealed within the single collection.

World range. *Tortula hoppeana* has long been considered to be a strictly Holarctic moss, until Ochyra (2004*d*) recorded it recently from the Antarctic and established it as a bipolar species (Fig. 161). In the Northern Hemisphere it has a circumpolar, strongly dissected arctic-alpine range. In the Arctic it is relatively rare and scattered in the Canadian Arctic Archipelago including Banks I. and Victoria I. (Vitt *et al.* 1988), Baffin I. (Brassard *et al.* 1979), Devon I. (Bryhn 1906; Steere 1946) and northern Ellesmere I. (Brassard 1971), widespread throughout Greenland (Holmen 1960), rare and localised on Spitsbergen (Kuc 1973*b*) and

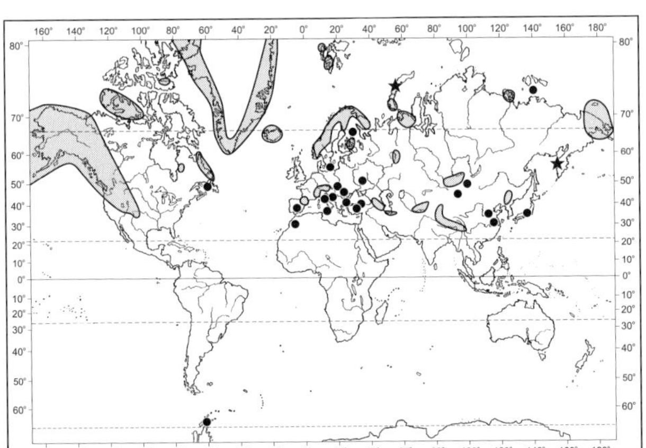

Figure 161. Global distribution of *Tortula hoppeana*.

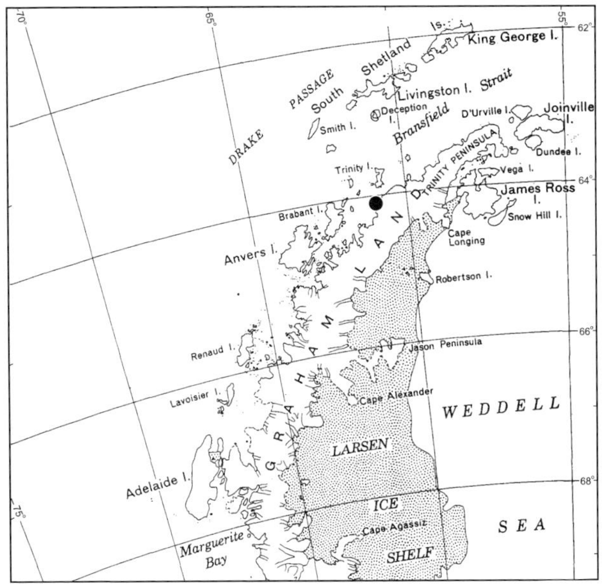

Figure 162. Distribution of *Tortula hoppeana* in Antarctica.

scattered in the Russian Arctic including the Polar Urals, Novaya Zemlya, Yakutia and Chukotka (Afonina & Czernyadjeva 1995). On mainland North America *T. hoppeana* occurs predominantly in the western part of the continent, ranging from Alaska, along the Rocky Mountains south to central California, Arizona and New Mexico (Vitt *et al.* 1988). In Eastern North America the species is rare and localised in Labrador and Quebec (Favreau & Brassard 1988) and in Newfoundland (Brassard 1983*a*). In Europe *T. hoppeana* is bicentric in distribution, being widespread on Iceland (Jóhannsson 1992), on the Kola Peninsula (Shlyakov & Konstantinova 1982), throughout the Scandes of Sweden and Norway (Söderström 1996) and rare in eastern Fennoscandia (Vaarama 1939). Elsewhere in Europe the species is known from all high mountains from the Pyrenees and Sierra Nevada in the west to the Ural Mountains (Ignatov & Ignatova 2003) and the Caucasus (Manakyan 1988) in the east, and extending southwards to Sicily (Cortini Pedrotti 2001*a,b*) and Turkey (Henderson & Prentice 1969), with an isolated station in the High Atlas of Morocco (Jelenc 1955). In Asia outside the Arctic *T. hoppeana* is known from Iran (Schiffner 1908; Ghahreman *et al.* 2003), Central Asia (Mamatkulov 1990), Central Siberia (Bardunov 1974), Mongolia (Tsegmed 2001*a,b*), China including Taiwan (Li *et al.* 2001) and central Honshu in Japan (Saito 1973*b*).

Distribution in Antarctica. *Tortula hoppeana* is known from a single locality at Cierva Point on the Danco Coast on the western Antarctic Peninsula at lat. 64°09′ S (Fig. 162). There is no precise information on the elevation of the locality, except for the general statement on the label that the moss was collected at 5–270 m. Also, no information on the size of the population and its local distribution is available.

Specimens examined. WEST ANTARCTICA. WEST ANTARCTIC PENINSULA. **DANCO COAST.** Cierva Pt., *OB 2* (AAS, KRAM).

←

Figure 160. *Tortula hoppeana*. — **1**. Plant with sporophyte. **2**. Portion of shoot, dry. **3**. Portion of shoot, wet. **4**. Cross-section of stem. **5–6**. Axillary hairs. **7–12**. Leaves. **13**. Leaf apex (papillae partly removed). **14**. Mid-leaf cells (papillae partly removed). **15**. Basal leaf cells. **16–19**. Cross-sections of leaves, a sequence from base to apex. **20, 23**. Deoperculate capsules. **21–22**. Operculate capsules. **24**. Exothecial cells in mid-urn. **25**. Exothecial cells at base of urn and stomata. **26**. Exothecial cells at orifice, peristome and spores. — [All from *OB 2*]. — Scale bars: a – 1 mm (2–3, 20–23); b – 100 µm (4–6); c – 1 mm (7–12) and 100 µm (24–26); d – 100 µm (13–19); e – 0.5 cm (1).

32. HENNEDIELLA Paris

Hennediella Paris, Index Bryol.: 557. 1896. ≡ *Hennedia* R.Br.bis, Trans. & Proc. New Zealand Inst. 25: 285. 1893, *hom. illeg.* [*non Hennedya* Harv., 1860]. — Lectotype (*vide* Zander 1993: p. 242): *Hennediella macrophylla* (R.Br.bis) Paris (*Hennedia macrophylla* R.Br.bis).

Beckettia Müll.Hal., Hedwigia 37: 77. 1898. ≡ *Hennediella* subgen. *Beckettia* (Müll.Hal.) G.Roth, Aussereur. Laubm.: 219. 1911. — Type: *Beckettia bruchioides* Müll.Hal. [= *Hennediella antarctica* (Ångstr.) Ochyra & Matteri].

Willia Müll.Hal. [unranked] *Schistidiella* Müll.Hal., Gen. Musc. Frond.: 424. 1900. ≡ *Willia* subgen. *Schistidiella* (Müll.Hal.) Broth. *in* Engl. & Prantl, Nat. Pflanzenfam. 1(3): 417. 1902. — Type: *Willia marginata* (Hook.f. & Wilson) Müll.Hal. (*Schistidium marginatum* Hook.f. & Wilson) [= *Hennediella marginata* (Hook.f. & Wilson) R.H.Zander].

Neobarbula Dusén, Bot. Not. 1905: 299. 1905. ≡ *Tortula* Hedw. sect. *Neobarbula* (Dusén) Cardot, Wiss. Erg. Schwed. Südpolar-Exp. 4(8): 95. 1908. — Type: *Neobarbula magellanica* Dusén [= *Hennediella densifolia* (Hook.f. & Wilson) R.H.Zander].

Plants small to moderate, loosely caespitose, densely tufted or gregarious, green to yellow-green above, brown to reddish-brown below. *Stems* short, erect, simple or irregularly branched, usually densely radiculose below, in cross-section elliptical to rounded-pentagonal, consisting of a small but distinct central strand surrounded by large, hyaline, thin-walled medullary cells, lacking a cortex and hyalodermis or with an indistinct sclerodermis composed of somewhat smaller, substereid, brownish cells; *axillary hairs* filiform, hyaline throughout, consisting of 5–8 cells, usually with 1–2 shorter and firm-walled basal cells. *Leaves* appressed, incurved or curled and often shrunken when dry, erect to patent and occasionally reflexed at mid-leaf when moist, ovate, oblong-lanceolate to ligulate or spathulate, plane or broadly concave distally, non-decurrent, not differentiated or ovate and somewhat sheathing at the base, acuminate to acute, occasionally obtuse, rounded, short-apiculate or fragile at the apex; *margins* plane or sometimes narrowly recurved, often dentate above, rarely entire, bordered by a few rows of short-rectangular to elongate, smooth or somewhat papillose, thick-walled and often yellowish or brownish cells forming 1–2(–3)-stratose border; *costa* percurrent to short-excurrent as a flattened, denticulate stout mucro, occasionally ending a few cells below the apex, usually inserted laterally with the laminal cells, covered by quadrate to short-rectangular, papillose epidermal cells on the ventral side in 4–5 rows of cells in mid-leaf and elongate, smooth or papillose cells on the dorsal side, in cross-section biconvex, consisting of a rounded or elliptical dorsal stereid band, 2–4 large guide cells in (1–)2 layers and a ventral epidermis of large cells usually forming a convex pad, dorsal epidermis absent or present, usually only laterally; *laminal cells* unistratose throughout, sometimes bi- or tristratose at margins, showing red coloration with KOH; *upper cells* large, quadrate, hexagonal or short-rectangular thin-walled or with evenly thickened walls, superficially plane to weakly convex, translucent to obscure, usually densely papillose with hollow, simple or bifid papillae, most often 6–8 per lumen, occasionally weakly papillose to smooth; *basal cells* rectangular, thin-walled, hyaline, lax, not differentiated across the base, gradually merging with the upper chlorophyllose cells, often distinctly ascending along the margins and passing into a distinct multiseriate marginal border. *Propagula* in the form of rhizoidal tubers, irregular or more or less ovoid, with protuberant cells or swollen rhizoidal segments sometimes present. *Autoicous*, *paroicous* and *dioicous*. *Perichaetial leaves* similar to the vegetative leaves, the latter only somewhat smaller. *Setae* singly per perichaetium, erect, straight, often fairly stout, yellow-brown to brown, dull, sinistrorse above. *Capsules* exserted or emergent, rarely immersed, ovoid to obloid-cylindrical from a short and inconspicuous neck, not or narrowed at the mouth, pachydermous, yellow-brown to dark brown, lustrous, smooth; *annulus* persistent, 1–4-seriate, composed of vesiculose cells; *operculum* obliquely long-rostrate from a convex or plano-convex base, with the cells arranged in the anticlockwise rows; *exothecial cells* rectangular below, quadrate to hexagonal above, thin- to thick-walled; *stomata* large in a single row at the base of the urn, phaneropore, with elliptical openings; *peristome teeth* absent, rudimentary or 16, deeply divided into 32 filamentous, terete, richly articulate, densely spiculose segments, twisted anticlockwise, arising from a high, spiculose basal membrane. *Spores* globose, small to large, yellowish, red to dark brown, smooth, densely and finely papillose to verrucose. *Calyptra* cucullate, smooth, sometimes much enlarged and covering the entire capsule.

Hennediella has long remained an obscure and poorly understood genus of the pottiaceous mosses, despite the considerable literature relating on it. The genus was established by Brown (1892) as *Hennedia* to accommodate three newly described New Zealand species, *H. macrophylla*, *H. intermedia* and *H. microphylla*. Because this generic name was an illegitimate younger homonym of the algal genus *Hennedya* Harv. of 1860, Paris (1896) renamed it as *Hennediella* and as such it was approved by Brotherus (1902, 1924*b*) and Dixon (1923*a*). The concept of the genus was essentially based on the very large, mitriform calyptra, confluent at the base, persistent when maturing and rupturing by lateral growth of the capsule. Dixon (1923*a*) considered the genus to be closely related to *Pottia* with which it was linked by *P. heimii* (Hedw.) Hampe and *P. grata* R.Br.bis, which he found to be identical to *Hennediella microphylla*. At the same time he suggested the close similarity and possible identity of the South Georgian *Pottia austrogeorgica* Cardot with *Hennediella microphylla*. Sainsbury (1952, 1955) discarded *Hennediella* as a distinct genus in favour of *Pottia* and suggested it may eventually constitute a separate section in this genus, distinguished by the dentate and bordered leaves and by the very large and mitriform calyptra. At the same time he included all species formerly placed in *Hennediella* within his concept of *Pottia macrophylla* (R.Br.bis) Sainsbury.

Blockeel (1990) resurrected *Hennediella* and emphasised the enlarged calyptrae coupled with the often short and stout setae, gymnostomous capsules and toothed leaf border of elongate cells to be a distinctive combination of features, sufficient to justify retention of *Hennediella* as a genus in its own right. Apart from *H. macrophylla*, he placed in it *H. austrogeorgica* (Cardot) Blockeel and *H. stanfordensis* (Steere) Blockeel. Finally, Zander (1993) presented a refined and expanded concept of *Hennediella* to which he transferred about 18 species previously positioned in *Pottia*, *Tortula*, *Desmatodon* and *Willia*. They are all united by the following suite of characters: (*1*) large and broadly lanceolate leaves; (*2*) distinct leaf border composed of elongate, thick-walled, coloured and epapillose cells; (*3*) serrate to dentate and mostly plane leaf margin; (*4*) large, thin-walled and strongly hollow-papillose upper laminal cells showing red KOH reaction; (*5*) elliptical, semicircular to rounded dorsal stereid band, usually with dorsal epidermis; and (*6*) enlarged and often inflated calyptra. *Hennediella* is doubtless closely related to *Tortula* with which it shares a similar leaf areolation and anatomy of the costa and a similar reduction series in sporophytes, but *Tortula* has generally broader leaves, more convex superficial surface of the laminal cells, lacks the marginal dentation and shows yellow KOH reaction of the upper laminal cells.

Zander (1993) placed in *Hennediella* 21 species but the real number is less if new reductions of names to synonymy presented here are considered, although this is partly compensated by descriptions of new species (Zander & Mahu 1999). *Hennediella* is a genus predominantly of the Southern Hemisphere where all its species occur, although three are bipolar taxa. The genus exhibits its greatest diversity in the central and northern Andes where eleven species, including eight local endemics, occur at high elevations from Bolivia to Colombia, occasionally extending northwards to the West Indies and Mexico (Delgadillo *et al.* 1995; Churchill *et al.* 2000).

Only two species have been recorded from South and East Africa and eight are widespread in the south-cool temperate and polar zone, five of which are endemic to the Subantarctic and southern South America. The genus has not yet been monographed, although valuable data on some species may be found in the old but still useful treatment of *Pottia s. lato* by Warnstorf (1916).

Two species of the genus are widely distributed in the Antarctic biome. They have been studied taxonomically by Savicz-Lyubitskaya & Smirnova (1963*b*, 1965*b*), Matteri (1977*a*), Kanda (1981*b*) and Ochyra (1998*a*) and, additionally, a valuable treatment of *Hennediella austrogeorgica* is available for South Georgia (Matteri 1977*b*).

1. Laminal cells more than 16 μm wide; leaf marginal border indistinct or absent; calyptra fugacious, covering only the upper half of the urn .... 1. ***H. heimii***
1. Laminal cells less than 16 μm wide; leaf marginal border prominent; calyptra often persistent, covering entire capsule .... 2. ***H. antarctica***

1. **Hennediella heimii** (Hedw.) R.H.Zander

Fig. 163

Hennediella heimii (Hedw.) R.H.Zander, Bull. Buffalo Soc. Nat. Sci. 32: 248. 1993. ≡ *Gymnostomum heimii* Hedw., Sp. Musc. Frond.: 32. 1801. ≡ *Bryum heimii* (Hedw.) Dicks., Fasc. Pl. Crypt. 4: index [i]. 1801. ≡ *Pottia heimii* (Hedw.) Hampe, Flora 20: 287. 1837. ≡ *Desmatodon heimii* (Hedw.) Mitt., J. Linn. Soc. Bot. 8: 28. 1864. ≡ *Tortula heimii* (Hedw.) Mitt., J. Linn. Soc. Bot. 12: 165. 1869. ≡ *Didymodon heimii* (Hedw.) Kindb., Eur. N. Am. Bryin. 2: 280. 1898. — TYPE CITATION: Prope Spandau primus invenit Heim, consilarius aulicus, Berolinensis. Circa Malchin ducatus Mecklenburgici clariass. Timm; Upsaliae ad palos vetustos Ol. Swarz, (in Austria D. Host, in Sudetibus Ludwig.); in addition, references to Hedwig (1785–1787) and Bridel (1798) are given. [Lectotype (*selected here*): The specimen marked 'a' on the sheet labelled 'Gymnostomum Heimii Hedw. St. Cr. 1. 80.t.30. a ab ipso inventore Cons. Aulico Heim acceptum (*manu Hedwig*) b. ex palis fontis in arce Upsal. c. Yarmouth (*manu Schwägrichen*)' – G-Hedwig/Schwägrichen!].

Bryum antarcticum Hook.f. & Wilson, Fl. Antarct. 1: 414, *pl. 153, f. 6.* 1847. — TYPE CITATION: Cockburn Island, lat. 64°S., long. 57°W. [Lectotype (*selected here*): 'Cockburn Island. N°. 6. Bryum antarcticum Hfil & Wils. Lat. 64°12′. Long. 56°56′. W' – BM-Hooker!; isotypes: BM-Hooker! (4 specimens), PC!]. First synonymised with *Pottia heimii* by Kanda (1981*b*: p. 102).

Pottia naumannii Müll.Hal., Bot. Jahrb. Syst. 5: 79. 1883 ['naumanni'], ***syn. nov.*** — TYPE CITATION: all species described as new by Müller (1883) are included in the section 'I. Bryologia Kerguelensis'; later Müller (1889) wrote: 'Insula Kerguelen, Betsy Cove (11.74)'. [Lectotype (*selected here*): 'Pottia Naumanni C. Müll. n. sp. Kerguelen. Betsy's Cove. Nov. 1874 l. Naumann' – HBG-Prahl!; isotype: PC-Cardot!].

Pottia oedipodioides Müll.Hal., Bot. Jahrb. Syst. 5: 79. 1883. ≡ *Hennediella oedipodioides* (Müll.Hal.) R.H.Zander, Bull. Buffalo Soc. Nat. Sci. 32: 248. 1993, ***syn. nov.*** — TYPE CITATION: all species described as new by Müller (1883) are included in the section 'I. Bryologia Kerguelensis'; later Müller (1889) wrote: 'Insula Kerguelen, Betsy Cove (11.74)'. [Lectotype (*selected here*): 'Ex Museo botanico Berolinensi. Pottia oedipodioides C. Müll. Kerguelen. Dr. Naumann 1874' – PC-Cardot!].

Pottia charcotii Cardot, Compt. Rend. Hebd. Séances Acad. Sci. 153: 602. 1911. ≡ *P. heimii* var. *charcotii* (Cardot) L.I.Savicz & Smirnova, Novosti Sist. Nizš. Rast. 2: 258. 1965 [*P. heimii* var. *charcotii* (Cardot) L.I.Savicz & Smirnova, Bot. Mater. Otd. Spor. Rast. Bot. Inst. Komarova Akad. Nauk SSSR 16: 193. 1963, *comb. inval. basion. non citat.*]. — TYPE CITATION: Terre de Graham: Cap des Trois-Pérez. [Lectotype (*selected here*): 'Herb. Mus. Paris 273^{c}. Pottia Charcotii Card. sp. nova. + Bryum sp. Terre du Graham: cap des Trois Pérez, fentes des rochers. 2ème expéd. Charcot. Leg. Gain, 6/3 1909' – PC-Cardot!; isotype: S!; syntype: 'Herb. Mus. Paris 272^{b}. Pottia Charcotii Card. sp. nova. Terre du Graham: cap des Trois Pérez, fentes des rochers. 2ème expéd. Charcot. Leg. Gain, 6/3 1909' – PC-Cardot!; isosyntype: H-Brotherus!]. First synonymised with *Pottia heimii* by Matteri (1977*a*: p. 141).

Pottia heimii var. *brevinervis* L.I.Savicz & Smirnova, Bot. Mater. Otd. Sporov. Rast. Inst. Bot. Komarova Akad. Nauk SSSR 16: 193. 1963. — TYPE CITATION: Antarctida Orientalis, pars centralis Banger Hills, ad pedes monticule 36 m s. m., sterilis (typus); in trajectu inter lacum angustias fjord Blizhnij dictas, ad marginem lacunae; ripa continentalis lacus Olgino, in terra humida; ad ripam rivuli dulcis, inter caespitulos muscorum lentissime fluentis (caulis unicus sterilis). [Holotype: not located at LE and possibly lost; neotype (*selected here*): 'Sectio Cryptogamica Instituti Botanici Academiae Scientiarum URSS № 10/410 Pottia heimii (Hedw.) Fuernr. var. brevinervis Lyd. Savicz et Z. Smirn. Station: Vost. Antarktida. Oasis Bunger Hills, r-n fjordov w sredney chasti, na perevale mezhdi ozerkom i fjordom 'Blizhnym', po krayu luzhitsy ['East Antarctica, Bunger Hills, the region of the fjords in the middle part between the lake 'Blizhnyy and the fjord, on the bank of a pool'] 1957.11.19 Leg. M. M. Hollerbach 1961 Det. L. I. Savicz i Z. Smirnova' – LE!;

isoneotypes: H!, KRAM!, LE!, NY!]. First synonymised with *Pottia heimii* by Kanda (1981*b*: p. 102).

Plants small to medium-sized, stiff and rigid or relatively soft, densely or loosely caespitose, sometimes gregarious or forming wide patches or mats, yellowish- or brownish-green to stramineous, reddish-brown, deep green or sordid-green above, light brown below, often one-coloured, yellow-brownish or green throughout, dull to lustrous. *Stems* erect, usually 0.4–1.2 or occasionally up to 2.5 cm tall, simple or sparsely, irregularly branched, often buried in soil and strongly radiculose at the base with the rhizoids often much longer than the stem, light to reddish-brown, lustrous, smooth, sparingly branched, in cross-section obscurely rounded-pentagonal to oval, with a small or large, distinct central strand, surrounded by 4–5-layered medulla of large, hyaline, thin-walled cells and an indistinct, 1–2-stratose outer cortex of somewhat smaller, thin-walled, brownish cells; *axillary hairs* filiform, hyaline throughout, composed of 5–7 mostly elongate cells and with 1–2 shorter basal cells. *Leaves* closely set, erect-appressed and imbricate, not or slightly altered or distant and contorted when dry, erect-spreading when moist, usually progressively larger towards the stem tip and sometimes rosulate, ovate, oblong-ovate, oblong-lanceolate, occasionally spathulate, short- to long-acuminate or acute, often with subtubulose to subcucullate apex on drying, 0.6–2.5 mm long, (0.2–)0.5–1.0 mm wide, broadly concave, non-decurrent; *margins* plane to erect, unistratose, entire throughout or crenulate in the distal part by papillose cells, sometimes more or less irregularly toothed or roughly serrate by projecting cells at the apex, not or slightly bordered by oblong to rhomboidal, less papillose, thicker-walled and yellowish cells in 1–2 rows; *costa* strong, reddish- to yellow-brown, sometimes pale yellowish and almost concolorous with the laminal cells, subpercurrent to short-excurrent as a cuspidate point, sometimes ceasing some distance below the leaf apex, 50–70 µm wide, covered by quadrate to short-rectangular, papillose epidermal cells on the ventral surface and elongate, smooth cells on the dorsal side, in cross-section plano- or biconvex, consisting of a prominent, rounded or elliptical dorsal stereid band, 2–4 large guide cells in 2 layers and a ventral epidermis of large cells, forming a flat or convex pad, dorsal epidermis absent or present, usually only laterally; *laminal cells* unistratose throughout, flat superficially; *upper cells* variable in shape, subquadrate, rounded-quadrate, elongate-hexagonal to oblong-rhomboidal, with thin to moderately, evenly thickened walls, (14–)16–25(–35) µm long, (13–)16–20(–25) µm wide, strongly papillose, especially on juvenile leaves, with simple or bifid, hollow, C- or O-shaped papillae, usually 6–8 per cell, but rather pellucid, often weakly papillose to entirely smooth, becoming sometimes oblong to rhomboidal, less papillose at the margin in 1–2 rows; *basal cells* rectangular, 40–80 µm long, 18–25 µm wide, thin-walled, lax, uniform throughout the base, rather imperceptibly merging into the upper chlorophyllose cells, often extending along the margins, forming an indistinct V-shaped basal region. *Asexual reproduction* sometimes present as swollen, linear, multicellular propagules arising from rhizoids attached to axillary shoots. *Dioicous, autoicous* or *synoicous*. *Gametangia* terminal. *Perichaetial leaves* usually larger than the vegetative leaves, 2.5–3.0 mm long, 0.7–1.0 mm wide, on small plants to 1.2 mm long and 0.5 mm wide, but otherwise similar in shape. *Setae* erect, straight, 5–10(–13) mm long, rather stout, yellowish but becoming brownish with age, sinistrorse above, sometimes dextrorse below. *Capsules* exserted, gymnostomous, systylious, from a short and inconspicuous neck ovoid, obloid to short-cylindrical, 0.9–1.6 mm long, 0.6–0.9 mm wide, straight or somewhat curved, not wide-mouthed when deoperculate, brown, smooth, lustrous, rather pachydermous; *annulus* persistent, 1–3-seriate, composed of small, rounded cells with strongly incrassate walls; *operculum* obliquely long-rostrate from a plano-convex to convex base, with a beak at least half or more the length of the urn, remaining attached to the columella after dehiscence and elevated above the capsule mouth owing to the elongation of the columella after the spores ripen; *exothecial cells* mostly rectangular, thin- to thicker-walled, 30–60 µm long, 15–20 µm wide, becoming shorter and wider, rounded or elliptical in 2–3 tiers at the orifice; *stomata* in one row at the neck of the urn, bicellular, phaneroporous, with a narrow, elliptical slit, pale, relatively large, 21–25 µm wide, 23–29 µm long; *peristome* absent. *Spores* short-ellipsoid, yellowish to brownish, densely papillose, 20–40 µm in diameter. *Calyptra* cucullate, smooth, early fugacious, covering about one-half the urn.

Chromosome numbers: $n = 26 + 2m$ – Ongul I., Enderby Land (Inoue 1976) and $n = 26$ (Przywara *et al.* 1984); additional five records of $n = 28$ from the Holarctic (Fritsch 1991).

Discussion. Apart from *Ceratodon purpureus* and *Bryum pseudotriquetrum*, *Hennediella heimii* is the most variable and polymorphous species of moss in the Antarctic, and its considerable morphological variability is coupled with its remarkable genetic diversity, at least in the Antarctic (Selkirk *et al.* 1998; Dale *et al.*

Figure 163. *Hennediella heimii*. — **1**. Plant with sporophytes, wet. **2**. Cross-section of stem. **3–4**. Axillary hairs. **5–12**. Leaves. **13–14**. Leaf apices (papillae partly removed). **15**. Leaf apex with epapillose cells. **16**. Mid-leaf cells. **17**. Lamina cells in transition zone to hyaline base. **18**. Basal leaf cells. **19–23**. Cross-sections of leaves. **24**. Perichaetial leaf. **25**. Systylious capsule, dry. **26**. Stomata and basal exothecial cells. **27**. Exothecial cells at mouth and spores. — [1, 6–7, 13, 17, 23–24, 26–28 from *RO 928/80*; 2, 4, 10–11 from *RO 2788/80*; 3, 8–9, 14, 20–21 from *Hollerbach 10/410*; 5 from *Gain 272b* (lectotype of *Pottia charcotii*), PC; 12, 15 from *Hooker 6* (holotype of *Bryum antarcticum*), BM; 16, 18–19 from *RO 2761/80*; 22 from *RO 1164/80*; 25 from *Kanda 780824-028*]. — Scale bars: a – 100 µm (26–27); b – 100 µm (2); c – 1 mm (25); d – 1 mm (24), 2 mm (1) and 100 µm (3–4, 13–23); e – 1 mm (5–12).

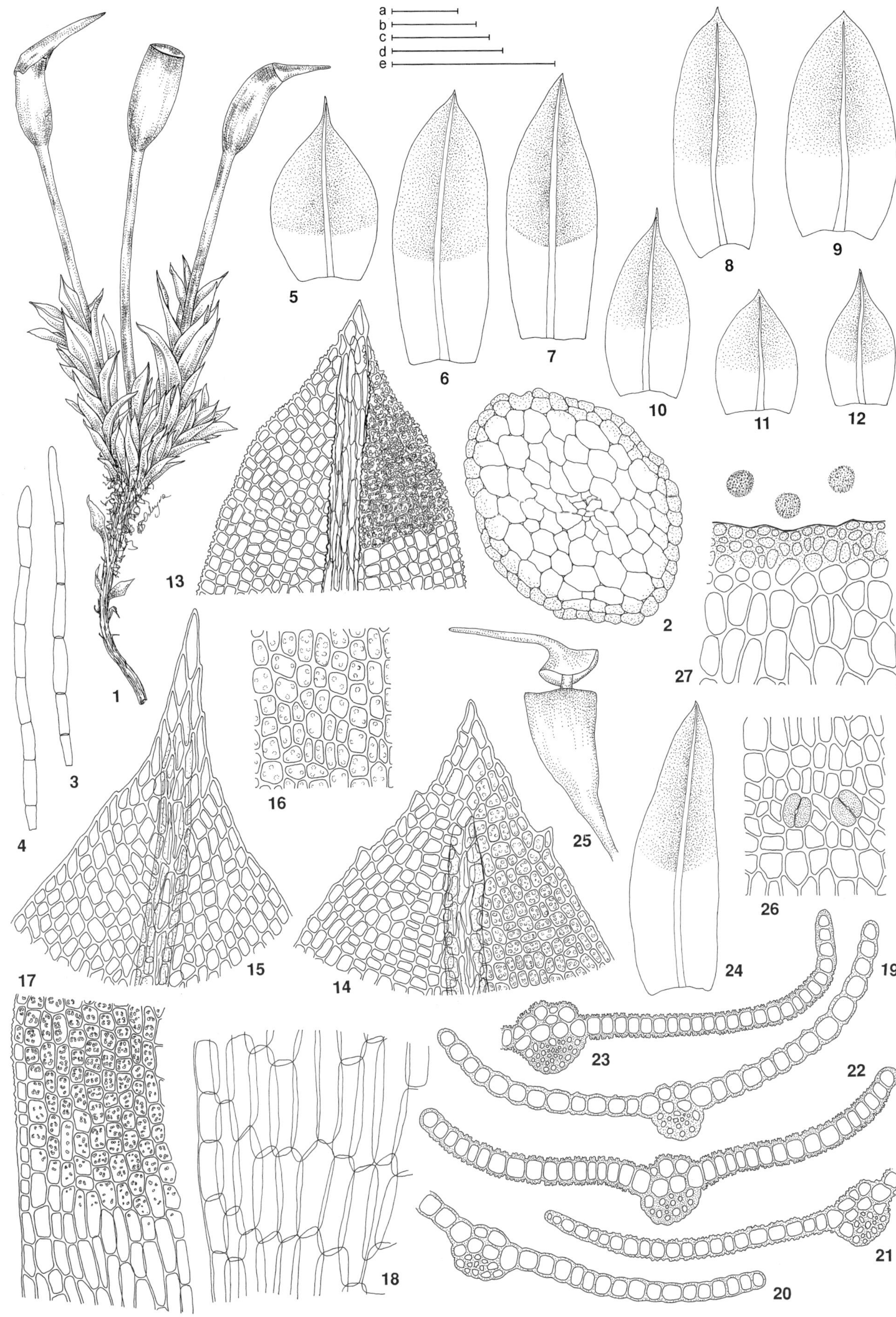
a
b
c
d
e
1
2
3
4
5
6
7
8
9
10
11
12
13
14
15
16
17
18
19
20
21
22
23
24
25
26
27

1999). In spite of variations, its typical expressions are readily recognised by their more or less roughly dentate or serrate leaf apex which may be acute to acuminate but scarcely hair-pointed, plane and indistinctly bordered leaf margins, densely papillose but rather translucent upper laminal cells which are usually mostly 16–35 μm long and 16–25 wide, and systylious capsules lacking a peristome. However, all the vegetative characters of this species are subject to considerable variation. The plants show much variability in height, shoot density, size, shape and density of leaves, but these are merely the result of local conditions of growth, exposure, wetness of habitat, etc. The Antarctic plants differ markedly in habit and generally two extreme phenotypes are observed. Many plants are small, rather stiff and rigid, with brown appressed and little altered leaves and grow in dense tufts which are often buried in soil. Such plants were described by Wilson & Hooker (1847) as *Bryum antarcticum* and are widespread in drier habitats throughout the Antarctic range of the species. The opposite extreme includes much larger, rather softer plants, with larger, spreading and somewhat contorted leaves which usually thrive in moist habitats. In the austral polar region such plants were described, among others, by Müller (1883, 1889) as *Pottia naumannii* and *P. oedipodioides* from Îles Kerguelen, but both names are here reduced to synonymy with *Hennediella heimii* because there are no discernible differences which separate them from this species.

The leaves of *Hennediella heimii* may vary in size and shape, the border is normally lacking or indistinct, the leaf margins are often entire in the upper part, the laminal cells may be papillose, and the costa may vanish in or below the leaf apex or may be excurrent. Some of these phenotypes were given separate taxonomic status, and Warnstorf (1916) enumerated 13 varieties and two forms, based principally on leaf form and shape. In the Antarctic some continental populations having the costae ceasing well below the apex were described as *Pottia heimii* var. *brevinervis* (Savicz-Lyubitskaya & Smirnova 1963*b*), whereas plants with epapillose laminal cells from the Graham Coast on the Antarctic Peninsula were described as a separate species, *Pottia charcotii*, by Cardot (1911*a*,*b*, 1913). This was subsequently reduced to a variety of *P. heimii* by Savicz-Lyubitskaya & Smirnova (1963*b*, 1965*b*). However, within the same population of this species it is possible to detect all stages intermediate between the extreme forms, and this clearly indicates that the variation of particular characters is continuous and, accordingly, taxa recognised on their basis do not merit separate taxonomic status.

Nyholm (1990) mentioned the occurrence of *Hennediella heimii* var. *arctica* (Lindb.) R.H.Zander in the Antarctic, but without citation of the specimens or their location. This is an arctic plant characterised by shorter, oblong-lanceolate and strongly concave leaves which are obtuse or shortly acute and entire to slightly toothed at the apex and the upper laminal cells which are opaque and markedly obscured by the densely arranged papillae and pale and smooth at the margins. In extreme cases the leaves are rounded-ovate or ovate-ligulate and broadly obtuse at the apex. In fact, such plants have not been discovered in Antarctic material and in the great majority of specimens examined the distal parts of the leaf laminae are rather pellucid and the leaf border is indistinct or wanting.

Hennediella heimii also exhibits remarkable variability of sporophyte characters, especially the length of the seta and the shape of the capsule. Normally, the capsule is relatively short, broad and straight, but many capsules are distinctly curved. However, variability of sporophytes is less obvious in the Antarctic because in many populations the capsules are immature and aborted.

In the field, *Hennediella heimii* may be readily mistaken for some morphs of *Ceratodon purpureus*, especially those described as *C. antarcticus*. The differences between these plants are discussed in detail under the latter taxon. Also, *H. heimii* may be confused with small plants of *Bryum pseudotriquetrum*, especially when both species grow together. However, the red coloration of the leaf base in *B. pseudotriquetrum*, together with the narrower leaf apex, the entire leaf margins and the anatomy of the costa which has the central stereid band encircled by the large ventral and dorsal epidermal cells should prevent misdetermination.

Hennediella heimii is readily distinguished from all other pottiaceous mosses in the Antarctic. Its congener, *H. antarctica*, is distinct on account of the much smaller distal laminal cells which are generally 10–15 μm wide and opaque due to the strong papillosity, the presence of a prominent marginal border from the apex to mid-leaf of elongate, less papillose or epapillose, thick-walled and yellow cells, the distinctly toothed upper leaf margin and the large, persistent calyptra which covers the entire capsule. *Tortula hoppeana* differs from *H. heimii* by its broadly acuminate to broadly acute, piliferous leaf apex and recurved to revolute leaf margins. *Syntrichia filaris* shares with *H. heimii* the dentate or obscurely denticulate leaf margin near the apex and the differentiated marginal border consisting of a row of the smaller and slightly papillose cells. However, it is generally a much larger plant with leaves 2.0–4.2 mm long and 0.7–1.0 mm wide, and the densely papillose distal chlorophyllose leaf cells which descend along the margins and sharply set off from the hyaline basal cells which exhibit a distinct fenestrate pattern. In addition, *S. filaris* lacks the stem central strand and the dorsal stereid band of the costa is distinctly crescent-shaped in transverse section.

Hennediella heimii is a Hedwigian species which was originally described as *Gymnostomum heimii* from European material (Hedwig 1785–1787, 1801). Because Hedwig (1801) cited several specimens in the protologue, none of which was indicated as the type, this species name is lectotypified here with a specimen from his personal herbarium which was doubtless seen and examined by that author.

Reproduction in Antarctica. *Hennediella heimii* readily produces sporophytes, sometimes in abundance, both in the maritime and continental Antarctic. In the former region plants with fully mature capsules have been observed in many populations, with the southernmost limit in Mars Oasis on eastern Alexander I. at lat. 71°49′S (*RILS 10440B*, AAS, KRAM). On the continent mature sporophytes have been recorded from the Sôya Coast, Enderby Land, at lat. 69°29′S (*Kanda 190177*, KRAM) and 69°40′36″S (*Kanda 190376*, KRAM), wheareas abundant young, immature sporophytes have been noted in southern Victoria Land at lat. 77°55′S (Seppelt *et al.* 1992). Rastorfer (1971) developed sporophytes in an artificial environment for what he called *Bryum antarcticum* collected in Miers Valley at lat. 78°06′34″S in the same area but, judging from the illustration of the capsule and operculum, the material refers rather to *Bryum pseudotriquetrum* (see also Kanda 1981*b*). Nonetheless, these are the farthest south locations for any moss sporophytes. In addition, *H. heimii* may propagate asexually by deciduous stem apices and axillary shoots, as well as by swollen rhizoidal segments arising from axillary shoots.

Habitat. *Hennediella heimii* is considered to be a halophytic moss characteristic of salt-marshes in the Northern Hemisphere (Adam 1976) or habitats rich in salts of sodium and potassium (Hübschmann 1960; Karczmarz 1961; Marstaller 1989; Oesau

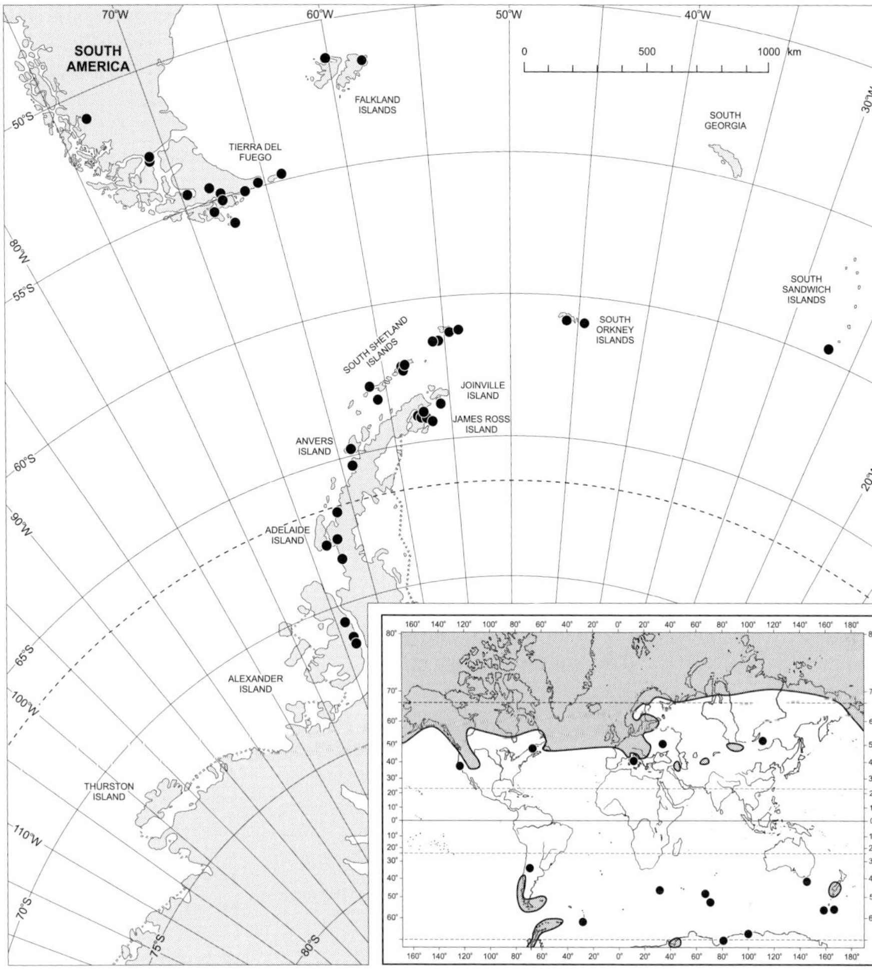

Figure 164. Distribution of *Hennediella heimii* in the Antarctic Peninsula and Fuegian regions. Inset: global distribution (localities in Dronning Maud Land and Victoria Land are not indicated).

2005). In the Antarctic it is a pioneer moss species colonising a variety of mineral substrates and habitats, especially near the sea coast, but also inland at places with no marine influence, growing in moist or dry situations in open and sunny or sheltered sites which are often unstable due to frost action or disturbance, as well as habitats enriched with nitrogen from birds or elephant seals. *H. heimii* thrives on muddy saline or alkaline soil, on gravel, fine clayey, silty or sandy soil, on moraines and marble detritus, at the margins of drainage channels, melt water runnels, seepage areas, ponds and streams, and below late snow beds; also on rock ledges, outcrops, scree, volcanic scoria and tephra rocks, rocky knolls used by birds as perches, in rock crevices and ledges. On Deception I. the species has been found inside old fuel tanks in the ruined whaling station. The species usually forms small pure tufts, but occasionally with an admixture of *Bryum pseudotriquetrum* and *Cephaloziella varians*; sometimes in continental habitats it is heavily colonised by cyanobacteria.

World range. A bipolar species in the strict sense, i.e. having no intermediate localities in the tropical region (Fig. 164, inset). In the Northern Hemisphere, *Hennediella heimii* has a continuous range throughout the Arctic and partly in the northern boreal zone, although generally the species is scattered and not abundant anywhere. In North America it extends southwards to central California (Norris & Shevock 2004) and New Mexico (Bartram 1931; Mahler 1981) in the west, and to Newfoundland (Brassard 1983*a*) and New Brunswick (Ireland 1982) in the eastern parts of the continent. Likewise, it is scattered around Greenland where it occurs mostly in its Arctic modification known as var. *arctica* (Holmen 1960). In Europe, the species is widespread in the northern regions including Svalbard (Kuc 1973*b*; Frisvoll & Elvebakk 1996), Iceland (Jóhannsson 1992), the British Isles and Ireland (Chamberlain & Preston 1992) and most parts of Scandinavia (Söderström 1996), and extends southwards to northern and central Italy (Cortini Pedrotti 2001*a*), Serbia (Pavletić 1955; Sabovljević & Stevanović 1999) and eastwards to Romania (Mohan 1998), the Ukraine (Bachurina & Mel'nichuk 1988), eastern Poland (Karczmarz 1961) and the Baltic countries (Söderström 1996), with some isolated occurrences in southern Russia (Ignatov & Ignatova 2003) and Azerbaydzhan (Lyubarskaya 1986). In Asia, *H. heimii* is scattered outside the Arctic in Middle Asia (Mamatkulov 1990), Central Siberia (Bardunov 1974) and northern Mongolia (Tsegmed 2001*a*,*b*). The record of the species from Japan (Sakurai 1954) is considered to be doubtful (Noguchi 1988), as is the case with its North African record (Ros *et al.* 1999). In the Southern Hemisphere, *H. heimii* has a wide range, predominantly in the Antarctic and Subantarctic including the Prince Edward Is. (Ochyra, personal observations), Îles Kerguelen (see list of synonyms above), Heard I. (Bergstrom & Seppelt 1988; Bergstrom & Selkirk 1997) and Macquarie I. (Seppelt 2004). Interestingly, the species has not hitherto been recorded from South Georgia, although it is widely distributed and locally abundant in the conterminous territories of South America and Antarctica. A single specimen from Maiviken in Cumberland West Bay on this island (*SWG 3464*, slide in AAS) which Matteri (1977*b*) considered as close to *H. heimii* because of its systylious capsules and entire leaf margins, is here interepreted as *H. antarctica* with which it shares the leaf areolation, systylious capsules are seldom observed in that species. It is known from a few stations on the South Island of New Zealand (Dixon 1923*a*; Sainsbury 1955), Campbell I. (Vitt 1974*a*), Tasmania (Weymouth & Rodway 1922; Sainsbury 1953; Dalton *et al.* 1991), but is unknown from south-east Australia as indicated on the map of its global distribution by Schofield (1974). It seems to be much more widespread in the temperate region of southern South America, extending from the Falkland Is. and Tierra del Fuego (Wilson & Hooker 1847; Dusén 1907) northwards to the Valdivian region in Chile (Greene 1986), with an isolated station in the Andes in Mendoza Province in north-western Argentina whence it was recorded as *Pottia guessfeldtii* Schlieph. (Schliephacke 1884).

Distribution in Antarctica. The geographical range of *Hennediella heimii* was formerly presented for this biome by Greene (1967*b* as *Bryum antarcticum*) and Kanda (1987*a* as *Pottia heimii*) when the species appeared to be quite rare and localised. However, examination of all available collections established it as a widely distributed but generally scattered pan-Antarctic species (Figs 164–165). In the maritime Antarctic, it extends from the South Sandwich Is. and South Orkney Is., where it is locally frequent and abundant on Signy I., through the South Shetland Is. where it is rather rare and infrequent, with the exception of Deception I., to Alexander I. where it has its southernmost locality in the West Antarctic at lat. 71°50′ S. On the east coast of the Antarctic Peninsula it is only relatively frequent on islands of the James Ross Island group where it was recorded for the first time in Antarctica on Cockburn I. as *Bryum antarcticum* (Wilson & Hooker 1847). Interestingly, the species has been recorded only rarely on mainland Antarctic Peninsula (e.g. Cape Pérez, Orford Cliff) although it is more frequent on the offshore islands. On the continent, *H. heimii* is widely scattered but not abundant in all sectors, although it constitutes a significant proportion of the terrestrial flora in the McMurdo Dry Valleys region of southern Victoria Land, where it reaches its southernmost station at lat. 78°09′ S in Miers Valley. Kanda (1987*e*) reported it from Vestfold Hills but his specimens have not been available for study. Nonetheless, the occurrence of the species in this area is confirmed by another collection (*RILS 6397*, AAS, KRAM). *Hennediella heimii* occurs predominantly at low elevations, from sea level to 150 m, and is much rarer at high elevations, reaching its highest station at 540 m on James Ross I.

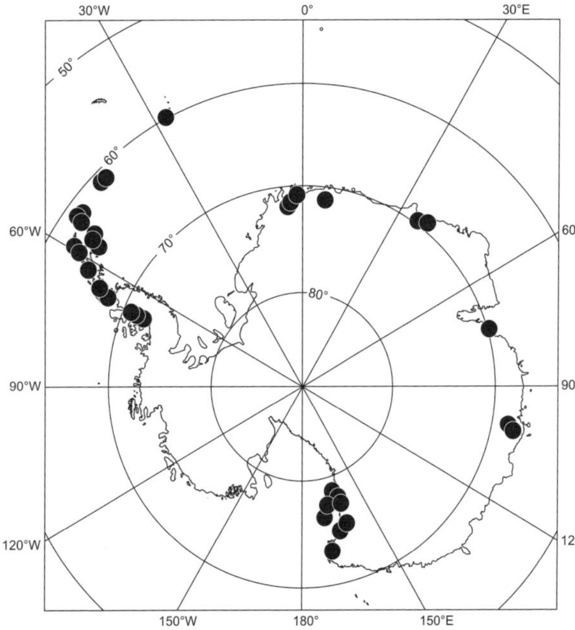

Figure 165. Distribution of *Hennediella heimii* in Antarctica.

Selected specimens examined. WEST ANTARCTICA. SOUTH SANDWICH IS. **VINDICATION I.** Braces Pt., *PC 105* (AAS, KRAM).

SOUTH ORKNEY IS. **CORONATION I.** Shingle Cove, 5 m, *RILS 453* (AAS, KRAM). **MATTHEWS I.** North-west coast, 15 m, *RILS 161* (AAS, KRAM); near The Divide, *ca* 35 m, *RILS 169B & 170B* (AAS, KRAM). **POWELL I.** Promontory on the south-east coast, 50 m, *RILS 215* (AAS, KRAM). **LAURIE I.** Martin Pt., *Gressitt 6300* (KRAM, NY). **SIGNY I.** Jebsen Pt., 65 m, *RILS 427* (AAS, KRAM); Moraine Valley, north end of Orwell Glacier, 30 m, *RILS 10825B* (AAS, KRAM); same locality, below Orwell Glacier, *ca* 15 m, *RILS 545 & 547* (AAS, KRAM); Factory Cove, behind BAS hut, 15 m, *REL 46* (AAS, KRAM) and 65 m, *REL 1054* (AAS, KRAM); Observation Bluff, 50 m, *RILS 325B* (AAS, KRAM); Polynesia Pt., 50 m, *RILS 10746* (AAS, KRAM); Gourlay Peninsula, *ca* 65 m, *RILS 496* (AAS, KRAM).

SOUTH SHETLAND IS. **ELEPHANT I.** *Sine loco*, Aug. 1916, *Shackleton Antarctic Expedition s.n.* (BM). **CLARENCE I.** *Sine loco*, 90 m, *Chuter 15, 32A, 36A & 37* (AAS, KRAM) and 95 m, *Chuter 61* (AAS, KRAM). **ASPLAND I.** West-facing sea cliff, 9 m, *JPB 74* (AAS, KRAM). **GIBBS I.** East of the Spit, 45–90 m, *JPB 181B* (AAS, KRAM) and 240 m, *JPB 163* (AAS, KRAM); near the Spit, 90 m, *JPB 142* (AAS, KRAM); west of the Spit, 110 m, *JPB 203* (AAS, KRAM). **KING GEORGE I.** ***Bransfield Strait***: Blue Dyke, 130 m, *RO 1164/80* (KRAM). ***Admiralty Bay***: Kasprowy Hill, 230 m, *RO 74/80* (KRAM); Breccia Crag, 140 m, *RO 928/80* (KRAM); Rembiszewski Nunataks, 190 m, *RO 2761/80* (KRAM). ***Potter Cove***: sea shore, 1 m, *Kanda 790609-092* (KRAM, NY), 2 m, *Kanda 790609-077,790609-080 & 790609-085* (KRAM, NY) and 5 m, *Kanda 790607-037* (KRAM). **LIVINGSTON I.** ***Byers Peninsula***: opposite Astor I., 8 m, *DCL 329* (AAS, KRAM). **DECEPTION I.** Neptunes Window, 35 m, *REL 30 & 31* (AAS, KRAM); Cathedral Crags, 35 m, *DM 89B & 91* (AAS, KRAM); Whalers Bay, whaling station, 2 m, *RILS 11230, 11231 & 11232* (AAS, KRAM) and 3–10 m, *RILS 3664 & 5791B* (AAS, KRAM); 100 m SE of whaling station, 10–20 m, *RILS 6546B* (AAS, KRAM); Ronald Hill, 15 m, *REL 1209* (AAS, KRAM), 25 m, *RILS 5778A* (AAS, KRAM) and *ca* 40 m, *RO 2788/80* (KRAM); Telefon Bay, 'New Island' hill, 20 m, *RILS 11290 & 11293B* (AAS, KRAM); Telefon Ridge, 60–180 m, *RILS 5796C & 5797B* (AAS, KRAM) and 210 m, *RILS 5744C* (AAS, KRAM); Kendall Terrace, 80 m, *RILS 5739D, 5740B & 5741A* (AAS, KRAM); Wensleydale Beacon, 2 m, *RILS 5600A* (AAS, KRAM), 3 m, *RILS 5601A* (AAS, KRAM) and 5–15 m, *RILS 6618* (AAS, KRAM); Fumarole Bay, south lagoon, 25 m, *RILS 11305B* (AAS, KRAM); Fumarole Bay, north end, 50 m, *RILS 11301* (AAS, KRAM); Vapour Col, 30 m, *DM 24B* (AAS, KRAM) and 100 m, *RILS 56116C* (AAS, KRAM); Collins Pt., 20–30 m, *RILS 3606A* (AAS, KRAM); west of Collins Pt., 155 m, *NJC 55* (AAS, KRAM); to east of Collins Pt., 25 m, *RILS 11344* (AAS, KRAM); 300 m north-west of South Pt., 50 m, *DM 137A* (AAS, KRAM).

WEST ANTARCTIC PENINSULA. **PALMER ARCHIPELAGO.** ***Melchior Archipelago***: Lysted I., *PAS 353.3 & 361* (FH). ***Anvers I.***: north of Biscoe Bay, *VK 159 & 636* (KRAM). **GRAHAM COAST.** Cape Pérez, *Gain 272B & 273C* (H, PC, S – type of *Pottia charcotii*). **LOUBET COAST.** Orford Cliff, *RILS 5277C, 5282 & 5288C* (AAS, KRAM). **FALLIÈRES COAST.** ***Adelaide I.***: Rothera Pt., 5 m, *Hughes 2 & 3A* (AAS, KRAM); Jenny I., 200 m, *RILS 4755* (AAS, KRAM). ***Marguerite Bay***: Blaiklock I., 8–15 m, *RILS 842B* (AAS, KRAM); Neny I., *Bryant 36* (NY). **GEORGE VI SOUND.** ***Alexander I.***: Lassus Mountains, 750 m, *BASM 208B* (AAS, KRAM); Ares Oasis, 30 m, *RILS 10440B* (AAS, KRAM) and 35 m, *RILS 10398* (AAS, KRAM); Mars Oasis, 35 m, *RILS 10425* (AAS, KRAM) and 175 m, *RILS 9430A* (AAS, KRAM).

EAST ANTARCTIC PENINSULA. **JOINVILLE ISLAND GROUP.** ***Paulet I.***: *sine loco*, *CS 458* (PC, S). **JAMES ROSS ISLAND GROUP.** ***Vega I.***: False Island Pt., 25–40 m, *RILS 7840* (AAS, KRAM). ***James Ross I.***: Cape Lachman, 30 m, *RILS 7334 & 7416* (AAS, KRAM); Andreassen Pt., 10 m, *RILS 7613* (AAS, KRAM); Fortress Hill, 150 m, *RILS 7750* (AAS, KRAM); valley SW of Fortress Hill, 25 m, *RILS 7809 & 7812* (AAS, KRAM); St. Martha Cove, 15 m, *RILS 11135A* (AAS, KRAM); valley south of St. Martha Cove, 50 m, *RILS 7470* (AAS, KRAM); hill south of St. Martha Cove, 60 m, *RILS 7617* (AAS, KRAM); Terrapin Hill, 175 m, *RILS 7737A* (AAS,

KRAM), 250 m, *RILS 7722B & 7725* (AAS, KRAM) and 540 m, *RILS 7714A* (AAS, KRAM); Blancmange Hill, 125 m, *RILS 7695 & 7705* (AAS, KRAM). ***Humps I.***: 2–3 m, *RILS 7398* (AAS, KRAM). ***Cockburn I.***: between foot of cliff and Adelie penguin colony, 120 m, *RILS 7916, 7918 & 7919A* (AAS, KRAM) and 150 m, *RILS 7939* (AAS, KRAM); toward of north end of plateau, *ca* 400 m, *RILS 7968* (AAS, KRAM); summit, *ca* 450 m, *RILS 7983 & 7985* (AAS, KRAM); *sine loco* [actually at the same site as *RILS 7916, 7918 & 7919A*], *Hooker 6* (BM, type of *Bryum antarcticum*). ***Seymour I.***: west of Cape Wiman, 25 m, *RILS 7888* (AAS, KRAM); *sine loco, CS 459* (PC, S).

EAST ANTARCTICA. DRONNING MAUD LAND. **KRONPRINSESSE MÄRTHA KYST.** Veten Nunatak, 72°37′S, 3°52′W, summer 1966–67, *Poole s.n.* (US); Robertskollen, 'Ice Axe Peak', 480 m, *Watkins 21A* (AAS, KRAM); Marsteinen Nunatak, 71°26′S, 1°42′W, summer 1966–67, *Poole s.n.* (US). **PRINSESSE ASTRID KYST.** Novolazarevskaya Station, *Meyer 4* (KRAM, NY).

ENDERBY LAND. **SÔYA COAST.** East Ongul I., *Matsuda 63* (KRAM, NY); Rundvågshetta, *Shimizu 751021023 & 751021024* (KRAM); Hamnenabben, *Shimizu 751113272* (KRAM); Breidvågnipa, *Shimizu 750519027* (KRAM). ***Skarvsnes***: Suribachi-ike, *Kanda 190177* (KRAM); Torinosu Bay, *Shimizu 750127004* (KRAM, NY); Dyako-ike, *Kanda 190318* (KRAM); Lake Oyako, *Nakanishi D-61* (AAS, KRAM). ***Skallen***: Magoke Pt., *Kanda 190371 & 190376* (KRAM, NY); Oo-ike, *Kanda 190339* (KRAM, NY). ***Langhovde***: *sine loco*, 20 m, *Kashiwadani 4006* (KRAM, LE) and *Shimizu 751114278* (KRAM); Hamna, *Shimizu 751125291* (KRAM); Naka-no-tani, *Shimizu 750516011* (KRAM, NY).

PRINCESS ELIZABETH LAND. **INGRID CHRISTENSEN COAST.** Vestfold Hills, Chernok Lake, close to Sordal Glacier, 10 m, *RILS 6397* (AAS, KRAM).

QUEEN MARY LAND. **KNOX COAST.** Bunger Oasis, west of the Station, *Korotkevich 10/410* (H, KRAM, LE, NY – type of *Pottia heimii* var. *brevinervis*) & *L6* (KRAM, LE); same locality, near the Polish Base by so-called 'Totalizator', *Różycki 5* (KRAM).

VICTORIA LAND. **BORCHGREVINK COAST.** Cape Hallett, 15 Dec 1963, *Schofield s.n* (US); moraine between Tinker Glacier and Burns Glacier, *RILS 9659* (AAS, KRAM); Edmondson Pt., 10 m, *DWHW 202A* (AAS, KRAM), 45 m, *RILS 9629* (AAS, KRAM) and 70 m, *RILS 10132* (AAS, KRAM). **SCOTT COAST.** ***Ross Sea***: Franklin Is., *ca* 65 m, 13 Jan 1965, *Shoup s.n.* (KRAM, NY). ***Granite Harbour Area***: Cuff Cape, 25 m, *WCS & SWG 65/54* (AAS, KRAM, NY); The Faltiron, *ca* 50 m, *WCS & SWG 65/48* (AAS, KRAM). ***McMurdo Sound Area***: McMurdo Sound, islet in old ice, Dec 1903, *Wilson s.n.* (*Discovery Expedition*) (PC); Gneiss Pt., *REL 4093* (AAS, KRAM); Spike Cape, *ca* 15 m, *WCS & SWG 65/42* (AAS, KRAM); Bay of Sails, *ca* 100 m, *WCS & SWG 65/40 & 65/65* (AAS, KRAM); Marble Pt., *ca* 15 m, *WCS & SWG 65/14 & 65/15* (KRAM, NY), 30 m, *Janetschek 5* (AAS, KRAM) and without elevation, *JRR MCM-5, 6, 7, 14, 15, 45 & 46* (US), 29 Dec 1965, *Spellerberg & Yeates s.n.* (AAS, KRAM) and 11 Dec 1963, *Fukushima HIRO-A.B. 68* (AAS, KRAM); Cape Bernacchi, *ca* 15 m, *WCS & SWG 65/11* (KRAM, NY) and *REL 4089* (AAS, KRAM); Taylor Valley, Lake Fryxell, *REL 4073* (AAS, KRAM); The Strand Moraines, 1908, *Priestley s.n.* (PC) and *REL 4070* (AAS, KRAM); south end of The Strand Moraines, *ca* 8 m, *WCS & SWG 65/6, 65/8 & 65/9* (KRAM, NY); near Hobbs Glacier, *REL 4081* (AAS, KRAM), *WCS & SWG 65/2* (AAS, KRAM) and *JRR MCM-38 & 42* (US); Miers Valley, *JRR MCM-29A, 51 & 55* (US) and *WCS 65/63* (KRAM, NY). ***Ross I.***: above Caughley Beach near Cape Bird, *REL 4039A & 4045A* (AAS, KRAM) and 300 m, *REL 4051* (AAS, KRAM); Cape Armitage, *REL 4013 & 4035* (AAS, KRAM). ***Ross Ice Shelf***: Bratina I., 5 m, *RILS 9990* (AAS, KRAM).

2. **Hennediella antarctica** (Ångstr.) Ochyra & Matteri Fig. 166, pl. 42

Hennediella antarctica (Ångstr.) Ochyra & Matteri, Fragm. Florist. Geobot. 41: 1006. 1996. ≡ *Gymnostomum antarcticum* Ångstr., Öfvers. Förh. Kongl. Svenska Vetensk.Akad. 29: 3. 1872. ≡ *Pottia antarctica* (Ångstr.) Müll.Hal., Flora 68: 413. 1885. — TYPE CITATION: »Sydamerikas kust« forsitan Porte Famine. [Lectotype (*vide* Ochyra & Matteri 1996*a*: p. 1006): 'P. antarctica »Sydamerikas kust« J. Ångstr. (Porte Famine) N. J. Andersson' – S-Ångström!; isotype: S!].

Pottia anderssonii Sauerb. *in* A.Jaeger & Sauerb., Ber. Thät. St. Gallischen Naturwiss. Ges. 1877–1878: 402. 1879, *nom. illeg. incl. spec. prior.* [Gymnostomum antarcticum Ångstr., 1872].

Pottia fuscomucronata Müll.Hal., Bot. Jahrb. Syst. 5: 79. 1883 ['fusco-mucronata'], ***syn. nov.*** — TYPE CITATION: all species described as new by Müller (1883) are included in the section 'I. Bryologia Kerguelensis'; later Müller (1889) wrote: 'Insula Kerguelen, Betsy Cove cum Pottia Naumanni, November 1874, c. fr. junioribus; Winter harbour (1.75) c. fr. maturis'. [Lectotype (*selected here*): 'Ex Museo botanico Berolinensi Pottia fusco-mucronata C. Müll. n. sp. Kerguelen. Dr. Naumann Nov. 1874' – S!; isotypes: BM!, PC-Cardot!, S-Roth!].

Hennedia macrophylla R.Br.bis, Trans. & Proc. New Zealand Inst. 25: 286, *pl. 31, f. 6–8.* 1893. ≡ *Hennediella macrophylla* (R.Br.bis) Paris, Index Bryol.: 557. 1896. ≡ *Pottia macrophylla* (R.Br.bis) Sainsbury, Rev. Bryol. Lichénol. 21: 217. 1952. — TYPE CITATION: Banks of the River Avon, near Christchurch. [Isotype: 'Hennediella macrophylla R. Br. Ter. Near Christchurch, New Zealand Leg. R. Brown ter.' – BM!]. First synonymised by Ochyra (1998*a*: p. 139).

Hennedia intermedia R.Br.bis, Trans. & Proc. New Zealand Inst. 25: 286, *pl. 32, f. 1–2.* 1893. ≡ *Hennediella intermedia* (R.Br.bis) Paris, Index Bryol.: 557. 1896, ***syn. nov.*** — TYPE CITATION: Damp banks, Port Hills, and plantations round the River Avon. [Lectotype (*selected here*): 'Musci Nova Zealandicae. Hennediella intermedia nova sp. Lyttelton Hills. Ch.[ristchur]ch. Coll. Robert Brown Herb. Robert Brown, Merivale, Christchurch' – BM-Dixon!; syntype: 'Herb. H. N. Dixon. Hennediella intermedia. Loc. Near Christchurch, New Zealand Coll. R. Brown ter. Comm. G. Brown Det. R. Brown ter.' – BM-Dixon!]. First synonymised with *Pottia macrophylla* by Sainsbury (1952: p. 217).

Hennedia microphylla R.Br.bis, Trans. & Proc. New Zealand Inst. 25: 286, *pl. 33, f. 1–3.* 1893. ≡ *Hennediella microphylla* (R.Br.bis) Paris, Index Bryol.: 557. 1896, ***syn. nov.*** — TYPE CITATION: Damp banks in the neighbourhood of Christchurch. Common. [Lectotype (*selected here*): 'Herb. H. N. Dixon. Hennediella microphylla Loc. Near R. Avon, Christchurch, New Zealand Coll. R. Brown ter Comm. G. Brown' – BM-Dixon!]. First synonymised with *Pottia macrophylla* by Sainsbury (1952: p. 217).

Pottia grata R.Br.bis, Trans. & Proc. New Zealand Inst. 26: 293, *pl. 33* (*Pottia grata*), *f. 1–3.* 1894, ***syn. nov.*** — TYPE CITATION: Damp banks, neighbourhood of Christchurch. Collected by R. B. [Isotype: 'Herb. H. N. Dixon. Pottia grata R. Br. ter. Loc. Near Christchurch, New Zealand Coll. R. Brown ter. Comm. G. Brown, Type' – BM-Dixon!]. First synonymised with *Pottia macrophylla* by Sainsbury (1952: p. 217).

Pottia austrogeorgica Cardot, Bull. Herb. Boissier Sér. 2, 6: 20. 1906 ['austro-georgica']. ≡ *Desmatodon austrogeorgicus* (Cardot) Ochyra *in* Ochyra, Vitt & D.G.Horton, Cryptog. Bryol. Lichénol. 7: 57. 1986. ≡ *Hennediella austrogeorgica* (Cardot) Blockeel, J. Bryol. 16: 191. 1990. — TYPE CITATION: [South Georgia] Royal bay: Moltke harbour. [Lectotype (*selected here*): 'Svenska Sydpolarexpeditionen 1901–03. N:r 290. Pottia austro-georgica Card. sp. nova South Georgia, Royal Bay, Moltke Harbour, 29/4 1902 Carl Skottsberg Det. J. Cardot' – PC!; isotypes: H-Brotherus!, LD!, UPS!]. First synonymised by Ochyra (1998*a*: p. 139).

Pottia acutidentata Cardot & Thér. *in* Cardot, Bull. Mus. Natl. Hist. Nat. 22: 338. 1916. ≡ *Hennediella acutidentata* (Cardot & Thér.) R.H.Zander, Bull. Buffalo Soc. Nat. Sci. 32: 248. 1993, ***syn. nov.*** — TYPE CITATION: [Kerguelen] (Bossière). [Lectotype (*selected here*): '19. Pottia acutidentata Card. & Thér. Kerguelen leg. R. Bossière 1913 – PC!; isotype: H-Brotherus!].

Pottia austrogeorgica var. *microphylla* Cardot & Broth., Kongl. Svenska Vetenskapsakad. Handl. 63(10): 21. 1923, ***syn. nov.*** — TYPE

CITATION: Georgia austr.: Bay of Isles, Rosita Harbour (29). [Lectotype (*selected here*): 'Expeditio suecica 1907–1909. 19. Pottia austrogeorgica Card. var. microphylla Card. var. nova Georgia australis, Bay of Isles, Rosita Harbour, ad rupes. 25.4 1909 Carl Skottsberg Det. J. Cardot' – PC-Cardot!; isotypes: H-Brotherus!, LD!, UPS!].

Plants small to medium, stiff and rigid to relatively soft, forming dense or looser tufts or patches, rarely more scattered and gregarious, dull, dark green, greenish- or yellowish-brown to brown. *Stems* erect, 0.4–1.0(–1.5) mm tall, simple or with 2–3 branches, usually buried in soil and with well developed rhizoidal system of long, lustrous, brown, smooth, sparingly branched rhizoids, in transverse section indistinctly rounded-pentagonal, consisting of a distinct central strand, 3–5-layered medulla of large, thin-walled, hyaline cells becoming smaller and somewhat brownish in the peripheral part and forming an indistinct cortex; *axillary hairs* filiform, hyaline throughout, 6–7-celled. *Leaves* crowded to somewhat distant, erect, closely appressed and imbricate to erect-spreading, incurved and twisted to little altered when dry, erecto-patent when moist, (1.0–)1.5–2.5(–3.2) mm long, (0.3–)0.4–0.7(–0.9) mm wide, broadly concave, non-decurrent, ovate to ovate- or oblong-lanceolate, occasionally spathulate, short- to long-acuminate or acute, usually with subtubulose apex when dry; *margins* plane, denticulate or serrate at the apex and minutely crenulate for some distance lower down by projections of the cell ends, unistratose or seldom with bistratose patches, distinctly bordered by 1–4 rows of thick-walled yellowish, rhombic, oblong-rhomboidal to rectangular, epapillose to weakly papillose cells, 10–30 μm long, 5–10 μm wide; *costa* strong, yellow- to red-brown, 50–80 μm wide at the base, gradually tapering upwards, percurrent to short-excurrent as a multicellular, toothed, serrate or crenate cusp composed of brown, elongate, thick-walled, smooth cells, covered by quadrate to short-rectangular, papillose epidermal cells on the ventral surface and elongate, smooth cells on the dorsal surface, in transverse section plano- or biconvex, consisting of a large, protuberant, rounded or elliptical dorsal stereid band lacking an epidermis, 2–4 large guide cells arranged in 2 layers and a ventral epidermis of large, papillose cells; *laminal cells* unistratose throughout, except for occasional bistratose patches at the margins, flat superficially; *upper cells* variable in shape, mostly isodiametric, rounded-quadrate, elliptical, hexagonal to short-rectangular, 10–15(–20) μm long, 9–14(–18) μm wide, with moderately, evenly thickened walls, distinctly obscured owing to the dense papillosity of simple or bifid, hollow, C- or O-shaped papillae, usually 6–9 per cell, becoming pellucid only at the marginal borders; *basal cells* rectangular, 30–80 μm long, 15–25 μm wide, thin-walled, hyaline, lax throughout the base or thicker-walled and long-rectangular at the margins forming distinct borders, usually becoming shorter and thicker-walled in the upper part of the base, gradually passing into papillose, chlorophyllose upper cells, extending along the margins and forming a distinct V-shaped basal region. *Autoicous* or *synoicous*. *Gametangia* terminal. *Perichaetial leaves* similar to vegetative leaves except for being larger and often more coarsely toothed at the apex. *Setae* erect, straight or somewhat bent, stout, 2–5(–8) mm long, yellow to yellowish-brown when young, becoming brown with age, twisted to the left when dry. *Capsules* exserted, erect, gymnostomous, ovoid to short-cylindrical, 1.0–1.5(–2.0) mm long, 0.5–0.9 mm wide, straight and symmetric or somewhat curved and asymmetric, usually widened at the mouth when dry and empty, gradually tapering to the seta, brown, glistening; *annulus* persistent, composed of 1–3 rows of rounded, thick-walled cells; *operculum* long-beaked from a convex base, with a stout, straight or slanted rostrum, not remaining attached to the columella after dehiscence; *exothecial cells* mostly rectangular, with thin to moderately thickened walls, 25–65 μm long, 15–20 μm wide, becoming shorter, rounded or elliptical in a few layers at the mouth; *stomata* 5–6 at the base of the urn, superficial, bicellular, with a narrow, elliptical slit; *peristome* absent. *Spores* ellipsoid, dark brown to reddish-brown, 27–39(–42) μm long, 26–38 μm wide, prominently verrucose. *Calyptra* large, 1.5–2.0 mm long, cucullate, smooth, naked, pale yellow with the brown apex, extending two-thirds down or enclosing the entire capsule, tightly embracing it at the base, not fugacious.

Chromosome number: $n = 26$ – Enderby Land (Inoue 1976 as *Desmatodon* sp.; *vide* Kanda 1981*b*).

Discussion. *Hennediella antarctica* is similar and closely related to *H. heimii*, but both species are readily separated by a set of both gametophyte and sporophyte characters; the differences between them are discussed under the latter species. *Hennediella antarctica* is distinguished by its rigid, erect to erect-spreading leaves that are distinctly bordered and coarsely toothed distally; small, densely papillose and obscure upper laminal cells that are mostly 10–14(–16) μm wide; and a very large calyptra that covers at least two thirds but most often the whole of the capsule and is not readily fugacious. In addition, the capsules are not systylious because the operculum is not persistently attached to the columella after dehiscence.

Hennediella antarctica is a very variable species throughout its broad geographical range and various local populations have been given taxonomic recognition. This species was originally described as *Gymnostomum antarcticum* by Ångström (1872)

→

Figure 166. *Hennediella antarctica*. — **1–3**. Plants with sporophytes, wet (1, 3) and dry (2). **4**. Cross-section of stem. **5–6**. Axillary hairs. **7–17**. Leaves. **18–20**. Leaf apices. **21–22**. Upper lamina cells at margin. **23**. Mid-leaf cells. **24**. Basal leaf cells. **25–27**. Cross-sections of leaves. **28**. Stoma and basal exothecial cells. **29**. Exothecial cells in middle of urn. **30**. Exothecial cells at mouth and spores. — [1, 4–6, 17, 24–25 from *RO 41/80*; 2–3, 9–10 from *RO 2351/80*; 7–8 from *Anderson s.n. & s.d.*, Port Famine (lectotype of *Gymnostomum antarcticum*), S; 11–12, 18, 21, 28 from *CS 290* (isotype of *Pottia austrogeorgica*), UPS; 13, 29–30 *CS 29* (isotype of *Pottia austrogeorgica* var. *microphylla*), UPS; 14–15 from *Shimizu 751114278*; 16, 19–20, 22–23, 26–27 from *RO 1284/80*]. — Scale bars: a – 1 mm (1, 3); b – 50 μm (21–23); c – 1 mm (2); d – 100 μm (4–6, 18–20, 24–30); e – 1 mm (7–17).

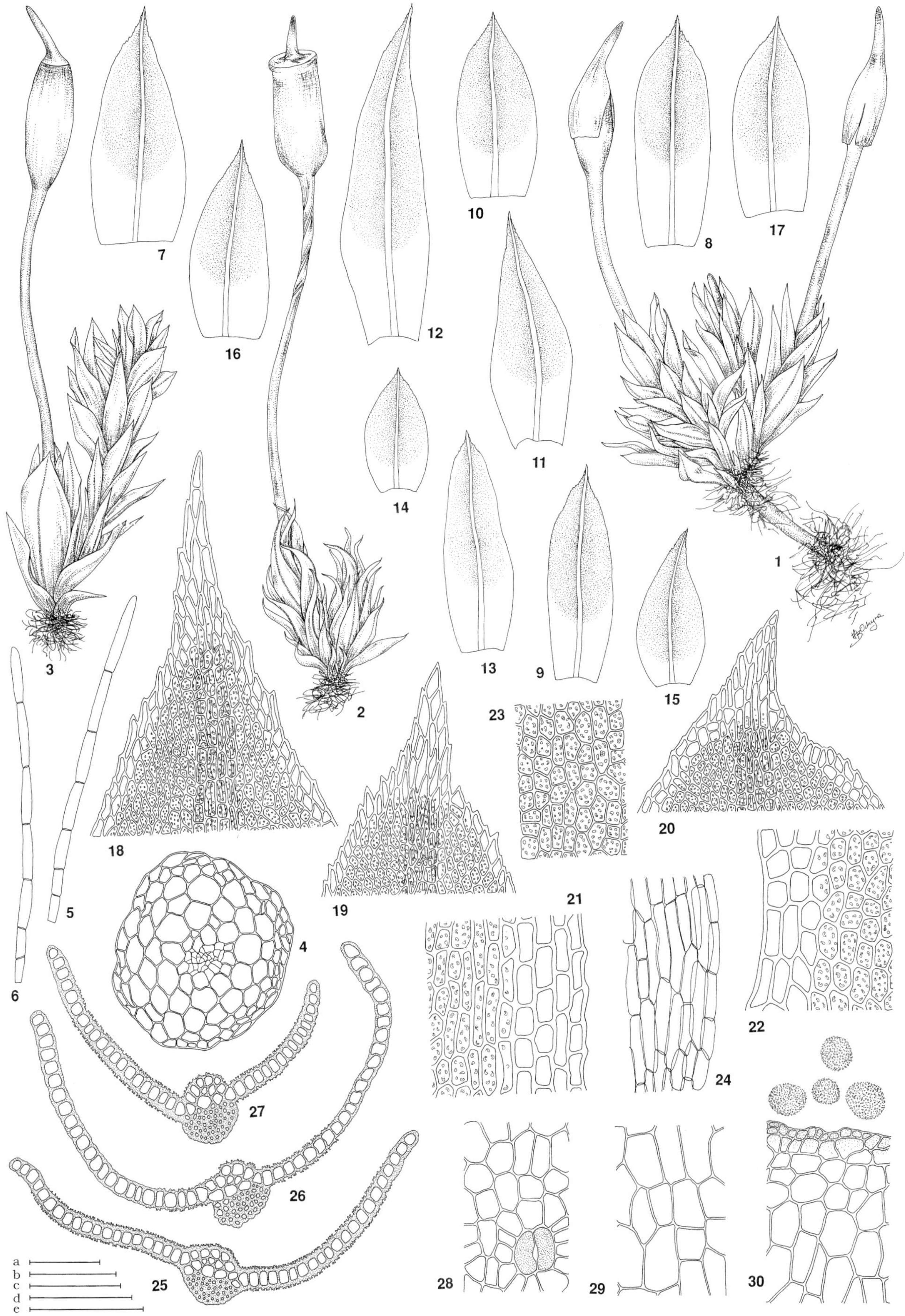
7
16
12
10
8
17
11
14
1
3
13
9
15
2
23
18
20
19
21
5
4
22
6
24
27
26
28
29
30
25
a
b
c
d
e

for a collection from Port Famine (=Puerto del Hambre) on the Brunswick Peninsula, southern Patagonia. However, this distinct species has been the source of considerable confusion, commencing with Jaeger (1873) who used the binomial *Pottia antarctica* Schimp. for a plant from 'Fretum Magellanicum', but this name was a *nomen nudum*. Jaeger & Sauerbeck (1879) transferred *Gymnostomum antarcticum* to *Pottia* and unnecessarily renamed it as *P. anderssonii* Sauerb. This error was corrected by Müller (1885) who rightly used the name *Pottia antarctica* (Ångstr.) Müll.Hal., although at the same time he cited some specimens under this name which clearly belonged within *H. heimii* (e.g. *Lechler 1091*, *1213*, BM). This generic placement was adopted by Cardot (1908*a*) who incorrectly ascribed the combination *P. antarctica* to Schimper, referring to the herbarium label of the plants collected in the Magellanian area by Lechler as No. 1199, although this material actually represents *H. heimii*. This error was subsequently repeated by Warnstorf (1916) who ascribed the combination *P. antarctica* to Brotherus (1902), and his concept of the species clearly referred to *H. heimii*. Finally, Ochyra & Matteri (1996*a*) transferred *Gymnostomum antarcticum* to *Hennediella*, and then Ochyra (1998*a*) synonymised *Hennediella macrophylla* and *Pottia austrogeorgica* with *H. antarctica*, two species that have long been considered as distinct in widely separated centres of the Southern Hemisphere. The type material of *H. antarctica* consists of medium-sized plants, with the leaves varying from ovate to oblong-lanceolate or lingulate, 1.5–2.5 mm long and the leaf border well developed in the apical part and less complete below but otherwise distinct, composed of 2–3 rows of thick-walled, yellowish, rhomboidal to elongate cells, and the upper laminal cells are rounded-quadrate to hexagonal, 10–15 μm wide and densely papillose. The seta is stout, 4–5 mm long and the capsule is short-cylindrical, 1.7–2.0 mm long.

Pottia austrogeorgica was described from subantarctic South Georgia (Cardot 1906*a*, 1908*a*) as a species closely related to *P. fuscomucronata* from Îles Kerguelen, and it was soon accepted by Warnstorf (1916). The species was rediscovered on South Georgia in 1909 and this material was recognised as var. *microphylla* (Cardot & Brotherus 1923) because it deviated from the type in leaf size. In the late 1960s and in the 1970s intensive bryological exploration of South Georgia and the conterminous maritime Antarctic yielded many additional specimens of *P. austrogeorgica*. These were thoroughly investigated by Matteri (1977*b*) who found that the species exhibited remarkable variation, especially in size of the plants and leaves. Accordingly, she stated that var. *microphylla* did not merit taxonomic recognition and the name was reduced to synonymy with *P. austrogeorgica*. Matteri (1977*a*) recorded this species from the South Sandwich Is. and South Orkney Is. in the northern maritime Antarctic and subsequently Kanda (1981*b*) reported it from continental Antarctica. Ochyra *et al.* (1986) extended the Antarctic range of *P. austrogeorgica* to the South Shetlands and at the same time transferred it to *Desmatodon*, but Blockeel (1990) later moved it to *Hennediella*. The original material and non-type collections of *H. austrogeorgica* fit perfectly the type of *H. antarctica* in all details of gametophyte and sporophyte features, including the leaf areolation, the form of the marginal leaf border, as well as the shape of leaves and capsules.

Hennediella macrophylla was originally described by Brown (1893) as *Hennedia macrophylla* from the South Island of New Zealand, along with two other species, *H. intermedia* and *H. microphylla*, but this author considered the last two doubtfully distinct from one another. The first of these has long been considered to be a New Zealand endemic, until Matteri (1988) recorded it from Tierra del Fuego. These species were recognised mostly on the basis of the seta length, *H. macrophylla* having the shortest seta, equal to the capsule in length, resulting in emergent, not exserted capsules as it was in the other two species. The taxonomic status and variability of these species, as well as *Pottia grata*, another species described by Brown (1894) from New Zealand, were discussed in detail by Dixon (1923*a*) who, having examined many New Zealand plants, stated that 'The series of plants mounted by Beckett shows ... a perplexing similarity between the two plants associated together, and I should not be at all surprised if it ultimately transpires that all forms of *Hennediella* belong to one variable species. I have found no structural differences, and the length of seta and degrees of exsertion of the capsule are most perplexingly variable'. This conclusion led Sainsbury (1952) to reduce these species' names to synonymy, retaining the name *Pottia macrophylla*. Blockeel (1990) confirmed the remarkable variability of this species in most gametophyte characters, including shape of leaf and leaf apex and form of the leaf border and, additionally, he found that the plants from England described as *Tortula brevis* by Whitehouse & Newton (1988) were conspecific with *H. macrophylla*. During the present work the type of *T. brevis* has not been examined, but investigation of material from the banks of the River Mole near Dorking (*H. & R. Ochyra 419/97*, KRAM) confirms the aptness of this decision.

Dixon (1923*a*) also showed the close similarity of *Hennediella microphylla* to *Pottia austrogeorgica* and *P. fuscomucronata* and stated that they are most probably inseparable from the former species. Blockeel (1990) recognised *Hennediella austrogeorgica* to be distinct from *H. macrophylla* 'in the less complete leaf border, in which the cells are ± quadrate, the stout red-brown nerve 60–80 μm wide, and the short ovoid-oblong capsule with a long-rostrate lid' and in the seta which tends to be more elongate and slender. However, all these characters are subject to great variation in the South Georgian and Antarctic populations designated as *H. austrogeorgica*, and retaining these species as distinct would be highly artificial. Therefore, Ochyra (1998*a*) considered them to be conspecific with *H. antarctica*.

Pottia fuscomucronata was described by Müller (1883, 1889) from Îles Kerguelen, and Brotherus (1906*a*) reported additional specimens from this archipelago which enabled evaluation of the local populations and their taxonomic status. This species was originally described as having the operculum adherent to the columella and, on this basis, Dixon (1923*a*) suggested its close relationship to *Hennediella heimii*. On the other hand, Matteri (1977*b*) maintained it as a good species showing additionally the lack of the distinctly toothed border as a diagnostic character of *P. fuscomucronata*. Capsules in the Kerguelen material are generally not systylious and only occasionally do some lids remain attached to the columella, but they are not markedly exserted over the capsule mouth nor seem to be permanently attached because the majority of the capsules are deoperculate. Careful examination of various populations of *Hennediella antarctica* from South Georgia and Antarctica revealed that this state of the capsules is not rare (e.g. *RILS 10824*, AAS, KRAM) and it can be interpreted as a brief phase of early dehiscence. Because the correspondence of all other characters of *P. fuscomucronata* and *H. antarctica* is perfect, including the leaf shape and areolation as well as sporophyte features, especially with phenotypes described as *Pottia austrogeorgica*, there are no reasons to maintain the former as a separate species, and it is considered here

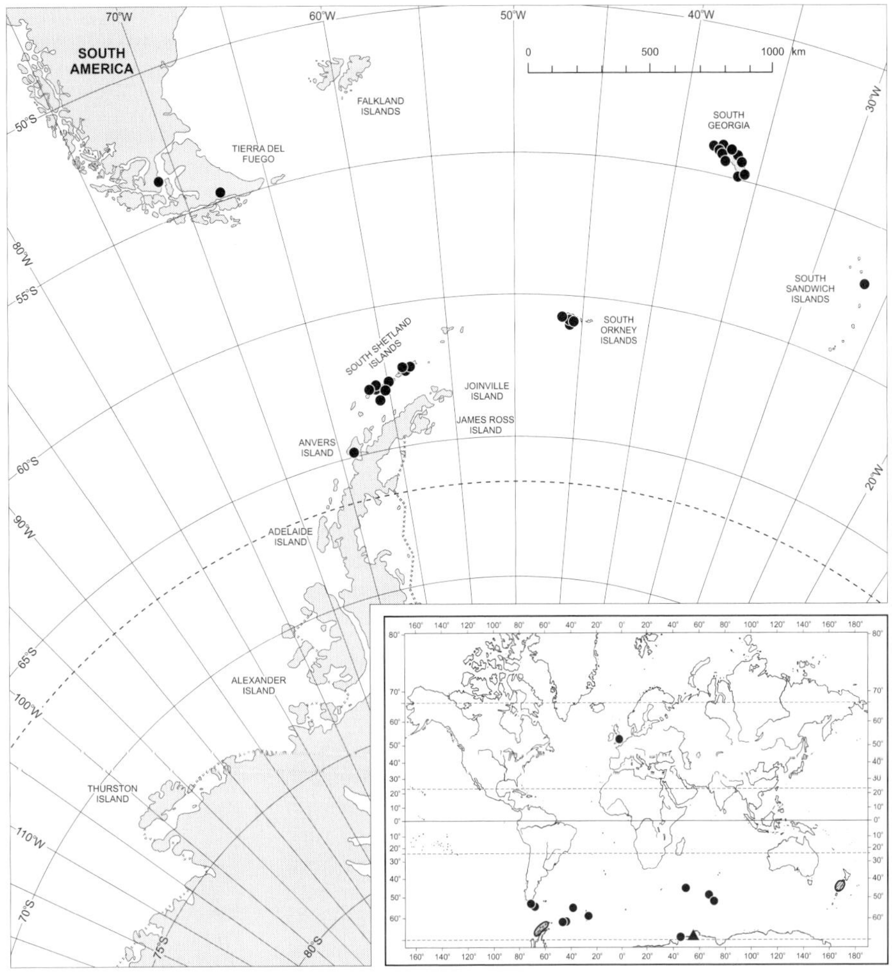

Figure 167. Distribution of *Hennediella antarctica* in the Antarctic Peninsula and Fuegian regions. Inset: global distribution. The literature record is marked by the triangle.

to be conspecific with *H. antarctica*. The same is also true for another species from Îles Kerguelen, *Hennediella acutidentata*, which was originally described from this archipelago as *Pottia acutidentata* (Cardot 1916). The type material of this species matches the local populations described as *P. fuscomucronata* and, accordingly, this species is also considered to be identical to *H. antarctica*.

Hennediella antarctica is one of the most prolific of all Antarctic mosses which, due to the profusion of fully mature sporophytes giving its populations a characteristic aspect, can hardly be confused with any other local moss species. In its lanceolate leaves with a shortly excurrent costa it approaches *Syntrichia saxicola*, but its leaves are entire and lack a marginal border and the plants are much larger. In contrast, *S. filaris* has denticulate leaf margins near the apex and has a somewhat differentiated marginal border, but it is a large plant, with the upper chlorophyllose leaf cells sharply demarcated from the basal hyaline cells descending along the margins. Additionally, in *S. filaris* the stem central duct is wanting and the costa is crescent-shaped in cross-section. In general, small plants of *Syntrichia* species may be distinguished by their spreading leaves with reflexed apices, in contrast to the erecto-patent leaves with straight apices in *H. antarctica*.

Reproduction in Antarctica. Most populations of *Hennediella antarctica* from the maritime Antarctic produce fully mature sporophytes in great profusion, whereas those from the continent are sterile. No means of asexual reproduction have been observed in the species.

Habitat. An epigean moss, occurring as a primary coloniser of bare soil which is usually subject to frost heaving and solifluction, or enriched by birds. It grows most often on base rich soil and gravel on moraines and recently exposed ground and on pockets of soil in scree, in rock crevices near the shore and on soil-covered ledges on sea cliffs, in moist or occasionally dry places, usually enriched by birds or mammals. In areas with recent volcanic activity it thrives on gravelly ash, including sites with some geothermal heating. On Signy I., *Hennediella antarctica* locally forms quite extensive stands on fine clayey or sandy soil littered with amphibolite, and these were described as a separate pioneer association within the fruticose lichen and moss cushion sub-formation (Lewis Smith 1972). Here, it forms almost pure stands with occasional admixtures of *Bryum pseudotriquetrum*, *Bartramia patens*, *Sanionia georgicouncinata* and *S. uncinata*, *Schistidium antarctici*, *Ceratodon purpureus*, *Syntrichia magellanica*, *S. saxicola* and *Brachythecium austrosalebrosum*.

World range. A bipolar species (Fig. 167, inset). It is widely distributed but scattered in the Southern Hemisphere where it is predominantly distributed on subantarctic South Georgia and in the northern maritime Antarctic but rare in continental Antarctica. Elsewhere in the Subantarctic it occurs on Îles Kerguelen (type of *Pottia fuscomucronata*; *Bell 3115*, AAS, KRAM – reported by Lightowlers (1986*a*) as *Tortula filaris*), Îles Crozet (Ochyra 2006) and Heard I. (*RDS 10655*, AAS, KRAM; this material was reported by Lightowlers (1986*a* as *Syntrichia filaris*). In addition, *Hennediella antarctica* is very rare in the Fuegian region, including the Brunswick Peninsula (*locus classicus* of *Gymnostomum antarcticum*) and Isla Grande de Tierra del Fuego (Matteri 1988, as *Pottia macrophylla*), and infrequent on the South Island of New Zealand (Dixon 1923*a*; Sainsbury 1955). The phytogeographical status of the species is difficult to define since it occurs in various climatic zones in the austral region. Because its maximum occurrence and abundance appears to be in the Subantarctic and Antarctic, especially on South Georgia, the South Orkneys and some South Shetlands, it is designated here as an amphiatlantic subantarctic species penetrating, on a limited scale, to the cool temperate zone in the Fuegian region and New Zealand. In the Northern Hemisphere *H. antarctica* is known only from southern England where it was originally described as *Tortula brevis* (Whitehouse & Newton 1988). Because the species is also known to occur on anthropogenic habitats such as cultivated flower beds and gardens, Blockeel (1990) considered this species to be introduced in the British Isles from New Zealand. The distribution of *H. antarctica* in Britain is centred on different river basins, and this may support the hypothesis that the species could have been introduced to several sites from which it has effectively dispersed along rivers and has colonised banks of the Rivers Thames and Tweed.

Distribution in Antarctica. *Hennediella antarctica* is a relatively rare but widely scattered species, having a bicentric distribution in the Antarctic (Fig. 167). It is most frequent in the northern maritime Antarctic, occurring predominantly in the South Orkney Is. (Signy I., Coronation I.) and South Shetland Is. (King George I., Livingston I., Deception I.). It is very rare elsewhere, being recorded only once in the South Sandwich Is. (Vindication I.) and on Anvers I., off the Danco Coast, where it reaches its southernmost limit of distribution at lat. 64°48′S. So far, it has not been recorded on the mainland Antarctic Peninsula but almost certainly occurs there. The second centre of its occurrence is continental Antarctica where it is infrequent on the Sôya Coast (Kanda 1981*b*) and in Amundsen Bay (Kanda 1987*e*), Enderby Land, growing near the sea shore. The information on its occurrence in Victoria Land (Ochyra 1998*a*) is erroneous. *Hennediella antarctica* is a species of low elevations, occurring often at sea level and predominantly at elevations below 100 m. Only occasionally it grows on nunataks, reaching its highest altitude of 265 m on King George I. (Ochyra 1998*a*).

Specimens examined. WEST ANTARCTICA. SOUTH SANDWICH IS. **VINDICATION I.** Towards northern end of the east side, 135 m, *REL 730B* (AAS, KRAM); Braces Pt., *PC 105* (AAS, KRAM).

SOUTH ORKNEY IS. **MONROE I.** South side, 80 m, *RILS 8000* (AAS, KRAM). **CORONATION I.** Meier Pt., 15–30 m, *RILS 586* (AAS, KRAM); Shingle Cove, 1.5–2 m, *RILS 8233* (AAS, KRAM); Olivine Pt., *RILS 127* (AAS, KRAM); Windscoop Buttress, 200 m, *RW 159A* (AAS, KRAM). **LYNCH I.** North side, 1.5–2 m, *RILS 8229* (AAS, KRAM). **SIGNY I.** Everson Ridge, 200 m, *RILS 8168* (AAS, KRAM); Thulla Pt., 30 m, *RILS 406* (AAS, KRAM); Foca Pt., 35 m, *RILS 336* (AAS, KRAM); near Moss Lake, 85 m, *RW 5* (AAS, KRAM); Factory Cove, 65 m, *RW 32* (AAS, KRAM); south-west of Factory Cove, 6 m, *RW 124* (AAS, KRAM) and 35 m, *RW 77 & 91* (AAS, KRAM); Observation Bluff, 10 m, *DCL 925B* (AAS, KRAM); between Observation Bluff and Polynesia Pt., *ca* 15 m, *REL 1085* (AAS, KRAM); north end of Orwell Glacier moraine, 2 m, *RILS 8093* (AAS, KRAM) and 30 m, *RILS 10824* (AAS, KRAM). **MOE I.** Above Landing Cove, 50–100 m, *RILS 5238B & 5247* (AAS, KRAM).

SOUTH SHETLAND IS. **KING GEORGE I.** ***Bransfield Strait***: Blue Dyke, 135 m, *RO 1154/80* (KRAM). ***Admiralty Bay***: Creeping Slopes, 120 m, *RO 1284/80* (KRAM); Siodło, 90 m, *RO 729/80* (KRAM); Sphinx Hill, 100 m, *254/80 & 277/80* (KRAM) and 130 m, *RO 211/80* (KRAM); moraines by the northern edge of Ecology Glacier, 20 m, *RO 906/80* (KRAM); Penguin Ridge, 140 m, *RO 1401/80* (KRAM); Point Thomas, 15 m, *Kanda 790619-017* (KRAM); Krzesanica, 130 m, *RO 2351/80* (KRAM) and 135 m, *1660/80* (KRAM); Ambona, 70 m, *RO 1643/80* (KRAM) and 80 m, *RO 1634/80* (KRAM); Petrified Forest Creek, 45 m, *RO 4836/79* (KRAM); Skua Cliff, 105 m, *RO 640/80* (KRAM); Jersak Hills, 175 m, *RO 5139/79* (KRAM); Italia Valley, 60 m, *RO 41/80* (KRAM) and 65 m, *RO 51/80* (KRAM); Cytadela, 5–20 m, *RO 942/80* (KRAM); Emerald Pt., 20 m, *RO 834/80* (KRAM); Stenhouse Bluff, 4 m, *RO 2611/80* (KRAM); Ullman Spur, 60 m, *RO 576/80* (KRAM); Szafer Ridge, 230 m, *RO 2577/80* (KRAM); Tern Nunatak, 265 m, *RO 2522/80* (KRAM); Smok, 10 m, *RO 2057/80 & 2082A/80* (KRAM) and 15 m, *RO 2053/80* (KRAM); Basalt Pt., 15 m, *RO 2231/80* (KRAM); Mt. Wawel, 20 m, *RO 2194/80* (KRAM). ***Drake Passage***: Atherton I., *BJ 113* (KRAM). ***Potter Peninsula***: Three Brothers Hill, 100 m, *Kanda 790609-074* (KRAM) and 120 m, *Kanda 790609-081 & 790690-101* (KRAM); Stranger Pt., 10 m, *Kanda 790609-086* (KRAM). **GREENWICH I.** Duff Pt., 15 m, *DCL 560* (AAS, KRAM). **LIVINGSTON I.** ***Byers Peninsula***: Pt. Smellie, 2 m, *KR 33B* (AAS, KRAM); Sayer Nunatak, 35 m, *RILS 5872* (AAS, KRAM); Robbery Beaches, 10 m, *KR 52B* (AAS, KRAM). ***Hurd Peninsula***: west side of Miers Bluff, *RMS 69-805* (US). **DECEPTION I.** Between Kroner Lake and Ronald Hill, 2–4 m, *RILS 11253* (AAS, KRAM); Wensleydale Beacon, 9 m, *RILS 5598A* (AAS, KRAM) and 30–100 m, *RILS 6675* (AAS, KRAM); Fumarole Bay, 100–200 m, *RILS 11437A* (AAS, KRAM); Stonethrow Ridge, 60 m, *RILS 5806E* (AAS, KRAM); entrance to Southern Bay in New Island Area, 1–2 m, *RILS 5677A* (AAS, KRAM).

WEST ANTARCTIC PENINSULA. **PALMER ARCHIPELAGO. *Anvers I.***: north of Biscoe Bay, *VK 636A* (KRAM).

EAST ANTARCTICA. ENDERBY LAND. **SÔYA COAST. *Skarvsnes***: Torinosu Cove, *Nakanishi E-2* (AAS, KRAM); Oyako-ike, *Kanda 190321 & 190322* (KRAM); *sine loco*, *Shimizu 7501032* (KRAM). ***Skallen***: Oo-ike, *Kanda 190337, 190338 & 190340* (KRAM). ***Langhovde***: *sine loco*, *Shimizu 750517042 & 751114278* (KRAM).

33. WILLIA MÜLL.HAL.

Willia Müll.Hal. *in* Neum., Int. Polarforsch. Deutsch. Exped. 2: 311. 1890. — Type: *Willia grimmioides* Müll.Hal. [= *Willia austroleucophaea* (Besch.) Broth. (*Grimmia austroleucophaea* Besch.)].

Plants small or sometimes medium-sized, slender or occasionally coarser, rigid, forming dull or somewhat glistening, rounded or hemispheric, compact cushions, hoary or not, dark green, brownish- or bluish-green, rufous to reddish-brown or brown above, blackish- or reddish-brown below, sometimes reddish-brown throughout. *Stems* erect, regularly dichotomously or fastigiately branched, rarely simple, densely rhizoidal at the base with long, light to reddish-brown, smooth, glistening rhizoids, in transverse section rounded-pentagonal, consisting of a distinct but weak central duct surrounded by a medullary tissue of large, thin-walled, hyaline cells, becoming somewhat smaller and brownish in 1–2 outermost rows but not forming a cortex; *axillary hairs* filiform, hyaline throughout, 5–10-celled, with 2–3 somewhat shorter and thicker-walled basal cells. *Leaves* crowded, conduplicate, loosely erect and twisted around the stem to contorted when dry, erect-spreading to erect-recurved when moist, lingulate to spathulate from a narrow, elliptical to rhomboidal, not or clearly sheathing base, weakly panduriform, non-decurrent, concave or broadly channelled, narrowly grooved along the costa, obtuse or broadly rounded to rounded-acute at the apex, abruptly changing into a long, hyaline, flexuose, weakly denticulate hair-point or a short, reddish-brown apiculus; *margins* plane, erect to weakly deflexed, unistratose throughout, entire or weakly denticulate and often bordered with 2–4 rows of thicker-walled, elongate-rhomboidal, hyaline cells near the apex; *costa* single, long-excurrent as a hyaline awn or short-excurrent as a strong, reddish-brown arista, inserted laterally or ventrally with the lamina, consisting of a ventral epidermis of quadrate to short-rectangular, papillose cells, in transverse section round to elliptical, consisting of a large, reniform to semicircular dorsal stereid band lacking an epidermis and covered with rectangular, smooth or slightly roughened cells, and 2 layers of guide cells including 4 large cells on the dorsal side and 2 smaller cells on the ventral side; *laminal cells* unistratose throughout, showing red reaction with KOH; *upper cells* subquadrate, small, thin to evenly thicker-walled, obscure due to the dense, solid or hollow, bifid papillae, 4–8 per lumen, abruptly demarcated from the basal cells and somewhat descending along the margins; *basal cells* rectangular, thin-walled, smooth, hyaline, lax, often becoming narrower and sturdier at the margins but not forming a distinct border. *Dioicous*. *Perigonia* terminal, bud-like. *Perichaetia* terminal; *inner perichaetial leaves* highly differentiated, loosely sheathing the seta, oblong-lanceolate to broadly lanceolate, long-piliferous with a hyaline or reddish-brown awn, with the areolation of rectangular to inflated rhomboidal cells in the lower half or throughout the leaf. *Setae* singly per perichaetium, short, stout, often fleshy, erect, straight to somewhat curved, yellow- to reddish-brown, dextrorse when dry. *Capsules* immersed, emergent to short-exserted, ovoid, ellipsoid to cylindrical, straight, erect to inclined; *annulus* persistent, often everted, 2–4-seriate, composed of highly vesiculose, thick-walled cells; *operculum* short- to long-conic; *exothecial cells* short- to long-rectangular, with thin to slightly thickened walls; *stomata* at the base of the urn, superficial, bicellular, with elliptical openings; *peristome* absent or well developed, consisting of 32 short, filamentous, low spiculose branched with several articulations and slightly twisted to the right, arising from a comparatively high, slightly spiculose basal membrane. *Spores* globose, light brown, smooth to papillose. *Calyptra* cucullate or long-mitrate, usually lobed below, naked, plicate below or smooth.

Willia is an oligotypic genus, closely related to *Syntrichia*, especially to plane-margined species of sect. *Aesiotortula* R.H.Zander, and shares with it gross characteristics. Zander (1993) suggested that it may ultimately be accepted as only a section of this genus. *Willia* is primarily characterised by sporophyte characteristics, namely by a tendency towards reduction of sporophyte and conspicuously differentiated perichaetial leaves. Moreover, the leaves are pandurate with plane margins which are indistinctly bordered by slightly thickened cells, the peristome teeth are wanting or short and arising from a high basal membrane, the annulus is persistent, multiseriate and sometimes everted on capsule dehiscensce, and the calyptra is large and often conic-mitrate. This combination of characteristics is largely unique in *Syntrichia*, although the African *S. cavallii* (Negri) Ochyra also has a conspicuously differentiated perichaetium and weakly pandurate leaves, although its leaf margins are recurved in the distal portion, and the plane leaf margins are characteristic of species of sect. *Aesiotortula*.

Willia is an austral genus of three species: *W. austroleucophaea* which is distributed mainly on subantarctic South Georgia (Bell 1974) and in the Fuegian region (Greene 1986); *W. calobolax* (Müll.Hal.) Lightowlers which is widespread in the Kerguelen Province of the Subantarctic and on Antipodes Island of New Zealand (Lightowlers 1985*a*), and *W. brachychaete* (Dusén) R.H.Zander which is endemic to western Patagonia (Dusén 1906; Kramer 1988; Zander 1993) (Fig. 201, inset). The genus was originally described by Müller (1890*a*) to accommodate a single species, *W. grimmioides*, and later Müller (1901) added to it *W. marginata* (Hook.f. & Wilson) Müll.Hal. from Îles Kerguelen, but Zander (1993) transferred this species to *Hennediella*. The generic name *Willia* commemorates Dr Hermann Will (1852–1930), a participant and naturalist on the German International Polar-Year Expedition 1882–1883 (Frahm & Eggers 2001). He and E. Mosthaff made a substantial collection of mosses on South Georgia (Greene 1972) which was subsequently studied and described by Müller (1890*a*) in his *Bryologia Austro-Georgiae*. Zander (1993) erroneously ascribed this eponymous name to Elie Jean François Guillou, a botanist to the French expedition on the ships *Astrolabe* and *Zélée* in 1837–1840 under the command of Captain M. J. Dumont d'Urville. Only *W. austroleucophaea* occurs in the Antarctic.

1. **Willia austroleucophaea** (Besch.) Broth. Fig. 168

Willia austroleucophaea (Besch.) Broth. *in* Engl. & Prantl, Nat. Pflanzenfam. 1(3): 453. 1902 ['austro-leucophaea']. ≡ *Grimmia austroleucophaea* Besch., Bull. Soc. Bot. France 32: lx. 1886 (1 Jan) ['1885']. ≡ *Guembelia austroleucophaea* (Besch.) Müll.Hal. *in* Paris, Index Bryol.: 540. 1896, *nom. inval. in synon.* — TYPE CITATION: Terre de Feu, île Horn (Hariot, n° 157). [Lectotype (*selected here*): 'Herb. Mus. Paris. Willia austro-leucophaea Broth. Grimmia austro-leucophaea (Besch.) sp. nov. Île Horn Fuegia! 1883 Hariot; the label inside packet reads: Grimmia austro-leucophaea Besch. 157 ~~Grimmia subleucophaea Besch.~~ sp. nov. Île Horn (T de Feu) 30 juin 83 H' – PC!; isotypes: BM!, PC!].

Willia grimmioides Müll.Hal. *in* Neumayer, Int. Polarforsch. Deutsch. Exped. 2: 311. 1890. — TYPE CITATION: Austro-Georgia, ad rupes des Köppenberges, 19. Majo 1883. [Lectotype (*selected here*): 'Ex Museo botanico Berolinensi Willia grimmioides C. Müll. Austro-Georgia Ad rupes Köppenberges 19. Majo 1883 Dr. Will' – PC!]. First synonymised by Bell (1974: p. 96).

Willia senilis Müll.Hal., Gen. Musc. Frond.: 424. 1901, *nom. nud.* [*Grimmia senilis* Besch. *in* Müll.Hal., Gen. Musc. Frond.: 424. 1901, *nom. nud. in synon.*], ***syn. nov.*** — ORIGINAL MATERIAL: vom Feuerlande [Chirotype: 'Herb. Mus. Paris Détroit de Magellan – Port Famine Voyage de l'Astrolabe et de la Zélée 1838–1840 n° 66 M. Jacquinot Hb. Bescherelle' – BM-Bescherelle!].

Plants small or moderate in size, slender, rigid, in low or deep, dense, rounded, typically hoary cushions, somewhat lustrous, green, yellowish-, greyish- to brownish-green above, dark brown or blackish-brown below. *Stems* erect, repeatedly forked or less often simple, fragile, loosely or tightly coherent below by dense, felt-like rhizoids, (0.5–)1.0–1.5(–2.5) cm tall, densely radiculose in the lower half with brown, smooth, glistening, branched rhizoids, in cross-section rounded-pentagonal, with a small but distinct central strand and 4–6-layered medulla of large, hyaline, thin-walled cells, with the outermost row of smaller, brownish cells, forming an indistinct cortex; *axillary hairs* filiform, uniseriate, hyaline throughout, composed of 8–10 cells, elongate distally and shorter at the base. *Leaves* densely set, folded, loosely erect and twisted around the stem on drying, erecto-patent with erect apices when moist, 1.2–1.8(–2.0) mm long (without a hair-point), (0.5–)0.6–0.8(–0.9) mm wide, lingulate to subspathulate from a narrow, elliptical, not sheathing base, more or less distinctly pandurate, non-decurrent, concave, obtuse or broadly rounded-acute at the apex, terminated by a long, hyaline, nearly smooth, flexuose hair-point, to 1 mm long on the uppermost leaves; *margins* plane, unistratose, entire or indistinctly denticulate or crenulate, often hyaline near the apex, unbordered or indistinctly bordered distally; *costa* single, strong, 60–70 µm wide, yellow-brown to dark brown, strongly convex dorsally, long-excurrent as a hyaline awn, covered with quadrate to short-rectangular, papillose epidermal cells on the ventral side and with elongate, smooth or somewhat roughened superficial cells on the dorsal side, in cross-section terete to elliptical, consisting of a salient, semicircular to reniform-shaped in outline dorsal stereid band, bistratose large guide cells and a ventral epidermis; *laminal cells* unistratose throughout; *upper cells* chlorophyllose, opaque, subquadrate to short-rectangular, (8–)10–15(–18) µm long, 8–15 µm wide, with thin- to moderately evenly thicker walls, densely papillose with solid or hollow, bifid papillae, 4–8 per cell, sharply set off from the basal cells, somewhat decurrent along the margins; *basal cells* rectangular, 25–70 µm long, 10–15 µm wide, hyaline, thin-walled, lax, becoming sturdier towards the margins but otherwise similar to inner cells, forming a typically fenestrate pattern. [*Dioicous*. *Perichetia* terminal. *Perichaetial leaves* markedly differentiated, loosely sheathing the seta, oblong-lanceolate to broadly lanceolate, 3.0–3.5 mm long, 1.0–1.3 mm wide, gradually long-acuminate, piliferous with a long, hyaline, nearly smooth, flexuose awn; *costa* single, strong, yellow-brown ending in the apex and long-excurrent as a hyaline awn; *areolation* of long-rectangular, yellowish-hyaline, thin-walled, smooth, lax cells in the lower half, becoming more sturdy, thicker-walled, shorter, rhomboidal to rectangular in the upper half and linear-flexuose, thick-walled in the apical part and forming a broad, long-decurrent border, entirely lacking quadrate, papillose cells in the innermost bracts or with a small elliptical area of quadrate, chlorophyllose, papillose cells below the apex near the costa in the median bracts. *Setae* straight to weakly curved, stout, fleshy, 2–3 mm long, reddish-brown, twisted clockwise when dry. *Capsules* immersed to emergent, erect to inclined, obloid to shortly cylindrical, 1.2–1.6 mm long, 0.5–0.6 mm wide, straight, symmetric, abruptly contracted to the seta, not narrowed at the mouth or appearing constricted at the mouth owing to the everted annulus, smooth, brown, slightly lustrous; *annulus* persistent, usually everted, composed of highly vesiculose, thick-walled cells in 3–4 rows; *operculum* stout, short-conical; *exothecial cells* mostly short rectangular to isodiametric in the middle of the urn, 25–50 µm long, 15–25 µm wide, becoming isodiametric, subquadrate to hexagonal towards the orifice, thin-walled; *stomata* few at the base of the urn, phaneroporous, bicellular, with small, elliptical pori; *peristome* absent. *Spores* spherical, 10–14 µm in diameter, brown, papillose. *Calyptra* long-mitrate, lobed and plicate at the base, smooth, naked, lustrous, dark brown at the tip, light brown to yellow-brownish below.]

Chromosome number: unknown.

Discussion. *Willia austroleucophaea* is a very distinct species when studied microscopically, although in the field in the Antarctic it may be mistaken for *Orthogrimmia sessitana*, *Grimmia plagiopodia*, *Syntrichia magellanica* and *S. caninervis* owing to the hoary appearance of its hemispheric compact cushions. Under the microscope, the strongly papillose upper laminal cells, as well as the subspathulate leaves with broadly rounded to rounded-acute apices will separate *W. austroleucophaea* from both grimmiaceous genera, while the weak but distinct panduration of the leaves and plane leaf margins will distinguish it from *S. magellanica* and *S. caninervis*. When fertile, the species is at once distinct from *Syntrichia* species in its immersed to emergent and gymnostomous capsules, as well as strongly differentiated perichaetial leaves, but this character cannot be used in the Antarctic as sporophytes are unknown. Antarctic plants of *Willia austroleucophaea* agree perfectly in the morphological and anatomical details of South Georgian and Fuegian populations, the only difference being their more gracile stature and smaller size of the leaves. They are certainly influenced by severe climatic conditions in the Antarctic, where the plants are smaller and less vigorous in comparison with those occurring in climatically more favourable northerly sites. *W. austroleucophaea* is easily distinguished from the two congeners. It shares the hyaline hair-point with *W. brachychaete* of western Patagonia but this epiphytic species is distinct in its well developed peristome. On the other hand, in the subantarctic *W. calobolax* the costa is shortly excurrent as a stout, reddish-brown, never hyaline arista.

Willia austroleucophaea was originally described as *Grimmia austroleucophaea* from sterile material collected on Horn I. in the Tierra del Fuego archipelago near Cape Horn by the French expedition of 1882–1883 (Bescherelle 1885, 1889).

→

Figure 168. *Willia austroleucophaea*. — **1**. Habit of sterile plant, dry. **2**. Plant with sporophytes, wet. **3**. Cross-section of stem. **4–5**. Axillary hairs. **6–8**. Lower leaves. **9–10**. Upper leaves. **11**. Leaf apex (papillae partly removed). **12**. Mid-leaf cells (papillae partly not indicated). **13**. Mid-leaf cells at margin (papillae partly removed). **14**. Cells in transition zone to hyaline base. **15**. Basal leaf cells. **16–19**. Cross-sections of leaf, a sequence from base to apex. **20**. Perichaetial leaf. **21**. Operculate capsule, dry. **22**. Deoperculate capsule, dry. **23**. Exothecial cells and stomata at base of urn. **24**. Exothecial cells in middle of urn. **25**. Exothecial cells at capsule mouth, annulus and spores. — [1, 3–9, 11–19 from *RILS 1896*; 2, 10, 20–25 from *SWG 3466*]. — Scale bars: a – 100 µm (11–15); b – 1 mm (1, 6–10, 20–22) and 100 µm (3–5, 16–19, 23–25); c – 1 mm (2).

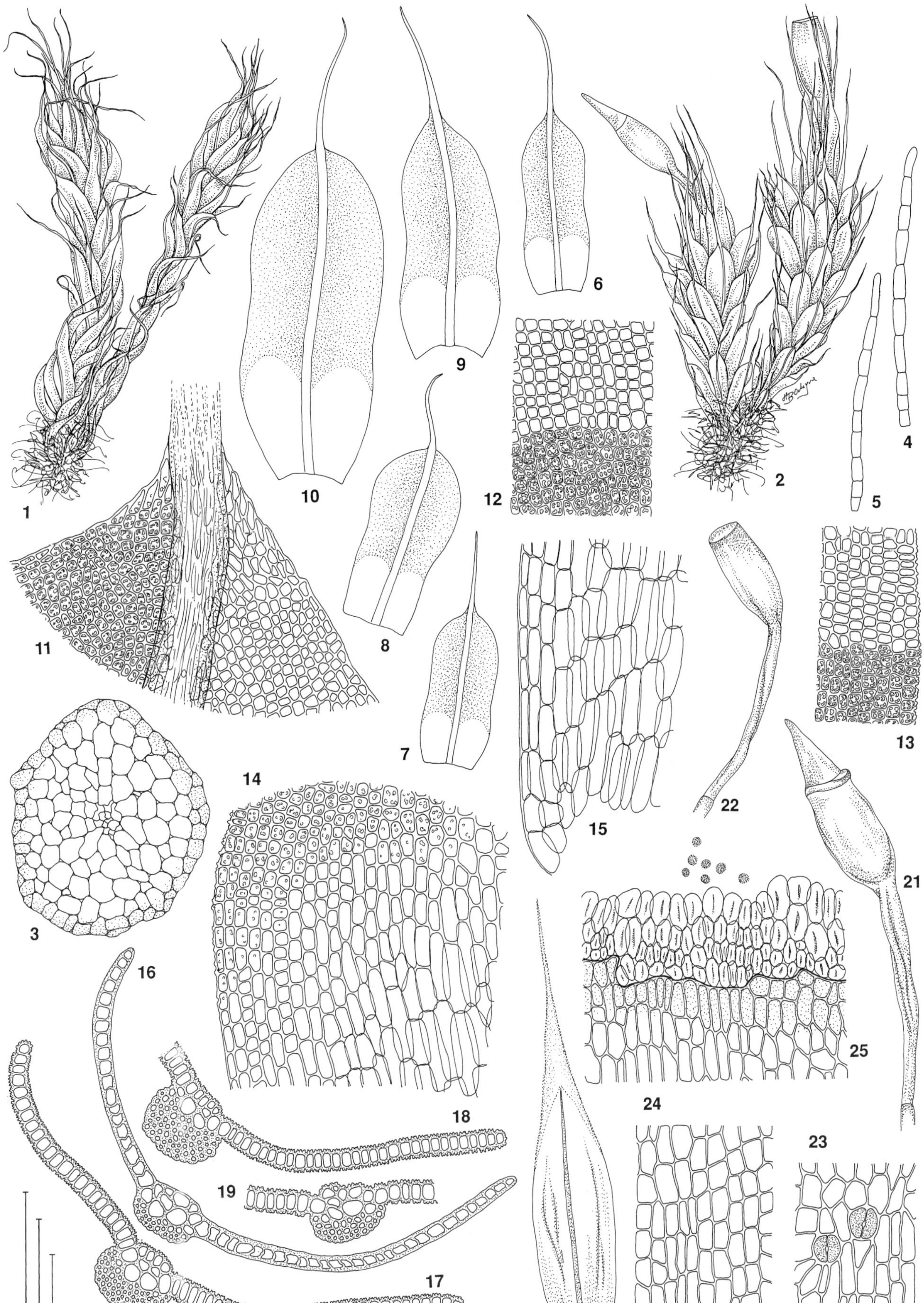
1
2
3
4
5
6
7
8
9
10
11
12
13
14
15
16
17
18
19
20
21
22
23
24
25
a b c

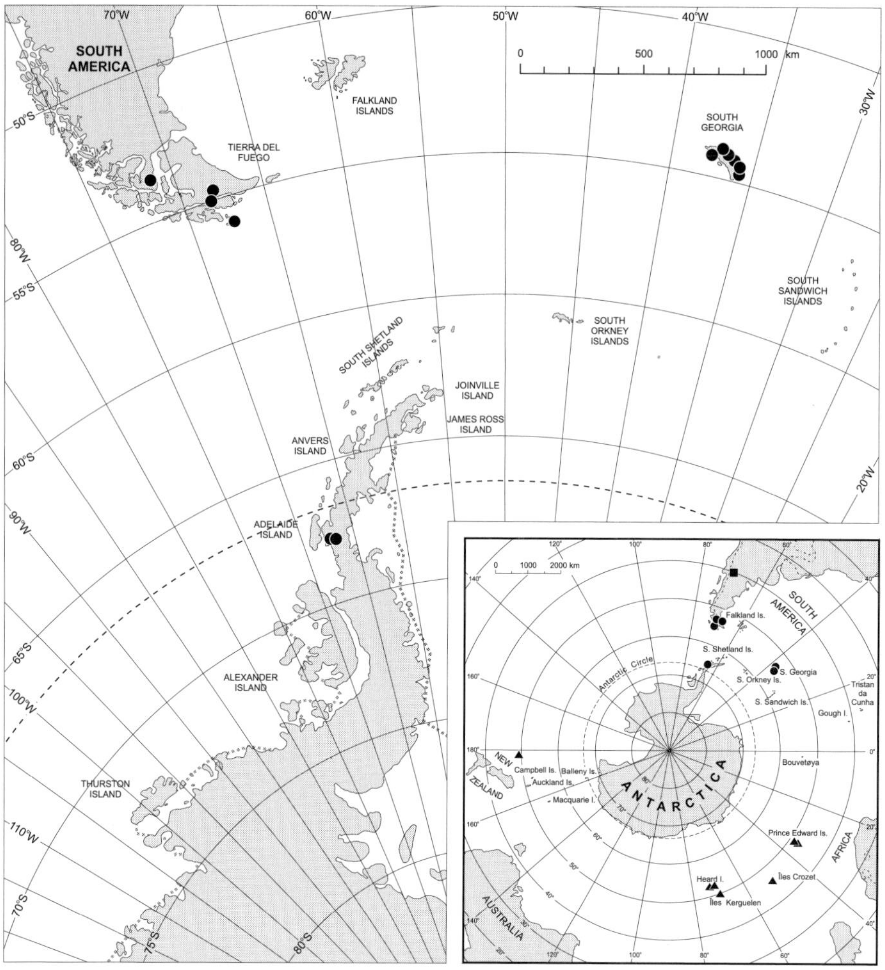

Figure 169. Global distribution of *Willia austroleucophaea*. Inset: distribution of the austral genus *Willia*: *W. austroleucophaea* (dots), *W. calobolax* (triangles) and *W. brachychaete* (square).

However, the species was collected much earlier in this region by M. Jacquinot, a naturalist with the French expedition on the ships *Astrolabe* and *Zélée* in 1837–1840. The specimen collected at Port Famine on the Brunswick Peninsula is not cited in the work of Montagne (1845*b*), which contains a report on the mosses collected by this expedition, but the voucher specimen is preserved in the Bescherelle herbarium at BM and this worker intended to describe it as *Grimmia senilis*. This name is mentioned by Müller (1901) as *Willia senilis* but, contrary to the suggestion of Wijk *et al.* (1969), it has not been validly published due to the lack of diagnosis and the original material represents typical *W. austroleucophaea*. Shortly after the description of the latter species by Bescherelle (1885, 1889), it was described once more from South Georgia as *Willia grimmioides* (Müller 1890*a*). Bell (1974) stated that the type material of this species name is preserved at HBG but, in fact, this material disagrees with the protologue in collecting site and date and is thus not eligible for typification. In fact, the only extant specimen of *W. grimmioides* whose label data perfectly concurs with data in the protologue is preserved in PC, and it is here selected as lectotype. Both *Willia austroleucophaea* and *W. grimmioides* were considered to be conspecific by Bell (1974) and this taxonomic conclusion is here confirmed upon examination of the relevant type collections.

Reproduction in Antarctica. Sterile.

Habitat. As elsewhere in its range, *Willia austroleucophaea* grows on pockets of dry sandy soil on rock ledges and in crevices, close to the seashore, forming pure cushions, lacking admixture of any other bryophyte species, but occasionally associated with some crustose lichens.

World range. A South American subantarctic species which appears to have maximum occurrence on South Georgia where it is widely distributed but never abundant around the island (Bell 1974), extending to a few isolated stations in the central maritime Antarctic and to the southernmost tip of South America (Fig. 169). Ochyra *et al.* (1998) designated *Willia austroleucophaea* as a temperate South American species, but in the southernmost tip of South America it is very rare and so far only a few collections are known, although it is probably undercollected there. Zanten (1971) recorded this species from the Prince Edward Is. in the Kerguelen Province of the Subantarctic, but these plants belong within another species of the genus, *W. calobolax* (Müll.Hal.) Lightowlers (Lightowlers 1985*a*; Ochyra 1998*d*).

Distribution in Antarctica. *Willia austroleucophaea* is known only from two sites in northern Marguerite Bay on the Fallières Coast on the western Antarctic Peninsula (Fig. 169), reaching its southernmost station at lat. 67°53′S on Lagotellerie I. where it was recorded at 300 m, this being an exceptional elevation for this species. The specimen from Horseshoe I. occurred on coastal rocks at the more typical altitude of 5–15 m.

Specimens examined. WEST ANTARCTICA. WEST ANTARCTIC PENINSULA. **Fallières Coast.** ***Marguerite Bay***: Horseshoe I., Lystad Bay, 5–8 m, *RILS 1753A* (AAS, KRAM) and 10–15 m, *RILS 1896* (AAS, KRAM); Lagotellerie I., 300 m, *RILS 4870* (AAS, KRAM).

34. SYNTRICHIA Brid.

Syntrichia Brid., J. Bot. (Schrader) 1800(1): 299. 1801. ≡ *Tortula* Hedw. sect. *Syntrichia* (Brid.) Lam. & DC., Syn. Pl. Fl. Gall. Descr.: 100. 1806. ≡ *Barbula* Hedw. sect. *Syntrichia* (Brid.) Müll.Hal., Syn. Musc. Frond. 1: 632. 1849. ≡ *Barbula* subgen. *Syntrichia* (Brid.) Kindb., Rev. Bryol. 36: 117. 1909. — Lectotype (*vide* Zander 1989: p. 432): *Syntrichia ruralis* (Hedw.) F.Weber & D.Mohr (*Barbula ruralis* Hedw.).

Tortula sect. *Rurales* De Not., Mem. Reale Accad. Sci. Torino 40: 286. 1936, *nom. illeg. incl. sect. prior.*

Sarconeurum Bryhn, Nyt Mag. Naturvidensk. 40: 204. 1902. — Type: *Sarconeurum antarcticum* Bryhn [= *Syntrichia sarconeurum* Ochyra & R.H.Zander]. First synonymised by Ochyra & Zander (2007: p. 210).

Plants from very small to very large, gracile to robust and coarse, rather stiff and rigid, usually in deep, compact or loose turves or tufts, occasionally scattered, dull to slightly lustrous, green, yellowish-, blackish-, olive- or glaucous-green, often with a tinge of red above, reddish-brown below. *Stems* erect, simple or branched, either dichotomously or by means of innovations, usually radiculose at the base or with a dense tomentum below, in transverse section rounded-pentagonal, with or without a central strand, consisting of a multilayered medulla of large, thin- to firm-walled, hyaline to yellowish-hyaline cells, without or with a 1–2-stratose cortex of substereid cells, lacking or with a hyalodermis, sometimes with collapsed outer walls; *rhizoids* often long and dense, brown to reddish-brown, smooth, glossy, branched; *axillary hairs* filiform, consisting of 3–17 cells, all hyaline or 1–3 basal cells yellowish, thicker-walled. *Leaves* often appreciably larger and more crowded at the stem apices, sometimes clustered in terminal rosettes, usually less crowded below, appressed to weakly spreading, straight or crisped, sometimes spirally twisted around the stem and folded when dry, erecto-patent, spreading to squarrose when moist, commonly widest above the middle, oblong- or lingulate-lanceolate, obovate, lingulate to broadly spathulate, rarely ovate, elliptical, ovate-lanceolate, oblong-ovate, sometimes lanceolate to broadly lanceolate, broadly channelled to carinate in the distal part, often narrowly grooved along the costa, with the base undifferentiated in shape to elliptical or oblong-elliptical and loosely sheathing, non-decurrent, abruptly narrowed to a broadly rounded, obtuse or broadly acute, occasionally narrowly acute or cucullate apex, sometimes long-acuminate or, very seldom, quickly contracting into a narrower, easily detached, acute, fragile propaguloid region, with a subula obtusely thickened, multistratose, solid and nearly cylindrical in the distal portion; *margins* plane or recurved at varying distances, sometimes broadly incurved, entire or sometimes serrulate or toothed distally, unbordered or sometimes with thick-walled, isodiametric or elongate and less papillose cells forming marginal or, less often, intramarginal border, unistratose or, rarely, bistratose; *costa* single, stout, yellow-brown to reddish-brown, usually strongly convex dorsally, subpercurrent to excurrent as a long, often hyaline, serrate to hispid, terete or rarely flattened awn or a shorter, concolorous apiculus, usually inserted at 45° with the lamina, covered with rounded-quadrate and papillose or occasionally elongate and smooth epidermal cells on the ventral side, elongate, papillose to spinose or nearly smooth superficial cells on the dorsal side, in cross-section round, elliptical to semicircular, consisting of a crescent-shaped or, very seldom, semicircular, 1–9-stratose dorsal stereid band usually lacking an epidermis, 2–4 large guide cells in 1–4 layers, with or without a hydroid strand; *laminal cells* uni- or, rarely, bi- to tristratose throughout or in patches, brick red with KOH; *upper cells* rounded-quadrate, short-rectangular to hexagonal, thin-walled, occasionally with collenchymatous thickenings, superficially strongly convex on both ventral and dorsal sides or flattened and not bulging, obscure or less often translucent, usually with (1–)4–8(–12) bifid or, sometimes, simple, solid or hollow papillae, very seldom smooth; *lower cells* usually highly differentiated across the base, rectangular or quadrate, thin-walled, hyaline, yellow or reddish, lax medially, becoming more sturdy, narrowly rectangular to linear, thicker-walled in several rows at the margins, usually sharply set off from the upper chlorophyllose cells, rising higher medially and forming characteristic fenestrate, pellucid groups on each side of the costa, occupying 15–50% of the total lamina. *Asexual reproduction* occasionally present as rhizoidal tubers, deciduous tips of the cauline leaves, sessile clusters of small, deciduous leaves or obovate, clavate or globose, multicellular, smooth brood bodies found on the ventral surface of the costa or in the leaf axils. *Dioicous* or *monoicous* (*autoicous* or *synoicous*). *Perichaetia* terminal; *inner perichaetial bracts* usually little differentiated in size and shape, very occasionally much enlarged and sheathing the seta. *Setae* singly per perichaetium or sometimes geminate, short to much elongated, erect, smooth, straight, red to reddish-brown, twisted anticlockwise when dry. *Capsules* exserted, ovoid to long-cylindrical, occasionally with a distinct neck, straight, symmetric or slightly curved, dark red or brown, becoming nearly black with age, smooth or striate when dry, mostly lustrous; *annulus* persistent, 2–4-seriate, composed of large, vesiculose cells with strongly incrassate walls; *operculum* conic-rostrate, rather large and stout, with a straight or somewhat oblique beak, sometimes equaling the urn in length; *exothecial cells* quadrate to rectangular, with thin to evenly thickened walls; *stomata* at the base of the urn, phaneroporous, with 2 guard cells and elliptical opening; *peristome* present, inserted below the mouth, consisting of a low or high basal membrane usually forming a high tesselated tube giving rise to 32 threadlike, spiculose teeth typically wound together to the left in 1 to several spirals, rarely basal membrane absent. *Spores* spherical, brownish, lightly papillose. *Calyptra* enlarged, cucullate, smooth, naked.

Although *Syntrichia* was recognised very early as a genus in its own right (Bridel 1801) it has not been universally accepted and most often it has been united with *Tortula*. Nonetheless, species which constitute *Syntrichia* have usually been considered as a separate group within *Tortula*, mostly as sect. *Syntrichia* (Brotherus 1902, 1924*b*) or sect. *Rurales* De Not. (Kramer 1980, 1988). Zander (1989, 1993) redefined *Syntrichia* and segregated it from the broadly understood *Tortula* on account of the following combination of characters: (*1*) brick red KOH reaction of the distal laminal cells; (*2*) absence of the upper marginal border of narrowly elongate cells; (*3*) the crescent-shaped or, less often, semicircular and markedly convex dorsal stereid band of the costa; (*4*) the lack of the differentiated dorsal epidermis.

Syntrichia is doubtless closely related to *Hennediella* with which it shares the red coloration of the upper laminal cells when treated with KOH, but it is distinguished by its ligulate to broadly spathulate and only very seldom lanceolate leaves, usually recurved and only occasionally toothed and/or bordered leaf margins, the lack of the dorsal epidermis and the crescentic to semicircular costal stereid band. This genus is also very near *Willia* which, however, differs from it in having the peristome distinctly reduced in length and complexity, strongly differentiated perichaetial leaves, short setae and distinctly panduriform cauline leaves.

As presently perceived, *Syntrichia* is a somewhat heterogeneous taxon and the placement of some species in this genus is still debatable and needs careful assessment. For example, *S. geheebiaeopsis* (Müll.Hal.) R.H.Zander, from the Subantarctic and southern South America, shares its leaf shape and areolation with *Bryoerythrophyllum* and *Mironia* R.H.Zander. However, it shows the typical costal morphology of *Syntrichia* including *S. cavalii* from East African mountains but the latter has markedly differentiated perichaetial leaves resembling those in the genus *Willia* but showing different sporophyte morphology. Zander (1993) divided the genus into three sections, namely the type section, sect. *Collotortula* R.H.Zander and sect. *Aesiotortula* R.H.Zander. The type section comprises most species of the genus (about 50) and is characterised by recurved leaf margins, upper chlorophyllose laminal cells with thin or evenly thickened walls lacking corner trigones, multiple papillae of the laminal cells and the lack of asexual reproduction. In contrast, sect. *Collotortula* comprises only 18 species which have distinct collenchymatous thickenings of the upper laminal cells, occasionally simple papillae, recurved leaf margins and club-shaped or ellipsoid propagula. Both sections are not sharply distinct because some species of sect. *Syntrichia* have weak triangular thickenings in the corners of the distal laminal cells.

Syntrichia sect. *Aesiotortula* is the smallest section, comprising only nine species, but it appears to be the most distinctive and well defined group within the genus. The plants are generally smaller than in other congeners, the leaf margins are plane all around, the costa is very often strongly thickened distally and semicircular in cross-section, and the propagula, if present, are leaf-like and borne terminally on the stem or as caducous leaf apices. Ochyra & Zander (2007) assessed the affinities of *Syntrichia pygmaea* (Dusén) R.H.Zander, a member of this section with an amphipacific south-temperate distribution pattern, and *Sarconeurum glaciale* (Müll.Hal.) Cardot & Bryhn, which is the only member of the monotypic genus *Sarconeurum* Bryhn and for a long time considered to be the only Antarctic endemic moss genus. This study has shown that both species are closely related and, because the former is doubtless a member of the genus *Syntrichia* as evidenced by its sporophyte characters, the notoriously sterile *Sarconeurum glaciale* is transferred to the former genus as *Syntrichia sarconeurum* Ochyra & R.H.Zander. Accordingly, the genus *Sarconeurum* has to be treated as congeneric with *Syntrichia*.

Syntrichia is a cosmopolitan genus, consisting of about 80 species. They are known from all continents but are predominantly distributed in temperate and cool regions of both hemispheres, whereas in the tropics they occur mostly on high mountains. The genus has gained much attention in various regional taxonomic treatments (e.g. Saito 1973*b*; Kramer 1980; Mishler 1985*a*,*b*, 1986, 1990, 1994; Gallego 2002, 2005; Gallego *et al.* 2002*a*,*b*; Werner *et al.* 2002, 2003, 2004), but no global revision is available. *Syntrichia* shows a remarkable diversity in cool temperate and polar regions in the Southern Hemisphere and, until recently, its species have been treated under *Tortula*.

There are several valuable treatments of *Syntrichia* species for southern South America (Dusén 1906, 1907), New Zealand (Lightowlers 1985*a*), South Georgia (Lightowlers 1984*a*, 1985*b*), Subantarctica (Lightowlers 1984*b*, 1986*a*) and Antarctica (Lightowlers 1987), as well as a preliminary taxonomic treatment and a synopsis of species for the whole south-temperate and austral polar region (Kramer 1988). In total, about a dozen species are known to occur in this vast area and, in some territories, they are among the most abundant bryophytes, for example at higher elevations on South Georgia (Lightowlers 1984*b*, 1985*b*). Of these, nine species have so far been recorded in the Subantarctic and five species and one variety in the Antarctic, following the generic concept adopted in this Flora. In the latter biome species of *Syntrichia* are mostly restricted to the northern Antarctic Peninsula region and only two species have wide pan-Antarctic distribution. Four Antarctic species represent sect. *Syntrichia* and one belongs within sect. *Aesiotortula*.

1. Cauline leaves with distinctive, solid, cylindrical, deciduous propaguloid apices; leaf margins plane; basal hyaline leaf cells gradually changing into chlorophyllose upper laminal cells along an ill-defined junction line at ± the same level across the leaf lamina .... 1. ***S. sarconeurum***

1. Cauline leaves lacking fragile, differentiated, caducous apices; leaf margins weakly to strongly recurved or revolute, rarely plane; basal hyaline cells abruptly changing into chlorophyllose upper cells along a well defined junction line at the much lower level at the leaf margins than at the costa, forming characteristic fenestrate, pellucid groups on each side of the costa 2

2. Costa percurrent or shortly excurrent into a mucro; leaves lanceolate, oblong-lanceolate, oblong-lingulate to lingulate-spathulate; leaf apex narrowly to broadly acute 3
2. Costa excurrent as a long or short, hyaline or reddish hair-point; leaves ovate, elliptical, oblong-elliptical to ovate-lingulate; leaf apex obtuse, rounded to rounded-acute 4
3. Leaves lanceolate to oblong-lanceolate, long-acuminate; leaf apex narrowly acute; leaf margin entire throughout; costa excurrent as a short cuspidate point .... 2. ***S. saxicola***
3. Leaves oblong-lanceolate to lanceolate or sublingulate, short-acuminate; leaf apex broadly acute; leaf margins dentate or serrate to entire, crenulate or sinuate at the apex; costa percurrent, ending in a distinct apical group of smooth, pellucid, rhomboidal cells .... 3. ***S. filaris***
4. Laminal cells mostly bistratose in the distal portion of leaf............ 4A. ***S. caninervis*** var. ***caninervis***
4. Laminal cells unistratose throughout or with occasional bistratose strands in the upper half of leaf............ 5
5. Costa on the dorsal side in the distal portion with tall, pedicellate and bifurcate papillae, 3–20 μm long; leaf lamina with scattered bistratose streaks above mid-leaf .... 4B. ***S. caninervis*** var. ***gypsophila***
5. Costa smooth or with low, less than 2.5 μm, simple or branched papillae on the dorsal side; leaf lamina unistratose .... 5. ***S. magellanica***

1. **Syntrichia sarconeurum** Ochyra & R.H.Zander
Fig. 170

Syntrichia sarconeurum Ochyra & R.H.Zander, Fragm. Florist. Geobot. Polonica 14: 210. 2007 [blocked by *Syntrichia glacialis* (Müll.Hal.) R.H.Zander and *S. antarctica* (Hampe) R.H.Zander]. ≡ *Didymodon glacialis* Hook.f. & Wilson, Fl. Antarct. 2: 408, *pl. 152, f. 6.* 1847, *hom. illeg.* [non *D. glacialis* (Brid.) Wallr., 1831]. ≡ *Leptotrichum glaciale* Müll.Hal., Syn. Musc. Frond. 2: 611. 1851. ≡ *Sarconeurum glaciale* (Müll.Hal.) Cardot & Bryhn, Natl. Antarct. Exped. Nat. Hist. 3 Musci: 3. 1907. — Type citation: Cockburn Island, lat. 64° S. 57° W (barren). [Lectotype (*selected here*): 'Cockburn Island N°. 4. Didymodon ? glacialis H.fil & Wils.' – BM-Wilson!; isotypes: BM!, NY-Mitten (4 specimens)!, PC!].

Sarconeurum antarcticum Bryhn, Nyt Mag. Naturvidensk. 40: 205, *pl. 1, f. 1–10, pl. 2, f. 1–12.* 1902. — Type citation: Habitat in terra antarctica '*Newnes-Land*' et in terra antarctica 'Geikie-Land', ubi clarus Carsten Borchcrevink [sic!], versus tractus austro-polares observandi causa peregrinatus, circiter sub parallelis 73° latitudinis australis et 170° longitudinis orientalis hunc muscum solum anno 1899 reperivit et specimina perpauca unde domum adtulit. Loco primo altitudine supra mare metrorum 6–7, in loco secundo altitudine circiter 100 m., locis ambobus ad rupes detritas, sociis speciebus lichenum mihi ignotis. [Lectotype (*selected here*): 'Sarconeurum antarcticum Bryhn In peninsula antarctica 'Newnes-Land' 20 pedes supra mare 1898–1900 leg. C. Borchgrevink' – O-Bryhn!; isolectotypes: H-Brotherus!, O-Bryhn (2 specimens)!, PC!; syntype: 'Sarconeurum antarcticum Bryhn In terra antarctica (100 m.) 'Gikie-Land' leg. E. Borchgrevink' – BM!, FH!, O-Bryhn!, O!]. First synonymised with *Sarconeurum glaciale* by Cardot (1907*b*: p. 3).

Plants small to rarely medium-sized, generally rigid and stiff, in compact or, occasionally, loose cushions or tufts, less commonly forming small detached spherical colonies or in continuous swards, sometimes gregarious, usually cohering below by dense, felt-like rhizoids, dull to slightly lustrous, dark or bright green, yellowish- to blackish-green above, tan, reddish-brown to stramineous-brown inside. *Stems* erect to ascending, (0.2–)0.5–1.5(–3.0) mm long, simple or sparingly branched, mostly slender and fragile, densely and almost uniformly foliated throughout, interlaced below with more or less dense tomentum of rhizoids, usually extending high among the leaves, in transverse section rounded-pentagonal, without or with a large central strand and a uniform parenchymatous tissue of large, almost uniform, thin-walled, somewhat collenchymatous, hyaline or yellowish-hyaline cells, lacking a differentiated cortex and hyalodermis or with 1 or 2 peripheral layers consisting of smaller and darker-coloured cells; *rhizoids* long, vigorous, greyish-brown to pale brown, smooth, lustrous, branched; *axillary hairs* filiform, hyaline throughout or with 2–3(–5) basal cells firm-walled and yellow, composed of 10–17 cells. *Leaves* crowded, somewhat smaller below and progressively but not appreciably larger at the stem apex, (0.8–)1.2–2.2(–2.7) mm long, (0.2–)0.4–0.6(–0.7) mm wide, non-decurrent, broadly channelled across the ventral surface, erect, somewhat shrunken and tightly or loosely appressed with incurved apices to erecto-patent when dry, spreading-recurved when moist, rather narrowly oblong- to lingulate-lanceolate or lanceolate, from a scarcely differentiated to ovate or oblong, slightly lustrous base, gradually tapering or quite rapidly contracted near or above mid-length into a long subula, channelled below, terete in the upper quarter, forming a cylindrical, flexuose, solid, fragile, multistratose, caducous, snout-like propagulum, to 0.7 mm long, 7–12 times as long as wide, obtuse or sharply apiculate with 1–3 smooth, pellucid cells; *margins* plane, entire, sometimes slightly undulate; *costa* single, prominent, well demarcated from the laminal cells, (25–)50–80(–95) μm wide below, brown, ending in the solid deciduous apex, distinctly convex dorsally, covered on both ventral and dorsal surfaces with quadrate to short-rectangular cells, in 3–4 rows of cells across the costa ventrally in the mid-leaf, in transverse section with one stereid or substereid band dorsally on large leaves, generally absent on small leaves, 2–3 large guide cells in 1 or very seldom 2 layers, without ventral stereid band or occasionally with small ventral stereid group near the leaf base on large leaves, with distinct ventral epidermis of large, papillose cells and lacking a dorsal epidermis; *laminal cells* unistratose throughout, showing the red coloration in the KOH reaction; *upper cells* quadrate to short-rectangular or transversely rectangular, especially at the margins, (12–)15–20(–27) μm wide, with walls evenly thickened or slightly collenchymatous, superficially flattened and not or slightly bulging, densely papillose with low, small, solid, conic or bifid papillae, several per lumen, resulting in obscure lamina; *basal cells* oblong to rectangular, 20–55 μm long, 15–20 μm wide, thin-walled, smooth, pellucid, hyaline, lax, bulging, becoming quadrate and sturdier in 2–4 marginal rows, gradually changing into the chlorophyllose cells along the ill-defined junction line at more or less the same level across the lamina or sometimes lower at the leaf margins than at the costa. *Asexual reproduction* common by deciduous, fragile propaguloid leaf apices, often being caducous late in the apical part of the stem. *Dioicous*. *Perigonia* terminal or seemingly lateral owing to subfloral innovations stretching in the same direction as that of the preceding stem, bud-like, to 2 mm long; *outer perigonial bracts* broadly ovate to ovate-lanceolate, gradually tapering to a subula, mostly with a caducous apex, 1.2–1.8 mm long, 0.8–0.9 mm wide, concave, distinctly costate, with the areolation similar to that of the vegetative leaves; *innermost perigonial bracts* broadly ovate, acuminate, to 1.2 mm long, 0.7–0.8 mm wide, deeply concave, bearing club-shaped, yellow-brown to brown antheridia, 5–8 per perigonium, intermixed with filiform, yellow to pale yellow-brownish paraphyses slightly exceeding the antheridia. *Perichaetia* terminal; *perichaetial leaves* undifferentiated and similar to the cauline leaves. *Sporophytes* unknown.

Chromosome number: unknown.

Discussion. *Syntrichia sarconeurum* is one of the most distinctive species of moss in the Antarctic, which may be distinguished readily from all other species, except for the two species of *Tortella*, by the characteristic fleshy and swollen, deciduous leaf apices forming propagules which, in spite of their fragile nature, are eventually caducous and often present on some leaves in the apical comal group. However, *Tortella fragilis* and *T. alpicola* are distinct from *S. sarconeurum* by their leaf areolation of the hyaline basal cells extending strongly up the margins and forming a V-shaped basal region that sharply demarcates from the upper chlorophyllose cells along the abrupt junction line. In contrast, the hyaline basal cells in *S. sarconeurum* are differentiated juxtacostally across the leaf base, and the junction line between them and the upper chlorophyllose cells is gradual and not sharply defined. Moreover, *Tortella* species have two, dorsal and ventral, stereid bands in the anatomical structure of the costa, whereas in *S. sarconeurum* only a dorsal stereid band is present.

Syntrichia sarconeurum is variable mainly in stature, coloration and size of plants and these features are doubtless conditioned by environmental conditions, plants growing in wetter and sheltered habitats being generally larger and more luxuriant. The variation of stature of the plants is coupled with the size of leaves whose length varies quite markedly, although the broken state of most mature leaves makes leaf dimensions difficult to assess. Generally, the propaguloid leaf apices constitute about one fourth or more of their total length. The leaves are mostly oblong- or ligulate-lanceolate and 1.2–2.2 mm long on average, but occasionally they are small, 0.8–1.2 mm long, 0.8 mm wide, broadly ovate to ovate-lanceolate and have recurved leaf apices on drying. The latter occur on male plants which are exceedingly rare and so far they have been observed at only two sites, on Camp Pt. (*RILS 4637A*, *4639* & *4687*, AAS, KRAM) and nearby Lagotellerie I. (*F.I.D.S. 473*, AAS, BM, KRAM, LE), both on the Fallières Coast on the Antarctic Peninsula. This may suggest the existence of sex dimorphism in *S. sarconeurum*, although the great rarity of male plants dictates caution in offering such an hypothesis. There is also some variation in the shape of the apical leaf constriction below the propagulum. In most plants the leaves are acuminate and not or only weakly constricted, but in some atypical leaves the apex is abruptly constricted below the propagulum. They occasionally occur on the same shoots with typically developed leaves.

The phenotypic plasticity of *Syntrichia sarconeurum* is associated with the quite conspicuous genetic variation in this species (Selkirk *et al.* 1997; Skotnicki *et al.* 1999, 2000, 2004). However, the results of genetic investigations, at least in the present phase, must be interpreted with caution. According to Skotnicki *et al.* (2004) the populations from the Vestfold Hills in Princess Elizabeth Land are isolated genetically, and distinct from those in Victoria Land and Dronning Maud Land. Conversely, the latter are genetically closely related to Chilean populations. However, the South American plants often designated as *Sarconeurum glaciale* (= *Syntrichia sarconeurum*) (Greene 1975; Matteri 1982, 1985*a*; Lightowlers 1985*c*) represent a distinct, though closely related species *S. lithophila* (Dusén) Ochyra & R.H.Zander (Ochyra & Zander 2007), whereas the Vestfold Hills populations do not deviate morphologically from other Antarctic populations of *S. sarconeurum*. It should be stressed that the voucher specimens used for molecular investigations by Skotnicki *et al.* (2004) have not been available during the present study.

→

Figure 170. *Syntrichia sarconeurum*. — **1**. Habit, male plant, dry. **2**. Habit, sterile plant, wet. **3**. Habit, sterile plant, dry. **4**. Cross-section of stem. **5**. Axillary hair. **6, 10–12, 14–16, 18-20**. Leaves with unbroken apices. **7–9, 13, 17, 21–22**. Leaves with caducous apices. **23**. Propaguloid leaf apex. **24**. Mid-leaf cells (papillae partly removed to show areolation). **25**. Mid-leaf cells and ventral supra-costal epidermal cells (papillae partly removed). **26**. Lamina cells in transition zone to hyaline base. **27**. Basal leaf cells. **28–31**. Cross-sections of leaves and costae. **32**. Cross-section of propaguloid apex. **33**. Perigonium. **34–36**. Perigonial bracts. **37**. Innermost perigonial bract with antheridia. — [1, 3, 18–19, 24 from *RILS 9973*; 2 from *RILS 4292*; 4–5, 7, 9, 16–17, 23, 25–27, 29–30 from *RO 2278/80*; 6, 8, 22, 28, 31 from *RILS 1967*; 10–12 from *RILS 9766*; 13–15 from *RILS 4629*; 20–21, 32–37 from *FIDSE 473*, BM]. — Scale bars: a – 100 μm (4–5); b – 1 mm (3); c – 0.5 cm (1) and 100 μm (23–32); d – 1 mm (10–22, 33–37) and 2 mm (2); e – 1 mm (6–9).

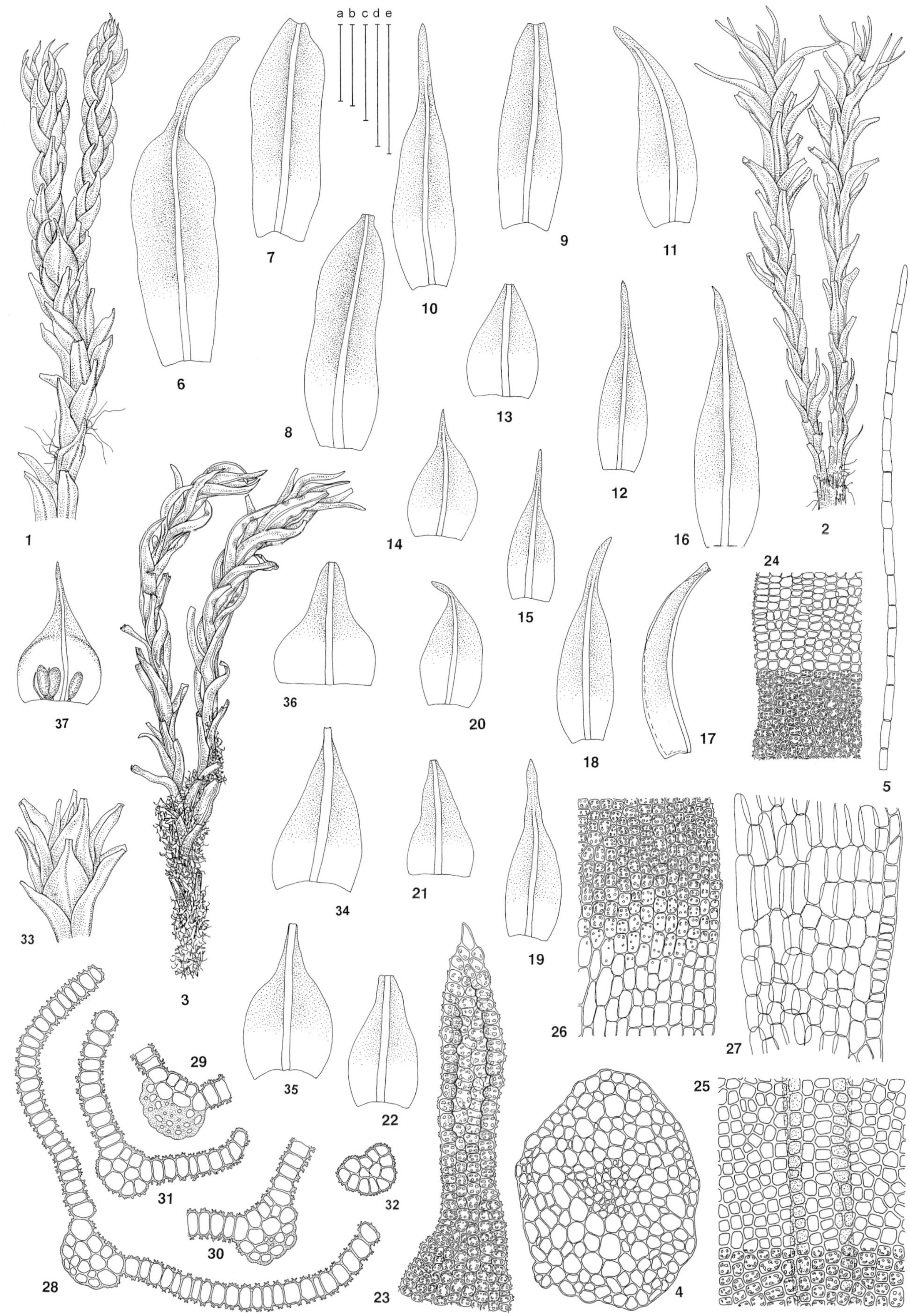
a b c d e
1
2
3
4
5
6
7
8
9
10
11
12
13
14
15
16
17
18
19
20
21
22
23
24
25
26
27
28
29
30
31
32
33
34
35
36
37

Syntrichia sarconeurum was described twice as a new species from material collected in Antarctica. The moss was first discovered by J. D. Hooker in January 1843 on Cockburn I. off the north-eastern coast of the Antarctic Peninsula during the British Antarctic Expedition 1839–1843, under the command of Captain James C. Ross. It was described and illustrated by Wilson & Hooker (1847) as a new species, *Didymodon glacialis*, but its affinities were not discussed and these authors placed it, with some hesitation, in *Didymodon*. In addition, the name proposed by Wilson & Hooker (1847) was illegitimate as a younger homonym of *D. glacialis* (Brid.) Wallr. of 1831, but Müller (1851) legitimised the epithet *glacialis* when transferring this species to *Leptotrichum* as *L. glaciale* Müll.Hal. Over a half century later this species was discovered in Victoria Land by C. Borchgrevink during the voyage of the British *Southern Cross* Expedition. It was reported by Gepp (1902) as *Barbula* sp., but Bryhn (1902) described this material as a new genus and species, *Sarconeurum antarcticum*. In 1902–1903 the species was rediscovered in Victoria Land by the British *Discovery* Expedition under the command of Robert F. Scott, and Cardot (1907*b*) confirmed the conspecificity of *Didymodon glacialis* and *Sarconeurum antarcticum*. He maintained *Sarconeurum* as a distinct genus, although failed to discuss its relationships, and made the relevant nomenclatural change, *Sarconeurum glaciale* (Müll.Hal.) Cardot & Bryhn upon transfer of *Didymodon glacialis* to *Sarconeurum*. This genus was subsequently accepted by Brotherus (1909, 1924) who placed it in the subfamily Trichostomoideae in the close proximity to the modern genus *Bryoerythrophyllum* and this placement has recently been accepted by Zander (2006). *Sarconeurum* was recognised as a monotypic genus until S. W. Greene added to it the second species, *S. tortelloides* (Greene *et al.* 1970) which was later reduced to a variety of *Tortella fragilis* (Zander & Hoe 1979). Its conspecificity with *T. alpicola* was indicated by Zander (1993; see p. 319–320).

Hilpert (1933) was the first to assess the taxonomic position and relationships of *Sarconeurum*. He exluded this genus from the Trichostomoideae and referred it to the Pottioideae, near *Tortula s. lato*, stressing the anatomical structure of the costa, leaf shape and areolation as the most remarkable indicators of the close relationships of these genera. In addition, this author pointed out the presence of the fleshy, fragile leaf apex in some species of *Tortula*, for example *T. atrovirens* (Sm.) Lindb. and considered this character as further evidence of the relationship of both *Sarconeurum* and *Tortula*. Actually, Hilpert (1933) maintained *Sarconeurum* as a genus in its own right only on account of its weak dorsal stereid band in the costa, composed of substereid cells and enlarged laminal cells. Had he known some austral species of *Tortula* bearing the ostensible similarity to *Sarconeurum glaciale*, his evaluation of the genus would probably have been more critical. Dusén (1905*c*) described *Tortula saxicola* from northern Tierra del Fuego but, because the epithet *saxicola* was already in use in this genus, he renamed this species as *T. lithophila* (Dusén 1906). A year later, another new species was described from southern Patagonia, *T. pygmaea* (Dusén 1907), and later Dixon (1923*a*) described a similar species from New Zealand, *T. abruptinervis*. All these species are characterised by having deciduous, fleshy, solid, propaguloid leaf apices and show great external similarity to *Sarconeurum glaciale*.

The presence of the propaguloid leaf apices in *Sarconeurum glaciale*, *Tortula lithophila* and *T. pygmaea* led Greene (1975) to the conclusion that these species are conspecific, the former name having priority. Accordingly the geographical range of *S. glaciale* was extended to southern South America. This concept was subsequently accepted by Matteri (1982) who detected fruiting plants designated as *S. glaciale* in Patagonia. On the basis of the structure of the peristome she confirmed Hilpert's (1933) suggestion on the close affinity of *Sarconeurum* to *Tortula* (and *Desmatodon*). However, the conspecificity of these species was not accepted by Zander (1978*c*) who showed a set of gametophyte characters that differentiated them. Pending further study of South American species, he tentatively used the name of the New Zealand species *Tortula abruptinervis*. These taxa were revisited by Lightowlers (1985*c*) who confirmed the identity of *Sarconeurum glaciale* and *Tortula lithophila* and their distinctness from *Tortula abruptinervis*. This concept was subsequently adopted by Zander (1993) who proposed the conspecificity of the latter species with *T. pygmaea*, and formally transferred this species to the reinstated genus *Syntrichia* as *S. pygmaea* (Dusén) R.H.Zander. The major problem with the taxonomic assessment of *Tortula lithophila* was the scarcity of the type material which was lost in the Dusén herbarium at S and only a small scrap was available in the Roth herbarium at S. The taxonomy of austral species of *Tortula* with caducous leaves has been re-examined by Ochyra & Zander (2007). Having studied all relevant type and numerous non-type collections of all species in question, they found that *T. abruptinervis*, *T. pygmaea* and *T. lithophila* are conspecific, the latter having priority. At the same time, a close alliance of this species to *Syntrichia* sect. *Aesiotortula* was demonstrated on account of the small statutre of plants, lingulate to spathulate leaves, plane leaf margins, bifid leaf papillae, semiterete costa in transverse section and the presence of leaf-like propagules, borne terminally on the stem or as caducous leaf apices and accordingly *T. lithophila* was transferred to *Syntrichia* as *S. lithophila* (Dusén) Ochyra & R.H.Zander.

Syntrichia lithophila is doubtless closely related to *Sarconeurum glaciale* but nonetheless the latter is clearly a distinct species on the basis of the following set of characters: (*1*) leaves oblong- to ovate-lanceolate, gradually tapering to the apex (vs. lingulate and abruptly narrowed at the apex); (*2*) leaf apex not or weakly constricted below the propagulum (vs. rapidly constricted below the propagulum, forming a distinct neck at the point of its attachment); (*3*) apical propagules late-caducous, linear, 7–12 times as long as wide (vs. early caducous, fusiform, 3–4 times as long as wide); (*4*) ventral leaf surface flat to broadly channelled along the costa (vs. narrowly grooved along the costa); (*5*) cells of the ventral surface of the costa quadrate to shortly rectangular (vs. long-rectangular); (*6*) upper laminal cells (12–)15–20(–27) µm wide (vs. 7–11(–14) µm wide).

The close similarity of both *Syntrichia lithophila* and *Sarconeurum glaciale* clearly indicates that recognition of *Sarconeurum* as a separate genus is unwarranted since its concept falls well within that of *Syntrichia* sect. *Aesiotortula* and, consequently, both taxa have been merged. The congenericity of *Sarconeurum* and *Syntrichia* affects the nomenclature of this well known Antarctic endemic because neither the epithet *glacialis* nor *antarctica* is available for this species in *Syntrichia*, being blocked by *S. glacialis* (Müll.Hal.) R.H.Zander and *S. antarctica* (Hampe) R.H.Zander, respectively. Therefore, a new name, *Syntrichia sarconeurum*, is proposed for this species retaining the former generic name which is firmly rooted in Antarctic botany, as a substantival epithet.

Reproduction in Antarctica. *Syntrichia sarconeurum* is invariably sterile. Despite the occasional formation of fully mature gametangia, sporophytes are unknown. A report of sporophytes in this species (Matteri 1982 as *Sarconeurum glaciale*) is incorrect because it was based on specimens from southern Patagonia that actually refer to *S. lithophila* (Zander 1993;

Ochyra & Zander 2007). Despite its sterility, *S. sarconeurum* propagates very effectively by asexual propagules that are produced in masses as readily detachable fragile leaf apices. Savicz-Lyubitskaya & Smirnova (1961, 1962, 1966) studied in detail modes of reproduction in continental populations of this species and described and illustrated the development of young plants from germinating propagules as well as from regenerating protonema arising from rhizoids.

Habitat. *Syntrichia sarconeurum* is a rupicolous and epigean moss, growing in various base-rich and acidic habitats, primarily in dry or moist, exposed and sheltered sites on rock faces as well as on ledges and in crevices and recesses, on scree and weathered rocks, sometimes in the proximity of nesting birds. It has often been collected from soil, silt, sand, and among fine rock fragments, sometimes being partially buried under a layer of fine gravel, while in areas with recent volcanic activity it grows on cinder, scoria and ash. In Victoria Land, in continental Antarctica, *S. sarconeurum* is often the principal component of the association of the short moss turf and cushion subformation in (Longton 1973). The species occurs as small, discrete colonies up to 5 cm wide and 4 cm deep amongst stones, particularly in seasonal drainage channels or, less often, as coalesced turves extending over several metres. The cushions are often moribund, with their surface covered by a thin growth of crustose lichens, particularly *Xanthoria mawsonii*, and occasionally cyanobacteria. The moss associates are infrequent and include *Bryum argenteum* (including var. *muticum*), *Syntrichia magellanica*, *Hennediella heimii*, *Ceratodon purpureus* and *Orthogrimmia sessitana*. In the maritime Antarctic, *S. sarconeurum* usually occurs as small cushions in various bryophyte and lichen assemblages of rock micro-habitats, for example on Signy I. in the South Orkneys, where it is rare and grows with the chasmophytic mosses *Hypnum revolutum*, *Distichium capillaceum*, *Didymodon brachyphyllus*, *Schistidium antarctici*, *Syntrichia saxicola* and *S. magellanica* (Lewis Smith 1972).

World range. *Syntrichia sarconeurum* is the most widespread of six strict Antarctic endemic moss species (Figs 171–172). Its reports from western Patagonia in southern South America (Greene 1975; Matteri 1982, 1985*a*; Greene 1986; Lightowlers 1985*c*; He 1998) are erroneous and are the result of the misinterpretation of the type collections of *Tortula pygmaea* and *T. lithophila*. Correctly, they refer to *Syntrichia lithophila* (Ochyra & Zander 2007).

Distribution in Antarctica. *Syntrichia sarconeurum* is a widespread and locally common, but seldom abundant, pan-Antarctic species (Figs 171–172). It is relatively rare and infrequent, though widely scattered, in the maritime Antarctic, ranging from the South Orkney Is. (Signy I.) and South Shetland Is. (Deception I.), where it is exceedingly rare, and along the western side of the Antarctic Peninsula through the Danco and Graham Coasts to northern Marguerite Bay, with one known locality on Alexander I. at lat. 70°52′S. The species is rare on the eastern coast of the Antarctic Peninsula, where it is relatively frequent only on James Ross I. and Cockburn I. and very rare and localised southwards, extending to the Wilkins Coast at lat. 69°28′S (Fig. 204). *Syntrichia sarconeurum* is widely distributed but scattered throughout both coastal and inland areas of the continent, with two apparent major centres of occurrence in Dronning Maud Land and Victoria Land (Fig. 203). It is very frequent in the latter region, particularly on Ross I., reaching its southernmost occurrence on the Dufek Coast at lat. 84°35′S, and this is the farthest south locality at which a moss has so far been discovered (Wise & Gressitt 1965). A single record from Enderby Land (Greene 1967*b*) was an error and later deleted (Greene *et al.* 1970). In addition, Seppelt & Green (1998) reported *S. sarconeurum* from the Prince Charles Mountains at Mac.Robertson Land but, unfortunately, the voucher collection has not been available for examination. *Syntrichia sarconeurum* occurs from near sea level to 1550 m. Its vertical range varies markedly in different parts of its broad geographical range and depends on local topography. All localities on the western coast of the Antarctic Peninsula in the maritime Antarctic are situated at low elevations, 2–60 m, whereas on the eastern coast of the Antarctic Peninsula the species occurs at higher elevations, including (5–)35–150 m on James Ross I., 150 m on the Oscar II Coast and 686 m on the Wilkins Coast. On the continent the species often occurs near sea level, but in Dronning Maud Land it ranges from 480 to 1550 m, whereas in Victoria Land it ranges from 10 to 1240 m. It appears to be well adapted to severe environments.

Selected specimens examined. WEST ANTARCTICA. SOUTH ORKNEY IS. **Signy I.** Ridge to south of North Pt., 50 m, *RILS 658 & 8055* (AAS, KRAM).

SOUTH SHETLAND IS. **Deception I.** Crags south of Neptunes Bellows, *ca* 50 m, *REL 24* (AAS, KRAM); Ronald Hill, *ca* 60 m, *RO 2778/80* (KRAM).

WEST ANTARCTIC PENINSULA. **Danco Coast.** Cuverville I., 20 m, *de Leeuw 31* (AAS, KRAM). ***Anvers I.***: Norsel Pt., 5–10 m, *RILS 1945* (BM); Hermit I., 6–8 m, *RILS 873* (AAS, KRAM); Litchfield I., 10–25 m, *RILS 1966* (AAS, KRAM) *& 1967* (BM). **Graham Coast.** Edge Hill, *ca* 35 m, *RWMC 660* (AAS, KRAM). **Wilhelm Archipelago.** ***Penola Strait***: Petermann I., *ca* 12 m, *RWMC 783* (AAS, KRAM) and 15 m, *REL 1351* (AAS, KRAM). ***Argentine Is.***: largest island between Irizar I. and Uruguay I., *ca* 15 m, *RWMC 437A* (AAS, KRAM); Uruguay I., 10–25 m, *RILS 4271* (AAS, KRAM) and *ca* 35 m, *RWMC 821A* (AAS, KRAM); Corner Is., 8 m, *REL 1367* (AAS, KRAM) and *ca* 10 m, *RWMC 569* (BM); island between Grotto I. and Corner Is., 3 m, *RWMC 574* (AAS, KRAM); Galindez I., Stella Creek, 5 m, *RWMC 404A* (AAS, KRAM); same island, Woozle Hill, ca 35 m, *REL 1329* (BM); Black I., 2–10 m, *RWMC 579* (AAS, KRAM) and *RILS 4292* (AAS, KRAM) and 13 m, *RWMC 832* (BM). ***Collins Bay***: Darboux I., 15 m, *RILS 3367* (AAS, KRAM). **Fallières Coast.** ***Adelaide I.***:

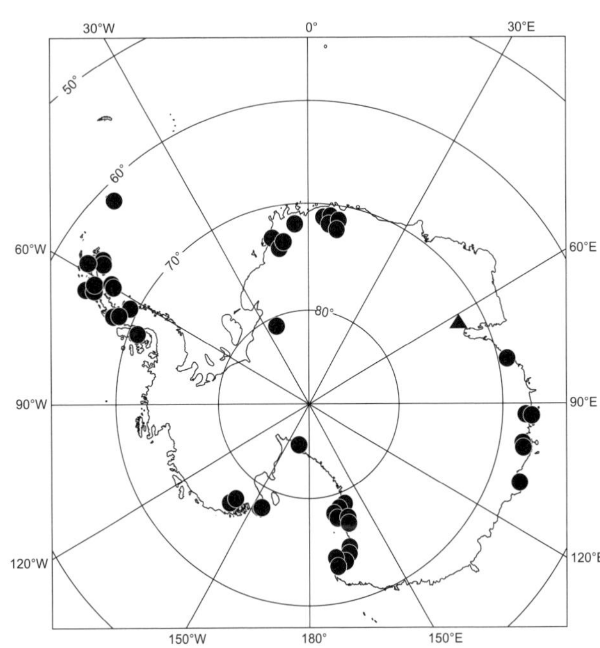

Figure 171. Distribution of *Syntrichia sarconeurum* in Antarctica. Literature record marked with the triangle.

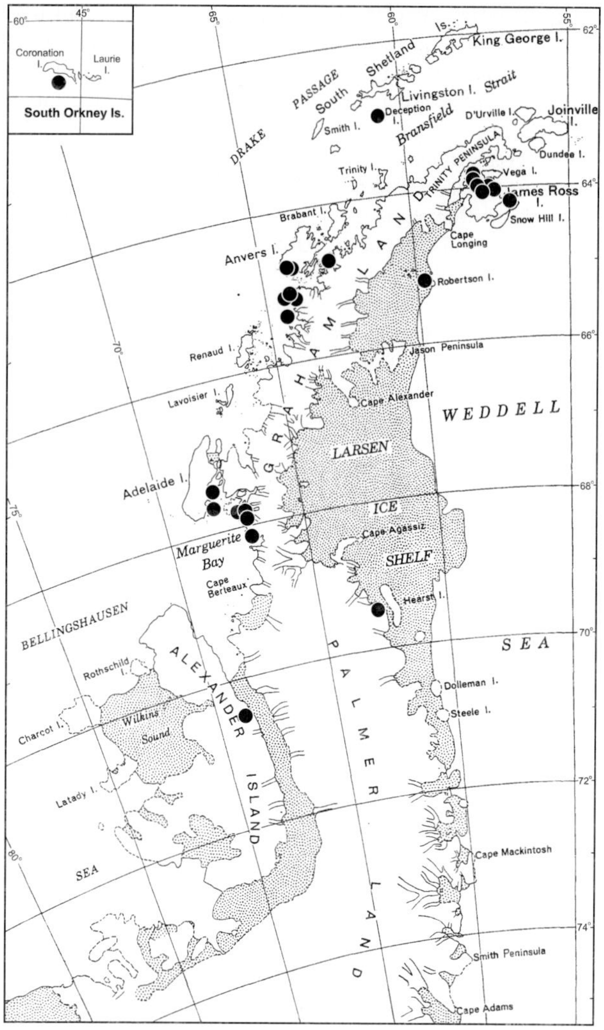

Figure 172. Distribution of *Syntrichia sarconeurum* in the Antarctic Peninsula region.

foot of Sighing Peak, *ca* 6 m, *FIDSE 453* (BM). ***Léonie Is.***: Léonie I., 20 m, *RILS 8995 & 10643* (AAS, KRAM) and 30 m, *RILS 8970A* (AAS, KRAM). ***Marguerite Bay***: Horshoe I., Lystad Bay, 5–8 m, *RILS 1753A & 1758* (AAS, KRAM); Lagotellerie I., 10 m, *FIDSE 473* (AAS, BM, KRAM, LE); Camp Pt., 10–15 m, *RILS 4623, 4629, 4637A, 4639 & 4687* (AAS, KRAM); Neny I., 3–5 m, *RILS 793A* (AAS, KRAM). **GEORGE VI SOUND.** ***Alexander I.***: Ganymede Heights, 100 m from King George VI Ice Shelf and Jupiter Glacier confluence, 10 m, *RILS 11563E* (AAS, KRAM).

EAST ANTARCTIC PENINSULA. **JAMES ROSS ISLAND GROUP.** ***James Ross I.***: east side of Lachman Crags, 35–40 m, *RILS 7338A* (AAS, KRAM); Andreassen Pt., 80 m, *RILS 7593* (AAS, KRAM); Fortress Hill, 150 m, *RILS 7753* (AAS, KRAM); col N of Martha Cove, 100 m, *RILS 7573* (AAS, KRAM); N side of Blancmange Hill, 125 m, *RILS 7706* (AAS, KRAM); NW of Dreadnought Pt., 5 m, *RILS 7504* (AAS, KRAM); Humps I., west peak, 50–100 m, *RILS 7400A* (AAS, KRAM). ***Cockburn I.***: between Adélie colony and foot of cliff on north side, 50–100 m, *RILS 7909B* (AAS, KRAM) and *Hooker 4* (BM, NY – type of *Didymodon glacialis*). **OSCAR II COAST.** Seal Nunataks, Hertha Nunatak, 150 m, *RILS 6425A* (AAS, KRAM). **WILKINS COAST.** Briesemeister Peak, 686 m, *MJC 42* (AAS, KRAM).

EAST ANTARCTICA. COATS LAND. **SHACKLETON RANGE.** North side of ridge extending east from Mt. Provender, *RILS 10000* (AAS, KRAM).

DRONNING MAUD LAND. **KRONPRINSESSE MÄRTHA KYST.** ***Vestfjella***: Kjabebeinet, 600 m, *Sømme 5* (AAS, KRAM); 'Lokalitet B & 4F', Nov 1968–Jan 1969, *Hjelle s.n.* (KRAM, O). ***Heimerfrontfjella***: Tottanfjella, Station 91a, *ca* 1550 m, *GTB Z91A* (AAS, KRAM) and station 92, 1667 m, *GTB Z92A* (AAS, BM, KRAM, NY, PC). ***Ahlmannryggen***: Robertskollen, 'Ice Axe Peak', 480 m, *Watkins 24A* (AAS, KRAM). **PRINSESSE ASTRID KYST.** Henriksenskjera, SW ridge, 1200 m, *RILS 10092A* (AAS, KRAM); same locality, 'spot 1328', *ca* 1300 m, *RILS 10081A* (AAS, KRAM); same locality, 'Twin Crown', 1363 m, *RILS 11139C* (AAS, KRAM); Schirmacher Oasis, *Korotkevich 11 & 25* (KRAM, LE); Drygalskifjella, nunatak 'Yunye', 1400–1500 m, 12.1.1960, *Konovalov 1B* (KRAM, LE); Novolazarevskaya Station, *Meyer 6* (AAS, KRAM, NY); Vasskilsata, 940 m, *RILS 10100A* (AAS, KRAM); 'Ekberget', St. 46F, lat. 72°16′S, 30 Jan 1951, *Norwegian-British-Swedish Antarctic Expedition 6* (NY, S).

PRINCESS ELIZABETH LAND. **INGRID CHRISTENSEN COAST.** Vestfold Hills, Mossel Hut area, 30 m, *RILS 6378* (AAS, KRAM); near Watts Hut, 10–30 m, *Kennedy 50* (AAS, KRAM).

QUEEN MARY LAND. **WILHELM II COAST.** West of Mirnyy Station, *Markov 1* (AAS, KRAM, LE, NY); Haswell Is., *Hollerbach 3/173* (AAS, KRAM, LE) and *Mawson's Expedition 1A* (BM). **KNOX COAST.** ***Bunger Oasis***: north of fjord 'Yuzhnyy', *Korotkevich 15* (KRAM, LE); Lake Figurnoe, *Korotkevich 26* (KRAM, LE, NY); north-west part of the oasis, *Hollerbach 12bis/562* (KRAM, LE); central part of the oasis, 'Bolshoy Razlom' valley near Lake Figurnoe, *Hollerbach 11/314* (KRAM, LE) & *11/312* (AAS, KRAM).

WILKES LAND. **BUDD COAST.** Clark Peninsula, Vincennes Bay, *GAL 2828A* (NY).

VICTORIA LAND. **PENNELL COAST.** Newnes Glacier ('Newnes Land') dropping from Adare Saddle to Protection Cove at the head of Robertson Bay, lat. 71°41′S, long. 170°14′E, 1898–1900, *Borchgrevink s.n.* (H, O – lectotype of *Sarconeurum antarcticum*); Geikie Ridge ('Geikie Land'), lat. 71°44′S, long. 169°36′E, 1898–1900, *Borchgrevink s.n.* (BM, FH, O – syntype of *Sarconeurum antarcticum*). **BORCHGREVINK COAST.** Cape Christie, *Strang MCM-10* (US); Edisto Inlet, Luther Peak, 18 Nov 1964, *Wise s.n.* (NY); west side of Whitehall Glacier, south of Crater Cirque, 100 m, *RILS 9901* (AAS, KRAM) and *DWHW 245* (AAS, KRAM) and 250–400 m, 11 Nov 1964, *Wise s.n.* (NY); Daniell Peninsula, 810 m, *DWHD 230* (AAS, KRAM) and 1000 m, *RILS 9923* (AAS, KRAM); Cape Phillips, *Bargagli M39* (AAS, KRAM), 200 m, *DWHW 236* (AAS, KRAM) and 250 m, *RILS 10168* (AAS, KRAM); Apostrophe I., 25 m, *RILS 9795 & 9821* (AAS, KRAM); ridge above Tinker and Burns Glaciers, 650 m, *RILS 9670, 9686 & 10300* (AAS, KRAM); Mt. McGee, 1540 m, *RILS 9651* (AAS, KRAM); Wood Bay, Kay I., 80–100 m, *RILS 9581* (AAS, KRAM), 100–130 m, *RILS 10200* (AAS, KRAM), 130 m, *RILS 10221* (AAS, KRAM) and 150 m, *RILS 9973* (AAS, KRAM); Harrow Peaks, 40 m, *RILS 9858* (AAS, KRAM) and 85 m, *RILS 9854 & 9885* (AAS, KRAM); Edmondson Pt., Adelie Lake, 20 m, *RILS 9766* (AAS, KRAM); Priestley Glacier, Lowry Bluff, 1200 m, *DWHW 224* (AAS, KRAM); Inexpressible I., 5–10 m, *RILS 9713* (AAS, KRAM) and 30 m, *DWHW 203 & 204A* (AAS, KRAM); Tarn Flat, 10 m, *RILS 10117* (AAS, KRAM); Lamplugh I., *Bargagli M52* (AAS, KRAM); Prior I., *Bargagli M22* and *RILS 9954A* (AAS, KRAM). **SCOTT COAST.** ***Granite Harbour Area***: Cuff Cape, *ca* 15 m, *WCS & SWG 65/55* (KRAM, NY); The Flatiron, *Schofield AA-5* (NY) and *ca* 50 m, *WCS & SWG 65/18* (KRAM, LD, NY) & *65/45 & 65/46* (KRAM, NY); Mt. Suess, 1000 m, *HJ 13B-C* (AAS, KRAM) and 1240 m, *HJ 15* (AAS, KRAM); Mt. England, 1040 m, *HJ 8A–D, 9 & 10* (AAS, KRAM) and 1170 m, *HJ 11* (AAS, KRAM). ***McMurdo Sound Area***: Gneiss Pt., *GAL 2196A & 2198A* (NY); Walcott Glacier, Lake Penny, *Schofield AA-8* (NY); same area, Lake Teardrop, 29 Jan 1964, *Schofield s.n.* (US). ***Ross I.***: Horseshoe Bay, lat. 77°35′S, *WCS & SWG 65/29* (KRAM, NY) and *ca* 35 m, *REL 4056* (AAS, KRAM, NY); Mt. Terror, 21 Jan 1902, *Discovery Expedition s.n.* (BM, PC); Hut Point Peninsula, *GAL 2193A* (KRAM, NY) and Nov 1908, *Joyce s.n.* (PC); Pram Pt., 80–100 m, *Lewis HP-10* (NY); New Zealand's Aurora Station, *HJ 4A-F* (AAS, KRAM); Crater Hill, *GAL 2177D, 2211 & 2217* (KRAM, NY), *Holm-Hansen 27* (NY), *ca* 130 m, *GAL 2217A* (KRAM, NY), 150 m, *REL 4018* (AAS, KRAM, NY) and *WCS & SWG 65/17* (KRAM, NY), 200 m, *HJ 3* (AAS, KRAM) and 250 m, *REL 4011* (AAS, KRAM, NY); Observation Hill, 100 m, *HJ 1* (AAS, KRAM) and 200 m, *HJ 18 & 19* (AAS, KRAM) and *REL 4005 & 4026* (AAS, KRAM, NY); Cape Armitage, *REL 4015 & 4024* (AAS, KRAM, NY); Scott Base, *Fukushima 805* (KRAM, LE), *JRR MCM-22* (US), 21 Jan 1962, *Rudolph s.n.* (NY) and 120 m, *HJ 2* (AAS, KRAM); Cape Mackay, 30 Jan 1962, *Little s.n.* (NY); Cape Crozier, north slope of Pat's Peak, *Rudolph 64024* (NY). **DUFEK COAST.** East side of Barrett Glacier, *ca* 500 m, lat. 84°35′S, *Wise & Shoup 16* (NY) and *Wise 18* (AAS, KRAM, NY).

MARIE BYRD LAND. **Edward VII Peninsula.** ***Rockerfeller Mountains***: Mt. Patterson, *Fitzsimmons 276* (NY). **Ruppert Coast.** ***Ford Range***: Skua Gull Peak, *PAS 7A* (US); Lichen Peak, *PAS G1, 46* (BM, NY) *& 51* (NY); Mt. Corey, *PAS 43A* (NY); peak No. 244, *Perkins 224A* (NY).

2. **Syntrichia saxicola** (Cardot) R.H.Zander

Fig. 173

Syntrichia saxicola (Cardot) R.H.Zander, Bull. Buffalo Soc. Nat. Sci. 32: 270, *pl. 111 f. 9–12.* 1993. ≡ *Tortula saxicola* Cardot, Bull. Herb. Boissier Sér 2, 5: 1002. 1905. — Type citation: Terre-de-Feu: Ushuaia. [Lectotype (*vide* Ochyra 1997*c*: p. 149): 'Svenska Sydpolarexpeditionen 1901–03. N:r 62. Tortula saxicola Card. spec. nov. Tierra del Fuego: Ushuaia, ad saxa Leg. C. Skottsberg 6.10.1902 Det. J. Cardot' – PC-Cardot!; isolectotypes: S!, UPS! Syntype: exactly same label data only Nr 63 – PC-Cardot!; isosyntype: S!, UPS!].

Tortula fuscoviridis Cardot, Bull. Herb. Boissier Sér 2, 6: 21. 1906. ≡ *Syntrichia saxicola* (Cardot) R.H.Zander, Bull. Buffalo Soc. Nat. Sci. 32: 269. 1993. — Type citation: [Géorgie du Sud] Royal bay: mt. Krokisius. [Lectotype (*selected here*): 'Svenska Sydpolarexpeditionen 1901–03. N:r 298. Tortula fuscoviridis Card. spec. nova South Georgia, Royal Bay, Mt. Krokisius usque 500 m Leg. C. Skottsberg 24.4.1902 Det. J. Cardot' – PC-Cardot!; isotypes: S!, UPS!]. First synonymised with *Tortula saxicola* by Lightowlers (1985*b*: p. 72, 75).

Tortula didymodontoides Broth. *in* Dryg., Deutsche Südpolar-Exped. 1901–1903, 8: 86, *pl. 7, f. 3.* 1906. ≡ *Syntrichia didymodontoides* (Broth.) R.H.Zander, Bull. Buffalo Soc. Nat. Sci. 32: 269. 1993. — Type citation: Keguelen: Felswand an der Südmire (Werth). [Lectotype (*selected here*) {or perhaps holotype}: 'Tortula didymodontoides Broth. n. sp. Kerguelen, 18/2/1903 leg. E. Werth' – H-Brotherus!]. First synonymised by Ochyra (1997*c*: p. 149).

Plants medium-sized to moderately large, stiff and rigid, densely caespitose, dull to somewhat glistening, brown to blackish-brown below, brown or orange-brown to yellowish- or greenish-, rarely blackish-brown above. *Stems* erect, (0.5–)1.0–4.0 cm tall, rather slender, sparingly dichotomously branched, sparsely to moderately radiculose at the base with reddish-brown, glossy, branched, smooth rhizoids, in transverse section obtusely rounded-pentagonal, without or with a weak central strand, 3–5-layered medulla of large, hyaline or yellowish-hyaline, thin-walled cells with clear corner thickenings and an outer 1–2-stratose cortex of smaller, substereid, brown cells, often with collapsed outer walls; *axillary hairs* filiform, hyaline throughout, long, composed of 15–18 cells, usually somewhat shorter at the base but otherwise similar to the distal cells. *Leaves* rather densely set, closely appressed, incurled to somewhat twisted when dry, erect-spreading to patent when moist, lanceolate to oblong-lanceolate from an oblong, hyaline base, sheathing the stem for a quarter to a third of the leaf length, gradually tapering to a long and finely acuminate apex, (1.5–)2.2–3.3(–4.0) mm long, (0.3–)0.5–0.8(–1.0) mm wide, non-decurrent, broadly to narrowly keeled; *margins* entire all around, recurved to revolute on both sides below, flat for down to ⅓ from the leaf apex; *costa* single, sharply demarcated from the laminal cells, percurrent to short excurrent as a reddish- or yellowish-brown, smooth, cuspidate point, 0.1–0.3 mm long, reddish-brown to brown, (40–)60–75(–90) μm wide, strongly convex and smooth or scabrous below with simple verrucate papillae on the dorsal side, flat or convex on the ventral side, in transverse section semiterete, consisting of a large and thick dorsal stereid band of 4–6 rows of small, sclerenchymatous cells, 1–2 rows of large guide cells and ventral epidermis of large, papillose cells; *laminal cells* unistratose throughout, red in KOH reaction; *upper cells* strongly chlorophyllose, irregularly rounded-quadrate to rounded-hexagonal, 6–14 μm wide, densely papillose on both ventral and dorsal faces with bifid, solid or hollow, C-shaped papillae, 4–8 per lumen, markedly obscuring the lamina and making it opaque, spreading down the margins of the upper basal part, separated from the basal cells along the sharp junction line; *basal cells* rectangular to linear, 40–120 μm long, 8–20 μm wide, hyaline, lax and inflated in the inner part of the base, becoming sturdier towards the margins and forming a ± distinct border. *Dioicous. Perigonia* terminal, bud-like, to 1.3 mm long, with numerous antheridia and yellowish paraphyses, slightly exceeding the antheridia. *Perichaetia* terminal; *perichaetial leaves* subconvolute, sheathing the seta, larger but otherwise similar to the cauline leaves. *Setae* erect, straight, 5–13 mm long, smooth, reddish-brown, dextrorse when dry. *Capsules* long-exserted, erect, symmetric, obloid to cylindrical, 1.9–2.3 mm long, brown, somewhat lustrous, smooth; *annulus* persistent, 2–4-seriate, composed of large, vesiculose cells with strongly incrassate walls; *operculum* conic-rostrate, straight, nearly half the length of the urn; *exothecial cells* thick-walled, rectangular, 30–80 μm long, 15–20 μm wide, becoming shorter and deeper brown in 3–4 rows below the mouth; *stomata* at the base of the urn in a single row, bicellular, superficial, with oval openings; *peristome* inserted below the mouth, to 1.2 mm high, yellow-brown, consisting of a high basal membrane forming a high tesselated tube giving rise to 32 threadlike, spiculose teeth, once spirally twisted counterclockwise. *Spores* spherical, yellowish to brownish, 10–13 μm in diameter, lightly papillose. *Calyptra* enlarged, cucullate, brown, smooth, naked.

Chromosome number: $n = 12$ – King George I. (Przywara *et al.* 1984 as *Tortula fuscoviridis*).

Discussion. *Syntrichia saxicola* is a distinct species which is unlikely to be confused with other species. In the field it is readily known by its brown to orange-brown plants growing mostly in dense tufts, and microscopically by its characteristic lanceolate, finely long-acuminate leaves with the entire leaf margins which are immediately diagnostic. Also, the cell arrangement at the leaf base is unique in *Syntrichia* because the isodiametric cells at the margins of the leaf limb descend along the margins towards the leaf base, whereas in other species of the genus they are elongate, mostly long-rectangular, and similar to the inner basal cells. In the Antarctic the species may be confused with *Syntrichia filaris* and *Hennediella antarctica*, but variously dentate to obscurely denticulate or crenulate to irregular leaf margins at the apex should distinguish these species. *Syntrichia saxicola* exhibits little variability in Antarctica, and most often populations differ mainly in size and coloration of the plants. In some populations the extreme tips of the cuspidate leaf point, and sometimes also the upper part of the leaf apex, become hyaline. However, such 'hyalinisation' of the lamina is occasionally found in other Antarctic species that typically lack hyaline hair-points, and probably this is an effect of the plants growing in harsh climatic conditions. There is also some variation in the leaf acumination, but generally the leaves are typically lanceolate to oblong-lanceolate and gradually taper to a shorter or longer acumen.

Syntrichia saxicola was originally described from Tierra del Fuego as *Tortula saxicola* (Cardot 1905*a*) and subsequently the species was described once again as *T. fuscoviridis* from South Georgia (Cardot 1906*a*). Lightowlers (1985*b*) found these two species inseparable, considering more slender stems and less acuminate leaves and larger, more papillose laminal cells of the latter as within the range of variation of the species. Brotherus (1906*a*) also described this species from Îles Kerguelen as *Tortula didymodontoides* but Ochyra (1997*c*) found it to be inseparable from *Syntrichia saxicola* and relegated this name to synonymy with this species name.

Reproduction in Antarctica. Very few Antarctic plants possess either sexual organs or sporophytes, but a few fertile specimens are known from Signy I. and Deception I. (Lightowlers 1986*b*; Convey & Lewis Smith 1993; Lewis Smith & Convey 2002). *Syntrichia saxicola* also produces sporophytes very rarely in the Fuegian region, and so far they have not been detected on South Georgia (Lightlowlers 1985*b*).

Habitat. *Syntrichia saxicola* occupies a relatively narrow range of habitats, growing in open or sheltered habitats on sandy or gravelly soil, morainic debris, scree, rock surfaces, ledges, in rock crevices and on ledges, in volcanic areas on scoria, lava cliffs and ash, sometimes entirely buried in it, and occasionally also on heated ground associated with fumaroles. It usually grows in dry habitats, although infrequently it also occurs in wet flushes, in seepage areas, on moist soil amongst *Deschampsia antarctica* and at melt stream margins, as well as in habitats disturbed by frost action, forming polygons and frost boils. On Signy I. in the South Orkney Is. the species grows on base-rich substrata, forming a distinct association in which it is often dominant, with *Syntrichia magellanica* and *Schistidium antarctici*, and frequently an admixture of *Brachythecium austro-salebrosum*, *Bryum pseudotriquetrum*, *Cephaloziella varians*, *Distichium capillaceum*, *Encalypta rhaptocarpa*, *Marchantia berteroana*, *Pohlia cruda* and *Sanionia uncinata*, and several species of calcicolous lichens (Lewis Smith 1972).

World range. An amphiatlantic subantarctic species (Fig. 174, inset), having maximum occurrence in the northern Antarctic Peninsula region (Lightowlers 1986*b*; Ochyra 1998*a*), and on subantarctic South Georgia (Lightowlers 1985*b*) and Îles Kerguelen (Lightowlers 1985*b*; Ochyra 1997*c* and personal observations). Outside this region it occurs rarely on East Falkland I. (Matteri 1986), on Isla Grande de Tierra del Fuego and Isla Dawson in the Fuegian region (Ochyra 1997*c*), in Cerro Cono in western Patagonia at an altitude of *ca* 1000 m (*Greene TBPA 254*, AAS, KRAM) and in the Río Aysén province (Seki 1974). It extends northward to the southern Valdivian province at Lago Espejo at lat. *ca* 41° S in Neuquén Province in Argentina (Bizot 1967) where it has also been collected at about 1000 m. Because most extra-subantarctic and antarctic localities are situated at high elevations, *S. saxicola* is considered as an austral polar rather then temperate species.

Distribution in Antarctica. *Syntrichia saxicola* is restricted in its distribution to the northern maritime Antarctic, having been recorded most frequently on Signy I. in the South Orkney Is. and King George I. and Deception I. in the South Shetland Is. It appears to be absent from the Antarctic Peninsula, but occurs on James Ross I. on the north-eastern coast of the Antarctic Peninsula, where its most southerly locality is at lat. 64° S (Fig. 174). The species occurs mainly from sea level to *ca* 150 m, but on King George I. it reaches 450 m.

Selected specimens examined. WEST ANTARCTICA. SOUTH ORKNEY IS. **CORONATION I.** Meier Pt., 25–35 m, *RILS 581* (BM); Windscoop Buttress, east of Sunshine Glacier, *ca* 200 m, *RW 170* (BM); Olivine Pt., *ca* 15 m, *RILS 150A* (AAS, KRAM). **POWELL I.** Promontory on SE coast, 35–65 m, *RILS 213A*AAS, KRAM); NE coast, *ca* 300 m, *RILS* (AAS, KRAM); NW coast, 300 m, *RILS 237* (BM). **MATTHEWS I.** NW coast, 15–35 m, *RILS 182B* (BM). **SIGNY I.** Cummings Cove, *RILS 1815B, E* (AAS, KRAM); Jebsen Pt., 200 m, *RILS 595* (AAS, KRAM); Snow Hill, *ca* 220 m, *RILS 601* (AAS, KRAM); Thulla Pt., 50 m, *MWH 767B* (AAS, KRAM); Spindrift Rocks, *ca* 65 m, *REL 1145* (BM); North Pt., *ca* 50 m, *RILS 660A* (BM); Jane Col, 150 m, *RILS 3157* (BM); south end of Three Lakes Valley, *REL 1188B* (AAS, KRAM); Knife Pt., 65 m, *RW 22B* (AAS, KRAM); SE shore of Factory Cove, *REL & RILS 833* (AAS, KRAM); Shagnasty I., 15 m, *RILS 543* (AAS, KRAM).

SOUTH SHETLAND IS. **ELEPHANT I.** 2 km SW of Endurance Glacier, 152 m, *JSA 44B* (AAS, KRAM); Walker Pt., 223 m, *JSA 192* (AAS, KRAM); Stinker Pt., 10 m, *JSA* (AAS, KRAM). Walker Pt., 92 m, *JSA 219* (BM). **CLARENCE I.** Southern end, 152 m, *JSA 2K* (BM). **ASPLAND I.** 30 m, *JPB 67C* (AAS, KRAM). **O'BRIEN I.** NE coast, 90 m, *JPB 9A & 14B* (AAS, KRAM). **GIBBS I.** East of the Spit, 45–90 m, *JPB 175A & 178* (AAS, BM, KRAM); The Spit, 30–120 m, *JPB 216* (BM). **KING GEORGE I.** ***Bransfield Strait***: Blue Dyke, 130 m, *RO 1165/80* (KRAM). ***Admiralty Bay***: Vauréal Peak, 80 m, *RO 5239/79* (KRAM); Cape Syrezol, 15 m, *RO 1864/80 & 1865/80* (KRAM); Harnasie Hill, 190 m, *RO 1838/80* (KRAM); Martins Head, 210 m, *1870/80* (KRAM); Creeping Slopes, 60 m, *RO 1230/80* (KRAM); Agat Pt., 6 m, *RO 413/80* (KRAM); Sphinx Hill, 6 m, *RO 411/80* (KRAM); Rescuers Hills, 100 m, *RO 172/80* (KRAM); Hala, 35 m, *RO 2394/80* (KRAM); Krzesanica, 65 m, *RO4978/79* (KRAM); Skua Cliff, 105 m, *RO 634/80* (KRAM); Ubocz, 180 m, *RO 5123/79* (KRAM); Krokiew, 165 m, *RO 1692/80* (KRAM); south-western branch of Panorama Ridge, 160 m, *RO 1706/80* (KRAM). ***Ezcurra Inlet***: Italia Valley, 90 m, *RO 121/80* (KRAM); Breccia Crag, 145 m, *RO 925/80* (KRAM); Cytadela, 10 m, *RO 962/80* (KRAM); Scalpel Pt., 160 m, *RO 1804/80* (KRAM); Barrel Pt., 3 m, *RO 893/80* (KRAM); Pond Hill, 40 m, *RO 861/80 & 1828/80* (KRAM); Emerald Pt., 10 m, *RO 831/80* (KRAM); Urbanek Crag, 4 m, *RO 1769/80* (KRAM); Dufayel I., Gdynia Pt., 10 m, *RO 1836a/80* (KRAM). ***MacKellar Inlet***: Misty Nunatak, 220 m, *RO 2307/80* (KRAM). ***Keller Peninsula***: Ore Pt., 90 m, *RO 447/80* (KRAM); Round Hill, 20 m, *RO 548/80* (KRAM); British Pt., 5 m, *RO 455/80* (KRAM); Moraine Pt., 40 m,

→

Figure 173. *Syntrichia saxicola*. — **1**. Plant with sporophytes. **2**. Portion of shoot, wet. **3**. Portion of shoot, dry. **4**. Cross-section of stem. **5–6**. Axillary hairs. **7–12**. Leaves. **13–14**. Leaf apices (papillae partly removed). **15**. Upper lamina cells at margin (papillae partly removed). **16**. Mid-leaf cells (papillae partly removed). **17**. Lamina cells in transition zone to hyaline base. **18**. Basal leaf cells. **19–21**. Cross-sections of leaves, a sequence from base to apex. **22**. Deoperculate capsule, dry. **23**. Operculate capsule, wet. **24**. Calyptra. **25**. Exothecial cells at orifice. **26**. Exothecial cells in middle of urn. **27**. Exothecial cells at base of urn and stomata. **28**. Peristome. **29**. Spores. — [1, 25–29 from *Roivainen 982*; 2, 9 from *RO 468/80*; 3, 22–24 from *CS 63* (syntype of *Tortula saxicola*), s; 4–8, 18–21 from *RO 957/80*; 10, 13–15 from *CS 298* (isotype of *Tortula fuscoviridis*), s; 11 from *RO 121/80*; 12, 16 from *RO 411/80*; 17 from *CS 62* (isolectotype of *Tortula saxicola*), s]. — Scale bars: a – 50 µm (15–16), 100 µm (5–6) and 1 mm (3, 22–24); b – 1 mm (7–12); c – 100 µm (4, 13–14, 17–18, 25–27) and 2 mm (2); d – 100 µm (19–21); e – 100 µm (29) and 0.5 cm (1); f – 1 mm (28).

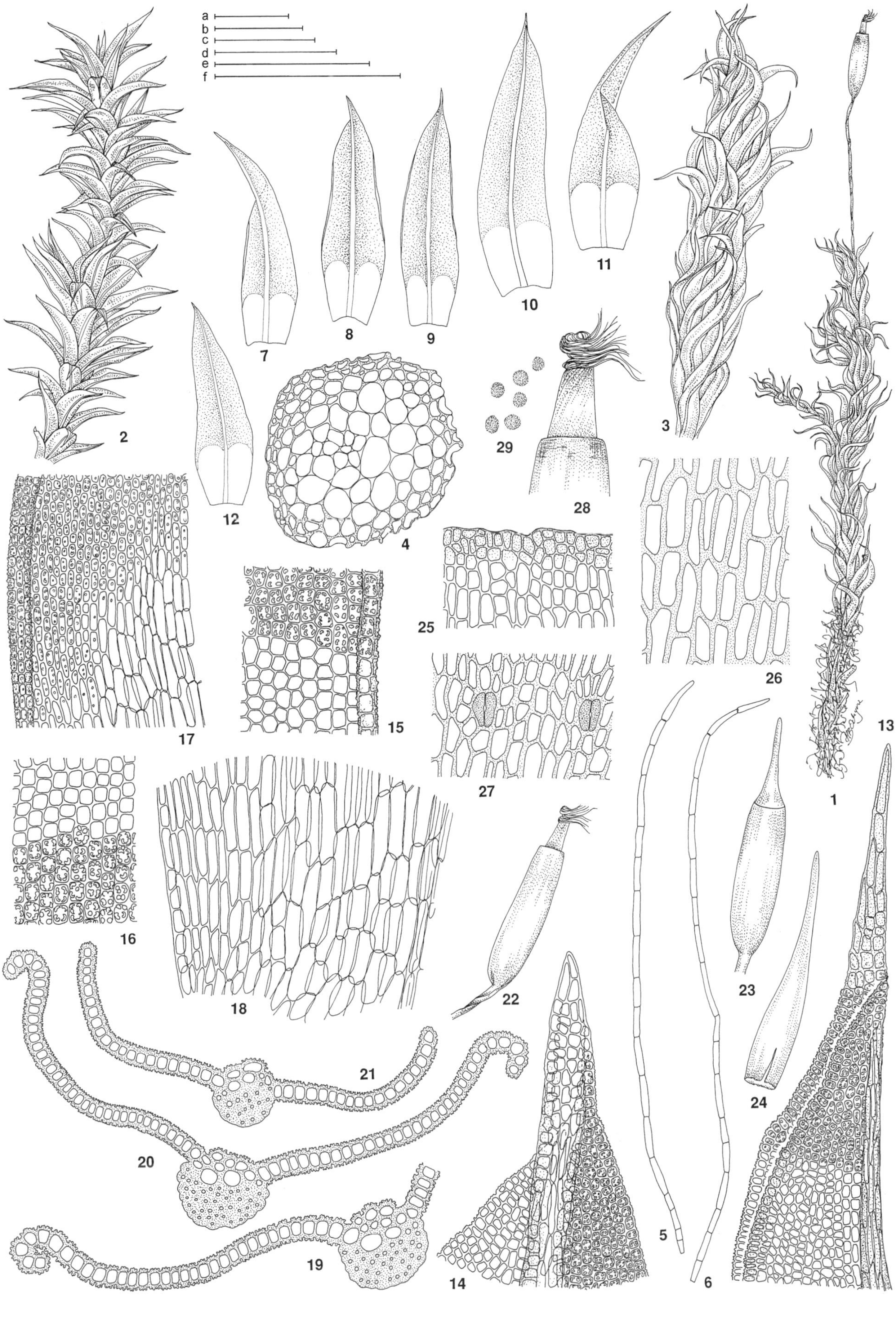
a
b
c
d
e
f
1
2
3
4
5
6
7
8
9
10
11
12
13
14
15
16
17
18
19
20
21
22
23
24
25
26
27
28
29

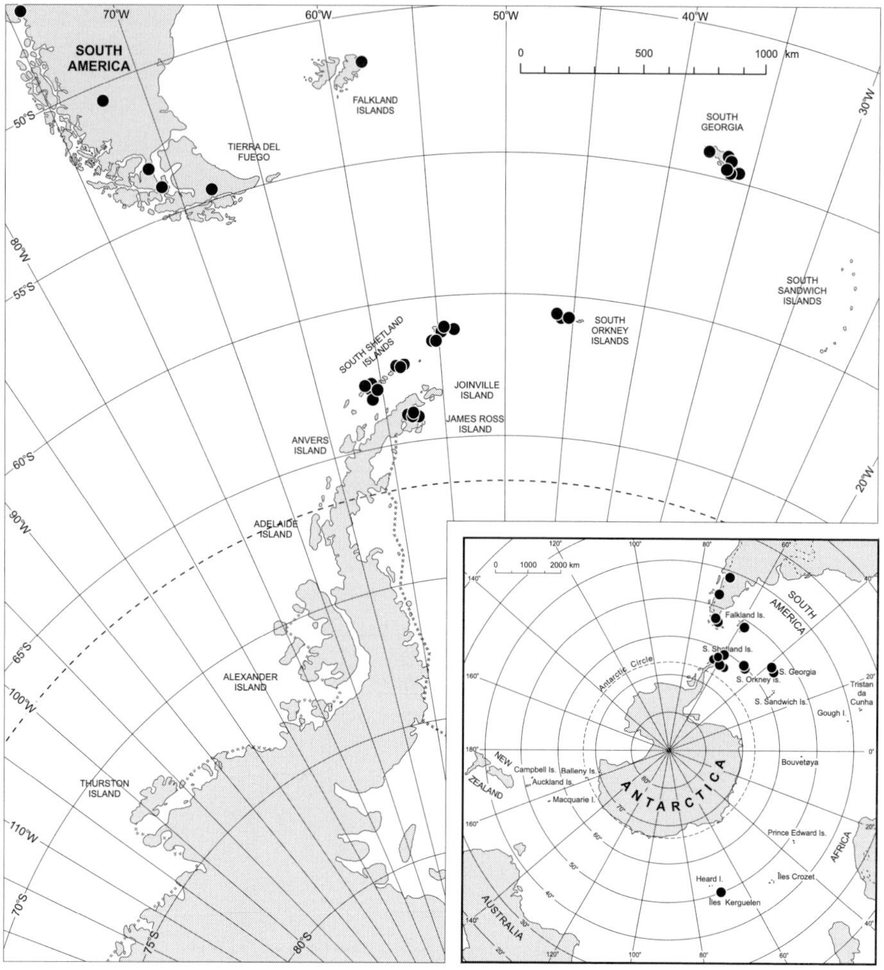

Figure 174. Distribution of *Syntrichia saxicola* in the Fuegian region and Antarctica. Inset: global distribution.

RO 468/80 (KRAM); Yellow Pt., 30 m, *RO 493/80* (KRAM); Tyrrell Ridge, 200 m, *RO 436/80* (KRAM). ***Martel Inlet***: Shark Fin, 145 m, *RO 2714/80* (KRAM); Ullman Spur, 12 m, *RO 562/80* (KRAM); Ternyck Needle, 430 m, *RO 1836/80* (KRAM); Tern Nunatak, 260 m, *RO 2502/80* (KRAM); Point Hennequin, 10 m, *RO 2247/80* (KRAM); Mt. Wawel, 120 m, *RO 2147/80 & 2195/80* (KRAM). ***King George Bay***: Lions Rump, *BJ 94* (KRAM); Penguin Island, *BJ 73* (KRAM). ***Fildes Peninsula***: Gemmel Peaks, *RO 2474/80* (KRAM); Suffield Pt., 50 m, *RO 2445/80* (KRAM). ***Marian Cove***: North Spit, *BJ 149* (KRAM). ***Barton Peninsula***: Winship Pt., *BJ 189* (KRAM). ***Potter Peninsula***: Three Brothers Hill, *BJ 166* (KRAM). **Livingston I.** ***Byers Peninsula***: between Chester Cove and Sealers Cove, 25–30 m, *RILS 3816B* (AAS, KRAM); west end opposite Astor I., *DCL 339* (AAS, KRAM); Ocoa Pt., 2 m, *KR 41B* (AAS, KRAM); Laager Pt., 5–15 m, *RILS 3804* (BM); Devils Pt., 20 m, *KR 24F* (AAS, KRAM); Robbery Beaches, 10 m, *KR 51D, 52A & 53F* (AAS, KRAM); Smellie Pt., 10 m, *KR 35C* (AAS, KRAM); bay off New Plymouth, 6 m, *DCL 321* (BM). ***Hurd Peninsula***: Johnsons Dock, 20 m, *Guzman 117A* (AAS, KRAM); Miers Bluff, *RMS 69-805* (US). **Deception I.** Cathedral Crags near Fildes Pt., *ca* 15 m, *NJC 88B & 133C* (AAS, KRAM), 65 m, *JK 29B* (AAS, KRAM) and 50–150 m, *RILS 6600* (AAS, KRAM); Neptunes Window, 15–35 m, *AC & PK 30A* (AAS, KRAM); Whalers Bay, 15–200 m, *RILS 6572 & 6585* (AAS, KRAM); Mt. Pond, 250 M, *RILS 81* (AAS, KRAM); Telefon Ridge, 150 m, *RILS 5811A* (AAS, KRAM); Cross Hill, *ca* 100 m, *NJC 11* (AAS, KRAM); col between Telefon Ridge and Stonethrow Ridge, 150 m, *RILS 5734A & 5735C* (AAS, KRAM); Stonethrow Ridge, 60 m, *RILS 5806D* (AAS, KRAM); Stonethrow Ridge, 30–60 m, *RILS 5760B* (AAS, KRAM); 150 m, *RILS 5811A* (AAS, KRAM); slopes to west of Wensleydale Beacon, 30–100 m, *RILS 6674C, 6677A, 6680A & 6681* (AAS, KRAM); above Fumarole Bay, 3–10 m, *RILS 3649* (AAS, KRAM) abd 10–30 m, *RILS 6629B & 6632A* (AAS, KRAM); Mt. Kirkwood, 170–250 m, *RILS 5764C & 5765C* (AAS, KRAM); Vapour Col, *ca* 175 m, *RILS 11269* (AAS, KRAM) and *NJC 23* (AAS, KRAM) and 400 m, *DM 35* (AAS, KRAM); Kendall Terrace, 65–85 m, *DM 101A* (AAS, KRAM); Crater Lake, 15 m, *RILS 5661A* (AAS, KRAM) and 25 m, *RILS 11428* (AAS, KRAM).

EAST ANTARCTIC PENINSULA. **James Ross Island group.** ***James Ross I.***: Abernethy Flats, 100 m, *RILS 7633* (AAS, KRAM); Lachman Crags, 15–50 m, *RILS 7336B, D & 7337B* (AAS, KRAM); Holluschickie Bay, Kotick Pt., 100 m, *RILS 5472* (AAS, KRAM); Humps I., 50–100 m, *RILS 7400B* (AAS, KRAM); Fortress Hill, 150 m, *RILS 7780,7792B & 7793* (AAS, KRAM); Blancmange Hill, 12–15 m, *RILS 7673* (AAS, KRAM) and 300 m, *RILS 7824A* (AAS, KRAM); St. Martha Cove, 150 m, *RILS 7539* (AAS, KRAM).

3. **Syntrichia filaris** (Müll.Hal.) R.H.Zander

Fig. 175

Syntrichia filaris (Müll.Hal.) R.H.Zander, Bull. Buffalo Soc. Nat. Sci. 32: 269. 1993. ≡ *Barbula filaris* Müll.Hal. *in* Neumayer, Int. Polarforsch. Deutsch. Exp. 2: 309. 1890. ≡ *Tortula filaris* (Müll.Hal.) Broth. *in* Engl. & Prantl, Nat. Pflanzenfam. 1(3): 433. 1902. — Type citation: Austro-Georgia, oberes Whalerthal, in rupium fissuris, 20. Martio 1883. [Lectotype (*vide* Lightowlers 1985*b*: p. 51): ‘Barbula (Syntrichia) filaris C.Müll. nsp. Austro-Georgia, oberes Whalerthal, in Felsspalten, 20.III.1883 Südpolarexpedition’ – HBG!; isotypes: H!, M!, PC!].

Tortula excelsa Cardot, Bull. Herb. Boissier Sér 2, 6: 30. 1906. — Type citation: [Antarctica] Shetland du Sud: île Nelson, Harmony Cove. [Lectotype (*vide* Lightowlers 1985*b*: p. 53): ‘Herb. J. Cardot. Tortula excelsa Card. sp. nova. Shetland du Sud: île Nelson. Leg. Skottsberg, 1902. Svenska Sydpol. Exped. n° 447’ – BM!; isotypes: BM!, PC! S!, S-Möller (2 specimens)!, UPS!]. First synonymised by Lightowlers (1985*b*: p. 49, 53).

Plants large and relatively robust to medium-sized, in tall, loose or compact cushions or turves, often coalescing and forming extensive mats and patches, rather rigid and stiff, dull or somewhat glistening owing to glossy hyaline leaf bases and backs of the nerves, bright yellow-, golden- or orange- to reddish-brown, light brown to brownish-green or light green throughout or only above, often brown to tan below. *Stems* erect to ascending, (0.8–)1.5–7.5 or, occasionally, to 10.0 cm tall, simple or sparingly dichotomously branched, with elongate, erect, fastigiate branches, sparcely to moderately radiculose at the base or in the lower half with long, brown to reddish-brown, shiny, branched rhizoids, in cross-section obscurely rounded-pentagonal, without or with a weakly developed, small central strand, composed of parenchymatous tissue of large, thin-walled, hyaline to yellowish-hyaline cells becoming not noticeably smaller and brownish in 1–2 outer layers and forming an indistinct cortex; *axillary hairs* filiform, hyaline throughout, composed of 9–11 elongate cells. *Leaves* evenly spaced, rather crowded, 1.0–4.4 mm long, 0.7–1.0 mm wide, erect-flexuose, curled or twisted when dry, patent to arcuately wide-spreading when moist, lanceolate to oblong-lanceolate or, occasionally, sublingulate, shortly acuminate, acute at the apex, differentiated into a smooth, hyaline, shiny, vaginate base, sheathing the stem for a fifth to a third of the leaf length and a chlorophyllose, papillose limb; *margins* plane to weakly recurved near the middle, entire below, dentate to obscurely denticulate or subserrulate or only crenulate to irregularly sinuate, sometimes entire at the apex; *costa* prominent, single, sharply demarcated from the laminal cells, gradually tapering upwards, 70–110 μm wide at the base, subpercurrent to percurrent and ending in a well defined group of relatively large, smooth, rhomboidal cells, yellow- to reddish-brown or brown, smooth on the dorsal side throughout or roughened with low, simple, verrucate papillae, in transverse section semi-terete to elliptical, strongly convex on the dorsal side, flat to weakly convex on the adaxial side, consisting of a relatively weak, lunate in outline, 2–3-stratose dorsal stereid band composed of stereid or substereid cells, 1–2 rows of large guide cells and a ventral, papillose epidermis, occasionally with 1–2 substereid cells in the proximal part of the costa, forming an imperfect ventral stereid group; *laminal cells* unistratose throughout, showing red KOH reaction; *upper cells* rounded-quadrate, subquadrate to subhexagonal, 6–20 μm long, 9–15 μm wide, firm-walled and weakly trigonous, sparsely pluripapillose with 2–6 hollow, C-shaped, branched papillae, making the limb moderately opaque, often becoming slightly smaller and less papillose, oblate, thicker-walled and yellowish at the marginal row forming an indistinct border; *basal cells* rectangular to linear, 40–110 μm long, 10–25 μm wide, smooth, sometimes lax and inflated, hyaline or yellowish-hyaline at the margins and near the costa, becoming narrower towards the margins and forming a yellowish-hyaline border, gradually passing into the chlorophyllose upper cells, with shortly rectangular, sparsely papillose but otherwise transparent cells at the junction line. [*Dioicous*. *Gametangia* terminal. *Perichaetial leaves* similar to the cauline leaves, but somewhat larger and longer acuminate, with margins recurved below and somewhat excurrent costae. *Setae* mostly singly per perichaetium, sometimes plants polysetous, erect, 10–20 mm long, smooth, red, dextrorse above, sinistrorse below. *Capsules* long-exserted, erect, cylindrical, straight or somewhat arcuate, 1.9–2.7 mm long, 0.3–0.6 mm wide, brown, somewhat glossy; *annulus* persistent, consisting of large, vesiculose cells with strongly incrassate walls in 2–4 rows; *operculum* subulate, straight to slightly oblique, about half the length of the urn; *exothecial cells* thin-walled, rectangular, in regular lengthwise rows, 35–85 μm long, 15–20 μm wide, becoming shorter and deeper brown in 3–4 rows below the mouth; *stomata* 5–6 at the extreme base of the urn in a single row, bicellular, superficial, with a narrowly elliptical slit; *peristome* inserted below the mouth, yellow-brown, to 1.6 mm high, consisting of 32 filiform, spiculose teeth, once spirally twisted anticlockwise, arising from a basal membrane forming a high tesselated tube. *Spores* globose, brownish, 10–14 μm in diameter, delicately papillose. *Calyptra* enlarged, cucullate, brown, smooth, naked.]

Chromosome number: $n = 12$ King George I. (Przywara *et al*. 1984 as *Tortula excelsa*).

Discussion. *Syntrichia filaris* is the largest and most robust representative of the Pottiaceae in the Antarctic biome and, in contrast to other members of the family, it is associated with wet habitats. The species is readily distinguished in the field by its bright brown coloration of the plants with yellowish, reddish or orange tinges. Apart from the stature and habit, its most distinctive characters include (*a*) lanceolate to oblong-lanceolate or, rarely, sublingulate leaves that are only short-acuminate; (*b*) plane to weakly recurved leaf margins that are typically denticulate near the apex, although the latter feature is subject to much variation in the Antarctic; (*c*) percurrent costa, ceasing in a distinct apical group of smooth, pellucid, rhomboidal cells; (*d*) relatively large and sparsely papillose laminal cells forming an indistinct marginal border in the distal portion. In many Antarctic populations the leaves are entire to subentire or only crenulate to sinuate on the upper margins, and this variation may cause some confusion with *S. saxicola*, another regional non-hair-pointed species of the genus. However, the latter species can be distinguished by its lanceolate and gradually long-acuminate leaves and a costa shortly excurrent as a smooth cusp, 0.1–03 mm long, in contrast to the oblong-lanceolate, short-acuminate leaves with a percurrent costa in *S. filaris*. In addition, the costa in the latter is broader and it has a weaker and thinner dorsal stereid band and the leaf margins are plane to only weakly recurved in mid-leaf. Moreover, *S. filaris* differs in its leaf areolation of larger and less densely papillose cells making the limbs more transparent and the presence of indistinct marginal borders. *Syntrichia andersonii* is unlikely to be mistaken for *S. filaris*, as its reddish, crenate to denticulate leaf mucro as well as the oblong-lingulate and weakly pandurate leaves are diagnostic. In leaf shape *S. filaris* resembles a miniature of *Hennediella antarctica*, but both species differ in their robustness, and other distinguishing characters are discussed under the latter species.

Syntrichia filaris was discovered in Antarctica on Nelson I. in the South Shetland Is. by C. Skottsberg in 1902 and Cardot (1906*a*, 1908*a*) described it as a new species, *Tortula excelsa*.

He compared it with South Georgian *T. filaris*, a species originally described from this island by Müller (1890*a*) as *Barbula filaris*, and found that both are very similar except for the robustness of the plants and leaf size. Lightowlers (1985*b*) noted that the Antarctic species, besides being larger and more robust, does not differ materially from the South Georgian species and, accordingly, reduced *T. excelsa* to synonymy with *T. filaris*. *Tortula robustula*, a species described by Cardot (1905*a*) from Tierra del Fuego and western Patagonia, was also reduced by Lightowlers (1985*b*) to *T. filaris*, and this taxonomic conclusion has been confirmed during the present study.

Reproduction in Antarctica. Consistently sterile.

Habitat. *Syntrichia filaris* is a wetland moss that grows in a variety of habitats including moist or wet rock surfaces, ledges and crevices, often below or near breeding seabird colonies, at stream sides, along melt-water runnels, in seepage areas traversed by rills, on soil in flushes, generally in open and wet sites, predominantly on base-rich, but also acidic substrata. It is one of the principal components of the moss hummock subformation in which it reaches the highest frequency and abundance in an association co-dominated by robust cushions of *Bryum pseudotriquetrum* and *Sanionia georgicouncinata*. It occupies flushed habitats and develops a shallow peat below the living moss, and in the wetter situations, where *S. filaris* is most abundant it is usually a reddish-, orange or yellow-brown colour, whereas in the drier sites the moss is bright green. Apart from the principal species of the association, the most frequent accompanying mosses are *Brachythecium austrosalebrosum*, *Syntrichia magellanica*, *Warnstorfia sarmentosa*, the liverwort *Cephaloziella varians*, and sometimes the vascular plants *Deschampsia antarctica* and *Colobanthus quitensis*.

World range. An amphiatlantic Subantarctic species, having maximum occurrence in austral polar regions including subantarctic South Georgia, the northern Antarctic Peninsula region and the isolated Bouvetøya, with northward extensions to the Falkland Is., Tierra del Fuego and southern Patagonia (Fig. 176). Lightowlers (1986*a*) reported *Syntrichia filaris* from the Kerguelen Provice of the Subantarctic, but these records are based upon misdetermination and the voucher collections from Îles Kerguelen (*Bell 3115*, AAS, KRAM) and MacDonald I. (*RDS 10655*, AAS, KRAM) represent *Hennediella antarctica*. However, *S. filaris* does occur on Îles Kerguelen and it was once recorded on Île aux Cochons in Golfe du Morbihan (*RO 440/07*, KRAM).

Distribution in Antarctica. *Syntrichia filaris* occurs most commonly in the South Orkneys and South Shetlands in the northern maritime Antarctic, being particularly frequent and abundant on Signy I. in the former archipelago, and on King George I. and Deception I. in the latter. In addition, there is a single record from Vega I. off the eastern coast of the Trinity Peninsula where it reaches its southernmost locality at lat. 63°59′ S, and several collections are known from the isolated island of Bouvetøya in the South Atlantic Ocean (Fig. 176). The records of *S. filaris* from the South Sandwich Is. (Convey *et al.* 2001) are based upon misdeterminations and the voucher specimens are actually *Hennediella heimii* and *Syntrichia magellanica*. Most localities of the species are at low elevations, from sea level to 100 m, and only sporadically has it been collected at higher sites up to 340 m.

Specimens examined. WEST ANTARCTICA. BOUVETØYA. Kapp Circoncision, *ca* 2 m, *MWH 863B* (AAS, KRAM); *ca* 15 m, *MWH 867A, C & 868C* (AAS, KRAM) and 20 m, *MWH 866* (AAS, KRAM); Nyrøysa, *TE 11, 96.1–3 & 206* (BG), 25 m, *TE 76* (BG), 30 m, *TE 73.1–2* (BG) and 177 m, *TE 235 & 242* (BG); Rustadkollen, 300–340 m, *TE 246* (BG); ridge above Nordstranda (Westwind Beach), 170 m, *TE 74.1–2* (BG) and 175 m, *TE 67* (BG); Westwind Beach, *Schoeman 25* (AAS, KRAM).

SOUTH ORKNEY IS. **Coronation I.** Meier Pt., 10 m, *RILS 8021 & 8024A* (AAS, KRAM), *ca* 15 m, *MMM 54* (AAS, KRAM) and 15–25 m, *RILS 583* (AAS, KRAM); Cape Vik, 100 m, *RILS 559* (AAS, KRAM); Shingle Cove *ca* 15 m, *RILS 441* (AAS, KRAM); Saunders Pt., *ca* 5 m, *RILS 39* (AAS, KRAM); Olivine Pt., *ca* 15 m, *RILS 150B* (BM). **Gosling I.** Northern part, 3–15 m, *MMM 29* (AAS, KRAM) and *ca* 15 m, *MMM 43* (AAS, KRAM). **Lynch I.** *Ca* 13 m, *DCL 990A* (AAS, KRAM). **Matthews I.** Near The Divide, *RILS 169A* (AAS, KRAM); NW coast, *RILS 182A* (BM). **Laurie I.** Mackenzie Peninsula, 150 m, *Bellisio 11691, 11694 & 11697* (KRAM, NY); Cerro Moneta, *Bellisio 11755* (KRAM, NY); Paso de la Bellavista, 120 m, *Bellisio 11786 & 11789* (KRAM, NY). **Signy I.** Thulla Pt., *ca* 15 m, *MWH 766C* (AAS, KRAM) and *ca* 65 m. *MWH 155B* (AAS, KRAM); Foca Cove, *RILS 1443, 1444 & 1445* (AAS, KRAM); near Spirndrift Rocks, ca 35 m, *MWH 766B* (AAS, KRAM); between Sprindrift Rocks and North Pt., *ca* 15 m, *MWH 82* (AAS, KRAM); North Pt., *ca* 15 m, *REL 856* (AAS, KRAM); Sombre Lake, 5 m, *RILS 1818* (AAS, KRAM); Stygian Cove, 50 m, *RILS 616* (AAS, KRAM); Waterpipe Beach, *ca* 13 m, *RILS 403* (AAS, KRAM); Elephant Flats, *ca* 15 m, *RILS 617* (AAS, KRAM); Factory Cove, *ca* 35 m, *MWH 749G* (AAS, KRAM); Paal Harbour, ca 65 m, *RILS 646B* (BM); Observation Bluff, 10 m, *MWH 69* (AAS, KRAM) and *ca* 65 m, *MWH 112* (AAS, KRAM). **Moe I.** Landing Cove, 30–50 m, *RILS 5232* (AAS, KRAM).

SOUTH SHETLAND IS. **Elephant I.** Cove 0.5 km south of Stinker Pt., 31 m, *JSA 104* (BM); 3.5 km east of Muckle Bluff, 183 m, *JSA 208A* (BM). **Clarence I.** Southern end, 152 m, *JSA 2H* (AAS, KRAM). **O'Brien I.** NE coast, 90 m, *JPB 20* (AAS, KRAM). **King George I.** ***Bransfield Strait***: Blue Dyke, 4 m, *RO 1212/80 & 1214/80* (KRAM). ***Admiralty Bay***: Vauréal Peak, 36 m, *RO 5251/79* (KRAM); Bastion, 240 m, *RO 896/80 & 1007/80* (KRAM); Uchatka Pt., 20 m, *RO 1052/80* (KRAM); Demay Pt., 15 m, *RO 1383/80* (KRAM); Creeping Slopes, 100 m, *RO 1252/80*. Sphinx Hill, 4 m, *RO 410/80* (KRAM); Rescuers Hills, 60 m, *RO 4875/79* (KRAM); Penguin Ridge, 30 m, *RO 1396/80* (KRAM); Geographers Creek, 95 m, *RO 1601/80* (KRAM); Ubocz, 110 m, *RO 4939/79* (KRAM); Jersak Hills, 135 m, *RO 5062/79* (KRAM). ***Ezcurra Inlet***: Kasprowy Hill, 190 m, *RO 69/80* (KRAM); Italia Valley, 5 m, *RO 5076/79* (KRAM); Emerald Pt., 25 m, *RO 827/80 & 857/80* (KRAM). ***Keller Peninsula***: Ore Pt., 10 m, *RO 525/80* (KRAM); Harpoon Pt., 60 m, *RO 552/80* (KRAM); British Pt., 3 m, *RO 458/80* (KRAM). ***Martel Inlet***: Ullman Spur, 2 m, *RO 600/80* (KRAM); Smok, 10 m, *RO 2051/80* (KRAM). ***Legru Bay***. Low Head, *BJ 85* (KRAM); Chopin Ridge, 10 m, *BJ 88* (KRAM); Lions Rump, *BJ 93* (KRAM). ***King George Bay***: Turret Pt., *BJ 133* (KRAM). ***Drake Passage***: Atherton I., *BJ 114* (KRAM). ***Fildes Peninsula***: Lake Kitezh, 15 m, *RO 2455/80* (KRAM); Durant Pt., *VK 582* (KRAM); Nebles Pt., *BJ 24 & 55* (KRAM). ***Marian Cove***: North Spit, *BJ 143* (KRAM). ***Barton Peninsula***: Winship Pt., *BJ 186* (KRAM). ***Potter Peninsula***: Three Brothers Hill, *BJ 164* (KRAM). **Nelson I.** Moss Bay, 26 Feb 1989, *Komárek s.n.* (KRAM); Harmony Cove, *CS 447* (PC, S, UPS – type of *Tortula excelsa*). **Robert I.** Coppermine Cove, 15–35 m, *RILS*

→

Figure 175. *Syntrichia filaris.* — **1**. Plant with sporophytes. **2**. Portion of shoot, dry. **3**. Portion of shoot, wet. **4**. Cross-section of stem. **5**. Axillary hair. **6–12**. Leaves. **13–14**. Leaf apices (papillae partly removed in 13). **15**. Upper lamina cells at margin. **16**. Mid-leaf cells. **17**. Lamina cells in the transition zone to the hyaline base. **18**. Basal leaf cells. **19–24**. Cross-sections of leaves, a sequence from base to apex. **25**. Capsule. **26**. Exothecial cells at orifice. **27**. Exothecial cells in middle of urn. **28**. Exothecial cells at base of urn and stoma. **29**. Peristome. **30**. Spores. — [1–2, 25–30 from *GCSC & SWG 626*; 3–4, 14–17, 19–21, 23 from *RO 5251/79*; 5 from *VK 446*; 6, 18 from *RO 600/80*; 7 from *RO 1383/80*; 22, 24 from *RO 1720/80*; 8 from *CS 447* (isotype of *Tortula excelsa*), S; 9, 13 from *Will s.n.*, 23.3.1883 (lectotype of *Barbula filaris*), HBG; 10 from *RO 4875/79*; 11 from *RO 5222/79*; 12 from *RILS 403*]. — Scale bars: a – 1 mm (3) and 100 µm (4, 26–28, 30); b – 1 mm (6–12); c – 0.5 mm (1), 1 mm (2, 25) & 100 µm (5, 19–24); d – 0.5 mm (29) and 100 µm (13–18).

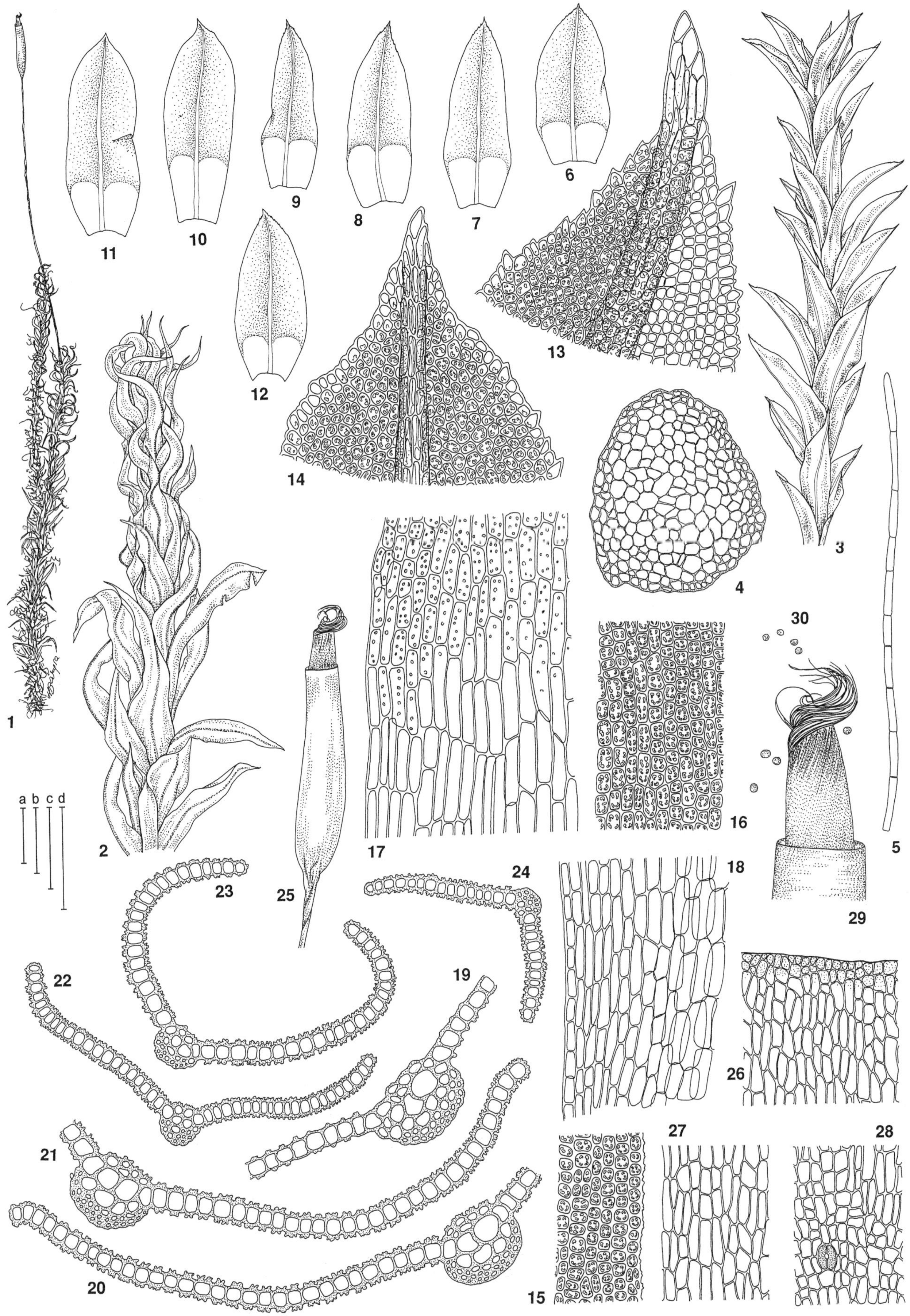
1
2
3
4
5
6
7
8
9
10
11
12
13
14
15
16
17
18
19
20
21
22
23
24
25
26
27
28
29
30
a b c d

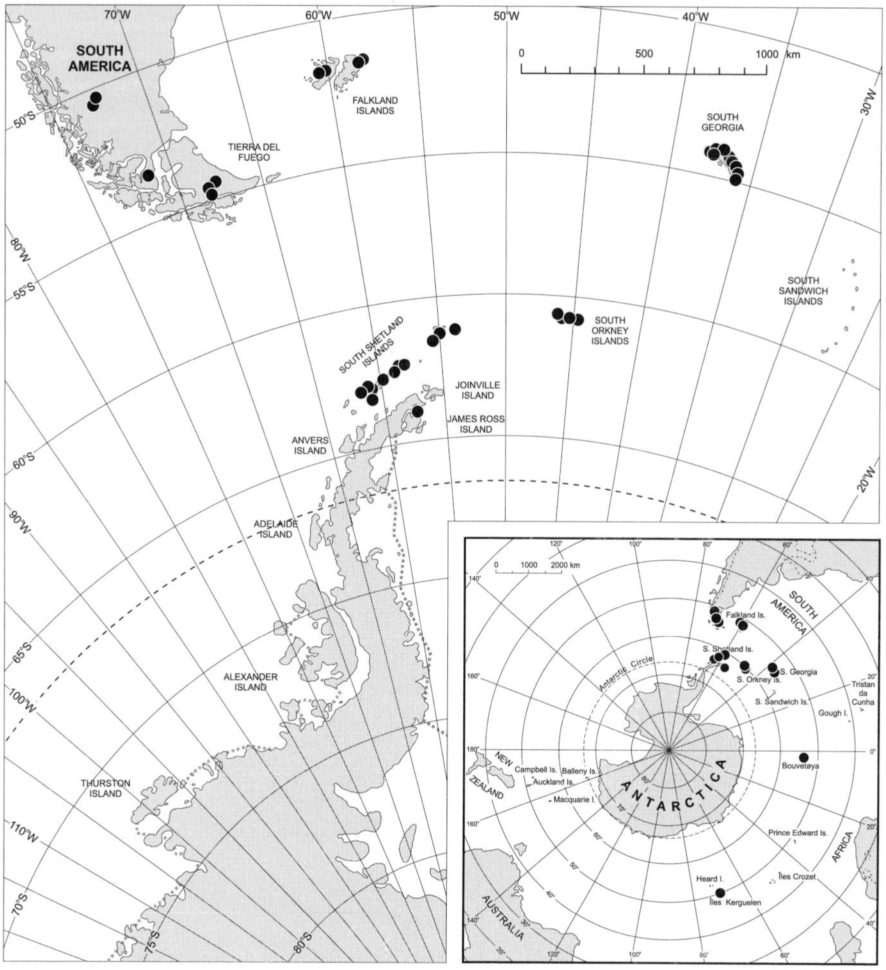

Figure 176. Distribution of *Syntrichia filaris* in the Fuegian region and Antarctica. Inset: global distribution.

914B (AAS, KRAM) and 15–50 m, *RILS 916* (BM). LIVINGSTON I. ***Byers Peninsula***: Laager Pt., 20 m, *KR 40D* (AAS, KRAM); Richards Cove, 20 m, *KR 50A* (AAS, KRAM); South Beaches, 6 m, *DCL 91* (AAS, KRAM) and 10 m, *DCL 452* (AAS, KRAM); above Hell Gates, 30 m, *DCL 266* (AAS, KRAM); Charity Glacier, 35 m, *DCL 55A* (BM); opposite Astor I., 5 m, *DCL 346* (AAS, KRAM); Smellie Pt., *KR 38B* (AAS, KRAM); Sayer Nunatak, 30 m, *RILS 5886A* (AAS, KRAM); Inott Pt., 8 m, *RILS 5890B* (AAS, KRAM). ***Hurd Peninsula***: west side of Miers Bluff, *RMS 69-800, 69-808 & 69-809* (US); Johnsons Dock, 20 m, *Guzmán 107* (AAS, KRAM), *Monroy 108* (AAS, KRAM) and *Redon 106B* (AAS, KRAM); near Spanish Base Camp 'Juan Carlos I', *Acuña 7003* (KRAM). DECEPTION I. South East Pt., 40–50 m, *RILS 11407* (AAS, KRAM) and *ca* 135 m, *RILS 737* (AAS, KRAM); Baily Head, 75 m, *DM 70B* (AAS, KRAM); Whalers Bay, *ca* 10 m, *RILS 80* (AAS, KRAM); Fumarole Bay, 8–30 m, *RILS 3648* (AAS, KRAM) and 50 m, *RILS 11351* (AAS, KRAM); between Stonethrow Ridge and Wensleydale, 7 m, *NJC 18B* (AAS, KRAM); Kendall Terrace, 60 m, *RILS 5742A* (AAS, KRAM); Vapour Col, 130 m, *RILS 5631A,5634C & 5636D* (AAS, KRAM) and 300 m, *MA 22A* (AAS, KRAM); Crater Lake, 12 m, *RILS 5769B* (AAS, KRAM) and 70–220 m, *RILS 6851B & 6853* (AAS, KRAM); Collins Pt., *ca* 35 m, *REL 876* (AAS, KRAM); Entrance Pt., 25 m, *DM 52A* (AAS, KRAM).

EAST ANTARCTIC PENINSULA. **JAMES ROSS ISLAND GROUP.** ***Vega I.***: False Island Pt., 25–30 m, *RILS 7831* (AAS, KRAM).

4. Syntrichia caninervis Mitt.

Syntrichia caninervis Mitt., J. Proc. Linn. Soc. Bot. Suppl. 1: 39. 1859. ≡ *Barbula caninervis* (Mitt.) A.Jaeger, Ber. Thätigk. St. Gallischen Naturw. Ges. 1871–1872: 453. 1873. ≡ *Tortula caninervis* (Mitt.) Broth. *in* Engl. & Prantl, Nat. Pflanzenfam 1(3): 435. 1902. — TYPE CITATION: In Tibet occid. reg. temp., ad rupes? T. Thomson ! (No. 174). [Lectotype (*vide* Gallego 2005: p. 77): 'W.119 Herb. Ind. Or. Hook. fil. & Thomson 174 Syntrichia caninervis Mitt. Hab. Rondu Tibet Regio temp Alt. 6.000 ped. Coll. T. T.' – BM-Hooker!].

Plants small, in dense tufts or patches, seldom gregarious, stiff and rigid, dull, olive-green, greenish- or greyish-brown to whitish above owing to hyalinised apical parts of the leaves, brown below, not or hoary due to long hyaline hair-points. *Stems* erect, 4–12 mm tall, not tightly clumped, simple or sparingly branched, radiculose at the base or tomentose throughout with clusters of long, branched, smooth, brown or reddish-brown, shining rhizoids, in transverse section rounded, without or with an indistinct central strand, consisting of a medulla of large, thin-walled, brownish- to yellowish-hyaline cells becoming somewhat smaller, brown in 1–2 outer rows; *axillary hairs* filamentous, uniseriate, consisting of 13–15 cells, hyaline throughout or with 3–5(–7) yellowish basal cells, elongate in the distal portion, shorter proximally. *Leaves* evenly spaced, densely set, appressed and imbricate, straight to slightly contorted-curved when dry, erect-spreading to patent when moist, ovate, ovate-lingulate to oblong-ovate, (0.8–)1.0–2.0 mm long, 0.6–1.0 mm wide, weakly sheathing at the base, rounded-obtuse to broadly acute, cucullate at the apex, piliferous, concave, broadly channelled in the distal part; *margins* recurved from the base near to the apex, in the upper half or third infolded, entire or crenulate-papillose by the projecting papillae, unistratose throughout or bistratose in places in the distal part, lacking a differentiated border; *costa* single, stout, yellow-brown, 70–110 μm wide at the base, not tapering upwards, strongly convex on the dorsal side, flat or convex ventrally, long-excurrent as a hyaline or greyish-hyaline, strongly spinose hair-point, 0.6–2.0 mm long, often as long as or longer than the leaf lamina, on the dorsal surface in the upper third strongly papillose with simple, bifurcate or branched and pedicellate papillae, 3–20 μm tall, becoming shorter, 2–5 μm towards the base, in cross-section with (1–)2(–3) layers of large guide cells, more or less distinct hydroids and (1–)2–4(–5)-stratose dorsal band of substereids or rarely stereids; *laminal cells* unistratose in the proximal half, regularly or irregularly bistratose in the distal half, usually except for 1–7(–10) marginal rows, sometimes unistratose throughout; *upper cells* quadrate, rounded-quadrate, hexagonal to transversely rectangular, especially towards the margins, 7–11 μm in diameter, thin- to moderately thick-walled, typically densely papillose with low, hollow, branched papillae, 4–6 per cell, obscuring the lumina and making them opaque, but usually becoming strongly eroded in plants buried in fine soil, making the leaf laminae transparent, distinctly descending down the leaf margins towards the base; *basal cells* forming differentiated, fenestrate areas on both sides of the costa, occupying ⅓–½ of the lamina and sharply demarcated from the upper area of the chlorophyllose cells, smooth, hyaline to yellowish-hyaline, thin- to firm-walled, 20–80 μm long, 10–25 μm wide near the costa, becoming narrower and shorter towards the base and forming ± distinct, multiseriate, yellowish border. [*Dioicous*. *Perigonia* terminal but becoming lateral by innovations, gemmiform, to 1 mm long; *inner perigonial bracts* ovate, epilose, concave, to 0.8 mm long, comprising 8–12 antheridia interspersed between numerous paraphyses. *Perichaetia* terminal, with outer perichaetial leaves similar to the cauline leaves and narrower inner perichaetial bracts. *Setae* erect, straight, 6–14 mm long, reddish-brown, smooth, dextrorse above, sinistrorse below. *Capsules* exserted, erect, straight to slightly curved, ovoid-cylindrical to cylindrical, 2.0–3.2 mm long, 0.6–0.8 mm wide, reddish-brown; *annulus* persistent, 2–3-seriate, composed of the outer row of vesiculose cells with strongly incrassate walls; *operculum* long-conical, narrow, blunt at the apex, slanted, 1.5–2.0 mm long; *exothecial cells* variable in shape but mostly oblong to rectangular, mixed with irregular, arranged in ± regular longitudinal rows, with strongly incrassate walls, (15–)20–60 μm long, 6–12 μm wide; *stomata* 5–6 at the base of the urn, bicellular, pale, superficial 40–50 μm long, 20–25 μm wide, with a narrow, elliptical slit; *peristome teeth* 32, filamentous, yellowish-orange to orange-reddish, strongly papillose, 700–800 μm long, once spirally twisted clockwise, arising from a basal membrane forming a pale yellowish, obliquely tesselated tube, 300–400 μm tall. *Spores* globose, brownish, papillose, 7.5–12.5(–17.5) μm in diameter. *Calyptra* cucullate, conic-cylindrical, brown, smooth, naked, covering the entire capsule or only the upper half of the urn.]

Chromosome number: unknown.

Discussion. *Syntrichia caninervis* is a core species of the complex consisting of three species and several varieties that is restricted in its occurrence to arid and semiarid regions of the Northern Hemisphere including the Mediterranean, North Africa, the Near and Middle East and Central Asia, with the disjunction in western North America from where it was originally described as *Tortula bistratosa* Flowers (Flowers 1951). Thus, this group represents a typical Circum-Tethyan element (Frey & Kürschner 1988). Apart from the central species *S. caninervis*, it comprises *S. rigescens* (Broth. & Geh.) Ochyra and *S. handelii* (Schiffn.) S.Agnew & Vondr. but, owing to remarkable plasticity and variability of the plants, the group is highly complex taxonomically and its taxa are not always easy to distinguish (Kramer 1978, 1980; Gallego *et al.* 2002*a*). All these taxa are distinct xerophytes and they are united primarily by having variously bistratose laminal cells in the distal half of the leaf. Agnew & Vondraček (1975) have shown that a very high degree of intergradation occurs between the several species of *Syntrichia* which have bistratose leaves in Asia and the Near East. Kramer (1978) suggested that this character is an adaptation for living in very dry habitats and arid climates, and has a limited taxonomic value because the bistratosity of the laminae exhibits its great variability in different populations. However, Gallego *et al.* (2002*a*) showed convincingly that, although in extreme cases it may range from entirely bistratose to entirely unistratose laminae, this character is relatively stable within the group and is of importance for the differentiation of taxa. The bistratosity of the lamina is coupled with a set of other characters including ovate, ovate-lingulate or ovate-lanceolate leaves that are mostly erect-spreading, patent to recurved on wetting; long and strongly spinose hyaline hair-points; and the anatomical structure of the costa that possesses hydroids and 1–3-stratose dorsal band of stereid or substereid cells that are clearly visible in transverse section. The combination of these characters makes this complex distinct and well defined.

Syntrichia caninervis is superficially similar to the common Antarctic species *S. magellanica* with which it shares the presence of the hyaline leaf hair-points. However, these two taxa are unlikely to be mistaken, and the bistratose laminal cells and the papillose costa on the dorsal surface in the former clearly distinguish *S. caninervis*. In addition, the leaf margin in this species is recurved to the apex and the costa is distinctly papillose on the dorsal side with simple and bifurcate papillae or pedicellate and branched papillae. Conversely, in *S. magellanica* the margins are plane or recurved to ⅔ way up the leaf and the abaxial surface of the costa is smooth or sometimes verrucate or obscurely dentate towards the apex. The latter character is particularly important for distinguishing the Antarctic plants of *S. caninervis* var. *gypsophila* that have the unistratose or nearly so leaf laminae. However, they are distinguished from *S. magellanica* by the spinose dorsal surface of the costa owing to the tall, pedicellate and branched papillae.

Syntrichia caninervis is a protean species which exhibits considerable variability in most characters including the extent of the stratosity of the laminal cells, the presence and length of the hair-point, the papillosity of the abaxial surface of the costa, the size of the plants and laminal cells, as well as the robustness of the costa. Accordingly, various phenotypic expressions of *S. caninervis* were given taxonomic recognition as separate species and Gallego *et al.* (2002*a*) gave nine specific names as its synonyms. In addition, these authors divided the species into four varieties, and Ignatov *et al.* (2002) added a fifth, var. *astrakhanica* Ignatov, Ignatova & Suragina, from Russia. The latter is remarkable as it produces gemmae on long branched filaments that are clustered in great abundance at the shoot tips.

Syntrichia caninervis is here reported for the first time from the Southern Hemisphere. The species is exceedingly rare and occasional and so far only two populations have been discovered on the eastern coast of the Antarctic Peninsula which, in contrast to the western coast, is drier and colder. The Antarctic plants are very variable and both populations can be assigned to two varieties, var. *caninervis* and var. *gypsophila*.

A. var. **caninervis** Fig. 177

Plants small, densely caespitose, almost entirely buried in fine, friable soil, dull, olive-green, greyish-brown or whitish above, brown below, somewhat hoary. *Stems* erect, 0.4–1.2 cm tall, simple or sparingly branched, covered throughout with clumps of long, branched, brown rhizoids, in cross-section without a central duct, consisting of an almost uniform medullary tissue of large, thin-walled, yellowish-hyaline cells, becoming somewhat smaller, thicker-walled and brown in 1–2 outer rows. *Leaves* crowded, imbricate, slightly curled or curved when dry, erect-spreading to spreading when moist, ovate to oblong-ovate, 0.8–1.8 mm long (without the hair-point), 0.6–0.9 mm wide, rounded-obtuse to broadly acute, cucullate at the apex, terminated with a sharply spinose hair-point, usually as long or nearly so as leaf lamina, occasionally shorter to reduced, not or somewhat recurved when wet; *margins* broadly recurved on both sides from the base to the apex or nearly so, entire; *costa* stout, yellow-brown, usually of the same width in most of its length, 70–110 μm wide, prominently convex dorsally, in transverse section semi-terete, consisting of (1–)2(–3) rows of enlarged guide cells and (1–)2–4(–5) layers of dorsal substereid or, rarely, stereid cells, with hydroids, lacking a dorsal epidermis, on the dorsal side with simple or branched papillae, 2–6 μm tall, in the distal half; *laminal cells* mostly regularly bistratose in the upper half, usually except for 1–7(–10) marginal rows, very seldom unistratose with occasional bistratose patches; *upper cells* rounded-quadrate, quadrate to hexagonal, becoming oblate towards the margins, 7–11 μm wide, with moderately thickened walls, rather densely papillose with 4–6 low, simple or hollow, branched papillae to smooth or nearly so on older leaves; *basal cells* rectangular, smooth, hyaline, thin-walled, 20–80 μm long, 10–25 μm wide near the costa, becoming shorter, narrower and sturdier towards the margins and forming yellowish, multiseriate borders, sharply demarcated from the upper chlorophyllose cells. Sterile.

Discussion. As elsewhere in its wide geographical range, the type variety of *Syntrichia caninervis* in the Antarctic is primarily distinguished by regularly bistratose laminal cells in the distal half of the leaf, although most often the marginal cells are unistratose and form 1–7(–10)-seriate border. However, the distal part of the leaf is very seldom entirely unistratose, only with occasional bistratose streaks, but such leaves occur on the same shoots that bear the leaves with bistratose laminal cells and therefore this may be considered as an aberration in the bistratosity of the leaf lamina. The Antarctic plants fall well within the range of variation of this taxon in the Northern Hemisphere and deviate only in having more cucullate leaves, a less papillose costa on the dorsal surface and in the anatomy of the costa. The dorsal part of the costa is composed mostly of the more or less homogeneous substereid cells forming a band that is not markedly demarcated from the guide cells. The costa is dorsally nearly smooth or papillose with low, simple to branched papillae that are 2–7 μm tall but are subject to considerable variation in this taxon. The leaves are generally somewhat smaller in Antarctic plants, but this is unlikely to cause problems with identification.

Reproduction in Antarctica. Sterile.

Habitat. Forming small patches on dry fine, friable soil stabilised by mosses.

World range. *Syntrichia caninervis* var. *caninervis* is a bipolar taxon, having a disjunct Circum-Tethyan range in the Northern Hemisphere (Fig. 178). It is widely distributed in the arid regions of the North American Southwest, ranging from Alberta, Montana and Washington via Oregon, Wyoming, Idaho, Utah, Nevada and Colorado to southern California, New Mexico and Arizona, with some isolated stations in Arctic Alaska (Steere 1978). In Eurasia, this taxon has its main centre of occurrence in the eastern Mediterranean region of the south-eastern Ukraine (Bachurina & Mel'nichuk 1988; Gallego 2005), the Caucasus (Manakyan 1989), Turkey, Lebanon and Jordan, in the Caspian Sea and Irano-Turanian regions and Central Asia, extending eastwards to the Altai Mountains in Central Siberia, Mongolia and central China, with some isolated stations in the western Mediterranean region of the eastern and central Iberian Peninsula of Spain and in the Saharo-Arabian region of Algeria and Morocco (Kramer 1980, 1988; Gallego *et al.* 2002*a*, 2005). In the Southern Hemisphere this taxon is known only from the arid region of the east coast of the Antarctic Peninsula.

The occurrence of *S. caninervis*, as well as other low growing bulbiform xerophytic taxa of the Pottiaceae, in this part of Antarctica supports the hypothesis of the Xerothermic Pangaean element related to the Permo-Triassic configuration of the continents (Schuster 1981, 1983; Frey & Kürschner 1988). Acceptance of this hypothesis would confirm the ancient origin of the Pottiaceae and indicate that the present-day occurrence of *S. caninervis* in the Antarctic has not been established as a result of recent immigration, but may represent an ancient, relictual pre-angiospermous distribution pattern, similar to that exhibited by some marchantioid and riccioid hepatics (Schuster 1981, 1983).

⟶

Figure 177. *Syntrichia caninervis* var. *caninervis*. — **1**. Habit. **2**. Portion of shoot, wet. **3**. Portion of shoot, dry. **4**. Cross-section of stem. **5–6**. Axillary hairs. **7–12**. Leaves. **13**. Leaf apex. **14**. Apex of hyaline hair-point. **15**. Mid-leaf cells. **16**. Cells in the transition zone between median and basal parts of leaf. **17**. Basal leaf cells. **18–23**. Cross-sections of leaves, a sequence from base to apex. — [1–2, 4 from *RILS 7891*; 3 from *RILS 7892*; 5–6, 10–23 from *RILS 7893B*; 7–9 from *RILS 7889A*]. — Scale bars: a – 1 mm (2–3); b – 100 μm (5–6); c – 1 mm (7–12) and 100 μm (18–23); d – 0.25 cm (1) and 100 μm (4, 13–17).

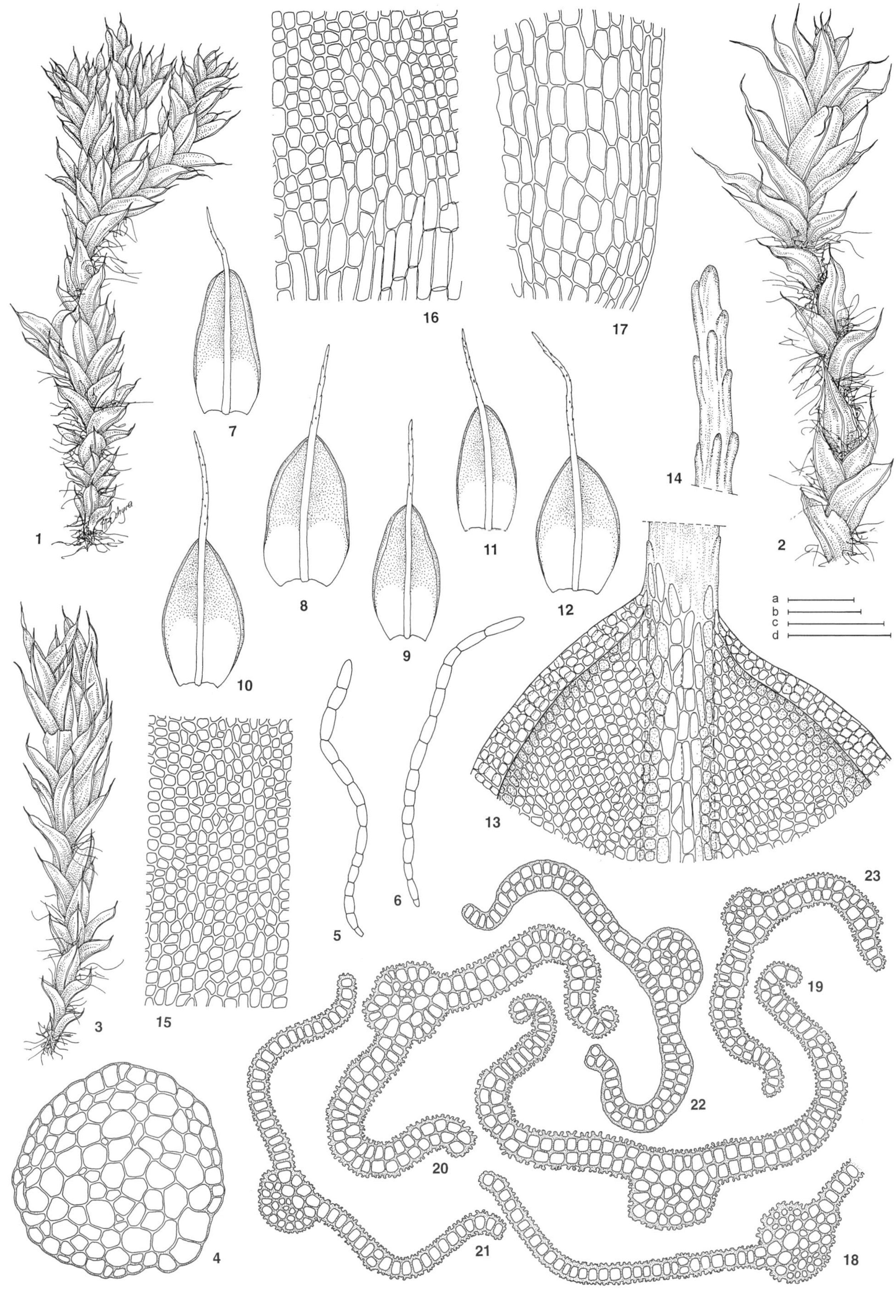
1
2
3
4
5
6
7
8
9
10
11
12
13
14
15
16
17
18
19
20
21
22
23
a
b
c
d

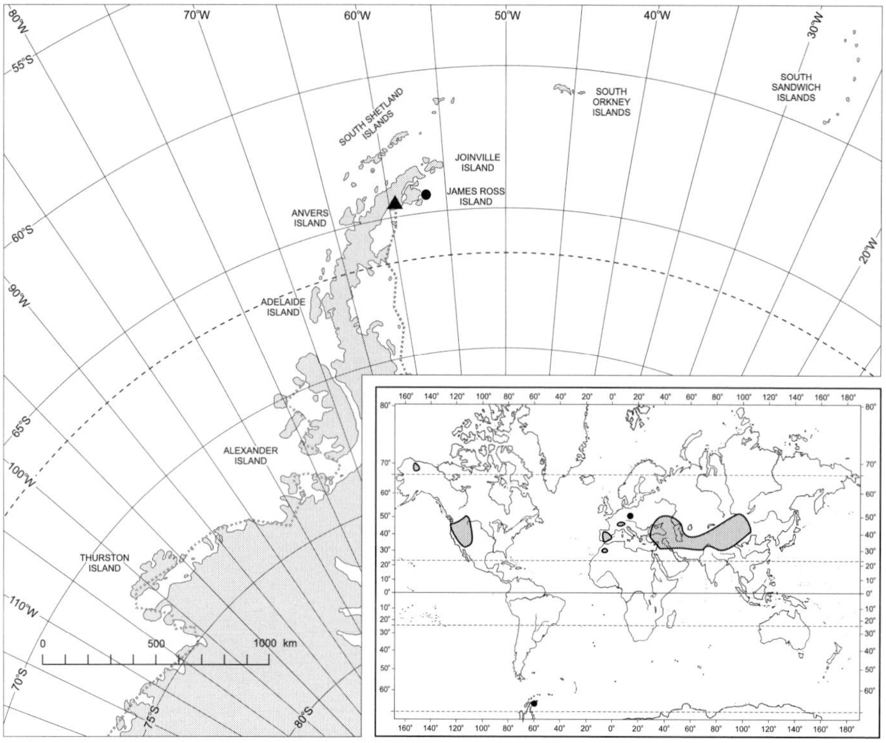

Figure 178. Distribution of *Syntrichia caninervis* var. *caninervis* (dot) and var. *gypsophila* (triangle) in Antarctica. Inset: global distribution of *S. caninervis*.

Distribution in Antarctica. So far, this taxon has been discovered only on Seymour I. off the east coast of the Antarctic Peninsula at lat. 64°13′S at elevation of 10–65 m (Fig. 210).

Specimens examined. WEST ANTARCTICA. EAST ANTARCTIC PENINSULA. **JAMES ROSS ISLAND GROUP.** ***Seymour I.***: 1 km west of Cape Wiman, 10 m, *RILS 7886A, 7889A, 7891, 7892 & 7893B* (AAS, KRAM); 1.5 km west of Cape Wiman, 65 m, *RILS 7886A* (AAS, KRAM).

B. var. **gypsophila** (J.J.Amann *ex* G.Roth) Ochyra Fig. 179

Syntrichia caninervis Mitt. var. *gypsophila* (J.J.Amann *ex* G.Roth) Ochyra, Fragm. Florist. Geobot. 37: 212. 1992. ≡ *Tortula ruralis* (Hedw.) P.Gaertn., B.Mey. & Scherb. var. *gypsophila* J.J.Amann *ex* G.Roth, Hedwigia 57: 135, *pl. 4, f. 7a–b.* 1915. ≡ *T. gypsophila* J.J.Amann *ex* G.Roth, Hedwigia 57: 135. 1915, *non rite publ.* ≡ *Syntrichia ruralis* (Hedw.) F.Weber & D.Mohr var. *gypsophila* (J.J.Amann *ex* G.Roth) J.J.Amann, Fl. Mouss. Suisse 2: 118. 1918. ≡ *Tortula caninervis* (Mitt.) Broth. var. *gypsophila* (J.J.Amann *ex* G.Roth) W.A.Kramer, Bryophyt. Biblioth. 21: 106. 1980. — TYPE CITATION: Im Wallis in der Schweiz, Sion-Montagne, in 550 bis 600 m von J. Amann gesammelt ..... 8. Dezember 1912. [Lectotype (*selected here*): 'J. Amann, Musci helvetici Tortula spuria mihi Kant. Valais: Murs de vignes et rochers à Sion-Montorge, 550–600 m 8. Dez. 1912 Leg. Amann 81. 2^{b}.12' – S-Roth (No. B104076)!; isotype: ZT!].

Plants small, densely tufted, stiff and rigid, dull, brown throughout or whitish at tips. *Stems* erect, 0.5–1.1 cm high, branched, densely radiculose at the base with long, brown, smooth, shining rhizoids. *Leaves* erect and appressed when dry, erect-spreading when moist, ovate, oblong-ovate to ovate-lingulate, 1.1–2.0 mm long, 0.6–1.0 mm wide, rounded-obtuse, cucullate and sometimes hyaline at the apex, ending with a long or short, strongly spinose hyaline hair-point, straight or recurved when wet; *margins* recurved from the base to near the apex, occasionally only to ⅔ way up the leaf; *costa* strong, 60–80 μm wide, yellow-brown, markedly convex dorsally, in transverse section with 1–2 rows of large guide cells, 2–4 layers of substereids or stereids and hydroids, on the dorsal side covered with tall, pedicellate, bifurcate papillae, 3–20 μm high in the upper third, becoming simple and branched, 2–5 μm high below; *laminal cells* unistratose throughout or with occasional bistratose streaks in the upper third; *upper cells* rounded-quadrate, rounded to hexagonal, moderately thick-walled, (8–)10–14(–16) μm wide, papillose with low, branched papillae, 3–6 per lumen, but not markedly obscuring the lamina, or nearly smooth on older leaves; *basal cells* rectangular, 30–60 μm long, 10–20 μm wide, thin-walled, hyaline near the costa, becoming narrower towards the margins, forming 4–7-seriate yellow border, sharply demarcated from the upper chlorophyllose cells. Sterile.

Discussion. *Syntrichia caninervis* var. *gypsophila* is superficially similar to the type variety of this species, but it differs in having irregularly bistratose laminal cells in the distal half of

⟶

Figure 179. *Syntrichia caninervis* var. *gypsophila*. — **1**. Habit. **2**. Portion of shoot with perichaetia, wet. **3**. Cross-section of stem. **4–6**. Axillary hairs. **7–12**. Leaves. **13**. Leaf apex with base of hyaline hair-point and showing papillosity of costa on dorsal side. **14**. Apex of hyaline hair-point. **15**. Mid-leaf cells (papillae partly removed). **16**. Cells in the transition zone between median and basal parts of leaf and basal leaf cells. **17–22**. Cross-sections of leaves, a sequence from base to apex. — [All from *CGB 10*]. — Scale bars: a – 1 mm (7–12); b – 100 μm (3, 13–14); c – 100 μm (15, 17–22); d – 1 mm (1–2) and 100 μm (4–6, 16).

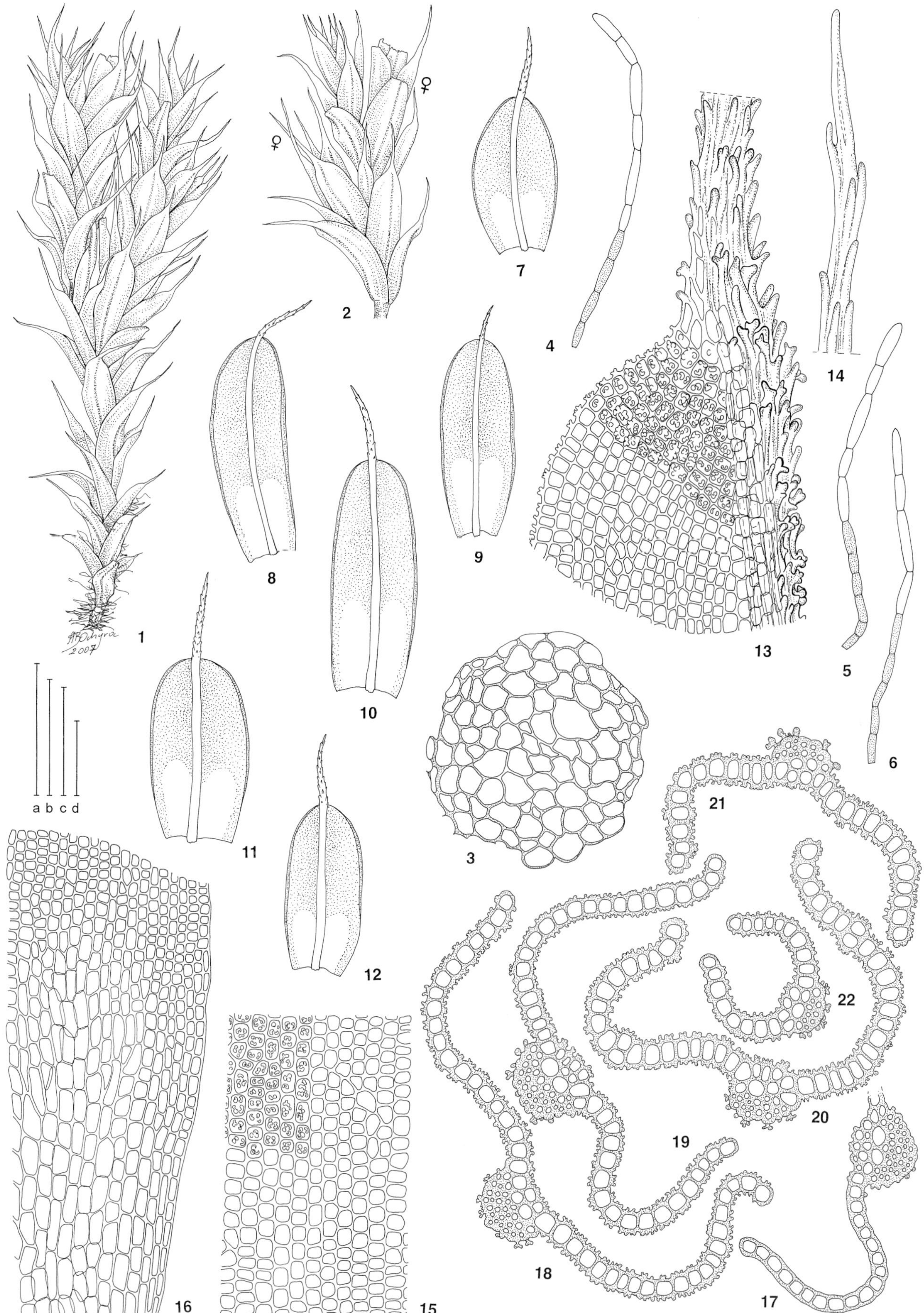
1
2
♀
♀
3
4
5
6
7
8
9
10
11
12
13
14
15
16
17
18
19
20
21
22
a b c d
A.Ochyra 2007

the leaves. In addition, the upper leaf cells are larger, 10–14 μm wide, and thinner-walled. The Antarctic population of this taxon is also recognised by its strongly papillose costa on the dorsal side in the upper third with high, pedicellate and bifurcate papillae that are 3–20 μm tall. A peculiar feature of the Antarctic plants of this variety is the entirely unistatose laminal cells which only occasionally exhibit bistratose strands. The extent of the bistratosity of the laminal cells in var. *gypsophila* varies markedly and sometimes it is reduced only to a few bistratose patches, but they are never completely unistratose in the plants from the Northern Hemisphere (M. Teresa Gallego, personal communication). It would be unwise to recognize the Antarctic population as a separate taxon, especially owing to the scantiness of the material, and the unistratosity of the laminal cells in the Antarctic plants should rather be considered as an extreme form in a series of various intermediate stages. In fact, the unistratose laminal cells, with only incidental bistratose patches, are typical of *S. caninervis* var. *astrakhanica*, a variety known only from the Caspian Sea area of Russia (Ignatov *et al.* 2002). This variety differs from *S. caninervis* var. *gypsophila* in having 2–4-celled, red-brown, spherical gemmae that are produced on the tips of the branched filaments in the leaf axils. The plants of this variety with abundant gemmae are markedly modified and have the leaves with percurrent or shortly excurrent costae. In contrast, the plants with less profuse gemmae have the leaves with typically developed hyaline hair-points, although the costa is usually smooth on the dorsal surface. No gemmae have so far been discovered in the Antarctic plants, and therefore they are determined as *S. caninervis* var. *gypsophila*, since they fit well the type of *Tortula ruralis* var. *gypsophila* from Switzerland (Roth 1915), except for the unistratose leaf laminae. In the *S. caninervis* complex gemmae are produced very rarely and, until recently, they have been known only in *S. rigescens*, whereas in *S. caninervis* itself they have hitherto been unknown (Gallego *et al.* 2002*a*). Gallego (2002) selected the specimen in ZT as lectotype of *Tortula ruralis* var. *gypsophila* but this typification has to be superseded. Roth (1915) described this taxon on the basis of the specimen preserved in his herbarium (now in S) and, because there is no evidence that he also used the specimen housed in ZT for description, the specimen in S must be designated as lectotype, whereas the specimen in ZT is simply an isotype.

Reproduction in Antarctica. Sterile.

Habitat. The single specimen was collected from a shaded vertical rock face on the south side of a mountain summit.

World range. *Syntrichia caninervis* var. *gypsophila* is currently known only from Eurasia and North Africa and its range is here sympatric with the range of the type variety of this species, but generally it appears to be a rarer and more scattered taxon. It is widely distributed but scattered in the Mediterranean region, including the eastern and central part of the Iberian Peninsula of Spain (Gallego 2002, 2005), with some scattered stations in the Alps of France, Switzerland and Italy, and with an isolated locality in Moravia in the Czech Republic, then recurring in the Ukraine, Turkey and in the Near and Middle East, including Iraq, Iran and Afghanistan. Two localities of this taxon are also known from Morocco in North Africa (Kramer 1980; Gallego *et al.* 2002*a*). A single, highly disjunct station is known from the Antarctic in the Southern Hemisphere.

Distribution in Antarctica. *Syntrichia caninervis* var. *gypsophila* is known only from a small nunatak at Larsen Inlet on the Nordenskjöld Coast on the eastern Antarctic Peninsula at lat. 64°22′S, at an altitude of about 830 m (Fig. 178).

Specimens examined. WEST ANTARCTICA. EAST ANTARCTIC PENINSULA. **NORDENSKJÖLD COAST.** Small nunatak to south of Tower Peak, between Sjögren Glacier and Larsen Inlet, *ca* 830 m, *CGB 10* (AAS, KRAM).

5. **Syntrichia magellanica** (Mont.) R.H.Zander Fig. 180

Syntrichia magellanica (Mont.) R.H.Zander, Bull. Buffalo Soc. Nat. Sci. 32: 269. 1993. ≡ *Tortula magellanica* Mont. *in* Gay, Fl. Chil. 7: 145. 1850. ≡ *Tortula princeps* De Not. var. *magellanica* (Mont.) Lightowlers, Brit. Antarct. Surv. Bull. 67: 61. 1985. — TYPE CITATION: Esta especie la cojió el Sr. Jacquinot en el puerto del Hambre. [Lectotype (*selected here*): 'Tortula hyperborea e Montag Port-Famine' – PC-Montagne!; isotype: BM-Schimper!].

Syntrichia fuegiana Mitt., J. Linn. Soc. Bot. 4: 71. 1859. ≡ *Tortula fuegiana* (Mitt.) Mitt., J. Linn. Soc. Bot. 12: 174. 1869. ≡ *Barbula fuegiana* (Mitt.) A.Jaeger, Ber. Thätigk. St. Gallischen Naturwiss. Ges. 1871–1872: 453. 1873. — TYPE CITATION: Falkland Islands, on sand-hills in Uranie Bay, *Dr. J. D. Hooker*; Capo Negro, Magellan, *Lechler*, 1088. [Lectotype (*vide* Lightowlers 1986*b*: p. 66): '1088. Syntrichia ~~erythropoda~~ fuegiana m. Musci Lechler Cabo Negro fret. magell. Oct. 52' – BM-Hooker! Syntype (2 labels): 'Tortula Sandhills Uranie Bay Sept. 1842' and 'Tortula Falkland W. 194 Barbula laevipila var?' – BM-Hooker! (2 specimens) = *Syntrichia antarctica*]. First synonymised with *Tortula princeps* var. *magellanica* by Lightowlers (1986*b*: p. 66).

Tortula monoica Cardot, Bull. Herb. Boissier Sér 2, 5: 1003. 1905. — TYPE CITATION: Iles Falkland: Port-Louis. [Lectotype (*vide* Lightowlers 1985*b*: p. 66): 'Svenska Sydpolarexpedition 1901–03. N:r 223. Tortula monoica Card. sp. nova + Ceratodon purpureus Brid. et Bryum argenteum H. Falkland Islands, Port Louis Greenpatch, in rupibus marit. D. 25/7 1902 C. Skottsberg. Det. J Cardot' – PC!; isolectotypes: H-Brotherus!, S! (4 specimens). Syntype: 'Svenska Sydpolarexpedition 1901–03 Nr 224 Tortula monoica Card. sp. nova + Ceratodon purpureus Brid. Falkland Islands, Port Louis D. 3/8 1902 C. Skottsberg. Det. J Cardot' – PC!; isosyntype: S!]. First synonymised with *Tortula princeps* var. *magellanica* by Lightowlers (1986*b*: p. 61).

Tortula grossiretis Cardot, Bull. Herb. Boissier Sér 2, 6: 21. 1906, ***syn. nov.*** — TYPE CITATION: [South Georgia] Cumberland bay: Pot Harbour. … Punta Arenas, détroit de Magellan (Racovitza, no 62 e). [Lectotype (*vide* Lightowlers 1985*b*: p. 64): 'Herb. J. Cardot Tortula grossiretis Card. sp. nova Géorgie du Sud: Cumberland Bay leg. Skottsberg, 1902, nº 299' – S!; isolectotype: PC!, S!; syntype: 'Herb. J. Cardot. Mousses du voyage de la Belgica nº 62e Barbula ruralis Hedw. forma Détroit de Magellan: environ de Punta Arenas. Leg. Racovitza, 9 Déc^{bre} 1897' – PC-Cardot!; isosyntype: BR!].

→

Figure 180. *Syntrichia magellanica*. — **1**. Plant with sporophytes. **2**. Portion of shoot, wet. **3**. Portion of shoot, dry. **4**. Cross-section of stem. **5–6**. Axillary hairs. **7–13**. Leaves. **14**. Leaf apex (papillae partly removed). **15–16**. Mid-leaf cells (papillae partly removed). **17**. Lamina cells in transition zone to hyaline base. **18**. Basal leaf cells. **19–22**. Cross-sections of leaves, a sequence from base to apex. **23**. Operculate capsule, wet. **24**. Calyptra. **25**. Peristome. **26**. Exothecial cells at orifice. **27**. Exothecial cells in middle of urn. **28**. Exothecial cells at base of urn and stoma. **29**. Spores. — [1, 8, 23–29 from *RO 2690/80*; 2–3 from *RO 963/80*; 11, 14–15, 17 from *RO 5227/79*; 4, 12–13 from *Gain 211* (lectotype of *Tortula heteroneura*), PC; 5–6, 22 from *VK 514*; 7, 19–21 from *RO 1831/80*; 9–10, 16, 18 from *CS 299* (isotype of *Tortula grossiretis*), S]. — Scale bars: a – 1 mm (2–3); b – 100 μm (4–6, 14–18, 26–29); c – 100 μm (19–22), 1 mm (7–13, 23–24) and 0.5 cm (1); d – 1 mm (25).

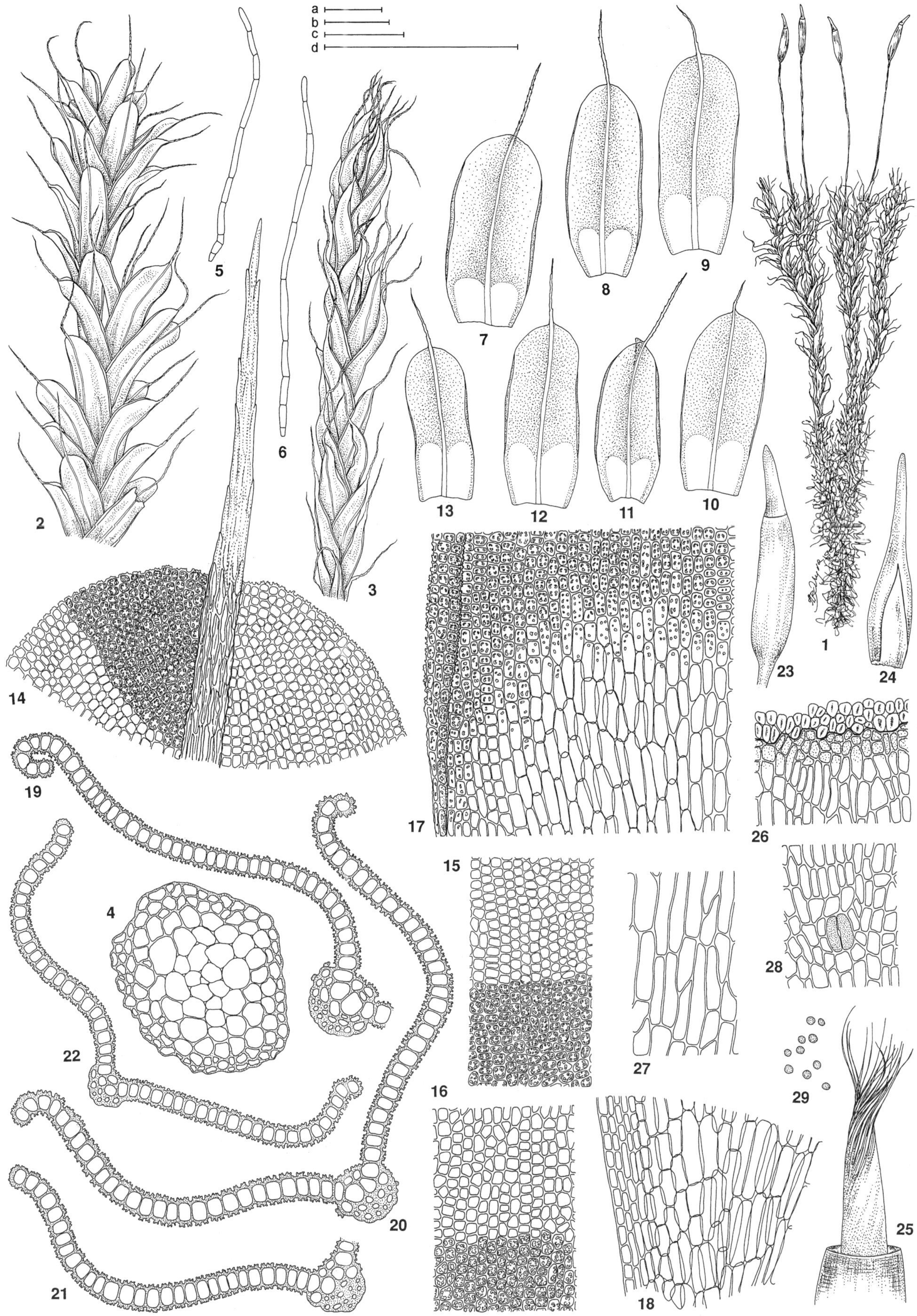
a
b
c
d
1
2
3
4
5
6
7
8
9
10
11
12
13
14
15
16
17
18
19
20
21
22
23
24
25
26
27
28
29

Tortula grossiretis var. *atrata* Cardot, Bull. Herb. Boissier Sér 2, 6: 21. 1906, ***syn. nov.*** — TYPE CITATION: [South Georgia] Cumberland bay: Pot Harbour. [Lectotype (*selected here*): 'Svenska Sydpolarexpedition 1901–03. N:r 300 Tortula grossiretis Card. sp. nova var. atrata Card. var. nova South Georgia, Cumberland Bay, Pot Harbour Leg. C. Skottsberg, 18.5.1902' – PC!; isotype: s!].

Tortula heteroneura Cardot, Compt. Rend. Hebd. Séances Acad. Sci. 153: 602. 1911, ***syn. nov.*** — TYPE CITATION: Ile Petermann. [Lectotype (*vide* Lightowlers 1985*b*: p. 65): 'Herb. Mus. Paris 211. Tortula heteroneura Card. sp. nova. Ile Petermann. sur la terre humide, entre les rochers. 2ème expéd. Charcot. Leg. Gain, 5/1 1909' – PC! isotypes: BM!, H-Brotherus! Syntypes: three specimens bearing exactly the same label data except for collecting numbers (*vide* Cardot 1913: p. 17) – nº 168ᵇ, 172 and 175 – PC!].

Tortula conferta E.B.Bartram, Bryologist 60: 140. 1957. ≡ *Tortula princeps* De Not. var. *conferta* (E.B.Bartram) Lightowlers, Brit. Antarct. Surv. Bull. 67: 61. 1985. ≡ *Syntrichia conferta* (E.B.Bartram) R.H.Zander, Bull. Buffalo Soc Nat. Sci. 32: 269. 1993, ***syn. nov.*** — TYPE CITATION: Melchior Archipelago: Lysted Island, 335.12, type, mixed with *Brachythecium antarcticum*. Lambda Island, NE. Landing, 345.1, mixed with *Pohlia cruda* var. *imbricata*. [Lectotype (*selected here*) {or perhaps holotype}: 'Plants of Antarctica. United States Antarctic Service Expedition of 1940–41. No. 335.12 Tortula conferta Bartr. sp. nov. Lysted Island + Brachythecium antarcticum Card. Melchior Archipelago. P. A. Siple, Collector, March 4, 1941' – FH-Bartram! Paratype: 'Plants of Antarctica. United States Antarctic Service Expedition of 1940–41. No. 345.1 Tortula conferta Bartr. sp. nov. + Pohlia cruda var. imbricata. Lambda Island, N. E. Landing, Melchior Archipelago. P. A. Siple, Collector, March 3, 1941' – FH-Bartram!].

Plants small, medium-sized to fairly large, slender to relatively robust, in loose or dense, dull, often hoary tufts, turves or patches, green, rusty-, yellow-, brownish- or olive-green to reddish- or blackish-brown when dry, typically dark green above and reddish-brown to brown below when moist. *Stems* erect, gracile to rather stiff and rigid, (0.2–)0.5–5.0 cm tall, sometimes as much as 8.0 cm, sparingly branched to simple, usually radiculose at the base with reddish-brown, smooth, glistening and densely branched rhizoids, or rhizoids scattered throughout the stem to almost lacking, in transverse section rounded-pentagonal, without a central strand, consisting of a 3–5-layered medullary tissue of large, thin-walled, hyaline cells, becoming smaller, thicker-walled, brown in 1–2 outer layers but not forming a distinct cortex; axillary hairs filiform, uniseriate, 8–12 cells long, all hyaline. *Leaves* crowded, evenly spaced, occasionally in interrupted tufts on the stem or subrosulate at the stem tips, appressed and slightly curled to flexuose-twisted when dry, erecto-patent to widely spreading and often distinctly recurved, rarely erect and straight to weakly recurved when moist, oblong- or oval-lingulate, elliptical or ovate, not or rarely weakly constricted at the middle, (0.5–)0.9–3.1(–3.5) mm long (excluding a hair-point), (0.3–)0.5–1.1(–1.3) mm wide, broadly concave, subacute to obtuse or rounded, recurved or cucullate at the apex, abruptly producing a hair-point, rarely reduced to an apiculus or absent; *margins* entire, papillose-crenulate above, plane near the base, recurved to revolute at mid-leaf up to two thirds of the leaf length, occasionally recurved nearly to the apex or sometimes plane throughout; *costa* single, abruptly demarcated from the laminal cells, mostly stout, 40–100 μm wide, percurrent or becoming obscure and ceasing below the apex, usually excurrent as a hyaline hair-point, smooth or verrucose to obscurely dentate towards the apex on the abaxial surface, covered with quadrate, papillose, epidermal cells on the adaxial surface, in transverse section semi-terete, strongly convex dorsally, lacking hydroids, consisting of 2–4-stratose, lunate in shape dorsal stereid or sometimes substereid band, 1–2(–3) layers of large guide cells and ventral epidermal cells papillose on the outer surface; *hair-point* 0.1–1.6 mm long, rarely absent, flexuose, hyaline throughout to reddish-brown at the base, mostly flat at the base, terete above, smooth to obsoletely serrate, seldom more sharply denticulate; *laminal cells* unistratose throughout, strongly differentiated into upper chlorophyllose and obscure cells and smooth, transparent basal cells; *upper cells* irregularly quadrate, oblate, or short-rectangular, not differentiated towards the margin, (8–)11–17(–25) μm long, (8–)10–15(–20) μm wide, firm-walled, densely and thickly pluripapillose with branched or hollow C-shaped papillae, 4–7 per lumen, spreading down the margins of the upper basal region of the lamina; *basal cells* sharply demarcated from the upper chlorophyllose cells along a well defined, arched junction line, seldom the transitional zone indistinct, rectangular to linear, 40–135 μm long, 10–30 μm wide, usually 3–10 times as long as wide, smooth, hyaline, thin-walled, sometimes lax and inflated, becoming short-rectangular and narrower near the margins and forming a distinct, multiseriate yellow border. *Monoicous* or *dioicous*. *Perigonia* terminal, gemmiform, to 1.3 mm long; inner perigonial bracts broadly ovate, concave, epilose, weakly costate, with numerous antheridia and yellowish paraphyses, slightly exceeding the antheridia. *Perichaetia* terminal; perichaetial leaves somewhat larger and more acute at the apex than the cauline leaves, subconvolute, sheathing the seta. *Setae* erect, straight, single in perichaetium or sometimes perichaetia polysetous, 5–15 mm long, smooth, reddish-brown below, pale brown above, dextrorse above, slightly sinistrorse at the base when dry. *Capsules* exserted, straight to somewhat arcuate, symmetric, erect, (1.5–)1.8–3.2 mm long, 0.35–0.8 mm wide, obloid to cylindrical, brown, somewhat lustrous, smooth; *annulus* persistent, composed of 2–3 rows of large, vesiculose cells with strongly incrassate walls; *operculum* subulate to conic-rostrate, straight to somewhat slanted, nearly half the length of the urn; *exothecial cells* thin-walled, rectangular, 70–140 μm long, 15–40 μm wide, becoming shorter, mostly hexagonal to rounded in 3–5 rows below the orifice; *stomata* at the base of the urn in a single row, bicellular, superficial, with oval openings; *peristome* inserted below the mouth, 0.6–1.7 mm high, orange to salmon-pink, consisting of a basal membrane forming a high tessellated tube, pale or fainty tinged with orange and becoming white with age, 0.4–1.0 mm tall, giving rise to 32 filiform, spiculose teeth, once spirally twisted into a dextrorse helix. *Spores* spherical, yellowish to brownish, 10–14(–18) μm in diameter, papillose. *Calyptra* enlarged, cucullate, brown, smooth, naked.

Chromosome numbers: *n* = 24, *n* = 36 and *n* = *ca* 40 – King George I. (Przywara *et al.* 1984 as *Tortula grossiretis*).

Discussion. The identity and taxonomy of the Antarctic populations of *Syntrichia* with hyaline hair-points have been the source of considerable confusion and misunderstanding, resulting in serious discrepancies among bryologists with regard to their taxonomic interpretation. They have been considered as species of *Tortula* and reported from this biome under several names, including *T. fuegiana* (Mitt.) Mitt. (Gepp 1902), *T. monoica* (Cardot 1906*a*, 1908*a*), *T. heteroneura* (Cardot 1911*a*,*b*, 1913), *T. grossiretis* (Dixon 1920) and *T. conferta* (Bartram 1957). All these names have been reduced to synonymy with *T. princeps* De Not. (Lightowlers 1985*a*,*b*, 1986*b*) or *Syntrichia princeps* (De Not.) Mitt. (Ochyra 1998*a*). *T. princeps* was originally described by De Notaris (1836) from Italy: 'ad truncos arborum et ad rupes in Sardinia – in Sicilia BALSAMO'. [Lectotype (*selected here*): 'Tortula princeps DNtrs ad truncos p. S[ta] Barbara m. Sardin. auste. Aprili 1835' – RO-De Notaris!; isolectotypes: RO-De Notaris!, RO-Fiorini Mazzanti! Syntype: 'Tortula princeps DNTrs in Aethna leg. Balsamo 1832' – RO-De Notaris!] and the species immediately gained wide acceptance, although its nomenclatural and taxonomic history has been somewhat chequered. Bruch *et al.* (1842) changed the name of the species for *Barbula muelleri* claiming that the moss was collected much earlier in Sardinia by Fr. A. Müller in 1825 and named by Ph. Bruch as *Syntrichia muelleri* in Müller's (1829) work on Sardinian mosses. In fact, this name is missing from that work, and the name *Barbula muelleri* Bruch & Schimp. was published for the first time by De Notaris (1838) as a *nomen nudum* in synonymy with *Tortula princeps*. *Barbula muelleri* is therefore illegitimate since Bruch *et al.* (1842) placed *Tortula princeps* in its synonymy. Wilson & Hooker (1847) transferred *B. muelleri* to *Tortula* but the name *T. muelleri* Wilson & Hook.f. is also illegitimate.

Syntrichia princeps is widely distributed in the arid regions in the Northern Hemisphere, including western North America and north Mexico, the Mediterranean region of Europe and North Africa, Macaronesia, the British Isles, Scandinavia, as well as the Near and Middle East. The species was reported quite early from the Southern Hemisphere, including South America (Wilson & Hooker 1847; Mitten 1869), New Zealand (Wilson 1854; Mitten 1867), Australia (Mitten 1856), Tasmania (Wilson 1859), and South Africa (Sim 1926), and consequently firmly established as a bipolar moss. However, detailed taxonomic studies of the austral taxa of *Syntrichia* with hyaline hair-points (Kramer 1988) revealed that *S. princeps* does not, in fact, occur in the Southern Hemisphere where it was replaced by a number of different species, including *S. antarctica* (Hampe) R.H.Zander and *S. magellanica*. Of these, *S. antarctica* is very close to *S. princeps* and both species share the presence of hydroids in the anatomical structure of the costa, but the former has smaller laminal cells, hydroid homologues in the costa and a more or less distinct marginal border in the leaf lamina. However, *S. magellanica* and its close relatives (*S. conferta* and *Tortula grossiretis*) are distinct from *S. princeps* and *S. antarctica* in lacking hydroids and hydroid homologues in the costa.

Species of the *Syntrichia magellanica* complex are very closely related taxa which differ primarily in leaf size, recurvature of leaf margins and length, denticulation and colour of hair-points. Lightowlers (1986*b*) initially recognised these taxa as varieties within the broadly understood and polymorphic *S. princeps*, namely var. *magellanica* and var. *conferta*, and placing *Tortula grossiretis* as the type variety. The latter consists of relatively large and robust plants, 0.5–8.0 cm tall, with leaves 1.2–3.5 mm long and 0.6–1.3 mm wide, plane or recurved in profile at the apex, pointed with a hyaline or reddish, smooth to denticulate awn, 0.2–1.6 mm long and with margins recurved or revolute to two third of the leaf length. In contrast, var. *magellanica* is a smaller plant, 0.5–1.5 cm tall, with smaller leaves, (1.0–)1.4–2.9 mm long and 0.5–1.0 mm wide that are cucullate to plane at the apex, have leaf margins recurved for about one third to a half of the leaf length and hair-points 0.3–1.0 mm long, hyaline and smooth to obscurely denticulate. Finally, var. *conferta* is a small and slender plant, 0.2–1.5 cm tall, with small leaves, 0.5–1.4 mm long and 0.3–0.75 mm wide, plane to weakly recurved margins, cucullate, plane to recurved leaf apex and a short, smooth or crenulate, hyaline hair-point, 0.1–0.4 mm long, often reduced to a short apiculus or absent. The dorsal stereid band in the costa in this variety is poorly developed and consists of substereid cells only, in contrast to two former varieties in which typical stereids are usually present.

Subsequently, Lightowlers (1986*b*) analysed in detail variation in the Antarctic plants of this species and found that in most cases these varieties cannot be reliably distinguished from one another and the delimitation of them is arbitrary. Therefore, he formally merged var. *conferta* with the type variety and suggested that all the Antarctic material represents a single, highly protean taxon. This appears to be the most practical and sensible solution because variation of all taxonomically important characters in this complex is continuous and no correlation is observed between the form of the leaves and the size of the plants. Consequently, most populations could be only randomly placed in either variety. This type of variation is typical not only for the plants from the Antarctic biome but also from more northerly latitudes. Therefore, the species that constitute the *Syntrichia magellanica* group are here considered conspecific, and *S. magellanica* is the oldest available name for this species. It was described for the first time as *Tortula magellanica* by Montagne (1850) from a specimen collected on the Brunswick Peninsula during the French south polar expedition of 1837–1840 on the ships *Astrolabe* and *Zélée* under the command of captain M. J. Dumont d'Urville, and initially determined as *T. hyperborea* (Brid.) Mont. (Montagne 1845*b*).

Kramer (1988) recognised five heterotypic synonyms of *Syntrichia magellanica* and, in the present treatment, four more synonyms are added. *Tortula grossiretis* and *T. grossiretis* var. *atrata* were described from subantarctic South Georgia (Cardot 1906*a*) and the former was subsequently reported from the Antarctic (Dixon 1920). The other two synonyms, *T. heteroneura* and *T. conferta*, refer to species described from Antarctica by, respectively, Cardot (1911*a*,*b*, 1913) and Bartram (1957). Lightowlers (1985*b*) considered *Syntrichia fuegiana* to be a synonym of *Tortula princeps* var. *magellanica* but Kramer (1988) questioned this decision and suggested that it is a distinct species recognised by the longer recurvature of leaf margins extending nearly to the rounded leaf apex. *Syntrichia fuegiana* was described by Mitten (1859) on the basis of two specimens: one collected at Cabo Negro on the Brunswick Peninsula on mainland South America, and another in Uranie Bay in the Falkland Is. Examination of the original collections of this species revealed the heterogeneity of the type material. The specimen from Cabo Negro (*Lechler 1088*, BM) clearly represents *Syntrichia magellanica* and, because it was chosen by Lightowlers (1985*b*) as lectotype, *S. fuegiana* must automatically be considered synonymous with this name. The specimen from the Falkland Is. (*Hooker 194*, BM) was originally determined by Wilson & Hooker (1847) as an un-named variety of *Tortula laevipila* (Brid.) Schwägr. and it clearly represents *Syntrichia antarctica*. Another specimen from the same locality (*Hooker 190*, BM) was described by E. Hampe (in Müller 1849) as *Barbula antarctica* (=*Syntrichia antarctica*).

Syntrichia magellanica was first recorded in Antarctica from Cockburn I. off Trinity Peninsula and reported as an unnamed variety of *Tortula laevipila* (Brid.) Schwägr. (Wilson & Hooker 1847). It is a distinct and well defined species that is diagnosed as follows: (*1*) hydroids and hydroid homologues absent; (*2*) stem central strand absent; (*3*) leaf margin plane to recurved to two third way up the leaf; (*4*) hair-point smooth to obscurely denticulate; (*5*) costa smooth or somewhat verrucate to weakly dentate towards the apex on the dorsal side. The species is related to *S. ruralis* (Hedw.) F.Weber & D.Mohr and *S. calcicola* J.J.Amann of the Northern Hemisphere, with which it shares the lack of hydroids and hydroid homologues in the costa. However, these species are distinct in having strongly recurved to squarrose leaves that end with moderately to sharply dentate hyaline hair-points, stronger dorsal stereid band of the costa with 3–6 layers of stereid cells, densely papillose dorsal surface of the costa and dioicous sex condition. However, in the Antarctic *S. magellanica* is a distinct and easily distinguished species by its oblong- or oval-lingulate, elliptical or ovate and piliferous leaves. It may be mistaken for *S. caninervis* and *Willia austroleucophaea*, but these species are exceedingly rare in this area and the probability of confusion is minimal. The former is distinct in its bistratose leaf laminae or, if unistratose, in its distinctly papillose costa on the dorsal surface and sharply dentate hair-points, while the latter is distinct in having pandurate leaves and plane leaf margins.

Reproduction in Antarctica. Normal male and female gametangia have often been observed and sporophytes have been detected in many populations of *Syntrichia magellanica* in the maritime Antarctic, although the capsules are often underdeveloped and bear aborted spores. No sporophytes have been observed in the continental populations.

Habitat. *Syntrichia magellanica* occurs in a wide range of habitats, both in dry or moist and open or sheltered situations and growing on both acidic and base-rich substrates. It usually grows on soil on rock ledges and in rock depressions, on pockets of fine or coarse gravelly, sandy and silty soil amongst rocks or boulders, volcanic debris, ash, scoria or lapilli, on scree, in crevices and on faces of various rock types, on peat or soil on rock ledges and on moraines, on alluvial soil, gravel and stones on sides of melt pools, at the margin of late snow beds and runnels, in seepage or flushed areas. It is often found in bird-influenced sites on maritime rocks and cliffs, near or in seabird colonies, in bird perching sites and below nesting ledges. In such habitats small forms of *S. magellanica* representing the *conferta* phenotypes predominate and, in some areas, for example on the Argentine Is., it forms a distinct community with *Bryum pseudotriquetrum*, *Ceratodon purpureus*, *Brachythecium austrosalebrosum* and *Sanionia georgicouncinata* on deposits of limpet shells near gull nests (Lewis Smith & Corner 1973). Occasionally, this species is found on whalebones on beaches, on decaying bird corpses and on mortar associated with buildings. *Syntrichia magellanica* is a widespread and frequent, but generally not abundant, associate of a variety of communities of the Antarctic tundra, especially the fruticose lichen and moss cushion sub-formation and moss hummock sub-formation. It is also a frequent associate of communities of the Antarctic herb turndra formation which is dominated by two native vascular plants, *Deschampsia antarctica* and *Colobanthus quitensis*.

World range. The main centre of occurrence of *Syntrichia magellanica* is in the Antarctic biome, with high frequency in the Antarctic Peninsula region and its offshore archipelagoes, and in northern Victoria Land on the Antarctic continent, as well as on some peri-Antarctic islands, namely Îles Kerguelen and South Georgia (Fig. 181). The species is relatively rare and infrequent on the Falkland Is. and in southern South America, including Tierra del Fuego and western Patagonia, but does not seem to extend north of lat. 50° S. Therefore, with some caution, *S. magellanica* is here designated as an Antarctic–Subantarctic species slightly penetrating into the cool temperate zone. It is evidently a polar counterpart of *S. antarctica* which is widely distributed in more northerly latitudes and shows a typical pan-south-temperate distribution pattern.

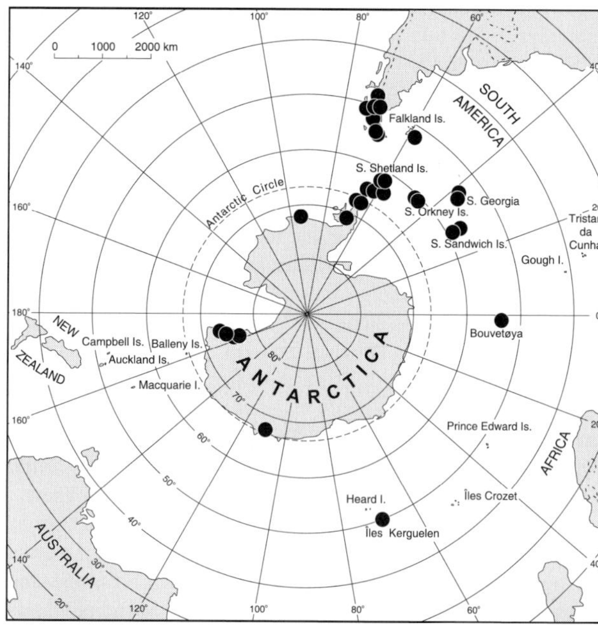

Figure 181. Global distribution of *Syntrichia magellanica*.

Distribution in Antarctica. *Syntrichia magellanica* is a pan-Antarctic species, although it has a very uneven and irregular distribution in this biome (Figs 182–183). It is widely distributed in the maritime Antarctic where it is locally one of the most frequent and abundant species of moss, for example on King George I. in the South Shetland Is. (Ochyra, 1998*a*). On Signy I., South Orkney Is., it locally forms closed stands of about 10 sq. m on soil associated with marble outcrops (Lewis Smith 1972). Its range extends from Bouvetøya, the South Sandwich Is., the South Orkney Is. and the South Shetland Is., and the west coast of the Antarctic Peninsula where it has three distinct centres of occurrence on the Danco and Graham Coasts, including offshore islands of Brabant, Anvers and Argentine Is., then appearing in northern Marguerite Bay and on southern Adelaide I. and, after a large disjunction, recurring on Alexander I. where it reaches lat. 72°05′ S. On the east coast of the Antarctic Peninsula the species appears to be less frequent, although it is widely scattered from the Trinity Peninsula and Joinville I. through James Ross I. south to Francis I. at lat. 67°37′ S (Fig. 182). On the continent, *S. magellanica* is locally frequent and widely distributed only on

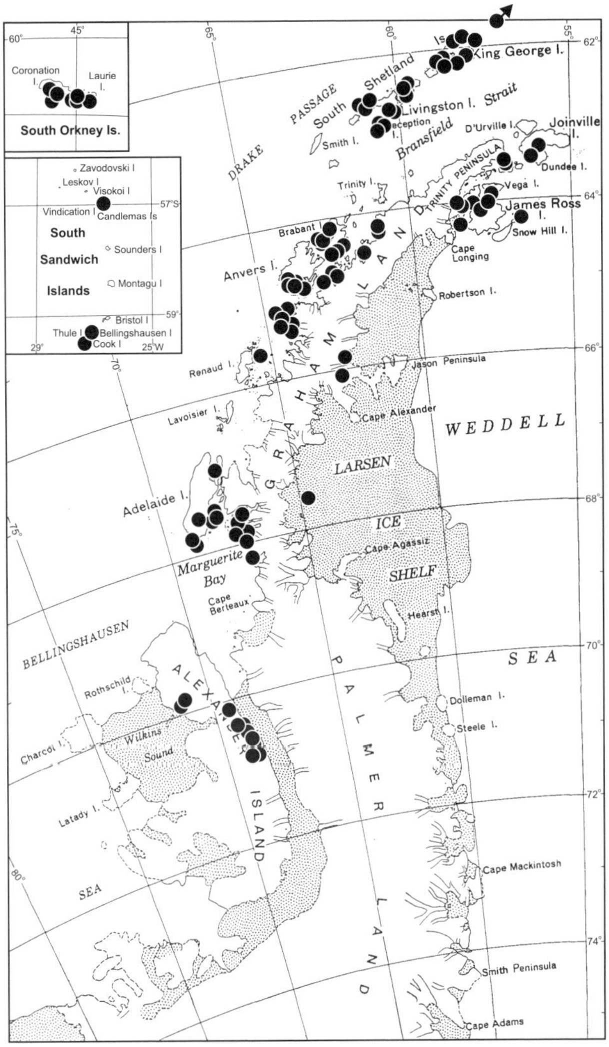

Figure 182. Distribution of *Syntrichia magellanica* in the Antarctic Peninsula region. The localities in the northern group of the South Shetland Is. are indicated by an arrow.

northern Victoria Land between lat. 72°38′ S and 75°41′ S, with solitary isolated records in Wilkes Land and Ellsworth Land (Fig. 183). The species grows predominantly at lower elevations, from sea level to 250 m but rarely extends to higher altitudes, for example on Alexander I. where it reaches *ca* 1060 m on a peak south of Bartok Glacier, and in northern Victoria Land on Mt. McGee in the Southern Cross Mountains where it was recorded at its highest elevation of 1100 m.

Specimens examined. WEST ANTARCTICA. BOUVETØYA. Kapp Circoncision, 40 m, *MWH 864A* (AAS, KRAM); Nyrøysa, *TE 93.2* (BG).

SOUTH SANDWICH IS. **Candlemas I.** Demon Pt., *PC 122D* (AAS, KRAM). **Freezeland Rock.** West face, 30 m, *MWH 808* (AAS, KRAM). **Bellingshausen I.** North side of north peak of crater rim, *ca* 250 m, *MWH 823D* (AAS, KRAM). **Southern Thule I.** Cape Flannery, 50 m, *PC 10* (AAS, KRAM).

SOUTH ORKNEY IS. **Coronation I.** Meier Pt., 10 m, *RILS 8024B* (AAS, KRAM) and 25 m, *RILS 582* (AAS, KRAM); Shingle Cove, 3 m, *RILS 442* (AAS, KRAM), 15 m, *NJC 122* (AAS, KRAM) and 65 m, *RILS 443* (AAS, KRAM). **Lynch I.** Below cliff, 15 m, *RILS 706* (AAS, KRAM). **Matthews I.** Near the Divide, 35 m, *RILS 170A* (BM). **Powell I.** John Peaks, *ca* 265 m, *RILS 265* (AAS, KRAM); south-east coast, 35–65 m, *RILS 213B* (AAS, KRAM). **Fredriksen I.** Central ridge, *ca* 115 m, *Edwards 1B* (AAS, KRAM). **Laurie I.** Cabo Destacamanto, 100 m, *Bellisio 11796* (KRAM, NY). **Signy I.** Cummings Cove, *RILS 1815D* (AAS, KRAM); Thulla Pt., 65 m, *MWH 155C* (AAS, KRAM); Foca Cove, 5 m, *RILS 8051* (AAS, KRAM); Spindrift Rocks, 85 m, *MWH 765F* (AAS, KRAM); North Pt., 75 m, *RILS 3167* (AAS, KRAM); between North Pt. and Robin Peak, 100 m, *RILS 3170* (BM); Stygian Cove, 20–30 m, *RILS 8146* (AAS, KRAM); Waterpipe Beach, *ca* 13 m, *RILS 381* (BM); Jane Col, 150 m, *RILS 3149* (AAS, KRAM); Snow Hill, 200 m, *RILS 600* (BM); Moraine Valley, below Orwell Glacier, 30 m, *RILS 10822* (AAS, KRAM); Factory Cove, *ca* 3 m, *RILS 666* (AAS, KRAM); Knife Pt., 6 m, *RILS 625* (AAS, KRAM); Berntsen Pt., *ca* 15 m, *REL 841* (BM); Paal Harbour, *ca* 85 m, *RILS 669* (AAS, KRAM); Pageant Pt., *ca* 15 m, *MWH 164A* (BM). **Moe I.** Landing Cove, 50–75 m, *RILS 5233* (AAS, KRAM).

SOUTH SHETLAND IS. **Elephant I.** 3 km south-west of Walker Pt., 92 m, *JSA 218* (AAS, KRAM); 0.5 km south-west of Endurance Glacier, 61 m, *JSA 71* (AAS, KRAM); 3.5 km east of Muckle Bluff, *JSA 208B* (AAS, KRAM). **Clarence I.** *Sine loco*, 95 m, *Chuter 44, 47, 51 & 59* (AAS, KRAM); southern end, 152 m, *JSA 2G* (AAS, KRAM). **Gibbs I.** East of the Spit, 105 m, *JPB 128* (AAS, KRAM). **Aspland I.** 30 m, *JPB 67A* (AAS, KRAM). **King George I.** ***Bransfield Strait***: Red Hill, 100 m, *RO 1092/80* (KRAM); Patelnia, 10 m, *RO 1135/80* (KRAM) and 12 m, *RO 1136/80* (KRAM). ***Admiralty Bay***: Vauréal Peak, 80 m, *RO 5238/79* (KRAM); Chabrier Rock, 6 m, *RO 1873/80* (KRAM); Cape Syrezol, 10 m, *RO 1851/80* (KRAM); Martins Head, 20 m, *RO 1866/80* (KRAM); Bastion, 240 m, *RO 1005/80* (KRAM); Brama, 200 m, *RO 785/80 & 786/80* (KRAM); Uchatka Pt., 12 m, *RO 1067/80* (KRAM); Demay Pt., 40 m, *RO 1332/80* (KRAM); Sphinx Hill, 95 m, *RO 236/80* (KRAM); Czajkowski Needle, 230 m, *RO 682/80* (KRAM); Rescurs Hills, 3 m, *RO 4904/79* (KRAM), 10 m, *RO 4885/79* (KRAM) and 110 m, *RO 167/80* (KRAM); Llano Pt., 10 m, *RO 4847/79* (KRAM), 20 m, *RO 4886/79* (KRAM), 15 m, *RO 4847/79* (KRAM) and 25 m, *RO 4884/79* (KRAM); Jasnorzewski Gardens, 5 m, *RO 1467/80 & 1480/80* (KRAM); Shag Pt. and Latarnia Rocks, 4 m, *RO 1554/80 & 1556/80* (KRAM); Observatory Creek, 50 m, *RO 4984/79* (KRAM); Ambona, 85 m, *RO 1628/80* (KRAM); Skua Cliff, 100 m, *RO 613/80* (KRAM). ***Ezcurra Inlet***: Kasprowy Hill, 160 m, *RO 163/80* (KRAM) and 250 m, *RO 162/80* (KRAM); Italia Valley, 5 m, *RO 5075/79* (KRAM); Breccia Crag, 1 m, *RO 921/80* (KRAM), 2 m, *RO 930/80* (KRAM) and 50 m, *RO 918/80* (KRAM); Emerald Pt., 15 m, *RO 846/80* (KRAM); Dufayel I., Gdynia Pt., 8 m, *RO 1778/80* (KRAM). ***MacKellar Inlet***: Klekowski Crag, 110 m, *RO 2280/80* (KRAM) and 250 m, *RO 2255/80* (KRAM); Komandor Peak, 230 m, *RO 1931/80* (KRAM). ***Keller Peninsula***: Ore Pt., 10 m, *RO 508/80* (KRAM); Moraine Pt., 50 m, *RO 467/80* (KRAM); Yellow Pt., 40 m, *RO 500/80* (KRAM). ***Martel Inlet***: Stenhouse Bluff, 4 m, *RO 2615/80* (KRAM) and 30 m, *RO 2590/80* (KRAM); Ternyck Needle, 430 m, *RO 1835/80* (KRAM); Szafer Ridge, 230 m, *RO 2566/80 & 2574/80* (KRAM); Tern Nunatak,

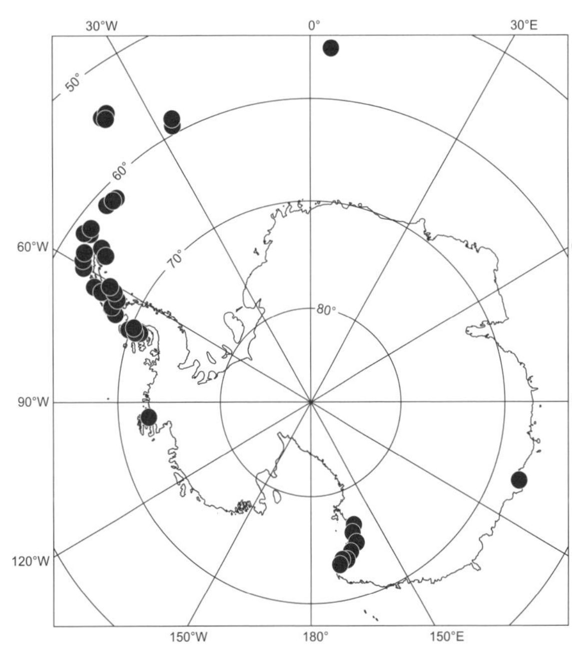

Figure 183. Distribution of *Syntrichia magellanica* in Antarctica.

255 m, *RO 2513/80* (KRAM; Warkocz 250 m, *RO 2547/80* (KRAM); Smok, 45 m, *RO 2070/80* (KRAM), 50 m, *RO 2069/80* (KRAM) and 80 m, *RO 2690/80* (KRAM); Basalt Pt., 15 m, *RO 2206/80* (KRAM); Mt. Wawel, 25 m, *RO 2128/80* (KRAM). ***Legru Bay***: Malczewski Pt., *BJ 13* (KRAM); Cinder Spur, 30 m, *BJ 2* (KRAM); Chopin Ridge, *BJ 90* (KRAM). ***King George Bay***: Lions Rump, *BJ 96 & 99* (KRAM); Turret Pt., *BJ 131 & BJ 135* (KRAM); Penguin I., *BJ 65 & 70* (KRAM). ***Sherrat Bay***: Three Sisters Pt., *BJ 121 & 124* (KRAM). ***Destruction Bay***: Trowbridge I., *BJ 9* (KRAM). ***Drake Passage***: Gam Pt., *BJ 10* (KRAM). ***Fildes Peninsula***: Suffield Pt., 50 m, *RO 2446/80* (KRAM); Nebles Pt., *BJ 25, 51 & 56* (KRAM); Green Pt., *BJ 21, 22 & 42* (KRAM). ***Marian Cove***: North Spit, *BJ 155* (KRAM). ***Potter Peninsula***: Stranger Pt., *BJ 109* (KRAM); Three Brothers Hill, *BJ 162* (KRAM); Florence Nunatak, 340 m, *RO 1312/80* (KRAM). **Robert I.** Coppermine Cove, 3–10 m, *RILS 915 & 934* (AAS, KRAM). **Aitcho I.** 5–50 m, *RILS 11456* (AAS, KRAM). **Greenwich I.** Discovery Bay, Ash Pt., 10 m, *DCL 703* (AAS, KRAM); Yankee Harbour, Spit Pt., 25 m, *DCL 543* (AAS, KRAM). **Livingston I.** ***Byers Peninsula***: South Beaches, 10 m, *DCL 170* (BM); Charity Glacier, 35 m, *DCL 55B* (AAS, KRAM); Barnard Pt., 15 m, *DCL 21* (BM); Clark Nunatak, 17 m, *DCL 115B* (AAS, KRAM); Dragon Cove, 12 m, *RILS 5859* (AAS, KRAM); Devils Pt., 100 m, *DCL 300* (AAS, KRAM); opposite Astor I., *DCL 347*(AAS, KRAM); between Chester Cove and Sealers Hill, 25–30 m, *RILS 3816A* (BM, KRAM). ***Hurd Peninsula***: Johnsons Dock, 20 m, *Monroe 116B* (AAS, KRAM); Miers Bluff, *RMS 69-805 & 69-810* (US); east side of False Bay, *RMS 69-579, 69-585 &69-586* (US). **Deception I.** Whalers Bay, 1 m, *RILS 3632* (AAS, KRAM) and 20 m, *RILS 5785A* (AAS, KRAM); Kroner Lake, *ca* 10 m, *RILS 82* (AAS, KRAM); Ronald Hill, 35 m, *BJT 252A* (AAS, KRAM) and 60 m, *RO 2774/80* (KRAM); Mt. Pond, 500 m, *RILS 11179* (AAS, KRAM); Crimson Hill, 10 m, *RILS 11525* (AAS, KRAM); Pendulum Cove, 10 m, *RO 2794/80* (KRAM); Telefon Ridge, 210 m, *RILS 5744F* (AAS, KRAM); Cross Hill, 60 m, *RILS 5795D* (AAS, KRAM); Kendall Terrace, 80 m, *RILS 5739C* (AAS, KRAM) and 170 m, *DM 109B* (AAS, KRAM); the col between Stonethrow Ridge and Telefon Ridge, 150 m, *RILS 5736C* (AAS, KRAM); Wensleydale Beacon, 165 m, *JBK75B* (AAS, KRAM); Fumarole Bay, 10–30 m, *RILS 6628 & 6640B* (AAS, KRAM); east face of Stonethrow Ridge, 60 m, *RILS 5804F* (AAS, KRAM); Crater Lake, 70–200 m, *RILS 6846C* (AAS, KRAM); Vapour Col, 100 m, *RILS 5625C* (AAS, KRAM) and 300 m, *DM 19* (AAS, KRAM); Mt. Kirkwood, 250 m, *RILS 5714B* (AAS, KRAM); Collins Pt., 20–30 m, *RILS 3603* (AAS, KRAM); Entrance Pt., 30–150 m, *RILS 6864* (AAS, KRAM).

WEST ANTARCTIC PENINSULA. **Danco Coast.** Cierva Pt., *VK 221, 244, 511, 514, 607 & 611* (KRAM); Spring Pt., *VK 1179* (KRAM); Charlotte Bay, Andrée I., 20–50 m, *RILS 4010* (AAS, KRAM); Cuverville I., *RILS 4191* (AAS, KRAM); Birdsend Bluff, 5–35 m, *RILS 4227* (AAS, KRAM); Beneden Head, 15–35 m, *AC & PK 55* (AAS, KRAM); Coughtrey Peninsula, Paradise Harbour, 100 m, *RILS 11616* (AAS, KRAM). **Palmer Archipelago.** ***Brabant I.***: Astrolabe Needle, *VK 890* (KRAM); Pinel Pt., 300 m, *AM 94B* (AAS, KRAM); Koch Glacier, 400 m, *AM 113* (AAS, KRAM); Mt. Bulcke, 300 m, *AM 105C* (AAS, KRAM). ***Melchior Archipelago***: Lysted I., *PAS 335.12* (FH, type of *Tortula conferta*); Lambda I., *PAS 345.1* (FH, NY); Omega I., *RILS 4128* (AAS, KRAM); Alpha I., *RILS 4106* (BM). ***Anvers I.***: Norsel Pt., *VK 150B* (KRAM) and 10–20 m, *RILS 1955A* (AAS, KRAM); Torgersen I., *RW 231* (AAS, KRAM); largest of Jubin Is., 5–10 m, *RILS 2009* (AAS, KRAM); Dream I., *VK 962 & 1026* (KRAM); Litchfield I., 10 m, *RWMC 384* (AAS, KRAM) and 10–25 m, *RILS 1965 & 1970* (AAS, KRAM); Hermit I., 5–15 m, *RILS 871* (AAS, KRAM); Laggard I., *ca* 15 m, *REL 1312* (AAS, KRAM); Stepping Stones, *VK 306* (KRAM); Cormorant I., 20 m, *Scott 39* (AAS, KRAM); Goudier I., Port Lockroy, 6–8 m, *RILS 731* (BM). **Graham Coast.** Cape Tuxen, 30–300 m, *RILS 4907* (AAS, KRAM); Cape Pérez, *ca* 7 m, *RWMC 635B* (AAS, KRAM). **Wilhelm Archipelago.** ***Penola Strait***: Petermann I., *Gain 168B, 172, 175 & 211* (PC, type of *Tortula heteroneura*); same island, Pt. Circoncision, 5–25 m, *RILS 4255* (AAS, KRAM). ***Argentine Is.***: Uruguay I., north-east side, *ca* 7 m, *RWMC 596* (BM) and 10–25 m, *RILS 4270* (AAS, KRAM); Galindez I., Corner Pt., *REL 1244* (AAS, KRAM); same Island, Woozle Hill, *ca* 12 m, *BJT 423A* (AAS, KRAM) and 25 m, *RILS 3291* (AAS, KRAM); same island, Stella Creek, 15 m, *REL 1324* (AAS, KRAM); Black I., 2–10 m, *RILS 4291 & 4298* (AAS, KRAM). ***Collins Bay***: Darboux I., 10–15 m, *RILS 3320* (AAS, KRAM). **Biscoe Is.** Armstrong Reef off Renauld I., 5 m, *RILS 5341B* (AAS, KRAM). **Loubet Coast.** ***Bennett Is.***: Pfaff I., *JK 178B* (AAS, KRAM). ***Laubeuf Fjord***: Webb I., 100–135 m, *DWB 3A* (AAS, KRAM). **Fallières Coast.** ***Adelaide I.***: Rothera Pt., 20 m, *JF 70A* (AAS, KRAM) and 20–30 m, *RILS 1864* (AAS, KRAM); Reptile Ridge, 140 m, *RILS 10598C* (AAS, KRAM); Mt. Gaudry, *ca* 465 m, *JK 143B* (BM); Jenny I., 200 m, *RILS 4760* (AAS, KRAM); Avian I., 15 m, *REL & RILS 936* (BM). ***Léonie Is.***: Lagoon I., 5 m, *RILS 9150 & 9160* (AAS, KRAM); Léonie I., 5–25 m, *RILS 8985A* (AAS, KRAM) and 150–175 m, *RILS 9135* (AAS, KRAM); Anchorage I., *RILS 9219* (AAS, KRAM). ***Dion Is.***: Consort I., 5–25 m, *RILS 3587* (AAS, KRAM). ***Marguerite Bay***: Piñero I., 10–25 m, *RILS 4829* (AAS, KRAM); Pourquoi Pas I., Lainez Pt., *Poncet 19* (KRAM); Horseshoe I., Lystad Bay, 10–15 m, *RILS 1900* (AAS, KRAM); Lagotellerie I., 50 m, *RILS 4839 & 4843* (AAS, KRAM); Camp Pt., 5–20 m, *RILS 4628, 4677 & 4686* (AAS, KRAM); Neny I., 7–10 m, *RILS 793B* (AAS, KRAM). **George VI Sound.** ***Alexander I.***: peak south of Bartok Glacier, *ca* 1060 m, *BASM 206* (AAS, KRAM); Debussy Heights, 400 m, *BASM 207B* (AAS, KRAM) and 500–700 m, *RILS 10004* (AAS, KRAM); Zebra Ridge, 600 m, *RILS 10024* (AAS, KRAM); Ablation Pt., *JJL & RBH 80 & 88* (AAS, KRAM); Ablation Valley, 2–3 m, *RILS 9275A* (AAS, KRAM), 75 m, *RILS 9292* (AAS, KRAM) and 400 m, *JJL & RBH 23 & 24* (AAS, KRAM); Ganymede Heights, *RILS 11563F* (AAS, KRAM); Moutonnée Valley, 183 m, *RILS 11082A & 1124A* (AAS, KRAM) and *ca* 200 m, *RILS 11120C* (AAS, KRAM); Middle Flatiron Valley, 315 m, *RILS 11099C* (AAS, KRAM), *ca* 420 m, *RILS 11100* (AAS, KRAM); Upper Flatiron Valley, *ca* 500 m, *RILS 11084, 11087B, 11088B & 11108* (AAS, KRAM); High Flatiron Valley, 532 m, *RILS 11093 & 11095* (AAS, KRAM); Probe Ridge, *DDWW 72E* (AAS, KRAM); Natal Ridge, *PC 696C* (AAS, KRAM) and *RILS 8850B* (AAS, KRAM); Viking Valley, 300 m, *RILS 9469* (AAS, KRAM); Ares Oasis, 50 m, *RILS 9461* (AAS, KRAM); Mars Oasis, 75 m, *RILS 10428* (AAS, KRAM), 80 m, *RILS 10379* (AAS, KRAM) and 175 m, *RILS 9431 & 9440* (AAS, KRAM); Herschel Heights, *RILS 8854 & 8855C* (AAS, KRAM); Horrocks Block, *RILS 8853A* (AAS, KRAM); Coal Nunatak, 100 m, *RILS 10318A & 10448* (AAS, KRAM).

EAST ANTARCTIC PENINSULA. **Trinity Peninsula.** Hope Bay, *REL 42* (AAS, KRAM) and *CS 427* (H, PC, S); Scar Hills, *ca* 50 m, *CGB 30B & 39* (BM); Mt. Flora. 25–50 m, *RILS 3841* (AAS, KRAM). **Joinville Island group.** ***Joinville I.***: north end of Active Sound, 200 m, *RILS 3733* (AAS, KRAM); Dundee I., Welchness Pt., 30–50 m, *RILS 5302A* (AAS, KRAM). **James Ross Island group.** ***Prince Gustav Channel***: Alectoria I., *ca* 50 m, *CGB 14* (AAS, KRAM); Persson I., 10 m, *RILS 5774* (AAS, KRAM). ***James Ross I.***: Abernethy Flats, 100 m, *RILS 7634A* (AAS, KRAM); Lachmann Crags, 5–35 m, *RILS 7326* (AAS, KRAM); Brandy Bay, 250 m, *RILS 5385A* (AAS, KRAM); Holluschickie Bay, 200 m, *RILS 5974A* (AAS, KRAM); Andreassen Pt., 80 m, *RILS 7594* (AAS, KRAM); Fortress Hill, 25 m, *RILS 7814* (AAS, KRAM) and 150 m, *RILS 7754 & 7792A* (AAS, KRAM); San José Pass, 165 m, *RILS 7528* (AAS, KRAM); St. Martha Cove, 15 m, *RILS 7480* (AAS, KRAM) and 80 m, *RILS 7650* (AAS, KRAM); Terrapin Hill, 250 m, *RILS 7724* (AAS, KRAM); Hidden Lake, 180 m, *RILS 5941B* (AAS, KRAM); Tumbledown Cliffs, 20 m, *RILS 5487B* (AAS, KRAM); Blancmange Hill, 125 m, *RILS 7694A* (AAS, KRAM). ***Vega I.***: False Island Pt., 25–30 m, *RILS 7832A* (AAS, KRAM). ***Cockburn I.***: *Hooker 2* (BM) and 50–150 m, *RILS 7909A & 7915* (AAS, KRAM); plateau, *ca* 400 m, *RILS 7966* (AAS, KRAM) and summit, *ca* 450 m, *RILS 7984* (AAS, KRAM). **Oscar II Coast.** 1 km north of Flask Glacier, 600 m, *RILS 5983* (AAS, KRAM); un-named nunatak on Jason Peninsula between Scar Inlet and Adie Inlet, *ca* 400 m, *RILS 2754D & 2759* (AAS, KRAM). **Bowman Coast.** Francis I., *ca* 325 m, *RILS 2775A & 2776A* (AAS, KRAM).

ELLSWORTH LAND. **Eights Coast.** Thurston Peninsula, Gneiss Islet, *Starr 84* (NY).

EAST ANTARCTICA. WILKES LAND. **Budd Coast.** Wilkes Station, *GAL 2828D* (NY).

VICTORIA LAND. **Borchgrevink Coast.** Crater Cirque, 30 m, *DWHW 247B* (AAS, KRAM) and 100 m, *RILS 9898, 9955 & 10173A* (AAS, KRAM); Edisto Inlet, Luther Peak, 18 Nov 1964, *Wise s.n.* (NY); west side of Whitehall Glacier, 270–430 m, 14 Nov 1964, *Wise s.n.* (NY); Apostrophe I., 25 m, *RILS 9791A & 9792* (AAS, KRAM); Cape King, 100–125 m, *RILS 10254, 10256A, 10259, 10260 & 10261A* (AAS, KRAM); Mt. McGee, 1100 m, *Bargagli 47* (AAS, KRAM); moraine of Tinker Glacier–Burns Glacier, *RILS 9657 & 9669* (AAS, KRAM); Harrow Peaks, 40 m, *RILS 9857* (AAS, KRAM), 85 m, *RILS 9849, 9850 & 9864B* (AAS, KRAM) and 150–200 m, *RILS 9560* (AAS, KRAM); Wood Bay, Kay I., 80–100 m, *RILS 9580* (AAS, KRAM), 100–130 m, *RILS 10198 & 10199A* (AAS, KRAM) and 150 m, *RILS 9960 & 10242A* (AAS, KRAM); 'Gondwana' Station area, 70 m, *RILS 9722 & 9723* (AAS, KRAM) and 150 m, *RILS 9742* (AAS, KRAM); Prior I., *RILS 9950B & 9951* (AAS, KRAM).

Excluded species of the Pottiaceae

Barbula unguiculata Hedw. — This species was reported by Wu & Hu (1990) from Fildes Peninsula on King George I. The voucher material (*Wei 8412*, KRAM) is a typical expression of *Syntrichia saxicola*. Accordingly, *B. unguiculata* must be withdrawn from the Antarctic moss flora. This species is essentially pan-Holarctic in distribution, having some localities in South America in Uruguay (Greene 1986).

Didymodon acutus (Brid.) K.Saito. — This species was recorded from Cockburn I. on the east coast of the Antarctic Peninsula by Wilson & Hooker (1847) as *Tortula gracilis* Hook. & Grev. The voucher material (*Hooker 3*, BM-Hooker) represents *Didymodon brachyphyllus*.

Hennediella densifolia (Hook.f. & Wilson) R.H.Zander. — This species is indicated for Antarctica in *Index muscorum* (Wijk *et al.* 1969 as *Tortula densifolia* Hook. f. & Wilson) but this report appears to be a bibliographic error. *Hennediella densifolia* is a southern South American and Falkland Is. species and no earlier references have been traced in the Antarctic botanical literature, nor are any specimens of this species from Antarctica known.

Syntrichia geheebiaeopsis (Müll.Hal.) R.H.Zander. — This species was reported from two localities on Fildes Peninsula, King George I. (Chen *et al.* 1995) as *Tortula geheebiaeopsis* (Müll.Hal.) Broth. However, the voucher collections represent *Syntrichia saxicola*, so the species must be withdrawn from the Antarctic flora. *Syntrichia geheebiaeopsis* is a rare subantarctic species, known from Îles Kerguelen, South Georgia and Isla de los Estados (Lightowlers 1986*b*).

S. laevipila Brid. — This species was recorded from Cockburn I. on the east coast of the Antarctic Peninsula by Wilson & Hooker (1847) as *Tortula laevipila* var. 4 *foliis obtusis areolis majusculis subrotundis*. This unnamed variety (*Hooker 2*, BM-Hooker, 3 specimens) represents the *conferta* expression of *S. magellanica*.

S. princeps (De Not.) Mitt. — The Antarctic plants of *Syntrichia* with hyaline hair-points were interpreted by Lightowlers (1986*b*) as *Tortula princeps* and this concept was subsequently adopted in the Antarctic botanical literature (e.g. Putzke & Pereira 1990, 2001; Chen *et al.* 1993, 1995; Lewis Smith 1996; Ochyra 1998*a*; Ochyra *et al.* 1998). These plants are considered in the present account as *Syntrichia magellanica* (see discussion under this species).

S. rubra (Mitt.) R.H.Zander. — This species was recorded from Bouvetøya by Zanten (1971) but the voucher material actually represents *Syntrichia filaris*.

S. ruralis (Hedw.) F.Weber & D.Mohr. — Wu & Hu (1990) reported this species from the Fildes Peninsula on King George I. as *Tortula ruralis* (Hedw.) P.Gaertn., B.Mey. & Scherb. This record is based on misdetermination, and examination of the voucher collection (*Wei 8086A*, KRAM) showed it to be a typical expression of *S. magellanica*. *Syntrichia ruralis* is essentially the Northern Hemisphere species (Gallego *et al.* 2002*b*) that is only rarely found in Southern Africa and Australia (Kramer 1988).

Tortula tenella Broth. — This species was reported from King George I. (Pizarro & Sáiz 1977) but this record is based on false synonymy of this species name with *Tortula monoica* suggested by Dixon (1923*a*). *Tortula tenella* is conspecific with *Syntrichia rubella* (Hook.f. & Wilson) R.H.Zander, an Australasian–South African species (Kramer 1988).

Subclass C4. **ORTHOTRICHIDAE** Ochyra

Orthotrichidae Ochyra *in* Ochyra, Żarnowiec & Bednarek-Ochyra, Cens. Cat. Polish Mosses: 31, 155. 2003.
≡ Diplolepideae order Orthotrichales Dixon *in* Verd., Man. Bryol.: 409. 1932.

Orthotrichidae Vitt, Goffinet & Hedd. *in* Bates, Ashton & Duckett, Bryol. Twenty-First Cent.: 119. 1998, *nom. nud.*

The Orthotrichidae comprise acrocarpous or cladocarpous mosses with the diplolepidous peristomes. They have usually been recognised as a distinct group in various classification systems of mosses, but Vitt *et al.* (1998) were the first to suggest recognition of the orthotrichalean mosses as a separate subclass, although the name Orthotrichidae was not then validly published. Ochyra (in Ochyra *et al.* 2003) formally validated this name when elevating the order Orthotrichales to the rank of subclass. The distinctness of the orthotrichalean mosses is primarily associated with peculiarities in the structure of the peristome which represents the separate *Orthotrichum*-type (Vitt 1981, 1984; Lewinsky 1989; Shaw 1986; Goffinet *et al.* 1999).

The orthotrichaceous peristome has basically a 4 : 2 : 4 or 4 : 2 : 2 peristomial formula, but occasionally a modified formula 4 : 2 : 3 is also observed and this has led to speculation on a possible relationship between the *Orthotrichum*-type and *Dicranum*-type of the peristomes with the latter derived from the former (Lewinsky 1989). In such interpretation the orthotrichaceous peristome should be considered as possibly pivotal in the evolution of the dicranaceous and bryaceous types. The *Orthotrichum*-type nearly approaches the conventional *Bryum*-type peristome but it is much reduced and highly modified, with the cellular pattern of both the exostome and/or the endostome strongly disorganised and difficult to interpret. It is mostly because of the clearly asymmetric anticlinal divisions occurring in the 8-celled IPL layer, and the lack of alignment of adjacent anticlinal IPL and PPL walls resulting from the post-divisural lateral displacement of the IPL walls caused by the swelling of the PPL walls (Goffinet *et al.* 1999). Additionally, the exostome teeth have a much more thickened dorsal surface than the ventral side, and the endostome has no or a highly reduced, or delicate, at times lacerated, basal membrane, stout and unkeeled segments alternating with the exostome teeth, and lack cilia. Vitt (1981) described opposite endostomial segments in some species of *Schlotheimia*, but this seems to represent only a minor peristome variation since the peristomial pattern in these species is the same as that found in species with alternating segments (Shaw & Rohrer 1984; Shaw 1986). Although none of these character-states is unique to the Orthotrichidae, and occasionally some of these pecularities of the peristome are found in other groups of bryalean and hypnalean mosses, they are associated in this group with unique plant morphologies and this certainly warrants its distinction as a separate subclass.

The orthotrichalean mosses are a rather isolated group which, in the early classification systems, was associated with the acrocarps and usually placed in the close proximity of grimmiaceous mosses (Müller 1848–1849; Schimper 1860, 1876; Limpricht 1876; Braithwaite 1889; Brotherus 1902–1903; Fleischer 1904). On the other hand, Fleischer (1920) considered them as pleurocarps and placed as a suborder in the order Isobryales. This placement was adopted by Brotherus (1924*b*) in the second edition of his treatment of Musci in *Die Natürlichen Pflanzenfamilien*. However, Dixon (1932*a*) reinstated the affinity of the orthotrichaceous mosses to the acrocarps and nested them as a separate order Orthotrichales in the Diplolepideae, close to the Funariales and Bryales. This placement has gained wide acceptance in modern classifications of mosses (Walther 1983; Vitt 1984; Buck & Goffinet 2000; Ochyra *et al.* 2003; Goffinet & Buck 2004).

The Orthotrichidae consist of two orders, Orthotrichales and Hedwigiales (Vitt *et al.* 1998; Ochyra *et al.* 2003). Because a peristome is lacking in all taxa belonging within the latter order, it is difficult to select the most plausible hypothesis of their true affinities. Even the growth form of the hedwigiacean mosses has been variously described as acrocarpous, pseudopleurocarpous, or pleurocarpous, but De Luna (1990*a*) provided developmental evidence of acrocarpy in *Hedwigia ciliata*. Also, the presence of the globular protonema is unique in the Hedwigiales (De Luna 1990*b*). On the basis of morphological analysis De Luna (1995) indicated that this otherwise bizarre group shows the closest affinities with the Orthotrichales, although the presence of the papillose pseudoparaphyllia is difficult to rationalise within the Orthotrichidae. Molecular data currently do not resolve the position of the hedwigiacean mosses and they are placed either within the *Orthotrichum* clade (De Luna *et al.* 1999) or the Bryidae clade (Newton *et al.* 2000; Goffinet *et al.* 2001).

Order VIII. **ORTHOTRICHALES** DIXON

Orthotrichales Dixon *in* Verd., Man. Bryol.: 409. 1932.

The orthotrichalean mosses were first recognised as a suprafamilial taxon by Fleischer (1920) who distinguished the suborder ['Unterreihe'] Orthotrichineae within the order Isobryales [= Leucodontales] to accommodate two families, Erpodiaceae and Orthotrichaceae. Brotherus (1924*b*) accepted this suborder and added to it the third family Ptychomitriaceae. This familial composition was adopted by Dixon (1932) when describing the order Orthotrichales. Finally, Miller & Harrington (1977) and Vitt (1982*a*) placed in this order the monotypic families Microtheciellaceae and Helicophyllaceae, respectively. As the result of subsequent studies all these families have been relegated from the Orthotrichales, leaving this order monotypic.

The Ptychomitriaceae are closely related to the Grimmiaceae on account of the *Racomitrium*-type peristome (Edwards 1979; Deguchi 1979; Churchill 1981; Tsubota *et al.* 2003). The Rhachitheciaceae also have close affinity to the Dicranidae (Goffinet 1997), whereas the Microtheciellaceae are placed near the Neckeraceae in the Hypnidae (Goffinet 1998). There have been various proposals with regard to the affinities of the Erpodiaceae. Goffinet *et al.* (1998) placed this family in the Haplolepideae, La-Farge *et al.* (2000) placed it in the Pottiales, whereas Buck & Goffinet (2000) and Goffinet & Buck (2004) transferred the Erpodiaceae to the Dicranales. However, Pursell & Allen (2002) found that the type of peristome reduction occurring in this family is probably analogous to that in the Bryaceae, and has no counterpart in the Haplolepideae. It argues for positioning the Erpodiaceae in the Diplolepideae. This hypothesis also seems to confirm the presence of the rhizoidal initials clustered at the abaxial leaf insertion and this is a characteristic feature of diplolepidean mosses (Hedenäs 1987*a*). Finally, the Helicophyllaceae are currently nested in the Hedwigiales (Buck & Goffinet 2000; Goffinet & Buck 2004), although the sole species of this family, *Helicophyllum torquatum*, has a set of unusual sporophyte features (immersed capsule, lacking peristome and stomata, flat operculum inserted within the thickened and inrolled capsule mouth, possibly the lack of calyptra) that isolate it from most other mosses.

Additionally, the genus *Drummondia* Hook., which was traditionally placed in the Orthotrichaceae and recognised as a separate subfamily Drummondioideae (Vitt 1972), also proved to be haplolepidous and was transferred to the Dicranidae as a member of the separate family Drummondiaceae (Crum 1987; Buck & Goffinet 2000; Goffinet & Buck 2004). Accordingly, the order Orthotrichales is monotypic and diagnosed by the characters of its only family Orthotrichaceae.

Family X. **ORTHOTRICHACEAE** Arn.

Orthotrichaceae Arn., Disp. Méth. Mousses: 13. 1825 [1826] ['Orthotrichoideae']. — Type: *Orthotrichum* Hedw.

Zygodontaceae Schimp., Coroll. Bryol. Eur.: 39. 1856. — Type: *Zygodon* Hook. & Taylor.

Macromitriaceae S.P.Churchill *in* S.P.Churchill, & Linares & G.M.Gonzáles Bibliot. José Jerónimo Triana 12: 588. 1995. — Type: *Macromitrium* Brid.

The Orthotrichaceae are an extremely large family that exhibits broad morphological heterogeneity, making it very complex and difficult to circumscribe. Gametophytically, orthotrichalean mosses resemble pottiaceous mosses, but the peristome is diplolepidous. They are small to modest-sized saxicolous or epiphytic plants, mostly associated with xerophytic habitats, erect to erect-ascending and occurring in cushions or tufts or prostrate with erect-ascending lateral branches and often with a mat-like growth form, sparsely or densely tomentose. The leaves throughout the family are crowded, often hygroscopic, erect-appressed to crispate or spirally twisted when dry, spreading to squarrose when moist, keeled, broadly concave to plane above, sometimes rugose or undulate, lanceolate, ovate-lanceolate, oblong-lanceolate, linear-lanceolate, ovate, elliptical, ligulate, or lingulate, without or with weakly, very seldom strongly differentiated alar cells, acute, acuminate or obtuse, rarely retuse. The margins are plane or revolute, sometimes incurved, generally entire but often serrate or denticulate, unbordered or bordered, occasionally with multistratose swollen limbidia. The costa is single, strong, ending below the apex, percurrent or sometimes excurrent, occasionally as a hyaline hair-point. The laminal cells are mostly unistratose or occasionally bistratose distally. The upper cells are small, isodiametric, thick-walled and variously papillose or smooth, while the lower cells are mainly elongate and hyaline or yellowish, tuberculate or smooth. The primary character of the family is the orthotrichaceous peristome consisting of 16 teeth and segments (Lewinsky 1989). It is of the arthrodontous type and diplolepidous with the endostome segments alternating or, rarely, appearing opposite the exostome teeth. The preperistome is absent, or present but reduced at the base of the teeth. The exostome teeth are 16, free or fused into 8 pairs, or occasionally united as a continuous irregular membrane, lanceolate or truncate, erect, spreading to reflexed or recurved when dry, with the dorsal plates thicker than the ventral ones. The endostome segments are 8 or 16, linear and hyaline, non-perforate, unkeeled, free or united at the base with a low basal membrane, and lack cilia. The peristome is sometimes absent or single and the teeth and segments are usually variously ornamented. The annulus is typically absent or poorly differentiated. The sporophytes are borne terminally on stems or erect branches, singly in perichaetia which have generally undifferentiated perichaetial leaves. The capsules are exserted, emergent or immersed, cylindrical to globose, not or contracted below the orifice, and smooth or ribbed. The stomata are bicellular, with elongate, superficial or immersed pori and with the subsidiary cells weakly or strongly covering the guard cells. The calyptra is mostly mitriform, less often cucullate, narrowly cylindrical, broadly campanulate to conic, smooth or plicate, naked or hairy and the capsules are immersed to shortly exserted. The operculum is flat and mostly conic, with a short or long, straight or oblique beak. The spores are isomorphic to dimorphic, granulose, reticulate or pitted, mono- or sometimes multicellular. Asexual reproduction is relatively frequent in the Orthotrichaceae, usually by multicellular ovoid or ellipsoid gemmae produced on leaves or, less often, rhizoids.

The morphology of orthotrichalean mosses sharply separates them from all other mosses and consequently the Orthotrichaceae are one of the earliest recognised families, although originally Arnott (1825) also placed in it the genera *Tetraphis*, *Octoblepharum* and *Calymperes*. The family is cosmopolitan in distribution and, with about 840 species (Crosby *et al.* 2000), it is one of the largest and most diverse families of mosses, although further reductions of species' names to synonymy are to be expected. It is much more abundant in tropical and warm temperate than cool and cold regions of the world. Most orthotrichlean species belong within five genera, namely *Macromitrium* (368), *Orthotrichum* (146), *Schlotheimia* (121), *Zygodon* (92), and *Ulota* (57). The remaining species are classified into small satellite genera comprising 1–4 species, but in two, *Groutiella* and *Macrocoma*, there are 13 and 10 species, respectively. Depending on taxonomic interpretations, the number of genera in the Orthotrichaceae varies from 23 (Goffinet & Vitt 1998; Crosby *et al.* 2000) to 20 (Goffinet *et al.* 2004).

Considerable morphological diversity in orthotrichalean mosses has resulted in various infrafamilial classifications of the Orthotrichaceae and, in extreme cases, for some genera separate families have been recognised, namely Zygodontaceae (Schimper 1856) and Macromitriaceae (Churchill *et al.* 1995) which, however, have not been widely accepted. Traditionally, genera of the Orthotrichaceae have been divided into four subfamilies (Brotherus 1925; Reimers 1954; Walther 1983). Vitt (1972, 1982*b*) also recognised four subfamilies, one of which, Drummondioideae, was established as new, whereas the Desmothecoideae was merged with the Macromitrioideae. Later, Vitt *et al.* (1993) suggested a refinement of this classification by the transfer of *Schlotheimia* and *Cardotiella* into a separate subfamily, but this proposal has not been substantiated.

Recently, Goffinet & Vitt (1998) presented a new classification of the Orthotrichaceae that is based on molecular phylogeny as well as comparative morphology. The family was divided into two sufamilies each composed of two tribes: the Zygodonteae and Orthotricheae (Orthotrichoideae) and the Macromitrieae and Schlotheimieae (Macromitrioideae). The genera of orthotrichalean species are disposed more or less evenly in both subfamilies, but the Macromitrioideae is richer and consist of about 430 species, although both of the largest genera of this subfamily, *Macromitrium* and *Schlotheimia*, still need taxonomic revision in some regions which should reduce the number of species.

In contrast to the northern polar region, the Orthotrichaceae are poorly represented in the southern counterpart. Only five genera and nine species have been recorded from within the Subantarctic and Antarctic, namely *Zygodon*, *Macromitrium*, *Ulota* and *Muelleriella*, each with two species, and *Orthotrichum* with one species. *Muelleriella* and *Orthotrichum* occur within the Antarctic botanical zone where they are infrequent components of the moss flora in the northern maritime Antarctic (Ochyra 1986). The Orthotrichaceae are much better represented in the cool temperate zone in the Southern Hemisphere, especially in southern South America, where three endemic genera are known, *Pleurorthotrichum* (Brotherus 1905; Lewinsky-Haapasaari 1994), *Matteria* and *Pentastichella* (Goffinet 1998). The forth genus, *Muelleriella*, is endemic to the islands of the Southern Ocean and southernmost tip of South America, and the South Island, New Zealand. Additionally, many species of *Zygodon*, *Ulota*, *Orthotrichum*, and *Macromitrium* have been reported or described from this biome, although most of them still await critical taxonomic revision.

KEY TO THE GENERA OF THE ANTARCTIC ORTHOTRICHACEAE

1. Laminal cells entirely smooth or pseudopapillose, 2–3-stratose distally; leaf margin 2–4-stratose, plane, often forming strong, swollen limbidium from about mid-leaf or below to the apex .... 35. ***Muelleriella***
1. Laminal cells densely papillose on both surfaces, unistratose with frequent bistratose strands above; leaf margins revolute almost throughout, unistratose to variously bistratose, not bulging .... 36. ***Orthotrichum***

35. MUELLERIELLA DUSÉN, *nom. cons.*

Muelleriella Dusén, Bot. Notiser 1905: 304. 1905, *nom. cons.* ≡ *Orthotrichum* Hedw. subgen. *Muelleriella* (Dusén) Cardot, Wiss. Erg. Schwed. Südpolar-Exp. 4(8): 121. 1908. — Lectotype (*vide* Vitt 1970: p. 298): *Muelleriella crassifolia* (Hook.f. & Wilson) Dusén (*Orthotrichum crassifolium* Hook.f. & Wilson).

Orthotrichum sect. *Brachysteleum* Mitt. *in* Hook.f., Handb. N. Zeal. Fl. 2: 432. 1867. — Lectotype (*vide* Ochyra 1997*d*: p. 174): *Muelleriella crassifolia* (Hook.f. & Wilson) Dusén (*Orthotrichum crassifolium* Hook.f. & Wilson).

Orthotrichum subgen. *Crassifolium* Lewinsky, J. Hattori Bot. Lab. 43: 55. 1977, *nom. illeg. incl. subgen. prior.*

Plants small, rigid and stiff, in loose or compact tufts or cushions, sometimes forming small mats, blackish-green, black to dark green or rarely brownish-black to brownish. *Stems* erect, unbranched to sparingly branched, densely radiculose at the base with reddish-brown, smooth, freely branched rhizoids, in transverse section rounded, lacking central strand and consisting of large, hyaline medullary cells with firm, orange walls and a poorly differentiated cortex of somewhat smaller, reddish-brown and thicker-walled cells in 1 or in places 2 layers; *axillary hairs* filiform, 3–6-celled, with 2 short, pigmented basal cells and 1–4 elongate, hyaline upper cells. *Leaves* loosely erect to tightly appressed, incurved, secund or falcate-secund, narrowly lanceolate to broadly subulate, ligulate-lanceolate to ligulate from an ovate or obovate base or ovate to oblong-ovate, obtuse, subacute to acute at the apex, concave, non-decurrent; *margins* plane, 2–4(–5)-stratose, not thickened or forming swollen limbidia extending from mid-leaf or below to near the apex; *costa* single, strong, ending below the apex, in transverse section biconvex, consisting of uniform stereid or substereid cells usually covered entirely or partially on both dorsal and ventral surfaces by larger epidermal cells, occasionally without enlarged epidermal cells; *laminal cells* smooth or pseudopapillose owing to cuticular thickenings, thick-walled, 2–3-stratose distally, unistratose in the proximal portion;

upper cells irregularly rounded to rounded-hexagonal; *basal cells* elongate- or short-rectangular to quadrate, often becoming slightly nodulose towards the costa, usually quadrate at the margins and sometimes forming a distinct border. *Autoicous*. *Setae* short, thin to thick and stout, straight to curved up to 90°. *Capsules* immersed to long exserted, ovoid, obloid to shortly cylindrical, sometimes subglobose, gradually or abruptly contracted to the seta, not or somewhat constricted under the mouth when dry and empty, smooth; *annulus* absent; *operculum* conical, shortly beaked; *exothecial cells* quadrate to rectangular, usually with strongly incrassate walls; *stomata* immersed, with strongly differentiated subsidiary cells covering half to whole guide cells; *peristome* diplolepidous, single; *exostome teeth* 16, erect, spreading to recurved when dry, densely papillose to irregularly papillose-striate; *endostome* absent; *preperistome* present, 1–4 cells high, sometimes up to half the length of the exostome. *Spores* multicellular, spherical, cubic to cubic-rectangular, 70–90 µm long in longest direction, 30–50 µm wide, appearing smooth or nearly so under the light microscope, but minutely papillose when viewed by SEM. *Calyptra* conic, mitriform, strongly plicate, glabrous or seldom with 1–2 multicellular hairs, sometimes roughened at the tip, covering half to almost the whole capsule.

Muelleriella is a small genus of four species and one subspecies, closely allied to *Orthotrichum* which is distinguished by the possession of an unusual combination of characters, including (*a*) single, yet diplolepidous, peristome with an exostome of 16 papillose, erect to recurved teeth; (*b*) well developed preperistome; (*c*) multicellular spores; (*d*) multistratose, smooth to pseudopapillose upper laminal cells; (*e*) plane, multistratose leaf margins, often forming swollen limbidia; (*f*) smooth capsules, and (*g*) blackish coloration of plants. Moreover, all species of *Muelleriella* are halophilous which is an unusual adaptation in mosses and is not found in any species of the Orthotrichaceae. Although none of these characters uniquely defines the genus, this suite of features is unknown in any other group of *Orthotrichum* species. Vitt (1979*b*) emphasised multicellular spores to be the most significant difference between *Muelleriella* and *Orthotrichum* but, in fact, such spores are occasionally found in the latter genus, for example in *O. steerei* Lewinsky from Ecuador (Lewinsky 1987), as well as in Chinese *O. dasymitrium* Lewinsky (Koponen & Luo 1992), *O. jetteae* B.H.Allen and *O. schofieldii* (B.C.Tan & J. Yu) B.H.Allen. The latter two species were originally placed in the separate genus *Orthomitrium* (Lewinsky-Haapasaari & Crosby 1996; Tan & Yu 1997), which was erected on a combination of unique characters including the puckered mouth of the capsule which was considered a unique trait. However, Allen (2002) pointed out that this character is scattered within the family Orthotrichaceae and should not serve to diagnose a distinct genus. Nonetheless, considering a unique combination of morphological characters, unusual ecological requirements, and restricted geographical distribution, the separate generic status of *Muelleriella* is retained in the present treatment.

Muelleriella was recognised as a genus in its own right by Dusén (1905*c*) to accommodate two species, *M. crassifolia* and *M. angustifolia* (Hook.f. & Wilson) Dusén, two of the earliest species described in this genus. However, these species were segregated much earlier as a section within *Orthotrichum*, sect. *Brachysteleum*, by Mitten (1867). Cardot (1908*a*) reduced *Muelleriella* to a subgenus within *Orthotrichum*, but the generic distinction of this taxon was supported by Vitt (1976*b*, 1979*b*) based on the aforementioned suite of characters. Lewinsky (1977*a*) initially rejected the generic distinction of *Muelleriella* and considered it as a subgenus within *Orthotrichum*, subgen. *Crassifolium*, but this name is nomenclaturally illegitimate as pointed out by Ochyra (1997*d*). Lewinsky's (1977*a*) concept of this taxon was based on reasoning that immersed stomata evolved only once in the Orthotrichaceae, so all taxa having such stomata, including *Muelleriella*, should be retained in the single genus *Orthotrichum*. This evolutionary hypothesis was defended by Vitt (1972, 1979*b*) who viewed *Muelleriella* as evolving along separate routes, the latter being confined to rocks while species of *Orthotrichum* evolved to more specialised substrata of either rock surfaces or trees. This concept was subsequently accepted by Lewinsky (1993) who reinstated *Muelleriella* as a separate genus, although she remained sceptical, particularly following cladistic analysis of *Orthotrichum* (Lewinsky-Haapasaari & Hedenäs 1998).

Muelleriella is one of few moss genera which are strictly subantarctic in distribution. Its range covers all subantarctic islands and extends into the southern cool temperate zone and the northern maritime Antarctic (Clarke & Lightowlers 1983; Ochyra 1986). Three species are narrow endemics of isolated islands, including Auckland Is. (*M. aucklandica*), Campbell I. (*M. angustifolia*) and Îles Kerguelen (*M. atrata*), and only one species (*M. crassifolia*) has a wide circumpolar range. A worldwide monograph of *Muelleriella* has been published by Vitt (1976*b*).

1. **Muelleriella crassifolia** (Hook.f. & Wilson) Dusén — Fig. 184

Muelleriella crassifolia (Hook.f. & Wilson) Dusén., Bot. Notiser 1905: 304. 1905. ≡ *Orthotrichum crassifolium* Hook. f. & Wilson, J. Bot. (Hooker) 3: 546. 1844. ≡ *Ulota crassifolia* (Hook.f. & Wilson) Hook.f. & Wilson *ex* Paris, Ind. Bryol. 3: 326. 1905, *nom. inval. in synon.* — Type citation: Lord Auckland's group and Campbell's Island. [Lectotype (*vide* Vitt 1976*b*: p. 100): 'Nº 35. Orthotrichum crassifolium H.fil. & Wils. Campbell Island Antarct. Exped. 1839–1843 J.D.H.' – BM-Hooker! Syntype: 'Nº 54. Orthotrichum crassifolium H.fil. & Wils. Lord Auckland's Islands Antarct. Exped. 1839–1843 J.D.H.' – BM-Hooker!; isosyntypes: BM-Hooker!, BR!, MANCH!].

Plants 5–15 mm tall, in small or medium-sized, dense or loose blackish- or brownish-green, dark green, rufous to black cushions. *Stems* erect, simple or sparsely branched, densely radiculose below with smooth, brown, branched rhizoids, in transverse section with 1(–2) layers of smaller, thicker-walled, brown cortical cells surrounding 3–4 layers of larger, hyaline or yellowish medullary cells, central strand lacking; *axillary hairs* numerous, 3–5-celled, filiform, with 2 short brownish basal cells and 1–3 elongate hyaline distal cells. *Leaves* crowded, erect when dry, erect-spreading to spreading when moist, 1.3–2.0 mm long, 0.4–0.6 mm wide, concave, ligulate-lanceolate to ligulate from an ovate or oblong-ovate base, acute at the apex, bistratose distally; *margins* plane, entire, 2–4-stratose, often forming swollen, fleshy limbidia extending from ⅓–¼ above the leaf base to the apex; *costa* single, gradually merging into lamina cells above, 55–70 µm wide, subpercurrent, in cross-section 4–6-stratose, composed of almost uniform cells or in the upper half central cells somewhat smaller and surrounded by a layer of larger laminal cells; *laminal cells* smooth or occasionally pseudopapillose due to longitudinal cuticular ridges, giving the leaves a papillose appearance in cross-section, thick-walled; *upper cells* rounded to rounded-hexagonal, 2–3-stratose, (6.0–)7.0–13.5 µm wide; *basal cells* unistratose, quadrate to shortly rectangular, about twice as long as wide, thick-walled, becoming elongate and nodulose towards the costa. *Autoicous*. *Perichaetia* terminal; *perichaetial leaves* similar to vegetative leaves but somewhat larger. *Setae* (1.5–)1.7–2.0 mm long, thick, straight or slightly curved. *Capsules* exserted, erect, ovoid to ovoid-cylindrical, 1.5–1.8(–2.0) mm long, 0.8–1.0 mm wide, brown, smooth, slightly constricted below the orifice when dry and empty; *annulus* none; *operculum* conic, with a short, straight rostrum; *exothecial cells* irregularly rectangular in the middle of the urn, 15–45 µm long, with strongly incrassate longitudinal walls, becoming shorter and irregular towards the base and transversely rectangular under the mouth; *stomata* numerous in the lower half of the urn, cryptoporous with subsidiary cells strongly differentiated and covering half to entire stoma; *peristome* single consisting of 16 exostome teeth; *teeth* spreading when mature and dry, becoming recurved with age, coarsely and irregularly papillose; *preperistome* present, well developed, consisting of 1–4 cells. *Spores* multicellular, globose (30–)40–90 µm in diameter. *Calyptra* conic, mitriform, strongly plicate, glabrous.

Chromosome number: unknown.

Discussion. *Muelleriella crassifolia* is unlikely to be confused with any moss in the Antarctic, even when sterile. Although superficially it may resemble species of *Schistidium* or *Orthotrichum*, it can be readily recognised by its distally bistratose leaves. Also, swollen and fleshy, multistratose limbidia in the upper half of the leaves are unknown in any other Antarctic moss, except for *Schistidium praemorsum*, which is distinguished by its keeled leaves. Fertile plants have distinctive exserted capsules which are covered with large, conic, strongly plicate and glabrous calyptrae. Additionally, multicellular spores are an unusual feature of this species. In contrast, species of *Schistidium* have capsules immersed in enlarged perichaetial leaves, and small unicellular spores.

Muelleriella crassifolia is a somewhat variable species throughout its wide geographical range, especially in capsule and seta length and leaf size. Vitt (1976*b*) recognised three intergrading, yet distinct populations, centring around southern South America, Kerguelen Province and the Australasian sector. The plants from the latter region are robust and have long setae and capsules, whereas the South American plants are small with shorter capsules and setae. The Antarctic populations are most similar to Fuegian plants in their small size, slender leaves and shorter capsules and setae and this supports the well substantiated theory that much of the maritime Antarctic bryophyte flora migrated from southern South America after retreat of glaciations.

Reproduction in Antarctica. The majority of the Antarctic populations are sterile, but mature sporophytes with normally developed spores are known from the South Orkney Is. (Lewis Smith 1972; Webb 1973; Convey 1994).

Habitat. *Muelleriella crassifolia* is a unique species in the Antarctic moss flora as it can tolerate high concentrations of marine salts (Lewis Smith 1978). It is an epipetric moss growing on coastal rocks in the supratidal and salt spray zones. It usually thrives on dry or wet, often nitrogen-enriched rocks which are colonised by an assemblage of nitrophilous and hygrohaline crustose lichens of the genera *Amandinea*, *Buellia*, *Caloplaca* and *Lecidea* and also associated with *Acarospora* spp., *Staurothele gelida* and *Xanthoria elegans*. They form a distinct zone on coastal rocks 1–2 m above the high-water mark which may be inundated with sea-spray. Occasionally, *M. crassifolia* also colonises the *Verrucaria* zone which extends from slightly below the high-water mark to about 5 m (Lewis Smith 1972).

Muelleriella crassifolia is the only species growing exclusively in such habitats in the Antarctic and it is one of very few bryophyte species which is known for its very high salinity tolerances. Apart from four species of *Muelleriella* there are few mosses that are able to tolerate salt water. These include *Schistidium maritimum* (Turner) Bruch & Schimp., which seems to be the ecological counterpart of *M. crassifolia* growing in identical habitats in the Northern Hemisphere. Also, the bipolar *Ulota phyllantha* Brid. and austral *Calyptopogon mnioides* (Schwägr.) Broth. are known to occur in the salt spray zone. In general, liverworts and mosses are unable to tolerate salt water, although some weedy, polytypic and polystenic species occur rarely in areas subject to tidal action or even in the salt spray zone (Engel & Schuster 1973; Vitt 1976*b*).

World range. *Muelleriella crassifolia* is a circumsubantarctic species widely distributed and locally common throughout all subantarctic islands (on Îles Crozet, Îles Kerguelen and Heard I. as subsp. *acuta* (Müll.Hal.) Vitt), as well as on Campbell I. and Auckland Is., penetrating northwards to the cool temperate region at the southernmost tip of the South Island, New Zealand, including offshore islands, and the southernmost part of South America, including the Falkland Is. and southwards to the northern maritime Antarctic (Fig. 185, inset).

Distribution in Antarctica. The Antarctic distribution of *Muelleriella crassifolia* was summarised by Clarke & Lightowlers (1983) and Ochyra (1986). The species is restricted in its distribution to the northern maritime Antarctic including the South Orkney Is. and South Shetland Is. and the western coast of the Antarctic Peninsula, where it extends from Astrolabe I. off Trinity Peninsula, to Berthelot I. off the Graham Coast at lat. 65°20′S

→

Figure 184. *Muelleriella crassifolia*. – **1**. Habit. **2**. Cross-section of the stem. **3–5**. Axillary hairs. **6–9**. Vegetative leaves. **10**. Leaf apex. **11**. Upper leaf cells. **12**. Mid-leaf cells at the margins. **13**. Basal leaf cells. **14–21**. Cross sections of the leaves, a sequence from the apex to the base. **22–23**. Perichaetial leaves. **24**. Young operculate capsule, wet. **25**. Old deoperculate capsule, wet. **26**. Old deoperculate capsule, dry. **27**. Exothecial cells above the middle of the urn. **28**. Lower exothecial cells and stoma. **29**. Portion of the peristome and exothecial cells below the mouth. **30–31**. Spores. – [1, 22–31 from *Matteri 1286*; 2, 5–7, 10–13, 16, 18–19 from *RO 367/80*; 3–4, 8–9, 14–15, 17, 20–21 from *BJ M115*]. – Scale bars: a – 100 µm (29); b – 1 mm (1, 22–26); c – 100 µm (2–5, 10–21, 27–28, 30–31); d – 1 mm (6–9).

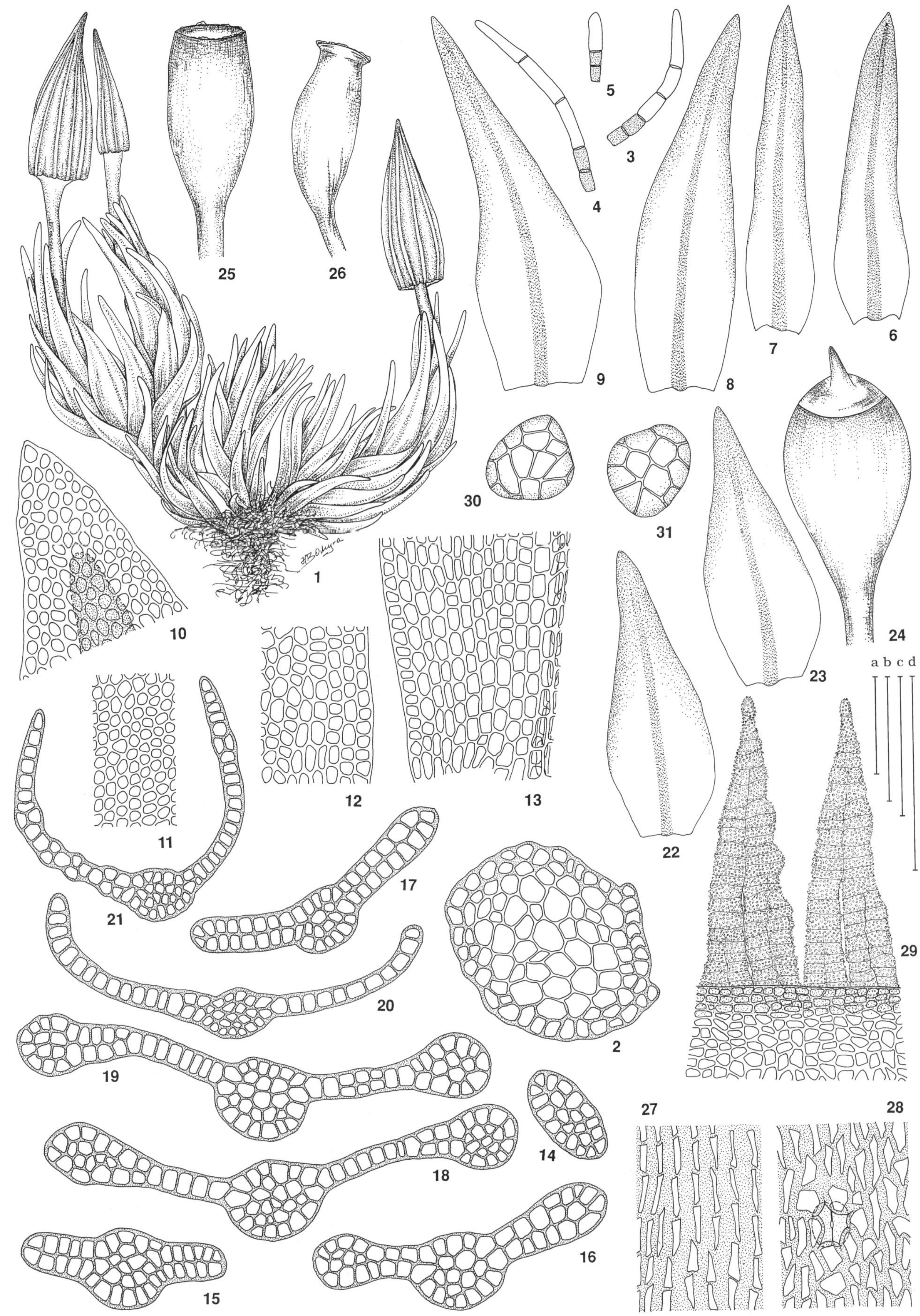
1
2
3
4
5
6
7
8
9
10
11
12
13
14
15
16
17
18
19
20
21
22
23
24
25
26
27
28
29
30
31
a b c d

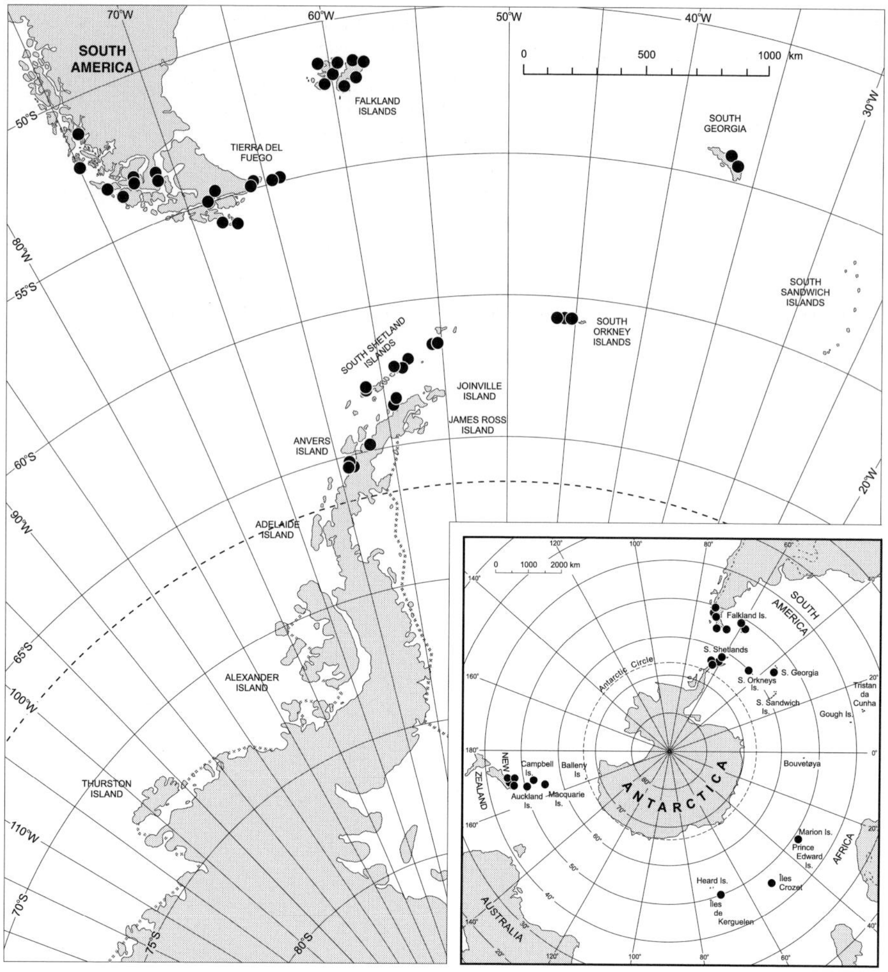

Figure 185. Distribution of *Muelleriella crassifolia* in the Fuegian region and Antarctica. Inset: global distribution.

(Fig. 185). It grows at low elevations, from sea level to 15 m, and only occasionally grows higher at 50–100 m on maritime cliffs subjected to the sea spray, and generally is scattered and seldom abundant.

Specimens examined. WEST ANTARCTICA. SOUTH ORKNEY IS. **Coronation I.** Saunders Pt., 2–3 m, *RILS 37* (AAS, KRAM); Shingle Cove, 1.5 m, *RILS 8230* (AAS, KRAM) and 5 m, *RILS 451* (AAS). **Lynch I.** Cliffs by shore, 2–5 m, *RILS 506* (AAS, KRAM); north side, 1.5 m, *RILS 8231* (AAS, KRAM). **Matthews I.** North of Coffer I., *RILS 192* (BM, S). **Powell I.** West of John Peaks, 8 m, *RILS 255* (AAS, KRAM); west of Falkland Harbour, 15 m, *RILS 281* (BM). **Signy I.** Tern Cove, 2–3 m, *RILS 404* (AAS, BM, KRAM, LE); Factory Cove, 10 m, *RILS 10* (BM); Observation Bluff, 2–3 m, *RILS 8080* (AAS, KRAM); between Observation Bluff and Polynesia Pt., 15 m, *REL 1083* (AAS, KRAM, PC); Polynesia Pt., 10–13 m, *MWH 664B* (AAS, KRAM) and 5–15 m, *RILS 2000, 2004 & 5060* (AAS, KRAM); Rusty Bluff above Paal Harbour, 100 m, *RILS 350* (AAS, KRAM); Berntsen Pt., 4 m, *RW 87* (AAS, KRAM); Lenton Pt., 3 m, *RILS 8077* (AAS, KRAM).

SOUTH SHETLAND IS. **Aspland I.** 9 m, *JPB 75* (BM). **Eadie I.** *JSA 231* (AAS, KRAM). **King George I.** ***Admiralty Bay***: Agat Pt., 7–8 m, *RO 357/80 & 367/80* (KRAM); Demay Pt., 5–6 m, *RO 1393/80* (AAS, KRAM). ***Fildes Peninsula***: Atherton I., *BJ 115* (KRAM). ***Venus Bay***: Brimstone Peak, 10 m, *BJ 7* (KRAM). **Livingston I.** ***Byers Peninsula***: Laager Pt., 20 m, *KR 40D* (AAS, KRAM).

WEST ANTARCTIC PENINSULA. **Trinity Peninsula.** Astrolabe I., *RILS 4056 & 4074* (AAS, KRAM); Cape Roquemaurel, 10–50 m, *RILS 3929* (AAS, KRAM). **Danco Coast.** Charlotte Bay, Andrée I., *RILS 4000* (BM, PC, S). **Graham Coast.** Cape Tuxen, *RWMC 670* (AAS, BM, KRAM). **Wilhelm Archipelago.** ***Argentine Is.***: Uruguay I., *RWMC 596A* (AAS, KRAM, LE); the largest island between Irizar I. and Uruguay I., *RWMC 708* (BM). ***Collins Bay***: Berthelot Is., *RWMC 618* (AAS, KRAM).

36. ORTHOTRICHUM Hedw.

Orthotrichum Hedw., Spec. Musc. Frond.: 162. 1801. — Lectotype (*vide* Grout 1935: p. 106): *Orthotrichum anomalum* Hedw.

Brachytrichum Röhl., Deutsch. Fl. Ed. 2, Kryptog. Gew. 3: 47. 1813, *nom. illeg. incl. gen. prior.*

Dorcadion Adans. *ex* Lindb., Utkast Eur. Bladmoss.: 7, 35, 39. 1878, *nom. illeg. incl. lectotyp. gen.*

Orthomitrium Lewinsky-Haapasaari & Crosby, Novon 6: 2. 1996. — Type: *Orthomitrium tuberculatum* Lewinsky-Haapasaari & Crosby [= *Orthotrichum jetteae* B.H.Allen]. First synonymised by Allen (2002: p. 628).

Plants small to fairly large and robust, typically in small, dense or more or less loose tufts, cushions or mats, dull, stiff and rigid, olive-green, dark green, yellowish-green, brownish to black. *Stems* erect or ascending, forked or fastigiately branched, sparsely radiculose at the base, in transverse section without central strand and with a distinct, 1–2-stratose cortex of small, red-brown cells. *Leaves* erect-appressed to imbricate, contorted-flexuose or crisped when dry, spreading to recurved when moist, ovate, elliptical, lanceolate, ovate-, oblong- to linear-lanceolate, or lingulate, gradually acuminate, acute, subacute to rounded or rarely obtuse, occasionally aristate, piliferous, cuspidate or truncate, plane or channelled at the apex, plane to keeled, rarely undulate, not or decurrent; *margins* mostly reflexed, recurved to revolute, entire or rarely denticulate or crenulate near the apex, unistratose or rarely bistratose; *costa* single, strong, subpercurrent to percurrent, with no marked internal differentiation; *laminal cells* unistratose to partially or completely bistratose in the distal portion, opaque or translucent; *upper cells* isodiametric, mostly rounded to rounded-hexagonal, thick-walled, with 1–3 simple or branched papillae per lumen, rarely smooth; *basal cells* elongate, rectangular to linear, hyaline and pellucid, smooth, with moderately to strongly thickened and nodulose walls, becoming shorter towards the margins but not forming distinct borders. *Asexual reproduction* occasionally present as cylindric, club-shaped or rarely globular septate gemmae arising from leaves. *Gonio-* or *autoicous*, occasionally *dioicous*. *Setae* short to elongate, erect, straight, slender to stout, sinistrorse. *Capsules* immersed, emergent to exserted, erect, straight, symmetrical or, occasionally, slightly asymmetrical, ovoid to cylindrical, rarely globose, smooth or rarely 8- or 16-ribbed, often becoming deeply furrowed and sometimes strongly constricted below the mouth when dry, smooth or rarely puckered at the mouth, more or less gradually tapering to seta; *annulus* none or poorly differentiated and persistent; *operculum* stoutly rostrate from a convex or conic base; *exothecial cells* variable in shape, mostly elongate, thick-walled, often differentiated in coloured longitudinal bands, 3–8 cells wide corresponding to the ribs; *stomata* superficial or immersed in central and basal parts of the urn; *peristome* double, more rarely single or absent; *exostome teeth* 16 or fused in 8 pairs, broadly lanceolate to triangular, sometimes rudimentary, acute, obtuse or truncate at the apex, erect, recurved to reflexed when dry, finely to coarsely papillose or ridged-striate; *endostome segments* 8 or 16, or sometimes lacking, as tall as the teeth or shorter, filiform to broadly lanceolate, acute or obtuse, often with irregular margins, hyaline to yellowish, smooth to variously ornamented. *Spores* spherical, nearly always unicellular, more or less papillose. *Calyptra* campanulate-mitriform, covering most of the urn, usually plicate, smooth or rarely papillose, hairy or glabrous.

Orthotrichum is one of the most distinctive and readily recognisable of all moss genera, its immersed to short-exserted and often ribbed capsules with large, campanulate-mitrate and usually hairy calyptrae covering the entire capsules being a major characteristic of this genus, differentiating it from most other genera. The plants have orthotropic stems that are mostly sympodially branched and they most often exhibit typically xerophytic features, including small, isodiametric and mostly papillose cells with incrassate walls. This is associated with the ecological preferences of species of *Orthotrichum* which generally grow in open, sunny habitats, either epiphytically on bark of deciduous trees or epipetrically on rock faces and boulders. Various subgenera and sections of this genus have their own particular habitat preferences to which the species are more or less restricted (Vitt 1971*b*).

Orthotrichum is one of the largest of all moss genera, currently consisting of about 150 taxa, although originally about 250 species were given names in *Orthotrichum*. A worldwide synopsis of species and infrageneric taxa of *Orthotrichum* has been published by Lewinsky (1993) who recognised 116 species and 11 varieties distributed in seven subgenera. Since then, 26 new species were described in, or transferred to, *Orthotrichum* (Crosby & Margill 1997, 2004; Crosby 2000; Allen 2002; Norris *et al.* 2004; Erdağ *et al.* 2004; Lara *et al.* 2006; Garilleti *et al.* 2006; Goffinet *et al.* 2007).

The genus is cosmopolitan in distribution, with particularly high concentrations of species in cool and temperate areas of both hemispheres, whereas in the tropics species of *Orthotrichum* avoid very wet and desert lowland regions. *Orthotrichum* has attracted much attention and recently it has been revised taxonomically in most regions of the world, including North America (Vitt 1973), Greenland (Lewinsky 1977*b*), Svalbard (Frisvoll & Lewinsky 1981), Central and South America (Lewinsky 1985, 1987), Africa (Lewinsky 1978), Australasia (Lewinsky 1984), and Southeast Asia (Lewinsky 1992). Also, an illustrated key was provided for European species of the genus (Lewinsky-Haapasaari 1995).

In the moss flora of Antarctica *Orthotrichum* is represented by a single cosmopolitan species, *O. rupestre*. It belongs within the small subgenus *Phaneroporum* Delogn., comprising only six species and two varieties which are primarily saxicoles characterised by having erect to spreading peristome teeth and superficial stomata.

1. **Orthotrichum rupestre** Schwägr. Fig. 186

Orthotrichum rupestre Schwägr., Spec. Musc. Suppl. 1(2): 27, *t. 54*. 1816. ≡ *Dorcadion rupestre* (Schwägr.) Lindb., Musci Scand.: 29. 1879. — TYPE CITATION: Legit in Helvetiae rupibus Schleicher, in Carinthiae alpibus (Pasterze) ego ipse, in Gallia simillimum Persoonus. [Lectotype (*vide* Lewinsky 1983: p. 56): 'Orthotr. rupestre Pasterze, in Carinthiae alpibus' – G-Hedwig/Schwägrichen (*non vidi*)].

Orthotrichum rupicola Müll.Hal, Bot. Jahrb. Syst. 5: 80. 1883 ['-um']. — TYPE CITATION: all new species described by Müller (1883) under the heading 'Bryologia kerguelensis'; later Müller (1889) wrote: Ins. Kerguelen, Foundery branch, in rupibus siccis cum Orthotricho crassifolio et Orthotricho angustifolio; in addition, the specimen of *Orthotrichum rupestre* collected on Îles Kerguelen by A. E. Eaton and reported by Mitten (1876*b*: p. 194) is cited. [Lectotype (*vide* Ochyra 1986: p. 144): 'Ex Museo botanico Berolinensi. Orthotrichum rupicolum C. Müll. Kerguelen Islands, Gazellen-Expedition Dr. Naumann 1874' – BM-Bescherelle!; isotypes: BM!, H!, PC!]. First synonymised by Lewinsky (1978: p. 63).

Orthotrichum antarcticum Cardot, Rev. Bryol. 27: 43. 1900. — TYPE CITATION: [Antarctica] IXe débarquement; later, Cardot (1901) gave more detailed locality data: Détroit de Gerlache: IXème débarquement, Terre de Danco, près du cap Anna Osterrieth, sur les petites terrasses humides de la falaise. (N° 205^{c}; 29 janvier 1898). [Lectotype (*selected here*): 'Herb. J. Cardot. Mousses du voyage de la Belgica. N° 205c. Orthotrichum antarcticum Card. sp. nova. Canal de la Belgica: sur les petites terrasses détrampées de la falaise du IXème débarquement. Leg. Racovitza, 29 janvier 1898' – PC!; isotypes: BR!, S-Möller! S-Roth!]. First synonymised by Ochyra (1986: p. 144).

Plants medium-sized, 1.0–2.5(–3.0) cm tall, in dull, dense, rigid cushions or looser tufts, yellowish- or olive-green to yellowish-brown above, dark brown to blackish-brown below. *Stems* brown, fasciculately branched, densely radiculose below with tufts of brown, branched, smooth rhizoids, in transverse section rounded, with 1(–2) layers of smaller, thick- and brown-walled cortical cells surrounding 4–6 layers of larger, hyaline or yellowish medullary cells, central duct absent; *axillary hairs* numerous, rather short, 3–5-celled, with 1 or 2 brown basal cells. *Leaves* erect, appressed when dry, widely spreading to slightly recurved when wet, (2.3–)2.5–3.0(–3.2) mm long, 0.7–1.0 mm wide, slightly decurrent, lanceolate from an ovate base, long or short acute to acuminate, concave, keeled in the median part, unistratose with common bistratose strands in the upper portion; *margins* entire, recurved to revolute almost throughout, unistratose or intermittently bistratose in 1 or few rows of cells above; *costa* single, relatively strong, (55–)65–70(–80) μm wide at base, yellowish-brown, vanishing just below the apex, convex dorsally, canaliculate ventrally, 2–3-stratose in the upper and median part, 4-stratose below, in transverse section composed of more or less homogeneous cells; *laminal cells* partially bistratose distally, unistratose in the proximal portion; *upper cells* isodiametric to slightly elongate, rounded-quadrate to rhomboidal, 6–8(–10) μm wide, 6–13 μm long, with incrassate walls and 1–3 forked papillae per cell; *basal cells* elongate, mostly rectangular, 5.5–10.5 μm long, (20–)30–70 μm wide, thick-walled, nodose to almost straight-walled, yellowish- to orange-brown in older leaves; *alar cells* differentiated, thick- or thin-walled, isodiametric to elongate, forming somewhat inflated, hyaline to orange-brown decurrencies. *Gonioautoicous*. *Perichaetial leaves* not differentiated. *Setae* short, up to 1.7 mm long, straight, erect. [*Capsules* immersed to emergent, short cylindric, 8-ribbed in the upper part, not contracted below the mouth; *operculum* conic with a short, straight lid; *exothecial cells* slightly differentiated into stripes of the upper part of the capsule, subquadrate to rectangular, thick-walled, becoming transversely rectangular below the orifice; *stomata* numerous in the central part of the urn, superficial; *peristome* double; *exostome teeth* 16, entire or perforate along the middle line, yellow-orange, erect to slightly recurved when dry, densely covered with small papillae on the outer surface; *endostome segments* 8, short, one third of the height of the exostome, of one row of cells, hyaline, smooth. *Spores* globose, rather roughly papillose, 20–25 μm in diameter.] *Calyptra* oblong-conic, yellow, smooth, with numerous straight hairs.

Chromosome numbers: unknown in the Antarctic and Subantarctic; $n = 6$ (4 records) and $n = 12$ (1 record) from the Holarctic (Fritsch 1991).

Discussion. When fertile, *Orthotrichum rupestre* is readily recognisable by its immersed to emergent capsules, superficial stomata and erect to spreading peristome teeth in the dry condition. In Antarctica mature sporophytes are unknown, and therefore the species may be confused with other taxa, especially *Muelleriella crassifolia* or species of *Schistidium*, but the smooth laminal cells in the latter distinguish them from *O. rupestre*. The papillose leaves of *O. rupestre* approach those of the Antarctic species of *Syntrichia*, especially *S. saxicola* and *S. filaris* with which it additionally shares the general habit. However, the percurrent or excurrent costae and smooth, hyaline basal cells abruptly separated from the upper chlorophyllose cells readily distinguish these species.

Orthotrichum rupestre was first collected in Antarctica by E. Racovitza during the *Belgica* Expedition and described by Cardot (1900, 1901) as *O. antarcticum*. However, Ochyra (1986) reduced the latter name to synonymy with *O. rupestre*. Lightowlers (1987) reported this species from Cuverville I., a small island situated in the same region from which the type of *O. antarcticum* was collected.

Orthotrichum rupestre was described by Schwägrichen (1816) but the name itself was published a decade earlier as a *nomen nudum* in *Plantae cryptogamicae helveticae*, a series of the exsiccati distributed by Schleicher (1805, 1807*a*,*b*). It has been common practice to ascribe the authority of *O. rupestre* to 'Schleich. *ex* Schwägr.' but it is contrary to Art. 46.3 and 46.4 of the current Code (McNeill *et al*. 2006) because the citation of a name in a list of synonyms cannot be considered ascription. Therefore, the name *O. rupestre* should be ascribed to Schwägrichen alone.

Reproduction in Antarctica. Plants with immature sporophytes have been found only once, in the population from Darboux I., off the Graham Coast, at the southernmost locality of the species. All other plants are consistently sterile.

Habitat. The habitats occupied by *Orthotrichum rupestre* in Antarctica are similar to those elsewhere in its range. It grows on rock faces, crevices and ledges in open and dry, usually shaded situations which are occupied by communities of the fruticose lichen and moss cushion subformation. It forms dense and compact cushions, more rarely looser mats which are practically pure, but occasionally associated with *Sanionia uncinata*, *Pohlia cruda* and *Bryum pseudotriquetrum*. Like *Muelleriella crassifolia* it occurs close to the shore and in some sites also occurs within the spray zone.

World range. *Orthotrichum rupestre* is a cosmopolitan oreophyte having the widest, though quite disjunct, geographical range of all species of the genus (Fig. 187, inset). It is widely scattered throughout mountainous regions of the Holarctic including North America (Vitt 1991; Crum & Anderson 1981), the southernmost tip of Greenland (Lewinsky 1977*b*), most regions

→

Figure 186. *Orthotrichum rupestre*. – **1**. Habit. **2**. Portion of branch with sporophyte. **3**. Sterile branch with perigonia. **4**. Portion of cross-section of stem. **5–9**. Axillary hairs. **10–14**. Vegetative leaves. **15**. Leaf apex. **16**. Upper leaf cells at margin. **17**. Mid-leaf cells. **18**. Alar cells. **19**. Basal juxtacostal cells. **20–22**. Cross sections of the leaves, a sequence from the apex to the base. **23**. Deoperculate capsule, dry. **24**. Stoma. **25**. Exothecial cells above middle of urn. **26**. Portion of peristome, spores and exothecial cells below mouth. – [1–2, 23–26 from *Naumann s.n.*, 1874 (type of *O. rupicola*), BM; 3–9, 11–12, 14–19 from *VK 599*; 21–22 from *VK 519*; 10, 13, 20 from *ER 205C* (type of *O. antarcticum*), BR]. – Scale bars: a – 0.5 cm (1); b – 100 μm (4–9, 15–22, 24–26); c – 1 mm (10–14, 23); d – 1 mm (2–3).

1
2
3
4
5
6
7
8
9
10
11
12
13
14
15
16
17
18
19
20
21
22
23
24
25
26
a b c d

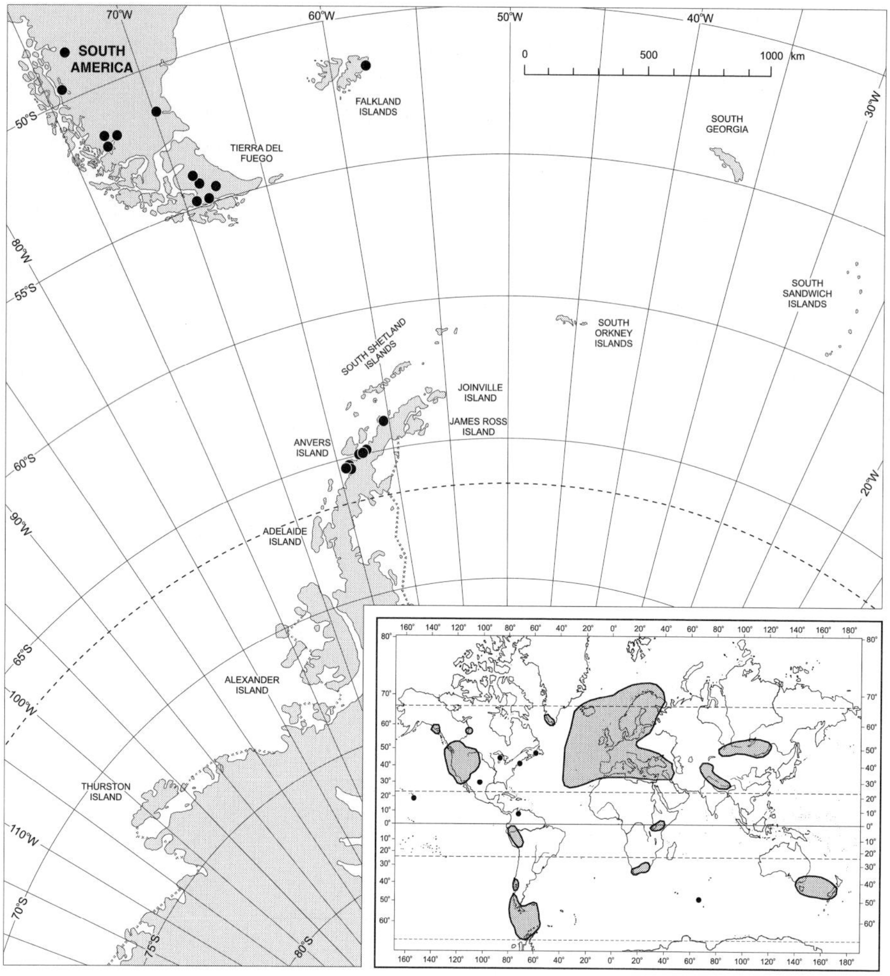

Figure 187. Distribution of *Orthotrichum rupestre* in the Fuegian region and Antarctica. Inset: global distribution.

of Europe eastwards to the Caucasus, Central Siberia (Ignatov & Lewinsky-Haapasaari 1994) and Mongolia (Abramova & Abramov 1983; Tsegmed 2001*a*,*b*), as well as Northwest India and West China (Lewinsky 1992). Outside the Holarctic the species occurs in South America extending along the Andean chain from Venezuela to Tierra del Fuego (Lewinsky 1987), in East and South African mountains (Lewinsky 1978) and Australasia (Lewinsky 1984). In the austral polar region it is known only from subantarctic Îles Kerguelen, from where it was originally described as *O. rupicola* (Müller 1883, 1889), and in the northern Antarctic Peninsula region.

Distribution in Antarctica. *Orthotrichum rupestre* is currently known from a very restricted area on the Danco and Graham Coasts on the western side of the Antarctic Peninsula, extending from 64°09′S at Cierva Pt. to 65°24′S on Darboux I. (Fig. 187). The species grows at low elevations, from near sea level to about 25 m, occasionally ascending to 100 m.

Specimens examined. WEST ANTARCTICA. WEST ANTARCTIC PENINSULA. DANCO COAST. Cierva Pt., Primavera Station, *VK 184, 488 & 599* (KRAM); Cape Anna, *ER 205C* (BR, PC, S – type of *Orthotrichum antarcticum*) & *151D* (BR, PC); Cuverville I., 2–20 m, *RILS 4170* (AAS), *4180* (AAS, BM, KRAM) *& 4181* (AAS, KRAM); Coughtrey Peninsula, Paradise Harbour, Almirante Brown Station, *VK 395* (KRAM). **GRAHAM COAST.** Cape Tuxen, 100 m, *REL 1356* (AAS, KRAM). **WILHELM ARCHIPELAGO.** ***Penola Strait***: Petermann I., 13 m, *RWMC 729* (AAS, KRAM) and 25 m, *RWMC 730B* (AAS, KRAM). ***Collins Bay***: Berthelot Is., 23 m, *RWMC 617* (AAS, KRAM); Darboux I., 10–15 m, *RILS 3315* (AAS, KRAM).

Subclass C5. **BRYIDAE** ENGL.

Bryidae Engl., Syllabus Grosse Ausg.: 47. 1892 ['Unterklasse Bryales'].

The subclass Bryidae is roughly equivalent to the group of mosses that was described by Philibert (1884) as 'Diplolépidées' (Taylor 1962). Fleischer (1904, 1908) subsequently latinised this term as the Diplolepideae and since then it has been firmly established in bryology for the designation of taxa having a double peristome, rather than a single one as in the two other major groups of the Arthrodontei, namely Haplolepideae and Heterolepideae. The Diplolepideae *sensu* Fleischer (1904, 1908) and Dixon (1932*a*) are a heterogeneous group since these authors neglected the basic differences in the structure of the diplolepidous peristomes and failed to distinguish the *Funaria*, *Orthotrichum* and *Bryum* peristome-types.

The Bryidae comprise taxa that are characterised

by having the *Bryum* peristome-type, i.e. diplolepidous alternate peristome which is the most complicated and best elaborated of the diplolepidous type. The exostome teeth are more strongly thickened, sturdy and distinctly trabeculate on the ventral (inner) surface and most have a reticulate pattern of ornamentation on the outer (dorsal) surface. The endostome typically has a well developed high basal membrane that is divided above into 16 broad segments that lie alternate to the exostome teeth and often develops one to three or even four cilia between each segment which are positioned opposite the outer teeth. The basal membrane is typically keeled with the 16 keels continuing upward as processes along the median lines of the segments. The segments are often split or widely perforated along the keel. In well developed endostomes the cilia may be nodulose to more or less appendiculate. In less well developed endostomes, the cilia are fewer and never appendiculate or may be absent, the segments are narrow and without a keel, often approaching the cilia of well developed endostomes in appearance, and the basal membrane is very low or virtually absent.

In the *Bryum* peristome-type a pattern 4:2:6 is common, but a variable number of cell divisions in the IPL results in peristomial formulas ranging from 4:2:8 14 in the most intricate peristomes to 4:2:4–6 or even 4:2:2 in simpler types (Edwards 1979). There is a large range of variations and reductions in this type of peristome, particularly in the presence and number of cilia and height of the basal membrane, but sometimes the endostome or exostome is entirely lacking, resulting in the single exostomial or endostomial peristome in some taxa.

The core of the subclass Bryidae constitutes taxa that are characterised by their acrocarpous growth form, with perichaetia produced terminally on main shoots or on short lateral branches. However, some recent classifications (e.g. Buck & Goffinet 2000; Goffinet & Buck 2004) also incorporate some families with creeping stems and lateral sporophytes, or erect stems and dendroid growth form that have palmate, pinnate or umbellate branching pattern, namely the Spiridentaceae, Cyrtopodaceae, Pterobryellaceae, Racopilaceae, and Hypnodendraceae. Yet, the position of some of these families in classifications is inconsistent, and in various systems they have been placed either in the Bryidae (e.g. Fleischer 1920, 1923; Brotherus 1924*b*; Buck & Goffinet 2000; Goffinet & Buck 2004), in the orders Leucodontales (e.g. Touw 1971; Vitt 1982*a*, 1984; Buck & Vitt 1986; Norris & Koponen 1996) or Hypnodendrales (Bell *et al.* 2007) of the pleurocarpous subclass Hypnidae.

The placement of these pleurocarpous families in the Bryidae is based mostly on molecular data which, unfortunately, currently do not fully resolve the problem of the relationships of these pleurocarpous families (Buck *et al.* 2000*a*,*b*; De Luna *et al.* 2000; Tsubota *et al.* 2002). Therefore, in the present treatment, the Bryidae are interpreted as including only acrocarpous and cladocarpous mosses with a ciliate alternate diplolepidous peristome-type, whereas the pleurocarpous families with the bryalean peristomes are dispersed within orders of the subclass Hypnidae as suggested by Vitt *et al.* (1998).

Goffinet & Buck (2004) placed five orders in the Bryidae which have subsequently been included in two superorders. These are the Splachnales, Orthotrichales, Hedwigiales, and Bryales which constitute the superorder Bryananae, and the Rhizogoniales which are placed in the monotypic superorder Rhizogonianae. Following the concept of Vitt *et al.* (1998) and Ochyra *et al.* (2003), the Orthotrichales and Hedwigiales are placed in the separate subclass Orthotrichidae. Because molecular data place members of the Splachnaceae close to the orthotrichalean clade (Vitt *et al.* 1998), the subclass Splachnidae[1] is here recognised to accommodate them.

The placement of the Splachnales in a separate subclass is also strongly supported by morphological characters, in particular by peculiar features of the sporophyte in splachnalean mosses, primarily by the unique structure of the peristome. Although highly modified and aberrant, it is of the diplolepidous opposite type, so altogether the splachnalean peristome has more in common with that of the *Funaria*-type than the *Bryum*-type (Koponen 1978, 1982; Edwards 1979; Vitt 1984). Basically, this peristome is double, with the endostome either adhering and fused to the exostome (in *Splachnum*) or, most often, lacking, this making the peristome single and exostomial. In some taxa (*Moseniella* and *Voitia*) the peristome is absent. The basic number of the exostome teeth in the Splachnales is 16, but these can be fused into 8 tooth-pairs or split into 32 divisions. There is also great variation in the thickness of deposition on the peristomial layers which causes the teeth to be erect, reflexed, or inflexed. The OPL is often much more thickened than the PPL and its outer wall is mostly baculate. Outstanding features of splachnalean mosses are the often highly differentiated neck of the capsule which may be greatly expanded and skirt- or umbrella-shaped or strongly swollen and bladder-shaped, as well as a high nitrogen requirement. Additionally, odours emanating from capsules of *Splachnum* and *Tetraplodon* appear to attract flies, resulting in entomophily in these genera (Koponen & Koponen 1978).

Of the remaining two orders of the Bryidae, the Rhizogoniales appear to comprise a heterogeneous assemblage of families and genera, since Goffinet & Buck (2004) placed in this order, both the typically acrocarpous families Rhizogoniaceae and Calomniaceae, as well as typically pleurocarpous families with the bryalean peristome-type, such as the Cyrtopodaceae, Hypnodendraceae, and Pterobryellaceae. By exclusion of these families and transferring them to the Hypnidae, recognition of the Rhizogoniales as a separate order has a weak foundation and appears to be redundant. This is because taxa of the Rhizogoniaceae

[1] Subclass Splachnidae Ochyra, *stat. nov.* Based on Eubryales suborder Splachnineae M.Fleisch., Hedwigia 61: 394 1920 ['Unterreihe Splachnineae'] [Subclass Splachnidae Vitt, Goffinet & Hedd. *in* Bates, Ashton & Duckett, Bryol. Twenty-First Cent.: 119. 1998, *nom. nud.*]

and Calomniaceae perfectly fit the concept of the Bryales, and some genera of the former family, namely *Leptotheca*, *Cryptopodium*, *Pyrrhobryum*, and *Hymenodontopsis*, have been nested in the Mniaceae (Koponen 1988) which is a typically bryalean moss family. Therefore, the Rhizogoniales are here united with the Bryales.

Consequently, the subclass Bryidae is here interpreted as a monotypic taxon, consisting of the single order Bryales, which is significantly diverse gametophytically and sporophytically and therefore nearly impossible to characterise. This is a very large group that comprises about 2000 species distributed worldwide, making it the third largest subclass of true mosses, after the Hypnidae and Dicranidae.

Order IX. **BRYALES** LIMPR.

Bryales Limpr. *in* Cohn, Krypt.-Fl. Schlesien 1(1): 55. 1876 ['Ordnung Bryineae'].

Rhizogoniales Goffinet & W.R.Buck *in* Goffinet, Hollowell & Magill, Monogr. Syst. Bot. Missouri Bot. Gard. 98: 235. 2004, ***syn. nov.*** Validly published by reference to Bryales suborder Rhizogoniineae M.Fleisch., Musci Buitenzorg 4: xvi. 1923.

Taxa of the Bryales are characterised by having the classic diplolepidous alternate peristome which is designated as the *Bryum*-type. In spite of this common developmental feature, there is great variation in the form, complexity, and ornamentation of mature peristomes within families and genera, and even sometimes within otherwise well defined species. The peristome is basically double and ciliate, but occasionally some taxa have a single exostomial or endostomial peristome due to reductions of the endostome or exostome, respectively. Sometimes the peristome is entirely absent, and in relatively few taxa of the Bryales the capsules are cleistocarpous.

Gametophytically, bryalean mosses are significantly variable and all main features of leaf shapes and areolation are present in this group in different combinations. The leaf areolation is lax or firm and the leaf margins are bordered or unbordered. The laminal cells vary from isodiametric to linear and, in general, they are predominately smooth, while papillose, mammillose or prorulose cells are rare. Likewise, alar cells and hyaline hair-points are very infrequent. The leaves are variable in size and shape, ranging from elliptic, rounded or ovate, through lanceolate or linear-lanceolate to setaceous. The foliage is generally terete but sometimes the leaves are conspicuously ranked or, occasionally, complanate. The costa is single and usually strong, mostly percurrent to excurrent, and usually internally differentiated into well developed guide cells and stereid cells. Asexual reproduction is relatively frequent in some genera by multicellular propagules produced in leaf axils, terminal clusters or as rhizoidal tubers. Sporophytes in the Bryales also exhibit remarkable variability. They are produced terminally or seldom laterally, and the capsules are generally exserted and only seldom immersed. They are erect, inclined, horizontal to pendulous, globose to cylindrical, smooth to furrowed and often have a differentiated apophysis.

The Bryales are the second largest order of the diplolepideous mosses, after the Hypnales, which in the number of species (about 1500) is similar to the Pottiales, the second largest order of the haplolepideous mosses after the Dicranales. Depending on taxonomic concepts its species are classified into 13–15, mostly mono- or oligogeneric families and 55–60 genera. These numbers are not definite since the two largest families of the Bryales, Bartramiaceae and Bryaceae, are not well studied taxonomically and very probably many reductions of species names to synonymy, as well as the fragmentation of some large genera, are to be expected. In the Antarctic the order is represented by two families, Bryaceae and Meesiaceae, that comprise 16 species belonging within five genera.

Family XI. **BRYACEAE** SCHWÄGR.

Bryaceae Schwägr. *in* Willd., Spec. Pl. Ed. 4, 5(2): 47. 1830 ['Fam. Bryi']. — Type: *Bryum* Hedw.

Mielichhoferiaceae Schimp., Coroll. Bryol. Eur.: 62. 1856. — Type: *Mielichhoferia* Nees & Hornsch.

Pleurobryaceae Schimp., Syn. Musc. Eur.: 325. 1860, *nom. illeg.*

Rhodobryaceae Nyholm, Ill. Fl. Nord. Mosses 3: 220. 1993, *non rite publ. descr. angl.*

Orthodontiaceae Goffinet *in* W.R.Buck & Goffinet *in* A.J.Shaw & Goffinet, Bryoph. Biol.: 104. 2000, ***syn. nov.*** Validly published by reference to Bryaceae subfam. Orthodontioideae Broth. *in* Engl., Nat. Pflanzenfam. Ed. 2, 11: 347. 1924. — Type: *Orthodontium* Schwägr.

The Bryaceae are a family of small to medium, less often large and robust, perennial acrocarps, usually associated in tufts and sometimes with subterranean stolons giving rise to erect stems. With few exceptions, for example the genera *Leptobryum* and *Orthodontium* with long-subulate, linear-lanceolate to setaceous leaves and *Epipterygium* with complanate and dimorphic leaves, gametophytes of bryalean mosses are constructed according to an established pattern and consequently lack distinct taxonomic features. The stems are mostly short and erect, simple or sparsely branching, often by subfloral innovations, with terete foliage of leaves arranged in many rows, usually matted with rhizoids in the lower part and in cross-section normally with a central strand. The leaves are often small, scale-like and distant below, progressively increasing in size upwards and often becoming crowded above, sometimes in comose tufts. They are erect to erect-spreading, rarely altered when dry, lanceolate, linear-lanceolate, ovate, obovate or ovate-lanceolate, rarely oval, elliptic to round, mostly acute and sometimes piliferous or hair-pointed, plane or concave, decurrent or non-decurrent. The margins are unbordered or bordered by narrow, elongate and thick-walled cells, unistratose or, very seldom, bi- to multistratose and forming a swollen limbidium, entire or toothed, plane or recurved. The costa is always present, single and mostly strong, percurrent or excurrent, less often ceasing well below the apex, usually with internal differentiation into well developed guide and stereid cells. The laminal cells are firm or lax, smooth and typically translucent, unistratose, relatively large, rhombic to rhomboidal-hexagonal or sometimes narrower and vermicular in the distal portion, rectangular to quadrate and sometimes pigmented in the proximal part.

The inflorescences in bryalean mosses are terminal on the main shoots or sometimes on short basal branches. The setae are elongate, erect, straight or curved near the tip, single or sometimes multiple in the perichaetium. There is extensive variation in capsule shape and orientation but, nevertheless, they best characterise the family. They are mostly elongate with a well developed apophysis that is wrinkled when dry and gradually tapered towards the seta. The capsules are frequently cernuous, inclined to horizontal, less often suberect to erect, symmetric or somewhat curved and usually smooth. Their shape varies broadly and ranges from clavate or pyriform through ovoid, obloid or globose to cylindrical or linear-cylindrical. The lids are convex to short-conic, umbonate to apiculate, or sometimes short-rostrate. The annulus is usually present, large and revoluble, or sometimes not revoluble. The peristome in the Bryaceae is remarkably variable and the degree of variation is greater than in any other family of arthrodontous mosses. The traditional generic limits in this family are often based on its and other sporophyte characters (Shaw & Crum 1982, 1984; Shaw & Fife 1984; Shaw 1984*b*, 1985*b*,*d*). The peristome varies from the perfect diplolepidous *Bryum*-type with 16 narrowly triangular exostome teeth and the endostome with a high basal membrane, broad and perforate segments and well developed, nodose or appendiculate cilia, through any conceivable type of reduction of the exostome and/or endostome to entirely absent. There is also extensive variability in the ornamentation pattern of the dorsal exostomial surface, ranging from the most common reticulate-papillose type through papillose or striate to entirely smooth (Shaw 1985*b*).

The Bryaceae were established relatively early as a separate family by Schwägrichen (1830), but it was then a heterogenerous taxon comprising, apart from the truly bryalean genera (*Bryum*, *Pohlia*, *Plagiobryum* and *Acidodontium*), also species of *Timmia* (Timmiaceae), *Meesia* and *Amblyodon* (Meesiaceae). The broadly conceived Bryaceae, including taxa now classified in various families of the Bryales such as the Mniaceae, Aulacomniaceae, Bartramiaceae and Meesiaceae, were accepted in all major pre-1880 treatments of mosses (Schimper 1856, 1860, 1876; Milde 1869; Limpricht 1876). The modern concept of this family based upon European taxa was outlined by Lindberg (1879), Braithwaite (1890–1893) and Limpricht (1891–1892), whereas Brotherus (1903–1904, 1924*b*) described the family on a worldwide scale.

The Bryaceae were divided into three subfamilies (Brotherus 1924*b*), the Orthodontioideae (lacking basal membrane and with linear segments of the peristome), the Mielichhoferioideae (with lateral inflorescences and exostome, endostome, or both, reduced or none), and the Bryoideae (with well developed exostome and endostome). Andrews (1935) subdivided the latter into the Pohlioideae, having long, narrow cells, and the Bryoideae in the strict sense, having shorter, rhombic cells, but they were designated as 'groups', not subfamilies, as claimed by some authors (e.g. Ochi 1959; Shaw 1985*b*; Griffin 1994*a*). However, the 'group' Pohlioideae has been adopted as a subfamily by Ochi (1959), but this name is not validly published due to the lack of the Latin diagnosis or a reference to such because Andrews (1935) gave only a brief English description of the 'group' Pohlioideae. The distinction between the subfamilies are not absolute and similarities in gametophyte characters relate genera of various subfamilies and markedly obscure limits between them, in spite of differences in peristomial structure (Pedersen 2000).

A considerable number of phylogenetic analyses based upon molecular studies has been published recently on the Bryaceae (e.g. Cox & Hedderson 1999, 2003; Cox *et al.* 2000; Pedersen & Hedenäs 2002*a*, 2005; Pedersen *et al.* 2003, 2007; Holyoak & Pedersen 2007). They showed that the family is a heterogeneous taxon and, as a result, some typically bryalean genera have been transferred elsewhere, for example *Leptobryum* to the Meesiaceae (Newton *et al.* 2000) and *Pohlia* and *Schizymenium* to the Mniaceae (Buck & Goffinet 2000; Goffinet *et al.* 2001; Goffinet & Buck 2004). However, phylogenetic analyses based on morphological and anatomical characters give topologies that are incongruent with the results of the molecular studies (Pedersen 2000; Pedersen & Hedenäs 2002*a*; Holyoak & Pedersen 2007). Nonetheless, remarkable changes in the taxonomic concepts of some traditional genera have been proposed based exclusively on chloroplast DNA evidence (Pedersen & Hedenäs 2005; Pedersen 2005; Holyoak & Pedersen 2007). These new ideas appear to be inconclusive because of the low confidence levels of some gene analyses and, in many cases, morphologically-based cladograms are more reliable than those based on molecular analyses (Zander 1998*b*, 2001, 2003, 2004; Yoshiyuki *et al.* 2002; Douady *et al.* 2003). Until more convincing evidence for reclassification of the Bryaceae is available, the traditional circumscription of the bryalen genera is retained here for practical reasons.

The Bryaceae are the largest family of the order Bryales, consisting of about 1000 species which are classified into 16–18 genera, although it is suggested that some large and heterogeneous genera should be split into smaller and more natural taxa (Spence 1987, 1996, 2005; Pedersen & Hedenäs 2005; Pedersen 2005). This family is taxonomically one of the most difficult and identification at the species and genus levels often requires good sporophyte material. Sterile plants sometimes cannot be determined even to genus level without extensive prior knowledge of the family. This is particularly so in the Antarctic biome where many bryalean mosses seldom produce mature sporophytes and, at the same time, some species are dominant constituents of some plant communities.

In the Antarctic botanical zone 15 species of the Bryaceae are known and these belong to four genera. A similar number of species is known from the Subantarctic but the family remains poorly studied taxonomically in this region, so the final number of taxa may change as taxonomic investigations and phytogeographical exploration progress.

KEY TO THE GENERA OF THE ANTARCTIC BRYACEAE

1. Leaves dimorphic, slenderly lanceolate below and setaceous from a broad, clasping base above; costa broad occupying about ⅓ width of the leaf base .... 37. ***Leptobryum***

1. Leaves undifferentiated, relatively broad, ovate, ovate-lanceolate to orbicular, lacking a broad, clasping base; costa narrow occupying ¼ or less width of the leaf base 2

  2. Perichaetia on short, lateral branchlets; laminal cells fusiform with spindle-like ends, 6–8(–10) µm wide; peristome absent or sometimes only rudimentary endostome present; leaves narrow, 0.3–0.6 mm wide, strongly lustrous, usually with a reddish hue .... 38. ***Schizymenium***

  2. Perichaetia terminal on main shoots; laminal cells rhomboidal to oblong-rhomboidal, rarely elongate and sinuose, with oblique, truncate or rounded apices, 9–14(–24) µm wide; peristome double; leaves broader, 0.4–1.3 mm wide, dull or lustrous, greenish to brownish 3

3. Leaves not distinctly bordered; laminal cells oblong-rhomboidal to linear-flexuose, relatively long, 4 : 1 or longer .... 39. ***Pohlia***

3. Leaves distinctly bordered, if unbordered then the laminal cells rather broad and short, rhomboidal-hexagonal, generally less than 4 : 1 .... 40. ***Bryum***

37. LEPTOBRYUM (SCHIMP.) WILSON

Leptobryum (Schimp.) Wilson, Bryol. Brit.: 219. 1855. ≡ *Bryum* Hedw. [unranked] *Leptobryum* Schimp. *in* Bruch, Schimp. & W.Gümbel, Bryol. Eur. 4: 1. 1851 [Fasc. 46–47 Consp. Vol. 4: iii]. ≡ *Bryum* subgen. *Leptobryum* (Schimp.) Kindb., Rev. Bryol. 36: 117. 1909. — Type: *Leptobryum pyriforme* (Hedw.) Wilson (*Webera pyriformis* Hedw.).

Plants small and slender, loosely or densely caespitose, occasionally gregarious, soft, lustrous, light- or yellow-green. *Stems* erect, simple, thin, radiculose at the base or in the lower half with finely papillose, branched, pale reddish-brown to purplish rhizoids, usually bearing tubers, in transverse section rounded-triangular, consisting of a large central strand, 2–3 layers of large, hyaline thin-walled medullary cells with distinct corner thickenings and a 1–2-stratose cortex of small, substereid cells with thick, dark yellow- to reddish-brown walls; *axillary hairs* 3–4-cellular, with 2 rectangular, maroon-red basal cells and 1–2 oblong, hyaline and readily deciduous distal cells. *Leaves* dimorphic, non-decurrent, canaliculate; *lower leaves* distant, narrowly lanceolate, acute to acuminate, erect-spreading from the insertion on wetting, becoming abruptly larger and crowded on the stem tips and forming a comal tuft, erect- to wide-spreading, sometimes erect to secund, flexuose on drying, setaceous to long-subulate from a broader, oblong, clasping base, gradually long-acuminate, acute at the apex; *margins* plane, entire below, serrulate to subentire above, unbordered, uni- or occasionally bistratose at places in the distal portion; *costa* strong, not sharply demarcated from the laminal cells, broad, filling about ⅓–⅔ the leaf base and most of the subula, percurrent to shortly excurrent, in transverse section depressed to crescent-shaped, weakly differentiated distally, with 2–5 large guide cells, 2–3-layered band of substereid cells on the dorsal side and lacking a ventral stereid group and with a layer of larger epidermal cells on the ventral surface; *laminal cells* smooth, thin-walled, unistratose except for occasional bistratose marginal patches above, long-linear in the subula, long-rectangular in the base, not differentiated in the basal angles. *Asexual reproduction* by abundant multicellular, obloid-ellopsoid to globose, reddish to dark brown tubers with non-protuberant cells, occurring on main rhizoids or sometimes on rhizoidal stalks in the axils of the lower leaves. *Synoicous*. *Perichaetial leaves* undifferentiated, similar to the upper vegetative leaves. *Setae* terminal, elongate, slender, flexuose-curved, smooth, reddish-yellow. *Capsules* long-exserted, strongly inclined to pendulous, pear-shaped from a long, narrow apophysis, narrowed at the mouth, smooth, leptodermous, lustrous, brownish-yellow, becoming reddish-brown with age; *annulus* biseriate, revoluble, composed of large, thick-walled, vesiculose cells; *operculum* conic-convex, mammillate; *exothecial cells* thin-walled, rectangular, becoming quadrate to oblate in several rows at the orifice; *stomata* numerous in the neck, superficial, bicellular, long-pored, slightly raised; *peristome* double; *exostome teeth* 16, lanceolate, yellowish, densely papillose on the outer surface, trabeculate on the inner side; *endostome* about as long as the exostome teeth, yellowish-hyaline, finely papillose, with a high basal membrane about half the length of the exostome teeth, well developed, keeled, broadly perforate segments and (1–)2–4 appendiculate cilia. *Spores* spherical, finely papillose. *Calyptra* cucullate, smooth, naked.

Leptobryum is here interpreted as a monotypic genus that consists of a single species, *L. pyriforme*. Arts (2001) placed in this genus a second species, *L. wilsonii* (Mitt.) Broth., but it is now transferred to *Pohlia* (see discussion under this species). Such interpretation makes *Leptobryum* one of the most characteristic and easiest to recognise of all bryalean mosses, both when sterile and fertile. Gametophytically, in its slenderly lanceolate to setaceous leaves that strongly clasp the stem, *Leptobryum* resembles some dicranaceous mosses of the genus *Dicranella*, but the shiny pyriform capsules with long and slender apophyses and perfect diplolepideous alternate and ciliate peristomes of the *Bryum*-type are identical to those found in many bryalean mosses. This combination of characters defines *Leptobryum* as a distinct genus. Because the only species of the genus, *L. pyriforme*, is synoicous, the fully mature sporophytes are commonly produced throughout its wide geographical range, including Antarctica.

The safest character to distinguish the sterile plants of *Leptobryum* and *Dicranella* are axillary hairs. They consist of 3–4 cells in the former, of which the lower two are rectangular, persistent and dark red or maroon-red pigmented, whereas the upper one or two cells are oblong to claviform, hyaline and mostly fragile and deciduous. In contrast, the axillary hairs in *Dicranella* are longer and filiform, 3–7-celled with two brown-pigmented basal cells. Axillary hairs similar to those in *Leptobryum* are typically found in other unrelated families of mosses, including the Meesiaceae, Splachnaceae, Funariaceae, as well as in the Mniaceae and Aulacomniaceae. Nonetheless, Zolotov & Ignatov (2001) considered the axillary hairs to support the alignment of *Leptobryum* within the Meesiaceae, as was implied by molecular data (Newton *et al.* 2000; Cox *et al.* 2000). However, it does not seem that the axillary hairs alone would be a reliable character deciding the relationships of genera and families. Hedenäs (1990) studied the axillary hairs in many pleurocarpous mosses and showed they are not specific to families in the Hypnidae.

Further, molecular data show the close alliance of the Meesiaceae with the Splachnidae. Such relationship has weak foundations and no morphological support since both splachnalean and bryalean mosses have completely different peristome-types. The former have diplolepidous-opposite peristomes, whereas the latter are characterised by having diplolepidous-alternate peristomes and it is difficult to accept a close affinity of taxa with a distinctly different peristome structure and development.

Even if the Meesiaceae are retained in the Bryales, *Leptobryum* is a discordant element in this family, and the only characters that unite this genus with it are the capsules with distinct necks and axillary hairs with club-shaped distal cells. On the other hand, the Meesiaceae have an unusual double peristome which is basically bryalean but highly modified. The exostome teeth are short, usually less than half of the endostome length, slightly horizontally striate on the dorsal surface, and the endostome has a short basal membrane, narrow segments with vertically elongate dorsal lamellae and secondary thickenings in irregular, often discontinuous, yellow patches, as well as no or rudimentary cilia. The peristomial features imply a closer affinity of *Leptobryum* to the Bryaceae than the Meesiaceae, and this traditional placement of this genus is retained here.

Leptobryum was recognised as a separate taxon in a little obscure situation by Schimper (1851) as an unranked subdivision of *Bryum* in the conspectus to genera and species published in the fourth volume of *Bryologia europaea*. The reference to plate 355 constitutes a reference to the short description published in the treatment of *Bryum* by Bruch *et al.* (1839). However, Schimper alone should be considered as the author of the name *Leptobryum* since P. Bruch, the co-author of taxonomic treatments in volumes I–IV of this opus, died in 1847 (Barnhart 1944). *Leptobryum* was recognised as a genus in its own right almost simultaneously, by Wilson (1855) and Schimper (1856). The former provided a short description of the genus and albeit he gave no direct reference to Schimper's (1851) conspectus, the authority of this generic name has to be '(Schimp.) Wilson' under Art. 33.3 of the current Code (McNeill *et al.* 2006). Furthermore, Wilson (1855) cited *Bryum* (*Leptobryum*) *pyriforme* in synonymy of *Leptobryum pyriforme*, with a reference to the conspectus in *Bryologia europaea*, and this is a direct link to the basionym. Schimper (1856) also ascribed the generic name *Leptobryum* to himself.

1. **Leptobryum pyriforme** (Hedw.) Wilson

Fig. 188

Leptobryum pyriforme (Hedw.) Wilson, Bryol. Brit.: 219. 1855. ≡ *Webera pyriformis* Hedw., Sp. Musc. Frond.: 169. 1801. ≡ *Mnium pyriforme* (Hedw.) Funck, Crypt. Gew. Fichtelgeb. 3: no. 61. 1802. ≡ *Bryum pyriforme* (Hedw.) P.Gaertn., B.Mey. & Scherb., Oekon. Fl. Wetterau 3(2): 105. 1802. ≡ *Hypnum pyriforme* (Hedw.) F.Weber & D.Mohr, Bot. Taschenbuch: 274. 1807. ≡ *Pohlia pyriformis* (Hedw.) Müll.Hal., Deutschl. Moose: 507. 1853 ['pyriforme'], *nom. inval. in synon.* — Type citation: Locis lapidosis, sabulosis, subhumidis, in Germania; also references are given to Dillenius (1741: t. 50, f. 60), Linnaeus (1763: p. 1112), Schreber (1771: p. 81), Murray (1784: p. 947), and Hedwig (1785–1787: p. 8); in the latter the following specimen is cited: 'Eadem circumstantia me quoque seduxit olim, cum uberior huius speciei disquisitio Chemnitii non concederetur ob rarissimum eius in illa regione proventum, ut pro Bryo vero haberem venditaremque in mea Muscorum frondosorum Historia Naturali P. II. p. 94.'), and Ehrhart (1791: no. 165). [Lectotype (*selected here*): the upper left specimen on the sheet bearing the label 'Webera pyriformis Hedw. St. Cr. 1 p. 5. t. 3 Bryum Dill. H. M. 60 Mnium pyriforme Linn. Chemnitzii partim, partim Lipsiae lecta' – G-Hedwig/Schwägrichen!].

Plants small, densely tufted or scattered, soft and slender, strongly glistening, yellowish-green. *Stems* unbranched, erect, slender, 0.5–1.5 cm tall, laxly foliate, dark reddish-brown, strongly lustrous, in cross-section obtusely triangular, with a distinct, large central duct and a 1–2-layered cortex of small, thick-walled, reddish-brown cells; *axillary hairs* short, composed of 3–4 cells, with 2 brownish-red or maroon-red basal cells and 1–2 hyaline, deciduous and often club-shaped distal cells; *rhizoids* clustered at the base or in the lower part of the stem, pale reddish-brown to purplish, bearing numerous tubers. *Leaves* distant below, crowded at the stem tips; *lower leaves* lanceolate, 1.0–1.5 mm long, 0.18–0.25 mm wide at the base, erecto-patent; *upper leaves* long-subulate to setaceous from an oblong, clasping base, 4–5 mm long, 0.25–0.4 mm wide at the base, flexuose when dry; *margins* plane, denticulate near the apex to almost entire, with occasional bistratose patches above, unbordered; *costa* broad, crescent-shaped to flattened, occupying ⅓–⅔ of the leaf base and filling most of the subula, subpercurrent to short-excurrent; *laminal cells* smooth, thin-walled, transparent, long-rectangular to linear in the subula, 70–100 μm long, 4–10 μm wide, becoming shorter and wider, often lax below, oblong-hexagonal to short-rectangular, 40–80 μm long, 8–15 μm wide. *Vegetative propagules* common and numerous, subterranean on main rhizoids or in the axils of the lower leaves, ellipsoid to globose, multicellular, dark purplish to dark reddish-brown, (90–)150–300(–350) μm long, 90–150 μm wide. *Synoicous*. *Setae* terminal, erect, elongate, 15–25 mm long, flexuose-curved to almost straight, smooth, yellowish- or orange-brown, strongly lustrous. *Capsules* long-exserted, horizontal to pendulous, pyriform with a distinct, slender, strongly wrinkled neck on drying, 1.5–1.8 mm long, brownish-yellow to orange-brown with age, leptodermous; *annulus* biseriate, deciduous; *operculum* conic-convex, mammillate; *exothecial cells* thin-walled, mostly rectangular, 30–65 μm long, 20–30 μm wide, becoming quadrate to oblate and smaller in several rows at the orifice; *stomata* numerous in the neck, phaneroporous; *peristome* double; *exostome teeth* 16, lanceolate, yellow, densely papillose, 360–380 μm long; *endostome* yellowish-hyaline, finely papillose, equalling the exostome teeth in height, with a basal membrane about ⅓ of the exostome teeth length, broadly perforate segments and 1–2 appendiculate cilia. *Spores* spherical, finely papillose, 14–18 μm in diameter. *Calyptra* cucullate, smooth, naked.

Chromosome numbers: unknown in the Antarctic and Subantarctic; $n = 10$ (1 record), $n = 20$ (9 records), $n = 20 + 2$m (1 record), $n = 20 + 4$ m (1 record), $n = 21$ (2 records) and $n = 33$ (1 record) from the Holarctic (Fritsch 1991).

Discussion. When in fruit the regularly pear-shaped, pendulous, strongly lustrous and brownish-yellow to orange-brown capsules of *Leptobryum pyriforme* readily separate this species from all other Antarctic mosses. Superficially, it can be confused with some species of *Dicranella* and *Campylopus pyriformis* with which it shares setaceous leaves as well as the frequent occurrence of propagules. However, these species differ from *L. pyriforme* in having short-rectangular to subquadrate laminal cells, and in *Dicranella hookeri* and *D. campylophylla* they are additionally highly mammillose giving, in the latter species, a papillose appearance to the leaf subula. Additionally, these species have much longer, filiform axillary hairs, composed of 4–7 cells which, in *Campylopus pyriformis*, are hyaline throughout, and in *Dicranella* species the basal cells are brownish, not reddish-brown as those in *L. pyriforme*. The latter also produces in great abundance large, ellipsoid and purplish-brown rhizoidal and axillary tubers that are composed of several, not protuberant cells, whereas those in *Dicranella campylophylla* are smaller, globose and consist of many slightly protuberant cells. Rhizoidal tubers also occur in *Campylopus pyriformis* but they are large, pale brownish, multicellular, ovoid and long-cylindrical to vermiform. *Leptobryum pyriforme* also shares a broad costa with *C. pyriformis*, but the costa of the latter has a characteristic internal anatomy with large ventral hyalocysts and dorsal stereids and/or substereids which excludes any alliance of this species with *L. pyriforme*.

Reproduction in Antarctica. The population of *Leptobryum pyriforme* from Deception I. produced sporophytes in profusion, although the majority remained immature at the beginning of the austral autumn on 25 March 1981 when the moss was collected. Nonetheless, a few mature capsules with more or less well developed peristomes and mature spores have been observed. The populations from Galindez I. are sterile, but the plants from both sites produce propagules in abundance. They arise as tubers from the main rhizoids but also occur in the leaf axils. Their size and shape are similar to those found in this species in other parts of its wide geographical range (Correns 1899).

Habitat. On the volcanically active Deception I. *Leptobryum pyriforme* grew at a geothermal site forming extensive patches on heated, moist, cindery ground in association with *Bryum archangelicum*, *B. dichotomum*, *B. orbiculatifolium*, *Ceratodon purpureus*, *Didymodon brachyphyllus*, *Ditrichum conicum*, *Philonotis polymorpha*, *Polytrichum juniperinum*, *Syntrichia filaris*, *S. magellanica*, and *Marchantia berteroana* (Lewis Smith 1984*b*,*c*, 2005*a*,*b*,*c*). However, it was not seen when the site was revisited in January 2002 by RILS. On the other hand, on Galindez I. off the Graham Coast on the Antarctic Peninsula, the species was found in small quantity intermingled in tufts of larger mosses such as *Chorisodontium aciphyllum*, *Polytrichum strictum*, *P. juniperinum*, *Sanionia georgicouncinata* and *Pohlia nutans* and the shoots of *L. pyriforme* were discovered in samples of these species cultured in the Ukraine and may be contaminations (Ochyra & Tyshchenko 2006). The associated species are principal constituents of the *Polytrichum strictum*-*Chorisodontium aciphyllum* sociation within the moss turf sub-formation

→

Figure 188. *Leptobryum pyriforme*. — **1**. Plants with mature capsule, dry. **2**. Sterile plant, dry. **3**. Cross-section of stem. **4–6**. Axillary hairs. **7**. Rhizoidal tubers. **8–12**. Leaves. **13–14**. Leaf apices. **15–16**. Cell in leaf subula. **17**. Mid-leaf cells. **18–19**. Basal cells. **20–25**. Cross-sections of leaf, a sequence from base to apex. **26–27**. Perichaetial leaves. **28**. Deoperculate capsule, dry. **29**. Operculate capsules, dry. **30**. Exothecial cells in middle of urn. **31**. Lower exothecial cells and stomata. **32**. Portion of peristome and spores. — [All from *RILS 3644A*]. — Scale bars: a – 1 mm (1–2), 100 μm (7) and 300 μm (4–6); b – 100 μm (30–32) and 200 μm (13–19); c – 1 mm (26–29); d – 100 μm (3); e – 1 mm (8–12); f – 100 μm (20–25).

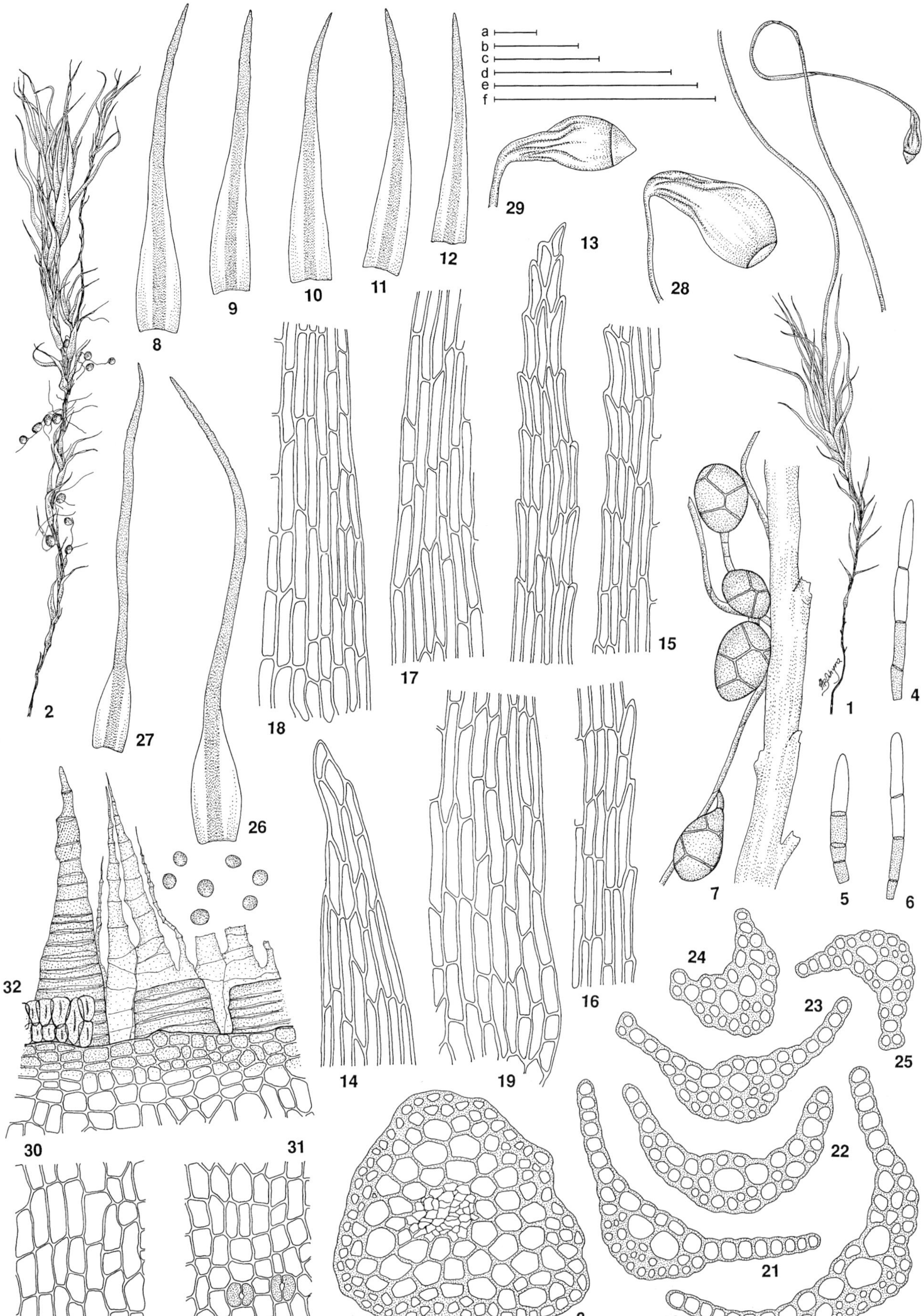
a
b
c
d
e
f
1
2
3
4
5
6
7
8
9
10
11
12
13
14
15
16
17
18
19
20
21
22
23
24
25
26
27
28
29
30
31
32

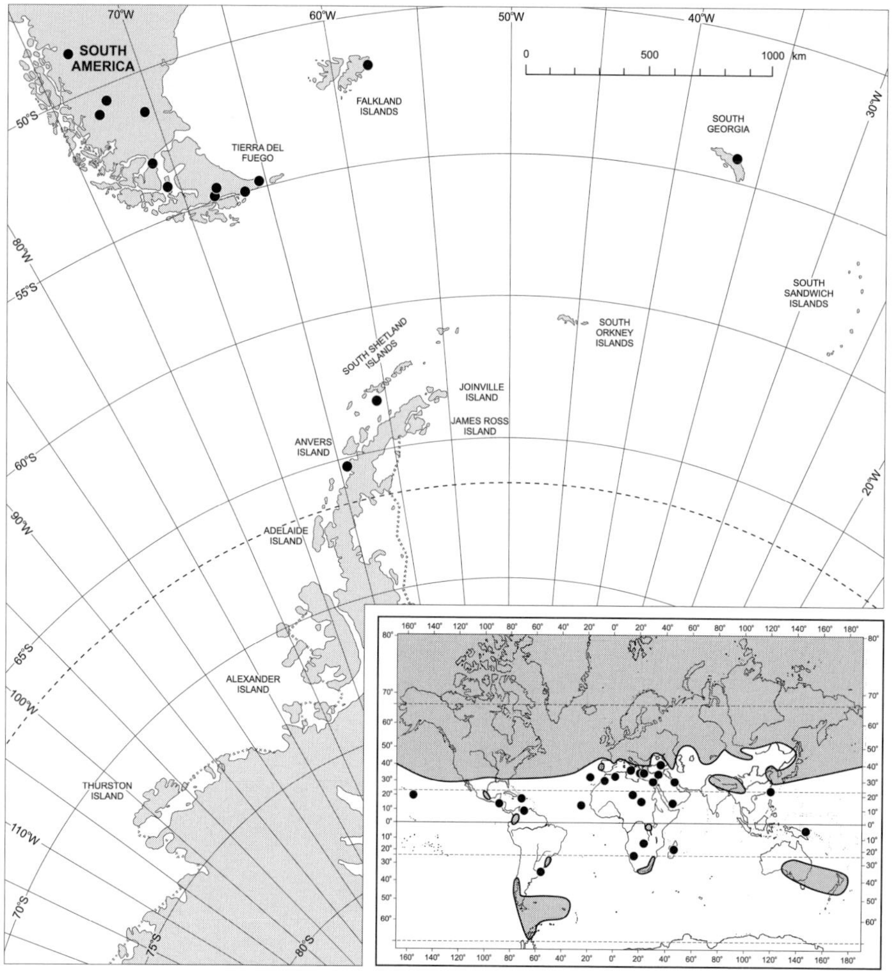

Figure 189. Distribution of *Leptobryum pyriforme* in the Fuegian region and Antarctica. Inset: global distribution.

which forms characteristic peat banks on Galindez I. (Lewis Smith & Corner 1973; Fenton & Lewis Smith 1982). Lewis Smith (1987) reported the occurrence of *L.* cf. *pyriforme* cultured in a soil incubation experiment on Signy I.

World range. *Leptobryum pyriforme* is usually considered to be a cosmopolitan species although, in fact, outside the Holarctic it occurs largely as an annual or short-lived perennial of cultivated land or burnt ground and a frequent weed in glasshouses. It seems to be more typically a bipolar species with many intermediate localities at montane elevations in the tropics (Fig. 189, inset). In the Northern Hemisphere it has a wide pan-Holarctic range and extending southwards to California, Arizona, New Mexico and the Gulf of Mexico in North America (Crum & Anderson 1981), the Mediterranean region and the Caucasus (Manakyan 1989) in Europe, northern and south-western Mongolia (Tsegmed 2001*a*), Central Asia, Central Siberia, Russian Far East (Ignatov & Afonina 1992), north-eastern China (Redfearn *et al.* 1996), Korea and Japan (Ochi 1959), with some isolated stations on Madeira in Macaronesia (Hübschmann 1973; Sjögren 2001), in the Iberian Peninsula (Casares Gil 1915; Sérgio & Carvalho 2003), Sicily (Dia *et al.* 1987), Greece (Duell 1995), Morocco and Algeria (Jelenc 1955) and Egypt (Ros *et al.* 1999) in North Africa, Lebanon (Henderson & Muirhead 1956) and Israel (Herrnstadt & Heyn 2004), Kuwait (El-Saadawi 1976) and the Arabian Peninsula (Kürschner 2000) in the Near East, the Himalayan region, Yunnan (Zhang & Li 2002), and Taiwan (Chuang 1973). In the Southern Hemisphere *L. pyriforme* is widely distributed in western and southern Patagonia (Greene 1986) and scattered and localised on the Falkland Is. (Matteri 1986), South Georgia (Ochyra *et al.* 2002) and in the maritime Antarctic, then recurring in southern Australia (Streimann & Curnow 1989; Streimann & Klanzenga 2002), on Tasmania (Dalton *et al.* 1991) and in New Zealand (Beever *et al.* 1992). In the tropics the species is widely scattered in the Neotropics, from Mexico (Crum 1994*e*), Honduras (Allen 2002) and Haiti (Crum 1965) to Venezuela (Griffin 1975), Colombia (Churchill *et al.* 1995) to Ecuador (Steere 1948) and south-eastern Brazil (Yano 1981), in sub-Saharan Africa from the Cape Verde Is. and Chad to the Cape Province and Madagascar (Rooy & Magill 1987; O'Shea 2003), and in Asia Eddy (1996) reported *L. pyriforme* to be 'scattered throughout Malesia, occurring spasmodically at all altitudes'. It has been reported from Papua New Guinea (Koponen & Norris 1985), but not from other Malesian islands, for instance Borneo (Touw 1978), the Lesser Sunda Is. (Touw 1992), or the Philippines (Iwatsuki & Tan 1979; Tan & Iwatsuki 1991). Also, the species is exceedingly rare in the Pacific region, being recorded only once from Maui in the Hawaiian Is. (Brotherus 1927; Miller *et al.* 1978).

Distribution in Antarctica. An exceedingly rare species, known only from two localities in the maritime Antarctic, on Deception I. in the South Shetland Is. and on Galindez I. in the Argentine Is. archipelago on the west coast of the Antarctic Peninsula where it has its southernmost occurrence at lat. *ca* 65°15′ S (Fig. 189). *Leptobryum pyrirforme* had been reported several times from the Antarctic continent (Imura *et al.* 1992; Kanda & Mochida 1992; Kanda & Okada 1993; Okada

& Kanda 1994) but these records correctly refer to *Pohlia wilsonii*. The locality on Deception I. no longer appears to exist since a geothermal gully on the southern slope of the Pendulum Cove ridge, where the species had been recorded in 1981, was possibly destroyed by erosion as the species was not seen in 2002 (Lewis Smith 2005*a*,*b*). *Leptobryum pyriforme* was not discovered on the Argentine Is. during botanical surveys of the archipelago during the 1960s to 1980s (Longton 1966; Lewis Smith & Corner 1973; Lewis Smith, unpublished) but the slender shoots of this moss may be easily overlooked, especially when growing intermingled amongst the shoots of other mosses. All Antarctic collections of *L. pyriforme* are from low elevations, from sea level to 50 m.

Specimens examined. WEST ANTARCTICA. SOUTH SHETLAND IS. **Deception I.** Pendulum Cove, beyond ridge behind the former Chilean base, *ca* 16 m, *RILS 3644A* (AAS, KRAM).

WEST ANTARCTIC PENINSULA. **Wilhelm Archipelago.** ***Argentine Is.***: Galindez I., Meek Channel, north-east of the Vernadsky (formerly Faraday) Station, *Manilo 3* (KRAM); between Meek Channel and Woozle Hill, 50 m, *Manilo 7* (KRAM).

38. SCHIZYMENIUM Harv.

Schizymenium Harv. *in* Hook., Icon. Pl. 3: *tab. 202*. 1839. — Lectotype (*vide* Wijk *et al.* 1967: p. 362): *Schizymenium bryoides* Harv.

Schizhymenium Harv. *ex* Müll.Hal., Bot. Zeit. (Berlin) 1: 604. 1843, *nom. illeg. orthogr. pro* Schizymenium.

Schisthymenium Harv. *ex* Saut., Flora 28: 156. 1845, *nom. illeg. orthogr. pro* Schizymenium.

Schishymenium Harv. *ex* Paris, Index Bryol. Ed. 2, 4: 227, *nom. illeg. orthogr. pro* Schizymenium.

Plants small and slender to medium-sized, lustrous or dull, loosely or densely caespitose, light green, green, whitish-green, yellowish to brownish. *Stems* erect, simple or sparingly and irregularly forked when sterile and bearing subfloral innovations when fertile, red throughout or only below, yellow above, radiculose at the base with reddish-orange to reddish-brownish, branched, papillose-roughened rhizoids, in transverse section obtusely pentagonal, with a small to large central strand, surrounded by 3–4-layered medulla of abruptly large, hyaline, thin-walled cells and with a 1–3-stratose cortex of small cells with incrassate, red to reddish-brown cells; *axillary hairs* short, filiform, 5–8-celled, with 3–4 short, quadrate, reddish-brown basal cells and 3–4 elongate, hyaline distal cells. *Leaves* mostly distant below, crowded above, less often evenly spaced, soft, appressed and erect, not or scarcely altered on drying, erect-spreading when wet, lanceolate, ovate- to linear-lanceolate, gradually long-acuminate to acute, plane, not or weakly decurrent; *margins* plane, erect or narrowly recurved, unbordered to indistinctly bordered with longer and narrower cells, serrate to serrulate or sinuolate to sinuolate-serrulate above, entire or serrulate below; *costa* single, ending at or below the apex, in transverse section terete, strongly convex dorsally, without or with 2 large guide cells often accompanied by hydroids, large ventral epidermal cells and 2–4-stratose dorsal stereid group, usually with a row of larger epidermal cells; *laminal cells* unistratose, smooth, transparent, thin- to moderately thick-walled, rhomboidal, linear-rhomboidal to linear-vermicular, usually longer and narrower towards the margins above, usually becoming weakly differentiated at the base, laxer, short-rectangular to subquadrate, less often similar to upper cells. *Paroicous* or, occasionally *dioicous* or *synoicous*. *Perigonia* lateral, bud-like, without or with few filiform paraphyses. *Perichaetia* lateral or, rarely, terminal on small stems at the base of vegetative plants; *perichaetial leaves* not strongly differentiated, similar to the vegetative leaves but somewhat smaller. *Setae* erect, flexuose, slender, yellow-brown to yellowish-orange. *Capsules* exserted, erect, inclined to horizontal or pendent, symmetric or slightly curved, pyriform to obloid-cylindrical, tapered to a short or fairly long neck, smooth or sometimes wrinkled-sulcate at the apophysis when dry; *annulus* large and revoluble, 2–3-seriate, composed of large, vesiculose cells with strongly incrassate walls; *operculum* convex-conic, apiculate or hemispheric and blunt; *exothecial cells* irregular, elongate, rectangular to angular, with thin, straight or sinuouse walls; *stomata* many or sparse in the neck, phaneroporous, superficial or somewhat sunken, bicellular, with narrow, elongate pori; *peristome* slender and delicate, single and endostomial, rarely double or absent; *exostome teeth*, when present, 16, usually rudimentary, short and blunt, rarely exceeding above the annulus, hyaline or pale yellowish, smooth or, rarely, finely papillose distally, unbordered, weakly trabeculate on the inner surface; *endostome* imperfect, consisting of a very low to relatively high basal membrane giving rise to narrow, filiform, often irregular in outline and blunt, hyaline to yellowish, smooth or granulose-papillose, flat or keeled, not or narrowly perforate, often anastomosing segments, without or with rudimentary cilia. *Spores* globose, lightly or densely papillose to nearly smooth. *Calyptra* small, cucullate, fugacious, smooth and naked.

Schizymenium is gametophytically extremely similar to *Mielichhoferia*, and soon after its inception (Harvey 1839), these genera were united by Müller (1849), the latter generic name having priority. However, thorough and detailed studies of peristome homology in some bryaceous mosses (Shaw & Crum 1982, 1984) resulted in the redefinition of some bryalean genera, and consequently after nearly a century *Schizymenium* was resurrected (Shaw 1985*d*). The genus is now circumscribed as having either a single endostomial peristome, or a double peristome in which the endostome is well developed and the exostome teeth exhibit various degrees of reduction, from rudimentary to nearly well developed, but are shorter than the endostome and smooth or seldom finely papillose near the apex on the outer surface. In a very few species the peristome is lacking or the endostome is very rudimentary. Almost all species of *Schizymenium* are paroicous and only a few are dioicous or synoicous. In contrast, species of *Mielichhoferia* have either a single exostomial peristome, or a double peristome in which the exostome teeth are well developed and the endostome is highly reduced to a rudimentary basal membrane. All but one species of *Mielichhoferia* are dioicous, except for one, *M. paroica*, from Costa Rica, which is paroicous (Shaw *et al.* 1998).

Schizymenium and *Mielichhoferia*, together with the monotypic Mexican genus *Synthetodontium*, have traditionally been classified in the separate subfamily Mielichhoferioideae within the Bryaceae (Brotherus 1903, 1924*b*). However, this taxon was distinguished much earlier as a separate family, Mielichhoferiaceae, by Schimper (1856) who later changed this legitimate name into the Pleurobryaceae ['Pleurobryeae'], which is an illegitimate and forgotten name (Schimper 1860, 1876). This concept did not gain wide acceptance, and soon the Mielichhoferiaceae were treated as an unranked group within the Bryaceae (Lindberg 1879; Limpricht 1891), and Brotherus (1903) gave it the status of subfamily.

The Mielichhoferioideae are characterised by variously reduced types of peristome structure and highly branched plants often bearing inflorescences in a lateral position, although inflorescences that appear to be lateral are, in some species, terminal on short stems that become overtopped by innovations and displaced to a pseudolateral position. It seems that the branching pattern is overvalued as a diagnostic trait at the subfamily level, since the degree of lateral displacement of inflorescences is determined by timing (Meusel 1935). In taxa where the innovations develop early in the season and therefore have time for considerable growth, the inflorescences are borne on short side branches. Conversely, in taxa which develop the subfloral innovations late, for example in *Bryum* and *Pohlia*, they have the appearance of secondary branches.

Gametophytically, *Schizymenium* is very similar to *Pohlia*, although the plants from both genera differ in branching pattern and, in the latter genus, innovations are common but bifurcation of the stem is rare, yet distinct, in some species. The differences between the highly branched plants of *Schizymenium* and simple or sparsely branched plants of *Pohlia* are a matter of degree, albeit with widely different extremes. Also, the differences in the development of the peristome are not sharply defined between the two genera. Although most species of *Pohlia* have the perfect double *Bryum*-type peristome, there are also species in this genus with variously reduced exostome and endostome teeth. Conversely, in a few species of *Schizymenium*, for example *S. pseudopohlia* (Shaw 1987), the exostome teeth are well developed and morphologically similar to those in *Pohlia*. The differences in the ornamentation of the outer surface of the exostome teeth in both genera (densely papillose-striate in *Pohlia* versus weakly papillose or smooth in *Schizymenium*) are also a matter of degree rather than quality. In *Pohlia* the outer perichaetial leaves are generally larger, with more sharply serrate margins, robust costae and longer cells, whereas in *Schizymenium* the perichaetial bracts are smaller than the cauline leaves.

Schizymenium was originally described by Harvey (1839) who placed in this genus two species, namely *S. bryoides* that was described as new from the Cape Province in South Africa, and *Weissia campylocarpa* from the Andes which was not given formally a name under this generic name. The placement of the latter species was generally overlooked by bryologists and *S. bryoides* was considered the holotype of *Schizymenium*. Nevertheless, because two species were originally included in this genus, lectotypification of this generic name was necessary and Wijk *et al.* (1967) selected *S. bryoides* as the lectotype of *Schizymenium*.

Schizymenium is a medium-sized genus of about 50 species, although the final number will probably be much lower because many species have not yet been critically assessed and studied since their description. The centre of its diversity is in South America where about 30 species are known; six species occur in Central America (Allen 2002), nine in Africa (O'Shea 2003), three in Malesia (Eddy 1996), three on the Hawaiian Is. (Staples *et al.* 2004), and one each in Australasia (*S. bryoides*, Fife 1995; Streimann & Klanzenga 2002), North America (California – *S. shevockii*, Shaw 2000), and Europe (Iberian Peninsula – *S. ponteverdense*, Sérgio 1980). Hitherto, three species have been discovered in austral polar regions, namely *S. bryoides* on Macquarie I. (Seppelt 2004), *S. campylocarpum* in the Kerguelen Province (Mitten 1876*a*,*b*; Hébrard 1970*a*; Zanten 1971), and *S. pusillum* on South Georgia (Clarke 1973*a*) and in the maritime Antarctic (Clarke & Lightowlers 1983).

1. **Schizymenium pusillum** (Hook. & Wilson) A.J.Shaw — Fig. 190

Schizymenium pusillum (Hook. & Wilson) A.J.Shaw, Bryologist 88: 30. 1985. ≡ *Leptostomum pusillum* Hook. & Wilson, London J. Bot. 3: 154. 1844. ≡ *Mielichhoferia pusilla* (Hook. & Wilson) Mitt., J. Linn. Soc. Bot. 12: 321. 1869. — TYPE CITATION: [Brazil] Summit of the Organ Mountains. [Lectotype (*selected here*): '30. Summit of the Organ Mt. March 1841 Herb. G. Gardiner Leptostomum ? pusillum' – BM-Hooker!; isotypes: BM!, BM-Hooker!].

Mielichhoferia microstoma Hampe, Ann. Sci. Nat. Bot. Sér. 5, 4: 335. 1865. — TYPE CITATION: [Colombia] Bogota, Guadalupe, 3100 metr., in montosis, aug. 1863, leg. A. Lindig. [Lectotype (*selected here*): 'Herb. Lindig Mielichhoferia microstoma Hp Nova Granata, Bogota, Guadalupe Alt. 3100 metr. – 1863 Aug.' – BM-Hampe!; isotypes: BM-Hampe! (2 specimens), H-SOL!]. First synonymised with *Mielichhoferia pusilla* by Mitten (1869: p. 321).

Mielichhoferia austrogeorgica Müll.Hal. *in* Neumayer, Int. Polarforsch. Deutsch. Exp. 2: 293. 1890 ['austro-georgica']. ≡ *Schizymenium austrogeorgicum* (Müll.Hal.) A.J.Shaw, Bryologist 88: 30. 1985, ***syn. nov.*** — TYPE CITATION: Austro-Georgia, ad rupes des Vexirberges der Ostseite, 17. Februario 1882, cum Blindiis caespites parvos pulvinatos sistens. [Lectotype (*selected here*): 'Mielichhoferia austrogeorgica C. Muell. Will Ostseite der Vexirberges beim Blindiis' – HBG!; isotype: M!].

Plants small to medium-sized, generally gracile, soft, forming compact or occasionally looser, usually deep tufts or turves, lustrous, light green, yellowish or golden-green, often reddish-tinged, mostly tomentose below with reddish-

Figure 190. *Schizymenium pusillum*. — **1.** Plant with mature capsule, dry. **2.** Plant with perigonia, dry. **3.** Cross-section of stem. **4–6.** Axillary hairs. **7–12.** Leaves. **13–14.** Leaf apices. **15.** Mid-leaf cells. **16–17.** Basal leaf cells. **18–21.** Cross-sections of leaves. **22–23.** Perichaetial leaves. **24.** Deoperculate capsule, dry. **25–26.** Operculate capsules, dry. **27.** Lower exothecial cells and stomata. **28.** Exothecial cells in middle of urn. **29.** Exothecial cells at orifice, rudimentary peristome, annulus and spores. **30.** Eperistomate capsule mouth. — [1, 10–11, 13, 15–16, 24, 26–28, 30 from *RWMC 451*; 2 from *GCSC & SWG 494*; 3–6, 8, 14, 17–23, 29 from *REL 440*; 7 from *Will s.n.*, Ostseite des Vexirberges (lectotype of *Mielichhoferia austrogeorgica*), HBG; 9 from *SWG 3480*; 12 from *Schulz 101*; 25 from *RILS 3283*]. — Scale bars: a – 1 mm (24–26); b – 100 µm (13–17); c – 2 mm (2) and 100 µm (3, 18–21, 27–30); d – 100 µm (4–6); e – 0.5 cm (1) and 1 mm (7–12, 22–23).

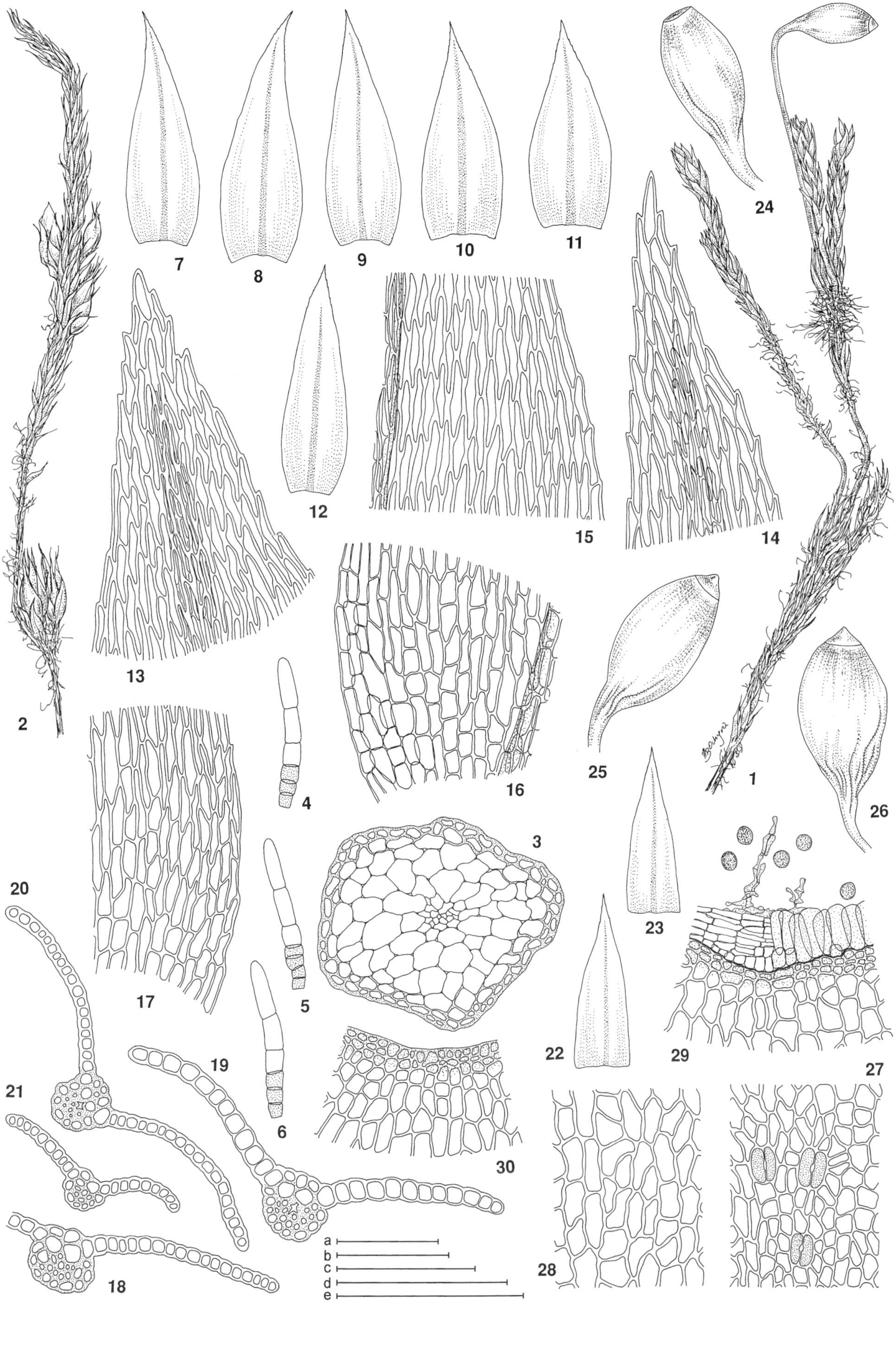
1
2
3
4
5
6
7
8
9
10
11
12
13
14
15
16
17
18
19
20
21
22
23
24
25
26
27
28
29
30
a
b
c
d
e

brown, papillose-roughened, branched, glossy rhizoids. *Stems* erect, 1–3 or rarely up to 6 cm tall, irregularly branched by innovations, rarely simple, uniformly foliate, red, glistening, in cross-section rounded-pentagonal, with a distinct central strand, 3–4-stratose medulla of large, thin-walled, hyaline cells and a 1–2-layered cortex of small, thick-walled, reddish cells; *axillary hairs* filiform, short, composed of 6–8 cells, with 3–4 short, quadrate, reddish-brown basal cells and 3–5 elongate, hyaline distal cells. *Leaves* of innovations evenly spaced, crowded above, distant below, erect to erect-spreading, not much altered on wetting, ovate-lanceolate to ovate, often slightly asymmetric, 0.7–1.5(–2.1) mm long, 0.3–0.5 mm wide, plane to somewhat carinate, acuminate to sharply acute, not or slightly decurrent; *margins* plane, unbordered, serrate to serulate-sinuolate above mid-leaf, entire to bluntly and remotely serrulate below; *costa* single, strong, yellowish-brown to reddish-brown below, sharply demarcated from the laminal cells, ceasing below the apex or seldom percurrent, 45–60 μm wide near the base, gradually tapering upwards, usually flexuose above, in transverse section strongly convex dorsally, consisting of 2 large guide cells usually accompanied by hydroids, large, 3–4-stratose dorsal stereid band with a row of larger dorsal epidermal cells and a row of large ventral epidermal cells; *laminal cells* unistratose throughout, smooth, pellucid, narrowly linear-rhomboidal to linear-vermicular, mostly fusiform, 25–85 μm long, 6–8 μm wide, becoming wider and shorter, rhomboidal, rectangular to subquadrate near the base, 20–60 μm long, 12–20 μm wide, not differentiated in the basal angles. *Synoicous* or *paroicous*. *Perigonia* lateral, gemmiform, to 1 mm long, with ovate, acute, concave bracts and many club-shaped, brown to reddish-brown antheridia, without or with sparse, yellowish paraphyses slightly exceeding the antheridia. *Perichaetia* lateral, at or near the base of the stems; *perichaetial leaves* somewhat smaller but otherwise similar in shape to the vegetative leaves. *Setae* erect, flexuose, 5–15 mm long, smooth, lustrous, yellow-brownish, light brown to orange-brown. *Capsules* inclined, horizontal to pendent, pyriform to obloid-cylindrical, 1.5–2.1 mm long, 0.7–1.0 mm wide, yellow-brown to brown, lustrous, symmetric to somewhat gibbous, with a short or fairly long neck, smooth or wrinkled-sulcate on drying; *annulus* biseriate, revoluble, composed of large, vesiculose, thick-walled cells; *operculum* convex-conic, apiculate; *exothecial cells* variable in shape, subquadrate, rectangular, hexagonal to multiangular, 20–40 μm long, 12–25 μm wide, with moderately thickened, straight or somewhat wavy walls, becoming much smaller, elliptical in 1–3 rows at the mouth and forming a dark yellow or orange-brownish capsule rim; *stomata* numerous in several rows in the apophysis, phaneroporous, bicellular, with long and narrow slit; *peristome* none or rudimentary and fugacious, consisting of very small, hyaline, smooth, irregular segments arising from a relatively high basal membrane. *Spores* spherical, opaque, brownish, papillose, 18–24 μm in diameter.

Chromosome number: unknown.

Discussion. The taxonomy and distribution of the moss genus *Schizymenium* has been revised for subantarctic South Georgia by Clarke (1973*a*) and for Antarctica by Clarke & Lightowlers (1983). Accordingly, identification of *Schizymenium* specimens from this region should be straightforward since the possibility of confusion with other regional bryalean mosses has been elucidated. Fertile plants of *S. pusillum* are easily recognisable owing to the formation of gemmiform, synoicous or paroicous gametangia and sporophytes on short side branches and eperistomate capsules, although sometimes the endostomial peristome is present but very rudimentary, delicate and fugacious. Gametophytically *S. pusillum* is very much like a *Pohlia*, but the species fruits regularly in Antarctica, so confusion with that genus and other bryalean mosses is minimal.

Schizymenium pusillum is most likely to be mistaken for *Pohlia nutans*, the commonest species of the genus in the Antarctic biome, and the differences between these species were summarised by Clarke & Lightowlers (1983). In the field, its light green, yellowish or golden-green colour, often with a reddish cast should aid its recognition. Sterile plants of both species differ in leaf areolation. *Schizymenium pusillum* has mostly fusiform and narrow cells, 6–8 μm wide, with thin or firm walls and spindle-like ends. In contrast, *P. nutans* has broader cells, 9–14 μm wide, with thick walls, mostly more than 3 μm, and truncate or oblique ends. Moreover, the leaves in *S. pusillum* are often slightly asymmetric. Some plants of *S. pusillum* with a reddish or glaucous tinge may be confused with *Pohlia wahlenbergii*, but the broad and lax laminal cells, 16–24 μm wide, as well as broader leaf apex, immediately separate them.

If *Schizymenium pusillum* is fairly constant in morphological and anatomical details of the leaves, it exhibits much variation in the size of the plants. Sometimes they grow in deep turves, to 6 cm tall, which have distinct annual growth bands, but usually the plants are smaller and do not exceed 3 cm. The type material of *S. pusillum* from Brazil (W. J. Hooker & Wilson 1844) is a small plant, about 1 cm tall, and the type material of *Mielichhoferia microstoma* from Colombia (Hampe 1865) is even smaller. The same is true of the original collection of *M. austrogeorgica* from South Georgia consisted of small sterile plants (Müller 1890*a*), although many subsequent collections of this species from this subantarctic island are larger and more luxuriant and commonly produce sporophytes. Nonetheless, because correspondence in critical morphological and anatomical characteristics of gametophytes and sporophytes between these species is high, *M. austrogeorgica* is here reduced to synonymy with *S. pusillum*. In the genus *Schizymenium* the only other eperistomate species is *S. schiedeanum* (Müll.Hal.) A.J.Shaw from Central America, but this species has much longer, linear-fusiform laminal cells, 60–130 μm long, and obloid-cylindrical capsules, 1.5–2.5 mm long.

Reproduction in Antarctica. *Schizymenium pusillum* is able to complete its life cycle at high latitudes, and the majority of specimens from the Antarctic produce mature sporophytes, some in great profusion, most probably owing to the frequent occurrence of synoicous or paroicous inflorescences.

Habitat. A rupicolous moss growing in dry or moist rock crevices, on rock faces and ledges, on humus in cracks of rocks and beneath overhangs, usually in sheltered and shaded, less often in exposed, situations on metamorphic and volcanic rocks. Most populations are near sea level, 2–25 m altitude. It forms pure, monospecific, compacts tufts or turves, with an occasional admixture of solitary shoots of *Bartramia patens*, *Polytrichum juniperinum*, *Barbilophozia hatcheri* and *Lophozia excisa*.

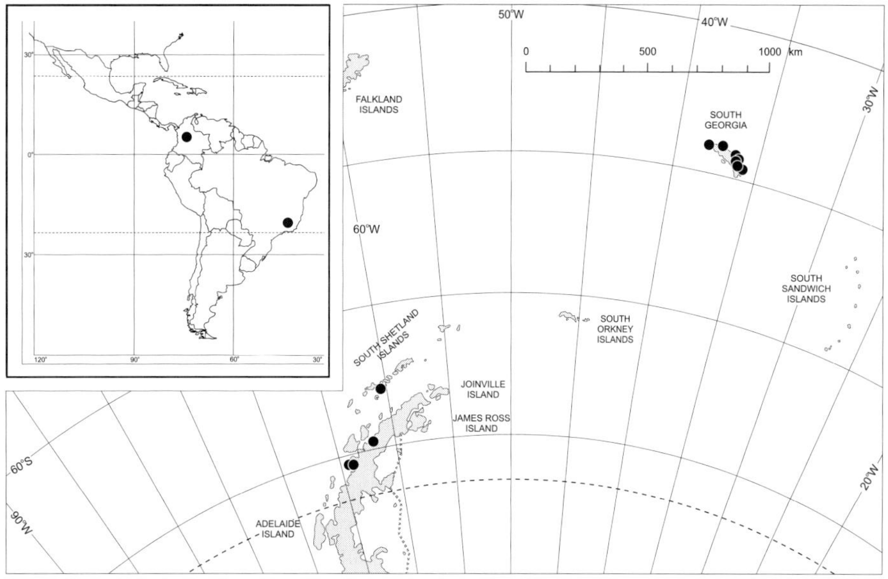

Figure 191. Distribution of *Schizymenium pusillum* in the Subantarctic and Antarctic. Inset: distribution in South America.

World range. A South American subantarctic species, having the main centre of its occurrence on South Georgia, extending southwards to the Antarctic Peninsula region and occasionally penetrating northwards into the tropics at high elevations in the northern Andes of Colombia and Serra dos Orgãos in south-eastern Brazil (Fig. 191, inset). The Brazilian Highlands in this region are well known as a refugium of many austral-Antarctic cool-adapted mosses, such as *Dicranoloma billardierei*, *Lepyrodon tomentosus*, *Vittia pachyloma*, *Orthodontium pellucens*, *Thuidiopsis furfurosa*, *Chrysoblastella chilensis*, and *Ptychomnium cygnisetum* (Herzog 1925; Sehnem 1955).

Distribution in Antarctica. A very rare and localised species, occurring at three known centres in the maritime Antarctic (Fig. 191). It is exceedingly rare on Livingston I. in the South Shetland Is. (Sancho *et al.* 1999), but more frequent and locally abundant on Andrée I. in Charlotte Bay on the Danco Coast, and on Rasmussen I. and the Argentine Is. off the Graham Coast (Clarke & Lightowlers 1983) where it reaches its southernmost occurrence at lat. 65°15′S. The records of *Schizymenium pusillum* from Bellingshausen I. in the South Sandwich Is. (Convey *et al.* 2001) are based upon misdetermination of specimens which correctly represent *Pohlia nutans* and *P. drummondii*. The species occurs at low elevations, from 2 to 25 m.

Specimens examined. WEST ANTARCTICA. SOUTH SHETLAND IS. **Livingston I.** ***Hurd Peninsula***: Johnsons Dock, 25 m, *Schulz 101* (KRAM).

WEST ANTARCTIC PENINSULA. **Danco Coast.** Charlotte Bay, Andrée I., 15–20 m, *RILS 4003*, *4019 & 4020* (BM) and 25 m, *RILS 4017* (AAS, KRAM) *&* *4036* (BM). **Graham Coast.** Rasmussen I., *RILS 1924* (AAS, KRAM, LE, PC). **Wilhelm Archipelago.** ***Argentine Is.***: Galindez I., 5 m, *BGLE 1114A* (AAS, BM, KRAM) and *RILS 938* (AAS, KRAM); same island, cliff overlooking Meek Channel, *ca* 7 m, *RWMC 451* (AAS, KRAM, LE); same island, Woozle Hill, 25 m, *RILS 3283* (AAS, KRAM); same island, Windscoop, *ca* 16 m, *RWMC 561* (AAS, KRAM).

39. **POHLIA** Hedw.

Pohlia Hedw., Spec. Musc. Frond.: 171. 1801. ≡ *Bryum* Hedw. subgen. *Pohlia* (Hedw.) Bruch & Schimp. *in* Bruch, Schimp. & W.Gümbel, Bryol. Eur. 4: 91. 1839 [Fasc. 6–9 Mon.: 21]. ≡ *Webera* Hedw. sect. *Pohlia* (Hedw.) Schimp., Coroll. Bryol. Eur.: 64. 1856. ≡ *Webera* subgen. *Pohlia* (Hedw.) Milde, Bryol. Siles.: 200. 1869. — Type: *Pohlia elongata* Hedw.

Webera Hedw., Spec. Musc. Frond.: 168. 1801, *hom. illeg.* ≡ *Bryum* sect. *Webera* Relh., Fl. Cantabr. Ed. 2: 425. 1802. ≡ *Bryum* subgen. *Webera* (Relh.) Bruch & Schimp. *in* Bruch, Schimp. & W.Gümbel, Bryol. Eur. 4: 102. 1839 [Fasc. 6–9 Mon.: 32]. ≡ *Pohlia* sect. *Webera* (Relh.) Margad., Lindbergia 1: 129. 1972. — Lectotype (*vide* Margadant 1972: p. 129): *Webera nutans* Hedw. [*Pohlia nutans* (Hedw.) Lindb.].

Lamprophyllum Lindb., Öfvers. Förh. Kongl. Svenska Vetensk.-Akad. 20: 394. 1863, *hom. illeg.* ≡ *Pohlia* [unranked] *Lamprophyllum* Lindb., Musci Scand.: 17. 1879. ≡ *Webera* sect. *Lamprophyllum* (Lindb.) Sakurai, Bot. Mag. (Tokyo) 53: 288. 1939. ≡ *Pohlia* sect. *Lamprophyllum* (Lindb.) Ochi, Rev. Bry. Japan: 23. 1959. — Lectotype (*vide* Margadant 1972: p. 129): *Lamprophyllum nutans* (Hedw.) Lindb. [*Pohlia nutans* (Hedw.) Lindb.].

Kaurinia Lindb. *in* Bryhn., Nyt Mag. Naturvidensk. 32: 124. 1891, *nom. nud.*

Mniobryum Schimp. *ex* Limpr., Laubm. Deutschl. 2: 272. 1892, *nom. cons.* [*Bryum* Hedw. (unranked) *Mniobryum* Schimp. *in* Bruch, Schimp. & W.Gümbel, Bryol. Eur. 4: 1. 1851 (Fasc. 46–47 Consp. Vol. 4: iii), *nom. nud.*]. ≡ *Webera* subgen. *Mniobryum* (Limpr.) J.J.Amann, Rev. Bryol. 20: 43. 1893. ≡ *Webera* sect. *Mniobryum* (Limpr.) Cardot, Wiss. Erg. Schwed. Südpolar-Exp. 4(8): 132. 1908. ≡ *Pohlia* sect. *Mniobryum* (Limpr.) Ochi, Rev. Bry. Japan: 40. 1959, *nom. illeg. incl. lectotyp. sect. prior.* [*Pohlia* sect. *Mniobryum* (Limpr.) Nyholm, Ill. Moss Fl. Fennoscandia. II. Musci 3: 195. 1958, *non rite publ. basion. non citat.*]. ≡ *Pohlia* subgen. *Mniobryum* (Limpr.) Ochyra *in* Ochyra, Żarnowiec & Bednarek-Ochyra, Cens. Cat. Polish Mosses: 158. 2003. — Lectotype (*vide* Lanjouw *et al.* 1956: p. 216): *Mniobryum carneum* (F.Weber & D.Mohr) Limpr., *nom. illeg.* [= *Pohlia melanodon* (Brid.) A.J.Shaw]. First synonymised by Andrews (1935: p. 188).

Wollnya Herzog, Beih. Bot. Centralbl. 26: 69. 1910, ***syn. nov.*** — Type: *Wollnya stellata* Herzog [= *Pohlia wilsonii* (Mitt.) Ochyra].

Plants small and slender to rather large and robust, loosely to densely tufted or gregarious, soft to fairly stiff and rigid, dull or glossy, sometimes opalescent, green, yellow, yellowish- or reddish-green, yellow- or reddish-brown, brownish-green, golden or seldom whitish to glaucous-green. *Stems* erect, simple or forked at the base or sparingly and irregularly branched by innovation when fertile, sparsely radiculose or densely matted at the base or in the lower part with pale orange-red to brown, branched, papillose rhizoids, in transverse section rounded to obtusely pentagonal, with more or less distinct central strand, 3–7-stratose medulla of large, hyaline, thin- to thick-walled cells and a 1–4(–5)-layered cortex of small, thick-walled, red-brown cells; *axillary hairs* short, filiform, 3–7-celled, with 2–5 short, quadrate, reddish-brown pigmented basal cells and 2–4 hyaline, elongate distal cells. *Leaves* crowded or spaced, erect and imbricate to erect-spreading, little altered on drying, soft or stiff, ovate, lanceolate to ovate- or linear-lanceolate, acuminate, acute or rarely obtuse at the apex, symmetric or somewhat asymmetric, unbordered, plane to broadly keeled, not or weakly to strongly decurrent; *margins* plane to narrowly recurved, serrate to serrulate to entire near the apex, entire to serrulate below; *costa* single, generally not strong, sharply set off from the laminal cells, ending above mid-leaf to percurrent, rarely shortly excurrent, in transverse section terete, usually strongly convex dorsally, with 2–4 median guide cells, often accompanied by hydroids, with a more or less well developed dorsal stereid band, without or with a few ventral stereids and a row of enlarged, thin- to firm-walled dorsal and ventral epidermal cells; *laminal cells* unistratose throughout, smooth, pellucid, with thin to thick walls, long and narrow, oblong-hexagonal to linear or rarely vermicular above, undifferentiated at the margins or somewhat longer and narrower than in the interior, becoming shorter and laxer, rectangular near the base, often quadrate at the insertion. *Asexual reproduction* occasionally by multicellular rhizoidal tubers or, more frequently, by propagules arising singly or in clusters of 3 to many in the upper leaf axils, consisting of a multicellular body and 0–4 apical or subapical leaf primordia. *Dioicous* or *paroicous*. *Perigonia* terminal on main shoots, occasionally on innovations; *perigonial bracts* ovate, concave, shortly acute to long-acuminate, erect to wide-spreading or squarrose. *Perichaetia* on the top of main stems; *perichaetial leaves* scarcely to strongly differentiated, with recurved margins. *Setae* erect, elongate, slender to fairly stout, flexuous and twisted when dry, generally curved or hooked at the apex, smooth, lustrous, singly or, rarely, 2 to several per perichaetium. *Capsules* small to large, exserted, erect, inclined to pendent, claviform, pyriform, ovoid to linear-cylindric, sometimes urceolate, with a short or long apophysis, symmetric or nearly so; *annulus* none to well developed, 1–3-seriate, removable or separating irregularly; *operculum* convex-conic, mammilate or apiculate to short-rostrate; *exothecial cells* variable in shape, isodiametric to elongate, with thin to firm, straight or sinuose walls, sometimes collenchymatous, becoming smaller and thicker-walled in several rows at the orifice; *stomata* many in the neck, bicellular, with 2 guide cells and elliptical to elongate pori, superficial to immersed; *peristome* double; *exostome teeth* 16, lanceolate, acute to acuminate, seldom blunt at the apex, yellowish to brownish, papillose throughout on the outer surface, weakly to strongly trabeculate on the inner side; *endostome* well developed or incomplete, hyaline to yellow-brown, consisting of a low or tall basal membrane, elongate, broad, keeled and widely perforate or narrow, flat and entire segments, and long and nodulose or, occasionally, appendiculate, rudimentary or no cilia. *Spores* globose, finely roughened to coarsely papillose. *Calyptra* cucullate, smooth, naked.

Pohlia is a large and very variable genus of bryalean mosses that is notoriously difficult taxonomically owing to its commonly nondescript and often technical characters. On the whole it is characterised by usually having a perfectly developed, double peristome with nodulose cilia; lanceolate to ovate-lanceolate decurrent leaves with areolation of long and narrow cells; unbordered, plane to narrowly recurved or reflexed margins that are serrate to serrulate distally; and subpercurrent to percurrent costae. However, these character states are not absolute and may appear in various combinations, and consequently *Pohlia* grades in nearly every trait into *Bryum*. In addition, leaves in different parts of single plants may be morphologically dissimilar and this contributes to taxonomic difficulties in some species of the genus and the merging of the generic limits between *Pohlia*, *Bryum* and some other bryalean genera. Therefore, sterile plants often cannot be readily assigned to the appropriate genus without prior knowledge of individual species.

The difficulties with the accurate circumscription of *Pohlia* are reflected in the chequered taxonomic and nomenclatural history of this genus. In the 19th century species were often placed in *Bryum*, including *Bryologia europaea*, an opus regarded as the oracle as far as taxonomic and nomenclatural questions of European mosses were concerned (Bruch *et al.* 1839). Species which currently belong within this genus were divided by Hedwig (1801) into two genera, namely the monotypic *Pohlia* (for *P. elongata*) and *Webera* (for *W. nutans*, *W. longicollis* and *W. pyriformis*). Both generic names were in common use until the 1880s, the latter even more frequently, although the name *Webera* was a younger illegitimate homonym and Lindberg (1863*b*, 1868, 1871) replaced it with *Lamprophyllum*, but this name also proved to be a later illegitimate homonym. Finally, Lindberg (1878, 1879) presented an amended concept of *Pohlia* and merged with it *Lamprophyllum*, and thus, by inference, with *Webera*. Since then there was wide acceptance of the generic name *Pohlia*, which was incorporated in most major works (e.g. Braithwaite 1890–1893; Brotherus 1903, 1924*b*; Warnstorf 1906; Mönkemeyer 1927; Andrews 1935), although some authors continued to use *Webera* (e.g. Limpricht 1892; Dixon & Jameson 1896, 1924; Loeske 1903; Roth 1905; Jensen 1939). Nowadays, *Pohlia* has gained universal acceptance and its status as a genus in its own right is not questioned.

Apart from the two aforementioned competing names *Pohlia* and *Webera*, a third name is associated with pohlioid species, namely *Mniobryum*. It was introduced by Schimper (1851) as a *nomen nudum* for an unranked subdivision of *Bryum*. Limpricht (1892) used this name for a new genus consisting of three species recognised by a combination of features including small to very small size, shortly pyriform to urceolate capsules that are generally little longer than wide on drying, immersed to emergent stomata, and a lack of differentiated annular cells. Although reasons exist for treating *Mniobryum* as a separate genus (Loeske 1910), it is very close to *Pohlia* in a strict sense and some species could be placed in either. Therefore, soon after its inception, it was reduced to a subgenus (Amann 1893) or section (Cardot 1908*a*) within *Webera*. Andrews (1935) merged both genera and subsequent authors considered it either a section (Nyholm 1958, 1993; Ochi 1959) or subgenus (Shaw 1981*a*, 1982*a*) within *Pohlia*. There have been some nomenclatural problems regarding the appropriate name for these infrageneric taxa. These have been resolved by Ochyra *et al.* (2003) who showed that *Pohlia* subgen. *Mniobryum* (Limpr.) Ochyra and sect. *Apalodictyon* (Müll.Hal.) Ochyra should be used, respectively, as subgeneric and sectional names for species classified in the genus *Mniobryum*. In the present treatment, the monotypic genus *Wollnya* is evaluated and it is also considered to be congeneric with *Pohlia*. The status of this genus is discussed in detail under *P. wilsonii*.

Pohlia is the second largest genus of the Bryaceae and is distributed worldwide. There are about 150 species that are still residual in this genus, but well over half of them, mostly from tropical and austral areas, have not been taxonomically assessed (Crosby *et al.* 2000), so it is likely that many of their names will be reduced to synonymy. Morphologically, the genus is very diverse and its species are grouped into three subgenera (Shaw 1982*a*): (*1*) subgen. *Pohlia*, including species that have elongate capsules; straight-walled exothecial cells; superficial stomata; compound and revoluble annuli; well developed peristomes with reddish-orange to brownish exostome teeth that are bordered, strongly trabeculate and coarsely papillose above and finely papillose below, and hyaline or very rarely yellow endostome that has a low or high basal membrane, broad keeled and variously perforated and long and nodose to short and rudimentary cilia; and generally lack propagules; (*2*) subgen. *Nyholmiella*, consisting of species that have elongate or isodiametric capsules; straight or wavy walls of the exothecial cells; superficial or immersed stomata; compound and removable annuli; reduced peristomes with whitish to whitish-yellow exostome teeth that are narrow, unbordered, weakly trabeculate and densely and evenly papillose throughout, and linear, flat or weakly keeled and not or weakly perforate segments of the endostome arising from a low or tall basal membrane, and no or rudimentary cilia; and often with axillary propagules; (*3*) subgen. *Mniobryum* comprising species with broadly pyriform to urceolate capsules that are exannulate or with poorly developed and irregularly separating annuli and have superficial or sunken stomata and strongly wavy exothecial cells; well developed peristomes with reddish-orange to brownish exostome, unbordered, distinctly trabeculate, and coarsely papillose above and finely papillose below the exostome teeth, and hyaline to yellow-brown endostome with a high or low basal membrane, broad, keeled and broadly perforate segments and long and nodose cilia; with or without axillary gemmae and rhizoidal tubers. The latter subgenus consists of two sections: (*1*) sect. *Cacodon* that consists of small to moderately large plants often bearing axillary gemmae or rhizoidal tubers and having larger capsules, often more than 2 mm, with irregularly separating, 1–3-seriate annuli, superficial or slightly immersed stomata, short-rectangular exothecial cells with evenly thickened walls and well developed, hyaline endostomes; (*2*) sect. *Apalodictyon* which includes small plants lacking propagules and having small, less than 2 mm, exannulate capsules with collenchymatous exothecial cells and usually sunken stomata and well developed, yellow to yellow-brown endostomes.

The genus *Pohlia* is represented on every continent, with the maximum concentration of species in cool and temperate regions of the Northern Hemisphere but penetrating the tropics at high elevations. Species of *Pohlia* occupy a wide range of habitats, but most taxa occur on acidic soil, often in disturbed sites, but also in undisturbed habitats on stream banks, in soil-filled rock fissures, amongst rocks in tundra and on exposed sites in the alpine zone (Shaw 1981*b*). Taxonomically, the genus is unevenly studied and so far it has been revised only for Japan and conterminous regions (Ochi 1959), North and Central America (Shaw 1982*b*), and Australia (Shaw 2006). On the other hand, many studies have been devoted to the propaguliferous species of the genus belonging within sect. *Cacodon* (Wilczek & Demaret 1970; Lewis & Smith 1977, 1978; Demaret & Wilczek 1978, 1979; Shaw 1981*a*,*c*,*d*,*e*; Karczmarz & Żarnowiec 1989; Arts 1990; Demaret & Arts 1993; Czernyadjeva 1997, 1999).

In the polar and cool temperate regions of the Southern Hemisphere *Pohlia* does not exhibit high species diversity and, hitherto, about a dozen species have been recorded from southern South America, New Zealand, Australia and Antarctica (Greene 1986; Fife 1995; Shaw 2006), of which about half are bipolar species that are widespread in the Northern Hemisphere. The number of truly austral species remains unknown until a revision of the genus in this region is completed. So far, local taxonomic treatments of *Pohlia* are available only for South Georgia (Clarke 1973*a*) and Antarctica (Greene *et al.* 1970; Lightowlers 1983; Ochyra 1998*a*). At present five species of the genus are known to occur in the Antarctic, belonging within the type subgenus (*P. cruda* and *P. nutans*) and subgen. *Mniobryum* (*P. wahlenbergii* – sect. *Apalodictyon*; *P. drummondii* and *P. wilsonii* – sect. *Cacodon*).

1. Leaf areolation predominantly lax; majority of median laminal cells 15 μm or more wide 2
1. Leaf areolation mostly sturdy; majority of median laminal cells less than 15 μm wide 3
   2. Propagules absent; leaf margin toothed to serrulate in the distal portion; plants whitish to glaucous-green, often with pinkish tinge .... 1. ***P. wahlenbergii***
   2. Multicellular tubers abundant on matted rhizoids; leaf margin entire to weakly crenulate near the apex; plants yellow-green .... 2. ***P. wilsonii***
3. Leaves long-decurrent; axillary gemmae usually present .... 3. ***P. drummondii***
3. Leaves not or weakly decurrent; propagules always absent 4
   4. Plants with a conspicuous bluish sheen; cell walls thin, less than 2 μm; upper laminal cells vermicular, 70–120 μm long; costa red to reddish-brown near the base .... 4. ***P. cruda***

4. Plants dull to slightly lustrous; cell walls thick, more than 3 µm; upper laminal cells elongate-hexagonal, 30–80 µm long; costa dark green to brownish at the base .... 5. ***P. nutans***

1. **Pohlia wahlenbergii** (F.Weber & D.Mohr) A.L.Andrews Fig. 192

Pohlia wahlenbergii (F.Weber & D.Mohr) A.L.Andrews *in* Grout, Moss Fl. N. Amer. 2: 203. 1935. ≡ *Hypnum wahlenbergii* F.Weber & D.Mohr, Bot. Taschenbuch: 280. 1807. ≡ *Bryum wahlenbergii* (F.Weber & D.Mohr) Schwägr., Sp. Musc. Frond. Suppl. 1: 92, *f. 70*. 1816. ≡ *Webera wahlenbergii* (F.Weber & D.Mohr) Fürnr., Flora 12 Ergänzungsbl. 2: 35. 1829. ≡ *Mniobryum wahlenbergii* (F.Weber & D.Mohr) Jenn., Man. Mosses W. Pennsylvania: 146. 1913. — TYPE CITATION: *Bryum albicans* Wahlenb. Fl. lappon. msr. ['Hab. ubi aqua fontana per terram sterilem glareosam diffluit praecipue in Nordlandia inferalpina passim, sed rarissime capsuliferam ex. gr. ad Storstensnäs Tromsöae lectum' (Wahlenberg 1812: p. 353)]. In arena insull. *Danubii* pr. *Ratisb.* Funck (ante 12 annos). [Lectotype (*selected here*): 'Nordlandia Norvegia; in margine terra sabuloso madefacto juxta mare ad Storstensnäs juxtam Tromsöam lectum d. 16 Juli 1800 Mnium albicans' – UPS!].

Plants small to medium-sized, slender, flaccid to fairly firm, forming dense or loose turves or tufts, dull or slightly glistening, whitish, whitish- to glaucous-green above or throughout, reddish, reddish- or pinkish-green below or throughout. *Stems* erect, simple or occasionally forking, 0.5–2.5 cm tall, red, sparsely or densely radiculose at the base or densely matted with rhizoids in the lower part, in cross-section obtusely pentagonal, consisting of a relatively large and distinct central strand, 3–4-stratose medulla of large, hyaline, thin-walled cells and a 1–2-layered cortex of small, thick-walled, red cells; *rhizoids* branching, reddish to reddish-brown, papillose, shining; *axillary hairs* filiform, 5–6-celled, consisting of 2 short, quadrate, reddish-brown basal cells and 3–4 elongate, hyaline distal cells. *Leaves* remote to fairly close, erect to erect-spreading, somewhat shrunken when dry, erecto-patent to patent when wet, 0.7–1.1(–2.1) mm long, 0.4–0.5(–0.7) mm wide, ovate, ovate-lanceolate to narrowly lanceolate, acute or acuminate to almost obtuse, plane, not or weakly decurrent; *margins* plane, unbordered, sharply toothed with projecting cell ends to serrulate in the upper half, entire to bluntly serrulate below, occasionally crenulate to nearly smooth throughout; *costa* single, ending several cells below the apex, slender, 45–70 µm wide at the base, gradually tapering upwards, straight or flexuose above, reddish to reddish-brown, in cross-section convex dorsally, consisting of 2 large guide cells, a weak dorsal stereid band and a row of dorsal and ventral epidermal cells; *laminal cells* smooth, pellucid, unistratose throughout, thin-walled, lax to relatively sturdy, oblong-rhomboidal, becoming broadly rhomboidal near the apex and somewhat narrower towards the margins, 55–110 µm long, (12–)15–25 µm wide, not much differentiated in the leaf base. [*Dioicous*. *Perigonia* terminal on simple stems, conspicuous, swollen; *outer perigonial bracts* broadly acute, wide-spreading. *Perichaetia* terminal on simple or innovating stems; *perichaetial leaves* scarcely differentiated, lanceolate, to 2 mm long, with lax areolation. *Setae* erect, slender, flexuose, 2.5–4.0 mm long, reddish. *Capsules* inclined to pendulous, shortly and broadly pyriform to urceolate, short-necked, 1.5–2.5 mm long, brown, lustrous; *annulus* none; *operculum* low conic, apiculate; *exothecial cells* ± isodiametric, irregular, 20–50 µm wide, thick- and sinuose-walled, weakly collenchymatous, becoming smaller and flattened in 4–5 rows at the mouth; *stomata* sunken, bicellular, with elongate pori; *peristome* double; *exostome teeth* lanceolate, acute, brown, reddish- or yellow-brown, coarsely papillose distally, more finely so below, bordered, trabeculate on the inner surface; *endostome* hyaline to pale yellowish, with a basal membrane about half the length of the exostome teeth, keeled and broadly perforate segments and nodulose cilia grouped in 2s or 3s. *Spores* globose, brownish-yellow, 13–24 µm in diameter, finely roughened. *Calyptra* cucullate, smooth, naked.]

Chromosome numbers: unknown in the Antarctic and Subantarctic; the chromosomal record of $n = 11$ from King George I. (Przywara *et al.* 1984) correctly refers to *Pohlia cruda*; additionally, $n = 10$ (1 record) and $n = 11$ (2 records) from the Holarctic (Fritsch 1991).

Discussion. When typically developed *Pohlia wahlenbergii* is easy to distinguish by its rather dull, whitish, bluish- or reddish-green to reddish leaves which are usually well spaced and sharply contrast with the dark red stems. It usually grows in damp or wet seepy habitats where the plants may be luxuriant, tall and soft. Microscopically, the species is recognised by its characteristic, lax leaf areolation of broad cells, mostly 15–25 µm wide. The Antarctic populations occurring at ecologically and geographically marginal sites comprise small and slender plants, with scarcely any superficial resemblance to the well developed forms from temperate regions. Yet, they retain all typical anatomical features of the species. The leaves are small, mostly about 1 mm long, not or somewhat decurrent and dull to slightly glistening. The leaf margins are distinctly dentate in the distal half by projecting cell ends, although sometimes the dentation is obscure and the margins are only crenate. The laminal cells are very broad, mostly 15–20 µm but often to 25 µm wide, oblong-rhomboidal throughout the lamina, becoming only narrower towards the margins and shorter and broadly rhomboidal to rhombic near the apex, and they exhibit remarkable cell laxity due to a bulging and displacement of cell walls perpendicular to the

→

Figure 192. *Pohlia wahlenbergii*. — **1–2**. Habit, dry. **3**. Sterile plant, dry. **4**. Cross-section of stem. **5–6**. Axillary hairs. **7–16**. Leaves. **17–19**. Leaf apices. **20–21**. Upper leaf cells at margin. **22**. Basal cells. **23–25**. Cross-sections of leaves. — [1, 5–7, 9–10, 15 from *RO 922/80*, 3–4, 8, 11, 16, 19, 22–25 from *RO 5086/79*, 20 from *RO 5080/79* and 21 from *RO 5079/79*; 2, 14, 18 from *RILS 5832B*, 12 from *RILS 5824E* and 13, 17 from *RILS 5825B*]. — Scale bars: a – 100 µm (4, 17–22); b – 50 µm (5–6); c – 100 µm (23–25); d – 1 cm (1), 4 mm (2), 2 mm (3) and 1 mm (7–16).

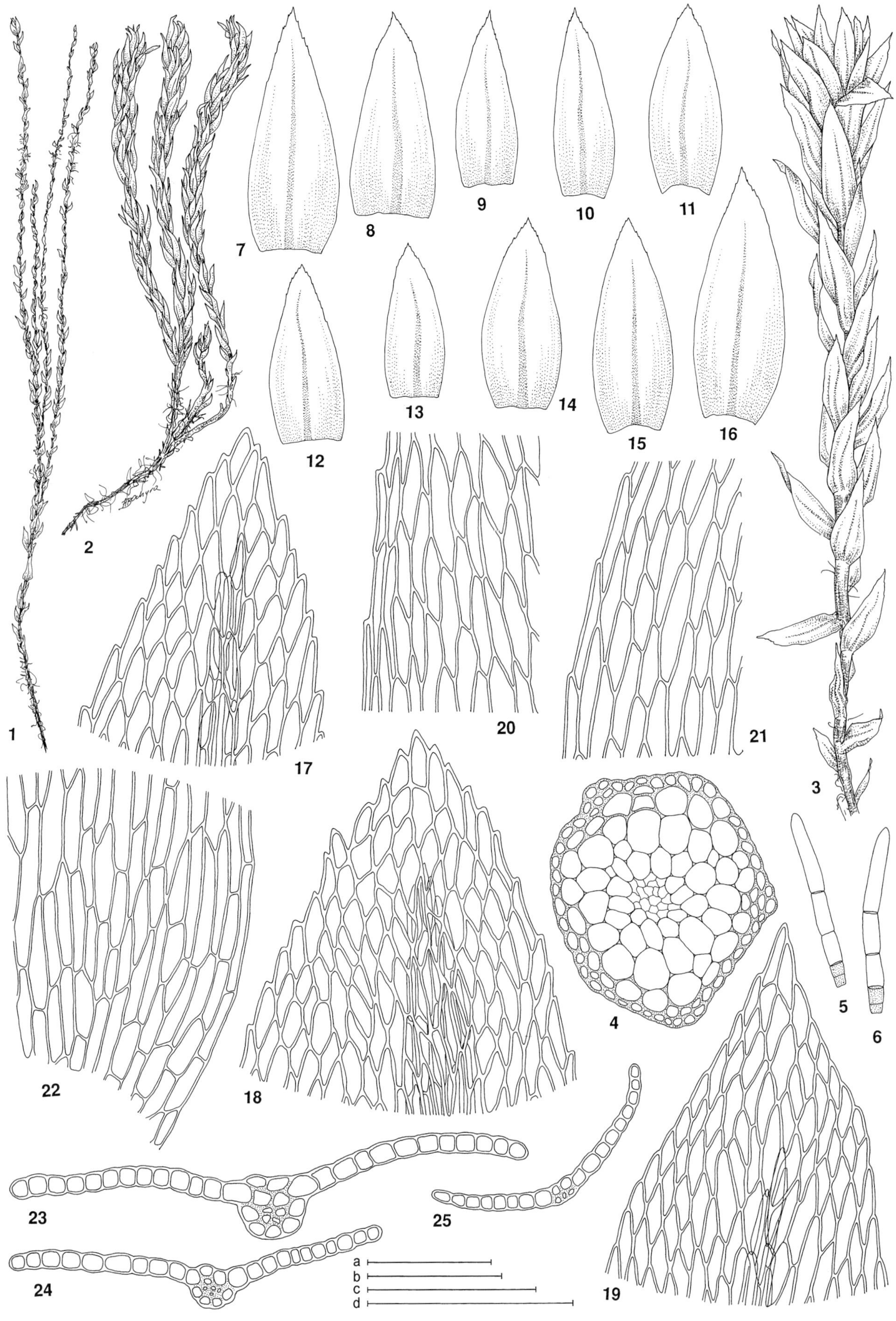
1
2
3
4
5
6
7
8
9
10
11
12
13
14
15
16
17
18
19
20
21
22
23
24
25
a
b
c
d

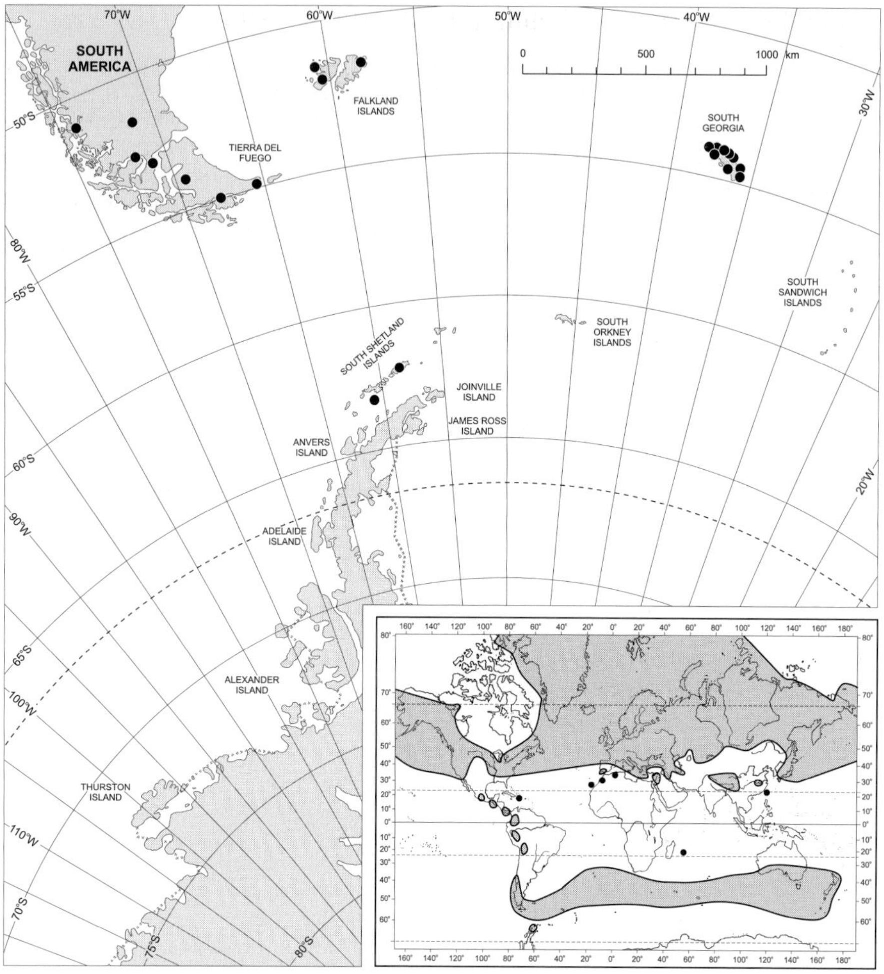

Figure 193. Distribution of *Pohlia wahlenbergii* in the Fuegian region and Antarctica. Inset: global distribution.

plane of the leaf. The lax leaf areolation is generally considered to be clear-cut and safe taxonomic character in *P. wahlenbergii* in temperate latitudes, but at higher latitudes it may be a little misleading. This is because some bryalean mosses, including *Pohlia cruda*, *P. nutans*, *P. drummondii* and most species of *Bryum*, show a greater tendency to cell laxity in the severe and harsh climatic conditions in Antarctica (Lightowlers 1983).

Pohlia cruda may be confused with *P. wahlenbergii* as the stem apices and young shoots of the former species are often pale green or whitish with a characteristic bluish sheen and the laminal cells may show varying degrees of laxity. Such plants have been detected in various parts of the maritime Antarctic, for example on Brabant I. (*AM 56B*, AAS, KRAM) and, additionally, they are small with their habit unlike typical expressions of *P. cruda*. Nonetheless, the laminal cells of *P. cruda* are long-rhomboidal with a sigmoid curve and they are, at most, 18 μm wide, whereas those in *P. wahlenbergii* are always straight and usually wider than 15 μm. Moreover, the leaves in *P. cruda* are ovate to elliptic and narrowed at the insertion, giving them a cordate appearance, with the widest point at about one third of the leaf length from the leaf base. In contrast, the leaves in *P. wahlenbergii* are ovate to ovate-lanceolate, not distinctly narrowed at the insertion, and not cordate, gradually narrowed upwards from the widest point at about one fifth of the leaf length above the leaf base. However, the possibility of confusion between these two species is minimal, because *P. cruda* can be distinguished visually by the light green, strongly lustrous leaves with a strong red costa, whereas those of *P. wahlenbergii* are dull to weakly lustrous, varying between reddish or pinkish and green in colour.

Cell laxity is also observed in some Antarctic populations of *Pohlia drummondii* (e.g. *RO 873/80*, KRAM), and this character coupled with the pale green coloration of the plants may also suggest similarity to *P. wahlenbergii*. In such plants the majority of the laminal cells are narrower, to 14 μm wide, the leaves are long-decurrent and distinctly carinate, and they often produce axillary bulbils.

Pohlia wahlenbergii is generally considered as a species in which asexual reproduction is unknown. Only Nyholm (1958, 1993) provided information that the 'vegetative propagation sometimes occurs by means of dark gemmae or short branchlets found in the leaf axils' in this species. Also, Seppelt (2004) stated that 'axillary bulbils have been found in some collections' from subantarctic Macquarie I. and considered them to be a key character for *P. wahlenbergii*. Unfortunately, neither author gives a detailed description or illustrations of these propagules, although this unusual discovery would certainly deserve more attention.

Pohlia wahlenbergii was described as a new species, *Hypnum wahlenbergii*, by Weber & Mohr (1807) on the basis of a specimen collected by H. Ch. Funck in Bavaria in southern Germany, but they also cited the specimen(s) named *Bryum albicans* by G. Wahlenberg for his then unpublished *Flora lapponica*. The original herbarium of these authors is largely unknown and many collections from their herbarium preserved at KIEL were destroyed during World War II (Isoviita 1979). Also, the specimen gathered by Funck near Regensburg could not be located at M or other herbaria consulted. Therefore, the specimen collected in 1800 by G. Wahlenberg in Lapponia, Norway, and cited in his *Flora lapponica* under *Mnium albicans* (Wahlenberg

1812) is here selected as lectotype. This material was certainly studied by F. Weber and D. Mohr during their visit to Sweden in 1803 (Weber & Mohr 1804).

Reproduction in Antarctica. Sterile.

Habitat. On the volcanic Deception I. *Pohlia wahlenbergii* is associated with heated ground in fumarole areas, growing on cemented soil, ash and scoria and in cracks in rocks in wet places in association with *Cryptochila grandiflora* and *Ditrichum conicum*. On King George I. the species thrives at the margin of melt water channels and in wet crevices in andesite rocks.

World range. *Pohlia wahlenbergii* is generally considered a cosmopolitan species, although it should appropriately be designated as a bipolar species with some intermediate stations in the tropics (Fig. 193, inset). It has a continuous range in the Northern Hemisphere reaching its northernmost stations at lat. *ca* 82°N on northern Ellesmere I. in the Nearctic (Brassard 1971), Peary Land in northern Greenland (Holmen 1960) and Spitsbergen (Kuc 1973*b*), but appears to be absent from the Canadian Arctic Archipelago (Shaw 1982*b*). In the southern Holarctic the species extends to California, Arizona, Arkansas, Louisiana and North Carolina (Shaw 1982*b*) in North America (Shaw 1982*b*), the Mediterranean region, northern Turkey and the Caucasus in Europe, the Dead Sea (Herrnstadt & Heyn 2004), central Iraq (Agnew & Vondraček 1975), northern Iran (Frey & Kürschner 1977), Afghanistan (Froehlich 1955), northern Mongolia (Tsegmed 2001*a*,*b*) and Japan (Ochi 1959), with an isolated station in the Canary Is. (Duell 1980), Morocco and Algeria in North Africa (Jelenc 1955), in the Himalayan region (Gangulee 1974), Sichuan, Yunnan and central east China (Redfearn *et al.* 1996) and Taiwan (Ochi 1959; Chuang 1973). In the Southern Hemisphere *P. wahlenbergii* is widely scattered but rather infrequent in the temperate zone, including the *Nothofagus* zone on the western fringes of southern South America from the Valdivian Province to Tierra del Fuego (Greene 1986; Ochi 1990*a*), on the Falkland Is. (Matteri 1986), Tristan da Cunha (Dixon 1960; Wace & Dickson 1965), Gough I. (Dixon 1934*b*; Wace & Dickson 1965), in southern Australia (Streimann & Klanzenga 2002) and in New Zealand and offshore islands (Sainsbury 1955; Vitt 1974*a*, 1979*a*), on some subantarctic islands, including South Georgia (Clarke 1973*a*), Marion I. (Zanten 1971), Îles Kerguelen (Hébrard 1970*a*) and Macquarie I. (Seppelt 2004), as well as rarely penetrating the northern maritime Antarctic. In the tropics *P. wahlenbergii* is exceedingly rare and is confined in its occurrence to the Neotropics where it occurs at altimontane elevations along the Andean Cordillera from southern Mexico to Bolivia, with an isolated station in the Dominican Republic (Shaw 1982*b*; Allen 2002; Churchill *et al.* 2000). It is absent from the Palaeotropics, except for one unspecified record from Mauritius (O'Shea 2003).

Distribution in Antarctica. A very rare and infrequent species, known only from King George I. and Deception I. in the South Shetland Is. (Fig. 193), reaching its southernmost occurrence at lat. 62°59′S on the latter island. The moss occurs from near sea level to about 500 m.

Specimens examined. WEST ANTARCTICA. SOUTH SHETLAND IS. **King Goerge I.** ***Admiralty Bay***: Ezcurra Inlet, Dutkiewicz Cliff, 5 m, *RO 5079/79*, 10 m, *RO 5080/79* and 20 m, *RO 5086/79* (KRAM); Breccia Crag, 30 m, *RO 922/80* (KRAM). **Deception I.** Hill 1.5 km NNW of Mt. Pond, 410–525 m, *RILS 6904* (AAS, KRAM); ridge north of Mt. Pond, 485–500 M, *RILS 11161B*,, *11163B & 11167* (AAS, KRAM); north summit of Mt. Pond, 480 m, *RILS 5824B, E, 5825B, 5831C & 5832B* (AAS, KRAM); Wensleydale Cliff, 60 m, *RILS 5603* (AAS, KRAM); south end of Fumarole Bay, *ca* 125 m, *RILS 11274A, 11281 & 11283A* (AAS, KRAM); 2 km north of Vapour Col, *ca* 175 m, *RILS 11261C, 11264 & 11266D* (AAS, KRAM).

2. **Pohlia wilsonii** (Mitt.) Ochyra Fig. 194

Pohlia wilsonii (Mitt.) Ochyra, *comb. nov.* ≡ *Bryum wilsonii* Mitt., Hooker's J. Bot. Kew Gard. Misc. 3: 55. 1851. ≡ *Brachymenium wilsonii* (Mitt.) A.Jaeger, Ber. Thätigk. St. Gallischen Naturw. Ges. 1873–1874: 112. 1875. ≡ *Webera wilsonii* (Mitt.) A.Jaeger, Ber. Thätigk. St. Gallischen Naturw. Ges. 1873–1874: 126. 1875. ≡ *Leptobryum wilsonii* (Mitt.) Broth. *in* Engl. & Prantl, Nat. Pflanzenfam. 1(3): 546. 1903. ≡ *Wollnya wilsonii* (Mitt.) Herzog, Beih. Bot. Centralbl. 28: 271. 1911. — Type citation: All new species described by Mitten (1851) were collected 'by Professor W. Jameson in the vicinity of Quito'. [Lectotype (*selected here*): '36. Bryum Wilsonii Mitten Mss Quito Jameson' – NY-Mitten!; isotypes: '36 Brachym. pyriforme Wils. B. Wilsonii Mitten in Jameson's mosses Quito Jameson' – BM-Hooker! (2 specimens), NY-Mitten! (2 specimens); syntype: 'Brachymenium pyriforme Cerro Pelado S. America' – NY-Mitten!].

Wollnya stellata Herzog, Beih. Bot. Centralbl. 26: 69, *t. 3a–d*. 1910. ≡ *Leptobryum stellatum* (Herzog) Broth. *in* Engl., Nat. Pflanzenfam. Ed. 2, 10: 374. 1924. — Type citation: Bolivia; An feuchten Felsen im Llavatal bei Cochabamba ca. 3800 m; Januar, 08. [Lectotype (*selected here*): 'Herbarium Haussknecht, Jena Flora von Bolivia. Willnya stellata nov. gen. nov. spec. An Felsen neben dem Weg im Valle de Llave bei Cochabamba ca. 3600 . leg. Th. Herzog Januar, 08' – JE-Herzog!; isotype: H-Brotherus!, JE-Herzog! (2 specimens)]. First synonymised with *Leptobryum wilsonii* by Arts (2001: p. 327).

Plants aquatic, submerged, forming loose tufts, relatively stiff and fragile, lustrous, pale yellowish to pale yellowish-green above, blackish-brown below. *Stems* erect, simple or sparingly branched, to 13 cm tall, deeply red to reddish-brown, strongly shining, densely matted throughout with a tomentum of reddish-brown to brown, branched, glistening, smooth or finely roughened rhizoids, in transverse section oval to obtusely pentagonal, consisting of a large central strand, 2–3-layered medulla of large, hyaline, thin-walled cells, with undifferentiated or a 1(–2)-stratose cortex of somewhat smaller, reddish-brown cells similar to the medullary cells; *axillary hairs* filiform, 3–4-celled, consisting of 2–3 short, quadrate to short-rectangular, brownish basal cells and 1–2 elongate, hyaline distal cells. *Leaves* widely spaced, soft and delicate, erect-spreading to widely patent or squarrose, flexuose when dry, oblong-lanceolate, lanceolate to ovate, 1.0–1.3 mm long, 0.25–0.35 mm wide, acute to subacute, plane, non-decurrent; *margins* plane, unistratose, unbordered, entire to sinuolate above; *costa* single, slender, ceasing far below the apex, 20–40 μm wide at the base, yellow-brownish, usually weakly differentiated from the laminal cells and sometimes subconcolorous, in transverse section elliptical, scarcely convex on the dorsal side, with a dorsal and ventral row of large epidermal cells, 2–3 large guide cells and small dorsal stereid or substereid band; *laminal cells* unistratose, smooth, transparent, thin-walled and lax, oblong-rhomboidal, oblong-hexagonal, rectangular to linear-rectangular, 50–200 μm long, 14–20(–24) μm wide, not differentiated in the basal angles. *Asexual reproduction* by means of spherical to ellipsoid, brown to dark brown, 150–200 μm long, 120–150 μm wide tubers, composed of 10–15 non-bulging cells, produced on short stalks on rhizoids and in the leaf axils. [*Dioicous*. *Perigonia* terminal, subcapitate-disciform; *perigonial bracts* 3–4 mm long, from an ovate, concave base abruptly narrowed to a lorate-subulate, flexuose, horizontally spreading to recurved acumen, with a broad, diffusing costa and lax areolation and numerous, club-shaped, orange-brown antheridia intermixed with filiform, hyaline to pale yellowish-hyaline paraphyses slightly exceeding the antheridia. *Perichaetia* terminal; *perichaetial leaves* linear-lanceolate, subulate, 2.5–4.0 mm long, straight, erect, with a relatively broad, indistinct and subconcolorous costa, ending below the apex, subentire to bluntly serrulate-sinuolate, with a lax areolation of elongate-hexagonal to linear-rectangular cells and many archegonia intermingled with hyaline to pale yellowish-hyaline paraphyses, slightly exceeding the archegonia. *Setae* erect, elongate, 1.3–3.5 cm long, fairly slender, flexuose, reddish- or yellowish-brown to pale brown. *Capsules* long-exserted, erect to inclined, pyriform from a long, narrow neck, 1.5–3.0 mm long, narrowed at the mouth, brown, lustrous; *annulus* not revoluble, 2–3-seriate, composed of large, vesiculose, thick-walled cells; *operculum* convex-conic, apiculate; *exothecial cells* variable in size and shape, isodiametric to elongate, varying from quadrate, irregular, oblong-hexagonal to rectangular, 25–75 μm long, 10–20 μm wide, with moderately thickened, wavy, collenchymatous walls, becoming smaller, rounded to oblate in 2–3 tiers at the orifice; *stomata* numerous in the neck, superficial, bicellular, with elongate, narrow pori; *peristome* double; *exostome teeth* 16, pale yellowish, lanceolate, acute, unbordered, 350–400 μm long, 80–85 μm wide at the base, finely papillose throughout, coarser at the apices, trabeculate on the inner side; *endostome* as tall as the exostome, pale yellowish to hyaline, finely roughened, consisting of a basal membrane about half the height of the exostome, keeled, fairly broad, narrowly perforate segments and no or 1–2 reduced cilia. *Spores* spherical, pale brownish, finely roughened, 18–35 μm in diameter. *Calyptra* cucullate, naked, smooth.]

Chromosome number: $n = 23$ (Kanda & Okada 1993, as *Dicranella* sp.).

Discussion. The above description of gametophytes of *Pohlia wilsonii* is based on the Antarctic plants which are morphologically highly modified owing to submergence in water. Typically, this species grows in loose or dense tufts that are yellow-green and strongly lustrous and the stems in the lower part are conspicuously tomentose due to the dense brown rhizoids. Yet the leaves, except for being only remotely disposed on the stem, retain the shape and areolation typical of this species which make it very distinct and difficult to mistake for any other moss species in Antarctica. Some submerged plants of *Bryum pseudotriquetrum* are morphologically similar to *P. wilsonii* and may be mistaken when growing occasionally in mixed stands (Kanda & Mochida 1992). However, *B. pseudotriquetrum* never produces rhizoidal tubers which are found in great abundance in *P. wilsonii* (Imura & Kanda 1986) and, additionally, they differ in leaf shape and areolation. The leaves in *P. wilsonii* are oblong-lanceolate, lanceolate, seldom ovate, acute or subacute at the apex and the laminal cells are oblong-rhomboidal, oblong-hexagonal, rectangular to linear-rectangular, whereas the inundated plants of *B. pseudotriquetrum* have lanceolate and long-acuminate or ovate to elliptical leaves, often with obtuse apices and the laminal cells are short, rhomboidal to hexagonal.

The discovery of *Pohlia wilsonii* in Antarctica has a long and complex history. This moss was discovered in the Sôya Station area by Nakanishi (1977) who suggested that it was superficially similar to *Bryum korotkevicziae* or its variety, var. *hollerbachii* (= *B. pseudotriquetrum*). Ochi (1979) and Imura & Kanda (1986) treated it as *Bryum* sp., Kanda & Iwatsuki (1989) suggested it was a *Dicranella* but without a specific designation, whereas Imura *et al.* (1992) and Kanda & Mochida (1992) assigned this puzzling moss to *Leptobryum pyriforme*. Finally, Ochyra *et al.* (1998) identified it as *Leptobryum wilsonii* and this identity was confirmed by Arts (2001). Although the Antarctic material perfectly fits the type and numerous non-type collections of *P. wilsonii* from South America, this identity was rejected by Imura *et al.* (2003) because in this normally dioicous species there have been found synoicous inflorescences on shoots from cultured tubers (Imura *et al.* 1992). This does not seem to be a strong evidence because sexual condition often varies in species of *Pohlia* and in every species such deviations are possible, especially in cultured material.

Pohlia wilsonii was described as *Bryum wilsonii* from material collected in the Andes of Ecuador by Mitten (1851) who, inspired by the colour of its leaves and the capsule shape, showed the close affinity of this species to *Leptobryum pyriforme* and half a century later Brotherus (1903) formally transferred it to *Leptobryum*. Since then the species has been firmly established as a member of this genus, and such placement was approved by Arts (2001) in his survey of *Leptobryum* as one of two species belonging to it. Although *P. wilsonii* is not very variable in its basic diagnostic traits, the species has been described several times from South America, and the authors suggested various generic placements for these species, including *Leptobryum* (Dusén 1903*b* as *L. pottiaceum*; Thériot 1922 as *L. escomelii*), *Webera* Cardot 1911*d*) and *Mniobryum* (Bartram 1933*b* as *M. bracteatum*). Herzog (1910) described the monotypic genus *Wollnya* from Bolivia with a single species, *W. stellata* and, with some reservation, placed it in the Bryaceae. A year later he revisited this genus (Herzog 1911) and found that *Leptobryum wilsonii* was closely related to, though specifically distinct from, *Wollnya stellata* and transferred it to *Wollnya*. Also, he transferred the latter genus to the Bryaceae. Arts (2001) considered both species of *Wollnya* to be conspecific and returned *W. wilsonii* to *Leptobryum*, thus inferring the congenericity of *Leptobryum* and *Wollnya*.

Leptobryum is certainly not the appropriate genus for *L. wilsonii*, since this species is gametophytically completely unlike *L. pyriforme*, the generitype of *Leptobryum*. The slenderly lanceolate lower leaves and setaceous upper leaves with a broad costa that occupy about one third of the leaf base width, as well as the narrow, long-linear, sturdy laminal cells in *L. pyriforme*, are quite unlike the oblong-lanceolate, delicate leaves with a narrow costa and broad, oblong-hexagonal to linear-rectangular laminal cells in *L. wilsonii*. On the other hand, both species share

→

Figure 194. *Pohlia wilsonii*. — **1**. Sterile plant, dry. **2**. Plant with mature capsule, wet. **3**. Cross-section of stem. **4–6**. Axillary hairs. **7–15**. Leaves. **16–17**. Leaf apices. **18**. Mid-leaf cells. **19**. Basal cells. **20–24**. Cross-sections of leaf, a sequence from apex to base. **25–29**. Rhizoidal tubers. — [1, 3, 7–13, 16–29 from *Nakanishi 800710-033*, NIPR; 2, 4–6, 14–15 from *Herzog 4831*, JE]. — Scale bars: a – 100 μm (3) and 200 μm (25–29); b – 2 mm (1–2), 100 μm (4–6, 20–24) and 200 μm (16–19); c – 1 mm (7–15).

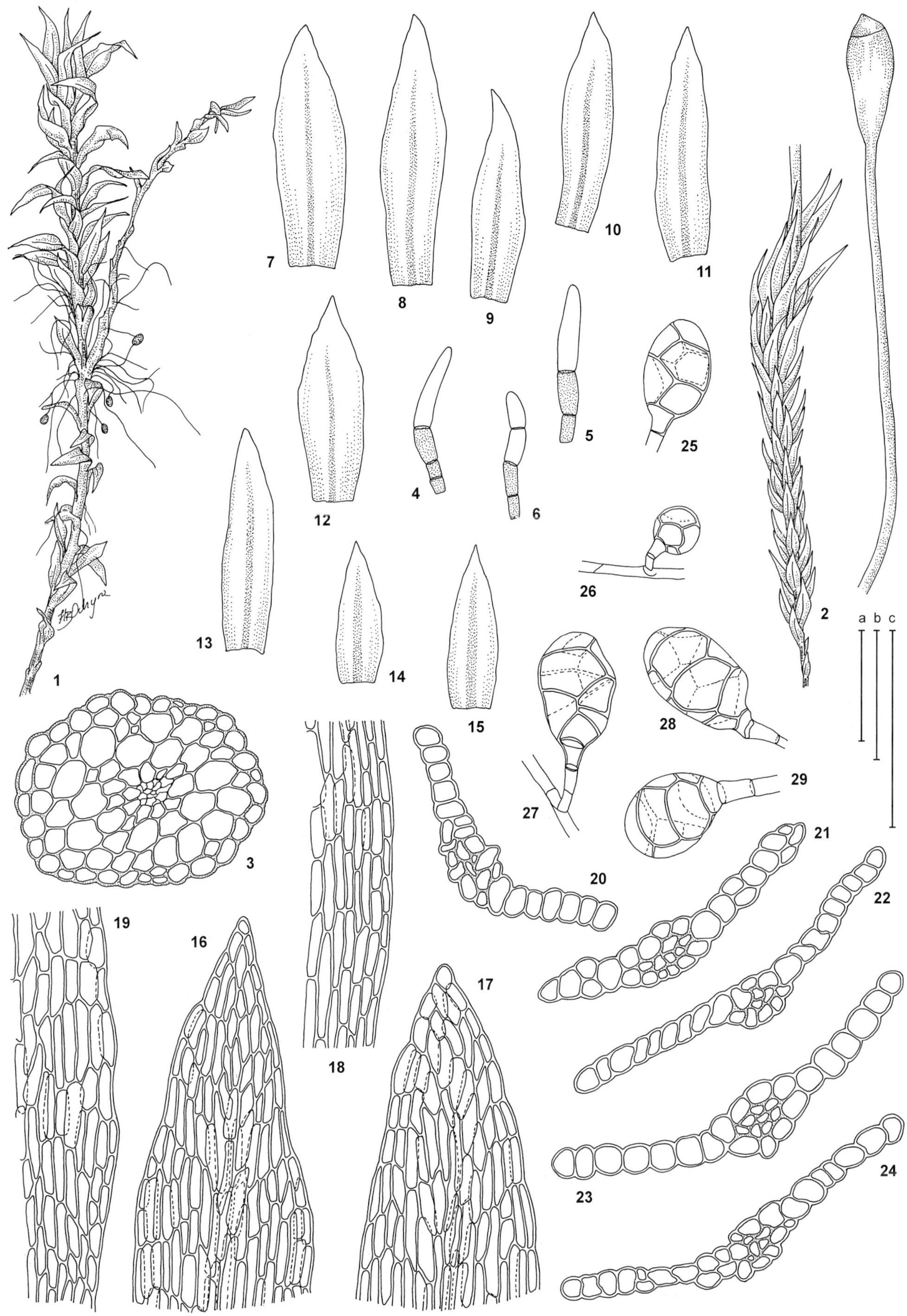
1
2
3
4
5
6
7
8
9
10
11
12
13
14
15
16
17
18
19
20
21
22
23
24
25
26
27
28
29
a b c

nearly identical capsule shape with the perfect *Bryum* peristome-type, although in *L. pyriforme* the cilia are well developed and appendiculate in pairs to tetrads and the annulus is revoluble, whereas in *L. wilsonii* they are absent or rudimentary and the annulus is irregularly separated. These species differ essentially in the shape of the exothecial cells which have straight walls in *L. pyriforme* and strongly wavy walls, especially in the apophysis, in *L. wilsonii*. This character, together with the areolation and shape of the vegetative leaves, clearly indicates the relationship of *L. wilsonii* to *Pohlia* subgen. *Mniobryum* sect. *Cacodon* and, accordingly, this species is here transferred to this genus as *Pohlia wilsonii*, and the genus *Wollnya* is considered congeneric with *Pohlia*. Such a position of this species was inferred by Shaw (1982*b*) who transferred *Webera integra* to *Pohlia*. However, this author was unaware of either the conspecificity of this species with *P. wilsonii* (Arts 2001) or the sporophyte characters of this species, which are critical for establishing the present placement of this species.

Cytological data also seem to confirm the present generic placement of *Pohlia wilsonii*. Kanda & Okada (1993) reported the chromosome number of the Antarctic plants to be $n = 22 + m$. They are evidently diploids since the basic chromosome number is $x = 11$ in the genus *Pohlia*, whereas it is $x = 10$ in *Leptobryum* (Steere 1972; Fritsch 1991).

Reproduction in Antarctica. *Pohlia wilsonii* produces sporophytes abundantly in most South American stations but the submerged Antarctic plants are entirely sterile. In contrast, they form gemmae in great profusion on rhizoids and in the leaf axils which readily germinate in culture and develop plants with synoicous inflorescences (Imura *et al.* 1992).

Habitat. In Antarctica *Pohlia wilsonii* grows totally submerged on the bottom of lakes at depths of 2 to 5 m. Here it forms columnar growth, with *Bryum pseudotriquetrum* and algae, up to 50–75 cm tall (Lewis Smith 2000). This is an unusual habitat for this species which typically is a terrestrial moss growing on soil, clay or in rock crevices, but often in moist or wet places in seepage areas or on stream banks. There are numerous examples of mosses and liverworts that are capable of growing successfully deeply submerged (e.g. Persson 1942; Light 1975; Light & Lewis Smith 1976; Malme 1978; Ignatov & Kurbatova 1990; Winton & Beever 2004). This phenomenon is frequently observed in the Antarctic and several species of mosses and liverworts, including *Bryum pseudotriquetrum*, *Campylium polygamum*, *Drepanocladus longifolius*, *Plagiothecium orthocarpum*, *Warnstorfia sarmentosa*, and *Pachyglossa dissitifolia* have been found at various depths, from 2 to 33 m, in lakes (e.g. Savicz-Lyubitskaya & Smirnova 1959, 1960, 1964*a,b,c*, 1965; Light & Heywood 1973, 1975; Priddle 1979; Kaspar *et al.* 1982; Seppelt 1983; Kanda & Ohtani 1991; Imura *et al.* 1992, 2003; Kanda & Mochida 1992; Tewari & Pant 1996). Such plants are usually highly modified, yet the diagnostic features of each species are retained.

World range. An amphiatlantic south-temperate species, with the maximum occurrence in South America where it ranges along the Andean Cordillera from Tierra del Fuego to Ecuador, occurring in the Central and North Andes at high elevations only, from 1300 m in Peru to 4600 m in Bolivia (Churchill *et al.* 2000), with a highly isolated, disjunct station in central Mexico (Shaw 1982*b*), and once discovered in Lesotho in southern Africa (Arts 2001) and on the Antarctic continent (Fig. 195).

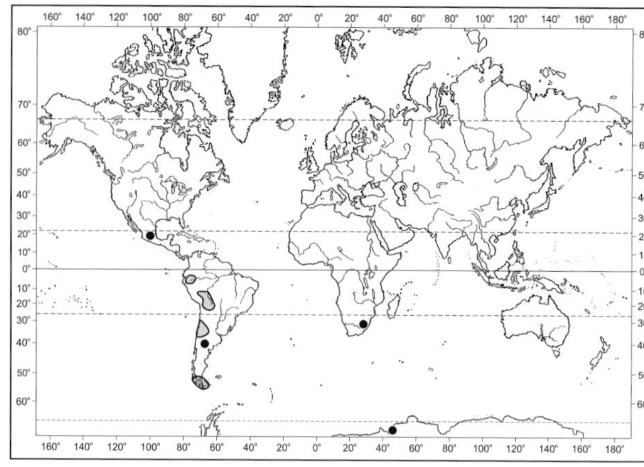

Figure 195. Global distribution of *Pohlia wilsonii*.

Distribution in Antarctica. Known only from a few lakes in the Skarvsnes region on the Sôya Coast, Enderby Land, in continental Antarctica at lat. 69°29′20″ S (Kanda & Mochida 1992) (Fig. 195). Tewari & Pant (1996) reported an unnamed *Leptobryum* from Zub Lake in Schirmacher Oasis at lat. 70°45′ S in Dronning Maud Land but the voucher specimen has not been available for study. However, these plants apper to be *Pohlia wilsonii*, at least judging from the illustrations of the leaves and gemmae presented by these authors.

Specimens examined. EAST ANTARCTICA. ENDERBY LAND. **SÔYA COAST.** Skarvsnes, Mt. Suribachi, Kokesumi Lake, 70 m, *Nakanishi C-87 & C-88* (AAS, KRAM, NIPR).

3. **Pohlia drummondii** (Müll.Hal.) A.L.Andrews Fig. 196

Pohlia drummondii (Müll.Hal.) A.L.Andrews *in* Grout, Moss Fl. N. Amer. 2: 196. 1935. ≡ *Bryum drummondii* Müll.Hal., Bot. Zeit. (Berlin) 20: 328. 1862. ≡ *Webera drummondii* (Müll.Hal.) A.Jaeger, Ber. Thätigk. St. Gallischen Naturw. Ges. 1874–1875: 137. 1876. — TYPE CITATION: In summitatibus mont. Rocky Mountains Americae septentrionalis, locis humidis: Drummond specimina pulcherrima legit. [Lectotype (*vide* Shaw 1981*c*: p. 27): 'Bryum Drummondii mihi nsp. Rocky Mountains: Drummond N. Amer. N° 263' – BM-Hampe!].

Bryum inflexum Müll.Hal. *in* Neumayer, Int. Polarforsch. Deutsch. Exp. 2: 296. 1890. ≡ *Webera inflexa* (Müll.Hal.) Kindb., Enum. Bryin. Exot. Suppl. 2: 108. 1891. ≡ *Pohlia inflexa* (Müll.Hal.) Wijk & Margad., Taxon 14: 197. 1965. — TYPE CITATION: Austro-Georgia, 'Bach-Grund am Ausgange des Brocken-Thales' inter Psilopilum tapes n. sp., 23. Januario 1883. [Holotype: B-Müller Hallensis†; neotype (*selected here*): 'British Antarctic Survey Flora of South Georgia Pohlia inflexa (C. Muell.) Wijk & Marg. Wet rocks by waterfall. Alt. c. 75 ft. South-west side of valley, behind north-west shore of Moltke Harbour,

→

Figure 196. *Pohlia drummondii.* — **1**. Habit, dry. **2**. Sterile shoot, dry. **3**. Sterile shoot, wet. **4**. Cross-section of stem. **5–6**. Axillary hairs. **7–13**. Leaves. **14–15**. Leaf apices. **16**. Mid-leaf cells. **17**. Mid-leaf cells at margin. **18**. Basal cells. **19–21**. Cross-sections of leaves. **22**. Cross-section of costa. **23–26**. Gemmae. — [1–4, 9, 19–22 from *RILS 3986*; 5–6, 10–11, 14–18, 23–25 from *RO 873/80*; 7 from *RO 2392/80*; 8 from *RO 4915/79*; 12 from *RO 867/80*; 13 from *AM 54A*; 26 from *RILS 5861A*]. — Scale bars: a – 1 mm (2–3); b – 0.5 cm (1) and 100 μm (4); c – 100 μm (14–22); d – 1 mm (7–13); e – 1 mm (23–26) and 100 μm (5–6).

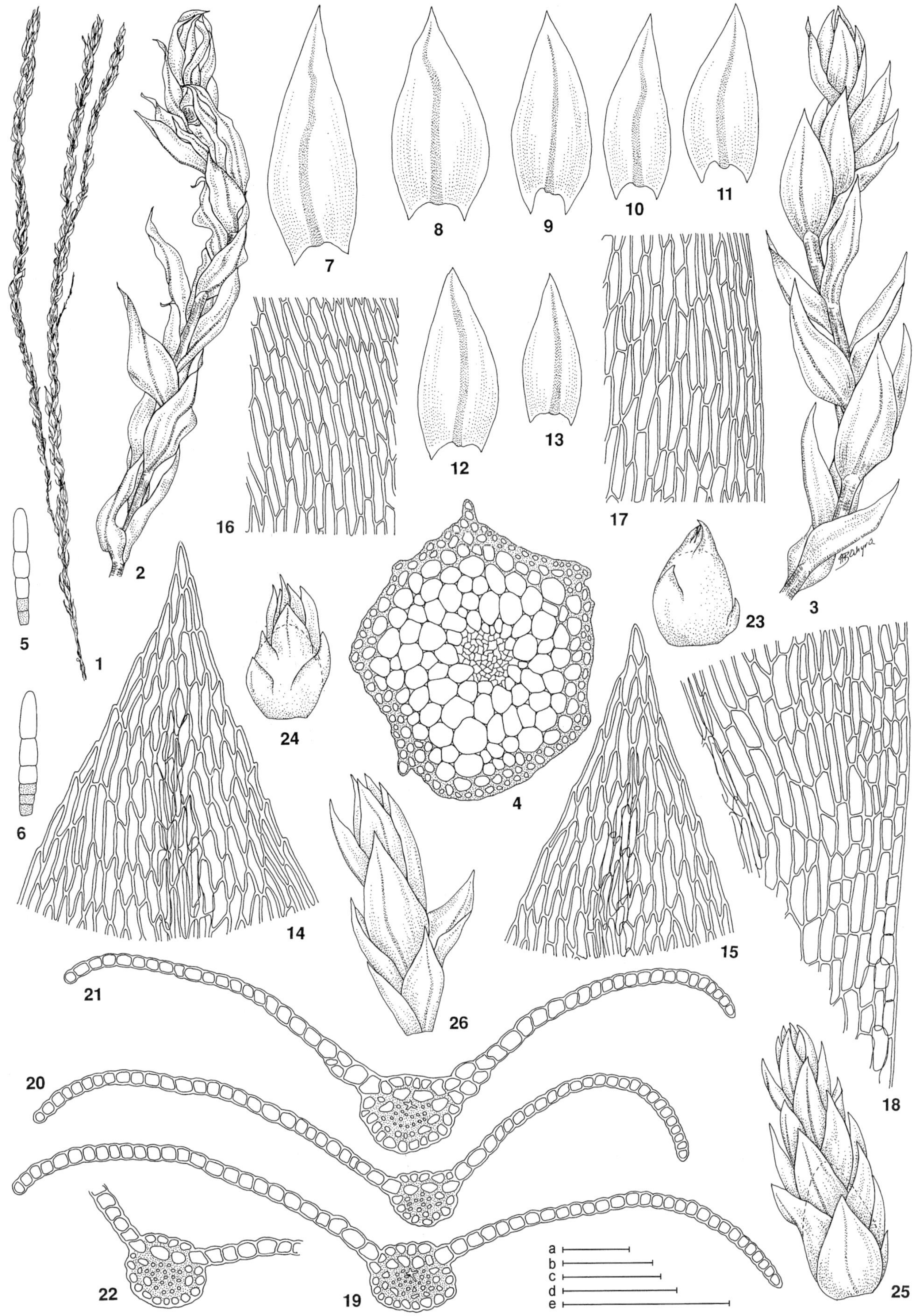
1
2
3
4
5
6
7
8
9
10
11
12
13
14
15
16
17
18
19
20
21
22
23
24
25
26
a
b
c
d
e

Royal Bay. GR 159098 Leg. S. W. Greene 22.2.1961 Det. C. G. S. Clarke, February 1970 No. Greene 2376' – KRAM!]. First synonymised by Ochyra (1998*a*: p. 201).

Plants small to medium-sized, slender, forming fairly loose tufts or extensive turves, glossy, yellow, yellow-green or yellow-brown, occasionally with a reddish hue. *Stems* erect, (0.5–)1.5–3.5(–4.5) cm high, simple or sparsely branched, red to blackish-red, except at the apex, often matted with reddish-brown, lustrous, finely papillose rhizoids in the lower part, in transverse section pentagonal, with a large and distinct central strand, 3–5-layered medulla of large, hyaline, thin-walled cells and (1–)2(–3)-stratose cortex of small, thick-walled, red to red-brown cells; *axillary hairs* short, filiform, 5–6 celled, with 2–3 quadrate, reddish-brown basal cells and 2–3 hyaline, short-rectangular distal cells. *Leaves* evenly and rather widely spaced, erect-appressed, catenulate, folded and somewhat twisted on drying, erect-spreading when moist, oblong-lanceolate, ovate to ovate-lanceolate, 0.9–2.3(–2.8) mm long, 0.5–0.9 mm wide, acute to broadly acute, carinate, long-decurrent; *margins* serrulate to crenate near the apex, entire below, plane to deflexed above, broadly recurved below; *costa* single, percurrent or ending below the apex, relatively strong, 80–100 μm wide at the base, gradually tapering upwards, flexuose above, red at the base, reddish-brown above, in transverse section strongly convex dorsally, consisting of 3–5 large guide cells, a strong dorsal stereid band and a row of ventral and dorsal epidermal cells; *laminal cells* unistratose throughout, smooth, pellucid, thin-walled, occasionally lax, 40–90(–120) μm long, 10–14(–15) μm wide, oblong-hexagonal or oblong-rhomboidal near the apex, becoming linear-rhomboidal to linear-flexuose towards the base, undifferentiated at the margins or slightly narrower in a single row. *Asexual reproduction* by ovoid to obloid gemmae occasionally present, arising singly in upper leaf axils, 300–600 μm long, green when young, red-brown with age, with 3–8, conspicuous, triangular-laminate leaf primordia scattered on the bulbiform gemma body as well as at the apex. [*Dioicous*. *Perigonia* terminal, bulbiform; *perigonial bracts* concave, ovate, broadly acute to long-acuminate. *Perichaetia* terminal, with perichaetial bracts not much differentiated from the vegetative leaves. *Setae* 1.5–3.5 cm tall, erect, flexuose, orange-red to red-brown. *Capsules* pendulous, broadly pyriform, 1.8–2.5 mm long, with a short to elongate neck; *annulus* irregularly separating, consisting of 2–3 rows of large, vesiculose cells with strongly incrassate walls; *operculum* low-convex, apiculate; *exothecial cells* isodiametric to hexagonal, 15–40 μm long, 10–20 μm wide, with thick, sinuose walls; *stomata* many in the neck, superficial, bicellular, with narrow, elongate pori; *peristome* double; *exostome teeth* 16, lanceolate, acute, yellow to reddish-yellow, unbordered, finely papillose below, more strongly so at the apex, trabeculate on the inner surface; *endostome* hyaline, finely papillose to glabrous, with broad, narrowly perforate segments arising from a low basal membrane, one third as high as the exostome teeth, and 1–2 nodulose cilia. *Spores* globose, 18–22 μm in diameter, finely roughened. *Calyptra* cucullate, smooth, naked.]

Chromosome number: unknown in the Antarctic and Subantarctic; n = 11 – a single record from the British Isles (Ramsay 1969).

Discussion. *Pohlia drummondii* is the only gemmiferous species of the genus in Antarctica, although the propagules are only occasionally present. The reddish-brown gemmae are enclosed within the leaves farther down the stem from the apex and are not easy to see until the shoots are studied under the microscope.

The non-propaguliferous plants can be distinguished from other local species of *Pohlia* by having strongly carinate leaves when dry which give them a catenulate appearance. In its strongly lustrous leaves *Pohlia drummondii* is similar to *P. cruda*, but the long leaf decurrencies, red costa, as well as shorter and straight laminal cells should prevent confusion, since the latter has non-decurrent leaves and vermicular laminal cells with a distinct sigmoid curve. *Pohlia drummondii* is also unlikely to be mistaken for *P. wahlenbergii*, which has flat to weakly concave, non-decurrent leaves and mostly dull to slightly glossy leaves, and the laminal cells are very broad, more than 15 μm, and form a very lax areolation. However, in some populations of *P. drummondii* growing in wet conditions, there is a strong tendency to cell laxity (Lightowlers 1983), but, nonetheless, the width of the cells does not exceed 15 μm. *Pohlia nutans* is separated from *P. drummondii* by is thick-walled laminal cells.

Pohlia drummondii was described from the Rocky Mountains by Müller (1862) as *Bryum drummondii* and, for a long time, it was considered to be a North American endemic until Andrews (1935) showed that European *P. commutata* Lindb. was conspecific with it. However, this taxonomic conclusion was questioned by Demaret & Wilczek (1978), who did not accept the separate status of these species, and Shaw (1982*b*) has since confirmed the identity of North American and European populations. Cardot & Brotherus (1923) reported this species from southern South America (as *P. commutata*), and Ochyra (1998*a*) recorded it from King George I. in the maritime Antarctic and, additionally, considered the South Georgian *P. inflexa* to be conspecific with it. The latter species was described from this subantarctic island by Müller (1890*a* as *Bryum inflexum*) and, although the type material of this species name has apparently been destroyed (Clarke 1973*b*), numerous non-type collections clearly confirm the identity of this moss.

Propaguliferous species of *Pohlia* are infrequent in the austral region and, so far, only two endemic species have been found in the temperate zone of the Australasian sector, namely *P. ochii* Vitt on Campbell I. (Vitt 1971*a*) with linear-vermicular gemmae and *P. australis* A.J.Shaw & Fife from the South Island, New Zealand (Shaw & Fife 1985) with spherical to obloid gemmae. The latter species is similar to *P. drummondii* in the shape of the propagules, but it has longer, linear and vermicular laminal cells, 100–150 μm long, and strongly serrate leaves near the apex. In addition, the gemmiferous Holarctic *P. camptotrachela* (Renauld & Cardot) Broth. is known from New Zealand (Fife 1995) and *P. bulbifera* (Warnst.) Warnst. from southern Australia (Streimann & Klanzenga 2002).

Reproduction in Antarctica. Inflorescences and sporophytes are unknown in the Antarctic, but bulbiferous axillary gemmae are often present in the axils of the upper leaves and they provide a means of vegetative propagation.

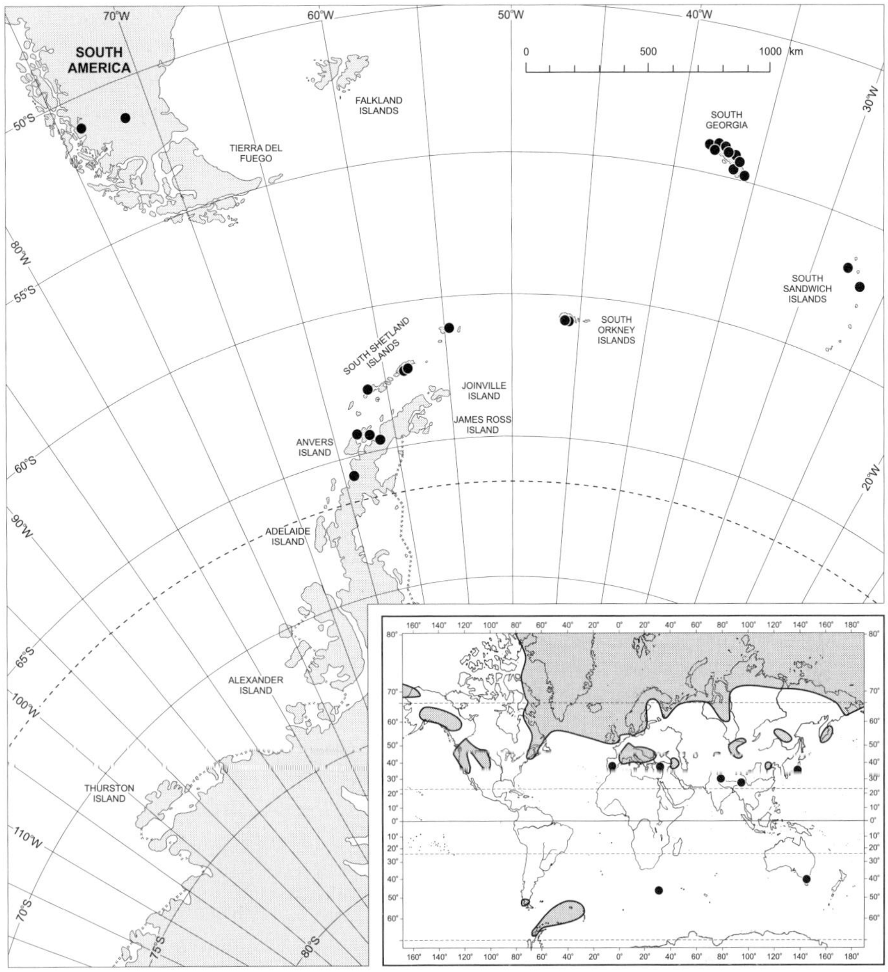

Figure 197. Distribution of *Pohlia drummondii* in the Fuegian region and Antarctica. Inset: global distribution.

Habitat. *Pohlia drummondii* typically grows on sandy, gravelly or glacial soil in wet or moist habitats by stream sides and melt water runnels, as well as in wet flushes, in damp or moist rock cracks or crevices, and once it was found submerged in a temporary shallow pool on King George I. On the volcanic South Sandwich Is. it grows on geothermal ground and on moist rocks near fumarole vents. Its most frequent associates include *Sanionia uncinata*, *Polytrichastrum alpinum*, *Warnstorfia fontinaliopsis*, *Ditrichum hyalinum*, *Meesia uliginosa*, *Cephaloziella varians*, and *Lophozia excisa*.

World range. *Pohlia drummondii* is a bipolar species in the strict sense, i.e. not having intermediate occurrences in the tropics (Fig. 197, inset). In the Northern Hemisphere it has an arctic-boreal-montane range which is continuous throughout northern regions of Eurasia, whereas in North America it is bicentric in distribution, being almost absent in the Nearctic (Shaw 1982*b*). In the southern part of the Holarctic it occurs in the mountains, extending southwards to California and Colorado in western North America and to New Hampshire in the eastern part of the continent (Shaw 1981*c*). In Europe, it is scattered throughout Iceland (Jóhannsson 1995), the British Isles (Wigginton 1994), Scandinavia (Söderström 1998), and the Kola Peninsula (Shlyakov & Konstantinova 1982), and appearing in the Central European mountains from the Pyrenees and Sierra Nevada to the Caucasus. In Asia, the species is widely scattered in Arctic Siberia and recurring in the mountains of Central Siberia and the southern Russian Far East (Czernyadjeva 1999), northern Mongolia (Tsegmed 2001*a*,*b*), north-eastern China (Redfearn *et al.* 1996), Honshu in Japan (Akiyama 1996), Yunnan (Zhang & Li 2002) and Kashmir (Brotherus 1898). In the Southern Hemisphere, *P. drummondii* is predominatly a circumsubantarctic species, being most frequent on South Georgia (Clarke 1973*a*) and penetrating the northern maritime Antarctic, as well as southern South America (Cardot & Brotherus 1923; Matteri 1985*a*) and south-eastern Australia and Tasmania (Shaw 2006, as *Pohlia inflexa*) where it occurs at high elevations. Moreover, it occurs also on Marion I. in the Subantarctic (*Huntley 963D*, GRO, KRAM, PRE).

Distribution in Antarctica. A rare and widely scattered species, restricted in its distribution to the northern maritime Antarctic (Fig. 197). It extends from the South Sandwich Is., through the South Orkney Is. and South Shetland Is. to lat. 65°32′S on the Graham Coast on the western Antarctic Peninsula. It occurs predominantly at low elevations, from near sea level to 100 m, but has been found at 238 m on Elephant I.

Specimens examined. WEST ANTARCTICA. SOUTH SANDWICH IS. **Leskov I.** Summit ridge, 100 m, *PC 245M*, *247E*, *249D & 254B* (AAS, KRAM). **Bellingshausen I.** Southern crater lip access point, 150 m, *PC 46B* (AAS, KRAM).

SOUTH ORKNEY IS. **Coronation I.** Cape Hansen, 20–200 m, *RILS 3228A* (AAS, KRAM, PC). **Lynch I.** North facing slope, *ca* 13 m, *DCL 946B* (AAS, KRAM). **Signy I.** Factory Cove, *RILS 1784* (AAS, KRAM); Paal Harbour, *ca* 80 m, *RILS 424* (AAS, KRAM).

SOUTH SHETLAND IS. **Elephant I.** 0.5 km south-west of Endurance Glacier, 238 m, *JSA 171B* (AAS, KRAM). **King George I.** ***Admiralty Bay***: Rescuers Hills, 15 m, *RO 4915/79* (KRAM); Jasnorzewski Gardens, 4 m, *RO 1495/80* (KRAM); Hala, 18 m, *RO 5216/79* (KRAM), 20 m, *RO 5209/79* (KRAM) & *RO 5212/79* (KRAM), 25 m, *RO 2392/80* (KRAM) and

30 m, *RO 5219/79* (KRAM); Ezcurra Inlet, Pond Hill, 70 m, *RO 874/80* (KRAM) and 140 m, *RO 867/80 & 873/80* (KRAM); MacKellar Inlet, Crépin Pt., 10 m, *RO 2048/80* (KRAM). **LIVINGSTON I.** ***Byers Peninsula***: Dragon Cove, 12 m, *RILS 5861A* (AAS, KRAM).

WEST ANTARCTIC PENINSULA. **DANCO COAST.** Charlotte Bay, Recess Cove, 2–10 m, *RILS 3986* (AAS, KRAM) & *3987* (AAS, BM, KRAM, NY). **PALMER ARCHIPELAGO.** ***Brabant I.***: Hill Bay, 5 m, *AM 54A & 56B* (AAS, KRAM). ***Melchior Is.***: Gamma I., 5–10 m, *RILS 4157A* (AAS, KRAM). **WILHELM ARCHIPELAGO.** ***Grandidier Channel***: Lahille I., 2–5 m, *RILS 4322* (AAS, KRAM).

4. **Pohlia cruda** (Hedw.) Lindb Fig. 198

Pohlia cruda (Hedw.) Lindb., Musci Scand.: 18. 1879. ≡ *Mnium crudum* Hedw., Sp. Musc. Frond.: 189. 1801. ≡ *Hypnum crudum* (Hedw.) F.Weber & D.Mohr, Index Mus. Pl. Crypt.: sheet 2 [4]. 1803. ≡ *Bryum crudum* (Hedw.) Sm., Fl. Brit. 3: 1361. 1804. ≡ *Polla cruda* (Hedw.) Brid., Bryol. Univ. 2: 817. 1827, *comb. non rite publ.* ≡ *Webera cruda* (Hedw.) Fürnr., Flora 12 Ergänzungsbl. 2: 35. 1829. ≡ *Lamprophyllum crudum* (Hedw.) Lindb., Acta Soc. Sci. Fenn. 10: 87. 1871. — TYPE CITATION: Locis udis, montosis ad saxa in Germania passit; also references to Dillenius (1741: t. 51, f. 76), Linnaeus (1763: p. 1112), Murray (1784: p. 947), and Hedwig (1785–1787: p. 99) are provided; in the latter the following specimens are cited: 'ad Liebichenstein Penigii, Lichtenwalde, die rotlufer Schlucht, molendium circa Chemnitium in Saxonia, haud infrequens mihi obviam fuit'. [Lectotype (*selected here*): the first right specimen on the sheet bearing the label 'Bryum pendulum hornum molle foliis et lanceolatis et gramineis Dill. H. M. 401. t. 51 f. 76 Mnium crudum antheris pendulis, calyptra recurvata foliis pellucidis Linn. Spec. pl. 2. p. 1576. n. ii' – G-Hedwig/Schwägrichen!].

Webera cruda var. *imbricata* Cardot, Rev. Bryol. 27: 43. 1900. ≡ *Pohlia cruda* var. *imbricata* (Cardot) E.B.Bartram, Bryologist 60: 141. 1957. — TYPE CITATION: [Antarctica] Xme, XIme et XXme débarquements; later, Cardot (1901: p. 34) gave details of the specimens: 'Détroit de Gerlache: XIème débarquement, Terre de Danco, cap Van Beneden, sur les corniches de la falaise, dans les endroits détrempés. (N° 233^{a}; 1er février 1898). XXème débarquement, Terre de Danco, à l'entrée du chenal de Lemaire, sur des roches isolées au milieu d'un glacier, à 50 mètres au-dessus de niveau de la mer; stérile. (N° 270^{c}; 12er février 1898). Xème débarquement, île Brabant, sur des roches complètement entourées de glace, à 350 mètres au-dessus de niveau de la mer (forme rabougrie); stérile. (N° 252^{c}; 30 janvier 1898). [Lectotype (*vide* Ochyra 1998*a*: p. 200): 'Herb. J. Cardot. Mousses du voyage de la Belgica. n° 270c. Webera cruda Sch. var. imbricata Card. Canal de Gerlache XXème débarquement, Terre de Danco, à l'entrée du chenal de Lemaire, sur des roches isolées au milieu d'un glacier, à 50 mètres au-dessus de niveau de la mer; stérile. 12 février 1898' – PC!; isolectotype: BR!]. First synonymised by Ochyra (1998*a*: p. 198).

Plants medium-sized or, occasionally small, and slender to quite large and fairly robust, densely or loosely caespitose or occurring intermingled with shoots of other mosses, soft, light green verging to whitish-, glaucous- or yellowish-green, very strongly glossy with a metallic lustre and opalescent. *Stems* erect or spreading, simple or seldom forked, red and lustrous, often appearing ± julaceous due to the imbricate foliage, sparsely or densely radiculose below with reddish- or orange-brown, branching, glistening and finely papillose rhizoids or rhizoids extending up along the older stems in scattered clusters, in transverse section obtusely pentagonal, consisting of a medium but distinct central strand, 2–4-stratose medulla of large, hyaline and thin-walled cells and 1–2-layered cortex of small cells with red, incrassate walls. *Leaves* smaller and distant below, becoming progressively larger and crowded above, erect-appressed and imbricate to erect-spreading, slightly shrunken when dry, ovate, elliptical to ovate-lanceolate, often cordate from a narrow insertion, (1.2–)1.5–2.5(–3.0) mm long, (0.6–)0.8–1.1(–1.4) mm wide, acute to shortly acuminate, not or shortly decurrent, slightly to strongly concave, somewhat carinate; *margins* plane to narrowly reflexed, unbordered, serrulate to crenate in the distal half, entire below; *costa* single, strong, 100–140 µm wide at the base, gradually tapering upwards, often flexuose above, ending some distance below the apex, red near the base, reddish-brown to brown above, in cross-section terete, strongly convex on the dorsal side, consisting of 2–4 large guide cells accompanied by hydroids, a large dorsal stereid band and a row of ventral and dorsal, relatively small epidermal cells; *laminal cells* smooth, translucent, unistratose throughout, thin-walled, sturdy or somewhat lax, oblong- to linear-hexagonal or linear, often somewhat vermicular with a distinct sigmoid curve, rounded at the apices, (60–)80–130(–160) µm long, (8–)9–12(–18) µm wide, becoming somewhat narrowed towards the margins but not forming a distinct border, shorter near the base and often red tinged at the insertion. *Dioicous* or rarely *synoicous*. *Perigonia* terminal, gemmiform, with long-acuminate bracts. *Perichaetia* terminal; *perichaetial leaves* usually from a reddish base, long and narrow, elongate linear-lanceolate, 3.5–4.5 mm long, 0.5–0.7 mm wide, long-acuminate, patent, recurved and flexuose, with margins recurved and distinctly toothed above. *Setae* erect, 1–2 cm long, fairly strong, slightly flexuose, smooth, orange- to reddish-brown. *Capsules* exserted, inclined to horizontal or cernuous, obloid-cylindrical to somewhat clavate, slightly gibbous, with a short and rather indistinct neck, 2.5–3.5 mm long, brownish; *annulus* revoluble, 2–3-seriate, consisting of large, vesiculose cells with strongly incrassate walls; *operculum* convex-conic, umbonate to apiculate; *exothecial cells* variable in size and shape, rounded, rectangular, rhomboidal to oblong-hexagonal, 30–100 µm long, 20–30 µm wide, collenchymatous, with slightly thickened and wavy walls, especially in the neck, becoming smaller, rounded but concolorous with lower cells in 2–4 tiers at the orifice; *stomata* fairly numerous in several rows in the neck, phaneropore, bicellular, concolorous with exothecial cells,

→

Figure 198. *Pohlia cruda*. — **1**. Sterile plant, dry. **2**. Portion of stem with perichaetium, dry. **3**. Plant with mature capsule. **4**. Portion of cross-section of stem. **5**. Axillary hair. **6–11**. Leaves. **12–13**. Leaf apices. **14–15**. Mid-leaf cells at margin. **16**. Mid-leaf cells. **17**. Basal cells. **18**. Cross-sections of mid-leaf. **19–22**. Cross-sections of costa, a sequence from leaf base to leaf apex. **23–24**. Perichaetial leaves. **25–26**. Operculate capsules, wet. **27**. Lower exothecial cells and stoma. **28**. Exothecial cells in middle of urn. **29**. Portion of the peristome and spores. — [1, 6–7, 14, 16–17 from *RO 555/80*; 2, 25–29 from *GCSC & SWG 675*; 3, 8, 23–24 from *REL 2311*; 4–5, 10–11, 18–22 from *RO 1613/80*; 9, 12–13, 15 from *RO 4958/79*]. — Scale bars: a – 1 mm (1–3, 25–26); b – 100 µm (12–17); c – 100 µm (4, 18–22); d – 1 mm (6–11, 23–24), 100 µm (27–29) and 200 µm (5).

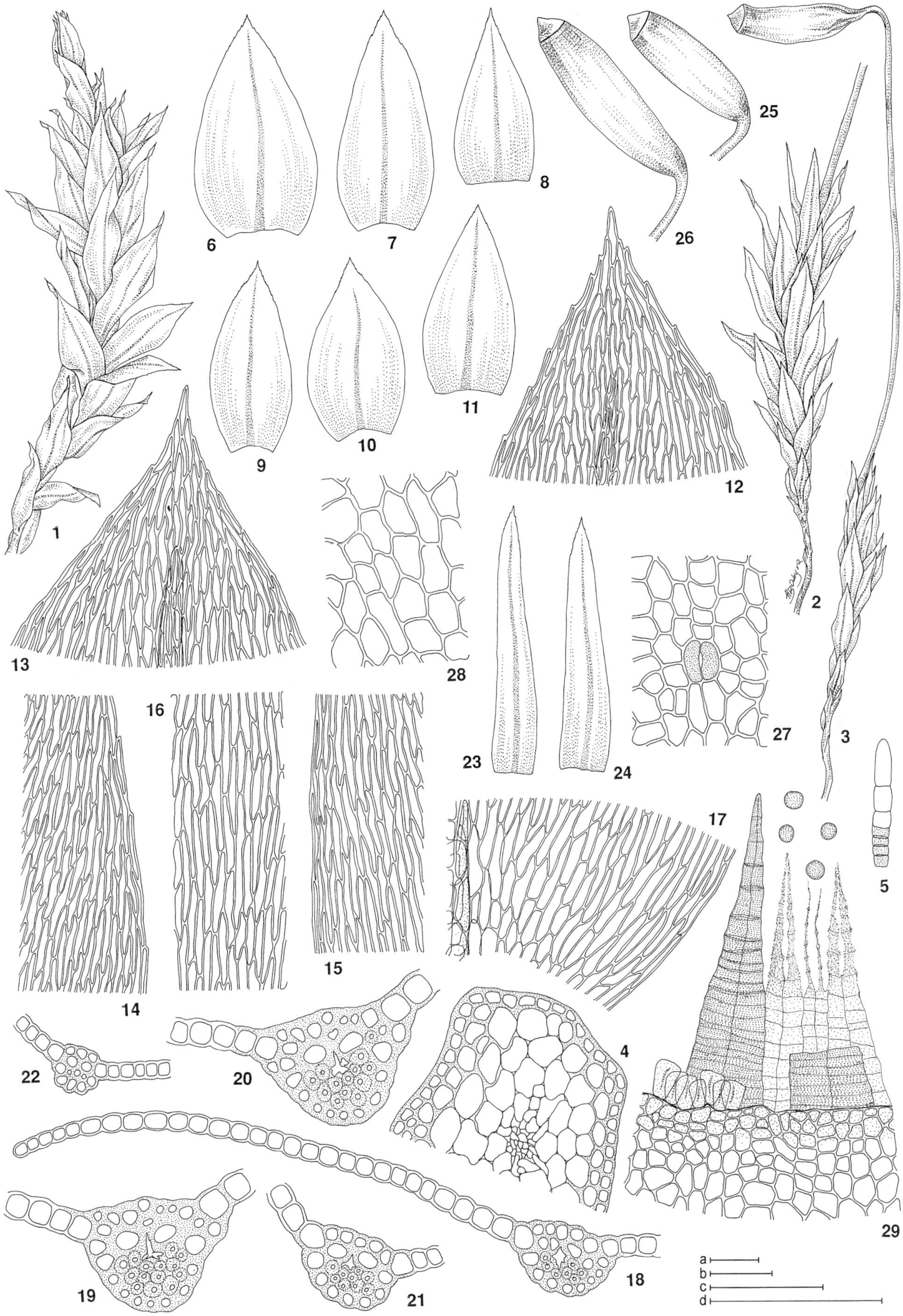
1
2
3
4
5
6
7
8
9
10
11
12
13
14
15
16
17
18
19
20
21
22
23
24
25
26
27
28
29
a
b
c
d

with narrow, elongate slits; *peristome* double; *exostome teeth* 16, lanceolate, yellowish or brownish-yellow, finely papillose, narrowly margined, strongly trabeculate on the inner side; *endostome* hyaline, finely papillose, equal in height with the exostome, consisting of a high basal membrane giving rise to broad, keeled segments, widely perforate at keel or nearly split into two, and well developed, slightly nodulose cilia in pairs, rarely singly or in triples. *Spores* spherical, brown, 18–25 μm in diameter, papillose. *Calyptra* cucullate, smooth, naked.

Chromosome numbers: $n = 11$ – South Georgia (Newton 1972); $n = 11$ (Przywara *et al.* 1984 as *P. wahlenbergii*); in addition, $n = 10$ (1 record), $n = 10 + 4m$ (1 record), $n = 11$ (1 record), $n = 22$ (4 records) and $n = 40$ (1 record) from the Holarctic (Fritsch 1991).

Discussion. *Pohlia cruda* is a handsome species with its gracefully slender appearance and a fresh, strongly metallic sheen which make it very conspicuous and readily recognisable in the field. The leaves are usually wide and shortly pointed, except the comal ones which are often narrowed and much more finely tapered. The species is not very variable in leaf shape and areolation, but exhibits greater variation in its robustness. The Antarctic plants usually grow in short or deep dense tufts and, in comparison with those from temperate regions, are more or less distinctly julaceous due to the strongly concave and closely imbricate leaves. This prompted Cardot (1900, 1901) to recognize them as a separate taxon, *Webera cruda* var. *imbricata*, which was later given a name under *Pohlia cruda* (Bartram 1957). However, this variety seems to be no more than a habitat modification in response to the harsh climatic and environmental conditions. Similar expressions have often been recorded from high elevations in Holarctic mountains and in the Arctic, and variously recognised as *Webera cruda* var. *compacta* Velen., var. *alpina* I.Hagen or var. *densa* Thér. The similarity of the latter variety to var. *imbricata* was emphasised by Cardot (1901). Such phenotypes are linked with normal expressions of *P. cruda* by numerous intermediates, especially on subantarctic South Georgia (Clarke 1973*a*), and the retention of the Antarctic populations as a distinct taxon is unjustified because they represent adaptations to the rigours of the polar environment. Consequently, Ochyra (1998*a*) reduced *Webera cruda* var. *imbricata* to a synonym of *P. cruda*.

Vegetative shoots of *Pohlia cruda* could sometimes be mistaken for *P. wahlenbergii* and *P. drummondii*, especially because some Antarctic populations of *P. drummondii* and *P. cruda* exhibit a marked tendency to laxity in leaf areolation. The differences between these species and possible sources of confusion are discussed in detail under the respective species.

Pohlia cruda is a species known from the earliest days of bryology, and Hedwig (1801) validated its name as *Mnium crudum*. In the Hedwig herbarium at G there are two sheets with his handwritten labels, comprising permanently glued single shoots as well as larger tufts of the moss. One of these sheets, bearing two different plants was annotated by A. J. Shaw as the lectotype (Shaw 1982*b*). However, this 'lectotypification' has to be superseded because no definite plant was selected as lectotype, and in the present treatment one of two specimens on this sheet is chosen as lectotype.

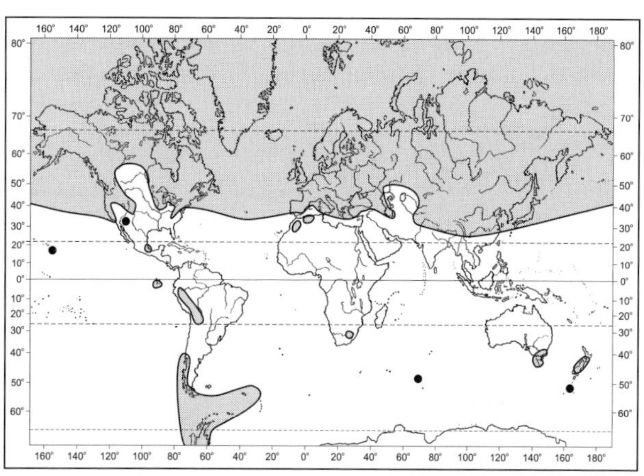

Figure 199. Global distribution of *Pohlia cruda*.

Reproduction in Antarctica. Predominantly sterile, but very rarely producing sporophytes on Galindez I. in the Argentine Is. archipelago off the Graham Coast, most of which remain immature. Inflorescences, particularly female, are not uncommon, but gametangial maturation is abnormal and usually unsuccessful, the situation being analogous to that with South Georgian populations of the species (Clarke & Greene 1970).

Habitat. *Pohlia cruda* is primarily a chasmophytic moss, occurring in a diversity of acidic and base-rich habitats, commonly in sheltered and shaded crevices in rocks faces, cliffs and on ledges, under rock overhangs, often together with species such as *Bartramia patens*, *Bryum pseudotriquetrum*, *Distichium capillaceum*, *Encalypta rhaptocarpa*, *Cratoneuropsis relaxa* subsp. *minor*, *Platydictya jungermannioides*, and *Schistidium antarctici*. It also thrives on sandy, gravelly, and fine glacial soil in wet or damp situations in seepage areas below snow patches, rocky ledges moistened by melt water, at the margins of melt runnels, fine alluvial soil adjacent to streams, and on flushed soil on scree and moraines, as well as in dry situations on pockets of soil between boulders, stones on moraines and scree slopes, and on fine gravel between boulders. Occasionally, it occurs in dark crevices in moss peat banks, or fissures filled with limpet shells near gull nests, and in areas with recent volcanic activity on ashy soil and between scoria blocks.

World range. *Pohlia cruda* is often designated a cosmopolitan or subcosmopolitan species (Dierßen 2001), but it is a classical bipolar species with not many intermediate localities in the tropics (Fig. 199). In the Northern Hemisphere it has a continuous pan-Holarctic range, extending southwards to California, Arizona, Tennessee and North Carolina (Shaw 1982*b*), the Azores (Eggers 1982), the Mediterranean region, central Turkey (Henderson & Prentice 1969), northern Iraq (Agnew & Vondraček 1975) and Iran (Schiffner 1908), Pakistan (Higuchi & Nishimura 2003), northern India (Gangulee 1974) and southern China (Redfearn *et al.* 1986), extending to Morocco and Algeria in North Africa (Ros *et al.* 1999). In the tropics, it is very rare in Mexico, and reports from other areas in Central America are erroneous (Shaw 1982*b*; Allen 2002); it is known in the Hawaiian Is. (Bartram 1952*b*), the Galapagos Is. (Gradstein & Weber 1982), and at high elevations in the Andes from Ecuador to Bolivia (Churchill *et al.* 2000), as well as Lesotho in southern Africa (Rooy & Magill 1987). In temperate and cool regions of the Southern Hemisphere *P. cruda* is known from the *Nothofagus* zone on the western fringes of southern South America from the Valdivian province to Tierra del Fuego (Greene 1986; Ochi 1990*a*), subantarctic South Georgia (Clarke 1973*a*) and the maritime Antarctic, then appearing on Îles Kerguelen (Brotherus 1906*a*) and in south-eastern Australia and Tasmania (Shaw 2006), New Zealand (Sainsbury 1955) and the Auckland Is. (Vitt 1979*a*).

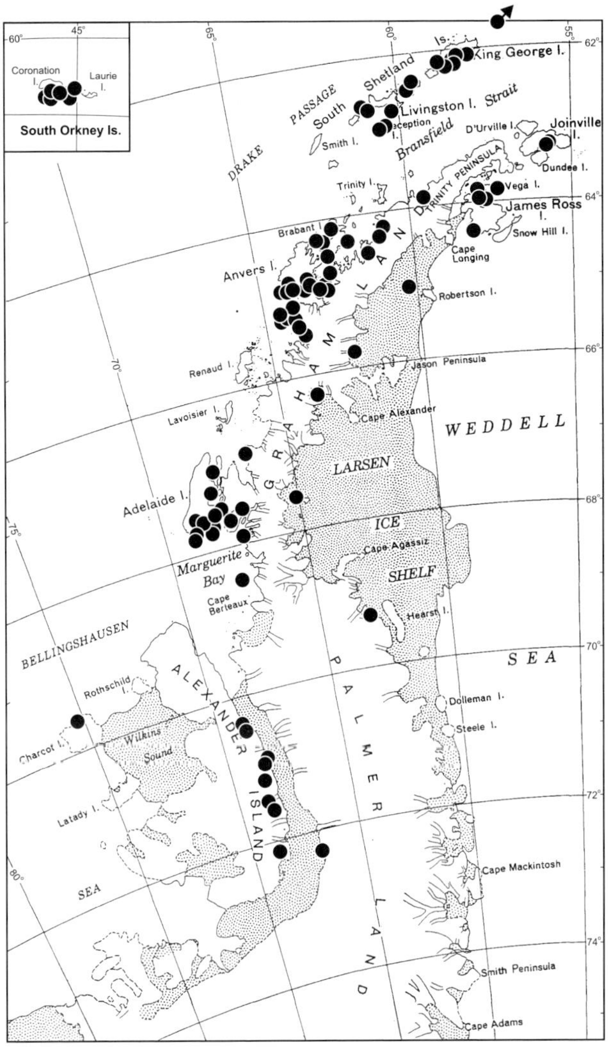

Figure 200. Distribution of *Pohlia cruda* in Antarctica. Localities in the northern group of the South Shetland Is. (Elephant I. and Gibbs I.) are indicated by an arrow.

Distribution in Antarctica. *Pohlia cruda* is a widespread and locally frequent species, occurring throughout the maritime Antarctic, from the South Orkneys and South Shetlands along the western coast of the Antarctic Peninsula to Alexander I. and the Rymill Coast where it reaches its southernmost station at lat. 72°21′S on the east coast of George VI Sound. In addition, it is rare and widely scattered on the eastern coast of the Antarctic Peninsula, ranging from Joinville I. to Finley Heights on the Wilkins Coast at lat. 69°14′S (Fig. 200). The species does not show special elevational preferences and occurs from sea level to 830 m on Adelaide I., although most stations are concentrated at lower elevations below 200 m.

Selected specimens examined. WEST ANTARCTICA. SOUTH ORKNEY IS. **Coronation I.** Cape Vik, 100 m, *RILS 573A-B* (BM); Laws Glacier, *ca* 15 m, *RW 135* (AAS, KRAM); Olivine Pt., *ca* 15 m, *RILS 143* (AAS, KRAM, PC); Saunders Pt., 5 m, *RILS 38* (AAS, KRAM); Windscoop Buttress, 200 m, *RW 163, 167 & 169* (AAS, KRAM). **Matthews I.** West side, *ca* 65 m, *RILS 168* (AAS, KRAM). **Powell I.** South-east coast, *ca* 115 m, *RILS 209* (AAS, KRAM); west side, 300 m, *RILS 234 & 235A* (AAS, KRAM). **Signy I.** Thulla Pt., 70 m, *RILS 8047D* (AAS, KRAM); hill betweem Spindrift Rocks and North Pt., *ca* 15 m, *MWH 184A* (AAS, KRAM); Three Lakes Valley, 15 m, *REL 1185* (BM, KRAM, LE, PC); Moss Lake, *ca* 95 m, *RW 143B* (AAS, KRAM); between Jane Peak and Snow Hill, 150 m, *REL 1193* (AAS, KRAM); Factory Cove, 35 m, *MWH 749A* (AAS, KRAM); Orwell Glacier, 30 m, *RILS 10821* (AAS, KRAM). **Moe I.** 100–150 m, *RILS 5252* (AAS, KRAM).

SOUTH SHETLAND IS. **Elephant I.** 2 km south-west of Endurance Glacier, 152 m, *JSA 49* (AAS, KRAM). **Gibbs I.** West of the Spit, 120 m, *JPB 189* (AAS, KRAM). **King George I.** ***Bransfield Strait***: Red Hill, 110 m, *RO 1115/80* (KRAM). ***Admiralty Bay***: Creeping Slopes, 60 m, *RO 1237/80* (KRAM); Czajkowski Needle, 240 m, *RO 685/80* (KRAM); Jardine Peak, 275 m, *RO 5197/79* (KRAM); Breccia Crag, 140 m, *RO 927/80* (KRAM); Keller Peninsula, Harpoon Pt., 60 m, *RO 555/80* (KRAM); Martel Inlet, Shark Fin, 160 m, *RO 2656/80* (KRAM); Mt. Wawel, 40 m, *RO 2156/80* (KRAM) and 120 m, *RO 2178/80* (KRAM); Rembiszewski Nunataks, 200 m, *RO 2743/80* (KRAM). ***King George Bay***: Lions Rump, *DCL 799A* (AAS, KRAM). ***Fildes Peninsula***: Suffield Pt., *RO 2438/80* (KRAM). ***Barton Peninsula***: Noel Hill, *BJ 199* (KRAM). **Robert I.** Coppermine Cove, *ca* 35 m, *RILS 924* (AAS, KRAM). **Greenwich I.** Cerro Lopez Glacier, 20 m, 3 Feb 1953, *Frödin s.n.* (S). **Livingston I.** ***Byers Peninsula***: Ocoa Pt., 2 m, *KR 41A* (AAS, KRAM); Devils Pt., 20 m, *KR 24G* (AAS, KRAM); Robbery Beaches, 10 m, *KR 52D* (AAS, KRAM); Clark Nunatak, 30 m, *DCL 125* (AAS, KRAM, LE, PC); Sayer Nunatak, 35 m, *RILS 5866* (AAS, KRAM). ***Hurd Peninsula***: north of the Bulgarian Station, *Pandurski 102* (KRAM); Johnsons Dock, 35 m, DC*L 74* (AAS, KRAM). **Deception I.** Hill NNW of Mt. Pond, 410–525 m, *RILS 6903* (AAS, KRAM); Fumarole Bay, 10–30 m, *RILS 6629A* (AAS, KRAM); Stonethrow Ridge 20 m, *NJC 20A* (AAS, KRAM); Crater Lake, 110 m, *RILS 5726A* (AAS, KRAM).

WEST ANTARCTIC PENINSULA. **Davis Coast.** Almond Pt., 5–20 m, *AC & PK 74A* (AAS, KRAM). **Danco Coast.** Cierva Pt., *VK 485, 600, 633 & 666* (KRAM); Spring Pt., *VK 914, 969 & 1250* (KRAM); Charlotte Bay, Recess Cove, 2–10 m, *RILS 3985* (AAS, KRAM); Cuverville I., 10–20 m, *RILS 11040 & 11957* (AAS, KRAM); Birdsend Bluff, 5–35 m, *RILS 4220 & 4222* (AAS, KRAM); Beneden Head, *ER 233A* (BR, PC); Neko Harbour, 15–20 m, *RILS 11935* (AAS, KRAM); Waterboat Pt., Gonzales Videla Base, 4 Jan 1961, *Leech s.n.* (KRAM); Coughtrey Peninsula, Paradise Harbour, Almirante Brown Station, *VK 83* (KRAM). **Palmer Archipelago.** ***Brabant I.***: Metchnikoff Pt., 100 m, *WH 31A* (AAS, KRAM); Koch Glacier, 400 m, *AM 111* (AAS, KRAM); east coast, 350 m, *ER 252C* (BR, PC). ***Melchior Archipelago***: Lambda I., Sevilla Landing, *PAS 348* (FH, LE); Omega I., 2–10 m, *RILS 4119* (AAS, KRAM); Gamma I., 5–10 m, *RILS 4146* (AAS, KRAM). ***Anvers I.***: Norsel Pt., 5–10 m, *RILS 1949* (BM) and 15 m, *REL 1227* (AAS, KRAM); Litchfield I., 10 m, *RWMC 370A* (AAS, KRAM); Hermit I., 23 m, *RILS 878A* (BM); Shortcut I., *VK 58* (KRAM); Laggard I., 15 m, *REL 1315* (AAS, KRAM). ***Wiencke I.***: Goudier I., Port Lockroy, *BJT 110* (BM, KRAM, PC) *& 177* (AAS, KRAM). **Graham Coast.** Edge Hill, 65 m, *RWMC 659A* (AAS, KRAM); Rasmussen I., 6 m, *AS 80* (AAS, KRAM); Cape Tuxen, *Gain 193A & 208* (PC, S); Cape Pérez, 15 m, *RWMC 637A* (AAS, KRAM). **Wilhelm Archipelago.** ***Lemaire Channel***: entrance to the channel, *ER 270C* (BR, PC). ***Penola Strait***: Petermann I., 13 m, *RWMC 724A* (AAS, KRAM) and 60 m, *AS 123* (AAS, KRAM). ***Argentine Is.***: Irizar I., 6 m, *RWMC 604A* (AAS, KRAM); Uruguay I., 13 m, *RWMC 718B* (AAS, KRAM); Corner Is., 6 m, *RWMC 567* (AAS, KRAM); Galindez I., Meek Channel, 6 m, *RWMC 700A* (AAS, KRAM); Skua I., 4 m, *RWMC 453* (AAS, KRAM); Black I., 6 m, *RWMC 577* (AAS, KRAM); The Barchans, largest island, 6 m, *RWMC 748* (AAS, BM, KRAM, NY). ***Collins Bay***: Green I., *BGLE 1083A* (BM); Berthelot Is., largest island, 65 m, *RWMC 626A* (AAS, KRAM); Darboux I., 10–15 m, *RILS 3311* (BM). ***Grandidier Channel***: Lahille I., 2–5 m, *RILS 4331* (AAS, KRAM); second largest island off Takaki Pt., 5–20 m, *RILS 4364* (AAS, KRAM); south arm of Beascochea Bay, *BGLE 1195A* (BM, LE, NY, PC). **Loubet Coast.** ***Lallemand Fjord***: Orford Cliff, *RILS 5275C, 5279, 5283A & 5286A* (AAS, KRAM). ***Bennett Is.***: Pfaff I., *JK 177* (AAS, KRAM). ***Laubeuf Fjord***: Webb I., Ryder Bay, *JK 258* (AAS, KRAM). **Fallières Coast.** ***Adelaide I.***: Mt. Bouvier, *JK 155* (BM); point east of Mt. Bouvier, 15 m, *JK 234* (AAS, KRAM); Mt. Mangin, 270 m, *JK 148A* (AAS, KRAM); Rothera Pt., 15 m, *JF 2* (AAS, KRAM); Reptile Ridge, 175 m, *RILS 10631* (AAS, KRAM); Stork Ridge, *PC 270* (AAS, KRAM); Mt. Gaundry, *ca* 330 m, *JK 142B* (BM) and *ca* 470 m, *JK 144* (BM); Mt. Liotard, *ca* 830 m, *JK 189B* (AAS, KRAM); Jenny I., 10 m, *JK 117* (BM, KRAM, NY) and *RILS 4757* (AAS, KRAM); Avian I., *ca* 6 m, *RILS 774B* (BM). ***Léonie Is.***: Lagoon I., 5 m, *RILS 9149* (AAS, KRAM); Léonie I., *RILS 8833A* (AAS, KRAM), 30 m, *RILS 8954D, 8958 & 8976* (AAS, KRAM). ***Marguerite Bay***: Blaiklock I., *Poncet 6B* (AAS, KRAM) and 8–15 m, *RILS 838* (AAS, KRAM, PC); Pourquoi Pas I., Dalgleish Bay, 5–20 m, *RILS 4774* (AAS, KRAM); Camp Pt., 5–20 m, *RILS 4634, 4678 & 4685* (AAS, KRAM);

Mushroom I., 60 m, *Poncet 1B* (AAS, KRAM). **RYMILL COAST.** Buttress Nunataks, 125 m, *FIDS 33* (BM). **GEORGE VI SOUND.** ***Charcot I.***: central north facing slope, 85 m, *RILS 10482A* (AAS, KRAM). ***Alexander I.***: Ablation Valley, Ablation Lake, 2–3 m, *RILS 9224B, 9225 & 9242* (AAS, KRAM), *ca* 5 m, *RILS 2709B* (AAS, KRAM); Himalia Ridge near Jupiter Glacier, 360 m, *RILS 1126A* (AAS, KRAM); Moutonnée Valley, *ca* 200 m, *RILS 11120D* (AAS, KRAM); Fossil Bluff, *ca* 370 m, *BJT 509A* (AAS, KRAM); Gateway Pass, Astarte Horn, *DDWW 114* (AAS, KRAM); Ares Ridge, 150 m, *RILS 9494* (AAS, KRAM); Mars Oasis, 70–75 m, *RILS 9412* (AAS, KRAM), 80 m, *RILS 10378* (AAS, KRAM), 125 m, *RILS 9417* (AAS, KRAM); Coal Nunatak, 100 m, *RILS 10442* (AAS, KRAM); Eros Glacier, *ca* 65 m, *BJT 523* (AAS, BM, KRAM, NY, PC).

EAST ANTARCTIC PENINSULA. **JOINVILLE ISLAND GROUP.** ***Joinville I.***: Mt. Alexander, 60–75 m, *RILS 3671* (AAS, KRAM); north-east of Mt. Alexander, 10–15 m, *RILS 3709* (AAS, KRAM). **JAMES ROSS ISLAND GROUP.** ***James Ross I.***: Abernethy Flats, 100 m, *RILS 7623* (AAS, KRAM); Lachmann Crags, 20–35 m, *RILS 7321B & 7379* (AAS, KRAM); St. Martha Cove, 50 m, *RILS 7511* (AAS, KRAM), 75 m, *RILS 7434A* (AAS, KRAM) and 80 m, *RILS 7649* (AAS, KRAM); Nygren Pt., 30 m, *RILS 5479* (AAS, KRAM). ***Vega I.***: Pastorizo Bay, 50 m, *RILS 7856* (AAS, KRAM). **OSCAR II COAST.** Seal Nunataks, Bruce Nunatak, 50 m, *RILS 6415A* (AAS, KRAM); Bildad Peak, *RILS 2768B* (AAS, KRAM); Cape Casey, Cabinet Inlet, *RT 3B* (AAS, KRAM). **BOWMAN COAST.** Francis I., *ca* 325 m, *RILS 2775B & 2776B* (AAS, KRAM). **WILKINS COAST.** Finley Heights, *ca* 235 m, *MJC 5A* (AAS, KRAM).

5. **Pohlia nutans** (Hedw.) Lindb — Fig. 201

Pohlia nutans (Hedw.) Lindb., Musci Scand.: 18. 1879. ≡ *Webera nutans* Hedw., Sp. Musc. Frond.: 168. 1801. ≡ *Hypnum nutans* (Hedw.) F.Weber & D.Mohr, Index Mus. Pl. Crypt.: sheet 2 [4]. 1803. ≡ *Bryum nutans* (Hedw.) Sm., Fl. Brit. 3: 1347. 1804. ≡ *Lamprophyllum nutans* (Hedw.) Lindb., Acta Soc. Sci. Fenn. 10: 25. 1871. — TYPE CITATION: Locis siccioribus, sterilibus, arenosis, apertis et umbrosis gregatim, nec non in uliginosis turfosis sparsim et omni Germania frequens; solo pingui rarior; also references to Dillenius (1741: t. 50, f. 61), Weiss (1770: p. 153), Schreber (1771: p. 81), and Hedwig (1785–1787: p. 9) is provided; in the latter the following specimens are cited: 'Chemnitzii, et per omnem circumsitam montosam regionem, ubiuis fere abunde mihi iste Muscus obvenit. Nemque tamen planiorem nostram Lipsicam respuit. Habitat enim hic loci cum in silva Universitati propria tum ad vallem lacus, qui est in pineto lindenthaliensi'. [Lectotype (*selected here*): [icon in] Dillenius (1741): f. 61 on pl. 50 of *Bryum trichodes laete virens, capitulis cernuis oblongis*; epitype (*selected here*): 'Böhmen: Isergebirge, Raspenau, an einen Waldwege beim Eisenbahndurchlasse, 360 m s. m., 18. Juli 1911 legit Jos. Blumrich' (E. Bauer, *Musci europaei exsiccati* No. 909) – KRAM!; isoepitypes: BM, FH, G, NY, PC and apparently in many other herbaria to which these exsiccati were distributed (*vide* Sayre 1971)].

Webera racovitzae Cardot, Rev. Bryol. 27: 44. 1900. ≡ *Pohlia racovitzae* (Cardot) Broth. *in* Engl. & Prantl, Nat. Pflanzenfam. 1(3): 549. 1903. — TYPE CITATION: [Antarctica] Xme et XIIme débarquements; later, Cardot (1901: p. 35) gave details of the specimens: 'Détroit de Gerlache: Xème débarquement, île Brabant, sur des roches complètement entourées de glaces, à 350 mètres au-dessus de niveau de la mer (forme rabougrie); stérile. (N° 252^{a}; 30 janvier 1898). XIIme débarquement, île de Cavelier de Cuverville, autour des touffes de graminées (*Aira antarctica*), sur une corniche de la falaise (N° 244; 2 février 1898). [Lectotype (*selected here*): 'Herb. J. Cardot. Mousses du voyage de la Belgica. n° 244. Webera Racovitzae Card. sp. nova Canal de Gerlache XIIème débarquement, sur une corniche de la falaise, autour des touffes de graminées. Leg. Racovitza 2 févr. 1898' – PC!; isolectotypes: BR!, S-Roth!]. First synonymised by Greene *et al.* (1970: p. 19).

Plants small and slender but often elongate to medium-sized, in dense or loose tufts or scattered singly or in small groups amongst other bryophytes, somewhat lustrous to dull, green to yellow- or olive-green above, pale brown to yellow-brown below. *Stems* erect, simple or sparsely branched, (0.5–)1.0–4.0(–7.5) cm tall, ± black to blackish-red, usually abundantly radiculose below with brown, branched, lustrous, finely papillose rhizoids, in transverse section obscurely pentagonal to rounded, with a large and distinct central strand, 3–5-layered medulla of large, thin-walled and hyaline cells and 1–2-stratose cortex of small cells with blackish-red to black, incrassate walls; *axillary hairs* short, filiform, 4–6-celled, with 2–3 short-rectangular to quadrate, reddish-brownish cells and 2–4 elongate, hyaline distal cells. *Leaves* distant to crowded, usually smaller below and increasing in size towards the stem apex, often forming a ± distinct comal group, erect, appressed-imbricate and often twisted around the stem on drying, erect to erecto-patent when moist, (0.8–)1.1–1.7(–2.5) mm long, (0.3–)0.5–0.8(–1.1) mm wide, gradually acuminate to acute, distinctly keeled-concave, not or weakly decurrent; *margins* plane to reflexed, at least in the lower part, serrate to weakly denticulate or crenulate in the distal half, entire below, unbordered; *costa* single, rather stout, subpercurrent to percurrent, 80–130 µm wide at the base, gradually narrowed upwards, often flexuose above, dark brown to blackish-brown below, brown, yellow-brown to dark green-brown above, in transverse section semiterete, strongly convex dorsally, flat to weakly convex on the ventral side, comprising 4–5 large, guide cells with hydroids, a large dorsal stereid band and a row of relatively large ventral and rather smaller dorsal epidermal cells; *laminal cells* smooth, pellucid, unistratose throughout, sturdy, thick-walled, occasionally laxer, narrowly hexagonal to oblong-hexagonal above, elongate- to linear-hexagonal below, (30–)35–70(–90) µm long, 7–11(–18) µm wide, straight-walled, rounded at the apices, usually somewhat narrowed and elongate at the margins but not forming a distinct border, undifferentiated at the basal angles. *Paroicous*, occasionally *synoicous*. *Gametangia* terminal; *perichaetial leaves* narrowly elongate-lanceolate, 2.2–2.6 mm long, 0.3–0.5 mm wide, finely acuminate, distinctly toothed, with percurrent costae. *Setae* erect, elongate, 0.5–1.5(–2.0) mm long, rather slender, straight to flexuose, yellow- or orange-brown, smooth. *Capsules* exserted, horizontal to pendent, pyriform, with a short and

→

Figure 201. *Pohlia nutans*. — **1–2**. Plants with mature capsules. **3**. Sterile plant, dry. **4**. Cross-section of stem. **5–7**. Axillary hairs. **8–14**. Leaves. **15–16**. Leaf apices. **17–18**. Mid-leaf cells at margin. **19**. Mid-leaf cells. **20**. Basal cells. **21**. Cross-sections of costa. **22–23**. Cross-sections of leaves. **24–25**. Perichaetial leaves. **26**. Deoperculate capsule, dry. **27–28**. Operculate capsules, dry. **29**. Lower exothecial cells and stomata. **30**. Exothecial cells in middle of urn. **31**. Portion of peristome and spores. — [1, 4–7, 9, 14, 21–24, 29–30 from *RILS 3425*; 2, 12–13, 25–28, 31 from *REL 1224*; 3 from *RO 722/80*; 8, 19–20 from *Scott 134a*; 10–11, 15–18 from *RO 707/80*]. — Scale bars: a – 100 µm (15–20); b – 1mm (3) and 100 µm (29–30); c – 100 µm (5–7, 31); d – 1 mm (8–14, 24–25); e – 2 mm (1–2, 26–28) and 100 µm (4, 21–23).

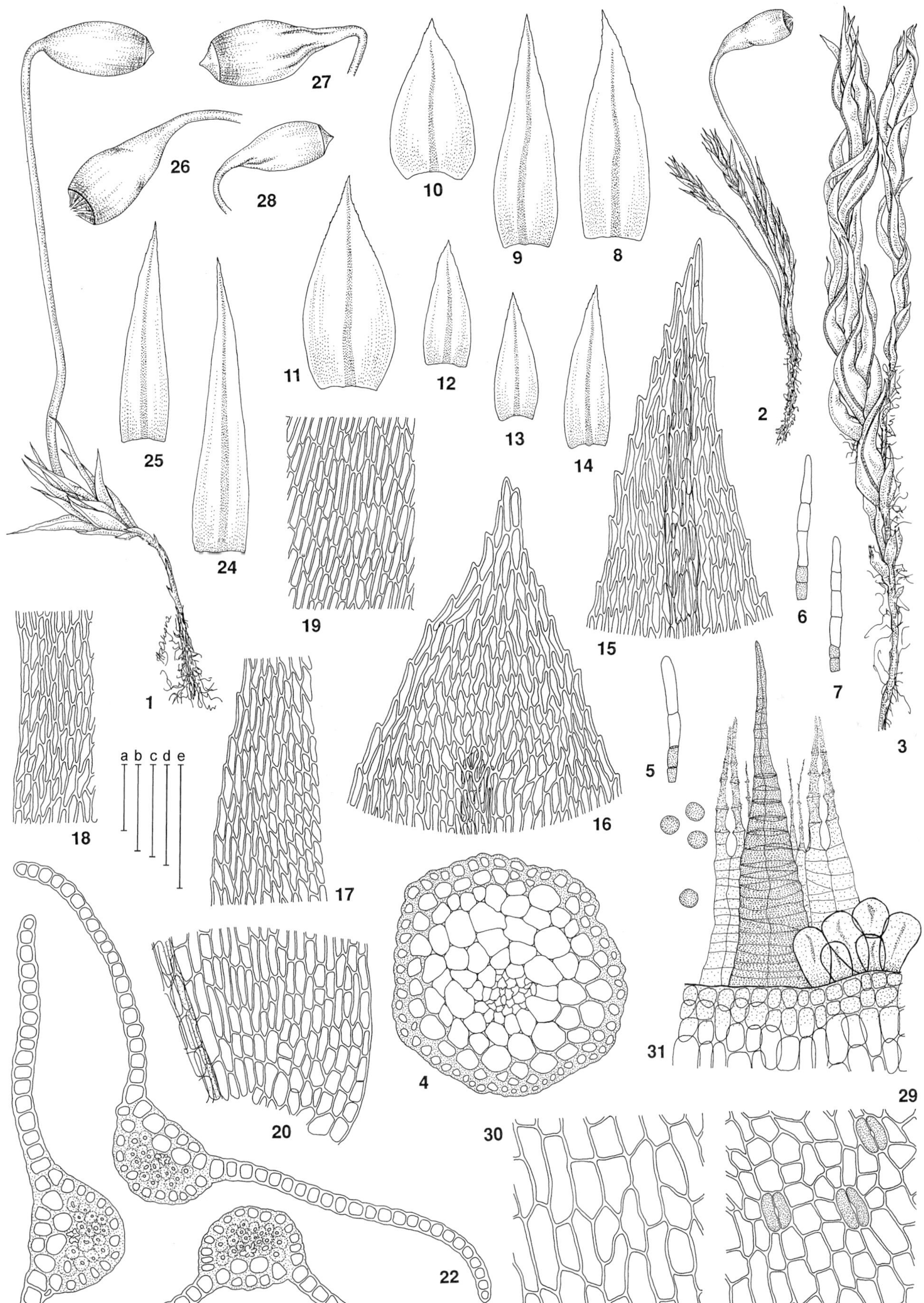
1
2
3
4
5
6
7
8
9
10
11
12
13
14
15
16
17
18
19
20
21
22
23
24
25
26
27
28
29
30
31
a b c d e

noticeably distinct neck, 2.5–3.5 mm long, symmetrical to slightly curved, light brown; *annulus* revoluble, consisting of 2–3 rows of large, vesiculose, thick-walled cells; *operculum* convex-conic, bluntly apiculate; *exothecial cells* elongate-rectangular, 35–100 µm long, 20–35(–40) µm wide, with thin, evenly thickened and slightly wavy walls, becoming isodiametric but otherwise concolorous in 2–3 tiers at the orifice; *stomata* fairly numerous in several rows in the neck, phaneroporous, bicellular, with narrow, elongate slit-like pori; *peristome* double; *exostome teeth* 16, lanceolate, 340–370 µm long, acute, pale yellow to yellow brownish, finely papillose, narrowly margined; *endostome* light yellowish to hyaline, finely roughened, with a high basal membrane and well developed, broad, keeled and perforate segments alternating with 2 well developed, nodulose cilia. *Spores* globose, brownish, 16–22 µm in diameter, finely to moderately papillose. *Calyptra* cucullate, naked, smooth.

Chromosome numbers: $n = 33$ – South Georgia (Newton 1972) and King George I. (Kuta *et al.* 1982); moreover, various chromosomal numbers, including $n = 11$ (2 records), $n = 20$ (1 record), $n = 20 + \mathrm{m}$ (1 record), $n = 21 + \mathrm{m}$ (1 record), $n = 22$ (17 records), $n = ca\ 27$ (1 record), $n = 31 + 2\mathrm{m}$ (1 record), $n = 32 + \mathrm{m}$ (1 record) and $n = 33$ (3 records), from the Holarctic were detected (Fritsch 1991).

Discussion. When typically developed, *Pohlia nutans* does not pose problems with identification, the thick-walled laminal cells and paroicous inflorescences being immediately diagnostic. The species is distinct and usually easily recognised with its slightly glossy to dull, green tufts and crowded, elongate, erect-appressed upper leaves that are often more or less twisted around the stem. Often it can be confused with *Bryum* species. However, it is separated from all local species of this genus by its unbordered leaves which are dentate or denticulate in their upper part.

As elsewhere, the species is exceedingly variable in Antarctica, in particular in its size and growth form, but the leaf areolation remains a relatively stable character, although in some populations a tendency to cell laxity is observed. In Antarctica plants predominantly form dense short turves or cushions and have simple, unbranched stems that are often densely radiculose below the apex and within the compact tufts. They are very similar to the plants occurring in the Arctic or in the alpine zone in the Northern Hemisphere, and were once recognised as distinct species or varieties of *P. nutans*: var. *bicolor* (Hoppe & Hornsch.) Hult (≡ *Webera bicolor* Hoppe & Hornsch.) and var. *subdenticulata* (Brid.) Delogne (≡ *Bryum subdenticulatum* Brid.). Now, they are not accepted as distinct taxa but are considered to be merely environmental variants that are conditioned by severe climatic conditions.

The possibility of confusion of *P. nutans* with other congeners in Antarctica is minimal. *Pohlia cruda* is readily distinguished by its pale, whitish-green colour and strong metallic or bluish gloss, as well as by its elliptic, ovate to ovate-lanceolate leaves. *Pohlia drummondii* is also a strongly lustrous plant with distinctly and long-decurrent leaves. *Pohlia wilsonii* and *P. wahlenbergii* have pale yellowish, whitish- to glaucous-green or reddish- to pinkish-green plants with a lax leaf areolation of very broad and thin-walled laminal cells.

Pohlia nutans was recorded for the first time from Antarctica by Cardot (1901) on the basis of specimens collected by E. Racovitza, naturalist on the *Belgica* Expedition (1897–1899) in the Antarctic Peninsula region. He compared it to the European *P. nutans* var. *bicolor* and var. *subdenticulata* which are altimontane expressions of the species growing in the alpine zone, thus in environmental conditions approaching those in the Antarctic. Apart from *P. nutans* Cardot (1900, 1901) described from the same collection a new and closely related species, *Webera racovitzae*, to which he assigned plants having no distinct comal leaves as well as tall and unbranched stems. Except for a slightly different habit, these plants perfectly match *P. nutans* in all anatomical details of the leaves, and Greene *et al.* (1970) considered both species to be identical taxa. A few years later, Cardot (1906*b*, 1907*a*) described a new variety within this species, *W. racovitzae* var. *laxiretis*, from specimens collected on Wiencke I., Danco Coast, by the First French Antarctic Expedition (1903–1905), but this taxon is considered to be identical to *Bryum pseudotriquetrum*.

Pohlia nutans is a species known from the early days of bryology, and Hedwig (1801) validated its name as *Webera nutans*. Shaw (1982*b*) showed that the lectotype of this species name is preserved in the herbarium of Hedwig and Schwägrichen at Geneva. Indeed, it holds a single sheet with the following label handwritten by J. Hedwig 'Webera nutans Hedw. St. Cr. 1. 9. t. 4. Specimina Lipsiae in Universitaetsholz inter Sphagn. palustre'. In addition, another specimen marked with 'b' is glued to this sheet and is annotated in Schwägrichen's handwriting as follows: 'b. a Richardson mare polar. Plerique caules simplices, quidem uno alterno rameo versus apicem caulis'. J. Richardson was a surgeon in the British Royal Navy and this specimen was collected during J. Franklin's Polar Sea journey of 1819–1822 (Sayre 1971). Thus, this specimen is not eligible for lectotypification of *Webera nutans*.

Because the sheet with the original material of *Pohlia nutans* at G bears no indication of lectotype, the lectotypification in Shaw's (1982*b*) work has to be superseded because there is no evidence that this author examined the Hedwigian plants from Leipzig. This specimen is not directly cited by Hedwig (1801) but he gave an indirect reference to it by citation of his earlier work (Hedwig 1785–1787) where this specimen is quoted in detail. This specimen consists of seven shoots which are glued on the sheet. The plants are dioicous and have small spores, 11–14 µm in diameter, but they evidently represent what is currently designated as *Pohlia sphagnicola* (Bruch & Schimp.) Lindb. & Arnell. Thus, selecting this specimen as the lectotype of *Webera nutans* would have disastrous consequences for the stabilisation of moss nomenclature.

The only other element that is eligible for lectotypification of *Webera nutans* is the illustration in *Historia muscorum* of Dillenius (1741: fig. 61 on pl. 50) of the species designated by the phrase-name as *Bryum trichodes laete virens, capitulis cernuis oblongis*. It was identified by Lindberg (1883; see also De Sloover 1995) as *Pohlia nutans*. This illustration is hereby formally designated as lectotype of this species' name, but it is not adequate for unmistakable identification of the species since it lacks many microscopic details which are crucial for the correct circumscription of this species, especially the anatomy, length and shape of the costa, the leaf areolation and the shape of the leaf apex. Therefore, the specimen collected in continental Europe and widely distributed in Bauer's *Musci europaei exsiccati* as No. 909 is here selected as epitype in order to preserve current usage of *Pohlia nutans*.

Reproduction in Antarctica. Inflorescences are present in many populations of *Pohlia nutans*, but sporophytes are not common, although fruiting specimens are known in most parts of the maritime Antarctic, from the South Sandwich Is. to Marguerite Bay in the south-western Antarctic Peninsula. Savicz-Lyubitskaya & Smirnova (1971, 1973) described the vegetative reproduction of *Pohlia nutans* in continental Antarctica by means of brittle terminal buds and brittle axillary flagellae. However, this observation cannot be confirmed since, of five specimens recorded by these authors from the Bunger Hills on the Knox Coast, Queen Mary Land, only one was located at LE, and it lacked any propagules.

Habitat. *Pohlia nutans* is a polyedaphic species, showing a broad ecological amplitude and growing in a wide range of habitats in Antarctica, preferably on wet, moist or damp, less commonly on dry, sheltered or exposed, rock ledges and crevices on hillsides, outcrops, scree, cliffs, in rock crevices on mineral soil or humus, glacial detritus and stony ground between rocks, stones, pebbles and gravel, in hollows on moss peat banks, in seepage areas, at margins of melt streams and runnels, in wet flushes, sometimes as a colonizer of bare soil, near bird nesting sites and on whale bones. The species is a constant component of communities of the moss carpet subformation, being the principal associate of *Brachythecium austrosalebrosum*, *Sanionia georgicouncinata*, *Warnstorfia fontinaliopsis*, *W. sarmentosa*, and *Cephaloziella varians*. It is also present in associations of the moss turf subformation, growing together with *Chorisodontium aciphyllum* and *Polytrichum strictum*, the principal components of moss peat banks which are a characteristic feature of the landscape in many parts of the northern maritime Antarctic (Fenton & Lewis Smith 1982). Also, *P. nutans* is a constant associate in stands of *Polytrichastrum alpinum* in low lying coastal areas that develop around the margins of areas disturbed by wallowing elephant seals, breeding colonies of giant petrels and sometimes around the margins of penguin rookeries. It is a frequent constituent of the Antarctic herb tundra formation, growing amongst *Deschampsia antarctica* and *Colobanthus quitensis*, together with *Sanionia georgicouncinata*, *Polytrichastrum alpinum*, *Syntrichia magellanica*, *S. filaris*, *Brachythecium austrosalebrosum*, *Bryum pseudotriquetrum*, *Ceratodon purpureus*, and several other moss species. On the volcanic South Sandwich Is., Deception I. (South Shetland Is.) and Mt. Rittmann in northern Victoria Land, *P. nutans* grows most often on heated ground in the fumarolic areas. Here, it usually grows on ash and scoria and cinder deposits on moraines, and is a coloniser of young lava flow. In the South Sandwich Is. it is a common constituent of vegetation around and inside of fumarole vents, often very active

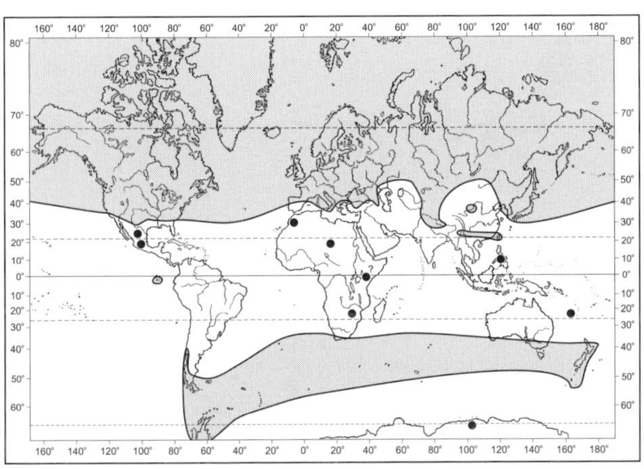

Figure 202. Global distribution of *Pohlia nutans*. The occurrence in Victoria Land is not indicated.

and emitting sulphurous steam forming luxuriant and closed turves. In such situations *P. nutans* grows with *Brachythecium austrosalebrosum*, *Campylopus introflexus*, *Cephaloziella varians*, *Cryptochila grandiflora*, *Lophozia excisa*, *Marchantia berteroana*, *Polytrichastrum alpinum*, and *Riccardia georgiensis* (Longton & Holdgate 1979). On Mt. Rittmann it thrives as small circular colonies on fine compacted heated clay around fumaroles where it is the only plant present. Here, it develops a unique growth form, with each plant comprising a rosette of short yellow-green shoots (Bargagli *et al.* 1996; Lewis Smith 2000; Skotnicki *et al.* 2002).

World range. *Pohlia nutans* is commonly considered as a (sub-) cosmopolitan species (e.g. Zanten 1971; Eddy 1996; Dierßen 2001), but it is a classical bipolar species with very occasional occurrences in the tropics (Fig. 202). It is very common in the Northern Hemisphere, having a continuous range throughout the arctic, boreal and temperate zones and extending in the south to California, Arizona, New Mexico, Arkansas and Georgia in North America (Shaw 1982*b*), to the Azores (Eggers 1982), Morocco (Jelenc 1955), central Portugal (Sérgio & Carvalho 2003), northern Spain (Casares Gil 1915), Sicily (Cortini Pedrotti 2001*a*,*b*), northern Greece (Duell 1995), northern Turkey (Robinson & Godfrey 1960; Henderson & Prentice 1969), the Caucasus (Brotherus 1893), northern Iran (Nicholson 1920), Pakistan (Higuchi & Nishimura 2003), northern India (Gangulee 1974), Yunnan (Zhang & Li 2002), Taiwan (Chuang 1973) and Japan (Iwatsuki 2004). In the tropics the species is rare and occasional and, so far, it has been recorded only from Mexico (Shaw 1982*b*), the Galapagos Is. (Gradstein & Weber 1982), the Philippines (Bartram 1929) and New Caledonia (Mitten 1873; Bescherelle 1873). In tropical Africa it is known only from Chad (Jelenc 1955), Zimbabwe (Best 1990; Manyanga & Perold 2004) and an uncertain record has been provided from the Kilimanjaro Mts. in Tanzania (Bizot *et al.* 1979). In the temperate, cool and cold regions of the Southern Hemisphere, *P. nutans* is scattered in the *Nothofagus* zone on the western fringes of southern South America, and is recorded from the Valdivian province to Tierra del Fuego (Greene 1986), Falkland Is. (Matteri 1986), Tristan da Cunha (Dixon 1960), Gough I. (Cardot 1911*c*), the Cape region of South Africa (Rooy & Magill 1987), St. Paul I. (Bescherelle 1875) and Amsterdam I. (Bescherelle 1875; Tixier 1980), then appearing in south-eastern Australia and Tasmania (Streimann & Klanzenga 2002) and New Zealand (Sainsbury 1955), as well as Auckland Is. (Vitt 1979*a*) and Campbell I. (Vitt 1974*a*). In Subantarctica the species is not uncommon on South Georgia (Clarke 1973*a*), rare on the Prince Edward Is. (Zanten 1971), Île de la Possession in the Îles Crozet archipelago (*Bell 3011*, AAS, KRAM) and Îles Kerguelen (Mitten 1876*a*,*b*). It is common in the maritime Antarctic but very rare in continental Antarctica.

Distribution in Antarctica. A pan-Antarctic species, common and locally abundant in the maritime Antarctic (Fig. 203) but rare on the continent (Fig. 204). In West Antarctica *Pohlia nutans* occurs on all peri-Antarctic archipelagoes (South Sandwich Is., South Orkney Is. and South Shetland Is.), on the isolated Bouvetøya, and along the west coast of the Antarctic Peninsula, from Trinity Peninsula to northern Marguerite Bay. It is much less frequent farther south on islands in southern Marguerite Bay, Charcot I., Alexander I. and on the Rymill Coast on the eastern coast of George VI Sound where it reaches its southernmost locality in West Antarctica at lat. 72°21′ S. On the east coast of the Antarctic Peninsula *P. nutans* is widely scattered on isolated nunataks, reaching its southernmost station on Engel Peaks, at 1015 m altitude, on the Wilkins Coast at lat.

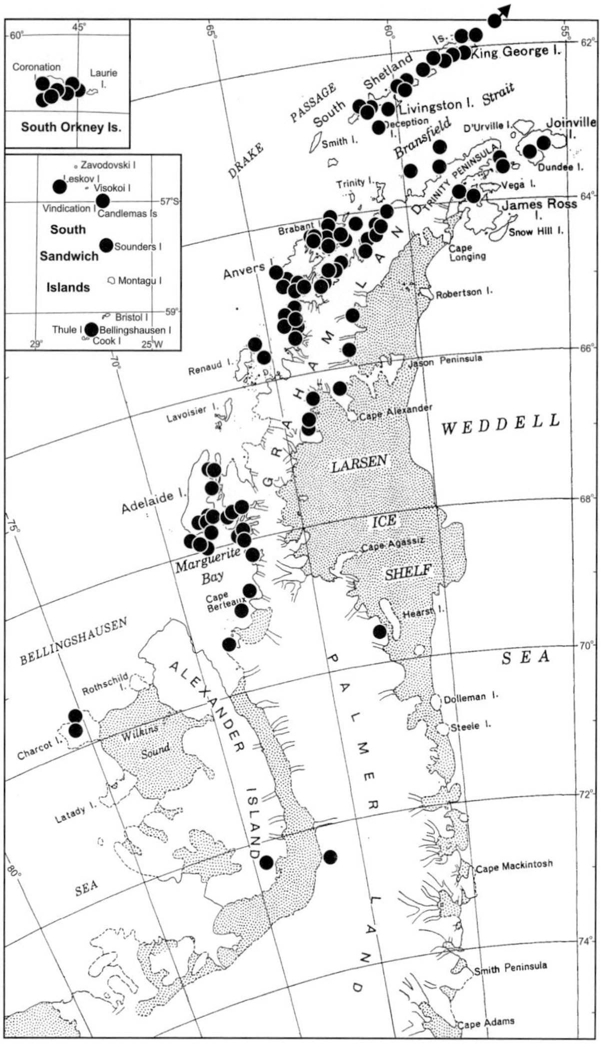

Figure 203. Distribution of *Pohlia nutans* in the Antarctic Peninsula region. Localities on Elephant I. in the northern group of the South Shetland Is. are indicated by an arrow.

69°32′ S. On the continent, the species is exceedingly rare in Queen Mary Land, northern Victoria Land and Ellsworth Land, reaching its southernmost occurrence on the actively volcanic Mt. Rittmann at lat. 73°28′ S (Bargagli *et al.* 1996; Skotnicki *et al.* 2002). The species also exhibits a very broad altitudinal range, occurring from sea level to 2800 m on Mt. Rittmann, but with the greatest concentration of sites at lower elevations, from near sea level to 400 m.

Selected specimens examined. WEST ANTARCTICA. BOUVETØYA. Moseryggen, 260–287, *TE 310 & 402* (BG).

SOUTH SANDWICH IS. **LESKOV I.** Summit ridge, 100 m, *PC 246G & 249G* (AAS, KRAM), *ca* 225 m, *MWH 833A* (AAS, KRAM, LE, PC) and 300 m, *REL 508A* (AAS, KRAM, LE, NY). **CANDLEMAS I.** North of Sea Serpent Cove, *KA 5B & 6* (AAS, KRAM); Lucifer Hill, *PC 107E, 108C & 111* (AAS, KRAM); Demon Pt., *PC 122A* (AAS, KRAM); Clinker Gulch, 50 m below summit, *PC 129B & 131* (AAS, KRAM); Sarcophagus Pt., *PC 167A* (AAS, KRAM); above Medusa Pool and Chimera Flats, *PC 172A* (AAS, KRAM); Gorgon Pool, *PC 173* (AAS, KRAM); Breakbones Plateau, *PC 201C, 209C & 212C* (AAS, KRAM). **SAUNDERS I.** Yellowstone Crags, 10 m, *PC 82, 84 & 86* (AAS, KRAM); Cordelia Bay, *MWH* (AAS, KRAM, LE, PC); 1.5 km south of Harper Pt., 15 m, *MWH 847* (AAS, BM, KRAM, LE, PC, S). **BELLINGSHAUSEN I.** Low point of south crater rim, 150 m, *PC 23* (AAS, KRAM); southern slope outside main crater, 100 m, *PC 36B & 78C* (AAS, KRAM); south slopes of main cone, 100 m, *MWH 407* (AAS, KRAM).

SOUTH ORKNEY IS. **MONROE I.** West coast, 30–100 m, *MMM 80* (AAS, KRAM); 80 m, *RILS 8001* (AAS, KRAM). **CORONATION I.** Mansfield Pt., 35 m, *RILS 293* (AAS, KRAM, PC); Cape Vik, 35–100 m, *RILS 570* (BM, S); Cape Hansen, 20–200 m, *RILS 3238* (AAS, KRAM); Shingle Cove, 25–75 m, *RILS 4966* (AAS, KRAM); Olivine Pt., 15 m, *RILS 120A* (AAS, KRAM, LE) and 35 m, *RILS 138, 145 & 149* (AAS, KRAM); Saunders Pt., *ca* 15 m, *RILS 19A* (BM). **LYNCH I.** North facing bank, *ca* 13 m, *DCL 942B* (AAS, KRAM). **MATTHEWS I.** North-west coast, 65 m, *RILS 157C* (AAS, KRAM); north-east coast, 15–30 m, *RILS 190* (AAS, KRAM). **COFFER I.** South side, 65 m, *RILS 181* (AAS, KRAM, LE, NY, PC). **CHRISTOFFERSEN I.** Southern side, 50 m, *KR 188A* (AAS, KRAM). **POWELL I.** Falkland Harbour, *ca* 15 m, *RILS 56* (BM); south-east coast, 135 m, *RILS 195* (AAS, KRAM). **MICHELSEN I.** East side, *ca* 15 m, *RILS 45* (BM). **FREDRIKSEN I.** Western face, 50 m, *DI 8D* (BM). **LAURIE I.** Cape Dundas, 65 m, *MMM 179* (AAS, KRAM). **SADDLE I.** Eastern part, 50–150 m, *MMM 170* (AAS, KRAM). **SIGNY I.** Jebsen Pt., 15 m, *MWH 162A* (AAS, KRAM); Foca Pt., 50 m, *REL 1124* (AAS, KRAM, LE, NY, PC); Starfish Cove, 15 m, *MWH 124A& 125A* (AAS, KRAM); between Jane Peak and Snow Hill, 150 m, *REL 1196* (AAS, KRAM, LE); north Elephant Flats, 15 m, *RILS 371* (AAS, KRAM); Borge Bay, 15 m, *FIDS 68C* (AAS, KRAM, LE, PC); Mooring Pt., *ca* 15 m, *MWH 97* (BM, KRAM, PC, S); Factory Cove, 15 m, *MWH 716A* (AAS, KRAM); Berntsen Pt., 15 m, *MWH 668A* (BM); Observation Bluff, 115 m, *DCL 929* (AAS, KRAM); Rusty Bluff, 220 m, *MWH 740B* (AAS, KRAM, S); Gneiss Hill, 175 m, *RILS 8109* (AAS, KRAM); Clowes Bay, 10–15 m, *MWH 694A & 703A* (AAS, KRAM). **MOE I.** Snipe Peak, 75 m, *KR 609* (AAS, KRAM); 10–200 m, *RILS 5250* (AAS, KRAM).

SOUTH SHETLAND IS. **ELEPHANT I.** 2 km south-west of Endurance Glacier, 152 m, *JSA 32D & 36B* (AAS, KRAM); 0.5 km south of Strinker Pt., *JSA 88B* (AAS, KRAM). **KING GEORGE I.** ***Admiralty Bay***: Creeping Slopes, 60 m, *RO 1238/80* (KRAM); Penguin Ridge, 45 m, *RO 1431/80* (KRAM); Ornithologists Creek, 18 m, *RO 718/80* (KRAM); MacKellar Inlet, Crépin Pt., 5 m, *RO 2036/80* (KRAM); Keller Peninsula, Ore Pt., 15 m, *RO 512/80* (KRAM); Martel Inlet, Tern Nunatak, 260 m, *RO 2520/80* (KRAM). ***Sherrat Bay***: Three Sisters Pt., *BJ 123* (KRAM). ***Drake Passage***: Pyrites I., 10 m, *BJ 32* (KRAM). ***Venus Bay***: North Foreland, *BJ 103* (KRAM). ***Fildes Peninsula***: Ardley I., *RO 2485/80* (KRAM); Bellinghausen Station, 15 m, *RO 2424/80 & 2432/80* (KRAM). ***Barton Peninsula***: Winship Pt., *BJ 188* (KRAM). ***Potter Peninsula***: Stranger Pt., *BJ 111* (KRAM). **NELSON I.** Harmony Cove, *CS 451* (PC, S). **ROBERT I.** Edward Pt., 50 m, *DCL 608* (AAS, KRAM); Coppermine Cove, 15–30 m, 35 m, *RILS 923B* (AAS, KRAM). **GREENWICH I.** Yankee Harbour, Spit Pt., 25 M, *DCL 534* (AAS, KRAM); Discovery Bay, Spark Pt., *DCL 644* (AAS, KRAM). **HALF MOON I.** Hill south of Huts, 50 m, *DCL 591A* (BM). **LIVINGSTON I.**

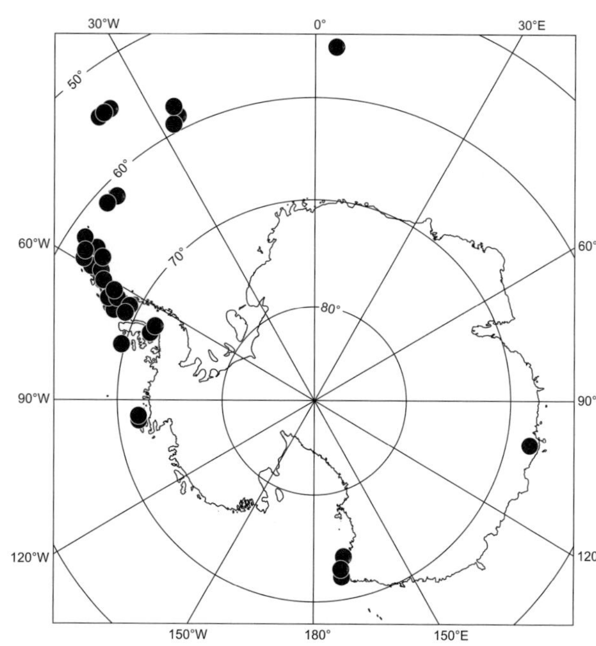

Figure 204. Distribution of *Pohlia nutans* in Antarctica.

Byers Peninsula: Clark Nunatak, 60 m, *DCL 163* (AAS, KRAM); South Beaches, 17 m, *DCL 438* (AAS, KRAM); Ocoa Pt., 20–75 m, *KR44G* (AAS, KRAM); between Chester Cone and Sealers Hill, 25–30 m, *RILS 3826* (AAS, KRAM). **DECEPTION I.** Whalers Bay, 5 m, *RILS 11226* (AAS, KRAM); between Kroner Lake and Ronald Hill, 2–4 m, *RILS 11240* (AAS, KRAM) and 15 m, *REL 1211* (AAS, KRAM); 1.5 km north of Mt. Pond, 500 m, *DM 39B* (AAS, KRAM) and *RILS 1159* (AAS, KRAM); Pendulum Cove, 100 m, *RILS 5749B & 5752A* (AAS, KRAM); Crater Lake, 15 m, *RILS 5651D* (AAS, KRAM).

WEST ANTARCTIC PENINSULA. **TRINITY PENINSULA.** Astrolabe I., 50–70 m, *RILS 4096 & 4097* (AAS, KRAM); Cape Roquemaurel, 10–50 m, *RILS 3931* (AAS, KRAM). **DAVIS COAST.** Trinity I., *RILS 700B* (AAS, KRAM); Cape Andreas, *CS 457* (PC, S). **DANCO COAST.** Cierva Pt., *VK 631 & 708A* (KRAM); Moss I., *CS 453 & 456* (PC, S); Spring Pt., *VK 1169 & 1261* (KRAM); Valdivia I., *CS 454* (PC, S); Charlotte Bay, Recess Cove, 2–10 m, *RILS 3939* (AAS, KRAM); Useful I., *RMS 69-343* (US); Cuverville I., *ER 244 & 474* (BR, PC); Beneden Head, *ER 473* (BR, PC); Coughtrey Peninsula, Paradise Harbour, Almirante Brown Station, *VK 344A* (KRAM); Bryde I., 100 m, *JS 179B* (AAS, KRAM). **PALMER ARCHIPELAGO.** ***Brabant I.***: Metchnikoff Pt., 50 m, *AM73C* (AAS, KRAM) and 100 m, *WH 19C* (AAS, KRAM); Claude Pt., 30 m, *WH 124B*(AAS, KRAM) and 200 m, *WH 204B* (AAS, KRAM); Minot Pt., 100 m, *AM 128B* (AAS, KRAM); Mt. Bulcke, 300 m, *AM 98 & 100B* (AAS, KRAM); Pinel Pt., 350 m, *AM 46B & 48C* (AAS, KRAM); Hunt I., 400 m, *AM 51B* (AAS, KRAM); east coast, 350 m, *ER 252A* (BR, PC); Two Hummock I., *RILS 704B* (AAS, KRAM). ***Melchior Archipelago***: Lystad I., *PAS 360.8* (FH); Omega I., 3 m, *IML 7935* (AAS, KRAM, NY). ***Anvers I.***: Cape Monaco, *VK 819* (KRAM); Norsel Pt., *VK 208, 1064 & 1073* (KRAM); Joubin Js., *JRR & JMG PAL-20* (US); Dream I, *VK 700 & 705* (KRAM); Litchfield I., *VK 1077, 1123, 1136, 1148 & 1157* (KRAM); Shortcut I., *VK 1055* (KRAM); Hermit I., *ca* 30 m, *RILS 889B* (AAS, KRAM); Cormorant I., 19 m, *AS 14* (AAS, KRAM). ***Wiencke I.***: Doumer I., 65 m, *FIDS 6* (AAS, KRAM); Goudier I., Port Lockroy, *IML 1761* (AAS, BM, KRAM). **GRAHAM COAST.** Edge Hill, 65 m, *RWMC 655* (AAS, KRAM); Rassmussen I., *Gain 280A, 281 & 282* (PC); Cape Tuxen, 13 m, *RWMC 673* (AAS, KRAM) and 330 m, *RWMC 684A* (AAS, KRAM); Mt. Demaria, 400 m, *RILS 10659* (AAS, KRAM); Cape Pérez, *Gain 272C* (PC); Prospect Pt., *PC 792B* (AAS, KRAM). **WILHELM ARCHIPELAGO.** ***Lemaire Channel***: Booth-Wandel I., *Gain 134A* (PC). ***Penola Strait***: Petermann I., 15 m, *REL 01359* (AAS, KRAM) and 65 m, *RWMC 839B* (AAS, KRAM). ***Argentine Is.***: Irizar I., 8 m, *RWMC 715* (AAS, KRAM); largest island between Irizar I. and Uruguay I., 15 m, *RWMC* (AAS, KRAM); Uruguay I., 23 m, *RWMC 440* (AAS, KRAM); Corner Is., 10 m, *RWMC 580* (AAS, KRAM); Galindez I., Stella Creek, 35 m, *RWMC 416B & 430* (AAS, KRAM); Winter I., 3 m, *BJT 127* (BM, PC); Skua I., 6 m, *RWMC 778B* (BM, KRAM, LE, NY); The Barchans, largest island, 6 m, *RWMC 749B* (AAS, KRAM). ***Collins Bay***: Green I., 20–60 m, *RILS 3373* (BM); Berthelot Is., largest island, 65 m, *RWMC 622B* (AAS, KRAM); Sommerville I., 25 m, *BJT 122B* (AAS, KRAM, NY); Darboux I., 10–15 m, *RILS 3312* (AAS, KRAM). ***Grandidier Channel***: Lahille I., 2–5 m, *RILS 4318* (AAS, KRAM); second largest island off Takaki Pt., 10–35 m, *RILS 4344 & 4345* (AAS, KRAM); south arm of Beascochea Bay, *BGLE 1195B* (AAS, BM, KRAM, LE, NY, PC, S); Hook I., *RMS 69-343* (US). **BISCOE IS.** Thomsen Is., 3 m, *RILS 5326B* (AAS, KRAM). **LOUBET COAST.** ***Bennett Is.***: Jona I., *JK 160A* (AAS, KRAM); Megaw I., *JK 165A* (AAS, KRAM). ***Laubeuf Fjord***: Webb I., Ryder Bay, *JK 255* (AAS, KRAM). **FALLIÈRES COAST.** ***Adelaide I.***: Rothera Pt., 17 m, *JF 30A & 31B* (AAS, KRAM) and 25 m, *JF 46* (AAS, KRAM); Stork Ridge, *PC 266B* (AAS, KRAM); point east of Mt. Bouvier, 15 m, *JK 245* (AAS, BM, KRAM, S); 7 km north of Cape Alexandra, *JK 202* (BM); Jenny I., 10 m, *JK 119A* (AAS, KRAM); Avian I., 50 m, *JK 136* (AAS, KRAM); Guébriant Is., 55 m, *BJT* (AAS, KRAM). ***Léonie Is.***: Lagoon I., *VK 273* (KRAM); Léonie I., 30 m, *RILS 8953* (AAS, KRAM); Anchorage Is., 25–35 m, *RILS 9208B* (AAS, KRAM). ***Dion Is.***: Courtier I., 5–25 m, *RILS 3588* (AAS, KRAM). ***Dismal Is.***: Faure I., 10–25 m, *RILS 3475* (AAS, KRAM). ***Marguerite Bay***: Blaiklock I., 5 m, *DWB 13C & 14A* (AAS, KRAM); Pourquoi Pas I., *Poncet 12* (KRAM); Horsehoe I., Lystad Bay, 10–15 m, *RILS 1895* (AAS, KRAM); Lagotellerie I., 50 m, *RILS 4840* (AAS, KRAM); Camp Pt., 10–15 m, *RILS 4622B* (AAS, KRAM); Neny I., 8–15 m, *RILS 797* (AAS, KRAM); Refuge Is., 10–25 m, *RILS 4525 & 4526* (AAS, KRAM); Stonington I., East Base, *Perkins 201* (US); Mushroom I., 60 m, *Poncet 1E* (AAS, KRAM). ***Terra Firma Is.***: Alamode I., 50 m, *PC 449B & 450B* (AAS, KRAM). ***Mica Is.***: the most southerly island, *FIDS 19A* (BM). **RYMILL COAST.** Batterbee Mts., 200 m, *RILS 2687* (AAS, KRAM). **GEORGE VI SOUND.** ***Charcot I.***: central north facing slope, 85 m, *PC 344B* (AAS, KRAM); summit area, 125 m, *RILS 10477* (AAS, KRAM). ***Alexander I.***: Stephenson Nunatak, *Fuchs E616/9* (BM).

EAST ANTARCTIC PENINSULA. **TRINITY PENINSULA.** Hope Bay, Base Hut, *RWMC 870C* (AAS, KRAM); Mt. Flora, 35–65 m, *FIDS 18R* (BM) and 20–25 m, *RILS 3845* (AAS, KRAM). **JOINVILLE ISLAND GROUP.** ***Joinville I.***: Mt. Alexander, 85 m, *RILS 3765* (AAS, KRAM); north end of Active Sound, 200 m, *RILS 3732* (AAS, KRAM). **JAMES ROSS ISLAND GROUP.** ***Prince Gustav Channel***: west side, *ca* 750 m, *CGB 215* (AAS, KRAM). ***James Ross I.***: Andreassen Pt., 10 m, *RILS 7606* (AAS, KRAM). **OSCAR II COAST.** Starbuck Glacier, 155 m, *PK 34* (AAS, KRAM); Gulliver Nunatak, *ca* 265 m, *PK 11* (AAS, KRAM). **FOYN COAST.** Cabinet Inlet, Cape Casey, 300–650 m, *RT 23 & 40* (AAS, KRAM); Mt. Denucé, *ca* 125 m, *RILS 2769A* (AAS, KRAM); Mt. Hayes, *ca* 185 m, *RILS 2766B* (AAS, KRAM). **WILKINS COAST.** Engel Peaks, 1000–1150 m, *MJC 63A* (AAS, KRAM).

ELLSWORTH LAND. **EIGHTS COAST.** ***Thurston Peninsula***: Mt. Dowling, *RILS 5416A* (AAS, KRAM); Gneiss Islet, *Starr s.n.* (AAS, KRAM, NY).

EAST ANTARCTICA. QUEEN MARY LAND. **KNOX COAST.** Bunger Oasis, near valley of seven lakes, 10 Jan 1957, *Korotkevich s.n.* (KRAM, LE).

VICTORIA LAND. **PENNELL COAST.** Cape North, 12 Feb 1963, *Orton s.n.* (AAS, KRAM, NY). **BORCHGREVINK COAST.** Crater Cirque, 11 Jan 1958, *Croll et al. s.n.* (AAS, KRAM); Mt. Rittmann, *ca* 2500 m, *RILS 9959, 10110 & 11765* (AAS, KRAM) and 2800 m, *RILS 11764* (AAS, KRAM).

40. BRYUM HEDW.

Bryum Hedw., Sp. Musc. Frond.: 178. 1801. — Lectotype (*vide* Britton 1918: p. 442): *Bryum argenteum* Hedw.

Ptychostomum Hornsch., Flora 5: 62. 1822. ≡ *Bryum* subgen. *Ptychostomum* (Hornsch.) Bruch & Schimp. *in* Bruch, Schimp. & W.Gümbel, Bryol. Eur. 4: 84. 1839 [Fasc. 6–9 Mon.: 14]. — Lectotype (*vide* Spence 2005: p. 19): *Ptychostomum cernuum* (Hedw.) Hornsch. (*Cynodontium cernuum* Hedw.) [= *Bryum uliginosum* (Brid.) Bruch. & Schimp.]. First synonymised by Bruch *et al.* (1839: p. 86).

Hemisynapsium Brid., Bryol. Univ. 1: 604. 1826. ≡ *Bryum* sect. *Hemisynapsium* (Brid.) Limpr., Laubm. Deutschl. 2: 285. 1892. — Type: *Hemisynapsium arcticum* (R. Br.) Brid. (*Pohlia arctica* R.Br.) [= *Bryum arcticum* (R. Br.) Bruch & Schimp.]. First synonymised by Limpricht (1892: p. 285).

Cladodium Brid., Bryol. Univ. 1: 620. 1826. ≡ *Pohlia* Hedw. [unranked] *Cladodium* (Brid.) Huebener, Muscol. Germ.: 475. 1833. ≡ *Bryum* subgen. *Cladodium* (Brid.) Bruch. & Schimp. *in* Bruch, Schimp. & W.Gümbel, Bryol. Eur. 4: 86. 1839 [Fasc. 6–9 Mon.: 16]. ≡ *Bryum* sect. *Cladodium* (Brid.) Bruch & Schimp. *in* Bruch, Schimp. & W.Gümbel, Bryol. Eur. 4: 153. 1846 (Fasc. 32 Mon.: 1). ≡ *Bryum* subsect. *Cladodium* (Brid.) I.Hagen, Tromsø Mus. Aarsh. 21–22: 119. 1901. — Lectotype (*vide* Pfeiffer 1873: p. 770): *Cladodium inclinatum* (Brid.) Brid. (*Pohlia inclinata* Sw. *ex* Brid.) [= *Bryum archangelicum* Bruch & Schimp.]. First synonymised by Bruch *et al.* (1839: p. 86).

Hemisinapsium Brid. *ex* Endl., Gen. Pl.: 52. 1836, *orthogr. pro* Hemisynapsium Brid.

Argyro-bryum Hampe, Linnaea 40: 312. 1876 ['Argyro-Bryum']. — Type: *Argyro-bryum subrotundum* Hampe [≡ *Bryum subrotundifolium* A.Jaeger]. First synonymised by Jaeger & Sauerbeck (1880: p. 443).

Argyrobryum (Müll.Hal.) Kindb., Bih. Kongl. Svenska Vetensk.-Akad. Handl. 6(19): 15. 1882. ≡ *Bryum* sect. *Argyrobryum* Müll.Hal., Syn. Musc. Frond. 1: 313. 1849. ≡ *Bryum* subgen. *Argyrobryum* (Müll.Hal.) J.J.Amann, Rev. Bryol. 20: 44. 1893. ≡ *Bryum* [unranked] *Argyrobryum* (Müll.Hal.) Broth. *in* Engl. & Prantl, Nat. Pflanzenfam. 1(3): 584. 1904, *nom. illeg. incl. lectotyp. gen.* [Lectotype (*vide* Mönkemeyer 1927: p. 484): *Bryum argenteum* Hedw.].

Rosulabryum J.R.Spence, Bryologist 99: 222. 1896. — Type: *Rosulabryum albolimbatum* (Hampe) J.R.Spence (*Rhodobryum albolimbatum* Hampe). First synonymised by Allen (2002: p. 327).

Imbribryum N.Pedersen, Bryologist 108: 449. 2005 [*Imbribryum* N. Pedersen & Hedenäs, Bryologist 108: 127. 2004, *non rite publ.*], ***syn. nov.*** — Type: *Imbribryum alpinum* (With.) N.Pedersen (*Bryum alpinum* With.) (18 Aug).

Gemmabryum J.R.Spence & H.P.Ramsay, Phytologia 87: 63. 2005 (14 Sep), ***syn. nov.*** — Type: *Gemmabryum pachytheca* (Müll.Hal.) J.R.Spence & H.P.Ramsay ['pachythecum'] (*Bryum pachytheca* Müll.Hal.).

Ochiobryum J.R.Spence & H.P.Ramsay, Phytologia 87: 69. 2005, ***syn. nov.*** — Type: *Ochiobryum blandum* (Hook.f. & Wilson) J.R.Spence & H.P.Ramsay (*Bryum blandum* Hook.f. & Wilson).

Plants perennial, small and slender to large and robust, densely or loosely caespitose, sometimes gregarious or intermixed with other mosses, stiff and rigid to soft, dull or shiny, principally green, yellow, orange and brown, less often red, pinkish, whitish or silvery-green, usually with all possible combinations and casts of these colours. *Stems* erect, mostly branching, especially by innovation, less often simple, mostly radiculose near the base or throughout except for the apex with brown to reddish-brown, branching, lustrous, smooth to papillose rhizoids, in transverse section pentagonal to rounded or obtusely angular, consisting of a small to large central strand, several layers of large, thin-walled, hyaline medullary cells and usually distinct 1–3-stratose cortex of small, thick-walled and pigmented cells, sometimes less prominent, composed of only somewhat smaller, firm-walled and coloured cells; *axillary hairs* filiform, consisting of 3–8 cells, including 1–3 short, quadrate, reddish-brown basal cells and 2–6 elongate, hyaline upper cells. *Leaves* generally smaller and distant below, larger and crowded above, sometimes forming a rosette at the stem apex, less often equally spaced along the elongate stem, erect to erect-spreading, sometimes erect-appressed and imbricate, usually not much altered but sometimes twisted or shrunken when dry, erecto-patent when moist, ovate, ovate-lanceolate, lanceolate, oblong-lanceolate, oval, elliptic to orbicular, rarely spathulate, gradually long acuminate to acute, less often rounded, obtuse to obtuse-apiculate at the apex, often awned, not or decurrent, concave to broadly carinate; *margins* plane, erect, recurved to revolute, entire to serrulate near the apex, not or distinctly bordered, uni- or occasionally bistratose; *costa* single, strong to relatively weak, subpercurrent, percurrent to long-excurrent, mostly prominently convex dorsally, in cross-section terete, consisting of several guide cells, often hydroids and a large or small dorsal stereid band, lacking ventral stereids and with a single row of large ventral and larger or smaller dorsal epidermal cells; *laminal cells* smooth, pellucid, unistratose throughout, except sometimes at margins, rhombic, rhomboid- or oblong-hexagonal, occasionally linear-rhomboidal above, rectangular to quadrate below, thin- to thick-walled, not differentiated at the basal angles, often narrow and linear at the margins, forming a prominent border, green, hyaline or red at the base. *Asexual reproduction* occasional to frequent by means of multicellular, globose or ellipsoid, red, brown, purplish, violet or crimson rhizoidal tubers, smooth or with protuberant cells or filiform or gemmiform, ovoid to subcylindrical axillary gemmae. *Dioicous*, *autoicous* or *synoicous*. *Inflorescences* terminal; *perichaetial leaves* ± triangular, narrower than the cauline leaves, inner ones small. *Setae* elongate, straight to flexuose, smooth, red to brownish-red, curved or hooked at the tip. *Capsules* exserted, inclined, horizontal to mostly pendulous, clavate or pyriform, sometimes ovoid to subglobose, with a prominent neck, brown, smooth, symmetric or slightly curved; *annulus* always present, deciduous, consisting of 2–4 rows of large, vesiculose cells with strongly incrassate walls; *operculum* convex-conic to hemispheric, minutely apiculate to umbonate; *exothecial cells* variable in size and shape, ranging from ± isodiametric and irregular to rectangular, thin-walled, becoming smaller, oblate to quadrate in 2–4 tiers at the orifice; *stomata* numerous in several rows in the neck, phaneroporous, bicellular, with long and narrow pori; *peristome* double; *exostome teeth* 16, lanceolate, acuminate, yellow to brown, densely finely dotted below, hyaline and papillose above, usually bordered, strongly trabeculate on the inner surface; *endostome* hyaline, finely papillose, with a short or tall basal membrane giving rise to lanceolate, ± keeled, narrowly or broadly perforate segments, alternating with 1–4, long or short, appendiculate or nodulose cilia, rarely lacking or none. *Spores* spherical, papillose to smooth. *Calyptra* cucullate, fugacious, naked, smooth.

Bryum is one of the largest and most polymorphous of moss genera, comprising species with widely divergent morphologies. Taxonomically, the genus is exceedingly difficult owing to its generally nondescript characters, the most important of which are the leaf areolation, especially the length/width ratio of the upper laminal cells, thickness of cell walls, colour of the basal leaf cells, costa length, development of the marginal border and stance of the leaf margin, sexuality of the plants, leaf decurrency, as well as the capsule shape and the development of the peristome. Typically, the leaves in *Bryum* are bordered with narrowly linear, thick-walled cells, the upper laminal cells are large, rhomboidal to oblong-rhomboidal, the costa is excurrent to percurrent, the capsules are pendulous and more or less pyriform, and the peristome is well developed, usually of the perfect *Bryum*-type, with cilia, high basal membrane and broad and perforate segments. Within the Bryaceae, the genus can be confused with *Pohlia* which has narrower, oblong-lanceolate and elimbate leaves with long and narrow laminal cells and percurrent costae. However, these features fall well within the range of variability of character states in *Bryum*, so sterile material may cause problems with determination.

Considerable morphological variation within the genus *Bryum* prompted various early bryologists to elaborate different classification systems of its species. Two concepts were developed. In the first of these, some authors circumscribed *Bryum* very broadly, placing in it most species of the Bryoideae in the modern sense, making it a typical catch-all genus. It was subsequently subdivided into many infrageneric taxa of various, often not precisely defined, ranks (e.g. Müller 1848–1849; Bruch *et al.* 1839; Schimper 1856, 1860, 1876; Podpěra 1954), resulting in considerable nomenclatural confusion. Even in the most recent treatments of Ochi (1970, 1972, 1973*a*,*b*, 1980, 1981, 1992) *Bryum* is broadly understood and encompases the genera *Anomobryum* and *Rhodobryum* which have gained wide acceptance. Ochi (1992) presented an infrageneric classification of the genus *Bryum* which was divided into three subgenera, namely subgen. *Anomobryum*, subgen. *Bryum* and subgen. *Rhodobryum*, and the type subgenus, i.e. *Bryum s. str.*, was subsequently subdivided into seven sections and 13 subsections. Isoviita (1992) attempted to clarify some nomenclatural issues associated with accepted infrageneric taxa of *Bryum*.

The second concept in the taxonomy of *Bryum* is characterised by segregation of various groups of species into separate genera. Some of these genera have not been widely accepted and have fallen into obsolescence, namely *Ptychostomum* Hornsch. (Hornschuch 1822), *Cladodium* Brid. and *Hemisynapsium* Brid. (Bridel 1826), as well as *Argyro-bryum* Hampe (Hampe 1876) and *Argyrobryum* (Müll.Hal.) Kindb. (Kindberg 1882, 1883). Conversely, four genera, namely *Epipterygium* Lindb. and *Plagiobryum* Lindb. (Lindberg 1863*a*) [the latter superseding *Zieria* Schimp., *hom. illeg.* (Schimper 1856)], *Anomobryum* Schimp. (Schimper 1860) and *Rhodobryum* (Schimp.) Limpr. (Limpricht 1892), have gained wide acceptance and are used in most Floras, checklists and catalogues of mosses.

As presently understood, *Bryum* is a very large, heterogeneous and impractical genus which canot realistically be defined and defended. Spence (1987) proposed a preliminary revised classification of the genus and recognised three informal groups of species within it, based primarily on gametophyte traits because sporophyte characters proved to vary markedly even in closely related species, especially with respect to the development of the peristome. Later, Spence (1996) formally distinguished one of these groups comprising robust species with a strongly rosulate habit as the separate genus *Rosulabryum*, and recently Spence (2005) resurrected the genus *Ptychostomum* for species corresponding to the informal group '*Cernuibryum*' of Spence (1987) and described a new monotypic genus *Plagiobryoides* for a strange Central American species, *P. incrassatolimbata*, characterised by strongly limbate leaves (Ochi & Salazar Allen 1990). Finally, Spence & Ramsay (2005, 2006) recognised two more new genera, *Gemmabryum* and *Ochiobryum*. The former accommodates those species of *Bryum* and *Brachymenium* that have elongate distal and median laminal cells, short and wide proximal laminal cells, imbricate and elimbate leaves and frequently produce axillary gemmae or rhizoidal tubers. The latter genus comprises only two species, from China and Australasia, and is characterised by having complanate leaves, a unique character in the genus *Bryum s. lato*.

However, this relatively clearly defined situation was upset by the application of molecular methods in the taxonomy of *Bryum* (Pedersen & Hedenäs 2002*a*, 2003, 2005; Pedersen *et al.* 2003). Based exclusively on evidence from chloroplast DNA and ignoring morphology, Pedersen & Hedenäs (2005) expanded the concept of the genus *Plagiobryum* that had hitherto been considered to be a very distinct and well defined genus both in terms of sporophyte and gametophyte characters, instantly rendering it an obscure taxon that is totally undefinable in morphological categories. In addition, the new genus *Imbribryum* was established for *Bryum alpinum*, *B. blandum* Hook.f. & Wilson and *B. muehlenbeckii* Bruch & Schimp. It too was without a clear description and designation of diagnostic characters, and the Latin diagnosis provides only some general traits which can also be applied to many species of *Bryum* which have only distant morphological affinities. This generic name was initially invalidly published because the generitype was not indicated, but later this name was validated and *Imbribryum alpinum* designated as its type (Pedersen 2005). Although not clearly defined, nomenclaturally it is a correct name, having priority over *Gemmabryum* of Spence & Ramsay (2005).

The taxonomic and classificational problems associated with segregates of *Bryum* are additionally complicated by strictly nomenclatural issues. It was unfortunate that Britton (1918) selected *Bryum argenteum* as lectotype of this generic name since, morphologically, this species is very close to *Anomobryum* and could easily be accommodated in that genus (Spence & Ramsay 2002). Unfortunately, the proposal to conserve *Bryum* with the conserved type *B. caespiticium* Hedw. was rejected by the Committee for Bryophyta (Zijlstra 2002) because 'Several members think that *B. argenteum* and *B. caespiticium* might be closely related. Some recent investigations (not yet published) support this, and moreover it seems that these two species are not closely related with *Anomobryum*'. Rejection of this proposal has resulted in the introduction of many new names for species transferred to *Ptychostomum*. This is the largest segregate of *Bryum* (Spence 2005) and this decision has not helped to stabilise the nomenclature. However, on morphological grounds, *B. argenteum* is very close to *Anomobryum* and both taxa share the lack of guide cells in the anatomical structure of the costa and the strongly thickened walls of exothecial cells that are markedly differentiated in several rows at the orifice. In addition, at least two species of *Anomobryum*, *A. hyalinum* T.J.Kop. & D.H.Norris and *A. ochii* T.J.Kop. & D.H.Norris, from New Guinea (Koponen & Norris 1984; Eddy 1996) are also silver in coloration. Also, the presence of rounded and obtuse leaf apices cannot be regarded as a character to exclude the close relationship of *B. argenteum* with *Anomobryum* because *B. argenteum* var. *muticum* constitutes a perfect link between these taxa. In contrast, *B. caespiticium* with its well developed guide cells in the costa is certainly distantly related to *B. argenteum* and clearly belongs within *Ptychostomum*, the largest segregate of the broadly conceived genus *Bryum*.

In the present treatment the traditional concept of *Bryum* is adopted since the new ideas regarding the taxonomy and classification of the genus are not yet fully resolved, and a thorough and detailed discussion of these issues exceeds the scope of the present Flora. In the traditional circumscription *Bryum* is one of the largest of all moss genera and, according to Crosby *et al.* (2000), there are still 448 species, although nearly 250 of these are poorly known and have mostly not been critically examined since being described. Conversely, Ochi (1992) reduced the genus to approximately 150 species and this number is probably the most realistic since the level of species inflation had previously been very high in *Bryum*. Many of the species appear to have been published without reference to earlier literature, especially if they came from previously uninvestigated areas.

Some authors apparently felt that the same species could not occur in geographically widely separated areas. Moreover, lack of illustrations of many species, brief diagnoses and a failure to compare original collections made their interpretation difficult. In the absence of definitive distinguishing characters, new species have often been described on the basis of characters that are particularly subject to variation, such as the mode of branching, size and coloration of the plants, leaf and capsule shape, and costa structure. Early bryologists often distinguished every environmental variant as a separate taxon, and this resulted in the proposal of innumerable names for infraspecific taxa at all ranks. The best examples of this situation is the synopsis of Eurasiatic taxa of *Bryum* presented in a series of publications by Podpěra (1942, 1943, 1944, 1945, 1950, 1951*a,b*, 1952, 1953, 1955, 1956, 1957, 1958, 1959, 1960, 1962, 1973*a,b*).

The genus *Bryum* has been intensively investigated in many parts of the world by Ochi (1959, 1970, 1972, 1973*a,b*, 1979, 1980, 1981, 1982, 1985, 1990*a,b*) and these studies resulted in many reductions of superfluously described species, especially from tropical and austral regions. In addition, many taxonomic revisions have been published of various groups and complexes within *Bryum*, for instance *B. erythrocarpum* (Crundwell & Nyholm 1964), *B. capillare* (Syed 1973), *B. billardierei* (Mohamed 1979), and *B. bicolor* (Wilczek & Demaret 1976; Smith & Whitehouse 1978; Holyoak 2003).

In Antarctica several species of *Bryum* are very common and locally abundant and are often important constituents of plant communities, especially in wetter places. Many Antarctic populations are sterile and, because sporophyte characters are usually critical for determination of *Bryum* species, numerous plants cannot be determined with full certainty. Therefore, sterile material is often attributed to a species mostly by default. The genus *Bryum* is difficult due to its generally nondescript, often technical characters, and the tendency of leaves taken from different parts of single plants to have dissimilar morphologies. Some species show great variation that is induced by different habitat conditions. As a result of these difficulties numerous species have been described from Antarctica, often from sterile material, which do not merit taxonomic recognition.

In the polar and cool temperate regions of the Southern Hemisphere *Bryum* exhibits moderate species richness and about 50 species have been recorded from the Subantarctic and Antarctic. Some of them, especially local endemics, have been studied and their names reduced to synonymy with names of species described earlier (Ochi 1972, 1979, 1982; Seppelt 1983). A lot of old records from this biome are still in need of verification, including a revision of voucher collections. At present seven species and one variety of *Bryum* are recognised in the Antarctic. These belong within four sections *sensu* Ochi (1992), with consideration of the correct names for them as suggested by Isoviita (1992) and Ochyra *et al.* (2003). These are: sect. *Bryum* (*B. argenteum* incl. var. *muticum*), sect. *Doliolidium* (Müll.Hal.) Müll.Hal. (*B. dichotomum*), sect. *Amblyophyllum* Müll.Hal. (*B. nivale* and *B. orbiculatifolium*) and sect. *Cladodium* (Brid.) Husn. (*B. pseudotriquetrum*, *B. archangelicum* and *B. pallescens*).

1. Plants whitish-silvery, occasionally pale yellowish- or greenish-whitish; upper parts of all leaves echlorophyllose, hyaline, lacking coloration; costa weak, ending well below the apex, lacking guide cells 2
1. Plants brown or green, not silvery, at least below; at least lower leaves chlorophyllose throughout; costa relatively strong, excurrent, percurrent or sometimes ceasing a few cells below the apex, with guide cells 3
    2. Most leaves apiculate; apiculus typically more than 10 μm long, usually much longer .... 1A. ***B. argenteum*** var. ***argenteum***
    2. Most leaves rounded at the apex; apiculus absent or rarely present, very short, less than 5 μm long .... 1B. ***B. argenteum*** var. ***muticum***
3. Bulbils present in the axils of the upper leaves on sterile shoots .... 2. ***B. dichotomum***
3. Axillary bulbils or gemmae lacking 4
    4. Leaf base green, brownish to yellowish-brown 5
    4. Leaf base red 6
5. Leaf base brownish to yellowish-brown; basal cells lax, sharply demarcated from the upper cells .... 3. ***B. nivale***
5. Leaf base green; basal cells sturdy or laxer, not markedly different from the upper cells .... 4. ***B. orbiculatifolium***
    6. Costa percurrent to short-excurrent; leaves ovate- to oblong-lanceolate, acute to short-acuminate, often cuspidate, somewhat decurrent or sometimes suborbicular to broadly ovate or oblong-lingulate, strongly concave on innovations .... 5. ***B. pseudotriquetrum***
    6. Costa long-excurrent; leaves oblong-lanceolate to ovate-lanceolate, mostly gradually long-acuminate, non-decurrent 7
7. Leaves straight, not altered when dry, ovate- to oblong-lanceolate; most upper laminal cells 3 or more times as long as wide; cilia of the endostome lacking or rudimentary and reduced to short stubs .... 6. ***B. archangelicum***
7. Leaves distinctly shrunken when dry, predominantly elongate, oblong-lanceolate; most upper laminal cells shorter, less than 3 times as long as wide; cilia of the endostome present, appendiculate .... 7. ***B. pallescens***

1. **Bryum argenteum** Hedw.

Bryum argenteum Hedw., Sp. Musc. Frond.: 181. 1801. ≡ *Hypnum argenteum* (Hedw.) F.Weber & D.Mohr, Index Mus. Pl. Crypt.: sheet 2 [4]. 1803. ≡ *Mnium argenteum* (Hedw.) Hoffm. *ex* P.Beauv., Prodr. Aethéogam: 74. 1805. ≡ *Argyrobryum argenteum* (Hedw.) Kindb., Bih. Kongl. Svenska Vetensk.-Akad. Handl. 6(19): 15. 1882. — Type citation: Ad muros, tecta, rupes Europae; also references to Morison (1699, p. 629, tab. 6, fig. 17), Vaillant (1727: p. 134, t. 26, f. 3), Dillenius (1741: p. 392, t. 50, f. 62), Linnaeus (1763: p. 1569), Haller (1768: p. 47), Weiss (1770: p. 199) and Murray (1784: p. 947). [Lectotype (*selected here*): the central specimen glued to the sheet between the specimens marked 'b' and 'a' on the sheet bearing the label 'Bryum argenteum antheris pendulis surculis cylindricis imbricatis laevibus Linn. Sp. pl. 2. p. 1586 n. 27 Bryum pendulum iulaceum argenteum et sericeum Dill. H. Musc. 392. t. 50. f. 62' – G-Hedwig/Schwägrichen!].

Plants small, occasionally moderately sized and slender, soft, in close flat or cushion-shaped tufts or patches, sometimes more loosely spreading to gregarious, whitish, silvery, pale green, pale yellowish-green or pale yellowish, shining and often silky on drying, occasionally with pale reddish or brownish tinge below. *Stems* erect, (0.2–)0.4–1.0(–3.0) cm tall, slender, filiform, fragile, simple or fastigiately branching, terete and julaceous, red, usually radiculose at the base or with clusters of rhizoids scattered throughout except for the stem tips, in transverse section round, comprising a distinct central strand, 3–5-layered medulla of large, thin-walled, hyaline cells and 1–2-stratose cortex of smaller cells with thin and red walls; rhizoids lightly fuscous to reddish-brown, branched, finely papillose; *axillary hairs* short, filiform, 2–4-celled, with 1–2 short, reddish-brown basal cells and 1–2 hyaline, elongate distal cells. *Leaves* crowded and equally spaced throughout the stem, or crowded and comose at the stem tip and distant below, erect-imbricate, both dry and moist, 0.3–0.8(–1.2) mm long, 0.25–0.6 mm wide, ovate, widely ovate, obovate, orbicular to transversely elliptical, widest in the lower half or in mid-leaf, strongly concave to cochleariform, acute to filiform-acuminate or broadly rounded, obtuse to apiculate at the apex, occasionally cucullate, non-decurrent, echlorophyllose in the distal part or throughout, chlorophyllose or sometimes lower leaves pale reddish-tinged at the base; *margins* entire, plane or slightly reflexed on larger leaves below, unbordered; *costa* single, usually vanishing considerably below the apex or reaching the base of the acumen or sometimes ending a few cells below the apex, sometimes very indistinct, obsolete to absent on some leaves, slender, (20–)40–55(–65) µm wide at the base, gradually tapering upwards, occasionally spurred distally, yellowish-brown and subconcolorous with the laminal cells, sometimes reddish-brown near the base, in transverse section hardly convex on the dorsal side, lacking guide cells and with a weak central band of stereid or substereid cells and ventral and dorsal rows of larger epidermal cells; *laminal cells* unistratose throughout, rhomboidal, oblong- to linear-rhomboidal to rectangular or occasionally hexagonal to oblong-hexagonal in the distal half and mid-leaf, (25–)35–70(–100) µm long, (8–)12–18(–24) µm wide, thin- to thick-walled, frequently thickened at the corners, becoming short-rectangular to quadrate at the base, 18–30 µm long, 13–18(–25) µm wide. *Asexual reproduction* very common by short, compact, readily deciduous branchlets or swollen, filamentous, rhizoidal gemmae. [*Dioicous*. *Perigonia* terminal, bud-like; perigonial bracts ovate, short-acuminate, concave, with a costa ceasing below the apex. *Perichaetia* terminal, with inner perichaetial leaves triangular, shortly costate. *Setae* erect, slender, 1–2 cm tall, red to red-brown, smooth, curved at the apex. *Capsules* exserted, pendent, obloid to obloid-cylindrical from a short neck, 2.0–2.6 mm long, 0.7–1.0 mm wide, symmetric, somewhat constricted below the mouth when dry, dark red to blackish-red, orange to red below the mouth, glossy or dull; *annulus* deciduous, triseriate, composed of large, vesiculose, thick-walled cells; *operculum* low conic to high conic-convex, apiculate, lustrous, usually long-persistent on the capsule; *exothecial cells* with strongly incrassate walls, usually 8–10 µm wide, predominantly rectangular, less often isodiametric in mid-urn, 15–50 µm long, 8–14 µm wide, becoming mostly isodiametric and irregular in the neck and oblate, transversely hexagonal, short-rectangular to quadrate, thin- to firm-walled in 4–10 tiers at the mouth; *stomata* numerous in many rows in the neck, phaneroporous, bicellular, with elongate, very narrow pori; *peristome* double; *exostome teeth* 16, lanceolate, 350–450 µm long, yellow, yellowish-brown to orange below, hyaline at the apex, finely papillose, narrowly bordered, trabeculate on the inner side; *endostome* as tall as the exostome teeth, yellowish to hyaline, finely papillose, consisting of a basal membrane, about half or more the length of the exostome, and broad, perforate, keeled processes alternating with 2–3 appendiculate cilia. *Spores* spherical, brownish, 10–16 µm in diameter, faintly papillose. *Calyptra* cucullate, 1.2–1.5 mm long, naked, smooth.]

Chromosome numbers: unknown in the Antarctic and Subantarctic and the chromosome number n = 10 from Enderby Land (Tatuno 1963; Inoue 1976) refers to *Bryum pseudotriquetrum*; n = 10 (24 records), n = 10 + m (2 records), n = 11 (6 records) and n = 20 (1 record) from different parts of the world have been detected (Fritsch 1991).

Discussion. *Bryum argenteum* is one of the most distinctive species of bryalean mosses and is easily recognised by its small to very small size, julaceous stems, and leaves with a weak costa lacking guide cells and smooth margins, as well as by its exothecial cells with strongly incrassate walls that greatly exceed the width of the lumina and become markedly different, thinner-walled and flattened to quadrate in 4–10 tiers at the orifice. However, the most characteristic and readily observable feature of this species that is emphasised in all Floras and taxonomic treatments of the genus, and from which is coined its common name 'silver moss', is the whitish-silvery to occasionally pale whitish-yellowish tint of the plants. It is due to the extreme thinness of the texture of the leaves which are usually devoid of chlorophyll in the distal part and are only faintly coloured in the proximal half.

Throughout its worldwide range *Bryum argenteum* exhibits considerable variation in the robustness of plants, leaf size, acumen length and costa length and some of this variability is also observed in Antarctic populations. In general, the shape of leaves of *B. argenteum* in Antarctica ranges from broadly ovate or oblong-ovate and tapering to an acute or filiform apex to orbicular or transversely elliptical with rounded and obtuse apices. These extreme morphotypes are unlike each other and their morphology may suggest specific distinctness, but they are linked by all possible intergradations within the same population or even on the same shoot. For convenience they are here recognised as varieties, var. *argenteum* and var. *muticum*, although some collections can only be randomly classified into one of these taxa since they are morphologically perfectly intermediate. Interestingly, phenotypes designated as *B. argenteum* var. *lanatum* (P.Beauv.) Hampe that are typical of dry, sunny and exposed habitats have not so far been found in Antarctica. This variety is a relatively slender and more silvery-white plant with leaves narrowed to a long-subulate apex and a costa extending into the acumen or becoming long-excurrent as a toothed awn. In some Antarctic plants the leaves are long-acuminate and filiform at the apex but the costa ceases well below it or occasionally extends to the base of the acumen as an ill-defined strip.

A. var. **argenteum** Fig. 205

Bryum argenteum var. *subamblyolepis* Cardot & Broth., Kongl. Svenska Vetenskapsakad. Handl. 63(10): 38. 1923, ***syn. nov.*** — TYPE CITATION: W. Falkland, Westpoint Island (589). [Lectotype (*selected here*): 'Expeditio suecica 1907–1909. 589 Bryum argenteum (L.) Hedw. var. subamlyolepis Card. var. nov. Falkland, Westpoint Island ad terram argillosam 6.12.1907 Leg. T. Halle Det. J. Cardot' – PC!; isotypes: LD!, S!].

Bryum siplei E.B.Bartram, Ann. Missouri Bot. Gard. 25: 723. 1938. — TYPE CITATION: [Antarctica] King Edward VII Land, Rockefeller Mts., Mt. Helen Washington, P. Siple, F. A. Wade, S. Corey & O. D. Stancliff 100. [Lectotype (*selected here*) {or perhaps holotype}: 'Byrd Antarctic Expedition II Mosses of King Edward VII Land, Antarctica No. 100 Bryum siplei Bartr. sp. nova. Locality: Mt. Helen Washington Collected by: Paul A. Siple Determined by: Edwin B. Bartram' – FH-Bartram!; isotype: AAS!]. First synonymised by Greene (1967*b*: p. 12).

Plants small, slender, julaceous, mostly in dense or loose tufts or patches, less often gregarious, pale green, pale yellowish or yellowish-green or whitish, silvery and shining on drying. *Stems* erect, to 10 mm high, fragile, branching below, red, red-brown to yellowish-green, frequently radiculose below with red-brown, lustrous, papillose, branching rhizoids or with clusters of rhizoids scattered along the stem, in cross-section with a distinct central strand, 3–5-layered medullary tissue of large, thin-walled, hyaline cells and 1–2-layered cortex of somewhat smaller cells with thicker, red-walls; *axillary hairs* short, filiform, 2–4-cellular, with 1–2 short, quadrate, reddish-brown basal cells and 1–2 elongate, hyaline, distal cells. *Leaves* crowded, occasionally somewhat distant, erect, imbricate wet or dry, concave, broadly ovate to oval or broadly oblong to obovate, 0.3–0.75(–1.2) mm long, 0.25–0.6 mm wide, non-decurrent, tapering to an acuminate, acute, subacute, apiculate or filiform and often recurved apex, pellucid, the lower very slightly chlorophyllose, the upper quite colourless and hyaline for upper half of the leaf, green, yellowish-green below, often reddish at the base; *margins* entire, plane to somewhat reflexed below, unbordered; *costa* single, variable in length, usually vanishing well below the apex and extending ½–¾ way up the leaf, but sometimes ill-defined above and running into the apiculus, slender, often faint, (35–)40–55(–65) μm wide at the base, gradually narrowed upwards, sometimes spurred in the upper part or forked at the tip, yellowish-brown to pale brownish and subconcolorous with the laminal cells, especially distally, sometimes reddish to reddish-brown near the base, in transverse section lacking guide cells, with a row of thin-walled, large epidermal cells on the ventral and dorsal sides and a central band of stereid or substereid cells; *laminal cells* pellucid, unistratose throughout, variable in length and size; *upper cells* hyaline, thin- to firm-walled, short to elongate, rhombic, irregularly rhomboidal to rectangular, becoming chlorophyllose, firm- to thick-walled, rectangular to elongate-rhomboidal, occasionally flexuose, obtuse or obliquely truncate at the ends in mid-leaf, (25–)35–70(–100) μm long, (10–)12–18(–25) μm wide, becoming slightly narrower towards the margins but not forming distinct borders; *basal cells* short-rectangular to quadrate, thick-walled, sometimes weakly reddish pigmented. *Asexual reproduction* by short, compact, ovoid to obloid, green, leafy, deciduous branchlets, arising from main shoot, fragile and readily detached tips of shoots or swollen, filamentous, rhizoidal gemmae. Sterile.

Discussion. *Bryum argenteum* was known long before it was formally designated with this binomial by Linnaeus (1753, 1763) and a full list of its early synonyms was presented by Dillenius (1741). The Linnaean binomial was validated by Hedwig (1801) and the specimen from his personal herbarium is here selected as lectotype. It represents a typical expression of *B. argenteum* that is characterised by leaves that are broadly ovate, imbricate, concave and quite shortly tapering to an acuminate apex, hyaline and chlorophyllose in the distal half, with a slender costa extending to ¾ way up the leaf, and elongate-rhomboidal laminal cells with moderately thickened walls, becoming quadrate to short-rectangular and green near the base.

Bryum argenteum var. *argenteum* is extremely variable in gametophyte features such as leaf shape, leaf apex, costa length as well as sporophytes including capsule shape, size and orientation, seta length and development of the peristome (Tan & Koponen 1989). Longton & MacIver (1977) and Longton (1981) showed experimentally that it exhibits well marked variation in morphological and physiological traits, and Skotnicki *et al.* (1998) described remarkable genetic variability between populations from widely separated geographical regions and observed

⟶

Figure 205. *Bryum argenteum* var. *argenteum*. — **1**. Habit, dry. **2**. Portion of stem, dry. **3**. Cross-section of stem. **4–6**. Axillary hairs. **7–20**. Leaves. **21–22**. Leaf apices. **23–24**. Mid-leaf cells. **25–26**. Basal cells. **27–29**. Cross-sections of leaves. **30–31**. Gemmae. — [1–2, 13–15, 25 from *PK 41A*; 3–6, 18–21, 23, 27–29 from *Seppelt 19601*; 7–12, 22, 24, 26 from *RO 972/80*; 16–17, 30–31 from *BAS Misc 208B*]. — Scale bars: a – 50 μm (4–6) and 100 μm (3, 21–29); b – 1 mm (2); c – 1 mm (7–20, 30–31) and 2 mm (1).

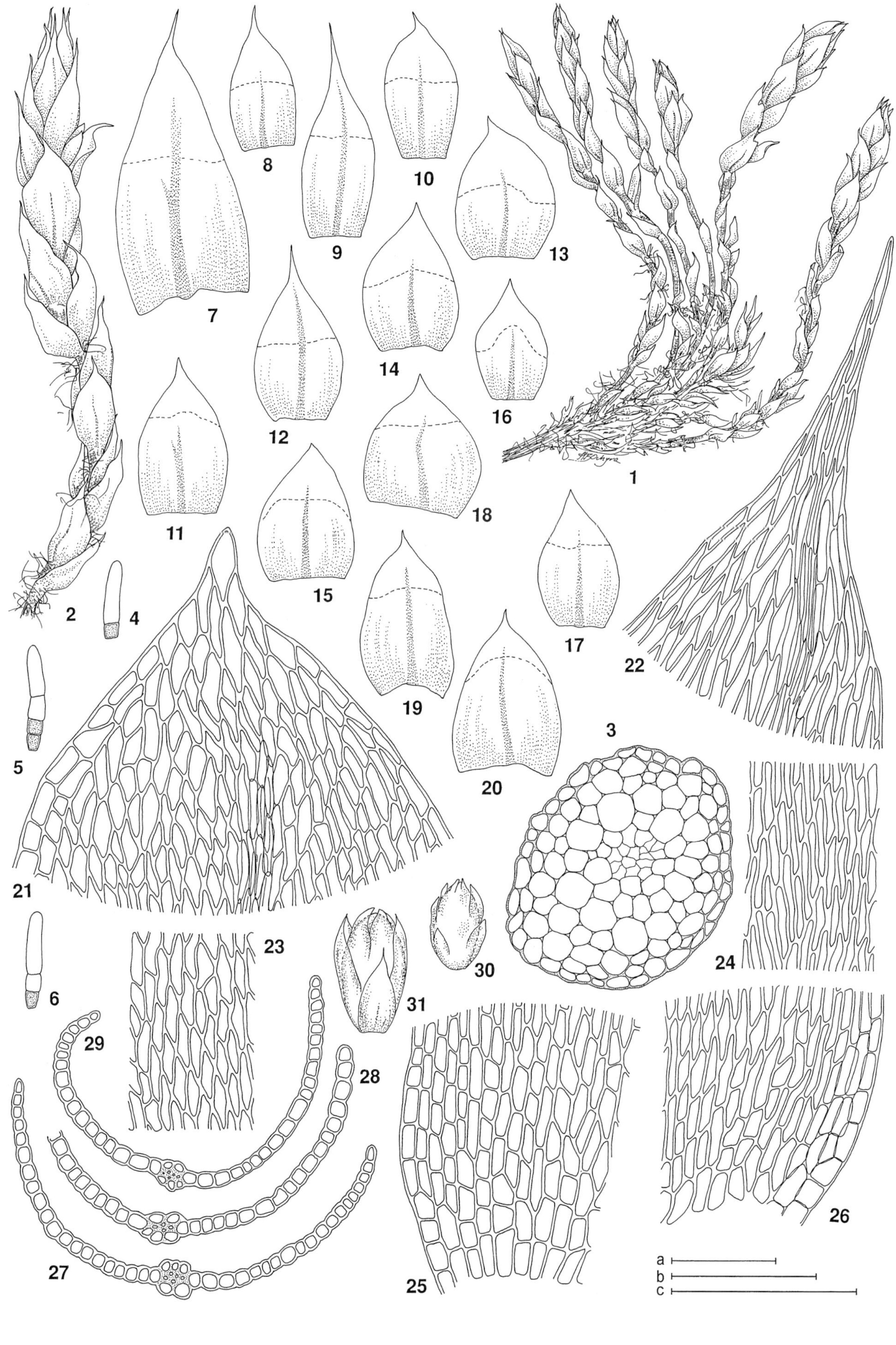

1
2
3
4
5
6
7
8
9
10
11
12
13
14
15
16
17
18
19
20
21
22
23
24
25
26
27
28
29
30
31
a
b
c

some marked genetic deviations in Antarctic plants in comparison with northern plants. It is likely that these differences may be partly due to misdetermination of the Antarctic material. For example, material from Cape Armitage on Ross I. in southern Victoria Land (*REL 4016*, AAS, KRAM) which was used for the study of morphological and physiological variability by Longton & MacIver (1977) and Longton (1981) actually represents *B. pseudotriquetrum*.

Considerable variation in plant and leaf size, leaf shape, acumen length and costa length is present in Antarctic plants of *Bryum argenteum* var. *argenteum*. For example, the costa may end below mid-leaf, extend to ¾ the leaf length or become subpercurrent to percurrent. A continual gradation in costa length was observed in Antarctic plants and the length of the costa varies among plants in a population or even leaves on the same plant. The same is also true for the leaf shape and acumination which gives a spectacular effect in the overall appearance of the plants. Numerous Antarctic populations, especially in wetter sites, tend to have much shorter apiculi to the leaves which may disappear on lower leaves but are present on apical leaves. Nonetheless, such plants have been placed in the type variety but separation of typical *B. argenteum* var. *argenteum* from var. *muticum* seems to be very subtle and ill-defined and at least partly habitat-determined. A tendency of the leaves of *B. argenteum* to become shorter and more obtuse has been clearly demonstrated experimentally in cultured material from subantarctic Macquarie I. (Seppelt & Selkirk 1984). Deciduous shoots cultured at 4°C under continuous light produced new shoots with orbicular, ecostate leaves with rounded apices and all cells chlorophyllous, while at 21°C and in continuous light the leaves developed an acute apex and costa and the cells in the distal ½–⅓ became hyaline and resembled typical leaves of specimens from temperate regions.

In the polar regions the type variety of *Bryum argenteum* may be confused with some phenotypes of *B. pseudotriquetrum*. This is due to the tendency of *B. pseudotriquetrum* plants living in extremely harsh climatic conditions, as well as those growing submerged in lakes and pools, to 'hyalinisation' of the uppermost leaves on the shoots which gives them a silvery appearance. Such plants usually have ovate, oblong-ovate or obovate leaves that are mostly 0.5–1.6 mm long and are obtuse, subacute to acute at the apex and the costa is percurrent to subpercurrent, only occasionally vanishing some distance below the apex. Also, they have a tendency to be chlorophyllose at the apex which is exacerbated in the harsh climate of continental Antarctica, but the lower leaves are normally chlorophyllose, brownish, reddish or greenish-brown. They are not infrequent on the Antarctic continent and Brotherus (1906*a*) described such phenotypes as a separate species, *B. filicaule*, from Gaussberg on the King Leopold and Queen Astrid Coast on Kaiser Wilhelm II Land. Additionally, they have been recorded as *B. argenteum* from Queen Mary Land and Wilkes Land (Savicz-Lyubitskaya & Smirnova 1964*d*), as well as from Enderby Land (Horikawa & Ando 1961, 1967*a*,*b*), but examination of voucher collections revealed that they all represent the *filicaule* phenotypes of *B. pseudotriquetrum*. Unfortunately, we have not had access to numerous specimens reported by Seppelt & Kanda (1986) from the Sôya Coast in Enderby Land, so we cannot confirm the identity of their material. Judging from the detailed drawings of the leaf areolation which is composed of predominantly very broad, thin-walled and hexagonal cells, it is possible that these plants may represent stunted and depauperate forms of *B. pseudotriquetrum*.

The *filicaule* phenotypes of *Bryum pseudotriquetrum* are usually small to very small plants with whitish uppermost leaves which perfectly imitate *B. argenteum*, for example the specimen from Seabee Hook at Cape Hallett in northern Victoria Land (*Schofield s.n.*, 17 Dec 1963, US). Sometimes they have a relatively strong costa, at least on some leaves, that frequently reaches the apex or ends shortly below it and has large guide cells visible in transverse section (see Fig. 2.4–6 in Savicz-Lyubitskaya & Smirnova 1964*d*). These are missing from *B. argenteum*, but also they cannot be detected in poorly developed costae in small plants of *B. pseudotriquetrum*. The lower leaves in such plants are mostly finely textured and filmy because of the very lax areolation of thin-walled, broad, short, hexagonal to oblong-hexagonal cells. This areolation is quite unlike that in *B. argenteum* var. *argenteum* in which the median, chlorophyllose cells are rectangular, oblong- to linear-hexagonal, narrow and generally firm- to thick-walled.

The conclusion based upon morphological and anatomical grounds is also clearly supported by cytological data. Tatuno (1963) established the chromosome number $n = 20$ for *Bryum pseudotriquetrum* plants from the Sôya Coast in Enderby Land. This count is typical of this species but was also detected once in *B. argenteum* amongst over 30 chromosomal records for this species (Fritsch 1991). The voucher specimens studied chromosomally (*Matsuda A.B. 64*, KRAM, LE) truly represent *B. pseudotriquetrum*. On the other hand, Inoue (1976) obtained a chromosome number $n = 10$ in plants of *B. argenteum* from the same region but the voucher material has not been available for examination. However, it should be noted that both chromosomal numbers, $n = 20$ and $n = 10$, are known to occur in *B. pseudotriquetrum* with equal frequency, whereas the latter number evidently predominates in *B. argenteum* (Fritsch 1991).

Gemmiparous plants *Bryum argenteum* var. *argenteum* that produce abundantly axillary bulbils in the upper part of shoots, and which serve as vegetative propagation, were described by Bartram (1938) as a separate species, *B. siplei*, from material from Marie Byrd Land. As stated in the original description, except for the propagules, the differences between this species and *B. argenteum* are slight and of little significance. Consequently, Greene (1967*b*) considered both species to be conspecific and this taxonomic conclusion was confirmed by Robinson (1972*a*) and Ochi (1979), and it is maintained here after examination of the original collections. Some plants of *B. argenteum* with propagules may be mistaken for *B. dichotomum*, but the strong, percurrent to excurrent costa safely separates the latter species.

Reproduction in Antarctica. Sterile, although female plants have been occasionally seen from Antarctica. Vegetative propagation is relatively frequent by abundant bulbils that are compact, abbreviated fragile branchlets produced in the upper part of shoots, and readily detachable tips of shoots and resemble those observed in this taxon in other parts of its cosmopolitan range (Clare & Terry 1960). Seppelt & Green (1998) described and illustrated filamentous, anastomosing propagules composed of many swollen cells developing terminally on rhizoids in plants from southern Victoria Land. They designated them as *Bryum subrotundifolium*, but here they are interpreted as *B. argenteum* var. *argenteum*.

Habitat. The type variety of *Bryum argenteum* grows in a wide range of habitats, but it is most abundant in open situations in the moist margins of drainage channels and melt pools, on seepage slopes, on scree, morainic detritus, beneath rock overhangs, and on sand, clay or gravel of outwash fans. It also thrives in dry, open or sheltered sites in rock crevices, on pockets of soil between stones and boulders, on rock ledges and outcrops, marble knolls, raised marine beaches and on soil boils amongst patterned ground. It often grows in bird nesting areas, on moist rock slabs below nests enriched with nitrogen and, in actively volcanic areas, it has been recorded from heated soil associated with fumaroles, on ash slopes and close to the mouth of quiescent vents. In Victoria Land *B. argenteum* var. *argenteum* often forms pure stands of 10 m^2 or more. Here, it is locally the principal component of the moss sociation developed on sandy soil covering steep, north-facing, seepage slopes in which it provides about 75% of the cover, with an additional 10% contributed by the cyanobacterium *Nostoc* sp. (Longton 1973). In the maritime Antarctic it occurs as scattered patches usually associated with *Ceratodon purpureus*, *Dicranella hookeri* (in volcanic sites), *Hennediella heimii*, *Pohlia nutans*, *Schistidium antarctici*, and *Syntrichia magellanica*.

World range. A truly cosmopolitan, ubiquitous and weedy moss, occurring throughout the world, often in anthropogenically influenced habitats including concrete, brick, thatch, paths and bonfire sites, and common even in large cities owing to its apparent ability to tolerate extreme atmospheric pollution.

Distribution in Antarctica. A widely distributed, scattered and infrequent taxon, although very locally common and abundant. In the maritime Antarctic it is rare but locally frequent on the South Sandwich Is., rare in the South Orkneys and South Shetland Is., and scattered on the west coast of the Antarctic Peninsula from Anvers I. to Two Step Moraine on Alexander I. at lat. 71°53′S. On the east coast of the Antarctic Peninsula *Bryum argenteum* var. *argenteum* is rare on the Trinity Peninsula and in the Joinville and James Ross Islands groups, with a very isolated site on the Oscar II Coast at lat. 65°37′S (Fig. 206). The continental distribution was reviewed and mapped by Greene (1967*b*) and Kanda (1987*a*). The latter author indicated the occurrence of the *B. argenteum s. lato* in the coastal area of continental Antarctica, with the most frequent distribution in Victoria Land. Because of the unavailability of voucher collections the distribution map (Fig. 207) is based only upon verified collections and, for continental Antarctica, it shows the occurrence of *B. argenteum* var. *argenteum* only in Victoria Land, where the taxon extends almost continuously from Cape Hallett on the Borchgrevink Coast to Black I. in the McMurdo Sound area at lat. 78°12′S. In addition, it is known from a single collection from Marie Byrd Land whence it was described as *B. siplei* (Bartram 1938). The localities from Queen Mary Land reported by Savicz-Lyubitskaya & Smirnova (1964*d*) are based upon misdetermined specimens. Also, this taxon has not been recorded from the Vestfold Hills (Seppelt 1984, 1986), and its occurrence on the Sôya Coast in Enderby Land needs confirmation. We examined the voucher collections of *B. argenteum* reported from this area by Horikawa & Ando (1961, 1967*a*,*b*) and they proved to be *B. pseudotriquetrum*. Additional specimens from this region reported by Seppelt & Kanda (1986) as well as from Casey Station in Wilkes Land (Seppelt & Selkirk 1984) were unavailable for study. This taxon was also reported from Dronning Maud Land (Kanda 1987*a*; Richter 1990, 1995) but no voucher material was located. *B. argenteum* var. *argenteum* most often occurs at elevations from sea level to 100 m, occasionally to 200 m, and very rarely above 400 m, with the highest record at 750 m in the Lassus Mountains on Alexander I. and 1200 m in Victoria Land.

Selected specimens examined. WEST ANTARCTICA. SOUTH SANDWICH IS. **Candlemas I.** Lucifer Hill, *119F*, *137D & 241B* (AAS, KRAM); Breakbones Plateau, *PC 202B*, *203C*, *204A & 207A* (AAS, KRAM); between lagoons and most southerly lava ridge, 65–100 m, *REL 555* (AAS, KRAM). **Bellingsausen I.** South coast, 20 m, *PC 75C* (AAS, KRAM).

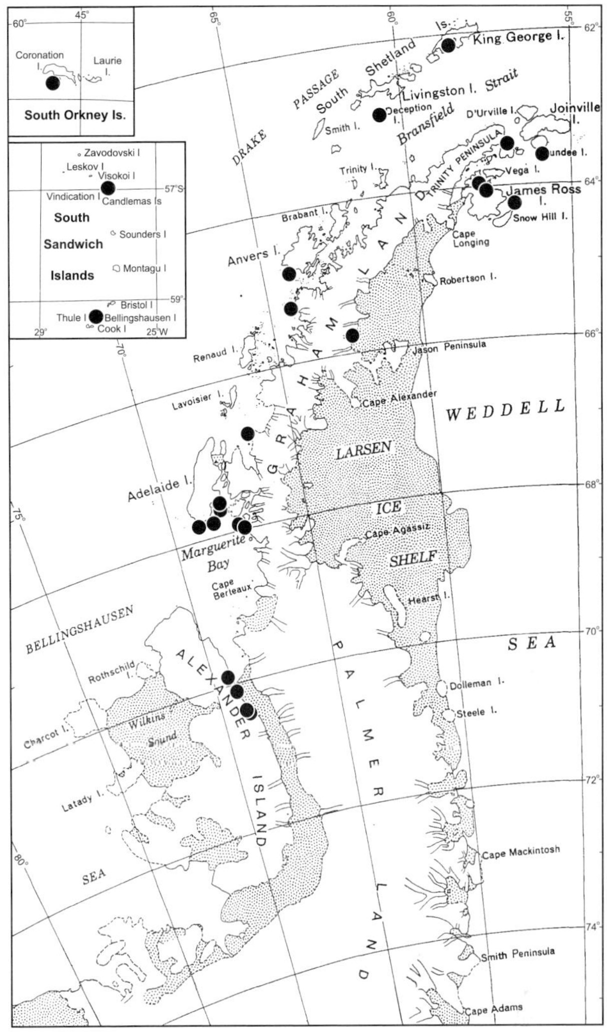

Figure 206. Distribution map for *Bryum argenteum* var. *argenteum* in the Antarctic Peninsula region.

SOUTH ORKNEY IS. **Signy I.** South of Waterpipe Beach, 7 m, *RILS 301*(AAS, KRAM); west of Waterpipe Beach, 8 m, *RILS 493* (AAS, KRAM); south of Waterpipe Beach, 6 m, *RILS 301* (AAS, KRAM); north of Spindrift Rocks, 3–6 m, *RILS 8062A* (AAS, KRAM).

SOUTH SHETLAND IS. **King George I.** ***Admiralty Bay***: Ezcurra Inlet, Golden Cove, Cytadela, 10 m, *RO 972/80* (KRAM). **Deception I.** Mt. Pond, 500 m, *RILS 11150A* (AAS, KRAM); Collins Pt., 50–70 m, *RILS 6758* (AAS, KRAM).

WEST ANTARCTIC PENINSULA. **Palmer Archipelago.** ***Anvers I.***: Hermit I., *ca* 15 m, *RILS 884* (AAS, KRAM). **Wilhelm Archipelago.** ***Collins Bay***: Darboux I., 10–15 m, *RILS 3325A* (AAS, KRAM). **Loubet Coast.** Orford Cliff, *RILS 5276C* (AAS, KRAM). **Fallières Coast.** ***Adelaide I.***: Rothera Pt., 20–30 m, *RILS 1867A* (AAS, KRAM); Jenny I., 15–50 m, *RILS 4713A* (AAS, KRAM) and 200 m, *RILS 4759* (AAS, KRAM); Avian I., *ca* 25 m, *RILS 769BB* (AAS, KRAM). ***Léonie Is.***: Lagoon I., 6 m, *RILS 9163* (AAS, KRAM); Léonie I., 15–20 m, *RILS 8996* (AAS, KRAM). ***Marguerite Bay***: Horseshoe I., Lystad Bay, 10–15 m, *RILS 1892* (AAS, KRAM) and 20 m, *RILS 8255* (AAS, KRAM); Camp Pt., 10–20 m, *RILS 4633 & 4683* (AAS, KRAM). **George VI Sound.** ***Alexander I.***: Lassus Mts., 750 m, *BAS Misc. 208B* (AAS, KRAM); Zebra Ridge, 600 m, *RILS 10023B* (AAS, KRAM); Ablation Valley, 75 m, *RILS 9294A* (AAS, KRAM); Mars Oasis, 70–75 m, *RILS 9408* (AAS, KRAM), 80 m, *RILS 10381* (AAS, KRAM) and 175 m, *RILS 9448* (AAS, KRAM); Two Step Moraine, 30 m, *RILS 9490* (AAS, KRAM).

EAST ANTARCTIC PENINSULA. **Trinity Peninsula.** Hope Bay, Scar Hills, *ca* 50 m, *CGB 30A* (AAS, KRAM). **Joinville Island group.** ***Dundee I.***: Welchness Pt., *RILS 5301B* (AAS, KRAM). ***Paulet I.***: without

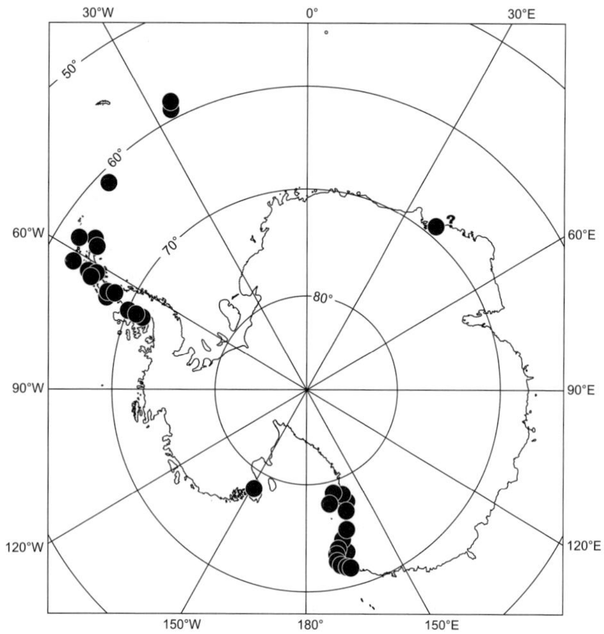

Figure 207. Distribution map for *Bryum argenteum* var. *argenteum* in Antarctica. Literature record marked with a question mark.

closer data, *CS 458 p.p.* (PC, S). **JAMES ROSS ISLAND GROUP.** ***James Ross I.***: St. Martha Cove, 50 m, *RILS 7461A & 7462B* (AAS, KRAM); Crame Col, 100 m, *RILS 7407* (AAS, KRAM); Terrapin Hill, 250 m, *RILS 7729* (AAS, KRAM). ***Cockburn I.***: without locality, *Hooker 5* (BM); north side of island above Adélie colony, 15–150 m, *RILS 7900* (AAS, KRAM); between Adélie colony and foot of cliff, 120 m, *RILS 7917* (AAS, KRAM); plateau, *ca* 400 m, *RILS 7969* (AAS, KRAM). **OSCAR II COAST.** Starbuck Glacier, *ca* 400 m, *PK 41A* (AAS, KRAM).

EAST ANTARCTICA. VICTORIA LAND. **BORCHGREVINK COAST.** Cape Hallett, Seabee Hook, *ca* 50 m, *SWG 65/58A & 65/59* (AAS, KRAM); Hallett Base, *Strang MCM-11* (US); Luther Peak, *ca* 80 m, *SWG 65/61* (AAS, KRAM); Crater Hill, *ca* 30 m, *DWHW 247A & 249B* (AAS, KRAM) and 100 m, *RILS 10173B* (AAS, KRAM); Crater Cirque, 100 m, *RILS 9894* (AAS, KRAM); Cape Phillips, 200 m, *DWHW 242* (AAS, KRAM); Intersection Ridge of Tinker-Burns Glaciers, 650 m, *RILS 9687A* (AAS, KRAM); moraine of Tinker-Burns Glacier, 300–500 m, *RILS 9668* (AAS, KRAM); ridge to north of Mt. McGee, 1200 m, *RILS 10169* (AAS, KRAM); Kay I., *DWHW 209A, 210A & 211A* (AAS, KRAM) and 100–130 m, *RILS 10201* (AAS, KRAM); Kay I., Wood Bay, 50 m, *RILS 10193* (AAS, KRAM); Harrow Peaks, 150–200 m, *RILS 9556* (AAS, KRAM); Baker Rocks, 1.5 m, *RILS 9685*(AAS, KRAM); Edmondson Pt., 10 m, *DWHW 202B* (AAS, KRAM) and 45 m, *RILS 9630* (AAS, KRAM); Cape Washington, 150 m, *RILS 9835* (AAS, KRAM); Cape King, 100–125 m, *RILS 10253 & 10261B* (AAS, KRAM); Gerlache Inlet, Gondwana Station, 70 m, *RILS 9724* (AAS, KRAM) and 150 m, *RILS 9745* (AAS, KRAM); Inexpressible I., 5–10 m, *RILS 9712 & 11775* (AAS, KRAM); Starr Nunatak, *DWHW 206A* (AAS, KRAM); Whitmer Peninsula, *Bargagli 54* (AAS, KRAM). **SCOTT COAST.** ***Granite Harbour Area***: Gregory I., *RDS 19601* (KRAM); Cuff Cape, *ca* 25 m, *WCS & SWG 65/51* (KRAM, NY); The Flatiron, *ca* 8 m, *WCS & SWG 65/19* (AAS, KRAM). ***McMurdo Sound Area***: Spike Cape, *ca* 15 m, *WCS & SWG 65/43* (KRAM, NY); Bay of Sails, *ca* 15 m, *WCS & SWG 65/39* (AAS, KRAM) and *ca* 100 m, *WCS & SWG 65/67*(AAS, KRAM); Marble Pt., *JRR MCM-13, 47 & 48* (US) and *ca* 15 m, *WCS & SWG 65/13* (AAS, KRAM); Cape Bernacchi, *ca* 15 m, *WCS & SWG 65/10* (AAS, KRAM, NY); Rocky Pt., *Spellerberg & Yeates 27270* (AAS, KRAM); Blue Glacier, *ca* 8 m, *WCS & SWG 65/5* (AAS, KRAM); Dailey I., *HJ 6C* (AAS, KRAM); Hobbs Glacier, *JRR MCM-43* (US); Miers Valley, *WCS & SWG 65/64* (AAS, KRAM, NY); Black I., *JRR MCM-43* (US). ***Ross I.***: Horsehoe Bay, *Standtman MCM-34* (US); north of Cape Royds, ca 65 m, *WCS & SWG 65/28* (AAS, KRAM); Mt. Erebus, *ca* 85 m, *WCS & SWG 65/31* (AAS, KRAM); Turks Head, *ca* 80 m, *WCS & SWG 65/33* (KRAM, NY); Arrival Heights, *ca* 200 m, *WCS & SWG 65/36* (AAS, KRAM); Observation Hill, 23 Dec 1963, *Schofield s.n.* (US).

MARIE BYRD LAND. **EDWARD VII PENINSULA.** ***Rockefeller Mountains***: Mt. Helen Washington, *PAS, Wade, Corey & Stancliff 100* (AAS, FH – type of *Bryum siplei*).

B. var. **muticum** Brid. Fig. 208

Bryum argenteum var. *muticum* Brid., Bryol. Univ. 1: 846. 1826. — TYPE CITATION: In Andibus. Humboldt et Bonpland. [Lectotype (*selected here*): the specimen on the sheet of *Bryum argenteum* 'Nº 83 Humboldt.' with a note written by W. Wilson '83 Humboldt perhaps a distinct species with the leaves not acuminate, lax, elliptico-rotund' – BM-Hooker! (No. 918165); isotypes: BM-Wilson! (4 specimens: Nos. 918157, 918159, 918160), G-Hedwig/Schwägrichen!].

Bryum argyroides Brid., Bryol. Univ. 1: 846. 1826, *nom. nud. in synon.*

Bryum argenteum var. *obtusifolium* Broth. *in* Dryg., Deutsche Südpolar-Exped. 1901–1903, 8: 89. 1906, ***syn. nov.*** — TYPE CITATION: Kerguelen: auf torfigen Substrat dicht am Strande der Pinguinbucht (Werth). Steril. Crozet-Gruppe: Possession-Insel, in unmittelbarer Nähe des Strandes (Werth). Steril. [Lectotype (*selected here*): 'Bryum argenteum L. f. fol. obtusis Crozet-Gruppe, Possession Insel 25/12/1901 leg. E. Werth' – H-Brotherus!; syntype: 'Bryum argenteum L. f. fol. obtusis Kerguelen, Enterbucht 8/11/1902 leg. E. Werth' – H-Brotherus!].

Argyro-bryum subrotundum Hampe, Linnaea 40: 312. 1876. ≡ *Bryum subrotundifolium* A.Jaeger *in* A.Jaeger & Sauerb., Ber. Thätigk. St. Gallischen Naturw. Ges. 1877–1878: 443. 1880 [*non Bryum subrotundum* Brid. 1817]. ≡ *Argyrobryum subrotundum* (Hampe) Wijk, Margad. & Florsch., Ind. Musc. 1: 95. 1959. ≡ *Anomobryum subrotundifolium* (A.Jaeger) J.R.Spence & H.P.Ramsay, Telopea 9: 787. 2002, ***syn. nov.*** — TYPE CITATION: In Monte Ararat Australiae leg. W. Sullivan. [Lectotype (*selected here*) {or perhaps holotype}: 'Argyro Bryum subrotundum Hp. Mt. Ararat Australiae Jul 1875 Sullivan' – BM-Hampe!; isotype: MEL (not seen, *vide* Ramsay & Seur 1994: p. 115)].

Bryum maceratum Müll.Hal., Linnaea 43: 385. 1882, ***syn. nov.*** — TYPE CITATION: Argentinia subtropica, in declivi Cordillerarum Boliviano, 1873, cum *Bryo ganophyllo* et *Funaria linearidente* mihi arcte consociatum. [Lectotype (*selected here*): 'Bryum (Argyrobryum) maceratum C. Müll. In declivi Cordillerarum Boliviano, 1873' – BM-Bescherelle!].

Bryum ellipsifolium Müll.Hal., Flora 73: 477. 1890, ***syn. nov.*** — TYPE CITATION: Africa or. trop., monte Kilima-Ndsaharo, alt. 3000–4800 m; Dr. Hans Meyer 1889. [Lectotype (*selected here*): 'Bryum (Argentobryum) ellipsifolium mihi Kilimandcharo, 3000–4800 m Dr Hans Meyer lg Hb. C. Müll.' – PC!].

Bryum myurella Dusén, Bot. Not. 1905: 304. 1905, ***hom. illeg., syn. nov.*** — TYPE CITATION: Fuegia australis in litore arenoso. [Isotypes: 'Museum Botanicum Upsaliense 202 Bryum Myurella Dus.

Figure 208. *Bryum argenteum* var. *muticum*. — **1–2**. Sterile plants, dry. **3**. Portion of stem, wet. **4**. Cross-section of stem. **5–7**. Axillary hairs. **8–17**. Leaves. **18**. Leaf apex. **19**. Mid-leaf cells. **20**. Mid-leaf cells at margin. **21**. Basal cells. **22**. Cross-section of leaf in middle part. **23**. Cross-section of leaf near apex. — [1, 10 from *CS 460* (isotype *of Bryum cephalozioides*), S; 2, 17 from *ER 229b* (isotype of *Bryum amblyolepis*), BR; 3, 11 from *RILS 4013*; 4–9, 18–20, 22–23 from *OB 9*; 12, 21 from *REL 1210*; 13–14 from *Rudolph s.n.*, 2 Nov 1961; 15–16 from *Sullivan s.n.*, Jul 1875 (holotype of *Argyro-bryum subrotundum*), BM]. — Scale bars: a – 1 mm (1) and 100 µm (18–21); b – 2 mm (2) and 100 µm (4, 22–23); c – 1 mm (3); d – 1 mm (8–17) and 100 µm (5–7).

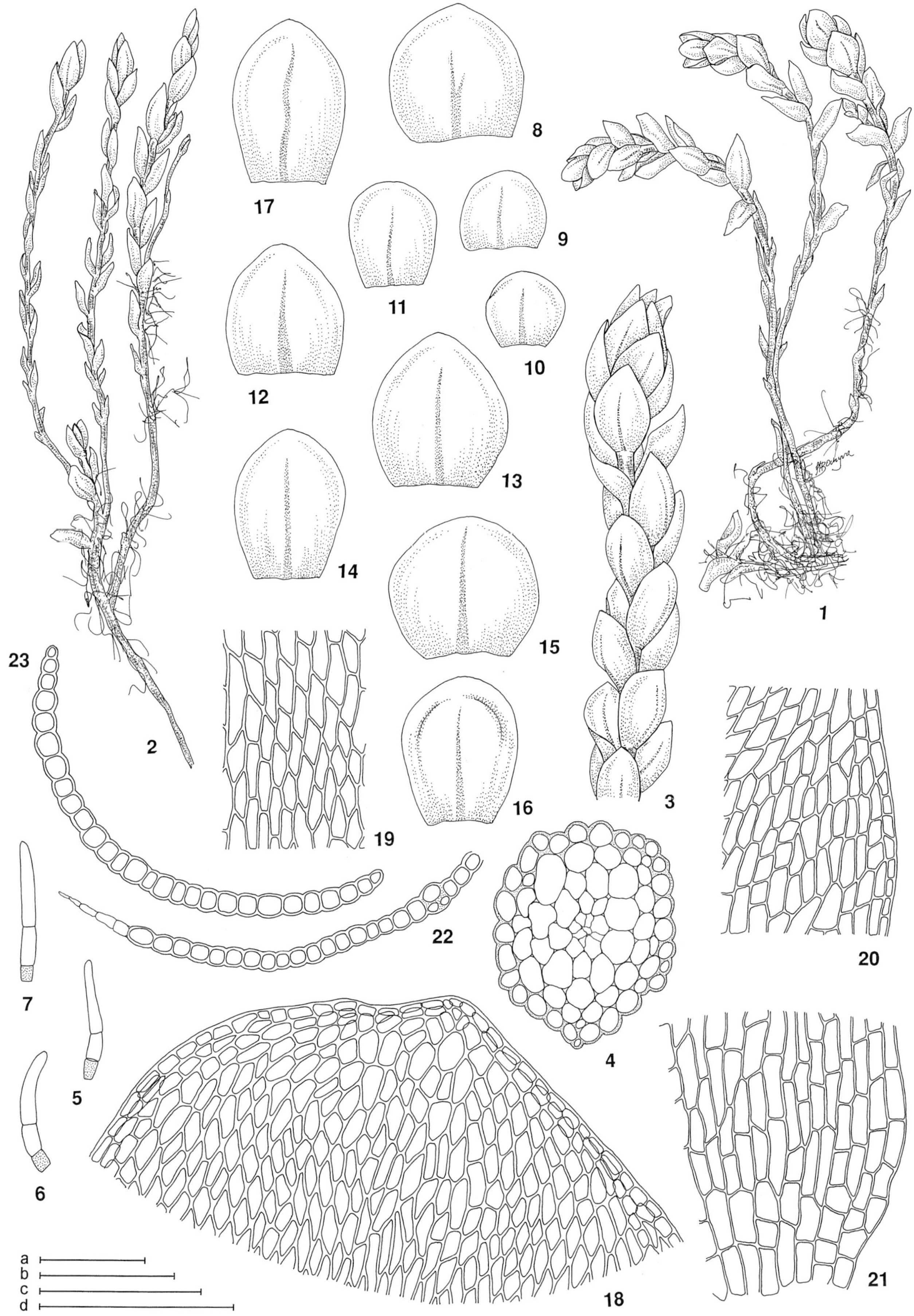

1
2
3
4
5
6
7
8
9
10
11
12
13
14
15
16
17
18
19
20
21
22
23
a
b
c
d

Tierra del Fuego. Rio Azopardo, på marken 1/3 1896 P. Dusén' – BM!, LD!].

Bryum amblyolepis Cardot, Rev. Bryol. 27: 44. 1900, ***syn. nov.*** — TYPE CITATION: [Antarctica] XIme débarquement; later, Cardot (1901: p. 37) gave details of the specimen: 'Détroit de Gerlache: XIème débarquement, Terre de Danco, cap Van Beneden, sur les corniches de la falaise. (N° 229^{b}; 1^{r} février 1898). [Lectotype (*selected here*): 'Herb. J. Cardot. Mousses du voyage de la Belgica. n° 229b. Bryum amblyolepis Card. sp. nova. Canal de Belgica: XIème débarquement, sur les corniches de la falaise. Leg. Racovitza, 1 févr. 1898' – PC!; isolectotype: BR!].

Bryum cephalozioides Cardot, Bull. Herb. Boissier Sér. 2, 6: 31. 1906, ***syn. nov.*** — TYPE CITATION: Archipel de Graham: île Paulet. [Lectotype (*selected here*): 'Svenska Sydpolarexpeditionen 1901–03. N:r 460. Bryum cephalozioides Card. sp. nova. Graham Archipelago Paulet Island /11 1902 Carl Skottsberg Det. J. Cardot' – PC-Cardot!; isotypes: S!, UPS!].

Plants small to medium-sized, gracile, soft, densely caespitose, sometimes distinctly layered because of repeated annual growth from innovations, occasionally scattered amongst other mosses as single gemmiform shoots or small clumps, pale yellowish-green, light green, whitish-green, whitish-yellow, pale brownish, light yellowish to whitish or silvery throughout or only above, sometimes pale reddish or brownish below. *Stems* erect, slender, filiform, fragile, 0.4–2.5(–3.0) cm tall, simple or fastigiately branching, julaceous, turgid, usually gemmiform at the tips, red to yellowish-brown, densely radiculose at the base or with clusters of brownish, branching, glossy, lightly papillose rhizoids scattered along the stem, in transverse section round, with a small but distinct central strand, 2–3 layers of large, hyaline, thin-walled cells forming a medullary tissue and 1–2 rows of somewhat smaller cortical cells with firm and red walls; *axillary cells* short, filiform, 3–4-cellular, with 1–2 short, quadrate, reddish-brown basal cells and 2–3 hyaline, elongate distal cells. *Leaves* somewhat distant below, becoming appreciably larger and crowded at the stem apex and usually forming a distinct capitulum, mostly laxly erect-imbricate below, closely erect-imbricate above, both dry and wet, 0.3–0.8 mm long, 0.3–0.8 mm wide, broadly ovate to oblong-ovate, orbicular to cochleariform-concave, sometimes transversely elliptical, rounded and obtuse at the apex, sometimes with a short, broad, blunt apiculus, less than 5 μm long, strongly concave and often cucullate, non-decurrent, usually hyaline in the distal half, sometimes hyaline cells forming a distinct, narrow, scarious strip descending well below mid-leaf along the margins, light green, yellowish-green to pale yellowish-brownish below, often reddish at the base, sometimes colourless throughout; *margins* entire, plane, lacking a border; *costa* single, sometimes spurred distally or forked at the tip, ceasing in mid-leaf to subpercurrent, quite often very weak, obsolete, indistinct and ending below mid-leaf to entirely absent, (20–)30–40(–50) μm wide at the base, pale yellowish-brownish and almost concolorous with the laminal cells or often reddish-brownish below, in cross-section terete, weakly convex on the dorsal and ventral sides, consisting of large dorsal and ventral epidermal cells and a small, highly reduced central group of a few substereid cells; *laminal cells* unistratose throughout, rhombic to rhomboidal, shorter, thin- to firm-walled, hyaline in the distal half, becoming chlorophyllose, thin- to thick-walled, sturdy or lax, 30–80 μm long, 14–18(–25) μm wide, in mid-leaf, quadrate to short-rectangular, generally wider, often with a reddish tint at the base. *Asexual reproduction* occasional by short, axillary, easily detachable short branchlets. Sterile.

Discussion. Morphologically, *Bryum argenteum* var. *muticum* is a distinct taxon that is recognisable by its julaceous habit, whitish, whitish-green, pale yellowish-green to light green coloration, and broadly ovate to orbicular, deeply concave and imbricate leaves that are rounded and obtuse at the apex and comose at the stem tips, giving the shoots a gemmiform appearance. Microscopically, it is characterised by its laxer leaf areolation of broader cells and relatively slender and obsolete costae, usually extending ¾ way up the leaf but often vanishing below mid-leaf or, in the extreme cases, entirely absent. This variety is a morphological and ecological counterpart of *B. argenteum* var. *lanatum* which often grows in wet habitats and has a totally reduced leaf hair-point or apiculus. It is connected to the type variety by a continual series of intermediates which prevent the successful determination of many populations. Therefore, recognition of these three phenotypes as distinct species, as recently proposed by Spence & Ramsay (2002, 2006), seems to be unreasonable because no distinct differences exist between them.

Here, *Bryum argenteum* var. *muticum* is interpreted as including plants with all leaves having rounded and obtuse apices which are devoid of any apiculi. This character is quite well correlated with relatively broad, thin-walled and sometimes lax laminal cells. Therefore *B. argenteum* var. *subamblyolepis* from the Falkland Is. (Cardot & Brotherus 1923) is here considered to be identical to the type variety rather than var. *muticum*. Despite the presence of rounded and obtuse leaves in the lower part of the stem, the upper leaves are distinctly acute. This interpretation agrees with the original concept of *B. argenteum* var. *muticum* presented by Bridel (1826) who, in the discussion of this taxon, mentioned plants from Chamblonde near Lausanne in Switzerland which had variable leaf apices and excluded them from this variety. Examination of the voucher collection in the Bridel herbarium (B) confirms this observation and, on single shoots, there is a gradual change from perfectly obtuse to distinctly apiculate leaves. Such plants frequently occur in Antarctica and, additionally, plants at the more exposed raised centre of large tufts or turves usually represent epilose phenotypes whereas peripheral plants or those growing in small depressions of mats have distinctly acute and apiculate leaves.

There has been considerable controversy regarding the identity and taxonomic status of Antarctic plants which are here referred to as *Bryum argenteum* var. *muticum*. Seppelt & Green (1998) accepted the specific distinctness of this taxon and named it *B. subrotundifolium*, indicating the yellowish-green coloration of the plants to be the essential difference between this species and the whitish-green or silvery plants of *B. argenteum*. However, the colour of plants in the *B. argenteum* complex is subject to remarkable variation and cannot validly be used as a satisfactory discriminating character. Coloration in the *B. argenteum* complex has been found to be partially a physiological response to the degree of shading or exposure to solar irradiance and UV-B rather than correlated only with morphological expressions of various characters. In general, the exceptional phenotypic plasticity within *B. argenteum s. lato* is almost certainly coupled with the high level of genetic variation in populations of this species in polar and temperate regions (Skotnicki *et al.* 1998).

The name *Bryum subrotundifolium* itself merits some comment. It was introduced by A. Jaeger (in Jaeger & Sauerbeck 1880) as a replacement for *Argyro-bryum subrotundum*, an Australian species described by Hampe (1876) from the state of Victoria, upon its transfer to *Bryum*. Under this generic name the epithet *subrotundum* is blocked by *B. subrotundum* Brid. of 1817. *B. subrotundifolium* was initially placed in the newly described genus *Argyro-bryum* whose name is solely validated, under Art. 42.1 of the Code (McNeill *et al.* 2006), by the description of its single species, *Argyro-bryum subrotundum*. This hyphenated generic name is very similar to the non-hyphenated sectional name *Bryum* sect. *Argyrobryum* Müll.Hal. which was formally elevated to generic rank by Kindberg (1882) and lectotypified with *B. argenteum* by Mönkemeyer (1927). Despite the ostensible similarity, these two generic names remain heterotypic and they cannot be considered orthographic variants as defined in Art. 61.1. The compilers of *Index muscorum* introduced serious confusion regarding the interpretation of the name *Argyro-bryum subrotundum* by considering this combination as invalidly published and neglecting the generic name *Argyro-bryum*. In fact, these authors transferred this species to *Argyrobryum* and proposed a new combination, *Argyrobryum subrotundum*.

Bryum subrotundifolium perfectly fits the type of *B. argenteum* var. *muticum* and is here reduced to synonymy with this varietal name. The latter taxon was described by Bridel (1826) from specimens collected in the Andes by A. von Humbold and A. J. A. G. Bonpland during their expedition to the Neotropics (Stearn 1968). According to Kunth (1822) the material was collected in Ecuador, then a part of the Viceroyalty of Nueva Granada, and represented *B. argenteum* with broadly ovate, minutely apiculate and yellowish leaves. Similar phenotypes were given specific or varietal names in other parts of the world, for example *B. maceratum* in the Central Andes (Müller 1882), *B. ellipsifolium* in the Kilimanjaro Mountains in East Africa (Müller 1890*b*), *B. myurella* in Tierra del Fuego (Dusén 1905*c*), *B. argenteum* var. *obtusifolium* from Îles Kerguelen and Crozet (Brotherus 1906*a*), as well as *B. amblyolepis* and *B. cephalozioides* in West Antarctica (Cardot 1900, 1901, 1906*a*, 1908*a*). Because they all match perfectly the type of *B. argenteum* var. *muticum*, their names are synonymised with the latter. The Antarctic plants represent the extreme phenotypes with regard to the size of the plants, *B. amblyolepis* being a relatively large moss and *B. cephalozioides* being a delicate and minute plant with small and often ecostate leaves, but no other clear-cut differences are observed. Both species have been considered conspecific with the broadly understood *B. argenteum* by Ochi (1979) and Robinson (1972*a*), respectively.

In Antarctica, *Bryum argenteum* var. *muticum* is a distinct moss. The only other moss species in this biome with leaves of comparable size and shape are *Stegonia latifolia* and *Bryum orbiculatifolium*. It shares the whitish to silvery coloration of the plants with *S. latifolia* which, however, is a short, gemmiform plant and has different leaf areolation of sturdy, thick-walled, rhombic, rhomboidal to oblong-hexagonal upper laminal cells that are distinctly bulging on the ventral side and remarkably thick-walled on the dorsal surface, and the costa has a markedly different anatomy, with a protuberant ventral epidermis, two large guide cells and a dorsal stereid band lacking an epidermis in the distal portion. *B. orbiculatifolium* differs in its entirely chlorophyllose laminal cells, green basal leaf cells and the presence of guide cells in the costa.

Reproduction in Antarctica. Sterile but occasionally propagating asexually by readily detatched axillary branchlets. This variety is generally known in the barren state and only the type material from Ecuador was collected in fine fruiting condition.

Habitat. *Bryum argenteum* var. *muticum* generally occurs in similar habitats to those of the type variety and often both grow together in mixed stands, although it seems that this variety is associated with wetter sites. It is frequently found on wet, mineral, gravelly or silty soil in seepage areas on slopes, often below late lying snow patches, at margins of melt-water runnels, streams and pools, at edges of frost boils, as well as in wet rock crevices and on ledges and outcrops, on moist soil among stones, under shaded and moist overhangs, frequently in bird nesting and perching areas, occasionally on decomposed and mummified seal carcasses. It usually grows in compact tufts or turves which are devoid of any accompanying species, but occasionally in more open patches *Syntrichia magellanica*, *S. sarconeurum*, *Ceratodon purpureus* and *Hennediella heimii* grow as admixtures. However, at some sites in Victoria Land, this variety forms extensive pure, sometimes closed, stands of 25 sq. metres or more, for example at Edmondson Pt. (Lewis Smith 1999 as *B. argenteum*). An exceptional stand occupying about 2.5 ha, much of it affording 100% cover, exists on Beaufort I., Ross Sea (Seppelt *et al.* 1999 as *B. subrotundifolium*).

World range. The global distribution of *Bryum argenteum* var. *muticum* is poorly known but most probably it is also a cosmopolitan taxon. It appears to be a rarer than var. *argenteum* and, at least in the Southern Hemisphere, restricted to polar regions and the alpine zone of tropical mountains, where it is common in wet habitats, for example in the Kilimanjaro Mountains (Ochyra & Bednarek-Ochyra, personal observations). It is rare in Australia (Spence & Ramsay 2002, 2006), common on some Subantarctic islands, for example the Prince Edward Is., Îles Crozet and Kerguelen (Ochyra, personal observations) but rare on South Georgia (Ochyra *et al.* 2002), and only a few collections have been made in southern South America. In the Northern Hemisphere, forms of *B. argenteum* with rounded and obtuse leaves are not infrequent and they have been variously named. Podpěra (1956) designated them as *B. argenteum* var. *julaceum* 'Schrad. *ex* Bland.' of 1809, but this name was nomenclaturally incorrect because *B. julaceum* of Schrader (1794) was validated by Gaertner *et al.* (1802). This name also referred to a species known as *Anomobryum julaceum* (Schrad. *ex* P.Gaertn., B.Mey. & Scherb.) Schimp. Furthermore, the name *B. argenteum* var. *majus* Schwägr. cannot be used for this taxon because it is illegitimate and includes the type of *B. argenteum* var. *julaceum* (Schrad. *ex* P.Gaertn., B.Mey. & Scherb.) Blandow.

Distribution in Antarctica. The Antarctic distribution of *Bryum argenteum* var. *muticum* (Figs 209–210) is similar to that of the type variety. This taxon is widely scattered along the west coast of the Antarctic Peninsula, from the Danco Coast to Alexander I. where it reaches its southernmost site at lat. 71°50′S, but it is exceedingly rare on the peri-Antarctic archipelagoes of the South Sandwich Is. and South Shetland Is. and apparently absent from the South Orkney Is. On the east coast of the Antarctic Peninsula it is restricted to the northernmost part, including Trinity Peninsula, Dundee I. and James Ross I., with an isolated station on the Foyn Coast at lat. 66°22′S. On the continent it is widespread only in Victoria Land, especially in the southern part, extending to lat. 77°53′S on Dailey I. It appears to be absent from the other sectors of continental Antarctica, except for Dronning Maud Land (Ochyra & Singh 2008) and Enderby Land, from where a single collection lacking any closer locality data is known. *B. argenteum* var. *muticum* occurs predominantly at low elevations and has the greatest concentration of localities from near sea level to about 100 m, with a few records at higher altitudes, and exceptionally it occurs at 1000–1050 m in southern Victoria Land.

Selected specimens examined. WEST ANTARCTICA. SOUTH SANDWICH IS. **Candlemas I.** Lucifer Hill, *PC 230E* (AAS, KRAM).

SOUTH SHETLAND IS. **Deception I.** Ronald Hill, 15 m, *REL 1210* (AAS, KRAM) and *RILS 733A* (AAS, KRAM), 25 m, *RILS 72 & 5778E* (AAS, KRAM) and 60 m, *RO 2782A/80* (KRAM).

WEST ANTARCTIC PENINSULA. **Danco Coast.** Cierva Pt., *VK 185* (KRAM) and 15 m, *OB 9* (AAS, KRAM); Charlotte Bay, Andrée I., 25 m, *RILS 4013* (AAS, KRAM); Cuverville I., 2–20 m, *RILS 4176, 11954A & 11955* (AAS, KRAM); Birdsend Bluff, 5–35 m, *RILS 4221* (AAS, KRAM); Beneden Head, *ER 229B* (BR, PC – type of *Bryum amblyolepis*); Coughtrey Peninsula, Paradise Harbour, Almirante Brown Station, *VK 339* (KRAM) and *Schmitt 213* (KRAM, NY). **Fallières Coast.** ***Adelaide I.***: Rothera Pt., 2 m, *RILS 8922* (AAS, KRAM); Reptile Ridge, 140 m, *RILS 10598A* (AAS, KRAM); Jenny I., *Gain 258B* (PC); Avian I., *ca* 25 m, *RILS 769B* (AAS, KRAM). ***Marguerite Bay***: Lagotellerie I., *ca* 25 m, *RILS 1777* (AAS, KRAM) and 165 m, *RILS 4865* (AAS, KRAM); Neny I., *ca* 15 m, *RILS 788* (AAS, KRAM); Refuge Is., largest island, 10–25 m, *RILS 4523* (AAS, KRAM). **George VI Sound.** ***Alexander I.***: Zebra Ridge, 600 m, *RILS 10023A* (AAS, KRAM); Ares Oasis, 30 m, *RILS 10440D* (AAS, KRAM); Mars Oasis, 70–75 m, *RILS 10405* (AAS, KRAM).

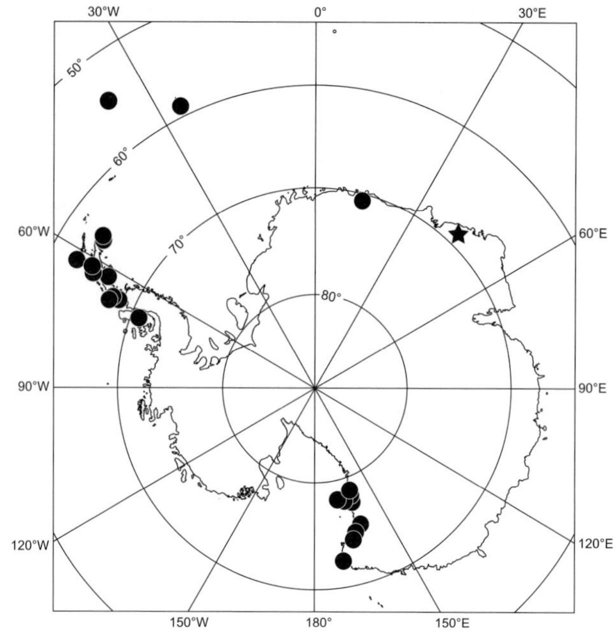

Figure 210. Distribution of *Bryum argenteum* var. *muticum* in Antarctica. Inadequately localised station in Enderby Land indicated by the star.

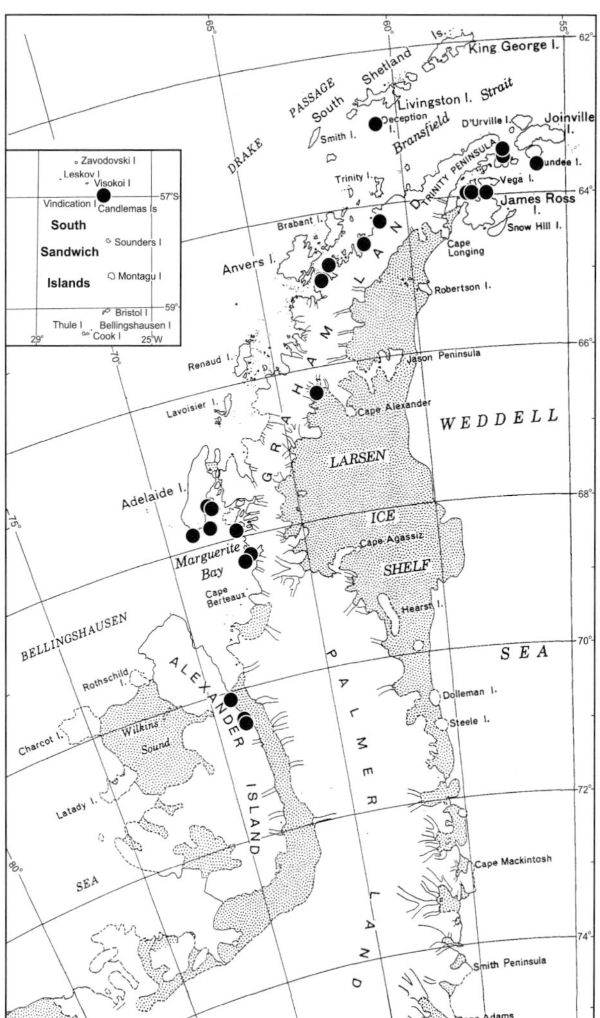

Figure 209. Distribution of *Bryum argenteum* var. *muticum* in the Antarctic Peninsula region.

EAST ANTARCTIC PENINSULA. **Trinity Peninsula.** Mt. Flora, 25–50 m, *RILS 3859* (AAS, KRAM) and *ca* 85 m, *CGB 7* (AAS, KRAM); Lake Boeckella, *ca* 35 m, *CGB 6* (AAS, KRAM). **Joinville Island group.** ***Paulet I.***: without closer data, *CS 460* (PC, S – type of *Bryum cephalozioides*). **James Ross Island group.** ***James Ross I.***: Lachman Crags, 20–30 m, *RILS 7328A* (AAS, KRAM); San José Pass, 165 m, *RILS 7523A & 7530* (AAS, KRAM); St. Martha Cove, 15 m, *RILS 7616* (AAS, KRAM); Blancmange Hill, 12–15 m, *RILS 7678A* (AAS, KRAM) and 125 m, *RILS 7694B & 7696* (AAS, KRAM). **Foyn Coast.** Cabinet Inlet, Cape Casey, 335–670 m, *RT 4* (AAS, KRAM).

EAST ANTARCTICA. DRONNING MAUD LAND. ***Schirmacher Oasis***: central part, 39 m, *Singh 9 & 50* (KRAM) and 78 m, *Singh 51* (KRAM).

ENDERBY LAND. Without closer data, *Ream 1596* (AAS, KRAM).

VICTORIA LAND. **Borchgrevink Coast.** Cape Hallett, Seabee Hook, *ca* 15 m, *SWG 65/57* (KRAM, NY); Hallett Base, 2 Nov 1961, *Rudolph s.n.* (AAS, KRAM); Crater Cirque, 100 m, *RILS 9896B* (AAS, KRAM); Daniell Peninsula, 1000 m, *RILS 9924* (AAS, KRAM); Apostrophe I., 25 m, *RILS 9788* (AAS, KRAM); Harrow Peaks, 75 m, *RILS 9644B* (AAS, KRAM); Baker Rocks, 750 m, *RILS 9537* (AAS, KRAM); Edmondson Pt., *DWHW 256A* (AAS, KRAM); Inexpressible I., 30 m, *DWHW 204B* (AAS, KRAM); Prior I., *RILS 9950A* (AAS, KRAM). **Scott Coast.** ***Granite Harbour Area***: Mt. Suess, 1000 m, *HJ 13A* (AAS, KRAM); Mt. England, 1050 m, *HJ 12A–B* (AAS, KRAM). ***McMurdo Sound Area***: Bartley Glacier, Wright Valley, *JRR MCM-31* (US); Strand Moraines, north of Blue Glacier, *WCS & SWG 65/7* (AAS, KRAM, LD, NY) and 1908, *Priestley s.n.* (PC); Dailey I., 90 m. *HJ 6A–B* (AAS, KRAM). ***Ross I.***: Horseshoe Bay, *Strandtman MCM-32* (US); Cape Royds, Jan 1909, *Murray s.n.* (PC); Mt. Terror, 21 Jan 1902, *Discovery Expedition s.n.* (PC); New Zealand's Aurora Station, *HJ 4G* (AAS, KRAM); Crater Hill, *NZ Geological Survey Party 1957–1958* (AAS, KRAM); Cape Crozier, 160 m, *HJ 7* (AAS, KRAM).

2. **Bryum dichotomum** Hedw. Fig. 211

Bryum dichotomum Hedw., Sp. Musc. Frond.: 183, *pl. 42, f. 8–12*, 1801. ≡ *Mnium dichotomum* (Hedw.) P.Beauv., Prodr. Aethéogam: 74. 1805. — Type citation: Nova Seelandia. [Lectotype (*selected here*): the third shoot from the left on the sheet labelled 'Bryum dichotomum Spec. Musc. p. 183. tab. 42. f. 8–12 aus Neu Seeland' – G-Hedwig/Schwägrichen!].

Bryum urbanskyi Broth. *in* Dryg., Deutsche Südpolar-Exped. 1901–1903, 8: 89, *pl. 8, f. 1.* 1906. — TYPE CITATION: Kerguelen: am Schwarzen See (Urbansky). [Lectotype (*selected here*): 'Bryum (Doliolid.) Urbanskyi Broth. n. sp. Kerguelen, am Schwarzen See 30/1 1903 leg. Urbansky.' – H-Brotherus!; isotype: PC-Cardot!]. First synonymised by Ochyra (1998*a*: p. 208).

Bryum anomobryoides Cardot & Broth., Kongl. Svenska Vetenskapsakad. Handl. 63(10): 38. 1923, ***syn. nov.*** — TYPE CITATION: Patagonia occ.: Archip. Adelaide, Isla Atalaya, Puerto Cuarenta Dias (557). Fuegia: Lago Fagnano: ad saxa riparia irrorata. [Lectotype (*selected here*): 'Herb. J. Cardot Bryum anomobryoides Card. sp. nova Patagonie: île Atakaya, rochers maritimes Leg. C. Skottsberg, 1908' – PC-Cardot!; isolectotypes: H-Brotherus!, S!].

Bryum laetevirens Cardot & Broth., Kongl. Svenska Vetenskapsakad. Handl. 63(10): 41, *pl. 3, f. 3a–e*. 1923, ***syn. nov.*** — TYPE CITATION: Patagonia occ.: Peel Inlet, Puerto Témpanos, ad rupes litoreas (555). [Lectotype (*selected here*): 'Expeditio suecica 1907–1909. 555 Bryum laetevirens Card. nova spec. Patag. occid. Peel Inlet Pto Témpanos 17.6.1908 Det. J. Cardot. Leg. ~~T. Halle et~~ C. Skottsberg.' – PC-Cardot!; isotypes: BM!, H-Brotherus!, LD! S-Möller!].

Plants small to occasionally larger, in dense or loose tufts, sometimes gregarious and growing with other mosses, slender, scarcely to distinctly lustrous, green, yellowish- or brownish-green to yellow-brownish above, brownish below. *Stems* erect, slender, simple or branching by innovating shoots, 0.2–1.0(–2.5) cm tall, reddish, radiculose at the base and often matted together with long, pale reddish-brown to reddish, papillose, branched rhizoids in clusters scattered along the stem, in cross-section rounded to obtusely pentagonal, consisting of a distinct, rather large central strand, 2–3-layered medullary tissue of large, hyaline, thin-walled cells and 2(–3)-stratose cortex of smaller cells with thick, red walls; *axillary hairs* short, filiform, 3–4-celled, consisting of a short, quadrate, reddish-brown basal cell and 2–3 hyaline, elongate distal cells. *Leaves* closely spaced throughout or somewhat distant below, often smaller below and larger and more crowded above, forming comose tufts at the stem tips, erect-imbricate to somewhat erect-spreading, firm and scarcely altered on drying, slightly more erecto-patent when moist, 0.7–1.5 mm long, 0.5–0.8 mm wide, ovate, ovate-lanceolate to lanceolate, short-acuminate to broadly acute, narrowed at the base, non-decurrent, concave to carinate-concave, green throughout or reddish at the base or in the proximal half; *margins* entire to slightly serrulate or irregularly notched towards the apex on young leaves, plane to recurved below, not or indistinctly bordered; *costa* single, rather strong, percurrent to long-excurrent as a rigid, smooth or slightly denticulate, aristate awn, 40–70 μm wide, yellow, yellow-or reddish-brown to reddish below, sharply demarcated from the laminal cells, in transverse section round, usually distinctly convex on the dorsal side towards the base, consisting of distinct ventral and dorsal rows of epidermal cells, 2–3 large guide cells in the lower half and a small, central band of stereid or substereid cells; *laminal cells* unistratose throughout, translucent, firm to moderately thick-walled, rhomboid to rhomboid- or oblong-hexagonal above, 25–60 μm long, 8–15 μm wide, becoming narrower and elongate in 1–3 rows at the margins but not forming a noticeable border, short-rectangular to quadrate towards the base. *Asexual reproduction* by bulbils frequently produced in the axils of the upper leaves, 1–2 per axil, yellowish- or brownish-green to green, ovoid to subcylindrical, (150–)220–450(–500) μm long, with acute to subacute or obtuse to rarely rounded, ecostate leaf primordia, arising from the upper ½–⅔ of bulbil, occasionally from near the base, and reaching half or less of its total length. [*Dioicous*. *Setae* erect, 0.5–1.5 cm long, slender, often flexuose, hooked at the apex, red or purple, smooth. *Capsules* exserted, horizontal to pendulous, small, 0.75–1.5 mm long, obovoid to obloid-cylindrical, slightly contracted below the mouth when dry, brown, reddish-brown to dark red when mature, with a short, broad neck abruptly narrowed to the seta, wrinkled when dry; *annulus* compound, revoluble, of 2–3 rows of large, vesiculose, incrassate cells; *operculum* convex-conic, acute or minutely apiculate; *exothecial cells* variable in shape, irregular, rounded-quadrate to oblong in mid-urn, firm- to thick-walled, 15–35 μm long, 6–15 μm wide, becoming predominantly transversely rectangular in the neck and oblate in 3–4 tiers at the orifice; *stomata* numerous in the lower part of the neck, small, superficial, bicellular, with narrow, elongate pori; *peristome* double, inserted close to the mouth, 340–370 μm long; *exostome teeth* 16, lanceolate, yellow-brown, brown to yellow below, paler yellow above, hyaline-yellowish at the tips, finely papillose below and more coarsely papillose at the tips, noticeably bordered with irregular margins, distinctly trabeculate on the inner side; *endostome* yellowish-hyaline, lightly papillose, consisting of a basal membrane to about half of the exostome length, well developed, rather narrow, keeled processes, with narrow or ovate perforations, and 2–3 long, nodose to appendiculate cilia. *Spores* globose, lightly yellow, 12–16 μm in diameter. *Calyptra* cucullate, small, fugacious, smooth, naked.]

Chromosome numbers: unknown in the Antarctic and Subantarctic; $n = 10$ (5 records) and $n = 24$ from Australia, the British Isles and India (Fritch 1991 as *B. bicolor*).

Discussion. In Antarctica *Bryum dichotomum* is a distinctive and unmistakable species belonging within sect. *Doliolidium* (Müll.Hal.) Müll.Hal. that is characterised by its small size, ovate to ovate-lanceolate leaves with yellow, yellowish-brown to reddish or reddish-brown, usually strongly aristate costae, absence of a leaf border, and large leafy bulbils that are produced in the upper leaf axils on sterile shoots. Gemmae, also inserted in the leaf axils, are often copiously produced and their formation seems to be a response to unfavourable environmental factors, such as severe drought (Sarla 1987; El-Saadawi & Zanaty 1990), or to the presence of certain chemicals, for example chelates and salicylic acid (Chopra & Sarla 1985). Bulbils exhibit remarkable morphological variability in shape, number per leaf axil and size, and they have been used as an important diagnostic character for the determination of species. *Bryum* species with axillary gemmae form a complex of closely related taxa which is usually designated as the *B. bicolor* aggregate (Wilczek & Demaret 1976; Smith & Whitehouse 1978).

There has been considerable uncertainty and ambiguity about the concept of some species in the *Bryum bicolor* complex, especially those that produce single or geminate bulbils in the leaf axils. The number of bulbils on a single plant seems to depend on environmental conditions (Shaw 1990), but there does not appear to be any evidence that the number of bulbils per leaf axil is susceptible to modification under different growth conditions. The key problem in the taxonomy of the *B. bicolor* complex is the status of the Southern Hemisphere *B. dichotomum* and the Northern Hemisphere *B. bicolor*. Demaret & Wilczek (1980) studied the type material of both species names and found that they refer to distinct species which differ mainly in bulbil morphology: in *B. dichotomum* the leaf primordia occur in the distal half of the bulbil, whereas in *B. bicolor* they arise from the base of the bulbil. Accordingly, the name *B. bicolor* continued to be applied to the northern plants until recently (e.g. Smith 1978; Corley *et al.* 1981; Crum & Anderson 1981; Noguchi 1988; Nyholm 1993; Demaret 1993; Lo Giudice 1996; Zolotov 2000; Vanderpoorten & Zartman 2002; Ochyra *et al.* 2003), although some workers stressed that further studies were needed to elucidate the interrelationships between both species.

An alternative concept of *Bryum dichotomum* and *B. bicolor* was presented by Ochi (1980) who considered them to be identical taxa (Ochi 1970, 1972). Morphology of bulbils is variable in this complex. In austral populations bulbils with leaf primordia restricted to their distal portions are predominant, but sometimes the leaf primordia may commence from near the base, and such gemmae have occasionally been observed in Antarctic plants. However, similar bulbils have also been described and illustrated in Holarctic plants (e.g. Smith & Whitehouse 1978: f. 1f–g, i; Vanderpoorten & Zartman 2002; Smith 2004: f. 187). Because morphology of the axillary gemmae in *B. bicolor* intergrades with *B. dichotomum* to such an extent that recognition of some populations could be arbitrary, the conspecificity of both species is here accepted, the latter name having priority. This concept has occasionally been adopted in the Northern Hemisphere (e.g. Spence 1988; Anderson *et al.* 1990; Ignatov & Afonina 1992; Redfearn *et al.* 1996; Holyoak 2003; Sérgio & Carvalho 2003; Rykovsky & Maslovsky 2004), and recently it was accepted in the new edition of *The Moss Flora of Britain and Ireland* (Smith 2004). For the sake of accuracy it should be noted that sheet 39 in the Dickson herbarium at BM, which corresponds to the description of this species (Dickson 1801), consists of two different groups of shoots which appear superficially to be distinct collections (L. T. Ellis, personal communication, 25 Nov 2005). Therefore, Demaret & Wilczek's (1980) examination of the type material cannot be considered to be tantamount to lectotypification of *B. bicolor*.

Other characters of *Bryum dichotomum* such as spore and capsule size, leaf morphology and costa length are too variable to be used as reliable taxonomic characters within the complex. This is also true of Antarctic material in which the costa may be percurrent or even subpercurrent to long-excurrent. This wide variation of gametophyte characters led various authors to describe many morphotypes of *B. dichotomum*, especially in the Southern Hemisphere, as distinct species. However, many names of these species have been reduced to synonymy (Dixon 1926; Ochi 1970, 1972, 1980, 1982; Ochyra 1998*a*), and here two more synonyms are added, *B. anomobryoides* and *B. laetevirens*, which were used for species described from Patagonia (Cardot & Brotherus 1923). These species were considered by Ochi (1982) to be conspecific with *B. urbanskyi* from Îles Kerguelen which, in turn, was treated as a problematic species, but Ochyra (1998*a*) reduced it to synonymy with *B. dichotomum*.

Some species described from southern South America which belong within the *Bryum bicolor* complex have never been taxonomically assessed, for example *B. gemmatum* Müll.Hal. from Tierra del Fuego (Müller 1885), *B. miserum* Cardot from the Falkland Is., *B. delitescens* Cardot from Isla de los Estados and *B. tenuirete* Cardot from western Patagonia (Cardot 1905*a*, 1908*a*), as well as *B. pluricolor* Dusén *ex* Thér. from southern Chile (Thériot 1917). These may represent distinct species or may be closely related or even identical to recently described new species in this complex from Europe. Therefore, a worldwide revision of this complex is needed because it should yield a more precise circumscription of particular species and provide important information on their distribution.

Bryum dichotomum is most likely to be mistaken for the bulbiferous *Pohlia drummondii* but the narrowly recurved and usually entire leaf margins, percurrent to long-excurrent costa and relatively short laminal cells should distinguish the former species. The species was reported for the first time from Antarctica from King George I. in the South Shetlands (Ochi & Ochyra 1986). Paradoxically, this first report in this biome was based upon misnamed specimens which correctly belonged to *B. archangelicum* as shown by Ochyra (1998*a*) who, at the same time, provided new Antarctic records of *B. dichotomum* from Deception I. and Alexander I.

Reproduction in Antarctica. Sterile, but usually producing caducous axillary gemmae in masses which provide a means for dispersal and propagation.

Habitat. A colonist in a variety of habitats, growing on moist stony or sandy soil, fine gravel, marble detritus and in crevices of rocks, and in areas with recent volcanic activity on cindery and compacted fine soil and scoria, usually along fissures on heated ground associated with fumaroles. It usually forms small pure stands, but in looser tufts *Brachythecium austrosalebrosum*, *Campylopus introflexus*, *Cryptochila grandiflora*, *Dicranella hookeri*, *Encalypta procera*, *Pohlia nutans*, and *Syntrichia magellanica* have been recorded as occasional associates.

World range. The geographical range of *Bryum dichotomum* is uncertain because of the current state of its taxonomy, and the global distribution map (Fig. 212) is rather the range of the whole *B. dichotomum* complex. Notwithstanding the specific name, it was often considered a cosmopolitan species (e.g. Gangulee 1974; Rooy & Magill 1987; Eddy 1996; Dierßen 2001), although it has not apparently been recorded from major regions of the world, for example from the Arctic and continental Antarctic, and it is rare and incidental in the tropics. Therefore, *B. dichotomum* should be interpreted as a bipolar temperate species, occasionally extending into the tropics, mostly at high elevations, or as a weed of cultivated land. In the Northern

Figure 211. *Bryum dichotomum*. — **1**. Sterile plant, wet. **2**. Sterile plant, dry. **3**. Portion of sterile shoot with brood bodies in axils of upper leaves. **4**. Cross-section of stem. **5–7**. Axillary hairs. **8–16**. Leaves. **17–19**. Leaf apices. **20–21**. Mid-leaf cells at margin. **22**. Upper leaf cells at margin. **23**. Mid-leaf cells. **24**. Basal cells. **25–28**. Cross-sections of leaves. **29–33**. Gemmae. — [1, 3, 13–15, 17, 21, 23–24, 30 from *RILS 4004A*; 2, 4–10, 18, 22, 25–29, 31, 33 from *RILS 5751B*; 11, 20 from *RILS 5752E*; 12 from *RILS 9338*; 16, 19, 32 from *RILS 3663*]. — Scale bars: a – 0.5 mm (1–2) & 100 µm (29–33); b – 50 µm (5–7 & 100 µm (17–24); c – 0.5 mm (8–16) & 100 µm (4, 25–28); d – 1 mm (3).

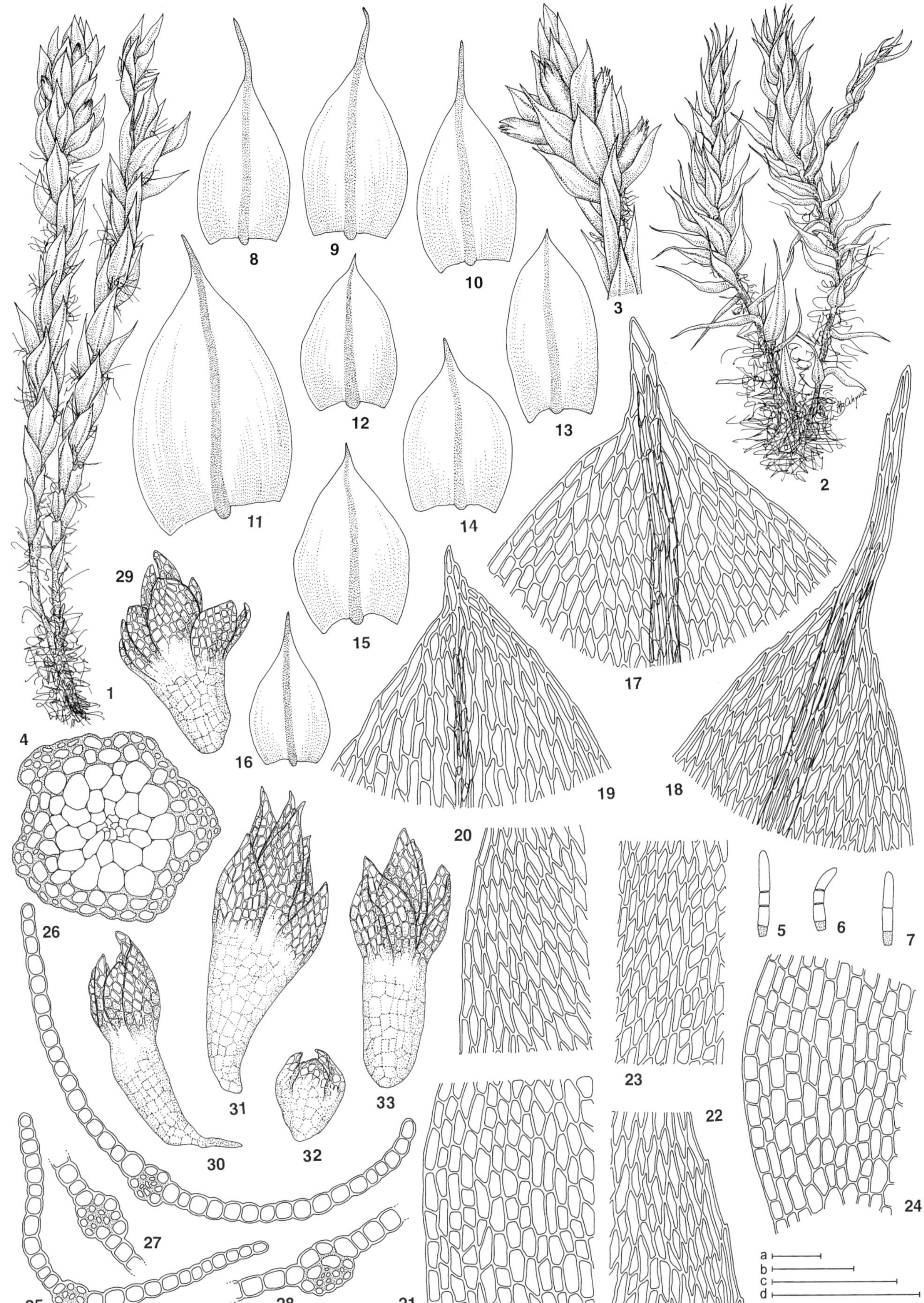
1
2
3
4
5
6
7
8
9
10
11
12
13
14
15
16
17
18
19
20
21
22
23
24
25
26
27
28
29
30
31
32
33
a
b
c
d

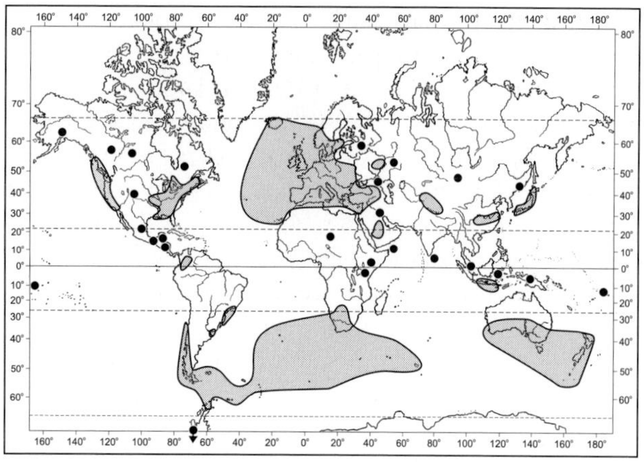

Figure 212. Global distribution of *Bryum dichotomum*.

Hemisphere *B. dichotomum* has a pan-Holarctic, though strongly dissected range, with the main centre of occurrence in Europe where it extends from Iceland and southern Scandinavia to the Mediterranean region and North Africa (Ros *et al.* 1999, 2000), Macaronesia in the west (Eggers 1982) and the Caucasus and Ural Mountains in the east (Zolotov 2000). In Asia, the species is infrequent and occasional in the Near East including the Arabian Peninsula (Frey & Kürschner 1991; Kürschner 2000; Herrnstadt & Heyn 2004), Central Asia and the Himalayan region (Noguchi *et al.* 1966; Gangulee 1974; Ochi 1990*b,c*; Kattel & Adhikari 1992), Mongolia (Ignatov *et al.* 2004), Russian Far East (Bardunov & Cherdantseva 1982), China and Taiwan (Redfearn *et al.* 1996) and Japan (Ochi 1959, 1980). In North America it is bicentric in distribution, occurring in the western part of the continent from British Columbia to southern California and Colorado, with isolated stations in southern Alaska, Alberta and Saskatchewan, and in the eastern part from Ontario, Quebec and New Brunswick south to Florida and Louisiana and west to Kansas (Vanderpoorten & Zartman 2002).

In the Southern Hemisphere *Bryum dichotomum* has a wide, pan-temperate range (Fig. 213, inset) which covers southern Australia and Tasmania (Ochi 1970; Streimann & Klanzenga 2002), New Zealand (Dixon 1926; Beever *et al.* 1992) and the islands of Auckland (Vitt 1979*a*) and Campbell (Vitt 1974*a*), southern Africa (Rooy & Magill 1987) including Tristan da Cunha and Gough I. (Ochi 1972), and the *Nothofagus* zone along the western fringe of southern South America, from the Valdivian region to Tierra del Fuego and the Falkland Is., then appearing on the Subantarctic Prince Edward Is. archipelago (Zanten 1971), Îles Crozet (Hébrard 1970*a*), Îles Kerguelen (Ochyra, personal observations) and Macquarie I. (Seppelt 2004) and extending to the peri-Antarctic islands of Bouvetøya, South Sandwich Is., South Orkneys and South Shetland Is. and to the western coast of the Antarctic Peninsula southwards to Alexander I. In the tropics *B. dichotomum* is rather infrequent in the Neotropics, extending from Mexico, Belize and Nicaragua in Central America (Ochi 1994; Allen 2002) to the northern Andes of Colombia and Ecuador and south-eastern Brazil and the River Plate area in South America (Ochi 1980, 1982), in Chad, Tanzania, Kenya and Socotra in Africa (O'Shea 2003), Sri Lanka

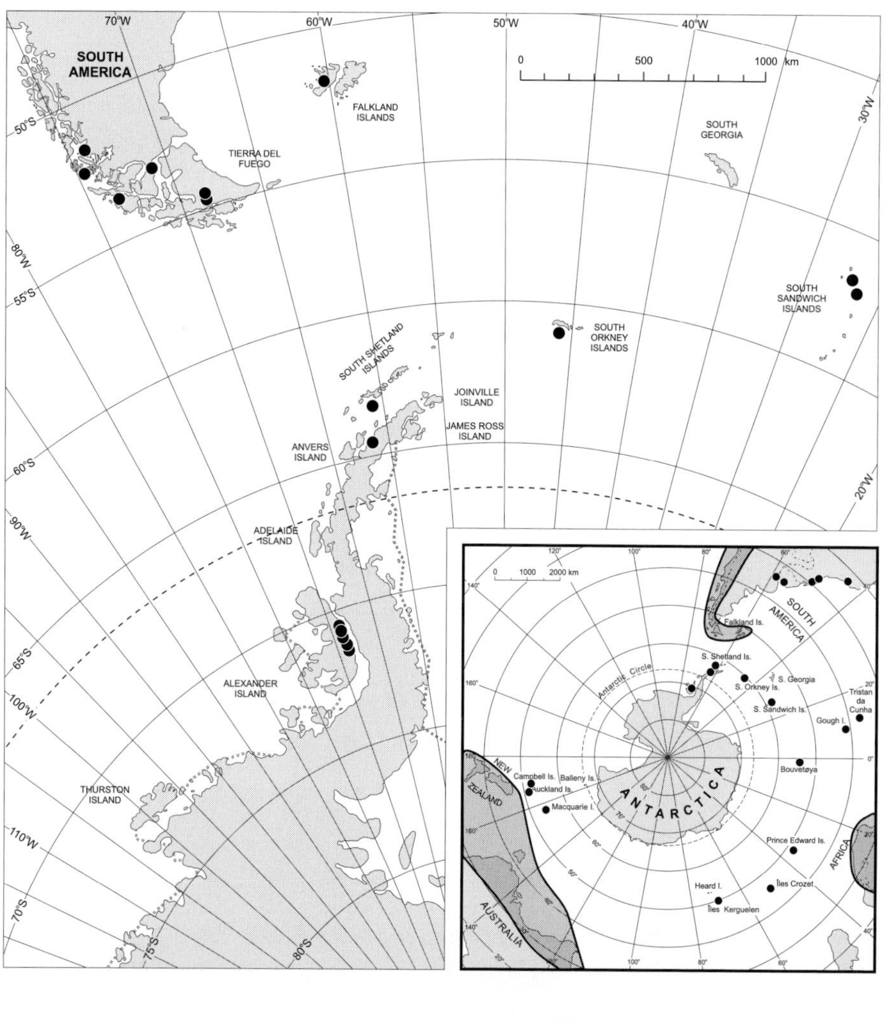

Figure 213. Distribution of *Bryum dichotomum* in Antarctica. Inset: Distribution in temperate and cold regions of the Southern Hemisphere.

(Ochi 1980, 1985) and Malesia (Eddy 1996), as well as from Samoa and Fiji in Melanesia in the Pacific region (Schultze-Motel 1974; Miller *et al.* 1978).

Distribution in Antarctica. *Bryum dichotomum* is widely distributed but scattered in the maritime Antarctic (Fig. 213). It occurs mainly in areas with volcanic activity, namely on two islands in the South Sandwich Is., Deception I. and Bouvetøya, and it is relatively frequent on Alexander I. where it extends as far south as lat. 71°42′ S. In addition, it is known from solitary stations on Signy I. in the South Orkneys and the Danco Coast on the western coast of the Antarctic Peninsula. Its altitudinal range extends from sea level to about 600 m on Alexander I. This species was reported as *Bryum atropurpureum* Bruch & Schimp., with a question mark, from Schirmacher Oasis in Dronning Maud Land in continental Antarctica (Singh & Semwal 2000) but no voucher specimen was available for study.

Selected specimens examined. WEST ANTARCTICA. BOUVETØYA. Nyrøysa, 30 m, *TE 70 & 287* (BG) and 0–51 m, 22 Feb 1979, *TE s.n.* (BG).

SOUTH SANDWICH IS. **Visokoi I.** Finger Pt., *ca* 35 m, *REL 754* (AAS, KRAM). **Candlemas I.** Lucifer Hill, south-west crater, *PC 107D* (AAS, KRAM); same locality, summit of west ridge above Clinker Gulch, *PC 231C* (AAS, KRAM).

SOUTH ORKNEY IS. **Signy I.** Gourlay Peninsula, *ca* 65 m, *RILS 496* (AAS, KRAM).

SOUTH SHETLAND IS. **Deception I.** Whalers Bay, 5 m, *RILS 11222, 11223 & 11225* (AAS, KRAM); Whalers Bay, 'Perchuč Cone', 240 m, *RILS 11235, 11236 & 11237* (AAS, KRAM); Ronald Hill, 80–100 m, *RILS 6872 & 6873A* (AAS, KRAM); Mt. Pond, 480 m, *RILS 5819B* (AAS, KRAM), 485–500 m, *RILS 11164A* (AAS, KRAM) and 500 m, *RILS 11150B & 11151A* (AAS, KRAM); Pendulum Cove, 2–4 m, *RILS 3280, 3663, 5745A, 5746C & 11313B* (AAS, KRAM) and 100 m, *RILS 5751B, 5752E & 5754B* (AAS, KRAM); Fumarole Bay, 100 m, *RILS 5664B & 5666A* (AAS, KRAM) and 125 m, *RILS 11272B, 11273A, 11279B, 11285 & 11289* (AAS, KRAM); Lake Irizar, 100–125 m, *RILS 6909B* (AAS, KRAM); 2 km north of Vapour Col, *ca* 175 m, *RILS 11261B & 11266F* (AAS, KRAM).

WEST ANTARCTIC PENINSULA. **Danco Coast.** Charlotte Bay, Andrée I., 250 m, *RILS 4004A* (AAS, KRAM). **George VI Sound. *Alexander I.***: Ablation Col, *RILS 11116 & 11117* (AAS, KRAM); Ablation Valley, *DDWW 36A* (AAS, KRAM) and 75 m, *RILS 9294A* (AAS, KRAM); Ablation Lake, 1–3 m, *RILS 9245A & 9246* (AAS, KRAM); High Flatiron Valley, 532 m, *RILS 11069* (AAS, KRAM); below Khufu Peak, 300–400 m, *RILS 9321 & 9330* (AAS, KRAM) and 500–600 m, *RILS 9338* (AAS, KRAM); Scarab Bluff, 490 m, *RILS 9366A* (AAS, KRAM); Mars Glacier, *RILS 8849A* (AAS, KRAM); Two Step Moraine, 30 m, *RILS 9491* (AAS, KRAM).

3. **Bryum nivale** Müll.Hal. Fig. 214

Bryum nivale Müll.Hal., Syn. Musc. Frond. 1: 262. 1848. — Type citation: Chile australis in rivulis niveis in jugis Andium inter montem ignivomum Antuco et la Sierra velada. Muscus omnium altissime in Andium crescens: Pöppig, 16. Febr. 1829. Hb. Kunzeanum. [Lectotype (*selected here*): 'Bryum nivale mihi Br. cyclophyllo prox.hermaphroditum. fruct. ignot. Chile austr. in rivulis niveis in jugis Andinum inter montem igniv. Antuco et la Serra velada, muscus omnium altissime crescens. Pöppig lgt.' – BM-Hampe! (No. 663430); isotype: BM-Hampe!].

Plants medium-sized, forming loose tufts or growing intermingled with other mosses, stiff and fairly rigid, scarcely glistening, green, yellowish-green, brownish-green to reddish-brown. *Stems* erect to ascending, 1–6 cm tall, simple or seldom forked because of innovations, reddish-brown and lustrous, evenly foliate to slightly comose, sparsely radiculose at the base with pale brownish, branched, finely roughened, lustrous rhizoids, in transverse section obtusely pentagonal, consisting of a large central strand, 3–4-stratose medulla of large, hyaline and thin-walled cells and 1(–2)-layered cortex of somewhat smaller cells with thin to somewhat thicker, reddish-brown walls. *Leaves* moderately crowded to distant, sometimes crowded at the apex and forming a coma, slightly altered on drying, erect-spreading to widely patent, often with a squarrose-recurved limb from a widespreading base, ovate, broadly ovate to suborbicular, 1.3–1.9 mm long, 0.9–1.5 mm wide, deeply concave, cordate, semi-vaginant and often loosely clasping, pale brownish to greenish, shortly decurrent at the base, broadly acute or rounded and shortly apiculate at the apex; *margins* entire, plane, lacking a distinct border; *costa* ceasing below the apex, clearly demarcated from the laminal cells, greenish- or yellowish-brown to brown, 60–80 μm wide near the base, gradually narrowed upwards, in cross-section distinctly convex dorsally, consisting of a prominent central stereid group, 3–4 enlarged guide cells and distinct epidermis on ventral and dorsal surfaces; *laminal cells* unistratose throughout, transparent; *upper* and *median leaves* rhombic to shortly oblong-rhomboidal, 30–80 μm long, 20–30 μm wide, thin-walled, sometimes lax on young leaves, often becoming narrower, rectangular to elongate-hexagonal at the margins but not forming a distinct border; *basal cells* oblong-hexagonal to rectangular, 45–85 μm long, 15–20(–25) μm wide, thin-walled, lax, turgid, brownish, sharply demarcated from the upper cells and forming prominent auricles on the mature leaves. *Asexual reproduction* not seen. *Dioicous*. *Perigonia* terminal; *outer perigonial bracts* squarrose, 2.0–2.5 mm long, 1.0–1.3 mm wide, from a broad, ovate, clasping base rather abruptly narrowed to a broad and fairly long acumen, with a strong and percurrent costa and areolation similar to that in the vegetative leaves; *inner perigonial bracts* smaller, 1.0–1.2 mm long, 0.9–1.0 mm wide, broadly ovate, rapidly narrowed to a short and broad acumen, with a strong and percurrent costa; *antheridia* 10–15 per perigonium, large, pale violet, intermixed with filiform, yellow to yellow-brown paraphyses, distinctly exceeding the antheridia. *Female plants* and *sporophytes* unknown.

Chromosome number: unknown.

Discussion. *Bryum nivale* is a distinct moss which may be distinguished by its deeply concave, broadly ovate to suborbicular leaves that are broadly acute and blunt or obtuse and shortly apiculate at the apex. This leaf shape is coupled with its characteristic areolation in the proximal portion which consists of rectangular to oblong-hexagonal, thin-walled, very lax, bulging and brownish cells which form prominent, shortly decurrent auricles on mature leaves and sharply set off from median and upper laminal cells. The latter are short, rhombic to shortly oblong-rhomboidal and thin- but otherwise firm-walled. Such sharp differentiation of the leaf areolation in *B. nivale* is unique in *Bryum* and in most species of this genus the differentiated basal leaf cells gradually pass into the upper cells. Only occasionally basal cells form distinct 2–3-seriate auricles in some populations of the Asian *B. handelii* Broth. and were described as a separate species, *B. pulchroalare* Broth. (Brotherus 1929).

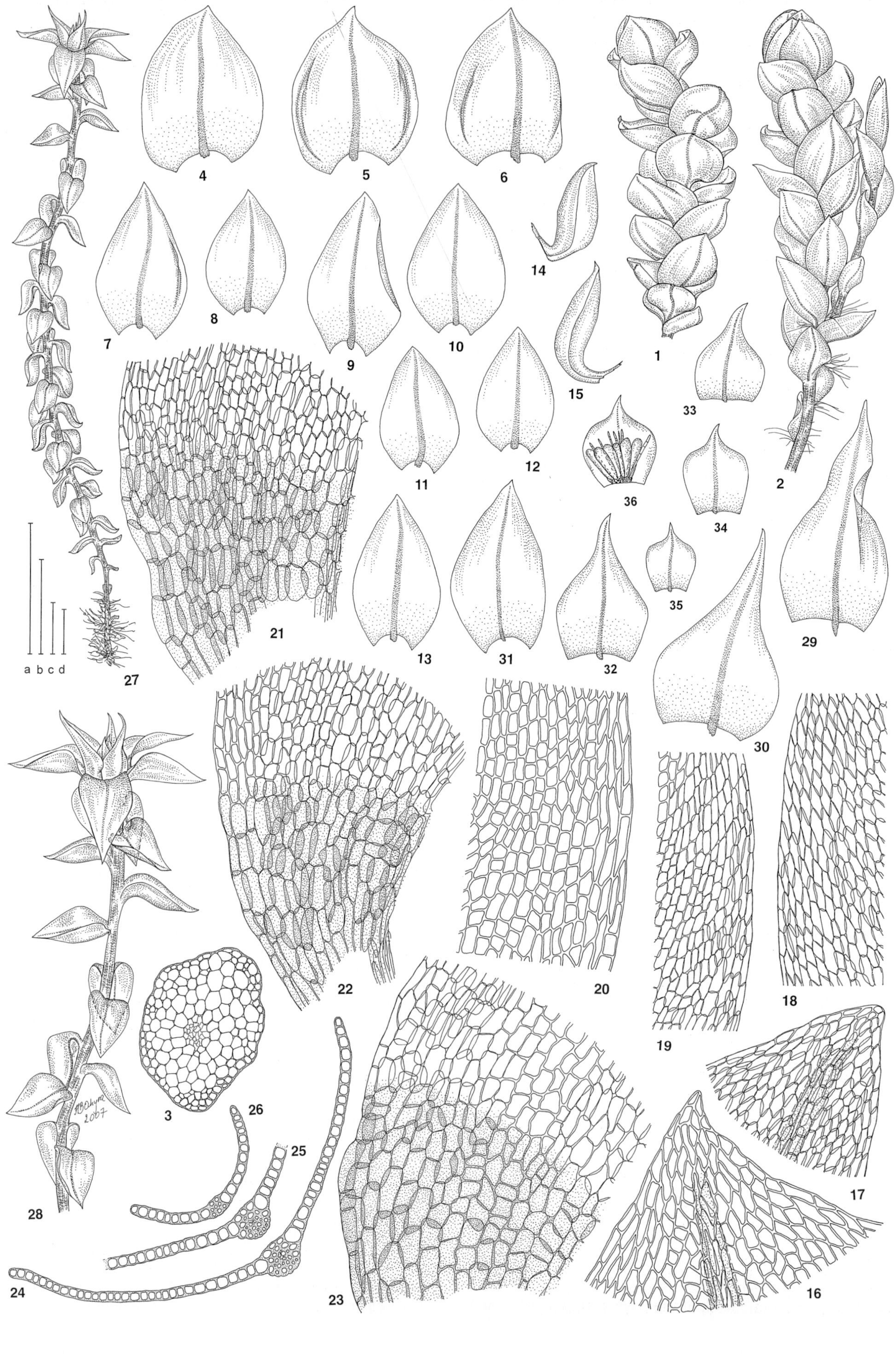
4
5
6
7
8
9
10
11
12
13
14
15
1
2
33
34
35
36
21
27
a b c d
31
32
29
30
22
20
19
18
3
26
25
24
23
28
17
16

The differences in the areolation result in the abrupt differentiation of the leaf lamina into the distinct, cordate, semi-vaginant base loosely clasping the stem and erecto-patent to widespreading or squarrose-recurved limb which is visible on plants with the distant foliage. Interestingly, such leaf areolation is also present on the outer perigonial bracts.

Bryum nivale belongs within sect. *Amblyophyllum* which is diagnosed by well spaced, concave and elimbate leaves that are ovate, oval, elliptic to orbicular with a rounded, obtuse or apiculate apex and decurrent base and rather short, rhombic to rhomboidal and thin-walled laminal cells. This section consists of about 20 species that are separated into three subsections (Ochi 1992). *Bryum nivale* exhibits some overall similarity to the Holarctic *B. cyclophyllum* (Schwägr.) Bruch & Schimp. (which was stressed by Müller on the herbarium label of the type material) and *B. cryophilum* Mårtensson and neotropical *B. renauldii* Renauld & Cardot which, however, are distinct in having weakly differentiated basal cells (Ochyra 1991*a*). In the Antarctic the leaf shape of *B. nivale* is similar to that of *B. orbiculatifolium* and the *filicaule* phenotypes of *B. pseudotriquetrum*. The former is a closely related species to *B. nivale* but it is a much smaller plant with very small leaves, less than 1 mm long, having undifferentiated, green basal cells. In contrast, the *filicaule* phenotypes of *B. pseudotriquetrum* have leaves of comparable size and shape but the proximal cells are red and do not form a distinct base with an areolation sharply differentiated from that in the upper portion of the leaf.

Bryum nivale is a very rare and poorly known species which has hitherto been known only from the type collection from a single locality in Bio-Bio region of central Chile (Müller 1848). Ochi (1982) suggested that the species may be 'an extremely modified form of *B. turbinatum*' from which it differs in its synoicous sex condition and leaf shape. In fact, *B. nivale* seems to be much closer to *B. schleicheri* DC. than to *B. turbinatum* (Hedw.) Turner, as emphasised in the original description by Müller (1848). The only difference between these species is the sex condition, *B. schleicheri* being a dioicous species, but the correspondence in other characters is perfect. The Antarctic plants from Alexander I. deviate also from the type material of *B. nivale* in their dioicy and poorly developed or absent leaf margin, but the latter may be owing to permanent or seasonal submersion and occurrence in extreme climatic conditions.

A parallel situation is observed in other species of the genus in Antarctica, especially in *B. pseudotriquetrum*, in which plants living in similar habitats are devoid of a marginal border and have been described as a separate species, *B. korotkevicziae*. Initially, the Antarctic specimens were intended to be described as a new species but the scantiness of the material and its variable development dictates caution in recognition of a new species on such weak grounds. The Antarctic specimens are very variable and the submerged plants growing in the melt pool have remotely foliated stems, whereas the terrestrial plants growing in otherwise wet habitats have more closely spaced leaves; similar variation is also observed in *B. schleicheri* in the Northern Hemisphere. The species is in need of more detailed study which would benefit from further collections which should contribute to its better understanding and circumscription. It is very likely that the species occurs beyond the Antarctic but careful examination of the type specimens of all species described from South America and Subantarctica has not revealed any similar taxon. However, the species does occur rarely on Marion I. in the Subantarctic (Ochyra & Hertel 1990 as *B. amblyolepis*; Ochyra, personal observations). Interestingly, the plants from Marion I. were found on a steep cliff with water trickling down from the nesting site of grey-headed albatrosses, and these plants also have poorly developed marginal border.

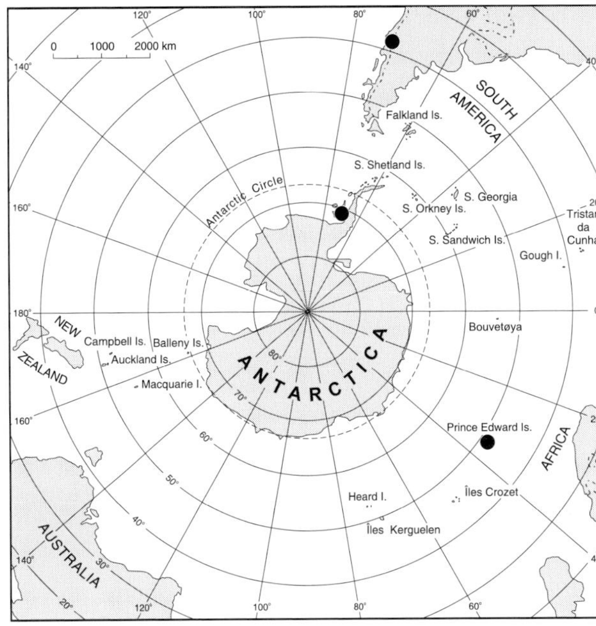

Figure 215. Global distribution of *Bryum nivale*.

Reproduction in Antarctica. Sterile, but male plants with perigonia are known.

Habitat. On silt and fine gravelly soil in a shallow melt pool, forming pure loose tufts or growing intermixed with *Campylium polygamum*.

World range. So far, the species is known from three widely separated areas in the Southern Hemisphere, including Alexander I. in Antarctica, Marion I. in Subantarctica and the Andes of Central Chile where it was found in the alpine zone (Fig. 215). Therefore, *Bryum nivale* may be provisionally designated as a subantarctic species extending to the Antarctic and alpine zone in the Andes.

Distribution in Antarctica. Known only from two collections made in one site in Striation Valley which is situated at lat. 70°53′S north of Jupiter Glacier on the east side of Alexander I. in the West Antarctic (Fig. 215).

Specimens examined. WEST ANTARCTICA. WEST ANTARCTIC PENINSULA. **George VI Sound.** ***Alexander I.***: Striation Valley, *RILS 11129B & 11130* (AAS, KRAM).

←

Figure 214. *Bryum nivale*. — **1–2**. Habit, wet. **3**. Cross-section of stem. **4–15**. Leaves. **16**. Leaf apex of terrestrial plant. **17**. Leaf apex of submerged plant. **18–19**. Mid-leaf cells at margin of submerged plant. **20**. Mid-leaf cells at margin of terrestrial plant. **21–22**. Basal cells of submerged plant. **23**. Basal cells of terrestrial plant. **24–26**. Cross-sections of leaf. **27**. Habit of male submerged plant. **28**. Portion of shoot of submerged plant with perigonium. **29–32**. Outer perigonial bracts. **33–35**. Inner perigonial bracts. **36**. Innermost perigonial bract with antheridia and paraphyses. — [1-2, 4-6, 16, 20, 23 from *RILS 11129B*; 3, 7-15, 17-19, 21-22, 24-36 from *RILS 11130*]. — Scale bars: a - 0.5 cm (27); b - 1 mm (4-15, 29-36); c - 100 μm (3, 16-26); d - 1 mm (1-2, 28).

4. **Bryum orbiculatifolium** Cardot & Broth.

Fig. 216

Bryum orbiculatifolium Cardot & Broth., Kongl. Svenska Vetenskapsakad. Handl. 63(10): 42. 1923. — TYPE CITATION: Patagonia occ.: Peel Inlet, ad terram glareosam, saepe irrigatam prope glaciem (540). [Lectotype (*selected here*): 'Expeditio suecica 1907–1909. 212. Campylopus guaitecae Dus. + Bryum orbiculatifolium Card. nova spec. Patagonia occid. Peel Inlet. am grossen Gletscher 16.6.1908 Leg. C. Skottsberg.' – PC-Cardot!; isotype: S!].

Bryum cymbifoliellum Dixon *in* Christoph., Results Norweg. Sci. Exped. Tristan da Cunha 48: 32. 1960. — TYPE CITATION: Tristan da Cunha: *Empetrum* heath, above Burntwood, ca 1200 m, Nos. 784c, 785c), type. Crater, 2000 m., No. 736. [Holotype: 'Herb. H. N. Dixon. Ref. No. 784c, 785 Bryum cymbifoliellum Dix. sp. nov. Empetrum heath, above Burntwood, c. 1200 m, Tristan da Cunha Coll. Christophersen & Mejland Date 8 Jan. 1938 Comm. Bot. Mus., Oslo type' – BM-Dixon!; isotypes: FH-Bartram!, O!]. First synonymised by Ochi (1982: p. 25).

Plants small, densely or loosely caespitose to, occasionally, gregarious, gracile but fairly rigid, somewhat glistening, pale to dark green or less often yellowish- to brownish-green above, reddish-brown to brown below. *Stems* erect, slender, simple or branched by innovating shoots overtopping the main stem and giving the plants a layered appearance, 0.2–1.0 mm tall, julaceous, turgid and markedly bud-like at the tips, reddish to reddish-brown, densely radiculose at the base and along the stems with clusters of long, branched, glossy, brown, finely papillose rhizoids, in cross-section round to obscurely round-pentagonal, consisting of a small central strand, 3–4 layers of large, hyaline, thin-walled medullary cells and a row of somewhat smaller cortical cells with thicker, reddish to reddish-brown walls; *axillary hairs* short, composed of 3–4 cells, with a short, quadrate, reddish-brown basal cell and elongate distal cells. *Leaves* smaller and somewhat distant below, becoming larger and crowded upwards, usually forming a coma at the stem tips, fairly rigid, erect, imbricate, both dry and wet, 0.4–0.8(–1.1) mm long, 0.4–0.8 mm wide, orbicular, cochleariform-concave to broadly ovate, sometimes transversely elliptical, oblong-ovate on young innovations, rounded and obtuse or sometimes with a short, broad, blunt apiculus, strongly concave and often cucullate, non-decurrent, green throughout or occasionally pale brownish at the base; *margins* entire, plane, erect to weakly involute, unbordered; *costa* single, weak or sometimes obsolete, subpercurrent to ceasing well below the apex, 30–40 µm wide at the base, brownish to pale yellowish-green and almost concolorous with the laminal cells on the leaves from innovations, in transverse section consisting of 2–3 larger guide cells, a small dorsal stereid band without differentiated epidermal cells, and 2–3 large ventral epidermal cells; *laminal cells* unistratose throughout, transparent, thin- to moderately thick-walled, rhombic, rhomboidal to oblong-hexagonal distally, 30–60 µm long, 15–20(–25) µm wide, becoming somewhat shorter and wider at the apex and rectangular and narrower at the margins but not forming a border, rectangular to oblong-hexagonal, green or occasionally faintly brownish, 40–70 µm long, 12–25 µm wide, at the base. *Asexual reproduction* unknown. [*Dioicous*. *Setae* erect, 1.5–3.0 cm long, slender, brown, weakly flexuose, smooth, glossy. *Capsules* long-exserted, horizontal to pendent, 2–3 mm long, 1.0–1.2 mm wide, pyriform, not contracted below the mouth, brown, with a short, broad neck, somewhat wrinkled when dry, abruptly narrowed into the seta; *annulus* and *operculum* not seen; *exothecial cells* very variable in shape, mostly irregular, rounded to oblong in mid-urn, thick-walled, 10–35 µm long, 5–20 µm wide, becoming oblate, deeply brown in 3–4 tiers at the capsule rim; *peristome* double, inserted close to the mouth, 500–550 µm long; *exostome teeth* 16, lanceolate, yellow to yellow-brownish, finely dotted below, pale yellowish to hyaline, finely papillose at the tips, unbordered, distinctly trabeculate on the inner side; *endostome* hyaline to hyaline-yellowish, smooth, consisting of a high basal membrane to about ¾ of the exostome length and rather broad, keeled, widely perforate processes and 2–3 nodose or appendiculate cilia. *Spores* spherical, pale yellowish, 8–10 µm in diameter. *Calyptra* not seen.]

Chromosome number: unknown.

Discussion. *Bryum orbiculatifolium* is a small moss typically with deeply concave and imbricate leaves that are rounded and obtuse or finely apiculate at the apex. The leaves are usually crowded distally, giving the plants a gemmiform habit. In size and aspect the species is nearly identical to the *muticum* form of *B. argenteum* but it differs in having entirely chlorophyllose laminal cells resulting in the pale to dark green or yellowish- or brownish-green coloration of the plants, green leaf base, stouter costa usually ending a short distance from the leaf apex with 2–3 guide cells as clearly seen in transverse section. In addition, the species is distinct from *B. argenteum* var. *muticum* in having large, pyriform capsules with a prominent apophysis that are exserted on exceptionally long setae which are 5 to 7 times longer than the minute leafy gametophytes. However, fertile plants with fully developed sporophytes have hitherto been found only once, in Chile. *Bryum orbiculatifolium* exhibits some variability with regard to the robustness of the plants and leaf shape. This is closely associated with its ecology in the Antarctic where it usually grows on unstable ground with its shoots often buried by fine volcanic ash. They continue their growth by repeated annual

→

Figure 216. *Bryum orbiculatifolium*. — **1**. Fertile plant with mature deoperculate capsule, wet. **2–3**. Habit, wet. **4**. Innovation. **5**. Cross-section of stem. **6–9**. Axillary hairs. **10–17**. Stem leaves. **18–19**. Leaves from innovations. **20–21**. Leaf apices. **22**. Mid-leaf cells at margin. **23–24**. Mid-leaf cells. **25**. Basal cells. **26–28**. Cross-sections of leaves. **29**. Portion of peristome, spores and exothecial cells at mouth. **30**. Exothecial cells in mid-urn. **31**. Exothecial cells in neck and stoma. — [1, 13–14, 29–31 from *Schwabe 36/b*, JE; 2, 16–18 from *RILS 11239*; 3–11, 20–28 from *PC 137E*; 12 from *Christophersen & Mejland 784C* – type of *B. cymbifoliellum*, BM; 15 from *CS 212* – type of *B. orbiculatifolium*, PC; 19 from *RILS 11166*]. — Scale bars: a – 1 mm (10–19) and 100 µm (6–9); b – 1 mm (4); c – 100 µm (30–31); d – 100 µm (5, 20–28); e – 1 mm (2): f – 1 mm (1, 3) and 100 µm (29).

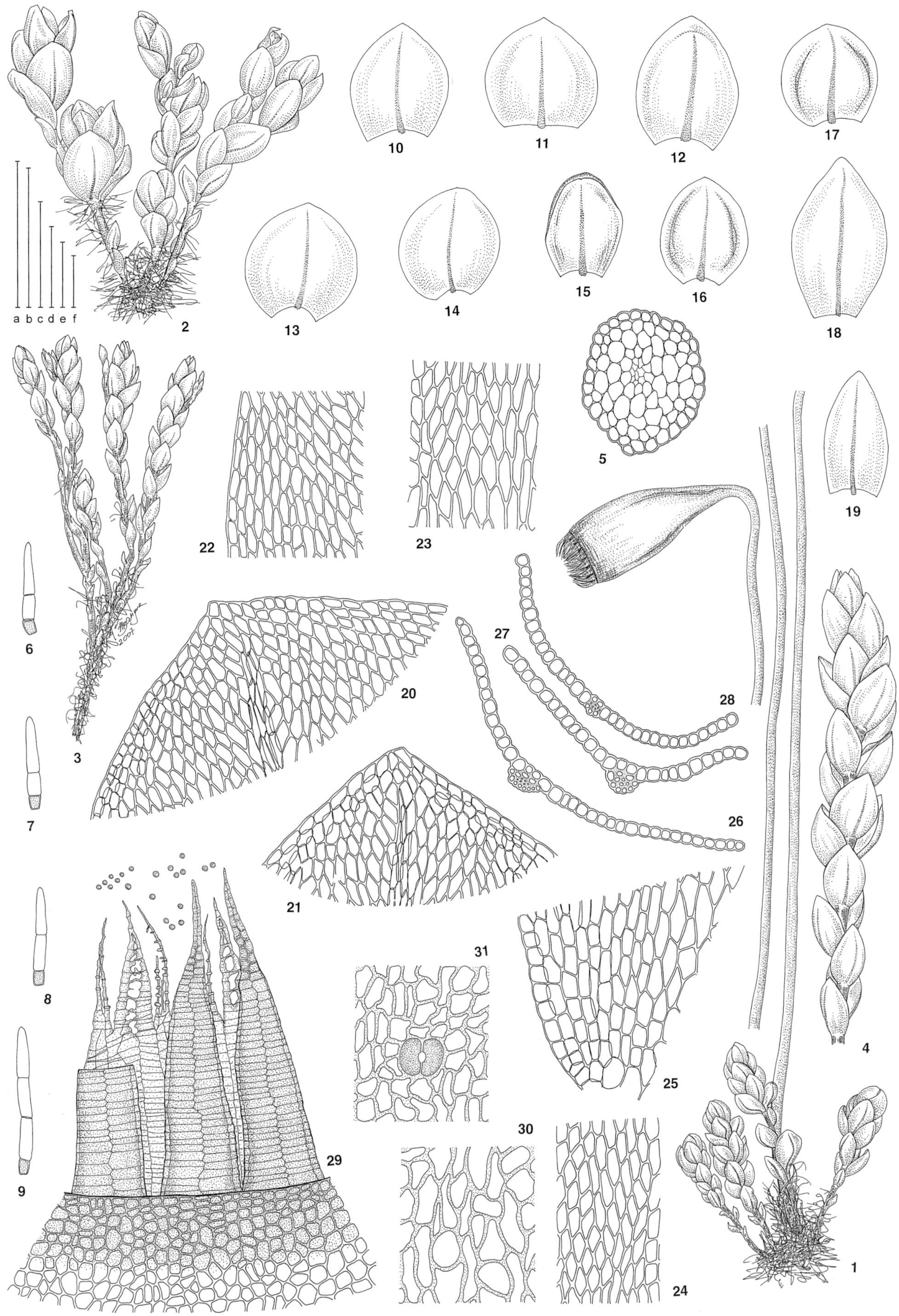
a b c d e f
1
2
3
4
5
6
7
8
9
10
11
12
13
14
15
16
17
18
19
20
21
22
23
24
25
26
27
28
29
30
31

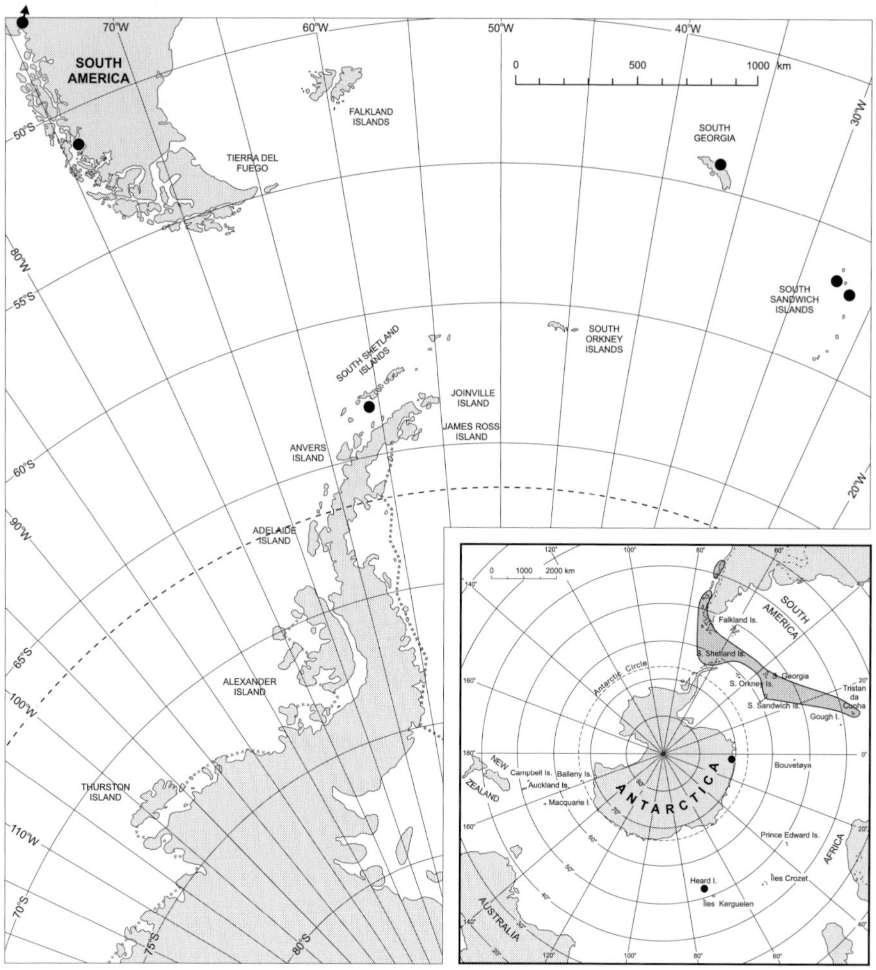

Figure 217. Distribution of *Bryum orbiculatifolium* in the Subantarctic and Antarctic. Inset: global distribution. Localities in Bio-Bio and Llaniquihue Provinces of Chile are not indicated.

innovations which give the plants a layered appearance. The young leaves of these innovations are oblong-ovate to lingulate and less concave, and only with age do they become orbicular and cochleariform. Such offshoots are occasionally found in the type material of *B. cymbifoliellum* from Tristan da Cunha (Dixon 1960) which is considered to be conspecific with *B. orbiculatifolium* (Ochi 1972, 1982), but they are absent from the type specimen of the latter. Nevertheless, the Antarctic plants fit well with the concept of this species, and the correspondence of the shape of the mature leaves and their areolation is perfect.

When describing *Bryum orbiculatifolium*, Cardot & Brotherus (1923) compared it with the Holarctic *B. cyclophyllum* which, however, is a larger, dull, laxly tufted and distantly foliated plant. The leaves of this species are soft, ovate, rounded-ovate to oblong-ovate, obtuse and not apiculate at the apex, generally less concave and distinctly bordered with narrow, elongate, partly bistratose cells. Nonetheless, both species share the green colour of the basal leaf cells and concave leaves that are rounded and obtuse or shortly apiculate at the apex. These traits warrant the placement of *B. cyclophyllum* and *B. orbiculatifolium* in subsect. *Euleucodontium* within sect. *Amblyophyllum*, although the latter is better known as sect. *Leucodontium* (J.J.Amann *ex* Broth.) Grout (Isoviita 1992) which consists of ten species from various part of the world (Ochi 1992).

In the Antarctic *Bryum orbiculatifolium* has been mistaken for the *filicaule* phenotypes of *B. pseudotriquetrum* and Dixon (1960) compared his species *B. cymbifoliellum* newly described from Tristan da Cunha, with the Antarctic *B. filicaule*. This similarity is so ostensible that, in fact, the first record of *B. orbiculatifolium* from Antarctica (Ochyra & Ochi 1986*a*) was based on plants that correctly refer to the *filicaule* phenotypes of *B. pseudotriquetrum*. Although they share a similar leaf shape, the *filicaule* phenotypes of *B. pseudotriquetrum* are distinct in having larger leaves with reddish to intensively red basal leaf cells, as well as only weakly concave leaves that are broadly acute at the apex.

Reproduction in Antarctica. Sterile.

Habitat. In the Antarctic, *Bryum orbiculatifolium* grows exclusively in volcanic areas, mostly on heat-cemented soil, geothermal ground, dry scoria and ashy slopes associated with fumaroles, and only occasionally has it been collected from unheated yellowish tufa rock outcrops, but immediately adjacent to fumaroles. It grows mostly in monospecific tufts and only occasionally species such as *Hymenoloma grimmiaceum*, *Ceratodon purpureus* and *Bryum pseudotriquetrum* occur as associates.

World range. A very rare and occasional, possibly undercollected, amphiatlantic south-temperate species (Fig. 217, inset), known from western Patagonia (Ochi 1982), subantarctic South Georgia (Ochyra *et al*. 2002) and Heard I. (Ochyra & Seppelt 2008*a*), Tristan da Cunha (Ochi 1972), and extending to the northern maritime and continental Antarctic.

Distribution in Antarctica. The species occurs primarily on some volcanic islands in the South Sandwich Is. and South Shetland Is. (Fig. 217), usually forming small, inconspicuous populations from near sea level to 500 m on Deception I., and once it was found on the Antarctic continent (Ochyra & Singh 2008).

Selected specimens examined. WEST ANTARCTICA. SOUTH SANDWICH IS. **Leskov I.** Summit ridge, 100 m, *PC 245H* (AAS, KRAM). **Candlemas I.** Lucifer Hill, *PC 137E* (AAS, KRAM).

SOUTH SHETLAND IS. **Deception I.** Whalers Bay, 75 m, *DM 10* (AAS, KRAM); between Kroner Lake and Ronald Hill, 2–4 m, *RILS 11239* (AAS, KRAM); geothermal area ENE of Ronald Hill, 80–100 m, *RILS 6875, 6877, 6878 & 6879* (AAS, KRAM); ridge north of Mt. Pond, 485–500 m, *RILS 11158B, 11164B & 11166* (AAS, KRAM); same locality, 500 m, *RILS 11155, 11156 & 11181* (AAS, KRAM) and *DM 40* (AAS, KRAM); east of Collins Pt., 15–25 m, *RILS 6861* (AAS, KRAM).

EAST ANTARCTICA. DRONNING MAUD LAND. ***Schirmacher Oasis***: eastern part, 60 m, *Singh 49 & 52* (KRAM).

5. **Bryum pseudotriquetrum** (Hedw.) P.Gaertn., B.Mey. & Scherb. Figs 218–219

Bryum pseudotriquetrum (Hedw.) P.Gaertn., B.Mey. & Scherb., Oekon. Fl. Wetterau 3: 102. 1802. ≡ *Mnium pseudotriquetrum* Hedw., Sp. Musc. Frond.: 190. 1801. ≡ *Hypnum pseudotriquetrum* (Hedw.) F.Weber & D.Mohr, Bot. Taschenb.: 288. 1807. ≡ *Bryum bimum* (Brid.) Turner subsp. *pseudotriquetrum* (Hedw.) Dixon, Stud. Handb. Brit. Mosses: 329. 1896, *nom. illeg. prior ut spec.* ≡ *Ptychostomum pseudotriquetrum* (Brid.) J.R.Spence *ex* Holyoak & N.Pedersen, J. Bryol. 29: 120. 2007 [*P. pseudotriquetrum* (Brid.) J.R.Spence, Phytologia 87: 23. 2005, *comb. inval. basion. non citat.*]. ≡ *Plagiobryum pseudotriquetrum* (Bruch & Schimp.) Pedersen *in* Pedersen & Hedenäs, Bryologist 108: 127. 2005. — Type citation: Locis apricis udis paludosis Europae; also, reference to pl. 7 in Hedwig (1792), pl. 51, f. 72 & 73 in Dillenius (1741) and pl. 24 f. 2 in Vaillant (1727) are given. [Lectotype (*vide* Demaret & Empain 1985: p. 280): 'Herb. Duby junct. Herb. Boissier 1886. Br. pstrq. Meesia longiseta Hostio Austr. 99. Simile Timmiae austriacae s differt folio supremis ovatis, l. ovatolanceolatis margine revolutis, quae integra videtur, siccitate conduplicatis, caulinis per id, quod margine valde revolvuntur lanceolatis, omnibis deest latiusculis basis. Habet aliquum similitudinem cum br. orbitale sed folia tortilia pedunculi flexuosi n rigidi distinguum. Sub nomine Meesiae longisetae misit Host (austriacus) AD 1799 ad Schwaegrichen vel ad Hedwigium. Bryum pseudotriquetrum Hedw. Paris, Index bryologicus, p. 201' – G-Hedwig/Schwägrichen!].

Webera gerlachei Cardot, Rev. Bryol. 27: 44. 1900. ≡ *Bryum gerlachei* (Cardot) Cardot, Voy. Belgica Mousses: 36. 1901. — Type citation: [Détroit du Gerlache] Ier, IXme, XIme et XVIIIme débarquements [later Cardot (1901) wrote: Détroit du Gerlache: Ier débarquement, île Auguste, à la surface de l'argile mèlée de guano. (N°s 215^b^, 215^c^; 26 janvier 1898). IXème débarquement, Terre de Danco, près du Cap Anna Osterrieth, sur les petites terrasses humides de la falaise. (N° 205^a^; 29 janvier 1898). XIème débarquement, Terre de Danco, cap Van Beneden, sur les corniches de la falaise. (N°s 229^a^; 1er février 1898). XVIIIème débarquement, île Banck, sur du sable quartzeux. (N° 264; février 1898). [Lectotype (*selected here*): 'Herb. J. Cardot. Voyage de la Belgica n° 264. Webera Gerlachei Card. sp. nova. Canal de Gerlache: XVIIIème débarquement, sur du sable quartzeux. Leg. Racovitza, 10 févr. 1898' – PC!; isolectotype: S!; syntypes: (1) 'Herb. J. Cardot. Voyage de la Belgica n° 205a. Webera Gerlachei Card. sp. nova. Canal de Gerlache: IXème débarquement, sur les petites terrasses humides de la falaise. Leg. Racovitza, 29 janv. 1898.' – PC!; (2) 'Herb. J. Cardot. Voyage de la Belgica n° 215b. Webera Gerlachei Card. sp. nova. Canal de Gerlache: Ier débarquement, à la surface de l'argile mèlée de guano. Leg. Racovitza, 29 janv. 1898.' – PC!; (3) same label data, No. 215c – PC!; (4) 'Herb. J. Cardot. Voyage de la Belgica n° 229a. Webera Gerlachei Card. sp. nova. Canal de Gerlache: XIème débarquement, sur les corniches de la falaise. Leg. Racovitza, 1 févr. 1898.' – PC!]. First synonymised by Ochi (1970: p. 44).

Bryum austropolare Cardot, Rev. Bryol. 27: 45. 1900 ['austro-polare']. — Type citation: [Détroit du Gerlache] IXme et XIme débarquements [later Cardot (1901) wrote: Détroit du Gerlache: IXème débarquement, Terre de Danco, près du Cap Anna Osterrieth, sur les roches et les petites terrasses humides de la falaise. (N°s 151^b^, 151^c^, 205^a^, 29 janvier 1898). XIème débarquement, Terre de Danco, cap Van Beneden, sur les corniches de la falaise. (N°s 233^b^, 234^b^; 1er février 1898). [Lectotype (*selected here*): 'Herb. J. Cardot. Mousses du voyage de la Belgica N° 233^b^ Bryum austro-polare Card. sp. nova. Canal de la Belgica: sur des corniches de la falaise, XIème débarquement. Leg. Racovitza, 1 févr. 98' – PC!; isolectotype: BR!; syntypes: (1) same label data, No. 234^b^ – BR!, PC!; (2) 'Herb. J. Cardot. Mousses du voyage de la Belgica N° 151^b^ Bryum austro-polare Card. sp. nova. Canal de la Belgica: IXème débarquement, sur les roches humides. Leg. Racovitza, 29 janv. 98' – PC!; isolectotype: BR!; (3) same label data, No. 151^c^ – BR!, PC!; (4): same label data, No. 205^a^ – BR!, PC!]. First synonymised by Ochi (1970: p. 44).

Bryum filicaule Broth. *in* Dryg., Deutsche Südpolar-Exped. 1901–1903, 8: 91, *f. 4.* 1906, ***syn. nov.*** — Type citation: [Antarctica] Gaussberg: an Felsen (Vanhöffen). [Lectotype (*designated here*) {or perhaps holotype}: 'Moos Bryum filicaule Broth. n. sp. vom Gaussberg 66°49′S in 89°39′O (Basalt!) der Kaiser Wilhelm II-Küste Deutsch Südpolar Expedition E. v. Drygalski 1902/03 leg. Prof. E. Vanhöffen, Kiel' – H-Brotherus!].

Webera racovitzae Cardot var. *laxiretis* Cardot, Rev. Bryol. 33: 34. 1906. ≡ *Pohlia racovitzae* (Cardot.) Broth. var. *laxiretis* (Cardot) Steere, Natl. Res. Council Publ. 839: 27. 1961, *comb. inval. basion. non citat.* — Type citation: Ile Wiencke. [Lectotype (*selected here*): 'Herb. J. Cardot. Webera Racovitzae Card. var. laxiretis Card. var. nova. Détroit de Gerlache: île Wiencke. 5/2 1905. Expéd. Charcot. n° 405^a^' – PC!; isolectotypes: H-Brotherus!, PC!, S!]; syntypes: (1) Herb. Mus. Paris. Webera Racovitzae Card. var. laxirete Card. var. nova. Ile Wiencke. 7 février 1905. Exped. Charcot n° 401^c^ Det. J. Cardot' – PC!; (2) Herb. Mus. Paris. Webera Racovitzae Card. var. laxirete Card. var. nova (+ Hypnum uncinatum Hedw.) Expéd. Charcot (Terres australes) n° 463 Determ. J. Cardot' – PC! (3 specimens). First synonymised by Ochyra (1998*a*: p. 194).

Bryum algens Cardot, Nat. Antarct. Exped. 1901–04, Nat. Hist. 4: 5, *pl. 2, fig. 1–21.* 1907. — Type citation: [Antarctica, southern Victoria Land] Granite Harbour, Jan. 20, 1902, c. flor. islet in old ice, MacMurdo Sound, Dec. 1903; ster. [Lectotype (*vide* Ochyra 1998*a*: p. 196): 'Herb. J. Cardot. Bryum algens Card. sp.nova. Terre Victoria: Granite Harbour, l. S. 77°. Expéd. de la Discovery, 20/1 1902' – PC!: isotypes: PC! (4 specimens); syntype: 'Herb. J. Cardot. Bryum algens Card. sp.nova. f. rabougrie. Antarctide: from one of islets in old ice, MacMurdo strait, leg. Dr Wilson. Exped. de la Discovery, Dec. 1903' – PC! (2 specimens). First synonymised by Ochyra (1998*a*: p. 194).

Bryum perangustidens Cardot, Compt. Rend. Hebd. Séances Acad. Sci. 153: 603. 1911. — Type citation: [Antarctica] Ile Petermann and Terre de Graham: Baie Marguerite, île Jenny; later, Cardot (1913) wrote: Ile Petermann, formant de vastez gazons sur de la terre humide, 5 janvier 1909 (n° 176); baie Marguerite: île Jenny, sur la terre humide, entre les cailloux, dans les éboulis, 15 et 30 janvier 1909 (n°s 218, 219, 231^b^, 243, 245^a^, 246^a^, 248^b^, 255, 256, 260^b^). Fert. [Lectotype (*selected here*): 'Herb. Mus. Paris. 176. Bryum perangustidens Card. sp. nova. Ile Petermann, sur le terre humide. 2ème expéd. Charcot. Leg. Gain, 5/1 1909' – PC-Cardot! syntypes: all syntypes: *Gain 218, 219, 231^b^, 243, 245^a^, 246^a^, 248^b^, 255, 256 & 260^b^* – PC! belong within *Bryum archgangelicum*]. First synonymised by Ochi (1979: p. 77).

Bryum korotkevicziae L.I.Savicz & Smirnova, Inform. Bjull. Sovetsk. Antark. Eksped. 7: 34, *f. 1–3.* 1959. — Type citation: Antarctida orientalis. Lacus 'Figurnoje' in oasi Bangeri 4 I 1957. In aqua, 33–36 m alt., in lacus fundo saxoso, detritu tecto. [Holotype: 'Sectio Cryptogamica Instituti Botanici Academiae Sc. URSS Bryum Korotkevicziae L. Savicz et Z. Smirn. Antarctida orientalis. Lacus 'Figurnoje' in oasis 'Bunger Hills'. In lacus, 33–36 m alt., fundo saxoso, detritu tecto. 4.I.1957. Leg. V. S. Korotkevicz Det. L. Savicz et Z. Smirnova' – LE!; isotype: KRAM!]. First synonymised by Ochyra (1998*a*: p. 194).

B. korotkevicziae var. *hollerbachii* L.I.Savicz & Smirnova, Inform. Bjull. Sovetsk. Antark. Eksped. 17: 25, *f. 1.* 1960. — TYPE CITATION: [Antarctica] 'Oasis minor' (in 15 km ad boreo-occidentem ab Bangeri Hills). In ripa orientali arenosa lacus haud profundi, ad glaciei liquaefactae marginem in aqua inventum est. In lacus fundo crescit et undis maximus pro parte ad ripam ejicitur. M. M. Hollerbach 10/II 1957 legit. [Holotype: 'Sectio Cryptogamica Instituti Botanici Academiae Sc. URSS № 8/529 Bryum Korotkevicziae L. Savicz et Z. Smirn. var. Hollerbachii L. Savicz et Z. Smirn. 'Malyy oasis', vybrosy na peschanom plazhe vostochnogo berega melkovodnogo ozera, u kromki lda, v taloy vode. Sobran 10 fevralya 1957 g. M. M. Gollerbakhom' (Little oasis (15 km to NW of Bunger Hills) thrown out ashore a non-deep lake and in the thawed water. Collected on 10 February 1957 by M. M. Hollerbach) – LE!; isotypes: AAS!, KRAM!]. First synonymised by Ochyra (1998*a*: p. 194).

B. inconnexum var. *fragile* Horik. & Ando, Hikobia 2: 166, *f. 3G–I.* 1961. — TYPE CITATION: Ongul Is.: East Ongul, 300 m west of Showa Base, on sandy place of dry valley (Feb. 1957, Leg. T. Kaji, A.B. No. 11 in HIRO). [Holotype: 'Herbarium of Hiroshima University, Hiroshima, Japan. Bryum inconnexum Card. var. fragile Horikawa et Ando type East Ongul, 30 m west of the Showa Base, on sandy place of dry valley (ca. 69° S, 39°30′ E), Feb. 2, 1957. Leg. T. Kaji (HIRO–A.B. No. 11)' – HIRO!; isotype: PC!]. First synonymised by Ochyra (1998*a*: p. 196).

B. ongulense Horik. & Ando, Hikobia 2: 166, *f. 3G–I.* 1961. — TYPE CITATION: Ongul Is.: East Ongul, 300 m west of Showa Base, on sandy place of dry valley (Feb. 1957, Leg. T. Kaji, A.B. No. 12 in HIRO). [Holotype: 'Herbarium of Horishima University, Hiroshima, Japan. Bryum ongulense Horikawa et Ando type East Ongul, 30 m west of the Showa Base, on sandy place of dry valley (ca. 69° S, 39°30′ E), Feb. 2, 1957. Leg. T. Kaji (HIRO–A.B. No. 12)' – HIRO!; isotype: AAS!]. First synonymised by Ochyra (1998*a*: p. 194).

B. confusum Ochyra *ex* Lewis Smith, J. Bryol. 27: 200. 2005, *nom. nud.*, ***syn. nov.*** — CHIROTYPE: [Antarctica] South Orkney Islands. Margin of melt water stream by marble outcrops. Alt. 50 ft. South of hut at Foca Point, Signy Island [60°42′ S, 45°40′ W] Leg. R. I. Lewis Smith 638A, 23–1–1966 – AAS!, KRAM!

Plants medium-sized to large, occasionally small, gracile and filiform to fairly robust, moderately stiff to pliant, in dense or loose, often very compact and tightly coherent tufts, turves or patches, sparsely radiculose to densely matted with a tomentum of rhizoids below, lustrous, green, yellow, yellow-, reddish- or brownish-green above, brown, blackish-brown, black or decolorate below. *Stems* erect or ascending from a decumbent base, (0.3–)1.0–7.5 or, occasionally in submerged plants, to 30 or more cm long, evenly foliated, sometimes julaceous, filiform and fragile to thick, simple or sparingly branched, in transverse section obscurely bluntly pentagonal, consisting of a large central strand, 3–5-stratose medulla of large, hyaline, thin-walled cells and 1–2(–3)-layered cortex of small cells with thick, red to red-brown walls; *rhizoids* long, branched, pale brown to reddish-brown, papillose; *axillary hairs* filiform, consisting of 5–7 cells, with 1–2 short, reddish to reddish-brown basal cells and 3–6 hyaline, elongate distal cells. *Leaves* evenly arranged along the stem or sometimes rather distant below and more crowded and somewhat larger above, occasionally distant in submerged plants, closely imbricate-appressed, erect to erecto-patent, shrunken and flexuose or weakly to not altered when dry, erect-spreading when moist, ovate- or oblong-lanceolate, broadly ovate to suborbicular, (0.8–)1.2–2.3(–2.8) mm long, (0.4–)0.6–1.0(–1.1) mm wide, narrowed at the insertion and shortly decurrent to non-decurrent, shortly acuminate, acute to broadly acute and often cuspidate, rarely broadly rounded, obtuse or apiculate at the apex, moderately to strongly concave; *margins* entire or weakly serrulate at the apex, narrowly to broadly recurved to varying distances or plane, distinctly bordered with 2–3 rows of elongate, unistratose cells to almost unbordered; *costa* robust to fairly weak, percurrent to short-excurrent, seldom ceasing well below the apex, red to reddish-brown below, 80–150 µm wide at the base, in transverse section prominent on the dorsal side, terete, consisting of a large central stereid band with a single layer of central and dorsal large epidermal cells and a row of large guide cells immediately below the ventral epidermis; *laminal cells* unistratose throughout, transparent, reddish at the base; *upper cells* hexagonal to rhomboidal-hexagonal to rhomboidal, thin- to firm-walled, sometimes porose, 30–60 µm long, 15–20(–25) µm wide, sturdy or sometimes lax, becoming narrowed and elongate, firm-walled at the margin and forming a distinct border or only somewhat longer than the inner cells and forming no or indistinct uniseriate border; *basal cells* short- to long-rectangular, oblong-hexagonal, rarely subquadrate, (30–)40–80(–100) µm long, 15–25 µm wide, thin-walled, often lax, red. *Dioicous* or *synoicous*. *Setae* erect to flexuose when dry, 5–20 mm long, pale brown, smooth. *Capsules* horizontal to pendent, clavate to cylindrical with a distinct neck, 1.5–2.5(–3.5) mm long, 0.6–1.0 mm wide, straight and symmetrical, pale brown, somewhat darker at the mouth; *annulus* compound, revoluble, 2–3-seriate, composed of large, vesiculose cells with strongly incrassate walls; *operculum* conic-convex, apiculate; *exothecial cells* variable in size and shape, isodiametric and irregular to oblong, 30–80 µm long, 20–40 µm wide, firm-walled, becoming smaller, quadrate to oblate in 2–3 rows at the orifice; *stomata* numerous in several rows in the neck, phaneroporous, bicellular, with elongate pori;

Figure 218. *Bryum pseudotriquetrum*. — **1**. Fertile plant, dry. **2.** Sterile plants, wet. **3**. Portion of sterile shoot, dry. **4.** Cross-section of the stem. **5–8.** Axillary hairs. **9–17**. Leaves. **18–20**. Leaf apices. **21–23**. Upper leaf cells at the margin. **24–25**. Mid-leaf cells. **26**. Basal cells. **27–30.** Cross-sections of the leaves. **31**. Deoperculate capsules, dry. **32–33**. Operculate capsules, wet. **34**. Lower exothecial cells and stoma. **35**. Exothecial cells in the middle of the urn. **36**. Exothecial cells at the orifice, portion of the peristome, annulus and spores. — [1, 31–33 from *RO 4889/79*; 17, 34–36 from *Gain 176* (lectotype of *B. perangustidens*), PC; 2–9, 18, 23–24, 26–30 from *RO 364/80*; 12 from *RO 2773/80*; 10, 22 from *ER 233B* (lectotype of *B. austropolare*), PC; 15 from *ER 264* (lectotype of *B. gerlachei*), PC; 11 from *Vanhöffen s.n.*, 1902/03 (type of *B. filicaule*), H; 13, 19, 25 from *Hollerbach 4/149*, LE; 14 from *Exped. Charcot*, 1905 (type of *Webera racovitzae* var. *laxiretis*), PC; 16, 20–21 from *Discovery Exped.*, 20/1 1902 (lectotype of *B. algens*), PC]. — Scale bars: a – 1 mm (9–17), 4 mm (2) and 200 µm (27–30); b – 2 mm (1, 3, 31–33) and 100 µm (5–8); c – 100 µm (4, 34–36); d – 100 µm (18–26).

a
b
c
d
1
2
3
4
5
6
7
8
9
10
11
12
13
14
15
16
17
18
19
20
21
22
23
24
25
26
27
28
29
30
31
32
33
34
35
36

peristome double; *exostome teeth* lanceolate, 350–500 µm long, pale yellow proximally, hyaline distally, papillose, bordered, consisting 15–20 transverse bars, strongly trabeculate on the inner surface; *endostome* hyaline to pale yellowish-hyaline, consisting of a basal membrane to ½ or less of the exostome length, giving rise to broad, perforate processes and 1–2 appendiculate cilia. *Spores* spherical, brownish-green, 11–16 µm in diameter, smooth or faintly papillose. *Calyptra* cucullate.

Chromosome numbers: n = 10 – Ongul I., Enderby Land (Inoue 1976 as *Bryum argenteum*), n = 20 – South Georgia (Newton 1980 as *B. algens*) and Ongul I., Enderby Land (Tatuno 1963 as *B. inconnexum*; the same chromosome number and n = 11 have been found in the Holarctic (Fritsch 1991).

Discussion. *Bryum pseudotriquetrum* is characterised by having shortly pointed, broad and long-decurrent leaves, short-excurrent costa, short distal laminal cells and reddish leaf base. Yet, in the extreme environmental conditions in Antarctica, the species exhibits extensive morphological plasticity which led to considerable taxonomic confusion. It is therefore hardly surprising that various phenotypes of this protean species have been recognised as distinct taxa, not only in the early days of Antarctic exploration, but even in the 1960s. Jules Cardot, who monopolised the examination of almost all moss collections made by the early Antarctic expeditions, described several taxa which clearly belonged within the *B. pseudotriquetrum* complex. These are: *Webera gerlachei* and *Bryum austropolare* (Cardot 1900, 1901), *Webera racovitzae* var. *laxiretis* (Cardot 1906*b*, 1907*a*), *Bryum algens* (Cardot 1907*b*) and *B. perangustidens* (Cardot 1911*a*,*b*, 1913). In addition, Brotherus (1906*a*) described another new species, *B. filicaule*, from Wilhelm II Land in continental Antarctica. Hugh N. Dixon (in Dixon & Watts 1918) was the first to express the opinion that these taxa were hardly distinct from one another and represented a single species which he then called, incorrectly, *Bryum antarcticum*, because this species is actually conspecific with *Hennediella heimii*. Following subsequent taxonomic investigations of Antarctic species of *Bryum*, Horikawa & Ando (1961) reduced *B. austropolare*, *B. inconnexum* var. *tomentosum* and *Webera racovitzae* var. *laxiretis* to synonymy with *B. inconnexum*, another species described by Cardot (1900, 1901) from the northern maritime Antarctic, but this conclusion is not supported by examination of the relevant type collections. At the same time, two new taxa were described, *B. inconnexum* var. *fragile* and *B. ongulense* (Horikawa & Ando 1961), but they were soon considered to be merely morphological variants of *B. inconnexum* (Horikawa & Ando 1967*a*). A very important study of the variation of plants then designated as *B. inconnexum* was conducted by Nakanishi (1979) who, on the basis of accurate and precise observations, concluded that the leaves on innovations in this species are subject to considerable variation in size and shape as well as in the length and the stoutness of the costa, and these features appeared to be strictly correlated with water conditions in the habitat. These observations have subsequently been confirmed by Seppelt & Kanda (1986). It is very likely that the remarkable morphological variability in this species is a response of the plants to the wide-ranging environmental conditions that it tolerates. Not surprisingly, perhaps, *B. pseudotriquetrum* exhibits considerable genetic variation in Antarctica (Skotnicki *et al.* 1998).

Bryum pseudotriquetrum exhibits a very similar pattern of variability in the Northern Hemisphere where the plants having leaves that are rather obtuse than acute, more or less widely ovate rather than ovate-lanceolate, concave rather than almost flat, not or only shortly decurrent rather than markedly decurrent, and with plane rather than recurved margins were given taxonomic recognition as *B. neodamense* Itzigs., *B. ovatum* Jur. and *B. subneodamense* Kindb. in various parts of its wide geographical range in the Holarctic. Despite their striking morphological characters, these species are considered to be inconstant phenotypes of *B. pseudotriquetrum* which occur locally in calcareous habitats that are subject to periodic inundation. This conclusion based upon assessment of morphological traits is also confirmed by phylogenetic analyses of DNA sequences in various phenotypes of this species (Holyoak & Hedenäs 2006). Also, an analysis of the flavonoid composition of *B. pseudotriquetrum* (as *B. algens*) by Webby *et al.* (1996) demonstrated a number of chemotypes in this species. Fresh and herbarium specimens were sampled from habitats ranging from dry fellfield to wet acid and calcareous flushes, and from sites on Signy I., James Ross I. and continental Antarctica. Such chemotypes may be expected in many other moss species with a wide ecological amplitude.

Ochi (1979) presented a detailed and thorough discussion of the Antarctic material of *Bryum* and concluded that all non-silvery plants of this genus represented the *bimum* phenotype of *B. pseudotriquetrum*. This taxonomic conclusion was followed by reduction of several Antarctic taxon names to synonymy with the latter name, including *B. inconnexum*, *B. crateris*, *B. perangustidens* and *B. ongulense*, although in most cases he examined only very fragmentary and incomplete original material of the taxa concerned. For example, Cardot (1911*a*,*b*, 1913) based the description of *B. perangustidens* on the basis of eleven specimens collected on two remote islands in two different regions of the western Antarctic Peninsula, namely on Petermann I. on the Graham Coast (one specimen) and Jenny I. on the Fallierères Coast in Marguerite Bay (ten specimens). On the basis of examination of the single specimen from Petermann I. (actually only a duplicate from BM which was incorrectly designated isotype), *B. perangustidens* was reduced to synonymy with *B. pseudotriquetrum*. Having examined all the syntypes of *B. perangustidens* we found that all specimens from Jenny I., which are in fine fruiting condition, in fact represent *B. archangelicum*, whereas the single specimen from Petermann I., also with sporophytes, truly belonged within *B. pseudotriquetrum*. In order to preserve the current taxonomic status of *B. perangustidens*, i.e. its conspecificity with *B. pseudotriquetrum* as suggested by Ochi (1979), the material from Petermann I. was selected as lectotype.

Although the concept of *Bryum pseudotriquetrum* proposed by Ochi (1979) for Antarctica might have appeared attractive, especially for non-bryologists, lumping all non-silvery plants of *Bryum* with this species was certainly an unrealistic simplification since some populations, which additionally were collected in fine fruiting condition, appeared to represent different species, such as *Bryum amblyodon* (now *B. archangelicum*) (Ochyra

→

Figure 219. *Bryum pseudotriquetrum*. — **1–3**. Habit, wet. **4–34**. Leaves. **35**. Leaf apex. **36**. Mid-leaf cells at margin. **37**. Basal cells. **38**. Cross-section of stem. **39–44**. Cross-sections of leaves. — [1, 16–20, 43–44 from *Li 40*; 2, 4–12, 35–36, 38, 40–42 from *RO 2456/80*; 3, 28–30, 37, 39 from *Li 7*; 13–15 from *Hollerbach 1/33* (type of *B. korotkevicziae*); 21–27 from *Vanhöffen s.n.*, 1902/03 (type of *B. filicaule*), H; 31–34 from *RILS 638A*]. — Scale bars: a – 1 mm (1–3) and 100 µm (35–37); b – 100 µm (39–44); c – 100 µm (38); d – 1 mm (4–34).

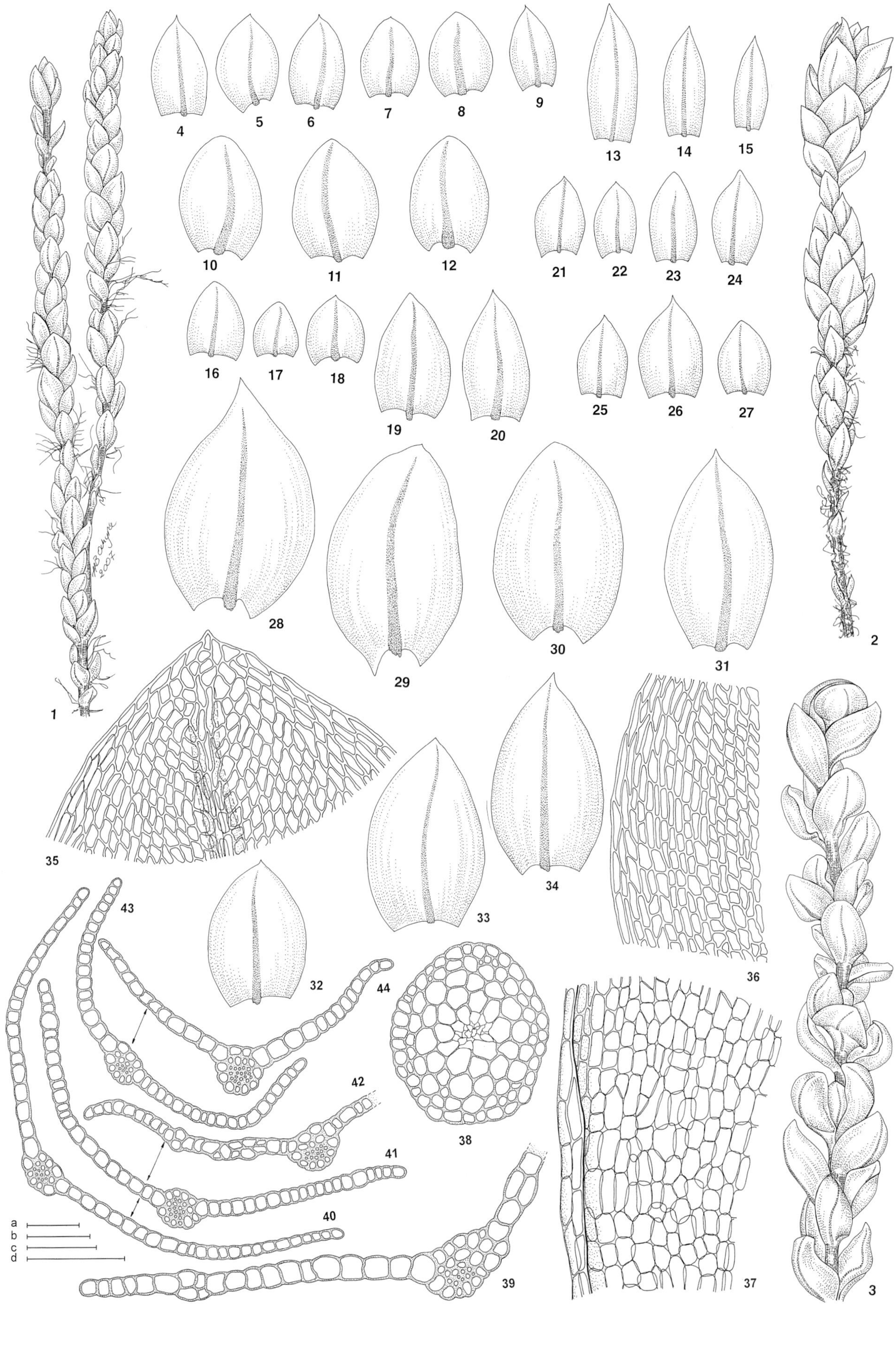
1
2
3
4
5
6
7
8
9
10
11
12
13
14
15
16
17
18
19
20
21
22
23
24
25
26
27
28
29
30
31
32
33
34
35
36
37
38
39
40
41
42
43
44
a
b
c
d

& Ochi 1986; Kanda & Ochi 1986) and *B. pallescens* (Ochi & Kanda 1991). Examination of over 900 specimens of the *B. pseudotriquetrum* complex from Antarctica, including all relevant type collections, led to the conclusion that the most sensible and realistic solution is to consider as *B. pseudotriquetrum* only those populations which have short-acuminate leaves, at least on the older parts of the stem. Sometimes the leaves on the uppermost parts of innovations are markedly modified and dissimilar to the older leaves in the lower part of the stem and such plants may imitate other species. This is best examplified by material which was originally reported from King George I. as *B. orbiculatifolium* (*Ochyra 2456/80*, KRAM) (Ochyra & Ochi 1986). The uppermost leaves of these plants are suborbicular, cochleariform, with obtuse or shortly apiculate apex and almost identical to those in *B. orbiculatifolium*, except for being somewhat larger, 1.0–1.3 mm long (Fig. 251.4–12). However, they have distinctly red basal cells, whereas the true *B. orbiculatifolium* has a green leaf base, and the older leaves are short-acuminate with a broadly acute apex. In some populations the upper leaves on the innovations are similar in shape but markedly larger, 1.6–2.1 mm long (e.g. *Li 7*, AAS, KRAM). Similar phenotypes of *B. pseudotriquetrum* were described from the continental Antarctic as *B. filicaule* and are characterised by having filiform shoots with imbricate leaves that are ovate and short-acuminate with a rounded to shortly apiculate apex (Fig. 251.16–20). They can sometimes be mistaken for *B. argenteum* and they are discussed under that species (p. 456).

In numerous populations of *Bryum pseudotriquetrum*, especially from continental Antarctica, the uppermost leaves on the innovations are distinctly long-acuminate and at first glance may appear to represent either *B. archangelicum* or *B. pallescens*. However, the lower leaves on the older parts of the stems are short-acuminate with percurrent to short-excurrent costae, typical of *B. pseudotriquetrum*. In contrast, in *B. archangelicum* and *B. pallescens* the lower leaves are distinctly long-acuminate and the costa is long-excurrent. Such plants were described as separate species, *B. algens* and *B. ongulense*, and possibly also the specimen reported from Enderby Land as *B. amblyodon* may belong here (Kanda & Ochi 1986), but the material was not available for examination. However, the peristome in this plant is described and illustrated as having reduced or no cilia, a feature typical of *B. archangelicum*, but in most Antarctic fruiting populations of *B. pseudotriquetrum* the peristomes are variously reduced and malformed. For example, this is clearly visible on the illustrations (and SEM micrographs submitted by R. D. Seppelt) of the specimen reported as *B. algens* from Kemp Land (Filson & Willis 1975).

Some populations of *Bryum pseudotriquetrum* in the Antarctic grow submerged in lakes, sometimes at considerable depth. The shoots of such plants are etiolated and their leaves are much smaller and distantly arranged on the stems. These features are a direct response to the restricted light availability in the deep-water habitats, one of the main factors controlling the occurrence of aquatic mosses (Winton & Beever 2004). The availability of light in Antarctic lakes is generally strongly dependent on the season, ice and snow cover, and water chemistry. Such strongly modified plants were described in Antarctica as separate taxa, *Bryum korotkevicziae* (Savicz-Lyubitskaya & Smirnova 1959) and *B. korotkevicziae* var. *hollerbachii* (Savicz-Lyubitskaya & Smirnova 1960). The status of these taxa was discussed in detail by Seppelt (1983) and they were considered to be identical to *B. algens*, and therefore to *B. pseudotriquetrum* because the latter name was reduced to synonymy with *B. pseudotriquetrum* (Ochyra 1998*a*).

Reproduction in Antarctica. Fertile plants of *Bryum pseudotriquetrum* occur quite often throughout the maritime and continental Antarctic, but mostly these are either male or female plants with antheridia and archegonia, respectively. Fruiting plants are fairly rare because the severe climatic conditions usually restrict or prevent development of mature sporophytes and viable spores. Even if mature sporophytes are produced, the majority are aborted due to frost or windblown soil abrasion (Rastorfer 1971), although, in sites with a favourable microclimate, *B. pseudotriquetrum* fruits prolifically. However, the species often propagates asexually by the juvenile shoot apices that may easily break off and serve as vegetative propagules which are capable of becoming established in moist soil (Kanda 1987*d*; Imura *et al.* 1993; Lewis Smith 1999; Wagner & Seppelt 2006). Furthermore, Kanda (1987*d*) reported rhizoidal tubers in aquatic plants of *B. pseudotriquetrum*, but this material was unavailable for study. Such tubers have never been observed in this species and none have been found in other Antarctic populations.

Habitat. As commented by Savicz-Lyubitskaya & Smirnova (1972) and confirmed by subsequent collections and field observations, *Bryum pseudotriquetrum* is probably the most ubiquitous moss species, at least in continental Antarctica. It grows, often in abundance, in a wide range of habitats, both wet and xeric, exposed and sheltered, in bird perching or nesting areas enriched with nitrogen and in sites not affected by animals, cold or heated ground on volcanic islands, including the margins of pools and lakes, seepage areas where its tufts coalesce and form bright green slightly undulating turves covering many square metres, snow-bed depressions, margins of melt-water channels and runnels, springs and low lying flushed areas. It is also common on stony and gravelly ground, rocky knolls, cliffs, rock ledges, outcrops and scree, bare patterned ground, sandy, rocky, silty and clayey soil as well as silt, debris and detritus between rocks and stones, rock crevices, and, occasionally, mortar of old buildings and other man-made structures. In volcanically active areas it is often associated with fumaroles and grows on coarse scoria, lapilli and volcanic ash, often being repeatedly buried in it, and also along stream banks. Occasionally, *B. pseudotriquetrum* grows in freshwater lakes (Lewis Smith 2000), sometimes at considerable depth, for example at 81 m in Radok Lake in the Amery Oasis in Mac.Robertson Land in continental Antarctica (Wagner & Seppelt 2006). It has been found in such habitats in several sites in Antarctica, including Figurnoye Lake in the Bunger Oasis in Quen Mary Land, whence it was described as a separate species, *B. korotkevicziae* (Savicz-Lyubitskaya & Smirnova 1959, 1960), Sôya Coast (Kanda & Ohtani 1991), Ablation Lake on Alexander I. and Lake Vanda in southern Victoria Land (Kaspar *et al.* 1982).

In the maritime Antarctic *Bryum pseudotriquetrum* is a constant component of communities of the moss hummock subformation which occur chiefly in situations with a permanent supply of moving water, usually enriched by cations derived from nearby rocks and soil (Lewis Smith 1972), and may develop a shallow peat beneath the living moss. Apart from *Brachythecium austrosalebrosum*, *Sanionia georgicouncinata* and *Syntrichia filaris*, *B. pseudotriquetrum* is the principal component of these communities, and *Warnstorfia sarmentosa*, *W. fontinaliopsis* and *Cephaloziella varians* usually occur as small admixtures. In the most favourable habitats *B. pseudotriquetrum* may form turves about 7–8 cm thick, with living leaves in the uppermost 0.5–1.5 cm. Moreover, the species is a frequent constituent of the Antarctic herb tundra formation which is dominated by the only two species of vascular plants native to this biome, *Deschampsia antarctica* and *Colobanthus quitensis*, and occupies well drained, sheltered habitats, usually on sloping ground at low elevations in coastal areas. It is also a frequent associate in various fellfield communities of the fruticose lichen and cushion moss subformation (Lewis Smith 1972, 1993*c*, 1996 as *B. algens*).

In continental Antarctica, with its extremely harsh environmental conditions of low temperature, short growing season, dry soil and strong wind, *Bryum pseudotriquetrum* often occurs sparsely as very short xeric tufts on dry gravelly soil, often largely covered by a thin film of fine soil. New growth of the moss is quickly buried and in such habitats it spends much of its existence subpsammically, with few leaves and very little stem being produced, but with an extensive rhizoidal system attaining up to 20 mm in length. This plays an important role in stabilising the soil in which the plants are established, and also in retaining moisture. In such habitats *B. pseudotriquetrum* usually coexists with *Ceratodon purpureus*, *Coscinodon lawianus* and *Hennediella heimii* forming communities classified in the short moss turf and cushion subformation (Longton 1973; Lewis Smith 1990; Okitsu *et al.* 2003, 2004). Additionally, the species also occurs in moist soil and seepage areas where it forms large closed stands of several square metres in which it is the only moss or forms more extensive mixed turves with *Schistidium antarctici*, *Ceratodon purpureus*, *Cephaloziella varians* and cyanobacteria. In such habitats its colonies are composed of densely packed shoots that are bound together by a dense tomentum of rhizoids (Matsuda 1968; Nakanishi 1977; Seppelt & Ashton 1978; Nakatsubo & Ohtani 1992; Imura *et al.* 1994; Lewis Smith 1999).

World range. Although *Bryum pseudotriquetrum* is generally considered to be a subcosmopolitan or cosmopolitan species, it is actually a bipolar moss with intermediate stations at high elevations in the Neotropics and tropical Africa (Fig. 220). In the Northern Hemisphere it has a continuous pan-Holarctic range, reaching its southern limits in California, Arizona, east Texas and Florida in North America (Crum & Anderson 1981) and in the Mediterranean region in Europe and North Africa, including Morocco, Algeria and Tunisia (Ros *et al.* 1999), Sinai (Ochi 1985), northern Iraq (Agnew & Vondraček 1975), central Iran (Ochi 1985), the Himalayas and central China (Zhang *et al.* 2007) and Japan (Ochi 1959), with some isolated stations in Yemen on the Arabian Peninsula (Kürschner & Ochyra 1999) and Taiwan (Kuo & Chiang 1987). In the Southern Hemisphere the species occurs on almost all subantarctic islands (except the Prince Edward Islands), in southern Australia and New Zealand (Ochi 1970), Tristan da Cunha (Ochi 1972), southern South America extending to Central Chile (Ochi 1982, 1990*a*) and extends to the Antarctic. In the tropics it is absent from Oceania and tropical Asia and Australia, and in the Palaeotropics it is scattered only in Africa, from Jebel Marra in Sudan (Townsend 1985) and Ethiopia to the Cape (O'Shea 2006), with an isolated station on Réunion (Müller 2000*b*) and Madagascar (Crosby *et al.* 1983 as *B. austroventricosum* Renauld). Likewise, it is widely distributed but generally scattered and infrequent in tropical America, including Mexico (Ochi 1994), Guatemala (Allen 2002) and the Dominican Republic (Ochi 1980), as well as the Andes where it ranges from Venezuela to Bolivia (Ochi 1980) and northern Argentina (Matteri 2003), with a few isolated stations in south-eastern Brazil and Buenos Aires Province in Argentina (Ochi 1980).

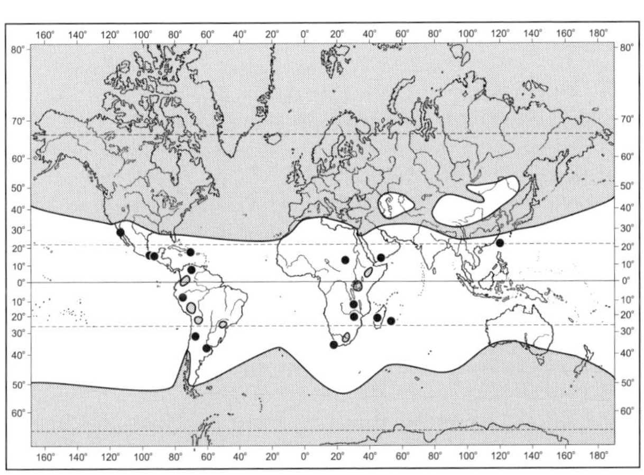

Figure 220. Global distribution of *Bryum pseudotriquetrum*.

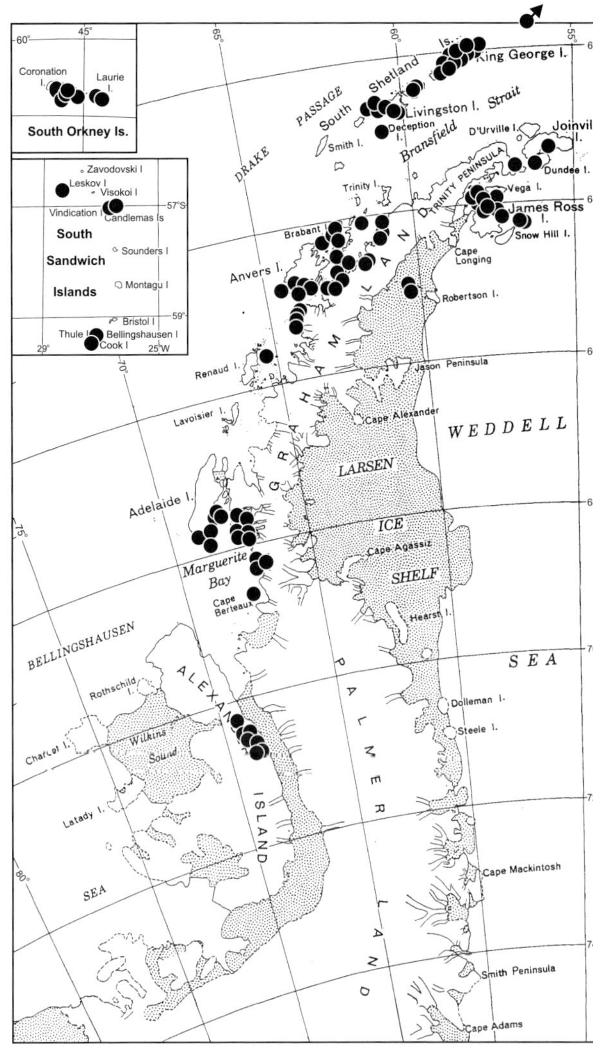

Figure 221. Distribution of *Bryum pseudotriquetrum* in the Antarctic Peninsula region. Locality on Clarence I. in the northern group of the South Shetland Is. is indicated by an arrow.

Distribution in Antarctica. *Bryum pseudotriquetrum* is one of the most widespread and common species of moss, both in terms of frequency and cover, in Antarctica (Figs 221–222). It is fairly frequent on Bouvetøya and in some islands of the South Sandwich Is., common throughout the South Orkney Is. and South Shetland Is. archipelagoes, and along the Antarctic Peninsula, although somewhat unevenly in the latter region (Fig. 221). On its western coast it appears to be absent from Trinity Peninsula

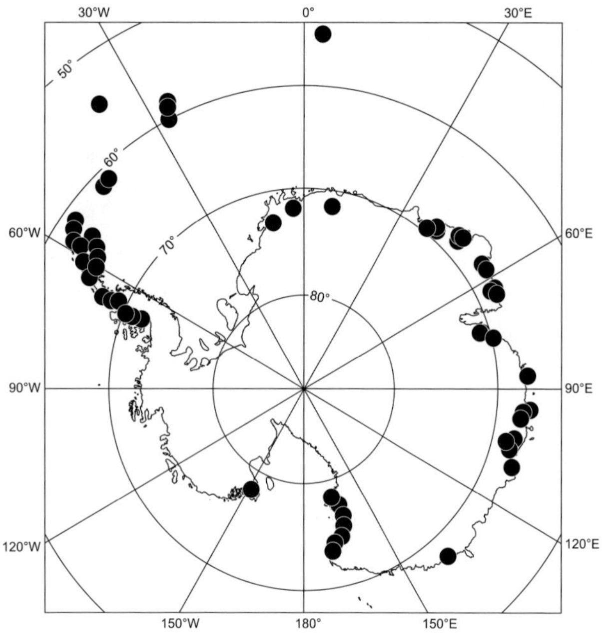

Figure 222. Distribution of *Bryum pseudotriquetrum* in Antarctica.

and scarce on the Davis Coast, but it is common on the Danco Coast and some islands of the adjoining Palmer Archipelago and becoming less frequent on the Graham Coast and absent from the Loubet Coast in the central part of the Antarctic Peninsula. However, it recurs in southern Adelaide I., Marguerite Bay and on the Fallières Coast, extending southwards to the Tierra Firma Is. A quite disjunct and isolated centre of its distribution occurs on the eastern coast of Alexander I., from Ablation Pt. to Coal Nunatak, where, at lat. 72°01′S, it has its southernmost locality in West Antarctica. On the eastern coast of the Antarctic Peninsula, the species is very rare on the mainland, being known only from Tabarin Peninsula, the northernmost part of the Antarctic Peninsula, and in the Joinville Island group, but much more frequent in the James Ross Island group, with some isolated stations on Seal Nunataks on the Nordenskjöld Coast at lat. 65°10′S. In continental Antarctica *B. pseudotriquetrum* has a wide range, though dissected owing to the discontinuity of ice-free areas (Fig. 222). It extends from western Dronning Maud Land to Wilkes Land and then recurs in Victoria Land, where it is very widespread and common, extending as far south at lat. 77°32′S in Lake Vanda (Kasper *et al.* 1982) and in the Canada Glacier area at lat. 77°37′S in (Schwarz *et al.* 1992). Some solitary records of the species are known from the eastern coast of the Ross Sea on Edward VII Peninsula and from Adélie Land, where it is one of only two species of moss recorded in this part of the continent (by the Australian Antarctic Expedition 1911–1914; Seppelt & Ochyra, in preparation).

Selected specimens examined. WEST ANTARCTICA. BOUVETØYA. Nyrøysa, *TE 207, 236, 263, 321, 345 & 365* (BG); same locality, 10 m, *TE 72* (BG) and 20 m, *TE 71 & 95.2* (BG).

SOUTH SANDWICH IS. **Leskov I.** Northern end, *ca*, 120 m, *REL 519* (AAS, KRAM). **Cadlemas I.** Sea Serpent Cove, *KA 7* (AAS, KRAM); Demon Pt., *PC 122B & 123B* (AAS, KRAM); Medusa Pool, *PC 109* (AAS, KRAM). **Vindication I.** *ca* 100 m, *REL 718* (AAS, KRAM). **Bellingshausen I.** 20 m, *PC 75B* (AAS, KRAM). **Cook I.** Resolution Pt., 50 m, *PC 19B* (AAS, KRAM).

SOUTH ORKNEY IS. **Coronation I.** Meier Pt., 15–25 m, *RILS 574 & 578* (AAS, KRAM); Cape Vik, *ca* 100 m, *RILS 572* (AAS, KRAM); Shingle Cove, 50 m, *RILS 493B* (AAS, KRAM); Olivine Pt., *ca* 35 m, *RILS 148* (AAS, KRAM); Saunders Pt., *ca* 15 m, *RILS 19B & 29B* (AAS, KRAM). **Lynch I.** 8 m, *RILS 699* (AAS, KRAM). **Matthews I.** Near The Divide, 35 m, *RILS 164* (AAS, KRAM). **Laurie I.** Scotia Bay, Cabo Moneta, *Bellisio 11757* (KRAM, NY). **Signy I.** Thula Pt., 35 m, *RILS 637* (AAS, KRAM); Moss Lake, 55 m, *RILS 10761 & 10762* (AAS, KRAM); Elephant Flats, 10 m, *RILS 327* (AAS, KRAM); Limestone Valley, *RILS 10900* (AAS, KRAM); below Orwell Glacier, *ca* 35 m, *RILS 634* (AAS, KRAM); north end of Moraine Valley, 100 m, *RILS 635* (AAS, KRAM); Knife Pt., *ca* 3 m, *RILS 691B* (AAS, KRAM); Gourlay Peninsula, 15 m, *RILS 469* (AAS, KRAM). **Moe I.** Landing Cove, 20–50 m, *RILS 5234* (AAS, KRAM).

SOUTH SHETLAND IS. **Clarence I.** 90 m, *Chuter 36* (AAS, KRAM). **King George I.** ***Bransfield Strait***: Red Hill, 80 m, *RO 1120/80* (KRAM); Blue Dyke, 5 m, *RO 1208/80* (KRAM). ***Admiralty Bay***: Uchatka Pt., 10 m, *RO 1329/80* (KRAM); Demay Pt., *VK 383, 410, 417, 424, 429, 448 & 457* (KRAM); Creeping Slopes, 100 m, *RO 1256/80* (KRAM); Zamek, 330 m, *RO 26/80* (KRAM); Sphinx Hill, 130 m, *RO 197/80* (KRAM); Penguin Ridge, 25 m, *RO 1405/80* (KRAM); Hala, 35 m, *RO 5221/79* (KRAM); Geographers Creek, 90 m, *RO 1614/80* (KRAM). ***Ezcurra Inlet***: Kasprowy Hill, 200 m, *RO 68/80* (KRAM); Italia Valley, 100 m, *RO 120/80* (KRAM); Breccia Crag, 140 m, *RO 917/80* (KRAM); Dufayel Island, Gdynia Pt., 8 m, *RO 1759/80* (KRAM). ***MacKellar Inlet***: Crépin Pt., 15 m, *RO 2045/80* (KRAM). ***Keller Peninsula***: Speil Pt., 4 m, *RO 545/80* (KRAM); British Pt., 3 m, *RO 457/80 & 459/80* (KRAM). ***Martel Inlet***: Stenhouse Bluff, 4 m, *RO 2617/80* (KRAM); Ullman Spur, 70 m, *RO 563/80* (KRAM); Mt. Wawel, 130 m, *RO 2181/80* (KRAM). ***Viéville Glacier***: Rembiszewski Nunataks, 200 m, *RO 2742/80* (KRAM); Vauréal Peak, 50 m, *RO 5255/79* (KRAM). ***Legru Bay***: Low Head, *BJ 84* (KRAM). ***King George Bay***: Lions Rump, *BJ 9* (KRAM); Turret Pt., *BJ 136 & 140* (KRAM); Penguin I., *BJ 61* (KRAM). ***Sherrat Bay***: Three Sisters Pt., *BJ 125* (KRAM). ***Destruction Bay***: Cape Melville, *BJ 79* (KRAM). ***Drake Passage***: Pyrites I., *BJ 33* (KRAM); False Round, *BJ 75* (KRAM); Pottinger Pt., 10 m, *BJ 30* (KRAM); Stigant Pt., *BJ 15* (KRAM); Tartar Pt., *BJ 5* (KRAM); Davey Pt., *BJ 50* (KRAM). ***Fildes Peninsula***: Bellinghausen Station, 15 m, *RO 2421/80* (KRAM); Lake Kitezh, 15 m, *RO 2469/80* (KRAM); Suffield Pt., 30 m, *RO 2435/80 & 2439/80* (KRAM); Green Pt., *BJ 19* (KRAM). ***Marian Cove***. North Spit, *BJ 145* (KRAM). ***Barton Peninsula***: Winship Point, *BJ 191* (KRAM). **Nelson I.** Moss Bay, 20 Jan 1989, *Komárek s.n.* (KRAM). **Greenwich I.** English Sound, 8 m, *Frödin 2* (KRAM, S); Discovery Bay, *RMS 69-984* (US). **Aitcho I.** 3–80 m, *RILS 11457* (AAS, KRAM). **Livingston I.** ***Byers Peninsula***: Sayer Nunatak, 30 m, *RILS 5886B* (AAS, KRAM); Pt. Smellie, *KR 38G* (AAS, KRAM); Devils Pt., 20 m, *KR 24H* (AAS, KRAM) and 100 m, *KR 23D* (AAS, KRAM); South Beaches, Vietor Rock, 8 m, *DCL 461* (AAS, KRAM); Robbery Beach, 5 m, *KR 51C* (AAS, KRAM); Lair Pt., 5 m, *KR 65C* (AAS, KRAM). ***Hurd Peninsula***: Miers Bluff, *RMS 69-808 & 69-825* (US); False Bay, *RMS 69-419 & 69-422* (US). **Deception I.** Macaroni Pt., 80 m, *DM 129* (AAS, KRAM); Bailey Head, 45 m, *Linskens 5* (AAS, KRAM); Cathedral Crags, 15 m, *NJC 133B* (AAS, KRAM), 50 m, *RILS 10406* (AAS, KRAM) and 65 m, *RILS 71* (AAS, KRAM); Whalers Bay, 2 m, *RILS 11233* (AAS, KRAM), 8 m, *RILS 5791A* (AAS, KRAM) and 75 m, *DM 94B & 95* (AAS, KRAM); Ronald Hill, 25 m, *RILS 5778C* (AAS, KRAM); Mt. Pond, 50 m, *RILS 5772B* (AAS, KRAM); Pendulum Cove, 5 m, *RILS 11189* (AAS, KRAM); Telefon Bay, 3 m, *DM 9* (AAS, KRAM); Telefon Ridge, 60–180 m, *RILS 5796A* (AAS, KRAM); Kendall Terrace, 80 m, *RILS 5741B* (AAS, KRAM); Wensleydale Beacon, 30–100 m, *RILS 6668* (AAS, KRAM); Stonethrow Ridge, 105–240 m, *RILS 5802C* (AAS, KRAM); Fumarole Bay, 100 m, *RILS 11504* (AAS, KRAM); Crater Lake, 110 m, *RILS 5729A* (AAS, KRAM); Vapour Col, 130 m, *RILS 5634E* (AAS, KRAM); Collins Pt., 25 m, *RILS 11335B & 11340* (AAS, KRAM) and 20–130 m, *RILS 6698* (AAS, KRAM); Entrance Pt., 50–70 m, *RILS 6792B & 6797* (AAS, KRAM).

WEST ANTARCTIC PENINSULA. **Trinity Peninsula.** Astrolabe I., *RILS 6975* (AAS, KRAM). **Davis Coast.** Cape Herschel, 20 m, *RILS 11642B* (AAS, KRAM). **Danco Coast.** Cierva Pt., *VK 665, 666A & 708* (KRAM); Spring Pt., *VK 30, 901, 915, 1178, 1274 & 1278* (KRAM); Challenger I., *CS 452* (BM, LD, PC, S); Charlotte Bay, 5–10 m, *RILS 3967B* (AAS, KRAM); Cape Anna, *ER 151B, 151C & 205A* (BR, PC – syntype of *Bryum austropolare*); Cuverville I., 25 m, *RILS 8809* (AAS, KRAM), 40 m, *RILS 8810* (AAS, KRAM) and 100 m, *Leeuw 15* (AAS, KRAM); Danco I., *RILS 7160* (AAS, KRAM); Mt. Banck, *ER 264* (PC, S – lectotype of *Webera gerlachei*); Birdsend Bluff, 5–35 m, *RILS 4219B* (AAS, KRAM); Beneden Head, *ER 233B & 234B* (BR, PC – lectotype of *Bryum austropolare*); Andvord Bay, Neko Harbour, 15–20 m, *RILS 11932 & 11934* (AAS,

KRAM); Paradise Harbour, Coughtrey Peninsula, Almirante Brown Station, *VK 334, 376 & 1116* (KRAM). **PALMER ARCHIPELAGO. *Christiania Is.***: Intercurrence I., 7 m, *DWB 1B, 1D & 2A* (AAS, KRAM). ***Brabant I.***: Metchnikoff Pt., 100 m, *WH 18B* (AAS, KRAM); Claude Pt., 200 m, *WH 205A* (AAS, KRAM); Astrolabe Needle, *VK 393* (KRAM); Minot Pt., *VK 572* (KRAM); Chiriguano Bay, 3 m, *AM 34* (AAS, KRAM) and 5 m *AM 95* (AAS, KRAM). ***Melchior Is.***: Melchior Harbour, *IML 7962, 7971, 7973, 7974 & 7977* (KRAM, NY). ***Anvers I.***: Norsel Pt., *ca* 10 m, *RILS 2372* (AAS, KRAM) and *VK 203 & 981* (KRAM); Biscoe Pt., *VK 158, 257, 276 & 334* (KRAM); Copper Peak, 2–3 m, *RILS 1908* (AAS, KRAM); Joubin Is., *JRS & JMG PAL-21* (US); Dream I., *VK 688 & 1000* (KRAM); Hermit I., *ca* 13 m, *RILS 883* (AAS, KRAM); Stepping Stones, *VK 52 & 55* (KRAM). ***Wiencke I.***: Goudier I., Port Lockroy, *VK 109, 111 & 118* (KRAM) and *Gain 101 & 104* (PC). **GRAHAM COAST.** Cape Pérez, *Poncet 5D* (AAS, KRAM). **BISCOE IS.**: Renaud I., Armstrong Reef, 4 m, *RILS 5343A* (AAS, KRAM). **WILHELM ARCHIPELAGO. *Lemaire Channel***: Wandel I., *Turquet 12B, 230B, 244B & 671* (PC). ***Penola Strait***: Petermann I., *Gain 168A* (PC). ***Collins Bay***: Green I., *BGLE 1083C* (BM). **LOUBET COAST. *Laubeuf Fjord***: Piñero I., 10–25 m, *RILS 4826* (AAS, KRAM). **FALLIÈRES COAST. *Adelaide I.***: Rothera Pt., 8 m, *RILS 87A & 95 70A* (AAS, KRAM); Avian I., 20 m, *RILS 769A* (AAS, KRAM). ***Léonie Is.***: Lagoon I., *BGLE 1476* (BM); Léonie I., 30 m, *RILS 8956* (AAS, KRAM). ***Marguerite Bay***: Jenny I., 15–50 m, *RILS 4713B* (AAS, KRAM); Guébériant Is., 40 m, *DWB 7C* (AAS, KRAM); Dion Is., Emperor I., 30 m, *DWB 17C* (AAS, KRAM); Dion Is., Courtier Is., 5–25 m, *RILS 3586* (AAS, KRAM); Blaiklock I., 8–15 m, *RILS 842A* (AAS, KRAM); Pourquoi Pas I., 5–20 m, *RILS 4773* (AAS, KRAM); Lagotellerie I., 50–75 m, *RILS 1776* (AAS, KRAM); Horseshoe I., 10–15 m, *RILS 1894* (AAS, KRAM); Line Is., *ca* 5 m, *RILS 4603* (AAS, KRAM); Camp Pt., 10–20 m, *RILS 4682* (AAS, KRAM); Faure Is., Dismal I., 5–25 m, *RILS 3473 & 3530* (AAS, KRAM); Neny I., 15 m, *RILS 788* (AAS, KRAM); Millerand I., 5–15 m, *RILS 3762A*(AAS, KRAM); Debenham Is., *BGLE 1559* (BM); Roman Four Promontory, 10–30 m, *RILS 4567A* (AAS, KRAM); Red Rock Ridge, 5 m, *RILS 4529* (AAS, KRAM); Refuge Is., 10–25 m, *RILS 4524* (AAS, KRAM); Terra Firma Is., *BASM 127 & 217* (AAS, KRAM); Flyspot Rocks, *Poncet 2A* (AAS, KRAM). **GEORGE VI SOUND. *Alexander I.***: Ablation Lake, 2–3 m, *9316A* (AAS, KRAM); Ablation Valley, 75 m, *RILS 9293A* (AAS, KRAM); Ablation Peak ridge, 717 m, *RILS 11115* (AAS, KRAM); Ganymede Heights, 5 m, *RILS 11109* (AAS, KRAM); Moutonnée Valley, 200 m, *RILS 11120A* (AAS, KRAM); Striation Valley, *RILS 11118* (AAS, KRAM); Middle Flatiron Valley, *ca* 315 m, *RILS 11101 & 11106* (AAS, KRAM); Upper Flatiron Valley, *ca* 500 m, *RILS 11107* (AAS, KRAM); High Flatiron, 532 m, *RILS 11071 & 11090A* (AAS, KRAM); Horrocks Block, *RILS 8853B* (AAS, KRAM); Ares Oasis, 30 m, *RILS 10441* (AAS, KRAM); Natal Ridge, 650 m, *RILS 9485* (AAS, KRAM); Mars Oasis, 70–75 m, *RILS 9401 & 9410* (AAS, KRAM), 175 m, *RILS 9428A & 9449* (AAS, KRAM); Viking Valley, 300 m, *RILS 9468* (AAS, KRAM); Coal Nunatak, 100 m, *RILS 10444 & 10455* (AAS, KRAM).

EAST ANTARCTIC PENINSULA. **TRINITY PENINSULA.** Tabarin Peninsula, Brown Bluff, 6 m, *RILS 11929* (AAS, KRAM). **JOINVILLE ISLAND GROUP. *Joinville I.***: Mt. Alexander, 15–50 m, *RILS 3707* (AAS, KRAM); Dundee I., Welchness Pt., *RILS 5300 & 5301A* (AAS, KRAM). **JAMES ROSS ISLAND GROUP. *Vega I.***: False Island Pt., 25–40 m, *RILS 7825, 7829B, 7832B & 7839B* (AAS, KRAM). ***James Ross I.***: Cape Lachman, 30 m, *RILS 7374B* (AAS, KRAM); Abernethy Flats, 100 m, *RILS 7832A* (AAS, KRAM); Andreassen Pt., 10 m, *RILS 7610B* (AAS, KRAM); Fortress Hill, 150 m, *RILS 7774* (AAS, KRAM); St. Martha Cove, 15 m, *RILS 11137B* (AAS, KRAM); hill south of St. Martha Cove, 60 m, *RILS 7617* (AAS, KRAM); San José Pass, 165 m, *RILS 7523B & 7531A* (AAS, KRAM); Dreadnought Pt., 10 m, *RILS 7499* (AAS, KRAM); Terrapin Hill, 175 m, *RILS 7737B* (AAS, KRAM) and 250 m, *RILS 7716 & 7730* (AAS, KRAM); Blancmange Hill, 12–15 m, *RILS 7672 & 7688* (AAS, KRAM). ***Humps I.***: 2–3 m, *RILS 7397* (AAS, KRAM). ***Cockburn I.***: between foot of cliff and Adelie penguin colony, 120 m, *RILS 7923* (AAS, KRAM) and *ca* 400 m, *RILS 7967A* (AAS, KRAM). ***Seymour I.***: west of Cape Wiman, 10 m, *RILS 7890* (AAS, KRAM) and 20 m, *RILS 7875* (AAS, KRAM). **NORDENSKJÖLD COAST. *Seal Nunataks***: Oceania Nunatak, 100 m, *RILS 6431* (AAS, KRAM); Castor Nunatak, 150 m, *RILS 6432* (AAS, KRAM).

EAST ANTARCTICA. DRONNING MAUD LAND. **KRONPRINSESSE MÄRTHA KYST. *Vestfjella***: Audunfjell, 300 m, *Sømme 3* (AAS, KRAM). ***Ahlmannryggen***: Robertskollen, 'Ice Axe Peak', 480 m, *Watkins 21b & 22* (AAS, KRAM). **PRINSESSE ASTRID KYST.** Novolazarevskaya Station, *Meyer 1 & 5* (KRAM, NY).

ENDERBY LAND. **SÔYA COAST.** East Ongul I., *Matsuda 56* (HIRO); West Ongul I., 27 m, 26 Jan 1968, *Kashiwadani s.n.* (HIRO); Rundvågshetta, *Shimizu 751021025* (KRAM). ***Skarvsnes***: Torinosu Bay, *Shimizu 750127014* (KRAM, NY); Mt. Tenpyo, *Shimizu 751104194* (KRAM); Mt. Suribati, *Shimizu 751104215* (KRAM). ***Skallen***: Skallevikhalsen Hills, *Shimizu 751029136* (KRAM, NY). **PRINCE OLAV COAST. *Molodezhnaya Station***: Mt. Garnet, Jan 1968, *MacNamara n.n.* (US); Alasheeva Cove, 100 m *Konovalov 15* (KRAM, LE). ***Amundsen Bay***: unnamed island, *Korotkevich 11 & 50(4/192)* (KRAM, LE).

KEMP LAND. Øygarden Group, 0–10 m, *ADK 40* (AAS, KRAM); Sheehan Nunatak, 27 Feb 1936, *Scoresby a & b* (BM); Stillwell Hills, 30 m, *ADK Z4* (AAS, KRAM); Kemp Peak, 250 m, *ADK Z9* (AAS, KRAM).

MAC.ROBERTSON LAND. **MAWSON COAST.** Fold I., *RDS 9608* (AAS, KRAM); Davis Vestfold Hills, 1961, *Hay s.n.* (AAS, KRAM); Snow Ridge, Chapman Ridge, *Filson 4288* (AAS, KRAM); Forbes Glacier, *RDS 9603* (AAS, KRAM); Mawson Station, *Gwynn AB54* (AAS, KRAM).

PRINCESS ELIZABETH LAND. **INGRID CHRISTENSEN COAST. *Larsemann Hills***: without closer data, *RILS 10322, 10323, 10324, 10331, 1033210341, 10342 &10363* (AAS, KRAM). ***Vestfold Hills***: Chelnok Lake close to Sorsdal Glacier, 10 m, *RILS 6397* (AAS, KRAM); Taleg Gorge, 30 m, *RILS 6391A* (AAS, KRAM); Mossel Hut, 10 m, *RILS 6375* (AAS, KRAM); NE of Znezhnyy Bay, 45 m, 20 Sep 1994, *Gibson s.n.* (KRAM).

WILHELM II LAND. Gaussberg, 1902–1903, *Vanhöffen s.n.* (H – type of *Bryum filicaule*), *Korotkevich 430* (KRAM, LE), 23 Jan 1957, *Dolgyskin s.n.* (KRAM, LE), 1 Jan 1957, *Kuksin 65* (KRAM, LE), 20 Jan 1957, *Eveteev 6/203* (AAS, KRAM, LE).

QUEEN MARY LAND. **PRAVDA COAST.** Stroiteley Is., *Markov 17* (KRAM, LE); Haswell Is., *Meyer 15* (KRAM, NY), *Korotkevich 394 & 402* (KRAM, LE), *Hollerbach 1/33, 3/179 & 4/149* (KRAM, LE); Mirnyy Station, *Markov 1* (KRAM, LE); Chugunov I., *Korotkevich 74* (KRAM, LE). **KNOX COAST. *Bunger Oasis***: Lake Figurnoe, *Hollerbach 11/315 & 12/382* (KRAM, LE); Mt. Vialov, *Hollerbach 12/594* (KRAM, LE) and *Korotkevicz 3* (KRAM, LE); Polish 'Dobrowolski' Base, *Różycki 1 & 2* (KRAM); same locality, near the Polish 'Dobrowolski' Base by so-called 'Totalizator', *Różycki 3* (KRAM). ***Grierson Is.***: Ptichyy I., *Korotkevich 198* (KRAM, LE).

WILKES LAND. **BUDD COAST. *Windmill Is.***: Clark Peninsula, Whitney Pt., 25 m, *RILS 6198A* (AAS, KRAM); Browning Peninsula, 50 m, *RILS 6349 & 6361* (AAS, KRAM); Bailey Peninsula, *ca* 30 m, *RILS 6232, 6242 & 6254A* (AAS, KRAM) and 50 m, *RILS 6153* (AAS, KRAM); Odbert I., *RDS 14008* (KRAM); Peterson I., *RILS 6324 & 6332* (AAS, KRAM); Shirley I., *RILS 6307* (KRAM); Herring I., *RDS 14214* (KRAM).

TERRE ADÉLIE. Without details, 1912, *Bauer s.n.* (*ex herb. L. Rodway 386, W. A. Weymouth, Tasmanian mosses 2786*) (BM).

VICTORIA LAND. **BORCHGREVINK COAST.** Cape Hallett, Seabee Hook, 17 Dec 1963, *Schofield s.n* (US) and 50 m, *SWG 65/58B* (AAS, KRAM); Crater Cirque, 100 m, *RILS 9893 & 9900* (AAS, KRAM); Apostrophe I., 25 m, *RILS 9789* (AAS, KRAM); Cape King, 75 m, *RILS 10130* (AAS, KRAM) and 100–125 m, *RILS 10251* (AAS, KRAM); Edmondson Pt., *DWHW 201B* (AAS, KRAM), 5 m, *RILS 9838* (AAS, KRAM); North Valley, 25 m, *RILS 9648* (AAS, KRAM) and 45 m, *RILS 9631* (AAS, KRAM); Kay I., Wood Bay, *RILS 10191* (AAS, KRAM), 125 m, *RILS 10211* (AAS, KRAM) and 150 m, *RILS 9962 & 9974* (AAS, KRAM); Harrow Peaks, 75 m, *RILS 9864A* (AAS, KRAM) and 75 m, *RILS 9646* (AAS, KRAM); Mt. McGee, 680 m, *Bargagli 47D* (AAS, KRAM); 'Gondwana' Station, 70 m, *RILS 9726A* (AAS, KRAM) and 150 m, *RILS 9738* (AAS, KRAM); Prior I., *RILS 9953* (AAS, KRAM); Starr Nunatak, 150 m, *Bargagli 23* (AAS, KRAM). **SCOTT COAST. *Granite Harbour Area***: Kar Plateau, *Ugolini 43* (US); The Flatiron, *Ugolini 36a–b* (US); Granite Harbour, 20 Jan 1902. *Discovery Expedition s.n.* (PC – type of *Bryum algens*). ***McMurdo Sound Area***: Cape Roberts, *Pittard MCM-20* (US) and *Strandtman MCM-19* (US); Marble Pt., *Ugolini 33* (US) and *Fukushima 808* (KRAM); Alcott Glacier, Lake Penny, lat. 78°10′S, *Schofield AA-8* (NY); same area, Lake Teardrop, 29 Jan 1964, *Schofield s.n.* (US); Taylor Valley, Lake Fryxell, *REL 4072* (AAS, KRAM); Blue Glacier, 60 m, *Janetschek 17* (AAS, KRAM); Hobbs Glacier, 8 m, *WCS & SWG 65/8* (KRAM, LE) and 50–100 m, *REL 4082* (AAS, KRAM); Miers Valley, *JRR MCM-29, 44 & 54* (US). ***Ross I.***: Horseshoe Bay, *Strandtman MCM-35* (US); Cape Armitage, *REL 4016 & 4028* (AAS, KRAM, NY); Turks Head, *ca* 85 m, *REL 4067* (AAS, KRAM);

MARIE BYRD LAND. **EDWARD VII PENINSULA. RUPPERT COAST. *Ford Range***: Skua Gull Peak, *PAS 72* (US).

6. **Bryum archangelicum** Bruch & Schimp.

Fig. 223

Bryum archangelicum Bruch & Schimp. *in* Bruch, Schimp. & W.Gümbel, Bryol. Eur. 4: 153, *f. 333*. 1846 [Fasc. 32 Mon Suppl. 1: 1, *f. 1*]. ≡ *Plagiobryum archangelicum* (Bruch & Schimp.) Pedersen *in* Pedersen & Hedenäs, Bryologist 108: 125. 2005. — TYPE CITATION: Ad Mare-Album prope Kipina ditionis Archangelicae am. Angström detexit et nobiscum communicavit. [Lectotype (*selected here*): 'Bryum Clad. archangelicum Bryol. eur. Angström l. & d. 44 Kipina dit. Archangel' – BM-Schimper!; isotype: BM-Schimper!].

Pohlia inclinata Sw. *ex* Brid., Muscol. Recent. Suppl. 2(3): 144. 1803. ≡ *Bryum inclinatum* (Brid.) Turton, Gen. Syst. Nat. 2: 1727. 1806, *hom. illeg.* ≡ *Leskea inclinata* (Brid.) F.Weber & D.Mohr, Ind. Mus. Pl. Crypt.: 4. 1803. ≡ *Cladodium inclinatum* (Brid.) Brid., Bryol. Univ. 1: 621. 1826. ≡ *Mnium turbinatum* Hedw. var. *inclinatum* (Brid.) Myr., Coroll. Fl. Upsal.: 61. 1833. ≡ *Ptychostomum inclinatum* (Brid.) J.R.Spence, Phytologia 87: 21. 2005, *comb. inval. basion. non citat.* — TYPE CITATION: In subhumidis Sueciae, ad palos putrescentes fontis arcis Upsaliensis inter *Hypnum riparium*, alibique confertim viventem clar. Swarz reperit; also, a reference to fig. 11 on pl. 5 in Swartz (1799) is given. [Lectotype (*selected here*): [icon in] Swartz (1799): f. 11 on pl. 5; epitype (*selected here*): 'Rabenhorst, Bryotheca europaea. 1087. Bryum inclinatum (Sw.). Cf. 1026. Bärwalde (Neumark) {now Mieszkowice 24 km NW of Kostrzyń in Pomerania, NW Poland}, an sonnigen Seeufer häufig. R. Ruthe.' – KRAM!]. First synonymised by Holyoak (2004: p. 250).

Bryum inclinatum var. *magellanicum* Cardot, Rev. Bryol. 27: 41. 1900, ***syn. nov.*** — TYPE CITATION: Détroit de Magellan: environs de Punta Areanas. [Lectotype (*vide* Ochyra & Ochi 1986*b*: p. 216): 'Herb. J. Cardot. Mousses du voyage de la Belgica N° 59 Bryum inclinatum Bryol. eur. var. magellanicum Card. var. nov. Détroit de Magellan: environs de Punta-Arenas. Leg. Racovitza, 9 décembre 1897' – PC-Cardot!; isolectotypes: BR!, PC-Cardot!; syntypes, exactly the same label data, No. 62c, 63 & 64 – BR!, PC-Cardot!].

Bryum imperfectum Cardot, Rev. Bryol. 27: 41. 1900, ***syn. nov.*** — TYPE CITATION: [Détroit du Gerlache] XXme débarquement; later Cardot (1901) wrote: Détroit du Gerlache: XXème débarquement, Terre de Danco, á l'entrée du chenal de Lemaire, sur des roches isolées au milieu d'un glacier, à 50 mètres au dessus niveau de la mer. (N° 268b; 12 février 1898). [Lectotype (*selected here*): 'Herb. J. Cardot. Mousses du voyage de la Belgica N° 268b Bryum imperfectum Card. sp. nova. Canal de la Belgica: XXème débarquement, sur des roches isolées au milieu d'un glacier. Leg. Racovitza, 12 févr. 1898' – PC!; isotype: BR!].

Bryum crateris Dixon, Bryologist 23: 67. 1920, ***syn. nov.*** — TYPE CITATION: [Deception Island, collected by Mr James Robins] No. 451; 452; the latter sterile. [Lectotype (*selected here*): 'Herb. H. N. Dixon Ref. No. 451 Bryum crateris Dixon sp. nov. {Bryologist, xxiii, 67 (1920)} On volcanic detritus, Deception I., 63° lat. 60°30′ long. Coll. James C. Robins 1918 Comm. Rev. D. Lillie' – BM-Dixon!; isolectotype: H-Brotherus!; syntype: the same label data, No. 452 – BM-Dixon!].

Plants medium-sized, forming short to relatively tall, dense tufts or patches, fairly stiff and rigid, scarcely radiculose to densely tomentose below, dull to slightly lustrous, green, yellow- or brownish-green above, brownish below. *Stems* 0.5–2.0 or occasionally 2.5 cm high, repeatedly branched by innovations, sometimes rosulate, red, usually densely matted in the lower part with a tomentum of reddish-brown, branched, papillose rhizoids, in transverse section obscurely rounded-pentagonal, consisting of a distinct central strand, 3–4-stratose medulla of large, hyaline, thin-walled cells and 1–2-layered cortex of small cells with red, thick-walls; *axillary hairs* filiform, short, 4–6-celled, with 2–3 short, quadrate, brownish-red basal cells and 2–4 elongate, hyaline distal cells. *Leaves* crowded above and often forming comal tufts at the stem apex, erect-appressed, straight not or only weakly contorted when dry, erect-spreading when moist, 1.5–3.5 mm long, 0.8–1.2 mm wide, ovate to lanceolate or oblong-lanceolate, acuminate to longly acuminate, non-decurrent, concave to keeled-concave, reddish at the base; *margins* entire or obscurely denticulate towards the apex, with a broad unistratose border of elongate and narrow cells, recurved to revolute throughout; *costa* stout, 80–110 μm at the base, short- to long-excurrent as a rigid, smooth or weakly denticulate awn, red to red-brownish in the proximal part, in transverse section strongly convex dorsally, terete, consisting of a central stereid band covered with a dorsal epidermis and with a row of guide cells immediately below the ventral epidermis; *laminal cells* unistratose throughout, transparent; *upper cells* elongate-rhomboidal to fusiform, firm-walled, 50–100 μm long, 10–12 μm wide near the apex, becoming rhomboidal to oblong-hexagonal, 40–60 μm, 12–15 μm wide, in mid-leaf, very narrow, elongate and thick-walled in 3–4 rows at the margin forming a distinct, unistratose border; *basal cells* rectangular to elongate-rectangular, 40–110 μm long, 12–25 μm wide, reddish at the base, often somewhat swollen in the basal angles. *Synoicous*. *Setae* reddish-brown, smooth, 8–30 mm long, erect, straight to flexuose when dry. *Capsules* long-exserted, cernuous to horizontal, straight and symmetric, elongate-pyriform to elongate-ovoid, with a well developed neck abruptly narrowed to the seta, 2–3 mm long, pale brownish with a darker brown strip at the mouth; *annulus* compound, revoluble, composed of 3 rows of large, incrassate, vesiculose cells; *operculum* conic-convex, apiculate to mammilate, to 0.5 mm long; *exothecial cells* isodiametric to rectangular, fairly irregular in shape, 40–100 μm long, 20–35 μm wide, firm-walled, in rather regular longitudinal rows, becoming smaller and oblate in 2–3 rows at the orifice, not forming a distinct colourful strip and concolorous with exothecial cells in the middle of the urn; *stomata* numerous

→

Figure 223. *Bryum archangelicum*. — **1**. Fertile plant, dry. **2.** Plant with mature capsule, wet. **3.** Cross-section of the stem. **4–6.** Axillary hairs. **7–15**. Leaves. **16–18**. Leaf apices. **19–20**. Mid-leaf cells at the margin. **21**. Mid-leaf cells. **22**. Basal cells. **23–26.** Cross-sections of the leaves. **27–28**. Deoperculate capsules, dry. **29**. Operculate capsule, wet. **30**. Lower exothecial cells and stomata. **31**. Exothecial cells in the middle of the urn. **32**. Exothecial cells at the orifice, portion of the peristome, annulus and spores. — [1–6, 11–12, 18, 20–31 from *RO 967/80*; 7–8, 16, 19 from *Prance 28295*; 9–10 from *ER 268B* (type of *B. imperfectum*), PC; 13, 17 from *Robbins 451* (type of *B. crateris*), BM; 14 from *RO 958/800*; 15 from *BJ 91*]. — Scale bars: a – 100 μm (3–6, 23–26, 32) and 200 μm (17); b – 1 mm (7–15); c – 100 μm (16, 18–22, 30–31); d – 1 mm (2, 27–29); e – 1 mm (1).

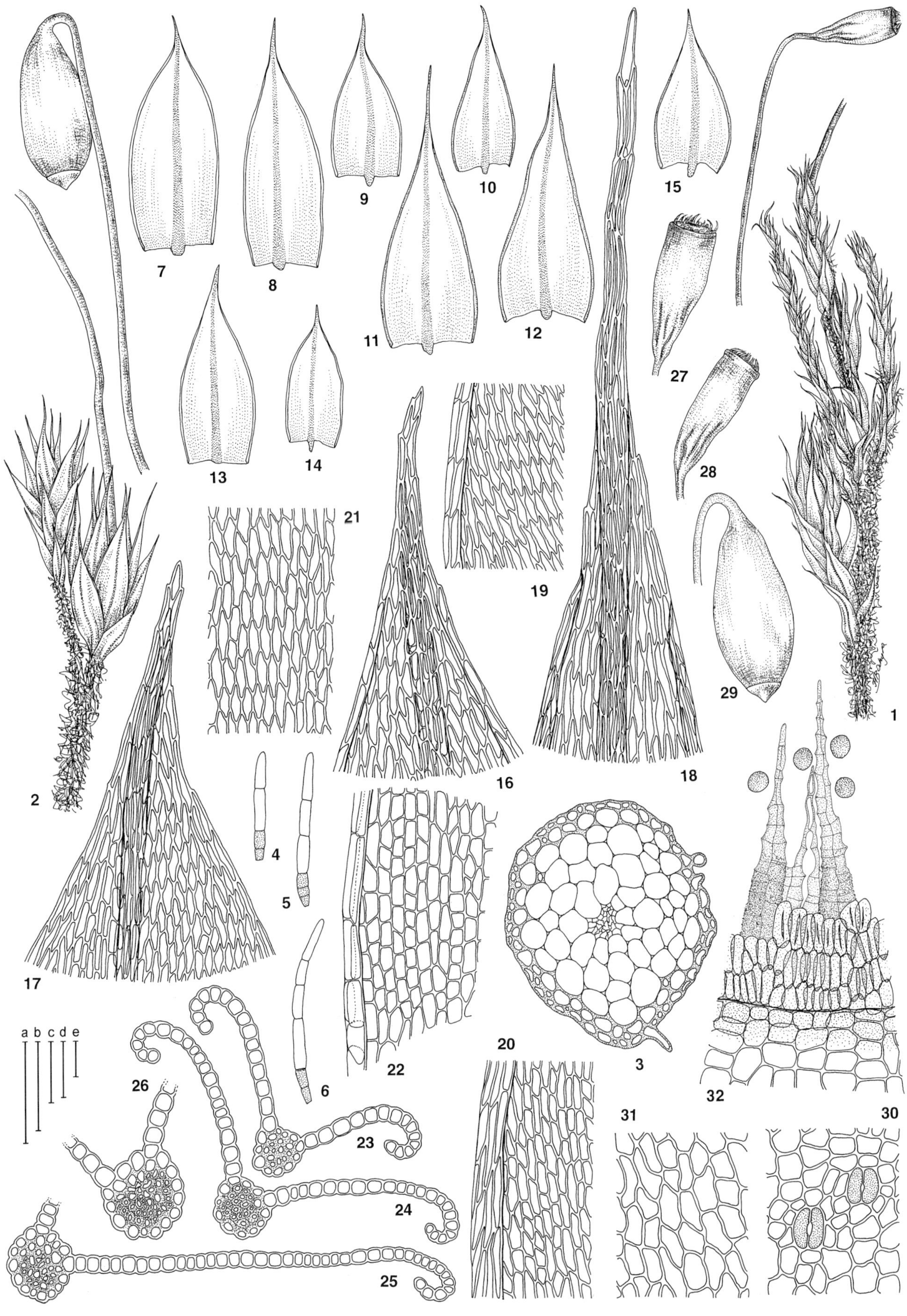
1
2
3
4
5
6
7
8
9
10
11
12
13
14
15
16
17
18
19
20
21
22
23
24
25
26
27
28
29
30
31
32
a b c d e

in 4–5 rows in the neck, phaneroporous, bicellular, with narrow, elongate slit; *peristome* double; *exostome teeth* lanceolate, 300–370 µm long, pale yelowish to hyaline above, orange-yellowish to deeply yellow below, regularly finely papillose throughout, with 10–12 transverse bars, narrowly hyaline bordered, trabeculate on the inner side; *endostome* pale yellowish to hyaline, with a basal membrane about half the length of the exostome teeth and broadly perforate processes, smooth to finely papillose, with or without rudimentary stubby cilia. *Spores* spherical, granulose, 18–30 µm in diameter. *Calyptra* falling early, cucullate.

Chromosome numbers: unknown in the Antarctic and Subantarctic; n = 10, n = 20 and n = 30 have been found in the Northern Hemisphere (Fritsch 1991 as *Bryum inclinatum*).

Discussion. When fertile, *Bryum archangelicum* can be identified by having rudimentary stubby or no cilia of the endostome. This species frequently produces sporophytes in Antarctica, although they are often aborted and the peristomes are reduced. In general, the gametophytes of *B. archangelicum* are nondescript and in other parts of its wide geographical range sterile plants are difficult to distinguish from other species, for example *B. caespiticium* Hedw. However, in the Antarctic *B. archangelicum* differs from other local species in its ovate-lanceolate, closely appressed leaves that are straight, not or only slightly contorted on drying, and have a robust costa that is long-excurrent as a stout, straight, smooth or denticulate awn. This character should separate Antarctic populations of *B. archangelicum* from *B. pallescens* in which the leaves are predominantly oblong-lanceolate and distinctly contorted when dry. Additionally, the laminal cells near the apex in *B. archangelicum* are elongate, mostly fusiform to elongate-rhomboidal, and at least three times as long as wide, whereas in *B. pallescens* and *B. pseudotriquetrum* they are short, rhomboidal, only 2–3 times as long as wide and never spindle-shaped.

Bryum archangelicum has a relatively complicated and chequered taxonomic and nomenclatural history. It was described as a separate species, *Pohlia inclinata*, by Swartz (1799) and this pre-1801 name was subsequently validated by Bridel (1803). Turton (1806) transferred this species to *Bryum*, but under this generic name the epithet *inclinatum* is blocked by *Bryum inclinatum* (Hedw.) Dicks. of 1801 [=*Distichium inclinatum* (Hedw.) Bruch & Schimp.]. As a result, various names have subsequently been suggested as replacements of the nomenclaturally unavailable epithet *inclinatum*. Andrews (1940) found that the North American *B. stenotrichum* Müll.Hal. (Müller 1887) is identical to *B. inclinatum* (Brid.) Turton and this name was used in some recent Floras (e.g. Crum & Anderson 1981; Nyholm 1993). Later, Ochi (1980) suggested that the older name for this species is *B. amblyodon* Müll.Hal., the name used for an Andean species described by Müller (1879) from northern Argentina. This name gained wide acceptance and it has also been used in the Antarctic literature (e.g. Kanda & Ochi 1986; Ochi & Ochyra 1986*b*; Ochyra *et al.* 1998; Ochyra 1998*a*). However, closer examination of the original material of *B. amblyodon* revealed that this species is identical to *B. pseudotriquetrum*. Geissler (1984) suggested *B. imbricatum* (Hedw.) Bruch & Schimp. to be the proper name for this species, but Demaret & Geissler (1990) showed that this is a distinct species differing from *B. inclinatum* in its dioicous sex condition and indistinct leaf border. They concluded that *B. amblyodon* was the earliest valid name that should be referred to the species previously known as *B. inclinatum*. Finally, Holyoak (2004) proved that *B. archangelicum*, a species described from northern Russia (Bruch *et al.* 1839), is identical to what has commonly been named *B. inclinatum* and consequently its name represents the oldest valid name for *B. inclinatum*. It is worth noting that also Ochi (1959) suggested the close similarity of these species. On the other hand, many authors (e.g. Savicz-Lyubitskaya 1954; Nyholm 1993; Zolotov 2006) still consider *B. archangelicum* to be specifically distinct from *B. amblyodon sensu* recent authors (i.e. *B. inclinatum*) in having the aulacodontous exostome teeth (i.e. with the trabeculae on the inner surface being narrower in the middle part resulting in slight perforations or less ornamented areas along the median line that are seen as pellucid spots on the outer surface of the teeth), ferruginous spores in the reflexed light, very long excurrent, aristate costae and inconspicuous leaf borders. These features are present in the type material of *B. archangelicum* but they appear to be poorly developed and are impractical when examining sterile material or immature capsules. Nevertheless, notwithstanding the final taxonomic concept, the epithet *inclinatum* is still available for this species when it is considered a member of *Ptychostomum*, one of several segregates of the large and doubtless heterogeneous genus *Bryum*. Spence (2005) used the epithet *inclinatum* under *Ptychostomum*, but this new combination is invalidly published since the basionym is not cited (he incorrectly quoted only the combination of this name under *Cladodium*).

In the Antarctic *Bryum archangelicum* was described twice as a separate taxon, namely *B. imperfectum* from material collected on the Graham Coast of the Antarctic Peninsula (Cardot 1900, 1901) and *B. crateris* (Dixon 1920) from the volcanic Deception I. in the South Shetland Is. These species were considered conspecific with *B. amblyodon* by Ochi & Ochyra (1986*b*) and Ochyra (1998*a*), respectively, and their names are here reduced to synonymy with *B. archangelicum*. In addition, Bartram (1957) reported *B. inclinatum* var. *magellanicum* from the Melchior Islands archipelago but this variety, which had originally been described from southern South America (Cardot 1900, 1901), was considered to be identical to *B. amblyodon* by Ochi & Ochyra (1986*b*) and in the present treatment its name is reduced to synonymy with *B. archangelicum*. It is also worth noting that all syntypes of *B. perangustidens* originating from Jenny I. in northern Marguerite Bay belong within *B. archangelicum*.

Reproduction in Antarctica. Frequently fertile, though frequently with aborted and malformed capsules having variously reduced peristome teeth.

Habitat. *Bryum archangelicum* is a pioneer moss growing on bare soil or rocks in dry or moist habitats, on mineral, sandy or gravelly soil, in crevices or depressions of rocks, on scree, rock ledges and outcrops, on pockets of soil amongst boulders and stones, sometimes on patches of moss peat, on stony margins of solifluction lobes, in flushes and on sorted circles on raised marine beaches, and in volcanic areas it often grows on heated ground in geothermal fissures, on volcanic ash and detritus, between scoria and lapilli, on ashy slopes, and, occasionally, on bricks and mortar of the foundations of ruined buildings (Deception I.), as well as in bird nesting areas. It usually forms monospecific tufts, occasionally with an admixture of species such as *Cephaloziella varians*, *Ceratodon purpureus*, *Didymodon brachyphyllus*, *Sanionia georgicouncinata*, *Syntrichia saxicola* and *S. magellanica*.

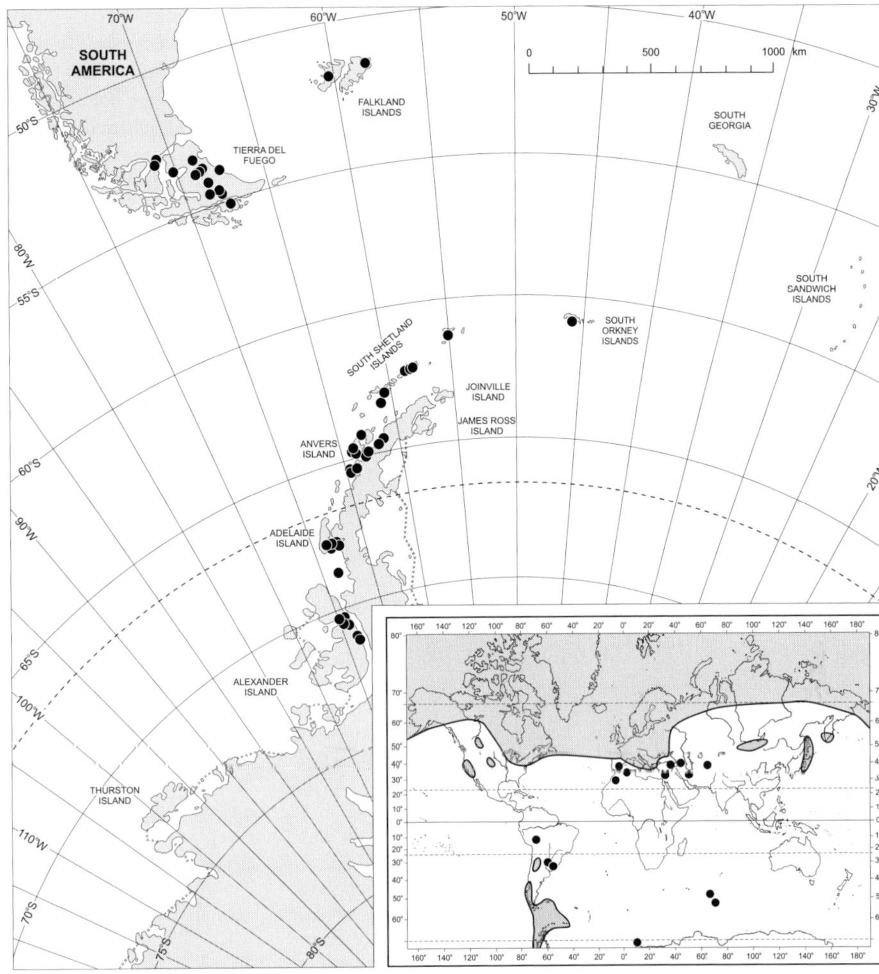

Figure 224. Distribution of *Bryum archangelicum* in the Antarctic Peninsula region. Inset: global distribution.

World range. A bipolar species with some intermediate localities in the tropical and subtropical regions in South America (Fig. 224, inset). In the Northern Hemisphere, *Bryum archangelicum* has a continuous range throughout the arctic and boreal zones, extending southwards to California, Colorado, Minnesota, Michigan and Maine in North America (Crum & Anderson 1981; Norris & Shevock 2004) and to the Mediterranean region in Europe, with some isolated stations in Morocco and Algeria in North Africa (Jelenc 1955), the Caucasus (Zolotov 2006) and Turkey, Cyprus, Iran and Afghanistan in the Near East (Frey & Kürschner 1991), Central Siberia, Kamchatka and Sakhalin in the Russian Far East (Zolotov 2006) and Hokkaido and central Honshu in Japan (Ochi 1959). In the Southern Hemisphere the species is scattered in western Patagonia, the Falkland Is. and Tierra del Fuego archipelago in southern South America, with some stations in the Andes of northern Argentina and Bolivia, Paraguay and Uruguay (Ochi 1981, 1982). It extends into the Antarctic along the western coast of the Antarctic Peninsula to central Alexander I. An isolated station of *B. archangelicum* is known from Îles Kerguelen (Ochi 1972) and Heard I. (Ochyra & Seppelt 2008*b*).

Distribution in Antarctica. *Bryum archangelicum* is a rather infrequent, though locally common and abundant species, for example on Deception I., that is confined in its distribution to the maritime Antarctic (Fig. 224). It is scattered on the peri-Antarctic archipelagoes of the South Orkney and South Shetland Is., as well as on the western coast of the Antarctic Peninsula, ranging from Charlotte Bay on the Danco Coast to central Alexander I. where it reaches its southernmost station at lat. 71°52′ S. Kanda & Ochi (1986) and Kanda (1987*a*) reported the species from a number of localities in continental Antarctica, including Enderby Land, Princess Elizabeth Land and Victoria Land but no voucher collections have been available for investigation. However, judging from the illustrations presented by these authors it is likely that the specimens so-named represent *B. pseudotriquetrum* since the leaves have short-excurrent costae. A single specimen from the Sôya Coast was illustrated in fine fruiting condition and its peristome described as having rudimentary cilia. However, the sporophytes of continental plants of *B. pseudotriquetrum*, for example as reported by Filson & Willis (1975) from Kemp Land, frequently have a strongly reduced inner peristome and this is probably a response to the harsh climatic conditions. Nonetheless, *B. archangelicum* does occur on the Antarctic continent and three specimens collected from the Schirmacher Oasis in Dronning Maud Land certainly belong within this species. They are apparently sterile but the correspondence in the gametophyte characters with the specimens of this species known from the maritime Antarctic is perfect. The altitudinal range of *B. archangelicum* extends in Antarctica from sea level to 320 m.

Selected specimens examined. WEST ANTARCTICA. SOUTH ORKNEY IS. **Signy I.** Foca Pt., *ca* 35 m, *RW 125* (AAS, KRAM); North Pt., *ca* 50 m, *RILS 656* (AAS, KRAM); Changing Lake, 50 m, *RILS 10766* (AAS, KRAM); Moraine Valley, 50 m, *RILS 10777* (AAS, KRAM); Orwell Glacier moraine, 30 m, *RILS 10820* (AAS, KRAM).

SOUTH SHETLAND IS. **Elephant I.** 500 m south-west of Endurance Glacier, 183 m, *JSA 146A* (AAS, KRAM). **King George I.** ***Admiralty***

Bay: Rembiszewski Nunataks, 200 m, *RO 2756/80* (KRAM); Upłaz, 45 m, *RO 1532/80* (KRAM); Ubocz, 90 m, *4949/79*. ***Ezcurra Inlet***: Krzymiński Pt, 6 m, *RO 944/80* (KRAM); Cytadela, 50 m, *RO R967/80* (KRAM); Pond Hill, 70 m, *RO 882/80* (KRAM). ***Dufayel I.***: Gdynia Pt., 8 m, *RO 1779/80* (KRAM). ***MacKellar Inlet***: Klekowski Crag, 110 m, *RO 2281/80* (KRAM). ***Martel Inlet***: Point Hennequin, 15 m, *RO 2239/80* (KRAM). ***King George Bay***: Lions Rump, *BJ 91* (KRAM). **LIVINGSTON I.** ***Hurd Peninsula***: east side of False Bay, *RMS 69-406* (US). **DECEPTION I.** Baily Head, 35 m, *DM 79A* (AAS, KRAM); Neptunes Bellows, 50–150 m, *RILS 6601A* (AAS, KRAM); Whalers Bay, 4 m, *RO 2767/80* (KRAM); Ronald Hill, 40 m, *RO 2787/80* (KRAM); Crater Lake, 70–200 m, *RILS 6846A & 6829A* (AAS, KRAM); Pendulum Cove, Crimson Hill, 10 m, *RILS 11331* (AAS, KRAM); Telefon Bay, 2–5 m, *RILS 6654* (AAS, KRAM); Kendall Terrace, 65–85 m, *DM 105 & 107* (AAS, KRAM); Fumarole Bay, 125 m, *RILS 11275* (AAS, KRAM); Vapour Col, 130 m, *RILS 5640B* (AAS, KRAM); Mt. Kirkwood, 170–250 m, *RILS 5765F* (AAS, KRAM); Collins Pt., 50–70 m, *RILS 6696A* (AAS, KRAM); Entrance Pt., 25 m, *DM 52C* (AAS, KRAM).

WEST ANTARCTIC PENINSULA. **DANCO COAST.** Charlotte Bay, 5–10 m, *RILS 3967A & 3968* (AAS, KRAM); Andrée I., 20–50 m, *RILS 4009* (AAS, KRAM); Cuverville I., 10–20 m, *RILS 11958 & 11959A* (AAS, KRAM) and 30–40 m, *RILS 8808* (AAS, KRAM); Andvord Bay, Neko Harbour, 15–20 m, *RILS 11931B* (AAS, KRAM); Coughtrey Peninsula, Paradise Harbour, *VC 1022* (KRAM). **PALMER ARCHIPELAGO.** ***Melchior Is.***: Lambda I., *PAS 347.3* (LD); Omega I., *Popovici 8392, 9205 & 10676* (AAS, KRAM). ***Anvers I.***: Bonaparte Pt., *RW 234 & 236* (AAS, KRAM); Norsel Pt., *RMS 69-833* (US) and 10–20 m, *RILS 1957* (AAS, KRAM); Hermit I., 10–25 m, *RILS 882* (AAS, KRAM); Shortcut I., *Prance 28595* (KRAM, NY); Goudier I., Port Lockroy, 2–10 m, *RILS 732B & 5291A* (AAS, KRAM). **GRAHAM COAST.** Cape Tuxen, Waddington Bay area, 20 m, *RILS 10656* (AAS, KRAM) and 80 m, *RILS 10669* (AAS, KRAM). **WILHELM ARCHIPELAGO.** ***Lemaire Channel***: without details, *ER 268B* (BR, PC – type of *Bryum imperfectum*). ***Argentine Is.***: Galindez I., 8 m, *RWMC 429A* (AAS, KRAM); Black I., 2–10 m, *RILS 4294* (AAS, KRAM). ***Collins Bay***: Darboux I., *Poncet 8D* (AAS, KRAM). **FALLIÈRES COAST.** ***Adelaide I.***: Rothera Pt., *RILS 8926* (AAS, KRAM) and 20–30 m, *RILS 1867B* (AAS, KRAM); Reptile Ridge, 140 m, *RILS 10598B* (AAS, KRAM); Jenny I., *Gain 218, 219, 231B, 243, 245A, 246A, 248B, 255, 256 & 260B* (PC – syntypes of *Bryum perangustidens*). ***Léonie Is.***: Lagoon I., 6 m, *RILS 9164A* (AAS, KRAM); Léonie I., 150–175 m, *RILS 9132* (AAS, KRAM). ***Marguerite Bay***: Pourquoi Pas I., *DDWW 154 & 157* (AAS, KRAM) and *Poncet 7C* (AAS, KRAM); Lagotellerie I., 50 m, *RILS 1765 & 4841* (AAS, KRAM); Camp Pt., 10–20 m, *RILS 4682* (AAS, KRAM); Millerand I., 5–15 m, *RILS 3762C* (AAS, KRAM); Terra Firma Is., Barn Rock, *Poncet 4* (KRAM). **GEORGE VI SOUND.** ***Alexander I.***: Ablation Valley delta area, *RILS 11113* (AAS, KRAM); Ablation Valley, 5 m, *RILS 11128* (AAS, KRAM) and 220 m, *RILS 9300* (AAS, KRAM); Fossil Bluff, plateau SW of Scarab Bluff, 400 m, *RILS 9356* (AAS, KRAM); Mars Glacier, Syrtis Hill, *RILS 8846, 8847 & 8848* (AAS, KRAM); Probe Ridge, *DDWW 73A* (AAS, KRAM); Two Step Moraine, 30 m, *RILS 9489* (AAS, KRAM).

EAST ANTARCTICA. DRONNING MAUD LAND. ***Schirmacher Oasis***: central part, 126 m, *Singh 48* (KRAM), 133 m, *Singh 42* (KRAM) and 146 m, *Singh 17* (KRAM). ***Baalsrudfjellet***: nutatak to the east of the Schirmacher Oasis, *Singh 24* (KRAM).

7. **Bryum pallescens** Schwägr. Fig. 225

Bryum pallescens Schwägr., Sp. Musc. Frond. Suppl. 1(2): 107, *pl. 75*. 1816, *nom. cons. prop.* ≡ *B. pseudotriquetrum* (Hedw.) P.Gaertn., B.Mey. & Scherb. subsp. *pallescens* (Schwägr.) Dixon, Stud. Handb. Brit. Mosses Ed. 3: 364. 1924. ≡ *B. pseudotriquetrum* var. *pallescens* (Schwägr.) Gangulee, Mosses E. India 2(4): 993. 1974. — TYPE CITATION: In subalpinis Sudetum et Helvetiae legerunt Schleicher et Car. Ludwig, Salisburgi Flörke, in alpinis graniticis Styriae ego, vere et aestate florens. [Lectotype (*selected here*): '[Switzerland] 153.811 Schl. Mn pallescens? facie duo minora inter singulum ... [illeg.] laciniae maior [illeg] in [illeg.] Ludw Scripsit Schwaegrichen cui misit Schleicher AD. 1811 teste G. Colomb-Duplan' (G-Hedwig/Schwägrichen!; isolectotype: BM-Turner/Hooker!].

Bryum inconnexum Cardot, Rev. Bryol. 27: 44. 1900, ***syn. nov.*** — TYPE CITATION: [Détroit du Gerlache] IXme et XXme débarquement [later Cardot (1901) wrote: Détroit du Gerlache: IXème débarquement, Terre de Danco, près du Cap Anna Osterrieth, sur les petites terrasses humides de la falaise. (N° 250^{b}; 29 janvier 1898). XXème débarquement, Terre de Danco, á l'entrée du chenal de Lemaire, sur des roches isolées au milieu d'un glacier, à 50 mètres au dessus niveau de la mer. (N° 268^{a}; 12 février 1898). [Lectotype (*selected here*): 'Herb. J. Cardot. Voyage de la Belgica n° 268^{a} Bryum inconnexum Card. sp. nova. Canal de Gerlache: XXème débarquement, sur des roches isolées au milieu d'un glacier, alt. 50 m. Leg. Racovitza, 12 févr. 1898' – PC!; isolectotypes: BR!, S-Dusén!; syntype: 'Herb. J. Cardot. Voyage de la Belgica n° 205^{b} Bryum inconnexum Card. sp. nova. Canal de Gerlache: IXème débarquement, sur les petites terrasses detrampes de la falaise. Leg. Racovitza, 29 janvier 1898' – PC!; isosyntype: BR!].

Bryum inconnexum var. *tomentosum* Cardot, Rev. Bryol. 33: 34. 1906. ≡ *B. inconnexum* fo. *tomentosum* (Cardot) Kuc, Rev. Bryol. Lichénol. 36: 667. 1969, ***syn. nov.*** — TYPE CITATION: Ile Wandel. [Lectotype (*selected here*): 'Herb. Mus. Paris. Bryum inconnexum Card. var. tomentosum Card. var. nova. Ile Wandel. 26 Nov^{re} 1904. Exped. Charcot n° 244^{a}' – PC-Cardot!; isotype: S-Möller!; syntypes: (1) 'Herb. J. Cardot. Bryum inconnexum Card. var. tomentosum Card. Ile Wandel 23/11 1904 Exped. Charcot n° 234^{b}' – PC!; (2) 'Herb. J. Cardot. Bryum inconnexum Card. var. tomentosum Card. Ile Wandel 25/11 1904 Exped. Charcot n° 230^{a}' – PC!].

Plants medium-sized to relatively large but slender, in short to tall, occasionally deep, dense or fairly loose tufts or sometimes forming extensive patches, usually densely tomentose below, lustrous, green to yellowish-green, sometimes tinged with red above, brown to reddish-brown below. *Stems* 1–4 or occasionally 6.5 cm high, strongly branched by innovations, usually with a brownish tomentum below of long, branched, brownish, papillose rhizoids, in cross-section obscurely pentagonal, with a prominent central strand, 3–4-layered medulla of large, hyaline, thin-walled cells and 2–3(–4)-stratose cortex of small cells with red, thick walls; *axillary hairs* relatively short, 3–6-celled, with 1–2 short, quadrate, red to reddish-brown basal cells and 2–4 hyaline, elongate distal cells. *Leaves* evenly set or crowded at the stem apex, spirally twisted and appressed when dry, erect-spreading when moist, 1.5–3.2 mm long, 0.7–1.2 mm wide, ovate- or oblong-lanceolate to triangular or nearly so, non-decurrent, gradually long-acuminate, concave to carinate-concave distally, reddish at the base; *margins* entire, recurved to revolute throughout, with a prominent unistratose border of narrow, elongate cells; *costa* robust, red in the proximal portion, 80–120 μm wide

⟶

Figure 225. *Bryum pallescens*. — **1**. Sterile plant, dry. **2.** Plant with mature capsule, wet. **3.** Cross-section of the stem. **4–6.** Axillary hairs. **7–13**. Leaves. **14–15**. Leaf apices. **16**. Mid-leaf cells. **17**. Mid-leaf cells at the margin. **18**. Basal cells. **19–22.** Cross-sections of the leaves. **23**. Deoperculate capsule, wet. **24**. Operculate capsule, wet. **25**. Deoperculate capsule, dry. **26**. Lower exothecial cells and stoma. **27**. Exothecial cells in the middle of the urn. **28.** Exothecial cells at the orifice, portion of the peristome and spores. — [1, 10–11 from *Exped. Charcot 244A* (type of *B. inconnexum* var. *tomentosum*), PC; 2–9, 14, 17, 19–28 from *VK 601*; 12–13, 15, 18 from *ER 268A* (type of *B. inconnexum*), PC]. — Scale bars: a – 100 μm (3, 19–22); b – 1 mm (1); c – 1 mm (7–13) and 100 μm (4–6); d – 1 mm (23–25) and 100 μm (14–18, 26–28); e – 1 mm (2).

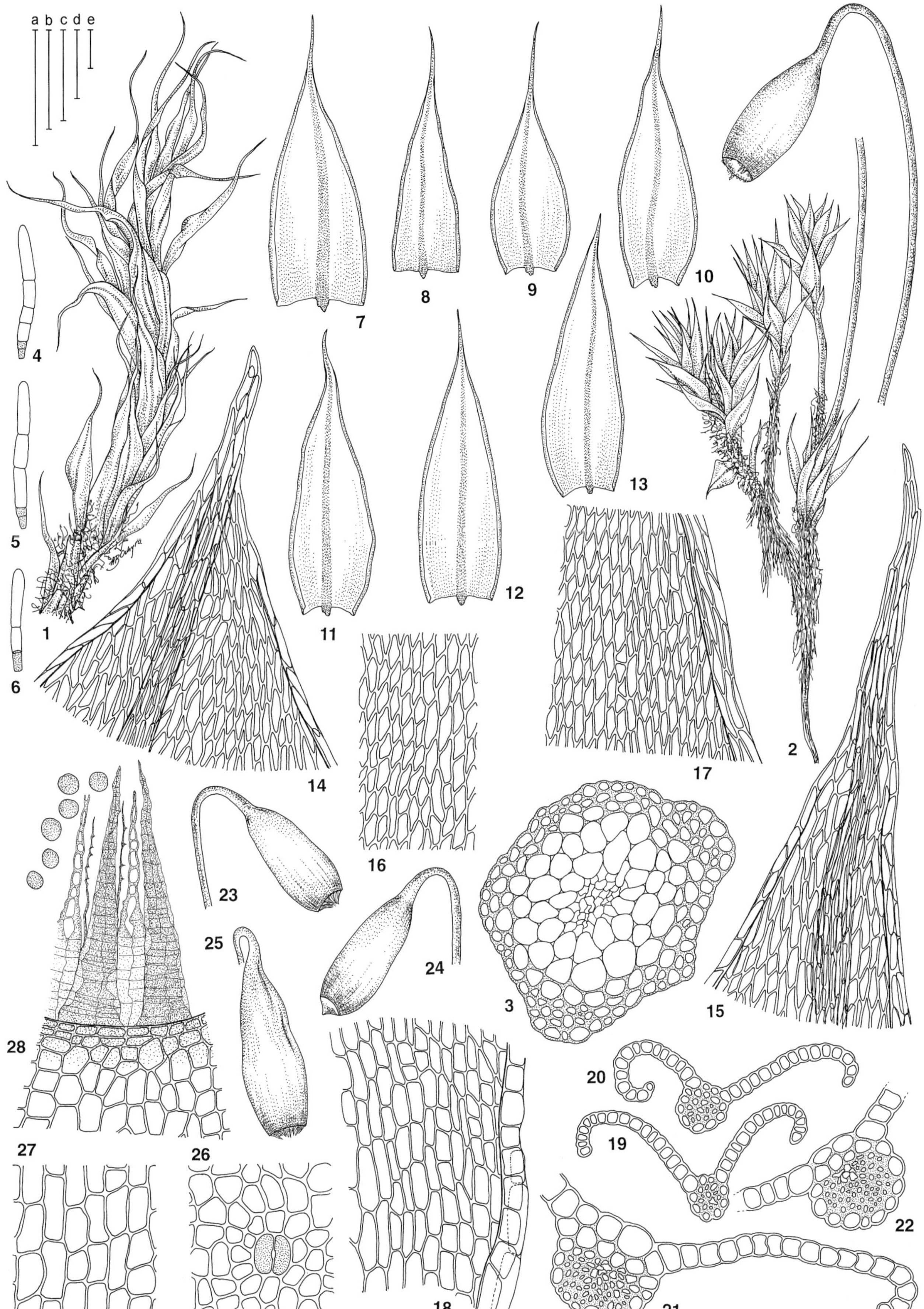
a b c d e
1
2
3
4
5
6
7
8
9
10
11
12
13
14
15
16
17
18
19
20
21
22
23
24
25
26
27
28

at the base, noticeably narrowed distally, long-excurrent as an entire or weakly denticulate awn, in transverse section terete, consisting of a strong stereid band covered with a distinct epidermis on the dorsal side and a row of large guide cells immediately below the ventral epidermis; *laminal cells* unistratose throughout, translucent; *upper cells* oblong- or rhomboidal-hexagonal to rhomboidal, 35–50(–60) µm long, 14–18(–20) µm wide, firm-walled, becoming narrow, long-rectangular to linear, thick-walled at the margin, forming a well marked, 3–5-seriate border, usually obscured because of the revolute leaf margin; *basal cells* reddish, rectangular, firm- to thin-walled, 50–90 µm long, 20–25 µm wide, becoming short-rectangular to subquadrate, 30–50 µm long, 20–25 µm wide near the margins, sometimes slightly turgid in the basal angles. *Autoicous* or *synoicous*. *Setae* red-brown to reddish, erect, smooth, 1–2 cm long, somewhat flexuose when dry. *Capsules* inclined to pendulous, pale brown, ochraceous to cinnamon-brown, straight and symmetric, 3–4 mm long, clavate to narrowly pyriform, with a distinct apophysis rapidly narrowed to the seta; *annulus* revoluble, compound, 3–4-seriate; *operculum* convex, mammilate; *exothecial cells* variable in size and shape, mostly oblong to rectangular to irregularly isodiametric, 30–100 µm long, 25–40 µm wide, thin-walled, becoming quadrate to oblate in 2–3 darker brown rows at the mouth; *stomata* numerous in many rows in the neck, superficial, bicellular, with long and narrow pori, large, 50–60 µm in diameter; *peristome* double; *exostome teeth* lanceolate, 400–450 µm long, yellow below, hyaline and papillose distally, narrowly hyaline bordered, with 20–25 transverse bars; *endostome* hyaline, nearly smooth, with a basal membrane about half the length of the teeth and fairly broad segments with wide, ovate perforations, cilia 1–3, nodose. *Spores* brownish-green, finely papillose, 15–21 µm in diameter. *Calyptra* cucullate.

Chromosome numbers: unknown in the Antarctic and Subantarctic; $n = 10$ (3 records), $n = 11$ (1 record), $n = 12$ (1 record) and $n = 20$ (2 records) from the Holarctic (Fritsch 1991).

Discussion. Although *Bryum pallescens* is predominantly sterile in Antarctica, it is rather easy to distinguish from all other congeners on the basis of its leaves being oblong-lanceolate and very long-acuminate with a costa excurrent as a long awn. In addition, the leaves are distinctly spirally twisted on drying and this feature should separate *B. pallescens* from *B. archangelicum* which has fairly rigid, straight or only slightly altered leaves when dry. Moreover, the latter species has mostly elongate-rhomboidal to fusiform laminal cells near the apex that are usually 3–6 times as long as wide, whereas *B. pallescens* has short, rhomboidal to oblong-rhomboidal apical cells, mostly less than 3 times as long as wide. Fertile plants of *B. pallescens* are distinct from *B. archangelicum* in their synoicous sex condition and the lack of cilia in the peristome, whereas *B. pallescens* is an autoicous plant and only occasionally are some plants polygamous, i.e. autoicous with male and female buds at the tips of separate branches, but the perichaetia often include a few antheridia. However, this is an impractical character for identifying *B. pallescens* in the Antarctic as the plants are usually sterile, but the sexual condition can be checked in most populations of *B. archangelicum* as it is often fertile in this biome.

Bryum pallescens is unlikely to be mistaken for *B. pseudotriquetrum*, the commonest and most variable species of the genus in Antarctica. In the latter species the leaves are mostly unaltered when dry, and are short-acuminate, with a short-excurrent or percurrent costa, and plants are dioicous or synoicous. These species are distantly related and belong within different subsections of sect. *Cladodium* in the hierarchical classification of the genus *Bryum* proposed by Ochi (1992), namely subsect. *Caespiticia* (C.J.Hartm.) J.J.Amann (*B. pseudotriquetrum*) and subsect. *Cernuiformia* (Kindb.) Broth. (*B. pallescens*), although the nomenclature of the subsection names is still provisional and needs careful evaluation (Isoviita 1992). Some authors considered *B. pallescens* as a subspecies (Dixon & Jameson 1924) or variety (Gangulee 1974) of *B. pseudotriquetrum*.

Bryum pallescens was described by Schwägrichen (1816) but the name was actually published a decade earlier as a *nomen nudum* in *Plantae cryptogamicae helveticae*, a series of exsiccati distributed by Schleicher (1807*a,b*) who ascribed this name to S. E. Bridel. Nowadays, it is a common practice to ascribe the authority of *B. pallescens* to 'Schleich. *ex* Schwägr.' but it is contrary to the rules of the current Code (McNeill *et al.* 2006). There is no basis under Art. 46.4 (and the definition of ascription in Art. 46.3) for any *ex* citation and Art. 46.3 makes it clear that appearance of a citation in a list of synonyms is not ascription, so the traditional 'Schleich. *ex* Schwägr.' has no basis under the current Art. 46 and the name *B. pallescens* should be ascribed to Schwägrichen alone. It is also worth noting that Demaret (1986) incorrectly designated the original specimen from the Hedwig/Schwägrichen herbarium at G as holotype of *B. pallescens*. In fact, since Schwägrichen (1816) cited in the protologue no fewer than five specimens, two of which are available in his personal herbarium, lectotypification was necessary.

However, *Bryum pallescens* was described much earlier by Weber & Mohr (1807) as *Hypnum boreale* and this species was soon given the name under *Bryum* by Funck (1808). The latter name has remained totally neglected since Bruch *et al.* (1839) reduced this species to variety of *B. pallescens*. Although it cannot be accepted from the point of view of the rules of nomenclature, this decision has gained wide acceptance because *Bryologia europaea* was at that time considered to be the oracle as far as taxonomic and nomenclatural concepts regarding European mosses were concerned. Consequently, the older name *Bryum boreale* was rejected and the use of *B. pallescens* became common practice, without exception, in all Floras and catalogues of mosses. The conspecificity of *B. boreale* and *B. pallescens* was confirmed by Ochyra & Bednarek-Ochyra (in preparation) and, to stabilise the nomenclature, *B. pallescens* will be proposed to be conserved against *B. boreale* (Ochyra & Bednarek-Ochyra, in preparation).

Bryum pallescens was recorded for the first time under this name from the Antarctic by Ochi & Kanda (1991) on the basis of specimens from Anvers I. and Deception I. The voucher collection from Anvers I. was originally determined by W. C. Steere as *B. stenotrichum* Müll.Hal., which name has long been used for the species currently named *B. archangelicum*. The duplicate of this collection at KRAM is richly fruiting and clearly represents *B. archangelicum* on account of the peristome that lacks cilia and elongate upper laminal cells, as can also be seen in the illustration by Ochi & Kanda (1991). Unfortunately, the specimen collected on Deception I. by R. Covarrubias (No. 21968 in the Mahu herbarium) was not available for study, but another specimen

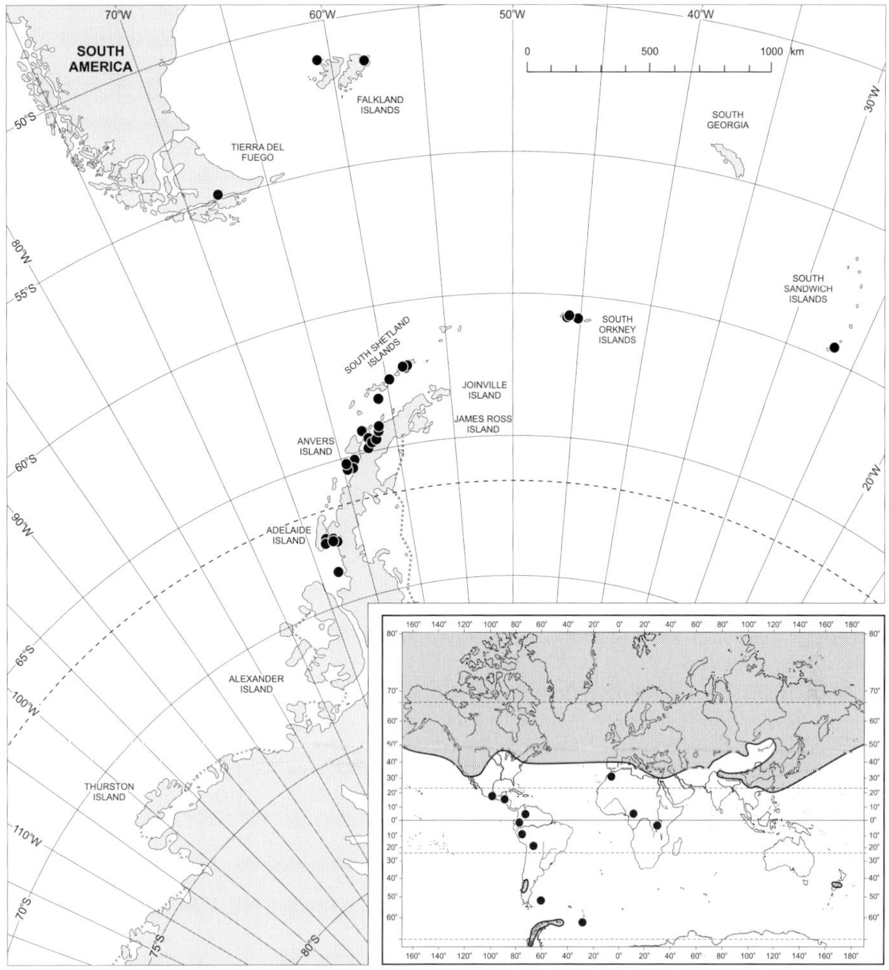

Figure 226. Distribution of *Bryum pallescens* in the Antarctic Peninsula region. Inset: global distribution.

from the same site (leg. R. Covarrubias, *Mahu 21934*, AAS, KRAM) was examined and it clearly represents *B. pallescens*. This species was collected for the first time in the Antarctic Peninsula region by E. Racovitza, naturalist to the *Belgica* expedition, and Cardot (1900, 1901) described it as a separate species, *B. inconnexum*. This name was subsequently used for designation of the specimens from various parts of the Antarctic (e.g. Horikawa & Ando 1961, 1967*a*; Tatuno 1963; Matsuda 1964, 1968; Kuc 1969; Robinson 1972*a*; Longton 1979; Nakanishi 1979), but all these plants correctly refer to *B. pseudotriquetrum*. However, the type material of *B. inconnexum*, though sterile, truly represents *B. pallescens* and Cardot (1900, 1901) actually compared the newly described Antarctic species to it. He showed the major differences between these species in the habit of the plants, the development of the rhizoidal tomentum and the weaker leaf margin in *B. inconnexum*, but these features are subject to considerable variation in *B. pallescens*. Because the correspondence in other characters of both species is excellent, they are here considered to be conspecific taxa. The same is also true for *B. inconnexum* var. *tomentosum*, a variety described by Cardot (1906*b*, 1907*a*) from the same region, which differs from the type variety in having narrower leaves and a better developed tomentum of rhozoids. It should be noted that Ochi (1979) considered these taxa to be identical to *B. pseudotriquetrum* but this species has a distinct short-excurrent costa and short-acuminate leaves.

Reproduction in Antarctica. Mostly sterile, occasionally with sporophytes which usually have aborted capsules with variously reduced peristomes.

Habitat. In dry or damp, sheltered or open sites, on clayey, sandy or rocky soil, gravel, on humus and debris in rock crevices and fissures on cliffs, rock outcrops and ledges, sometimes moistened by melt water, on pockets of soil between stones, scree and volcanic ash and scoria. It is often a colonist in fellfield communities and on recently deglaciated terrain. Occasionally, *Bryum pallescens* forms luxuriant extensive stands along the margins of runnels, melt water streams and in calcareous flushes where it grows in various communities of the moss carpet and moss hummock subformations (notably on north-west Signy I.). It was sometimes found in typically anthropogenic habitats including bricks and mortar of ruined buildings and inside old whaling tanks (Deception I.). It often forms pure coalesced cushions in communities of the moss carpet subformation with an admixture of species such as *Brachythecium austrosalebrosum*, *Bryum pseudotriquetrum*, *Cephaloziella varians*, *Sanionia georgicouncinata*, *Warnstorfia sarmentosa* and *W. fontinaliopsis*.

World range. *Bryum pallescens* is a bipolar species with some intermediate occurrences at high elevations in the tropical mountains (Fig. 226, inset). In the Northern Hemisphere it has a continuous pan-Holarctic range in the polar, boreal and temperate zones, extending southwards to California (Norris & Shevock 2004) and Colorado (Weber 1973; Weber & Wittmann 2007) and the northern United States (Crum & Anderson 1981) in North America, to the Mediterranean region in Europe and Morocco in North Africa (Jelenc 1955; Ros *et al.* 1999), the Near East and Central Asia (Frey & Kürschner 1991), northern Mongolia (Tsegmed 2001*a*,*b*), southern China and Taiwan (Zhang *et al.*

2007) and Japan (Ochi 1959). The species is very rare in the neotropics, in southern Mexico (Ochi 1994) and Guatemala (Allen 2002) in Central America and in the northern and central Andes of Colombia, Ecuador, Peru and Bolivia (Ochi 1981) and in Africa including the Cameroon Mountains (Mitten 1863) and the Ruwenzori Mountains (Demaret 1940). In the temperate zone of the Southern Hemisphere *B. pallescens* is exceedingly rare in western Patagonia, Tierra del Fuego and the Falkland Is. (Ochi 1981) and on the South Island of New Zealand (Ochi 1970, 1984), and extends to the maritime Antarctic.

Distribution in Antarctica. Despite its wide range throughout the maritime Antarctic (Fig. 226), *Bryum pallescens* is generally an infrequent species, and only in a few areas does it occur in abundance (e.g. Signy I.). It extends from the southern South Sandwich Is., through the South Orkney and South Shetland Is. along the western coast of the Antarctic Peninsula to Marguerite Bay, reaching its southernmost occurrence in the Terra Firma Is. at lat. 68°43′ S. It grows from sea level to 300 m, although most occurrences are below 100 m.

Selected specimens examined. WEST ANTARCTICA. SOUTH SANDWICH IS. **BELLINGSHAUSEN I.** Within main crater on west side, 135 m, *MWH 429D* (AAS, KRAM).

SOUTH ORKNEY IS. **CORONATION I.** Island opposite Saunders Pt., 8 m, *RILS 31B* (AAS, KRAM). **POWELL I.** West of John Peaks, 15 m, *RILS 252* (AAS, KRAM). **SIGNY I.** Foca Cove, 5 m, *RILS 8052A* (AAS, KRAM); Foca Pt., 15 m, *RILS 397* (AAS, KRAM) and 35 m, *RILS 344* (AAS, KRAM); Moss Braes, 25–50 m, *RILS 10911* (AAS, KRAM); Robin Peak, 100 m, *RILS 5162* (AAS, KRAM); Mirounga Flats, 10 m, *RILS 10800* (AAS, KRAM) and 13 m, *RILS 387* (AAS, KRAM); below Orwell Glacier, *ca* 35 m, *RILS 632* (AAS, KRAM).

SOUTH SHETLAND IS. **KING GEORGE I.** ***Fildes Peninsula***: Ardley I., *Kühnemann 92A* (AAS, KRAM); without closer data, *Li 35* (AAS, KRAM). ***Admiralty Bay***: Agat Pt., 15 m, *RO 378/80 RILS 31B* (KRAM); between Wróbel Glacier and Ornithologists Creek, 120 m, *RO 645/80 RILS 31B* (KRAM). **NELSON I.** Without closer data, *Li 4* (AAS, KRAM). **ROBERT I.** Coppermine Cove, 6–50 m, *RILS 911* (AAS, KRAM). **DECEPTION I.** Whaling Station, 10–20 m, *RILS 6546A* (AAS, KRAM); Crater Lake, 15 m, *RILS 5661C* (AAS, KRAM); Pendulum Cove, 5 m, *RILS 11190* (AAS, KRAM); between Pendulum Cove and Telefon Bay, *Mahu 21934* (AAS, KRAM); Fumarole Bay, 50 m, *RILS 11528* (AAS, KRAM); Wensleydale Beacon, 2 m, *RILS 5600B* (AAS, KRAM).

WEST ANTARCTIC PENINSULA. **DAVIS COAST.** Island southeast of Cape Herschell, 20 m, *RILS 11642C* (AAS, KRAM). **DANCO COAST.** Cierva Pt., *VK 601* (KRAM); Spring Pt., *VK 902 & 970* (KRAM); Charlotte Bay, Andrée I., 25 m, *RILS 4007* (AAS, KRAM); Birdsend Bluff, 15–20 m, *RILS 4219C* (AAS, KRAM); Andvord Bay, Neko Harbour, 15–20 m, *RILS 11993* (AAS, KRAM). **PALMER ARCHIPELAGO.** ***Brabant I.*** Claude Pt., 200 m, *WH 206A & 207B* (AAS, KRAM); Pinel Pt., 300 m, *AM 94A* (AAS, KRAM). **GRAHAM COAST.** Cape Tuxen, *Fowbert 28* (AAS, KRAM). ***Lemaire Channel***: without details, *ER 268A* (BR, PC, S – type of *Bryum inconnexum*); Wandel I., *Turquet 230A, 234B & 244A* (PC – type of *Bryum inconnexum* var. *tomentosum*) *& 164B & 234A* (PC). ***Argentine Is.***: Black I., 7 m, *RWMC 578* (AAS, KRAM). ***Collins Bay***: Green I., *BGLE 1083D* (BM). **LOUBET COAST.** Orford Cliff, *RILS 5271A, 5272, 5274, 5276B, 5277B, 5278A & 5287D* (AAS, KRAM). **FALLIÈRES COAST.** ***Adelaide I.***: Rothera Pt., 8 m, *RILS 87B* (AAS, KRAM); Reptile Ridge, 300 m, *RILS 3466* (AAS, KRAM); Jenny I., *Gain 244* (PC). ***Léonie Is.***: Léonie I., 30 m, *RILS 8978* (AAS, KRAM). ***Marguerite Bay***: Horseshoe I., *RILS 1893* (AAS, KRAM); Lagotellerie I., 50 m, *RILS 1767* (AAS, KRAM); Terra Firma Is., Alamode I., *Poncet 3D* (AAS, KRAM).

EXCLUDED SPECIES OF THE BRYACEAE

Bryum billardierei Schwägr. — The species was reported from the South Orkney Is. by Haji Mohamed (1979) as follows: 'Inaccessible Island: 1873, Challenger Expedition, Moseley (BM)'. This is clearly an error because the *Challenger* Expedition of 1872–1876 never operated in the Antarctic. Inaccessible I. belongs within the Tristan da Cunha archipelago and N. H. Moseley, a naturalist on this expedition, intensively botanised on the island on 17 October 1873 (Moseley 1874). Inaccessible Is. is a group of three main islands and offlying rocks situated west of Coronation I. in the South Orkney Islands archipelago (Hattersley-Smith 1991) but hitherto no plant collections have been made there.

B. bornholmense Wink. & R.Ruthe. — This provisionally named species was reported from Schirmacher Oasis in Dronning Maud Land, continental Antarctica (Singh & Semwal 2000), but no material was available for examination. *Bryum bornholmense* is a Euro-American species of the *B. erythrocarpum* complex which is characterised by the presence of rhizoidal tubers. This species has never been recorded in the Southern Hemisphere and Singh & Semwal (2000) made no mention of the propagules in the Antarctic material, although this is the most characteristic feature of *B. bornholmense*.

B. gemmiferum R.Wilczek & Demaret. — This species was named, with a question mark, from the geothermal ground on the volcanic Deception I. in the South Shetland Is. (Lewis Smith 1991) but the voucher collections correctly belong within *B. dichotomum*.

B. mildeanum Jur. — This species was reported, with a question mark, from Schirmacher Oasis in Dronning Maud Land (Singh & Semwal 2000), but no material was available for examination. *Bryum mildeanum* is a Eurasian species found also in North Africa and Macaronesia, but has never been recorded beyond the Holarctic.

B. muehlenbeckii Bruch & Schimp. — This species was recorded from King George I. and Nelson I., South Shetland Is. (Chen *et al.* 1993, 1995; Li *et al.* 1998). Some voucher collections were examined (*Li 7, 25 & 40*, AAS, KRAM) and they clearly belong within *B. pseudotriquetrum*.

B. pseudomicron Müll.Hal. — This name was provisionally used for a specimen (*REL 754*, AAS, KRAM) from Visokoi I., South Sandwich Is., and subsequently published by Convey *et al.* (2001). Correctly, this material belongs within *B. dichotomum*.

B. urbanskyi Broth. — The specimens reported from King George I., South Shetland Is., as *Bryum urbanskyi* (Ochyra & Ochi 1986) are, in fact, *B. pseudotriquetrum*. The leaves are red at base and recurved at the margin and the gemmae are absent. The type material of this species from Îles Kerguelen represents *B. dichotomum* (Ochyra 1998*a*).

Family XII. **MEESIACEAE** Schimp.

Meesiaceae Schimp., Coroll. Bryol. Eur.: 82. 1856. — Type: *Meesia* Hedw., *nom. cons.*

The Meesiaceae are a small family of acrocarpous mosses consisting of four genera with eight or nine species. The genera of this family, three of which are monotypic, show wide divergence in gametophyte characters but close similarity in the sporophyte. It is the erect-pyriform capsule with the urn curved and inclined, sometimes slightly so to cylindric, above an erect, long and straight apophysis which most clearly unifies this family. The capsules are large compared with the size of the gametophyte and, additionally, they are situated on remarkably long, slender and flexuose setae, this giving a characteristic appearance to the plants. The peristome is generally present (only *Neomeesia* Deguchi has a gymnostomous capsule) and variously developed but diplolepidous in structure. The exostome consists of 16 teeth which are densely and finely papillose or faintly transversely striate on the dorsal surface and they show various stages of reduction. In *Paludella* Brid. the teeth are nearly equal in height to the endostome segments but in *Meesia* Hedw. and *Amblyodon* P.Beauv. they are much shorter than the endostome and blunt or truncate at the apex. The endostome consists of a low basal membrane, closely adherent to the exostome teeth, and narrow, long, linear and keeled segments which are narrowly perforated along the median line. The latter are usually irregularly lobed at the margins and sometimes united at the tips, and often with irregular, discontinuous thickenings on the dorsal surface. Cilia are lacking or vestigial, but represented by irregular chains of nodose deposits on the membrane because of the incomplete resorption of the inner peristomial layer. The operculum is bluntly conic and the calyptra is cucullate.

Gametophytically, genera of the Meesiaceae are strongly diversified with regard to the leaf shape and areolation. The leaves are unicostate, keeled or concave, arranged in 3–5 ranks, erect or erect-clasping at the base, erect-spreading, wide-spreading, squarrose or squarrose-recurved above and vary from ovate, lanceolate, oblong- or ovate-lanceolate, lingulate to obovate. The upper laminal cells are thick- to thin-walled, smooth or mammillose and uni- to bistratose, whereas the basal cells are enlarged and bulging. The costa is strong, ceasing well below the apex or subpercurrent, in cross-section with small, ventrally exposed guide cells and a well developed dorsal stereid band or with large, substereid median and small outer cells. The leaf margins are entire or denticulate to serrate distally, recurved to plane or nearly so. The stems are usually densely tomentose below with rhizoids arising from lenticel-shaped clusters of initials. The axillary hairs are peculiar and consist of a single, markedly elongate, cylindrical or club-shaped, hyaline distal cell and 1–6 short, quadrate to short-rectangular proximal cells that are lilac, pale purplish or brown, but not brick-red as sometimes described (Griffin & Buck 1989).

The essential differences in the gametophyte and sporophyte characters between the genera of the Meesiaceae prompted Abramova (1956) to subdivide the family into three subfamilies, although the names of these taxa were not validly published. This classification was subsequently developed by Matteri & Ochyra (1999) and the family was divided into four monogeneric subfamilies: Meesioideae, Paludelloideae Matteri & Ochyra, Neomeesioideae Matteri & Ochyra and Amblyodontoideae Matteri & Ochyra. On the basis of molecular evidence, Buck & Goffinet (2000) transferred the bryalean genus *Leptobryum* to the Meesiaceae and Ochyra *et al.* (2003) recognised the separate subfamily Leptobryoideae to accommodate this otherwise morphologically discordant element in this family. Although *Leptobryum* and the typical meesialean genera share a pyriform capsule with a well developed neck, more or less similar costal anatomy and the shape of the axillary hairs, the genus is here excluded from the Meesiaceae and returned to the Bryaceae on account of its perfect bryoid peristome.

The Meesiaceae are predominantly a Northern Hemisphere family comprising large perennial mosses associated primarily with wet habitats. The family consists of one oligotypic genus *Meesia* Hedw. with five species and three monotypic genera, *Paludella*, *Amblyodon* and *Neomeesia*. In southern South America the Meesiaceae are represented by three species (Matteri & Ochyra 1999), namely the Fuegian-Patagonian endemic *Neomeesia paludella* (Besch.) Deguchi (Deguchi 1983) and the bipolar *Amblyodon dealbatus* (Hedw.) P.Beauv. (Ochyra 1992*b*) and *Meesia uliginosa* Hedw. which extends to the northern maritime Antarctic (Ochyra & Lewis Smith 1999).

41. **MEESIA** Hedw., *nom. cons.*

Sp. Musc. Frond.: 173. 1801. — Lectotype (*vide* Grout 1935: p. 181): *Meesia longiseta* Hedw.

Plants medium-sized to large, loosely or densely tufted, often densely tomentose below, dark green or yellowish to brown or blackish. *Stems* erect, simple or sparingly branched by innovations, in cross-section obtusely rounded-triangular to rounded-pentagonal, comprising a distinct central strand, 3–6 layers of large, thin-walled, hyaline cells and 2–5-stratose cortex of small, stereid cells with incrassate walls or somewhat larger, substereid cells with moderately thickened walls, sometimes with a distinct hyalodermis of large, hyaline cells with thin walls; *rhizoids* extensively but irregularly branched, red-brown, densely papillose, arranged on almost orbicular macronematal initials in a median axillary position relative to the leaves; *axillary hairs* filiform, 3–10-celled, consisting of an elongate, cylindrical to claviform, hyaline terminal cell and 2–4 short-rectangular to subquadrate, lilac to purplish basal cells. *Leaves* moderately crowded, not much to strongly contorted on drying, erect, erecto-patent, squarrose to squarrose-recurved, multiranked, sometimes trifarious, not or distinctly decurrent, oblong-lanceolate, oblong-ovate to lance-ligulate or linear-lingulate, not or distinctly keeled above the shoulders, rounded or acute at the apex or gradually narrowed to a slenderly acute apex; *margins* entire or distinctly serrulate above the shoulders, plane to recurved or revolute; *costa* single, strong, broader at the base, vanishing near or well below the apex, in transverse section covered on the ventral and dorsal side with a distinct epidermis of large, firm-walled cells and consisting of a large central group of ± homogeneous, fairly large substereid or small stereid cells with strongly incrassate walls and a row of distinct or indistinct guide cells immediately below the ventral epidermis; *laminal cells* unistratose throughout, transparent, shortly rectangular, smooth or slightly mammillose, rather thick-walled in the distal portion, elongate, rectangular, bulging and hyaline near the base, not differentiated in the basal angles. *Autoicous* or *dioicous*. *Setae* usually very long and slender, flexuose when dry, smooth, reddish, yellow- or red-brown. *Capsules* elongate, pear-shaped, curved and asymmetric from an erect, elongate apophysis comprising up to ½ of the capsule length, inclined to horizontal, smooth, brown; *annulus* narrow and poorly differentiated; *operculum* bluntly conic to hemispheric *exothecial cells* thick-walled, sometimes with collenchymatous thickenings, isodiametric, quadrate to hexagonal on the ventral side, rectangular on the dorsal side of the capsule, becoming oblate in 5–10 rows at the orifice; *stomata* numerous in the neck, superficial, bicellular, with oval pori; *peristome* double; *exostome teeth* about 2–4 times shorter than the segments, inserted far below the mouth, brownish to hyaline, with prominent ventral trabeculae, rounded or truncate at the apex, minutely papillose or cross-striolate; *endostome* smooth, consisting of narrow, keeled and narrowly slit segments joined at the base in a low basal membrane, often connected by transverse nodulose filaments, with no or rudimentary cilia. *Spores* large, irregularly spherical, finely papillose. *Calyptra* small, cucullate, fugacious.

Meesia is easily recognised when sporophytes are present because no other moss genus has such a long seta with a strongly curved, asymmetric capsule above the distinct neck and with the peristome including strongly reduced exostome teeth that are greatly exceeded by the longer endostome. When sterile, the genus is difficult to circumscribe because of differing facies associated with the leaf shape and arrangement. The enlarged and bulging proximal and small and firm-walled distal laminal cells coupled with a fairly broad costa lacking guide cells or with a row of larger deuters situated immediately below the ventral epidermis and characteristic axillary hairs with large club-shaped terminal cells roughly characterise the genus. *Meesia* has not been monographed on a worldwide scale but most individual species placed in it have been taxonomically assessed and the superbly illustrated revision of the genus (and the family Meesiaceae) in the former Soviet Union is still useful (Abramova 1956).

Meesia is a small genus of five or six species. Four of these are widely distributed from the arctic to temperate regions in the Northern Hemisphere (Ochyra *et al.* 1988*a,b,c,d*), of which two occur at isolated disjunct stations beyond the Holarctic, namely *M. triquetra* (Jolycl.) Ångstr. in New Guinea (Norris *et al.* 1999) and south-eastern Australia (Willis 1955; Streimann & Touw 1981) and *M. uliginosa* in southern South America (Matteri & Ochyra 1999) and Antarctica (Ochyra & Lewis Smith 1999). There have been several species described in this genus from outside the Holarctic, but all of them, except for one or two, are identical to other species of the genus, for instance *M. hymenostoma* Cardot & Broth. is conspecific with *M. uliginosa* (Matteri & Ochyra 1999), or correctly belong within other genera, for example *Meesia kenyae* P.de la Varde from tropical East Africa is conspecific with *Ceratodon purpureus* (Ochyra 2001*a*), *M. novae-zealandiae* Dixon & Sainsbury from New Zealand is identical to *Philonotis pyriformis* (R.Br.bis) Wijk & Margad. (Fife 1995), *M. patagonica* Dusén is inseparable from *Dicranella hookeri* (Matteri & Ochyra 1999) and *M. lavardei* Thér. from the Andes of Colombia correctly belongs within *Barbula* (Churchill *et al.* 1995). One species, *Meesia austrogeorgica* Müll.Hal., was described from subantarctic South Georgia but it is conspecific with *Philonotis polymorpha* (Clarke 1973*a*; Ochyra 1998*a*). The genus has been reviewed in Antarctica and one species, *M. uliginosa*, is known to occur in the northern maritime Antarctic (Ochyra & Lewis Smith 1999).

1. **Meesia uliginosa** Hedw. Fig. 227

Meesia uliginosa Hedw., Spec. Musc. Frond. 173. 1801. — TYPE CITATION: In pratis turfosis paludosis Hessiae, Saxoniae, aliisque Germaniae locis, in rupibus alpestribus calcareis Austriae vulgaris; also references to Dillenius (1741: p. 389, pl. 49, f. 58), Linnaeus (1763: p. 1585), Murray (1784: p. 949), Hedwig (1787: p. 1, pl. 1–2) and Dickson (1790: p. 7). [Lectotype (*selected here*): the specimen glued to the sheet and marked 'a' bearing the label 'Meesia uliginosa Hedw. Cryptog. p. 1. t. 1. f. 1–2 Bryum trichodes Linn. {manu Hedwig} b. angustifolia alp Helvet. c. longirostris Brid. mt. d. stricta a Bridel alp Carinth {manu Schwägrichen} – G-Hedwig/ Schwaegrichen!].

→

Figure 227. *Meesia uliginosa*. — **1**. Habit. **2**. Portion of shoot, dry. **3**. Portion of shoot, wet. **4**. Male plant with two perigonia. **5**. Cross-section of stem. **6–10**. Axillary hairs. **11–16**. Leaves. **17–18**. Leaf apices. **19**. Upper lamina cells below apex. **20**. Mid-leaf cells. **21**. Basal leaf cells. **22–27**. Cross-sections of leaves, a sequence from apex to base. — [1, 5–6, 8–11, 18, 21–22, 25–27 from *RO 1157/80*; 2–3, 13–17, 20 from *RO 1240/80*; 4, 7, 23–24 from *RO 1625/80*; 12, 19 from *RILS 7610A*]. — Scale bars: a – 1 mm (11–16) and 200 µm (6–10, 17–21); b – 100 µm (22–27); c – 1 mm (1–4); d – 50 µm (5).

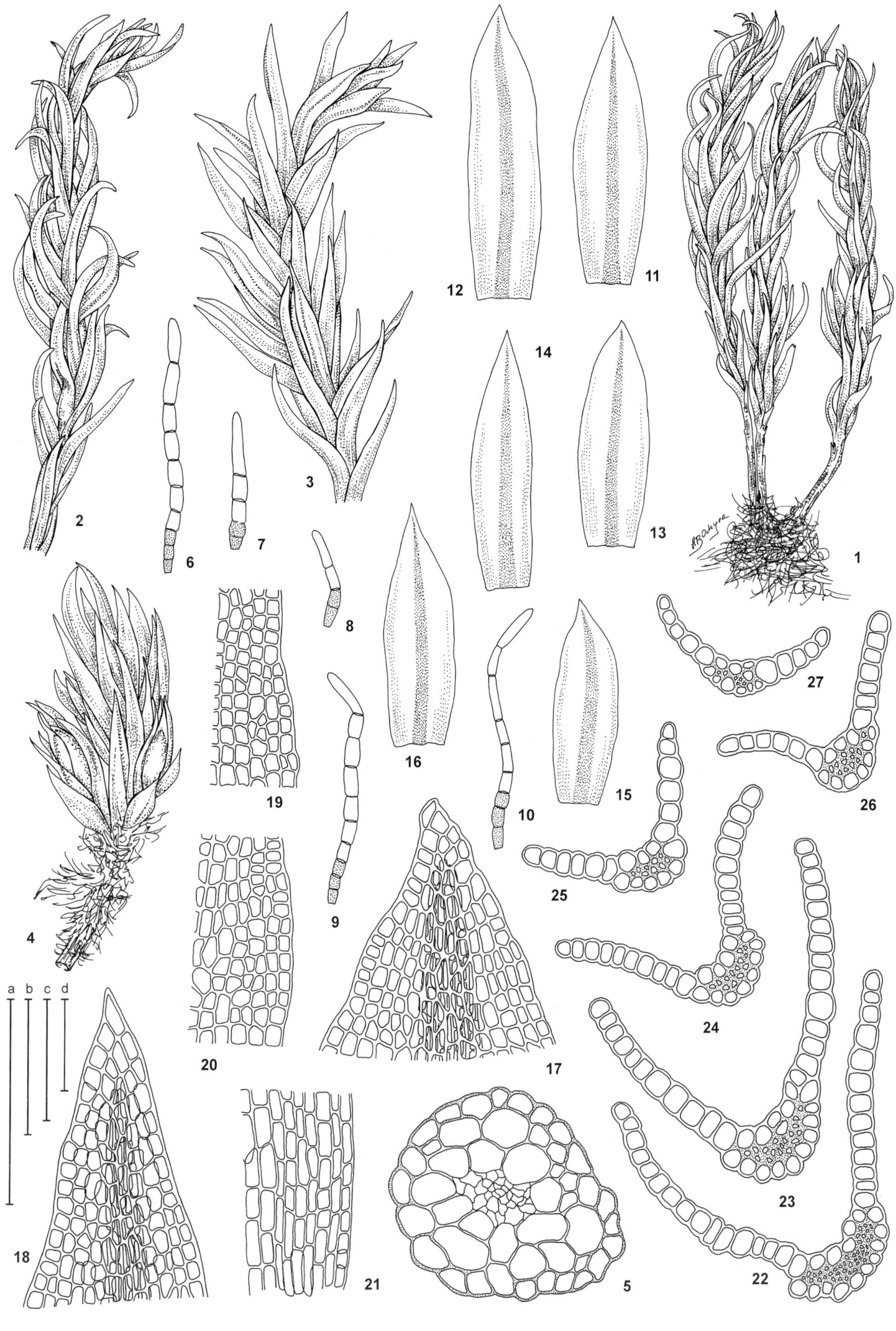

1
2
3
4
5
6
7
8
9
10
11
12
13
14
15
16
17
18
19
20
21
22
23
24
25
26
27
a b c d

Ceratodon kinggeorgicus Kanda, Hikobia 9: 324, *f. 3.* 1986. — TYPE CITATION: [Antarctica, South Shetland Islands] King George Island, Potter Cove (70 m alt., coll. H. Kanda 120. [Holotype: 'Herbarium of National Institute of Polar Research Tokyo. Japan 790608–054 Ceratodon kinggeorgicus Kanda Am 6 S. Shetland Islands Potter Cove, King George Island Sp: sterile 621400S 0584000W 70 m Coll. H. Kanda 120 790130 Det. H. Kanda Field No. 120' – NIPR!]. First synonymised by Ochyra & Lewis Smith (1998: p. 6).

Plants small, in compact, somewhat lustrous tufts, sordid-, yellowish- or blackish-green to dark green above, light brown to brownish-green below, rather sparsely radiculose below with light brown, branched, slightly papillose rhizoids. *Stems* erect, 0.5–1.5 cm tall, slender, simple or forked, brown, in transverse section oval, consisting of a relatively large central strand, 2–3 layers of large, thin-walled medullary cells and undifferentiated epidermis with brownish cell walls; *axillary hairs* numerous, filiform, slender, 4–10-cellular, with 2–3 brown to reddish-brown basal cells and hyaline upper cells. *Leaves* arranged in many rows, densely set, appressed, straight or somewhat contorted when dry, erect-spreading when wet, oblong-lanceolate to ligulate, mostly shortly acute to obtuse, 1.0–1.5 mm long, (0.2–)0.3–0.4 mm wide, not decurrent, often keeled; *margins* flat, entire; *costa* stout, 90–110 µm wide at base, occupying about ⅓–½ the width of the leaf base, vanishing just below the apex, in transverse section with small, nearly homogeneous cells with strongly incrassate walls, forming the central stereid band, surrounded by a row of large epidermal cells on the ventral and dorsal surfaces; *laminal cells* unistratose throughout, translucent, smooth, not or slightly bulging in transverse section, subquadrate to shortly rectangular distally, 18–25 µm long, 13–20 µm wide, firm- to thick-walled; *basal cells* rectangular, bulging, thin-walled, up to 50 µm long, 15–20 µm wide. *Asexual reproduction* unknown. *Autoicous* or *dioicous*. *Perichaetial leaves* larger than the vegetative leaves, ovate-lanceolate, rather abruptly narrowed into the short apex. [*Setae* 2–4(–7) cm long, thin, flexuose, reddish. *Capsules* light brown, lustrous, asymmetric, narrowly pyriform, 1.5–3.5 mm long, curved with an erect, straight neck; *operculum* hemispheric, flat-topped; *exothecial cells* thick-walled on the ventral side; *exostome teeth* short, truncate, smooth, hyaline with yellowish-brown lamellae; *segments* of the endostome long, smooth, yellowish-brown. *Spores* 40–50 µm in diameter, yellowish-green, finely papillose. *Calyptra* small, cucullate, fugacious.]

Chromosome numbers: unknown in the Antarctic and Subantarctic; $n = 13$ (1 record), $n = 13 + \text{m}$ (1 record), $n = 14$ (2 records) and $n = 14 + \text{m}$ (1 record) from the Holarctic (Fritsch 1991).

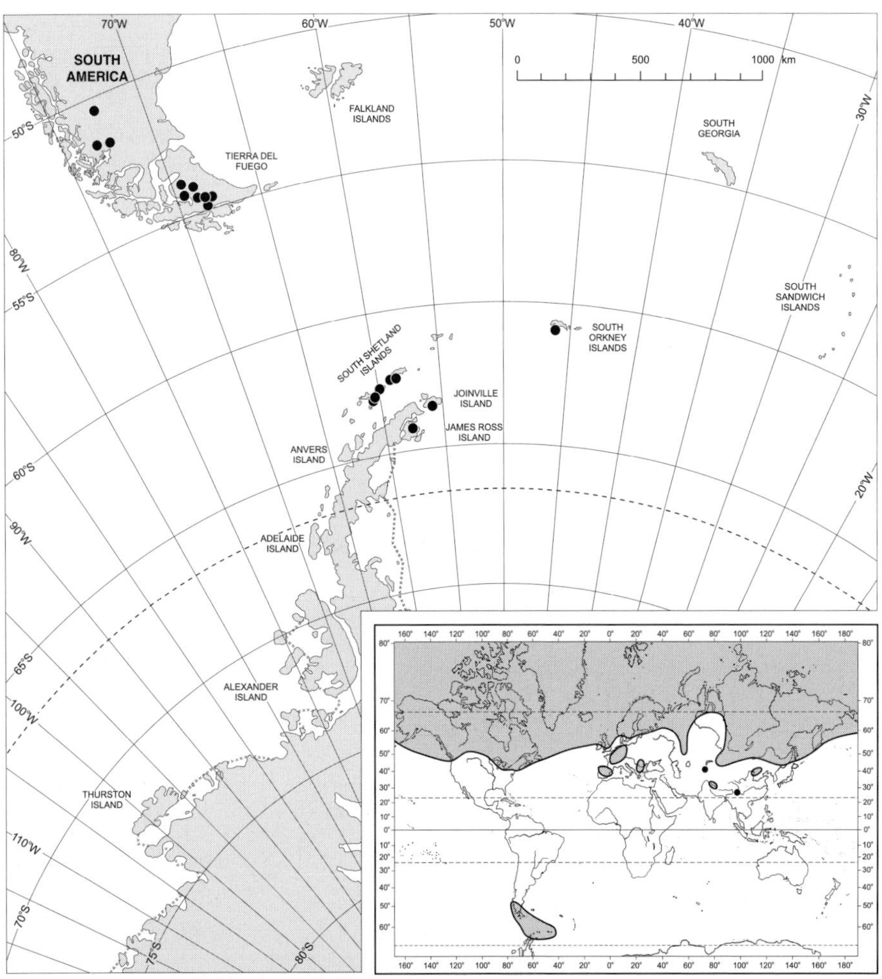

Figure 228. Distribution of *Meesia uliginosa* in the Fuegian region and Antarctica. Inset: global distribution.

Discussion. When sterile the species is rather inconspicuous and, *in situ*, it can be easily overlooked or mistaken for some bryalean mosses, for instance *Pohlia nutans*, or various phenotypes of *Ceratodon purpureus* with blunt-tipped leaves. The external similarity in habit and leaf areolation to the latter species is so striking that Kanda (1986*a*) described material from King George Island as a new species, *C. kinggeorgicus*, which in fact is inseparable from *Meesia uliginosa* (Ochyra & Lewis Smith 1999). This species is distinguished by its erect, oblong-lanceolate or ligulate, shortly apiculate to obtuse leaves with a very broad costa which occupies about half of the leaf base. In cross-section the costa shows nearly homogeneous, small substereid or stereid inner cells surrounded by an outer layer of large, thinner-walled epidermal cells and this character allows the species to be safely distinguished from *Ceratodon*, *Bryum* and *Pohlia* in which the costa has a distinct row of large guide cells.

Despite its distinctiveness *Meesia uliginosa* varies considerably in size, density of shoots, leaf shape and length of seta, but Antarctic populations represent a remarkably uniform phenotype which is characterised by smaller-sized plants, growing in low compact turves, with shortly acute leaves. In the Northern Hemisphere such plants are typical of high-alpine habitats and are often recognised as *M. uliginosa* var. *minor* (Brid.) F.Weber & D.Mohr. However, these seem to be merely an ecological variant of no taxonomic value because they are linked with taller plants growing in subalpine or lowland rich fens by a continuous series of intermediates.

Reproduction in Antarctica. Sporophytes unknown, but sexual organs are present on some plants and this suggests that plants with sporophytes may exist here. Fully mature sporophytes are frequent in southern South America (Matteri & Ochyra 1999).

Habitat. Forming small dark green to blackish-green compact turves in moist, acidic or basic sites, frequently on gravelly, sandy, marble or schist soil, or in soil-filled crevices in rocks, less frequently on wet or damp ground at the margins of melt water channels below late snow patches, occasionally also amongst *Deschampsia antarctica* and *Polytrichastrum alpinum* or in mats of *Sanionia georgicouncinata*. Turves of *Meesia uliginosa* are very often interwoven with shoots of *Cephaloziella varians* and mosses including *Pohlia cruda*, *Bartramia patens* and *Bryum* species.

World range. A bipolar species having a continuous circumpolar arctic-boreal-montane range in the Northern Hemisphere (Ochyra *et al.* 1988*a*) (Fig. 228, inset). It is common throughout the Arctic, with decreasing frequency in the southern part of the Holarctic where it extends as far south as Sikkim in the Himalayas and Sichuan Province in China (Li & He 2007). In the Southern Hemisphere the species is known from a few stations in southern South America including Santa Cruz Province in Argentina, Tierra del Fuego and the offshore archipelagoes of the northern Antarctic Peninsula.

Distribution in Antarctica. The species is locally frequent only on King George I. in the South Shetland Is., and rare in the South Orkney Is. (Signy I.) and on James Ross I. on the north-eastern coast of the Antarctic Peninsula where it reaches its southernmost station at lat. 63°54′S (Fig. 228). Its altitudinal range extends from near sea level to 200 m.

Specimens examined. WEST ANTARCTICA. SOUTH ORKNEY IS. **Signy I.** Thulla Pt., *ca* 65 m, *MWH 155D* (AAS, KRAM); Foca Cove, 35 m, *RILS 5009* (AAS, KRAM) and 45 m, *RILS 3174* (AAS, KRAM); Stygian Cove, 25 m, *RILS 10819* (AAS, KRAM); hill SW of Mirounga Cove, 35 m, *RILS 8156* (AAS, KRAM); Marble Knolls, 10 m, *RILS 10797A* (AAS, KRAM); Mooring Pt., 50 m, *RILS 5082* (AAS, KRAM); Observation Bluff, *ca* 40 m, *RILS 3120* (AAS, KRAM).

SOUTH SHETLAND IS. **King George I.** ***Admiralty Bay***: Red Hill, 100 m, *RO 1091/80* (KRAM); Blue Dyke, 120–130 m, *RO 1157/80 & 1171/80* (KRAM); Creeping Slopes, 140 m, *RO 1240/80* (KRAM); Sphinx Hill, 95 m, *RO 250/80* (KRAM); Llano Pt., 50 m, *RO 4901/79* (KRAM); Hala, 30–35 m, *RO 5202/79 & 2386/80* (KRAM); Krzesanica, 130 m, *RO 1661/80* (KRAM); Ambona, 75–80 m, *RO 1625/80 & 1639/80* (KRAM); Skua Cliff, 105 m, *RO 613/80* (KRAM); Ubocz, 95–125 m, *RO 4956/79, 2372/80 & 2378/80* (KRAM); Jersak Hills, 180 m, *RO 5144/79* (KRAM). ***Martel Inlet***: Smok Hill, 5 m, *RO 2056/80* (KRAM); Basalt Pt., 10 m, *RO 2210/80* (KRAM); Mt. Wawel, 40 m, *RO 2116/80* (KRAM). ***Viéville Glacier***: Rembiszewski Nunataks, 200 m, *RO 2725/80* (KRAM). ***Fildes Peninsula***: Bellingshausen Station, 15 m, *RO 2420/80* (KRAM); Fildes Strait, *Li 890102* (AAS, KRAM). ***Potter Peninsula***: Stranger Pt., 5–45 m, *Kanda 112, 113, 116 & 117* (NIPR – paratypes of *Ceratodon kinggeorgicus*); Three Brothers Hill, 70 m, *Kanda 120* (NIPR – holotype of *Ceratodon kinggeorgicus*). **Robert I.** Coppermine Cove, 25–35 m, *RILS 927* (AAS, KRAM). **Livingston I.** ***Hurd Peninsula***: Spanish Base, *Acuña 7004* (KRAM); False Bay, *RMS 69-415* (US); Johnsons Dock, 20 m, *Monroy 116A* (AAS, KRAM).

EAST ANTARCTIC PENINSULA. **Joinville Island group.** ***Joinville I.***: NE slope of Mt. Alexander, 15–50 m, *RILS 3705* (AAS, KRAM). **James Ross Island group.** ***James Ross I.***: north of Andreassen Pt., 10 m, *RILS 7610A & 7611C* (AAS, KRAM).

Family XIII. **BARTRAMIACEAE** Schwägr.

Bartramiaceae Schwägr. *in* Willd., Spec. Pl. **5**(2): 90. 1830 [‘Bartramiae’]. — Type: *Bartramia* Hedw., *nom. cons.*

The Bartramiaceae are a large and one of the most conspicuous and clear-cut of all moss families. The apple-shaped capsule which is usually erect or inclined to pendulous, symmetric or not, often oblique-mouthed and mostly longitudinally furrowed when dry, is an outstanding characteristic of the family. Less frequently, they are irregularly wrinkled, warty or rugose, to smooth on drying. The peristome, if present, is variously developed but diplolepidous in structure. In its best development it is double and bryoid, but frequently there is a strong tendency for its reduction and a few species are cleistocarpous. The exostome teeth are smooth or variously ornamented, yellow-brown to red-yellow, mostly with high trabeculae at the back, sometimes fused at the tips. The endostome is sometimes rudimentary and occasionally it is adherent to the exostome. It is somewhat shorter than the exostome, keeled and consists of a fairly high basal membrane, lanceolate segments and 1–3 short, never appendiculate, cilia which are sometimes lacking. Each segment is split along the median line nearly to the base of the basal membrane, and each half segment diverges outward toward the cilia, where they often meet the diverging half of the adjacent segment, thus framing the cilia. The setae vary from elongate and bearing exserted capsules to very short with subsessile capsules. They are smooth, straight or, occasionally, flexuose to arcuate. The operculum is typically low convex-umbonate to high conic or rarely rostrate. The spores are large, spherical to reniform and often densely coarsely papillose (Hirohama 1977).

A prominent feature of the gametophyte in this family is the *Climacium*-type of branch development which is characterised by the presence of dormant branch buds (Akiyama 1990; Akiyama & Nishimura 1993). This is very unusual in acrocarpous mosses in which, with a few exceptions only, *Bryum*-type branch development dominates. The Bartramiaceae have a broader diversity of morphological variation in axillary hairs which are 2- to multicelled and often consistent within a genus. The leaves tend to be narrow and often their base is differentiated. They are mostly lanceolate-acuminate, often with a sheathing base, rarely ovate and obtuse, imbricate to widespreading or recurved, straight, flexuose or crispate. The leaf margins are variously serrate with two-ranked teeth or with cell ends projecting as mammillae with contiguous cells having mammillae more or less fused. The costa is single, well or ill defined, often scabrous or spinose on the dorsal surface, subpercurrent to long-excurrent as a brown or hyaline awn. The laminal cells tend to be sharply angled, narrow, and strongly prorate or with papillae over the lumina, rarely smooth or mammillose. The stems are typically erect or occasionally decumbent, unbranched or with subfloral whorls of innovations and, in transverse section, they consist of a distinct central strand and very often an outer hyalodermis of large, thin-walled cells (Kawai 1982). The stems are often densely matted below with a reddish tomentum of strongly papillose or, rarely, smooth rhizoids (Hirohama & Iwatsuki, 1980).

Griffin & Buck (1989) subdivided the Bartramiaceae into three subfamilies, mainly on axillary hair morphology, which show a broader diversity than in any of the other related families of the Bryales. In the type subfamily Bartramioideae, the axillary hairs are composed of 3–10 largely elongate cells, whereas in the subfamily Breutelioideae they are generally short and are composed of short cells and terminated by a globose hyaline cell. The third subfamily Conostomoideae has somewhat intermediate axillary hairs and they are short and composed of 2–3 short cells, with a claviform or undifferentiated terminal cell.

The Bartramiaceae are a medium-sized family consisting of ten genera with about 280 species (Crosby *et al.* 2000), but realistically this number could be markedly reduced when some large and under-investigated genera such as *Philonotis*, *Breutelia* and *Bartramia* are fully taxonomically revised. The family has a cosmopolitan distribution and some species are important constituents of the flora and vegetation in polar regions. It is well represented in the cool temperate zone in the Southern Hemisphere and in the Subantarctic (Matteri 1968, 1973*a*, 1984, 1985*b*; Newton 1973, 1974*b*; Bell 1973*a*; Fransén 2004*b*, 2005), but in the Antarctic only four species and three genera have so far been recorded, namely *Conostomum* (Conostomoideae), *Philonotis* (Breutelioideae) and *Bartramia* (Bartramioideae).

KEY TO THE ANTARCTIC GENERA OF THE BARTRAMIACEAE

1. Leaves closely appressed in 5 distinct ranks; rhizoids smooth; operculum obliquely rostrate; peristome teeth fused at the tips, forming a perforated cone ...... 42. ***Conostomum***
1. Leaves not ranked, variously erect to spreading; rhizoids papillose; operculum conic to umbonate; peristome teeth not fused at the tips 2
    2. Leaves from an ovate, not clasping base lanceolate, erect-spreading; axillary hairs with rounded, short, more or less inflated terminal cell; leaves unistratose .... 43. ***Philonotis***
    2. Leaves from an oblong-ovate, clasping base narrowly lanceolate to subulate; axillary hairs with elongate terminal cell; leaves bistratose .... 44. ***Bartramia***

42. CONOSTOMUM Sw. *ex* F.Weber & D.Mohr

Conostomum Sw. *ex* F.Weber & D.Mohr, Naturh. Reise Schwedens: 121. 1804. ≡ *Bartramia* Hedw. sect. *Conostomum* (Sw. *ex* F.Weber & D.Mohr) Bruch & Schimp. *in* Bruch, Schimp. & W.Gümbel, Bryol. Eur. 4: 1. 1851 [Fasc. 46–47 Consp. Vol. 4: iii]. — Type: *Conostomum arcticum* Sw. *ex* F.Weber & D.Mohr, *nom. illeg.* [*Mnium tetragonum* Dicks. *ex* Hedw.; *Conostomum tetragonum* (Dicks. *ex* Hedw.) Lindb.].

Plants small to medium-sized, densely tufted, glaucous, green or yellow-green above, red-brown below. *Stems* erect, simple or irregularly branched, often with a few subperichaetial innovations, densely tomentose below with smooth, reddish-brown, branched rhizoids, in transverse section rounded-pentagonal, consisting a large central strand and a distinct hyalodermis of thin-walled, often collapsed cells on the outer surface; *axillary hairs* short, filiform, 2–3-celled, composed of short cells, hyaline throughout or with a reddish-brown basal cell. *Leaves* crowded, stiff, imbricate, erect-appressed when dry and wet, in five distinct, spiral ranks, not plicate, triangular to lanceolate or oblong-lanceolate, rarely ovate or oblong-ovate to lingulate, slenderly acuminate or rounded-obtuse at the apex; *margins* flat or narrowly reflexed above, sharply serrate to subentire in the upper part; *costa* long excurrent to subpercurrent, narrow to very broad and occupying about ½ to ¾ the width of the leaf lamina, mammillose-toothed on the dorsal side; *laminal cells* unistratose to variously bistratose near the costa; *upper cells* firm-walled, short-rhomboidal to rectangular, smooth or prorate from the upper ends on both surfaces towards the apex; *basal cells* rectangular, somewhat broader, undifferentiated in the alar region or quadrate, enlarged and bulging in small groups in the basal angles. *Autoicous* or *dioicous*. *Perichaetial leaves* larger than the vegetative leaves, semi-sheathing at the base, elongate-lanceolate, with a thinner costa and longer cells. *Setae* short or elongate, straight or curved immediately below the capsule, smooth, yellow-brown, untwisted. *Capsules* broadly obovoid to obloid with a short, swollen neck, horizontal to cernuous and asymmetric, sometimes erect and straight, wrinkled to furrowed when dry, small-mouthed, stegocarpous or cleistocarpous; *operculum* obliquely stoutly conic-rostrate; *annulus* none; *exothecial cells* ± isodiametric, hexagonal, firm- to thick-walled, sometimes in indistinct bands of differing cell wall thickness; *peristome* single or occasionally lacking, with 16 narrowly triangular to linear-lanceolate, orange to red-brown, smooth or finely papillose to faintly striolate at the base, narrowly bordered teeth united at the apex. *Spores* large, spherical to reniform, coarsely but lowly baculate-insulate. *Calyptra* cucullate, naked, fugacious.

Conostomum is a very distinct genus that is defined by 5-ranked, erect-appressed and imbricate leaves, giving the stem a ridged appearance, smooth rhizoids and 3-celled axillary hairs which are composed of cells about the same size, with a not inflated apical cell, hyaline throughout or with a brownish pigmented basal cell. Sporophytically, the most characteristic feature of the genus is the single peristome with connate apically teeth which are erect and form a cone preventing spore escape (the Latin name of the genus alludes to this character). The lower parts of the teeth become reflexed on drying but the cohesion of the teeth causes the apical lattice to be drawn closer to the capsule mouth forming gaps between the teeth which allow gradual dispersal of the spores under dry conditions in a manner similar to that in *Funaria hygrometrica*.

The genus is readily related to the Bartramiaceae by the glaucous colour of the plants, stem with a hyalodermis and a well developed large central strand, prorate upper laminal cells, as well as globose and sulcate or wrinkled capsules with a conical-rostrate operculum and coarsely papillose spores. Owing to its distinctness, *Conostomum* is, besides *Bartramia*, the oldest genus recognised in the Bartramiaceae. This name was introduced by O. Swartz but it was validly published by Weber & Mohr (1804) who studied the material in his herbarium during their visit to Sweden in 1803. Only two years later Swartz (1806) thoroughly re-described the genus *Conostomum* and superbly illustrated two species then included in it, namely the northern *C. boreale* (= *C. tetragonum*) and the southern *C. australe* (= *C. pentastichum*).

Conostomum is a small genus of seven species, six of which occur in the temperate or montane regions in the Southern Hemisphere (Matteri 1984; Fife 1998). Two of these, *Conostomum magellanicum* and *C. pentastichum*, extend to the northern maritime Antarctic and the latter slightly crosses the equator and reaches southern Mexico (Griffin 1994*b*). The seventh species, *C. tetragonum*, has a wide, though highly dissected, circumpolar, arctic-boreal-montane range in the Northern Hemisphere. A worldwide revision of *Conostomum* has been published by Frahm *et al.* (1996) and in South Georgia the genus was surveyed by Bell (1973*a*).

1. Leaves lanceolate to linear- or ovate-lanceolate, acuminate to aristate; costa narrow, fairly well-defined, at least distally, long-excurrent; leaf margin narrowly reflexed above ... . 1. ***C. pentastichum***

1. Leaves ovate to oblong-ovate, rounded-obtuse; costa very broad, ill-defined, vanishing below the apex; leaf margin flat .... 2. ***C. magellanicum***

1. **Conostomum pentastichum** (Brid.) Lindb. Fig. 229

Conostomum pentastichum (Brid.) Lindb., Öfvers. Förh. Kongl. Svenska Vetensk.-Akad. 20: 392. 1863. ≡ *Bartramia pentasticha* Brid., Musc. Rec. 2: 134, *t. 1*, *f. 8*. 1803. — Type citation: In Freto magellanico, unde a Commersono relata est mihique a cel. Jussieu communicata, habitat. [Lectotype (*selected here*) {or perhaps holotype}: 'Conostomum australe Sw. Barthramia pentasticha Nobis E freto Magellanico Comersonus retulit Paris 1797' – B-Bridel!].

Conostomum australe Sw., Neues J. f. Bot. 1(3): 31, *f. 6*. 1806, *nom. illeg. incl. spec. prior.*

Bryum angulosum Menzies *in* Sw., Neues J. f. Bot. 1(3): 32. 1806, *nom. inval. in synon.*

Conostomum antarcticum Sw. *ex* F.Weber & D.Mohr, Naturh. Reise Schwedens: 122. 1804, *nom. nud.*, ***syn. nov.***

Conostomum rhynchostegium Müll.Hal. *in* Neumayer, Int. Polarforsch. Deutsch. Exped. 2: 303. 1890. — Type citation: South Georgia, Hoch-Plateau and Whaler Bay, leg. Will, 1882–1883. [Lectotype (*vide* Bell 1973*a*: p. 46): 'Conostomum rhynchostegium C.Müll. Fundort: Quelle auf dem Hochplateau, in dicht verfiltzten Polstern, findet sich außerdem nicht selten an Bachufern (Kl. Wasserfall, Gletcherthal) Süd-Georgien, Januar 1883 Will' – M!; isolectotype: HBG!; syntypes: H!, HBG!, M!]. First synonymised with *Conostomum australe* by Cardot (1906*a*: p. 1004) and subsequently with *C. pentastichum* by Bell (1973*a*: p. 45).

Plants small to medium-sized, densely caespitose, forming large cushions or extensive turves, yellow-green to glaucous, somewhat lustrous and silky above, brownish below. *Stems* erect, 2–5 cm tall, irregularly branched by subperichaetial innovations, reddish-brown, moderately tomentose below with yellow- or reddish-brown to brownish, branched, smooth rhizoids, in cross-section rounded to rounded-pentagonal, with a large central strand, 5–6 layers of large, thin-walled medullary cells, 2–3-stratose cortex of substereid, small, thick-walled cells and unistratose hyalodermis of large, thin-walled cells, often with eroding outer walls, giving the stem a toothed wheel appearance; *axillary hairs* 3-celled, filiform, hyaline or with a brown-pigmented basal cell. *Leaves* crowded, stiffly erect-appressed when moist, not altered when dry, in 5 distinct ranks, lanceolate to oblong- or linear-lanceolate, acuminate to aristate, (1.4–)1.7–2.5(–3.5) mm long, 0.3–0.5 mm wide; *margins* unistratose, generally narrowly recurved on both sides in the upper ½ or more, plane below, serrate near the apex, with teeth frequently paired; *costa* single, yellow-brown, 90–120 µm wide at base, often not sharply demarcated from the lamina cells in the lower half, short- to long-excurrent as a smooth hair-point, (0.3–) 0.4–0.8 mm long, often becoming hyaline at the tips, strongly mammillose-toothed dorsally, weakly so ventrally, because of the projection of the distal ends of the superficial cells, in cross-section with not well defined guide cells, a large inner group of small stereid cells surrounded by a row of large, mammillose outer epidermal cells; *laminal cells* bistratose juxtacostally from the base to the apex with 4–10 rows of unistratose cells at margins, with strongly thickened adjoining walls and thin and bulging outer walls; *upper cells* variable in shape, mostly irregularly rhomboidal to narrowly rectangular, (15–)17–38(–60) µm long, 5–15(–20) µm wide, smooth or papillose at the upper ends on both surfaces towards the apex; *lower cells* thin-walled, lax, smooth, rectangular to hexagonal, up to 100 µm; *alar cells* not differentiated. *Autoicous*. *Inflorescences* terminal; *perigonia* large, cupulate; *perichaetial leaves* similar to the vegetative leaves but larger, 2.5–4.0 mm long, 0.5–0.6 mm wide, with linear, thick-walled laminal cells in the distal portion. [*Setae* 1–2 cm long, erect, yellow-red to yellow-brown, untwisted. *Capsules* suberect to horizontal, obloid to ellipsoid, gibbous, 1.5–3.0 cm long, 1.5–1.8 mm wide, sulcate when dry, distinctly narrowed at the mouth, 0.7–0.9 mm in diameter; *operculum* rostrate, slanted, to 1.2 mm long, asymmetric; *exothecial cells* hexagonal to oblong-hexagonal, thick-walled, becoming smaller, subquadrate, reddish-brown in 3–4 rows at the orifice; *stomata* phaneroporous, bicellular with narrow, elongate pori, numerous at the base of the urn; *peristome* single, fragile, consisting of 16 narrowly lanceolate to linear, yellow-reddish to reddish-brown teeth, to 900 µm long, inserted below the capsule rim, faintly striolated in the lower half, nearly smooth distally, conspicuously trabeculate on the dorsal surface, slit along the middle and fused at the tips. *Spores* 50–65 µm wide, oval-reniform, reddish-brown, with bulging hexagonal plates at the distal surface. *Calyptra* cucullate, fugacious, smooth.]

Chromosome number: $n = 8$ – South Georgia (Newton 1980).

Discussion. The pale green to yellow-green plants, sometimes with a bluish or whitish hue, with the unusual rigid ranking of the leaves along the stem and the costa excurrent as an aristate awn of varying length which is usually hyaline at the tip, make this species almost unmistakable for any other moss in Antarctica. Additionally, the strongly mammillose-toothed upper laminal cells are a useful microscopic character for the safe recognition of the species. In its 5-ranked leaves it is similar to *Conostomum magellanicum*, which differs in possessing a broadly rounded-obtuse leaf apex without hairpoint and a broad, ill-defined costa ending well below the apex. Frahm *et al.* (1996) considered *C. pentastichum* to be conspecific with the northern *C. tetragonum*, thus making the latter a bipolar species. This taxonomic conclusion cannot be accepted because *C. tetragonum* appears to be distinct in its entirely unistratose laminal cells, sharply delimited costa, not imperceptibly merging with the laminal cells, dioicous sex condition and smaller spores that are 40–50 µm in diameter.

Conostomum pentastichum has an interesting taxonomic and nomenclatural history. The species was first collected in the Magellan Strait area by Ph. Commerson, a naturalist to the French expedition which circumnavigated the world, under the command of L. A. Bougainville, in 1766–1769. The material was subsequently donated by A. L. Jussieu to various European botanists and was described as a new species, *Bartramia pentasticha* by Bridel (1803) and *Conostomum australe* by Swartz (1806), but the latter name is illegitimate because it includes the type of *B. pentasticha*. The latter author also cited a specimen collected in 1795 by A. Menzies, collected on Isla de los Estados near Tierra del Fuego during his voyage with Captain George Vancouver on the *Discovery* (1791–1795). Swartz (1806) intended to describe it as a new species, *Bryum angulosum*, and this name is cited in synonymy with *Conostomum australe*. He initially intended to use the name *C. antarcticum* for the austral species, and this name is cited by Weber & Mohr (1804).

Reproduction in Antarctica. Sporophytes are unknown, although both male and female inflorescences are present on a specimen from Coronation I. (*RW 133A*, AAS, KRAM), while a specimen from adjacent Signy I. (*RILS 457*, AAS, KRAM) bears perigonia. The species regularly and abundantly fruits on South Georgia, in southern South America and elsewhere throughout its wide geographical range in the Southern Hemisphere.

Habitat. It is a species of generally damp or wet gravelly soil, growing most often on flushed, north- or west-facing slopes, on exposed ledges and in rock crevices. The most frequent associates are *Andreaea regularis* and *Polytrichastrum alpinum* and its tufts are often intermingled with hepatics, for example *Cephaloziella varians* and *Lophozia excisa.*

→

Figure 229. *Conostomum pentastichum.* — **1**. Plant with mature capsules. **2**. Portion of stem, wet. **3**. Portion of cross-section of stem. **4–5**. Axillary hairs. **6–14**. Leaves. **15–16**. Upper lamina cells at margin. **17**. Mid-leaf cells at margin. **18**. Basal leaf cells. **19–21**. Cross-sections of leaves, a sequence from base to apex. **22–23**. Perichaetial leaves. **24**. Upper lamina cells of inner perichaetial leaf. **25–26**. Operculate capsule, dry. **27**. Deoperculate capsule, dry. **28**. Calyptra. **29**. Exothecial cells in middle of urn. **30**. Lower exothecial cells and stomata. **31**. Peristome teeth and rim cells of capsule. **32–37**. Spores. — [1, 8–9, 26 from *Poelt s.n.*, 14.i.1989; 2–3, 10–11, 19–21 from *RILS 460*; 4–5, 13–15, 17 from *RILS 1833*; 6, 22–25, 29–31, 35–37 from *Will s.n.*, 10.v.1883 (lectotype of *C. rhynchostegium*), M; 7 from *REL 1089*; 12, 16, 18 from *RILS 457*; 27 from *RDS 15613*; 28, 32–34 from *SWG 1313*]. — Scale bars: a – 1 mm (2, 22–23, 25–28) and 50 µm (19–21); b – 50 µm (17, 24); c – 100 µm (3, 18, 29–37); d – 1 mm (6–14), 500 µm (15–16) and 50 µm (4–5); e – 0.5 cm (1).

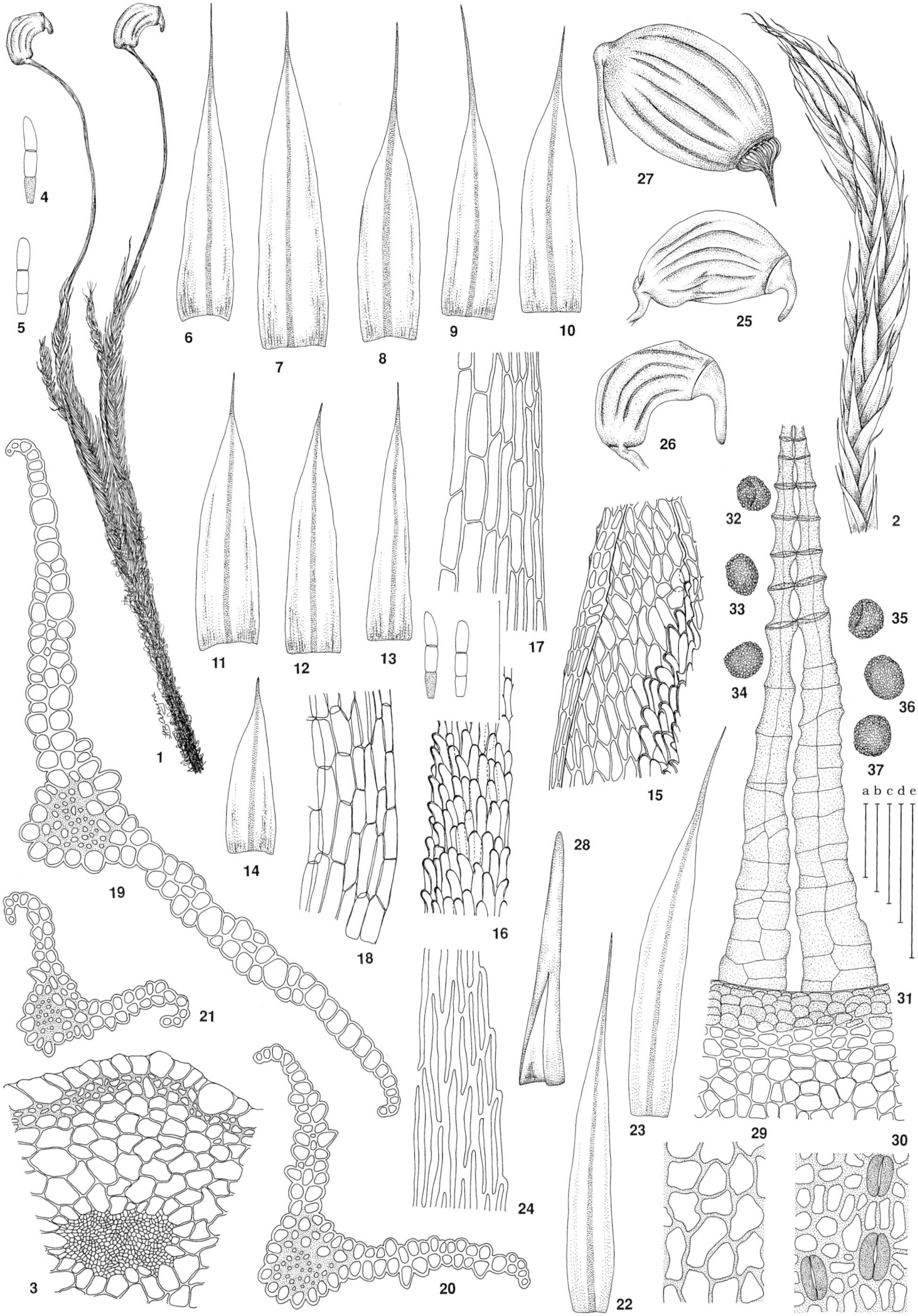
1
2
3
4
5
6
7
8
9
10
11
12
13
14
15
16
17
18
19
20
21
22
23
24
25
26
27
28
29
30
31
32
33
34
35
36
37
a b c d e

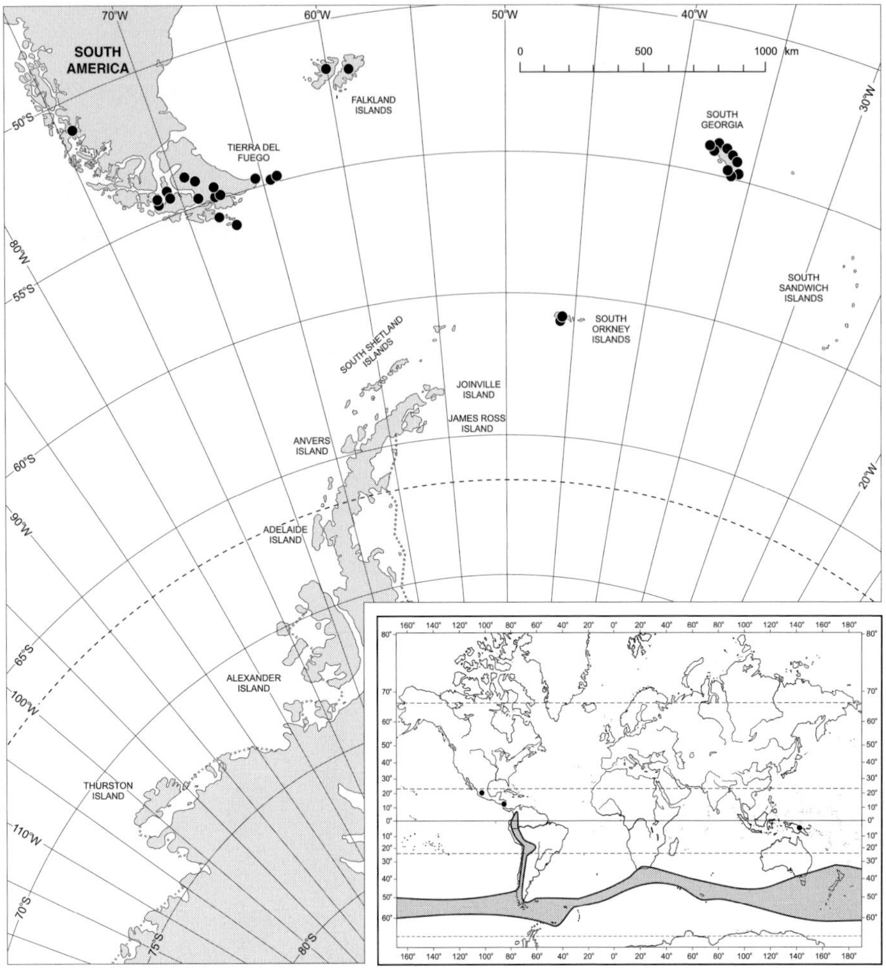

Figure 230. Distribution of *Conostomum pentastichum* in the Fuegian region and Antarctica. Inset: global distribution.

World range. A pan-south-temperate species widely distributed in New Zealand (Beever *et al.* 1992; Fife 1998), including the offshore Auckland Is. (Vitt 1979*a*) and Campbell I. (Vitt 1974*a*), south-eastern Australia (Streimann & Curnow 1989; Streimann & Klanzenga 2002), Tasmania (Dalton *et al.* 1991), South Africa (Magill 1987*b*) and southern South America (Matteri 1984, 1985*b*), as well as on the subantarctic islands of Macquarie (Seppelt 2004), Kerguelen (Mitten 1879; Hemsley 1884 both as *Philonotis australis*) and South Georgia (Bell 1973*a*), extending to the South Orkney Is. in Antarctica (Bell 1973*b*) (Fig. 230, inset). In addition, it penetrates the tropics occurring at high elevations in New Guinea (Virtanen 1999), and extends along the Andean Cordillera from Tierra del Fuego as far north as Costa Rica and Oaxaca Province in southern Mexico (Griffin 1994*b*; Delgadillo *et al.* 1995).

Distribution in Antarctica. *Conostomum pentastichum* is currently known exclusively from Coronation I. and numerous sites on Signy I. in the South Orkney Is. archipelago (Fig. 230). It occurs at an elevation of about 20–200 m.

Specimens examined. WEST ANTARCTICA. SOUTH ORKNEY IS. CORONATION I. Cape Hansen, *RILS 1832 & 1833* (BM, KRAM), 20–200 m, *RILS 3225* (AAS, KRAM) and 85 m, *RILS 5185* (AAS, KRAM); E side of Laws Glacier, *RW 113A* (AAS, KRAM) and *ca* 100 m, *RW 133A* (AAS, KRAM). SIGNY I. Rusty Bluff, *RILS 457* (AAS, H, KRAM, LE) *& 460* (BM); between Observation Bluff and Polynesia Pt., *REL 1087* (AAS, KRAM, NY, PC, S); Moraine Valley, *ca* 35 m, *REL 1089* (BM, KRAM, NY, PRE, S) and 100 m, *RILS 460* (BM, KRAM); Porteous Pt., 50 m, *RILS 8676A* (AAS, KRAM).

2. Conostomum magellanicum Sull. Fig. 231

Conostomum magellanicum Sull., Kew J. Bot. **2**: 316. 1850. — TYPE CITATION: all new species reported by Sullivant (1850) came 'from the vicinity of Orange Harbour, Tierra del Fuego' and were 'a part of the botanical collections made by the 'United States Exploring Expedition', under the command of Captain Charles Wilkes'. [Lectotype (*selected here*) {or perhaps holotype}: 'U.S. A. Ex. Wilkes 1838–1842 Conostomum magellanicum Sulliv. Orange Harbour, Fuegia' – FH-Sullivant!; isotypes: BM-Hooker!, NY!, PC!, PC-Cardot!].

Conostomum perangulatum Cardot, Bull. Herb. Boissier Sér. 2,

Figure 231. *Conostomum magellanicum*. — 1. Plant with mature capsules. 2. Portion of stem, wet. 3. Portion of cross-section of stem. 4–5. Axillary hairs. 6–13. Leaves. 14. Leaf apex. 15. Mid-leaf cells at margin. 16. Upper lamina cells at margin. 17. Basal leaf cells. 18–20. Cross-sections of leaves, a sequence from apex to base. 21–22. Perichaetial leaves. 23. Upper lamina cells of inner perichaetial leaf. 24. Operculate capsule, dry. 25. Lower exothecial cells and stomata. 26. Exothecial cells in middle of urn. 27. Peristome teeth and rim cells of capsule. 28–30. Spores. — [1, 24–30 from *SWG 2359*; 2 from *RO 559/80*; 3, 8, 10–11, 14, 16–17 from *RILS 6*; 4–5 from *RILS*; 6, 15, 21–23 from *CS 109* (type of *C. perangulatum*), PC; 7 from *U. S. Ant. Exp. Wilkes 1838–1842 s.n.*, Orange Harbour (isotype of *C. magellanicum*), BM; 9, 12–13, 18–20 from *RO 1638/80*]. — Scale bars: a – 1 mm (24); b – 1 mm (2) and 50 µm (15–16, 18–20); c – 100 µm (25–30); d – 1 mm (21–22); e – 100 µm (3, 14, 17, 23); f – 0.5 cm (1); g – 1 mm (6–13) and 50 µm (4–5).

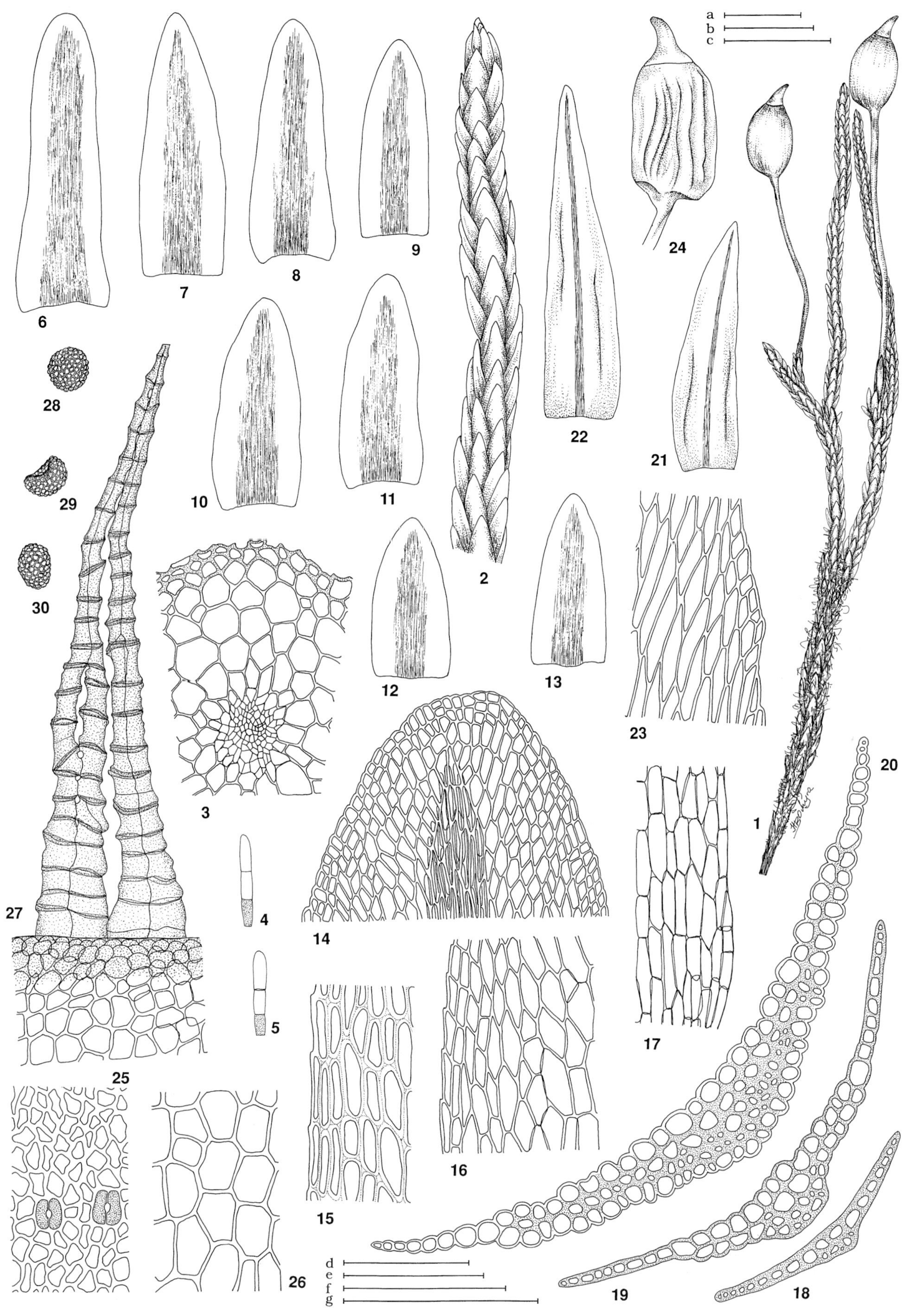
a
b
c
6
7
8
9
24
22
21
28
29
30
10
11
2
12
13
23
3
20
1
27
4
14
5
17
25
15
16
26
19
18
d
e
f
g

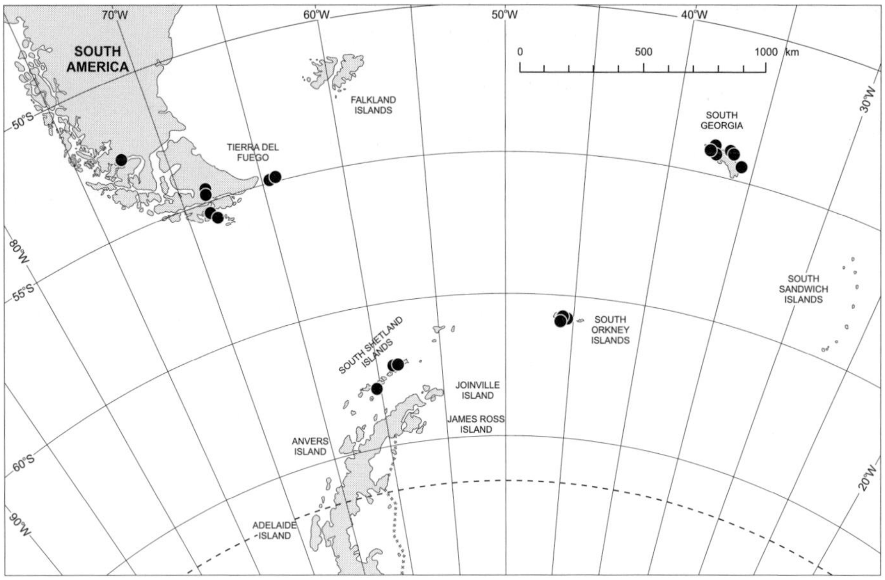

Figure 232. Global distribution of *Conostomum magellanicum*.

5: 1006. 1905. — TYPE CITATION: Terre-de-Feu, Bahia Tekenika. [Lectotype (*selected here*): 'Svenska Sydpolarexpeditionen 1901–03. N:r 109. Conostomum perangulatum Card. sp. nova. Ile Hoste: Baie Tekenika. 5 et 6/11 1902. Carl Skottsberg Det. J. Cardot.' – PC-Cardot!; isotype: S!]. First synonymised by Bell (1973*a*: p. 42).

Plants small to medium-sized, occasionally large, forming compact turves, sometimes scattered among other mosses, yellow- or brown-green to pale yellow or whitish, somewhat glossy above, brownish and densely tomentose below with brown, smooth rhizoids. *Stems* filiform, rigid but fragile, erect, simple or sparingly branched, 1–3 or, occasionally, 6 cm tall, brownish, in cross-section rounded to obtusely 5-angled, with a large central strand, 4–5-stratose medulla of large, thin-walled cells, 1–2-stratose cortex of substereid, smaller and thicker-walled cells and unistratose hyalodermis of large, thin-walled cells, usually with collapsed outer walls; *axillary hairs* filiform, short, consisting of 3 short cells, hyaline throughout or with 1 or 2 brownish basal cells. *Leaves* crowded, erect, closely imbricate in 5 distinct, spiral ranks, ovate, oblong-ovate to lingulate, broadly rounded-obtuse, concave, 0.8–1.5(–2.3) mm long, 0.4–0.6 mm wide; *margins* flat, often crenulate towards the apex; *costa* single, brownish-yellow, very broad and occupying about ½–⅔ the width of the leaf lamina, 100–200 μm wide at base, not clearly demarcated and gradually merging into the laminal cells, vanishing well below the apex, smooth or weakly mammillose on both surfaces because of the projecting upper cell ends, in transverse section consisting of a row of ill defined guide cells and a large group of small, thick-walled stereid cells surrounded on both ventral and dorsal surfaces by a row of larger, thin-walled and bulging epidermal cells; *laminal cells* unistratose near the margins, gradually passing into the bistratose costa, variable in shape, mostly with bulging outer walls; *upper cells* rhomboidal, subquadrate, oblong-hexagonal to short-rectangular, (10–)15–35(–50) μm long, 5–10(–15) μm wide, smooth or weakly prorate near the costa, becoming narrower, thicker-walled at the margins and forming a narrow 3–5-seriate border; *lower cells* thin-walled, lax, smooth, rectangular, 20–50 μm long, 10–20 μm wide. *Autoicous. Male inflorescences* terminal, inconspicuous, gemmiform; [*perichaetial leaves* oblong, acuminate, obtuse at the apex, with a narrow, well defined costa ending below the apex and oblong-rhomboidal, thin-walled laminal cells. *Setae* erect, 0.8–1.5 cm long, yellow-reddish, untwisted, smooth. *Capsules* erect, obloid, almost symmetric, furrowed when dry, brown, to 2 mm long, narrowed at the mouth; *operculum* obliquely conic, stout; *exothecial cells* hexagonal, thin-walled, becoming smaller and thicker-walled near the urn base, weakly differentiated at the mouth; *stomata* phaneroporous, bicellular with narrow, oval pori, numerous at the base of the urn; *peristome* single, fragile, consisting of 16 linear-lanceolate, orange-brown, smooth, strongly trabeculate on the dorsal surface teeth, slit along the median line and fused at the tips. *Spores* reniform, yellow-brown, verrucose 50–70 μm in diameter.]

Chromosome number: $n = 8$ – King George I. (Kuta *et al*. 1982).

Discussion. *Conostomum magellanicum* is one of the most distinctive of all the Antarctic mosses, the ovate to oblong-ovate, rounded-obtuse and closely imbricate leaves arranged in five distinct ranks enabling its recognition in the field with a hand-lens. Microscopically, the species is readily known by its broad and ill-defined costa which occupies about half of the leaf lamina. The leaf shape remains a constant character despite considerable variation in plant size. Plants from the South Shetland Is. are generally smaller and rather delicate in contrast to those from the South Orkney Is. or South Georgia, but the correspondence in all taxonomically important characters is perfect.

With its unique rounded-obtuse leaves and subpercurrent costa, *Conostomum magellanicum* is an isolated and quite discordant species in the genus, all other species having distinctly acuminate and aristate leaves. Yet, the species bears some similarity to the very rare southern Patagonian and Fuegian species, *C. perpusillum* Cardot & Broth., in which the upper leaves are shortly acuminate and cuspidate but the lower ones are shortly acute at the apex and have subpercurrent costae. Likewise, in the rare Bolivian species, *C. macrotheca* Herzog, the costa vanishes below the apex which, however, is narrowly acute. *Conostomum magellanicum* was described by Sullivant (1850, 1859) from material collected by the United States Antarctic Expedition of 1838–1842 under the command of Captain Charles Wilkes in Orange Harbour on Isla Hoste in the Tierra del Fuego archipelago. For about 40 years it was known only from the type collection, until it was rediscovered on Isla de los Estados by Spegazzini (Müller 1885) and later on Isla Hoste in Bahia Tekenika by C. Skottsberg during the Swedish South Polar Expedition of 1901–1903. Cardot (1905*a*, 1908*a*) described this material as a new species, *C. perangulatum*, which does not differ materially from *C. magellanicum* and, accordingly, Bell (1973*a*) considered both to be conspecific.

Reproduction in Antarctica. Sporophytes have not been seen on any Antarctic material and only one specimen, from Signy I. (*RILS 605* AAS, KRAM), bears perigonia. The species fruits rarely on South Georgia and in Tierra del Fuego.

Habitat. The species is predominantly associated with wet sites, thriving most often on bare ground on rock ledges, among boulders, on scree and on gravelly soil on cliffs. It often grows where snow lies long into the summer together with species such as *Pachyglossa dissitifolia*, *Anthelia juratzkana*, *Cephaloziella varians*, *Herzogobryum teres*, and *Warnstorfia sarmentosa*, as well as at the margins of melt water runnels.

World range. A Fuegian–Antarctic species occurring in Magallanes Province of Chile and in Tierra del Fuego (Matteri 1985), on South Georgia (Bell 1973*a*) and in the northern maritime Antarctic (Bell 1973*b*) (Fig. 232).

Distribution in Antarctica. A rare species known only from the South Orkney and South Shetland Is. and extending to lat. 62°37′S on Livingston I. (Fig. 232). It was reported for the first time from this region by Lewis Smith (1972 as *C. perangulatum*) and subsequently by Bell (1973*b*). The species is usually confined to higher elevations, mostly above 80 m and rarely descends to about 15 m. Its highest station is at 275 m on King George I.

Specimens examined. **WEST ANTARCTICA.** SOUTH ORKNEY IS. **CORONATION I.** Olivine Pt., 2 m, *RILS 8683* (AAS, KRAM), *ca* 15 m, *RILS 132* (AAS, BM, KRAM) and 25 m, *RILS 8685* (AAS, KRAM); Cape Hansen, *RILS 1831* (BM) and 85 m, *RILS 5188* (AAS, KRAM); island opposite Saunders Pt., 5–7 m, *RILS 27* (AAS, KRAM). **POWELL I.** Promontory to west of John Peaks, 50 m, *RILS 253* (AAS, KRAM, NY, PC, S). **SIGNY I.** Waterpipe Beach, *ca* 15 m, *RILS 505B* (AAS, KRAM); Factory Cove, 50 m, *REL 1068 & 1077* (AAS, KRAM) and *ca* 65 m, *RILS 6* (AAS, BM, KRAM); Moraine Valley, *ca* 35 m, *RILS 695G* (AAS, KRAM), 50 m, *RILS 544* (AAS, KRAM) and *ca* 65 m, *RILS 304* (AAS, KRAM, LE); Rusty Bluff, *RILS 605* (AAS, KRAM); between Moraine Valley and Rusty Bluff, 80 m, *RILS 5154* (AAS, KRAM); above Paal Harbour, 75 m, *RILS 10870 & 10872B* (AAS, KRAM).

SOUTH SHETLAND IS. **KING GEORGE I.** ***Bransfield Strait***: Red Hill, 100 m, *RO 1108/80* (KRAM). ***Admiralty Bay***: Ambona, 85 m, *RO 1638/80* (KRAM); Ubocz, 95 m, *RO 4852/79* (KRAM); unnamed hill between Jersak Hills and Krokiew, 3 Jan 1980, *RO s.n.* (KRAM); Jardine Peak, 275 m, *RO 5196/79* (KRAM); Italia Valley, 100 m, *RO 114/80* (KRAM); British Pt., *ca* 13 m, *RILS 712* (BM, KRAM); Ullman Spur, 90 m, *RO 559/8* (KRAM); Mt. Wawel, 20 m, *RO 2150/80* (KRAM) and 40 m, *RO 2121/80* (KRAM). **LIVINGSTON I.** ***Hurd Peninsula***: Spanish Base, *Schulz 112* (KRAM).

43. PHILONOTIS Brid.

Philonotis Brid., Bryol. Univ. 2: 15. 1827. ≡ *Bartramia* Hedw. sect. *Philonotis* (Brid.) Müll.Hal., Syn. Musc. Frond. 1: 471. 1848. — Lectotype (*vide* Flowers 1935: 164): *Philonotis fontana* (Hedw.) Brid. (*Mnium fontanum* Hedw.)

Plants small and slender, medium-sized to large and robust, loosely or densely caespitose, pliant or rigid, dull or glossy, green, yellow-green or sometimes bluish-green to glaucous, often whitish or bluish-white when growing. *Stems* erect or ascending, occasionally prostrate, with subfloral whorls of branches near the tops of the primary stems, moderately to densely interwoven with a ferruginous tomentum of lightly papillose to smooth, much branched, reddish-brown or brown rhizoids, in cross-section with a large central duct, several rows of large, thin-walled medullary cells, 2–4-stratose corted of small, stereid or substereid cells, with thick, reddish-brown walls and an outer hyalodermis of large, thin-walled, turgid and fragile cells; *axillary hairs* short, 2- to multi-celled, terminated with a large, globular, hyaline cell. *Leaves* often dimorphic, erect-appressed, erect-spreading to wide-spreading, straight or falcate, sometimes secund or spirally arranged, lanceolate, ovate- or oblong-lanceolate to narrowly triangular-lanceolate, acute to slenderly long-acuminate, sometimes obtuse to rounded at the apex, lacking a differentiated basis, not plicate; *margins* plane to narrowly revolute, singly or doubly, sharply or bluntly serrulate nearly throughout, sometimes entire or crenulate, occasionally distinctly bordered; *costa* subpercurrent, percurrent to long-excurrent as a spinose to weakly toothed arista, in cross-section with a row of guide cells and 1 or 2 stereid bands; *laminal cells* unistratose, transparent; *upper cells* quadrate, oblong to linear, prorate at one or both ends, sometimes centrally papillose or smooth; *basal cells* enlarged and thin-walled, often laxer at the angles but not forming distinct auricles, occasionally enlarged, thin-walled and forming a distinct hyaline marginal border. *Asexual reproduction* occasionally by deciduous brood branches or leafy axillary propagules. *Dioicous* or, rarely, *autoicous* or *synoicous*. *Perigonia* terminal and discoid or lateral and budlike; *perigonial leaves* broadly ovate, narrowed into a sharp or obtuse point. *Perichaetia* terminal; *perichaetial leaves* differing markedly in shape from the vegetative leaves or not differentiated. *Setae* elongate, erect, straight to flexuose, smooth, red to purple. *Capsules* inclined to cernuous, subglobose to ovoid, arcuate or straight, longitudinally ribbed, rugose or smooth when dry; *operculum* low-conic, mammillate to bluntly apiculate; *annulus* none; *exothecial cells* short, variable in shape, thin- to thick-walled, becoming smaller in several rows at the orifice; *peristome* double; *exostome teeth* 16, lanceolate to narrowly triangular, yellow to red-brown, strongly trabeculate, finely papillose in the proximal part, strongly so distally; *endostome segments* arising from a high basal membrane, nearly as long as the exostome, keeled, yellowish or pale brownish, finely papillose-striate below, split along the median line and each half diverging toward cilia; *cilia* 2–3, well developed or rudimentary. *Spores* rounded to reniform, coarsely papillose, brown to orange-brown. *Calyptra* cucullate.

The generic limits of *Philonotis* are reasonably well established and it can be easily separated from other genera of the Bartramiaceae by the combination of the following characters: axillary hairs short terminated with a large, globose cell, not plicate and mostly lanceolate leaves, laminal cells prominently papillose, stems with a distinct hyalodermis, long exserted capsules and stems with frequent subfloral whorls of branches. The generic name *Philonotis*, meaning wetness-loving, accurately reflects the ecological preferences of its species which are mostly associated with wet habitats.

Philonotis is a cosmopolitan genus comprising about 170 species, but many of them doubtless do not deserve any taxonomic recognition and may be considered conspecific with species described earlier. As a result, the real number of species is much lower and probably fluctuates between 120 and 130 but, unfortunately, the genus has been taxonomically assessed only for some regions, for example in East Asia and the West Pacific (Kabiersch 1938; Ochi 1962, 1963), New Guinea (Koponen & Norris 1996) and southern South America (Matteri 1968, 1985*b*). The excessive number of described species is primarily due to the unusually large amount of gametophyte variation exhibited by some species, a phenomenon typical of mosses growing in wet sites. A detailed account of the vegetative variations in *Philonotis* was discussed by Flowers (1935, 1973).

The genus is represented in southern South America by ten species (Matteri 1968, 1985), three of which occur on South Georgia (Clarke 1973*a*). One species, *Ph. gourdonii*, was described from Antarctica (Cardot 1911*a,b*, 1913) from a very restricted area on the volcanic Deception I. but it is here considered to be conspecific with *Ph. polymorpha.*

1. **Philonotis polymorpha** (Müll.Hal.) Kindb. Fig. 233

Philonotis polymorpha (Müll.Hal.) Kindb., Enum. Bryin. Exot.: 93. 1889. ≡ *Bartramia polymorpha* Müll.Hal., Bot. Jahrb. Syst. 5: 79. 1883. — TYPE CITATION: all new species described by Müller (1883) are included in the section 'I. Bryologia Kerguelensis'; later Müller (1889) provided details of the type collection as follows: 'Ins. Kerguelen, Betsy Cove, in rupibus ut videtur inundatis (12. 74)'. [Lectotype (*selected here*): 'Philonotis polymorpha (CM) Kerguelen leg. Naumann' – H-Brotherus!].

Bartramia acicularis Müll.Hal. *in* Neumayer, Int. Polarforsch. Deutsch. Exped. 2: 307. 1890. ≡ *Philonotis acicularis* (Müll.Hal.) Kindb., Enum. Bryin. Exot.: 108. 1891, ***syn. nov.*** — TYPE CITATION: Austro-Georgia, Hochplateau, 2. Majo 1883. [Lectotype (*selected here*): 'Philonotis acicularis C. Müller n. sp. Fundort: Hochplateau Süd-Georgien 2/V 83 Will' – M!; isotypes: H-Brotherus!, HBG!].

Meesia austrogeorgica Müll.Hal. *in* Neumayer, Int. Polarforsch. Deutsch. Exped. 2: 307. 1890 ['austro-georgica'], ***syn. nov.*** — TYPE CITATION: In fonte des Hochplateaus, 16. Novembri 1882. [Lectotype (*selected here*): 'Meesia austro-georgica C. Müller Fundort: Hochplateau, Quelle Süd-Georgien 83. 2V Will']. First synonymised with *Philonotis acicularis* by Clarke (1973*a*: p. 55).

Philonotis varians Cardot, Bull. Herb. Boissier Sér. 2, 6: 24. 1906, ***syn. nov.*** — TYPE CITATION: Cumberland Bay: Jason harbour, Pot harbour, Bore valley. [Lectotype (*selected here*): 'Herb. J. Cardot. Philonotis varians Card. sp. nova. Géorgie du Sud: Cumberland Bay. Leg. Skottsberg, 1902. Svenska Sydpol. Exped. nº 358' – PC!; isolectotypes: S!, S-Möller! (3 specimens); syntypes: Skottsberg 357, 360, 361 & 422 – PC!, S!].

Philonotis varians fo. *robusta* Cardot, Wiss. Erg. Schwed. Südpolar-Exp. 4(8): 231. 1908, *nom. nud.*, ***syn. nov.*** — ORIGINAL MATERIAL: Baie Cumberland: Pot-Harbour (Skottsberg, nº 356) [Chirotypes: 'Svenska Sydpolarexpeditionen 1901–03. N:r 356. Philonotis ~~crassula~~ varians Card. sp. nova. f. robusta. South Georgia, Cumberland Bay, 18/5 1902 Pot-Habour Carl Skottsberg Det. J. Cardot' – PC!, S!].

Philonotis varians fo. *gracilis* Cardot, Wiss. Erg. Schwed. Südpolar-Exp. 4(8): 231. 1908, *nom. nud.*, ***syn. nov.*** — ORIGINAL MATERIAL: Baie Cumberland: Jason-Harbour (Skottsberg, nº 359) [Chirotypes: 'Svenska Sydpolarexpeditionen 1901–03. N:r 356. Philonotis varians Card. sp. nova. + Webera ampliretis (C.M.) Par. South Georgia, Cumberland Bay, 23/4 1902 Jason Habour Carl Skottsberg Det. J. Cardot' – PC!, S!].

Philonotis elegantissima Broth. *ex* Dism., Mém. Soc. Bot. France 17: 32. 1910, *nom. illeg. incl. spec. prior.*

Philonotis gourdonii Cardot, Compt. Rend. Ac. Sci. Paris 153: 603. 1911, ***syn. nov.*** — TYPE CITATION: Shetland méridionales: île Deception. [Lectotype (*selected here*): 'Herb. Mus. Paris 297. Philonotis Gourdonii Card. sp. nova. Shetland du Sud: île Deception, sommet du mont Pond, dépôts calcaires et sulfureux des sources chaudes, 450 m. 2ème expéd. Charcot. Leg. Gourdon, 28/12/1909.' – PC!; isotypes: BM! (2 specimens), PC! (4 specimens)].

Philonotis luteola E.B.Bartram, Farlowia 2: 315. 1946, *hom. illeg.* [non *Ph. luteola* Cardot 1911]. ≡ *Ph. bartramii* Matteri, Anales Inst. Patagonia 4: 273. 1973, ***syn. nov.*** — TYPE CITATION: Fuegia media: Estncia Vicuna, Puesto 20, Cerro Fuentes, reg. alp., No. 734. [ISOTYPES: 'Philonotis luteola Bartram n.sp. Chile, Fuegia media, Estancia Vicuña, Puesto 20, Cerro Fuentes, in riculo alpino, 600 m s.m. 3.1.1929 H. Roivainen' – H!, KRAM!]. First synonymised with *Philonotis acicularis* by Matteri (1985*b*).

Plants small and slender to medium-sized or fairly large forming dense or loose, extensive turves, soft or stiff and rigid, dull, sordid- to yellow-green above, brownish below, occasionally blackish-brown throughout. *Stems* erect to ascending, 1–6 cm tall, with few to many slender and elongate branches produced from the apical region of the previous year's shoots, the latter becoming prostrate,

Figure 233. *Philonotis polymorpha.* — **1**. Plant with mature capsule. **2–3**. Sterile plants, dry. **4**. Sterile plant, wet. **5**. Portion of cross-section of stem. **6–8**. Axillary hairs. **9–18**. Leaves. **19–20**. Leaf apices. **21**. Upper lamina cells at margin. **22**. Mid-leaf cells. **23**. Basal leaf cells. **24–26**. Cross-sections of leaves, a sequence from base to apex. **27**. Plant with perigonium. **28–29**. Perigonial bracts. **30**. Deoperculate capsule, dry. **31**. Operculate capsule, dry. **32**. Lower exothecial cells and stomata. **33**. Exothecial cells in middle of urn. **34**. Peristome teeth, spores and rim cells of capsule. — [1, 6–8, 17–19, 30–34 from *GCSC & SWG 187*; 2, 13–15 from *Gourdon 297* (isotype of *Ph. gourdonii*), PC; 3–5, 10–12, 20–26 from *RILS 5835*; 9 from *Will s.n.*, 2.v.1883 (type of *Bartramia acicularis*), HBG; 16 from *Will s.n.*, 2.v.1883 type of *Meesia austrogeorgica*), HBG; 27–29 from *GCSC & SWG 486*]. — Scale bars: a – 1 mm (30–31); b – 50 μm (21–26), 1 mm (3–4, 28–29) and 2 mm (2); c – 0.5 cm (1); d – 50 μm (6–8), 100 μm (5, 19–20, 32–34), 1 mm (9–18) and 2 mm (27).

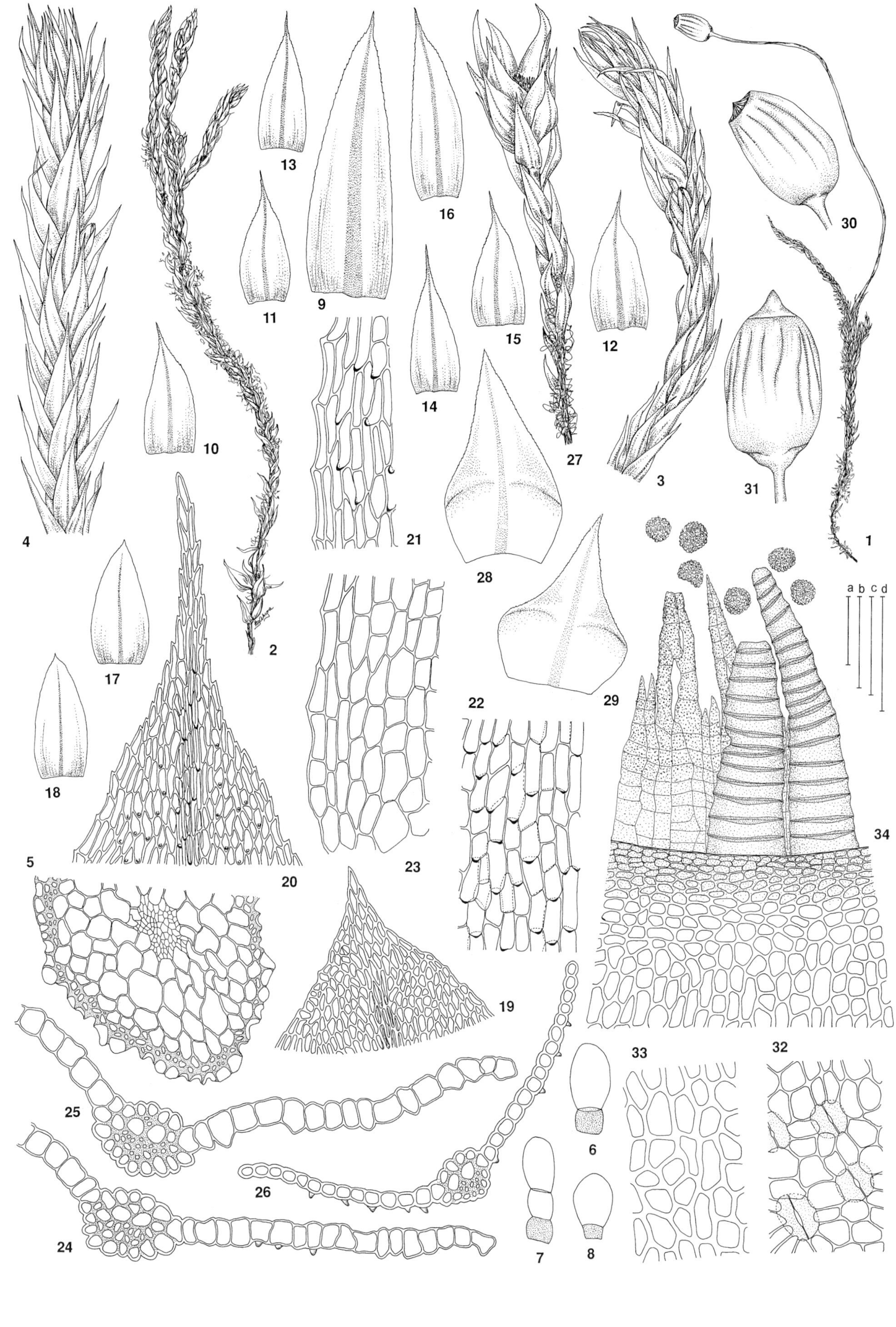

4
13
11
9
16
15
14
12
27
3
30
31
1
10
2
21
28
a b c d
17
18
5
20
22
29
23
34
19
25
26
24
6
7
8
33
32

moderately tomentose, in transverse section with a large central duct, 4–5-layered medulla of large, thin-walled cells, 1–2-layered cortex of small, thick-walled, bright brown cells and an outer hyalodermis; *axillary hairs* 2–3-celled, with a brown basal cell. *Leaves* straight, distant to densely set, erect to erect-flexuose when dry, erecto-patent to homomallous when wet, concave, strongly carinate, (0.8–)1.0–1.2(–2.0) mm long, (0.3–)0.4–0.6(–0.8) mm wide, variable in shape, dimorphous, at base of stems ovate to oblong-ovate, acute to obtuse, with a subpercurrent costa, becoming oblong-lanceolate, short-acuminate to aristate with an excurrent costa higher up the stem; *margins* plane throughout or somewhat recurved in the lower half, entire below, variously singly serrate above; *costa* intensely yellow to yellow-brownish, sharply set off from the laminal cells, 50–70(–80) µm wide at base, papillosely toothed on dorsal side above, in cross-section with papillose-mammillose epidermis, few conductive cells and two or only dorsal stereid bands; *laminal cells* unistratose, transparent; *upper cells* rectangular to oblong-hexagonal, 15–35(–40) µm long, 7–10(–15) µm wide, weakly papillose by projecting lower cell ends, more prominently so on the dorsal surface, becoming narrower and elongate near the margins and forming an indistinct border; *lower cells* laxer and larger, subquadrate to short-rectangular, up to 20 µm wide and to 45 µm long. [*Dioicous*. *Perigonia* enlarged, discoid; *perigonial bracts* erecto-patent, broadly ovate, shortly acuminate, obtuse or broadly acute at the apex, with a subpercurrent costa. *Perichaetia* terminal; *perichaetial leaves* aristate. *Setae* 1–2 cm long, straight, brown; *capsules* erect, ovoid, subsymmetric, sulcate whole of its length when dry, *ca* 2 mm long, 1.5 mm wide; *operculum* conic-convex with a short blunt point; *exothecial cells* thick-walled, subquadrate, short-rectangular to hexagonal, 10–20 µm wide, 20–50 µm long; *stomata* numerous at base of the urn, bicellular, variously oriented, with narrowly elongate slits; *peristome* double; *exostome teeth* 16, lanceolate, yellowish to yellowish-red, papillose, blunt at the tips; *endostome* pale, consisting of segments with acuminate apices, as long as the exostome or longer, arising from a basal membrane of about half of the endostome length and with 2–3 short cilia. *Spores* rounded to reniform, 22–30 µm in diameter, strongly verrucose.]

Chromosome number: unknown.

Discussion. As the specific epithet implies, *Philonotis polymorpha* is identified by its dimorphous or polymorphous leaves which are ovate to oblong-ovate, acute to obtuse, with a subpercurrent costa in the lower part of the stem and oblong-lanceolate, short-acuminate to aristate with an excurrent costa in the upper half. The species is, however, very variable, especially in leaf shape and costa excurrence, and seasonal variation may contribute much to its confusion. In the early spring phenotypes the leaf apices are rather obtuse and the costae vanish below them. The leaves become progressively more acute in the summertime and at the end of the growing season they show a distinct gradation between these two extremes. The species is likely to be mistaken in the field for some bryaceous mosses, especially *Pohlia drummondii* which grows also in wet sites. Both species are difficult to distinguish without microscopic examination. The latter species is easily recognised microscopically by its entirely smooth laminal cells, always percurrent costa and entire leaf margins which are only slightly crenate at the apex.

The considerable morphological plasticity of *Philonotis polymorpha* resulted in much taxonomic confusion because various phenotypes were named as separate species, especially if the specimens originated from remote geographical areas. It was first described by Müller (1883, 1889) from Îles Kerguelen on the basis of material collected in 1874 by F. C. Naumann during the *Gazelle* expedition. Later, the same author described it twice from South Georgia as *Bartramia acicularis* and *Meesia austrogeorgica* (Müller 1890*a*), but the former name was soon transferred to *Philonotis* (Kindberg 1891) and was usually used as the correct name for this species by subsequent authors. The species was recollected on South Georgia by C. Skottsberg during the Swedish South Polar Expedition of 1901–1903, and Cardot (1906*a*, 1908*a*) described it as a new species, *Ph. varians*. Cardot (1908*a*) indicated it as a protean species and illustrated a spectrum of its leaves showing wide variation in their shape, but also indicated considerable differences in the stature of the plants, though robust and gracile forms remained formally undescribed. Dismier (1910) claimed *Ph. elegantissima* to be an older name for *Ph. varians*, but this name was used by V. F. Brotherus in the herbarium labels only for a specimen collected by Kurtz in Mendoza Province of Argentina and it has never been validly published. A few years later Cardot (1911*a*,*b*, 1913) described this species as *Ph. gourdonii* from a single collection from a geothermal area on Deception I. in Antarctica and, finally, Bartram (1946) described it from Isla Grande de Tierra del Fuego as another new species, *Ph. luteola*. Because this name was a later homonym of *Ph. luteola* described by Cardot (1911*f*) from Mexico, Matteri (1973*b*) changed it to *Ph. bartramii*, but subsequently reduced it to synonymy with *Ph. acicularis* (Matteri 1985*b*). Here, all these species are considered conspecific with *Ph. polymorpha* because they perfectly match the range of variation of this species as observed on the islands in the Kerguelen Province of the Subantarctic (Ochyra, unpublished).

As is the case with other large moss genera, *Philonotis* has a well developed infrageneric classification (Brotherus 1924*b*). The genus is divided into several sections, the largest of which, sect. *Philonotula* (Schimp.) Broth., consists of mostly exotic dioicous species with smaller and gemmiform perigonia and the second largest section, sect. *Philonotis*, comprises dioicous, robust species having enlarged, discoid perigonia and mammillae at the lower end or between the middle and the lower end of the laminal cells. To this section belongs Antarctic *Ph. polymorpha*. The third section, sect. *Heteromorphae* (Kindb.) Ochyra, consists of dioicous but slender species with mammillae at or near the upper end of the laminal cells. Finally, the two monotypic sections, sect. *Catenularia* (Müll.Hal.) Paris and sect. *Pseudomniobryum* Broth., contain exotic species, with very peculiar features of the gametophytes. The former consists of austral *Ph. scabrifolia* (Hook.f. & Wilson) Braithw. and Mexican *Ph. corticata* H.A.Crum & D.G.Griffin (Crum & Griffin 1981) and is characterised by the laminal cells having central papillae projecting on both surfaces, whereas the latter consists of the Afro-American temperate *Ph. vagans* (Hook.f. & Wilson) Mitt. which has distinctly limbate leaves with a multiseriate hyaline border in the proximal part.

Reproduction in Antarctica. Neither sexual organs nor sporophytes are known in Antarctic plants. Likewise, inflorescences are very local on South Georgia and sporophytes are correspondingly infrequent.

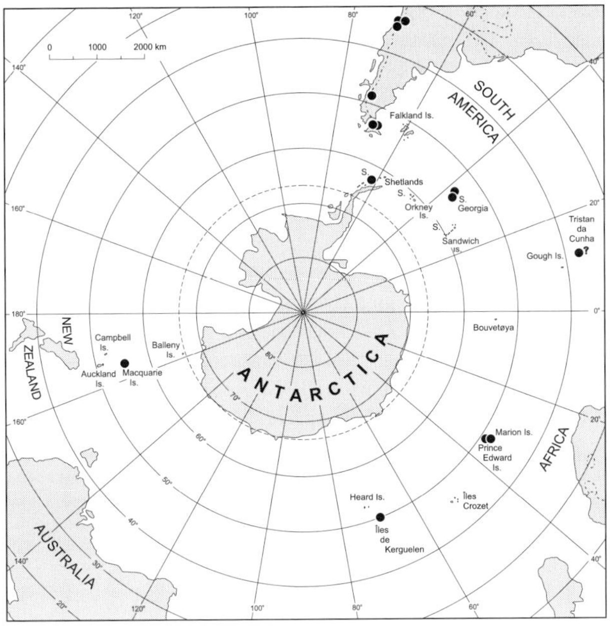

Figure 234. Global distribution of *Philonotis polymorpha*.

Habitat. In the Antarctic *Philonotis polymorpha* is known only on Deception I., South Shetland Is., where it grows in a few very restricted areas on heated ground with active fumarolic fissures emitting steam, and on neighbouring moist ashy soil which is diffusely heated. At its main site at Mt. Pond, it forms a unique community, together with *Dicranella hookeri*, which dominates large stands up to about 100 sq. m, as well as smaller isolated patches. Outside Antarctica *Ph. polymorpha* grows mostly on stream banks and in all wet habitats, occasionally forming swards of several square metres, and plants reach a considerable length, to 30 cm, for example on Marion I. in Kerguelen Province (Ochyra, personal observations).

World distribution. A circumsubantarctic species (Fig. 234) that is common and abundant on South Georgia (Clarke 1973*a*), the Prince Edward Is. (Zanten 1971 as *Ph.* cf. *angustifolia*; Ochyra, personal observations), Îles Kerguelen (Brotherus 1906*a*; Cardot 1916; Hébrard 1970*a*,*b*) and Macquarie I. (Ochyra *et al.* 2008). On the latter island, at least some specimens named *Aulacomnium palustre* (Hedw.) Schwägr. belong to this species (e.g. *RDS 6453*, *7472*, *10302*, *11522*, *11600* & *12109* – ADT, KRAM) and, actually, the otherwise excellent drawings of this species in Seppelt (2004) and Bell & Catcheside (2006) illustrate *Ph. polymorpha.* In addition, the species is scattered at high elevations in southern South America, from Tierra del Fuego to Mendoza Province in Argentina and Region Metropolitana in Chile (He 1998), and Dixon (1937) reported it, with some reservation, from Tristan da Cunha.

Distribution in Antarctica. The species is known only from a few very restricted areas on the volcanic Deception I. (Fig. 234) where it was first recorded in 1909 and described by Cardot (1911*a*, 1913) as *Philonotis gourdonii*. It still occurs at this locality despite three eruptions of the volcano between 1967 and 1970 (Collins 1969; Young & Kläy 1971; Lewis Smith 1984*b*,*c*, 1988*d*, 2005*a*,*b*,*c*). It is currently known from three widely separated geothermal areas on this island, the main site being near the summit of Mt. Pond between 480 and 500 m altitude. It is less frequent at sites near Vapour Col at an elevation of 175 m and Telefon Bay near sea level. In 1964 it was found in very small quantity 'in interior of covered whale boat embedded in beach with warm fumaroles' in Whalers Bay. However, this area was destroyed by a lahar caused by the 1970 eruption, and *Ph. polymorpha* has not been recorded in this area since then.

Specimens examined. WEST ANTARCTICA. SOUTH SHETLAND IS. **Deception I.** Mt. Pond, 450 m, *Gourdon 297* (PC – type of *Ph. gourdonii*); Mt. Pond, east face of northern summit, 480 m, *RILS 5826C*, *5827B 5829A*, *5823C*, *5824D*, *5831A*, *5830A*, *5833B*, *5834B* & *5835* (AAS, KRAM); hill 1.5 km NNW of Mt. Pond, 410–525 m, *RILS 6891*, *6892*, *6896* & *6900* (AAS, KRAM); ridge north of Mt. Pond, 485–500 m, *RILS 11144*, *11145*, *11154B* & *11160* (AAS, KRAM); Telefon Bay, above small lake near 'New Island' Hill, 20 m, *RILS 11291* & *11294* (AAS, KRAM); 2 km north of Vapour Col, *ca* 175 m, *RILS 11256B* & *11266E* (AAS, KRAM); north side of Vapour Col, 400 m, *DM 30*, *33A* & *36C* (AAS, KRAM); without closer locality, in beach with warm fumaroles, 18 Feb 1964, *Watson s.n.* (US).

44. BARTRAMIA Hedw., *nom. cons.*

Bartramia Hedw., Sp. Musc. Frond.: 164. 1801. — Lectotype (*vide* Flowers 1935: p. 158): *Bartramia halleriana* Hedw.

Cephaloxis P.Beauv., Prodr. Aethéogam.: 32, 50. 1805, *nom. inval. in synon.*

Plants small to large and robust, often rigid, densely or loosely caespitose, dull, green, yellow-green or bluish above, yellow-brown and sparsely to densely tomentose below. *Stems* erect, simple or often forked, not whorled, red to reddish-brown, in cross-section rounded, with a large central strand, 1–4-stratose cortex of small, thick-walled cells and unistratose hyalodermis of hyaline, thin-walled cells, usually with collapsed outer walls; *rhizoids* arising from lenticel-shaped clusters of initials on the stem and around branch primordia, reddish-brown, much branched, smooth or papillose; *axillary hairs* 3- to multicelled, relatively long, composed of elongate cells, with cross-walls intensively pigmented and the basal cells sometimes brown. *Leaves* crowded or distantly spaced, not ranked, erect, erect-appressed to spreading, straight and rigid, flexuose or sometimes crisped when dry, linear to lanceolate from a broad, non-plicate, sheathing base or lacking a vaginant base; *margins* flat or narrowly recurved, sharply serrate above, with the teeth single or paired, uni-, bi- or sometimes tristratose in several rows of cells; *costa* strong, prominent at back, subpercurrent to excurrent as a brown or rarely hyaline arista, sharply toothed on the back distally, in cross-section with a row of guide cells and large dorsal stereid band; *laminal cells* bi- or unistratose, transparent or opaque distally, pellucid in the sheathing base; *upper cells* subquadrate to long-rectangular, thick-walled, strongly prorate to mammillose by projecting lower and upper cell ends; *basal cells* elongate-rectangular to linear, pale, thin-walled, smooth; alar cells weakly differentiated, yellow-reddish, subquadrate to rectangular, bulging. *Synoicous, autoicous* or *dioicous*. *Perigonia* gemmiform. *Perichaetial leaves* similar to the vegetative leaves or sometimes differentiated, ending in long capillary points. *Setae* variable, very short to long, straight or arcuate, smooth. *Capsules* erect, inclined or pendulous, stegocarpous or occasionally cleistocarpous, mostly asymmetric and obliquely mouthed, subglobose to ovoid, furrowed or rarely wrinkled when dry; *operculum* plano-conical or conical-mammilate, bluntly pointed; *annulus* absent; *exothecial cells* variable and mostly irregular, hexagonal, oblong or quadrate, thin- or thick-walled; *stomata* phaneroporous, bicellular in the neck; *peristome* double, single or sometimes lacking; *exostome teeth* 16, narrowly triangular, smooth or papillose, often cleft above; *endostome* consisting of keeled segments arising from a high or low basal membrane and often poorly developed cilia. *Spores* reniform to spherical, coarsely papillose. *Calyptra* small, cucullate.

Bartramia is a difficult genus to circumscribe and characterise because its species exhibit an array of gross morphological features which correspond to three distinct groups: one has strongly crispate leaves with recurved margins in the distal portion and transparent distal cells in the limb which arise from a semi-vaginant, subclasping and otherwise not so highly differentiated base; another has straight and rigid leaves with abruptly expanded, tightly sheathing base, plane or weakly recurved leaf margins and obscure upper laminal cells; the third group has neither of these sets of characteristics. These groups are recognised as sections, sect. *Bartramia*, sect. *Pyridium* Müll.Hal., which is often incorrectly named sect. *Vaginella* Müll.Hal., and sect. *Strictidium* Müll.Hal., respectively. Additionally, the type section consists of autoicous species, whereas the last two sections comprise dioicous or synoicous taxa. In general, the plants usually have a characteristic rigid appearance which can be credited to a combination of leaf characters. The leaves of species of the type section and sect. *Pyridium* consist of echlorophyllose, non-plicate bases, closely clasping the stem and abruptly constricted to lanceolate, linear or subulate, photosynthetic limbs with serrate, plane margins and strongly papillose cells because of the projecting ends. A smaller group of species belonging within sect. *Strictidium* has the leaves without sheathing bases. In general, the plants of the genus *Bartramia* are without whorled, subfloral innovations and the axillary hairs terminate in elongate cells and the cross-walls of their basal cells are strongly brown-coloured, although the cells themselves are hyaline. The exserted capsules and differentiated hyalodermis in the stem separate *Bartramia* from the closely related genus *Leiomela* (Mitt.) Broth.

Some species of *Bartramia* are very variable morphologically and this resulted in recognition of a great number of species in the genus. However, recent taxonomic investigations of sect. *Bartramia* and sect. *Pyridium* (Fransén 1995, 2004*a*,*b*, 2005) greatly reduced the number of species in the genus *Bartramia*. Currently the former section consists of only three species while the latter section contains 22 species with some infraspecific taxa. Sect. *Strictidium* has not hitherto been taxonomically assessed, but it consists of two unquestionably accepted species, *B. stricta* Brid. and *B. compacta* Hornsch., and four insufficiently known species. Thus, *Bartramia* as a whole contains about 30 species which are distributed worldwide, although in the tropical region they are confined to high elevations.

Many species have been described from southern South America and the subantarctic islands, but most appeared to be identical to other species, so at present only five or six species are known from this region (Newton 1973, 1974*b*; Matteri 1984, 1985*b*, Fransén 2004*a*,*b*). Four species and one variety have been reported from Antarctica by Cardot (1906*b*, 1907*a*, 1908*a*, 1911*a*,*b*, 1913) but Robinson (1972*a*) proved that the relevant specimens represent a single species, *Bartramia patens*, which is widespread in the maritime Antarctic. A detailed account of the taxonomy and variation of this species in South Georgia was presented by Newton (1973, 1974*b*).

1. **Bartramia patens** Brid. Fig. 235, pl. 37

Bartramia patens Brid., Musc. Rec. 2(3): 134, *t. 1*, *f. 7*. 1803. ≡ *B. ithyphylla* Brid. subsp. *patens* (Brid.) Fransén, Lindbergia 29: 93. 1995. — TYPE CITATION: In freto Magellanico in imo sinu Baie de Bougainville dicto a Commersono lectam clar. De Iussieu benigne communicavit. [Lectotype (*selected here*) {or perhaps holotype}: 'Barthramia patens Bryol. Univ. e Freto Magellanico Commerson retulit Tortain & Jussieu dedit 1803 1797' – B-Bridel!; isotype: 'Decembre 1767 leg. Commerson' – PC!].

Bartramia diminutiva Müll.Hal., Bot. Jahrb. 5: 79. 1883. — TYPE CITATION: all new species described by Müller (1883) are included in the section 'I. Bryologia Kerguelensis'; later Müller (1889) provided details of the type collection as follows: 'Ins. Kerguelen, Winter harbour (13. 1. 75)'. [Lectotype (*selected here*): 'Bartr. diminutiva C.M. Kerguelen leg. Naumann' – H-Brotherus!]. First synonymised by Robinson (1972*a*: p. 174).

Bartramia pycnocolea Müll.Hal. *in* Neumayer, Int. Polarforsch. Deutsch. Exped. 2: 304. 1890 ['pycnocoleos']. — TYPE CITATION: Austro-Georgia, ad rupes im Hochthale über dem oberen Whaler-Thale, 18. Martio 1883. [Lectotype (*selected here*.): 'Bartramia pycnocoleos C. Müll. n.sp. Felsen im Hochthal über dem oberen Whalerthal, 160 mtr. über dem Whalerthal Süd-Georgien 18/III 83 Will' – HBG!; isotypes: H-Brotherus!, M!]. First synonymised by Robinson (1972*a*: p. 174).

Bartramia subpatens Müll.Hal. *in* Neumayer, Int. Polarforsch. Deutsch. Exped. 2: 305. 1890, *hom. illeg.* [non *B. subpatens* Hampe 1865]. ≡ *B. austrogeorgica* Paris, Ind. Bryol.: 104. 1894. ≡ *B. patens* fo. *austrogeorgica* (Paris) Cardot, Wiss. Erg. Schwed. Südpolar-Exp. 4(8): 224. 1908. — TYPE CITATION: Austro-Georgia, Whaler-Bay, 30. Vivbr. 1882. [Lectotype (*selected here*): № 46 Bartramia subpatens C. Müll. nsp. Fundort: Whalerbay N-Seite des Pirnerberges Süd-Georgien 19/XI 83 Will' – HBG!; isotypes: H-Brotherus!, M!]. First synonymised by Cardot (1906*a*: p. 23).

Bartramia oreadella Müll.Hal. *in* Neumayer, Int. Polarforsch. Deutsch. Exped. 2: 305. 1890. — TYPE CITATION: Austro-Georgia, in fissuris rupium des oberen Whaler-Thales, 23. Martio 1883. [Lectotype

→

Figure 235. *Bartramia patens*. — **1**. Plant with mature capsule. **2**. Portion of stem, wet. **3**. Portion of stem, dry. **4**. Portion of cross-section of stem. **5–7**. Axillary hairs. **8–11**. Leaves. **12**. Leaf apex. **13**. Cells at margin in subula. **14**. Upper lamina cells at leaf shoulder. **15**. Mid-sheath cells. **16–21**. Cross-sections of leaves, a sequence from apex to base. **22**. Calyptra. **23**. Deoperculate capsule, dry. **24**. Lower exothecial cells and stoma. **25**. Exothecial cells in middle of urn. **26**. Peristome teeth, spores and rim cells of capsule. — [1, 4–8, 11–21, 24–26 from *RO 2619/80*; 2 from *RO 2554/80*); 3 from *RO 916/80*; 9–10 from *RO 924/80*; 22 from *RO 735/80*; 23 from *Gain 247B*, PC]. — Scale bars: a – 1 mm (1–3, 22–23); b – 50 µm (4–7, 12–21); c – 100 µm (24–26); d – 1 mm (8–11).

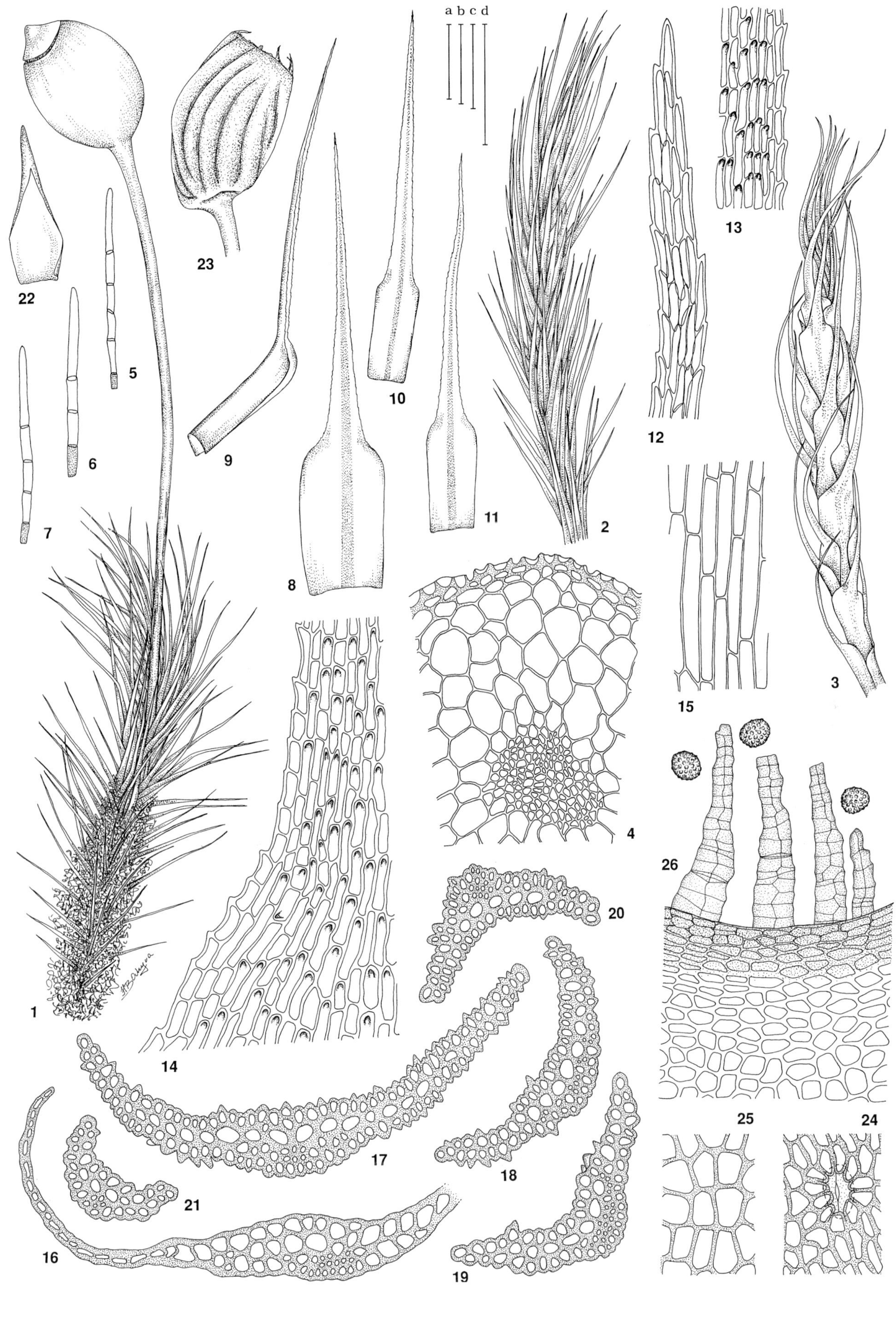
a b c d
1
2
3
4
5
6
7
8
9
10
11
12
13
14
15
16
17
18
19
20
21
22
23
24
25
26

(*selected here*): 'Bartramia oreadella C. Müll. nsp. Fundort: oberes Whalerthal in Felsspalten Süd-Georgien 20/III 83 Will' – HBG!; isotype: M!]. First synonymised by Robinson (1972*a*: p. 174).

Bartramia oreadella var. *microphylla* Cardot, Bull. Herb. Boissier, Sér. 2, 5: 1006. 1906. — TYPE CITATION: Terre de Feu: Ushuaia. [Lectotype (*selected here*): 'Svenska Sydpolarexpeditionen 1901–03. N:r 15 in parte. Bartramia oreadella C. Müll. var. microphylla Card. var. nova In monte supra Ushuaia reg. alpina 11/3 1902 Carl Skottsberg Det. J. Cardot' – PC!; isotype: S!]. First synonymised by Newton (1973: p. 13).

Plants of moderate size to fairly large, in dense, dull tufts, (0.5–) 1.0–4.5(–7.5) cm tall, bright to dark green, blue- or yellowish-green to glaucous. *Stems* erect, sparingly dichotomously branched by innovations, rarely simple, dark brown, frequently radiculose with brown- or reddish-brown, branched, smooth to faintly papillose radiculose, in cross-section oval to obscurely rounded-pentagonal, with a large central strand, 5–7-layered medulla of large, hyaline, thin-walled cells and 1–2-stratose cortex of small, thick-walled cells and a unistratose hyalodemis of thin-walled, hyaline cells. *Leaves* crowded, rigid, strict and imbricate, often flexuose and secund, erect when dry, erect-spreading when wet, (2.0–)2.5–6.5(–8.0) mm long (including base), consisting of pale, square or oblong clasping base, (0.5–)0.8–1.5(–2.0) mm long, 0.5–1.0(–1.5) mm wide, abruptly constricted into a linear-subulate limb, (1.5–)2.0–5.0(–6.5) mm long, 0.2–0.6 mm wide, usually intact or sometimes with broken tips, in cross-section 2–3-stratose except at the margin for a few rows of cells; *margins* plane, serrate to the shoulders, entire below; *costa* strong, percurrent to excurrent, 100–200 µm wide, indistinct above and appearing to fill most of the limb, in cross-section plano-convex to biconvex, with a distinct row of large guide cells, large dorsal stereid band and none or rudimentary ventral band; *upper laminal cells* oblong to short-rectangular, green and obscure, (7–)10–35(–55) µm long, 7–10 µm wide, papillose singly or doubly by projecting ends; *basal cells* transparent, hyaline, thin-walled, linear, 60–150 µm long, 8–15 µm wide. *Synoicous* or rarely *dioicous*. *Perichaetial leaves* shorter than or as long as the vegetative leaves. *Setae* erect, straight or flexuose, reddish-brown, (0.5–)1.0–3.0(–3.5) mm long. *Capsules* inclined to suberect, subglobose, (1.4–)2.0–2.5 mm wide, asymmetric, narrow-mouthed, light brown to reddish-brown, sulcate when dry; *operculum* low-conic, shortly apiculate; *annulus* none; *exothecial cells* subquadrate to hexagonal, thick-walled, usually with collenchymatous thickenings or with thicker longitudinal than transverse walls, (20–)30–70(–85) µm long, (15–)20–40(–60) µm wide; *stomata* many at the base of the urn, bicellular, with narrow elongate pori; *peristome* double; *exostome teeth* lanceolate, (230–)350–400(–460) µm long, yellow- or red-brown, finely papillose below, smooth or nearly so distally, strongly trabeculate on the inner surface; *endostome segments* arising from a low basal membrane, pale yellowish, smooth, irregularly lanceolate, rudimentary, shorter than the teeth and adhering to them. *Spores* reniform, 30–40 µm wide along the equatorial axis and 25–35 µm in the polar axis, brown, usually with slightly prominent, large irregular papillae, occasionally finely papillose throughout. *Calyptra* cucullate.

Chromosome numbers: $n = 11 + m$ – Signy I. (Newton 1980) and $n = 16$ – South Georgia and King George I. (Newton 1980; Kuta *et al.* 1982).

Discussion. This is one of the most distinctive of all Antarctic mosses being recognisable on sight by its glaucous coloration of the plants and rigidly erect leaves which are abruptly subulate from a white, sheathing base. In leaf shape *Bartramia patens* is similar to species of *Polytrichum* or *Polytrichastrum* but the lack of the longitudinal lamellae on the ventral surface of the limbs immediately precludes any possibility of confusion. Moreover, the upper portion of the leaf in *B. patens* is often brittle and the limbs frequently break off when touched, a feature that is not observed in the Antarctic polytrichalean genera. Microscopically, it is very easily known by its 2–3-stratose leaves and papillose laminal cells. In addition, the subglobose capsules, which are deeply furrowed, when dry are characteristic of this moss.

Despite its distinctiveness the species is polymorphous and very variable in the shape and size of the leaves and this gave raise to recognition of various phenotypes as species on their own right. Antarctic plants share exactly the same variation pattern which was accurately and thoroughly described by Newton (1973) for its South Georgian populations. Because continuous variation is observed between the extreme phenotypes, this clearly indicates that their taxonomic recognition is unjustified and supports the reduction of many taxon names to synonymy with *Bartramia patens* throughout its fairly wide geographical range.

Bartramia patens is closely related to *B. ithyphylla* Brid., a species which is considered to be bipolar in distribution (Ochyra *et al.* 1992). However, Fransén (2004*b*) recognised the latter as a strictly Holarctic species and reduced *B. patens* to a subspecies of *B. ithyphylla* because it was impossible to differentiate southern and northern populations on morphological grounds alone. In practice, *B. patens* may be considered merely as a geographical race of *B. ithyphylla*. Yet, *B. patens* is here accepted as a species in its own right because, though closely related, these species can be differentiated on the basis of the anatomy of the leaf lamina. *Bartramia patens* has a distinctly tristratose limb in the median part, with a rather flat costa on the dorsal side because of a small dorsal stereid band which consists of less than 15 cells. In contrast, *B. ithyphylla* has a bistratose limb in the middle part, with a costa prominently convex dorsally owing to the large dorsal stereid band which usually consists of 20–30 cells.

Antarctic specimens of the genus *Bartramia* are quite easy for interpretation and all clearly belong within *B. patens*, but the genus poses many more taxonomic problems in the Subantarctic and the cool temperate zone where, besides *B. patens*, three other species are known, namely *B. subsymmetrica* Cardot, *B. robusta* Hook.f. & Wilson and *B. ithyphylloides* Müll.Hal. When fertile, these species are easy to distinguish, the latter having double peristome and the former two having single peristomes, but sterile plants are much more difficult to distinguish. Fortunately, fruiting populations are abundant for all species of the genus, and in doubtful cases careful examination of the costal anatomy may be helpful.

Reproduction in Antarctica. This is one of few species commonly and abundantly producing fully mature sporophytes throughout this region.

Habitat. The species grows in a wide range of exposed and sheltered habitats including boulders, damp rock crevices, rock outcrops, overhangs and ledges, stony ground, scree, particularly in places where there is some accumulation of soil or organic matter, both on basic and acidic substrata. It is a common constituent of various associations of the fruticose lichen and moss cushion subformation, being especially frequent in the *Andreaea–Usnea* association.

World distribution. An amphiatlantic temperate species common in the *Nothofagus* zone at the western fringes of southern South America from the Juan Fernandez Islands (Robinson 1975) and Neuquén Province in Argentina at lat. *ca* 40° S (Matteri 1984) to Tierra del Fuego and Isla de los Estados (Matteri 1985*b*) and the Falkland Is. (Matteri 1986). It extends to some subantarctic islands including South Georgia (Newton 1974*b*), the Prince Edward Is. (Ochyra & Hertel 1990), Îles Kerguelen (Hébrard 1970*a*) and Heard I. (Bergstrom & Selkirk 1997). Furthermore, it occurs on Tristan da Cunha in the cool temperate zone (Dixon 1960) and extends to the maritime Antarctic (Fig. 236).

Distribution in Antarctica. A widespread and locally common species in the maritime Antarctic (Fig. 237). It is very rare and occasional in the South Sandwich Is. and fairly common in the South Orkney and South Shetland Is. and on the western coast of the Antarctic Peninsula, from Astrolabe I. off Trinity Peninsula southwards to central Marguerite Bay reaching its southernmost locality at a highly isolated station, Rils Nunatak, on the northern coast of Charcot I. at lat. 69°45′ S. On the eastern coast of the Antarctic Peninsula *B. patens* is very rare and occasional on Joinville I. and James Ross I., with a single locality on the Oscar II Coast at lat. 65°46′ S. Its altitudinal range is from near sea level to 700 m.

Selected specimens examined. WEST ANTARCTICA. SOUTH SANDWICH IS. **Leskov I.** Eastern slope of the summit ridge, 100 m, *PC 247A* (AAS, KRAM) and *ca* 200 m, *REL 490C & 492* (AAS, KRAM). **Candlemas I.** Breakbones Plateau, *PC 200A* (AAS, KRAM).

SOUTH ORKNEY IS. **Coronation I.** Mansfield Pt., *ca* 35 m, *RILS 296* (AAS, KRAM); Cape Vik, *RILS 573G* (AAS, KRAM); Cape Hansen, 20–200 m, *RILS 3230* (AAS, KRAM); Shingle Cove, *ca* 85 m, *RILS 452* (AAS, KRAM); Windscoop Buttress, 200 m, *RW 156 & 162* (AAS, KRAM); Olivine Pt., 17–65 m, *RILS 139* (AAS, KRAM); Saunders Pt., 5–7 m, *RILS 27* (AAS, KRAM). **Lynch I.** *Ca* 15 m, *DCL 962* (AAS, KRAM). **Mathews I.** The Divide, 65 m, *RILS 176* (AAS, KRAM) and 100 m, *RILS 178B* (AAS, KRAM); north-west coast, *ca* 35 m, *RILS 158* (AAS, KRAM). **Powell I.** John Peaks, *ca* 265 m, *RILS 269* (AAS, KRAM). **Signy I.** Thulla Pt., 65 m, *MWH 156* (AAS, KRAM); east of Spindrift Rocks, 85 m, *MWH 765E* (AAS, KRAM); Knob Lake, 35 m, *RILS 623* (AAS, KRAM); between Limestone Valley and Jane Peak, 100 m, *RILS 5143A* (AAS, KRAM); between Factory Cove and Paal Harbour, *MWH 39* (AAS, KRAM); Observation Bluff, *MWH 778A* (AAS, KRAM) and 50 m, *RILS 318* (AAS, KRAM); Moraine Valley, Rusty Bluff, 65 m, *RILS 415* (AAS, KRAM); Berntsen Pt., 30 m, *BJT 362* (AAS, KRAM). **Moe I.** 20–200 m, *RILS 5255* (AAS, KRAM).

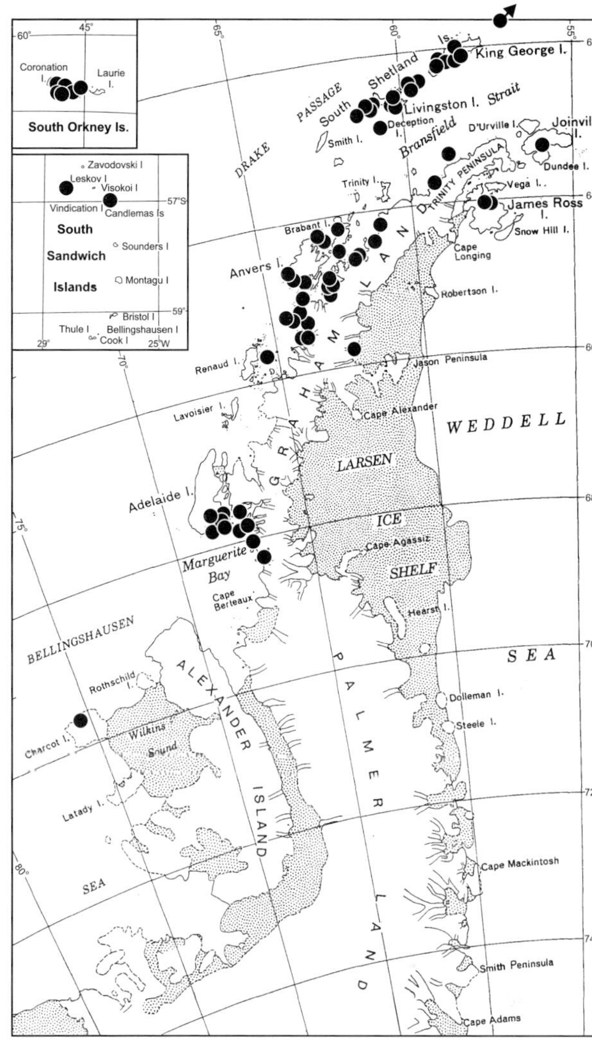

Figure 237. Distribution of *Bartramia patens* in Antarctica. The occurrence on Elephant I. in the northern group of the South Shetland Is. is indicated by an arrow.

SOUTH SHETLAND IS. **Elephant I.** Walker Pt., 150 m, *JPB 259* (AAS, KRAM); 2 km south-west of Endurance Glacier, 152 m, *JSA 30* (AAS, KRAM); Emma Cove, 138 m, *JSA 121* (AAS, KRAM). **King George I.** ***Bransfield Strait***: Red Hill, 100 m, *RO 1095/80* (KRAM); Blue Dyke, 130 m, *RO 1150/80* (KRAM). ***Admiralty Bay***: Rembiszewski Nunataks, 200 m, *RO 2746/80* (KRAM); Zamek, 330 m, *RO 16/80* (KRAM); Sphinx Hill, 95 m, *RO 226/80* (KRAM); Czajkowski Needle, 230 m, *RO 683/80* (KRAM); Rescuers Hills, 25 m, *RO 4890/79* (KRAM). ***Ezcurra Inlet***: Jersak Hills, 170 m, *RO 5117/79* (KRAM), 185 m, *RO 5128/79* (KRAM) and 200 m, *RO 5089/79* (KRAM); Jardine Peak, 275 m, *RO 5171/79* (KRAM); Italia Valley, 5 m, *RO 5077/79* (KRAM), 60 m, *RO 46/80* (KRAM) and 200 m, *RO 65/80* (KRAM); Dutkiewicz Cliff, 15 m, *RO 5084/79* (KRAM); Dufayel I., Gdynia Pt., 10 m, *RO 1777/80* (KRAM). ***MacKellar Inlet***: Klekowski Crag, 230 m, *RO 2270/80* (KRAM); Komandor Peak, 290 m, *RO 1954/80* (KRAM); Crépin Pt., 10 m, *RO 2996/80* (KRAM) and 20 m, *RO 2049/80* (KRAM). ***Keller Peninsula***: Harpoon Pt., 5 m, *RO 554/80* (KRAM); Moraine Pt., 30 m, *RO 487/80* (KRAM) and 35 m, *RO 466/80* (KRAM); Piasecki Pass, 200 m, *RO 431/80* (KRAM). ***Martel Inlet***: Shark Fin, 180 m, *RO 2638/80* (KRAM); Stenhouse Bluff, 25 m, *RO 2619/80* (KRAM); Mt. Wawel, 40 m, *RO 2136/80* (KRAM). ***Drake Passage***: Davey Pt., *BJ 48* (KRAM). ***Fildes Peninsula***: Lake Kitezh, *RO 2454/80* (KRAM). ***Barton Peninsula***: Winship Pt., *BJ 182*

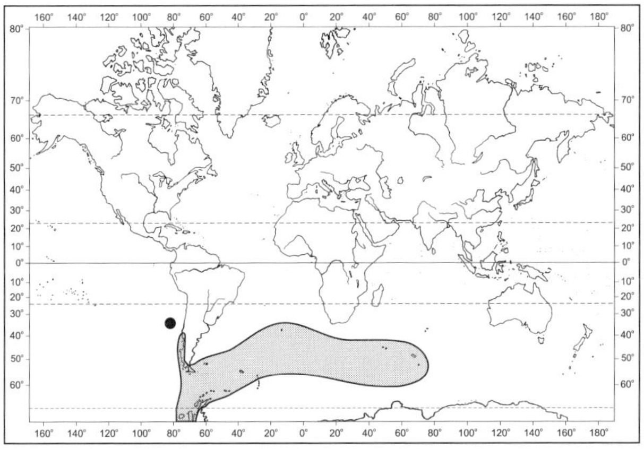

Figure 236. Global distribution of *Bartramia patens*.

(KRAM). ***Potter Cove***: Three Brothers Hill, *BJ 157* (KRAM). **NELSON I.** Rip Pt., 80 m, *DCL 795* (AAS, KRAM). **ROBERT I.** Coppermine Cove, 15–35 m, *RILS 910* (AAS, KRAM). **GREENWICH I.** Yankee Harbour, Spit Pt., 40 m, *DCL 546* (AAS, KRAM); Duff Pt., 15 m, *DCL 562* (AAS, KRAM). **LIVINGSTON I.** ***Byers Peninsula***: Cutler Stack, 3 m, *KR 66C* (AAS, KRAM); Robbery Beaches, 10 m, *KR 54B & 56F* (AAS, KRAM) and 15 m, *KR 53E* (AAS, KRAM); Smellie Pt., *KR 38A* (AAS, KRAM); Devils Pt., 20 m, *KR 24C* (AAS, KRAM) and 100 m, *KR 23G* (AAS, KRAM); Ocoa Pt., *KR 41D & 42C* (AAS, KRAM); South Beaches, *DCL 490* (AAS, KRAM); Clark Nunatak, 180 m, *DCL 219B* (AAS, KRAM); Charity Glacier, 35 m, *DCL 56* (AAS, KRAM). ***Hurd Peninsula***: Spanish Station, *Acuña 7005* (KRAM); north of the Bulgarian Base, *Pandurski 102A* (KRAM); Johnsons Dock, *Redon 109* (AAS, KRAM); False Bay, *RMS 69-406a, 421, 532, 546, 551, 581, 599a & 693* (US); Miers Bluff, *RMS 69-828* (US). ***Moon Bay***: Inott Pt., *RILS 5890A* (AAS, KRAM). **SNOW I.** President Head, *KR 2B* (AAS, KRAM). **DECEPTION I.** Cathedral Crags, 50–150 m, *RILS 6596A* (AAS, KRAM); Ronald Hill, *ca* 12 m, *RILS 63* (AAS, KRAM); between Kroner Lake and Ronald Hill, 2–4 m, *RILS 11245* (AAS, KRAM); East Cross Hill, 60 m, *RILS 5795E* (AAS, KRAM); Telefon Ridge, 30–60 m, *RILS 5760C* (AAS, KRAM); Kendall Terrace, 80 m, *RILS 5740A* (AAS, KRAM); Fumarole Bay, 50 m, *RILS 11353* (AAS, KRAM); Stonethrow Ridge, 60 m, *RILS 5806B* (AAS, KRAM); Wensleydale Gap, 300 m, *NJC 29* (AAS, KRAM); Wensleydale Beacon, 30–100 m, *RILS 6674B* (AAS, KRAM); north-east of Mt. Kirkwood, 200 m, *RILS 5721A* (AAS, KRAM); Mt. Kirkwood, 170–250 m, *RILS 5764E* (AAS, KRAM); Vapour Col, *ca* 175 m, *RILS 11265* (AAS, KRAM); W of Crater Lake, 50 m, *RILS 5768A* (AAS, KRAM); Entrance Pt., 50–70 m, *RILS 6793* (AAS, KRAM).

WEST ANTARCTIC PENINSULA. **TRINITY PENINSULA.** Astrolabe I., 10–30 m, *RILS 4060* (AAS, KRAM). **DAVIS COAST.** Cape Kjellman, *CS 450* (PC, S). **DANCO COAST.** Cierva Pt., *VK 524, 525, 602, 613 & 931*a (KRAM) and 5–270 m, *OB 1* (AAS, KRAM); Spring Pt., *VK 470, 898, 973, 1180 & 1252* (KRAM); Andrée I., 20–50 m, *RILS 4002* (AAS, KRAM); Cuverville I., 10–20 m, *RILS11941* (AAS, KRAM) and 15 m, *Leeuw 17A* (AAS, KRAM); Birdsend Bluff, 5–35 m, *RILS 4228* (AAS, KRAM); Coughtrey Peninsula, Paradise Harbour, *VC 1018 & 1025* (KRAM); Bryde I., *RMS 69-395 & 399* (US). **PALMER ARCHIPELAGO.** ***Brabant I.***: S of Minot Pt., *VK 573* (KRAM); Koch Glacier, 400 m, *AM 112* (AAS, KRAM). ***Melchior Archipelago***: Lysted I., Channel Pt., *PAS 378.3* (FH); same island, Chain Pt., *PAS 377.1* (FH); Lambda I., Sevilla Landing, *PAS 347.1* (FH); Kappa I., *Popovici 8393* (AAS, KRAM); Gamma I., 5–10 m, *RILS 4156* (AAS, KRAM); Omega I., 2–10 m, *RILS 4127* (AAS, KRAM); Alpha I., 2–5 m, *RILS 4104* (AAS, KRAM). ***Anvers I.***: Joubin Is., *JRR & JMG PAL-22* (US); Cape Monaco, *VK 807* (KRAM); Norsel Pt., *RMS 69-841, 844 & 868* (US); Bonaparte Pt., *VK 134* (KRAM); Cormorant I., *AS 58* (AAS, KRAM); Dream I., *VK 930* (KRAM); Litchfield I., *RWMC 378* (AAS, KRAM); Hermit I., *ca* 15 m, *RILS 886* (AAS, KRAM); Laggard I., 15 m, *REL 1314* (AAS, KRAM). **GRAHAM COAST.** Cape Tuxen, 20 m, *RILS 10662* (AAS, KRAM) and 65 m, *RWMC 850B* (AAS, KRAM); Rasmussen I., *Gain 280b & 281* (PC) and *RILS 1916A* (AAS, KRAM); Edge Hill, 65 m, *RWMC 659B* (AAS, KRAM). **WILHELM ARCHIPELAGO.** ***Lemaire Channel***: Booth I., *Turquet 230* (PC). ***Penola Strait***: Petermann I., 15 m, *REL 1343A* (AAS, KRAM), 50 m, *AS 127A & 128* (AAS, KRAM) and 60 m, *AS 62* (AAS, KRAM). ***Argentine Is.***: Irizar I., 8 m, *RWMC 698A* (AAS, KRAM); Uruguay I., 23 m, *RWMC 439B* (AAS, KRAM); Galindez I., 8 m, *RWMC 429A* (AAS, KRAM), 25 m, *RWMC 423* (AAS, KRAM) and 2–40 m, *RILS 3308* (AAS, KRAM); Skua I., 15 m, *RWMC 461B & 506C* (AAS, KRAM); Black I., 2–10 m, *RILS 4297* (AAS, KRAM). ***Collins Bay***: Berthelot Is., Green I., *RILS 3368C* (AAS, KRAM); Darboux I., 10–15 m, *RILS 3333* (AAS, KRAM). ***Grandidier Channel***: Lahille I., 2–5 m, *RILS 4323* (AAS, KRAM); Takaki Pt., 5–20 m, *RILS 4389* (AAS, KRAM) and 10–35 m, *RILS 4340* (AAS, KRAM). **BISCOE IS.**: Hook I., *RMS 69-505A* (US). **LOUBET COAST.** ***Bennett Is.***: Pfaff I., *JK 176B* (AAS, KRAM). ***Laubeuf Fjord***: Piñero I., 10–25 m, *RILS 4828* (AAS, KRAM). **FALLIÈRES COAST.** ***Adelaide I.***: Rothera Pt., 10 m, *JF 33A* (AAS, KRAM), 15 m, *RILS 93* (AAS, KRAM) and 30 m, *JF 58A* (AAS, KRAM); Stork Ridge, *PC 261B* (AAS, KRAM) and 300 m, *RILS 1879, 3423 & 3466* (AAS, KRAM); Reptile Ridge, 140 m, *RILS 10581A* (AAS, KRAM); and 175 m, *RILS 10633* (AAS, KRAM); Mt. Mangin, 245 m, *JK 148B* (AAS, KRAM) and 550 m, *JK 150A* (AAS, KRAM). ***Léonie Is.***: Lagoon I., 10–15 m, *RILS 9185* (AAS, KRAM); Léonie I., *Gain 234A* (PC) and 10–15 m, *RILS 8975* (AAS, KRAM), 5–50 m, *RILS 8987B* (AAS, KRAM) and 150–175 m, *RILS 9122 & 9128* (AAS, KRAM); Anchorage I. 25–35 m, *RILS 9207* (AAS, KRAM). ***Marguerite Bay***: Jenny I., *Gain 221, 222a, 223, 231A, 247A, 246B, 247B, 248B & 259B* (PC), 15–50 m, *RILS 4714* (AAS, KRAM) and 200 m, *RILS 4756* (AAS, KRAM); Blaiklock I., 8–15 m, *RILS 849* (AAS, KRAM); Pourquoi Pas I., *DDWW 156B* (AAS, KRAM); Lagotellerie I., 50–75 m, *RILS 1779* (AAS, KRAM); Horseshoe I., Lystad Bay, 5–8 m, *RILS 1755* (AAS, KRAM); Camp Pt., 5–15 m, *RILS 4674* (AAS, KRAM) and 10–15 m, *RILS 4635* (AAS, KRAM); Neny I., 50 m, *RILS 789* (AAS, KRAM). **GEORGE VI SOUND.** ***Charcot I.***: Rils Nunatak, 85 m, *RILS 10494* (AAS, KRAM) and 125 m, *RILS 10539 & 10711A* (AAS, KRAM).

WEST ANTARCTIC PENINSULA. **JOINVILLE ISLAND GROUP.** ***Joinville I.***: Mt. Alexander, 15–50 m, *RILS 3702* (AAS, KRAM) and 60–75 m, *RILS 3670* (AAS, KRAM). **JAMES ROSS ISLAND GROUP.** ***James Ross I.***: Lachman Crags, 20–35 m, *RILS 7321A* (AAS, KRAM); valley north of St. Martha Cove, 50 m, *RILS 7510B* (AAS, KRAM) and 80 m, *RILS 7517* (AAS, KRAM); hill south of St. Martha Cove, 75 m, *RILS 7433* (AAS, KRAM) and *ca* 150 m, *RILS 7429* (AAS, KRAM). **OSCAR II COAST.** Flask Glacier, 700 m, *RILS 5989A* (AAS, KRAM).

Subclass C6. **HYPNIDAE** W.R.BUCK, GOFFINET & A.J.SHAW

Hypnidae W.R.Buck, Goffinet & A.J.Shaw, Molec. Phylogen. Evol. 16: 195. 2000.

The Hypnidae comprise pleurocarpous mosses all of which have diplolepidous alternate peristomes. These are well defined by having two alternate rows of teeth. The outer row (exostome) consists of 16 robust triangular or lanceolate teeth each with inner trabeculae derived from the cross-walls of a single column of cells and two columns of plates separated by a zig-zag median line on the outer face. The inner row (endostome) consists of a delicate basal membrane bearing a network of cells on each side which is divided in the distal portion into 16 often perforated segments that alternate with the exostome teeth, and between these inner teeth there are 16 groups of 2–4 fine cilia which lie inside the outer teeth. Although it is commonly designated as the *Hypnum* peristome-type or hypnoid peristome (Hedenäs 1994*a*) this is, in fact, a typical *Bryum* peristome-type; the former term is used only because of its presence in pleurocarpous mosses, not owing to the essential structural differences. This prompted some bryologists to place bryalean and hypnalean mosses in one order Bryales (Robinson 1971) or subclass Bryidae (Crosby 1980). In most cases these peristomes are virtually indistinguishable from each other and the only difference which is emphasised as differentiating bryoid and hypnoid peristomes is the ornamentation of the outer face of the exostome. It is papillose throughout in the former, whereas in the latter it is transversely striate in the proximal portion and papillose distally.

As with the bryoid peristome, the hypnoid peristome exhibits a large range of modifications resulting in various kinds of highly specialised peristomes, for example neckeroid, leskeoid and hookerioid which are characteristic of various families of pleurocarpous mosses and present various peculiarities in their morphologies (Vitt 1981; Hedenäs 1999, 2001*a*; Huttunen *et al*. 2004). Moreover, even within the perfect hypnoid peristomes there exists much variation and they differ in characters such as exostome colour and border, perforation of the endostome processes and number and kind of cilia.

The Hypnidae are generally considered to be a monophyletic group which comprises exclusively pleurocarpous mosses and its monophyly is strongly supported by nuclear, chloroplast and mitochondrial DNA sequences (De Luna *et al*. 1999; Newton *et al*. 2000; Buck *et al*. 2000*a*; O'Brien 2007). They are characterised by extensive branching and lateral sporophyte position. The term 'pleurocarpy' was introduced by Nees *et al*. (1823) to designate the lateral position of the female gametangia (i.e. perichaetia) and it contrasts with 'acrocarpy' and 'cladocarpy', situations in which the perichaetia are produced terminally on main shoots or lateral branches, respectively (LaFarge-England 1996).

Apart from the pleurocarpous growth form, the subclass is defined by the presence of pseudoparaphyllia which are small, unistratose, foliose or filamentous structures occurring only around branch primordia or branch bases (Bonnot 1967; Ireland 1971; Berthier 1972; Ignatov 1999; Ignatov & Hedenäs 2007), as well as paraphyllia which are small, green, filiform, lanceolate or leaf-like, sometimes branched structures often produced on stems and branches (Noguchi 1972; Schofield & Hébant 1984). Homogeneous anatomy of the costa is another feature typical of pleurocarpous mosses (Hedenäs 1994*b*). However, the latter two features are not satisfactory to circumscribe the Hypnidae because pseudoparaphyllia and paraphyllia are missing from some pleurocarps and costae are absent in many taxa.

The classification of pleurocarpous mosses has undergone considerable changes over the past two centuries and excellent reviews of various systems in the historical perspective was outlined by Buck (1991, 2007). Until the early 1980s the classificational system proposed by Fleischer (1904–1923, 1920) and subsequently popularised by Brotherus (1925) in the second edition of the Musci in *Die Naturlichen Pflanzenfamilien* exclusively dominated moss systematics. But in the last two decades of the 20th century there was an increasing interest in the study of morphological traits in both gametophyte and sporophyte and, accordingly, many new suggestions and proposals have been presented regarding the reclassification of pleurocarpous mosses as a whole (Vitt 1982*a*, 1984; Buck & Vitt 1986; Hedenäs 1995, 1998*c*) or of various groups within them, for instance Hookeriales (Crosby 1974; Buck 1987, 1988; Hedenäs 1996*a*), Entodontaceae (Buck 1980), Leskeaceae and Thuidiaceae (Buck & Crum 1990; Touw 2001), Amblystegiaceae (Vanderpoorten *et al*. 2002*a*,*b*; Ignatov & Ignatova 2004; Hedenäs *et al*. 2005; Hedenäs & Vanderpoorten 2007), Vittiaceae (Vanderpoorten *et al*. 2003), Brachytheciaceae (Ignatov & Huttunen 2002; Huttunen & Ignatov 2004), Sematophyllaceae (Hedenäs 1996*d*), Hypnaceae (Ignatov & Ignatova 2004; Gardiner *et al*. 2005), Meteoriaceae (Buck 1994) and Plagiotheciaceae (Buck & Ireland 1985; Hedenäs 1987*a*; Pedersen & Hedenäs 2002*b*).

The advent of DNA sequencing in moss taxonomy in the last decade provided some important insights into evolutionary relationships that previously had not been discerned on the basis of morphological studies. As a result, a lot of changes, sometimes revolutionary ones, have been proposed at various levels of the taxonomic hierarchy and novel hypotheses drawn from analyses of molecular data were integrated into a new classification system of the pleurocarps (Buck & Goffinet 2000; Goffinet & Buck 2004). This trend in taxonomic and phylogenetic research on mosses has continued and an excellent review of the current problems in this field is presented by Newton & Tangney (2007).

Traditionally, the pleurocarpous mosses of the subclass Hypnidae have been divided into three orders, Hookeriales, Isobryales (= Leucodontales) and Hypnobryales (= Hypnales) (Fleischer 1920; Brotherus 1925). This ordinal classification rested on characters such as branching mode and architecture of the peristome. The Hookeriales are characterised by a unique architecture of the endostome in which the horizontal walls separating its inner cells are left in place and form baffle-like walls; molecular data confirm the monophyly of this order (Buck *et al*. 2004).

In contrast, the Leucodontales comprise mainly epiphytic taxa with sympodial branching and highly specialised peristomes, whereas the Hypnales are primarily terricoles with monopodial branching and well developed peristomes. However, numerous molecular data (e.g. Buck *et al*. 2000*a*; Quandt *et al*. 2000; Goffinet *et al*. 2001; Tsubota *et al*. 2002; Arikawa & Higuchi 2003; Gardiner *et al*. 2005) provided evidence that taxa belonging within these orders are interspersed on the phylogenetic tree. In practice they are inseparable from each other and have been formally merged by Buck *et al*. (2000*a*). Thus, the Hypnidae are now considered to consist of two orders, Hookeriales and Hypnales.

The Hypnidae are the largest subclass of the Bryophyta and comprise approximately half of all extant moss taxa, including about 50 families, 470 genera and over 5200 species (Crosby *et al*. 2000), but species richness is highly asymmetric. Only five families, 41 genera and about 720 species belong within the Hookeriales and the remaining taxa are the Hypnales which is the largest order of mosses. The former order comprises mostly tropical and subtropical taxa. It is absent from the Antarctic and very poorly represented in some regions of the Subantarctic, namely on Macquarie I. where two genera each with a single species (Seppelt 2004) and on the Prince Edward Is. where two genera and three species (Ochyra 2008) have been recorded. In general, the pleurocarpous mosses of the subclass Hypnidae have nearly similar diversity in tropical and temperate areas (Hedenäs 2007).

Order X. **HYPNALES** W.R.Buck & Vitt

Hypnales W.R.Buck & Vitt, Taxon 35: 33. 1986 [validly published by reference to suborder Hypnineae M.Fleisch., Hedwigia 61: 399. 1920].

Metacranales M.Fleisch. [Untergruppe] Hypnobryineae M.Fleisch., Musci Fl. Buitenzorg 3: xxi. 1908, *nom. illeg.*

Isobryales M.Fleisch., Hedwigia 61: 396. 1920, *nom. illeg.*

Hypnobryales M.Fleisch., Hedwigia 61: 398. 1920, *nom. illeg.*

Leucodontales W.R.Buck & Vitt, Taxon 35: 34. 1986 [validly published by reference to suborder Leucodontineae M.Fleisch., Hedwigia 61: 397. 1920]. First synonymised by Buck *et al.* (2000*a*: p. 197).

When considered as a single taxon encompassing former isobryalean and hypnobryalean taxa, the order is exceedingly morphologically variable. The stature of the plants ranges from small and slender to robust. The stems are usually interwoven, creeping and prostrate, ascending to erect, mostly profusely and variously branched, often irregularly so to frondose, pinnate or dendroid. The paraphyllia and/or pseudoparaphyllia are present or absent. The leaf areolation is extremely variable, the upper laminal cells being isodiametric to elongate, smooth or papillose and thin- or thick-walled. The basal angular cells are often differentiated and form distinct auricles. The leaves are spirally arranged or complanate, linear-lanceolate to orbicular, plane or cochleariform, rounded to long-acuminate at the apex, sometimes awned to aristate due to an excurrent costa. The branch leaves are similar to and smaller than the stem leaves, or differ in shape. The costa is long or short, single and entire or branched to spurred, double or absent. Asexual reproduction is occasionally caused by dwarf deciduous axillary branches or axillary gemmae. The sporophytes are also considerably variable. The setae are short or elongate, resulting in immersed or exserted capsules. The capsules are erect, cylindrical and symmetric to inclined or horizontal, curved and asymmetric. The peristome is perfect to variously reduced, with narrow or broad, solid or perforate processes of the endostome, arising from a low or high basal membrane or sometimes free.

Despite the extremely low level of molecular variability in the Hypnales as compared to the other mosses (Shaw *et al.* 2003), the order may have undergone rapid radiation and reached a high rate of diversification, and now it is the largest order of mosses, consisting of about 4400 species placed in about 430 genera and 45 families. The limits between some families at present are very problematic, as are limits between some of the constituent genera. The real number of species in the Hypnales is unknown because many genera are still in need of taxonomic assessment, especially in extra-Holarctic areas where many reductions of superfluous species names to synonymy are to be expected. Likewise, there is a continuous tendency to split large and heterogeneous genera into smaller but better defined and more natural ones. In the Antarctic the Hypnales are not well represented and only four families, eleven genera and 17 species are known from this biome.

The supremacy of acrocarpous mosses over pleurocarpous ones is an interesting feature of the polar and cold as well as dry regions throughout the world and the tendency for high numbers of acrocarps and low numbers of pleurocarps is evident in both hemispheres. Of the 113 taxa (111 species and 2 varieties) comprising the Antarctic moss flora, only 17 are pleurocarps, giving a ratio 6.65 : 1 for the two growth forms, and this is the most extreme value for any biome. All belong to the maritime Antarctic and none is known in continental Antarctica. On subantarctic Marion and Prince Edward Islands is it 3.17 : 1 and on Macquarie I. 2.3 : 1. The ratio continues to decrease northwards; for example, on the Auckland Is. it is 1.86 : 1, and on Juan Fernandez Is. 1.37 : 1. For comparison, in Devon I. in the Canadian High Arctic Archipelago at lat. *ca* 75° N this ratio is 2.6 : 1, more than three times lower than that at a similar latitude on Alexander I. in Antarctica. Yet, despite their poor species diversity, some pleurocarpous mosses are locally very abundant and play an important role in the Antarctic tundra vegetation.

Family XIV. **AMBLYSTEGIACEAE** Kindb.

Amblystegiaceae Kindb, Rev. Bryol. 12: 26. 1885. — Type: *Amblystegium* Schimp.

Cratoneuraceae Mönk. *in* Pascher, Süsswasserflora 14: 116. 1914. — Type: *Cratoneuron* (Sull.) Spruce.

Hypnobartlettiaceae Ochyra, Lindbergia 11: 7. 1985. — Type: *Hypnobartlettia* Ochyra.

Vittiaceae Ochyra, J. Hattori Bot. Lab. 62: 409. 1987. — Type: *Vittia* Ochyra.

Campyliaceae W.R.Buck *in* A.J.Shaw & Goffinet, Bryoph. Biol.: 110. 2000. — Type: *Campylium* (Sull.) Mitt.

Calliergonaceae Vanderpoorten, Hedenäs, C.J.Cox & A.J.Shaw, Taxon 51: 120. 2002, ***syn. nov.*** — Type: *Calliergon* (Sull.) Kindb.

Scorpidiaceae Ignatov & Ignatova, Moss Fl. Middle Eur. Russia 2: 830, 942. 2004, ***syn. nov.*** — Type: *Scorpidium* (Schimp.) Limpr.

The Amblystegiaceae are a rather small family of pleurocarpous mosses consisting of approximately 60–70 species, but many species are very prominent constituents of the bryophyte flora, especially in cold and temperate regions of the earth. They are mainly associated with wet or otherwise moist habitats, although some species grow in dry situations on rocks, bark or soil. Although the Amblystegiaceae were recognised as a separate family in the late 1800s (Kindberg 1885; Roth 1899), they are difficult to circumscribe and still serve as a catch-all family for various aquatic taxa whose affinities are difficult to evaluate because of the sterile condition of the plants. Various attempts have recently been undertaken to rearrange taxa belonging within this family, using both morphological and molecular data. Accordingly, the genera traditionally placed in the Amblystegiaceae were realigned into several smaller and seemingly better defined families, including the Hypnobartlettiaceae and Donrichardsiaceae (Ochyra 1985*a*,*b*; Váňa 1986), Vittiaceae (Ochyra 1987*c*; Vanderpoorten *et al.* 2003), Campyliaceae (Buck & Goffinet 2000), Calliergonaceae (Vanderpoorten *et al.* 2002*a*,*b*; Hedenäs *et al.* 2005) and Scorpidiaceae (Ignatov & Ignatova 2004). None of these appear to be satisfactory because the phylogeny of these taxa, as reconstructed by nuclear and chloroplast sequence data, and classifications based on their highly variable morphologies, do not seem to overlap at all well (Hedenäs & Vanderpoorten 2007).

The Amblystegiaceae are closely related to the Hypnaceae and Thuidiaceae and the limits between them are sometimes obscure and unclear (Hedenäs 1998*a*). This may explain the relatively late recognition of the Amblystegiaceae as a family in its own right (Ochyra 2003*d*) and earlier most of its members were placed in the Hypnaceae (Schimper 1856). Both families share a perfect hypnaceous peristome with yellow-brown exostome teeth which are cross-striolate below and widely bordered at the transition zone, stomata with narrow and elongate pori, curved and asymmetric capsules which are usually strongly contracted below the orifice and plicate inner perichaetial leaves with a single, long and well developed costa and straight apices. The leaves of most species of the Amblystegiaceae are unicostate, although the costae are often variously branched, spurred or double, short or long. In many species the angular leaf cells are strongly differentiated and form inflated and well marked auricles. The paraphyllia are mostly lacking but in a few species they are sparse or abundant. They are mostly uniseriate and filiform, and only rarely foliose.

Likewise, the generic limits in the Amblystegiaceae are often controversial and many species can be arranged in a linear series of integrades. In general, there is now a tendency for recognition of smaller but more homogeneous genera and therefore some large genera have been split into several segregates, for example *Drepanocladus* (Müll.Hal.) G.Roth, *Calliergon* (Sull.) Kindb., *Campylium* (Sull.) Mitt., *Sciaromium* (Mitt.) Mitt., *Cratoneuron* (Sull.) Spruce or *Amblystegium* Schimp. have been split into several segregates (Loeske 1907*a*,*b*, 1910; Tuomikoski & Koponen 1979; Ochyra 1987*b*,*c*,*d*, 1989*b*; Hedenäs 1989*b*, 1993*a*,*b*, 1997*a*, 2006). Hence, the number of the accepted genera in this family varies from 20 to 30.

There have been several subfamilies proposed within the Amblystegiaceae (Szafran 1961; Kanda 1975, 1976) but their circumscription is not precise because it is difficult to find reliable consistent characters for them. Therefore, the final infrafamilial classification of the Amblystegiaceae must await further studies.

Most members of the Amblystegiaceae are confined in their occurrence to the Northern Hemisphere, especially to the cool and temperate regions where they are the main constituents of various types of mire vegetation. Outside the Holarctic species of this family are less prominent. In the tropics they are largely restricted to higher elevations in the mountains, but they are quite frequent in temperate and cool areas in the Southern Hemisphere and some species are here widespread and common peatland mosses.

In southern South America, South Georgia and Antarctica the Amblystegiaceae are well represented, both in terms of the number of species as well as their abundance and these play an important role in the vegetation cover, with species of *Sanionia* being amongst the commonest mosses in the maritime Antarctic. So far, seven species and five genera have been discovered in this biome. Except for *Warnstorfia fontinaliopsis* and *Cratoneuropsis relaxa* subsp. *minor*, all are circumpolar Holarctic species occurring in arctic, boreal and temperate zones. In contrast to the taxonomy of genera of amblystegialean mosses, species of this family are mostly well circumscribed and defined, and therefore various northern Floras are useful for determination of Antarctic members of this family (e.g. Hedenäs 1993*a*, 2003*a*), together with the taxonomic surveys of the Amblystegiaceae for the Neotropics (Hedenäs 2003*b*) and Tierra del Fuego (Ochyra & Matteri 2001).

KEY TO THE GENERA OF THE ANTARCTIC AMBLYSTEGIACEAE

1. Paraphyllia present, uniseriate to narrowly lanceolate; angular cells diffusely delimited, never auriculate; small plants .... 45. ***Cratoneuropsis***
1. Paraphyllia lacking; angular cells forming distinct, ± well delimited and often auriculate groups; medium-sized to large plants 2
   2. Leaves erect-spreading, straight, broadly lanceolate, gradually tapering into a fine, channelled acumen; costa single, often branched from the base, or short and double .... 46. ***Campylium***

2. Leaves erect, straight or falcato-secund (at least at the tips of stems and branches), lanceolate, ovate or oblong, gradually narrowed to a plane acumen or rarely rounded and cucullate at the apex; costa single 3

3. Stem with a perfect unistratose ring of hyalodermis; leaves usually distinctly plicate ... . 47. ***Sanionia***

3. Stem without or with an incomplete hyalodermis; stem leaves not or indistinctly plicate 4

4. Costa long excurrent; nematogon cells at the leaf apex absent; plants distichously branched; reddish colour always absent .... 48. ***Drepanocladus***

4. Costa ending well below the leaf apex; nematogon cells present at the apex; plants radially branched; reddish colour often present .... 49. ***Warnstorfia***

45. CRATONEUROPSIS (BROTH.) M.FLEISCH.

Cratoneuropsis (Broth.) M.Fleisch., Musci Buitenzorg 4: 1535. 1923. ≡ *Campylium* (Sull.) Mitt. sect. *Cratoneuropsis* Broth. *in* Engl. & Prantl, Nat. Pflanzenfam. 1(3): 1043. 1908. — Lectotype (*selected here*): *Cratoneuropsis relaxa* (Hook.f. & Wilson) M.Fleisch. *ex* Broth. (*Hypnum relaxum* Hook.f. & Wilson).

Orthotheciella (Müll.Hal.) Ochyra, Moss Fl. King George Island Antarctica: 224. 1998. ≡ *Hypnum* Hedw. sect. *Orthotheciella* Müll.Hal. *in* Engl., Forschungsr. Gazelle 4 (Bot.): 36. 1889. ≡ *Pseudoleskea* Schimp. sect. *Orthotheciella* (Müll.Hal.) Broth. *in* Engl. & Prantl, Nat. Pflanzenfam. 1(3): 1002. 1907, ***syn. nov.*** — Lectotype (*vide* Ochyra 1998*a*: p. 224): *Hypnum filum* Müll.Hal. [= *Cratoneuropsis relaxa* (Hook.f. & Wilson) Broth. subsp. *minor* (Wilson & Hook.f.) Ochyra (*Hypnum filicinum* Hedw. var. *minus* Wilson & Hook.f.)].

Plants very variable in size and stature, ranging from very small and slender through medium-sized to very large and robust, mostly rigid and stiff, in loose or dense, large patches, intricate mats or deep tufts, dull or lustrous, green, yellow-green, yellow or golden-brown to brown, occasionally blackish-brown to black. *Stems* yellowish-green to brown, filiform and slender or stout, prostrate and straggling, ascending or crowded and erect, sparingly or richly, freely to subpinnately branched, sometimes pinnately branched and giving the plant a feather-like appearance, in transverse section with a 1–3(–4)-stratose cortex of relatively small and thick-walled, yellowish-green to brown cells gradually passing into 4–7-layered medulla of large, thin-walled, hyaline cells, with a small but distinct central strand; *rhizoids* absent or sparsely scattered in clusters along the stem, occasionally rich along the stem facing the substrate, light brown, smooth, branched; *pseudoparaphyllia* foliose to filamentous; *paraphyllia* filiform to narrowly lanceolate, consisting of a single row of cells or 2–3 rows of cells in the basal part and uniseriate distally, infrequent to very many on the surface of the stems and stronger branches, sometimes missing on slender and delicate plants; *axillary hairs* infrequent, 2–4-celled, with 1–2 short brownish basal cells and 2–3 elongate, hyaline upper cells. *Stem* and *branch leaves* similar, the latter somewhat smaller, crowded to distant, very variable in shape and stance, plane or weakly to deeply concave, smooth, striate or plicate, not or shortly decurrent, straight, falcate and secund or homomallous, wide-spreading to strongly squarrose-recurved, ovate, ovate-lanceolate to lanceolate, often from a cordate or ovate, erect-spreading and loosely subclasping base short-acuminate or tapering to a long, slender and folded acumen; *margins* plane, serrulate or crenulate all around or subentire to entire; *costa* single, strong, subpercurrent to percurrent, occasionally vanishing below the apex or long-excurrent, yellow and sharply set off from the laminal cells, often curved above, terete; *laminal cells* unistratose, transparent to scarcely pellucid, smooth or prorulose; *upper* and *median cells* variable, mostly not uniform in size and shape, shortly elliptic, rhomboidal, oblong-rhomboidal, oblong-hexagonal to linear,with truncate or spindle-like apices, thin-, firm- to thick-walled, with un-pitted walls; *basal cells* usually somewhat larger, oblong to rectangular, often yellowish or brownish in 1–2 rows at the insertion; *angular cells* undifferentiated or subquadrate to short-rectangular, with thick and often yellow walls or somewhat enlarged, subhyaline but not forming distinct auricles and gradually merging into the laminal cells. *Dioicous. Inner perichaetial leaves* hyaline to pale yellow, elongate lanceolate, gradually long-acuminate, plicate, with a single, long-excurrent costa, serrate margins and oblong-hexagonal to linear laminal cells; *outer perichaetial leaves* smaller, rather suddenly acuminate, with a weaker costa. *Setae* elongate, straight or flexuose when dry, smooth, brown to reddish-brown. *Capsules* reddish-brown to brown, obloid to cylindrical, arcuate, asymmetric, inclined to horizontal, sometimes suberect, not or strongly contracted below the mouth when dry and empty; *operculum* conic-convex, apiculate; *annulus* present, separating, 1–3-seriate; *exothecial* cells isodiametric below to oblong or rectangular above, becoming short and subquadrate in 3–4 tiers below the orifice, thin- to firm-walled; *stomata* few, bicellular, superficial, long-pored in the neck; *peristome* double; *exostome teeth* yellow, strongly trabeculate ventrally, cross-striolate below, becoming striolate-dotted in the transitional zone and distinctly papillose, bordered and dentate at the margins distally; *endostome teeth* well developed, pale yellowish, finely papillose throughout, basal membrane high, processes not or narrowly cleft, *cilia* 2–3, well developed, nodose, somewhat shorter than the processes. *Spores* globose, finely papillose. *Calyptra* cucullate, naked.

Cratoneuropsis is a distinct and natural genus which shares some similarity with the northern genus *Cratoneuron* in having paraphyllia on the stems and branches, robust, percurrent to excurrent costae, a leaf areolation of predominantly short, rhomboidal, oblong-rhomboidal and oblong-hexagonal cells, as well as the dioicous sex condition. This remarkable similarity is possibly a result of convergence associated with the predilections of both genera to wet habitats. However, the paraphyllia in *Cratoneuropsis* are filiform, composed of a single row of cells or two cell rows in the proximal portion and one row in the distal part, whereas in *Cratoneuron* the paraphyllia are broadly foliose. In addition, *Cratoneuropsis* has no or only weakly differentiated angular leaf cells which do not form sharply distinct auricles, whereas in *Cratoneuron* the alar cells are large, typically hyaline and thin-walled and form prominent, inflated and decurrent auricles which occupy the whole leaf insertion and are abruptly delimited from the adjacent laminal cells.

As presently conceived, *Cratoneuropsis* is a monotypic genus, consisting of a single species, *C. relaxa* (Hook.f. & Wilson) Broth. which exhibits an incredible range of variability in the size and branching of the plants, as well as in the shape and set of the leaves. The genus has so far been considered to be an Australasian endemic, occurring in south-eastern Australia and Tasmania (Streimann & Klanzenga 2002), New Zealand (Sainsbury 1955) and its offshore islands in the cool-south-temperate zone (Vitt 1974*a*, 1979*a*) and extending to subantarctic Macquarie I. (Seppelt 2004). Here, the geographical range of *Cratoneuropsis* is markedly extended and the genus is considered to be a pan-south-temperate taxon because the genus *Orthotheciella* (Müll.Hal.) Ochyra is included in it. The latter genus was initially recognised as an unranked group within *Hypnum* to accommodate four species described from Îles Kerguelen (Müller 1889). Brotherus (1907) gave it the status of a separate section, sect. *Orthotheciella*, within the genus *Pseudoleskea* Schimp. and Ochyra (1998*a*) raised it to genus. *Orthotheciella* is distinguished by its filiform, uniseriate or partially biseriate paraphyllia sparsely occurring on the surface of the stem and larger branches which are exactly of the same shape as those in *Cratoneuropsis*. Moreover, *Orthotheciella* shares with *Cratoneuropsis* an identical leaf areolation, lack of distinct alar cells and strong percurrent to excurrent costa and these characters warrant their congenericity. The only real difference between the generitypes of *Cratoneuropsis* and *Orthotheciella* is the stature of the plants, a criterion which is of minor importance in phylogenetic considerations, especially because it is subject to remarkable variability in taxa belonging within both genera.

The presence of the paraphyllia and the dioicous sex condition are the fundamental characters separating *Cratoneuropsis*, especially those taxa which were placed in *Orthotheciella*, from the morphologically similar genus *Hygroamblystegium* Loeske which is also frequently associated with moist to wet habitats. The ostensible external similarity of the plants in having stout and flexuose costae, short laminal cells, not markedly differentiated alar cells and ovate-lanceolate, acuminate to bluntly acute leaves led Reimers (1926) and Bartram (1946) to the placement of southern South American species, which are now considered to be identical to *Cratoneuropsis relaxa* subsp. *minor*, in the genus *Hygroamblystegium*. On the other hand, these species have often quite thick-walled laminal cells and this suggests their affinity to *Pseudoleskea* and, in fact, Cardot (1900, 1901, 1906*a*,*b*, 1908*a*), and Brotherus (1907, 1925) placed in this genus a number of newly described species from the Southern Hemisphere.

The genus *Orthotheciella* was lectotypified with *Hypnum filum* and this name was erroneously synonymised with *Hygroamblystegium varium* (Hedw.) Mönk (Ochyra 1998*a*), a species having a predominantly Holarctic distribution. Though morphologically similar, this species differs from *Cratoneuropsis relaxa* subsp. *minor* in having a narrower costa ending in the leaf acumen but well below the apex, a usually quite well delimited, ovate to broadly ovate alar cell group along the basal margin of the leaf and an autoicous sex condition. However, this unfortunate taxonomic conclusion has no nomenclatural consequences and *Orthotheciella* is now considered to be congeneric with *Cratoneuropsis*. This genus was originally recognised by Brotherus (1907) as a section of *Campylium* because of the similarity of one of numerous habitat expression of *Cratoneuropsis relaxa*, known as *C. decussata* (Hook.f. & Wilson) Broth., to species of *Campylium* owing to its strongly squarrose-recurved leaves. This section was subsequently elevated to genus (Fleischer 1923) which has gained wide and universal acceptance.

1. **Cratoneuropsis relaxa** (Hook.f. & Wilson) M.Fleisch. *ex* Broth. subsp. **minor** (Wilson & Hook.f.) Ochyra Fig. 238

Cratoneuropsis relaxa (Hook.f. & Wilson) M.Fleisch. *ex* Broth. subsp. *minor* (Wilson & Hook.f.) Ochyra, *stat. et comb. nov.* ≡ *Hypnum filicinum* Hedw. var. *minus* Wilson & Hook.f. *in* Hook.f., Fl. Antarct. 1: 141. 1845 ['minor']. ≡ *Cratoneuropsis relaxa* (Hook.f. & Wilson) M.Fleisch. subsp. *minor* (Wilson & Hook.f.) Myrcha, Ochyra & Tatur *in* Klekowski & Opaliński, First Pol.-Sov. Antarct. Symp.: 163. 1991, *comb. inval. basion. non citat.* — TYPE CITATION: Lord Auckland's group; in woods by the banks of streams (Dr. Lyall.). [Lectotype (*vide* Ochyra 1989*b*: 235): 'Lord Aucklands Islands D. Lyall 26 {handwritten label} No. Hypnum filicinum L. var. β minor Lord Aucklands Islands Antarct. Exp. 1839–1843. ~~J. D. H.~~ Dr Lyall' – BM!; isolectotypes: BM-Hooker!, BM-Wilson! (3 specimens), BR!, E!, H-Lindberg!, MANCH-Spruce!, MICH!; syntypes: (1) same locality, Lyall nº 33 – BM!, BM-Hooker!, BM-Wilson!; (2) same locality, Lyall nº 37 – BM-Hooker!, BM-Wilson!].

Amblystegium chilense Lorentz, Bot. Zeitung (Berlin) 24: 188. 1866. ≡ *Hygroamblystegium chilense* (Lorentz) Reimers, Hedwigia 66: 70. 1926. ≡ *Pseudoleskea chilensis* (Lorentz) Ochyra, J. Bryol. 14: 459. 1987, ***syn. nov.*** — TYPE CITATION: all species reported by Lorentz (1866) were collected 'in Chile prope Valdiviam et prope Corral'. [Lectotype (*vide* Ochyra 1987*b*: p. 459): 'In ligno aqueductus prope Corral, puerto de Valdivia, leg. Dr Kl. Krause' – GOET!; isotypes: BM-Schimper!, M!, NY-Jaeger! (2 specimens), PC!].

Amblystegium kerguelense Mitt., Phil. Trans. 168 (extra vol.): 37. 1879. ≡ *Hygroamblystegium kerguelense* (Mitt.) Broth. *in* Engl. & Prantl, Nat. Pflanzenfam. 1(3): 1019. 1908. ≡ *Cratoneuron kerguelense* (Mitt.) Broth. *in* Engl., Nat. Pflanzenfam. Ed. 2, 11: 334. 1925, ***syn. nov.*** — TYPE CITATION: [Îles Kerguelen] Christmas Harbour, Hooker. Near Swain's Bay, barren, Eaton. [Lectotype (*vide* Ochyra 1989*b*: p. 235): 'Amblystegium kerguelense Mitt. Near Swain's Bay, Kerguelen Rev. E. A. Eaton Transit Venus Exp. 22.xii.1874' – BM!; syntypes: (1): 'No. 266. Hypnum filicinum var. 2 Kerguelen's Land. Antarct. Exp. 1839–1843. J. D. H. – BM-Hooker!; (2) 'Hypnum serpens var. Kerguelen's Land 752 Antarct. Exp. 1839–1843 J. D. H. 267' – BM-Hampe!, BM-Hooker! (2 specimens), BM-Wilson!; (2) 'Hypnum serpens var. 2 Kerguelen's Land 753 Antarct. Exp. 1839–1843 J. D. H. 268' – BM!, BM-Hampe!, BM-Hooker! (2 specimens), NY-Mitten!]. First synonymised with *Pseudoleskea chilensis* by Ochyra (1989*b*: p. 235).

Hypnum chalarocladum Müll.Hal., Bot. Jahrb. Syst. 5: 82. 1883. ≡ *Pseudoleskea chalaroclada* (Müll.Hal.) Paris, Index Bryol.: 1034. 1898. ≡ *Hygroamblystegium chalarocladum* (Müll.Hal.) Reimers, Hedwigia 66: 71. 1926, ***syn. nov.*** — Type citation: all new species described by Müller (1883) are included in the section 'I. Bryologia Kerguelensis'; later Müller (1889: p. 36) wrote: 'Ins. Kerguelen, Betsy Cove (11. 74)'. [Lectotype (*selected here*): 'Hypnum (Pseudoleskea) chalarocladum C. Müll.Kerguelenland, Gazelle-Expedition, Dr Naumann 1874' – BM!; isotypes: FH-Fleischer!, NY!, PC!, S!, W!].

Hypnum desmiocladum Müll.Hal., Bot. Jahrb. Syst. 5: 83. 1883. ≡ *Pseudoleskea desmioclada* (Müll.Hal.) Paris, Index Bryol.: 1034. 1898. ≡ *Hygroamblystegium desmiocladum* (Müll.Hal.) Reimers, Hedwigia 66: 71. 1926, ***syn. nov.*** — Type citation: all new species described by Müller (1883) are included in the section 'I. Bryologia Kerguelensis'; later Müller (1889: p. 35) wrote: 'Ins. Kerguelen, Port Palliser, Betsy Cove (2. 75)'. [Lectotype (*selected here*): 'Hypnum (Pseudoleskea) desmioclada C. Müll. Kerguelenland, Betsy Cove, Nov. 1874 Dr Naumann' – S!; isotypes: BM! (2 specimens)!, E!, S! (2 specimens)].

Hypnum filum Müll.Hal., Bot. Jahrb. Syst. 5: 83. 1883. ≡ *Pseudoleskea filum* (Müll.Hal.) Paris, Index Bryol.: 1035. 1898. ≡ *Hygroamblystegium filum* (Müll.Hal.) Reimers, Hedwigia 66: 71. 1926, ***syn. nov.*** — Type citation: all new species described by Müller (1883) are included in the section 'I. Bryologia Kerguelensis'; later Müller (1889: p. 35) wrote: 'Ins. Kerguelen, Winter harbour (1. 75)'. [Lectotype (*selected here*): 'Hypnum filum C. Müll. n. sp. (Sectio nova: Orthotheciella C. M.) Kerguelen, Winter Harb. Jan. 1875 I. Naumann' – HBG!; isotype: JE-Herzog!].

Hypnum austrocatenulatum Müll.Hal, Bot. Jahrb. Syst. 5: 83. 1883 ['austro-catenulatum']. ≡ *Pseudoleskea austrocatenulata* (Müll. Hal.) Paris, Index Bryol.: 1034. 1898. ≡ *Hygroamblystegium austrocatenulatum* (Müll.Hal.) Reimers, Hedwigia 66: 71. 1926, ***syn. nov.*** — Type citation: all new species described by Müller (1883) are included in the section 'I. Bryologia Kerguelensis'; later Müller (1889: p. 35) wrote: 'Ins. Kerguelen, Betsy Cove (11. 74)'. [Lectotype (*selected here*): 'Hypnum austro-catenulatum C. Müll. n. sp. Kerguelenland, Gazellen-Expedition Dr. Naumann 1874' – BM!; isotype: S!].

Leskea fuegiana Besch. *in* Hariot, Miss. Sc. Cape Horn Bot.: 298. 1889. ≡ *Pseudoleskea fuegiana* (Besch.) Cardot, Bull. Herb. Boissier Sér. 2, 5: 1009. 1905. ≡ *Hygroamblystegium fuegianum* (Besch.) Reimers, Hedwigia 66: 71. 1926, ***syn. nov.*** — Type citation: Terre de Feu: île Wollaston. – Otarie (Hariot). [Lectotype (*selected here*): 'Herb. Émil Bescherelle: 1900. Leskea fuegiana Besch. Terre de Feu: île Wollaston legit Hariot n° 84 21 juin 1883' – BM-Bescherelle!; syntype: 'Herb. Émil Bescherelle: 1900. Leskea fuegiana Besch. Île Otarie (Wollaston) legit Hariot n° 92' – BM-Bescherelle!].

Hypnum austrofluviatile Müll.Hal. *in* Neumayer, Int. Polarforsch. Deutsch. Exp. 2: 321. 1890 ['austro-fluviatile']. ≡ *Amblystegium austrofluviatile* (Müll.Hal.) Paris, Index Bryol.: 12. 1894. ≡ *Hygroamblystegium austrofluviatile* (Müll.Hal.) Broth. *in* Engl. & Prantl, Nat. Pflanzenfam. 1(3): 1028. 1908, ***syn. nov.*** — Type citation: Austro-Georgia, 'Quelle auf dem Hochplateau', 14. Julio 1883, sterile. [Lectotype (*selected here*): 'N° 10 Hypnum austro-fluviatile C. Müll. n. sp. Standort: Quellen auf dem Hochplateau (orig. Müll.) Südgeorgien 14/VII 83 Will' – M!].

Pseudoleskea antarctica Cardot, Rev. Bryol. 27: 45. 1900. ≡ *Hygroamblystegium antarcticum* (Cardot) Reimers, Hedwigia 66: 71. 1926, ***syn. nov.*** — Type citation: [Détroit du Gerlache] XIme débarquement; later Cardot (1901: p. 41) wrote: Détroit de Gerlache: XIème débarquement, Terre de Danco, cap Van Beneden, sur les corniches de la falaise. (N°s 231^{a} et 232^{d}; 1^{r} février 1898). [Lectotype (*vide* Ochyra 1898*a*: p. 226): 'Herb. J. Cardot. Mousses du voyage de la Belgica N° 231^{a} Pseudoleskea antarctica Card. sp. nova. Canal de la Belgica. XIème débarquement sur des corniches de la falaise leg. Racovitza, 1 févr. 1898' – PC!; isolectotypes: BR!, FH!, H-Brotherus!, S! (3 specimens); syntypes: same label data, No. 232d – BR!, PC!].

Sciaromium gracile Dusén *in* W.B.Scott, Rep. Princeton Univ. Exp. Patagonia 8(1) Botany: 102, *f. 25 & pl. 11 f. 6.* 1903, ***syn. nov.*** — Type citation: Fuegia australis, ubi ad Lapataia in consortio Sciaromi depasti lecta est. [Lectotype (*vide* Ochyra 1987*b*: p. 459): 'Princeton Scientific Expedition to Patagonia 1896–97. P. 2. Sciaromium gracile Dusén. S. Patagonia, J. B. Hatcher' – S!; isotypes: C!, FH!, NY!, PC!]. First synonymised with *Pseudoleskea chilensis* by Ochyra (1987*b*: p. 459).

Amblystegium philonotis Dusén *ex* Paris, Index Bryol. Ed. 2, 5: 146. 1906. *nom. nud.*, ***syn. nov.*** — Original material: Am. merid.: Patagon. occid. Dusén M. Amer. austr. n. 494. [Chirotypes: '494. Amblystegium philonotis Dus. Patagonia occ. in valle fl. Aysen in truncis putridis. Jan. 28 a. 1897 P. Dusén' – BM!, E!, PC!, S!, W!].

Pseudoleskea fuegiana (Besch.) Cardot var. *skottsbergii* Cardot, Bull. Herb. Boissier Sér. 2, 5: 1009. 1905. ≡ *Hygroamblystegium fuegianum* (Besch.) Reimers var. *skottsbergii* (Cardot) E.B.Bartram *in* Roiv., Ann. Bot. Soc. Zool.-Bot. 'Vanamo' 5(10): 36. 1934, ***syn. nov.*** — Type citation: Terre-de-Feu: Ushuaia, Rio Olivia. [Lectotype (*selected here*): 'Svenska Sydpolarexpeditionen 1901–03. N:r 184. Pseudoleskea fuegiana (Besch.) Card. var. Skottsbergii var. nova. Terre de Feu: Ushuaia, Rio Olivia 11/3 1902 Carl Skottsberg Det. J. Cardot' – PC-Cardot!; isotypes: H-Brotherus!, S!].

Pseudoleskea lurida Cardot, Bull. Herb. Boissier Sér. 2, 5: 1009. 1905. ≡ *Hygroamblystegium luridum* (Cardot) Reimers, Hedwigia 66: 71. 1926, ***syn. nov.*** — Type citation: Terre-de-Feu: Rio Olivia. [Lectotype (*selected here*): 'Svenska Sydpolarexpeditionen 1901–03. N:r 185. Pseudoleskea lurida Card. sp. nova. Terre de Feu: Rio Olivia 10/10 1902 Carl Skottsberg Det. J. Cardot' – PC-Cardot!; isotype: S!].

Pseudoleskea platyphylla Cardot, Bull. Herb. Boissier Sér. 2, 6: 1007. 1906. ≡ *Hygroamblystegium platyphyllum* (Cardot) Reimers, Hedwigia 66: 71. 1926, ***syn. nov.*** — Type citation: [South Georgia] Cumberland bay: Pot harbour. [Lectotype (*selected here*): 'Herb. J. Cardot. Svenska Sydpolarexped. 1901–03 n° 417 in parte Pseudoleskea platyphylla Card. sp. nova. Géorgie du Sud: Baie Cumberland, supra Pot Harbour Leg. C. Skottsberg 21/5 1902' – PC-Cardot!; isotypes: S!, S-Roth!].

Pseudoleskea calochroa Cardot, Bull. Herb. Boissier Sér. 2, 6: 1007. 1906. ≡ *Hygroamblystegium calochroum* (Cardot) Reimers, Hedwigia 66: 71. 1926, ***syn. nov.*** — Type citation: [Géorgie du Sud] Cumberland Bay: May harbour. [Lectotype (*selected here*): 'Svenska Sydpolarexpeditionen 1901–03. N:r 403 Pseudoleskea calochroa Card. sp. nova South Georgia May Harbour Carl Skottsberg D. 2/5 1902 Det. J. Cardot' – PC-Cardot!; isotypes: H-Brotherus!, S!].

Pseudoleskea strictula Cardot, Bull. Herb. Boissier Sér. 2, 6: 1007. 1906. ≡ *Hygroamblystegium strictulum* (Cardot) Reimers, Hedwigia 66: 71. 1926, ***syn. nov.*** — Type citation: [Géorgie du Sud] Cumberland Bay: Pot harbour. [Lectotype (*selected here*): 'Svenska Sydpolarexpeditionen 1901–03. N:r 417 Pseudoleskea strictula Card. sp. nova South Georgia, Cumberland Bay Carl Skottsberg D. 21/5 1902 Det. J. Cardot' – PC-Cardot!; isotypes: FH-Fleischer!, H-Brotherus!, S! (3 specimens), S-Roth!, UPS!].

Pseudoleskea fuegiana (Besch.) Cardot var. *gracilis* Cardot & Broth., Kongl. Svenska Vetenskapsakad. Handl. 63(10): 61. 1923. ≡ *Hygroamblystegium fuegianum* (Besch.) Reimers var. *gracilis* (Cardot & Broth.) E.B.Bartram, Farlowia 2: 317. 1946, ***syn. nov.*** — Type citation: Islas Guiatecas: Melinca, ad truncos (806). [Lectotype (*selected here*): 'Herb. J. Cardot Pseudoleskea fuegiana (Besch.) Card. var. gracilis Card. var. nova. Ins. Guaitecas, Melinca, ad truncos leg. Skottsberg 1908, n. 806' – PC!; isotypes: H-Brotherus!, UPS!].

Pseudoleskea sordidoviridis Cardot & Broth., Kongl. Svenska Vetenskapsakad. Handl. 63(10): 61. 1923. ≡ *Hygroamblystegium sordidoviride* (Cardot & Broth.) Reimers, Hedwigia 66: 71. 1926, ***syn. nov.*** — TYPE CITATION: E. Falkland: North Arm (821); Sparrow Cove. [Lectotype (*selected here*): 'Herb. J. Cardot Expeditio suecica 1907–1909 821. Pseudoleskea sordidoviridis Card. sp. nova. Falkland: île orientale, North Arm. 28.1.1908 C. Skottsberg' – PC!; isotypes: H-Brotherus!, UPS!; syntype: 'Herb. J. Cardot Expeditio suecica 1907–1909 820. Pseudoleskea sordidoviridis Card. sp. nova. Falkland: île orientale, Port William, Sparrow Cove, ruisseau. 10.1.1908 C. Skottsberg' – H-Brotherus!, LD!, PC!, S!].

Amblystegium tenellum Cardot & Broth., Kongl. Svenska Vetenskapsakad. Handl. 63(10): 62, *pl. 3. f. 13a–c.* 1923. ≡ *Hygroamblystegium tenellum* (Cardot & Broth.) Reimers, Hedwigia 66: 71. 1926, ***syn. nov.*** — TYPE CITATION: Patagonia andina: Ultima Esperanza, in caverna (896). [Lectotype (*selected here*): 'Herb. J. Cardot Amblystegium tenellum Card. sp. nova. Patagonia, Ultima Esperanza, Cueva Mylodre, 21.2.1908 leg. C. Skottsberg n. 896' – PC!; isotypes: H-Brotherus!, UPS!].

Amblystegium excurrens Cardot & Broth., Kongl. Svenska Vetenskapsakad. Handl. 63(10): 62, *pl. 3. f. 14a–b.* 1923. ≡ *Hygroamblystegium tenellum* (Cardot & Broth.) Reimers, Hedwigia 66: 71. 1926. ≡ *Hygroamblystegium fuegianum* (Besch.) Reimers var. *excurrens* (Cardot & Broth.) E.B.Bartram, Farlowia 2: 317. 1946, ***syn. nov.*** — TYPE CITATION: Patagonia australis: Skyring, Puerto Pinto (897) (1896). [Lectotype (*selected here*): 'Herb. J. Cardot Amblystegium excurrens Card. sp. nova. Patagonia, Skyring, Puerto Pinto pierres dans la forêt leg. Halle et Skottsberg 1908, n. 897' – PC!; isotypes: H-Brotherus!, S!, UPS!].

Hygroamblystegium fuegianum var. *compactum* E.B.Bartram, Farlowia 2: 317. 1946, ***syn. nov.*** — TYPE CITATION: Chile: Prov. de Magellanes, Fuegia media, Lago Linch, Cerro Chico, reg. alp., No. 1380 type. [Lectotype (*selected here*): 'Bryophyta fuegiana. 1380. Hygroamblystegium filum (C. Müller) Reimers var. compactum Bartram n. var. Fuegia media, Lago Linch, Cerro Chico, ad rupem phylliticam in reg. alp. ca 600 m s.m. 20.12.1928 leg. H. Roivainen' – FH-Bartram!; isotypes: F!, H!, S!].

Hygroamblystegium fuegianum var. *secundum* E.B.Bartram, Farlowia 2: 317. 1946, ***syn. nov.*** — TYPE CITATION: Chile: Prov. de Magellanes, Fuegia media, Lago Fagnano, Punta Pizzaro, No. 1213 type. [Lectotype (*selected here*): 'Bryophyta fuegiana. 1213. Hygroamblystegium fuegianum (Besch.) Reim. var. subsecundum Bartram n. var. Fuegia media, Lago Fagnano, Estancia Cameron, Punta Pizarro, ad rupem phylliticam 19.I.1929 Leg. H. Roivainen' – FH-Bartram!; isotypes: H!, S!].

Lescuraea chilensis Herzog, Rev. Bryol. Lichénol. 23: 87, *f. 24.* 1954, ***syn. nov.*** — TYPE CITATION: Mittelchile: Cuesta La Dormida, 1.030 m., auf Erde, leg. G. H. Schwabe, n° 140/b. [Lectotype (*selected here*) {or perhaps holotype}: 'Herbarium Haussknecht, Jena ~~Pseudoleskea~~ Lescuraea chilensis Herz. n. sp. Fl. v. Mittelchile Cuesta de la Dormida, 1030 m leg. G. H. Schwabe, 3. 11. 40 no. 140/b pp' – JE-Herzog!].

Plants small to medium-sized, in loose or dense, yellow-green, yellow or golden-brown to brown, rather rigid, dull or lustrous, usually intricate mats. *Stems* yellowish-green to brown, 1–4 cm long, prostrate to ascending, freely to subpinnately branched, not or sparsely radiculose with light brown, smooth, branched rhizoids, in transverse section with a 1–2(–3)-stratose cortex of relatively small and thick-walled, yellowish-green to brown cells surrounding 4–5 layers of large, thin-walled, hyaline medullary cells, central strand small but distinct; *pseudoparaphyllia* foliose to filamentous; *paraphyllia* infrequent to many on the surface of the stems and stronger branches, sometimes missing on slender and delicate plants, uniseriate and filiform to narrowly lanceolate with 2 rows of cells below; *axillary hairs* infrequent, 2–4-celled, with 1–2 short brownish basal cells and 2–3 elongate, hyaline upper cells; *branches* slender, elongate, straight, erect, subjulaceous. *Stem* and *branch leaves* similar, the latter somewhat smaller, 0.7–1.8 mm long, 0.3–0.8 mm wide, erect, erect to erect-spreading, crowded, straight to secund or homomallous at the branch and stem apices, subimbricate, closely appressed when dry, erect-spreading when wet, non-decurrent, ovate-lanceolate, gradually long-acuminate, acute, plane or weakly concave, smooth or plicate; *margins* plane, serrulate or crenulate all around, sometimes subentire to entire; *costa* single, strong, subpercurrent to percurrent, occasionally vanishing well below the apex and spurred or long-excurrent, 40–70 µm wide at base, yellow and sharply set off from the laminal cells, often curved above, terete; *laminal cells* unistratose, transparent; *upper* and *median cells* smooth, variable, mostly not uniform in size and shape, shortly elliptic, rhomboidal to oblong-rhomboidal, (10–)12–25(–30) µm long, (4–)5–7(–9) µm wide, firm to thick-walled, occasionally with thin walls; *basal cells* usually somewhat larger, yellowish in 1–2 rows at the insertion; *angular cells* subquadrate to short rectangular, with thick and often yellow walls, otherwise similar to adjacent laminal cells and not forming distinct groups. [*Dioicous. Inner perichaetial leaves* pale yellow, lanceolate, gradually long-acuminate, plicate, with a single, long-excurrent costa, serrate at the margins and with areolation of oblong-hexagonal to linear cells. *Setae* elongate, 1.0–2.5 cm long, flexuose when dry, smooth, reddish-brown. *Capsules* reddish-brown to light brown, cylindrical, with a distinct neck, 1.8–2.5 mm long, arcuate, asymmetric, inclined to horizontal, sometimes suberect, usually contracted below the mouth when dry and empty; *operculum* conic-convex, apiculate; *annulus* present, separating, 1–3-seriate; *exothecial* cells isodiametric below to oblong or rectangular above, becoming short and subquadrate in 3–4 tiers below the orifice, thin- to firm-walled; *stomata* few, bicellular, superficial, long-pored in the neck; *peristome* double; *exostome teeth* lanceolate, yellow, cross-striolate below, distinctly papillose and dentate at the margins above; *endostome teeth* well developed, pale yellowish, finely papillose throughout, basal membrane high, processes not or narrowly cleft, *cilia* 2–3, well developed, nodose, somewhat shorter than the processes. *Spores* globose, 10–16 µm in diameter, yellow, finely papillose. *Calyptra* cucullate, naked.]

Chromosome number: unknown.

Discussion. Despite its enormous phenotypic plasticity, *Cratoneuropsis relaxa* subsp. *minor* is very easy to distinguish and in the Antarctic it is unlikely to be confused with any other moss, the strong and percurrent to excurrent costa immediately separating this taxon from all other local pleurocarps, except for *Drepanocladus longifolius*. This is the only other pleurocarpous moss in this region with a strong, percurrent to excurrent costa but it is a much larger moss with lanceolate and long-acuminate leaves, distinct alar cells forming small, inflated, yellowish to hyaline auricles and linear-flexuose laminal cells.

The variation of the protean *Cratoneuropsis relaxa* was discussed by Sainsbury (1955) who found that the species is extremely plastic in all structural features and this plasticity is doubtless conditioned environmentally and is associated with the frequency and ability of the species to colonise a series of wet and moist habitats. Its various phenotypes differ markedly in the coloration, branching pattern and size of the plants, shape and stance of the leaves, stoutness of the costa, leaf areolation and abundance of paraphyllia. Typically, the Australasian populations have leaves varying considerably in their recurvature from straight to markedly squarrose-recurved, notwithstanding the stature of the plants. Plants growing in streams with fast-flowing water have markedly reduced paraphyllia and straight, erect, broadly ovate, shortly acuminate leaves which are not folded in the acumen. These were described as species of *Sciaromium*, namely *S. bellii* Broth., *S. elimbatum* Broth. & Watts and *S. forsythii* Broth., but all these names have been reduced to synonymy with *C. relaxa* (Sainsbury 1948*a*; Stone 1981; Ochyra 1987*b*; Ochyra & Streimann 2001).

Despite this random variation conditioned by local environmental changes, the morphology of *Cratoneuropsis relaxa* also varies over a latitudinal gradient. Although all of the individual characters exhibit a high degree of variation among and within sites, there are clear trends of decreasing plant size, leaf length and width, leaf acumination and abundance of paraphyllia and trends of increasing imbrication of the leaves to the stems and branches with increased latitude and climatic severity. Consequently, southern plants occurring on the Southern Ocean islands are often much smaller and possess leaves that are markedly appressed, less acuminate with straight apices and paraphyllia that are very sparse or absent. Both morphotypes of *C. relaxa* occur in abundance on the Auckland Is. and Campbell I. situated 400 and 700 km, respectively, south-east of New Zealand in the cool-south-temperate zone. However, on subantarctic islands in the Kerguelen Province, in South Georgia and the northern maritime Antarctic, as well as in southern South America, including the Falkland Is., large plants with spreading leaves and recurved to squarrose leaf apices are absent. Here, plants tend to have variously appressed leaves and straight or eventually secund leaf apices. A similar morphological pattern of variation over latitudinal gradients is exhibited by other mosses, for instance *Meesia triquetra*, *Polytrichum strictum* and *Hylocomium splendens* (Hedw.) Schimp. (Montagnes & Vitt 1991), and these morphological trends parallel those in vascular plants (Bliss 1962; Chapin 1983; Wilson 1966).

The small stature of the plants can be controlled genetically, through selective pressures and this apparently explains the different morphologies of the plants in southern South America and southern New Zealand which occur at similar latitudes. Recognition of the South American and subantarctic populations as a separate species, for which the oldest available name is *Amblystegium chilense* (Lorentz 1866), is very tempting but the continuous variation and imperceptible merging of the southern *chilensis* phenotypes into northern typical *relaxa* or *decussata* phenotypes on the Auckland Is. and Campbell Is. suggest that *Cratoneuropsis relaxa* should be interpreted as a single species. In fact, such an interpretation of plants collected by H. N. Moseley during the *Challenger* Expedition on subantarctic Marion I. and named *Amblystegium decussatum* (Hook.f. & Wilson) Mitt., was presented by Mitten (1876*a*, 1884) who commented on them: 'A slender state, unlike the larger New-Zealand forms; but the substance of the leaves is not different'. In a similar way Mitten (1879) commented on specimens from Îles Kerguelen collected by A. E. Eaton during the British Transit of Venus Expedition: 'All the specimens referred to this species [*Amblystegium decussatum*] have but little external resemblance to the complete state found in New Zealand, but they agree very closely in the areolation of their leaves, and it is probable they are only slender forms similar to those produced by *A. filicinum*'. Because these phenotypes are quite well separated geographically and meet only in a relatively narrow area on two oceanic archipelagoes, it seems quite reasonable to afford them taxonomic recognition as subspecies. Interestingly, *C. relaxa* was first recognised as a variety, *Hypnum filicinum* var. *minus* from material collected from the Auckland Is. (Wilson & Hooker 1845) and nine years later it was described as a species in its own right, *Hypnum relaxum*, from the North Island of New Zealand (Wilson 1854). In fact, at the same time another species was also described, *H. decussatum*, which represents a very handsome phenotype with strongly squarrose-recurved leaves. These species were considered conspecific by Sainsbury (1955) who choose the former as the final name for this taxon.

The Antarctic populations of *Cratoneuropsis relaxa* subsp. *minor* mostly resemble those from South Georgia and Tierra del Fuego which are characterised by having broad costae, 40–55 μm wide at the base, which are mostly percurrent to short-excurrent. A few plants from Signy I. in the South Orkney Is. were found to grow deeply submerged in lakes and they have much stronger and wider costae, reaching 60–70 μm at the base. On the Antarctic Peninsula this taxon has been referred to as *Hygroamblystegium austrofluviatile* (Lewis Smith 1996). There are only a few plants from Elephant I. and Gibbs I. in which the costae vanish well below the apex on most leaves and the plants from Elephant I. were reported by Darbishire (1923) as *Amblystegium subvarium*. This species was originally described from Îles Kerguelen (Brotherus 1906*a*) and is inseparable from *Hygroamblystegium varium*. Such phenotypes represent highly reduced filiform plants. These are quite common on all subantarctic islands and were usually reported as *Amblystegium serpens* (Hedw.) Schimp. (for illustration see Seppelt 2004). In general, *C. relaxa* subsp. *minor* is an exceedingly variable taxon producing a large number of phenotypic expressions, and Reimers (1926) compared its polymorphism to that of *Hypnum cupressiforme* Hedw. Over 20 phenotypes from the geographical range of *C. relaxa* subsp. *minor* were given taxonomic recognition as species or infraspecific taxa and numerous unpublished taxon names exist in various herbaria.

Reproduction in Antarctica. Always sterile both in Antarctica and South Georgia. Although it is common and abundant in southern South America, it produces sporophytes very seldom (Ochyra & Matteri 2001).

→

Figure 238. *Cratoneuropsis relaxa* subsp. *minor*. — **1–2**. Habit. **3–4**. Portions of branches, dry. **5**. Cross-section of the stem. **6–9**. Paraphyllia. **10–13**. Axillary hairs. **14–16**. Pseudoparaphyllia. **17–26**. Leaves. **27**. Leaf apex. **28–29**. Upper leaf cells at the margin. **30–31**. Mid-leaf cells. **32–33**. Basal leaf cells. — [1, 4, 10, 13, 17–20, 28, 30) from *REL &RILS 832*; 2–3, 5, 11–12, 23–27, 32–33 from *RILS 651*; 6–9 from *RILS 466*; 14–16 from *RO 4846/79*; 21–22, 29–31from *ER 231A* (lectotype of *Pseudoleskea antarctica*), PC]. — Scale bars: a – 100 μm (5); b – 1 mm (3–4); c – 0.5 cm (1–2) and 100 μm (6–16, 27–33); d – 1 mm (17–26).

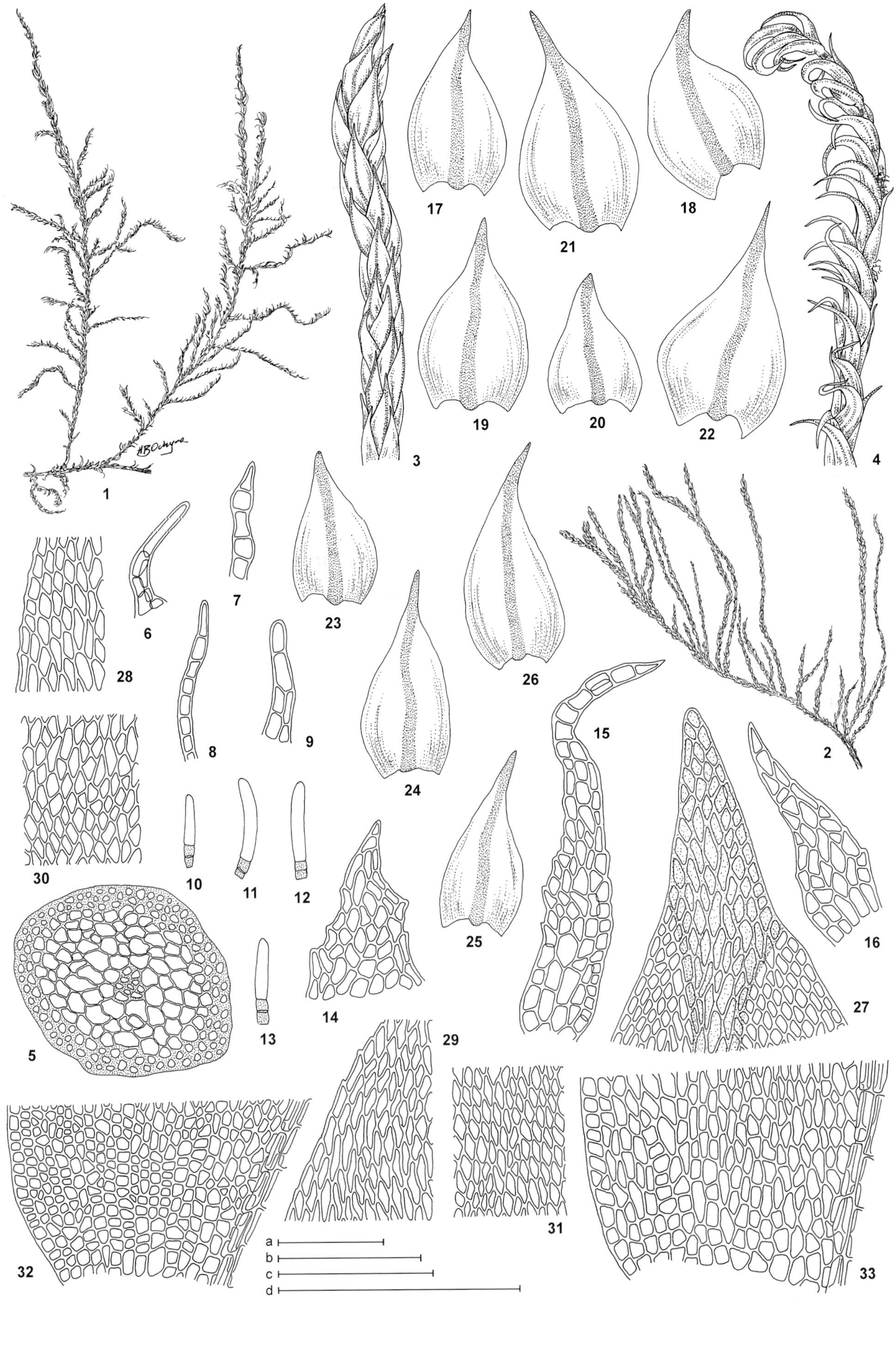
1
2
3
4
5
6
7
8
9
10
11
12
13
14
15
16
17
18
19
20
21
22
23
24
25
26
27
28
29
30
31
32
33
a
b
c
d

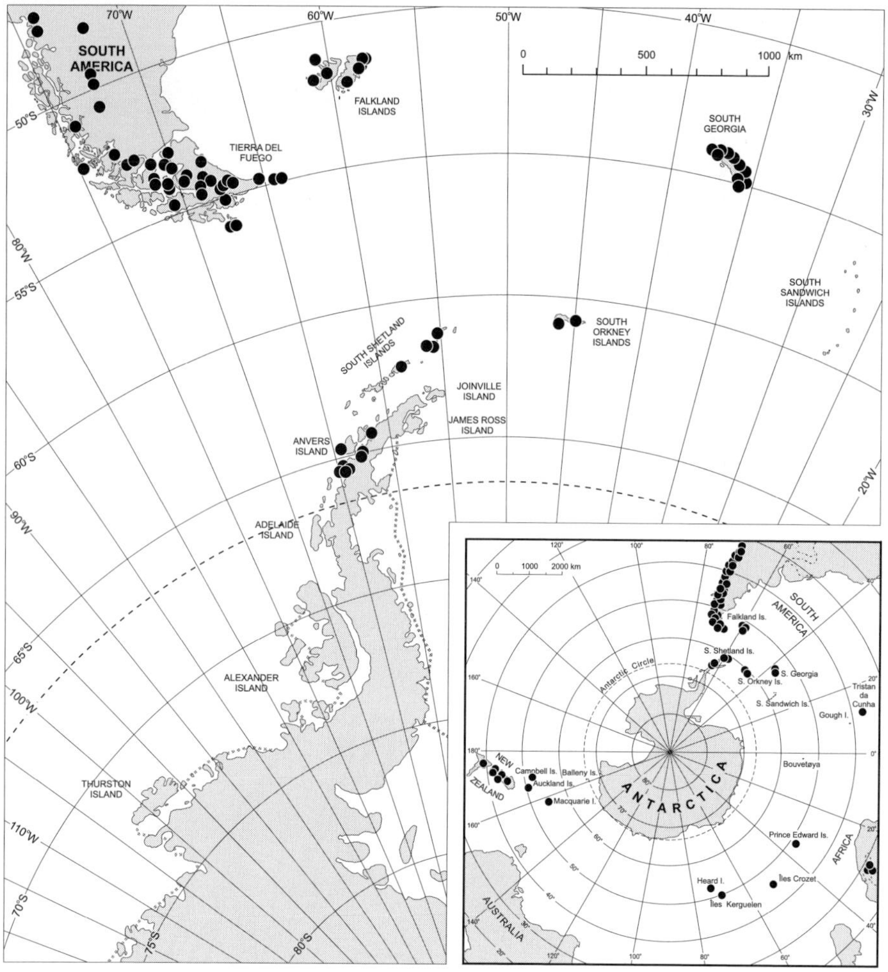

Figure 239. Distribution of *Cratoneuropsis relaxa* subsp. *minor* in the Fuegian region and Antarctica. Inset: global distribution.

Habitat. The species usually grows in damp or wet crevices and concavities in rocks and cliffs, on sheltered rock ledges and beneath overhanging rocks, on mineral soil, sometimes on fine scree. On Signy I., South Orkney Is., it is typically associated with moist calcareous rock. It has also been found submerged in lakes to a depth of 5 m.

World range. *Cratoneuropsis relaxa* subsp. *minor* is fundamentally a pan-south-temperate taxon (Fig. 239, inset) with maximum occurrence in southern South America where it is a weedy moss, very common and abundant along the western fringes of the continent from Cape Horn, Tierra del Fuego and Isla de los Estados to central Chile, and on the Southern Ocean islands from the Falkland Is. and South Georgia in the west to Macquarie I., Auckland Is. and Campbell I. in the east, extending to Tristan da Cunha and South Africa in the north and to the maritime Antarctic in the south. The taxon is exceedingly rare in New Zealand, where it is represented by only a few collections and, so far, it has not been detected on mainland Australia or Tasmania. The record of *Orthotheciella varia* from the latter area (Ochyra & Streimann 2001) refers to *Hygroamblystegium varium*.

Distribution in Antarctica. This taxon is infrequent in the South Orkney Is. and South Shetland Is. and scattered on the western coast of the Antarctic Peninsula on the Danco Coast and Graham Coast, extending as far south as lat. 65°30′S (Fig. 239). It grows usually at low elevations and only rarely above 100 m, the highest record being *ca* 500 m on Coronation I., South Orkney Is.

Specimens examined. **WEST ANTARCTICA.** SOUTH ORKNEY IS. **CORONATION I.** Knobs Ridge, *ca* 500 m, *RILS 1451* (AAS, KRAM). **POWELL I.** SE coast, ca. 65 m, *RILS 211* (AAS, KRAM). **SIGNY I.** Thulla Pt., *ca* 3.5 m, *RILS 432* (AAS, KRAM); Foca Cove, *ca* 15 m, *RILS 596 & 1811* (AAS, KRAM); Foca Pt., 35 m, *REL 1113* (AAS, KRAM); Spindrift Rocks, 3–6 m, *RILS 8059* (AAS, KRAM); between North Pt. and Spindrift Rocks, 65 m, *REL 1133* (AAS, KRAM); North Pt., *ca* 5 m, *RILS 466* (AAS, KRAM); Stygian Cove, *RILS 1805* (AAS, KRAM) and 25–30 m, *RILS 10788* (AAS, KRAM); Waterpipe Beach, 25 m, *RILS 651* (AAS, KRAM) and 35 m, *RILS 481* (AAS, KRAM); Knob Lake, *RILS 7006* (AAS, KRAM); Three Lakes Valley, *RBH & JJL 31* (AAS, KRAM) and *RILS 596* (AAS, KRAM); Factory Cove, 8–16 m, *REL & RILS 832* (AAS, KRAM); Limestone Valley, 15 m, *RILS 1790, 5139 & 5141* (AAS, KRAM).

SOUTH SHETLAND IS. **ELEPHANT I.** Without locality, 1916, *Clark s.n.* (BM). **ASPLAND I.** 9 m, *JPB 72* (AAS, KRAM). **GIBBS I.** The Spit, 120 m, *JPB 105A & 106* (AAS, KRAM). **KING GEORGE I.** ***Fildes Peninsula***: Ardley I., *Kühnemann 84* (AAS, KRAM). ***Admiralty Bay***: Creeping Slopes, 60 m, *RO 4846/79* (KRAM).

WEST ANTARCTIC PENINSULA. **DANCO COAST.** Charlotte Bay, Andrée I., 25 m, *RILS 4004B, 4012 & 4014* (AAS, KRAM); Cuverville I., 2–20 m, *RILS 4171, 4172 & 4186* (AAS, KRAM); Beneden Head, *ER 231A 232D* (BR, PC – type of *Pseudoleskea antarctica*); Paradise Harbour, *Popovici 9213* (BA, KRAM). **PALMER ARCHIPELAGO.** ***Anvers I.***: Hermit I., 17 m, *RILS 877A* (AAS, KRAM); Norsel Pt., 5–10 m, *RILS 1952 & 1953* (AAS, KRAM). **GRAHAM COAST.** Cape Pérez, *Gain 273A* (H, PC). **WILHELM ARCHIPELAGO.** ***Penola Strait***: Petermann I., 13 m, *RWMC 724B* (AAS, KRAM) and 17 m, *REL 1345B* (AAS, KRAM). ***Argentine Is.***: Corner Is., 8 m, *REL 1368* (AAS, KRAM); Galindez I., Stella Creek, 3.5 m, *RWMC 402* (AAS, KRAM) and *BGLE* (BM); Skua I., Skua Creek, *ca* 7 m, *RWMC 557* (AAS, KRAM) and *ca* 13 m, *RWMC 504A* (AAS, KRAM); Black I., *RWMC 576* (AAS, KRAM) and 2–10 m, *RILS 4290* (AAS, KRAM). ***Collins Bay***: Berthelot Is., *ca* 20 m, *RWMC 616B* (AAS, KRAM) and *ca* 35 m, *RWMC 627* (AAS, KRAM); Darboux I., *RILS 3310, 3334 & 3364A* (AAS, KRAM).

46. CAMPYLIUM (SULL.) MITT.

Campylium (Sull.) Mitt., J. Linn. Soc. Bot. 12: 631. 1869. ≡ *Hypnum* Hedw. sect. *Campylium* Sull. *in* A.Gray, Manual Ed. 2: 677. 1856. ≡ *Hypnum* subgen. *Campylium* (Sull.) A.A.Fisch.Waldh., Bull. Soc. Natural Moscow 37: 37. 1864. — Lectotype (*vide* Pfeiffer 1872: p. 571): *Campylium stellatum* (Hedw.) Lange & C. E. O. Jensen (*Hypnum stellatum* Hedw.).

Plants small, medium-sized or large, in loose or dense mats or tufts, more or less lustrous, green, yellow-green, golden or brownish. *Stems* prostrate to ascending, freely and irregularly to pinnately branched in one plane or almost unbranched, in cross section lacking a hyalodermis, with a 1–2(–4)-stratose cortex of small, thick-walled and brown cells, 4–6 layers of large, hyaline, thin-walled medullary cells and a small but distinct central strand; *rhizoids* inserted intercalary, slightly to profusely branched, reddish-papillose, smooth or weakly warty-papillose; *pseudoparaphyllia* large, foliose, mostly polymorphous; *paraphyllia* absent or occasionally present, uniseriate, narrowly triangular or ovate; *axillary hairs* 2–4(–5)-celled, hyaline above with a single quadrate, brownish basal cell. *Stem* and *branch leaves* similar, crowded or distant, erect-spreading to widely patent, sometimes squarrose, straight or falcate, not plicate, concave or strongly so, from a broadly ovate or ovate-cordate, erect to spreading base gradually or suddenly narrowed to a short or long, slender, subulate, channelled or rarely almost tubular acumen; *margins* plane, entire or weakly sinuate; *costa* single, slender and elongate, sometimes branched, extending to ¾ way up the leaf, or short and double, occasionally lacking; *laminal cells* unistratose, pellucid, linear-flexuose, thin- to thick-walled, porose or not, smooth; *alar cells* rectangular to quadrate, small or enlarged, thin-walled and hyaline, becoming thick-walled and brownish-coloured with age, forming a well marked, broadly ovate or rectangular, often decurrent group. *Autoicous* or *dioicous*. *Inner perichaetial leaves* elongate-lanceolate, gradually or rapidly narrowed into a long, often filiform acumen, plicate, with a slender or strong costa reaching mid-leaf. *Setae* elongate, red or reddish-yellow, smooth. *Capsules* subcylindrical, curved, strongly contracted below the mouth when dry and empty, horizontal or rarely inclined; *operculum* conic-convex with a blunt or acute beak; *annulus* deciduous, 2–3-seriate; *exothecial cells* mostly rectangular, firm-walled; *stomata* present in the neck, bicellular, superficial, long-pored; *exostome teeth* lance-subulate, yellow-brownish, cross-striolate below, papillose above, dentate at the margins; *endostome* fused in the high basal membrane, with keeled, narrowly cleft or entire segments and 1–3(–4) nodose or appendiculate cilia. *Spores* globose, delicately papillose. *Calyptra* cucullate, naked.

A small genus consisting of seven species widely distibuted and locally common and abundant in cold and temperate areas in the Northern Hemisphere and with some species occurring at bipolar stations in the Southern Hemisphere and/or at altimontane elevations in the tropics. In older treatments *Campylium* comprised a large assortment of various, mostly remotely related species characterised by having variable, single or mostly double costae. Hedenäs (1997*a*) presented a partial generic revision of this complex and dispersed most of its species to other genera, retaining in *Campylium s. str.* only four species. He transferred *C. polygamum* (and the other two arctic species) to *Drepanocladus* but we found that in general the alar cells, pseudoparaphyllia and bordering of the exostome teeth are essentially of the same type in these genera and the costa condition and channelled leaf apex strongly unify *C. polygamum* with the generitype *C. stellatum.*

1. **Campylium polygamum** (Schimp.) Lange & C.E.O.Jensen — Fig. 240

Campylium polygamum (Schimp.) Lange & C.E.O.Jensen, Medd. Grønland 3: 329. 1887. ≡ *Amblystegium polygamum* Schimp. *in* Bruch, Schimp. & W.Gümbel, Bryol. Eur. 6: 60, *pl. 572*. 1853 [Fasc. 55–56 Mon.: 16, fig. 10]. ≡ *Campyliadelphus polygamus* (Schimp.) Kanda, J. Sci. Hiroshima Univ. Ser. B, Div. 2 (Bot.) 15(2): 262. 1976 ['1975']. ≡ *Drepanocladus polygamus* (Schimp.) Hedenäs, Bryologist 100: 82. 1997. — TYPE CITATION: In stagno exsiccato Kirkeler-Weiher dicto prope Biponten, ubi Th. Gumbel anno 1843 detexit. [Lectotype (*fide* Ochyra *in* Hedenäs 1997*a*: p. 82): 'H. riparium gymnandrum foliis ramulos. squarrosis patulis Hypnum gymnandrum Kirkeler Weiher April 1843' – BM!].

Plants medium-sized to moderately robust, rather rigid, densely caespitose, lustrous, green or yellowish- to brownish-green. *Stems* creeping, ascending to erect, 2–7 cm long, sometimes to 15 cm in submerged forms, irregularly and sparsely or richly branched, in cross-section round or oval, consisting of 2–3 layers of small, yellow-brown to brown, thick-walled cortical cells and a medulla of 5–6 layers of large, thin-walled, hyaline cells as well as a small but distinct central duct; *branches* erect or ascending, crowded, up to 1 cm long; *rhizoids* very sparse, smooth, reddish-brown; *pseudoparaphyllia* foliose, mostly triangular; *axillary hairs* hyaline to brownish, 3–4-celled. *Stem* and *branch leaves* similar in shape, usually crowded, erect-spreading to widely spreading or recurved, often flexuose-twisted when dry, 2–3 mm long, 0.7–1.0 mm wide, broadly lanceolate from an ovate or rounded-triangular base, gradually tapering to a long, slender, often recurved and canaliculate acumen, markedly narrowed at the insertion; *margins* plane, entire all around; *costa* yellow, to 75 μm wide at the base, variable, single but frequently short and branched almost from the base, rarely unbranched and reaching ½–¾ way up the leaf; *laminal cells* unistratose, pellucid; *upper* and *median cells* elongate, linear-flexuose with attenuate, truncate or oblique ends, 5–8(–14) μm wide, (20–)50–130 μm long, 10–18 times as long as wide, thin- to thick-walled and often porose above the base; *basal cells* shorter and broader, 30–50 μm long, 10–15 μm wide, 3–5 times as long as wide, with incrassate and pitted walls; *alar cells* inflated and hyaline, becoming thick-walled and brownish with age, forming distinct, transversely triangular

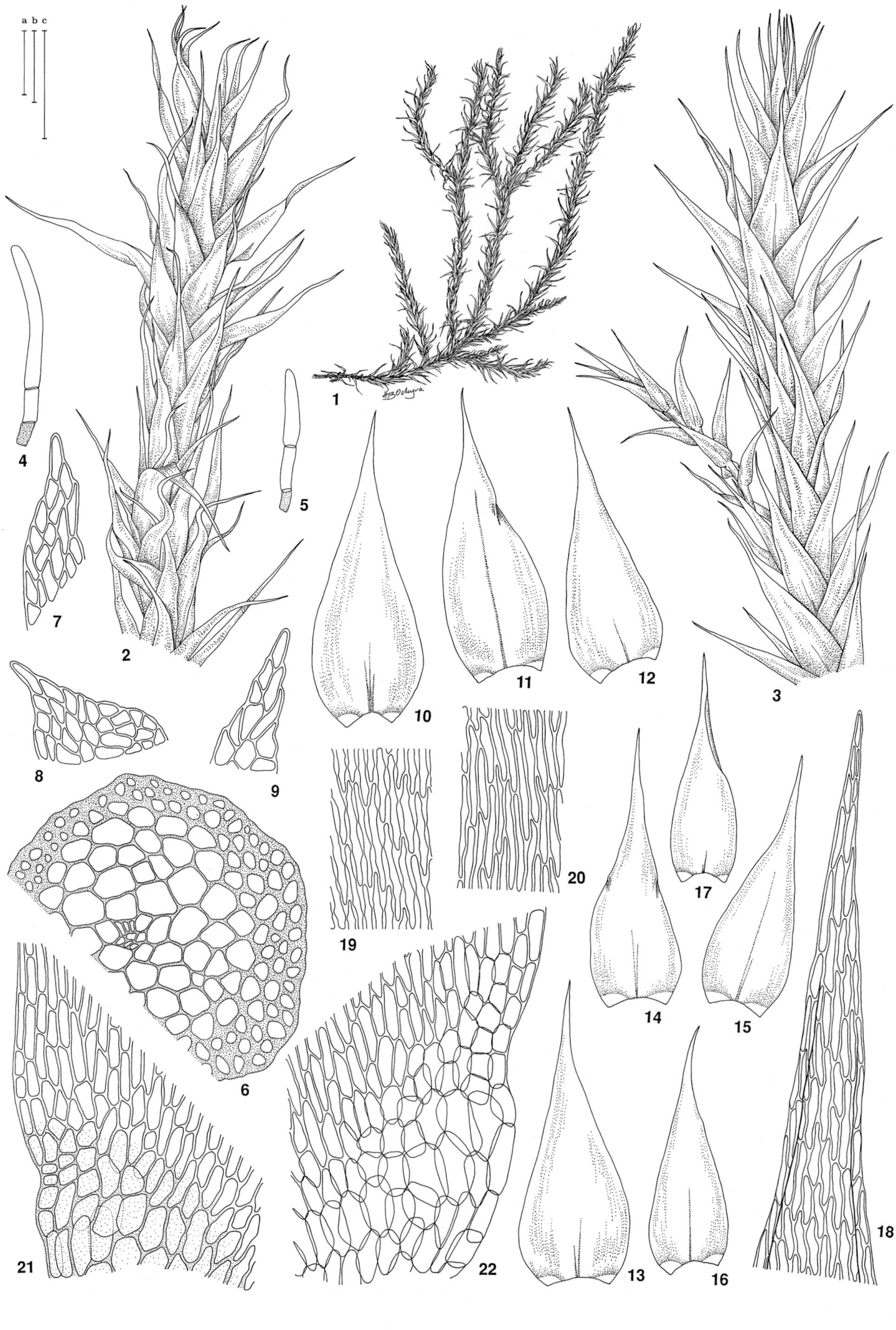
a b c
1
2
3
4
5
6
7
8
9
10
11
12
13
14
15
16
17
18
19
20
21
22

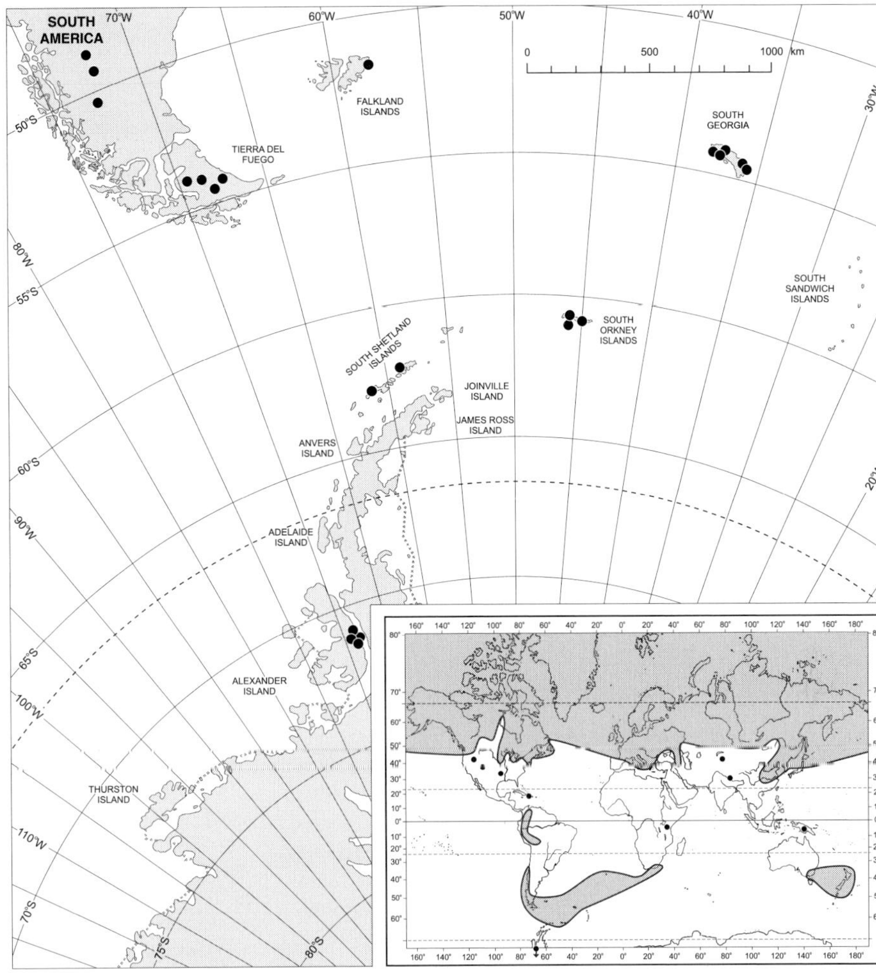

Figure 241. Distribution of *Campylium polygamum* in the Fuegian region and Antarctica. Inset: global distribution. Localities on Alexander I. are indicated by the arrow.

to broadly ovate, somewhat inflated auricles, extending from the margin to ⅔ of the leaf insertion or to the costa. [*Autoicous* or *polygamous. Inner perichaetial leaves* elongate-lanceolate, 3.0–3.5 mm long, 0.8–1.0 mm wide, gradually tapering into a long filiform acumen, plicate, with a strong costa reaching above the mid-leaf. *Setae* elongate, 3.5–4.0 cm long, red or reddish-brown, flexuose, smooth; *capsules* cylindrical to subcylindrical with a distinct neck, 2.0–2.5(–3.0) mm long, 0.7–0.9 mm wide, strongly curved to subsymmetric, slightly or strongly constricted below the mouth when dry and empty, brown, suberect to inclined; *operculum* conic-convex with a blunt, stright beak; *annulus* separating, biseriate; *exothecial cells* isodiametric to elongate, rather thin-walled; *stomata* present in the neck, bicellular, superficial, with long pori; *exostome teeth* lanceolate-subulate, brown-yellowish, cross-striolate below, roughly papillose above; *endostome teeth* consisting of a high, slightly papillose basal membrane giving rise to keeled, narrowly cleft or entire segments and 1–2 nodose cilia. *Spores* globose, 14–16 μm in diameter, minutely papillose. *Calyptra* cucullate, naked.]

Chromosome numbers: $n = 11$ – Signy I. (Newton 1980); $n = 18$ – King George I. (Przywara *et al.* 1984); $n = 10 + m$, 20, 22 – the other three records for European material (Fritsch 1991).

Discussion. In the Antarctic flora *Campylium polygamum* is unlikely to be confused with other species, except perhaps for some expressions of *Warnstorfia sarmentosa* which also possesses large, pellucid and inflated auricles sharply separated from the adjacent laminal cells. However, differences in shape of the leaves, particularly the obtuse and short apiculate apices in the latter should separate the two taxa. In addition, in *W. sarmentosa* red coloration of plants occurs commonly, whereas in *C. polygamum* red coloration is always lacking. The species also bears a superficial resemblance to some phenotypes of *Warnstorfia fontinaliopsis*, but in this species the alar cells gradually pass into the median laminal cells and never form large, pellucid and sharply demarcated groups, and the leaf apices are always plane, never canaliculate, and bear rhizoid initials.

The species shows considerable variation in the formation of its costa. It is either single and extends up to ¾ of the leaf length or short and double or branched, usually with one long branch

←

Figure 240. *Campylium polygamum*. — **1**. Habit. **2**. Portion of branch, dry. **3**. Portion of branch, wet. **4–5**. Axillary hairs. **6**. Portion of cross-section of the stem. **7–9**. Pseudoparaphyllia. **10–17**. Leaves. **18**. Leaf apex. **19–20**. Mid-leaf cells. **21–22**. Alar cells. — [1–5, 10–12, 18–20 from *RO 2458/80*; 6–9, 13, 15, 22 from *RO 2478/80*; 14, 16–17, 21 from *RILS 603*]. — Scale bars: a – 1 mm (2); b – 1 mm (10–17) and 50 μm (4–9, 18–22); c – 1 cm (1) and 1 mm (3).

and 1 or 2 short lateral branches. Interestingly, these various costae may occur on the same stem.

Despite its wide distribution, *Campylium polygamum* was recognised relatively late as a species in its own right. It was Schimper (1853*a*) who described it as *Amblystegium polygamum* from material collected in Germany but in, fact, it had been distinguished as a separate taxon a quarter-century earlier by Bridel (1827) as *Hypnum stellatum* Hedw. [var.] *terrae-novae* Brid. from Newfoundland (Ochyra 1995). Additionally, in the 1850s two descriptions of *C. polygamum* were made, ascribing to it full specific rank, in both cases from exotic areas, by Wilson (1854) in New Zealand as *Hypnum nodiflorum* Wilson and by Mitten (1859) in Tibet as *H. tibetanum* Mitt. (Ochyra 1998*e*).

Reproduction in Antarctica. The species is exclusively sterile in the region, although occasional sporophytes are known from Tierra del Fuego (Ochyra & Matteri 2001).

Habitat. A typical hydrophytic and aquatic moss growing on wet ground at the margins of lakes and pools, seepage areas, wet scree and soil amongst boulders, and occasionally floating, entirely submerged or with emerging shoots in lakes and pools. It has been found submerged at a depth of 9 m on Alexander I. (Light & Heywood 1973) and 10 m in Lake Kitezh on King George I. (Ochyra 1998*a*).

World range. *Campylium polygamum* is a bipolar species (Fig. 241, inset) having some transitional stations at altimontane elevations in the tropics, including New Guinea (Ochyra *et al.* 1991), East Africa (Ochyra & Pócs 1992), the northern and central Andes from Venezuela to Bolivia and the West Indies (Hedenäs 2003*b*; Quandt & Vanderpoorten 2006). In the Northern Hemisphere it has a continuous circumpolar range in the arctic, boreal and temperate zones, extending southwards to the northern United States in North America, the Mediterranean region and the Caucasus in Europe, Central Siberia, central east China and central Honshu in Japan (Kanda 1975). In the Southern Hemisphere it is widespread in Patagonia, Tierra del Fuego (Ochyra & Matteri 2001), the Falkland Is. (Matteri 1985), South Georgia (Newton 1983*a*), Tristan da Cunha (Dixon 1960), as well as in south-eastern Australia (Streimann & Curnow 1989) and New Zealand (Sainsbury 1955), extending to West Antarctica.

Distribution in Antarctica. An infrequent species, occasionally found on Signy I. in the South Orkney Is., where it locally forms stands of several square metres, and on King George I. and Livingston I. in the South Shetland Is., as well as on Alexander I. where it extends as far south as lat. 71°53′S (Fig. 241). It usually grows at low elevations but also occurs at 420 m on Alexander I.

Specimens examined. WEST ANTARCTICA. SOUTH ORKNEY IS. **SIGNY I.** Jebsen Pt., *RILS 594b* (AAS, KRAM); Everson Ridge, 200 m, *RILS 8170* (AAS, KRAM); Foca Pt., 35 m, *RILS 333* (AAS, KRAM); Stygian Cove, *RILS 490* (AAS, KRAM); Knob Lake, 35 m, *RILS 8161* (AAS, KRAM); Waterpipe Beach, *ca* 35 m, *RILS 480 & 591* (AAS, KRAM); Snow Hill, 217 m, *RILS 603* (AAS, KRAM); Orwell Glacier moraine, 30 m, *RILS 8577* (AAS, KRAM); Andreaea Plateau, 170 m, *RILS 10926* (AAS, KRAM); Rusty Bluff above Paal Harbour, 35 m, *RILS 354* (AAS, KRAM).

SOUTH SHETLAND IS. **KING GEORGE I. *Fildes Peninsula***: Gemmel Peaks, *ca* 15 m, *RO 2475/80 & 2478/80* (AAS, KRAM); Lake Kitezh, 15 m, *Mahu 20553 & 21898* (KRAM), *Zippel B-120* (KRAM) and *RO 2458/80* (AAS, KRAM), *2459/80 & 2462/80* (KRAM); without closer locality data, 14 Feb 1989, *Komárek s.n.* (KRAM) and *Li GWS10, NN16, WS17 & WS23* (AAS, KRAM). **LIVINGSTON I. *Byers Peninsula***: between Smellie Pt. and Chester Cone, 35–50 m, *RILS 3777* (AAS, KRAM).

WEST ANTARCTIC PENINSULA. **GEORGE VI SOUND. *Alexander I.***: Ablation Pt., *JJL & RBH 4, 5, 39, 43 & 82* (AAS, KRAM); Moutonnée Valley, *RILS 11112, 11121, 11125 & 11127* (AAS, KRAM); Ganymede Heights, *RILS 11077* (AAS, KRAM); Striation Valley, 1–2 m, *RILS 11080 & 11129A* (AAS, KRAM); Middle Flatiron Valley, 420 m, *RILS 11097C* (AAS, KRAM); Mars Oasis, *PC 699A* (AAS, KRAM) and 30 m, *RILS 10421* (AAS, KRAM), 50 m, *RILS 9466* (AAS, KRAM) and 70 m, *RILS 9414* (AAS, KRAM); Viking Valley, Secret Lake, *DDWW 1 & 60* (AAS, KRAM) and 200 m, *RILS 9488* (AAS, KRAM); Viking Valley bedrock, below Scarp Edge, *DDWW 65* (AAS, KRAM).

47. SANIONIA LOESKE

Sanionia Loeske, Verh. Bot. Vereins Prov. Brandenburg 49: 63. 1907. ≡ *Drepanocladus* (Müll.Hal.) G.Roth sect. *Sanionia* (Loeske) Broth. *in* Engl. & Prantl, Nat. Pflanzenfam. 1(3): 1033. 1908. ≡ *Drepanocladus* (Müll.Hal.) G.Roth subgen. *Sanionia* (Loeske) Podp., Consp. Musc. Eur.: 586. 1954. — Lectotype (*vide* Hedenäs 1989*b*: p. 403): *Sanionia uncinata* (Hedw.) Loeske (*Hypnum uncinatum* Hedw.).

Plants small and slender, medium-sized to large and robust, soft or rigid, forming extensive flat mats or deep tufts, somewhat glossy, green, yellowish-green or golden to brownish. *Stems* prostrate or ascending to erect, irregularly to pinnately branched in one plane, seldom almost unbranched, in transverse section with a distinct central strand, well developed, 2–4-stratose cortex of small, brown and incrassate cells and unistratose outer hyalodermis of thin-walled hyaline cells; *rhizoids* sparse, slightly branched, smooth, reddish-brown, inserted at or just below the leaf base; *pseudoparaphyllia* foliose; *axillary hairs* 1–8-celled, hyaline. *Stem* and *branch leaves* similar, lanceolate, mostly falcato-secund, from a rounded-triangular to broadly ovate base gradually tapering to a lanceolate-subulate, almost straight to circinate acumen, strongly to weakly plicate, concave, non-decurrent; *margins* finely to strongly serrate or denticulate all around, plane below, slightly recurved above; *costa* single, strong, ending in the acumen, plano-convex in cross-section; *laminal cells* unistratose, transparent, smooth; *upper cells* linear-flexuose, with thin and eporose or thicker and somewhat porose walls; *basal cells* shorter and broader, thicker-walled and more strongly porose; *angular cells* differentiated, few to many, hyaline, thin-walled, quadrate to rectangular, forming transversely triangular or rectangular, inflated groups; *supra-alar cells* differentiated, with smooth or strongly porose walls forming a group smaller or larger than the alars. *Autoicous. Inner perichaetial leaves* straight and erect, lanceolate to ovate-lanceolate, plicate, with denticulate to dentate margins and a single costa. *Setae* smooth, elongate, flexuose, red to reddish-brown. *Capsules* short-cylindrical to cylindrical, curved and asymmetric or cylindrical and symmetric, inclined to horizontal or erect; *operculum* conic-convex, short-rostrate; *annulus* present, separating; *exothecial cells* quadrate to short-rectangular, rather thin-walled, oblate to isodiametric in 3–7 rows just below the orifice; *stomata* many in the neck, bicellular, with narrow, long pori; *peristome* double; *exostome teeth* 16, yellowish-brown to brownish, lanceolate, cross-striolate below, papillose above, with dentate margins in the distal half; *endostome* well developed, with a high or low basal membrane and normally developed or reduced, entire or narrowly cleft segments and 1–3, nodose or rudimentary cilia or cilia lacking. *Spores* globose, minutely papillose. *Calyptra* cucullate, smooth, naked.

Sanionia is a small genus which was segregated from the catch-all *Drepanocladus* (Loeske 1907*a*,*b*). It is distinct from other genera of the *Calliergon–Scorpidium–Drepanocladus* complex by the combination of several characters including plicate leaves, the presence of a perfect hyalodermis in the stem, well developed alar cells but having the ontogeny of the *aduncus* type (Hedenäs 1987*b*), serrulate leaf margins, the lack of rhizoid initials at the leaf apices and cross-striolate rather than dotted lower portions of the exostome teeth. In the Antarctic the genus is unmistakable because of its strongly falcate and costate leaves which are unknown in any other regional taxon, except for the only species of *Hypnum*, which has short and double costae.

Sanionia consists of four species, three of which are bipolar in distribution, while *S. orthothecioides* (Lindb.) Loeske has a circumpolar, though strongly dissected range in the arctic and northern boreal zones in the Northern Hemisphere. Two of these, *S. uncinata* and *S. georgicouncinata*, are also widespread in the cool and cold regions in the Southern Hemisphere, including Antarctica, and the former of these occurs additionally in tropical mountains in Africa and South America. The fourth species, *S. symmetrica* (Renauld & Cardot) Wheldon occurs predominantly western North America, with a few records in southern Chile. A useful taxonomic revision of the Fennoscandian species of *Sanionia* was published by Hedenäs (1989*b*), and in the Southern Hemisphere the genus was revised in Tierra del Fuego (Ochyra & Matteri 2001) and in the Neotropics (Hedenäs 2003*b*).

1. Alar cells forming a transversely triangular group, abruptly separated from the small, quadrate or rectangular and mostly chlorophyllose supra-alar cells; stem leaves long to very long and slenderly acuminate and moderately serrulate at margins in the upper part; inner perichaetial leaves gradually long-acuminate; endostome with a high basal membrane .... 1. ***S. uncinata***
1. Alar cells forming an isodiametric group, gradually passing into large, rectangular to long-rectangular and often echlorophyllose supra-alar cells; stem leaves acuminate to long-acuminate with strongly serrate margins above; inner perichaetial leaves rather suddenly long-acuminate; endostome reduced .... 2. ***S. georgicouncinata***

1. **Sanionia uncinata** (Hedw.) Loeske Fig. 242

Sanionia uncinata (Hedw.) Loeske, Hedwigia 47: 195. 1908. ≡ *Hypnum uncinatum* Hedw., Sp. Musc. Frond.: 289. 1801. ≡ *Stereodon uncinatus* (Hedw.) Brid., Bryol. Univ. 2: 825. 1827. ≡ *Amblystegium uncinatum* (Hedw.) De Not., Cronac. Briol. Ital. 2: 24. 1867. ≡ *Harpidium uncinatum* (Hedw.) Lange & C.E.O.Jensen, Medd. Grønland 3: 323. 1887. ≡ *Drepanocladus uncinatus* (Hedw.) Warnst., Beih. Bot. Centralbl. 13: 402, 417. 1903. — TYPE CITATION: Locis montosis Saxoniae metalliferae, montis piniferi Franconiae, alpinis Austriae, Salisburgi, et in Suecia; also a reference to Hedwig (1797, p. 65, t. 25). [LECTOTYPE (*fide* Hedenäs 1989*b*: p. 404): the lower plant on the sheet bearing the label 'Hypnum uncinatum Hedw. St. Cr. 4. p. 65. t. 25. Chemnitzii Sax lect.' G-Hedwig/Schwaegrichen!].

Hypnum austroaduncum Müll.Hal., Bot. Jahrb. Syst. 5: 82. 1883 ['austro-aduncum']. ≡ *Drepanocladus austroaduncus* (Müll.Hal.) Broth. *ex* Paris, Coll. Nom. Broth.: 10. 1909, ***syn. nov.*** — TYPE CITATION: all new species described by Müller (1883) are included in the section 'I. Bryologia Kerguelensis'; later Müller (1889: p. 33) wrote: 'Ins. Kerguelen, Betsy Cove, cum Psilopilo antarctico, Decbr. 1874. Winter harbour, Whate Bay, Januario 1875'. [Lectotype (*selected here*): 'Herb. Émil Bescherelle: 1900. Hypnum (Harpidium) austro-aduncum C. Müller Kerguelen: Dr. F. Naumann Debr 1874 Hb. C. Müll.' – BM-Bescherelle!; isolectotype: H-Brotherus!; syntype: 'Hypnum (Drepanocladus) austro-aduncum C. Müll nsp. Kerguelen Christmas Harbour (à île Crozier) 12.I.75 l. Naumann' – HBG!].

Hypnum austrouncinatum Müll.Hal., Bot. Jahrb. Syst. 5: 82. 1883 ['austro-uncinatum']. — TYPE CITATION: all new species described by Müller (1883) are included in the section 'I. Bryologia Kerguelensis'. [Lectotype (*selected here*): 'Aus dem Herbarium A. Geheeb Hypnum austro-uncinatum C. Müll. n. sp. (1883) Flora der Kerguelen-Insel Leg. Dr. Naumann, 1873 Dr. Alex. Braun ad A. Geheeb misit 1876' – JE-Herzog!; isotypes: BM! (two specimens), BM-Dixon!, H-Brotherus!, NY, S-Froelich!, S-Kindberg!, S-Möller!]. First synonymised by Cardot (1901: p. 43) and Kaalaas (1912: p. 116).

Hypnum uncinatum Hedw. fo. *gracile* Renauld *in* Cardot, Voy. Belgica Mousses: 44. 1901, *nom. inval.* [Art. 34.1(b)].

Drepanocladus uncinatus (Hedw.) Warnst. fo. *gracillimus* Cardot, Deux. Exped. Antarct. Française Mousses: 28: 1913 ['gracillima'], *nom. nud.*, ***syn. nov.*** — ORIGINAL MATERIAL: [Antarctic Peninsula] Échatillons sans numéro ni localité. [Chirotype: 'Herb. Mus. Paris 120. Drepanocladus uncinatus (Hedw.) Warnst. var. subjulaceus (BS.) Warnst. forma gracillima. 2ème expéd. Charcot. Leg. Gain' – PC!].

Drepanocladus plicatus Dixon, Kongel. Norske Vidensk. Selsk. Forh. 7: 3. 1934. — TYPE CITATION: [South Georgia] West Cumberland Bay, 20 m alt., 25 November, 1931 (29). 'Dammen', Grytviken, 4 April 1933 (75). [Lectotype (*selected here*): 'Herb. H. N. Dixon Ref. № 29 Drepanocladus plicatus Dix. sp. nov. {K. Norske Vidensk. Selsk. Forh. VII, Nr 1, 3 (1934)} West Cumberland Bay, South Georgia 20 m alt. Coll. T. Tröim 15 Nov. 1931 type' – BM-Dixon!; isolectotype: TRH!; syntype: 'Herb. H. N. Dixon Ref. № 75 Drepanocladus plicatus Dix. 'Dammen' (The Dam), Grytviken, South Georgia Coll. T. Tröim 4 Apr. 1933 Com. Herb. Mus. Niederosis' – BM-Dixon!; isosyntype: TRH!]. First synonymised by Ochyra & Matteri (2001: p. 46].

Plants small and slender, medium-sized to relatively robust growing in extensive mats or deep tufts, green, yellow- to brownish-green. *Stems* creeping to ascending or erect, (1.5–)2.0–6.5(–10.0) cm long, usually ± pinnately branched, less often irregularly to subpinnately branched in one plane, in transverse section oval, consisting of a unistratose outer, perfectly developed hyalodermis of large, hyaline and thin-walled cells, 2–4-layered cortex of small, brownish, incrassate cells, 4–7-stratose medulla of large, thin-walled and hyaline cells and distinct central strand; *rhizoids* rather scanty, smooth, reddish-brown, inserted in or just below the leaf base; *pseudoparaphyllia* broad, foliose; *axillary hairs* large, abundant, 1–8-cellular, with a short brownish basal cell and elongate and hyaline upper cells. *Branch* and *stem leaves* similar, the latter only somewhat smaller and often more circinate, lanceolate, from a rounded-triangular to ovate base, gradually tapering to a long or filiform, strongly falcato-secund to circinate, rarely almost straight, acumen, 2–4 mm long, 0.3–1.1 mm wide, concave, usually strongly plicate, but sometimes almost smooth, not or hardly decurrent; *margins* denticulate to finely serrulate all around to almost smooth near the base, plane or slightly recurved above; *costa* single, rather strong, often situated in a deep furrow, extending into acumen, (30–)40–70 μm wide at the base, in transverse section plano-convex, 2–4-stratose, composed of ± homogeneous cells; *laminal cells* unistratose, pellucid, smooth or slightly prorate at distal ends near the apex; *upper cells* linear to slightly linear-flexuose, 25–110 μm long, 4–7(–8) μm wide, mostly thin-walled and eporose, rarely thicker-walled and pitted, rounded or shortly tapering at the ends; *basal cells* shorter and wider with more incrassate and porose walls, 20–45 μm long, 8–13 μm wide; *angular cells* hyaline, thin-walled, quadrate to rectangular forming a transversely triangular, inflated, not or slightly decurrent group; *supra-alar cells* rather sharply delimited from the alar cells, wider and shorter, eporose or slightly pitted, chlorophyllose, forming a smaller or similar size group as the alar cells. *Autoicous. Inner perichaetial leaves* lanceolate, gradually long-acuminate, 3.5–7.5 mm long, somewhat concave, plicate, with margins denticulate distally and a single costa ending in the acumen. *Setae* elongate, 1.5–3.0 mm long, smooth, red to reddish-brown, dextrorse below, untwisted or sinistrorse when dry above. *Capsules* cylindrical, 1.5–2.5 mm long, curved to symmetric, inclined to horizontal or erect, with a distinct neck; *operculum* convex-conic, short-beaked; *annulus* of 2–3 rows of cells, separating; *exothecial cells* short- or long-rectangular, with slightly thickened walls, isodiametric in 1–3 rows below the orifice; *stomata* bicellular, superficial, long-pored; *peristome* double; *exostome teeth* 16, yellowish-brown, with the outer surface cross-striolate below and papillose and dentate at margins above, trabeculate on the inner surface; *endostome* well developed, pale brownish to yellowish, finely papillose to almost smooth, with a high basal membrane reaching almost half of the total endostome height, entire or narrowly perforated processes and 1–3 nodose cilia. *Spores* spherical, (11–)13–15(–17) μm in diamater, finely papillose. *Calyptra* cucullate, naked.

Chromosome number: $n = 11$ – South Georgia (Newton 1972); $n = 9 + \mathrm{m}, 10, 11, 12, 20$ – the other 27 records for the Northern Hemisphere material (Fritsch 1991).

Differentiation. Some small plants of *Sanionia uncinata* may be mistaken in the field for *Hypnum revolutum*, but its unicostate leaves and distinct pellucid, hyaline alar cells contrast sharply with the bicostate leaves of the latter species in which the alar cells are composed of small, thick-walled and never pellucid and hyaline cells. In the field, it may be difficult to distinguish this species from *S. georgicouncinata*. Both species are morphologically quite similar, although *S. uncinata* is generally a smaller plant when growing in 'normal' conditions, but in the extreme climatic conditions in Antarctica the plants are often strongly reduced in size, so this feature may be very misleading.

The safest characters separating both Antarctic species relate, unfortunately, to the sporophyte but these have little practical value because plants are almost always sterile. Fertile plants of *S. uncinata* have a perfectly developed endostome with a high basal membrane and narrow perforations along the median line of the segments, and the inner perichaetial leaves are gradually acuminate to a very long filiform acumen. In contrast, *S. georgicouncinata* has a strongly reduced endostome which is irregularly perforated in all parts and the inner perichaetial leaves are suddenly short-acuminate. When sterile, *S. uncinata* is best distinguished by its alar cells which are sharply distinct from the supra-alar cells, whereas in *S. georgicouncinata* the transition to the supra-alar cells is gradual, and consequently both groups of cells form a large rectangular to ovate group extending along the basal leaf margin. In addition, the stem leaves are longer-acuminate and weaker serrate in *S. uncinata* and the basal cells in *S. georgicouncinata* are often thin-walled, markedly widened and very lax.

As elsewhere, *Sanionia uncinata* varies greatly in plant size and colour as well as falcateness and plicateness of the leaves. Plant size is closely related to the habitat and, in general, stunted plants grow in dry and exposed habitats, whereas large and robust plants thrive in wet and sheltered places. The orthophyllous phenotypes which were described as *Drepanocladus plicatus* from South Georgia (Dixon 1934*c*) are uncommon in Antarctica (Lewis Smith 1972, 1978; Lewis Smith & Gimingham 1976; Priddle 1979). They have a markedly different habit in comparison with 'normal' plants having strongly falcate leaves, but except for the leaf shape itself, all structural features of the plants are the same. These plants grow in calcareous flushes on Signy I. and are usually distinct from typical phenotypes of *S. uncinata*.

In contrast to the Northern Hemisphere where numerous infraspecific taxa were recognised on account of the remarkable variability of *Sanionia uncinata* (Hedenäs 1989*b*), in the Southern Hemisphere the species was surprisingly interpreted as a uniform taxon. It was described twice from Îles Kerguelen, as *Hypnum austrouncinatum* and *H. austroaduncum* (Müller 1883,

→

Figure 242. *Sanionia uncinata*. — **1**. Plant with sporophytes, dry. **2**. Portion of branch, dry. **3**. Cross-section of stem. **4–6**. Axillary hairs. **7**. Pseudoparaphyllium. **8–14**. Leaves. **15**. Upper cells at margin. **16–17**. Mid-leaf cells. **18**. Mid-leaf cells at margin. **19–21**. Angular cells. **22**. Basal cells. **23–24**. Outlines of cross-sections of stem leaves. **25–26**. Perichaetial leaves. **27**. Margin of perichaetial leaf at shoulder. **28**. Operculate capsule, dry. **29**. Deoperculate capsule, dry. **30**. Stoma and exothecial cells at base of urn. **31**. Exothecial cells in middle of urn. **32**. Portion of peristome, spores, annulus and exothecial cells at mouth. — [1–7, 11–16, 18–20, 22–27 from *Schulz 113*; 8–10 from *RO 5120/79*; 17, 21 from *RO 1283/80*; 28 from *Moreau & Guarrera 10629*; 29 from *Roivainen 5129*; 30–32 from *Roivainen s.n.*, 17 Jan 1929]. — Scale bars: a – 1 mm (2, 25–26); b – 1 mm (8–14) and 200 μm (23–24); c – 0.5 cm (1) and 2 mm (28–29); d – 100 μm (15–22, 27, 30–31) and 200 μm (3–7, 32).

1
2
3
4
5
6
7
8
9
10
11
12
13
14
15
16
17
18
19
20
21
22
23
24
25
26
27
28
29
30
31
32
a
b
c
d

1889), but these species perfectly match typical expressions of *S. uncinata*. In addition, two populations from subantarctic South Georgia comprising exceptionally robust plants were described as *Drepanocladus plicatus* (Dixon 1934*c*), but this species also does not differ materially from typical phenotypes of *S. uncinata* (Ochyra & Matteri 2001).

An interesting assessment of the variability of Antarctic plants of *Sanionia uncinata*, collected by E. Racovitza during the *Belgica* Expedition 1897–1899, was presented by F. Renauld who recognised two distinct morphotypes in this species (Cardot 1901). One of these was designated as a new form, *Hypnum uncinatum* var. *subjulaceum* Schimp. fo. *gracile*, corresponding to *H. austrouncinatum* from Îles Kerguelen, but it is no more than a delicate expression of the otherwise typical *Sanionia uncinata*. The second morphotype was recognised as another new form, *Hypnum uncinatum* var. *plumulosum* Schimp. fo. *glaciale*, corresponding to *H. georgicouncinatum* from South Georgia, but this was considered to be only an ecological variant of the species growing in sites enriched with nitrogen from sea birds. These observations were very accurate and currently these morphotypes refer to two distinct species, *S. uncinata* and *S. georgicouncinata*. Nevertheless, these two names, *Hypnum uncinatum* fo. *gracile* and *H. uncinatum* fo. *glaciale*, were invalidly published according to Art. 34.1(b) of the current Code (McNeill *et al.* 2006) since they must be considered as provisional name which were merely proposed in anticipation of the future acceptance of these taxa. Later, Cardot (1913) designated one collection from the Antarctic Peninsula region as *Drepanocladus uncinatus* var. *subjulaceus* (Schimp.) Warnst. fo. *gracillimus* but this taxon has never been described, and the original material matches well the slender phenotype of *S. uncinata*. It is worth noting that Berggren (1875) used the same epithet for a variety of *Hypnum uncinatum* from Svalbard, and this name is also reduced to synonymy with *S. uncinata* (Hedenäs 1989*b*). However, Cardot (1913) evidently intended to recognise a different taxon within the then widely accepted variety, *Hypnum uncinatum* var. *subjulaceum* (Warnstorf 1903) but failed to provide its diagnosis.

Reproduction in Antarctica. Fully mature sporophytes are produced very rarely, although sexual organs are occasionally observed. The species commonly and abundantly fruits in Tierra del Fuego and fairly often on some subantarctic islands, including the Prince Edward Is., Îles Crozet and Îles Kerguelen (Ochyra, personal observations) and South Georgia (Lewis Smith, personal observations).

Habitat. In the maritime Antarctic *Sanionia uncinata* is an ubiquitous and weedy moss growing in a wide range of dry and wet habitats. It is associated with most communities of both cryptogam and herb tundra formations and it seems to avoid only communities of dry habitats dominated by crustose lichens in which bryophytes occur only occasionally. The species grows on rock ledges and outcrops, scree, stony ground, as well as on mineral soil, usually in drier habitats, but it is also found at the margins of streams and lakes, in damp flushes, seepage areas and sometimes in lakes and pools, occasionally submerged to a depth of 5 m.

World range. A bipolar species (Fig. 243) having a continuous circumpolar arctic-boreal-temperate range in the Northern Hemisphere, extending southwards to the Canary Is. (Eggers 1982), and a pan-temperate distribution in the Southern Hemisphere including southern South America (Ochyra & Matteri 2001), KwaZulu-Natal in South Africa (Sim 1926), south-eastern

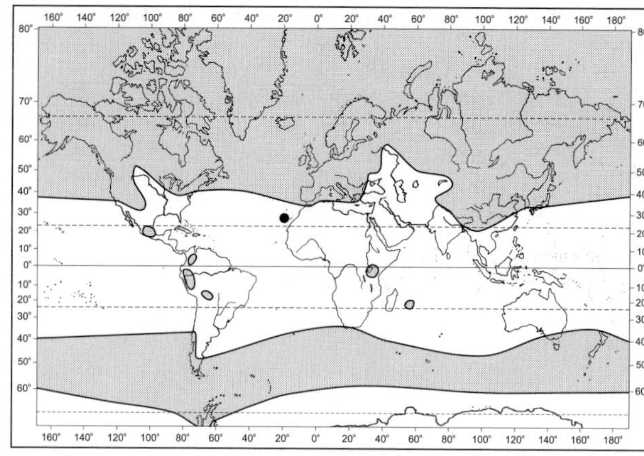

Figure 243. Global distribution of *Sanionia uncinata*.

Australia (Streimann & Curnow 1989), Tasmania (Dalton *et al.* 1991) and New Zealand (Sainsbury 1955), and all cool-temperate and subantarctic islands of the Southern Ocean and the maritime Antarctic with some transitional high mountain stations in the northern and central Andes (Hedenäs 2002*b*) and East Africa (O'Shea 2006) in the tropics.

Distribution in Antarctica. *Sanionia uncinata* is a widely distributed and often very common and abundant species on peri-Antarctic islands and archipelagoes, inluding Bouvetøya, the South Sandwich, South Orkney and South Shetland Is. and in the Antarctic Peninsula where it extends along its western coast from Tower I. off the southern Trinity Peninsula to the Graham Coast, with some isolated stations in Marguerite Bay, extending as far south the Terra Firma Is. at lat. 68°43′S (Fig. 244). It occurs predominantly at low elevetations, from sea level to about 100–200 m, occasionally extending to about 300 m and reaching its highest altitude of about 400 m on Vindication I. in the South Sandwich Is.

Selected specimens examined. WEST ANTARCTICA. BOUVETØYA. Nyrøysa, 0–51 m, *TE 367* (BG); Rustadkollen, 300–340 m, *TE 237, 247 & 261* (BG); nunatak east of Rustadkollen, *ca* 295 m, *TE 047.1-3* (BG).

SOUTH SANDWICH IS. **LESKOV I.** Summit ridge, 100 m, *PC 245F & 246C* (AAS, KRAM); east slope of summit ridge, 100 m, *PC 246C, 247C, 249A, 250F & 251A* (AAS, KRAM). **VINDICATION I.** East side of summit mass, *ca* 400 m, *MWH 852B* (AAS, KRAM); north-east corner of the island, *ca* 180 m, *MWH 440* (AAS, KRAM). **MONTAGU I.** Allen Pt., 150 m, *Vaughan 802* (AAS, KRAM). **BELLINGSHAUSEN I.** Southern slopes of crater, 130 m, *PC 62B* (AAS, KRAM); west of access point, 150 m, *PC 74A* (AAS, KRAM); north side of north peak of crater rim, *ca* 250 m, *MWH 823B* (AAS, KRAM); southern slopes of main cone, *ca* 80 m, *MWH 821C* (AAS, KRAM).

SOUTH ORKNEY IS. **CORONATION I.** Windscoop Buttress east of Sunshine Glacier, 175 m, *RW 63C* (AAS, KRAM); Saunders Pt., *RILS 26* (AAS, KRAM). **CHRISTOFFERSEN I.** Southern end, 50 m, *KR 243B* (AAS, KRAM). **MATTHEWS I.** 15–100 m, *RILS 174* (AAS, KRAM). **STEEPHOLM I.** 75 m, *RILS 3250* (AAS, KRAM). **POWELL I.** East Falkland Harbour, *RILS 55* (AAS, KRAM). **FREDRIKSEN I.** North-east side, 100 m, *NJC 12A* (AAS, KRAM). **LAURIE I.** Cabo Moneta, 150 m, *Bellisio 11765* (KRAM, NY); Cabo Destacamento, 100 m, *Bellisio 11800* (KRAM, NY). **SIGNY I.** Thulla Pt., 70 m, *RILS 8047B & 8048* (AAS, KRAM); Foca Cove, *RILS 1806* (AAS, KRAM); between North Pt. and Spindrift Rocks, *RILS 1787* (AAS, KRAM); Robin Peak, *RILS 1875* (AAS, KRAM); Stygian Cove, 20 m, *RILS 10785* (AAS, KRAM); Waterpipe Beach, 8 m, *RILS 467* (AAS, KRAM); Changing Lake, 40 m, *RILS 10794A* (AAS, KRAM); Sombre Lake, *RILS 1817 & 1823B* (AAS, KRAM); Knob Lake, 4 m, *RILS 1835* (AAS, KRAM); Factory Cove, *ca* 65 m, *RILS 5* (AAS, KRAM); Observation Bluff, 30 m, *RILS 10856* (AAS, KRAM) and 35 m, *RILS 624& 1782* (AAS, KRAM); Moraine Valley, 75 m, *RILS 10854* (AAS, KRAM).

SOUTH SHETLAND IS. **King George I.** ***Potter Peninsula***: Three Brothers Hill, *BJ 161 & 161A* (KRAM). ***Fildes Peninsula***. Suffield Pt., *RO 2453/80* (KRAM); Durant Pt., *VK 579* (KRAM); Ardley I., *VK 278 & 552* (KRAM). ***Destruction Bay***. Faraway Nunataks, *BJ 170* (KRAM). ***King George Bay***. Penguin I., *Leech 175* (KRAM, NY). ***Admiralty Bay***: The Tower, 340 m, *RO 816/80* (KRAM); Creeping Slopes, 130 m, *RO 1283/80* (KRAM); Czajkowski Needle, 240 m, *RO 676/80* (KRAM); Jersak Hills, 125 m, 200 m, *RO 5120/79* (KRAM). ***Ezcurra Inlet***: Italia Valley, 100 m, *RO 96/80* (KRAM); Pond Hill, 90 m, *RO 875/80* (KRAM); Urbanek Crag, 180 m, *RO 2326/80* (KRAM); Klekowski Crag, 250 m, *RO 2268/80* (KRAM). ***MacKellar Inlet***: Misty Nunatak, 250 m, *RO 2030/80* (KRAM); Admiralen Peak, 306 m, *RO 2017/80* (KRAM); Crépin Pt., 5 m, *RO 2044/80* (KRAM); Kapitan Peak, 210 m, *RO 1909/80* (KRAM). ***Martel Inlet***: Yellow Pt, 40 m, *RO 495/80* (KRAM); Shark Fin, 140 m, *RO 2633/80* (KRAM); Precious Peaks, 100 m, *RO 2661/80* (KRAM); Tern Nunatak, 266 m, *RO 2519/80* (KRAM); Point Hennequin, 10 m, *RO 2244/80* (KRAM); Basalt Pt., 25 m, *RO 2198/80* (KRAM). **Robert I.** Coppermine, 7 m, *Frödin 3B* (NY). **Greenwich I.** Discovery Bay, NE of Mt. Plymouth, *RMS 69–982* (US); Yankee Harbour, *Schmitt 217* (KRAM, NY). **Desolation I.** North-facing slope, *KR 68A* (AAS, KRAM). **Livingston I.** ***Byers Peninsula***: Lair Pt., 5 m, *KR 66D* (AAS, KRAM); Cutler Stack, 3 m, *KR 66D* (AAS, KRAM); between Smellie Pt. and Devils Pt., 5 m, *KR 28E* (AAS, KRAM); Laager Pt., 5–15 m, *RILS 3803 & 3803A* (AAS, KRAM) and 30 m, *KR 39C* (AAS, KRAM). ***Hurd Peninsula***: False Bay, *RMS 69-411, 69-543, 69-568, 69-692 & 69-699* (US). **Snow I.** President Head, *KR 2A, 4B & 7A* (AAS, KRAM). **Deception I.** Cathedral Crags, *Leech 169* (KRAM, NY); Whalers Bay, abandoned British Station, *RO 2765/80* (KRAM); above Lake Kroner, 40 m, *RO 2771/80* (KRAM); Mt. Pond, 60 m, *RILS 5775A* (AAS, KRAM); Pendulum Cove, east of Crimson Hill, 130 m, *RILS 11210B* (AAS, KRAM); Cross Hill, 60 m, *RILS 5795C* (AAS, KRAM); Fumarole Bay, 50 m, *RILS 11350* (AAS, KRAM), 100 m, *RILS 11494* (AAS, KRAM) and *ca* 125 m, *RILS 11277* (AAS, KRAM); Stonethrow Ridge, 45–230 m, *RILS 5804C & 5809D* (AAS, KRAM); Kendall Terrace, 60 m, *RILS 5742B* (AAS, KRAM); Mt. Kirkwood, 170–250 m, *RILS 5714A & 5764A* (AAS, KRAM); Vapour Col, 130 m, *RILS 5634B & 5636B* (AAS, KRAM); Crater Lake, 12 m, *RILS 5769A* (AAS, KRAM) and 30 m, *RILS 5766A & 5767D* (AAS, KRAM); opposite Låvebrua I., *NJC 67A* (AAS, KRAM).

WEST ANTARCTIC PENINSULA. **Davis Coast.** Tower Island, *NJC 97* (AAS, KRAM). **Danco Coast.** Cierva Pt., *OB 5B* (KRAM); Spring Pt., *VK 473, 1190, 1213, 1222, 1245, 1230, 1251 & 1262* (KRAM) and 10–30 m, *RILS 3876* (AAS, KRAM); Beneden Head, *ER 230A & 234D* (BR, PC); Paradise Harbour, Coughtrey Peninsula, Almirante Brown Station, *RMS 69-382* (US). **Palmer Archipelago.** ***Brabant Island***: Astrolabe Needle, *VK 885* (KRAM). ***Melchior Islands***: Lysted I., *PAS 335.5 & 360.9* (FH); Lysted I., 8–10 m, *PAS 371.3* (FH); Lambda I., Sevilla Landing, *PAS 347.2* (FH). ***Anvers I.***: Norsel Pt., *VK 955* (KRAM); Cape Monaco, *VK 781 & 824* (KRAM); Bonaparte Pt., *VK 46, 1039 & 1164* (KRAM); Dream I., *VK 641, 673 & 989* (KRAM). **Graham Coast.** Cape Tuxen, *Gain 203* (PC). ***Grandidier Channel***: Takaki Pt., 10–35 m, *RILS 4365B* (AAS, KRAM). **Wilhelm Archipelago.** ***Argentine Is.***: Galindez I., 3 m, *RILS 940* (AAS, KRAM). **Biscoe Is.**: Thomsen Is., 3 m, *RILS 5326BB* (AAS, KRAM). **Fallières Coast.** ***Adelaide I.***: Stork Ridge, *PC 265A* (AAS, KRAM); Avian Island, 20 Jan 1963, *Berg s.n.* (US). ***Marguerite Bay***: Neny I., 10–15 m, *RILS 791* (AAS, KRAM); Refuge Is., largest island, 10–25 m, *RILS 4521* (AAS, KRAM); Terra Firma Is., Barn Rock, *Poncet 7* (COLO, KRAM); same group, Alamode I., *Poncet 3C* (AAS, KRAM).

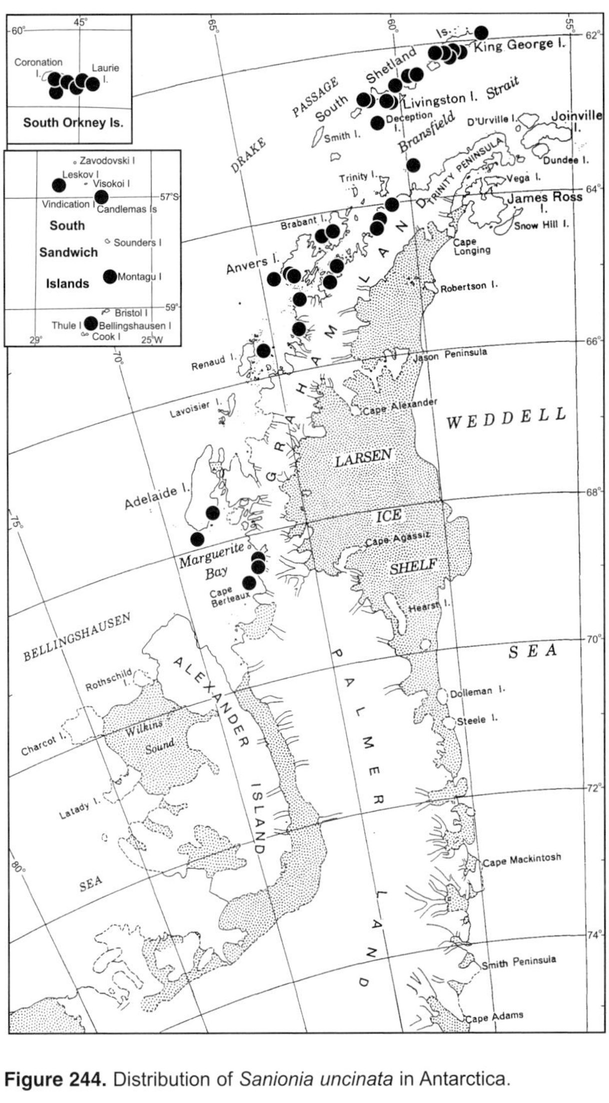

Figure 244. Distribution of *Sanionia uncinata* in Antarctica.

2. **Sanionia georgicouncinata** (Müll.Hal.) Ochyra & Hedenäs Fig. 245

Sanionia georgicouncinata (Müll.Hal.) Ochyra & Hedenäs *in* Ochyra, Moss Fl. King George Island Antarctica: 227. 1998. ≡ *Hypnum georgicouncinatum* Müll.Hal. *in* Neumayer, Int. Polarforsch. Deutsch. Exp. 2: 321. 1890 ['georgicouncinatum'] ≡ *Drepanocladus uncinatus* (Hedw.) Warnst. fo. *georgicouncinatus* (Müll.Hal.) Cardot *in* Cardot & Broth., Kongl. Svenska Vetenskapsakad. Handl. 63(10): 64. 1923. ≡ *Sanionia uncinata* (Hedw.) Loeske var. *georgicouncinata* (Müll.Hal.) Putzke & A.B.Pereira, Antarctic Mosses: 51. 2001, *comb. inval. basion. non citat.* — Type citation: Austro-Georgia, 'Köppenberg', Sumpf auf der Westseite', 18. I. 1883. [Lectotype (*vide* Ochyra 1998*a*: p. 227): 'No. 43. C. Müller. Hypnum georgicouncinatum C. Müller Fundort: Köppenberg. Sumpf auf der W-Seite. Süd-Georgien 18/I 83 Will' – M!; isotype: PC-Cardot!].

Hypnum uncinatum Hedw. var. *polare* Renauld *in* Cardot & Thér., Proc. Wash. Acad. Sci. 4: 340. 1902. [*H. uncinatum* fo. *polare* (Renauld *in* Cardot, Voy. Belgica Mousses: 44. 1901, *nom. inval.* {Art. 34.1(b)}]. ≡ *H. uncinatum* fo. *polare* (Renauld) Renauld *ex* Cardot, Rev. Bryol. 33: 35. 1906, ***syn. nov.*** — Type citation: From St. Matthew Island (Trelease, 2159), Plover Bay, Siberia (Trelease, 1978). [Lectotype (*selected here*): 'The Harriman Alaska Expedition Plants of Bering Strait collected by Wm. Trelease and De Alton Saunders. Hypnum uncinatum Hedw. var. polare Ren. Det. Renauld. No. 1978 Collector T. Plover Bay Siberia July 11, 1899' – PC-Cardot!].

Sanionia nivalis Hedenäs, Ann. Bot. Fenn. 26: 411, *f. 10–12.* 1989. — Type citation: Sweden, Torne Lappmark: Jukkasjärvi, S. of Nissuntjårro's peak 1738 m, 1200–1300 m a.s.l., 23.VII.1988, L. Hedenäs TL88-16. [Holotype: 'TL 88-16 Flora suecica Sanionia nivalis Hedenäs Torne Lappmark: Jukkasjärvi, S of Nissuntjårro's peak 1738 m; between boulders 200–300 m from remaining snow in late snow-bed, slope tow. E-N. Alt. 1200–1300 m a.s.l., 23.VII.1988, L. Hedenäs' – S!]. First synonymised by Ochyra (1998*a*: p. 227).

Plants medium-sized to fairly large and robust, in soft, loose or dense extensive carpets or deep tufts, somewhat lustrous, green, yellow, yellow- or brownish-green. *Stems* creeping or more crowded and ascending to erect, (1.5–)2.0–7.5(–14.0) cm long, irregularly to subpinnately branched, sometimes almost unbranched, in cross-section with a perfect, unistratose hyalodermis, 2–3-stratose cortex of small, brown, thick-walled cells, 4–7-layered medulla of large, hyaline, firm-walled cells and distinct central duct; *rhizoids* sparse, reddish-brown, branched; *pseudoparaphyllia* foliose, broad; *axillary hairs* filiform, abundant, consisting of 1–2 short, brown basal cells and (2–)4–6 elongate, hyaline upper cells. *Stem* and *branch leaves* similar, strongly falcato-secund, sometimes almost straight, lanceolate from a broadly ovate to ovate-triangular base, (1.5–)2.0–3.5(–4.0) mm long, 0.5–1.2(–1.4) mm wide, concave to deeply concave, smooth to weakly plicate; *margins* serrulate throughout to almost entire below, plane or weakly recurved above; *costa* single, ceasing in the acumen, (30–)35–70(–80) µm wide at base, plano-convex in cross-section, situated in a shallow, wide-angled furrow; *mid-leaf cells* linear to linear-flexuose, (20–)25–100(–115) µm long, (4–)5–9(–11) µm wide, thin-walled, not pitted, with rounded or shortly tapering apices; *basal cells* shorter and wider, with more incrassate and pitted walls, 15–40 µm long, 10–16 µm wide; *angular cells* forming a large, ovate group extending along the basal margins, composed of quadrate to rectangular, hyaline, thin-walled alar cells, forming an inflated, isodiametric group, and rectangular, thin-walled, mostly echlorophyllose supra-alar cells, poorly differentiated from and forming a group of similar size or larger than the alar cells. *Autoicous*. *Inner perichaetial leaves* ovate-lanceolate, 2.5–5.5 mm long, abruptly narrowed to acute to short-acuminate apex, with strongly denticulate upper margins. *Setae* elongate, 2.0–3.5 mm long, smooth, red to reddish-brown, not twisted. *Capsules* short-cylindrical, 1.5–2.5 mm long, straight or slightly curved, erect to slighly inclined; *operculum* convex-conic, short-beaked; *annulus* of 2–3 rows of cells, separating; *exothecial cells* quadrate to short-rectangular, with thin to slightly incrassate walls, isodiametric in 1–3 rows below the orifice; stomata bicellular, superficial, long-pored; *peristome* double; [*exostome teeth* 16, yellowish to brownish, with the outer surface cross-striolate below and very weakly papillose and often irregularly striolate above, trabeculate on the inner surface; *endostome teeth* strongly reduced, hyaline to slightly yellowish, finely papillose throughout or only in the upper part, strongly perforated, with a low basal membrane reaching to ⅓ of the total endostome height, narrow processes and no or rudimentary cilia. *Spores* spherical, (11–)14–21 µm in diameter, finely papillose.] *Calyptra* cucullate, smooth, naked.

Chromosome numbers: n = 20, 30 – King George I. (Kuta *et al.* 1982 as *Sanionia uncinata*).

Discussion. *Sanionia georgicouncinata* was described relatively early from South Georgia (Müller 1890*a*), but subsequently this species fell into taxonomic obsolescence being considered as a form of the then all-encompassing *S. uncinata* (Cardot & Brotherus 1923). Ferdinand Renauld drew attention to the unusual form of the basal leaf cells in the Antarctic plants collected by E. Racovitza during the 'Belgica' Expedition and suggested recognising them as a separate form of *S. uncinata*, namely *Hypnum uncinatum* fo. *polare* (Cardot 1901). Further collections of this form were made by the Swedish South Polar Expedition of 1901–1903 (Cardot 1906*a*) and French Antarctic Expedition of 1903–1905 (Cardot 1906*b*). However, at about the same time, an analogous form of *Sanionia uncinata* was discovered in 1899 in the Bering Sea area by the Harriman Alaska Expedition. Ferdinand Renauld studied the material and designated it as *Hypnum uncinatum* var. *polare* (Cardot & Thériot 1902). It was the first valid description of this taxon because *H. uncinatum* fo. *polare* was only a provisional name referred to in a letter from Renauld to Cardot and, as such, it was invalidly published according to Art. 34.1(b) of the current Code (McNeill *et al.* 2006). Examination of the original material of *H. uncinatum* var. *polare* from the Bering Sea confirms perfect correspondence of Antarctic and Arctic specimens and the identity of *Hypnum uncinatum* fo. *polare* with *Sanionia georgicouncinata*. It is noteworthy that the specimen from Plover Bay (*Trelease 1978*, PC), which is here chosen as a lectotype of this form name, represents the first record of *S. georgicouncinata* from Chukotka (Afonina 2004). Finally, Hedenäs (1989*b*) recognised similar plants from Scandinavia as the separate species *S. nivalis*. However, detailed comparison of this species with the type of *Hypnum georgicouncinatum* revealed that they are inseparable from each other and that they constituted one species (Ochyra 1998*a*). The Antarctic plants are generally smaller and more slender in comparison to the northern ones, but the correspondence in critical characters is excellent. For the differences between *S. georgicouncinata* and *S. uncinata* see comment under the latter species.

Reproduction in Antarctica. The species has only once been found with immature sporophytes on King George I., and once with juvenile setae in Charlotte Bay on the Danco Coast on the western Antarctic Peninsula.

Habitat. *Sanionia georgicouncinata* grows predominantly in wet habitats in various communities of the moss carpet subformation. It is especially frequent at the drier margins of swampy areas, forming extensive pure stands in moist hollows, in late snow patches and along streams and melt runnels as well as on almost barren, frost-disturbed stony ground. In addition, the species is a frequent associate of the herb tundra dominated by *Deschampsia antarctica* and *Colobanthus quitensis*, in particular in sheltered habitats on rock ledges in coastal areas.

→

Figure 245. *Sanionia georgicouncinata*. — **1**. Plant with sporophyte. **2**. Sterile plant. **3**. Portion of branch, dry. **4**. Cross-section of stem. **5**. Pseudoparaphyllium. **6–8**. Axillary hairs. **9–17**. Leaves. **18**. Upper cells at margin. **19–21**. Mid-leaf cells. **22**. Basal cells. **23–26**. Angular cells. **27–28**. Outlines of cross-sections of stem leaves. **29–31**. Perichaetial leaves. **32**. Margin of perichaetial leaf at shoulder. **33**. Capsule, wet. **34**. Stomata and exothecial cells at base of urn. **35**. Exothecial cells in middle of urn. **36**. Portion of peristome, annulus, exothecial cells at mouth and spores. — [1, 3, 11–12, 25, 29–35 from *RO 2630/80*; 2, 15–16 from *RO 5667/79*; 4–8, 17–23, 27–28 from *RO 646/80*; 9–10, 26 from *Trelease 1978* – lectotype of *Hypnum uncinatum* var. *polare*, PC; 13–14 from *RO 1258/80*; 24 from *RO 2633/80*; 36 from *Hedenäs TL 88-33* – paratype of *Sanionia nivalis*)]. Scale bars: a – 0.5 cm (2); b – 1 mm (9–17, 29–31), 100 µm (4–8, 23–26) and 200 µm (27–28); c – 0.5 cm (1) and 200 µm (34–36); d – 2 mm (3, 33) and 100 µm (18–22, 32).

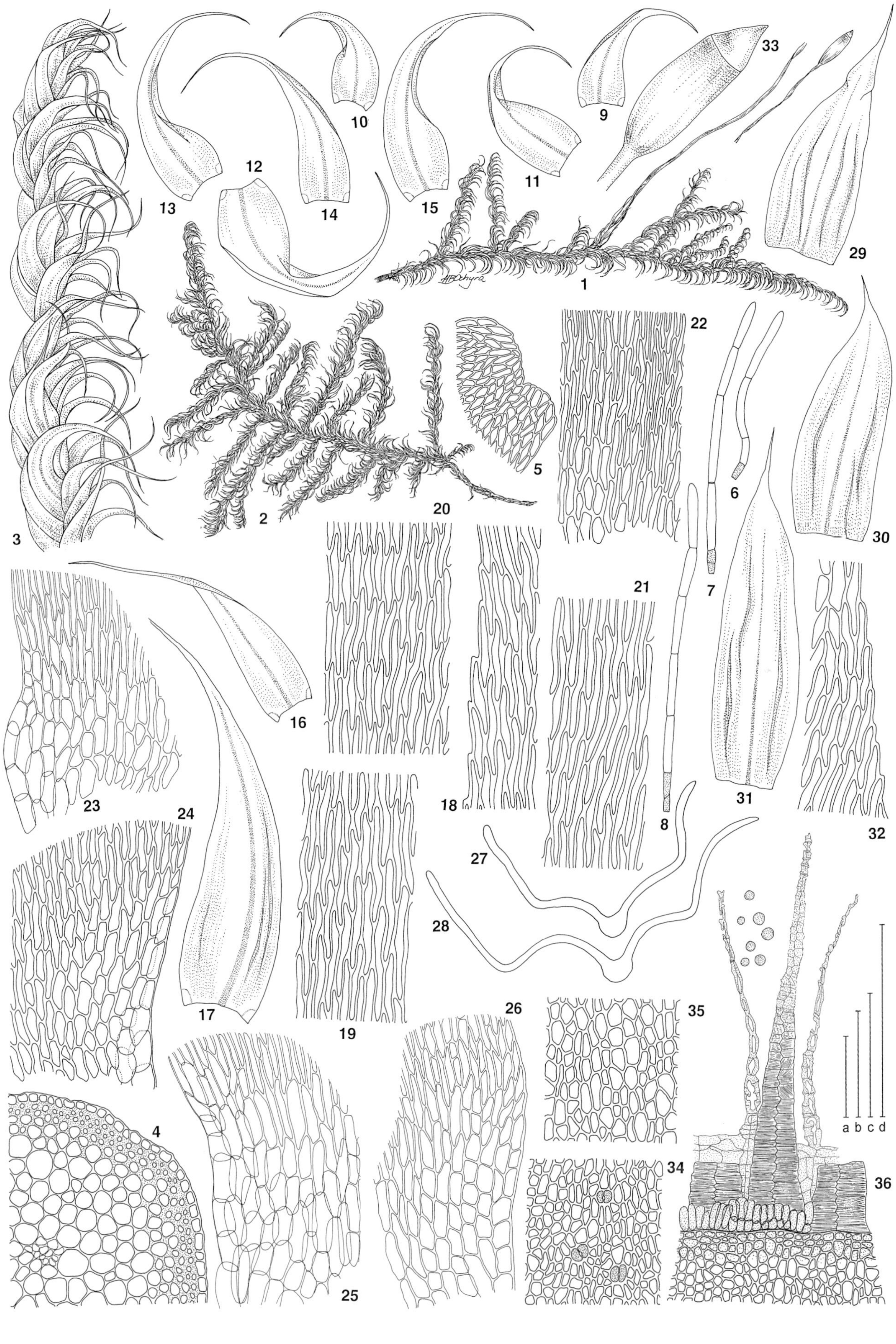
1
2
3
4
5
6
7
8
9
10
11
12
13
14
15
16
17
18
19
20
21
22
23
24
25
26
27
28
29
30
31
32
33
34
35
36
a b c d

World range. A bipolar species (Fig. 246). Its distribution in the Northern Hemisphere is still imperfectly known, but the present data indicate that it is an arctic-boreal-montane species having a strongly dissected circumpolar range. So far, it has been frequently collected in Fennoscandia (Hedenäs 1989*b*), and rarely in Iceland (Jóhannsson 1998), Spitsbergen (Frisvoll & Elvebakk 1996) and occasionally in East Greenland, Alaska and the Northwest Territories of Canada and the Rocky Mountains of Colorado, as well as very occasionally in Arctic Siberia (Ignatov *et al.* 2006). In the Southern Hemisphere it is restricted to South Georgia and the maritime Antarctic but, so far, no records are known from Tierra del Fuego (Ochyra & Matteri 2001).

Distribution in Antarctica. A very widespread and common species, ranging from Bouvetøya and the South Sandwich Is. along the western coast of the Antarctic Peninsula to southern Marguerite Bay, where it reaches its southernmost occurrence at lat. 68°43′ S in Terra Firma Is. In addition, it occurs rarely in the northern part of the eastern coast of the Antarctic Peninsula on Joinville and James Ross Islands groups (Fig. 247). It occurs at low elevations, from sea level to about 100 m, and only occasionally is it found at higher altitudes, to 340 m on Bouvetøya and 300 m on King George I.

Selected specimens examined. **WEST ANTARCTICA.** BOUVETØYA. Kapp Valdivia, 60–90 m, *TE 26.1-2, 284, 302 & 304* (BG); Nyrøysa, 20–35 m, *TE 206, 295, 366.1-2* (BG); south-western slope of Rustadkollen, 300–340 m, *TE 230, 248, 256, 257 & 360* (BG); Rustadkollen, top plateau, 300–340 m, *TE 223* (BG); nunatak east of Rustadkollen, *ca* 275 m, *TE 50* (BG).

SOUTH SANDWICH IS. **LESKOV I.** Summit ridge, *ca* 200 m, *REL 527* (AAS, KRAM); western slopes of island below summit ridge, 70 m, *PC 255B* (AAS, KRAM). **CANDLEMAS I.** Western shore, *ca* ¼ mile south of lagoons, 35 m, *REL 587* (AAS, KRAM); Lucifer Hill just above Clinker, *PC 147B* (AAS, KRAM); above Medusa Pool and Chimaera Flats, *PC 171 & 172D* (AAS, KRAM); Breakbones Plateau, *PC 205B* (AAS, KRAM). **MONTAGU I.** Scarlett Pt., 50 m, *PC 98A & 99B* (AAS, KRAM); Allen Pt., 70 m, *PC 100D* (AAS, KRAM). **BRISTOL I.** Wilson Rock, *PC 80* (AAS, KRAM). **SOUTHERN THULE.** Cape Flannery, *PC 8* (AAS, KRAM). **COOK I.** Reef Pt., 30 m, *PC 14A* (AAS, KRAM) and 50 m, *PC 13A* (AAS, KRAM); Resolution Pt., 50 m, *PC 19A* (AAS, KRAM). **BELLINGSHAUSEN I.** Crater rim, 100 m, *PC 24C* (AAS, KRAM), 150 m, *PC 39F, 46D & 56E* (AAS, KRAM) and 170 m, *PC 32D* (AAS, KRAM); wouth-western slope of island, 40 m, *PC 77* (AAS, KRAM); western coast, 40 m, *PC 79* (AAS, KRAM).

SOUTH ORKNEY IS. **MONROE I.** Slopes and plateau , 2–200 m, *RILS 5248* (AAS, KRAM). **CORONATION I.** Cape Hansen, *NJC 119* (AAS, KRAM). **LYNCH I.** North side, 3–30 m, *RILS 507* (AAS, KRAM). **POWELL I.** Michelsen I., *RILS 43* (AAS, KRAM); south-eastern coast, *ca* 130 m, *RILS 201* (AAS, KRAM); promontory opposite and E of Christoffersen I., *RILS 48* (AAS, KRAM). **CHRISTOFFERSEN I.** Falkland Harbour, 60 m, *KR 221C* (AAS, KRAM). **MICHELSEN I.** Eastern side, *RILS 43* (AAS, KRAM). **FREDRIKSEN I.** Central ridge, *ca* 115 m, *Edwards 1A* (AAS, KRAM). **LAURIE I.** Cape Moneta, 150 m, *Bellisio 11749* (KRAM, NY). **SADDLE I.** *Rudmose Brown s.n.* 4.2.1903 (E, PC). **SIGNY I.** Foca Cove, 25 m, *RILS 10908* (AAS, KRAM); Stygian Cove, 20 m, *RILS 10786* (AAS, KRAM); Borge Bay, *ca* 110 m, *Discovery Exped. 8* (BM); Drying Pt., *ca* 3 m, *RILS 691E* (AAS, KRAM). **MOE I.** 2–200 m, *RILS 5248* (AAS, KRAM).

SOUTH SHETLAND IS. **ELEPHANT I.** Point Wild, *KA 20A* (AAS, KRAM). **GIBBS I.** Near the Spit, 90 m, *JPB 139* (AAS, KRAM). **KING GEORGE I.** ***Potter Peninsula***: Stranger Pt., *BJ 105* (KRAM); Three Brothers Hill, *BJ 161A* (KRAM). ***Barton Peninsula***: Winship Pt., *BJ 181* (KRAM); Noel Hill, *BJ 197* (KRAM); Narębski Pt., *BJ 178* (KRAM). ***Marian Cove***: North Spit, *BJ 144* (KRAM). ***Fildes Peninsula***: Green Pt., *BJ 17 & 39* (KRAM); Nebles Pt., *BJ 52* (KRAM); Ardley I., 5 m, *RO 2482/80* (KRAM); Gemmel Peaks, 15 m, *RO 2471/80* (KRAM). ***Drake Passage***: Atherton I., *BJ 112* (KRAM); Bell I., *BJ 3* (KRAM); Davey Pt., 60 m, *BJ* 47 (KRAM); Tartar Pt., 10 m, *BJ 6* (KRAM); Pottinger Pt., 10 m, *BJ 31* (KRAM); Ridley I., *BJ 28* (KRAM); False Round, *BJ 75* (KRAM); Gam Pt., *BJ 11* (KRAM); Pyrites I., *BJ 35* (KRAM). ***Destruction Bay***: North Foreland, *BJ 100* (KRAM); Jenny Buttress, *BJ 168* (KRAM); Wrona Buttress, *BJ 175* (KRAM). ***Sherrat Bay***: Three Sisters Pt., *BJ 116* (KRAM). ***King George Bay***. Penguin I., *BJ 64 & BJ 68* (KRAM); Turret Pt., *BJ 126* (KRAM); Lions Rump, *BJ 92* (KRAM). ***Legru Bay***. Low Head, *BJ 82* (KRAM). ***Admiralty Bay***: Patelnia, 12 m, *RO 1132/80* (KRAM); Bastion, 250 m, *RO 1000/80* (KRAM) and 280 m, *RO 898/80* (KRAM); Zamek, 330 m, *RO 24/80* (KRAM); Demay Pt., 40 m, *RO 1384/80* (KRAM); Creeping Slopes, 100 m, *RO 1258/80* (KRAM); Llano Pt., 10 m, *RO 4851/79 & RO 4880/79* (KRAM); Rakusa Pt., 8 m, *RO 303/80* (KRAM); Ambona, 85 m, *RO 1636/80* (KRAM); Ubocz, 85 m, *RO 4964/79* (KRAM); Vauréal Peak, 35 m, *RO 5246/79* (KRAM); Cape Syrezol, 15 m, *RO 1850/80* (KRAM). ***Ezcurra Inlet***: Jardine Peak, 275 m, *RO 5167/79* (KRAM); Belweder, 120 m, *RO 1598/80* (KRAM); Emerald Pt., 3 m, *RO 858/80* (KRAM); Dufayel I., Gdynia Pt., 4 m, *RO 1767/80* (KRAM). ***MacKellar Inlet***: Crépin Pt., 20 m, *RO 2038/80* (KRAM); Keller Peninsula, Ore Pt., 7 m, *RO 518/80* (KRAM) and 30 m, *RO 522/80* (KRAM). ***Martel Inlet***: Moraine Pt., 20 m, *RO 463/80* (KRAM); Mt. Flagstaff, 140 m, *RO 422/80* (KRAM); Stenhouse Bluff, 30 m, *RO 2583/80* (KRAM); Ullman Spur, 90 m, *RO 561/80* (KRAM); Warkocz, 300 m, *RO 2540/80* (KRAM); Mt. Wawel, 40 m, *RO 2131/80* (KRAM). **NELSON I.** Harmony Cove, *CS 479* (PC, S); Moss Bay, 26 Feb 1989, *Komárek s.n.* (KRAM). **ROBERT I.** Coppermine Cove, 3–50 m, *RILS 920* (AAS, KRAM). **AITCHO I.** 5–80 m, *RILS 11450* (AAS, KRAM). **GREENWICH I.** Discovery Bay, NE of Mt. Plymouth, *RMS 69-981 & 69-985* (US). **HALFMOON I.** 22 Jan 1967, *Keith s.n.* (NY). **LIVINGSTON I.** ***Byers Peninsula***: Ocoa Pt., 50 m, *KR 42B* (AAS, KRAM); Lair Pt., 2 m, *KR 61B* (AAS, KRAM); Robbery Beaches, 10 m, *KR 52F, 53C & 56E* (AAS, KRAM); Smellie Pt., *KR 38E* (AAS, KRAM); Devils Pt., 100 m, *KR 23A* (AAS, KRAM). ***Hurd Peninsula***: Spanish base 'Juan Carlos I', *Acuña 7001* (KRAM); False Bay, *RMS 69-582* (US); Miers Bluff, *RMS 69-823* (US); Barnard Pt., 15 m, *DCL 22* (AAS, KRAM). ***Moon Bay***: Inott Pt., *RILS 5890C* (AAS, KRAM). **SNOW I.** President Head, *KR 1A* (AAS, KRAM). **DECEPTION I.** Whalers Bay, 5 m, *RILS 11229* (AAS, KRAM) and 12 m, *RILS 5792A* (AAS, KRAM); between Kroner Lake and Whalers Bay, 3–5 m, *CGB 42* (AAS, KRAM); Mt. Pond, 50 m, *RILS 5771B & 5772A* (AAS, KRAM); Telefon Ridge, 30–60 m, *RILS 5758A & 5761C* (AAS, KRAM); Fumarole Bay, 100–200 m, *RILS 11436* (AAS, KRAM); Wensleydale Beacon, 105–240 m, *RILS 5802D* (AAS, KRAM); Crater Lake, 70–200 m, *RILS 6833C* (AAS, KRAM); Entrance Pt., 50–70 m, *RILS 6792A* (AAS, KRAM).

WEST ANTARCTIC PENINSULA. **TRINITY PENINSULA.** Astrolabe I., *ca* 25 m, *NJC 91* (AAS, KRAM) and 10–30 m, *RILS 4051 & 4093* (AAS, KRAM); Mt. Flora, 25–50 m, *RILS 3844* (AAS, KRAM) and *ca* 85 m, *CGB 22* (AAS, KRAM); between Mt. Flora and Scar Hills, *ca* 85 m, *CGB 17* (AAS, KRAM). **DAVIS COAST.** Cape Herschel, 20 m, *RILS 11642A* (AAS, KRAM); Tower I., *RILS 701B & 703* (AAS, KRAM); Cape Kjellman, *CS 473 & 482* (PC, S); Cape Andreas, *CS 480 & 481* (PC, S); Cape Roquemaurel, 10–50 m, *RILS 3920* (AAS, KRAM). **DANCO COAST.** Young Pt., Bone Bay, 5–15 m, *RILS 3902* (AAS, KRAM); Diamonen I., *ER 158 & 200* (BR, PC); Two Hummock I., *Wiley & R. Smith 704B* (AAS, KRAM); Cierva Pt., *VK 260, 519, 588, 609, 630, 657 & 663* (KRAM); Moss Island, *CS 483* (PC, S); Spring Pt., *VK 910, 1177, 1197, 1249 & 1275* (KRAM); Valdivia Pt., *CS 484* (PC, S); Challenger I., *CS 452 pp.* (PC, S); Charlotte Bay, Recess Cove, *RILS 3978* (AAS, KRAM); Andrée I., 25 m, *RILS 4006 & 4038* (AAS, KRAM); Cuverville I., 60 m, *RILS 8802* (AAS, KRAM); Useful I., *RMS*

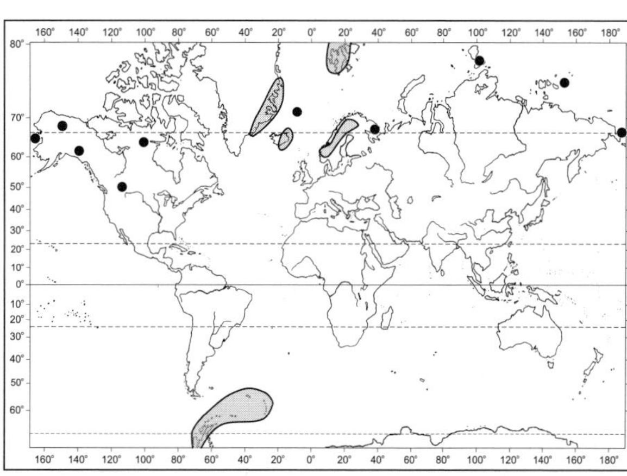

Figure 246. Global distribution of *Sanionia georgicouncinata*.

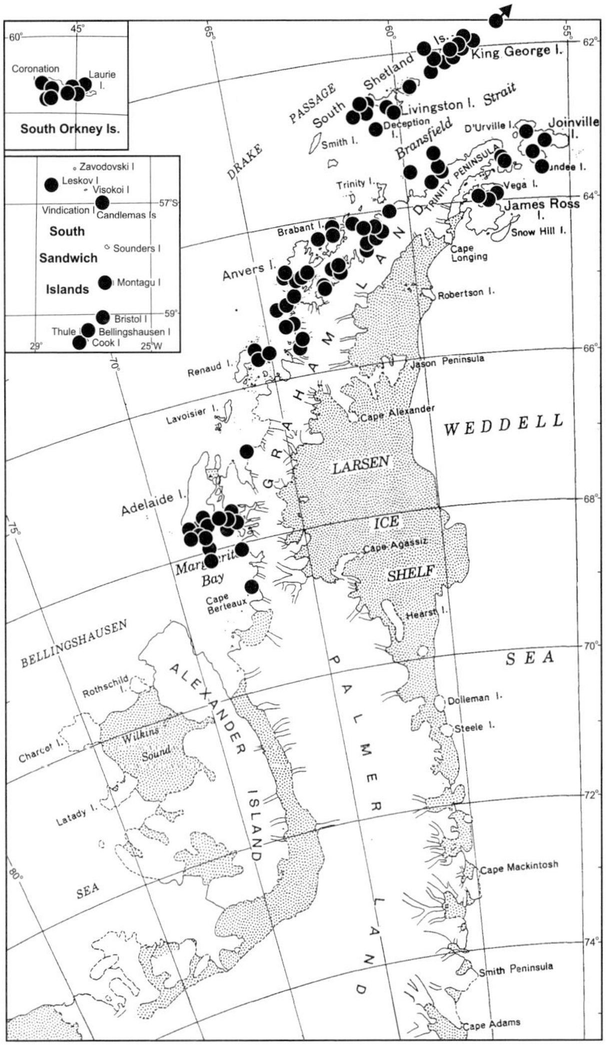

Figure 247. Distribution of *Sanionia georgicouncinata* in Antarctica. Localities on Elephant I. and Gibbs I. in the northern group of the South Shetland Is. are indicated by the arrow.

69-342 (US); Birdsend Bluff, 5–35 m, *RILS 4217A* (AAS, KRAM); Beneden Head, *ER 229C, 230B & 230C* (BR, PC); Bryde I., *RMS 69-364 & 64-393* (US); Paradise Harbour, Coughtrey Peninsula, Almirante Brown Station, *VK 1019* (KRAM), 7 Jan 1985, *Birkenmajer s.n.* (KRAM) and *Leech 420* (NY). **Palmer Archipelago.** ***Christiania Is.***: Intercurrence I., 7 m, *DWB 1C, 2C* (AAS, KRAM). ***Brabant I.***: Cape Roux, 10 m, *AM 10A* (AAS, KRAM); 'Easter Island' north-east of Metchnikoff Pt., 80 m, *AM 71* (AAS, KRAM); Astrolabe Needle, *VK 324 & 394* (KRAM); Driencourt Pt., 10 m, *AM 14B* (AAS, KRAM); Minot Pt., *VK 570* (KRAM). ***Melchior Is.***: Sidders I., *IML 7970, 7978 & 7979* (NY); Lysted I., 6–10 m, *PAS 360.7 & 371.9* (FH); Lambda I., Sevilla Landing, *PAS 348.5* (FH); Bailey I., *PAS 350* (FH); Eta I., Point Andersen, *Pujals 20131* (TENN). ***Anvers I.***: Norsel P., *Lippert 6, 8 & 13* (NY) and *VK 149, 211, 230, 532, 903 & 945a* (KRAM); Cape Monaco, *VK 536* (KRAM); Bonaparte Pt., *VK 30, 170 & 1162* (KRAM); Biscoe Pt., *JRR & JMG PAL–3, 4, 82, 84, 88 & 89* (US); Arthur Harbour, Palmer Station, *IML 8003A & 8074A* (NY); Joubin Is., *JRR & JMG PAL–16, 17 & 19* (US); Litchfield I., *VK 1104* (KRAM); Shortcut I., *VK 60 & 517* (KRAM); Dream I., *VK 575, 675, 699, 704, 906, 924, 704, 935, 944 & 964* (KRAM); Stepping Stone I., *VK 48 & 57* (KRAM); Goudier I., Port Lockroy, *Gain 96* (BM, PC), *97* (PC, S), *99, 100 & 109* (PC), *VK 95* (KRAM) and *RILS 726* (AAS, KRAM); Point Damoy, 10–50 m, 8 Feb 1985, *Birkenmajer s.n.* (KRAM); Bob I., *ER 476* (BR, PC); Wiencke I., *Turquet 338A, 401a, 403b & 405b* (PC); Neumayer Channel, *ER 258A & 259B* (BR, PC). **Graham Coast.** Cape Tuxen, *Gain 199* (PC), *KA 14 & 15* (AAS, KRAM) and 30–300 m, *RILS 4886* (AAS, KRAM); Cape Pérez, *Poncet 5C* (AAS, KRAM). **Wilhelm Archipelago.** ***Lemaire Channel***: entrance to the channel, *ER 268D & 270D* (BR, PC); Booth I., *Turquet 252B* (PC) and *Gain 127* (PC, S), *130B, 133, 137 & 150* (PC), *131* (BM, PC, S) *& 136* (BM, PC, S); Wandel I., *Turquet 12B, 230B, 244B & 671* (PC). ***Penola Strait***: Petermann I., *Lionville 145 & 146* (PC, S) and *Gain 156* (BM, PC), *157, 163, 164A* (PC, S), *166B* (BM, PC, S) *& 175B* (PC); Port Circoncision, *Schmitt 207a* (NY); Penguin I. north-west of Yalour I., *Schmitt 209* (NY). ***Argentine Is***: Easternmost island of the Corner Is., *JRR & JMG PAL–52, 58, 100, 109, 114, 116 & 120* (US); Galindez I., *ca* 3 m, *RILS 940* (AAS, KRAM). ***Collins Bay***: Green I., 5–70 m, *RILS 3369* (AAS, KRAM); Berthelot Is., *Gourdon 180B & 183* (PC); Darboux I., *Poncet 8C* (AAS, KRAM). ***Grandidier Channel***: Takaki Pt., 10–35 m, *RILS 4336 & 4385* (AAS, KRAM); Lahille I., 2–5 m, *RILS 4310* (AAS, KRAM). **Biscoe Is.** Hook I., 15–40 m, *RMS 69-873* (US); Thomsen Is., 3 m, *RILS 5326A* (AAS, KRAM); Pitt Is., 10 m, *RILS 5322A,B* (AAS, KRAM); Trundle I., 35 m, *RILS 5337A* (AAS, KRAM); Snodgrass I., 3 m, *RILS 5344A* (AAS, KRAM); Armstrong Reef of Renauld I., 5 m, *RILS 5341A* (AAS, KRAM); between Watkins I. and Belding I., 15 m, *RILS 5320B* (AAS, KRAM). **Loubet Coast.** Orford Cliff, *RILS 5281A* (AAS, KRAM). ***Laubeuf Fjord***: Webb I., Ryder Bay, *JK 254* (AAS, KRAM). **Fallières Coast.** ***Adelaide I.***: Rothera Pt., *Schmitt 199D* (NY) and 29 Dec 1984, *Birkenmajer s.n.* (KRAM); Reptile Ridge, 205 m, 30 Dec 1984, *Birkenmajer s.n.* (KRAM); Avian I., *JRR & JMG PAL–125, 127, 137, 141, 146, 147, 148, 149 & 150* (US) and *ca* 12 m, *RILS 776* (AAS, KRAM); Cone I., 10 m, *RILS 4885* (AAS, KRAM). ***Léonie Is.***: Léonie I., *Gain 234B* (PC); Lagoon I., *VK 272* (KRAM) and 10 m, *RILS 3597A & 8828B* (AAS, KRAM). ***Marguerite Bay***: Limpet I., *DWB 11* (AAS, KRAM); Jenny I., *Gain 224* (BM, PC, S) and 15–50 m, *RILS 4719* (AAS, KRAM); Guébriant Is., 40 m, *DWB 7B & 8* (AAS, KRAM); Dion Is., Emperor I., 30 m, *DWB 17A* (AAS, KRAM); Piñero I., 10–25 m, *RILS 4825* (AAS, KRAM); Blaiklock I., 8–15 m, *RILS 846* (AAS, KRAM); Pourquoi Pas I., *Poncet 15* (COLO, KRAM); Lagotellerie I., *ca* 50 m, *RILS 1771* (AAS, KRAM); Neny I., 8–17 m, *RILS 791* (AAS, KRAM); Dismal I., 10–25 m, *RILS 3478 & 3529* (AAS, KRAM); Faure Is., island in north-west part, 5–20 m, *RILS 3508* (AAS, KRAM); Millerand I., *RILS 3763* (AAS, KRAM). Terra Firma Is., 12 m, *Borne 218* (AAS, KRAM).

EAST ANTARCTIC PENINSULA. **Joinville Island group.** ***d'Urville I.***: 10 m, *CGB 8* (AAS, KRAM). ***Joinville I.***: Mt. Alexander, 15–50 m, *RILS 3701* (AAS, KRAM); Active Sound, 200 m, *RILS 3734* (AAS, KRAM). ***Paulet I.***: *CS 485* (PC, S). **James Ross Island group.** ***Vega I.***: False Island Pt., 25–30 m, *RILS 7827* (AAS, KRAM). ***James Ross I.***: Lachman Craggs, 35 m, *RILS 7319* (AAS, KRAM); valley north of St. Martha Cove, 70 m, *RILS 7516* (AAS, KRAM).

48. DREPANOCLADUS (Müll.Hal.) G.Roth, *nom. cons.*

Drepanocladus (Müll.Hal.) G.Roth, Hedwigia 38 (Beibl.): 6. 1899. ≡ *Hypnum* Hedw. subsect. *Drepanocladus* Müll.Hal., Syn. Musc. Frond. 2: 321. 1851. ≡ *Hypnum* subgen. *Drepanocladus* (Müll. Hal.) Limpr., Laubm. Deutschl. 3: 362. 1897, *nom. illeg. incl. subgen. prior.* — Type: *Drepanocladus aduncus* (Hedw.) Warnst. (*Hypnum aduncum* Hedw.), *typ. cons.*

Plants variable in size, medium-sized to relatively large and robust, in loose or dense, soft to fairly stiff mats or tufts, light or dark green, yellow-green, yellowish or brownish. *Stems* prostrate and spreading or crowded and ascending to suberect, sparingly and irregularly to regularly subpinnately branched in ± one plane, in transverse section lacking a hyalodermis, with a small but distinct central strand, medulla of large, hyaline and thin-walled cells and cortex of small and thick-walled cells; *rhizoids* scattered on the stem at or just below the leaf insertion, brownish, slightly or sometimes profusely branched, smooth; *pseudoparaphyllia* small, foliose, triangular to broadly ovate; *paraphyllia* absent; *axillary hairs* well developed, abundant, 1–2(–4)-celled, hyaline or yellowish-orange to pale brownish with age, composed of a short, quadrate to short-rectangular, mostly hyaline or yellow basal cell and elongate upper cells. *Stem* and *branch leaves* similar, the latter only somewhat smaller, crowded or spaced, straight to falcate-secund, erect to erecto-patent, concave, not or hardly decurrent, non-plicate or striolate when dry, narrowly lanceolate to ovate-lanceolate or ovate, from a broadly ovate, ovate-triangular, cordate-ovate to oblong-lanceolate base gradually or abruptly tapering to a short or long, sometimes squarrose acumen; *margins* plane, entire or somewhat sinuate below; *costa* single, relatively weak to strong, ending in the mid-leaf to long-excurrent, biconvex in cross-section; *laminal cells* smooth, unistratose or occasionally with bistratose streaks, transparent; *upper* and *median cells* linear-flexuose to oblong-rhomboidal, with thin to relatively thick, smooth or pitted walls; *angular cells* clearly differentiated, quadrate or rectangular, hyaline and thin-walled to yellowish and thick-walled forming a distinct, inflated, not or hardly decurrent, transversely triangular to ovate-triangular group, extending to the costa or occupying ½–⅓ the width of the leaf base. *Dioicous* or occasionally *autoicous*. *Inner perichaetial leaves* straight and erect, ovate or ovate-lanceolate, plicate, unicostate, with entire margins. *Setae* elongate, smooth, red to reddish-brown becoming light brown with age, straight or flexuose, untwisted or laxly sinistrorse above. *Capsules* cylindrical, curved, inclined to horizontal, not or constricted below the mouth when dry and empty; *operculum* convex-conic, with a short, straight beak; *annulus* present, 2–3-seriate, separating; *exothecial cells* mostly rectangular to oblong-hexagonal, moderately thick-walled, becoming oblate, darker coloured in 3–4 tiers at the orifice; *stomata* numerous in the base of the urn, phaneroporous, composed of two reniform cells surrounding the elongate pore; *peristome* double; *exostome teeth* 16, yellowish-brown, cross-striolate or dotted-striolate below, papillose above, with entire or weakly dentate margins; *endostome* with a high, smooth or minutely papillose, hyaline or yellowish basal membrane, giving rise to entire or narrowly perforated processes and (2–)3–4 nodose and finely verrucose cilia. *Spores* globose, yellow-brown to pale brown, slightly and minutely papillose. *Calyptra* cucullate, naked, smooth.

In older bryological treatments *Drepanocladus* has been interpreted as a large genus encompassing a number of hydrophytic species with more or less falcato-secund leaves (Warnstorf 1903). As such, it was an artificial genus comprising several unrelated complexes of species. Loeske (1907*a*,*b*, 1910) introduced the basis for the division of the 'old' *Drepanocladus* into the five genera *Sanionia* Loeske, *Limprichtia* Loeske, *Pseudocalliergon* (Limpr.) Loeske, *Warnstorfia* Loeske and *Drepanocladus s. str.*, but his taxonomic concepts have long been neglected by bryologists. Later, this concept was further expanded and refined by Paul (1916), Tuomikoski & Koponen (1979) and Hedenäs (1989*a*,*b*, 1992, 1993*b*, 2003*a*, 2006) and another three genera have been added to this complex, namely *Loeskypnum* H.K.G.Paul, *Sarmentypnum* Tuom. & T.J.Kop. and *Hamatocaulis* Hedenäs. Most of these are smaller but better defined segregates and now seem to have gained wide acceptance. Because of the great uniformity of sporophyte characters in the Hypnales, they differ mostly in major gametophyte features including shape of leaves and alars cells, branching pattern, plant coloration, axillary hairs and leaf margins. It should be noted that *Drepanocladus* (Müll.Hal.) G.Roth was proposed for conservation by Isoviita & Ochyra (1990) because it was antedated by two earlier names, *Drepano-hypnym* Hampe and *Drepanocladus* Müll.Hal., the latter name being synonymous with *Sanionia.*

Drepanocladus s. str. consists of seven species centered around *D. aduncus* (Hedw.) Warnst. and *D. sendtneri* (Schimp.) Warnst. which correspond to two sections recognised within the genus, namely sect. *Drepanocladus* which was monographed by Żarnowiec (2001) and sect. *Sendtneri* (G.Roth) Ochyra & Żarnowiec which was overviewed by Hedenäs (1998*b*). They are widely distributed throughout much of the Northern Hemisphere with numerous disjunct and isolated outposts in the tropics and Southern Hemisphere. In the Antarctic the genus is represented by a single species, *D. longifolius*, whereas in Tierra del Fuego another species, *D. polycarpos* (Voit) Warnst., is occasionally found (Ochyra & Matteri 1996*b*, 2001). The latter is also known from the Prince Edward Is. and Macquarie I. and a third species, *D. aduncus*, has been recorded from Îles Kerguelen and Macquarie I. (Żarnowiec 2001).

1. **Drepanocladus longifolius** (Wilson *ex* Mitt.) Broth. *ex* Paris — Fig. 248

Drepanocladus longifolius (Wilson *ex* Mitt.) Broth. *ex* Paris., Coll. Nom. Broth.: 10. 1909. ≡ *Hypnum longifolium* Wilson, Bryol. Brit.: 391. 1855, *hom. illeg.* [non *H. longifolium* Schultz 1806]. ≡ *Amblystegium longifolium* Wilson *ex* Mitt., J. Linn. Soc. Bot. **12**: 571. 1869. — TYPE CITATION: Ins. Falkland Islands, J. D. Hooker, Lechler, n. 96. [Lectotype (*vide* Ochyra *in* Sastre-De Jesús 1992: p. 57): 'H. longifolium Wils. in part Falkland Islands Dr Hooker H. falcatum H. longifol. Falklands Lyall № 47/W. 228 Hypnum aduncum ~~uncinatum~~ H. near falcatum Brid. H. longifolium Mitt.' – NY-Mitten!; isolectotypes: BM-Hampe!, BM-Wilson! (5 specimens), E!, H-Lindberg!, MANCH-Spruce!; syntype: '96. Musci Hypnum fluitans Hedw. Ost Falkland Inseln Port William Sept. 50' – BM-Hooker!].

Plants medium-sized, in relatively rigid, lustrous, dense, interwoven and intricate, floating mats, dark or yellowish-green to golden-brownish. *Stems* prostrate, elongate, wiry, 5–15 or to 20 cm, irregularly and remotely branched, often appearing bristly because of the denuded, stout costae, in transverse section rounded with

Figure 248. *Drepanocladus longifolius.* — **1**. Habit. **2–3**. Portions of branches, dry. **4**. Portion of cross-section of stem. **5–9**. Axillary hairs. **10–15**. Leaves. **16–17**. Pseudoparaphyllia. **18**. Lamina cells in leaf subula. **19**. Upper cells at margin. **20**. Mid-leaf cells. **21–22**. Alar cells. **23**. Cross-section of costa at base. **24–26**. Leaf cross-sections. — [1, 4-9, 13–20, 23–26 from *RILS 3830B*; 2–3, 10–12, 21–22 from *RBH 28*]. — Scale bars: a – 2 mm (10–15) & 100 µm (5–9); b – 100 µm (23–26); c – 100 µm (4, 18–22); d – 0.5 cm (1–2); e – 100 µm (16–17); f – 1 mm (3).

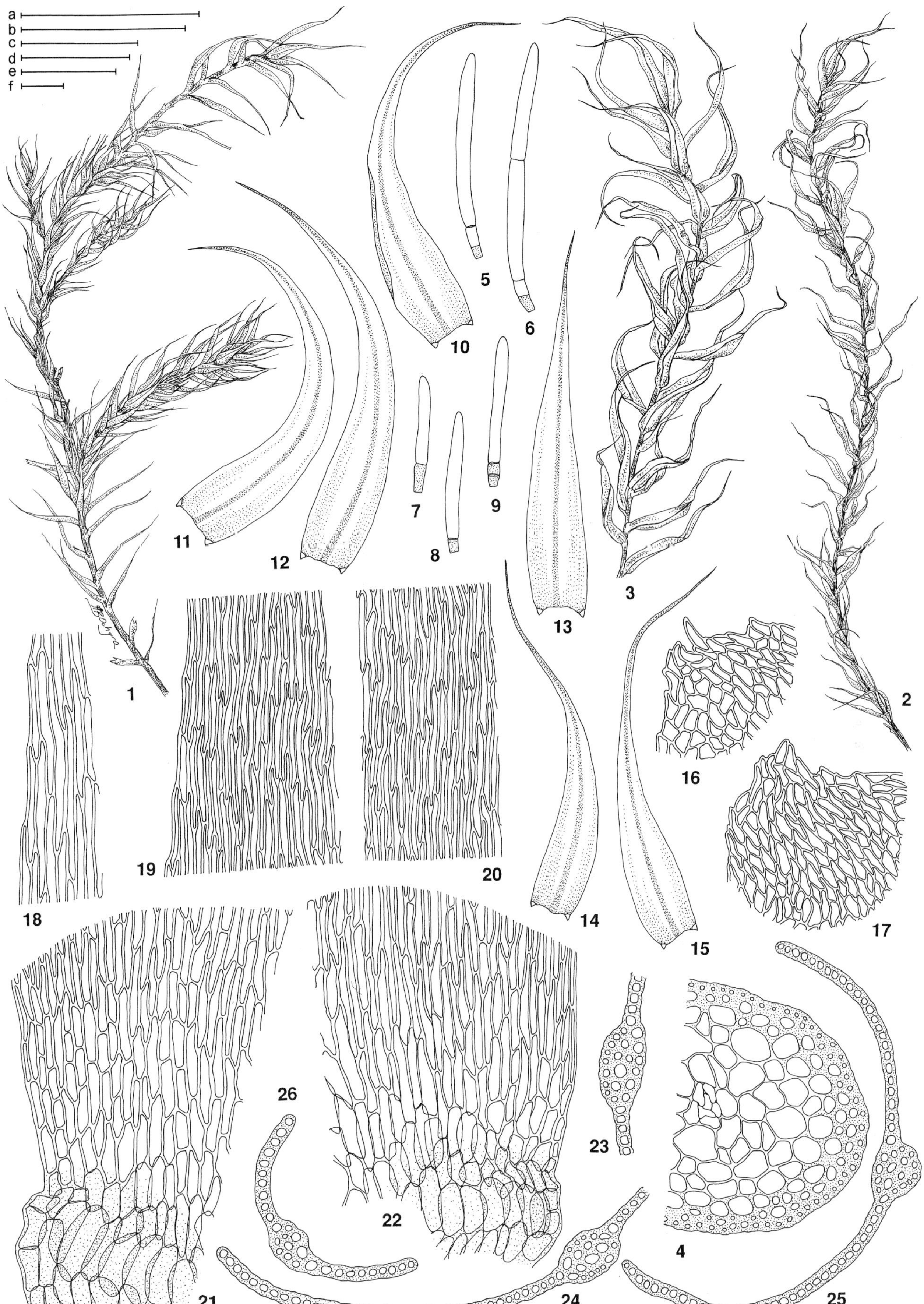
a
b
c
d
e
f
1
2
3
4
5
6
7
8
9
10
11
12
13
14
15
16
17
18
19
20
21
22
23
24
25
26

a small central strand, 4–6-stratose medulla of large, hyaline and firm-walled cells and 2–3-stratose cortex of small, thick-walled, brown cells; *rhizoids* not observed in floating forms, normally rather infrequent, mostly in the lower part of stem, brownish, smooth, branched; *pseudoparaphyllia* small, foliose, triangular to irregular in outline; *axillary hairs* abundant, 2–4-celled with 1–2 small, quadrate, basal cells and elongate, hyaline or light brownish upper cells. *Stem* and *branch leaves* similar, spaced, straight to falcate or strongly so, often altered when dry, not striolate, concave, non-decurrent, narrowly lanceolate, gradually narrowed to a long-acuminate or filiform acumen, erecto-patent to patent, 3.5–5.0 mm long, 0.6–0.9 mm wide; *margins* plane, entire or sinuate above the base; *costa* strong, intensely yellow, biconvex in cross-section, short to long excurrent, 70–100(–110) μm wide at the base; *laminal cells* unistratose or sometimes with bistratose streaks (beyond Antarctica), transparent, smooth; *upper* and *median cells* linear to linear-flexuose, tapering at both ends, rarely oblique, 4–7(–9) μm wide, 60–120(–150) μm long, thin- to moderately thick-walled with smooth or pitted walls; *basal cells* shorter and wider, oblong-rhomboidal to rectangular, 30–50 μm long, 8–15 μm wide, mostly with thick, yellowish and slightly porose walls; *angular cells* rectangular, with thin to thick walls, forming small, inflated, yellowish to hyaline, ovate to ovate-triangular, not or hardly decurrent groups, clearly separated from the adjacent cells, occupying about ⅓ or less of the leaf base. [*Dioicous. Inner perichaetial leaves* erect and straight, ovate-lanceolate, gradually or rather abruptly long acuminate, often filiform, 3–5 mm long, plicate, with a single and excurrent costa and entire margins. *Setae* elongate, 3–6 cm long, reddish-brown, smooth, straight to flexuose. *Capsules* cylindrical, 3–5 mm long, asymmetric and curved to symmetric and almost straight, inclined to erect, slightly constricted below the mouth when dry and empty, brown, with a distinct neck; *operculum* convex-conic, with a short, stout, slanted beak; *exothecial cells* rectangular, 20–75 μm long, 12–30 μm wide, relatively thick-walled, becoming subquadrate below the orifice; *stomata* in the lower part of the urn, superficial, bicellular, long-pored; *exostome teeth* 16, lanceolate, yellowish-brown, dotted-striolate below, papillose above; *endostome* with a high, hyaline to yellowish, smooth to finely papillose basal membrane, giving rise to narrow, entire to narrowly cleft processes separated by 1–2 nodose cilia. *Spores* spherical, finely papillose, 13–17 μm wide. *Calyptra* cucullate, naked.]

Chromosome number: unknown.

Discussion. *Drepanocladus longifolius* is a very distinct species in the Antarctic being the only pleurocarpous moss with lanceolate and long-acuminate leaves having a strong and percurrent to long-excurrent costa. Because of deep submergence, the local plants are morphologically highly modified in comparison with the typical phenotypes from the South American mainland which grow in swampy habitats, but only seldom submerged. They have strongly falcato-secund leaves which are often circinate at the tips of the stem and branches. However, the submerged plants retain the characteristic shape of the alar cells which form small, convex, rounded to triangular groups occupying about ⅓ or less of the leaf base. In the Antarctic *D. longifolius* grows in intricate mats together with highly modified plants of *Warnstorfia sarmentosa* characterised by having oblong-lanceolate and acuminate leaves. Plants with these modifications show external convergence, but in *W. sarmentosa* the costa always ceases well below the apex.

The discovery of *Drepanocladus longifolius* is generally attributed to Mitten (1869) who described it as *Amblystegium longifolium*. However, the species was actually recognised for the first time by Wilson (1855) as *Hypnum longifolium* on the basis of the specimen collected on the Falkland Is. during the British Antarctic voyage of 1839–1843 and reported by Wilson & Hooker (1847) as *Hypnum falcatum* Brid. Because *H. longifolium* Wilson is a later illegitimate homonym, this name could not stand, but Mitten (1869) adopted the epithet *longifolium* for this species under the generic name *Amblystegium* and cited the specimen collected by J. D. Hooker from the Falkland Is., adding one more specimen from the same archipelago gathered by W. Lechler. The species was subsequently shifted to *Drepanocladus* by Paris (1909), although Williams (1916) overlooked this transfer and superfluously proposed the new combination *D. longifolius* (Mitt.) R.S.Williams.

Dixon (1929) assessed *Drepanocladus longifolius* taxonomically and considered it to be identical to the New Zealand *D. brachiatus* (Mitt.) Paris, and the conspecificity of these species was subsequently adopted by Smirnova (1956). However, this conclusion is unacceptable because the two species differ markedly in the form of their costae, *D. brachiatus* having the costa ceasing far below the apex, whereas in *D. longifolius* the costa is excurrent as a stout point.

Hedenäs (1997*b*, 2003*b*) accepted *Drepanocladus longifolius* as a distinct species, but at the same time, markedly modified its concept by merging with it *D. capillifolius* (Warnst.) Warnst. from the Northern Hemisphere, and this concept has gained wide acceptance (Hill *et al.* 2006). Unfortunately, this concept cannot be defended because the two species differ essentially in a number of characters (Ochyra & Matteri 2001; Żarnowiec 2001), the most important of which is the form of the alar cells. In *D. longifolius* they form a small oval auriculate group, gradually separated from the adjacent basal cells and which occupies about ⅓ the leaf insertion, clearly indicating the relationship of the species to sect. *Sendtneri* of the genus. In contrast, in *D. capillifolius* the alar cells form a large transversely triangular and sharply demarcated group reaching the costa or nearly so, and this character clearly indicates the relationship of this species with the type section of *Drepanocladus*. In addition, the two species differ totally in their habits, *D. longifolius* being a large plant with strongly falcato-secund leaves, whereas *D. capillifolius* is usually a medium-sized plant with straight or nearly straight leaves. It is unlikely that these two species will be confused in the field. The Antarctic plants are a little atypical and have straight leaves, but this modification is certainly a response to permanent submergence.

Reproduction in Antarctica. All Antarctic plants are sterile, but in general this species fruits very seldom and hitherto only a few plants with mature sporophytes have been detected in Tierra del Fuego (Ochyra & Matteri 2001) and in the Falkland Is. (type material).

Habitat. In Antarctica *Drepanocladus longifolius* grows exclusively submerged in water at depths of 4–10 m forming intricate mats over silt and mud in oligotrophic lakes on Signy I.

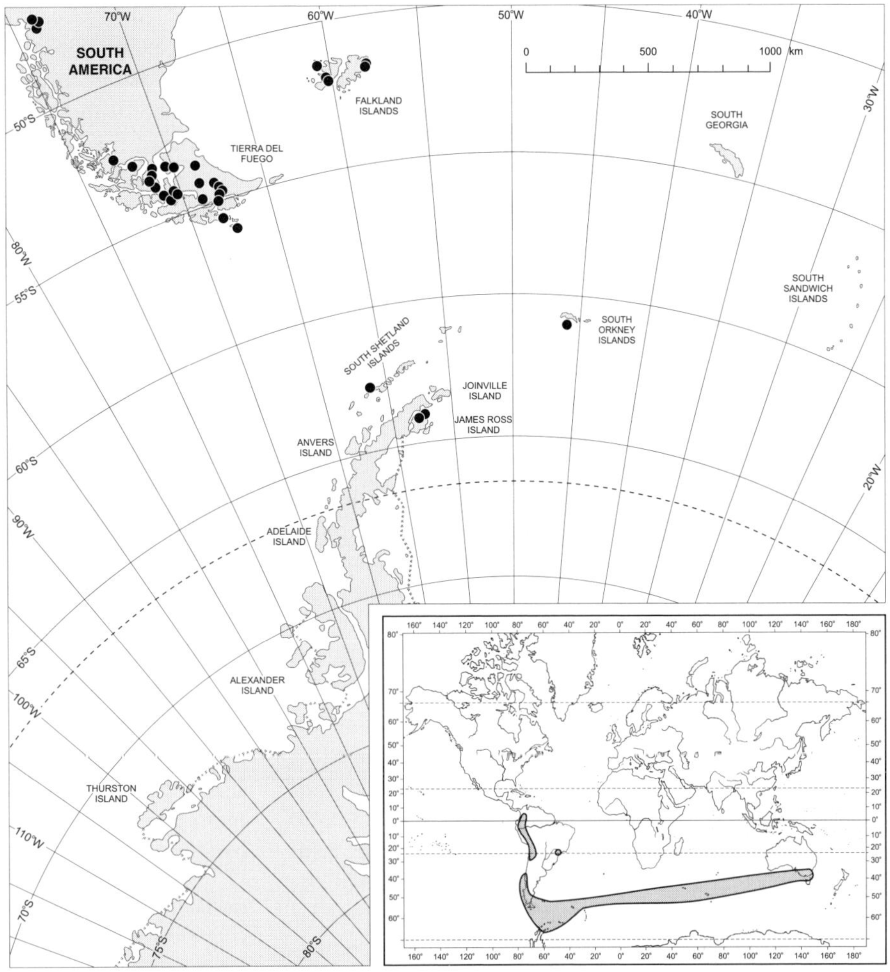

Figure 249. Distribution of *Drepanocladus longifolius* in the Fuegian region and Antarctica. Inset: global distribution.

(Heywood 1967, 1968; Priddle 1979). It grows in a similar lake on Livingston I. where it also occurs in masses washed up along the shorelines. It is most often associated with *Warnstorfia sarmentosa* and, on Livingston I., it also grows with the liverwort *Pachyglossa dissitifolia* Herzog & Grolle. On James Ross I. the species was found as a sub-fossil, with a radiocarbon age of 9525 years B.P. in a deep sedimentary deposit of a former lake (Ingólfsson *et al.* 1992 as *Drepanocladus* cf. *aduncus*).

World range. An austral pan-temperate species with the main centre of occurrence in southern South America, including western and southern Patagonia, Tierra del Fuego and the Falkland Is. (Ochyra & Matteri 2001), and extending along the Andean chain at high elevations from Bolivia to Colombia (Hedenäs 2003*b*), then appearing on some subantarctic islands including South Georgia (Ochyra *et al.* 2002), Îles Crozet (Ochyra, personal observations) and Îles Kerguelen (Hedenäs 1997*b*; Ochyra, personal observations) and extending to south-eastern Australia and the northern maritime Antarctic (Fig. 249, inset).

Distribution in Antarctica. Very rare and occasional in a few lakes on Signy I. in the South Orkney Is., on Livingston I. in the South Shetland Is. and Vega I. off the eastern coast of Trinity Peninsula; also recorded from sub-fossil deposits on James Ross I. on the eastern coast of the Antarctic Peninsula (Fig. 249).

Specimens examined. WEST ANTARCTICA. SOUTH ORKNEY IS. **Signy I.** Moss Lake, *BAS Misc. 28* (AAS, KRAM) and *RILS 1820, 1821 & 1824* (AAS, KRAM); Heywood Lake (sediment), *RILS 10365B, C, E & F* (AAS).

SOUTH SHETLAND IS. **Livingston I.** ***Byers Peninsula***: Chester Cone, 30 m, *RILS 3830B & 3831B* (AAS, KRAM).

EAST ANTARCTIC PENINSULA. **James Ross Island group.** ***Vega I.***: Pastorizo Bay, 50 m, 0–3 m depth, *RILS 7311* (AAS, KRAM). ***James Ross I.***: Cape Lachman, 10 m, *RILS 7309* (AAS, KRAM) [sub-fossil collection]; 2 km south-west of Cape Lachman, 30 m, *RILS 7308* (AAS, KRAM).

49. WARNSTORFIA Loeske

Warnstorfia Loeske, Hedwigia 46: 310. 1907. ≡ *Drepanocladus* (Müll.Hal.) G.Roth sect. *Warnstorfia* (Loeske) Broth. *in* Engl. & Prantl, Nat. Pflanzenfam. 1(3): 1034. 1908. ≡ *Drepanocladus* subgen. *Warnstorfia* (Loeske) Tuom., Ann. Bot. Soc. Fenn. Vanamo 23: 10. 1949. — Lectotype (*vide* Ochyra *et al.* 1991: p. 101): *Warnstorfia fluitans* (Hedw.) Loeske (*Hypnum fluitans* Hedw.).

Sarmentypnum Tuom. & T.J.Kop., Ann. Bot. Fenn. 16: 223. 1979. ≡ *Calliergon* (Sull.) Kindb. subgen. *Sarmentypnum* (Tuom. & T.J.Kop.) Zanten & J.K.Bartlett, Cryptog. Bryol. Lichénol. 7: 32. 1986. ≡ *Warnstorfia* Loeske subgen. *Sarmentypnum* (Tuom. & T.J.Kop.,) Ochyra, Cens. Cat. Polish Mosses: 180. 2003. — Type: *Sarmentypnum sarmentosum* (Wahlenb.) Tuom. & T.J.Kop. (*Hypnum sarmentosum* Wahlenb.). First synonymised by Hedenäs (1993*b*: p. 464).

Plants medium-sized to rather large and robust in soft, shiny mats, green, yellow-green, yellowish, brownish, reddish- or yellow-brown or often clearly red. *Stems* prostrate to erect-spreading, sparsely to richly, irregularly to regularly, often pinnately, radially branched, rarely almost unbranched, in cross-section with a small central strand, medulla of large, hyaline, thin-walled cells and well or poorly developed cortex of moderately thick-walled cells, sometimes with an imperfectly developed hyalodermis; *rhizoids* smooth, reddish-brown, slightly branched, scattered along the stems, sometimes inserted at various points on the leaves, especially near the apex or along the upper part of the costa; *pseudoparaphyllia* foliose, narrow to broad; *paraphyllia* lacking; *axillary hairs* abundant, strong or weak, with 1–7 elongate, hyaline upper cells and 1–2 short, brown basal cells. *Stem* and *branch leaves* similar, straight or falcate, triangular, ovate, ovate-lanceolate, lanceolate to oblong-lanceolate, gradually long-acuminate or abruptly tapering to a shortly acuminate or broadly rounded and mostly recurved-apiculate, plane or cucullate apex, concave or deeply concave to almost plane, not plicate, not or distinctly decurrent; *margins* plane, sharply or bluntly serrulate to denticulate or entire to sinuose; *costa* single, weak or robust, ending just above mid-leaf to longly excurrent, simple or occasionally branched; *laminal cells* unistratose throughout, transparent, smooth; *upper* and *median cells* linear, straight to flexuose, with thin- or thick and often pitted walls; *nematogon cells* present just below the apex, hyaline, shorter and broader than the surrounding laminal cells, rhomboidal to oblong-rhomboidal; *basal cells* wider and shorter, more strongly incrassate and porose; *angular cells* quadrate or elongate, mostly inflated to strongly inflated, hyaline and becoming incrassate with age forming transversely triangular, quadrate or ovate groups reaching the costa or extending half the distance between the margin and the costa, not or sharply set off from the adjacent lamina cells; *supra-alar cells* undifferentiated or well differentiated. *Autoicous* or *dioicous*. *Inner perichaetial leaves* straight and erect, lanceolate to ovate-lanceolate, gradually or suddenly narrowed to a long or short acumen, not plicate, with a single costa extending to the acumen or long-excurrent. *Setae* elongate, reddish, smooth, straight or flexuose. *Capsules* cylindrical, curved, inclined to horizontal; *operculum* convex-conic, short-beaked; *annulus* persistent; *exothecial cells* short to elongate-rectangular with more or less incrassate walls; *stomata* with long pores, bicellular, superficial, placed in the neck of the capsule; *exostome teeth* 16, yellow-brown, dotted and reticulate below, papillose above on the dorsal surface, strongly trabeculate on the ventral face, dentate at the margins; *endostome* well developed, papillose or finely papillose above, more finely papillose below, with entire or narrowly perforate processes and 2–3(–5) nodose cilia arising from a high basal membrane. *Spores* spherical, finely papillose. *Calyptra* cucullate, naked, smooth.

Warnstorfia is here interpreted in the broad sense as outlined by Hedenäs (1993*b*) to include *W. sarmentosa*, a species which was segregated by Tuomikoski & Koponen (1979) into the monotypic genus *Sarmentypnum*. In its typical expression this species differs from other congeners in its entire leaf margins and rounded leaf apex, but aquatic and submerged expressions of this moss have distantly minutely serrulate leaf margins and the leaf apices become more long-acuminate. As a whole *Warnstorfia* is very diversified gametophytically and somewhat heterogeneous and difficult to circumscribe clearly, but the red colours occurring in most species under normal growth conditions, the serrulate leaf margins, triangular, ovate to lanceolate, acuminate to rounded-apiculate leaves and the distinct alar cells forming triangular or ovate, mostly decurrent groups are distinctive characters of this genus.

Recently, based on molecular data, Hedenäs (2006) resurrected the genus *Sarmentypnum* and transferred to it seven species which have traditionally been associated with *Warnstorfia*, leaving the latter as a small genus consisting of three autoicous species, namely *W. fluitans*, *W. fontinaliopsis* (Müll.Hal.) Ochyra and *W. pseudostraminea* (Müll.Hal.) Tuom. & T.J.Kop. Segregation of these two genera is weakly supported morphologically since the sex condition alone does not seem to be a strong character at the generic level. There are many examples of moss genera comprising both autoicous and dioicous species, sometimes forming pairs of sibling taxa.

As interpreted here, *Warnstorfia* is a small genus consisting of nine species which grow predominantly in wet or aquatic habitats in the Northern Hemisphere. Eight of these are Holarctic taxa and only one, *W. fontinaliopsis* is an austral species. Three of the northern species are bipolar in distribution and additionally occur in tropical mountains at high elevations. Two species are widely distributed in the maritime Antarctic.

1. Leaves oblong-ovate to oblong-lanceolate; leaf apex rounded, normally shortly apiculate and cucullate or ± abruptly or shortly acuminate; alar cells hyaline or reddish-tinged forming sharply delimited, ovate to transversely triangular, decurrent groups .... 1. ***W. sarmentosa***
1. Leaves broadly ovate to ovate-lanceolate, gradually long acuminate; leaf apex acute plane; alar cells chlorophyllose forming indistinct, non-decurrent groups gradually passing into the laminal cells .... 2. ***W. fontinaliopsis***

1. **Warnstorfia sarmentosa** (Wahlenb.) Hedenäs

Fig. 250

Warnstorfia sarmentosa (Wahlenb.) Hedenäs, J. Bryol. 17: 470. 1993. ≡ *Hypnum sarmentosum* Wahlenb., Fl. Lapp.: 380. 1812. ≡ *H. trifarium* F.Weber & D.Mohr subsp. *sarmentosum* (Wahlenb.) Hampe, Flora 20: 275. 1837. ≡ *H. trifarium* F.Weber & D.Mohr var. *sarmentosum* (Wahlenb.) Rabenh, Deutschl. Krypt. Fl. 2(3): 290. 1848. ≡ *Amblystegium sarmentosum* (Wahlenb.) De Not., Cronac. Briol. Ital. 2: 23. 1867. ≡ *Calliergon sarmentosum* (Wahlenb.) Kindb.,

Can. Rec. Sc. 6(2): 72. 1894. ≡ *Acrocladium sarmentosum* (Wahlenb.) Rich. & Wall., Trans. Br. Bryol. Soc. 1(4): 25. 1950. ≡ *Sarmentypnum sarmentosum* (Wahlenb.) Tuom. & T.J.Kop., Ann. Bot. Fennici 16: 223. 1979. — TYPE CITATION: In rupibus alpinis uni aqua nivalis defluit passim. [Lectotype (*vide* Hedenäs 1993*b*: 470): 'Hypnum sarmentosum Flor. Lapp. Norska Nordland pa fjället Krönan vid Rorstad. Wahlenberg scripsit et misit, Juni 1807' – S-Swartz!; isotypes: S-Swartz! (2 specimens)].

Calliergon luipichense R.S.Williams, Bull. New York Bot. Gard. 6: 259. 1909. ≡ *Warnstorfia luipichensis* (R.S.Williams) Hedenäs, J. Bryol. 17: 473. 1993. ≡ *Sarmentypnum luipichense* (R.S.Williams) Hedenäs, J. Hattori Bot. Lab. 100: 132. 2006. — TYPE CITATION: [Bolivia] Growing in a small pond at 4300 meters elevation near the Luipichi Pass, Sept. 9, 1901 (2021). [Lectotype (*selected here*): 'Plants of Bolivia collected by R. S. Williams 1901–1902 Calliergon luipichense sp. nov. RSW. Near Luipichio Pass, 4300 metres, in pond No 2021 Sept. 9. 1901' – NY!; isotypes: H-Brotherus!, JE-Herzog!]. First synonymised with *Calliergon sarmentosum* by Karczmarz (1971: p. 126)].

Plants medium-sized to relatively large and robust, lustrous, in compact or loose mats or tufts, pale or yellow-green, yellow-brown, dark brown, orange to purplish-red. *Stems* erect or prostrate, flexible, 5–15 cm long, occasionally in floating and submerged forms to 25 cm, simple or sparingly, freely and irregularly branched, often terete, in transverse section composed of a 2–3-stratose cortex of smaller, thick-walled, brown cells surrounding 3–5 layers of large, hyaline, thin-walled medullary cells, central strand small but distinct; *rhizoids* lacking or few, light brown, smooth; *pseudoparaphyllia* foliose; *axillary hairs* filiform, 4–5-celled, consisting of 1–2 short, brown basal cells and 2–3 hyaline, elongate upper cells. *Stem* and *branch leaves* similar, crowded or spaced in submerged form, (1.0–)1.5–2.5(–3.8) mm long, 0.8-1.0 mm wide, patent, erecto-patent or loosely appressed, becoming closely imbricate at the tips of the stems and branches, resulting in their julaceous and cuspidate appearance, concave, auriculate and decurrent, oblong-ovate, oblong to oblong-lanceolate, rounded and cucullate at the apex and usually shortly reflexed-apiculate or the apiculus lacking on the older leaves, sometimes gradually narrowed towards the acuminate or acute apiculate to rounded-apiculate, convolute apex; *margins* entire or weakly serrulate near the apex, sometimes minutely serrulate below in submerged forms; *costa* single, well-defined, 60–75 µm wide at base, extending ¾ or more way up the leaf; *laminal cells* unistratose throughout, pellucid, smooth; *upper* and *median* cells linear-flexuose, mostly terete at the apices, (30)45–90(–120) µm long, 4–6(–8) µm wide, 6–13 times as long as wide, firm- to thick-walled, often porose; *nematogon cells* few, usually present below the apex, colourless, quadrate or rhombic; *basal cells* wider and shorter, 20–35 µm long, 8–10 µm wide, often with incrassate and porose walls and with yellow or reddish-brown tinge; *angular cells* thin-walled, large, hyaline or reddish-tinged, forming decurrent, inflated, elongate-elliptic decurrencies sharply set off from the adjacent basal cells and extending to the middle of the base, sometimes the cells between the costa and auricles thiner-walled and non-porose, giving an impression of the alars extending to the costa. [*Dioicous. Perichaetial leaves* imbricate and closely appressed, concave, sheathing the seta, oblong, abruptly acuminate or apiculate, entire. *Setae* elongate, 3.5–4.5 cm long, reddish-orange below, paler above, glabrous throughout, straight or flexuose. *Capsules* cylindrical, 2.5–3.2 mm long, *ca* 1 mm wide, inclined to suberect, curved, asymmetric, brown, contracted below the mouth; *operculum* conic-convex, rostrate, exannulate; *exothecial cells* hexagonal, firm-walled, somewhat collenchymatous, becoming short and isodiametric in 3–4 rows below the orifice; *stomata* bicellular, superficial, with a long pore; *exostome teeth* 16, yellow, cross-striolate and reticulate below, papillose and dentate above; *endostome* with a high, pale yellowish basal membrane giving rise to narrow, entire or narrowly cleft segments and 1–3 nodose cilia. *Spores* globose, 16–20 µm wide, green, finely papillose. *Calyptra* cucullate, naked.]

Chromosome number: $n = 11$ – King George I. (Kuta *et al.* 1982).

Discussion. *Warnstorfia sarmentosa* is readily recognised because plants are frequently tinged with orange, brown or red and have cuspidate apices to the stems and branches and oblong-lanceolate leaves which are mostly cucullate and abruptly narrowed to rounded- or acute-apiculate apices. Like all mosses associated with wet habitats *W. sarmentosa* is very variable in size and habit. For this reason, modern moss taxonomy is prejudiced against recognition of multitudinous expressions at infraspecific levels. Karczmarz (1971) accepted nine varieties and forms of this species, none of which was based on characters that would be clearly independent from environmental influences (degree of spacing of leaves, their width, shape of leaf apices, branching pattern, etc.) and their scattered distributions over a very wide geographical range give no hint of evolutionary distinction.

One environmental expression from the Bolivian Andes was described as a separate species, *Calliergon luipichense* (Williams 1909). Although this species name was reduced to synonymy with *C. sarmentosum* (Karczmarz 1971), the plants so-named are still considered as a distinct species by Hedenäs (1993*b*, 2003*b*, 2006) who maintains that it differs from the closely related *Warnstorfia sarmentosa* in its leaves gradually or rapidly tapering to acuminate, acute-apiculate or rarely rounded-apiculate apices that have at least partially denticulate leaf margins. The leaf apices vary considerably in *W. sarmentosa*, even on the same shoot, from rounded- or acute-apiculate to long-acuminate, and phenotypes with distinctly acuminate leaves were given taxonomic recognition as varieties (*Hypnum sarmentosum* var. *beringianum* Cardot & Thér., var. *acuminatum* Bryhn and var. *fontinaloides* Berggr.) which are not accepted by modern moss taxonomy. Likewise, minutely and remotely serrulate leaf margins are observed in various populations of *W. sarmentosa*, including those in the Antarctic (Fig. 282, f. 25–26, 31). However, as with other species of the genus *Warnstorfia*, there is a remarkable tendency to reduction of the serrulation of leaf margins in *W. sarmentosa*. Therefore, recognition of a species using such highly plastic and easily modified characters is very hazardous, and certainly *W. luipichensis* does not deserve specific distinction, as already suggested by Karczmarz (1971), and the identity of this species with *W. sarmentosa* is here confirmed.

The majority of Antarctic plants are rather slender and pale- to yellowish-green, but reddish-tinged forms are sometimes

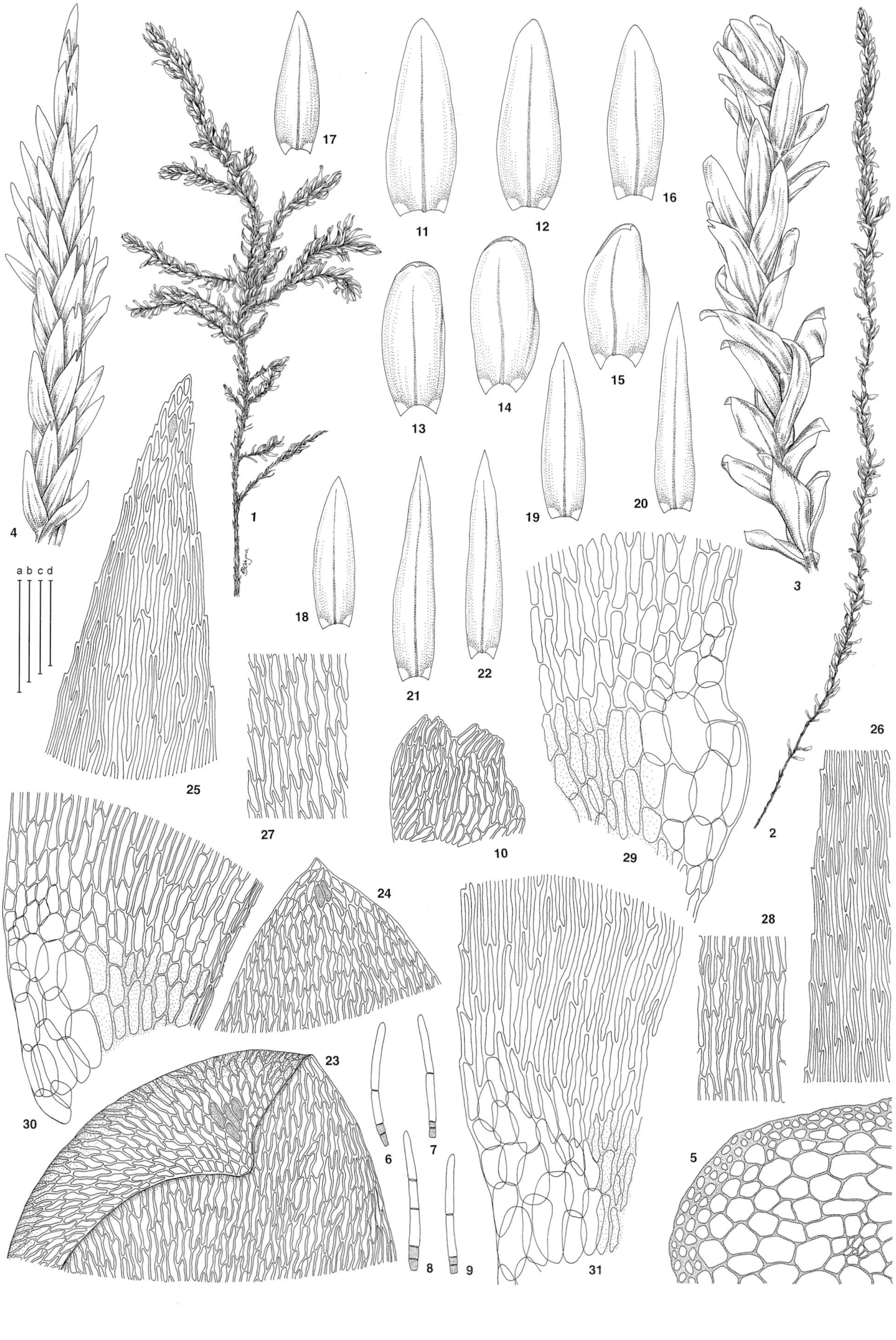
1
2
3
4
5
6
7
8
9
10
11
12
13
14
15
16
17
18
19
20
21
22
23
24
25
26
27
28
29
30
31
a b c d

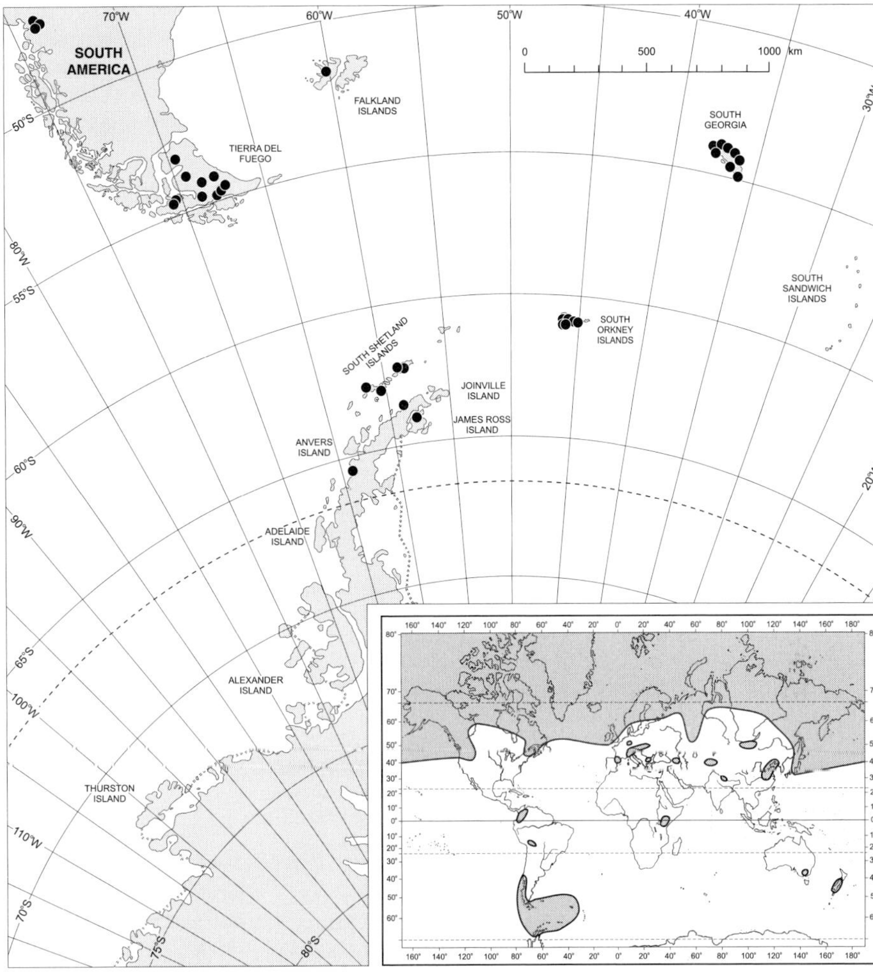

Figure 251. Distribution of *Warnstorfia sarmentosa* in the Fuegian region and Antarctica. Inset: global distribution.

found. A few populations occur at considerable depth in lakes, and these plants have more acuminate leaf apices which become additionally convolute on drying and the auricles extend to the costa because of the thin-walled and non-porose basal cells.

In the Antarctic *Warnstorfia sarmentosa* grows frequently with *W. fontinaliopsis*, and they adopt a similar overall appearance. However, they can be readily distinguished from each other in the field with use of a hand-lens, the former having oblong-lanceolate leaves, rounded and cucullate at the apices and frequently shortly reflexed-apiculate, although the points are often missing from the older leaves, whereas the latter has ovate to ovate-lanceolate and long acuminate leaves, acute or obtuse but not broadly rounded at the apex.

Reproduction in Antarctica. Always sterile both in the Antarctic and subantarctic South Georgia. It fruits only rarely in Tierra del Fuego (Ochyra & Matteri 2001).

Habitat. A typically hydrophilous moss growing in a wide range of wet and often flushed habitats, forming spongy carpets in well drained swamps on level or sloping ground, on wet rock faces and boulders, along the margins of small or temporary melt runnels, at the margin of lakes and ponds, sometimes floating in water and submerged to a depth of 10 m on Signy I. (Priddle & Dartnall 1978; Priddle 1979). Together with *Warnstorfia fontinaliopsis* and *Sanionia georgicouncinata* it is the major constituent of various associations of the moss carpet sub-formation which develop chiefly in situations with a permanent supply of standing water. Additional frequent associates include *Cephaloziella varians* and *Pohlia nutans*. *Warnstorfia sarmentosa* is also a frequent constituent of communities of the moss hummock subformation which occur in habitats with moving water and are dominated by *Brachythecium austrosalebrosum*, *Syntrichia filaris*, *Bryum pseudotriquetrum* and *Sanionia georgicouncinata*.

World range. A bipolar species having a continuous circumpolar arctic-boreal-alpine range throughout the Northern Hemisphere and scattered in the Southern Hemisphere, including southern South America and the Falkland Islands (Ochyra & Matteri 2001), South Georgia (Newton 1979*a*) and the northern Antarctic Peninsula region in the Western Hemisphere and south-eastern Australia and the South Island of New Zealand in the Eastern Hemisphere (Karczmarz 1971). Some scattered stations of *Warnstorfia sarmentosa* occur in the mountains in the tropics, including New Guinea (Ochyra *et al.* 1991), East Africa (Ochyra

←

Figure 250. *Warnstorfia sarmentosa*. — **1–2**. Habit. **3**. Portion of branch, dry. **4**. Portion of branch, wet. **5**. Portion of stem cross-section. **6–9**. Axillary hairs. **10**. Pseudoparaphyllium. **11–22**. Leaves. **23–24**. Leaf apices in terrestrial plants. **25**. Leaf apex in submerged plant. **26**. Mid-leaf cells at margin in submerged plant. **27–28**. Mid-leaf cells. **29–31**. Alar cells. — [1, 5 from *RO 2365/80*; 2 from *RO 1012/80*; 3, 13, 15, from *RO 1185/80*; 4, 11–12, 16 from *RO 1492/80*; 6–10, 27–28 from *RO 1047/80*; 17–22, 25–26, 29, 31 from *RILS 3830A*; 23 from *RO 5104/79*; 24, 30 from *RO 2186/80*]. — Scale bars: a – 100 µm (5–9, 23–31); b – 1 mm (3–4, 11–22); c – 100 µm (10); d – 0.5 cm (1–2).

1990*b*) and the central and northern Andes of Bolivia, Ecuador, Colombia and Venezuela (Hedenäs 2003*b*) (Fig. 251, inset).

Distribution in Antarctica. *Warnstorfia sarmentosa* is widely distributed and locally common only in the South Orkney Is. and South Shetland Is. in the northern maritime Antarctic. On the Antarctic Peninsula it is exceedingly rare and, so far, it has been recorded from only three widely separated locations on the Davis and Graham Coasts on the western Antarctic Peninsula, extending as far south as Lahille I. at lat. 65°15′ S, and a single record for James Ross I. on the north-eastern coast of the Antarctic Peninula (Fig. 251). The species occurs predominantly at low elevations and the highest concentration of its sites is from sea level to about 100 m, but it occasionally extends to 250 m on Signy I., South Orkney Is., and 275 m on King George I., South Shetland Is.

Selected specimens examined. WEST ANTARCTICA. SOUTH ORKNEY IS. **CORONATION I.** Meier Pt., 8–25 m, *RILS 576* (AAS, KRAM); Shingle Cove, 15 m, *RILS 435 & 436* (AAS, KRAM); Olivine Pt., 15 m, *RILS 137* (AAS, KRAM); Saunders Pt., *ca* 15 m, *RILS 25A* (AAS, KRAM); island opposite Saunders Pt., *ca* 15 m, *RILS 24* (AAS, KRAM). **POWELL I.** John Peaks, *ca* 15 m, *RILS 245* (AAS, KRAM). **SIGNY I.** Thulla Pt., 35 m, *REL 1131* (AAS, KRAM); Foca Pt., 35 m, *REL 857* (AAS, KRAM); Spindrift Rocks, *ca* 35 m, *MWH 776C* (AAS, KRAM); Cummings Cove, Emerald Lake, 30 m, *RILS 1857 & 8138* (AAS, KRAM); Robin Plateau, 190 m, *RILS 7019* (AAS, KRAM); Stygian Cove, 8 m, *REL 1170* (AAS, KRAM); Sombre Lake, *RILS 1816 & 1823A* (AAS, KRAM); Moss Lake, *RBH & JJL 28A* (AAS, KRAM) and *RILS 1819* (AAS, KRAM); Mirounga Flats, 7 m, *RILS 383* (AAS, KRAM); Factory Cove, 65 m, *RILS 4* (AAS, KRAM); Moraine Valley, 10 m, *RILS 375B* (AAS, KRAM) and *ca* 40 m, *RILS 331B* (AAS, KRAM); Rusty Bluff above Paal Harbour, 100 m, (AAS, KRAM); Gourlay Peninsula, 35 m, *RILS 359* (AAS, KRAM). **MOE I.** 10–20 m, *RILS 5262* (AAS, KRAM).

SOUTH SHETLAND IS. **ELEPHANT I.** Walker Pt., 120 m, *JPB* 256 (AAS, KRAM). **KING GEORGE I.** ***Potter Cove***: without closer locality, *Schmitt 226a* (KRAM, NY). ***Marian Cove***: North Spit, *BJ 152* (KRAM). ***Fildes Peninsula***: Nebles Pt., 5 m, *BJ 59* (KRAM); Suffield Pt., 40 m, *RO 2437/80* (KRAM); Ardley I., 15 m, *RO 2489/80* (KRAM); Gemmel Peaks, 15 m, *RO 2464/80 & 2479/80* (KRAM); Lake Kitezh, *ca* 15 m, *RO 2462/80 & 2465/80* (KRAM). ***Admiralty Bay***: Blue Dyke, 40 m, *RO 1185/80* (KRAM); Bastion, 240 m, *RO 1012/80* (KRAM); Uchatka Pt., 20 m, *RO 1047/80* (KRAM); Demay Pt., 65 m, *RO 1373/80* (KRAM); Creeping Slopes, 100 m, *RO 1253/80* (KRAM); Rescurs Hills, 3 m, *RO 4918/79* (KRAM); Jasnorzewski Gardens, 4 m, *RO 1492/80* (KRAM); Upłaz, 80 m, *RO 4965/79* (KRAM); Petrified Forest Creek, 180 m, *RO 4837/79* (KRAM); Ubocz, 95 m, *RO 4993/79* (KRAM) and 125 m, *RO 2365/80* (KRAM); Jersak Hills, 170 m, *RO 5104/79* (KRAM). ***Ezcurra Inlet***: Jardine Peak, 275 m, *RO 5165/79* (KRAM); Italia Valley, 70 m, *RO 58/80* (KRAM) and 100 m, *RO 90/80* (KRAM). ***Keller Peninsula***: Speil Pt., 6 m, *RO 527/80* (KRAM); British Pt., 3 m, *RO 456/80* (KRAM). ***Martel Inlet***: Ullman Spur, 50 m, *RO 572/80* (KRAM); Point Hennequin, 5 m, *RO 2240/80* (KRAM) and 10 m, *RO 2248/80* (KRAM); Mt. Wawel, 20 m, *RO 2186/80* (KRAM) and 25 m, *RO 2190/80* (KRAM). **NELSON I.** Moss Bay, *Komárek 88* (KRAM). **ROBERT I.** Coppermine Cove, 5 m, *Frödin 5 & 11* (AAS, KRAM, S) and 8–15 m, *RILS 919* (AAS, KRAM). **LIVINGSTON I.** ***Byers Peninsula***: Robbery Beaches, 10 m, *KR 52H* (AAS, KRAM); between Smellie Pt. and Chester Cove, 35–50 m, *RILS 3778* (AAS, KRAM); above Smellie Pt., *KR 38H* (AAS, KRAM); large lake south of Chester Cone, 30 m, *RILS 3830A & 3831B* (AAS, KRAM); Clark Nunatak, 180 m, *DCL 236* (AAS, KRAM); Negro Hill, 100 m, *KR 16B* (AAS, KRAM).

WEST ANTARCTIC PENINSULA. **DAVIS COAST.** Cape Kjellman, *CS 473* (PC, S). **GRAHAM COAST.** ***Grandidier Channel***: Lahille I., 2–5 m, *RILS 4315* (AAS, KRAM).

EAST ANTARCTIC PENINSULA. **JAMES ROSS ISLAND GROUP.** ***James Ross I.***: Andreassen Pt., 70 m, *RILS 7310* (AAS, KRAM).

2. **Warnstorfia fontinaliopsis** (Müll.Hal.) Ochyra Fig. 252

Warnstorfia fontinaliopsis (Müll.Hal.) Ochyra *in* Ochyra & Matteri *in* Guarrera, Gamundi & Matteri, Fl. Cript. Tierra del Fuego 14(10): 72. 2001. ≡ *Hypnum fontinaliopsis* Müll.Hal., Bot. Jahrb. Syst. 5: 82. 1883. ≡ *Drepanocladus fontinaliopsis* (Müll.Hal.) Broth. *ex* Paris, Coll. Nom. Broth.: 10. 1909. — TYPE CITATION: all new species described by Müller (1883) are included in the section 'I. Bryologia Kerguelensis'; later Müller (1889: p. 34) wrote: 'Ins. Kerguelen, Betsy Cove, in aquosis'. [Lectotype (*selected here*): 'Ex Museo botanico Berolinensi Hypnum (Drepanocl.) fontinaliopsis C. Müll. Kerguelensland: Betsy Cove Dr Naumann 1875' – HBG!; isotype: PC!].

Hypnum laculosum Müll.Hal., Flora 68: 425. 1885. ≡ *H. fluitans* Hedw. var. *laculosum* (Müll.Hal) Paris, Ind. Bryol.: 641. 1896. ≡ *Drepanocladus laculosus* (Müll.Hal.) Broth. *in* Cardot & Broth., Kongl. Svenska Vetenskapsakad. Handl. 63(10): 64. 1923. ≡ *Warnstorfia laculosa* (Müll.Hal.) Ochyra & Matteri, Fragm. Florist. Geobot. 42: 181. 1997. — TYPE CITATION: Fuegia, Staten Island, ad laculos alpinos inter Port Cook et Port S. John, Martio 1882: Spegazzini. [Lectotype (*vide* Ochyra & Matteri 1997: p. 181): 'Hypnum laculosum C. Mueller Loc. Pto. Cook In loculis alpinis intra P. Cook et P. S. Johnn, S. I. 2/82 Spegazzini 114' – LPS! (No. 41294)]. First synonymised by Ochyra & Matteri (2001: p. 72).

Hypnum austrostramineum Müll.Hal. *in* Neumayer, Int. Polarforsch. Deutsch. Exp. 2: 319. 1890 ['austro-stramincum']. ≡ *Drepanocladus austrostramineus* (Müll.Hal.) Broth. *ex* Paris, Coll. Nom. Broth.: 10. 1909. ≡ *Calliergidium austrostramineum* (Müll.Hal.) E.B.Bartram, Farlowia 2: 317. 1946. ≡ *Brachythecium austrostramineum* (Müll.Hal.) Ochyra *in* Ochyra, Vitt & Horton, Cryptog. Bryol. Lichénol. 7: 57. 1986. ≡ *Warnstorfia austrostraminea* (Müll.Hal.) Ochyra, Fragm. Florist. Geobot. 41: 480. 1996. — TYPE CITATION: Austro-Georgia, in locis paludosis 'der Landzuge', 25. Januario 1883. [Lectotype (*vide* Ochyra 1996*g*: p. 481): 'N° 41. Hypnum austro-stramineum C. Müll. Fundort: Landzuge an sumpfiger Quellen Süd-Georgien. 25/I 83 Will – HBG!; isotype: PC!]. First synonymised by Ochyra & Matteri (2001: p. 72)].

Hypnum austrostramineum var. *gracillimum* Müll.Hal. *in* Neumayer, Int. Polarforsch. Deutsch. Exp. 2: 319. 1890. ≡ *Drepanocladus austrostramineus* (Müll.Hal.) Broth. *ex* Paris var. *gracillimus* (Müll.Hal.) Cardot & Broth., Kongl. Svenska Vetenskapsakad. Handl. 63(10): 65. 1923. ≡ *Warnstorfia austrostraminea* (Müll.Hal.) Ochyra var. *gracillima* (Müll.Hal.) Ochyra, Fragm. Florist. Geobot. 41: 481. 1996, ***syn. nov.*** — TYPE CITATION: [South Georgia] Eodem loco in paludibus profundis, 7. Jan. 1883. [Lectotype (*vide* Ochyra 1996*g*: p. 481): 'Ex Museo botanico Berolinensi. Hypnum (Drepanocladus) austro-stramineum var. gracillimum C.Müll. Austro-Georgia Fundort: in locis valde paludosis Dr. Will' – PC!; isotype: PC!].

Hypnum austrostramineum var. *subfluitans* Müll.Hal. *in* Neumayer, Int. Polarforsch. Deutsch. Exp. 2: 320. 1890, ***syn. nov.*** — TYPE

Figure 252. *Warnstorfia fontinaliopsis.* — **1**. Habit. **2–5**. Portions of shoots, dry. **6**. Portion of cross-section of stem. **7–10**. Axillary hairs. **11–13**. Pseudoparaphyllia. **14–26**. Leaves. **27–29**. Leaf apices. **30–32**. Upper lamina cells at margin. **33–34**. Mid-leaf cells. **35**. Alar cells. **36**. Cross-section of leaf. — [1, 15, 32 from *RO 2488/80*; 2, 6–14, 35–36 from *VK 1196*; 3–4, 26 from *RO 702/80*; 5, 25 from *Turquet 252A* (lectotype of *Brachythecium turquetii*), PC; 16 from *RO 4818/79*; 17 from *RO 1418/80*; 18 from *VK 81*; 19, 27 from *Turquet 339A* (lectotype of *Hypnum austrostramineum* var. *minus*), PC; 20, 30 from *Will 41* (lectotype of *Hypnum austrostramineum*), HBG; 21 from *CS 64* (type of *Drepanocladus austrostramineus* var. *tenuissimus*), PC; 22 from *Will s.n.*, 7.i.1883 (lectotype of *Hypnum austrostramineum* var. *gracillimum*), PC; 23, 28, 34 from *Will 40* (lectotype of *H. austrostramineum* var. *subfluitans*), HBG; 24, 29, 31, 33 from *RO 1483/80*]. — Scale bars: a – 1 mm (4); b – 1 mm (2–3, 5); c – 1 mm (14–26); d – 100 µm (6–13, 27–36); e – 0.5 cm (1).

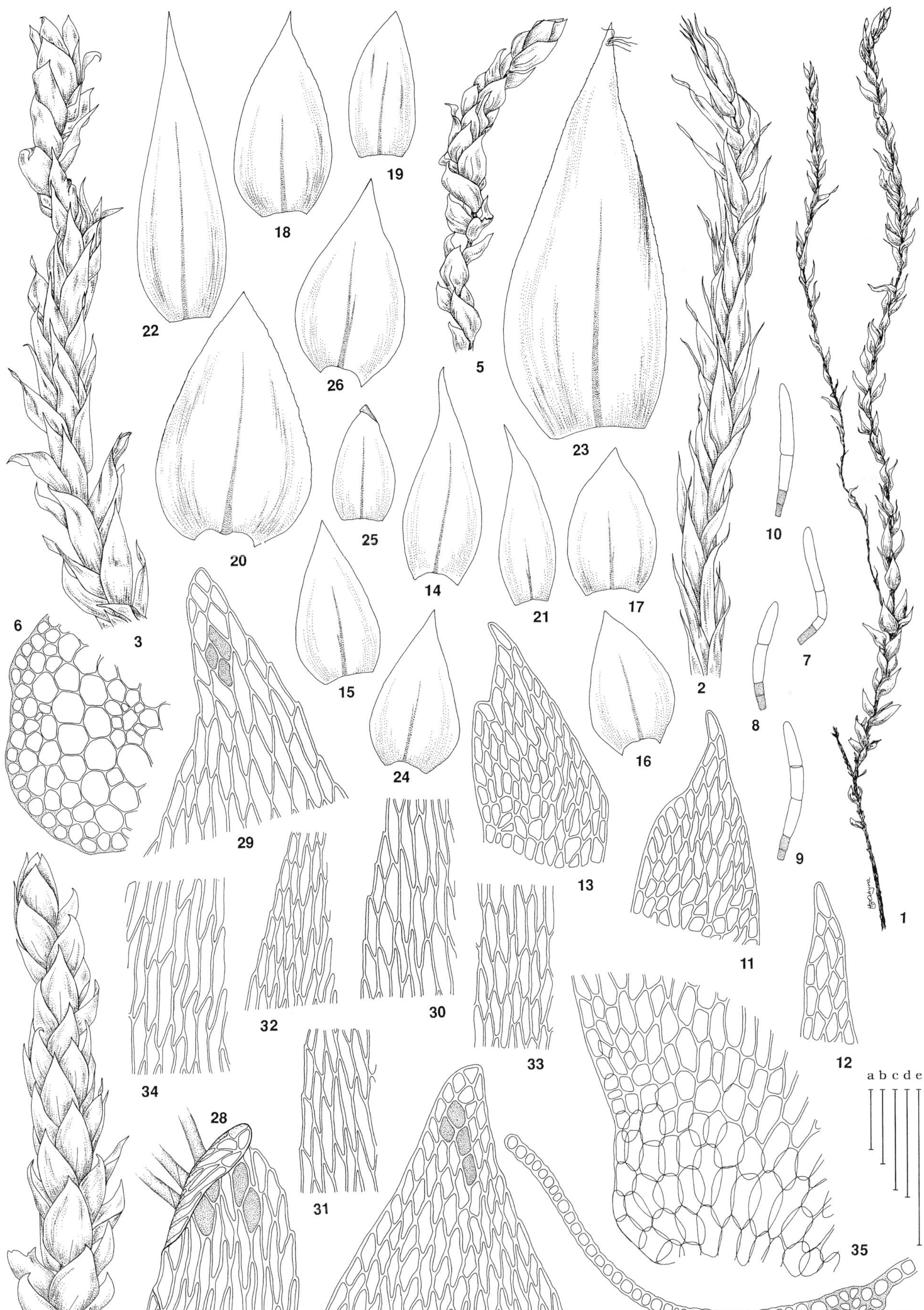
a b c d e

CITATION: Eodem loco in locis profundis prope 'dem grossen Teich', 14. Januario 1883. [Lectotype (*vide* Ochyra 1996*g*: p. 481): 'N° 40. Hypnum austro-stramineum var. subfluitans C. Müll. Fundort: Landzuge An sehr sumpfiger Orten in der Nähe des grossen Teiches Süd-Georgien. 14/I 83 Will – HBG].

Brachythecium turquetii Cardot, Rev. Bryol. 33: 34. 1906, ***syn. nov***. — TYPE CITATION: Île Wandel. [Lectotype (*vide* Ochyra 1998*f*: p. 288): 'Herb. Mus. Paris. Brachythecium Turquetii Card. sp. nova. Ile Wandel 23 nov[bre] 1904 Exped[n] Charcot aux terres australes no. 292[a]. Determ. J. Cardot' – PC!; syntypes: (1) same label, No. 240a – PC!, S-Roth!; (2) same label, without number – H!, PC!]. First synonymised with *Calliergidium austrostramineum* by Robinson (1972*a*: p. 176) and *Warnstorfia laculosa* by by Ochyra (1998*f*: p. 288)].

Hypnum austrostramineum var. *minus* Cardot, Rev. Bryol. 33: 35. 1906. ≡ *Warnstorfia austrostraminea* var. *minor* (Cardot) Ochyra, Fragm. Florist. Geobot. 41: 481. 1996. — TYPE CITATION: Île Wiencke. [Lectotype (*vide* Ochyra 1996*g*: p. 481): 'Herb. J. Cardot. Hypnum austro-stramineum C. M. var. minus Card. var. nov. Detr. de Gerlache: île Wiencke 29/II 1904 Expéd. Charcot n° 339[a]' – PC! First synonymised by Ochyra & Matteri (2001: p. 72)].

Drepanocladus austrostramineus var. *tenuissimus* Cardot & Broth., Kongl. Svenska Vetenskapsakad. Handl. 63(10): 65. 1923. ≡ *Warnstorfia austrostraminea* var. *tenuissima* (Cardot & Broth.) Ochyra, Fragm. Florist. Geobot. 41: 481. 1996, ***syn. nov***. — TYPE CITATION: Georgia austr.: Cumberland Bay, Moraine Fjord (64). [Lectotype (*selected here*): 'Expeditio suecica 1907–1909. 64. Drepanocladus austrostramineus (C. M.) Broth. var. tenuissimus Card. var. nov. Georgia australis: Cumberland Bay, Moraine fiord. 18.4.1909. Carl Skottsberg' – PC-Cardot!; isotype: H-Brotherus!].

Plants varying in size from small and slender to large and fairly robust, usually medium-sized, densely or loosely, often deeply caespitose or forming soft, extensive mats or carpets, sometimes floating in water, somewhat shiny, pale yellowish- to sordid-green above, light brown below, less frequently tinged with brown or orange, occasionally straw-coloured thoughout. *Stems* erect-ascending, sometimes creeping, usually pale brownish, simple or irregularly sparsely branched, elongate, (1.5–)2.0–6.5 (–14.0) cm tall, often terete and julaceous, with straight and bud-like apices because of crowded and imbricate leaves, in cross-section round, consisting of a medium-sized central strand of relatively large cells, 3–5-stratose medulla of large, thin-walled, hyaline cells and 1–2-stratose cortex of somewhat smaller but otherwise larger, thicker-walled cells; *rhizoids* sparse, mostly at base of stems, light brown, smooth, weakly branched, inserted intercalary on the stem or arising from the nematogon cells below the leaf apex; *pseudoparaphyllia* abundant, relatively large, lanceolate to triangular or ovate-triangular, mostly entire; *axillary hairs* short, comprising 1–2 short, brown basal cells and 2–3 elongate, hyaline upper cells. *Stem* and *branch leaves* similar in size and shape, crowded and loosely imbricate, becoming closely appressed above resulting in bud-like tips to the shoots, erect-spreading, sometimes remote and spreading, straight or occasionally falcato-secund at least at the tips of stems and branches, (0.9–)1.2–3.2(–4.0) mm long, (0.5–)0.6–1.0(–1.3) mm wide, ovate or broadly ovate, ovate-lanceolate to lanceolate, gradually narrowed to a short or long, rounded-obtuse or acute apex, sometimes bearing radicles at back of the tip arising from the rhizoid initials, concave to strongly concave, non-striolate or plicate, narrowed at the insertion, not to shortly decurrent but not noticeably auriculate; *margins* entire to remotely serrulate or sinuate, plane; *costa* single, sometimes branched or spurred, pale and almost concolorous with the lamina cells to intensely yellow and distinct, 60–80 μm wide at base, gradually tapering upwards, extending ½–¾ way up the leaf, plano-convex in cross-section and composed of homogeneous cells; *laminal cells* unistratose throughout, transparent, smooth; *upper cells* rhomboidal or oblong-rhomboidal to linear, straight, with truncate or shortly fusiformly narrowed cell-ends, (40–)50–100(–160) μm long, (8–)10–14(–16) μm wide, with thin- to firm, smooth or slightly pitted walls, often lax, becoming shorter towards the apex; *rhizoid initials* common at the apex, consisting of 3–5 rhombic, hyaline cells, sometimes producing bundles of rhizoids; *basal cells* quadrate to short-rectangular, 20–70 μm long, 20–25 μm wide, lax, thin-walled, not noticeably differentiated in the basal angles, gradually passing into the lamina cells. [*Autoicous*. *Perigonia* bud-like, *ca* 1 mm long; *perigonial bracts* ovate, 0.6–0.7 mm long, abruptly short acuminate, concave, ecostate, with entire margins; *antheridia* 7–9 per perigonium, claviform, brown; *paraphyses* few, yellowish as long as or slightly exceeding the antheridia. *Perichaetia* cylindric; *inner perichaetial leaves* straight and erect, oblong-ovate, suddenly narrowed to a shortly acuminate apex, 2.0–2.5 mm long, 0.8–1.0 mm wide, non-plicate, with a weak costa reaching the mid-leaf. *Setae* yellow-reddish, 4.5–6.0 cm long, flexuose, not twisted, smooth. *Capsules* suberect to horizontal, curved, asymmetric, brown; *operculum* conic, blunt at the apex; *exothecial cells* rounded to quadrate, thick-walled, in regular longitudinal rows, 15–40 μm wide, becoming elliptical towards the base of the urn; *stomata* many near base of the urn, superficial, bicellular, round-pored; *peristome* double; *exostome teeth* lanceolate, dotted-striolate on the dorsal side below, papillose distally, with dentate, narrow margin in the upper half; *endostome* pale yellowish to hyaline, consisting of a fairly low basal membrane, about 40% of the endostome length, giving rise to narrowly lanceolate processes with narrow perforations and 2–3 short, entire cilia. *Spores* globose, brownish, papillose, 20–22 μm in diameter. *Calyptra* cucullate, naked, smooth.]

Chromosome number: *n* = 22 – King George I. (Kuta *et al.* 1982).

Discussion. Although *Warnstorfia fontinaliopsis* is a notoriously variable and protean species, it is usually easily recognised and unlikely to be confused with any other moss in Antarctica, except possibly *Brachythecium austrosalebrosum*. However, they can be readily separated when examined microscopically. The species is diagnosed by its terete and subjulaceous stems and branches, broadly ovate to ovate-lanceolate and deeply concave leaves with rounded-obtuse to acute apices, entire to subentire leaf margins, lax leaf areolation, especially at leaf bases, and almost always present hyaline, rhombic nematogon cells below the leaf apex. In its concave and broadly ovate leaves, as well as in the overall appearance of the plants, the species may be mistaken in the field for *Brachythecium austrosalebrosum* with which it shares a similar coloration and ecological preferences. Therefore, in early ecological accounts *W. fontinaliopsis* was sometimes erroneously referred to as this species (e.g. Gimingham & Lewis Smith 1970; Lewis Smith 1972; Allison & Lewis Smith 1973; Lewis Smith & Corner 1973). However, in *B. austrosalebrosum* rhizoid initials at the leaf apex are always absent, the leaf margins are usually sharply serrulate above and the basal cells are not so lax and thin-walled, but usually thicker-walled and porose. These characters separate it also from other species of *Brachythecium*. *Warnstorfia fontinaliospsis* frequently grows with *W. sarmentosa* and *Sanionia georgicouncinata* and may be confused with the former, but its broadly rounded leaf apices and the large alar cells are the safest characters with which to separate them. The lack of sharply demarcated alar cells distinguish it also from *Campylium polygamum*.

In the Antarctic botanical literature the species is usually known as *Calliergidium austrostramineum*, a species originally described as *Hypnum austrostramineum* from South Georgia (Müller 1890*a*). The taxonomic position of this species was unclear and puzzled bryologists since was first described, and consequently it was transferred from genus to genus including *Drepanocladus* (Brotherus 1908; Paris 1909), *Calliergidium* (Bartram 1946) and *Brachythecium* (Ochyra *et al.* 1986). However, Ochyra (1996*g*) showed that the best placement for it is in *Warnstorfia* where it bears some distant affinities to *W. fluitans* (Hedw.) Loeske. Subsequently, Ochyra & Matteri (1997) reported an older name for it, *Hypnum laculosum*, which was used for the species described from Isla de los Estados (Müller 1885). Finally, it was named *Warnstorfia fontinaliopsis* by Ochyra (in Ochyra & Matteri 2001). This poorly known species was described from scanty material collected from Îles Kerguelen (Müller 1883, 1889) and it was initially considered to be identical with *Drepanocladus fluitans* (Hedw.) Warnst. (Ochyra & Bartlett 1986). However, subsequent re-examination of the type material and availability of better developed specimens from Prince Edward I. and Îles Kerguelen (Ochyra, personal observations) revealed that the plants from Kerguelen Province are inseparable from Fuegian, South Georgian and Antarctic plants, and that *Hypnum fontinaliopsis* was the oldest available name for this otherwise distinct species.

Warnstorfia laculosa is an exceedingly variable species, differing greatly in the robustness and coloration of the plants and leaf size and shape according to habitats. This polymorphism is largely induced by differences in environment, particularly in water supply, as well as by harsh climatic conditions in the far south. Müller (1890*a*) was the first to observe the remarkable variability in this species and recognised, besides the type variety, two other varieties of slender and floating plants. Two further varieties were later recognised by Cardot & Brotherus (1923) and Cardot (1906*a*) from South Georgia and Antarctica, respectively. Additionally, Cardot (1906*b*) described some plants from the Graham Coast region on the west Antarctic Peninsula as *Brachythecium turquetii*. Although this taxon showed convergence with brachythecialean mosses it is inseparable from *Warnstorfia fontinaliospsis* (Ochyra 1998*f*). Because of the range of environmental variation, recognition of taxonomically valid varieties from the practical point of view is almost impossible and *W. fontinaliospsis* has to be treated as a single, polymorphous species, as with its northern counterpart, *W. fluitans*.

Reproduction in Antarctica. Always sterile in the Antarctic but plants with sporophytes have been recorded on South Georgia.

Habitat. The moss thrives in permanently moist or wet habitats, locally forming almost pure stands of more than 100 sq. m,

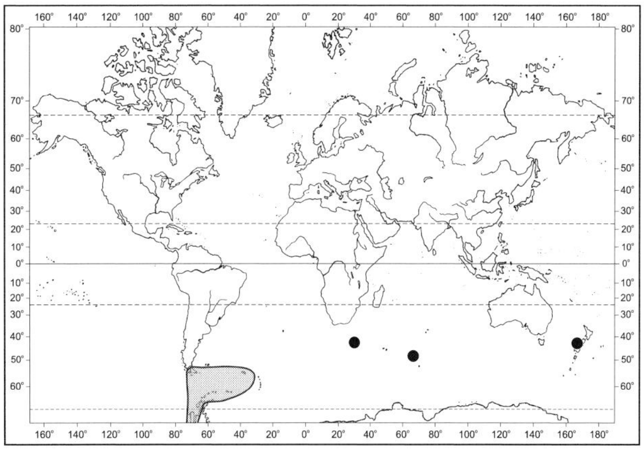

Figure 253. Global distribution of *Warnstorfia fontinaliopsis*.

on level or gently sloping ground, in swamps, and along the margins of streams and melt water channels, in wet hollows with late snow patches, on dripping rocks, frequently in water at the margins of sluggish streams, pools and lakes, rarely submerged to 4 m, occcasionally on dry rock ledges and on dry soil in rock depressions. In hollows with late lying snow, notably on Signy I., an extremely fine slender form of *Warnstorfia fontinaliopsis* forms carpets of several square metres. Its most frequent associates are *Warnstorfia sarmentosa*, *Sanionia georgicouncinata* and *Cephaloziella varians*. *Warnstorfia fontinaliopsis* is one of the major constituents of the swampy vegetation belonging within the moss carpet sub-formation.

World range. *Warnstorfia fontinaliopsis* may be designated, with some reservation, as a subantarctic species, having a disjunct circumpolar range in the Southern Hemisphere (Fig. 253), with maximum occurrence on subantarctic South Georgia and in the northern maritime Antarctic. In addition, it is exceedingly rare on Prince Edward I. and on Îles Kerguelen (Ochyra, personal observations) in Subantarctica and infrequent in southern South America, including Isla de los Estados, Isla Grande de Tierra del Fuego and western Patagonia (Ochyra & Matteri 2001) and very rare on the South Island of New Zealand (Bartlett 1984).

Distribution in Antarctica. A widespread and locally common species in the South Orkney Is. and South Shetland Is., scattered and localised in the northern part of the west Antarctic Peninsula and fairly common in its central part on the Graham Coast, extending southwards to northern Marguerite Bay at lat. 67°55′S (Fig. 254). It occurs most often at low elevations, from sea level to 50 m, and only seldom ascends to 100 m on Signy I.

Selected specimens examined. **WEST ANTARCTICA.** SOUTH ORKNEY IS. **Monroe I.** Plateau and slopes, 80 m, *RILS 8002* (AAS, KRAM). **Coronation I.** Meier Pt., 17–30 m, *RILS 575* (AAS, KRAM); Mansfield Pt., 15 m, *RILS 291* (AAS, KRAM); Cape Hansen, 20–200 m, *RILS 3220 & 3233* (AAS, KRAM); Shingle Cove, 15 m, *RILS 437* (AAS, KRAM); Olivine Pt., 15 m, *RILS 131 & 140* (AAS, KRAM); island opposite Saunders Point, *ca* 15 m, *RILS 24* (AAS, KRAM). **Lynch I.** 10–17 m, *RILS 517 & 518* (AAS, KRAM). **Matthews I.** The Divide, 15–100 m, *RILS 174B* (AAS, KRAM). **Coffer I.** 15–50 m, *RILS 183A* (AAS, KRAM). **Powell I.** Promontory opposite and east of Christoffersen Island, *RILS 48* (AAS, KRAM); west of John Peaks, 35 m, *RILS 246* (AAS, KRAM); Falkland Harbour, *ca* 35 m, *RILS 276* (AAS, KRAM). **Signy I.** Foca Pt., 7 m, *RILS 341* (AAS, KRAM); NW coast, *RILS 10909* (AAS, KRAM); North Pt., 80 m, *RILS 8600* (AAS, KRAM); Robin Peak, *RILS 1785* (AAS, KRAM); Stygian Cove, 25 m, *RILS 7022* (AAS, KRAM); Three Lakes Valley, *ca* 15 m, *RW 77* (AAS, KRAM); Sombre Lake, *RILS 1822* (AAS, KRAM); Heywood Lake, 6 m, *RILS 10365D* (AAS, KRAM); Knife Pt., 20 m, *RILS 8583 & 8584* (AAS, KRAM); Factory Cove,

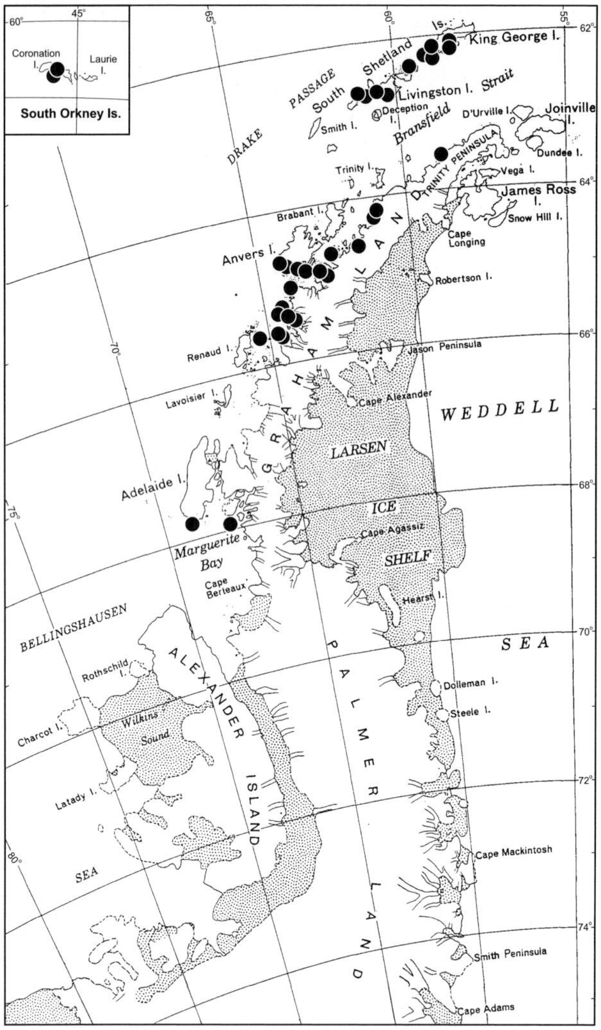

Figure 254. Distribution of *Warnstorfia fontinaliopsis* in Antarctica.

3 m, *RILS 5049* (AAS, KRAM), 65 m, *RILS 4* (AAS, KRAM); Station, 70 m *RILS 7015* (AAS, KRAM); Moraine Valley, *ca* 35 m, *RILS 611 & 1807* (AAS, KRAM), 65 m, *RILS 10897* (AAS, KRAM) and 100 m, *RILS 10843* (AAS, KRAM); Observation Bluff, 2–3 m, *RILS 694C* (AAS, KRAM) and 20 m, *RILS 8081* (AAS, KRAM); Gourlay Peninsula, *ca* 35 m, *RILS 361, 362, 364 & 365* (AAS, KRAM); Hillier Moss, 15 m, *RILS 7016* (AAS, KRAM), 50 m, *RILS 7021 & 7085* (AAS, KRAM) and 80 m, *RILS 7017* (AAS, KRAM); between Lenton Pt. and Gourlay Peninsula, 25 m, *RILS 592* (AAS, KRAM); Lenton Pt., 25 m, *RILS 694AB* (AAS, KRAM). **MOE I.** 10–100 m, *RILS 5264* (AAS, KRAM).

SOUTH SHETLAND IS. **KING GEORGE I.** ***Fildes Peninsula***: Ardley I., 10 m, *RO 2488/80* (AAS, KRAM); Bukhta Korablya, 28 Mar 1976, *Petrov s.n.* (KRAM). ***Drake Passage***: Pyrites I., *BJ 38* (KRAM). ***King George Bay***: Turret Pt., 30 m, *DCL 818* (AAS, KRAM). ***Bransfield Strait***: Blue Dyke, 40 m, *RO 1193/80* (AAS, KRAM). ***Admiralty Bay***: Agat Pt., 12 m, *RO 375/80 & 376/80* (AAS, KRAM); Rescuers Hills, 30 m, *RO 4818/79* (AAS, KRAM); Llano Pt., 55 m, *RO 4917/79* (KRAM) and 60 m, *RO 179/80* (KRAM); Penguin Ridge, 20 m, *RO 1411/80* (KRAM), 30 m, *RO 1444/80* (KRAM), 40 m, *RO 1440/80* (KRAM), and 50 m, *RO 1408/80 & 1428/80* (AAS, KRAM); Jasnorzewski Gardens, 4 m, *RO 1483/80, 1484/80, 1487/80 & 1493/80* (KRAM). **NELSON I.** Moss Bay, *VK 89* (KRAM); Harmony Cove, *Montiel CM4755 & 4780* (KRAM). **ROBERT I.** Coppermine Cove, 10 m, *RILS 913* (AAS, KRAM). **AITCHO IS.** 5–6 m, *RILS 11450 & 11451* (AAS, KRAM). **LIVINGSTON I.** ***Byers Peninsula***: South Beaches, 6 m, *DCL 413* (AAS, KRAM); Hell Gates, *DCL 285* (AAS, KRAM); between Smellie Pt. and Chester Cone, 35–50 m, *RILS 3779* (AAS, KRAM). ***Hurd Peninsula***: False Bay, *RMS 69-507* (US).

WEST ANTARCTIC PENINSULA. **DAVIS COAST.** Cape Herschel, 15 m, *RILS 11640B* (AAS, KRAM). **DANCO COAST.** Cierva Pt., *OB 19* (AAS, KRAM); Spring Pt., *VK 1196* (KRAM) and 10–30 m, *RILS 3877, 3878 & 3879* (AAS, KRAM); Charlotte Bay, Recess Cove, 2–10 m, *RILS 3980 & 3981* (AAS, KRAM); Cuverville I., *Racovitza 240A* (BR, PC), 10–15 m, *Leeuw 37, 41 & 42* (AAS, KRAM), 10–20 m, *RILS 4177, 11948 & 11959B* (AAS, KRAM) and 45 m, *RILS 8807* (AAS, KRAM); Bryde I., *RMS 69-354* (US); Paradise Harbour, Coughtrey Peninsula, Almirante Brown Station, *VK 226A* (KRAM), 7 Oct 1985, *Birkenmajer s.n.* (KRAM) and 85 m, *Linskens 13B* (AAS, KRAM). **PALMER ARCHIPELAGO. *Anvers I.***: Norsel Pt., *VK 945 & 1066* (KRAM), *REL 1233 & 1234* (AAS, KRAM), *RILS 2375* (AAS, KRAM) and *RMS 69-850 & 851* (US); Palmer Station, *RILS 5292* (AAS, KRAM); Bonaparte Pt., 24 Feb 1975, *Hormel Institute 8-1-0-5* (AAS, KRAM); Dream I., *VK 642, 674 & 983* (KRAM); Shortcut I., *VK 1048 & 1049* (KRAM); Litchfield I., *JRRr & JMG PAL-157* (US) and *VK 81 & 499* (KRAM); Hermit Island, *RILS 879* (AAS, KRAM); Laggard I., 35 m, *REL 1294* (AAS, KRAM); Wiencke I., *ER 259A* (BR, PC, S) and *Turquet 339A, 344B & 401B* (PC); same island, Lécuyer Pt,, 10 m, *RILS 725* (AAS, KRAM). **GRAHAM COAST.** Cape Rasmussen, *RILS 1914* (AAS, KRAM); Cape Tuxen, Waddington Bay area, 25 m, *RILS 10664* (AAS, KRAM); Cape Tuxen, 10–100 m, *RILS 4889* (AAS, KRAM). **WILHELM ARCHIPELAGO.** ***Lemaire Channel***: Fish Is., *S. & J. Poncet 1* (KRAM); Booth I., *Turquet 240A* (H, PC – syntype of *Brachythecium turquetii*) *& 292A* (PC – lectotype of *B. turquetii*). ***Argentine Is.***: Uruguay I., 10–25 m, *RILS 4267* (AAS, KRAM); Corner Is., *JRR PAL-80* (US) and *JRR & JMG PAL-106* (US); Galindez I., *RMS 69-216 & 255* (US), 15 m, *RILS 749* (AAS, KRAM) and *REL 1236, 1253 & 1254* (AAS, KRAM), 25 m, *RILS 750* (AAS, KRAM) and 35 m *REL 1238* (AAS, KRAM); same island, Woozle Hill, 25 m, *RILS 3286* (AAS, KRAM); Black I. 2–10 m, *RILS 4289* (AAS, KRAM). ***Collins Bay***: Berthelot Is., Green I., *BGLE 1049* (BM) and *RILS 3370* (AAS, KRAM). ***Grandidier Channel***: Takaki Pt., *RILS 4337B, 4338 & 4339* (AAS, KRAM); Lahille I., 2–5 m, *RILS 4312, 4313 & 4317* (AAS, KRAM). **BISCOE IS.** Hook I., *RMS 69-889A* (US). **FALLIÈRES COAST. *Adelaide I.***: Avian I., *ca* 15 m, *RILS 775B* (AAS, KRAM) and *REL 1258 & 1259* (AAS, KRAM). ***Marguerite Bay***: Line Is., 5 m, *RILS 4606* (AAS, KRAM) and 15 m, *RILS 775B* (AAS, KRAM).

EXCLUDED SPECIES OF THE AMBLYSTEGIACEAE

Drepanocladus aduncus (Hedw.) Warnst. — The material so named, with some reservation, was reported by Light & Heywood (1973) and Priddle (1979) from Signy I. The specimens are characterised by their long excurrent costae and represent *Drepanocladus longifolius*, and all voucher specimens are cited under this species. In typical *D. aduncus* the costa is much weaker and vanishes well below the leaf apex. Additionally, this species was reported from Fildes Peninsula on King George I. by Chen *et al.* (1993, 1995). Examination of two specimens (*Li S1 & AN19*, AAS, KRAM) showed that they represented a small but otherwise typical and frequently found expression of *Sanionia georgicouncinata.* The almost entire leaf margins and lax basal cells might indeed suggest *Drepanocladus aduncus*, but the stems have a perfect hyalodermis and the angular cells form elongate groups along the proximal leaf margins.

Drepanocladus brachiatus (Mitt.) Broth. *ex* Paris — The species was reported from Wiencke I. near the Antarctic Peninsula by Dixon (1935). Examination of the original collection (Port Lockroy, 23 Jan 1931, *Discovery Expedition 14*, BM-Dixon) shows that the material represents *Sanionia georgicouncinata*. *Drepanocladus brachiatus* is a New Zealand species which is closely related or possibly conspecific with *D. sordidus* (Müll.Hal.) Hedenäs.

Hygrohypnum polare (Lindb.) Loeske — This taxon was reported from Victoria Land by Linskens *et al.* (1991) as *Hygrohypnum palustre* Loeske var. *polare* (Lindb.) Mönk. but no voucher specimens are available. Since no pleurocarpous moss has been so far recorded in continental Antarctica, this report is very doubtful, especially because it comes from an area well collected bryologically. *Hygrohypnum polare* is an arctic-alpine species which has never been collected outside the Northern Hemisphere.

Straminergon stramineum (Brid.) Hedenäs — This species was reported as *Calliergon stramineum* (Brid.) Kindb. from King George I. (Wu & Hu 1990; Chen *et al.*, 1993, 1995) and Signy I. (Gimingham & Smith 1971). These specimens represent *Warnstorfia sarmentosa.* The Antarctic populations of the latter species often lack red coloration and sometimes also a recurved apiculus at the leaf apices but are otherwise quite similar to *Straminergon stramineum*; it is a Northern Hemisphere species with some extra-Holarctic stations in the Andes and New Zealand. In *S. stramineum* the alar cells form narrow, oval groups extending upwards along the proximal margins and the axillary hairs are few, very weak, with 1–2 shorter upper cells, whereas in *W. sarmentosa* alar cells form broader, often transversely triangular groups and the axillary hairs are more abundant and larger, with 2–3 elongate upper cells. The King George I. specimens (*Wei 8308*, KRAM and *Li K12 & As16*, AAS, KRAM) represent typical expressions of *W. sarmentosa* with some reddish coloration and apiculate leaf apices.

Warnstorfia exannulata (Schimp.) Loeske — Chen *et al.* (1993, 1995) reported this species from Ardley I. off Fildes Peninsula on King George I. The specimen examined (*Li AN11*, AAS, KRAM) revealed it to be a typical orthophyllous plant of *Sanionia uncinata.* It differs from *Warnstorfia exannulata* in having a perfect hyalodermis to the stem, transversely rectangular alar cells not reaching the costa and a lack of nematogon cells.

Family XV. BRACHYTHECIACEAE Schimp.

Brachytheciaceae Schimp., Syn. Musc. Eur. Ed. 2, 1: 'xcv' [cxv]. 1876 ['Brachythecieae']. — Type: *Brachythecium* Schimp.

Camptotheciaceae Schimp., Syn. Musc. Eur. Ed. 2, 1: 'xcv' [cxv]. 1876 ['Camptothecieae']. — Type: *Camptothecium* Schimp.

Donrichardsiaceae Ochyra, Lindbergia 11: 7. 1985. — Type: *Donrichardsia* H.A.Crum & L.E.Anderson.

Apart from the Sematophyllaceae and Hypnaceae, the Brachytheciaceae are the largest family in the Hypnales. It has a worldwide distribution and comprises about 350–400 species classified into about 40 genera. As with other families of pleurocarpous mosses, the real number of species is difficult to evaluate because many of its genera have never been revised taxonomically on a global scale. The family is best characterised in terms of the sporophyte characters. The inner perichaetial leaves are always non-plicate, abruptly long-acuminate from an erect and broad base, usually with a reflexed acumen and ecostate or with short and very weak costae. They sharply contrast with the perichaetial leaves of the Amblystegiaceae which are gametophytically very similar to the Brachytheciaceae but have plicate and costate inner perichaetial leaves with straight apices. The capsules are relatively short, mostly inclined and asymmetric, not completely contracted and have a flaring mouth when dry and with a perfect hypnaceous peristome. They are more or less strongly convex on their dorsal side and not or only slightly convex on the ventral side. The exostome teeth are most often red-brown, brown-red or red, cross-striolate on the outer surface and their border is gradually narrowed upwards from the transition zone, whereas the endostome segments are usually widely perforated (Ignatov *et al.* 1998). The important features of the family are round-pored stomata and setae that are often roughened by papillae. The operculum is convex-conic to long-rostrate, usually slanted. The setae are variously mammillose throughout (or only above), though sometimes smooth throughout, whereas in all members of the Amblystegiaceae, except *Amblystegium pseudosubtile* Hedenäs from Haiti in the West Indies (Hedenäs 2003*b*), the setae are always smooth.

Gametophytically, the Brachytheciaceae are strongly diversified and difficult to characterise except in generalities. The leaves are usually narrow and elongate, often strongly concave and plicate, mostly erect and appressed and not becoming squarrose on wetting, but in some species they are falcate. The leaves are usually gradually long-acuminate and only rarely shortly acuminate, blunt or rounded at the apex which is usually plane, although sometimes it is markedly twisted. The costa is single and only rarely branched, often projected as a dorsal spine, mostly ending well below the apex and only rarely is it percurrent to excurrent. The laminal cells are always smooth, rarely prorate at the ends, but never centrally papillose, mostly elongate, rhomboidal to linear-flexuose. The alar cells are usually differentiated but not sharply demarcated from the adjacent laminal cells, small, quadrate and sturdy to lax and inflated, sometimes forming decurrencies or, rarely prominent auricles.

In contrast to the Amblystegiaceae, species the Brachytheciaceae are mostly associated with drier habitats, though some of them are also true aquatics. Until recently, the traditional morphological classification of the family was adopted as outlined by Brotherus (1925) who classified over three quarters of the known species into four large genera, namely *Brachythecium*, *Rhynchostegium* Schimp., *Euryhynchium* Schimp. and *Rhynchostegiella* (Schimp.) Limpr., whereas the remaining species were placed in about 35 small, oligotypic genera. Robinson (1962) proposed a revised generic classification of the family for North America but it has not been widely adopted. Based upon combined analysis of morphological and molecular data sets, these four large and polyphyletic genera were split into a number of segregates which are smaller and better defined genera (Ignatov & Huttunen 2002, Huttunen & Ignatov 2004; Vanderpoorten *et al.* 2005; Huttunen *et al.* 2007). At the same time the Brachytheciaceae were divided into four subfamilies, namely Eurhynchioideae Milde, Brachythecioideae, Helicodontioideae M.Fleisch. and Homalothecioideae Ignatov & Huttunen. These classifications are still provisional and the status and relationships of many taxa still require clarification.

Since the Brachytheciaceae are mostly temperate in distribution, they are not well represented in the cool and cold regions in the Southern Hemisphere, at least in terms of the number of taxa. In practice, the family is represented in the Subantarctic and Antarctic by seven species of *Brachythecium*, two of *Sciuro-hypnum* and two genera, *Brachytheciastrum* Ignatov & Huttunen and *Platyhypnidium* M.Fleisch., each with a single species in this region. In more northerly latitudes, in the cool and warm temperate zones, genera such as *Rhynchostegium*, *Eurhynchium* and *Eriodon* Mont. are also found infrequently. So far, the family has been rather under-studied taxonomically, and the most valuable treatments of the Brachytheciaceae have been completed for Japan and East Asia (Takaki 1955*a*,*b*, 1956) and New Guinea (Ignatov *et al.* 1999). In the Southern Hemisphere the entire family was studied for Australia (Hedenäs 1996*b*, 2002).

KEY TO THE GENERA OF THE ANTARCTIC BRACHYTHECIACEAE

1. Leaves distinctly vertically plicate, not or scarcely decurrent; stem leaves ovate-lanceolate, broadly ovate to elongate-triangular, gradually or, seldom, more abruptly acuminate .... 50. ***Brachythecium***
1. Leaves smooth or hardly plicate, longly or shortly decurrent with tapering auricles; stem leaves, cordate-triangular to broadly ovate-cordate abruptly narrowed to a filiform acumen .... 51. ***Sciuro-hypnum***

50. **BRACHYTHECIUM** SCHIMP.

Brachythecium Schimp. *in* Bruch, Schimp. & W.Gümbel, Bryol. Eur. 6: 5. 1853 [Fasc. 52–54 Mon.: 1]. ≡ *Hypnum* Hedw. sect. *Brachythecium* (Schimp.) Sull. *in* A.Gray, Manual Ed. 2: 675. 1856. ≡ *Hypnum* subgen. *Brachythecium* (Schimp.) Hobk., Syn. Brit. Mosses: 148. 1873. — Lectotype (*vide* Robinson 1962: p. 103): *Brachythecium rivulare* Schimp.

Paramyurium (Limpr.) Warnst., Krypt.-Fl. Brandenburg Laubm.: 763. 1905. ≡ *Eurhynchium* Schimp. subgen. *Paramyurium* Limpr., Laubm. Deutschl. 3: 173. 1896. — Lectotype (*vide* Ignatov & Huttunen 2002: p. 265): *Eurhynchium tommasinii* (Boulay) Molendo (*Hypnum tommasinii* Boulay) [= *Brachythecium tommasinii* (Boulay) Ignatov & Huttunen]. First synonymised by Ignatov & Huttunen (2002: p. 265).

Chamberlainia Grout, Moss Fl. N. Amer. 3: 27. 1928. ≡ *Brachythecium* subgen. *Chamberlainia* (Grout) Vohra, Rec. Bot. Surv. India 23: 215. 1983. — Lectotype (*vide* Robinson 1963: p. 103): *Chamberlainia acuminata* (Hedw.) Grout (*Leskea acuminata* J. Hedwig). First synonymised by Vohra (1983: p. 215) and Anderson *et al.* (1990: p. 477).

Cryptoneurum Thér. & P.de la Varde, Trav. Bryol. 2 (Rev. Bryol. Lichénol. 13–14): 22. 1944. — Type: *Cryptoneurum acuminatum* Thér. & P.de la Varde [= *Brachythecium ruderale* (Brid.) W. R. Buck]. First synonymised by Buck (1998: p. 239].

Plants small and slender to rather large and robust, often soft and ± glistening, forming loose or dense mats, patches, wefts or deep tufts, green, yellowish, golden or whitish-green to stramineous with age. *Stems* prostrate, decumbent or ascending, sometimes erect, freely and irregularly to subpinnately branched, with straight or curved branches, terete and julaceous to complanate-foliate, in transverse section comprising a 1–2(–3)-stratose cortex of small, thick-walled, substereid cells surrounding several rows of larger, thin-walled, hyaline medullary cells and a small central strand; *paraphyllia* lacking; *pseudoparaphyllia* foliose to lanceolate, acuminate to acute; *axillary hairs* short, 2–4-celled, with 1–2 short, brown basal cells and elongate distal cell, often gradually tapering to the distal end; *rhizoids* often scanty, slightly branched, smooth, brown to pale brownish. *Stem* and *branch leaves* somewhat differentiated with the stem leaves mostly larger and longer-acuminate and with a stronger costa, generally crowded, erect, imbricate or erect-spreading to spreading, plane or concave, deeply longitudinally plicate to smooth, ovate to lanceolate, long-acuminate to acute, sometimes twisted at the apex, not or somewhat decurrent; *margins* plane to recurved all around or only partially, entire to serrulate throughout or only distally; *costa* single, rarely branched or spurred, ending near the mid-leaf to subpercurrent, without or sometimes with a spine projecting at the back; *laminal cells* smooth, unistratose, transparent, elongate-rhomboidal to oblong-linear or linear-fusiform, becoming shorter and laxer and often porose at base and shorter at the extreme apex; *angular cells* subquadrate to oblong-hexagonal or inflated-rectangular, usually gradually differentiated, sometimes abruptly so in conspicuous oval alar group, occasionally forming broad, noticeable decurrencies. *Asexual reproduction* none. *Autoicous*, *dioicous* or, rarely, *polyoicous*. *Perichaetial leaves* ecostate or rarely with weak, short costae, slenderly long-acuminate, usually with reflexed apices. *Setae* elongate, mostly straight and twisted when dry, smooth or scabrous above or throughout, reddish. *Capsules* erect, inclined to horizontal, ovoid, obloid to short-cylindrical, gibbous or arcuate, usually dark brown; *operculum* conic to convex-conic, with a blunt or acute beak, occasionally short-rostrate; *annulus* undifferentiated, persistent or separating by fragments; *exothecial cells* isodiametric to oblong, thick-walled, not collenchymatous; *stomata* superficial, bicellular, round-pored; *peristome* double, xero- or hydrocastique; *exostome teeth* 16, lance-subulate, shouldered, reddish-brown, trabeculate at the back, cross-striolate below, papillose above, bordered; *endostome* comprising a high basal membrane giving rise to keeled, narrowly perforate to gaping segments, as high as the teeth, separated by 1–3 nodose to appendiculate cilia, shorter than the segments, rarely vestigial or lacking cilia. *Spores* globose, minutely roughened to smooth. *Calyptra* cucullate, smooth, naked.

Before splitting, *Brachythecium* was the largest genus of the Brachytheciaceae consisting of approximately 175 described species (Crosby *et al.* 2000), some of which are very widespread and weedy species. At present, the more narrowly defined genus consists of about 50–60 species (Ignatov & Huttunen 2002), but further reductions of names of exotic species to synonymy are still expected. *Brachythecium* is a rather nondescript genus that is best identified and defined by absence of traits than by their presence. Plants are assigned to *Brachythecium* on account of their sporophyte and gametophyte characters, including conic-convex lids rather than long-beaked ones, smooth or finely roughened setae rather than distinctly scabrous ones, plicate, ovate-lanceolate, gradually acuminate and non-decurrent leaves rather than smooth, cordate-triangular, abruptly acuminate and prominently decurrent ones, and the costa not ending in a strong spine on the dorsal side. In some species, for example *B. rivulare* Schimp. and *B. subplicatum* Hampe the alar cells form auriculate groups of large, hyaline and inflated cells, not small and sturdy like those in *Sciuro-hypnum*.

In general, *Brachythecium s. str.* corresponds to sect. *Brachythecium*, sect. *Albicantia* Ochyra and sect. *Salebrosa* Broth. of the broadly conceived genus. The type section of *Brachythecium* has traditionally been designated sect. *Rutabula* Broth., but Robinson (1962) lectotypified this generic name with *B. rivulare* Schimp. which only just belongs within the type section. McFarland (1988) questioned the aptness of this lectotypification and argued that Schimper (1853*b*), in his discussion on the philosophy used in the segregation of *Brachythecium* from the then all-encompassing *Hypnum*, stated that he based his concept of this genus upon the four 'Hauptgrundformen' species, namely *B. populeum* (Hedw.) Schimp., *B. velutinum* (Hedw.) Schimp., *B. reflexum* (Starke) Schimp. and *B. rutabulum* (Hedw.) Schimp., but none of these was indicated as the type. Even if the selection of *B. rivulare* as the lectotype is not particularly satisfactory because the species is rather atypical in the genus in having large, pellucid and broadly decurrent auricles, this selection of this species is not in serious conflict with the protologue and cannot be rejected. The change of the lectotype is not particularly justified because *B. rutabulum*, a suggested new lectotype, is a species closely related to *B. rivulare*, so the change has no practical consequences and does not affect the current nomenclature.

In the austral polar regions species of *Brachythecium* are prominent constituents of the flora and vegetation, both in terms of frequency and cover, and so far six species have been reported from this biome, two of which are widely distributed in the maritime Antarctic. *Brachythecium* is taxonomically one of the most difficult of all moss genera and distinctions between some species are very blurred. Only some local revisions of the broadly understood genus are available, for example for Japan and conterminous territories (Takaki 1955*a,b*, 1956), the Himalayas (Vohra 1983) and Latin America including also Antarctic and Subantarctic taxa (McFarland 1988). Newton (1979*a,b*) provided a thorough revision of *Brachythecium* species in South Georgia. *Brachythecium* has been divided into several sections, but no satisfactory infrageneric classification of the genus has hitherto been presented. All Antarctic species are here assigned to sect. *Salebrosa*, although McFarland (1988) suggested sect. *Albicantia* as the proper place for *B. austroglareosum* and sect. *Reflexa* Broth. for *B. subpilosum*.

1. Leaf margins intermittently recurved all around .... 1. ***B. austroglareosum***
1. Leaf margins plane or recurved only at the extreme base 2
    2. Leaves broadly ovate to ovate-lanceolate, gradually or suddenly short-acuminate, non-decurrent; stems erect to ascending; seta glabrous .... 2. ***B. austrosalebrosum***
    2. Leaves elongate- to ovate-triangular, suddenly long-acuminate, not or scarcely decurrent; stems prostrate; seta rough .... 3. ***B. subpilosum***

1. **Brachythecium austroglareosum** (Müll.Hal.) Kindb. Fig. 255

Brachythecium austroglareosum (Müll.Hal.) Kindb., Enum. Bryin. Exot. Suppl. 2: 97. 1891. ≡ *Hypnum austroglareosum* Müll.Hal., Bot. Jahrb. Syst. 5: 82. 1883 ['austro-glareosum']. — Type citation: all new species described by Müller (1883) are included in the section 'I. Bryologia Kerguelensis'; later Müller (1889: p. 32) wrote: 'Ins. Kerguelen, Betsy Cove, cum Hypno paradoxo associatum, Novbr. 1874'. [Lectotype (*selected here*): 'Ex Museo botanico Berolinensi. Hypnum (Brachythecium) austro-glareosum Kerguelen, Betsy's Cove Dr Naumann, 1874' – PC-Cardot!; isotypes: BM!, H-Brotherus!, LE!, S!].

Brachythecium austroglareosum var. *diffusum* Cardot, Compt. Rend. Ac. Sci. Paris 153: 603. 1911. — TYPE CITATION: Baie Marguerite: île Jenny. [Lectotype (*selected here*): 'Herbarium J. Cardot. Brachythecium austroglareosum (C.M.) Par. var. diffusum Card. var. nova. Baie Marguerite: île Jenny, dans une grotte humide. – 2ème exped. Charcot. Leg. Gain, 1909. n. 249' – H-Brotherus!; isotypes: BM!, PC (unavailable, on loan since 29 Apr 1981 to Dr D. K. Smith, Knoxville). First synonymised by Newton (1979*a*: p. 137; 1979*b*: p. 129).

Brachythecium gramontii Cardot, Bull. Mus. Natl. Hist. Nat. 22: 341. 1916. — TYPE CITATION: Kerguelen (Rallier du Baty). [Lectotype (*selected here*): 'Herb. Mus. Paris Brachythecium Gramontii Card. sp. nova Kerguelen Exped. Rallier du Baty, 1913–14 Det. J. Cardot 11/1915' — PC!; isotypes: BM!, PC!]. First synonymised by Newton (1979*a*: p. 137; 1979*b*: p. 129).

Plants small to medium-sized, rarely fairly large, generally gracile, sericeous, somewhat lustrous, pale green, yellow-green or yellowish, rarely golden above, light to whitish-brown or brown below, in dense mats or deep tufts. *Stems* prostrate, decumbent or erect, slender, julaceous, 2–6 cm tall, simple or irregularly sparingly branched, with tapering tips, in transverse section with a small but distinct central strand, 2–3-layered cortex of small, brownish, thicker-walled cells surrounding larger, thin-walled, hyaline medullary cells disposed in 3–5 layers; *branches* short, unequal, erect, tapering at the tips; *pseudoparaphyllia* foliose, orbicular, broadly ovate to ovate, acute to acuminate; *rhizoids* reddish-brown to pale brown, weakly branched, sparsely scattered along the stem in clusters below the leaf insertion; *axillary hairs* 2–3-celled, short, filiform, consisting of a long, hyaline apical cell and 1–2 short, brownish basal cells. *Leaves* crowded, occasionally somewhat spaced, fairly stiff, erect-imbricate to erect-spreading on drying, little altered on wetting, 1.4–2.6(–2.8) mm long, 0.4–0.8(–0.9) mm wide, narrowly ovate-lanceolate to elongate-triangular, gradually long-acuminate, with long, slender and usually twisted apices, slightly concave, distinctly plicate, not or scarcely decurrent; *margins* intermittently recurved all around, entire to minutely serrulate in the distal portion; *costa* broad at the base, gradually narrowed upwards, 40–60 µm wide at the base, vanishing ⅔–¾ way up the leaf, occasionally shorter and branched; *laminal cells* unistratose throughout, transparent, smooth; *upper cells* fusiform to linear-flexuose, (45–)50–90(–120) µm long, 5–11 µm wide, with moderately thickened and weakly pitted walls, about 5–12 times as long as wide; *basal cells* shorter and wider, oblong-rhomboidal, 20–70 µm long, 7–17 µm wide, with thickened and porose walls; *alar cells* quadrate to rectangular, thick-walled, regularly seriate, 4–10 in marginal rows. [*Autoicous*. *Perigonia* bud-like. *Perichaetial leaves* long, slenderly long-acuminate ecostate. *Setae* erect, straight, smooth, 1.9–2.5 cm long, red-brown to blackish-brown. *Capsules* obloid, inclined to horizontal, asymmetric, 1.4–1.6 mm long, 0.8–0.9 mm wide, red-brown; *opeculum* conic-convex, strongly apiculate, 0.5–0.8 mm long; *annulus* biseriate, fragmentarily deciduous, composed of enlarged, vesiculose, thick-walled cells; *exothecial cells* subquadrate to short-rectangular, moderately thick-walled; *stomata* fairly numerous in the urn base, superficial, bicellular, round-pored; *peristome* double; *exostome teeth* 16, lanceolate, reddish-brown, cross-striolate in the proximal portion, hyaline and papillose distally, bordered; *endostome* as high as the exostome, composed of a high basal membrane about half the peristome length and keeled, perforate segments alternating with 1–2 nodose cilia, shorter than the segments. *Spores* globose, finely papillose, 12–20 µm in diameter. *Calyptra* cucullate, smooth, naked.]

Chromosome number: $n = 13$ – South Georgia (Newton 1979*b*).

Discussion. In its julaceous stems and branches *Brachythecium austroglareosum* is similar to *B. austrosalebrosum* but it is a smaller and slender, sericeous plant with finely long-acuminate, markedly longitudinally plicate leaves and pitted walls to the laminal cells, while its distinctly, though intermittently, recurved leaf margins along the whole length separate the species from all congeners in the region. The species shows little variation in Antarctica though Cardot (1911*a*,*b*, 1913) recognised a plant collected in a cave and forming looser tufts with more widely spaced leaves as var. *diffusum*. However, this is evidently a habitat modification, which does not deserve taxonomic recognition, and Newton (1979*a*,*b*) rightly discarded its recognition as a taxon in its own right and placed this name amongst synonyms of *B. austroglareousm*.

Brachythecium austroglareosum was originally described from Îles Kerguelen as *Hypnum austroglareosum* (Müller 1883, 1889) and soon it was transferred to *Brachythecium* by Kindberg (1891). The species was subsequently rediscovered in this archipelago (Brotherus 1906*a*) and later Cardot (1916) described it again as *B. gramontii*. The latter species is represented by exceptionally large, luxuriant though slender plants with the stems reaching 15 cm in length, but otherwise perfectly matching the type of *B. austroglareosum*.

McFarland (1988) assigned *Brachythecium austroglareosum* to sect. *Albicantia* which is characterised by recurved leaf margins, ovate-lanceolate and deeply concave leaves with a prominent oval marginal group of basal angular cells which are longly decurrent in a broad band (Ochyra 1998*g*). Although *B. austroglareosum* shares some external resemblance as well as partially recurved leaf margins with other species of this section, including *B. groenlandicum* (C.E.O.Jensen) Schljakov, *B. glareosum* (Spruce) Schimp. and *B. erythrorrizon* Schimp., these latter species differ in having distinctly decurrent leaves, whereas *B. austroglareosum* has not or only weakly decurrent leaves. On the other hand, these features place it in sect. *Salebrosa* which is characterised by the lack of or very short leaf decurrencies,

→

Figure 255. *Brachythecium austroglareosum*. — **1**. Habit, dry. **2**. Portion of branch, dry. **3**. Portion of cross-section of stem. **4**. Pseudoparaphyllium. **5–10**. Axillary hairs. **11–16**. Leaves. **17**. Leaf apex. **18–19**. Upper leaf cells at margin. **20–21**. Mid-leaf cells. **22–23**. Basal juxtacostal cells. **24–25**. Alar cells. — [1–10, 15–16, 20, 23–24 from *Gain 253*, PC; 11 from *VK 1260*; 12–13, 17–19, 21–22, 25 from *Gain 241E*, PC; 14 from *Naumann s.n.*, 1879 (lectotype of *Hypnum austroglareosum*), PC]. — Scale bars: a – 50 µm (17–25); b – 1 mm (2); c – 1 mm (1, 11–16); d – 100 µm (4–10) and 200 µm (3).

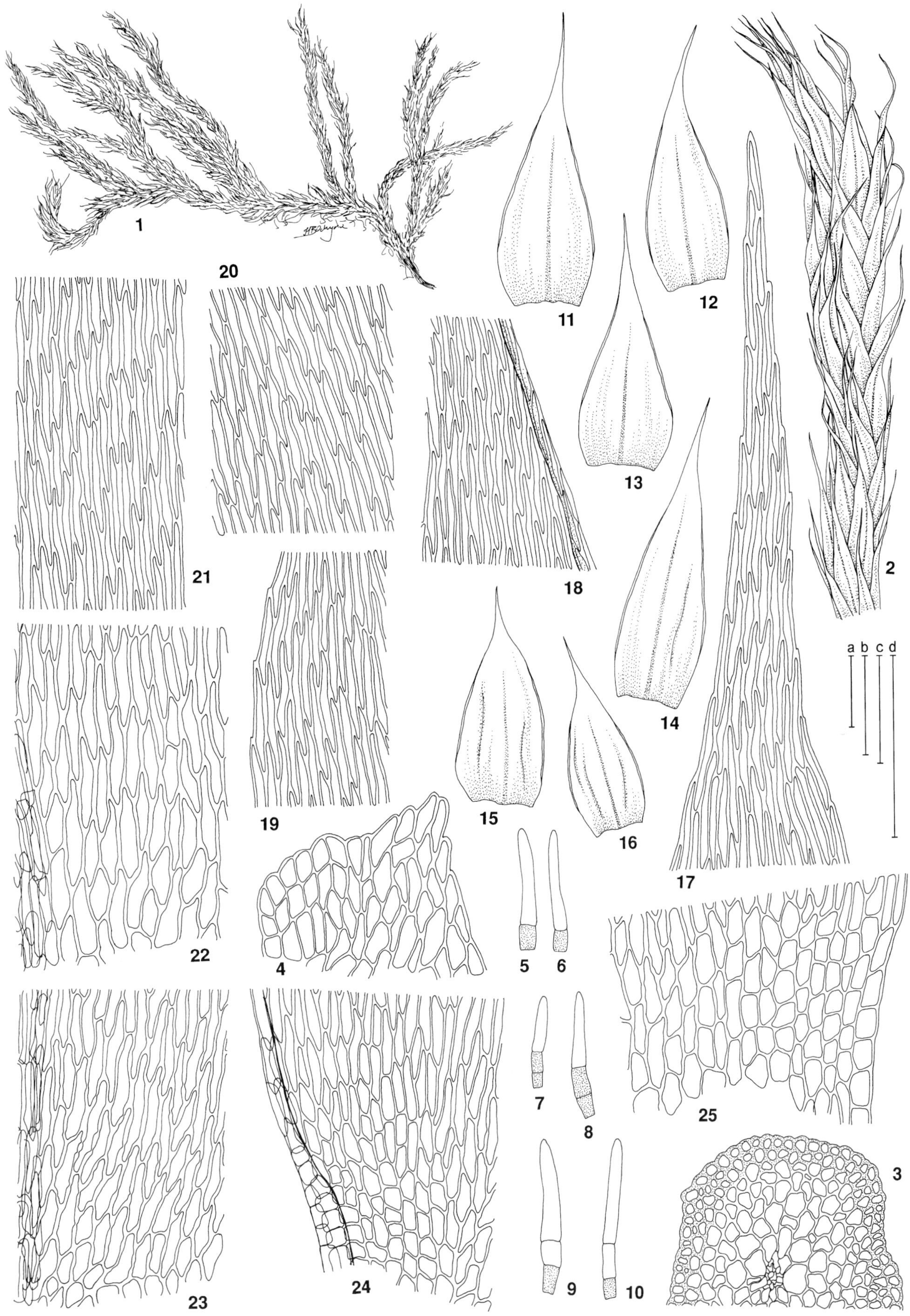
1
2
3
4
5
6
7
8
9
10
11
12
13
14
15
16
17
18
19
20
21
22
23
24
25
a
b
c
d

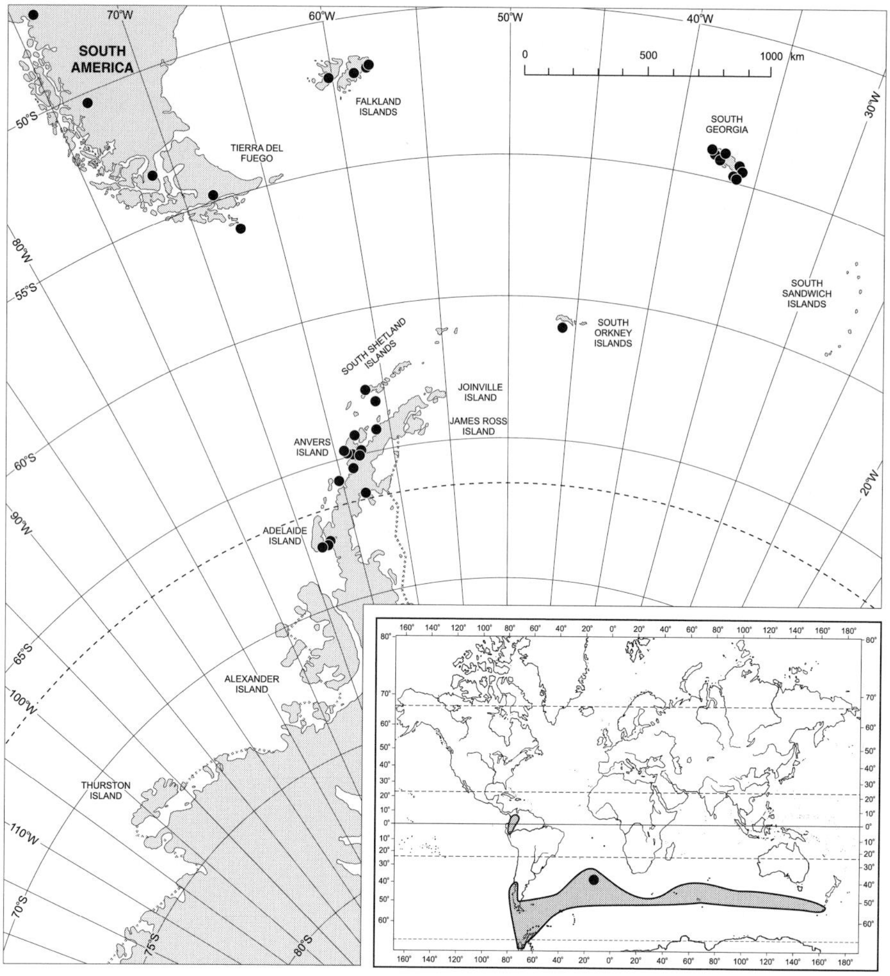

Figure 256. Distribution of *Brachythecium austroglareosum* in the Fuegian region and Antarctica. Inset: global distribution.

and therefore *B. austroglareosum* is positioned in this section following the original concept of Brotherus (1908, 1925). This placement is confirmed inferentially by McFarland (1988) who presented the hypothesis that *B. austroglareosum* is possibly conspecific with *B. turgidum* (Hartm.) Kindb., a pan-Holarctic, arctic-boreal-alpine species. This suggestion is valid because the coincidence of many gametophyte and sporophyte traits, habitat preferences and growth form between the two species is strikingly obvious and it is likely that detailed comparative studies will reveal *B. austroglareosum* to be a bipolar species.

Reproduction in Antarctica. Always sterile.

Habitat. *Brachythecium austroglareosum* typically grows on soil and humus in wet or moist rock crevices, on rock ledges, and sometimes in sheltered caves.

World range. A pan-south-temperate species (Fig. 256, inset) which is widely distributed in the *Nothofagus* zone in southern South America from the Valdivian region to Tierra del Fuego and the Falkland Is., and subsequently appearing in the maritime Antarctic and on the subantarctic islands of South Georgia, Îles Crozet and Îles Kerguelen, extending to Gough I. (Ochyra 1999*i*) in the north and to the Auckland Is. in the east. In addition, as with other antipodal cool-adapted mosses, *Brachythecium austroglareosum* recurs in the northern Andes of Peru, Ecuador and Colombia growing at high elevations of 3300–4100 m (McFarland 1988).

Distribution in Antarctica. *Brachythecium austroglareosum* is rare and occasional in the South Orkney Is. and South Shetland Is. and quite rare and widely scattered in the northern and central parts of the western Antarctic Peninsula, with a few highly isolated stations in northern Marguerite Bay where it extends as far south as lat. 67°34′ S; it is also occasionally found on the eastern coast of the Antarctic Peninsula (Fig. 256). It occurs at low elevations from near sea level to about 80 m, and only sporadically above 330 m on the Foyn Coast on the eastern Antarctic Peninsula.

Specimens examined. WEST ANTARCTICA. SOUTH ORKNEY IS. **SIGNY I.** Foca Cove, 5 m, *RILS 8050A* (AAS, KRAM).

SOUTH SHETLAND IS. **LIVINGSTON I.** ***Byers Peninsula***: Laager Pt., 30 m, *KR 39D* (AAS, KRAM). **DECEPTION I.** Punta Murature, *Kühnemann 281* (AAS, KRAM); Fumarole Bay, 100 m, *RILS 11493, 11509, 11510A & 11526* (AAS, KRAM).

WEST ANTARCTIC PENINSULA. **DANCO COAST.** Spring Pt., *VK 1260 & 1277* (KRAM); Andvord Bay, Neko Harbour, 15–20 m, *RILS 11930A* (AAS, KRAM); Paradise Harbour, Coughtrey Peninsula, Almirante Brown Station, opposite Bryde I., *RMS 69-365 & 69-402* (US). **PALMER ARCHIPELAGO.** ***Melchior Is.***: Omega I., *Popovici 16075* (KRAM, LE). ***Anvers I.***: Hermit I., 10–15 m, *RILS 881* (AAS, KRAM); Goudier I., Port Lockroy, 3–6 m, *RILS 727* (AAS, KRAM). **GRAHAM COAST.** Cape Pérez, *Gain 273B* (PC). **WILHELM ARCHIPELAGO.** ***Lemaire Channel***: Booth I., *Turquet 244C* (PC). ***Argentine Is.***: Uruguay I., *ca* 10 m, *RILS 926* (AAS, KRAM): Galindez I., Woozle Hill, 2 m, *RILS 3287 & 3289* (AAS, KRAM). **BISCOE IS.** Hook I., *RMS 69-705A* (US). **FALLIÈRES COAST.** ***Léonie Is.***: Léonie I., 20 m, *RILS 10649B* (AAS, KRAM). ***Marguerite Bay***: Jenny I., *Gain 241C, 242B,, 250B, 253 & 261* (PC) *& 249* (BM, H – type of *B. austroglareosum* var. *diffusum*); Blaiklock I., *Poncet 6A* (AAS, KRAM).

EAST ANTARCTIC PENINSULA. **FOYN COAST.** North shore of Cabinet Inlet, Eden Glacier, 330–660 m, *RT 28* (AAS, KRAM).

2. **Brachythecium austrosalebrosum** (Müll.Hal.) Kindb. Fig. 257

Brachythecium austrosalebrosum (Müll.Hal.) Kindb., Enum. Bryin. Exot. Suppl. 2: 97. 1891. ≡ *Hypnum austrosalebrosum* Müll.Hal., Bot. Jahrb. Syst. 5: 82. 1883 ['austro-salebrosum']. — Type citation: all new species described by Müller (1883) are included in the section 'I. Bryologia Kerguelensis'; later Müller (1889: p. 32) wrote: 'Ins. Kerguelen, sine loco natali (12. 1. 75'. [Lectotype (*selected here*): 'Ex Museo botanico Berolinensi. Brachythecium austro-salebrosum C. Müll. n. sp. Kerguelensland Gazellen Expedition Dr Naumann 1875' – le!; isolectotypes: je!, s!; syntype: 'Ex Museo botanico Berolinensi. Brachythecium austro-salebrosum C. Müll. n. sp. Kerguelensland Engl. Jahrb. V. p. 82. Gazellen-Expedition Dr. Naumann 1874' – bm!].

Hypnum euryodictyon Müll.Hal., Bot. Jahrb. Syst. 5: 82. 1883. ≡ *Brachythecium euryodictyon* (Müll.Hal.) Kindb., Enum. Bryin. Exot. Suppl. 2: 97. 1891. — Type citation: all new species described by Müller (1883) are included in the section 'I. Bryologia Kerguelensis'; later Müller (1889: p. 32) wrote: 'Ins. Kerguelen, Betsy Cove, Decbr. 1874'. [Lectotype (*selected here*): 'Ex Museo botanico Berolinensi. Brachythecium euryodictyon C. Müll. n. sp. Kerguelensland Gazellen-Expedition Dr. Naumann 1874' – bm!; isotypes: bm!, h-Brotherus!, pc!, s-Dusén!]. First synonymised by Newton (1979*a*: p. 138; 1979*b*: p. 127).

Hypnum georgicoglareosum Müll.Hal. *in* Neumayer, Int. Polarforsch. Deutsch. Exp. 2: 318. 1890 ['georgico glareosum']. ≡ *Brachythecium georgicoglareosum* (Müll.Hal.) Kindb., Enum. Bryin. Exot. Suppl. 2: 97. 1891. — Type citation: Austro-Georgia, 'Quelle auf dem Hochplateau' cum Hypno austro-fluviatili, 14. Julio 1883, sterile; 'Landzunge', rarius, 14. Januario 1883, sterile. [Lectotype (*selected here*): 'N° 3. Hypnum georgico-glareosum C. Müll. n.sp. Fundort: Quellen auf dem Hochplateau Süd-Georgien, 14/VII 83 Will Südpolarexpedition' – hbg!; syntype: 'N° 3. Hypnum georgico-glareosum CMüll. n.sp. Fundort: Landzunge nicht häufig Süd-Georgien, 14/I 83 Will Südpolarexpedition' – hbg!]. First synonymised by Newton (1979*a*: p. 138; 1979*b*: p. 127).

Brachythecium antarcticum Cardot, Rev. Bryol. 27: 46. 1900. — Type citation: [Détroit du Gerlache] XIme débarquement [later Cardot (1901: p. 42) wrote: Détroit de Gerlache: XIème débarquement, Terre de Danco, cap Van Beneden, sur les corniches de la falaise. (N°s 230^{a} in parte, 232^{a}, 232^{c}, 233^{c}, 234^{c}; 1^{r} février 1898). [Lectotype (*selected here*): 'Herb. J. Cardot. Mousses du voyage de la Belgica n° 232^{a} Brachythecium antarcticum Card. sp. nova. Canal de la Belgica. XIème débarquement sur des corniches de la falaise leg. Racovitza, 1 févr. 1898' – pc!; isolectotypes: br!, h-Brotherus!, s!; syntypes: same label data, No. 230a', 232c, 233c, 234c – br!, pc!]. First synonymised by Newton (1979*a*: p. 138; 1979*b*: p. 127).

Brachythecium antarcticum var. *cavifolium* Cardot, Rev. Bryol. 27: 46. 1900. — Type citation: [Détroit du Gerlache] Ier, IXme et XIme débarquement [later Cardot (1901: p. 42) wrote: Détroit de Gerlache: Ier débarquement, île Auguste, à la surface de l'argile mélée de guano. (N° 215^{a}; 26 janvier 1898). IXème débarquement, Terre de Danco, près du cap Anna Osterrieth, sur les petites terrasses humides de la falaise. (N°s 151^{a} et 205^{d}; 29 janvier 1898). XIème débarquement, Terre de Danco, cap Van Beneden, sur les corniches de la falaise. (N°s 232^{a} in parte; 1^{r} février 1898). [Lectotype (*selected here*): 'Herb. J. Cardot. Mousses du voyage de la Belgica n° 215^{a} Brachythecium antarcticum Card. sp. nova. var. cavifolium Card. var. nova. Canal de la Belgica. Ier débarquement, à la surface de l'argile mélée de guano. Leg. Racovitza, 26 janvier 1898' – pc!; isolectotype: br!; syntypes: same label data, No. 205d', 232a', 233a' – br!, pc!]. First synonymised by Newton (1979*a*: p. 138; 1979*b*: p. 127).

Brachythecium antarcticum fo. *robustum* Skottsb., Ymer 1905: 413. 1905. — Type citation: At Beatrice Point [Orleans Channel, Graham's Land] I found, at the foot of a steep mountain wall, a small plateau, leaning gently towards north and irrigated by the melting snow from the cliff above. [Lectotype (*selected here*): 'Svenska Sydpolarexpeditionen 1901–03 N:r 474. Brachythecium antarcticum Card. var. cavifolium Card. f. robusta. Terre Louis-Philippe: pointe Béatrice 29/11 1902 Carl Skottsberg Det. J. Cardot' – pc!; isotype: h-Brotherus!, s!]. First synonymised by Ochyra (1999*f*: p. 513).

Brachythecium skottsbergii Cardot, Bull. Herb. Boissier Sér. 2, 6: 1008. 1906. — Type citation: [Géorgie du Sud] Cumberland bay: May harbour, Pot harbour, Jason harbour, Moraine fiord. [Lectotype (*selected here*): 'Svenska Sydpolarexpeditionen 1901–03 N:r 405. Brachythecium Skottsbergii Card. sp. nova. South Georgia, Cumberland bay May harbour 2/5 1902 Carl Skottsberg Det. J. Cardot' – pc!; isolectotypes: h-Brotherus!, s!; syntypes: (1) 'Svenska Sydpolarexpeditionen 1901–03 N:r 406. Brachythecium Skottsbergii Card. sp. nova. South Georgia, Cumberland bay Moraine Fiord 20/5 1902 Carl Skottsberg Det. J. Cardot' – pc!, s!; (2) 'Svenska Sydpolarexpeditionen 1901–03 N:r 407. Brachythecium Skottsbergii Card. forma gracilior + Pseudoleskea strictula Card. Géorgie du Sud: baie Cumberland, Pot harbour 21/5 1902 Carl Skottsberg Det. J. Cardot' – pc!, s!]. First synonymised by Newton (1979*a*: p. 138; 1979*b*: p. 128).

Plants medium-sized to relatively large and robust, mostly soft, in dense or looser intertwining mats or often in deep coalescing tufts forming large hummocks, lustrous, light to yellow-green or, rarely, olive-green above, dark brown to whitish-brown below. *Stems* erect to ascending, sometimes prostrate with ascending branches, simple or sparsely branched, (1.5–)2.5–7.5(–25.0) cm tall, often julaceous, densely foliate in upper parts, often becoming denuded below, in cross-section round, with a small central strand and 1–2 layers of thin-walled, brownish outer cortical cells slightly smaller than the large, hyaline medullary cells; *pseudoparaphyllia* large, foliose, triangular to ovate-acute; *rhizoids* sparse, slightly branched, smooth, dark brown; *axillary hairs* short, filiform, 2–3(–4)-celled, with 1 or 2 quadrate, brown basal cells and an elongate apical cell. *Leaves* erect to erecto-patent, imbricate, tightly appressed at the stem and branch tips, (1.8–)2.3–3.4(–3.8) mm long, (0.5–)1.0–1.5(–1.7) mm wide, very seldom widely spaced, strongly or weakly concave, not or slightly and narrowly decurrent, deeply plicate on drying and weakly plicate on wetting, broadly ovate to ovate-lanceolate, suddenly or gradually short-acuminate, rarely long-acuminate; *margins* plane above, slightly recurved at the extreme base, entire to serrulate above; *costa* single, weak, almost concolorous with the laminal cells, widest at the base, 40–60 µm wide, gradually tapering upwards, extending ½–¾ way up the leaf; *laminal cells* unistratose, pellucid, smooth; *upper* and *median* cells oblong-rhomboidal to linear-fusiform, tapering at the apices, thin-walled, becoming shorter towards the leaf apex, (55–)70–115(–125) µm long, 7–12 µm wide, *ca* 8–11 times as long as wide; *basal cells* shorter and broader, rather lax, oblong-hexagonal to rectangular, with thickened and slightly porose walls, 25–85(–100) µm long, 12–25 µm wide; *alar cells* rounded, quadrate to short-rectangular, often collenchymatous, 15–45 µm long, 11–30 µm wide, lax but not inflated, forming poorly defined, small, flat, rounded auricles. *Autoicous. Setae* elongate, orange-yellow to dark red-brown, glabrous. *Capsules* obloid, inclined to horizontal, curved, asymmetric, 1.3–2.4 mm long, 0.6–1.0 mm wide, red-brown; *opeculum* conic-convex, strongly apiculate, 0.4–0.9 mm long; *annulus* 1–2-seriate, deciduous, composed of large, vesiculose, thick-walled cells; *exothecial cells* subquadrate to short-rectangular, moderately thick-walled; *stomata* many at the base of the urn, superficial, bicellular, round-pored; *peristome* double; *exostome teeth* 16, lanceolate, reddish-brown, cross-striolate below, papillose above, with dentate border; *endostome* of the same length as the exostome, consisting of a basal membrane of *ca* ½ the peristome length and keeled, perforate segments and 1–2 nodose cilia, shorter than the segments. *Spores* and *calyptra* not seen.]

Chromosome numbers: *n* = 10, 20 – King George I. (Kuta *et al.* 1982); *n* = 11 – Signy I. (Newton 1980); *n* = 10, 13 – South Georgia (Newton 1972, 1979*a*).

Discussion. *Brachythecium austrosalebrosum* is distinguished by its tufted growth form, broadly ovate to ovate-lanceolate and deeply concave leaves which are not or only shortly and narrowly decurrent and abruptly or gradually tapering to short apices. In the field the species can be easily recognised by its growth form, the plants mostly growing in tufts which coalesce to produce locally extensive undulating hummocks. Robust plants of *B. austrosalebrosum* may be confused with *Sciuro-hypnum fuegianum* but the very broad decurrencies of the leaves and thinly tufted or matted growth form of the latter should preclude any confusion. *Brachythecium austroglareosum* is a smaller plant with narrower leaves, longer and narrower median laminal cells with mostly pitted walls and recurved leaf margins, whereas *B. subpilosum* is also smaller and has narrowly triangular to ovate-triangular leaves which are abruptly long-acuminate and often shortly decurrent.

Brachythecium austrosalebrosum shows considerable variation in growth form and Antarctic plants are generally smaller in size and bear leaves with shorter and broader apices, and are always sterile in comparison with those from more northerly latitudes. These dwarf growth forms were recognised as separate taxa, *B. antarcticum* and *B. antarcticum* var. *cavifolium* (Cardot 1900, 1901), but they are ecological variants conditioned by harsh local climatic conditions and, as such, do not merit taxonomic recognition. Newton (1979*a,b*) reduced these names to synonymy with *B. austrosalebrosum*. On the other hand, plants growing in steam vents, moist hollows associated with fumaroles, and fissures with geothermal heating on Bouvetøya are very large and luxuriant with prostrate or pendent stems up to 25 cm long and widely spaced, plane or slightly concave leaves. The common phenotype of the species is characterised by having robust deep tufts with julaceous shoots and ovate-lanceolate, deeply concave and imbricate leaves abruptly contracted to a fine point.

The species was originally described from Îles Kerguelen under two names, *Hypnum austrosalebrosum* and *H. euryodictyon* (Müller 1883, 1889), but Newton (1979*a,b*) correctly considered them synonyms and accepted the former epithet as the final name for this species. It was perhaps an unfortunate choice because it suggests a close affinity of the austral species with its northern counterpart *Brachythecium salebrosum* (F.Weber & D.Mohr) Schimp. Müller (1883, 1889) indeed compared his newly described species to *B. salebrosum* and showed differences in the stature of the plants, leaf shape and areolation between them, the Kerguelen taxon having larger, rigid and compressed stems, entire leaf margins, short and cuspidate leaf apices and plane and impressed alar cells. Both species certainly belong within sect. *Salebrosa* but they actually differ sharply in leaf shape.

Reproduction in Antarctica. Always sterile in Antarctica, while on subantarctic South Georgia only gametangia have been found.

Habitat. *Brachythecium austrosalebrosum* is predominantly a hydrophytic species, although sometimes it also thrives in dry sites. It grows on gravelly or sandy soil at the margins of melt water streams and runnels, usually in places with some degree of flushing, on wet rocks, cliffs, rock ledges, stones and boulders subject to dripping water or moistened by melt water, on moraines, scree, moist debris, in depressions and hollows between boulders, seepage areas below late snow patches, and generally in sites with moving water. Occasionally, it has also been found on mortar of old buildings. On volcanic islands it grows in fumaroles, on fine or coarse ashy soil, cinder slopes covered with scoria and lapilli, ashy knolls, sometimes on rock outcrops with geothermal fissures. The species usually constitutes a separate association within the moss hummock subformation (Lewis Smith 1972; Longton 1979) which occupies flushed habitats along the margins of melt runnels and on wet sloping rocks. It forms coalescing hummocks producing locally extensive closed undulating stands covering several sqare metres. Associated species usually include *Sanionia georgicouncinata*, *Bryum pseudotriquetrum*, *Warnstorfia samentosa* and *Cephaloziella varians*. On Bouvetøya it forms a similar association in the fumarole area in which it cohabits with *Syntrichia filaris* (Bell & Blom 1986).

World range. A circumsubantarctic species (Fig. 258), widely distributed on all islands from South Georgia to Macquarie I., extending to the maritime Antarctic and to the Falkland Is., Fuegian region and western Patagonia in the southernmost part of South America (McFarland 1988), with some isolated stations at high elevation in the Andes of Bolivia (Ochyra 1999*j*) and Peru (Allen 1988). McFarland (1988) considered *Brachythecium austrosalebrosum* to be 'circum-subantarctic in distribution from Patagonia to Kerguelen Island' but, in fact, its distribution pattern suggests an amphiatlantic subantarctic species (Ochyra 1998*a*). However, McFarland (1988) reported *B. austrosalebrosum* from Macquarie I. on the basis of a specimen collected in 1930 by the British-Australian-New Zealand Antarctic Research Expedition of 1929–1931 (*BANZARE 996*, BM), and this island together with Île de la Possession of the Îles Crozet archipelago, was cited by this author in the entry 'Antarctica'. Seppelt (2004) overlooked this record and reported *B. salebrosum* from

⟶

Figure 257. *Brachythecium austrosalebrosum.* — **1**. Habit. **2–3**. Portion of branches, dry. **4**. Cross-section of stem. **5–8**. Axillary hairs. **9–11**. Pseudoparaphyllia. **12–20**. Leaves. **21–22**. Leaf apices. **23–24**. Mid-leaf cells. **25**. Alar cells. **26**. Basal leaf cells. — [1, 4, 9–11, 14–15, 23 from *RO 1382/80*; 2, 19, 21, 24, 26 from *RO 4916/79*; 3, 16 from *RO 644/80*; 5–8, 17 from *RO 5256/79*; 12–13 from *ER 215A* (type of *Brachythecium antarcticum* var. *cavifolium*), BR; 18, 22 from *ER 232A* (type of *B. antarcticum*), BR; 20, 25 from *Naumann s.n.*, 1874 (lectotype of *Hypnum austrosalebrosum*), LE]. — Scale bars: a – 1 mm (2–3, 12–20); b – 0.5 cm (1), 0.5 mm (9–11) and 100 µm (4, 21–26); c – 100 µm (5–8).

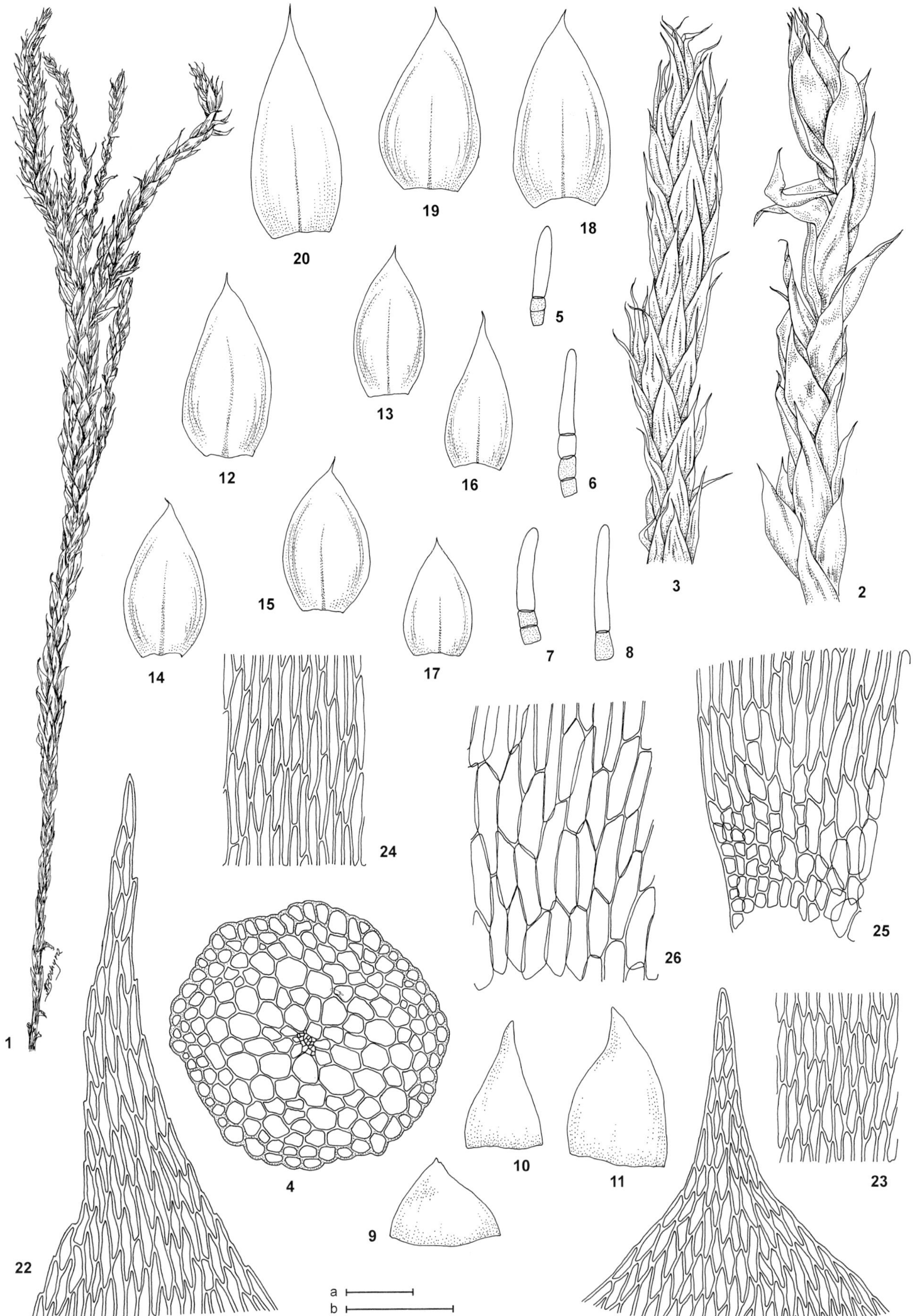

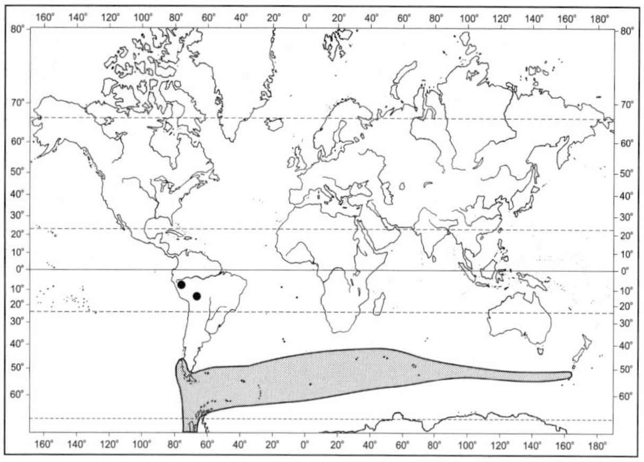

Figure 258. Global distribution of *Brachythecium austrosalebrosum*.

Macquarie I. but judging from his excellent illustration this material should correctly be referred to *B. austrosalebrosum*.

Distribution in Antarctica. A widely distributed and locally common and abundant species throughout the maritime Antarctic (Fig. 259), extending from Bouvetøya and the South Sandwich Is. through the South Orkney Is. to the South Shetland Is. and the western side of the Antarctic Peninsula. Here, *Brachythecium austrosalebrosum* is infrequent and rare on the Davis Coast, but rather common on the Danco Coast and offshore islands, decreasing in its frequency southwards, becoming infrequent on the Graham Coast and very rare on the Loubet Coast. It appears to be more abundant in northern Marguerite Bay, with a few highly isolated stations on Charcot I. (Convey *et al.* 2000), in the northern part of George VI Sound and reaching its southernmost station at lat. 70°03′ S in the Douglas Range on Alexander I. where it was found at its highest elevation of 600 m. Moreover, it occurs in the northern part of the east side of the Antarctic Peninsula in the Joinville Island and James Ross Island groups, extending southwards to lat. 64°08′ S. In general, *B. austrosalebrosum* is a species of lower elevations, reaching maximum frequency from sea level to about 100–150 m and only occasionally has it been recorded at elevations above 200 m.

Selected specimens examined. WEST ANTARCTICA. BOUVETØYA. Rustadkollen, 300–340 m, *TE 259* (BG); nunatak east of Rustadkollen, 295 m, *TE 69.1* (BG); north of Aagaardbreen, 80–100 m, *TE 370 & 371* (BG); Nordstranda, 170 m, *TE 54.1* (BG) and 175 m, *TE 66.1* (BG); Moseryggen, 260–287 m, *TE 251* (BG); Nyrøysa, 0–51 m, *TE 64, 97, 276, 312, 313, 346.1 & 358* (BG); Kapp Circoncision, 20 m, *MWH 865* (AAS, KRAM); south of Kapp Circoncision, 2 m, *MWH 863A* (AAS, KRAM) and 15 m, *MWH 867B* (AAS, KRAM).

SOUTH SANDWICH IS. **VINDICATION I.** North end of centre islet of trio close to east coast, *KA 3* (AAS, KRAM); south slope of the island, *ca* 200 m, *MWH 853 & 855* (AAS, KRAM); east facing cliff towards northern end of the island, 100 m, *REL 717* (AAS, KRAM); east side of the island, 135 m, *REL & MWH 735* (AAS, KRAM); north-east coast of the island, 100 m, *MWH 851B* (AAS, KRAM). **CANDLEMAS I.** South of lagoons, base of main cliff, 15 m, *MWH 445B* (AAS, KRAM) and 35 m, *REL 588* (AAS, KRAM); above penguin rookery by southern shore of eastern lagoon, *REL 593* (AAS, KRAM); under crag projecting through north end of ice cap, 50 m, *REL 685* (AAS, KRAM); Lucifer Hill, N-facing crags south of Demon Pt., *PC 118A, 122C & 123A* (AAS, KRAM); north-west of Lucifer Hill, *PC 230A & 241C* (AAS, KRAM); Clinker Gulch, *PC 159B & 160* (AAS, KRAM); above Medusa Pool and Chimaera Flats, *PC 172C* (AAS, KRAM); Gorgon Pool, *PC 00174* (AAS, KRAM); Demon Pt., *PC 187B* (AAS, KRAM); Breakbones Plateau, *PC 207B & 212A* (AAS, KRAM). **BELLINGSHAUSEN I.** South coast, 20 m, *PC 76A* (AAS, KRAM).

SOUTH ORKNEY IS. **CORONATION I.** Sandefjord Bay, *Discovery Exped. 5* (BM); Maier Pt., 5–50 m, *RILS 580* (AAS, KRAM); Mansfield Pt., 15 m, *RILS 290* (AAS, KRAM); Cape Vik, 35–100 m, *RILS 561* (AAS, KRAM); Cape Hansen, 20–200 m, *RILS 3227* (AAS, KRAM); Windscoop Buttress, 175 m, *RW 51A* (AAS, KRAM); Olivine Pt., 15–35 m, *RILS 128* (AAS, KRAM); Saunders Pt., 5 m, *RILS 28* (AAS, KRAM) and *ca* 7 m, *RILS 35A* (AAS, KRAM). **LYNCH I.** 5–15 m, *RILS 508* (AAS, KRAM). **MATTHEWS I.** The Divide, *ca* 50 m, *RILS 175* (AAS, KRAM). **POWELL I.** Michelsen I., 25 m, *RILS 40* (AAS, KRAM); Falkland Harbour, *RILS 51A* (AAS, KRAM); Ellefsen Harbour, *ca* 10 m, *RILS 47* (AAS, KRAM). **MICHELSEN I.** East side, *ca* 25 m, *RILS 40* (AAS, KRAM). **LAURIE I.** Scotia Bay, *Valette 12* (PC). **SIGNY I.** Foca Cove, 5 m, *RILS 8052B* (AAS, KRAM); Moss Lake, 50 m, *RILS 10760* (AAS, KRAM); Three Lakes Valley, north of Jane Peak, *ca* 85 m, *RILS 598* (AAS, KRAM); Mirounga Flats, 10 m, *RILS 386* (AAS, KRAM); Elephant Flats, 15 m, *RILS 649* (AAS, KRAM); Knife Pt., 3 m, *RILS 691C* (AAS, KRAM); Observation Bluff, 5 m, *RILS 8079B* (AAS, KRAM); Snow Hill, *ca* 215 m, *RILS 626* (AAS, KRAM); Paal Harbour, *MWH 753C* (AAS, KRAM). **MOE I.** 2–100 m, *RILS 5246* (AAS, KRAM) and 75 m, *KR 608* (AAS, KRAM).

SOUTH SHETLAND IS. **GIBBS I.** The Spit, 30–120 m, *JPB 210* (AAS, KRAM). **KING GEORGE I.** ***Fildes Peninsula***: Suffield Pt., 4 m, *RO 2441/80* (KRAM); Bellinghausen Station, 15 m, *RO 2423/80* (KRAM); Lake Kitezh, 16 m, *RO 2460/80* (KRAM); Flat Top Peninsula, 10 m, *John & Sudgen 16A* (AAS, KRAM). ***Drake Passage***: Ridley I., *BJ 29* (KRAM). ***Destruction Bay***: Cape Melville, *BJ 81* (KRAM). ***Sherrat Bay***: Three Sisters Pt., *BJ 122* (KRAM). ***King George Bay***: Turret Pt., *BJ 130* (KRAM); Lions Rump, *BJ 98* (KRAM). ***Admiralty Bay***: Vauréal Peak, 30 m, *RO 5257/79* (KRAM); Cape Vauréal, 35 m, *RO 5256/79* (KRAM); Blue Dyke, 10 m, *RO 1202/80* (KRAM); Bastion, 240 m, *RO 1008/80* (KRAM); Uchatka Pt., 20 m, *RO 1053/80* (AAS, KRAM); Demay Pt, 40 m, *RO 1382/80* (KRAM); Creeping Slopes, 60 m, *RO 1293/80 & RO 1295/80* (KRAM); Rescurs Hills, 35 m, *RO 4881/79* (AAS, KRAM); Llano Pt., 15 m, *RO 4916/* (AAS, KRAM); Rakusa Pt., 8 m, *RO 319/80* (KRAM); Penguin Ridge, 45 m, *RO 1432/80* (AAS, KRAM); Jasnorzewski Gardens, 3 m, *RO 1494/80* (KRAM); Hala, 18 m, *RO 5214/79* (KRAM); Upłaz, 50 m, *RO 2356/80* (KRAM); Ambona, 85 m, *RO 1646/80* (KRAM). ***Ezcurra Inlet***: Jardine Peak, 230 m, *RO 5195/79* (AAS, KRAM); Kasprowy Hill, 200 m, *RO 72/80* (KRAM). ***Keller Peninsula***: British Pt., 5 m, *RO 460/80* (KRAM). ***Martel Inlet***: Basalt Pt., 10 m, *RO 2237/80* (KRAM); Mt. Wawel, 20 m, *RO 2193/80* (KRAM). **NELSON I.** Moss Bay, 26 Feb 1989, *Komárek s.n.* (KRAM); Harmony Cove, *CS 469* (PC, S). **ROBERT I.** Coppermine Cove, 15–30 m, *RILS 912 & 914* (AAS, KRAM). **AITCHO I.** 3–80 m, *RILS 11452* (AAS, KRAM). **LIVINGSTON I.** ***Byers Peninsula***: Lair Pt., 5 m, *KR 65A* (AAS, KRAM); Pt. Smellie, 10 m, *KR 35F* (AAS, KRAM); between Pt. Smellie and Devils Pt., 5 m, *KR 291* (AAS, KRAM). ***Hurd Peninsula***: Spanish Station, *Schulz 136* (KRAM); Miers Bluff, *RMS 69-801, 806, 826 & 827* (US); Barnard Pt., 30 m, *DCL 7* (AAS, KRAM). **DECEPTION I.** Baily Head, 75 m, *DM 70A* (AAS, KRAM); Sewing-Machine Needles, *Bennett 9, 10 & 11* (BM); between South Pt. and Baily Head, 100 m, *REL 860* (AAS, KRAM); Whalers Bay, 150 m north-east of the former whaling station, 75 m, *DM 94A* (AAS, KRAM); Ronald Hill, 65–80 m, *REL 14* (AAS, KRAM); Telefon Bay, 3–5 m, *RILS 11371* (AAS, KRAM); Stonethrow Ridge between Argentine Base and Wensleydale Beacon, 7 m, *NJC 18A* (AAS, KRAM); Fumarole Bay, near the Argentine Base, 15 m, *RO 2797/80* (KRAM) and 100 m, *RILS 11497* (AAS, KRAM); Vapour Col, 130 m, *RILS 5633B* (AAS, KRAM); Crater Lake, 30 m, *RILS 11415* (AAS, KRAM); Collins Pt., 5–15 m, *RILS 58* (AAS, KRAM), 20–30 m, *RILS 3605B & 3609* (AAS, KRAM) and 65 m, *REL 881A* (AAS, KRAM); ridge running south-west of Entrance Pt., 80–170 m, *RILS 6784B* (AAS, KRAM); Entrance Pt., 100 m, *DM 6B* (AAS, KRAM).

WEST ANTARCTIC PENINSULA. **TRINITY PENINSULA.** Astrolabe I., 50–70 m, *RILS 4094* (AAS, KRAM). **DAVIS COAST.** Cape Kjellman, *CS 473* (PC, S); Cape Andreas, *CS 474* (PC, S); Tower I., 15–30 m, *NJC 93* (AAS, KRAM); Cape Roquemaurel, 10–50 m, *RILS 3921* (AAS, KRAM). **DANCO COAST.** Cierva Pt., *VK 485B, 492, 597, 608, 614, 664 & 711* (KRAM); Spring Pt., *VK 474, 911 & 974* (KRAM); Challenger I., *CS 470 & 471* (PC, S); Charlotte Bay, Recess Cove, 2–10 m, *RILS 3982* (AAS, KRAM); Cape Anna, *ER 151A & 205D* (BR, PC); Cuverville I., 20 m, *RILS 8806* (AAS, KRAM), 30 m, *RILS 8805* (AAS, KRAM) and 40 m, *RILS 8803 & 8804* (AAS, KRAM); Danco I., *RILS 7159* (AAS, KRAM); Birdsend Bluff, 5–35 m, *RILS 4216* (AAS, KRAM); Beneden Head, *ER 230A, 232A, 232C, 233C, 234C* (BR, PC); Andvord Bay, Neko Harbour, 15–20 m, *RILS 11930B & 11931A* (AAS, KRAM); Paradise Harbour, Coughtrey Peninsula, Almirante Brown Station, *VK 225, 338, 375, 376A, 479, 1014 & 1023*

(KRAM) and *Schmitt 214B* (KRAM, NY), 10 m, 7 Jan 1985, *Birkenmajer s.n.* (KRAM) and 100 m, *RILS 11614B & 11615* (AAS, KRAM); Waterfall Buttress, 20 m, 14 Jan 1985, *Birkenmajer s.n.* (KRAM). **PALMER ARCHIPELAGO.** ***Christiania Is.***: Intercurrence I., 8 m, *DWB 1A* (AAS, KRAM). ***Gerlache Strait***: Auguste I., *ER 215A* (BR, PC); Two Hummock I., *RILS 704D* (AAS, KRAM). ***Melchior Is.***: Lysted I., Channel Pt., *PAS 335.2 & 378* (FH); Omega I., 2–10 m, *RILS 4115* (AAS, KRAM); Gamma I. 5–10 m, *RILS 4113* (AAS, KRAM). ***Brabant I.***: Metchnikoff Pt., 100 m, *WH 19B* (AAS, KRAM); Claude Pt., *WH 207C* (AAS, KRAM), 200 m, *WH 206B* (AAS, KRAM) and 300 m, *AM 2B* (AAS, KRAM); south-west of Hill Bay, 5 m, *AM 54B* (AAS, KRAM). ***Anvers I.***: Norsel Pt., *VK 229* (KRAM) and *RILS 2374* (AAS, KRAM); Dream I., *VK 689 & 908* (KRAM); Hermit I., 15–25 m, *RILS 880 & 881* (AAS, KRAM); Goudier I., Port Lockroy, *RILS 727* (AAS, KRAM). **GRAHAM COAST.** Cape Tuxen, 30–300 m, *RILS 4887* (AAS, KRAM). ***Grandidier Channel***: Takaki Pt., 10–35 m, *RILS 4337A* (AAS, KRAM); Lahille I., 2–5 m, *RILS 4311* (AAS, KRAM). **WILHELM ARCHIPELAGO.** ***Lemaire Channel***: Booth I., *Turquet 239A, 239B & 245B* (PC). ***Penola Strait***: Petermann I., *Gain 166D, 166E & 212* (PC) and 5–25 m, *RILS 4254* (AAS, KRAM). ***Argentine Is.***: Uruguay I., 8 m, *RILS 946* (AAS, KRAM); Galindez I., 15 m, *RILS 758* (AAS, KRAM). ***Collins Bay***: Berthelot Is., Green I., *BGLE 1084* (BM) and 20–60 m, *RILS 3372* (AAS, KRAM); Darboux I., *RILS 3322* (AAS, KRAM). **BISCOE IS.** Hook I., *RMS 69-511* (US). **LOUBET COAST.** Orford Cliff, *RILS 5270B & 5281B* (AAS, KRAM). **FALLIÈRES COAST.** ***Adelaide I.***: Rothera Pt., 10 m, *RILS 770* (AAS, KRAM) and 15 m, *RILS 94* (AAS, KRAM); Avian I, *ca* 15 m, *RILS 99, 100 & 775A-B* (AAS, KRAM). ***Léonie Is.***: Lagoon I., 5 m, *RILS 9146* (AAS, KRAM); Léonie I., 30 m, *RILS 8927C, 8952 & 8977* (AAS, KRAM) and 150–175 m, *RILS 9134* (AAS, KRAM); Anchorage I., 25 m, *RILS 11841B* (AAS, KRAM). ***Marguerite Bay***: Guébriant Is., 40 m, *DWB 7A* (AAS, KRAM); Jenny I., 15–50 m, *RILS 4718* (AAS, KRAM); Limpet I., 20 m, *DWB 9 & 10* (AAS, KRAM); Blaiklock I., 15–20 m, *RILS 845* (AAS, KRAM); Pourquoi Pas I., 5–20 m, *RILS 4771* (AAS, KRAM); Lagotellerie I., *ca* 50 m, *RILS 1768, 1769 & 1770* (AAS, KRAM). **RYMILL COAST.** Rhyolite Is., below 50 m, *PC 639 & 640* (AAS, KRAM). **GEORGE VI SOUND.** ***Charcot I.***: central north-facing slope, *RILS 10525 & 10709* (AAS) and 85 m, *RILS 10507C* (AAS, KRAM). ***Alexander I.***: Zebra Ridge, *ca* 155 m, *DDWW 150* (AAS, KRAM) and 600 m, *RILS 10028* (AAS, KRAM).

EAST ANTARCTIC PENINSULA. **JOINVILLE ISLAND GROUP.** ***Paulet I.***: *CS 476, 477 & 478* (PC, S). **JAMES ROSS ISLAND GROUP.** ***Vega I.***: Cape Lamb, 200 m, *RILS 5387B* (AAS, KRAM); False Island Pt., 25–30 m, *RILS 7826* (AAS, KRAM). ***James Ross I.***: Fortress Hill, 25–150 m, *RILS 7808 & 7771* (AAS, KRAM); Blancmange Hill, 12–15 m, *RILS 7685* (AAS, KRAM); St. Martha Cove, 100 m, *RILS 5376* (AAS, KRAM); valley south of St. Martha Cove, 15 m, *RILS 7622* (AAS, KRAM); Lachman Crags, 2–35 m, *RILS 7318* (AAS, KRAM) and 400 m, *RILS 5396* (AAS, KRAM); Brandy Bay, 250 m, *RILS 5385B* (AAS, KRAM); Tumbledown Cliffs, 60 m, *RILS 5380A* (AAS, KRAM) Röhss Bay, Cape Obelisk, 5 m, *RILS 5504A* (AAS, KRAM); Ula Pt., 30 m, *RILS 5378B* (AAS, KRAM).

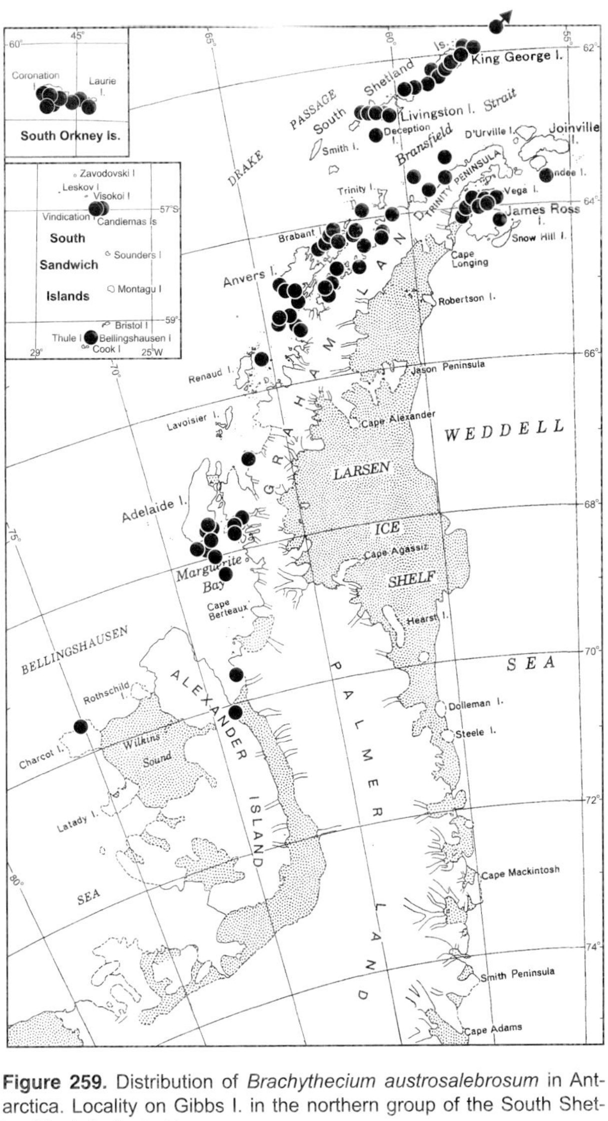

Figure 259. Distribution of *Brachythecium austrosalebrosum* in Antarctica. Locality on Gibbs I. in the northern group of the South Shetland Is. is indicated by an arrow.

3. **Brachythecium subpilosum** (Hook.f. & Wilson) A.Jaeger Fig. 260

Brachythecium subpilosum (Hook.f. & Wilson) A.Jaeger, Ber. Thät. St. Gall. Naturw. Ges. 1876-1877: 342. 1878. ≡ *Hypnum subpilosum* Hook.f. & Wilson, London J. Bot. 3: 553. 1844. — TYPE CITATION: Hermite Island, Cape Horn. [Lectotype (*selected here*): 'Hypnum subpilosum Cape Spencer (C. Horn) W. 165' – BM-Wilson!; isotypes: E!, PC-Cardot!].

Plants medium-sized, fairly stiff, glossy, bright- to yellow-green or yellow-brown, forming extensive, dense mats or shallow tufts. *Stems* creeping to ascending, 1.5–5.5(–10.5) cm long, freely branched, rigid, closely foliate, ± julaceous and terete, in cross-section with a small but distinct central strand, 5–6-stratose medulla of large, thin-walled and hyaline cells and a cortex consisting of small, brown, thick-walled cells in 2–3 layers; *pseudoparaphyllia* foliose, orbicular to broadly ovate-apiculate; *rhizoids* abundant, in fascicles on the ventral side of the stem facing the substrate, weakly branched, smooth, reddish-brown; *axillary hairs* short, filiform, 3–4-celled, with a short, brown basal cells and elongate, hyaline apical cell. *Leaves* imbricate, erect to erecto-patent when dry, little altered when moist, (1.2–)1.5–2.5(–3.0) mm long, (0.5–)0.7–1.3(–1.5) mm wide, shallowly to deeply concave, not or slightly decurrent, strongly plicate, triangular to broadly ovate-cordate or ovate-lanceolate, suddenly constricted into a long, subpiliferous acumen; *margins* recurved immediately above the base, plane above, serrulate all around, sometimes subentire; *costa* stout, gradually tapering upwards, 40–70 µm wide at base, reaching the mid-leaf or somewhat higher; *laminal cells* smooth, pellucid, unistratose; *upper* and *median cells* linear-flexuose to fusiform, thin-walled, (50–)55–115(–140) µm long, 6–11 µm wide, *ca* 9–16 times as long as wide, becoming shorter towards the apex; *basal cells* shorter and wider, 20–85 µm long, 8–17 µm wide, rectangular to oblong hexagonal, thicker-walled and slightly pitted; *alar cells* rounded, quadrate to short rectangular, thick-walled,

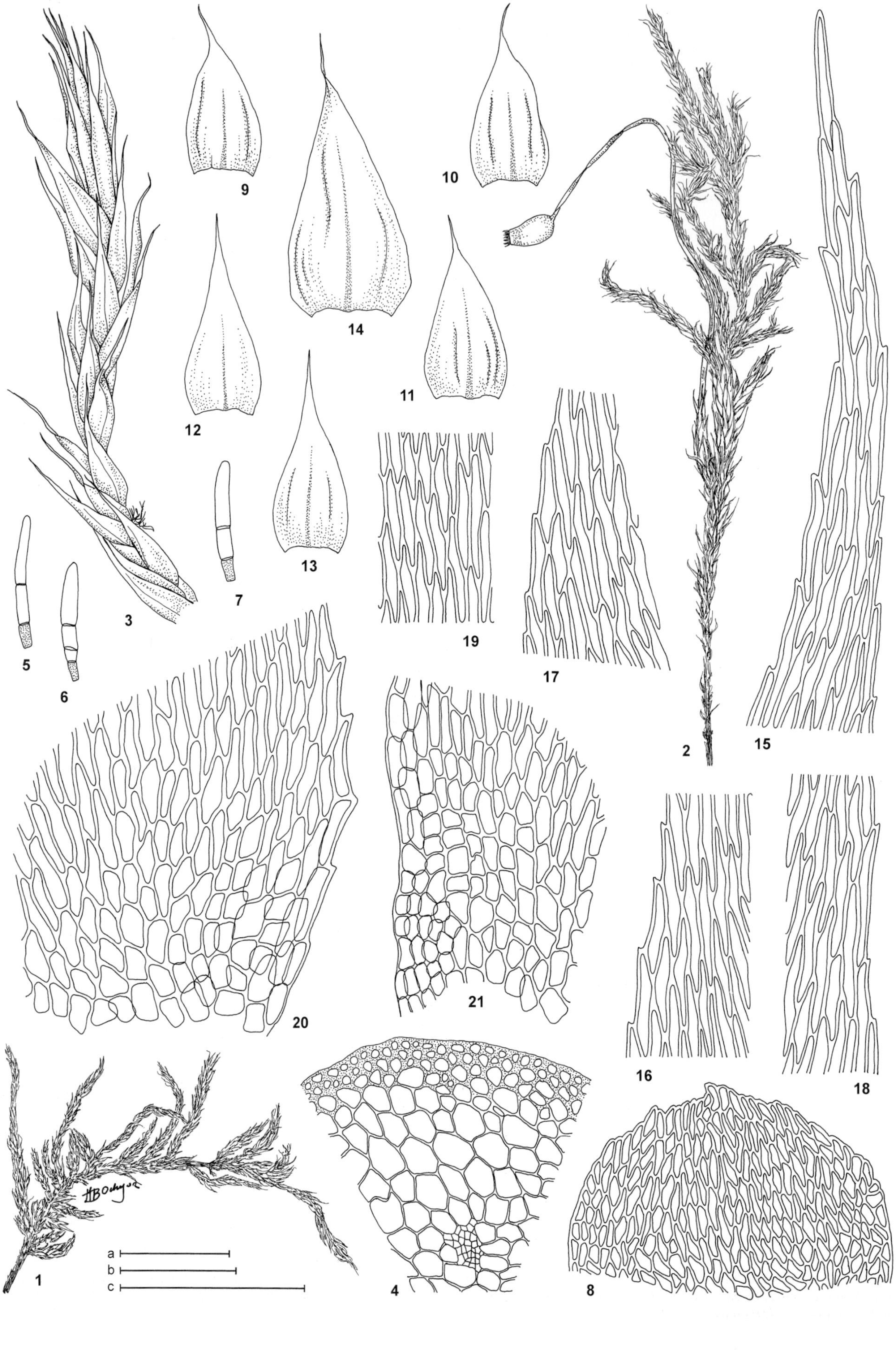
1
2
3
4
5
6
7
8
9
10
11
12
13
14
15
16
17
18
19
20
21
a
b
c

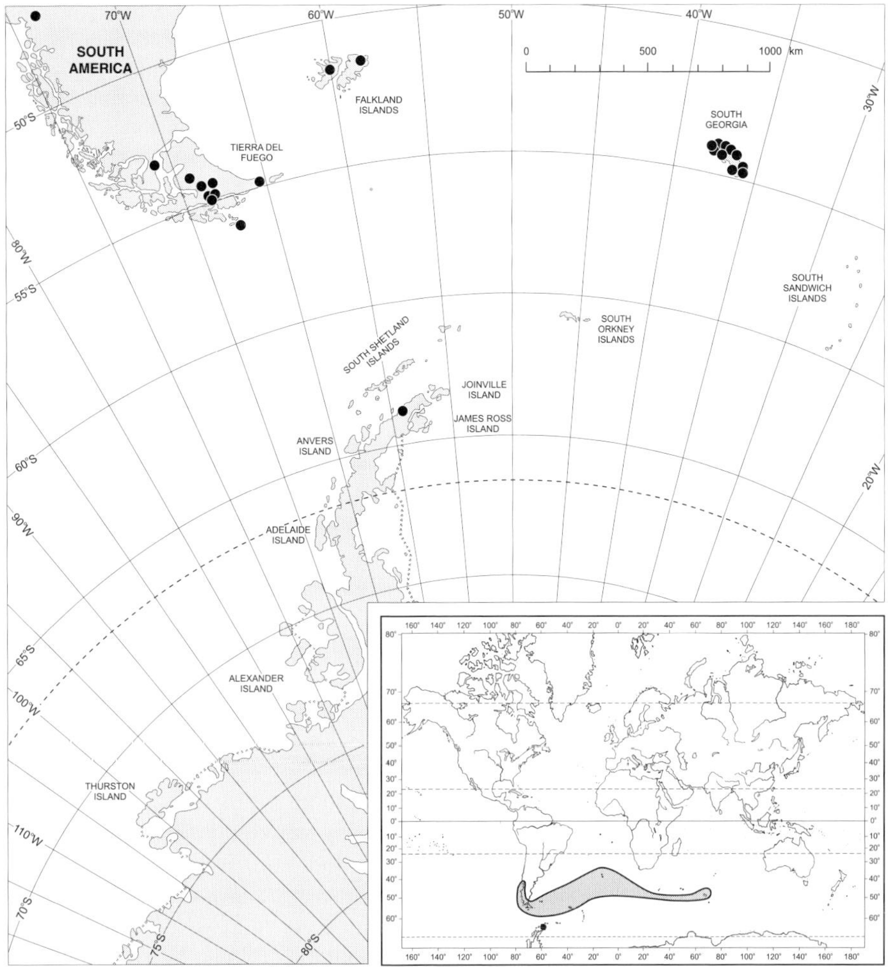

Figure 261. Distribution of *Brachythecium subpilosum* in the Fuegian region and Antarctica. Inset: global distribution.

somewhat opaque, seriate, forming distinct auricles, not or shortly decurrent. [*Autoicous*. *Inner perichaetial leaves* lanceolate, abruptly long-acuminate, ecostate, with plane margins, entire or serrulate at the apex. *Setae* red to orange-red, 1–2 cm long, scabrous throughout with low diffuse papillae, sometimes nearly smooth. *Capsules* inclined, obloid, curved, 1.3–2.5 mm long, 0.8–1.1 mm wide, red-brown; *operculum* conic, apiculate, 0.6–1.0 mm long; *annulus* biseriate, deciduous; *peristome* double; *exostome teeth* lanceolate, pale orange, transversely striolate below, paler and papillose above; *endostome* consisting of a high basal membrane giving rise to broadly perforated segments alternating with 2–3, nodose and appendiculate cilia. *Spores* globose, (13–)15–20(–24) µm in diameter, minutely roughened. *Calyptra* cucullate, naked.]

Chromosome number: $n = 13$ – South Georgia (Newton 1979*a*).

Discussion. *Brachythecium subpilosum* is a distinct species distinguished by its triangular, subpiliferous leaves. It may be confused in the field with *Sciuro-hypnum glaciale* but the plicate and not or scarcely decurrent leaves of *B. subpilosum* contrast with the smooth and broadly and long-decurrent leaves in the latter species and this should preclude misidentification of the two species. *Brachythecium austroglareosum* is distinct in its wholly recurved leaf margins and *B. austrosalebrosum* has abruptly short-acuminate leaves.

Apart from *Brachytheciastrum paradoxum* (Hook.f. & Wilson) Ignatov & Huttunen, *Brachythecium subpilosum* is the oldest species of a brachythecialean moss described from southern South America. It was discovered by J. D. Hooker on Hermite I. near Cape Horn during his Antarctic voyage of 1839–1843 and subsequently described by Hooker & Wilson (1844) as *Hypnum subpilosum*. Jaeger (1878) transferred it to the genus *Brachythecium* and Brotherus (1908, 1925) placed it in sect. *Salebrosa*, whereas McFarland (1988) transferred it to sect. *Reflexa* which is currently recognised as a separate genus, *Sciuro-hypnum* (Ignatov & Huttunen 2002). *Brachythecium subpilosum* shares with the latter taxon papillose setae and triangular and rapidly acuminate leaves, but these features are also found in other species of sect. *Salebrosa*. On the other hand, the lack of prominent decurrent auricles and strongly plicate leaves clearly place *B. subpilosum* in *Brachythecium* sect. *Salebrosa*.

←

Figure 260. *Brachythecium subpilosum*. — **1–2**. Habit. **3**. Portion of branch, dry. **4**. Portion of cross-section of stem. **5–7**. Axillary hairs. **8**. Pseudoparaphyllium. **9–14**. Leaves. **15**. Leaf apex. **16–17**. Upper leaf cells at margin. **18**. Mid-leaf cells at margin. **19**. Mid-leaf cells. **20–21**. Alar cells. — [1, 3, 9–11, 17–19, 21 from *CS 472*, PC; 2, 4–8, 12–16, 20 from *Bell 1052*]. — Scale bars: a – 0.5 cm (1–2) and 100 µm (8); b – 1 mm (3, 9–14); c – 50 µm (4) and 100 µm (5–7, 15–21).

Reproduction in Antarctica. The only Antarctic specimen is sterile, but the species frequently fruits on South Georgia and in Tierra del Fuego.

Habitat. No information on the habitat is available on the label for the only Antarctic collection of *Brachythecium subpilosum*, but Skottsberg (1912) wrote that the species was found 'zwischen den Felsen dicht oberhalb des Ufers' together with *Warnstorfia sarmentosa* and *Sanionia georgicouncinata*, and also *Cladonia pyxidata* and *Prasiola crispa*. This suggests a habitat similar to that in which the species grows elsewhere in the Subantarctic where it is usually found on moist rocks, stream banks, rock outcrops and ledges, and on whale bones. In Tierra del Fuego *B. subpilosum* is a weedy species thriving in a wide range of wet and dry habitats on rocks and soil in grasslands, rotting logs and stumps, and at the bases of trees and on the ground in woodland.

World range. An amphiatlantic temperate species common in the Magellanian region of western Patagonia and reaching to the Valdivian region at lat. *ca* 40° S, in Tierra del Fuego and on the Falkland Is., as well as on the subantarctic islands of South Georgia (Newton 1979*b*) and Îles Kerguelen (Brotherus 1906*a*), extending to the Tristan da Cunha group (Dixon 1960) and Gough I. (Cardot 1911*c*; Bartram 1959) and to the northern Antarctic Peninsula (Fig. 261).

Distribution in Antarctica. The species has been collected only once on the Davis Coast on the west side of the Antarctic Peninsula (Fig. 261). It is remarkable that the only specimen known from the Antarctic was collected over 100 years ago.

Specimen examined. WEST ANTARCTICA. WEST ANTARCTIC PENINSULA. **DAVIS COAST.** Cape Kjellman, *CS 472* (PC, S, UPS).

51. SCIURO-HYPNUM HAMPE

Sciuro-hypnum Hampe, Linnaea 38: 220. 1874. — Type: *Sciuro-hypnum borgenii* Hampe.

Chionobryum Głow., Oesterr. Bot. Z. 63: 279. 1913. — Type: *Chionobryum venturii* (De Not.) Głow. (*Bryum venturii* De Not.) [= *Sciuro-hypnum glaciale* (Schimp.) Ignatov & Huttunen]. First synonymised by Ignatov & Huttunen (2002: p. 269).

Cratoneurella H.Rob., Bryologist 65: 140. 1962. — Type: *Cratoneurella uncinifolia* (Broth. & Paris) H.Rob. (*Brachythecium uncinifolium* Broth. & Paris). First synonymised by Ignatov & Huttunen (2002: p. 269).

Plants small, medium-sized to large, slender to moderately robust, soft or fairly rigid and stiff, loosely or densely caespitose, or forming wefts, dense or loose extensive patches or tightly adhering mats, mostly glossy, dark or pale green, yellowish-green, yellow, brownish or yellow- or golden-brownish. *Stems* prostrate, ascending, arched to erect, slender and thread-like to fairly robust, turgid and terete, sparsely or closely, irregularly to pinnately branched, with branches often confined to one side of the stem, ascending, erect or sometimes curved, julaceous or with a complanate foliage, in transverse section with a distinct central strand, 2–3-stratose cortex of small, thick-walled substereid cells surrounding a 3–6-stratose medulla of large, parenchymatous, thin-walled, hyaline cells; *pseudoparaphyllia* foliose, ovate to triangular, obtuse to apiculate; *axillary hairs* filiform, 2–3(–5)-celled, consisting of 1–2 elongate, hyaline distal cells and 1–3 short, quadrate, brownish-pigmented basal cells. *Stem* and *branch leaves* usually differentiated, smooth or faintly striolate to indistinctly plicate, plane to strongly concave, long- or short-decurrent to non-decurrent; *stem leaves* crowded to more distant, erect-imbricate to erect-spreading, sometimes homomallous when dry, erect-spreading to patent when moist, cordate-triangular, broadly ovate-cordate, deltoid-ovate, lanceolate or ovate-lanceolate, gradually or abruptly narrowed to a short or long, channelled or plane, sometimes filiform, flexuose-spreading to reflexed, occasionally twisted acumen; *margins* strongly recurved above the base, plane above, serrulate to finely denticulate distally or entire throughout, sometimes with a few teeth at the base of the acumen, sometimes excavate at the basal angles and broadly decurrent to not or only scarcely decurrent; *costa* ceasing in or above the mid-leaf to subpercurrent, not ending as a dorsal spine, occasionally shorter and forked, smooth on the dorsal side or occasionally serrulate at the back at or near the tip; *branch leaves* usually shorter and narrower, often more crowded, more loosely erect to imbricate than the stem leaves, less decurrent or non-decurrent, with more strongly serrulate margins and costa ending in a spine; *laminal cells* unistratose, pellucid, smooth; *upper* and *median cells* oblong-hexagonal, shortly oblong-rhomboidal, linear-rhomboidal to linear-flexuose, thin- to moderately thick-walled; *basal cells* rhomboidal to rectangular, usually wider than the upper cells, often with pitted walls, sometimes opaque and brownish; *alar cells* quadrate, rounded-quadrate to shortly oblong, usually with thicker and porose walls, somewhat opaque, usually forming long or short decurrencies, occasionally non-decurrent. *Asexual reproduction* unknown. *Autoicous* or very rarely *dioicous*. *Inner perichaetial leaves* lanceolate to oblong-lanceolate, from an erect, broad sheathing base fairly suddenly narrowed into a short or elongate, often reflexed acumen, ecostate or with a faint single costa. *Setae* elongate, straight, often stout, red, purple, red- or orange-brown to red-yellow, scabrous throughout or only in the upper part, smooth below. *Capsules* obloid, ellipsoid or ovoid, suberect, inclined to horizontal, asymmetric, gibbous, not constricted below the mouth, brownish, becoming blackish with age; *operculum* convex-conic or high-conic, acute, mammillate to shortly apiculate, rarely rostrate; *annulus* occasionally totally indiscernible, most often consisting of 1–2 rows of large, vesiculose, thick-walled cells, fragmenting, sometimes persistent; *exothecial cells* variable in shape, rounded-quadrate, rectangular, rhomboidal to hexagonal, not or arranged in ± distinct rows, with evenly thickened walls or with the vertical walls more thickened than the horizontal ones; *stomata* bicellular, phaneroporous, with oval pori, numerous at the base of the urn; *peristome* double; *exostome teeth* lanceolate-subulate, reddish- or yellow-brown to red, cross-striolate below, becoming pale yellowish to hyaline, papillose above, distinctly margined, dentate near the apex; *endostome* yellow or pale orange, consisting of a high, finely papillose to nearly glabrous basal membrane, comprising about half of the peristome length, perforated processes and appendiculate or nodose cilia in groups of 1–3. *Spores* spherical, yellow, yellow-green to brown, finely papillose, rather small. *Calyptra* cucullate, smooth, naked.

Sciuro-hypnum is a rather medium-sized segregate of the broadly conceived genus *Brachythecium*, being roughly an equivalent of its sect. *Reflexa* and subgen. *Cirriphyllopsis* Broth. (Brotherus 1908, 1925). Species which constitute the core of the genus are characterised by having plane or concave, imbricate, smooth or faintly plicate leaves that are triangular-cordate to broadly ovate-cordate with the alar cells forming usually long and broad decurrencies. In addition, they have setae that are mostly papillose, at least in the distal portion. *Sciuro-hypnum* consists of about 20–25 species which are primarily widely distributed in the Northern Hemisphere, with some species occurring in Africa and South America. However, the real number of its species cannot be precisely established until some exotic taxa are taxonomically assessed.

Although *Sciuro-hypnum* has only recently been resurrected, it has gained some acceptance in the bryological literature (e.g. Ochyra *et al.* 2003; Ignatov & Ignatova 2004; Hill *et al.* 2006; Ignatov *et al.* 2006). In some of these treatments the incorrect orthography '*Sciurohypnum*' is used. When establishing this generic name, Hampe (1874) used the hyphenised version of this name and this original spelling has to be retained and deletion of the hyphen may only be effected by conservation.

In the cool-temperate and polar regions of the Southern Hemisphere *Sciuro-hypnum* is represented by three species occurring only in the American sector, namely *S. filirepens* (Dusén) Ochyra & Żarnowiec, *S. glaciale* and *S. fuegianum*. The first of these is endemic to the Valdivian and Magellanian provinces in western Patagonia of Chile (McFarland 1988), whereas the other two species occur on subantarctic South Georgia and in the northern maritime Antarctic.

1. Stem leaves broadly triangular-cordate abruptly narrowed into a long acumen; leaf apex mostly obtuse usually forming shoulders at the base of the acumen; margins with few large and sharp teeth at the base of the acumen; large plants .... 1. ***S. fuegianum***
1. Stem leaves ovate-cordate, gradually long- or short-acuminate; leaf apex not forming shoulders at the base of the acumen; margins extensively serrate to serrulate above but the teeth undifferentiated; smaller plants .... 2. ***S. glaciale***

1. Sciuro-hypnum fuegianum (Broth.) Ochyra & Żarnowiec Fig. 262

Sciuro-hypnum fuegianum (Broth.) Ochyra & Żarnowiec, Cens. Cat. Polish Mosses: 174. 2003. ≡ *Brachythecium fuegianum* Broth. *in* Engl. & Prantl, Nat. Pflanzenfam. 1(3): 1143. 1909. ≡ *Stereophyllum fuegianum* Besch. *in* Hariot, Miss. Sc. Cape Horn Bot.: 302, *t. 5, f. 19.* 1889, *hom. illeg.* ≡ *Brachythecium stereophyllum* Müll.Hal. *in* Besch. *in* Hariot, Miss. Sc. Cape Horn Bot.: 302. 1889, *hom. illeg.* — TYPE CITATION: Terre de Feu: île L'Hermite, baie Saint-Martin (Hariot). [Lectotype (*selected here*): 'Herb. Émil Bescherelle: 1900. ~~Hypnum~~ Stereophyllum fuegianum Besch. 138 Ile Hermite (T de Feu) Baie St Martin 1 julliet 83 dH.' – BM-Bescherelle!; isotype: BM-Bescherelle!, PC!, PC-Bizot!].

Brachythecium stereophyllum Müll.Hal. *in* Besch. *in* Hariot, Miss. Sc. Cape Horn Bot.: 302. 1889, *nom. inval. in synon.*

Brachythecium majusculum M.E.Newton, Brit. Antarct. Surv. Bull. 39: 45, *f. 1.* 1974 [*Brachythecium majusculum* Dusén *in* W.B.Scott, Rep. Princeton Univ. Exp. Patagonia 8(1) Botany: 125. 1903, *nom. nud.*]. ≡ *Sciuro-hypnum majusculum* (M.E.Newton) Ignatov & Huttunen, Arctoa 11: 270. 2003, ***syn. nov.*** — TYPE CITATION: Amongst wet *Deschampsia* sward, alt. *c.* 200 ft. (*c.* 61 m.), behind Jordan Cove, Bird Island, South Georgia, grid reference 030 154, Leg. S. W. Greene, 15.xii.1960, Greene 266. [Holotype: 'British Antarctic Survey Flora of South Georgia Brachythecium majusculum M. E. Newton Amongst wet *Deschampsia* sward. Alt. c. 200 ft. Behind Jordan Cove, Bird Island, GR 030 154 Leg. S. W. Greene 15.12.1960 Det. M. E. Newton, June 1973 No. Greene 266' – AAS!; isotypes: H!, KRAM!, NY!, S!].

Plants medium-sized to large and robust, fairly rigid, in loose, glossy mats or deep tufts, yellowish-green to bright green above, whitish to drab below. *Stems* creeping to ascending, 2–7 cm tall, julaceous, terete, densely foliate, sparsely irregularly branched with branches of various length confined to one side or subpinnate, in cross-section with a small but distinct central duct, 5–7 layers of large, thin-walled and hyaline medullary cells and 1–2-stratose cortex of small, intensely yellow to yellow-brownish, thicker-walled cells; *rhizoids* sparse, brown, smooth, branched; *pseudoparaphyllia* foliose, orbicular to ovate, apiculate; *axillary hairs* short, filiform, 2–3-celled, with an elongate apical cell and 1–2 brown basal cells. *Leaves* imbricate, erecto-patent, little altered when wet, (1.7–)2.0–3.0(–3.5) mm long, (0.9–)1.2–1.9(–2.2) mm wide, deeply concave, slightly plicate, long and broadly decurrent, broadly ovate-triangular, triangular-cordate to ovate-cordate, mostly obtuse at the apex and suddenly constricted into a short or long, often aristate, channelled acumen, 0.3–0.8 mm long, apices of larger leaves forming shoulders at the base of the acumen, rarely gradually acuminate on smaller stem and branch leaves; *margins* recurved at base below the widest point of the lamina, plane above, entire below, serrulate above, usually with a few sharp teeth at the base of the acumen, rarely entire throughout; *costa* short and weak, 40–70 μm wide at base, gradually tapering upwards, sometimes forked or with short lateral spurs, reaching ¼–½ way up the leaf, rarely higher; *branch leaves* smaller, 1.1–2.3 mm long, 0.7–1.4 mm wide, more gradually acuminate; *laminal cells* smooth, transparent; *upper* and *median cells* linear-flexuose to fusiform, thin-walled, faintly pitted, 65–120 µm long, 6–13 µm wide, *ca* 6–12 times as long as wide, becoming shorter, to 40 µm, near the apex; *basal cells* laxer, shorter and wider, 40–100 µm long, 11–28 µm wide, with incrassate and porose walls; *alar cells* quadrate to shortly rectangular, seriate, with slightly thickened and non-porose walls forming clearly defined, triangular, long and broadly decurrent auricles. [*Autoicous*. *Inner perichaetial leaves* oblong-lanceolate, from an erect, broad sheathing base abruptly narrowed into a elongate, reflexed acumen, ecostate. *Setae* elongate, straight, 1.0–2.5 cm long, orange- to reddish-brown, scabrous throughout. *Capsules* broadly obloid to shortly cylindrical, suberect to inclined, curved, 1.5–2.7 mm long, 0.5–1.2 mm wide, dark reddish-brown to blackish-brown; *operculum* conic, apiculate, 0.5–0.8 mm long; *annulus* biseriate, fragmentarily deciduous; *peristome* double; *exothecial cells* short, rounded-quadrate to short-rectangular or rhomboidal, thick-walled horizontal ones; *peristome* double; *exostome teeth* lance-subulate, yellow-brown, cross-striolate below, hyaline and densely papillose above; *endostome* yellow, almost smooth, composed of a high basal membrane, perforated processes and 2–3 nodose and appendiculate cilia. *Spores* globose, green, 17–21 µm wide, almost smooth. *Calyptra* not seen.]

Chromosome number: unknown.

Discussion. *Sciuro-hypnum fuegianum* is one of the largest species in the genus and one of most robust brachythecialean mosses in the Antarctic. In its large size and julaceous stems and branches it resembles some phenotypes of *Brachythecium austrosalebrosum* but its soft growth form, and non-decurrent leaves sharply contrast with the rigid growth form, broadly ovate-cordate and longly and broadly decurrent leaves in *S. fuegianum*. Also, the broadly ovate-triangular, triangular-cordate to ovate-cordate, not plicate and distinctly shouldered leaves of *S. fuegianum* are completely unlike the leaves of *B. austrosalebrosum* which are broadly ovate to ovate-lanceolate, suddenly or gradually short-acuminate and distinctly plicate. In the shape of the leaf decurrencies *S. fuegianum* approaches very much the closely related *S. glaciale* but this species is generally a smaller and less robust plant, with less rigid stems and branches and has ovate and gradually acuminate, less concave leaves without shoulders at the base of the acumen and with serrulate to serrate margins, and the leaf bases form narrower decurrent wings. *Brachythecium subpilosum* also shows some external similarity to *S. fuegianum* but it is at once distinct in its triangular leaves that are not or scarcely decurrent.

Sciuro-hypnum fuegianum was reported from the Antarctic by Lightowlers (1987) as *Brachythecium majusculum*. This species was described by Newton (1974*c*) from South Georgia, but a number of collections from Tierra del Fuego have also been mentioned in the protologue. The name of the species was used for the first time by Dusén (1903*b*) for a plant from Tierra del Fuego but, as a *nomen nudum*, it could not stand. However, it transpired that this species was described much earlier from the Fuegian region by Bescherelle (1885, 1889) as *Stereophyllum fuegianum*, although this author introduced some confusion concerning the interpretation of this name. Initially, Bescherelle (1885) reported under *S. fuegianum* two specimens collected from Hermite I. No description of the species was provided but *Brachythecium longidens* Müll.Hal. was cited as a synonym. The latter species was described a few months earlier, in August of 1885, by Müller (1885), so the name proposed by Bescherelle (1885) is validly published according to Art. 7.3, but at the same time this name is illegitimate according to Art. 52.2(e) because it includes the type of an earlier name (McNeill *et al.* 2006). Four years later, Bescherelle (1889) again published the name *Stereophyllum fuegianum* giving a direct reference to the same name published in 1885, but the synonymous name *Brachythecium longidens* was excluded. Although *S. fuegianum* of 1889 is validly published, this name is also illegitimate as the younger homonym of *S. fuegianum* of 1885. However, Brotherus (1909) legitimised the epithet *fuegianum* when adopting it for this species upon transferring it to the genus *Brachythecium*. The type material of *Brachythecium majusculum* perfectly matches the original material of *B. fuegianum* in all details and the two species have to be considered conspecific, the latter name having priority.

However, a question remains about the status of *Hypnum longidens* with which Bescherelle (1885) initially associated his *Stereophyllum fuegianum*. Unfortunately, this problem cannot be satisfactorily resolved at the moment until the type material of the former is located and examined. Bescherelle (1885) cited under his *Stereophyllum fuegianum* the specimens collected on Hermite I. by Hahn and Hariot No. 130, but his *S. fuegianum* of 1889 was based upon specimen No. 138 of Hariot alone and this may suggest his uncertainty about the status of this species. Also, a careful analysis of the protologue of *Hypnum longidens* does not give a definite answer regarding the status of this species. Müller (1885) compared this species to *Brachythecium rutabulum* from which it differs in having glabrous setae and rudimentary cilia in the peristome, and these features are unknown in any local species of *Brachythecium*. On the other hand, the long costa vanishing in the leaf acumen and the weakly scabrous seta with low and scattered papillae and cordate-ovate leaves are typical of *Sciuro-hypnum longidecurrens*, whereas the long and percurrent costa precluded an affinity of this species with *Brachythecium subpilosum*. The different identity of *Hypnum longidens* and *Stereophyllum fuegianum* was confirmed inferentially by Karl Müller of Halle, the author of the former species. He examined the plant, described by Bescherelle (1889) as a new species, and designated it as *Brachythecium stereophyllum* in June 1886. This name has never been validly published and Bescherelle (1889) placed it in synonymy of *Stereophyllum fuegianum*.

Reproduction in Antarctica. Sporophytes and sexual organs are unknown in Antarctica, but they are produced occasionally by plants from South Georgia and Tierra del Fuego, although mature and fully developed capsules are infrequent.

Habitat. The species grows on moist rock ledges and on soil and yellowish tufa boulders amongst coarse scoria on Deception I.,

→

Figure 262. *Sciuro-hypnum fuegianum*. — **1**. Habit. **2**. Portion of branch, dry. **3**. Portion of branch, wet. **4**. Portion of cross-section of stem. **5–8**. Axillary hairs. **9**. Pseudoparaphyllium. **10–15**. Leaves. **16–17**. Leaf apices. **18**. Upper leaf cells at margin. **19**. Mid-leaf cells. **20**. Basal cells. **21**. Alar cells. — [1, 4–11, 14–21 from *PC 66B*: 2–3, 12–13 from *SWG 266* (isotype of *Brachythecium majusculum*)]. — Scale bars: a – 100 µm (5–8) and 200 µm (9); b – 0.5 cm (1), 2 mm (2–3) and 100 µm (16–21); c – 1 mm (10–15) and 100 µm (4).

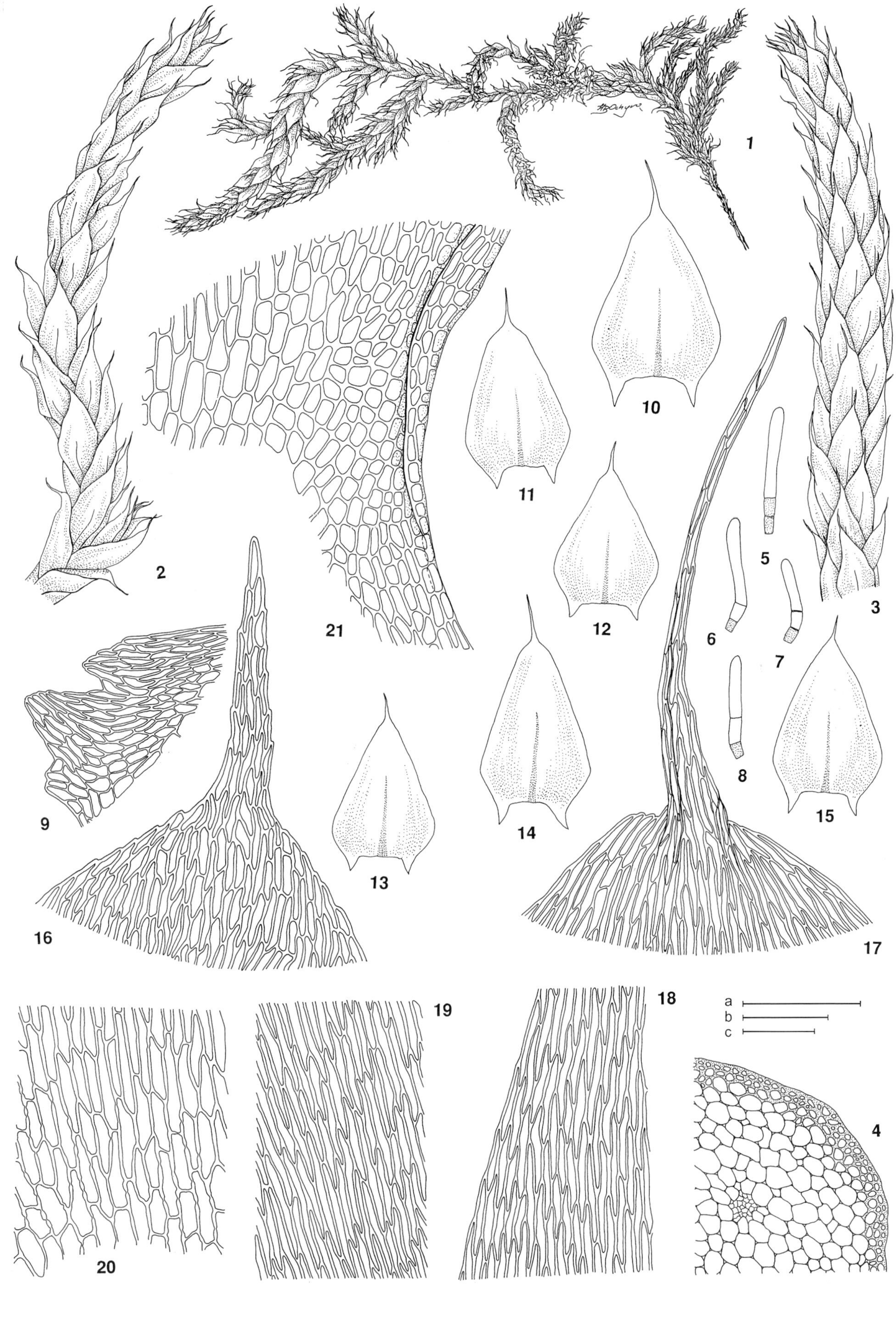
1
2
3
4
5
6
7
8
9
10
11
12
13
14
15
16
17
18
19
20
21
a
b
c

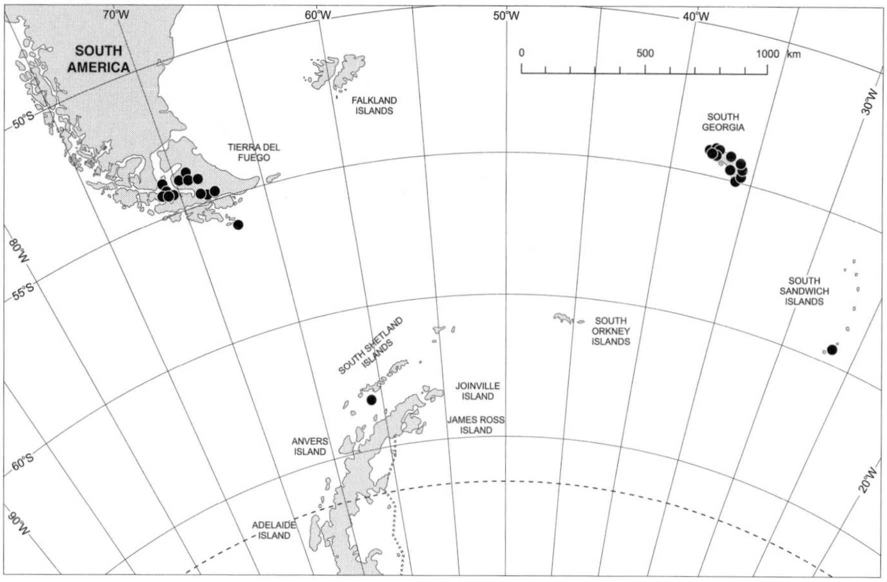

Figure 263. Global distribution of *Sciuro-hypnum fuegianum*.

and on Bellingshausen I. it thrives abundantly in hollows on moist and heated ground in the outer bryophyte zone around fumaroles, usually in association with *Ceratodon purpureus* and *Sanionia georgicouncinata*, under overhanging rocks and on moist buttresses with steam emerging from numerous fissures. On South Georgia *Sciuro-hypnum fuegianum* grows frequently on sandy soil and gravel in moist and wet sites by streams and waterfalls and boggy areas dominated by *Acaena* and *Deschampsia*, whereas in Tierra del Fuego it is also found at the bases of trees.

World range. A south-temperate American species known only from Tierra del Fuego, subantarctic South Georgia and Antarctica (Fig. 263).

Distribution in Antarctica. A very rare and occasional species on heated ground on the volcanic islands in the South Sandwich Is. and South Shetland Is. archipelagoes (Fig. 263). It is fairly frequent on Bellingshausen I. where it occurs at elevations of 75–200 m, whereas on Deception I. the species is exceedingly scarce at two sites, reaching its highest elevation in Antarctica on the summit of Mt. Kirkwood at 350 m. The specimen reported from the Danco Coast on the western Coast of the Antarctic Peninsula (Lightowlers 1987) is interpreted here as *Sciuro-hypnum glaciale*.

Specimens examined. WEST ANTARCTICA. SOUTH SANDWICH IS. **BELLINGSHAUSEN I.** South-east rim of crater, *ca* 200 m, *MWH 824D* (AAS, KRAM); crater rim, 100 m, *PC 27A* (AAS, KRAM), 120 m, *PC 66B* (AAS, KRAM) and 170 m, *PC 33* (AAS, KRAM); immediately to west of the southern access to crater lip, 150 m, *PC 59B* (AAS, KRAM); southern slopes of the island, *ca* 75 m, *MWH 815C* (AAS, KRAM) and 250 m, *MWH 420B* (AAS, KRAM); within the main crater on west side, *ca* 120 m, *MWH 426* (AAS, KRAM).

SOUTH SHETLAND IS. **DECEPTION I.** Pendulum Cove, lava ridge east of Crimson Hill, 130 m, *RILS 11210A* (AAS, KRAM); Mt. Kirkwood, 350 m, *DM 118C & 119C* (AAS, KRAM).

2. Sciuro-hypnum glaciale (Schimp.) Ignatov & Huttunen Fig. 264

Sciuro-hypnum glaciale (Schimp.) Ignatov & Huttunen, Arctoa 11: 270. 2003. ≡ *Brachythecium glaciale* Schimp. *in* Bruch, Schimp. & W.Gümbel, Bryol. Eur. 6: 15, *pl. 542*. 1853 [Fasc. 52–54 Mon.: 11, *pl. 8*]. — TYPE CITATION: In depressis nivos inter Alpes Faulhorn et Wildgerst Helvetiae altitudinae 8000' W. P. Schimper anno 1842 detexit, serius cl. Bryhn e Norvegiae Alpe Gustafjeld dedit et cl. Fr. Arnold e bavarica Alpe Karwendel misit. [Lectotype (*vide* Ignatov & Koponen 1996: p. 294): 'Hypnum glaciale Zwischen dem Faulhorn u. dem Wildgerst' – BM-Schimper!; syntype: 'Hypnum Starkii var. Gustafjeld Norweg occid. ex herb. Blytt litt. R.' – BM-Schimper!].

Plants small to medium-sized, fairly rigid, in glossy intricate mats or less often dense tufts, light green, brownish or yellowish-green above, brown below, occasionally brown throughout. *Stems* creeping or ascending to erect, 1–5 cm tall, irregularly to subpinnately branched, in transverse section with a small central strand, 4–6-stratose medulla of large, parenchymatous, thin-walled and hyaline cells and 2–3-layered cortex of small, thick-walled, brown cells; *rhizoids* sparse, brown, smooth, weakly branched, in clusters immediately below the leaf insertion; *pseudoparaphyllia* broad, orbicular to ovate-acuminate; *axillary hairs* 2–4-celled, with an elongate, hyaline apical cell and 1–3 short, quadrate, brownish basal cells. *Stem* and *branch leaves* similar, the latter narrower, loosely imbricate to erecto-patent, not or slightly altered when dry, (1.0–)1.2–1.9(–2.6) mm long, (0.5–)0.6–1.1 (–1.3) mm wide, concave, not or faintly plicate, long and broadly decurrent, ovate, cordate-ovate to rounded-triangular-cordate, without shoulder at the apex, gradually or suddenly narrowed to a slender, short or long, plane

Figure 264. *Sciuro-hypnum glaciale*. — **1**. Habit. **2–3**. Portion of branches, dry. **4**. Portion of cross-section of stem. **5–8**. Axillary hairs. **9–10**. Pseudoparaphyllia. **11**. Cell areolation of pseudoparaphyllium. **12–19**. Leaves. **20–21**. Leaf apices. **22**. Upper leaf cells at margin. **23–25**. Mid-leaf cells. **26**. Basal leaf cells. **27**. Alar cells. — [1, 3 from *RO 1533/80*; 2, 4–11, 13–14, 25–26 from *RO 1725/80*; 12, 20, 22 from *Recio 1916*; 15–16, 19 from *RO 478/80*; 17–18, 21, 27 from *RO 636/80*; 23 from *DM 23*; 24 from *RILS 4011*]. — Scale bars: a – 1 mm (2–3); b – 1 mm (1, 12–19) and 100 µm (4, 11, 20–27); c – 100 µm (5–8) and 0.5 mm (9–10).

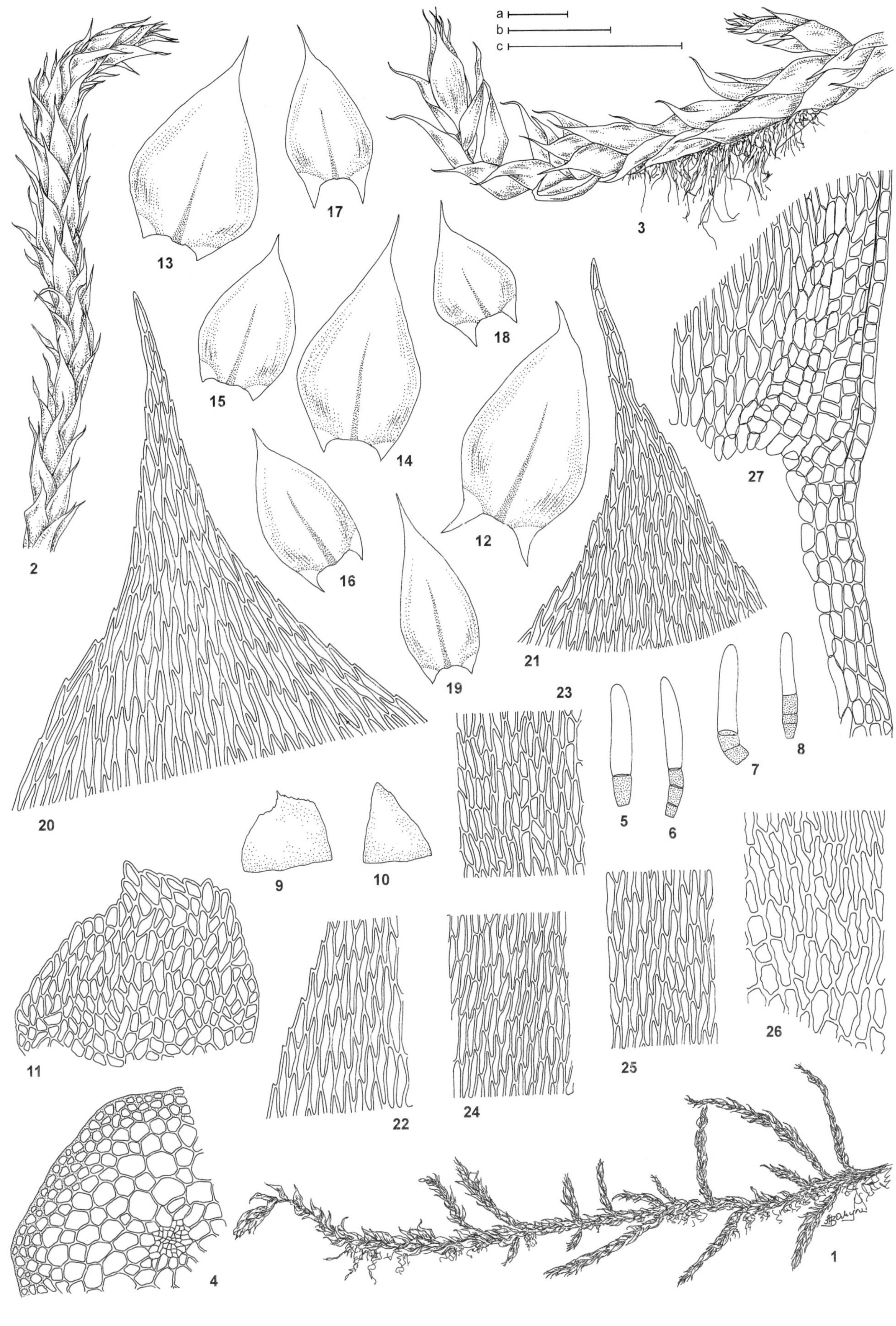
a
b
c
1
2
3
4
5
6
7
8
9
10
11
12
13
14
15
16
17
18
19
20
21
22
23
24
25
26
27

acumen; *margins* recurved below the widest point, plane above, entire below, serrulate in the upper half; *costa* single, well defined, gradually tapering, 40–80 µm wide at base, vanishing about half-way up the leaf; *laminal cells* unistratose, smooth, transparent; *upper* and *median cells* hexagonal, elongate-rhomboidal, elongate-hexagonal to linear, straight or flexuose, distinctly tapering at the ends, thin- to firm-walled, not porose, (30–)40–60(–90) µm long, 5–12 µm wide, *ca* 6–10 times as long as wide; *basal cells* shorter and wider, rectangular to oblong-hexagonal, 30–80 µm long, 12–26 µm wide, thicker-walled, porose; *alar cells* quadrate to short-rectangular, numerous, seriate, evenly thicker-walled above and becoming thin-walled in the decurrent part, forming broad and long decurrencies. [*Autoicous. Inner perichaetial leaves* oblong-ovate, fairly abruptly constricted to a short or long, recurved or flexuose acumen from a broad sheathing base, ecostate. *Setae* stout, 8–12 mm long, scabrous throughout, orange-reddish. *Capsules* inclined to horizontal, ovoid to ellipsoid, strongly gibbous, 1.5–2.2 mm long, 1.1–1.3 mm wide, pachydermous, red-brown to blackish; *operculum* high-conic, apiculate, to 0.7 mm long; *annulus* 2-seriate, long persistent, composed of small, thick-walled cells; *exothecial cells* quadrate, hexagonal to oblong, evenly thick-walled; *peristome* double; *exostome teeth* lanceolate, purple at the base, yellowish-brown, cross-striolate above, hyaline, vertically papillose-striate and broadly bordered at the apex; *endostome* yellow, almost smooth, consisting of a high basal membrane giving rise to lanceolate, keeled and perforate processes, alternating with nodose or appendiculate cilia in groups of 2. *Spores* globose, brownish-green, 12–16 µm in diameter, finely papillose to almost smooth. *Calyptra* cucullate, naked, smooth.]

Chromosome number: $n = 8$ – South Georgia (Newton 1979*a*).

Discussion. As a member of the genus *Sciuro-hypnum*, *S. glaciale* is readily known by its loosely imbricate, concave and prominently decurrent leaves. In the Northern Hemisphere the species is very variable and its several distinct phenotypes were taxonomically recognised as varieties or separate species. These are sometimes difficult to distinguish from the closely related congeners, including *S. reflexum*, *S. starkei* (Brid.) Ignatov & Huttunen and *S. oedipodium* (Mitt.) Ignatov & Huttunen, and there have been much confusion between them (Mårtensson 1956; Crundwell 1959; Nyholm 1965). The Antarctic populations are rather uniform morphologically, although the plants in some populations are fairly large and can be mistaken for *S. fuegianum* and *Brachythecium subpilosum*. The latter has not or scarcely decurrent, plicate, triangular-cordate or triangular leaves which sharply contrast with the non-plicate, broadly decurrent and ovate, cordate-ovate to rounded-triangular leaves of *S. glaciale*. In fact, the specimens from King George I. which were distributed in *Bryophyta Antarctica Exsiccata* Nos 72, 97, 145 and 174 (Ochyra 1984; Ochyra *et al.* 1986) as *B. subpilosum* correctly belong within *S. glaciale* (Lightowlers 1987; Ochyra 1998*a*).

Much more problematic may be distinction between *Sciuro-hypnum glaciale* and its sibling species *S. fuegianum* and sometimes very large specimens of the former are superficially very similar to the latter species. For example, the specimen reported from Andrée I. on the Danco Coast as *Brachythecium majusculum* (Lightowlers 1987) is here interpreted as a robust plant of *S. glaciale* and actually much smaller and typical plants of this species were collected on the same site. The stem leaves of *S. glaciale* are ovate, cordate-ovate to rounded-triangular-cordate, gradually tapering to the acumen, making the leaf apex acute. In contrast, the stem leaves of *S. fuegianum* are broadly triangular-cordate, sometimes broader than long, abruptly narrowed at the apex and forming distinct shoulders from which a long piliferous acumen usually emerges. For better evaluation of this character, a number of the leaves should be checked because some variation of this character in both species is observed. Newton (1974*c*, 1979*b*) emphasised the difference in the degree of serrulation of the leaf margins between the two species, namely *S. glaciale* has extensively and uniformly serrate to serrulate leaf margins, whereas *S. fuegianum* has a smooth leaf margin, with only a few large and sharp teeth at the base of the acumen. In fact, even the type specimens of *Sciuro-hypnum fuegianum* and *Brachythecium majusculum* have the leaf margins distinctly serrulate in the upper half on most leaves, but nevertheless the large teeth at the leaf shoulder are truly characteristic of this species.

According to McFarland (1988) the austral plants named *Sciuro-hypnum glaciale* are actually dwarf phenotypes of *S. fuegianum* because of a wide range of intergrading plant sizes, growth habits and leaf shapes amongst various populations of this species. They are a reflection of the microenvironmental conditions and a response to the short growing season and severe climate in which they occur. Antarctic plants are quite similar to those from northern localities. These were described as *Brachythecium gelidum* Bryhn and are now considered as a variety of *Sciuro-hypnum glaciale* var. *gelidum* (Bryhn) Ochyra & Żarnowiec. They are larger plants with elongate stems and longer laminal cells and some larger plants of this variety are quite similar to *S. fuegianum*. The problem of the taxonomic status of the small austral populations of *Sciuro-hypnum* needs a careful assessment in the broadest geographical scope, and it is very likely that both austral species of this genus, *S. fuegianum* and *S. glaciale*, should be merged with more broadly conceived *S. glaciale*.

Reproduction in Antarctica. Always sterile, but plants with gametangia and immature sporophytes are known from South Georgia.

Habitat. *Sciuro-hypnum glaciale* thrives in dry or wet situations on gravelly or stony ground and soil, on moraines, among boulders, on rock ledges and banks of melt water channels.

World range. A bipolar species having a strongly discontinuous arctic-boreal-montane range in the Northern Hemisphere (Fig. 265, inset), with its main centre of occurrence in northern Europe, including Iceland (Jóhannsson 1997), Svalbard (Frisvoll & Elvebakk 1996), Scotland (Crundwell 1959; Rothero 1994), Fennoscandia (Nyholm 1965) and northern regions of European Russia (Abramova *et al.* 1961; Ignatov & Ignatova 2004), rarely in the Pyrenees, Alps, Carpathians and Caucasus. In Asia it is known only from the Chukchi Peninsula (Afonina 1988, 2004), the Sayan Mts. (Bardunov 1974) and China (Ignatov & Koponen 1996; Wang & Hu 2005) and additionally it occurs in Greenland (Lange & Jensen 1887; Jensen 1898). It was erroneously reported from Japan (Noguchi 1991) and North America (Crum & Anderson 1981). In the southern Hemisphere the species occurs in Tierra del Fuego and western Patagonia in southern South America, on subantarctic South Georgia and in the northern maritime Antarctic.

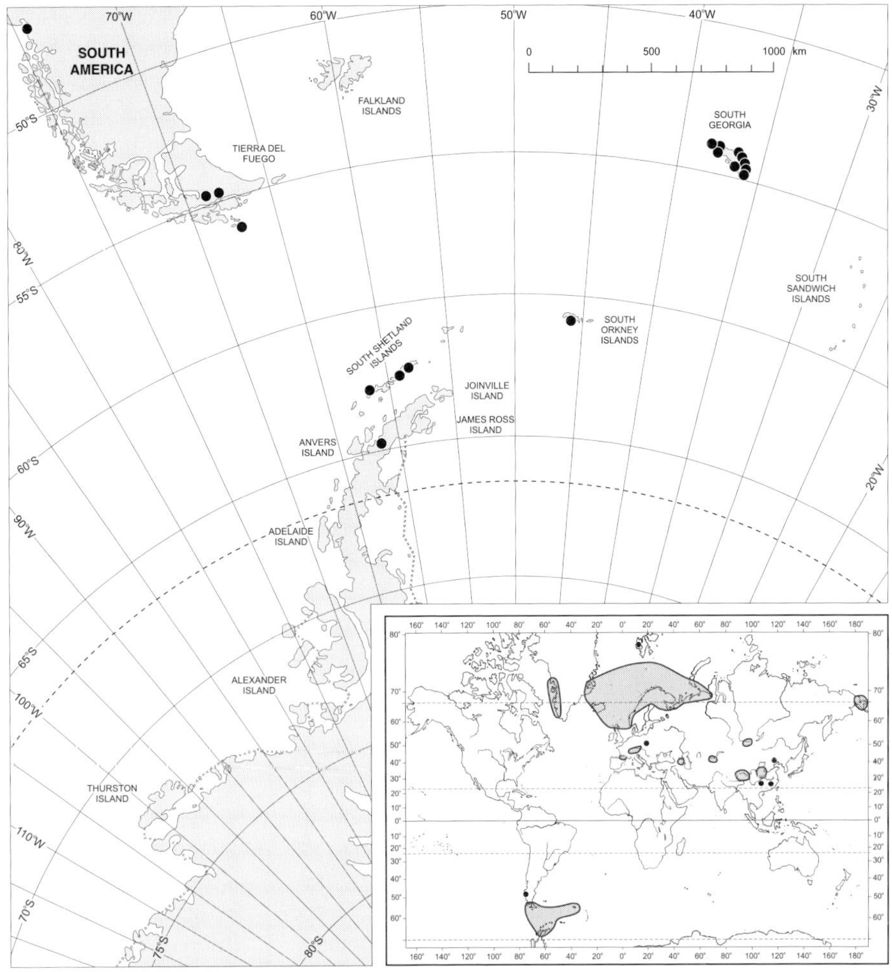

Figure 265. Distribution of *Sciuro-hypnum glaciale* in the Fuegian region and Antarctica. Inset: global distribution.

Distribution in Antarctica. A very rare and localised species, known from scattered localities in the South Orkney Is. (Signy I.), South Shetland Is. (King George I., Nelson I. and Livingston I.) and from a single locality on the western coast of the Antarctic Peninsula (Fig. 265). It grows at low elevations, from near sea level to 80 m.

Specimens examined. WEST ANTARCTICA. SOUTH ORKNEY IS. **Signy I.** Stygian Cove, 20 m, *RILS 10787* (AAS, KRAM).

SOUTH SHETLAND IS. **King George I.** ***Fildes Peninsula***: Bukhta Korablya between Suffield Pt. and Vent Pt., 26 Mar 1976, *Petrov s.n.* (KRAM). ***Admiralty Bay***: Demay Pt., 80 m, *RO 1309/80* (KRAM); Upłaz, 45 m, *RO 1546/80* (KRAM) and 100 m, *RO 1533/80* (KRAM); Skua Cliff, 30 m, *RO 636/80* (AAS, KRAM) and 80 m, *RO 161/80* (KRAM). ***Ezcurra Inlet***: North-east branch of Panorama Ridge, 4 m, *RO 1725/80* (AAS, KRAM); Italia Valley, 8 m, *RO 5078/79* (KRAM). ***Keller Peninsula***: Moraine Pt., 15 m, *RO 478/80* (AAS, KRAM). **Nelson I.** Cruls Refuge, *Recio 1916* (AAS, KRAM). **Livingston I.** ***Byers Peninsula***: Negro Hill, 5 m, *KR 18B* (AAS, KRAM).

WEST ANTARCTIC PENINSULA. **Danco Coast.** Andrée I., 25 m, *RILS 4011 & 4015* (AAS, KRAM).

Excluded species of the Brachytheciaceae

Brachythecium subplicatum (Hampe) A.Jaeger — This species was reported from several stations on Fildes Peninsula, King George I., and from one locality on Nelson I. in the South Shetland Is. archipelago. Two specimens examined (*Li FW7 & NN5*, AAS, KRAM) represent robust forms of *B. austrosalebrosum*. *Brachythecium subplicatum* is an austral pan-temperate species characterised by having small, rounded, decurrent and pellucid auricles and the leaves suddenly constricted to a short subpiliferous point. *Brachythecium* '*subplicatum*' was also mentioned by Gimingham (1967) and Gimingham & Lewis Smith (1970) from Signy I. as a moss referable to the 'moss hummock subformation'. Although no material collected by this author has been located, this provisionally named moss refers to *B. austrosalebrosum* which is a common constituent of this subformation on this island (Lewis Smith 1972) and robust plants of this species are quite similar to *B. subplicatum*.

Sciuro-hypnum populeum (Hedw.) Ignatov & Huttunen — This species was reported as *Brachythecium populeum* from a sub-fossil collection in a sediment core taken from a lake on Fildes Peninsula on King George I. in the South Shetland Is. It was described as being identical to the extant moss living in and close to this lake (Zhou *et al.* 1989). No material was available for examination but the occurrence of this species in the Antarctic is very dubious. *Sciuro-hypnum populeum* is a pan-Holarctic temperate species which has never been collected south of the Tropic of Cancer. The Antarctic moss is described as having 'leafs [sic!] of long needle shape with a central rib reaching the tip' which are 'distributed spirally and relatively sparse on the stem, densely on the top'. Judging from the description and rather crude micrographs it is likely that this is *Campylium polygamum* which often grows in lakes in this area and at lake margins (Chen *et al.* 1995; Ochyra 1998*a*), but it is also possible that this is a species of *Sanionia* which also sometimes grows submerged; the somewhat curved leaves indicated in the photographs make this assumption highy probable. *Drepanocladus longifolius* was found in lake sediments on James Ross I., as well as in lakes on Signy I. and Livingston I. but so far it has not been recorded from King George I. and, additionally, it has much longer leaves.

Family XVI. **PLAGIOTHECIACEAE** M.FLEISCH.

Plagiotheciaceae M.Fleisch., Nova Guinea 8: 748. 1912. Validly published by reference to subfam. Plagiothecioideae Broth. *in* Engl. & Prantl, Nat. Pflanzenfam. 1(3): 1021, 1078. 1908 ['Plagiothecieae']. — Type: *Plagiothecium* Schimp.

Since its inception, the family Plagiotheciaceae has been variously interpreted, usually as an assemblage of diverse and apparently remotely related moss genera with a perfect hypnoid peristome and more or less complanate foliage (Brotherus 1909; Fleischer 1912; Iwatsuki 1970, 1987; Lewinsky 1974; Noguchi 1994). All these discordant elements have been dispersed to other families of hypnalean mosses, leaving the Plagiotheciaceae monotypic (Buck & Ireland 1985, 1989; Buck 1998), consisting of the single genus *Plagiothecium.* However, as a result of extensive morphological studies and cladistic analyses (Hedenäs 1987*a*, 1989*c*, 1995, 1996*c*), supported by molecular evidence (Pedersen & Hedenäs 2001, 2002*b*), the concept of the Plagiotheciaceae was recently expanded and 14 genera of different morphologies were transferred to this family. These include, amongst other things, *Myurella* Schimp., *Struckia* Müll.Hal., *Rhizofabronia* (Broth.) M.Fleisch., *Orthothecium* Schimp., *Catagonium* Broth. and *Isopterygiopsis* Z.Iwats. The unity of such a defined family is warranted, among others, by the absence of pseudoparaphyllia, the purplish and granular-papillose rhizoids which are inserted in the leaf axils and the whitish-yellow exostome teeth. These features are not constant and do not occur in all genera, and it is likely that they have evolved irrespective of the degree of the relationships of the taxa concerned. For example, in many species of the type genus of the family itself, *Plagiothecium*, the rhizoids are intercalary and smooth, not axillary and papillose, but usually they arise at the insertion of the costa on the dorsal side or on the stem below the costa base and are smooth. Therefore, the Plagiotheciaceae as a multigeneric taxon cannot be defended as a natural taxon and, until more convincing evidence is provided, for the time being it seems sensible to consider this family as monogeneric, especially because other molecular studies suggest its monophyly (Tsubota *et al.* 2002).

The Plagiotheciaceae are primarily defined by a combination of gametophyte characters. The plants are mostly complanately foliate, although in some species the stems and branches are terete. The leaves are ovate, ovate- to oblong-lanceolate, often strongly asymmetrical, straight or weakly falcate but sometimes directed downwards, acute to acuminate, occasionally piliferous and strongly to scarcely decurrent, with entire to weakly denticulate margins at the apex. The most characteristic feature of the family is the shape of the leaf costae. They are usually described as double, but actually they are not formed by two branches independently arising from the leaf base. The costa is, in fact, single and consists of a short basal stump which is forked above with branches of varying length, but generally they are short and do not or scarcely exceed the middle of the leaf. The laminal cells are smooth, linear-flexuoxe and heavily chlorophyllose, often with a group of short and broad, echlorophyllose nematogon cells near the leaf apex from which arise rhizoids. The alar cells are differentiated, usually short and inflated, and form decurrent, triangular or often auriculate and oval groups. The stem has a poorly developed cortex, composed of only somewhat smaller, thin-walled cells which are clearly visible in transverse section, but not forming a true hyalodermis. Sporophytically, the Plagiotheciaceae are very similar to other hypnalean mosses. The peristome is double and perfectly hypnoid, consisting of the 16 exostome teeth which are whitish-yellow, cross-striolate below and papillose above, and the endostome has a medium to high basal membrane giving rise to the segments which are narrow or broad, keeled, unperforated or very narrowly cleft and shorter than or almost as long as the teeth. Cilia are present and well developed or lacking.

The characters of the Plagiotheciaceae concur to a high degree with most characters of the Hypnaceae and therefore these families seem to be scarcely distinct from each other. However, the complanate-foliate habit and straight or only weakly falcate leaves may serve as sufficient diagnostic characters at the family level. Therefore, some workers still regard the Plagiotheciaceae as a weak segregate and consider it as a subfamily within the broadly circumscribed Hypnaceae (Nishimura *et al.* 1984).

The Plagiotheciaceae are a fairly small family of hypnalean mosses which is predominantly associated with cool and temperate areas, while in the tropics its species occur mostly on mountains. In the austral polar regions the family is scarcely represented and its members are not prominent constituents of the tundra vegetation.

52. PLAGIOTHECIUM SCHIMP.

Plagiothecium Schimp. *in* Bruch, Schimp. & W.Gü,bel, Bryol. Eur. 5: 179. 1851 [Fasc. 48 Mon.: 1]. ≡ *Stereodon* (Brid.) Mitt. sect. *Plagiothecium* (Schimp.) Mitt., J. Linn. Soc. Bot. 4: 88. 1859. — Lectotype (*vide* Grout 1932: p. 155): *Plagiothecium denticulatum* (Hedw.) Schimp. (*Hypnum denticulatum* Hedw.).

Philoscia Berk., Handb. Brit. Mosses: 49, 146. 1863. — Type: *Philoscia latebricola* (Schimp.) Berk. (*Plagiothecium latebricola* Schimp.). First synonymised by Buck & Ireland (1989: p. 3).

Plagiotheciella M.Fleisch. *ex* Broth. *in* Engl., Nat. Pflanzenfam. Ed. 2, 11: 465. 1925, *nom. illeg. incl. gen. prior.*

Saviczia Abramova & I.I.Abramov, Novosti Syst. Nizš. Rast. [3]: 298. 1966. — Type: *Saviczia obtusissima* (Broth.) Abramova & I.I.Abramov (*Plagiothecium obtusissimum* Broth.). First synonymised by Iwatsuki (1970: p. 341).

Plants small to medium-sized, occasionally large, slender or stout, soft, dull or more often lustrous, in loose or compact, usually flat mats or less often in deep tufts, green to yellow-green or whitish-yellow-green. *Stems* prostrate or ascending to erect, simple or sparingly irregularly branched, complanate-foliate to julaceous, in transverse section with the cortex composed of fairly large and thin-walled cells in 1–3 layers, surrounding several rows of larger, thin-walled, hyaline medullary cells and distinct central strand; *rhizoids* papillose to smooth, purplish at least when young, intercalary, arising just below the leaf insertion or at leaf tips; *pseudoparaphyllia* absent; *axillary hairs* filiform, 4–10-celled, hyaline, with 2–3 short, brownish-pigmented basal cells. *Stem* and *branch leaves* mostly similar, crowded or distantly arranged, erect to erect-spreading, rarely widely patent, imbricate to strongly complanate, symmetric or often asymmetric, ovate, ovate- to oblong-lanceolate, occasionally elliptical, acute, acuminate, rarely piliferous, sometimes with apices curved downwards, concave, smooth or occasionally undulate, decurrent; *margins* plane to recurved, entire or serrate at the apex; *costa* branched from a short basal stalk, reaching ⅓–½ way up the leaf, usually with one branch longer and extending to the mid-leaf, occasionally one branch reduced making the costa seemingly single, sometimes almost absent; *laminal cells* smooth, transparent, heavily chlorophyllose, unistratose throughout, thin-walled; *upper* and *median cells* linear-flexuose to linear-rhomboidal, gradually tapering at the ends; *alar cells* quadrate to rectangular, forming short or long, triangular, ovate or auriculate decurrencies. *Asexual reproduction* often present by cylindrical or fusiform propagules composed of 2–7 cells in one row, clustered in the leaf axils, sometimes on leaf surfaces. *Autoicous* or *dioicous*. *Perichaetia* large and conspicuous; *inner perichaetial leaves* oblong to oblong-lanceolate, convolute at the base, shortly or longly acuminate. *Setae* elongate, straight or flexuose, smooth, reddish, distally twisted to the left, below to the right. *Capsules* erect to inclined, rarely cernuous, straight or arcuate, obloid, ovoid to cylindrical, smooth or striate to furrowed when dry, sometimes weakly constricted below the mouth; *operculum* conic to short-rostrate, straight or somewhat oblique; *annulus* 1–3-seriate, often tardily deciduous; *exothecial cells* isodiametric to elongate, thin- to moderately thick-walled; *stomata* in the base of the urn, bicellular, superficial, with rounded pori; *peristome* double; *exostome teeth* lanceolate-subulate, yellowish-whitish, hyaline bordered, cross-striolate below, papillose above, occasionally papillose throughout, trabeculate at the back; *endostome* hyaline or pale yellowish, smooth, consisting of a medium-high to high basal membrane, broad or narrow, keeled, not or narrowly perforate segments, shorter or almost as long as the exostome teeth and 1–3 appendiculate cilia, occasionally lacking or vestigial. *Spores* spherical, fairly small, smooth or minutely roughened. *Calyptra* cucullate, naked, smooth.

Plagiothecium is usually not difficult to distinguish by its characteristic complanate habit coupled with strongly asymmetric and often markedly falcato-recurved lateral leaves that are entire throughout or somewhat denticulate at the apex. Additional microscopic features enabling safe recognition of the genus include the lack of pseudoparaphyllia, short and distinctly branched costae, papillose and often purplish rhizoids arising just below the leaf insertion and the frequent presence of a group of hyaline nematogon cells in the leaf apex from which rhizoids often arise. These characters allow unmistakable recognition of the genus in the temperate regions of its range but in the tropics the genus may be much more easily confused with some hookerialean mosses. For instance, the most common species of the genus in the Neotropics, *P. drepanophyllum* Renauld & Cardot (Buck & Ireland 1989), proved to be identical to *Lepidopilum scabrisetum* (Schwägr.) Steere and consequently the plants so-named received the name *P. standleyi* E.B.Bartram (Ochyra & Buck 2002).

Plagiothecium is fairly diverse morphologically and some species or groups of species were segregated into small genera. One of the most obvious and distinct of these is *Buckiella* Ireland which accommodates the Holarctic *B. undulata* (Hedw.) Ireland and Hawaiian *B. draytonii* (Sull.) Ireland. It is characterised by having minute, granular cuticular papillae covering the leaf cells, and is unrelated to *Plagiothecium*. It is placed in the Hypnaceae (Ireland 2001) and the distant relationship of species belonging within *Buckiella* is also supported by molecular evidence (Arikawa & Higuchi 1999).

Plagiothecium is a medium-sized genus of about 40–45 species widely distributed in the temperate zone of both hemispheres, although it is also quite well represented in the tropics at high elevations. It has been revised taxonomically for Europe (Jedlička 1947, 1948, 1950, 1954, 1961), North America (Ireland 1969), tropical and temperate South America (Buck & Ireland 1989), Australasia (Ireland 1992) and Antarctica (Bednarek-Ochyra *et al.* 1999). Only four species have hitherto been recorded in the cool temperate and cold regions of the Southern Hemisphere, namely *P. lucidum* (Hook.f. & Wilson) Paris in Tierra del Fuego (Buck & Ireland 1989) which, in fact, is an Afro-American species (Ochyra *et al.* 2000),

P. falklandicum (Cardot & Broth.) M.E.Newton on the Falkland Is. and South Georgia (Newton 1983*b*) and in Tierra del Fuego (Ochyra 2001*b*), for which the correct name is *P. lamprostachys* (Hampe) A.Jaeger (Ochyra 2002*b*; Ochyra & Matcham 2005), as well as *P. ovalifolium* Cardot and *P. orthocarpum* Mitt. The last two species extend to the Antarctic.

1. Leaf apices broadly acute; median leaf cells 14–20 µm wide, 6–9 times as long as wide; margins recurved throughout; leaf decurrencies broad and short .... 1. ***P. ovalifolium***
1. Leaf apices narrowly acute to acuminate; median leaf cells 9–15 µm wide, 10–15 times as long as wide; margins plane; leaf decurrencies long and narrow .... 2. ***P. orthocarpum***

1. **Plagiothecium ovalifolium** Cardot Fig. 266

Plagiothecium ovalifolium Cardot, Bull. Herb. Boissier Sér. 2, 5: 1010. 1905. — TYPE CITATION: Terre-de-Feu: Ushuaia, Lapataia. [Lectotype (*vide* Buck & Ireland 1989: p. 15): 'Svenska Sydpolar-expeditionen 1901–03. N:r 178. Plagiothecium ovalifolium Card. sp. nov. Tierra del Fuego, in silva ad Bahia Lapataia 9/10 1902 Carl Skottsberg Det. J. Cardot' – PC!; isolectotype: S!; syntypes: 'Svenska Sydpolarexpeditionen 1901–03. N:r 177. Plagiothecium ovalifolium Card. sp. nov. Tierra del Fuego, in monte supra Ushuaia, 530 m.s.m. 3/3 1902 Carl Skottsberg Det. J. Cardot' – PC!, S!].

Plants small to medium-sized, in soft, lustrous, mostly dense deep tufts or extensive mats, green to whitish-green or yellow- to sordid-green above, yellowish to brownish inside. *Stems* prostrate to ascending or erect, to 7 cm long, usually complanate-foliate, simple or irregularly branched, in cross-section round, with a weak central strand, 1–3-layered cortex of smaller, brownish-yellow, thin-walled cortical cells rather sharply demarcated from larger, thin-walled, hyaline medullary cells. *Leaves* distant to crowded, patent, ovate to oblong-ovate, gradually and broadly acute, smooth, decurrent, little altered on drying, mostly symmetric or nearly so; *margins* entire or occasionally with a few small denticulations at the apex, narrowly recurved throughout; *costa* double, usually with one branch extending to ⅓–½ or more the leaf length, sometimes very short; *laminal cells* transparent, elongate-hexagonal, thin-walled, 85–130 µm long, (10–)14–20 µm wide, 6–9 times as long as wide; *alar cells* poorly differentiated, rectangular to quadrate, occasionally somewhat inflated and spherical, forming short and broad decurrencies, 3–6 cells wide, 5–10 cells long, mostly stripping off with dissection. [*Autoicous*. *Perichaetial leaves* broadly oblong-ovate, convolute. *Setae* straight, smooth, 1.8–2.5 cm long, reddish, sinistrorse above. *Capsules* inclined to horizontal, cylindrical, 2.0–2.5 mm long, brown, smooth, somewhat constricted at the mouth on drying; *operculum* conic convex, short, with a slanted beak; *annulus* 2–3-seriate, deciduous; *exothecial cells* variable in shape, isodiametric to shortly rectangular, firm-walled; *peristome* double; *exostome teeth* lanceolate, pale yellow, papillose-striate below, papillose distally; *endostome* hyaline yellowish, papillose-spiculose, with a high basal membrane, keeled and strongly perforate segments and mostly single, appendiculate cilia. *Spores* spherical, finely roughened, 10–12 µm in diameter. *Calyptra* cucullate, naked, smooth.]

Chromosome number: unknown.

Discussion. *Plagiothecium ovalifolium* is distinguished from all other Antarctic mosses by its flattened habit as well as ovate to oblong-ovate, symmetric or nearly so, widely-spaced and broadly acute leaves. It is externally similar to *P. orthocarpum* to which it seems to be most closely related, but differs in its entirely recurved leaf margins and shorter and broader leaf cells. In its broad leaf decurrencies *P. ovalifolium* resembles *P. lamprostachys*, but the julaceous shoots, closely imbricate, deeply concave plane leaf margins and narrow laminal cells, 5–11 µm wide, of the latter species at once preclude confusion between them.

The recurved leaf margins are rather an unusual feature in the genus *Plagiothecium* but occur, for example, in the Arctic *P. berggrenianum* Frisvoll. This species, originally described from Spitsbergen (Frisvoll 1981) and subsequently found in the Russian Arctic (Ukrainskaya 1996; Ignatov *et al.* 2001), is immediately recognised from *P. ovalifolium* by its julaceous shoots, hooked leaf apices and gradually tapering leaf decurrencies. In contrast, *P. ovalifolium* is a complanate-leaved plant with plane leaf apices and broad decurrencies.

Reproduction in Antarctica. Sterile.

Habitat. The species shows a very narrow ecological amplitude in Antarctica, growing almost exclusively in damp, shaded, sheltered fissures and on ledges of siliceous rocks. It usually forms small dense mats with an occasional admixture of other mosses, for instance *Bartramia patens*, *Sanionia georgicouncinata*, *Pohlia cruda*, *Andreaea gainii*, *Hymenoloma antarcticum*, and liverworts, for example *Barbilophozia hatcheri* and *Lophozia excisa*.

World range. *Plagiothecium ovalifolium* is an amphiatlantic south-temperate species (Fig. 267). It has a very narrow range in Tierra del Fuego, the Falkland Is. (Matteri 1986) and southern Patagonia, extending as far north as lat. *ca* 50° S near Lago Argentino (Buck & Ireland 1989). Additionally, it is known from some subantarctic islands including the Prince Edward Is. (Ochyra & Hertel 1990; Bednarek-Ochyra *et al.* 1999; Ochyra, personal observations), Îles Crozet (Hébrard 1970*a*; Bednarek-Ochyra *et al.* 1999; Ochyra, personal observations) and Îles Kerguelen (Ochyra, personal observations). All specimens

→

Figure 266. *Plagiothecium ovalifolium.* — **1**. Habit. **2**. Portion of branch, dry. **3**. Portion of branch, wet. **4**. Cross-section of stem. **5–7**. Axillary hairs. **8–12**. Leaves. **13–14**. Leaf apices. **15**. Mid-leaf cells. **16–17**. Alar cells. — [1–9, 14–16 from *RILS 5012*; 10–11, 13 from *RILS 8232*; 12, 17 from *CS 177* (lectotype of *P. ovalifolium*), PC]. — Scale bars: a – 1 mm (2–3); b – 100 µm (13–17); c – 0.5 cm (1) and 1 mm (8–12); d – 100 µm (4–7). (Reproduced from *Bryobrothera*, 5: p. 57, 1999, with permission.).

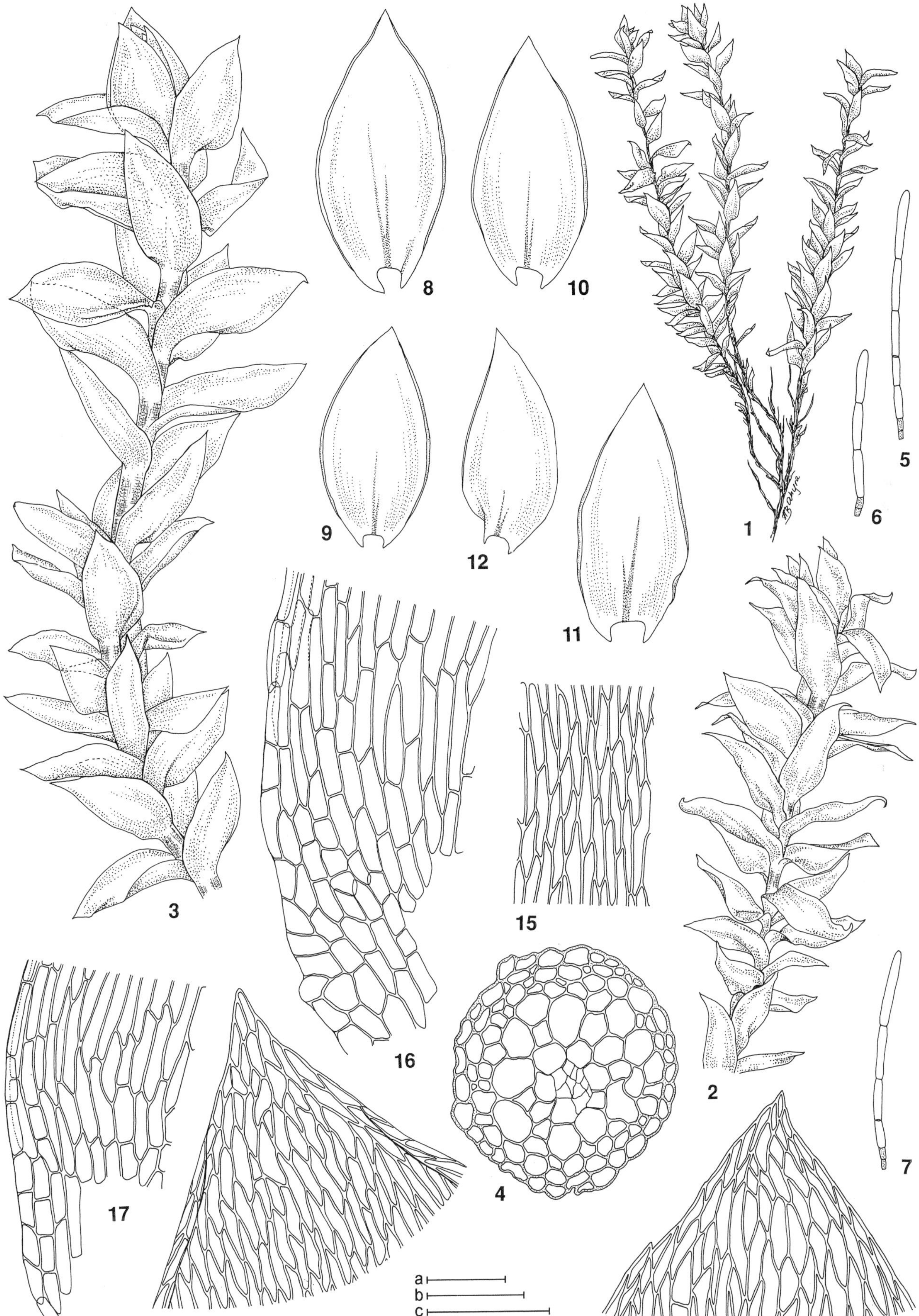
1
2
3
4
5
6
7
8
9
10
11
12
13
14
15
16
17
a
b
c
d

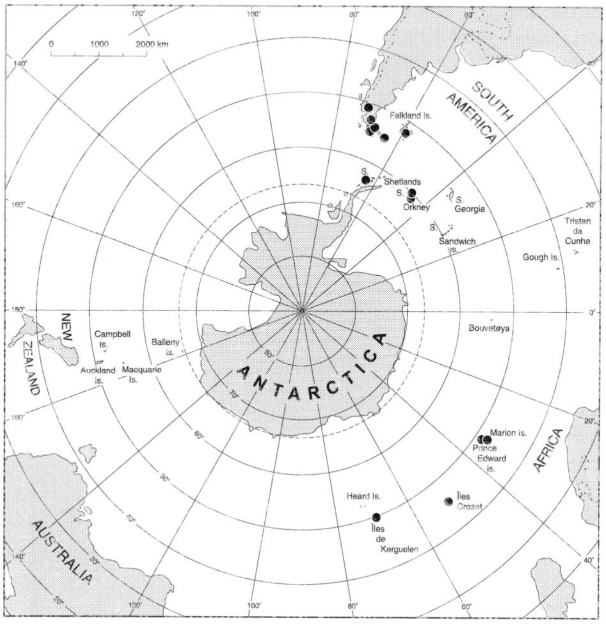

Figure 267. Global distribution of *Plagiothecium ovalifolium*. (Reproduced from *Bryobrothera*, 5: p. 58, 1999, with permission.)

reported by Zanten (1971) from Marion I. and Gremmen (1982) in the Prince Edward Is. as *P. platyphyllum* belong to the present species.

Antarctic distribution. *Plagiothecium ovalifolium* is very restricted in its Antarctic distribution and, so far, it has been recorded only from the South Orkney Is. and Petermann I. off the Graham Coast on the west side of the Antarctic Peninsula (Fig. 267). All Antarctic reports of *Plagiothecium* sp. (Lewis Smith 1972; Lewis Smith & Corner 1973) should be referred to this species.

Specimens examined. WEST ANTARCTICA. SOUTH ORKNEY IS. CORONATION **I.** Shingle Cove, 1–2 m, *RILS 8232* (AAS, KRAM). **LYNCH I.** 15 m, *RILS 523 & 1825* (AAS, KRAM). **SIGNY I.** Foca Pt., 50 m, *REL 1116* (AAS, KRAM); Knob Lake, 15 m, *RILS 5010* (AAS, KRAM); Cemetary Bay, 35 m, *RILS 5012* (AAS, KRAM); Observation Bluff, *ca* 85 m, *RILS 15* (AAS, KRAM); Rusty Bluff above Paal Harbour, *RILS 1804* (AAS, KRAM), 50 m, *RILS 8222* (AAS, KRAM) and 80 m, *RILS 5011* (AAS, KRAM); Paal Harbour, east side of Rusty Bluff, *ca* 100 m, *RILS 348 & 349* (AAS, KRAM).

WEST ANTARCTIC PENINSULA. **WILHELM ARCHIPELAGO.** ***Penola Strait***: Petermann I., *ca* 13 m, *RWMC 784B* (AAS, KRAM) and *ca* 17 m, *REL 1344* (AAS, KRAM).

2. **Plagiothecium orthocarpum** Mitt. Fig. 268

Plagiothecium orthocarpum Mitt., J. Linn. Soc. Bot. 12: 521. 1869. ≡ *Hypnum orthocarpon* Hornsch. *ex* Müll.Hal., Syn. Musc. Frond. 2: 253. 1851, *hom. illeg*. [*non H. orthocarpon* Ångstr. 1842]. [*H. orthocarpum* Hornsch., Bot. Zeitung (Regensburg) 3: 520. 1820, *nom. nud*.; *Isothecium orthocarpon* Brid., Bryol. Univ. 2: 387. 1827, *nom. nud*.]. — TYPE CITATION: Chile, ad Talcahuano: Adalbert de Chamisso 1816. [Lectotype (*selected here*): 'Herbarium Haussknecht, Jena Herbarium Hugo Dohl Schmalhalden Hypnum orthocarpum Brid. Chile' – JE!].

Hypnum georgicoantarcticum Müll.Hal. *in* Neumayer, Int. Polarforsch. Deutsch. Exp. 2: 321. 1890 ['georgico-antarcticum']. ≡ *Plagiothecium georgicoantarcticum* (Müll.Hal.) Kindb., Enum. Bryin. Exot. 102. 1899, ***syn. nov.*** — TYPE CITATION: Austro-Georgia, in rupium fissuris am Ausgange des Brockenthales, 24. Januario 1883; latere orientali des Vexirberges in rupium fissuris, 17. Febr. 1883. [Lectotype (*vide* Bednarek-Ochyra *et al.* 1999: p. 59): '№ 50 51. Plagiothecium georgico-antarcticum C. Müll. nsp. Fundort: In Spalten eines Felsens am Ausgang des Brockenthales Südgeorgien 24/I 83 Will Südpolarexpedition' – HBG!; isotype: PC!].

Plagiothecium conostegium Herzog, Biblioth. Bot. 87: 154. 1916, ***syn. nov.*** — TYPE CITATION: [Bolivia] An der Waldgrenze über Tablas, ca. 3400 m, No. 2497; im Nebelwald über Comarapa, ca. 2600 m, No. 4239; zwischen San Mateo und Sunchal No. 4513/a. [Lectotype (*vide* Buck & Ireland 1989: p. 13): 'Plantae in itinere secundo per Bolivian lectae. No. 2947. Plagiothecium conostegium H. n. sp. An der Waldgrenze über Tablas ca 3400 m leg. Th. Herzog Mai 1911' – JE!; isolectotype: S!; syntype: 'Plantae in itinere secundo per Bolivian lectae. No. 4513/a. Plagiothecium conostegium H. n. sp. Zwischen San Mateo und Sunchal. ca 2600 m? leg. Th. Herzog April 1911' – JE!].

Plagiothecium simonovii L.I.Savicz & Smirnova, Inform. Bjull. Sovetsk. Antark. Eksped. 49: 34, *f. 1*. 1964, ***syn. nov.*** — TYPE CITATION: Antarctida Orientalis, terra reginae Maud, Oasis Schirmacher, lacus Glubokoje, 32,3 m alt., 16/I 1962, leg. I. M. Simonov. [Holotype: 'Plagiothecium simonovii L. Savicz & Z. Smirn. Vost. Antarktida, Zemlya Korolevy Mod, Oasis Shirmakhera, Ozero Glubokoye, s glubi 32.3 m. 16.i.1962 Leg. I. M. Simonov ii.1964 Det. L. Savicz-Lyubitskaya & Z. Smirnova – LE!; isotypes: AAS!, KRAM!].

Plants totally submerged, soft, dull, pale green. *Stems* freely floating, 5–16 cm long, simple or sparingly branched, often with elongate, apical flagellae, in transverse section elliptic or obtusely angulate, consisting of a small central strand surrounded by 3–4 rows of very large, thin-walled, hyaline medullary cells and 2–3 rows of slightly smaller, thin-walled, somewhat brownish outer cortical cells; *rhizoids* lacking or sparse scattered along the stem, smooth, brownish. *Leaves* widely spaced or more crowded, concave, 1.5–2.5 mm long, 0.4–0.8 mm wide, narrowed at the insertion, decurrent; *lateral leaves* erecto-patent to spreading-recurved, oblong-lanceolate to ovate-lanceolate, gradually long-acuminate, apiculate, asymmetric, often keeled and conduplicate; *ventral* and *dorsal leaves* symmetric, appressed, ovate-lanceolate to broadly ovate, acuminate, abruptly apiculate; *margins* plane, entire or sparsely serrulate at the extreme apex; *costae* strong, branched, one branch usually reaching ⅓–½ the leaf length, sometimes branches arising from a relatively tall basal stalk; *laminal cells* transparent, smooth, thin-walled, linear-flexuose, 60–145 μm long, 9–15 μm wide, 10–15 times as long as wide, gradually narrowed towards the margins, 90–215 μm long, 6–11 μm wide; *basal cells* lax, shorter and wider, 105–165 μm long, 12–21 μm wide; *alar* cells hyaline, quadrate to elongate, enlarged and inflated forming small, indistinct decurrencies, 2–3 cells wide, to 8 cells long, extending down the stem

Figure 268. *Plagiothecium orthocarpum*. — **1.** Habit. **2.** Portion of the branch. **3.** Cross-section of stem. **4–8.** Leaves. **9.** Leaf from flagelliform offshoot. **10.** Leaf apex. **11.** Mid-leaf cells at margin. **12.** Mid-leaf cells. **13.** Basal leaf cells showing decurrency. — [All from *Simonov s.n.*, 16.i.1962 (type of *P. simonovii*)]. — Scale bars: a – 1 mm (2); b – 0.5 cm (1) and 1 mm (4–9); c – 100 μm (3, 10–13). (Reproduced from *Bryobrothera*, 5: p. 60, 1999, with permission.)

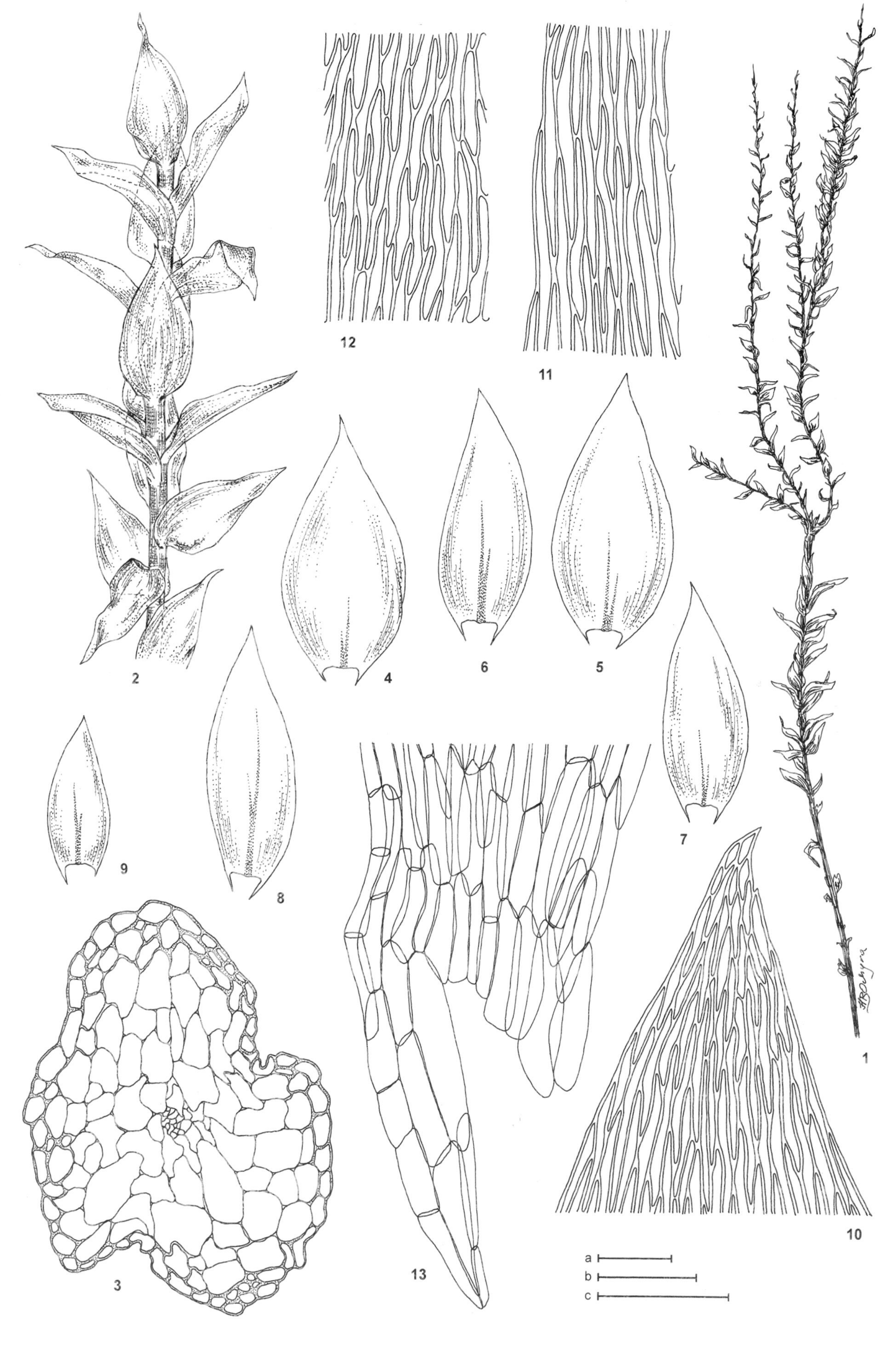
12
11
2
4
6
5
7
9
8
3
13
10
1
a
b
c

and often stripping off with dissection. [*Autoicous. Setae* straight, smooth, reddish, 1.5–2.5 cm long, twisted to the left above. *Capsules* suberect, inclined to horizontal, cylindrical, straight to slightly curved, 1.8–2.1 mm long, pale brown to olivaceous, usually striate or furrowed and somewhat constricted below the orifice when dry; *operculum* shortly rostrate, with a somewhat slanted, acute beak; *annulus* biseriate, deciduous; *exothecial cells* subquadrate to narrowly oblong-rhomboidal, moderately thick-walled; *peristome* double; *exostome teeth* lanceolate, pale yellowish, hyaline bordered, papillose-striate below, papillose above, strongly trabeculate on the ventral surface; *endostome* hyaline or pale yellowish, smooth, consisting of a high basal membrane, about half the length of the exostome teeth, giving rise to keeled, entire or very narrowly perforate segments, alternated with appendiculate cilia in groups of 3. *Spores* globose, finely roughened, 12–15 μm in diameter. *Calyptra* cucullate, glabrous, smooth.]

Chromosome number: unknown.

Discussion. The unique Antarctic deep-water population of *Plagiothecium* was recognised by Savicz-Lyubitskaya & Smirnova (1964*c*, 1965*c*) as a new species, *P. simonovii*, which was compared with the European *P. platyphyllum* Mönk. The Antarctic species differed in its larger size, angular stems, smaller and shorter acuminate leaves, longer laminal cells and elongate cells in leaf decurrencies. These characters distinctly separate both species and, additionally, in *P. platyphyllum* there is a group of rhombic, thin-walled and hyaline, sometimes eroded rhizoid initials in the leaf apex, and the leaf margins are sharply denticulate at the apex. Besides the unusual habit of the plants conditioned by their deep submergence, the leaf characters of *P. simonovii* including leaf shape, areolation and decurrencies agree perfectly with those of *P. georgicoantarcticum* from South Georgia and, intially, these species were considered to be conspecific, the latter having priority (Bednarek-Ochyra *et al.* 1999). However, subsequent studies revealed that the final name for this otherwise distinct and widespread species is *P. orthocarpum*.

This obscure and neglected species has a long and fairly chequered history. The original material was collected in 1816 in Chile by A. von Chamisso (1781–1838), botanist to the Russian scientific voyage around the world in the years 1815–1818, on the *Rurik* under the command of Otto von Kotzebue. This species was named as *Hypnum orthocarpum* by Hornschuch (1820), but he never published its description, and this name was subsequently placed by Steudel (1824) in his *Nomenclator botanicus*. Bridel (1827) mentioned this taxon as *Isothecium orthocarpon* but did not describe it since the moss was unknown to him ('Plura de hoc nobis ignoto musco a clarissimo Hornschuchio ex ejus promissis expectamus'). Finally, Müller (1851) provided the diagnosis of this species, although the name *Hypnum orthocarpon* was illegitimate because it was antedated by *H. orthocarpon* Ångstr. of 1842. However, this epithet was saved by Mitten (1869) upon the transfer of this species to *Plagiothecium*, although this species was apparently unknown to him and he only cited verbatim Müller's (1851) diagnosis. Buck & Ireland (1989) did not see any original material and suggested that *P. orthocarpum* may be an older name for *P. ovalifolium*. However, examination of a scrap of the apparently original material of *P. orthocarpum* revealed that this species is identical to *P. georgicoantarcticum* and its name has priority. Accordingly, *P. simonovii* and *P. conostegium*, which were formerly considered synonymous with *P. georgicoantarcticum* (Bednarek-Ochyra *et al.* 1999), are now reduced to synonymy with *P. orthocarpum*.

Plagiothecium orthocarpum is a practically unmistakable species in the cool-temperate and polar regions of the Southern Hemisphere. *Plagiothecium lamprostachys* and *P. ovalifolium* are distinguished by their large leaf decurrencies and *P. lucidum* has slenderly long-acuminate leaves and very narrow mid-leaf cells, 5–6(–8) μm wide.

Reproduction in Antarctica. Sterile.

Habitat. In normal conditions *Plagiothecium orthocarpum* is a terrestrial moss, growing on humus and soil in shaded sites, predominantly in humid forests. The only record of this species in the Antarctic is based on specimens which were collected in a lake at depths between 23 and 32 m. This is a remarkable habitat for *Plagiothecium*, all species of this genus being typically terrestrial plants associated with dry habitats and only occasionally growing in damp or wet sites, for instance European *P. platyphyllum*. However, bryological literature provides numerous examples of mosses and liverworts that are capable of growing successfully deeply submerged (Persson 1942; Light 1975; Light & Lewis Smith 1976; Malme 1978; Ignatov & Kurbatova 1990; Winton & Beever 2004). This phenomenon is frequently observed in the Antarctic and several species have been found at various depths in lakes (Savicz-Lyubitskaya & Smirnova 1959, 1960, 1964*a,b*; Light & Heywood 1973, 1975; Priddle & Dartnall 1978; Priddle 1979; Kaspar *et al.* 1982; Seppelt 1983; Kanda & Ohtani 1991; Kanda & Mochida 1992; Imura *et al.* 1992, 1999, 2003; Tewari & Pant 1996; Kudoh *et al.* 2003*a,b,c*; Wagner & Seppelt 2006). Such plants are usually highly modified, yet the diagnostic features of each species are retained.

The occurrence of this terrestrial species as a deep-water moss in a lake which is permanently covered by several metres of ice (minimum thickness *ca* 2.0–2.5 m in summer) in an extremely isolated site *ca* 2800 km from its nearest known locality (South Georgia) is remarkable. Simonov & Fedotov (1964) described Lake Glubokoye as rising by up to 4 m in depth during the summer and having reversed thermal stratification, with the bottom layer remaining at 4° C throughout the year while the surface does not exceed 0.2° C. Fragments of the moss were found in bottom sediments to a depth of 49 cm, indicating that the moss had occupied the lake bed for a very long time.

World range. An amphiatlantic south-temperate species, penetrating into the tropics along the Andean cordillera from Tierra

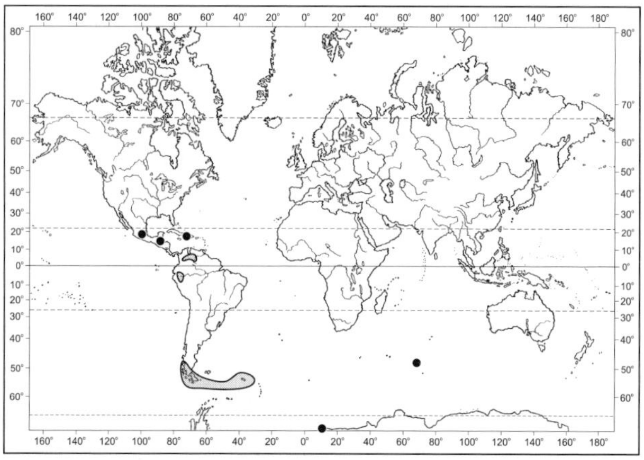

Figure 269. Global distribution of *Plagiothecium orthocarpum*.

del Fuego to southern Mexico (Buck & Ireland 1989) and the West Indies (Buck 1998), and recurring in continental Antarctica in the south (Fig. 269). In the Subantarctic *Plagiothecium orthocarpum* is infrequent in South Georgia and Îles Kerguelen (Bednarek-Ochyra *et al.* 1999) but has not been seen from the Prince Edward Is. and Îles Crozet (Ochyra, personal observations).

Distribution in Antarctica. *Plagiothecium orthocarpum* is known only from a single site in Ozero Glubokoye (= deep lake) in Schirmacher Oasis in Dronning Maud Land in continental Antarctica at lat. 70°45′S and long. 11°20′ – 11°55′W (Fig. 269). The species was found on several occasions in 1962 and 1963 at depths from 23 to 32.5 m. It has not been collected elsewhere in Antarctica since then, although Richter (1995) commented that this moss was also found in 1986 and 1989 in other deep lakes in the southern part of Schirmacher Oasis, for instance in lakes Pomornik, Zub and Karovoye, adding that 'it may be assumed that this species is distributed in all suitable lakes up to Lake Zapadnoye' at the western end of the oasis. All of these lakes are situated at elevations between 80 and 100 m. Unfortunately, no specimens mentioned by this author from other lakes in the oasis have been available for study. Material reported as *P. simonovii* by Campos *et al.* (1978) from Lake Kitezh on Fildes Peninsula, King George I. (*Mahu 20553*, KRAM) actually represents *Campylium polygamum* and specimens referred to as *Plagiothecium* sp. from Alexander I. (Lewis Smith 1988*c*) are *Hypnum revolutum*.

Specimens examined. EAST ANTARCTICA. DRONNING MAUD LAND. **KRONPRINSESSE MÄRTHA KYST.** Schirmacher Oasis, Lake Glubokoye, at a depth of 32.3 m, 16 Jan 1962, *Simonov s.n.* (AAS, KRAM, LE – type of *P. simonovii*); same locality, at a depth of 32 m, 15 Jan 1962, *Simonov s.n.* (KRAM, LE); same locality, at a depth of 32.5 m, 1 Feb 1962, *Simonov s.n.* (KRAM, LE); same locality, at a depth of 23.5 m, 20 Jan 1961, *Simonov s.n.* (LE).

Family XVII. **HYPNACEAE** Schimp.

Hypnaceae Schimp., Coroll. Bryol. Eur. 113. 1856. — Type: *Hypnum* Hedw., *nom. cons.*

Orthotheciaceae Schimp., Coroll. Bryol. Eur. 103. 1856. — Type: *Orthothecium* Schimp.

Pylaisiaceae Schimp., Syn. Musc. Eur. 241. 1860. — Type: *Pylaisia* Schimp., *nom. cons.*

Microthamniaceae Kindb., Bot. Centralbl. 77: 386. 1899, *nom. illeg.*

The delimitation of the Hypnaceae is one of the most difficult and controversial problems in the taxonomy of pleurocarpous mosses. As presently understood the family is a convenient repository for various taxa extracted from other families of the Hypnales, for example from the Sematophyllaceae (Seki 1968), Hylocomiaceae (Andrews 1954), Rhytidiaceae (Noguchi 1972), Entodontaceae (Buck 1980) and Plagiotheciaceae (Buck & Ireland 1985). Some recent attempts to redefine the family (e.g. Nishimura *et al.* 1984; Nishimura 1999) appear to be unsatisfactory and its exact circumscription must await further studies of taxa whose familial positions have so far not been thoroughly examined. An evaluation of the phylogenetic relationships among the Holarctic Hypnaceae, Amblystegiaceae and Thuidiaceae making use of the cladistic analysis (Hedenäs 1995, 1996*d*, 1998*a*) and molecular data (Tsubota *et al.* 2002; Gardiner *et al.* 2005) showed only that the Hypnaceae is phylogenetically a very heterogeneous, catch-all family, and no satisfactory proposals for its reclassification have been presented. Therefore, the old classification of the Hypnaceae proposed by Brotherus (1925) is still in use, with some subsequent modifications (Nishimura 1999). He placed in the Hypnaceae 31 genera, but since then more than twice that number of genera have been transferred to or recognised as new within it. For obvious reasons the total number of species in this family cannot be precisely enumerated at present, but it is estimated it may comprise about 850 species classified into 65 genera (Crosby *et al.* 2000). Of these, about two thirds of the genera are mono- or oligotypic taxa, consisting of 1–5 species. Whereas the five largest genera encompass about three quarters of the known species of the Hypnaceae, namely *Ectropothecium* Mitt. (202), *Isopterygium* Mitt. (151), *Vesicularia* (Müll. Hal.) Müll.Hal. (116), *Mittenothamnium* Henn. (82) and *Hypnum* (81), except for the last genus, they have not yet been taxonomically revised on a global scale but, in general, considerable reductions of the number of species are expected owing to the high level of species inflation in tropical regions (Touw 1974).

The Hypnaceae are defined by a perfect hypnoid peristome consisting of broadly bordered exostome teeth which are cross-striolated on the front surface and highly trabeculate at the back and have a distinct median zig-zag line and endostome segments arising from a high basal membrane alternating with cilia. This 'ideal' peristome is found mostly in taxa with curved, asymmetric and inclined to horizontal capsules, whereas in taxa with erect and symmetric capsules the endostome is mostly variously reduced. The leaves are commonly curved, homomallous or falcato-secund, occasionally complanate or julaceous, mostly wide-spreading to erect-spreading, usually uniform but sometimes somewhat differentiated. They are typically lanceolate-ovate, broadly oblong-ovate to ovate, sometimes cordate, often asymmetric, rarely plicate, gradually long-acuminate, with straight or sometimes recurved-squarrose apices, flat or concave, smooth or rarely plicate and not or sometimes decurrent. The costa is short and double or often lacking. The laminal cells are usually thin-walled, prosenchymatous, mostly linear but sometimes shorter, oblong-hexagonal to oblong-rhomboidal, smooth or occasionally prorate or centrally papillose at the back. The alar cells are mostly differentiated, predominantly small and quadrate but rarely enlarged and inflated. The plants vary from very small and slender to robust and coarse, growing in loose or dense, often shiny mats, patches or wefts. The stems are predominantly prostrate, less often ascending or erect, irregularly to very regularly, pinnately branched and sometimes frondose, in transverse section with or without a central strand, sometimes with a unistratose hyalodermis. Pseudoparaphyllia are usually present, filiform to foliose, rarely absent and paraphyllia are always lacking.

The Hypnaceae are a worldwide family with the greatest concentration of taxa in the Northern Hemisphere and tropics. In the Subantarctic and Antarctic, as well as in the cool temperate zone in the Southern Hemisphere, it is poorly represented and its members are not prominent constituents of local floras. Only three genera, *Platydictya*, *Isopterygiopsis* and *Hypnum*, each with one species, are present in the Antarctic.

KEY TO THE ANTARCTIC GENERA OF THE HYPNACEAE

1. Leaves falcate-secund; alar cells well differentiated, small and subquadrate; rhizoids arising below the leaf insertion, smooth; capsules inclined to pendulous, asymmetric .... 53. ***Hypnum***
1. Leaves straight, occasionally somewhat secund; alar cells scarcely differentiated; rhizoids in the leaf axils, papillose; capsules ± erect, symmetric 2
    2. Plants very small; leaves minute, 0.15–0.5 mm long; pseudoparaphyllia present, filiform; laminal cells oblong-rhomboidal to shortly rhomboidal or rhombic; perichaetial leaves spinosely dentate at margins .... 54. ***Platydictya***
    2. Plants larger; leaves more than 0.5 mm long; pseudoparaphyllia lacking; laminal cells linear-flexuose to linear-fusiform; margins of the perichaetial leaves entire ... . 55. ***Isopterygiopsis***

53. **HYPNUM** HEDW., *nom. cons.*

Hypnum Hedw., Spec. Musc. Frond.: 236. 1801, *nom. cons.* — Type: *Hypnum cupressiforme* Hedw., *typ. cons.*

Stereodon Brid., Bryol. Univ. 1: xlv. 1826 and 2: 823. 1827, *nom. illeg. incl. lectotyp. gen. prior.*

Drepanium (Schimp.) Lange & C.E.O.Jensen, Meddel. Grønland 3: 326. 1887, *nom. illeg. incl. lectotyp. gen. prior.*

Plants small and slender to large and coarse, soft or stiff and rigid, often lustrous, in thinly to densely interwoven mats or tufts, green, golden, yellowish, brownish, golden-green to golden-brown. *Stems* creeping to ascending, rarely erect, irregularly to regularly pinnately branched, sometimes regularly frondose-pinnate, usually terete but sometimes complanate-foliate, in transverse section with a variously developed cortex of small, firm-to thick-walled cells, sometimes with a unistratose hyalodermis of enlarged, thin-walled cells surrounding large, thin-walled, usually hyaline medullary cells and a small or larger central strand, rarely lacking; *branches* spreading or ascending, rarely julaceous or complanately foliate, often hooked at the tips; *rhizoids* smooth, inserted on the stem below the leaf bases; *pseudoparaphyllia* present, foliose to subfilamentous; *axillary hairs* filiform, fairly short, consisting of 1–2 quadrate, brown basal cells and 2–5 elongate, hyaline distal cells. *Stem* and *branch leaves* similar, crowded, mostly falcate-secund, sometimes circinate or sometimes erect-spreading, symmetric, ovate, oblong-lanceolate to lanceolate-ovate, gradually short-acuminate to long-subulate, not or shortly decurrent, concave, smooth or rarely plicate; *margins* erect or reflexed in the proximal part, plane, rarely recurved to revolute, entire to serrulate distally; *costa* double, very short to comparatively long, sometimes lacking; *laminal cells* pellucid, smooth, oblong-rhomboidal to linear-flexuose, with thin to thick and sometimes porose walls, especially at the insertion; *alar cells* differentiated, small and quadrate, thick-walled and coloured, forming ± triangular groups or sometimes oblong, inflated, thin-walled and hyaline, forming distinct pellucid auricles in the basal angles. *Asexual reproduction* unknown. *Autoicous* or *dioicous*. *Perichaetial leaves* lanceolate to oblong-ovate, short- to slenderly long-acuminate, smooth or usually plicate, ecostate, subentire to serrulate at margins above, with undifferentiated alar cells. *Setae* elongate, smooth, light brown, yellowish-red to reddish-brown, dextrorse below and sinistrorse above when dry. *Capsules* suberect, inclined to horizontal, sometimes cernuous, usually asymmetric and arcuate, obloid to long-cylindrical, mostly smooth when dry; *operculum* conic, convex-conic, apiculate to mammillate, sometimes short-rostrate with an oblique beak; *annulus* mostly present, (1–)2–3-seriate, composed of large, thick-walled cells, usually partially deciduous; *exothecial cells* subquadrate to rectangular, thin- to thick-walled, sometimes slightly collenchymatous, thin- to firm-walled, sometimes with distinctly thickened longitudinal walls; *stomata* sparse to numerous in the neck of the urn, bicellular, superficial, long-pored; *peristome* double; *exostome teeth* subulate-acuminate, shouldered, bordered, yellowish to brownish, finely transversely striolate below, hyaline and papillose above, highly trabeculate; *endostome* consisting of keeled, pale or yellowish-pellucid, entire or narrowly cleft segments arising from a high basal membrane, alternating with long, rarely rudimentary, nodose and papillose, not appendiculate cilia in groups of (1–)2–3. *Spores* globose to ellipsoid, nearly smooth to finely papillose. *Calyptra* cucullate, naked, smooth.

Hypnum is one of the ancient generic moss names which is derived from a Greek word meaning 'sleep' or 'recumbent'. In the early days of bryology the majority of pleurocarpous mosses had been placed in this genus, making it an extremely heterogeneous and all-encompassing taxon. In modern moss taxonomy most of the species once included in *Hypnum* have been transferred to other genera, but in the present circumscription the genus still remains a large taxon encompassing an assortment of species, and is consequently difficult to define. In general, modern *Hypnum* is characterised by prostrate, regularly pinnate-branched plants with falcate-secund to circinate leaves, short and double costae, smooth and elongate laminal cells, variously differentiated alar cells, smooth and elongate setae, and inclined and curved capsules.

There are about 80 species which are still residual in *Hypnum* (Crosby *et al.* 2000). Although many exotic species have not been studied since their description, it seems that their real number in this genus is probably close to 40, at least judging from a series of excellent taxonomic treatments dealing with various sections of *Hypnum* by Ando (1972*a*,*b*, 1973, 1976, 1986, 1987, 1989, 1990, 1992, 1993, 1996). The genus was subdivided by Ando (1973) into 11 sections, although sect. *Polyptera* Ando, was subsequently raised to genus in its own right, *Caribaeohypnum* Ando & Higuchi (Ando & Higuchi 1984). There have been several attempts to split *Hypnum* into smaller segregates but these have not gained wider acceptance. Additionally, such names as *Stereodon* and *Drepanium* are illegitimate because they originally included *Hypnum cupressiforme* which is accepted as the conserved type of *Hypnum* (Dixon 1932*b*).

In the cold regions of the Southern Hemisphere the genus is poorly represented. Two species, *Hypnum cupressiforme* and *H. chrysogaster* Müll.Hal., occur on various subantarctic islands, while a third, *H. revolutum*, is widely distributed but scattered in the maritime Antarctic. It belongs within a small section *Revolutohypnum* Mönk. consisting of four Holarctic species.

1. **Hypnum revolutum** (Mitt.) Lindb. Fig. 270

Hypnum revolutum (Mitt.) Lindb., Oefv. K. Vet. Ak. Foerh. 23: 542. 1867. ≡ *Stereodon revolutus* Mitt., J. Linn. Soc. Bot. Suppl. 1: 97. 1859. — Type citation: In Tibet. occid. reg. alp., in summo montis Hera La, alt. 18,700 ped., H. Strachey! [Lectotype (*selected here*) {or perhaps holotype}: 'Top of Hera Lu 18.700 Tibet H. Strachey (Stereodon revolutus) – NY-Mitten!; isotype: BM-Hooker!].

Plants small to medium-sized, slender to fairly coarse, stiff and rigid, shiny in yellow-green to yellow- or golden-brown, dense mats or deep tufts. *Stems* procumbent or ascending to erect, 2–5 cm long, sometimes filiform, terete to julaceous, irregularly to evenly pinnately branched, in cross-section oval or round, with a small but distinct central duct, 3–4-stratose medulla of large, thin-walled, hyaline to yellowish-hyaline cells and 2–3-stratose cortex of small, yellow-brown, thick-walled cells; *branches* subjulaceous, 2–5 mm long, patent; *pseudoparaphyllia* foliose, lanceolate to ovate; *axillary hairs* filiform, hyaline throughout, 3–5-celled, consisting of a short basal cell and elongate distal cells. *Stem leaves* slightly curved to falcate-secund, often homomallous, sometimes almost straight, broadly ovate to oblong-lanceolate, gradually or abruptly tapering to an acumen of varying length, long-acuminate to short-acute, concave, slightly plicate, (1.0–)1.2–1.8(–2.0) mm long, 0.4–0.6 mm wide, non-decurrent; *branch leaves* somewhat smaller, similar to the stem leaves or narrower, ovate-lanceolate, more distinctly serrulate in the upper ⅓ or less; *margins* narrowly revolute from the base to the apex or sometimes plane on one or both sides, entire, sinuate-serrulate or serrulate at the apex; *costa* short and double, slender, usually distinct, rarely almost obsolete; *laminal cells* oblong- to linear-rhomboidal or fusiform, rather short and wide in the mid-leaf, 20–50 μm long, 4–5(–6) μm wide, thin- to thick-walled, becoming wider, with thicker and porose walls at the insertion; *alar cells* subquadrate, numerous, (6–)8–15 in the marginal row, thick-walled, forming clearly defined but not very conspicuous, ± triangular groups in the basal angles. [*Dioicous*. *Inner perichaetial leaves* oblong-lanceolate, plicate, shortly acuminate, entire, ecostate. *Setae* erect, straight, smooth, 1–2 cm long, yellowish- to reddish-brown, sinistrorse above, dextrorse below. *Capsules* inclined to horizontal, obloid-cylindrical to cylindrical, 2–3 mm long, 0.6–0.8 mm wide, curved, pale to yellowish-brown, somewhat striate and constricted below the mouth on drying; *operculum* convex-conic, apiculate; *annulus* 2(–3)-seriate, persistent; *cilia* 2–3, nodose. *Spores* almost smooth, globose, 10–15 μm in diameter. *Calyptra* cucullate, naked, smooth.]

Chromosome number: unknown in the Antarctic and Subantarctic; n = 14 from Canada (Anderson & Crum 1959).

Discussion. In the Antarctic, *Hypnum revolutum* is easy to recognise by its falcate-secund leaves which are unknown in any other local pleurocarps, except for species of *Sanionia*. Some dwarf, underdeveloped plants of the latter genus, especially *S. uncinata*, may occasionally be confused in the field with *H. revolutum* but when observed microscopically they are at once different in having long, single costae, large pellucid alar cells and always flat leaf margins. In contrast, the leaves of *H. revolutum* are rather shortly acuminate, with short, double costae and non-excavate alar regions, consisting of homogeneous, subquadrate cells, and the plants are small to medium-sized and often have julaceous, filiform stems and branches.

In its wide geographical range *Hypnum revolutum* exhibits great polymorphism in the robustness, mode of ramification and foliage of the plants, areolation and plication of the leaf lamina and recurvature of its margins. As in other parts of the range, Ando (1973) recognised in the Antarctic material two infraspecific taxa, namely *H. revolutum* var. *dolomiticum* (Milde) Mönk. and *H. revolutum* fo. *pumilum* (Husn.) Ando within the type variety in which he placed most of the Antarctic specimens. The former varietal name was later changed to *H. revolutum* var. *ravaudii* (Husn.) Ando (Ando & Matteri 1982). It comprises smaller plants with weakly plicate leaves and slightly recurved to plane margins, whereas the latter form is similar to it in overall appearance and its margins are variable in degree of revoluteness,

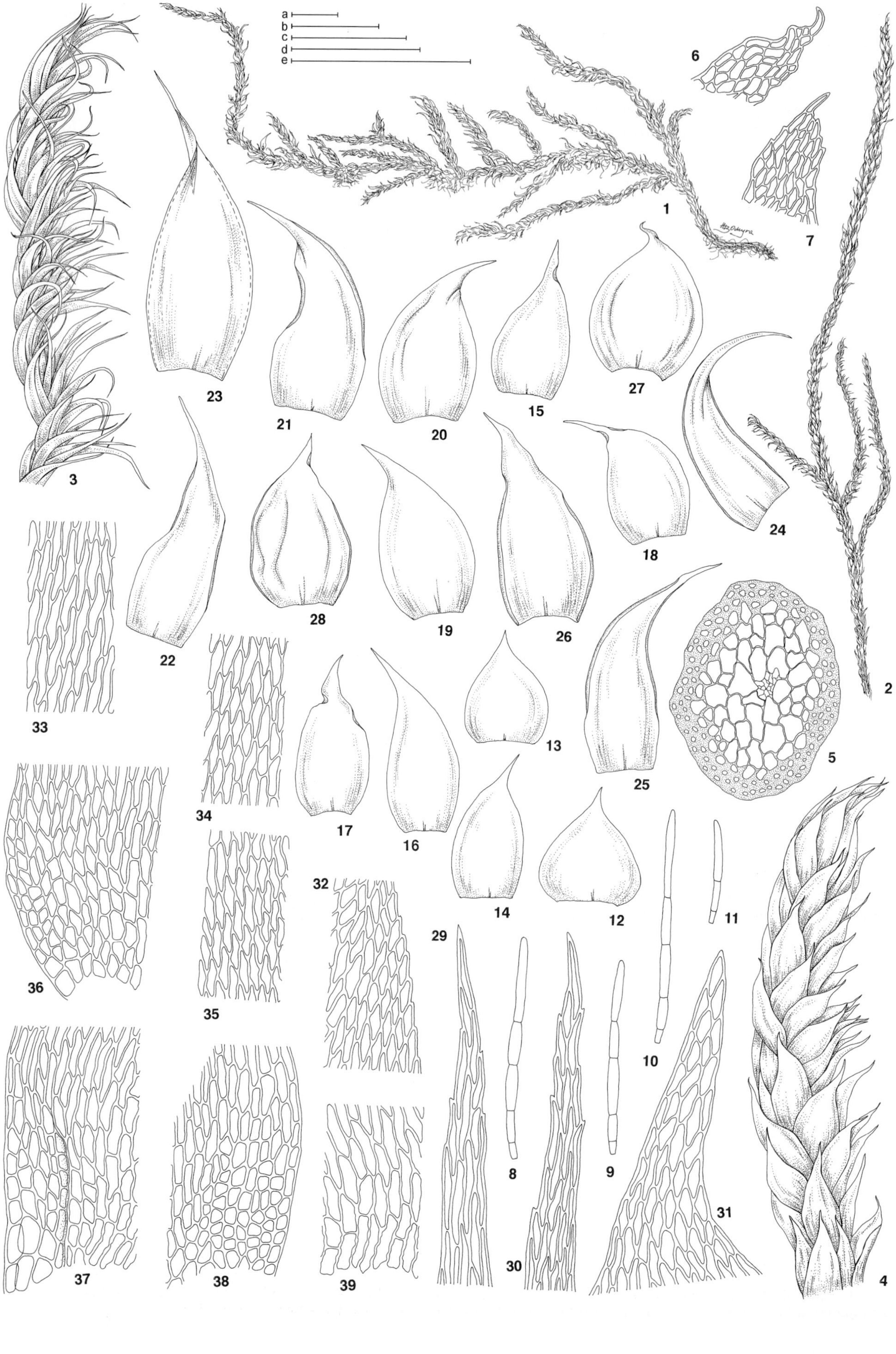
a
b
c
d
e
1
2
3
4
5
6
7
8
9
10
11
12
13
14
15
16
17
18
19
20
21
22
23
24
25
26
27
28
29
30
31
32
33
34
35
36
37
38
39

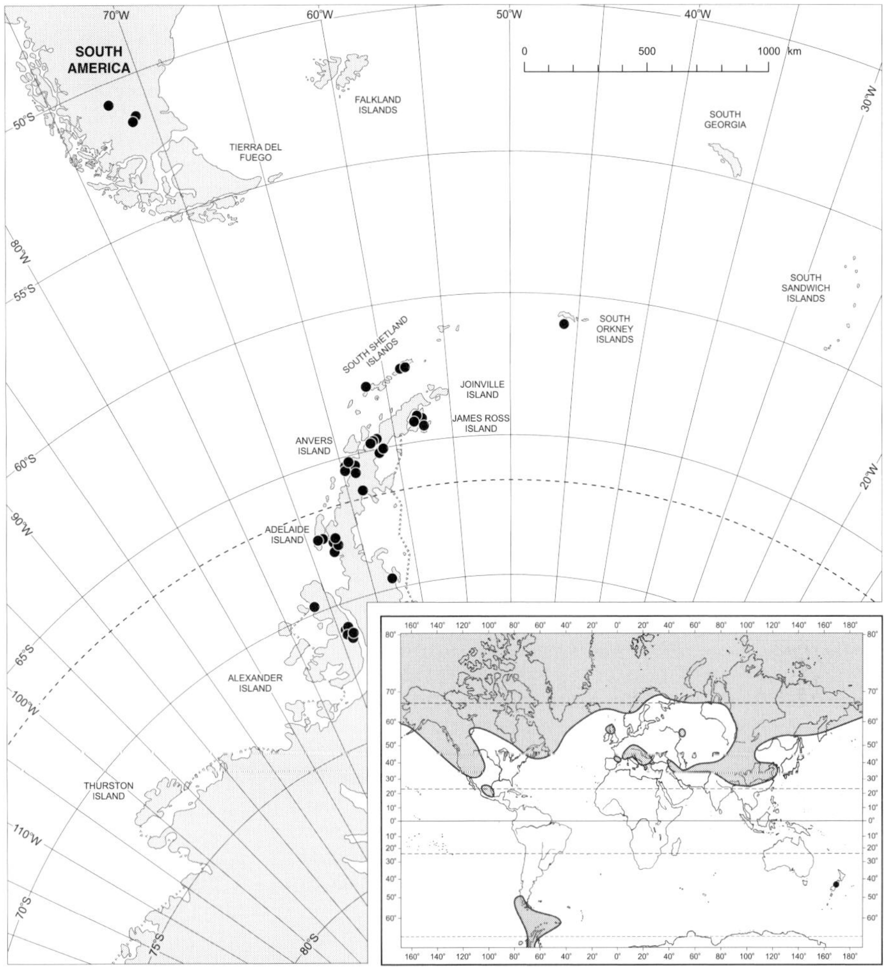

Figure 271. Distribution of *Hypnum revolutum* in the Fuegian region and Antarctica. Inset: global distribution.

from strongly revolute to almost plane on both sides and additionally it has broadly ovate, shorter-acuminate and less falcate leaves and shorter and wider lamina cells. These characters are all extremely variable and as such are not acceptable for the delimitation of separate taxa because they are linked by numerous intermediates and consequently named at random, especially because plants exhibiting characters of these taxa often can be found within the same mat. Therefore these infraspecific taxa are not recognised here and they are considered to be merely habitat expressions conditioned by severe climatic conditions.

Reproduction in Antarctica. Neither sexaul organs nor sporophytes have been found on Antarctic material.

Habitat. *Hypnum revolutum* grows mostly in dry and exposed, as well as in damp or wet and sheltered sites, forming small patches on rock outcrops and ledges, on pockets of mineral soil among boulders and large blocks, in rock crevices and windscoops, on stony and gravelly soil in melt runnels and at the margins of solifluction lobes. It thrives on both calcareous and acidic rocks, although it is much more frequent on basic substrates. Samples of *H. revolutum* are often free of other species, but sometimes it grows together with *Distichium capillaceum*, *Encalypta procera*, *Tortella alpicola*, *Syntrichia saxicola*, *Platydictya jungermannioides* and *Cratoneuropsis relaxa* subsp. *minor*. In general, the species is found in various associations of the fruticose lichen and moss cushion subformation, and sometimes its dry mats are encrusted with lichens.

World range. A bipolar species (Fig. 271, inset), arctic-boreal-montane in the Northern Hemisphere where it has a circumpolar, continuous range in the Arctic and disjunct occurrences in the mountains of the southern part of the Holarctic, including the Himalayas, Central Asia, the Caucasus and the Sierra Nevada in Europe, and southern Mexico in Central America (Ando 1972*a*, 1973). In the Southern Hemisphere it was first recorded from the Antarctic (Cardot 1900, 1901; Bartram 1957) and subsequently in southern Patagonia (Ando & Matteri 1982; Matteri 1985*a*) and on the South Island in New Zealand (Ochyra 1988*b*).

Distribution in Antarctica. *Hypnum revolutum* is widely but unevenly distributed in West Antarctica, ranging from the South Orkney Is. to Alexander I. where it reaches its southernmost stations at lat. 71°53′S (Fig. 271). The species is exceedingly rare in the South Orkney Is. and South Shetland Is., although

←

Figure 270. *Hypnum revolutum*. — **1–2**. Habit. **3–4**. Portions of branches, dry. 5. Cross-section of stem. **6–7**. Pseudoparaphyllia. **8–11**. Axillary hairs. **12–28**. Leaves. **29–31**. Leaf apices. **32**. Upper leaf cells at margin. **33–35**. Mid-leaf cells at margin. **36–38**. Alar cells. **39**. Basal leaf cells. — [1, 12–13, 31, 35–36 from *RO 2090/80*; 2, 8, 14–15 from *RO 2171/80*; 3, 5–7, 9–11, 23–25, 29, 37 from *RO 430/80*; 4, 17–18 from *RO 1148/80*; 16 from *RO 22267/80* ; 19–20 from *RO 442/80*; 21–22, 30, 33, 38 from *RO 2284/80*; 26, 39 from *RO 2584/80*; 27–28, 32, 34 from *RO 448/80*]. Scale bars: a – 1 mm (1–2); b – 1 mm (3) and 50 mm (29–39); c – 1 mm (4); d – 100 mm (5–11); e – 1 mm (12–28).

it is locally common and abundant on King George I. (Ochyra 1998*a*). It occurs sporadically along the western coast of the Antarctic Peninsula and has two distinct centres of occurrence, on the offshore islands of the Graham Coast and in northern Marguerite Bay, whereas in the northern part it is very rare on the Danco Coast and absent from Trinity Peninsula. Additionally, *H. revolutum* occurs occasionally on the eastern coast of the Antarctic Peninsula, where it is fairly frequent on James Ross I. and very rare southwards, extending to lat. 66°12′ S on the Foyn Coast. The species exhibits a remarkable elevational amplitude, occurring from near sea level to the summits of several nunataks, reaching its highest altitude of 515 m on Alexander I. and possibly on the Foyn Coast, where it was collected at elevations from '330–650' m.

Selected specimens examined. WEST ANTARCTICA. SOUTH ORKNEY IS. **SIGNY I.** Knife Pt., 25 m, *RILS 5067* (AAS, KRAM); Paal Harbour, *RILS 1450* (AAS, KRAM); east side of Rusty Bluff above Paal Harbour, *ca* 35 m, *RILS 354* (AAS, KRAM).

SOUTH SHETLAND IS. **KING GEORGE I.** ***Fildes Peninsula***: Two Summit I., *Li 890213* (AAS, KRAM); Ostrov Geologov, *Li 890124* (AAS, KRAM); Great Wall Station, *Li DTN2* (AAS, KRAM); Green Pt., *BJ 214* (KRAM). ***Bransfield Strait***: Blue Dyke, 130 m, *RO 1148/80* (AAS, KRAM). ***Admiralty Bay***: Brama, 200 m, *RO 791/80* (KRAM) and 210 m, *RO 802/80* (KRAM). ***Ezcurra Inlet***: Breccia Crag, 140 m, *RO 920/80* (KRAM); Klekowski Crag, 220 m, *RO 2267/80* (KRAM) and 250 m, *RO 2284/80* (AAS, KRAM). ***Keller Peninsula***: Ore Pt., 90 m, *RO 448/80* (AAS, KRAM); Tyrrell Ridge, 220 m, *RO 430/80* (KRAM); Mt. Birkenmajer, 280, *RO 442/80* (AAS, KRAM); without specific locality, *RMS 69-941 & 69-956* (KRAM). ***Martel Inlet***: Stenhouse Bluff, 30 m, *RO 2584/80* (AAS, KRAM); Szafer Ridge, 250 m, *RO 2578/80 & 2268/80* (KRAM); Tern Nunatak, 266 m, *RO 2518/80* (KRAM); Smok, 40 m, *RO 2090/80 & 2091/80* (KRAM); Mt. Wawel, 80 m, *RO 2171/80* (KRAM). **LIVINGSTON I.** ***Byers Peninsula***: bay off New Plymouth, *DCL 318B* (AAS, KRAM).

WEST ANTARCTIC PENINSULA. **DANCO COAST.** Cape Anna, *ER 202B* (BR, PC); Cuverville I., 2–20 m, *RILS 4168* (AAS, KRAM); Birdsend Bluff, 5–35 m, *RILS 2414 & 4215* (AAS, KRAM); Coughtrey Peninsula, Paradise Harbour, Almirante Brown Station, *VK 398* (KRAM). **GRAHAM COAST.** Edge Hill, *ca* 20 m, *RWMC 654A* (AAS, HIRO, KRAM); Cape Tuxen, 3–300 m, *RILS 4888* (AAS, KRAM); Cape Pérez, *ca.* 7 m, *RWMC 635A* (AAS, HIRO, KRAM). **WILHELM ARCHIPELAGO.** ***Penola Strait***: Petermann I., 15 m, *REL 1345AA & 1345C* (AAS, KRAM) and 20 m, *RWMC 796* (AAS, KRAM). ***Argentine Is.***: Corner Is., *ca* 8 m, *REL 1368* (AAS, KRAM); most westerly of the Corner Is., 8 m, *RWMC 666* (AAS, HIRO, KRAM) and *ca* 10 m, *RWMC 667* (AAS, KRAM); Galindez I., Stella Creek, *REL 1321* (AAS, KRAM) and 20 m, *RWMC 409* (AAS, KRAM); same island, Woozle Hill, 20 m, *RILS 937* (AAS, KRAM). ***Collins Bay***: largest of Berthelot Is., *ca* 35 m, *RWMC 628B* (AAS, HIRO, KRAM); Darboux I., 10–15 m, *RILS 3334A & 3335* (AAS, KRAM). **FALLIÈRES COAST.** ***Adelaide I.***: Rothera Pt., 6 m, *RILS 8887* (AAS, KRAM) and 20 m, *JF 69A, 69B & 70B* (AAS, KRAM); Reptile Ridge, 140 m, *RILS 10580* (AAS, KRAM); Mt. Gaudry, *ca* 465 m, *JK 143A* (AAS, KRAM). ***Léonie Is.***: Lagoon I., *BGLE 1476* (BM); Léonie I., 30 m, *RILS 8927F* (AAS, KRAM) and 150–175 m, *RILS 9129* (AAS, KRAM). ***Marguerite Bay***: Jenny I., *ca* 15 m, *BJT 439A* (AAS, KRAM) and 200 m, *RILS 4753* (AAS, KRAM); Pourquoi-Pas I., Lainez Pt., *S. & J. Poncet 14* (KRAM); Lagotellerie I., 300 m, *RILS 4868* (AAS, KRAM); Camp Pt., 5–20 m, *RILS 4636, 4637, 4638, 4679, 4680 & 4681* (AAS, KRAM). **GEORGE VI SOUND.** ***Alexander I.***: Debussy Heights, 400 m, *BAS Misc. 207A* (AAS, KRAM) and *RILS 6022A* (AAS, KRAM); Ablation Valley, 2–3 m, *RILS 9220B, 9250, 9255B, 9257B & 9276* (AAS, KRAM), 5 m, *RILS 2708A & 2710* (AAS, KRAM), 75 m, *RILS 9290* (AAS, KRAM); 155 m, *RILS 2716* (AAS, KRAM) and 400 m, *JJL & RBH 23A* (AAS, KRAM); Middle Flatiron Valley, 315 m, *RILS 11103B* (AAS, KRAM) and 420 m, *RILS 11096B & 11104* (AAS, KRAM); High Flatiron Valley, 510 m, *RILS 11073B* (AAS, KRAM) and 515 m, *RILS 11070 & 11092* (AAS, KRAM); Probe Ridge, *DDWW 72A* (AAS, KRAM); Natal Ridge, *PC 693, 694, 695A & 696A* (AAS, KRAM); Viking Valley, 300 m, *RILS 9471* (AAS, KRAM); Mars Oasis, 75–80 m, *RILS 10380 & 10427* (AAS, KRAM).

EAST ANTARCTIC PENINSULA. **JAMES ROSS ISLAND GROUP.** ***Vega I.***: False Island Pt., 25–40 m, *RILS 7853* (AAS, KRAM). ***James Ross I.***: Abernethy Flats, 100 m, *RILS 7634B* (AAS, KRAM); Blancmange Hill, 12–15 m, *RILS 7680* (AAS, KRAM) and 300 m, *RILS 7821, 7822 & 7824B* (AAS, KRAM); west side of point 536, 300 m, *RILS 5956A* (AAS, KRAM); south of St. Martha Cove, 35 m, *RILS 7455B, 7456 & 7457* (AAS, KRAM), 120 m, *RILS 7620* (AAS, KRAM) and 150 m, *RILS 7535, 7536 & 7541B* (AAS, KRAM); north of St. Martha Cove, 80 m, *RILS 7647* (AAS, KRAM); Lachmann Crags, 20–35 m, *RILS 7320, 7325, 7333C, 7339 & 7380A* (AAS, KRAM) and 400 m, *RILS 5393A* (AAS, KRAM) ; Fortress Hill, 150 m, *RILS 7758 & 7772* (AAS, KRAM); Andreassen Pt., 10 m, *RILS 7612* (AAS, KRAM); Crame Col, 100 m, *RILS 7409* (AAS, KRAM); Terrapin Hill, 85 m, *RILS 7740* (AAS, KRAM), 250 m, *RILS 7726* (AAS, KRAM) and 540 m, *RILS 7712* (AAS, KRAM); Dreadnought Pt., 10 m, *RILS 7498* (AAS, KRAM); Cape Obelisk, 60 m, *RILS 5951* (AAS, KRAM); Hamilton Pt., 100 m, *RILS 11648* (AAS, KRAM). **OSCAR II COAST.** ***Exasperation Inlet***: north-east of Caution Pt., *PK 19* (AAS, KRAM); Starbuck Glacier, *ca* 370 m, *PK 42* (AAS, KRAM). **FOYN COAST.** ***Cabinet Inlet***: Eden Glacier, 330–650 m, *RT 29 & 31B* (AAS, KRAM). **WILKINS COAST.** Large nunatak, lat. 69°30′ S, long. 62°41′ W, 610–760 m, *MJC 51* (AAS, HIRO, KRAM).

54. PLATYDICTYA BERK.

Platydictya Berk., Handb. Brit. Mosses: 145. 1863. — Type: *Platydictya sprucei* (Bruch ex Spruce) Berk. (*Leskea sprucei* Bruch *ex* Spruce) [= *Platydictya jungermannioides* (Brid.) H.A.Crum (*Hypnum jungermannioides* Brid.)].

Apterygium Kindb., Rev. Bryol. 12: 29. 1885, *nom. illeg. incl. typ. gen. prior.*

Amblystegiella Loeske, Moosfl. Harz.: 295. 1903, *nom. illeg. incl. typ. gen. prior.*

Plants minute and slender, soft, dull or somewhat lustrous, forming silky, pale or dark green, brownish- or yellowish-green or brownish patches or thin tufts. *Stems* prostrate, filiform, freely and irregularly branched, with erect-ascending branches, pale brownish, fragile, in cross-section with 1–2-stratose cortex of somewhat smaller, thicker-walled cells surrounding large, thin-walled, hyaline medullary cells and without a central strand; *rhizoids* in the leaf axils, purple-brown, densely granulose-papillose; *pseudoparaphyllia* small, filiform. *Leaves* very small, rarely exceeding 0.5 mm, erect-spreading, sometimes subsecund to distinctly secund, ovate-lanceolate to linear-lanceolate, straight to curved, slenderly acute, not or slightly narrowed at the base, slightly concave, non-decurrent; *margins* flat, finely serrulate; *costa* lacking or very slightly indicated; *laminal cells* pellucid, smooth, oblong-rhomboidal to shortly rhomboidal or rhombic, firm-walled; *alar cells* almost undifferentiated, subquadrate to short-rectangular, forming inconspicuous groups in the basal angles. *Asexual reproduction* small propagules clustered in leaf axils on stems and branches. *Dioicous* or *autoicous*. *Inner perichaetial leaves* small, non-plicate, erect, ovate-lanceolate, long acuminate, ecostate, spinosely dentate at margins. *Setae* elongate, comparatively long, smooth, red below, becoming yellowish upwards. *Capsules* erect and symmetric or inclined to horizontal and slightly asymmetric, obloid-cylindrical to obovoid, constricted below the mouth; *operculum* convex, mammillate to conic, apiculate; *annulus* present, consisting of 1–2 rows of vesiculose cells, deciduous; *exothecial cells* thin-walled, quadrate to short-rectangular; *stomata* few at the base of the urn, bicellular, round, pale, superficial; *peristome* double; *exostome teeth* lanceolate, yellow to yellowish-brown, cross-striolate below, pale and papillose above; *endostome* pale, finely papillose. comprising keeled, narrowly perforated segments arising from a moderately high basal membrane and 1–3 fragile, sometimes rudimentary or lacking cilia. *Spores* small, finely papillose to almost smooth. *Calyptra* cucullate, naked, smooth.

A small genus of two species, one of which, *Platydictya minutissima* Sull., is a narrow endemic in eastern North America (Crum & Anderson 1981), whereas the second one, *P. jungermannioides*, is a bipolar species occurring also in the Antarctic. For a long time *Platydictya* was known as *Amblystegiella* Loeske but Crum (1964) showed that this is a nomenclaturally illegitimate name. Until recently, the genus comprised at least seven species, but all, except for the above mentioned two species, were transferred to the separate genus *Serpoleskea* (Limpr.) Warnst. in which rhizoids are smooth and intercalary, arising from the stem below leaf bases (Hedenäs 1987*a*).

Platydictya has traditionally been placed in the Amblystegiaceae, but now the genus, is considered to be a member of the Hypnaceae (Grout 1932; Crum & Anderson 1981; Vitt 1984). In its granular-papillose rhizoids arising in the leaf axils, brownish-yellow exostome teeth with borders strongly widened at the transition zone as well as in its small and smooth perichaetial leaves, *Platydictya* is a discordant element in the Hypnaceae. For this reason Hedenäs (1987*a*) positioned it in the Plagiotheciaceae. Nevertheless, at present the placement of *Platydictya*, along with *Isopterygiopsis* with which it shares a similar type of rhizoids, in the Hypnaceae seems to be the best familial position for this genus, since these genera share the general morphology and many structural characters with other hypnalean mosses.

1. **Platydictya jungermannioides** (Brid.) H.A.Crum Fig. 272

Platydictya jungermannioides (Brid.) H.A.Crum, Michigan Bot. 3: 60. 1964. ≡ *Hypnum jungermannioides* Brid., Musc. Rec. 2: 255. 1812. ≡ *Amblystegiella jungermannioides* (Brid.) Giacom., Atti Ist. Bot. Univ. Pavia Ser. 5, 4: 262. 1947. — Type citation: In introitu speluncae Baumannianae in Hercynia legit Weberus. [Neotype (*vide* Isoviita 1979: p. 281): 'Hypnum jungermannioides Brid. Am Eingange in die Baumannshöhle 24 Sept. 97 Flörke' – JE!].

Amblystegium densissimum Cardot, Rev. Bryol. 27: 46. 1900. ≡ *Amblystegiella densissima* (Cardot) Broth. *in* Engl. & Prantl, Nat. Pflanzenfam. 1(3): 1025. 1908. ≡ *Platydictya densissima* (Cardot) H.Rob. *in* Llano, Antarct. Res. Ser. Amer. Geophys. Union 20: 170. 1972. — Type citation: [Antarctica] détroit de Gerlache, IXme débarquement; later Cardot (1901) wrote: Terre de Danco, près du cap Anna Osterrieth, sur les petites terrasses humides de la falaise. (N° 205e; 29 janvier 1898). [Lectotype (*selected here*): 'Herb. J. Cardot. Mousses du voyage de la Belgica N° 205e Amblystegium densissimum Card. sp. nova. Canal de la Belgica. IXème débarquement, sur les petites terrasses humides de la falaise leg. Racovitza, 29 janvier 1898' – PC!; isotype: BR!]. First synonymised by Ochyra (1999*k*: p. 562)].

Plants very small, thread-like, in dense, soft, silky mats, light green above, brownish below, sometimes in deep, to 2 cm, tufts. *Stems* filiform, fragile, erect, 5–10(–20) mm long, subjulaceous, sparingly or profusely, irregularly branched, with branches easily detached, in cross-section without central strand, consisting of large, thin-walled, hyaline medullary cells and 1–2-layered cortex of smaller, thicker-walled, pale brownish cells; *rhizoids* arising from leaf axils, purple-brown, granular-papillose; *pseudoparaphyllia* small, filiform; *axillary hairs* short, filiform, 2–3-celled, hyaline throughout. *Leaves* crowded, erect to erect-spreading, 0.2–0.3(–0.35) mm long, 0.1–0.17 mm wide, somewhat concave, straight or weakly curved, ovate-lanceolate, slenderly acuminate, somewhat narrowed at the insertion, non-decurrent; *margin* plane, serrulate to sinuate; *costa* lacking or sometimes very slightly marked; *laminal cells* transparent, unistratose, smooth, firm-walled, shortly oblong-rhomboidal, (15–)18–27(–30) µm long, 6–7 µm wide; *alar cells* subquadrate to short-rectangular, forming inconspicuous groups of 3–5 cells along the margins in the basal angles. *Gemmae* often clustered in leaf axils on stems and branches, smooth, elongate, obcuneate bodies consisting of 2–4 cells in a single row. [*Dioicous*. *Inner perichaetial leaves* elongate from an ovate base long-acuminate, sharply denticulate all around, ecostate. *Setae* erect, smooth, 5–11 mm long, purple, sinistrorse above on drying. *Capsules* erect to slightly inclined, obloid to short-cylindrical, 0.5–1.0 mm long, 0.4–0.5 mm wide, symmetric or nearly so, smooth, brownish, constricted at the mouth, gradually tapering to the seta; *operculum* conic-convex, stoutly mammillate, to shortly rostellate, 0.3–0.4 mm long; *annulus* 1(–2)-seriate, deciduous; *exothecial cells* thin-walled, quadrate to short-rectangular; *stomata* rather few at the base of the urn, large, pale, bicellular, phaneroporous, with oval pori; *peristome* double, inserted below the capsule rim; *exostome teeth* lanceolate, pale yellowish to hyaline, bordered, dentate at the margins, cross-striolate below, papillose above on the dorsal surface, weakly trabeculate on the ventral side; *endostome* colourless, with a fairly high basal membrane, narrow and entire processes and with no or 1–2 rudimentary to ± well developed, nodose cilia. *Spores* globose, 11–13 µm in diamater, almost smooth. *Calyptra* cucullate, smooth, naked.]

Chromosome number: unknown.

Discussion. *Platydictya jungermannioiedes* is the smallest of all the mosses occurring in the Antarctic. It can be easily distinguished in the field with the naked eye by its characteristic habit of very slender, filiform and fragile plants thriving in dense, dark or light green tufts. This moss resembles tiny liverworts and hence its specific name alludes to characteristic habit and stature of the plants. Microscopically, the species has very small leaves which are ecostate and have the margins serrulate all around, although the species is very variable in the degree of leaf serrulation. In addition, it is the only pleurocarpous moss in the Antarctic which produces abundant propagules.

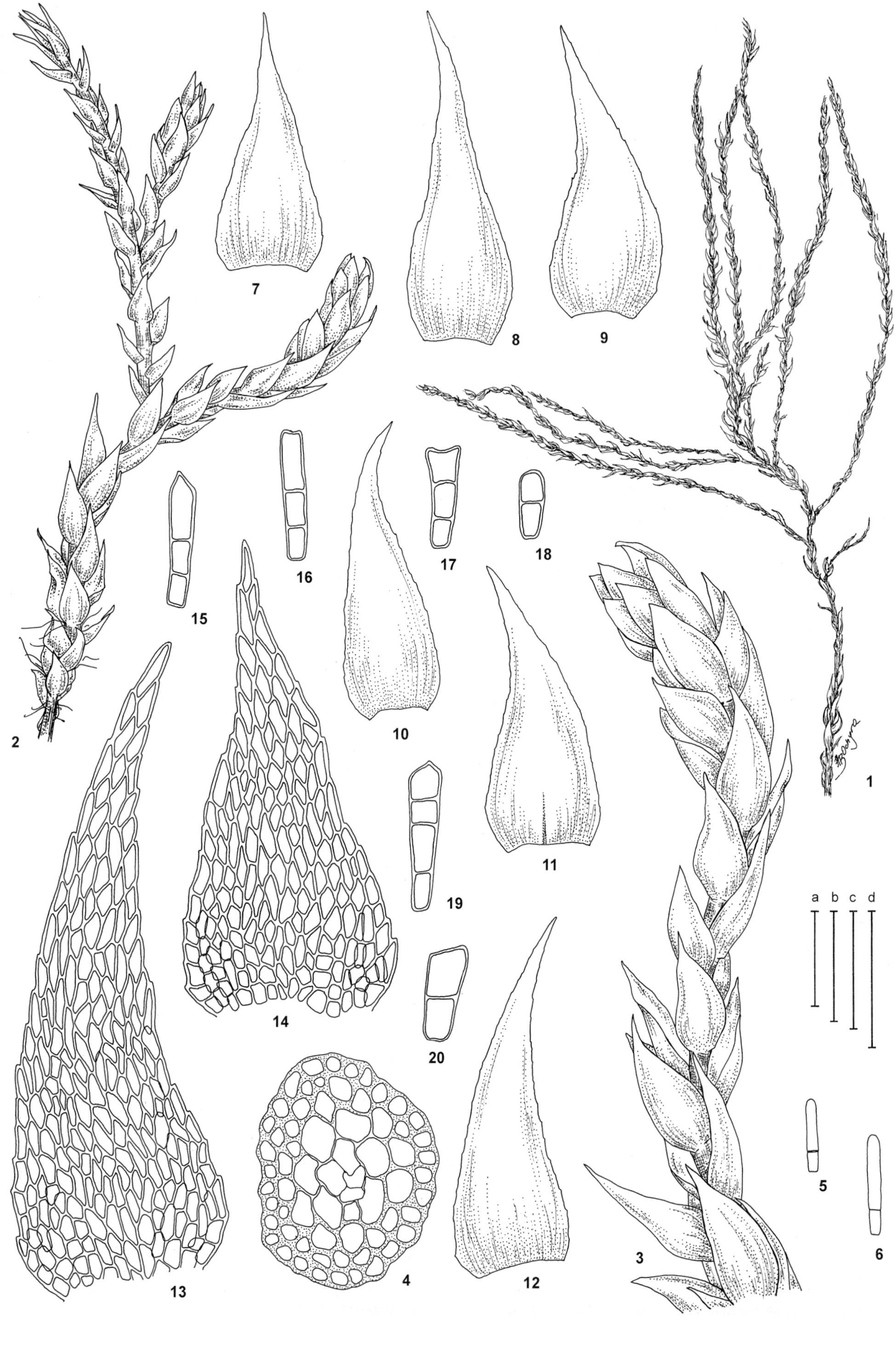
1
2
3
4
5
6
7
8
9
10
11
12
13
14
15
16
17
18
19
20
a b c d

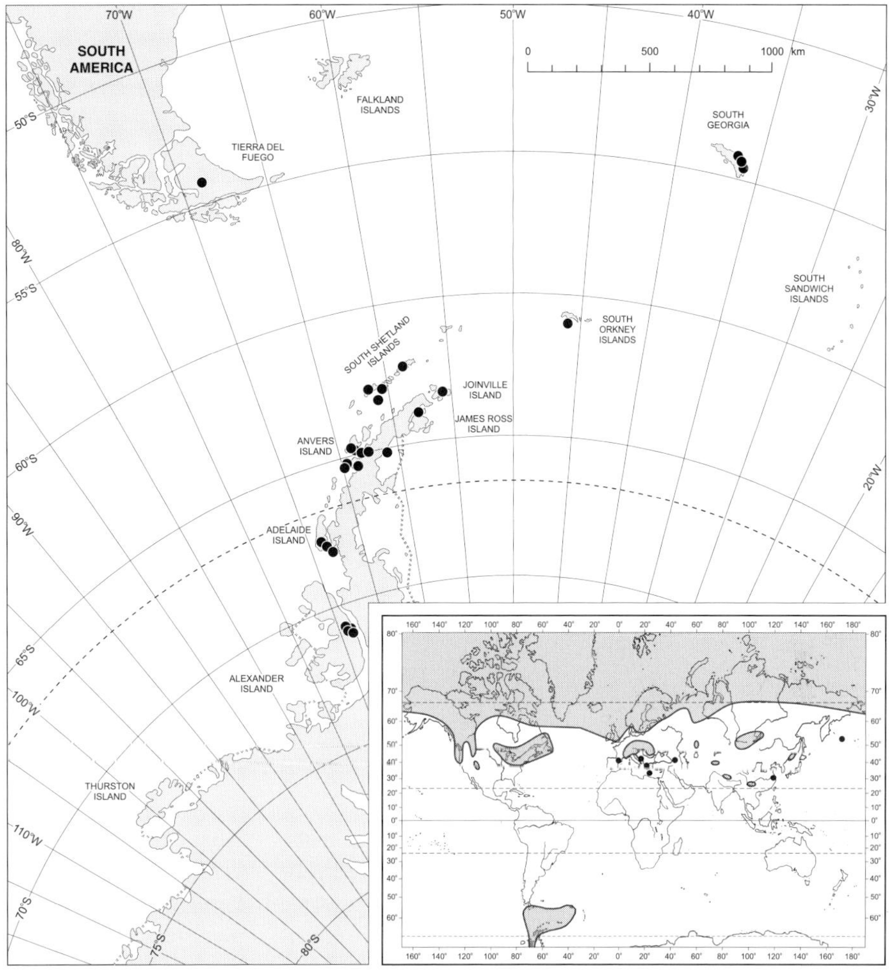

Figure 273. Distribution of *Platydictya jungermannioides* in the Fuegian region and Antarctica. Inset: global distribution. (Reproduced from *Annales Botanici Fennici*, 36: p. 56, 1998, with permission.)

Cardot (1900, 1901) recognised Antarctic plants as a separate species, *Amblystegium densissimum*, on account of their very compact tufts, subjulaceous branches and stems and concave, shorter acute and densely arranged leaves. Many Antarctic populations grow in very dense tufts, but the structural characters of the leaves do not differ materially from those found in plants from high latitudes in the Northern Hemisphere. Therefore, they do not merit taxonomic recognition and this Antarctic species is considered conspecific with *P. jungermannioides* (Ochyra 1999*k*).

Reproduction in Antarctica. Always sterile and without sexual organs, but gemmae in the leaf axils or arising from marginal cells near the leaf apex are present in most specimens and their effectiveness in dispersal seems to be well reflected in the wide geographical distribution of *Platydictya jungermannioides.*

Habitat. The species is predominantly associated with calcareous habitats and only occasionally grows on acidic sites. It thrives on damp or wet mineral soil, gravel or humus in deep and sheltered rock crevices, in rock concavities, often under overhanging rocks on cliffs, outcrops and knolls of marble and amphibolite. On Signy I. *Platydictya jungermannioides* is a constituent of the association dominated by *Syntrichia saxicola* and *Schistidium antarctici* in which it cohabits, apart from the dominant species, with *Cephaloziella varians*, *Marchantia berteroana*, *Syntrichia magellanica*, *Cratoneuropsis relaxa* subsp. *minor* and *Pohlia cruda.*

World range. A bipolar species (Fig. 273, inset), widely distributed in the Northern Hemisphere where it has a continuous circumpolar range throughout the Arctic and boreal zone, becoming scattered in the mountains in the southern parts of the Holarctic (Mogensen 1986). In the Southern Hemisphere it has the main centre of its distribution in the Antarctic and it is very rare in Tierra del Fuego (Ochyra & Matteri 2001) and South Georgia (Ochyra *et al.* 2002).

Distribution in Antarctica. *Platydictya jungermannioides* is widely distributed but scattered throughout the maritime Antarctic, ranging from the South Orkney Is. to Alexander I. where it reaches its southernmost locality at lat 71°31′S (Fig. 273). It is fairly frequent on Signy I. in the South Orkney Is., rare and localised on the South Shetland Is. and on the western coast of the Antarctic Peninsula, with a concentration of localities in the northern part of the Graham Coast and lower frequency in the Danco Coast and in northern Marguerite May. In addition, the species occurs occasionally on the eastern coast of the Antarctic Peninsula, from Joinville I. to the northern Oscar II Coast.

◄——

Figure 272. *Platydictya jungermannioides.* — **1–2**. Habit. **3**. Portion of shoot, dry. **4**. Cross-section of the stem. **5–6**. Axillary hairs. **7–12**. Leaves. **13–14**. Leaf areolation. **15–20**. Propagules. — [1, 4, 8–13, 15–20 from *RO 926/80*; 2–3 from *JK 146*; 5–7, 14 from *ER 205E* (type of *Amblystegium densissimum*), BR]. — Scale bars: a – 25 µm (5–6) and 50 µm (13–14); b – 100 µm (7–12); c – 1 mm (1), 0.5 mm (2) and 0.25 mm (3); d – 50 µm (4, 15–20). (Reproduced from *Annales Botanici Fennici*, 36: p. 54, 1999, with permission.)

P. jungermannioides occurs predominantly at low elevations, from near sea level to about 200 m but it has been found at higher elevations, reaching 600 m on Alexander I.

Specimens examined. WEST ANTARCTICA. SOUTH ORKNEY IS. **SIGNY I.** Foca Pt., 5 m, *RILS 8050A* (AAS, KRAM), *ca* 12 m, *RILS 3173* (AAS, KRAM), 17 m, *RILS 366 & 8050A* (AAS, KRAM) and *ca* 37 m, *RILS 502A & 5134* (AAS, KRAM); Stygian Cove, 25–30 m, *RILS 10789* (AAS, KRAM); Marble Knolls, 20 m, *RILS 10894* (AAS, KRAM); Knob Lake, 15 m, *RILS 5132B* (AAS, KRAM); Elephant Flats, *ca* 17 m, *RILS 609* (AAS, KRAM) and *ca* 115 m, *RILS 627* (AAS, KRAM); Limestone Valley, *RILS 1809 & 1810* (AAS, KRAM) and 15 m, *RILS 5139* (AAS, KRAM); Snow Hill, *ca* 215 m, *RILS 602 & 603A* (AAS, KRAM); Knife Pt., 25 m, *RILS 5066 & 5067B* (AAS, KRAM); Factory Cove, *ca* 35 m, *REL 1063* (AAS, KRAM); between Observation Bluff and Polynesia Pt., *ca* 15 m, *REL 1084* (AAS, KRAM); above Paal Harbour, 75 m, *RILS 10867 & 10868B* (AAS, KRAM), 80 m, *RILS 5122* (AAS, KRAM) and *ca* 85 m, *RILS 676A* (AAS, KRAM); Rusty Bluff, 100 m, *RILS 346* (AAS, KRAM).

SOUTH SHETLAND IS. **KING GEORGE I.** ***Admiralty Bay***: Breccia Crag, 140 m, *RO 926/80* (KRAM). **LIVINGSTON I.** ***Byers Peninsula***: 20 m, *KR 24E* (AAS, KRAM). ***Hurd Peninsula***: South Bay, north-east of the Spanish base 'Juan Carlos I', 20 m, *BASM 254* (AAS, KRAM). **DECEPTION I.** Fumarole Bay, 100 m, *RILS 11508* (AAS, KRAM).

WEST ANTARCTIC PENINSULA. **DANCO COAST.** Cape Anna, *ER 205E* (BR, PC – type of *Amblystegium densissimum*); Cuverville I., 10–20 m, *RILS 11946A* (AAS, KRAM). **PALMER ARCHIPELAGO.** ***Anvers I.***: Norsel Pt, 5–10 m, *RILS 1951* (AAS, KRAM); Arthur Harbour, Hermit I., *ca ca* 23 m, *RILS 878B* (AAS, KRAM). **GRAHAM COAST.** Edge Hill, 10 m, *RILS 651* (AAS, KRAM). **WILHELM ARCHIPELAGO.** ***Penola Strait***: Peterman I., *ca* 13 m, *RWMC 721 & 724B* (AAS, KRAM) and *ca* 17 m, *REL 1345B, 1346 & 1347B* (AAS, KRAM). ***Argentine Is.***: Irizar I., *ca* 7 m, *RWMC 605* (AAS, KRAM); between Irizar I. and Uruguay I., *ca* 16 m, *RWMC 432* (AAS, KRAM); Galindez I., overlooking Stella Creek, *ca* 3 m, *RWMC 402A & 555* (AAS, KRAM) and 5 m, *RWMC 469* (AAS, KRAM). **FALLIÈRES COAST.** ***Adelaide I.***: Mt. Gaudry, *ca* 465 m, *JK 146* (AAS, KRAM). ***Léonie Is.***: Léonie I., 20 m, *RILS 10645* (AAS, KRAM) and 25 m, *RILS 8981A* (AAS, KRAM). ***Marguerite Bay***: Lagotellerie I., 35–80 m, *RILS 1780 & 2194A* (AAS, KRAM). **GEORGE VI SOUND.** ***Alexander I.***: Zebra Ridge, 600 m, *RILS 10029* (AAS, KRAM); Succession Cliffs, 20 m, *BJT 533* (AAS, KRAM); Waitabit Cliffs, *BJT 518B* (AAS, KRAM).

EAST ANTARCTIC PENINSULA. **JOINVILLE ISLAND GROUP.** ***Joinville I.***: Mt. Alexander, 60–75 m, *RILS 3667A* (AAS, KRAM). **JAMES ROSS ISLAND GROUP.** ***James Ross I.***: Lachmann Crags 35 m, *RILS 7382A & 7383* (AAS, KRAM). **OSCAR II COAST.** ***Exasperation Inlet***: Starbuck Glacier, *ca* 370 m, *PK 40* (AAS, KRAM).

55. ISOPTERYGIOPSIS Z.IWATS.

Isopterygiopsis Z.Iwats., J. Hattori Bot. Lab. 33: 379. 1970. — Type: *Isopterygiopsis muelleriana* (Schimp.) Iwats. (*Plagiothecium muellerianum* Schimp.).

Plants small and slender, forming thin, flat, dense mats or deep tufts with distinct metallic lustre, pale green, yellowish, yellowish-green to brownish-green above, brownish below. *Stems* creeping, sometimes stoloniform, simple or sparsely irregularly branched, with erect or spreading, complanate-foliate or terete branches, in cross-section rounded, with an outer hyalodermis of large, thin-walled cells or epidermal cells small and thick-walled in 1–2 layers, surrounding large, thin-walled, colourless medullary cells and ± distinct central strand; *rhizoids* clustered in the leaf axils, branched, purple, granular-papillose; *pseudoparaphyllia* none; *axillary hairs* filiform, 2–3-celled with a long hyaline distal cell and 1–2 short, brownish basal cells. *Stem* and *branch leaves* similar, distant or usually crowded and pectinate in arrangement, not or rigidly distichous-complanate, imbricate, erect-spreading, sometimes weakly falcate, flat or concave, symmetric, non-decurrent, lanceolate to oblong-lanceolate, rather abruptly narrowed to a long, filiform acumen; *margins* plane to erect, entire, slightly serrulate or with some blunt teeth near the basal angles; *costa* thin, short and double, or almost lacking; *laminal cells* pellucid, prosenchymatous, linear-fusiform, smooth, narrow; *alar cells* none or slightly differentiated, green, quadrate to short-rectangular, not inflated or decurrent. *Asexual reproduction* sometimes present by means of pale, filamentous, cylindric to fusiform, 2–6-celled propagules present in the leaf axils. *Autoicous* or *dioicous*. *Inner perichaetial leaves* oblong, sheathing at the base, gradually or abruptly tapering to a fairly short, straight acumen. *Setae* smooth, elongate, straight, reddish or orange-yellow. *Capsules* erect to inclined, obloid to obloid-cylindrical, straight to somewhat curved, smooth, constricted below the mouth; *operculum* stout, straight or obliquely rostrate from a conic base; *annulus* biseriate, deciduous in fragments or long persistent; *exothecial cells* quadrate to rectangular, with thick and often wavy walls; *stomata* at the base of the urn, rounded, bicellular, phaneroporous, with rounded pori; *peristome* double; *exostome teeth* lanceolate, yellowish-hyaline and cross-striolate below, hyaline and papillose distally, narrowly hyaline bordered, trabeculate on the ventral side; *endostome* hyaline, finely papillose, with a fairly high basal membrane giving rise to narrow, lanceolate, keeled, narrowly perforate segments alternating with delicate, nodose cilia in groups 1–2, sometimes rudimentary or lacking. *Spores* spherical, smooth to minutely papillose, dark green. *Calyptra* cucullate, naked, smooth.

Isopterygiopsis is one of several segregates which were split off from the catch-all genus *Isopterygium* Mitt. (Iwatsuki 1970). This genus, along with another segregate *Herzogiella* Broth., is characterised by having axillary, granular-papillose and purplish rhizoids and lacking pseudoparaphyllia. This rare form of rhizoids prompted Hedenäs (1987*a*) to transfer *Isopterygiopsis* to the Plagiotheciaceae but here the genus is retained in its traditional placement in the Hypnaceae since it shares gross morphology and anatomical details with other members of this family. Initially, *Isopterygiopsis* was a monotypic genus, comprising a single species *I. muelleriana* which is characterised by the presence of a unistratose hyalodermis in the stem anatomy (Ochyra 1976), but later two other species have been added to the genus, *I. alpicola* (Lindb. & Arnell) Hedenäs (Hedenäs 1988) and *I. pulchella* (Iwatsuki 1987). They are widely distributed but rather infrequent arctic-boreal-montane species in the Northern Hemisphere. *Isopterygiopsis pulchella* is bipolar in distribution and *I. muelleriana* occurs occasionally in East African mountains (Hedenäs 2001*b*).

1. **Isopterygiopsis pulchella** (Hedw.) Z.Iwats. Fig. 274

Isopterygiopsis pulchella (Hedw.) Z.Iwats., J. Hattori Bot. Lab. 63: 450. 1987. ≡ *Leskea pulchella* Hedw., Spec. Musc. Frond.: 220, *pl. 55, f. 7–12*. 1801. ≡ *Hypnum pulchellum* (Hedw.) Dicks. *ex* Brid., Muscol. Recent. 2(2): 101. 1801, *hom. illeg.* ≡ *Plagiothecium pulchellum* (Hedw.) Schimp. *in* Bruch, Schimp. & W.Gümbel, Bryol. Eur. 5: 187, *pl. 497*. 1851 [Fasc. 48 Mon.: 9, *pl. 4*]. ≡ *Isopterygium pulchellum* (Hedw.) A.Jaeger, Ber. Thätigk. St. Gallischen Naturwiss. Ges. 1876–1877: 441. 1878. — Type citation: In silvis umbrosis Scotiae; also, a reference to fig. 6 on pl. 5 in Dickson (1790) is given. [Lectotype (*selected here*): [icon in] Hedwig (1801): f. 7–12 on pl. 55; epitype (*selected here*): 'Rabenhorst, Bryotheca europaea. 16. Plagiothecium pulchellum (Hedw.) Br. Sch. et Gümb. Coroll. p. 114. Br. E. V. T. 497. Hypnum pulchellum Hedw. Spec. Musc. II. 225 Musc. Fr. IV. T. 25. Rabenh. Handb. II. 3. p. 275. Am Untersberg bei Salzburg ges. v. Dr. Schwartz' – KRAM!].

Plagiothecium antarcticum Mitt., J. Linn. Soc. Bot. 15: 71. 1876. ≡ *Isopterygium antarcticum* (Mitt.) Cardot, Bull. Mus. Natl. Hist. Nat. 22: 341. 1916. ≡ *I. pulchellum* var. *antarcticum* (Mitt.) Zanten, Marion & Prince Edward Isl.: 216. 1971, *comb. inval. basion. non citat.*, ***syn. nov.*** — Type citation: Marion Island. [Lectotype (*selected here*): 'P. antarcticum m. Marion Is[d]. Challenger Exp.' – NY!; isotype: BM!].

Hypnum austropulchellum Müll.Hal., Bot. Jahrb. Syst. 5: 82. 1883 ['austro pulchellum']. = *Plagiothecium austropulchellum* (Müll.Hal.) Kindb., Enum. Bryin. Exot. Suppl. 2: 102. 1891, ***syn. nov.*** — Type citation: all new species described by Müller (1883) are included in the section 'I. Bryologia Kerguelensis'; later Müller (1889: p. 35) wrote: 'Insula Kerguelen, Betsy Cove, cum Bryo austro-crudo et Bartramia subexigua associatum, 12. Januario 1875. Quoque in Foundery branch caespitibus latis'. [Lectotype (*selected here*): 'Ex Museo botanico Berolinensi. Hypnum (Plagioth.) austro-pulchellum C. Müll. n. sp. Kerguelensl. Foundery branch. Dr. Naumann 1874' – S-Dusén!; syntypes: (1) 'Ex Museo botanico Berolinensi. Plagiothecium austropulchellum C. Müll. n. sp. Kerguelensland Gazellen-Expedition. Dr. Naumann. 1874' – S!; (2)): 'Herb. Émil Bescherelle: 1900. Plagiothecium antarcticum C. Müll Kerguelen, Betsy's Cove, Dec. 1874: Dr. Fr. Naumann Hb. C. Müll' – BM-Bescherelle! (3 specimens)]. First synonymised with *Isopterygium antarcticum* by Cardot (1916: p. 341).

Hypnum magellanicum Müll.Hal., Flora 68: 425. 1885, *hom. illeg.* [non *Hypnum magellanicum* P.Beauv. 1805]. ≡ *Isopterygium fuegianum* Besch. *in* Hariot, Miss. Sc. Cape Horn Bot.: 301. 1889. ≡ *Plagiothecium magellanicum* Paris, Index Bryol.: 707, 963. 1897, *nom. illeg. incl. spec. prior.* ≡ *Plagiothecium fuegianum* (Besch.) Dusén *in* W.B.Scott, Rep. Princeton Univ. Exp. Patagonia 8(1) Botany: 101. 1903. — Type citation: Clarence Island, S. Holl Bay: Hariot No. 63. [Lectotype (*selected here*): 'Herb. Émil Bescherelle: 1900. ~~Isopterygium fuegianum,~~ Besch. = (Hyp. Donianum Mitt. non C. Müller) S. Holl bay – Ile Clarence Plagiothecium magellanicum C.M. 20 Jan No. 63 Hariot' – BM-Bescherelle!; syntype: 'Herb. Émil Bescherelle: 1900. Isopter. fuegianum Besch. S Holl Bay Ile Clarence Terre de Feu Nariot n° 63 9 Mai 83' – BM-Bescherelle!]. First synonymised by Ireland (1991: p. 267).

Plagiothecium leptoplumulosum Dusén *in* W.B.Scott, Rep. Princeton Univ. Exp. Patagonia 8(1) Botany: 100, *f. 23 & pl. 9, f. 3–5*. 1903. ≡ *Isopterygium leptoplumulosum* (Dusén) Cardot, Wiss. Erg. Schwed. Südpolar-Exp. 4(8): 178. 1908. — Type citation: Hab. speciei Patagonia australi, ubi verisimilariter in territorio fontinali fluminis Rio Chico lecta est. [Lectotype (*selected here*):): 'Princeton Scientific Expedition to Patagonia 1896–97. P. 15. Plagiothecium leptoplumulosum Dusén Locality Patag. austr. [verisimilariter] In territorio Rio Chico Collected by J. B. Hatcher. Det P. Dusén Princeton University, Princeton, N. J.' – S!; isotypes: NY! (2 specimens), S!]. First synonymised by Ireland (1991: p. 267).

Plants small and slender, in soft, lustrous, dense mats or deep tufts, bright-green to light yellowish-green, yellowish or yellowish-brown with age. *Stems* filiform, creeping, to 2 cm long, irregularly, richly branched, branches ascending, terete to complanate-foliate, in transverse section round or oval, with a small, distinct or indistinct central strand and an outer cortex of 1–2 rows of smaller, moderately thick-walled, brownish cells sharply contrasting with large, thin-walled and hyaline medullary cells; *rhizoids* clustered in the leaf axils, branched, brown, purplish-brownish, coarsely papillose; *pseudoparaphyllia* lacking; *axillary hairs* filiform, 2–3-celled, with elongate, hyaline apical cell and 1–2 short, brown basal cells. *Leaves* crowded to rather distant, erect to erect-spreading, straight to falcate, often homomallous and ± secund at the tips, symmetric, elongate-triangular to narrowly lanceolate, slenderly long-acuminate, 0.7–0.9(–1.2) mm long, (0.25–)0.3–0.35(–0.4) mm wide, plane to slightly concave, non-decurrent; *margins* plane to erect, entire or with a few blunt dentations near the base; *costa* short and double or lacking; *laminal cells* smooth, transparent, firm-walled, linear-flexuose to linear-fusiform, 60–100 μm long, 5–6 μm wide, 10–15 times as long as wide, becoming shorter and wider at the insertion, with thick and somewhat pitted walls; *alar cells* scarcely differentiated, consisting of 1–3 short-rectangular cells along the margins. *Autoicous*. *Inner perichaetial leaves* ovate to oblong, abruptly narrowed to a straight, filiform apex. *Setae* erect, straight to somewhat curved, elongate, 8–15 mm long, smooth, yellow- to orange-reddish, twisted to the left above. *Capsules* suberect, nearly symmetric to inclined, obloid-cylindrical, 1.4–1.6 mm long, ca 0.5 mm wide, brown, smooth, with a distinct neck wrinkled when dry, contracted below the mouth on drying; *operculum* bluntly low-conic, to 0.5 mm long, with a straight or slightly oblique beak; *annulus* deciduous, consisting of 2 rows of cells, with strongly incrassate walls; *exothecial cells* quadrate, rounded-quadrate to short-rectangular, in regular longitudinal rows, with firm or thick, usually wavy walls and collenchymatous thickenings, becoming somewhat smaller towards the base of the urn and the mouth; *stomata* few at the base of the urn, bicellular, superficial, with elliptical pori; *exostome teeth* lanceolate, hyaline-yellowish and cross-striolate below, hyaline and papillose above, hyaline bordered, dentate at the margins, trabeculate on the ventral side; *endostome* with a high basal membrane, keeled, entire to narrowly slit segments, and short, 1–2 cilia or occasionally rudimentary to lacking. *Spores* globose, small, 10–12 μm wide, minutely roughened.

Chromosome numbers: unknown in the Antarctic and Subantarctic; $n = 10 + 2m$ (1 record), $n = 11$ (5 records) and $n = 22$ (3 records) from the Holarctic (Fritsch 1991).

Discussion. The species is distinct with its non-decurrent, slightly concave, usually falcate and homomallous, narrowly lanceolate and long-acuminate leaves with narrow, linear-flexuose to linear-fusiform cells throughout the lamina. Antarctic plants are generally smaller than those from the Subantarctic or Tierra del

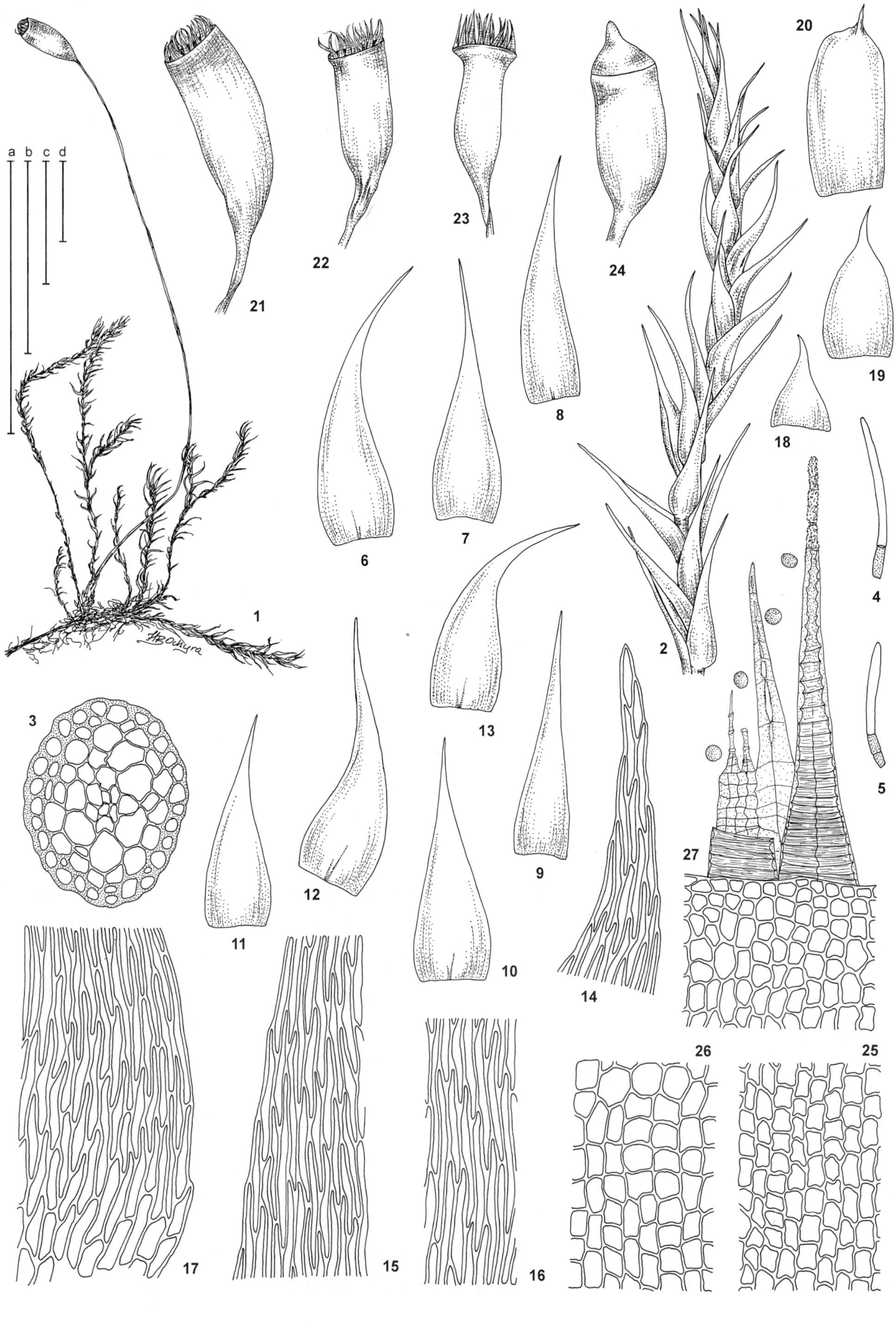
a
b
c
d
1
2
3
4
5
6
7
8
9
10
11
12
13
14
15
16
17
18
19
20
21
22
23
24
25
26
27

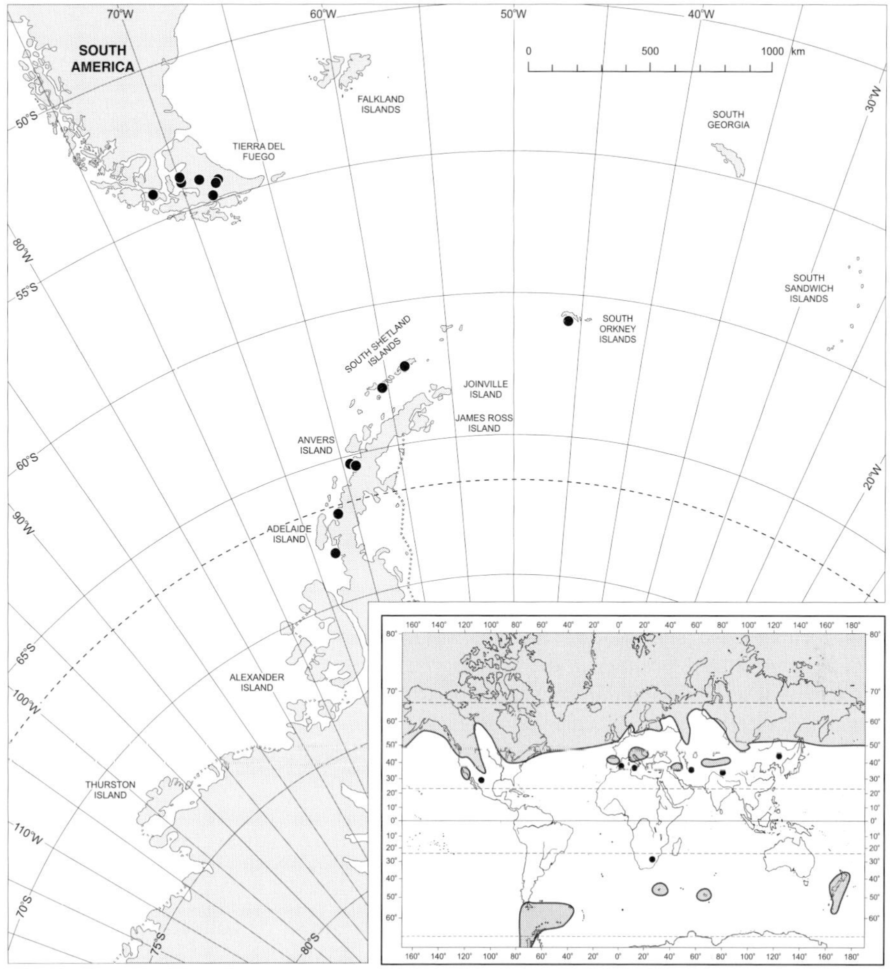

Figure 275. Distribution of *Isopterygiopsis pulchella* in the Fuegian region and Antarctica. Inset: global distribution.

Fuego and they may sometimes be confused with *Platydictya jungermannioides*. However, the plants are larger and their characteristic metallic lustre makes *Isopterygiopsis pulchella* easy to distinguish, even in the field. Additionally, the leaves are always much larger than those in *P. jungermannioides* and they are never shorter than 0.5 mm. The species is not very variable in Antarctica and only the leaves may vary to some extent in their acumination and falcateness. Some Antarctic plants of *I. pulchella* are of similar size and stature to tiny and filiform expressions of *Hypnum revolutum* but the latter has ovate or oblong-lanceolate leaves that are rather abruptly narrowed to a short, usually distinctly serrulate acumen. In addition, they differ markedly in the areolation of usually shorter, oblong-rhomboidal cells and, primarily in having distinct group of subquadrate, thick-walled cells in the basal angles.

Some morphological expressions of *Isopterygiopsis pulchella*, from its wide geographical range in southern South America and Subantarctica, were described as species in their own rights, namely *Plagiothecium antarcticum* from Marion I. (Mitten 1876*a*), *Hypnum austropulchellum* from Îles Kerguelen (Müller 1883, 1889) as well as *Hypnum magellanicum* and *Plagiothecium leptoplumulosum* from Tierra del Fuego (Müller 1885; Dusén 1903*b*). Examination of the original collections of all of these species names showed that they represent minor morphological variants of otherwise typical plants of *I. pulchella* and all these names are reduced to synonymy, confirming suggestions by some earlier authors regarding the identity of these species.

Isopterygiopsis pulchella was discovered for the first time in Scotland and described by Dickson (1790) as *Hypnum pulchellum*. This pre-1801 name was subsequently validated as *Leskea pulchella* by Hedwig (1801). No original material matching the protologue is available in the personal Hedwig herbarium at G (Price 2005), therefore the typification of this species name by Ireland (1969) and Hedenäs & Geissler (1999) has to be superseded. The latter authors showed the specimen marked 'b. a Turnero' as the lectotype of *Leskea pulchella* but this material is not cited in the protologue and certainly was not studied by J. Hedwig. In the absence of the original material in the Hedwig herbarium which could be eligible for typification of *Leskea pulchella*, the plate in *Species muscorum frondosorum*,

←

Figure 274. *Isopterygiopsis pulchella*. — **1**. Habit. **2**. Portion of shoot. **3**. Cross-section of stem. **4–5**. Axillary hairs. **6–13**. Leaves. **14**. Leaf apex. **15**. Upper lamina cells at margin. **16**. Mid-leaf cells. **17**. Basal leaf cells. **18**. Outer perichaetial leaf. **19–20**. Inner perichaetial leaves. **21**. Deoperculate capsule, wet. **22–23**. Deoperculate capsule, dry. **24**. Operculate capsule, wet. **25**. Exothecial cells at base of urn. **26**. Exothecial cells in middle of urn. **27**. Portion of peristome, spores and cells at mouth of capsule. — [2–5, 11–13, 16 from *RILS 676*; 1, 8, 18–20, 23–27 from *Naumann s.n.*, 1874 (lectotype of *Hypnum austropulchellum*), S; 6–7, 14–15, 17 from *RO 1785/80*; 9 from *Hariot 63* (lectotype of *Isopterygium fuegianum*), BM; 10 from *Moseley s.n.*, Marion Island (isotype of *Plagiothecium antarcticum*), BM; 21–22 from *RO 528/95*]. — Scale bars: a – 1 mm (6–13, 18–20); b – 0.5 cm (1) and 1 mm (2); c – 1 mm (21–24) and 100 µm (25–27); d – 50 µm (3–5, 14–17).

which accurately depicts the habit, leaf shape, capsule and the peristome teeth, is here chosen as lectotype and the specimen widely distributed by L. Rabenhorst as No. 16 in his *Bryotheca europaea* is here chosen as epitype. It is worth noting that Dickson (1790) presented an incorrect illustration of the leaf which is shown as having a prominent percurrent costa which is unknown in *I. pulchella*. Hence, this illustration is rejected as iconotype.

Reproduction in Antarctica. Three populations of *Isopterygiopsis pulchella* from Livingston I., Cape Tuxen and Neny I. were found with sporophytes, but all other collections are sterile. In the latter site fully mature sporophytes are abundant. However, gemmae have not been observed in Antarctic plants, although they are frequent in the Northern Hemisphere.

Habitat. The species grows on mineral soil as well as on soil rich in humus, in dry or damp, sheltered habitats such as crevices in rocks and cliffs and under overhanging rocks. It is most common on calcareous substrata and very rarely grows on acidic rocks, forming almost pure tufts, usually without associated species, but it has occasionally been found with *Distichium capillaceum* and *Pohlia nutans*.

World range. A bipolar species (Fig. 275, inset), practically without intermediate localities in the tropical zone, except for one record at high elevation in Lesotho in South Africa (Hodgetts *et al.* 1999). It has a circumpolar continuous range in the arctic and boreal zone in the Northern Hemisphere, decreasing in frequency southwards and occurring mostly in the mountains in the southern part of the Holarctic. Likewise, *Isopterygiopsis pulchella* is widely distributed in the Southern Hemisphere in the temperate zone, including New Zealand and its offshore islands, southern South America and some subantarctic islands, including South Georgia (Ochyra *et al.* 2002), the Prince Edward Is. (Zanten 1971; Ochyra personal observations) and Îles Kerguelen (Cardot 1916; Ochyra, personal observations) and then recurring in the maritime Antarctic.

Distribution in Antarctica. A rare, widely scattered and localised species, known from the South Orkney Is. and South Shetland Is. and three centres on the western coast of the Antarctic Peninsula, extending to Neny I. in northern Marguerite Bay at lat. 68°12′S (Fig. 275). Surprisingly, the southernmost population on Neny I. produced fully mature sporophytes in great profusion. The species grows at low elevations, from near sea level to about 85 m.

Specimens examined. WEST ANTARCTICA. SOUTH ORKNEY IS. **SIGNY I.** Above Paal Harbour, *ca* 85 m, *RILS 676B* (AAS, KRAM).

SOUTH SHETLAND IS. **KING GEORGE I.** ***Admiralty Bay***: Dufayel I., Gdynia Pt., 40 m, *RO 1785/80* (KRAM). **LIVINGSTON I.** ***Hurd Peninsula***: near Bulgarian base, *Schulz 200* (KRAM).

WEST ANTARCTIC PENINSULA. **GRAHAM COAST.** Cape Tuxen, *RWMC 847* (AAS, KRAM); Cape Tuxen, Waddington Bay, 20 m, *RILS 10660* (AAS, KRAM). **WILHELM ARCHIPELAGO.** ***Argentine Is.***: Galindez I., overlooking Skua Creek, *ca* 7 m, *RWMC 587* (AAS, KRAM). **LOUBET COAST.** Orford Cliff, *RILS 5269B* (aas, kram). **FALLIÈRES COAST.** ***Marguerite Bay***: Neny I., 50 m, *RILS 792* (AAS, KRAM).

EXCLUDED SPECIES OF THE HYPNACEAE

Hypnum recurvatum (Lindb. & Arnell) Kindb. fo. ***strictum*** (Meyl.) Ando — This taxon was reported from Fildes Peninsula on King George I. by Chen *et al.* (1995) and it was the first record of this boreal-montane Holarctic species in the Southern Hemisphere. This record is based upon misidentification and examination of the herbarium collections showed that they are actually *H. revolutum* and more exactly the habitat expression called by Ando (1973) fo. *pumilum*. It resembles *H. revolutum* in leaf shape and areolation, but differs in having a very small alar group consisting of only a few subquadrate cells in the marginal row.

Glossary

This glossary is based primarily on *Glossarium polyglottum bryologiae* (Magill 1990), *Mosses and other bryophytes, an illustrated glossary* (Malcolm & Malcolm 2006) and glossaries in the Floras of Crum (2004) and Smith (2004). In order to facilitate comparison of terms we have listed others that have the same (*synonyms*) and opposite (*antonyms*) meanings (abbreviated to 'syn.' and 'ant.', respectively), as well as those with similar meaning to or otherwise associated with a given term (under the heading 'see also').

A

abaxial – facing away from the stem or axis; see also *dorsal*, used in reference of leaves; ant.: *adaxial* or *ventral*

acidophile – growing in acidic habitats; ant.: *basiphile*

acidophilous – acid-loving; ant.: *basiphilous*

acrocarpous – terminal fruited; mosses in which the sporophytes are usually borne terminally at the apex of the stems or main branches with further vegetative growth continued in the same direction as the preceding stem by lateral branch(es) or innovation(s); acrocarpous mosses usually grow erect in tufts or cushions or occur as scattered individuals and are not or sparsely branched; see also *cladocarpous*; ant.: *pleurocarpous*

acropetal – developing or maturing from the base to the apex, hence with youngest organs or tissue at the apex; ant.: *basipetal*

acumen – a tapering point, with an angle at the apex less than 15°

acuminate – having a gradually slenderly diminishing point, with a terminal angle less than 15°; longer than acute

acute – distinctly and sharply pointed, but not drawn out, with an angle of 15° at the apex

adaxial – facing towards the stem or axis; see also *dorsal*, used in reference to leaves; ant.: *adaxial*

adherent – a close union of two separate parts or organs but which are not fused; see also *coherent*; ant.: *adnate*

adnate – having two separate parts or organs fused together; see also *coherent*; ant.: *adherent*

adventive – an introduced plant, not original to the area; ant.: *indigenous* and *native*

alar cells – cells at the basal angles of a leaf, usually differentiated in size, shape and colour from other leaf cells; see also *angular cells*

alternate – said of leaves or branches that are staggered along a stem first on one side and then the other; also referring to a peristome with exostome teeth alternate in lock-step with the endostome segments; ant.: *opposite*

altimontane – pertaining to mountains and high upland regions above the timber line

amphithecium (*pl.* **amphithecia**) – the peripheral layer of cells in the capsule surrounding the endothecium in the embryonic stage of its development that gives rise to all tissues from the epidermis to the outer spore sac and, in *Sphagnum* also producing the spore sac; see also *endothecium*

andreaealean referring to mosses of the class Andreaeopsida; syn.: *granite* or *lantern mosses*

androecium (*pl.* **androecia**) – the male gametoecium or male inflorescence consisting of antheridia and the surrounding bracts; syn.: *perigonium*

angulate – angled, having angles or sharp corners

angular cells – cells at the basal angles of a leaf, usually referred to as alar cells

annual – living and completing the life cycle in one year; see also *biennial* and *ephemeral*; ant.: *perennial*

annular – formed or shaped like a ring, used of any organs disposed in a circle

annulate – having an annulus; ant.: *exannulate*

annulus (*pl.* **annuli**) – a ring of variously differentiated cells between the mouth of the capsule and the operculum, aiding in the opening of the capsule

antheridium (*pl.* **antheridia**) – the male gametagium; a fusiform to cylindrical, stalked structure containing spermatozoids; ant.: *archegonium*

anticlinal – perpendicular to the surface; see also *radial*; ant.: *periclinal*

antrorse – directed upwards, turned up or forward; ant.: *retrorse*

apical – at the tip of any structure

apiculate – abruptly ending in a short point, somewhat longer than mucronate

apiculus (*pl.* **apiculi**) – a short, abrupt point, somewhat longer than a mucro

apophysis (*pl.* **apophyses**) – a strongly differentiated, sterile neck at the base of a capsule between the urn and the seta; syn.: *hypophysis*

appendiculate – with short, thin transverse bars or projections

appressed – lying close or flat for the whole length of the organ, leaves closely applied to the stem

approximate – drawn close together but not united

aquatic – living in and completing the life cycle in water; ant.: *terrestrial*

archegonium (*pl.* **archegonia**) – the female gametangium or sexual organ; a flask-shaped structure consisting of a stalk, a swollen basal part or venter and a tubular neck, containing the egg cell which after fertilisation develops the sporophyte; ant.: *perigonium*

archesporium (*pl.* **archesporia**) – the cells within the embryonic sporangium, ultimately giving rise to spores by meiosis

arcuate – bent like a bow, curved

areolation – the arrangement of the leaf cells

arista – an awn

aristate – awned, with a bristle-like hair-point

arthrodontous – with reference to a peristome type consisting of one or two rings of triangular teeth or segments composed essentially of differentially thickened periclinal wall-pairs; the teeth and segments are separated as a result of erosion or splitting of unthickened anticlinal walls, and are exposed after dehiscence of the operculum; transverse wall-pairs are usually thickened and remain as articulations or trabeculae up the face of the teeth; ant.: *nematodontous*

articulate – with thickened joints or articulations between the segments of the peristome teeth; syn.: *trabeculate*

articulation – a thickened projecting cross-bar formed from horizontal walls of either face of arthrodontous peristome teeth; syn.: *trabecula*

ascending – growing obliquely upwards from the substrate; see also *decumbent*

asexual reproduction – regeneration by means of specialised structures such as gemmae, caducous leaves or branches

attenuate – slenderly narrowed or tapered

auricle – an ear-like group of enlarged or otherwise differentiated alar cells at each corner of a leaf base forming a distinctive lobe, often sharply delimited from other leaf cells

auriculate – with auricles

austral – southern; applied to plants which are native to the Southern Hemisphere, particularly the cool temperate and polar regions

autoicous – having antheridia and archegonia borne on different branches of the same plant; see also *cladautoicous*, *gonioautoicous* and *pseudautoicous*

awn – a bristle-like hair-point, usually formed by an excurrent costa; see also *arista*

axil – the upper angle formed between the axis and any organ which arises from it, especially of a leaf or branch

axillary hair – uniseriate hair, usually found in the axils of the uppermost leaves

axis (*pl.* **axes**) – the main stem; an imaginary line round which organs are developed

B

basal – located at the base; syn.: *proximal*; ant.: *apical*, *terminal* and *distal*

basal cell – cells at the base or bottom; in leaves, usually differentiated cells of the lower ¼–⅓ of a leaf

basal membrane – a short or long cylindrical or tubular membrane at the base of the endostome, commonly bearing processes and cilia or at the base of the exostome and giving rise to the teeth

basipetal – developing or maturing from the apex to the base, hence with oldest organs or tissue at the apex; ant.: *actopetal*

basiphile – growing in basic habitats, such as limestone, dolomite or chalk; see also *calciphile, calcicole*; ant.: *acidophile*

basiphilophus – base-loving; ant.: *acidophilous*

beak – the prolonged projection from the operculum of a capsule; see also *rostrum*

bi – prefix meaning 'two'

bicellular – composed of two cells

biconvex – bulging out on both surfaces

bicostate – with two costae

biennial – living and completing the life cycle in two years or growing seasons; see also *annual*, *ephemeral* and *perennial*

bifarious – arranged in two rows; syn.: *distichous*

bifid – forked or divided into two ± equal parts; syn.: *bifurcate*

biflagellate – equipped with two flagella; the sperm of all bryophytes are biflagellate

bifurcate – divided into two branches or prongs, twice-forked; syn.: *bifid*

bilateral – arranged on opposite sides

bilobed – divided into two lobes

binate – arranged in two parts or in pairs: syn.: *geminate*

bipinnate – with both primary and secondary stems pinnate, twice-pinnately branched

biseriate – in two rows

bistratose – composed of two cell layers

blade – the flat green portion of a leaf, differentiated or not from the basal part of a leaf

border – margins differentiated in structure or cell shape, size, colour, or thickness of cells from the rest of the leaf lamina; or peristome teeth, having a narrow pale margin; see also *limbidium*

bordered – having a border

boreal – northern; especially pertaining to the zone of the coniferous forests between the treeless tundra and the deciduous forests of temperate latitudes

bract – a leaf-like structure, similar to or modified from the vegetative leaves surrounding and protecting a gametangium or gemma-cup

bracteate – provided with bracts

bracteolate – having branchlets

branch – a lateral division of the stem

brevipilose – bearing fairly short and soft hairs

brood body – any structures that serve in the vegetative reproduction, such as gemmae, bulbils, tubers and surculi; syn.: *propagulum*

bryoid – applied to mosses that have a *Bryum*-like peristome with a well developed exostome and endostome or having a *Bryum*-like appearance or leaves with large, thin-walled rhombic or hexagonal cells

bulbiform – bulb-shaped; a rosette of leaves imbricate to form a bud-like plant

bulbil – a small bulb-shaped axillary bud, often with apical and sometimes with lateral projections or 'leaf primordia' or a rhizoidal gemma, serving in the vegetative reproduction; see also *brood body* and *tuber*

C

caducous – dropping off early, readily falling off; referring to vegetative propagules in the form of leaflets, branchlets or leaf apices

caespitose – tufted; growing in dense patches, cushions, tufts or sods

calcareous – any material containing calcium carbonate, such as limestone or dolomite

calcicole, **calciphile** – a plant addicted to a calcareous substrate; see also *basiphile*; ant.: *acidophile*

calyptra (*pl.* **calyptrae**) – the cap or hood developed from the venter of the archegonium covering a developing capsule or at least the operculum of older capsules, falling or not as capsule matures

campanulate – bell-shaped; applied to an elongate and cylindrical calyptra

canaliculate – with a U-shaped longitudinal groove formed by upturned margins; syn. *channelled*

canescent – appearing frosted because of numerous hyaline hair-points; syn.: *hoary*

capillary – slender, comparable with a hair, hair-like

capsule – the terminal spore-producing portion of a sporophyte, usually differentiated into an apical operculum, central urn or sporangium and a sterile basal neck or apophysis

carinate – shaped like a keel of a boat; with a narrow longitudinal channel V-shaped in cross-section

carpet – growth-form in which the living moss shoots are vertically aligned parallel to each other and arise from a horizontal matrix of decaying or moribund shoots, resembling the pile of a carpet

castaneous – dark brown, chestnut-coloured

catenulate – shaped like a chain; formed of parts united or linked as in a chain

cauline – belonging to the stem or arising from it

central strand – a small group of elongate cells forming a central axis of some stems, typically made up of hydroids

cell – the microscopic basic structural unit of all living organisms, in plants containing protoplasm with a single nucleus enclosed by a cellulose wall

cernuous – nodding or drooping at an angle between horizontal and vertically downwards

channelled – with a broad U-shaped longitudinal groove formed by upturned margins; hollowed out like a gutter and semicircular in cross-section; syn.: *canaliculate*

chasmophyte – a plant which grows in rock crevices or fissures

chirotype – the specimen on which a manuscript or herbarium name or a *nomen nudum* is based

chlorophyllose – having chlorophyll and therefore green; ant.: *echlorophyllose*

cilium (*pl.* **cilia**) – a uniseriate hair-like structure alternating with processes in the inner peristome; a delicate hair- or thread-like structure

ciliate – having fine hairs along a margin or on a surface

circinate – coiled into a ring or partially so; curved in a circle, with the tip or apex innermost

cirrate – curved or wavy like a tendril; ending with a curled or wavy appendage, usually an extension of the costa

cladautoicous – monoicous, with the male inflorescence on a separate branch; see also *autoicous*, *gonioautoicous* and *pseudautoicous*

cladocarpous – mosses in which sporophytes are borne terminally on short lateral branches; see also *acrocarpous* and *pleurocarpous*

clavate – club-shaped; lenghened and thickened towards the tip

cleistocarpous – pertaining to capsules that do not dehisce regularly owing to the lack of an operculum and the spores are released by rupture or decay of the capsule wall; syn.: *inoperculate*

coalesced – fused together

cochleariform – shaped like a spoon

coherent – stiking together of different parts or organs but not fusing; ant.: *adnate*

collenchymatous – with walls more heavily thickened at the angles than elsewhere; syn.: *corner thickening*

columella – the central axis of sterile tissue in a sporangium in most mosses, surrounding by a spore-producing tissue

comal tuft – a cluster of crowded leaves at the apex of a stem or branch

***combinatio nova* (*comb. nov.*)** – see *new combination*

comose – tufted, having tufts of large, crowded leaves at the stem apex

complanate – flattened into one plane

compound – similar parts aggregated into a common whole; ant.: *simple*

compressed – flattened lengthwise, squeezed together, or strongly condensed

concave – hollow, as the inside of a saucer

concolorous – uniform in tint or colour

conduplicate – folded lengthwise so sharply that the two sides are nearly parallel

confluent – merging, blended into one; passing by degrees one into the other

congener – another member of the same genus

congeneric – belonging to the same genus

congenerity – the condition of belonging to the same genus

conic, **conical** – cone-shaped; having the figure of a cone

connate – united, fused with a structure or organ of the same kind

connivent – coming into contact or converging; directed or pointing together

conspecific – belonging within the same species

conspecificity – the condition of belonging to the same species

conserved name – a name of a family, genus, or species ruled as legitimate and with precedence over the other specified names even though it may have been illegitimate when published or lack priority; its type, orthography, or gender has been fixed by the conservation process; see also *nomen conservandum*

constricted – drawn together, tightened, abruptly narrowed; see also *contracted*

contiguous – adjoining, when neighbouring parts are in contact without fusion

contorted – irregularly twisted, bent into irregular curves

contracted – abruptly narrowed or shortened; see also *constricted*

convolute – when one part is wholly rolled up in another; rolled together and forming a sheath; spindled, enfolded or rolled to form a cylinder or sheath; syn.: *sheathing*

cordate – heart-shaped; applied to leaves with the large, rounded ends at the base or point of attachement

coriaceous – leathery or tough

corner thickening – cell walls more heavily thickened at the corners

correct name – a name of a taxon with a particular circumscription, position, and rank that must be adopted in accordance with the rules

cortex (*pl.* **cortices**) – the external tissue of a stem or branch, surrounding a medulla, usually composed of several layers of much differentiated stereid or substereid cells; sometimes used to denote the internal tissue located between the central strand and epidermis (medulla)

cortical – referring or belonging to the cortex

cosmopolitan – distributed throughout the world

costa (*pl.* **costae**) – a thickened vein in the centre of a leaf; syn.: *midrib*

costate – having a costa or midrib

crassate – thick, dense or coarse

creeping – running along or under the ground and rooting at intervals; syn.: *prostrate*; see also *procumbent*

crenate – scalloped along the edge with coarse or tiny rounded teeth, usually bulging walls of individual cells

crenulate – scalloped along the edge with minute rounded teeth, usually bulging cells

cribrose – pierced like a sieve; resembling a sieve with fine perforations

crisped – strongly curled, twisted, or wavy like crisp bacon

crispate – wavy; often used more loosely to mean variously curled, twisted or contorted

cristate – crested; with a pale serrate margin resembling a crest; crested with a ridge resembling a rooster's comb

cryptoporous – applied to stoma with its guard cells and pore that are sunken in a depression or overarched by the adjacent exothecial cells; ant.: *phaneroporous*

C-shaped papilla – a papilla appearing crescent-shaped when viewed from above with an ordinary light microscope

cubic – cube-shaped, like a solid body having six equal square faces

cucullate – hooded or hood-like in shape; said of a calyptra that has a single long split up one side; also used to described strongly concave leaves, hooded at the apex by erect or incurved margins

cuneate – wedge-shaped

cupulate, **cupuliform** – shaped as a cup with a somewhat flared mouth; see also *cyathiform*

cushion – a growth-form of densely-arranged branched shoots radiating from a central point and forming dense hemispherical or rounded colonies

cusp – a sharp, rigid point

cuspidate – abruptly tipped with or gradually tapering to a sharp, rigid and stout point; referring to a shortly excurrent costa or a stem or branch apex formed by imbricate concave leaves

cuticle – the non-cellular covering of leaves or capsules, often variously roughened

cyathiform – cup-shaped with a mouth that is only somewhat flared

cygneous – curved downwards like a swan's neck

cylindrical – shaped like cylinder, elongate with a circular cross-section

cymbiform – shaped like a broad and shallow boat

D

deciduous – falling off, not persistent, or lost at maturity; mostly applied to the annulus and operculum

declinate – bent or curved downward or forward

decolorate – colourless, discharged of colour

decumbent – prostrate toward the base but with ascending or erect tips

decurrent – running down, as when leaves are prolonged beyond their insertion

decurved – curved downwards

deflexed – bent abruptly downwards; see also *reflexed*; ant.: *inflexed*

defoliate – leaves removed or lost, having cast its leaves; ant.: *foliate*

dehiscent – splitting open or rupturing; applied to capsules opening by an operculum or valves

deltoid – triangular, resembling a capital delta, the fourth letter in the Greek alphabet

dendroid, **dendritic** – branched or tree-like in form, resembling a miniature tree

dentate – having sharp, coarse, salient teeth directed forward, usually composed of two or more cells; see also *serrate*

denticulate – minutely toothed, the teeth usually composed of only the projecting tips of cells; see also *serrulate*

denudate – made bare, left naked, stripped; syn.: *denuded*

denuded – eroded or torn away, stripped; syn.: *denudate*

deoperculate – having lost the operculum, referring to a capsule after the operculum has fallen

depauperate – poorly developed, stunted

depressed – flattened as viewed from above

descending – tending gradually downward

descriptive name – a name of a taxon above the rank of family not based on a generic name

deuters – the thin-walled and highly vacuolated cells with very large lumina stretching across the leaf costa in many mosses as viewed in its cross-section; syn.: *guide cells*

dextrorse – turned to the right, or twisting upwards in an anti-clockwise spiral when viewed from above, as in a right-hand standard screw thread; referring to the torsion of a seta; ant.: *sinistrorse*

diagnosis – a statement of that which in the opinion of its author distinguishes the taxon from other taxa

diagonal – slantwise, traversing obliquely

diaphanous – clear and glass-like, colourless; syn.: *transparent*

dichotomous – separated in pairs, division of a structure to yield two, more or less equal, parts; see also *bifurcate*

diffuse – widely or loosely spreading

dimorphic – occurring under two different forms

dioicous – with antheridia and archegonia borne on different plants; ant.: *monoicous*

diploid – applied to the state of an organism that contains two complete sets of chromosomes (denoted by 2*n*); the typical chromosome level of the sporophyte generation in mosses; see also *haploid* and *polyploid*

diplolepidous – with a double articulated peristome, i.e. having two concentric circles of plates forming the articulated peristome teeth; ant.: *haplolepidous*

disciform – saucer-shaped, flat and circular, orbicular

discoid – flattened and disc-like and thickest in the middle

dissected – deeply divided or cut in many segments

distal – located away from the base or point of attachment; syn.: *terminal*; ant.: *proximal*

distant – well spaced, scattered, widely separated, not close together; syn.: *remote*

distichous – arranged in two vertical rows or ranks on opposite sides of a stem; syn.: *bifarious*

divergent – turned in opposite directions

divisural line – a longitudinal line, usually zig-zag, found on either or both surfaces of a peristome tooth; syn.: *median line*

dorsal – relating to the back, the surface turned away from the axis which in the case of a leaf is the lower surface; syn.: *abaxial*; ant.: *ventral* or *adaxial*

dorsal plate – the outer surface of an arthrodontous peristome tooth, each plate representing the inner periclinal wall of one cell

dorsiventral – flattened, with distinct upper and lower surfaces

dotted-striolate – finely striate, marked with fine or delicate longitudinal lines of minute pustules

double peristome – a peristome with two rings of teeth; ant.: *single peristome*

doubly serrate – with teeth overlapping and joined in pairs

E

e-, ex- – prefix meaning 'not' or 'without'

ecad – a morphological form resulting from the adaptation of a species to a particular habitat

echinulate – roughened by minute spines

echlorophyllose – lacking chlorophyll and therefore colourless; ant.: *chlorophyllose*

ecostate – lacking a costa or midrib; ant.: *costate*

edentate – lacking teeth; ant.: *dentate*

elamellate, elamellose – lacking lamellae

elimbate – lacking a differentiated border; ant.: *limbate*

ellipsoid – a solid that has an elliptical outline

elliptical – shaped like an ellipse, oblong with convex sides or ends

elongate – stretched out, drawn out in length

emarginate – having a deeper notch or indentation cut out at a broad rounded apex; see also *retuse*

emergent – partially exposed; used of capsules which rise slightly above the tips of the perichaetial leaves; see also *exserted* and *immersed*

endemic – a distribution pattern in which a species or other taxon is confined to a given area, as an island, country or a floristic area

endogenous – arising from deep-seated tissue; ant.: *exogenous*

endostome – the inner ring of teeth of a diplolepidous peristome, formed from contiguous periclinal wall-pairs of the inner and primary peristomial layers; homologous with the haplolepidous peristome; syn.: *inner peristome*

endothecium (*pl.* **endothecia**) – the inner tissue of the embryonic capsule giving rise to all tissue interior to the spore sac, including the columella and spore sac in most mosses and only the columella in *Sphagnum*; see also *amphithecium*

entire – without toothing or division, more or less smooth on the margin

eperistomate – without a peristome; syn.: *gymnostomous*; ant.: *peristomate*

ephemeral – a short-lived plant that completes its life cycle within only a single growing season; see also *annual*, *biennial* and *perennial*

epidermis (*pl.* **epidermides**) – the outermost layer of cells in an organ, such as a stem or costa; often included with the cortex

epigean – growing upon the ground; on land as opposed to water

epilithic – growing on rocks; syn.: *petrophilous*, *rupestral*, *rupicolous* and *saxicolous*

epiphragm (*pl.* **epiphragmata**) – a circular membrane at the mouth of a capsule formed by the expanded tip of the columella and attached to the tips of the peristome teeth or the mouth of the capsule

epiphytic – growing on another plant but not parasitising

epithet – the second word in a binary combination, referring to the specific name following a generic name

epitype – a specimen or illustration selected to serve as an interpretative type when the holotype, lectotype, or previously designated neotype, or all original material associated with a validly published name cannot be identified for the purpose of precise application of the name of a taxon

equidistant – uniformly spaced

erect – said of anything that is pointing upwards; applied to leaves directed towards the stem apex but not pressed against the stem; with capsules straight, not curved; with leaf margins curved upward

erect-patent – pointing upwards at an angle of about 20–45°, for example the adaxial angle between a leaf and stem; syn.: *erect-spreading*

erect-spreading – pointing upwards at an angle of about 20–45°, for example the adaxial angle between a leaf and stem; syn.: *erect-patent*

eroded – worn away

erose – irregularly notched or ragged, as though bitten or gnawed

estomatose – lacking stomata

exannulate – without an annulus; ant.: *annulate*

excurrent – running through to the apex and projecting beyond as an awn

exine – the outermost wall layer of a spore that is produced by a spore, usually containing the highly waterproof, rot-resistant, and UV-absorbing carotenoid polymer sporopollenin; see also *intine*

exogenous – arising from superficial tissue; ant.: *endogenous*

exostome – outer circle of teeth of the diplolepidous peristome; in genera with a single peristome it is homologous with the endostome of diplolepidous mosses; syn.: *outer peristome*

exothecium (*pl.* **exothecia**) – the superficial outermost layer of cells of a capsule

exserted – protruded beyond, projecting and exposed; applied to capsules carried well above the perichaetial leaves; ant.: *immersed*

exsiccata (*pl.* **exsiccatae**) – a set of dried plant specimens, usually numbered and with printed labels, accompanied by a booklet of labels, distributed to herbaria or collectors and intended to be reliably identified reference collections of species

F

falcate – curved like the blade of a sickle; syn.: *falciform*

falcato-secund – with leaves that are both curved like a sickle and turned toward one side of the stem or branch

falciform – sickle-shaped; syn.: *falcate*

fastigiate – with branches that are erect, almost parallel and reach the same length

female inflorescence – the female sex organs with the cluster of modified leaves surrounding them; syn.: *perichaetium*

fenestrate – pierced with broad openings resembling windows

ferrugineous – rust-coloured

fertile – producing sex organs; ant.: *sterile*

fibrillose – furnished with fine spiral or annular fibre-like wall thickenings

filament – any thread-like, long and sometimes branched structure, composed of a row of cells attached end-to-end

filamentous – thread-like, formed of filaments; syn.: *filiform*

filiform – thread-shaped, long and slender; syn.: *filamentous*

fimbriate – with the margin bordered by long slender processes; syn.: *fringed*

flabellate – shaped as a fan, dilated in a wedge-shaped, sometimes plaited

flaccid – soft, flabby or limp; syn.: *flexible* and *pliant*

flagelliform branch – a slender, long and gradually tapering shoot resembling a runner or whip, with leaves getting steadily smaller towards the tip, functioning as a soil-anchor or propagule

flagellum (*pl.* **flagella**) – a whip-like structure of bryophyte spermatozoids, functioning in locomotion

fleshy – thick, pulpy; syn.: *succulent*

flexible – capable of being bent but elastic enough to be able to resume its original shape and figure; syn.: *flaccid* and *pliant*

flexuose – bent alternately in opposite directions, wavy, twisted

fluted – with a shallow rounded grooves or furrows

foliage – the leafy covering

foliate – leafy, closed with leaves; ant.: *defoliate*

foliose – leafy or leaf-like, closely covered with leaves

foot – the basal portion of a sporophyte, embedded in the gametophyte and serving as both an organ of attachment and for absorption of water and mineral nutrients

forked – branched, divided into parts; syn.: *furcate*

fringed – margined with a hair-like or ciliate appendages; syn.: *fimbriate*

fruit – a colloquial term for the capsule or, more loosely, the sporophyte

fugacious – soon perishing, vanishing or readily falling away

fulvous – tawny, dull yellow-brown

furcate – branched, divided into parts; syn.: *forked*

furrowed – with a lengthwise groove; see also *sulcate* and *striate*

fuscous – dull brown, often tinged with black

fusiform – shaped like a spindle, long and narrow and tapered at both ends

G

gametangium (*pl.* **gametangia**) – a sex organ that forms gametes, either an antheridium that produces sperms or an archegonium that produces an egg

gametoecium – gametangia together with surrounding bracts; syn.: *inflorescence*

gametophore – a gametangium-bearing stalk; loosely used for the leafy stem of the mature gametophyte plant developed from a protonema

gametophyte – the haploid generation which bears the sexual organs and produces gametes, in turn giving rise to the sporophyte; in mosses the dominant generation, consisting of green, leafy plants

geminate – arranged or combined in pairs, doubled; syn.: *binate*

gemma (*pl.* **gemmae**) – a type of vegetative propagule, brood body borne on various parts of the gametophyte plant

gemma cup – a cup-shaped structure, formed from a rosette of modified leaves that functions as an aid in the distribution of gemmae by water splash; syn.: *splash-cup*

gemmiferous – producing and bearing gemmae

gemmiform – bud-shaped, applied to plants or structures that are bud-like

gibbous – bulging or swollen; ant.: *ventricose*

glabrous – lacking papillae, verrucae or hairs; syn.: *smooth* or *naked*; ant.: *papillose*, *verrusose* or *hairy*

glaucous – having a whitish, bluish or greyish hue or colour, sea-green

globose – spherical or nearly so

globular – spheroidal in shape

gonioautoicous – having a bud-like male inflorescence borne axillary on the same stem or branch as the female inflorescence; see also *autoicous*, *cladautoicous* and *pseudautoicous*

granite mosses – a common name for andreaealean mosses which refers to their typical habitat of siliceous rocks; syn.: *andreaealean* or *lantern mosses*

granulose – minutely roughened with blunt projections

gregarious – growing in company, associated but in loose tufts or mixed mats, not densely matted

grimmialean – referring to mosses of the family Grimmiaceae

grooved – with a lengthwise furrow, furrowed; see also *fluted*

growth-form – the genetically determined overall morphology of an individual moss plant

guard cells – the epidermal cells which surround the opening of a stoma, usually paired and kindney-shaped, occasionally a single cell with a central slit or pore, which open or close the stoma according to their state of turgidity

guide cells – the large, elongate, empty, rather thin-walled and highly vacuolated cells stretching across the costa in many mosses as viewed in its cross-section; syn.: *deuters*

gymnostomous – lacking a peristome; syn.: *eperistomate*; ant.: *ochyrostomous*

gynoecium (*pl.* **gynoecia**) – the female gametoecium or female inflorescence consisting of archegonia and the surrounding bracts; syn.: *perichaetium*

H

habit – the general appearance or aspect of a plant

habitat – the ecological niche or micro-environment in which a plant grows

hair-point – a fine, weak, hair-like leaf tip, usually whitish when dry, formed by an excurrent costa; see also *awn* and *arista*

hairy – covered with hairs or producing hairs

halophyte – a plant which grows within the influence of salt water, such as coastal dunes and sea cliffs that are regularly exposed to or inundated by salt spray or tidal flooding

hamate – hook-shaped, more abruptly curved than falcate; syn.: *uncinate*

haploid – applied to the state of an organism that contains one complete set of chromosomes (denoted by n); the normal chromosome level of the gametophyte generation in mosses; see also *diploid* and *polyploid*

haplolepidous – with a single articulated peristome, i.e. having a single circle of articulated teeth; ant.: *diplolepidous*

heliophilous – thriving in a sunny habitat, exposed to full sunlight; syn.: *photophilous*; ant.: *sciophilous*

heliophyte – a plant adapted to full sunlight; syn.: *photophyte*; ant.: *sciophyte*

hemispherical – flat on one side and bulging out on the other side, in reference to a solid in three-dimensional shapes

heterogeneous – dissimilar in nature, not uniform in kind, made up of contrasting or unrelated parts; ant.: *homogeneous*

heteroicous – with several form of gametoecia on the same plant or various plants of the same species; see also *polygamous* or *polyoicous*

heterophyllous – having different stem and branch leaves ; ant.: *isophyllous*

heterotypic synonym – a synonym based on a type different from that of accepted name, designated by the symbol of the algebraic identity (=); syn.: *taxonomic synonym*; ant.: *homotypic* or *nomenclatural synonym*

hispid – beset with short, rough, stiff hairs or bristles; bristly

hoary – whitish or greyish and appearing frosted from massed hyaline hair-points; syn.: *canescent*

holantarctic – a pattern of distribution restricted to the floral kingdom Antarctis (called also Holantarctis) in the Southern Hemisphere which includes most areas of the world south of 40° S latitude, namely the continent of Antarctica, Patagonia (southern Chile, southern Argentina, Tierra del Fuego), most of New Zealand and all islands of the Southern Ocean; Tasmania is excluded since its flora is more closely related to that found in the Australian floral kingdom

holarctic – a pattern of distribution restricted to the largest floral kingdom Holarctis in the northern part of the Northern Hemisphere

holotype – the one specimen or illustration used by the author or designated by the author as the nomenclatural type

homogeneous – uniform, of the same kind or nature, composed of similar parts; ant.: *heterogeneous*

homonym – a name spelled exactly like another name published for a taxon of the same rank based on a different type

homotypic synonym – a synonym based on the same type as that of another name in the same rank, designated by the symbol of congruence (≡); syn.: *nomenclatural synonym*; ant.: *heterotypic* or *taxonomic synonym*

hummock – a growth-form resembling that of the cushion growth-form but with taller and more loosely arranged shoots, usually arising from a matrix of decaying shoots

hyaline – colourless or transparent when moist, usually white when dry

hyalocyst – large, empty, water-storage, hyaline cell as in the leaves of species of the Sphagnaceae, Calymperaceae, Leucobryaceae and Dicranaceae; syn.: *leucocyst*

hyalodermis (*pl.* **hyalodermides**) – an outer layer of the epidermis of a stem composed of differentiated, usually enlarged, thin-walled and colourless cells

hydric – very wet, pertaining to water, with reference to habitat; ant.: *xeric*

hydro- – prefix referring to an aquatic, generally submerged, existence

hydroid – a tracheid-like conductive cell in the central strand and sometimes also in the costa of some mosses

hydrome – a specialised hydroid tissue in the central strand, costa, and seta of some mosses that transports water, analogous to the xylem of vascular plants

hydromorphic – structurally adapted for living in true aquatic habitats; ant.: *xeromorphic*

hydrophile – a plant which grows in an aquatic habitat; ant.: *xerophile*

hydrophilous – preferring an aquatic habitat; ant.: *xerophilous*

hydrophyte – an aquatic plant, partially or wholly submerged or floating; ant.: *xerophyte*

hygro- – prefix referring to a wet but not aquatic existence

hygrocastique – capsules opening under moist conditions; ant.: *xerocastique*

hygromorphic – structurally adapted for living in wet but not true aquatic habitats; ant.: *xeromorphic*

hygrophile – a moisture-loving plant; ant.: *xerophile*

hygrophilous – preferring a wet but not aquatic habitat; ant.: *xerophilous*

hygrophyte – a plant which needs a large supply of moisture for its growth; ant.: *xerophyte*

hygroscopic – readily absorbing moisture; applied to peristome teeth that bend in or out in response to subtle changes in humidity

hypnalean – referring to mosses of the family Hypnaceae

hypnoid peristome – a complete double peristome, with an outer row of 16 lanceolate exostome teeth which are cross-striolate at the base and trabeculate at the back and a well developed inner endostome consisting of a high basal membrane and 16 lanceolate, keeled segments alternating with one or more cilia

hypophysis (*pl.* **hypophyses**) – a markedly differentiated, conspicuous swelling or expansion of the sterile basal part of a capsule, between the seta and the urn; syn.: *apophysis*

I

icon (*pl.* **icones**) – a plate, drawing or engraving

illegitimate name – a validly published name that is not in accordance with one or more rules of the Code, principally those on superfluity and homonymy; see also *nomen illegetimum*

imbricate – overlapping as the tiles on a roof, closely appressed and overlapping

immersed – submerged or below the surface; applied to capsules that are completely overtopped by the perichaetial leaves; see also *emergent*; ant.: *exserted*; also referring to stomata with guard cells sunken below the level of the exothecial cells; syn.: *cryptoporous*

imperfect – said of a structure in which certain parts usually present are not developed

incised – dissected or cut into sharp divisions separated by a narrow sinus

inclined – falling away from the horizontal direction; referring to capsules that are between the erect and horizontal positions

included – enclosed or nearly so by surrounding organs or structures

incoherent – loosely tufted

incrassate – thickened, especially applied to cell wall thickness

incurved – curved upwards and inwards; referring to leaf tips or margins; ant.: *recurved*

indehiscent – pertaining to a capsule that does not open by distinct mechanisms such as an operculum or valves

indigenous – not introduced, original the area, native; syn.: *native*; ant.: *adventive*

inflated – swollen, bladder-like

inflexed – turned abruptly or bent inwards or upwards; referring to leaf margins or leaves on a stem; ant.: *reflexed*

inflorescence – a cluster of sex organs with surrounding specialised leaves; syn.: *gametoecium*

infraspecific rank – a taxonomic category within the species: the subspecies (abbreviated subsp.), variety (var.) and form (fo.)

inner peristome – the inner ring of teeth of a diplolepidous peristome, formed from contiguous periclinal wall-pairs of the inner and primary peristomial layers; homologous with the haplolepidous peristome; syn.: *endostome*

inner peristomial layer (IPL) – the innermost layer of the peristomial cylinder that contributes in the formation of arthrodontous peristomes; see also *outer* and *primary peristomial layer*

innovation – a newly formed shoot, usually arising below an inflorescence after the formation of sex organs

inoperculate – wanting an operculum; applied to a moss capsule that has no regular mechanism for opening; syn.: *cleistocarpous*

inrolled – loosely rolled inward; applied to a leaf margin; syn.: *involute*

insertion – the place where one organ is attached to its support; referring to the line or point of attachment of a leaf or branch to a stem

intercalary – inserted between; applied to branches arising below the shoot apex or to rhizoids inserted between the leaves on a stem

intine – the innermost wall layer of a spore; see also *exine*

intricate – interwoven or entangled

involute – having the edges of the leaves tightly rolled inward; syn.: *inrolled*; ant.: *revolute*

irregular – not uniform, wanting in regularity of form, as in the random spacing of teeth along a leaf margin or the arrangement of cells in a leaf blade

iso- – prefix meaning 'equal' or denoting the uniformity

isodiametric – of equal dimensions; applied to cells that are about as broad as they are long, regardless of their shape

isophyllous – having all stem and branch leaves similar; ant.: *heterophyllous*

isopolar – referring to spores with similar extremities

isosyntype – a duplicate of a syntype

isotype – a duplicate specimen of the holotype

J

julaceous – resembling a catkin; referring to smoothly cylindric stems and branches with crowded and imbricate leaves

juvenile – applied to the early forms of a moss arising from the protonema and giving rise to the adult plant

juxtacostal – located near or next to the costa

K

keel – a ridge like the keel of a boat formed along a sharp fold of a leaf

keeled – sharply folded along the middle, like the keel of a boat, V-shaped in transverse section; syn.: *carinate*

KOH reaction – a change of the colour in cell walls, usually to yellow, orange or red, after soaking in a 2–10% solution of potassium hydroxide; with many pottialean mosses the resultant colorations are consistent enough to be relied on as diagnostic characters

L

lacerate – deeply and irregularly torn or slashed

laciniate – slashed or finely and deeply dissected into narrow lobes

lacunose – perforated with holes or when the surface is covered with depressions, pertaining to cells which have irregular or seemingly empty lumina because their walls are deeply porose or strongly thickened

lamella (*pl.* **lamellae**) – flat, parallel, unistratose, photosynthetic plates, ridges or filamentous outgrowths along a costa or leaf blade in the Polytrichaceae as well as some pottialean and grimmialean mosses

lamellate, **lamellose** – with photosynthetic plates, ridges or filamentous outgrowths

lamina (*pl.* **laminae**) – the flattened, uni- or polystratose and green part of the leaf blade as distinct from the costa as well as the border and/or differentiated basal part, if present

laminal cells – cells of the lamina; see also *areolation*

lanceolate – shaped like the head of a spear, 3–6 times longer than wide, widest below the middle, and with slightly convex sides tapering gradually towards the apex (narrower than ovate)

lantern mosses – a common name for mosses of the family Andreaeaceae inspired by their capsules which look like a traditional Chinese lantern when they are dry and fully open; syn.: *andreaealean* or *granite mosses*

lateral – fixed on or near the side of an organ

latticed – cross-barred

lax – loose or soft; applied to large, bulging and thin-walled cells as well as to plants with shoots loosely arranged in a tuft and to spacing leaves on a stem

leaf – a photosynthetic lateral outgrowth borne by the stem or axis, in mosses generally consisting of a mostly unistratose lamina with or without polystratose costa

lectotype – a specimen or illustration (or a species in the case of generic names) designated from the original material as the nomenclatural type if no holotype was indicated at the time of publication, or if it is missing, or if it is found to belong to more than one taxon

legitimate name – a validly published name that is in accordance with all rules of nomenclature

lenticular – shaped like a double-convex lens

leptoid – a specialised conductive cell present in the central strand of stems, setae and costae of some mosses, analogous to the sieve tubes of vascular plants in form and function

leptodermous – thin-walled, especially pertaining to leaf or exothecial cells; ant.: *pachydermous*

leptome – a specialised tissue in the central strand, costae and setae of some mosses that transports dissolved energy-rich products of photosynthesis, analogous to the phloem of vascular plants

leucocyst – large, empty, water-storage, hyaline cell as in the leaves of species of the Sphagnaceae, Calymperaceae, Leucobryaceae and Dicranaceae; syn.: *hyalocyst*

libriform – referring to a narrow, thick-walled cell of woody tissue resembling bast, wood-fibre

lid – the dehiscent cover of a capsule; syn.: *operculum*

ligulate – resembling a strap, narrow, parallel-sided and fairly short

limb – the upper part of a leaf when it is differentiated into the green upper part and usually enlarged, sheathing and often hyaline basal part

limbate – with a well developed border of distinct, elongate and thick-walled cells

limbidium (*pl.* **limbidia**) – a well developed, often bi- to polystratose border of a leaf sharply differentiated from the rest of the leaf

linear – very narrow and several times longer than wide with nearly parallel sides, referring to leaves and cells; narrower than ligulate

lingulate – resembling a tongue, broad and parallel-sided with a slightly broadened apex

lobate – dividing into or bearing lobes

lobe – any larger segment of a divided leaf or other organ

lobule – a small lobe

longitudinal – arranged or placed lengthwise, parallel to the long axis

lower – towards the base, well below the middle; applied to the position of leaves on a stem, cells of a leaf; syn.: *proximal*

lumen (*pl.* **lumina**) – the cavity of a cell inside the cell walls

lunate – half-moon shaped

lunulate – shaped like a crescent moon

lustrous – having a sheen or gloss that readily reflects light

M

macrostomous – with a large capsule mouth; ant.: *microstomous*

male inflorescence – the male sex organs with the cluster of modified leaves surrounding them; syn.: *perigonium*

mammilla (*pl.* **mammillae**) – a bulge on the cell surface having a uniformly thickened walls and a nipple-like tip into which the cell lumen projects; a low, rounded projection resembling a nipple or teat on an operculum

mammillate – convex or hemispherical with a short, blunt central protuberance

mammillose – having mammillae

margin – the edge of a structure, usually of a leaf

marginal – referring to the edge of a structure, usually of a leaf

mat – a growth form with tightly interwoven stems and branches forming a flattened or complanate patch adhering to the substrate with rhizoids

m-chromosome – smallest chromosome, less than half the length of any other member of the complement, symbolised by 'm'

median line – a lengthwise line, usually zig-zag and sometimes recessed in a furrow, found on either or both surfaces of a peristome tooth; the line represents the contiguous anticlinal walls of two cell columns from one of the layers of the peristomial cylinder; syn.: *divisural line*

medulla (*pl.* **medullae**) – the internal stem tissue when the term cortex is used for the epidermis

mesic – moist, intermediate between very wet (*hydric*) and very dry (*xeric*)

micrometre – one-millionth of a metre or one-thousandth of a millimetre, symbolised by 'μm'

microphyllous – bearing minute leaves, small-leaved; pertaining to flagelliform branches that produce smaller than normal leaves

microstomous – with a small and narrow capsule mouth; ant.: *macrostomous*

midrib – a vein in the centre of a leaf; syn.: *costa*

mid-leaf – middle third of a leaf between the upper leaf and lower leaf; referring to median leaf cells between the margins and the costa

minute – very small, inconspicuous

mitrate – shaped like a bishop's cap (mitre), conical with an undivided or equally lobed base; referring to a calyptra; syn.: *mitriform*; ant.: *cucullate*

mitriform – mitrate, pertaining to calyptrae; syn.: *mitrate*

mono- – prefix meaning 'one'

moniliform – necklace-shaped, like a string of beads

monoicous – having male and female gametangia on the same plant, bisexual; ant. *dioicous*

monopodial – a growth form with a single and continuous axis with unlimited growth, the situation typical of pleurocarpous mosses; ant.: *sympodial*

monotypic – with only one taxon; for example, a monotypic genus consists of a single species: see also *oligotypic*

montane – pertaining to mountains, cool moist upland regions below the timber line

morphology – a general appearance, outward form or structure of a plant

mucilage papilla – a specialised cell that emits a clear, gel-like secretion

mucro – an abrupt, relatively broad and short, straight point at a leaf apex, somewhat shorter than an apiculus

mucronate – possessing a short, fairly broad and short, straight point, usually caused by a shortly excurrent costa

multi- – prefix meaning 'many'; syn.: *poly-* and *pluri-*

multicellular – composed of many cells

multifid – cleft into many segments or lobes

muticous – without a point or awn

N

n – the haploid chromosome number of an organism or tissue

naked – lacking hairs, cilia, rhizoids or ornamentation; syn.: *smooth*

native – not introduced, original to the area; syn.: *indigenous*; ant.: *adventive*

navicular – boat-shaped; syn.: *cymbiform*

nearctic – a pattern of distribution restricted to the American Arctic

neck – the sterile basal portion of a capsule between the top of the seta and the urn, occasionally markedly differentiated; syn.: *apophysis* and *hypophysis*

nematocyst – a rhizoid initial cell in the leaf lamina; syn.: *nematogon*

nematodontous – with reference to a peristome type in the Polytrichopsida, consisting essentially of whole dead cells with more or less evenly thickened walls; ant.: *arthrodontous*

nematogon – a differentiated leaf cell, usually short and thin-walled, from which rhizoids arise; syn.: *nematocyst*

neophyte – a plant that is not native to an area and was introduced recently

neotropic – a pattern of distribution restricted to the American tropics

neotype – a specimen or illustration selected to serve as nomenclatural type if no original material is extant or as long as it is missing

nerve – the midrib of a leaf; see *costa*

new combination (***comb. nov.***) – a combination formed from a previously published legitimate name and employing the same final epithet (or employing the name itself if formed from a generic, subgeneric or sectional name)

nitrophilous – referring to plants preferring substrates rich in nitrogenous compounds; ant.: *nitrophobous*

nitrophobous – referring to plants which avoid nitrogen-rich habitats; ant.: *nitrophilous*

nodose – with coarse knob- or bead-like thickenings; referring to longitudinal cell walls or cilia with bead-like thickenings corresponding in position to the articulations of the processes

nodulose – with minute knob- or bead-like thickenings

nomenclatural novelties – new names and descriptions or diagnoses of a new taxa

nomenclatural synonym – see *homotypic synonym*

nomenclatural type – the element to which the name of a taxon is permanently attached

***nomen conservandum* (*nom. cons.*)** – see *conserved name*

***nomen illegitimum* (*nom. illeg.*)** – see *illegitimate name*

***nomen novum* (*nom. nov.*)** – see *replacement name*

***nomen nudum* (*pl. nomina nuda*) (*nom. nud.*)** – a name of a new taxon published without description or diagnosis or a reference to a description or diagnosis

***nomen rejiciendum* (*nom. rej.*)** – a name rejected in favour of a name conserved under Art. 14 or a name ruled as rejected under Art. 56 of the Internacional Code of Botanical Nomenclature (ICBN); syn.: *rejected name*

non-septate – lacking cross-walls

nutant – nodding or drooping more than 90° from vertical

O

ob- – prefix meaning 'inversely' or 'reversely'

obconical – shaped like an upside-down cone, conical with the narrow end at the base

obcordate – inversely cordate or heart-shaped, the notch being apical

obcuneate – inversely cuneate or wedge-shaped

oblate – flattened, wider than long; ant.: *prolate*

oblique – slanted, pertaining to a capsule mouth or a beak of the operculum

obloid – a solid with an oblong profile, a three-dimensional equivalent of oblong, pertaining to capsules with rounded edges and corners

oblong – rectangular with rounded or obtuse corners or ends

oblong-hexagonal – approaching a rhombic outline but longer, with the lateral angles obtuse; syn.: *rhomboidal*

obovate – reversed ovate, with the broader distal portion rather than the base

obovoid – an obovate solid, a three dimensional equivalent of obovate

obscure – dark, indistinct or hard to see, as with leaf cells obscured by dense surface papillae

obsolete – barely visible or almost absent, used of an organ which is scarcely apparent or has vanished, as with a costa that is thin or is almost the same colour as the cells around it

obtriangular – shaped like an upside-down triangle

obtuse – bluntly pointed, with the sides making an angle of more than 90°

ochraceous – brownish-yellow, ochre-coloured

ochyrostomous – applied by Hedwig to the capsule bearing a peristome; syn.: *peristomate*; ant.: *gymnostomous*

oligotypic – with several taxa; for example, an oligotypic genus consists of 3–5 species; see also *monotypic*

olivaceous – olive-green, yellowish-green darkened with black

opaque – dull, matt, lustreless or not shining; not allowing any light through; ant.: *transparent* and *translucent*

operculum (*pl.* **opercula**) – the lid covering the mouth of a capsule which separates by a transverse line at maturity to release the spores

opposite – of leaves or branches arranged along a stem directly across from each other; ant.: *alternate*

orbicular – circular in outline or nearly so; syn.: *rotund*

oreophyte, orophyte – a plant dwelling in mountainous regions

orifice – a mouth of the capsule; an opening through which spores are released from the capsule

O-shaped papilla – a papilla appearing like a ring when viewed from above with an ordinary light microscope

outer peristome – the outer ring of teeth of a diplolepidous peristome, formed from contiguous periclinal wall-pairs of the outer and primary peristomial layers; missing or rudimentary in haplolepidous peristomes; syn.: *exostome*

outer peristomial layer (OPL) – the outermost layer of the peristomial cylinder that contributes to the formation of arthrodontous peristomes; see also see also *inner* and *primary peristomial layer*

oval – short-elliptic

ovate – shaped like a longitudinal section of a hen's egg, broadest below the middle, in reference to a plane

ovoid – an egg-shaped solid, a three dimensional equivalent of ovate

P

pachydermous – thick-skinned; applied to very thickened and firm cells of moss capsules or stems; ant.: *leptodermous*

paired – associated in groups of two; applied to peristome teeth and marginal teeth on leaves

paludicolous – growing in marshy places

pandurate – fiddle-shaped, drawn in at the middle; syn.: *panduriform*

panduriform – fiddle-shaped, shaped line a body of violin, obovate with a sinus on either side; syn.: *pandurate*

papilla (*pl.* **papillae**) – a solid superficial protuberance on a cell surface

papillate – roughened or ornamented with papillae; syn.: *papillose*

papillose – covered with papillae; syn.: *papillate*

paraphyllium (*pl.* **paraphyllia**) – a small, green, leafy or filamentous interfoliar appendage produced randomly on the stems or branches of many pleurocarpous mosses

paraphysis (*pl.* **paraphyses**) – a sterile, filiform, usually uniseriate hair associated with antheridia and archegonia in mosses

paratype – a specimen cited in the protologue that is neither the holotype nor an isotype, nor one of syntypes if two or more specimens were simultaneously designated as types

parenchymatous – applied to thin-walled and isodiametric cells, with non-overlapping end-walls; ant.: *prosenchymatous*

paroicous – having antheridia in the axils of the leaf immediately below the perichaetium

patent – making an angle of 45° or more, for example the adaxial angle between a leaf and stem

patulous – standing open, making an angle of 45–90°, for example the adaxial angle between a leaf and stem; syn.: *spreading*

pectinate – resembling a comb, divided in narrow segments set close like the teeth of a comb

pedicellate – borne on a stalk

pellucid – wholly or partially transparent; see also *transparent* and *translucent*; ant.: *opaque*

pendent – drooping or hanging down from its support; syn.: *pendulous*

pendulous – hanging downward; referring to capsules drooping and inclined beyond horizontal and to stems and branches that hang; syn.: *pendent*

penicillate – pencil-shaped, ending in a tuft like a brush

pennate – arranged like a barbs on a feather; syn.: *pinnate*

pentagonal – five-angled

pentastichous – in five rows or ranks

percurrent – extending throughout the entire length; referring to a costa extending to the leaf apex

perennial – a plant which lasts several years; most moss gametophytes are perennials, whereas most sporophytes are annuals; see also *biennial* and *ephemeral*; ant.: *annual*

perfect peristome – a complete double peristome with an outer ring of 16 exostome teeth and a well developed inner endostome of 16 keeled segments alternating with one or more thin cilia

perforate – pierced through or having translucent dots that look like little holes

perichaetial leaf – the modified leaf enclosing the archegonia

perichaetium (*pl.* **perichaetia**) – the female inflorescence, consisting of leaves or bracts surrounding the archegonia; see also *male inflorescence* or *perigonium*

periclinal – a plane of division oriented parallel to the surface; ant.: *anticlinal*

perigonial leaf – the modified leaf enclosing the antheridia

perigonium (*pl.* **perigonia**) – the male inflorescence, consisting of a cluster of leaves or bracts surrounding the antheridia; see also *female inflorescence* or *perichaetium*

peristomate – provided with a peristome; syn.: *ochyrostomous*; ant.: *eperistomate*

peristome – the single or double ring of teeth inside the mouth of a capsule revealed after fall of the operculum

peristomial formula – a descriptive equation listing the peristomial number from the OPL to the IPL, indicating relative degree of wall thickenings, any lateral displacements of the IPL and prostomial development

persistent – long lasting, remaining till the part which bears it is wholly matured, not falling or non-deciduous

petrophilous – rock dwelling, preferring rocks

phaneroporous – applied to stomata with its guard cells and pores that are in the same level as the adjacent exothecial cells; ant.: *cryptoporous*

phenotype – observable characteristics of an individual resulting from the interaction of its genotype with the environment

photophilous – a sun-loving plant, thriving in full sunlight; syn.: *heliophilous*; ant.: *sciophilous*

photophyte – a plant thriving in sites exposed to full sunlight; syn.: *heliophyte*; ant.: *sciophyte*

photosynthetic – containing the pigment chlorophyll plus the methabolic pathways for converting energy-poor carbon dioxide into energy-rich carbohydrates using the energy of sunlight

piliferous – bearing hair-points

pilose – bearing long and soft hairs

pinnate – with many spreading branches on opposite sides of the axis, arranged like the barbs on a feather, feathery; syn.: *pennate*; see also *plumose*

pit – a small depression or cavity in the cell wall; syn.: *pore*

pitted – having small cavities, openings, holes or depressions in the cell walls; syn.: *porous*

plane – flate or level, lying only in one plane; applied to leaf margins or blade

plagiotropic – having the direction of growth at an oblique angle or horizontal

plano-convex – flat on one side and bulging on the other side; syn.: *semicircular*

pleated – folded or furrowed lengthwise; syn.: *plicate*

pleurocarpous – lateral fruited; mosses in which the sporophytes are usually borne laterally from a perichaetial bud or a short specialised branch rather than at the tips of the stems or main branches; pleurocarpous mosses are usually freely branched and grow creeping or prostrate in mats, patches or wefts; see also *cladocarpous*; ant.: *acrocarpous*

pliant – bending easily, soft; syn.: *flaccid* and *flexible*

plica (*pl.* **plicae**) – a lengthwise pleat or folding

plicate – longitudinally folded; syn.: *pleated*

plumose – with numerous close and regular, spreading branches on opposite sides of the axis, thus feathery in appearance; see also *pinnate* and *pennate*

pluri- – prefix meaning 'several' or 'many'; syn.: *multi-* and *poly-*

pluripapillose – with several papillae per cell

pluriseriate – arranged in several or many rows

pluristratose – composed of several or many cell layers

poikilohydric – referring to an organism that lacks a mechanism to prevent water loss or desiccation; bryophytes cannot actively control their internal water content as they lack or have only rudimentary tissues that are specialised for internal conduction, most of them tend to be poikilohydric

poly- – prefix meaning 'many' or 'numerous'; syn.: *multi-* and *pluri-*

polygamous – with several form of gametoecia on the same plant or various plants of the same species; see also *heteroicous* or *polyoicous*

polygonal – a plane figure having many sides

polymorphic – very variable, with several or various diversities of forms

polyoicous – with several form of gametoecia on the same plant or various plants of the same species; see also *heteroicous* or *polygamous*

polyploid – applied to the state of an organism that contains more than two complete sets of chromosomes; see also *haploid* and *diploid*

polysetous – with more than one sporophyte produced from a single perichaetium

polysety – the production of more than one sporophyte from a single perichaetium

pore – any small aperture; a hole in a cell wall; syn.: *pit*

porose – pierced with small holes, having cell wall pits; syn.: *pitted*

pottialean – referring to mosses of the order Phascales or Pottiales

preperistome – a vestigial structure outside, and usually adhering to, the main peristome teeth; syn.: *prostome*

primary leaf – the initial leaf on a juvenile plant, generally different morphologically from the later produced leaves
primary peristomial layer (PPL) – the middle layer of the peristomial cylinder that contributes in the formation of arthrodontous peristomes; see *inner* and *outer peristomial layers*
primary stem – a main stem, usually creeping and prostrate or rhizome-like, with either scale-like or strongly reduced leaves
primordium (*pl.* **primordia**) – an organ in a very early stage of development
priority – a right to precedence established by the date of valid publication of a legitimate name or of an illegitimate earlier homonym or by the date of designation of a type
prismatic – prism-shaped, with flat faces separated by angles
process – a single tooth-like division of the endostome of the diplolepidous peristome; syn.: *segment*
procumbent – spreading, lying flat on the ground; see also *creeping* and *prostrate*
prolate – lengthened out, longer than wide; ant.: *oblate*
proliferation – repeated production of new shoots or innovations or a structure formed by proliferating
proliferous – bearing progeny as offshoots, continuing growth by the production of new stems or innovations
prolific – fertile, fruitful
propagulum (*pl.* **propagula**), **propagule** – any structures that reproduce a plant vegetatively, such as gemmae, bulbils, tubers and surculi; syn.: *brood body*
prorate – having a papillose surface by the projecting ends of prosenchymatous cells
prorulate – having a delicately and minutely roughened or coarse surface by protruding tips of prosenchymatous cells; syn.: *scindulose*
prosenchymatous – applied to narrow, elongate cells with tapering overlapping end walls; ant.: *parenchymatous*
prostome – a rudimentary structure of irregular thickenings or projections outside, and usually adhering to, the main peristome teeth; syn.: *preperistome*
prostrate – closely appressed to the ground, spreading; syn.: *creeping*; see also *procumbent*
protean – exceedingly variable
protologue – everything associated with a name at its valid publication, i.e. description or diagnosis, illustrations, references, synonymy, geographical data, citation of specimens, discussion and comments
protonema (*pl.* **protonemata**) – the first growth stage of a bryophyte being a green, branched filamentous, globose or thalloid structure, produced from a germinating spore on which the conspicuous plant is developed
protuberant – bulging out
proximal – the part nearest the base or point of attachment; syn.: *basal*; ant.: *terminal* and *distal*
psammic, **psammophilous** – sand-loving
psammophile – a plant confined to sandy habitats
pseudautoicous – with dwarf male plants growing epiphytically in the female; see also *autoicous*, *cladautoicous* and *gonioautoicous*
pseudo- – prefix denoting ‘false’
pseudopapillose – referring to lengthwise cuticular ridges or thickenings over the surface of the laminal cells and stems which give them a papillose appearance in transverse section
pseudoparaphyllium (*pl.* **pseudoparaphyllia**) – a small, unistratose, foliose, lanceolate or filiform structure superficially resembling paraphyllium; often found in pleurocarpous mosses clustered around the bases of branches or branch primordia
pseudopodium (*pl.* **pseudopodia**) – a stalk-like extension of the stem tip bearing clusters of gemmae or a sporangium in *Andreaea* and *Sphagnum*
pulvinate – cushion-shaped
punctate – marked with readily visible dots or depressions, resembling pin-pricks; usually referring to spore markings
punctulate – covered with minute dots or depressions
pyriform – resembling a pear in shape, pear-shaped

Q

quadrangular – four-cornered
quadrate – square, with four sides of ± equal length
quinquefarious – arranged in five ranks

R

radial – radiating, as from a centre; arranged symmetrically in three or more rows around a central axis
radicle – a slender filament attaching a moss to its substratum; syn.: *rhizoid*
radiculose – covered with rhizoids
ramification – branching
ramose – much-branched, having many branches
ramulose – having many branchlets
rank – a vertical row of similar structures such as leaves on a stem or teeth in a peristome; a level in a classification hierarchy
ranked – arranged in vertical rows
rectangular – of the form of rectangle, right-angled, with two sides 2–3 times longer than wide
recurved – curved backward and downward; referring to leaves curved outward from the stem, or leaf margins curved downward and inward, leaf tips and marginal teeth; in peristome teeth curved outward and ± downward; ant.: *incurved*
reduced – not fully developed, smaller in size than typical or expected; simplified during evolution, structurally simpler than in ancestral stock
reflexed – abruptly bent or turned downward or backward; referring to leaf margins or leaves on a stem; syn.: *deflexed*; ant.: *inflexed*
regular – symmetric
rejected name – a name the use of which is prohibited, either by formal action under Art. 14 or 56 of the International Code of Botanical Nomenclature (ICBN) overriding other provisions of the Code or because it was nomenclaturally superfluous when published or a later homonym; syn.: *nomen rejiciendum*
remote – scattered, widely separated, not close together; syn.: *distant*
reniform – kidney-shaped
repand – slightly wavy or lumpy, with slightly uneven margin; syn.: *sinuolate*
replacement name – a name proposed as a substitute for a previously published name; syn.: *nomen novum*
reticulate – forming a network, netted, as the network pattern produced by cell wall thickenings

retrorse – directed or turned backward or downward; ant.: *antrorse*

retuse – with a shallow notch or indentation at a broadly rounded apex; see also *emarginate*

revoluble – falling away as a ring or separating in a curled row; referring to an annulus that falls away in a ring

revolute – strongly rolled backward and downward from the margin or apex; ant.: *involute*

rheophilous – thriving in a habitat of fast-flowing water

rheophyte – a plant growing in fast-moving water

rhizoid – root-like; a delicate filament on the stem or leaf of a moss, usually uniseriate multicellular, simple or branched structure, generally with oblique end-walls that function in absorption and anchorage

rhizoid initial – a rhizoid-generating cell, usually differentiating in size and shape from neighbouring cells

rhizoidal – referring to rhizoids

rhizoidal gemma – a gemma borne on rhizoids; syn.: *tuber*

rhizomatous – having a slender, horizontal and creeping subterranean stem

rhizome – a slender, horizontal stem, prostrating on or under ground, sending off erect secondary stems

rhombic – shaped like rhomb, an equilateral oblique-angled figure; diamond-shaped, in the traditional playing-card sense

rhomboidal – approaching a rhombic outline but longer, with the lateral angles obtuse; syn.: *oblong-hexagonal*

rib – rised vertical line of cells on the walls of capsules

robust – large

rosette – a compact cluster or whorl of leaves in a circular form surrounding the stem

rostellate – the diminutive of rostrate, with a short beak

rostrate – with a beak, narrowed into a long slender tip or point

rostrum – any beak-like extension

rosulate – collected into a rosette; referring to leaves enlarged and crowded at tops of stems

rotund – round; syn.: *orbicular*

ruderal – a plant that thrives on disturbed or waste ground

rudimentary – poorly developed; syn.: *vestigial*

rufescent – turning red or red-brown

rufous – reddish or brownish-red

rugose – irregularly wavy, wrinkled, or undulate crosswise; see also *undulate*

rugulose – minutely or somewhat rugose; syn.: *wrinkled*

rupestral – growing among rocks; syn.: *epilithic*, *petrophilous*, *rupicolous* and *saxicolous*

rupicolous – dwelling among rocks; syn.: *epilithic*, *petrophilous*, *rupestral* and *saxicolous*

S

saxicolous – growing on rocks; syn.: *epilithic*, *petrophilous*, *rupestral* and *rupicolous*

scabrous – rough to the touch, usually because of covering of papillae

scale leaf – small, vestigial or scale-like leaf, usually on underground shoots, often with little chlorophyll

scarious – thin and dry, papery or membranous in texture, not green

scindula (*pl.* **scindulae**) – tiny papillae formed by the protruding tips of cells overlapping at their ends

scindulose – having a minutely papillose appearance caused by the protruding tips of cells overlapping at their ends; syn.: *prorulate*

sciophilous – thriving in shady habitats, shade-loving; ant.: *heliophilous* or *photophilous*

sciophyte – a plant of the shade; ant.: *heliophyte* or *photophyte*

sclerenchymatous – composed of cells with strongly incrassate walls; ant.: *parenchymatous*

scleroderm – an internal tissue of small cells with heavily thickened walls in the stem and branches

secund – turned or bent towards one side

segment – one of the divisions into which a plant organ, as a leaf, may be cleft; a piece cut off; a single tooth-like division of the endostome of the diplolepidous peristome; syn.: *process*

semi- – prefix meaning 'half'

semicircular – flat on one side and bulging out on the other side, in reference to a plane in two-dimensional shapes; syn.: *plano-convex*

semiterete – rounded on one side, flat on the other; see also *plano-convex*

sensu lato – a Latin phrase meaning 'in the broad sense' and either written out in full or abbreviated *s. lato*, *s. lat.* or *s. l.*; ant.: *sensu stricto*

sensu stricto – a Latin phrase meaning 'in the narrow sense' and either written out in full or abbreviated *s. str.* or *s. s.*; ant.: *sensu lato*

septate – divided by a partition, having cell walls

seriate – disposed in series of rows or layers, either transverse or longitudinal

sericeous – silky, clothed with close-pressed, fine, soft and straight hairs

serrate – saw-toothed; beset with regular antrorse teeth on the margin that are composed of one or more cells; see also *dentate*

serrulate – serrate, but the teeth minute; see also *denticulate*

sesqui- – prefix meaning one and a half

sesquilepidous – a periastome having a 2 : 3 peristomial number, i.e. two cell columns in the primary peristomial layer opposite three in the inner peristome layer; characteristic of the haplolepidous peristome

sessile – destitute of a stalk or seta, as though sitting close

seta (*pl.* **setae**) – the stalk of a sporophyte between the foot and the capsule

setaceous – bristle-like, less than subulate

sheath – a tubular or enrolled part or organ, as the expanded and usually modified base of the leaf clasping the stem or surrounding the lower part of the seta

sheathing – surrounding or closely clasping a stem or seta; syn.: *convolute*

shoot – a leafy stem

shoulder – the distal portion of the basal part of a leaf abruptly narrowed into the upper part of the leaf

sigmoid – S-shaped, doubly curved in opposite directions

simple – of one piece or series; ant.: *compound*

single peristome – a peristome with only a single ring of teeth; ant.: *double peristome*

sinistrorse – turned to the left, or twisting upwards in a clockwise spiral when viewed from above, as in a left-hand non-standard screw thread; referring to the torsion of a seta; ant.: *dextrorse*

sinuate – having a wavy or uneven wall or margin; syn.: *sinuose*

sinuolate – having a minutely wavy or uneven wall or margin; syn.: *repand*

sinuose – deeply waved; syn.: *sinuate*

smooth – lacking any surface irregularities such as papillae, mammillae, spines, hairs or lamellae; syn.: *naked*

sordid – having a dirty or discoloured hue

spathulate – tapering toward the base from a flat, broad, rounded apex, resembling a spatula

spiculose – covered with sharp and minute teeth or papillae

spinose – having sharp, pointed teeth or very high cell papillae or mammillae

splash-cup – a gemma-producing structure shaped like a cup, formed from a rosette of modified leaves; syn.: *gemma cup*

sporangium – the portion of a sporophyte being a spore-containing cavity in the capsule that surrounds the columella or, more loosely, the entire capsule

spore – a minute, mostly spherical and nearly always unicellular body produced in the capsule as a result of reduction division, giving rise on germination to protonemata

spore sac – a spore-containing cavity in the capsule that surrounds the columella in most mosses, but overarches it in species of *Andreaea* and *Sphagnum*

sporophyte – the asexual diploid spore-producing generation which remains attached to the gametophyte and partially dependent on it, typically consisting of a foot embedded in the gametophyte, seta and capsule which produces spores following meiosis

spreading – making an angle of 45–90°, for example the adaxial angle between a leaf and stem; syn.: *patulous*

squamose – scaly or scale-like, covered with or composed of scales

squarrose – rough or scurfy with spreading and outstanding processes, as the tips of leaves curved away at ± right angles

squarrose-recurved – rough or scurfy with spreading and outstanding processes, as the tips of leaves strongly curved downwards

stance – the manner in which the leaves are held in relation to the stem

stegocarpous – referring to a moss capsule that dehisces by means of an operculum; ant.: *cleistocarpous*

stellate – star-shaped or radiating as like the points of a star

stem – the main axis of a moss gametophyte, growing by means of a single apical cell and bearing leaves and sexual organs

stenotypic – applied to a taxon which exhibits a narrow range of variation; ant.: *polymorphic*

stereid – the element of a strengthening tissue, a slender, elongate, fibre-like cell with strongly thickened walls, occurring in groups in the costa of the leaves or forming the cortex of the stems of some mosses, appearing in the transverse section as almost solid with a spot in the middle representing the lumen

stereid band – a group of stereid cells

stereid cell – see *stereid*

sterile – barren, lacking reproductive organs; generally referring to absence of sexual structures but can also mean absence of asexual propagules

stipe – the unbranched basal part of an erect stem in a dendroid or frondose moss

stoloniferous – with a long, slender and spreading branch or stem with reduced and scale-like leaves throughout which arches away from its parent plant like a runner of a strawberry plant

stoma (*pl.* **stomata**) – a breathing pore for gas exchange in the capsule wall, usually surrounded by two guard cells that control its opening

stomatose – having stomata

stramineous – straw-coloured or straw-like

stratose – layered

striate – marked with longitudinal parallel lines, as grooves, furrows or ridges

strict – close or very straight and upright

striolate – finely striate, marked with fine or delicate longitudinal ridges

struma (*pl.* **strumae**) – gall-like swelling on an organ, as on one side of a moss capsule

strumulose – having a small gall-like swelling on one side of a moss capsule

sub- – prefix meaning 'nearly', 'almost', or 'somewhat', as in subacute, subentire, subpercurrent, subquadrate, or subtubulose; also used to mean 'under', as in submedian

subfloral innovation – a branch arising just below a perichaetium; syn.: *subperichaetial innovation*

subpercurrent – applied to a moss costa that does not extend to the leaf apex but ceases somewhat below

subperichaetial innovation – a branch that arises immediately below a perichaetium; syn.: *subfloral innovation*

subpsammic – buried or partially buried by sand; with reference to the ephemeral burial of some mosses in habitats of sandy or silty soil

substereid – almost with the characteristics of a stereid, but with walls not as strongly thickened

subterranean – underground

subula (*pl.* **subulae**) – a sharp, fine, long and slender point

subulate – slenderly long-acuminate, needle-like or awl-shaped

succulent – having thick fleshy leaves; syn.: *fleshy*

sulcate – strongly folded lengthwise or with deep longitudinal furrows or grooves

superficial – on the surface of an organ; applied to stomata with pores that are in the same plane as the surface of the exothecium; syn.: *phaneroporous*; ant.: *cryptoporous*

supra-alar – referring to cells or area located immediately above the alar cells

suprabasal – located just above the base

surculus (*pl.* **surculi**) – a reduced branch or leaf that functions as a vegetative brood body or an erect secondary stem growing from a horizontal primary stem

sympodial – a growth form with the main axis of limited growth, further growth continuing by means of secondary axes that arise as subfloral innovations, the situation typical of acrocarpous mosses; ant.: *monopodial*

synonym – a name considered to apply to the same taxon as the accepted name

syntype – any specimen cited in the protologue when no holotype was designated, or any of two or more specimens simultaneously designated as types

systylius – with the operculum fixed to the columella and elevated above the capsule mouth after the capsule has opened

synoicous – having antheridia and archegonia mixed together in the same gametoecium or inflorescence

T

tangential – at right angles to the radial or medullary rays
tapering – narrowing towards one end
taxon (*pl.* **taxa**) – a taxonomic group of any rank
taxonomic synonym – see *heterotypic synonym*
teeth (*sing.* **tooth**) – a division of the haplolepidous peristome or the outer peristome of a diplolepidous peristome; also applied to sharp projections at the margin or face of leaves
terete – circular in transverse section, cylindric and usually tapering
terminal – at the apex, tip or distal end; ant.: *basal*
terrestrial – growing on dry land; ant.: *aquatic*
terricolous – growing on or preferring a soil substrate
tessellated – resembling a mosaic of square tiles, applied to the basal membrane of the peristome of *Syntrichia*
tetrad – a set or group of four, referring to a group of four developing spores
tetragonal – four-angled
tetrahedral apical cell – the four-faced cell at the tip of a bryophyte stem or thallus dividing on its three upper faces to form tissue
thallose – composed of a flat plate of tissue
theca – the spore-bearing portion of a moss capsule; syn.: *urn*; see also *apophysis* or *neck*
tier – transverse or horizontal row
tiered – arranged in transverse or horizontal rows; see also *seriate*
tomentose – densely radiculose, woolly; covered with a tomentum
tomentum – a felt-like covering of abundant long rhizoids on the stem or rarely leaves
tortuose – bent or twisted in different directions
torus – the thickening of the closing membrane in bordered pits
trabecula (*pl.* **trabeculae**) – projecting cross-bars formed from horizontal walls on either face of the teeth of an arthrodontous peristome
trabeculate – cross-barred; syn.: *articulate*
translucent – allowing some light through, but not clear; ant.: *opaque*
transparent – clear and glass-like, allowing all light through; syn.: *diaphanous*; ant.: *opaque*
transverse – across, at right angles to the long axis of a plant or organ
tri- – prefix meaning 'three'; see also *bi-* and *uni-*
triangular – shaped like a triangle, with straight or somewhat concave sides
trifarious – arranged in three ranks; syn.: *tristichous*
triseriate – arranged in three rows
tristichous – arranged in three ranks; syn.: *trifarious*
tristratose – made up of three layers
truncate – abruptly cut off or squarred off at the apex
tuber – gemma borne on rhizoids; syn.: *rhizoidal gemma*
tubular – cylindrical and apparently hollow
tubulose – inrolled and overlapping to form a tube, usually referring to leaves
tuft – a growth-form with stems erect but radiating at the edges
tumid – inflated, swollen; see also *turgid*
turbinate – shaped like an inverted cone or pear
turf – a growth-form in which the shoots are short or tall, crowded, erect and parallel, generally arising from a vertically aligned matrix of decaying or moribund shoots
turgid – swollen but not with air; see also *tumid*
type – see *nomenclatural type*

U

ubiquitous – occurring everywhere
umbonate – convex with an abrupt, rounded boss in the centre
uncinate – shaped like a hook; syn.: *hamate*
undulate – somewhat wavy in one direction; see also *rugose*
unequal – dissimilar; applied to objects or structures of different sizes or shapes
uni- – prefix meaning 'one'
unipapillose – bearing a single papilla per cell
uniseriate – arranged in one row; applied to hair-like structures comprised of a single row of cells
unistratose – one-layered; comprising a single cell layer
upper – away from the base or point of attachment; towards the apex of a leaf or stem; syn.: *distal*
urceolate – urn-shape; applied to capsules narrowed somewhat below the wide mouth and again strongly narrowed at the base
urn – the spore-bearing portion of a capsule, as opposed to the sterile apophysis; syn.: *theca*; see also *apophysis* or *neck*

V

vaginant – sheathing or wrapping round
vaginate – sheathed
vaginula – the sheath or ring enveloping the base of a seta remaining after the separation of the calyptra and derived from the venter of the archegonium
validly published name – a name effectively published and in accordance with rules of the Code
valve – the segment into which the capsule in *Andreaea* naturally separates at maturity
venter – the swollen basal portion of an archegonium, containing the egg
ventral – the upper surface of a leaf; syn.: *adaxial*; ant.: *dorsal*
ventricose – bulging or swollen like a stomach on one side toward the base, resembling a belly; ant.: *gibbous*
vermicular – shaped like a worm, long, narrow and somewhat curved, commonly with rounded ends; usually applied to cells
verruca (*pl.* **verrucae**) – a small elevation or swelling on a surface
verrucose – coarsely roughened or warty; usually applied to the surface ornamentation of a mature spore
verruculose – delicately or irregularly roughened or warty; usually applied to the surface ornamentation of a mature spore
verticillate – arranged in a circle or ring; syn.: *whorled*
vesiculose – inflated like a bladder
vestigial – reduced to only a non-functioning trace or remnant; syn.: *rudimentary*
vide – a Latin word meaning 'see', for referring readers to some other source of information
voucher – a specimen preserved as a sample of plant material studied or illustrated for a publication

W

wart – a hard or firm protuberance or excrescence; see also *papilla* and *verruca*

warty – covered with warts or verrucae

whorled – arranged in a circle or ring; syn.: *verticillate*

wide-spreading – held at ± right angles to the stem; see also *patent*, *erect-patent* and *erect-spreading*

wing – a thin, flat membranous expansion or appendage

wrinkled – creased; syn.: *rugose*

X

x – basic or ancestral gametophyte chromosome number of a taxon; see also *n*

xer- – prefix referring to a dry existence

xeric – very dry, with reference to habitat; ant.: *hydric*

xerocastique – capsules opening under dry conditions; ant.: *hygrocastique*

xeromorphic – with morphological structures adapted for surviving in very dry habitats; ant.: *hydromorphic* and *hygromorphic*

xerophile – a plant which grows in a dry situation; ant.: *hydrophile* and *hygrophile*

xerophilous – preferring a dry habitat; ant.: *hydrophilous* and *hygrophilous*

xerophyte – a plant which can subsist with a small amount of moisture; ant.: *hydrophyte* and *hygrophyte*

Z

zig-zag – having short bends or angles from side to side; a median line on diplolepidous exostome teeth

References

A

Abolin', A. A. 1985: [*Polytrichum strictum* (Polytrichaceae) – an original species or a modificant *P. juniperinum*?]. *Botanicheskiy Zhurnal*, **70:** 1503–1511 (in Russian).

Abramov, I. I. & L. I. Savicz-Lyubitskaya. 1963: [Division Bryopsida. Bryophytes]. *In*: Yu. A. Orlov (ed.), [*Principles of palaeobotany. Guide for palaeontologists and geologists of the U.S.S.R.*]. Moskva, Izdatel'stvo Akademiy Nauk SSSR: pp. 344–414 (in Russian).

Abramova, A. L. 1956: [A monographic survey of the family Meesiaceae in the USSR]. *Trudy Botanicheskogo Instituta im. V. L. Komarova Akademiy Nauk SSSR, Seria 2 (Sporovye Rastenya)*, **10:** 393–489 (in Russian).

——— & I. I. Abramov. 1983: [Conspectus of the moss flora of the Mongolian People's Republic]. *Biologicheskye Resursy i Prirodnye Uslovya Mongol'skoy Narodnoy Respubliki*, **17:** 1–221 (in Russian).

———, K. I. Ladyzhenskaya & L. I. Savicz-Lyubitskaya. 1954: [Andreaeales nec non Bryales (Tetraphidales, Polytrichales, Buxbaumiales, Schistostegales)]. *In*: V. P. Savicz (ed.), *Flora cryptogamarum URSS*. **3**. *Musci frondosi (2)*. Moskva and Leningrad, typis et impensis Academiae Scientiarum URSS: 331 pp. (in Russian).

———, L. I. Savicz-Lyubitskaya & Z. N. Smirnova. 1961: [*Handbook of the mosses of the Arctic USSR*]. Moskva and Leningrad, Izdatel'stvo Akademiy Nauk SSSR: 714 pp. (in Russian).

Adam, P. 1976: The occurrence of bryophytes on British saltmarshes. *Journal of Bryology*, **9:** 265–274.

Affolter, J. 1980: The 'Antarctic Flora': researches of Charles Darwin and Joseph Hooker. *Contributions from the University of Michigan Herbarium*, **14:** 1–9.

Afonina, O. M. 1988: [*Leafy mosses of the Chukotka Peninsula. Part II. Fam. Bryaceae – Hypnaceae*]. Magadan, Akademia Nauk SSSR, Dal'nevostochnoe Otdelenye, Institut Biologicheskikh Problem Severa: 47 pp. (in Russian).

———. 2004: [*Conspectus of the moss flora of Chukotka*]. Sankt Petersburg, Botanicheskiy Institut im. Komarova RAN, Rossiyskaya Akademia Nauk: 259 pp. (in Russian).

——— & I. V. Czernyadjeva. 1995: Mosses of the Russian Arctic: check-list and bibliography. *Arctoa, a Journal of Bryology*, **5:** 99–142.

Agnew, S. & M. Vondraček. 1975: A moss flora of Iraq. *Feddes Repertorium*, **86:** 341–489.

Akiyama, H. 1990: Morphology and taxonomic significance of dormant branch primordia, dormant buds, and vegetative reproductive organs in the suborder Leucodontineae and Neckerineae (Musci, Isobryales). *The Bryologist*, **93:** 395–408.

———. 1996: [Taxonomical and ecological notes on Asian bryophytes, 6. *Pohlia drummondii* var. *drummondii*, new to Japan]. *Acta Phytotaxonomica et Geobotanica*, **47:** 125–126 (in Japanese).

——— & N. Nishimura. 1993: The *Climacium*-type branch development in the Bartramiaceae and its taxonomic significance. *The Bryologist*, **96:** 185–191.

Allen, B. 1987*a*: A revision of the Dicnemonaceae. *Journal of the Hattori Botanical Laboratory*, **62:** 1–100.

———. 1987*b*: A revision of the genus *Mesotus* (Musci: Dicranaceae). *Journal of Bryology*, **14:** 441–452.

———. 1988: A contribution to the moss flora of Peru, including a new species of *Dicranum*. *The Bryologist*, **91:** 91–94.

———. 1994*a*: Moss Flora of Central America. Part 1. Sphagnaceae – Calymperaceae. *Monographs in Systematic Botany from the Missouri Botanical Garden*, **49:** 1–242.

———. 1994*b*: The genus *Blindia* (Musci: Seligeriaceae) in Maine. *Evansia*, **11:** 88–92.

———. 1998: The genus *Timmia* (Musci: Timmiaceae) in Maine. *Evansia*, **15:** 115–121.

———. 1999: A revision of the moss genus *Lepyrodon* (Leucodontales, Lepyrodontaceae). *In*: J. Enroth, J. Hyvönen & S. Piippo (eds), *From Finland to Japan to New Guinea to China – that is Timo Koponen, bryologist*. *Bryobrothera*, **5:** 23–48.

———. 2000: The genus *Dicranoweisia* (Musci: Dicranaceae) in Maine. *Evansia*, **17:** 26–30.

———. 2001: The genus *Seligeria* (Musci: Seligeriaceae) in Maine. *Evansia*, **18:** 137–142.

———. 2002: Moss Flora of Central America. Part 2. Encalyptaceae – Orthotrichaceae. *Monographs in Systematic Botany from the Missouri Botanical Garden*, **90:** 1–699.

———. 2005 ['2006']: Maine mosses. Sphagnaceae – Timmiaceae. *Memoirs of the New York Botanical Garden*, **93:** i–xi + 1–419.

Allison, J. S. & R. I. Lewis Smith. 1973: The vegetation of Elephant Island, South Shetland Islands. *British Antarctic Survey Bulletin*, **33–34:** 185–212.

Allison, K. W. 1971: Mosses discovered in New Zealand since the publication of Sainsbury's 'A Handbook of New Zealand Mosses' in 1955. *New Zealand Journal of Botany*, **9:** 672–673.

Amann, J. 1893: Etudes sur le genre *Bryum* (2e article). *Revue Bryologique*, **20:** 39–46.

——— (with collaboration of C. Meylan & P. Culmann). 1918: *Flore des mousses de la Suisse. Deuxième partie. Bryogéographie de la Suisse.* Lausanne, Imprimeries Réunies S. A.: 414 pp. + 12 pls.

Anderson, L. E. & H. Crum. 1959 ['1958']: Cytotaxonomic studies on mosses of the Canadian Rocky Mountains. *National Museum of Canada Bulletin*, **160** *Contributions to Botany*: 1–89.

———, ——— & W. R. Buck. 1990: List of the mosses of North America north of Mexico. *The Bryologist*, **93:** 448–499.

Ando, H. 1972*a*: Distribution and speciation in the genus *Hypnum* in the circum-Pacific region. *Journal of the Hattori Botanical Laboratory*, **35:** 68–98.

———. 1972*b*: Studies on the genus *Hypnum* Hedw. (I). *Journal of Science of the Hiroshima University Series B, Div. 2 (Botany)*, **14:** 53–73.

———. 1973: Studies on the genus *Hypnum* Hedw. (II). *Journal of Science of the Hiroshima University Series B, Div. 2 (Botany)*, **14:** 165–207.

———. 1976: Studies on the genus *Hypnum* Hedw. (III). *Journal of Science of the Hiroshima University Series B, Div. 2 (Botany)*, **16:** 1–46.

———. 1979: Ecology of terrestrial plants in the Antarctic with particular reference to bryophytes. *Memoirs of National Institute of Polar Research Special Issue*, **11:** 81–103.

———. 1986: Studies on the genus *Hypnum* Hedw. (IV). *Hikobia*, **9:** 467–484.

———. 1987: Studies on the genus *Hypnum* Hedw. (V). *Hikobia*, **10:** 43–54.

———. 1989: Studies on the genus *Hypnum* Hedw. (VI). *Hikobia*, **10:** 269–291.

———. 1990: Studies on the genus *Hypnum* Hedw. (VII). *Hikobia*, **10:** 409–417.

———. 1992: Studies on the genus *Hypnum* Hedw. (VIII). *Hikobia*, **11:** 111–123.

———. 1993: Studies on the genus *Hypnum* Hedw. (IX). *Hikobia*, **11:** 265–275.

———. 1996: Studies on the genus *Hypnum* Hedw. (X). *Hikobia*, **12:** 9–17.

——— & M. Higuchi. 1984: *Caribaeohypnum*, a new genus for a northern Latin American moss, *Hypnum polypterum* (Mitt.) Broth. *Cryptogamie, Bryologie – Lichénologie*, **5:** 7–14.

——— & C. M. Matteri. 1982: Patagonian bryophytes 7. *Hypnum revolutum* (Mitt.) Lindb. new to South America, with notes on the *Hypnum* species known from the Transecta zone, S Argentina. *Lindbergia*, **8:** 60–64.

Andrews, A. L. 1922: The status of *Gyroweisia* in North America. *The Bryologist*, **25:** 97–100.

———. 1935: Bryaceae. *In*: A. J. Grout (ed.), *Moss flora of North America north of Mexico*. **2**(3). Newfane, Vermont, privately published: pp. 184–210 + pls. 65A, 71–83.

———. 1940: Bryaceae (II). *In*: A. J. Grout (ed.), *Moss flora of North America north of Mexico*. **2**(4). Newfane, Vermont, privately published: pp. 211–242 + pls. 65A, 84–98.

———. 1954: Taxonomic notes. XII. The families Rhytidiaceae and Hylocomiaceae. *The Bryologist*, **57:** 251–261.

——— & F. J. Hermann. 1959: A natural hybrid in the Ditrichaceae. *The Bryologist*, **62:** 119–122.

Anonymous. 1821: New Shetland. *Literary Gazette and Journal of Belles Lettres, Arts, Science, etc. (London)*, **5:** 712–713 [10 Nov].

———. 1946: Edward Bransfield's Antarctic voyage. 1819–20, and the discovery of the Antarctic continent. *Polar Record*, **4**(32): 385–393.

———. 1952: Richard Sherratt's chart of the South Shetland Islands, 1821. *Polar Record*, **6**(43): 362–365 + 1 map.

———. 1961: The Antarctic Treaty [Washington, December 1, 1959]. *Treaty Series*. **97**. Command 1535. London, Her Majesty's Stationery Office: 7 pp.

———. 1993: *Protocol on environmental protection to the Antarctic Treaty*. London, Her Majesty's Stationery Office: 38 pp.

———. 1997: *List of protected areas in Antarctica*. London, Foreign and Commonwealth Office: 33 pp.

Antarctic Protected Areas Information Archive. 2006 (with regular updates): http://cep.ats.aq/cep/apa/index.html

Arikawa, T. & M. Higuchi. 1999: Phylogenetic analysis of the Plagiotheciaceae (Musci) and its relatives based on *rbc*L gene sequences. *Cryptogamie, Bryologie*, **20:** 231–245.

——— & ———. 2003: Preliminary phylogenetic analysis of *Pylaisia* (Hypnaceae, Musci) and its relatives based on *rbc*L gene sequences. *Journal of the Hattori Botanical Laboratory*, **94:** 87–106.

Arnott, G. A. W. 1825: *Disposition méthodique des espèces des mousses*. Paris, Imprimerie de J. Tastu: iv + 72 pp.

Arts, Th. 1986: The occurrence of tubers in *Campylopus pyriformis*. *Lindbergia*, **12:** 125–128.

———. 1989 ['1990']: The occurrence of rhizoidal tubers in the genus *Campylopus*. *Lindbergia*, **15:** 60–64.

———. 1990: A reappraisal of the propaguliferous species of *Pohlia annotina*-complex in Belgium and the Grand Duchy de Luxembourg. *In*: P. Compère, A. R. Perry, E. Robbrecht & H. Stieperaere (eds), *Bryology and lichenology in Belgium. A symposium celebrating the tenth anniversary of the 'Vlaamse Werkgroep Bryologie' Meise 26 November 1988. Mémoires de la Société Royale de Botanique de Belgie*, **12:** 20–37.

———. 1993: Rhizoidal tubers in the genus *Campylopus* and their systematic relevance. *Lindbergia*, **18:** 31–40.

———. 1994: Rhizoidal tubers and protonemal gemmae in European *Ditrichum* species. *Journal of Bryology*, **18:** 43–61.

———. 2001: The moss genus *Leptobryum* and the identity of *Pohlia integra*. *Journal of Bryology*, **23:** 325–330.

——— & J.-P. Frahm. 1990: *Campylopus pyriformis* new to North America. *The Bryologist*, **93:** 290–294.

Arzeni, C. B. 1965: Bryophytes of the cedar groves of Lebanon. *The Bryologist*, **68:** 109–113.

Ayukawa, E., S. Imura & H. Kanda. 2001: Bryophyte propagule bank in the Yukidori Valley, Langhovde, Antarctica. *Antarctic Record*, **45:** 320–328.

Å

Ångström, J. 1872: Förteckning och beskrifning öfver mossor, samlade af Professor N. J. Andersson under Fregatten Eugenies vedrldomsegling åren 1851–53. *Öfversigt af Förhandlingar: Kongliga Svenska Vetenskaps-Akademien*, **29:** 3–29.

B

Bachurina, G. F. & V. M. Mel'nichuk. 1987–1988: [*The moss flora of the Ukrainian SSR – Andreaeidae, Bryidae*]. Kiev, Naukova Dumka: 180 pp. (part 1, 1987); 180 pp. (part 2, 1988) (in Ukrainian).

Baker, J. H. 1972: The rate of production and decomposition of *Chorisodontium aciphyllum* (Hook. f. & Wils.) Broth. *British Antarctic Survey Bulletin*, **27:** 123–129.

Bardunov, L. V. 1961: [Mosses of the shores and mountains of northern Lake Baykal]. *Trudy Vostochno-Sibirskogo Filyala Academiy Nauk SSSR Seriya Biologicheskaya*, **42:** 1–120 (in Russian).

———. 1965: [*Mosses of the Eastern Sayan*]. Moskva and Leningrad, Izdatel'stvo Nauka, Leningradskoye Otdelenye: 161 pp. (in Russian).

———. 1969: [*The handbook of the mosses of Central Siberia*]. Leningrad, Izdatel'stvo Nauka, Leningradskoye Otdelenye: 330 pp. (in Russian).

———. 1974: [*Mosses of the Altai and Sayan*]. Novosibirsk, Izdatel'stvo Nauka, Sibirskoe Otdelenye: 168 pp. (in Russian)

——— & V. Ya. Cherdantseva. 1982: [*Mosses of the Southern Primorye*]. Novosibirsk, Izdatel'stvo Nauka, Sibirskoe Otdelenye: 207 pp. (in Russian)

Bargagli, R., P. A. Broady & D. W. H. Walton. 1996: Preliminary investigation on the thermal biosystem of Mount Rittmann fumaroles (northern Victoria Land, Antarctica). *Antarctic Science*, **8:** 121–126.

Barnhart, J. H. 1944: Dates of the 'Bryologia europaea'. *The Bryologist*, **47:** 97–108.

Bartlett, J. K. 1984: New or interesting records of mosses from New Zealand. *National Museum of New Zealand Records*, **2:** 181–189.

———. 1985: New records of some New Zealand mosses. *New Zealand Journal of Botany*, **23:** 171–177.

——— & J.-P. Frahm. 1983: Notes on *Campylopus* and *Chorisodontium* from New Zealand. *Journal of Bryology*, **12:** 365–382.

——— & D. H. Vitt. 1986: A survey of species in the genus *Blindia* (Bryopsida, Seligeriaceae). *New Zealand Journal of Botany*, **24:** 203–246.

Bartram, E. B. 1928: Mosses of Jamaica. *Jamaica Naturalist*, **1:** 5–6, 15–20.

———. 1929: Mosses of the Philippines. *The Philippine Journal of Science*, **68:** 1–437 + pls. 1–29.

———. 1931: Mosses of New Mexico collected by Bro. Arsène. *Annales de Cryptogamie Exotique*, **4:** 153–160.

———. 1933*a*: Manual of Hawaiian mosses. *Bernice P. Bishop Museum Bulletin*, **101:** 1–275.

———. 1933*b*: Mosses of the River Napo, Ecuador. *Revue Bryologique et Lichénologique, Nouvelle Série*, **6:** 9–18.

———. 1938: Mosses. *In*: *The Second Byrd Antarctic Expedition – Botany. Annals of the Missouri Botanical Garden*, **25:** 719–724.

———. 1946: New mosses from Tierra del Fuego. *Farlowia*, **2:** 309–319.

———. 1949: Mosses of Guatemala. *Fieldiana: Botany*, **25:** 1–442.

———. 1952*a*: Mosses of Chile and Argentina mainly collected by R. Santenson. *Svensk Botanisk Tidskrift*, **46:** 242–253.

———. 1952*b*: High altitude mosses from Mauna Kea, Island of Hawaii. *Occasional Papers of Bernice P. Bishop Museum*, **20:** 297–300.

———. 1957: Mosses from the United States Antarctic Service Expedition, 1940–41. *The Bryologist*, **60:** 139–143.

———. 1959: Contribution to the mosses of Gough Island. *Revue Bryologique et Lichénologique, Nouvelle Série*, **28:** 208–210.

Bauer, E. 1908: *Musci europaei exsiccati. Schedae und Bemerkungen zur sechsten Serie*. Prag, Carl Bellmann: 14 pp.

Bauhin, C. 1623: *Pinax theatri botanici sive index in Theophrasti Dioscoridis Plinii et botanicorum qui à seculo scripserunt opera plantarum circiter sex millium ab ipsis exhibitarum nomina cum earundum synonymiis & differentiis methodicè secundum earum & genera & species proponens. Opus xl. Hactenus non editum summopere expetitum & ad auctoris intelligendus plurimùm facicus*. Sumptibus & typis Ludovici Regis: xxiv + 522 + [1–22, index] pp.

Bayfield, N. B. 1973: Notes on water relations of *Polytrichum commune* Hedw. *Journal of Bryology*, **7:** 607–617.

Bednarek-Ochyra, H. 1993*a*: Three species of mosses new to Svalbard. *Fragmenta Floristica et Geobotanica*, **38:** 322–323.

———. 1993*b*: The identity of two neglected species of *Racomitrium* (Musci, Grimmiaceae) from Patagonia. *In*: R. Ochyra & L. Stuchlik (eds), *Botanostephane kornasiana. Botanical contributions presented to Jan Kornaœ in celebration of his 70th birthday. Fragmenta Floristica et Geobotanica Supplementum* **2:** 83–90.

———. 1995*a*: Rodzaj *Racomitrium* (Musci, Grimmiaceae) w Polsce: taksonomia, ekologia i fitogeografia [The genus *Racomitrium* (Musci, Grimmiaceae) in Poland: taxonomy, ecology and phytogeography]. *Fragmenta Floristica et Geobotanica Series Polonica*, **2:** 1–307 (in Polish with extensive English summary).

———. 1995*b*: A note on *Racomitrium tortipilum* (Musci, Grimmiaceae) from Brazil. *Fragmenta Floristica et Geobotanica*, **40:** 177–180.

———. 1996: *Grimmia hartmanii* (Musci, Grimmiaceae), an addition to the bryoflora of the Crimea. *Fragmenta Floristica et Geobotanica*, **41:** 482–484.

———. 1999: *Racomitrium aduncoides* (Musci, Grimmiaceae), a remarkable new species from eastern North America. *Fragmenta Floristica et Geobotanica*, **44:** 277–285.

———. 2000: *Racomitrium ryszardii* (Musci, Grimmiaceae), a new hydrophilous species from the Pacific Northwest with comments on *Racomitrium aquaticum* in North America. *Cryptogamie, Bryologie*, **21:** 275–284.

———. 2001: *Racomitrium subsecundum* (Hook. & Grev. ex Harv. in Hook.) Wilson in Mitt. & Wilson. Ecuador and Bolivia. *In*: T. Blockeel (ed.), *New national and regional bryophyte records, 4. Journal of Bryology*, **23:** 151.

———. 2004*a*: *Codriophorus corrugatus* (Bryopsida, Grimmiaceae), a new species from East Asia and Southern Alaska. *The Bryologist*, **107:** 377–384.

———. 2004*b*: On the occurrence of *Racomitrium aquaticum* (Schrad.) Brid. (Grimmiaceae, Musci) in Asiatic Russia. *Cryptogamie, Bryologie*, **25:** 83–89.

———. 2006: *A taxonomic monograph of the moss genus* Codriophorus *P. Beauv. (Grimmiaceae)*. Kraków, W. Szafer Institute of Botany, Polish Academy of Sciences: 276 pp.

——— & **R. Ochyra. 1992:** A major range extension of *Racomitrium laevigatum*. *Fragmenta Floristica et Geobotanica*, **37:** 7–12.

——— & ———. **1994:** *Racomitrium lamprocarpum* (Musci, Grimmiaceae) in southern South America. *Fragmenta Floristica et Geobotanica*, **39:** 361–367

——— & ———. **1996:** *Racomitrium curiosissimum* (Musci, Grimmiaceae) an exquisite new species from New Zealand. *Fragmenta Floristica et Geobotanica*, **41:** 973–984.

——— & ———. **1998:** *Racomitrium lamprocarpum* (Müll. Hal.) Jaeg. – an addition to the moss flora of Îles Kerguelen and the Subantarctic. *Journal of Bryology*, **20:** 525–528.

——— & ———. **2000:** *Racomitrium norrisii* sp. nov. (Grimmiaceae, Bryopsida) – endemic of western North America. *Annales Botanici Fennici*, **37:** 235–241.

——— & ———. **2003:** *Racomitrium patagonicum*, a new moss species from southern South America. *Journal of Bryology*, **25:** 181–187.

——— & ———. **2007:** *Bucklandiella lamprocarpa* (Müll.Hal.) Bednarek-Ochyra & Ochyra. Macquarie Island. *In*: T. Blockeel (ed.), *New national and regional bryophyte records, 15. Journal of Bryology*, **29:** 139–140.

——— & ———. **2008*a*:** *Bucklandiella pachydictyon* (Cardot) Bednarek-Ochyra & Ochyra. Prince Edward Islands and Îles Crozet. *In*: T. Blockeel (ed.), *New national and regional bryophyte records, 18. Journal of Bryology*, **30**: in press.

——— & ———. **2008*b*:** *Bucklandiella didyma* (Mont.) Bednarek-Ochyra & Ochyra. Australia and New Zealand. *In*: T. Blockeel (ed.), *New national and regional bryophyte records, 18. Journal of Bryology*, **30**: in press.

———, **R. I. Lewis Smith & R. Ochyra. 1999:** The genus *Plagiothecium* (Plagiotheciaceae, Bryopsida) in Antarctica. *In*: J. Enroth, J. Hyvönen & S. Piippo (eds), *From Finland to Japan to New Guinea to China – that is Timo Koponen, bryologist. Bryobrothera*, **5:** 55–64.

———, **C. M. Matteri & R. Ochyra. 1996:** A major range extension of *Racomitrium lamprocarpum* (Musci, Grimmiaceae) in South America. *Fragmenta Floristica et Geobotanica*, **41:** 995–1000.

———, **R. Ochyra & W. R. Buck. 1999:** The genus *Racomitrium* (Grimmiaceae) in Brazil, with the first report of *R. subsecundum* in South America. *Brittonia*, **51:** 93–105.

———, ——— **& P. Szmajda. 1990:** M. 665. *Racomitrium sudeticum* (Funck) Bruch & Schimp. *In*: R. Ochyra & P. Szmajda (eds), *Atlas of the geographical distribution of spore plants in Poland. Series V. Mosses (Musci)*. **6**. Kraków, W. Szafer Institute of Botany and Poznań, Adam Mickiewicz University: pp. 41–46 + 1 map.

———, **J. Váňa, R. Ochyra & R. I. Lewis Smith. 2000:** *The liverwort flora of Antarctica*. Cracow, Polish Academy of Sciences, Institute of Botany: xvi + 236 + [2] pp.

Beever, J., K. W. Allison & J. Child. 1992: *The mosses of New Zealand*. Ed. 2. Dunedin, University of Otago Press: x + 214 pp.

———, **B. Malcolm & N. Malcolm. 2002:** *The moss genus* Fissidens *in New Zealand – an illustrated key*. Nelson, Micro-Optics: iv + 91 pp.

Bell, B. G. 1973*a*: A synoptic flora of South Georgian mosses: II. *Chorisodontium*, *Dicranoloma*, *Dicranum*, *Platyneurum* and *Conostomum*. *British Antarctic Survey Bulletin*, **37:** 33–52.

———. **1973*b*:** Notes on Antarctic bryophytes: I. New records from the Antarctic botanical zone. *British Antarctic Survey Bulletin*, **36:** 131–132.

———. **1973*c*:** Notes on Antarctic bryophytes: II. New records of *Racomitria* from the Antarctic botanical zone. *British Antarctic Survey Bulletin*, **37:** 91–94

———. **1974:** A synoptic flora of South Georgian mosses: V. *Willia* and *Racomitrium*. *British Antarctic Survey Bulletin*, **38**: 73–101.

———. **1976:** Notes on Antarctic bryophytes: VI. The genus *Dicranoweisia* on Signy Island, South Orkney Islands. *British Antarctic Survey Bulletin*, **44:** 97–100.

———. **1977:** Notes on Antarctic bryophytes: IX. A previously unreported botanical collection made on South Georgia during the German International Polar-Year Expedition, 1882–83. *British Antarctic Survey Bulletin*, **46:** 137–139.

———. **1983:** The type specimen of the moss *Dicranum oleodictyon* Dix. *British Antarctic Survey Bulletin*, **61:** 74.

———. **1984:** A synoptic flora of South Georgian mosses: *Grimmia* and *Schistidium*. *British Antarctic Survey Bulletin*, **63:** 71–109.

——— **& H. H. Blom. 1986:** Bryophytes of Bouvetøya. *Norsk Polarinstitutt Skrifter*, **185:** 11–22.

——— **& S. W. Greene. 1975:** Additions and corrections to the distribution records in A synoptic Flora of South Georgian Mosses, Parts I–V. *British Antarctic Survey Bulletin*, **40:** 65–68.

Bell, G. H. & D. G. Catcheside. 2006: Aulacomniaceae. *In*: P. M. McCarthy (ed.), *Flora of Australia*. **51**. *Mosses (1)*. Canberra, Australian Biological Resources Study: pp. 245–247.

Bell, N. E., D. Quandt, T. J. O'Brien & A. E. Newton. 2007: Taxonomy and phylogeny in the earliest diverging pleurocarps: square holes and bifurcating pegs. *The Bryologist*, **110:** 533–560.

Bellingshausen, Th. Th. 1831: [*The twofold expedition in the Southern Ocean and the voyage round the world, during 1819, 20 and 21, accomplished by the sloops* Vostok *and* Mirnyy *under the command of Captain Bellingshausen, the Commander of the sloop* Vostok. *The sloop* Mirnyy *was commanded by Lieutenant Lazarev*]. **1–2**. St. Petersburg, v Tipografii Ivana Glazunova: 398 pp. (Vol. 1) and 327 pp. (Vol. 2) (in Russian).

Bergeron, T. 1944: On some meteorological conditions for the dissemination of spores, pollen etc., and a supposed wind transport of *Aloina* spores from the region of lower Yenisey to southwestern Finland in July 1936. *Svensk Botanisk Tidskrift*, **38:** 269–292.

Berggren, S. 1875: Musci et Hepaticae spetsbergenses. Bericht über die Untersuchung der Moosflora Spitzbergens und Beeren-Eilands während der Schwedischen Expedition 1864 und 1868, und Verzeichnis der dort gesammelten Arten. *Kongliga Svenska Vetenskapsakademiens Handlingar*, **13(7):** 1–103.

Bergstrom, D. M. & P. Selkirk. 1997: Distribution of bryophytes on Subantarctic Heard Island. *The Bryologist*, **100:** 349–355.

——— **& R. D. Seppelt. 1988 ['1989']:** The moss flora of Heard Island: revised checklist, annotations and phytogeographical considerations. *Lindbergia*, **14:** 184–190.

Berthier, J. 1972: Recherches sur la structure et le dévelopment de l'apex du gamétophyte feuillé des mousses. *Revue Bryologique et Lichénologique, Nouvelle Série*, **38:** 421–551.

Bescherelle, É. 1873: Florule bryologique de la Nouvelle-Calédonie. *Annales des Sciences Naturelles Botanique Série 5*, **18:** 184–245.

———. 1875: Note sur les mousses des îles Saint-Paul et d'Amsterdam. *Comptes Rendus Hebdomadaires des Séances de l'Académie des Sciences*, **81:** 720–725.

———. 1885 ['1886']: Mousses nouvelles de l'Amérique australe. *Bulletin de la Société Botanique de France*, **32** (sess. extraord.): liv–lxix.

———. 1889: Mousses. *In*: *Mission scientifique du Cap Horn 1882–1883*. **5**. Botanique. Paris, Gauthier-Villars et Fils: pp. 253–312 + pls. 1–6.

Best, E. B. 1990: The Bryophyta of Zimbabwe – an annotated check-list. *Kirkia*, **13:** 293–318.

Beyer, L. & M. Bölter (eds). 2002: *Geoecology of Antarctic ice-free coastal landscapes. Ecological Studies*. **154**. Berlin – Heidelberg, Springer-Verlag: xxiii + 427 pp.

Bijlsma, R. J., G. M. Dirkse & H. van Melick. 1985: *Aloina brevirostris* (Hook. & Grev.) Kindb. in Nederland. *Lindbergia*, **11:** 165–166.

Birkenmajer, K., R. Ochyra, I. U. Olsson & L. Stuchlik. 1985: Mid-Holocene radiocarbon-dated peat at Admiralty Bay, King George Island (South Shetland Islands, West Antarctica). *Bulletin of the Polish Academy of Sciences, Earth Sciences*, **33:** 7–13.

Bizot, M. 1967: Quelques mousses africaines et américaines. *Bulletin de la Société Botanique de France*, **114:** 423–427.

——— & G. Kilbertus. 1979: Les *Campylopus* africains subg. *Lucidus* Biz. et Kilb. *Revue Bryologique et Lichénologique, Nouvelle Série*, **45:** 61–95.

———, T. Pócs & A. J. Sharp. 1979: Results of a bryogeographical expedition to East Africa in 1968, II. *Journal of the Hattori Botanical Laboratory*, **45:** 145–165.

Björck, S., N. Malmer, C. Hjort, P. Sandgren, Ó. Ingolfsson, B. Wallén, R. I. Lewis Smith & B. L. Jónsson. 1991: Stratigraphic and paleoclimatic studies of a 5500-year old moss bank on Elephant Island, Antarctica. *Arctic and Alpine Research*, **23:** 361–374.

Blagodatskikh, A. S. 1984: [*Mosses of the Kolyma Highland*]. Magadan, Akademia Nauk SSSR, Dal'nevostochnyy Nauchnyy Tsentr, Institut Biologicheskikh Problem Severa: 45 pp. (in Russian).

Bliss, L. C. 1962: Adaptations of arctic and alpine plants to environmental conditions. *Arctic*, **15:** 117–145.

———. 1979: Vascular plant vegetation of the southern circumpolar region in relation to antarctic, alpine, and arctic vegetation. *Canadian Journal of Botany*, **57:** 2167–2178.

Blockeel, T. L. 1990: The genus *Hennediella* Par.: a note on the affinities of *Tortula brevis* Whitehouse & Newton and *T. stanfordensis* Steere. *Journal of Bryology*, **16:** 187–192.

———. 1992*a*: *Tortella fragilis*. *In*: M. O. Hill, C. D. Preston & A. J. E. Smith (eds), *Atlas of the bryophytes of Britain and Ireland*. **2**. Mosses (except Diplolepideae). Colchester, Harley Books: p. 319.

———. 1992*b*: *Stegonia latifolia*. *In*: M. O. Hill, C. D. Preston & A. J. E. Smith (eds), *Atlas of the bryophytes of Britain and Ireland*. **2**. Mosses (except Diplolepideae). Colchester, Harley Books: p. 240.

———. 1992*c*: *Pterygoneurum ovatum*. *In*: M. O. Hill, C. D. Preston & A. J. E. Smith (eds), *Atlas of the bryophytes of Britain and Ireland*. **2**. Mosses (except Diplolepideae). Colchester, Harley Books: p. 238.

———. 1992*d*: *Aloina brevirostris*. *In*: M. O. Hill, C. D. Preston & A. J. E. Smith (eds), *Atlas of the bryophytes of Britain and Ireland*. **2**. Mosses (except Diplolepideae). Colchester, Harley Books: p. 231.

———. 1994: *Amphidium lapponicum*. *In*: M. O. Hill, C. D. Preston & A. J. E. Smith (eds), *Atlas of the bryophytes of Britain and Ireland*. **3**. Mosses (Diplolepideae). Colchester, Harley Books: p. 176.

——— & A. J. E. Smith. 1998: *Pottiopsis* gen. nov. and notes on other taxa of British and Irish Pottiaceae. *Journal of Bryology*, **20:** 65–68.

Blom, H. H. 1996: A revision of the *Schistidium apocarpum* complex in Norway and Sweden. *Bryophytorum Bibliotheca*, **49:** 1–333.

———. 1998: *Schistidium*. *In*: E. Nyholm, *Illustrated flora of Nordic mosses*. **4**. Copenhagen and Lund, Nordic Bryological Society: pp. 287–330.

Bonner, W. N. & R. I. Lewis Smith (eds). 1985: *Conservation areas in the Antarctic*. Cambridge, Scientific Committee on Antarctic Research, International Council of Scientific Union: viii + 299 pp.

Bonnot, E. J. 1967: Sur la valeur et la signification des paraphylles chez les Bryales. *Bulletin de la Société Botanique de France Mémoires*, **1967:** 236–248.

Bopp, M. & I. Capesius. 1996: New aspects of bryophyte taxonomy provided by a molecular approach. *Botanica Acta*, **109:** 1–5.

——— & ———. 1998: A molecular approach to bryophyte systematics. *In*: J. W. Bates, N. W. Ashton & J. G. Duckett (eds), *Bryology for the twenty-first century*. Leeds, Maney Publishing and the British Bryological Society: pp. 79–88.

Boros, A. 1968: *Bryogeographie und Bryoflora Ungarns*. Budapest, Akadémiai Kiadó: 466 pp.

Bower, F. O. 1935: *Primitive land plants also known as the Archegoniatae*. London, Macmillan: 658 pp.

Bowra, G. T., M. W. Holdgate & P. J. Tilbrook. 1966: Biological investigations in Tottanfjella and central Heimefrontfjella. *British Antarctic Survey Bulletin*, **9:** 63–70.

Braithwaite, R. 1887: *The British moss-flora. Part X. Fam. VIII.—Tortulaceae*. London, published by the author: pp. 245–287 + pls. xxxvii–xlii.

———. 1889: *The British moss-flora. Part XII. Fam. X. Grimmiaceae. II*. London, published by the author: pp. 57–98 + pls. liv–lix.

———. 1890–1893: *The British moss-flora. Parts XIII–XV. Fam. XV. Bryaceae*. London, published by the author: pp. 139–197 + pls. lxvi–lxxv.

Brassard, G. R. 1971: The mosses of northern Ellesmere Island, Arctic Canada. II. Annotated list of the taxa. *The Bryologist*, **74:** 282–311.

———. 1983*a*: Checklist of the mosses of the island of Newfoundland, Canada. *The Bryologist*, **86:** 54–63.

———. 1983*b*: Bryogeography, with special reference to mosses. *In*: G. R. South (ed.), *Biogeography and ecology of the island of Newfoundland*. The Hague, Dr. W. Junk Publishers: pp. 361–384.

——, A. J. Fife & J. Weber. 1979: Mosses from Baffin Island, Arctic Canada. *Lindbergia*, **5:** 99–104.

Braun, A. 1864: *Uebersicht des natürlichen Systems nach der Anordnung desselben. In*: P. Ascherson, *Flora der Provinz Brandenburg, der Altmark und des Herzogsthums Magdeburg. Zum Gebrauche in Schulen und auf Exkursionen.* **1**(2). Berlin, August Hirschwald: pp. 22–146.

Braunmiller, H., J. Poelt & W. Schultze-Motel. 1971: Über die Verbreitung einiger Arten der Laubmoosgattung *Tortella* in Mitteleuropa. *Archiv für Naturschutz und Landschaftsforschung*, **11:** 169–178.

Bremer, B. 1980: A taxonomic revision of *Schistidium* (Grimmiaceae, Bryophyta) 1. *Lindbergia*, **6:** 1–16.

——. **1981 ['1980']:** A taxonomic revision of *Schistidium* (Grimmiaceae, Bryophyta) 2. *Lindbergia*, **6:** 89–117.

——. **1982 ['1981']:** A taxonomic revision of *Schistidium* (Grimmiaceae, Bryophyta) 3. *Lindbergia*, **7:** 73–90.

Bridel, S. E. 1798: *Muscologia recentiorum seu analysis, historia, et descriptio methodica omnium muscorum frondosorum, hucusque cognitorum, ad normam Hedwigii.* Gothae, apud Carolum Guil. Ettingerum and Parisiis, apud Barrois Iuniorim: x + 222 pp. + tab. i–vi.

——. **1801:** Animadversiones in Muscologiae recentiorum tomum secundum. *Journal für die Botanik*, **1800(2):** 268–299.

——. **1803:** *Muscologia recentiorum seu analysis, historia et descriptio methodica omnium muscorum frondosorum hucusque cognitorum ad normam Hedwigii.* **2**(3). Gothae, apud Carolum Guil. Ettingerum and Parisiis, apud Barrois Iuniorim: viii + 9–178 [+ 6, ind.] pp. + tab. i–ii.

——. **1819:** *Methodus nova muscorum ad naturae normam melius instituta et Muscologiae recentiorum accommodata.* Gothae, apud A. Ukertum: xviii + 220 pp.

——. **1826–1827:** *Bryologia universa seu systematica·ad novam methodum dispositio, historia et descriptio omnium muscorum et descriptio omnium muscorum frondosorum hucusque cognitorum cum synonymia ex auctoribus probatissimis.* **1–2**. Lipsiae, sumtibus Joan. Ambros. Barth: [ii] + 856 pp. (Vol. 1, 1826–1827) and xlvi + 848 + [2] pp. + x + iii pls. (Vol. 2, 1827).

Britton, E. G. 1918: Class 1. Musci. Mosses. *In*: N. L. Britton (ed.), *Flora of Bermuda (illustrated).* New York, Charles Scribner's Sons: pp. 430–448.

—— **& J. T. Emerson. 1913:** Andreaeaceae. *In*: N. L. Britton & L. M. Underwood (eds), *North American Flora.* **15**(1). New York, New York Botanical Garden: pp. 35–39.

Broady, P. 1984: Taxonomic and ecological investigations of algae on steam-warmed soil on Mt Erebus, Ross Island, Antarctica. *Phycologia*, **23:** 257–271.

——, **D. R. Given, L. Greenfield & K. Thomson. 1987:** The biota and environment of fumaroles on Mt Melbourne, Northern Victoria Land. *Polar Biology*, **7:** 97–113.

Brotherus, V. F. 1893: Enumeratio muscorum Caucasi. *Acta Societatis Scientiarum Fennicae*, **19(12):** i–v + 1–170.

——. **1898:** Contributions to the bryological flora of northwestern Himalaya. *Acta Societatis Scientiarum Fennicae*, **24:** 1–46.

——. **1901*a*:** Andreaeaceae. *In*: A. Engler & K. Prantl (eds), *Die Natürlichen Pflanzenfamilien nebst ihren Gattungen und wichtigeren Arten inbesondere den Nutzpflanzen*, **1**(3[2]) – *Musci (Laubmoose).* Leipzig, Wilhelm Engelmann: pp. 265–268.

——. **1901*b*–1909:** III. Unterklasse Bryales. II. Spezieller Teil. *In*: A. Engler & K. Prantl (eds), *Die Natürlichen Pflanzenfamilien nebst ihren Gattungen und wichtigeren Arten inbesondere den Nutzpflanzen.* **1**(3) – *Musci (Laubmoose).* Leipzig, Wilhelm Engelmann: pp. 277–1246.

——. **1905:** *Pleurorthotrichum*, eine neue Laubmoosgattung aus Chile. *Oefversigt af Förhandlingar, Finska Vetenskaps-Societeten*, **47(15):** 1–4 + 1 Tafel.

——. **1906*a*:** Die Laubmoose der Deutschen Südpolar-Expedition 1901–1903. *In*: E. von Drygalski (ed.), *Deutsche Südpolar-Expedition 1901–1903 im Auftrage des Reichsamtes des Innern.* **8** (Botanik). Berlin, Georg Reimer: pp. 83–96.

——. **1906*b*:** Lieutnant Olufsen's second Pamir-Expedition. *Botanisk Tidsskrift*, **27:** 203–208.

——. **1923:** Die Laubmoose Fennoskandia. *Flora Fennica*, **1:** i–xiii + 1–635.

——. **1924*a*:** Musci (Laubmoose). Andreaeaceae. *In*: A. Engler (ed.), *Die Natürlichen Pflanzenfamilien nebst ihren Gattungen und wichtigeren Arten inbesondere den Nutzpflanzen.* **10**. Leipzig, Wilhelm Engelmann: pp. 129–131.

——. **1924*b*:** Musci (Laubmoose). Unterklasse Bryales. II. Spezieller Teil. *In*: A. Engler (ed.), *Die Natürlichen Pflanzenfamilien nebst ihren Gattungen und wichtigeren Arten inbesondere den Nutzpflanzen.* **10**. Leipzig, Wilhelm Engelmann: pp. 143–478.

——. **1924*c*:** Musci Insulae-Paschalis. *In*: C. Skottsberg (ed.), *The natural history of Juan Fernandez and Easter Island.* **2** (Botany). Uppsala, Almquist & Wiksells Boktryckerei AB: pp. 241–246 + pls. 21–23.

——. **1925:** Musci (Laubmoose). *In*: A. Engler (ed.), *Die Natürlichen Pflanzenfamilien nebst ihren Gattungen und wichtigeren Arten inbesondere den Nutzpflanzen.* **11**. Leipzig, Wilhelm Engelmann: iv + 542 pp.

——. **1927:** Hawaiian mosses. *Bernice P. Bishop Museum Bulletin*, **40:** 1–37 + pls. i–viii.

——. **1929:** Musci novi asiatici. *Revue Bryologique, Nouvelle Série*, **2**: 1–16 + pl. i.

Brown, R. 1893 ['1891']: Notes on a proposed new genus of New Zealand mosses; together with a description of three new species. *Transactions and Proceedings of the New Zealand Institute*, **25:** 285–287 + pls. xxxi(5–8)–xxxiii.

——. **1894:** Notes on New Zealand mosses: genus *Pottia.* *Transactions and Proceedings of the New Zealand Institute*, **26:** 288–296 + pls. xxxi–xxxiv.

Bruch, P. & W. P. Schimper. 1843: Corrections and remarks upon Drummond's first (Arctic and Canadian) collection of North American mosses. *London Journal of Botany*, **2:** 663–670.

——, —— **& W. T. Gümbel. 1839:** *Bryum. In*: W. P. Schimper (ed.), *Bryologia europaea seu genera muscorum europaeorum monographicae illustrata.* **2**. Stuttgartiae, sumptibus Librariae E. Schweizerbart: pp. 71–152 + pls. 331–332, 334, 339–348, 351–356, 358–360, 363–376, 378–380, 382, 384 [Fasc. 6–9: 82 pp. + 41 pls.].

——, —— & ——. **1842:** *Barbula. In*: W. P. Schimper (ed.), *Bryologia europaea seu genera muscorum europaeorum monographicae illustrata.* **2**. Stuttgartiae, sumptibus Librariae E. Schweizerbart: pp. 63–108 + pls. 137–145, 147–151, 153–166, 168–169 [Fasc. 13–15: 46 pp. + 29 pls.].

——, —— & ——. **1843:** *Pottia* and *Anacalypta*. *In*: W. P. Schimper (ed.), *Bryologia europaea seu genera muscorum europaeorum monographicae illustrata.* **2**. Stuttgartiae, sumptibus Librariae E. Schweizerbart: pp. 29–41, 45–50 + Tab. 117–128 [Fasc. 18–20: 13 & 6 pp. + 7 & 4 pls.].

——, —— & ——. **1845*a*:** *Grimmia*. *In*: W. P. Schimper (ed.), *Bryologia europaea seu genera muscorum europaeorum monographicae illustrata.* **3**. Stuttgartiae, sumptibus Librariae E. Schweizerbart: pp. 103–132 + Tab. 236–252, 254–261 [Fasc. 25–28: 30 pp. + 24 pls.].

——, —— & ——. **1845*b*:** *Racomitrium*. *In*: W. P. Schimper (ed.), *Bryologia europaea seu genera muscorum europaeorum monographicae illustrata.* **3**. Stuttgartiae, sumptibus Librariae E. Schweizerbart: pp. 135–147 + Tab. 266–271 [Fasc. 25–28: 13 pp. + 2 & 8 (*Dryptodon*) pls.].

——, —— & ——. **1846:** *Didymodon*. *In*: W. P. Schimper (ed.), *Bryologia europaea seu genera muscorum europaeorum monographicae illustrata.* **2**. Stuttgartiae, sumptibus Librariae E. Schweizerbart: pp. 135–141 + Tab. 185–188 [Fasc. 29–30: 7 pp. + 4 pls.].

Brummitt, R. K. & C. E. Powell (eds). 1992: *Authors of plant names. A list of authors of scientific names of plants, with recommended standard forms of their names, including abbreviations*. Kew, Royal Botanic Gardens: 732 pp.

Bryan, V. S. 1956: Chromosomes and systematic position of the inoperculate mosses, *Pleuridium* and *Bruchia*. *American Journal of Botany*, **43:** 460–468.

Bryhn, N. 1902: *Sarconeurum*, genus muscorum novum. *Sarconeurum antarcticum* sp. nov. *In*: N. Wille (ed.), *Mittheilungen über einige von C. Borchgrevink auf dem antarctischen Festlande gesammelte Pflanzen. Nyt Magazin for Naturvidenskaberne*, **40:** 204–208 + Tab. 1–2.

——. **1906:** Bryophyta in itinere polari norvagorum [sic!] secundo collecta. *In*: *Report of the second Norwegian Arctic Expedition in the 'Fram' 1898–1902*. **11**. Kristiania, A. Brøgger: 260 pp. + pls i–ii.

Buck, W. R. 1979: A re-evaluation of the Bruchiaceae with the description of a new genus. *Brittonia*, **31:** 469–473.

——. **1980:** A generic revision of the Entodontaceae. *Journal of the Hattori Botanical Laboratory*, **48:** 71–159.

——. **1987:** Taxonomic and nomenclatural rearrangements in the Hookeriales with notes on West Indian taxa. *Brittonia*, **39:** 210–224.

——. **1988:** Another view of familial delimitation in the Hookeriales. *Journal of the Hattori Botanical Laboratory*, **64:** 29–36.

——. **1991:** The basis for familial classification of pleurocarpous mosses. *Advances in Bryology*, **4:** 169–185.

——. **1994:** A new attempt at understanding the Meteoriaceae. *Journal of the Hattori Botanical Laboratory*, **75:** 51–72.

——. **1997:** A new Brazilian species of *Racomitrium* (Grimmiaceae). *Brittonia*, **49:** 463–465.

——. **1998:** Pleurocarpous mosses of the West Indies. *Memoirs of the New York Botanical Gardens*, **82:** 1–400.

——. **2007:** The history of pleurocarp classification: two steps forward, one step back. *In*: A. E. Newton & R. S. Tangney (eds), *Pleurocarpous mosses. Systematics and evolution. The Systematics Association Special Volume Series*. **71**. Boca Raton, London, New York, CRC Press, Taylor & Francis Groups: pp. 1–18.

—— **& H. Crum. 1990:** An evaluation of familial limits among the genera traditionally aligned with the Thuidiaceae and Leskeaceae. *Contributions from the University of Michigan Herbarium*, **17:** 55–69.

—— **& B. Goffinet. 2000:** Morphology and classification of mosses. *In*: A. J. Shaw & B. Goffinet (eds), *Bryophyte biology*. Cambridge, Cambridge University Press: pp. 71–123.

—— **& R. R. Ireland. 1985:** A reclassification of the Plagiotheciaceae. *Nova Hedwigia*, **41:** 89–124.

—— & ——. **1989:** Plagiotheciaceae. *Flora Neotropica Monograph*, **50:** 1–22.

—— **& W. C. Steere. 1983:** Un listado preliminar de los musgos de la Española. *Moscocoa*, **2:** 28–53.

—— **& D. H. Vitt. 1986:** Suggestions for a new familial classification of pleurocarpous mosses. *Taxon*, **35:** 21–60.

——, —— **& J. Shaw. 2000*a*:** Testing morphological concepts of orders of pleurocarpous mosses (Bryophyta) using phylogenetic reconstructions based on *trn*L-*trn*F and *rps*4 sequences. *Molecular Phylogenetics and Evolution*, **16:** 180–198.

——, —— & ——. **2000*b*:** Novel relationships in pleurocarpous mosses as revealed by cpDNA sequences. *The Bryologist*, **103:** 774–789.

——**, C. J. Cox, J. A. Shaw & B. Goffinet. 2004:** Ordinal relationships of pleurocarpous mosses, with special emphasis on the Hookeriales. *Systematics and Biodiversity*, **2:** 121–145.

Burley, J. S. & N. M. Pritchard. 1990: Revision of the genus *Ceratodon* (Bryophyta). *Harvard Papers in Botany*, **2:** 17–76.

Buxbaum, J. Ch. 1728: *Plantarum minus cognitarum centuria I. complectens plantas circa Byzantium et in oriente observatas*. Petropoli, ex Typographia Academiae: xii +48 pp. + lxv pls.

Buysson, R. Du. 1883: Clef analytique des mousses de la famille des Grimmiées. *Feuille des Jeunes Naturalistes*, **13:** 63–65.

C

Calabrese, G. M. 2006: A taxonomic revision of *Zygodon* (Orthotrichaceae) in southern South America. *The Bryologist*, **109:** 453–509.

—— **& C. M. Matteri. 1999:** Los musgos (Bryophyta) de Puerto Blest y alrededores (Parque Nacional Nahuel Huapi, Argentina) I. Lista comentada de la especies. *Revista del Museo Argentino de Ciencias Naturales, Nueva Serie*, **1:** 23–48.

Cameron, R. E., G. H. Lacy, F. A. Morelli & J. B. Marsh. 1971: Farthest south soil microbial and ecological investigations. *Antarctic Journal of the United States*, **6:** 105–106.

Campos, H., J. Arenas & W. Steffen. 1978: Antecedentes y observaciones limnológicas en los principales logos de la Isla Rey Jorge, Shetland del Sur, Antárctica. *Serie Científica Instituto Antártico Chileno*, **24:** 11–19.

Campbell, D. H. 1895: *The structure and development of the mosses and ferns (Archegoniatae)*. London – New York, Macmillan and Co.: viii + 544 pp.

——. **1905:** *The structure and development of the mosses and ferns (Archegoniatae)*. Ed. 2. London – New York, Macmillan and Co.: vii + 657 pp.

———. **1918:** *The structure and development of the mosses and ferns (Archegoniatae).* Ed. 3. New York and London, The Macmillan Company: ix + 708 pp.

Cano, M. J. & M. T. Gallego. 2003: Lectotypification of twenty names of taxa referable to *Tortula* Hedw. (Pottiaceae, Bryophyta). *Taxon*, **52:** 611–618.

——— **& J. Guerra & R. M. Ros. 1993:** A revision of the moss genus *Crossidium* (Pottiaceae) with the description of the new genus *Microcrossidium. Plant Systematics and Evolution*, **188:** 213–235.

———, ——— **&** ———. **1994:** *Pterygoneurum compactum* sp. nov. (Musci: Pottiaceae) from Spain. *The Bryologist*, **97:** 412–415.

———, **R. M. Ros, T. M. Gallego, J. A. Jiménez & J. Guerra. 2002:** Contribution to the bryophyte flora of Morocco: the Anti-Atlas catalogue. *Cryptogamie, Bryologie,* **23:** 249–262.

Cantrill, D. J. 1997: Hepatophytes from the Early Cretaceous of Alexander Island, Antarctica: systematics and palaeoecology. *International Journal of Plant Science*, **158:** 476–488.

———. **2000:** A Cretaceous (Aptian) flora from President Head, Snow Island, Antarctica. *Palaeontographica, Abteilung B*, **253:** 153–191.

Cao, T. & Ch. Gao. 1995: A revised taxonomic account of the genus *Andreaea* (Andreaeaceae, Musci) in China. Contributions to the Bryoflora of China 6. *Harvard Papers in Botany*, **7:** 11–24.

——— **&** ———. **1999:** Andreaeaceae. *In*: Ch. Gao & M. R. Crosby (eds), *Moss Flora of China. English version.* **1**. *Sphagnaceae – Leucobryaceae*. Beijing – New York, Science Press and St. Louis, Missouri Botanical Garden: pp. 49–56.

——— **& S. He. 1999:** Ditrichaceae. *In*: Ch. Gao & M. R. Crosby (eds), *Moss flora of China, English version.* **1**. *Sphagnaceae – Leucobryaceae*. Beijing – New York, Science Press and St. Louis, Missouri Botanical Garden: pp. 58–82.

——— **& D. H. Vitt. 1986:** A taxonomic revision and phylogenetic analysis of *Grimmia* and *Schistidium* (Bryopsida; Grimmiaceae) in China. *Journal of the Hattori Botanical Laboratory*, **61:** 123–247.

———, **S. He & D. H. Vitt. 2003:** Grimmiaceae. *In*: Ch. Gao, M. R. Crosby & S. He (eds), *Moss Flora of China. English version.* **3**. *Grimmiaceae – Tetraphidaceae*. Beijing – New York, Science Press and St. Louis, Missouri Botanical Garden: pp. 3–76.

———, **D. G. Horton & Ch. Gao. 1992:** A revision of the genus *Encalypta* (Encalyptaceae, Musci) in China. *In*: T. Koponen & J. Hyvönen (eds), *Proceedings of the Congress of East Asiatic bryology, Helsinki, August 12–19, 1990. Bryobrothera*, **1:** 251–268.

Capesius, I. & M. Stech. 1997: Molecular relationships within mosses based on 18S rRNA gene sequences. *Nova Hedwigia*, **64:** 525–533.

Cardot, J. 1900: Note préliminaire sur les Mousses recueillies par l'Expédition antarctique belge. *Revue Bryologique,* **27:** 38–46.

———. **1901:** Mousses et coup d'oeil sur la flore bryologique des Terres Magellaniques. *In*: *Expédition Antactique Belge. Résultats du Voyage du S. Y. Belgica en 1897–1898–1899 sous le commandement de A. de Gerlache de Gomery. Rapports Scientifiques. Botanique.* Anvers, Imprimerie J.-E. Buschmann: 48 pp. + pls. i–xiv.

———. **1905*a*:** Notice préliminaire sur les mousses recueillies par l'Expédition antarctique suédoise. I. Espèces de la région magellanique. *Bulletin de l'Herbier Boissier, Série 2,* **5:** 997–1011.

———. **1905*b*:** Deux genres nouveaux de mousses acrocarpes. *Revue Bryologique,* **32:** 45–47.

———. **1906*a*:** Notice préliminaire sur les mousses recueillies par l'Expédition antarctique suédoise. II. Espèces de la Géorgie du Sud. III. Espèces de l'Antarctide. *Bulletin de l'Herbier Boissier, Série 2,* **6:** 16–32.

———. **1906*b*:** Les mousses de l'Expédition Charcot. *Revue Bryologique,* **33:** 33–35.

———. **1906*c*:** Note sur la végétation bryologique de l'Antarctide. *Comptes Rendus Hebdomadaires des Séances de l'Académie des Sciences*, **42:** 456–458.

———. **1907*a*:** Mousses. *In*: *Expédition antarctique française (1903–1905) commandée par le Dr Jean Charcot. Sciences Naturelles: Documents Scientifiques. Botanique.* Paris, Masson et Cie: 9 pp. + 2 maps.

———. **1907*b*:** Musci. *In*: *National Antarctic Expedition. 1901–1904. Natural History.* **3**. Zoology and Botany. London, Trustees of the British Museum: 6 pp. + pls. i–ii.

———. **1908*a*:** La flore bryologique des terres magellaniques, de la Géorgie du Sud et de l'Antarctide. *In*: *Wissenschaftliche Ergebnisse der Schwedischen Südpolar-Expedition 1901–1903 unter Leitung von Dr. Otto Nordenskjöld.* **4**(8). Stockholm, Lithographisches Institut des Generalstabs: 298 pp. + 11 pls.

———. **1908*b*:** Note sur la flore de l'Antarctide. *In*: *Comptes rendus de l'Association Française pour l'Advancement des Sciences. Congrès de Reims. – 1907.* Paris, Imprimerie Chaix: pp. 452–460.

———. **1910*a*:** Musci. *In*: J. Briquet (ed.), *Recueil des documents destinés a servir de base aux débats de la section de nomenclature systématique du Congrès International de Botanique de Bruxelles 1910 présenté au nom du Bureau permanent de nomenclature et des Commissions de nomenclature cryptogamique et paléobotanique.* Berlin, Comité d'Organisation de Congrès International de Botanique: pp. 49–50.

———. **1910*b*:** Musci. *In*: J. Murray (ed.), *British Antarctic Expedition 1907–9 under the command of Sir E. H. Shackleton, C. V. O. Reports on Scientific investigations.* **1**. Biology. Part 4. London, William Heinemann: pp. 77–79.

———. **1911*a*:** Sur les mousses rapportées par l'Expédition antarctique du Pourquoi-Pas? *Comptes Rendus Hebdomadaires des Séances de l'Académie des Sciences*, **153:** 599–603 [October – beginning of November].

———. **1911*b*:** Note sur les mousses rapportées par la seconde expédition antarctique française. *Revue Bryologique,* **38**: 124–127 [end of November].

———. **1911*c*:** Les mousses de l'Expédition nationale antarctique écossaise. *Transactions of the Royal Society of Edinburgh*, **48:** 67–82 + pls i–iii.

———. **1911*d*:** Diagnoses préliminaires de mousses mexicaines (8e article). *Revue Bryologique*, **38:** 1–9.

———. **1911*e*:** Diagnoses préliminaires de mousses mexicaines (9e article). *Revue Bryologique*, **38:** 33–43.

———. **1913:** Mousses. *In*: *Deuxiéme Expédition Antarctique Française (1908–1910) commandée par le Dr Jean Charcot.*

Sciences Naturelles: Documents scientifiques. Paris, Masson et Cie: 31 pp. + 5 pls.

———. **1914:** *Philibertiella* Card., genre nouveau de la tribu Ditrichées. *Revue Bryologique*, **41:** 37–38.

———. **1916:** Note sur des mousses de Kerguelen. *Bulletin du Muséum d'Histoire Naturelle*, **22:** 336–341.

——— **& V. F. Brotherus. 1923:** Botanische Ergebnisse der Schwedischen Expedition nach Patagonien und dem Feuerlande 1907–1909. X. Les mousses. *Kongliga Svenska Vetenskapsakademiens Handlingar*, **63(10):** 1–74 + pls. 1–4.

——— **& I. Thériot. 1902:** Papers from the Harriman Alaska Expedition. XXIX. The mosses of Alaska. *Proceedings of the Washington Academy of Sciences*, **4:** 293–372 + pls. xiii–xxiii.

Carrión, J. S., M. J. Cano & J. Guerra. 1995: Spore morphology in the moss genus *Pterygoneurum* Jur. (Pottiaceae). *Nova Hedwigia*, **61:** 481–496.

Casares Gil, A. 1915: Enumeración y distribución geográfica de las muscíneas de la Península Ibérica. *Trabajos del Museo Nacional de Ciencias Naturales Serie Botánica*, **8:** 1–179.

Casas Sicart, C. 1986: Catálogo de los briófitos de la verniente española del Pirineo Central y de Andorra. *Collectanea Botanica*, **16:** 255–321.

Casas, C., M. Brugués & R. M. Cros. 2001: *Flora dels briòfits des Països Catalans*. **1**. *Molses*. Barcelona, Institut d'Estudis Catalans: 278 pp. (in Catalonian).

———**, M. Brugués, R. M. Cros & C. Sérgio. 1989:** *Cartografia de briòfits. Peninsula Ibèrica i les Illes Belears, Canàries, Açores i Madeira.* **2:** 51–100. Barcelona, Institut d'Estudis Catalans, Universitat Autònoma de Barcelona: 158 pp. + 50 maps folio.

Catcheside, D. G. 1980: *Mosses of South Australia*. Adelaide, D. J. Woolman: 364 pp.

——— **& J.-P. Frahm. 1985:** Additions to the *Campylopus* flora of Australia. *Journal of Bryology*, **13:** 359–367.

Cavers, F. 1911: The inter-relationship of the Bryophyta. *New Phytologist Reprint*, **4:** i–vii + 1–203.

Chalmers, M. O., M. A. Harper & W. A. Marshall. 1996: *An illustrated catalogue of airborne microbiota from the maritime Antarctic*. Cambridge, British Antarctic Survey: 175 pp.

Chałubiński, T. 1886: *Enumeratio muscorum frondosorum tatrensium, hucusque cognitorum*. Warszawa, E. Skikawski: viii + 208 pp.

Chamberlain, D. F. & C. D. Preston. 1992: *Pottia heimii*. *In*: M. O. Hill, C. D. Preston & A. J. E. Smith (eds), *Atlas of the bryophytes of Britain and Ireland*. **2**. Mosses (except Diplolepideae). Colchester, Harley Books: p. 250.

Chapin, F. S. III. 1983: Direct and indirect effects of temperature on arctic plants. *Polar Biology*, **2:** 47–52.

Chen, F-D., X.-D. Li & H. Kanda. 1993: [An investigation of the bryophytes in the Fildes Peninsula, King George Island, Antarctica]. *Antarctic Research (Chinese Edition),* **5:** 46–50 (in Chinese with English summary.)

———**,** ——— **& J.-X. Ziu. 1995:** [*Handbook of Fildes Peninsula Antarctica mosses*]. Ocean Press, Beijing: [iii] + [56 unnumbered plates] + 157 pp. (in Chinese).

Chen, P. C. 1941: Studien über die ostasiatischen Arten der Pottiaceae, I–II. *Hedwigia*, **80:** 1–76, 141–322.

Chiang, T.-Y. 1998: Taxonomic revision of *Andreaea* (Mosses, Andreaeaceae) of Taiwan. *Botanical Bulletin of Academia Sinica*, **39:** 57–68.

Chopra, R. N. & P. K. Kumra. 1988: *Biology of bryophytes*. New Delhi, Wiley Eastern Limited: xi + 350 pp.

——— **& Sarla. 1985:** Induction of buds in the moss *Bryum pallescens* Schleich. ex Schwaegr. by a metal chelate, Fe-EDTA. *Journal of Bryology*, **13:** 423–428.

Chopra, R. S. 1967: Relationships between liverworts and mosses. *Phytomorphology*, **17:** 70–78.

———. **1975:** Taxonomy of Indian mosses (An introduction). *Botanical Monograph*, **10:** i–xl + 1–631.

Chuang, C.-C. 1973: A moss flora of Taiwan exclusive of essentially pleurocarpous families. *Journal of the Hattori Botanical Laboratory*, **37:** 419–509.

Churchill, S. P. 1981: A phylogenetic analysis, classification and synopsis of the genera of the Grimmiaceae (Musci). *In*: A. A. Funk & D. R. Brooks (eds), *Advances in cladistics. Proceedings of the First Meeting of the Willi Hennig Society*. New York, The New York Botanical Garden: pp. 127–144.

———. **1985:** A synopsis of the Kansas mosses, with keys and distribution maps. *University of Kansas Scientific Bulletin*, **53:** 1–64.

———. **1987:** Systematics and biogeography of *Jaffueliobryum* (Grimmiaceae). *In*: W. R. Buck (ed.), *Bryostephane steereana: a collection of bryological papers presented to William Campbell Steere on the occasion of his 80th birthday. Memoirs of the New York Botanical Garden*, **45:** 691–708.

———. **1998:** Catalog of Amazonian mosses. *Journal of the Hattori Botanical Laboratory*, **85:** 191–238.

——— **& W. R. Buck. 1982:** A taxonomic investigation of *Leptotheca* (Rhizogoniaceae). *Brittonia*, **34:** 1–11.

———**, D. Griffin III & J. Muñoz. 2000:** A checklist of the mosses of the tropical Andean countries. *Ruizia*, **17:** 1–203.

———**, C. E. L. Linares & G. M. Gonzáles (illustrations). 1995:** Prodromus bryologiae novo-granatensis. Introduccion a la flora de musgos de Colombia. **1** (Adelotheciaceae a Funariaceae) and **2** (Grimmiaceae a Trachypodaceae). *Biblioteca 'Jose Jeronimo Triana'*, **12:** 1–8 [unnumbered] + 1–453 + i–xxvi (part 1) and 1–8 [unnumbered] + 455–924 (part 2).

Clare, D. & T. B. Terry. 1960: Dispersal of *Bryum argenteum*. *Transactions of the British Bryological Society*, **3:** 748.

Clark, M. R. & P. R. Dingwall. 1985: *Conservation of islands in the Southern Ocean. A review of the protected areas of Insulantarctica*. Gland – Cambridge, International Union for Conservation Nature and Natural Resources: vii + 188 [+ 5] pp.

Clarke, G. C. S. 1973*a*: A synoptic flora of South Georgian mosses: III. *Leptotheca*, *Philonotis*, *Mielichhoferia* and *Pohlia*. *British Antarctic Survey Bulletin*, **37:** 53–79.

———. **1973*b*:** Notes on Antarctic bryophytes: III. The type specimen of *Pohlia inflexa* (C. Muell.) Wijk & Marg. *British Antarctic Survey Bulletin*, **37:** 97–98.

——— **& S. W. Greene. 1970:** Reproductive performance of two species of *Pohlia* at widely separated stations. *Transactions of the British Bryological Society*, **6:** 114–128.

——— **& P. J. Lightowlers. 1983:** Notes on Antarctic bryophytes: XI. *Mielichhoferia austro-georgica* and *Muelleriella crassifolia*. *British Antarctic Survey Bulletin*, **59:** 35–39.

Clifford, H. T. 1955: On the distribution of *Rhacomitrium crispulum* (H. f. & W.) H. f. & W. *The Bryologist*, **58:** 330–334.

Collins, N. J. 1969: The effects of volcanic activity on the vegetation of Deception Island. *British Antarctic Survey Bulletin*, **21:** 79–94.

———. 1973: Productivity of selected bryophyte communities in the maritime Antarctic. *In*: L. C. Bliss & F. E. Wielgolaski (eds), *Primary production and production processes. Tundra biome*. Edmonton, University of Alberta Printing Services: pp. 177–183.

———. 1976*a*: The development of moss-peat banks in relation to changing climate and ice cover on Signy Island in the maritime Antarctic. *British Antarctic Survey Bulletin*, **43:** 85–102.

———. 1976*b*: Growth and population dynamics of the moss *Polytrichum alpestre* in the maritime Antarctic. *Oikos*, **27:** 389–401.

———. 1977: The growth of mosses in two contrasting communities in the maritime Antarctic: measurement and prediction of net annual production. *In*: G. A. Llano (ed.), *Adaptations within Antarctic ecosystems. The Proceedings of the Third SCAR Symposium on Antarctic biology*. Washington, D.C., Smithsonian Institution: pp. 921–933.

Convey, P. 1994: Modelling reproductive effort in sub- and maritime Antarctic mosses. *Oecologia*, **100:** 45–53

———. 2003: Maritime Antarctic climate change: signals from terrestrial biology. *In*: E. Domack, A. Leventner, A. Burnett, R. Bindschadler, P. Convey & M. Kirby (eds), *Antarctic Peninsula climate variability. Historical and paleoenvironmental perspectives.* Antarctic Research Series. **79**. Washington, D.C., American Geophysical Union: pp. 145–158.

——— & R. I. Lewis Smith. 1993: Investment in sexual reproduction by Antarctic mosses. *Oikos*, **68:** 293–302.

——— & ———. 2006: Geothermal bryophyte habitats in the South Shetland Islands, maritime Antarctic. *Journal of Vegetation Science*, **17:** 529–538.

———, ———, D. A. Hodgson & H. J. Peat. 2001: The flora of the South Sandwich Islands, with particular reference to the influence of geothermal heating. *Journal of Biogeography*, **27:** 1279–1295.

———, ———, H. J. Peat & P. J. A. Pugh. 2000: The terrestrial biota of Charcot Island, eastern Bellingshausen Sea, Antarctica: an example of extreme isolation. *Antarctic Science*, **12:** 406–413.

Corbière, L. 1913: Nouvelle contribution à la flore bryologique du Maroc d'après les récoltes du Lieutenant Mouret. *Revue Bryologique*, **40:** 51–57.

Corley, M. F. V. 1976: The taxonomy of *Campylopus pyriformis* (Schultz) Brid. and related species. *Journal of Bryology*, **9:** 193–212.

———. 1992: *Campylopus introflexus*. *In*: M. O. Hill, C. D. Preston & A. J. E. Smith (eds), *Atlas of the bryophytes of Britain and Ireland*. **2**. Mosses (except Diplolepideae). Colchester, Harley Books: p. 180.

——— & J.-P. Frahm. 1982: Taxonomy and world distribution of *Campylopus pyriformis* (Schultz) Brid. *Journal of Bryology*, **12:** 187–190.

———, A. C. Crundwell, R. Düll, M. O. Hill & A. J .E. Smith. 1981 ['1982']: Mosses of Europe and the Azores; an annotated list of species, with synonyms from the recent literature. *Journal of Bryology*, **11:** 609–689.

Correns, C. 1899: Untersuchungen über die Vermehrung der Laubmoose durch Brutorgane und Stecklinge. Jena, Verlag von Gustav Fischer: xxiv + 472 pp.

Cortini Pedrotti, C. 2001*a*: New check-list of the mosses of Italy. *Flora Mediterranea*, **11:** 23–107.

———. 2001*b*: *Flora dei muschi d'Italia. Sphagnopsida, Andreaeopsida, Bryopsida (I parte)*. Roma, Antonio Delfino Editore Medicina-Science: xiii + 817 pp.

Cox, C. J. & T. A. J. Hedderson. 1999: Phylogenetic relationships among the ciliate arthrodontous mosses: evidence from chloroplast and nuclear DNA sequences. *Plant Systematics and Evolution*, **215:** 119–139.

——— & ———. 2003: Phylogenetic relationships within the moss family Bryaceae based on chloroplast DNA evidence. *Journal of Bryology*, **25:** 31–40.

———, B. Goffinet, A. E. Newton, A. J. Shaw & T. A. J. Hedderson. 2000: Phylogenetic relationships among the diplolepideous-alternate mosses (Bryidae) inferred from nuclear and chloroplast DNA sequences. *The Bryologist*, **103:** 224–241.

Craig, E. J. 1939: *Aloina*. *In*: A. J. Grout (ed.), *Moss flora of North America north of Mexico*. **1**(4). Newfane, Vermont, privately published: pp. 211–215 + pl. 92A, E–F.

Crandall-Stotler, B. 1984: Musci, hepatics and anthocerotes – an essay on analogues. *In*: R. M. Schuster (ed.), *New manual of bryology*. **2**. Nichinan, Hattori Botanical Laboratory: pp. 1093–1129.

——— & R. E. Stotler. 2000: Morphology and classification of the Marchantiophyta. *In*: A. J. Shaw & B. Goffinet (eds), *Bryophyte biology*. Cambridge, Cambridge University Press: pp. 21–70.

Crosby, M. R. 1974: Toward a revised classification of the Hookeriales. *Journal of the Hattori Botanical Laboratory*, **38:** 129–141.

———. 1980: The diversity and relationships of mosses. *In*: R. J. Taylor & A. E. Leviton (eds), *The mosses of North America*. San Francisco, The Pacific Division of the American Association for the Advencement of Science: pp. 115–129.

———. 2000: Index of mosses 1996–1998. *Monographs in Systematic Botany from the Missouri Botanical Garden*, **80:** 1–65.

———. 2003: *Vade mecum bryologiae I. Abbreviations for bryological nomenclatural literature, with full titles and other useful information*. St. Louis, Missouri Botanical Garden: 165 pp.

——— & R. E. Magill. 1981: *A dictionary of mosses*. Third printing. *Monographs in Systematic Botany from the Missouri Botanical Garden*, **3:** 1–43.

——— & ———. 1997: Index of mosses 1993–1995. *Monographs in Systematic Botany from the Missouri Botanical Garden*, **62:** 1–106.

——— & ———. 2004: Index of mosses 1999–2001. *Monographs in Systematic Botany from the Missouri Botanical Garden*, **96:** 1–45.

———, ——— & Ch. R. Bauer. 1992: Index of mosses 1963–1989. *Monographs in Systematic Botany from the Missouri Botanical Garden*, **42:** 1–646.

———, ———, B. Allen & S. He. 2000: *A checklist of the mosses*. St. Louis, Missouri Botanical Garden: 320 pp.

———, U. Schultze-Motel & W. Schultze-Motel. 1983: Katalog der Laubmoose von Madagaskar und dem umliegenden Inseln. *Willdenowia*, 13: 187–255.

Crum, H. 1964: Mosses of the Douglas Lake region of Michigan. *Michigan Botanist*, 3: 3–12, 48–63.

———. **1965:** New moss records from Haiti. *The Bryologist*, **68:** 232–233.

———. **1987:** *Leratia* and its place in the Orthotrichaceae, subfamily Macromitrioideae. *In*: W. R. Buck (ed.), *Bryostephane steereana: a collection of bryological papers presented to William Campbell Steere on the occasion of his 80th birthday. Memoirs of the New York Botanical Garden*, **45:** 602–606.

———. **1994*a*:** *Dicranella*. *In*: A. J. Sharp, H. Crum & P. M. Eckel (eds.), *The moss flora of Mexico. Part one Sphagnales to Bryales. Memoirs of the New York Botanical Garden*, **69:** 112–116.

———. **1994*b*:** Seligeriales. *In*: A. J. Sharp, H. Crum & P. M. Eckel (eds.), *The moss flora of Mexico. Part one Sphagnales to Bryales. Memoirs of the New York Botanical Garden*, **69:** 415–418.

———. **1994*c*:** Grimmiales. *In*: A. J. Sharp, H. Crum & P. M. Eckel (eds), *The moss flora of Mexico. Part one Sphagnales to Bryales. Memoirs of the New York Botanical Garden*, **69:** 386–415.

———. **1994*d*:** *Pterygoneurum*. *In*: A. J. Sharp, H. Crum & P. M. Eckel (eds.), *The moss flora of Mexico. Part one Sphagnales to Bryales. Memoirs of the New York Botanical Garden*, **69:** 368–372.

———. **1994*e*:** *Leptobryum*. *In*: A. J. Sharp, H. Crum & P. M. Eckel (eds.), *The moss flora of Mexico. Part one Sphagnales to Bryales. Memoirs of the New York Botanical Garden*, **69:** 503–505.

———. **2001:** *Structural diversity of bryophytes*. Ann Arbor, The University of Michigan Herbarium: 379 pp.

———. **2004:** *Mosses of Great Lakes Forest*. Ed. 4. Ann Arbor, The University of Michigan Herbarium: vii + 592 pp.

———. **2007:** *Dicranella*. *In*: R. H. Zander (Lead ed.), *Flora of North America north of Mexico*. **27**. *Bryophytes: mosses (1)*. New York – Oxford, Oxford University Press: pp. 386–393.

——— **& L. E. Anderson. 1981:** *Mosses of eastern North America*. **1–2**. New York, Columbia University Press: 1328 pp.

——— & ———. **1989:** New names for some North American mosses. *The Bryologists*, **92:** 533.

——— **& E. B. Bartram. 1958:** A survey of the moss flora of Jamaica. *Bulletin of the Institute of Jamaica Science Series*, **8:** 1–90.

——— **& D. Griffin, III. 1981:** *Philonotis corticata*, new from Mexico. *The Bryologist*, **84:** 399–401.

——— **& W. C. Steere. 1958:** A contribution to the bryology of Haiti. *The American Midland Naturalist*, **60:** 1–51.

Crundwell, A. C. 1959: A revision of the British material of *Brachythecium glaciale* and *B. starkei*. *Transactions of the British Bryological Society*, **3:** 565–567.

———. **1970:** Notes on the nomenclature of British mosses. I. *Transactions of the British Bryological Society*, **6:** 133–138.

———. **1971:** Notes on the nomenclature of British mosses. II. *Transactions of the British Bryological Society*, **6:** 323–326.

——— **& E. Nyholm. 1962:** Notes on the genus *Tortella*. I. *T. inclinata*, *T. densa*, *T. flavovirens* and *T. glareicola*. *Transactions of the British Bryological Society*, **4:** 187–193.

——— & ———. **1963:** Notes on the genus *Tortella*. II. *Tortella arctica*. *The Bryologist*, **66:** 184–191.

——— & ———. **1964:** The European species of the *Bryum erythrocarpum* complex. *Transactions of the British Bryological Society*, **4:** 597–637.

——— & ———. **1977:** *Dicranella howei* Ren. & Card. and its relationship to *D. varia* (Hedw.) Schimp. *Lindbergia*, **4:** 35–38.

——— **& H. L. K. Whitehouse. 1978:** *Bryum rubens* in North America. *The Bryologist*, **81:** 323–324.

Czernyadjeva, I. V. 1997: [The species of the genus *Pohlia* (Musci) with propagula]. *Botanicheskiy Zhurnal*, **82:** 102–122 (in Russian with English summary).

———. **1999:** On the distribution of propaguliferous species of *Pohlia* (Bryaceae, Musci) in Russia. *Arctoa, a Journal of Bryology*, **8:** 51–56.

———. **2001:** Moss flora of Yamal Peninsula (West Siberian Arctic). *Arctoa, a Journal of Bryology*, **10:** 121–150.

D

Dale, T. M., M. L. Skotnicki, K. D. Adam & P. M. Selkirk. 1999: Genetic diversity in the moss *Hennediella heimii* in Miers Valley, southern Victoria Land, Antarctica. *Polar Biology*, **21:** 228–233.

Dalton, P. J., R. D. Seppelt & A. M. Buchanan. 1991: An annotated checklist of Tasmanian mosses. *In*: M. R. Banks, S. J. Smith, A. E. Orchard & G. Kantvilas (eds), *Aspects of Tasmanian botany. A tribute to Winifred Curtis*. Hobart, Royal Society of Tasmania: pp. 15–32.

Darbishire, O. V. 1923: Cryptogams from the Antarctic. *The Journal of Botany British and Foreign*, **61:** 105–107.

Davis, R. C. & P. M. Harrisson. 1982: Prediction of photosynthesis in maritime Antarctic mosses. *Comité National Français des Recherches Antarctiques*, **51:** 241–247.

Debenham, F. (ed.). 1945: *The Voyage of Captain Bellingshausen's to the Antarctic Seas, 1819–1821*. **1–2**. Translated from Russian. The Hakluyt Society Publications, Second Series. **91–92**. London, The Hakluyt Society: xxx + 259 pp. (Vol. 1); viii + pp. 261–474 pp. (Vol. 2).

Deguchi, H. 1977: [Small male-branches of *Ptychomitrium* (Grimmiaceae) arised from the base of vaginula inside the perichaetial leaf circle]. *Miscellanea Bryologica et Lichenologica*, **7:** 177–179 (in Japanese).

———. **1979 ['1978']:** A revision of the genera *Grimmia*, *Schistidium* and *Coscinodon* (Musci) of Japan. *Journal of Science of the Hiroshima University, Series B, Div. 2 (Botany)*, **16:** 121–256.

———. **1983:** *Neomeesia* Deguchi, a new genus of the family Meesiaceae from southern South America. *Bulletin of the National Science Museum, Series B (Botany)*, **9:** 143–148.

———. **1984:** Studies on some Patagonian species of Grimmiaceae (Musci, Bryophyta). *In*: H. Inoue (ed.), *Studies on cryptogams in southern Chile*. Tokyo, Kenseisha: pp. 17–72.

———. **1987:** Studies on some Peruvian species of the Grimmiaceae (Musci, Bryophyta). *In*: H. Inoue (ed.), *Studies on cryptogams in southern Peru*. Tokyo, Tokai University Press: pp. 19–74.

———. **1991*a*:** A list of moss collections made during the expeditions to Chile in 1981 and 1987 (1). *Bulletin of the National Science Museum, Series B (Botany)*, **17:** 15–34.

———. **1991*b*:** A list of moss collections made during the expeditions to Chile in 1981 and 1987 (2). *Bulletin of the National Science Museum, Series B (Botany)*, **17:** 67–83.

Delgadillo, C. M. 1973*a*: A new species, nomenclatural changes, and generic limits in *Aloina*, *Aloinella*, and *Crossidium* (Musci). *The Bryologist*, **76:** 271–277.

———. **1973*b*:** A quantitative study of *Aloina*, *Aloinella*, and *Crossidium* (Musci). *The Bryologist*, **76:** 301–305.

———. **1975:** Taxonomic revision of *Aloina*, *Aloinella*, and *Crossidium* (Musci). *The Bryologist*, **78:** 245–303.

———. **2002:** The distinction between *Grimmia mexicana* and *G. ochyriana*. *The Bryologist*, **103:** 762–764

——— **& A. S. Cárdenas. 1979:** Musgos de Zacatecas, Mexico. I. *Boletin de la Sociedad Botanica de México*, **38:** 1–6.

——— **& M. M. Schiavone. 2004:** *Aloina* and *Aloinella* (Bryopsida, Pottiaceae) in northern Argentina. *Brittonia*, **56:** 291–293.

——— **& J. L. Villaseñor R. 2002:** The status of South American *Grimmia herzogii* Broth. (Musci). *Taxon*, **51:** 123–129.

———**, B. Bello & A. S. Cárdenas. 1995:** LATMOSS. A catalogue of the neotropical mosses. *Monographs in Systematic Botany from the Missouri Botanical Garden*, **56**: 1–191.

De Luna, E. 1990*a*: Developmental evidence of acrocarpy in *Hedwigia ciliata* (Musci: Hedwigiaceae). *Tropical Bryology*, **2:** 53–60.

———. **1990*b*:** Protonemal development in the Hedwigiaceae (Musci) and its systematic significance. *Systematic Botany*, **15:** 192–204.

———. **1995:** The circumscription and phylogenetic relationships of the Hedwigiaceae (Musci). *Systematic Botany*, **20:** 347–373.

———**, A. E. Newton, A. Withey, D. Gonzáles & B. D. Mishler. 1999:** The transition to pleurocarpy: a phylogenetic analysis of the main diplolepidous lineages based on *rbc*L sequences and morphology. *The Bryologist*, **102:** 634–650.

———**, W. R. Buck, H. Akiyama, T. Arikawa, H. Tsubota, D. Gonzáles, A. E. Newton & A. J. Shaw. 2000:** Ordinal phylogeny within the hypnobryalean pleurocarpous mosses inferred from cladistic analyses of three chloroplast DNA sequence data sets: *trn*L-F, *rps*4, and *rbc*L. *The Bryologist*, **103:** 242–256.

Demaret, F. 1940: Prodrome des bryophytes du Congo belge et du Ruanda-Urundi. *Bulletin Jardin Botanique de l'État*, **16:** 21–104.

———. **1986:** Etude du matériel-type de trois taxons de *Bryum*. *Bulletin Jardin Botanique National de Belgique*, **56:** 305–314.

———. **1993:** *Bryum*. *In*: *Flore générale de Belgique. Bryophytes.* **3**(2). Meise, Ministère de l'Agriculture, Jardin Botanique National de Belgique: pp. 152–258.

——— **& Th. Arts. 1993:** *Pohlia*. *In*: *Flore générale de Belgique. Bryophytes.* **3**(2). Meise, Ministère de l'Agriculture, Jardin Botanique National de Belgique: pp. 113–145.

——— **& A. Empain. 1985:** *Bryum pseudotriquetrum* (Hedw.) Gaert., Meyer & Scherb. et sa variété *bimum* (Schreb.) Lilj. *Bulletin Jardin Botanique National de Belgique*, **55:** 275–290.

——— **& P. Geissler. 1990:** L'identité de *Bryum inclinatum* (Brid.) Turton homonyme illégitime et les espèces voisines. *Bulletin Jardin Botanique National de Belgique*, **60:** 223–228.

——— **& R. Wilczek. 1978:** *Pohlia drummondii* (C. Müll.) Andrews et les taxons confondus avec lui. *Bulletin Jardin Botanique National de Belgique*, **48:** 447–463.

——— **&** ———**. 1979:** Les *Pohlia* de Belgique. *Bulletin Jardin Botanique National de Belgique*, **49:** 149–158.

——— **&** ———**. 1980:** La valeur spécifique de *Bryum dichotomum* Hedw., *B. bulbillosum* Mont., *B. bicolor* Dicks. et *B. barnesii* Wood. *Bulletin Jardin Botanique National de Belgique*, **50:** 187–196.

De Notaris, J. 1836: Specimen de Tortulis italicis. *Memorie della Reale Accademia delle Scienze di Torino*, **40:** 281–326.

———. **1838:** *Syllabus muscorum in Italia et in insulis circumstantibus hucusque cognitorum.* Taurini, ex Typographia Cantari sumptibus auctoris: xx + 331 pp.

———. **1862:** *Musci italici*. **1**. Genova, presso l'autore and Torino, Ermanno Loescher: 69 pp. + 35 pls.

De Sloover, J. L. 1977: Note de bryologie africaine IX. *Andreaea, Racomitrium, Gymnostomiella, Thuidium. Bulletin du Jardin Botanique National de Belgique*, **47:** 155–181.

———. **1986:** Note de bryologie africaine XIII. Polytrichaceae. *Bulletin du Jardin Botanique National de Belgique*, **56:** 241–300.

———. **1995:** Les muscinées illustrées par J. J. Dillenius en 1741et determinées par S. O. Lindberg en 1883. Sciences du vivant – Botanique N° 1. Namur, Press Universitaires de Namur: 95 pp.

Dia, M. G. & F. M. Raimondo. 1984: Nomenclatural and taxonomic observations on «*Racomitrium marginatum*» Lojac. *Webbia*, **38:** 687–693.

———**, G. Miceli & F. M. Raimondo. 1987:** Check-list dei muschi noti in Sicilia. *Webbia*, **61:** 61–123.

Dickson, J. 1790: *Fasciculus secundus plantarum cryptogamicarum Britanniae.* Londini, apud auctorem et G. Nicol: 31 pp. + pls. iv–vi.

———. **1793:** *Fasciculus tertius plantarum cryptogamicarum Britanniae.* Londini, apud auctorem et G. Nicol: 24 pp. + pls. vii–ix.

———. **1801:** *Fasciculus quartus plantarum cryptogamicarum Britanniae.* Londini, apud auctorem et G. Nicol: 28 pp. + 4 unnumbered pages of Index + pls. x–xii.

Dierßen, K. 2001: Distribution, ecological amplitude and phytosociological characterization of European bryophytes. *Bryophytorum Bibliotheca*, **56:** 1–289.

Dillenius, J. J. 1741 ['1742']: *Historia muscorum in qua circiter sexcentae species veteres et novae ad sua genera relatae describuntur et iconibus genuinis illustrantur: cum appendice et indice synonymorum.* Oxonii, e Theatro Sheldoniano: xvi + 576 pp. + lxxxv pls.

Dismier, M. G. 1910: Revision des *Philonotis* de l'Amerique. *Mémoires de la Société Botanique de France*, **17:** 1–37.

Dixon, H. N. 1912: On some mosses from New Zealand. *Journal of the Linnean Society, Botany*, **40:** 433–459.

———. **1913:** Studies in the bryology of New Zealand, with special reference to the herbarium of Robert Brown, of Christchurch, New Zealand. Part I. *New Zealand Institute Bulletin*, **3:** 1–29 + pls. i–iv.

———. **1914:** Studies in the bryology of New Zealand, with special reference to the herbarium of Robert Brown, of Christchurch, New Zealand. Part II. *New Zealand Institute Bulletin*, **3:** 31–74 + pls. v–vi.

———. **1920:** Contribution to the Antarctic bryology. *The Bryologist*, **23:** 65–71.

———. **1923*a*:** Studies in the bryology of New Zealand, with special reference to the herbarium of Robert Brown, of Christchurch, New Zealand. Part III. *New Zealand Institute Bulletin*, **3:** 75–152 + pls. vii–viii.

———. **1923*b*:** Dr. Stirton's new British mosses revised. *Journal of Botany, British and Foreign*, **61:** 10–17, 46–52, 69–75.

———. **1926:** Studies in the bryology of New Zealand, with special reference to the herbarium of Robert Brown, of Christchurch, New Zealand. Part IV. *New Zealand Institute Bulletin*, **3:** 153–238 + pl. ix.

———. **1927:** Studies in the bryology of New Zealand, with special reference to the herbarium of Robert Brown, of Christchurch, New Zealand. Part V. *New Zealand Institute Bulletin*, **3:** 239–298.

———. **1929:** Studies in the bryology of New Zealand, with special reference to the herbarium of Robert Brown, of Christchurch, New Zealand. Part VI. *New Zealand Institute Bulletin*, **3:** 299–372 + pl. x.

———. **1930:** Additions to the moss flora of the north-western Himalayas. *Annales Bryologici*, **3:** 51–70.

———. **1932*a*:** Classification of mosses. *In*: F. Verdoorn (ed.), *Manual of bryology*. The Hague, Martinus Nijhoff: pp. 397–412.

———. **1932*b* ['1931']:** Proposals on bryological nomenclature. *Revue Bryologique, Nouvelle Série*, **4:** 119–122.

———. **1934*a* ['1933]:** The nomenclature of the *Species Muscorum*. *Revue Bryologique, Nouvelle Série*, **6:** 93–115.

———. **1934*b*:** Bryophyta. *In*: E. Christophersen (ed.), *Plants of Gough Island (Diego Alvarez). Scientific Results of the Norwegian Antarctic Expeditions 1927–1928 et sqq., Instituted and Financed by Consul Lars Christensen*. **13**. Oslo, I Kommisjon hos Jacob Dybwad: pp. 9–10.

———. **1934*c*:** More mosses from South Georgia. *Det Kongelige Norske Videnskabers Selskab Forhandlinger*, **7:** 1–3.

———. **1935:** The mosses. *In*: J. W. S. Marr (ed.), *The South Orkney Islands. Discovery Reports*, **10:** 388–382.

———. **1937:** Musci. *In*: E. Christophersen, *Plants of Tristan da Cunha. Scientific results of the Norwegian Antarctic Expeditions 1927–1928 et sqq., instituted and financed by consul Lars Christensen*. **16**. Oslo, I Kommisjon hos Jacob Dybwad: pp. 12–13.

———. **1939:** Mosses. *In*: R. A. Dyer, *The flora of Tristan da Cunha: H.M.S. Carlisle Expedition, 1937. Bothalia*, **3:** 598–601.

———. **1942:** Madagascar mosses. *Journal of Botany, British and Foreign*, **80:** 41–50 + pl. 627.

———. **1960:** Mosses of Tristan da Cunha. *In*: E. Christophersen (ed.), *Results of the Norwegian Scientific Expedition to Tristan da Cunha 1937–1938*. **48**. Oslo, Kommissjon hos H. Aschehoug & Co. (W. Nygaard): 49 pp.

——— **& H. G. Jameson. 1896:** *The student's handbook of British mosses*. Eastbourne, V. T. Sumfield and London, John Wheldon & Co.: xlvi + 520 pp. + pls. i–lx.

——— **&** ———. **1924:** *The student's handbook of British mosses*. Eastbourne, Sumfield & Day Ltd and London, Wheldon & Wesley: xlvi + 582 pp. + pls. i–lxiii.

——— **& W. W. Watts. 1918:** Mosses. *In*: *Australasian Antarctic Expedition 1911–14. under the leadership of Sir Douglas Mawson, D.Sc., B.E. Scientific Reports. Series C. – Zoology and Botany*. **7**(1). Sydney, William Applegate Gullick: 9 pp.

Douady, C. J., F. Delsuc, Y. Boucher, W. F. Doolittle & E. J. Douzery. 2003: Comparison of Bayesian and maximum likelihood bootstrap measures of phylogenetic reliability. *Molecular Biology and Evolution*, **20:** 248–254.

Doweld, A. B. 2001: *Prosyllabus Tracheophytorum. Tentamen systematis plantarum vascularium (Tracheophyta)*. Geos, Moscow. 110 + lxxx pp.

Drummond, Th. 1828: *Musci americani; or, specimens of the mosses collected in British North America, and chiefly among the Rocky Mountains during the second land arctic expedition under the command of Captain Franklin, R. N. by Thomas Drummond, assistant naturalist to the expedition*. **1**. Numbers 1–158; **2**. Numbers 159–286. Glasgow.

Duarte Bello, P. P. 1997: Musgos de Cuba. *Fontqueria*, **47:** i–xxii + 1–717.

Duell, R. 1980: Bryoflora und Bryogeographie der Insel La Palma, Canaren. *Cryptogamie, Bryologie – Lichénologie*, **1:** 151–188.

———. **1984:** Distribution of the European and Macaronesian mosses (Bryophytina). Part I. *Bryologische Beitraege*, **4:** 1–113.

———. **1994:** *Deutschlands Moose. Die Verbreitung der deutschen Moose in der Bundesrepublik Deutschland in der heutigen Grenzen, ihre vertikale und zonale Verbreitung, ihre Arealtypen, Sporophytehäufigkeit, sowie Angaben zum Rückgang der Arten und zu ihrer Gefährdung. 2. Teil Grimmiales – Orthotrichales.* Bad Münstereifel – Ohlerath: 211 pp.

———. **1995:** Übersicht der Moose Griechenland. *Bryologische Beitraege*, **10:** 1–125.

——— **& L. Meinunger. 1989:** *Deutschlands Moose. Die Verbreitung der deutschen Moose in der Bundesrepublik Deutschland in der heutigen Grenzen, ihre vertikale und zonale Verbreitung, ihre Arealtypen, Sporophytehäufigkeit, sowie Angaben zum Rückgang der Arten und zu ihrer Gefährdung. 1. Teil: Anthocerotae, Marchantiatae, Bryatae: Sphagnidae, Andreaeidae, Bryidae: Tetraphidales – Pottiales.* Bad Münstereifel – Ohlerath, IDH-Verlag: 368 pp.

Du Rietz, G. E. 1940: Problems of bipolar plant distribution. *Acta Phytogeographica Suecica*, **13:** 215–282.

Dusén, P. 1903*a*: Beiträge zur Bryologie der Magellansländer, von Westpatagonien und Südchile. 1. *Arkiv för Botanik*, **1:** 441–465 + pls. 18–28.

———. **1903*b*:** Patagonian and Fuegian mosses. *In*: W. B. Scott (ed.), *Reports of the Princeton University expeditions to Patagonia, 1896–1899*. **8**(1) – *Botany*. The University, Princeton and Schweizerbart'sche Verlagshandlung (E. Nägele), Stuttgart: pp. 63–125 + pls vii–xi.

———. **1905*a*:** Beiträge zur Bryologie der Magellansländer, von Westpatagonien und Südchile. 2. *Arkiv för Botanik*, **4(1):** 1–45 + pls. 1–11.

———. **1905*b*:** Beiträge zur Bryologie der Magellansländer, von Westpatagonien und Südchile. 3. (Continuatio Dicranacearum.). *Arkiv för Botanik*, **4(13):** 1–24 + pls. 1–8.

———. **1905c:** Musci nonnuli novi e Fuegia et Patagonia reportati. *Botaniska Notiser*, **1905:** 299–310.

———. **1906:** Beiträge zur Bryologie der Magellansländer, von Westpatagonien und Südchile. IV. *Arkiv för Botanik*, **6(8):** 1–40 + pls. 1–12.

———. **1907:** Beiträge zur Bryologie der Magellansländer, von Westpatagonien und Südchile. V. *Arkiv för Botanik*, **6(10):** 1–32 + pls. 1–6.

E

Eckel, P. M. 1991: *Tortella tortelloides* (Musci: Pottiaceae) new to North America. *The Bryologist*, **94:** 84–87.

———. **1997a:** *Tortella knightii*, an Australasian species new to South America in Tierra del Fuego. *The Bryologist*, **100:** 210–211.

———. **1997b:** The moss *Tortella alpicola* Dix. new to Alberta and the Yukon Territory with a discussion of its range and comments on related species. *The Canadian Field-Naturalist*, **111:** 320–322.

———. **1998:** Re-evaluation of *Tortella* (Musci, Pottiaceae) in conterminous U.S.A. and Canada with a treatment of the European species *Tortella nitida*. *Bulletin of the Buffalo Society of Natural Sciences*, **36:** 117–191.

———. **1999:** *Tortella alpicola* Dixon, new to the moss flora of South Dakota. *Evansia*, **16:** 167.

Eddy, A. 1988: *A handbook of Malesian mosses*. **1**. *Sphagnales to Dicranales*. London, The Natural History Museum: [iii] + 204 pp.

———. **1990:** *A handbook of Malesian mosses*. **2**. *Leucobryaceae to Buxbaumiaceae*. London, The Natural History Museum: [vi] + 256 pp.

———. **1996:** *A handbook of Malesian mosses*. **3**. *Splachnobryaceae to Leptostomataceae*. London, The Natural History Museum: [vi] + 277 pp.

Edwards, S. R. 1977: Taxonomic implications of cell patterns in haplolepidous moss peristomes. *Bulletin of the British Bryological Society*, **31:** 13–14.

———. **1979:** Taxonomic implications of cell patterns in haplolepidous moss peristomes. *In*: G. C. S. Clarke & J. G. Duckett (eds), *Bryophyte systematics*. The Systematics Association Special Volume. **14**. London, Academic Press: pp. 317–346.

———. **1984:** Homologies and inter-relationships of moss peristomes. *In*: R. M. Schuster (ed.), *New manual of bryology*. **2**. Nichinan, The Hattori Botanical Laboratory: pp. 658–695.

Eggers, J. 1982: Artenliste der Moose Makaronesiens. *Cryptogamie, Bryologie – Lichénologie*, **3:** 283–335.

Ehrhart, J. F. 1787: *Beiträge zur Naturkunde, und den damit werwandten Wissenschaften, besonders der Botanik, Chemie, Haus- und Landwirtschaft, Arzneigelehrtheit und Apothekerkunst*. **1**. Hannover and Osnabrück, Schmidtische Buchhandlung: viii + 192 pp.

———. **1791:** *Plantae cryptogamae Linneae, quas in locis earum natalibus collegit et exsiccavit Fridericus Ehrhart*. Decades 17–24, Nos. 161–240. Hannoverae.

Eights, J. 1833: Description of a new crustaceous animal found on the shores of the South Shetland Islands, with remarks on their natural history. *Transactions of the Albany Institute*, **2:** 53–69 + pls 1–2.

El-Saadawi, W. E. 1976: Some mosses from Kuwait. *The Bryologist*, **79:** 515–518.

——— **& M. S. Zanaty. 1990:** *Bryum bicolor* Dicks. and *Funaria hygrometrica* Hedw. develop remarkable persisting structures in extreme environment. *Journal of the Hattori Botanical Laboratory*, **68:** 285–291.

Elvebakk, A. & H. Hertel. 1996: Part 6. Lichens. *In*: A. Elvebakk & P. Prestrud (eds), *A catalogue of Svalbard plants, fungi, algae and cyanobacteria. Norsk Polarinstitutt Skrifter*, **198:** 271–359.

Elven, R. & A. Elvebakk. 1996: Part 1. Vascular plants. *In*: A. Elvebakk & P. Prestrud (eds), *A catalogue of Svalbard plants, fungi, algae and cyanobacteria. Norsk Polarinstitutt Skrifter*, **198:** 9–55.

Engel, J. J. 1978: A taxonomic and phytogeographic study of Brunswick Peninsula (Strait of Magellan). Hepaticae and Anthocerotae. *Fieldiana Botany*, **41:** 1–319.

———. **1980:** A monograph of *Clasmatocolea* (Hepaticae). *Fieldiana Botany, New Series*, **3:** i–viii + 1–229.

———. **1990:** Falkland Islands (Islas Malvinas). Hepaticae and Anthocerotophyta: a taxonomic and phytogeographic study. *Fieldiana Botany, New Series*, **25:** 1–209.

——— **& R. M. Schuster. 1973:** On some tidal zone Hepaticae from South Chile, with comments on maritime dispersal. *Bulletin of the Torrey Botanical Club*, **100:** 29–35.

——— **& G. L. Smith Merrill. 2004:** Austral Hepaticae. 35. A taxonomic and phylogenetic study of *Telaranea* (Lepidoziaceae), with a monograph of the genus in temperate Australasia and commentary on extra-Australasian taxa. *Fieldiana Botany, New Series*, **30:** 1–265.

Engelskjøn, T. 1981: Terrestrial vegetation of Bouvetøya: a preliminary account. *Norsk Polarinstitutt Skrifter*, **175:** 17–28.

———. **1985:** Botany of Gjelskvikfjella and Mühlig-Hofmanfjella, Dronning Maud Land. *Norsk Polainstitutt Rapport Serie*, **22:** 43–54.

———. **1986:** Botany of two Antarctic mountain ranges: Gjelsvikfjella and Mühlig-Hofmannfjella, Dronning Maud Land. I. General ecology and development of the Antarctic cold desert cryptogam formation. *Polar Research*, **4:** 205–224.

———. **1987:** Botany of Bouvetøya, South Atlantic Ocean. II. The terrestrial vegetation of Bouvetøya. *Polar Research*, **5:** 129–163.

——— **& P. M. Jørgensen. 1986:** Phytogeographical relations of the cryptogamic flora of Bouvetøya. *Norsk Polarinstittut Skrifter*, **185:** 71–79.

Enroth, J. 1991: Bryophyte Flora of the Huon Peninsula, Papua New Guinea. XLIII. Funariaceae (Musci). *In*: T. Koponen (ed.), *Bryophyte flora of the Huon Peninsula, Papua New Guinea*. XXXIX–XLVII. *Acta Botanica Fennica*, **143:** 57–69.

Ensminger, J. T., L. N. McCold & J. W. Webb. 1999: Environmental impact assessment under the National Environmental Policy Act and the Protocol on Environmental Protection to the Antarctic Treaty. *Environmental Management*, **24:** 13–23.

Erdağ, A., H. Kürschner & G. Parolly. 2004: *Orthotrichum leblebicii* sp. nov. (Orthotrichaceae, Bryopsida), and two further new epiphytic *Orthotrichum* records from southern Turkey. *Nova Hedwigia*, **78:** 517–526.

F

Fabiszewski, J. & B. Wojtuń. 1997: The occurrence and development of peat mounds on King George Island (maritime Antarctic). *Acta Societatis Botanicorum Poloniae*, **66:** 223–229.

Farr, E. R., J. A. Leussink & F. A. Stafleu (eds.). 1979: *Index Nominum Genericorum (Plantarum)*. **1** (Aa–*Epochnium*). Utrecht, Bohn, Scheltema & Holkema and The Hague, dr. W. Junk b.v., Publishers: xxvi + 630 pp.

Favreau, M. & G. Brassard. 1988: Catalogue bibliographique des bryophytes du Quebec et du Labrador. *Memorial University of Newfoundland Occasional Papers in Biology*, **12:** 1–114.

Fenton, J. H. C. 1980: The rate of peat accumulation in Antarctic moss banks. *Journal of Ecology*, **68:** 211–228.

——. **1982:** The formation of vertical edges on Antarctic moss peat banks. *Arctic and Alpine Research*, **14:** 21–26.

—— **& R. I. Lewis Smith. 1982:** Distribution, composition and general characteristics of the moss banks of the maritime Antarctic. *British Antarctic Survey Bulletin*, **51:** 215–236.

Fife, A. J. 1982: Taxonomic and nomenclatural observations on the Funariaceae. 1. *Physcomitrium*, *Physcomitrella*, and *Goniomitrium* in New Zealand. *Lindbergia*, **8:** 96–104.

——. **1984:** Records of new or otherwise interesting mosses in New Zealand, including a new species. *New Zealand Journal of Botany*, **22:** 1–6.

——. **1985:** A generic revision of the Funariaceae (Bryophyta: Musci). Part I. *Journal of the Hattori Botanical Laboratory*, **58:** 149–196.

——. **1986:** Taxonomic and nomenclatural observations on the Funariaceae. 4. A review of *Entosthodon laxus* with incidental notes on *E. obtusifolius*. *The Bryologist*, **89:** 302–309.

——. **1987:** Taxonomic and nomenclatural observations on the Funariaceae. 5. A revision of the Andean species of *Entosthodon*. *In*: W. R. Buck (ed.), *Bryostephane steereana: a collection of bryological papers presented to William Campbell Steere on the occasion of his 80th birthday. Memoirs of the New York Botanical Garden*, **45:** 301–325.

——. **1995:** Checklist of the mosses of New Zealand. *The Bryologist*, **98:** 313–337.

——. **1996:** A synopsis of New Zealand *Sphagna*, with a description of *S. simplex* sp. nov. *New Zealand Journal of Botany*, **34:** 309–328.

——. **1998:** A synopsis of the New Zealand representatives of *Conostomum* (Musci: Bartramiaceae). *New Zealand Journal of Botany*, **36:** 605–615.

——. **2000:** A synopsis of the New Zealand species of *Schistidium* (Grimmiaceae; Musci), with observations on a little known species of *Racomitrium*. *New Zealand Journal of Botany*, **38:** 191–204.

—— **& P. J. Dalton. 2005:** A reconsideration of *Pleurophascum* (Musci: Pleurophascaceae) and specific status for a New Zealand endemic, *Pleurophascum ovalifolium* stat. et nom. nov. *New Zealand Journal of Botany*, **43:** 871–884.

—— **& R. D. Seppelt. 2001:** A revision of the family Funariaceae (Musci) in Australia. *Hikobia*, **13:** 473–490.

Filson, R. B. 1966: The lichens and mosses of Mac.Robertson Land. *ANARE Scientific Reports, Series B (II) Botany*, **82:** [i–vii] + 1–169.

—— **& J. H. Willis. 1975:** A fruiting occurrence of *Bryum algens* Card. in East Antarctica. *Muelleria*, **3:** 112–116.

Fleischer, M. 1904–1923: *Die Musci der Flora von Buitenzorg zugleisch Laubmoosflora von Java mit Berücksichtigung aller Familien und Gattungen der gesamtem Laubmooswelt*. Leiden, E. J. Brill: i–xxxi + 1–379 (vol. 1, 1904); i–xviii + 381–643 (vol. 2, 1904); i–xxiv + 645–1103 (vol. 3, 1908); i–xxxi + 1105–1729 (vol. 4, 1923).

——. **1912:** Laubmoose. *In*: *Nova Guinea. Résultats de l'Expédition scientifique néerlandaise à la Nouvelle-Guinée*. **8**(4). Leide, E. J. Brill: pp. 735–753 + Tab. cxix–cxxiv.

——. **1920:** Natürliches System der Laubmoose. *Hedwigia*, **61:** 390–400.

Fleming, W. L. S., A. Q. Stevenson, B. B. Roberts & G. C. L. Bertram. 1938: Notes on the scientific work of the British Graham Land Expedition, 1934–1937. *The Geographical Journal*, **91:** 508–532.

Florschütz, P. A. & J. Florschütz-de Waard. 1974: Studies on Colombia cryptogams. I. Variation of characters in South American species of *Campylopus*. *Journal of the Hattori Botanical Laboratory*, **38:** 111–114.

Florschütz-de Waard, J. & P. A. Florschütz. 1979: Estudios sobre criptógamas colombianas III. Lista comentada de los musgos de Colombia. *The Bryologist*, **82:** 215–259.

Flowers, S. 1935: Family Bartramiaceae. *In*: A. J. Grout (ed.), *Moss flora of North America north of Mexico*. **2**(3). Newfane, Vermont, privately published: pp. 152–180 + pls. 67–70.

——. **1951:** A new species of *Tortula* from Utah and Arizona. *The Bryologist*, **54:** 278–282.

——. **1956:** *Dicranoweisia crispula* and *D. cirrhata*. *The Bryologist*, **59:** 239–244.

——. **1973:** *Mosses: Utah and the West*. Provo, Brigham Young University Press: xii + 567 pp.

Fojcik, B. & A. Gumieniak. 1999: [A locality for *Campylopus introflexus* (Musci, Dicranaceae) in Płaskowyż Kolbuszowski (Kotlina Sandomierska)]. *Fragmenta Floristica et Geobotanica Series Polonica*, **6:** 288–290 (in Polish with English summary).

Fox, A. J. & A. P. R. Cooper. 1994: Measured properties of the Antarctic ice sheet derived from the SCAR Antarctic digital database. *Polar Record*, **30:** 201–206.

Frahm, J.-P. 1972: Die Ausbreitung von *Campylopus introflexus* (Hedw.) Brid. in Mitteleuropa. *Herzogia*, **2:** 317–330.

——. **1974:** Zur Unterscheidung und Verbreitung von *Campylopus introflexus* (Hedw.) Brid. und *C. polytrichoides* De Not. *Revue Bryologique et Lichénologique, Nouvelle Série*, **40:** 33–44.

——. **1975:** Conspectus der mittel- und südamerikanischen *Campylopus*-Arten (Dicranaceae). *Bryophytorum Bibliotheca*, **5:** 1–143 + Tafelanhang with Figs 1–14.

——. **1976:** Zur *Campylopus*-Flora von Chile. *Herzogia*, **4:** 141–160.

——. **1978:** Taxonomische Notizen zur Gattung *Campylopus* VI. *Nova Hedwigia*, **30:** 835–849.

——. **1980:** Synopsis of the genus *Campylopus* in North America north of Mexico. *The Bryologist*, **83:** 570–588.

——. **1981:** Bestimmungsschlüssel und Illustrationen zu Gattung *Chorisodontium* Broth. *Herzogia*, **5:** 499–516.

——. **1982:** Problems of infrageneric classification in *Campylopus In*: P. Geissler & S. W. Greene (eds), *Bryophyte taxonomy. Methods, practices and floristic exploration.*

Proceedings of the International Association of Bryologists Taxonomic Workshop Meeting, Genève 27 Aug – 2 Sept 1979. Beihefte zur Nova Hedwigia, **71:** 219–224.

———. **1983:** A new infrageneric classification in the genus *Campylopus* Brid. *Journal of the Hattori Botanical Laboratory*, **54:** 207–225.

———. **1984:** Phytogeography of European *Campylopus* species. *In*: J. Váňa (ed.), *Proceedings of the Third Meeting of the Bryologists from Central and East Europe*. Praha, Univerzita Karlova: pp. 191–212.

———. **1985:** Afrikanische *Campylopus*-Arten (Dicranaceae, Musci). *Bryophytorum Bibliotheca*, **31:** 1–216.

———. **1986:** Ecosystematics and geotaxonomy of the Campylopodioideae. *The Bryological Times*, **36:** 1–5.

———. **1987*a*:** A revised list of the *Campylopus* species of the world. *Bryologische Beitraege*, **7:** 1–117.

———. **1987*b*:** A survey of the *Campylopus* species of Australia. *Journal of Bryology*, **14:** 701–727.

———. **1988:** The subantarctic and southern hemispheric species of *Campylopus* (Dicranaceae), with contributions to the origin and speciation of the genus. *Journal of the Hattori Botanical Laboratory*, **64:** 367–387.

———. **1989:** The genus *Chorisodontium* (Dicranaceae, Musci) in the Neotropics. *Tropical Bryology*, **1:** 22–24.

———. **1990*a*:** *Campylopus*, a modern and successful genus!? *Tropical Bryology*, **2:** 91–101.

———. **1990*b*:** A short survey of the *Campylopus*-flora of New Caledonia. *Cryptogamie, Bryologie – Lichénologie*, **11:** 369–375.

———. **1991*a*:** A phenetic and cladisitic study of the Campylopodioideae. *Journal of the Hattori Botanical Laboratory*, **69:** 65–78.

———. **1991*b*:** Dicranaceae: Campylopodioideae, Paraleucobryoideae. *Flora Neotropica Monograph*, **54:** 1–238.

———. **1992:** A revision of the East-Asian species of *Campylopus*. *Journal of the Hattori Botanical Laboratory*, **71:** 133–164.

———. **1994*a*:** A new synopsis of the *Campylopus* species from Australia. *Journal of Bryology*, **18:** 311–327.

———. **1994*b*:** Taxonomische Notizen zur Gattung *Campylopus* XVI. *Nova Hedwigia*, **59:** 147–155.

———. **1995:** Laubmoose. *In*: W. Frey, J.-P. Frahm, E. Fischer & W. Lobin, *Die Moos- und Farnpflanzen Europas. Kleine Kryptogamenflora begründed von Helmut Gams*. **4**. Stuttgart, Jena and New York, Gustav Fischer Verlag: pp. 121–318.

———. **1996:** Revision der Gattung *Rhacocarpus* Lindb. (Musci). *Cryptogamie, Bryologie – Lichénologie*, **17:** 39–65.

———. **1999:** A type catalogue of Campylopodioideae and Paraleucobryoideae (Musci, Dicranaceae), Part II, *Campylopus*. *Tropical Bryology*, **16:** 17–102.

———. **2005:** A contribution to the *Campylopus* flora of Chile. *Tropical Bryology*, **26:** 39–43.

——— **& J. Eggers. 2001:** *Lexikon deutschsprachiger Bryologen*. Norderstedt, Books on Demand GmbH: 672 pp.

———**, T. Klöcker, R. Schmidt & Ch. Schöter. 2000:** Revision der Gattung *Amphidium* (Musci, Dicranaceae). *Tropical Bryology*, **18:** 171–184.

———**, H. Börner, N. Streiber, B. Wallau & S. Weitkus. 1996:** Revision der Gattung *Conostomum* (Musci, Bartramiaceae). *Tropical Bryology*, **12:** 97–114.

Fransén, S. 1995: A taxonomic revision of Neotropical *Bartramia* section *Vaginella* C. Müll. *Lindbergia*, **20:** 147–179.

———. **2004*a*:** A taxonomic revision of extra-Neotropical *Bartramia* section *Vaginella* C. Müll. *Lindbergia*, **29:** 73–107.

———. **2004*b*:** A taxonomic revision of *Bartramia* section *Bartramia*. *Lindbergia*, **29:** 113–122.

———. **2005:** Taxonomic revision of the moss genus *Bartramia* Hedw. sections *Bartramia* and *Vaginella* C. Müll. Göteborg, Göteborg University, Faculty of Science: 122 pp.

Frenot, Y., S. L. Chown, J. Whinam, P. M. Selkirk, P. Convey, M. Skotnicki & D. M. Bergstrom. 2005: Biological invasions in the Antarctic: extent, impacts and implications. *Biological Reviews*, **80:** 45–72.

Frey, W. 1977: Neue Vorstellungen über die Verwandschaftsgruppen und die Stammegeschichte der Laubmoose. *In*: W. Frey, H. Hurka & F. Oberwinkler (eds), *Beiträge zur Biologie der niederen Pflanzen. Systematik – Stammgeschichte – Ökologie*. Stuttgart and New York, Gustav Fischer Verlag: pp. 117–139.

——— **& J.-P. Frahm. 1987:** The twist mechanism in the cygneous setae of the genus *Campylopus*. Morphology, structure and function. *Nova Hedwigia*, **44:** 291–304.

——— **& H. Kürschner. 1977:** Studies on the bryophyte flora and vegetation of the Mohammed Reza Shah National Park, N. Iran. *Iranian Journal of Botany*, **1:** 137–153.

——— **& ———. 1988:** Bryophytes of the Arabian Peninsula and Socotra. Floristics, phytogeography and definition of the Xerothermic Pangaean element. Studies on Arabian bryophytes 12. *Nova Hedwigia*, **46:** 37–120.

——— **& ———. 1991:** Conspectus bryophytorum orientalium et arabicorum. An annotated catalogue of the bryophytes of Southwest Asia. *Bryophytorum Bibliotheca*, **39:** 1–181.

———**, I. Herrnstadt & H. Kürschner. 1990:** *Pterygoneurum crossidioides* (Pottiaceae, Musci), a new species to the desert flora of the Dead Sea. *Nova Hedwigia*, **50:** 239–244.

Frisvoll, A. A. 1981: Fifteen bryophytes new to Svalbard, including notes on some rare and interesting species. *Lindbergia*, **7:** 91–102.

———. **1982:** Revision of Svalbard bryophytes. II. The genus *Racomitrium*. *Lindbergia*, **9:** 41–52.

———. **1983*a*:** A taxonomic revision of the *Racomitrium canescens* group (Bryophyta, Grimmiales). *Gunneria*, **41:** 1–181.

———. **1983*b*:** Bryophytes from Jan Mayen, including twenty-five species new to the island. *The Bryologist*, **86:** 332–341.

———. **1984*a*:** Taxonomic note on *Racomitrium crispulum* (Hook. f. et Wils.) Hook. f. et Wils. *Journal of Bryology*, **13:** 285–290.

———. **1984*b*:** Lectotypification of *Racomitrium affine* (Web. & Mohr) Lindb., *R. heterostichum* (Hedw.) Brid., *R. microcarpon* (Hedw.) Brid., *R. obtusum* (Brid.) Brid., and *R. sudeticum* (Funck) Bruch & Schimp. *Journal of the Hattori Botanical Laboratory*, **57:** 299–318.

———. **1986:** Southern hemisphere synonyms of *Racomitrium sudeticum* (Funck) Bruch et Schimp. *Journal of Bryology*, **14:** 339–346.

———. **1987 ['1986']:** Lectotypification of *Racomitrium lanuginosum* (Hedw.) Brid. *Lindbergia*, **12:** 83–86.

———. **1988:** A taxonomic revision of the *Racomitrium heterostichum* group (Bryophyta, Grimmiaceae) in N. and C. America, Europe and Asia. *Gunneria*, **59:** 1–289.

——— **& A. Elvebakk. 1996:** Part 2. Bryophytes. *In*: A. Elvebakk & P. Prestrud (eds), *A catalogue of Svalbard plants, fungi, algae and cyanobacteria. Norsk Polarinstitutt Skrifter*, **198:** 57–172.

——— **& J. Lewinsky. 1981:** The genus *Orthotrichum* in Svalbard. *Lindbergia*, **7:** 2–9.

Fritsch, R. 1991: Index to bryophyte chromosome counts. *Bryophytorum Bibliotheca*, **40:** 1–352.

Froehlich, J. 1955: Die von Dr. A. Gilli in den Jahren 1949 bis 1951 in Zentral-, Ost- und Nordost-Afghanistan gesammelten Bryophyten. *Mitteilungen der Thüringischen Botanischen Gesellschaft*, **1:** 59–70.

———. **1964:** Bryophyten aus Afghanistan und Nordwest-Pakistan. *Annalen des Naturhistorichen Museum Wien*, **67:** 149–158.

Fudali, E. 1992: Further spreading of *Campylopus introflexus* (Musci, Dicranaceae) in Poland. *Fragmenta Floristica et Geobotanica*, **37:** 503–506.

Fulford, M. H. 1963: Continental drift and the distribution patterns in leafy Hepaticae. *In*: *Polar wandering and continental drift*. Society of Economic Paleontologists and Mineralogists. A Division of the American Association of Petroleum Geologists: pp. 140–145.

Funck, H. Ch. 1808: *Cryptogamische Gewächse besonders des Fichtelgebirg's*. **11**. Leipzig, in Commission bey Iohann Ambrosius Barth: 4 pp.

———. **1820:** *Deutschlands Mooses. Ein Taschenherbarium zum Gebrauch auf botanischen Excursionen*. Baireuth, F. C. Birner: 71 pp.

Furmańczyk, K. & R. Ochyra. 1982: Plant communities of the Admiralty Bay region (King George Island, South Shetland Islands, Antarctic). I. Jasnorzewski Gardens. *Polish Polar Research*, **3:** 25–39.

Fürnrohr, H. Ch. 1829*a*: [Review of] Sam. El. a Bridel-Brideri, a consil. secret. legation. Saxo-Gothanae etc. *Bryologia universa seu systematica ad novam methodum dispositio, historia et descriptio omnium muscorum hucusque cognitorum cum synonymia ex auctoribus probatissimis*. Lipsiae sumtibus Joan. Ambros. Barth. 8. max. Vol. primum 1826. XLVI et 856 pag. Vol. secundum. 1827. 848 pag. *Ergänzungsblätter zur Flora oder Botanische Zeitung*, **2:** 1–60.

———. **1829*b*:** Noch ein Wort über *Barbula obtusifolia* und *Bryum erythrocarpon*; nebst einigen anderen bryologischen Bemerkungen. *Flora oder Botanische Zeitung*, **12:** 577–585; 593–601.

G

Gaertner, P. G., B. Meyer & J. Scherbius. 1802: *Oekonomisch-technische Flora der Wetterau*. **3**(2). Frankfurt am Main, Philipp Heinrich Guilhauman: 388 + 32 pp.

Gallego, M. T. 2002: Pottiaceae: *Syntrichia*. *In*: J. Guerra & R. M. Ros (eds), *Flora briofítica ibérica*. Murcia, Sociedad Española de Briología: 31 pp.

———. **2005:** A taxonomic study of the genus *Syntrichia* Brid. (Pottiaceae, Musci) in the Mediterranean region and Macaronesia. *Journal of the Hattori Botanical Laboratory*, **98:** 47–122.

——— **& M. J. Cano. 1998:** *Aloina brevirostris* (Hook. & Grev.) Kindb., new for Greece, the Iberian Peninsula and the Canary Islands. *Journal of Bryology*, **20:** 245–246.

——— **&** ———. **2000:** Género modelo *Aloina* (Pottiaceae). *In*: J. Guerra & R. M. Ros (eds), *Flora briofítica ibérica*. Murcia, Sociedad Española de Briología: 16 pp.

———, ———, **R. M. Ros & J. Guerra. 1998:** *Aloina humilis* sp. nov. (Bryopsida, Pottiaceae) from the Canary Islands. *Nova Hedwigia*, **67:** 119–124.

———, ———, ——— **&** ———. **1999:** The genus *Aloina* (Pottiaceae, Musci) in the Mediterranean region and neighbouring areas. *Nova Hedwigia*, **69:** 173–194.

———, ———, ——— **&** ———. **2002*a*:** New data on a Circum-Tethyan group of *Syntrichia* (Pottiaceae, Bryophyta): the *S. caninervis* complex. *Systematic Botany*, **27:** 643–653.

———, ———, ——— **&** ———. **2002*b*:** An overview of *Syntrichia ruralis* complex (Pottiaceae: Musci) in the Mediterranean region and neighbouring areas. *Botanical Journal of the Linnean Society*, **138:** 209–224.

Gangulee, H. C. 1969–1972: *Mosses of eastern India and adjacent regions – a monograph*. **1**. Calcutta, Privately published: pp. i–iv + 1–170 (Fascicle 1, 1969); pp. v–xiii + 171–566 (Fascicle 2, 1971); pp. xiv–xix + 567–830 (Fascicle 3, 1972).

———. **1974:** *Mosses of eastern India and adjacent regions – a monograph*. **2**(4). Calcutta, Privately published: pp. xx–xxvi + 831–1134.

Gao, Ch., D. H. Vitt & S. He. 1999: Dicranaceae. *In*: Ch. Gao, M. R. Crosby & S. He (eds), *Moss Flora of China. English version*. **1**. *Sphagnaceae – Leucobryaceae*. Beijing – New York, Science Press and St. Louis, Missouri Botanical Garden: pp. 90–241.

Gardiner, A., M. Ignatov, S. Huttunen & A. Troitsky. 2005: On resurrection of the families Pseudoleskeaceae Schimp. and Pylaisiaceae Schimp. (Musci, Hypnales). *Taxon*, **54:** 651–663.

Garilleti, R., F. Lata & V. Mazimpaka. 2006: *Orthotrichum anodon* (Orthotrichaceae), a new species from California, and its relationships to other *Orthotricha* with puckered mouth. *The Bryologist*, **109:** 188–196.

Gee, C. T. 1989: Revision of the Late Jurassic/Early Cretaceous flora from Hope Bay, Antarctica. *Palaeontographica, Abteilung B*, **213:** 149–214.

Geheeb, A. 1878: Mousses nouvelles de l'Afrique méridionale. *Revue Bryologique*, **5:** 68–72.

———. **1910:** Bryologia atlantica. Die Laubmoose der atlantischen Inseln (unter Ausschluß der europäischen und arktischen Gebiete). *Bibliotheca Botanica*, **73:** 1–73 + pls. i–xx.

Geissler, P. 1984: Notulae bryofloristicae helveticae. *Candollea*, **39:** 641–646.

——— **& E. Maier. 1995:** Lectotypifications of Central European *Grimmia* species (Musci, Grimmiaceae). *Candollea*, **50:** 495–514.

Gepp, A. 1902: Cryptogamia. Musci. *In*: *Report on the collections of natural history made in Antarctic regions during the voyage of the 'Southern Cross'*. London, British Museum (Natural History): p. 319.

Ghahreman, A., K. Mehdigholi, T. Nejad Sattari & F. Attar. 2003: A preliminary study of moss flora of Golestan Province and comparison with Iranian mosses. *The Iranian Journal of Botany*, **10:** 63–81.

Giacomini, V. 1955: Sull' autonomia specifica e sul ciclo di forme di *Campylopus polytrichoides* De Not. *Atti dell'Istituto Botanico dell'Università di Pavia, Seria 5*, **13:** 45–83.

Giese, M. & J.-P. Frahm. 1985: A revision of *Campylopodium* (C. Müll.) Besch. *Lindbergia*, **11:** 125–133.

Gilbert, G. E., T. J. Early & H. T. King, H.T. 1969: Biological survey of Ellsworth Land. *Antarctic Journal of the United States*, **4:** 95–96.

Gimingham, C. H. 1967: Quantitative community analysis of bryophyte ecology on Signy Island. *In*: J. E. Smith (organiser), *A discussion on the Antarctic terrestrial ecosystem. Philosophical Transactions of the Royal Society of London, Series B, Biological Sciences*, **252:** 251–259.

——— & R. I. Lewis Smith. 1970: Bryophyte and lichen communities in the maritime Antarctic. *In*: M. W. Holdgate (ed.), *Antarctic ecology*. **2**. London, Academic Press: pp. 752–785.

——— & ———. 1971: Growth form and water relations of mosses in the maritime Antarctic. *British Antarctic Survey Bulletin*, **25:** 1–21.

Given, D. R., C. D. Meurk & M. Foggo. 1991: *Checklist of the flora of Terra Nova Bay and Wood Bay, North Victoria Land and southwestern Ross Island, Ross Dependency, Antarctica*. Christchurch, DSIR Land Resources and Lower Hutt, Central Institute of Technology: 7 pp.

Gmelin, J. F. 1791: *Caroli à Linné, ... Systema naturae per regna tria naturae, secundum classes, ordines, genera, species, cum characteribus differentiis*. Editio 13. **2**(2). Georg Emanuel Beer, Leipzig: pp. 885–1661.

Godley, E. J. 1960: The botany of southern Chile in relation to New Zealand and the Subantarctic. *In*: C. F. A. Pantin (ed.), *A discussion on the biology of the southern cold temperate zone. Proceedings of the Royal Society of London, Series B*, **152:** 457–475.

Goebel, K. I. E. 1930: *Organographie der Pflanzen inbesondere der Archegoniaten und Samenpflanzen*. **2**. Ed. 3. Jena, Gustav Fischer: pp. i–x + 643–1378.

Goffinet, B. 1997: The Rhachitheciaceae: revised circumscription and ordinal affinities. *The Bryologist*, **100:** 425–439.

———. 1998: A reconsideration of the affinities of *Kleioweisiopsis*, *Pleurozygodontopsis*, *Trigonodictyon*, and the Microtheciellaceae (Bryopsida, Orthotrichales). *Journal of Bryology*, **20:** 69–81.

——— & W. R. Buck. 2004: Systematics of the Bryophyta (mosses): from molecules to a revised classification. *In*: B. Goffinet, V. Hollowell & R. Magill (eds), *Molecular systematics of bryophytes. Monographs in Systematic Botany from the Missouri Botanical Garden*, **98:** 205–239.

——— & D. H. Vitt. 1998: Revised generic classification of the Orthotrichaceae based on a molecular phylogeny and comparative morphology. *In*: J. W. Bates, N. W. Ashton & J. G. Duckett (eds), *Bryology for the twenty-first century*. Leeds, Maney Publishing and the British Bryological Society: pp. 143–159.

———, R. J. Bayer & D. H. Vitt. 1998: Circumscription and phylogeny of the Orthotrichales (Bryopsida) inferred from *rbc*L sequence analyses. *American Journal of Botany*, **85:** 1324–1337.

———, W. R. Buck & M. A. Wall. 2007: *Orthotrichum freyanum* (Orthotrichaceae), a new epiphytic moss from Chile. *In*: H. Kürschner, T. Pfeiffer & M. Stech (eds.), *Festschrift to Wolfgang Frey – a scientist's life between deserts and rain forest. Nova Hedwigia Beiheft*, **131:** 1–11.

———, C. J. Cox, A. J. Shaw & T. A. J. Hedderson. 2001: The Bryophyta (mosses): systematic and evolutionary inferences from an *rps4* gene (cpDNA) phylogeny. *Annals of Botany*, **87:** 191–208.

———, J. Shaw, L. E. Anderson & B. D. Mishler. 1999: Peristome development in mosses in relation to systematic and evolution. V. Diplolepideae: Orthotrichaceae. *The Bryologist*, **102:** 581–594.

———, ———, C. J. Cox, N. J. Wickett & S. B. Boles. 2004: Phylogenetic inferences in the Orthotrichoideae (Orthotrichaceae, Bryophyta) based on variation in four loci from all genomes. *In*: B. Goffinet, V. Hollowell & R. Magill (eds), *Molecular systematics of bryophytes. Monographs in Systematic Botany from the Missouri Botanical Garden*, **98:** 270–289.

Gradstein, S. R. & H. J. M. Sipman. 1978: Taxonomy and world distribution of *Campylopus introflexus* and *C. pilifer* (= *C. polytrichoides*): a new synthesis. *The Bryologist*, **81:** 114–121.

——— & W. A. Weber. 1982: Bryogeography of the Galapagos Islands. *Journal of the Hattori Botanical Laboratory*, **52:** 127–152.

———, S. P. Churchill & N. Salazar-Allen. 2001: Guide to the bryophytes of tropical America. *Memoirs of the New York Botanical Garden*, **86:** i–viii + 1–577.

Gray, N. F., C. H. E. Wyborn & R. I. Lewis Smith. 1982: Nematophagous fungi from the maritime Antarctic. *Oikos*, **38:** 194–201.

Green, T. G. A., D. Kulle, S. Pannewitz, L. G. Sancho & B. Schroeter, B. 2005: UV-A protection in mosses growing in continental Antarctica. *Polar Biology*, **28:** 822–827.

Greene, D. M. 1972: The herbarium of the British Antarctic Survey. *British Antarctic Survey Bulletin*, **31:** 107–109.

———. 1986: *A conspectus of the mosses of Antarctica, South Georgia, the Falkland Islands and southern South America*. Cambridge, British Antarctic Survey: 314 pp.

——— & S. W. Greene. 1975: The data bank of the British Antarctic Survey's botanical section. *In*: J. P. Brenen, R. Ross & J. T. Williams (eds), *Computers in botanical collections*. New York, Plenum Publishing Corporation: pp. 79–87.

Greene, S. W. 1964*a*: Plants of the land. *In*: R. E. Priestley, R. J. Adie & G. de Q. Robin (eds), *Antarctic research. A review of British scientific achievements in Antarctica*. London, Butterworth: pp. 239–253.

———. 1964*b*: Problems and progress in Antarctic bryology. *In*: R. Carrick, M. W. Holdgate & J. Prévost (eds), *Biologie antarctique. Proceedings of the First Symposium on Antarctic Biology, Paris, 2–8 Sept. 1962*. Paris, Hermann: pp. 173–179.

———. 1967*a*: The changing pattern of Antarctic botanical studies. *In*: *Japanese Antarctic Research Expedition Scientific Reports, Special Issue*. **1**. *Proceedings of the Symposium on Pacific–Antarctic Sciences*. Tokyo, National Science Museum, Department of Polar Research: pp. 236–244.

———. 1967*b*: Bryophyte distribution. *In*: S. W. Greene, J. L. Gressitt, D. Koob, G. A. Llano, E. D. Rudolph, R. Singer, W. C. Steere & F. C. Ugolini (eds), *Antarctic map folio series*. **5**. *Terrestrial life in Antarctica*. New York, American Geographical Society: pp. 11–13 + pl. 5.

———. **1968*a*:** Studies in the Antarctic bryology I. – A basic check list for mosses. *Revue Bryologique et Lichénologique, Nouvelle Série*, **36:** 132–138.

———. **1968*b*:** Studies in Antarctic bryology. II. – *Andreaea, Neuroloma. Revue Bryologique et Lichénologique, Nouvelle Série*, **36:** 139–146.

———. **1970:** Studies on *Colobanthus quitensis* (Kunth) Bartl. and *Deschampsia antarctica* Desv.: I. Introduction. *British Antarctic Survey Bulletin*, **23:** 19–24.

———. **1973 ['1974']:** A synoptic flora of South Georgian mosses: I. *Dendroligotrichum*, *Polytrichum* and *Psilopilum*. *British Antarctic Survey Bulletin*, **36:** 1–32.

———. **1975:** The Antarctic moss *Sarconeurum glaciale* (C. Muell.) Card. et Bryhn in southern South America. *British Antarctic Survey Bulletin*, **41–42:** 187–191.

——— **& D. M. Greene. 1963:** Check list of the sub-Antarctic and Antarctic vascular flora. *Polar Record*, **11:** 411–418.

——— **& A. Holtom. 1971:** Studies on *Colobanthus quitensis* (Kunth) Bartl. and *Deschampsia antarctica* Desv.: III. Distribution, habitats and performance in the Antarctic botanical zone. *British Antarctic Survey Bulletin*, **26:** 1–29.

——— **& D. W. H. Walton. 1975:** An annotated check list of the sub-Antarctic and Antarctic vascular flora. *Polar Record*, **17:** 473–484.

———**, D. M. Greene, P. D. Brown & J. M. Pacey. 1970:** Antarctic moss flora I. The genera *Andreaea*, *Pohlia*, *Polytrichum*, *Psilopilum* and *Sarconeurum*. *British Antarctic Survey Scientific Reports*, **64:** 1–118.

Greenfield, L. G. 2000: Some properties of water soluble substances from antarctic and arctic mosses and lichens. *In*: B. Schroeter, M. Schlensog & T. G. A. Green (eds), *New aspects in cryptogamic Research. Contributions in honour of Ludger Kappen. Bibliotheca Lichenologica*, **75:** 71–78.

Gremmen, N. J. M. 1982: *The vegetation of the Subantarctic islands Marion and Prince Edward.* The Hague – Boston – London, Dr. W. Junk Publishers: x + 149 pp.

Greuter, W., R. K. Brummitt, E. Farr, N. Kilian, P. M. Kirk & P. C. Silva (eds). 1993: *Names in current use for extant plant genera.* Königstein, Koeltz Scientific Books: xxvii + 1464 pp.

———**, H. M. Burdet, W. G. Chaloner, V. Demoulin, R. Grolle, D. L. Hawksworth, D. H. Nicolson, P. C. Silva, F. A. Stafleu, E. G. Voss & J. McNeill. 1988:** *International Code of Botanical Nomenclature adopted by the Fourteenth International Botanical Congress, Berlin, July–August 1987. Regnum Vegetabile.* **118**. Königstein, Koeltz Scientific Books: xiv + 328 pp.

Greven, H. C. 1995: Grimmia *Hedw. (Grimmiaceae, Musci) in Europe*. Leiden, Backhuys Publishers: 160 pp.

———. **1998*a*:** Synopsis of *Grimmia* Hedw. in New Zealand, including *Grimmia wilsonii*, sp. nov. *Journal of Bryology*, **20:** 389–402.

———. **1998*b*:** *Grimmia* Hedw. in Iceland, including *Grimmia grisea* Cardot, new to Europe. *Lindbergia*, **23:** 91–93.

———. **1999:** Key to *Grimmia*, *Coscinodon* and *Hydrogrimmia* in Europe. *Journal of Bryology*, **20:** 3117–121.

———. **2000*a*:** *Grimmia herzogii*, a misunderstood South American endemic. *The Bryologist*, **103:** 760–761.

———. **2000*b*:** Synopsis of *Grimmia* in Australia. *Journal of Bryology*, **22:** 217–222

———. **2002*a*:** *Grimmia nevadense* [sic!], a new species from California. *The Bryologist*, **105:** 273–275.

———. **2002*b*:** *Grimmia limprichtii*: a distinct taxon. *Herzogia*, **14:** 211–212.

———. **2002*c*:** *Grimmia herzogii* Broth., a misunderstood South American endemic. *The Bryologist*, **103:** 760–761.

———. **2003:** *Grimmias of the world.* Leiden, Backhuys Publishers: 247 pp.

Greville, R. K. & G. A. W. Arnott. 1822–1826: *Tentamen methodi muscorum; or, a new arrangement of the genera of mosses, with characters, and observations on their distribution, history and structure*. Edinburgh: iv + 127 pp.

Griffin, D. III. 1975: Additions to the moss flora of Venezuela. *The Bryologist*, **78:** 212–215.

———. **1994*a*:** Key to the families and genera of the Bryales. *In*: A. J. Sharp, H. Crum & P. M. Eckel (eds.), *The moss flora of Mexico. Part one Sphagnales to Bryales. Memoirs of the New York Botanical Garden*, **69:** 452–453.

———. **1994*b*:** Bartramiaceae. *In*: A. J. Sharp, H. Crum & P. M. Eckel (eds.), *The moss flora of Mexico. Part one Sphagnales to Bryales. Memoirs of the New York Botanical Garden*, **69:** 537–574.

——— **& W. R. Buck. 1989:** Taxonomic and phylogenetic studies on the Bartramiaceae. *The Bryologist*, **92:** 368–380.

Grims, F. (with collaboration of H. Köckinger, R. Krisai, A. Schriebl, M. Suanjak, H. Zechmeister & F. Ehrendorfer). 1999: Die Laubmoose Österreichs. Catalogus Florae Austriae, II. Teil, Bryophyten (Moose), Heft 1, Musci (Laubmoose). *Biosystematik and Ecology Series, Österreichische Akademie der Wissenschaften*, **15:** i–vii + 1–418.

Grolle, R. 1963: Monographie der Lebermoosgattung *Leptoscyphus* Mitt. *Nova Acta Leopoldina, Neue Folge*, **161:** 1–143.

———. **1965:** Die Lebermoosgattungen *Blepharidophyllum* Ångstr. und *Krunodiplophyllum* nov. gen. (Scapaniaceae). *Journal of the Hattori Botanical Laboratory*, **28:** 55–74.

———. **1966:** *Herzogobryum* – eine beblätterte Lebermoosgattung mit dorsal verzanten Merophyten. *Österreichische Botanische Zeitschrift*, **113:** 220–234.

———. **1969:** Grossdisjunktionen in artarealen lateinamerikanischer Lebermoose. *In*: E. J. Fittkau, J. Illies, H. Klinge, G. H. Schwabe & H. Sioli (eds), *Biogeography and ecology in South America*. **2**. The Hague, Dr. W. Junk N.V: pp. 562–582.

———. **2002:** The Hepaticae and Anthocerotae of the subantarctic and temperate islands in the eastern Southern Hemisphere (90° E to 0°): an annotated catalogue. *Journal of Bryology*, **24:** 57–80.

Grout, A. J. 1932: Subfamily Hypneae and subfamily Stereophylleae. *In*: A. J. Grout (ed.), *Moss flora of North America north of Mexico*. **3**(3). Newfane, Vermont, privately published: pp. 121–166 + pls 32–43.

———. **1935:** Family Meesiaceae. *In*: A. J. Grout (ed.), *Moss flora of North America north of Mexico*. **2**(3). Newfane, Vermont, privately published: pp. 180–183 + pls. 63–64, 65B.

———. **1939:** *Phascum*. *In*: A. J. Grout (ed.), *Moss flora of North America north of Mexico*. **1**(4). Newfane, Vermont, privately published: pp. 195–197 + pl. 92A.1a–1g.

Guerra, J. & M. J. Cano. 2000: A taxonomic contribution on the European cleistocarpous species of Pottiaceae (Musci). *Journal of Bryology*, **22:** 91–97.

———, ——— & R. M. Ros. 1995a: On the identity of *Pterygoneurum macleanum* and *P. kemsleyi* (Pottiaceae, Musci). *The Bryologist*, **98:** 45–59.

———, ——— & ———. 1995b: El género *Pterygoneurum* Jur. (Pottiaceae, Musci) en la Península Ibérica. *Cryptogamie, Bryologie – Lichénologie*, **16:** 165–175.

H

Haberlandt, G. 1886: Beiträge zur Anatomie und Physiologie der Laubmoose. *Jahrbücher für Wissenschaftliche Botanik*, **17:** 359–498 + pls. 21–27.

Hagen, I. 1909: Forarbejder til en Norsk løvmosflora. IX. Grimmiaceae. *Det Kgl. Norske Videnskabers Selskabs Skrifter*, **1909(5):** 1–94.

———. 1915: Forarbejder til en Norsk løvmosflora. XX. Dicranaceae. *Det Kgl. Norske Videnskabers Selskabs Skrifter*, **1914(1):** 1–192.

Haji Mohamed, M. A. 1979: A taxonomic study of *Bryum billardieri* Schwaegr. and related species. *Journal of Bryology*, **10:** 401–465.

Halle, T. G. 1913: The Mesozoic flora of Graham Land. *In*: *Wissenschaftliche Ergebnisse der Schwedischen Südpolar-Expedition 1901–1903 unter Leitung von Dr. Otto Nordenskjöld*. **3**(14). Stockholm, Lithographisches Institut des Generalstabs: 123 pp. + 9 pls.

Haller, A. von. 1768: *Historia stirpium indigenarum Helvetiae inchoata*. **3**. Bernae, sumptibus Societatis Typographicae: 204 pp. + pls. xlv–xlviii.

———. 1742: *Enumeratio methodica stirpium Helvetiae indigenarum. Qua omnium brevis descriptio et synonymia compendium virium medicarum dubiarum declaratio novarum et rariorum uberior historia et icones continentur*. **1–2**. Gottingae, ex Officina Academica Abrami Vandenhoek: 794 pp. + 24 pls.

Hampe, E. 1846: Ueber *Hyophila javanica* Brid. *Botanische Zeitung (Berlin)*, **4:** 266–269.

———. 1847: Ein Referat über die Columbischen Moose, welche von Herrn Moritz gesammelt wurden. *Linnaea*, **20:** 65–98.

———. 1853: Ueber Classification der Moose. *Botanische Zeitung (Berlin)*, **11:** 297–309, 321–332.

———. 1865: Musci. *In*: J. Triana & J. E. Planchon (eds), *Prodromus florae novo-granatensis ou énumération des plantes de la Nouvelle-Grenade avec description des espèces nouvelles. Annales des Sciences Naturelles Botanique, Série 5*, **4:** 324–378.

———. 1867: *Ditrichum* Timm. statt *Leptotrichum* Hpe. *Flora*, **50:** 181–182.

———. 1876: Musci novi Musei Melbournei. Continuatio. *Linnaea*, **40:** 301–326.

Haring, I. M. 1938: *Tortella*. *In*: A. J. Grout (ed.), *Moss flora of North America north of Mexico*. **1**(3). Newfane, Vermont, privately published: pp. 165–170 + pl. 78A.

Harvey, W. H. 1836: Musci indici. *In*: W. J. Hooker, *Icones plantarum; or figures, with brief descriptive characters and remarks of new or rare plants*. **1**(1). London, Longman, Rees, Orme, Brown, Green, & Longman: tab. 17–24.

———. 1839: *Schizymenium bryoides*. *In*: W. J. Hooker, *Icones plantarum; or figures, with brief descriptive characters and remarks of new or rare plants*. **3**(1). London, Longman, Rees, Orme, Brown, Green, & Longman: tab. 202.

——— & J. D. Hooker. 1840: Musci indici; or list of mosses collected in the East Indies by Dr Wallich; with references to the figures of the new species published in Hooker's Icones plantarum, vol. I. tabs. XVII–XXIV; by the honorable W. H. Harvey: to which are added those collected by Dr Royle in the northern part of India, by J. D. Hooker. *London Journal of Botany*, **2:** 1–21.

Hastings, R. I. 1996: The genus *Coscinodon* (Bryopsida, Grimmiaceae) in South America, including a new species. *The Bryologist*, **99:** 418–427.

———. 1999: Taxonomy and biogeography of the genus *Coscinodon* (Bryopsida, Grimmiaceae) in North America, including a new species. *The Bryologist*, **102:** 265–286.

——— & H. Deguchi. 1997: Taxonomy of *Coscinodon arctolimnius* (Bryopsida, Grimmiaceae) including a new subspecies from Asia. *The Bryologist*, **100:** 50–55.

——— & H. C. Greven. 2007a: *Grimmia milleri* sp. nov. (Grimmiaceae) from northeastern North America and the status of *Grimmia afroincurva*. *The Bryologist*, **110:** 500–505.

——— & ———. 2007b: *Grimmia*. *In*: R. H. Zander (Lead ed.), *Flora of North America north of Mexico*. **27**. *Bryophytes: mosses (1)*. New York – Oxford, Oxford University Press: pp. 225–258.

Hässel de Menendez, G. G., S. W. Greene & C. M. Matteri. 1984: The occurrence and distribution of bryophytes in southern Patagonia between latitude 51° and 52° S. *Journal of the Hattori Botanical Laboratory*, **55:** 45–64.

Hattersley-Smith, M. A. 1991: The history of place-names in the British Antarctic Territory. *British Antarctic Survey Scientific Reports*, **113:** 1–670.

He, S. 1998: A checklist of the mosses of Chile. *Journal of the Hattori Botanical Laboratory*, **85:** 103–189.

Headland, R. K. 1989: *Chronological list of Antarctic expeditions and related historical events*. Cambridge – Sydney, Cambridge University Press: ix + 730 pp.

Heap, J. 1994: The Antarctic Treaty System: Introduction. *In*: J. Heap (ed.), *Handbook of the Antarctic Treaty system*. **1**. *General measures*. Washington, D.C., U.S. Department of State: pp. 8–19.

Hébant, Ch. 1974: Studies on the development of the conducting tissue-system in the gametophytes of some Polytrichales II. Development and structure at maturity of the hydroids of the central strand. *Journal of the Hattori Botanical Laboratory*, **38:** 565–607.

———. 1977: The conducting tissues of bryophytes. *Bryophytorum Bibliotheca*, **10:** i–xi + 1–157 + pls. 1–80.

Hébrard, J. P. 1970a: Muscinées récoltées aux Iles Kerguelen et Crozet pendant la campagne d'été des Terres Australes et Antarctiques Françaises (1969). *Revue Bryologique et Lichénologique, Nouvelle Série*, **37:** 135–162.

———. 1970b: Contribution à l'étude de quelques groupments muscinaux des Iles Kerguelen. *Revue Bryologique et Lichénologique, Nouvelle Série*, **37:** 315–343.

Hedderson, T. A., R. Chapman & C. J. Cox. 1998: Bryophytes and the origins and diversification of land plants: new evidence from molecules. *In*: J. W. Bates, N. W. Ashton & J. G. Duckett (eds), *Bryology for the twenty-first century*. Leeds, Maney Publishing and the British Bryological Society: pp. 65–77.

———, ——— & **W. L. Rootes. 1996:** Phylogenetic relationships of bryophytes inferred from small-subunit rRNA gene sequences. *Plant Systematics and Evolution*, **200:** 213–224.

———, **C. J. Cox & J. G. Gibbins. 1999:** Phylogenetic relationships of the Wardiaceae (Musci); evidence from 18s rRNA and rps4 gene sequences. *The Bryologist*, **102:** 26–31.

Hedenäs, L. 1987*a*: North European mosses with axillary rhizoids, a taxonomic study. *Journal of Bryology*, **14:** 429–439.

———. **1987*b*:** On the ontogeny of alar cells in *Drepanocladus aduncus*, *D. exannulatus* and some other species. *Journal of Bryology*, **14:** 753–759.

———. **1988:** *Isopterygiopsis alpicola* (Lindb. & Arn.) Hedenäs comb. nov. *Journal of Bryology*, **15:** 495.

———. **1989*a*:** The genera *Scorpidium* and *Hamatocaulis*, gen. nov., in northern Europe. *Lindbergia*, **15:** 8–36.

———. **1989*b*:** The genus *Sanionia* (Musci) in northwestern Europe, a taxonomic revision. *Annales Botanici Fennici*, **26:** 399–419.

———. **1989*c*:** Some neglected character distribution patterns among the pleurocarpous mosses. *The Bryologist*, **92:** 157–163.

———. **1990:** Axillary hairs in pleurocarpous mosses – a comparative study. *Lindbergia*, **15:** 166–180.

———. **1992 ['1990']:** The genus *Pseudocalliergon* in northern Europe. *Lindbergia*, **16:** 80–99.

———. **1993*a*:** Field and microscope keys to the Fennoscandian species of the *Calliergon–Scorpidium–Drepanocladus* complex, including some related or similar species. Märsta, Biodetektor: 79 pp.

———. **1993*b*:** A generic revision of the *Calliergon–Scorpidium–Drepanocladus* group. *Journal of Bryology*, **17:** 447–479.

———. **1994*a*:** What is a hypnoid peristome? *The Bryological Times*, **78:** 3.

———. **1994*b*:** The basal pleurocarpous diplolepidous mosses – a cladistic approach. *The Bryologist*, **97:** 225–243.

———. **1995:** Higher taxonomic level relationships among diplolepidous pleurocarpous mosses – a cladistic overview. *Journal of Bryology*, **18:** 723–781.

———. **1996*a*:** A cladistic overview of the 'Hookeriales'. *Lindbergia*, **21:** 107–143.

———. **1996*b*:** Taxonomic and nomenclatural notes on Australian Brachytheciaceae. *Nova Hedwigia*, **62:** 451–462.

———. **1996*c*:** On the taxonomic position of *Struckia* C. Müll. *Journal of the Hattori Botanical Laboratory*, **80:** 241–245.

———. **1996*d*:** A cladistic evaluation of relationships between the Hookeriales, the Sematophyllaceae and some other genera. *Lindbergia*, **21:** 49–82.

———. **1997*a*:** A partial generic revision of *Campylium* (Musci). *The Bryologist*, **100:** 65–88.

———. **1997*b*:** The *Drepanocladus s. str.* species with excurrent costae (Musci, Amblystegiaceae). *Nova Hedwigia*, **64:** 535–547.

———. **1998*a*:** An evaluation of phylogenetic relationships among the Thuidiaceae, the Amblystegiaceae, and the temperate members of the Hypnaceae. *Lindbergia*, **22:** 101–133.

———. **1998*b*:** An overview of the *Drepanocladus sendtneri* complex. *Journal of Bryology*, **20:** 83–102.

———. **1998*c*:** Cladistic studies on pleurocarpous mosses: research needs, and use of results. *In*: J. W. Bates, N. W. Ashton & J. G. Duckett (eds), *Bryology for the twenty-first century*. Leeds, Maney Publishing and the British Bryological Society: pp. 125–141.

———. **1999:** How important is phylogenetic history in explaining character states in pleurocarpous mosses? *Canadian Journal of Botany*, **77:** 1723–1742.

———. **2001*a*:** Environmental factors potentially affecting character states in pleurocarpous mosses. *The Bryologist*, **104:** 72–91.

———. **2001*b*:** The identity of *Isopterygium sericifolium* Dixon. *Journal of Bryology*, **23:** 145–146.

———. **2002:** An overview of the family Brachytheciaceae (Bryophyta) in Australia. *Journal of the Hattori Botanical Laboratory*, **92:** 51–90.

———. **2003*a*:** The European species of the *Calliergon–Scorpidium–Drepanocladus* complex, including some related or similar species. *Meylania*, **28:** 1–116.

———. **2003*b*:** Amblystegiaceae (Musci). *Flora Neotropica Monograph*, **89:** i–iv + 1–107.

———. **2006:** Additional insights into the phylogeny of *Calliergon*, *Loeskypnum*, *Straminergon*, and *Warnstorfia* (Bryophyta: Calliergonaceae). *Journal of the Hattori Botanical Laboratory*, **100:** 125–134.

———. **2007:** Global diversity among pleurocarpous mosses. *The Bryologist*, **110:** 319–331.

——— **& P. Geissler. 1999:** Lectotypification of Hedwig names: holarctic pleurocarpous mosses. *Candollea*, **54:** 417–432.

——— **& A. Vanderpoorten. 2007:** The Amblystegiaceae and Calliergonaceae. *In*: A. E. Newton & R. S. Tangney (eds), *Pleurocarpous mosses. Systematics and evolution. The Systematics Association Special Volume Series*. **71**. Boca Raton, London, New York, CRC Press, Taylor & Francis Groups: pp. 163–176.

———, **G. Oliván & P. Eldenäs. 2005:** Phylogeny of the Calliergonaceae (Bryophyta) based on molecular and morphological data. *Plant Systematics and Evolution*, **252:** 49–61.

Hedgpeth, J. W. 1971: James Eights of the Antarctic (1798–1882). *In*: L. O. Quam (ed.), *Research in the Antarctic. American Association for the Advancement of Science Publication*. **93**. Washington, D.C., American Association for the Advancement of Science: pp. 3–45.

Hedwig, J. 1782: *Fundamentum historiae naturalis muscorum frondosorum concernens eorum flores, fructus, seminalem propagationem adiecta generum dispositione methodica iconibus illustratis. Pars I–II.* Lipsiae, apud Siegfried Lebrecht Crusium: xxviii + 112 pp. + x pls. (Pars I); xi + 107 + x pls. (Pars II).

———. **1784:** *Theoria generationis et fructificationis plantarum cryptogamicarum Linnaei mere propriis observationibus et experimentis superstructa*. Petropoli, Typis Academiae Imper. Scientiarum: 164 pp. + pls. i–xxxvii.

———. **1785–1787:** *Descriptio et adumbratio microscopico-analytica muscorum frondosorum nec non aliorum vegetatntium e classe cryptogamica Linnaei novorum dubiisque vexatorum*. **1**. Lipsiae, in bibliopolio I. G. Mülleriano: xviii + 109 pp. + Tab. i–xl.

———. **1789:** *Descriptio et adumbratio microscopico-analytica muscorum frondosorum nec non aliorum vegetatntium e classe cryptogamica Linnaei novorum dubiisque vexatorum*. **2**.

Lipsiae, in bibliopolio I. G. Mülleriano: i + 112 pp. + Tab. i–xl.

———. **1791–1792:** *Descriptio et adumbratio microscopico-analytica muscorum frondosorum nec non aliorum vegetatntium e classe cryptogamica Linnaei novorum dubiisque vexatorum*. **3**. Lipsiae, in bibliopolio I. G. Mülleriano: xiv + 100 pp. + Tab. i–xl.

———. **1797:** *Descriptio et adumbratio microscopico-analytica muscorum frondosorum nec non aliorum vegetatntium e classe cryptogamica Linnaei novorum dubiisque vexatorum*. **4**. Lipsiae, in bibliopolio I. G. Mülleriano: iv + 106 pp. + Tab. i–xl.

———. **1801:** *Species muscorum frondosorum descriptae et tabulis aeneis LXXVII coloratis illustratae*. Lipsiae, sumtu Joannis Ambrosii Barthii, Parisiis, apud Amand Koenig: vi + 353 pp. + pls. i–lxxvii.

Hemsley, W. B. 1884: Kerguelen Island. Musci. *In*: J. Murray (ed.), *Report on the scientific results of the voyage of H.M.S. Challenger during the years 1873–76 under the command of Captain George S. Nares, R.N., F.R.S. and the late Captain Frank Tourle Thomson, R.N. Botany*. **1**(2). London, Longmans & Co.; John Murry; Macmillan & Co.; Simpkin, Marshall, & Co.; Trübner & Co.; E. Stranford; J. D. Potter; Kegan Paul, Trench, & Co.; and Bernard Quaritch; Edinburgh, Adam & Charles Black and Douglas & Foulis; Dublin, A. Thom & Co. and Hodges, Figgis, & Co.: pp. 219–229.

Henderson, D. M. & C. W. Muirhead. 1956: Eastern Mediterranean bryophytes. *Notes from the Royal Botanic Garden Edinburgh*, **22:** 121–129.

——— **& H. T. Prentice. 1969:** Contributions to the bryophyte flora of Turkey: VIII. *Notes from the Royal Botanic Garden Edinburgh*, **29:** 235–262.

Herrnstadt, I. & C. C. Heyn. 2004: Bryopsida (mosses). *In*: C. C. Heyn & I. Herrnstadt (eds), *Flora palaestina. The bryophyte flora of Israel and adjacent regions*. Jerusalem, The Israel Academy of Sciences and Humanities: pp. 11–520 + pls. i–xiv & 1–7.

Herzog, Th. 1910: Beiträge zur Laubmoosflora von Bolivia. *Beihefte zum Botanischen Centralblatt*, **26:** 45–102.

———. **1911:** Bemerkungen zu der neuen Laubmoos-Gattung *Wollnya*. *Beihefte zum Botanischen Centralblatt*, **28:** 268–271 + pl. i.

———. **1916:** Die Bryophyten meiner zweiten Reise durch Bolivia. *Bibliotheca Botanica*, **87:** 1–347 + pls. A, i-viii.

———. **1925:** Die Moose Südbrasiliens als pflanzengeographische Zeugen. *Veröffentlichungen des Geobotanischen Institut Rübel in Zürich*, **3:** 519–539.

———. **1937:** Neue Bryophyten vom Ruwenzori and aus dem Patagonischen Inlandeisgebiet. *Repertorium Specierum Novarum Regni Vegetabilis,* **41:** 285–293 + pls. ccxxxiv–ccxxxvi.

———. **1938:** Contribución al conocimiento de la flora briofita del sur de Chile. Parte sístemática. *Archivos de la Escuela de Farmacia de la Facultad de Ciencias Médicas de Córdoba (R.A.) Sección Científica*, **7:** 1–56.

———. **1939:** Verzeichnis der gesammelten Bryophyten. *In*: Th. Herzog, G. H. Schwabe & E. Schwabe, *Zur Bryophytenflora Südchiles. (Aus dem südchilenischen Küstengebiet. Beitrag 7.). Beihefte zum Botanischen Centralblat, Abteilung B*, **60:** 15–51.

Hesselbo, A. 1923: Mosses collected on the north-coast of Greenland by the late Dr. Th. Wulff. *In*: *Den II. Thule Ekspedition til Grønland nordkyst 1916–18*. **10**. *Meddelelser om Grønland*, **64:** 271–277.

Heywood, R. B. 1967: Ecology of the freshwater lakes of Signy Island, South Orkney Islands. I. Catchment areas, drainage systems and lake morphology. *British Antarctic Survey Bulletin*, **14:** 25–43.

———. **1968:** Ecology of the freshwater lakes of Signy Island, South Orkney Islands. 2. Physical and chemical properties of the lakes. *British Antarctic Survey Bulletin*, **18:** 11–44.

Higuchi, M. & N. Nishimura. 2003: Mosses of Pakistan. *Journal of the Hattori Botanical Laboratory*, **93:** 273–291.

Hill, M. O., N. Bell, M. A. Bruggemann-Nannenga, M. Brugués, M. J. Cano, J. Enroth, K. I. Flatberg, J.-P. Frahm, M. T. Gallego, R. Garilleti, J. Guerra, L. Hedenäs, D. T. Holyoak, J. Hyvönen, M. S. Ignatov, F. Lara, V. Mazimpaka, J. Muñoz & L. Söderström. 2006: An annotated checklist of the mosses of Europe and Macaronesia. *Journal of Bryology*, **28:** 198–267.

Hilpert, F. 1933: Studien zur Systematic der Trichostomaceen. *Beihefte zum Botanische Centralblatt*, **50:** 585–706.

Hirohama, T. 1977: Spore morphology of bryophytes observed by scanning electron microscope, II. Bartramiaceae. *Bulletin of the National Science Museum, Series B (Botany)*, **3:** 37–44 + pls. 1–6 (in Japanese with English summary).

———. **1978:** Spore morphology of bryophytes observed by scanning electron microscope, IV. Grimmiaceae. *Bulletin of the National Science Museum, Series B (Botany)*, **4:** 33–42 + pls. 1–8 (in Japanese with English summary).

——— **& Z. Iwatsuki. 1980:** Surface ornamentation of rhizoids of the species of Bartramiaceae (Musci). *Journal of the Hattori Botanical Laboratory*, **48:** 259–275.

Hodgetts, N. G., H. W. Matcham & J. G. Duckett. 1999: Bryophytes collected in Lesotho, the Natal Drakensberg and the Orange Free State, southern Africa. *Journal of Bryology*, **21:** 133–155.

Hoe, W. J. 1974: Annotated checklist of Hawaiian mosses. *Lyonia*, **1:** 1–45.

Hoffmann, G. F. 1796: *Deutschlands Flora oder Botanisches Taschenbuch. Zweyter Theil für das Jahr 1795. Cryptogamie*. Erlangae, Johan Jacob Palm: lxviii [unnumbered] + 200 pp. + xiv pls.

Holdgate, M. W. 1964: Terrestrial ecology in the maritime Antarctic. *In*: R. Carrick, M. W. Holdgate & J. Prévost (eds), *Biologie antarctique. Proceedings of the First Symposium on Antarctic Biology, Paris, 2–8 Sept. 1962*. Paris, Hermann: pp. 181–194.

———. **1977:** Terrestrial ecosystems in the Antarctic. *In*: V. Fuchs & R. M. Laws (eds), *Scientific research in Antarctica. Philosophical Transactions of the Royal Society of London, Series B*, **279:** 5–25.

Holmen, K. 1960: The mosses of Peary Land North Greenland. A list of the species collected between Victoria Fjord and Danmark Fjord. *Meddelelser om Grønland*, **163(2):** 1–96.

Holmgren, P. K., W. Keuken & B. K. Schofield. 1990: *Index herbariorum. Part I. Herbaria of the world*. Ed. 8. Regnum Vegetabile. **120**. New York, New York Botanical Garden: x + 693 pp.

Holyoak, D. T. 2003: A taxonomic review of some British coastal species of the *Bryum bicolor* complex, with a description of *Bryum dyffrynense* sp. nov. *Journal of Bryology*, **25:** 107–113.

———. **2004:** Taxonomic notes on some European species of *Bryum* (Bryopsida: Bryaceae). *Journal of Bryology*, **26:** 247–264.

——— **& L. Hedenäs. 2006:** Morphological, ecological and molecular studies on the intergrading taxa *Bryum neodamense* and *B. pseudotriquetrum* (Bryopsida: Brayaceae). *Journal of Bryology*, **28:** 299–311.

——— **& N. Pedersen. 2007:** Conflicting molecular and morphological evidence of evolution within the Bryaceae (Bryopsida) and its implications for generic taxonomy. *Journal of Bryology*, **29:** 111–124.

Hooker, J. D. 1847: *The botany of the Antarctic voyage of H. M. Discovery ships* Erebus *and* Terror *in the years 1839–43, under the command of Captain Sir James Clark Ross, Kt., R. N., F. R. S.* Vol. **1**. *Flora antarctica*. London, Reeve Brothers: xii + 574 pp. + 198 pls.

——— **& W. Wilson. 1844:** Musci antarctici; being characters with brief descriptions of the new species of mosses discovered during the voyage of H. M. Discovery ships, *Erebus* and *Terror*, in the southern circumpolar regions, together with those of Tasmania. *London Journal of Botany*, **3:** 533–556.

Hooker, W. J. 1818: *Musci exotici; containing figures and descriptions of new or little known foreign mosses and other cryptogamic subjects*. **1**. London, Richard and Arthur Taylor for Longman, Hurst, Rees, Orme, and Brown: viii + 96 pls.

———. **1834:** Mr. Drummond's collections. *London Journal of Botany*, **1:** 50–60, 183–202.

———. **1835–1836:** Notice concerning Drummond's collections made chiefly in the southern and western parts of the United States. *Companion to the Botanical Magazine*, **1:** 21–26, 39–49, 95–101, 170–177 (1835); **2:** 60–64 (1836)

———. **& R. K. Greville. 1824:** On the genus *Tortula*, of the order Musci. *The Edinburgh Journal of Science*, **1:** 287–302 + pl. xii.

——— **& W. Wilson. 1844:** Enumeration of the mosses and Hepaticae, collected in Brazil by George Gardner, Esq. *London Journal of Botany*, **6:** 149–167.

——— **&** ———. **1847:** Brief characters of some new mosses, collected in New Granada by W. Purdie. *London Journal of Botany*, **6:** 289–292.

Hoppe, D. H. 1801: Nachtrag zu der Abhandlung über die in Deutschland nachsenden Arten des Widerthons (*Polytrichum*). *Botanisches Taschenbuch*, **1801:** 196–199.

Horikawa, Y. & H. Ando. 1961: Mosses of the Ongul Islands collected during the 1957–1960 Japanese Antarctic Research Expedition. *Hikobia,* **2:** 160–178.

——— **&** ———. **1963:** A review of the Antarctic species of *Ceratodon* described by Cardot. *Hikobia*, **3:** 275–280.

——— **&** ———. **1967*a*:** The mosses of the Ongul Islands and adjoining coastal areas of the Antarctic Continent. *In*: *Japanese Antarctic Research Expedition Scientific Reports*, *Special Issue*. **1**. *Proceedings of the Symposium on Pacific–Antarctic Sciences*. Tokyo, National Science Museum, Department of Polar Research: pp. 245–252.

——— **&** ———. **1967*b*:** [Mosses of Antarctica]. *Polar News*, **6:** 12–19 (in Japanese).

Hornschuch, F. 1819*a*: Neue Laubmoose. *Flora oder Botanische Zeitung (Regensburg)*, **2:** 81–111.

———. **1819*b*:** Berichtigung. *Flora oder Botanische Zeitung (Regensburg)*, **2:** 442–444.

———. **1820:** Ueber die von Chamisso und Bergius gesammelten Moose. *Flora oder Botanische Zeitung (Regensburg)*, **3:** 513–522.

———. **1822:** *Ptychostomum* novum muscorum frondosorum genus. *Flora*, **5:** 62–64.

Horton, D. G. 1982*a*: A revision of the Encalyptaceae (Musci), with particular reference to the North American taxa. Part I. *Journal of the Hattori Botanical Laboratory*, **53:** 365–418.

———. **1982*b*:** Geographical isolation and historical factors in relation to specific and infraspecific categories in mosses. *In*: P. Geissler & S. W. Greene (eds), *Bryophyte taxonomy: methods, practices and floristic exploration. Beihefte zur Nova Hedwigia*, **71:** 21–29.

———. **1983:** A revision of the Encalyptaceae (Musci), with particular reference to the North American taxa. Part II. *Journal of the Hattori Botanical Laboratory*, **54:** 353–532.

———. **1994:** Encalyptales. *In*: A. J. Sharp, H. Crum & P. M. Eckel (eds), *The moss flora of Mexico. Part one: Sphagnales to Bryaales. Memoirs of the New York Botanical Garden*, **69:** 418–422.

Host, N. Th. 1831: *Flora austriaca*. **2**. Viennae, sumptibus Frider. Beck Bibliop. Univer.: [ii] + 768 pp.

Howe, M. A. 1899: The Hepaticae and Anthocerotes of California. *Memoirs of the Torrey Botanical Club*, **7:** 1–208.

Hu, S.-S. 1998: Moss community types and species diversity of southern Fildes Peninsula (King George Island, South Shetland Islands), Antarctica. *Journal of the Hattori Botanical Laboratory*, **84:** 187–198.

——— **& R. D. Seppelt. 1989:** Moss communities and their characteristics in ice-free areas of the Windmill Islands, Wilkes Land, continental Antarctica. *In*: *Proceedings of the International Symposium on Antarctic Research, Hangzou, China, 8–12 May 1989*. Tianjin, Ocean Press: pp. 317–324.

Hudson, W. 1762: *Flora anglica, exhibens plantas per regnum anglae sponte crescentes distributas secundum systema sexuale: cum differentiis specierum, synonimis auctorum, nominibus incolarum, solo locorum, tempore florendi, officinalibus pharmacopoeorum*. Londini, J. Nourse et C. Moran: viii + [8] + 506 + [22, index].

Huntley, B. J. 1971: Vegetation. *In*: E. M. van Zinderen Bakker, Sr., J. M. Winterbottom & R. A. Dyer (eds), *Marion and Prince Edward Islands. Report on the South African Biological and Geological Expedition 1965–1966*. Cape Town, A. A. Balkema: pp. 98–160.

Husnot, P. T. 1884–1890: *Muscologia gallica. Descriptions et figures des mousses de France et de contrées voisines*. **1**. Cahan, T. Husnot and Paris, F. Savy: viii + 284 pp. + pls. i–lxxix.

Huttunen, S. & M. S. Ignatov. 2004: Phylogeny of Brachytheciaceae (Bryophyta), based on morphology and sequence level data. *Cladistics*, **20:** 151–183.

———**, A. A. Gardiner & M. S. Ignatov. 2007:** Advances in knowledge of the Brachytheciaceae (Bryophyta). *In*: A. E. Newton & R. S. Tangney (eds), *Pleurocarpous mosses. Systematics and evolution*. The Systematics Association Special Volume Series. **71**. Boca Raton, London, New York, CRC Press, Taylor & Francis Groups: pp. 117–143.

———**, M. S. Ignatov, K. Müller & D. Quandt. 2004:** Phylogeny and evolution of epiphytism in the three moss families Meteoriaceae, Brachytheciaceae and Lembophyllaceae. *In*: B. Goffinet, V. Hollowell & R. Magill (eds), *Molecular* *systematics of bryophytes*. *Monographs in Systematic Botany from the Missouri Botanical Garden*, **98:** 328–361.

Hübener, J. W. P. 1833: *Muscologia germanica oder Beschreibung der deutschen Laubmoose. Im erweiterten Umfange nach dem jetzigen Stande der Wissenschaft, nebst Erörterung der Standörter und ihrer Entdecker, der Synonyme seit Hoffmann und Roth, mit erläuternden Anmerkungen*. Leipzig, bei Friedrich Hofmeister: xviii + 722 pp.

Hübschmann, A. von. 1960: Das *Pottietum hemii*, eine salzliebende Kleinmoosgesellschaft. *Mitteilungen der Floristisch-soziologischen Arbeitsgemeinschaft, Neue Folge*, **8:** 124–126.

———. 1973: Bryologische Studien auf der Azoreninsel São Miguel. *Revista de Facultade de Ciências de Lisboa 2.ª Série C*, **17:** 627–702.

Hyvönen, J. 1986: Bryophyte flora of the Huon Peninsula, Papua New Guinea. XVIII. Polytrichaceae and Buxbaumiaceae (Musci). *In*: T. Koponen (ed.), *Bryophyte flora of the Huon Peninsula, Papua New Guinea. XIV–XIX. Acta Botanica Fennica*, **133:** 107–149.

———. 1991*a*: *Chorisodontium* (Dicranaceae, Musci) in southern South America. *Annales Botanici Fennici*, **28:** 247–258.

———. 1991*b*: *Tortella fragilis* (Pottiaceae) reported for southern South America. *The Bryologist*, **94:** 416–418.

———. 2006: Polytrichaceae. *In*: P. M. McCarthy (ed.), *Flora of Australia*. **51**. *Mosses (1)*. Canberra, Australian Biological Resources Study and Melbourne, CSIRO Publishing: pp. 124–143 (except *Dawsonia*).

———, T. A. Hedderson, G. L. Smith Merrill. J. G. Gibbings & S. Koskinen. 1998: On phylogeny of the Polytrichales. *The Bryologist*, **101:** 489–504.

———, S. Koskinen, G. L. Smith Merrill T. A. Hedderson & S. Stenroos. 2004: Phylogeny of the Polytrichales (Bryophyta) based on simultaneous analysis of molecular and morphological data. *Molecular Phylogenetics and Evolution*, **31:** 915–928.

I

Ignatov, M. S. 1994: Bryophytes of Altai Mountains. I. Study area and history of its bryological exploration. *Arctoa, a Journal of Bryology*, **3:** 13–27 (with Russian abstract).

———. 1999: Bryophyte flora of the Huon Peninsula, Papua New Guinea. LXIII. On the pseudoparaphyllia in Brachytheciaceae and Meteoriaceae (Musci). *In*: S. Piippo (ed.), *Bryophyte flora of the Huon Peninsula, Papua New Guinea. LX–LXIV. Acta Botanica Fennica*, **165:** 73–83.

——— & O. M. Afonina (eds). 1992: Check-list of mosses of the former USSR. *Arctoa, a Journal of Bryology*, **1:** 1–85.

——— & T. Cao. 1994: Bryophytes of Altai Mountains. VI. The family Grimmiaceae (Musci). *Arctoa, a Journal of Bryology*, **3:** 67–122.

——— & L. Hedenäs. 2007: Homologies of stem structures in pleurocarpous mosses, especially of pseudoparaphyllia and similar structures. *In*: A. E. Newton & R. S. Tangney (eds), *Pleurocarpous mosses. Systematics and evolution*. The Systematics Association Special Volume Series. **71**. Boca Raton, London, New York, CRC Press, Taylor & Francis Groups: pp. 269–286.

——— & S. Huttunen. 2002 ['2003']: Brachytheciaceae (Bryophyta) – a family of sibling genera. *Arctoa, a Journal of Bryology*, **11:** 245–296.

——— & E.A. Ignatova. 2003: [Moss flora of the Middle European Russia. Vol. 1: Sphagnaceae – Hedwigiaceae]. *Arctoa, a Journal of Bryology*, **11** *Suppl.* **1:** 1–608 (in Russian).

——— & ———. 2004: [Moss flora of the Middle European Russia. Vol. 2: Fontinalaceae – Amblystegiaceae]. *Arctoa, a Journal of Bryology*, **11** *Suppl.* **2:** 610–960 (in Russian).

——— & T. Koponen. 1996: On the taxonomy of some East Asian *Brachythecium* (Brachytheciaceae, Musci). *Annales Botanici Fennici*, **33:** 285–301.

——— & N. B. Kurbatova. 1990: A review of deep-water bryophytes with new records from USSR. *Hikobia*, **10:** 393–401.

——— & J. Lewinsky-Haapasaari. 1994: Bryophytes of Altai Mountains. II. The genera *Amphidium* Schimp., *Orthotrichum* Hedw. and *Zygodon* Hook. & Tayl. (Orthotrichaceae, Musci). *Arctoa, a Journal of Bryology*, **3:** 29–57.

——— & G. L. Smith Merrill. 1995: Bryophytes of Altai Mountains. VI. The family Polytrichaceae (Musci). *Arctoa, a Journal of Bryology*, **5:** 61–97.

———, O. M. Afonina & E. A. Ignatova. 2006: Check-list of mosses of East Europe and North Asia. *Arctoa, a Journal of Bryology*, **15:** 1–130.

———, E.A. Ignatova & S. A. Sugarina. 2002: A new variety of *Syntrichia caninervis* (Pottiaceae, Musci). *Arctoa, a Journal of Bryology*, **11:** 333–336.

———, T. Koponen & D. H. Norris. 1999: Bryophyte flora of the Huon Peninsula, Papua New Guinea. LXII. Brachytheciaceae (Musci), excluding *Homalothecium* and *Palamocladium*. *In*: S. Piippo (ed.), *Bryophyte flora of the Huon Peninsula, Papua New Guinea. LX–LXIV. Acta Botanica Fennica*, **165:** 23–72.

———, H. Robinson & E. A. Ignatova. 1998: On the exostome ornamentation of Brachytheciaceae. *Arctoa, a Journal of Bryology*, **7:** 153–188.

———, E. I. Ivanova, E. A. Ignatova & K. K. Krivoshapkin. 2001: On the moss flora of Ust-Maya District (Republic Sakha/Yakutia, East Siberia). *Arctoa, a Journal of Bryology*, **10:** 165–184.

———, Ts. Tsegmed, B. Tan, X.-L. Bai & V. Zolotov. 2004: Mosses of Gobi in Mongolia. *Journal of the Hattori Botanical Laboratory*, **96:** 183–210.

Imura, S. & H. Kanda. 1986: The gemmae of the mosses collected from the Syowa Station area, Antarctica. *Memoirs of National Institute of Polar Research Special Issue*, **44:** 241–246.

———, M. Higuchi, H. Kanda & Z. Iwatsuki. 1992: Culture of rhizoidal tubers on an aquatic moss in the lakes near the Syowa Station area, Antarctica. *Proceedings of the NIPR Symposium on Polar Biology*, **5:** 114–117.

———, ———, ——— & ———. 1993: Vegetative reproduction of mosses in soil around the Antarctic moss community (extended abstract). *Proceedings of the NIPR Symposium on Polar Biology*, **6:** 179–181.

———, ———, ——— & ———. 1994: Structure of moss colonies in the Syowa Station area, Antarctica. *Proceedings of the NIPR Symposium on Polar Biology*, **7:** 232–236.

———, T. Bando, S. Saito, K. Seto & H. Kanda. 1999: Benthic moss pillars in Antarctic lakes. *Polar Biology*, **22:** 137–140.

———, T. Bando, K. Seto, S. Ohtani, S. Kudoh & H. Kanda. 2003: Distribution of aquatic mosses in the Sôya Coast region, East Antarctica. *Polar Bioscience*, **16:** 1–10.

Ingerpuu, N., A. Kalda, L. Kannukene, H. Krall, M. Leis & K. Vellak. 1994: List of the Estonian bryophytes. *The Naturalist's Notebook*, **94:** 1–175.

Ingólfsson, Ó., C. Hjort, S. Björck & R. I. Lewis Smith. 1992: Lake Pleistocene and Holocene glacial history of James Ross Island, Antarctic Peninsula. *Boreas*, **21:** 209–222.

Inoue, S. 1976: Chromosome studies on five species of Antarctic mosses. *Kumamoto Journal of Science, Biology*, **13:** 1–5.

———. 1979: A preliminary report on moss chromosome numbers. *Miscellanea Bryologica et Lichenologica*, **8:** 109–113.

Ireland, R. R. 1960: Notes on *Pogonatum alpinum* in eastern United States and Canada. *The Bryologist*, **63:** 46–50.

———. 1969: A taxonomic revision of the genus *Plagiothecium* for North America north of Mexico. *National Museum of Natural Sciences Publications in Botany*, **1:** i–viii + 1–118.

———. 1971: Moss pseudoparaphyllia. *The Bryologist*, **74:** 312–238.

———. 1982: Moss flora of the maritime provinces. *National Museums of Canada Publications in Botany*, **13:** 1–738.

———. 1991: A preliminary study of the genus *Isopterygium* in Latin America. *Caldasia*, **16:** 265–276.

———. 1992: Studies of the genus *Plagiothecium* in Australasia. *The Bryologist*, **95:** 221–224.

———. 2001: *Buckiella* a new genus in the Hypnaceae (Musci). *Novon*, **11:** 55–62.

——— & L. M. Ley. 1992: Atlas of Ontario mosses. *Syllogeus*, **70:** i–v + 1–138.

———, G. Bellolio, R. Rodríguez & J. Larraín. 2006: Studies on the moss flora of the Bío-Bío Region of Chile. *Tropical Bryology*, **28:** 63–77.

———, G. R. Brassard, W. B. Schofield & D. H. Vitt. 1987: Checklist of the mosses of Canada II. *Lindbergia*, **13:** 1–62.

Isoviita, P. 1979 ['1980']: *Platydictya jungermannioides* and *Jungermannia setacea* – a problem of typification. *Annales Botanici Fennici*, **16:** 275–284.

———. 1992: Notes on the sectional nomenclature of the Bryoideae. *In*: H. Ochi, *A revised infrageneric classification of the genus* Bryum *and related genera (Bryaceae, Musci). Appendix. In*: T. Koponen & J. Hyvönen (eds), *Proceedings of the Congress of East Asiatic Bryology, Helsinki, August 12–19, 1990. Bryobrothera*, **1:** 231–244.

——— & R. Ochyra. 1990: Proposal to conserve *Drepanocladus* (Musci, Amblystegiaceae). *Taxon*, **39:** 348–351.

Iwatsuki, Z. 1970: A revision of *Plagiothecium* and its related genera from Japan and her adjacent areas. I. *Journal of the Hattori Botanical Laboratory*, **33:** 331–380.

———. 1987: Notes on *Isopterygium* Mitt. (Plagiotheciaceae). *Journal of the Hattori Botanical Laboratory*, **63:** 445–451.

———. 2004: New catalog of the mosses of Japan. *Journal of the Hattori Botanical Laboratory*, **96:** 1–182.

——— & B. C. Tan. 1979: Checklist of Philippine mosses. *Kalikasan*, **8:** 179–210.

J

Jaeger, A. 1873: Genera et species muscorum systematice disposita seu Adumbratio florae muscorum totius orbis terrarum. Pars III. *Bericht über die Thätigkeit der St. Gallischen Naturwissenschaftlichen Gesellschaft*, **1871–1872:** 309–490.

——— & F. Sauerbeck. 1879: Genera et species muscorum systematice disposita seu Adumbratio florae muscorum totius orbis terrarum. Pars IX. *Bericht über die Thätigkeit der St. Gallischen Naturwissenschaftlichen Gesellschaft*, **1877–1878:** 257–361.

——— & ———. 1880: Supplementum ad adumbrationem muscorum et conspectus systematis. *Bericht über die Thätigkeit der St. Gallischen Naturwissenschaftlichen Gesellschaft*, **1877–1878:** 362–514.

Jalas, J. 1955: *Racomitrium lanuginosum* (Hedw.) Brid. als Klimaindikatorin Ostfennoskandien. *Archivum Societatis Zoologicae Botanicae Fennicae 'Vanamo'*, **9** Suppl.: 73–88.

Járai-Komlódi, M. & S. Orbán. 1975: Spore morphological studies on recent European *Encalypta* species. *Acta Botanica Academiae Scientiarum Hungaricae*, **21:** 305–345.

Jedlička, J. 1947: Clavis specierum et formarum generis *Plagiothecium* Br. eur. s. str. *Sbornik Klubu Přirodovědeckého v Brně*, **28:** 34–44.

———. 1948: Monographia specierum europaearum gen. *Plagiothecium* s. s. Partis specialis I. Summarium. *Spisy Vydávané Přirodovědeckou Fakultou Masarykovy University*, **308** *Řada L*, **2:** 1–45.

———. 1950: Monographia specierum europaearum gen. *Plagiothecium* s. s. Icones. *Spisy Vydávané Přirodovědeckou Fakultou Masarykovy University*, **318** *Řada L*, **4:** tab. 1–7 (photographs) + tab. i–xxxiii (line drawings).

———. 1954: Ad monographiam generis *Plagiothecium* Bryol. eur. s. str. addimentum. *Spisy Vydávané Přirodovědeckou Fakultou Masarykovy University*, **356** *Řada L*, **8:** 175–236.

———. 1961: Monographia specierum europaearum gen. *Plagiothecium* Bryol. eur. s. str. Partis specialis II. Distributio geographica specialis. *Spisy Přirodovědecké Fakulty University J. E. Purkyné*, **422** *Řada L*, **16:** 173–223.

Jelenc, F. 1955: *Muscinées de Afrique du Nord (Algérie, Tunisie, Maroc, Sahara)*. Oran, L. Fouque: 152 pp.

———. 1967: Muscinées de l'Afrique du Nord (Supplément). *Revue Bryologique et Lichénologique, Nouvelle Série*, **35:** 186–215.

Jenkin, J. F. 1972: Macquarie Island, subantarctic. *In*: T. Rosswall & O.W. Heal (eds), *Structure and function of tundra ecosystems. Ecological Bulletins*, **20:** 375–397.

Jennings, O. E. 1913: *A manual of the mosses of Western Pennsylvania*. Pittsburgh, privately published: v + 429 pp.

Jensen, C. 1898: Mosser fra Øst-Grønland. *Meddelelser om Grønland*, **15:** 365–443.

———. 1939: *Skandinaviens bladmossflora*. København, Ejnar Munksgaard: 535 pp.

Jiménez, J. A., R. M. Ros, M. J. Cano & J. Guerra. 2002: Contribution to the bryophyte flora of Morocco: terricolous and saxicolous bryophytes of the Jbel Bouhalla. *Journal of Bryology*, **24:** 243–250.

Jóhannsson, B. 1990: [Icelandic bryophytes. Archidiaceae, Funariaceae, Splachnaceae and Orthotrichaceae]. *Fjölrit Náttúrufræðistofnunar*, **15:** 1–80 (in Icelandic with English summary).

———. 1991: [Icelandic bryophytes. Dicranaceae]. *Fjölrit Náttúrufræðistofnunar*, **19:** 1–119 (in Icelandic with English summary).

———. 1992: [Icelandic bryophytes. Pottiaceae]. *Fjölrit Náttúrufræðistofnunar*, **21:** 1–122 (in Icelandic with English summary).

———. **1995:** [Icelandic bryophytes. Bryaceae]. *Fjölrit Náttúrufræðistofnunar*, **27:** 1–162 (in Icelandic with English summary).

———. **1997:** [Icelandic bryophytes. Brachytheciaceae]. *Fjölrit Náttúrufræðistofnunar*, **33:** 1–83 (in Icelandic with English summary).

———. **1998:** [Icelandic bryophytes. Amblystegiaceae]. *Fjölrit Náttúrufræðistofnunar*, **34:** 1–126 (in Icelandic with English summary).

———. **2003:** [Icelandic bryophytes. Lists and additions]. *Fjölrit Náttúrufræðistofnunar*, **44:** 1–135 (in Icelandic with English summary).

Johnson, P. N. & D. J. Campbell. 1975: Vascular flora of the Auckland Islands. *New Zealand Journal of Botany*, **13:** 665–720.

Jones, G. N. 1933: Family Grimmiaceae. *In*: A. J. Grout (ed.), *Moss flora of North America north of Mexico*. **2**(1). Newfane, Vermont, privately published: pp. 1–65 + pls. 1–25.

Jules, E. S. & A. J. Shaw. 1994: Adaptation to metal-contaminated soils in populations of the moss, *Ceratodon purpureus*: vegetative growth and reproductive expression. *American Journal of Botany*, **81:** 791–797.

Juratzka, J. 1882: *Die Laubmoosflora von Oesterreich-Ungarn. Handschriftlicher Nachlass Jakob Juratzka's enthaltend die Beschreibung der in Oesterreich-Ungarn wachsenden Laubmoose mit Ausnahme der Leskeaceae, Hypnaceae, der Andreaeaceae und der Sphagnaceae*. Wien, W. Braumüller and Leipzig, F. A. Brockhaus: viii + 385 pp.

K

Kaalaas, B. 1912: Bryophyten aus den Crozetinseln. II. *Nyt Magazin for Naturvidenskaberne*, **50:** 98–119.

Kabiersch, W. 1938: Studien über die ostasiatischen Arten einiger Laubmoosfamilien II (Rhizogoniaceae, Bartramiaceae, Aulacomniaceae, Meesiaceae). *Hedwigia*, **77:** 71–136.

Kanda, H. 1975: A revision of the family Amblystegiaceae of Japan. I. *Journal of Science of the Hiroshima University, Series B, Div. 2*, **15:** 201–276.

———. **1976:** A revision of the family Amblystegiaceae of Japan. II. *Journal of Science of the Hiroshima University, Series B, Div. 2*, **16:** 47–119.

———. **1979:** Regenerative development in culture of Antarctic palnts of *Ceratodon purpureus* (Hedw.) Brid. *Memoirs of National Institute of Polar Research, Special Issue*, **11:** 58–69.

———. **1981*a*:** Flora and vegetation of mosses in ice-free areas of Soya Coast and Prince Olav Coast, East Antarctica. *Hikobia, Supplement,* **1:** 91–100.

———. **1981*b*:** Two moss species of the genus *Pottia* collected from the vicinity of Syowa Station, East Antarctica. *Antarctic Record*, **71:** 96–108.

———. **1986*a*:** Antarctic species of *Ceratodon* collected from King George Island, South Shetland Islands, Antarctica. *Hikobia*, **9:** 319–325.

———. **1986*b*:** Moss communities in some ice-free areas along the Soya Coast, East Antarctica. *Memoirs of National Institute of Polar Research, Special Issue*, **44:** 229–240.

———. **1987*a*:** [*Handbook of Antarctic mosses*]. Tokyo, National Polar Institute: [v] + 83 pp. (in Japanese).

———. **1987*b*:** *Catalog of moss specimens from Antarctica and adjacent regions housed mainly in the herbarium of the National Institute of Polar Research*. Tokyo, National Polar Institute: iii + 186 pp.

———. **1987*c*:** Moss vegetation in the Yukidori Valley, Langhovde, East Antarctica. *In*: *Papers on Plant Ecology and Taxonomy to the Memory of Dr. Satoshi Nakanishi*. Kobe, The Kobe Geobotanical Society: pp. 197–204.

———. **1987*d*:** Ecology of the moss vegetation in the Syowa Station area, Enderby Land, Antarctica. *In*: T. Pócs, T. Simon, Z. Tuba & J. Podani (eds), *Proceedings of the IAB Conference of Bryology, Budapest–Vácrátót, Hungary, 5–10 August 1985. Symposia Biologica Hungarica*. **35**. Part A. Budapest, Akadémiai Kiadó: 259–267.

———. **1987*e*:** New localities and taxonomic remarks for some mosses in continental Antarctica. *Proceedings of the NIPR Symposium on Polar Biology*, **1:** 265–272.

——— **& Z. Iwatsuki: 1989:** Two aquatic mosses in the lakes near Syowa Station, continental Antarctica. *Hikobia*, **10:** 293–297.

——— **& V. Komárková. 1997:** Antarctic terrestrial ecosystems. *In*: F. E. Wielgolaski (ed.), *Ecosystems of the World*. **3**. *Polar and alpine tundra*. Amsterdam, Elsevier: pp. 721–761.

——— **& Y. Mochida. 1992:** Aquatic mosses found in lakes of the Skarvsnes region, Syowa Station area, Antarctica (Extended Abstract). *Proceedings of the NIPR Symposium on Polar Biology*, **4:** 177–179.

——— **& H. Ochi. 1986:** Fruiting plants of *Bryum* found in the vicinity of Suowa Station, Antarctica. *Memoirs of National Institute of Polar Research Special Issue*, **44:** 220–228.

——— **& S. Ohtani. 1991:** Morphology of the aquatic mosses collected in Lake Yukidori, Langhovde, Antarctica. *Proceedings of the NIPR Symposium on Polar Biology*, **4:** 114–122.

——— **& H. Okada. 1993:** Chromosome study on the submerged moss collected from Antarctic lakes. *Proceedings of the NIPR Symposium on Polar Biology*, **6:** 121–125.

———**, S. Ohtani & S. Imura. 2002:** Plant communities at Dronning Maud Land. *In*: L. Beyer & M. Bölter (eds), *Geoecology of Antarctic ice-free coastal landscapes. Ecological Studies*, **154:** 249–264.

Kannukene, L. & N. Matveyeva. 1996: Mosses from the Arctic tundra of the Taimyr Peninsula, Siberia. *Proceedings of the Estonian Academy of Sciences Biology*, **45:** 51–67.

Kappen, L. 1985: Vegetation and ecology of ice-free areas of northern Victoria Land, Antarctica. 1. The lichen vegetation of Birthday Ridge and an inland mountain. *Polar Biology,* **4:** 213–225.

———. **2004:** The diversity of lichens in Antarctica, a review and comments. *In*: P. Döbbeler & G. Rambold (eds), *Contributions to lichenology. Festschrift in honour of Hannes Hertel. Bibliotheca Lichenologica*, **88:** 331–343.

——— **& B. Schroeter. 2002:** Plants and lichens in the Antarctic, their way of life and their relevance to soil formation. *In*: L. Beyer & M. Bölter (eds), *Geoecology of Antarctic ice-free coastal landscapes. Ecological Studies*, **154:** 327–373.

——— **& H. Straka. 1988:** Pollen and spore transport into Antarctica. *Polar Biology*, **8:** 173–180.

——— **& F. Valladares. 1999:** Opportunistic growth and desiccation tolerance: the ecological success of poikilohydrous autotrophs. *In*: F. I. Pugnaire & F. Valladares (eds),

Handbook of functional plant ecology. New York, Marcel Dekker: pp. 9–80.

———, **R. I. Lewis Smith & M. Meyer. 1989:** Carbon dioxide exchange of two ecodemes of *Schistidium antarctici* in continental Antarctica. *Polar Biology*, **9:** 415–422.

Karczmarz, K. 1961: [*Pottia Heimii* (Hedw.) Br. eur. (*Desmatodon Heimii* Lazar.) in Poland]. *Ekologia Polska, Seria B*, **7:** 297–302 (in Polish with English summary).

———. **1971 ['1972']:** A monograph of the genus *Calliergon* (Sull.) Kindb. *Monographiae Botanicae*, **34:** 1–209 + pls. i–xx.

——— **& J. Żarnowiec. 1989:** Studies on propaguliferous species of *Pohlia* section *Pohliella* in Poland. Taxonomy and distribution. *Annales Universitatis Mariae Curie-Skłodowska, Sectio C*, **44:** 143–164.

Kaspar, M., G. M. Jr., Simmons, B. C. Parker, K. G. Seaburg, R. A. Jr. Wharton & R. I. Lewis Smith. 1982: *Bryum* Hedw. from Lake Vanda, Antarctica. *The Bryologist*, **85:** 424–430.

Kattel, L. P. & M. K. Adhikari. 1992: *Mosses of Nepal (List and references)*. Kathmandu, Natural History Society of Nepal: 107 pp.

Kawai, I. 1965: Studies on the genus *Grimmia*, with reference to the affinity of the gametophyte. *The Science Reports of Kanazawa University*, **10:** 79–132 + pls. i–lxiv.

———. **1968:** Taxonomic studies on the midrib in Musci. (1) Significance of the midrib in systematic botany. *The Science Reports of Kanazawa University*, **13:** 127–157.

———. **1969:** Studies on the affinity regarding the conducting tissue of the sporophyte in the Musci (2). On the seta in some species of the family Polytrichaceae. *The Science Reports of Kanazawa University*, **14:** 39–51 + pls. i–vi.

———. **1982:** Systematic studies on the conducting tissue of the gametophyte in Musci. (12) Anatomical characteristics of stems in some species of Bartramiaceae. *The Science Reports of Kanazawa University*, **26(2):** 31–50 + pls. i–vi.

Kennedy, A. D. 1993: Photosynthetic response of the Antarctic moss *Polytrichum alpestre* Hoppe to low temperatures and freeze-thaw stress. *Polar Biology*, **13:** 271–279.

Kilbertus, G. & M. Bizot. 1971: Contribution à la connaissance de la flore des muscinés d'Afghanistan. *Bulletin de l'Académie et de la Société Lorraine des Sciences*, **10:** 67–75.

Kim, J. H., I.-Y. Ahn, K. S. Lee, H. Chung & H.-G. Choi. 2007: Vegetation of Barton Peninsula in the neighbourhood of King Sejong Station (King George Island, maritime Antarctic). *Polar Biology*, **30:** 903–916.

Kindberg, N. C. 1882: Die Familien und Gattungen der Laubmoose (Bryineae) Schwedens und Norwegens hauptsächlich nach dem Lindbergschen Systeme. *Bihang till Kongliga Svenska Vetenskaps-Akademiens Handligar*, **6(19):** 1–25.

———. **1883:** Die Arten der Laubmoose (Bryineae) Schwedens und Norwegens. *Bihang till Kongliga Svenska Vetenskaps-Akademiens Handligar*, **7(9):** 1–167.

———. **1885:** Table analytique des mousses pleurocarpes européennes. *Revue Bryologique*, **12:** 24–31.

———. **1891:** *Enumeratio Bryinearum exoticarum. Suppl.* **2**. Linkoeping, Officina Corresp. Ostrogoth.: pp. 97–108.

———. **1898:** *European and N. American Bryineae (Mosses)*. **2**. Linköping, Linköpings Lithografiska Akiebolag: pp. [i–iii] + 153–410.

King, J. C. & Turner, J. 1997: *Antarctic meteorology and climatology*. Cambridge, Cambridge University Press: 425 pp.

———, **J. Turner, G. J. Marshall, W. M. Connally & T. A. Lachlan-Cope. 2003:** Antarctic Peninsula climate variability and its causes as revealed by analysis of instrumental records. *In*: E. Domack, A. Leventner, A. Burnett, R. Bindschadler, P. Convey & M. Kirby (eds), *Antarctic Peninsula climate variability. Historical and paleoenvironmental perspectives*. Antarctic Research Series. **79**. Washington, D.C., American Geophysical Union: pp. 17–30.

Kis, G. 1985: *Mosses of south-east tropical Africa. An annotated list with distributional data*. Vácrátót, Institute of Ecology and Botany of the Hungarian Academy of Sciences: 170 pp.

Klanzenga, N. 2003: A revision of the Australasian species of *Dicranoloma* (Bryophyta, Dicranaceae). *Australian Systematic Botany*, **16:** 427–471.

Koch, L. 1950: Mosses of California: an annotated list of species. *Leaflets of Western Botany*, **6:** 1–40.

———. **1955:** Mosses of California V. *Pterigoneurum ovatum* and *P. subsessile*. *Madroño*, **13:** 100–101.

Komárková, V. 1984: Studies on plant communities of the Antarctic Peninsula near Palmer Station. *Antarctic Journal of the United States*, **19:** 180–182.

———, **S. Poncet & J. Poncet. 1985:** Two native Antarctic vascular plants, *Deschampsia antarctica* and *Colobanthus quitensis*: a new southernmost locality and other localities in the Antarctic Peninsula area. *Arctic and Alpine Research*, **17:** 401–416.

———, ——— **&** ———. **1990:** Additional and revisited localities of vascular plants *Deschampsia antarctica* Desv. and *Colobanthus quitensis* (Kunth) Bartl. in the Antarctic Peninsula area. *Arctic and Alpine Research*, **22:** 108–113.

Koperski, M. 1989: Die *Racomitrium heterostichum*-Gruppe in der nordwestdeutschen Tiefebene. *Drosera*, **1989:** 95–100.

Koponen, A. 1978: The peristome and spores in Splachnaceae and their evolutionary significance. *In*: C. Suire (ed.), *Congrès International de Bryologie, Bordeaux 21–23 Novembre 1977. Comptes Rendus. Bryophytorum Bibliotheca*, **13:** 535–567.

———. **1982:** On the structure and function of the peristome in Splachnaceae. *Journal of the Hattori Botanical Laboratory*, **53:** 73–98.

——— **& T. Koponen. 1978:** Evidence of entomophily in Splachnaceae (Bryophyta). *In*: C. Suire (ed.), *Congrès International de Bryologie, Bordeaux 21–23 Novembre 1977. Comptes Rendus. Bryophytorum Bibliotheca*, **13:** 569–577.

Koponen, T. 1988: The phylogeny and classification of Mniaceae and Rhizogoniaceae (Musci). *Journal of the Hattori Botanical Laboratory*, **64:** 37–46.

——— **& J.-X. Luo. 1992:** Moss flora of Wo-long Nature Reserve, Sichuan Province, China. *Bryobrothera*, **1:** 161–175.

——— **& D. H. Norris. 1984:** Bryophyte flora of the Huon Peninsula, Papua New Guinea. IV. *Anomobryum*, *Bryum* and *Rhodobryum* (Bryaceae, Musci). *Annales Botanici Fennici*, **21:** 265–290.

——— **&** ———. **1985:** Bryophyte flora of the Huon Peninsula, Papua New Guinea. XI. *Brachymenium*, *Epipterygium*, *Leptobryum*, *Mielichhoferia*, *Orthodontium* and *Pohlia* (Bryaceae), and Leptostomataceae (Musci). *In*: T. Koponen (ed.),

Bryophyte flora of the Huon Peninsula, Papua New Guinea. LVII–LIXI. Acta Botanica Fennica, **131:** 99–127.

——— & ———. **1996:** Bryophyte flora of the Huon Peninsula, Papua New Guinea. LVII. *Fleischerobryum* and *Philonotis* (Bartramiaceae, Musci). *In*: T. Koponen (ed.), *Bryophyte flora of the Huon Peninsula, Papua New Guinea. VII–XIII. Acta Botanica Fennica*, **156:** 1–21.

——— **V. Oittinen. 1967:** *Aloina brevirostris* (Hook. & Grev.) Kindb. found in Finland. *Memoranda Societatis pro Fauna et Flora Fennica*, **43:** 5–7.

Koppe, F. 1976: Beobachtungen über die Moosvegetation und Moosflora von Zypern. *Herzogia*, **4:** 95–136.

Kramer, W. 1980: *Tortula* Hedw. sect. *Rurales* De Not. (Pottiaceae, Musci) in der östlciehn Holarktis. *Bryophytorum Bibliotheca*, **21:** 1–165 + pls. 1–29.

———. **1988:** Beiträge zur Systematik und Bryogeographie einiger Sippen von *Tortula* Hedw. sect. *Rurales* De Not. (Pottiaceae, Musci) unter besonderer Berücksichtigung der Südhemisphäre. *Journal of the Hattori Botanical Laboratory*, **65:** 81–144.

Kruijer, J. D. 2002: Hypopterygiaceae of the world. *Blumea Supplement*, **13:** 1–388.

Kuc, M. 1963: Flora of mosses and their distribution on the north coast of Hornsund (S.W.-Svalbard). *Fragmenta Floristica et Geobotanica*, **9:** 291–366.

———. **1964:** Briogeografia wyżyn południowych Polski [Bryogeography of the southern uplands of Poland]. *Monographiae Botanicae*, **17:** 1–212 (in Polish with English summary).

———. **1969:** Some mosses from the Antarctic oasis. *Revue Bryologique et Lichénologique, Nouvelle Série*, **36:** 655–672.

———. **1973*a*:** Bryogeography of expedition area, Axel Heiberg Island, N. W. T., Canada. *Bryophytorum Bibliotheca*, **2:** 1–120.

———. **1973*b*:** A review of the mosses of Svalbard. *Revue Bryologique et Lichénologique, Nouvelle Série*, **39:** 401–472.

Kucyniak, J. 1958: Une mousse inattendue por le Québec: *Pterigoneurum ovatum*. *Naturaliste Canadien*, **85:** 217–224.

Kudoh, S., Y. Kashino & S. Imura. 2003*a*: Ecological studies of aquatic moss pillars in Antarctic lakes 3. Light response and chilling and heat sensitivity of photosynthesis. *Polar Bioscience*, **16:** 33–42.

———, **K. Watanabe & S. Imura. 2003*b*:** Ecological studies of aquatic moss pillars in Antarctic lakes 2. Temperature and light environment at the moss habitat. *Polar Bioscience*, **16:** 23–32.

———. **Y. Tsuchiya, E. Ayukawa, S. Imura & H. Kanda. 2003*c*:** Ecological studies of aquatic moss pillars in Antarctic lakes 1. Macro structure and carbon, nitrogen and chlorophyll *a* contents. *Polar Bioscience*, **16:** 11–22.

Kühnemann, O. 1938: Catalogo de los musgos Argentinos. *Lilloa*, **2:** 37–183.

——— **& M. F. Gonçalves-Carralves. 1975:** Monografia preliminar para la flora criptogámica fueguina: los musgos de la familia Ptychomniaceae (Bryophyta). *Darwiniana*, **19:** 583–617.

Kunth, C. S. 1822: *Synopsis plantarum, in itinere ad plagam aequinoctialem orbis novi, collegerunt Al. de Humbold et Am. Bonpland*. **1**. Parisis and Argentorati, apud F. G. Levrault: [vii] + iv + 491 pp.

Kuo, Ch.-M. & T.-Y. Chiang. 1987: Index of Taiwan mosses. *Taiwania*, **32:** 119–207.

Kürschner, H. 2000: Bryophyte flora of the Arabian Peninsula and Socotra. *Bryophytorum Bibliotheca*, **55:** 1–131.

——— **& R. Ochyra. 1999:** Two remarkable hygrophytic mosses new to the bryophyte flora of Yemen (Studies in Arabian bryophytes 25). *Fragmenta Floristica et Geobotanica*, **44:** 269–275.

Kuta, E., L. Przywara & R. Ochyra. 1982: Karyological studies on Antarctic mosses I. *The Bryologist*, **85:** 131–138.

L

La Farge-England, C. 1996: Growth from, branching pattern, and perichaetial position in mosses: cladocarpy and pleurocarpy redefined. *The Bryologist*, **99:** 170–186.

———, **B. D. Mishler, J. A. Wheeler, D. P. Wall, K. Johannes, S. Schaffer & A. J. Shaw. 2000:** Phylogenetic relationships within the haplolepidous mosses. *The Bryologist*, **103:** 257–276.

Lal, J. 2004: Bryophytes of Schirmacher Oasis. *In*: *Nineteenth Indian Expedition to Antarctica, Scientific Report, Technical Publication*. **17**. New Delhi, Department of Ocean Development: pp. 165–171.

Lange, J. & C. E. O. Jensen. 1887: *Oversigt over Grønlands mosser*. Kjøbenhavn, Bianco Lunos Kgl. Hof-Bogtrykkeri (F. Dreyer): 110 pp.

Lanjouw, J. (chairman), Ch. Baehni, W. Robyns, R. C. Rollins, R. Ross, J. Rousseau, G. M. Schulze, A. C. Smith, R. de Vilmorin (members) & F. A. Stafleu (secretary). 1956: *International Code of Botanical Nomenclature adopted by the Eighth International Botanical Congress, Paris, July 1954*. Regnum Vegetabile. **8**. Utrecht, Kemink en Zoon N. V.: 338 pp.

Lara, F., R. Garilleti & V. Mazimpaka. 2006: *Orthotrichum galiciae* (Orthotrichaceae, Bryopsida), a new species from Ethiopia. *Nova Hedwigia*, **83:** 199–205.

Lawton, E. 1971: *Moss flora of the Pacific Northwest*. Nichinan, The Hattori Botanical Laboratory: xiii + 362 pp. + 195 pls.

———. **1972:** The genus *Rhacomitrium* in America and Japan. *Journal of the Hattori Botanical Laboratory*, **35:** 252–262.

———. **1973:** *Rhacomitrium crispulum* and some related species. *Bulletin of the Torrey Botanical Club*, **100:** 230–235.

Lazarenko, A. S. 1938: [Data on the Middle-Asia bryoflora]. *Journal de l'Institut Botanique de l'Académie des Sciences de la RSS d'Ukraine* **26–27:** 191–216 (in Ukrainian with English summary).

Leeuw, C. de, A. Aptroot & B. van Zanten. 1998: The lichen and bryophyte vegetation of Cuverville Island, Antarctica. *Nova Hedwigia*, **67:** 469–480.

Lewinsky, J. 1974: The family Plagiotheciaceae in Denmark. *Lindbergia*, **2:** 185–217.

———. **1976:** On the taxonomic position of *Amphidium* Schimp. *Lindbergia*, **3:** 227–231.

———. **1977*a*:** The moss genus *Orthotrichum*. Morphological studies and evolutionary remarks. *Journal of the Hattori Botanical Laboratory*, **43:** 31–61.

———. **1977*b*:** The family Orthotrichaceae in Greenland. A taxonomic revision. *Lindbergia*, **4:** 57–103.

———. **1978:** The genus *Orthotrichum* Hedw. (Musci) in Africa south of the Tropic of Cancer. *Botanisk Tidsskrift*, **72:** 61–85.

———. **1983:** Lectotypification of *Orthotrichum rupestre* Schleich. ex Schwaegr. *Lindbergia*, **9:** 53–56.

———. **1984:** The genus *Orthotrichum* Hedw. (Musci) in Australasia. A taxonomic revision. *Journal of the Hattori Botanical Laboratory*, **56:** 369–460.

———. **1985:** *Orthotrichum* (Orthotrichaceae) in South America 1. Introduction and taxonomic revision of taxa with immersed stomata. *Lindbergia*, **10:** 65–94.

———. **1987:** *Orthotrichum* (Orthotrichaceae) in South America 2. Taxonomic revision of taxa with superficial stomata. *In*: W. R. Buck (ed.), *Bryostephane steereana: a collection of bryological papers presented to William Campbell Steere on the occasion of his 80th birthday. Memoirs of the New York Botanical Garden*, **45:** 326–370.

———. **1989:** Does the orthotrichaceous peristome exist? A study of peristome evolution in the genus *Orthotrichum* Hedw. with a possible derivation of the haplolepidous peristome. *Journal of the Hattori Botanical Laboratory*, **67:** 335–363.

———. **1992:** The genus *Orthotrichum* Hedw. (Orthotrichaceae, Musci) in Southeast Asia. A taxonomic revision. *Journal of the Hattori Botanical Laboratory*, **72:** 1–88.

———. **1993:** A synopsis of the genus *Orthotrichum* Hedw. (Musci, Orthotrichaceae). *Bryobrothera*, **2:** 1–59.

——— **& G. S. Mogensen. 1977:** Distribution maps of bryophytes in Greenland 4. *Lindbergia*, **4:** 135–142.

Lewinsky-Haapasaari, J. 1994: The genus *Pleurorthotrichum* Broth. *Lindbergia*, **19:** 11–24.

———. **1995:** Illustrierter Bestimmungsschlüssel zu den europäischen *Orthotrichum*-Arten. *Meylania*, **9:** 3–56.

——— **& M. R. Crosby. 1996:** *Orthomitrium tuberculatum* (Orthotrichaceae), a new genus and species from Guizhou, China. *Novon*, **6:** 1–5.

——— **& L. Hedenäs. 1998:** A cladistic analysis of the moss genus *Orthotrichum*. *The Bryologist*, **101:** 519–555.

Lewis, K. & A. J. E. Smith. 1977: Studies on some bulbiferous species of *Pohlia* section *Pohliella* I. Experimental investigations. *Journal of Bryology*, **9:** 539–556.

——— **&** ———. **1978:** Studies on some bulbiferous species of *Pohlia* section *Pohliella* II. Taxonomy. *Journal of Bryology*, **10:** 9–27.

Lewis, L. A., B. D. Mishler & R. Vilgalys. 1997: Phylogenetic relationships of the liverworts (Hepaticae), a basal embryophyte lineage, inferred from nucleotide sequence data of the chloroplast gene *rbc*L. *Molecular Phylogenetics and Evolution*, **7:** 377–393.

Lewis Smith, R. I. 1972: Vegetation of the South Orkney Islands with particular reference to Signy Island. *British Antarctic Survey Scientific Reports*, **68:** 1–124 + pls. 1–6.

———. **1978:** Summer and winter concentrations of sodium, potassium and calcium in some maritime Antarctic cryptogams. *Journal of Ecology*, **66:** 891–909.

———. **1979:** Peat forming vegetation in the Antarctic. *In*: E. Kivinen, L. Heikurainen & P. Pakarinen (organisers), *Proceedings of the International Symposium on Classification of Peat and Peatland*. Helsinki, International Peat Society: pp. 58–67.

———. **1981:** Plant succession and re-exposed moss banks on a deglaciated headland in Arthur Harbour, Anvers Island. *British Antarctic Survey Bulletin*, **51:** 193–199.

———. **1982:** Farthest south and highest occurrences of vascular plants in the Antarctic. *Polar Record*, **21**(131): 170–173.

———. **1984*a*:** Terrestrial plant biology of the sub-Antarctic and Antarctic. *In*: R. M. Laws (ed.), *Antarctic ecology*. **1**. London, Academic Press (Harcourt Brace Jovanovich, Publishers): pp. 61–162.

———. **1984*b*:** Colonization and recovery by cryptogams following recent volcanic activity on Deception Island, South Shetland Islands. *British Antarctic Survey Bulletin*, **62:** 25–51.

———. **1984*c*:** Colonization by bryophytes following recent volcanic activity on an Antarctic island. *Journal of the Hattori Botanical Laboratory*, **56:** 53–63.

———. **1985:** A unique community of pioneer mosses dominated by *Pterygoneurum* cf. *ovatum* in the Antarctic. *Journal of Bryology*, **13:** 509–514.

———. **1987:** The bryophyte propagule bank of Antarctic fellfield soils. *In*: T. Pócs, T. Simon, Z. Tuba & J. Podani (eds), *Proceedings of the IAB Conference of Bryology, Budapest–Vácrátót, Hungary, 5–10 August 1985. Symposia Biologica Hungarica*. **35**. Part A. Budapest, Akadémiai Kiadó: pp. 233–245.

———. **1988*a*:** Recording bryophyte microclimates in remote and severe environments. *In*: J. M. Glime (ed.), *Methods in bryology*. Nichinan, Hattori Botanical Laboratory: pp. 275–284.

———. **1988*b*:** Classification and ordination of cryptogamic communities in Wilkes Land, continental Antarctic. *Vegetatio*, **76:** 155–166.

———. **1988*c*:** Bryophyte oases in Ablation Valley on Alexander Island, Antarctica. *The Bryologist*, **91:** 45–50.

———. **1988*d*:** Botanical survey of Deception Island. *British Antarctic Survey Bulletin*, **80:** 129–136.

———. **1988*e*:** Destruction of Antarctic terrestrial ecosystems by a rapidly increasing fur seal population. *Biological Conservation*, **45:** 55–72.

———. **1989:** Aspects of cryptogam water relations at a continental Antarctic site. *Polarforschung*, **58:** 139–153.

———. **1990*a*:** Plant community dynamics in Wilkes Land, Antarctica. *Proceedings of the NIPR Symposium on Polar Biology*, **3:** 229–244.

———. **1990*b*:** Signy Island as a paradigm of biological and environmental change in Antarctic terrestrial ecosystems. *In*: K. R. Kerry & G. Hempel (eds), *Antarctic ecosystems: ecological change and conservation*. Berlin, Springer-Verlag: pp. 32–50.

———. **1991:** Exotic sporomorpha as indicators of potential immigrant colonists in Antarctica. *Grana*, **30:** 313–324.

———. **1993*a*:** The role of bryophyte propagule banks in primary succession: case-study of an Antarctic fellfield soil. *In*: J. Miles & D. W. H. Walton (eds), *Primary succession on land*. The British Ecological Society Special Publication. **12**. Oxford, Blackwell Scientific Publications: pp. 55–78.

———. **1993*b*:** The vegetation of Cockburn Island, Antarctica. *Polar Biology*, **13:** 535–542.

———. **1993*c*:** Dry coastal ecosystems of Antarctica. *In*: E. van der Maarel (ed.), *Ecosystems of the World*. **2A**. *Dry coastal* *ecosystems. Polar regions and Europe*. Amsterdam, Elsevier: pp. 51–71.

———. **1993*d*:** Dry coastal ecosystems on sub-Antarctic islands. *In*: E. van der Maarel (ed.), *Ecosystems of the World*. **2A**. *Dry coastal ecosystems. Polar regions and Europe*. Amsterdam, Elsevier: pp. 73–93.

———. **1996:** Terrestrial and freshwater biotic components of the western Antarctic Peninsula. *In*: R. Ross, E. Hofmann & L. Quetin (eds), *Foundations for ecological research west of the Antarctic Peninsula*. Antarctic Research Series. **70**. Washington, D.C., American Geophysical Union: pp. 15–59.

———. **1997*a*:** Impact of an increasing fur seal population on Antarctic plant communities: resilience and recovery. *In*: B. Battaglia, J. Valencia & D. W. H. Walton (eds), *Antarctic communities. Species, structure and survival*. Cambridge, Cambridge University Press: pp. 432–436.

———. **1997*b*:** Oases as centres of high plant diversity and dispersal in Antarctica. *In*: W. B. Lyons, C. Howard-Williams & I. Hawes (eds), *Ecosystem processes in Antarctic ice-free landscapes*. Rotterdam, Balkema: pp. 119–128.

———. **1999:** Biological and environmental characteristics of three cosmopolitan mosses dominant in continental Antarctica. *Journal of Vegetation Science*, **10:** 231–242.

———. **2000:** Plants of extreme habitats in Antarctica. *In*: B. Schroeter, M. Schlensog & T. G. A. Green (eds), *New aspects in cryptogamic research. Contributions in honour of Ludger Kappen. Bibliotheca Lichenologica*, **75:** 405–419.

———. **2001:** Plant colonisation response to climate change in the Antarctic. *In*: P. Prošek & K. Láska (eds), *Ecology of the Antarctic coastal oasis. Folia Facultatis Scientiarium Naturalium Universitatis Masarykianae Brunensis Geographia*, **25:** 19–33.

———. **2005*a*:** The bryophyte flora of geothermal habitats on Deception Island, Antarctica. *Journal of the Hattori Botanical Laboratory*, **97:** 233–284.

———. **2005*b*:** The thermophilic bryoflora of Deception Island: unique plant communities as a criterion for designating an Antarctic Specially Protected Area. *Antarctic Science*, **17:** 17–27.

———. **2005*c*:** Bryophyte diversity and ecology of two geologically contrasting Antarctic islands. *Journal of Bryology*, **27:** 195–206.

——— **& P. Convey. 2002:** Enhanced sexual reproduction in bryophytes at high latitudes in the maritime Antarctic. *Journal of Bryology*, **24:** 107–117.

——— **& R. W. M. Corner. 1973:** Vegetation of the Arthur Harbour-Argentine Islands region of the Antarctic Peninsula. *British Antarctic Survey Bulletin*, **33–34:** 89–122.

——— **& A. M. Coupar. 1987:** The colonization potential of bryophyte propagules in Antarctic fellfield sites. *Comité National Français Recherches Antarctiques*, **58:** 189–204.

——— **& C. H. Gimingham. 1976:** Classification of cryptogamic communities in the maritime Antarctic. *British Antarctic Survey Bulletin*, **43:** 25–47.

——— **& R. Ochyra. 2006:** Antarctic soil propagule banks reveal a potential source of exotic plant colonists. *Journal of the Hattori Botanical Laboratory*, **100:** 325–331.

——— **& S. Poncet. 1985:** New southernmost record for Antarctic flowering plants. *Polar Record*, **22:** 425–427.

——— **& ———. 1987:** *Deschampsia antarctica* and *Colobanthus quitensis* in the Terra Firma Islands. *British Antarctic Survey Bulletin*, **74:** 31–35.

Li, X.-D., J.-X. Liu & F.-D. Chen. 1998: A preliminary study on bryoflora of the Fildes Peninsula, King George Island, Antarctica. *Journal of the Hattori Botanical Laboratory*, **84:** 199–202.

Li, X-J. (ed.). 1985: [Bryoflora of Xizang]. *In*: [*The Series of the Scientific Expedition to Quinghai-Xizang Plateau*]. Beijing, Science Press: x + 581 pp. (in Chinese).

——— **& S. He. 2007:** Meesiaceae. *In*: X.-J. Li, M. R. Crosby & S. He (eds), *Moss Flora of China. English version*. **4**. *Bryaceae – Timmiaceae*. Beijing – New York, Science Press and St. Louis, Missouri Botanical Garden: pp. 157–160.

———, ——— **& Z. Iwatsuki. 2001.** Pottiaceae. *In*: X.-J. Li, M. R. Crosby & S. He (eds), *Moss Flora of China. English version*. **2**. *Fissidentaceae – Ptychomitriaceae*. Beijing – New York, Science Press and St. Louis, Missouri Botanical Garden: pp. 114–249.

Light, J. J. 1975: Clear lakes and aquatic bryophytes in the mountains of Scotland. *Journal of Ecology*, **63:** 937–943.

——— **& R. B. Heywood. 1973:** Deep-water mosses in Antarctic lakes. *Nature*, **242:** 535–536.

——— **& ———. 1975:** Is the vegetation of continental Antarctica predominantly aquatic? *Nature*, **256:** 199–200.

——— **& Lewis Smith, R. I. 1976:** Deep-water bryophytes from the highest Scottish lochs. *Journal of Bryology*, **9:** 55–62.

Lightowlers, P. J. 1983: Notes on the Antarctic mosses *Pohlia cruda* and *P. inflexa*. *British Antarctic Survey Bulletin*, **60:** 83–87

———. **1984*a*:** A new variety of the moss *Tortula robusta* from South Georgia. *British Antarctic Survey Bulletin*, **64:** 63–66.

———. **1984*b*:** Taxonomy of sub-antarctic mosses – *Tortula*. *Institute of Terrestrial Ecology Annual Report*, **1983:** 70–71.

———. **1985*a*:** Taxonomic notes on New Zealand species of *Tortula*. *Journal of Bryology*, **13:** 369–375.

———. **1985*b*:** A synoptic flora of South Georgian mosses: *Tortula*. *British Antarctic Survey Bulletin*, **67:** 41–77.

———. **1985*c*:** *Tortula lithophila* Dus. is *Sarconeurum glaciale* (C. Muell.) Card. et Bryhn. *The Bryologist*, **88:** 365–366.

———. **1986*a*:** Taxonomy and distribution of the subantarctic species of *Tortula*. *Journal of Bryology*, **14:** 281–295.

———. **1986*b*:** The moss genus *Tortula* from the Antarctic botanical zone. *British Antarctic Survey Bulletin*, **72:** 71–76.

———. **1987:** New or noteworthy mosses from the Antarctic Peninsula region. *British Antarctic Survey Bulletin*, **76:** 91–94.

———. **1988:** The Antarctic Plant Database at the BAS Herbarium. *Taxon*, **37:** 378–380.

Limpricht, K. G. 1876: Laubmoose. *In*: F. Cohn (ed.), *Kryptogamen-Flora von Schlesien*. **1**. Breslau, J. U. Kern's Verlag (Max Müller): pp. 27–224.

———. **1885–1889:** Die Laubmoose Deutschlands, Oesterreichs und der Schweiz. *In*: *Dr L. Rabenhorst's Kryptogamen-Flora von Deutschland, Oesterreich und der Schweiz*. 2 Aufl. **4**(1). *Sphagnaceae, Andreaeaceae, Archidiaceae, Bryineae (Cleistocarpae, Stegocarpae [Acrocarpae])*. Leipzig, Eduard Kummer: viii + 836 pp.

———. **1890–1895:** Die Laubmoose Deutschlands, Oesterreichs und der Schweiz. *In*: *Dr L. Rabenhorst's Kryptogamen-Flora von Deutschland, Oesterreich und der Schweiz*. 2 Aufl. **4**(2).

Bryineae (Stegocarpae [Acrocarpae, Pleurocarpae excl. Hypnaceae]). Leipzig, Eduard Kummer: [7] + 853 pp.

Lindberg, S. O. 1863*a*: Om ett nytt slägte, *Epipterygium*, bland bladmossorna. *Öfversigt af Förhandlingar: Kongl. Svenska Vetenskaps-Akademien*, **19:** 599–609.

———**. 1863*b*:** Bidrag till mossornas synonymi. *Öfversigt af Förhandlingar: Kongliga Svenska Vetenskaps-Akademien*, **20:** 385–418.

———**. 1864:** De Tortulis et ceteris Trichostomeis europaeis. *Öfversigt af Förhandlingar: Kongliga Svenska Vetenskaps-Akademien*, **21:** 213–254.

———**. 1868:** Observationes de Mniaceis europaeis. *Notiser ur Sällskapets pro Fauna et Flora Fennica Förhandlingar*, **9:** 39–88.

———**. 1871:** Revisio critica iconum in opere Flora Danica muscos illustrantium. *Acta Societatis Scientiarum Fennicae*, **10:** 1–118.

———**. 1878:** *Utkast till en naturlig gruppering af Europas bladmossor med toppsittande frukt (Bryinae acrocarpae).* Helsingfors, J. C. Frenckell & Sons tryckeri: i + 39 pp.

———**. 1879:** *Musci scandinavici in systemate novo naturali dispositi.* Upsaliae, ex Officina Iesaiae Edquist: 50 pp.

———**. 1883:** *Kritisk granskning af mossorna uti Dillenii Historia muscorum.* Helsingfors, J. C. Frenckell & Son: 59 pp.

——— **& H. W. Arnell. 1890:** Musci Asiae borealis. Beschreibung der von den Schwedischen Expeditionen nach Sibirien in den Jahren 1875 und 1876 gesammelten Moose mit Berücksichtigung aller früheren bryologischen Angaben für das russische Nord-Asien. Zweiter Theil Laubmoose. *Kongliga Svenska Vetenskaps-Akademiens Handlingar*, **23(10):** 1–163.

Lindsay, D. C. 1971: Vegetation of the South Shetland Islands. *British Antarctic Survey Bulletin*, **25:** 59–83.

Linnaeus, C. 1753: *Species plantarum, exhibentes plantas rite cognitas, ad genera relatas, cum differentiis specificis, nominibus trivialibus, synonymis selectis, locis natalibus, secundum systema sexuale digestas.* **1–2**. Holmiae, Impensis Laurentii Salvii: xii + 1200 pp. [+ 30 pp., index].

———**. 1763:** *Species plantarum, exhibentes plantas rite cognitas, ad genera relatas, cum differentiis specificis, nominibus trivialibus, synonymis selectis, locis natalibus, secundum systema sexuale digestas.* Ed. 2. **2**. Holmiae, Impensis Direct. Laurentii Salvii, Holmiae: pp. i + 785–1684 [+ 1–64, indexes].

Linskens, H. F., R. Bargagli, M. Cresti & S. Focardi. 1993: Entrapment of long-distance transported pollen grains by various moss species in coastal Victoria Land, Antarctica. *Polar Biology*, **13:** 81–87.

———**,** ———**, S. Focardii & M. Cresti. 1991:** Antarctic moss turf as pollen traps. *Proceedings of the Koninklijke Nederlandse Akademie van Wetenschappen*, **94:** 233–241.

Liu, J. & X. Li. 1998*a*: Studies on the plant morphology of *Andreaea* Hedw. from Fildes Peninsula, Antarctica. *Chenia*, **5:** 227–230.

——— **&** ———**. 1998*b*:** [SEM studies on the plant morphology of *Schistidium* Brid. from Fildes Peninsula, Antarctica]. *Chinese Journal of Polar Research*, **10:** 217–221 (in Chinese with English summary).

Loeske, L. 1903: *Moosflora des Harzes. Hilfsbuch für die bryologische Forschng im Harze und dessem Umgebung mit Verbreitungsangaben und Bestimmungstabellen.* Leipzig, Verlag von Gebrüder Borntraeger: xx + 350 pp.

———**. 1907*a*:** *Drepanocladus*, eine biologische Mischgattung. *Hedwigia*, **46:** 300–321.

———**. 1907*b*:** Bryologische Beobachtungen aus den Algäuer Alpen. *Verhandlungen des Botanischen vereins der Provinz Brandenburg*, **49:** 30–65.

———**. 1908:** Die Moose des Arlberggebietes. *Hedwigia*, **47:** 156–199.

———**. 1910:** *Studien zur vergleichenden Morphologie und phylogenetischen Systematik der Laubmoose.* Berlin, Max Lande: 224 pp.

———**. 1913:** *Die Laubmoose Europas. I. Grimmiaceae.* Berlin-Schöneberg, Max Lande (Hoffmann & Campe's Verlag): xvi + 207 pp.

———**. 1914–1929:** Die Laubmoose Europas. II. Funariaceae. *Repertorium Specierum Novarum Regni Vegetabilis Sonderbeiheft* **B:** [i–iii] + ii–viii + 1–120

———**. 1930:** Monographie der europäischen Grimmiaceen. *Bibliotheca Botanica*, **101:** i–ix + 1–236.

———**. 1931:** *Ceratodon purpureus* mit Brutfäden. *Nederlandsch Kruidkundig Archief*, **1931**: 178–181.

Lo Giudice, R. 1996: On the *Bryum bicolor* complex in Italy including *Bryum gemmilucens* Wilcz. & Dem., new species to the Italian bryological flora. *Flora Mediterranea*, **6:** 61–67.

Long, D. G. 1985: Polytrichaceae. *In*: G. S. Mogensen (ed.), *Illustrated moss flora of Arctic North America.* **1**. *Meddelelser om Grønland Bioscense*, **17:** 1–57.

———**. 1995:** The *Musci Indici*: its authors, types and localities. *Botanical Journal of the Linnean Society*, **119:** 1–33.

Longton, R. E. 1966: Botanical studies in the Antarctic during the 1963–64 and 1964–65 seasons. *British Antarctic Survey Bulletin*, **10:** 85–95.

———**. 1967:** Vegetation in the maritime Antarctic. *In*: J. E. Smith (organiser), *A discussion on the Antarctic terrestrial ecosystem. Philosophical Transactions of the Royal Society of London, Series B, Biological Sciences*, **252:** 213–235.

———**. 1972:** Reproduction of Antarctic mosses in the genera *Polytrichum* and *Psilopilum* with particular reference to temperature. *British Antarctic Survey Bulletin*, **27:** 51–96.

———**. 1973:** A classification of terrestrial vegetation near McMurdo Sound, continental Antarctica. *Canadian Journal of Botany*, **51:** 2339–2346.

———**. 1979:** Vegetation ecology and classification in the Antarctic Zone. *Canadian Journal of Botany*, **57**: 2264–2278.

———**. 1980:** Physiological ecology of mosses. *In*: R. J. Taylor & A. E. Leviton (eds), *The mosses of North America.* San Francisco, The Pacific Division of the American Association for the Advencement of Science: pp. 77–113.

———**. 1981:** Inter-population variation in morphology and physiology in the cosmopolitan moss *Bryum argenteum* Hedw. *Journal of Bryology*, **11:** 501–520.

———**. 1988:** *Biology of polar bryophytes and lichens.* Cambridge, Cambridge University Press: 391 pp.

——— **& M. W. Holdgate. 1967:** Temperature relationships of Antarctic vegetation. *In*: J. E. Smith (organiser), *A discussion on the Antarctic terrestrial ecosystem. Philosophical Transactions of the Royal Society of London, Series B, Biological Sciences*, **252:** 237–250.

——— & ———. **1979:** The South Sandwich Islands: IV. Botany. *British Antarctic Survey Scientific Reports*, **94:** 1–53 + pl. 1.

——— **& A. MacIver. 1977:** Climatic relationshops in Antarctic and Northern Hemisphere populations of a cosmopolitan moss, *Bryum argenteum* Hedw. *In*: G. A. Llano (ed.), *Adaptations within Antarctic ecosystems. The Proceedings of the Third SCAR Symposium on Antarctic biology*. Washington, D.C., Smithsonian Institution: pp. 899–919.

——— **& R. M. Schuster. 1983:** Reproductive biology. *In*: R. M. Schuster (ed.), *New manual of bryology*. **1**. Nichinan, The Hattori Botanical Laboratory: pp. 386–462.

Lorch, W. 1908: Die Polytrichaceen. Eine biologische Monographie. *Abhandlungen der Bayerischen Akademie der Wissenschaften, Mathematisch-Naturwissenschaftliche Klasse*, **23:** 445–546.

———. **1931:** *Anatomie der Laubmoose*. *In*: K. Lindbauer (ed.), *Handbuch der Pflanzenanatomie*. **7**(1). *Bryophyten*. Berlin, Gebrüder Borntraeger: viii + 358 pp.

Lorentz, P. G. 1866: Musci frondosi in Chile prope Valdiviam et prope Corral lecti per Dr. Krause. *Botanische Zeitung (Berlin)*, **24:** 185–189.

Lyon, G. L. & W. F. Giggenbach. 1974: Geothermal activity in Victoria Land, Antarctica. *New Zealand Journal of Geology and Geophysics*, **17:** 511–521.

Lyubarskaya, L. B. 1986: [*Conspect of the moss flora of Azerbaydzhan*]. Baku, Institut Botaniki AN Azerbaydzhanskoy SSR: 178 pp. (Manuscript deposited in VINITI Order No. 9882) (in Russian).

M

Mackenzie Lamb, I. 1970: Antarctic terrestrial plants and their ecology. *In*: M. W. Holdgate (ed.), *Antarctic ecology*. **2**. London, Academic Press: pp. 733–751.

Mägdefrau, K. 1982: Life-forms of bryophytes. *In*: A. J. E. Smith (ed.), *Bryophyte ecology*. London, Chapman & Hall: 45–58.

Magill, R. M. 1977: Towards the conservation of four family names in Musci. *Taxon*, **26:** 597–598.

———. **1981:** Sphagnaceaeae – Grimmiaceae. *In*: E. O. Leistner (ed.), *Flora of southern Africa which deals with the territories of South Africa, Transkei, Lesotho, Swaziland, Bophuthatswana, South West Africa/Namibia, Botswana and Venda. Bryophyta. Part 1. Mosses*. **1**. Pretoria, Botanical Research Institute: pp. i–xv + 1–291.

———. **1987*a*:** On the endostomial nature of the *Dichelodontium* (Ptychomniaceae) peristome. *In*: W. R. Buck (ed.), *Bryostephane steereana: a collection of bryological papers presented to William Campbell Steere on the occasion of his 80th birthday. Memoirs of the New York Botanical Garden*, **45:** 71–86.

———. **1987*b***: Gigaspermaceae – Bartramiaceae. *In*: E. O. Leistner (ed.), *Flora of southern Africa which deals with the territories of South Africa, Transkei, Lesotho, Swaziland, Bophuthatswana, South West Africa/Namibia, Botswana and Venda. Bryophyta. Part 1. Mosses*. **2**. Pretoria, Botanical Research Institute: pp. i–ix + 293–443.

——— **(ed.). 1990:** Glossarium polyglottum bryologiae. A multilingual glossary for bryology. *Monographs in Systematic Botany from the Missouri Botanical Garden*, **33:** [1–9] + 1–297.

———. **1993:** Conserved names for mosses: a brief history. *Taxon*, **42:** 5–15.

———. **1998:** Erpodiaceae – Hookeriaceae. *In*: E. O. Leistner (ed.), *Flora of southern Africa which deals with the territories of South Africa, Transkei, Lesotho, Swaziland, Bophuthatswana, South West Africa/Namibia, Botswana and Venda. Bryophyta. Part 1. Mosses*. **3**. Pretoria, Botanical Research Institute: pp. i–vii + 445–622.

Mahler, W. F. 1981: Preliminary checklist of mosses of New Mexico. *The Bryologist*, **81:** 593–599.

Maier, E. 1998: Zur systematischen Stellung von *Grimmia pitardii* Corb. (Musci). *Candollea*, **53:** 301–308.

———. **1999:** Auf den Spuren von Lantzius-Beninga – Die Bedeutung der Peristommerkmale bei Laubmoosen (Bryophyta). *Stuttgarter Beiträge zur Naturkunde Serie A (Biologie)*, **591:** 1–13.

———. **2002:** The genus *Grimmia* (Musci, Grimmiaceae) in the Himalaya. *Candollea*, **57:** 143–238.

——— **& P. Geissler. 1995:** *Grimmia* in Mitteleuropa: ein Bestimmungsschlüssel. *Herzogia*, **11:** 1–80.

Malcolm, B. & N. Malcolm. 2006: *Mosses and other bryophytes, an illustrated glossary*. Ed. 2. Nelson, New Zealand, Micro-Optics Press: iv + 336 pp.

Malme, L. 1978: Floristic and ecological studies of bryophytes in some Norwegian inland lakes. *Norwegian Journal of Botany*, **25:** 271–279.

Malta, N. 1926: Die Gattung *Zygodon* Hook. et Tayl. Eine monographische Studie. *Latvijas Universitates Botaniskā Dārza Darbi*, **1:** 1–185.

Mamatkulov, U. K. 1990: [*The flora of the bryophytes of the Tadjik SSR*. **1**. *Sphagnaceae – Grimmiaceae*]. Dushanbe, Donish: 234 pp. (in Russian).

Manakyan, V. A. 1989: [*The mosses of south-eastern Armenia*]. Erevan, Izdatel'stvo Armyanskoy Akademiy Nauk SSR: 313 pp. (in Russian)

Manyanga, Ph. & S. M. Perold. 2004: A checklist of Zimbabwean bryophytes. *Southern African Botanical Diversity Network Report*, **21:** 1–22.

Margadant, W. D. 1972: Notes on the nomenclature of Musci. *Lindbergia*, **1:** 121–129.

——— **& P. Geissler. 1995:** Seventeen proposals concerning nomina conservanda for genera of Musci. *Taxon*, **44:** 613–624.

Marshall, W. A. 1996: Biological particles over Antarctica. *Nature*, **383:** 680.

——— **& P. Convey. 1997:** Dispersal of moss propagules on Signy Island, maritime Antarctic. *Polar Biology*, **18:** 376–383.

Marstaller, R. 1989: Die Moosgesellschaften der Ordnung *Funarietalia hygrometricae* v. Hübschmann 1957. *Gleditschia*, **17:** 237–250.

Mårtensson, O. 1956: Bryophytes of the Torneträsk area, northern Swedish Lappland. II. Musci. *Kungliga Svenska Vetenskkapsakademiens Avhandlingar i Naturskyddsärenden*, **14:** 1–321.

Matcham, H. W. & J. G. Duckett. 2008*a*: *Pterygoneurum ovatum* (Hedw.) Dixon. *In*: T. Blockeel (ed.), *New national and regional bryophyte records, 18. Journal of Bryology*, **30:** in press.

——— & ———. **2008*b*:** *Aloina brevirostris* (Hook. & Grev.) Kindb. Chile. *In*: T. Blockeel (ed.), *New national and regional bryophyte records, 18. Journal of Bryology*, **30:** in press.

Matsuda, T. 1964: Ecological studies on the community of mosses at Langhovde region, Antarctica. *The Antarctic Record*, **21:** 25–38.

———. **1968:** Ecological study of the moss community and microorganisms in the vicinity of Syowa Station, Antarctica. *In*: *Japanese Antarctic Research Expedition Scientific Reports, Series E*. **29**. *Biology*. Tokyo, National Science Museum, Department of Polar Research: pp. 1–58.

Matsui, T. & Iwatsuki, Z. 1990: A taxonomic revision of the family Ditrichaceae (Musci) of Japan, Korea and Taiwan. *Journal of the Hattori Botanical Laboratory*, **68:** 317–366.

Matteri, C. M. 1968: Las especies de «*Philonotis* (Bartramiaceae)» del sur de Argentina. *Revista del Museo Argentino de Ciencias Naturales «Bernardino Rivadavia» e Instituto Nacional de Investigacion de la Ciencias Naturales Botanica*, **3:** 183–234.

———. **1973*a*:** El genero «*Breutelia*» (Bartramiaceae, Musci) en la region andino-patagonica. *Revista del Museo Argentino de Ciencias Naturales «Bernardino Rivadavia» e Instituto Nacional de Investigacion de la Ciencias Naturales Botanica*, **4:** 319–360.

———. **1973*b*:** Notas briologicas (Musci) I. *Anales del Instituto de la Patagonia*, **4:** 273.

———. **1973*c*:** Revision de la Hypopterygiaceae (Musci) austroamericanas. *Boletín de la Sociedad de Botánica*, **15:** 229–250.

———. **1975:** Orden Hookeriales. *In*: S.A. Guerrera, I. Gamundi de Amos & D. Rabinovich de Halperin (eds), *Flora criptogámica de Tierra del Fuego*. **14**(9). Buenos Aires, Consejo Nacional de Investigaciones Científicas y Técnicas de la Republica Argentina: 60 pp.

———. **1977*a*:** Notes on Antarctic bryophytes: X. The genus *Pottia* from the Antarctic botanical zone. *British Antarctic Survey Bulletin*, **46:** 140–143.

———. **1977*b*:** A synoptic flora of South Georgian mosses: VII. *Pottia*. *British Antarctic Survey Bulletin*, **46:** 23–28.

———. **1982:** Patagonian bryophytes 6. Fruiting *Sarconeurum glaciale* (C. Muell.) Card. et Bryhn newly found in southern Patagonia. *Lindbergia*, **8:** 105–109.

———. **1984:** Sinopsis de las especies andino-patagonicas, antarticas y subantarticas de los generos *Bartramia*, *Bartramidula* y *Conostomum* (Bartramiaceae, Musci). *Darwiniana*, **25:** 143–162.

———. **1985*a*:** Catalogo de los musgos. *In*: O. Boelcke, D. M. Moore & F. A. Roig (eds), *Transecta botanica de la Patagonia austral*. Buenos Aires, Consejo Nacional de Investigaciones Cientificas y Tecnicas (Argentina), Instituto de la Patagonia (Chile) and Royal Society (Gran Bretaña): pp. 265–297.

———. **1985*b*:** Bartramiaceae. *In*: S.A. Guerrera, I. Gamundi de Amos & D. Rabinovich de Halperin (eds), *Flora criptogámica de Tierra del Fuego*. **14**(7). Buenos Aires, Consejo Nacional de Investigaciones Científicas y Técnicas de la Republica Argentina: 62 pp.

———. **1986:** Los Musci (Bryophyta) de las Islas Malvinas, su habitat y distribución. *Nova Hedwigia*, **43:** 159–189.

———. **1988:** New records for the Fuegian moss flora and the status of *Oedipodium fuegianum* Kühn. et Gonç.-Carr. *Lindbergia*, **14:** 63–66.

———. **1991:** The genus *Leptostomum* R. Brown (Musci) in southern South America. *Journal of the Hattori Botanical Laboratory*, **69:** 257–264.

———. **1996:** *Ditrichum brevirostrum* (R. Brown ter.) Broth. newly discovered in northern Patagonia and the Fuegian region. *Lindbergia*, **21:** 3–8.

———. **2003:** Los musgos (Bryophyta) de Argentina. *Tropical Bryology*, **24:** 33–100.

——— **& G. M. Calabrese. 1999:** Taxonomic notes and distribution of *Matteria papillosula* and *Microdus euchlorus*, from northern Patagonia (Argentina), with a new combination in *Campylopodium* (Bryopsida). *Annales Botanici Fennici*, **36:** 43–49.

——— **& R. M. Farias. 1999:** Lectotypification of *Andreaea pachyphylla* (Musci, Andreaeopsida). *Lindbergia*, **24:** 11–14.

——— **& R. Ochyra. 1999:** The Meesiaceae (Musci) in southern South America, with notes on the subdivision of the family. *In*: H. Manitz (ed.), *Riclef-Grolle-Festschrift. Haussknechtia, Beiheft*, **9:** 225–242.

——— **& M. M. Schiavone. 1991:** La vegetación muscinal de Bahía Buen Suceso, Tierra del Fuego, Argentina. *The Bryologist*, **94:** 368–376.

——— & ———. **2002:** Catálogo de los musgos (Bryophyta) de la Région Fueguina en Argentina y Chile. *Revista del Museo Argentino de Ciencias Naturales, Nueva Serie*, **4:** 111–138.

——— & ———. **2004:** Preliminary survey of the mosses (Bryophyta) from Los Glaciares National Park, Santa Cruz Province, Argentina. *Revista del Museo Argentino de Ciencias Naturales, Nueva Serie*, **6:** 9–24.

May, R. 1986: Notes on some Macaronesian *Tortella* species. *Bryologische Beiträge*, **6:** 58–66.

McCarthy, P. M. (ed.). 2006: *Flora of Australia*. **51**. *Mosses (1)*. Canberra, Australian Biological Resources Study and Melbourne, CSIRO Publishing: x + 450 pp.

McFarland, K. D. 1988: Revision of *Brachythecium* (Musci) for Mexico, Central America, South America, Antarctica, and circum-subantarctic islands. Mss. of Ph.D. thesis, The University of Tennessee, Knoxville; printed in 1994 by xerographic processes by UMI Dissertation Services, Ann Arbor: xii + 537 pp.

McIntosh, T. T. 2007: *Schistidium*. *In*: R. H. Zander (Lead ed.), *Flora of North America north of Mexico*. **27**. *Bryophytes: mosses (1)*. New York – Oxford, Oxford University Press: pp. 207–225.

McNeill, J. (Chairman), F. R. Barrie, H. M. Burdet, V. Demoulin, D. L. Hawksworth, K. Marhold, D. H. Nicolson, J. Prado, P. C. Solfva, J. E. Skog, J. H. Wiersema (members) & N. J. Turland (secretary). 2006: *International Code of Botanical Nomenclature (Vienna Code)*. Regnum Vegetabile. **146**. Ruggel, Lichtenstein, A. R. G. Gantner Verlag: xviii + 568 pp.

Meinunger, L. & W. Schröder. 2007: *Verbreitungsatlas der Moose Deutschlands*. **2**. Regensburg, Verlag von Regensburgische Botanische Gesellschaft: 699 pp.

Melick, D. R. & R. D. Seppelt. 1997: Vegetation patterns in relation to climatic and endogenous changes in Wilkes Land, continental Antarctica. *Journal of Ecology*, **85:** 43–56.

———, M. J. Hovenden & R. D. Seppelt. 1994: Phytogeography of bryophyte and lichen vegetation in the Windmill Islands, Wilkes Land, continental Antarctica. *Vegetatio*, **111:** 71–87.

Mel'nichuk, V. M. 1970: [*Handbook of the mosses of the central region and the south of the European part of the U.S.S.R.*]. Kiev, Naukova Dumka: 442 pp. (in Russian).

Menzel M. 1986: Beitrag zur andinen Laubmoosflora von Peru. *Willdenowia*, **15:** 529–555.

Menzies, A. 1798: A new arrangement of the genus *Polytrichum*. *Transactions of the Linnaean Society*, **4:** 63–84 + tab. 7–8.

Meurck, C. D. 1989: Subantarctic islands and Antarctica. *In*: *Triennial Report of Botany Division Christchurch – New Zealand*. Christchurch, Department of Scientific and Industrial Research: p. 46.

Meusel, H. 1935: Wuchsformen und Wuchstypen der europäischen Laubmoose. *Nova Acta Leopoldina, Neue Folge*, **3:** 123–277.

Mikulášková, E. 2006: [Development in distribution of the neophytic moss *Campylopus introflexus* in the Czech Republic]. *Bryonora*, **38:** 1–6 (in Czech with English abstract).

Milde, J. 1869: *Bryologia silesiaca. Laubmoos-Flora von Nord- und Mittel-Deutschland unter besonderer Berücksichtigung Schlesiens und mit Hinzunahme der Floren von Jütland, Holland, der Rheinpflaz, von Baden, Franken, Böhmen, Mähren und der Umgegend von München.* Leipzig, Arthur Felix: ix + 410 pp.

Miller, H. A. 1979: The phylogeny and distribution of the Musci. *In*: G. C. S. Clarke & J. G. Duckett (eds), *Bryophyte systematics*. The Systematics Association Special Volume. **14**. London, Academic Press: pp. 12–39.

——— & A. J. Harrington. 1977: Microtheciellaceae, fam. nov. *Journal of Bryology*, **9:** 519–524.

———, H. O. Whittier & B. A. Whittier. 1978: Prodromus florae muscorum Polynesiae with a key to genera. *Bryophytorum Bibliotheca*, **16:** 1–334.

Mishler, B. D. 1985*a*: The phylogenetic relationships of *Tortula*: an SEM survey and a preliminary cladistic analysis. *The Bryologist*, **88:** 388–403.

———. 1985*b*: Biosystematic studies of the *Tortula ruralis* complex. I. Variation of taxonomic characters in culture. *Journal of the Hattori Botanical Laboratory*, **58:** 225–253.

———. 1986: Ontogeny and phylogeny in *Tortula* (Musci: Pottiaceae). *Systematic Botany*, **11:** 189–208.

———. 1990: Reproductive biology and species distinctions in the moss genus *Tortula*, as represented in Mexico. *Systematic Botany*, **15:** 86–97.

———. 1994: *Tortula*. *In*: A. J. Sharp, H. Crum & P. M. Eckel (eds.), *The moss flora of Mexico. Part one Sphagnales to Bryales. Memoirs of the New York Botanical Garden*, **69:** 319–350.

——— & S. P. Churchill. 1984: A cladistic approach to the phylogeny of the 'bryophytes'. *Brittonia*, **36:** 406–424.

———, L. A. Lewis, M. A. Buchheim, K. S. Renzaglia, D. J. Garbary, C. F. Delwiche, F. W. Zechman, T. S. Kantz & R. L. Chapman. 1994: Phylogenetic relationships of the 'green algae' and 'bryophytes'. *Annals of the Missouri Botanical Garden*, **81:** 451–483.

Mitten, W. 1851: Catalogue of cryptogamic plants collected by Professor W. Jameson in the vicinity of Quito. *Hooker's Journal of Botany and Kew Garden Miscellany*, **3:** 49–57; 351–361.

———. 1856: A list of the Musci and Hepaticae collected in Victoria, Australia, by Dr. F. Mueller. *Hooker's Journal of Botany and Kew Garden Miscellany*, **8:** 257–266 + pl. xii.

———. 1859: Musci Indiae Orientalis; an enumeration of the mosses of the East Indies. *Journal of the Proceedings of the Linnean Society Supplement to Botany*, **1:** 1–171.

———. 1863: On the Musci and Hepaticae from the Cameroon Mountains and from the River Niger. *The Journal of the Linnean Society Botany*, **7:** 147–169.

———. 1867: Musci. *In*: J. D. Hooker (ed.), *Handbook of the New Zealand flora: a systematic description of the native plants of New Zealand and the Chatham, Kermadec's Lord Auckland's Campbell's and Macquarrie's islands*. **2**. London, Reeve and Co.: pp. 393–497.

———. 1869: Musci austro-americani, sive Enumeratio muscorum omnium austro-americanorum mihihucusque cognitorum, eorum praecipue in terris amazonicis andinisque Ricardo Spruceo lectorum. *The Journal of the Linnean Society Botany*, **12:** 1–659.

———. 1873: Musci. *In*: B. Seeman (ed.), *Flora vitiensis: a description of the plants of the Viti or Fiji Islands with an account of their history, use, and properties*. London, L. Reeve and Co.: pp. 378–404 + pls. xcvii–xcviii.

———. 1875: Musci (Mosses). *In*: J. Ch. Melliss (ed.), *St. Helena: a physical, historical, and topographical description of the island, including its geology, fauna, flora, and meteorology*. London, L. Reeve & Co.: pp. 357–366 + pl. 56.

———. 1876*a*: The Musci and Hepaticae collected by H. N. Moseley, M. A., naturalist to H. M. S. 'Challenger'. *The Journal of the Linnean Society Botany*, **15:** 59–73.

———. 1876*b*: A list of the Musci and Hepaticae collected in Kerguelen's Island by Rev. A. E Eaton, A. M. *The Journal of the Linnean Society Botany*, **15:** 193–197.

———. 1879: Musci. *In*: *An account of the petrological, botanical, and zoological collections made in Kerguelen's Land and Rodriguez during the Transit of Venus Expedition carried out by order of Her Majesty's government in the years 1874–75. Philosophical Transactions of the Royal Society of London*, **168** (extra volume): 24–39 + pl. iii.

———. 1883: Australian mosses. *Transactions and Proceedings of Royal Society of Victoria*, **19:** 50–96.

———. 1884: Prince Edward Group. Marion Island. Musci. Kerguelen Island. Musci. *In*: J. Murray (ed.), *Report on the scientific results of the voyage of H.M.S. Challenger during the years 1873–76 under the command of Captain George S. Nares, R.N., F.R.S. and the late Captain Frank Tourle Thomson, R.N. Botany*. **1**(2). London, Longmans & Co.; John Murray; Macmillan & Co.; Simpkin, Marshall, & Co.; Trübner & Co.; E. Stranford; J. D. Potter; Kegan Paul, Trench, & Co.; and Bernard Quaritch; Edinburgh, Adam & Charles Black and Douglas & Foulis; Dublin, A. Thom & Co. and Hodges, Figgis, & Co.: pp. 197–201.

Mogensen, G. S. 1986: Taxonomy and distribution of Greenland mosses. II. *Platydictya* Berk. (Musci: Amblystegiaceae). *Lindbergia*, **12:** 139–143.

——— & J. Lewinsky. 1987 ['1986']: Distribution maps of bryophytes in Greenland 11. *Lindbergia*, **12:** 149–152.

——— & R. H. Zander. 1999: Four moss species new to Greenland: *Barbula amplexifolia*, *Didymodon brachyphyllus*,

D. michiganensis, and *Gyroweisia tenuis* (Pottiaceae, Musci). *Lindbergia*, **24:** 77–83.

Mohamed, M. A. H. 1979: A taxonomic study of *Bryum billardieri* Schwaegr. and related species. *Journal of Bryology*, **10:** 401–465.

Mohan, Gh. 1975: *Studiul familiei Grimmiaceae din brioflora Romaniei*. Bucureşti, Universitatea Bucureşti, Facultatea de Biologie: 76 pp. + 31 pls.

———. 1998: Catalogul briofitelor din România. *Acta Botanica Horti Bucurestensis*, **1998:** 1–432.

Mönkemeyer, W. 1927: Die Laubmoose Europas. Handbuch zur Einführung in die Mooskunde, Bestimmungstabellen und Beschreibung der europäischen Gattungen und Arten, nebst kurzer Darstellung aller Famillien und ihrer wichtigsten Gattungen. — *In*: *Dr L. Rabenhorst's Kryptogamen-Flora von Deutschland, Oesterreich und der Schweiz*. **4**. *Ergänzungsband: Andreaeales – Bryales*. Leipzig, Akademische Verlagsgesellschaft m. b. H.: x + 960 pp.

Montagne, C. 1845*a*: Cinquième centurie de plantes cellulaires exotiques nouvelles. *Annales des Sciences Naturelles Botanique, Série 3*, **4:** 86–123 + pl. 5.

———. 1845*b*: Familia II. Musci Dill. Linn. *In*: M. M. Hombron & M. Jacquinot (eds), *Voyage au pole sud et dans l'Océanie sur les corvettes l'Astrolabe et la Zélée, exécuté par ordre du roi pendant les années 1837–1838–1839–1840, sous le commandement de M. J. Dumont-s-Urville, Captaine de vaisseau. Botanique*. **1**. *Plantes cellulaires*. Paris, Gide et Cie: pp. 281–335 + pl. 20.

———. 1850: Flora chilena. Plantas cellulares. I. Musgos. *In*: C. Gay (ed.), *Historia física y politica de Chile según documentos adquiridos en esta república durante doce años de residencia en ella y publicada bajo los auspicios del supremo gobierno. Botánica*. **7**. Paris, en casa del autor and Santiago, Museo de Historia Natural: pp. 1–202.

Montagnes, R. J. S. & D. H. Vitt. 1991: Patterns of morphological variation in *Meesia triquetra* (Bryopsida: Meesiaceae) over an arctic-boreal gradient. *Systematic Botany*, **16:** 726–735.

Morison, R. 1699: *Plantarum historiae universalis oxoniensis pars tertia seu herbarium distributio nova per tabulas cognationis & affinitatis ex libro naturae observata & detecta*. **3**. Oxonii, e Theatro Sheldoniano: xxiv + 659 + [9] pp.

Moseley, H. N. 1874: Notes on the plants collected in the islands of the Tristan d'Acunha group. *Journal of the Linnean Society Botany*, **14:** 377–384.

Müller, C. [A.] 1894–1900: Musci (Laubmoose). *In*: A. Engler & K. Prantl (eds), *Die Natürlichen Pflanzenfamilien nebst ihren Gattungen und wichtigeren Arten inbesondere den Nutzpflanzen*, **1**(3[1]). Leipzig, Wilhelm Engelmann: pp. 142–202.

Müller, F. 2000*a*. *Amphidium lapponicum* (Hedw.) Schimp. Chile. *In*: T. Blockeel (ed.), *New national and regional bryophyte records, 3. Journal of Bryology*, **22:** 304–305.

———. 2000*b*: Additions to the bryophyte flora of Réunion (East African islands). *Tropical Bryology*, **18:** 91–95.

———. 2002: New records and new synonyms for the southern South American moss flora. *Nova Hedwigia*, **74:** 445–450.

——— & R. A. Pursell. 2003: The genus *Fissidens* (Musci, Fissidentaceae) in Chile. *Journal of the Hattori Botanical Laboratory*, **93:** 117–139.

——— & A. Schäfer-Verwimp. 1999: New bryophyte taxon records from tropical countries III. *Tropical Bryology*, **16:** 195–201.

Müller, Fa. A. 1829: Erstes Verzeichniss sardinischer Laubmoose, wie auch derjenigen welche von meinen Freunde Hrn. Fleischer bei Smyrna aufgefunded worden sind, nebst Beschreibungen und Abbildungen einiger neuen Arten. *Flora*, **12:** 385–396; 401–410.

Müller, K. 1845: Beiträge zu einer Flora der Aequinoctial-Gegenden. *Linnaea*, **18:** 667–709 + Taf. ix–x.

———. 1846: Bryologische Notizen. Aus einem Briefe des Herrn. E. Hampe. *Botanische Zeitung (Berlin)*, **4:** 124–125.

———. 1848–1851: *Synopsis muscorum frondosorum omnium hucusque cognitorum*. **1–2**. Berolini, sumptibus Alb. Foerstner: viii + 812 (Pars prima – Musci vegetationis acrocarpicae; 1848–1849); 772 pp. (Pars secunda – Musci vegetationis pleurocarpicae; 1850–1851).

———. 1862: Additamenta ad Synopsin Muscorum nova. *Botanische Zeitung (Berlin)*, **20:** 327–329, 337–339, 348–350, 361–362, 373–374, 382–382, 392–393.

———. 1879: Prodromus bryologiae argentinicae I. *Linnaea*, **42:** 217–460.

———. 1882: Prodromus bryologiae argentinicae II, seu Musci lorentziani argentinici. *Linnaea*, **43:** 341–486.

———. 1883: Die auf der Expedition S. M. S. »Gazelle« von Dr. Naumann gesammelten Laubmoose. *Botanische Jahrbücher für Systematik, Pflanzengeschichte und Pflanzengeographie*, **5:** 76–88.

———. 1885: Bryologia fuegiana. *Flora*, **68:** 391–429.

———. 1887: Beiträge zur Bryologie Nord-Amerika's. *Flora*, **70:** 219–225.

———. 1889: Laubmoose (Musci frondosi). *In*: A. Engler (ed.), *Die Forschungsreise S.M.S. 'Gazelle' in den Jahren 1874 bis 1876 unter Kommando des Kapitän zur See Freiherrn von Schleinitz*. **4**. Botanik. Berlin, Ernst Siegfried Mittler und Sohn: 64 pp.

———. 1890*a*: Bryologia Austro-Georgiae. *In*: G. Neumayer (ed.), *Die Internationale Polarforschung 1882–83. Die Deutschen Expeditionen und ihre Ergebnisse*. **2**. Berlin, A. Asher: pp. 279–322.

———. 1890*b*: Die Moose von vier Kilimandscharo-Expeditionen. *Flora*, **73:** 465–499.

———. 1901: *Genera muscorum frondosorum. Classes schistocarporum, cleistocarporum, stegocarporum complectentia, exceptis orthotrichaceis et pleurocarpis. Gattungen und Gruppen der Laubmoose in historischer und systematischer Beziehung, sowie nach ihrer geographischen Verbreitung unter Berücksichtigung der Arten*. Leipzig, Verlag von Eduard Kummer: vii + 474 pp.

Muñoz, J. 1998*a*: Materials toward a revision of *Grimmia* (Musci: Grimmiaceae): nomenclature and taxonomy of *Grimmia longirostris*. *Annals of the Missouri Botanical Garden*, **85:** 352–363.

———. 1998*b*: Amended description of *Coscinodon calyptratus* (Grimmiaceae) and major range extensions into Eurasia and New Zealand. *The Bryologist*, **101:** 89–92.

———. 1998*c*: A taxonomic revision of *Grimmia* subgenus *Orthogrimmia* (Musci, Grimmiaceae). *Annals of the Missouri Botanical Garden*, **85:** 367–403.

——. **1999:** A revision of *Grimmia* (Musci, Grimmiaceae) in the Americas. 1: Latin America. *Annals of the Missouri Botanical Garden*, **86:** 118–191.

——. **2002:** *Grimmia exquisita* (Musci, Grimmiaceae), a new species from central Asia. *Journal of Bryology* **24:** 315–318,

—— **& F. Pando. 2000:** A world synopsis of the genus *Grimmia* (Musci, Grimmiaceae). *Monographs in Systematic Botany from the Missouri Botanical Garden*, **83:** 1–133.

——**, J. R. Shevock & D. Toren. 2002:** *Grimmia serrana* (Bryopsida, Grimmiaceae), a new species from California, U.S.A. *Journal of Bryology*, **24:** 143–146.

——**, Á. M. Felicísimo, F. Cabezas. A. R. Burgaz & I. Martínez. 2004:** Wind as a long-distance dispersal vehicle in the Southern Hemisphere. *Science*, **304:** 1144–1147.

Murray, B. M. 1982: Andreaeopsida. *In*: S. P. Parker (ed.), *Synopsis and classification of living organisms*. **1**. New York, McGrew-Hill Book Company: pp. 305–307.

——. **1984:** A revision of the monotypic genera *Indusiella*, *Aligrimmia* and *Coscinodontella* (Musci: Grimmiaceae), with comments on convergent xeromorphological features. *The Bryologist*, **87:** 24–36.

——. **1986:** *Andreaea sinuosa*, sp. nov. (Musci: Andreaeaceae), from Alaska, British Columbia and Scotland. *The Bryologist*, **89:** 189–194.

——. **1987*a*:** Andreaeobryaceae. *In*: G. S. Mogensen (ed.), *Illustrated moss flora of Arctic North America. 3. Andreaeobryaceae – Tetraphidaceae. Meddelelser om Grønland Bioscense*, **23:** 3–6.

——. **1987*b*:** Andreaeaceae. *In*: G. S. Mogensen (ed.), *Illustrated moss flora of Arctic North America. 3. Andreaeobryaceae – Tetraphidaceae. Bioscense*, **23:** 6–24.

——. **1988*a*:** Systematics of the Andreaeopsida (Bryophyta): two orders with links to *Takakia*. *In*: J. Engel & S. Hattori (eds), *Bryological contributions presented in celebration of the distinguished scholarship of Rudolf M. Schuster*. *Beihefte zur Nova Hedwigia*, **90:** 289–336.

——. **1988*b*:** The genus *Andreaea* in Britain and Ireland. *Journal of Bryology*, **15:** 17–82.

——. **1992:** Bryophyte flora of the Alaskan steppes. *Bryobrothera*, **1:** 9–33.

——. **2006:** Andreaeaceae. *In*: P. M. McCarthy (ed.), *Flora of Australia*. **51**. *Mosses (1)*. Canberra, Australian Biological Resources Study and Melbourne, CSIRO Publishing: pp. 108–123.

Murray, J. A. 1784: *Caroli à Linné equitis Systema vegetabilium secundum classes ordines genera species cum characteribus et differentiis. Editio decima quarta.* Gottingae, typis et impensis Jo. Christ. Dieterich: xx + 987 + [988–1004, index] pp.

Myrcha, A., R. Ochyra & A. Tatur. 1991: Site of Special Scientific Interest No. 8 – Western Shores of Admiralty Bay, King George Island, South Shetland Islands. *In*: R. Z. Klekowski & K. W. Opaliński (eds), *The First Polish-Soviet Antarctic Symposium 'Arctowski '85'*. Polish Academy of Sciences, Institute of Ecology, Warszawa: pp. 157–168.

N

Nakanishi, S. 1977: Ecological studies of the moss and lichen communities in the ice-free areas near Syowa Station, Antarctica. *Antarctic Record*, **59:** 68–96.

——. **1979:** On the variation of leaf characters of an Antarctic moss, *Bryum inconnexum*. *Memoirs of National Institute of Polar Research Special Issue, Proceedings of the Symposium on Terrestrial Ecosystem in the Syowa Station Area*, **11:** 47–57.

Nakatsubo, T. & S. Ohtani. 1992: Note on the structure of moss colonies composed of two species on King George Island, the South Shetland Islands. *Antarctic Record*, **36:** 285–293.

Necker, N. J. 1771: *Methodus muscorum per classes, ordines, genera ad species cum synonymis, nominibus trivialibus, locis natalibus, observationibus digestorum, aeneisque figuris illustratorum*. Mannheim: Typograph. Academ. Elect. Scient.: xvii + 296 pp. + i pl.

Nees von Esenbeck, Ch. G., F. Hornschuch & J. Sturm. 1823: *Bryologia germanica, oder Beschreibung der in Deutschland und in der Schweiz wachsenden Laubmoose*. **1**. Nürnberg, bey Jacob Sturm: [xvi] + cliii + 206 pp. + xii pls.

Newsham, K. K. 2003: UVB radiation arising from stratospheric ozone depletion influences the pigmentation of the Antarctic moss *Andreaea regularis*. *Oecologia*, **135:** 327–331.

Newton, A. E. 2000: Evolution of the major moss lineages: phylogenetic analysis based on multiple gene sequences and morphology. *The Bryologist*, **103:** 187–211.

—— **& R. S. Tangney. 2007:** *Pleurocarpous mosses. Systematics and evolution*. The Systematics Association Special Volume Series. **71**. Boca Raton, London, New York, CRC Press, Taylor & Francis Groups: [18] + 434 pp.

——**, C. J. Cox, J. G. Duckett, J. A. Wheeler, B. Goffinet, T. A. J. Hedderson & B. D. Mishler. 2000:** Evolution of the major moss lineages: phylogenetic analyses based on multiple gene sequences and morphology. *The Bryologist*, **103:** 187–211.

Newton, M. E. 1972: Chromosome studies in some South Georgian bryophytes. *British Antarctic Survey Bulletin*, **33:** 41–49.

——. **1973:** A taxonomic assessment of *Bartramia*, *Breutelia* and *Exodokidium* on South Georgia. *British Antarctic Survey Bulletin*, **32:** 1–14.

——. **1974*a*:** Notes on Antarctic bryophytes: IV. *Encalypta* Hedw. *British Antarctic Survey Bulletin*, **39:** 1–6.

——. **1974*b*:** A synoptic flora of South Georgian mosses: IV. *Bartramia* and *Breutelia*. *British Antarctic Survey Bulletin*, **38:** 59–71.

——. **1974*c*:** Notes on Antarctic bryophytes: V. *Brachythecium majusculum* M. E. Newton, sp. nov. *British Antarctic Survey Bulletin*, **39:** 45–48.

——. **1977*a*:** A synoptic flora of South Georgian mosses: VI. *Cheilothela*, *Dicranella*, *Distichium*, *Myurella* and *Catagonium*. *British Antarctic Survey Bulletin*, **46:** 1–21.

——. **1977*b*:** Notes on Antarctic bryophytes: VII. The occurrence of *Distichium* B. S. G. and *Dicranella* (C. Muell.) Schimp. in the Antarctic botanical zone. *British Antarctic Survey Bulletin*, **46:** 131–135.

——. **1979*a*:** A synoptic flora of South Georgian mosses: VIII. *Calliergon* and *Brachythecium*. *British Antarctic Survey Bulletin*, **48:** 133–157.

——. **1979*b*:** A taxonomic assessment of *Brachythecium* on South Georgia. *British Antarctic Survey Bulletin*, **48:** 119–132.

———. **1980:** Chromosome studies in some Antarctic and sub-Antarctic bryophytes. *British Antarctic Survey Bulletin,* **50:** 77–86.

———. **1983*a*:** A synoptic flora of South Georgian mosses: *Campylium. British Antarctic Survey Bulletin,* **61:** 53–58.

———. **1983*b*:** A new species of the moss genus *Plagiothecium* from South Georgia. *British Antarctic Survey Bulletin,* **60:** 63–67.

Nicholson, W. E. 1920: Mosses from the Caspian and Black Sea regions. *The Bryologist*, **23:** 90–91.

Nishimura, N. 1999: A historical review of Hypnaceae. *In*: J. Enroth, J. Hyvönen & S. Piippo (eds), *From Finland to Japan to New Guinea to China – that is Timo Koponen, bryologist. Bryobrothera*, **5:** 283–298.

———**, M. Higuchi, T. Seki & H. Ando. 1984:** Delimitation and the subdivision of the moss family Hypnaceae. *Journal of the Hattori Botanical Laboratory*, **55:** 227–234.

Noakes, T. D. & R. E. Longton. 1988: Pathways of water movement in relation to structure and micromorphology in Antarctic mosses. *Polarfoschung*, **58:** 125–138.

——— **&** ———**. 1989:** Studies on water relations in mosses from the cold-Antarctic. *In*: R. B. Heywood (ed.), *Proceedings of British Antarctic Special Topic Award Scheme Symposium 9–10 November 1988.* Cambridge, British Antarctic Survey: pp. 103–116.

Noguchi, A. 1972: On the delimitation of the genera of Hylocomiaceae and Rhytidiaceae. *Journal of the Hattori Botanical Laboratory*, **35:** 155–168.

———. **1974:** Musci japonici. X. The genus *Racomitrium*. *Journal of the Hattori Botanical Laboratory*, **38:** 337–369.

———. **1987–1994:** *Illustrated moss flora of Japan*. **1–5**. Nichinan, The Hattori Botanical Laboratory: pp. 1–3 + 1–242 + i–vi (part 1, 1987); pp. 243–490 + i–vi (part 2, 1988); pp. 491–742 + i–vi (part 3, 1989); pp. 743–1012 + i–ix (part 4, 1991); pp. 1013–1253 (part 5, 1994).

———**, N. Takaki & H. Inoue. 1966:** Bryophytes collected by Dr. K. Yoda in eastern Nepal. *Bulletin of the National Science Museum*, **9:** 359–386.

Norris, D. H. & Koponen, T. 1988: Bryophyte flora of the Huon Peninsula, Papua New Guinea. XXIV. Andreaeaceae, Archidiaceae, Seligeriaceae, and Ditrichaceae (Musci). *Annales Botanici Fennici*, **25:** 165–177.

——— **&** ———**. 1989:** Bryophyte flora of the Huon Peninsula, Papua New Guinea. XXVIII. Pottiaceae (Musci). *In*: T. Koponen (ed.), *Bryophyte flora of the Huon Peninsula, Papua New Guinea. XXVI–XXVIII. Acta Botanica Fennica* **137:** 81–138.

——— **&** ———**. 1990:** Bryophyte flora of the Huon Peninsula, Papua New Guinea. XXXV. Dicranaceae and Dicnemonaceae (Musci). *In*: T. Koponen (ed.), *Bryophyte flora of the Huon Peninsula, Papua New Guinea. XXXV–XXXVI. Acta Botanica Fennica*, **139:** 1–64.

——— **&** ———**. 1996:** Bryophyte flora of the Huon Peninsula, Papua New Guinea. LVIII. *Hypnodendron* (Hypnodendraceae, Musci). *In*: J. Enroth (ed.), *Bryophyte flora of the Huon Peninsula, Papua New Guinea. LVII–LIX. Acta Botanica Fennica*, **156:** 23–48.

——— **& J. R. Shevock. 2004:** Contributions toward a bryoflora of California: I. A specimen-based catalogue of mosses. *Madroño*, **51:** 1–131.

———**, T. Koponen & S. Piippo. 1999:** Bryophyte flora of the Huon Peninsula, Papua New Guinea. LXVI. Meesiaceae (Musci), with lists of boreal to temperate disjunct, boreal, and widely spread species in New Guinea. *Annales Botanici Fennici*, **36:** 257–263.

———**, J. R. Shevock & B. Goffinet. 2004:** *Orthotrichum kellmanii* (Bryopsida, Orthotrichaceae), a remarkable new species from the central coast of California. *The Bryologist*, **107:** 209–214.

Nyholm, E. 1954–1965: *Illustrated moss flora of Fennoscandia. II. Musci.* Lund, CWK Gleerup: pp. 1–87 + 1 map (part 1, 1954); pp. 87–189 (part 2, 1956); pp. 189–288 (part 3, 1958); pp. 408–647 (part 5, 1965).

———. **1987:** *Illustrated flora of Nordic mosses.* **1**. *Fissidentaceae – Seligeriaceae*. Copenhagen – Lund, Nordic Bryological Society: pp. 1–72.

———. **1990:** *Illustrated flora of Nordic mosses.* **2**. *Pottiaceae – Splachnaceae – Schistostegaceae*. Copenhagen – Lund, Nordic Bryological Society: pp. 73–141.

———. **1993:** *Illustrated flora of Nordic mosses.* **3**. *Bryaceae – Rhodobryaceae – Mniaceae – Cinclidiaceae – Plagiomniaceae*. Copenhagen – Lund, Nordic Bryological Society: pp. 143–244.

———. **1998:** *Illustrated flora of Nordic mosses.* **4**. *Aulacomniaceae – Meesiaceae – Catoscopiaceae – Bartramiaceae – Timmiaceae – Encalyptaceae – Grimmiaceae – Ptychomitriaceae – Hedwigiaceae – Orthotrichaceae*. Copenhagen – Lund, Nordic Bryological Society: pp. 249–405.

O

O'Brien, T. J. 2007: The phylogenetic distribution of pleurocarpous mosses: evidence from cpDNA sequences. *In*: A. E. Newton & R. S. Tangney (eds), *Pleurocarpous mosses. Systematics and evolution*. The Systematics Association Special Volume Series. **71**. Boca Raton, London, New York, CRC Press, Taylor & Francis Groups: pp. 19–40.

Ochi, H. 1959: *A revision of the Bryaceae in Japan and adjacent regions*. Tottori, The Biological Institute, Faculty of Liberal Arts, Tottori University: 121 pp.

———. **1962:** Contribution to the mosses of Bartramiaceae in Japan and the adjacent regions (1). *Nova Hedwigia*, **4:** 87–108 + pls. 38–45.

———. **1963:** Contribution to the mosses of Bartramiaceae in Japan and the adjacent regions (2). *Nova Hedwigia*, **5:** 91–115 + pls. 7–24.

———. **1970:** A revision of the subfamily Bryoideae in Australia, Tasmania, New Zealand and the adjacent islands. *The Journal of the Faculty of Education Tottori University Natural Sciences*, **21:** 7–67.

———. **1972:** A revision of African Bryoideae, Musci (First part). *The Journal of the Faculty of Education Tottori University Natural Sciences*, **23:** 7–126.

———. **1973*a*:** A revision of African Bryoideae, Musci (Second part). *The Journal of the Faculty of Education Tottori University Natural Sciences*, **24:** 23–50.

———. **1973*b*:** Supplement to the subfamily Bryoideae (Musci) in Australia and New Zealand. *Hikobia*, **6:** 217–223.

———. **1979:** A taxonomic review of the genus *Bryum*, Musci in Antarctica. *Memoirs of National Institute of Polar Research Special Issue*, **11:** 70–80.

———. **1980:** A revision of the neotropical Bryoideae, Musci (First part). *The Journal of the Faculty of Education Tottori University Natural Sciences*, **29:** 49–154.

———. **1981:** A revision of the neotropical Bryoideae, Musci (Second part). *The Journal of the Faculty of Education Tottori University Natural Sciences*, **30:** 21–55.

———. **1982:** A revision of the Bryoideae (Musci) in southern South America. *The Journal of the Faculty of Education Tottori University Natural Sciences*, **31:** 11–47.

———. **1984:** Revision of the Australasian Bryoideae (Musci) Supplement II. Studies on the *Bryum* type specimens of R. Brown ter. *New Zealand Journal of Botany*, **22:** 179–182.

———. **1985:** An annotated list of the subfamily Bryoideae in South, Southeast and East Asia. *The Journal of the Faculty of Education Tottori University Natural Sciences*, **34:** 41–96.

———. **1990*a*:** Bryaceous mosses as the source of distributional records from the Andean region. *Bulletin of the National Science Museum Series B*, **16:** 61–71.

———. **1990*b*:** Bryaceae. *In*: M. Higuchi & N. Takaki (eds.), *Mosses from Nepal collected by Botanical Expedition of National Science Museum, Tokyo. In*: M. Watanabe & S. B. Malla (eds), *Cryptogams of the Himalayas*. **2**. *Central and eastern Nepal*. Tsukuba, National Science Museum: pp. 132–139.

———. **1990*c*:** Noteworthy additional bryaceous mosses (Bryaceae) to the Central Himalayan (Nepal) alpine belt. *Hikobia*, **10:** 419–422.

———. **1992:** A revised infrageneric classification of the genus *Bryum* and related genera (Bryaceae, Musci). *In*: T. Koponen & J. Hyvönen (eds), *Proceedings of the Congress of East Asiatic Bryology, Helsinki, August 12–19, 1990. Bryobrothera*, **1:** 231–241(–244).

———. **1994:** *Bryum*. *In*: A. J. Sharp, H. Crum & P. M. Eckel (eds.), *The moss flora of Mexico. Part one Sphagnales to Bryales. Memoirs of the New York Botanical Garden*, **69:** 454–489.

——— **& H. Kanda. 1991:** New additions of bryaceous mosses (Musci, Bryaceae) to the Antarctic flora. *Proceedings of the NIPR Symposium on Polar Biology*, **4:** 123–126.

——— **& R. Ochyra. 1986:** *Bryum dichotomum* Hedw., a species new to the Antarctic. *Lindbergia*, **11:** 157–160.

——— **& N. Salazar Allen. 1990:** *Bryum incrassatolimbatum* new to Panama, with first description of its sporophyte. *The Bryologist*, **93:** 286–287.

Ochyra, R. 1976: Taxonomic position and geographical distribution of *Isopterygiopsis muelleriana* (Schimp.) Iwats. *Fragmenta Floristica et Geobotanica*, **22:** 129–135 + 1 map.

———. **1983:** [Adventive bryophytes]. *Wiadomości Botaniczne*, **27:** 31–44 (in Polish).

———. **1984:** *Bryophyta Antarctica Exsiccata. Numbers 1–200.* Edmonton, University of Alberta: 27 [unnumbered] pp.

———. **1985*a*:** *Hypnobartlettia fontana* gen. et sp. nov. (Musci: Hypnobartlettiaceae fam. nov.), a unique moss from New Zealand. *Lindbergia*, **11:** 2–8.

———. **1985*b*:** *Koponenia*, a new pleurocarpous moss genus from Bolivia. *Journal of Bryology*, **13:** 479–486.

———. **1986 ['1985']:** On the Antarctic species of the family Orthotrichaceae. *Lindbergia*, **11:** 141–146.

———. **1987*a*:** *Schistidium steerei* (Grimmiaceae), a remarkable new species from the Antarctic, with a note on *S. obtusifolium*. *In*: W. R. Buck (ed.), *Bryostephane steereana: a collection of bryological papers presented to William Campbell Steere on the occasion of his 80th birthday. Memoirs of the New York Botanical Garden*, **45:** 607–614.

———. **1987*b*:** A revision of the moss genus *Sciaromium* (Mitt.) Mitt. I. General remarks and the section *Aloma* Dusén. *Journal of Bryology*, **14:** 453–464.

———. **1987*c*:** A revision of the moss genus *Sciaromium* (Mitt.) Mitt. II. The section *Limbidium* Dusén, with a description of *Vittia* gen. nov. (Vittiaceae fam. nov.). *Journal of the Hattori Botanical Laboratory*, **62:** 387–415.

———. **1987*d*:** A revision of the moss genus *Sciaromium* (Mitt.) Mitt. III. The section *Platyloma* Broth. *Journal of the Hattori Botanical Laboratory*, **63:** 107–132.

———. **1988*a*:** [*Dicranella staphylina* H. Whiteh., a new species in the moss flora of Poland]. *Badania Fizjograficzne nad Polską Zachodnią Seria B – Botanika*, **37:** 177–185 (in Polish with English summary).

———. **1988*b*:** *Hypnum revolutum* (Mitt.) Lindb. new to New Zealand. *Journal of Bryology*, **15:** 236–237.

———. **1988*c*:** What is *Dichelyma antarcticum* C. Muell.? *Polish Polar Research*, **8:** 403–410.

———. **1989*a*:** Lectotypification of *Schistidium pulvinatum* (Hedw.) Brid. (Musci: Grimmiaceae) and its consequences. *Nova Hedwigia*, **48:** 85–106.

———. **1989*b*:** Animadversions on the moss genus *Cratoneuron* (Sull.) Spruce. *Journal of the Hattori Botanical Laboratory*, **67:** 203–242.

———. **1989*c*:** The alleged occurrence of *Bryum perlimbatum* Card. (Musci, Bryaceae) in South Africa. *Journal of Bryology*, **15:** 551–553.

———. **1990*a*:** The discovery of the South Georgian endemic species *Schistidium urnulaceum* (C. Muell.) B. G. Bell (Musci: Grimmiaceae) in the Antarctic. *Polish Polar Research*, **11:** 133–146.

———. **1990*b*:** The identity of *Calliergon keniae* (Musci). *Bulletin du Jardin Botanique National de Belgique*, **60:** 111–117.

———. **1990*c*:** *Streimannia turgida*, a new genus and species of Brachytheciaceae from Australia. *Journal of Bryology*, **16:** 223–229.

———. **1990*d*:** *Epipterygium* (Musci, Bryaceae), a genus new to Australasia. *Polish Botanical Studies*, **1:** 51–57.

———. **1991*a*:** *Hygrohypnum ellipticum* is *Bryum renauldii* (Musci). *Fragmenta Floristica et Geobotanica*, **35:** 71–75

———. **1991*b*:** *Bryostreimannia* nom. nov. replaces *Streimannia* Ochyra, hom. illeg. *Journal of Bryology*, **16**: 643–644.

———. **1992*a*:** A new synonym of *Blindia magellanica* (Musci, Seligeriaceae). *Fragmenta Floristica et Geobotanica*, **37:** 221–225.

———. **1992*b*:** *Amblyodon dealbatus* (Musci, Meesiaceae) – a bipolar disjunct. *Fragmenta Floristica et Geobotanica*, **37:** 251–259.

———. **1993*a*:** The taxonomic status of *Blindia skottsbergii* (Musci, Seligeriaceae). *Fragmenta Floristica et Geobotanica*, **38:** 503–508.

———. **1993*b*:** Antipodal mosses: I. A revision of the genus *Holodontium* (Seligeriaceae). *Fragmenta Floristica et Geobotanica*, **38:** 75–98.

——. **1993*c*:** *Grimmia plagiopodia* (Musci, Grimmiaceae) in the Southern Hemisphere. *Fragmenta Floristica et Geobotanica*, **38:** 21–27.

——. **1993*d*:** Taxonomic results of the BRYOTROP Expedition to Zaire and Rwanda. 20. Grimmiaceae, Funariaceae, Bartramiaceae (*Philonotis*), Amblystegiaceae, Plagiotheciaceae. *Tropical Bryology*, **8:** 181–187.

——. **1993*e*:** New synonyms of *Racomitrium lamprocarpum* (Musci, Grimmiaceae). *Fragmenta Floristica et Geobotanica*, **38:** 738–741.

——. **1993*f*:** Two new species of *Grimmia* (Musci, Grimmiaceae) from New Zealand. *In*: R. Ochyra & L. Stuchlik (eds), *Botanostephane kornasiana. Botanical contributions presented to Jan Kornaś in celebration of his 70th birthday. Fragmenta Floristica et Geobotanica Supplementum* **2:** 219–227.

——. **1994*a*:** Antipodal mosses: II. The identity of *Ditrichum perporodictyon* (Musci, Ditrichaceae). *Fragmenta Floristica et Geobotanica*, **39:** 113–116.

——. **1994*b*:** Nomenclatural notes on mosses: 4. Seligeriaceae subfam. Dicranoweisioideae. *Fragmenta Floristica et Geobotanica*, **39:** 314.

——. **1994*c*:** Antipodal mosses: III. Taxonomic status of the genus *Monocranum*. *Fragmenta Floristica et Geobotanica*, **39:** 453–460.

——. **1995:** On some Bridelian taxa of amblystegiaceous mosses. *Fragmenta Floristica et Geobotanica*, **40:** 899–904.

——. **1996*a*:** Antipodal mosses: IV. A note on *Andreaea hookeri* (Andreaeaceae). *Fragmenta Floristica et Geobotanica*, **41:** 287–290.

——. **1996*b*:** Nomenclatural notes on mosses: 10. *Andreaea gainii* and *A. paralella* [sic] (Andreaeaceae) from Antarctica. *Fragmenta Floristica et Geobotanica*, **41:** 479–480.

——. **1996*c*:** Antipodal mosses: V. On *Distichium austrogeorgicum* (Ditrichaceae). *Fragmenta Floristica et Geobotanica*, **41:** 335–338.

——. **1996*d*:** Rhizoid gemmae in *Ditrichum brotherusii* (Musci, Ditrichaceae) from Antarctica. *Fragmenta Floristica et Geobotanica*, **41:** 1019–1021.

——. **1996*e*:** *Ditrichum lewis-smithii* (Ditrichaceae, Bryopsida), a new species from Antarctica. *Annales Botanici Fennici*, **33:** 303–309.

——. **1996*f*:** Nomenclatural notes on mosses: 9. *Austrophilibertiella* nom. nov. (Ditrichaceae). *Fragmenta Floristica et Geobotanica*, **41:** 477–478.

——. **1996*g*:** Nomenclatural notes on mosses: 11. *Warnstorfia austro-straminea* comb. nov. (Amblystegiaceae). *Fragmenta Floristica et Geobotanica*, **41:** 480–482.

——. **1996*h*:** One more synonym of *Holodontium strictum* (Musci, Seligeriaceae). *Fragmenta Floristica et Geobotanica*, **41:** 1011–1014.

——. **1996*i*:** The southernmost locality for *Limprichtia revolvens* (Musci, Amblystegiaceae). *Fragmenta Floristica et Geobotanica*, **41:** 1014–1016.

——. **1997*a*:** Nomenclatural notes on mosses: 15. Typification of *Polytrichum subpiliferum* (Polytrichaceae). *Fragmenta Floristica et Geobotanica*, **42:** 175–176.

——. **1997*b*:** *Leptotrichella* replaces *Microdus* (Musci, Dicranaceae). *Fragmenta Floristica et Geobotanica*, **42:** 559–565.

——. **1997*c*:** Antipodal mosses: VII. A reappraisal of *Tortula didymodontoides* (Musci, Pottiaceae) from Îles Kerguelen. *Fragmenta Floristica et Geobotanica*, **42:** 147–152.

——. **1997*d*:** Nomenclatural notes on mosses: 14. Synonyms of *Muelleriella* (Orthotrichaceae). *Fragmenta Floristica et Geobotanica*, **42:** 173–175.

——. **1997*e*:** The oldest synonym of *Blindia torrentium* (Musci, Seligeriaceae) from the Falkland Islands. *Fragmenta Floristica et Geobotanica*, **42:** 575–578.

——. **1998*a*:** *The moss flora of King George Island, Antarctica*. Cracow, Polish Academy of Sciences, W. Szafer Institute of Botany: xxiv + 278 pp.

——. **1998*b*:** New names and combinations in *Schistidium* (Musci, Grimmiaceae). *Fragmenta Floristica et Geobotanica*, **43:** 103–108.

——. **1998*c*:** *Schistidium halinae* (Grimmiaceae, Bryopsida), a new moss species from the Antarctic. *Annales Botanici Fennici*, **35:** 267–273.

——. **1998*d*:** The type of *Barbula calobolax* var. *angustinervia* (Musci, Pottiaceae). *Fragmenta Floristica et Geobotanica*, **43:** 285–287.

——. **1998*e*:** The two oldest recognitions of *Campylium polygamum* (Musci, Amblystegiaceae). *Fragmenta Floristica et Geobotanica*, **43:** 279–283.

——. **1998*f*:** The taxonomic status of *Brachythecium turquetii* (Musci, Brachytheciaceae). *Fragmenta Floristica et Geobotanica*, **43:** 287–289.

——. **1998*g*:** Nomenclatural notes on mosses: 17. Validation of *Brachythecium* sect. *Albicantia* (Brachytheciaceae). *Fragmenta Floristica et Geobotanica*, **43:** 289–290.

——. **1999*a*:** *Ditrichum austrogeorgicum*: a synonym of *D. hyalinum* (Ditrichaceae, Bryopsida). *Cryptogamie, Bryologie*, **20:** 247–253.

——. **1999*b*:** Antipodal mosses: XII. A reconsideration of the Chilean endemic *Ditrichum fontanum* (Musci, Ditrichaceae). *Fragmenta Floristica et Geobotanica*, **44:** 239–244.

——. **1999*c*:** The taxonomic status of two varieties of *Ditrichum* (Musci, Ditrichaceae) from southern South America. *Fragmenta Floristica et Geobotanica*, **44:** 504–507.

——. **1999*d*:** *Dicranella gremmenii* (Musci, Dicranaceae) sp. nov. from Subantarctic Marion Island. *Fragmenta Floristica et Geobotanica*, **44:** 219–225.

——. **1999*e*:** Antipodal mosses: XI. Additional new synonyms of *Anisothecium hookeri* (Dicranaceae). *Fragmenta Floristica et Geobotanica*, **44:** 233–238.

——. **1999*f*:** The identities of some forms of Antarctic mosses. *Fragmenta Floristica et Geobotanica*, **44:** 511–514.

——. **1999*g*:** Nomenclatural notes on mosses: XIX. *Verrucidens* (Seligeriaceae). *Fragmenta Floristica et Geobotanica*, **44:** 499–502.

——. **1999*h*:** New combinations in neotropical mosses. *Fragmenta Floristica et Geobotanica*, **44:** 255–259.

——. **1999*i*:** *Brachythecium austroglareosum* (Musci, Brachytheciaceae), an addition to the moss flora of Gough Island, Tristan da Cunha group. *Fragmenta Floristica et Geobotanica*, **44:** 514–517.

——. **1999*j*:** *Brachythecium austrosalebrosum* (Musci, Brachytheciaceae) new to Bolivia. *Fragmenta Floristica et Geobotanica*, **44:** 517–519.

——. **1999*k*:** Antipodal mosses: IX. *Platydictya* (Bryopsida, Hypnaceae). *Annales Botanici Fennici*, **36:** 51–58.

——. **1999*l*:** Antipodal mosses: X. The status of *Dichodontium brasiliense* and *D. opacifolium* (Dicranaceae). *Fragmenta Floristica et Geobotanica*, **44:** 227–232.

——. **1999*m*:** Antipodal mosses: XIII. A neotypification and the taxonomic status of *Barbula validinervia* (Pottiaceae), with a note on muscological research on Îles Kerguelen. *Fragmenta Floristica et Geobotanica*, **44:** 245–253.

——. **1999*n*:** Nomenclatural notes on mosses: XX. *Brachydontium microcarpum, comb. nov.* (Seligeriaceae). *Fragmenta Floristica et Geobotanica*, **44:** 502–504.

——. **2001*a*:** Does the family Meesiaceae (Musci) occur in Africa? *Cryptogamie, Bryologie*, **22:** 23–28.

——. **2001*b*:** *Plagiothecium falklandicum* (Cardot & Broth.) M.E.Newton. Argentina: Tierra del Fuego. *In*: T. Blockeel (ed.), *New national and regional bryophyte records, 4. Journal of Bryology*, **23**: 150–151.

——. **2002*a*:** Nomenclatural changes in subfamilies of the Dicranaceae. *Cryptogamie, Bryologie*, **23:** 345–349.

——. **2002*b*:** *Plagiothecium lamprostachys* (Hampe) A.Jaeger, a forgotten name for an Australasian moss. *Journal of Bryology*, **24:** 85–86.

——. **2002*c*:** [Review of] J. Putzke and A. B. Pereira (2001): *The Antarctic mosses, with special reference to the South Shetland Islands. Polish Polar Research*, **23:** 293–298.

——. **2003*a*:** *Campylopus pyriformis* (Schultz) Brid. The Falkland Islands. *In*: T. Blockeel (ed.), *New national and regional bryophyte records, 7. Journal of Bryology*, **25**: 141–142.

——. **2003*b*:** *Schistidium lewis-smithii* (Bryopsida, Grimmiaceae), a new species from the maritime Antarctic. *Nova Hedwigia*, **77:** 363–372.

——. **2003*c*:** Antipodal mosses: VIII. *Valdonia* gen. nov. (Seligeriaceae) from the Kerguelen Province in the Subantarctic. *Cryptogamie, Bryologie*, **24:** 87–102.

——. **2003*d*:** The first recognition of the family Amblystegiaceae. *Journal of Bryology*, **25:** 135–140.

——. **2003*e*:** A nomenclatural comment on the northern Andean *Chorisodontium* (Dicranaceae) and citation of Hooker names. *Journal of Bryology*, **25:** 66–68.

——. **2004*a*:** Antipodal mosses: XIV. On the taxonomic status of *Grimmia lawiana* (Bryopsida: Grimmiaceae) from the continental Antarctic. *Polish Polar Research*, **25:** 111–122.

——. **2004*b*:** *Schistidium leptoneurum* species nova from the South Shetland Islands, Antarctica. *Cryptogamie, Bryologie*, **25:** 125–130.

——. **2004*c*:** Antipodal mosses: XV. Taxonomy and distribution of *Schistidium cupulare* (Bryopsida: Grimmiaceae). *Polish Polar Research*, **25**: 123–133.

——. **2004*d*:** *Tortula hoppeana*, the correct name for *Desmatodon latifolius* in *Tortula* (Bryopsida, Pottiaceae). *The Bryologist*, **107:** 497–500.

——. **2005:** *Didymodon brachyphyllus* (Sull.) R.H.Zander. South Georgia. *In*: T. Blockeel (ed.), *New national and regional bryophyte records, 11. Journal of Bryology*, **27:** 164.

——. **2006:** *Hennediella antarctica* (Ångstr.) Ochyra & Matteri. Îles Crozet. *In*: T. Blockeel (ed.), *New national and regional bryophyte records, 12. Journal of Bryology*, **28:** 69.

——. **2008:** Mosses. *In*: S. L. Chown & P. W. Froneman (eds), *The Prince Edward Islands: land-sea interactions in a changing ecosystem*. Stellenbosch, African SunMedia: pp. 383–389.

—— **& O. M. Afonina. 1986:** The taxonomic position and geographical distribution of *Grimmia andreaeopsis* C. Muell. (Grimmiaceae, Musci). *Polish Polar Research* **7:** 319–332.

—— **& J. K. Bartlett. 1986:** The identity of *Drepanocladus fontinaliopsis* (C. Muel;l.) Broth. ex Par. (Bryopsida, Amblystegiaceae). *New Zealand Journal of Botany*, **24:** 361–368.

—— **& H. Bednarek-Ochyra. 1994:** A reconsideration of *Grimmia brandegei* (Musci, Grimmiaceae). *Fragmenta Floristica et Geobotanica*, **39:** 667–670.

—— **& ——. 1997:** The type specimen of *Leptotrichum praealtum* (Musci, Ditrichaceae). *Fragmenta Floristica et Geobotanica* **42:** 567–570.

—— **& ——. 1999:** *Racomitrium valdon-smithii* (Musci, Grimmiaceae) sp. nov. from subantarctic Marion Island. *Fragmenta Floristica et Geobotanica*, **44:** 209–217.

—— **& ——. 2004:** *Streptocolea atrata* (Bryopsida: Grimmiaceae), newly found in western North America, with a review of its global distribution. *The Bryologist* **107:** 542–549.

—— **& ——. 2007*a*:** *Didymodon australasiae* Cardot & Broth. Heard Island. *In*: T. Blockeel (ed.), *New national and regional bryophyte records, 17. Journal of Bryology*, **29:** 277–278.

—— **& ——. 2007*b*:** *Bucklandiella*. *In*: R. H. Zander (Lead ed.), *Flora of North America north of Mexico*. **27**. *Bryophytes: mosses (1)*. New York – Oxford, Oxford University Press: pp. 267–285.

—— **& B. G. Bell. 1984:** A record of *Schistidium falcatum* (Bryophyte: Musci) from the Antarctic. *British Antarctic Survey Bulletin*, **64:** 77–79.

—— **& D. A. Broughton. 2004:** Additions to the moss flora of the Falkland Islands. *Journal of Bryology*, **26:** 232–236.

—— **& W. R. Buck. 2002:** A re-appraisal of the type material of *Plagiothecium drepanophyllum. The Bryologist*, **105:** 641–644.

—— **& ——. 2003:** *Arctoa fulvella*, new to Tierra del Fuego, with notes on trans-American bipolar bryogeography. *The Bryologist*, **106:** 532–538.

—— **& H. Deguchi. 1999:** *Ditrichum gemmiferum* (Musci, Ditrichaceae) in Chile. *Fragmenta Floristica et Geobotanica*, **44:** 509–511.

——**& H. Hertel. 1990 ['1991']:** Contribution to the moss flora of the subantarctic island Marion. *Polish Botanical Studies*, **1:** 19–34.

—— **& P. Isoviita. 1989:** Proposal to conserve *Schistidium* (Musci, Grimmiaceae) with a new type. *Taxon*, **38:** 665–669.

—— **& E. Kuta. 1990 ['1992']:** Chromosome studies on Polish bryophytes, VI. *Acta Biologica Cracoviensia Series Botanica*, **32:** 197–218.

—— **& R. I. Lewis Smith. 1996:** Antipodal mosses: VI. *Stegonia latifolia* (Musci, Pottiaceae) in the Antarctic. *Fragmenta Floristica et Geobotanica*, **41:** 985–994.

—— **& ——. 1998:** Antarctic species in the genus *Ditrichum* (Ditrichaceae, Bryopsida), with a description of *D. gemmiferum* sp. nov. *Annales Botanici Fennici*, **35:** 33–53.

—— **& ——. 1999:** *Meesia uliginosa* Hedw. (Musci, Meesiaceae) in Antarctica. *Cryptogamie, Bryologie*, **20:** 5–10.

——— & D. Lamy. 2004: New names for mosses proposed by Palisot de Beauvois in his '*Muscologie*' of 1822. *Cryptogamie, Bryologie*, **25:** 179–196.

——— & P. J. Lightowlers. 1988: The South Georgian moss flora: *Vittia*. *British Antarctic Survey Bulletin*, **80:** 121–127.

——— & M. Mahu. 1990: *Cochlearidium aymaranum*, a strange new genus and species of Cochlearidiaceae (fam. Nov.) from Chile. *Journal of Bryology*, **16:** 215–222.

——— & H. W. Matcham. 2005: *Plagiothecium lamprostachys* (Hampe) A.Jaeger. South Africa and Lesotho. *In*: T. Blockeel (ed.), *New national and regional bryophyte records, 4. Journal of Bryology*, **27:** 165–166.

——— & C. M. Matteri. 1996*a*: Nomenclatural changes in South American and sub-Antarctic mosses. *Fragmenta Floristica et Geobotanica*, **41:** 1005–1010.

——— & ———. 1996*b*: *Drepanocladus polycarpos* (Musci, Amblystegiaceae), an addition to the moss flora of Tierra del Fuego. *Fragmenta Floristica et Geobotanica*, **41:** 1001–1004.

——— & ———. 1997: An earlier name for *Warnstorfia austro-straminea* (Musci, Amblystegiaceae). *Fragmenta Floristica et Geobotanica*, **42:** 179–182.

——— & ———. 2001: Amblystegiaceae. *In*: S. A. Guerrera, I. J. Gamundi de Amos & C. M. Matteri (eds), *Flora criptogámica de Tierra del Fuego*. **14**(10). Buenos Aires, Consejo Nacional de Investigaciones Científicas y Técnicas de la Republica Argentina: 96 pp.

——— & ———. 2007: *Encalypta procera* Bruch. Argentina. *In*: T. Blockeel (ed.), *New national and regional bryophyte records, 15. Journal of Bryology*, **29**: 140.

——— & J. Muñoz. 2000: Further nomenclatural changes and corrections in *Schistidium*. *Journal of Bryology*, **22:** 141–142.

——— & M. E. Newton. 1985: The taxonomy and distribution of *Dicranella cardotii* (R. Br. ter.) Dix., an addition to the moss flora of Antarctica. *Lindbergia*, **11:** 94–98.

——— & H. Ochi. 1986: New or otherwise interesting species of the genus *Bryum* (Musci, Bryaceae) in the Antarctic. *Acta Botanica Hungarica*, **32:** 209–219.

——— & T. Pócs. 1987 ['1985']: East African bryophytes, XI. Dr. L. Péntek's collection from Mozambique. *Abstracta Botanica*, **9** *Supplementum*, **2:** 63–86.

——— & ———. 1992: *Bryophyta Africana Selecta*: a new exsiccata from Africa. *Fragmenta Floristica et Geobotanica*, **37:** 379–388.

——— & R. S. Poulsen. 2003: Four new moss records from Îles Kerguelen. *Journal of Bryology*, **25:** 136–138.

——— & R. D. Seppelt. 2008*a*: *Bryum orbiculatifolium* Cardot & Broth. Heard Island. *In*: T. Blockeel (ed.), *New national and regional bryophyte records, 18. Journal of Bryology*, **30:** in press.

——— & R. D. Seppelt. 2008*b*: *Bryum archangelicum* Bruch & Schimp. Heard Island. *In*: T. Blockeel (ed.), *New national and regional bryophyte records, 18. Journal of Bryology*, **30:** in press.

——— & S. M. Singh. 2008: Three remarkable moss records from Dronning Maud Land, continental Antarctica. *Nova Hedwigia*, **86:** 497–506.

——— & V. R. Smith. 2003: *Entosthodon productus* Mitt. (Funariaceae) on Marion Island – the first record in the Subantarctic. *Cryptogamie, Bryologie*, **24:** 173–180.

——— & H. Streimann. 2001: Studies on austral Amblystegiaceae: I. Australian taxa. *Nova Hedwigia*, **73:** 453–460.

——— & P. Szmajda. 1978: An annotated list of Polish mosses. *Fragmenta Floristica et Geobotanica* **24:** 93–145.

——— & ———. 1983: *Amphidium lapponicum*. *In*: J. Szweykowski & T. Wojterski (eds), *Atlas of geographical distribution of spore-plants in Poland. Series V. Mosses (Musci)*. **1**. Warszawa – Poznań, Państwowe Wydawnictwo Naukowe: p. 17 + 1 map.

——— & O. Tyshchenko. 2006: *Leptobryum pyriforme* (Hedw.) Wilson. Antarctic Peninsula. *In*: T. Blockeel (ed.), *New national and regional bryophyte records, 13. Journal of Bryology*, **28:** 151–152.

——— & J. Váňa. 1989: The hepatics reported from the Antarctic and an outline of their phytogeography. *Polish Polar Research*, **10:** 211–229.

——— & R. H. Zander. 2002: The genera *Didymodon* and *Bryoerythrophyllum* (Pottiaceae) in Antarctica. *Journal of Bryology*, **24:** 33–44.

——— & ———. 2007: [Is *Tortula lithophila* conspecific with *Sarconeurum glaciale* (Bryopsida: Pottiaceae)?]. *Fragmenta Floristica et Geobotanica Polonica*, **14:** 209–212 (in Polish with extensive English summary).

———, H. Bednarek-Ochyra & R. I. Lewis Smith. 1998: 170 years of research of the Antarctic moss flora. *In*: P. Głowacki & J. Bednarek (eds), *Polish Polar Studies. 25th Polar Symposium. The 100th anniversary of Prof. Henryk Arctowski's and Prof. Antoni Bolesław Dobrowolski's participation in the Belgica expedition to the Antarctic in 1887–1889* [sic!], *Warszawa*, 1998. Institute of Geophysics of the Polish Academy of Sciences, Warszawa: pp. 159–177.

———, ——— & ———. 2002: New and rare moss species from subantarctic South Georgia. *Nova Hedwigia*, **74:** 121–147.

———, ——— & ———. 2003*a*: *Schistidium deceptionense*, another new endemic moss species from the Antarctic. *The Bryologist*, **106:** 569–574.

———, ——— & ———. 2003*b*: *Andreaea mutabilis* Hook. f. & Wilson. South Georgia. *In*: T. Blockeel (ed.), *New national and regional bryophyte records, 7. Journal of Bryology*, **25:** 141.

———, ——— & R. D. Seppelt. 2008: *Philonotis polymorpha* (Müll.Hal.) Broth. Macquarie Island. *In*: T. Blockeel (ed.), *New national and regional bryophyte records, 18. Journal of Bryology*, **30:** in press.

———, ——— & P. Szmajda. 1990: M. 271. *Racomitrium lanuginosum* (Hedw.) Brid. *In*: R. Ochyra & P. Szmajda (eds), *Atlas of the geographical distribution of spore plants in Poland. Series V. Mosses (Musci)*. **5**. Kraków, W. Szafer Institute of Botany and Poznań, Adam Mickiewicz University: pp. 29–33 + 1 map.

———, R. Kempa & W. R. Buck. 2000: *Plagiothecium lucidum* (Hook.f. & Wils.) Par. in tropical Africa. *Tropical Bryology*, **18:** 147–152.

———, T. Koponen & D. H. Norris. 1991: Bryophyte flora of the Huon Peninsula, Papua New Guinea. XLVI. Amblystegiaceae (Musci). *In*: T. Koponen (ed.), *Bryophyte flora of the Huon Peninsula, Papua New Guinea. XL–XLVIII. Acta Botanica Fennica* **143:** 91–106.

———, C. Sérgio & R. Schumacker. 1988: *Racomitrium lamprocarpum* (C. Muell.) Jaeg., an austral moss disjunct in

Portugal, with taxonomic and phytogeographic notes. *Bulletin du Jardin Botanique National de Belgique*, **58:** 225–258.

———, **V. R. Smith & N. J. M. Gremmen. 2003:** *Thuidium delicatulum* (Hedw.) Schimp. (Thuidiaceae) – another bipolar moss disjunct from Subantarctic Marion Island. *Cryptogamie, Bryologie*, **24:** 253–263.

———, **P. Szmajda & H. Bednarek-Ochyra. 1992:** M. 393. *Bartramia ithyphylla* Brid. *In*: R. Ochyra & P. Szmajda (eds), *Atlas of the geographical distribution of mosses in Poland.* **8**. Kraków, W. Szafer Institute of Botany and Poznań, Adam Mickiewicz University: pp. 25–34 + 1 map.

———, **D. H. Vitt & D. G. Horton. 1986:** An annotated guide to *Bryophyta Antarctica Exsiccata*. *Cryptogamie, Bryologie – Lichénologie*, **7:** 53–62.

———, **J. Żarnowiec & H. Bednarek-Ochyra. 2003:** *Census catalogue of Polish mosses*. Kraków, Polish Academy of Sciences, Institute of Botany: 372 pp.

———, **H. Bednarek-Ochyra, Th. Arts & R. I. Lewis Smith. 2000:** Occurrence of the neotropical moss *Dicranella hilariana* (Mont.) Mitt. in the Antarctic. *Tropical Bryology*, **18:** 153–160.

———, **P. Isoviita, J. Żarnowiec & H. Bednarek-Ochyra. 2006:** Correct author citations for class and some subclass names of the Bryophyta. *Annales Botanici Fennici*, **43:** 118–122.

———, **P. Szmajda, H. Bednarek & W. Bocheński. 1988*a*:** M. 386. *Meesia uliginosa* Hedw. *In*: Z. Tobolewski & T. Wojterski (eds), *Atlas of geographical distribution of spore-plants in Poland. Series V. Mosses (Musci)*. **3**. Warszawa – Poznań, Państwowe Wydawnictwo Naukowe: p. 11–13 + 1 map.

———, **P. Szmajda, H. Bednarek & W. Bocheński. 1988*b*:** M. 387. *Meesia hexasticha* (Funck) Bruch. *In*: Z. Tobolewski & T. Wojterski (eds), *Atlas of geographical distribution of spore-plants in Poland. Series V. Mosses (Musci)*. **3**. Warszawa – Poznań, Państwowe Wydawnictwo Naukowe: p. 15–16 + 1 map.

———, **P. Szmajda, H. Bednarek & W. Bocheński. 1988*c*:** M. 388. *Meesia longiseta* Hedw. *In*: Z. Tobolewski & T. Wojterski (eds), *Atlas of geographical distribution of spore-plants in Poland. Series V. Mosses (Musci)*. **3**. Warszawa – Poznań, Państwowe Wydawnictwo Naukowe: p. 17–19 + 1 map.

———, **P. Szmajda, H. Bednarek & W. Bocheński. 1988*d*:** M. 389. *Meesia triquetra* (Richt.) Aongstr. *In*: Z. Tobolewski & T. Wojterski (eds), *Atlas of geographical distribution of spore-plants in Poland. Series V. Mosses (Musci)*. **3**. Warszawa – Poznań, Państwowe Wydawnictwo Naukowe: p. 21–26 + 1 map.

Ociepa, A. M. 2007: Jurassic liverworts from Mount Flora, Hope Bay, Antarctic Peninsula. *Polish Polar Research*, **28:** 31–36.

Oeder, G. Ch. 1766: [*Flora danica*] *Icones plantarum sponte nascentium in regnis Daniae et Norvegiae, in ducatibus Slesvici et Holsaticae, et in comitatibus Oldenburgi et Delmenhorstiae; ad illustrandum opus de iisdem plantis, regio jussu exarandum, Florae danicae nomine inscriptum*. **2(5)**. Kiøbenhavn, Trykt hos hans Kongl. Majests. Hof. Bogtrykker Nicolaus Møller: 10 pp. + pls. 241–300.

Oesau, A. 2003: *Pterygoneurum papillosum* (Bryopsida: Pottiaceae), a new moss species from Germany. *Journal of Bryology*, **25:** 247–252.

———**. 2005:** Zur Verbreitung und Vergesellschaftung des Salzmooses *Desmatodon heimii* (Hedw.) Mitt. in Rheinhessen (Rheinland-Pfalz). *Limprichtia*, **26:** 1–12.

Okada, H. & H. Kanda. 1994: Cytotaxonomical aspects of Antarctic mosses, with special references to the proportion of polyploidy. *Nankyoku Shiryô (Antarctic Record)*, **38:** 54–62.

Okitsu, S., S. Imura & E. Ayukawa. 2003: Structure and dynamics of the *Ceratodon purpureus–Bryum pseudotriquetrum* community in the Yukidori Valley, Langhovde, continental Antarctica. *Polar Bioscience*, **16:** 49–60.

———, ——— **&** ———**. 2004:** Micro-relief distribution of major mosses in ice-free areas along the Sôya Coast, the Syowa Station area, East Antarctica. *Polar Bioscience*, **17:** 69–82.

Olech, M. 2002: Plant communities on King George Island. *In*: L. Beyer & M. Bölter (eds), *Geoecology of Antarctic ice-free coastal landscapes. Ecological Studies*, **154:** 215–231.

Osada, T. 1965: Japanese Polytrichaceae. I. Introduction and the genus *Pogonatum*. *Journal of the Hattori Botanical Laboratory*, **28:** 171–201.

———**. 1966:** Japanese Polytrichaceae. II. The genera *Polytrichum*, *Oligotrichum*, *Bartramiopsis* and *Atrichum* and phytogeography. *Journal of the Hattori Botanical Laboratory*, **29:** 1–52.

O'Shea, B. J. 1999: Checklist of the mosses of sub-Saharan Africa (version 3, 11/99). *Tropical Bryology Research Reports*, **1:** 1–133.

———**. 2003:** Checklist of the mosses of sub-Saharan Africa (version 4, 12/99). *Tropical Bryology Research Reports*, **4:** 1–182.

———**. 2006:** Checklist of the mosses of sub-Saharan Africa (version 5, 12/06). *Tropical Bryology Research Reports*, **6:** 1–252.

Otnyukova, T. N., E. A. Ignatova, M. S. Ignatov & V. E. Fedosov. 2004: New records of *Tortella alpicola* Dix. in Eurasia. *Arctoa, a Journal of Bryology*, **13:** 197–201.

Ø

Øvstedal, D. O. & R. I. Lewis Smith. 2001: *Lichens of Antarctica and South Georgia. A guide to their identification and ecology*. Cambridge, Cambridge University Press: xii + 411 pp.

P

Papp, C. 1933: Contribution à la monographie du *Polytrichum juniperinum* Willd. *Revue Bryologique et Lichénologique, Nouvelle Série*, **6:** 154–170 + pls. iii–iv.

Parihar, N. S. 1965: *An introduction to Embryophyta*. **1**. *Bryophyta*. Ed. 5. Allahabad, Central Book Depot: vii + 377 pp

Paris, É. G. 1894–1898: *Index bryologicus sive enumeratio muscorum hucusque cognitorum adjunctis synonyma distributioneque geographica locupletissimis*. Parisiis, apud Paul Klinksieck: vi + 1380 pp.

———**. 1909:** *Collatio nominum brotherianorum et Indicis bryologicis*. Parisiis, Libraria J.-B. Baillière et Filii: [3] + 37 pp.

Paul, H. 1916: Ueber *Hypnum turgescens* T. Jensen. Eine systematisch-geographische Studie. *Bryologische Zeitschrift*, **1:** 145–160.

Pavletić, Z. 1955: *Prodromus of the bryophyte flora of Yugoslavia.* Zagreb, Jugoslavenska Akademija Znanosti i Umjetnosti: 577 pp. (in Serbo-Croatian).

Peat, H. J. 1998: The Antarctic Plant Database: a specimen and literature based information system. *Taxon*, **47:** 85–93.

———, A. Clarke & P. Convey. 2006: Diversity and biogeography of the Antarctic flora. *Journal of Biogeography*, **34:** 132–146.

Pedersen, N. 2000: A cladistic overview of the Bryaceae (Musci) based on morphological and anatomical data and with emphasis on the genus *Bryum*. *Journal of Bryology*, **22:** 193–206.

———. 2005: Validation of *Imbribryum* (Bryaceae). *The Bryologist*, **108:** 449.

——— & L. Hedenäs. 2001: Phylogenetic relationships within the Plagiotheciaceae. *Lindbergia*, **26:** 62–76.

——— & ———. 2002*a*: Phylogenetic relationships between *Bryum* and supposedly closely related genera. *Journal of Bryology*, **24:** 277–289.

——— & ———. 2002*b*: Phylogeny of the Plagiotheciaceae based on molecular and morphological evidence. *The Bryologist*, **105:** 310–324.

——— & ———. 2003: Phylogenetic investigations of a well supported clade within the Bryaceae: evidence from seven chloroplast sequences and morphology. *Plant Systematics and Evolution*, **240:** 115–132.

——— & ———. 2005: Taxonomic and nomenclatural implications of phylogenetic studies of the Bryaceae based on molecular data and morphology. *The Bryologist*, **108:** 123–128.

———, C. J. Cox & L. Hedenäs. 2003: Phylogeny of the moss family Bryaceae inferred from chloroplast DNA sequences and morphology. *Systematic Botany*, **28:** 471–482.

———, D. T. Holyoak & A. E. Newton. 2007: Systematics and morphological evolution within the moss family Bryaceae: a comparison between parsimony and Bayesian methods for reconstruction of ancestral character states. *Molecular Phylogenetics and Evolution*, **43:** 891–907.

Pereira, A. B. & J. Putzke, J. 1994: Floristic composition of Stinker Point, Elephant Island, Antarctica. *Korean Journal of Polar Research*, **5:** 37–47.

Persson, H. 1942: Bryophytes from the bottom of some lakes in north Sweden. *Botaniska Notiser*, **1942:** 308–324.

———. 1944: On some species of *Aloina*, with special reference to their dispersal by the wind. *Svensk Botanisk Tidskrift*, **38:** 260–268.

———. 1952: New or otherwise interesting bryophytes from Alaska-Yukon. *The Bryologist*, **55:** 1–25; 88–116.

———. 1963: Bryophytes of Alaska and Yukon Territory collected by Handford T. Shacklette. *The Bryologist*, **66:** 1–26.

Petrov, S. 1975: [*Bryophyta bulgarica clavis diagnostica*]. Sofia, in aedibus Academiae Scientiarum Bulgaricae: 535 pp. (in Bulgarian).

Pettersson, B. 1940: Experimentelle untersuchungen über die euanemochore Verbreitung der Sporenpflanzen. *Acta Botanica Fennica*, **25:** 1–102.

Pfeiffer, L. K. G. 1871–1875: *Nomenclator botanicus. Nomimum ad finem anni 1858 publici juris factorum, classes, ordines, tribus, families, divisiones, genera, subgenera vel sectiones designantium enumeratio alphabetica. Adjectis auctoribus, temporibus, locis systematicis apud varios, notis literaris atque etymologicis et synonymis.* **1**. Cassellis, sumptibus Theodori Fischeri: pp. i–vi + 1–808 (part 1(1), 1871–1873); pp. i–vi + 809–1876 (part 1(2), 1874–1875).

Philibert, H. 1884: De l'importance du péristome pour les affinités naturelles des mousses. 2e article. *Revue Bryologique*, **11:** 65–72.

———. 1889: Etudes sur le péristome. Huitiemè article. *Revue Bryologique*, **16:** 39–44, 67–77.

Pickard, J. 1986: Spatial relations of the vegetation of the Vestfold Hills. *In*: J. Pickard (ed.), *Antarctic Oasis. Terrestrial environments and history of the Vestfold Hills*. Sydney, Academic Press: pp. 275–308.

——— & R. D. Seppelt. 1984*a*: Phytogeography of Antarctica. *Journal of Biogeography*, **11:** 83–102.

——— & ———. 1984*b*: Holocene occurrence of the moss *Bryum algens* Card. in the Vestfold Hills, Antarctica. *Journal of Bryology*, **13:** 209–217.

Pitkin, P. H. 1992: *Distichium inclinatum*. *In*: M. O. Hill, C. D. Preston & A. J. E. Smith (eds), *Atlas of the bryophytes of Britain and Ireland.* **2**. *Mosses (except Diplolepideae).* Colchester, Harley Books: p. 100.

Pizarro, C. & F. Sáiz. 1977: Estudio de la taxocenosis muscinal de la península Fildes (isla Rey Jorge, Shetland del Sur). *Serie Científica Instituto Antártico Chileno*, **5:** 81–96.

Podpěra, J. 1942: *Bryum* generis monographiae prodromus. 1. Species Eurasiae septentrionalis. Pars 1. *Práce Moravské Přírodovědecké Společnosti*, **13(2):** 1–58.

———. 1943: *Bryum* generis monographiae prodromus. 1. Species Eurasiae septentrionalis. Pars 2. *Práce Moravské Přírodovědecké Společnosti*, **15(4):** 1–52.

———. 1944: *Bryum* generis monographiae prodromus. 1. Species Eurasiae septentrionalis. Pars 3. *Práce Moravské Přírodovědecké Společnosti*, **16(6):** 1–46.

———. 1945: *Bryum* generis monographiae prodromus. 1. Species Eurasiae septentrionalis. Pars 4. *Práce Moravské Přírodovědecké Společnosti*, **17(4):** 1–104.

———. 1950: *Bryum* generis monographiae prodromus. 1. Species Eurasiae septentrionalis. Pars. 5. *Práce Moravskoslezské Akademie Věd Přírodních*, **22:** 357–472.

———. 1951*a*: *Bryum* generis monographiae prodromus. 1. Species Eurasiae septentrionalis. Pars. 5. *Práce Moravskoslezské Akademie Věd Přírodních*, **23:** 1–40.

———. 1951*b*: *Bryum* generis monographiae prodromus. 1. Species Eurasiae septentrionalis. Pars 6. systematica. *Práce Moravskoslezské Akademie Věd Přírodních*, **23:** 451–465.

———. 1952: *Bryum* generis monographiae prodromus. 1. Species Eurasiae septentrionalis. Pars 7. systematica. *Práce Moravskoslezské Akademie Věd Přírodních*, **24:** 267–306.

———. 1953: *Bryum* generis monographiae prodromus. 1. Species Eurasiae septentrionalis. Pars 8. systematica. *Práce Moravskoslezské Akademie Věd Přírodních*, **25:** 53–120.

———. 1954: *Conspectus muscorum europaeorum*. Praha, Nakladatvelství Československé Akademie Věd: 697 pp.

———. 1955: *Bryum* generis monographiae prodromus. 1. Species Eurasiae septentrionalis. Pars 9. systematica. *Práce Brněnské Základny Československé Akademie Věd*, **27:** 105–156.

———. 1956: *Bryum* generis monographiae prodromus. 1. Species Eurasiae septentrionalis. Pars 10. systematica. *Práce Brněnské Základny Československé Akademie Věd*, **28:** 147–261.

———. 1957: *Bryum* generis monographiae prodromus. 1. Species Eurasiae septentrionalis. Pars 11. systematica. *Práce Brněnské Základny Československé Akademie Věd*, **29:** 85–188.

———. **1958:** *Bryum* generis monographiae prodromus. 1. Species Eurasiae septentrionalis. Pars 12. systematica. *Práce Brněnské Základny Československé Akademie Věd*, **30:** 89–130.

———. **1959:** *Bryum* generis monographiae prodromus. 1. Species Eurasiae septentrionalis. Pars 13. systematica. *Práce Brněnské Základny Československé Akademie Věd*, **31:** 253–320.

———. **1960:** *Bryum* generis monographiae prodromus. 1. Species Eurasiae septentrionalis. Pars 14. systematica. *Práce Brněnské Základny Československé Akademie Věd*, **32:** 301–352 + tab. 1–12.

———. **1962:** *Bryum* generis monographiae prodromus. 1. Species Eurasiae septentrionalis. Pars 15. systematica. *Práce Brněnské Základny Československé Akademie Věd*, **34:** 49–102 + tab. i–vii.

———. **1973*a*:** *Bryum* generis monographiae prodromus. I. 1. Species Eurasiae septentrionalis. Pars 16. systematica. Prague, Academia, Publishing House of the Czechoslovak Academy of Sciences: 257 pp.

———. **1973*b*:** *Bryum* generis monographiae prodromus. II. 1. Species Eurasiae septentrionalis. Pars 17. systematica. Prague, Academia, Publishing House of the Czechoslovak Academy of Sciences: 470 pp.

Polunin, N. 1959: *Circumpolar Arctic flora*. Oxford, Oxford University Press: 514 pp.

Porsild, A. E. 1964: Illustrated flora of the Canadian Arctic Archipelago. *National Museums of Canada Bulletin*, **146** *Biological Series*, **50:** i–iv + 1–218.

Post, A. 1990: Photoprotective pigment as an adaptative strategy in the Antarctic moss *Ceratodon purpureus*. *Polar Biology*, **10:** 241–245.

Potier de la Varde, R. 1955: Mousses récoltées par M. le Dr. Olov Hedberg, en Afrique orientale, au cours de la mission suédoise de 1948. *Arkiv för Botanik, Serie 2*, **3(8):** 125–204.

Price, M. J. 2005: Catalogue of the Hedwig-Schwägrichen herbarium (G). Part 1. Type material and a review of typifications for the Hedwig moss names. *Boissiera*, **61:** 3–388.

Priddle, J. 1979: Morphology and adaptation of aquatic mosses in an Antarctic lake. *Journal of Bryology*, **10:** 517–529.

——— **& H. J. G. Dartnall. 1978:** The biology of an Antarctic aquatic moss community. *Freshwater Biology*, **8:** 469–480.

Proctor, M. C. F. 1979: Structure and eco-physiological adaptation in bryophytes. *In*: G. C. S. Clarke & J. G. Duckett (eds), *Bryophyte systematics*. The Systematics Association Special Volume. **14**. London, Academic Press: pp. 479–509.

———. **1980:** Diffusion resistance in bryophytes. *In*: J. Grace, E. D. Ford & P. G. Jarvis (eds), *Plants and their atmospheric environment*. Oxford, Blackwell: pp. 219–229.

———. **1981:** Physiological ecology of bryophytes. *Advances in Bryology*, **1:** 79–166.

———. **1982:** Physiological ecology: water relations, light and temperature responses, carbon balance. *In*: A. J. E. Smith (ed.), *Bryophyte ecology*. London, Chapman & Hall: pp. 333–381.

———. **2005:** Why do Polytrichaceae have lamellae? *Journal of Bryology*, **27:** 221–229.

Przywara, L., E. Kuta & R. Ochyra. 1984: Cytological studies on Antarctic mosses II. *Journal of the Hattori Botanical Laboratory*, **57:** 127–137.

———**, R. Ochyra & E. Kuta. 1984 ['1983']:** Chromosome studies on Polish bryophytes II. *Lindbergia*, **9:** 178–185.

Puri, P. 1985: Bryophytes – a broad perspective. Delhi – Lucknow, Atma Ram & Sons: viii + 562 pp.

Pursell, R. A. & B. Allen. 2002: Erpodiaceae. *In*: B. Allen (ed.), Moss Flora of Central America. Part 2. Encalyptaceae – Orthotrichaceae. *Monographs in Systematic Botany from the Missouri Botanical Garden*, **90:** 523–531.

Putzke, J. & A. B. Pereira. 1990: Mosses of King George Island, Antarctica. *Pesquisa Antártica Brasileira*, **2:** 17–71.

——— **&** ———. **2001:** *The Antarctic mosses with special reference to the South Shetland Islands*. Editora da ULBRA, Canoas: 196 pp.

———**, A. B. Pereira & M. T. L. Putzke. 1998:** Moss communities of Rip Point in northern Nelson Island, South Shetland Islands, Antarctica. *Pesquisa Antártica Brasileira*, **3:** 103–115.

Q

Quandt, D. & A. Vanderpoorten. 2006: *Drepanocladus polygamus* (Schimp.) Hedenäs. Venezuela. *In*: T. Blockeel (ed.), *New national and regional bryophyte records, 12. Journal of Bryology*, **28:** 68–69.

———**, R. Tangney, J.-P. Frahm & W. Frey. 2000:** A molecular contribution for understanding the Lembophyllaceae (Bryopsida) based on noncoding chloroplast region (cpDNA) and ITS2 (nrDNA) sequence data. *Journal of the Hattori Botanical Laboratory*, **89:** 71–92.

R

Raffaelli, M. 1973: Il genere «Andreaea» Hedw. in Italia. Distribuzione, caratteri distintive e chiave analitica delle specie. *Webbia*, **28:** 1–30.

Rams, S., R. M. Ros & O. Werner. 2006: *Tortella alpicola* (Pottiaceae) from Spain, new to western Europe. *The Bryologist*, **109:** 404–407.

Ramsay, H. P. 1969: Cytological studies on some mosses from the British Isles. *Botanical Journal of the Linnean Society*, **62:** 85–121.

———. **1974:** Cytological studies on Australian mosses. *Australian Journal of Botany*, **22:** 293–348.

——— **& J. Seur. 1994:** Register of type specimens of mosses in Australian herbaria. *Flora of Australia Supplementary Series*, **2:** i–viii + 1–142.

Rastorfer, J. R. 1971: Vegetative regeneration and sporophyte development of *Bryum antarcticum* in an artificial environment. *Journal of the Hattori Botanical Laboratory*, **34:** 391–397.

Raven, P. H. & D. I. Axelrod. 1974: Angiosperm biogeography and past continental movements. *Annals of the Missouri Botanical Garden*, **61:** 539–673.

Ray, J. J. 1696: *Synopsis methodica stirpium britannicarum, in qua tum notae generum characteristicae traduntur, tum species singulae breviter describuntur; ducentae quinquaginta plus minus novae species partim suis locis inseruntur, partim in appendice seorsim exhibentur. Cum indice et virum epitome*. Ed. 2. Londini: impensis S. Smith & B. Walford,: xl + 346 [+ 347–365] pp.

Razgulyaeva, L. V., M. G. Napreenko, Ch. Wolfram & M. S. Ignatov. 2001: *Campylopus introflexus* (Dicranaceae, Musci) – an addition to the moss flora of Russia. *Arctoa, a Journal of Bryology,* **10:** 185–188.

Redfearn, P. L. Jr., B. C. Tan & S. He. 1996: A newly updated and annotated checklist of Chinese mosses. *Journal of the Hattori Botanical Laboratory*, **79:** 163–357.

Redhead, S. A. 1973: Observations of the rhizoids of *Andreaea*. *The Bryologist*, **76:** 185–187.

Rees, P. M. & C. J. Cleal. 2004: Lower Jurassic floras from Hope Bay and Botany Bay, Antarctica. *Special Papers in Palaeontology*, **72:** 5–89.

Reese, W. D. 1984: *Mosses of the Gulf south from the Rio Grande to the Apalachicola*. Baton Rouge – London, Louisiana University Press: xvi + 252 pp.

——— & W. R. Buck. 1991: *Weissiodicranum insularum* gen. et sp. nov. *The Bryologist*, **94:** 308–310.

——— & R. H. Zander. 1987: Serpotortellaceae, a new family of mosses (Musci) from Madagascar and Réunion. *The Bryologist*, **90:** 230–236.

Reimers, H. 1926: Beiträge zur Bryophytenflora Südamerikas. I. u. II. *Hedwigia*, **66:** 27–78.

———. 1936: Über die Laubmoosgattungen *Bryoporteria* Thér., *Camptodontium* Dus., *Hymenoloma* Dus., *Verrucidens* Card. und ihre systematische Stellung. *Notizblatt des Botanischen Gartens und Museums zu Berlin-Dahlem*, **13:** 36–52.

———. 1954: Bryophyta. Moose. *In*: H. Melchior & E. Werdermann (eds), *A. Engler's Syllabus der Pflanzenfamilien mit besonderer Berücksichtigung der Nutzpflanzen nebst einer Übersicht über die Florenreiche und Florengebiete der Erde*. Ed. 12. **1**. *Allgemeiner Teil. Bakterien bis Gymnospermen*. Berlin Nikolassee, Gebrüder Borntraeger: pp. 218–268.

Renzanglia, K. S. & K. C. Vaughn. 2000: Anatomy, development and classification of hornworts. *In*: A. J. Shaw & B. Goffinet (eds), *Bryophyte biology*. Cambridge, Cambridge University Press: pp. 1–20.

Richards, E. W. 1963: *Campylopus introflexus* (Hedw.) Brid. and *C. polytrichoides* De Not. in the British Isles; a preliminary account. *Transactions of the British Bryological Society*, **4:** 404–412 + pl. 1.

——— & A. J. E. Smith. 1975: A progress report on *Campylopus introflexus* (Hedw.) Brid. and *C. polytrichoides* De Not. in Britain and Ireland. *Journal of Bryology*, **8:** 293–298.

Richter, W. 1990: The bryoflora of the Schirmacher Oasis (East Antarctica) in relations to hydrosphere and cryosphere. *Geodätische und geophysikalische Veröffentlichungen Reihe I*, **16:** 489–494.

———. 1995: Bryophytes in the Schirmacher Oasis. *In*: P. Bormann & D. Fritzsche (eds), *The Schirmacher Oasis, Queen Maud Land, East Antarctica, and its surroundings. Petermanns Geographische Mitteilungen*. **289**. Gotha, Justus Perthes Verlag: pp. 341–342 + map 9.

Risse, S. 1985: Rhizoidal tubers on *Ditrichum heteromallum* (Hedw.) Britt. *Journal of Bryology*, **13:** 527–531.

———. 1986: Beobachtungen an Rhizoidengemmen von *Dicranella rufescens* (With.) Schimp. mit Bemerkungen zur Ontogenese und morphologischen Einteilung der Rhizoidengemmen. *Lindbergia*, **12:** 9–14.

———. 1987: Rhizoid gemmae in mosses. *Lindbergia*, **13:** 111–126.

———. 1988: Rhizoid gemmae in *Campylopus pyriformis* from the southern hemisphere. *Journal of Bryology*, **15:** 496–497.

Robinson, H. 1962: Generic revisions of North American Brachytheciaceae. *The Bryologist*, **65:** 73–146.

———. 1967: Preliminary studies on the bryophytes of Colombia. *The Bryologist*, **70:** 1–61.

———. 1970: A revision of the moss genus *Trichostomopsis*. *Phytologia*, **20:** 184–191.

———. 1971: A revised classification for the orders and families of mosses. *Phytologia*, **21:** 289–293.

———. 1972*a*: Observations on the origin and taxonomy of the Antarctic moss flora. *In*: G. A. Llano (ed.), *Antarctic terrestrial biology*. Antarctic Research Series. **20**. Washington, D. C., American Geophysical Union: pp. 163–177.

———. 1972*b*: Scanning electron microscope studies on moss leaves and peristomes. *The Bryologist*, **74:** 473–483.

———. 1974: Notes on the mosses of Juan Fernandez and southern South America. *Phytologia*, **29:** 116–119.

———. 1975: The mosses of Juan Fernandez. *Smithsonian Contributions to Botany*, **27**: i–iv + 1–88.

———. 1986: Notes on bryogeography of Venezuela. *The Bryologist*, **89:** 8–12.

——— & R. K. Godfrey. 1960: Contribution to the bryophyte flora of Turkey. *Revue Bryologique et Lichénologique, Nouvelle Série*, **29:** 244–253.

Robinson, S. A., J. Wasley & A. K. Tobin. 2003: Living on the edge – plants and global change in continental and maritime Antarctica. *Global Change Biology*, **9:** 1681–1717.

Rodway, L. 1914: *Tasmanian Bryophyta*. **1**. *Mosses*. Hobart, The Royal Society of Tasmania: 163 pp.

Roivainen, H. 1955*a*: Contribution to the Fuegian species of the genus *Rhacomitrium* Bridel. *Archivum Societatis Zoologicae Botanicae Fennicae 'Vanamo'*, **9(2):** 85–98.

———. 1955*b*: *Bucklandia Bartramii*, a new genus and species of leaf moss from Tierra del Fuego. *Archivum Societatis Zoologicae Botanicae Fennicae 'Vanamo'*, **9(2):** 98–100.

———. 1972: *Bucklandiella* Roiv., nomen novum (Musci). *Annales Botanici Fennici*, **9:** 116.

——— & E. B. Bartram. 1937: Bryological investigations in Tierra del Fuego. *Annales Botanici Societatis Zoologicae-Botanicae Fennicae Vanamo* **9(2):** i–x + 1–58.

Rooy, J. van. 1992: The genus *Amphidium* in southern Africa. *Lindbergia*, **17:** 59–63.

——— & R. E. Magill. 1987: Bryaceae. *In*: E. O. Leistner (ed.), *Flora of southern Africa which deals with the territories of South Africa, Transkei, Lesotho, Swaziland, Bophuthatswana, South West Africa/Namibia, Botswana and Venda. Bryophyta. Part 1. Mosses*. **2**. Pretoria, Botanical Research Institute: pp. 335–393.

Ros, R. M., M. J. Cano & J. Guerra. 1999: Bryophyte checklist of Northern Africa. *Journal of Bryology*, **21:** 207–244.

———, ———, J. Muñoz & J. Guerra. 2000: Contribution to the bryophyte flora of Morocco: the Jbel Toubkal. *Journal of Bryology*, **22:** 283–289.

Roth, G. 1899: Uebersichr über die Familie der Hypnaceen. *Hedwigia*, **38** *Beiblatt* **1:** 3–8.

———. 1904: *Die europäischen Laubmoose*. **1**. *Kleistokarpische und Akrokarpische Moose bis zu den Bryaceen*. Leipzig, Verlag von Wilhelm Engelmann: xiii + 598 pp. + pls. i–lii.

———. **1905:** *Die europäischen Laubmoose*. **1**. *Schluss der akrokarpischen Mooses und pleurokarpische Mooses*. Leipzig, Verlag von Wilhelm Engelmann: xvi + 733 pp. + pls. i–lxii.

———. **1911:** *Die aussereuropäischen Laubmoose*. **1**. *Enthaltend die Andreaeaceae, Archidiaceae, Cleistocarpae und Trematodonteae*. Dresden, C. Heinrich: x + 331 + pls. i–xxviii.

———. **1913:** *Die aussereuropäischen Laubmoose. Nachtrag I zu Band I*. Dresden, C. Heinrich: 18 pp. + pls. i–ii.

———. **1915:** Neuere and noch weniger bekannte europäische Laubmoose. X. *Hedwigia*, **57:** 132–140.

———. **1916:** Nachtrag III zu Band I der aussereuropäischen Laubmoose von 1910/11. *Hedwigia*, **57:** 257–262 + pl. v.

Rothero, G. P. 1994: *Brachythecium glaciale* Br. Eur. *In*: M. O. Hill, C. D. Preston & A. J. E. Smith (eds), *Atlas of the bryophytes of Britain and Ireland*. **3**. Mosses (Diplolepideae). Colchester, Harley Books: p. 322.

Rothmaler, W. 1951: Die Abteilungen und Klassen der Pflanzen. *Feddes Repertorium Specierum Novarum Regni Vegetabilis*, **54:** 256–266.

Rudolph, E. D. 1970: Local dissemination of plant propagues in Antarctica. *In*: M. W. Holdgate (ed.), *Antarctic ecology*. **2**. London, Academic Press: pp. 812–817.

Ruhland, W. 1924*a*: Musci. Allgemeiner Teil. *In*: A. Engler (ed.), *Die Natürlichen Pflanzenfamilien nebst ihren Gattungen und wichtigeren Arten inbesondere den Nutzpflanzen*. **10**. Leipzig, Wilhelm Engelmann: pp. 1–100.

———. **1924*b*:** Unterklasse Sphagnales. Allgemeine Verhältnisse. *In*: A. Engler (ed.), *Die Natürlichen Pflanzenfamilien nebst ihren Gattungen und wichtigeren Arten inbesondere den Nutzpflanzen*, **10**. Leipzig, Wilhelm Engelmann: pp. 101–105.

———. **1924*c*:** Unterklasse Andreaeales. Allgemeine Verhältnisse. *In*: A. Engler (ed.), *Die Natürlichen Pflanzenfamilien nebst ihren Gattungen und wichtigeren Arten inbesondere den Nutzpflanzen*, **10**. Leipzig, Wilhelm Engelmann: pp. 126–128.

Rushing, A. E. & Snider, J. A. 1980: Observations on sporophyte development in *Lorentziella imbricata* (Mitt.) Broth. *Journal of the Hattori Botanical Laboratory*, **47:** 35–44.

Ryan, P. G., B. P. Watkins, R. I. Lewis Smith, H. Dastych, A. Eicker, W. Foissner, H. Heatwole, W. R. Miller & G. Thompson. 1989: Biological survey of Robertskollen, western Dronning Maud Land: area description and preliminary species lists. *South African Journal of Antarctic Research*, **19:** 10–20.

Rykovsky, G. F. & O. M. Maslovsky. 2004: [*Bryophyta*. **1**. Andreaeopsida – Bryopsida]. *In*: V. I. Parfenov (ed.), [*Flora of Belarus*]. Minsk, Technalohija: 437 pp. [in Russian with ecological data in English].

S

Sabovljević, M. & V. Stevanović. 1999: Moss conspectus of the Federal Republic of Yugoslavia. *Flora Mediterranea*, **9:** 65–95.

Sainsbury, G. O. K. 1948*a*: Synonyms of some New Zealand mosses. *Revue Bryologique et Lichénologique, Nouvelle Série*, **17:** 79–85.

———. **1948*b*:** Bryobartramiaceae, a new moss family. *The Bryologist*, **51:** 9–13.

———. **1952:** Critical New Zealand mosses. *Revue Bryologique et Lichénologique, Nouvelle Série*, **21:** 213–225.

———. **1953:** Notes on Tasmanian mosses from Rodway's Herbarium. *Papers and Proceedings of the Royal Society of Tasmania*, **87:** 83–91.

———. **1955:** A handbook of New Zealand mosses. *Royal Society of New Zealand Bulletin*, **5:** 1–490.

——— **& K. W. Allison. 1962:** Mosses from the Auckland Islands and Campbell Island. *Records of the Dominion Museum*, **4:** 133–142.

Saito, K. 1973*a*: Notes on the Pottiaceae (3). *Journal of Japanese Botany*, **48:** 161–167.

———. **1973*b*:** Memoir of the Japanese Pottiaceae (I). *Bulletin of the National Science Museum Tokyo*, **16:** 61–91.

———. **1975:** A monograph of Japanese Pottiaceae (Musci). *Journal of the Hattori Botanical Laboratory*, **39:** 373–537.

Sakurai, K. 1954: *Muscologia japonica*. Tokyo: 245 pp. + 70 pls.

Salazar Allen, N. 1993: A revision of the pantropical moss genus *Leucophanes* Brid. *Bryophytorum Bibliotheca*, **46:** 1–281.

Sancho, L. G., F. Schulz, B. Schroeter & L. Kappen. 1999: Bryophyte and lichen flora of South Bay (Livingston Island: South Shetland Islands, Antarctica). *Nova Hedwigia*, **68:** 301–337.

Sarla. 1987: In vitro formation and behaviour of gemmae in *Bryum capillare* Hedw. *Phytomorphology*, **37:** 173–180.

Sastre-De Jesús, I. 1992: Central America, South America and the West Indies. *In*: B. M. Thiers (project ccordinator), *Indices to the species of mosses and lichens described by William Mitten. Memoirs of the New York Botanical Garden*, **68:** 57–88.

Savicz-Lyubitskaya, L. I. 1954: [A survey of the genus *Bryum* Hedw. in the USSR]. *Trudy Botanicheskogo Instituta im. V. L. Komarova Akademii Nauk SSSR, Seria II, Sporovye Rastenya*, **9:** 495–634 (in Russian).

——— **& Z. N. Smirnova. 1959:** [A new species of the genus *Bryum* Hedw. from Bunger's Oasis]. *Informatsionnyy Bjulletin Sovetskoy Antarkticheskoy Ekspeditsiy*, **7:** 34–39 (in Russian).

——— **&** ———. **1960:** [New variety of *Bryum korotkevicziae* Sav.-Ljub. et Z. Smirn.]. *Informatsionnyy Bjulletin Sovetskoy Antarkticheskoy Ekspeditsiy*, **17:** 25–27 (in Russian).

——— **&** ———. **1961:** On the modes of reproduction of *Sarconeurum glaciale* (Hook. fil. et Wils.) Card. et Bryhn, an endemic moss of the Antarctica. *Revue Bryologique et Lichénologique, Nouvelle Série*, **30:** 216–222.

——— **&** ———. **1962:** [The endemic moss of the Antarctic continent – *Sarconeurum glaciale* (Hook. fil. et Wils.) Card. et Bryhn]. *In*: A. P. Andryashev & P. V. Ushakov (eds), [*Biological reports of the Soviet Antarctic Expedition (1955–1958)*]. **1**. Moskva – Leningrad, Izdatel'stvo Akademiy Nauk SSSR: pp. 295–309 (in Russian).

——— **&** ———. **1963*a*:** [A contribution to the biology and geography of *Bryoerythrophyllum recurvirostre* (Hedw.) Chen – a new species in the bryoflora of the Antarctica]. *Botanicheskiy Zhurnal*, **48:** 350–361 (in Russian with English summary).

——— **&** ———. **1963*b*:** [De speciebus *Pottiae* Fuernr. in Antarctida inventis]. *Botanicheskye Materialy Otdela Sporovykh Rasteniy Botanicheskogo Instituta Akademiy Nauk SSSR*, **16:** 188–195 (in Russian).

——— & ———. **1964*a*:** New species of the *Bryum* Hedw. from the Bunger Hills. *Soviet Antarctic Expedition Information Bulletin*, **1:** 308–313.

——— & ———. **1964*b*:** New variety of *Bryum korotkevicziae* Sav.-Ljub. et Z. Smirn. *Soviet Antarctic Expedition Information Bulletin*, **2:** 206–208.

——— & ———. **1964*c*:** [A deep-water member of the genus *Plagiothecium* Br. et Sch. in Antarctica]. *Informatsionnyy Bjulletin Sovetskoy Antarkticheskoy Ekspeditsiy*, **49:** 33–39 (in Russian).

——— & ———. **1964*d*:** [Notula de Bryo argenteo Hedw. ex Antarctida]. *Novosti Sistematiki Nizshikh Rasteniy*, **1:** 292–301 (in Russian).

——— & ———. **1965*a*:** [De *Grimmia antarctici* Card. ex Antarctide Orientalis]. *Novosti Sistematiki Nizshikh Rasteniy*, **2:** 258–259 (in Russian).

——— & ———. **1965*b*:** [De speciebus Pottiae Fuernr. in Antarctide inventis supplementa]. *Novosti Sistematiki Nizshikh Rasteniy*, **2:** 251–258 (in Russian).

——— & ———. **1965*c*:** A deep-water member of the genus *Plagiothecium* Br. et Sch. in Antarctica. *Soviet Antarctic Expedition Information Bulletin*, **5:** 240–243.

———. **1966:** The endemic moss of the Antarctic continent – *Sarconeurum glaciale* (Hook. fil. et Wils.) Card. et Bryhn. *In*: A. P. Andryashev & P. V. Ushakov (eds), Biological reports of the Soviet Antarctic Expedition (1955–1958). **1**. Jerusalem, Israel Program for Scientific Translations: pp. 301–316.

——— & ———. **1970*a*:** [*Grimmia plagiopodia* Hedw. from East Antarctica]. *Bjulletin Sovetskoy Antarkticheskoy Ekspeditsiy*, **80:** 94–97 (in Russian).

——— & ———. **1970*b*:** [*Grimmia donniana* Sm. ex Antarctida Orientali]. *Novosti Sistematiki Nizshikh Rasteniy*, **6:** 249–253 (in Russian).

——— & ———. **1970*c*:** [*The handbook of the mosses of the U.S.S.R.*]. Leningrad, Izdatel'stvo 'Nauka', Leningradskoe Otdelenye: 824 pp. (in Russian).

——— & ———. **1971:** [The moss *Pohlia nutans* (Hedw.) Lindb. in East Antarctica]. *Bjulletin Sovetskoy Antarkticheskoy Ekspeditsiy*, **82:** 72–75 (in Russian).

——— & ———. **1972:** [*Bryum algens* Card. – the most widespread moss in East Antarctica]. *Trudy Sovetskoy Antarkticheskoy Ekspeditsiy*, **60:** 328–345 (in Russian).

——— & ———. **1973:** The moss *Pohlia nutans* (Hedw.) Lindb. in East Antarctica. *Soviet Antarctic Expedition Information Bulletin*, **8:** 222–224.

——— & ———. **1978:** [Leafy mosses of Eastern Antarctic region]. *Botanicheskiy Zhurnal*, **63:** 272–290 (in Russian).

Sayre, G. 1959: *Dates of publications describing Musci, 1801–1821*. Troy: 102 pp.

———. **1971:** Cryptogamae exsiccatae. An annotated bibliography of published exsiccatae of Algae, Lichenes, Hepaticae, and Musci. *Memoirs of the New York Botanical Garden*, **19(2):** 175–276.

Schaffner, J. H. 1909: Classification of plants, IV. *The Ohio Naturalist*, **9:** 446–455.

Schiavone, M. M. 1993: Orden Polytrichinales. *In*: S. A. Guerrera, I. Gamundi de Amos & C. M. Matteri (eds), *Flora criptogámica de Tierra del Fuego*. **14**(12). Buenos Aires, Consejo Nacional de Investigaciones Científicas y Técnicas de la Republica Argentina: 61 pp.

——— **& J. Hyvönen. 1991:** Distribution maps of mosses from Tierra del Fuego, Argentina. 3. *Lindbergia*, **17:** 47–49.

——— & ———. **1993:** Distribution maps of mosses from Tierra del Fuego, Argentina. 4. *Lindbergia*, **18:** 113–115.

Schiffner, V. 1908: Beiträge zur Kenntnis der Bryophyten von Persien und Lydien. *Oesterreichische Botanische Zeitschrift*, **58:** 225–231, 304–318, 341–351 + Taf. vii–ix.

Schimper, W. Ph. 1851: Conspectus generum et specierum que in Vol. IV continentur. *In*: W. P. Schimper (ed.), *Bryologia europaea seu genera muscorum europaeorum monographicae illustrata*. **4**. Stuttgartiae, sumptibus Librariae E. Schweizerbart: pp. iii–iv.

———. **1853*a*:** *Amblystegium*. *In*: W. P. Schimper (ed.), *Bryologia europaea seu genera muscorum europaeorum monographicae illustrata*. **6**. Stuttgartiae, sumptibus Librariae E. Schweizerbart: pp. 45–61 + Tab. 561–566, 568–573 [Fasc. 45–61: 17 pp. + 11 pls.].

———. **1853*b*:** *Brachythecium*. *In*: W. P. Schimper (ed.), *Bryologia europaea seu genera muscorum europaeorum monographicae illustrata*. **6**. Stuttgartiae, sumptibus Librariae E. Schweizerbart: pp. 5–26 + Tab. 535–555 [Fasc. 52–54: 22 pp. + 21 pls.].

———. **1855:** *Barbula*. Supplementum IV. *In*: W. P. Schimper (ed.), *Bryologia europaea seu genera muscorum europaeorum monographicae illustrata*. **6**. Stuttgartiae, sumptibus Librariae E. Schweizerbart: pp. 159–160 + pl. 639 [Fasc. 62–64: 2 pp. + pl. 4].

———. **1856:** *Corollarium bryologiae europaeae, conspectum diagnosticum familiarum, generum et specierum, adnotationes novas atque emendationes complectens*. Stuttgartiae, sumptibus librariae E. Schweizerbart: 140 pp.

———. **1860:** *Synopsis muscorum europaeorum praemissa introductione de elementis bryologicis tractante*. Stuttgart, sumptibus Librariae E. Schweizerbart. vi + lix + 733 pp. + 1 map + pls. 1–8.

———. **1876:** *Synopsis muscorum europaeorum praemissa introductione de elementis bryologicis tractante*. **1**. Introductio. Stuttgartiae, sumptibus Librariae E. Schweizerbart (E. Koch). cxxx pp. + viii tab.

———. **1879:** *Palaeophytologie*. *In*: K. A. Zittel (ed.), *Handbuch der Palaeontologie*. II. Abtheilung. **1**. München – Leipzig, R. Oldenbourg: pp. 1–152.

Schleicher, J. Ch. 1805 ['1806']: Platae cryptogamicae helveticae, quas in locis earum natalibus collegit et exsiccavit Schleicher. Centuria tertia. *Neuer Journal für Botanik*, **1(2):** 198–200.

———. **1807*a*:** Platae cryptogamicae helveticae. Centuria quinta. *Neuer Journal für Botanik*, **2(1):** 170–172.

———. **1807*b*:** *Catalogus hucusque absolutus omnium plantarum in Helvetia cis et transalpina sponte nascentium. Quas continuis itineribus in usum botanophilorum collegit, nomine genuino destinxit ac collatione cum celebriorum auctorum descriptionibus et iconibus rite facta summo studio novissime*. Bex, in pago helveto-vaudensi: i + 39 pp.

Schlensog, M. & B. Schroeter. 2000: Poikilohydry in Antarctic cryptogams and its role for photosynthetic performance in mesic and xeric habitats. *In*: W. Davison, C. Howard-Williams & P. Broady (eds), *Antarctic ecosystems: models for wider ecological understanding*. Christchurch, New Zealand Natural Science: pp. 175–182.

Schliephacke, K. 1884: *Pottia Güssfeldti*, ein neues Laubmoos. *Berichte der Deutschen Botanischen Gesellschaft*, **2:** 461–462.

Schofield, W. B. 1974: Bipolar disjunctive mosses in the Southern Hemisphere, with particular reference to New Zealand. *Journal of the Hattori Botanical Laboratory*, **38:** 13–32.

———. 1976: Bryophytes of British Columbia III. Habitat and distributional information for selected mosses. *Syesis*, **9:** 317–354.

———. 1985: *Introduction to bryology*. New York, Macmillan Publishing Company and London, Collier Macmillan Publishers: xvi + 431 pp.

———. 2007: *Dicranoweisia*. *In*: R. H. Zander (Lead ed.), *Flora of North America north of Mexico*. **27**. *Bryophytes: mosses (1)*. New York – Oxford, Oxford University Press: pp. 395–397.

——— & H. A. Crum. 1972: Disjunctions in bryophytes. *Annals of the Missouri Botanical Garden*, **59:** 174–202.

——— & Ch. Hébant. 1984: The morphology and anatomy of the moss gametophore. *In*: R. M. Schuster (ed.), *New manual of bryology*. **2**. Nichinan, The Hattori Botanical Laboratory: pp. 627–657.

Schrader, H. A. 1794: *Spicilegium florae Germaniae*. Hannoverae, impensis Christiani Ritscheri: viii + 194 pp. + 4 pls.

Schreber, J. Ch. D. 1771: *Spicilegium florae lipsicae*. Lipsiae, prostat in bibliopolo Dykiano: xii + 148 + [32, indexes] pp.

Schroeter, B., L. Kappen & F. Schultz. 1997: Hydration-related spatial and temporal variation of photosynthetic activity in Antarctic lichens. *In*: B. Battaglia, J. Valencia & D. W. H. Walton (eds), *Antarctic communities. Species, structure and survival*. Cambridge, Cambridge University Press: pp. 221–225.

Schultze-Motel, W. 1970: Monographie der Laubmoosgattung *Andreaea* I. Die costaten Arten. *Willdenowia*, **6:** 25–110.

———. 1974: Die Moose der Samoa-Inseln. *Willdenowia*, **7:** 333–408.

Schuster, R. M. 1963: Studies on antipodal Hepaticae I. Annotated keys to the genera of antipodal Hepaticae with special reference to New Zealand and Tasmania. *Journal of the Hattori Botanical Laboratory*, **26:** 185–309.

———. 1969: Problems of antipodal distribution in lower land plants. *Taxon,* **18:** 46–91.

———. 1972: Continental movements, 'Wallace's Line', and Indomalayan-Australasian dispersal of land plants: some eclectic concepts. *Botanical Review*, **38:** 3–86.

———. 1976: Plate tectonics and its bearing on the geographical origin and dispersal of angiosperms. *In*: C. B. Beck (ed.), *Origin and early evolution of angiosperms*. New York, Columbia University Press: pp. 48–138.

———. 1977: The evolution and early diversification of the Hepaticae and Anthocerotae. *In*: W. Frey, H. Hurka & F. Oberwinkler (eds), *Beiträge zur Biologie der niederen Pflanzen*. Stuttgart and New York, Fischer Verlag: pp. 107–115.

———. 1979: On the persistence and dispersal of transantarctic Hepaticae. *Canadian Journal of Botany,* **57:** 2179–2225.

———. 1981: Paleoecology, origin, distribution through time, and evolution of Hepaticae and Anthocerotae. *In*: K. J. Niklas (ed.), *Paleobotany, paleoecology, and evolution*. **2**. New York, Praeger Publishers: pp. 129–191.

———. 1982: Generic and familial endemism in the hepatic flora of Gondwanaland: origin and causes. *Journal of the Hattori Botanical Laboratory,* **52:** 3–35.

———. 1983*a*: Phytogeography of the Bryophyta. *In*: R. M. Schuster (ed.), *New manual of bryology*. **1**. Nichinan, Hattori Botanical Laboratory: pp. 463–626.

——— (ed.). 1983*b*–1984: *New manual of bryology*. **1–2**. Nichinan, The Hattori Botanical Laboratory: vii + 626 pp. (1983*b*, Vol. 1); [1–7] + 627–1295 pp. (1984, Vol. 2).

———. 1989: Studies on the hepatic flora of the Prince Edward Islands. I. Aneuraceae. *Journal of the Hattori Botanical Laboratory*, **67:** 59–108.

———. 2000: Austral Hepaticae Part I. *Nova Hedwigia Beiheft*, **118:** i–vii + 1–524.

———. 2002: Austral Hepaticae Part II. *Nova Hedwigia Beiheft*, **119:** i–viii + 1–606.

Schwägrichen, Ch. F. 1804: Musci novi detecti et descripti. *In*: J. A. Schultes, *Reise auf den Glockner*. **2**. Wien, J. W. Degen: pp. 362–366.

———. 1816: *Species muscorum frondosorum descriptae et tabulis aeneis LXXVII coloratis illustratae opus postumum. Supplementum primum. Sectio posterior*. Lipsiae, sumtu Joannis Ambrosii Barthii, Parisiis, apud Treutel et Würz: vi + 373 pp. + pls. l–c.

———. 1830: Species muscorum frondosorum. Pars I. *In*: C. L. Willdenow (ed.), *Caroli a Linné Species plantarum exhibentes plantas rite cognitas ad genera relatas cum differentiis specificis, nominbus trivialibus, synonymis selectis, locis natalibus secundum systema sexuale digestas*. Ed. 4. **5**(2[1]). Berolini, impensis G. C. Nauck: xiv + 122 pp.

Schwarz, A. M. J., T. G. A. Green & R. D. Seppelt. 1992: Terrestrial vegetation at Canada Glacier, Southern Victoria Land, Antarctica. *Polar Biology*, **12:** 397–404.

Schwarz, O. M. 1994: The development of the peristome-forming layers in the Funariaceae. *International Journal of Plant Science*, **155:** 640–657.

Scott, G. A. M. & I. G. Stone (with illustrations by C. Roser). 1976: *The mosses of southern Australia*. London – New York – San Francisco, Academic Press: xv + 495 pp.

Segarra, J.-G., F. Puche, W. Frey & H. Kürschner. 1998: *Pterygoneurum squamosum* (Pottiaceae, Musci) a new species from Spain. *Nova Hedwigia*, **67:** 511–515.

Sehnem, A. 1955: Vegetationsbild der Laubmoose von Rio Grande do Sul, Brasilien. *Mitteilungen der Thüringischen Botanischen Gesellschaft*, **1:** 208–221.

Seki, T. 1968: A revision of the family Sematophyllaceae of Japan with special reference to a statistical demarcation of the family. *Journal of Science of the Hiroshima University, Series B, Div. 2 (Botany)*, **12:** 1–80.

———. **1974:** A moss flora of Provincia de Aisén, Chile. Results of the Second Scientific Expedition to Patagonia by Hokkaido and Hiroshima Universities, 1967. *Journal of Science of the Hiroshima University, Series B, Div. 2 (Botany)*, **15:** 9–101 + pls. i–iv.

Selkirk, P. M., M. L. Skotnicki, J. Ninham, M. B. Connett & J. Armstrong. 1998: Genetic variation and dispersal of *Bryum argenteum* and *Hennediella heimii* populations in the Garwood Valley, southern Victoria Land, Antarctica. *Antarctic Science*, **10:** 423–430.

———, ———, K. D. Adam, M. B. Connett, T. Dale, T. W. Joe & J. Armstrong. 1997: Genetic variation in

Antarctic populations of the moss *Sarconeurum glaciale*. *Polar Biology*, **18:** 344–350.

Senchina, D. S. 2005: A historical survey of botanical exploration in Antarctica. *Huntia*, **12:** 31–69.

Seppelt, R. D. 1980: The taxonomic status of *Pseudodistichium* Card. (Ditrichaceae). *Lindbergia*, **6:** 126–128.

———. **1982:** A monographic revision of the genus *Ditrichum* (Musci: Ditrichaceae). I. Australian and New Zealand taxa. *Journal of the Hattori Botanical Laboratory*, **51:** 99–150.

———. **1983:** The status of the Antarctic moss *Bryum korotkevicziae*. *Lindbergia*, **9:** 21–26.

———. **1984:** The bryoflora of the Vestfold Hills and Ingrid Christensen Coast, Antarctica. *ANARE Research Notes*, **20:** 1–30.

———. **1986:** Bryophytes of the Vestfold Hills. *In*: J. Pickard (ed.), *Antarctic oasis. Terrestrial environments and history of the Vestfold Hills*. Sydney, Academic Press: pp. 221–245.

———. **1996:** A re-examination of *Ditrichum immersum* van Zanten. *Journal of Bryology*, **19:** 267–277.

———. **2002:** Plant communities at Wilkes Land. *In*: L. Beyer & M. Bölter (eds), *Geoecology of Antarctic ice-free coastal landscapes. Ecological Studies*, **154:** 233–248.

———. **2004:** *The moss flora of Macquarie Island*. Australian Antarctic Division, Kingston: [14] + 328 pp.

———. **2007:** *Distichium*. *In*: R. H. Zander (Lead ed.), *Flora of North America north of Mexico*. **27**. *Bryophytes: mosses (1)*. New York – Oxford, Oxford University Press: pp. 448–449.

——— **& D. H. Ashton. 1978:** Studies on the ecology of the vegetation at Mawson Station, Antarctica. *Australian Journal of Ecology*, **3:** 373–388.

——— **& T. G. A. Green. 1998:** A bryophyte flora for Southern Victoria Land, Antarctica. *New Zealand Journal of Botany*, **36:** 617–635.

——— **& R. J. Hancock. 1991:** Growth responses of *Ceratodon purpureus* in culture. *Journal of the Hattori Botanical Laboratory*, **69:** 133–145.

——— **& H. Kanda. 1986:** Morphological variation and taxonomic interpretation in the moss genus *Bryum* in Antarctica. *Memoirs of the National Institute of Polar Research, Series E, Biology and Medical Science*, **37:** 27–42.

——— **& P. M. Selkirk. 1984:** Effects of submersion on morphology and the implications of induced environmental modification on the taxonomic interpretation of selected Antarctic moss species. *Journal of the Hattori Botanical Laboratory*, **55:** 273–279.

———**, T. G. A. Green & M. Skotnicki. 1999:** Notes on the flora, invertebrate fauna and biological significance of Beaufort Island, Ross Sea, Antarctica. *Polarforschung*, **66:** 53–59.

———**, R. I. Lewis Smith & H. Kanda. 1998:** Antarctic bryology: past achievements and new perspectives. *Journal of the Hattori Botanical Laboratory*, **84:** 203–239.

———**, T. G. A. Green, A.-M. J. Schwarz & A. Frost. 1992:** Extreme southern locations for moss sporophytes in Antarctica. *Antarctic Science*, **4:** 37–39.

Sérgio, C. 1980: Sur la présence de *Mielichhoferia ponteverdensis* Luis. au Portugal. *Cryptogamie, Bryologie – Lichénologie*, **1:** 91–94.

——— **& S. Carvalho. 2003:** Annotated catalogue of Portuguese bryophytes. *Portugaliae Acta Biologica*, **21:** 5–230.

Shaw, A. J. 1981*a*: Taxonomic studies on *Pohlia* subg. *Mniobryum*, including *P. brevinervis* Lindb. & Arn., new for North America. *The Bryologist*, **84:** 505–514.

———. **1981*b*:** Ecological diversification among nine species of *Pohlia* (Musci) in western North America. *Canadian Journal of Botany*, **59:** 2359–2378.

———. **1981*c*:** A taxonomic revision of the propaguliferous species of *Pohlia* (Musci) in North America. *Journal of the Hattori Botanical Laboratory*, **50:** 1–81.

———. **1981*d*:** The nomenclature of *Pohlia annotina* (Hedw.) Lindb. and related species. *Lindbergia*, **7:** 10–22.

———. **1981*e*:** *Pohlia andrewsii* and *P. tundrae*, two new arctic-alpine propaguliferous species from North America. *The Bryologist*, **84:** 65–74.

———. **1982*a*:** Character analysis, phylogeny, and classification of the moss genus *Pohlia*. *Canadian Journal of Botany*, **62:** 219–229.

———. **1982*b*:** *Pohlia* (Musci) in North and Central America and the West Indies. *Contributions from the University of Michigan Herbarium*, **15:** 219–295.

———. **1984*a*:** A reinterpretation of peristome structure in *Pseudoditrichum mirabile* Steere & Iwats. (Pseudoditrichaceae). *The Bryologist*, **87:** 314–318.

———. **1984*b*:** Character analysis, phylogeny, and classification of the moss genus *Pohlia*. *Canadian Journal of Botany*, **62:** 219–229.

———. **1985*a*:** Peristome structure in the Mitteniales (ord. nov.: Musci), a neglected novelty. *Systematic Botany*, **10:** 224–233.

———. **1985*b*:** The correlation between taxonomy and peristome structure in the Bryaceae. *Journal of the Hattori Botanical Laboratory*, **59:** 79–100.

———. **1985*c*:** Exostomial ornamentation in the Bryales. *Journal of the Hattori Botanical Laboratory*, **59:** 303–324.

———. **1985*d*:** Nomenclatural changes in the Bryaceae subfamily Mielichhoferioideae. *The Bryologist*, **88:** 28–30.

———. **1986:** Peristome structure in the Orthotrichaceae. *Journal of the Hattori Botanical Laboratory*, **60:** 119–136.

———. **1987:** Systematic studies on the Bryaceae. *In*: W. R. Buck (ed.), *Bryostephane steereana: a collection of bryological papers presented to William Campbell Steere on the occasion of his 80th birthday. Memoirs of the New York Botanical Garden*, **45:** 682–690.

———. **2000:** *Schizymenium shevockii* (Bryaceae), a new species of moss from California, based on morphological and molecular evidence. *Systematic Botany*, **25:** 188–196.

———. **1988:** Peristome development in mosses in relation to systematics and evolution. II. *Tetraphis pellucida* (Tetraphidaceae). *American Journal of Botany*, **75:** 1019–1032.

———. **1990:** Genetic and environmental effects on morphology and asexual reproduction in the moss, *Bryum bicolor*. *The Bryologist*, **93:** 1–6.

———. **2006:** A revision of the moss genus *Pohlia* Hedw. (Mniaceae) in Australia. *Systematic Botany*, **31:** 247–257.

——— **& B. H. Allen. 1985:** Anatomy and morphology of the peristome in *Discelium nudum* (Musci: Disceliaceae). *The Bryologist*, **88:** 263–267.

——— **& L. E. Anderson. 1986:** Morphology and homology of the peristome teeth in *Hymenodon* and *Hymenodontopsis* (Rhizogoniaceae: Musci). *Systematic Botany*, **11:** 446–454.

——— & **S. C. Beer. 1999:** Life history variation in gametophyte populations of the moss *Ceratodon purpureus* (Ditrichaceae). *American Journal of Botany*, **86:** 512–521.

——— & **H. Crum. 1982:** Comments on the Mielichhoferioideae of Mexico with the description of a new species of *Synthetodontium*. *Contributions from the University of Michigan Herbarium*, **15:** 209–217.

——— & ———. **1984:** Peristome homology in *Mielichhoferia* and a taxonomic account of North American species. *Journal of the Hattori Botanical Laboratory*, **57:** 363–381.

——— & **A. J. Fife. 1984:** The evolutionary and taxonomic significance of peristome morphology in *Anomobryum* (Bryaceae, Musci). *Journal of the Hattori Botanical Laboratory*, **57:** 285–298.

——— & ———. **1985:** *Pohlia australis* sp. nov. (Musci) from New Zealand with notes on some other austral Pohlias. *New Zealand Journal of Botany*, **23:** 183–186.

——— & **B. Goffinet (eds). 2000:** *Bryophyte biology*. Cambridge, Cambridge University Press: x + 476 pp.

——— & **H. Robinson. 1984:** On the development, evolution, and function of peristomes in mosses. *Journal of the Hattori Botanical Laboratory*, **57:** 319–335.

——— & **J. R. Rohrer. 1984:** Endostomial architecture in diplolepidous mosses. *Journal of the Hattori Botanical Laboratory*, **57:** 41–61.

———, **B. Allen & W. R. Buck. 1998:** New species of Bryaceae (*Mielichhoferia*, *Brachymenium*) from Costa Rica and Réunion. *The Bryologist*, **101:** 600–604.

———, **L. E. Anderson & B. D. Mishler. 1987:** Peristome development in mosses in relation to systematics and evolution. I. *Diphyscium foliosum* (Diphysciaceae). *In*: W. R. Buck (ed.), *Bryostephane steereana: a collection of bryological papers presented to William Campbell Steere on the occasion of his 80th birthday. Memoirs of the New York Botanical Garden*, **45:** 55–70.

———, ——— & ———. **1989*a*:** Peristome development in mosses in relation to systematics and evolution. III. *Funaria hygrometrica*, *Bryum pseudocapillare*, and *B. bicolor*. *Systematic Botany*, **14:** 24–36.

———, ——— & ———. **2000:** Paedomorphic sporophyte development in *Bruchia flexuosa* (Bruchiaceae). *The Bryologist*, **103:** 147–155.

———, **C. J. Cox & B. Goffinet. 2005:** Global patterns of moss diversity: taxonomic and molecular inferences. *Taxon*, **54:** 337–352.

———, **B. D. Mishler & L. E. Anderson. 1989*b*:** Peristome development in mosses in relation to systematics and evolution. IV. Haplolepideae: Ditrichaceae and Dicranaceae. *The Bryologist*, **92:** 314–325.

———, **C. J. Cox, B. Goffinet, W. R. Buck & S. B. Boles. 2003:** Phylogenetic evidence of a rapid radiation of pleurocarpous mosses (Bryophyta). *Evolution*, **57:** 2226–2241.

Sherratt, R. 1821: Observations on South Shetland. *Imperial Magazine (London)*, **1821:** columns 1214–1218.

Shevock, J. R. & D. Toren. 2001: A catalogue of mosses for the city and county of San Francisco, California. *Madroño*, **48:** 1–16.

Showers, D. 1982: The mosses of Lassen Volcanic National Park, California. *The Bryologist*, **85:** 324–328.

Shlyakov, R. N. & N. A. Konstantinova. 1982: [*Conspectus of the bryophyte flora of the Murmansk region*]. Apatity, Akademia Nauk SSSR Ordena Lenina, Kol'skiy Filial im. S. M. Kirova Ordena 'Znak Pocheta', Polyarno-Alpiyskiy Botanicheskiy Sad-Institut: 227 pp. (in Russian).

Sim, T. R. 1926: The Bryophyta of South Africa comprising Sphaerocarpales, Marchantiales, Jungermanniales, Anthocerotales, Sphagnales, Andreaeales, Polytrichales, Bryales. *Transactions of the Royal Society of South Africa*, **15:** 1–475.

Simonov, I. M. & V. I. Fedotov. 1964: [Lakes in the Schirmacher Ponds]. *Soviet Antarctic Expedition Information Bulletin*, **5:** 160–163 (in Russian).

Singh, D. K. & R. C. Semwal. 2000: Bryoflora of Schirmacher Oasis, East Antarctica: a preliminary study. *In*: *Sixteenth Indian Expedition to Antarctica, Scientific Report, Technical Publication*. **14**. New Delhi, Department of Ocean Development: pp. 173–185.

Siple, P. A. 1938: The Second Byrd Antarctic Expedition – Botany. I. Ecology and geographical distribution. *Annals of the Missouri Botanical Garden*, **25:** 467–514 + pls 32–37.

Sjögren, E. 2001: Distribution of Azorean bryophytes up to 1999, their island distribution and information on their presence elsewhere, including Madeira and the Canary Islands. *Boletim do Museu Municipal do Funchal (História Natural) Suplemento*, **7:** 1–89.

Skotnicki, M. L., R. Bargagli & J. A. Ninham. 2002: Genetic diversity in the moss *Pohlia nutans* on geothermal ground of Mt. Rittmann, Victoria Land, Antarctica. *Polar Biology*, **25:** 771–777.

———, **A. Mackenzie & P. M. Selkirk. 2004:** Mosses surviving on the edge: origins, genetic diversity, and mutations in Antarctica. *In*: B. Goffinet, V. Hollowell & R. Magill (eds), *Molecular systematics of bryophytes. Monographs in Systematic Botany from the Missouri Botanical Garden*, **98:** 388–403.

———, **J. A. Ninham & P. M. Selkirk. 1998:** High levels of RAPD diversity in the moss *Bryum argenteum* in Australia, New Zealand, and Antarctica. *The Bryologist*, **101:** 412–421.

———, ——— & ———. **1999:** Genetic diversity and dispersal of the moss *Sarconeurum glaciale* on Ross Island, East Antarctica. *Molecular Ecology*, **8:** 753–762.

———, ——— & ———. **2000:** Genetic diversity, mutagenesis and dispersal of Antarctic mosses – a review of progress with molecular studies. *Antarctic Science*, **12:** 363–373.

———, **P. M. Selkirk & C. Beard. 1998:** RAPD profiling of genetic diversity in two populations of the moss *Ceratodon purpureus* in Victoria Land, Antarctica. *Polar Biology*, **19:** 172–176.

———, ——— & **J. A. Ninham. 1998:** RAPD analysis of genetic variation and dispersal of the moss *Bryum pseudotriquetrum* from Southern Victoria Land, Antarctica. *Polar Biology*, **20:** 121–126.

———, **P. M. Selkirk, P. Broady, K. D. Adam & J. A. Ninham. 2001:** Dispersal of the moss *Campylopus pyriformis* on geothermal ground near the summits of Mount Erebus and Mount Melbourne, Victoria Land, Antarctica. *Antarctic Science*, **13:** 280–285.

Skottsberg, C. 1905: Some remarks upon the geographical distribution of vegetation in the colder Southern Hemisphere. *Ymer*, **1905:** 402–427 + Taf. 7–8.

———. **1912:** Einige Bemerkungen über die Vegetationsverhältnisse des Graham Landes. *In*: *Wissenschaftliche Ergebnisse* *der Schwedischen Südpolar-Expedition 1901–1903*. **4**(13). Stockholm, Lithographisches Institut des Generalstabs: 16 pp. + 3 pls.

———. **1960:** Remarks on the plant geography of the southern cold temperate zone. *In*: C. F. A. Pantin (ed.), *A discussion on the biology of the southern cold temperate zone. Proceedings of the Royal Society of London, Series B,* **152:** 447–457.

Slavick, A. D. 1992: Africa. *In*: B. M. Thiers (project coordinator), *Indices to the species of mosses and lichens described by William Mitten. Memoirs of the New York Botanical Garden*, **68:** 9–17.

Smirnova, Z. N. 1956: [De identificatione *Drepanocladi brachiati* (Mitt.) Dix. et *D. longifolii* (Wils.) Williams]. *Botanicheskye Materialy Otdela Sporovykh Rasteniy Botanicheskogo Instituta Akademiy Nauk SSSR*, **11:** 219–228 (in Russian).

Smith, A. J. E. 1978: *The moss flora of Britain and Ireland*. Cambridge, Cambridge University Press: viii + 706 pp.

———. **2004:** *The moss flora of Britain and Ireland*. Ed. 2. Cambridge, Cambridge University Press: xii + 1012 pp.

——— **& M. E. Newton. 1968:** Chromosome studies on some British and Irish mosses. III. *Transactions of the British Bryological Society*, **5:** 463–522.

——— **& H. L. K. Whitehouse, 1978:** An account of the British species of the *Bryum bicolor* complex including *B. dunense* sp. nov. *Journal of Bryology*, **10:** 29–47.

Smith, G. L. 1969: On *Atrichopsis*, with notes on some austral *Psilopilum* species (Polytrichaceae). *Bulletin of the Torrey Botanical Club*, **96:** 60–69.

———. **1971:** A conspectus of the genera of Polytrichaceae. *Memoirs of the New York Botanical Garden*, **21(3):** 1–83.

———. **1972:** Continental drift and the distribution of Polytrichaceae. *Journal of the Hattori Botanical Laboratory*, **35:** 41–49.

———. **1974:** New developments in the taxonomy of Polytrichaceae: epiphragm structure and spore morphology as generic characters. *Journal of the Hattori Botanical Laboratory*, **38:** 143–150.

Smith, G. M. 1955: *Cryptogamic botany*. **2**. *Bryophytes and pteridophytes*. Ed. 2. New York and London, McGraw-Hill Book Company: vii + 399 pp.

Smith Merrill, G. L. 1994: Subclass Polytrichidae. *In*: A. J. Sharp, H. Crum & P. M. Eckel (eds), *The moss flora of Mexico. Part two: Orthotrichales to Polytrichales. Memoirs of the New York Botanical Garden*, **69:** 1068–1092.

———. **2007:** Polytrichaceae. *In*: R. H. Zander (Lead ed.), *Flora of North America north of Mexico*. **27**. *Bryophytes: mosses (1)*. New York – Oxford, Oxford University Press: pp. 121–161.

Smoot, E. L. & Th. N. Taylor. 1986: Structurally preserved fossil plants from Antarctica: II. A Permian moss from the Transantarctic Mountains. *American Journal of Botany*, **73:** 1683–1691.

Snider, J. A. 1975*a*: Sporophyte development in the genus *Archidium* (Musci). *Journal of the Hattori Botanical Laboratory*, **39:** 85–104.

———. **1975*b*:** A revision of the genus *Archidium* (Musci). *Journal of the Hattori Botanical Laboratory*, **39:** 105–201.

———. **1994*a*:** Subclass Archidiidae. *In*: A. J. Sharp, H. Crum & P. M. Eckel (eds), *The moss flora of Mexico. Part one: Sphagnales to Bryales. Memoirs of the New York Botanical Garden*, **69:** 24–27.

———. **1994*b*:** *Distichium*. *In*: A. J. Sharp, H. Crum & P. M. Eckel (eds), *The moss flora of Mexico. Part one: Sphagnales to Bryales. Memoirs of the New York Botanical Garden*, **69:** 104–105.

Söderström, L. (ed.). 1996: *Preliminary distribution maps of bryophytes in northwestern Europe*. **2**. Musci (A–I). Trondheim, Mossornas Vänner: 72 pp.

———. **1998:** *Preliminary distribution maps of bryophytes in northwestern Europe*. **3**. Musci (J–Z). Trondheim, Mossornas Vänner: 68 pp.

Soldan, Z. 1996: [Distribution of neophytic mosses *Campylopus introflexus* and *Orthodontium lineare* in the Czech Republic]. *Bryonora*, **18:** 10–19 (in Czech with English summary).

Sollman, Ph. 2000: A taxonomic revision of *Pseudosymblepharis* Broth. (Musci: Pottiaceae) in Asia. *Tropical Bryology*, **18:** 129–145.

Spagnuolo, V., P. Caputo, S. Cozzolino, R. Castaldo & P. De Luca. 1999: Patterns of relationships in Trichostomoideae (Pottiaceae, Musci). *Plant Systematics and Evolution*, **216:** 69–79.

Spence, J. R. 1987: A proposed reclassification of *Bryum*, *Anomobryum* and *Brachymenium* (Musci, Bryaceae). *Journal of Bryology*, **14:** 659–676.

———. **1988:** *Bryum* Hedw. (Bryaceae) in western North America. *The Bryologist*, **91:** 73–85.

———. **1996:** *Rosulabryum* genus novum (Bryaceae). *The Bryologist*, **99:** 221–225.

———. **2005:** New genera and combinations in Bryaceae (Bryales, Musci) for North America. *Phytologia*, **87:** 15–28.

——— **& H. P. Ramsay. 1999:** Proposal to conserve the name *Bryum* (Musci, Bryaceae) with a conserved type. *Taxon*, **48:** 827–828.

——— **& ———. 2002:** The genus *Anomobryum* Schimp. (Bryopsida, Bryaceae) in Australia. *Telopea*, **9:** 777–792,

——— **& ———. 2005:** New genera and combinations in the Bryaceae (Bryales, Musci) for Australia. *Phytologia*, **87:** 61–72.

——— **& ———. 2006:** Bryaceae. *In*: P. M. McCarthy (ed.), *Flora of Australia Volume* **51**. *Mosses (1)*. Canberra, Australian Biological Resources Study and Melbourne, CSIRO Publishing: pp. 274–348.

Staples, G. W., C. T. Imada, W. J. Hoe & C. W. Smith. 2004: A revised checklist of Hawaiian mosses. *Tropical Bryology*, **25:** 35–69.

Stearn, W. T. (ed.). 1968: *Humboldt, Bonpland, Kunth and tropical American botany. A miscellany on the 'Nova genera et species plantarum'*. Lehre, Verlag von J. Cramer: [7] + 159 pp.

Stebel, A. 1995: *Campylopus introflexus* (Musci, Dicranaceae) found in the Carpathians. *Fragmenta Floristica et Geobotanica*, **40:** 909–911.

Stech, M. 1999*a*: A reclassification of Dicranaceae (Bryopsida) based on non-coding cpDNA sequence data. *Journal of the Hattori Botanical Laboratory*, **86:** 137–159.

———. **1999*b*:** A molecular systematic contribution to the position of *Amphidium* Schimp. (Rhabdoweisiaceae, Bryopsida). *Nova Hedwigia*, **68:** 291–300.

Steere, W. C. 1938: *Didymodon*. *In*: A. J. Grout (ed.), *Moss flora of North America north of Mexico*. **1**(3). Newfane, Vermont, privately published: pp. 185–190 + pls. 89C–90.

———. **1946:** Musci. *In*: N. Polunin (ed.), *Botany of the Canadian Eastern Arctic. Part II, Thallophyta and Bryophyta. National Museums of Canada Bulletin*, **97** *Biological Series*, **26:** 370–490.

———. **1947:** A consideration of the concept of genus in Musci. *The Bryologist*, **50:** 247–258.

———. **1948:** A contribution to the bryogeography of Ecuador. I. A review of the species of Musci previously reported. *The Bryologist*, **51:** 65–167.

———. **1951:** Bryophyta of Arctic America. IV. The mosses of Corwallis Island. *The Bryologist*, **54:** 181–202.

———. **1954:** Chromosome number and behavior in Antarctic mosses. *Botanical Gazette*, **116:** 93–133.

———. **1961:** A preliminary review of the bryophytes of Antarctica. *In*: *Science in Antarctica. Part I: The life sciences in Antarctica. National Academy of Sciences, National Research Council, Washington, DC, Publication*, **839:** 20–33.

———. **1965:** Antarctic Bryophyta. *BioScience*, **15:** 283–285.

———. **1972:** Chromosome numbers in bryophytes. *Journal of the Hattori Botanical Laboratory*, **35:** 99–125.

———. **1977:** Bryophytes from Great Bear Lake and Copper Mine, Northwest Territories, Canada. *Journal of the Hattori Botanical Laboratory*, **42:** 425–465.

———. **1978:** The mosses of Arctic Alaska. *Bryophytorum Bibliotheca*, **14:** i–x + 1–508.

Stepanova, N. A. 1986: [*Conspectus of the moss flora of the Yakutian tundras*]. Yakutsk, Akademia Nauk SSSR, Sibirskoe Otdelenye, Yakutskiy Filial: 119 pp. (in Rusian).

Steudel, E. 1824: *Nomenclator botanicus enumerans ordine alphabetico nomina atque synonyma tum generica tum specifica a Linnaeo et recentioribus de re botanica scriptoribus plantis cryptogamis imposita*. Stuttgartiae et Tubingae, sumtibus J. G. Cottae: xviii + 450 pp.

Stewart, W. G. 1988: A study of the distribution and ecology of the saxicolous mosses *Grimmia anodon*, *G. plagiopodia* and *G. pulvinata* in southern Ontario. *The Michigan Botanist*, **27:** 3–11.

Stone, I. G. 1981: *Sciaromium elimbatum* Broth. et Watts and *S. forsythii* Broth. reduced to synonymy in *Cratoneuropsis relaxa* (Hook. f. et Wils.) Fleisch. in Broth. *Journal of Bryology*, **11:** 844–845.

Størmer, P. 1954: Mosses from Tirich Mir. A contribution to the bryology of NW Pakistan. *In*: *Scientific results of the Norwegian Expedition to Tirich Mir 1950*. **3**. *Nytt Magasin for Botanik*, **3:** 213–226.

———. **1963:** Iranian plants collected by Per Wendelbo in 1959. VI. Mosses (Musci). *Årbokfor Universitetet i Bergen, Matematisk-Naturvitenskapelig Serie*, **1963(11):** 1–34.

Streiff, A. 2006: Phylogenetic study of *Grimmia* (Grimmiaceae) based on plastid DNA sequences (*trn*L-*trn*F and *rps*4) and on morphological characters. *The Bryologist*, **109:** 224–235.

Streimann, H. 1997: Taxonomic studies on Australian Hookeriaceae (Musci). 1. Introduction, and the genera *Achrophyllum*, *Callicostella*, *Chaetomitrium* and *Cyclodictyon*. *Journal of the Hattori Botanical Laboratory*, **82:** 281–304.

———. **1999:** Taxonomic studies on Australian Hookeriaceae (Musci). 2. The genera *Distichophyllum* and *Bryobrothera*. *Journal of the Hattori Botanical Laboratory*, **86:** 89–119.

———. **2000:** Taxonomic studies on Australian Hookeriaceae (Musci). 3. The genera *Calyptrochaeta*, *Daltonia*, *Hookeriopsis* and *Sauloma*. *Journal of the Hattori Botanical Laboratory*, **88:** 101–138.

———. **2001:** Taxonomic studies on Australian Hookeriaceae (Musci). 4. Summary and bryogeographic notes. *Journal of the Hattori Botanical Laboratory*, **90:** 211–220.

——— **& J. Curnow. 1989:** Catalogue of mosses of Australia and its external territories. *Australian Flora and Fauna Series*, **10**: i–viii + 1–479.

——— **& N. Klanzenga. 2002:** Catalogue of Australian mosses. *Flora of Australia Supplementary Series*, **17:** [1–5] + 1–259.

——— **& A. Touw. 1981:** New records for some Australian mosses. *Journal of the Hattori Botanical Laboratory*, **49:** 261–271.

Sudgen, D. E. & C. M. Clapperton. 1977: The maximum ice extent on island group in the Scotia Sea, Antarctica. *Quaternary Research*, **7:** 268–282.

Sullivant, W. S. 1850: Notices of several new species of mosses and Hepaticae from Tierra del Fuego. *Hooker's Journal of Botany and Kew Garden Miscellany*, **2:** 315–318.

———. **1856*a* ['1857']:** Descriptions of the mosses and liverworts. *In*: A. W. Whipple & J. C. Ives (eds), *Explorations and surveys for a railroad route from the Mississippi River to the Pacific Ocean*. **4**. *Route near the thirty-fifth parallel explored by Lieutnant A. W. Whipple, topographical engineers, in 1853 and 1854. Part 2. Report on the botany of the expedition*. Washington, D.C., War Department: pp. 185–193 + pls. i–x.

———. **1856*b* ['1857']:** *Musci boreali-americani* (Text). Cantibrigiae, privately published. *Musci boreali- americani, sive specimina exsiccata muscorum in Americae rebuspublicis foederatis detectorum*. 355 + 58 specimens (The exsiccata). Columbi Ohioensum.

———. **1859:** *United States Exploring Expedition. During the years 1838, 1839, 1840, 1841, 1842. Under the command of Charles Wilkes, U.S.N. Botany. Musci*. Philadelphia, C. Sherman & Son: 32 pp. + 26 pls.

Swartz, O. 1799: *Dispositio systematica muscorum frondosorum Sueciae. Adjectis descriptionibus et iconibus novarum specierum*. Erlangae: apud Ioannem Iacobum Palm: 112 pp. + 9 pls.

———. **1806:** *Conostomum*, eine neue Moos-Gattung. *Neues Journal für die Botanik*, **1(3):** 24–36 + pls. v–vi.

Syed, H. 1973: A taxonomic study of *Bryum capillare* Hedw. and related species. *Journal of Bryology*, **7:** 265–326.

Szafran, B. 1957: [Mosses (Musci)]. *In*: [*Polish Flora. Spore plants of Poland and adjacent territories*]. **1**. Warszawa, Państwowe Wydawnictwo Naukowe: 449 pp. (in Polish).

———. **1961:** [Mosses (Musci)]. *In*: [*Polish Flora. Spore plants of Poland and adjacent territories*]. **2**. Warszawa, Państwowe Wydawnictwo Naukowe: 407 pp. (in Polish).

Š

Šmarda, J. 1948: [The mosses of Slovakia (ČSR)]. *Časopis Zemského Musea v Brně*, **32:** 6–80 + 4 unnumbered pages with photographs (in Slovak with English summary).

T

Takaki, N. 1955*a*: Researches on the Brachytheciaceae of Japan and its adjacent areas (I). *Journal of the Hattori Botanical Laboratory*, **14:** 1–28.

———. **1955*b*:** Researches on the Brachytheciaceae of Japan and its adjacent areas. II. *Journal of the Hattori Botanical Laboratory*, **15:** 1–69.

———. **1956:** Researches on the Brachytheciaceae of Japan and its adjacent areas. III. *Journal of the Hattori Botanical Laboratory*, **16:** 1–71.

Tallis, J. H. 1958: Studies in the biology and ecology of *Racomitrium lanuginosum* Brid. I. Distribution and ecology. *Journal of Ecology*, **46:** 271–288.

———. **1959:** Studies in the biology and ecology of *Racomitrium lanuginosum* Brid. II. Growth, reproduction and physiology. *Journal of Ecology*, **47:** 325–350.

Tan, B. C. & Z. Iwatsuki. 1991: A new annotated Philippine moss checklist. *Harvard Papers in Botany*, **3:** 1–64.

——— **& T. Koponen. 1989:** Additions and corrections for Philippine moss flora. *Cryptogamie, Bryologie – Lichénologie*, **10:** 235–245.

——— **& J. Yu. 1997:** Mosses of Quinghai-Tibetan Plateau, China. *Journal of the Hattori Botanical Laboratory*, **82:** 305–320.

Tangney, R. S. 1997: A taxonomic revision of the genus *Camptochaete* Reichdt., Lembophyllaceae (Musci). *Journal of the Hattori Botanical Laboratory*, **81:** 53–121.

Tatuno, S. 1963: Zytologische Untersuchungen über die Laubmoose von Antarktis. *Hikobia*, **3:** 268–273 (in Japanese with German summary).

Taylor, E. C. Sr. 1962: The Philibert peristome articles. An abridged translation. *The Bryologist*, **65:** 175–212.

Taylor, Th. 1847: Descriptions of new Musci and Hepaticae, collected by Professor William Jameson on Pichincha, near Quito. *London Journal of Botany*, **6:** 328–342.

———. **1848:** On the specific characters of certain new cryptogamic plants, lately received from, and collected by, Professor William Jameson, on Pichincha, near Quito. *London Journal of Botany*, **7:** 278–285.

Tewari, S. D. & G. Pant. 1996: Some moss collections from Dakshin Gangotri, Antarctica. *The Bryological Times*, **91:** 7.

Thal, J. 1588: *Sylva hercynia, sive catalogus plantarum sponte nascentium in montibus, et locis vicinis Hercyniae, quae respicit Saxoniam, conscriptus singulari studio*. Francofurto ad Moenum, Joh. Feyerabend: 133 pp. + ix pls.

Thériot, I. 1917: Contribution à la flore bryologique du Chili. *Revista Chilena de Historia Natural*, **21:** 6–37.

———. **1922:** *Leptobryum escomelii* Ther. sp. nov. *The Bryologist*, **25:** 90–92.

———. **1924:** Mousses de Kerguelen rapportées par M. Etienne Peau (1923–1924). *Bulletin Mensuel de la Société Linnéenne de la Seine Maritime*, **11:** 102–106.

———. **1934:** Contribution à la flore bryologique du Chili (11e article). *Revue Bryologique et Lichénologique, Nouvelle Série*, **7:** 167–183.

———. **1935:** Contribution à la flore bryologique du Chili (12.e article) (1). *Revista Chilena de Historia Natural*, **39:** 16–21.

Thomas, B. A. 1972: A probable moss from the Lower Carboniferous of the Forest of Dean, Gloucestershire. *Annals of Botany*, **36:** 155–161.

Thor, G. 1995: Reassessment of the first lichen and moss collections from Heimefrontfjella, Dronning Maud Land. *Antarctic Science*, **7:** 261–264.

Tingey, R. J. 1991: *The geology of Antarctica*. *In*: *Oxford Monographs on Geology and Geophysics*. **17**. Oxford, Clarendon Press: 680 pp. + 1 colour map.

Tixier, P. 1980: Bryophyta exotica. VI. Données muscologiques nouvelles sur l'île d'Amsterdam (Océan Indien du Sud). *Nova Hedwigia*, **32:** 483–491.

Tournefort, J. P. de. 1700: *Institutiones rei herbariae. Editio altera*. Parisiis, e Typografia regia: xx + 697 pp. + 476 pls.

Touw, A. 1971: A taxonomic revision of the Hypnodendraceae (Musci). *Blumea*, **19:** 211–354.

———. **1974:** Some notes on taxonomic and floristic research on exotic mosses. *Journal of the Hattori Botanical Laboratory*, **38:** 123–128.

———. **1978:** The mosses reported from Borneo. *Journal of the Hattori Botanical Laboratory*, **44:** 147–176.

———. **1992:** A survey of the mosses of the Lesser Sunda Islands (Nusa Tenggara), Indonesia. *Journal of the Hattori Botanical Laboratory*, **71:** 289–366.

———. **2001:** A taxonomic revision of the Thuidiaceae (Musci) of tropical Asia, the western Pacific, and Hawaii. *Journal of the Hattori Botanical Laboratory*, **91:** 1–136.

——— **& W. V. Rubers. 1989:** De Nederlandse bladmossen. Flora en verspreidingsatlas van de Nederlandse Musci (*Sphagnum* uitgezonderd). *In*: H. D. van Behemen, D. A. G. Buizer & A. Littel (eds), Natuurhistorische Bibliothek van de KNNV. **50**. Utrecht, Stichting Uitgeverij Koninklijke Nedelandse Natuurhistorische Vereniging: 532 pp.

Townsend, C. C. 1985 ['1984']: A small collection of mosses from Jebel Marra, Sudan Republic. *Lindbergia*, **10:** 175–180.

Triggs, G. D. (ed.). 1987: *The Antarctic Treaty regime. Law, environment and resources*. Cambridge, Cambridge University Press: xxii + 239 pp.

Tsegmed, T. 2001*a*: Checklist and distribution of mosses in Mongolia. *Arctoa, a Journal of Bryology,* **10:** 1–18.

———. **2001*b*:** [*Handbook of mosses of Mongolia*]. Ulaanbatar, Soyombo Printing: 473 pp. (in Mongolian).

Tsubota, H., T. Arikawa, H. Akiyama, E. De Luna, D. Gonzáles, M. Higuchi & H. Deguchi. 2002: Molecular phylogeny of hypnobryalean mosses as inferred from a large-scale dataset of chloroplasts *rbc*L, with special reference to the Hypnaceae and possibly related families. *Hikobia*, **13:** 645–665.

———**, Y. Ageno, B. Estébanez, T. Yamaguchi & H. Deguchi. 2003:** Molecular phylogeny of the Grimmiales (Musci) based on chloroplast *rbc*L sequences. *Hikobia*, **14:** 55–70.

Tuomikoski, R. & T. Koponen. 1979: On the generic taxonomy of *Calliergon* and *Drepanocladus* (Musci, Amblystegiaceae). *Annales Botanici Fennici*, **16:** 213–227.

———, ——— **& T. Ahti. 1973:** The mosses of the island of Newfoundland. *Annales Botanici Fennici*, **10:** 217–264.

Turrill, W. B. 1953: *Pioneer plant geography: the phytogeographical research of Sir Joseph Dalton Hooker*. The Hague, Martinus Nijhoff: x + 267 pp.

Turton 1806: *A general system of nature, thgrough the three grand kingdoms of animals, vegetables, and minerals; systematically divided into their several classes, orders, genera, species, and varieties with their habitations, manners, economy, structure and peculiarities. Vegetable Kingdom*. **2**. London, Andrews, Lambert, & c. (printed for Lackington, Allen, and Co.): i + pp. 889–1851 [+ 1852–1880, index].

U

Ukrainskaya, G. Ya. 1996: [*Plagiothecium berggrenianum* (Plagiotheciaceae, Musci) in Russia]. *Botanicheskiy Zhurnal*, **81:** 87–91 (in Russian with English summary).

UNEP. 1992: *Convention on Biological Diversity*. United Nations Environment Programme, Environmental Law and Institutions Programme Activity Centre: 52 pp.

V

Vaarama, A. 1939: *Desmatodon latifolius* (Hedw.) Bryol. eur. in Laatokka-Karelien. *Annales Botanici Societatis Zoologicae-Botanicae Fennicae 'Vanamo'*, **11:** 44–50.

Vaillant, S. 1727: *Botanicon parisiense ou dénombrement par ordre alphabétique des plantes, qui se trouvent aux environs de Paris compris dans la carte de la prevoté & de l'élection de la dite ville par le sieur Danet Gendre année MDCCXXII. Avec plusieurs descriptions des plantes, leurs synonymes, le terms de fleurir & de grainer et une critique des auteurs de botanique par feu monsieur Sebastien Vaillant, ... Enriché de plus de trois cents figures, dessinées par le sieur Claude Aubriet, peintre du cabinet du roy*. Leide & Amsterdam, chez Jean & Herman Verbeek et Balthazar Lakeman: xxxvi + xvi + 205 + [49] pp. + pls. i–xxxiii.

Váňa, J. 1986: *Ochyraea tatrensis* gen. et sp. nov., a remarkable pleurocarpous moss from Czechoslovakia. *Journal of Bryology*, **14:** 261–267.

——— & N. Gremmen. 2005: Hepatics of Heard Island. *Cryptogamie, Bryologie*, **26:** 79–90.

——— & ———. 2006: Checklist of the hepatic flora of sub-Antarctic Îles Kerguelen. *Cryptogamie, Bryologie*, **27:** 131–139.

Vanderpoorten, A. & Ch. E. Zartman. 2002: The *Bryum bicolor* complex in North America. *The Bryologist*, **105:** 128–139.

———, L. Hedenäs, C. J. Cox & A. J. Shaw. 2002*a*: Circumscription, classification, and taxonomy of Amblystegiaceae (Bryopsida) inferred from nuclear and chloroplast DNA sequence data and morphology. *Taxon*, **51:** 115–122.

———, ———, ——— & ———. 2002*b*: Phylogeny and morphological evolution of the Amblystegiaceae. *Molecular Phylogenetics and Evolution*, **23:** 1–21.

———, M. Ignatov, S. Huttunen & B. Goffinet. 2005: A molecular and morphological recircumscription of *Brachytheciastrum* (Brachytheciaceae, Bryopsida). *Taxon*, **51:** 369–376.

———, B. Goffinet, L. Hedenäs, C. J. Cox & A. J. Shaw. 2003: A taxonomic reassessment of the Vittiaceae (Hypnales, Bryopsida): evidence from phylogenetic analyses of combined chloroplast and nuclear sequence data. *Plant Systematics and Evolution*, **241:** 1–12.

Venturi, G. 1868: Il *Desmatodon griseus* di Juratzka. *Commentario della Fauna, Flore e Gea del Veneto e del Trentino*, **1:** 119–124.

———. 1883: De la *Pottia latifolia* Schpr. *Revue Bryologique*, **10:** 95–96.

Verdoorn, F. 1932: *Manual of bryology*. The Hague, Martinus Nijhoff: ix + 486 pp.

Vilhelm, J. 1922: Variabilité du genre *Schistidium* en Tchécoslovaquie. *Acta Botanica Bohemica*, **1:** 43–55.

———. 1924: [Variabilité du genre *Grimmia* en Tchécoslovaquie]. *Věstnik Královské České Společnosti Nauk, Třida Matematicko-Přirodovědecké*, **1924:** 1–46 (in Czech with French summary).

Virtanen, V. 1999: Bryophyte flora of the Huon Peninsula, Papua New Guinea. LX. *Bartramia*, *Conostomum* and *Leiomela* (Bartramiaceae, Musci). *In*: E. Piippo (ed.), *Bryophyte flora of the Huon Peninsula, Papua New Guinea. LX–LXIV. Acta Botanica Fennica*, **165:** 1–15.

Vital, D. M. 1984: Notes on *Campylopus* and *Microcampylopus*. *Cryptogamie, Bryologie – Lichénologie*, **5:** 15–26.

Vitt, D. H. 1970: Proposal to conserve the generic name *Muelleriella* Dusén, Bot. Not. 1905: 304. 1905 (Musci, Orthotrichaceae). *Taxon*, **19:** 298.

———. 1971*a*: New species of mosses from Campbell Island, New Zealand. *The Bryologist*, **74:** 464–472.

———. 1971*b*: The infrageneric evolution, phylogeny, and taxonomy of the genus *Orthotrichum* (Musci) in North America. *Nova Hedwigia*, **21:** 683–711.

———. 1972: A monograph of the genus *Drummondia*. *Canadian Journal of Botany*, **50:** 1191–1208.

———. 1973: A revision of the genus *Orthotrichum* in North America, north of Mexico. *Bryophytorum Bibliotheca*, **1:** 1–208 + pls. 1–60.

———. 1974*a*: A key and synopsis of the mosses of Campbell Island, New Zealand. *New Zealand Journal of Botany*, **12:** 185–210.

———. 1974*b*: The mosses reported from the Yukon Territory, Canada. *Journal of the Hattori Botanical Laboratory*, **38:** 299–312.

———. 1976*a*: The genus *Seligeria* in North America. *Lindbergia*, **3:** 241–275.

———. 1976*b*: A monograph of the genus *Muelleriella* Dusén. *Journal of the Hattori Botanical Laboratory*, **40:** 91–113.

———. 1977: A taxonomic study of the genus *Crosbya* (=*Bellia*; Musci). *Canadian Journal of Botany*, **55:** 2079–2089.

———. 1979*a*: The moss flora of the Auckland Islands, New Zealand, with a consideration of habitats, origins, and adaptations. *Canadian Journal of Botany*, **57:** 2226–2263.

———. 1979*b*: The status of *Muelleriella* nom. cons., with *Muelleriella atrata* (Mitten) Vitt, an earlier name for *M. engelii* Vitt from Kerguélen Island. *Lindbergia*, **5:** 93–95.

———. 1980: A comparative study of *Andreaea acutifolia*, *A. mitabilis*, and *A. rupestris*. *New Zealand Journal of Botany*, **18:** 367–377.

———. 1981: Adaptative modes of the moss sporophyte. *The Bryologist*, **84:** 166–186.

———. 1982*a*: Bryopsida. *In*: S. P. Parker (ed.), *Synopsis and classification of living organisms*. **1**. New York, McGraw-Hill Book Company: pp. 307–336 + pls. 31–34.

———. 1982*b*: The genera of Orthotrichaceae. *In*: P. Geissler & S. W. Greene (eds), *Bryophyte taxonomy: methods, practices and floristic exploration. Proceedings of the International Association of Bryologists Taxonomic Workshop Meeting, Genève, 27 Aug – 2 Sept 1979. Beihefte zur Nova Hedwigia*, **71:** 261–268.

———. 1984: Classification of the Bryopsida. *In*: R. M. Schuster (ed.), *New manual of bryology*. **2**. Nichinan, The Hattori Botanical Laboratory: pp. 696–759.

———. 1991: The distribution of North American bryophytes. *Orthotrichum rupestre* Schwaegr. *Evansia*, **8:** 78–80.

———. **2000:** The classification of mosses: two-hundred years after Hedwig. *Nova Hedwigia*, **70:** 25–36.

——— **& J. K. Bartlett. 1983:** The genus *Seligeria* in New Zealand. *The Bryologist* **86:** 106–113.

——— **& C. D. Hamilton. 1974:** A scanning electron microscopy study of the spores and selected peristomes of the North American Encalyptaceae (Musci). *Canadian Journal of Botany*, **52:** 1973–1981.

——— **& C. Marsh. 1988:** Population variation and phytogeography of *Racomitrium lanuginosum* and *R. pruinosum*. *In*: J. J. Engel & S. Hattori (eds), *Bryological contributions presented in celebration of the distinguished scholarship of Rudolf M. Schuster. Beihefte zur Nova Hedwigia*, **90:** 235–260.

——— **& H. P. Ramsay. 1985*a*:** The *Macromitrium* complex in Australasia (Orthotrichaceae: Bryopsida). Part I. Taxonomy and phylogenetic relationships. *Journal of the Hattori Botanical Laboratory*, **59:** 325–451.

——— **& ———. 1985*b*:** The *Macromitrium* complex in Australasia (Orthotrichaceae: Bryopsida). Part II. Distribution, ecology, and paleography. *Journal of the Hattori Botanical Laboratory*, **59:** 453–468.

——— **& ———. 1986:** The *Macromitrium* complex in Australasia (Orthotrichaceae: Bryopsida). Part III. Cytotaxonomy. *Journal of the Hattori Botanical Laboratory*, **61:** 1–43.

———**, B. Goffinet & T. Hedderson. 1998:** The ordinal classification of the mosses: questions and answers for the 1990s. *In*: J. W. Bates, N. W. Ashton & J. G. Duckett (eds), *Bryology for the twenty-first century*. Leeds, Maney Publishing and the British Bryological Society: pp. 113–123.

———**, T. Koponen & D. H. Norris. 1993:** Bryophyte flora of the Huon Peninsula, Papua New Guinea. LIII. *Ulota* and *Schlotheimia* (Orthotrichaceae, Musci). *Acta Botanica Fennica*, **148:** 5–25.

———**, J. E. Marsh & R. B. Bovey. 1988:** *Mosses, lichens and ferns of Northwest North America*. Edmonton, Lone Pine Publishing: 296 pp.

Vohra, J. N. 1983: *Hypnobryales suborder Leskeineae (Musci) of the Himalayas*. Records of the Botanical Survey of India. **23**. Calcutta, Botanical Survey of India: [7] + iii + 336 pp.

Voss, E. G. (Chairman), H. M. Burdet, W. G. Chaloner, V. Demoulin, P. Hiepko, J. McNeill, R. D. Meikle, D. H. Nicolson, R. C. Rollins, P. C. Silva (members) & W. Greuter (secretary). 1983: *International Code of Botanical Nomenclature adopted by the Thirteenth International Botanical Congress, Sydney, August 1981*. Regnum Vegetabile. **111.** Utrecht – Antwerpen, Bohn, Scheltema & Holkema and The Hague – Boston, Dr. W. Junk, Publishers: xv + 472 pp.

Vysotska, O. I. 1975: [New data on chromosome numbers of Bryopsida in the Ukraine]. *Ukrainskiy Botanicheskiy Zhurnal*, **32:** 498–503 (in Ukrainian with English summary).

W

Wace, N. M. 1960: The botany of the southern oceanic islands. *In*: C. F. A. Pantin (ed.), *A discussion on the biology of the southern cold temperate zone*. *Proceedings of the Royal Society of London, Series B*, **152:** 447–457.

———. **1965:** Vascular plants. *In*: J. van Mieghem & P. van Oye (eds), *Biogeography and ecology in Antarctica*. Monographiae Biologicae. **15**. The Hague, Dr. W. Junk Publishers: pp. 201–266.

———. **1969:** The terrestrial biogeography of Antarctica. *In*: E. M. van Zinderen Bakker (ed.), *Palaeoecology of Africa and the surrounding islands and Antarctica*. **5**. Cape Town, A. A. Balkema: pp. 213–238.

——— **& J. H. Dickson. 1965:** The terrestrial botany of the Tristan da Cunha Islands. *In*: *The biological report of the Royal Society expedition to Tristan da Cunha, 1962. Part II. Philosophical Transactions of the Royal Society of London, Series B, Biological Sciences*, **249(759):** 273–360.

——— **& M. W. Holdgate. 1958:** The vegetation of Tristan da Cunha. *The Journal of Ecology*, **46:** 593–620.

Wagner, B. & R. D. Seppelt. 2006: Deep-water occurrence of the moss *Bryum pseudotriquetrum* in Radok Lake, Amery Oasis, East Antarctica. *Polar Biology*, **29:** 791–795.

Wahlenberg, G. 1812: *Flora lapponica exhibens plantas geographice et botanice consideratas, in Lapponiis Suecis scilicet Umensi, Pitensi, Lulensi, Tornensi et Kemensi scilicet Nordlandia et Finmarkia utraque indigenas, itineribus annorum 1800, 1802, 1807 et denuo investigata*. Berolini, in Taberna Libraria Scholae realis: lxv + 550 pp. + pl. 1–30 + chart to p. lxv + 1 coloured map.

Walther, K. 1934: Untersuchungen über die Variabilität innerhalb des Formenkreises von *Polytrichum juniperinum* Willd. *Annales Bryologici*, **7:** 121–156.

———. **1940:** Revision des Verwandschaftskreises von *Polytrichum piliferum* Schreb. *Repertorium Specierum Novarum Regni Vegetabilis*, **48:** 251–255 + pl. cccxix.

———. **1967:** Beiträge zur Moosflora Westanatoliens I. *Mitteilungen aus dem Staatsinstitut für Allgemeine Botanik in Hamburg*, **12:** 129–186.

———. **1983:** Bryophytina. Laubmoose. *In*: J. Gerloff & J. Poelt (eds), *A. Engler's Syllabus der Pflanzenfamilien mit besonderer Berücksichtigung der Nutzpflanzen nebst einer Übersicht über die Florechreiche und Florengebiete der Erde*. Ed. 13. **5**(2). Berlin – Stuttgart, Gebrüder Borntraeger: x + 108 pp.

Wang, J.-F. & R.-L. Hu. 2005: [Brachytheciaceae]. *In*: R.-L. Hu & J.-F. Wang (eds), [*Flora Bryophytorum Sinicorum*. **7**. *Hypnobryales*]. Beijing, Science Press: pp. 82–182 (in Chinese).

Wang, M.-Z., P.-Ch. Wu, W.-X. Xu & R.-L. Xiong. 2000: [Polytrichaceae]. *In*: Z.-G. Wu (ed.), [*Flora yunnanica*. **19.** (*Bryophyta: Musci*)]. Beijing, Science Press: 589–631 (in Chinese).

Wareham, R. T. 1939: *Pottia*. *In*: A. J. Grout (ed.), *Moss flora of North America north of Mexico*. **1**(4). Newfane, Vermont, privately published: pp. 197–208 + pls. 93–97 & 103x.

Warnstorf, C. 1903: Die europäischen Harpidien. *Beiheft zum Botanischen Centralblatt*, **13:** 388–430 + Taf. 13–14.

———. **1906:** *Kryptogamenflora der Mark Brandenburg und angrenzender Gebiete*. **2**. *Laubmoose*. Leipzig, Verlag von Gebrüder Bornträger: xii + 1160 pp.

———. **1913:** Zur Bryo-Geographie des Russischen Reiches. Eine Erinnerung an Dr. E. Zickendrath. *Hedwigia*, **53:** 184–320.

———. **1916:** *Pottia*-Studien als Vorarbeiten zu einer Monographie des Genus '*Pottia* Ehrh.' sens. str. *Hedwigia*, **58:** 35–152.

Watson, E. V. 1964: An annotated list of bryophytes of Jan Mayen Island. *Nytt Magasin for Botanik*, **11:** 151–212.

———. **1967:** *The structure and life of bryophytes*. London, Hutchinson University Library: 192 pp.

Watson, W. 1914: Xerophytic adaptations of bryophytes in relation to habitat. *New Phytologist*, **13:** 149–169; 181–190.

Watts, W. W. & Th. Whiteledge. 1902: Census muscorum australensium. A classified catalogue of the frondose mosses of Australia and Tasmania, collated from publications and herbaria records. *Proceedings of the Linnean Society of New South Wales*, **23(3)** *Supplement*: 1–90.

——— & ———. 1906: Census muscorum australensium. A classified catalogue of the frondose mosses of Australia and Tasmania, collated from publications and herbaria records. *Proceedings of the Linnean Society of New South Wales*, **30(4)** *Supplement*: 91–163.

Webb, R. 1973: Reproductive behaviour of mosses on Signy Island, South Orkney Islands. *British Antarctic Survey Bulletin*, **36:** 61–77.

Webby, R. F., K. R. Markham & R. I. Lewis Smith. 1996: Chemotypes of the Antarctric moss *Bryum algens* delineated by their flavonoid constituents. *Biochemical Systematics and Ecology*, **24:** 469–475.

Weber, F. & D. M. H. Mohr. 1804: *Naturhistorische Reise Schwedens*. Göttingen, bei Heinrich Dieterich: 207 pp. + 3 pls.

——— & ———. 1807: *Botanisches Taschenbuch auf das Jahr 1807. Deutschland's Kryptogamische Gewächse. Erste Abtheilung. Filices, Musci frondosi et Hepatici*. Kiel, in der akademischen Buchhandlung: xlvi + 509 pp. + xii pls.

Weber, W. A. 1973: Guide to the mosses of Colorado. Keys and ecological notes based on field and herbarium studies. *Institute of Arctic and Alpine Research, University of Colorado, Occasional Paper*, **6:** i–iii + 1–48.

——— & R. C. Wittmann. 2007: *Bryophytes of Colorado: mosses, liverworts, and hornworts*. Santa Fe, Pilgrims Process, Inc.: 238 pp.

Webster, W. H. B. 1834: *Narrative of a voyage to the southern Atlantic Ocean, in the years 1828, 29, 30, performed in H. M. sloop* Chanticleer, *under the command of the late Captain Henry Foster, F.R.S. & c. by order of the Lords Commissioners of the Admiralty*. London, Richard Bentley: xii + 399 pp. (Vol. 1); viii + 398 pp. (Vol. 2).

Wegener, C. 1993: *Racomitrium sudeticum* new to Spitsbergen. *Polarflokken*, **17:** 3–6.

Weiss, F. W. 1770: *Plantae cryptogamicae florae gottingensis*. Göttingae, litteris Joannis Henrici Schulzii: [vi] + xii + 333 pp.

Weitz, S. & C. C. Heyn. 1981: Intra-specific differentiation within the cosmopolitan moss species *Funaria hygrometrica* Hedw. *The Bryologist*, **84:** 315–334.

Welch, W. H. 1943: The systematic position of the genera *Wardia*, *Hydropogon*, and *Hydropogonella*. *The Bryologist*, **46:** 25–46.

Werner, O., R. M. Ros & B. Goffinet. 2007: A reconsideration of the systematic position of *Goniomitrium* (Funariaceae) based on chloroplast sequence markers. *The Bryologist*, **110:** 108–114.

———, ——— & M. Grundmann. 2005: Molecular phylogeny of Trichostomoideae (Pottiaceae, Bryophyta) based on nrITS sequence data. *Taxon*, **54:** 361–368.

———, ———, M. J. Cano & J. Guerra. 2002: *Tortula* and some related genera (Pottiaceae, Musci): phylogenetic relationships based on chloroplast *rps*4 sequences. *Plant Systematics and Evolution*, **235:** 197–207.

———, ———, ——— & ———. 2003: On the systematic position of *Tortula inermis* and *Tortula bolanderi* (Pottiaceae, Musci) based on chloroplast *rps4* sequences. *Nova Hedwigia*, **76:** 137–145.

———, ———, ——— & ———. 2004: Molecular phylogeny of Pottiaceae (Musci) based on chloroplast *rps*4 sequence data. *Plant Systematics and Evolution*, **243:** 147–164.

Wettstein, F. von. 1932: Genetik. *In*: F. Verdoorn (ed.), *Manual of bryology*. The Hague, Martinus Nijhoff: pp. 233–272

Weymouth, W. A. & L. Rodway. 1922: Bryophyte notes. *Papers and Proceedings of the Royal Society of Tasmania*, **1921:** 173–175.

Whitehouse, H. L. K. 1966: The occurrence of tubers in European mosses. *Transactions of the British Bryological Society*, **5:** 103–116.

———. 1969: *Dicranella staphylina*, a new European species. *Transactions of the British Bryological Society*, **5:** 757–765.

——— & M. E. Newton. 1988: *Tortula brevis* sp. nov. and *T. stanfordensis* Steere: morphology, cytology and geographical distribution. *Journal of Bryology*, **15:** 83–99.

Wiggins, I. L. & J. H. Thomas. 1962: *A flora of the Alaskan – Arctic slope*. Arctic Institute of North America. Special Publication. **4**. Toronto, University of Toronto Press: vii + 425 pp.

Wigginton, M. J. 1994: *Pohlia drummondii* (C. Müll.) Andrews. *In*: M. O. Hill, C. D. Preston & A. J. E. Smith (eds), *Atlas of the bryophytes of Britain and Ireland*. **3**. Mosses (Diplolepideae). Colchester, Harley Books: p. 62.

Wijk, R. van der. 1929: Über den Bau und die Entwicklung der Peristomzähne bei *Polytrichum*. *Recueil des Travaux Botaniques Néerlandais*, **26:** 289–395.

———, W. D. Margadant & P. A. Florschütz. 1959: *Index muscorum*. **1** (A–C). Regnum Vegetabile. **17**. Utrecht, Kemink en Zoon N.V.: xxviii + 548 pp.

———, ——— & ———. 1962: *Index muscorum*. **2** (D–*Hypno*). Reg. Veg. **26**. Utrecht, Kemink en Zoon N.V.: 535 pp.

———, ——— & ———. 1967: *Index muscorum*. **4** (P–S). Reg. Veg. **48**. Utrecht, Kemink en Zoon N.V.: 604 pp.

———, ——— & ———. 1969: *Index muscorum*. **5** (T–Z, Appendix). Reg. Veg. **65**. Utrecht, Kemink en Zoon N.V.: xii + 922 pp.

Wilczek, R. & F. Demaret. 1970: Les *Pohlia* propagilifères de Belgique (Bryaceae). *Bulletin Jardin Botanique National de Belgique*, **40:** 405–422.

——— & ———. 1976: Les espèces du complexe *Bryum bicolor* (Musci). *Bulletin Jardin Botanique National de Belgique*, **46:** 511–541.

Willdenow, C. L. 1787: *Florae berolinensis prodromus secundum systema Linneanum ab illustr, viro ac Eq. C. P. Thunbergio emendatum conscriptus*. Berolini, impensis Wilhelmi Viewegii: xvi + 439 pp. + vii pls.

Williams, H. 1959: *Pterigoneurum* [sic] *ovatum* in Ontario. *The Bryologist*, **62:** 156–158.

Williams, R. S. 1909: Bolivian mosses. Part II. *Bulletin of the New York Botanical Garden*, **6:** 227–261.

———. 1913*a*: Dicranaceae, Leucobryaceae. *In*: *North American Flora*. **15**(2). New York, The New York Botanical Garden: pp. 77–166.

———. 1913*b*: The genus *Husnotiella* Cardot. *The Bryologist*, **16:** 25.

———. **1916:** Peruvian mosses. *Bulletin of the Torrey Botanical Club*, **43:** 323–334 + pls. 17–20.

Willis, J. H. 1955: New and interesting moss records for Australia. *Victorian Naturalist*, **72:** 73–78.

Wilson, J. W. 1966: An analysis of plant growth and its control in arctic environments. *Annals of Botany*, **30:** 383–402.

Wilson, W. 1841: Remarks on Drummond's Musci Americani, collected in British North America, during the second Land Arctic Expedition, under the command of Sir John Franklin, R. N. – Glasgow, 1828. *London Journal of Botany*, **3:** 343–349.

———. **1854:** Nat. Ord. C. Musci, Juss. *In*: J. D. Hooker, *The botany of the Antarctic voyage of H. M. Discovery ships* Erebus *and* Terror *in the years 1839–43, under the command of Captain Sir James Clark Ross, Kt., R. N., F. R. S.* Vol. **1**. Flora Novae-Zelandiae. Part II. Flowerless plants. London, Lovell Reeve: pp. 160–221 + pls. clxxi–clxxvii.

———. **1855:** *Bryologia Britannica; containing the mosses of Great Britain and Ireland, systematically arranged and described according to the method of Bruch and Schimper, with illustrative plates: being a new (third) edition, with many additions and alterations, of the Muscologia Britannica of Messrs. Hooker and Taylor*. London, Longman, Brown, Green, and Longmans: xx + 445 pp. + lxi pls.

———. **1859:** Nat. Ord. C. Musci, Juss. *In*: J. D. Hooker, *The botany of the Antarctic voyage of H. M. Discovery ships* Erebus *and* Terror *in the years 1839–43, under the command of Captain Sir James Clark Ross, Kt., R. N., F. R. S.* **1**. *Flora Tasmaniae. Part II. Flowerless plants*. London, Lovell Reeve: pp. 160–210 + pls. clxxi–clxxvi.

——— **& J. D. Hooker. 1845:** Musci. *In*: J. D. Hooker, *The botany of the Antractic voyage of H. M. Discovery ships* Erebus *and* Terror *in the years 1839–43, under the command of Captain Sir James Clark Ross, Kt., R. N., F. R. S.* **1**. *Flora antarctica. Part I., Botany of Lord Auckland's group and Campbell's Island*. London, Reeve Brothers: pp. 117–143 + pls. lvii–lxii.

——— **& J. D. Hooker. 1847:** Musci. *In*: J. D. Hooker, *The botany of the Antractic voyage of H. M. Discovery ships* Erebus *and* Terror *in the years 1839–43, under the command of Captain Sir James Clark Ross, Kt., R. N., F. R. S.* **1**. *Flora antarctica. Part. II., Botany of Fuegia, the Falklands, Kerguelen's Land, etc.* London, Reeve Brothers: pp. 395–423, 550–551 + pls. cli–clv.

——— **& W. J. Hooker. 1841:** *Musci americani; or, specimens of mosses, Jungermanniae & c. collected by the late Thomas Drummond, in the southern states of North America arranged and named by W. Wilson and Sir W. J. Hooker. Numbers 1–180*. Warrington.

Winkworth, R. C., S. J. Wagstaff, D. Glenny & P. J. Lockhart. 2002: Plant dispersal N.E.W.S. from New Zealand. *Trends in Ecology and Evolution*, **17:** 514–520.

Winton, M. D. de & J. E. Beever. 2004: Deep-water bryophyte records from New Zealand lakes. *New Zealand Journal of Marine and Freshwater Research*, **38:** 329–340.

Wise, K. A. J. & Gressitt, J. L. 1965: Far southern animals and plants. *Nature*, **207(4992):** 101–102.

Wright, C. H. 1905*a*: The mosses of Gough Island. *In*: R. N. Rudmose Brown, *The botany of Gough Island.–II. Journal of the Linnean Society Botany*, **37:** 264–265.

———. **1905*b*:** The mosses of the South Orkneys. *In*: R. N. Rudmose Brown, C. H. Wright & O. V. Darbishire, *The botany of the South Orkneys. Transaction and Proceedings of the Botanical Society of Edinburgh*, **23:** 103–104.

Wu, P.-C. & S.-S. Hu. 1990: [The bryophytes of the King George Island, Antarctica]. *Acta Botanica Sinica,* **32**: 477–482 (in Chinese with English summary).

——— **& M.-Z. Wang. 2005:** Polytrichaceae. *In*: Ch. Gao, M. R. Crosby & S. He (eds), *Moss Flora of China. English version.* **8**. *Sematophyllaceae – Polytrichaceae*. Beijing – New York, Science Press and St. Louis, Missouri Botanical Garden: pp. 306–366.

Y

Yano, O. 1981: A checklist of Brazilian mosses. *Journal of the Hattori Botanical Laboratory*, **50:** 279–456.

Yli-Rekola, M. 1980: Infraspecific variation of *Polytrichastrum alpinum* (Musci, Polytrichaceae). I. Comparison of multivariate methods. *Annales Botanici Fennici*, **17:** 277–291.

Yoshiyuki, S., G. V. Glazko & M. Nei. 2002: Overcredibility of molecular phylogenies obtained by Bayesian phylogenetics. *Proceedings of the National Academy of Sciences*, **99:** 16138–16143.

Young, A. 1821: Notice of the voyage of Edward Barnsfield [sic!], master of His Majesty's ship *Andromache* to New South Shetland. *Edinburgh Philosophical Journal*, **3:** 345–348.

Young, S. B. & J.-R. Kläy. 1971: Bryophytes in the 1969 crater of Deception Island, Antarctica: an apparent case of rapid long-distance dispersal. *Ohio Journal of Science*, **71:** 358–362.

Z

Zander, R. H. 1978*a*: A synopsis of *Bryoerythrophyllum* and *Morinia* (Pottiaceae) in the New World. *The Bryologist*, **81:** 539–560.

———. **1978*b*:** New combinations in *Didymodon* (Musci) and a key to the taxa in North America north of Mexico. *Phytologia*, **41:** 11–32.

———. **1978*c*:** *Didymodon sinuosus* in Alaska and a key to species of Pottiaceae with propaguloid leaf apices. *The Bryologist*, **81:** 572–575.

———. **1980:** Acid-base color reactions: the status of *Triquetrella ferruginea*, *Barbula inaequifolia* and *B. calcarea*. *The Bryologist*, **83:** 228–233.

———. **1981*a*:** Descriptions and illustrations of *Barbula*, *Pseudocrossidium* and *Bryoerythrophyllum* (p.p.) of Mexico. *Cryptogamie, Bryologie – Lichénologie*, **2:** 1–22.

———. **1981*b*:** *Didymodon* (Pottiaceae) in Mexico and California: taxonomy and nomenclature of discontinuous and non-discontinuous taxa. *Cryptogamie, Bryologie – Lichénologie*, **2:** 379–422.

———. **1986:** Notes on *Bryoerythrophyllum* (Musci). *The Bryologist*, **89:** 13–16.

———. **1989:** Seven new genera in Pottiaceae (Musci) and a lectotype for *Syntrichia*. *Phytologia*, **65:** 424–436.

———. **1993:** Genera of the Pottiaceae: mosses of harsh environment. *Bulletin of the Buffalo Society of Natural Sciences*, **32:** i–vi + 1–378.

———. **1994*a*:** *Bryoerythrophyllum*. *In*: A. J. Sharp, H. Crum & P. M. Eckel (eds.), *The moss flora of Mexico. Part one*

Sphagnales to Bryales. Memoirs of the New York Botanical Garden, **69:** 273–283.

———. **1994*b*:** *Didymodon*. *In*: A. J. Sharp, H. Crum & P. M. Eckel (eds.), *The moss flora of Mexico. Part one Sphagnales to Bryales. Memoirs of the New York Botanical Garden*, **69:** 299–319.

———. **1994*c*:** *Phascum*. *In*: A. J. Sharp, H. Crum & P. M. Eckel (eds.), *The moss flora of Mexico. Part one Sphagnales to Bryales. Memoirs of the New York Botanical Garden*, **69:** 377–380.

———. **1995:** Phylogenetic relationships of *Hyophiladelphus* gen. nov. (Pottiaceae, Musci) and a perspective on the cladistic method. *The Bryologist*, **98:** 363–374.

———. **1998*a*:** A phylogrammatic evolutionary analysis of the moss genus *Didymodon* in North America north of Mexico. *Bulletin of the Buffalo Society of Natural Sciences*, **36:** 81–115.

———. **1998*b*:** Phylogenetic reconstruction, a critique. *Taxon*, **47:** 681–693.

———. **2001:** A conditional probability of recostruction measure for internal cladogram branches. *Systematic Biology*, **50:** 425–437.

———. **2003:** Reliable phylogenetic resolution of morphological data can be better than that of molecular data. *Taxon*, **52:** 109–112.

———. **2004:** Minimal values for reliability of bootstrap and jackknife proportions, decay index, and Bayesian posterior probability. *PhyloInformatics*, **2:** 1–13.

———. **2006:** The Pottiaceae s. str. as an evolutionary lazarus taxon. *Journal of the Hattori Botanical Laboratory*, **100:** 581–602.

———. **2007:** *Bryoerythrophyllum*. *In*: R. H. Zander (Lead ed.), *Flora of North America north of Mexico*. **27**. *Bryophytes: mosses (1)*. New York – Oxford, Oxford University Press: pp. 565–569.

——— **& H. J. During. 1999:** *Neophoenix* (Pottiaceae), a new African moss genus found through soil diaspore bank analysis. *Taxon*, **48:** 657–662.

——— **& W. J. Hoe. 1979:** Geographic disjunction and heterophylly in *Tortella fragilis* var. *tortelloides* (= *Sarconeurum tortelloides*). *The Bryologist*, **82:** 84–87.

——— **& R. R. Ireland. 1979:** Propaguliferous *Ceratodon purpureus* in riparian environments. *The Bryologist*, **82:** 474–478.

——— **& M. Mahu. 1999:** A new species of *Hennediella* from Chile. *The Bryologist*, **102:** 349–351.

——— **& R. Ochyra. 2001:** *Didymodon tectorum* and. *D. brachyphyllus* (Musci, Pottiaceae) in North America. *The Bryologist*, **104:** 372–377.

Zanten, B. O. van. 1971: Musci. *In*: E. M. van Zinderen Bakker, Sr., J. M. Winterbottom & R. A. Dyer (eds), *Marion and Prince Edward Islands. Report on the South African Biological and Geological Expedition 1965–1966*. Cape Town, A. A. Balkema: pp. 173–227.

———. **1976:** Preliminary report on germination experiments designed to estimate the survival chances of moss spores during aerial transoceanic long-distance dispersal in the Southern Hemisphere with special reference to New Zealand. *Journal of the Hattori Botanical Laboratory*, **41:** 133–140.

———. **1978:** Experimental studies on transoceanic long-range dispersal of moss species in the Southern Hemisphere. *Journal of the Hattori Botanical Laboratory*, **44:** 455–482.

———. **1983:** Possibilities of long-range dispersal in bryophytes with special reference to the Southern Hemisphere. *In*: K. Kubitzki (ed.), *Dispersal and distribution: An International Symposium. Sondernbände des Naturwissenschaftlichen Vereins in Hamburg*, **7:** 49–64.

———. **1984:** Some considerations on the feasibility of long-distance transport in bryophytes. *Acta Botanica Neerlandica*, **33:** 231–232.

——— **& S. R. Gradstein. 1988:** Experimental dispersal geography of Neotropical liverworts. *In*: J. J. Engel & S. Hattori (eds), *Bryological contributions presented in celebration of the distinguished scholarship of Rudolf M. Schuster. Beihefte zur Nova Hedwigia*, **90:** 41–94.

——— **& T. Pócs. 1981:** Distribution and dispersal of bryophytes. *Advances in Bryology*, **1:** 479–562.

Zhang, D.-C. & X.-J. Li. 2002: [Bryaceae]. *In*: X.-J. Li (ed.), [*Flora yunnanica*. **18** (Bryophyta: Musci)]. Beijing, Science Press: pp. 382–448 (in Chinese).

——— , ——— **& S. He. 2007:** Bryaceae. *In*: X.-J. Li, M. R. Crosby & S. He (eds), *Moss Flora of China. English version*. **4**. *Bryaceae – Timmiaceae*. Beijing – New York, Science Press and St. Louis, Missouri Botanical Garden: pp. 3–92.

Zhou, J.-L., Z.-W. Zhang, J.-H. Cao, J.-H. Ji & P.-Ch. Wu. 1995: Chromosome observation on 3 moss in the King George Island, Antarctica. *Antarctic Research (Chinese Edition)*, **7:** 80–82 (in Chinese with English summary).

Zhou, S.-Y., Y.-L. Zhou & J.-L. Zhao. 1989: Electron microscopic observation and electron microprobe analysis of the moss in the water and bed sediments of the West Lake of the Great Wall Station in the south pole area. *In*: G. Kun (ed.), *Proceedings of the International Symposium on Antarctic Research. Chinese Committee on Antarctic Research*. Tianjin, China Ocean Press: pp. 350–355.

Zijlstra, G. 2002: Report of the Committee for Bryophyta: 7. *Taxon*, **51:** 793–794.

Zolotov, V. I. 2000: The genus *Bryum* (Bryaceae, Musci) in Middle European Russia. *Arctoa, a Journal of Bryology,* **9:** 155–232.

———. **2006:** On the systematics and distribution of some species of *Bryum* (Bryaceae, Bryophyta) in Russia. *Arctoa, a Journal of Bryology,* **15:** 155–162.

——— **& M. S. Ignatov. 2001:** On the axillary hairs of *Leptobryum* Meesiaceae, Musci) and some other acrocarpous mosses. *Arctoa, a Journal of Bryology*, **10:** 189–200.

Zomlefer, W. B. 1993: A revision of *Rigodium* (Musci: Rigodiaceae). *The Bryologist*, **96:** 1–72.

Ż

Żarnowiec, J. 2001: *A taxonomic monograph of the* Drepanocladus aduncus *group* (Bryopsida: Amblystegiaceae). Bielsko-Biała, Łódź Technical University, Bielsko Biała Branch: 248 pp.

Index to Latin Plant Names

Roman type is used for accepted scientific names for plants treated as established members of the Flora and names in *italics* are synonyms or plants not established in the Flora. New scientific names and combinations are in **boldface**. Page numbers in **boldface** indicate the primary reference page for a taxon; page numbers in *italics* indicate that the name is considered a synonym on that page; page number with an asterisk (*) indicates a map and with a plus (+) shows an illustration.

B

C

D

H

T

X

Z